Standard Atomic Weights of the Elements 2007, IUPAC

Based on Relative Atomic Mass of $^{12}C = 12$, where ^{12}C is a neutral atom in its nuclear and electronic ground state.†

Name	Symbol	Atomic Number	Atomic Weight
Actinium*	Ac	89	(227)
Aluminum	Al	13	26.9815386(8)
Americium*	Am	95	(243)
Antimony	Sb	51	121.760(1)
Argon	Ar	18	39.948(1)
Arsenic	As	33	74.92160(2)
Astatine*	At	85	(210)
Barium	Ba	56	137.327(7)
Berkelium*	Bk	97	(247)
Beryllium	Be	4	9.012182(3)
Bismuth	Bi	83	208.98040(1)
Bohrium*	Bh	107	(264)
Boron	B	5	10.811(7)
Bromine	Br	35	79.904(1)
Cadmium	Cd	48	112.411(8)
Calcium	Ca	20	40.078(4)
Californium*	Cf	98	(251)
Carbon	C	6	12.0107(8)
Cerium	Ce	58	140.116(1)
Cesium	Cs	55	132.9054519(2)
Chlorine	Cl	17	35.453(2)
Chromium	Cr	24	51.9961(6)
Cobalt	Co	27	58.933195(5)
Copper	Cu	29	63.546(3)
Curium*	Cm	96	(247)
Darmstadtium*	Ds	110	(281)
Dubnium*	Db	105	(262)
Dysprosium	Dy	66	162.500(1)
Einsteinium*	Es	99	(252)
Erbium	Er	68	167.259(3)
Europium	Eu	63	151.964(1)
Fermium*	Fm	100	(257)
Fluorine	F	9	18.9984032(5)
Francium*	Fr	87	(223)
Gadolinium	Gd	64	157.25(3)
Gallium	Ga	31	69.723(1)
Germanium	Ge	32	72.64(1)
Gold	Au	79	196.966569(4)
Hafnium	Hf	72	178.49(2)
Hassium*	Hs	108	(277)
Helium	He	2	4.002602(2)
Holmium	Ho	67	164.93032(2)
Hydrogen	H	1	1.00794(7)
Indium	In	49	114.818(3)
Iodine	I	53	126.90447(3)
Iridium	Ir	77	192.217(3)
Iron	Fe	26	55.845(2)
Krypton	Kr	36	83.798(2)
Lanthanum	La	57	138.90547(7)
Lawrencium*	Lr	103	(262)
Lead	Pb	82	207.2(1)
Lithium	Li	3	[6.941(2)]†
Lutetium	Lu	71	174.9668(1)
Magnesium	Mg	12	24.3050(6)
Manganese	Mn	25	54.938045(5)
Meitnerium*	Mt	109	(268)
Mendelevium*	Md	101	(258)
Mercury	Hg	80	200.59(2)
Molybdenum	Mo	42	95.96(2)
Neodymium	Nd	60	144.242(3)
Neon	Ne	10	20.1797(6)
Neptunium*	Np	93	(237)
Nickel	Ni	28	58.6934(4)
Niobium	Nb	41	92.90638(2)
Nitrogen	N	7	14.0067(2)
Nobelium*	No		(259)
Osmium	Os		
Oxygen	O		
Palladium	Pd		
Phosphorus	P		
Platinum	Pt		
Plutonium*	Pu		
Polonium*	Po	84	
Potassium	K	19	39.098[illegible](1)
Praseodymium	Pr	59	140.90765(2)
Promethium*	Pm	61	(145)
Protactinium*	Pa	91	231.03588(2)
Radium*	Ra	88	(226)
Radon*	Rn	86	(222)
Rhenium	Re	75	186.207(1)
Rhodium	Rh	45	102.90550(2)
Roentgenium*	Rg	111	(272)
Rubidium	Rb	37	85.4678(3)
Ruthenium	Ru	44	101.07(2)
Rutherfordium*	Rf	104	(261)
Samarium	Sm	62	150.36(2)
Scandium	Sc	21	44.955912(6)
Seaborgium*	Sg	106	(266)
Selenium	Se	34	78.96(3)
Silicon	Si	14	28.0855(3)
Silver	Ag	47	107.8682(2)
Sodium	Na	11	22.98976928(2)
Strontium	Sr	38	87.62(1)
Sulfur	S	16	32.065(5)
Tantalum	Ta	73	180.94788(2)
Technetium*	Tc	43	(98)
Tellurium	Te	52	127.60(3)
Terbium	Tb	65	158.92535(2)
Thallium	Tl	81	204.3833(2)
Thorium*	Th	90	232.03806(2)
Thulium	Tm	69	168.93421(2)
Tin	Sn	50	118.710(7)
Titanium	Ti	22	47.867(1)
Tungsten	W	74	183.84(1)
Uranium*	U	92	238.02891(3)
Vanadium	V	23	50.9415(1)
Xenon	Xe	54	131.293(6)
Ytterbium	Yb	70	173.054(5)
Yttrium	Y	39	88.90585(2)
Zinc	Zn	30	65.38(2)
Zirconium	Zr	40	91.224(2)
—‡*		112	(285)
—‡*		113	(284)
—‡*		114	(289)
—‡*		115	(288)
—‡*		116	(293)
—‡*		118	(294)

† The atomic weights of many elements vary depending on the origin and treatment of the sample. This is particularly true for Li; commercially available lithium-containing materials have Li atomic weights in the range of 6.939 and 6.996. Uncertainties are given in parentheses following the last significant figure to which they are attributed.

* Elements with no stable nuclide; the value given in parentheses is the atomic mass number of the isotope of longest known half-life. However, three such elements (Th, Pa, and U) have a characteristic terrestrial isotopic composition, and the atomic weight is tabulated for these.

‡ Not yet named.

Instant Access to OWL (two semesters) ISBN-10: 0-495-05099-7 • ISBN-13: 978-0-495-05099-5
Instant Access to OWL e-Book (two semesters) ISBN-10: 0-495-39160-3 • ISBN-13: 978-0-495-39160-9

Developed at the University of Massachusetts, Amherst, and class tested by tens of thousands of chemistry students, **OWL** is a fully customizable and flexible web-based learning system. **OWL** supports mastery learning and offers numerical, chemical, and contextual parameterization to produce thousands of problems correlated to this text. The **OWL** system also features a database of simulations, tutorials, and exercises, as well as end-of-chapter problems from the text. With **OWL**, you get the most widely used online learning system available for chemistry with unsurpassed reliability and dedicated training and support.

Features

- Interactive simulations of chemical systems are accompanied by guiding questions that lead you through an exploration of the simulation. These concept-building tools guide you to your own discovery of chemical concepts and relationships.
- Interactive problem-solving tutors ask questions and then give feedback that helps you solve the problem.
- Explorations of animations, movies, and graphic images help you examine the chemical principles behind multimedia presentations of chemical events.
- Re-try questions over and over, covering the same concept, but using different numerical values and chemical systems until you get it right.
- The optional **e-Book in OWL** includes the complete electronic version of the text, fully integrated and linked to **OWL** homework problems.

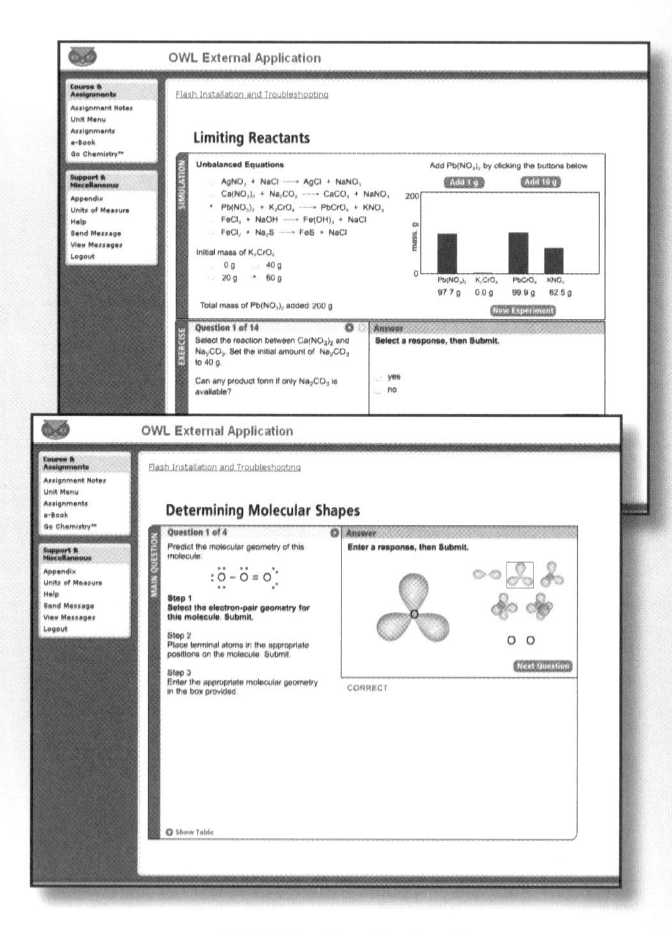

For this textbook, **OWL** includes parameterized end-of-chapter questions from the text (marked in the text with a blue square ■) and tutorials based on the Estimation boxes in the text.

Go Chemistry™

Go Chemistry™ is a set of easy-to-use essential videos that can be downloaded to your video iPod or portable video player—ideal for students on the go! Mini video lectures include animations and problems for quick summaries of key concepts. E-flashcards briefly introduce a key concept, and then test your understanding of the basics with a series of questions. **Go Chemistry** also plays on QuickTime, Windows Media Player, iTunes™, and iPhones™.

OWL Quick Prep

This online short course has proved to be very effective at helping students like you review key chemistry concepts and hit the ground running in first-term introductory chemistry. **Quick Prep** is available 24 hours a day, is self-paced and self-contained, and is the perfect solution to help students prepare for their introductory chemistry course.

If an access card came packaged with your text, you may be able to use these assets immediately. If not, visit **www.ichapters.com**, our preferred online store, to purchase access to **Go Chemistry** and other exciting products from Cengage Learning.

PRINCIPLES OF

Chemistry

THE MOLECULAR SCIENCE

John W. Moore
University of Wisconsin—Madison

Conrad L. Stanitski
Franklin and Marshall College

Peter C. Jurs
Pennsylvania State University

Australia • Brazil • Japan • Korea • Mexico • Singapore • Spain • United Kingdom • United States

Principles of Chemistry: The Molecular Science
John W. Moore, Conrad L. Stanitski, Peter C. Jurs

Publisher: Mary Finch
Senior Acquisitions Editor: Lisa Lockwood
Senior Development Editor: Peter McGahey
Associate Development Editor: Brandi Kirksey
Assistant Editor: Elizabeth Woods
Editorial Assistant: Jon Olafsson
Senior Media Editor: Lisa Weber
Marketing Manager: Nicole Hamm
Marketing Coordinator: Kevin Carroll
Marketing Communications Manager: Linda Yip
Project Manager, Editorial Production: Jennifer Risden
Creative Director: Rob Hugel
Art Director: John Walker
Print Buyer: Becky Cross
Permissions Editors: John Hill, Margaret Chamberlain-Gaston
Production Service: Graphic World Inc.
Text Designer: tani hasegawa
Photo Researcher: Sue C. Howard
Copy Editor: Graphic World Inc.
Illustrators: Graphic World Inc.; Greg Gambino
OWL Producers: Stephen Battisti, Cindy Stein, and David Hart in the Center for Educational Software Development at the University of Massachusetts, Amherst, and Cow Town Productions
Cover Designer: William Stanton
Cover Image: © Whit Bronaugh
Compositor: Graphic World Inc.

Library of Congress Control Number: 2008938348

ISBN-13: 978-0-495-39079-4

ISBN-10: 0-495-39079-8

Brooks/Cole
10 Davis Drive
Belmont, CA 94002-3098
USA

Cengage Learning is a leading provider of customized learning solutions with office locations around the globe, including Singapore, the United Kingdom, Australia, Mexico, Brazil, and Japan. Locate your local office at **www.cengage.com/international.**

Cengage Learning products are represented in Canada by Nelson Education, Ltd.

To learn more about Brooks/Cole, visit **www.cengage.com/brookscole**

Purchase any of our products at your local college store or at our preferred online store **www.ichapters.com.**

About the Cover
As you look carefully at the photo on the front cover of this book, you will notice that each spherical water droplet hanging from a blade of grass acts as a lens and shows an image of the background of the photo, a blue flower. By studying chemistry, you will develop a new way of perceiving and thinking about the fascinating physical world around you. Chemistry concepts explain observable phenomena that occur on the macroscale (water droplets are spherical, ice floats) by describing what is happening at the molecular scale (nanoscale). In the case of water droplets, what we observe has to do with the shape and orientation of water molecules. Water droplets are spherical because water molecules stick together in specific orientations, forming interconnected puckered hexagonal rings. This gives water other interesting properties, such as the hexagonal shapes of snowflakes and the fact that ice floats in liquid water. As you use this book, let it help you to understand more fully the vast range of natural phenomena—from the cosmos to a cell's nucleus to atoms and molecules—in which chemistry plays a major role.

Printed in the United States of America
1 2 3 4 5 6 7 12 11 10 09

To our wives—Betty (JWM), Barbara (CLS), and Elaine (PCJ)—for your patience, support, understanding, and love.

About the Authors

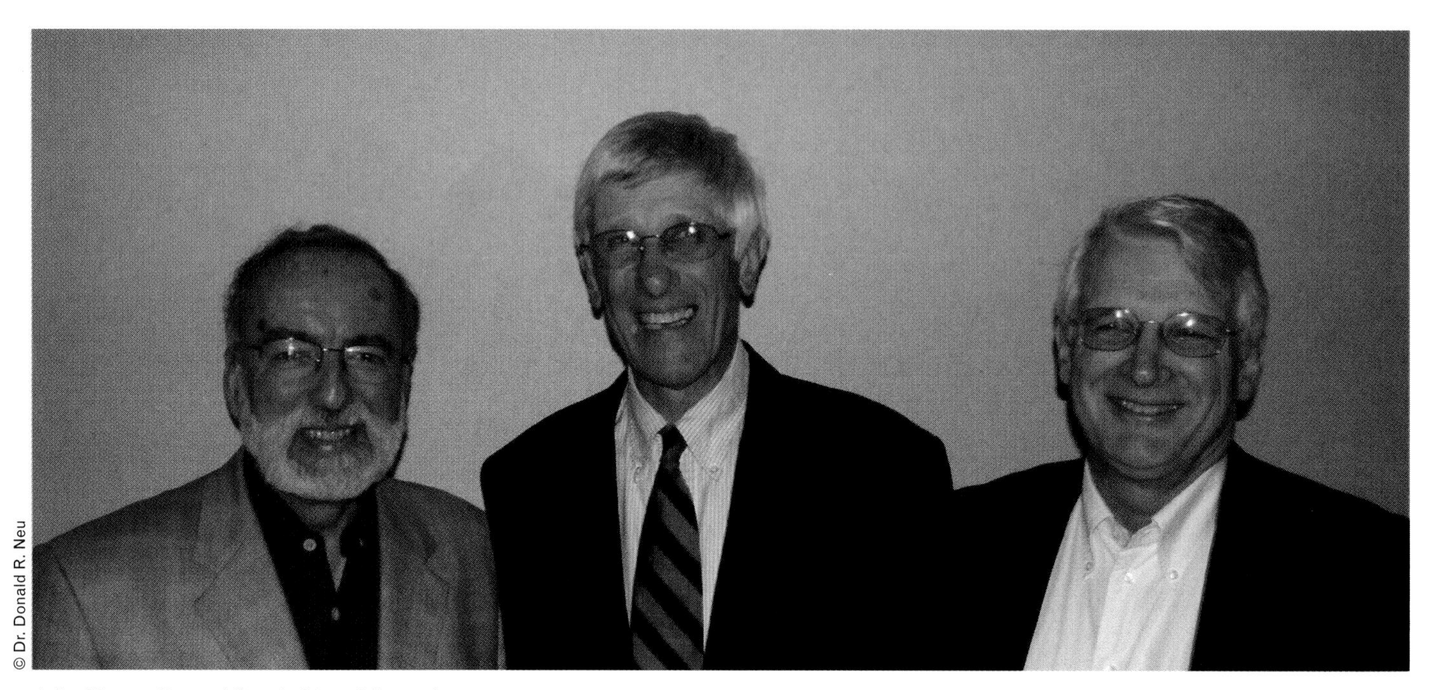

© Dr. Donald R. Neu

John Moore, Conrad Stanitski, and Peter Jurs.

John W. Moore received an A.B. magna cum laude from Franklin and Marshall College and a Ph.D. from Northwestern University. He held a National Science Foundation (NSF) postdoctoral fellowship at the University of Copenhagen and taught at Indiana University and Eastern Michigan University before joining the faculty of the University of Wisconsin–Madison in 1989. At the University of Wisconsin, Dr. Moore is W. T. Lippincott Professor of Chemistry and Director of the Institute for Chemical Education. He has been Editor of the *Journal of Chemical Education (JCE)* since 1996. He has won the American Chemical Society (ACS) George C. Pimentel Award in Chemical Education and the James Flack Norris Award for Excellence in Teaching Chemistry. In 2003, he won the Benjamin Smith Reynolds Award at the University of Wisconsin–Madison in recognition of his excellence in teaching chemistry to engineering students. Dr. Moore has recently received the second of two major grants from the NSF to support development of a chemistry pathway for the NSF-supported National Science Digital Library.

Conrad L. Stanitski is Distinguished Emeritus Professor of Chemistry at the University of Central Arkansas and is currently Visiting Professor at Franklin and Marshall College. He received his B.S. in Science Education from Bloomsburg State College, M.A. in Chemical Education from the University of Northern Iowa, and Ph.D. in Inorganic Chemistry from the University of Connecticut. He has co-authored chemistry textbooks for science majors, allied health science students, nonscience majors, and high school chemistry students. Dr. Stanitski has won many teaching awards, including the CMA CATALYST National Award for Excellence in Chemistry Teaching, the Gustav Ohaus–National Science Teachers Association Award for Creative Innovations in College Science Teaching, the Thomas R. Branch Award for Teaching Excellence and the Samuel Nelson Gray Distinguished Professor Award from Randolph-Macon College, and the 2002 Western Connecticut ACS Section Visiting Scientist Award. He was Chair of the American Chemical Society Division of Chemical Education (2001) and has been an elected Councilor for that division. An instrumental and vocal performer, he also enjoys jogging, tennis, rowing, and reading.

Peter C. Jurs is Professor Emeritus of Chemistry at the Pennsylvania State University. Dr. Jurs earned his B.S. in Chemistry from Stanford University and his Ph.D. in Chemistry from the University of Washington. He then joined the faculty of Pennsylvania State University, where he has been Professor of Chemistry since 1978. Dr. Jurs's research interests have focused on the application of computational methods to chemical and biological problems, including the development of models linking molecular structure to chemical or biological properties (drug design). For this work he was awarded the ACS Award for Computers in Chemistry in 1990. Dr. Jurs has been Assistant Head for Undergraduate Education at Penn State, and he works with the Chemical Education Interest Group to enhance and improve the undergraduate program. In 1995, he was awarded the C. I. Noll Award for Outstanding Undergraduate Teaching. Dr. Jurs serves as an elected Councilor for the American Chemical Society Computer Division.

Contents Overview

Detailed Contents

© Paul Steel/Photolibrary

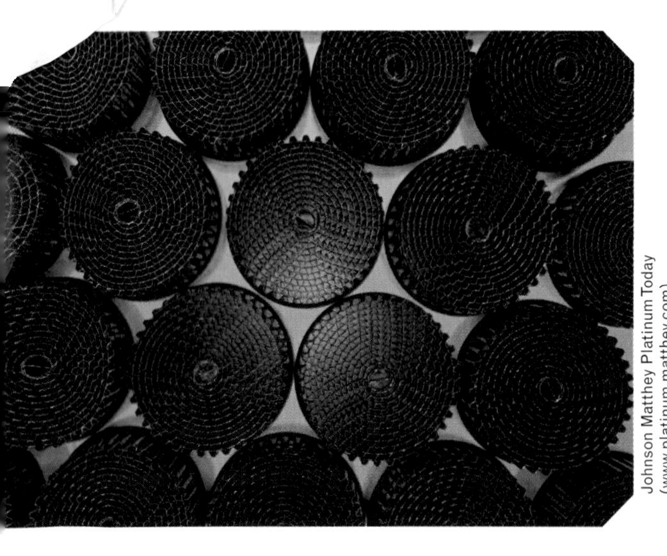
Johnson Matthey Platinum Today (www.platinum.matthey.com)

George D. Lepp/Corbis

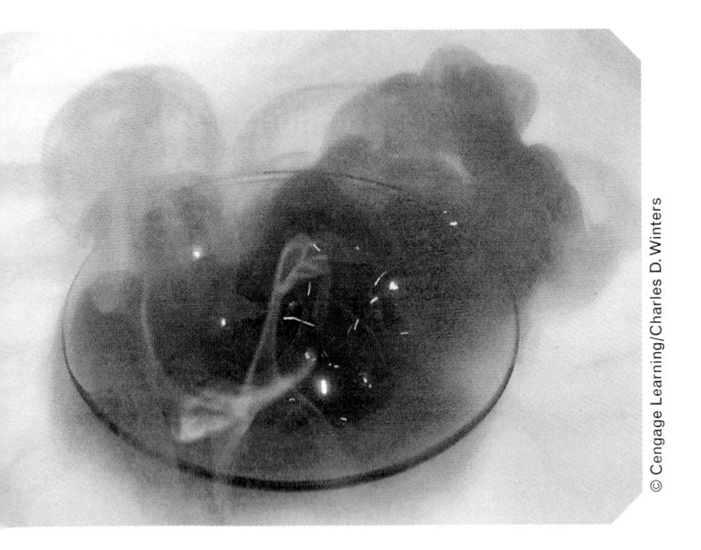

© Cengage Learning/Charles D. Winters

Preface

Students take a two-semester general chemistry course for science majors for a variety of reasons, largely those associated with using the course as a pre- or co-requisite for other science-related courses. At the very core of this textbook are two assertions: (1) Such students need to recognize the relevance of chemistry to solving important problems and the important contributions chemistry makes to other disciplines; and (2) It is essential that such students gain a working knowledge of how chemistry principles are applied to solve problems in a broad spectrum of applications. Examples of such applications are creating new and improving existing chemical pathways that lead to the more efficient synthesis of new pharmaceuticals; developing a deeper understanding of alternative energy sources to mitigate global warming; and understanding how new, more efficient catalysts can be developed that could help to decrease air pollution and to minimize production of chemical waste from industrial processes. Knowledge of chemistry provides a way of interpreting macroscale phenomena at the molecular level that can be applied to many critical 21st century problems, including those just given.

Why a Principles Textbook?

This book is slimmer than a typical bulky general chemistry textbook. This textbook is our direct response to concerns that many general chemistry textbooks contain more material, sometimes much more, than can be addressed meaningfully in the required two-semester course. Students find such large books too cumbersome to bring regularly to class or to their out-of-class study places. In addition, a growing unease exists nationally about the increasing costs associated with higher education, among them the cost of textbooks, including large general chemistry books.

Principles of Chemistry: The Molecular Science was written to address these concerns. Its title highlights the fact that the chemical principles included in the book are limited to those likely to be covered in most general chemistry courses for science majors. These chemical concepts and principles are discussed in this book at the level of depth and rigor expected in standard general chemistry textbooks. We have spent considerable effort to ensure that such is the case; the depth of coverage and the rigor have not been compromised.

In writing *Principles of Chemistry: The Molecular Science,* we used our other textbook *Chemistry: The Molecular Science,* third edition, as a starting point because of its clear presentation of major chemical principles. To assist us, we sought and received significant feedback from more than two dozen professors of general chemistry regarding what material was of secondary importance and could be deleted while retaining the major principles. Additionally, we carefully reviewed the American Chemical Society's general chemistry examinations to determine the major chemical concepts assessed in those exams and we retained the chemical principles associated with them. Consequently, to achieve the reduced size desired of the *Principles* book, some topics from the third edition of *Chemistry: The Molecular Science* were reduced in scope (e.g., organic functional groups, coordination compounds) or omitted (e.g., water purification and surfactants, stratospheric ozone depletion, ceramics and cement, noncovalent forces in living cells, aromatic compounds).

To stay within the guidelines to reduce the textbook's size, the *Tools of Chemistry* and the *Chemistry You Can Do* features from the third edition do not appear in the *Principles* textbook. The *Summary Problems* have all been moved to the student companion website. *Chemistry in the News* features have been retained, as were *Estimation* boxes, popular items of our larger textbook. Through judicious choices, the variety of the chapter-end Questions for Review and Thought is retained while the number of such questions has decreased a bit to approximately 80 per chapter, still a sufficient number.

Goals

Our overarching goal in this *Principles* textbook, as in *Chemistry: The Molecular Science,* third edition, remains to involve science and engineering students in active study of what modern chemistry is, how it applies to a broad range of disciplines, and what effects it has on their own lives. Throughout this *Principles* book, a high level of rigor is maintained so that students in mainstream general chemistry courses for science majors and engineers will learn the chemical principles and concepts and develop the problem-solving skills essential to their future ability to use chemical ideas effectively. We have selected and carefully refined the book's many unique features in support of this goal.

More specifically, we intend that this textbook will help students develop:

- A broad overview of chemistry and chemical reactions;
- An understanding of the most important principles, concepts, and models used by chemists and those in chemistry-related fields;
- The ability to apply the facts, concepts, models, and principles of chemistry appropriately to new situations in chemistry, to other sciences and engineering, and to other disciplines;
- An appreciation of the many ways that chemistry affects the daily lives of all people, students included; and
- Motivation to study in ways that help all students achieve real learning that results in long-term retention of facts, concepts, and principles, and how to apply them.

Modern chemistry is inextricably entwined with many other disciplines and so we use, as appropriate, some organic chemistry and industrial chemistry as contexts to discuss applications in these areas in conjunction with the chemical principles on which they are based. This contextual approach serves to motivate students whose interests lie in related disciplines and also provides a more accurate picture of the multidisciplinary collaborations prevalent in contemporary chemical research and modern industrial chemistry.

Audience

We have written *Principles of Chemistry: The Molecular Science* with the intention for it to be used for mainstream, two-semester general chemistry courses designed for students who expect to pursue further study in science, engineering, or science-related disciplines. Those planning to major in chemistry, biochemistry, biological sciences, engineering, geological and environmental sciences, agricultural sciences, materials science, physics, and many related areas will benefit from this book and its approach. We assume that the students who use this book have a basic foundation in mathematics (algebra and geometry) and in general science. Almost all will also have had a chemistry course before coming to college.

Features

We strongly encourage students to understand chemical concepts and principles, and to learn to apply them to problem solving. We believe that such understanding is essential if students are to be able to use what they learn in subsequent courses and in their future careers. All too often we hear professors in courses for which general chemistry is a prerequisite complain that students have not retained what was taught. This book is unique in its thoughtful choice of features that address this issue and help students achieve long-term retention of the material.

Problem Solving

Problem solving is introduced in Chapter 1, and a framework is built there that is followed throughout the book. Each chapter contains many worked-out **Problem-Solving Examples**—a total of 205 in the book as a whole. Most consist of five parts: a Question (problem); an Answer, stated briefly; a Strategy and Explanation section that outlines one approach to solving the problem and provides significant help for students whose answer did not agree with ours; a Reasonable Answer Check section marked with a ☑ that indicates how a student could check whether a result is reasonable; and an associated **Problem-Solving Practice** that provides a similar question or questions, with answers appearing only in an appendix. We explicitly encourage students first to define the problem, develop a plan, and work out an answer without looking at either the Answer or the Explanation, and only then to compare their answer with ours. If their answer did not agree with ours, students are asked to repeat their work. Only then do we suggest that they look at the Strategy and Explanation, which is couched in conceptual as well as numeric terms so that it will improve students' understanding, not just their ability to answer an identical question on an exam. The Reasonable Answer Check section helps students learn how to use estimated results and other criteria to decide whether an answer is reasonable, an ability that will serve them well in the future. By providing similar practice problems that are answered in the back of the book, we encourage students to immediately consolidate their thinking and improve their ability to apply their new understanding to related problems.

Enhancing students' abilities to estimate results is the goal of the **Estimation** boxes found in many chapters. These are a unique feature of this book. Each Estimation box poses a problem that relates to the content of the chapter in which it appears and for which a rough calculation suffices. Students gain knowledge of various means of approximation, such as back-of-the-envelope calculations, and are encouraged to use diverse sources of information, such as encyclopedias, handbooks, and the Internet.

To further ensure that students do not merely memorize algorithmic solutions to specific problems, we provide 234 **Exercises,** which immediately follow introduction of new concepts *within* each chapter. Often the results that students obtain from a numeric Exercise provide insights into the concepts. Most Exercises are thought provoking and require that students apply conceptual thinking. Exercises that are more conceptual than mathematical are clearly designated.

Examples, Problem-Solving Examples and Practice Problems, Estimation boxes, and Exercises are all intended to stimulate active thinking and participation by students as they read the text and to help them hone their understanding of concepts. The grand total of more than 600 of these **active-learning items** is an exceptionally high number for this kind of textbook.

Conceptual Understanding

We believe that a sound conceptual foundation is the best means by which students can approach and solve a wide variety of real-world problems. This approach is supported by considerable evidence in the literature: Students learn better and retain what they learn longer when they have mastered fundamental concepts. To understand chemistry requires familiarity with at least three conceptual levels:

- **Macroscale** (laboratory and real-world phenomena)
- **Nanoscale** (models involving particles: atoms, molecules, and ions)
- **Symbolic** (chemical formulas and equations)

These three conceptual levels are explicitly defined and illustrated in Chapter 1. This chapter emphasizes the value of the chemist's unique nanoscale perspective on science and the world with a specific example of solving a real-world problem by application of chemical thinking—using fundamental chemical principles, a cost-effective method was developed to successfully decrease toxic levels of naturally occurring arsenic in drinking water sources in Bangladesh. This theme of conceptual understanding and its application to problems continues throughout the book. Many of the problem-solving features already mentioned have been specifically designed to support conceptual understanding.

Units are introduced on a need-to-know basis at the first point in the book where they contribute to the discussion. Units for length and mass are defined in Chapter 2, in conjunction with the discussion of the sizes and masses of atoms and subatomic particles. Energy units are defined in Chapter 6, where they are first needed to deal with kinetic and potential energy, work, and heat. In each case, introducing units at the time when the need for them can be made clear allows relationships that would otherwise appear pointless and arbitrary to support the development of closely related concepts.

Whenever possible, both in the text and in the end-of-chapter questions, **we use real chemical systems in examples and problems.** In the kinetics chapter, for example, the text and problems utilize real reactions and real data from which to determine reaction rates or reaction orders. Instead of $A + B \rightarrow C + D$, students will find $I^- + CH_3Br \rightarrow CH_3I + Br^-$. Some data have been taken from the recent research literature. The same approach is employed in many other chapters, where real chemical systems are used as examples.

Most important, we provide **clear, direct, thorough, and understandable explanations** of all topics, including those such as kinetics, thermodynamics, aqueous equilibria, and electrochemistry that many students find daunting. The methods of science and concepts such as chemical and physical properties; purification and separation; the relation of macroscale,

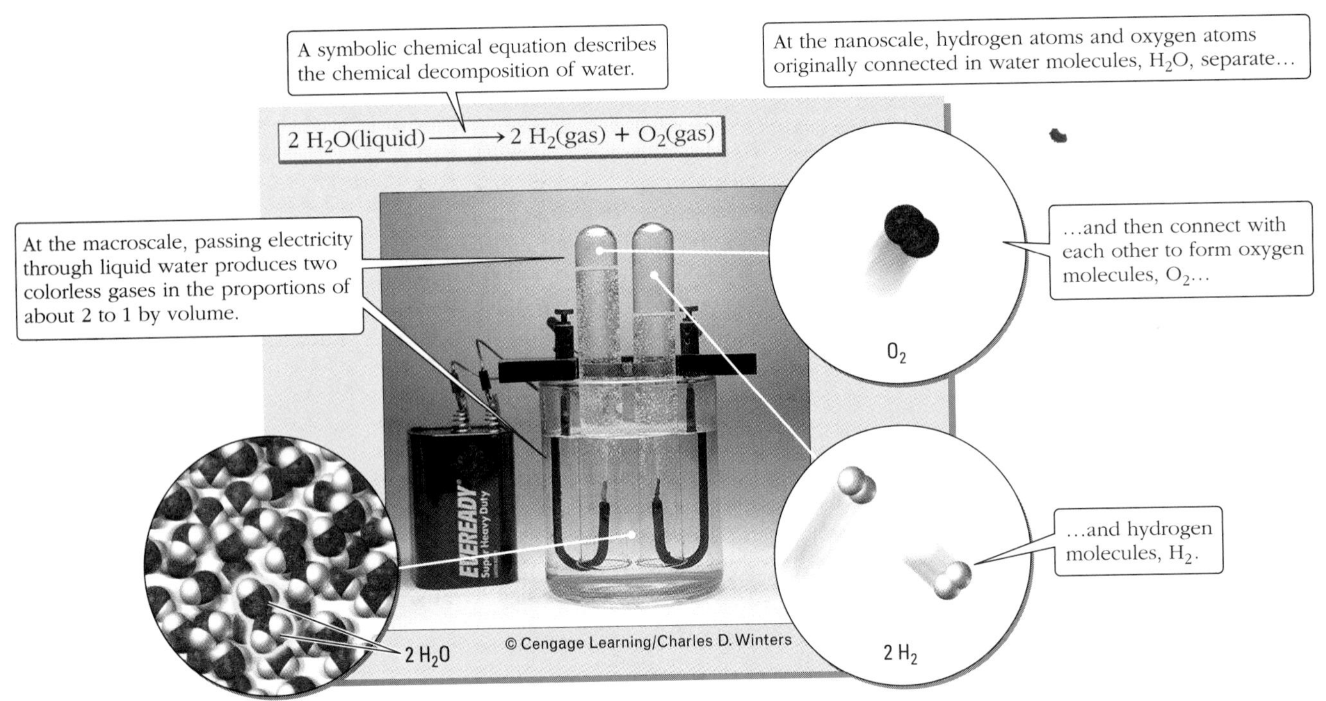

A symbolic chemical equation describes the chemical decomposition of water.
At the nanoscale, hydrogen atoms and oxygen atoms originally connected in water molecules, H_2O, separate...
$2\ H_2O(\text{liquid}) \longrightarrow 2\ H_2(\text{gas}) + O_2(\text{gas})$
At the macroscale, passing electricity through liquid water produces two colorless gases in the proportions of about 2 to 1 by volume.
...and then connect with each other to form oxygen molecules, O_2...
O_2
...and hydrogen molecules, H_2.
EVEREADY
Super Heavy Duty
$2\ H_2O$
© Cengage Learning/Charles D. Winters
$2\ H_2$

Figure 1 As seen in this figure from Chapter 1, figures throughout the book often include symbolic formulas and chemical equations along with nanoscale views of atoms, molecules, and ions for an effective illustration of how chemists think about the nanoscale world.

nanoscale, and symbolic representations; elements and compounds; and kinetic-molecular theory are introduced in Chapter 1 so that they can be used throughout the later discussion. Rather than being bogged down with discussions of units and nomenclature, students begin this book with an overview of what real chemistry is about—together with fundamental ideas that they will need to understand it.

Visualization for Understanding

Illustrations help students to visualize atoms and molecules and to make connections among macroscale observations, nanoscale models, and symbolic representations of chemistry. The **illustrations** in *Principles of Chemistry: The Molecular Science* are designed to engage today's visually oriented students. The success of the illustration program is exemplified by the fact that the first edition of our larger textbook, *Chemistry: The Molecular Science,* was awarded the coveted Talbot prize for visual excellence by the Society of Academic Authors. The illustrations in this *Principles* book continue that level of excellence. High-quality color photographs of substances and reactions, many by Charles D. Winters, are presented together with greatly magnified illustrations of the atoms, molecules, and/or ions involved created originally by J/B Woolsey Associates LLC. New drawings for this new book have been created by Graphic World Inc. Often these illustrations are accompanied by the symbolic formula for a substance or equation for a reaction, as in Figure 1. These **nanoscale views of atoms, molecules, and ions** have been generated with molecular modeling software and then combined by a skilled artist with the photographs and formulas or equations. Similar illustrations appear in exercises, examples, and end-of-chapter problems, thereby ensuring that students are tested on the ideas they represent. The result provides an exceptionally effective way for students to learn how chemists think about the nanoscale world of atoms, molecules, and ions.

Often the story is carried solely by an illustration and accompanying text that points out the most important parts of the figure. An example is the visual story of molecular structure shown in Figure 2. In other cases, text in balloons explains the operation of instruments, apparatus, and experiments; clarifies the development of a mathematical derivation; or points out salient features of graphs or nanoscale pictures. Throughout the book visual interest is high, and visualizations of many kinds are used to support conceptual development.

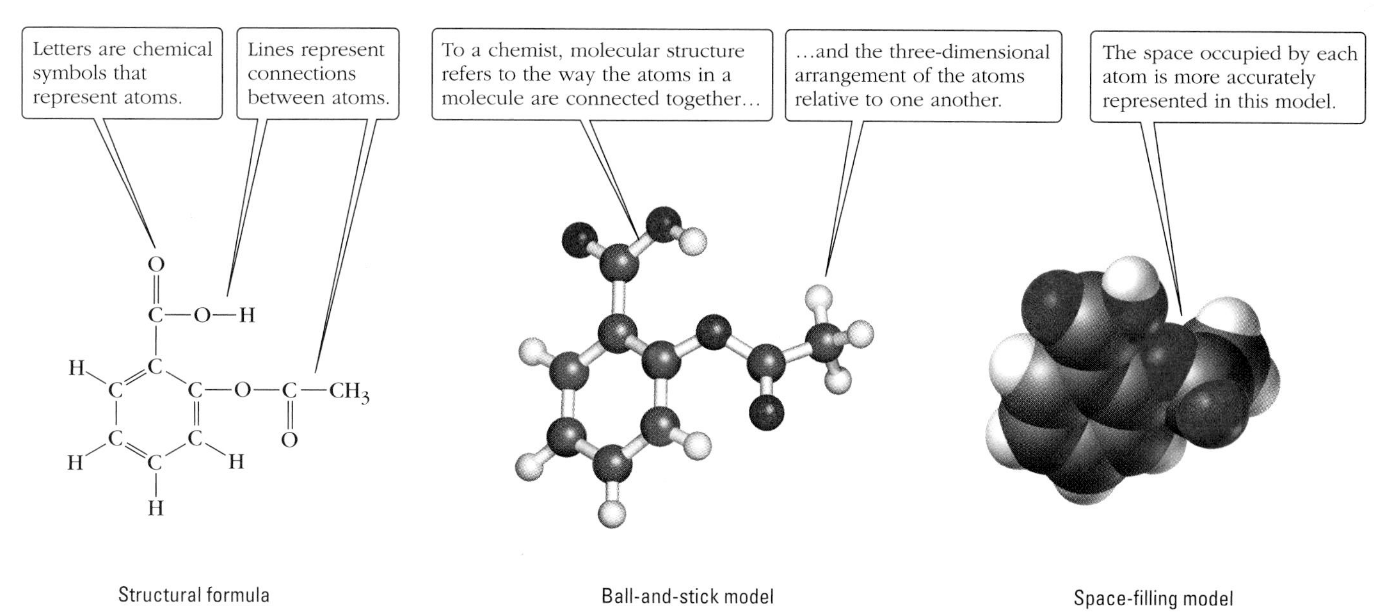

Figure 2 Thoughout the book illustrations like this are often used to tell the story through art and accompanying text balloons.

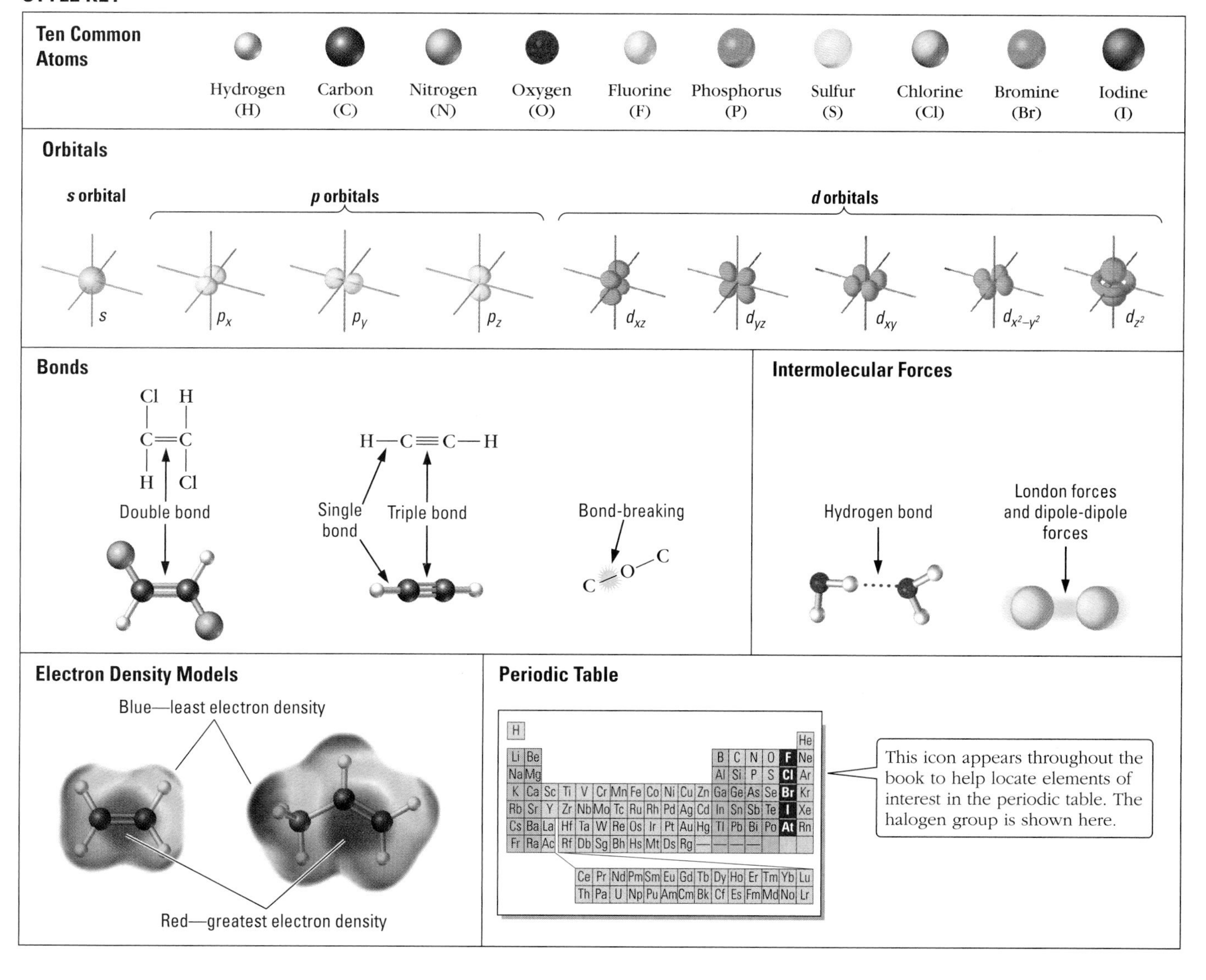

Integrated Media

Advances in learning technologies, such as web-based tools, have proven effective in assisting general chemistry students to gain a deeper understanding of major concepts and principles. In that regard, we integrate into this *Principles* textbook's presentations several digital-based study tools:

- Active Figures on the companion website provide an animated version of designated in-text figures that illustrate important concepts accompanied by consolidating exercises. Active Figures are indicated by an annotation next to the in-text figures.
- Estimation boxes from the text are expanded into modules on the companion website to allow expanded development of approximation skills.
- Many chapter-end Questions for Review and Thought, those indicated by the ■ icon, are assignable in a parameterized version of the OWL homework system.
- To assist further study, the *In Closing* section at the end of each chapter is a listing of benchmarks correlated with questions assignable in OWL.
- For users of OWL, an integrated e-Book of the text is available to allow for easy review of relevant text sections when attempting a problem.

Interdisciplinary Applications

Whenever possible we directly include practical applications, especially those applications that students will revisit when they study other natural science and engineering disciplines. Applications have been integrated where they are relevant, rather than being relegated to isolated chapters and separated from the principles and facts on which they are based. We intend that students should see that chemistry is a lively, relevant subject that is fundamental to a broad range of disciplines and that it can help solve important, real-world problems.

Simple organic compounds provide excellent examples that can be used to advantage in areas such as stoichiometry and molecular formulas. To take advantage of this synergy, we incorporate, where appropriate, a limited number of basic organic structure and bonding concepts into the text, although not as extensively as in the third edition of *Chemistry: The Molecular Science*. In this *Principles* book, the discussion of the properties of molecular compounds, for example, presents the key concepts of structural formulas, functional groups, and structural isomers due to their fundamental importance in structure/property relationships. A limited number of functional groups (alcohols, aldehydes, ketones, carboxylic acids) and *cis/trans* isomerism are introduced. A large percentage of students in most general chemistry courses are planning careers in biological or medical areas that make constant use of biochemistry. Many of the chemical principles that students encounter in general chemistry are directly applicable to biochemistry. After all, Nature doesn't use one set of chemical principles for general chemistry and different ones for organic chemistry and biochemistry; they are all the same. For this reason, we have chosen to show, in carefully selected examples, how general chemistry principles underlie selected fundamental biochemical phenomena.

Other Features

Chemistry in the News boxes bring the latest discoveries in chemistry and applications of chemistry to the attention of students, making clear that chemistry is continually changing and developing—it is not merely a static compendium of items to memorize. Like any other human pursuit, chemistry depends on people, so we include in nearly every chapter a portrait of a scientist, **biographical sketches** of men and women who have advanced our understanding or applied chemistry imaginatively to important problems.

End-of-Chapter Study Aids

At the end of each chapter, students will find many ways to test their understanding and to consolidate their learning. **Summary Problems,** available at the student companion website, bring together concepts and problem-solving skills from throughout each chapter (except for Chapter 1). Students are challenged in these problems to answer a multifaceted question that builds on and is relevant to the chapter's content. The **In Closing** section highlights the learning goals for the chapter, provides references to the sections in the chapter that address each goal, and includes references to end-of-chapter questions available in the OWL homework system. **Key Terms** are listed, with references to the sections where they are defined.

A broad range of chapter-end **Questions for Review and Thought** is provided to serve as a basis for homework or in-class problem solving. Among these initially are **Review Questions,** ones not answered in the back of the book, that test vocabulary and simple concepts. Next are **Topical Questions,** those keyed to the major topics in the chapter and listed under headings that correspond with each section in the chapter. These questions are often accompanied by photographs, graphs, and diagrams that make the situations described more concrete and realistic. Usually a question that is answered at the end of the book is paired with a similar one that is not. There are also many **General Questions** that are not explicitly keyed by section headings. Often these General Questions require students to integrate several principles and concepts. **Applying Concepts** includes questions specifically designed to test conceptual learning. Many of these questions include diagrams of atoms, molecules, or ions and require students to relate macroscopic observations, nanoscale models, and symbolic formulas and equations. **More Challenging Questions** require students to apply more thought and to better integrate multiple concepts than do typical end-of-chapter questions. **Conceptual Challenge Problems,** most of which were written by H. Graden Kirksey of the University of Memphis, are especially important in helping students assess and improve their conceptual thinking ability. Designed for group work, the Conceptual Challenge Problems are rigorous and thought provoking. Much effective learning can be induced by dividing a class into groups of three or four students and then assigning these groups to work collaboratively on these problems.

Organization

The order of chapters reflects the most common division of content between the first and second semesters of a typical general chemistry course. The first few chapters briefly review basic material that most students should have encountered in high school. Next, the book develops the ideas of chemical reactions, stoichiometry, and energy transfers during reactions. We then deal with the electronic structure of atoms, bonding and molecular structures, and the way in which structure affects properties. To finish the first-semester course, there are adjacent chapters on gases and on liquids and solids.

The next chapters deal with kinetics and chemical equilibrium principles that establish fundamental understanding of how fast reactions will occur and what concentrations of reactants and products will remain when equilibrium is reached. These fundamental concepts are then applied in chapters related to solutions, as well as to acid-base equilibria, buffers, and solubility equilibria in aqueous solutions. A chapter on thermodynamics and Gibbs free energy is followed by one on electrochemistry, which makes use of thermodynamic ideas. The book finishes with a chapter on nuclear chemistry.

To help students connect chemical ideas that are closely related but are presented in different chapters, we have included **numerous cross references (indicated by the ⇐ symbol).** The cross references will help students link a concept being developed in the chapter they are currently reading with an earlier, related principle or fact. They also provide many opportunities for students to review material encountered earlier.

Principles of Chemistry: The Molecular Science can be divided into several sections, each of which treats an important aspect of chemistry:

Fundamental Ideas of Chemistry

Chapter 1, The Nature of Chemistry, is designed to capture students' interest from the start by concentrating on chemistry (not on math, units, and significant figures, which are treated comprehensively in an appendix) and how modern science works. The chapter begins with the question, Why Care About Chemistry? and then describes the use of chemical principles to develop a successful, low-cost method that reduces toxic levels of naturally occurring arsenic in well water in Bangladesh. The chapter also introduces major concepts that bear on all of chemistry, emphasizing the three conceptual levels with which students must be facile to understand chemical phenomena: macroscale, nanoscale, and symbolic.

Chapter 2, Atoms and Elements, introduces units and dimensional analysis on a need-to-know basis in the context of the sizes of atoms. It concentrates on thorough, understandable treatment of the concepts of atomic structure, atomic weight, and moles of elements, making the connections among them clear. It concludes by introducing the periodic table and highlighting the periodicity of properties of elements.

Chapter 3, Chemical Compounds, distinguishes ionic compounds from molecular compounds, and illustrates molecular compounds with the simplest alkanes. The important theme of structure is reinforced by showing several carefully selected examples of structural isomerism in short-chain alkanes. Charges of monoatomic ions are related to the periodic table. Molar masses of compounds and determining formulas fit logically into the chapter's structure.

Chemical Reactions

Chapter 4, Quantities of Reactants and Products, begins a three-chapter sequence that treats chemical reactions qualitatively and quantitatively. Students learn how to balance equations and to use four typical inorganic reaction patterns to predict reaction products. A single stepwise method is provided for solving all stoichiometry problems, and 15 Problem-Solving Examples demonstrate a broad range of stoichiometry calculations.

Chapter 5, Chemical Reactions, has a strong descriptive chemistry focus, dealing with exchange reactions, acid-base reactions, and oxidation-reduction reactions in aqueous solutions. It includes real-world occurrences of each type of reaction. Students learn how to recognize a redox reaction from the chemical nature of the reactants (not just by using oxidation numbers) and how to do aqueous acid-base titration calculations.

Chapter 6, Energy and Chemical Reactions, begins with a thorough and straightforward introduction to forms of energy, conservation of energy, heat and work, system and surroundings, and exothermic and endothermic processes. Carefully designed illustrations help students to understand thermodynamic principles. Heat capacity, heats of changes of state, and heats of reactions are clearly explained, as are calorimetry, Hess's Law, and standard enthalpy changes.

Electrons, Bonding, and Structure

Chapter 7, Electron Configurations and the Periodic Table, introduces atomic spectra, quantum theory, and quantum numbers, using color-coded illustrations to visualize the different energy levels of *s, p, d,* and *f* orbitals. The *s*-, *p*-, *d*-, and *f*-block locations in the periodic table are used to predict electron configurations. Periodic trends in atomic and ionic radii, ionization energy, and electron affinity are presented. The chapter ties together these concepts with a discussion and application of the Born-Haber cycle.

Chapter 8, Covalent Bonding, provides simple stepwise guidelines for writing Lewis structures, with many examples of how to use them. The role of single and multiple bonds in hydrocarbons is smoothly integrated with the introduction to covalent bonding. Bond properties—bond length, bond energy, and electronegativity and bond polarity—are described. The discussion of polar bonds is enhanced by illustrations using molecular models that show variations in electron density. Formal charge is discussed as a method to assess resonance structures. Molecular orbital theory is introduced as well.

Chapter 9, Molecular Structures, provides a thorough presentation of valence-shell electron-pair repulsion (VSEPR) theory and orbital hybridization. Molecular geometry and polarity are extensively illustrated with computer-generated models, and the relation of molecular structure, polarity, and hydrogen bonding to attractions among molecules is clearly developed and illustrated in solved problems. The importance of noncovalent interactions and forces between molecules is emphasized.

States of Matter

Chapter 10, Gases and the Atmosphere, uses kinetic-molecular theory to interpret the behavior of gases and then describes each of the individual gas laws. Mathematical problem solving uses the ideal gas law or the combined gas law, and many conceptual Exercises throughout the chapter emphasize qualitative understanding of gas properties. Gas stoichiometry is presented in a uniquely concise and clear manner. Then, the properties of gases are applied to the greenhouse effect and enhanced global warming.

Chapter 11, Liquids, Solids, and Materials, begins by discussing the properties of liquids and the nature of phase changes. The unique and vitally important properties of water are treated thoroughly. The principles of crystal structure are introduced using cubic unit cells only. The fact that much current chemical research involves materials is illustrated by the discussions of network solids, metals, semiconductors, and insulators.

Reactions: How Fast and How Far?

Chapter 12, Chemical Kinetics: Rates of Reactions, presents one of the most difficult topics in the course with extraordinary clarity. Defining reaction rate, finding rate laws from initial rates and integrated rate laws, and using the Arrhenius equation are thoroughly developed. How molecular changes during unimolecular and bimolecular elementary reactions relate to activation energy initiates the treatment of reaction mechanisms. Catalysis is shown to involve changing a reaction mechanism. Both enzymes and industrial catalysts are described using concepts developed earlier in the chapter.

Chapter 13, Chemical Equilibrium, emphasizes equally a qualitative understanding of the nature of equilibrium and the solving of related mathematical problems. That equilibrium results from equal but opposite reaction rates is fully explained. Both Le Chatelier's principle and the reaction quotient, Q, are used to predict shifts in equilibria. A unique section on equilibrium at the nanoscale introduces briefly and qualitatively how enthalpy changes and entropy changes affect equilibria. Optimizing the yield of the Haber-Bosch ammonia synthesis elegantly illustrates how kinetics, equilibrium, and enthalpy and entropy changes control the outcome of a chemical reaction.

Reactions in Aqueous Solution

Chapter 14, The Chemistry of Solutes and Solutions, builds on principles previously introduced, showing the influence of enthalpy and entropy on solution properties. Understanding of solubility, Henry's Law, concentration units (including ppm and ppb), and colligative properties (including osmosis) is reinforced by numerous Exercises and 13 Problem-Solving Examples.

Chapter 15, Acids and Bases, concentrates initially on the Brønsted-Lowry acid-base concept, clearly delineating proton transfers using color-coded illustrations and molecular models. In addition to a full exploration of pH and the meaning of and calculations associated with K_a and K_b, acid strength is related to molecular structure. Student interest is enhanced by the use of common acids and bases. The hydrolysis of salts is described together with related problems. Lewis acids and bases are defined and illustrated using practical examples.

Chapter 16, Additional Aqueous Equilibria, extends the treatment of acid-base and solubility equilibria to buffers, titration, and precipitation. The Henderson-Hasselbalch equation is applied to buffer pH calculations. Calculations of points on titration curves are shown, and the interpretation of several types of titration curves provides conceptual understanding. The final section deals with the various factors that affect solubility (pH, common ions, complex ions, and amphoterism), and with selective precipitation.

Thermodynamics, Electrochemistry, and Nuclear Chemistry

Chapter 17, Thermodynamics: Directionality of Chemical Reactions, explores the nature and significance of entropy, both qualitatively and quantitatively. The signs of Gibbs free energy changes are related to the easily understood classification of reactions as reactant- or product-favored, with the discussion deliberately avoiding the often-misinterpreted term "spontaneous." The thermodynamic significance of coupling one reaction with another is illustrated using industrial examples. Energy conservation is defined thermodynamically. A closing section reinforces the important distinction between thermodynamic and kinetic stability.

Chapter 18, Electrochemistry and Its Applications, defines redox reactions and uses half-reactions to balance redox equations. Electrochemical cells, cell voltage, standard cell potentials, the relation of cell potential to Gibbs free energy, and the effect of concentrations on cell potential are all explored. Practical applications include batteries, fuel cells, electrolysis, and corrosion.

Chapter 19, Nuclear Chemistry, deals with radioactivity, nuclear reactions, nuclear stability, and rates of nuclear decay reactions. Also provided are a thorough description of nuclear fission and nuclear fusion, and a discussion of radiation units, background radiation, and nuclear radiation effects.

Supporting Materials

For the Instructor

Supporting instructor materials are available to qualified adopters. Please consult your local Cengage Learning Brooks/Cole representative for details.

Visit **www.cengage.com/chemistry/moore** to

- See samples of materials
- Request a desk copy
- Locate your local representative
- Download electronic files of the *Instructor's Solutions Manual,* the *Test Bank,* and other helpful materials for instructors and students

PowerLecture™

PowerLecture with ExamView® Instructor's CD-ROM. ISBN-10: 0-495-39155-7, ISBN-13: 978-0-495-39155-5

PowerLecture is a one-stop digital library and presentation tool that includes

- Prepared Microsoft® PowerPoint® Lecture Slides authored by Stephen C. Foster of Mississippi State University that cover all key points from the text in a convenient format that you can enhance with your own materials or with additional interactive video and animations on the CD-ROM for personalized, media-enhanced lectures.
- Image libraries in PowerPoint and JPEG formats that contain electronic files for all text art, most photographs, and all numbered tables in the text. These files can be used to create your own transparencies or PowerPoint lectures.
- Electronic files for the complete *Instructor's Solutions Manual* and *Test Bank.*
- Sample chapters from the *Student Solutions Manual* and *Study Guide.*
- ExamView testing software, with all the test items from the *Test Bank* in electronic format, enables you to create customized tests of up to 250 items in print or online.

Instructor's Solutions Manual by Judy L. Ozment, Pennsylvania State University.
Contains fully worked-out solutions to all end-of-chapter questions and Conceptual Challenge Problems. Solutions match the problem-solving strategies used in the text. Available for download on the Faculty Companion Site at **www.cengage.com/chemistry/moore** and on the instructor's PowerLecture CD-ROM.

OWL: Online Web-based Learning by Roberta Day and Beatrice Botch of the University of Massachusetts, Amherst, and William Vining of the State University of New York at Oneonta.
Instant Access to OWL (two semesters) ISBN-10: 0-495-05099-7, ISBN-13: 978-0-495-05099-5
Instant Access to OWL e-Book (two semesters) ISBN-10: 0-495-39160-3, ISBN-13: 978-0-495-39160-9
Developed at the University of Massachusetts, Amherst, and class tested by tens of thousands of chemistry students, OWL is a fully customizable and flexible web-based learning system. OWL supports mastery learning and offers numerical, chemical, and contextual parameterization to produce thousands of problems correlated to this text. The OWL system also features a database of simulations, tutorials, and exercises, as well as end-of-chapter problems from the text. With OWL, you get the most widely used online learning system available for chemistry with unsurpassed reliability and dedicated training and support. And now OWL for General Chemistry includes Go Chemistry™—mini video lectures covering key chemistry concepts that students can view onscreen or download to their portable video player to study on the go! Within OWL, students can download five free modules! For *Principles of Chemistry: The Molecular Science,* first edition, OWL includes parameterized end-of-chapter questions from the text (marked in the text with ■) and tutorials based on the Estimation boxes in the text.

The optional **e-Book in OWL** includes the complete electronic version of the text, fully integrated and linked to OWL homework problems. Most e-Books in OWL are interactive and offer highlighting, notetaking, and bookmarking features that can all be saved. To view an OWL demo and for more information, visit **www.cengage.com/owl** or contact your Cengage Learning Brooks/Cole representative.

Test Bank by Marcia Gillette, Indiana University Kokomo, and Jason Holland, University of Central Missouri
Containing more than 1000 questions carefully matched to the corresponding text sections, the *Test Bank* is available as digital files and in ExamView format on the instructor's PowerLecture CD-ROM.

Faculty Companion Website. This site contains the *Instructor's Manual* and other tools. Accessible from **www.cengage.com/moore.**

Cengage Learning Custom Solutions develops personalized text solutions to meet your course needs. Match your learning materials to your syllabus and create the perfect learning solution—your customized text will contain the same thought-provoking, scientifically sound content, superior authorship, and stunning art that you've come to expect from Cengage Learning Brooks/Cole texts, yet in a more flexible format. Visit **www.cengage.com/custom.com** to start building your book today.

Chemistry Comes Alive! Video Series. A broad range of videos and animations suitable for use in lecture presentations, for independent study, or for incorporation into the instructor's own tutorials is available separately from JCE Software on eight CD-ROMs (**http://www.jce.divched.org/JCESoft/Programs/VideoCD/CCA/index.html**) or online via JCE Web Software (**http://www.jce.divched.org/JCESoft/jcesoftSubscriber.html**).

***JCE* QBank** (available separately from the *Journal of Chemical Education;* see **http://www.jce.divched.org/JCEDLib/QBank/index.html**).
Contains more than 3500 homework and quiz questions suitable for delivery via WebCT, Desire2Learn, or Moodle course management systems, hundreds of ConcepTest questions that can be used with "clickers" to make lectures more interactive, and a collection of conceptual questions together with a discussion of how to write conceptual questions. Available to all *JCE* subscribers.

For the Student

Visit the *Principles of Chemistry: The Molecular Science* website at **www.cengage.com/chemistry/moore** to see samples of select student supplements. Students can purchase any Cengage Learning product at your local college store or at our preferred online store **www.ichapters.com.**

OWL for General Chemistry.
See the above description in the instructor support materials section.

OWL Quick Prep for General Chemistry (90 Day Access Code) by Beatrice Botch and Roberta Day, University of Massachusetts, Amherst.
ISBN-10: 0-495-11042-6, ISBN-13: 978-0-495-11042-2
Quick Prep is a self-paced, online short course that helps students succeed in general chemistry. Students who completed Quick Prep through an organized class or self-study averaged almost a full letter grade higher in their subsequent general chemistry courses than those who did not. Intended to be taken prior to the start of the semester, Quick Prep is appropriate for both underprepared students and for students who seek a review of basic skills and concepts. Quick Prep is an approximately 20-hour commitment delivered through the online learning system OWL with no textbook required and can be completed at any time on the student's schedule. To view an OWL Quick Prep demonstration and for more information, visit **www.cengage.com/chemistry/quickprep** or contact your Cengage Learning Brooks/Cole representative. Instant Access Codes are available at **www.ichapters.com.**

Go Chemistry™ for General Chemistry.
ISBN-10: 0-495-38228-0, ISBN-13: 978-0-495-38228-7

Go Chemistry™ is a set of 27 easy-to-use essential videos that can be downloaded to your video iPod, iPhone, or portable video player—ideal for the student on the go! Developed by textbook author John Kotz, these new electronic tools are designed to help students quickly review essential chemistry topics. Mini video lectures include animations and problems for a quick summary of key concepts. Selected Go Chemistry modules have e-flashcards to briefly introduce a key concept and then test student understanding of the basics with a series of questions. Go Chemistry also plays on iTunes, Windows Media Player, and QuickTime. To purchase Go Chemistry, enter ISBN 0-495-38228-0 at **www.ichapters.com.**

Student Solutions Manual by Judy L. Ozment, Pennsylvania State University.
ISBN-10 0-495-39158-1, ISBN-13: 978-0-495-39158-6
Contains fully worked-out solutions to end-of-chapter questions that have blue, boldfaced numbers. Solutions match the problem-solving strategies used in the main text. A sample is available on the Student Companion Website at **www.cengage.com/chemistry/moore.**

Study Guide by Michael J. Sanger, Middle Tennessee State University.
ISBN-10: 0-495-39118-2, ISBN-13: 978-0-495-39157-9
Contains learning tools such as brief notes on chapter sections with examples, reviews of key terms, and practice tests with answers provided. A sample is available on the Student Companion Website at **www.cengage.com/chemistry/moore.**

Student Companion Website.
Accessible from **www.cengage.com/chemistry/moore,** this site provides online study tools including Active Figure interactive versions of key pieces of art from the text, interactive modules based on the text's Estimation boxes, and Summary Problems—excellent exam preparation aides—that bring together concepts and problem-solving skills from throughout each chapter.

Survival Guide for General Chemistry with Math Review and Proficiency Questions, Second Edition by Charles H. Atwood, University of Georgia.
ISBN-10: 0-495-38751-7, ISBN-13 978-0-495-38751-0
Intended to help you practice for exams, this survival guide shows you how to solve difficult problems by dissecting them into manageable chunks. The guide includes three levels of proficiency questions—A, B, and minimal—to quickly build confidence as you master the knowledge you need to succeed in your course.

Essential Algebra for Chemistry Students, Second Edition by David W. Ball, Cleveland State University.
ISBN-10: 0-495-01327-7, ISBN-13 978-0-495-01327-3
This short book is intended for students who lack confidence and/or competency in their essential mathematics skills necessary to succeed in general chemistry. Each chapter focuses on a specific type of skill and has worked-out examples to show how these skills translate to chemical problem solving. Includes references to OWL, our web-based tutorial program, offering students access to online algebra skills exercises.

General Chemistry: Guided Explorations by David Hanson, State University of New York at Stony Brook.
ISBN-10: 0-195-11599-1, ISBN-13: 978-0-495-11599-1
This student workbook is designed to support Process Oriented Guided Inquiry Learning (POGIL) with activities that promote a student-focused active classroom. It is an excellent ancillary to *Principles of Chemistry: The Molecular Science* or any other general chemistry text.

General Chemistry Collection CD-ROM (available separately from *JCE Software;* see **http://jce.divched.org/JCESoft/Programs/GCC/index.html**).
Contains many software programs, animations, and videos that correlate with the content of this book. Arrangements can be made to make this item available to students at very low cost. Call (800) 991-5534 for more information about *JCE* products.

For the Laboratory

Laboratory Handbook for General Chemistry, Third Edition by Conrad L. Stanitski, Franklin and Marshall College, Norman E. Griswold, Nebraska Wesleyan College, H. A. Neidig, Lebanon Valley College, and James N. Spencer, Franklin and Marshall College.
ISBN-10: 0-495-01890-2, ISBN-13: 978-0-495-01890-2

This "how-to" guide containing specific information about the basic equipment, techniques, and operations necessary for successful laboratory experiments helps students perform their laboratory work more effectively, efficiently, and safely. The third edition includes video demonstrations of a number of common laboratory techniques.

Cengage Learning Brooks/Cole Lab Manuals.
We offer a variety of printed manuals to meet all your general chemistry laboratory needs. Instructors can visit the chemistry site at **www.cengage.com/chemistry** for a full listing and description of these laboratory manuals, laboratory notebooks, and laboratory handbooks. All Cengage Learning laboratory manuals can be customized for your specific needs.

Signature Labs . . . for the customized laboratory.
Signature Labs is Cengage Learning's digital library of tried-and-true labs that help you take the guesswork out of running your chemistry laboratory. Select just the experiments you want from hundreds of options and approaches. Provide your students with only the experiments they will conduct and know you will get the results you seek. Visit **www.signaturelabs.com** to begin building your manual today.

ChemPages Laboratory CD-ROM (available separately from *JCE Software* or as part of *JCE Web Software;* see **http://jce.divched.org/JCESoft/Programs/CPL/index.html** or **http://www.jce.divched.org/JCESoft/jcesoftSubscriber.html**).
A collection of videos with voiceover and text showing how to perform the most common laboratory techniques used by students in first-year chemistry courses.

Note: Unless otherwise noted, the website domain names (URLs) provided here are not published by Cengage Learning Brooks/Cole and the Publisher can accept no responsibility or liability for these sites' content. Because of the dynamic nature of the Internet, Cengage Learning Brooks/Cole cannot in any case guarantee the continued availability of third-party websites.

Reviewers

Reviewers play a critical role in the preparation of any textbook. The individuals listed below helped to shape this textbook into one that is not merely accurate and up to date, but a valuable practical resource for teaching and testing students.

Reviewers for *Principles of Chemistry: The Molecular Science,* First Edition

Sheila W. Armentrout, *East Tennessee State University*
Michael E. Clay, *College of San Mateo*
Renée S. Cole, *University of Central Missouri*
S. Michael Condren, *Christian Brothers University*
Vincent Giannamore, *Nicholls State University*
Marcia L. Gillette, *Indiana University Kokomo*
Pamela J. Hagrman, *Onondaga Community College*
Lynn G. Hartshorn, *University of Saint Thomas*
John Hawes, *Lake City Community College*
Andrew B. Helms, *Jacksonville State University*
Booker Juma, *Fayetteville State University*
Dennis McMinn, *Gonzaga University*
Ray Mohseni, *East Tennessee State University*
Al Nichols, *Jacksonville State University*
Stephen J. Paddison, *University of Alabama-Huntsville*
Morgan Ponder, *Samford University*
William Quintana, *New Mexico State University*
D. Paul Rillema, *Wichita State University*
Jason E. Ritchie, *University of Mississippi*
René Rodriguez, *Idaho State University*
Amar S. Tung, *Lincoln University*
Lin Zhu, *Indiana University-Purdue University at Indianapolis*

We also thank these people who were dedicated to checking the accuracy of the text and art: Marcia L. Gillette, Indiana University Kokomo, conducted a continuity review, and René Rodriguez, Idaho State University, conducted an accuracy review.

Acknowledgments

Although at times it may seem that way to an author, writing and producing a textbook is not a solitary endeavor. We have had very high quality assistance in all of the many aspects in bringing this book to its final form. For that we extend heartfelt thanks to everyone who contributed to the project.

Principles of Chemistry: The Molecular Science is the second book this author team has had the pleasure of working on with Lisa Lockwood, senior acquisitions editor. We undertook this project at Lisa's suggestion, and she has supported and overseen the entire project with zest and good humor. The result is an excellent, rigorous, mainstream general chemistry textbook.

Peter McGahey, in his role of senior development editor, is someone whose many skills give added dimension to the term "multitasking." He provided stellar advice and active support throughout the project. Available even when on "vacation," Peter went the extra mile when things needed to be provided, sometimes with very short response time. He assembled an excellent group of expert reviewers, obtained reviews from them in timely fashion, and provided invaluable feedback based on their comments. His calm, caring, and thoughtful demeanor provided a seamless connection between the authors and the many other members of the production staff. Thanks, Peter, again!

Jennifer Risden, content project manager, made the invaluable contribution of helping to keep the authors on track and providing timely queries and suggestions regarding editing, layout, and appearance of the book. Lisa Weber served as senior media editor; her ability to organize all the multimedia elements was essential. Liz Woods, assistant editor, kept materials flowing to the authors in a timely manner. Brandi Kirksey, associate development editor, has ably handled all of the ancillary print materials.

To succeed in reaching and influencing large numbers of students, a book such as this one needs to be adopted and used. Nicole Hamm, marketing manager, directs the marketing and sales programs, and many local sales representatives throughout the country have helped and will help get this book to students who can benefit from it.

Authors (and users) care about what is inside a book and also how the book looks. The illustration program for this book is exceptional, carefully done to support student learning in every possible way. The many photographs of Charles D. Winters of Oneonta, New York, provide students with close-up views of chemistry in action. We thank Charlie again for his artist's eye in bringing to life what could be rather mundane phenomena through someone else's lens. Our thanks go to Sue C. Howard, photo researcher, for finding new photos that complemented the illustration program, and to John Walker, art director.

Copy editing, layout, and production of the book were done by Graphic World Publishing Services under the superb direction of Julie Ninnis, production editor. Julie's calm efficiency in overseeing all the manifold aspects of production helped to ensure that this book is of the highest possible quality. We appreciate the work of Maryalice Ditzler as copy editor for her close reading of the manuscript. Our sincere thanks to all of the Graphic World Inc. staff who contributed to this project.

The active-learning, conceptual approach of this book has been greatly influenced by the systemic curriculum enhancement project, *Establishing New Traditions: Revitalizing the Curriculum,* funded by the National Science Foundation, Directorate for Education and Human Resources, Division of Undergraduate Education, grant DUE-9455928.

We also thank the many teachers, colleagues, students, and others who have contributed to our knowledge of chemistry and helped us devise better ways to help others learn it. Collectively, the authors of this book have many years of experience teaching and learning, and we have tried to incorporate as much of that as possible into our presentation of chemistry.

Finally, we thank our families and friends who have supported all of our efforts—and who can reasonably expect more of our time and attention now that this project is complete.

We hope that using this book conveys to both faculty and students the excitement that chemistry offers and results in a lively and productive experience for them.

John W. Moore
Madison, Wisconsin

Conrad L. Stanitski
Lancaster, Pennsylvania

Peter C. Jurs
State College, Pennsylvania

1

The Nature of Chemistry

Rafiqur Rahman/Reuters/Landov

Chemistry can help to identify and solve many problems faced by people throughout the world. The photo shows a Bangladeshi boy collecting drinking water from a well at a village near Jessore, 275 kilometers (172 miles) southwest of Dhaka, the capital of Bangladesh. The pump is painted green to indicate that the water is free of contamination by arsenic and is therefore safe to drink. Analytical tools developed by chemists allowed the problem of arsenic contamination to be identified, and chemistry also contributed substantially to finding methods for removing arsenic from contaminated water. This is only one of thousands of examples of chemistry's crucial contributions to modern society.

Welcome to the world of chemical science! This chapter describes how modern chemical research is done and how it can be applied to questions and problems that affect our daily lives. It also provides an overview of the methods of science and the fundamental ideas of chemistry. These ideas are extremely important and very powerful. They will be applied over and over throughout your study of chemistry and of many other sciences.

1.1 Why Care About Chemistry?

Very human accounts of how fascinating—even romantic—chemistry can be are provided by Primo Levi in his autobiography, *The Periodic Table* (New York: Schocken Books, 1984), and by Oliver Sacks in *Uncle Tungsten: Memories of a Chemical Boyhood* (New York: Knopf, 2001). Levi was sentenced to a death camp during World War II but survived because the Nazis found his chemistry skills useful; those same skills made him a special kind of writer. Sacks describes how his mother and other relatives encouraged his interest in metals, diamonds, magnets, medicines, and other chemicals and how he learned that "science is a territory of freedom and friendship in the midst of tyranny and hatred."

Atoms are the extremely small particles that are the building blocks of all matter (Section 1.9). In molecules, atoms combine to give the smallest particles with the properties of a particular substance (Section 1.10).

Why study chemistry? There are many good reasons. **Chemistry** is the science of matter and its transformations from one form to another. **Matter** is anything that has mass and occupies space. Consequently, chemistry has an enormous impact on our daily lives, on other sciences, and even on areas as diverse as art, music, cooking, and recreation. Chemical transformations happen all the time, everywhere. Chemistry is intimately involved in the air we breathe and the reasons we need to breathe it; in purifying the water we drink; in growing, cooking, and digesting the food we eat; and in the discovery and production of medicines to help maintain health. Chemists continually provide new ways of transforming matter into different forms with useful properties. Examples include the plastic disks used in CD and DVD players; the microchips and batteries in cell phones or computers; and the steel, aluminum, rubber, plastic, and other components of automobiles.

Chemists are people who are fascinated by matter and its transformations—as you are likely to be after seeing and experiencing chemistry in action. Chemists have a unique and spectacularly successful way of thinking about and interpreting the material world around them—an atomic and molecular perspective. Knowledge and understanding of chemistry are crucial in biology, pharmacology, medicine, geology, materials science, many branches of engineering, and other sciences. Modern research is often done by teams of scientists whose members represent several of these different disciplines. In such teams, ability to communicate and collaborate is just as important as knowledge in a single field. Studying chemistry can help you learn how chemists think about the world and solve problems, which in turn can lead to effective collaborations. Such knowledge will be useful in many career paths and will help you become a better-informed citizen in a world that is becoming technologically more and more complex—and interesting.

1.2 Cleaning Drinking Water

"The whole of science is nothing more than a refinement of everyday thinking." —Albert Einstein

How modern science works and why the chemist's unique perspective is so valuable can be seen through an example. (At this point you need not fully understand the science, so don't worry if some words or ideas are unfamiliar.) From many possibilities, we have chosen to discuss an issue that directly affects more than a hundred million people worldwide—the toxic effects of arsenic in drinking water. The story of what has been called the largest mass poisoning in history has unfolded in the delta of the Ganges River in Bangladesh and northeastern India during the past 30 years.

The element arsenic has been known for more than 800 years, and its compounds have been known much longer than that. Orpiment, As_2S_3, a bright yellow mineral known to the ancient Greeks as *arsenikon,* probably gave the element its name. Ancient Chinese writings describe the toxicity of arsenic compounds and their use as pesticides. They also indicate that those who worked with the arsenical pesticides for two years or more began to develop symptoms of poisoning. Arsenic is a popular poison in detective stories because arsenic was often used by real-life killers. Arsenic compounds were readily available as a weed or insect killer and could easily be added surreptitiously to the victim's food or drink. How then could it be that arsenic would accidentally be present in the drinking water supply of large numbers of people?

In the 1960s the United Nations Children's Fund (UNICEF) began a campaign to reduce the number of children (and adults) in Bangladesh dying from diseases such as dysentery and cholera caused by drinking water contaminated by bacteria. Most drinking water came from rivers, shallow hand-dug wells, and other surface sources. Ineffective sewage treatment and annual flooding caused by monsoon rains almost guaranteed that the water would be contaminated. Therefore, UNICEF and other aid agencies funded and encouraged installation of wells from which people could obtain uncontaminated water. Called tube wells because a steel tube was driven 20 to 100 feet into the ground and a hand pump installed, these wells halved the mortality rate for children under five, saving 125,000 children every year.

At the time the wells were installed, nobody checked the water for contaminants other than bacteria, apparently on the assumption that groundwater would be pure enough to drink. In many parts of the Ganges delta there are high concentrations of arsenic compounds in sediments just below the surface. Groundwater can dissolve these compounds from the sediments, producing dangerous concentrations of arsenic (above 50 ppb, parts per billion; 1 ppb is one gram of arsenic per billion grams of water) in one of every five wells. Wells contaminated with more than 50 ppb arsenic serve about 20 million people. It is estimated that more than 100 million people in Bangladesh, India, and surrounding countries are drinking water that contains a greater concentration of arsenic than 10 ppb, the maximum safe level recommended by the World Health Organization.

In the early 1980s scientists A. K. Chakraborty and K. C. Saha at the School of Tropical Medicine in Kolkata (Calcutta) began to see many patients who had spots on their skin that changed from white to brown to black. There was thickening and eventual cracking of the skin, as well as other clinical symptoms of arsenic poisoning. Chakraborty and Saha were aware of water analyses carried out as early as 1978 that revealed high concentrations of arsenic. In 1987 they published in the *Indian Journal of Medical Research* their conclusions that people in both India and Bangladesh were being poisoned by arsenic in the water they were drinking. It was not until the latter part of the 1990s, however, that the full scale of the problem was recognized and a major program of testing water from the tube wells began.

According to UNICEF there are nearly ten million tube wells in Bangladesh. A little more than half had been tested for arsenic by 2007. If the concentration of arsenic was below 50 ppb, then the well's pump was painted green to indicate that the water was reasonably safe to drink. If a well tested above 50 ppb, then it was painted red. About 15% of the pumps were painted red. In more than 8300 villages 80% or more of the pumps were red, leaving the inhabitants with very few sources of clean water to drink. Once they knew which wells were contaminated, nearly a third of the people switched to a clean well, but more than half of the people with contaminated water have done nothing and are still exposed to unacceptable levels of arsenic.

STRDEL/AFP/Getty Images

Figure 1.1 A SONO filter in a village hut in Bangladesh. The red container holds the composite iron matrix. The aqua bucket holds the sand and charcoal filters.

This problem was so severe that in 2005 the U.S. National Academy of Engineering announced the Grainger Challenge Prize for Sustainable Development, which would award $1 million for development of an inexpensive system to reduce arsenic to safe levels in drinking water. In February 2007 the National Academy announced a winner: Abul Hussam, a professor of chemistry and biochemistry at George Mason University. Hussam invented the SONO filter, which effectively removes arsenic from drinking water, lasts for many years, and binds the arsenic so tightly that there is little or no hazard associated with disposing of used filters. A photograph of the SONO filter is shown in Figure 1.1. Although it looks crude, a simple device is exactly what is needed in a country where the average annual income is $480 per person.

The chemistry of the SONO filter is interesting, but most of the details are proprietary (that is, the inventor is keeping them secret). In groundwater the pH is between 6.5 and 7.5 and arsenic is present in both the 3+ oxidation state (as H_3AsO_3) and the 5+ oxidation state (as $H_2AsO_4^-$ or $HAsO_4^{2-}$). The filter must remove all three chemical species. Its primary active material is called a composite iron matrix (CIM). The CIM is made from cast-iron chips by a special process. The upper bucket of the filter contains three layers: 10 kg coarse river sand on top, then 5 to 10 kg CIM, and 10 kg brick chips and coarse river sand at the bottom. Water passes slowly through this mixture and drains into the lower bucket, which contains 10 kg coarse river sand on top, wood charcoal to adsorb organic material, 9 kg fine river sand, and 3.5 kg brick chips. Water is tapped from the bottom of the lower bucket. To ensure that the water remains in contact with the contents of both buckets for

One kilogram (kg) weighs 2.2 pounds.

long enough to remove the dissolved arsenic species, the taps are designed to keep the flow of water slow.

In the upper bucket, arsenic in the 3+ oxidation state is oxidized to the 5+ state, by oxygen in the air. Manganese, which constitutes 1% to 2% by weight of the CIM, catalyzes the oxidation. The negative ions containing arsenic ($H_2AsO_4^-$ or $HAsO_4^{2-}$) become tightly attached to iron atoms at the surfaces of iron particles in the CIM, replacing hydroxide ions (OH^-) that are present on the iron surfaces. The chemical reactions are

$$=FeOH + H_2AsO_4^- \longrightarrow =FeHAsO_4^- + H_2O$$

and

$$=FeOH + HAsO_4^{2-} \longrightarrow =FeAsO_4^{2-} + H_2O$$

where =Fe represents an iron atom on the surface of an iron turning. The arsenic species are so tightly attached to the iron atoms that they do not come off, no matter how much water passes through the filter.

Thus, chemistry has made major contributions to help alleviate the problem of contaminated drinking water in Bangladesh. Analytical chemists have devised tests for arsenic that are sensitive enough to detect one gram of arsenic in a billion grams of water (1 ppb)—a quantity smaller than a needle in a haystack. And a chemist has devised a way to remove arsenic from the water that works well, does not further pollute the environment, and costs very little. All of this effort depends on applying the methods of science to improve everyday life.

1.3 How Science Is Done

How science is done is dealt with in *Oxygen,* a play written by chemists Carl Djerassi and Roald Hoffmann that premiered in 2001. By revisiting the discovery of oxygen, the play provides many insights regarding the process of science and the people who make science their life's work.

The story of arsenic in drinking water illustrates many aspects of how people do science and how scientific knowledge changes and improves over time. In antiquity it was known that compounds of arsenic were poisons, even though the element itself had not been discovered. For example, the Chinese used As_2O_3 to rid their fields of pests and they knew that workers who produced the poison usually lost their hair and became ill with less than two years of exposure. This led to the reasonable hypothesis that "pee-song" (as the poison was known in Chinese) is harmful to humans. A **hypothesis** is an idea that is tentatively proposed as an explanation for some observation and provides a basis for experimentation. The hypothesis that pee-song or "white arsenic" (as it was known in Western countries) could be used as a poison was amply verified by the Borgias and others who poisoned their rivals to gain power or money.

Testing a hypothesis may involve collecting qualitative data, quantitative data, or both. Qualitative data, the observation that people drinking water from tube wells in Bangladesh developed spots on their skin that changed color and the skin eventually became hard and cracked, provided a clue that arsenic poisoning might be occurring. **Qualitative** data do not involve numbers. The qualitative data led to quantitative studies showing that one in five of the wells tested had arsenic levels greater than 50 ppb. These **quantitative** studies reported numbers to indicate how much arsenic was present. Other quantitative clinical studies had already demonstrated that levels of arsenic above 50 ppb would lead to poisoning and possibly cancer.

A scientific **law** is a statement that summarizes and explains a wide range of experimental results and has not been contradicted by experiments. *A law can predict unknown results and also can be disproved or falsified by new experiments.* When the results of a new experiment contradict a law, that's exciting to a scientist. If enough scientists repeat the experiment and get the same contradictory result, then the law must be modified to account for the new results—or even discarded altogether.

A well-tested hypothesis is designated as a **theory**—a unifying principle that explains a body of facts and the laws based on them. (Notice that in everyday speech the word "theory" is often used to designate what a scientist means by "hypothesis.") A theory usually suggests new hypotheses and experiments, and, like a law, it may have to be modified or even discarded if contradicted by new experimental results. A **model** makes a theory more concrete, often in a physical or a mathematical form. Models of molecules and ions, for example, tell us how the atoms are connected in arsenic species (such as H_3AsO_3) and give clues to why arsenic could be tightly bound to iron atoms at the surface of a cast-iron chip. Molecular models can be constructed by using spheres to represent atoms and sticks to represent the connections between the atoms. Or a computer can be used to calculate the locations of the atoms and display model molecular structures on a screen (as was done to create the models of arsenic-containing species in the margin). The theories that matter is made of atoms and molecules, that atoms are arranged in specific molecular structures, and that the properties of matter depend on those structures are fundamental to chemists' unique atomic/molecular perspective on the world and to nearly everything modern chemists do. Clearly it is important that you become as familiar as you can with these theories and with models based on them.

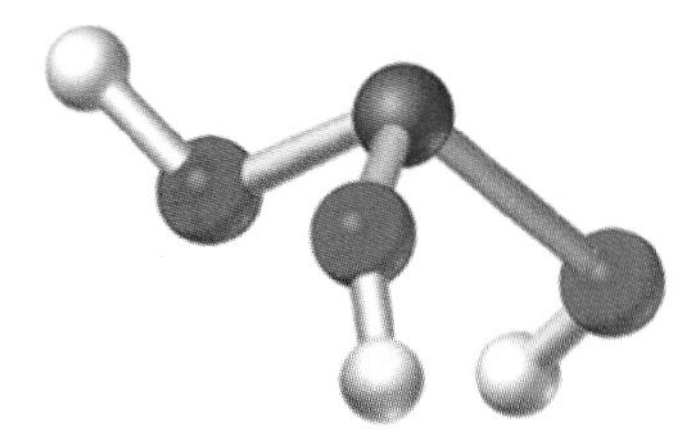

H_3AsO_3

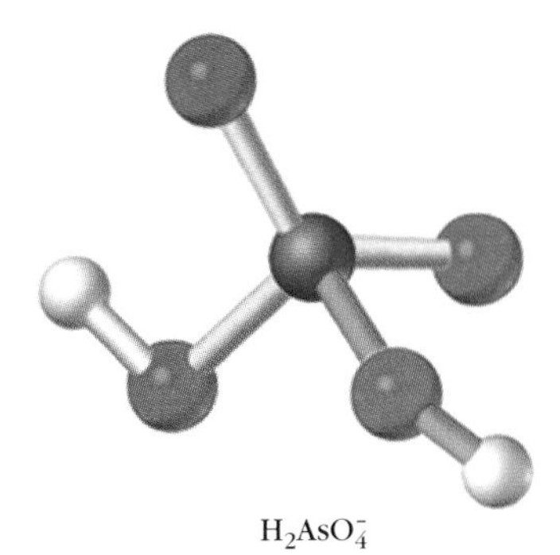

$H_2AsO_4^-$

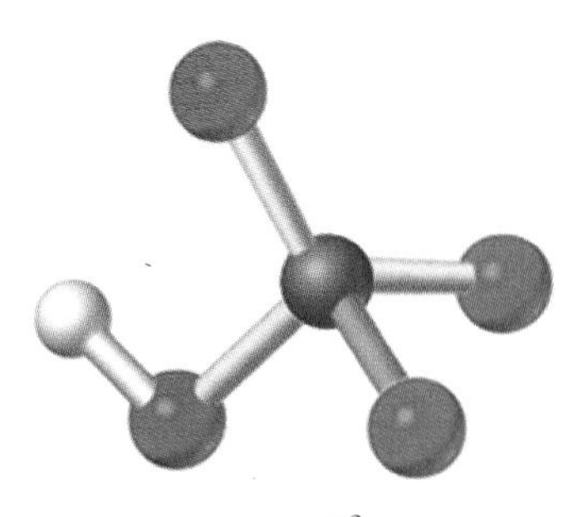

$HAsO_4^{2-}$

Structures of arsenic-containing species found in drinking water in Bangladesh.

Another important aspect of the way science is done involves communication. Science is based on experiments and on hypotheses, laws, and theories that can be contradicted by experiments. Therefore, it is essential that experimental results be communicated to all scientists working in any specific area of research as quickly and accurately as possible. Scientific communication allows contributions to be made by scientists in different parts of the world and greatly enhances the rapidity with which science can develop. In addition, communication among members of scientific research teams is crucial to their success. Poor communication (and failure of some experts to read and act on published information) kept the full enormity of the arsenic problem from becoming known for almost 10 years. Had communication been better, many fewer residents of Bangladesh and India would be ill today. The importance of scientific communication is emphasized by the fact that the Internet was created not by commercial interests, but by scientists who saw its great potential for communicating scientific information.

In the remainder of this chapter we discuss fundamental concepts of chemistry that have been revealed by applying the processes of science to the study of matter. We begin by considering how matter can be classified according to characteristic properties.

1.4 Identifying Matter: Physical Properties

One type of matter can be distinguished from another by observing the properties of samples of matter and classifying the matter according to those properties. A **substance** is a type of matter that has the same properties and the same composition throughout a sample. Each substance has characteristic properties that are different from the properties of any other substance (Figure 1.2). In addition, one sample of a substance has the same composition as every other sample of that substance—it consists of the same stuff in the same proportions.

You can distinguish sugar from water because you know that sugar consists of small white particles of solid, while water is a colorless liquid. Metals can be recognized as a class of substances because they usually are solids, have high densities, feel cold to the touch, and have shiny surfaces. These properties can be observed and measured without changing the composition of a substance. They are called **physical properties.**

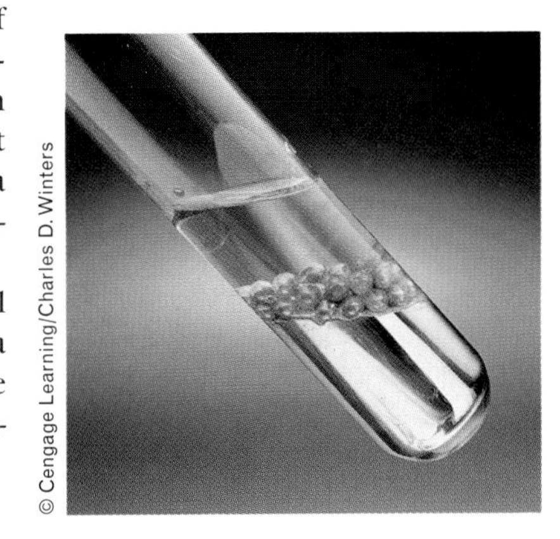

Figure 1.2 Substances have characteristic physical properties. The test tube contains silvery liquid mercury in the bottom, small spheres of solid orange copper in the middle, and colorless liquid water above the copper. Each of these substances has characteristic properties that differentiate it from the others.

Some Physical Properties

Temperature	Melting point	Hardness, brittleness
Pressure	Boiling point	Heat capacity
Mass	Density	Thermal conductivity
Volume	Color	Electrical conductivity
State (solid, liquid, gas)	Shape of solid crystals	

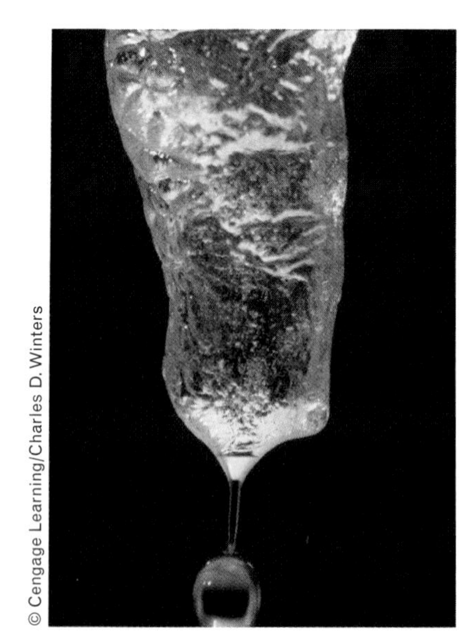

Figure 1.3 Physical change. When ice melts, it changes—physically—from a solid to a liquid, but it is still water.

Physical Change

As a substance's temperature or pressure changes, or if it is mechanically manipulated, some of its physical properties may change. Changes in the physical properties of a substance are called **physical changes.** The same substance is present before and after a physical change, but the substance's physical state or the gross size and shape of its pieces may have changed. Examples are melting a solid (Figure 1.3), boiling a liquid, hammering a copper wire into a flat shape, and grinding sugar into a fine powder.

Melting and Boiling Point

An important way to help identify a substance is to measure the temperature at which the solid melts (the substance's **melting point**) and at which the liquid boils (its **boiling point**). If two or more substances are in a mixture, the melting point depends on how much of each is present, but for a single substance the melting point is always the same. This is also true of the boiling point (as long as the pressure on the boiling liquid is the same). In addition, the melting point of a pure crystalline sample of a substance is sharp—there is almost no change in temperature as the sample melts. When a mixture of two or more substances melts, the temperature when liquid first appears can be quite different from the temperature when the last of the solid is gone.

Temperature is the property of matter that determines whether there can be heat energy transfer from one object to another. It is represented by the symbol T. Energy transfers of its own accord from an object at a higher temperature to a cooler object. In the United States, everyday temperatures are reported using the Fahrenheit temperature scale. On this scale the freezing point of water is by definition 32 °F and the boiling point is 212 °F. The **Celsius temperature scale** is used in most countries of the world and in science. On this scale 0 °C is the freezing point and 100 °C is the boiling point of pure water at a pressure of one atmosphere. The number of units between the freezing and boiling points of water is 180 Fahrenheit degrees and 100 Celsius degrees. This means that the Celsius degree is almost twice as large as the Fahrenheit degree. It takes only 5 Celsius degrees to cover the same temperature range as 9 Fahrenheit degrees, and this relationship can be used to calculate a temperature on one scale from a temperature on the other (see Appendix B.2).

If you need to convert from the Fahrenheit to the Celsius scale or from Celsius to Fahrenheit, an explanation of how to do so is in Appendix B.2.

Because temperatures in scientific studies are usually measured in Celsius units, there is little need to make conversions to and from the Fahrenheit scale, but it is quite helpful to be familiar with how large various Celsius temperatures are. For example, it is useful to know that water freezes at 0 °C and boils at 100 °C, a comfortable room temperature is about 22 °C, your body temperature is 37 °C, and the hottest water you could put your hand into without serious burns is about 60 °C.

Exercises that are labeled conceptual are designed to test your understanding of a concept.

Answers to exercises are provided near the end of the book in a section with color on the edges of the pages.

CONCEPTUAL
EXERCISE 1.1 Temperature

(a) Which is the higher temperature, 110 °C or 180 °F?
(b) Which is the lower temperature, 36 °C or 100 °F?
(c) The melting point of gallium is 29.8 °C. If you hold a sample of gallium in your hand, will it melt?

The density of a substance varies depending on the temperature and the pressure. Densities of liquids and solids change very little as pressure changes, and they change less with temperature than do densities of gases. Because the volume of a gas varies significantly with temperature and pressure, the density of a gas can help identify the gas only if the temperature and pressure are specified.

Density

Another property that is often used to help identify a substance is **density,** the ratio of the mass of a sample to its volume. If you have ten pounds of sugar, it occupies ten times the volume that one pound of sugar does. In mathematical terms, a substance's volume is directly proportional to its mass. This means that a substance's density has the same value regardless of how big the sample is.

$$\text{Density} = \frac{\text{mass}}{\text{volume}} \qquad d = \frac{m}{V}$$

Even if they look similar, you can tell a sample of aluminum from a sample of lead by picking each up. Your brain will automatically estimate which sample has greater mass for the same volume, telling you which is the lead. At 20 °C, aluminum has a density of 2.70 g/mL, placing it among the least dense metals. Lead's density is 11.34 g/mL, so a sample of lead is much heavier than a sample of aluminum of the same size.

Suppose that you are trying to identify a liquid. Your hypothesis is that the liquid is ethanol (ethyl alcohol). By determining its density you could test the hypothesis. You could weigh a clean, dry graduated cylinder and then add some of the liquid to it. Suppose that, from the markings on the cylinder, you read the volume of liquid to be 8.30 mL (at 20 °C). You could then weigh the cylinder with the liquid and subtract the mass of the empty cylinder to obtain the mass of liquid. Suppose the liquid mass is 6.544 g. The density can then be calculated as

$$d = \frac{m}{V} = \frac{6.544 \text{ g}}{8.30 \text{ mL}} = 0.788 \text{ g/mL}$$

From a table of physical properties of various substances you find that the density of ethanol is 0.789 g/mL, which helps confirm your hypothesis that the substance is ethanol.

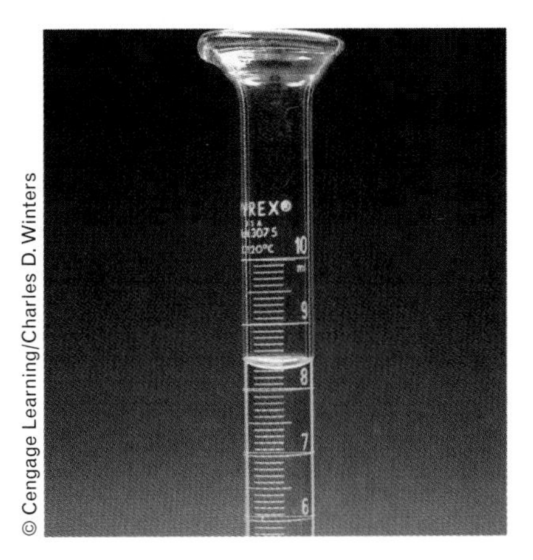

Graduated cylinder containing 8.30 mL of liquid.

If you divide 6.544 by 8.30 on a scientific calculator, the answer might come up as 0.788433735. This displays more digits than are meaningful, and we have rounded the result to only three significant digits. Rules for deciding how many digits should be reported in the result of a calculation and procedures for rounding numbers are introduced in Section 2.4 and Appendix A.3.

A useful source of data on densities and other physical properties of substances is the *CRC Handbook of Chemistry and Physics,* published by the CRC Press. Information is also available via the Internet—for example, the National Institute for Standards and Technology's WebBook at **http://webbook.nist.gov**.

CONCEPTUAL
EXERCISE 1.2 Density of Liquids

When 5.0 mL each of vegetable oil, water, and kerosene are put into a large test tube, they form three layers, as shown in the photo.

(a) List the three liquids in order of increasing density (smallest density first, largest density last).
(b) If an additional 5.0 mL of vegetable oil is poured into the test tube, what will happen? Describe the appearance of the tube.
(c) If 5.0 mL of kerosene is added to the test tube with the 5.0 mL of vegetable oil in part (b), will there be a permanent change in the order of liquids from top to bottom of the tube? Why or why not?

EXERCISE 1.3 Physical Properties and Changes

Identify each physical property and physical change mentioned in each of these statements. Also identify the qualitative and the quantitative information given in each statement.

(a) The blue chemical compound azulene melts at 99 °C.
(b) The white crystals of table salt are cubic.
(c) A sample of lead has a mass of 0.123 g and melts at 327 °C.
(d) Ethanol is a colorless liquid that vaporizes easily; it boils at 78 °C and its density is 0.789 g/mL.

Liquid densities. Kerosene *(top layer)*, vegetable oil *(middle layer)*, and water *(bottom layer)* have different densities.

Measurements and Calculations: Dimensional Analysis

Determining a property such as density requires scientific measurements and calculations. The result of a measurement, such as 6.544 g or 8.30 mL, usually consists of a number and a unit. Both the number and the unit should be included in calculations. For example, the densities in Table 1.1 have units of grams per milliliter (g/mL), because density is defined as the mass of a sample divided by its volume. When a mass is divided by a volume, the units (g for the mass and mL for the volume) are also divided. The result is grams divided by milliliters (g/mL). That is, both numbers *and*

Table 1.1 Densities of Some Substances at 20 °C

Substance	Density (g/mL)	Substance	Density (g/mL)
Butane	0.579	Titanium	4.50
Ethanol	0.789	Zinc	7.14
Benzene	0.880	Iron	7.86
Water	0.998	Nickel	8.90
Bromobenzene	1.49	Copper	8.93
Magnesium	1.74	Lead	11.34
Sodium chloride	2.16	Mercury	13.55
Aluminum	2.70	Gold	19.32

units follow the rules of algebra. This is an example of **dimensional analysis,** a method of using units in calculations to check for correctness. More detailed descriptions of dimensional analysis are given in Section 2.3 and Appendix A.2. We will use this technique for problem solving throughout the book.

Suppose that you want to know whether you could lift a gallon (3784 mL) of the liquid metal mercury. To answer the question, calculate the mass of the mercury using its density, 13.55 g/mL, obtained from Table 1.1. One way to do this is to use the equation that defines density, $d = m/V$. Then solve algebraically for m, and calculate the result:

Because mercury and mercury vapor are poisonous, carrying a gallon of it around is not a good idea unless it is in a sealed container.

$$m = V \times d = 3784\ \cancel{\text{mL}} \times \frac{13.55\ \text{g}}{1\ \cancel{\text{mL}}} = 51{,}270\ \text{g} \qquad (1.1)$$

This equation emphasizes the fact that mass is proportional to volume, because the volume is multiplied by a proportionality constant, the density. Notice also that the units of volume (mL) appeared once in the denominator of a fraction and once in the numerator, thereby dividing out (canceling) and leaving only mass units (g). (The result, 51,270 g, is more than 100 pounds, so you could probably lift the mercury, but not easily.)

In Equation 1.1, a known quantity (the volume) was multiplied by a proportionality factor (the density), and the units canceled, giving an answer (the mass) with appropriate units. A general approach to this kind of problem is to recognize that the quantity you want to calculate (the mass) is proportional to a quantity whose value you know (the volume). Then use a proportionality factor that relates the two quantities, setting things up so that the units cancel.

$$\text{Known } \cancel{\text{quantity units}} \times \frac{\text{desired quantity units}}{\text{known } \cancel{\text{quantity units}}} = \text{desired quantity units}$$

proportionality (conversion) factor

$$3784\ \cancel{\text{mL}} \times \frac{13.55\ \text{g}}{1\ \cancel{\text{mL}}} = 51{,}270\ \text{g}$$

A **proportionality factor** is a ratio (fraction) whose numerator and denominator have different units but refer to the same thing. In the preceding example, the proportionality factor is the density, which relates the mass and volume of the same sample of mercury. A proportionality factor is often called a **conversion factor** because it enables us to convert from one kind of unit to a different kind of unit.

In this book, units and dimensional analysis techniques are introduced at the first point where you need to know them. Appendices A and B provide all of this information in one place.

Because a conversion factor is a fraction, every conversion factor can be expressed in two ways. The conversion factor in the example just given could be expressed either as the density or as its reciprocal:

$$\frac{13.55\ \text{g}}{1\ \text{mL}} \quad \text{or} \quad \frac{1\ \text{mL}}{13.55\ \text{g}}$$

The first fraction enables conversion from volume units (mL) to mass units (g). The second allows mass units to be converted to volume units. Which conversion factor to use depends on which units are in the known quantity and which units are in the quantity that we want to calculate. Setting up the calculation so that the units cancel ensures that we are using the appropriate conversion factor. (See Appendix A.2 for more examples.)

PROBLEM-SOLVING EXAMPLE 1.1 Density

In an old movie, thieves are shown running off with pieces of gold bullion that are about a foot long and have a square cross section of about six inches. The volume of each piece of gold is 7000 mL. Calculate the mass of gold and express the result in pounds (lb). Based on your result, is what the movie shows physically possible? (1 lb = 454 g)

Answer 1.4×10^5 g; 300 lb; probably not

Strategy and Explanation A good approach to problem solving is to (1) define the problem, (2) develop a plan, (3) execute the plan, and (4) check your result to see whether it is reasonable. (These four steps are described in more detail in Appendix A.1.)

Step 1: *Define the problem.* You are asked to calculate the mass of the gold, and you know the volume.

Step 2: *Develop a plan.* Density relates mass and volume and is the appropriate proportionality factor, so look up the density in a table. Mass is proportional to volume, so the volume either has to be multiplied by the density or divided by the density. Use the units to decide which.

Step 3: *Execute the plan.* According to Table 1.1, the density of gold is 19.32 g/mL. Setting up the calculation so that the unit (milliliter) cancels gives

$$7000\ \cancel{\text{mL}} \times \frac{19.32\ \text{g}}{1\ \cancel{\text{mL}}} = 1.35 \times 10^5\ \text{g}$$

This can be converted to pounds

$$1.35 \times 10^5\ \cancel{\text{g}} \times \frac{1\ \text{lb}}{454\ \cancel{\text{g}}} = 300\ \text{lb}$$

Notice that the result is expressed to one significant figure, because the volume was given to only one significant figure and only multiplications and divisions were done.

☑ **Reasonable Answer Check** Gold is nearly 20 times denser than water. A liter (1000 mL) of water is about a quart and a quart of water (2 pints) weighs about two pounds. Seven liters (7000 mL) of water should weigh 14 lb, and 20 times 14 gives 280 lb, so the answer is reasonable. The movie scene is not—few people could run while carrying a 300-lb object!

PROBLEM-SOLVING PRACTICE 1.1

Find the volume occupied by a 4.33-g sample of benzene.

Rules for assigning the appropriate number of significant figures to a result are given in Appendix A.3.

The checkmark symbol accompanied by the words "Reasonable Answer Check" will be used throughout this book to indicate how to check the answer to a problem to make certain a reasonable result has been obtained.

Answers to Problem-Solving Practice problems are given near the end of the book in a section with color on the edges of the pages.

This book includes many examples, like Problem-Solving Example 1.1, that illustrate general problem-solving techniques and ways to approach specific types of problems. Usually, each of these examples states a problem; gives the answer; explains one way to analyze the problem, plan a solution, and execute the plan; and describes a way to check that the result is reasonable. We urge you to first try to solve the problem on your own. Then check to see whether your answer matches the one given. If it does not match, try again before reading the explanation. After you have tried twice, read the explanation to find out why your reasoning differs from that given. If your answer is correct, but your reasoning differs from the explanation, you may have discovered an alternative solution to the problem. Finally, work out the Problem-Solving Practice that accompanies the example. It relates to the same concept and allows you to improve your problem-solving skills.

1.5 Chemical Changes and Chemical Properties

Another way to identify a substance is to observe how it reacts chemically. For example, if you heat a white, granular solid carefully and it caramelizes (turns brown and becomes a syrupy liquid—see Figure 1.4), it is a good bet that the white solid is ordinary table sugar (sucrose). When heated gently, sucrose decomposes to give water and other new substances. If you heat sucrose very hot, it will char, leaving behind a black residue that is mainly carbon (and is hard to clean up). If you drip some water onto a sample of sodium metal, the sodium will react violently with the water, producing a solution of lye (sodium hydroxide) and a flammable gas, hydrogen (Figure 1.5). These are examples of **chemical changes** or **chemical reactions.** In a chemical reaction, one or more substances (the **reactants**) are transformed into one or more different substances (the **products**). Reactant substances are replaced by product substances as the reaction occurs. This process is indicated by writing the reactants, an arrow, and then the products:

$$\underset{\text{Reactant}}{\text{Sucrose}} \xrightarrow[\text{changes to}]{} \underbrace{\text{Carbon} + \text{Water}}_{\text{Products}}$$

Chemical reactions make chemistry interesting, exciting, and valuable. If you know how, you can make arsenic-containing species stick to iron surfaces, clothing from crude petroleum, or even a silk purse from a sow's ear (it has been done). This is a very empowering idea, and human society has gained a great deal from it. Our way of life is greatly enhanced by our ability to use and control chemical reactions. And life itself is based on chemical reactions. Biological cells are filled with water-based solutions in which thousands of chemical reactions are happening all the time.

Sucrose (Reactant) changes to Carbon + Water (Products)
Water
Carbon
1 When table sugar (sucrose) is heated . . .
2 . . . it caramelizes, turning brown.
3 Heating to a higher temperature causes further decomposition (charring) to carbon and water vapor.
© Cengage Learning/Charles D. Winters

Figure 1.4 Chemical change. Heat can caramelize or char sugar.

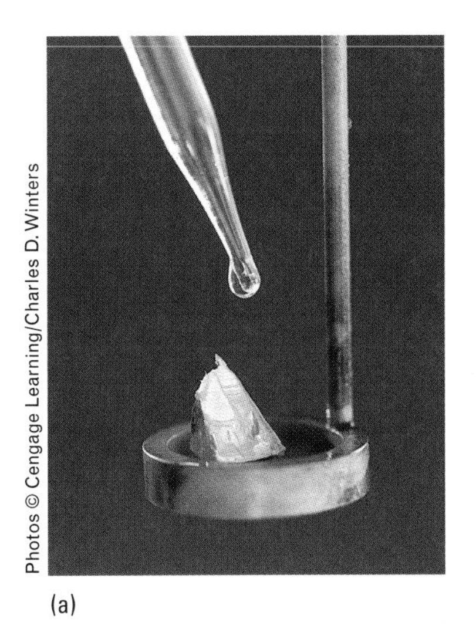

(a) (b)

Figure 1.5 Chemical change. When a drop of water (a) hits a piece of sodium, the resulting violent reaction (b) produces flammable hydrogen gas and a solution of sodium hydroxide (lye). Production of motion, heat, and light when substances are mixed is evidence that a chemical reaction is occurring.

Figure 1.6 Chemical change. Vinegar, which is an acid, has been added to an egg, causing colorless carbon dioxide gas to bubble away from the eggshell, which consists mainly of calcium carbonate. Production of gas bubbles when substances come into contact is one kind of evidence that a chemical reaction is occurring.

Chemical Properties

A substance's **chemical properties** describe the kinds of chemical reactions the substance can undergo. One chemical property of metallic sodium is that it reacts rapidly with water to produce hydrogen and a solution of sodium hydroxide (Figure 1.5). Because it also reacts rapidly with air and a number of other substances, sodium is also said to have a more general chemical property: It is highly reactive. A chemical property of substances known as metal carbonates is that they produce carbon dioxide when treated with an acid (Figure 1.6). Fuels are substances that have the chemical property of reacting with oxygen or air and at the same time transferring large quantities of energy to their surroundings. An example is natural gas (mainly methane), which is shown reacting with oxygen from the air in a gas stove in Figure 1.7. A substance's chemical properties tell us how it will behave when it contacts air or water, when it is heated or cooled, when it is exposed to sunlight, or when it is mixed with another substance. Such knowledge is very useful to chemists, biochemists, geologists, chemical engineers, and many other kinds of scientists.

Energy

Chemical reactions are usually accompanied by transfers of energy. (Physical changes also involve energy transfers, but usually they are smaller than those for chemical changes.) **Energy** is defined as the capacity to do work—that is, to make something happen. Combustion of a fuel, as in Figure 1.7, transforms energy stored in chemical bonds in the fuel molecules and oxygen molecules into motion of the product molecules and of other nearby molecules. This corresponds to a higher temperature in the vicinity of the flame. The chemical reaction in a light stick transforms energy stored in molecules into light energy, with only a little heat transfer (Figure 1.8). A chemical reaction in a battery makes a calculator work by forcing electrons to flow through an electric circuit.

Figure 1.7 Combustion of natural gas. Natural gas, which in the United States consists mostly of methane, burns in air, transferring energy that raises the temperature of its surroundings.

Energy supplied from somewhere else can cause chemical reactions to occur. For example, photosynthesis takes place when sunlight illuminates green plants. Some of the sunlight's energy is stored in carbohydrate molecules and oxygen molecules that

Figure 1.8 Transforming energy. In each of these light sticks a chemical reaction transforms energy stored in molecules into light. Unlike an electric light bulb, the light sticks do not get hot.

are produced from carbon dioxide and water by photosynthesis. Aluminum, which you may have used as foil to wrap and store food, is produced by passing electricity through a molten, aluminum-containing ore. You consume and metabolize food, using the energy stored in food molecules to cause chemical reactions to occur in the cells of your body. The relation between chemical changes and energy is an important theme of chemistry.

CONCEPTUAL
EXERCISE 1.4 Chemical and Physical Changes

Identify the chemical and physical changes that are described in this statement: Propane gas burns, and the heat of the combustion reaction is used to hard-boil an egg.

1.6 Classifying Matter: Substances and Mixtures

Once its chemical and physical properties are known, a sample of matter can be classified on the basis of those properties. Most of the matter we encounter every day is like the composite iron matrix in a SONO filter, concrete, or the carbon fiber composite frame of a high-tech bicycle—not uniform throughout. There are variations in color, hardness, and other properties from one part of a sample to another. This makes these materials complicated, but also interesting. A major advance in chemistry occurred when it was realized that it was possible to separate several component substances from such nonuniform samples. For example, the fact that groundwater can separate (dissolve) arsenic-containing substances from soil is the means by which arsenic gets into drinking water in Bangladesh.

Often, as in the case of soil or concrete, we can easily see that one part of a sample is different from another part. In other cases a sample may appear completely uniform to the unaided eye, but a microscope can reveal that it is not. For example, blood appears smooth in texture, but magnification reveals red and white cells within the liquid (Figure 1.9). The same is true of milk. A mixture in which the uneven texture of the material can be seen with the naked eye or with a microscope is classified as a **heterogeneous mixture.** Properties in one region are different from the properties in another region.

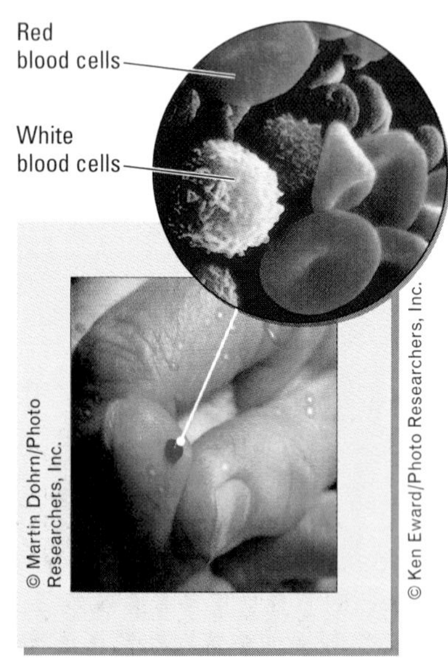

Figure 1.9 A heterogeneous mixture. Blood appears to be uniform to the unaided eye, but a microscope reveals that it is not homogeneous. The properties of red blood cells differ from the properties of the surrounding blood plasma, for example.

A **homogeneous mixture,** or **solution,** is completely uniform and consists of two or more substances in the same phase—solid, liquid, or gas (Figure 1.10). No amount of optical magnification will reveal different properties in one region of a solution compared with those in another. Heterogeneity exists in a solution only at the scale of atoms and molecules, which are too small to be seen with visible light. Examples of solutions are clear air (mostly a mixture of nitrogen and oxygen gases), sugar water, and some brass alloys (which are homogeneous mixtures of copper and zinc). The properties of a homogeneous mixture are the same everywhere in any particular sample, but they can vary from one sample to another depending on how much of one component is present relative to another component.

Separation and Purification

Earlier in this chapter we stated that a substance has characteristic properties that distinguish it from all other substances. However, for those characteristic properties to be observed, the substance must be separated from all other substances; that is, it must be purified. The melting point of an impure substance is different from that of the purified substance. The color and appearance of a mixture may also differ from those of a pure substance. Therefore, when we talk about the properties of a substance, it is assumed that we are referring to a pure substance—one from which all other substances have been separated.

Purification usually has to be done in several repeated steps and monitored by observing some property of the substance being purified. For example, iron can be separated from a heterogeneous mixture of iron and sulfur with a magnet, as shown in Figure 1.11. In this example, color, which depends on the relative quantities of iron and sulfur, indicates purity. The bright yellow color of sulfur is assumed to indicate that all the iron has been removed.

Concluding that a substance is pure on the basis of a single property of the mixture could be misleading because other methods of purification might change some other properties of the sample. It is safe to call sulfur pure when a variety of methods of purification fail to change its physical and chemical properties. Purification is important because it allows us to attribute properties (such as suitability of water for humans to drink) to specific substances and then to study systematically which kinds of substances have properties that we find useful. In some cases, insufficient purification of a substance has led scientists to attribute to that substance properties that were actually due to a tiny trace of another substance.

Only a few substances occur in nature in pure form. Gold, diamonds, and silicon dioxide (quartz) are examples. We live in a world of mixtures; all living things, the air and food on which we depend, and many products of technology are mixtures. Much of what we know about chemistry, however, is based on separating and purifying the components of those mixtures and then determining their properties. To date, more than 34 million substances have been reported, and many more are being discovered or synthesized by chemists every year. When pure, each of these substances has its own particular composition and its own characteristic properties.

A good example of the importance of purification is the high-purity silicon needed to produce transistors and computer chips. In one billion grams (about 1000 tons) of highly pure silicon there has to be less than one gram of impurity. Once the silicon has been purified, small but accurately known quantities of specific substances, such as boron or arsenic, can be introduced to give the electronic chip the desired properties.

Figure 1.10 A solution. When solid salt (sodium chloride) is stirred into liquid water, it dissolves to form a homogeneous liquid mixture. Each portion of the solution has exactly the same saltiness as every other portion, and other properties are also the same throughout the solution.

High-purity silicon.

Detection and Analysis

The example of arsenic in drinking water shows that it is very important to know whether a substance is present in a sample and to be able to find out how much of it is there. Does an ore contain enough of a valuable metal to make it worthwhile to mine the ore? Is there enough mercury in a sample of fish to make it unsafe for humans to eat the fish?

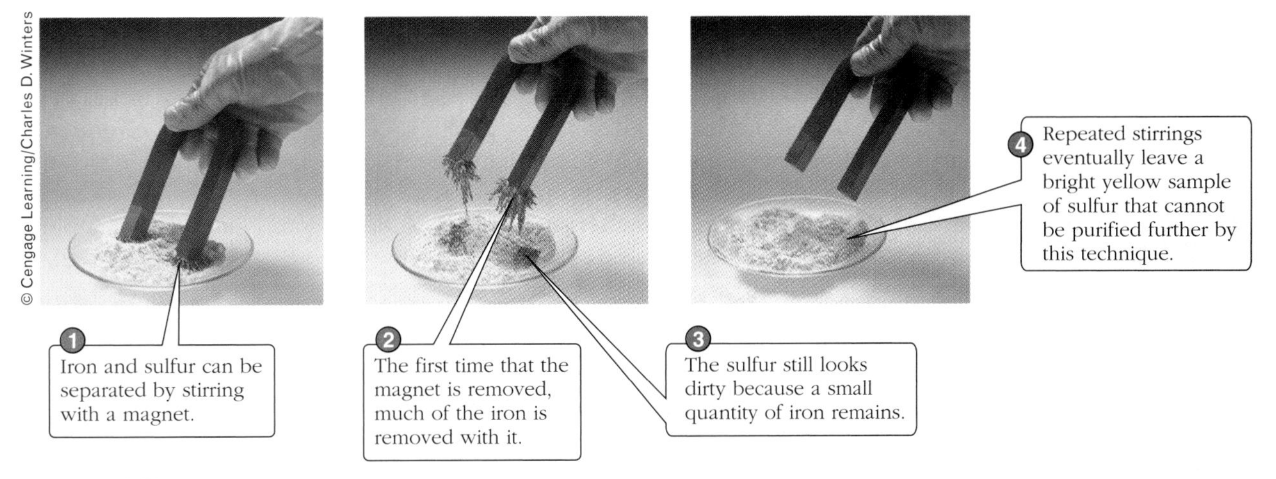

1 Iron and sulfur can be separated by stirring with a magnet.
2 The first time that the magnet is removed, much of the iron is removed with it.
3 The sulfur still looks dirty because a small quantity of iron remains.
4 Repeated stirrings eventually leave a bright yellow sample of sulfur that cannot be purified further by this technique.

Figure 1.11 Separating a mixture: iron and sulfur.

One part per million (ppm) means we can find one gram of a substance in one million grams of total sample. That corresponds to one tenth of a drop of water in a bucket of water. One part per billion corresponds to a drop in a swimming pool, and one part per trillion corresponds to a drop in a large supermarket.

Absence of evidence is not evidence of absence.

Answering questions like these is the job of *analytical chemists,* and they improve their methods every year. For example, in 1960 mercury could be detected at a concentration of one part per million, in 1970 the detection limit was one part per billion, and by 1980 the limit had dropped to one part per trillion. Thus, in 20 years the ability to detect small concentrations of mercury had increased by a factor of one million. This improvement has an important effect. Because we can detect smaller and smaller concentrations of contaminants, such contaminants can be found in many more samples. A few decades ago, toxic substances were usually not found when food, air, or water was tested, but that did not mean they were not there. It just meant that our analytical methods were unable to detect them. Today, with much better methods, toxic substances can be detected in most samples, which prompts demands that concentrations of such substances should be reduced to zero.

Although we expect that chemistry will push detection limits lower and lower, there will always be a limit below which an impurity will be undetectable. Proving that there are no contaminants in a sample will never be possible. This is a specific instance of the general rule that it is impossible to prove a negative. To put this idea another way, it will never be possible to prove that we have produced a completely pure sample of a substance, and, therefore, it is unproductive to legislate that there should be zero contamination in food or other substances. It is more important to use chemical analysis to determine a safe level of a toxin than to try to prove that the toxin is completely absent. In some cases, very small concentrations of a substance are beneficial but larger concentrations are toxic. (Examples of this are selenium and arsenic in the human diet.) Analytical chemistry can help us to determine the optimal ranges of concentration.

1.7 Classifying Matter: Elements and Compounds

In 1661 Robert Boyle was the first to propose that elements could be defined by the fact that they could not be decomposed into two or more simpler substances.

Most of the substances separated from mixtures can be converted to two or more simpler substances by chemical reactions—a process called *decomposition.* Substances are often decomposed by heating them, illuminating them with sunlight, or passing electricity through them. For example, table sugar (sucrose) can be separated from sugar cane and purified. When heated it decomposes via a complex series of chemical changes (caramelization—shown earlier in Figure 1.4) that produces the brown color and flavor of caramel candy. If heated for a longer time at a high enough temperature, sucrose is converted completely to two other substances, carbon and water. Furthermore, if the water is collected, it can be decomposed still further to pure hydrogen and oxygen by passing an electric current through it. However, nobody has found a way to decompose carbon, hydrogen, or oxygen.

Figure 1.12 A compound and its elements. Table sugar, sucrose (a), is composed of the elements carbon (b), oxygen (c), and hydrogen (d). When elements are combined in a compound, the properties of the elements are no longer evident. Only the properties of the compound can be observed.

Substances like carbon, hydrogen, and oxygen that cannot be changed by chemical reactions into two or more new substances are called **chemical elements** (or just elements). Substances that can be decomposed, like sucrose and water, are **chemical compounds** (or just compounds). When elements are chemically combined in a compound, their original characteristic properties—such as color, hardness, and melting point—are replaced by the characteristic properties of the compound. For example, sucrose is composed of these three elements:

- Carbon, which is usually a black powder, but is also commonly seen in the form of diamonds
- Hydrogen, a colorless, flammable gas with the lowest density known
- Oxygen, a colorless gas necessary for human respiration

As you know from experience, sucrose is a white, crystalline powder that is completely unlike any of these three elements (Figure 1.12).

If a compound consists of two or more different elements, how is it different from a mixture? There are two ways: (1) A compound has specific composition and (2) a compound has specific properties. Both the composition and the properties of a mixture can vary. A solution of sugar in water can be very sweet or only a little sweet, depending on how much sugar has been dissolved. There is no particular composition of a sugar solution that is favored over any other, and each different composition has its own set of properties. On the other hand, 100.0 g pure water always contains 11.2 g hydrogen and 88.8 g oxygen. Pure water always melts at 0.0 °C and boils at 100 °C (at one atmosphere pressure), and it is always a colorless liquid at room temperature.

PROBLEM-SOLVING EXAMPLE 1.2 Elements and Compounds

A shiny, hard solid (substance A) is heated in the presence of carbon dioxide gas. After a few minutes, a white solid (substance B) and a black solid (substance C) are formed. No other substances are found. When the black solid is heated in the presence of pure oxygen, carbon dioxide is formed. Decide whether each substance (A, B, and C) is an element or a compound, and give a reason for your choice in each case. If there is insufficient evidence to decide, say so.

Answer A, insufficient evidence; B, compound; C, element

Explanation Substance C must be an element, because it combines with oxygen to form a compound, carbon dioxide, that contains only two elements; in fact, substance C must be carbon. Substance B must be a compound, because it must contain oxygen (from the carbon dioxide) and at least one other element (from substance A). There is not enough evidence to decide whether substance A is an element or a compound. If substance A is an element, then substance B must be an oxide of that element. However, there could be two or more elements in substance A (that is, it could be a compound), and the compound could still combine with oxygen from carbon dioxide to form a new compound.

☑ **Reasonable Answer Check** Substance C is black, and carbon (graphite) is black. You could test experimentally to see whether substance A could be decomposed by heating it in a vacuum; if two or more new substances were formed, then substance A would have to be a compound. If there was no change, substance A could be assumed to be an element.

PROBLEM-SOLVING PRACTICE 1.2

A student grinds an unknown sample (A) to a fine powder and attempts to dissolve the sample in 100 mL pure water. Some solid (B) remains undissolved. When the water is separated from the solid and allowed to evaporate, a white powder (C) forms. The dry white powder (C) is found to weigh 0.034 g. All of sample C can be dissolved in 25 mL pure water. Can you say whether each sample A, B, and C is an element, a compound, or a mixture? Explain briefly.

Types of Matter

What we have just said about separating mixtures to obtain elements or compounds and decomposing compounds to obtain elements leads to a useful way to classify matter (Figure 1.13). Heterogeneous mixtures such as iron with sulfur can be separated using simple manipulation—such as a magnet. Homogeneous mixtures are somewhat more difficult to separate, but physical processes will serve. For example, salt water can be purified for drinking by distilling: heating to evaporate the water and cooling to condense the water vapor back to liquid. When enough water has evaporated, salt crystals will form and they can be separated from the solution. Most difficult of all is separation of the elements that are combined in a compound. Such a separation requires a chemical change, which may involve reactions with other substances or sizable inputs of energy.

Desalinization of water (removing salt) could provide drinking water for large numbers of people who live in dry climates near the ocean. However, distillation requires a lot of energy resources and, therefore, is expensive. When solar energy can be used to evaporate the water, desalinization is less costly.

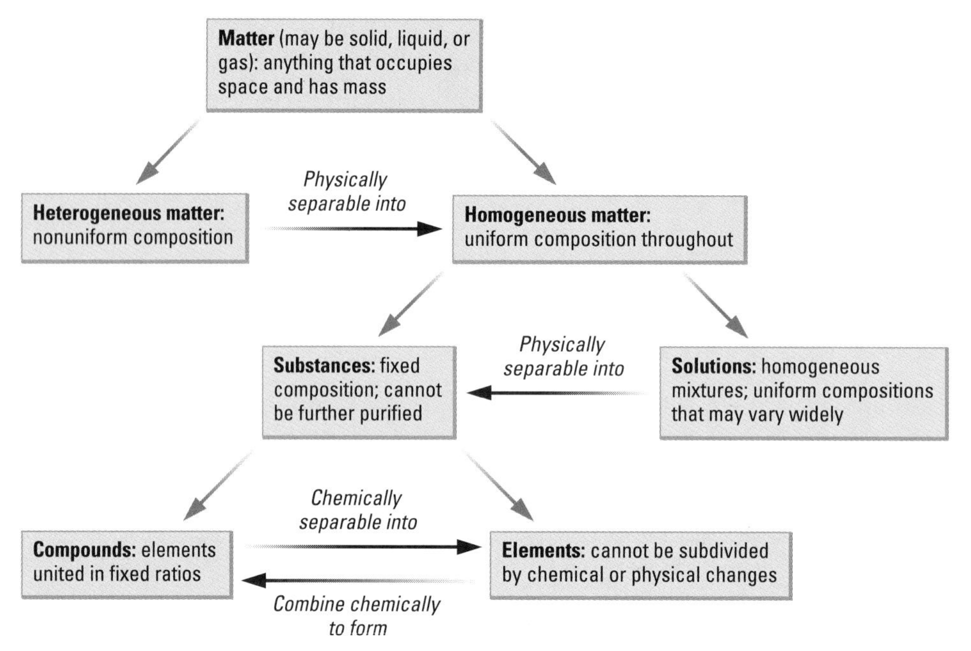

Figure 1.13 A scheme for classifying matter.

CONCEPTUAL
EXERCISE 1.5 Classifying Matter

Classify each of these with regard to the type of matter described:
(a) Sugar dissolved in water
(b) The soda pop in a can of carbonated beverage
(c) Used motor oil freshly drained from a car
(d) The diamond in a piece of jewelry
(e) A 25-cent coin
(f) A single crystal of sugar

1.8 Nanoscale Theories and Models

To further illustrate how the methods of science are applied to matter, we now consider how a theory based on atoms and molecules can account for the physical properties, chemical properties, and classification scheme that we have just described. Physical and chemical properties can be observed by the unaided human senses and refer to samples of matter large enough to be seen, measured, and handled. Such samples are macroscopic; their size places them at the **macroscale.** By contrast, samples of matter so small that they have to be viewed with a microscope are **microscale** samples. Blood cells and bacteria, for example, are matter at the microscale. The matter that really interests chemists, however, is at the **nanoscale.** The term is based on the prefix "nano," which comes from the International System of Units (SI units) and indicates something one billion times smaller than something else. (See Table 1.2 for some important SI prefixes and length units.) For example, a line that is one billion (1×10^9) times shorter than 1 meter is 1 *nano*meter (1×10^{-9} m) long. The sizes of atoms and molecules are at the nanoscale. An average-sized atom such as a sulfur atom has a diameter of two tenths of a nanometer (0.2 nm = 2×10^{-10} m), a water molecule is about the same size, and an aspirin molecule is about three quarters of a nanometer (0.75 nm = 7.5×10^{-10} m) across. Figure 1.14 indicates the relative sizes of various objects at the macroscale, microscale, and nanoscale.

The International System of Units is the modern version of the metric system. It is described in more detail in Appendix B.

Using 1×10^9 to represent 1,000,000,000 or one billion is called scientific notation. It is reviewed in Appendix A.5.

Table 1.2 Some SI (Metric) Prefixes and Units for Length

Prefix	Abbreviation	Meaning	Example
kilo	k	10^3	1 kilometer (km) = 1×10^3 meter (m)
deci	d	10^{-1}	1 decimeter (dm) = 1×10^{-1} m = 0.1 m
centi	c	10^{-2}	1 centimeter (cm) = 1×10^{-2} m = 0.01 m
milli	m	10^{-3}	1 millimeter (mm) = 1×10^{-3} m = 0.001 m
micro	μ	10^{-6}	1 micrometer (μm) = 1×10^{-6} m
nano	n	10^{-9}	1 nanometer (nm) = 1×10^{-9} m
pico	p	10^{-12}	1 picometer (pm) = 1×10^{-12} m

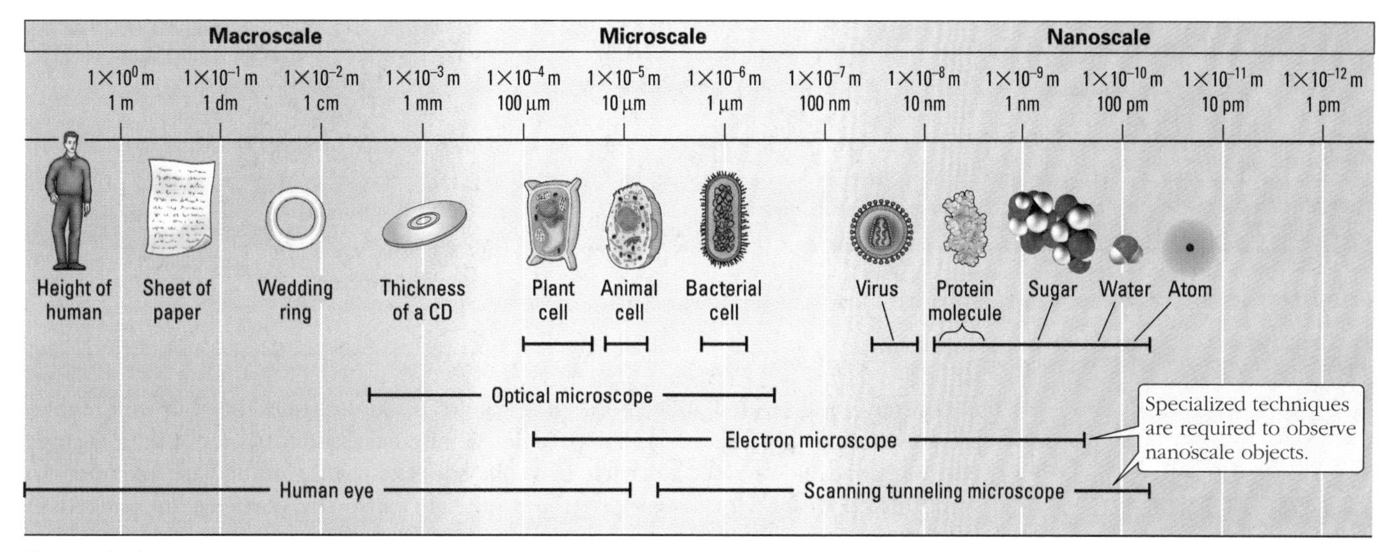

Figure 1.14 **Macroscale, microscale, and nanoscale.**

Earlier, we described the chemist's unique atomic and molecular perspective. It is a fundamental idea of chemistry that matter is the way it is because of the nature of its constituent atoms and molecules. Those atoms and molecules are very, very tiny. Therefore, we need to use imagination creatively to discover useful theories that connect the behavior of tiny nanoscale constituents to the observed behavior of chemical substances at the macroscale. Learning chemistry enables you to "see" in the things all around you nanoscale structure that cannot be seen with your eyes.

Jacob Bronowski, in a television series and book titled *The Ascent of Man,* had this to say about the importance of imagination: "There are many gifts that are unique in man; but at the center of them all, the root from which all knowledge grows, lies the ability to draw conclusions from what we see to what we do not see."

States of Matter: Solids, Liquids, and Gases

An easily observed and very useful property of matter is its physical state. Is it a solid, liquid, or gas? A **solid** can be recognized because it has a rigid shape and a fixed volume that changes very little as temperature and pressure change (Figure 1.15). Like a solid, a **liquid** has a fixed volume, but a liquid is fluid—it takes on the shape of its container and has no definite form of its own. **Gases** are also fluid, but gases expand to fill whatever containers they occupy and their volumes vary considerably with temperature and pressure. For most substances, when compared at the same conditions, the volume of the solid is slightly less than the volume of the same mass of liquid, but the volume of the same mass of gas is much, much larger. As the temperature is raised, most solids melt to form liquids; eventually, if the temperature is raised enough, most liquids boil to form gases.

A theory that deals with matter at the nanoscale is the **kinetic-molecular theory.** It states that all matter consists of extremely tiny particles (atoms or molecules) that are

Figure 1.15 **Quartz crystal.** Quartz, like any solid, has a rigid shape. Its volume changes very little with changes in temperature or pressure.

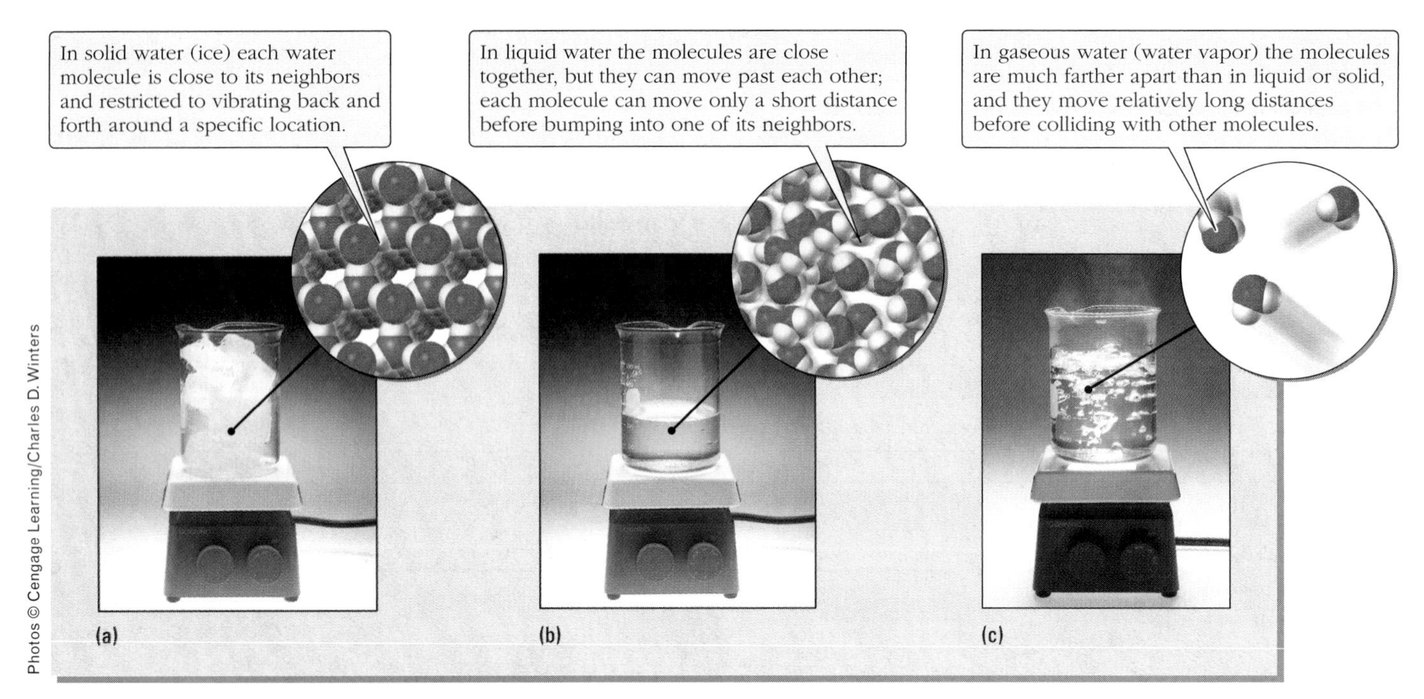

In solid water (ice) each water molecule is close to its neighbors and restricted to vibrating back and forth around a specific location.
In liquid water the molecules are close together, but they can move past each other; each molecule can move only a short distance before bumping into one of its neighbors.
In gaseous water (water vapor) the molecules are much farther apart than in liquid or solid, and they move relatively long distances before colliding with other molecules.
(a)
(b)
(c)
Photos © Cengage Learning/Charles D. Winters

Figure 1.16 Nanoscale representation of three states of matter.

in constant motion. In a solid these particles are packed closely together in a regular array, as shown in Figure 1.16a. The particles vibrate back and forth about their average positions, but seldom does a particle in a solid squeeze past its immediate neighbors to come into contact with a new set of particles. Because the particles are packed so tightly and in such a regular arrangement, a solid is rigid, its volume is fixed, and the volume of a given mass is small. The external shape of a solid often reflects the internal arrangement of its particles. This relation between the observable structure of the solid and the arrangement of the particles from which it is made is one reason that scientists have long been fascinated by the shapes of crystals and minerals (Figure 1.17).

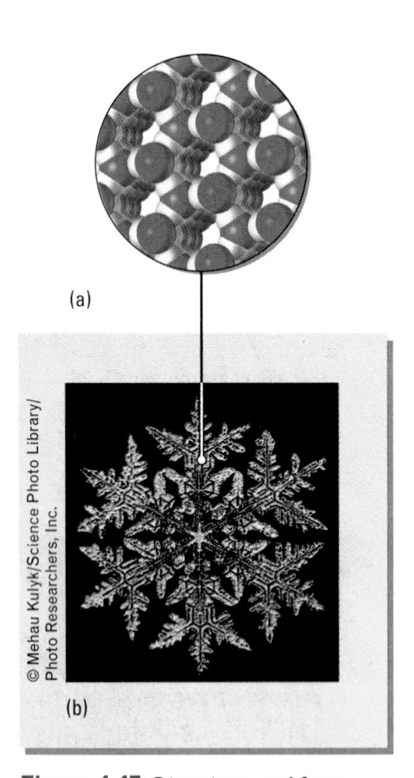

Figure 1.17 Structure and form. (a) In the nanoscale structure of ice, each water molecule occupies a position in a regular array or lattice. (b) The form of a snowflake reflects the hexagonal symmetry of the nanoscale structure of ice.

The kinetic-molecular theory of matter can also be used to interpret the properties of liquids, as shown in Figure 1.16b. Liquids are fluid because the atoms or molecules are arranged more haphazardly than in solids. Particles are not confined to specific locations but rather can move past one another. No particle goes very far without bumping into another—the particles in a liquid interact with their neighbors continually. Because the particles are usually a little farther apart in a liquid than in the corresponding solid, the volume is usually a little bigger. (Ice and liquid water, which are shown in Figure 1.16, are an important exception to this last generality. As you can see from the figure, the water molecules in ice are arranged so that there are empty hexagonal channels. When ice melts, these channels become partially filled by water molecules, accounting for the slightly smaller volume of the same mass of liquid water.)

Like liquids, gases are fluid because their nanoscale particles can easily move past one another. As shown in Figure 1.16c, the particles fly about to fill any container they are in; hence, a gas has no fixed shape or volume. In a gas the particles are much farther apart than in a solid or a liquid. They move significant distances before hitting other particles or the walls of the container. The particles also move quite rapidly. In air at room temperature, for example, the average molecule is going faster than 1000 miles per hour. A particle hits another particle every so often, but most of the time each is quite far away from all the others. Consequently, the nature of the particles is much less important in determining the properties of a gas.

Temperature can also be interpreted using the kinetic-molecular theory. The higher the temperature is, the more active the nanoscale particles are. A solid melts when its temperature is raised to the point where the particles vibrate fast enough and far enough to push each other out of the way and move out of their regularly spaced positions. The substance becomes a liquid because the particles are now behaving as they do in a liquid, bumping into one another and pushing past their neighbors. As the temperature goes higher, the particles move even faster, until finally they can escape the clutches of their comrades and become independent; the substance becomes a gas. Increasing temperature corresponds to faster and faster motions of atoms and molecules. This is a general rule that you will find useful in many future discussions of chemistry (Figure 1.18).

Using the kinetic-molecular theory to interpret the properties of solids, liquids, and gases and the effect of changing temperature provides a very simple example of how chemists use nanoscale theories and models to interpret and explain macroscale observations. In the remainder of this chapter and throughout your study of chemistry, you should try to imagine how the atoms and molecules are arranged and what they are doing whenever you consider a macroscale sample of matter. That is, you should try to develop the chemist's special perspective on the relation of nanoscale structure to macroscale behavior.

Because the nature of the particles is relatively unimportant in determining the behavior of gases, all gases can be described fairly accurately by the ideal gas law, which is introduced in Chapter 10.

The late Richard Feynmann, a Nobel laureate in physics, said, "If in some cataclysm all of scientific knowledge were to be destroyed, and only one sentence passed on to the next generation of creatures, what statement would contain the most information in the fewest words? I believe it is the atomic hypothesis, that all things are made of atoms, little particles that move around in perpetual motion."

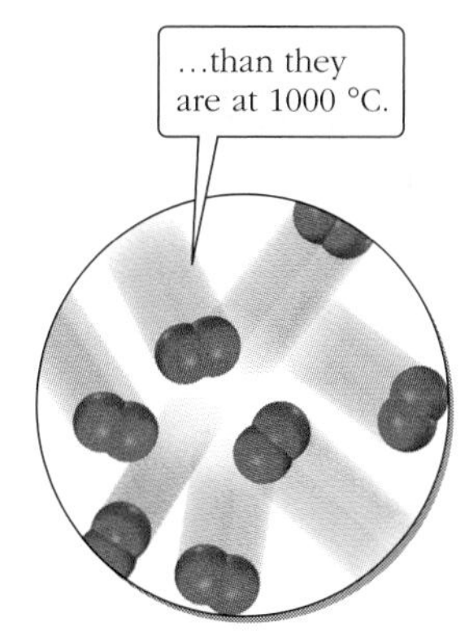
On average, gas molecules are moving much slower at 25 °C...
...than they are at 1000 °C.

Figure 1.18 Molecular speed and temperature.

CONCEPTUAL
EXERCISE 1.6 Kinetic-Molecular Theory

Use the idea that matter consists of tiny particles in motion to interpret each observation.

(a) An ice cube sitting in the sun slowly melts, and the liquid water eventually evaporates.
(b) Wet clothes hung on a line eventually dry.
(c) Moisture appears on the outside of a glass of ice water.
(d) Evaporation of a solution of sugar in water forms crystals.

1.9 The Atomic Theory

The existence of elements can be explained by a nanoscale model involving particles, just as the properties of solids, liquids, and gases can be. This model, which is closely related to the kinetic-molecular theory, is called the **atomic theory.** It was proposed in 1803 by John Dalton. According to Dalton's theory, an element cannot be decomposed into two or more new substances because at the nanoscale it consists of one and only one kind of atom and because atoms are indivisible under the conditions of chemical reactions. An **atom** is the smallest particle of an element that embodies the chemical properties of that element. An element, such as the sample of copper in Figure 1.19, is made up entirely of atoms of the same kind.

The fact that a compound can be decomposed into two or more different substances can be explained by saying that each compound must contain two or more different kinds of atoms. The process of decomposition involves separating at least one type of atom from atoms of the other kind(s). For example, charring of sugar corresponds to separating atoms of carbon from atoms of oxygen and atoms of hydrogen.

Dalton also said that each kind of atom must have its own properties—in particular, a characteristic mass. This idea allowed his theory to account for the masses of different elements that combine in chemical reactions to form compounds. An important success of Dalton's ideas was that they could be used to interpret known chemical facts quantitatively.

Two laws known in Dalton's time could be explained by the atomic theory. One was based on experiments in which the reactants were carefully weighed before a chemical reaction, and the reaction products were carefully collected and weighed afterward. The results led to the **law of conservation of mass** (also called the law of

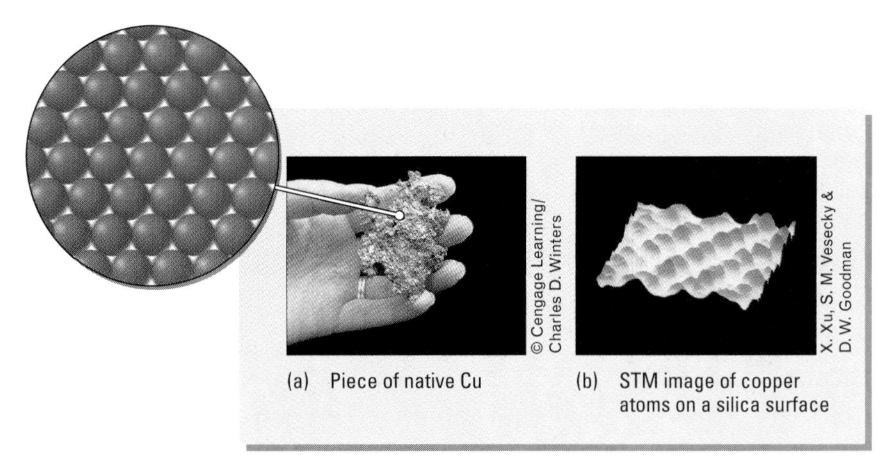

Figure 1.19 Elements, atoms, and the nanoscale world of chemistry. (a) A macroscopic sample of naturally occurring copper metal with a nanoscale, magnified representation of a tiny portion of its surface. It is clear that all the atoms in the sample of copper are the same kind of atoms. (b) A scanning tunneling microscopy (STM) image, enhanced by a computer, of a layer of copper atoms on the surface of silica (a compound of silicon and oxygen). The section of the layer shown is 1.7 nm square and the rows of atoms are separated by about 0.44 nm.

conservation of matter): *There is no detectable change in mass during an ordinary chemical reaction.* The atomic theory says that mass is conserved because the same number of atoms of each kind is present before and after a reaction, and each of those kinds of atoms has its same characteristic mass before and after the reaction.

Our bodies are made up of atoms from the distant past—atoms from other people and other things. Some of the carbon, hydrogen, and oxygen atoms in our carbohydrates have come from the breaths (first and last) of both famous and ordinary persons of the past.

The other law was based on the observation that in a chemical compound the proportions of the elements by mass are always the same. Water always contains 1 g hydrogen for every 8 g oxygen, and carbon monoxide always contains 4 g oxygen for every 3 g carbon. The **law of constant composition** summarizes such observations: *A chemical compound always contains the same elements in the same proportions by mass.* The atomic theory explains this observation by saying that atoms of different elements always combine in the same ratio in a compound. For example, in carbon monoxide there is always one carbon atom for each oxygen atom. If the mass of an oxygen atom is $\frac{4}{3}$ times the mass of a carbon atom, then the ratio of mass of oxygen to mass of carbon in carbon monoxide will always be 4:3.

Dalton's theory has been modified to account for discoveries since his time. The *modern* atomic theory is based on these assumptions:

All matter is composed of atoms, which are extremely tiny. Interactions among atoms account for the properties of matter.

All atoms of a given element have the same chemical properties. Atoms of different elements have different chemical properties.

According to the modern theory, atoms of the same element have the same chemical properties but are not necessarily identical in all respects. The discussion of isotopes in Chapter 2 shows how atoms of the same element can differ in mass.

Compounds are formed by the chemical combination of two or more different kinds of atoms. Atoms usually combine in the ratio of small whole numbers. For example, in a carbon monoxide molecule there is one carbon atom and one oxygen atom; a carbon dioxide molecule consists of one carbon atom and two oxygen atoms.

A chemical reaction involves joining, separating, or rearranging atoms. Atoms in the reactant substances form new combinations in the product substances. Atoms are not created, destroyed, or converted into other kinds of atoms during a chemical reaction.

The hallmark of a good theory is that it suggests new experiments, and this was true of the atomic theory. Dalton realized that it predicted a law that had not yet been discovered. If compounds are formed by combining atoms of different elements on the

ESTIMATION How Tiny Are Atoms and Molecules?

It is often useful to estimate an approximate value for something. Usually this estimation can be done quickly, and often it can be done without a calculator. The idea is to pick round numbers that you can work with in your head, or to use some other method that allows a quick estimate. If you really need an accurate value, an estimate is still useful to check whether the accurate value is in the right ballpark. Often an estimate is referred to as a "back-of-the-envelope" calculation, because estimates might be done over lunch on any piece of paper that is at hand. Some estimates are referred to as "order-of-magnitude calculations" because only the power of ten (the order of magnitude) in the answer is obtained. To help you develop estimation skills, most chapters in this book will provide you with an example of estimating something.

To get a more intuitive feeling for how small atoms and molecules are, estimate how many hydrogen atoms could fit inside a 12-oz (355-mL) soft-drink can. Make the same estimate for protein molecules. Use the approximate sizes given in Figure 1.14.

Because 1 mL is the same volume as a cube 1 cm on each side (1 cm^3), the volume of the can is the same as the volume of 355 cubes 1 cm on each side. Therefore, we can first estimate how many atoms would fit into a 1-cm cube and then multiply that number by 355.

According to Figure 1.14, a typical atom has a diameter slightly less than 100 pm. Because this is an estimate, and to make the numbers easy to handle, assume that we are dealing with an atom that is 100 pm in diameter. Then the atom's diameter is 100×10^{-12} m = 1×10^{-10} m, and it will require 10^{10} of these atoms lined up in a row to make a length of 1 m. Since 1 cm is $\frac{1}{100}$ (10^{-2}) of a meter, only $10^{-2} \times 10^{10} = 10^8$ atoms would fit in 1 cm.

In three dimensions, there could be 10^8 atoms along each of the three perpendicular edges of a 1-cm cube (the *x*, *y*, and *z* directions). The one row along the *x*-axis could be repeated 10^8 times along the *y*-axis, and then that layer of atoms could be repeated 10^8 times along the *z*-axis. Therefore, the number of atoms that we estimate would fit inside the cube is $10^8 \times 10^8 \times 10^8 = 10^{24}$ atoms. Multiplying this by 355 gives $355 \times 10^{24} = 3.6 \times 10^{26}$ atoms in the soft-drink can.

This estimate is a bit low. A hydrogen atom's diameter is less than 100 pm, so more hydrogen atoms would fit inside the can. Also, atoms are usually thought of as spheres, so they could pack together more closely than they would if just lined up in rows. Therefore, an even larger number of atoms than 3.6×10^{26} could fit inside the can.

For a typical protein molecule, Figure 1.14 indicates a diameter on the order of 5 nm = 5000 pm. That is 50 times bigger than the 100-pm diameter we used for the hydrogen atom. Thus, there would be 50 times fewer protein molecules in the *x* direction, 50 times fewer in the *y* direction, and 50 times fewer in the *z* direction. Therefore, the number of protein molecules would be fewer by $50 \times 50 \times 50$ = 125,000. The number of protein molecules can thus be estimated as $(3.6 \times 10^{26})/(1.25 \times 10^5)$. Because 3.6 is roughly three times 1.25, and because we are estimating, not calculating accurately, we can take the result to be 3×10^{21} protein molecules. That's still a whole lot of molecules!

Visit this book's companion website at **www.cengage.com/chemistry/moore** to work an interactive module based on this material.

nanoscale, then in some cases there might be more than a single combination. An example is carbon monoxide and carbon dioxide. In carbon monoxide there is one oxygen atom for each carbon atom, while in carbon dioxide there are two oxygen atoms per carbon atom. Therefore, in carbon dioxide the mass of oxygen per gram of carbon ought to be twice as great as it is in carbon monoxide (because twice as many oxygen atoms will weigh twice as much). Dalton called this the **law of multiple proportions,** and he carried out quantitative experiments seeking data to confirm or deny it. Dalton and others obtained data consistent with the law of multiple proportions, thereby enhancing acceptance of the atomic theory.

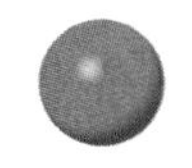

oxygen atom

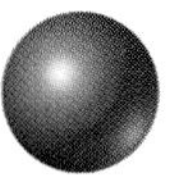

carbon atom

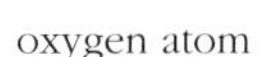

carbon monoxide

carbon dioxide

1.10 The Chemical Elements

Every element has been given a unique *name* and a *symbol* derived from the name. These names and symbols are listed in the periodic table inside the front cover of the book. The first letter of each symbol is capitalized; the second letter, if there is one, is lowercase, as in He, the symbol for helium. Elements discovered a long time ago have names and symbols with Latin or other origins, such as Au for gold (from *aurum,* meaning "bright dawn") and Fe for iron (from *ferrum*). The names of more

Table 1.3 The Names of Some Chemical Elements

Element	Symbol	Date Discovered	Discoverer	Derivation of Name/Symbol
Carbon	C	Ancient	Ancient	Latin, *carbo* (charcoal)
Curium	Cm	1944	G. Seaborg, et al.	Honoring Marie and Pierre Curie, Nobel Prize winners for discovery of radioactive elements
Hydrogen	H	1766	H. Cavendish	Greek, *hydro* (water) and *genes* (generator)
Meitnerium	Mt	1982	P. Armbruster, et al.	Honoring Lise Meitner, codiscoverer of nuclear fission
Mendelevium	Md	1955	G. Seaborg, et al.	Honoring Dmitri Mendeleev, who devised the periodic table
Mercury	Hg	Ancient	Ancient	For Mercury, messenger of the gods, because it flows quickly; symbol from Greek *hydrargyrum,* liquid silver
Polonium	Po	1898	M. Curie and P. Curie	In honor of Poland, Marie Curie's native country
Seaborgium	Sg	1974	G. Seaborg, et al.	Honoring Glenn Seaborg, Nobel Prize winner for synthesis of new elements
Sodium	Na	1807	H. Davy	Latin, *soda* (sodium carbonate); symbol from Latin *natrium*
Tin	Sn	Ancient	Ancient	German, *Zinn;* symbol from Latin, *stannum*

recently discovered elements are derived from their place of discovery or from a person or place of significance (Table 1.3).

Elements are being synthesized even now. Element 115 was made in 2004, but only a few atoms were observed.

Ancient people knew of nine elements—gold (Au), silver (Ag), copper (Cu), tin (Sn), lead (Pb), mercury (Hg), iron (Fe), sulfur (S), and carbon (C). Most of the other naturally occurring elements were discovered during the 1800s, as one by one they were separated from minerals in the earth's crust or from the earth's oceans or atmosphere. Currently, more than 110 elements are known, but only 90 occur in nature. Elements such as technetium (Tc), neptunium (Np), mendelevium (Md), seaborgium (Sg), and meitnerium (Mt) have been made using nuclear reactions, beginning in the 1930s.

Figure 1.20 Some metallic elements—iron, aluminum, copper, and gold. The steel ball bearing is principally iron. The rod is made of aluminum. The inner coil is gold and the other one is copper. Metals are malleable, ductile, and conduct electricity.

Types of Elements

The vast majority of the elements are **metals**—only 24 are not. You are probably familiar with many properties of metals. At room temperature they are solids (except for mercury, which is a liquid), they conduct electricity (and conduct better as the temperature decreases), they are ductile (can be drawn into wires), they are malleable (can be rolled into sheets), and they can form alloys (solutions of one or more metals in another metal). In a solid metal, individual metal atoms are packed close to each other, so metals usually have fairly high densities. Figure 1.20 shows some common metals. Iron (Fe) and aluminum (Al) are used in automobile parts because of their ductility, malleability, and relatively low cost. Copper (Cu) is used in electrical wiring because it conducts electricity better than most metals. Gold (Au) is used for the vital electrical contacts in automobile air bags and in some computers because it does not corrode and is an excellent electrical conductor.

On the periodic table inside the front cover of this book, the metals, nonmetals, and metalloids are color-coded: gray and blue for metals, lavender for nonmetals, and orange for metalloids.

In contrast, **nonmetals** do not conduct electricity (with a few exceptions, such as graphite, one form of carbon). Nonmetals are more diverse in their physical properties than are metals (Figure 1.21). At room temperature some nonmetals are solids (such as phosphorus, sulfur, and iodine), bromine is a liquid, and others are gases (such as hydrogen, nitrogen, and chlorine). The nonmetals helium (He), neon (Ne), argon (Ar), krypton (Kr), xenon (Xe), and radon (Rn) are gases that consist of individual atoms.

A few elements—boron, silicon, germanium, arsenic, antimony, and tellurium—are classified as **metalloids.** Some properties of metalloids are typical of metals and other properties are characteristic of nonmetals. For example, some metalloids are shiny like metals, but they do not conduct electricity as well as metals. Many of the metalloids are semiconductors and are essential for the electronics industry.

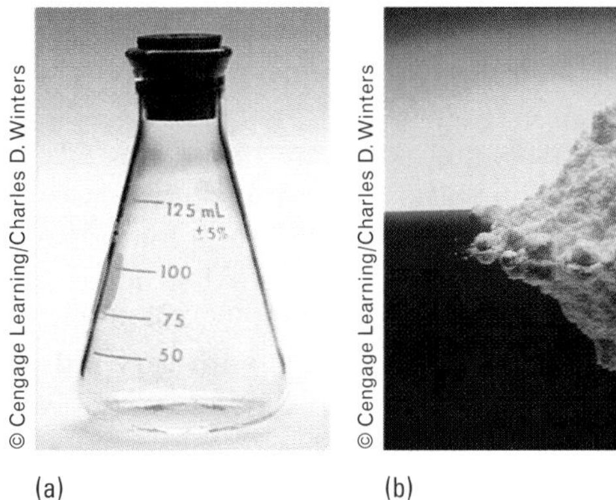

© Cengage Learning/Charles D. Winters

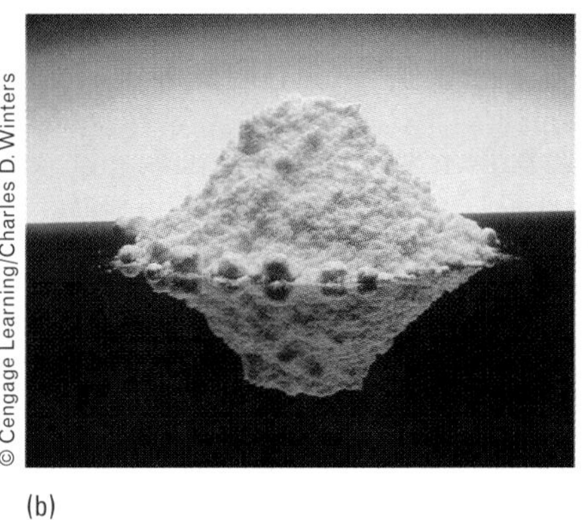

© Cengage Learning/Charles D. Winters

© Cengage Learning/Larry Cameron

(a) (b) (c) (d)

Figure 1.21 Some nonmetallic elements—(a) chlorine, (b) sulfur, (c) bromine, (d) iodine. Nonmetals occur as solids, liquids, and gases and have very low electrical conductivities. Bromine is the only nonmetal that is a liquid at room temperature.

EXERCISE 1.7 Elements

Use Table 1.3, the periodic table inside the front cover, and/or the list of elements inside the front cover to answer these questions.

(a) Four elements are named for planets in our solar system (including the ex-planet Pluto). Give their names and symbols.

(b) One element is named for a state in the United States. Name the element and give its symbol.

(c) Two elements are named in honor of women. What are their names and symbols?

(d) Several elements are named for countries or regions of the world. Find at least four of these and give names and symbols.

(e) List the symbols of all elements that are nonmetals.

Chemical bonds are strong attractions that hold atoms together. Bonding is discussed in detail in Chapter 8.

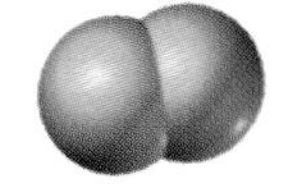

Space-filling model of Cl_2

Elements that consist of diatomic molecules are H_2, N_2, O_2, F_2, Cl_2, Br_2, and I_2. You need to remember that these elements consist of diatomic molecules, because they will be encountered frequently. Most of these elements are close together in the periodic table, which makes it easier to remember them.

Elements That Consist of Molecules

Most elements that are nonmetals consist of molecules on the nanoscale. A **molecule** is a unit of matter in which two or more atoms are chemically bonded together. For example, a chlorine molecule contains two chlorine atoms and can be represented by the chemical formula Cl_2. A **chemical formula** uses the symbols for the elements to represent the atomic composition of a substance. In gaseous chlorine, Cl_2 molecules are the particles that fly about and collide with each other and the container walls. Molecules, like Cl_2, that consist of two atoms are called **diatomic molecules.** Oxygen, O_2, and nitrogen, N_2, also exist as diatomic molecules, as do hydrogen, H_2; fluorine, F_2; bromine, Br_2; and iodine, I_2.

EXERCISE 1.8 Elements That Consist of Diatomic Molecules or Are Metalloids

On a copy of the periodic table, circle the symbols of the elements that

(a) Consist of diatomic molecules; (b) Are metalloids.

Devise rules related to the periodic table that will help you to remember which elements these are.

Allotropes

Oxygen and carbon are among the elements that exist as **allotropes,** different forms of the same element in the same physical state at the same temperature and pressure. Allotropes are possible because the same kind of atoms can be connected in different ways when they form molecules. For example, the allotropes of oxygen are O_2, sometimes called dioxygen, and O_3, ozone. Dioxygen, a major component of earth's atmos-

oxygen molecule, O_2

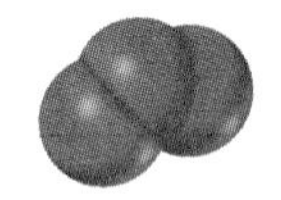

ozone molecule, O_3

A C_{60} molecule's size compared with a soccer ball is almost the same as a soccer ball's size compared with planet Earth.

For an up-to-date description of research on nanotubes, see "The Nanotube Site," **http://www.pa.msu.edu/cmp/csc/nanotube.html.**

phere, is by far the more common allotropic form. Ozone is a highly reactive, pale blue gas first detected by its characteristic pungent odor. Its name comes from *ozein,* a Greek word meaning "to smell."

Diamond and graphite, known for centuries, have quite different properties. Diamond is a hard, colorless solid and graphite is a soft, black solid, but both consist entirely of carbon atoms. This makes them allotropes of carbon, and for a long time they were thought to be the only allotropes of carbon with well-defined structures. Therefore, it was a surprise in the 1980s when another carbon allotrope was discovered in soot produced when carbon-containing materials are burned with very little oxygen. The new allotrope consists of 60-carbon atom cages and represents a new class of molecules. The C_{60} molecule resembles a soccer ball with a carbon atom at each corner of each of the black pentagons in Figure 1.22b. Each five-membered ring of carbon atoms is surrounded by five six-membered rings. This molecular structure of carbon pentagons and hexagons reminded its discoverers of a geodesic dome (Figure 1.22a), a structure popularized years ago by the innovative American philosopher and engineer R. Buckminster Fuller. Therefore, the official name of the C_{60} allotrope is buckminsterfullerene. Chemists often call it simply a "buckyball." C_{60} buckyballs belong to a larger molecular family of even-numbered carbon cages that is collectively called fullerenes.

Carbon atoms can also form concentric tubes that resemble rolled-up chicken wire. These single- and multi-walled *nanotubes* of only carbon atoms are excellent electrical conductors and extremely strong. Imagine the exciting applications for such properties, including making buckyfibers that could substitute for the metal wires now used to transmit electrical power. Dozens of uses have been proposed for fullerenes, buckytubes, and buckyfibers, among them microscopic ball bearings, lightweight batteries, new lubricants, nanoscale electric switches, new plastics, and antitumor therapy for cancer patients (by enclosing a radioactive atom within the cage). All these applications await an inexpensive way of making buckyballs and other fullerenes. Currently buckyballs, the cheapest fullerene, are more expensive than gold.

EXERCISE 1.9 Allotropes

A student says that tin and lead are allotropes because they are both dull gray metals. Why is the statement wrong?

Grant Heilman Photography

(a)

© Cengage Learning/Charles D. Winters

(b)

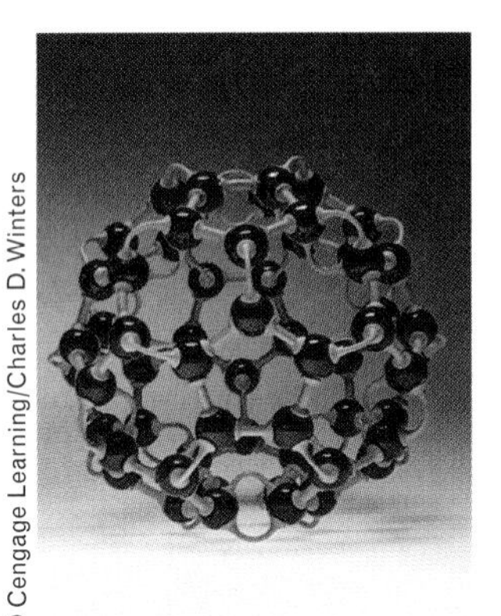

© Cengage Learning/Charles D. Winters

(c)

Figure 1.22 Models for fullerenes. (a) Geodesic domes at Elmira College, Elmira, New York. Geodesic domes, such as those designed originally by R. Buckminster Fuller, contain linked hexagons and pentagons. (b) A soccer ball is a model for the C_{60} structure. (c) The C_{60} fullerene molecule, which is made up of five-membered rings (black rings on the soccer ball) and six-membered rings (white rings on the ball).

1.11 Communicating Chemistry: Symbolism

Chemical symbols—such as Na, I, or Mt—are a shorthand way of indicating what kind of atoms we are talking about. Chemical formulas tell us how many atoms of an element are combined in a molecule and in what ratios atoms are combined in compounds. For example, the formula Cl_2 tells us that there are two chlorine atoms in a chlorine molecule. The formulas CO and CO_2 tell us that carbon and oxygen form two different compounds—one that has equal numbers of C and O atoms and one that has twice as many O atoms as C atoms. In other words, chemical symbols and formulas symbolize the nanoscale composition of each substance.

Chemical symbols and formulas also represent the macroscale properties of elements and compounds. That is, the symbol Na brings to mind a highly reactive metal, and the formula H_2O represents a colorless liquid that freezes at 0 °C, boils at 100 °C, and reacts violently with Na. Because chemists are familiar with both the nanoscale and macroscale characteristics of substances, they usually use symbols to abbreviate their representations of both. Symbols are also useful for representing chemical reactions. For example, the charring of sucrose mentioned earlier is represented by

Sucrose	$\longrightarrow$	Carbon	+	Water
$C_{12}H_{22}O_{11}$	$\longrightarrow$	12 C	+	11 H_2O
Reactant	changes to		Products	

The symbolic aspect of chemistry is the third part of the chemist's special view of the world. It is important that you become familiar and comfortable with using chemical symbols and formulas to represent chemical substances and their reactions. Figure 1.23 shows how chemical symbolism can be applied to the process of decomposing water with electricity (electrolysis of water).

AP Photo/Michael Scate

Sir Harold Kroto 1939–

Along with the late Richard E. Smalley and Robert Curl, Harold Kroto received the Nobel Prize in Chemistry in 1996 for discovering fullerenes. In the same year Kroto, who is British, received a knighthood for his work. At the time of the discovery, Smalley assembled paper hexagons and pentagons to make a model of the C_{60} molecule. He asked Kroto, "Who was the architect who worked with big domes?" When Kroto replied, "Buckminster Fuller," the two shouted with glee, "It's Buckminster Fuller—ene!" The structure's name had been coined.

PROBLEM-SOLVING EXAMPLE 1.3 Macroscale, Nanoscale, and Symbolic Representations

The figure shows a sample of water boiling. In spaces labeled A, indicate whether the macroscale or the nanoscale is represented. In spaces labeled B, draw the molecules that would be present with appropriate distances between them. One of the circles represents a bubble of gas within the liquid. The other represents the liquid. In space C, write a symbolic representation of the boiling process.

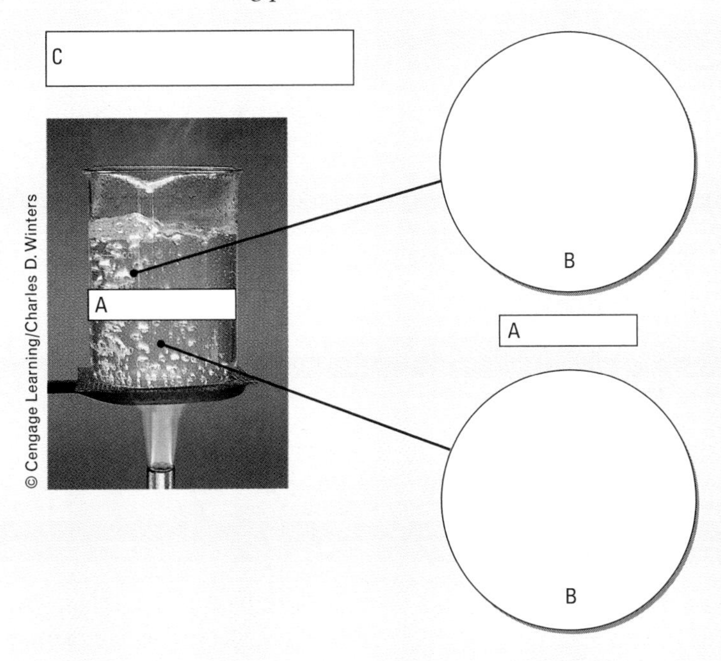

© Cengage Learning/Charles D. Winters

Answer

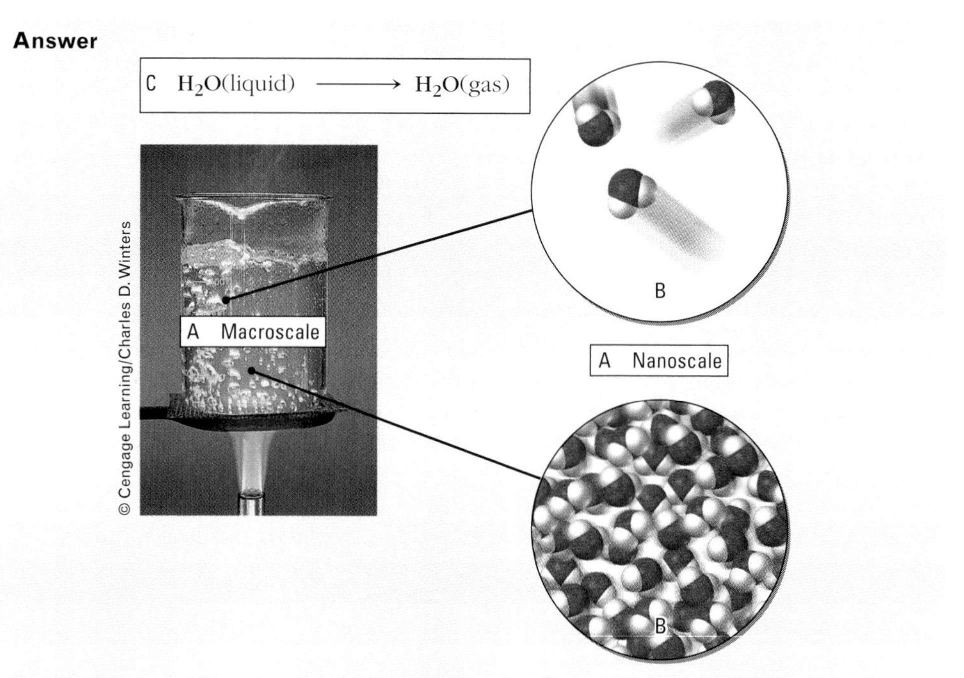

C H_2O(liquid) ⟶ H_2O(gas)
A Macroscale
B
A Nanoscale
B
© Cengage Learning/Charles D. Winters

Explanation Each water molecule consists of two hydrogen atoms and one oxygen atom. In liquid water the molecules are close together and oriented in various directions. In a bubble of gaseous water the molecules are much farther apart—there are fewer of them per unit volume. The symbolic representation is the equation

$$H_2O(\text{liquid}) \longrightarrow H_2O(\text{gas})$$

PROBLEM-SOLVING PRACTICE 1.3

Draw a nanoscale representation and a symbolic representation for both allotropes of oxygen. Describe the properties of each allotrope at the macroscale.

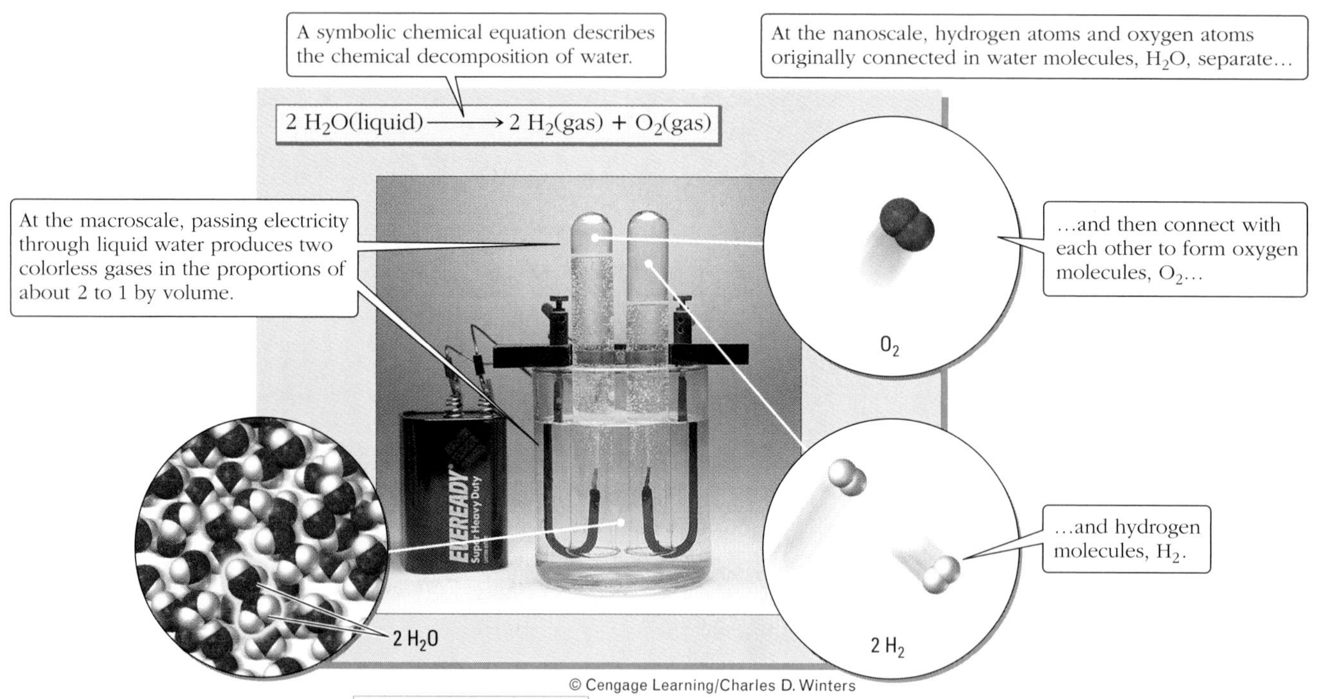

A symbolic chemical equation describes the chemical decomposition of water.
At the nanoscale, hydrogen atoms and oxygen atoms originally connected in water molecules, H_2O, separate...
2 H_2O(liquid) ⟶ 2 H_2(gas) + O_2(gas)
At the macroscale, passing electricity through liquid water produces two colorless gases in the proportions of about 2 to 1 by volume.
...and then connect with each other to form oxygen molecules, O_2...
O_2
EVEREADY
Super Heavy Duty
...and hydrogen molecules, H_2.
2 H_2O
2 H_2

© Cengage Learning/Charles D. Winters

Active Figure 1.23 **Symbolic, macroscale, and nanoscale representations of a chemical reaction.** Visit this book's companion website at **www.cengage.com/chemistry/moore** to test your understanding of the concepts in this figure.

1.12 Modern Chemical Sciences

Our goal in this chapter has been to make clear many of the reasons you should care about chemistry. Chemistry is fundamental to understanding many other sciences and to understanding how the material world around us works. Chemistry provides a unique, nanoscale perspective that has been highly successful in stimulating scientific inquiry and in the development of high-tech materials, modern medicines, and many other advances that benefit us every day. Chemistry is happening all around us and within us all the time. Knowledge of chemistry is a key to understanding and making the most of our internal and external environments and to making better decisions about how we live our lives and structure our economy and society. Finally, chemistry—the properties of elements and compounds, the nanoscale theories and models that interpret those properties, and the changes of one kind of substance into another—is just plain interesting and fun. Chemistry presents an intellectual challenge and provides ways to satisfy intellectual curiosity while helping us to better understand the world in which we live.

Modern chemistry overlaps more and more with biology, medicine, engineering, and other sciences. Because it is central to understanding matter and its transformations, chemistry becomes continually more important in a world that relies on chemical knowledge to produce the materials and energy required for a comfortable and productive way of life. The breadth of chemistry is recognized by the term "chemical sciences," which includes chemistry, chemical engineering, chemical biology, materials chemistry, industrial chemistry, environmental chemistry, medicinal chemistry, and many other fields. Practitioners of the chemical sciences produce new plastics, medicines, superconductors, composite materials, and electronic devices. The chemical sciences enable better understanding of the mechanisms of biological processes, how proteins function, and how to imitate the action of organisms that carry out important functions. The chemical sciences enable us to measure tiny concentrations of substances, separate one substance from another, and determine whether a given substance is helpful or harmful to humans. Practitioners of the chemical sciences create new industrial processes and products that are less hazardous, produce less pollution, and generate smaller quantities of wastes. The enthusiasm of chemists for research in all of these areas and the many discoveries that are made every day offer ample evidence that chemistry is an energetic and exciting science. We hope that this excitement is evident in this chapter and in the rest of the book.

IN CLOSING

Having studied this chapter, you should be able to . . .

- Appreciate the power of chemistry to answer intriguing questions (Section 1.1).
- Describe the approach used by scientists in solving problems (Sections 1.2, 1.3).
- Understand the differences among a hypothesis, a theory, and a law (Section 1.3).
- Define quantitative and qualitative observations (Section 1.3).
- Identify the physical properties of matter or physical changes occurring in a sample of matter (Section 1.4). ■ End-of-chapter Question assignable in OWL: 14
- Estimate Celsius temperatures for commonly encountered situations (Section 1.4).
- Calculate mass, volume, or density, given any two of the three (Section 1.4).
- Identify the chemical properties of matter or chemical changes occurring in a sample of matter (Section 1.5). ■ Question 20
- Explain the difference between homogeneous and heterogeneous mixtures (Section 1.6).
- Describe the importance of separation, purification, and analysis (Section 1.6).

Selected end-of-chapter Questions may be assigned in OWL.

For quick review, download Go Chemistry mini-lecture flashcard modules (or purchase them at **www.ichapters.com**).

- Understand the difference between a chemical element and a chemical compound (Sections 1.7, 1.9). ■ Questions 26, 28
- Classify matter (Section 1.7, Figure 1.13).
- Describe characteristic properties of the three states of matter—gases, liquids, and solids (Section 1.8).
- Identify relative sizes at the macroscale, microscale, and nanoscale levels (Section 1.8).
- Describe the kinetic-molecular theory at the nanoscale level (Section 1.8).
- Use the postulates of modern atomic theory to explain macroscopic observations about elements, compounds, conservation of mass, constant composition, and multiple proportions (Section 1.9). ■ Questions 37, 46, 52
- Distinguish metals, nonmetals, and metalloids according to their properties (Section 1.10). ■ Questions 44, 46, 64
- Identify elements that consist of molecules, and define allotropes (Section 1.10).
- Distinguish among macroscale, nanoscale, and symbolic representations of substances and chemical processes (Section 1.11).

KEY TERMS

The following terms were defined and given in boldface type in this chapter. Be sure to understand each of these terms and the concepts with which they are associated. (The number of the section where each term is introduced is given in parentheses.)

allotrope *(Section 1.10)*
atom *(1.9)*
atomic theory *(1.9)*
boiling point *(1.4)*
Celsius temperature scale *(1.4)*
chemical change *(1.5)*
chemical compound *(1.7)*
chemical element *(1.7)*
chemical formula *(1.10)*
chemical property *(1.5)*
chemical reaction *(1.5)*
chemistry *(1.1)*
conservation of mass, law of *(1.9)*
constant composition, law of *(1.9)*
conversion factor *(1.4)*
density *(1.4)*
diatomic molecule *(1.10)*
dimensional analysis *(1.4)*
energy *(1.5)*
gas *(1.8)*
heterogeneous mixture *(1.6)*
homogeneous mixture *(1.6)*
hypothesis *(1.3)*
kinetic-molecular theory *(1.8)*
law *(1.3)*
liquid *(1.8)*
macroscale *(1.8)*
matter *(1.1)*
melting point *(1.4)*
metal *(1.10)*
metalloid *(1.10)*
microscale *(1.8)*
model *(1.3)*
molecule *(1.10)*
multiple proportions, law of *(1.9)*
nanoscale *(1.8)*
nonmetal *(1.10)*
physical changes *(1.4)*
physical properties *(1.4)*
product *(1.5)*
proportionality factor *(1.4)*
qualitative *(1.3)*
quantitative *(1.3)*
reactant *(1.5)*
solid *(1.8)*
solution *(1.6)*
substance *(1.4)*
temperature *(1.4)*
theory *(1.3)*

QUESTIONS FOR REVIEW AND THOUGHT

■ denotes questions assignable in OWL.

Blue-numbered questions have short answers at the back of this book and fully worked solutions in the *Student Solutions Manual.*

Review Questions

These questions test vocabulary and simple concepts.

1. Choose an object in your room, such as a CD player or television set. Write down five qualitative observations and five quantitative observations regarding the object you chose.
2. What are three important characteristics of a scientific law? Name two laws that were mentioned in this chapter. State each of the laws that you named.
3. How does a scientific theory differ from a law? How are theories and models related?
4. What is the unique perspective that chemists use to make sense out of the material world? Give at least one example of how that perspective can be applied to a significant problem.
5. Give two examples of situations in which purity of a chemical substance is important.

Topical Questions

These questions are keyed to the major topics in the chapter. Usually a question that is answered at the end of this book is paired with a similar one that is not.

Why Care About Chemistry?

6. Make a list of at least four issues faced by our society that require scientific studies and scientific data before a democratic society can make informed, rational decisions. Exchange lists with another student and evaluate the quality of each other's choices.
7. Make a list of at least four questions you have wondered about that may involve chemistry. Compare your list with a list from another student taking the same chemistry course. Evaluate the quality of each other's questions and decide how "chemical" they are.

How Science Is Done

8. Identify the information in each sentence as qualitative or quantitative.
   (a) The element gallium melts at 29.8 °C.
   (b) A chemical compound containing cobalt and chlorine is blue.
   (c) Aluminum metal is a conductor of electricity.
   (d) The chemical compound ethanol boils at 79 °C.
   (e) A chemical compound containing lead and sulfur forms shiny, plate-like, yellow crystals.
9. Make as many qualitative and quantitative observations as you can regarding what is shown in the photograph.

© Cengage Learning/Charles D. Winters

Identifying Matter: Physical Properties

10. The elements sulfur and bromine are shown in the photograph. Based on the photograph, describe as many properties of each sample as you can. Are any properties the same? Which properties are different?

© Cengage Learning/Charles D. Winters

Sulfur and bromine. The sulfur is on the flat dish; the bromine is in a closed flask supported on a cork ring.

11. In the accompanying photo, you see a crystal of the mineral calcite surrounded by piles of calcium and carbon, two of the elements that combine to make the mineral. (The other element combined in calcite is oxygen.) Based on the photo, describe some of the physical properties of the elements and the mineral. Are any properties the same? Are any properties different?

© Cengage Learning/Charles D. Winters

Calcite (the transparent, cube-like crystal) and two of its constituent elements, calcium (chips) and carbon (black grains). The calcium chips are covered with a thin film of calcium oxide.

12. The boiling point of a liquid is 20 °C. If you hold a sample of the substance in your hand, will it boil? Explain briefly.
13. Dry Ice (solid carbon dioxide) sublimes (changes from solid to gas without forming liquid) at −78.6 °C. Suppose you had a sample of gaseous carbon dioxide and the temperature was 30° F below zero (a very cold day). Would solid carbon dioxide form? Explain briefly how you answered the question.
14. ■ A 105.5-g sample of a metal was placed into water in a graduated cylinder, and it completely submerged. The water level rose from 25.4 mL to 37.2 mL. Use data in Table 1.1 to identify the metal.
15. An irregularly shaped piece of lead weighs 10.0 g. It is carefully lowered into a graduated cylinder containing 30.0 mL ethanol, and it sinks to the bottom of the cylinder. To what volume reading does the ethanol rise?
16. An unknown sample of a metal is 1.0 cm thick, 2.0 cm wide, and 10.0 cm long. Its mass is 54.0 g. Use data in Table 1.1 to identify the metal. (Remember that 1 cm^3 = 1 mL.)
17. Find the volume occupied by a 4.33-g sample of benzene.

Chemical Changes and Chemical Properties

18. In each case, identify the underlined property as a physical or chemical property. Give a reason for your choice.
    (a) The normal <u>color</u> of the element bromine is red-orange.
    (b) Iron is <u>transformed into rust</u> in the presence of air and water.
    (c) Dynamite can <u>explode</u>.
    (d) Aluminum metal, the <u>shiny</u> "foil" used in the kitchen, <u>melts</u> at 660 °C.
19. In each case, identify the underlined property as a physical or a chemical property. Give a reason for your choice.
    (a) Dry Ice <u>sublimes</u> (changes directly from a solid to a gas) at −78.6 °C.
    (b) Methanol (methyl alcohol) <u>burns in air</u> with a colorless flame.
    (c) Sugar is <u>soluble in water</u>.
    (d) Hydrogen peroxide, H_2O_2, <u>decomposes to form oxygen, O_2, and water, H_2O</u>.
20. ■ In each case, describe the change as a chemical or physical change. Give a reason for your choice.
    (a) A cup of household bleach changes the color of your favorite T-shirt from purple to pink.
    (b) The fuels in the space shuttle (hydrogen and oxygen) combine to give water and provide the energy to lift the shuttle into space.
    (c) An ice cube in your glass of lemonade melts.
21. In each case, describe the change as a chemical or physical change. Give a reason for your choice.
    (a) Salt dissolves when you add it to water.
    (b) Food is digested and metabolized in your body.
    (c) Crystalline sugar is ground into a fine powder.
    (d) When potassium is added to water there is a purplish-pink flame and the water becomes basic (alkaline).

Classifying Matter: Substances and Mixtures

22. Small chips of iron are mixed with sand (see photo). Is this a homogeneous or heterogeneous mixture? Suggest a way to separate the iron and sand from each other.

Layers of sand, iron, and sand.

23. Suppose that you have a solution of sugar in water. Is this a homogeneous or heterogeneous mixture? Describe an experimental procedure by which you can separate the two substances.
24. Identify each of these as a homogeneous or a heterogeneous mixture.
    (a) Vodka (b) Blood
    (c) Cowhide (d) Bread
25. Identify each of these as a homogeneous or a heterogeneous mixture.
    (a) An asphalt (blacktop) road (b) Clear ocean water
    (c) Iced tea with ice cubes (d) Filtered apple cider

Classifying Matter: Elements and Compounds

26. ■ For each of the changes described, decide whether two or more elements formed a compound or if a compound decomposed (to form elements or other compounds). Explain your reasoning in each case.
    (a) Upon heating, a blue powder turned white and lost mass.
    (b) A white solid forms three different gases when heated. The total mass of the gases is the same as that of the solid.
27. For each of the changes described, decide whether two or more elements formed a compound or if a compound decomposed (to form elements or other compounds). Explain your reasoning in each case.
    (a) After a reddish-colored metal is placed in a flame, it turns black and has a higher mass.
    (b) A white solid is heated in oxygen and forms two gases. The mass of the gases is the same as the masses of the solid and the oxygen.
28. Classify each of these with regard to the type of matter (element, compound, heterogeneous mixture, or homogeneous mixture). Explain your choice in each case.
    (a) A piece of newspaper
    (b) Solid, granulated sugar
    (c) Freshly squeezed orange juice
    (d) Gold jewelry

29. Classify each of these as an element, a compound, a heterogeneous mixture, or a homogeneous mixture. Explain your choice in each case.
    (a) Table salt (sodium chloride)
    (b) Methane (which burns in pure oxygen to form only carbon dioxide and water)
    (c) Chocolate chip cookie
    (d) Silicon
30. A black powder is placed in a long glass tube. Hydrogen gas is passed into the tube so that the hydrogen sweeps out all other gases. The powder is then heated with a Bunsen burner. The powder turns red-orange, and water vapor can be seen condensing at the unheated far end of the tube. The red-orange color remains after the tube cools.
    (a) Was the original black substance an element? Explain briefly.
    (b) Is the new red-orange substance an element? Explain briefly.
31. A finely divided black substance is placed in a glass tube filled with air. When the tube is heated with a Bunsen burner, the black substance turns red-orange. The total mass of the red-orange substance is greater than that of the black substance.
    (a) Can you conclude that the black substance is an element? Explain briefly.
    (b) Can you conclude that the red-orange substance is a compound? Explain briefly.

Nanoscale Theories and Models

32. ■ The accompanying photo shows a crystal of the mineral halite, a form of ordinary salt. Are these crystals at the macroscale, microscale, or nanoscale? How would you describe the shape of these crystals? What might this tell you about the arrangement of the atoms deep inside the crystal?

A halite (sodium chloride) crystal.

33. The photograph shows an end-on view of tiny wires made from nickel metal by special processing. The scale bar is 1 μm long. Are these wires at the macroscale, microscale, or nanoscale?

Nickel wires. (The scale bar is 1 μm long.)

34. When you open a can of soft drink, the carbon dioxide gas inside expands rapidly as it rushes from the can. Describe this process in terms of the kinetic-molecular theory.
35. After you wash your clothes, you hang them on a line in the sun to dry. Describe the change or changes that occur in terms of the kinetic-molecular theory. Are the changes that occur physical or chemical changes?
36. Sucrose has to be heated to a high temperature before it caramelizes. Use the kinetic-molecular theory to explain why sugar caramelizes only at high temperatures.
37. ■ Give a nanoscale interpretation of the fact that at the melting point the density of solid mercury is greater than the density of liquid mercury, and at the boiling point the density of liquid mercury is greater than the density of gaseous mercury.
38. ■ Make these unit conversions, using the prefixes in Table 1.2.
    (a) 32.75 km to meters (b) 0.0342 mm to nanometers
    (c) 1.21×10^{-12} km to micrometers
39. Make these unit conversions, using the prefixes in Table 1.2.
    (a) 0.00572 kg to grams (b) 8.347×10^{7} nL to liters
    (c) 423.7 g to kilograms

The Atomic Theory

40. Explain in your own words, by writing a short paragraph, how the atomic theory explains conservation of mass during a chemical reaction and during a physical change.
41. Explain in your own words, by writing a short paragraph, how the atomic theory explains constant composition of chemical compounds.
42. Explain in your own words, by writing a short paragraph, how the atomic theory predicts the law of multiple proportions.
43. The element chromium forms three different oxides (that contain only chromium and oxygen). The percentage of chromium (number of grams of chromium in 100 g oxide) in these compounds is 52.0%, 68.4%, and 76.5%. Do these data conform to the law of multiple proportions? Explain why or why not.

The Chemical Elements

44. ■ Name and give the symbols for two elements that
    (a) Are metals (b) Are nonmetals
    (c) Are metalloids (d) Consist of diatomic molecules
45. ■ Name and give the symbols for two elements that
    (a) Are gases at room temperature
    (b) Are solids at room temperature
    (c) Do not consist of molecules
    (d) Have different allotropic forms

Communicating Chemistry: Symbolism

46. ■ Write a chemical formula for each substance, and draw a nanoscale picture of how the molecules are arranged at room temperature.
    (a) Water, a liquid whose molecules contain two hydrogen atoms and one oxygen atom each
    (b) Nitrogen, a gas that consists of diatomic molecules
    (c) Neon
    (d) Chlorine
47. Write a chemical formula for each substance and draw a nanoscale picture of how the molecules are arranged at room temperature.
    (a) Iodine, a solid that consists of diatomic molecules
    (b) Ozone
    (c) Helium
    (d) Carbon dioxide
48. Write a nanoscale representation and a symbolic representation and describe what happens at the macroscale when hydrogen reacts chemically with oxygen to form water vapor.
49. Write a nanoscale representation and a symbolic representation and describe what happens at the macroscale when bromine evaporates to form bromine vapor.

General Questions

These questions are not explicitly keyed to chapter topics; many require integration of several concepts.

50. Classify the information in each of these statements as quantitative or qualitative and as relating to a physical or chemical property.
    (a) A white chemical compound has a mass of 1.456 g. When placed in water containing a dye, it causes the red color of the dye to fade to colorless.
    (b) A sample of lithium metal, with a mass of 0.6 g, was placed in water. The metal reacted with the water to produce the compound lithium hydroxide and the element hydrogen.
51. Classify the information in each of these statements as quantitative or qualitative and as relating to a physical or chemical property.
    (a) A liter of water, colored with a purple dye, was passed through a charcoal filter. The charcoal adsorbed the dye, and colorless water came through. Later, the purple dye was removed from the charcoal and retained its color.
    (b) When a white powder dissolved in a test tube of water, the test tube felt cold. Hydrochloric acid was then added, and a white solid formed.
52. ■ The density of solid potassium is 0.86 g/mL. The density of solid calcium is 1.55 g/mL, almost twice as great. However, the mass of a potassium atom is only slightly less than the mass of a calcium atom. Provide a nanoscale explanation of these facts.
53. The density of gaseous helium at 25 °C and normal atmospheric pressure is 1.64×10^{-4} g/mL. At the same temperature and pressure the density of argon gas is 1.63×10^{-3} g/mL. The mass of an atom of argon is almost exactly ten times the mass of an atom of helium. Provide a nanoscale explanation of why the densities differ as they do.

Applying Concepts

These questions test conceptual learning.

54. ■ Using Table 1.1, but without using your calculator, decide which has the larger mass:
    (a) 20. mL butane or 20. mL bromobenzene
    (b) 10. mL benzene or 1.0 mL gold
    (c) 0.732 mL copper or 0.732 mL lead
55. Using Table 1.1, but without using your calculator, decide which has the larger volume:
    (a) 1.0 g ethanol or 1.0 g bromobenzene
    (b) 10. g aluminum or 12. g water
    (c) 20 g gold or 40 g magnesium
56. ■ At 25 °C the density of water is 0.997 g/mL, whereas the density of ice at −10 °C is 0.917 g/mL.
    (a) If a plastic soft-drink bottle (volume = 250 mL) is completely filled with pure water, capped, and then frozen at −10 °C, what volume will the solid occupy?
    (b) What will the bottle look like when you take it out of the freezer?
57. When water alone (instead of engine coolant, which contains water and other substances) was used in automobile radiators to cool cast-iron engine blocks, it sometimes happened in winter that the engine block would crack, ruining the engine. Cast iron is not pure iron and is relatively hard and brittle. Explain in your own words how the engine block in a car might crack in cold weather.
58. Bromobenzene does not dissolve in either water or ethanol.
    (a) If you pour 2 mL water into a test tube that contains 2 mL bromobenzene, which liquid will be on top?
    (b) If you pour 2 mL ethanol carefully into the test tube with bromobenzene and water described in part (a) without shaking or mixing the liquids, what will happen?
    (c) What will happen if you thoroughly stir the mixture in part (b)?
59. Water does not mix with either benzene or bromobenzene when it is stirred together with either of them, but benzene and bromobenzene do mix.
    (a) If you pour 2 mL bromobenzene into a test tube, then add 2 mL water and stir, what would the test tube look like a few minutes later?
    (b) Suppose you add 2 mL benzene to the test tube in part (a), pouring the benzene carefully down the side of the tube so that the liquids do not mix together. Describe the appearance of the test tube now.
    (c) If the test tube containing all three liquids is thoroughly shaken and then allowed to stand for 5 minutes, what will the tube look like?

60. The figure shows a nanoscale view of the atoms of mercury in a thermometer registering 10 °C.

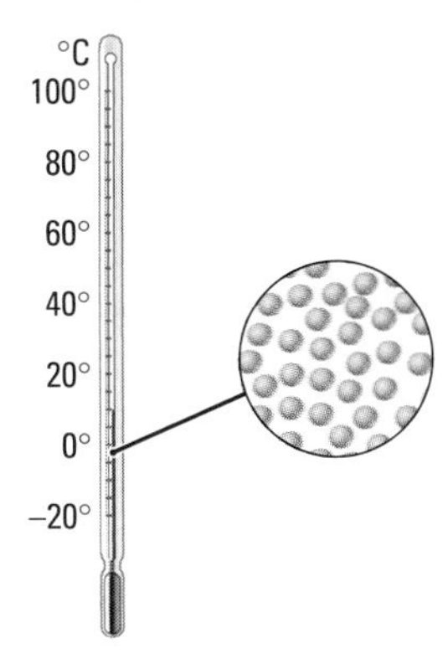

Which nanoscale drawing best represents the atoms in the liquid in this same thermometer at 90 °C? (Assume that the same volume of liquid is shown in each nanoscale drawing.)

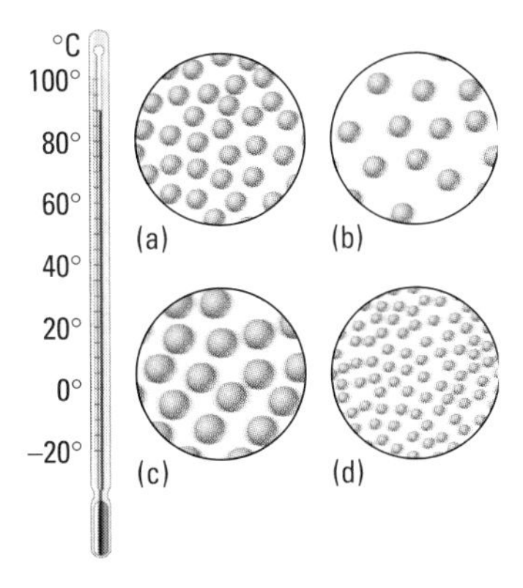

61. Answer these questions using figures (a) through (i). (Each question may have more than one answer.)

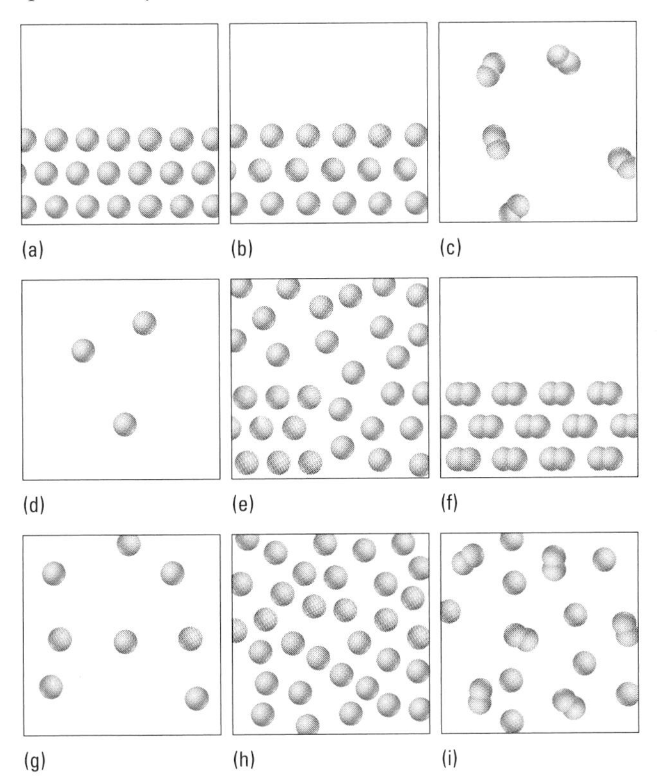

(a) Which represents nanoscale particles in a sample of solid?
(b) Which represents nanoscale particles in a sample of liquid?
(c) Which represents nanoscale particles in a sample of gas?
(d) Which represents nanoscale particles in a sample of an element?
(e) Which represents nanoscale particles in a sample of a compound?
(f) Which represents nanoscale particles in a sample of a pure substance?
(g) Which represents nanoscale particles in a sample of a mixture?

More Challenging Questions

These questions require more thought and integrate several concepts.

62. ■ The element platinum has a solid-state structure in which platinum atoms are arranged in a cubic shape that repeats throughout the solid. If the length of an edge of the cube is 392 pm (1 pm = 1×10^{-12} m), what is the volume of the cube in cubic meters?

63. The compound sodium chloride has a solid-state structure in which there is a repeating cubic arrangement of sodium ions and chloride ions. If the volume of the cube is 1.81×10^{-22} cm^3, what is the length of an edge of the cube in pm (1 pm = 1×10^{-12} m)?

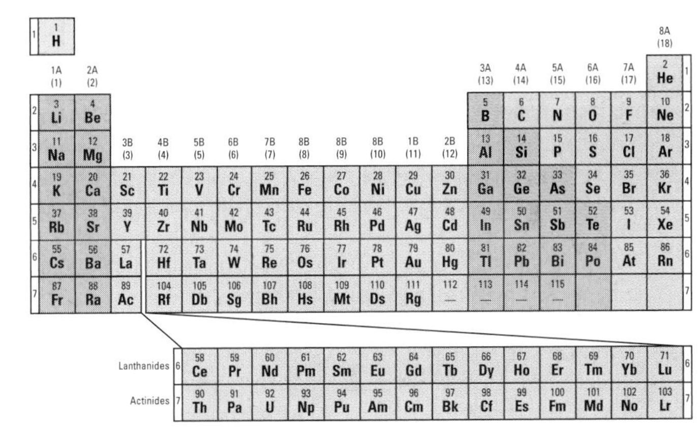

64. ■ The periodic table shown here is color coded gray, blue, orange, and lavender. Identify the color of the area (or colors of the areas) in which you would expect to find each type of element.
(a) A metal
(b) A nonmetal
(c) A metalloid

65. The periodic table shown here is color coded gray, blue, orange, and lavender. Identify the color of the area (or colors of the areas) in which you would expect to find each type of element.
(a) A shiny solid that conducts electricity
(b) A gas whose molecules consist of single atoms
(c) An element that is a semiconductor
(d) A yellow solid that has very low electrical conductivity

66. ■ When someone discovers a new substance, it is relatively easy to show that the substance is not an element, but it is quite difficult to prove that the substance is an element. Explain why this is so, and relate your explanation to the discussion of scientific laws and theories in Section 1.3.

67. Soap can be made by mixing animal or vegetable fat with a concentrated solution of lye and heating it in a large vat. Suppose that 3.24 kg vegetable fat is placed in a large iron vat and then 50.0 L water and 5.0 kg lye (sodium hydroxide, NaOH) are added. The vat is placed over a fire and heated for two hours, and soap forms.
    (a) Classify each of the materials identified in the soap-making process as a substance or a mixture. For each substance, indicate whether it is an element or a compound. For each mixture, indicate whether it is homogeneous or heterogeneous.
    (b) Assuming that the fat and lye are completely converted into soap, what mass of soap is produced?
    (c) What physical and chemical processes occur as the soap is made?

68. The densities of several elements are given in Table 1.1.
    (a) Of the elements nickel, gold, lead, and magnesium, which will float on liquid mercury at 20 °C?
    (b) Of the elements titanium, copper, iron, and gold, which will float highest on the mercury? That is, which element will have the smallest fraction of its volume below the surface of the liquid?

69. You have some metal shot (small spheres about 5 mm in diameter), and you want to identify the metal. You have a flask that is known to contain exactly 100.0 mL when filled with liquid to a mark in the flask's neck. When the flask is filled with water at 20 °C, the mass of flask and water is 122.3 g. The water is emptied from the flask and 20 of the small spheres of metal are carefully placed in the flask. The 20 small spheres had a mass of 42.3 g. The flask is again filled to the mark with water at 20 °C and weighed. This time the mass is 159.9 g.
    (a) What metal is in the spheres? (Assume that the spheres are all the same and consist of pure metal.)
    (b) What volume would 500 spheres occupy?

70. Suppose you are trying to get lemon juice and you have no juicer. Some people say that you can get more juice from a lemon if you roll it on a hard surface, applying pressure with the palm of your hand before you cut it and squeeze out the juice. Others claim that you will get more juice if you first heat the lemon in a microwave and then cut and squeeze it. Apply the methods of science to arrive at a technique that will give the most juice from a lemon. Carry out experiments and draw conclusions based on them. Try to generate a hypothesis to explain your results.

71. If you drink orange juice soon after you brush your teeth, the orange juice tastes quite different. Apply the methods of science to find what causes this effect. Carry out experiments and draw conclusions based on them.

Conceptual Challenge Problems

These rigorous, thought-provoking problems integrate conceptual learning with problem solving and are suitable for group work.

CP1.A (Section 1.3) Some people use expressions such as "a rolling stone gathers no moss" and "where there is no light there is no life." Why do you believe these are "laws of nature"?

CP1.B (Section 1.3) Parents teach their children to wash their hands before eating. (a) Do all parents accept the germ theory of disease? (b) Are all diseases caused by germs?

CP1.C (Section 1.8) In Section 1.8 you read that, on an atomic scale, all matter is in constant motion. (For example, the average speed of a molecule of nitrogen or oxygen in the air is greater than 1000 miles per hour at room temperature.) (a) What evidence can you put forward that supports the kinetic-molecular theory? (b) Suppose you accept the notion that molecules of air are moving at speeds near 1000 miles per hour. What can you then reason about the paths that these molecules take when moving at this speed?

CP1.D (Section 1.8) Some scientists think there are living things smaller than bacteria (*New York Times,* January 18, 2000; p. D1). Called "nanobes," they are roughly cylindrical and range from 20 to 150 nm long and about 10 nm in diameter. One approach to determining whether nanobes are living is to estimate how many atoms and molecules could make up a nanobe. If the number is too small, then there would not be enough DNA, protein, and other biological molecules to carry out life processes. To test this method, estimate an upper limit for the number of atoms that could be in a nanobe. (Use a small atom, such as hydrogen.) Also estimate how many protein molecules could fit inside a nanobe. Do your estimates rule out the possibility that a nanobe could be living? Explain why or why not.

CP1.E (Section 1.12) The life expectancy of U.S. citizens in 1992 was 76 years. In 1916 the life expectancy was only 52 years. This is an increase of 46% in a lifetime. (a) Could this astonishing increase occur again? (b) To what single source would you attribute this noteworthy increase in life expectancy? Why did you identify this one source as being most influential?

CP1.F Helium-filled balloons rise and will fly away unless tethered by a string. Use the kinetic-molecular theory to explain why a helium-filled balloon is "lighter than air."

2

Atoms and Elements

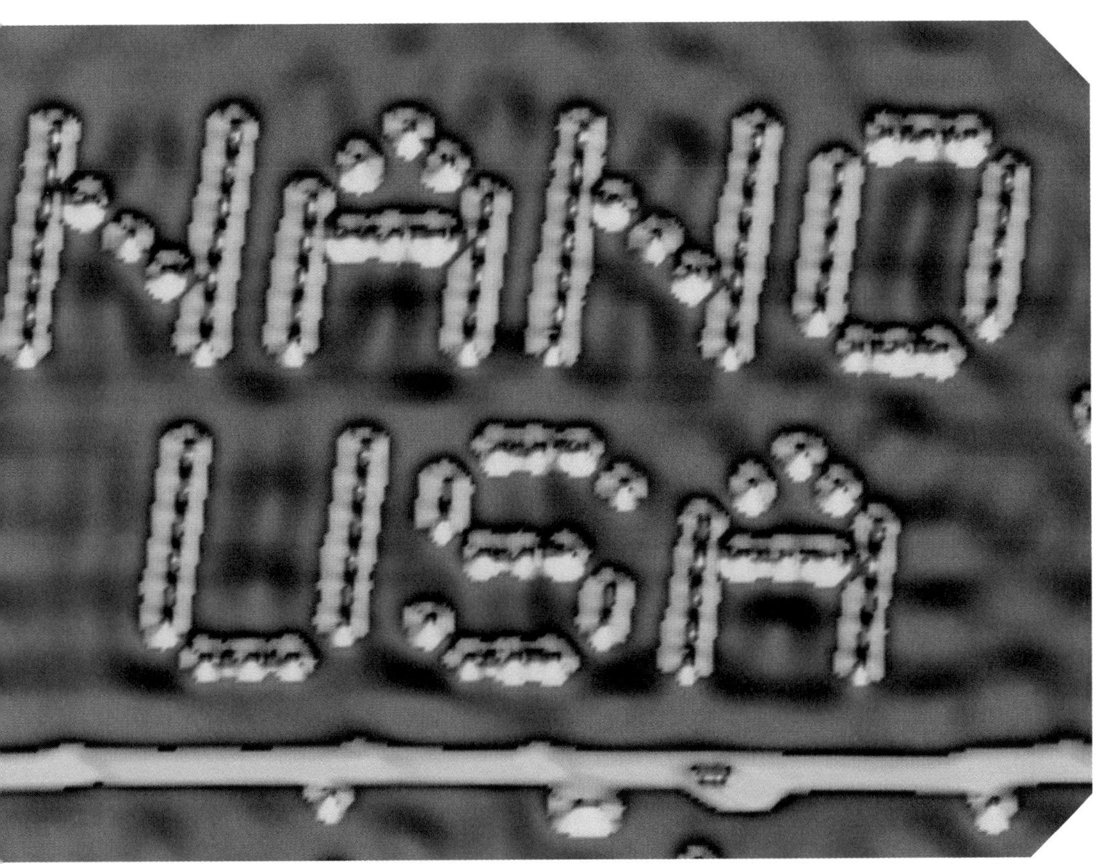

Image reproduced by permission of IBM Research, Almaden Research Center. Unauthorized use not permitted.

The blue bumps making up the letters in the photograph are images of 112 individual CO molecules on a copper surface. Each letter is 4 nm high and 3 nm wide. The overall image was generated by a scanning tunneling microscope (STM), which can detect individual atoms or molecules, allowing us to make images of nanoscale atomic arrangements.

To study chemistry, we need to start with atoms—the basic building blocks of matter. Early theories of the atom considered atoms to be indivisible, but we know now that atoms are more complicated than that. Elements differ from one another because of differences in the internal structure of their atoms. Under the right conditions, smaller particles within atoms—known as *subatomic particles*—can be removed or rearranged. The term **atomic structure** refers to the identity and arrangement of these subatomic particles in the atom. An understanding of the details of atomic structure aids in the understanding of how elements combine to form compounds and how atoms are rearranged in chemical reactions. Atomic structure also accounts for the properties of materials. The next few sections describe how experiments support the idea that atoms are composed of smaller (subatomic) particles.

2.1 Atomic Structure and Subatomic Particles

Electrical charges played an important role in many of the experiments from which the theory of atomic structure was derived. Two types of electrical charge exist: positive and negative. ***Electrical charges of the same type repel one another, and charges of the opposite type attract one another.*** A positively charged particle repels another positively charged particle. Likewise, two negatively charged particles repel each other. In contrast, two particles with opposite signs attract each other.

CONCEPTUAL
EXERCISE 2.1 Electrical Charge

When you comb your hair on a dry day, your hair sticks to the comb. How could you explain this behavior in terms of a nanoscale model in which atoms contain positive and negative charges?

Radioactivity

In 1896 Henri Becquerel discovered that a sample of a uranium ore emitted rays that darkened a photographic plate, even though the plate was covered by a protective black paper. In 1898 Marie and Pierre Curie isolated the new elements polonium and radium, which emitted the same kind of rays. Marie suggested that atoms of such elements spontaneously emit these rays and named the phenomenon **radioactivity.**

Atoms of radioactive elements can emit three types of radiation: alpha (α), beta (β), and gamma (γ) rays. These radiations behave differently when passed between electrically charged plates (Figure 2.1). Alpha and beta rays are deflected, but gamma rays are not. These events occur because alpha rays and beta rays are composed of charged particles that come from within the radioactive atom. Alpha rays have a 2+ charge, and

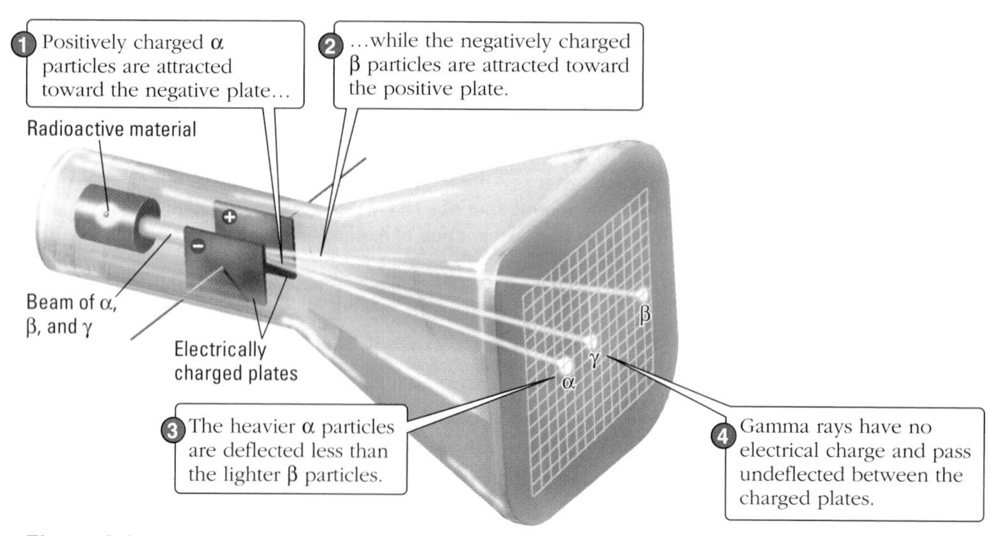
1 Positively charged α particles are attracted toward the negative plate…
2 …while the negatively charged β particles are attracted toward the positive plate.
Radioactive material
Beam of α, β, and γ
Electrically charged plates
β
γ
α
3 The heavier α particles are deflected less than the lighter β particles.
4 Gamma rays have no electrical charge and pass undeflected between the charged plates.

Figure 2.1 The α, β, and γ rays from a radioactive sample are separated by an electrical field.

beta rays have a 1− charge. Alpha rays and beta rays are particles because they have mass—they are matter. In the experiment shown in Figure 2.1, alpha particles are deflected less and so must be heavier than beta particles. Gamma rays have no detectable charge or mass; they behave like light rays. If radioactive atoms can break apart to produce subatomic alpha and beta particles, then there must be something smaller inside the atoms.

Electrons

Further evidence that atoms are composed of subatomic particles came from experiments with specially constructed glass tubes called cathode-ray tubes. Most of the air has been removed from these tubes and a metal electrode sealed into each end. When a sufficiently high voltage is applied to the electrodes, a beam of rays flows from the negatively charged electrode (the *cathode*) to the positively charged electrode (the *anode*). These rays, known as cathode rays, come directly from the metal atoms of the cathode. The cathode rays travel in straight lines, are attracted toward positively charged plates, can be deflected by a magnetic field, can cast sharp shadows, can heat metal objects red hot, and can cause gases and fluorescent materials to glow. When cathode rays strike a fluorescent screen, the energy transferred causes light to be given off as tiny flashes. Thus, the properties of a cathode ray are those of a beam of negatively charged particles, each of which produces a light flash when it hits a fluorescent screen. Sir Joseph John Thomson suggested that cathode rays consist of the same particles that had earlier been named **electrons** and had been suggested to be the carriers of electricity. He also observed that cathode rays were produced from electrodes made of different metals. This implied that electrons are constituents of the atoms of those different elements.

In 1897 Thomson used a specially designed cathode-ray tube to simultaneously apply electric and magnetic fields to a beam of cathode rays. By balancing the electric field against the magnetic field and using the basic laws of electricity and magnetism, Thomson calculated the *ratio* of mass to charge for the electrons in the cathode-ray beam: 5.60×10^{-9} grams per coulomb (g/C). (The coulomb, C, is a fundamental unit of electrical charge.)

Fourteen years later, Robert Millikan used a cleverly devised experiment to measure the charge of an electron (Figure 2.2). Tiny oil droplets were sprayed into a chamber. As they settled slowly through the air, the droplets were exposed to x-rays, which caused electrons to be transferred from gas molecules in the air to the droplets. Using a small microscope to observe individual droplets, Millikan adjusted the electrical charge of plates above and below the droplets so that the electrostatic attraction just balanced the gravitational attraction. In this way he could suspend a single droplet motionless. From equations describing these forces, Millikan calculated the charge on the suspended droplets. Different droplets had different charges, but Millikan found that each was an integer multiple of the smallest charge. The smallest charge was 1.60×10^{-19} C. Millikan assumed this to be the fundamental quantity of charge, the charge on an electron. Given this value and the mass-to-charge ratio determined by Thomson, the mass of an electron could be computed: $(1.60 \times 10^{-19}\text{ C})(5.60 \times 10^{-9}\text{ g/C}) = 8.96 \times 10^{-28}$ g. The currently accepted most accurate value for the electron's mass is $9.10938215 \times 10^{-28}$ g, and the currently accepted most accurate value for the electron's charge in coulombs is $-1.602176487 \times 10^{-19}$ C. This quantity is called the *electronic charge*. For convenience, the charges on subatomic particles are given in multiples of electronic charge rather than in coulombs. So the charge on an electron is 1−.

See Appendix A.2 for a review of scientific notation, which is used to represent very small or very large numbers as powers of 10. For example, 0.000001 is 1×10^{-6} (the decimal point moves six places to the left to give the −6 exponent) and 2,000,000 is 2×10^{6} (the decimal point moves six places to the right to give the +6 exponent).

Electron
Charge = 1−
Mass = 9.10938×10^{-28} g

Other experiments provided further evidence that the electron is a *fundamental* particle of matter; that is, it is present in *all* matter. The beta particles emitted by radioactive elements were found to have the same properties as cathode rays, which are streams of electrons.

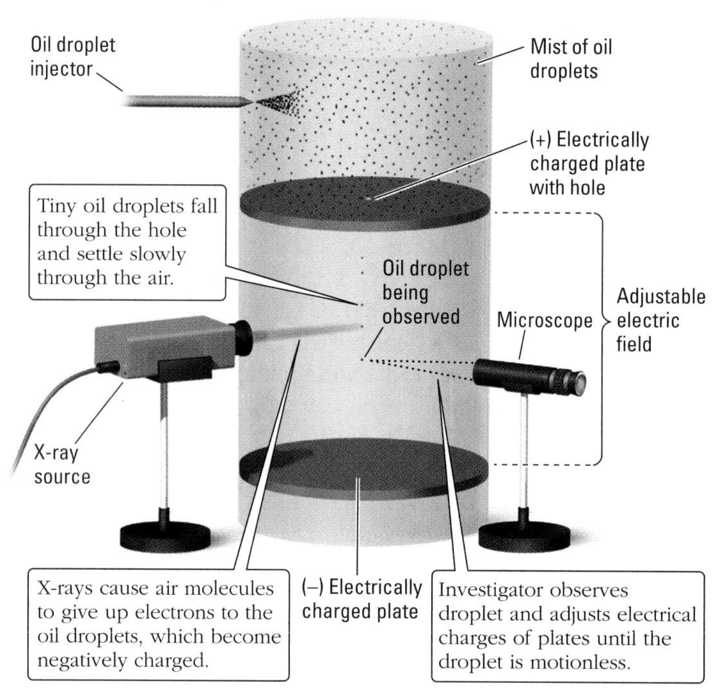

Figure 2.2 Millikan oil-drop experiment. From the known mass of the droplets and the applied voltage at which the charged droplets were held stationary, Millikan could calculate the charges on the droplets.

Protons

When atoms lose electrons, the atoms become positively charged. When atoms gain electrons, the atoms become negatively charged. Such charged atoms, or similarly charged groups of atoms, are known as **ions.** From experiments with positive ions, formed by knocking electrons out of atoms, the existence of a positively charged, fundamental particle was deduced. Positively charged particles with different mass-to-charge ratios were formed by atoms of different elements. The variation in masses showed that atoms of different elements must contain different numbers of positive particles. Those from hydrogen atoms had the smallest mass-to-charge ratio, indicating that they are the fundamental positively charged particles of atomic structure. Such particles are called **protons.** The mass of a proton is known from experiment to be $1.672621637 \times 10^{-24}$ g, which is about 1800 times the mass of an electron. The charge on a proton is $1.602176487 \times 10^{-19}$ C, equal in size, but opposite in sign, to the charge on an electron. The proton's charge is represented by 1+. Thus, an atom that has lost two electrons has a charge of 2+.

As mass increases, mass-to-charge ratio increases for a given amount of charge. For a fixed charge, doubling the mass will double the mass-to-charge ratio. For a fixed mass, doubling the charge will halve the mass-to-charge ratio.

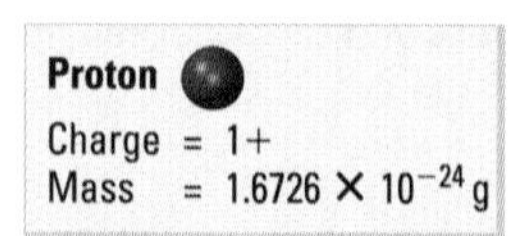

2.2 The Nuclear Atom

The Nucleus

Once it was known that there were subatomic particles, the next question scientists wanted to answer was, How are these particles arranged in an atom? Between 1909 and 1911 Ernest Rutherford reported experiments (Figure 2.3) that led to a better understanding of atomic structure. Alpha particles (which have the same mass as helium atoms and a 2+ charge) were allowed to hit a very thin sheet of gold foil. Almost all of the alpha particles passed through undeflected. However, a very few alpha particles were deflected through large angles, and some came almost straight

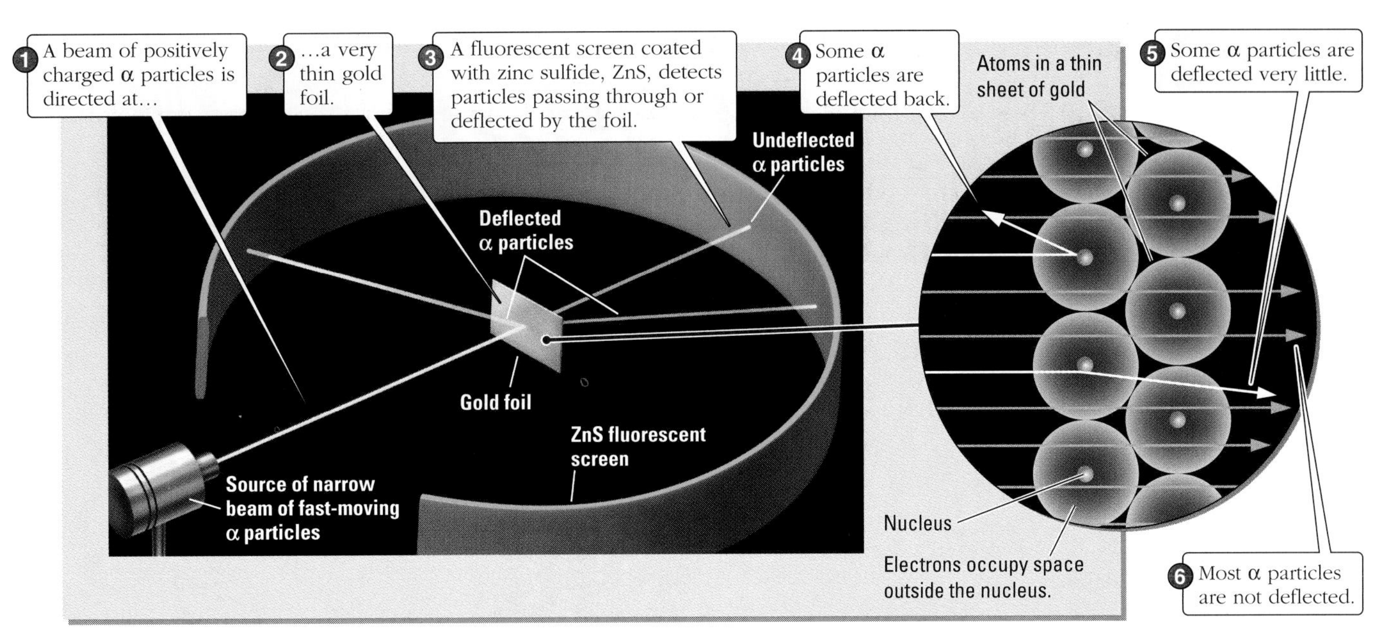

Figure 2.3 The Rutherford experiment and its interpretation.

back toward the source. Rutherford described this unexpected result by saying, "It was about as credible as if you had fired a 15-inch [artillery] shell at a piece of paper and it came back and hit you."

The only way to account for the observations was to conclude that all of the positive charge and most of the mass of the atom are concentrated in a very small region (Figure 2.3). Rutherford called this tiny atomic core the **nucleus.** Only such a region could be sufficiently dense and highly charged to repel an alpha particle. From their results, Rutherford and his associates calculated values for the charge and radius of the gold nucleus. The currently accepted values are a charge of 79+ and a radius of approximately 1×10^{-13} cm. This makes the nucleus about 10,000 times *smaller* than the atom. Most of the volume of the atom is occupied by the electrons. Somehow the space outside the nucleus is occupied by the negatively charged electrons, but their arrangement was unknown to Rutherford and other scientists of the time. The arrangement of electrons in atoms is now well understood and is described in Chapter 7.

Alpha particles are four times heavier than the lightest atoms, which are hydrogen atoms.

In 1920 Ernest Rutherford proposed that the nucleus might contain an uncharged particle whose mass approximated that of a proton.

Neutrons

Atoms are electrically neutral (no net charge), so they must contain equal numbers of protons and electrons. However, most neutral atoms have masses greater than the sum of the masses of their protons and electrons. The additional mass indicates that subatomic particles with mass but no charge must also be present. Because they have no charge, these particles are more difficult to detect experimentally. In 1932 James Chadwick devised a clever experiment that detected the neutral particles by having them knock protons out of atoms and then detecting the protons. The neutral subatomic particles are called **neutrons.** They have no electrical charge and a mass of $1.674927211 \times 10^{-24}$ g, nearly the same as the mass of a proton.

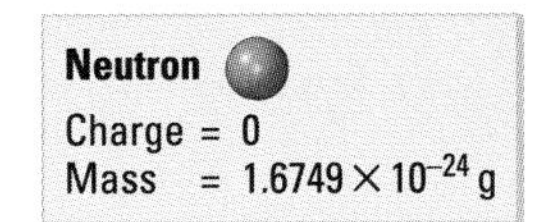

In summary, there are three primary subatomic particles: protons, neutrons, and electrons.

- Protons and neutrons make up the nucleus, providing most of the atom's mass; the protons provide all of its positive charge.
- The nuclear radius is approximately 10,000 times smaller than the radius of the entire atom.

Ernest Rutherford 1871–1937

Born on a farm in New Zealand, Rutherford earned his Ph.D. in physics from Cambridge University in 1895. He discovered alpha and beta radiation and coined the term "half-life." For proving that alpha radiation is composed of helium nuclei and that beta radiation consists of electrons, Rutherford received the Nobel Prize in Chemistry in 1908. As a professor at Cambridge University, he guided the work of no fewer than ten future Nobel Prize recipients. Element 104 is named in Rutherford's honor.

- Negatively charged electrons outside the nucleus occupy most of the volume of the atom, but contribute very little mass.
- A neutral atom has no net electrical charge because the number of electrons outside the nucleus equals the number of protons inside the nucleus.

To chemists, the electrons are the most important subatomic particles because they are the first part of the atom to contact another atom. The electrons in atoms largely determine how elements combine to form chemical compounds.

CONCEPTUAL **EXERCISE** **2.2 Describing Atoms**

If an atom had a radius of 100 m, it would approximately fill a football stadium.
(a) What would the approximate radius of the nucleus of such an atom be?
(b) What common object is about that size?

2.3 The Sizes of Atoms and the Units Used to Represent Them

Atoms are extremely small. One teaspoon of water contains about three times as many atoms as the Atlantic Ocean contains teaspoons of water. To do quantitative calculations in chemistry, it is important to understand the units used to express the sizes of very large and very small quantities.

To state the size of an object on the macroscale in the United States (for example, yourself), we would give your weight in pounds and your height in feet and inches. Pounds, feet, and inches are part of the measurement system used in the United States, but almost nowhere else in the world. Most of the world uses the **metric system** of units for recording and reporting measurements. The metric system is a decimal system that adjusts the size of its basic units by multiplying or dividing them by multiples of 10.

The International System of units (or SI units) is the officially recognized measurement system of science. It is derived from the metric system and is described in Appendix B. The units for mass, length, and volume are introduced here. Other units are introduced as they are needed in later chapters.

In the metric system, your weight (really, your mass) would be given in kilograms. The **mass** of an object is a fundamental measure of the quantity of matter in that object. The metric units for mass are *grams* or multiples or fractions of grams. The prefixes listed in Table 2.1 are used with all metric units. A *kilo*gram, for example, is equal to 1000 grams and is a convenient size for measuring the mass of a person.

Strictly speaking, the pound is a unit of weight rather than mass. The weight of an object depends on the local force of gravity. For measurements made at the Earth's surface, the distinction between mass and weight is not generally useful.

Table 2.1 Some Prefixes Used in the SI and Metric Systems

Prefix	Abbreviation	Meaning	Example
mega	M	10^6	1 megaton = 1×10^6 tons
kilo	k	10^3	1 kilometer (km) = 1×10^3 meter (m) 1 kilogram (kg) = 1×10^3 gram (g)
deci	d	10^{-1}	1 decimeter (dm) = 1×10^{-1} m 1 deciliter (dL) = 1×10^{-1} liter (L)
centi	c	10^{-2}	1 centimeter (cm) = 1×10^{-2} m
milli	m	10^{-3}	1 milligram (mg) = 1×10^{-3} g
micro	μ	10^{-6}	1 micrometer (μm) = 1×10^{-6} m
nano	n	10^{-9}	1 nanometer (nm) = 1×10^{-9} m 1 nanogram (ng) = 1×10^{-9} g
pico	p	10^{-12}	1 picometer (pm) = 1×10^{-12} m
femto	f	10^{-15}	1 femtogram (fg) = 1×10^{-15} g

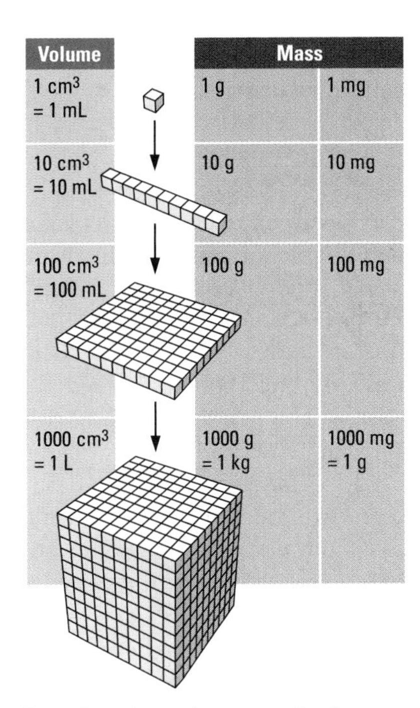

Relative sizes of mass and volume units.

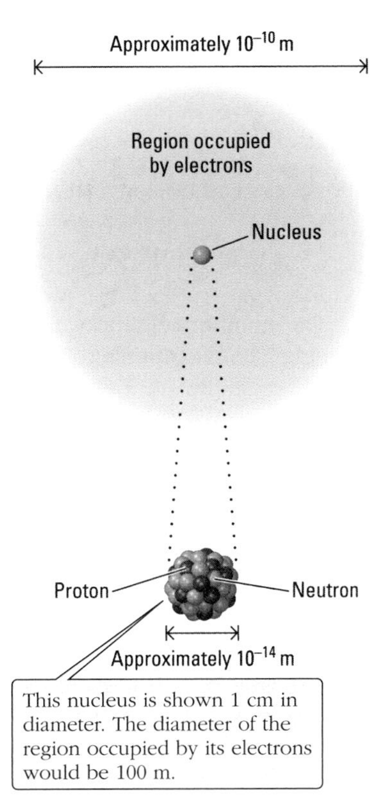

Relative sizes of the atomic nucleus and an atom (not to scale).

For objects much smaller than people, prefixes that represent negative powers of 10 are used. For example, 1 *milli*gram equals 1×10^{-3} g.

$$1 \text{ milligram (mg)} = \frac{1}{1000} \times 1 \text{ g} = 0.001 \text{ g} = 1 \times 10^{-3} \text{ g}$$

Individual atoms are too small to be weighed directly; their masses can be measured only by indirect experiments. An atom's mass is on the order of 1×10^{-22} g. For example, a sample of copper that weighs one *nano*gram (1 ng = 1×10^{-9} g) contains about 9×10^{12} copper atoms. The most sensitive laboratory balances can weigh samples of about 0.0000001 g (1×10^{-7} g = 0.1 microgram, μg).

Your height in metric units would be given in *meters,* the metric unit for length. Six feet is equivalent to 1.83 m. Atoms aren't nearly this big. The sizes of atoms are reported in *pico*meters (1 pm = 1×10^{-12} m), and the radius of a typical atom is very small—between 30 and 300 pm. For example, the radius of a copper atom is 128 pm (128×10^{-12} m).

To get a feeling for these dimensions, consider how many copper atoms it would take to form a single file of copper atoms across a U.S. penny with a diameter of 1.90×10^{-2} m. This distance can be expressed in picometers by using a conversion factor ***(⟵ p. 8)*** based on 1 pm = 1×10^{-12} m.

Conversion factors are the basis for dimensional analysis, a commonly used problem-solving technique. It is described in detail in Appendix A.2.

$$1.90 \times 10^{-2} \text{ m} \times \underbrace{\frac{1 \text{ pm}}{1 \times 10^{-12} \text{ m}}}_{\text{conversion factor}} = 1.90 \times 10^{10} \text{ pm}$$

Note that the units m (for meters) cancel, leaving the answer in pm, the units we want. A penny is 1.90×10^{10} pm in diameter.

Every conversion factor can be used in two ways. We just converted meters to picometers by using

$$\frac{1 \text{ pm}}{1 \times 10^{-12} \text{ m}}$$

Picometers can be converted to meters by inverting this conversion factor:

$$8.70 \times 10^{10} \text{ pm} \times \frac{1 \times 10^{-12} \text{ m}}{1 \text{ pm}} = 8.70 \times 10^{-2} \text{ m} = 0.0870 \text{ m}$$

The number of copper atoms needed to stretch across a penny can be calculated by using a conversion factor linking the penny's diameter in picometers with the diameter of a single copper atom. The diameter of a Cu atom is twice the radius, 2×128 pm = 256 pm. Therefore, the conversion factor is 1 Cu atom per 256 pm, and

Notice how "Cu atom" is included in the conversion to keep track of what kind of atom we are interested in.

$$1.90 \times 10^{10} \text{ pm} \times \frac{1 \text{ Cu atom}}{256 \text{ pm}} = 7.42 \times 10^{7} \text{ Cu atoms, or } 74{,}200{,}000 \text{ Cu atoms}$$

Thus, it takes more than 74 million copper atoms to reach across the penny's diameter. Atoms are indeed tiny.

In chemistry, the most commonly used length units are the centimeter, the millimeter, the nanometer, and the picometer. The most commonly used mass units are the kilogram, gram, and milligram. The relationships among these units and some other units are given inside the back cover.

Problem-Solving Examples 2.1 and 2.2 illustrate the use of dimensional analysis in unit conversion problems. Notice that in these examples, and throughout the book, the answers are given before the strategy and explanation of how the answers are found. We urge you to first try to answer the problem on your own. Then check to see whether your answer is correct. If it does not match, try again. Finally, read the explanation, which usually also includes the strategy for solving the problem. If your answer is correct, but your reasoning differs from the explanation, you might have discovered an alternative way to solve the problem.

PROBLEM-SOLVING EXAMPLE 2.1 Conversion of Units

A medium-sized paperback book has a mass of 530. g. What is the book's mass in kilograms and in pounds?

Answer 0.530 kg and 1.17 lb

Strategy and Explanation Our strategy is to use conversion factors that relate what we know to what we are trying to calculate. For this problem, we use conversion factors from inside the back cover and metric prefixes from Table 2.1 to calculate the answer, being sure to set up the calculation so that only the desired unit remains. The relationship between grams and kilograms is 1 kg = 10^3 g, so 530. g is

$$530. \text{ g} \times \frac{1 \text{ kg}}{10^3 \text{ g}} = 530. \times 10^{-3} \text{ kg} = 0.530 \text{ kg}$$

One pound equals 453.6 g, so we convert grams to pounds as follows:

$$530. \text{ g} \times \frac{1 \text{ lb}}{453.6 \text{ g}} = 1.17 \text{ lb}$$

☑ **Reasonable Answer Check** There are about 2.2 lb per kg, and the book's mass is about ½ kg. Thus, its mass in pounds should be about 2.2/2 = 1.1, which is close to our more accurate answer.

On September 23, 1999, the NASA Mars Climate Orbiter spacecraft approached too close and burned up in Mars's atmosphere because of a navigational error due to a failed translation of English units into metric units by the spacecraft's software program.

PROBLEM-SOLVING PRACTICE 2.1

(a) A car requires 10 gallons to fill its gas tank. How many liters is this?

(b) An American football field is 100 yards long. How many meters is this?

PROBLEM-SOLVING EXAMPLE 2.2 Nanoscale Distances

The individual letters in the STM image shown as the Chapter Opener photograph are 4.0 nm high and 3.0 nm wide. How many letters could fit in a distance of 1.00 mm (about the width of the head of a pin)?

Answer 3.3×10^5 letters

Strategy and Explanation Our strategy uses conversion factors to relate what we know to what we want to calculate. A nanometer is 1×10^{-9} m and a millimeter is 1×10^{-3} m. The width of a letter in meters is

$$3.0\ \cancel{\text{nm}} \times \frac{1 \times 10^{-9}\ \text{m}}{1\ \cancel{\text{nm}}} = 3.0 \times 10^{-9}\ \text{m}$$

Therefore, the number of 3.0-nm-wide letters that could fit into 1.00 mm is

$$1.0 \times 10^{-3}\ \cancel{\text{m}} \times \frac{1\ \text{letter}}{3.0 \times 10^{-9}\ \cancel{\text{m}}} = 3.3 \times 10^5\ \text{letters}$$

☑ **Reasonable Answer Check** If the letters are 3.0 nm wide, then about one third of 10^9 could fit into 1 m. One mm is 1/1000 of a meter, and multiplying these two estimates gives $(0.33 \times 10^9)(1/1000) = 0.33 \times 10^6$, which is another way of writing the answer we calculated.

PROBLEM-SOLVING PRACTICE 2.2

Do the following conversions using factors based on the equalities inside the back cover.

(a) How many grams of sugar are in a 5-lb bag of sugar?

(b) Over a period of time, a donor gives 3 pints of blood. How many milliliters (mL) has the donor given?

(c) The same donor's 160-lb body contains approximately 5 L of blood. Considering that 1 L is nearly equal to 1 quart, estimate the percentage of the donor's blood that has been donated in all.

The liter (L) and milliliter (mL) are the most common volume units of chemistry. There are 1000 mL in 1 L. One liter is a bit larger than a quart, and a teaspoon of water has a volume of about 5 mL. Chemists often use the terms *milliliter* and *cubic centimeter* (cm^3 or sometimes cc) interchangeably because they are equivalent ($1\ mL = 1\ cm^3$).

As illustrated in Problem-Solving Example 2.3, two or more steps of a calculation using dimensional analysis are best written in a single setup and entered into a calculator as a single calculation.

PROBLEM-SOLVING EXAMPLE 2.3 Volume Units

A chemist uses 50. μL (microliters) of a sample for her analysis. What is the volume in mL? In cm^3? In L?

Answer 5.0×10^{-2} mL; $5.0 \times 10^{-2}\ cm^3$; 5.0×10^{-5} L

Strategy and Explanation Use the conversion factors in Table 2.1 that involve micro and milli: $1\ \mu L = 1 \times 10^{-6}$ L and 1 L = 1000 mL. Multiply the conversion factors to cancel μL and L, leaving only mL.

$$50.\ \cancel{\mu\text{L}} \times \frac{1 \times 10^{-6}\ \cancel{\text{L}}}{1\ \cancel{\mu\text{L}}} \times \frac{10^3\ \text{mL}}{1\ \cancel{\text{L}}} = 5.0 \times 10^{-2}\ \text{mL}$$

Because 1 mL and 1 cm^3 are equivalent, the sample size can also be expressed as $5.0 \times 10^{-2}\ cm^3$. Since there are 1000 mL in 1 L, the sample size expressed in liters is 5.0×10^{-5} L.

☑ **Reasonable Answer Check** There is a factor of 1000 between μL and mL, and the final number is a factor of 1000 smaller than the original volume, so the answer is reasonable.

PROBLEM-SOLVING PRACTICE 2.3

A patient's blood cholesterol level measures 165 mg/dL. Express this value in g/L.

Glassware for measuring the volume of liquids.

2.4 Uncertainty and Significant Figures

Measurements always include some degree of uncertainty, because we can never measure quantities exactly or know the value of a quantity with absolute certainty. Scientists have adopted standardized ways of expressing uncertainty in numerical results of measurements.

When the result of a measurement is expressed numerically, the final digit reported is uncertain. The digits we write down from a measurement—both the certain ones and the final, uncertain one—are called **significant figures.** For example, the number 5.025 has four significant figures and the number 4.0 has two significant figures.

To determine the number of significant figures in a measurement, read the number from left to right and count all digits, starting with the first digit that is not zero. All the digits in the number are significant except any zeros that are used only to position the decimal point.

Appendix A.3 discusses precision, accuracy, and significant figures in detail.

In Section 2.5, we discuss the masses of atoms and begin calculating with numbers that involve uncertainty. Significant figures provide a simple means for keeping track of these uncertainties.

Example	Number of Significant Figures
1.23 g	3
0.00123 g	3; the zeros to the left of the 1 simply locate the decimal point. The number of significant figures is more obvious if you write numbers in scientific notation. Thus, $0.00123 = 1.23 \times 10^{-3}$.
2.0 g and 0.020 g	2; both have two significant figures. When a number is greater than 1, *all zeros to the right of the decimal point are significant.* For a number less than 1, only zeros to the right of the first significant figure are significant.
100 g	1; in numbers that do not contain a decimal point, trailing zeros may or may not be significant. To eliminate possible confusion, the practice followed in this book is to include a decimal point if the zeros are significant. Thus, 100. has three significant figures, while 100 has only one. Alternatively, we can write in scientific notation 1.00×10^2 (three significant figures) or 1×10^2 (one significant figure). For a number written in scientific notation, all digits are significant.
100 cm/m	Infinite number of significant figures, because this is a defined quantity. There are *exactly* 100 centimeters in one meter.
$\pi = 3.1415926\ldots$	The value of π is known to a greater number of significant figures than any data you will ever use in a calculation.

PROBLEM-SOLVING EXAMPLE 2.4 Significant Figures

How many significant figures are present in each of these numbers?

(a) 0.0001171 m (b) 26.94 mL (c) 207 cm
(d) 0.7011 g (e) 0.0010 L (f) 12,400. s

Answer

(a) Four (b) Four (c) Three
(d) Four (e) Two (f) Five

Strategy and Explanation Apply the principles of significant figures given in the preceding examples.

(a) The leading zeros do not count, so there are four significant figures.
(b) All four of the digits are significant.
(c) All three digits are significant figures.
(d) Four digits follow the decimal, and all are significant figures.
(e) The leading zeros do not count, so there are two significant figures.
(f) Since there is a decimal point, all five of the digits are significant.

PROBLEM-SOLVING PRACTICE 2.4

Determine the number of significant figures in these numbers: (a) 0.00602 g; (b) 22.871 mg; (c) 344. °C; (d) 100.0 mL; (e) 0.00042 m; (f) 0.002001 L.

Significant Figures in Calculations

When numbers are combined in a calculation, the number of significant figures in the result is determined by the number of significant figures in the starting numbers and the nature of the arithmetic operation being performed.

Addition and Subtraction: *In addition or subtraction, the number of decimal places in the answer equals the number of decimal places in the number with the fewest decimal places.* Suppose you add these three numbers:

0.12	2 significant figures	2 decimal places
1.6	2 significant figures	1 decimal place
10.976	5 significant figures	3 decimal places
12.696 rounds to 12.7		

This sum should be reported as 12.7, a number with one decimal place, because 1.6 has only one decimal place.

Multiplication and Division: *In multiplication or division, the number of significant figures in the answer is the same as that in the quantity with the fewest significant figures.*

$$\frac{0.7608}{0.0546} = 13.9 \qquad \text{or, in scientific notation, } 1.39 \times 10^1$$

The numerator, 0.7608, has four significant figures, but the denominator has only three, so the result must be reported with three significant figures.

Rules for Rounding

The numerical result obtained in many calculations must be rounded to retain the proper number of significant figures. When you round a number to reduce the number of digits, follow these rules:

- The last digit retained is increased by 1 if the following digit is greater than 5 or a 5 followed by other digits.
- The last digit retained is left unchanged if the following digit is less than 5.
- If the digit to be removed is 5, and there are no digits to the right of the 5, then the last digit retained is increased by 1 if that digit is odd, and the last digit retained is left unchanged if that digit is even.

Full Number	Number Rounded to Three Significant Digits
12.696	12.7
16.249	16.2
18.35	18.4
18.45	18.4
24.752	24.7
18.351	18.4

One last word regarding significant figures, rounding, and calculations. In working problems on a calculator, you should do the calculation using all the digits allowed by the calculator and round only at the end of the problem. Rounding in the middle of a calculation sequence can introduce small errors that can accumulate later in the calculation. If your answers do not quite agree with those in the appendices of this book, this practice may be the source of the disagreement.

PROBLEM-SOLVING EXAMPLE 2.5 Rounding Significant Figures

Do these calculations and round the result to the proper number of significant figures.

(a) $15.80 + 0.0060 - 2.0 + 0.081$

(b) $\frac{55.0}{12.34}$

(c) $\frac{12.7732 - 2.3317}{5.007}$

(d) $2.16 \times 10^3 + 4.01 \times 10^2$

Answer

(a) 13.9 (b) 4.46 (c) 2.085 (d) 2.56×10^3

Strategy and Explanation In each case, the calculation is done with no rounding. Then the rules for rounding are applied to the answer.

(a) 13.887 is rounded to 13.9 with one decimal place because 2.0 has one decimal place.

(b) 4.457 is rounded to 4.46 with three significant figures just as in 55.0, which also has three significant figures.

(c) The number of significant figures in the result is governed by the least number of significant figures in the quotient. The denominator, 5.007, has four significant figures, so the calculator result, 2.08538 . . . , is properly expressed with four significant figures as 2.085.

(d) We change 4.01×10^2 to 0.401×10^3 and then sum 2.16×10^3 plus 0.401×10^3 and round to three significant figures to get 2.56×10^3.

PROBLEM-SOLVING PRACTICE 2.5

Do these calculations and round the result to the proper number of significant figures.

(a) $244.2 + 0.1732$ (b) 6.19×5.2222 (c) $\frac{7.2234 - 11.3851}{4.22}$

2.5 Atomic Numbers and Mass Numbers

Each element has a unique atomic number.

The periodic table is organized by atomic number; this is discussed in Section 2.9.

Experiments done early in the twentieth century found that *atoms of the same element have the same numbers of protons in their nuclei.* This number is called the **atomic number** and is given the symbol *Z*. In the periodic table on the inside front cover of this book, the atomic number for each element is written above the element's symbol. For example, a copper atom has a nucleus containing 29 protons, so its atomic number is 29 ($Z = 29$). A lead atom (Pb) has 82 protons in its nucleus, so the atomic number for lead is 82.

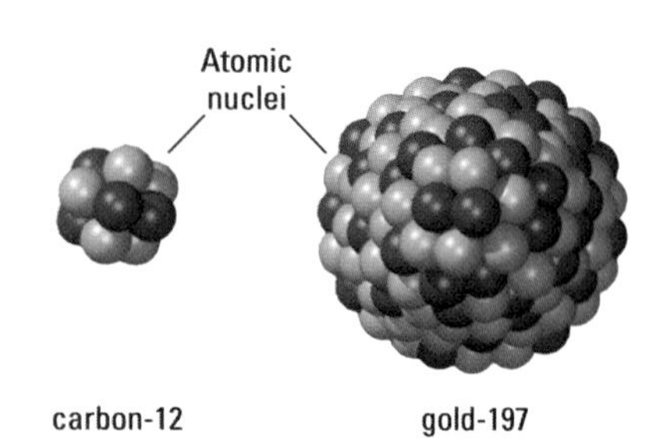

The scale of atomic masses is defined relative to a standard, the mass of a carbon atom that has 6 protons and 6 neutrons in its nucleus. The masses of atoms of every other element are established relative to the mass of this carbon atom, which is defined as having a mass of exactly 12 atomic mass units. In terms of macroscale mass units, 1 **atomic mass unit (amu)** = 1.66054×10^{-24} g. For example, when an experiment shows that a gold atom, on average, is 16.4 times as massive as the standard carbon atom, we then know mass of the gold atom in amu and grams.

$$16.4 \times 12 \text{ amu} = 197 \text{ amu}$$

$$197 \text{ amu} \times \frac{1.66054 \times 10^{-24} \text{ g}}{1 \text{ amu}} = 3.27 \times 10^{-22} \text{ g}$$

1 atomic mass unit (amu) = $\frac{1}{12}$ the mass of a carbon atom having 6 protons and 6 neutrons in the nucleus.

The masses of the fundamental subatomic particles in atomic mass units have been determined experimentally. The proton and the neutron have masses very close to 1 amu, whereas the electron's mass is approximately 1800 times smaller.

Particle	Mass (grams)	Mass (atomic mass units)	Charge
Electron	$9.10938215 \times 10^{-28}$	0.000548579	1−
Proton	$1.6726216 \times 10^{-24}$	1.00728	1+
Neutron	$1.67492721 \times 10^{-24}$	1.00866	0

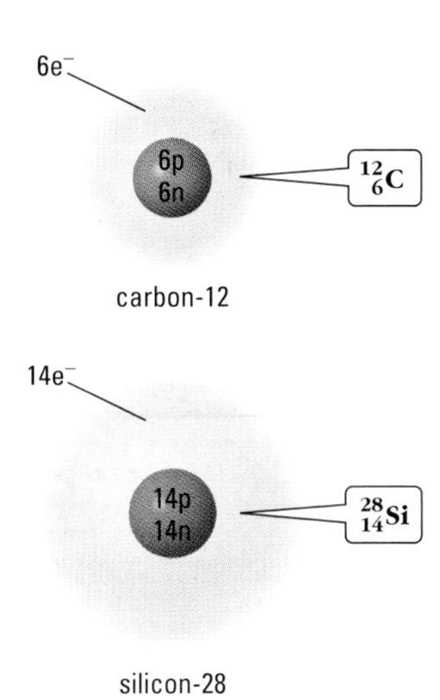

Once a relative scale of atomic masses has been established, we can estimate the mass of any atom whose nuclear composition is known. The proton and neutron have masses so close to 1 amu that the difference can be ignored in an estimate. Electrons have much less mass than protons or neutrons. Even though the number of electrons in an atom must equal the number of protons, the mass of the electrons is so small that they never affect the atomic mass by more than 0.1%, so the electrons' mass need not be considered. To estimate an atom's mass, we add up its number of protons and neutrons. This sum, called the **mass number** of that particular atom, is given the symbol A. For example, a copper atom that has 29 protons and 34 neutrons in its nucleus has a mass number, A, of 63. A lead atom that has 82 protons and 126 neutrons has $A = 208$.

With this information, an atom of known composition, such as a lead-208 atom, can be represented as follows:

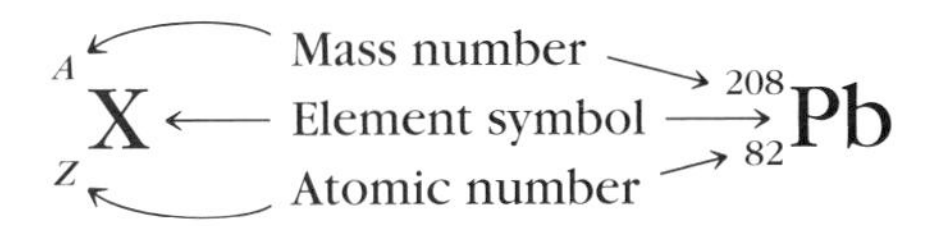

Each element has its own unique one- or two-letter symbol. Because each element is defined by the number of protons its atoms contain, knowing which element you are dealing with means you automatically know the number of protons its atoms have. Thus, the Z part of the notation is redundant. For example, the lead atom might be represented by the symbol ^{208}Pb because the Pb tells us the element is lead, and lead by definition always contains 82 protons. Whether we use the symbol or the alternative notation lead-208, we simply say "lead-208."

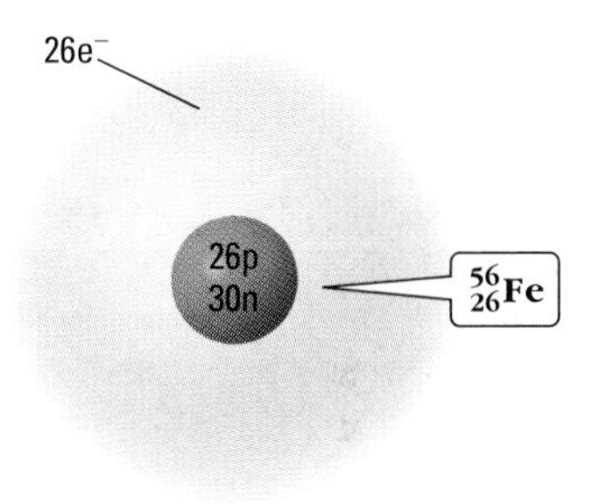

PROBLEM-SOLVING EXAMPLE 2.6 Atomic Nuclei

Iodine-131 is used in medicine to assess thyroid gland function. How many protons and neutrons are present in an iodine-131 atom?

Answer 53 protons and 78 neutrons

Strategy and Explanation The periodic table inside the front cover of this book shows that the atomic number of iodine (I) is 53. Therefore, the atom has 53 protons in its nucleus. Because the mass number of the atom is the sum of the number of protons and neutrons in the nucleus,

$$\text{Mass number} = \text{number of protons} + \text{number of neutrons}$$

$$131 = 53 + \text{number of neutrons}$$

$$\text{Number of neutrons} = 131 - 53 = 78$$

PROBLEM-SOLVING PRACTICE 2.6

(a) What is the mass number of a phosphorus atom with 16 neutrons?
(b) How many protons, neutrons, and electrons are there in a neon-22 atom?
(c) Write the symbol for the atom with 82 protons and 125 neutrons.

The actual mass of an atom is slightly less than the sum of the masses of its protons, neutrons, and electrons. The difference, known as the mass defect, is related to the energy that binds nuclear particles together, a topic discussed in Chapter 19.

Two elements can't have the same atomic number. If two atoms differ in their number of protons, they are atoms of different elements. If only their number of neutrons differs, they are isotopes of a single element, such as neon-20, neon-21, and neon-22.

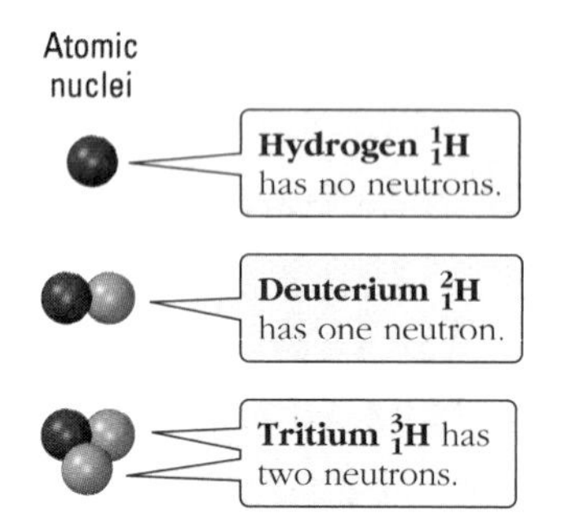
Atomic nuclei
Hydrogen $^{1}_{1}H$ has no neutrons.
Deuterium $^{2}_{1}H$ has one neutron.
Tritium $^{3}_{1}H$ has two neutrons.

Hydrogen isotopes. Hydrogen, deuterium, and tritium each contain one proton. Hydrogen has no neutrons; deuterium and tritium have one and two neutrons, respectively.

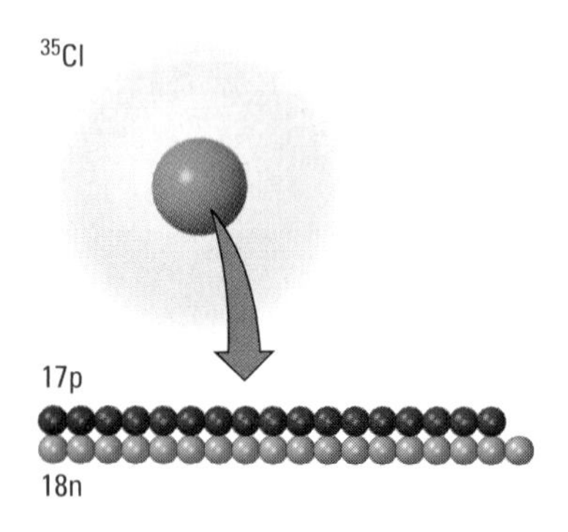
^{35}Cl
17p
18n

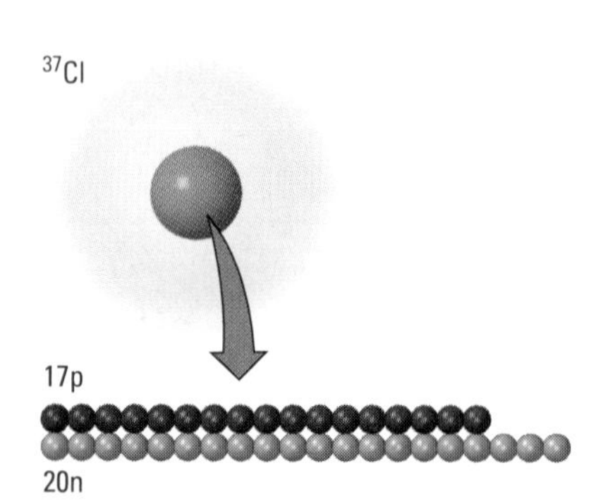
^{37}Cl
17p
20n

Chlorine isotopes. Chlorine-35 and chlorine-37 atoms each contain 17 protons; chlorine-35 atoms have 18 neutrons, and chlorine-37 atoms contain 20 neutrons.

Although an atom's mass approximately equals its mass number, the actual mass is not an integral number. For example, the actual mass of a gold-196 atom is 195.9231 amu, slightly less than the mass number 196. The masses of atoms are determined experimentally using mass spectrometers.

Mass spectrometric analysis of most naturally occurring elements reveals that not all atoms of an element have the same mass. For example, all silicon atoms have 14 protons, but some silicon nuclei have 14 neutrons, others have 15, and others have 16. Thus, naturally occurring silicon (atomic number 14) is always a mixture of silicon-28, silicon-29, and silicon-30 atoms. Such different atoms of the *same* element are called isotopes. **Isotopes** are atoms with the *same* atomic number (Z) but different mass numbers (A). All the silicon atoms have 14 protons; that is what makes them silicon atoms. But these isotopes differ in their mass numbers because they have different numbers of neutrons.

PROBLEM-SOLVING EXAMPLE 2.7 Isotopes

Boron has two isotopes, one with five neutrons and the other with six neutrons. What are the mass numbers and symbols of these isotopes?

Answer The mass numbers are 10 and 11. The symbols are $^{10}_{5}B$ and $^{11}_{5}B$.

Strategy and Explanation We use the entry for boron in the periodic table to find the answer. Boron has an atomic number of 5, so it has five protons in its nucleus. Therefore, the mass numbers of the two isotopes are given by the sum of their numbers of protons and neutrons:

$$\text{Isotope 1: B} = 5 \text{ protons} + 5 \text{ neutrons} = 10 \text{ (boron-10)}$$

$$\text{Isotope 2: B} = 5 \text{ protons} + 6 \text{ neutrons} = 11 \text{ (boron-11)}$$

Placing the atomic number at the bottom left and the mass number at the top left gives the symbols $^{10}_{5}B$ and $^{11}_{5}B$.

PROBLEM-SOLVING PRACTICE 2.7

Naturally occurring magnesium has three isotopes with 12, 13, and 14 neutrons. What are the mass numbers and symbols of these three isotopes?

We usually refer to a particular isotope by giving its mass number. For example, $^{238}_{92}U$ is referred to as uranium-238. But a few isotopes have distinctive names and symbols because of their importance, such as the isotopes of hydrogen. All hydrogen isotopes have just one proton. When the single proton is the only nuclear particle, the element is simply called hydrogen. With one neutron as well as one proton present, the isotope $^{2}_{1}H$ is called either deuterium or heavy hydrogen (symbol D). When two neutrons are present, the isotope $^{3}_{1}H$ is called tritium (symbol T).

CONCEPTUAL EXERCISE 2.3 Isotopes

A student in your chemistry class tells you that nitrogen-14 and nitrogen-15 are not isotopes because they have the same number of protons. How would you refute this statement?

2.6 Isotopes and Atomic Weight

Copper has two naturally occurring isotopes, copper-63 and copper-65, with atomic masses of 62.9296 amu and 64.9278 amu, respectively. In a macroscopic collection of naturally occurring copper atoms, the average mass of the atoms is neither 63 (all copper-63) nor 65 (all copper-65). Rather, the average atomic mass will fall between 63 and 65, with its exact value depending on the proportion of each isotope in the mixture. The proportion of atoms of each isotope in a natural sample of an element is called the **percent abundance,** the percentage of atoms of a *particular* isotope.

The concept of percent is widely used in chemistry, and it is worth briefly reviewing here. For example, earth's atmosphere contains approximately 78% nitrogen, 21% oxygen, and 1% argon. U.S. pennies minted after 1982 contain 2.4% copper; the remainder is zinc.

PROBLEM-SOLVING EXAMPLE 2.8 Applying Percent

The U.S. Mint issued state quarters over the 10-year period 1999–2008. The quarters each weigh 5.670 g and contain 8.33% nickel and the remainder copper. What mass of each element is contained in each quarter?

Answer 0.472 g nickel and 5.20 g copper

Strategy and Explanation We need the mass of each element in each quarter. We start by calculating the mass of nickel. Its percentage, 8.33%, means that every 100. g of coin contains 8.33 g nickel.

$$5.670\ \text{g quarter} \times \frac{8.33\ \text{g nickel}}{100.\ \text{g quarter}} = 0.472\ \text{g nickel}$$

The mass of copper is found the same way, using the conversion factor 91.67 g copper per 100. g of quarter.

$$5.670\ \text{g quarter} \times \frac{91.67\ \text{g copper}}{100.\ \text{g quarter}} = 5.198\ \text{g copper}$$

We could have obtained this value directly by recognizing that the masses of nickel and copper must sum to the mass of the quarter. Therefore,

$$5.670\ \text{g quarter} = 0.472\ \text{g nickel} + x\ \text{g copper}$$

Solving for x, the mass of copper, gives

$$5.670\ \text{g} - 0.472\ \text{g} = 5.198\ \text{g copper}$$

☑ **Reasonable Answer Check** The ratio of nickel to copper is about 11:1 (91.67/8.33 = 11), and the ratio of the masses calculated is also about 11:1 (5.198/0.472 = 11), so the answer is reasonable.

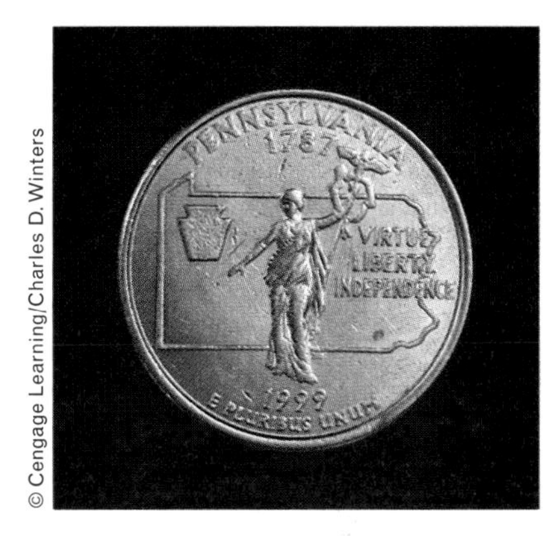

© Cengage Learning/Charles D. Winters

The Pennsylvania quarter. It shows the statue "Commonwealth," an outline of the state, the state motto, and a keystone.

The sum of the percentages for the composition of a sample must be 100.

PROBLEM-SOLVING PRACTICE 2.8

Many heating devices such as hair dryers contain nichrome wire, an alloy containing 80.% nickel and 20.% chromium, which gets hot when an electric current passes through it. If a heating device contains 75 g nichrome wire, how many grams of nickel and how many grams of chromium does the wire contain?

The percent abundance of each isotope in a sample of an element is given as follows:

$$\text{Percent abundance} = \frac{\text{number of atoms of a given isotope}}{\text{total number of atoms of all isotopes of that element}} \times 100\%$$

Table 2.2 gives information about the percent abundance for naturally occurring isotopes of hydrogen, boron, and bromine. The percent abundance and isotopic mass of each isotope can be used to find the average mass of atoms of that element, and this average mass is called the atomic weight of the element. The **atomic weight** of an element is the average mass of a representative sample of atoms of the element, expressed in atomic mass units.

Table 2.2 Isotopic Masses of the Stable Isotopes of Hydrogen, Boron, and Bromine

Element	Symbol	Atomic Weight (amu)	Mass Number	Isotopic Mass (amu)	Percent Abundance
Hydrogen	H	1.00794	1	1.007825	99.9855
	D		2	2.0141022	0.0145
Boron	B	10.811	10	10.012939	19.91
			11	11.009305	80.09
Bromine	Br	79.904	79	78.918336	50.69
			81	80.916289	49.31

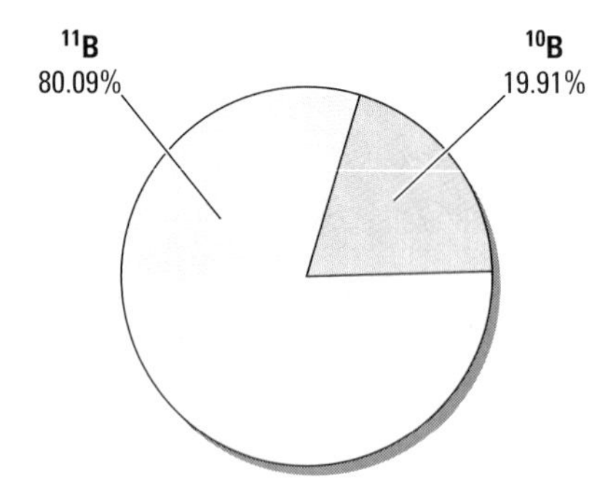

Percent abundance of boron-10 and boron-11.

Boron, for example, is a relatively rare element present in compounds used in laundry detergents, mild antiseptics, and Pyrex cookware. It has two naturally occurring isotopes: boron-10, with a mass of 10.0129 amu and 19.91% abundance, and boron-11, with a mass of 11.0093 amu and 80.09% abundance. Since the abundances are approximately 20% and 80%, respectively, you can estimate the atomic weight of boron: 20 atoms out of every 100, or 2 atoms out of every 10, are boron-10. If you then add up the masses of 10 atoms, you have 2 atoms with a mass of about 10 amu and 8 atoms with a mass of about 11 amu, so the sum is 108 amu, and the average is 108 amu/10 = 10.8 amu. This approximation is about right when you consider that the mass numbers of the boron isotopes are 10 and 11 and that boron is 80% boron-11 and only 20% boron-10. Therefore, the atomic weight should be about two tenths of the way down from 11 to 10, or 10.8. Thus, each atomic weight is a weighted average that accounts for the proportion of each isotope, not just the usual arithmetic average in which the values are simply summed and divided by the number of values.

In general, the atomic weight of an element is found from the percent abundance data as shown by the following more exact calculation for boron. The mass of each isotope is multiplied by its fractional abundance, the percent abundance expressed as a decimal, to calculate the weighted average, the atomic weight.

$$\begin{aligned}\text{Atomic weight} &= [(\text{fractional abundance } ^{10}\text{B})(\text{isotopic mass } ^{10}\text{B}) \\ &\quad + (\text{fractional abundance } ^{11}\text{B})(\text{isotopic mass } ^{11}\text{B})] \\ &= (0.1991)(10.0129 \text{ amu}) + (0.8009)(11.0093 \text{ amu}) \\ &= 10.81 \text{ amu}\end{aligned}$$

1 amu = $\frac{1}{12}$ mass of carbon-12 atom

The term "atomic weight" is so commonly used that it has become accepted, even though it is really a mass rather than a weight.

Our earlier estimate was quite close to the more exact result. The arithmetic average of the isotopic masses of boron is (10.0129 + 11.093)/2 = 10.55, which is quite different from the actual atomic weight.

The atomic weight of each stable (nonradioactive) element has been determined; these values appear in the periodic table in the inside front cover of this book. For most elements, the abundances of the isotopes are the same no matter where a sample is collected. Therefore, the atomic weights in the periodic table are used whenever an atomic weight is needed. In the periodic table, each element's box contains the atomic number, the symbol, and the atomic weight. For example, the periodic table entry for zinc is

30 ⟵	Atomic number
Zn ⟵	Symbol
65.382 ⟵	Atomic weight

Elemental zinc.

EXERCISE 2.4 Atomic Weight

Verify that the atomic weight of lithium is 6.941 amu, given this information:

$^{6}_{3}$Li mass = 6.015121 amu and percent abundance = 7.500%

$^{7}_{3}$Li mass = 7.016003 amu and percent abundance = 92.50%

CONCEPTUAL EXERCISE 2.5 Isotopic Abundance

Naturally occurring magnesium contains three isotopes: ^{24}Mg (78.70%), ^{25}Mg (10.13%), and ^{26}Mg (11.17%). Estimate the atomic weight of Mg and compare your estimate with the atomic weight calculated by finding the arithmetic average of the atomic masses. Which value is larger? Why is it larger?

CONCEPTUAL EXERCISE 2.6 Percent Abundance

Gallium has two abundant isotopes, and its atomic weight is 69.72 amu. If you knew only this value and not the percent abundance of the isotopes, make the case that the percent abundance of each of the two gallium isotopes cannot be 50%.

2.7 Amounts of Substances: The Mole

As noted earlier, atoms are much too small to be seen directly or weighed individually on the most sensitive balance. However, when working with chemical reagents, it is essential to know how many atoms, molecules, or other nanoscale units of an element or compound you have. To connect the macroscale world, where chemicals can be measured, weighed, and manipulated, to the nanoscale world of individual atoms or molecules, chemists have defined a convenient unit of matter that contains a known number of particles. This chemical counting unit is the **mole (mol),** defined as the amount of substance that contains as many atoms, molecules, ions, or other nanoscale units as there are atoms in *exactly* 12 g of carbon-12.

One mole of carbon has a mass of 12.01 g, not exactly 12 g, because naturally occurring carbon contains both carbon-12 (98.89%) and carbon-13 (1.11%). By definition, one mole of carbon-12 has a mass of exactly 12 g.

The term "mole" is derived from the Latin word *moles* meaning a "heap" or "pile."

When used with a number, mole is abbreviated mol, for example, 0.5 mol.

The essential point to understand about moles is that ***one mole always contains the same number of particles,*** no matter what substance or what kind of particles we are talking about. The number of particles in a mole is

$$1 \text{ mol} = 6.02214179 \times 10^{23} \text{ particles}$$

The mole is the connection between the macroscale and nanoscale worlds, the visible and the not directly visible.

The number of particles in a mole is known as **Avogadro's number** after Amadeo Avogadro (1776–1856), an Italian physicist who conceived the basic idea but never experimentally determined the number, which came later. It is important to realize that the value of Avogadro's number is a definition tied to the number of atoms in 12 g carbon-12.

$$\text{Avogadro's number} = 6.02214179 \times 10^{23} \text{ per mole} = 6.02214179 \times 10^{23} \text{ mol}^{-1}$$

One difficulty in comprehending Avogadro's number is its sheer size. Writing it out fully yields

$$6.02214179 \times 10^{23} = 602{,}214{,}179{,}000{,}000{,}000{,}000{,}000$$

or 602,214.179 × 1 million × 1 million × 1 million. Although Avogadro's number is known to nine significant figures, we will most often use it rounded to 6.022×10^{23}.

Figure 2.4 One-mole quantities of six elements. S, Mg, Cr (lower, left to right); Cu, Al, Pb (upper, left to right).

Atomic weights are given in the periodic table of elements in the inside front cover of this book.

There are many analogies used to try to give a feeling for the size of this number. If you poured Avogadro's number of marshmallows over the continental United States, the marshmallows would cover the country to a depth of approximately 650 miles. Or, if one mole of pennies were divided evenly among every man, woman, and child in the United States, your share alone would pay off the national debt (about \$9.4 trillion, or $\$9.4 \times 10^{12}$) 2.1 times.

You can think of the mole simply as a counting unit, analogous to the counting units we use for ordinary items such as doughnuts or bagels by the dozen, shoes by the pair, or sheets of paper by the ream (500 sheets). Atoms, molecules, and other particles in chemistry are counted by the mole. The different masses of the elements shown in Figure 2.4 each contain one mole of atoms. For each element in the figure, the mass in grams (the macroscale) is numerically equal to the atomic weight in atomic mass units (the nanoscale). The **molar mass** of any substance is the mass, in grams, of one mole of that substance. Molar mass has the units of grams per mole (g/mol).

For example,

$$\begin{aligned}\text{molar mass of copper (Cu)} &= \text{mass of 1 mol Cu atoms}\\ &= \text{mass of } 6.022 \times 10^{23} \text{ Cu atoms}\\ &= 63.546 \text{ g/mol}\\ \text{molar mass of aluminum (Al)} &= \text{mass of 1 mol Al atoms}\\ &= \text{mass of } 6.022 \times 10^{23} \text{ Al atoms}\\ &= 26.9815 \text{ g/mol}\end{aligned}$$

Each molar mass of copper or aluminum contains Avogadro's number of atoms. *Molar mass differs from one element to the next because the atoms of different*

ESTIMATION The Size of Avogadro's Number

Chemists and other scientists often use estimates in place of exact calculations when they want to know the approximate value of a quantity. Analogies to help us understand the extremely large value of Avogadro's number are an example.

If 1 mol of green peas were spread evenly over the continental United States, how deep would the layer of peas be? The surface area of the continental United States is about 3.0×10^6 square miles (mi^2). There are 5280 feet per mile.

Let's start with an estimate of a green pea's size: $\frac{1}{4}$-inch diameter. Then 4 peas would fit along a 1-inch line, and 48 would fit along a 1-foot line, and $48^3 = 110{,}592$ would fit into 1 cubic foot (ft^3). Since we are estimating, we will approximate by saying 1×10^5 peas per cubic foot.

Now, let's estimate how many cubic feet of peas are in 1 mol of peas:

$$\frac{6.022 \times 10^{23} \text{ peas}}{1 \text{ mol peas}} \times \frac{1 \text{ ft}^3}{1 \times 10^5 \text{ peas}} \cong \frac{6.0 \times 10^{18} \text{ ft}^3}{1 \text{ mol peas}}$$

(The "approximately equal" sign, $\cong$, is an indicator of these approximations.) The surface area of the continental United States is 3.0×10^6 square miles (mi^2), which is about

$$3.0 \times 10^6 \text{ mi}^2 \times \left(\frac{5280 \text{ ft}}{1 \text{ mi}}\right)^2 \cong 8.4 \times 10^{13} \text{ ft}^2$$

so 1 mol of peas spread evenly over this area will have a depth of

$$\frac{6.0 \times 10^{18} \text{ ft}^3}{1 \text{ mol peas}} \times \frac{1}{8.4 \times 10^{13} \text{ ft}^2} \cong \frac{7.1 \times 10^4 \text{ ft}}{1 \text{ mol peas}}$$

or

$$\frac{7.1 \times 10^4 \text{ ft}}{1 \text{ mol peas}} \times \frac{1 \text{ mi}}{5280 \text{ ft}} \cong \frac{14 \text{ mi}}{1 \text{ mol peas}}$$

Note that in many parts of the estimate, we rounded or used fewer significant figures than we could have used. Our purpose was to *estimate* the final answer, not to compute it exactly. The final answer, 14 miles, is not particularly accurate, but it is a valid estimate. The depth would be more than 10 miles but less than 20 miles. It would not be 6 inches or even 6 feet. Estimating served the overall purpose of developing the analogy for understanding the size of Avogadro's number.

Visit this book's companion website at **www.cengage.com/chemistry/moore** to work an interactive module based on this material.

elements have different masses. Think of a mole as analogous to a dozen. We could have a dozen golf balls, a dozen baseballs, or a dozen bowling balls, 12 items in each case. The dozen items do not weigh the same, however, because the individual items do not weigh the same: 45 g per golf ball, 134 g per baseball, and 7200 g per bowling ball. In a similar way, the mass of Avogadro's number of atoms of one element is different from the mass of Avogadro's number of atoms of another element because the atoms of different elements differ in mass.

Items can be counted by weighing. Knowing the mass of one nail, we can estimate the number of nails in this 5-lb box.

2.8 Molar Mass and Problem Solving

Understanding the idea of a mole and applying it properly are *essential* to doing quantitative chemistry. In particular, it is *absolutely necessary* to be able to make two basic conversions: *moles* → *mass* and *mass* → *moles*. To do these and many other calculations in chemistry, it is most helpful to use dimensional analysis in the same way it is used in unit conversions. Along with calculating the final answer, write the units with all quantities in a calculation and cancel the units. If the problem is set up properly, the answer will have the desired units.

Let's see how these concepts apply to converting mass to moles or moles to mass. *In either case, the conversion factor is provided by the molar mass of the substance, the number of grams in one mole, that is, grams per mole (g/mol).*

Mass ⇄ moles conversions for substance A

Mass A ⟶ moles A

$$\text{Grams A} \times \underbrace{\frac{1 \text{ mol A}}{\text{grams A}}}_{\frac{1}{\text{molar mass}}} = \text{moles A}$$

Moles A ⟶ mass A

$$\text{Moles A} \times \underbrace{\frac{\text{grams A}}{1 \text{ mol A}}}_{\text{molar mass}} = \text{grams A}$$

Suppose you need 0.250 mol Cu for an experiment. How many grams of Cu should you use? The atomic weight of Cu is 63.546 amu, so the molar mass of Cu is 63.546 g/mol. To calculate the mass of 0.250 mol Cu, you need the conversion factor 63.546 g Cu/1 mol Cu.

$$0.250 \text{ } \cancel{\text{mol Cu}} \times \frac{63.55 \text{ g Cu}}{1 \text{ } \cancel{\text{mol Cu}}} = 15.9 \text{ g Cu}$$

In this book we will, when possible, *use one more significant figure in the molar mass than in any of the other data in the problem.* In the problem just completed, note that we used four significant figures in the molar mass of Cu when three were given in the number of moles. Using one more significant figure in the molar mass guarantees that its precision is greater than that of the other numbers and does not limit the precision of the result of the computation.

EXERCISE 2.7 Grams, Moles, and Avogadro's Number

You have a 10.00-g sample of lithium and a 10.00-g sample of iridium. How many atoms are in each sample, and how many more atoms are in the lithium sample than in the iridium sample?

Frequently, a problem requires converting a mass to the equivalent number of moles, such as calculating the number of moles of bromine in 10.00 g bromine. Because bromine is a diatomic element, it consists of Br_2 molecules. Therefore, there are 2 mol Br atoms in 1 mol Br_2 molecules. The molar mass of Br_2 is twice its atomic mass, 2×79.904 g/mol = 159.81 g/mol. To calculate the moles of bromine in 10.00 g of Br_2, use the molar mass of Br_2 as the conversion factor, 1 mol Br_2/159.81 g Br_2.

$$10.00 \text{ g } Br_2 \times \frac{1 \text{ mol } Br_2}{159.81 \text{ g } Br_2} = 6.257 \times 10^{-2} \text{ mol } Br_2$$

Titanium and aluminum are metals used in modern airplane manufacturing.

PROBLEM-SOLVING EXAMPLE 2.9 Mass and Moles

(a) Titanium (Ti) is a metal used to build airplanes. How many moles of Ti are in a 100. g sample of the pure metal?

(b) Aluminum is also used in airplane manufacturing. A piece of Al contains 2.16 mol Al. Is the mass of Al greater or less than the mass of Ti in part (a)?

Answer (a) 2.09 mol Ti (b) 58.3 g Al, which is less than the mass of Ti

Strategy and Explanation

(a) This is a mass-to-moles conversion that is solved using the molar mass of Ti as the conversion factor 1 mol Ti/47.87 g Ti.

$$100. \text{ g Ti} \times \frac{1 \text{ mol Ti}}{47.87 \text{ g Ti}} = 2.09 \text{ mol Ti}$$

(b) This is a moles-to-mass conversion that requires the conversion factor 26.98 g Al/1 mol Al.

$$2.16 \text{ mol Al} \times \frac{26.98 \text{ g Al}}{1 \text{ mol Al}} = 58.3 \text{ g Al}$$

This mass is less than the 100. g mass of titanium.

PROBLEM-SOLVING PRACTICE 2.9

Calculate (a) the number of moles in 1.00 mg molybdenum (Mo) and (b) the number of grams in 5.00×10^{-3} mol gold (Au).

Dmitri Mendeleev 1834–1907

Originally from Siberia, Mendeleev spent most of his life in St. Petersburg. He taught at the University of St. Petersburg, where he wrote books and published his concept of chemical periodicity, which helped systematize inorganic chemistry. Later in life he moved on to other interests, including studying the natural resources of Russia and their commercial applications.

2.9 The Periodic Table

You have already used the periodic table inside the front cover of this book to obtain atomic numbers and atomic weights of elements. But it is much more valuable than this. The periodic table is an exceptionally useful tool in chemistry. It allows us to organize and interrelate the chemical and physical properties of the elements. For example, elements can be classified as metals, nonmetals, or metalloids by their positions in the periodic table. You should become familiar with its main features and terminology.

Dmitri Mendeleev (1834–1907), while a professor at the University of St. Petersburg, Russia, realized that listing the elements in order of increasing atomic weight revealed a periodic repetition of the properties. He summarized his findings in the table that has come to be called the periodic table. By lining up the elements in horizontal rows in order of increasing atomic weight and starting a new row when he came to an element with properties similar to one already in the previous row, he saw that the resulting columns contained elements with similar properties. Mendeleev found that some positions in his table were not filled, he predicted that new elements would be found that filled the gaps, and he predicted properties of the undiscovered elements. Two of the missing elements—gallium (Ga) and germanium (Ge)—were soon discovered, with properties very close to those Mendeleev had predicted.

It gives a useful perspective to realize that Mendeleev developed the periodic table nearly a half-century before electrons, protons, and neutrons were known.

Later experiments by H. G. J. Moseley demonstrated that elements in the periodic table should be ordered by atomic numbers rather than atomic weights. Arranging the elements in order of increasing atomic number gives the **law of**

CHEMISTRY IN THE NEWS

Periodic Table Stamp

This striking 0.30-euro stamp was issued by the Spanish Post Office on February 2, 2007, to commemorate the 100th anniversary of the death of Mendeleev, who proposed the periodic table in 1869. The stamp shows an artistic representation in which the blue and yellow areas denote the main group elements, the red area denotes the transition elements, and the green area denotes the lanthanides and actinides. The sizes of the four colored areas are in the correct proportions regarding the number of periods and groups of elements in each region. The small white squares embedded in the red and yellow regions show the locations of four elements—gallium, germanium, scandium, and technetium—not yet discovered in 1869. Mendeleev used the periodicity of the table to correctly predict properties for gallium, germanium, and scandium, elements that were discovered in 1875, 1886, and 1878, respectively. Technetium, the first artificially synthesized element, was produced in 1937.

Spanish postage stamp of the periodic table.

chemical periodicity: ***The properties of the elements are periodic functions of their atomic numbers (numbers of protons).***

Periodic Table Features

Elements in the **periodic table** are arranged according to atomic number so that *elements with similar chemical properties occur in vertical columns* called **groups.** The table commonly used in the United States has groups numbered 1 through 8 (Figure 2.5), with each number followed by either an A or a B. The A groups (Groups 1A and 2A on the left of the table and Groups 3A through 8A at the right) are collectively known as **main group elements.** The B groups (in the middle of the table) are called **transition elements.**

The horizontal rows of the table are called **periods,** and they are numbered beginning with 1 for the period containing only H and He. Sodium (Na) is, for example, in Group 1A and is the first element in the third period. Silver (Ag) is in Group 1B and is in the fifth period.

The table in Figure 2.5 and inside the front cover helps us to recognize that most elements are metals (gray and blue), far fewer elements are nonmetals (lavender), and even fewer are metalloids (orange). Elements usually become less metallic from left to right across a period, and eventually one or more nonmetals are found in each period. The six metalloids (B, Si, Ge, As, Sb, and Te) fall along a zigzag line passing between Al and Si, Ge and As, and Sb and Te.

Periodicity of piano keys.

Many interactive periodic tables are available on the web. Two that you may want to explore are at **http://www.chemeddl.org/collections/ptl** and **http://periodictable.com.**

Important Regions of the Periodic Table

The elements (except hydrogen) in the leftmost column (Group 1A) are called **alkali metals.** Elements in Group 2A are known as the **alkaline earth metals.** The alkali metals and alkaline earth metals are very reactive and are found in nature only combined with other elements in compounds. The **halogens** (Group 7A) consist of diatomic molecules and are highly reactive. Group 8A, on the far right of the periodic table, contains the **noble gases,** the least reactive elements.

The **transition elements,** or transition metals, fill the center of the periodic table in Periods 4 through 7. The **lanthanides** and **actinides** are metallic elements listed separately in two rows at the bottom of the periodic table.

Periodicity of the elements means a recurrence of similar properties at regular intervals when the elements are arranged in the correct order.

An alternative convention for numbering the groups in the periodic table uses the numbers 1 through 18, with no letters.

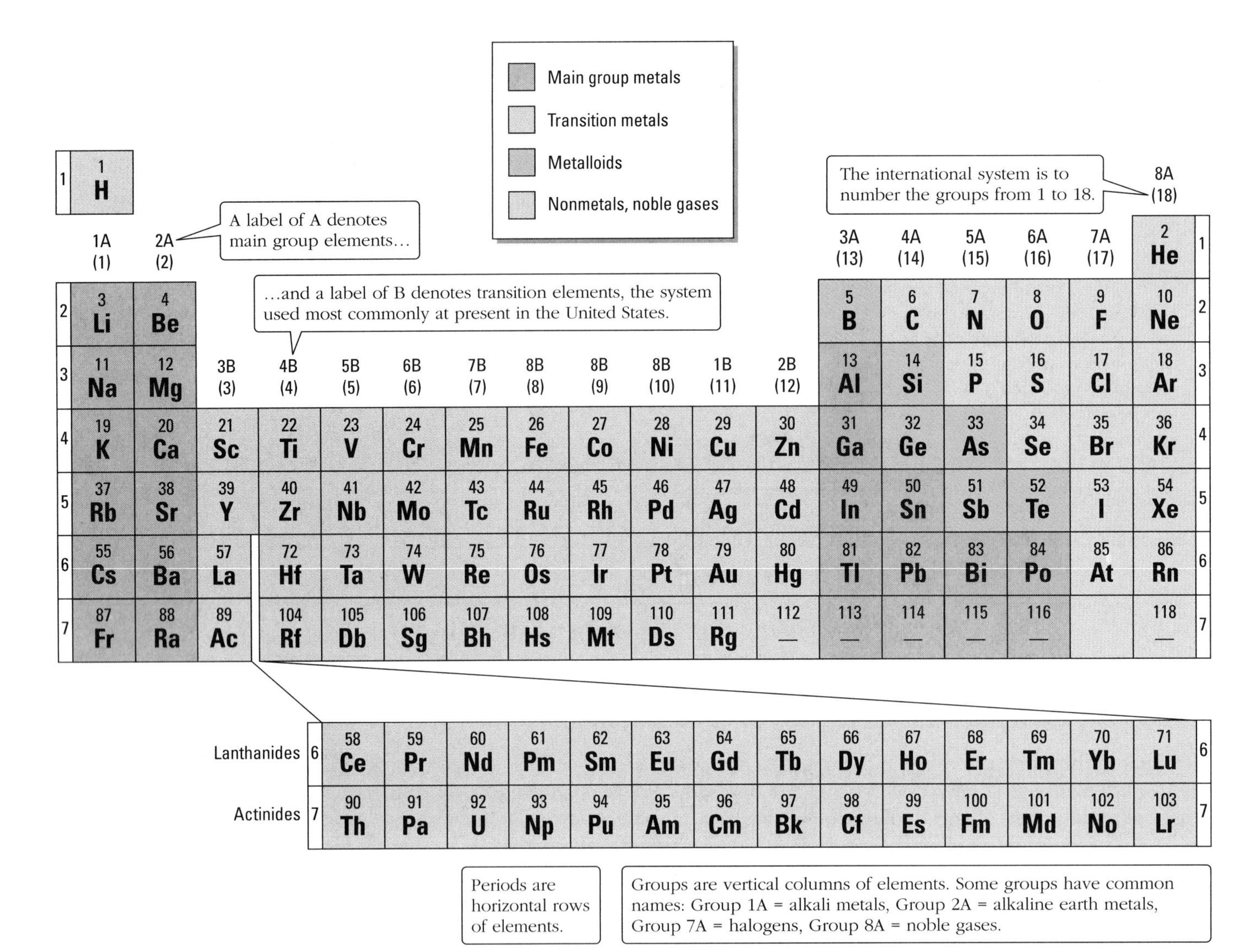

Figure 2.5 Modern periodic table of the elements. Elements are listed across the periods in ascending order of atomic number.

IN CLOSING

Having studied this chapter, you should be able to . . .

- Describe radioactivity, electrons, protons, and neutrons and the general structure of the atom (Sections 2.1, 2.2).
- Use conversion factors for the units for mass, volume, and length common in chemistry (Section 2.3). ■ End-of-chapter Questions assignable in OWL: 10, 65
- Identify the correct number of significant figures in a number and carry significant figures through calculations (Section 2.4). ■ Questions 13, 17
- Define isotope and give the mass number and number of neutrons for a specific isotope (Section 2.5). ■ Question 29
- Calculate the atomic weight of an element from isotopic abundances (Section 2.6). ■ Questions 35, 37, 62

Selected end-of-chapter Questions may be assigned in OWL.

For quick review, download Go Chemistry mini-lecture flashcard modules (or purchase them at **www.ichapters.com**).

- Explain the difference between the atomic number and the atomic weight of an element and find this information for any element (Sections 2.5, 2.6).
- Relate masses of elements to the mole, Avogadro's number, and molar mass (Section 2.7). ■ Questions 47, 69
- Do gram-mole and mole-gram conversions for elements (Section 2.8). ■ Questions 41, 42
- Identify the periodic table location of groups, periods, alkali metals, alkaline earth metals, halogens, noble gases, transition elements, lanthanides, and actinides (Section 2.9). ■ Questions 51, 53

Prepare for an exam with a **Summary Problem** that brings together concepts and problem-solving skills from throughout the chapter. Go to **www.cengage.com/chemistry/moore** to download this chapter's Summary Problem from the Student Companion Site.

KEY TERMS

actinides *(Section 2.9)*
alkali metals *(2.9)*
alkaline earth metals *(2.9)*
atomic mass unit (amu) *(2.5)*
atomic number *(2.5)*
atomic structure *(Introduction)*
atomic weight *(2.6)*
Avogadro's number *(2.7)*
chemical periodicity, law of *(2.9)*
electron *(2.1)*
group *(2.9)*
halogens *(2.9)*
ion *(2.1)*
isotope *(2.5)*
lanthanides *(2.9)*
main group elements *(2.9)*
mass *(2.3)*
mass number *(2.5)*
metric system *(2.3)*
molar mass *(2.7)*
mole (mol) *(2.7)*
neutron *(2.2)*
noble gases *(2.9)*
nucleus *(2.2)*
percent abundance *(2.6)*
period *(2.9)*
periodic table *(2.9)*
proton *(2.1)*
radioactivity *(2.1)*
significant figures *(2.4)*
transition elements *(2.9)*

QUESTIONS FOR REVIEW AND THOUGHT

■ denotes questions assignable in OWL.

Blue-numbered questions have short answers at the back of this book and fully worked solutions in the *Student Solutions Manual.*

Review Questions

1. What is the fundamental unit of electrical charge?
2. The positively charged particle in an atom is called the proton.
   (a) How much heavier is a proton than an electron?
   (b) What is the difference in the charge on a proton and an electron?
3. Ernest Rutherford's famous gold-foil experiment examined the structure of atoms.
   (a) What surprising result was observed?
   (b) The results of the gold-foil experiment enabled Rutherford to calculate that the nucleus is much smaller than the atom. How much smaller?
4. In any given *neutral* atom, how many protons are there compared with the number of electrons?

Topical Questions

Units and Unit Conversions

5. If the nucleus of an atom were the size of a golf ball (4 cm in diameter), what would be the diameter of the atom?
6. If a sheet of business paper is exactly 11 inches high, what is its height in centimeters? In millimeters? In meters?
7. The maximum speed limit in many states is 65 miles per hour. What is this speed in kilometers per hour?
8. A Volkswagen engine has a displacement of 120. in^3. What is this volume in cubic centimeters? In liters?
9. Calculate how many square inches there are in one square meter.
10. ■ One square mile contains exactly 640 acres. How many square meters are in one acre?
11. On May 18, 1980, Mt. St. Helens in Washington erupted. The 9677-ft-high summit was lowered by 1314 ft by the eruption. Approximately 0.67 cubic mile of debris was released into the atmosphere. How many cubic meters of debris was released?

12. Suppose a room is 18 ft long and 15 ft wide, and the distance from floor to ceiling is 8 ft, 6 in. You need to know the volume of the room in metric units for some scientific calculations. What is the room's volume in cubic meters? In liters?

Significant Figures

13. ■ How many significant figures are present in these measured quantities?
    (a) 1374 kg
    (b) 0.00348 s
    (c) 5.619 mm
    (d) 2.475×10^{-3} cm
    (e) 33.1 mL
14. How many significant figures are present in these measured quantities?
    (a) 1.022×10^{2} km
    (b) 34 m^2
    (c) 0.042 L
    (d) 28.2 °C
    (e) 323. mg
15. For each of these numbers, round to three significant digits and write the result in scientific notation.
    (a) 0.0004332
    (b) 44.7337
    (c) 22.4555
    (d) 0.0088418
16. For each of these numbers, round to four significant digits and write the result in scientific notation.
    (a) 247.583
    (b) 100,578
    (c) 0.0000348719
    (d) 0.004003881
17. ■ Perform these calculations and express the result with the proper number of significant figures.
    (a) $\dfrac{4.850 \text{ g} - 2.34 \text{ g}}{1.3 \text{ mL}}$
    (b) $V = \pi r^3$, where $r = 4.112$ cm
    (c) $(4.66 \times 10^{-3}) \times 4.666$
    (d) $\dfrac{0.003400}{65.2}$
18. Perform these calculations and express the result with the proper number of significant figures.
    (a) $2221.05 - \dfrac{3256.5}{3.20}$
    (b) $343.2 \times (2.01 \times 10^{-3})$
    (c) $S = 4\pi r^2$, where $r = 2.55$ cm
    (d) $\dfrac{2802}{15} - (0.0025 \times 10{,}000.)$

Percent

19. Silver jewelry is actually a mixture of silver and copper. If a bracelet with a mass of 17.6 g contains 14.1 g silver, what is the percentage of silver? Of copper?
20. The solder once used by plumbers to fasten copper pipes together consists of 67% lead and 33% tin. What is the mass of lead (in grams) in a 1.00-lb block of solder? What is the mass of tin?
21. Automobile batteries are filled with sulfuric acid. What is the mass of the acid (in grams) in 500. mL of the battery acid solution if the density of the solution is 1.285 g/cm^3 and the solution is 38.08% sulfuric acid by mass?
22. When popcorn pops, it loses water explosively. If a kernel of corn weighing 0.125 g before popping weighs 0.106 g afterward, what percentage of its mass did it lose on popping?

Isotopes

23. Are these statements true or false? Explain why in each case.
    (a) Atoms of the same element always have the same mass number.
    (b) Atoms of the same element can have different atomic numbers.
24. What is the difference between the mass number and the atomic number of an atom?
25. How many electrons, protons, and neutrons are present in an atom of cobalt-60?
26. The artificial radioactive element technetium is used in many medical studies. Give the number of electrons, protons, and neutrons in an atom of technetium-99.
27. The atomic weight of bromine is 79.904. The natural abundance of ^{81}Br is 49.31%. What is the atomic weight of the only other natural isotope of bromine?
28. The atomic weight of boron is 10.811. The natural abundance of ^{10}B is 19.91%. What is the atomic weight of the only other natural isotope of boron?
29. ■ How many electrons, protons, and neutrons are there in an atom of (a) calcium-40, $^{40}_{20}$Ca, (b) tin-119, $^{119}_{50}$Sn, and (c) plutonium-244, $^{244}_{94}$Pu?
30. How many electrons, protons, and neutrons are there in an atom of (a) carbon-13, $^{13}_{6}$C, (b) chromium-50, $^{50}_{24}$Cr, and (c) bismuth-205, $^{205}_{83}$Bi?
31. Fill in this table:

Z	*A*	Number of Neutrons	Element
35	81	______	______
______	______	62	Pd
77	______	115	______
______	151	______	Eu

32. Fill in this table:

Z	*A*	Number of Neutrons	Element
60	144	______	______
______	______	12	Mg
64	______	94	______
______	37	______	Cl

33. Which of these are isotopes of element X, whose atomic number is 9: $^{18}_{9}X$, $^{20}_{9}X$, $^{9}_{4}X$, $^{15}_{9}X$?
34. Which of these species are isotopes of the same element: $^{20}_{10}X$, $^{20}_{11}X$, $^{21}_{10}X$, $^{20}_{12}X$? Explain.

Atomic Weight

35. ■ Verify that the atomic weight of lithium is 6.941 amu, given this information:

    ^{6}Li, exact mass = 6.015121 amu

    percent abundance = 7.500%

    ^{7}Li, exact mass = 7.016003 amu

    percent abundance = 92.50%
36. Verify that the atomic weight of magnesium is 24.3050 amu, given this information:

    ^{24}Mg, exact mass = 23.985042 amu

    percent abundance = 78.99%

    ^{25}Mg, exact mass = 24.98537 amu

    percent abundance = 10.00%

    ^{26}Mg, exact mass = 25.982593 amu

    percent abundance = 11.01%
37. ■ Gallium has two naturally occurring isotopes, ^{69}Ga and ^{71}Ga, with masses of 68.9257 amu and 70.9249 amu, respectively. Calculate the abundances of these isotopes of gallium.
38. Silver has two stable isotopes, ^{107}Ag and ^{109}Ag, with masses of 106.90509 amu and 108.90476 amu, respectively. Calculate the abundances of these isotopes of silver.
39. Lithium has two stable isotopes, ^{6}Li and ^{7}Li. Since the atomic weight of lithium is 6.941, which is the more abundant isotope?
40. Argon has three naturally occurring isotopes: 0.337% ^{36}Ar, 0.063% ^{38}Ar, and 99.60% ^{40}Ar. Estimate the atomic weight of argon. If the masses of the isotopes are 35.968, 37.963, and 39.962, respectively, what is the atomic weight of natural argon?

The Mole

41. ■ Calculate the number of grams in
    (a) 2.5 mol boron
    (b) 0.015 mol O_2
    (c) 1.25×10^{-3} mol iron
    (d) 653 mol helium
42. Calculate the number of grams in
    (a) 6.03 mol gold
    (b) 0.045 mol uranium
    (c) 15.6 mol Ne
    (d) 3.63×10^{-4} mol plutonium
43. ■ Calculate the number of moles represented by each of these:
    (a) 127.08 g Cu
    (b) 20.0 g calcium
    (c) 16.75 g Al
    (d) 0.012 g potassium
    (e) 5.0 mg americium
44. Calculate the number of moles represented by each of these:
    (a) 16.0 g Na
    (b) 0.0034 g platinum
    (c) 1.54 g P
    (d) 0.876 g arsenic
    (e) 0.983 g Xe
45. If you have a 35.67-g piece of chromium metal on your car, how many atoms of chromium do you have?
46. If you have a ring that contains 1.94 g of gold, how many atoms of gold are in the ring?
47. ■ What is the average mass in grams of one copper atom?
48. What is the average mass in grams of one atom of titanium?

The Periodic Table

49. Name and give symbols for (a) three elements that are metals, (b) four elements that are nonmetals, and (c) two elements that are metalloids. In each case, also locate the element in the periodic table by giving the group and period in which the element is found.
50. Name and give symbols for three transition metals in the fourth period. Look up each of your choices in a dictionary, a book such as *The Handbook of Chemistry and Physics*, or on the Internet, and make a list of their properties. Also list the uses of each element.
51. ■ How many elements are there in Group 4A of the periodic table? Give the name and symbol of each of these elements. Tell whether each is a metal, nonmetal, or metalloid.
52. How many elements are there in the fourth period of the periodic table? Give the name and symbol of each of these elements. Tell whether each is a metal, metalloid, or nonmetal.
53. ■ The symbols for the four elements whose names begin with the letter I are In, I, Ir, and Fe. Match each symbol with one of the statements below.
    (a) a halogen
    (b) a main group metal
    (c) a transition metal in Period 6
    (d) a transition metal in Period 4
54. The symbols for four of the eight elements whose names begin with the letter S are Si, Ag, Na, and S. Match each symbol with one of the statements below.
    (a) a solid nonmetal
    (b) an alkali metal
    (c) a transition metal
    (d) a metalloid
55. Use the periodic table to identify these elements:
    (a) Name an element in Group 2A.
    (b) Name an element in the third period.
    (c) What element is in the second period in Group 4A?
    (d) What element is in the third period in Group 6A?
    (e) What halogen is in the fifth period?
    (f) What alkaline earth element is in the third period?
    (g) What noble gas element is in the fourth period?
    (h) What nonmetal is in Group 6A and the second period?
    (i) Name a metalloid in the fourth period.
56. Use the periodic table to identify these elements:
    (a) Name an element in Group 2B.
    (b) Name an element in the fifth period.
    (c) What element is in the sixth period in Group 4A?
    (d) What element is in the third period in Group 5A?
    (e) What alkali metal is in the third period?
    (f) What noble gas is in the fifth period?
    (g) Name the element in Group 6A and the fourth period. Is it a metal, nonmetal, or metalloid?
    (h) Name a metalloid in Group 5A.

General Questions

57. The density of a solution of sulfuric acid is 1.285 g/cm^3, and it is 38.08% acid by mass. What volume of the acid solution (in mL) do you need to supply 125 g sulfuric acid?
58. In addition to the metric units of nm and pm, a commonly used unit is the angstrom, where 1 Å = 1×10^{-10} m. If the distance between the Pt atom and the N atom in a compound is 1.97 Å, what is the distance in nm? In pm?
59. The smallest repeating unit of a crystal of common salt is a cube with an edge length of 0.563 nm. What is the volume of this cube in nm^3? In cm^3?
60. ■ The cancer drug cisplatin contains 65.0% platinum. If you have 1.53 g of the compound, how many grams of platinum does this sample contain?
61. The fluoridation of city water supplies has been practiced in the United States for several decades because it is believed that fluoride prevents tooth decay, especially in young children. This is done by continuously adding sodium fluoride to water as it comes from a reservoir. Assume you live in a medium-sized city of 150,000 people and that each person uses 175 gal of water per day. How many tons of sodium fluoride must you add to the water supply each year (365 days) to have the required fluoride concentration of 1 part per million (that is, 1 ton of fluoride per million tons of water)? (Sodium fluoride is 45.0% fluoride, and one U.S. gallon of water has a mass of 8.34 lb.)
62. ■ Potassium has three stable isotopes, ^{39}K, ^{40}K, and ^{41}K, but ^{40}K has a very low natural abundance. Which of the other two is the more abundant?
63. Which one of these symbols conveys more information about the atom: ^{37}Cl or $_{17}Cl$? Explain.
64. Gems and precious stones are measured in carats, a weight unit equivalent to 200 mg. If you have a 2.3-carat diamond in a ring, how many moles of carbon do you have?
65. ■ Gold prices fluctuate, depending on the international situation. If gold is selling for $800 per troy ounce, how much must you spend to purchase 1.00 mol gold (1 troy ounce is equivalent to 31.1 g)?
66. A piece of copper wire is 25 ft long and has a diameter of 2.0 mm. Copper has a density of 8.92 g/cm^3. How many moles of copper and how many atoms of copper are there in the piece of wire?

Applying Concepts

67. Which sets of values are possible? Why are the others not possible?

	Mass Number	Atomic Number	Number of Protons	Number of Neutrons
(a)	19	42	19	23
(b)	235	92	92	143
(c)	53	131	131	79
(d)	32	15	15	15
(e)	14	7	7	7
(f)	40	18	18	40

68. Which sets of values are possible? Why are the others not possible?

	Mass Number	Atomic Number	Number of Protons	Number of Neutrons
(a)	53	25	25	29
(b)	195	78	195	117
(c)	33	16	16	16
(d)	52	24	24	28
(e)	35	17	18	17

69. ■ Which member of each pair has the greater number of atoms? Explain why.
    (a) 1 mol Cl or 1 mol Cl_2
    (b) 1 molecule of O_2 or 1 mol O_2
    (c) 1 nitrogen atom or 1 nitrogen molecule
    (d) 6.022×10^{23} fluorine molecules or 1 mol fluorine molecules
    (e) 20.2 g Ne or 1 mol Ne
    (f) 1 molecule of Br_2 or 159.8 g Br_2
    (g) 107.9 g Ag or 6.9 g Li
    (h) 58.9 g Co or 58.9 g Cu
    (i) 1 g calcium or 6.022×10^{23} calcium atoms
    (j) 1 g chlorine atoms or 1 g chlorine molecules
70. Which member of each pair has the greater mass? Explain why.
    (a) 1 mol iron or 1 mol aluminum
    (b) 6.022×10^{23} lead atoms or 1 mol lead
    (c) 1 copper atom or 1 mol copper
    (d) 1 mol Cl or 1 mol Cl_2
    (e) 1 g oxygen atoms or 1 g oxygen molecules
    (f) 24.3 g Mg or 1 mol Mg
    (g) 1 mol Na or 1 g Na
    (h) 4.0 g He or 6.022×10^{23} He atoms
    (i) 1 molecule of I_2 or 1 mol I_2
    (j) 1 oxygen molecule or 1 oxygen atom

More Challenging Questions

71. At 25 °C, the density of water is 0.997 g/cm^3, whereas the density of ice at −10 °C is 0.917 g/cm^3.
    (a) If a plastic soft-drink bottle (volume = 250 mL) is filled with pure water, capped, and then frozen at −10 °C, what volume will the solid occupy?
    (b) Could the ice be contained within the bottle?
72. A high-quality analytical balance can weigh accurately to the nearest 1.0×10^{-4} g. How many carbon atoms are present in 1.000 mg carbon, which could be weighed by such a balance? Given the precision of the balance, what are the high and low limits on the number of atoms present in the 1.000-mg sample?
73. Eleven of the elements in the periodic table are found in nature as gases at room temperature. List them. Where are they located in the periodic table?
74. Air mostly consists of diatomic molecules of nitrogen (about 80%) and oxygen (about 20%). Draw a nanoscale picture of a sample of air that contains a total of 10 molecules.

75. Identify the element that satisfies each of these descriptions:
    (a) A member of the same group as oxygen whose atoms contain 34 electrons
    (b) A member of the alkali metal group whose atoms contain 20 neutrons
    (c) A halogen whose atoms contain 35 protons and 44 neutrons
    (d) A noble gas whose atoms contain 10 protons and 10 neutrons

Conceptual Challenge Problems

CP2.A (Section 2.1) Suppose you are faced with a problem similar to the one faced by Robert Millikan when he analyzed data from his oil-drop experiment. Below are the masses of three stacks of dimes. What do you conclude to be the mass of a dime, and what is your argument?

Stack 1 = 9.12 g

Stack 2 = 15.96 g

Stack 3 = 27.36 g

CP2.B (Section 2.3) The age of the universe is unknown, but some conclude from measuring Hubble's constant that it is about 18 billion years old, which is about four times the age of the earth. If so, what is the age of the universe in seconds? If you had a sample of carbon with the same number of carbon atoms as there have been seconds since the universe began, could you measure this sample on a laboratory balance that can detect masses as small as 0.1 mg?

3

Chemical Compounds

Stalactites, natural stone formations hanging from the ceilings of underground caves, are formed by the interaction of water and limestone (calcite), which is largely $CaCO_3$. The water is weakly acidic due to dissolved carbon dioxide and dissolves $CaCO_3$. When the solution is exposed once again to air, the CO_2 escapes, and solid calcite is deposited. Over long time periods, fantastic shapes of solid stone are formed.

One of the most important things chemists do is synthesize new chemical compounds, substances that on the nanoscale consist of new, unique combinations of atoms. These compounds may have properties similar to those of existing compounds, or they may be very different. Often chemists can custom-design a new compound to have desirable properties. All compounds contain at least two elements, and most compounds contain more than two elements. This chapter deals with two major, general types of chemical compounds—those consisting of individual molecules, and those made of the positively and negatively charged atoms or groups of atoms called ions. We will now examine how compounds are represented by symbols, formulas, and names and how formulas represent the macroscale masses and compositions of compounds.

3.1 Molecular Compounds

In a **molecular compound** at the nanoscale level, atoms of two or more different elements are combined into the independent units known as molecules (← *p. 23*). Every day we inhale, exhale, metabolize, and in other ways use thousands of molecular compounds. Water, carbon dioxide, sucrose (table sugar), and caffeine, as well as carbohydrates, proteins, and fats, are among the many common molecular compounds in our bodies.

Metabolism is a general term for all of the chemical reactions that act to keep a living thing functioning. We *metabolize* food molecules to extract energy and produce other molecules needed by our bodies. Our *metabolic* reactions are controlled by enzymes (discussed in Section 12.9) and other kinds of molecules.

Molecular Formulas

The composition of a molecular compound is represented in writing by its **molecular formula,** in which the number and kinds of atoms combined to make one molecule of the compound are indicated by subscripts and elemental symbols. For example, the molecular formula for water, H_2O, shows that there are three atoms per molecule—two hydrogen atoms and one oxygen atom. The subscript to the right of each element's symbol indicates the number of atoms of that element present in the molecule. If the subscript is omitted, it is understood to be 1, as for the O in H_2O. These same principles apply to the molecular formulas of all molecules.

H_2O H—O—H

Space-filling model

Ball-and-stick model

Some molecules are classified as **inorganic compounds** because they do not contain carbon—for example, sulfur dioxide, SO_2, an air pollutant, or ammonia, NH_3, which, dissolved in water, is used as a household cleaning agent. (Many inorganic compounds are ionic compounds, which are described in Section 3.5 of this chapter.) The majority of **organic compounds** are composed of molecules. Organic compounds invariably contain carbon, usually contain hydrogen, and may also contain oxygen, nitrogen, sulfur, phosphorus, or halogens. Such compounds are of great interest because they are the basis for the clothes we wear, the food we eat, the fuels we burn, and the living organisms in our environment. For example, ethanol, C_2H_6O, is the organic compound familiar as a component of "alcoholic" beverages, and methane, CH_4, is the organic compound that is the major component of natural gas.

Some elements are also composed of molecules. In oxygen, for example, two oxygen atoms are joined in an O_2 molecule (← *p. 23*).

The formula of a molecular compound, especially an organic compound, can be written in several different ways. The molecular formula given previously for ethanol, C_2H_6O, is one example. For an organic compound, the symbols of the elements other than carbon are frequently written in alphabetical order, and each has a subscript indicating the total number of atoms of that type in the molecule, as illustrated by C_2H_6O. Because of the huge number of organic compounds, this formula may not give sufficient information to indicate what compound is represented. Such identification requires more information about how the atoms are connected to each other. A **structural formula** shows exactly how atoms are connected. In ethanol, for example, the first carbon atom is connected to three hydrogen atoms, and the second carbon atom is connected to two hydrogen atoms and an —OH group.

Lines represent bonds (chemical connections) between atoms.

```
    H   H
    |   |
H — C — C — O — H
    |   |
    H   H
```

The lines in structural formulas represent chemical bonds between atoms.

The formula can also be written in a modified form to show how the atoms are grouped together in the molecule. Such formulas, called **condensed formulas,** emphasize the atoms or groups of atoms connected to each carbon atom. For ethanol, the condensed formula is CH_3CH_2OH. If you compare this to the structural formula for ethanol, you can easily see that they represent the same structure. The —OH attached to the C atom is a distinctive grouping of atoms that characterizes the group of organic compounds known as alcohols. Such groups distinctive to the various classes of organic compounds are known as **functional groups.**

To summarize, three different ways of writing formulas are shown here for ethanol:

Structural formula	Condensed formula	Molecular formula
H H | | H—C—C—O—H | | H H	CH_3CH_2OH	C_2H_6O

As illustrated earlier for molecular elements (*p. 23*), molecular compounds can also be represented by ball-and-stick and space-filling models.

H H
| |
H—C—C—O—H
| |
H H

Structural formula

Ball-and-stick model

Space-filling model

Atom colors in molecular models: H, light gray; C, dark gray; N, blue; O, red; S, yellow.

Some additional examples of ball-and-stick molecular models are given in Table 3.1.

Table 3.1 Examples of Simple Molecular Compounds

Name	Molecular Formula	Number and Kind of Atoms	Molecular Model
Carbon dioxide	CO_2	3 total: 1 carbon, 2 oxygen	
Ammonia	NH_3	4 total: 1 nitrogen, 3 hydrogen	
Nitrogen dioxide	NO_2	3 total: 1 nitrogen, 2 oxygen	
Carbon tetrachloride	CCl_4	5 total: 1 carbon, 4 chlorine	
Pentane	C_5H_{12}	17 total: 5 carbon, 12 hydrogen	

PROBLEM-SOLVING EXAMPLE 3.1 Condensed and Molecular Formulas

(a) Write the molecular formulas for these molecules:

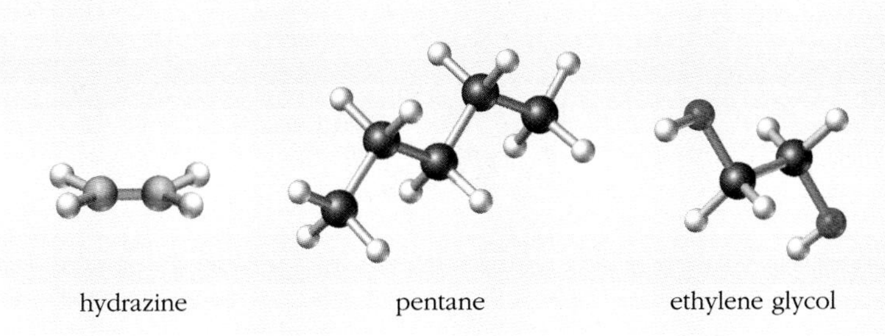

hydrazine pentane ethylene glycol

(b) Write the condensed formulas for hydrazine, pentane, and ethylene glycol.
(c) Write the molecular formula for this molecule:

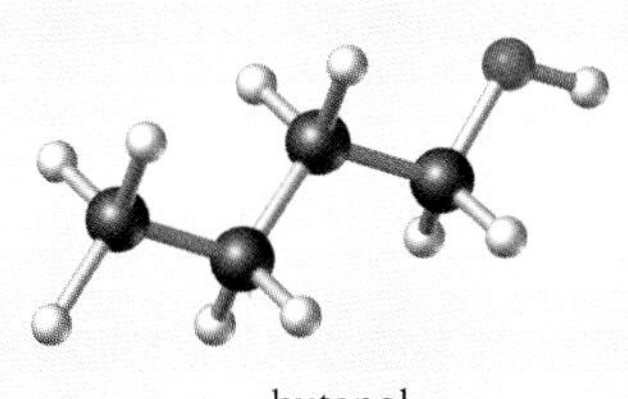

butanol

Answer

(a) hydrazine, N_2H_4; pentane, C_5H_{12}; ethylene glycol, $C_2H_6O_2$

(b) H_2NNH_2 (hydrazine) $CH_3CH_2CH_2CH_2CH_3$ (pentane) $HOCH_2CH_2OH$ (ethylene glycol)

(c) $C_4H_{10}O$

Strategy and Explanation

(a) Simply count the atoms of each type in each molecule to obtain the molecular formulas. Then write the symbols with their subscripts in alphabetical order.
(b) In condensed formulas, each carbon atom and its hydrogen atoms are written without connecting lines (CH_3, CH_2, or CH). Other groups are usually written on the same line with the carbon and hydrogen atoms if the groups are at the beginning or end of the molecule. Otherwise, they are connected above or below the line by straight lines to the respective carbon atoms. Condensed formulas emphasize important groups in molecules, such as the —OH groups in ethylene glycol.
(c) Count the atoms of each type in the molecule to obtain the molecular formula.

PROBLEM-SOLVING PRACTICE 3.1

Write the molecular formulas for these compounds.
(a) Adenosine triphosphate (ATP), an energy source in biochemical reactions, contains 10 carbon, 11 hydrogen, 13 oxygen, 5 nitrogen, and 3 phosphorus atoms per molecule
(b) Capsaicin, the active ingredient in chili peppers, has 18 carbon, 27 hydrogen, 3 oxygen, and 1 nitrogen atoms per molecule
(c) Oxalic acid, has the condensed formula HOOCCOOH and is found in rhubarb.

An automotive antifreeze that contains propylene glycol.

EXERCISE 3.1 Structural, Condensed, and Molecular Formulas

A molecular model of propylene glycol, used in some "environmentally friendly" antifreezes, looks like this:

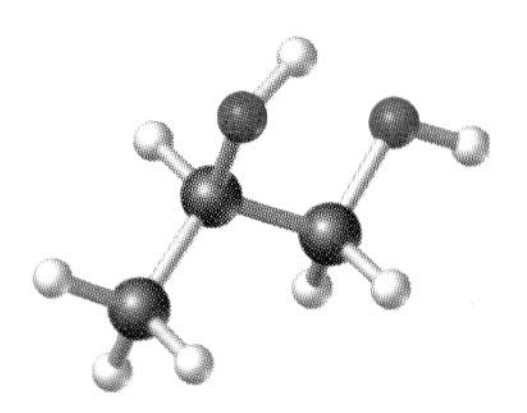

propylene glycol

Write the structural formula, the condensed formula, and the molecular formula for propylene glycol.

3.2 Naming Binary Inorganic Compounds

Each element has a unique name, as does each chemical compound. The names of compounds are assigned in a systematic way based on well-established rules. We will begin by applying the rules used to name simple binary molecular compounds. We will introduce other naming rules as we need them.

Hydrogen compounds with carbon are discussed in the next section.

Binary molecular compounds consist of molecules that contain atoms of only two elements. For hydrogen compounds containing oxygen, sulfur, the halogens, and most other nonmetals, the hydrogen is written first in the formula and named first. The other nonmetal is then named, with the nonmetal's name changed to end in *-ide*. For example, HCl is named hydrogen chloride. There is a binary compound of hydrogen with every nonmetal except the noble gases.

H_2O is written with H before O, as are the hydrogen compounds of Groups 6A and 7A: H_2S, H_2Se, and HF, HCl, HBr, and HI. Other H-containing compounds are usually written with the H atom after the other atom.

Formula	Name
HCl	Hydrogen chloride
HBr	Hydrogen bromide
HI	Hydrogen iodide
H_2Se	Hydrogen selenide

hydrogen chloride

hydrogen bromide

hydrogen iodide

Many binary molecular compounds contain nonmetallic elements from Groups 4A, 5A, 6A, and 7A of the periodic table. In these compounds the elements are listed in formulas and names in the order of the group numbers, and prefixes are used to designate the number of a particular kind of atom. The prefixes are listed in Table 3.2. Table 3.3 illustrates how these prefixes are applied.

Table 3.2 Prefixes Used in Naming Chemical Compounds

Prefix	Number
Mono-	1
Di-	2
Tri-	3
Tetra-	4
Penta-	5
Hexa-	6
Hepta-	7
Octa-	8
Nona-	9
Deca-	10

A number of binary nonmetal compounds were discovered and named years ago, before systematic naming rules were developed. Such *common names* are still used today and must simply be learned.

Formula	Common Name	Formula	Common Name
H_2O	Water	NO	Nitric oxide
NH_3	Ammonia	N_2O	Nitrous oxide ("laughing gas")
N_2H_4	Hydrazine	PH_3	Phosphine

Table 3.3 Examples of Binary Compounds

Molecular Formula	Name	Use
CO	Carbon monoxide	Steel manufacturing
NO_2	Nitrogen dioxide	Preparation of nitric acid
N_2O	Dinitrogen oxide	Anesthetic; spray can propellant
N_2O_5	Dinitrogen pentaoxide	Forms nitric acid
PBr_3	Phosphorus tribromide	Forms phosphorous acid
PBr_5	Phosphorus pentabromide	Forms phosphoric acid
SF_6	Sulfur hexafluoride	Transformer insulator
P_4O_{10}	Tetraphosphorus decaoxide	Drying agent

PROBLEM-SOLVING EXAMPLE 3.2 Naming Binary Inorganic Compounds

Name these compounds: (a) CO_2, (b) SiO_2, (c) SO_3, (d) N_2O_4, (e) PCl_5.

Answer

(a) Carbon dioxide
(b) Silicon dioxide
(c) Sulfur trioxide
(d) Dinitrogen tetraoxide
(e) Phosphorus pentachloride

Strategy and Explanation These compounds consist entirely of nonmetals, so they are all molecular compounds. The prefixes in Table 3.2 are used as necessary.

(a) Use *di-* to represent the two oxygen atoms.
(b) Use *di-* for the two oxygen atoms.
(c) Use *tri-* for the three oxygen atoms.
(d) Use *di-* for the two nitrogen atoms and *tetra-* for the four oxygen atoms.
(e) Use *penta-* for the five chlorine atoms.

PROBLEM-SOLVING PRACTICE 3.2

Name these compounds: (a) SO_2, (b) BF_3, (c) CCl_4.

EXERCISE 3.2 Names and Formulas of Compounds

Give the formula for each of these binary nonmetal compounds:

(a) Carbon disulfide
(b) Phosphorus trichloride
(c) Sulfur dibromide
(d) Selenium dioxide
(e) Oxygen difluoride
(f) Xenon trioxide

3.3 Hydrocarbons

Hydrocarbons are organic compounds composed of only carbon and hydrogen; they are the simplest class of organic compounds. Millions of organic compounds, including hydrocarbons, are known. They vary enormously in structure and function, ranging from the simple molecule methane (CH_4, the major constituent of natural gas) to large, complex biochemical molecules such as proteins, which often contain hundreds or thousands of atoms. Organic compounds are the main constituents of living matter. In organic compounds the carbon atoms are nearly always bonded to other carbon atoms and to hydrogen atoms. Among the reasons for the enormous variety of organic compounds is the characteristic property of carbon atoms to form strong, stable bonds with up to four other carbon atoms. A **chemical bond** is an attractive force between

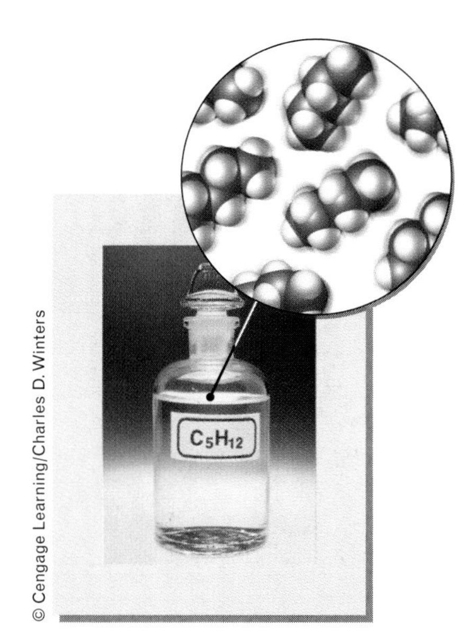

Pentane is an alkane used as a solvent.

Table 3.4 The First Ten Alkane Hydrocarbons, C_nH_{2n+2}

Molecular Formula	Name	Boiling Point (°C)	Physical State at Room Temperature
CH_4	Methane	−161.6	Gas
C_2H_6	Ethane	−88.6	Gas
C_3H_8	Propane	−42.1	Gas
C_4H_{10}	Butane	−0.5	Gas
C_5H_{12}	Pentane	36.1	Liquid
C_6H_{14}	Hexane	68.7	Liquid
C_7H_{16}	Heptane	98.4	Liquid
C_8H_{18}	Octane	125.7	Liquid
C_9H_{20}	Nonane	150.8	Liquid
$C_{10}H_{22}$	Decane	174.0	Liquid

two atoms holding them together. Through their carbon-carbon bonds, carbon atoms can form chains, branched chains, rings, and other more complicated structures. With such a large number of compounds, dividing them into classes is necessary to make organic chemistry manageable.

The simplest major class of hydrocarbons is the **alkanes,** which are economically important fuels and lubricants. The simplest alkane is methane, CH_4, which has a central carbon atom with four bonds joining it to four H atoms. The general formula for non-cyclic alkanes is C_nH_{2n+2}, where n is an integer. Table 3.4 provides some information about the first ten alkanes. The first four (methane, ethane, propane, butane) have common names that must be memorized. For $n = 5$ or greater, the names are systematic. The prefixes of Table 3.2 indicate the number of carbon atoms in the molecule, and the ending *-ane* indicates that the compound is an alk*ane.* For example, the five-carbon alkane is *pentane.*

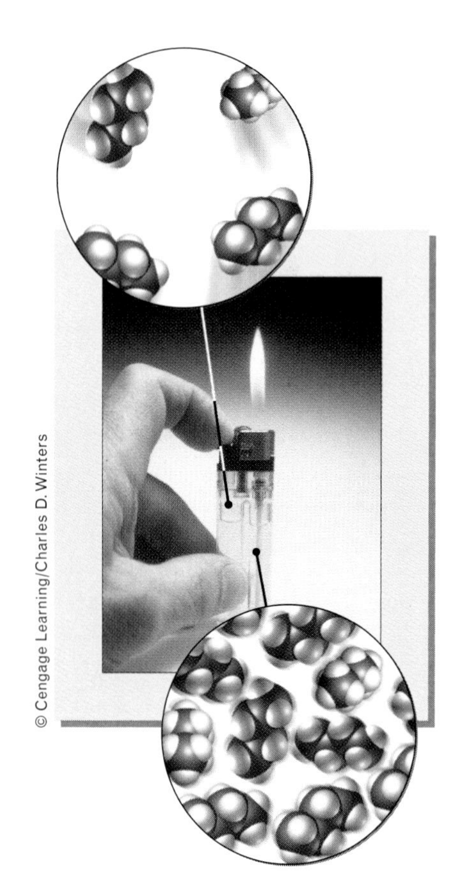

Butane, $CH_3CH_2CH_2CH_3$, is the fuel in this lighter. Butane molecules are present in the liquid and gaseous states in the lighter.

Methane, the simplest alkane, makes up about 85% of natural gas in the United States. Methane is also known to be one of the greenhouse gases (Section 10.9), meaning that it is one of the chemicals implicated in the problem of global warming. Ethane, propane, and butane are used as heating fuel for homes and in industry. In these simple alkanes, the carbon atoms are connected in unbranched (straight) chains, and each carbon atom is connected to either two or three hydrogen atoms.

```
    H            H  H             H  H  H              H  H  H  H
    |            |  |             |  |  |              |  |  |  |
 H—C—H       H—C—C—H,       H—C—C—C—H         H—C—C—C—C—H
    |            |  |             |  |  |              |  |  |  |
    H            H  H             H  H  H              H  H  H  H
 methane       ethane          propane              butane
```

Larger alkanes have longer chains of carbon atoms with hydrogens attached to each carbon.

EXERCISE 3.3 Alkane Molecular Formulas

(a) Using the general formula for non-cyclic alkanes, C_nH_{2n+2}, write the molecular formulas for the alkanes containing 16 and 28 carbon atoms.
(b) How many hydrogen atoms are present in tetradecane, which has 14 carbon atoms?
(c) Verify that each of these formulas corresponds to the general formula for non-cyclic alkanes.

The molecular structures of hydrocarbons provide the framework for the discussion of the structures of all other organic compounds. If a different atom or combination of atoms replaces one of the hydrogens in the molecular structure of an alkane, a compound with different properties results. A hydrogen atom in an alkane can be replaced by a single atom such as a halogen, for example. In this way ethane, CH_3CH_3, becomes chloroethane, CH_3CH_2Cl. The replacement can be a combination of atoms such as an oxygen bonded to a hydrogen, —OH, so ethane, CH_3CH_3, can be changed to ethanol, CH_3CH_2OH.

The molecular structures of organic compounds determine their properties. For example, comparing ethane, CH_3CH_3, to ethanol, CH_3CH_2OH, where the —OH group is substituted for one of the hydrogens, the boiling point changes from −88.6 °C to 78.5 °C. This is due to the types of intermolecular interactions that are present; these effects will be fully explained in Section 9.6.

PROBLEM-SOLVING EXAMPLE 3.3 Alkanes

Table 3.4 gives the boiling points for the first ten alkane hydrocarbons.
(a) Is the change in boiling point constant from one alkane to the next in the series?
(b) What do you propose as the explanation for the manner in which the boiling point changes from one alkane to the next?

Answer
(a) No. The increment is large between methane and ethane and gets progressively smaller as the molecules get larger. It is only 23 °C between nonane and decane.
(b) Larger molecules interact more strongly and therefore require a higher temperature to move them apart to become gases.

Strategy and Explanation We analyze the data given in Table 3.4.
(a) The boiling point differences between successive alkanes get smaller as the alkanes get larger.
(b) The larger molecules require a higher temperature to overcome their attraction to one another and to cease being liquid and become gaseous (***p. 19***).

PROBLEM-SOLVING PRACTICE 3.3

Consider a series of molecules formed from the alkanes by substituting one of the hydrogen atoms with a chlorine atom. Would you expect a similar trend in changes in boiling points among this set of compounds as you observed with the alkanes themselves?

3.4 Alkanes and Their Isomers

Two or more compounds having the same molecular formula but different arrangements of atoms are called **isomers.** Isomers differ from one another in one or more physical properties, such as boiling point, color, and solubility; chemical reactivity differs as well. Several types of isomerism are possible, particularly in organic compounds. **Constitutional isomers** (also called structural isomers) are compounds with the same molecular formula that differ in the order in which their atoms are bonded.

In this context, "straight chain" means a chain of carbon atoms with no branches to other carbon atoms; the carbon atoms are in an unbranched sequence. As you can see from the molecular model of butane, the chain is not actually straight, but rather is a zigzag.

Straight-Chain and Branched-Chain Isomers of Alkanes

The first three alkanes—methane, ethane, and propane—have only one possible structural arrangement. When we come to the alkane with four carbon atoms, C_4H_{10}, there are two possible arrangements—a *straight* chain of four carbons (butane) or a *branched* chain of three carbons with the fourth carbon attached to the central atom of the chain of three (methylpropane), as shown in the table.

Butane and methylpropane are constitutional isomers because they have the same molecular formula, but they are different compounds with different properties. Two constitutional isomers are different from each other in the same sense that two different structures built with identical Lego blocks are different from each other.

Molecular Formula	Condensed Formula	Structural Formula	Molecular Model
Butane C_4H_{10}	$CH_3CH_2CH_2CH_3$ Melting point −138 °C Boiling point −0.5 °C	H H H H │ │ │ │ H—C—C—C—C—H │ │ │ │ H H H H	
Methylpropane C_4H_{10}	CH_3 │ CH_3—CH—CH_3 Melting point −145 °C Boiling point −11.6 °C	H │ H—C—H H │ H │ │ │ H—C—C—C—H │ │ │ H H H	

Historically, straight-chain hydrocarbons were referred to as *normal* hydrocarbons, and *n*- was used as a prefix in their names. The current practice is not to use *n*-. If a hydrocarbon's name is given without indication that it is a branched-chain compound, assume it is a straight-chain hydrocarbon.

Methylpropane, the branched isomer of butane, has a *methyl* group (CH_3—) bonded to the central carbon atom. A methyl group is the simplest example of an **alkyl group,** the fragment of the molecule that remains when a hydrogen atom is removed from an alkane. Addition of one hydrogen to a methyl group gives methane:

```
   H         H
   |         |
H—C—H     H—C—      CH3—     methyl
   |         |                group
   H         H
```

Addition of one hydrogen to an ethyl group gives ethane:

```
   H H         H H
   | |         | |
H—C—C—H     H—C—C—      CH3CH2—     ethyl
   | |         | |                   group
   H H         H H
```

When we consider an alkyl group with three carbons, there are two possibilities:

```
   H H H           H H H
   | | |           | | |
H—C—C—C—H       H—C—C—C—H
   | | |           | | |
   H H H           H H H

                     |
CH3CH2CH2—       CH3CHCH3

  propyl          isopropyl
  group           group
```

Table 3.5 Some Common Alkyl Groups

Name	Condensed Structural Representation
Methyl	CH_3—
Ethyl	CH_3CH_2—
Propyl	$CH_3CH_2CH_2$—
Isopropyl	CH_3CH— │ CH_3 or $(CH_3)_2CH$—

Alkyl groups are named by dropping *-ane* from the parent alkane name and adding *-yl.* Theoretically, an alkyl group can be derived from any alkane. Some of the more common examples of alkyl groups are given in Table 3.5.

The number of alkane constitutional isomers grows rapidly as the number of carbon atoms increases because of the possibility of chain branching. Table 3.6 shows the number of isomers for some alkanes. Chain branching is another reason for the extraordinarily large number of possible organic compounds. Because each isomer has its own specific properties, organic compounds can also have a range of properties.

Table 3.6 Alkane Isomers

Molecular Formula	Number of Isomers
C_4H_{10}	2
C_5H_{12}	3
C_6H_{14}	5
C_7H_{16}	9
C_8H_{18}	18
C_9H_{20}	35
$C_{10}H_{22}$	75
$C_{12}H_{26}$	355
$C_{15}H_{32}$	4347
$C_{20}H_{42}$	366,319

CONCEPTUAL **EXERCISE** **3.4 Straight-Chain and Branched-Chain Isomers**

Three constitutional isomers are possible for pentane. Write structural and condensed formulas for these isomers.

3.5 Ions and Ionic Compounds

Not all compounds are molecular. A compound whose nanoscale composition consists of positive and negative ions is classified as an **ionic compound.** Many common substances, such as table salt, NaCl; lime, CaO; lye, NaOH; and baking soda, $NaHCO_3$, are ionic compounds.

When metals react with nonmetals, the ***metal atoms typically lose electrons to form positive ions.*** Any positive ion is referred to as a **cation** (pronounced CAT-ion). Cations *always* have *fewer* electrons than protons. For example, Figure 3.1 shows that

The terms "cation" and "anion" are derived from the Greek words *ion* (traveling), *cat* (down), and *an* (up).

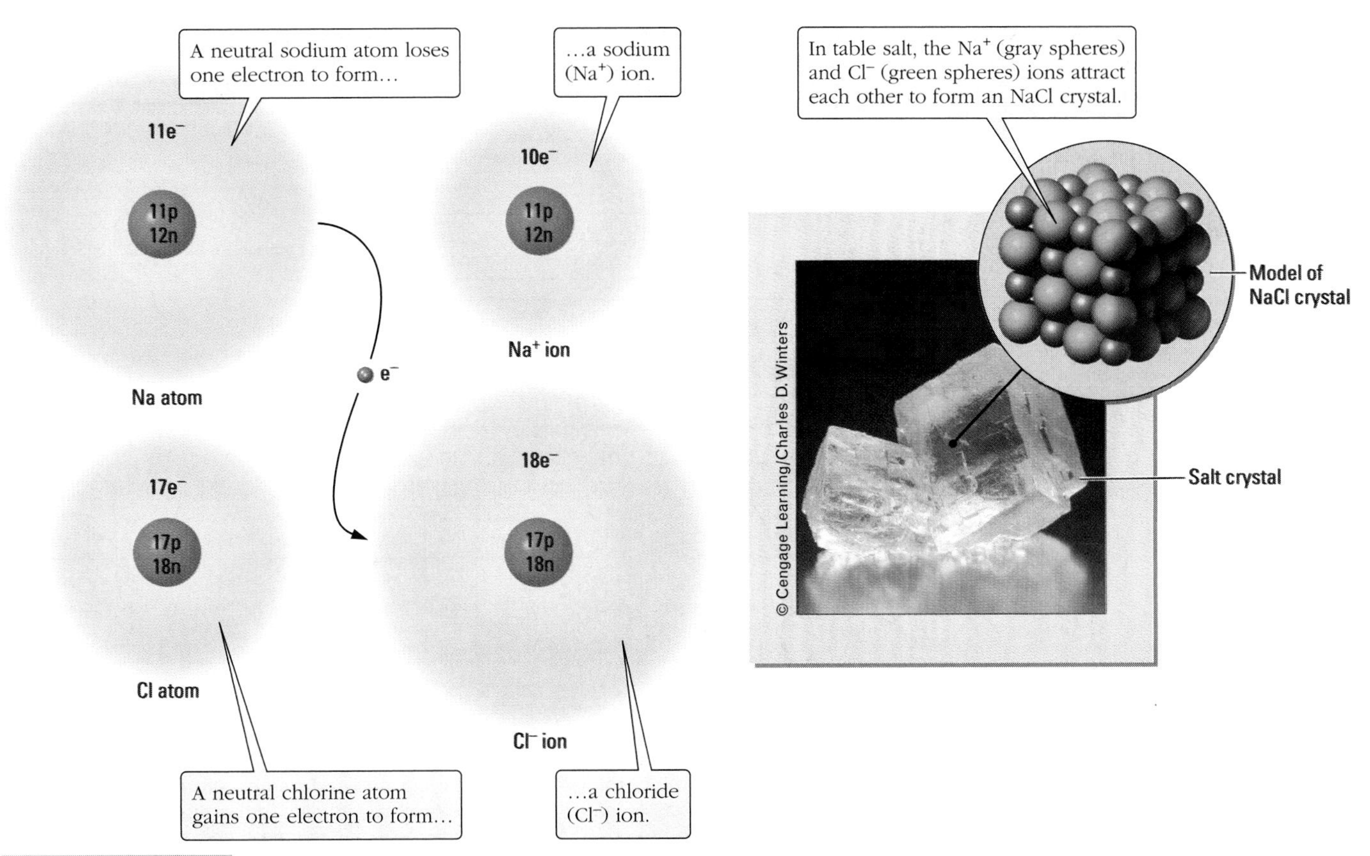

Active Figure 3.1 **Formation of the ionic compound NaCl.** Visit this book's companion website at **www.cengage.com/chemistry/moore** to test your understanding of the concepts in this figure.

Ionic compounds. Red iron(III) oxide, black copper(II) bromide, CaF_2 (front crystal), and NaCl (rear crystal).

an electrically neutral sodium atom, which has 11 protons [11+] and 11 electrons [11−], can lose one electron to become a sodium cation, which, with 11 protons but only 10 electrons, has a net 1+ charge and is symbolized as Na^+. ***The quantity of positive charge on a cation equals the number of electrons lost by the neutral metal atom.*** For example, when a neutral magnesium atom loses two electrons, it forms a 2+ magnesium ion, Mg^{2+}.

Conversely, when nonmetals react with metals, the ***nonmetal atoms typically gain electrons to form negatively charged ions.*** Any negative ion is an **anion** (pronounced ANN-ion). Anions *always* have *more* electrons than protons. Figure 3.1 shows that a neutral chlorine atom (17 protons, 17 electrons) can gain an electron to form a chlor*ide* ion, Cl^-. With 17 protons and 18 electrons, the chloride ion has a *net* 1− charge. ***The quantity of negative charge on a nonmetal anion equals the number of electrons gained by the neutral nonmetal atom.*** For example, a neutral sulfur atom that gains two electrons forms a sulf*ide* ion, S^{2-}.

Monatomic Ions

A **monatomic ion** is a single atom that has lost or gained electrons. The charges of the common monatomic ions are given in Figure 3.2. Notice that metals of Groups 1A, 2A, and 3A form monatomic ions with charges *equal to the A group number.* For example,

Group	Neutral Metal Atom	Electrons Lost = (A-Group Number)	Metal Ion
1A	K (19 protons, 19 electrons)	1	K^+ (19 protons, 18 electrons)
2A	Mg (12 protons, 12 electrons)	2	Mg^{2+} (12 protons, 10 electrons)
3A	Al (13 protons, 13 electrons)	3	Al^{3+} (13 protons, 10 electrons)

It is extremely important that you know the ions commonly formed by the elements shown in Figure 3.2 so that you can recognize ionic compounds and their formulas and write their formulas as reaction products (Section 5.1).

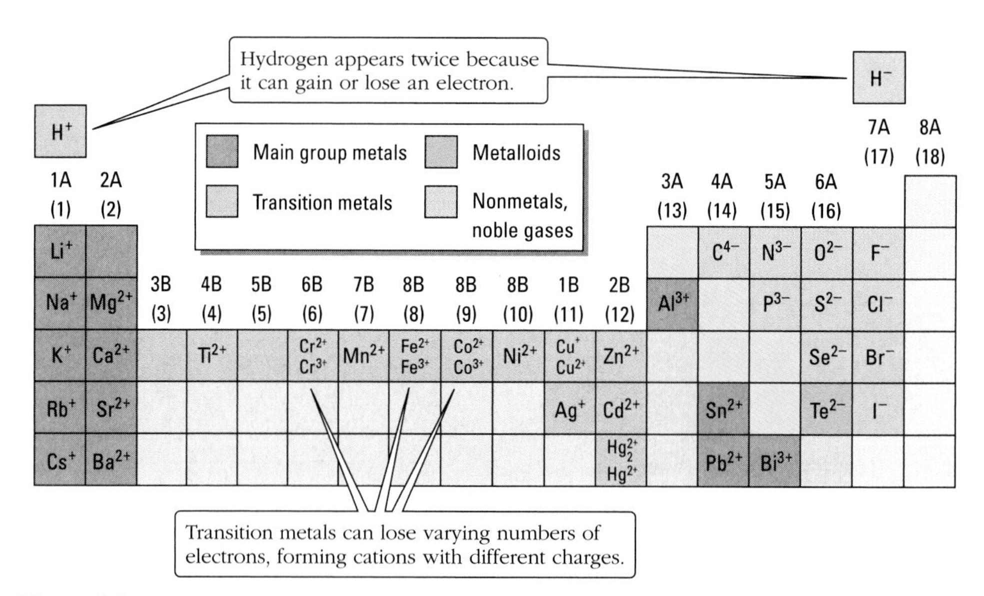

Figure 3.2 Charges on some common monatomic cations and anions. Note that metals generally form cations. The cation charge is given by the group number in the case of the main group elements of Groups 1A, 2A, and 3A *(gray).* For transition elements *(blue),* the positive charge is variable, and other ions in addition to those illustrated are possible. Nonmetals *(lavender)* generally form anions that have a charge equal to 8 minus the A group number.

Nonmetals of Groups 5A, 6A, and 7A form monatomic ions that have a negative charge *usually equal to 8 minus the A group number.* For example,

In Section 8.2 we explain the basis for the (8 − group number) relationship for nonmetals.

Group	Neutral Nonmetal Atom	Electrons Gained = 8 − (A-Group Number)	Nonmetal Ion
5A	N (7 protons, 7 electrons)	3 = (8 − 5)	N^{3-} (7 protons, 10 electrons)
6A	S (16 protons, 16 electrons)	2 = (8 − 6)	S^{2-} (16 protons, 18 electrons)
7A	F (9 protons, 9 electrons)	1 = (8 − 7)	F^{-} (9 protons, 10 electrons)

You might have noticed in Figure 3.2 that hydrogen appears at two locations in the periodic table. This is because a hydrogen atom can either gain or lose an electron. When it loses an electron, it forms a hydrogen ion, H^+ (1 proton, 0 electrons). When it gains an electron, it forms a hydride ion, H^- (1 proton, 2 electrons).

Noble gas atoms do not easily lose or gain electrons and have no common ions to list in Figure 3.2.

Transition metals form cations, but these metals can lose varying numbers of electrons, thus forming ions of different charges (Figure 3.2). Therefore, the group number is not an accurate guide to charges in these cases. It is important to learn which ions are formed most frequently by these transition metals. Many transition metals form 2+ and 3+ ions. For example, iron atoms can lose two or three electrons to form Fe^{2+} (26 protons, 24 electrons) or Fe^{3+} (26 protons, 23 electrons), respectively.

An older naming system for distinguishing between metal ions of different charges uses the ending *-ic* for the ion of higher charge and *-ous* for the ion of lower charge. These endings are combined with the element's name—for example, Fe^{2+} (ferrous) and Fe^{3+} (ferric) or Cu^+ (cuprous) and Cu^{2+} (cupric). We will not use these names in this book, but you might encounter them elsewhere.

PROBLEM-SOLVING EXAMPLE 3.4 Predicting Ion Charges

Using a periodic table, predict the charges on ions of aluminum, calcium, and phosphorus, and write symbols for these ions.

Answer Al^{3+}, Ca^{2+}, P^{3-}

Strategy and Explanation We find each element in the periodic table and use its position to answer the questions. Aluminum is a Group 3A metal, so it loses three electrons to give the Al^{3+} cation.

$$Al \longrightarrow Al^{3+} + 3\,e^-$$

Calcium is a Group 2A metal, so it loses two electrons to give the Ca^{2+} cation.

$$Ca \longrightarrow Ca^{2+} + 2\,e^-$$

Phosphorus is a Group 5A nonmetal, so it gains 8 − 5 = 3 electrons to give the P^{3-} anion.

$$P + 3\,e^- \longrightarrow P^{3-}$$

PROBLEM-SOLVING PRACTICE 3.4

For each of the ions listed below, explain whether it is likely to be found in an ionic compound.

(a) Ca^{4+} (b) Cr^{2+} (c) Sr^-

Polyatomic Ions

A **polyatomic ion** is a *unit* of two or more atoms that bears a net electrical charge. Table 3.7 lists some common polyatomic ions. Polyatomic ions are found in many places—oceans, minerals, living cells, and foods. For example, hydrogen carbonate (bicarbonate) ion, HCO_3^-, is present in rain water, sea water, blood, and baking soda. It consists of one carbon atom, three oxygen atoms, and one hydrogen atom, with one

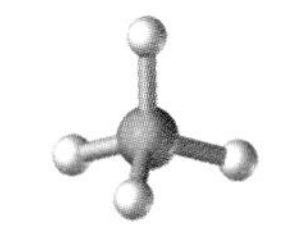
NH_4^+
ammonium ion

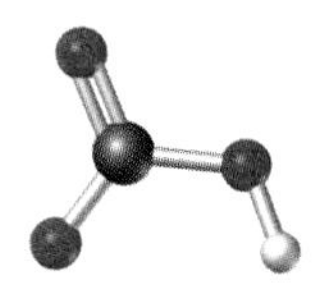
HCO_3^-
hydrogen carbonate ion

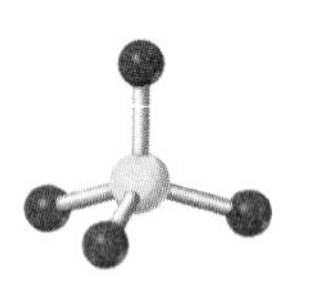
SO_4^{2-}
sulfate ion

Table 3.7 Common Polyatomic Ions

Formula	Name	Formula	Name
Cation (1+)			
NH_4^+	Ammonium		
Anions (1−)			
OH^-	Hydroxide	NO_2^-	Nitrite
HSO_4^-	Hydrogen sulfate	NO_3^-	Nitrate
CH_3COO^-	Acetate	MnO_4^-	Permanganate
ClO^-	Hypochlorite	$H_2PO_4^-$	Dihydrogen phosphate
ClO_2^-	Chlorite	CN^-	Cyanide
ClO_3^-	Chlorate	HCO_3^-	Hydrogen carbonate (bicarbonate)
ClO_4^-	Perchlorate		
Anions (2−)			
CO_3^{2-}	Carbonate	SO_3^{2-}	Sulfite
HPO_4^{2-}	Monohydrogen phosphate	SO_4^{2-}	Sulfate
$Cr_2O_7^{2-}$	Dichromate	$C_2O_4^{2-}$	Oxalate
$S_2O_3^{2-}$	Thiosulfate		
Anion (3−)			
PO_4^{3-}	Phosphate		

unit of negative charge spread over the group of five atoms. The polyatomic sulfate ion, SO_4^{2-}, consists of one sulfur atom and four oxygen atoms and has an overall charge of 2−. One of the most common polyatomic cations is NH_4^+, the ammonium ion. In this case, four hydrogen atoms are connected to a nitrogen atom, and the group bears a net 1+ charge. (We discuss the naming of polyatomic ions containing oxygen atoms in Section 3.6.)

In many chemical reactions the polyatomic ion unit remains intact. *It is important to know the names, formulas, and charges of the common polyatomic ions listed in Table 3.7.*

Ionic Compounds

In ionic compounds, cations and anions are attracted to each other by electrostatic forces, the forces of attraction between positive and negative charges. This is called **ionic bonding.** The strength of the electrostatic force dictates many of the properties of ionic compounds.

The attraction between oppositely charged ions increases with charge and decreases with the distance between the ions. The force between two charged particles is given quantitatively by **Coulomb's law:**

$$F = k\frac{Q_1Q_2}{d^2}$$

where Q_1 and Q_2 are the magnitudes of the charges on the two interacting particles, d is the distance between the two particles, and k is a constant. For ions separated by the same distance, the attractive force between 2+ and 2− ions is four times greater than that between 1+ and 1− ions. The attractive force also increases as the distance between the centers of the ions decreases. Thus, a small cation and a small anion will attract each other more strongly than will larger ions. (We discuss the sizes of ions in Section 7.10.)

Classifying compounds as ionic or molecular is very useful because these two types of compounds have quite different properties and thus different uses. The following generalizations will help you to predict whether a compound is ionic.

1. When a compound is composed of a metal cation (the elements in the gray and blue areas in Figure 3.2) and a nonmetal anion (the elements in the lavender area of Figure 3.2), it is an ionic compound, especially if the metal atom is combined with just one or two nonmetal atoms. Examples of such compounds include NaCl, $CaCl_2$, and KI.
2. When a compound is composed of two or more nonmetals, it is likely to be a molecular compound. Examples of such compounds include H_2O, NH_3, and CCl_4. Organic compounds, which contain carbon and hydrogen and possibly also oxygen, nitrogen, or halogens, are molecular compounds. Examples include acetic acid, CH_3COOH, and urea, CH_4N_2O.
3. If several nonmetal atoms are combined into a polyatomic anion, such as SO_4^{2-}, and this anion is combined with a metal ion, such as Ca^{2+}, the compound will be ionic. Examples include $CaSO_4$, $NaNO_3$, and KOH. If several nonmetal atoms are combined into a polyatomic cation, such as NH_4^+, this cation can be combined with an anion to form an ionic compound. Examples include NH_4Cl and $(NH_4)_2SO_4$.
4. Metal atoms can be part of polyatomic anions (MnO_4^- or $Cr_2O_7^{2-}$). When such polyatomic anions are combined with a metal ion, an ionic compound results—for example, $K_2Cr_2O_7$.
5. Metalloids (the elements in the orange area of Figure 3.2) can be incorporated into either ionic or molecular compounds. For example, the metalloid boron can combine with the nonmetal chlorine to form the molecular compound BCl_3. The metalloid arsenic is found in the arsenate ion in the ionic compound K_3AsO_4.

Don't confuse NH_4^+, ammonium ion, a polyatomic ion, with NH_3, which is a molecular compound.

Potassium dichromate, $K_2Cr_2O_7$. This beautiful orange-red compound contains potassium ions, K^+, and dichromate ions, $Cr_2O_7^{2-}$.

PROBLEM-SOLVING EXAMPLE 3.5 Ionic and Molecular Compounds

Predict whether each compound is likely to be ionic or molecular:

(a) Li_2CO_3 (b) C_3H_8 (c) $Fe_2(SO_4)_3$
(d) N_2H_4 (e) Na_2S (f) CO_2

Answer

(a) Ionic (b) Molecular (c) Ionic
(d) Molecular (e) Ionic (f) Molecular

Strategy and Explanation For each compound we use the location of its elements in the periodic table and our knowledge of polyatomic ions to answer the question.

(a) Lithium is a metal, so the lithium ion, Li^+, and the polyatomic carbonate ion, CO_3^{2-}, form an ionic compound, lithium carbonate.
(b) This compound is composed entirely of nonmetal atoms, so it is molecular. It is propane, an alkane hydrocarbon.
(c) Iron is a metal, so the Fe^{3+} ion and the polyatomic SO_4^{2-} sulfate ion form an ionic compound, iron(III) sulfate.
(d) This compound is composed entirely of nonmetal atoms, so it is molecular. It is called hydrazine.
(e) Sodium is a metal and sulfur is a nonmetal, so they form an ionic compound of Na^+ ions and S^{2-} ions, called sodium sulfide.
(f) This compound is composed entirely of nonmetal atoms, so it is molecular. It is called carbon dioxide.

PROBLEM-SOLVING PRACTICE 3.5

Predict whether each of these compounds is likely to be ionic or molecular:

(a) CH_4 (b) $CaBr_2$ (c) $MgCl_2$ (d) PCl_3 (e) KCl

Writing Formulas for Ionic Compounds

As with the formulas for molecular compounds, a subscript of 1 in formulas of ionic compounds is understood to be there and is not written.

All compounds are electrically neutral. Therefore, when cations and anions combine to form an ionic compound, there must be zero *net* charge. The *total* positive charge of all the cations must equal the *total* negative charge of all the anions. For example, consider the ionic compound formed when potassium reacts with sulfur. Potassium is a Group 1A metal, so a potassium atom loses one electron to become a K^+ ion. Sulfur is a Group 6A nonmetal, so a sulfur atom gains two electrons to become an S^{2-} ion. To make the compound electrically neutral, two K^+ ions (total charge 2+) are needed for each S^{2-} ion. Consequently, the compound has the formula K_2S. The subscripts in an ionic compound formula show the numbers of ions included in the simplest formula unit. In this case, the subscript 2 indicates two K^+ ions for every S^{2-} ion.

Similarly, aluminum oxide, a combination of Al^{3+} and O^{2-} ions, has the formula Al_2O_3: 2 Al^{3+} gives 6+ charge; 3 O^{2-} gives 6− charge; total charge = 0.

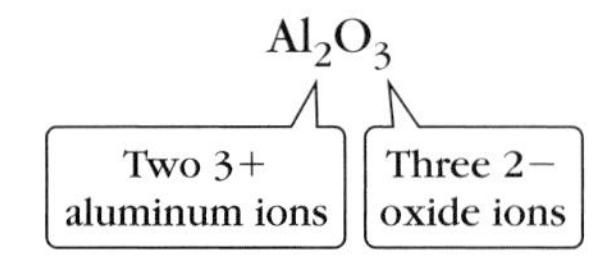

Notice that in writing the formulas for ionic compounds, *the cation symbol is written first, followed by the anion symbol.* The charges of the ions are *not* included in the formulas of ionic compounds.

Let's now consider several ionic compounds of magnesium, a Group 2A metal that forms Mg^{2+} ions.

Combining Ions	Overall Charge	Formula
Mg^{2+} and Br^-	(2+) + 2(1−) = 0	$MgBr_2$
Mg^{2+} and SO_4^{2-}	(2+) + (2−) = 0	$MgSO_4$
Mg^{2+} and OH^-	(2+) + 2(1−) = 0	$Mg(OH)_2$
Mg^{2+} and PO_4^{3-}	3(2+) + 2(3−) = 0	$Mg_3(PO_4)_2$

Notice in the latter two cases that when a polyatomic ion occurs more than once in a formula, the polyatomic ion's formula is put in parentheses followed by the necessary subscript.

$Mg_3(PO_4)_2$

Three 2+ magnesium ions

Two 3− phosphate ions

PROBLEM-SOLVING EXAMPLE 3.6 Ions in Ionic Compounds

For each compound, give the symbol or formula of each ion present and indicate how many of each ion are represented in the formula:

(a) Li_2S (b) Na_2SO_3 (c) $Ca(CH_3COO)_2$ (d) $Al_2(SO_4)_3$

Answer

(a) Two Li^+, one S^{2-} (b) Two Na^+, one SO_3^{2-}

(c) One Ca^{2+}, two CH_3COO^- (d) Two Al^{3+}, three SO_4^{2-}

Strategy and Explanation

(a) Lithium is a Group 1A element and *always* forms 1+ ions. The S^{2-} ion is formed from sulfur, a Group 6A element, by gaining two electrons (8 − 6 = 2). To maintain electrical neutrality, there must be two Li^+ ions for each S^{2-} ion.

(b) Sodium is a Group 1A element and therefore forms Na^+. Two Na^+ ions offset the 2− charge of the single polyatomic sulfite ion, SO_3^{2-} (see Table 3.7).

(c) Calcium is a Group 2A element that always forms 2+ ions. To have electrical neutrality, the two acetate ions with their 1− charge are needed to offset the 2+ charge on the Ca^{2+}.

(d) Aluminum, a Group 3A element, forms Al^{3+} ions. A 2 : 3 combination of two Al^{3+} ions ($2 \times 3+ = 6+$) with three 2− sulfate ions, SO_4^{2-} ($3 \times 2- = 6-$), gives electrical neutrality.

PROBLEM-SOLVING PRACTICE 3.6

Determine how many ions and how many atoms there are in each of these formulas.

(a) $In_2(SO_3)_3$ (b) $(NH_4)_3PO_4$

PROBLEM-SOLVING EXAMPLE 3.7 Formulas of Ionic Compounds

Write the correct formulas for ionic compounds composed of (a) calcium and fluoride ions, (b) barium and phosphate ions, (c) Fe^{3+} and nitrate ions, and (d) sodium and carbonate ions.

Answer

(a) CaF_2 (b) $Ba_3(PO_4)_2$ (c) $Fe(NO_3)_3$ (d) Na_2CO_3

Strategy and Explanation We use the location of each element in the periodic table and the charges on polyatomic anions to answer the questions.

(a) Calcium is a Group 2A metal, so it forms 2+ ions. Fluorine is a Group 7A nonmetal that forms 1− ions. Therefore, we need two 1− F^- ions for every Ca^{2+} ion to make CaF_2.

(b) Barium is a Group 2A metal, so it forms 2+ ions. Phosphate is a 3− polyatomic ion. Therefore, we need three Ba^{2+} ions and two PO_4^{3-} ions to form $Ba_3(PO_4)_2$. Because there is more than one polyatomic phosphate ion, it is enclosed in parentheses followed by the proper subscript.

(c) Iron is in its Fe^{3+} state. Nitrate is a 1− polyatomic ion. Therefore, we need three nitrate ions for each Fe^{3+} ion to form $Fe(NO_3)_3$. The polyatomic nitrate ion is enclosed in parentheses followed by the proper subscript.

(d) Carbonate is a 2− polyatomic ion that combines with two Na^+ ions to form Na_2CO_3. The polyatomic carbonate ion is not enclosed in parentheses since the formula contains only one carbonate.

PROBLEM-SOLVING PRACTICE 3.7

For each of these ionic compounds, write a list of which ions and how many of each are present.

(a) $MgBr_2$ (b) Li_2CO_3 (c) NH_4Cl (d) $Fe_2(SO_4)_3$

(e) Copper is a transition element that can form two compounds with bromine containing either Cu^+ or Cu^{2+}. Write the formulas for these compounds.

3.6 Naming Ions and Ionic Compounds

Ionic compounds can be named unambiguously by using the rules given in this section. You should learn these rules thoroughly.

Naming Positive Ions

Virtually all cations used in this book are metal ions that can be named by the rules given below. The ammonium ion (NH_4^+) is the major exception; it is a polyatomic ion composed of nonmetal atoms.

An unusual cation that you will see on occasion is Hg_2^{2+}, the mercury(I) ion. The Roman numeral (I) is used to show that the ion is composed of two Hg^+ ions bonded together, giving an overall 2+ charge.

1a. ***For metals that form only one kind of cation, the name is simply the name of the metal plus the word "ion."*** For example, Mg^{2+} is the magnesium ion.

The Stock system is named after Alfred Stock (1876–1946), a German chemist famous for his work on the hydrogen compounds of boron and silicon.

1b. ***For metals that can form more than one kind of cation, the name of each ion must indicate its charge. The charge is indicated by a Roman numeral in parentheses immediately following the ion's name (the Stock system).*** For example, Cu^{2+} is the copper(II) ion and Cu^{+} is the copper(I) ion.

© Cengage Learning/Charles D. Winters

Copper(I) oxide *(left)* and copper(II) oxide. The different copper ion charges result in different colors.

Naming Negative Ions

These rules apply to naming anions.

2a. ***A monatomic anion is named by adding* -ide *to the stem of the name of the nonmetal element from which the ion is derived.*** For example, a *phosph*orus atom gives a *phosph*ide ion, and a *chlor*ine atom forms a *chlor*ide ion. Anions of Group 7A elements, the halogens, are collectively called **halide ions.**

2b. ***The names of the most common polyatomic ions are given in Table 3.7 (p. 74).*** Most must simply be memorized. However, some guidelines can help, especially for **oxoanions,** which are polyatomic ions containing oxygen.

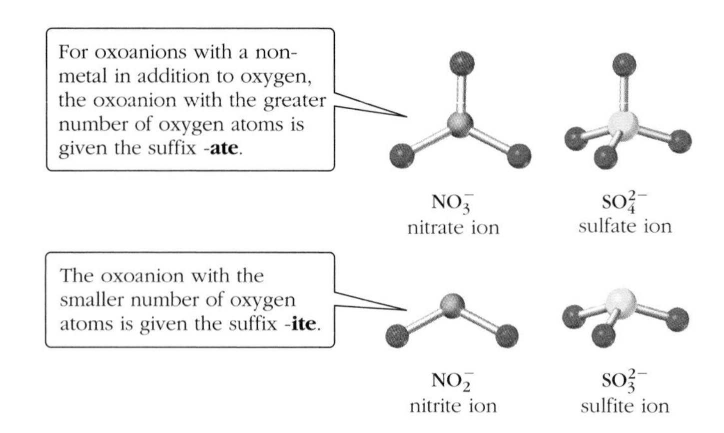

For oxoanions with a nonmetal in addition to oxygen, the oxoanion with the greater number of oxygen atoms is given the suffix -ate.
NO_3^-
nitrate ion
SO_4^{2-}
sulfate ion
The oxoanion with the smaller number of oxygen atoms is given the suffix -ite.
NO_2^-
nitrite ion
SO_3^{2-}
sulfite ion

Note that *-ate* and *-ite* suffixes do not relate to the ion's charge, but to the relative number of oxygen atoms: *-ate* ions contain one more oxygen atom than *-ite* ions (NO_3^- vs. NO_2^-).

When more than two different oxoanions of a given nonmetal exist, a more extended naming scheme must be used. When there are four oxoanions involved, the two middle ones are named according to the *-ate* and *-ite* endings; then the ion containing the largest number of oxygen atoms is given the prefix *per-* and the suffix *-ate,* and the one containing the smallest number is given the prefix *hypo-* and the suffix *-ite.* The oxoanions of chlorine are good examples:

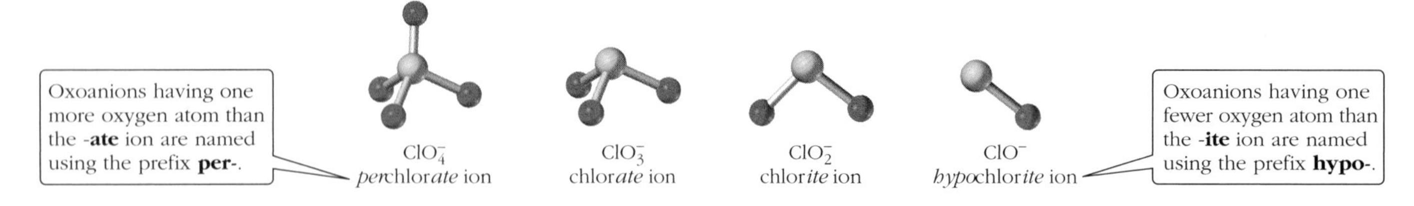

Oxoanions having one more oxygen atom than the -ate ion are named using the prefix per-.
ClO_4^-
*per*chlor*ate* ion
ClO_3^-
chlor*ate* ion
ClO_2^-
chlor*ite* ion
ClO^-
*hypo*chlor*ite* ion
Oxoanions having one fewer oxygen atom than the -ite ion are named using the prefix hypo-.

The same naming rules also apply to the oxoanions of bromine and iodine.

Note that the negative charge of the polyatomic ion decreases by one for each hydrogen added.

Oxoanions containing hydrogen are named simply by adding the word "hydrogen" before the name of the oxoanion, for example, hydrogen sulfate ion, HSO_4^-. When an oxoanion of a given nonmetal can combine with different numbers of hydrogen atoms, we must use prefixes to indicate which ion we are talking about: *di*hydrogen phosphate for $H_2PO_4^-$ and *mono*hydrogen phosphate for HPO_4^{2-}. Because many hydrogen-containing oxoanions have common names that are used often, you should know them. For example, the hydrogen carbonate ion, HCO_3^-, is often called the bicarbonate ion.

Table 3.8 Names of Some Common Ionic Compounds

Common Name	Systematic Name	Formula
Baking soda	Sodium hydrogen carbonate	$NaHCO_3$
Lime	Calcium oxide	CaO
Milk of magnesia	Magnesium hydroxide	$Mg(OH)_2$
Table salt	Sodium chloride	$NaCl$
Smelling salts	Ammonium carbonate	$(NH_4)_2CO_3$
Lye	Sodium hydroxide	$NaOH$
Blue vitriol	Copper(II) sulfate pentahydrate	$CuSO_4 \cdot 5\ H_2O$

Sodium chloride, NaCl. This common ionic compound contains sodium ions (Na^+) and chloride ions (Cl^-).

Naming Ionic Compounds

Table 3.8 lists a number of common ionic compounds. We will use these compounds to demonstrate the rules for systematically naming ionic compounds. One basic naming rule is by now probably apparent—***the name of the cation comes first, then the name of the anion.*** Also, in naming a compound, the word "ion" is not used with the metal name.

Notice these examples from Table 3.8:

- Calcium oxide, CaO, is named from calcium for Ca^{2+} (Rule 1a) and oxide for O^{2-} (Rule 2a). Likewise, sodium chloride is derived from sodium (Na^+, Rule 1a) and chloride (Cl^-, Rule 2a).
- Ammonium carbonate, $(NH_4)_2CO_3$, contains two polyatomic ions that are named in Table 3.7 ***(⇐ p. 74).***
- In the name copper(II) sulfate, the (II) indicates that Cu^{2+} is present, not Cu^+, the other possibility.

PROBLEM-SOLVING EXAMPLE 3.8 Using Formulas to Name Ionic Compounds

Name each of these ionic compounds.

(a) KCl (b) $Ca(OH)_2$ (c) $Fe_3(PO_4)_2$
(d) $Al(NO_3)_3$ (e) $(NH_4)_2SO_4$

Answer

(a) Potassium chloride (b) Calcium hydroxide (c) Iron(II) phosphate
(d) Aluminum nitrate (e) Ammonium sulfate

Strategy and Explanation

(a) The potassium ion, K^+, and the chloride ion, Cl^-, combine to form potassium chloride.
(b) The calcium ion, Ca^{2+}, and the hydroxide ion, OH^-, combine to form calcium hydroxide.
(c) The iron(II) ion, Fe^{2+}, and the phosphate ion, PO_4^{3-}, combine to give iron(II) phosphate.
(d) The aluminum ion, Al^{3+}, combines with the nitrate ion, NO_3^-, to form aluminum nitrate.
(e) The ammonium ion, NH_4^+, and the sulfate ion, SO_4^{2-}, combine to form ammonium sulfate.

PROBLEM-SOLVING PRACTICE 3.8

Name each of these ionic compounds:

(a) KNO_2 (b) $NaHSO_3$ (c) $Mn(OH)_2$
(d) $Mn_2(SO_4)_3$ (e) Ba_3N_2 (f) LiH

PROBLEM-SOLVING EXAMPLE 3.9 Using Names to Write Formulas of Ionic Compounds

Write the correct formula for each of these ionic compounds:
(a) Ammonium sulfide (b) Potassium sulfate
(c) Copper(II) nitrate (d) Iron(II) chloride

Answer
(a) $(NH_4)_2S$ (b) K_2SO_4 (c) $Cu(NO_3)_2$ (d) $FeCl_2$

Strategy and Explanation We use our knowledge of ion charges and the given charges for the metals to write the formulas.

(a) The ammonium cation is NH_4^+ and the sulfide ion is S^{2-}, so two ammonium ions are needed for one sulfide ion to make the neutral compound.
(b) The potassium cation is K^+ and the sulfate anion is SO_4^{2-}, so two potassium ions are needed.
(c) The copper(II) cation is Cu^{2+} and the anion is NO_3^-, so two nitrate ions are needed.
(d) The iron(II) ion is Fe^{2+} and the chloride ion is Cl^-, so two chloride ions are needed.

PROBLEM-SOLVING PRACTICE 3.9

Write the correct formula for each of these ionic compounds:
(a) Potassium dihydrogen phosphate (b) Copper(I) hydroxide
(c) Sodium hypochlorite (d) Ammonium perchlorate
(e) Chromium(III) chloride (f) Iron(II) sulfite

3.7 Bonding in and Properties of Ionic Compounds

The book *Salt: A World History,* by Mark Kurlansky, is a compelling account of table salt's importance through the ages.

As is true for all compounds, the properties of an ionic compound differ significantly from those of its component elements. Consider the familiar ionic compound, table salt (sodium chloride, NaCl), composed of Na^+ and Cl^- ions. Sodium chloride is a white, crystalline, water-soluble solid, very different from its component elements, metallic sodium and gaseous chlorine. Sodium is an extremely reactive metal that reacts violently with water. Chlorine is a diatomic, toxic gas that reacts with water. Sodium *ions* and chloride *ions* do not undergo such reactions, and NaCl dissolves uneventfully in water.

In ionic solids, cations and anions are held by ionic bonding in an orderly array called a **crystal lattice,** in which each cation is surrounded by anions and each anion is surrounded by cations. Such an arrangement maximizes the attraction between cations and anions and minimizes the repulsion between ions of like charge. In sodium chloride, as shown in Figure 3.3, six chloride ions surround each sodium ion, and six sodium ions surround each chloride ion. As indicated in the formula, there is one sodium ion for each chloride ion.

The formula of an ionic compound indicates only the smallest whole-number ratio of the number of cations to the number of anions in the compound. In NaCl that ratio is 1:1. An Na^+Cl^- pair is referred to as a **formula unit** of sodium chloride. Note that the formula unit of an ionic compound has no independent existence outside of the crystal, which is different from the individual molecules of a molecular compound such as H_2O.

The regular array of ions in a crystal lattice gives ionic compounds two of their characteristic properties—high melting points and distinctive crystalline shapes. The melting points are related to the charges and sizes of the ions. For ions of similar size, such as O^{2-} and F^-, *the larger the charges, the higher the melting point,* because of the greater attraction between ions of higher charge (← ***p. 74***). For example, CaO

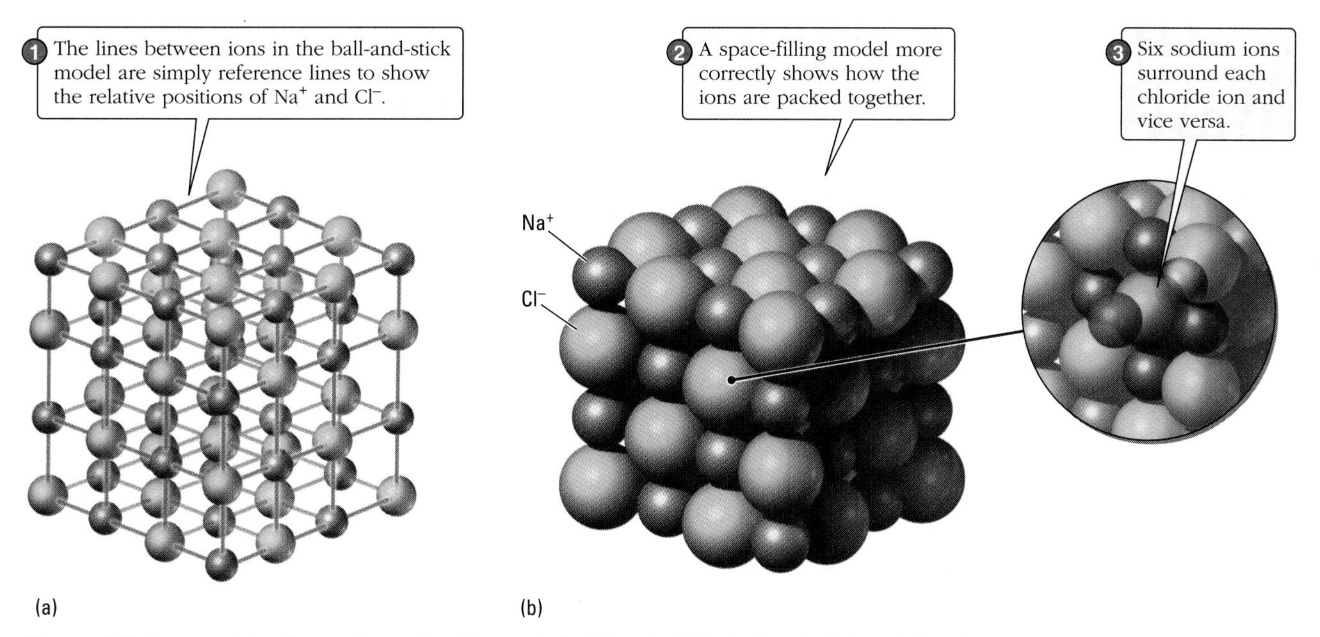

Figure 3.3 Two models of a sodium chloride crystal lattice. (a) This ball-and-stick model illustrates clearly how the ions are arranged, although it shows the ions too far apart. (b) Although a space-filling model shows how the ions are packed, it is difficult to see the locations of ions other than those on the faces of the crystal lattice.

(composed of doubly charged Ca^{2+} and O^{2-} ions) melts at 2572 °C, whereas NaF (composed of singly charged Na^+ and F^- ions) melts at 993 °C.

The crystals of ionic solids have characteristic shapes because the ions are held rather rigidly in position by strong attractive forces. Such alignment creates planes of ions within the crystals. Ionic crystals can be cleaved by an outside force that causes the planes of ions to shift slightly, bringing ions of like charge closer together (Figure 3.4). The resulting repulsion causes the layers on opposite sides of the cleavage plane to separate, and the crystal splits apart.

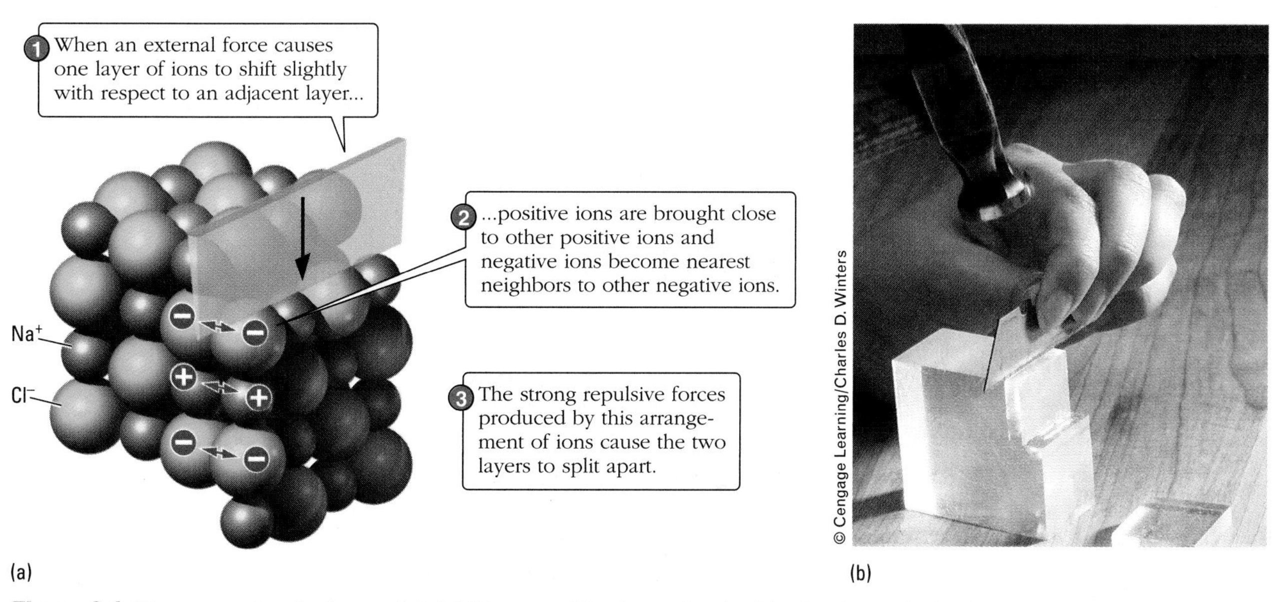

Figure 3.4 Cleavage of an ionic crystal. (a) Diagram of the forces involved in cleaving an ionic crystal. (b) A sharp blow on a knife edge lying along a plane of a salt crystal causes the crystal to split.

Figure 3.5 A molten ionic compound conducts an electric current. When an ionic compound is melted, ions are freed from the crystal lattice and migrate to the electrodes dipping into the melt. An electric current flows, and the light bulb illuminates, showing a complete circuit.

Table 3.9 Properties of Molecular and Ionic Compounds

Molecular Compounds	Ionic Compounds
Many are formed by combination of nonmetals with other nonmetals or with some metals	Formed by combination of reactive metals with reactive nonmetals
Gases, liquids, solids	Crystalline solids
Brittle and weak or soft and waxy solids	Hard and brittle solids
Low melting points	High melting points
Low boiling points (−250 to 600 °C)	High boiling points (700 to 3500 °C)
Poor conductors of electricity	Good conductors of electricity when molten; poor conductors of electricity when solid
Poor conductors of heat	Poor conductors of heat
Many insoluble in water but soluble in organic solvents	Many soluble in water
Examples: hydrocarbons, H_2O, CO_2	Examples: NaCl, CaF_2, NH_4NO_3

An electric current is the movement of charged particles from one place to another.

Because the ions in a crystal can only vibrate about fixed positions, ionic solids do not conduct electricity. However, when an ionic solid melts, as shown in Figure 3.5, the ions are free to move and conduct an electric current. Cations move toward the negative electrode and anions move toward the positive electrode, which results in an electric current.

The general properties of molecular and ionic compounds are summarized in Table 3.9. In particular, note the differences in physical state, electrical conductivity, melting point, and water solubility.

CONCEPTUAL EXERCISE **3.5 Properties of Molecular and Ionic Compounds**

Is a compound that is solid at room temperature and soluble in water likely to be a molecular or ionic compound? Why?

Ionic Compounds in Aqueous Solution: Electrolytes

Many ionic compounds are soluble in water. As a result, the oceans, rivers, lakes, and even the tap water in our residences contain many kinds of ions in solution. This makes the solubilities of ionic compounds and the properties of ions in solution of great practical interest.

When an ionic compound dissolves in water, it **dissociates**—the oppositely charged ions separate from one another. For example, when solid NaCl dissolves in water, it dissociates into Na^+ and Cl^- ions that become uniformly mixed with water molecules and dispersed throughout the solution.

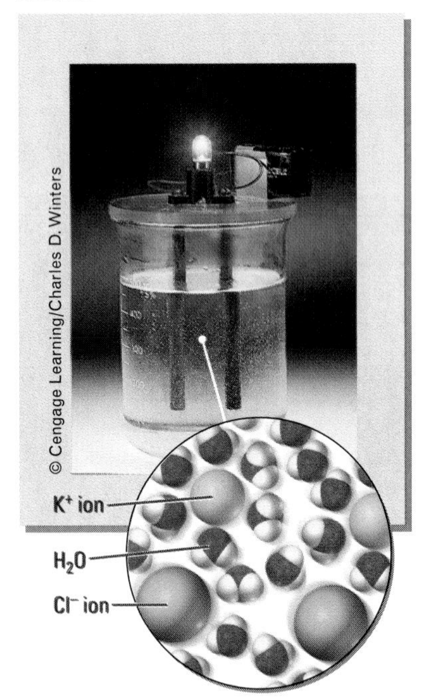

Figure 3.6 Electrical conductivity of an ionic compound solution. When an electrolyte, such as KCl, is dissolved in water and provides ions that move about, the electrical circuit is completed and the light bulb in the circuit glows. The ions of every KCl unit have dissociated: K^+ and Cl^-. The Cl^- ions move toward the positive electrode, and the K^+ ions move toward the negative electrode, transporting electrical charge through the solution.

Aqueous solutions of ionic compounds conduct electricity because the ions are free to move about (Figure 3.6). (This is the same mechanism of conductivity as for molten ionic compounds.) Substances that conduct electricity when dissolved in water are called **electrolytes.**

Most molecular compounds that are water-soluble continue to exist as molecules in solution; table sugar (chemical name, sucrose) is an example. Substances such as sucrose whose solutions do not conduct electricity are called **nonelectrolytes.** Be sure you understand the difference between these two important properties of a compound: its solubility and its ability to release ions in solution. In Section 5.1 we will provide a much more detailed discussion of solubility and the dissociation of ions from electrolytes in solution.

We use the term "dissociate" for ionic compounds that separate into their constituent ions in water. The term "ionize" is used for molecular compounds whose molecules react with water to form ions.

3.8 Moles of Compounds

In talking about compounds in quantities large enough to manipulate conveniently, we deal with moles of these compounds.

Molar Mass of Molecular Compounds

The most recognizable molecular formula, H_2O, shows us that there are two H atoms for every O atom in a water molecule. In two water molecules, therefore, there are four H atoms and two O atoms; in a dozen water molecules, there are two dozen H atoms and one dozen O atoms. We can extend this until we have one mole of water molecules (Avogadro's number of molecules, 6.022×10^{23}), each containing two moles of hydrogen atoms and one mole of oxygen atoms ***(← p. 51)***. We can also say that in 1.000 mol water there are 2.000 mol H atoms and 1.000 mol O atoms:

One mole of a molecular compound means 6.022×10^{23} molecules, not one molecule.

H_2O	H	O
6.022×10^{23} water molecules	$2(6.022 \times 10^{23}$ H atoms)	6.022×10^{23} O atoms
1.000 mol H_2O molecules	2.000 mol H atoms	1.000 mol O atoms
18.0153 g H_2O	2(1.0079 g H) = 2.0158 g H	15.9994 g O

The mass of one mole of water molecules—the *molar mass*—is the sum of the masses of two moles of H atoms and one mole of O atoms: 2.0158 g H + 15.9994 g O = 18.0152 g water in a mole of water. For chemical compounds, the *molar mass,* in grams per mole, is *numerically the same* as the **molecular weight,** the sum of the atomic weights (in amu) of *all* the atoms in the compound's formula. The molar masses of several molecular compounds are shown in the following table.

Compound	Structural Formula	Molecular Weight	Molar Mass
Ammonia NH_3	H—N—H | H	14.01 amu, N + 3(1.01 amu, H) = 17.04 amu	17.04 g/mol
Trifluoromethane CHF_3	F | F—C—F | H	12.01 amu, C + 1.01 amu, H + 3(19.00 amu, F) = 70.02 amu	70.02 g/mol
Sulfur dioxide SO_2	O=S—O	32.07 amu, S + 2(16.00 amu, O) = 64.07 amu	64.07 g/mol
Glycerol $C_3H_8O_3$	CH_2OH | CHOH | CH_2OH	3(12.01 amu, C) + 8(1.01 amu, H) + 3(16.00 amu, O) = 92.11 amu	92.11 g/mol

Molar Mass of Ionic Compounds

Because ionic compounds do not contain individual molecules, the term "formula weight" is sometimes used for ionic compounds instead of "molecular weight." As with molecular weight, an ionic compound's **formula weight** is the sum of the atomic weights of all the atoms in the compound's formula. The molar mass of an ionic compound, expressed in grams per mole (g/mol), is numerically equivalent to its formula weight. The term "molar mass" is used for both molecular and ionic compounds.

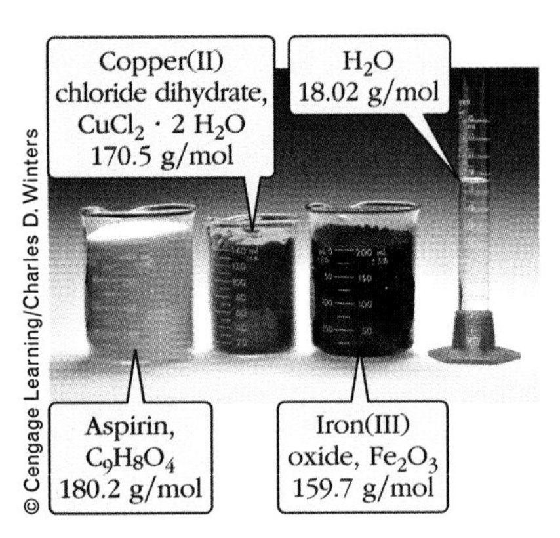

One-mole quantities of four compounds.

Compound	Formula Weight	Molar Mass
Sodium chloride NaCl	22.99 amu, Na + 35.45 amu, Cl = 58.44 amu	58.44 g/mol
Magnesium oxide MgO	24.31 amu, Mg + 16.00 amu, O = 40.31 amu	40.31 g/mol
Potassium sulfide K_2S	2(39.10 amu, K) + 32.07 amu, S = 110.27 amu	110.27 g/mol
Calcium nitrate $Ca(NO_3)_2$	40.08 amu, Ca + 2(14.01 amu, N) + 6(16.00 amu, O) = 164.10 amu	164.10 g/mol
Magnesium phosphate $Mg_3(PO_4)_2$	3(24.31 amu, Mg) + 2(30.97 amu, P) + 8(16.00 amu, O) = 262.87 amu	262.87 g/mol

Notice that $Mg_3(PO_4)_2$ has 2 P atoms and $2 \times 4 = 8$ O atoms because there are two PO_4^{3-} ions in the formula.

EXERCISE 3.6 Molar Masses

Calculate the molar mass of each of these compounds:

(a) K_2HPO_4 (b) $C_{27}H_{46}O$ (cholesterol)
(c) $Mn_2(SO_4)_3$ (d) $C_8H_{10}N_4O_2$ (caffeine)

Gram–Mole Conversions

As you might expect, it is essential to be able to do gram-mole conversions for compounds, just as we did for elements (← ***p. 53***). Here also, the key to such conversions is using molar mass as a conversion factor.

PROBLEM-SOLVING EXAMPLE 3.10 Grams to Moles

Calcium phosphate, $Ca_3(PO_4)_2$, is an ionic compound that is the main constituent of bone. How many moles of $Ca_3(PO_4)_2$ are in 10.0 g of the compound?

Answer 3.22×10^{-2} mol $Ca_3(PO_4)_2$

Strategy and Explanation To convert from grams to moles, we first find calcium phosphate's molar mass, which is the sum of the molar masses of the atoms in the formula.

$$3 \text{ mol Ca } (40.08 \text{ g Ca/mol Ca}) + 2 \text{ mol P } (30.97 \text{ g P/mol P}) + 8 \text{ mol O } (16.00 \text{ g O/mol O})$$
$$= 310.18 \text{ g/mol } Ca_3(PO_4)_2$$

This molar mass is used to convert mass to moles:

$$10.0 \text{ g } \cancel{Ca_3(PO_4)_2} \times \frac{1 \text{ mol } Ca_3(PO_4)_2}{310.18 \text{ g } \cancel{Ca_3(PO_4)_2}} = 3.22 \times 10^{-2} \text{ mol } Ca_3(PO_4)_2$$

☑ **Reasonable Answer Check** We started with 10.0 g calcium phosphate, which has a molar mass of about 300 g/mol, so 10/300 is about 1/30, which is close to the more accurate answer we calculated.

PROBLEM-SOLVING PRACTICE 3.10

Calculate the number of moles in 12.5 g of each of these ionic compounds:
(a) $K_2Cr_2O_7$ (b) $KMnO_4$ (c) $(NH_4)_2CO_3$

PROBLEM-SOLVING EXAMPLE 3.11 Moles to Grams

Cortisone, $C_{21}H_{28}O_5$, is an anti-inflammatory steroid. How many grams of cortisone are in 5.0 mmol cortisone? (1 mmol = 10^{-3} mol)

Answer 1.8 g

Strategy and Explanation First, calculate the molar mass of cortisone, which is 360.46 g/mol. Then use the molar mass to convert from moles to grams:

$$5.0 \times 10^{-3}\ \cancel{\text{mol cortisone}} \times \frac{360.46\ \text{g cortisone}}{1\ \cancel{\text{mol cortisone}}} = 1.8\ \text{g cortisone}$$

PROBLEM-SOLVING PRACTICE 3.11

(a) How many grams are in 5.0 mmol sucrose, $C_{12}H_{22}O_{11}$?
(b) How many milligrams are in 3.0 μmol adrenocorticotropic hormone (ACTH), which has a molar mass of approximately 4600 g/mol? (1 μmol = 10^{-6} mol)

PROBLEM-SOLVING EXAMPLE 3.12 Grams and Moles

Aspartame is a widely used artificial sweetener (NutraSweet) that is almost 200 times sweeter than sucrose. One sample of aspartame, $C_{14}H_{18}N_2O_5$, has a mass of 1.80 g; another contains 0.220 mol aspartame. To see which sample is larger, answer these questions:
(a) What is the molar mass of aspartame?
(b) How many moles of aspartame are in the 1.80-g sample?
(c) How many grams of aspartame are in the 0.220-mol sample?

Answer
(a) 294.3 g/mol (b) 6.12×10^{-3} mol (c) 64.7 g
The 0.220-mol sample is larger.

Strategy and Explanation
(a) Calculate the molar mass from the molecular formula.

$$[14\ \text{mol C} \times (12.011\ \text{g C/mol C})] + [18\ \text{mol H} \times (1.008\ \text{g H/mol H})]$$
$$+ [2\ \text{mol N} \times (14.007\ \text{g N/mol N})] + [5\ \text{mol O} \times (15.999\ \text{g O/mol O})]$$
$$= 294.3\ \text{g}$$

Thus, the molar mass of aspartame is 294.3 g/mol.
(b) Use the molar mass to calculate moles from grams:

$$1.80\ \cancel{\text{g aspartame}} \times \frac{1\ \text{mol aspartame}}{294.3\ \cancel{\text{g aspartame}}} = 6.12 \times 10^{-3}\ \text{mol aspartame}$$

(c) Converting moles to grams is the opposite of part (b):

$$0.220\ \cancel{\text{mol aspartame}} \times \frac{294.3\ \text{g aspartame}}{1\ \cancel{\text{mol aspartame}}} = 64.7\ \text{g aspartame}$$

The 0.220-mol sample of aspartame is the larger one.

PROBLEM-SOLVING PRACTICE 3.12

(a) Calculate the number of moles in 10.0 g $C_{27}H_{46}O$ (cholesterol) and 10.0 g $Mn_2(SO_4)_3$.
(b) Calculate the number of grams in 0.25 mol K_2HPO_4 and 0.25 mol $C_8H_{10}N_4O_2$ (caffeine).

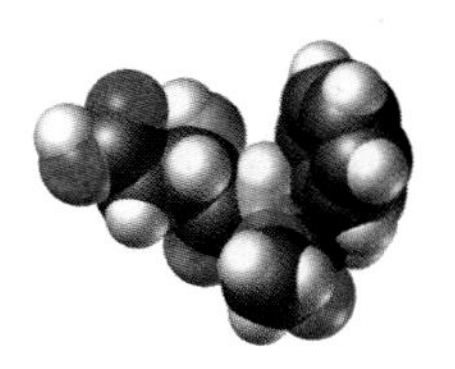
aspartame

CONCEPTUAL EXERCISE **3.7 Moles and Formulas**

Is this statement true? "Two different compounds have the same formula. Therefore, 100 g of each compound contains the same number of moles." Justify your answer.

$CuSO_4 \cdot 5\ H_2O$.

Moles of Ionic Hydrates

Many ionic compounds, known as **ionic hydrates** or hydrated compounds, have water molecules trapped within the crystal lattice. The associated water is called the **water of hydration.** For example, the formula for a beautiful deep-blue compound named copper(II) sulfate pentahydrate is $CuSO_4 \cdot 5\ H_2O$. The $\cdot 5\ H_2O$ and the term "pentahydrate" indicate five moles of water associated with every mole of copper(II) sulfate. The molar mass of a hydrate includes the mass of the water of hydration. Thus, the molar mass of $CuSO_4 \cdot 5\ H_2O$ is 249.7 g: 159.6 g $CuSO_4$ + 90.1 g (for 5 mol H_2O) = 249.7 g. There are many ionic hydrates, including the frequently encountered ones listed in Table 3.10.

One commonly used hydrate may well be in the walls of your room. Plasterboard (sometimes called wallboard or gypsum board) contains hydrated calcium sulfate, or gypsum, $CaSO_4 \cdot 2\ H_2O$, as well as unhydrated $CaSO_4$, sandwiched between two thicknesses of paper. Gypsum is a natural mineral that can be mined. It is also formed when sulfur dioxide is removed from electric power plant exhaust gases by reacting SO_2 with calcium oxide.

Calcium sulfate *hemi*hydrate contains one water molecule per two $CaSO_4$ units. The prefix *hemi-* refers to $\frac{1}{2}$ just as in the familiar word "hemisphere."

Heating gypsum to 180 °C drives off some of the water of hydration to form calcium sulfate hemihydrate, $CaSO_4 \cdot \frac{1}{2}\ H_2O$, commonly called Plaster of Paris. This compound is widely used in casts for broken limbs. When water is added to it, it forms a thick slurry that can be poured into a mold or spread over a part of the body. As the slurry hardens, it takes on additional water of hydration and its volume increases, forming a rigid protective cast.

EXERCISE **3.8 Moles of an Ionic Hydrate**

A home remedy calls for 2 teaspoons (20 g) Epsom salt (see Table 3.10). Calculate the number of moles of the hydrate represented by this mass.

Gypsum in its crystalline form. Gypsum is hydrated calcium sulfate, $CaSO_4 \cdot 2\ H_2O$.

Table 3.10 Some Common Hydrated Ionic Compounds

Formula	Systematic Name	Common Name	Uses
$Na_2CO_3 \cdot 10\ H_2O$	Sodium carbonate decahydrate	Washing soda	Water softener
$MgSO_4 \cdot 7\ H_2O$	Magnesium sulfate heptahydrate	Epsom salt	Dyeing and tanning
$CaSO_4 \cdot 2\ H_2O$	Calcium sulfate dihydrate	Gypsum	Wallboard
$CaSO_4 \cdot \frac{1}{2}\ H_2O$	Calcium sulfate hemihydrate	Plaster of Paris	Casts, molds
$CuSO_4 \cdot 5\ H_2O$	Copper(II) sulfate pentahydrate	Blue vitriol	Algicide, root killer

3.9 Percent Composition

You saw in the previous section that the composition of any compound can be expressed as either *(1) the number of atoms of each type per molecule or formula unit* or *(2) the mass of each element in a mole of the compound.* The latter relationship provides the information needed to find the **percent composition by mass** of each element of the compound (also called the *mass percent*).

It is important to recognize that the percent composition of a compound by mass is independent of the quantity of the compound. The percent composition by mass remains the same whether a sample contains 1 mg, 1 g, or 1 kg of the compound.

PROBLEM-SOLVING EXAMPLE 3.13 Percent Composition by Mass

Propane, C_3H_8, is the fuel used in gas grills. Calculate the percentages of carbon and hydrogen in propane.

Answer 81.72% carbon and 18.28% hydrogen

Strategy and Explanation First, we calculate the molar mass of propane: 3(12.01 g C) + 8(1.0079 g H) = 44.09 g C_3H_8/mol. We can then calculate the percentages of each element from the mass of each element in 1 mol C_3H_8 divided by the molar mass of propane.

$$\% \text{ C} = \frac{\text{mass of C in 1 mol } C_3H_8}{\text{mass of } C_3H_8 \text{ in 1 mol } C_3H_8} \times 100\%$$

$$= \frac{3 \times 12.01 \text{ g C}}{44.09 \text{ g } C_3H_8} \times 100\% = 81.72\% \text{ C}$$

$$\% \text{ H} = \frac{\text{mass of H in 1 mol } C_3H_8}{\text{mass of } C_3H_8 \text{ in 1 mol } C_3H_8} \times 100\%$$

$$= \frac{8 \times 1.0079 \text{ g H}}{44.09 \text{ g } C_3H_8} \times 100\% = 18.28\% \text{ H}$$

These answers can also be expressed as 81.70 g C per 100.0 g C_3H_8 and 18.28 g H per 100.0 g C_3H_8.

☑ **Reasonable Answer Check** Each carbon atom has twelve times the mass of a hydrogen atom. Propane has approximately 12 × 3 = 36 amu of carbon and approximately 1 × 8 = 8 amu of hydrogen. So the percentage of carbon should be about 36/8 = 4.5 times as large as the percentage of hydrogen. This agrees with our more carefully calculated answer.

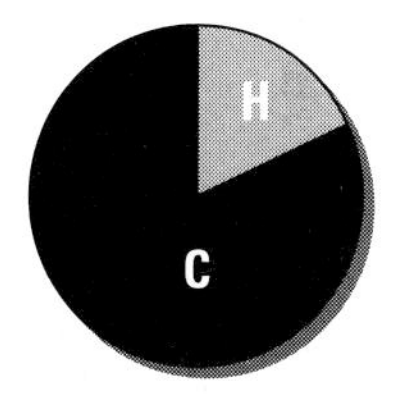

Mass percent carbon and hydrogen in propane, C_3H_8.

PROBLEM-SOLVING PRACTICE 3.13

What is the percentage of each element in silicon dioxide, SiO_2?

Note that the percentages calculated in Problem-Solving Example 3.13 add up to 100%. Therefore, once we calculated the percentage of carbon in propane, we also could have determined the percentage of hydrogen simply by subtracting: 100% − 81.7% C = 18.3% H. Calculating all percentages and adding them to confirm that they give 100% is a good way to check for errors.

In many circumstances, we do not need to know the percent composition of a compound with many significant figures, but rather we are interested in having an estimate of the percentage. For example, our objective could be to estimate whether the percentage of oxygen in a compound is about half or closer to two thirds. The labels on garden fertilizers show the approximate percent composition of the product and allow the consumer to compare approximate compositions of products.

PROBLEM-SOLVING EXAMPLE 3.14 Percent Composition of Hydrated Salt

Epsom salt is $MgSO_4 \cdot 7\,H_2O$. (a) What is the percent by mass of water in Epsom salt? (b) What are the percentages of each element in Epsom salt?

Epsom salt.

Answer

(a) 51.16% water (b) 9.862% Mg; 13.01% S; 71.40% O; 5.724% H

Strategy and Explanation

(a) We first find the molar mass of Epsom salt, which is the sum of the molar masses of the atoms in the chemical formula:

$$1 \text{ mol Mg } (24.31 \text{ g/mol}) + 1 \text{ mol S } (32.07 \text{ g/mol}) + 4 \text{ mol O } (16.00 \text{ g/mol})$$

$$+ 14 \text{ mol H } (1.008 \text{ g/mol}) + 7 \text{ mol O } (16.00 \text{ g/mol}) = 246.49 \text{ g/mol Epsom salt}$$

Because 1 mol Epsom salt contains 7 mol H_2O, the mass of water in 1 mol Epsom salt is $7 \times (18.015) = 126.11$ g

$$\% \, H_2O = \frac{126.11 \text{ g water per mol Epsom salt}}{246.49 \text{ g/mol Epsom salt}} \times 100\% = 51.16\% \, H_2O$$

(b) We calculate the percentage of magnesium from the ratio of the mass of magnesium in 1 mol Epsom salt to the mass of Epsom salt in 1 mol:

$$\% \, Mg = \frac{\text{mass of Mg in 1 mol Epsom salt}}{\text{mass of Epsom salt in 1 mol}} = \frac{24.31 \text{ g Mg}}{246.49 \text{ g/mol Epsom salt}} \times 100\% = 9.862\% \, Mg$$

We calculate the percentages for the remaining elements in the same way:

$$\% \, S = \frac{32.07 \text{ g S}}{246.49 \text{ g/mol Epsom salt}} \times 100\% = 13.01\% \, S$$

$$\% \, O = \frac{(64.00 + 112.00) \text{ g O}}{246.49 \text{ g/mol Epsom salt}} \times 100\% = 71.40\% \, O$$

$$\% \, H = \frac{14.11 \text{ g H}}{246.49 \text{ g/mol Epsom salt}} \times 100\% = 5.724\% \, H$$

☑ **Reasonable Answer Check** In the formula of the hydrated salt, there are seven waters with a combined mass of $7 \times 18 = 126$ g, and there are six other atoms with molar masses ranging between 16 and 32 that total to 120 g. Thus, the hydrated salt should be about 50% water by weight, and it is. There are 11 oxygen atoms in the formula, so oxygen should have the largest percent by weight, and it does. The percentages sum to 99.98% due to rounding.

PROBLEM-SOLVING PRACTICE 3.14

What is the mass percent of each element in hydrated nickel(II) chloride, $NiCl_2 \cdot 6\,H_2O$?

EXERCISE 3.9 Percent Composition

Express the composition of each compound first as the mass of each element in 1.000 mol of the compound, and then as the mass percent of each element:

(a) SF_6 (b) $C_{12}H_{22}O_{11}$
(c) $Al_2(SO_4)_3$ (d) $U(OTeF_5)_6$

3.10 Determining Empirical and Molecular Formulas

A formula can be used to derive the percent composition by mass of a compound, and the reverse process also works—we can determine the formula of a compound from mass percent data. In doing so, keep in mind that the subscripts in a formula indicate the relative numbers of moles of each element in one mole of that compound.

We can apply this method to finding the formula of diborane, a compound consisting of boron and hydrogen. Experiments show that diborane is 78.13% B and

21.87% H. Based on these percentages, a 100.0-g diborane sample contains 78.13 g B and 21.87 g H. From this information we can calculate the number of moles of each element in the sample:

$$78.13 \text{ g B} \times \frac{1 \text{ mol B}}{10.811 \text{ g B}} = 7.227 \text{ mol B}$$

$$21.87 \text{ g H} \times \frac{1 \text{ mol H}}{1.0079 \text{ g H}} = 21.70 \text{ mol H}$$

To determine the formula from these data, we next need to find the number of moles of each element *relative to the other element*—in this case, the ratio of moles of hydrogen to moles of boron. Looking at the numbers reveals that there are about three times as many moles of H atoms as there are moles of B atoms. To calculate the ratio exactly, we divide the larger number of moles by the smaller number of moles. For diborane that ratio is

$$\frac{21.70 \text{ mol H}}{7.227 \text{ mol B}} = \frac{3.003 \text{ mol H}}{1.000 \text{ mol B}}$$

This ratio confirms that there are three moles of H atoms for every one mole of B atoms and that there are three hydrogen atoms for each boron atom. This information gives the formula BH_3, which may or may not be the molecular formula of diborane.

For a molecular compound such as diborane, the molecular formula must also accurately reflect the *total number of atoms in a molecule of the compound.* The calculation we have done gives the *simplest possible ratio of atoms in the molecule,* and BH_3 is the simplest formula for diborane. A formula that reports the simplest possible ratio of atoms in the molecule is called an **empirical formula.** Multiples of the simplest formula are possible, such as B_2H_6, B_3H_9, and so on.

To determine the actual molecular formula from the empirical formula requires that we *experimentally determine* the molar mass of the compound and then compare our result with the molar mass predicted by the empirical formula. If the two molar masses are the same, the empirical and molecular formulas are the same. However, if the experimentally determined molar mass is some multiple of the value predicted by the empirical formula, the molecular formula is that multiple of the empirical formula. In the case of diborane, experiments indicate that the molar mass is 27.67 g/mol. This compares with the molar mass of 13.84 g/mol for BH_3, and so the molecular formula is a multiple of the empirical formula. That multiple is 27.67/13.84 = 2.00. Thus, the molecular formula of diborane is B_2H_6, two times BH_3.

The molar mass of the molecular formula divided by the molar mass of the empirical formula always yields a small integer. If the integer is 1, then the molecular formula and the empirical formula are the same.

PROBLEM-SOLVING EXAMPLE 3.15 Molecular Formula from Percent Composition by Mass Data

When oxygen reacts with phosphorus, two possible oxides can form. One contains 56.34% P and 43.66% O, and its experimentally determined molar mass is 219.90 g/mol. Determine its molecular formula.

Answer P_4O_6

Strategy and Explanation The first step in finding a molecular formula from percent composition by mass and molar mass is to calculate the relative number of moles of each element and then determine the empirical formula. Considering the percent composition, we know that a 100.0-g sample of this phosphorus oxide contains 56.34 g P and 43.66 g O, so the numbers of moles of each element are

$$56.34 \text{ g P} \times \frac{1 \text{ mol P}}{30.97 \text{ g P}} = 1.819 \text{ mol P}$$

$$43.66 \text{ g O} \times \frac{1 \text{ mol O}}{16.00 \text{ g O}} = 2.729 \text{ mol O}$$

Thus, the mole ratio (and atom ratio) is

$$\frac{2.729 \text{ mol O}}{1.819 \text{ mol P}} = \frac{1.500 \text{ mol O}}{1.000 \text{ mol P}}$$

Because we can't have partial atoms, we double the numbers to convert to whole numbers, which gives us three moles of oxygen atoms for every two moles of phosphorus atoms. This gives an empirical formula of P_2O_3. The molar mass corresponding to this empirical formula is

$$\text{a}2 \cancel{\text{mol P}} \times \frac{30.97 \text{ g P}}{1 \cancel{\text{mol P}}}\text{b} + \text{a}3 \cancel{\text{mol O}} \times \frac{16.00 \text{ g O}}{1 \cancel{\text{mol O}}}\text{b} = 109.95 \text{ g } P_2O_3 \text{ per mole } P_2O_3$$

compared with a known molar mass of 219.90 g/mol. The known molar mass is $\frac{219.90}{109.95} = 2$ times the molar mass predicted by the empirical formula, so the molecular formula is P_4O_6, twice the empirical formula.

☑ **Reasonable Answer Check** The molar mass of P is about 31 g, so we should have $31 \text{ g} \times 4 = 124$ g P in 1 mol P_4O_6. This would give a P percent by mass of 124/220 = 56%, which is just about the percent given in the problem.

PROBLEM-SOLVING PRACTICE 3.15

The other phosphorus oxide contains 43.64% P and 56.36% O, and its experimentally determined molar mass is 283.89 g/mol. Determine its empirical and molecular formulas.

aspirin

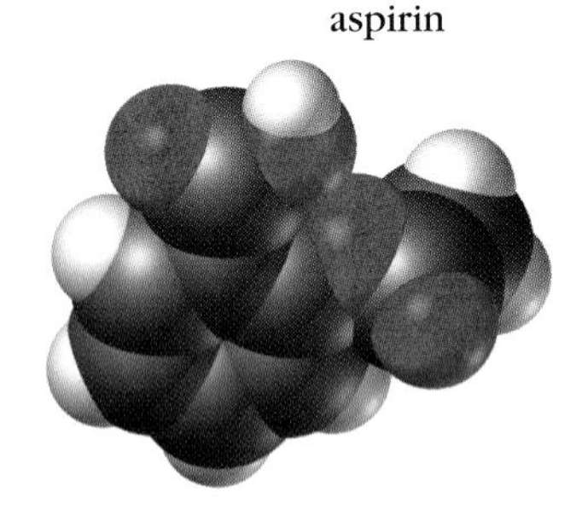

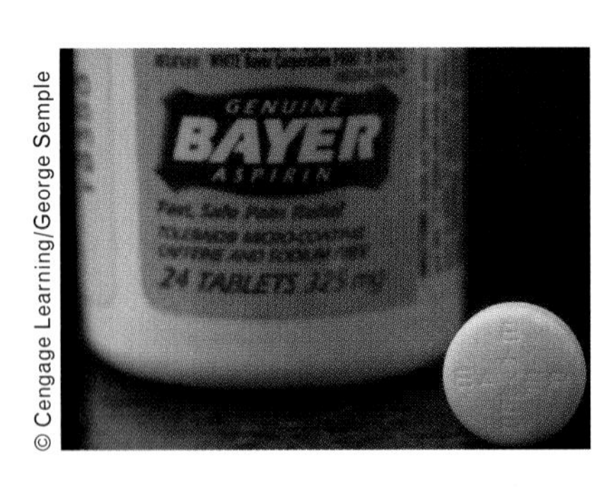

PROBLEM-SOLVING EXAMPLE 3.16 Molecular Formula from Percent Composition by Mass Data

Aspirin, a commonly used analgesic, has a molar mass of 180.15 g/mol. It contains 60.00% C, 4.4756% H, and the rest is oxygen. What are its empirical and molecular formulas?

Answer Both the empirical and molecular formulas are $C_9H_8O_4$.

Strategy and Explanation First find the number of moles of each element in 100.0 g of the compound:

$$60.00 \cancel{\text{g C}} \times \frac{1 \text{ mol C}}{12.01 \cancel{\text{g C}}} = 4.996 \text{ mol C}$$

$$4.4756 \cancel{\text{g H}} \times \frac{1 \text{ mol H}}{1.0079 \cancel{\text{g H}}} = 4.441 \text{ mol H}$$

That leaves moles of oxygen to be calculated. The mass of oxygen in the sample must be

$$100.0 \text{ g sample} - 60.00 \text{ g C} - 4.4756 \text{ g H} = 35.52 \text{ g O}$$

Converting this to moles of oxygen gives

$$35.52 \cancel{\text{g O}} \times \frac{1 \text{ mol O}}{15.9994 \cancel{\text{g O}}} = 2.220 \text{ mol O}$$

Base the mole ratio on the smallest number of moles present, in this case, moles of oxygen:

$$\frac{4.496 \text{ mol C}}{2.220 \text{ mol O}} = \frac{2.25 \text{ mol O}}{1.00 \text{ mol O}}$$

$$\frac{4.441 \text{ mol H}}{2.220 \text{ mol O}} = \frac{2.00 \text{ mol H}}{1.00 \text{ mol O}}$$

Therefore, the empirical formula has an atom ratio of 2.25 C to 1.00 O and 2.00 H to 1.00 O. To get small whole numbers for empirical formula subscripts, we multiply each molar number by 4, so the empirical formula is $C_9H_8O_4$. The molar mass predicted by this empirical formula is 180.15 g/mol, the same as the experimentally determined molar mass, indicating that the molecular formula is the same as the empirical formula.

☑ **Reasonable Answer Check** The molar mass of C is 12 g/mol, so we should have $12 \text{ g/mol} \times 9 \text{ mol} = 108$ g C in 1 mol $C_9H_8O_4$. This would give a mass percent C of 108/180 = 60%, which is the percent given in the problem.

PROBLEM-SOLVING PRACTICE 3.16

Vitamin C (ascorbic acid) contains 40.9% C, 4.58% H, and 54.5% O and has an experimentally determined molar mass of 176.13 g/mol. Determine its empirical and molecular formulas.

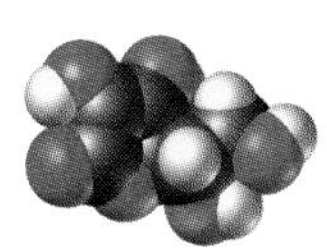

3.11 The Biological Periodic Table

Most of the more than 100 known elements are not directly involved with our personal health and well-being. However, more than 30 of the elements, shown in Figure 3.7, are absolutely essential to human life. Among these essential elements are metals, nonmetals, and metalloids from across the periodic table. All are necessary as part of a well-balanced diet.

If you weigh 150 lb, about 90 lb (60%) is water, 30 lb is fat, and the remaining 30 lb is a combination of proteins, carbohydrates, and calcium, phosphorus, and other dietary minerals.

Table 3.11 lists the building block elements and major minerals in order of their relative abundances per million atoms in the body, showing the preeminence of four of the nonmetals—oxygen, carbon, hydrogen, and nitrogen. These four nonmetals, the building block elements, contribute most of the atoms in the biologically significant chemicals—

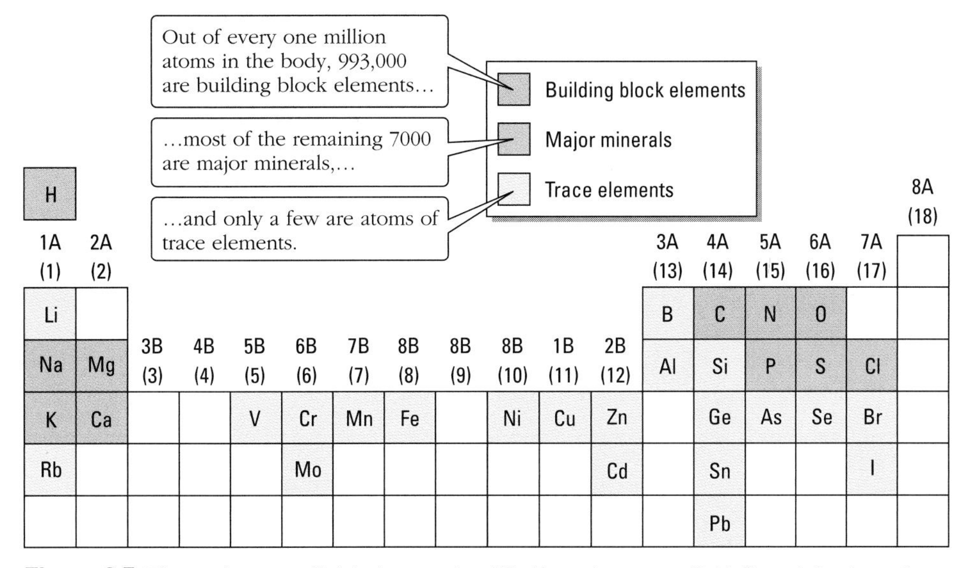

Figure 3.7 Elements essential to human health. Four elements—C, H, N, and O—form the many organic compounds that make up living organisms. The major minerals are required in relatively large amounts; trace elements are required in lesser amounts.

Table 3.11 Major Elements of the Human Body

Element	Symbol	Relative Abundance (Atoms/Million Atoms in the Body)	Element	Symbol	Relative Abundance (Atoms/Million Atoms in the Body)
Hydrogen	H	630,000	Chlorine	Cl	570
Oxygen	O	255,000	Sulfur	S	490
Carbon	C	94,500	Sodium	Na	410
Nitrogen	N	13,500	Potassium	K	260
Calcium	Ca	3100	Magnesium	Mg	130
Phosphorus	P	2200			

the biochemicals—composing all plants and animals. With few exceptions, a major one being water, the biochemicals that these nonmetals form are organic compounds.

Nonmetals are also present as anions in body fluids, including chloride ion, Cl^-; phosphorus in three ionic forms, PO_4^{3-}, HPO_4^{2-}, and $H_2PO_4^-$; and H and carbon as hydrogen carbonate ion, HCO_3^-, and carbonate ion, CO_3^{2-}. Metals are present in the body as cations in solution (for example, Na^+, K^+) and in solids (Ca^{2+} in bones and teeth). Metals are also incorporated into large biomolecules (for example, Fe^{2+} in hemoglobin and Co^{3+} in vitamin B-12).

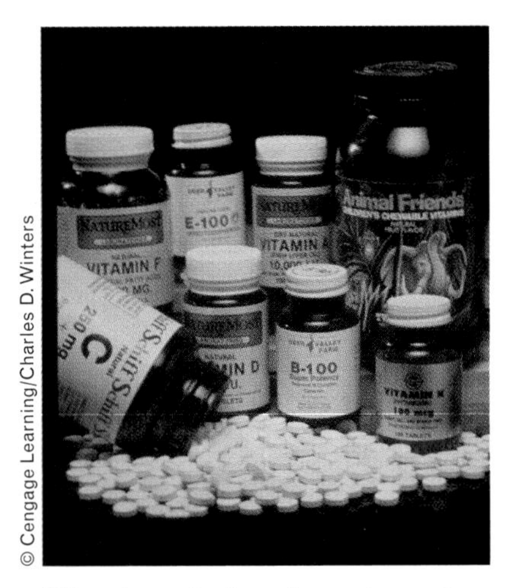

Vitamin and mineral supplements.

EXERCISE 3.10 Essential Elements

Using Figure 3.7, identify (a) the essential nonmetals, (b) the essential alkaline earth metals, (c) the essential halide ions, and (d) four essential transition metals.

The Dietary Minerals

The general term **dietary minerals** refers to the essential elements other than carbon, hydrogen, oxygen, or nitrogen. The dietary necessity and effects of these elements go far beyond that implied by their collective presence as only about 4% of our body weight. They exemplify the old saying, "Good things come in small packages." Because the body uses them efficiently, recycling them through many reactions, dietary minerals are required in only small amounts, but their absence from your diet can cause significant health problems.

The 28 dietary minerals indicated in Figure 3.7 are classified into the relatively more abundant **major minerals** and the less plentiful **trace elements.** Major minerals are those present in quantities greater than 0.01% of body mass (100 mg per kg)—for example, more than 6 g for a 60-kg (132-lb) individual. Trace elements are present in smaller (sometimes far smaller) amounts. For example, the necessary daily total intake of iodine is only 150 μg. In the context of nutrition, major minerals and trace elements usually refer to ions in soluble ionic compounds in the diet.

IN CLOSING

Having studied this chapter, you should be able to . . .

- Interpret the meaning of molecular formulas, condensed formulas, and structural formulas (Section 3.1).
- Name binary molecular compounds, including straight-chain alkanes (Sections 3.2, 3.3). ■ End-of-chapter Question assignable in OWL: 5
- Write structural formulas for and identify straight- and branched-chain alkane constitutional isomers (Section 3.4).
- Predict the charges on monatomic ions of metals and nonmetals (Section 3.5).
- Know the names and formulas of polyatomic ions (Section 3.5).
- Describe the properties of ionic compounds and compare them with the properties of molecular compounds (Sections 3.5, 3.7).
- Given their names, write the formulas of ionic compounds (Section 3.5).
- Given their formulas, name ionic compounds (Section 3.6). ■ Question 19
- Describe electrolytes in aqueous solution and summarize the differences between electrolytes and nonelectrolytes (Section 3.7). ■ Questions 27, 28
- Thoroughly explain the use of the mole concept for chemical compounds (Section 3.8).

Selected end-of-chapter Questions may be assigned in OWL.

For quick review, download Go Chemistry mini-lecture flashcard modules (or purchase them at **www.ichapters.com**).

Prepare for an exam with a **Summary Problem** that brings together concepts and problem-solving skills from throughout the chapter. Go to **www.cengage.com/chemistry/moore** to download this chapter's Summary Problem from the Student Companion Site.

- Calculate the molar mass of a compound (Section 3.8). ■ Question 33
- Calculate the number of moles of a compound given the mass, and vice versa (Section 3.8). ■ Questions 35, 37
- Explain the formula of a hydrated ionic compound and calculate its molar mass (Section 3.8). ■ Question 56
- Express molecular composition in terms of percent composition (Section 3.9). ■ Question 41
- Use percent composition and molar mass to determine the empirical and molecular formulas of a compound (Section 3.10). ■ Question 50
- Identify biologically important elements (Section 3.11).

KEY TERMS

alkane *(Section 3.3)*
alkyl group *(3.4)*
anion *(3.5)*
binary molecular compound *(3.2)*
cation *(3.5)*
chemical bond *(3.3)*
condensed formula *(3.1)*
constitutional isomer *(3.4)*
Coulomb's law *(3.5)*
crystal lattice *(3.7)*
dietary mineral *(3.11)*
dissociation *(3.7)*
electrolyte *(3.7)*
empirical formula *(3.10)*
formula unit *(3.7)*
formula weight *(3.8)*
functional groups *(3.1)*
halide ion *(3.6)*
hydrocarbon *(3.3)*
inorganic compound *(3.1)*
ionic bonding *(3.5)*
ionic compound *(3.5)*
ionic hydrate *(3.8)*
isomer *(3.4)*
major mineral *(3.11)*
molecular compound *(3.1)*
molecular formula *(3.1)*
molecular weight *(3.8)*
monatomic ion *(3.5)*
nonelectrolyte *(3.7)*
organic compound *(3.1)*
oxoanion *(3.6)*
percent composition by mass *(3.9)*
polyatomic ion *(3.5)*
structural formula *(3.1)*
trace element *(3.11)*
water of hydration *(3.8)*

QUESTIONS FOR REVIEW AND THOUGHT

■ denotes questions assignable in OWL.

Blue-numbered questions have short answers at the back of this book and fully worked solutions in the *Student Solutions Manual.*

Review Questions

1. For each of these structural formulas, write the molecular formula and condensed formula.

```
        H  H          H               H  H  O
        |  |         /                |  |  ‖
(a) H—C—C—C=C             (b) H—C—C—C—OH
        |  |  |      \                |  |
        H  H  H       H               H  H

        H  H  H  O
        |  |  |  ‖
(c) H—N—C—C—C—OH
           |  |
           H  H
```

2. Given these condensed formulas, write the structural and molecular formulas.
   (a) CH_3OH
   (b) $CH_3CH_2NH_2$
   (c) $CH_3CH_2SCH_2CH_3$
3. Give the name for each of these binary nonmetal compounds.
   (a) NF_3 (b) HI
   (c) BBr_3 (d) C_6H_{14}
4. Give the formula for each of these binary nonmetal compounds.
   (a) Sulfur trioxide
   (b) Dinitrogen pentaoxide
   (c) Phosphorous pentachloride
   (d) Silicon tetrachloride
   (e) Diboron trioxide (commonly called boric oxide)

Topical Questions

Molecular and Structural Formulas

5. ■ Give the formula for each of these nonmetal compounds.
   (a) Bromine trifluoride
   (b) Xenon difluoride
   (c) Diphosphorus tetrafluoride
   (d) Pentadecane
   (e) Hydrazine
6. Write structural formulas for these alkanes.
   (a) Butane (b) Nonane
   (c) Hexane (d) Octane
   (e) Octadecane
7. Write the molecular formula of each of these compounds.
   (a) Benzene (a liquid hydrocarbon), with 6 carbon atoms and 6 hydrogen atoms per molecule
   (b) Vitamin C, with 6 carbon atoms, 8 hydrogen atoms, and 6 oxygen atoms per molecule
8. Write the formula for each molecule.
   (a) A molecule of the hydrocarbon heptane, which has 7 carbon atoms and 16 hydrogen atoms
   (b) A molecule of acrylonitrile (the basis of Orlon and Acrilan fibers), which has 3 carbon atoms, 3 hydrogen atoms, and 1 nitrogen atom
   (c) A molecule of Fenclorac (an anti-inflammatory drug), which has 14 carbon atoms, 16 hydrogen atoms, 2 chlorine atoms, and 2 oxygen atoms
9. Give the total number of atoms of each element in one formula unit of each of these compounds.
   (a) CaC_2O_4 (b) $C_6H_5CHCH_2$
   (c) $(NH_4)_2SO_4$ (d) $Pt(NH_3)_2Cl_2$
   (e) $K_4Fe(CN)_6$
10. Give the total number of atoms of each element in each of these molecules.
    (a) $C_6H_5COOC_2H_5$ (b) $HOOCCH_2CH_2COOH$
    (c) $NH_2CH_2CH_2COOH$ (d) $C_{10}H_9NH_2Fe$
    (e) $C_6H_2CH_3(NO_2)_3$

Constitutional Isomers

11. Consider two molecules that are constitutional isomers.
    (a) What is the same on the molecular level between these two molecules?
    (b) What is different on the molecular level between these two molecules?
12. Draw structural formulas for the five constitutional isomers of C_6H_{14}.

Predicting Ion Charges

13. Predict the charges for ions of these elements.
    (a) Magnesium (b) Zinc
    (c) Iron (d) Gallium
14. Predict the charges for ions of these elements.
    (a) Selenium (b) Fluorine
    (c) Silver (d) Nitrogen
15. Which of these are the correct formulas of compounds? For those that are not, give the correct formula.
    (a) AlCl (b) NaF_2
    (c) Ga_2O_3 (d) MgS
16. Which of these are the correct formulas of compounds? For those that are not, give the correct formula.
    (a) Ca_2O (b) $SrCl_2$
    (c) Fe_2O_5 (d) K_2O

Polyatomic Ions

17. For each of these compounds, tell what ions are present and how many there are per formula unit.
    (a) $Pb(NO_3)_2$ (b) $NiCO_3$
    (c) $(NH_4)_3PO_4$ (d) K_2SO_4
18. For each of these compounds, tell what ions are present and how many there are per formula unit.
    (a) $Ca(CH_3CO_2)_2$ (b) $Co_2(SO_4)_3$
    (c) $Al(OH)_3$ (d) $(NH_4)_2CO_3$
19. ■ Write the chemical formulas for these compounds.
    (a) Nickel(II) nitrate (b) Sodium bicarbonate
    (c) Lithium hypochlorite (d) Magnesium chlorate
    (e) Calcium sulfite
20. Write the chemical formulas for these compounds.
    (a) Iron(III) nitrate (b) Potassium carbonate
    (c) Sodium phosphate (d) Calcium chlorite
    (e) Sodium sulfate

Ionic Compounds

21. ■ Identify which of these substances are ionic.
    (a) CF_4 (b) $SrBr_2$
    (c) $Co(NO_3)_3$ (d) SiO_2
    (e) KCN (f) SCl_2
22. Which of these substances are ionic? Write the formula for each.
    (a) Methane (b) Dinitrogen pentaoxide
    (c) Ammonium sulfide (d) Hydrogen selenide
    (e) Sodium perchlorate
23. Give the correct formula for each of these ionic compounds.
    (a) Ammonium carbonate (b) Calcium iodide
    (c) Copper(II) bromide (d) Aluminum phosphate
24. Give the correct formula for each of these ionic compounds.
    (a) Calcium hydrogen carbonate
    (b) Potassium permanganate
    (c) Magnesium perchlorate
    (d) Ammonium monohydrogen phosphate
25. Correctly name each of these ionic compounds.
    (a) K_2S (b) $NiSO_4$
    (c) $(NH_4)_3PO_4$ (d) $Al(OH)_3$
    (e) $Co_2(SO_4)_3$
26. Correctly name each of these ionic compounds.
    (a) KH_2PO_4 (b) $CuSO_4$
    (c) $CrCl_3$ (d) $Ca(CH_3COO)_2$
    (e) $Fe_2(SO_4)_3$

Electrolytes

27. ■ For each of these electrolytes, what ions will be present in an aqueous solution?
    (a) KOH (b) K_2SO_4
    (c) $NaNO_3$ (d) NH_4Cl
28. For each of these electrolytes, what ions will be present in an aqueous solution?
    (a) CaI_2 (b) $Mg_3(PO_4)_2$
    (c) NiS (d) $MgBr_2$

29. Which of these substances would conduct electricity when dissolved in water?
   (a) NaCl (b) $CH_3CH_2CH_3$ (propane)
   (c) CH_3OH (methanol) (d) $Ca(NO_3)_2$
30. Which of these substances would conduct electricity when dissolved in water?
   (a) NH_4Cl (b) $CH_3CH_2CH_2CH_3$ (butane)
   (c) $C_{12}H_{22}O_{11}$ (table sugar) (d) $Ba(NO_3)_2$

Moles of Compounds

31. Fill in this table for one mol methanol, CH_3OH.

	CH_3OH	Carbon	Hydrogen	Oxygen
Number of moles				
Number of molecules or atoms				
Molar mass				

32. Fill in this table for one mol glucose, $C_6H_{12}O_6$.

	$C_6H_{12}O_6$	Carbon	Hydrogen	Oxygen
Number of moles				
Number of molecules or atoms				
Molar mass				

33. ■ Calculate the molar mass of each of these compounds.
   (a) Iron(III) oxide
   (b) Boron trifluoride
   (c) Dinitrogen oxide (laughing gas)
   (d) Manganese(II) chloride tetrahydrate
   (e) $C_6H_8O_6$, ascorbic acid
34. Calculate the molar mass of each of these compounds.
   (a) $B_{10}H_{14}$, a boron hydride once considered as a rocket fuel
   (b) $C_6H_2(CH_3)(NO_2)_3$, TNT, an explosive
   (c) $PtCl_2(NH_3)_2$, a cancer chemotherapy agent called cisplatin
   (d) $CH_3(CH_2)_3SH$, butyl mercaptan, a compound with a skunklike odor
   (e) $C_{20}H_{24}N_2O_2$, quinine, used as an antimalarial drug
35. ■ How many moles are represented by 1.00 g of each of these compounds?
   (a) CH_3OH, methanol
   (b) Cl_2CO, phosgene, a poisonous gas
   (c) Ammonium nitrate
   (d) Magnesium sulfate heptahydrate (Epsom salt)
   (e) $Ag(CH_3COO)$, silver acetate
36. How many moles are present in 0.250 g of each of these compounds?
   (a) $C_7H_5NO_3S$, saccharin, an artificial sweetener
   (b) $C_{13}H_{20}N_2O_2$, procaine, a painkiller used by dentists
   (c) $C_{20}H_{14}O_4$, phenolphthalein, a dye
37. ■ Acetaminophen, an analgesic, has the molecular formula $C_8H_9O_2N$.
   (a) What is the molar mass of acetaminophen?
   (b) How many moles are present in 5.32 g acetaminophen?
   (c) How many grams are present in 0.166 mol acetaminophen?
38. An Alka-Seltzer tablet contains 324 mg aspirin, $C_9H_8O_4$; 1904 mg $NaHCO_3$; and 1000. mg citric acid, $C_6H_8O_7$. (The last two compounds react with each other to provide the "fizz," bubbles of CO_2, when the tablet is put into water.)
   (a) Calculate the number of moles of each substance in the tablet.
   (b) If you take one tablet, how many molecules of aspirin are you consuming?
39. The use of CFCs (chlorofluorocarbons) has been curtailed because there is strong evidence that they cause environmental damage. If a spray can contains 250 g of one of these compounds, CCl_2F_2, how many molecules of this CFC are you releasing to the air when you empty the can?
40. CFCs (chlorofluorocarbons) are implicated in decreasing the ozone concentration in the stratosphere. A CFC substitute is CF_3CH_2F.
   (a) If you have 25.5 g of this compound, how many moles does this represent?
   (b) How many atoms of fluorine are contained in 25.5 g of the compound?

Percent Composition

41. ■ Calculate the molar mass of each of these compounds and the mass percent of each element.
   (a) PbS, lead(II) sulfide, galena
   (b) C_2H_6, ethane, a hydrocarbon fuel
   (c) CH_3COOH, acetic acid, an important ingredient in vinegar
   (d) NH_4NO_3, ammonium nitrate, a fertilizer
42. Calculate the molar mass of each of these compounds and the mass percent of each element.
   (a) $MgCO_3$, magnesium carbonate
   (b) C_6H_5OH, phenol, an organic compound used in some cleaners
   (c) $C_2H_3O_5N$, peroxyacetyl nitrate, an objectionable compound in photochemical smog
   (d) $C_4H_{10}O_3NPS$, acephate, an insecticide
43. The copper-containing compound $Cu(NH_3)_4SO_4 \cdot H_2O$ is a beautiful blue solid. Calculate the molar mass of the compound and the mass percent of each element.
44. Sucrose, table sugar, is $C_{12}H_{22}O_{11}$. When sucrose is heated, water is driven off. How many grams of pure carbon can be obtained from exactly one pound of sugar?
45. Carbonic anhydrase, an important enzyme in mammalian respiration, is a large zinc-containing protein with a molar mass of 3.00×10^4 g/mol. The zinc is 0.218% by mass of the protein. How many zinc atoms does each carbonic anhydrase molecule contain?
46. Nitrogen fixation in the root nodules of peas and other legumes occurs with a reaction involving a molybdenum-containing enzyme named *nitrogenase*. This enzyme contains two Mo atoms per molecule and is 0.0872% Mo by mass. What is the molar mass of the enzyme?
47. Disilane, Si_2H_x, contains 90.28% silicon by mass. What is the value of x in this compound?

Empirical and Molecular Formulas

48. The molecular formula of ascorbic acid (vitamin C) is $C_6H_8O_6$. What is its empirical formula?
49. The empirical formula of maleic acid is CHO. Its molar mass is 116.1 g/mol. What is its molecular formula?
50. ■ A compound with a molar mass of 100.0 g/mol has an elemental composition of 24.0% C, 3.0% H, 16.0% O, and 57.0% F. What is the molecular formula of the compound?
51. Acetylene is a colorless gas that is used as a fuel in welding torches, among other things. It is 92.26% C and 7.74% H. Its molar mass is 26.02 g/mol. Calculate the empirical and molecular formulas.
52. A compound contains 38.67% K, 13.85% N, and 47.47% O by mass. What is the empirical formula of the compound?
53. A compound contains 36.76% Fe, 21.11% S, and 42.13% O by mass. What is the empirical formula of the compound?
54. Cacodyl, a compound containing arsenic, was reported in 1842 by the German chemist Bunsen. It has an almost intolerable garliclike odor. Its molar mass is 210. g/mol, and it is 22.88% C, 5.76% H, and 71.36% As. Determine its empirical and molecular formulas.
55. The action of bacteria on meat and fish produces a poisonous and stinky compound called cadaverine. It is 58.77% C, 13.81% H, and 27.42% N. Its molar mass is 102.2 g/mol. Determine the molecular formula of cadaverine.
56. ■ If Epsom salt, $MgSO_4 \cdot x\ H_2O$, is heated to 250 °C, all of the water of hydration is lost. After a 1.687-g sample of the hydrate is heated, 0.824 g $MgSO_4$ remains. How many molecules of water are there per formula unit of $MgSO_4$?
57. The alum used in cooking is potassium aluminum sulfate hydrate, $KAl(SO_4)_2 \cdot x\ H_2O$. To find the value of x, you can heat a sample of the compound to drive off all the water and leave only $KAl(SO_4)_2$. Assume that you heat 4.74 g of the hydrated compound and that it loses 2.16 g water. What is the value of x?

Biological Periodic Table

58. Make a list of the top ten most abundant essential elements needed by the human body.
59. Which types of compounds contain the majority of the oxygen found in the human body?
60. (a) How are metals found in the body, as atoms or as ions?
    (b) What are two uses for metals in the human body?

General Questions

61. (a) Name each of these compounds.
    (b) Tell which ones are best described as ionic.
    (i) $ClBr_3$ (ii) NCl_3
    (iii) $CaSO_4$ (iv) C_7H_{16}
    (v) XeF_4 (vi) OF_2
    (vii) NaI (viii) Al_2S_3
    (ix) PCl_5 (x) K_3PO_4
62. (a) Write the formula for each of these compounds.
    (b) Tell which ones are best described as ionic.
    (i) Sodium hypochlorite
    (ii) Aluminum perchlorate
    (iii) Potassium permanganate
    (iv) Potassium dihydrogen phosphate
    (v) Chlorine trifluoride
    (vi) Boron tribromide
    (vii) Calcium acetate
    (viii) Sodium sulfite
    (ix) Disulfur tetrachloride
    (x) Phosphorus trifluoride
63. Elemental analysis of fluorocarbonyl hypofluorite gave 14.6% C, 39.0% O, and 46.3% F. If the molar mass of the compound is 82.0 g/mol, determine the (a) empirical and (b) molecular formulas of the compound.
64. Azulene, a beautiful blue hydrocarbon, is 93.71% C and has a molar mass of 128.16 g/mol. What are the (a) empirical and (b) molecular formulas of azulene?
65. Direct reaction of iodine, I_2, and chlorine, Cl_2, produces an iodine chloride, I_xCl_y, a bright yellow solid.
    (a) If you completely react 0.678 g iodine to produce 1.246 g I_xCl_y, what is the empirical formula of the compound?
    (b) A later experiment shows that the molar mass of I_xCl_y is 467 g/mol. What is the molecular formula of the compound?
66. Iron pyrite, often called "fool's gold," has the formula FeS_2. If you could convert 15.8 kg iron pyrite to iron metal and remove the sulfur, how many kilograms of the metal could you obtain?
67. Draw a diagram to indicate the arrangement of nanoscale particles of each substance. Consider each drawing to hold a very tiny portion of each substance. Each drawing should contain at least 16 particles, and it need not be three-dimensional.

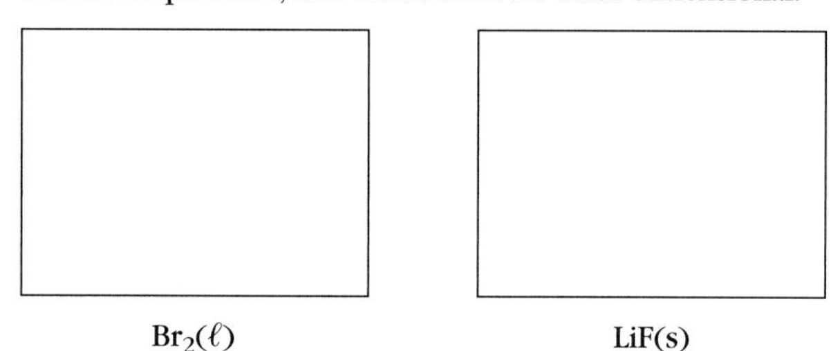

68. Draw diagrams of each nanoscale situation below. Represent atoms or monatomic ions as circles; represent molecules or polyatomic ions by overlapping circles for the atoms that make up the molecule or ion; and distinguish among different kinds of atoms by labeling or shading the circles. In each case draw representations of at least five nanoscale particles. Your diagrams can be two-dimensional.
    (a) A crystal of solid sodium chloride
    (b) The sodium chloride from part (a) after it has been melted
69. Draw diagrams of each nanoscale situation below. Represent atoms or monatomic ions as circles; represent molecules or polyatomic ions by overlapping circles for the atoms that make up the molecule or ion; and distinguish among different kinds of atoms by labeling or shading the circles. In each case draw representations of at least five nanoscale particles. Your diagrams can be two-dimensional.
    (a) A sample of solid lithium nitrate, $LiNO_3$
    (b) A sample of molten lithium nitrate
    (c) A molten sample of lithium nitrate after electrodes have been placed into it and a direct current applied to the electrodes

Applying Concepts

70. When asked to draw all the possible constitutional isomers for C_3H_8O, a student drew these structures. The student's instructor said some of the structures were identical.
    (a) How many actual isomers are there?
    (b) Which structures are identical?

$CH_3—CH_2—CH_2—OH$ $\quad$ $CH_3—CH_2—O—CH_3$

$CH_3—O—CH_2—CH_3$ $\quad$ $HO—CH_2—CH_2$ (with CH_3 on the second CH_2)

$CH_3—CH—CH_3$ (with OH on CH) $\quad$ $HO—CH—CH_3$ (with CH_3 on CH)

71. The formula for thallium nitrate is $TlNO_3$. Based on this information, what would be the formulas for thallium carbonate and thallium sulfate?
72. The name given with each of these formulas is incorrect. What are the correct names?
    (a) CaF_2, calcium difluoride
    (b) CuO, copper oxide
    (c) $NaNO_3$, sodium nitroxide
    (d) NI_3, nitrogen iodide
    (e) $FeCl_3$, iron(I) chloride
    (f) Li_2SO_4, dilithium sulfate
73. Based on the guidelines for naming oxoanions in a series, how would you name these species?
    (a) BrO_4^-, BrO_3^-, BrO_2^-, BrO^-
    (b) SeO_4^{2-}, SeO_3^{2-}
74. Which illustration best represents $CaCl_2$ dissolved in water?

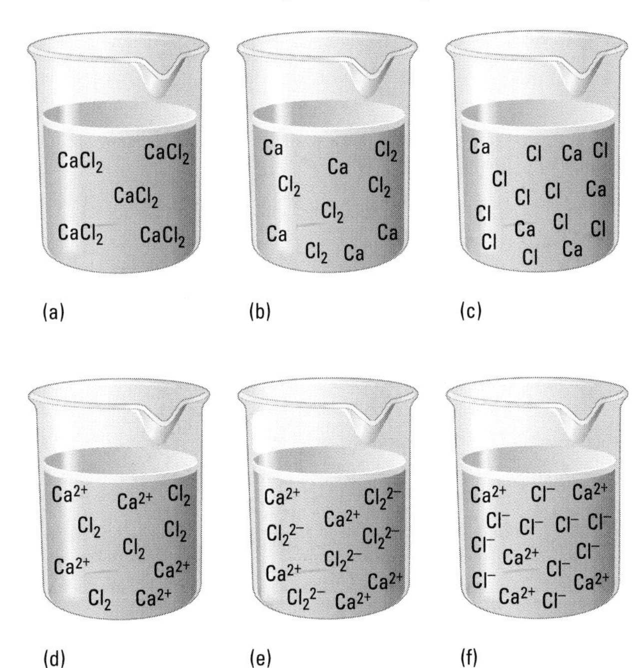

75. Which sample has the largest amount of NH_3?
    (a) 6.022×10^{24} molecules of NH_3
    (b) 0.1 mol NH_3
    (c) 17.03 g NH_3

More Challenging Questions

76. ■ A piece of nickel foil, 0.550 mm thick and 1.25 cm square, was allowed to react with fluorine, F_2, to give a nickel fluoride. (The density of nickel is 8.908 g/cm^3.)
    (a) How many moles of nickel foil were used?
    (b) If you isolate 1.261 g nickel fluoride, what is its formula?
    (c) What is its name?
77. One molecule of an unknown compound has a mass of 7.308×10^{-23} g, and 27.3% of that mass is due to carbon; the rest is oxygen. What is the compound?
78. What are the empirical formulas of these compounds?
    (a) A hydrate of Fe(III) thiocyanate, $Fe(SCN)_3$, that contains 19.0% water
    (b) A mineral hydrate of zinc sulfate that contains 43.86% water
    (c) A sodium aluminum silicate hydrate that contains 12.10% Na, 14.19% Al, 22.14% Si, 42.09% O, and 9.48% H_2O
79. The active ingredients in lawn and garden fertilizer are nitrogen, phosphorus, and potassium. Bags of fertilizer usually carry three numbers, as in 5-10-5 fertilizer (a typical fertilizer for flowers). The first number is the mass percent N; the second is the mass percent P_2O_5; the third is the mass percent K_2O. Thus, the active ingredients in a 5-10-5 product are equivalent to 5% N, 10% P_2O_5, and 5% K_2O, by weight. (The reporting of fertilizer ingredients in this way is a convention agreed on by fertilizer manufacturers.) What is the mass in pounds of each of these three elements (N, P, K) in a 100-lb bag of fertilizer?

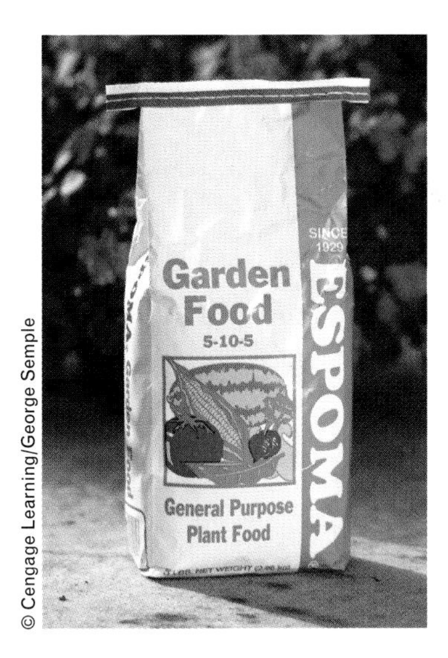

80. Four common ingredients in fertilizers are ammonium nitrate NH_4NO_3, ammonium sulfate $(NH_4)_2SO_4$, urea $(NH_4)_2CO$, and ammonium hydrogen phosphate $(NH_4)_2HPO_4$. On the basis of mass, which of these compounds has the largest mass percent nitrogen? What is the mass percent of nitrogen in this compound?

Conceptual Challenge Problems

CP3.A (Section 3.9) A chemist analyzes three compounds and reports these data for the percent by mass of the elements Ex, Ey, and Ez in each compound.

Compound	% Ex	% Ey	% Ez
A	37.485	12.583	49.931
B	40.002	6.7142	53.284
C	40.685	5.1216	54.193

Assume that you accept the notion that the numbers of atoms of the elements in compounds are in small whole-number ratios and that the number of atoms in a sample of any element is directly proportional to that sample's mass. What is possible for you to know about the empirical formulas for these three compounds?

CP3.B (Section 3.10) The following table displays on each horizontal row an empirical formula for one of the three compounds noted in CP3.A.

Compound A	Compound B	Compound C
________	$ExEy_2Ez$	________
$Ex_6Ey_8Ez_3$	________	________
________	________	Ex_3Ey_2Ez
________	$Ex_9Ey_2Ez_6$	________
________	________	$ExEy_2Ez_3$
$Ex_3Ey_8Ez_3$	________	________

Based only on what was learned in that problem, what is the empirical formula for the other two compounds in that row?

CP3.C (Section 3.10)

(a) Suppose that a chemist now determines that the ratio of the masses of equal numbers of atoms of Ez and Ex atoms is 1.3320 g Ez/1 g Ex. With this added information, what can now be known about the formulas for compounds A, B, and C in Problem CP3.A?

(b) Suppose that this chemist further determines that the ratio of the masses of equal numbers of atoms of Ex and Ey is 11.916 g Ex/1 g Ey. What is the ratio of the masses of equal numbers of Ez and Ey atoms?

(c) If the mass ratios of equal numbers of atoms of Ex, Ey, and Ez are known, what can be known about the formulas of the three compounds A, B, and C?

4

Quantities of Reactants and Products

NASA

Launch of a NASA Space Shuttle and its booster rockets. The shuttle engines are powered by the reaction of hydrogen with oxygen. How do the scientists and engineers designing the engine know how much of each reactant to use? Using a balanced chemical equation allows accurate calculations of the masses of reactants needed.

A major emphasis of chemistry is understanding chemical reactions. To work with chemical reactions requires knowing the correct formula of each reactant and product and the relative molar amounts of each involved in the reaction. This information is contained in balanced chemical equations—equations that are consistent with the law of conservation of mass and the existence of atoms. This chapter begins with a discussion of the nature of chemical equations. It is followed by a brief introduction to some general types of chemical reactions. Many of the very large number of known chemical reactions can be assigned to a few categories: combination, decomposition, displacement, and exchange. Next comes a description of how to write balanced chemical equations. After that, we will look at how the balanced chemical equation can be used to move from an understanding of what the reactants and products are to how much of each is involved under various conditions. Finally, we will introduce methods used to find the formulas of chemical compounds.

While the fundamentals of chemical reactions must be understood at the atomic and molecular levels (the nanoscale), the chemical reactions that we will describe are observable at the macroscale in the everyday world around us. Understanding and applying the quantitative relationships in chemical reactions are essential skills. You should know how to calculate the quantity of product that a reaction will generate from a given quantity of reactants. Facility with these quantitative calculations connects the nanoscale world of the chemical reaction to the macroscale world of laboratory and industrial manipulations of measurable quantities of chemicals.

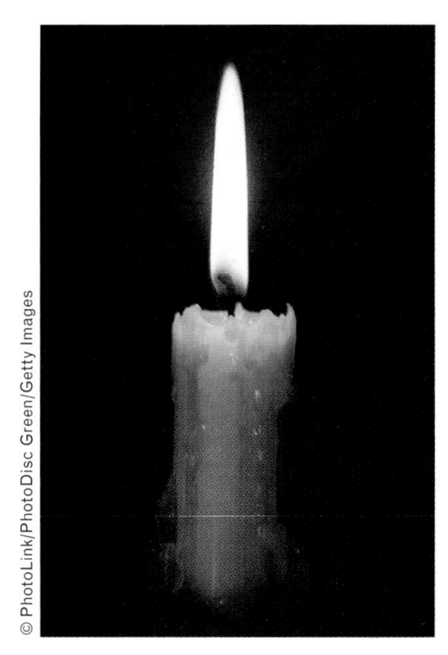

Combustion of hydrocarbons from candle wax produces a candle flame.

4.1 Chemical Equations

A candle flame can create a mood as well as provide light. It also is the result of a chemical reaction, a process in which reactants are converted into products ***(⇐ p. 10).*** The reactants and products can be elements, compounds, or both. In equation form we write

$$\text{Reactant(s)} \longrightarrow \text{product(s)}$$

where the arrow means "forms" or "yields" or "changes to."

In a burning candle, the reactants are hydrocarbons from the candle wax and oxygen from the air. Such reactions, in which an element or compound burns in air or oxygen, are called **combustion reactions.** The products of the complete combustion of hydrocarbons ***(⇐ p. 67)*** are always carbon dioxide and water.

$$\underset{\text{a hydrocarbon in candle wax}}{C_{25}H_{52}(s)} + \underset{\text{oxygen}}{38\ O_2(g)} \longrightarrow \underset{\text{carbon dioxide}}{25\ CO_2(g)} + \underset{\text{water}}{26\ H_2O(g)}$$

This **balanced chemical equation** indicates the relative amounts of reactants and products required so that the number of atoms of each element in the reactants equals the number of atoms of the same element in the products. In the next section we discuss how to write balanced equations.

Usually the physical states of the reactants and products are indicated in a chemical equation by placing one of these symbols after each reactant and product: (s) for solid, (ℓ) for liquid, and (g) for gas. The symbol (aq) is used to represent an **aqueous solution,** a substance dissolved in water. This is illustrated by the equation for the reaction of zinc metal with hydrochloric acid, an aqueous solution of hydrogen chloride. The products are hydrogen gas and an aqueous solution of zinc chloride, a soluble ionic compound ***(⇐ p. 71).***

$$\underbrace{Zn(s) + 2\ HCl(aq)}_{\text{Reactants}} \longrightarrow \underbrace{H_2(g) + ZnCl_2(aq)}_{\text{Products}}$$

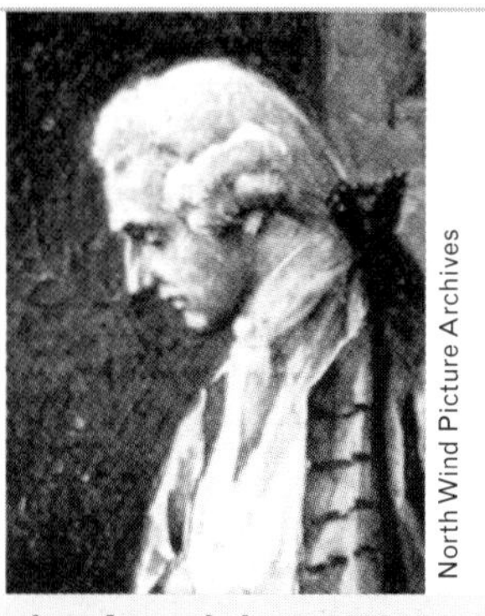

Antoine Lavoisier 1743–1794

One of the first to recognize the importance of exact scientific measurements and of carefully planned experiments, Lavoisier introduced principles for naming chemical substances that are still in use today. Further, he wrote a textbook, *Elements of Chemistry* (1789), in which he applied for the first time the principle of conservation of mass to chemistry and used the idea to write early versions of chemical equations. His life was cut short during the Reign of Terror of the French Revolution.

In the 18th century, the great French scientist Antoine Lavoisier introduced the law of conservation of mass, which later became part of Dalton's atomic theory. Lavoisier showed that mass is neither created nor destroyed in chemical reactions. Therefore, if you use 5 g of reactants they will form 5 g of products if the reaction is complete; if you use 500 mg of reactants they will form 500 mg of products; and so on. Combined with Dalton's atomic theory ***(⇐ p. 19),*** this also means that if there are 1000 atoms of a particular element in the reactants, then those 1000 atoms must appear in the products.

Consider, for example, the reaction between gaseous hydrogen and chlorine to produce hydrogen chloride gas:

$$H_2(g) + Cl_2(g) \longrightarrow 2\ HCl(g)$$

When applied to this reaction, the law of conservation of mass means that one diatomic molecule of H_2 (two atoms of hydrogen) and one diatomic molecule of Cl_2 (two atoms of Cl) must produce two molecules of HCl. The numbers in front of the formulas—the **coefficients**—in balanced equations show how matter is conserved. The 2 HCl indicates that two HCl molecules are formed, each containing one hydrogen atom and one chlorine atom. Note how the symbol of an element or the formula of a compound is multiplied through by the coefficient that precedes it. The equality of the number of atoms of each kind in the reactants and in the products is what makes the equation "balanced."

There must be an accounting for *all* atoms in a chemical reaction.

Recall that there are 6.022×10^{23} atoms in a mole of any atomic element (*p. 51*) or 6.022×10^{23} molecules in a mole of any molecular element or compound (*p. 83*).

Multiplying all the coefficients by the same factor gives the relative amounts of reactants and products at any scale. For example, 4 H_2 molecules will react with 4 Cl_2 molecules to produce 8 HCl molecules (Figure 4.1). If we continue to scale up the reaction, we can use Avogadro's number as the common factor. Thus, 1 mol H_2 molecules reacting with 1 mol Cl_2 molecules will produce 2 mol HCl molecules. As demanded by the conservation of mass, the number of atoms of each type in the reactants and the products is the same.

With the numbers of atoms balanced, the masses represented by the equation are also balanced. The molar masses show that 1.000 mol H_2 is equivalent to 2.016 g H_2 and that 1.000 mol Cl_2 is equivalent to 70.90 g Cl_2, so the total mass of reactants must be 2.016 g + 70.90 g = 72.92 g when 1.000 mol each of H_2 and Cl_2 are used. Conservation of mass demands that the same mass, 72.92 g HCl, must result from the reaction, and it does.

$$2.000 \text{ mol HCl} \times \frac{36.45 \text{ g HCl}}{1 \text{ mol HCl}} = 72.90 \text{ g HCl}$$

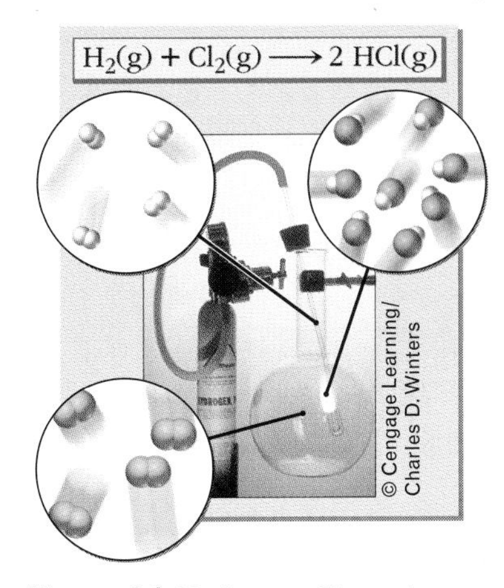

Figure 4.1 Hydrogen, H_2, and chlorine, Cl_2, react to form hydrogen chloride, HCl. Two molecules of HCl are formed when one H_2 molecule reacts with one Cl_2 molecule. This ratio is maintained when the reaction is carried out on a larger scale.

Relations among the masses of chemical reactants and products are called **stoichiometry** (stoy-key-AHM-uh-tree), and the coefficients (the multiplying numbers) in a balanced equation are called **stoichiometric coefficients** (or just coefficients).

EXERCISE 4.1 Chemical Equations

When methane burns this reaction is occurring:

$$CH_4(g) + 2\,O_2(g) \longrightarrow CO_2(g) + 2\,H_2O(g)$$

Write out in words the meaning of this chemical equation.

EXERCISE 4.2 Stoichiometric Coefficients

Heating iron metal in oxygen forms iron(III) oxide, Fe_2O_3 (Figure 4.2):

$$4\,Fe(s) + 3\,O_2(g) \longrightarrow 2\,Fe_2O_3(s)$$

(a) If 2.50 g Fe_2O_3 is formed by this reaction, what is the maximum *total* mass of iron metal and oxygen that reacted?
(b) Identify the stoichiometric coefficients in this equation.
(c) If 10,000 oxygen atoms reacted, how many Fe atoms were needed to react with this amount of oxygen?

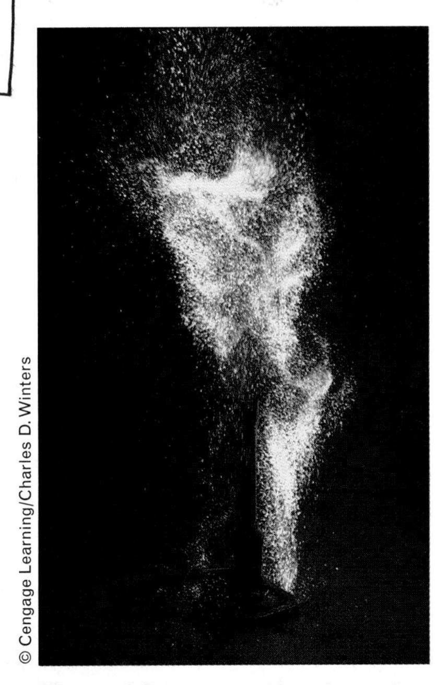

Figure 4.2 Powdered iron burns in air to form the iron oxide Fe_2O_3. The energy released during the reaction heats the particles to incandescence.

4.2 Patterns of Chemical Reactions

Many simple chemical reactions fall into one of the reaction patterns illustrated in Figure 4.3. Learning to recognize these reaction patterns is useful because they serve as a guide to predict what might happen when chemicals are mixed together or heated.

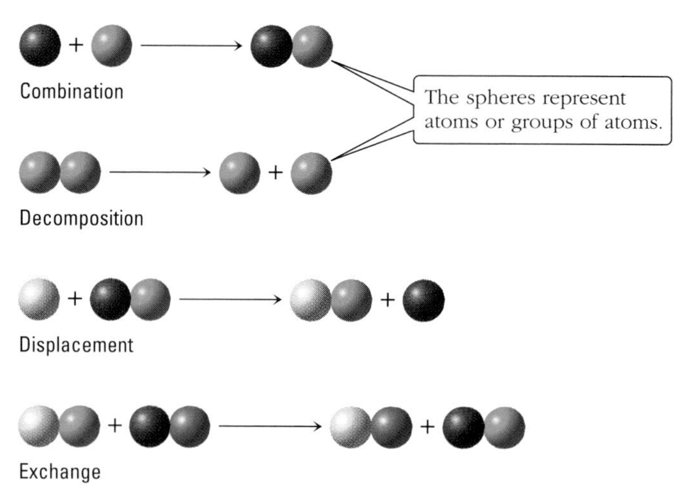

Figure 4.3 Four general types of chemical reactions. This classification of reaction patterns applies mainly to elements and inorganic compounds.

Take a look at the equation below. What does it mean to you at this stage in your study of chemistry?

$$Cl_2(g) + 2\ KBr(aq) \longrightarrow 2\ KCl(aq) + Br_2(\ell)$$

It's easy to let your eye slide by an equation on the printed page, but *don't* do that; ***read the equation.*** In this case it shows you that gaseous diatomic chlorine mixed with an aqueous solution of potassium bromide reacts to produce an aqueous solution of potassium chloride plus liquid diatomic bromine. After you learn to recognize reaction patterns, you will see that this is a *displacement* reaction—chlorine has displaced bromine so that the resulting compound in solution is KCl instead of the KBr originally present. Because chlorine displaces bromine, chlorine is what chemists describe as "more active" than bromine. The occurrence of this reaction implies that when chlorine is mixed with a solution of a different ionic bromide compound, displacement might also take place.

Throughout the rest of this chapter and the rest of this book, you should read and interpret chemical equations as we have just illustrated. Note the physical states of reactants and products. Mentally classify the reactions as described in the next few sections and look for what can be learned from each example. Most importantly, *don't* think that the equations must be memorized. There are far too many chemical reactions for that. Instead look for patterns, classes of reactions and the kinds of substances that undergo them, and information that can be applied in other situations. Doing so will give you insight into how chemistry is used every day in a wide variety of applications.

Combination Reactions

In a **combination reaction,** two or more substances react to form a single product.

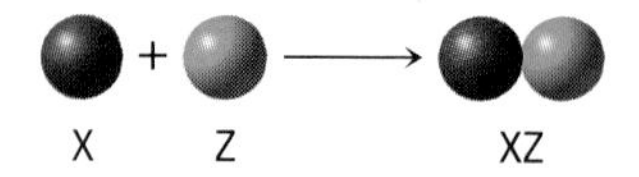

Recall that the halogens are F_2, Cl_2, Br_2, and I_2.

The halogens (Group 7A) and oxygen are such reactive elements that they undergo combination reactions with most other elements. Thus, if one of two possible reactants is oxygen or a halogen and the other reactant is another element, it is reasonable to expect that a combination reaction will occur.

The combination reaction of a metal with oxygen produces an *ionic compound,* a metal oxide. Like any ionic compound, the metal oxide must be electrically neutral. Because you can predict a reasonable positive charge for the metal ion by knowing its position in the periodic table and by using the guidelines in Section 3.5 (⟵ *p. 71*) you can determine the formula of the metal oxide. For example, when aluminum (Al, Group 3A), which forms Al^{3+} ions, reacts with O_2, which forms O^{2-} ions, the product must be aluminum oxide Al_2O_3 (2 Al^{3+} and 3 O^{2-} ions), a compound also known as *alumina,* or *corundum.*

$$4\,Al(s) + 3\,O_2(g) \longrightarrow \underset{\text{aluminum oxide}}{2\,Al_2O_3(s)}$$

The halogens also combine with metals to form ionic compounds with formulas that are predictable based on the charges of the ions formed. The halogens form 1− ions in simple ionic compounds. For example, sodium combines with chlorine, and zinc combines with iodine, to form sodium chloride and zinc iodide, respectively (Figure 4.4).

$$2\,Na(s) + Cl_2(g) \longrightarrow 2\,NaCl(s)$$

$$Zn(s) + I_2(s) \longrightarrow ZnI_2(s)$$

(a)

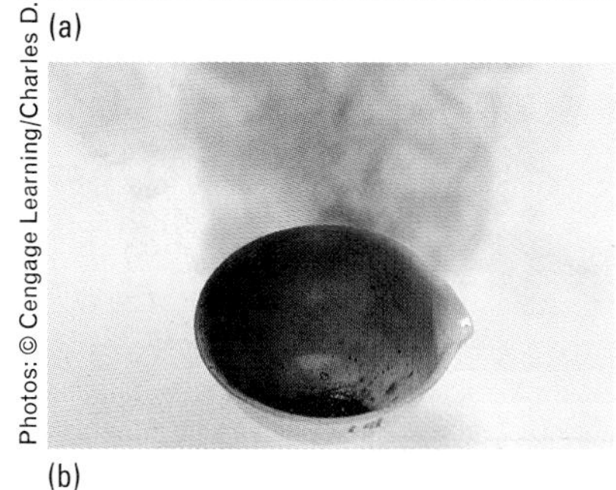

(b)

Photos: © Cengage Learning/Charles D. Winters

Figure 4.4 Combination of zinc and iodine. (a) The reactants: dark gray iodine crystals *(left)* and gray powdered zinc metal *(right).* (b) The reaction. In a vigorous combination reaction, zinc atoms react with diatomic iodine molecules to form zinc iodide, an ionic compound, and the heat of the reaction is great enough that excess iodine forms a purple vapor.

When nonmetals combine with oxygen or chlorine, the compounds formed are not ionic but are molecular, composed of *molecules.* For example, sulfur, the Group 6A neighbor of oxygen, combines with oxygen to form two oxides, SO_2 and SO_3, in reactions of great environmental and industrial significance.

$$S_8(s) + 8\,O_2(g) \longrightarrow \underset{\substack{\text{sulfur dioxide} \\ \text{(colorless; choking odor)}}}{8\,SO_2(g)}$$

$$2\,SO_2(g) + O_2(g) \longrightarrow \underset{\substack{\text{sulfur trioxide} \\ \text{(colorless; even more choking odor)}}}{2\,SO_3(g)}$$

At room temperature, sulfur is a bright yellow solid. At the nanoscale it consists of molecules with eight-membered rings, S_8.

Sulfur dioxide enters the atmosphere both from natural sources and from human activities. The eruption of Mount St. Helens in May 1980 injected millions of tons of SO_2 into the atmosphere, for example. But about 75% of the sulfur oxides in the atmosphere come from human activities, such as burning coal in power plants. All coal contains sulfur, usually from 1% to 4% by weight.

Yet another example of a combination reaction is that between an organic molecule such as ethylene, C_2H_4, and bromine, Br_2, to form dibromoethane:

$$C_2H_4(g) + Br_2(\ell) \longrightarrow C_2H_4Br_2(g)$$

EXERCISE 4.3 Combination Reactions

Indicate whether each equation for a combination reaction is balanced, and if it is not, why not.

(a) $Cu + O_2 \rightarrow CuO$ (b) $Cr + Br_2 \rightarrow CrBr_3$ (c) $S_8 + 3\,F_2 \rightarrow SF_6$

© Steve Terrill/Photolibrary

Mount St. Helens erupting.

CONCEPTUAL
EXERCISE 4.4 Combination Reactions

(a) What information is needed to predict the product of a combination reaction between two elements? (b) What specific information is needed to predict the product of a combination reaction between calcium and fluorine? (c) What is the product formed by this reaction?

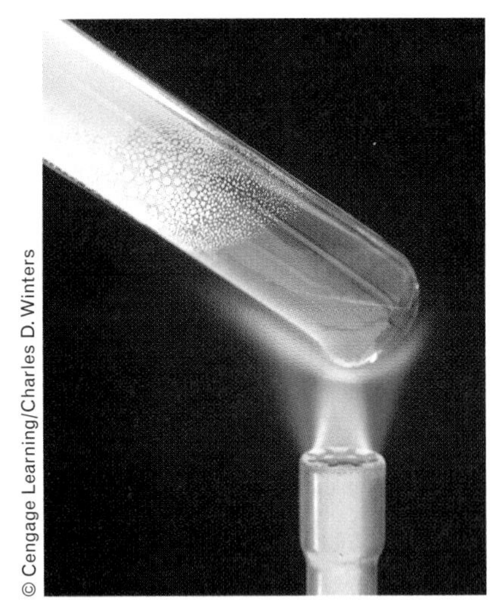

Decomposition of HgO. When heated, red mercury(II) oxide decomposes into liquid mercury and oxygen gas.

Decomposition Reactions

Decomposition reactions can be considered the opposite of combination reactions. In a **decomposition reaction**, one substance decomposes to form two or more products. The general reaction is

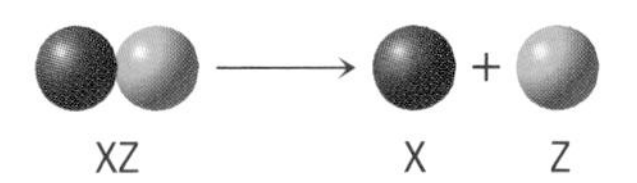

Many compounds that we would describe as "stable" because they exist without change under normal conditions of temperature and pressure undergo decomposition when the temperature is raised, a process known as *thermal decomposition*. For example, a few metal oxides decompose upon heating to give the metal and oxygen gas, the reverse of combination reactions. One of the best-known metal oxide decomposition reactions is the reaction by which Joseph Priestley discovered oxygen in 1774:

$$2\,HgO(s) \xrightarrow{\text{heat}} 2\,Hg(\ell) + O_2(g)$$

A very common and important type of decomposition reaction is illustrated by the chemistry of *metal carbonates* and calcium carbonate in particular. Many metal carbonates decompose to give metal oxides plus carbon dioxide:

$$\underset{\text{calcium carbonate}}{CaCO_3(s)} \xrightarrow{800-1000\ ^\circ C} \underset{\text{calcium oxide}}{CaO(s)} + CO_2(g)$$

Sea shells are composed largely of calcium carbonate.

Calcium is the fifth most abundant element in the earth's crust and the third most abundant metal (after Al and Fe). Naturally occurring calcium is mostly in the form of calcium carbonate from the fossilized remains of early marine life. Limestone, a form of calcium carbonate, is one of the basic raw materials of industry. Lime (calcium oxide), made by the decomposition reaction just discussed, is a raw material used for the manufacture of chemicals, in water treatment, and in the paper industry.

Some compounds are sufficiently unstable that their decomposition reactions are explosive. Nitroglycerin, $C_3H_5(NO_3)_3$, is such a compound. The formula for nitroglycerin contains parentheses around the NO_3 groups; they must be accounted for when balancing chemical equations. Nitroglycerin, a molecular organic compound, is very sensitive to vibrations and bumps and can decompose violently.

The NO_3 groups in nitroglycerin are not nitrate ions.

$$4\,C_3H_5(NO_3)_3(\ell) \longrightarrow 12\,CO_2(g) + 10\,H_2O(g) + 6\,N_2(g) + O_2(g)$$

Dynamite contains nitroglycerin.

Water, by contrast, is such a stable compound that it can be decomposed to hydrogen and oxygen only at a very high temperature or by using a direct electric current, a process called electrolysis (Figure 4.5).

$$2\,H_2O(\ell) \xrightarrow{\text{direct current}} 2\,H_2(g) + O_2(g)$$

PROBLEM-SOLVING EXAMPLE 4.1 Combination and Decomposition Reactions

Predict the reaction type and the formula of the missing species for each of these reactions:

(a) $2\,Fe(s) + 3$ ________ $(g) \longrightarrow 2\,FeCl_3(s)$

(b) $Cu(OH)_2(s) \longrightarrow CuO(s) +$ ________ (ℓ)

(c) $P_4(s) + 5\,O_2(g) \longrightarrow$ ________ (s)

(d) $CaSO_3(s) \longrightarrow$ ________ $(s) + SO_2(g)$

Answer

(a) Combination: Cl_2
(b) Decomposition: H_2O
(c) Combination: P_4O_{10}
(d) Decomposition: CaO

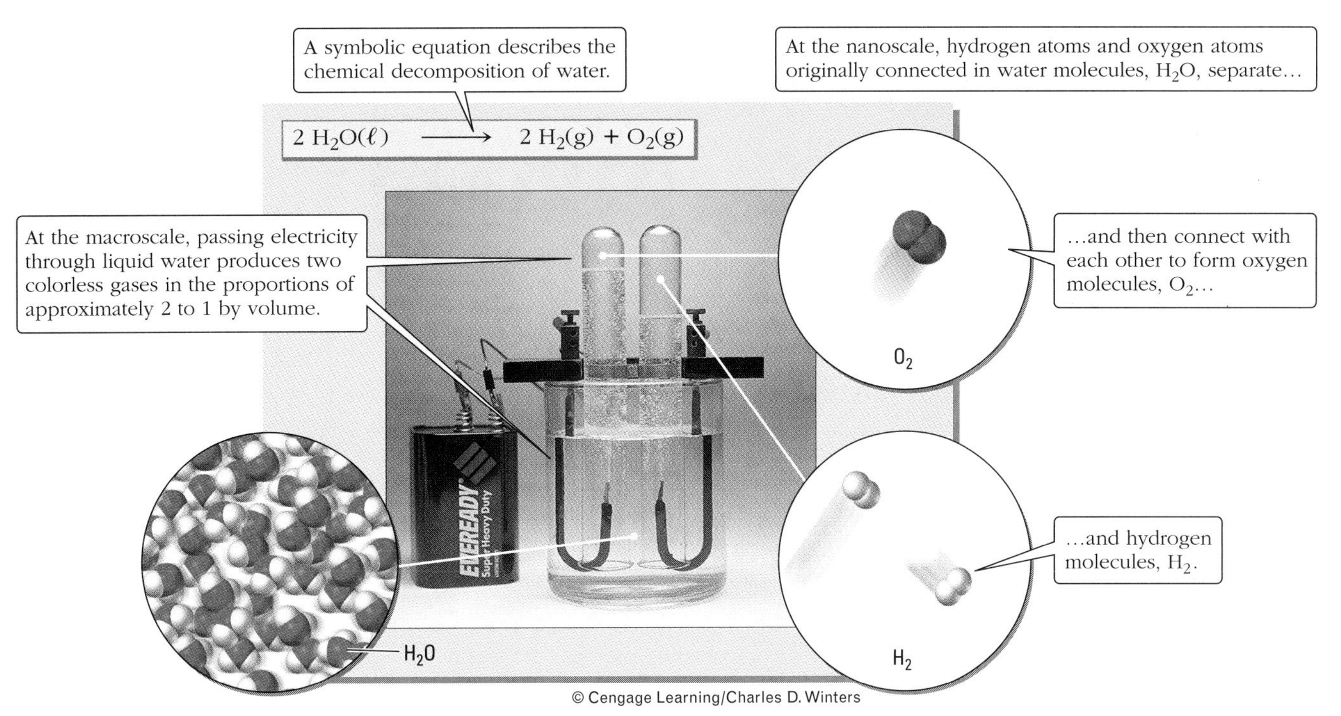

A symbolic equation describes the chemical decomposition of water.
At the nanoscale, hydrogen atoms and oxygen atoms originally connected in water molecules, H_2O, separate...
$2\ H_2O(\ell) \longrightarrow 2\ H_2(g) + O_2(g)$
At the macroscale, passing electricity through liquid water produces two colorless gases in the proportions of approximately 2 to 1 by volume.
...and then connect with each other to form oxygen molecules, O_2...
O_2
EVEREADY Super Heavy Duty
...and hydrogen molecules, H_2.
H_2O
H_2

© Cengage Learning/Charles D. Winters

Active Figure 4.5 **Decomposition of water.** A direct electric current decomposes water into gaseous hydrogen (H_2) and oxygen (O_2). Visit this book's companion website at **www.cengage.com/chemistry/moore** to test your understanding of the concepts in this figure.

Diatomic molecules.
(gasses) H_2 F_2 Cl_2 N_2 Br_2 O_2 I_2

Strategy and Explanation Use the fact that decomposition reactions have one reactant and combination reactions have one product. (a) Combination of Fe(s) and Cl_2(g) produces $FeCl_3$(s). (b) Decomposition of $Cu(OH)_2$(s) produces CuO(s) and $H_2O(\ell)$. (c) Combination of P_4(s) and O_2(g) produces P_4O_{10}(s). (d) Decomposition of $CaSO_3$(s) produces CaO(s) and SO_2(g).

PROBLEM-SOLVING PRACTICE 4.1

Predict the reaction type and the missing substance for each of these reactions:

(a) N_2 (g) + 2 O_2(g) ⟶ 2 NO_2(g)
(b) 4 Fe(s) + 3 O_2 (g) ⟶ 2 Fe_2O_3(s)
(c) 2 NaN_3(s) ⟶ 2 Na(s) + 3 N_2 (g)

EXERCISE 4.5 Combination and Decomposition Reactions

Predict the products formed by these reactions:

(a) Magnesium with chlorine
(b) The thermal decomposition of magnesium carbonate

The Granger Collection, NY.

Alfred Nobel 1833–1896

A Swedish chemist and engineer, Nobel discovered how to mix nitroglycerin (a liquid explosive that is extremely sensitive to light and heat) with diatomaceous earth to make dynamite, which could be handled and shipped safely. Nobel's talent as an entrepreneur combined with his many inventions (he held 355 patents) made him a very rich man. He never married and left his fortune to establish the Nobel Prizes, awarded annually to individuals who "have conferred the greatest benefits on mankind in the fields of physics, chemistry, physiology or medicine, literature and peace."

Displacement Reactions

Displacement reactions are those in which one element reacts with a compound to form a new compound and release a different element. The element released is said to have been displaced. The general equation for a **displacement reaction** is

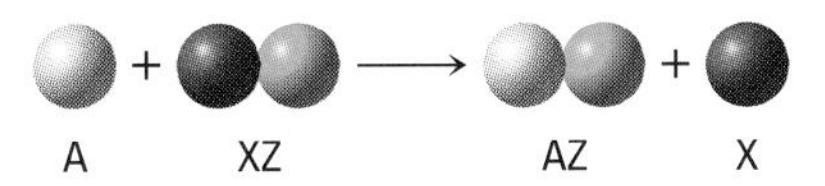

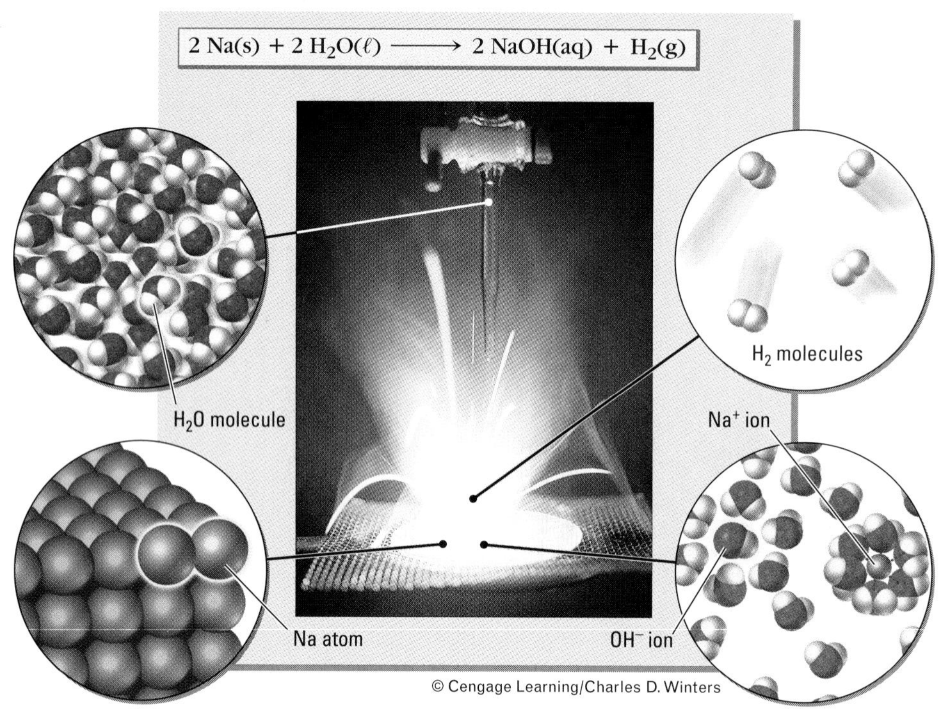

Figure 4.6 A displacement reaction. When liquid water drips from a buret onto a sample of solid sodium metal, the sodium displaces hydrogen gas from the water, and an aqueous solution of sodium hydroxide is formed. The hydrogen gas burns, producing the flame shown in the photograph. In the nanoscale pictures, the numbers of atoms, molecules, and ions that appear in the balanced equation are shown with yellow highlights.

The reaction of metallic sodium with water is such a reaction.

$$2\,Na(s) + 2\,H_2O(\ell) \longrightarrow 2\,NaOH(aq) + H_2(g)$$

If you think of water, H_2O, as H—O—H, it is easier to see that the Na displaces H from H_2O.

Here sodium displaces hydrogen from water (Figure 4.6). All the alkali metals, Group 1A, which are very reactive elements, react in this way when exposed to water.

Another example is the displacement reaction that occurs between metallic copper and an aqueous solution of silver nitrate.

$$Cu(s) + 2\,AgNO_3(aq) \longrightarrow Cu(NO_3)_2(aq) + 2\,Ag(s)$$

In this case, one metal displaces another. As you will see in Chapter 5, the metals can be arranged in a series from most reactive to least reactive (Section 5.5). This activity series can be used to predict the outcome of displacement reactions.

Exchange Reactions

Exchange reactions are also called metathesis or double-displacement reactions.

In an **exchange reaction**, there is an interchange of partners between two compounds. In general:

AD + XZ ⟶ AZ + XD

Mixing aqueous solutions of lead(II) nitrate and potassium chromate, for example, illustrates an exchange reaction in which an insoluble product is formed. The aqueous Pb^{2+} ions and K^+ ions exchange partners to form insoluble lead(II) chromate and water-soluble potassium nitrate:

$$\underset{\text{lead(II) nitrate}}{Pb(NO_3)_2(aq)} + \underset{\text{potassium chromate}}{K_2CrO_4(aq)} \longrightarrow \underset{\text{lead(II) chromate}}{PbCrO_4(s)} + \underset{\text{potassium nitrate}}{2\,KNO_3(aq)}$$

Such reactions (further discussed in Chapter 5) include several kinds of reactions that take place between reactants that are ionic compounds dissolved in water. They occur when reactant ions are removed from solution by the formation of one of three types of product: (1) an insoluble solid, (2) a molecular compound, or (3) a gas.

PROBLEM-SOLVING EXAMPLE 4.2 Classifying Reactions by Type

Classify each of these reactions as one of the four general types discussed in this section.

(a) $2\,Al(s) + 3\,Br_2(\ell) \longrightarrow Al_2Br_6(s)$
(b) $2\,K(s) + 2\,H_2O(\ell) \longrightarrow 2\,KOH(aq) + H_2(g)$
(c) $AgNO_3(aq) + NaCl(aq) \longrightarrow AgCl(s) + NaNO_3(aq)$
(d) $NH_4NO_3(s) \longrightarrow N_2O(g) + 2\,H_2O(g)$

Answer

(a) Combination (b) Displacement (c) Exchange (d) Decomposition

Strategy and Explanation Use the fact that in displacement reactions, one reactant is an element, and in exchange reactions, both reactants are compounds.

(a) With two reactants and a single product, this must be a combination reaction.
(b) The general equation for a displacement reaction, $A + XZ \longrightarrow AZ + X$, matches what occurs in the given reaction. Potassium (A) displaces hydrogen (X) from water (XZ) to form KOH (AZ) plus H_2 (X).
(c) The reactants exchange partners in this exchange reaction. Applying the general equation for an exchange reaction, $AD + XZ \longrightarrow AZ + XD$, to this case, we find A is Ag^+, D is NO_3^-, X is Na^+, and Z is Cl^-.
(d) In this reaction, a single substance, NH_4NO_3, decomposes to form two products, N_2O and H_2O.

PROBLEM-SOLVING PRACTICE 4.2

Classify each of these reactions as one of the four general reaction types described in this section.

(a) $2\,Al(OH)_3(s) \longrightarrow Al_2O_3(s) + 3\,H_2O(g)$
(b) $Na_2O(s) + H_2O(\ell) \longrightarrow 2\,NaOH(aq)$
(c) $S_8(s) + 24\,F_2(g) \longrightarrow 8\,SF_6(g)$
(d) $3\,NaOH(aq) + H_3PO_4(aq) \longrightarrow Na_3PO_4(aq) + 3\,H_2O(\ell)$
(e) $3\,C(s) + Fe_2O_3(s) \longrightarrow 3\,CO(g) + 2\,Fe(\ell)$

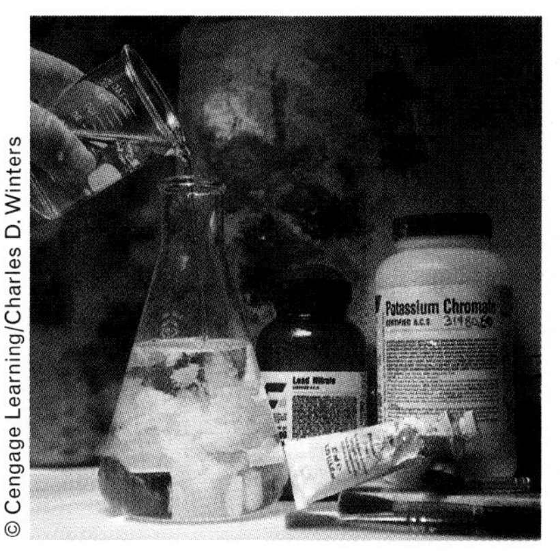

$Pb(NO_3)_2(aq) + K_2CrO_4(aq) \longrightarrow PbCrO_4(s) + 2\,KNO_3(aq)$. When lead(II) nitrate and potassium chromate solutions are mixed, a brilliant yellow precipitate of lead(II) chromate is formed.

4 types of rxns:
combination
Displacement
Exchange
Decomposition

4.3 Balancing Chemical Equations

Balancing a chemical equation means using coefficients so that the same number of atoms of each element appears on each side of the equation. We will begin with one of the general classes of reactions, the combination of reactants to produce a single product, to illustrate how to balance chemical equations by a largely trial-and-error process.

We will balance the equation for the formation of ammonia from nitrogen and hydrogen. Millions of tons of ammonia, NH_3, are manufactured worldwide annually by this reaction, using nitrogen extracted from air and hydrogen obtained from natural gas.

Step 1: Write an unbalanced equation containing the correct formulas of all reactants and products.

$$\text{(unbalanced equation)} \qquad N_2 + H_2 \longrightarrow NH_3$$

Clearly, both nitrogen and hydrogen are unbalanced. There are two nitrogen atoms on the left and only one on the right, and two hydrogen atoms on the left and three on the right.

Balancing an equation involves changing the coefficients, but the subscripts in the formulas *cannot* be changed. Changing subscripts would change the chemical species involved in the reaction. For example, in the ammonia synthesis, changing NH_3 to N_2H_2 would result in equal numbers of atoms of each kind on each side of the equation, but N_2H_2 is a completely different substance from NH_3, so the equation would be for a completely different reaction.

Step 2: ***Balance atoms of one of the elements.*** We start by using a coefficient of 2 on the right to balance the nitrogen atoms: 2 NH_3 indicates two ammonia molecules, each containing a nitrogen atom and three hydrogen atoms. On the right we now have two nitrogen atoms and six hydrogen atoms.

$$\text{(unbalanced equation)} \qquad N_2 + H_2 \longrightarrow 2\,NH_3$$

Step 3: ***Balance atoms of the remaining elements.*** To balance the six hydrogen atoms on the right, we use a coefficient of 3 for the H_2 on the left to furnish six hydrogen atoms.

$$\text{(balanced equation)} \qquad N_2 + 3\,H_2 \longrightarrow 2\,NH_3$$

Step 4: ***Verify that the number of atoms of each element is balanced.*** Do an atom count to check that the numbers of nitrogen and hydrogen atoms are the same on each side of the equation.

$$\begin{array}{lcccc} \text{(balanced equation)} & N_2 + & 3\,H_2 & \longrightarrow & 2\,NH_3 \\ \text{atom count:} & 2\,N + & (3 \times 2)\,H & = & 2\,N + (2 \times 3)\,H \\ & 2\,N + & 6\,H & = & 2\,N + 6\,H \end{array}$$

The physical states of the reactants and products are usually also included in the balanced equation. Thus, the final equation for ammonia formation is

$$N_2(g) + 3\,H_2(g) \longrightarrow 2\,NH_3(g)$$

PROBLEM-SOLVING EXAMPLE 4.3 Balancing a Chemical Equation

Ammonia gas reacts with oxygen gas to form gaseous nitrogen monoxide, NO, and water vapor at 1000 °C. Write the balanced equation for this reaction.

Answer $4\,NH_3(g) + 5\,O_2(g) \longrightarrow 4\,NO(g) + 6\,H_2O(g)$

Strategy and Explanation Use the stepwise approach to balancing chemical equations. Note that oxygen is in both products.

Step 1: *Write an unbalanced equation containing the correct formulas of all reactants and products.* The formula for ammonia is NH_3. The unbalanced equation is

$$\text{(unbalanced equation)} \qquad NH_3(g) + O_2(g) \longrightarrow NO(g) + H_2O(g)$$

Step 2: *Balance atoms of one of the elements.* Hydrogen is unbalanced since there are three hydrogen atoms on the left and two on the right. Whenever three and two atoms must be balanced, use coefficients to give six atoms on both sides of the equation. To do so, use a coefficient of 2 on the left and 3 on the right to have six hydrogens on each side.

$$\text{(unbalanced equation)} \qquad 2\,NH_3(g) + O_2(g) \longrightarrow NO(g) + 3\,H_2O(g)$$

Step 3: *Balance atoms of the remaining elements.* There are now two nitrogen atoms on the left and one on the right, so we balance nitrogen by using the coefficient 2 for the NO molecule.

$$\text{(unbalanced equation)} \qquad 2\,NH_3(g) + O_2(g) \longrightarrow 2\,NO(g) + 3\,H_2O(g)$$

Now there are two oxygen atoms on the left and five on the right. We use a coefficient of $\frac{5}{2}$ to balance the atoms of O_2.

$$\text{(balanced equation)} \qquad 2\,NH_3(g) + \tfrac{5}{2}O_2(g) \longrightarrow 2\,NO(g) + 3\,H_2O(g)$$

The equation is now balanced, but it is customary to use whole-number coefficients. Therefore, we multiply all the coefficients by 2 to get the final balanced equation.

$$\text{(balanced equation)} \qquad 4\,NH_3(g) + 5\,O_2(g) \longrightarrow 4\,NO(g) + 6\,H_2O(g)$$

Step 4: *Verify that the number of atoms of each element is balanced.*

(balanced equation)	$4\ NH_3(g)$	$+\ 5\ O_2(g)$	$\longrightarrow$	$4\ NO(g)$	$+\ 6\ H_2O(g)$
	4 N		=	4 N	
	12 H		=		12 H
		10 O	=	4 O	6 O

PROBLEM-SOLVING PRACTICE 4.3

Balance these equations:

(a) $Cr(s) + Cl_2(g) \longrightarrow CrCl_3(s)$ (b) $As_2O_3(s) + H_2(g) \longrightarrow As(s) + H_2O(\ell)$

We now turn to the combustion of propane, C_3H_8, to illustrate balancing a somewhat more complex chemical equation. We will assume that complete combustion occurs, meaning that the only products are carbon dioxide and water.

PROBLEM-SOLVING EXAMPLE 4.4 Balancing a Combustion Reaction Equation

Write a balanced equation for the complete combustion of propane, C_3H_8, the fuel used in gas grills.

Answer $C_3H_8(g) + 5\ O_2(g) \longrightarrow 3\ CO_2(g) + 4\ H_2O(\ell)$

Strategy and Explanation

Step 1: *Write an unbalanced equation containing the correct formulas of all reactants and products.* The initial equation is

$$\text{(unbalanced equation)} \quad C_3H_8 + O_2 \longrightarrow CO_2 + H_2O$$

Step 2: *Balance the atoms of one of the elements.* None of the elements are balanced, so we could start with C, H, or O. We will start with C because it appears in only one reactant and one product. The three carbon atoms in C_3H_8 will produce three CO_2 molecules.

$$\text{(unbalanced equation)} \quad C_3H_8 + O_2 \longrightarrow 3\ CO_2 + H_2O$$

Step 3: *Balance atoms of the remaining elements.* We next balance the H atoms. The eight H atoms in the reactants will combine with oxygen to produce four water molecules, each containing two H atoms.

$$\text{(unbalanced equation)} \quad C_3H_8 + O_2 \longrightarrow 3\ CO_2 + 4\ H_2O$$

Oxygen is the remaining element to balance. At this point, there are ten oxygen atoms in the products (3×2 in three CO_2 molecules, and 4×1 in four water molecules), but only two in O_2, a reactant. Therefore, O_2 in the reactants needs a coefficient of 5 to have ten oxygen atoms in the reactants.

$$C_3H_8(g) + 5\ O_2(g) \longrightarrow 3\ CO_2(g) + 4\ H_2O(\ell)$$

This combustion equation is now balanced.

Step 4: *Verify that the number of atoms of each element is balanced.*

$C_3H_8(g)$	$+\ 5\ O_2(g)$	$\longrightarrow$	$3\ CO_2(g)$		$+\ 4\ H_2O(\ell)$
3 C		=	3 C		
8 H		=			8 H
	10 O	=	6 O	+	4 O

It usually works best to first balance the element that appears in the fewest formulas; balance the element that appears in the most formulas last.

PROBLEM-SOLVING PRACTICE 4.4

Ethyl alcohol, C_2H_5OH, can be added to gasoline to create a cleaner-burning fuel. Write the balanced equation for:

(a) Complete combustion of ethyl alcohol to produce carbon dioxide and water

(b) Incomplete combustion of ethyl alcohol to produce carbon monoxide and water

When one or more polyatomic ions appear on both sides of a chemical equation, each one is treated as a whole during the balancing steps. When such an ion must have a subscript in the molecular formula, the polyatomic ion is enclosed in parentheses. For example, in the equation for the reaction between sodium phosphate and barium nitrate to produce barium phosphate and sodium nitrate,

$$2\ Na_3PO_4(aq) + 3\ Ba(NO_3)_2(aq) \longrightarrow Ba_3(PO_4)_2(s) + 6\ NaNO_3(aq)$$

the nitrate ions, NO_3^-, and phosphate ions, PO_4^{3-}, are kept together as units and are enclosed in parentheses when the polyatomic ion occurs more than once in a chemical formula.

4.4 The Mole and Chemical Reactions: The Macro–Nano Connection

When the molar mass—which links the number of atoms, molecules, or formula units with the mass of the atoms, molecules, or ionic compounds—is combined with a balanced chemical equation, the masses of the reactants and products can be calculated. In this way the nanoscale of chemical reactions is linked with the macroscale, at which we can measure masses of reactants and products by weighing.

We will explore these relationships using the combustion reaction between methane, CH_4, and oxygen, O_2, as an example (Figure 4.7). The balanced equation shows the number of molecules of the reactants, which are methane and oxygen, and of the products, which are carbon dioxide and water. The coefficients on each species can also be interpreted as the numbers of moles of each compound, and we can use the molar mass of each compound to calculate the mass of each reactant and product represented in the balanced equation.

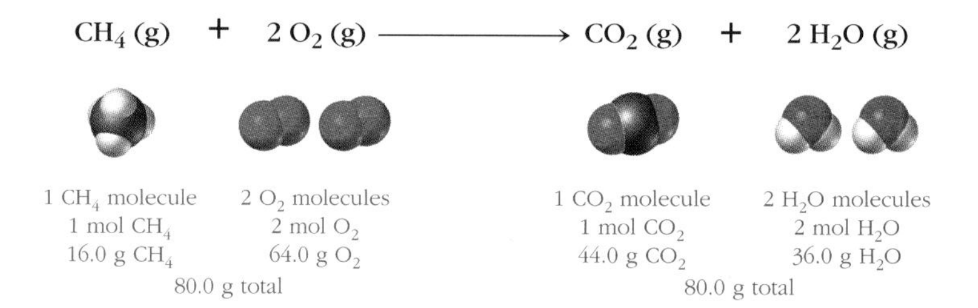

© Cengage Learning/Charles D. Winters

Figure 4.7 Combustion of methane with oxygen. Methane is the main component of natural gas, a primary fuel for industrial economies and the gas commonly used in laboratory Bunsen burners.

Notice that the total mass of reactants (16.0 g CH_4 + 64.0 g O_2 = 80.0 g reactants) equals the total mass of products (44.0 g CO_2 + 36.0 g H_2O = 80.0 g products), as must always be the case for a balanced equation.

The mole ratio is also known as the stoichiometric factor.

The stoichiometric coefficients in a balanced chemical equation provide the **mole ratios** that relate the numbers of moles of reactants and products to each other. These mole ratios are used in all quantitative calculations involving chemical reactions. Several of the mole ratios for the example equation are

$$\frac{2\ \text{mol}\ O_2}{1\ \text{mol}\ CH_4} \quad \text{or} \quad \frac{1\ \text{mol}\ CO_2}{1\ \text{mol}\ CH_4} \quad \text{or} \quad \frac{2\ \text{mol}\ H_2O}{2\ \text{mol}\ O_2}$$

These mole ratios tell us that 2 mol O_2 react with every 1 mol CH_4, or 1 mol CO_2 is formed for each 1 mol CH_4 that reacts, or 2 mol H_2O is formed for each 2 mol O_2 that reacts.

We can use the mole ratios in the balanced equation to calculate the molar amount of one reactant or product from the molar amount of another reactant or product. For example, we can calculate the number of moles of H_2O produced when 0.40 mol CH_4 is reacted fully with oxygen.

Moles of CH_4 ⟶ Moles of H_2O

$$0.40 \text{ mol } CH_4 \times \frac{2 \text{ mol } H_2O}{1 \text{ mol } CH_4} = 0.80 \text{ mol } H_2O$$

CONCEPTUAL **EXERCISE** **4.6 Mole Ratios**

Write all the possible mole ratios that can be obtained from the balanced equation for the reaction between Al and Br_2 to form Al_2Br_6.

The molar mass and the mole ratio, as illustrated in Figure 4.8, provide the links between masses and molar amounts of reactants and products. When the quantity of a reactant is given in grams, then we use its molar mass to convert to moles of reactant as a first step in using the balanced chemical equation. Then we use the balanced chemical equation to convert from moles of reactant to moles of product. Finally, we convert to grams of product if necessary by using its molar mass. Figure 4.8 illustrates this sequence of calculations.

PROBLEM-SOLVING EXAMPLE 4.5 Moles and Grams in Chemical Reactions

An iron ore named hematite, Fe_2O_3, can be reacted with carbon monoxide, CO, to form iron and carbon dioxide. How many moles and grams of iron are produced when 45.0 g hematite is reacted with sufficient CO?

Answer 0.564 mol and 31.5 g Fe

Strategy and Explanation Solving stoichiometry problems relies on the relationships illustrated in Figure 4.8 to connect masses and molar amounts of reactants or products. First, we need to write a balanced equation for the reaction.

$$Fe_2O_3(s) + 3\ CO(g) \longrightarrow 2\ Fe(s) + 3\ CO_2(g)$$

Now we can use the relationships connecting masses and moles of reactants or products, as illustrated in Figure 4.8. We know the mass of hematite, and to proceed we need to know the moles of hematite. The molar mass of hematite is 159.69 g/mol.

$$\text{Grams of hematite} \longrightarrow \text{moles of hematite}$$

$$\text{Amount of hematite reacted: } 45.0 \text{ g hematite} \times \frac{1 \text{ mol hematite}}{159.69 \text{ g hematite}} = 0.282 \text{ mol hematite}$$

We then use the mole ratio (stoichiometric factor) from the balanced reaction to convert moles of hematite reacted to moles of iron produced.

$$\text{Moles of hematite} \longrightarrow \text{moles of iron}$$

$$\text{Amount of iron formed: } 0.282 \text{ mol hematite} \times \frac{2 \text{ mol Fe}}{1 \text{ mol hematite}} = 0.564 \text{ mol Fe}$$

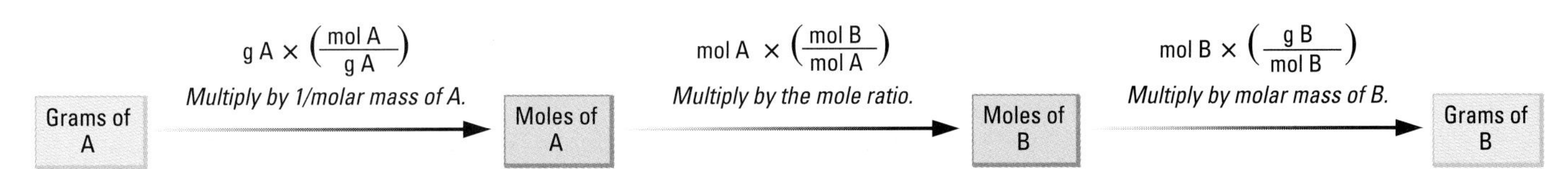

Figure 4.8 Stoichiometric relationships in a chemical reaction. The mass or molar amount of one reactant or product (A) is related to the mass or molar amount of another reactant or product (B) by the series of calculations shown.

We then multiply the number of moles of iron formed by the molar mass of iron to obtain the mass of iron formed.

$$0.564 \text{ mol Fe} \times \frac{55.845 \text{ g Fe}}{1 \text{ mol Fe}} = 31.5 \text{ g Fe}$$

☑ **Reasonable Answer Check** The balanced equation shows that for every one mole of hematite reacted, two moles of iron are produced. Therefore, approximately 0.3 mol hematite should produce twice that molar amount, or approximately 0.6 mol iron. The answer is reasonable.

PROBLEM-SOLVING PRACTICE 4.5

How many grams of carbon monoxide are required to react completely with 0.433 mol hematite?

We will illustrate the stoichiometric relationships of Figure 4.8 and a stepwise method for solving problems involving mass relations in chemical reactions to answer gram-to-gram conversion questions such as: How many grams of O_2 are needed to react completely with 5.0 g CH_4?

Step 1: ***Write the correct formulas for reactants and products and balance the chemical equation.*** The balanced equation is:

$$CH_4(g) + 2\,O_2(g) \longrightarrow CO_2(g) + 2\,H_2O(g)$$

Step 2: ***Decide what information about the problem is known and what is unknown. Map out a strategy for answering the question.*** In this example, you know the mass of CH_4 and you want to calculate the mass of O_2. You also know that you can use molar mass to convert mass of CH_4 to molar amount of CH_4. Then you can use the mole ratio from the balanced equation (2 mol O_2/1 mol CH_4) to calculate the moles of O_2 needed. Finally, you can use the molar mass of O_2 to convert the moles of O_2 to grams of O_2.

Step 3: ***Calculate moles from grams (if necessary).*** The known mass of CH_4 must be converted to molar amount because the coefficients of the balanced equation express mole relationships.

$$5.0 \text{ g } CH_4 \times \frac{1 \text{ mol } CH_4}{16.0 \text{ g } CH_4} = 0.313 \text{ mol } CH_4$$

In multistep calculations, remember to carry one additional significant figure in intermediate steps before rounding to the final value.

Step 4: ***Use the mole ratio to calculate the unknown number of moles, and then convert the number of moles to number of grams (if necessary).*** Calculate the number of moles and then grams of O_2.

$$0.313 \text{ mol } CH_4 \times \frac{2 \text{ mol } O_2}{1 \text{ mol } CH_4} = 0.626 \text{ mol } O_2$$

$$0.626 \text{ mol } O_2 \times \frac{32.0 \text{ g } O_2}{1 \text{ mol } O_2} = 20.\text{ g } O_2$$

Step 5: ***Check the answer to see whether it is reasonable.*** The starting mass of CH_4 of 5.0 g is about one third of a mole of CH_4. One third of a mole of CH_4 should react with two thirds of a mole of O_2 because the mole ratio is 1:2. The molar mass of O_2 is 32.0 g/mol, so two thirds of this amount is about 20. Therefore, the answer of 20. g O_2 is reasonable.

EXERCISE 4.7 Moles and Grams in Chemical Reactions

Verify that 10.8 g water is produced by the reaction of sufficient O_2 with 0.300 mol CH_4.

Problem-Solving Examples 4.6 and 4.7 illustrate further the application of the steps for solving problems involving mass relations in chemical reactions.

PROBLEM-SOLVING EXAMPLE 4.6 Moles and Grams in Chemical Reactions

When silicon dioxide, SiO_2, and carbon are reacted at high temperature, silicon carbide, SiC (also known as carborundum, an important industrial abrasive), and carbon monoxide, CO, are produced. Calculate the mass in grams of silicon carbide that will be formed by the complete reaction of 0.400 mol SiO_2 with sufficient carbon.

Answer 16.0 g SiC

Strategy and Explanation Use the stepwise approach to solve mass relations problems.

Step 1: *Write the correct formulas for reactants and products and balance the chemical equation.* SiO_2 and C are the reactants; SiC and CO are the products.

$$SiO_2(s) + 3\,C(s) \longrightarrow SiC(s) + 2\,CO(g)$$

Step 2: *Decide what information about the problem is known and what is unknown. Map out a strategy for answering the question.* We know how many moles of SiO_2 are available. Sufficient C is present, so there is enough C to react with all of the SiO_2. If we can calculate how many moles of SiC are formed, then the mass of SiC can be calculated.

Step 3: *Calculate moles from grams (if necessary).* We know there is 0.400 mol SiO_2.

Step 4: *Use the mole ratio to calculate the unknown number of moles, and then convert the number of moles to number of grams (if necessary).* Both steps are needed here to convert moles of SiO_2 to moles of SiC and then to grams of SiC. In a single setup the calculation is

$$0.400 \text{ mol } SiO_2 \times \frac{1 \text{ mol SiC}}{1 \text{ mol } SiO_2} \times \frac{40.10 \text{ g SiC}}{1 \text{ mol SiC}} = 16.0 \text{ g SiC}$$

Step 5: *Check the answer to see whether it is reasonable.* The answer, 16.0 g SiC, is reasonable because 1 mol SiO_2 would produce 1 mol SiC (40.10 g); therefore, four tenths of a mole of SiO_2 would produce four tenths of a mole of SiC (16.0 g).

☑ **Reasonable Answer Check** Step 5 verified that the answer is reasonable.

© Cengage Learning/Charles D. Winters

Silicon carbide, SiC. The grinding wheel *(left)* is coated with SiC. Naturally occurring silicon carbide *(right)* is also known as carborundum. It is one of the hardest substances known, making it valuable as an abrasive.

PROBLEM-SOLVING PRACTICE 4.6

Tin is extracted from its ore cassiterite, SnO_2, by reaction with carbon from coal.

$$SnO_2(s) + 2\,C(s) \rightarrow Sn(\ell) + 2\,CO(g)$$

(a) What mass of tin can be produced from 0.300 mol cassiterite?
(b) How many grams of carbon are required to produce this much tin?

PROBLEM-SOLVING EXAMPLE 4.7 Grams, Moles, and Grams

A popular candy bar contains 21.1 g sucrose (cane sugar), $C_{12}H_{22}O_{11}$. When the candy bar is eaten, the sucrose is metabolized according to the overall equation

$$\text{(unbalanced equation)} \quad C_{12}H_{22}O_{11}(s) + O_2(g) \longrightarrow CO_2(g) + H_2O(\ell)$$

Balance the chemical equation, and find the mass of O_2 consumed and the masses of CO_2 and H_2O produced by this reaction.

Answer $C_{12}H_{22}O_{11}(s) + 12\,O_2(g) \rightarrow 12\,CO_2(g) + 11\,H_2O(\ell)$
23.7 g O_2 is consumed; 32.6 g CO_2 and 12.2 g H_2O are produced.

Strategy and Explanation First, write the balanced chemical equation. Then use it to calculate the masses required. Coefficients of 12 and 11 for the CO_2 and H_2O products balance the C and H, giving 35 oxygen atoms in the products. In the reactants, these oxygen atoms are balanced by the 12 O_2 molecules plus the 11 oxygen atoms in sucrose.

$$C_{12}H_{22}O_{11}(s) + 12\ O_2(g) \longrightarrow 12\ CO_2(g) + 11\ H_2O(\ell)$$

Use sucrose's molar mass (342.3 g/mol) to convert grams of sucrose to moles of sucrose.

$$21.1\text{ g sucrose} \times \frac{1\text{ mol sucrose}}{342.3\text{ g sucrose}} = 0.06164\text{ mol sucrose}$$

Next, use the mole ratios and molar masses to calculate the masses of O_2, CO_2, and H_2O.

$$0.06164\text{ mol sucrose} \times \frac{12\text{ mol }O_2}{1\text{ mol sucrose}} \times \frac{31.99\text{ g }O_2}{1\text{ mol }O_2} = 23.7\text{ g }O_2$$

$$0.06164\text{ mol sucrose} \times \frac{12\text{ mol }CO_2}{1\text{ mol sucrose}} \times \frac{44.01\text{ g }CO_2}{1\text{ mol }CO_2} = 32.6\text{ g }CO_2$$

$$0.06164\text{ mol sucrose} \times \frac{11\text{ mol }H_2O}{1\text{ mol sucrose}} \times \frac{18.02\text{ g }H_2O}{1\text{ mol }H_2O} = 12.2\text{ g }H_2O$$

The mass of water could have been found by using the conservation of mass.

$$\text{Total mass of reactants} = \text{total mass of products}$$

$$\text{Total mass of reactants} = 21.1\text{ g sucrose} + 23.7\text{ g }O_2 = 44.8\text{ g}$$

$$\text{Total mass of products} = 32.6\text{ g }CO_2 + ?\text{ g }H_2O = 44.8\text{ g}$$

$$\text{Mass of }H_2O = 44.8\text{ g} - 32.6\text{ g} = 12.2\text{ g }H_2O$$

☑ **Reasonable Answer Check** These are reasonable answers because approximately 0.05 mol sucrose would require approximately 0.6 mol O_2 and would produce approximately 0.6 mol CO_2 and 0.5 mol H_2O. Therefore, the calculated masses should be somewhat smaller than the molar masses, and they are.

PROBLEM-SOLVING PRACTICE 4.7

A lump of coke (carbon) weighs 57 g.
(a) What mass of oxygen is required to burn it to carbon monoxide?
(b) How many grams of CO are produced?

An example of analytical chemistry is the quantitative determination of arsenic levels in drinking water in Bangladesh as described in Section 1.2.

To this point we have used the methods of stoichiometry to compute the quantity of products given the quantity of reactants. Now we turn to the reverse problem: Given the quantity of products, what quantitative information can we deduce about the reactants? Questions such as these are often confronted by *analytical chemistry,* a field in which chemists creatively identify pure substances and measure the quantities of components of mixtures. Although analytical chemistry is now largely done by instrumental methods, classic chemical reactions and stoichiometry still play a central role. The analysis of mixtures is often challenging. It can take a great deal of imagination to figure out how to use chemistry to determine what, and how much, is there.

PROBLEM-SOLVING EXAMPLE 4.8 Evaluating an Ore

The mass percent of titanium dioxide, TiO_2, in an ore can be evaluated by carrying out the reaction of the ore with bromine trifluoride and measuring the mass of oxygen gas evolved.

$$3\ TiO_2(s) + 4\ BrF_3(\ell) \longrightarrow 3\ TiF_4(s) + 2\ Br_2(\ell) + 3\ O_2(g)$$

If 2.376 g of a TiO_2-containing ore generates 0.143 g O_2, what is the mass percent of TiO_2 in the ore sample?

Answer 15.0% TiO_2

Strategy and Explanation The mass percent of TiO_2 is

$$\text{Mass percent } TiO_2 = \frac{\text{mass of } TiO_2}{\text{mass of ore sample}} \times 100\%$$

The mass of the sample is given, and the mass of TiO_2 can be determined from the known mass of oxygen and the balanced equation.

$$0.143 \text{ g } O_2 \times \frac{1 \text{ mol } O_2}{32.00 \text{ g } O_2} \times \frac{3 \text{ mol } TiO_2}{3 \text{ mol } O_2} \times \frac{79.88 \text{ g } TiO_2}{1 \text{ mol } TiO_2} = 0.3569 \text{ g } TiO_2$$

The mass percent of TiO_2 is

$$\frac{0.3569 \text{ g } TiO_2}{2.376 \text{ g sample}} \times 100\% = 15.0\%$$

Titanium dioxide is a valuable commercial product. It is so widely used in paints and pigments that an ore with only 15% TiO_2 can be mined profitably.

☑ **Reasonable Answer Check** According to the balanced equation, 2.4 g TiO_2 should produce $2.4 \times \frac{(3 \times 32)}{(3 \times 80)} \approx 1$ g of O_2 but this reaction actually produced about 0.14 g, which is close to 15% of the expected amount. The answer is reasonable.

PROBLEM-SOLVING PRACTICE 4.8

The purity of magnesium metal can be determined by reacting the metal with sufficient hydrochloric acid to form $MgCl_2$, evaporating the water from the resulting solution, and weighing the solid $MgCl_2$ formed.

$$Mg(s) + 2\,HCl(aq) \longrightarrow MgCl_2(aq) + H_2(g)$$

Calculate the percentage of magnesium in a 1.72-g sample that produced 6.46 g $MgCl_2$ when reacted with excess HCl.

4.5 Reactions with One Reactant in Limited Supply

In the previous section, we assumed that exactly stoichiometric amounts of reactants were present; each reactant was entirely consumed when the reaction was over. However, this is rarely the case when chemists carry out an actual synthesis, whether for small quantities in a laboratory or on a large scale in an industrial process. Usually, one reactant is more expensive or less readily available than others. The cheaper or more available reactants are used in excess to ensure that the more expensive material is completely converted to product.

The industrial production of methanol, CH_3OH, is such a case. Methanol, an important industrial product, is manufactured by the reaction of carbon monoxide and hydrogen.

$$CO(g) + 2\,H_2(g) \longrightarrow CH_3OH(\ell)$$

 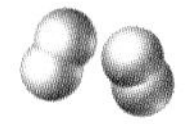 

Carbon monoxide is manufactured cheaply by burning coke (which is mostly carbon) in a limited supply of air so that there is insufficient oxygen to form carbon dioxide. Hydrogen is more expensive to manufacture. Therefore, methanol synthesis uses an excess of carbon monoxide, and the quantity of methanol produced is dictated by the quantity of hydrogen available. In this reaction, hydrogen acts as the limiting reactant.

A **limiting reactant** is the reactant that is completely converted to products during a reaction. Once the limiting reactant has been used up, no more product can form.

ESTIMATION How Much CO_2 Is Produced by Your Car?

Your car burns gasoline in a combustion reaction and produces water and carbon dioxide, CO_2, one of the major greenhouse gases, which is involved in global warming (discussed in detail in Section 10.9). For each gallon of gasoline that you burn in your car, how much CO_2 is produced? How much CO_2 is produced by your car per year?

To proceed with the estimation, we need to write a balanced chemical equation with the stoichiometric relationship between the reactant, gasoline, and the product of interest, CO_2. To write the chemical equation we need to make an assumption about the composition of gasoline. We will assume that the gasoline is octane, C_8H_{18}, so the reaction of interest is

$$2\, C_8H_{18} + 25\, O_2 \longrightarrow 16\, CO_2 + 18\, H_2O$$

One gallon equals 4 quarts, which equals

$$4 \text{ qt} \times \frac{1 \text{ L}}{1.057 \text{ qt}} = 3.78 \text{ L}$$

Gasoline floats on water, so its density must be less than that of water. Assume it is 0.80 g/mL, so

$$3.78 \text{ L} \times 0.80 \text{ g/mL} \times 10^3 \text{ mL/L} = 3.02 \times 10^3 \text{ g}$$

We now convert the grams of octane to moles using the molar mass of octane.

$$3020 \text{ g octane} \times \frac{1 \text{ mol}}{114.2 \text{ g}} = 26.4 \text{ mol octane}$$

To convert to moles of CO_2 we use the balanced equation, which shows that for every mole of octane consumed, eight moles of CO_2 are produced; thus, we have $26.4 \times 8 = 211$ mol CO_2. The molar mass of CO_2 is 44 g/mol, so 211 mol $\times$ 44 g/mol = 9280 g CO_2. Thus, for every gallon of gasoline burned, 9.3 kg (20.5 lb) CO_2 is produced.

If you drive your car 10,000 miles per year and get an average of 25 miles per gallon, you use about 400 gallons of gasoline per year. Burning this quantity of gasoline produces 400×9.3 kg = 3720 kg CO_2. That's 3720 kg $\times$ 2.2 lb/kg = 8200 lb, or more than 4 tons CO_2.

How much CO_2 is that? The 3720 kg CO_2 is about 85,000 mol CO_2. At room temperature and atmospheric pressure, that's about 2,080,000 L CO_2 or 2080 m^3 CO_2—enough to fill about 4000 1-m diameter balloons, or 11 such balloons each day of the year. In Section 10.9 we will discuss the effect that CO_2 is having on the earth's atmosphere and its link to global warming.

Visit this book's companion website at **www.cengage.com/chemistry/moore** to work an interactive module based on this material.

The limiting reactant must be used as the basis for calculating the maximum possible amount of product(s) because the limiting reactant limits the amount of product(s) that can be formed. ***The moles of product(s) formed are always determined by the starting number of moles of the limiting reactant.***

We can make an analogy to a chemistry "limiting reactant" in the assembling of cheese sandwiches. Each sandwich must have 2 slices of bread, 1 slice of cheese, and 1 lettuce leaf. Suppose we have available 40 slices of bread, 25 slices of cheese, and 30 lettuce leaves. How many complete cheese sandwiches can we assemble?

Each sandwich must have the ratio of 2 bread:1 cheese:1 lettuce (analogous to coefficients in a balanced chemical equation):

$$2 \text{ bread slices} + 1 \text{ cheese slice} + 1 \text{ lettuce leaf} = 1 \text{ cheese sandwich}$$

We have enough bread slices to make 20 sandwiches, enough cheese slices to make 25 sandwiches, and enough lettuce leaves to make 30 sandwiches. Thus, using the quantities of ingredients on hand and the 2:1 requirement for bread and cheese, we can assemble only 20 complete sandwiches. At that point, we run out of bread even though there are unused cheese slices and lettuce leaves. Thus, bread slices are the "limiting reactant" because they limit the number of sandwiches that can be made. The overall yield of sandwiches is limited by the bread slices, the "limiting reactant." Overall, the 20 sandwiches contain a total of 40 bread slices, 20 cheese slices, and 20 lettuce leaves. Five cheese slices and 10 lettuce leaves are unused, that is, in excess.

In determining the maximum number of cheese sandwiches that could be assembled, the "limiting reactant" was the bread slices. We ran out of bread slices before using up all the available cheese or lettuce. Similarly, the limiting reactant must be identified in a chemical reaction to determine how much product(s) will be produced if all the limiting reactant is converted to the desired product(s).

If we know which one of a set of reactants is the limiting reactant, we can use that information to solve a quantitative problem directly, as illustrated in Problem-Solving Example 4.9.

PROBLEM-SOLVING EXAMPLE 4.9 Moles of Product from Limiting Reactant

The organic compound urea, $(NH_2)_2CO$, can be prepared with this reaction between ammonia and carbon dioxide:

$$2\,NH_3(g) + CO_2(g) \longrightarrow (NH_2)_2CO(aq) + H_2O(\ell)$$

If 2.0 mol ammonia and 2.0 mol carbon dioxide are mixed, how many moles of urea are produced? Ammonia is the limiting reactant.

urea

You should be able to explain why NH_3 is the limiting reactant.

Answer 1.0 mol $(NH_2)_2CO$

Strategy and Explanation Start with the balanced equation and consider the stoichiometric coefficients. Concentrate on the NH_3 since it is given as the limiting reactant. The coefficients show that for every 2 mol NH_3 reacted, 1 mol $(NH_2)_2CO$ will be produced. Thus, we use this information to answer the question

$$2.0 \text{ mol } NH_3 \times \frac{1 \text{ mol } (NH_2)_2CO}{2 \text{ mol } NH_3} = 1.0 \text{ mol } (NH_2)_2CO$$

☑ **Reasonable Answer Check** The balanced equation shows that the number of moles of $(NH_2)_2CO$ produced must be one half the number of moles of NH_3 that reacted, and it is.

PROBLEM-SOLVING PRACTICE 4.9

If we reacted 2.0 mol NH_3 with 0.75 mol CO_2 (and CO_2 is now the limiting reactant), how many moles of $(NH_2)_2CO$ would be produced?

Next, we consider the case where the quantities of reactants are given, but the limiting reactant is not identified and therefore must be determined. There are two approaches to identifying the limiting reactant—the mole ratio method and the mass method. You can rely on whichever method works better for you.

Mole Ratio Method

In this approach, we start by calculating the number of moles of each reactant available. We then compare their mole ratio with the mole ratio from the stoichiometric coefficients of the balanced equation. From this comparison, we can figure out which reactant is limiting.

Mass Method

In this approach, we start by calculating the mass of product that would be produced from the available quantity of each reactant, assuming that an unlimited quantity of the other reactant were available. The limiting reactant is the one that produces the smaller mass of product.

Problem-Solving Example 4.10 illustrates these two methods for identifying a limiting reactant.

PROBLEM-SOLVING EXAMPLE 4.10 Limiting Reactant

Cisplatin is an anticancer drug used for treatment of solid tumors. It can be produced by reacting ammonia with potassium tetrachloroplatinate (ktcp).

$$\underset{\text{potassium tetrachloroplatinate}}{K_2PtCl_4(s)} + \underset{\text{ammonia}}{2\ NH_3(aq)} \longrightarrow \underset{\text{cisplatin}}{Pt(NH_3)_2Cl_2(s)} + 2\ KCl(aq)$$

If the reaction starts with 5.00 g ammonia and 50.0 g ktcp, (a) which is the limiting reactant? (b) How many grams of cisplatin are produced? Assume that all the limiting reactant is converted to cisplatin.

Answer

(a) Potassium tetrachloroplatinate (b) 36.0 g cisplatin

Strategy and Explanation

Method 1 (Mole Ratio Method) We start by finding the number of moles of each reactant that is available. We first calculate the molar mass of ktcp: 2(39.098) + (195.078) + 4(35.453) = 415.09 g/mol.

$$50.0\text{ g ktcp} \times \frac{1\text{ mol ktcp}}{415.09\text{ g ktcp}} = 0.120\text{ mol ktcp}$$

$$5.00\text{ g NH}_3 \times \frac{1\text{ mol NH}_3}{17.0\text{ g NH}_3} = 0.294\text{ mol NH}_3$$

From this information we can deduce which reactant is limiting. Looking at the stoichiometric coefficients of the balanced equation, we see that for every two moles of ammonia reacted there is one mole of ktcp reacted.

$$\text{Mole ratio} = \frac{2\text{ mol ammonia}}{1\text{ mol ktcp}}$$

There are two possibilities to consider. (a) Ammonia is limiting, which means that 0.294 mol ammonia would require half that number of moles of ktcp, or 0.147 mol ktcp. There is not this much ktcp available, so ktcp must be the limiting reactant. (b) Ktcp is limiting, which means that 0.120 mol ktcp would require twice as many moles of ammonia, or 0.240 mol ammonia. There is 0.294 mol ammonia available, which is more than enough; ammonia is in excess. Therefore, ktcp is the limiting reactant, and the quantity of cisplatin that can be produced by the reaction must be calculated based on the molar amount of the limiting reactant, 0.120 mol ktcp. We first calculate the molar mass of cisplatin: (195.078) + 2(14.0067) + 6(1.0079) + 2(35.453) = 300.04 g/mol. We then use the mole ratio to determine the mass of cisplatin that could be produced.

$$0.120\text{ mol ktcp} \times \frac{1\text{ mol cisplatin}}{1\text{ mol ktcp}} \times \frac{300.04\text{ g cisplatin}}{1\text{ mol cisplatin}} = 36.0\text{ g cisplatin}$$

Method 2 (Mass Method) We first calculate the mass of cisplatin that would be produced from 0.120 mol ktcp and sufficient ammonia. Alternatively, we calculate the mass of cisplatin formed from 0.294 mol ammonia and assume sufficient ktcp.

Mass of cisplatin produced from 0.120 mol ktcp and sufficient ammonia:

The mass method directly gives the maximum mass of product.

$$0.120\text{ mol ktcp} \times \frac{1\text{ mol cisplatin}}{1\text{ mol ktcp}} \times \frac{300.04\text{ g cisplatin}}{1\text{ mol cisplatin}} = 36.0\text{ g cisplatin}$$

Mass of cisplatin produced from 0.294 mol ammonia and sufficient ktcp:

$$0.294\text{ mol NH}_3 \times \frac{1\text{ mol cisplatin}}{2\text{ mol NH}_3} \times \frac{300.04\text{ g cisplatin}}{1\text{ mol cisplatin}} = 44.1\text{ g cisplatin}$$

This comparison shows that the amount of ktcp available would produce less cisplatin than that from the amount of ammonia available, providing proof that ktcp is the limiting reactant.

When cisplatin is produced industrially, ammonia is always provided in excess because it is much cheaper than potassium tetrachloroplatinate.

☑ **Reasonable Answer Check** The ratio of molar masses of cisplatin and ktcp is about three quarters (300/415), so we should have approximately three quarters as much cisplatin product as ktcp reactant (36/50), and we do.

PROBLEM-SOLVING PRACTICE 4.10

Carbon disulfide reacts with oxygen to form carbon dioxide and sulfur dioxide.

$$CS_2(\ell) + O_2(g) \longrightarrow CO_2(g) + SO_2(g)$$

A mixture of 3.5 g CS_2 and 17.5 g O_2 is reacted.
(a) Balance the equation.
(b) What is the limiting reactant?
(c) What is the maximum mass of sulfur dioxide that can be formed?

PROBLEM-SOLVING EXAMPLE 4.11 Limiting Reactant

Powdered aluminum can react with iron(III) oxide in the *thermite reaction* to form molten iron and aluminum oxide:

$$2\ Al(s) + Fe_2O_3(s) \longrightarrow 2\ Fe(\ell) + Al_2O_3(s)$$

Liquid iron is produced because the reaction releases so much energy that the temperature is very high. This liquid iron can be used to weld steel railroad rails. (a) What is the limiting reactant when a mixture of 100. g Al and 100. g Fe_2O_3 react? (b) What is the mass of liquid iron formed? (c) How many grams of the excess reactant remain after the reaction is complete?

Answer
(a) Fe_2O_3 (b) 69.9 g Fe (c) 66.4 g Al

Strategy and Explanation
(a) We begin by determining how many moles of each reactant are available.

$$100.\ g\ Al \times \frac{1\ mol\ Al}{26.98\ g\ Al} = 3.71\ mol\ Al$$

$$100.\ g\ Fe_2O_3 \times \frac{1\ mol\ Fe_2O_3}{159.69\ g\ Fe_2O_3} = 0.626\ mol\ Fe_2O_3$$

Next, using the mass method for this limiting reactant problem, we find the masses of iron produced, based on the available masses of each reactant.

$$3.71\ mol\ Al \times \frac{2\ mol\ Fe}{2\ mol\ Al} \times \frac{55.845\ g\ Fe}{1\ mol\ Fe} = 207.\ g\ Fe$$

$$0.626\ mol\ Fe_2O_3 \times \frac{2\ mol\ Fe}{1\ mol\ Fe_2O_3} \times \frac{55.845\ g\ Fe}{1\ mol\ Fe} = 69.9\ g\ Fe$$

Clearly, the mass of iron that can be formed using the given masses of aluminum and iron(III) oxide is controlled by the quantity of the iron(III) oxide. The iron(III) oxide is the limiting reactant because it produces less iron. Aluminum is in excess.
(b) The mass of iron formed is 69.9 g Fe.
(c) We can find the number of moles of Al that reacted from the moles of Fe_2O_3 used and the mole ratio of Al to Fe_2O_3.

$$0.626\ mol\ Fe_2O_3 \times \frac{2\ mol\ Al}{1\ mol\ Fe_2O_3} = 1.25\ mol\ Al$$

By subtracting 1.25 mol Al from the initial amount of Al (3.71 mol − 1.25 mol = 2.46 mol Al), we find that 2.46 mol Al remains unreacted. Therefore, the mass of unreacted Al is

$$2.46\ mol\ Al \times \frac{26.98\ g\ Al}{1\ mol\ Al} = 66.4\ g\ Al$$

Powdered aluminum reacts with iron(III) oxide extremely vigorously in the thermite reaction to form molten iron and aluminum oxide.

It is useful, although not necessary, to calculate the quantity of excess reactant remaining to verify that the reactant in excess is not the limiting reactant.

☑ **Reasonable Answer Check** The molar masses of Fe_2O_3 (160 g/mol) and Fe (56 g/mol) are in the ratio of approximately 3:1. One mole of Fe_2O_3 produces 2 mol Fe according to the balanced equation. So a given mass of Fe_2O_3 should produce about two thirds as much Fe, and this agrees with our more exact calculation.

PROBLEM-SOLVING PRACTICE 4.11

Preparation of the pure silicon used in silicon chips involves the reaction between purified liquid silicon tetrachloride and magnesium.

$$SiCl_4(\ell) + 2\,Mg(s) \longrightarrow Si(s) + 2\,MgCl_2(s)$$

If the reaction were run with 100. g each of $SiCl_4$ and Mg, which reactant would be limiting, and what mass of Si would be produced?

EXERCISE 4.8 Limiting Reactant

Urea is used as a fertilizer because it can react with water to release ammonia, which provides nitrogen to plants.

$$(NH_2)_2CO(s) + H_2O(\ell) \longrightarrow 2\,NH_3(aq) + CO_2(g)$$

(a) Determine the limiting reactant when 300. g urea and 100. g water are combined.
(b) How many grams of ammonia and how many grams of carbon dioxide form?
(c) What mass of the excess reactant remains after reaction?

CHEMISTRY IN THE NEWS

Smothering Fire—Water That Isn't Wet

Fire is a combustion reaction in which fuel and oxygen, O_2, combine, usually at high temperatures, to form water and carbon dioxide. Three factors are necessary for a fire: combustible fuel, oxygen, and a temperature above the ignition temperature of the fuel. Once the fire has started, it is self-supporting because it supplies the heat necessary to keep the temperature high (if sufficient fuel and oxygen are available). Quenching a fire requires removing the fuel, lowering the oxygen level, cooling the reaction mixture below the ignition temperature, or some combination of these.

An effective way to quench a fire is smothering, which reduces the amount of available oxygen below the level needed to support combustion. In other words, smothering decreases the amount of the limiting reactant. Foams, inert gas, and CO_2 are effective substances for smothering.

Developed by 3M, Novec 1230 is a new compound with very desirable fire-suppression properties. An organic compound with many carbon-fluorine bonds, it is a colorless liquid at room temperature (B.P. 49.2 °C) that feels like water. When placed on a fire, Novec vaporizes and smothers the fire. Novec 1230 is not an electrical conductor, so it can be used for electrical fires. In fact, in a demonstration on television, a laptop computer was immersed in Novec 1230 and continued to work. The compound is used in situations where water cannot be used, for example, clean rooms, hospitals, or museums. The new compound also does not affect the stratospheric ozone layer because its lifetime in the lower atmosphere is short, making it environmentally friendly.

$$CF_3{-}CF_2{-}\overset{\overset{\Large O}{\|}}{C}{-}\underset{\underset{\Large F}{|}}{\overset{\overset{\Large CF_3}{|}}{C}}{-}CF_3$$

Structural formula for Novec 1230.

Sources: New York Times, December 12, 2004; p. 103. http://solutions.3m.com/wps/portal/3M/en_US/Novec/Home/Product_Information/Fire_Protection/?WT.mc_id=www.3m.com/novec1230 http://en.wikipedia.org/wiki/Novec_1230

4.6 Evaluating the Success of a Synthesis: Percent Yield

A reaction that forms the maximum possible quantity of product is said to have a 100% yield. The percentage is based on the amount of limiting reactant used. This maximum possible quantity of product is called the **theoretical yield**. Often the **actual yield,** the quantity of desired product actually obtained from a synthesis in a laboratory or industrial chemical plant, is less than the theoretical yield.

The efficiency of a particular synthesis method is evaluated by calculating the **percent yield,** which is defined as

$$\text{Percent yield} = \frac{\text{actual yield}}{\text{theoretical yield}} \times 100\%$$

Percent yield can be applied, for example, to the synthesis of aspirin. Suppose a student carried out the synthesis and obtained 2.2 g aspirin rather than the theoretical yield of 2.6 g. What is the percent yield of this reaction?

$$\text{Percent yield} = \frac{\text{actual yield of product}}{\text{theoretical yield of product}} \times 100\% = \frac{2.2\text{ g}}{2.6\text{ g}} \times 100\% = 85\%$$

Although we hope to obtain as close to the theoretical yield as possible when carrying out a reaction, few reactions or experimental manipulations are 100% efficient, despite controlled conditions and careful laboratory techniques. Side reactions can occur that form products other than the desired one, and during the isolation and purification of the desired product, some of it may be lost. When chemists report the synthesis of a new compound or the development of a new synthesis, they also report the percent yield of the reaction or the overall series of reactions. Other chemists who wish to repeat the synthesis then have an idea of how much product can be expected from a certain amount of reactants.

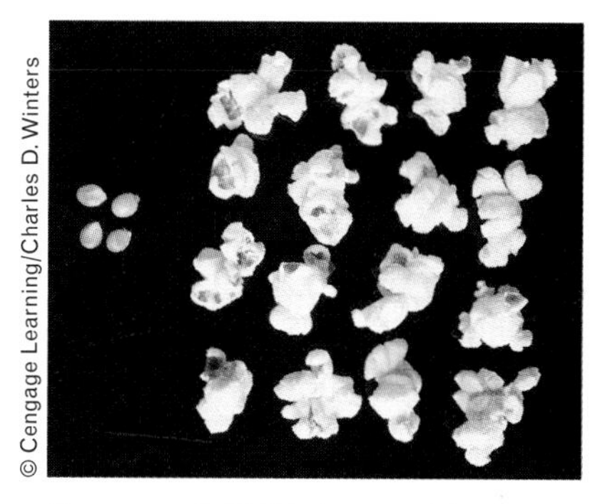

Popcorn yield. We began with 20 popcorn kernels, but only 16 of them popped. The percent yield of popcorn was (16/20) × 100% = 80%.

PROBLEM-SOLVING EXAMPLE 4.12 Calculating Percent Yield

Methanol, CH_3OH, is an excellent fuel, and it can be produced from carbon monoxide and hydrogen.

$$CO(g) + 2\,H_2(g) \longrightarrow CH_3OH(\ell)$$

If 500. g CO reacts with excess H_2 and 485. g CH_3OH is produced, what is the percent yield of the reaction?

Answer 85.0%

Strategy and Explanation To solve the problem we need to calculate the theoretical yield of CH_3OH. We start by calculating the number of moles of CO, the limiting reactant, that react.

$$500.\text{ g CO} \times \frac{1\text{ mol CO}}{28.0\text{ g CO}} = 17.86\text{ mol CO}$$

The coefficients of the balanced equation show that for every 1 mol CO reacted, 1 mol CH_3OH will be produced. Therefore, the maximum number of moles CH_3OH produced will be 17.86 mol CH_3OH. We convert this to the mass of CH_3OH, the theoretical yield:

$$17.86\text{ mol } CH_3OH \times \frac{32.0\text{ g } CH_3OH}{1\text{ mol } CH_3OH} = 571.\text{ g } CH_3OH$$

Thus, the theoretical yield is 571. g CH_3OH. The problem states that 485. g CH_3OH was produced. We calculate the percent yield:

$$\frac{485.\text{ g } CH_3OH \text{ (actual yield)}}{571.\text{ g } CH_3OH \text{ (theoretical yield)}} \times 100\% = 85.0\%$$

☑ **Reasonable Answer Check** The molar mass of CH_3OH is slightly greater than that of CO, and an equal number of moles of CO and CH_3OH are in the balanced equation. So *x* g CO should produce somewhat more than *x* g CH_3OH. Slightly less is actually produced, however, so a percent yield slightly less than 100% is about right.

PROBLEM-SOLVING PRACTICE 4.12

If the methanol synthesis reaction is run with an 85% yield, and you want to make 1.0 kg CH_3OH, how many grams of H_2 should you use if CO is available in excess?

PROBLEM-SOLVING EXAMPLE 4.13 Percent Yield

Ammonia can be produced from the reaction of a metal oxide such as calcium oxide with ammonium chloride:

$$CaO(s) + 2\ NH_4Cl(s) \rightarrow 2\ NH_3(g) + H_2O(g) + CaCl_2(s)$$

How many grams of calcium oxide would be needed to react with excess ammonium chloride to produce 1.00 g ammonia if the expected percent yield were 25%?

Answer 6.6 g CaO

Strategy and Explanation The expected percent yield expressed as a decimal is 0.25. So the theoretical yield is

$$\text{Theoretical yield} = \frac{\text{actual yield}}{\text{percent yield}} = \frac{1.00\text{ g } NH_3}{0.25} = 4.00\text{ g } NH_3$$

Ammonium chloride is in excess, so CaO is the limiting reactant and determines the amount of ammonia that will be produced. The mass of CaO needed can be calculated from the theoretical yield of ammonia and the 1:2 mole ratio for CaO and ammonia as given in the balanced equation.

$$4.00\text{ g } NH_3 \times \frac{1\text{ mol } NH_3}{17.03\text{ g } NH_3} \times \frac{1\text{ mol CaO}}{2\text{ mol } NH_3} \times \frac{56.077\text{ g CaO}}{1\text{ mol CaO}} = 6.6\text{ g CaO}$$

☑ **Reasonable Answer Check** To check the answer, we solve the problem a different way. The molar mass of CaO is approximately 56 g/mol, so 6.6 g CaO is approximately 0.12 mol CaO. The coefficients in the balanced equation tell us that this would produce twice as many moles of NH_3 or 0.24 mol NH_3, which is 0.24 × 17 g/mol = 4 g NH_3, if the yield were 100%. The actual yield is 25%, so the amount of ammonia expected is reduced to one fourth, approximately 1 g. This approximate calculation is consistent with our more accurate calculation, and the answer is reasonable.

PROBLEM-SOLVING PRACTICE 4.13

You heat 2.50 g copper with an excess of sulfur and synthesize 2.53 g copper(I) sulfide, Cu_2S:

$$16\ Cu(s) + S_8(s) \longrightarrow 8\ Cu_2S(s)$$

Your laboratory instructor expects students to have at least a 60% yield for this reaction. Did your synthesis meet this standard?

CONCEPTUAL
EXERCISE **4.9 Percent Yield**

Percent yield can be reduced by side reactions that produce undesired product(s) and by poor laboratory technique in isolating and purifying the desired product. Identify two other factors that could lead to a low percent yield.

Atom Economy—Another Approach to Tracing Starting Materials

Rather than concentrating simply on percent yield, the concept of **atom economy** focuses on the amounts of starting materials that are incorporated into the desired final product. The greater the fraction of starting atoms incorporated into the desired final product, the fewer waste by-products created. Of course, the objective is to devise syntheses that are as efficient as possible.

Whereas a high percent yield has often been the major goal of chemical synthesis, the concept of atom economy, quantified by the definition of percent atom economy, is becoming important.

$$\text{Percent atom economy} = \frac{\text{sum of atomic weight of atoms in the useful product}}{\text{sum of atomic weight of all atoms in reactants}} \times 100\%$$

Reactions for which all atoms in the reactants are found in the desired product have a percent atom economy of 100%. As an example, consider this combination reaction:

$$CO(g) + 2\,H_2(g) \longrightarrow CH_3OH(\ell)$$

The sum of atomic weights of all atoms in the reactants is $12.011 + 15.9994 + \{2 \times (2 \times 1.0079)\} = 32.042$ amu. The atomic weight of all the atoms in the product is $12.011 + \{3 \times (1.0079)\} + 15.9994 + 1.0079 = 32.042$ amu. The percent atom economy for this reaction is 100%.

Many other reactions in organic synthesis, however, generate other products in addition to the desired product. In such cases, the percent atom economy is far less than 100%. Devising strategies for synthesis of desired compounds with the least waste is a major goal of the current push toward "green chemistry."

Green chemistry aims to eliminate pollution by making chemical products that do not harm health or the environment. It encourages the development of production processes that reduce or eliminate hazardous chemicals. Green chemistry also aims to prevent pollution at its source rather than having to clean up problems after they occur. Each year since 1996 the U.S. Environmental Protection Agency has given Presidential Green Chemistry Challenge Awards for noteworthy green chemistry advances.

4.7 Percent Composition and Empirical Formulas

In Section 3.10, percent composition data were used to derive empirical and molecular formulas, but nothing was mentioned about how such data are obtained. One way to obtain such data is combustion analysis, which is often employed with organic compounds, most of which contain carbon and hydrogen. In **combustion analysis** a compound is burned in excess oxygen, which converts the carbon to carbon dioxide and the hydrogen to water. These combustion products are collected and weighed, and the masses are used to calculate the quantities of carbon and hydrogen in the original substance using the balanced combustion equation. A schematic diagram of the apparatus

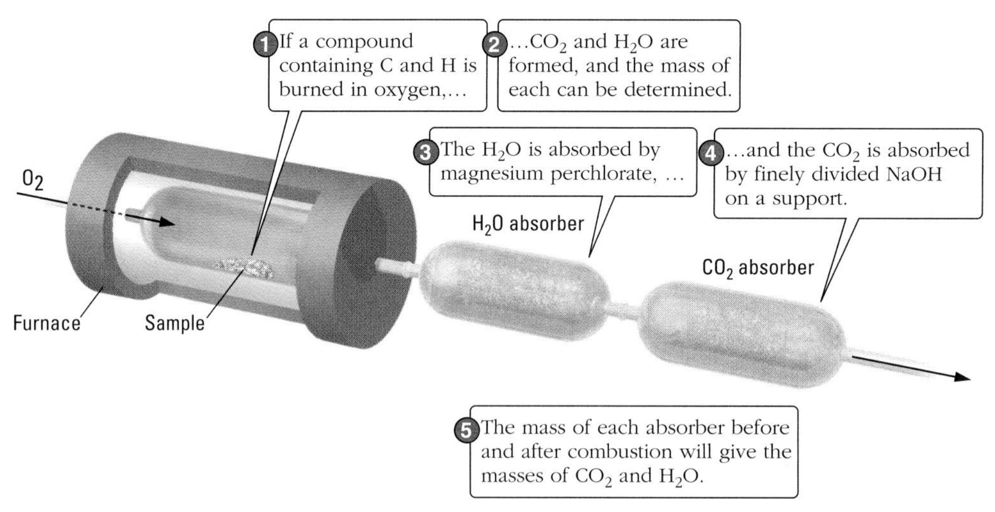

Figure 4.9 Combustion analysis. Schematic diagram of an apparatus for determining the empirical formula of an organic compound. Only a few milligrams of a combustible compound are needed for analysis.

is shown in Figure 4.9. Many organic compounds also contain oxygen. In such cases, the mass of oxygen in the sample can be determined simply by difference.

$$\text{Mass of oxygen} = \text{mass of sample} - (\text{mass of C} + \text{mass of H})$$

vitamin C

As an example, consider this problem. An analytical chemist used combustion analysis to determine the empirical formula of vitamin C, an organic compound containing only carbon, hydrogen, and oxygen. Combustion of 1.000 g of pure vitamin C produced 1.502 g CO_2 and 0.409 g H_2O. A different experiment determined that the molar mass of vitamin C is 176.12 g/mol.

The task is to determine the subscripts on C, H, and O in the empirical formula of vitamin C. Recall from Chapter 3 that the subscripts in a chemical formula tell how many moles of atoms of each element are in 1 mol of the compound. *All* of the carbon in the CO_2 and *all* of the hydrogen in the H_2O came from the vitamin C sample that was burned, so we can work backward to assess the composition of vitamin C.

First, we determine the masses of carbon and hydrogen in the original sample.

$$1.502 \text{ g } CO_2 \times \frac{1 \text{ mol } CO_2}{44.009 \text{ g } CO_2} \times \frac{1 \text{ mol C}}{1 \text{ mol } CO_2} \times \frac{12.011 \text{ g C}}{1 \text{ mol C}} = 0.4100 \text{ g C}$$

$$0.409 \text{ g } H_2O \times \frac{1 \text{ mol } H_2O}{18.015 \text{ g } H_2O} \times \frac{2 \text{ mol H}}{1 \text{ mol } H_2O} \times \frac{1.0079 \text{ g H}}{1 \text{ mol H}} = 0.04577 \text{ g H}$$

The mass of oxygen in the original sample can be calculated by difference.

$$1.000 \text{ g sample} - (0.4100 \text{ g C in sample} + 0.04577 \text{ g H in sample}) = 0.5442 \text{ g O in the sample}$$

From the mass data, we can now calculate how many moles of each element were in the sample.

Carrying an extra digit during the intermediate parts of a multistep problem and rounding at the end is good practice.

$$0.4100 \text{ g C} \times \frac{1 \text{ mol C}}{12.011 \text{ g C}} = 0.03414 \text{ mol C}$$

$$0.04577 \text{ g H} \times \frac{1 \text{ mol H}}{1.0079 \text{ g H}} = 0.04541 \text{ mol H}$$

$$0.5442 \text{ g O} \times \frac{1 \text{ mol O}}{15.999 \text{ g O}} = 0.03401 \text{ mol O}$$

Next, we find the mole ratios of the elements in the compound by dividing by the smallest number of moles.

$$\frac{0.04541 \text{ mol H}}{0.03401 \text{ mol O}} = \frac{1.335 \text{ mol H}}{1.000 \text{ mol O}}$$

$$\frac{0.03414 \text{ mol C}}{0.03401 \text{ mol O}} = \frac{1.004 \text{ mol C}}{1.000 \text{ mol O}}$$

The ratios are very close to 1.33 mol H to 1.00 mol O and a one-to-one ratio of C to O. Multiplying by 3 to get whole numbers gives the empirical formula of vitamin C as $C_3H_4O_3$. From this we can calculate an empirical molar mass of 88.06 g. Because the experimental molar mass is twice the calculated molar mass, the molecular formula of vitamin C is $C_6H_8O_6$, twice the empirical formula.

If after dividing by the smallest number of moles, the ratios are not whole numbers, multiply each subscript by a number that converts the fractions to whole numbers. For example, multiplying $NO_{2.5}$ by 2 changes it to N_2O_5.

For many organic compounds, the empirical and molecular formulas are the same. In addition, several different organic compounds can have the identical empirical and molecular formulas—they are isomers *(p. 69)*. In such cases, you must know the structural formula to fully describe the compound. Ethanol, CH_3CH_2OH, and dimethyl ether, CH_3OCH_3, for example, each have the same empirical formula and molecular formula, C_2H_6O, but they are different compounds with different properties.

CH_3CH_2OH ethanol

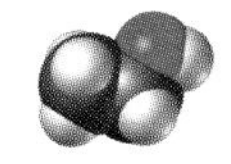

CH_3OCH_3 dimethyl ether

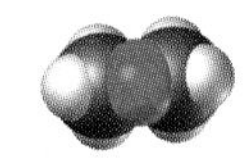

PROBLEM-SOLVING EXAMPLE 4.14 Empirical Formula from Combustion Analysis

Butyric acid, an organic compound with an extremely unpleasant odor, contains only carbon, hydrogen, and oxygen. When 1.20 g butyric acid was burned, 2.41 g CO_2 and 0.982 g H_2O were produced. Calculate the empirical formula of butyric acid. In a separate experiment, the molar mass of butyric acid was determined to be 88.1 g/mol. What is butyric acid's molecular formula?

Answer The empirical formula is C_2H_4O. The molecular formula is $C_4H_8O_2$.

Strategy and Explanation All of the carbon and hydrogen in the butyric acid are burned to form CO_2 and H_2O, respectively. Therefore, we use the masses of CO_2 and H_2O to calculate how many grams of C and H, respectively, were in the original butyric acid sample.

$$2.41 \text{ g CO}_2 \times \frac{1 \text{ mol CO}_2}{44.01 \text{ g CO}_2} \times \frac{1 \text{ mol C}}{1 \text{ mol CO}_2} \times \frac{12.01 \text{ g C}}{\text{mol C}} = 0.658 \text{ g C}$$

$$0.982 \text{ g H}_2\text{O} \times \frac{1 \text{ mol H}_2\text{O}}{18.02 \text{ g H}_2\text{O}} \times \frac{2 \text{ mol H}}{1 \text{ mol H}_2\text{O}} \times \frac{1.008 \text{ g H}}{1 \text{ mol H}} = 0.110 \text{ g H}$$

The remaining mass of the sample, 1.20 g − 0.658 g − 0.110 g = 0.432 g, must be oxygen.

$$0.658 \text{ g C} + 0.110 \text{ g H} + 0.432 \text{ g O} = 1.20 \text{ g sample}$$

We then find the number of moles of each element in the butyric acid sample.

$$0.658 \text{ g C} \times \frac{1 \text{ mol C}}{12.01 \text{ g C}} = 0.0548 \text{ mol C}$$

$$0.110 \text{ g H} \times \frac{1 \text{ mol H}}{1.008 \text{ g H}} = 0.109 \text{ mol H}$$

$$0.432 \text{ g O} \times \frac{1 \text{ mol O}}{16.00 \text{ g O}} = 0.0270 \text{ mol O}$$

To find the mole ratios of the elements, we divide the molar amount of each element by the smallest number of moles.

$$\frac{0.0548 \text{ mol C}}{0.0270 \text{ mol O}} = \frac{2.03 \text{ mol C}}{1.00 \text{ mol O}} \quad \text{and} \quad \frac{0.109 \text{ mol H}}{0.0270 \text{ mol O}} = \frac{4.03 \text{ mol H}}{1.00 \text{ mol O}}$$

The mole ratios show that for every oxygen atom in the molecule, there are two carbon atoms and four hydrogen atoms. Therefore, the empirical formula of butyric acid is C_2H_4O, which has a formula mass of 44.05 g/mol. The experimental molar mass is known to be twice this value, so the molecular formula of butyric acid is $C_4H_8O_2$, twice the empirical formula.

☑ **Reasonable Answer Check** The molar mass of butyric acid, $C_4H_8O_2$, is $(4 \times 12.01) + (8 \times 1.008) + (2 \times 15.9994) = 88.10$ g/mol, so the answer is reasonable.

PROBLEM-SOLVING PRACTICE 4.14

Phenol is a compound of carbon, hydrogen, and oxygen that is used commonly as a disinfectant. Combustion analysis of a 175-mg sample of phenol yielded 491 mg CO_2 and 100. mg H_2O.

(a) Calculate the empirical formula of phenol.
(b) What other information is necessary to determine whether the empirical formula is the actual molecular formula?

CONCEPTUAL EXERCISE 4.10 Formula from Combustion Analysis

Nicotine, a compound found in cigarettes, contains C, H, and N. Outline a method by which you could use combustion analysis to determine the empirical formula for nicotine.

Determining Formulas from Experimental Data

One technique to determine the formula of a binary compound formed by direct combination of its two elements is to measure the mass of reactants that is converted to the product compound.

PROBLEM-SOLVING EXAMPLE 4.15 Empirical Formula from Experimental Data

Solid red phosphorus reacts with liquid bromine to produce a phosphorus bromide.

$$P_4(s) + Br_2(\ell) \longrightarrow P_xBr_y(\ell)$$

If 0.347 g P_4 reacts with 0.860 mL Br_2, what is the empirical formula of the product? The density of bromine is 3.12 g/mL.

Answer PBr_3

Strategy and Explanation We start by calculating the moles of P_4 and Br_2 that combined:

$$0.347 \text{ g } P_4 \times \frac{1 \text{ mol } P_4}{123.90 \text{ g } P_4} = 2.801 \times 10^{-3} \text{ mol } P_4$$

To determine the moles of bromine, we first use the density of bromine to convert milliliters of bromine to grams:

$$0.860 \text{ mL } Br_2 \times \frac{3.12 \text{ g } Br_2}{1 \text{ mL } Br_2} = 2.68 \text{ g } Br_2$$

We then convert from grams to moles:

$$2.68 \text{ g } Br_2 \times \frac{1 \text{ mol } Br_2}{159.8 \text{ g } Br_2} = 1.677 \times 10^{-2} \text{ mol } Br_2$$

The mole ratio of bromine atoms to phosphorus atoms in a molecule of the product can be calculated from the moles of atoms of each element.

$$1.677 \times 10^{-2} \text{ mol } Br_2 \text{ molecules} \times \frac{2 \text{ mol Br atoms}}{1 \text{ mol } Br_2 \text{ molecules}} = 3.35 \times 10^{-2} \text{ mol Br atoms}$$

$$2.801 \times 10^{-3} \text{ mol } P_4 \text{ molecules} \times \frac{4 \text{ mol P atoms}}{1 \text{ mol } P_4 \text{ molecules}} = 1.12 \times 10^{-2} \text{ mol P atoms}$$

$$\frac{3.35 \times 10^{-2} \text{ mol Br atoms}}{1.12 \times 10^{-2} \text{ mol P atoms}} = \frac{2.99 \text{ mol Br atoms}}{1.00 \text{ mol P atoms}}$$

The mole ratio in the compound is 3.00 mol bromine atoms for 1.00 mol phosphorus atoms. The empirical formula is PBr_3. By other experimental methods, the known molar mass of this compound is found to be the same as the empirical formula molar mass. Given this additional information, we know that the molecular formula is also PBr_3.

☑ **Reasonable Answer Check** Phosphorus is a Group 5A element, so combining it with three bromines results in a reasonable molecular formula for such a combination of elements.

PROBLEM-SOLVING PRACTICE 4.15

The complete reaction of 0.569 g tin with 2.434 g iodine formed Sn_xI_y. What is the empirical formula of this tin iodide?

IN CLOSING

Having studied this chapter, you should be able to . . .

- Interpret the information conveyed by a balanced chemical equation (Section 4.1).
- Recognize the general reaction types: combination, decomposition, displacement, and exchange (Section 4.2). ■ End-of-chapter Questions assignable in OWL: 11, 13
- Balance simple chemical equations (Section 4.3). ■ Question 17
- Use mole ratios to calculate the number of moles or number of grams of one reactant or product from the number of moles or number of grams of another reactant or product by using the balanced chemical equation (Section 4.4). ■ Questions 23, 27, 54
- Use principles of stoichiometry in the chemical analysis of a mixture (Section 4.4). ■ Question 74
- Determine which of two reactants is the limiting reactant (Section 4.5). ■ Questions 37, 39
- Explain the differences among actual yield, theoretical yield, and percent yield, and calculate theoretical and percent yields (Section 4.6). ■ Question 44
- Use principles of stoichiometry to find the empirical formula of an unknown compound using combustion analysis or other mass data (Section 4.7). ■ Questions 50, 56

Selected end-of-chapter Questions may be assigned in OWL.

For quick review, download Go Chemistry mini-lecture flashcard modules (or purchase them at **www.ichapters.com**).

Prepare for an exam with a **Summary Problem** that brings together concepts and problem-solving skills from throughout the chapter. Go to **www.cengage.com/chemistry/moore** to download this chapter's Summary Problem from the Student Companion Site.

KEY TERMS

actual yield *(Section 4.6)*
aqueous solution *(4.1)*
atom economy *(4.6)*
balanced chemical equation *(4.1)*
coefficient *(4.1)*
combination reaction *(4.2)*
combustion analysis *(4.7)*
combustion reaction *(4.1)*
decomposition reaction *(4.2)*
displacement reaction *(4.2)*
exchange reaction *(4.2)*
limiting reactant *(4.5)*
mole ratio *(4.4)*
percent yield *(4.6)*
stoichiometric coefficient *(4.1)*
stoichiometry *(4.1)*
theoretical yield *(4.6)*

QUESTIONS FOR REVIEW AND THOUGHT

■ denotes questions assignable in OWL.

Blue-numbered questions have short answers at the back of this book and fully worked solutions in the *Student Solutions Manual.*

Review Questions

1. Complete the table for the reaction

$$3\ H_2(g) + N_2(g) \longrightarrow 2\ NH_3(g)$$

H_2	N_2	NH_3
_________ mol	1 mol	_________ mol
3 molecules	_________ molecules	_________ molecules
_________ g	_________ g	34.08 g

2. Write all the possible mole ratios for the reaction

$$3\ MgO(s) + 2\ Fe(s) \longrightarrow Fe_2O_3(s) + 3\ Mg(s)$$

3. If a 10.0-g mass of carbon is combined with an exact stoichiometric amount of oxygen (26.6 g) to make carbon dioxide, how many grams of CO_2 can be isolated?

4. Given the reaction

$$2\ Fe(s) + 3\ Cl_2(g) \longrightarrow 2\ FeCl_3(s)$$

fill in the missing conversion factors for the scheme

g Cl_2 —?→ g $FeCl_3$
|　　　　　|
?　　　　　?
↓　　　　　↓
mol Cl_2 —?→ mol $FeCl_3$

5. Does the limiting reactant determine the theoretical yield, actual yield, or both? Explain.

Topical Questions

Stoichiometry

6. For this reaction, fill in the table with the indicated quantities for the balanced equation.

$$4\ NH_3(g) + 5\ O_2(g) \longrightarrow 4\ NO(g) + 6\ H_2O(g)$$

	NH_3	O_2	NO	H_2O
No. of molecules				
No. of atoms				
No. of moles of molecules				
Mass				
Total mass of reactants				
Total mass of products				

7. For this reaction, fill in the table with the indicated quantities for the balanced equation.

$$2\ C_2H_6(g) + 7\ O_2(g) \longrightarrow 4\ CO_2(g) + 6\ H_2O(g)$$

	C_2H_6	O_2	CO_2	H_2O
No. of molecules				
No. of atoms				
No. of moles of molecules				
Mass				
Total mass of reactants				
Total mass of products				

8. The following diagram shows A (blue spheres) reacting with B (tan spheres). Which equation best describes the stoichiometry of the reaction depicted in this diagram?

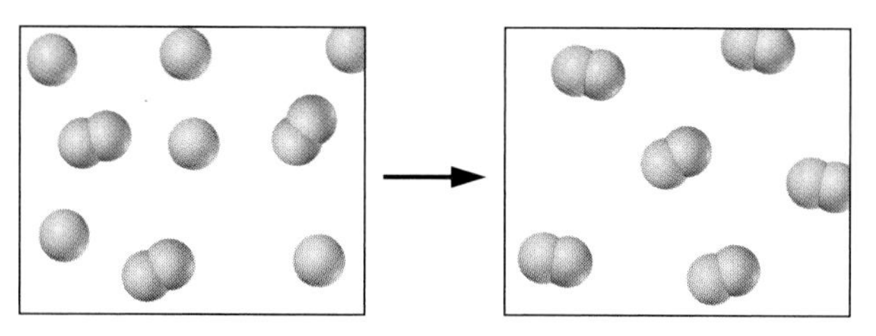

(a) $3\ A_2 + 6\ B \rightarrow 6\ AB$　(b) $A_2 + 2\ B \rightarrow 2\ AB$
(c) $2\ A + B \rightarrow AB$　(d) $3\ A + 6\ B \rightarrow 6\ AB$

9. The following diagram shows A (blue spheres) reacting with B (tan spheres). Write a balanced equation that describes the stoichiometry of the reaction shown in the diagram.

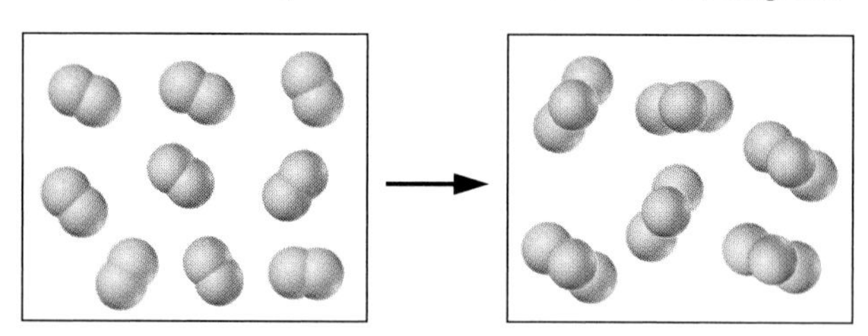

10. Balance this equation and determine which box represents reactants and which box represents products.

$$Sb(g) + Cl_2(g) \longrightarrow SbCl_3(g)$$

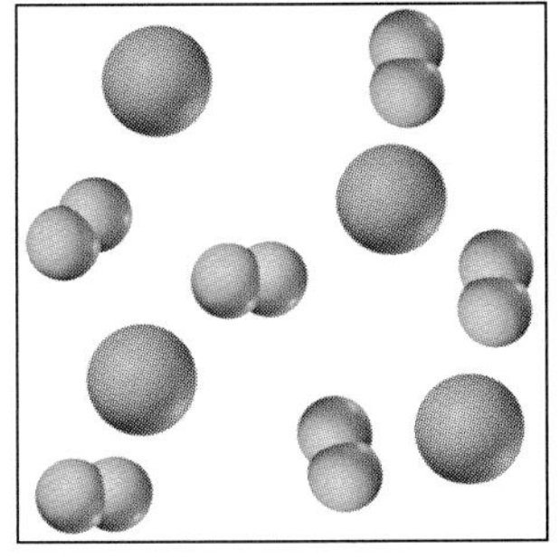

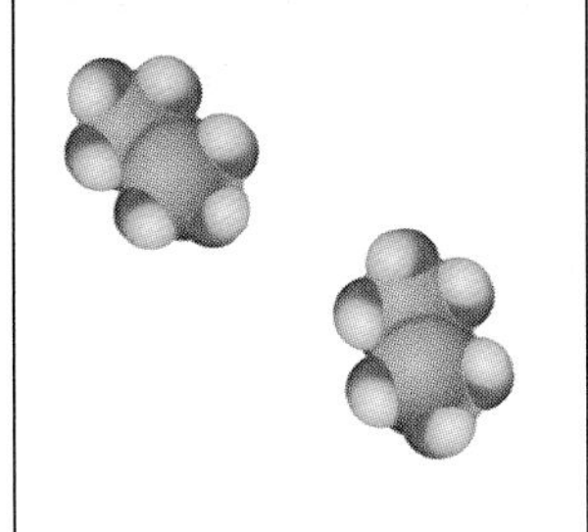

(a)

(b)

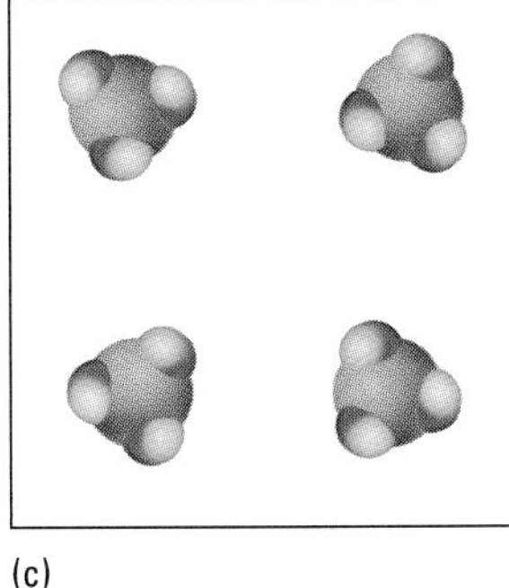

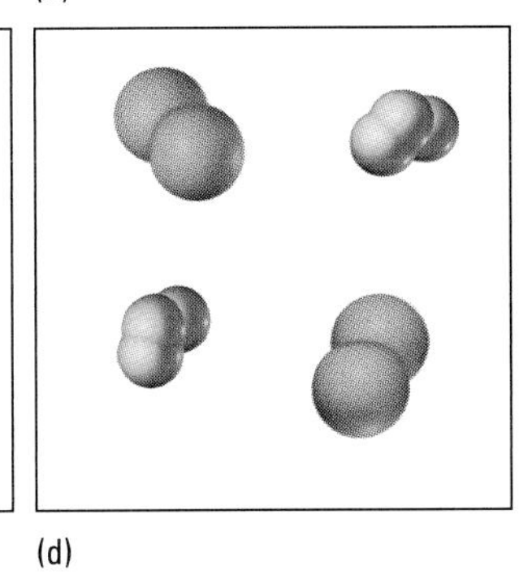

(c) (d)

Classification of Chemical Reactions

11. ■ Indicate whether each of these equations represents a combination, decomposition, displacement, or exchange reaction.
    (a) $Cu(s) + O_2(g) \longrightarrow 2\ CuO(s)$
    (b) $NH_4NO_3(s) \longrightarrow N_2O(g) + 2\ H_2O(\ell)$
    (c) $AgNO_3(aq) + KCl(aq) \longrightarrow AgCl(s) + KNO_3(aq)$
    (d) $Mg(s) + 2\ HCl(aq) \longrightarrow MgCl_2(aq) + H_2(g)$
12. Indicate whether each of these equations represents a combination, decomposition, displacement, or exchange reaction.
    (a) $C(s) + O_2(g) \longrightarrow CO_2(g)$
    (b) $2\ KClO_3(s) \longrightarrow 2\ KCl(s) + 3\ O_2(g)$
    (c) $BaCl_2(aq) + K_2SO_4(aq) \longrightarrow BaSO_4(s) + 2\ KCl(aq)$
    (d) $Mg(s) + CoSO_4(aq) \longrightarrow MgSO_4(aq) + Co(s)$
13. ■ Indicate whether each of these equations represents a combination, decomposition, displacement, or exchange reaction.
    (a) $PbCO_3(s) \longrightarrow PbO(s) + CO_2(g)$
    (b) $Cu(s) + 4\ HNO_3(aq) \longrightarrow$
    $Cu(NO_3)_2(aq) + 2\ H_2O(\ell) + 2\ NO_2(g)$
    (c) $2\ Zn(s) + O_2(g) \longrightarrow 2\ ZnO(s)$
    (d) $Pb(NO_3)_2(aq) + 2\ KI(aq) \longrightarrow PbI_2(s) + 2\ KNO_3(aq)$
14. Indicate whether each of these equations represents a combination, decomposition, displacement, or exchange reaction.
    (a) $Mg(s) + FeCl_2(aq) \longrightarrow MgCl_2(aq) + Fe(s)$
    (b) $ZnCO_3(s) \longrightarrow ZnO(s) + CO_2(g)$
    (c) $2\ C(s) + O_2(g) \longrightarrow 2\ CO(g)$
    (d) $CaCl_2(aq) + Na_2CO_3(aq) \longrightarrow CaCO_3(s) + 2\ NaCl(aq)$

Balancing Equations

15. Complete and balance these equations involving oxygen reacting with an element. Name the product in each case.
    (a) $Mg(s) + O_2(g) \longrightarrow$
    (b) $Ca(s) + O_2(g) \longrightarrow$
    (c) $In(s) + O_2(g) \longrightarrow$
16. Complete and balance these equations involving oxygen reacting with an element.
    (a) $Ti(s) + O_2(g) \longrightarrow$ titanium(IV) oxide
    (b) $S_8(s) + O_2(g) \longrightarrow$ sulfur dioxide
    (c) $Se(s) + O_2(g) \longrightarrow$ selenium dioxide
17. ■ Balance these equations.
    (a) $UO_2(s) + HF(\ell) \longrightarrow UF_4(s) + H_2O(\ell)$
    (b) $B_2O_3(s) + HF(\ell) \longrightarrow BF_3(g) + H_2O(\ell)$
    (c) $BF_3(g) + H_2O(\ell) \longrightarrow HF(\ell) + H_3BO_3(s)$
18. Balance these equations.
    (a) $MgO(s) + Fe(s) \longrightarrow Fe_2O_3(s) + Mg(s)$
    (b) $H_3BO_3(s) \longrightarrow B_2O_3(s) + H_2O(\ell)$
    (c) $NaNO_3(s) + H_2SO_4(aq) \longrightarrow Na_2SO_4(aq) + HNO_3(g)$
19. Balance these equations.
    (a) Reaction to produce hydrazine, N_2H_4:
    $$H_2NCl(aq) + NH_3(g) \longrightarrow NH_4Cl(aq) + N_2H_4(aq)$$
    (b) Reaction of the fuels (dimethylhydrazine and dinitrogen tetroxide) used in the Moon Lander and Space Shuttle:
    $$(CH_3)_2N_2H_2(\ell) + N_2O_4(g) \longrightarrow N_2(g) + H_2O(g) + CO_2(g)$$
    (c) Reaction of calcium carbide with water to produce acetylene, C_2H_2:
    $$CaC_2(s) + H_2O(\ell) \longrightarrow Ca(OH)_2(s) + C_2H_2(g)$$
20. Balance these equations.
    (a) Reaction of calcium cyanamide to produce ammonia:
    $$CaNCN(s) + H_2O(\ell) \longrightarrow CaCO_3(s) + NH_3(g)$$
    (b) Reaction to produce diborane, B_2H_6:
    $$NaBH_4(s) + H_2SO_4(aq) \longrightarrow B_2H_6(g) + H_2(g) + Na_2SO_4(aq)$$
    (c) Reaction to rid water of hydrogen sulfide, H_2S, a foul-smelling compound:
    $$H_2S(aq) + Cl_2(aq) \longrightarrow S_8(s) + HCl(aq)$$
21. Balance these combustion reactions.
    (a) $C_6H_{12}O_6 + O_2 \longrightarrow CO_2 + H_2O$
    (b) $C_5H_{12} + O_2 \longrightarrow CO_2 + H_2O$
    (c) $C_7H_{14}O_2 + O_2 \longrightarrow CO_2 + H_2O$
    (d) $C_2H_4O_2 + O_2 \longrightarrow CO_2 + H_2O$
22. Balance these equations.
    (a) $Mg + HNO_3 \longrightarrow H_2 + Mg(NO_3)_2$
    (b) $Al + Fe_2O_3 \longrightarrow Al_2O_3 + Fe$
    (c) $S_8 + O_2 \longrightarrow SO_3$
    (d) $SO_3 + H_2O \longrightarrow H_2SO_4$

The Mole and Chemical Reactions

23. ■ Chlorine can be produced in the laboratory by the reaction of hydrochloric acid with excess manganese(IV) oxide.

$$4\ HCl(aq) + MnO_2(s) \longrightarrow Cl_2(g) + 2\ H_2O(\ell) + MnCl_2(aq)$$

How many moles of HCl are needed to form 12.5 mol Cl_2?

24. Methane, CH_4, is the major component of natural gas. How many moles of oxygen are needed to burn 16.5 mol CH_4?

$$CH_4(g) + 2\ O_2(g) \longrightarrow CO_2(g) + 2\ H_2O(\ell)$$

25. Nitrogen monoxide is oxidized in air to give brown nitrogen dioxide.

$$2\ NO(g) + O_2(g) \longrightarrow 2\ NO_2(g)$$

Starting with 2.2 mol NO, how many moles and how many grams of O_2 are required for complete reaction? What mass of NO_2, in grams, is produced?

26. Aluminum reacts with oxygen to give aluminum oxide.

$$4\ Al(s) + 3\ O_2(g) \longrightarrow 2\ Al_2O_3(s)$$

If you have 6.0 mol Al, how many moles and how many grams of O_2 are needed for complete reaction? What mass of Al_2O_3, in grams, is produced?

27. ■ Many metals react with halogens to give metal halides. For example, iron reacts with chlorine to give iron(II) chloride, $FeCl_2$.

$$Fe(s) + Cl_2(g) \longrightarrow FeCl_2(s)$$

(a) Beginning with 10.0 g iron, what mass of Cl_2, in grams, is required for complete reaction?
(b) What quantity of $FeCl_2$, in moles and in grams, is expected?

28. Like many metals, manganese reacts with a halogen to give a metal halide.

$$2\ Mn(s) + 3\ F_2(g) \longrightarrow 2\ MnF_3(s)$$

(a) If you begin with 5.12 g Mn, what mass in grams of F_2 is required for complete reaction?
(b) What quantity in moles and in grams of the red solid MnF_3 is expected?

29. The final step in the manufacture of platinum metal (for use in automotive catalytic converters and other products) is the reaction

$$3\ (NH_4)_2PtCl_6(s) \longrightarrow 3\ Pt(s) + 2\ NH_4Cl(s) + 2\ N_2(g) + 16\ HCl(g)$$

Complete this table of reaction quantities for the reaction of 12.35 g $(NH_4)_2PtCl_6$.

$(NH_4)_2PtCl_6$	Pt	HCl
12.35 g	______ g	______ g
______ mol	______ mol	______ mol

30. Disulfur dichloride, S_2Cl_2, is used to vulcanize rubber. It can be made by treating molten sulfur with gaseous chlorine.

$$S_8(\ell) + 4\ Cl_2(g) \longrightarrow 4\ S_2Cl_2(g)$$

Complete this table of reaction quantities for the production of 103.5 g S_2Cl_2.

S_8	Cl_2	S_2Cl_2
______ g	______ g	103.5 g
______ mol	______ mol	______ mol

31. If 2.5 mol O_2 reacts with propane, C_3H_8, by combustion, how many moles of H_2O will be produced? How many grams of H_2O will be produced?

32. If you want to synthesize 1.45 g of the semiconducting material GaAs, what masses of Ga and of As, in grams, are required?

33. Careful decomposition of ammonium nitrate, NH_4NO_3, gives laughing gas (dinitrogen monoxide, N_2O) and water.
(a) Write a balanced equation for this reaction.
(b) Beginning with 10.0 g NH_4NO_3, what masses of N_2O and water are expected?

34. In making iron from iron ore, this reaction occurs.

$$Fe_2O_3(s) + 3\ CO(g) \longrightarrow 2\ Fe(s) + 3\ CO_2(g)$$

(a) How many grams of iron can be obtained from 1.00 kg iron(III) oxide?
(b) How many grams of CO are required?

Limiting Reactant

35. The reaction of Na_2SO_4 with $BaCl_2$ is

$$Na_2SO_4(aq) + BaCl_2(aq) \longrightarrow BaSO_4(s) + 2\ NaCl(aq)$$

If solutions containing exactly one gram of each reactant are mixed, which reactant is the limiting reactant, and how many grams of $BaSO_4$ are produced?

36. If a mixture of 100. g Al and 200. g MnO is reacted according to the reaction

$$2\ Al(s) + 3\ MnO(s) \longrightarrow Al_2O_3(s) + 3\ Mn(s)$$

which of the reactants is in excess and how many grams of it remain when the reaction is complete?

37. ■ Aluminum chloride, Al_2Cl_6, is an inexpensive reagent used in many industrial processes. It is made by treating scrap aluminum with chlorine according to the balanced equation

$$2\ Al(s) + 3\ Cl_2(g) \longrightarrow Al_2Cl_6(s)$$

(a) Which reactant is limiting if 2.70 g Al and 4.05 g Cl_2 are mixed?
(b) What mass of Al_2Cl_6 can be produced?
(c) What mass of the excess reactant will remain when the reaction is complete?

38. Methanol, CH_3OH, is a clean-burning, easily handled fuel. It can be made by the direct reaction of CO and H_2.

$$CO(g) + 2\,H_2(g) \longrightarrow CH_3OH(\ell)$$

(a) Starting with a mixture of 12.0 g H_2 and 74.5 g CO, which is the limiting reactant?
(b) What mass of the excess reactant, in grams, is left after reaction is complete?
(c) What mass of methanol can be obtained, in theory?

39. ■ Ammonia gas can be prepared by the reaction

$$CaO(s) + 2\,NH_4Cl(s) \longrightarrow 2\,NH_3(g) + H_2O(g) + CaCl_2(s)$$

If 112 g CaO reacts with 224 g NH_4Cl, how many moles of reactants and products are there when the reaction is complete?

40. This reaction between lithium hydroxide and carbon dioxide has been used to scrub CO_2 from spacecraft atmospheres:

$$2\,LiOH + CO_2 \longrightarrow Li_2CO_3 + H_2O$$

(a) If 0.500 kg LiOH were available, how many grams of CO_2 could be consumed?
(b) How many grams of water would be produced?

41. ■ Consider the chemical reaction $2\,S + 3\,O_2 \longrightarrow 2\,SO_3$. If the reaction is run by adding S indefinitely to a fixed amount of O_2, which of these graphs best represents the formation of SO_3? Explain your choice.

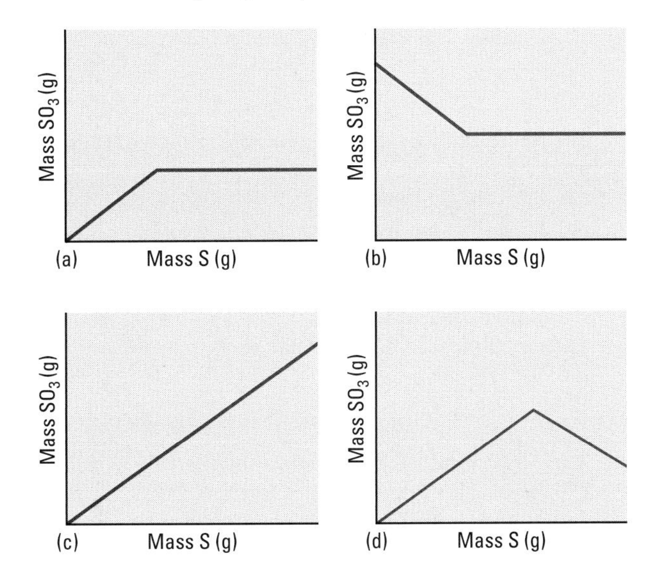

Percent Yield

42. Quicklime, CaO, is formed when calcium hydroxide is heated.

$$Ca(OH)_2(s) \longrightarrow CaO(s) + H_2O(\ell)$$

If the theoretical yield is 65.5 g but only 36.7 g quicklime is produced, what is the percent yield?

43. Diborane, B_2H_6, is valuable for the synthesis of new organic compounds. The boron compound can be made by the reaction

$$2\,NaBH_4(s) + I_2(s) \longrightarrow B_2H_6(g) + 2\,NaI(s) + H_2(g)$$

Suppose you use 1.203 g $NaBH_4$ and excess iodine, and you isolate 0.295 g B_2H_6. What is the percent yield of B_2H_6?

44. ■ Methanol, CH_3OH, is used in racing cars because it is a clean-burning fuel. It can be made by this reaction:

$$CO(g) + 2\,H_2(g) \longrightarrow CH_3OH(\ell)$$

What is the percent yield if 5.0×10^3 g H_2 reacts with excess CO to form 3.5×10^3 g CH_3OH?

45. If 3.7 g sodium metal and 4.3 g chlorine gas react to form NaCl, what is the theoretical yield? If 5.5 g NaCl was formed, what is the percent yield?

46. The ceramic silicon nitride, Si_3N_4, is made by heating silicon and nitrogen at an elevated temperature.

$$3\,Si(s) + 2\,N_2(g) \longrightarrow Si_3N_4(s)$$

How many grams of silicon must combine with excess N_2 to produce 1.0 kg Si_3N_4 if this process is 92% efficient?

47. Disulfur dichloride can be prepared by

$$3\,SCl_2 + 4\,NaF \longrightarrow SF_4 + S_2Cl_2 + 4\,NaCl$$

What is the percent yield of the reaction if 5.00 g SCl_2 reacts with excess NaF to produce 1.262 g S_2Cl_2?

Empirical Formulas

48. What is the empirical formula of a compound that contains 60.0% oxygen and 40.0% sulfur by mass?

49. A potassium salt was analyzed to have this percent composition: 26.57% K, 35.36% Cr, and 38.07% O. What is its empirical formula?

50. ■ Styrene, the building block of polystyrene, is a hydrocarbon. If 0.438 g of the compound is burned and produces 1.481 g CO_2 and 0.303 g H_2O, what is the empirical formula of the compound?

51. Mesitylene is a liquid hydrocarbon. If 0.115 g of the compound is burned in pure O_2 to give 0.379 g CO_2 and 0.1035 g H_2O, what is the empirical formula of the compound?

52. Propionic acid, an organic acid, contains only C, H, and O. If 0.236 g of the acid burns completely in O_2 and gives 0.421 g CO_2 and 0.172 g H_2O, what is the empirical formula of the acid?

53. Quinone, which is used in the dye industry and in photography, is an organic compound containing only C, H, and O. What is the empirical formula of the compound if 0.105 g of the compound gives 0.257 g CO_2 and 0.0350 g H_2O when burned completely?

General Questions

54. ■ Nitrogen gas can be prepared in the laboratory by the reaction of ammonia with copper(II) oxide according to this unbalanced equation:

$$NH_3(g) + CuO(s) \longrightarrow N_2(g) + Cu(s) + H_2O(g)$$

If 26.3 g of gaseous NH_3 is passed over a bed of solid CuO in stoichiometric excess, what mass, in grams, of N_2 can be isolated?

55. The overall chemical equation for the photosynthesis reaction in green plants is

$$6\ CO_2(g) + 6\ H_2O(\ell) \longrightarrow C_6H_{12}O_6(aq) + 6\ O_2(g)$$

How many grams of oxygen are produced by a plant when 50.0 g CO_2 is consumed?

56. ■ In an experiment, 1.056 g of a metal carbonate containing an unknown metal M was heated to give the metal oxide and 0.376 g CO_2.

$$MCO_3(s) \xrightarrow{\text{heat}} MO(s) + CO_2(g)$$

What is the identity of the metal M?

(a) Ni (b) Cu
(c) Zn (d) Ba

57. The cancer chemotherapy agent cisplatin is made by the reaction

$$(NH_4)_2PtCl_4(s) + 2\ NH_3(aq) \longrightarrow 2\ NH_4Cl(aq) + Pt(NH_3)_2Cl_2(s)$$

Assume that 15.5 g $(NH_4)_2PtCl_4$ is combined with 0.15 mol aqueous NH_3 to make cisplatin. What is the theoretical mass, in grams, of cisplatin that can be formed?

58. Silicon and hydrogen form a series of interesting compounds, Si_xH_y. To find the formula of one of them, a 6.22-g sample of the compound is burned in oxygen. All of the Si is converted to 11.64 g SiO_2 and all of the H to 6.980 g H_2O. What is the empirical formula of the silicon compound?

59. The Hargraves process is an industrial method for making sodium sulfate for use in papermaking.

$$4\ NaCl + 2\ SO_2 + 2\ H_2O + O_2 \longrightarrow 2\ Na_2SO_4 + 4\ HCl$$

(a) If you start with 10. mol of each reactant, which one will determine the amount of Na_2SO_4 produced?
(b) What if you start with 100. g of each reactant?

Applying Concepts

60. Ammonia can be formed by a direct reaction of nitrogen and hydrogen.

$$N_2(g) + 3\ H_2(g) \longrightarrow 2\ NH_3(g)$$

A tiny portion of the starting mixture is represented by this diagram, where the blue circles represent N and the white circles represent H.

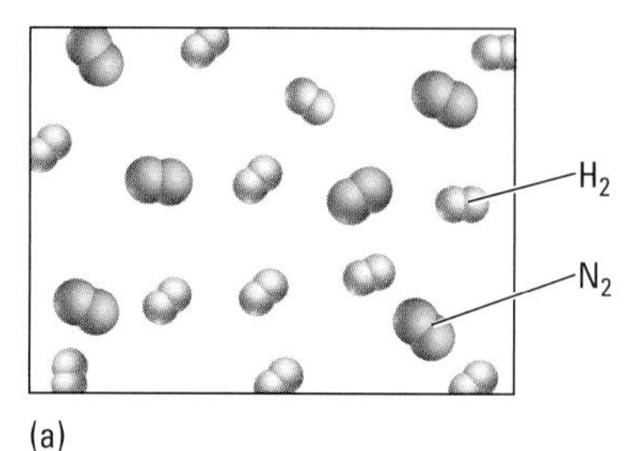

(a)

Which of these represents the product mixture?

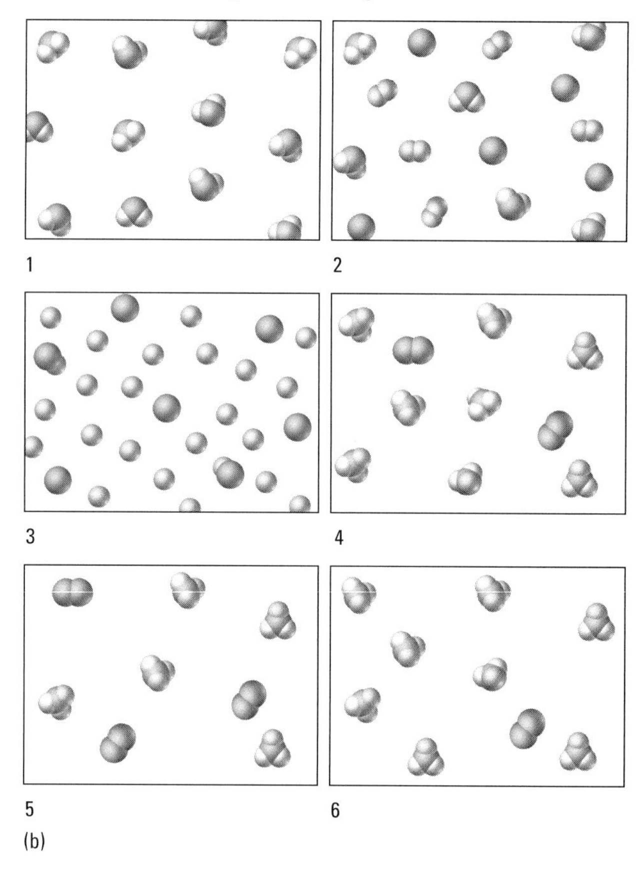

(b)

For the reaction of the given sample, which of these statements is true?

(a) N_2 is the limiting reactant.
(b) H_2 is the limiting reactant.
(c) NH_3 is the limiting reactant.
(d) No reactant is limiting; they are present in the correct stoichiometric ratio.

61. Carbon monoxide burns readily in oxygen to form carbon dioxide.

$$2\ CO(g) + O_2(g) \longrightarrow 2\ CO_2(g)$$

The box on the left represents a tiny portion of a mixture of CO and O_2. If these molecules react to form CO_2, what should the contents of the box on the right look like?

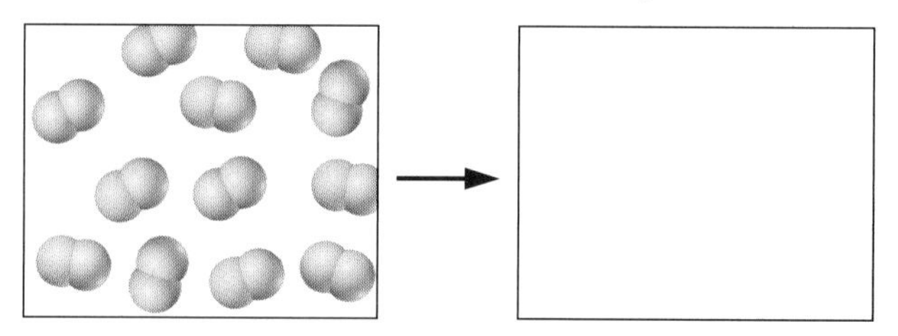

62. Which chemical equation best represents the reaction taking place in this illustration?
(a) $X_2 + Y_2 \longrightarrow n\ XY_3$
(b) $X_2 + 3\ Y_2 \longrightarrow 2\ XY_3$
(c) $6\ X_2 + 6\ Y_2 \longrightarrow 4\ XY_3 + 4\ X_2$
(d) $6\ X_2 + 6\ Y_2 \longrightarrow 4\ X_3Y + 4\ Y_2$

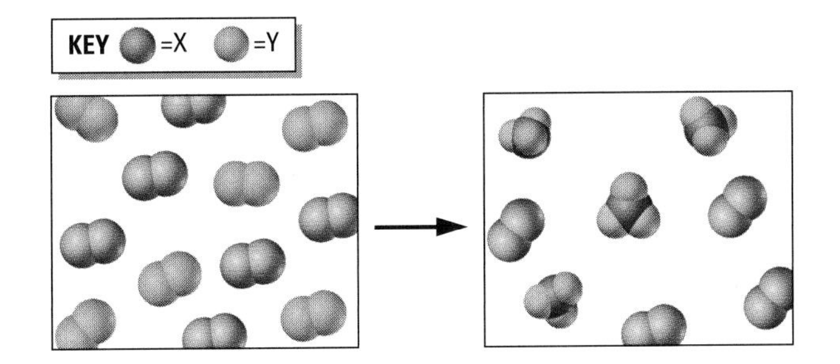

63. ■ A weighed sample of a metal is added to liquid bromine and allowed to react completely. The product substance is then separated from any leftover reactants and weighed. This experiment is repeated with several masses of the metal but with the same volume of bromine. This graph indicates the results. Explain why the graph has the shape that it does.

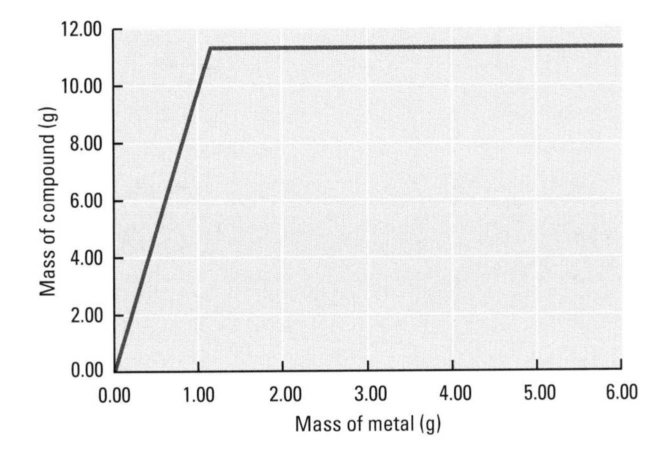

64. A series of experimental measurements like the ones described in Question 63 is carried out for iron reacting with bromine. This graph is obtained. What is the empirical formula of the compound formed by iron and bromine? Write a balanced equation for the reaction between iron and bromine. Name the product.

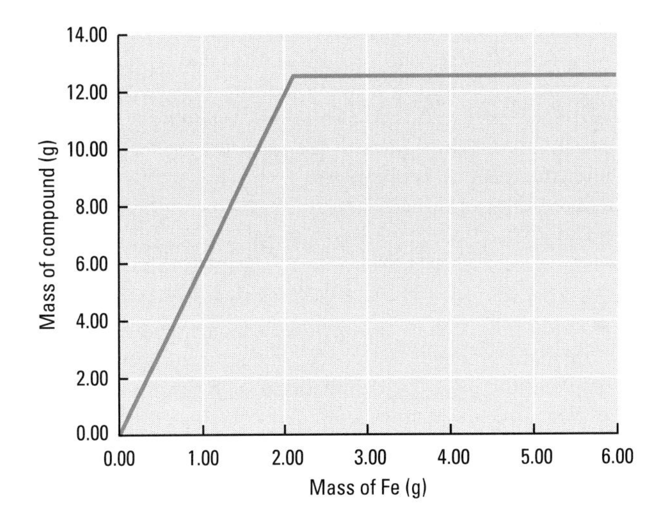

More Challenging Questions

65. ■ Hydrogen gas H_2(g) is reacted with a sample of Fe_2O_3(s) at 400 °C. Two products are formed: water vapor and a black solid compound that is 72.3% Fe and 27.7% O by mass. Write the balanced chemical equation for the reaction.

66. Write the balanced chemical equation for the complete combustion of malonic acid, an organic acid containing 34.62% C, 3.88% H, and the remainder O, by mass.

67. Aluminum bromide is a valuable laboratory chemical. What is the theoretical yield, in grams, of Al_2Br_6 if 25.0 mL liquid bromine (density = 3.12 g/mL) and excess aluminum metal are reacted?

$$2\ \text{Al(s)} + 3\ \text{Br}_2(\ell) \longrightarrow \text{Al}_2\text{Br}_6\text{(s)}$$

68. In a reaction, 1.2 g element A reacts with exactly 3.2 g oxygen to form an oxide, AO_x; 2.4 g element A reacts with exactly 3.2 g oxygen to form a second oxide, AO_y.
(a) What is the ratio x/y?
(b) If $x = 2$, what might be the identity of element A?

69. A copper ore contained Cu_2S and CuS plus 10% inert impurities. When 200.0 g of the ore was "roasted," it yielded 150.8 g of 90.0% pure copper and sulfur dioxide gas. What is the percentage of Cu_2S in the ore?

$$\text{Cu}_2\text{S} + \text{O}_2 \longrightarrow 2\ \text{Cu} + \text{SO}_2 \qquad \text{CuS} + \text{O}_2 \longrightarrow \text{Cu} + \text{SO}_2$$

70. A compound with the formula X_2S_3 has a mass of 10.00 g. It is then roasted (reacted with oxygen) to convert it to X_2O_3. After roasting, it weighs 7.410 g. What is the atomic mass of element X?

71. A metal carbonate decomposed to form its metal oxide and CO_2 when it was heated:

$$\text{MCO}_3\text{(s)} \longrightarrow \text{MO(s)} + \text{CO}_2\text{(g)} \qquad \text{(balanced)}$$

After the reaction, the metal oxide was found to have a mass 56.0% as large as the starting MCO_3. What metal was in the carbonate?

72. When solutions of silver nitrate and sodium carbonate are mixed, solid silver carbonate is formed and sodium nitrate remains in solution. If a solution containing 12.43 g sodium carbonate is mixed with a solution containing 8.37 g silver nitrate, how many grams of the four species are present after the reaction is complete?

73. This reaction produces sulfuric acid:

$$2\ SO_2 + O_2 + 2\ H_2O \longrightarrow 2\ H_2SO_4$$

If 200. g SO_2, 85 g O_2, and 66 g H_2O are mixed and the reaction proceeds to completion, which reactant is limiting, how many grams of H_2SO_4 are produced, and how many grams of the other two reactants are left over?

74. ■ You have an organic liquid that contains either ethyl alcohol, C_2H_5OH, or methyl alcohol, CH_3OH, or both. You burned a sample of the liquid weighing 0.280 g to form 0.385 g $CO_2(g)$. What was the composition of the sample of liquid?

75. L-Dopa is a drug used for the treatment of Parkinson's disease. Elemental analysis shows it to be 54.82% carbon, 7.10% nitrogen, 32.46% oxygen, and the remainder hydrogen.
   (a) What is L-dopa's empirical formula?
   (b) The molar mass of L-dopa is 197.19 g/mol; what is its molecular formula?

Conceptual Challenge Problems

CP4.A (Section 4.4) In Example 4.7 it was not possible to find the mass of O_2 directly from a knowledge of the mass of sucrose. Are there chemical reactions in which the mass of a product or another reactant can be known directly if you know the mass of a reactant? Cite a couple of these reactions.

CP4.B (Section 4.4) Glucose, $C_6H_{12}O_6$, a monosaccharide, and sucrose, $C_{12}H_{22}O_{11}$, a disaccharide, each undergo complete combustion with O_2 (metabolic conversion) to produce H_2O and CO_2.
(a) How many moles of O_2 are needed per mole of each sugar for the reaction to proceed?
(b) How many grams of O_2 are needed per mole of each sugar for the reaction to proceed?
(c) Which combustion reaction produces more H_2O per gram of sugar? How many grams of H_2O are produced per gram of each sugar?

5

Chemical Reactions

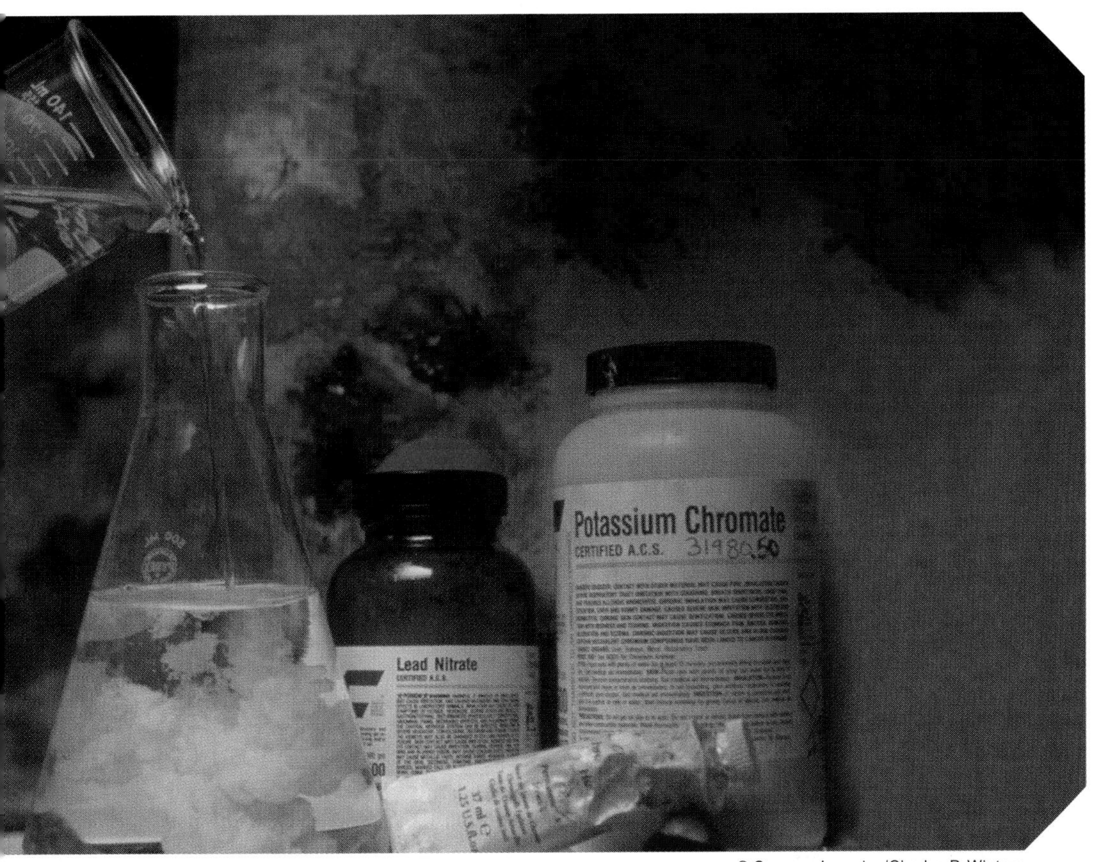

The brilliant yellow precipitate lead(II) chromate, $PbCrO_4$, is formed when lead(II) ions, Pb^{2+}, and chromate ions, CrO_4^{2-}, come together in an aqueous solution. The precipitation reaction occurs because the lead(II) chromate product is insoluble. Although the color of lead(II) chromate is so beautiful that it has been used as a pigment in paint, both lead and chromate are poisons, and the paint must be handled carefully.

Chemistry is concerned with how substances react and what products are formed when they react. A chemical compound can consist of molecules or oppositely charged ions, and often the compound's properties can be deduced from the behavior of these molecules or ions. The *chemical* properties of a compound are the transformations that the molecules or ions can undergo when the substance reacts. A central focus of chemistry is providing answers to questions such as these: When two substances are mixed, will a chemical reaction occur? If a chemical reaction occurs, what will the products be?

As you saw in Chapter 4 (⇐ *p. 101*), most reactions of simple ionic and molecular compounds can be assigned to a few general categories: combination, decomposition, displacement, and exchange. In this chapter we discuss chemical reactions in more detail, including oxidation-reduction reactions. The ability to recognize which type of reaction occurs for a particular set of reactants will allow you to predict the products.

Chemical reactions involving exchange of ions to form precipitates are discussed first, followed by net ionic equations, which focus on the active participants in such reactions. We then consider acid-base reactions, neutralization reactions, and reactions that form gases as products. Next comes a discussion of oxidation-reduction (redox) reactions, oxidation numbers as a means to organize our understanding of redox reactions, and the activity series of metals.

A great deal of chemistry—perhaps most—occurs in solution, and we introduce the means for quantitatively describing the concentrations of solutes in solutions. This discussion is followed by explorations of solution stoichiometry and finally aqueous titration, an analytical technique that is used to measure solute concentrations.

5.1 Exchange Reactions: Precipitation and Net Ionic Equations

Aqueous Solubility of Ionic Compounds

Many of the ionic compounds that you frequently encounter, such as table salt, baking soda, and household plant fertilizers, are soluble in water. It is therefore tempting to conclude that all ionic compounds are soluble in water, but such is not the case. Although many ionic compounds are water-soluble, some are only slightly soluble, and others dissolve hardly at all.

When an ionic compound dissolves in water, its ions separate and become surrounded by water molecules, as illustrated in Figure 5.1a. The process in which ions separate is called *dissociation.* Soluble ionic compounds are one type of strong electrolyte. Recall that an electrolyte is a substance whose aqueous solution contains ions

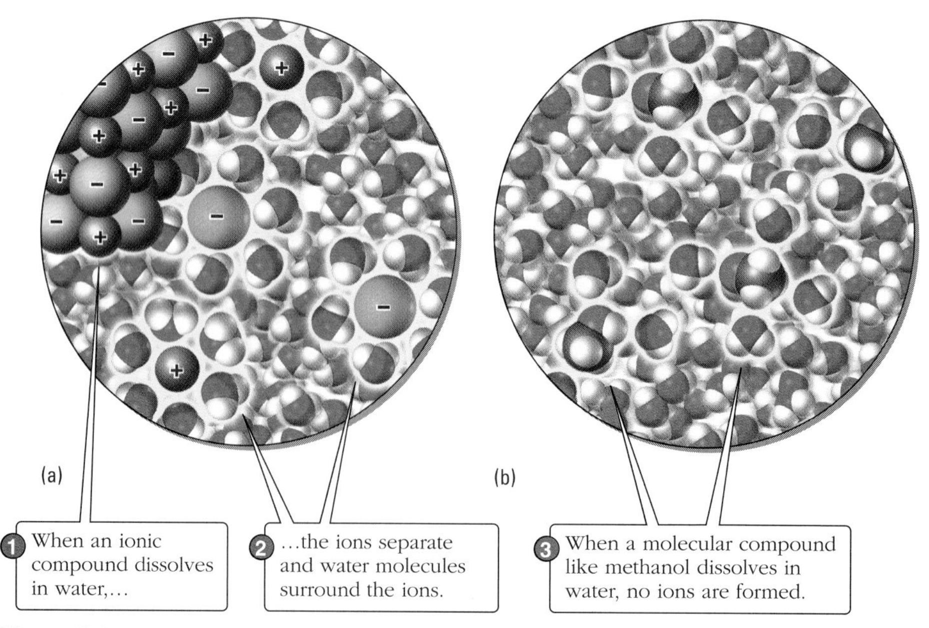
(a)
(b)
1 When an ionic compound dissolves in water,...
2 ...the ions separate and water molecules surround the ions.
3 When a molecular compound like methanol dissolves in water, no ions are formed.

Figure 5.1 Dissolution of (a) an ionic compound and (b) a molecular compound (methanol, CH_3OH) in water.

Table 5.1 Solubility Rules for Ionic Compounds

Usually Soluble	
Group 1A: Li^+, Na^+, K^+, Rb^+, Cs^+, ammonium NH_4^+:	All Group 1A (alkali metal) and ammonium salts are soluble.
Nitrates: NO_3^-	All nitrates are soluble.
Chlorides, bromides, iodides: Cl^-, Br^-, I^-	All common chlorides, bromides, and iodides are soluble except AgCl, Hg_2Cl_2, $PbCl_2$, AgBr, Hg_2Br_2, $PbBr_2$, AgI, Hg_2I_2, PbI_2.
Sulfates: SO_4^{2-}	Most sulfates are soluble; exceptions include $CaSO_4$, $SrSO_4$, $BaSO_4$, and $PbSO_4$.
Chlorates, ClO_3^-	All chlorates are soluble.
Perchlorates, ClO_4^-	All perchlorates are soluble.
Acetates, CH_3COO^-	All acetates are soluble.
Usually Insoluble	
Phosphates, PO_4^{3-}	All phosphates are insoluble except those of NH_4^+ and Group 1A elements (alkali metal cations).
Carbonates, CO_3^{2-}	All carbonates are insoluble except those of NH_4^+ and Group 1A elements (alkali metal cations).
Hydroxides, OH^-	All hydroxides are insoluble except those of NH_4^+ and Group 1A (alkali metal cations). $Sr(OH)_2$, $Ba(OH)_2$, and $Ca(OH)_2$ are slightly soluble.
Oxalates, $C_2O_4^{2-}$	All oxalates are insoluble except those of NH_4^+ and Group 1A (alkali metal cations).
Sulfides, S^{2-}	All sulfides are insoluble except those of NH_4^+, Group 1A (alkali metal cations) and Group 2A (MgS, CaS, and BaS are sparingly soluble).

and therefore conducts electricity. A **strong electrolyte** is completely converted to ions when it forms an aqueous solution. By contrast, most water-soluble molecular compounds do not ionize when they dissolve. This is illustrated in Figure 5.1b.

The solubility rules given in Table 5.1 are general guidelines for predicting the water solubilities of ionic compounds based on the ions they contain. *If a compound contains at least one of the ions indicated for soluble compounds in Table 5.1, then the compound is at least moderately soluble.*

Figure 5.2 shows examples illustrating the solubility rules for a few nitrates, hydroxides, and sulfides. Suppose you want to know whether $NiSO_4$ is soluble in water. $NiSO_4$ contains Ni^{2+} and SO_4^{2-} ions. Although Ni^{2+} is not mentioned in Table 5.1, substances containing SO_4^{2-} are described as soluble (except for $SrSO_4$, $BaSO_4$, and $PbSO_4$). Because $NiSO_4$ contains an ion, SO_4^{2-}, that indicates solubility and $NiSO_4$ is not one of the sulfate exceptions, it is predicted to be soluble.

PROBLEM-SOLVING EXAMPLE 5.1 Using Solubility Rules

Indicate what ions are present in each of these compounds, and then use Table 5.1 to predict whether each compound is water-soluble.

(a) $CaCl_2$ (b) $Fe(OH)_3$ (c) NH_4NO_3
(d) $CuCO_3$ (e) $Ni(ClO_3)_2$

Answer (a) Ca^{2+} and Cl^-, soluble. (b) Fe^{3+} and OH^-, insoluble. (c) NH_4^+ and NO_3^-, soluble. (d) Cu^{2+} and CO_3^{2-}, insoluble. (e) $Ni^{2+}$and ClO_3^-, soluble.

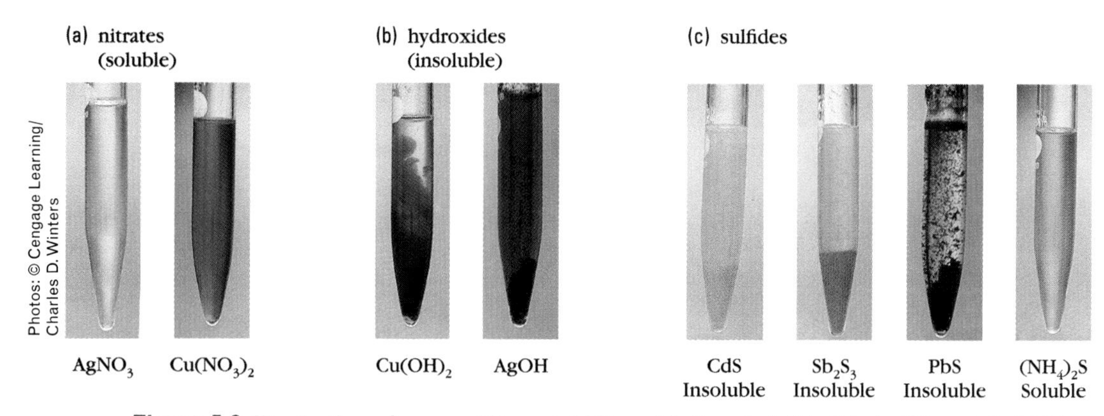

Figure 5.2 Illustration of some of the solubility guidelines in Table 5.1. $AgNO_3$ and $Cu(NO_3)_2$, like all nitrates, are soluble. $Cu(OH)_2$ and AgOH, like most hydroxides, are insoluble. CdS, Sb_2S_3, and PbS, like nearly all sulfides, are insoluble, but $(NH_4)_2S$ is an exception (it is soluble).

Strategy and Explanation The use of the solubility rules requires identifying the ions present and checking their aqueous solubility (Table 5.1).

(a) $CaCl_2$ contains Ca^{2+} and Cl^- ions. All chlorides are soluble, with a few exceptions for transition metals, so calcium chloride is soluble.

(b) $Fe(OH)_3$ contains Fe^{3+} and OH^- ions. As indicated in Table 5.1, all hydroxides are insoluble except alkali metals and a few other exceptions, so iron(III) hydroxide is insoluble.

(c) NH_4NO_3 contains NH_4^+ and NO_3^- ions. All ammonium salts are soluble, and all nitrates are soluble, so NH_4NO_3 is soluble.

(d) $CuCO_3$ contains Cu^{2+} and CO_3^{2-} ions. All carbonates are insoluable except those of ammonium and alkali metals, and copper is not an alkali metal, so $CuCO_3$ is insoluble.

(e) $Ni(ClO_3)_2$ contains Ni^{2+} and ClO_3^- ions. All chlorates are soluble, so $Ni(ClO_3)_2$ is soluble.

PROBLEM-SOLVING PRACTICE 5.1

Predict whether each of these compounds is likely to be water-soluble.

(a) NaF (b) $Ca(CH_3COO)_2$ (c) $SrCl_2$
(d) MgO (e) $PbCl_2$ (f) HgS

Recall that exchange reactions (p. 106) have this reaction pattern:

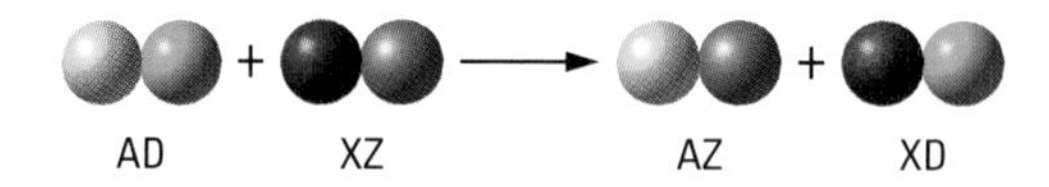

If all the reactants and the products of such a reaction are water-soluble ionic compounds, no overall reaction takes place. In such cases, mixing the solutions of AD and XZ just results in an aqueous solution containing the A, D, X, and Z ions.

What will happen when two aqueous solutions are mixed, one containing dissolved calcium nitrate, $Ca(NO_3)_2$, and the other containing dissolved sodium chloride, NaCl? Both are soluble ionic compounds (Table 5.1), so the resulting solution contains Ca^{2+}, NO_3^-, Na^+, and Cl^- ions. To decide whether a reaction will occur requires determining whether any two of these ions can react with each other to form a *new* compound. For an exchange reaction to occur, the calcium ion and the chloride ion would have to form calcium chloride, $CaCl_2$, and the sodium ion and the nitrate ion would have to form sodium nitrate, $NaNO_3$. Is either a possible chemical reaction? If either product is insoluble the answer is yes. Checking the solubility rules shows that both of these compounds are water-soluble. No reaction to remove the ions from solution is possible; therefore, when these two aqueous solutions are mixed together, no reaction occurs.

If, however, one or both of the potential products of the reaction remove ions from the solution, a reaction will occur. Three different kinds of products can cause an exchange reaction to occur in aqueous solution:

1. Formation of an *insoluble ionic compound:*

$$AgNO_3(aq) + KCl(aq) \longrightarrow KNO_3(aq) + AgCl(s)$$

2. Formation of a *molecular compound* that remains in solution. Most commonly this happens when water is produced in acid-base neutralization reactions:

$$H_2SO_4(aq) + 2\ NaOH(aq) \longrightarrow Na_2SO_4(aq) + 2\ H_2O(\ell)$$

3. Formation of a *gaseous molecular compound* that escapes from the solution:

$$2\ HCl(aq) + Na_2S(aq) \longrightarrow 2\ NaCl(aq) + H_2S(g)$$

© Cengage Learning/Charles D. Winters

Figure 5.3 Precipitation of barium sulfate. Mixing aqueous solutions of barium chloride, $BaCl_2$, and sodium sulfate, Na_2SO_4, forms a precipitate of barium sulfate, $BaSO_4$. Sodium chloride, NaCl, the other product of this exchange reaction, is water soluble, and Na^+ and Cl^- ions remain in solution.

Precipitation Reactions

Consider the possibility of an exchange reaction when aqueous solutions of barium chloride and sodium sulfate are mixed:

$$BaCl_2(aq) + Na_2SO_4(aq) \longrightarrow ? + ?$$

If the barium ions and sodium ions exchange partners to form $BaSO_4$ and NaCl, the equation will be

$$\underset{\text{barium chloride}}{BaCl_2(aq)} + \underset{\text{sodium sulfate}}{Na_2SO_4(aq)} \longrightarrow \underset{\text{barium sulfate}}{BaSO_4} + \underset{\text{sodium chloride}}{2\ NaCl}$$

Will a reaction occur? The answer is yes if an insoluble product—a **precipitate**—can form. Checking Table 5.1, we find that NaCl is soluble, but $BaSO_4$ is not soluble (sulfates of Ca^{2+}, Sr^{2+}, Ba^{2+}, and Pb^{2+} are insoluble). Therefore, an exchange reaction will occur, and solid barium sulfate will precipitate from the solution (Figure 5.3). Precipitate formation is indicated by an (s) next to the precipitate, a solid, in the overall equation. Because it is soluble, NaCl remains dissolved in solution, and we put (aq) next to NaCl in the equation.

$$BaCl_2(aq) + Na_2SO_4(aq) \longrightarrow BaSO_4(s) + 2\ NaCl(aq)$$

Ba^{2+} and Na^+ do not react with each other, and neither do Cl^- and SO_4^{2-}.

PROBLEM-SOLVING EXAMPLE 5.2 Exchange Reactions

For each of these pairs of ionic compounds, decide whether an exchange reaction will occur when their aqueous solutions are mixed, and write a balanced chemical equation for those reactions that will occur.

(a) $(NH_4)_2S$ and $Cu(NO_3)_2$ (b) $ZnCl_2$ and Na_2CO_3 (c) $CaCl_2$ and KNO_3

Answer

(a) CuS precipitates. $(NH_4)_2S(aq) + Cu(NO_3)_2(aq) \longrightarrow CuS(s) + 2\ NH_4NO_3(aq)$

(b) $ZnCO_3$ precipitates. $ZnCl_2(aq) + Na_2CO_3(aq) \longrightarrow ZnCO_3(s) + 2\ NaCl(aq)$

(c) No reaction occurs. Both possible products, $Ca(NO_3)_2$ and KCl, are soluble.

Strategy and Explanation In each case, consider which cation-anion combinations can form, and then decide whether the possible compounds will precipitate. You have to be careful to take into account any polyatomic ions, which must be kept together as a unit during the balancing of the chemical reaction.

(a) An exchange reaction between $(NH_4)_2S$ and $Cu(NO_3)_2$ forms CuS(s) and NH_4NO_3. Table 5.1 shows that all nitrates are soluble, so NH_4NO_3 remains in solution. CuS is not soluble and therefore precipitates.

(b) The exchange reaction between $ZnCl_2$ and Na_2CO_3 forms the insoluble product $ZnCO_3(s)$ and leaves soluble NaCl in solution.
(c) No precipitate forms when $CaCl_2$ and KNO_3 are mixed because each product, $Ca(NO_3)_2$ and KCl, is soluble. All four of the ions (Ca^{2+}, Cl^-, K^+, NO_3^-) remain in solution. No exchange reaction occurs because no product is formed that removes ions from the solution.

PROBLEM-SOLVING PRACTICE 5.2

Predict the products and write a balanced chemical equation for the exchange reaction in aqueous solution between each pair of ionic compounds. Use Table 5.1 to determine solubilities and indicate in the equation whether a precipitate forms.
(a) $NiCl_2$ and NaOH (b) K_2CO_3 and $CaBr_2$

Net Ionic Equations

In writing equations for exchange reactions in the preceding section, we used overall equations. There is another way to represent what happens, however. In each case in which a precipitate forms, the product that does not precipitate remains in solution. Therefore, its ions are in solution as reactants and remain there after the reaction. Such ions are commonly called **spectator ions** because, like the spectators at a play or game, they are present but are not involved directly in the real action. Consequently, the spectator ions can be left out of the equation that represents the chemical change that occurs. An equation that includes only the symbols or formulas of ions in solution or compounds that undergo change is called a **net ionic equation.** We will use the reaction of aqueous NaCl with $AgNO_3$ to form AgCl and $NaNO_3$ to illustrate the general steps for writing a net ionic equation.

Step 1: Write the overall balanced equation using the correct formulas for the reactants and products.

Overall chemical reaction:

$$\underset{\text{silver nitrate}}{AgNO_3} + \underset{\text{sodium chloride}}{NaCl} \longrightarrow \underset{\text{silver chloride}}{AgCl} + \underset{\text{sodium nitrate}}{NaNO_3}$$

Step 1 actually consists of two parts: first, write the unbalanced equation with the correct formulas for reactants and products; second, balance the equation.

Step 2: Use the general guidelines in Table 5.1 to determine the solubilities of reactants and products. In this case, the guidelines indicate that nitrates are soluble, so $AgNO_3$ and $NaNO_3$ are soluble. NaCl is water-soluble because almost all chlorides are soluble. However, AgCl is one of the insoluble chlorides (AgCl, Hg_2Cl_2, and $PbCl_2$). Using this information we can write

$$AgNO_3(aq) + NaCl(aq) \longrightarrow AgCl(s) + NaNO_3(aq)$$

Step 3: Recognize that all soluble ionic compounds dissociate into their component ions in aqueous solution. Therefore we have

$$AgNO_3(aq) \text{ consists of } Ag^+(aq) + NO_3^-(aq).$$

$$NaCl(aq) \text{ consists of } Na^+(aq) + Cl^-(aq).$$

$$NaNO_3(aq) \text{ consists of } Na^+(aq) + NO_3^-(aq).$$

Step 4: Use the ions from Step 3 to write a complete ionic equation with the ions in solution from each soluble compound shown separately.

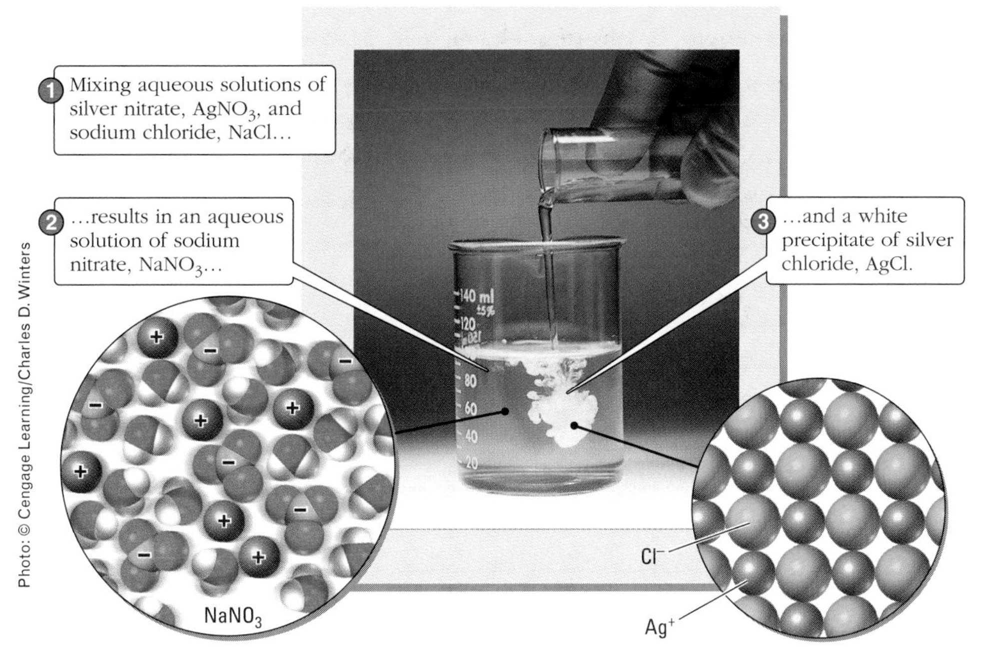

Active Figure 5.4 **Precipitation of silver chloride.** Visit this book's companion website at **www.cengage.com/chemistry/moore** to test your understanding of the concepts in this figure.

Complete ionic equation:

$$Ag^+(aq) + NO_3^-(aq) + Na^+(aq) + Cl^-(aq) \longrightarrow AgCl(s) + Na^+(aq) + NO_3^-(aq)$$

Note that the precipitate is represented by its complete formula (Figure 5.4).

Step 5: ***To obtain the net ionic equation, cancel out the spectator ions from each side of the complete ionic equation.*** Sodium ions and nitrate ions are the spectator ions in this example, and we cancel them from the complete ionic equation to give the net ionic equation.

Complete ionic equation:

$$Ag^+(aq) + \cancel{NO_3^-(aq)} + \cancel{Na^+(aq)} + Cl^-(aq) \longrightarrow AgCl(s) + \cancel{Na^+(aq)} + \cancel{NO_3^-(aq)}$$

Net ionic equation:

$$Ag^+(aq) + Cl^-(aq) \longrightarrow AgCl(s)$$

Step 6: ***Check that the sum of the charges is the same on each side of the net ionic equation.*** For the equation in Step 5 the sum of charges is zero on each side: $(1+) + (1-) = 0$ on the left; AgCl is an ionic compound with zero net charge on the right.

The charge must be the same on both sides of a balanced equation since electrons are neither created nor destroyed.

PROBLEM-SOLVING EXAMPLE 5.3 Net Ionic Equations

Write the net ionic equation that occurs when aqueous solutions of lead nitrate, $Pb(NO_3)_2$, and potassium iodide, KI, are mixed.

Answer $Pb^{2+}(aq) + 2\,I^-(aq) \longrightarrow PbI_2(s)$

Strategy and Explanation Use the stepwise procedure presented above.

Step 1: *Write the overall balanced equation using the correct formulas for the reactants and products.* This is an exchange reaction (***p. 106***).

$$Pb(NO_3)_2 + 2\ KI \longrightarrow PbI_2 + 2\ KNO_3$$

Step 2: *Determine the solubilities of reactants and products.* The solubility rules in Table 5.1 predict that all these reactants and products are soluble except PbI_2.

$$Pb(NO_3)_2(aq) + 2\ KI(aq) \longrightarrow PbI_2(s) + 2\ KNO_3(aq)$$

Step 3: *Identify the ions present when the soluble compounds dissociate in solution.*

1 mol $Pb(NO_3)_2$ contains 2 mol NO_3^- ions along with 1 mol Pb^{2+} ions.

$Pb(NO_3)_2(aq)$ consists of $Pb^{2+}(aq)$ and 2 NO_3^-.

$KI(aq)$ consists of $K^+(aq)$ and $I^-(aq)$.

$KNO_3(aq)$ consists of $K^+(aq)$ and $NO_3^-(aq)$.

Step 4: *Write the complete ionic equation.*

$$Pb^{2+}(aq) + 2\ NO_3^-(aq) + 2\ K^+(aq) + 2\ I^-(aq) \longrightarrow PbI_2(s) + 2\ K^+(aq) + 2\ NO_3^-(aq)$$

Steps 5 and 6: *Cancel spectator ions* [$K^+(aq)$ and $NO_3^-(aq)$] *to get the net ionic equation; check that charge is balanced.*

$$\text{Net ionic equation: } Pb^{2+}(aq) + 2\ I^-(aq) \longrightarrow PbI_2(s)$$

$$\text{Net charge} = (2+) + 2 \times (1-) = 0$$

☑ **Reasonable Answer Check** Each side of the net ionic equation has the same charge and types of atoms.

PROBLEM-SOLVING PRACTICE 5.3

Write a balanced equation for the reaction (if any) for each of these ionic compound pairs in aqueous solution. Then use the complete ionic equation to write their balanced net ionic equations.

(a) $BaCl_2$ and Na_2SO_4 (b) $(NH_4)_2S$ and $FeCl_2$

CONCEPTUAL EXERCISE 5.1 Net Ionic Equations

It is possible for an exchange reaction in which both products precipitate to occur in aqueous solution. Using Table 5.1, identify the reactants and products of an example of such a reaction.

If you live in an area with "hard water," you have probably noticed the scale that forms inside your teakettle or saucepans when you boil water in them. Hard water is mostly caused by the presence of the cations Ca^{2+}, Mg^{2+}, and also Fe^{2+} or Fe^{3+}. When the water also contains hydrogen carbonate ion, HCO_3^-, this reaction occurs when the water is heated:

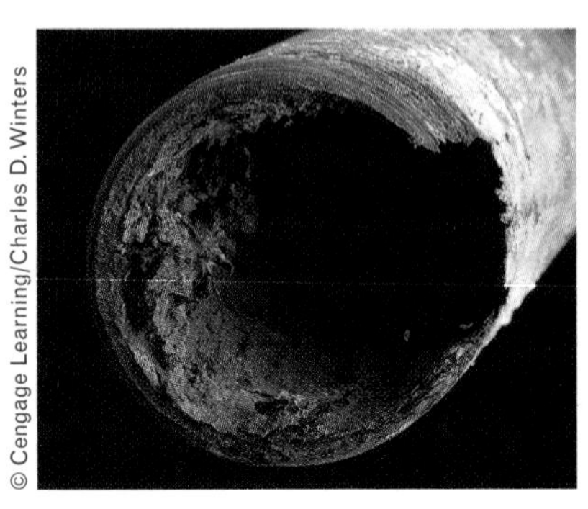

Boiler scale can form inside hot-water pipes.

$$2\ HCO_3^-(aq) \longrightarrow H_2O(\ell) + CO_2(g) + CO_3^{2-}(aq)$$

The carbon dioxide escapes from the hot water, and the hydrogen carbonate ions are slowly converted to carbonate ions. The carbonate ions can form precipitates with the calcium, magnesium, or iron ions to produce a scale that sticks to metal surfaces. In hot-water heating systems in areas with high calcium ion concentrations, the buildup of such *boiler scale* can plug the pipes.

$$Ca^{2+}(aq) + 2\ HCO_3^-(aq) \longrightarrow CaCO_3(s) + H_2O(\ell) + CO_2(g)$$

5.2 Acids, Bases, and Acid-Base Exchange Reactions

Acids and bases are two extremely important classes of compounds—so important that this book devotes two chapters to them later (Chapters 15 and 16). Here, we focus on a few general properties and consider how acids and bases react with each other. Acids have a number of properties in common, and so do bases. Some properties of acids are related to properties of bases. Solutions of acids change the colors of pigments in specific ways. For example, acids change the color of litmus from blue to red and cause the dye phenolphthalein to be colorless. In contrast, bases turn red litmus blue and make phenolphthalein pink. If an acid has made litmus red, adding a base will reverse the effect, making the litmus blue again. Thus, acids and bases seem to be opposites. A base can *neutralize* the effect of an acid, and an acid can neutralize the effect of a base.

Litmus is a dye derived from lichens. Phenolphthalein is a synthetic dye.

Acids have other characteristic properties. They taste sour, they produce bubbles of gas when reacting with limestone, and they dissolve many metals while producing a flammable gas. Although you should never taste substances in a chemistry laboratory, you have probably experienced the sour taste of at least one acid—vinegar, which is a dilute solution of acetic acid in water. Bases, in contrast, have a bitter taste. Soap, for example, contains a base. Rather than dissolving metals, bases often cause metal ions to form insoluble compounds that precipitate from solution. Such precipitates can be made to dissolve by adding an acid, another case in which an acid counteracts a property of a base.

Acids

The properties of acids can be explained by a common feature of acid molecules. An **acid** is any substance that increases the concentration of hydrogen ions, H^+, when dissolved in pure water. The H^+ ion is a hydrogen atom that has lost its one electron; the H^+ ion is just a proton. As a "naked" H^+ ion, it cannot exist by itself in water. Because H^+ is a very small, positively charged species, it interacts strongly with oxygen atoms of water molecules. Thus, H^+ combines with H_2O to form H_3O^+, known as the **hydronium ion.** Chapter 15 explores the importance of the hydronium ion to acid-base chemistry. For now, we represent the hydronium ion as $H^+(aq)$. The properties that acids have in common are those of hydrogen ions dissolved in water.

$$H^+ + H_2O \longrightarrow H_3O^+$$

Acids that are entirely converted to ions (completely ionized) when dissolved in water are *strong electrolytes* and are called **strong acids.** One of the most common strong acids is hydrochloric acid, which ionizes completely in aqueous solution to form hydrogen ions and chloride ions (Figure 5.5a).

$$HCl(aq) \longrightarrow H^+(aq) + Cl^-(aq)$$

The more complete, and proper, way to write an equation for the reaction is

$$HCl(aq) + H_2O(\ell) \longrightarrow H_3O^+(aq) + Cl^-(aq)$$

which explicitly shows the hydronium ion, H_3O^+. Table 5.2 lists some other common acids.

In contrast, acids and other substances that ionize only slightly are termed **weak electrolytes.** Acids that are only partially ionized in aqueous solution are termed **weak acids.** For example, when acetic acid dissolves in water, usually fewer than 5% of the acetic acid molecules are ionized at any time. The remainder of the acetic acid

(a) Strong acid (HCl)

(b) Weak acid (CH_3COOH)

KEY: water molecule; hydronium ion; chloride ion; acetate ion; acetic acid molecule

Figure 5.5 The ionization of acids in water. (a) A strong acid such as hydrochloric acid, HCl, is completely ionized in water; all the HCl molecules ionize to form $H_3O^+(aq)$ and $Cl^-(aq)$ ions. (b) Weak acids such as acetic acid, CH_3COOH, are only slightly ionized in water. Nonionized acetic acid molecules far outnumber aqueous H_3O^+ and CH_3COO^- ions formed by the ionization of acetic acid molecules.

The organic functional group —COOH is present in all organic carboxylic acids.

exists as nonionized molecules. Thus, because acetic acid is only slightly ionized in aqueous solution, it is a weak electrolyte and classified as a weak acid (Figure 5.5b).

$$CH_3COOH(aq) \rightleftharpoons H^+(aq) + CH_3COO^-(aq)$$

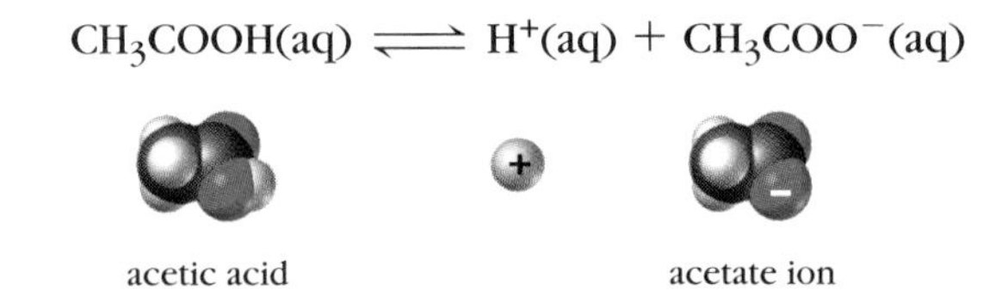

The double arrow in this equation for the ionization of acetic acid signifies a characteristic property of weak electrolytes. They establish a ***dynamic equilibrium*** in solution between the formation of the ions and their undissociated molecular form. In aqueous acetic acid, hydrogen ions and acetate ions recombine to form CH_3COOH molecules.

Some common acids, such as sulfuric acid, can provide more than 1 mol H^+ ions per mole of acid:

$$\underset{\text{sulfuric acid}}{H_2SO_4(aq)} \longrightarrow H^+(aq) + \underset{\text{hydrogen sulfate ion}}{HSO_4^-(aq)}$$

$$\underset{\text{hydrogen sulfate ion}}{HSO_4^-(aq)} \rightleftharpoons H^+(aq) + \underset{\text{sulfate ion}}{SO_4^{2-}(aq)}$$

The first ionization reaction is essentially complete, so sulfuric acid is considered a strong electrolyte (and a strong acid as well). However, the hydrogen sulfate ion, like acetic acid, is only partially ionized, so it is a weak electrolyte and also a weak acid.

CONCEPTUAL
EXERCISE 5.2 Dissociation of Acids

Phosphoric acid, H_3PO_4, has three protons that can ionize. Write the equations for its three ionization reactions, each of which is a dynamic equilibrium.

Table 5.2 Common Acids and Bases

Strong Acids (Strong Electrolytes)		Strong Bases (Strong Electrolytes)	
HCl	Hydrochloric acid	$LiOH$	Lithium hydroxide
HNO_3	Nitric acid	$NaOH$	Sodium hydroxide
H_2SO_4	Sulfuric acid	KOH	Potassium hydroxide
$HClO_4$	Perchloric acid	$Ca(OH)_2$	Calcium hydroxide‡
HBr	Hydrobromic acid	$Ba(OH)_2$	Barium hydroxide‡
HI	Hydroiodic acid	$Sr(OH)_2$	Strontium hydroxide‡
Weak Acids* (Weak Electrolytes)		**Weak Bases† (Weak Electrolytes)**	
H_3PO_4	Phosphoric acid	NH_3	Ammonia
CH_3COOH	Acetic acid	CH_3NH_2	Methylamine
H_2CO_3	Carbonic acid		
HCN	Hydrocyanic acid		
$HCOOH$	Formic acid		
C_6H_5COOH	Benzoic acid		

*Many organic acids are weak acids.
†Many organic amines (related to ammonia) are weak bases.
‡The hydroxides of calcium, barium, and strontium are only slightly soluble, but all that dissolves is dissociated into ions.

Bases

A **base** is a substance that increases the concentration of the **hydroxide ion,** OH^-, when dissolved in pure water. The properties that bases have in common are properties attributable to the aqueous hydroxide ion, $OH^-(aq)$. Compounds that contain hydroxide ions, such as sodium hydroxide or potassium hydroxide, are obvious bases. As ionic compounds they are strong electrolytes and **strong bases.**

$$NaOH(s) \xrightarrow{H_2O} Na^+(aq) + OH^-(aq)$$

A base that is slightly water-soluble, such as $Ca(OH)_2$, can still be a strong electrolyte if the amount of the compound that dissolves completely dissociates into ions.

Acids and bases that are strong electrolytes are strong acids and bases. Acids and bases that are weak electrolytes are weak acids and bases.

Ammonia, NH_3, is another very common base. Although the compound does not have an OH^- ion as part of its formula, it produces the ion by reaction with water.

$$NH_3(aq) + H_2O(\ell) \rightleftharpoons NH_4^+(aq) + OH^-(aq)$$

In the equilibrium between NH_3 and the NH_4^+ and OH^- ions, only a small concentration of the ions is present, so ammonia is a weak electrolyte (<5% ionized), and it is a **weak base.**

To summarize:

- Strong electrolytes are compounds that ionize completely in aqueous solutions. They can be ionic compounds (salts or strong bases) or molecular compounds that are strong acids.

- Weak electrolytes are molecular compounds that are weak acids or bases and establish equilibrium with water.
- Nonelectrolytes are molecular compounds that do not ionize in aqueous solution.

EXERCISE 5.3 Acids and Bases

(a) What ions are produced when perchloric acid, $HClO_4$, dissolves in water?
(b) Calcium hydroxide is only slightly soluble in water. What little does dissolve, however, is dissociated. What ions are produced? Write an equation for the dissociation of calcium hydroxide.

PROBLEM-SOLVING EXAMPLE 5.4 Strong Electrolytes, Weak Electrolytes, and Nonelectrolytes

Identify whether each of these substances in an aqueous solution will be a strong electrolyte, a weak electrolyte, or a nonelectrolyte: HBr, hydrogen bromide; LiOH, lithium hydroxide; HCOOH, formic acid; CH_3CH_2OH, ethanol.

Answer HBr is a strong electrolyte; LiOH is a strong electrolyte; HCOOH is a weak electrolyte; CH_3CH_2OH is a nonelectrolyte.

Strategy and Explanation For the acids and bases we refer to Table 5.2. Hydrogen bromide is a common strong acid and, therefore, is a strong electrolyte. Lithium hydroxide is a common strong base and completely dissociates into ions in aqueous solution, so it is a strong electrolyte. Formic acid is a weak acid because it only partially ionizes in aqueous solution, so it is a weak electrolyte. Ethanol is a molecular compound that does not dissociate into ions in aqueous solution, so it is a nonelectrolye.

PROBLEM-SOLVING PRACTICE 5.4

Look back through the discussion of electrolytes and Table 5.2 and identify at least one additional strong electrolyte, one additional weak electrolyte, and one additional nonelectrolyte beyond those discussed in Problem-Solving Example 5.4.

(a)

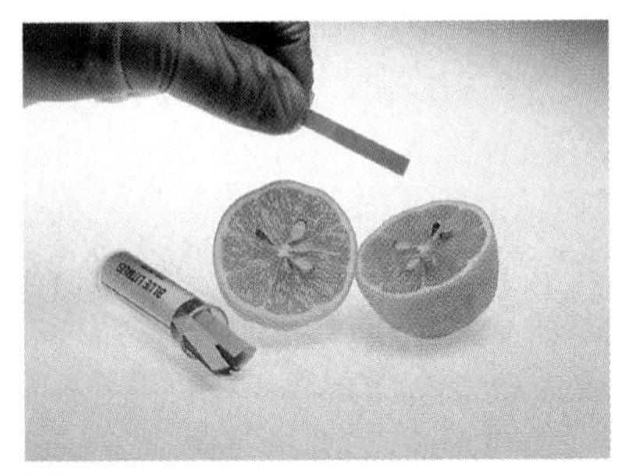

(b)

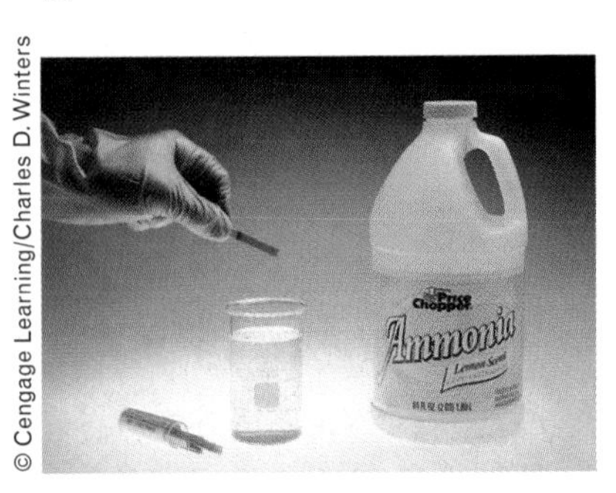

(c)

Acids and bases. (a) Many common foods and household products are acidic or basic. Citrus fruits contain citric acid, and household ammonia and oven cleaner are basic. (b) The acid in lemon juice turns blue litmus paper red, whereas (c) household ammonia turns red litmus paper blue.

Neutralization Reactions

When aqueous solutions of a strong acid (such as HCl) and a strong base (such as NaOH) are mixed, the ions in solution are the hydrogen ion and the anion from the acid, the metal cation, and the hydroxide ion from the base:

From hydrochloric acid: $H^+(aq)$, $Cl^-(aq)$
From sodium hydroxide: $Na^+(aq)$, $OH^-(aq)$

As in precipitation reactions, an exchange reaction will occur *if two of these ions can react with each other to form a compound that removes ions from solution*. In an acid-base reaction, that compound is water, formed by the combination of $H^+(aq)$ with $OH^-(aq)$.

When a strong acid and a strong base react, they neutralize each other. This happens because the hydrogen ions from the acid react with hydroxide ions from the base to form water. Water, the product, is a molecular compound. The other ions remain in the solution, which is an aqueous solution of a **salt** (an ionic compound whose cation comes from a base and whose anion comes from an acid). If the water were evaporated, the solid salt would remain. In the case of HCl plus NaOH, the salt is sodium chloride, NaCl.

$$\underset{\text{acid}}{HCl(aq)} + \underset{\text{base}}{NaOH(aq)} \longrightarrow \underset{\text{water}}{H\,OH(\ell)} + \underset{\text{salt}}{Na\,Cl(aq)}$$

The overall neutralization reaction can be written more generally as

$$\underset{\text{acid}}{HX(aq)} + \underset{\text{base}}{MOH(aq)} \longrightarrow \underset{\text{water}}{H\,OH(\ell)} + \underset{\text{salt}}{M\,X(aq)}$$

You should recognize this as an exchange reaction in which the $H^+(aq)$ ions from the aqueous acid and the $M^+(aq)$ ions from the metal hydroxide are exchange partners, as are the X^- and OH^- ions.

The salt that forms depends on the acid and base that react. Magnesium chloride, another salt, is formed when a commercial antacid containing magnesium hydroxide is swallowed to neutralize excess hydrochloric acid in the stomach.

$$\underset{\text{hydrochloric acid}}{2\,HCl(aq)} + \underset{\text{magnesium hydroxide}}{Mg(OH)_2(s)} \longrightarrow 2\,H_2O(\ell) + \underset{\text{magnesium chloride}}{MgCl_2(aq)}$$

Milk of magnesia consists of a suspension of finely divided particles of $Mg(OH)_2(s)$ in water.

Organic acids, such as acetic acid and propionic acid, which contain the acid functional group —COOH, also neutralize bases to form salts. The H in the —COOH functional group is the acidic proton. Its removal generates the $—COO^-$ anion. The reaction of propionic acid, CH_3CH_2COOH, and sodium hydroxide produces the salt sodium propionate, $NaCH_3CH_2COO$, containing sodium ions, Na^+, and propionate ions, $CH_3CH_2COO^-$. Sodium propionate is commonly used as a food preservative.

The —COOH combination, called the *acid functional group,* is present in organic acids and imparts acidic properties to compounds containing it.

$$\underset{\text{propionic acid}}{CH_3CH_2COOH(aq)} + NaOH(aq) \longrightarrow H_2O(\ell) + \underset{\text{sodium propionate}}{NaCH_3CH_2COO(aq)}$$

Although the propionic acid molecule contains a number of H atoms, it is only the H atom that is part of the acid functional group, —COOH, involved in this neutralization reaction.

Note that this reaction is analogous to the general reaction shown above on this page.

PROBLEM-SOLVING EXAMPLE 5.5 Balancing Neutralization Equations

Write a balanced chemical equation for the reaction of nitric acid, HNO_3, with calcium hydroxide, $Ca(OH)_2$, in aqueous solution.

Answer $2\,HNO_3(aq) + Ca(OH)_2(aq) \longrightarrow Ca(NO_3)_2(aq) + H_2O\,(\ell)$

Strategy and Explanation This is a neutralization reaction between an acid and a base, so the products are a salt and water. We begin by writing the unbalanced equation with all the substances.

$$\text{(unbalanced equation)} \quad HNO_3(aq) + Ca(OH)_2(aq) \longrightarrow Ca(NO_3)_2(aq) + H_2O\,(\ell)$$

It is generally a good idea to start with the ions and balance the hydrogen and oxygen atoms later. The calcium ions are in balance, but we need to add a coefficient of 2 to the nitric acid since two nitrate ions appear in the products.

$$\text{(unbalanced equation)} \quad 2\,HNO_3(aq) + Ca(OH)_2(aq) \longrightarrow Ca(NO_3)_2(aq) + H_2O\,(\ell)$$

All the coefficients are correct except for water. We count four hydrogen atoms in the reactants (two from nitric acid and two from calcium hydroxide), so we must put a coefficient of 2 in front of the water to balance the equation.

$$\text{(balanced equation)} \quad 2\,HNO_3(aq) + Ca(OH)_2(aq) \longrightarrow Ca(NO_3)_2(aq) + 2\,H_2O\,(\ell)$$

☑ **Reasonable Answer Check** We note that there are eight oxygen atoms in the reactants (six from nitric acid and two from calcium hydroxide) and there are eight oxygen atoms in the products (six from calcium nitrate and two from water).

PROBLEM-SOLVING PRACTICE 5.5

Write a balanced equation for the reaction of phosphoric acid, H_3PO_4, with sodium hydroxide, NaOH.

PROBLEM-SOLVING EXAMPLE 5.6 Acids, Bases, and Salts

Identify the acid and base used to form each of these salts: (a) $CaSO_4$, (b) $Mg(ClO_4)_2$. Write balanced equations for the formation of these compounds.

Answer (a) Calcium hydroxide, $Ca(OH)_2$, and sulfuric acid, H_2SO_4. (b) Magnesium hydroxide, $Mg(OH)_2$, and perchloric acid, $HClO_4$.

Strategy and Explanation A salt is formed from the cation of a base and the anion of an acid. (a) $CaSO_4$ contains calcium and sulfate ions. Ca^{2+} ions come from $Ca(OH)_2$, calcium hydroxide, and SO_4^{2-} ions come from H_2SO_4, sulfuric acid. The neutralization reaction between $Ca(OH)_2$ and H_2SO_4 produces $CaSO_4$ and water.

$$Ca(OH)_2(aq) + H_2SO_4(aq) \longrightarrow CaSO_4(s) + 2\ H_2O\ (\ell)$$

(b) Magnesium perchlorate contains magnesium and perchlorate ions, Mg^{2+} and ClO_4^-. Mg^{2+} ions could be derived from $Mg(OH)_2$, magnesium hydroxide, and ClO_4^- ions could be derived from $HClO_4$, perchloric acid. The neutralization reaction between $Mg(OH)_2$ and $HClO_4$ produces $Mg(ClO_4)_2$ and water.

$$Mg(OH)_2(aq) + 2\ HClO_4(aq) \longrightarrow Mg(ClO_4)_2(aq) + 2\ H_2O\ (\ell)$$

☑ **Reasonable Answer Check** The final neutralization equations each have the same types and numbers of atoms on each side.

PROBLEM-SOLVING PRACTICE 5.6

Identify the acid and the base that can react to form (a) $MgSO_4$ and (b) $SrCO_3$.

Net Ionic Equations for Acid-Base Reactions

Net ionic equations can be written for acid-base reactions as well as for precipitation reactions. This should not be surprising because precipitation and acid-base neutralization reactions are both exchange reactions.

Consider the reaction given earlier of magnesium hydroxide with hydrochloric acid to relieve excess stomach acid, HCl. The overall balanced equation is

$$2\ HCl(aq) + Mg(OH)_2(s) \longrightarrow 2\ H_2O(\ell) + MgCl_2(aq)$$

The acid and base furnish hydrogen ions and hydroxide ions, respectively.

$$2\ HCl(aq) \longrightarrow 2\ H^+(aq) + 2\ Cl^-(aq)$$

$$Mg(OH)_2(s) \rightleftharpoons Mg^{2+}(aq) + 2\ OH^-(aq)$$

Although magnesium hydroxide is not very soluble, the little that dissolves is completely dissociated.

Note that we retain the coefficients from the balanced overall equation (first step).

We now use this information to write a complete ionic equation. We use Table 5.1 to check the solubility of the product salt, $MgCl_2$. Magnesium chloride is soluble, so the Mg^{2+} and Cl^- ions remain in solution. The complete ionic equation is

$$\cancel{Mg^{2+}(aq)} + 2\ OH^-(aq) + 2\ H^+(aq) + \cancel{2\ Cl^-(aq)} \longrightarrow \cancel{Mg^{2+}(aq)} + \cancel{2\ Cl^-(aq)} + 2\ H_2O(\ell)$$

Canceling spectator ions from each side of the complete ionic equation yields the net ionic equation. In this case, magnesium ions and chloride ions are the spectator ions. Canceling them leaves us with this net ionic equation:

$$2\ H^+(aq) + 2\ OH^-(aq) \longrightarrow 2\ H_2O(\ell)$$

or simply

$$H^+(aq) + OH^-(aq) \longrightarrow H_2O(\ell)$$

This is the net ionic equation for the neutralization reaction between a strong acid and a strong base that yields a soluble salt. Note that, as always, there is *conservation of charge* in the net ionic equation. On the left, (1+) + (1−) = 0; on the right, water has zero net charge.

Next, consider a neutralization reaction between a weak acid, HCN, and a strong base, KOH.

$$HCN(aq) + KOH(aq) \longrightarrow KCN(aq) + H_2O(\ell)$$

The weak acid HCN is not completely ionized, so we leave it in the molecular form, but KOH and KCN are strong electrolytes. The complete ionic equation is

$$HCN(aq) + \cancel{K^+(aq)} + OH^-(aq) \longrightarrow \cancel{K^+(aq)} + CN^-(aq) + H_2O(\ell)$$

Canceling spectator ions yields

$$HCN(aq) + OH^-(aq) \longrightarrow CN^-(aq) + H_2O(\ell)$$

The net ionic equation for the neutralization of a weak acid with a strong base contains the molecular form of the acid and the anion of the salt. The net ionic equation shows that charge is conserved.

PROBLEM-SOLVING EXAMPLE 5.7 Neutralization Reaction with a Weak Acid

Write a balanced equation for the reaction of acetic acid, CH_3COOH, with calcium hydroxide, $Ca(OH)_2$. Then write the net ionic equation for this neutralization reaction.

Answer

Equation: $2\ CH_3COOH(aq) + Ca(OH)_2(aq) \longrightarrow Ca(CH_3COO)_2(aq) + 2\ H_2O(\ell)$

Net ionic equation: $CH_3COOH(aq) + OH^-(aq) \longrightarrow CH_3COO^-(aq) + H_2O(\ell)$

Strategy and Explanation We have been given the formula of an acid and a base that will react. The two products of the neutralization reaction are water and the salt calcium acetate, $Ca(CH_3COO)_2$, formed from the base's cation, Ca^{2+}, and the acid's anion, CH_3COO^-. Table 5.1 shows that $Ca(CH_3COO)_2$ is soluble.

We start by writing the four species involved in the reaction, not worrying for the moment about balancing the equation.

(unbalanced equation) $CH_3COOH(aq) + Ca(OH)_2(aq) \longrightarrow Ca(CH_3COO)_2(aq) + H_2O(\ell)$

To balance the equation, two hydrogen ions (H^+) from the acetic acid react with the two hydroxide ions (OH^-) of the calcium hydroxide. This also means that two water molecules will be produced by the reaction. We must put coefficients of 2 in front of the acetic acid and in front of water to balance the equation.

(balanced equation) $2\ CH_3COOH(aq) + Ca(OH)_2(aq) \longrightarrow Ca(CH_3COO)_2(aq) + 2\ H_2O(\ell)$

To write the net ionic equation, we must know whether the four substances involved in the reaction are strong or weak electrolytes. Acetic acid is a weak electrolyte (it is a weak acid). Calcium hydroxide is a strong electrolyte (strong base). Calcium acetate is a strong electrolyte (Table 5.1). Water is a molecular compound and a nonelectrolyte. We now write the complete ionic equation

$$2\ CH_3COOH(aq) + Ca^{2+}(aq) + 2\ OH^-(aq) \longrightarrow Ca^{2+}(aq) + 2\ CH_3COO^-(aq) + 2\ H_2O(\ell)$$

The calcium ions are spectator ions and are canceled to give the net ionic equation:

$$CH_3COOH(aq) + OH^-(aq) \longrightarrow CH_3COO^-(aq) + H_2O(\ell)$$

☑ **Reasonable Answer Check** Note that the spectator ions cancel and each side of the net ionic equation has the same charge (1−) and the same number and types of atoms.

PROBLEM-SOLVING PRACTICE 5.7

Write a balanced equation for the reaction of hydrocyanic acid, HCN, with calcium hydroxide, $Ca(OH)_2$. Then write the balanced complete ionic equation and the net ionic equation for this neutralization reaction.

> **EXERCISE 5.4 Neutralizations and Net Ionic Equations**
>
> Write balanced complete ionic equations and net ionic equations for the neutralization reactions of these acids and bases:
>
> (a) HCl and KOH
>
> (b) H_2SO_4 and $Ba(OH)_2$ (Remember that sulfuric acid can provide 2 mol $H^+(aq)$ per 1 mol sulfuric acid.)
>
> (c) CH_3COOH and NaOH

Figure 5.6 Reaction of calcium carbonate with an acid. A piece of coral that is largely calcium carbonate, $CaCO_3$, reacts readily with hydrochloric acid to give CO_2 gas and aqueous calcium chloride.

> **EXERCISE 5.5 Net Ionic Equations and Antacids**
>
> The commercial antacids Maalox, Di-Gel tablets, and Mylanta contain aluminum hydroxide or magnesium hydroxide that reacts with excess hydrochloric acid in the stomach. Write the balanced complete ionic equation and net ionic equation for the soothing neutralization reaction of aluminum hydroxide with HCl. Assume that dissolved aluminum hydroxide is completely dissociated.

Gas-Forming Exchange Reactions

The formation of a gas is the third way that exchange reactions can occur, since formation of the gas removes the molecular product from the solution. Escape of the gas from the solution removes ions from the solution. Acids are involved in many gas-forming exchange reactions.

The reaction of a metal carbonate with an acid is an excellent example of a gas-forming exchange reaction (Figure 5.6).

$$CaCO_3(s) + 2\,HCl(aq) \longrightarrow CaCl_2(aq) + \cancel{H_2CO_3(aq)}$$

$$\cancel{H_2CO_3(aq)} \longrightarrow H_2O(\ell) + CO_2(g)$$

$$\text{Overall reaction:}\quad CaCO_3(s) + 2\,HCl(aq) \longrightarrow CaCl_2(aq) + H_2O(\ell) + CO_2(g)$$

A salt and H_2CO_3 (carbonic acid) are always the products from an acid reacting with a metal carbonate, and their formation illustrates the exchange reaction pattern. Carbonic acid is unstable, however, and much of it is rapidly converted to water and CO_2 gas. If the reaction is done in an open container, most of the gas will bubble out of the solution.

Carbonates (which contain CO_3^{2-}) and hydrogen carbonates (which contain HCO_3^-) are bases because they react with protons (H^+ ions) in neutralization reactions. Carbon dioxide is always released when acids react with a metal carbonate or a metal hydrogen carbonate. For example, excess hydrochloric acid in the stomach is neutralized by ingesting commercial antacids such as Alka-Seltzer ($NaHCO_3$), Tums ($CaCO_3$), or Di-Gel liquid ($MgCO_3$). Taking an Alka-Seltzer or a Tums to relieve excess stomach acid produces these helpful reactions:

$$\text{Alka-Seltzer:}\quad NaHCO_3(aq) + HCl(aq) \longrightarrow NaCl(aq) + H_2O(\ell) + CO_2(g)$$

$$\text{Tums:}\quad CaCO_3(aq) + 2\,HCl(aq) \longrightarrow CaCl_2(aq) + H_2O(\ell) + CO_2(g)$$

The net ionic equations for these two reactions are

$$HCO_3^-(aq) + H^+(aq) \longrightarrow H_2O(\ell) + CO_2(g)$$

$$CO_3^{2-}(aq) + 2\,H^+(aq) \longrightarrow H_2O(\ell) + CO_2(g)$$

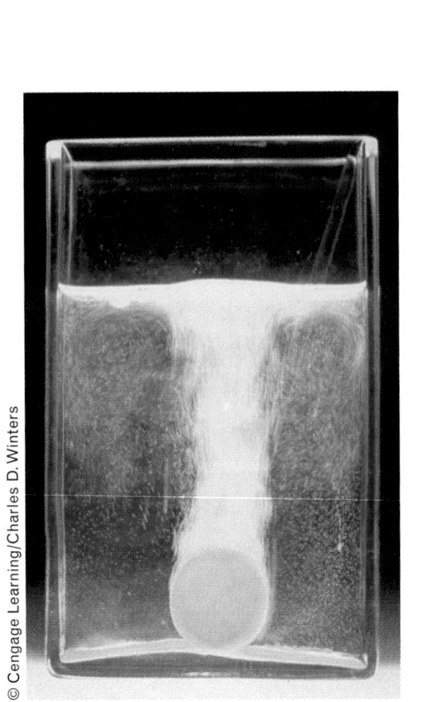

Antacid reacting with HCl.

Acids also react by exchange reactions with metal sulfites or sulfides to produce foul-smelling gaseous SO_2 or H_2S, respectively. With sulfites, the initial product is sulfurous acid, which, like carbonic acid, quickly decomposes.

$$CaSO_3(aq) + 2\,HCl(aq) \longrightarrow CaCl_2(aq) + \cancel{H_2SO_3(aq)}$$
$$\cancel{H_2SO_3(aq)} \longrightarrow H_2O(\ell) + SO_2(g)$$

Overall reaction: $$CaSO_3(aq) + 2\,HCl(aq) \longrightarrow CaCl_2(aq) + H_2O(\ell) + SO_2(g)$$

With sulfides, the gaseous product H_2S is formed directly.

$$Na_2S(aq) + 2\,HCl(aq) \longrightarrow 2\,NaCl(aq) + H_2S(g)$$

EXERCISE 5.6 Gas-Forming Reactions

Predict the products and write the balanced overall equation and the net ionic equation for each of these gas-generating reactions.

(a) $Na_2CO_3(aq) + H_2SO_4(aq) \longrightarrow$
(b) $FeS(s) + HCl(aq) \longrightarrow$
(c) $K_2SO_3(aq) + HCl(aq) \longrightarrow$

CONCEPTUAL EXERCISE 5.7 Exchange Reaction Classification

Identify each of these exchange reactions as a precipitation reaction, an acid-base reaction, or a gas-forming reaction. Predict the products of each reaction and write an overall balanced equation and net ionic equation for the reaction.

(a) $NiCO_3(s) + H_2SO_4(aq) \longrightarrow$
(b) $Sr(OH)_2(s) + HNO_3(aq) \longrightarrow$
(c) $BaCl_2(aq) + Na_2C_2O_4(aq) \longrightarrow$
(d) $PbCO_3(s) + H_2SO_4(aq) \longrightarrow$

5.3 Oxidation-Reduction Reactions

Now we turn to oxidation-reduction reactions, which are classified by what happens with electrons at the nanoscale level as a result of the reaction.

The terms "oxidation" and "reduction" come from reactions that have been known for centuries. Ancient civilizations learned how to change metal oxides and sulfides to the metal—that is, how to *reduce* ore to the metal. For example, cassiterite or tin(IV) oxide, SnO_2, is a tin ore discovered in Britain centuries ago. It is very easily reduced to tin by heating with carbon. In this reaction, tin is reduced from tin(IV) in the ore to tin metal.

SnO_2 loses oxygen and is reduced.

$$SnO_2(s) + 2\,C(s) \longrightarrow Sn(s) + 2\,CO(g)$$

When SnO_2 is reduced by carbon, oxygen is removed from the tin and added to the carbon, which is "oxidized" by the addition of oxygen. In fact, any process in which oxygen is added to another substance is an oxidation.

When magnesium burns in air, the magnesium is oxidized to magnesium oxide, MgO.

Mg combines with oxygen and is oxidized.

$$2\,Mg(s) + O_2(g) \longrightarrow 2\,MgO(s)$$

The experimental observations we have just outlined point to several fundamental conclusions:

- If one reactant is oxidized, another reactant in the same reaction must simultaneously be reduced. For this reason, we refer to such reactions as **oxidation-reduction reactions,** or **redox reactions** for short.
- Oxidation is the reverse of reduction.

Figure 5.7 Oxidation of copper metal by silver ion. A spiral of copper wire was immersed in an aqueous solution of silver nitrate, $AgNO_3$. With time, the copper reduces Ag^+ ions to silver metal crystals, and the copper metal is oxidized to Cu^{2+} ions. The blue color of the solution is due to the presence of aqueous copper(II) ion.

For example, the reactions we have just described show that addition of oxygen is oxidation and removal of oxygen is reduction. But oxidation and reduction are more than that, as we see next.

Redox Reactions and Electron Transfer

Oxidation and reduction reactions involve transfer of electrons from one reactant to another. When a substance *accepts electrons*, it is said to be **reduced.** The language is descriptive because in a **reduction** there is a decrease (reduction) in the real or apparent electric charge on an atom. For example, in this net ionic equation, Ag^+ ions are reduced to uncharged Ag atoms by accepting electrons from copper atoms (Figure 5.7).

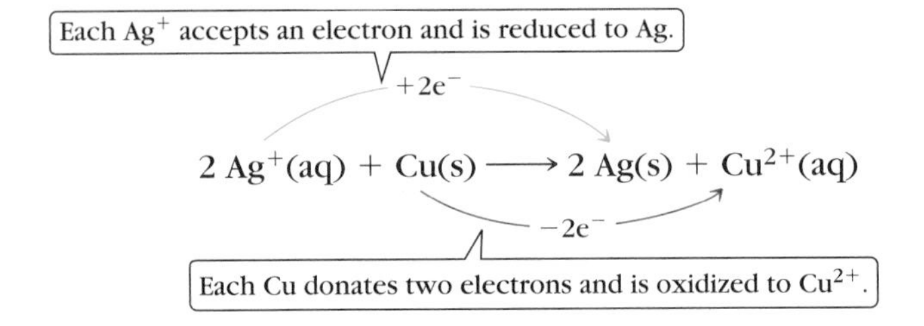

When a substance *loses electrons*, it is said to be **oxidized.** In **oxidation,** the real or apparent electrical charge on an atom of the substance increases when it gives up electrons. In our example, a copper metal atom releases two electrons forming Cu^{2+}; its electric charge has increased from zero to 2+, and the copper atom is said to have been oxidized. For this to happen, something must be available to take the electrons donated by the copper. In this case, Ag^+ is the electron acceptor. ***In every oxidation-reduction reaction, a reactant is reduced and a reactant is oxidized.***

Oxidation is the loss of electrons.

$$X \rightarrow X^+ + e^-$$

X loses one or more electrons and is oxidized.

Reduction is the gain of electrons.

$$Y + e^- \rightarrow Y^-$$

Y gains one or more electrons and is reduced.

In the reaction of magnesium with oxygen (Figure 5.8), oxygen gains electrons when converted to the oxide ion. The charge of each O atom changes from 0 to 2− as it is reduced.

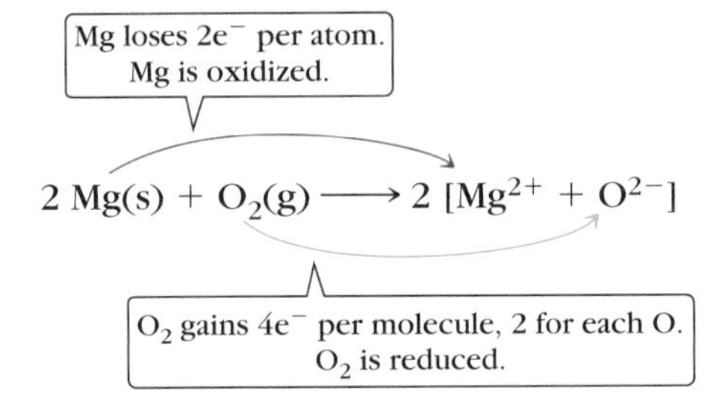

Each magnesium atom changes from 0 to 2+ as it is oxidized. All redox reactions can be analyzed in a similar manner.

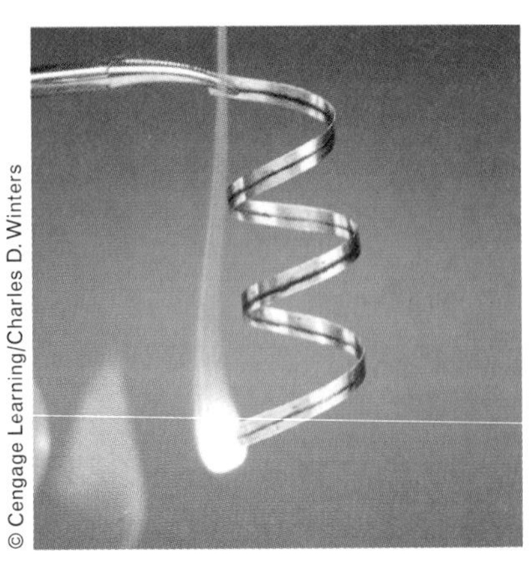

Figure 5.8 $Mg(s) + O_2(g)$. A piece of magnesium ribbon burns in air, oxidizing the metal to the white solid magnesium oxide, MgO.

Common Oxidizing and Reducing Agents

As stated above, in every redox reaction, a reactant is oxidized and a reactant is reduced. The species causing the oxidation (electron loss) is the **oxidizing agent,** and the species causing the reduction (electron gain) is the **reducing agent.** As a redox reaction proceeds, *the oxidizing agent is reduced and the reducing agent is oxidized.* In the reaction just described between Mg and O_2, Mg is oxidized, and it is the reducing agent; O_2 is reduced, and it is the oxidizing agent. Note that the oxidizing agent and the reducing agent are always reactants, not products. Figure 5.9 provides some guidelines to determine which species involved in a redox reaction is the oxidizing agent and which is the reducing agent.

Like oxygen, the halogens (F_2, Cl_2, Br_2, and I_2) are always oxidizing agents in their reactions with metals and most nonmetals. For example, consider the combination reaction of sodium metal with chlorine:

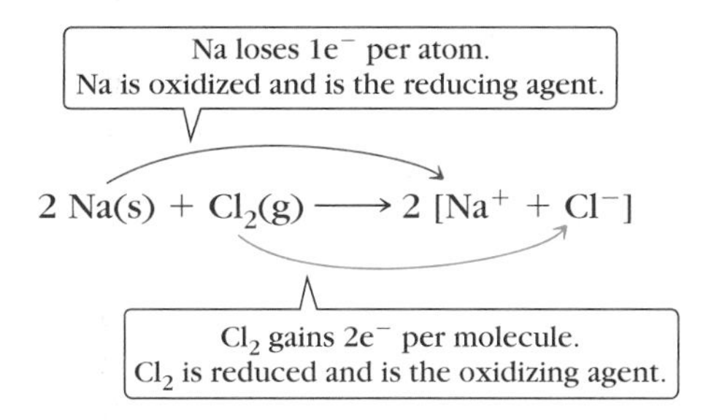

Na loses 1e⁻ per atom.
Na is oxidized and is the reducing agent.
2 Na(s) + Cl2(g) ⟶ 2 [Na+ + Cl−]
Cl2 gains 2e⁻ per molecule.
Cl2 is reduced and is the oxidizing agent.

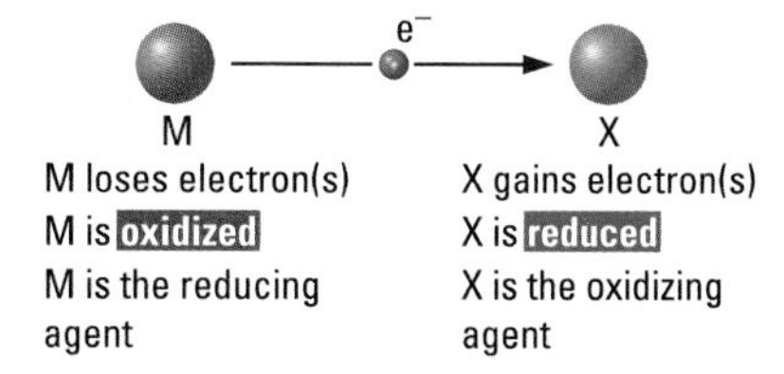

e⁻
M
X
M loses electron(s)
X gains electron(s)
M is oxidized
X is reduced
M is the reducing agent
X is the oxidizing agent

Figure 5.9 Oxidation-reduction relationships and electron transfer.

A useful memory aid for keeping the oxidation and reduction definitions straight is **OIL RIG (Oxidation Is Loss; Reduction Is Gain).**

Here sodium begins as the metallic element, but it ends up as the Na^+ ion after combining with chlorine. Thus, sodium is oxidized (loses electrons) and is the reducing agent. Chlorine ends up as Cl^-; Cl_2 has been reduced (gains electrons) and therefore is the oxidizing agent. The general reaction for halogen, X_2, reduction is

$$\text{Reduction reaction:} \quad \underset{\text{oxidizing agent}}{X_2} + 2e^- \longrightarrow 2\,X^-$$

That is, a halogen will always oxidize a metal to give a metal halide, and the formula of the product can be predicted from the charge on the metal ion and the charge of the halide. The halogens in decreasing order of oxidizing ability are as follows:

Note that the oxidizing agent is reduced, and the reducing agent is oxidized.

Oxidizing Agent	Usual Reduction Product
F_2 (strongest)	F^-
Cl_2	Cl^-
Br_2	Br^-
I_2 (weakest)	I^-

EXERCISE 5.8 Oxidizing and Reducing Agents

Identify which species is losing electrons and which is gaining electrons, which is oxidized and which is reduced, and which is the oxidizing agent and which is the reducing agent in this reaction:

$$2\,Ca(s) + O_2(g) \longrightarrow 2\,CaO(s)$$

EXERCISE 5.9 Redox Reactions

Write the chemical equation for chlorine gas undergoing a redox reaction with calcium metal. Which species is the oxidizing agent?

Chlorine is widely used as an oxidizing agent in water and sewage treatment. A common contaminant of water is hydrogen sulfide, H_2S, which gives a thoroughly unpleasant "rotten egg" odor to the water and may come from the decay of organic matter or from underground mineral deposits. Chlorine oxidizes H_2S to insoluble elemental sulfur, which is easily removed.

$$8\,Cl_2(g) + 8\,H_2S(aq) \longrightarrow S_8(s) + 16\,HCl(aq)$$

Oxidation and reduction occur readily when a strong oxidizing agent comes into contact with a strong reducing agent. Knowing the easily recognized oxidizing and reducing agents enables you to predict that a reaction will take place when they are combined and in some cases to predict what the products will be. Table 5.3 and the following points provide some guidelines.

- An element that has combined with oxygen has been oxidized. In the process each oxygen atom in oxygen, O_2, gains two electrons and becomes the oxide ion, O^{2-} (as in a metal oxide). Oxygen can also be combined in a molecule such as CO_2 or H_2O (as occurs in the combustion reaction of a hydrocarbon). Therefore, oxygen has been reduced. Since it has accepted electrons, oxygen is the oxidizing agent in such cases.
- An element that has combined with a halogen has been oxidized. In the process the halogen, X_2, is changed to halide ions, X^-, by adding an electron to each halogen atom. Therefore, the halogen atom has been reduced to the halide ion, and the halogen is the oxidizing agent. A halogen can also be combined in a molecule such as HCl. Among the halogens, fluorine and chlorine are particularly strong oxidizing agents.
- When an elemental metal combines with something to form a compound, the metal has been oxidized. In the process, it has lost electrons, usually to form a positive ion.

There are exceptions to the guideline that metals are always positively charged in compounds. However, you probably will not encounter these exceptions in introductory chemistry.

$$\text{Oxidation reaction:} \quad \underset{\text{reducing agent}}{M} \longrightarrow M^{n+} + ne^-$$

Therefore, the metal (an electron donor) has been oxidized and has functioned as a reducing agent. Most metals are reasonably good reducing agents, and metals such as sodium, magnesium, and aluminum from Groups 1A, 2A, and 3A are particularly good ones.

- Other common oxidizing and reducing agents are listed in Table 5.3, and some are described below. When one of these agents takes part in a reaction, it is reasonably certain that it is a redox reaction. (Nitric acid can be an exception. In addition to being a good oxidizing agent, it is an acid and functions only as an acid in reactions such as the decomposition of a metal carbonate, a non-redox reaction.)

Table 5.3 Common Oxidizing and Reducing Agents

Oxidizing Agent	Reaction Product	Reducing Agent	Reaction Product
O_2	O^{2-} or an oxygen-containing molecular compound	H_2 or hydrogen-containing molecular compound	H^+ or H combined in H_2O
H_2O_2 (hydrogen peroxide)	$H_2O(\ell)$	C used to reduce metal oxides	CO and CO_2
F_2, Cl_2, Br_2, or I_2 (halogens)	F^-, Cl^-, Br^-, or I^- (halide ions)	M, metals such as Na, K, Fe, or Al	M^{n+}, metal ions such as Na^+, K^+, Fe^{3+}, or Al^{3+}
HNO_3 (nitric acid)	Nitrogen oxides such as NO and NO_2		
$Cr_2O_7^{2-}$ (dichromate ion)	Cr^{3+} (chromium(III) ion), in acid solution		
MnO_4^- (permanganate ion)	Mn^{2+} (manganese(II) ion), in acid solution		

Figure 5.10 illustrates the action of concentrated nitric acid, HNO_3, as an oxidizing agent. Nitric acid oxidizes copper metal to give copper(II) nitrate, and copper metal reduces nitric acid to the brown gas NO_2. The net ionic equation is

$$\underset{\text{reducing agent}}{Cu(s)} + 4\,H^+(aq) + \underset{\text{oxidizing agent}}{2\,NO_3^-(aq)} \longrightarrow Cu^{2+}(aq) + 2\,NO_2(g) + 2\,H_2O(\ell)$$

© Cengage Learning/Charles D. Winters

Figure 5.10 Cu(s) + HNO_3(aq). Copper reacts vigorously with concentrated nitric acid to give brown NO_2 gas.

The metal is the reducing agent, since it is the substance oxidized. In fact, the most common reducing agents are metals. Some metal ions such as Fe^{2+} can also be reducing agents because they can be oxidized to ions of higher charge. Aqueous Fe^{2+} ion reacts readily with the strong oxidizing agent MnO_4^-, the permanganate ion. The Fe^{2+} ion is oxidized to Fe^{3+}, and the MnO_4^- ion is reduced to the Mn^{2+} ion.

$$5\,Fe^{2+}(aq) + MnO_4^-(aq) + 8\,H^+(aq) \longrightarrow 5\,Fe^{3+}(aq) + Mn^{2+}(aq) + 4\,H_2O(\ell)$$

It is important to be aware that it can be dangerous to mix a strong oxidizing agent with a strong reducing agent. A violent reaction, even an explosion, may take place. Chemicals should not be stored on laboratory shelves in alphabetical order, because such an ordering may place a strong oxidizing agent next to a strong reducing agent. In particular, swimming pool chemicals that contain chlorine and are strong oxidizing agents should not be stored in the hardware store or the garage next to easily oxidized materials such as ammonia.

EXERCISE 5.10 Oxidation-Reduction Reactions

Decide which of these reactions are oxidation-reduction reactions. In each case explain your choice and identify the oxidizing and reducing agents in the redox reactions.

(a) $NaOH(aq) + HNO_3(aq) \longrightarrow NaNO_3(aq) + H_2O(\ell)$
(b) $4\,Cr(s) + 3\,O_2(g) \longrightarrow 2\,Cr_2O_3(s)$
(c) $NiCO_3(s) + 2\,HCl(aq) \longrightarrow NiCl_2(aq) + H_2O(\ell) + CO_2(g)$
(d) $Cu(s) + Cl_2(g) \longrightarrow CuCl_2(s)$

5.4 Oxidation Numbers and Redox Reactions

An arbitrary bookkeeping system has been devised for keeping track of electrons in redox reactions. It extends the obvious oxidation and reduction case when neutral atoms become ions to reactions in which the changes are less obvious. The system is set up so that ***oxidation numbers always change in redox reactions.*** As a result, oxidation and reduction can be determined in the ways shown in Table 5.4.

How electrons participate in bonding atoms in molecules is the subject of Chapter 8.

Table 5.4 Recognizing Oxidation-Reduction Reactions

	Oxidation	Reduction
In terms of oxygen	Gain of oxygen	Loss of oxygen
In terms of halogen	Gain of halogen	Loss of halogen
In terms of hydrogen	Loss of hydrogen	Gain of hydrogen
In terms of electrons	Loss of electrons	Gain of electrons
In terms of oxidation numbers	Increase of oxidation number	Decrease of oxidation number

An **oxidation number** compares the charge of an uncombined atom with its actual charge or its relative charge in a compound. All neutral atoms have an equal number of protons and electrons and thus have no net charge. When sodium metal atoms (zero net charge) combine with chlorine atoms (zero net charge) to form sodium chloride, each sodium atom loses an electron to form a sodium ion, Na^+, and each chlorine atom gains an electron to form a chloride ion, Cl^-. Therefore, Na^+ has an oxidation number of +1 because it has one fewer electron than a sodium atom, and Cl^- has an oxidation number of −1 because it has one more electron than a chlorine atom. Oxidation numbers of atoms in molecular compounds are assigned as though electrons were completely transferred to form ions. In the molecular compound phosphorus trichloride, PCl_3, for example, chlorine is assigned an oxidation number of −1 even though it is not a Cl^- ion; the chlorine is directly bonded to the phosphorus. The chlorine atoms in PCl_3 are thought of as "possessing" more electrons than they have in Cl_2.

You can use this set of rules to determine oxidation numbers.

Oxidation numbers are also called oxidation states.

Rule 1: ***The oxidation number of an atom of a pure element is 0.*** When the atoms are not combined with those of any other element (for example, oxygen in O_2, sulfur in S_8, iron in metallic Fe, or chlorine in Cl_2), the oxidation number is 0.

In this book, oxidation numbers are written as +1, +2, etc., whereas charges on ions are written as 1+, 2+, etc.

Rule 2: ***The oxidation number of a monatomic ion equals its charge.*** Thus, the oxidation number of Cu^{2+} is +2; that of S^{2-} is −2.

Rule 3: ***Some elements have the same oxidation number in almost all their compounds and can be used as references for oxidation numbers of other elements in compounds.***

(a) Hydrogen has an oxidation number of +1 unless it is combined with a metal, in which case its oxidation number is −1.

(b) Fluorine has an oxidation number of −1 in all its compounds. Halogens other than fluorine have an oxidation number of −1 except when combined with a halogen above them in the periodic table or with oxygen.

(c) Oxygen has an oxidation number of −2 except in peroxides, such as hydrogen peroxide, H_2O_2, in which oxygen has an oxidation number of −1 (and hydrogen is +1).

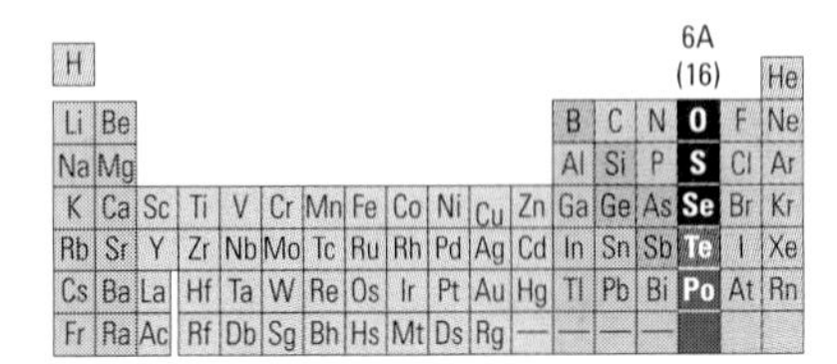

(d) In binary compounds (compounds of two elements), atoms of Group 6A elements (O, S, Se, Te) have an oxidation number of −2 except when combined with oxygen or halogens, in which case the Group 6A elements have positive oxidation numbers.

Rule 4: ***The sum of the oxidation numbers in a neutral compound is 0; the sum of the oxidation numbers in a polyatomic ion equals the charge on the ion.*** For example, in SO_2, the oxidation number of oxygen is −2, and with two O atoms, the total for oxygen is −4. Because the sum of the oxidation numbers must equal zero, the oxidation number of sulfur must be +4: $(+4) + 2(-2) = 0$. In the sulfite ion, SO_3^{2-}, the net charge is 2−. Because each oxygen is −2, the oxidation number of sulfur in sulfite must be +4: $(+4) + 3(-2) = 2-$.

$$\overset{+4}{S}\overset{-2}{O}{}_3^{2-}$$

Now, let's apply these rules to the equations for simple combination and displacement reactions involving sulfur and oxygen.

$$\text{Combination:} \quad \overset{0}{S_8}(s) + 8\,\overset{0}{O_2}(g) \longrightarrow 8\,\overset{+4}{S}\overset{-2}{O_2}(g)$$

$$\text{Combination:} \quad \overset{+2}{Zn}\overset{-2}{S}(s) + 2\,\overset{0}{O_2}(aq) \longrightarrow \overset{+2}{Zn}\overset{+6}{S}\overset{-2}{O_4}(aq)$$

$$\text{Displacement:} \quad \overset{+1}{Cu_2}\overset{-2}{S}(s) + \overset{0}{O_2}(g) \longrightarrow 2\,\overset{0}{Cu}(s) + \overset{+4}{S}\overset{-2}{O_2}(g)$$

These are all oxidation-reduction reactions, as shown by the fact that there has been a change in the oxidation numbers of atoms from reactants to products.

Every reaction in which an element becomes combined in a compound is a redox reaction. The oxidation number of the element must increase or decrease from its original value of zero. Combination reactions and displacement reactions in which one element displaces another are all redox reactions.

Those decomposition reactions in which elemental gases are produced are also redox reactions. Millions of tons of ammonium nitrate, NH_4NO_3, are used as fertilizer to supply nitrogen to crops. Ammonium nitrate is also used as an explosive that is decomposed by heating.

Ammonium nitrate was used in the 1995 bombing of the Federal Building in Oklahoma City, Oklahoma.

$$2\ NH_4NO_3(s) \longrightarrow 2\ N_2(g) + 4\ H_2O(g) + O_2(g)$$

Like a number of other explosives, ammonium nitrate contains an element with two different oxidation numbers, in effect having an oxidizing and reducing agent in the same compound.

$$\overset{-3\ +1}{NH_4^+} \qquad \overset{+5\ -2}{NO_3^-}$$

Note that nitrogen's oxidation number is -3 in the ammonium ion and $+5$ in the nitrate ion. Therefore, in the decomposition of ammonium nitrate to generate N_2, the N in the ammonium ion is oxidized from -3 to 0, and the ammonium ion is the reducing agent. The N in the nitrate ion is reduced from $+5$ to 0, and the nitrate ion is the oxidizing agent.

PROBLEM-SOLVING EXAMPLE 5.8 Applying Oxidation Numbers

Metallic copper and dilute nitric acid react according to this redox equation:

$$3\ Cu(s) + 8\ HNO_3(aq) \longrightarrow 3\ Cu(NO_3)_2(aq) + 2\ NO(g) + 4\ H_2O(\ell)$$

Assign oxidation numbers for each atom in the equation. Identify which element has been oxidized and which has been reduced.

Answer

$$\overset{0}{3\ Cu(s)} + \overset{+1+5-2}{8\ HNO_3(aq)} \longrightarrow \overset{+2+5-2}{3\ Cu(NO_3)_2(aq)} + \overset{+2-2}{2\ NO(g)} + \overset{+1-2}{4\ H_2O(\ell)}$$

Copper metal is oxidized. Nitrogen (in HNO_3) is reduced.

Strategy and Explanation Use the four rules introduced earlier and knowledge of the formulas of polyatomic ions to assign oxidation numbers. Copper is in its elemental state as a reactant, so its oxidation number is 0 (Rule 1).

For nitric acid we start by recognizing that it is a compound and has no net charge (Rule 4). Therefore, because the oxidation number of each oxygen is -2 (Rule 3c) for a total of -6 and the oxidation number of the hydrogen is $+1$ (Rule 3a), the oxidation number of nitrogen must be $+5$: $0 = 3(-2) + (+1) + (+5)$.

For the product combining copper and nitrate ions, we start by assigning the oxidation number of $+2$ to copper to balance the charges of the two nitrate ions. The oxidation numbers of the oxygen and nitrogen atoms within the nitrate anion are the same as in the reactant nitrate anions. For the NO molecules, we assign an oxidation number of -2 to oxygen (Rule 3c), so the oxidation number of the nitrogen in NO must be $+2$. The oxygen in water has oxidation number -2 and the hydrogen is $+1$.

Copper has changed from an oxidation number of 0 to an oxidation number of $+2$; it has been oxidized. Nitrogen has changed from an oxidation number of $+5$ to an oxidation number of $+2$; it has been reduced.

PROBLEM-SOLVING PRACTICE 5.8

Determine the oxidation number for each atom in this equation:

$$Sb_2S_3(s) + 3\ Fe(s) \longrightarrow 3\ FeS(s) + 2\ Sb(s)$$

Cite the oxidation number rule(s) you used to obtain your answers.

PROBLEM-SOLVING EXAMPLE 5.9 Oxidation-Reduction Reaction

Most metals we use are found in nature as cations in ores. The metal ion must be reduced to its elemental form, which is done with an appropriate oxidation-reduction reaction. The copper ore chalcocite, Cu_2S, is reacted with oxygen in a process called roasting to form metallic copper.

$$Cu_2S(s) + O_2(g) \longrightarrow 2\ Cu(s) + SO_2(g)$$

Identify the atoms that are oxidized and reduced, and name the oxidizing and reducing agents.

Answer Cu^+ ions and O_2 are reduced; S^{2-} ions are oxidized. O_2 and Cu^+ are the oxidizing agents; S^{2-} is the reducing agent.

Strategy and Explanation We first assign the oxidation numbers for all the atoms in reaction according to Rules 1 through 4.

$$\overset{+1\ -2}{Cu_2S(s)} + \overset{0}{O_2(g)} \longrightarrow 2\ \overset{0}{Cu(s)} + \overset{+4\ -2}{SO_2(g)}$$

The oxidation number of Cu^+ decreases from $+1$ to 0, so Cu^+ is reduced. The oxidation number of S increases from -2 to $+4$, so S^{2-} is oxidized. The oxidation number of oxygen decreases from 0 to -2, so oxygen is reduced. The oxidizing agents are Cu^+ and O_2, which accept electrons. The reducing agent is S^{2-} in Cu_2S, which donates electrons.

PROBLEM-SOLVING PRACTICE 5.9

Which are the oxidizing and reducing agents, and which atoms are oxidized and reduced in this reaction?

$$PbO(s) + CO(g) \longrightarrow Pb(s) + CO_2(g)$$

Exchange reactions of ionic compounds in aqueous solutions are not redox reactions because no change of oxidation numbers occurs. Consider, for example, the precipitation of barium sulfate when aqueous solutions of barium chloride and sulfuric acid are mixed.

$$Ba^{2+}(aq) + 2\ Cl^-(aq) + 2\ H^+(aq) + SO_4^{2-}(aq) \longrightarrow BaSO_4(s) + 2\ H^+(aq) + 2\ Cl^-(aq)$$

Net ionic equation: $$Ba^{2+}(aq) + SO_4^{2-}(aq) \longrightarrow BaSO_4(s)$$

The oxidation numbers of all atoms remain unchanged from the reactants to products, so this is not a redox reaction.

CONCEPTUAL EXERCISE 5.11 Redox in CFC Disposal

This redox reaction is used for the disposal of chlorofluorocarbons (CFCs) by their reaction with sodium oxalate, $Na_2C_2O_4$:

$$CF_2Cl_2(g) + 2\ Na_2C_2O_4(s) \longrightarrow 2\ NaF(s) + 2\ NaCl(s) + C(s) + 4\ CO_2(g)$$

(a) What is oxidized in this reaction?
(b) What is reduced?

5.5 Displacement Reactions, Redox, and the Activity Series

Recall that displacement reactions *(⟸ p. 105)* have this reaction pattern:

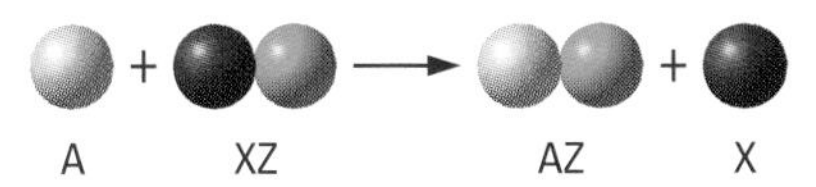

Displacement reactions, like combination reactions, are oxidation-reduction reactions. For example, in the reaction of hydrochloric acid with iron,

$$\mathrm{Fe(s) + 2\,HCl(aq) \longrightarrow FeCl_2(aq) + H_2(g)}$$

metallic iron is the reducing agent; it is oxidized from an oxidation number of 0 in Fe(s) to +2 in $FeCl_2$. Hydrogen ions, H^+, in hydrochloric acid are reduced to hydrogen gas (H_2), in which hydrogen has an oxidation number of 0.

Extensive studies with many metals have led to the development of a **metal activity series,** a ranking of relative reactivity of metals in displacement and other kinds of reactions (Table 5.5). The most reactive metals appear at the top of the series, and activity decreases going down the series. Metals at the top are powerful reducing agents and readily lose electrons to form cations. The metals at the lower end of the series are poor reducing agents. However, their cations (Au^+, Ag^+) are powerful oxidizing agents that readily gain an electron to form the free metal.

An element higher in the activity series will displace an element below it in the series from its compounds. For example, zinc displaces copper ions from copper(II) sulfate solution, and copper metal displaces silver ions from silver nitrate solution (Figure 5.11).

Table 5.5 Activity Series of Metals

Displace H_2 from $H_2O(\ell)$, steam, or acid	Li K Ba Sr Ca Na
Displace H_2 from steam or acid	Mg Al Mn Zn Cr
Displace H_2 from acid	Fe Ni Sn Pb
	H_2
Do not displace H_2 from $H_2O(\ell)$, steam, or acid	Sb Cu Hg Ag Pd Pt Au

Ease of oxidation increases (upward)

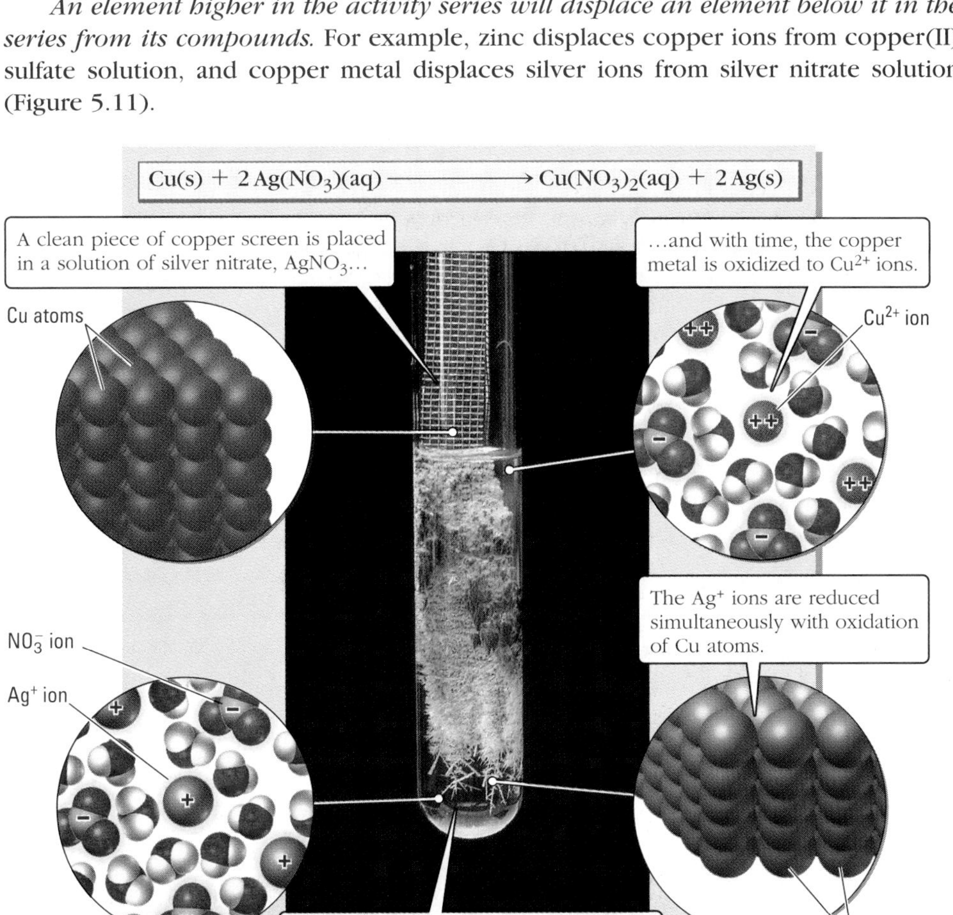

Active Figure 5.11 Metal + aqueous metal salt displacement reaction. The oxidation of copper metal by silver ion. (Atoms or ions that take part in the reaction have been highlighted in the nanoscale pictures.) Visit this book's companion website at **www.cengage.com/chemistry/moore** to test your understanding of the concepts in this figure.

Figure 5.12 Potassium, an active metal. When a drop of water falls onto a sample of potassium metal, it reacts vigorously to give hydrogen gas and a solution of potassium hydroxide.

$$Zn(s) + CuSO_4(aq) \longrightarrow ZnSO_4(aq) + Cu(s)$$

$$Cu(s) + 2\ AgNO_3(aq) \longrightarrow Cu(NO_3)_2(aq) + 2\ Ag(s)$$

In each case, the elemental metal (Zn, Cu) is the reducing agent and is oxidized; Cu^{2+} ions and Ag^+ ions are oxidizing agents and are reduced to Cu(s) and Ag(s), respectively.

Metals above hydrogen in the series react with acids whose anions are not oxidizing agents, such as hydrochloric acid, to form hydrogen, H_2, and the metal salt containing the cation of the metal and the anion of the acid. For example, $FeCl_2$ is formed from iron and hydrochloric acid, and $ZnBr_2$ is formed from zinc and hydrobromic acid.

$$Zn(s) + 2\ HBr(aq) \longrightarrow ZnBr_2(aq) + H_2(g)$$

Metals below hydrogen in the activity series do not displace hydrogen from acids in this way.

Very reactive metals—those at the top of the activity series, from lithium (Li) through sodium (Na)—can displace hydrogen from water. Some do so violently (Figure 5.12). Metals of intermediate activity (Mg through Cr) displace hydrogen from steam, but not from liquid water at room temperature.

Elements very low in the activity series are unreactive. Sometimes called noble metals (Au, Ag, Pt), they are prized for their nonreactivity. It is no accident that gold and silver have been used extensively for coinage since antiquity. These metals do not react with air, water, or even common acids, thus maintaining their luster (and value) for many years. Their low reactivity explains why they occur naturally as free metals and have been known as elements since antiquity.

PROBLEM-SOLVING EXAMPLE 5.10 Activity Series of Metals

Use the activity series found in Table 5.5 to predict which of these reactions will occur. Complete and balance the equations for those reactions that will occur.

(a) $Cr(s) + MgCl_2(aq) \longrightarrow$
(b) $Al(s) + Pb(NO_3)_2(aq) \longrightarrow$
(c) $Mg(s) +$ hydrochloric acid $\longrightarrow$

Answer

(a) No reaction
(b) $2\ Al(s) + 3\ Pb(NO_3)_2(aq) \longrightarrow 2\ Al(NO_3)_3(aq) + 3\ Pb(s)$
(c) $Mg(s) + 2\ HCl(aq) \longrightarrow MgCl_2(aq) + H_2(g)$

Strategy and Explanation

(a) Chromium is less active than magnesium, so it will not displace magnesium ions from magnesium chloride. Therefore, no reaction occurs.
(b) Aluminum is above lead in the activity series, so aluminum will displace lead ions from a solution of lead(II) nitrate to form metallic lead and Al^{3+} ions.

$$2\ Al(s) + 3\ Pb(NO_3)_2(aq) \longrightarrow 2\ Al(NO_3)_3(aq) + 3\ Pb(s)$$

(c) Magnesium is above hydrogen in the activity series, so it will displace hydrogen ions from HCl to form the metal salt $MgCl_2$ plus hydrogen gas.

$$Mg(s) + 2\ HCl(aq) \longrightarrow MgCl_2(aq) + H_2(g)$$

PROBLEM-SOLVING PRACTICE 5.10

Use Table 5.5 to predict whether each of these reactions will occur. If a reaction occurs, identify what has been oxidized or reduced and what the oxidizing agent and the reducing agent are.

(a) $2\ Al(s) + 3\ CuSO_4(aq) \longrightarrow Al_2(SO_4)_3(aq) + 3\ Cu(s)$
(b) $2\ Al(s) + Cr_2O_3(s) \longrightarrow Al_2O_3(s) + 2\ Cr(s)$
(c) $Pt(s) + 4\ HCl(aq) \longrightarrow PtCl_4(aq) + 2\ H_2(g)$
(d) $Au(s) + 3\ AgNO_3(aq) \longrightarrow Au(NO_3)_3(aq) + 3\ Ag(s)$

CONCEPTUAL
EXERCISE 5.12 Reaction Product Prediction

For these pairs of reactants, predict what kind of reaction would occur and what the products might be. Which reactions are redox reactions?

(a) Combustion of ethanol: $CH_3CH_2OH(\ell) + O_2(g) \longrightarrow ?$

(b) $Fe(s) + HNO_3(aq) \longrightarrow ?$

(c) $AgNO_3(aq) + KBr(aq) \longrightarrow ?$

5.6 Solution Concentration

Many of the chemicals in your body or in other living systems are dissolved in water—that is, they are in an aqueous solution. Like chemical reactions in living systems, many reactions studied in the chemical laboratory are carried out in solution. Frequently, this chemistry must be done quantitatively. For example, intravenous fluids administered to patients contain many compounds (salts, nutrients, drugs, and so on), and the concentration of each must be known accurately. To accomplish this task, we continue to use balanced equations and moles, but we measure volumes of solution rather than masses of solids, liquids, and gases.

A solution is a homogeneous mixture of a **solute,** the substance that has been dissolved, and the **solvent,** the substance in which the solute has been dissolved. To know the quantity of solute in a given volume of a liquid solution requires knowing the **concentration** of the solution—the relative quantities of solute and solvent. Molarity, which relates the amount of solute expressed in moles to the solution volume expressed in liters, is the most useful of the many ways of expressing solution concentration for studying chemical reactions in solution.

Molarity

The **molarity** of a solution is defined as the amount of solute expressed in moles per unit volume of solution, expressed in liters (mol/L).

$$\text{Molarity} = \frac{\text{moles of solute}}{\text{liters of solution}}$$

Note that the volume term in the denominator is liters of *solution,* not liters of solvent.

If, for example, 40.0 g (1.00 mol) NaOH is dissolved in sufficient water to produce a solution with a total volume of 1.00 L, the solution has a concentration of 1.00 mol NaOH/1.00 L of solution, which is a 1.00 molar solution. The molarity of this solution is reported as 1.00 M, where the capital M stands for moles/liter. Molarity is also represented by square brackets around the formula of a compound or ion, such as [NaOH] or $[OH^-]$. The brackets indicate amount (in moles) of the species (compound or ion) per unit volume (in liters) of solution.

molarity of NaOH solution > [NaOH] = 2 M < 2 moles per liter

A solution of known molarity can be made by adding the required amount of solute to a volumetric flask, adding some solvent to dissolve all the solute, and then adding sufficient solvent with continual mixing to fill the flask to the mark. As shown in Figure 5.13, the etched marking indicates the liquid level equal to the specified volume of the flask.

1 Combine ~240 mL distilled H_2O with 0.395 g (0.00250 mol) $KMnO_4$ in a 250.0-mL volumetric flask.

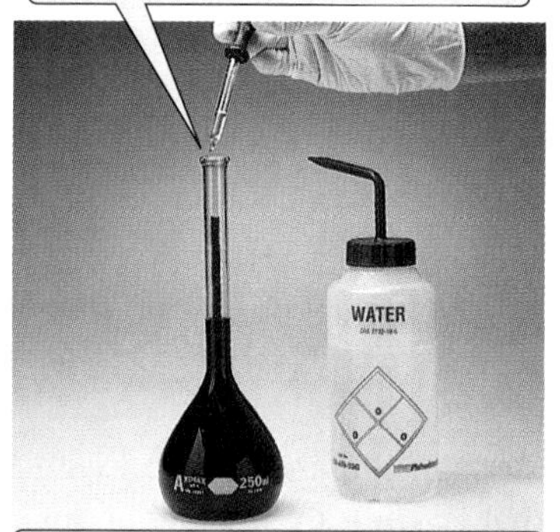

3 After the solid dissolves, add sufficient water to fill the flask to the mark etched in the neck, indicating a volume of 250.0 mL.
WATER
4 Shake the flask again to thoroughly mix its contents. The flask now contains 250.0 mL of 0.0100 M KMnO4 solution.

Figure 5.13 Solution preparation from a solid solute. Making a 0.0100 M aqueous solution of $KMnO_4$.

PROBLEM-SOLVING EXAMPLE 5.11 Molarity

Potassium permanganate, $KMnO_4$, is a strong oxidizing agent whose solutions are often used in laboratory experiments.

(a) If 0.433 g $KMnO_4$ is added to a 500.0-mL volumetric flask and water is added until the solution volume is exactly 500.0 mL, what is the molarity of the resulting solution?

(b) You need to prepare a 0.0250 M solution of $KMnO_4$ for an experiment. How many grams of $KMnO_4$ should be added with sufficient water to a 1.00-L volumetric flask to give the desired solution?

Answer

(a) 0.00548 M (b) 3.95 g

Strategy and Explanation

(a) To calculate the molarity of the solution, we need to calculate the moles of solute and the solution volume in liters. The volume was given as 500.0 mL, which is 0.5000 L. We use the molar mass of $KMnO_4$ (158.03 g/mol) to obtain the moles of solute.

$$0.433 \text{ g } KMnO_4 \times \frac{1 \text{ mol } KMnO_4}{158.03 \text{ g } KMnO_4} = 2.74 \times 10^{-3} \text{ mol } KMnO_4$$

We can now calculate the molarity.

$$\text{Molarity of } KMnO_4 = \frac{2.74 \times 10^{-3} \text{ mol } KMnO_4}{0.500 \text{ L solution}} = 5.48 \times 10^{-3} \text{ mol/L}$$

This can be expressed as 0.00548 M or in the notation $[KMnO_4] = 0.00548$ M.

(b) To make a 0.0250 M solution in a 1.00-L volumetric flask requires 0.0250 mol $KMnO_4$. We convert to grams using the molar mass of $KMnO_4$.

$$0.0250 \text{ mol } KMnO_4 \times \frac{158.03 \text{ g } KMnO_4}{1 \text{ mol } KMnO_4} = 3.95 \text{ g } KMnO_4$$

☑ **Reasonable Answer Check** (a) We have about a half gram of solute with a molar mass of about 160. We will put this solute into a half-liter flask, so it is as if we put about one gram into a one-liter flask. The molarity should be about 1/160 = 0.00625, which is close to our more exact answer. (b) We need a little more than 2/100 of a mole of $KMnO_4$, which has a molar mass of about 160. One one-hundredth of 160 is 1.6, so two one-hundredths is twice that, or 3.2, which is close to our more exact answer.

PROBLEM-SOLVING PRACTICE 5.11

Calculate the molarity of sodium sulfate in a solution that contains 36.0 g Na_2SO_4 in 750. mL solution.

EXERCISE 5.13 Cholesterol Molarity

A blood serum cholesterol level greater than 240 mg of cholesterol per deciliter (0.100 L) of blood usually indicates the need for medical intervention. Calculate this serum cholesterol level in molarity. Cholesterol's molecular formula is $C_{27}H_{46}O$.

Sometimes the molarity of a particular ion in a solution is required, a value that depends on the formula of the solute. For example, potassium chromate is a soluble ionic compound and a strong electrolyte that completely dissociates in solution to form 2 mol K^+ ions and 1 mol CrO_4^{2-} ions for each mole of K_2CrO_4 that dissolves:

$$\underset{\substack{1\text{ mol}\\100\%\text{ dissociation}}}{K_2CrO_4(aq)} \longrightarrow \underset{2\text{ mol}}{2\,K^+(aq)} + \underset{1\text{ mol}}{CrO_4^{2-}(aq)}$$

The K^+ concentration is twice the K_2CrO_4 concentration because each mole of K_2CrO_4 contains 2 mol K^+. Therefore, a 0.00283 M K_2CrO_4 solution has a K^+ concentration of 2×0.00283 M = 0.00566 M and a CrO_4^{2-} concentration of 0.00283 M.

EXERCISE 5.14 Molarity

A student dissolves 6.37 g aluminum nitrate in sufficient water to make 250.0 mL of solution. Calculate (a) the molarity of aluminum nitrate in this solution and (b) the molarity of aluminum ions and of nitrate ions in this solution.

CONCEPTUAL
EXERCISE 5.15 Molarity

When solutions are prepared, the final volume of solution can be different from the sum of the volumes of the solute and solvent because some expansion or contraction can occur. Why is it always better to describe solution preparation as "adding enough solvent" to make a certain volume of solution?

Preparing a Solution of Known Molarity by Diluting a More Concentrated One

Frequently, solutions of the same solute need to be available at several different molarities. For example, hydrochloric acid is often used at concentrations of 6.0 M, 1.0 M, and 0.050 M. To make these solutions, chemists often use a concentrated solution of known molarity and dilute samples of it with water to make solutions of lesser molarity. *The number of moles of solute in the sample that is diluted remains constant throughout the dilution operation.* Therefore, the number of moles of solute in the dilute solution must be the same as the number of moles of solute in the sample of the more concentrated solution. Diluting a solution does increase the volume, so the molarity of the solution is lowered by the dilution operation, even though the number of moles of solute remains unchanged.

The number of moles in each case is the same and a simple relationship applies:

$$Molarity(\text{conc.}) \times V(\text{conc.}) = Molarity(\text{dil}) \times V(\text{dil})$$

Consider two cases: A teaspoonful of sugar, $C_{12}H_{22}O_{11}$, is dissolved in a glass of water and a teaspoonful of sugar is dissolved in a swimming pool full of water. The swimming pool and the glass contain the same number of moles of sugar, but the concentration of sugar in the swimming pool is far less because the volume of solution in the pool is much greater than that in the glass.

where *Molarity*(conc.) and *V*(conc.) represent the molarity and the volume (in liters) of the concentrated solution, and *Molarity*(dil) and *V*(dil) represent the molarity and volume of the dilute solution. *Multiplying a volume in liters by a solution's molarity (moles/liter) yields the number of moles of solute.*

A quick and useful check on a dilution calculation is to make certain that the molarity of the diluted solution is lower than that of the concentrated solution.

We can calculate, for example, the concentration of a hydrochloric acid solution made by diluting 25.0 mL of 6.0 M HCl to 500. mL. In this case, we want to determine *Molarity*(dil) when *Molarity*(conc.) = 6.0 M, *V*(conc.) = 0.0250 L, and *V*(dil) = 0.500 L. We algebraically rearrange the relationship to get the concentration of the diluted HCl.

$$Molarity(\text{dil}) = \frac{Molarity(\text{conc.}) \times V(\text{conc.})}{V(\text{dil})}$$

$$= \frac{6.0 \text{ mol/L} \times 0.0250 \text{ L}}{0.500 \text{ L}} = 0.30 \text{ mol/L}$$

A diluted solution will always be less concentrated (lower molarity) than the more concentrated solution (Figure 5.14).

Use caution when diluting a concentrated acid. The more concentrated acid should be added slowly to the solvent (water) so that the heat released during the dilution is rapidly dissipated into a large volume of water. If water is added to the acid, the heat released by the dissolving could be sufficient to vaporize the solution, spraying the acid over you and anyone nearby.

EXERCISE 5.16 Moles of Solute in Solutions

Consider 100. mL of 6.0 M HCl solution, which is diluted with water to yield 500. mL of 1.20 M HCl. Show that 100. mL of the more concentrated solution contains the same number of moles of HCl as 500. mL of the more dilute solution.

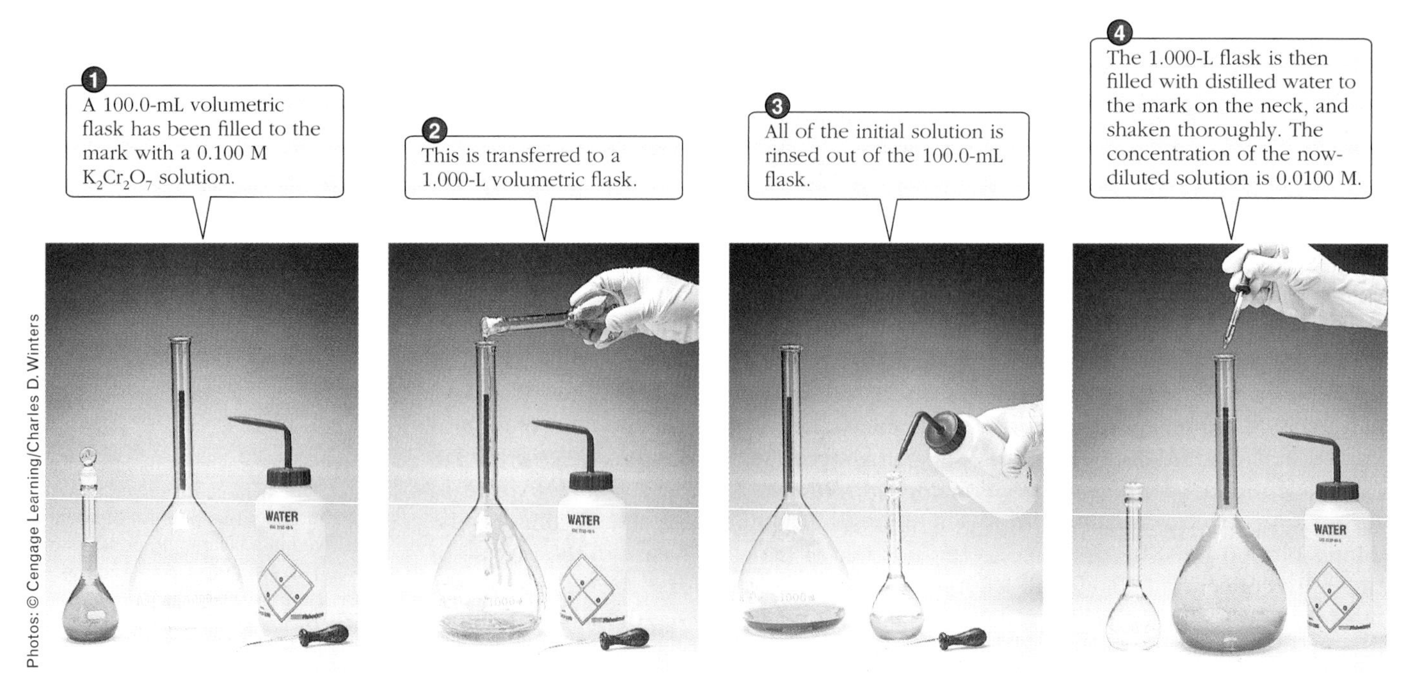

1
A 100.0-mL volumetric flask has been filled to the mark with a 0.100 M $K_2Cr_2O_7$ solution.
2
This is transferred to a 1.000-L volumetric flask.
3
All of the initial solution is rinsed out of the 100.0-mL flask.
4
The 1.000-L flask is then filled with distilled water to the mark on the neck, and shaken thoroughly. The concentration of the now-diluted solution is 0.0100 M.
WATER
WATER
WATER

Photos: © Cengage Learning/Charles D. Winters

Figure 5.14 Solution preparation by dilution.

PROBLEM-SOLVING EXAMPLE 5.12 Solution Concentration and Dilution

Describe how to prepare 500.0 mL of 1.00 M H_2SO_4 solution from a concentrated sulfuric acid solution that is 18.0 M.

Answer Add 27.8 mL of the concentrated sulfuric acid slowly and carefully to enough water to make up a total volume of 500.0 mL of solution.

Strategy and Explanation In this dilution problem, the concentrations of the concentrated (18.0 M) and less concentrated (1.00 M) solutions are given, as well as the volume of the diluted solution (500.0 mL). The volume of the concentrated sulfuric acid, *V*(conc.), to be diluted is needed, and can be calculated from this relationship:

$$Molarity(\text{conc.}) \times V(\text{conc.}) = Molarity(\text{dil}) \times V(\text{dil})$$

$$V(\text{conc.}) = \frac{Molarity(\text{dil}) \times V(\text{dil})}{Molarity(\text{conc.})}$$

$$= \frac{1.00 \text{ mol/L} \times 0.500 \text{ L}}{18.0 \text{ mol/L}} = 0.0278 \text{ L} = 27.8 \text{ mL}$$

Thus, 27.8 mL of concentrated sulfuric acid is added slowly, with stirring, to about 350 mL of distilled water. When the solution has cooled to room temperature, sufficient water is added to bring the final volume to 500.0 mL, resulting in a 1.00 M sulfuric acid solution.

☑ **Reasonable Answer Check** The ratio of molarities is 18:1, so the ratio of volumes should be 1:18, and it is.

PROBLEM-SOLVING PRACTICE 5.12

A laboratory procedure calls for 50.0 mL of 0.150 M NaOH. You have available 100. mL of 0.500 M NaOH. What volume of the more concentrated solution should be diluted to make the desired solution?

CONCEPTUAL EXERCISE 5.17 Solution Concentration

The molarity of a solution can be decreased by dilution. How could the molarity of a solution be increased without adding additional solute?

Preparing a Solution of Known Molarity from a Pure Solute

In Problem-Solving Example 5.11, we described finding the molarity of a $KMnO_4$ solution that was prepared from known quantities of solute and solution. More frequently, a solid or liquid solute (sometimes even a gas) must be used to make up a solution of known molarity. The problem becomes one of calculating what mass of solute to use to provide the proper number of moles.

Consider a laboratory experiment that requires 2.00 L of 0.750 M NH_4Cl solution. What mass of NH_4Cl must be dissolved in water to make 2.00 L of solution? The number of moles of NH_4Cl required can be calculated from the molarity.

$M \times L = \text{mol/L} \times L = \text{mol}$

$$0.750 \text{ mol/L } NH_4Cl \text{ solution} \times 2.00 \text{ L solution} = 1.500 \text{ mol } NH_4Cl$$

Then the molar mass can be used to calculate the number of grams of NH_4Cl needed.

$$1.500 \text{ mol } NH_4Cl \times 53.49 \text{ g/mol } NH_4Cl = 80.2 \text{ g } NH_4Cl$$

The solution is prepared by putting 80.2 g NH_4Cl into a beaker, dissolving it in pure water, rinsing all of the solution into a volumetric flask, and adding distilled water until the solution volume is 2.00 L, which results in a 0.750 M NH_4Cl solution.

PROBLEM-SOLVING EXAMPLE 5.13 Solute Mass and Molarity

Describe how to prepare 500.0 mL of 0.0250 M $K_2Cr_2O_7$ solution starting with solid potassium dichromate.

Answer Dissolve 3.68 g $K_2Cr_2O_7$ in water and add enough water to make 500.0 mL of solution.

Strategy and Explanation Use the definition of molarity and the molar mass of potassium dichromate to solve the problem. First, find the number of moles of the solute, $K_2Cr_2O_7$, in 500.0 mL of 0.0250 M $K_2Cr_2O_7$ solution by multiplying the volume in liters times the molarity of the solution.

$$0.500 \text{ L solution} \times \frac{0.0250 \text{ mol } K_2Cr_2O_7}{1 \text{ L solution}} = 1.25 \times 10^{-2} \text{ mol } K_2Cr_2O_7$$

From this calculate the number of grams of $K_2Cr_2O_7$.

$$1.25 \times 10^{-2} \text{ mol } K_2Cr_2O_7 \times \frac{294.2 \text{ g } K_2Cr_2O_7}{1 \text{ mol } K_2Cr_2O_7} = 3.68 \text{ g } K_2Cr_2O_7$$

The solution is prepared by putting 3.68 g $K_2Cr_2O_7$ into a 500-mL volumetric flask and adding enough distilled water to dissolve the solute and then additional water sufficient to bring the solution volume up to the mark of the flask. This results in 500.0 mL of 0.0250 M $K_2Cr_2O_7$ solution.

☑ **Reasonable Answer Check** The molar mass of $K_2Cr_2O_7$ is about 300 g/mol, and we want a 0.025 M solution, so we need about 300 g/mol × 0.025 mol/L = 7.5 g/L. But only one-half liter is required, so 0.5 L × 7.5 g/L = 3.75 g is needed, which agrees with our more accurate answer.

PROBLEM-SOLVING PRACTICE 5.13

Describe how you would prepare these solutions:
(a) 1.00 L of 0.125 M Na_2CO_3 from solid Na_2CO_3
(b) 100. mL of 0.0500 M Na_2CO_3 from a 0.125 M Na_2CO_3 solution
(c) 500. mL of 0.0215 M $KMnO_4$ from solid $KMnO_4$
(d) 250. mL of 0.00450 M $KMnO_4$ from 0.0215 M $KMnO_4$

5.7 Molarity and Reactions in Aqueous Solutions

Many kinds of reactions—acid-base (⇐ *p. 146*), precipitation (⇐ *p. 139*), and redox (⇐ *p. 151*)—occur in aqueous solutions. In such reactions, molarity is the concentration unit of choice because it quantitatively relates a volume of one reactant and the molar amount of that reactant contained in solution to the volume and corresponding molar amount of another reactant or product in solution. Molarity allows us to make conversions between volumes of solutions and moles of reactants and products as given by the stoichiometric coefficients. Molarity is used to link mass, amount (moles), and volume of solution (Figure 5.15).

PROBLEM-SOLVING EXAMPLE 5.14 Solution Reaction Stoichiometry

A major industrial use of hydrochloric acid is for "pickling," the removal of rust from steel by dipping the steel into very large baths of HCl. The acid reacts in an exchange reaction with rust, which is essentially Fe_2O_3, leaving behind a clean steel surface.

$$Fe_2O_3(s) + 6\,HCl(aq) \longrightarrow 2\,FeCl_3(aq) + 3\,H_2O(\ell)$$

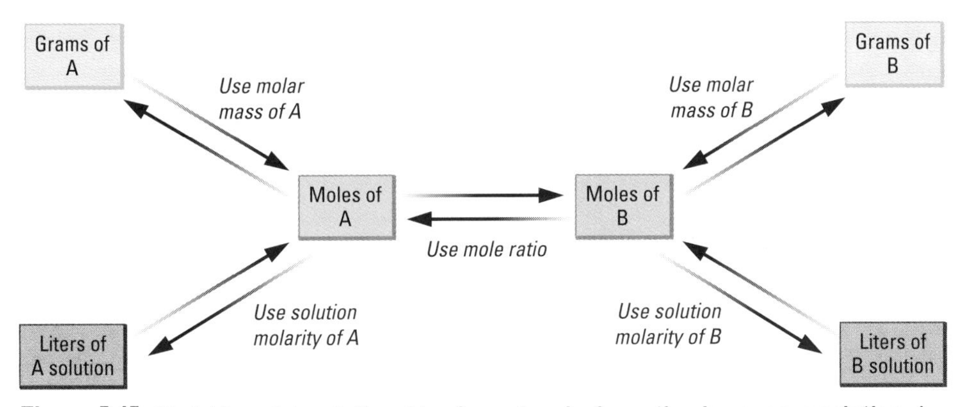

Figure 5.15 Stoichiometric relationships for a chemical reaction in aqueous solution. A mole ratio provides the connection between moles of a reactant or product and moles of another reactant or product.

Once the rust is taken off, the steel is removed from the acid bath and rinsed before the acid reacts significantly with the iron in the steel.

How many pounds of rust can be removed when rust-covered steel reacts with 800. L of 12.0 M HCl? Assume that only the rust reacts with the HCl (1.000 lb = 453.6 g).

Answer 563 lb Fe_2O_3

Strategy and Explanation We use the stoichiometric relationships in Figure 5.15. We must calculate the number of moles of HCl, then the number of moles of Fe_2O_3, and finally the mass of Fe_2O_3. First, calculate the number of moles of HCl available in 800. L of the solution.

$$800.\ \text{L HCl} \times \frac{12.0\ \text{mol HCl}}{1\ \text{L solution}} = 9.600 \times 10^3\ \text{mol HCl}$$

Then the mass of Fe_2O_3 can be determined.

$$(9.600 \times 10^3\ \text{mol HCl}) \times \frac{1\ \text{mol Fe}_2\text{O}_3}{6\ \text{mol HCl}} = 1.600 \times 10^3\ \text{mol Fe}_2\text{O}_3$$

$$(1.600 \times 10^3\ \text{mol Fe}_2\text{O}_3) \times \frac{159.7\ \text{g Fe}_2\text{O}_3}{1\ \text{mol Fe}_2\text{O}_3} = 2.555 \times 10^5\ \text{g Fe}_2\text{O}_3$$

$$(2.555 \times 10^5\ \text{g Fe}_2\text{O}_3) \times \frac{1\ \text{lb Fe}_2\text{O}_3}{453.6\ \text{g Fe}_2\text{O}_3} = 563\ \text{lb Fe}_2\text{O}_3$$

Placing all the conversion factors in the same mathematical setup gives

$$800.\ \text{L HCl} \times \frac{12.0\ \text{mol HCl}}{1\ \text{L solution}} \times \frac{1\ \text{mol Fe}_2\text{O}_3}{6\ \text{mol HCl}} \times \frac{159.7\ \text{g Fe}_2\text{O}_3}{1\ \text{mol Fe}_2\text{O}_3} \times \frac{1\ \text{lb Fe}_2\text{O}_3}{453.6\ \text{g Fe}_2\text{O}_3} = 563\ \text{lb Fe}_2\text{O}_3$$

The solution remaining in the acid bath presents a real disposal challenge because of the metal ions it contains.

☑ **Reasonable Answer Check** We have about 10,000 mol HCl. It will react with one sixth as many moles of rust, or about 1600 mol Fe_2O_3. Converting number of moles of Fe_2O_3 to number of pounds requires multiplying by the molar mass (160 g/mol) and then multiplying by the grams-to-pounds conversion factor (roughly 1 lb/500 g), which is equivalent to dividing by about 3. So $1600/3 \approx 500$, which checks with our more accurate answer.

PROBLEM-SOLVING PRACTICE 5.14

In a recent year, 1.2×10^{10} kg sodium hydroxide (NaOH) was produced in the United States by passing an electric current through brine, an aqueous solution of sodium chloride.

$$2\ NaCl(aq) + 2\ H_2O(\ell) \longrightarrow 2\ NaOH(aq) + Cl_2(g) + H_2(g)$$

What volume of brine is needed to produce this mass of NaOH? (*Note*: 1.0 L brine contains 360 g dissolved NaCl.)

EXERCISE 5.18 Molarity

Sodium chloride is used in intravenous solutions for medical applications. The NaCl concentration in such solutions must be accurately known and can be assessed by reacting the solution with an experimentally determined volume of $AgNO_3$ solution of known concentration. The net ionic equation is

$$Ag^+(aq) + Cl^-(aq) \longrightarrow AgCl(s)$$

Suppose that a chemical technician uses 19.3 mL of 0.200 M $AgNO_3$ to convert all the NaCl in a 25.0-mL sample of an intravenous solution to AgCl. Calculate the molarity of NaCl in the solution.

5.8 Aqueous Solution Titrations

One important quantitative use of aqueous solution reactions is to determine the unknown concentration of a reactant in a solution, such as the concentration of HCl in a solution of HCl. This is done with a titration using a **standard solution,** a solution whose concentration is known accurately. In a **titration,** a substance in the standard solution reacts with a known stoichiometry with the substance whose concentration is to be determined. When the stoichiometrically equivalent amount of standard solution has been added, the **equivalence point** is reached. At that point, the molar amount of reactant that has been added from the standard solution is exactly what is needed to react completely with the substance whose concentration is to be determined. The progress of the reaction is monitored by an indicator, a dye that changes color at the equivalence point, or through some other means with appropriate instruments. Phenolphthalein, for example, is commonly used as the indicator in strong acid–strong base titrations because it is colorless in acidic solutions and pink in basic solutions. The point at which the indicator is seen to change color is called the end point.

Acid-base titrations are described more extensively in Chapter 16.

A common example of a titration is the determination of the molarity of an acid by titration of the acid with a standard solution of a base. For example, we can use a standard solution of 0.100 M KOH to determine the concentration of an HCl solution. To carry out this titration, we use a carefully measured volume of the HCl solution and slowly add the standardized KOH solution until the equivalence point is reached (Figure 5.16). At that point, the number of moles of OH^- added to the HCl solution exactly matches the number of moles of H^+ that were in the original acid sample.

PROBLEM-SOLVING EXAMPLE 5.15 Acid-Base Titration

A student has an aqueous solution of calcium hydroxide that is approximately 0.10 M. She titrated a 50.0-mL sample of the calcium hydroxide solution with a standardized solution of 0.300 M $HNO_3(aq)$. To reach the end point, 41.4 mL of the HNO_3 solution was needed. What is the molarity of the calcium hydroxide solution?

Answer 0.124 M $Ca(OH)_2$

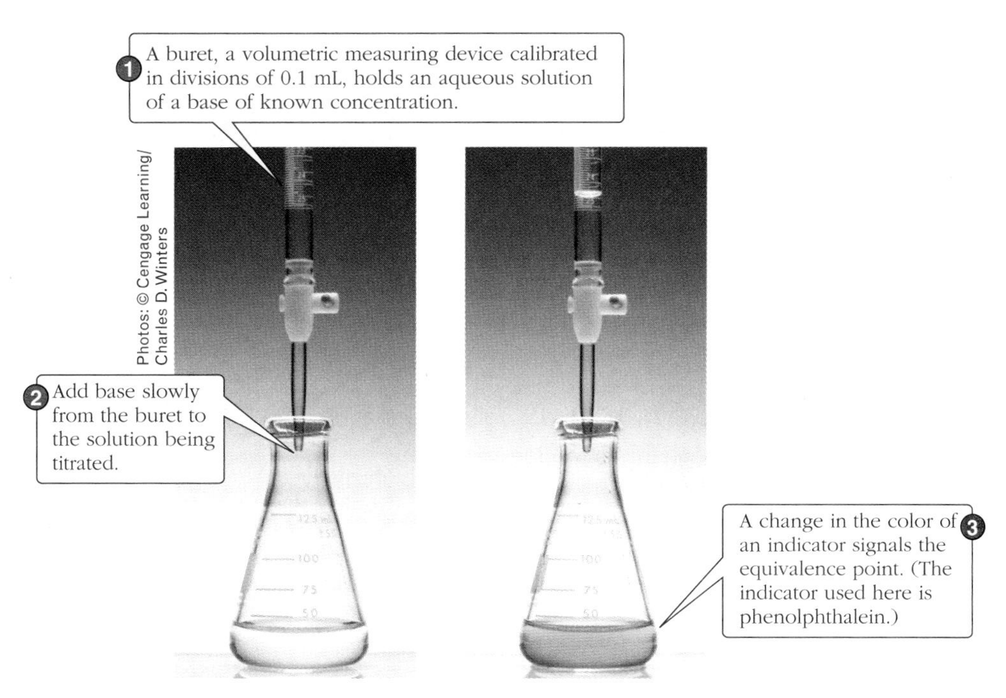

Figure 5.16 Titration of an acid in aqueous solution with a standard solution of base.

Strategy and Explanation Start by writing the balanced equation for this acid-base reaction.

$$2\ HNO_3(aq) + Ca(OH)_2(aq) \longrightarrow Ca(NO_3)_2(aq) + 2\ H_2O\ (\ell)$$

The net ionic equation is

$$H^+(aq) + OH^-(aq) \longrightarrow H_2O(\ell)$$

Then calculate the number of moles of HNO_3 consumed.

$$0.0414\ \text{L}\ HNO_3\ \text{solution} \times \frac{0.300\ \text{mol}\ HNO_3}{1.00\ \text{L}\ HNO_3\ \text{solution}} = 0.0124\ \text{mol}\ HNO_3$$

The balanced equation shows that for every 2 mol HNO_3 reacted, one mol $Ca(OH)_2$ is consumed. Therefore, if 1.24×10^{-2} mol HNO_3 was reacted, $(1.24 \times 10^{-2}$ mol $HNO_3)[(1$ mol $Ca(OH)_2)/(2$ mol $HNO_3)] = 6.21 \times 10^{-3}$ mol $Ca(OH)_2$ must have been consumed. From the number of moles of calcium hydroxide and the volume of the calcium hydroxide solution, calculate the molarity of the solution.

$$\frac{6.21 \times 10^{-3}\ \text{mol}\ Ca(OH)_2}{0.0500\ \text{L}\ Ca(OH)_2\ \text{solution}} = 0.124\ \text{M}\ Ca(OH)_2$$

☑ **Reasonable Answer Check** At the equivalence point, the number of moles of $H^+(aq)$ added and $OH^-(aq)$ in the initial sample must be equal. The number of moles of each reactant is its volume multiplied by its molarity. For the HNO_3 we have 0.0414 L × 0.300 M = 0.0124 mol HNO_3. For the $Ca(OH)_2$ we have 0.050 L × 0.124 M × 2 = 0.0124 mol $Ca(OH)_2$. The answer is reasonable.

PROBLEM-SOLVING PRACTICE 5.15

In a titration, a 20.0-mL sample of sulfuric acid, H_2SO_4, was titrated to the end point with 41.3 mL of 0.100 M NaOH. What is the molarity of the H_2SO_4 solution?

IN CLOSING

Having studied this chapter, you should be able to . . .

- Predict products of common types of chemical reactions: precipitation, acid-base, and gas-forming (Sections 5.1, 5.2, 5.3). ■ End-of-chapter Question assignable in OWL: 25
- Write a net ionic equation for a given reaction in aqueous solution (Section 5.1). ■ Questions 13, 15
- Recognize common acids and bases and predict when neutralization reactions will occur (Section 5.2). ■ Question 21
- Identify the acid and base used to form a specific salt (Section 5.2). ■ Question 23
- Recognize oxidation-reduction reactions and common oxidizing and reducing agents (Section 5.3). ■ Questions 32, 34
- Assign oxidation numbers to reactants and products in a redox reaction, identify what has been oxidized or reduced, and identify oxidizing agents and reducing agents (Section 5.4). ■ Question 27
- Use the activity series to predict products of displacement redox reactions (Section 5.5). ■ Question 40
- Define molarity and calculate molar concentrations (Section 5.6). ■ Questions 42, 44, 46, 48
- Determine how to prepare a solution of a given molarity from the solute and water or by dilution of a more concentrated solution (Section 5.6). ■ Question 48
- Solve stoichiometry problems by using solution molarities (Section 5.7). ■ Questions 51, 52
- Understand how aqueous solution titrations can be used to determine the concentration of an unknown solution (Section 5.8). ■ Question 56

Selected end-of-chapter Questions may be assigned in OWL.

For quick review, download Go Chemistry mini-lecture flashcard modules (or purchase them at **www.ichapters.com**).

Prepare for an exam with a **Summary Problem** that brings together concepts and problem-solving skills from throughout the chapter. Go to **www.cengage.com/chemistry/moore** to download this chapter's Summary Problem from the Student Companion Site.

KEY TERMS

acid *(Section 5.2)*
base *(5.2)*
concentration *(5.6)*
equivalence point *(5.8)*
hydronium ion *(5.2)*
hydroxide ion *(5.2)*
metal activity series *(5.5)*
molarity *(5.6)*
net ionic equation *(5.1)*
oxidation *(5.3)*
oxidation number *(5.4)*
oxidation-reduction reaction *(5.3)*
oxidized *(5.3)*
oxidizing agent *(5.3)*
precipitate *(5.1)*
redox reactions *(5.3)*
reduced *(5.3)*
reducing agent *(5.3)*
reduction *(5.3)*
salt *(5.2)*
solute *(5.6)*
solvent *(5.6)*
spectator ion *(5.1)*
standard solution *(5.8)*
strong acid *(5.2)*
strong base *(5.2)*
strong electrolyte *(5.1)*
titration *(5.8)*
weak acid *(5.2)*
weak base *(5.2)*
weak electrolyte *(5.2)*

QUESTIONS FOR REVIEW AND THOUGHT

■ denotes questions assignable in OWL.

Blue-numbered questions have short answers at the back of this book and fully worked solutions in the *Student Solutions Manual.*

Review Questions

1. Classify each of these reactions as a combination, decomposition, exchange, acid-base, or oxidation-reduction reaction.
   (a) $MgO(s) + 2\ HCl(aq) \longrightarrow MgCl_2(aq) + H_2O(\ell)$
   (b) $2\ NaHCO_3(s) \xrightarrow{heat} Na_2CO_3(s) + CO_2(g) + H_2O(g)$
   (c) $CaO(s) + SO_2(g) \longrightarrow CaSO_3(s)$
   (d) $3\ Cu(s) + 8\ HNO_3(aq) \longrightarrow 3\ Cu(NO_3)_2(aq) + 2\ NO(g) + 4\ H_2O(\ell)$
   (e) $2\ NO(g) + O_2(g) \longrightarrow 2\ NO_2(g)$
2. Find two examples in this chapter of the reaction of a metal with a halogen, write a balanced equation for each example, and name the product.
3. Find two examples of acid-base reactions in this chapter. Write balanced equations for these reactions, and name the reactants and products.
4. Find two examples of precipitation reactions in this chapter. Write balanced equations for these reactions, and name the reactants and products.
5. Explain the difference between oxidation and reduction. Give an example of each.
6. For each of the following, does the oxidation number increase or decrease in the course of a redox reaction?
   (a) An oxidizing agent
   (b) A reducing agent
   (c) A substance undergoing oxidation
   (d) A substance undergoing reduction

Topical Questions

Solubility

7. ■ Predict whether each of these compounds is likely to be water-soluble. Indicate which ions are present in solution for the water-soluble compounds.
   (a) $Fe(ClO_4)_2$ (b) Na_2SO_4
   (c) KBr (d) Na_2CO_3
8. Predict whether each of these compounds is likely to be water-soluble. For those compounds which are soluble, indicate which ions are present in solution.
   (a) $Ca(NO_3)_2$ (b) KCl
   (c) $CuSO_4$ (d) $FeCl_3$
9. Predict whether each of these compounds is likely to be water-soluble. Indicate which ions are present in solution for the water-soluble compounds.
   (a) Potassium monohydrogen phosphate
   (b) Sodium hypochlorite
   (c) Magnesium chloride
   (d) Calcium hydroxide
   (e) Aluminum bromide
10. Predict whether each of these compounds is likely to be water-soluble. Indicate which ions are present in solution for the water-soluble compounds.
   (a) Ammonium nitrate (b) Barium sulfate
   (c) Potassium acetate (d) Calcium carbonate
   (e) Sodium perchlorate

Exchange Reactions

11. ■ For each of these pairs of ionic compounds, write a balanced equation reflecting whether precipitation will occur in aqueous solution. For those combinations that do not produce a precipitate, write "NP."
   (a) $MnCl_2 + Na_2S$ (b) $HNO_3 + CuSO_4$
   (c) $NaOH + HClO_4$ (d) $Hg(NO_3)_2 + Na_2S$
   (e) $Pb(NO_3)_2 + HCl$ (f) $BaCl_2 + H_2SO_4$
12. For each of these pairs of ionic compounds, write a balanced equation reflecting whether precipitation will occur in aqueous solution. For those combinations that do not produce a precipitate, write "NP."
   (a) $HNO_3 + Na_3PO_4$ (b) $NaCl + Pb(CH_3COO)_2$
   (c) $(NH_4)_2S + NiCl_2$ (d) $K_2SO_4 + Cu(NO_3)_2$
   (e) $FeCl_3 + NaOH$ (f) $AgNO_3 + KCl$
13. Identify the water-insoluble product in each of these reactions. Write the net ionic equations for these reactions. Identify the spectator ions.
   (a) $CuCl_2(aq) + H_2S(aq) \longrightarrow CuS + 2\ HCl$
   (b) $CaCl_2(aq) + K_2CO_3(aq) \longrightarrow 2\ KCl + CaCO_3$
   (c) $AgNO_3(aq) + NaI(aq) \longrightarrow AgI + NaNO_3$
14. Identify the water-insoluble product in each of these reactions. Write the net ionic equations for these reactions. Identify the spectator ions.
   (a) $Pb(NO_3)_2(aq) + Na_2SO_4(aq) \longrightarrow PbSO_4 + NaNO_3$
   (b) $K_3PO_4(aq) + Mg(NO_3)_2(aq) \longrightarrow Mg_3(PO_4)_2 + KNO_3$
   (c) $(NH_4)_2SO_4(aq) + BaBr_2(aq) \longrightarrow BaSO_4 + NH_4Br$
15. ■ Balance each of these equations, and then write the complete ionic and net ionic equations.
   (a) $Zn(s) + HCl(aq) \longrightarrow H_2(g) + ZnCl_2(aq)$
   (b) $Mg(OH)_2(s) + HCl(aq) \longrightarrow MgCl_2(aq) + H_2O(\ell)$
   (c) $HNO_3(aq) + CaCO_3(s) \longrightarrow Ca(NO_3)_2(aq) + H_2O(\ell) + CO_2(g)$
   (d) $HCl(aq) + MnO_2(s) \longrightarrow MnCl_2(aq) + Cl_2(g) + H_2O(\ell)$
16. Balance each of these equations, and then write the complete ionic and net ionic equations.
   (a) $(NH_4)_2CO_3(aq) + Cu(NO_3)_2(aq) \longrightarrow CuCO_3(s) + NH_4NO_3(aq)$
   (b) $Pb(NO_3)_2(aq) + HCl(aq) \longrightarrow PbCl_2(s) + HNO_3(aq)$
   (c) $BaCO_3(s) + HCl(aq) \longrightarrow BaCl_2(aq) + H_2O(\ell) + CO_2(g)$
17. Balance each of these equations, and then write the complete ionic and net ionic equations. Refer to Tables 5.1 and 5.2 for information on solubility and on acids and bases. Show states (s, ℓ, g, aq) for all reactants and products.
   (a) $Ca(OH)_2 + HNO_3 \longrightarrow Ca(NO_3)_2 + H_2O$
   (b) $BaCl_2 + Na_2CO_3 \longrightarrow BaCO_3 + NaCl$
   (c) $Na_3PO_4 + Ni(NO_3)_2 \longrightarrow Ni_3(PO_4)_2 + NaNO_3$

18. Balance each of these equations, and then write the complete ionic and net ionic equations. Refer to Tables 5.1 and 5.2 for information on solubility and on acids and bases. Show states (s, ℓ, g, aq) for all reactants and products.
(a) $ZnCl_2 + KOH \longrightarrow KCl + Zn(OH)_2$
(b) $AgNO_3 + KI \longrightarrow AgI + KNO_3$
(c) $NaOH + FeCl_2 \longrightarrow Fe(OH)_2 + NaCl$

19. Balance the equation for this precipitation reaction, and then write the complete ionic and net ionic equations.

$$CdCl_2 + NaOH \longrightarrow Cd(OH)_2 + NaCl$$

20. Balance the equation for this precipitation reaction, and then write the complete ionic and net ionic equations.

$$Ni(NO_3)_2 + Na_2CO_3 \longrightarrow NiCO_3 + NaNO_3$$

21. ■ Classify each of these as an acid or a base. Which are strong and which are weak? What ions are produced when each is dissolved in water?
(a) KOH (b) $Mg(OH)_2$
(c) HClO (d) HBr
(e) LiOH (f) H_2SO_3

22. Classify each of these as an acid or a base. Which are strong and which are weak? What ions are produced when each is dissolved in water?
(a) HNO_3 (b) $Ca(OH)_2$
(c) NH_3 (d) H_3PO_4
(e) KOH (f) CH_3COOH

23. ■ Identify the acid and base used to form these salts, and write the overall neutralization reaction in both complete and net ionic form.
(a) $NaNO_2$ (b) $CaSO_4$
(c) NaI (d) $Mg_3(PO_4)_2$

24. Identify the acid and base used to form these salts, and write the overall neutralization reaction in both complete and net ionic form.
(a) $NaCH_3COO$ (b) $CaCl_2$
(c) LiBr (d) $Ba(NO_3)_2$

25. ■ Classify each of these exchange reactions as an acid-base reaction, a precipitation reaction, or a gas-forming reaction. Predict the products of the reaction, and then balance the completed equation.
(a) $MnCl_2(aq) + Na_2S(aq) \longrightarrow$
(b) $Na_2CO_3(aq) + ZnCl_2(aq) \longrightarrow$
(c) $K_2CO_3(aq) + HClO_4(aq) \longrightarrow$

26. Classify each of these exchange reactions as an acid-base reaction, a precipitation reaction, or a gas-forming reaction. Predict the products of the reaction, and then balance the completed equation.
(a) $Fe(OH)_3(s) + HNO_3(aq) \longrightarrow$
(b) $FeCO_3(s) + H_2SO_4(aq) \longrightarrow$
(c) $FeCl_2(aq) + (NH_4)_2S(aq) \longrightarrow$
(d) $Fe(NO_3)_2(aq) + Na_2CO_3(aq) \longrightarrow$

Oxidation-Reduction Reactions

27. ■ Assign oxidation numbers to each atom in these compounds.
(a) SO_3 (b) HNO_3
(c) $KMnO_4$ (d) H_2O
(e) LiOH (f) CH_2Cl_2

28. Assign oxidation numbers to each atom in these compounds.
(a) $Fe(OH)_3$ (b) $HClO_3$
(c) $CuCl_2$ (d) K_2CrO_4
(e) $Ni(OH)_2$ (f) N_2H_4

29. Assign oxidation numbers to each atom in these ions.
(a) SO_4^{2-} (b) NO_3^-
(c) MnO_4^- (d) $Cr(OH)_4^-$
(e) $H_2PO_4^-$ (f) $S_2O_3^{2-}$

30. What is the oxidation number of Mn in each of these species?
(a) $(MnF_6)^{3-}$ (b) Mn_2O_7
(c) MnO_4^- (d) $Mn(CN)_6^-$
(e) MnO_2

31. What is the oxidation number of S in each of these species?
(a) H_2SO_4 (b) H_2SO_3
(c) SO_2 (d) SO_3
(e) $H_2S_2O_7$ (f) $Na_2S_2O_3$

32. ■ Which of these reactions are oxidation-reduction reactions? Explain your answer briefly. Classify the remaining reactions.
(a) $CdCl_2(aq) + Na_2S(aq) \longrightarrow CdS(s) + 2\ NaCl(aq)$
(b) $2\ Ca(s) + O_2(g) \longrightarrow 2\ CaO(s)$
(c) $Ca(OH)_2(s) + 2\ HCl(aq) \longrightarrow CaCl_2(aq) + 2\ H_2O(\ell)$

33. Which of these reactions are oxidation-reduction reactions? Explain your answer briefly. Classify the remaining reactions.
(a) $Zn(s) + 2\ NO_3^-(aq) + 4\ H_3O^+(aq) \longrightarrow Zn^{2+}(aq) - 2\ NO_2(g) + 6\ H_2O(\ell)$
(b) $Zn(OH)_2(s) + H_2SO_4(aq) \longrightarrow ZnSO_4(aq) + 2\ H_2O(\ell)$
(c) $Ca(s) + 2\ H_2O(\ell) \longrightarrow Ca(OH)_2(s) + H_2(g)$

34. ■ Which of these substances are oxidizing agents?
(a) Zn (b) O_2
(c) HNO_3 (d) MnO_4^-
(e) H_2 (f) H^+

35. Which of these substances are reducing agents?
(a) Ca (b) Ca^{2+}
(c) $Cr_2O_7^{2-}$ (d) Al
(e) Br_2 (f) H_2

36. Identify the products of these redox combination reactions.
(a) $C(s) + O_2(g) \longrightarrow$
(b) $P_4(s) + Cl_2(g) \longrightarrow$
(c) $Ti(s) + Cl_2(g) \longrightarrow$
(d) $Mg(s) + N_2(g) \longrightarrow$
(e) $FeO(s) + O_2(g) \longrightarrow$
(f) $NO(g) + O_2(g) \longrightarrow$

37. Complete and balance these equations for redox displacement reactions.
(a) $K(s) + H_2O(\ell) \longrightarrow$
(b) $Mg(s) + HBr(aq) \longrightarrow$
(c) $NaBr(aq) + Cl_2(aq) \longrightarrow$
(d) $WO_3(s) + H_2(g) \longrightarrow$
(e) $H_2S(aq) + Cl_2(aq) \longrightarrow$

Activity Series

38. Give an example of a displacement reaction that is also a redox reaction and identify which species is (a) oxidized, (b) reduced, (c) the reducing agent, and (d) the oxidizing agent.

39. (a) In what groups of the periodic table are the most reactive metals found? Where do we find the least reactive metals?
    (b) Silver (Ag) does not react with 1 M HCl solution. Will Ag react with a solution of aluminum nitrate, $Al(NO_3)_3$? If so, write a chemical equation for the reaction.
    (c) Lead (Pb) will react very slowly with 1 M HCl solution. Aluminum will react with lead(II) sulfate solution, $PbSO_4$. Will Pb react with an $AgNO_3$ solution? If so, write a chemical equation for the reaction.
    (d) On the basis of the information obtained in answering parts (a), (b), and (c), arrange Ag, Al, and Pb in decreasing order of reactivity.
40. ■ Use the activity series of metals (Table 5.5) to predict the outcome of each of these reactions. If no reaction occurs, write "NR."
    (a) $Na^+(aq) + Zn(s) \longrightarrow$
    (b) $HCl(aq) + Pt(s) \longrightarrow$
    (c) $Ag^+(aq) + Au(s) \longrightarrow$
    (d) $Au^{3+}(aq) + Ag(s) \longrightarrow$
41. Using the activity series of metals (Table 5.5), predict whether these reactions will occur in aqueous solution.
    (a) $Mg(s) + Ca(s) \longrightarrow Mg^{2+}(aq) + Ca^{2+}(aq)$
    (b) $2\ Al^{3+}(aq) + 3\ Pb^{2+}(aq) \longrightarrow 2\ Al(s) + 3\ Pb(s)$
    (c) $H_2(g) + Zn^{2+}(aq) \longrightarrow 2\ H^+(aq) + Zn(s)$
    (d) $Mg(s) + Cu^{2+}(aq) \longrightarrow Mg^{2+}(aq) + Cu(s)$
    (e) $Pb(s) + 2\ H^+(aq) \longrightarrow H_2(g) + Pb^{2+}(aq)$
    (f) $2\ Ag^+(aq) + Cu(s) \longrightarrow 2\ Ag(s) + Cu^{2+}(aq)$
    (g) $2\ Al^{3+}(aq) + 3\ Zn(s) \longrightarrow 3\ Zn^{2+}(aq) + 2\ Al(s)$

Solution Concentrations

42. ■ Assume that 6.73 g Na_2CO_3 is dissolved in enough water to make 250. mL of solution.
    (a) What is the molarity of the sodium carbonate?
    (b) What are the concentrations of the Na^+ and CO_3^{2-} ions?
43. Some $K_2Cr_2O_7$, with a mass of 2.335 g, is dissolved in enough water to make 500. mL of solution.
    (a) What is the molarity of the potassium dichromate?
    (b) What are the concentrations of the K^+ and $Cr_2O_7^{2-}$ ions?
44. ■ What is the mass, in grams, of solute in 250. mL of a 0.0125 M solution of $KMnO_4$?
45. What is the mass, in grams, of solute in 100. mL of a 1.023×10^{-3} M solution of Na_3PO_4?
46. ■ What volume of 0.123 M NaOH, in milliliters, contains 25.0 g NaOH?
47. What volume of 2.06 M $KMnO_4$, in liters, contains 322 g solute?
48. ■ If 6.00 mL of 0.0250 M $CuSO_4$ is diluted to 10.0 mL with pure water, what is the concentration of copper(II) sulfate in the diluted solution?
49. If you dilute 25.0 mL of 1.50 M HCl to 500. mL, what is the molar concentration of the diluted HCl?

Calculations for Reactions in Solution

50. What mass, in grams, of Na_2CO_3 is required for complete reaction with 25.0 mL of 0.155 M HNO_3?

$$Na_2CO_3(aq) + 2\ HNO_3(aq) \longrightarrow 2\ NaNO_3(aq) + CO_2(g) + H_2O(\ell)$$

51. ■ Hydrazine, N_2H_4, a base like ammonia, can react with an acid such as sulfuric acid.

$$2\ N_2H_4(aq) + H_2SO_4(aq) \longrightarrow 2\ N_2H_5^+(aq) + SO_4^{-2}(aq)$$

What mass of hydrazine can react with 250. mL of 0.225 M H_2SO_4?

52. What volume, in milliliters, of 0.125 M HNO_3 is required to react completely with 1.30 g $Ba(OH)_2$?

$$2\ HNO_3(aq) + Ba(OH)_2(s) \longrightarrow Ba(NO_3)_2(aq) + 2\ H_2O(\ell)$$

53. Diborane, B_2H_6, can be produced by this reaction:

$$2\ NaBH_4(s) + H_2SO_4(aq) \longrightarrow 2\ H_2(g) + Na_2SO_4(aq) + B_2H_6(g)$$

What volume, in milliliters, of 0.0875 of M H_2SO_4 should be used to completely react with 1.35 g $NaBH_4$?

54. What is the maximum mass, in grams, of AgCl that can be precipitated by mixing 50.0 mL of 0.025 M $AgNO_3$ solution with 100.0 mL of 0.025 M NaCl solution? Which reactant is in excess? What is the concentration of the excess reactant remaining in solution after the AgCl has precipitated?
55. Suppose you mix 25.0 mL of 0.234 M $FeCl_3$ solution with 42.5 mL of 0.453 M NaOH.
    (a) What is the maximum mass, in grams, of $Fe(OH)_3$ that will precipitate?
    (b) Which reactant is in excess?
    (c) What is the concentration of the excess reactant remaining in solution after the maximum mass of $Fe(OH)_3$ has precipitated?
56. ■ If a volume of 32.45 mL HCl is used to completely neutralize 2.050 g Na_2CO_3 according to this equation, what is the molarity of the HCl?

$$Na_2CO_3(aq) + 2\ HCl(aq) \longrightarrow 2\ NaCl(aq) + CO_2(g) + H_2O(\ell)$$

57. Potassium acid phthalate, $KHC_8H_4O_4$, is used to standardize solutions of bases. The acidic anion reacts with bases according to this net ionic equation:

$$HC_8H_4O_4^-(aq) + OH^-(aq) \longrightarrow H_2O(\ell) + C_8H_4O_4^{2-}(aq)$$

If a 0.902-g sample of potassium acid phthalate requires 26.45 mL NaOH to react, what is the molarity of the NaOH?

General Questions

58. Magnesium metal reacts readily with HNO_3, as shown in this equation:

$$Mg(s) + HNO_3(aq) \longrightarrow Mg(NO_3)_2(aq) + NO_2(g) + H_2O(\ell)$$

    (a) Balance the equation.
    (b) Name each reactant and product.
    (c) Write the net ionic equation.
    (d) What type of reaction is this?
59. Aqueous solutions of $(NH_4)_2S$ and $Hg(NO_3)_2$ react to give HgS and NH_4NO_3.
    (a) Write the overall balanced equation. Indicate the state (s or aq) for each compound.
    (b) Name each compound.

(c) Write the net ionic equation.
(d) What type of reaction does this appear to be?

60. Classify these reactions and predict the products formed.
(a) $SO_3(g) + H_2O(\ell) \longrightarrow$
(b) $Sr(s) + H_2(g) \longrightarrow$
(c) $Mg(s) + H_2SO_4(aq, \text{dilute}) \longrightarrow$
(d) $Na_3PO_4(aq) + AgNO_3(aq) \longrightarrow$
(e) $Ca(HCO_3)_2(s) \xrightarrow{\text{heat}}$
(f) $Fe^{3+}(aq) + Sn^{2+}(aq) \longrightarrow$

61. What species (atoms, molecules, ions) are present in an aqueous solution of each of these compounds?
(a) NH_3 (b) CH_3COOH
(c) NaOH (d) HBr

62. Use the activity series to predict whether these reactions will occur.
(a) $Fe(s) + Mg^{2+}(aq) \longrightarrow Mg(s) + Fe^{2+}(aq)$
(b) $Ni(s) + Cu^{2+}(aq) \longrightarrow Ni^{2+}(aq) + Cu(s)$
(c) $Cu(s) + 2\,H^+(aq) \longrightarrow Cu^{2+}(aq) + H_2(g)$
(d) $Mg(s) + H_2O(g) \longrightarrow MgO(s) + H_2(g)$

63. Determine which of these are redox reactions. Identify the oxidizing and reducing agents in each of the redox reactions.
(a) $NaOH(aq) + H_3PO_4(aq) \rightarrow NaH_2PO_4(aq) + H_2O(\ell)$
(b) $NH_3(g) + CO_2(g) + H_2O(\ell) \rightarrow NH_4HCO_3(aq)$
(c) $TiCl_4(g) + 2\,Mg(\ell) \longrightarrow Ti(s) + 2\,MgCl_2(\ell)$
(d) $NaCl(s) + NaHSO_4(aq) \longrightarrow HCl(g) + Na_2SO_4(aq)$

Applying Concepts

64. When these pairs of reactants are combined in a beaker, (a) describe in words what the contents of the beaker would look like before and after any reaction occurs, (b) use different circles for atoms, molecules, and ions to draw a nanoscale (particulate-level) diagram of what the contents would look like, and (c) write a chemical equation to represent symbolically what the contents would look like.

$LiCl(aq)$ and $AgNO_3(aq)$

$NaOH(aq)$ and $HCl(aq)$

65. When these pairs of reactants are combined in a beaker, (a) describe in words what the contents of the beaker would look like before and after any reaction occurs, (b) use different circles for atoms, molecules, and ions to draw a particulate-level diagram of what the contents would look like, and (c) write a chemical equation to represent symbolically what the contents would look like.

$CaCO_3(s)$ and $HCl(aq)$

$NH_4NO_3(aq)$ and $KOH(aq)$

66. Explain how you could prepare barium sulfate by (a) an acid-base reaction, (b) a precipitation reaction, and (c) a gas-forming reaction. The materials you have to start with are $BaCO_3$, $Ba(OH)_2$, Na_2SO_4, and H_2SO_4.

67. An unknown solution contains either calcium ions or strontium ions, but not both. Which one of these solutions could you use to tell whether the ions present are Ca^{2+} or Sr^{2+}? Explain the reasoning behind your choice.

$NaOH(aq)$, $H_2SO_4(aq)$, $H_2S(aq)$

68. You prepared a NaCl solution by adding 58.44 g NaCl to a 1-L volumetric flask and then adding water to dissolve it. When you were finished, the final volume in your flask looked like this:

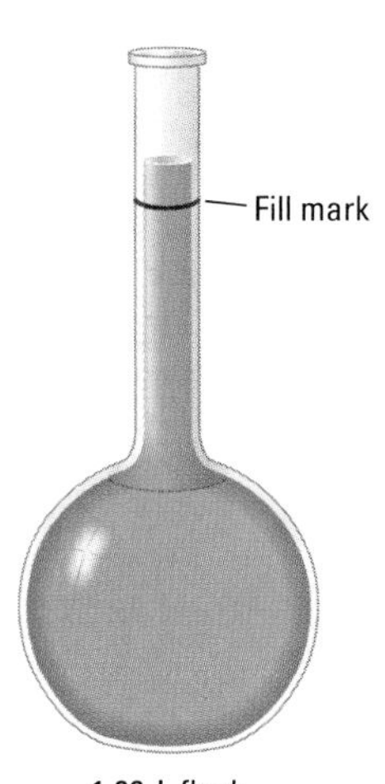

The solution you prepared is
(a) Greater than 1 M because you added more solvent than necessary.
(b) Less than 1 M because you added less solvent than necessary.
(c) Greater than 1 M because you added less solvent than necessary.
(d) Less than 1 M because you added more solvent than necessary.
(e) 1 M because the amount of solute, not solvent, determines the concentration.

69. These drawings represent beakers of aqueous solutions. Each orange circle represents a dissolved solute particle.

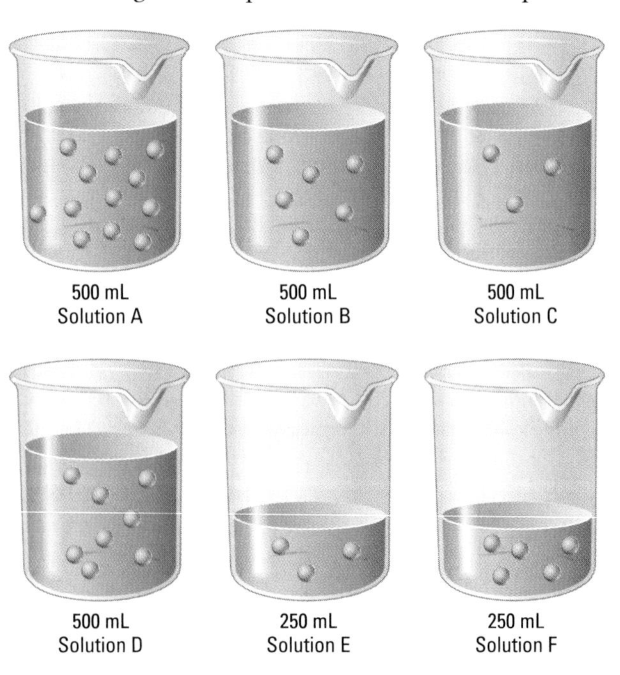

(a) Which solution is most concentrated?
(b) Which solution is least concentrated?

(c) Which two solutions have the same concentration?
(d) When solutions E and F are combined, the resulting solution has the same concentration as solution ________.
(e) When solutions B and E are combined, the resulting solution has the same concentration as solution ________.
(f) If you evaporate half of the water from solution B, the resulting solution will have the same concentration as solution ________.
(g) If you place half of solution A in another beaker and then add 250 mL water, the resulting solution will have the same concentration as solution ________.

70. Ten milliliters of a solution of an acid is mixed with 10 mL of a solution of a base. When the mixture was tested with litmus paper, the blue litmus turned red, and the red litmus remained red. Which of these interpretations is (are) correct?
(a) The mixture contains more hydrogen ions than hydroxide ions.
(b) The mixture contains more hydroxide ions than hydrogen ions.
(c) When an acid and a base react, water is formed, so the mixture cannot be acidic or basic.
(d) If the acid was HCl and the base was NaOH, the concentration of HCl in the initial acidic solution must have been greater than the concentration of NaOH in the initial basic solution.
(e) If the acid was H_2SO_4 and the base was NaOH, the concentration of H_2SO_4 in the initial acidic solution must have been greater than the concentration of NaOH in the initial basic solution.

71. Gold can be dissolved from gold-bearing rock by treating the rock with sodium cyanide in the presence of the oxygen in air.

$$4\ Au(s) + 8\ NaCN(aq) + O_2(g) + 2\ H_2O(\ell) \longrightarrow 4\ NaAu(CN)_2(aq) + 4\ NaOH(aq)$$

Once the gold is in solution in the form of the $Au(CN)_2^-$ ion, it can be precipitated as the metal according to the following unbalanced equation:

$$Au(CN)_2^-(aq) + Zn(s) \longrightarrow Zn^{2+}(aq) + Au(s) + CN^-(aq)$$

(a) Are the two reactions above acid-base or oxidation-reduction reactions? Briefly describe your reasoning.
(b) How many liters of 0.075 M NaCN will you need to extract the gold from 1000 kg of rock if the rock is 0.019% gold?
(c) How many kilograms of metallic zinc will you need to recover the gold from the $Au(CN)_2^-$ obtained from the gold in the rock?
(d) If the gold is recovered completely from the rock and the metal is made into a cylindrical rod 15.0 cm long, what is the diameter of the rod? (The density of gold is 19.3 g/cm^3.)

72. Four groups of students from an introductory chemistry laboratory are studying the reactions of solutions of alkali metal halides with aqueous silver nitrate, $AgNO_3$. They use these salts.

Group A: NaCl
Group B: KCl
Group C: NaBr
Group D: KBr

Each of the four groups dissolves 0.004 mol of their salt in some water. Each then adds various masses of silver nitrate, $AgNO_3$, to their solutions. After each group collects the precipitated silver halide, the mass of this product is plotted versus the mass of $AgNO_3$ added. The results are given on this graph.

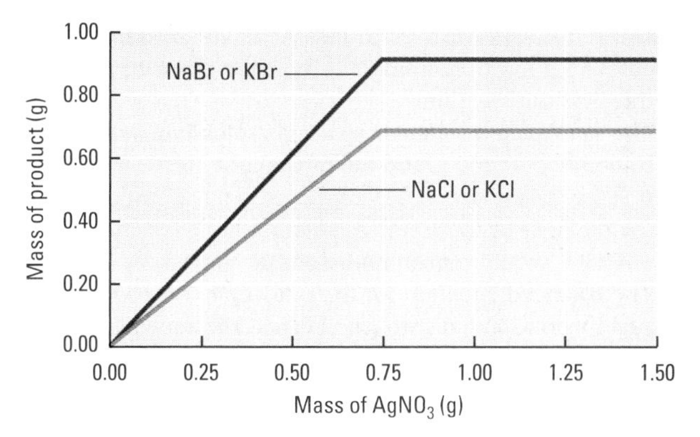

(a) Write the balanced net ionic equation for the reaction observed by each group.
(b) Explain why the data for groups A and B lie on the same line, whereas those for groups C and D lie on a different line.
(c) Explain the shape of the curve observed by each group. Why do they level off at the same mass of added $AgNO_3$ (0.75 g) but give different masses of product?

More Challenging Questions

73. You are given 0.954 g of an unknown acid, H_2A, which reacts with NaOH according to the balanced equation

$$H_2A(aq) + 2\ NaOH(aq) \longrightarrow Na_2A(aq) + 2\ H_2O(\ell)$$

If a volume of 36.04 mL 0.509 M NaOH is required to react with all of the acid, what is the molar mass of the acid?

74. You are given an acid and told only that it could be citric acid (molar mass = 192.1 g/mol) or tartaric acid (molar mass = 150.1 g/mol). To determine which acid you have, you react it with NaOH. The appropriate reactions are

Citric acid:

$$C_6H_8O_7(aq) + 3\ NaOH(aq) \longrightarrow Na_3C_6H_5O_7(aq) + 3\ H_2O(\ell)$$

Tartaric acid:

$$C_4H_6O_6(aq) + 2\ NaOH(aq) \longrightarrow Na_2C_4H_4O_6(aq) + 2\ H_2O(\ell)$$

You find that a 0.956-g sample requires 29.1 mL 0.513 M NaOH to reach the equivalence point. What is the unknown acid?

75. In the past, devices for testing a driver's breath for alcohol depended on this reaction:

$$\underset{\text{ethanol}}{3\ C_2H_5OH(aq)} + 2\ K_2Cr_2O_7(aq) + 8\ H_2SO_4(aq) \longrightarrow$$

$$\underset{\text{acetic acid}}{3\ CH_3COOH(aq)} + 2\ Cr_2(SO_4)_3(aq) + 2\ K_2SO_4(aq) + 11\ H_2O(\ell)$$

Write the net ionic equation for this reaction. What oxidation numbers are changing in the course of this reaction? Which substances are being oxidized and reduced? Which substance is the oxidizing agent and which is the reducing agent?

76. How much salt is in your chicken soup? A student added excess $AgNO_3(aq)$ to a 1-cup serving of regular chicken soup (240 mL) and got 5.55 g AgCl precipitate. How many grams of NaCl were in the regular chicken soup? Assume that all the chloride ions in the soup were from NaCl. In a second experiment, the same procedure was done with chicken soup advertised to have "less salt," and the student got 3.55 g AgCl precipitate. How many grams of NaCl are in the "less salt" version?

77. The salt calcium sulfate is sparingly soluble in water with a solubility of 0.209 g/100 mL of water at 30 °C. If you stirred 0.550 g $CaSO_4$ into 100.0 mL water at 30 °C, what would the molarity of the resulting solution be? How many grams of $CaSO_4$ would remain undissolved?

78. The balanced equation for the oxidation of ethanol to acetic acid by potassium dichromate in an acidic aqueous solution is

$$3\,C_2H_5OH(aq) + 2\,K_2Cr_2O_7(aq) + 16\,HCl(aq) \longrightarrow 3\,CH_3COOH(aq) + 4\,CrCl_3(aq) + 4\,KCl(aq) + 11\,H_2O(\ell)$$

What volume of a 0.600 M potassium dichromate solution is needed to generate 0.166 mol acetic acid (CH_3COOH) from a solution containing excess ethanol and HCl?

79. Dolomite, found in soil, is $CaMg(CO_3)_2$. If a 20.0-g sample of soil is titrated with 65.25 mL of 0.2500 M HCl, what is the mass percent of dolomite in the soil sample?

80. Vitamin C is ascorbic acid, $HC_6H_7O_6$, which can be titrated with a strong base.

$$HC_6H_7O_6(aq) + NaOH(aq) \longrightarrow NaC_6H_7O_6(aq) + H_2O(\ell)$$

In a laboratory experiment, a student dissolved a 500.0-mg vitamin C tablet in 200.0 mL water and then titrated it with 0.1250 M NaOH. It required 21.30 mL of the base to reach the equivalence point. What percentage of the mass of the tablet is impurity?

Conceptual Challenge Problems

CP5.A (Section 5.3) There is a conservation of the number of electrons exchanged during redox reactions, which is tantamount to stating that electrical charge is conserved during chemical reactions. The assignment of oxidation numbers is an arbitrary yet clever way to do the bookkeeping for these electrons. What makes it possible to assign the same oxidation number to all elements that are not bound to other elements not in chemical compounds?

CP5.B (Section 5.4) Consider these redox reactions:

$$HIO_3 + FeI_2 + HCl \longrightarrow FeCl_3 + ICl + H_2O$$

$$CuSCN + KIO_3 + HCl \longrightarrow CuSO_4 + KCl + HCN + ICl + H_2O$$

(a) Identify the species that have been oxidized or reduced in each of the reactions.

(b) After you have correctly identified the species that have been oxidized or reduced in each equation, you might like to try using oxidation numbers to balance each equation. This will be a challenge because, as you have discovered, more than one kind of atom is oxidized or reduced, although in all cases the product of the oxidation and reduction is unambiguous. Record the initial and final oxidation states of each kind of atom that is oxidized or reduced in each equation. Then decide on the coefficients that will equalize the oxidation number changes and satisfy any other atom balancing needed. Finally, balance the equation by adding the correct coefficients to it.

CP5.C (Section 5.5) A student was given four metals (A, B, C, and D) and solutions of their corresponding salts (AZ, BZ, CZ, and DZ). The student was asked to determine the relative reactivity of the four metals by reacting the metals with the solutions. The student's laboratory observations are indicated in the table. Arrange the four metals in order of decreasing activity.

Metal	AZ(aq)	BZ(aq)	CZ(aq)	DZ(aq)
A	No reaction	No reaction	No reaction	No reaction
B	Reaction	No reaction	Reaction	No reaction
C	Reaction	No reaction	No reaction	No reaction
D	Reaction	Reaction	Reaction	No reaction

CP5.D (Section 5.6) How would you prepare 1 L of 1.00×10^{-6} M NaCl (molar mass = 58.44 g/mol) solution by using a balance that can measure mass only to 0.01 g?

CP5.E (Section 5.8) How could you show that when baking soda reacts with the acetic acid, CH_3COOH, in vinegar, all of the carbon and oxygen atoms in the carbon dioxide produced come from the baking soda alone and none comes from the acetic acid in vinegar?

6

Energy and Chemical Reactions

© Royalty-Free/Corbis

Combustion of a fuel. Burning propane, C_3H_8, transfers a great deal of energy to anything in contact with the reactant and product molecules—in this case glass, boiling water, and soon-to-be-hard-boiled eggs. The energy transferred when a fuel such as propane burns can be transformed to provide many of the benefits of our technology-intensive society.

In our industrialized, high-technology, appliance-oriented society, the average use of energy per person is at nearly its highest point in history. The United States, with only 5% of the world's population, consumes about 30% of the world's energy resources. In every year since 1958 we have consumed more energy resources than have been produced within our borders. Most of the energy we use comes from chemical reactions: combustion of the fossil fuels coal, petroleum, and natural gas. The rest comes from hydroelectric power plants, nuclear power plants, solar energy and wind collectors, and burning wood and other plant material. Both U.S. and world energy use are growing rapidly.

"A theory is the more impressive the greater the simplicity of its premises is, the more different kinds of things it relates, and the more extended is its area of applicability. Therefore, the deep impression which classical thermodynamics made upon me. It is the only physical theory of universal content concerning which I am convinced that, within the framework of the applicability of its basic concepts, it will never be overthrown." (Albert Einstein, quoted in Schlipp, P. A. [ed.] "Albert Einstein: Philosopher-Scientist." In *The Library of Living Philosophers,* Vol. VII. Autobiographical notes, 3rd ed. LaSalle, IL: Open Court Publishing, 1969; p. 33.)

Chemical reactions involve transfers of energy. When a fuel burns, the energy of the products is less than the energy of the reactants. The leftover energy shows up in anything that is in contact with the reactants and products. For example, when propane burns in air, the carbon, hydrogen, and oxygen atoms in the C_3H_8 and O_2 reactant molecules rearrange to form CO_2 and H_2O product molecules.

$$C_3H_8(g) + 5\,O_2(g) \longrightarrow 3\,CO_2(g) + 4\,H_2O(g)$$

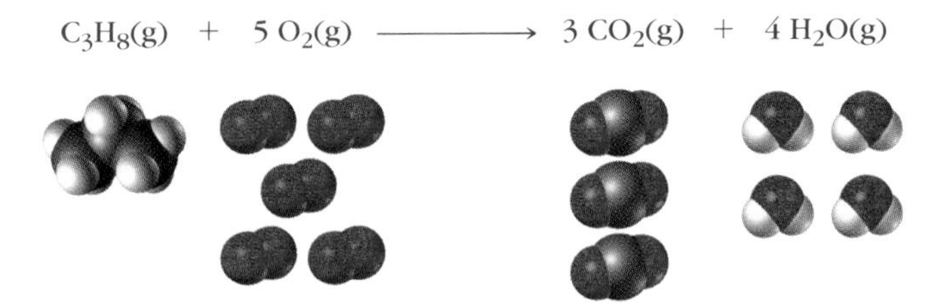

Because of the way their atoms are bonded together, the CO_2 and H_2O molecules have less total energy than the C_3H_8 and O_2 reactant molecules did. After the reaction, some energy that was in the reactants is not contained in the product molecules. That energy heats everything that is close to where the reaction takes place. To describe this effect, we say that the reaction transfers energy to its surroundings.

For the past hundred years or so, most of the energy society has used has come from combustion of fossil fuels, and this will continue to be true well into the future. Consequently, it is very important to understand how energy and chemical reactions are related and how chemistry might be used to alter our dependence on fossil fuels. This requires knowledge of **thermodynamics,** the science of heat, work, and transformations of one to the other. The fastest-growing new industries in the twenty-first century may well be those that capitalize on such knowledge and the new chemistry and chemical industries it spawns.

6.1 The Nature of Energy

What is energy? Where does the energy we use come from? And how can chemical reactions result in the transfer of energy to or from their surroundings? *Energy,* represented by E, was defined in Section 1.5 (*p. 14*) as the capacity to do work. If you climb a mountain or a staircase, you work against the force of gravity as you move upward, and your gravitational energy increases. The energy you use to do this work is released when food you have eaten is metabolized (undergoes chemical reactions) within your body. Energy from food enables you to work against the force of gravity as you climb, and it warms your body (climbing makes you hotter as well as higher). Therefore our study of the relations between energy and chemistry also needs to consider processes that involve work and processes that involve heat.

Energy can be classified as kinetic or potential. **Kinetic energy** is energy that something has because it is moving (Figure 6.1). Examples of kinetic energy are

- Energy of motion of a macroscale object, such as a moving baseball or automobile; this is often called *mechanical energy.*
- Energy of motion of nanoscale objects such as atoms, molecules, or ions; this is often called *thermal energy.*
- Energy of motion of electrons through an electrical conductor; this is often called *electrical energy.*
- Energy of periodic motion of nanoscale particles when a macroscale sample is alternately compressed and expanded (as when a sound wave passes through air).

Kinetic energy, E_k, can be calculated as $E_k = \frac{1}{2}mv^2$, where m represents the mass and v represents the velocity of a moving object.

Figure 6.1 Kinetic energy. As it speeds toward the player, a tennis ball has kinetic energy that depends on its mass and velocity.

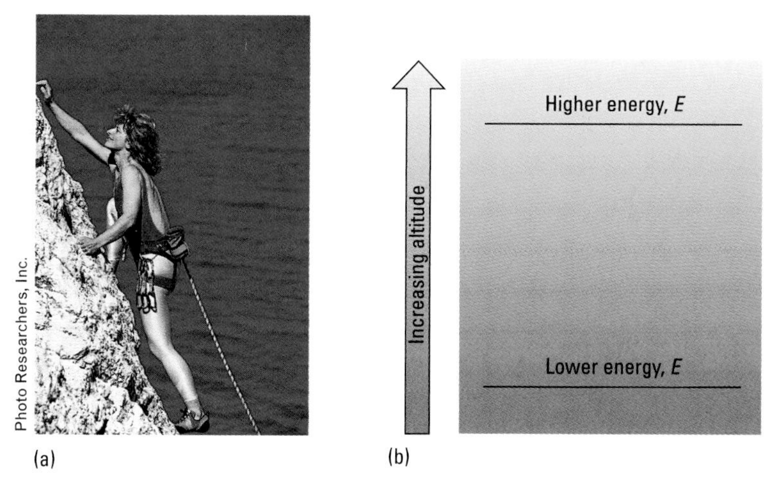

Rock climbing. (a) Climbing requires energy. (b) The higher the altitude, the greater the climber's gravitational energy.

Figure 6.2 Gravitational potential energy. Water on the brink of a waterfall has potential energy (stored energy that could be used to do work) because of its position relative to the earth; that energy could be used to generate electricity, for example, as in a hydroelectric power plant.

Potential energy is energy that something has as a result of its position and some force that is capable of changing that position. Examples include

- Energy that a ball held in your hand has because the force of gravity attracts it toward the floor; this is often called *gravitational energy*.
- Energy that charged particles have because they attract or repel each other; this is often called *electrostatic energy*. An example is the potential energy of positive and negative ions close together.
- Energy resulting from attractions and repulsions among electrons and atomic nuclei in molecules; this is often called *chemical potential energy* and is the kind of energy stored in foods and fuels.

Potential energy can be calculated in different ways, depending on the type of force that is involved. For example, near the surface of the earth, gravitational potential energy, E_p, can be calculated as $E_p = mgh$, where m is mass, g is the gravitational constant ($g = 9.8\ m/s^2$), and h is the height above the surface.

Potential energy can be converted to kinetic energy, and vice versa. As droplets of water fall over a waterfall (Figure 6.2), the potential energy they had at the top is converted to kinetic energy—they move faster and faster. Conversely, the kinetic energy of falling water could drive a water wheel to pump water to an elevated reservoir, where its potential energy would be higher.

James P. Joule 1818–1889

The energy unit joule is named for James P. Joule. The son of a brewer in Manchester, England, Joule was a student of John Dalton (← ***p. 19***). Joule established the idea that working and heating are both processes by which energy can be transferred from one sample of matter to another.

Energy Units

The SI unit of energy is the *joule* (rhymes with rule), symbol J. The joule is a derived unit, which means that it can be expressed as a combination of other more fundamental units: $1\ J = 1\ kg\ m^2/s^2$. If a 2.0-kg object (which weighs about $4\frac{1}{2}$ pounds) is moving with a velocity of 1.0 meter per second (roughly 2 miles per hour), its kinetic energy is

$$E_k = \tfrac{1}{2}mv^2 = \tfrac{1}{2} \times (2.0\ kg)(1.0\ m/s)^2 = 1.0\ kg\ m^2/s^2 = 1.0\ J$$

This is a relatively small quantity of energy. Because the joule is so small, we often use the kilojoule (1 kilojoule = 1 kJ = 1000 J) as a unit of energy.

Another energy unit is the *calorie*, symbol cal. By definition 1 cal = 4.184 J exactly. A calorie is very close to the quantity of energy required to raise the temperature of one gram of water by one degree Celsius. (The calorie was originally defined as the quan-

The joule is the unit of energy in the International System of units (SI units). SI units are described in Appendix B. A joule is approximately the quantity of energy required for one human heartbeat.

1 cal = 4.184 J exactly

1 Cal = 1 kcal = 1000 cal = 4.184 kJ = 4184 J

The food Calorie measures how much energy is released when a given quantity of food undergoes combustion with oxygen.

tity of energy required to raise the temperature of 1 g $H_2O(\ell)$ from 14.5 °C to 15.5 °C.) The "calorie" that you hear about in connection with nutrition and dieting is actually a kilocalorie (kcal) and is usually represented with a capital C. Thus, a breakfast cereal that gives you 100 Calories of nutritional energy actually provides 100 kcal = 100×10^3 cal. In many countries food energy is reported in kilojoules rather than in Calories. For example, the label on the packet of nonsugar sweetener shown in Figure 6.3 indicates that it provides 16 kJ of nutritional energy.

Figure 6.3 Food energy. A packet of artificial sweetener from Australia. As its label shows, the sweetener in the packet supplies 16 kJ of nutritional energy. It is equivalent in sweetness to 2 level teaspoonfuls of sugar, which would supply 140 kJ of nutritional energy.

PROBLEM-SOLVING EXAMPLE 6.1 Energy Units

A single Fritos snack chip has a food energy of 5.0 Cal. What is this energy in joules?

Answer 2.1×10^4 J

Strategy and Explanation To find the energy in joules, we use the fact that 1 Cal = 1 kcal, the definition of the prefix *kilo-* (= 1000), and the definition 1 cal = 4.184 J to generate appropriate proportionality factors (conversion factors).

$$E = 5.0\ \text{Cal} \times \frac{1\ \text{kcal}}{1\ \text{Cal}} \times \frac{1000\ \text{cal}}{1\ \text{kcal}} \times \frac{4.184\ \text{J}}{1\ \text{cal}} = 2.1 \times 10^4\ \text{J}$$

☑ **Reasonable Answer Check** 2.1×10^4 J is 21 kJ. Because 1 Cal = 1 kcal = 4.184 kJ, the result in kJ should be about four times the original 5 Cal (that is, about 20 kJ), which it is.

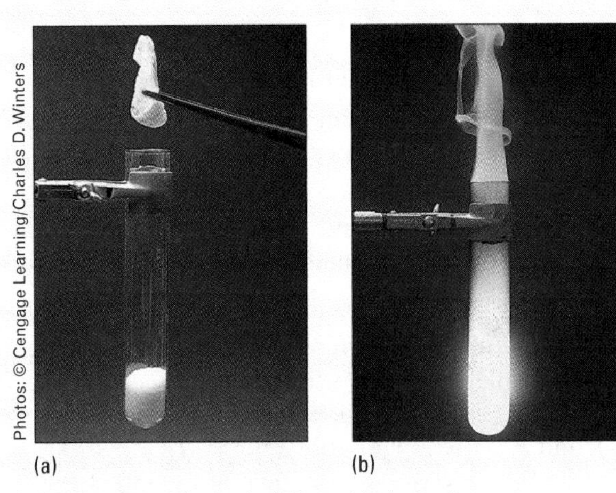

(a) (b)

Food energy. A single Fritos chip burns in oxygen generated by thermal decomposition of potassium chlorate.

PROBLEM-SOLVING PRACTICE 6.1

(a) If you eat a hot dog, it will provide 160 Calories of energy. Express this energy in joules.

(b) A watt (W) is a unit of power that corresponds to the transfer of one joule of energy in one second. The energy used by an x-watt light bulb operating for y seconds is $x \times y$ joules. If you turn on a 75-watt bulb for 3.0 hours, how many joules of electrical energy will be transformed into light and heat?

(c) The packet of nonsugar sweetener in Figure 6.3 provides 16 kJ of nutritional energy. Express this energy in kilocalories.

6.2 Conservation of Energy

In Section 1.8 (p. 16) the kinetic-molecular theory was described qualitatively. A corollary to this theory is that *molecules move faster, on average, as the temperature increases.*

When you dive from a diving board into a pool of water, several transformations of energy occur (Figure 6.4). Eventually, you float on the surface and the water becomes still. However, on average, the water molecules are moving a little faster in the vicin-

ESTIMATION Earth's Kinetic Energy

Estimate the earth's kinetic energy as it moves through space in orbit around the sun.

From an encyclopedia, a dictionary, or the Internet, you can obtain the facts that the earth's mass is about 3×10^{24} kg and its distance from the sun is about 150,000,000 km. Assume earth's orbit is a circle and calculate the distance traveled in a year as the circumference of this circle, $\pi d = 2\pi r = 2 \times 3.14 \times 1.5 \times 10^8$ km. Since $2 \times 3 = 6$, $1.5 \times 6 = 9$, and 3.14 is a bit more than 3, estimate the distance as 10×10^8 km. Earth's speed, then, is a bit less than 10×10^8 km/yr.

Because $1 \text{ J} = 1 \text{ kg m}^2/\text{s}^2$, convert the time unit from years to seconds. Estimate the number of seconds in 1 yr as $60 \text{ s/min} \times 60 \text{ min/h} \times 24 \text{ h/d} \times 365 \text{ d/yr} = 60 \times 60 \times 24 \times 365$ s/yr. To make the arithmetic easy, round 24 to 20 and 365 to 400, giving $60 \times 60 \times 20 \times 400 \text{ s/yr} = 6 \times 6 \times 2 \times 4 \times 10^5$ s/yr. This gives 288×10^5 s/yr, or 3×10^7 s/yr rounded to one significant figure. Therefore earth's speed is about $10/3 \times 10^8/10^7 \cong 30$ km/s or 3×10^4 m/s.

Now the equation for kinetic energy can be used.

$$E_k = \frac{1}{2}mv^2$$

$$= \frac{1}{2} \times (3 \times 10^{24} \text{ kg}) \times (3 \times 10^4 \text{ m/s})^2 \simeq 1 \times 10^{33} \text{ J}$$

Although the earth's speed is not high, its mass is very large. This results in an extraordinarily large kinetic energy—far more energy than has been involved in all of the hurricanes and typhoons that earth has ever experienced.

Visit this book's companion website at **www.cengage.com/chemistry/moore** to work an interactive module based on this material.

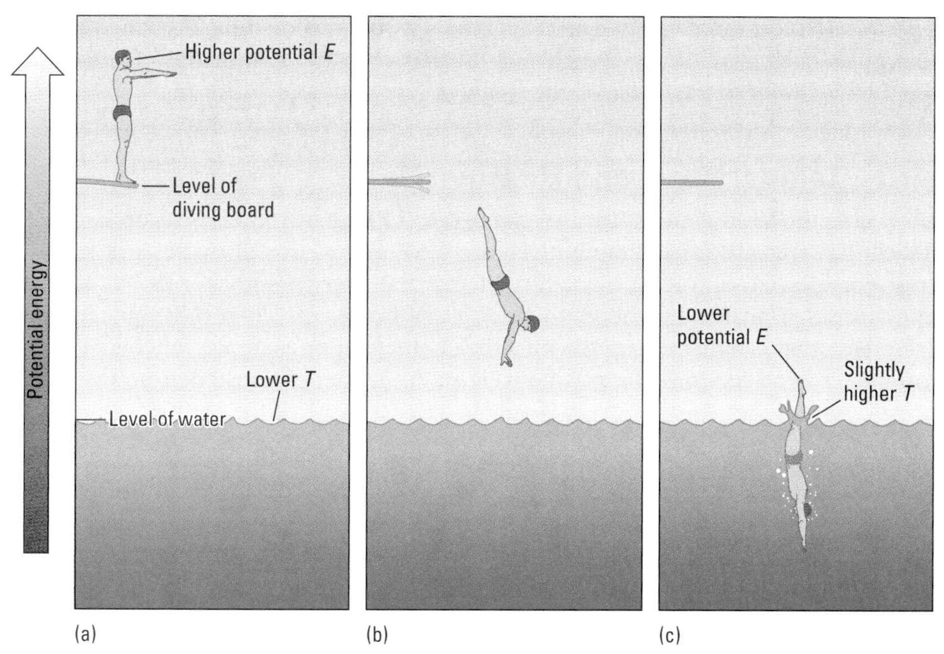

Figure 6.4 Energy transformations. Potential and kinetic energy are interconverted when someone dives into water. These interconversions are governed by the law of conservation of energy. (a) The diver has greater gravitational potential energy on the diving board than at the surface of the water, because the platform is higher above the earth. (b) Some of the potential energy has been converted into kinetic energy as the diver's altitude above the water decreases and velocity increases; maximum kinetic energy occurs just prior to impact with the water. (c) Upon impact, the diver works on the water, splashing it aside; eventually, the initial potential energy difference is converted into motion on the nanoscale—the temperature of the water has become slightly higher.

ity of your point of impact; that is, the temperature of the water is now a little higher. Energy has been transformed from potential to kinetic and from macroscale kinetic to nanoscale kinetic (that is, thermal). Nevertheless, the total quantity of energy, kinetic plus potential, is the same before and after the dive. In many, many experiments, the total energy has always been found to be the same before and after an event. These

The nature of scientific laws is discussed in Chapter 1 (p. 4).

experiments are summarized by the **law of conservation of energy,** which states that *energy can neither be created nor destroyed—the total energy of the universe is constant.* This is also called the **first law of thermodynamics.**

CONCEPTUAL EXERCISE **6.1 Energy Transfers**

You toss a rubber ball up into the air. It falls to the floor, bounces for a while, and eventually comes to rest. Several energy transfers are involved. Describe them and the changes they cause.

Work and *heat* refer to the quantity of energy transferred from one object or sample to another by working or heating *processes.* However, we often talk about work and heat as if they were forms of energy. Working and heating processes transfer energy from one form or one place to another. To emphasize this, we often will use the words *working* and *heating* where many people would use *work* and *heat.*

Energy and Working

When a force acts on an object and moves the object, the change in the object's kinetic energy is equal to the work done on the object. Work has to be done, for example, to accelerate a car from 0 to 60 miles per hour or to hit a baseball out of a stadium. Work is also required to increase the potential energy of an object. Thus, work has to be done to raise an object against the force of gravity (as in an elevator), to separate a sodium ion, Na^+, from a chloride ion, Cl^-, or to move an electron away from an atomic nucleus. The work done on an object corresponds to the quantity of energy transferred to that object; that is, doing **work** (or **working**) on an object is *a process that transfers energy to an object.* Conversely, if an object does work on something else, the quantity of energy associated with the object must decrease. In the rest of this chapter (and book), we will refer to a transfer of energy by doing work as a "work transfer."

Energy, Temperature, and Heating

According to the kinetic-molecular theory (p. 17), all matter consists of nanoscale particles that are in constant motion (Figure 6.5). Therefore, all matter has thermal energy. For a given sample, the quantity of thermal energy is greater the higher the temperature is. Transferring energy to a sample of matter usually results in a temperature increase that can be measured with a thermometer. For example, when a mercury thermometer is placed into warm water (Figure 6.6), energy transfers from the water

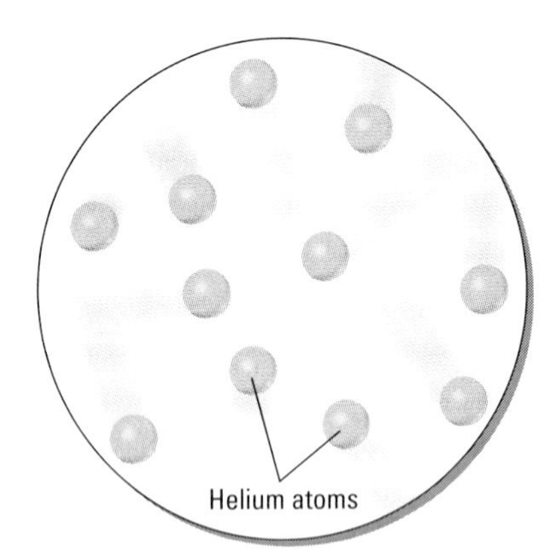

Figure 6.5 Thermal energy. According to the kinetic-molecular theory, nanoscale particles (atoms, molecules, and ions) are in constant motion. Here, atoms of gaseous helium are shown. Each atom has kinetic energy that depends on how fast it is moving (as indicated by the length of the "tail," which shows how far each atom travels per unit time). The thermal energy of the sample is the sum of the kinetic energies of all the helium atoms. The higher the temperature of the helium, the faster the average speed of the molecules, and therefore the greater the thermal energy.

Figure 6.6 Measuring temperature. The mercury in a thermometer expands because the mercury atoms are moving faster (have more energy) after the boiling water transfers energy to (heats) the mercury; the temperature and the volume of the mercury have both increased.

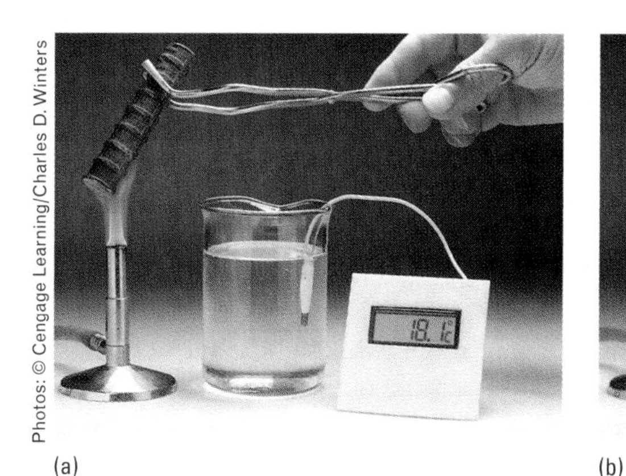

(a) (b)

Figure 6.7 Energy transfer by heating. Water in a beaker is heated when a hotter sample (a steel bar) is plunged into the water. There is a transfer of energy from the hotter metal bar to the cooler water. Eventually, enough energy is transferred so that the bar and the water reach the same temperature—that is, thermal equilibrium is achieved.

to the thermometer (the water heats the thermometer). The increased energy of the mercury atoms means that they move about more rapidly, which slightly increases the volume of the spaces between the atoms. Consequently, the mercury expands (as most substances do upon heating), and the column of mercury rises higher in the thermometer tube.

Transferring energy by heating is a process, but it is common to talk about that process as if heat were a form of energy. It is often said that one sample transfers heat to another, when what is meant is that one sample transfers energy by heating the other.

Heat (or **heating**) refers to *the energy transfer process that happens whenever two samples of matter at different temperatures are brought into contact.* ***Energy always transfers from the hotter to the cooler sample until both are at the same temperature.*** For example, a piece of metal at a high temperature in a Bunsen burner flame and a beaker of cold water (Figure 6.7a) are two samples of matter with different temperatures. When the hot metal is plunged into the cold water (Figure 6.7b), energy transfers from the metal to the water until the two samples reach the same temperature. Once that happens, the metal and water are said to be in **thermal equilibrium.** When thermal equilibrium is reached, the metal has heated the water (and the water has cooled the metal) to a common temperature. In the rest of this chapter (and book), we will refer to a transfer of energy by heating and cooling as a "heat transfer."

Usually most objects in a given region, such as your room, are at about the same temperature—at thermal equilibrium. A fresh cup of coffee, which is hotter than room temperature, transfers energy by heating the rest of the room until the coffee cools off (and the rest of the room warms up a bit). A can of cold soda, which is much cooler than its surroundings, receives energy from everything else until it warms up (and your room cools off a little). Because the total quantity of material in your room is very much greater than that in a cup of coffee or a can of soda, the room temperature changes only a tiny bit to reach thermal equilibrium, whereas the temperature of the coffee or the soda changes a lot.

A diagram such as Figure 6.8 can be used to show the energy transfer from a cup of hot coffee to your room. The upper horizontal line represents the energy of the hot coffee and the lower line represents the energy of the room-temperature coffee. Because the coffee started at a higher temperature (higher energy), the upper line is labeled the initial state. The lower line is the final state. During the change from initial to final state, energy transfers from the coffee to your room. Therefore, the energy of the coffee is lower in the final state than it was in the initial state.

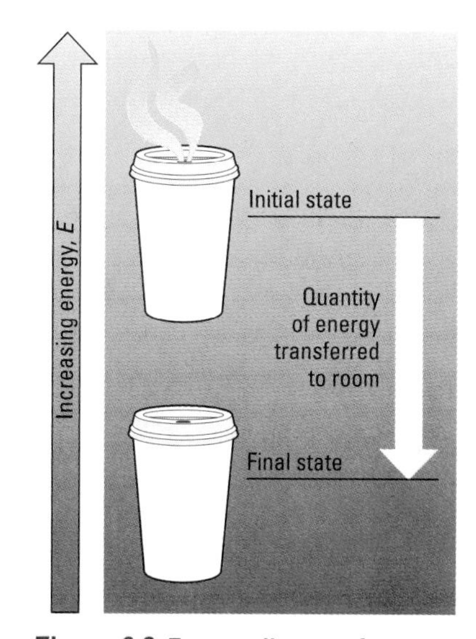

Figure 6.8 Energy diagram for a cup of hot coffee. The diagram compares the energy of a cup of hot coffee with the energy after the coffee has cooled to room temperature. The higher something is in the diagram, the more energy it has. As the coffee and cup cool to room temperature, energy is transferred to the surrounding matter in the room. According to the law of conservation of energy, the energy remaining in the coffee must be less after the change (in the final state) than it was before the change (in the initial state). The quantity of energy transferred is represented by the arrow from the initial to the final state.

EXERCISE 6.2 Energy Diagrams

(a) Draw an energy diagram like the one in Figure 6.8 for warming a can of cold soda to room temperature. Label the initial and final states and use an arrow to represent the change in energy of the can of soda.

(b) Draw a second energy diagram, to the same scale, to show the change in energy of the room as the can of cold soda warms to room temperature.

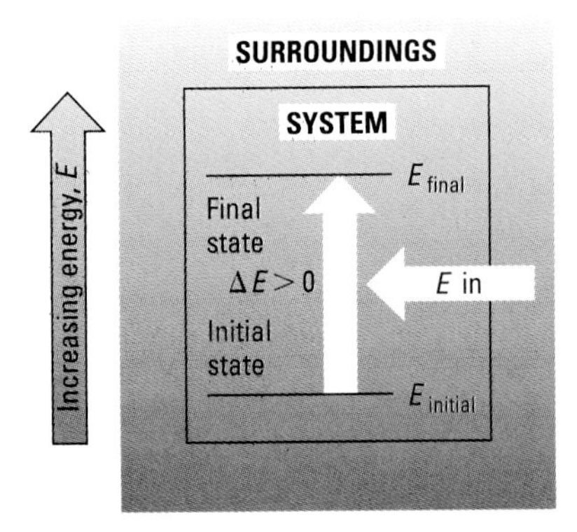

Δ*E* positive: Internal energy increases.

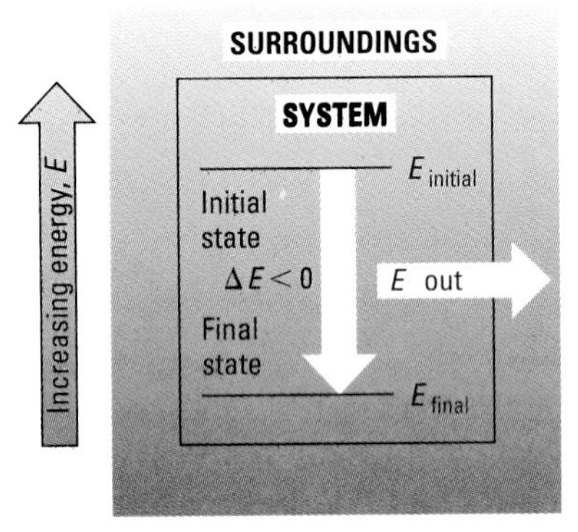

Δ*E* negative: Internal energy decreases.

System, Surroundings, and Internal Energy

In thermodynamics it is useful to define a *region of primary concern* as the **system.** Then we can decide whether energy transfers into or out of the system and keep an accounting of how much energy transfers in each direction. *Everything that can exchange energy with the system* is defined as the **surroundings.** A system may be delineated by an actual physical boundary, such as the inside surface of a flask or the membrane of a cell in your body. Or the boundary may be indistinct, as in the case of the solar system within its surroundings, the rest of the galaxy. In the case of a hot cup of coffee in your room, the cup and the coffee might be the system, and your room would be the surroundings. For a chemical reaction, the system is usually defined to be all of the atoms that make up the reactants. These same atoms will be bonded in a different way in the products after the reaction, and it is their energy before and after reaction that interests us most.

The **internal energy** of a system is the *sum of the individual energies (kinetic and potential) of all nanoscale particles (atoms, molecules, or ions)* in that system. Increasing the temperature increases the internal energy because it increases the average speed of motion of nanoscale particles. ***The total internal energy of a sample of matter depends on temperature, the type of particles, and the number of particles in the sample.*** For a given substance, internal energy depends on temperature and the size of the sample. Thus, despite being at a higher temperature, a cupful of boiling water contains less energy than a bathtub full of warm water.

Calculating Thermodynamic Changes

If we represent a system's internal energy by E, then the change in internal energy during any process is calculated as $E_{final} - E_{initial}$. That is, from the internal energy after the process is over subtract the internal energy before it began. Such a calculation is designated by using a Greek letter Δ (capital delta) before the quantity that changes. Thus, $E_{final} - E_{initial} = \Delta E$. *Whenever a change is indicated by Δ, a positive value indicates an increase and a negative value indicates a decrease.* Therefore, if the internal energy increases during a process, ΔE has a positive value ($\Delta E > 0$); if the internal energy decreases, ΔE is negative ($\Delta E < 0$).

A good analogy to this thermodynamic calculation is your bank account. Assume that in your account (the system) you have a balance B of \$260 ($B_{initial}$), and you withdraw \$60 in spending money. After the withdrawal the balance is \$200 (B_{final}). The change in your balance is

$$\text{Change in balance} = \Delta B = B_{final} - B_{initial} = \$200 - \$260 = -\$60$$

The negative sign on the \$60 indicates that money has been withdrawn from the account (system) and transferred to you (the surroundings). The cash itself is not negative, but during the process of withdrawing your money the balance in the bank went down, so ΔB was negative. Similarly, the *magnitude* of change of a thermodynamic quantity is a number with no sign. To indicate the *direction* of a change, we attach a negative sign (transferred out of the system) or a positive sign (transferred into the system).

CONCEPTUAL
EXERCISE 6.3 Direction of Energy Transfer

It takes about 1.5 kJ to raise the temperature of a can of Classic Coke from 25.0 °C to 26.0 °C. You put the can of Coke into a refrigerator to cool it from room temperature (25.0 °C) to 1.0 °C.

(a) What quantity of heat transfer is required? Express your answer in kilojoules.
(b) What is a reasonable choice of system for this situation?
(c) What constitutes the surroundings?
(d) What is the sign of ΔE for this situation? What is the calculated value of ΔE?
(e) Draw an energy diagram showing the system, the surroundings, the change in energy of the system, and the energy transfer between the system and the surroundings.

Conservation of Energy and Chemical Reactions

For many chemical reactions the only energy transfer processes are heating and doing work. If no other energy transfers (such as emitting light) take place, the law of conservation of energy for any system can be written as

$$\Delta E = q + w \qquad [6.1]$$

where q represents the quantity of energy transferred by heating the system, and w represents the quantity of energy transferred by doing work on the system. If energy is transferred into the system from the surroundings by heating, then q is positive; if energy is transferred into the system because the surroundings do work on the system, then w is positive. If energy is transferred out of the system by heating the surroundings, then q has a negative value; if energy is transferred out of the system because work is done on the surroundings, then w has a negative value. The *magnitudes* of q and w indicate the *quantities* of energy transferred, and the *signs* of q and w indicate the *direction* in which the energy is transferred. The relationships among ΔE, q, and w for a system are shown in Figure 6.9.

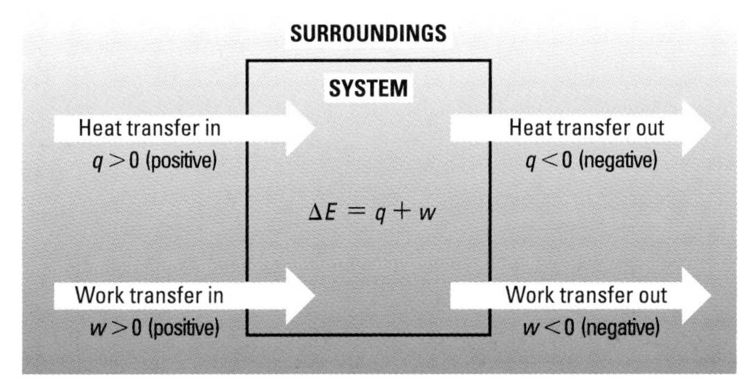

Figure 6.9 Internal energy, heat, and work. Schematic diagram showing energy transfers between a thermodynamic system and its surroundings.

PROBLEM-SOLVING EXAMPLE 6.2 Internal Energy, Heat, and Work

A fuel cell that operates on the reaction of hydrogen with oxygen powers a small automobile by running an electric motor. The motor draws 75.0 kilowatts (75.0 kJ/s) and runs for 2 minutes and 20 seconds. During this period, 5.0×10^3 kJ must be carried away from the fuel cell to prevent it from overheating. If the system is defined to be the hydrogen and oxygen that react, what is the change in the system's internal energy?

Answer -1.55×10^4 kJ

Strategy and Explanation Because the system is the reactants and products, the rest of the fuel cell is part of the surroundings. The problem states that the motor powers a car, which means that the system is doing work (to cause electric current to flow, which then makes the car move). Therefore energy is transferred out of the system and w must be negative. The work is

$$w = -75.0 \frac{\text{kJ}}{\text{s}} \times 140.\ \text{s} = -10.5 \times 10^3\ \text{kJ}$$

Energy is also transferred out of the system when the system heats its surroundings, which means that q must be negative, and $q = -5.0 \times 10^3$ kJ. Using Equation 6.1

$$\Delta E = q + w = (-5.0 \times 10^3\ \text{kJ}) + (-10.5 \times 10^3\ \text{kJ}) = -1.55 \times 10^4\ \text{kJ}$$

Thus the internal energy of the reaction product (water) is 15.5×10^3 kJ lower than the internal energy of the reactants (hydrogen and oxygen).

☑ **Reasonable Answer Check** ΔE is negative, which is reasonable. The internal energy of the reaction products should be lower than that of the reactants, because energy transferred from the reactions heats the surroundings and does work on the surroundings.

PROBLEM-SOLVING PRACTICE 6.2

Suppose that the internal energy decreases by 2400 J when a mixture of natural gas (methane) and oxygen is ignited and burns. If the surroundings are heated by 1.89 kJ, how much work was done by this system on the surroundings?

So far we have seen that

- energy transfers can occur either by heating or by working;
- it is convenient to define a system so that energy transfers into a system (positive) and out of a system (negative) can be accounted for; and
- the internal energy of a system can change as a result of heating or doing work on the system.

Our primary interest in this chapter is heat transfers (the "thermo" in thermodynamics). Heat transfers can take place between two objects at different temperatures. Heat transfers also accompany physical changes and chemical changes. The next three sections (6.3 to 6.5) show how quantitative measurements of heat transfers can be made, first for heating that results from a temperature difference and then for heating that accompanies a physical change.

6.3 Heat Capacity

The **heat capacity** of a sample of matter is *the quantity of energy required to increase the temperature of that sample by one degree.* Heat capacity depends on the mass of the sample and the substance of which it is made (or substances, if it is not pure). To determine the quantity of energy transferred by heating, we usually measure the change in temperature of a substance whose heat capacity is known. Often that substance is water.

Specific Heat Capacity

The notation $J\ g^{-1}\ °C^{-1}$ means that the units are joules divided by grams and divided by degrees Celsius; that is, $\frac{J}{g\ °C}$. We will use negative exponents to show unambiguously which units are in the denominator whenever the denominator includes two or more units.

To make useful comparisons among samples of different substances with different masses, the **specific heat capacity** (which is sometimes just called *specific heat*) is defined as the quantity of energy needed to increase the temperature of one gram of a substance by one degree Celsius. For water at 15 °C, the specific heat capacity is 1.00 cal $g^{-1}\ °C^{-1}$ or 4.184 J $g^{-1}\ °C^{-1}$; for common window glass, it is only about 0.8 J $g^{-1}\ °C^{-1}$. That is, it takes about five times as much heat to raise the temperature of a gram of water by 1 °C as it does for a gram of glass. Like density (⟵ p. 6), specific heat capacity is a property that can be used to distinguish one substance from another. It can also be used to distinguish a pure substance from a solution or mixture, because the specific heat capacity of a mixture will vary with the proportions of the mixture's components.

The specific heat capacity, c, of a substance can be determined experimentally by measuring the quantity of energy transferred to or from a known mass of the substance as its temperature rises or falls. We assume that there is no work transfer of energy to or from the sample and we treat the sample as a thermodynamic system, so $\Delta E = q$.

$$\text{Specific heat capacity} = \frac{\text{quantity of energy transferred by heating}}{\text{sample mass} \times \text{temperature change}}$$

or

$$c = \frac{q}{m \times \Delta T} \qquad [6.2]$$

Suppose that for a 25.0-g sample of ethylene glycol 90.7 J is required to change the temperature from 22.4 °C to 23.9 °C. (Ethylene glycol is used as a coolant in automobile engines.) Thus,

$$\Delta T = (23.9\ °\text{C} - 22.4\ °\text{C}) = 1.5\ °\text{C}$$

From Equation 6.2, the specific heat capacity of ethylene glycol is

$$c = \frac{q}{m \times \Delta T} = \frac{90.7\ \text{J}}{25.0\ \text{g} \times 1.5\ °\text{C}} = 2.4\ \text{J g}^{-1}\ °\text{C}^{-1}$$

© Don Mason/Corbis

Moderation of microclimate by water. In cities near bodies of water (such as Seattle, shown here), summertime temperatures are lower within a few hundred meters of the waterfront than they are a few kilometers away from the water. Wintertime temperatures are higher, unless the water freezes, in which case the moderating effect is less, because ice on the surface insulates the rest of the water from the air.

The specific heat capacities of many substances have been determined. A few values are listed in Table 6.1. Notice that water has one of the highest values. This is important because a high specific heat capacity means that a great deal of energy must be transferred to a large body of water to raise its temperature by just one degree. Conversely, a lot of energy must be transferred away from the water before its temperature falls by one degree. Thus, a lake or ocean can store an enormous quantity of energy and thereby moderate local temperatures. This has a profound influence on weather near lakes or oceans.

When the specific heat capacity of a substance is known, you can calculate the temperature change that should occur when a given quantity of energy is transferred to or from a sample of known mass. More important, by measuring the temperature change and the mass of a substance, you can calculate q, the quantity of energy transferred to or from it by heating. For these calculations it is convenient to rearrange Equation 6.2 algebraically as

$$\Delta T = \frac{q}{c \times m} \quad \text{or} \quad q = c \times m \times \Delta T \qquad [6.2']$$

The high specific heat capacity of water helps to keep your body temperature relatively constant. Water accounts for a large fraction of your body mass, and warming or cooling that water requires a lot of energy transfer.

PROBLEM-SOLVING EXAMPLE 6.3 Using Specific Heat Capacity

If 100.0 g water is cooled from 25.3 °C to 16.9 °C, what quantity of energy has been transferred away from the water?

Answer 3.5 kJ transferred away from the water

Strategy and Explanation Treat the water as a system. The quantity of energy is proportional to the specific heat capacity of water (Table 6.1), the mass of water, and the change in temperature. This is summarized in Equation 6.2′ as

$$\Delta E = q = c \times m \times \Delta T$$

$$= 4.184\ \text{J g}^{-1}\ °\text{C}^{-1} \times 100.0\ \text{g} \times (16.9\ °\text{C} - 25.3\ °\text{C}) = -3.5 \times 10^3\ \text{J} = -3.5\ \text{kJ}$$

☑ **Reasonable Answer Check** It requires about 4 J to heat 1 g water by 1 °C. In this case the temperature change is not quite 10 °C and we have 100 g water, so q should be about $4\ \text{J g}^{-1}\ °\text{C}^{-1} \times (-10\ °\text{C}) \times 100\ \text{g} = -4000\ \text{J} = -4\ \text{kJ}$, which it is. The sign is negative because energy is transferred out of the water as it cools.

PROBLEM-SOLVING PRACTICE 6.3

A piece of aluminum with a mass of 250. g is at an initial temperature of 5.0 °C. If 24.1 kJ is supplied to warm the Al, what is its final temperature? Obtain the specific heat capacity of Al from Table 6.1.

CONCEPTUAL EXERCISE 6.4 Specific Heat Capacity and Temperature Change

Suppose you put two 50-mL beakers in a refrigerator so energy is transferred out of each sample at the same constant rate. If one beaker contains 10. g pulverized glass and one contains 10. g carbon (graphite), which beaker has the lower temperature after 3 min in the refrigerator?

Table 6.1 Specific Heat Capacities for Some Elements, Compounds, and Common Solids

Substance	Specific Heat Capacity ($J\ g^{-1}\ °C^{-1}$)
Elements	
Aluminum, Al	0.902
Carbon (graphite), C	0.720
Iron, Fe	0.451
Copper, Cu	0.385
Gold, Au	0.128
Compounds	
Ammonia, $NH_3(\ell)$	4.70
Water (liquid), $H_2O(\ell)$	4.184
Ethanol, $C_2H_5OH(\ell)$	2.46
Ethylene glycol (antifreeze), $HOCH_2CH_2OH(\ell)$	2.42
Water (ice), $H_2O(s)$	2.06
Carbon tetrachloride, $CCl_4(\ell)$	0.861
A chlorofluorocarbon (CFC), $CCl_2F_2(\ell)$	0.598
Common solids	
Wood	1.76
Concrete	0.88
Glass	0.84
Granite	0.79

Samples of substances listed in Table 6.1: glass, water, copper, aluminum, graphite, iron.

Molar Heat Capacity

It is often useful to know the heat capacity of a sample in terms of the same number of particles instead of the same mass. For this purpose we use the **molar heat capacity,** symbol c_m. This is *the quantity of energy that must be transferred to increase the temperature of one mole of a substance by 1 °C.* The molar heat capacity is easily calculated from the specific heat capacity by using the molar mass of the substance. For example, the specific heat capacity of liquid ethanol is given in Table 6.1 as $2.46\ J\ g^{-1}\ °C^{-1}$. The molecular formula of ethanol is CH_3CH_2OH, so its molar mass is 46.07 g/mol. The molar heat capacity is

$$c_m = \frac{2.46\ J}{g\ °C} \times \frac{46.07\ g}{mol} = 113\ J\ mol^{-1}\ °C^{-1}$$

CONCEPTUAL **EXERCISE 6.5 Molar Heat Capacity**

Calculate the molar heat capacities of all the metals listed in Table 6.1. Compare these with the value just calculated for ethanol. Based on your results, suggest a way to predict the molar heat capacity of a metal. Can this same rule be applied to other kinds of substances?

As you should have found in Conceptual Exercise 6.5, molar heat capacities of metals are very similar. This can be explained if we consider what happens on the nanoscale when a metal is heated. The energy transferred by heating a solid makes the atoms vibrate more extensively about their average positions in the solid crystal lattice. Every metal consists of many, many atoms, all of the same kind and packed closely together; that is, the structures of all metals are very similar. As a consequence, the ways that the metal atoms can vibrate (and therefore the ways that their energies can be increased) are very similar. Thus, no matter what the metal, nearly the same quantity of energy must be transferred per metal atom to increase the temperature by 1 °C. The quantity of energy per mole is therefore very similar for all metals.

PROBLEM-SOLVING EXAMPLE **6.4 Transfer of Energy Between Samples by Heating**

Suppose that you have 100. mL H_2O at 20.0 °C and you add to the water 55.0 g iron pellets that had been heated to 425 °C. What is the temperature of both the water and the iron when thermal equilibrium is reached? (Assume that there is no energy transfer to the glass beaker or to the air or to anything else but the water. Assume also that no work is done, that no liquid water vaporizes, and that the density of water is 1.00 g/mL.)

Answer $T_{final} = 42.7\ °C$

Strategy and Explanation Thermal equilibrium means that the water and the iron pellets will have the same final temperature, which is what we want to calculate. Consider the iron to be the system and the water to be the surroundings. The energy transferred from the iron is the same energy that is transferred to the water. None of this energy goes anywhere other than to the water. Therefore $\Delta E_{water} = -\Delta E_{iron}$ and $q_{water} = -q_{iron}$. The quantity of energy transferred to the water and the quantity transferred from the iron are equal. They are opposite in algebraic sign because energy was transferred *from* the iron as its temperature dropped, and energy was transferred *to* the water to raise its temperature.

Specific heat capacities for iron and water are listed in Table 6.1. The mass of water is 100. mL × 1.00 g/mL = 100. g. $T_{initial}$ for the iron is 425 °C and $T_{initial}$ for the water is 20.0 °C.

$$q_{water} = -q_{iron}$$

$$c_{water} \times m_{water} \times \Delta T_{water} = -c_{iron} \times m_{iron} \times \Delta T_{iron}$$

$$(4.184\ J\ g^{-1}\ °C^{-1})(100.\ g)(T_{final} - 20.0\ °C) = -(0.451\ J\ g^{-1}\ °C^{-1})(55.0\ g)(T_{final} - 425\ °C)$$

$$(418.4 \text{ J °C}^{-1})T_{final} - (8.368 \times 10^3 \text{ J}) = -(24.80 \text{ J °C}^{-1})T_{final} + (1.054 \times 10^4 \text{ J})$$

$$(443.2 \text{ J °C}^{-1})T_{final} = 1.891 \times 10^4 \text{ J}$$

Solving, we find $T_{final} = 42.7$ °C. The iron has cooled a lot ($\Delta T_{iron} = -382$ °C) and the water has warmed a little ($\Delta T_{water} = 22.7$ °C).

The quantities of energy transferred have opposite signs because they take place from the iron (negative) to the water (positive).

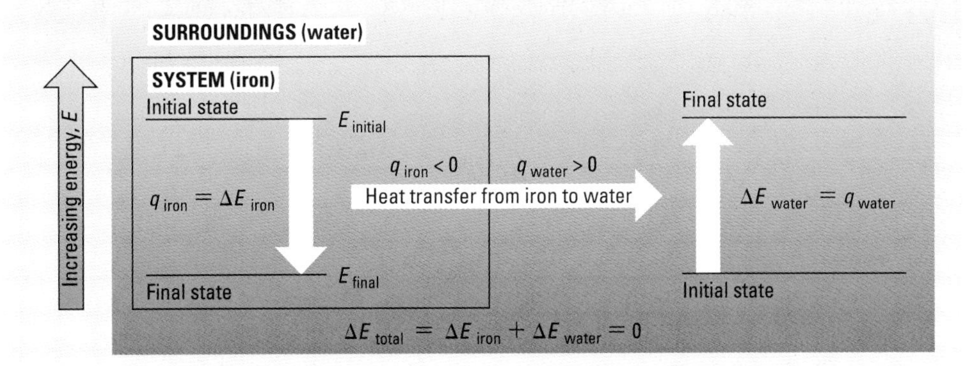

☑ **Reasonable Answer Check** As a check, note that the final temperature must be between the two initial values, which it is. Also, don't be concerned by the fact that transferring the same quantity of energy resulted in two very different values of ΔT; this difference arises because the specific heat capacities and masses of iron and water are different. There is much less iron and its specific heat capacity is smaller, so its temperature changes much more than the temperature of the water.

PROBLEM-SOLVING PRACTICE 6.4

A 400.-g bar of iron is heated in a flame and then immersed in 1000. g water in a beaker. The initial temperature of the water was 20.0 °C, and both the iron and the water are at 32.8 °C at the end of the experiment. What was the original temperature of the hot iron bar? (Assume that all energy transfer is between the water and the iron.)

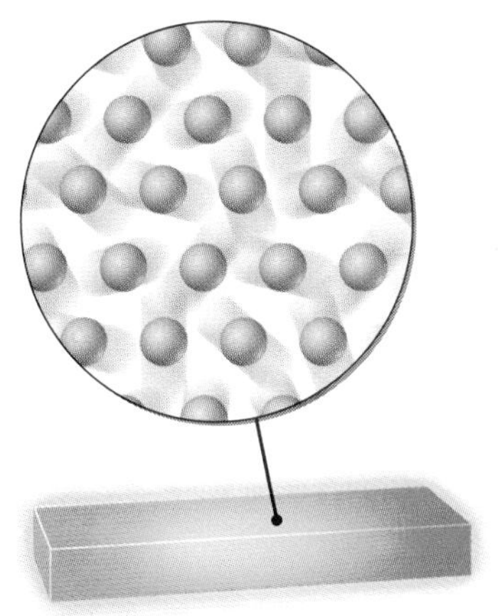

Hot iron bar

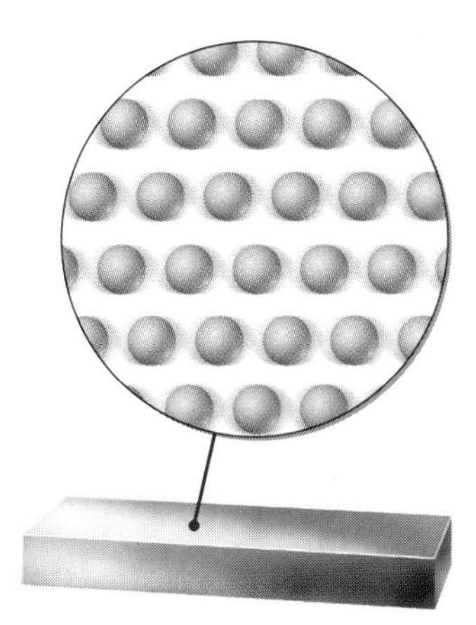

Cold iron bar

Hot and cold iron. On the nanoscale the atoms in the sample of hot iron are vibrating much farther from their average positions than those in the sample of room-temperature iron. The greater vibration of atoms in hot iron means harder collisions of iron atoms with water molecules. Such collisions transfer energy to the water molecules, heating the water.

6.4 Energy and Enthalpy

Using heat capacity we can account for transfers of energy between samples of matter as a result of temperature differences. But energy transfers also accompany physical or chemical changes, *even though there may be no change in temperature.* We will first consider the simpler case of physical change and then apply the same ideas to chemical changes.

Conservation of Energy and Changes of State

Consider a system that consists of water at its boiling temperature in a container with a balloon attached (Figure 6.10). The system is under a constant atmospheric pressure. If the water is heated, it will boil, the temperature will remain at 100 °C, and the steam produced by boiling the water will inflate the balloon (Figure 6.10b). If the heating stops, then the water will stop boiling, some of the steam will condense to liquid, and the volume of steam will decrease (Figure 6.10c). There will be heat transfer of energy to the surroundings. However, as long as steam is condensing to liquid water, the temperature will remain at 100 °C. In summary, transferring energy *into* the system produces more steam; transferring energy *out of* the system results in less steam. Both the boiling and condensing processes occur at the same temperature—the boiling point.

The boiling process can be represented by the equation

$$H_2O(\ell) \longrightarrow H_2O(g) \qquad \text{endothermic}$$

Changes of state (between solid and liquid, liquid and gas, or solid and gas) are described in more detail in Section 11.3. Because the temperature remains constant during a change of state, melting points and boiling points can be measured relatively easily and used to identify substances (⇐ p. 5).

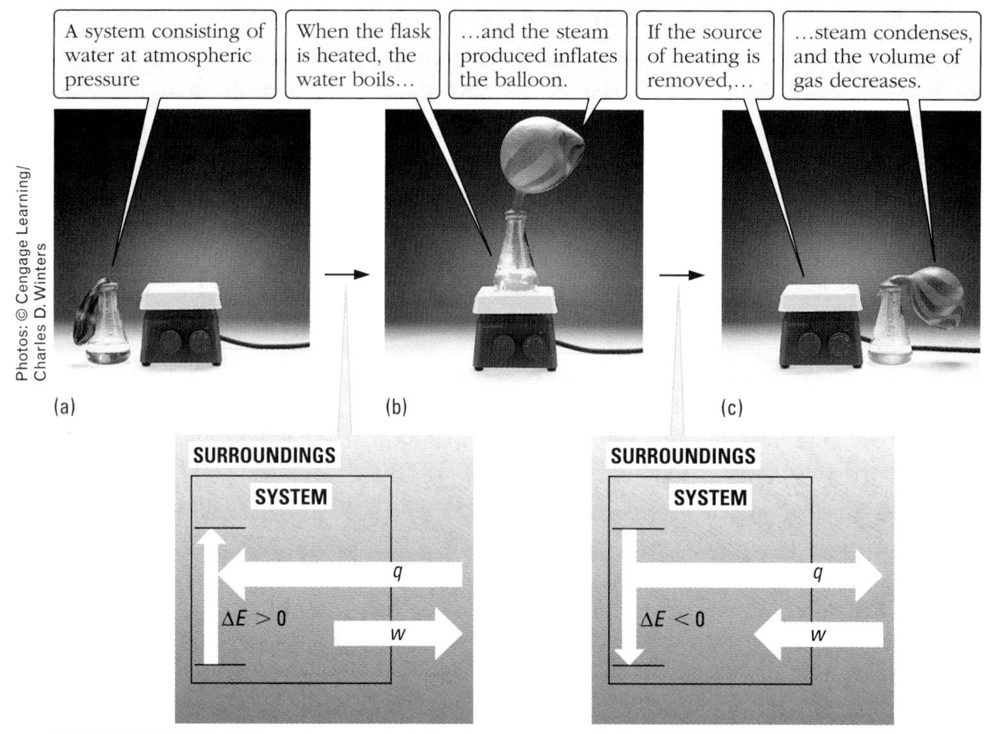

Active Figure 6.10 **Boiling water at constant pressure.** When water boils, the steam pushes against atmospheric pressure and does work on the atmosphere (which is part of the surroundings). The balloon allows the expansion of the steam to be seen; even if the balloon were not there, the steam would push back the surrounding air. In general, for any constant-pressure process, if a change in volume occurs, some work is done, either on the surroundings or on the system. Visit this book's companion website at **www.cengage.com/chemistry/moore** to test your understanding of the concepts in this figure.

Thermic or *thermo* comes from the Greek word *thermé,* meaning "heat." *Endo* comes from the Greek word *endon,* meaning "within or inside." *Endothermic* therefore indicates transfer of energy *into* the system.

We call this process **endothermic** because, as it occurs, energy must be transferred *into* the system to maintain constant temperature. If no energy transfer took place, the liquid water would get cooler. Evaporation of water (perspiration) from your skin, which occurs at a lower temperature than boiling, is an endothermic process that you are certainly familiar with. Energy must be transferred from your skin to the evaporating water, and this energy transfer cools your skin.

The opposite of boiling is condensation. It can be represented by the opposite equation,

$$H_2O(g) \longrightarrow H_2O(\ell) \qquad \text{exothermic}$$

Exo comes from the Greek word *exō,* meaning "out of." *Exothermic* indicates transfer of energy *out of* the system.

This process is said to be **exothermic** because energy must be transferred *out of* the system to maintain constant temperature. Because condensation of $H_2O(g)$ (steam) is exothermic, a burn from steam at 100 °C is much worse than a burn from liquid water at 100 °C. The steam heats the skin a lot more because there is a heat transfer due to the condensation as well as the difference in temperature between the water and your skin.

Phase Change	Direction of Energy Transfer	Sign of q	Type of Change
$H_2O(\ell) \rightarrow H_2O(g)$	Surroundings → system	Positive ($q > 0$)	*Endo*thermic
$H_2O(g) \rightarrow H_2O(\ell)$	System → surroundings	Negative ($q < 0$)	*Exo*thermic

The system in Figure 6.10 can be analyzed by using the law of conservation of energy, $\Delta E = q + w$. Vaporizing 1.0 g water requires heat transfer of 2260 J, so $q =$ 2260 J (a positive value because the transfer is from the surroundings to the system). At the same time, the expansion of the steam pushes back the atmosphere, doing work. The quantity of work is more difficult to calculate, but it is clear that w must be negative, because the system does work on the surroundings. Therefore, the internal energy of the system is increased by the quantity of heating and decreased by the quantity of work done.

Now suppose that the heating is stopped and the direction of heat transfer is reversed. The water stops boiling and some of the steam condenses to liquid water. The balloon deflates and the atmosphere pushes back the steam. If 1.0 g steam condenses, then $q = -2260$ J. Because the surrounding atmosphere pushes on the system, the surroundings have done work on the system, which makes w positive. As long as steam is condensing to liquid, the temperature remains at 100 °C. The internal energy of the system is increased by the work done on it and decreased by the heat transfer of energy to the surroundings.

The device described here is a crude example of a steam engine. Burning fuel boils water, and the steam does work. In a real steam engine the steam would drive a piston and then be allowed to escape, providing a means for the system to continually do work on its surroundings. Systems that convert heat into work are called heat engines. Another example is an internal combustion engine in an automobile, which converts heat from the combustion of fuel into work to move the car.

Enthalpy: Heat Transfer at Constant Pressure

In the previous section it was clear that work was done. When the balloon's volume increased, the balloon pushed aside air that had occupied the space the gas in the balloon expanded to fill. Work is done when a force moves something through some distance. If the flask containing the boiling water had been sealed, nothing would have moved and no work would have been done. Therefore, in a closed container where the system's volume is constant, $w = 0$, and

$$\Delta E = q + w = q + 0 = q_V$$

The subscript V indicates constant volume; that is, q_V is the heat transfer into a constant-volume system. This means that ***if a process is carried out in a closed container and the heat transfer is measured, ΔE has been determined***.

In plants, animals, laboratories, and the environment, physical processes and chemical reactions seldom take place in closed containers. Instead they are carried out in contact with the atmosphere. For example, the vaporization of water shown in Figure 6.10 took place under conditions of constant atmospheric pressure, and the expanding steam had to push back the atmosphere. In such a case,

$$\Delta E = q_P + w_{atm}$$

That is, ΔE differs from the heat transfer at constant pressure, q_P, by the work done to push back the atmosphere, w_{atm}.

To see how much work is required, consider the idealized system shown in the margin, which consists of a cylinder with a weightless piston. (The purpose of the piston is to distinguish the water system from the surrounding air.) When some of the water in the bottom of the cylinder boils, the volume of the system increases. The piston and the atmosphere are forced upward. The system (water and steam) does work to raise the surrounding air. The work can be calculated as $w_{atm} = -(\text{force} \times \text{distance}) = -(F \times d)$. (The negative sign indicates that the system is doing work on the surroundings.) The distance the piston moves is clear from the diagram. The force can be calculated from the pressure (P), which is defined as force per unit area (A). Since $P = F/A$, the force is $F = P \times A$, and the work is $w_{atm} = -F \times d = -P \times A \times d = -P\Delta V$. The change in the volume of the system, ΔV, is the volume of the cylinder through which the piston moves. This is calculated as the area of the base times the height, or $\Delta V = A \times d$. Thus the work is $w_{atm} = -P \times A \times d = -P\Delta V$. This means that the work of pushing back the atmosphere is always equal to the atmospheric pressure times the change in volume of the system.

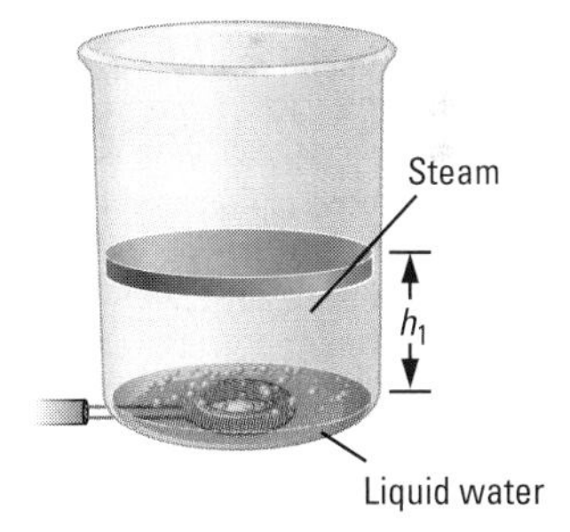

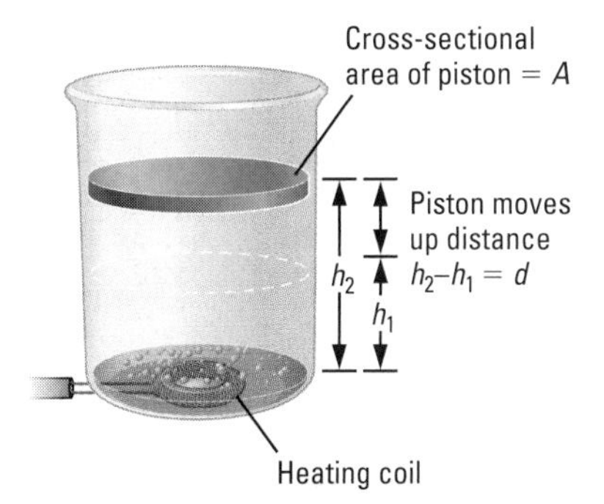

Vaporization of water. When a sample of water boils at constant pressure, energy must be supplied to expand the steam against atmospheric pressure.

The law of conservation of energy for a constant-pressure process can now be written as

$$\Delta E = q_P + w_{\text{atm}} \qquad \text{or} \qquad q_P = \Delta E - w_{\text{atm}} = \Delta E + P\Delta V$$

This equation says that when we carry out reactions in beakers or other containers open to the atmosphere, the heat transfer differs from the change in energy by an easily calculated term, $P\Delta V$. Therefore it is convenient to use q_P to characterize energy transfers in typical chemical and physical processes. *The quantity of thermal energy transferred into a system at constant pressure*, q_P, is called the **enthalpy change** of the system, symbolized by ΔH. Thus, $\Delta H = q_P$, and

$$\Delta H = \Delta E + P\Delta V$$

Because it is equal to the quantity of thermal energy transferred at constant pressure and because most chemical reactions are carried out at atmospheric (constant) pressure, the enthalpy change for a process is often called the *heat of that process.* For example, the enthalpy change for melting (fusion) is also called the heat of fusion.

ΔH accounts for all the energy transferred except the quantity that does the work of pushing back the atmosphere. For processes that do not involve gases, w_{atm} is very small. Even when gases are involved, w_{atm} is usually much smaller than q_P. That is, ΔH is closely related to the change in the internal energy of the system but is slightly different in magnitude. ***Whenever heat transfer is measured at constant pressure, it is ΔH that is determined.***

PROBLEM-SOLVING EXAMPLE 6.5 Changes of State, ΔH, and ΔE

Methanol, CH_3OH, boils at 65.0 °C. When 5.0 g methanol boils at 1 atm, the volume of $CH_3OH(g)$ is 4.32 L greater than the volume of the liquid. The heat transfer is 5865 J, and the process is endothermic. Calculate ΔH and ΔE. (The units 1 L × 1 atm = 101.3 J.)

Answer $\Delta H = 5865$ J; $\Delta E = 5427$ J

Strategy and Explanation The process takes place at constant pressure. By definition, $\Delta H = q_P$. Because thermal energy is transferred to the system, the sign of ΔH must be positive. Therefore $\Delta H = 5865$ J. Because the system expands, ΔV is positive. This makes the sign of w_{atm} negative, and

$$w_{\text{atm}} = -P\Delta V = -1 \text{ atm} \times 4.32 \text{ L} = -4.32 \text{ L atm} = -4.32 \times 101.3 \text{ J} = -438 \text{ J}$$

To calculate ΔE, add the expansion work to the enthalpy change.

$$\Delta E = q_P + w_{\text{atm}} = \Delta H + w_{\text{atm}} = 5865 \text{ J} - 438 \text{ J} = 5427 \text{ J}$$

☑ **Reasonable Answer Check** Boiling is an endothermic process, so ΔH must be positive. Because the system did work on the surroundings, the change in internal energy must be less than the enthalpy change, and it is.

The results of this example show that ΔE differs by less than 10% from ΔH—that is, by 440 J out of 5865 J, which is 7.5%. It is true for most physical and chemical processes that the work of pushing back the atmosphere is only a small fraction of the heat transfer of energy. Because ΔH is so close to ΔE, chemists often refer to enthalpy changes as energy changes.

PROBLEM-SOLVING PRACTICE 6.5

When potassium melts at atmospheric pressure, the heat transfer is 14.6 cal/g. The density of liquid potassium at its melting point is 0.82 g/mL and that of solid potassium is 0.86 g/mL. Given that a volume change of 1.00 mL at atmospheric pressure corresponds to 0.10 J, calculate ΔH and ΔE for melting 1.00 g potassium.

Freezing and Melting (Fusion)

Consider what happens when ice is heated at a slow, constant rate from −50 °C to +50 °C. A graph of temperature as a function of quantity of transferred energy is shown in Figure 6.11. When the temperature reaches 0 °C, it remains constant, despite the fact that energy is still being transferred to the sample. As long as ice is melting, thermal energy must be continually supplied to overcome forces that hold the water molecules in their regularly spaced positions in the nanoscale structure of solid ice. Overcoming these forces raises the potential energy of the water molecules and therefore requires a transfer of energy into the system.

Melting a solid is an example of a **change of state** or **phase change,** a physical process in which one state of matter is transformed into another. During a phase

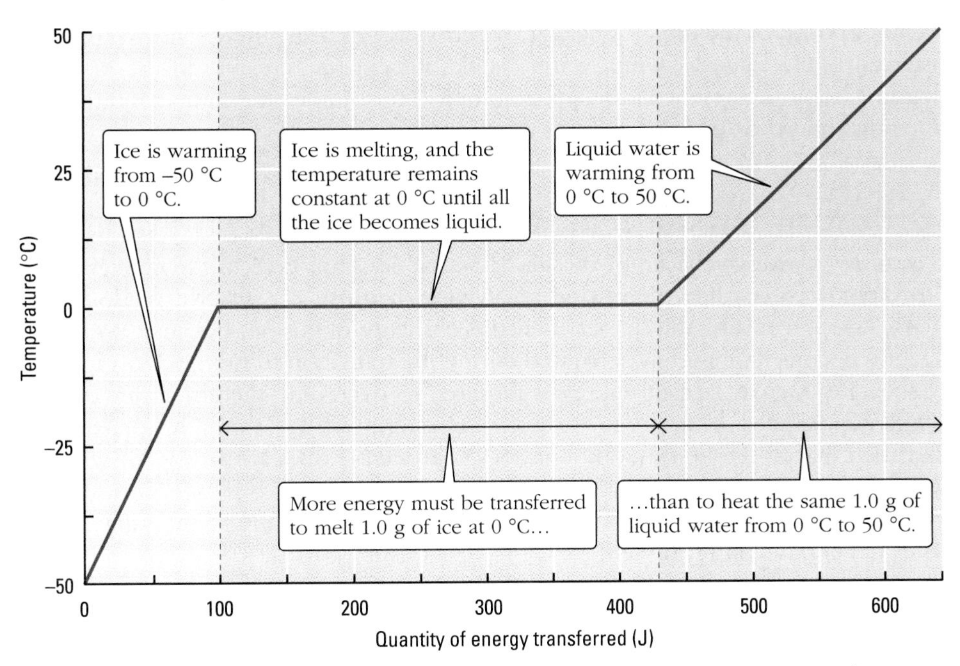

Figure 6.11 Heating graph. When a 1.0-g sample of ice is heated at a constant rate, the temperature does not always increase at a constant rate.

change, the temperature remains constant, but energy must be continually transferred into (melting, boiling) or out of (condensing, freezing) the system because the nanoscale particles have higher or lower potential energy after the phase change than they did before it. As shown in Figure 6.11, the quantity of energy transferred during a phase change is significant.

The quantity of thermal energy that must be transferred to a solid as it melts at constant pressure is called the **enthalpy of fusion.** For ice the enthalpy of fusion is 333 J/g at 0 °C. This same quantity of energy could raise the temperature of a 1.00-g block of iron from 0 °C to 738 °C (red hot), or it could melt 0.50 g ice and heat the liquid water from 0 °C to 80 °C. This is illustrated schematically in Figure 6.12.

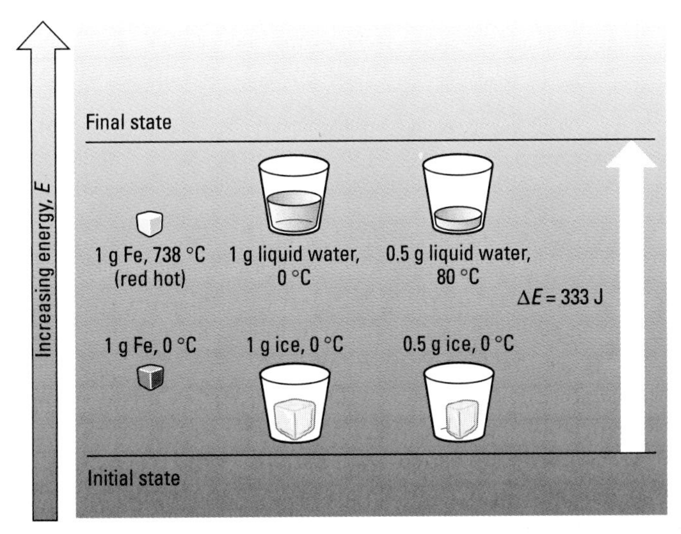

Figure 6.12 Heating, temperature change, and phase change. Heating a substance can cause a temperature change, a phase change, or both. Here, 333 J has been transferred to each of three samples: a 1-g block of iron at 0 °C; a 1-g block of ice at 0 °C; and a 0.5-g block of ice at 0 °C. The iron block becomes red hot; its temperature increases to 738 °C. The 1-g block of ice melts, resulting in 1 g of liquid water at 0 °C. The 0.5-g block of ice melts, and there is enough energy to heat the liquid water to 80 °C.

Tony Ranze/AFP/Getty Images

Protecting crops from freezing. Because heat is transferred to the surroundings as water freezes, one way to protect plants from freezing if the temperature drops just below the freezing point is to spray water on them. As the water freezes, energy transfer to the leaves, stems, and fruits keeps the plants themselves from freezing.

The opposite of melting is freezing. When water freezes, the quantity of energy transferred is the same as when water melts, but energy transfers in the opposite direction—from the system to the surroundings. Thus, under the same conditions of temperature and pressure, $\Delta H_{fusion} = -\Delta H_{freezing}$.

Vaporization and Condensation

The quantity of energy that must be transferred at constant pressure to convert a liquid to vapor (gas) is called the **enthalpy of vaporization.** For water it is 2260 J/g at 100 °C. This is considerably larger than the enthalpy of fusion, because the water molecules become completely separated during the transition from liquid to vapor. As they separate, a great deal of energy is required to overcome the attractions among the water molecules. Therefore, the potential energy of the vapor is considerably higher than that of the liquid. Although 333 J can melt 1.00 g ice at 0 °C, it will boil only 0.147 g water at 100 °C.

$$333\ \text{J} \times \frac{1.00\ \text{g water vaporized}}{2260\ \text{J}} = 0.147\ \text{g water vaporized}$$

The opposite of vaporization is condensation. Therefore, under the same conditions of temperature and pressure, $\Delta H_{vaporization} = -\Delta H_{condensation}$.

You experience cooling due to evaporation of water when you perspire. If you work up a real sweat, then lots of water evaporates from your skin, producing a much greater cooling effect. People who exercise in cool weather need to carry a sweatshirt or jacket. When they stop exercising they generate less body heat, but lots of perspiration remains on their skin. Its evaporation can cool the body enough to cause a chill.

CONCEPTUAL
EXERCISE 6.6 Heating and Cooling Graphs

(a) Assume that a 1.0-g sample of ice at −5 °C is heated at a uniform rate until the temperature is 105 °C. Draw a graph like the one in Figure 6.10 to show how temperature varies with energy transferred. Your graph should be drawn to approximately the correct scale.

(b) Assume that a 0.50-g sample of water is cooled [at the same uniform rate as the heating in part (a)] from 105 °C to −5 °C. Draw a cooling curve to show how temperature varies with energy transferred. Your graph should be drawn to the same scale as in part (a).

EXERCISE 6.7 Changes of State

Assume you have 1 cup of ice (237 g) at 0.0 °C. How much heating is required to melt the ice, warm the resulting water to 100.0 °C, and then boil the water to vapor at 100.0 °C? (*Hint:* Do three separate calculations and then add the results.)

State Functions and Path Independence

Both energy and enthalpy are **state functions,** *properties whose values are invariably the same if a system is in the same state.* A system's state is defined by its temperature, pressure, volume, mass, and composition. For the same initial and final states, *a change in a state function does not depend on the path by which the system changes from one state to another* (Figure 6.13). Returning to the bank account analogy (***p. 184***), your bank balance is independent of the path by which you change it. If you have \$1000 in the bank (initial state) and withdraw \$100, your balance will go down to \$900 (final state) and $\Delta B = -\$100$. If instead you had deposited \$500 and withdrawn \$600 you would have achieved the same change of $\Delta B = -\$100$ by a different pathway, and your final balance would still be \$900.

The fact that changes in a state function are independent of the sequence of events by which change occurs is important, because it allows us to apply laboratory measurements to real-life situations. For example, if you measure in the lab the heat transfer when 1.0 g glucose (dextrose sugar) burns in exactly the amount of oxygen required to convert it to carbon dioxide and water, you will find that $\Delta H = -15.5$ kJ. When you eat something that contains 1.0 g glucose and your body metabolizes the

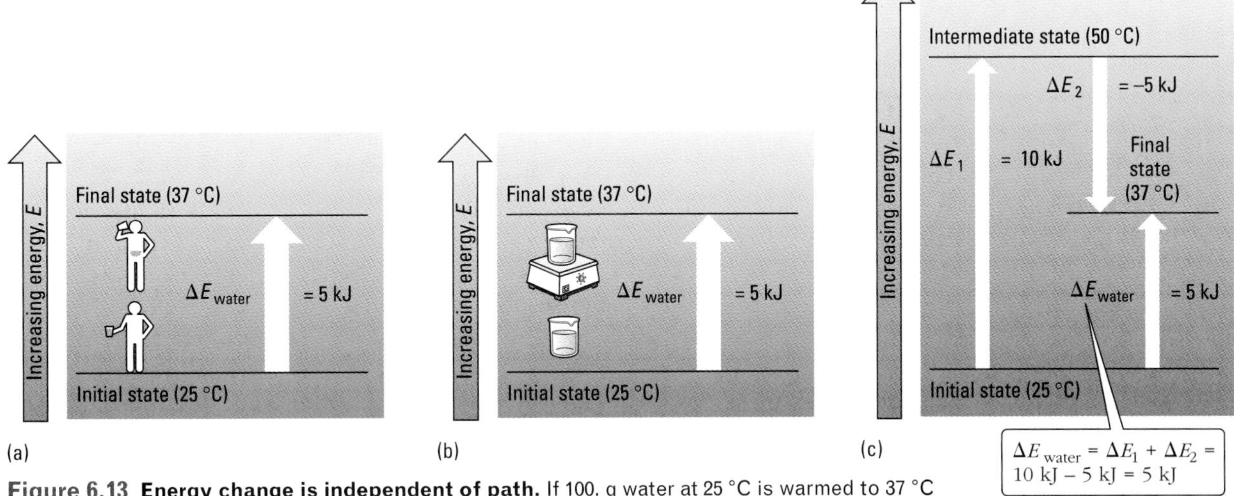

Figure 6.13 Energy change is independent of path. If 100. g water at 25 °C is warmed to 37 °C (body temperature) at atmospheric pressure, the change in energy of the water is the same whether (a) you drank the water and your body warmed it to 37 °C, (b) you put the water in a beaker and heated it with a hot plate, or (c) you heated the water to 50 °C and then cooled it to 37 °C.

glucose (producing the same products at the same temperature and pressure), there is the same change in enthalpy. Thus, laboratory measurements can be used to determine how much energy you can get from a given quantity of food, which is the basis for the caloric values listed on labels.

6.5 Thermochemical Expressions

To indicate the heat transfer that occurs when either a physical or chemical process takes place, we write a **thermochemical expression,** *a balanced chemical equation together with the corresponding value of the enthalpy change.* For evaporation of water near room temperature and at typical atmospheric pressure, this thermochemical expression can be written:

$$H_2O(\ell) \longrightarrow H_2O(g) \qquad \Delta H^\circ = +44.0 \text{ kJ} \qquad (25\ ^\circ\text{C, 1 bar})$$

In 1982 the International Union of Pure and Applied Chemistry chose a pressure of 1 bar as the standard for tabulating information for thermochemical expressions. This pressure is very close to the standard atmosphere: 1 bar = 0.98692 atm = 1×10^5 kg m^{-1} s^{-2}. (Pressure units are discussed further in Section 10.2.)

The symbol ΔH° (pronounced "delta-aitch-standard") represents the **standard enthalpy change,** which is defined as the *enthalpy change at the standard pressure of 1 bar and a specified temperature.* Because the value of the enthalpy change depends on the pressure at which the process is carried out, all enthalpy changes are reported at the same standard pressure, 1 bar. (The bar is a unit of pressure that is very close to the pressure of the earth's atmosphere at sea level; you may have heard this unit used in a weather report.) The value of the enthalpy change also varies slightly with temperature. For thermochemical expressions in this book, the temperature can be assumed to be 25 °C, unless some other temperature is specified.

Usually the surroundings contain far more matter than the system and hence have a much greater heat capacity. Consequently, the temperature of the surroundings often does not change significantly, even though energy transfer has occurred. For evaporation of water at 25 °C, the temperature of the surroundings would not drop much below 25 °C.

The thermochemical expression given above indicates that when *one mole* of liquid water (at 25 °C and 1 bar) evaporates to form *one mole* of water vapor (at 25 °C and 1 bar), 44.0 kJ of energy must be transferred from the surroundings to the system to maintain the temperature at 25 °C. The size of the enthalpy change depends on how much process (in this case evaporation) takes place. The more water that evaporates, the more the surroundings are cooled. If 2 mol $H_2O(\ell)$ is converted to 2 mol $H_2O(g)$, 88 kJ of energy is transferred; if 0.5 mol $H_2O(\ell)$ is converted to 0.5 mol $H_2O(g)$, only 22 kJ is required. The numerical value of ΔH° corresponds to the reaction as written, with the coefficients indicating moles of each reactant and moles of each product. For the thermochemical expression

$$2\,H_2O(\ell) \longrightarrow 2\,H_2O(g) \qquad \Delta H^\circ = +88.0\text{ kJ}$$

the process is evaporating 2 mol $H_2O(\ell)$ to form 2 mol $H_2O(g)$, both at 25 °C and 1 bar. For this process the enthalpy change is twice as great as for the case where there is a coefficient of 1 on each side of the equation.

Now consider water vapor condensing to form liquid. If 44.0 kJ of energy is required to do the work of separating the water molecules in 1 mol of the liquid as it vaporizes, the same quantity of energy will be released when the molecules move closer together as the vapor condenses to form liquid.

$$H_2O(g) \longrightarrow H_2O(\ell) \qquad \Delta H^\circ = -44.0\text{ kJ}$$

This thermochemical expression indicates that 44.0 kJ of energy is transferred to the surroundings *from* the system when 1 mol of water vapor condenses to liquid at 25 °C and 1 bar.

The idea here is similar to the example given earlier of water falling from top to bottom of a waterfall. The decrease in potential energy of the water when it falls from the top to the bottom of the waterfall is exactly equal to the increase in potential energy that would be required to take the same quantity of water from the bottom of the fall to the top. The signs are opposite because in one case potential energy is transferred *from* the water and in the other case it is transferred *to* the water.

CONCEPTUAL
EXERCISE 6.8 Interpreting Thermochemical Expressions

What part of the thermochemical expression for vaporization of water indicates that energy is transferred from the surroundings to the system when the evaporation process occurs?

CONCEPTUAL
EXERCISE 6.9 Thermochemical Expressions

Why is it essential to specify the state (s, ℓ, or g) of each reactant and each product in a thermochemical expression?

PROBLEM-SOLVING EXAMPLE 6.6 Changes of State and ΔH°

Calculate the energy transferred to the surroundings when water vapor in the air condenses at 25 °C to give rain in a thunderstorm. Suppose that one inch of rain falls over one square mile of ground, so that 6.6×10^{10} mL has fallen. (Assume $d_{H_2O(\ell)} = 1.0$ g/mL.)

Agronomists and meteorologists measure quantities of rainwater in units of acre-feet; an acre-foot is enough water to cover an acre of land to a depth of one foot.

Answer 1.6×10^{11} kJ

Strategy and Explanation The thermochemical expression for condensation of 1 mol water at 25 °C is

$$H_2O(g) \longrightarrow H_2O(\ell) \qquad \Delta H^\circ = -44.0\text{ kJ}$$

The standard enthalpy change tells how much heat transfer is required when 1 mol water condenses at constant pressure, so we first calculate how many moles of water condensed.

$$\text{Amount of water condensed} = 6.6 \times 10^{10}\text{ g water} \times \frac{1\text{ mol}}{18.0\text{ g}} = 3.66 \times 10^{9}\text{ mol water}$$

Next, calculate the quantity of energy transferred from the fact that 44.0 kJ is transferred per mole of water.

$$\text{Quantity of energy transferred} = 3.66 \times 10^{9}\text{ mol water} \times \frac{44.0\text{ kJ}}{1\text{ mol}} = 1.6 \times 10^{11}\text{ kJ}$$

The negative sign of ΔH° in the thermochemical expression indicates transfer of the 1.6×10^{11} kJ *from the water* (system) *to the surroundings.*

Since the explosion of 1000 tons of dynamite is equivalent to 4.2×10^9 kJ, the energy transferred by our hypothetical thunderstorm is about the same as that released when 38,000 tons of dynamite explodes! A great deal of energy can be stored in water vapor, which is one reason why storms can cause so much damage.

☑ **Reasonable Answer Check** The quantity of water is about 10^{11} g. The energy transfer is 44 kJ for 1 mol (18 g) water. Since 44 is about twice 18, this is about 2 kJ/g. Therefore, the number of kJ transferred in kJ should be about twice the number of grams, or about 2×10^{11} kJ, and it is.

Like all examples in this chapter, this one assumes that the temperature of the system remains constant, so that all the energy transfer associated with the phase change goes to or from the surroundings.

PROBLEM-SOLVING PRACTICE 6.6

The enthalpy change for sublimation of 1 mol solid iodine at 25 °C and 1 bar is 62.4 kJ. (Sublimation means changing directly from solid to gas.)

$$I_2(s) \longrightarrow I_2(g) \qquad \Delta H° = +62.4 \text{ kJ}$$

(a) What quantity of energy must be transferred to vaporize 10.0 g solid iodine?
(b) If 3.42 g iodine vapor changes to solid iodine, what quantity of energy is transferred?
(c) Is the process in part (b) exothermic or endothermic?

Richard Ramette

Iodine "thermometer." A glass sphere containing a few iodine crystals rests on the ground in desert sunshine. The hotter the flask, the more iodine sublimes, producing the beautiful violet color.

6.6 Enthalpy Changes for Chemical Reactions

Having developed methods for quantitative treatment of energy transfers as a result of temperature differences and as a result of phase changes, we are now ready to apply these ideas to energy transfers that accompany chemical reactions.

Like phase changes, chemical reactions can be exothermic or endothermic, but reactions usually involve much larger energy transfers than do phase changes. Indeed, a temperature change is one piece of evidence that a chemical reaction has taken place. The large energy transfers that occur during chemical reactions result from the breaking and forming of chemical bonds as reactants are converted into products. These energy transfers have important applications in living systems, in industrial processes, in heating or cooling your home, and in many other situations.

Hydrogen is an excellent fuel. It produces very little pollution when it burns in air, and its reaction with oxygen to form water is highly exothermic. It is used as a fuel in the Space Shuttle, for example. The thermochemical expression for formation of 1 mol water vapor from hydrogen and oxygen is

$$H_2(g) + \tfrac{1}{2}\,O_2(g) \longrightarrow H_2O(g) \qquad \Delta H° = -241.8 \text{ kJ} \qquad [6.3]$$

Like all thermochemical expressions, this one has four important characteristics:

- The *sign of* $\Delta H°$ indicates the direction of energy transfer.
- The magnitude of $\Delta H°$ depends on the *states of matter* of the reactants and products.
- The *balanced equation represents moles* of reactants and of products.
- The *quantity of energy transferred is proportional to the quantity of reaction* that occurs.

The relationship of reaction heat transfer and enthalpy change:

Reactant → product with transfer of thermal energy *from* system to surroundings. ΔH is negative; reaction is *exo*thermic.

Reactant → product with transfer of thermal energy *into* system from surroundings. ΔH is positive; reaction is *endo*thermic.

Sign of $\Delta H°$ Thermochemical Expression 6.3 tells us that this process is exothermic, because $\Delta H°$ is negative. Formation of 1 mol water vapor transfers 241.8 kJ of energy from the reacting chemicals to the surroundings. If 1 mol water vapor is decomposed to hydrogen and oxygen (the reverse process), the magnitude of $\Delta H°$ is the same, but the sign is opposite, indicating transfer of energy from the surroundings to the system:

$$H_2O(g) \longrightarrow H_2(g) + \tfrac{1}{2}\,O_2(g) \qquad \Delta H° = +241.8 \text{ kJ} \qquad [6.4]$$

The reverse of an exothermic process is endothermic. The magnitude of the energy transfer is the same, but the direction of transfer is opposite.

States of Matter If liquid water is involved instead of water vapor, the magnitude of $\Delta H°$ is different from that in Thermochemical Expression 6.3:

$$H_2(g) + \tfrac{1}{2}\,O_2(g) \longrightarrow H_2O(\ell) \qquad \Delta H° = -285.8 \text{ kJ} \qquad [6.5]$$

(a)

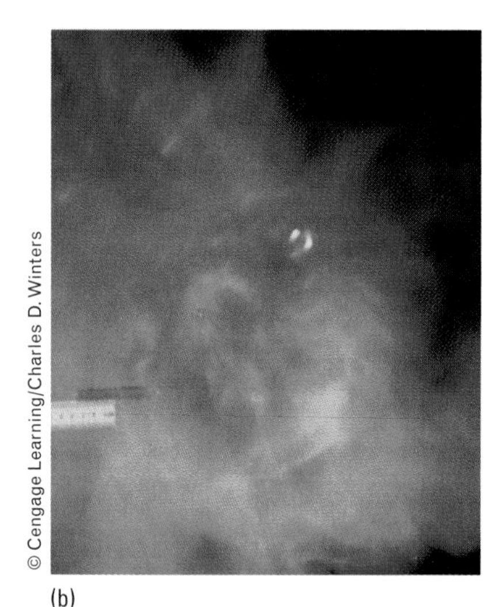

(b)

Combustion of hydrogen gas. The combination reaction of hydrogen and oxygen produces water vapor in a highly exothermic process.

The direct proportionality between quantity of reaction and quantity of heat transfer is in line with your everyday experience. Burning twice as much natural gas produces twice as much heating.

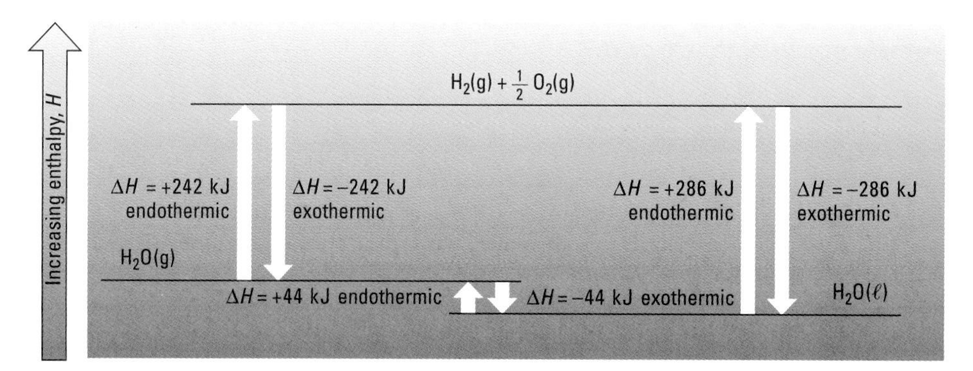

Figure 6.14 Enthalpy diagram. Water vapor [1 mol $H_2O(g)$], liquid water [1 mol $H_2O(\ell)$], and a stoichiometric mixture of hydrogen and oxygen gases [1 mol $H_2(g)$ and $\frac{1}{2}$ mol $O_2(g)$] all have different enthalpy values. The figure shows how these are related, with the highest enthalpy at the top.

Our discussion of phase changes *(p. 189)* showed that an enthalpy change occurs when a substance changes state. Vaporizing 1 mol $H_2O(\ell)$ requires 44.0 kJ. Forming 1 mol $H_2O(\ell)$ from $H_2(g)$ and $O_2(g)$ is 285.8 kJ − 241.8 kJ = 44.0 kJ more exothermic than is forming 1 mol $H_2O(g)$. Figure 6.14 shows the relationships among these quantities. The enthalpy of the reactants [$H_2(g)$ and $\frac{1}{2}$ $O_2(g)$] is greater than that of the product [$H_2O(g)$]. Because the system has less enthalpy after the reaction than before, the law of conservation of energy requires that 241.8 kJ must be transferred *to* the surroundings as the reaction takes place. $H_2O(\ell)$ has even less enthalpy than $H_2O(g)$, so when $H_2O(\ell)$ is formed, even more energy, 285.8 kJ, must be transferred to the surroundings.

Balanced Equation Represents Moles To write an equation for the formation of 1 mol H_2O it is necessary to use a fractional coefficient for O_2. This is acceptable in a thermochemical expression, because the coefficients mean moles, not molecules, and half a mole of O_2 is a perfectly reasonable quantity.

Quantity of Energy Is Proportional to Quantity of Reaction Thermochemical expressions obey the rules of stoichiometry *(p. 107)*. The more reaction there is, the more energy is transferred. Because the balanced equation represents moles, we can calculate how much heat transfer occurs from the number of moles of a reactant that is consumed or the number of moles of a product that is formed. If the thermochemical expression for combination of gaseous hydrogen and oxygen is written without the fractional coefficient, so that 2 mol $H_2O(g)$ is produced, then the energy transfer is twice as great; that is, 2(−241.8 kJ) = −483.6 kJ.

$$2\,H_2(g) + O_2(g) \longrightarrow 2\,H_2O(g) \qquad \Delta H^\circ = -483.6\text{ kJ} \qquad [6.6]$$

EXERCISE 6.10 Enthalpy Change and Stoichiometry

Calculate the change in enthalpy if 0.5000 mol $H_2(g)$ reacts with an excess of $O_2(g)$ to form water vapor at 25 °C.

PROBLEM-SOLVING EXAMPLE 6.7 Thermochemical Expressions

Given the thermochemical expression

$$2\,C_2H_6(g) + 7\,O_2(g) \longrightarrow 4\,CO_2(g) + 6\,H_2O(g) \qquad \Delta H^\circ = -2856\text{ kJ}$$

write a thermochemical expression for

(a) Formation of 1 mol $CO_2(g)$ by burning $C_2H_6(g)$
(b) Formation of 1 mol $C_2H_6(g)$ by reacting $CO_2(g)$ with $H_2O(g)$
(c) Combination of 1 mol $O_2(g)$ with a stoichiometric quantity of $C_2H_6(g)$

Answer

(a) $\frac{1}{2}\,C_2H_6(g) + \frac{7}{4}\,O_2(g) \rightarrow CO_2(g) + \frac{3}{2}\,H_2(g)$ $\Delta H° = -714.0$ kJ

(b) $2\,CO_2(g) + 3\,H_2O(g) \rightarrow C_2H_6(g) + \frac{7}{2}\,O_2(g)$ $\Delta H° = +1428$ kJ

(c) $\frac{2}{7}\,C_2H_6(g) + O_2(g) \rightarrow \frac{4}{7}\,CO_2(g) + \frac{6}{7}\,H_2O(g)$ $\Delta H° = -408.0$ kJ

Strategy and Explanation

(a) Producing 1 mol $CO_2(g)$ requires that one quarter the molar amount of each reactant and product be used and also makes the $\Delta H°$ value one quarter as big.

(b) Forming $C_2H_6(g)$ means that $C_2H_6(g)$ must be a product. This changes the direction of the reaction and the sign of $\Delta H°$; forming 1 mol $C_2H_6(g)$ requires that each coefficient be halved and this halves the size of $\Delta H°$.

(c) If 1 mol $O_2(g)$ reacts, only $\frac{2}{7}$ mol $C_2H_6(g)$ is required; each coefficient is one seventh its original value, and $\Delta H°$ is also one seventh the original value.

☑ **Reasonable Answer Check** In each case examine the coefficients, the direction of the chemical equation, and the sign of $\Delta H°$ to make certain that the appropriate quantity of reactant or product and the appropriate sign have been written.

PROBLEM-SOLVING PRACTICE 6.7

Given the thermochemical expression

$$BaO(s) + CO_2(g) \longrightarrow BaCO_3(s) \qquad \Delta H° = -662.8 \text{ kJ}$$

write the thermochemical expression for the production of 4 mol CO_2 by decomposition of solid barium carbonate.

In Section 4.4 we derived stoichiometric factors (mole ratios) from the coefficients in balanced chemical equations (← p. 108). Stoichiometric factors that relate quantity of energy transferred to quantity of reactant used up or quantity of product produced can be derived from a thermochemical expression. From the equation

$$2\,H_2(g) + O_2(g) \longrightarrow 2\,H_2O(g) \qquad \Delta H° = -483.6 \text{ kJ}$$

these factors (and their reciprocals) can be derived:

$$\frac{-483.6 \text{ kJ}}{2 \text{ mol } H_2 \text{ reacted}} \qquad \frac{-483.6 \text{ kJ}}{1 \text{ mol } O_2 \text{ reacted}} \qquad \frac{-483.6 \text{ kJ}}{2 \text{ mol } H_2O \text{ produced}}$$

The first factor says that 483.6 kJ of energy will transfer from the system to the surroundings whenever 2 mol H_2 is consumed in this reaction. The reciprocal of the second factor says that if the reaction transfers 483.6 kJ to the surroundings, then 1 mol of O_2 must have been used up. We shall refer to stoichiometric factors that include thermochemical information as *thermostoichiometric factors.*

EXERCISE 6.11 Thermostoichiometric Factors from Thermochemical Expressions

Write all of the thermostoichiometric factors (including their reciprocals) that can be derived from this expression:

$$N_2(g) + 3\,H_2(g) \longrightarrow 2\,NH_3(g) \qquad \Delta H° = -92.22 \text{ kJ}$$

CONCEPTUAL
EXERCISE 6.12 Hand Warmer

When the tightly sealed outer package is opened, the portable hand warmer shown in the margin transfers energy to its surroundings. In cold weather it can keep fingers or toes warm for several hours. Suggest a way that such a hand warmer could be designed. What chemicals might be used? Why is the tightly sealed package needed?

Portable hand warmer.

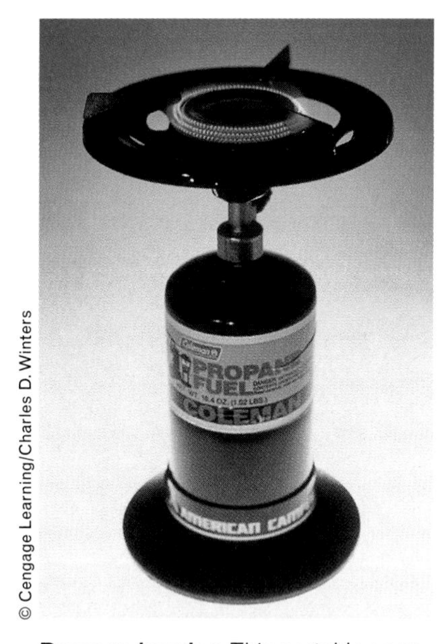

Propane burning. This portable camp stove burns propane fuel. Propane is a major component of liquified petroleum (LP) gas, which is used for heating some houses.

Enthalpy changes for reactions have many practical applications. For instance, when enthalpies of combustion are known, the quantity of energy transferred by the combustion of a given mass of fuel can be calculated. Suppose you are designing a heating system, and you want to know how much heating can be provided per pound (454 g) of propane, C_3H_8, burned in a furnace. The reaction that occurs is *exothermic* (which is not surprising, given that it is a combustion reaction).

$$C_3H_8(g) + 5\,O_2(g) \longrightarrow 3\,CO_2(g) + 4\,H_2O(\ell) \qquad \Delta H^\circ = -2220\text{ kJ}$$

According to this thermochemical expression, 2220 kJ of energy transfers to the surroundings for every 1 mol $C_3H_8(g)$ burned, for every 5 mol $O_2(g)$ consumed, for every 3 mol $CO_2(g)$ formed, and for every 4 mol $H_2O(\ell)$ produced. We know that 454 g $C_3H_8(g)$ has been burned, so we can calculate how many moles of propane that is.

$$\text{Amount of propane} = 454\text{ g} \times \frac{1\text{ mol } C_3H_8}{44.10\text{ g}} = 10.29\text{ mol } C_3H_8$$

Then we multiply by the appropriate thermostoichiometric factor to find the total energy transferred.

$$\text{Energy transferred} = 10.29\text{ mol } C_3H_8 \times \frac{-2220\text{ kJ}}{1\text{ mol } C_3H_8} = -22{,}900\text{ kJ}$$

Burning a pound of fuel such as propane releases a substantial quantity of energy.

PROBLEM-SOLVING EXAMPLE 6.8 Calculating Energy Transferred

The reaction of iron with oxygen from the air provides the energy transferred by the hot pack described in Conceptual Exercise 6.12. Assuming that the iron is converted to iron(III) oxide, how much heating can be provided by a hot pack that contains a tenth of a pound of iron? The thermochemical expression is

$$2\,Fe(s) + \tfrac{3}{2}\,O_2(g) \longrightarrow Fe_2O_3(s) \qquad \Delta H^\circ = -824.2\text{ kJ}$$

Answer −335 kJ

Strategy and Explanation Begin by calculating how many moles of iron are present. A pound is 454 g, so

$$\text{Amount of iron} = 0.100\text{ lb} \times \frac{454\text{ g}}{1\text{ lb}} \times \frac{1\text{ mol Fe}}{55.84\text{ g}} = 0.8130\text{ mol Fe}$$

Then use a thermostoichiometric factor to calculate the energy transferred. The appropriate factor is 824.2 kJ transferred to the surroundings per 2 mol Fe, so

$$\text{Energy transferred} = 0.8130\text{ mol Fe} \times \frac{-824.2\text{ kJ}}{2\text{ mol Fe}} = -335\text{ kJ}$$

Thus, 335 kJ is transferred by the reaction to heat your fingers.

☑ **Reasonable Answer Check** A tenth of a pound is about 45 g, which is a bit less than the molar mass of iron, so we are oxidizing less than a mole of iron. Two moles of iron gives about 800 kJ, so less than a mole should give less than 400 kJ, which makes 335 kJ a reasonable value. The sign should be negative because the enthalpy of the hand warmer (system) should go down when it transfers energy to your hand.

PROBLEM-SOLVING PRACTICE 6.8

How much thermal energy transfer is required to maintain constant temperature during decomposition of 12.6 g liquid water to the elements hydrogen and oxygen at 25.0 °C? In what direction does the energy transfer?

$$H_2O(\ell) \longrightarrow H_2(g) + \tfrac{1}{2}\,O_2(g) \qquad \Delta H^\circ = 285.8\text{ kJ}$$

6.7 Where Does the Energy Come From?

During melting or boiling, nanoscale particles (atoms, molecules, or ions) that attract each other are separated, which increases their potential energy. This requires transfer of energy from the surroundings to enable the particles to overcome their mutual attractions. During a chemical reaction, chemical compounds are created or broken down; that is, reactant molecules are converted into product molecules. Atoms in molecules are held together by chemical bonds. When existing chemical bonds are broken and new chemical bonds are formed, atomic nuclei and electrons move farther apart or closer together, and their energy increases or decreases. These energy differences are usually much greater than those for phase changes.

Consider the reaction of hydrogen gas with chlorine gas to form hydrogen chloride gas.

$$H_2(g) + Cl_2(g) \longrightarrow 2\ HCl(g) \qquad [6.7]$$

When this reaction occurs, the two hydrogen atoms in a H_2 molecule separate, as do the two chlorine atoms in a Cl_2 molecule. In the product the atoms are combined in a different way—as two HCl molecules. We can think of this change as involving two steps:

$$H_2(g) + Cl_2(g) \longrightarrow 2\ H(g) + 2\ Cl(g) \longrightarrow 2\ HCl(g)$$

The first step is to break all bonds in the reactant H_2 and Cl_2 molecules. The second step is to form the bonds in the two product HCl molecules. The net effect of these two steps is the same as for Equation 6.7: One hydrogen molecule and one chlorine molecule change into two hydrogen chloride molecules. The enthalpy changes for these two processes are shown in Figure 6.15.

The reaction of hydrogen with chlorine actually occurs by a complicated series of steps, but the details of how the atoms rearrange do not matter, because enthalpy is a state function and the initial and final states are the same. This means that we can concentrate on products and reactants and not worry about exactly what happens in between.

Another analogy for the enthalpy change for a reaction is the change in altitude when you climb a mountain. No matter which route you take to the summit (which atoms you separate or combine first), the difference in altitude between the summit and where you started to climb (the enthalpy difference between products and reactants) is the same.

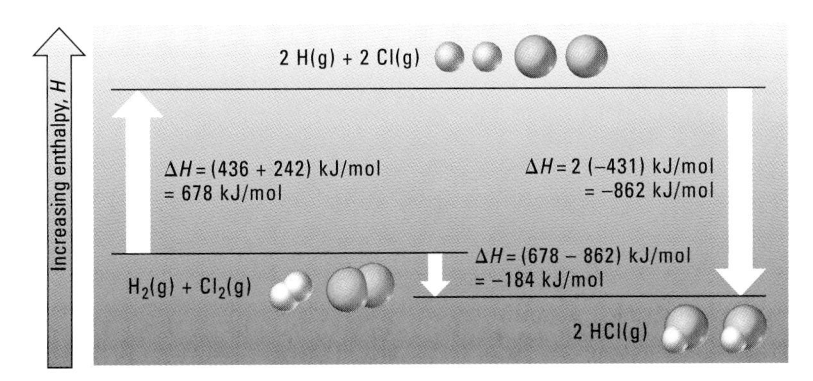

Figure 6.15 Stepwise energy changes in a reaction. Breaking a mole of H_2 molecules into H atoms requires 436 kJ. Breaking a mole of Cl_2 molecules into Cl atoms requires 242 kJ. Putting 2 mol H atoms together with 2 mol Cl atoms to form 2 mol HCl provides $2 \times (-431\ kJ) = -862$ kJ, so the reaction is exothermic. $\Delta H° = 436\ kJ + 242\ kJ - 862\ kJ = -184$ kJ. The relatively weak Cl—Cl bond in the reactants accounts for the fact that this reaction is exothermic.

CONCEPTUAL
EXERCISE 6.13 Reaction Pathways and Enthalpy Change

Suppose that the enthalpy change differed depending on the pathway a reaction took from reactants to products. For example, suppose that 190 kJ was released when a mole of hydrogen gas and a mole of chlorine gas combined to form two moles of hydrogen chloride (Equation 6.7), but that only 185 kJ was released when the same reactant molecules were broken into atoms and the atoms then recombined to form hydrogen chloride (Equation 6.8). Would this violate the first law of thermodynamics? Explain why or why not.

Bond Enthalpies

Bond enthalpy and bond energy differ because a volume change occurs when one molecule changes to two atoms at constant pressure. Therefore work is done on the surroundings (← **p. 191**) and $\Delta E \neq \Delta H$. For a more detailed discussion, see Treptow, R. S., *Journal of Chemical Education,* Vol. 72, 1995; p. 497.

Separating two atoms that are bonded together requires a transfer of energy into the system, because work must be done against the force holding the pair of atoms together. *The enthalpy change that occurs when two bonded atoms in a gas-phase molecule are separated completely at constant pressure* is called the **bond enthalpy** (or the **bond energy**—the two terms are often used interchangeably). The bond enthalpy is usually expressed per mole of bonds. For example, the bond enthalpy for a Cl_2 molecule is 242 kJ/mol, so we can write

$$Cl_2(g) \longrightarrow 2\ Cl(g) \qquad \Delta H^\circ = 242\ kJ \quad [6.8]$$

 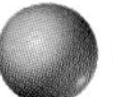 

Bond ***breaking*** is ***endo***thermic.

Bond enthalpies are always positive, and they range in magnitude from about 150 kJ/mol to a little more than 1000 kJ/mol. ***Bond breaking is always endothermic,*** because there is always a transfer of energy into the system (in this case, the mole of Cl_2 molecules) to separate pairs of bonded atoms. Conversely, when atoms come together to form a bond, energy will invariably be transferred to the surroundings because the potential energy of the atoms is lower when they are bonded together. Conservation of energy requires that if the system's energy goes down, the energy of the surroundings must go up. Thus, ***formation of bonds from separated atoms is always exothermic.*** How these generalizations apply to the reaction of hydrogen with chlorine to form hydrogen chloride is shown in Figure 6.15.

Bond ***making*** is ***exo***thermic.

Bond enthalpies provide a way to see what makes a process exothermic or endothermic. If, as in Figure 6.15, the total energy transferred out of the system when new bonds form is greater than the total energy transferred in to break all of the bonds in the reactants, then the reaction is exothermic. In terms of bond enthalpies there are two ways for an exothermic reaction to happen:

- Weaker bonds are broken, stronger bonds are formed, and the number of bonds is the same.
- Bonds in reactants and products are of about the same strength, but more bonds are formed than are broken.

An endothermic reaction involves breaking stronger bonds than are formed, breaking more bonds than are formed, or both.

CONCEPTUAL
EXERCISE 6.14 Enthalpy Change and Bond Enthalpies

Consider the endothermic reactions

(a) $2\ HF(g) \longrightarrow H_2(g) + F_2(g)$ (b) $2\ H_2O(g) \longrightarrow 2\ H_2(g) + O_2(g)$

      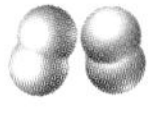 

In which case is formation of weaker bonds the more important factor in making the reaction endothermic? In which case is formation of fewer bonds more important?

6.8 Measuring Enthalpy Changes: Calorimetry

A thermochemical expression tells us how much energy is transferred as a chemical process occurs. This knowledge enables us to calculate the heat obtainable when a fuel is burned, as was done in the preceding section. Also, when reactions are carried out on a larger scale—say, in a chemical plant that manufactures sulfuric acid—the surroundings must have enough cooling capacity to prevent an exothermic reaction from overheating, speeding up, running out of control, and possibly damaging the plant. For these and many other reasons it is useful to know as many $\Delta H°$ values as possible.

For many reactions, direct experimental measurements can be made by using a **calorimeter,** *a device that measures heat transfers*. Calorimetric measurements can be made at constant volume or at constant pressure. Often, in finding heats of combustion or the caloric value of foods, where at least one of the reactants is a gas, the measurement is done at a constant volume in a *bomb calorimeter* (Figure 6.16). The "bomb" is a cylinder about the size of a large fruit juice can with heavy steel walls so that it can contain high pressures. A weighed sample of a combustible solid or liquid is placed in a dish inside the bomb. The bomb is then filled with pure $O_2(g)$ and placed in a water-filled container with well-insulated walls. The sample is ignited, usually by an electrical spark. When the sample burns, it warms the bomb and the water around it to the same temperature.

Figure 6.16 Combustion (bomb) calorimeter.

In this configuration, the oxygen and the compound represent the *system* and the bomb and the water around it are the *surroundings*. For the system, $\Delta E = q + w$. Because there is no change in volume of the sealed, rigid bomb, $w = 0$. Therefore, $\Delta E = q_V$. To calculate q_V and ΔE, we can sum the energy transfers from the reaction to the bomb and to the water. Because each of these is a transfer out of the system, each will be negative. For example, the energy transfer to heat the water can be calculated as $c_{\text{water}} \times m_{\text{water}} \times \Delta T_{\text{water}}$, where c_{water} is the specific heat capacity of water, m_{water} is the mass of water, and ΔT_{water} is the change in temperature of the water. Because this energy transfer is out of the system, the energy transfer from the system to the water is negative, that is, $-(c_{\text{water}} \times m_{\text{water}} \times \Delta T_{\text{water}})$. Problem-Solving Example 6.9 illustrates how this works.

PROBLEM-SOLVING EXAMPLE 6.9 Measuring Energy Change with a Bomb Calorimeter

A 3.30-g sample of the sugar glucose, $C_6H_{12}O_6(s)$, was placed in a bomb calorimeter, ignited, and burned to form carbon dioxide and water. The temperature of the water and the bomb changed from 22.4 °C to 34.1 °C. If the calorimeter contained 850. g water and had a heat capacity of 847 J/°C, what is ΔE for combustion of 1 mol glucose? (The heat capacity of the bomb is the energy transfer required to raise the bomb's temperature by 1 °C.)

Answer −2810 kJ

Strategy and Explanation When the glucose burns, it heats the calorimeter and the water. Calculate the heat transfer from the reaction to the calorimeter and the water from the temperature change and their heat capacities. (Look up the heat capacity of water in Table 6.1.) Use this result to calculate the heat transfer from the reaction. Then use a proportion to find the heat transfer for 1 mol glucose.

$$\Delta T = (34.1 - 22.4)°\text{C} = 11.7\ °\text{C}$$

$$\text{Energy transferred from system to bomb} = -\text{heat capacity of bomb} \times \Delta T$$

$$= -\frac{847\ \text{J}}{°\text{C}} \times 11.7\ °\text{C} = -9910\ \text{J} = -9.910\ \text{kJ}$$

$$\text{Energy transferred from system to water} = -(c \times m \times \Delta T)$$

$$= -\left(\frac{4.184\ \text{J}}{\text{g}\ °\text{C}} \times 850.\ \text{g} \times 11.7\ °\text{C}\right) = -41{,}610\ \text{J} = -41.61\ \text{kJ}$$

$$\Delta E = q_V = -9.910\ \text{kJ} - 41.61\ \text{kJ} = -51.52\ \text{kJ}$$

This quantity of energy transfer corresponds to burning 3.30 g glucose. To scale to 1 mol glucose, first calculate how many moles of glucose were burned.

$$3.30\ \text{g}\ C_6H_{12}O_6 \times \frac{1\ \text{mol}}{180.16\ \text{g}} = 1.832 \times 10^{-2}\ \text{mol}\ C_6H_{12}O_6$$

Then set up this proportion.

$$\frac{-51.52\ \text{kJ}}{1.832 \times 10^{-2}\ \text{mol}} = \frac{\Delta E}{1\ \text{mol}} \qquad \Delta E = 1\ \text{mol} \times \frac{-51.52\ \text{kJ}}{1.832 \times 10^{-2}\ \text{mol}} = -2.81 \times 10^3\ \text{kJ}$$

☑ **Reasonable Answer Check** The result is negative, which correctly reflects the fact that burning sugar is exothermic. A mole of glucose (180 g) is more than a third of a pound, and a third of a pound of sugar contains quite a bit of energy (many Calories), so it is reasonable that the magnitude of the answer is in the thousands of kilojoules.

(a)

(b)

Photos: © Jerold J. Jacobsen

Figure 6.17 Coffee cup calorimeter. (a) A simple constant-pressure calorimeter can be made from two coffee cups that are good thermal insulators, a cork or other insulating lid, a temperature probe, and a stirrer. (b) Close-up of the nested cups that make up the calorimeter. A reaction carried out in an aqueous solution within the calorimeter will change the temperature of the solution. Because the thermal insulation is extremely good, essentially no energy transfer can occur to or from anything outside the calorimeter. Therefore, the heat capacity of the solution and its change in temperature can be used to calculate q_P and ΔH.

PROBLEM-SOLVING PRACTICE 6.9

In Problem-Solving Example 6.1, a single Fritos chip was oxidized by potassium chlorate. Suppose that a single chip weighing 1.0 g is placed in a bomb calorimeter that has a heat capacity of 877 J/°C. The calorimeter contains 832 g water. When the bomb is filled with excess oxygen and the chip is ignited, the temperature rises from 20.64 °C to 25.43 °C. Use these data to verify the statement that the chip provides 5 Cal when metabolized.

CONCEPTUAL
EXERCISE 6.15 Comparing Enthalpy Change and Energy Change

Write a balanced equation for the combustion of glucose to form $CO_2(g)$ and $H_2O(\ell)$. Use what you already know about the volume of a mole of any gas at a given temperature and pressure (or look in Section 10.4) to predict whether ΔH would differ significantly from ΔE for the reaction in Problem-Solving Example 6.9.

When reactions take place in solution, it is much easier to use a calorimeter that is open to the atmosphere. An example, often encountered in introductory chemistry courses, is the *coffee cup calorimeter* shown in Figure 6.17. The nested coffee cups (which are made of expanded polystyrene) provide good thermal insulation; reactions can occur when solutions are poured together in the inner cup. Because a coffee cup calorimeter is a constant-pressure device, the measured heat transfer is q_P, which equals ΔH.

PROBLEM-SOLVING EXAMPLE 6.10 Measuring Enthalpy Change with a Coffee Cup Calorimeter

A coffee cup calorimeter is used to determine ΔH for the reaction

$$\text{NaOH(aq)} + \text{HCl(aq)} \longrightarrow H_2O(\ell) + \text{NaCl(aq)} \qquad \Delta H = ?$$

When 250. mL of 1.00 M NaOH was added to 250. mL of 1.00 M HCl at 1 bar, the temperature of the solution increased from 23.4 °C to 30.4 °C. Use this information to determine ΔH and complete the thermochemical expression. Assume that the heat capacities of the coffee cups, the temperature probe, and the stirrer are negligible, that the solution has the same density and the same specific heat capacity as water, and that there is no change in volume of the solutions upon mixing.

Answer $\Delta H = -59$ kJ

Strategy and Explanation Use the definition of specific heat capacity [Equation 6.2′ (← *p. 186*)] to calculate q_P, the heat transfer for the constant-pressure conditions (1 bar). Because the density of the solution is assumed to be the same as for water and the total volume is 500. mL, the mass of solution is 500. g. Because the reaction system heats the solution, q_P is negative and

$$q_P = -c \times m \times \Delta T$$
$$= -(4.184\ \text{J g}^{-1}\ °\text{C}^{-1})(500.\ \text{g})(30.4\ °\text{C} - 23.4\ °\text{C}) = -1.46 \times 10^4\ \text{J} = -14.6\ \text{kJ}$$

This quantity of heat transfer does not correspond to the equation as written, however; instead it corresponds to consumption of

$$250.\ \text{mL} \times \frac{1.00\ \text{mol}}{1000\ \text{mL}} = 0.250\ \text{mol HCl} \quad \text{and} \quad 250.\ \text{mL} \times \frac{1.00\ \text{mol}}{1000\ \text{mL}} = 0.250\ \text{mol NaOH}$$

From the balanced equation, 1 mol HCl is required for each 1 mol NaOH. Therefore the reactants are in the stoichiometric ratio and neither reactant is a limiting reactant. Because the chemical equation involves 1 mol HCl, the heat transfer must be scaled in proportion to this quantity of HCl.

$$\frac{\Delta H}{1\ \text{mol HCl}} = \frac{-14.6\ \text{kJ}}{0.250\ \text{mol HCl}}$$

$$\Delta H = 1\ \text{mol HCl} \times \frac{-14.6\ \text{kJ}}{0.250\ \text{mol HCl}} = -59\ \text{kJ}$$

(Note that because the reactants were in the stoichiometric ratio, the NaOH could also have been used in the preceding calculation.)

☑ **Reasonable Answer Check** The temperature of the surroundings increased, so the reaction is exothermic and $\Delta H°$ must be negative. The temperature of 500.-g solution went up 7.0 °C, so the heat transfer was about $(500 \times 7 \times 4)$ J = 14,000 J = 14 kJ. This corresponded to one-quarter mole of each reactant, so the heat transfer per mole must be about 4×14 kJ = 56 kJ. Therefore ΔH should be about −56 kJ, which it is.

PROBLEM-SOLVING PRACTICE 6.10

Suppose that 100. mL of 1.00 M HCl and 100. mL of 0.50 M NaOH, both at 20.4 °C, are mixed in a coffee cup calorimeter. Use the result from Problem-Solving Example 6.10 to predict what will be the highest temperature reached in the calorimeter after mixing the solutions. Make similar assumptions to those made in Problem-Solving Example 6.10.

When expanded polystyrene coffee cups are used to make a calorimeter, the masses of substances other than the solvent water are often so small that their heat capacities can be ignored; all of the energy of a reacton can be assumed to be transferred to the water.

CONCEPTUAL
EXERCISE 6.16 Calorimetry

In Problem-Solving Example 6.10 ΔT was observed to be 7.0 °C for mixing 250. mL of 1.00 M HCl and 250. mL of 1.00 M NaOH in a coffee cup calorimeter. Predict ΔT for mixing

(a) 200. mL of 1.0 M HCl and 200. mL of 1.0 M NaOH.
(b) 100. mL of 1.0 M H_2SO_4 and 100. mL of 1.0 M NaOH.

6.9 Hess's Law

Calorimetry works well for some reactions, but for many others it is difficult to use. Besides, it would be very time-consuming to measure values for every conceivable reaction, and it would take a great deal of space to tabulate so many values. Fortunately, there is a better way. It is based on **Hess's law,** which states that, *if the equation for a reaction is the sum of the equations for two or more other reactions, then $\Delta H°$ for the first reaction must be the sum of $\Delta H°$ values of the other reactions.* Hess's law is a corollary of the law of conservation of energy. It works even if

CHEMISTRY IN THE NEWS

How Small Can a Calorimeter Be?

With the development of nanotechnology, it has become important to make calorimetric measurements at the nanoscale. For example, when studying how reactions occur in an automobile catalytic converter, it is useful to be able to measure the heat transfer during catalytic dissociation of carbon monoxide molecules on a metal surface. Because the reaction occurs on a two-dimensional surface, a much smaller number of atoms and molecules are involved than would be present in a three-dimensional sample, and so the heat transfer is also much smaller. This calls for a small and highly sensitive calorimeter.

The ultimate in sensitivity and small size is a nanoscale calorimeter reported by W. Chung Fon, Keith C. Schwab, John M. Worlock, and Michael L. Roukes. A scanning electron micrograph of their calorimeter is shown in the photograph. The calorimeter was patterned from a 120-nm-thick layer of silicon nitride (SiN) on an area about 25 μm on a side. Four 8-μm-long and 600-nm-wide "beams" of SiN (the X shape in the photograph) suspend a thermometer and heater. The heater is made of gold and the thermometer is AuGe. Four thin films of niobium metal serve as electrical leads to the heater and thermometer. The device was prepared using special nanofabrication techniques.

The tiny volume of the metallic heater and thermometer ensure that their contribution to the heat capacity of the calorimeter is very small. This makes possible measurements of very small energy transfers. (The smaller the heat capacity is, the larger the temperature change is for a given energy transfer.)

When a pulse of electric current was applied to the gold heater, delivering 0.125 nW (0.125×10^{-9} J/s) at an initial temperature of 4.5 K, the calorimeter responded as shown in the graph. Unlike experiments involving coffee cup calorimeters that you may have performed, this experiment is over in about 75 μs (75×10^{-6} s)—a very short time. The graph shows an increase in temperature that levels off, which is similar to what would be observed in a coffee cup calorimeter. The 75-μs timescale, however, requires special electronics to measure the temperature change of less than 0.1 K (100 mK).

10 μm

1 μm

Nanocalorimeter

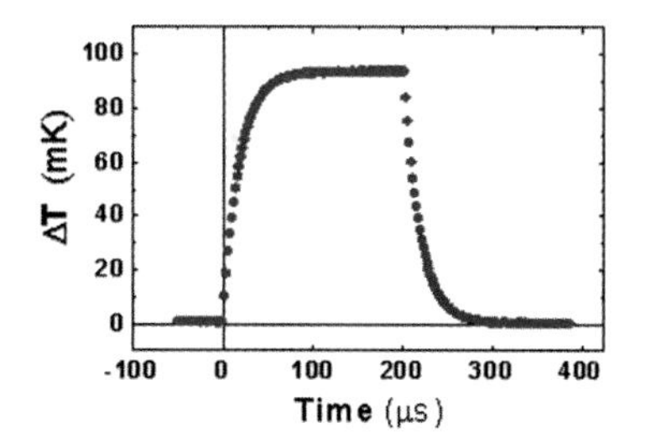

To check that the calorimeter could measure the heat capacity of a real substance, a thin layer of helium atoms was condensed on its surface—about 2 He atoms per square nanometer of surface. The heat capacity of this film was measured to be about 3 fJ/K (3×10^{-15} J/K). This result is similar to observations for thin layers of helium on other surfaces. Based on their experiments, the four scientists conclude that their nanocalorimeter could measure energy transfers as small as about 0.5 aJ/K (5×10^{-19} J/K)—very tiny energy transfers indeed!

Source: Nano Letters, Vol. 5, 2005; pp. 1968–1971.

the overall reaction does not actually occur by way of the separate equations that are summed.

For example, in Figure 6.14 (***p. 198***) we noted that the formation of liquid water from its elements $H_2(g)$ and $O_2(g)$ could be thought of as two successive changes: (a) formation of water vapor from the elements and (b) condensation of water vapor to liquid water. As shown below, the equation for formation of liquid water can be obtained by adding algebraically the chemical equations for these two steps. Therefore, according to Hess's law, the $\Delta H°$ value can be found by adding the $\Delta H°$ values for the two steps.

(a)	$H_2(g) + \frac{1}{2}O_2(g) \longrightarrow H_2O(g)$	$\Delta H_1° = -241.8$ kJ
(b)	$H_2O(g) \longrightarrow H_2O(\ell)$	$\Delta H_2° = -44.0$ kJ
(a) + (b)	$H_2(g) + \frac{1}{2}O_2(g) \longrightarrow H_2O(\ell)$	$\Delta H° = \Delta H_1° + \Delta H_2° = -285.8$ kJ

Hess's law is based on a fact we mentioned earlier (***p. 194***). A system's enthalpy will be the same no matter how the system is prepared. Therefore, at 25 °C and 1 bar, the initial system, $H_2(g) + \frac{1}{2}O_2(g)$, has a particular enthalpy value. The final system, $H_2O(\ell)$, also has a characteristic (but different) enthalpy. Whether we get from initial system to final system by a single step or by the two-step process of the chemical equations (a) and (b), the enthalpy change will be the same.

Here, 1 mol $H_2O(g)$ is a product of the first reaction and a reactant in the second. Thus, $H_2O(g)$ can be canceled out. This is similar to adding two algebraic equations: If the same quantity or term appears on both sides of the equation, it cancels. The net result is an equation for the overall reaction and its associated enthalpy change. This overall enthalpy change applies even if the liquid water is formed directly from hydrogen and oxygen.

Note that it takes 1 mol $H_2O(g)$ to cancel 1 mol $H_2O(g)$. If the coefficient of $H_2O(g)$ had been different on one side of the chemical reactions from the coefficient on the other side, $H_2O(g)$ could not have been completely canceled.

A useful approach to Hess's law is to analyze the equation whose $\Delta H°$ you are trying to calculate. Identify which reactants are desired in what quantities and also which products in what quantities. Then consider how the known thermochemical expressions could be changed to give reactants and products in appropriate quantities. For example, suppose you want the thermochemical expression for the reaction

$$\tfrac{1}{2}\,CH_4(g) + O_2(g) \longrightarrow \tfrac{1}{2}\,CO_2(g) + H_2O(\ell) \qquad \Delta H° = ?$$

and you already know the thermochemical expressions

$$\text{(a) } CH_4(g) + 2\,O_2(g) \longrightarrow CO_2(g) + 2\,H_2O(g) \qquad \Delta H_a° = -802.34 \text{ kJ}$$

and

$$\text{(b) } H_2O(\ell) \longrightarrow H_2O(g) \qquad \Delta H_b° = 44.01 \text{ kJ}$$

The target equation has only $\frac{1}{2}$ mol $CH_4(g)$ as a reactant; it also has $\frac{1}{2}$ mol $CO_2(g)$ and 1 mol $H_2O(\ell)$ as products. Equation (a) has the same reactants and products, but twice as many moles of each; also, water is in the gaseous state in equation (a). If we change each coefficient and the $\Delta H°$ value of expression (a) to one half their original values, we have the thermochemical expression

$$\text{(a}'\text{) } \tfrac{1}{2}\,CH_4(g) + O_2(g) \longrightarrow \tfrac{1}{2}\,CO_2(g) + H_2O(g) \qquad \Delta H_{a'}° = -401.17 \text{ kJ}$$

which differs from the target expression only in the phase of water. Expression (b) has liquid water on the left and gaseous water on the right, but our target expression has liquid water on the right. If the equation in (b) is reversed (which changes the sign of $\Delta H°$), the thermochemical expression becomes

$$\text{(b}'\text{) } H_2O(g) \longrightarrow H_2O(\ell) \qquad \Delta H_{b'}° = -44.01 \text{ kJ}$$

Summing the expressions (a′) and (b′) gives the target expression, from which $H_2O(g)$ has been canceled.

$$\text{(a}' + \text{b}'\text{) } \tfrac{1}{2}\,CH_4(g) + O_2(g) \longrightarrow \tfrac{1}{2}\,CO_2(g) + H_2O(\ell) \qquad \Delta H° = \Delta H_{a'}° + \Delta H_{b'}°$$

$$\Delta H° = (-401.17 \text{ kJ}) + (-44.01 \text{ kJ}) = -445.18 \text{ kJ}$$

PROBLEM-SOLVING EXAMPLE 6.11 Using Hess's Law

In designing a chemical plant for manufacturing the plastic polyethylene, you need to know the enthalpy change for the removal of H_2 from C_2H_6 (ethane) to give C_2H_4 (ethylene), a key step in the process.

$$C_2H_6(g) \longrightarrow C_2H_4(g) + H_2(g) \qquad \Delta H° = ?$$

From experiments you know these thermochemical expressions:

$$\text{(a) } 2\,C_2H_6(g) + 7\,O_2(g) \longrightarrow 4\,CO_2(g) + 6\,H_2O(\ell) \qquad \Delta H_a° = -3119.4 \text{ kJ}$$

$$\text{(b) } C_2H_4(g) + 3\,O_2(g) \longrightarrow 2\,CO_2(g) + 2\,H_2O(\ell) \qquad \Delta H_b° = -1410.9 \text{ kJ}$$

$$\text{(c) } 2\,H_2(g) + O_2(g) \longrightarrow 2\,H_2O(\ell) \qquad \Delta H_c° = -571.66 \text{ kJ}$$

Use this information to find the value of $\Delta H°$ for the formation of ethylene from ethane.

Answer $\Delta H° = 137.0$ kJ

Strategy and Explanation Analyze reactions (a), (b), and (c). Reaction (a) involves 2 mol ethane on the reactant side, but only 1 mol ethane is required in the desired reaction.

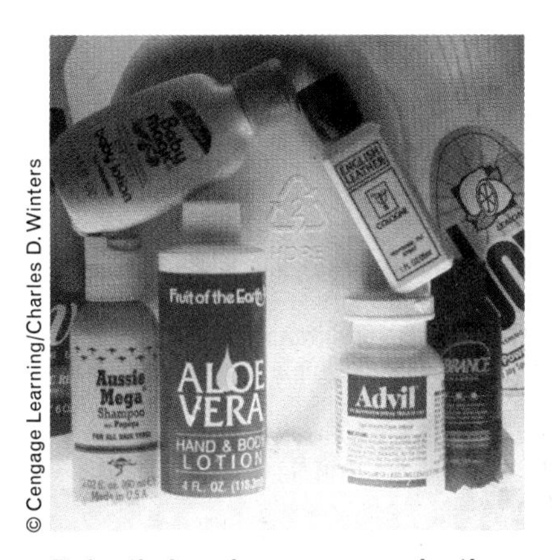

Polyethylene is a common plastic. Many products are packaged in polyethylene bottles.

Reaction (b) has C_2H_4 as a reactant, but C_2H_4 is a product in the desired reaction. Reaction (c) has 2 mol H_2 as a reactant, but 1 mol H_2 is a product in the desired reaction.

First, since the desired expression has only 1 mol ethane on the reactant side, we multiply expression (a) by $\frac{1}{2}$ to give an expression (a′) that also has 1 mol ethane on the reactant side. Halving the coefficients in the equation also halves the enthalpy change.

$$(a') = \tfrac{1}{2}(a) \qquad C_2H_6(g) + \tfrac{7}{2}O_2(g) \longrightarrow 2\,CO_2(g) + 3\,H_2O(\ell) \qquad \Delta H^\circ_{a'} = -1559.7\text{ kJ}$$

Next, we reverse expression (b) so that C_2H_4 is on the product side, giving expression (b′). This also reverses the sign of the enthalpy change.

$$(b') = -(b) \qquad 2\,CO_2(g) + 2\,H_2O(\ell) \longrightarrow C_2H_4(g) + 3\,O_2(g)$$
$$\Delta H^\circ_{b'} = -\Delta H^\circ_b = +1410.9\text{ kJ}$$

To get 1 mol $H_2(g)$ on the product side, we reverse expression (c) and multiply all coefficients by $\frac{1}{2}$. This changes the sign and halves the enthalpy change.

$$(c') = -\tfrac{1}{2}(c) \qquad H_2O(\ell) \longrightarrow H_2(g) + \tfrac{1}{2}O_2(g) \qquad \Delta H^\circ_{c'} = -\tfrac{1}{2}\Delta H^\circ_c = +285.83\text{ kJ}$$

Now it is possible to add expressions (a′), (b′), and (c′) to give the desired expression.

$$(a') \qquad C_2H_6(g) + \cancel{\tfrac{7}{2}O_2(g)} \longrightarrow \cancel{2\,CO_2(g)} + \cancel{3\,H_2O(\ell)} \qquad \Delta H^\circ_{a'} = -1559.7\text{ kJ}$$
$$(b') \qquad \cancel{2\,CO_2(g)} + \cancel{2\,H_2O(\ell)} \longrightarrow C_2H_4(g) + \cancel{3\,O_2(g)} \qquad \Delta H^\circ_{b'} = +1410.9\text{ kJ}$$
$$(c') \qquad \cancel{H_2O(\ell)} \longrightarrow H_2(g) + \cancel{\tfrac{1}{2}O_2(g)} \qquad \Delta H^\circ_{c'} = +285.83\text{ kJ}$$
$$\text{Net equation:} \qquad C_2H_6(g) \longrightarrow C_2H_4(g) + H_2(g) \qquad \Delta H^\circ_{net} = 137.0\text{ kJ}$$

When the chemical equations are added, there is $\frac{7}{2}$ mol $O_2(g)$ on the reactant side and $(3 + \frac{1}{2}) = \frac{7}{2}$ mol $O_2(g)$ on the product side. There is 3 mol $H_2O(\ell)$ on each side and 2 mol $CO_2(g)$ on each side. Therefore, $O_2(g)$, $CO_2(g)$, and $H_2O(\ell)$ all cancel, and the chemical equation for the conversion of ethane to ethylene and hydrogen remains.

☑ **Reasonable Answer Check** The overall process involves breaking a molecule apart into simpler molecules, which is likely to involve breaking bonds. Therefore it should be endothermic, and ΔH° should be positive.

PROBLEM-SOLVING PRACTICE 6.11

When iron is obtained from iron ore, an important reaction is conversion of $Fe_3O_4(s)$ to $FeO(s)$. Write a balanced equation for this reaction. Then use these thermochemical expressions to calculate ΔH° for the reaction.

$$3\,Fe(s) + 2\,O_2(g) \longrightarrow Fe_3O_4(s) \qquad \Delta H^\circ_1 = -1118.4\text{ kJ}$$
$$Fe(s) + \tfrac{1}{2}O_2(g) \longrightarrow FeO(s) \qquad \Delta H^\circ_2 = -272.0\text{ kJ}$$

6.10 Standard Molar Enthalpies of Formation

The word *molar* means "per mole." Thus, the standard molar enthalpy of formation is the standard enthalpy of formation per mole of compound formed.

Hess's law makes it possible to tabulate ΔH° values for a relatively few reactions and, by suitable combinations of these few reactions, to calculate ΔH° values for a great many other reactions. To make such a tabulation we use standard molar enthalpies of formation. The **standard molar enthalpy of formation,** H_f°, is the *standard enthalpy change for formation of one mole of a compound from its elements in their standard states.* The subscript *f* indicates *f*ormation of the compound. The **standard state** of an element or compound is the physical state in which it exists at 1 bar and a specified temperature. At 25 °C the standard state for hydrogen is $H_2(g)$ and for sodium chloride is $NaCl(s)$. For an element that can exist in several different allotropic forms (← p. 23) at 1 bar and 25 °C, the most stable form is usually selected as the standard state. For example, graphite, not diamond or buckminsterfullerene, is the standard state for carbon; $O_2(g)$, not $O_3(g)$, is the standard state for oxygen.

Some examples of thermochemical expressions involving standard molar enthalpies of formation are

$$H_2(g) + \frac{1}{2} O_2(g) \longrightarrow H_2O(\ell) \qquad \Delta H° = \Delta H_f°\{H_2O(\ell)\} = -285.8 \text{ kJ/mol}$$

$$2 \text{ C(graphite)} + 2 H_2(g) \longrightarrow C_2H_4(g) \qquad \Delta H° = \Delta H_f°\{C_2H_4(g)\} = 52.26 \text{ kJ/mol}$$

$$2 \text{ C(graphite)} + 3 H_2(g) + \frac{1}{2} O_2 \longrightarrow C_2H_5OH(\ell)$$

$$\Delta H° = \Delta H_f°\{C_2H_5OH(\ell)\} = -277.69 \text{ kJ/mol}$$

Notice that in each case *1 mol of a compound in its standard state is formed directly from appropriate amounts of elements in their standard states.*

Some examples of thermochemical expressions at 25 °C and 1 bar where $\Delta H°$ is *not* a standard molar enthalpy of formation (and the reason why it is not) are

$$MgO(s) + SO_3(g) \longrightarrow MgSO_4(s) \qquad \Delta H° = -287.5 \text{ kJ} \qquad [6.9]$$

(reactants are not elements)

and

$$P_4(s) + 6 Cl_2(g) \longrightarrow 4 PCl_3(\ell) \qquad \Delta H° = -1278.8 \text{ kJ} \qquad [6.10]$$

(4 mol product formed instead of 1 mol)

It is common to use the term "heat of formation" interchangeably with "enthalpy of formation." It is only the heat of reaction at constant pressure that is equivalent to the enthalpy change. If heat of reaction is measured under other conditions, it may not equal the enthalpy change. For example, when measured at constant volume in a bomb calorimeter, heat of reaction corresponds to the change of internal energy, not enthalpy.

PROBLEM-SOLVING EXAMPLE 6.12 Thermochemical Expressions for Standard Molar Enthalpies of Formation

Rewrite thermochemical Expressions 6.9 and 6.10 so that they represent standard molar enthalpies of formation of their products. (The standard molar enthalpy of formation value for $MgSO_4(s)$ is −1284.9 kJ/mol at 25 °C.)

Answer

$$Mg(s) + \frac{1}{8} S_8(s) + 2 O_2(g) \longrightarrow MgSO_4(s) \qquad \Delta H_f°\{MgSO_4(s)\} = -1284.9 \text{ kJ/mol}$$

and

$$\frac{1}{4} P_4(s) + \frac{3}{2} Cl_2(g) \longrightarrow PCl_3(\ell) \qquad \Delta H_f°\{PCl_3(\ell)\} = -319.7 \text{ kJ/mol}$$

Strategy and Explanation Expression 6.9 has 1 mol $MgSO_4(s)$ on the right side, but the reactants are not elements in their standard states. Write a new expression so that the left side contains the elements Mg(s), $S_8(s)$, and $O_2(g)$. For this expression $\Delta H°$ is the standard molar enthalpy of formation, −1284.9 kJ/mol. The new thermochemical expression is given in the Answer section above.

Expression 6.10 has elements in their standard states on the left side, but more than 1 mol of product is formed. Rewrite the expression so the right side involves only 1 mol $PCl_3(\ell)$, and reduce the coefficients of the elements on the left side in proportion—that is, divide all coefficients by 4. Then $\Delta H°$ must also be divided by 4 to obtain the second thermochemical expression in the Answer section.

☑ **Reasonable Answer Check** Check each expression carefully to make certain the substance whose standard enthalpy of formation you want is on the right side and has a coefficient of 1. For $PCl_3(\ell)$, $\Delta H_f°$ should be one fourth of about −1300 kJ, and it is.

© Cengage Learning/Charles D. Winters

Burning charcoal. Charcoal is mainly carbon, and it burns to form mainly carbon dioxide gas. The energy transfer from a charcoal grill could be estimated from the mass of charcoal and the standard molar enthalpy of formation of $CO_2(g)$.

PROBLEM-SOLVING PRACTICE 6.12

Write an appropriate thermochemical expression in each case. (You may need to use fractional coefficients.)

(a) The standard molar enthalpy of formation of $NH_3(g)$ at 25 °C is −46.11 kJ/mol.

(b) The standard molar enthalpy of formation of CO(g) at 25 °C is −110.525 kJ/mol.

CONCEPTUAL EXERCISE **6.17 Standard Molar Enthalpies of Formation of Elements**

Write the thermochemical expression that corresponds to the standard molar enthalpy of formation of $N_2(g)$.

(a) What process, if any, takes place in the chemical equation?
(b) What does this imply about the enthalpy change?

Table 6.2 and Appendix J list values of ΔH_f°, obtained from the National Institute for Standards and Technology (NIST), for many compounds. Notice that no values are listed in these tables for elements in their most stable forms, such as C(graphite) or $O_2(g)$. As you probably realized from Conceptual Exercise 6.17, *standard enthalpies of formation for the elements in their standard states are zero,* because forming an element in its standard state from the same element in its standard state involves no chemical or physical change.

Hess's law can be used to find the standard enthalpy change for any reaction if there is a set of reactions whose enthalpy changes are known and whose chemical equations, when added together, will give the equation for the desired reaction. For example, suppose you are a chemical engineer and want to know how much heating

Table 6.2 Selected Standard Molar Enthalpies of Formation at 25 °C*

Formula	Name	Standard Molar Enthalpy of Formation (kJ/mol)	Formula	Name	Standard Molar Enthalpy of Formation (kJ/mol)
$Al_2O_3(s)$	Aluminum oxide	−1675.7	HI(g)	Hydrogen iodide	26.48
$BaCO_3(s)$	Barium carbonate	−1216.3	KF(s)	Potassium fluoride	−567.27
$CaCO_3(s)$	Calcium carbonate	−1206.92	KCl(s)	Potassium chloride	−436.747
CaO(s)	Calcium oxide	−635.09	KBr(s)	Potassium bromide	−393.8
C(s, diamond)	Diamond	1.895	MgO(s)	Magnesium oxide	−601.70
$CCl_4(\ell)$	Carbon tetrachloride	−135.44	$MgSO_4(s)$	Magnesium sulfate	−1284.9
$CH_4(g)$	Methane	−74.81	$Mg(OH)_2(s)$	Magnesium hydroxide	−924.54
$C_2H_5OH(\ell)$	Ethyl alcohol	−277.69	NaF(s)	Sodium fluoride	−573.647
CO(g)	Carbon monoxide	−110.525	NaCl(s)	Sodium chloride	−411.153
$CO_2(g)$	Carbon dioxide	−393.509	NaBr(s)	Sodium bromide	−361.062
$C_2H_2(g)$	Acetylene (ethyne)	226.73	NaI(s)	Sodium iodide	−287.78
$C_2H_4(g)$	Ethylene (ethene)	52.26	$NH_3(g)$	Ammonia	−46.11
$C_2H_6(g)$	Ethane	−84.68	NO(g)	Nitrogen monoxide	90.25
$C_3H_8(g)$	Propane	−103.8	$NO_2(g)$	Nitrogen dioxide	33.18
$C_4H_{10}(g)$	Butane	−126.148	$O_3(g)$	Ozone	142.7
$C_6H_{12}O_6(s)$	α-D-Glucose	−1274.4	$PCl_3(\ell)$	Phosphorus trichloride	−319.7
$CuSO_4(s)$	Copper(II) sulfate	−771.36	$PCl_5(s)$	Phosphorus pentachloride	−443.5
$H_2O(g)$	Water vapor	−241.818	$SiO_2(s)$	Silicon dioxide (quartz)	−910.94
$H_2O(\ell)$	Liquid water	−285.830	$SnCl_2(s)$	Tin(II) chloride	−325.1
HF(g)	Hydrogen fluoride	−271.1	$SnCl_4(\ell)$	Tin(IV) chloride	−511.3
HCl(g)	Hydrogen chloride	−92.307	$SO_2(g)$	Sulfur dioxide	−296.830
HBr(g)	Hydrogen bromide	−36.40	$SO_3(g)$	Sulfur trioxide	−395.72

*From Wagman, D. D., Evans, W. H., Parker, V. B., Schuman, R. H., Halow, I., Bailey, S. M., Churney, K. L., and Nuttall, R. The NBS Tables of Chemical Thermodynamic Properties. *Journal of Physical and Chemical Reference Data,* Vol. 11, Suppl. 2, 1982. (NBS, the National Bureau of Standards, is now NIST, the National Institute for Standards and Technology.)

is required to decompose limestone (calcium carbonate) to lime (calcium oxide) and carbon dioxide.

$$\text{CaCO}_3(\text{s}) \longrightarrow \text{CaO(s)} + \text{CO}_2(\text{g}) \qquad \Delta H° = ?$$

As a first approximation you can assume that all substances are in their standard states at 25 °C and look up the standard molar enthalpy of formation of each substance in a table such as Table 6.2 or Appendix J. This gives the thermochemical expressions

$$\text{(a)} \quad \text{Ca(s)} + \text{C(graphite)} + \tfrac{3}{2}\,\text{O}_2(\text{g}) \longrightarrow \text{CaCO}_3(\text{s}) \qquad \Delta H_a° = -1206.9 \text{ kJ}$$

$$\text{(b)} \quad \text{Ca(s)} + \tfrac{1}{2}\,\text{O}_2(\text{g}) \longrightarrow \text{CaO(s)} \qquad \Delta H_b° = -635.1 \text{ kJ}$$

$$\text{(c)} \quad \text{C(graphite)} + \text{O}_2(\text{g}) \longrightarrow \text{CO}_2(\text{g}) \qquad \Delta H_c° = -393.5 \text{ kJ}$$

Now add the three chemical equations in such a way that the resulting equation is the one given above for the decomposition of limestone. In expression (a), $CaCO_3(s)$ is a product, but it must appear in the desired expression as a reactant. Therefore, the equation in (a) must be reversed, and the sign of $\Delta H_a°$ must also be reversed. On the other hand, CaO(s) and $CO_2(g)$ are products in the desired expression, so expressions (b) and (c) can be added with the same direction and sign of $\Delta H°$ as they have in the $\Delta H_f°$ equations:

$$\text{(a}') = -\text{(a)} \quad \text{CaCO}_3(\text{s}) \longrightarrow \text{Ca(s)} + \text{C(graphite)} + \tfrac{3}{2}\,\text{O}_2(\text{g}) \qquad \Delta H_{a'}° = +1206.9 \text{ kJ}$$

$$\text{(b)} \quad \text{Ca(s)} + \tfrac{1}{2}\,\text{O}_2(\text{g}) \longrightarrow \text{CaO(s)} \qquad \Delta H_b° = -635.1 \text{ kJ}$$

$$\text{(c)} \quad \text{C(graphite)} + \text{O}_2(\text{g}) \longrightarrow \text{CO}_2(\text{g}) \qquad \Delta H_c° = -393.5 \text{ kJ}$$

$$\text{CaCO}_3(\text{s}) \longrightarrow \text{CaO(s)} + \text{CO}_2(\text{g}) \qquad \Delta H° = +178.3 \text{ kJ}$$

When the expressions are added in this fashion, 1 mol each of C(graphite) and Ca(s) and $\frac{3}{2}$ mol $O_2(g)$ appear on opposite sides and so are canceled out. Thus, the sum of these chemical equations is the desired one for the decomposition of calcium carbonate, and the sum of the enthalpy changes of the three expressions gives that for the desired expression.

Another very useful conclusion can be drawn from this example. The calculation can be written mathematically as

$$\begin{aligned}\Delta H° &= \Delta H_f°\{\text{CaO(s)}\} + \Delta H_f°\{\text{CO}_2(\text{g})\} - \Delta H_f°\{\text{CaCO}_3(\text{s})\} \\ &= (-635.1 \text{ kJ}) + (-393.5 \text{ kJ}) - (-1206.9 \text{ kJ}) = 178.3 \text{ kJ}\end{aligned}$$

which involves adding the $\Delta H_f°$ values for the products of the reaction, CaO(s) and $CO_2(g)$, and subtracting the $\Delta H_f°$ value for the reactant, $CaCO_3(s)$. The mathematics of the problem can be summarized by the equation

$$\begin{aligned}\Delta H° = &\sum\{(\text{moles of product}) \times \Delta H_f°(\text{product})\} \\ &-\sum\{(\text{moles of reactant})\} \times \Delta H_f°(\text{reactant})\}\end{aligned} \qquad [6.11]$$

This equation says that to get the standard enthalpy change of the reaction you should (1) multiply the standard molar enthalpy of formation of each product by the number of moles of that product and then sum over all products, (2) multiply the standard molar enthalpy of formation of each reactant by the number of moles of that reactant and then sum over all reactants, and (3) subtract the sum for the reactants from the sum for the products. This is a useful shortcut to writing the thermochemical expressions for all appropriate formation reactions and applying Hess's law, as we did above.

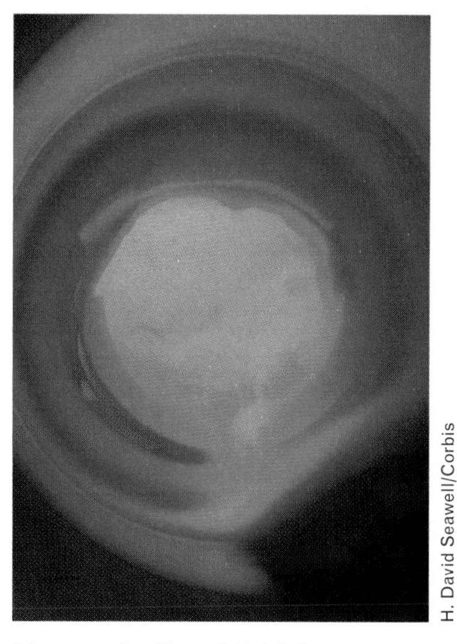

H. David Seawell/Corbis

Lime production. At high temperature in a lime kiln, calcium carbonate (limestone, $CaCO_3$) decomposes to calcium oxide (lime, CaO) and carbon dioxide (CO_2).

Reatha Clark King

Reatha Clark King 1938–

Reatha Clark was born in Georgia and married N. Judge King in 1961. She obtained degrees from Clark Atlanta University and the University of Chicago and began her career with the National Bureau of Standards (now the National Institute for Standards and Technology, NIST), where she determined enthalpies of formation of fluorine compounds that were important to the U.S. space program and NASA. She became a dean at York College, president of Metropolitan State University (Minneapolis), and served as president of the General Mills Foundation.

PROBLEM-SOLVING EXAMPLE 6.13 Using Standard Molar Enthalpies of Formation

Benzene, C_6H_6, is a commercially important hydrocarbon that is present in gasoline, where it enhances the octane rating. Calculate its enthalpy of combustion per mole; that is, find the value of $\Delta H°$ for the reaction

$$C_6H_6(\ell) + \frac{15}{2} O_2(g) \longrightarrow 6\, CO_2(g) + 3\, H_2O(\ell)$$

For benzene, $\Delta H_f°\{C_6H_6(\ell)\} = 49.0$ kJ/mol. Use Table 6.2 for any other values you may need.

Answer $\Delta H° = -3267.5$ kJ

Strategy and Explanation To calculate $\Delta H°$ you need standard molar enthalpies of formation for all compounds (and elements, if they are not in their standard states) involved in the reaction. (Since $O_2(g)$ is in its standard state, it is not included.) From Table 6.2,

$$C(\text{graphite}) + O_2(g) \longrightarrow CO_2(g) \qquad \Delta H_f° = -393.509 \text{ kJ/mol}$$

$$H_2(g) + \frac{1}{2} O_2(g) \longrightarrow H_2O(\ell) \qquad \Delta H_f° = -285.830 \text{ kJ/mol}$$

Using Equation 6.11,

$$\begin{aligned}\Delta H° &= [6 \text{ mol} \times \Delta H_f°\{CO_2(g)\} + 3 \text{ mol} \times \Delta H_f°\{H_2O(\ell)\}] - [1 \text{ mol } C_6H_6(\ell) \times \Delta H_f°\{C_6H_6(\ell)\}\\ &= [6 \text{ mol} \times (-393.509 \text{ kJ/mol}) + 3 \text{ mol} \times (-285.830 \text{ kJ/mol})]\\ &\quad -[1 \text{ mol} \times (49.0 \text{ kJ/mol})] = -3267.5 \text{ kJ}\end{aligned}$$

☑ **Reasonable Answer Check** As expected, the enthalpy change for combustion of a fuel is negative and large.

PROBLEM-SOLVING PRACTICE 6.13

Nitroglycerin is a powerful explosive because it decomposes exothermically and four different gases are formed.

$$2\, C_3H_5(NO_3)_3(\ell) \longrightarrow 3\, N_2(g) + \frac{1}{2} O_2(g) + 6\, CO_2(g) + 5\, H_2O(g)$$

For nitroglycerin, $\Delta H_f°\{C_3H_5(NO_3)_3(\ell)\} = -364$ kJ/mol. Using data from Table 6.2, calculate the energy transfer when 10.0 g nitroglycerin explodes.

When the enthalpy change for a reaction is known, it is possible to use that information to calculate $\Delta H_f°$ for one substance in the reaction provided that $\Delta H_f°$ values are known for all of the rest of the substances. Problem-Solving Example 6.14 indicates how to do this.

PROBLEM-SOLVING EXAMPLE 6.14 Standard Molar Enthalpy of Formation from Enthalpy of Combustion

Octane, C_8H_{18}, is a hydrocarbon that is present in gasoline. At 25 °C the enthalpy of combustion per mole for octane is −5116.0 kJ/mol. Use data from Table 6.2 to calculate the standard molar enthalpy of formation of octane. (Assume that water vapor is produced by the combustion reaction.)

Answer $\Delta H_f° = -208.4$ kJ

Strategy and Explanation Write a balanced equation for the target reaction whose $\Delta H°$ you want to calculate. Also write a balanced equation for combustion of octane, for which you know the standard enthalpy change. By studying these two equations, decide what additional information is needed to set up a Hess's law calculation that will yield the standard molar enthalpy of formation of octane.

(target reaction) $8\, C(s) + 9\, H_2(g) \longrightarrow C_8H_{18}(\ell) \qquad \Delta H_f° = ??$ kJ/mol

(a) $C_8H_{18}(\ell) + \frac{25}{2} O_2(g) \longrightarrow 8\, CO_2(g) + 9\, H_2O(g) \qquad \Delta H°_{\text{combustion}} = -5116.0$ kJ

Notice that the combustion equation involves 1 mol $C_8H_{18}(\ell)$ as a reactant and the target equation (for enthalpy of formation) involves 1 mol $C_8H_{18}(\ell)$ as a product. Therefore, it seems reasonable to reverse the combustion equation and see where that leads.

$$\text{(a')}\ 8\,CO_2(g) + 9\,H_2O(g) \longrightarrow C_8H_{18}(\ell) + \tfrac{25}{2}\,O_2(g) \qquad \Delta H^\circ = +5116.0\text{ kJ}$$

On the reactant side of the target equation we have 8 C(s) and 9 $H_2(g)$. These elements, combined with $O_2(g)$, are on the left side of equation (a′), so perhaps it would be reasonable to use the equations corresponding to standard molar enthalpies of formation of carbon dioxide and water. From Table 6.2, we have

$$\text{(b)}\ C(s) + O_2(g) \longrightarrow CO_2(g) \qquad \Delta H_f^\circ = -393.509\text{ kJ/mol}$$

$$\text{(c)}\ H_2(g) + \tfrac{1}{2}\,O_2(g) \longrightarrow H_2O(g) \qquad \Delta H_f^\circ = -241.818\text{ kJ/mol}$$

Multiplying equation (b) by 8 and equation (c) by 9 gives the correct number of moles of C(s) and of $H_2(g)$ on the reactant side of the target equation. This gives

$$\text{(a')}\ 8\,CO_2(g) + 9\,H_2O(g) \longrightarrow C_8H_{18}(\ell) + \tfrac{25}{2}\,O_2(g) \qquad \Delta H^\circ = +5116.0\text{ kJ}$$

$$\text{(b')}\ 8\,C(s) + 8\,O_2(g) \longrightarrow 8\,CO_2(g) \qquad \Delta H^\circ = -3148.072\text{ kJ}$$

$$\text{(c')}\ 9\,H_2(g) + \tfrac{9}{2}\,O_2(g) \longrightarrow 9\,H_2O(g) \qquad \Delta H^\circ = -2176.362\text{ kJ/mol}$$

$$8\,C(s) + 9\,H_2(g) \longrightarrow C_8H_{18}(\ell) \qquad \Delta H_f^\circ = \Delta H_{a'}^\circ + \Delta H_{b'}^\circ + \Delta H_{c'}^\circ = -208.4\text{ kJ/mol}$$

PROBLEM-SOLVING PRACTICE 6.14

Use data from Table 6.2 to calculate the molar heat of combustion of sulfur dioxide, $SO_2(g)$, to form sulfur trioxide, $SO_3(g)$.

IN CLOSING

Having studied this chapter, you should be able to . . .

- Understand the difference between kinetic energy and potential energy (Section 6.1).
- Be familiar with typical energy units and be able to convert from one unit to another (Section 6.1).
- Understand conservation of energy and energy transfer by heating and working (Section 6.2). ■ End-of-chapter Question assignable in OWL: 15
- Recognize and use thermodynamic terms: system, surroundings, heat, work, temperature, thermal equilibrium, exothermic, endothermic, and state function (Sections 6.2, 6.4).
- Use specific heat capacity and the sign conventions for transfer of energy (Section 6.3). ■ Question 29
- Distinguish between the change in internal energy and the change in enthalpy for a system (Section 6.4).
- Use thermochemical expressions and derive thermostoichiometric factors from them (Sections 6.5, 6.6). ■ Question 35
- Use the fact that the standard enthalpy change for a reaction, ΔH°, is proportional to the quantity of reactants consumed or products produced when the reaction occurs (Section 6.6). ■ Questions 37, 41, 43
- Understand the origin of the enthalpy change for a chemical reaction in terms of bond enthalpies (Section 6.7). ■ Questions 45–47
- Describe how calorimeters can measure the quantity of thermal energy transferred during a reaction (Section 6.8).

Selected end-of-chapter Questions may be assigned in OWL.

For quick review, download Go Chemistry mini-lecture flashcard modules (or purchase them at **www.ichapters.com**).

Prepare for an exam with a **Summary Problem** that brings together concepts and problem-solving skills from throughout the chapter. Go to **www.cengage.com/chemistry/moore** to download this chapter's Summary Problem from the Student Companion Site.

- Apply Hess's law to find the enthalpy change for a reaction (Sections 6.9, 6.10).
- Use standard molar enthalpies of formation to calculate the thermal energy transfer when a reaction takes place (Section 6.10). ■ Questions 59, 63

KEY TERMS

bond enthalpy (bond energy) *(Section 6.7)*

calorimeter *(6.8)*

change of state *(6.4)*

conservation of energy, law of *(6.2)*

endothermic *(6.4)*

enthalpy change *(6.4)*

enthalpy of fusion *(6.4)*

enthalpy of vaporization *(6.4)*

exothermic *(6.4)*

first law of thermodynamics *(6.2)*

heat/heating *(6.2)*

heat capacity *(6.3)*

Hess's law *(6.9)*

internal energy *(6.2)*

kinetic energy *(6.1)*

molar heat capacity *(6.3)*

phase change *(6.4)*

potential energy *(6.1)*

specific heat capacity *(6.3)*

standard enthalpy change *(6.5)*

standard molar enthalpy of formation *(6.10)*

standard state *(6.10)*

state function *(6.4)*

surroundings *(6.2)*

system *(6.2)*

thermal equilibrium *(6.2)*

thermochemical expression *(6.5)*

thermodynamics *(Introduction)*

work/working *(6.2)*

QUESTIONS FOR REVIEW AND THOUGHT

■ denotes questions assignable in OWL.

Blue-numbered questions have short answers at the back of this book and fully worked solutions in the *Student Solutions Manual.*

Review Questions

1. For each of the following, define a system and its surroundings and give the direction of heat transfer:
   (a) Propane is burning in a Bunsen burner in the laboratory.
   (b) After you have a swim, water droplets on your skin evaporate.
   (c) Water, originally at 25 °C, is placed in the freezing compartment of a refrigerator.
   (d) Two chemicals are mixed in a flask on a laboratory bench. A reaction occurs and heat is evolved.
2. Criticize each of these statements:
   (a) Enthalpy of formation refers to a reaction in which 1 mol of one or more reactants produces some quantity of product.
   (b) The standard enthalpy of formation of O_2 as a gas at 25 °C and a pressure of 1 atm is 15.0 kJ/mol.
3. Explain how a coffee cup calorimeter may be used to measure the enthalpy change of (a) a change in state and (b) a chemical reaction.
4. What is required for heat transfer of energy from one sample of matter to another to occur?
5. Name two exothermic processes and two endothermic processes that you encountered recently and that were not associated with your chemistry course.
6. Explain in your own words why it is useful in thermodynamics to distinguish a system from its surroundings.

Topical Questions

The Nature of Energy

7. (a) A 2-inch piece of two-layer chocolate cake with frosting provides 1670 kJ of energy. What is this in Cal?
   (b) If you were on a diet that calls for eating no more than 1200 Cal per day, how many joules would you consume per day?
8. Sulfur dioxide, SO_2, is found in wines and in polluted air. If a 32.1-g sample of sulfur is burned in the air to get 64.1 g SO_2, 297 kJ of energy is released. Express this energy in (a) joules, (b) calories, and (c) kilocalories.
9. ■ On a sunny day, solar energy reaches the earth at a rate of 4.0 J min^{-1} cm^{-2}. Suppose a house has a square, flat roof of dimensions 12 m by 12 m. How much solar energy reaches this roof in 1.0 h? (*Note:* This is why roofs painted with light-reflecting paint are better than black, unpainted roofs in warm climates: They reflect most of this energy rather than absorb it.)
10. A 100-W light bulb is left on for 14 h. How many joules of energy are used? With electricity at $0.09 per kWh, how much does it cost to leave the light on for 14 h?

Conservation of Energy

11. Describe how energy is changed from one form to another in these processes:
    (a) At a July 4th celebration, a match is lit and ignites the fuse of a rocket firecracker, which fires off and explodes at an altitude of 1000 ft.
    (b) A gallon of gasoline is pumped from an underground storage tank into the fuel tank of your car, and you use it up by driving 25 mi.
12. Analyze transfer of energy from one form to another in each situation below.
    (a) In a space shuttle, hydrogen and oxygen combine to form water, boosting the shuttle into orbit above the earth.
    (b) You eat a package of Fritos, go to class and listen to a lecture, walk back to your dorm, and climb the stairs to the fourth floor.
13. ■ Solid ammonium chloride is added to water in a beaker and dissolves. The beaker becomes cold to the touch.
    (a) Make an appropriate choice of system and surroundings and describe it unambiguously.
    (b) Explain why you chose the system and surroundings you did.
    (c) Identify transfers of energy and material into and out of the system that would be important for you to monitor in your study.
    (d) Is the process of dissolving $NH_4Cl(s)$ in water exothermic or endothermic?
14. A bar of Monel (an alloy of nickel, copper, iron, and manganese) is heated until it melts, poured into a mold, and solidifies.
    (a) Make an appropriate choice of system and surroundings and describe it unambiguously.
    (b) Explain why you chose the system and surroundings you did.
    (c) Identify transfers of energy and material into and out of the system that would be important for you to monitor in your study.
15. ■ If a system does 75.4 J of work on its surroundings and simultaneously there is 25.7 cal of heat transfer from the surroundings to the system, what is ΔE for the system?
16. A 20.0-g sample of water cools from 30 °C to 20.0 °C, which transfers 840 J to the surroundings. No work is done on the water. What is ΔE_{water}?

Heat Capacity

17. Which requires greater transfer of energy: (a) cooling 10.0 g water from 50 °C to 20 °C or (b) cooling 20.0 g Cu from 37 °C to 25 °C?
18. Which requires more energy: (a) warming 15.0 g water from 25 °C to 37 °C or (b) warming 60.0 g aluminum from 25 °C to 37 °C?
19. You hold a gram of copper in one hand and a gram of aluminum in the other. Each metal was originally at 0 °C. (Both metals are in the shape of a little ball that fits into your hand.) If they both are heated at the same rate, which will warm to your body temperature first?
20. Ethylene glycol, $(CH_2OH)_2$, is often used as an antifreeze in cars. Which requires greater transfer of thermal energy to warm from 25.0 °C to 100.0 °C, pure water or an equal mass of pure ethylene glycol?
21. If the cooling system in an automobile has a capacity of 5.00 quarts of liquid, compare the quantity of thermal energy transferred to the liquid in the system when its temperature is raised from 25.0 °C to 100.0 °C for water and ethylene glycol. The densities of water and ethylene glycol are 1.00 g/cm^3 and 1.113 g/cm^3, respectively. 1 quart = 0.946 L. Report your results in joules.
22. One way to cool a cup of coffee is to plunge an ice-cold piece of aluminum into it. Suppose a 20.0-g piece of aluminum is stored in the refrigerator at 32 °F (0.0 °C) and then put into a cup of coffee. The coffee's temperature drops from 90.0 °C to 75.0 °C. How much energy (in kilojoules) did the coffee transfer to the piece of aluminum?
23. An unknown metal requires 34.7 J to heat a 23.4-g sample of it from 17.3°C to 28.9 °C. Which of the metals in Table 6.1 is most likely to be the unknown?
24. An unknown metal requires 336.9 J to heat a 46.3-g sample of it from 24.3 °C to 43.2 °C. Which of the metals in Table 6.1 is most likely to be the unknown?
25. A chemical reaction occurs, and 20.7 J is transferred from the chemical system to its surroundings. (Assume that no work is done.)
    (a) What is the algebraic sign of $\Delta T_{surroundings}$?
    (b) What is the algebraic sign of ΔE_{system}?
26. A physical process called a phase transition occurs in a sample of an alloy, and 437 kJ transfers from the surroundings to the alloy. (Assume that no work is done.)
    (a) What is the algebraic sign of ΔT_{alloy}?
    (b) What is the algebraic sign of ΔE_{alloy}?

Energy and Enthalpy

27. The thermal energy transfer required to melt 1.00 g ice at 0 °C is 333 J. If one ice cube has a mass of 62.0 g, and a tray contains 20 ice cubes, how much energy is required to melt a tray of ice cubes at 0 °C?
28. Calculate the quantity of heating required to convert the water in four ice cubes (60.1 g each) from $H_2O(s)$ at 0 °C to $H_2O(g)$ at 100. °C. The enthalpy of fusion of ice at 0 °C is 333 J/g and the enthalpy of vaporization of liquid water at 100. °C is 2260 J/g.
29. ■ The hydrocarbon benzene, C_6H_6, boils at 80.1 °C. How much energy is required to heat 1.00 kg of this liquid from 20.0 °C to the boiling point and then change the liquid completely to a vapor at that temperature? The specific heat capacity of liquid C_6H_6 is 1.74 J g^{-1} °C^{-1}, and the enthalpy of vaporization is 395 J/g. Report your result in joules.
30. How much energy (in joules) would be required to raise the temperature of 1.00 lb lead (1.00 lb = 454 g) from room temperature (25 °C) to its melting point (327 °C), and then melt the lead completely at 327 °C? The specific heat capacity of lead is 0.159 J g^{-1} °C^{-1}, and its enthalpy of fusion is 24.7 J/g.
31. Draw a cooling graph for steam-to-water-to-ice.
32. Draw a heating graph for converting Dry Ice to carbon dioxide gas.

Thermochemical Expressions

33. Energy is stored in the body in adenosine triphosphate, ATP, which is formed by the reaction between adenosine diphosphate, ADP, and dihydrogen phosphate ions.

$$ADP^{3-}(aq) + H_2PO_4^{2-}(aq) \longrightarrow ATP^{4-}(aq) + H_2O(\ell) \quad \Delta H° = 38 \text{ kJ}$$

Is the reaction endothermic or exothermic?

34. Calcium carbide, CaC_2, is manufactured by reducing lime with carbon at high temperature. (The carbide is used in turn to make acetylene, an industrially important organic chemical.)

$$CaO(s) + 3\,C(s) \longrightarrow CaC_2(s) + CO(g) \quad \Delta H° = 464.8 \text{ kJ}$$

Is the reaction endothermic or exothermic?

35. ■ Given the thermochemical expression

$$H_2O(s) \longrightarrow H_2O(\ell) \quad \Delta H° = 6.0 \text{ kJ}$$

what quantity of energy is transferred to the surroundings when
(a) 34.2 mol liquid water freezes?
(b) 100.0 g liquid water freezes?

36. Given the thermochemical expression

$$CaO(s) + 3\,C(s) \longrightarrow CaC_2(s) + CO(g) \quad \Delta H° = 464.8 \text{ kJ}$$

what quantity of energy is transferred when
(a) 34.8 mol CO(g) is formed by this reaction?
(b) a metric ton (1000 kg) of $CaC_2(s)$ is manufactured?
(c) 0.432 mol carbon reacts with CaO(s)?

Enthalpy Changes for Chemical Reactions

37. ■ Given the thermochemical expression for combustion of isooctane (a component of gasoline),

$$2\,C_8H_{18}(\ell) + 25\,O_2(g) \longrightarrow 16\,CO_2(g) + 18\,H_2O(\ell) \quad \Delta H° = -10{,}992 \text{ kJ}$$

write a thermochemical expression for
(a) production of 4.00 mol $CO_2(g)$.
(b) combustion of 100. mol isooctane.
(c) combination of 1.00 mol isooctane with a stoichiometric quantity of air.

38. Given the thermochemical expression for combustion of benzene,

$$2\,C_6H_6(\ell) + 15\,O_2(g) \longrightarrow 12\,CO_2(g) + 6\,H_2O(\ell) \quad \Delta H° = -6534.8 \text{ kJ}$$

write a thermochemical expression for
(a) combustion of 0.50 mol benzene.
(b) consumption of 5 mol $O_2(g)$.
(c) production of 144 mol $CO_2(g)$.

39. Write all the thermostoichiometric factors that can be derived from the thermochemical expression

$$CaO(s) + 3\,C(s) \longrightarrow CaC_2(s) + CO(g) \quad \Delta H° = 464.8 \text{ kJ}$$

40. Write all the thermostoichiometric factors that can be derived from the thermochemical expression

$$2\,CH_3OH(\ell) + 3\,O_2(g) \longrightarrow 2\,CO_2(g) + 4\,H_2O(\ell) \quad \Delta H° = -1530 \text{ kJ}$$

41. ■ "Gasohol," a mixture of gasoline and ethanol, C_2H_5OH, is used as automobile fuel. The alcohol releases energy in a combustion reaction with O_2.

$$C_2H_5OH(\ell) + 3\,O_2(g) \longrightarrow 2\,CO_2(g) + 3\,H_2O(\ell)$$

If 0.115 g alcohol evolves 3.62 kJ when burned at constant pressure, what is the molar enthalpy of combustion for ethanol?

42. White phosphorus, P_4, ignites in air to produce P_4O_{10}.

$$P_4(s) + 5\,O_2(g) \longrightarrow P_4O_{10}(s)$$

When 3.56 g P_4 is burned, 85.8 kJ of thermal energy is evolved at constant pressure. What is the molar enthalpy of combustion of P_4?

43. ■ When wood is burned we may assume that the reaction is the combustion of cellulose (empirical formula, CH_2O).

$$CH_2O(s) + O_2(g) \longrightarrow CO_2(g) + H_2O(g) \quad \Delta H° = -425 \text{ kJ}$$

How much energy is released when a 10.0-lb wood log burns completely? (Assume the wood is 100% dry and burns via the reaction above.)

44. A plant takes CO_2 and H_2O from its surroundings and makes cellulose by the reverse of the reaction in the preceding problem. The energy provided for this process comes from the sun via photosynthesis. How much energy does it take for a plant to make 100 g of cellulose?

Where Does the Energy Come From?

Use these bond enthalpy values to answer the questions below.

Bond	Bond Enthalpy (kJ/mol)
H—F	566
H—Cl	431
H—Br	366
H—I	299
H—H	436
F—F	158
Cl—Cl	242
Br—Br	193
I—I	151

45. ■ Which of the four molecules HF, HCl, HBr, and HI has the strongest chemical bond?

46. ■ Which of the four molecules F_2, Cl_2, Br_2, and I_2 has the weakest chemical bond?

47. ■ For the reactions of molecular hydrogen with fluorine and with chlorine:
(a) Calculate the enthalpy change for breaking all the bonds in the reactants.
(b) Calculate the enthalpy change for forming all the bonds in the products.
(c) From the results in parts (a) and (b), calculate the enthalpy change for the reaction.
(d) Which reaction is most exothermic?

48. For the reactions of molecular hydrogen with bromine and with iodine:
   (a) Calculate the enthalpy change for breaking all the bonds in the reactants.
   (b) Calculate the enthalpy change for forming all the bonds in the products.
   (c) From the results in parts (a) and (b), calculate the enthalpy change for the reaction.
   (d) Which reaction is most exothermic?

Measuring Enthalpy Changes: Calorimetry

49. Suppose you add a small ice cube to room-temperature water in a coffee cup calorimeter. What is the final temperature when all of the ice is melted? Assume that you have 200. mL water at 25 °C and that the ice cube weighs 15.0 g and is at 0 °C before being added to the water.
50. A coffee cup calorimeter can be used to investigate the "cold pack reaction," the process that occurs when solid ammonium nitrate dissolves in water:

$$NH_4NO_3(s) \longrightarrow NH_4^+(aq) + NO_3^-(aq)$$

   Suppose 25.0 g solid NH_4NO_3 at 23.0 °C is added to 250. mL H_2O at the same temperature. After the solid is all dissolved, the temperature is measured to be 15.6 °C. Calculate the enthalpy change for the cold pack reaction. (*Hint:* Calculate the energy transferred per mole of NH_4NO_3.) Is the reaction endothermic or exothermic?
51. How much thermal energy is evolved by a reaction in a bomb calorimeter (Figure 6.16) in which the temperature of the bomb and water increases from 19.50 °C to 22.83 °C? The bomb has a heat capacity of 650 J/°C; the calorimeter contains 320. g water. Report your result in kilojoules.
52. Sulfur (2.56 g) was burned in a bomb calorimeter with excess $O_2(g)$. The temperature increased from 21.25 °C to 26.72 °C. The bomb had a heat capacity of 923 J $°C^{-1}$ and the calorimeter contained 815 g water. Calculate the heat transfer, per mole of SO_2 formed, in the course of the reaction

$$S(s) + O_2(g) \longrightarrow SO_2(g)$$

Hess's Law

53. Calculate the standard enthalpy change, $\Delta H°$, for the formation of 1 mol strontium carbonate (the material that gives the red color in fireworks) from its elements.

$$Sr(s) + C(\text{graphite}) + \tfrac{3}{2}O_2(g) \longrightarrow SrCO_3(s)$$

   The information available is

$$Sr(s) + \tfrac{1}{2}O_2(g) \longrightarrow SrO(s) \qquad \Delta H° = -592 \text{ kJ}$$
$$SrO(s) + CO_2(g) \longrightarrow SrCO_3(s) \qquad \Delta H° = -234 \text{ kJ}$$
$$C(\text{graphite}) + O_2(g) \longrightarrow CO_2(g) \qquad \Delta H° = -394 \text{ kJ}$$

54. What is the standard enthalpy change for the reaction of lead(II) chloride with chlorine to give lead(IV) chloride?

$$PbCl_2(s) + Cl_2(g) \longrightarrow PbCl_4(\ell)$$

   It is known that $PbCl_2(s)$ can be formed from the metal and $Cl_2(g)$,

$$Pb(s) + Cl_2(g) \longrightarrow PbCl_2(s) \qquad \Delta H° = -359.4 \text{ kJ}$$

   and that $PbCl_4(\ell)$ can be formed directly from the elements.

$$Pb(s) + 2\,Cl_2(g) \longrightarrow PbCl_4(\ell) \qquad \Delta H° = -329.3 \text{ kJ}$$

Standard Molar Enthalpies of Formation

55. For each compound below, write a balanced thermochemical expression depicting the formation of 1 mol of the compound. Standard molar enthalpies of formation are found in Appendix J.
   (a) $Al_2O_3(s)$ (b) $TiCl_4(\ell)$
   (c) $NH_4NO_3(s)$
56. The standard molar enthalpy of formation of glucose, $C_6H_{12}O_6(s)$, is −1274.4 kJ/mol.
   (a) Is the formation of glucose from its elements exothermic or endothermic?
   (b) Write a balanced equation depicting the formation of glucose from its elements and for which the enthalpy of reaction is −1274.4 kJ.
57. In photosynthesis, the sun's energy brings about the combination of CO_2 and H_2O to form O_2 and a carbon-containing compound such as a sugar. In its simplest form, the reaction could be written

$$6\,CO_2(g) + 6\,H_2O(\ell) \longrightarrow 6\,O_2(g) + C_6H_{12}O_6(s)$$

   Using the enthalpies of formation in Table 6.2, (a) calculate the enthalpy of reaction and (b) decide whether the reaction is exothermic or endothermic.
58. The first step in the production of nitric acid from ammonia involves the oxidation of NH_3.

$$4\,NH_3(g) + 5\,O_2(g) \longrightarrow 4\,NO(g) + 6\,H_2O(g)$$

   Use the information in Table 6.2 or Appendix J to calculate the enthalpy change for this reaction. Is the reaction exothermic or endothermic?
59. ■ Iron can react with oxygen to give iron(III) oxide. If 5.58 g Fe is heated in pure O_2 to give $Fe_2O_3(s)$, how much thermal energy is transferred out of this system (at constant pressure)?
60. The formation of aluminum oxide from its elements is highly exothermic. If 2.70 g Al metal is burned in pure O_2 to give Al_2O_3, how much thermal energy is evolved in the process (at constant pressure)?

General Questions

61. The specific heat capacity of copper is 0.385 J g^{-1} $°C^{-1}$, whereas it is 0.128 J g^{-1} $°C^{-1}$ for gold. Assume you place 100. g of each metal, originally at 25 °C, in a boiling water bath at 100 °C. If energy is transferred to each metal at the same rate, which piece of metal will reach 100 °C first?
62. Suppose you have 40.0 g ice at −25.0 °C and you add to it 60.0 g liquid water at 10.0 °C. What will be the composition and state (s or ℓ) that results? What will be the temperature? (Assume that there is no heat transfer of energy except that between the water and the ice.)
63. ■ The combustion of diborane, B_2H_6, proceeds according to the equation

$$B_2H_6(g) + 3\,O_2(g) \longrightarrow B_2O_3(s) + 3\,H_2O(\ell)$$

   and 2166 kJ is liberated per mole of $B_2H_6(g)$ (at constant pressure). Calculate the molar enthalpy of formation of $B_2H_6(g)$ using this information, the data in Table 6.2, and the fact that the standard molar enthalpy of formation for $B_2O_3(s)$ is −1273 kJ/mol.

64. In principle, copper could be used to generate valuable hydrogen gas from water.

$$Cu(s) + H_2O(g) \longrightarrow CuO(s) + H_2(g)$$

(a) Is the reaction exothermic or endothermic?
(b) If 2.00 g copper metal reacts with excess water vapor at constant pressure, how much thermal energy transfer is involved (either into or out of the system) in the reaction?

65. These two thermochemical expressions are known:

$$2\ C\ (\text{graphite}) + 2\ H_2(g) \longrightarrow C_2H_4(g) \qquad \Delta H° = 52.3\ kJ$$

$$C_2H_4Cl_2(\ell) \longrightarrow Cl_2(g) + C_2H_4(g) \qquad \Delta H° = 217.5\ kJ$$

Calculate the molar enthalpy of formation of $C_2H_4Cl_2(\ell)$.

66. Given the following information and the data in Table 6.2, calculate the standard molar enthalpy of formation for liquid hydrazine, N_2H_4.

$$N_2H_4(\ell) + O_2(g) \longrightarrow N_2(g) + 2\ H_2O(g) \qquad \Delta H° = -534\ kJ$$

67. In 1947 Texas City, Texas, was devastated by the explosion of a shipload of ammonium nitrate, a compound intended to be used as a fertilizer. When heated, ammonium nitrate can decompose exothermically to N_2O and water.

$$NH_4NO_3(s) \longrightarrow N_2O(g) + 2\ H_2O(g)$$

If the heat from this exothermic reaction is contained, higher temperatures are generated, at which point ammonium nitrate can decompose explosively to N_2, H_2O, and O_2.

$$2\ NH_4NO_3(s) \longrightarrow 2\ N_2(g) + 4\ H_2O(g) + O_2(g)$$

If oxidizable materials are present, fires can break out, as was the case at Texas City. Using the information in Appendix J, answer the following questions.
(a) How much thermal energy is evolved (at constant pressure and under standard conditions) by the first reaction?
(b) If 8.00 kg ammonium nitrate explodes (the second reaction), how much thermal energy is evolved (at constant pressure and under standard conditions)?

68. Uranium-235 is used as a fuel in nuclear power plants. Since natural uranium contains only a small amount of this isotope, the uranium must be enriched in uranium-235 before it can be used. To do this, uranium(IV) oxide is first converted to a gaseous compound, UF_6, and the isotopes are separated by a gaseous diffusion technique. Some key reactions are

$$UO_2(s) + 4\ HF(g) \longrightarrow UF_4(s) + 2\ H_2O(g)$$

$$UF_4(s) + F_2(g) \longrightarrow UF_6(g)$$

How much thermal energy transfer (at constant pressure) would be involved in producing 225 ton $UF_6(g)$ from $UO_2(s)$ (1 ton = 9.08×10^5 g)? Some necessary standard enthalpies of formation are

$$\Delta H_f°\{UO_2(s)\} = -1085\ kJ/mol$$

$$\Delta H_f°\{UF_4(s)\} = -1914\ kJ/mol$$

$$\Delta H_f°\{UF_6(g)\} = -2147\ kJ/mol$$

69. One method of producing H_2 on a large scale is this chemical cycle.

Step 1: $SO_2(g) + 2\ H_2O(g) + Br_2(g) \longrightarrow H_2SO_4(\ell) + 2\ HBr(g)$

Step 2: $H_2SO_4(\ell) \longrightarrow H_2O(g) + SO_2(g) + \frac{1}{2}O_2(g)$

Step 3: $2\ HBr(g) \longrightarrow H_2(g) + Br_2(g)$

Using the table of standard enthalpies of formation in Appendix J, calculate $\Delta H°$ for each step. What is the equation for the overall process, and what is its enthalpy change? Is the overall process exothermic or endothermic?

70. One reaction involved in the conversion of iron ore to the metal is

$$FeO(s) + CO(g) \longrightarrow Fe(s) + CO_2(g)$$

Calculate the standard enthalpy change for this reaction from these reactions of iron oxides with CO:

$$3\ Fe_2O_3(s) + CO(g) \longrightarrow 2\ Fe_3O_4(s) + CO_2(g) \qquad \Delta H° = -47\ kJ$$

$$Fe_2O_3(s) + 3\ CO(g) \longrightarrow 2\ Fe(s) + 3\ CO_2(g) \qquad \Delta H° = -25\ kJ$$

$$Fe_3O_4(s) + CO(g) \longrightarrow 3\ FeO(s) + CO_2(g) \qquad \Delta H° = 19\ kJ$$

Applying Concepts

71. ■ Consider this graph, which presents data for a 1.0-g mass of each of four substances, A, B, C, and D. Which substance has the highest specific heat capacity?

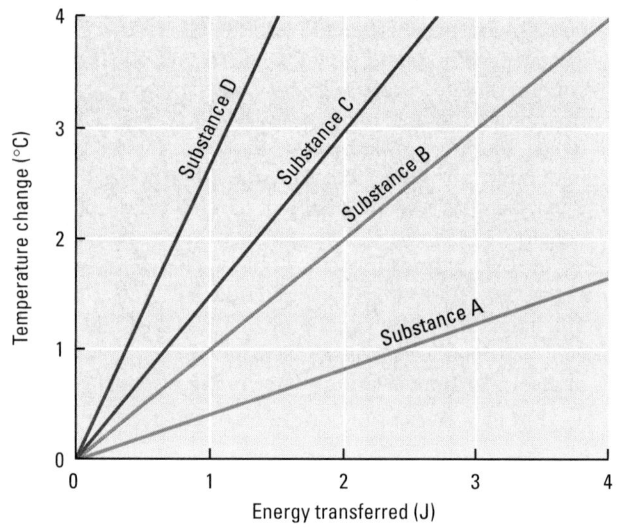

72. Based on the graph that accompanies Question 71, how much heat would need to be transferred to 10 g substance B to raise its temperature from 35 °C to 38 °C?

73. The sketch below shows two identical beakers with different volumes of water at the same temperature. Is the thermal energy content of beaker 1 greater than, less than, or equal to that of beaker 2? Explain your reasoning.

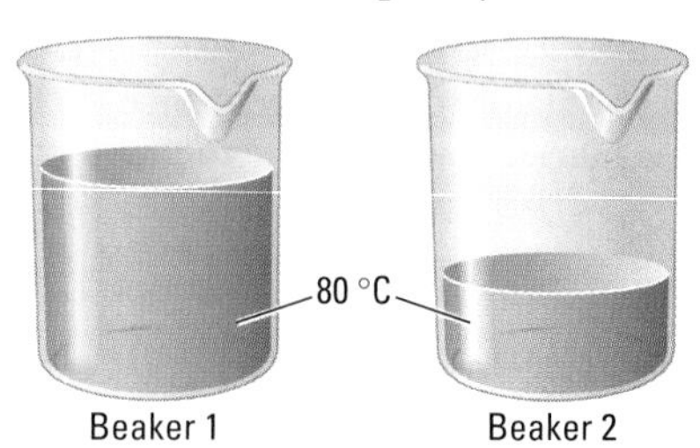

74. If the same quantity of thermal energy were transferred to each beaker in Question 73, would the temperature of beaker 1 be greater than, less than, or equal to that of beaker 2? Explain your reasoning.
75. Consider this thermochemical expression

$$2\ S(s) + 3\ O_2(g) \longrightarrow 2\ SO_3(g) \qquad \Delta H° = -791\ kJ$$

and the standard molar enthalpy of formation for $SO_3(g)$ listed in Table 6.2. Why are the enthalpy values different?
76. In this chapter, the symbols $\Delta H_f°$ and $\Delta H°$ were used to denote a change in enthalpy. What is similar and what is different about the enthalpy changes they represent?

More Challenging Questions

77. Oxygen difluoride, OF_2, is a colorless, very poisonous gas that reacts rapidly and exothermically with water vapor to produce O_2 and HF.

$$OF_2(g) + H_2O(g) \longrightarrow 2\ HF(g) + O_2(g) \qquad \Delta H° = -318\ kJ$$

Using this information and Table 6.2 or Appendix J, calculate the molar enthalpy of formation of $OF_2(g)$.
78. Suppose that 28.89 g $ClF_3(g)$ is reacted with 57.3 g Na(s) to form NaCl(s) and NaF(s). The reaction occurs at 1 bar and 25.0 °C. Calculate the quantity of energy transferred between the system and surroundings and describe in which direction the energy is transferred.
79. The four hydrocarbons of lowest molar mass are methane, ethane, propane, and butane. All are used extensively as fuels in our economy. Calculate the thermal energy transferred to the surroundings per gram (the fuel value) of each of these four fuels and rank them by this quantity.
80. For the four substances in the preceding question, calculate the energy density. (Energy density is the quantity of energy released per unit volume of fuel burned.) Assume that substances that are gases at room temperature have been forced to condense to liquids by applying pressure and, if necessary, by lowering temperature. Obtain needed density information from the *CRC Handbook of Chemistry and Physics,* 85th ed. (London: Taylor & Francis CRC Press; 2004) **http://crcpress.com**.
81. A student had five beakers, each containing 100. mL of 0.500 M NaOH(aq) and all at room temperature (20.0 °C). The student planned to add a carefully weighed quantity of solid ascorbic acid, $C_6H_8O_6$, to each beaker, stir until it dissolved, and measure the increase in temperature. After the fourth experiment, the student was interrupted and called away. The data table looked like this:

Experiment	Mass of Ascorbic Acid (g)	Final Temperature (°C)
1	2.20	21.7
2	4.40	23.3
3	8.81	26.7
4	13.22	26.6
5	17.62	—

(a) Predict the temperature the student would have observed in experiment 5. Explain why you predicted this temperature.
(b) For each experiment indicate which is the limiting reactant, sodium hydroxide or ascorbic acid.
(c) When ascorbic acid reacts with NaOH, how many hydrogen ions are involved? One, as in the case of HCl? Two, as in the case of H_2SO_4? Or three, as in the case of phosphoric acid, H_3PO_4? Explain clearly how you can tell, based on the student's calorimeter data.

82. In their home laboratory, two students do an experiment (a rather dangerous one—don't try it without proper safety precautions!) with drain cleaner (Drano, a solid) and toilet bowl cleaner (The Works, a liquid solution). The students measure 1 teaspoon (tsp) of Drano into each of four Styrofoam coffee cups and dissolve the solid in half a cup of water. Then they wash their hands and go have lunch. When they return, they measure the temperature of the solution in each of the four cups and find it to be 22.3 °C. Next they measure into separate small empty cups 1, 2, 3, and 4 tablespoons (Tbsp) of The Works. In each cup they add enough water to make the total volume 4 Tbsp. After a few minutes they measure the temperature of each cup and find it to be 22.3 °C. Finally the two students take each cup of The Works, pour it into a cup of Drano solution, and measure the temperature over a period of a few minutes. Their results are reported in the table below.

Experiment	Volume of The Works (Tbsp)	Highest Temperature (°C)
1	1	28.0
2	2	33.6
3	3	39.3
4	4	39.4

Discuss these results and interpret them in terms of the thermochemistry and stoichiometry of the reaction. Is the reaction exothermic or endothermic? Why is more energy transferred in some cases than others? For each experiment, which reactant, Drano or The Works, is limiting? Why are the final temperatures nearly the same in experiments 3 and 4? What can you conclude about the stoichiometric ratio between the two reactants?

Conceptual Challenge Problems

CP6.A (Section 6.2) Suppose a scientist discovered that energy was not conserved, but rather that 1×10^{-7}% of the energy transferred from one system vanishes before it enters another system. How would this affect electric utilities, thermochemical experiments in scientific laboratories, and scientific thinking?

CP6.B (Section 6.2) Suppose that someone were to tell your teacher during class that energy is not always conserved. This person states that he or she had previously learned that in the case of nuclear reactions, mass is converted into energy according to Einstein's equation $E = mc^2$. Hence, energy is continuously produced as mass is changed into energy. Your teacher quickly responds by giving the following assignment to the class: "Please write a paragraph or two to refute or clarify this student's thesis." What would you say?

CP6.C (Section 6.3) The specific heat capacities at 25 °C for three metals with widely differing molar masses are 3.6 J g^{-1} $°C^{-1}$ for Li, 0.25 J g^{-1} $°C^{-1}$ for Ag, and 0.11 J g^{-1} $°C^{-1}$ for Th. Suppose that you have three samples, one of each metal and each containing the same number of atoms.

(a) Is the energy transfer required to increase the temperature of each sample by 1 °C significantly different from one sample to the next?

(b) What interpretation can you make about temperature based on the result you found in part (a)?

CP6.D (Section 6.4) During one of your chemistry classes a student asks the professor, "Why does hot water freeze more quickly than cold water?"

(a) What do you expect the professor to say in answer to the student's question?

(b) In one experiment, two 100.-g samples of water were placed in identical containers on the same surface 1 decimeter apart in a room at −25 °C. One sample had an initial temperature of 78 °C, while the second was at 24 °C. The second sample took 151 min to freeze, and the first took 166 min (only 10% longer) to freeze. Clearly the cooler sample froze more quickly, but not nearly as quickly as one might have expected. How can this be so?

CP6.E (Section 6.10) Assume that glass has the same properties as pure SiO_2. The thermal conductivity (the rate at which heat transfer occurs through a substance) for aluminum is eight times that for SiO_2.

(a) Is it more efficient in time and energy to bake brownies in an aluminum pan or a glass pan?

(b) It is said that things cook more evenly in a glass pan than in an aluminum pan. Are there scientific data that indicate that this statement is reasonable?

7

Electron Configurations and the Periodic Table

© Paul Steel/Photolibrary

The spectacular colors of fireworks are due to transition of electrons among different energy levels in various metal ions. Electron transitions in Sr^{2+} ions give bright red color; Ca^{2+} orange; Ba^{2+} green; and Cu^{2+} blue. Mixtures of Sr^{2+} and Cu^{2+} produce purple. This chapter includes a discussion of the electron transitions that lead to emission of visible light.

The periodic table was created by Dmitri Mendeleev to summarize experimental observations ***(⇐ p. 54)***. He had no theory or model to explain why, for example, experimental data showed that all alkaline earth metals combine with oxygen in a 1:1 atom ratio—he just knew they did. In the early years of the twentieth century, however, it became evident that atoms contain electrons. As a result of these findings, explanations of periodic trends in physical and chemical properties began to be based on an understanding of the arrangement of electrons within atoms—on what we now call *electron configurations.* Studies of the interaction of light with atoms and molecules revealed that electrons in atoms are arranged in energy levels or shells. Soon it was understood that the occurrence of electrons in these shells is determined by the energies of the electrons. Electrons in the outermost shell are the *valence electrons;* the number of valence electrons and the shell in which they occur are the chief factors that determine chemical reactivity. This chapter describes the relationship of atomic electron configurations to atomic properties. Special emphasis is placed on the use of the periodic table to derive the electron configurations for atoms and ions.

7.1 Electromagnetic Radiation and Matter

Theories about the energy and arrangement of electrons in atoms are based on experimental studies of the interaction of matter with electromagnetic radiation, of which visible light is a familiar form. The human eye can distinguish the spectrum of colors that make up visible light. Interestingly, matter in some form is always associated with any color of light our eyes see. For example, the red glow of a neon sign comes from neon atoms excited by electricity, and fireworks displays are visible because of light emitted from metal ions excited by the heat of explosive reactions. How are these varied colors of light produced?

The speed of light through a substance (air, glass, or water, for example) depends on the chemical constitution of the substance and the wavelength of the light. This is the basis for using a glass prism to disperse light and is the explanation for rainbows.

Atoms that have more than the minimum quantity of energy are described as "excited." The extra energy can be released in the form of electromagnetic radiation, some of which is light in the visible region. Electromagnetic radiation and its applications are familiar to all of us; sunlight, automobile headlights, dental X-rays, microwave ovens, and cell phones all transmit or receive electromagnetic radiation (Figure 7.1). These kinds of radiation seem very different, but they are actually very similar. All **electromagnetic radiation** consists of oscillating perpendicular electric and magnetic fields that travel through space at the same rate (the "speed of light": 186,000 miles/second, or 2.998×10^8 m/s in a vacuum). Any of the various kinds of electromagnetic

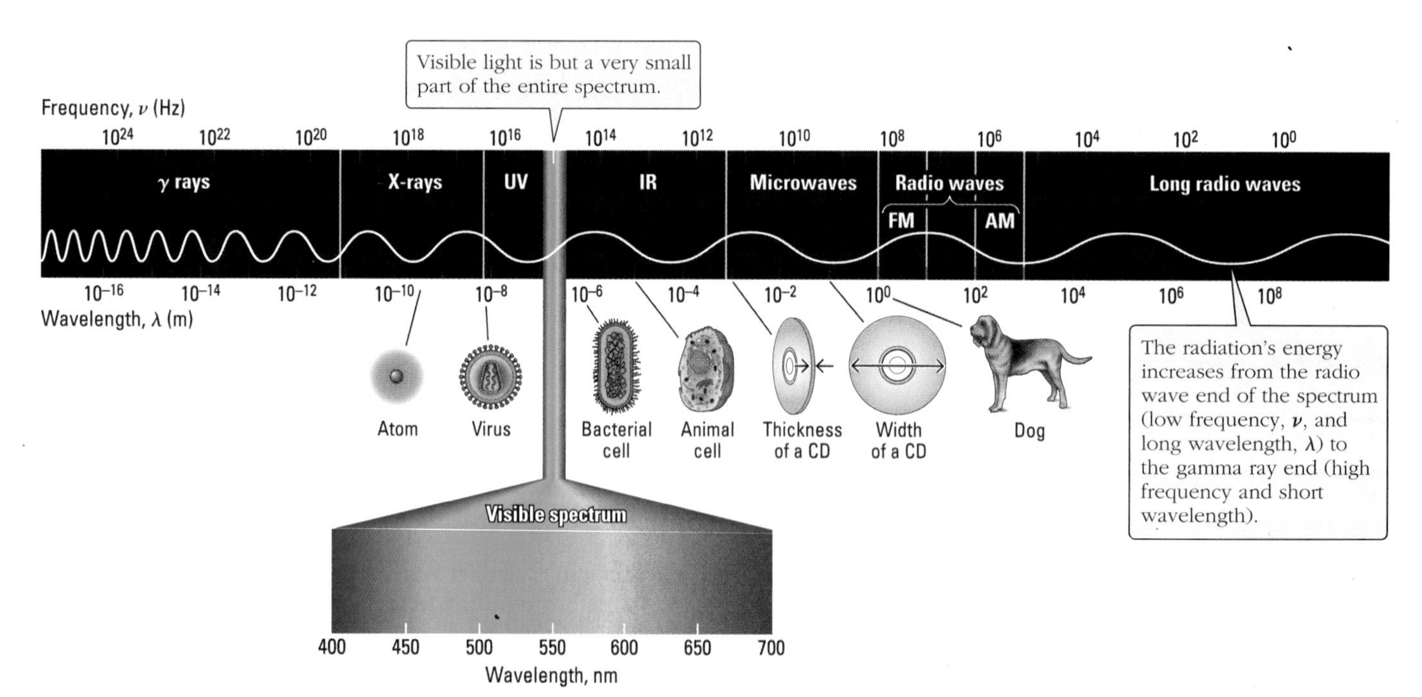

Figure 7.1 The electromagnetic spectrum. The size of wavelengths in the electromagnetic spectrum is compared with common objects. Visible light (enlarged section) is only a small part of the entire spectrum.

Table 7.1 Useful Wavelength Units for Different Regions of the Electromagnetic Spectrum

Wavelength Unit	Unit Length (m)	Radiation Type
Picometer, pm	10^{-12}	γ (gamma)
Ångström, Å	10^{-10}	X-ray
Nanometer, nm	10^{-9}	Ultraviolet, visible
Micrometer, μm	10^{-6}	Infrared
Millimeter, mm	10^{-3}	Infrared
Centimeter, cm	10^{-2}	Microwave
Meter, m	1	TV, radio

radiation can be described in terms of frequency (ν) and wavelength (λ). A **spectrum** is the distribution of intensities of wavelengths or frequencies of electromagnetic radiation emitted or absorbed by an object. Figure 7.1 gives wavelength and frequency values for several regions of the spectrum.

As illustrated in Figure 7.2, the **wavelength** is the distance between adjacent crests (or troughs) in a wave, and the **frequency** is the number of complete waves passing a point in a given period of time—that is, cycles per second or simply per second, 1/s. Reciprocal units such as 1/s are often represented in the negative exponent form, s^{-1}, which means "per second." A frequency of 4.80×10^{14} s^{-1} means that 4.80×10^{14} waves pass a fixed point every second. The unit s^{-1} is given the name *hertz* (Hz).

The hertz unit (cycles/s, or s^{-1}) was named in honor of Heinrich Hertz (1857–1894), a German physicist.

For wave phenomena such as sound, the intensity (loudness) is related to the *amplitude* (height of the wave crest) (Figure 7.2). The higher the amplitude, the more intense is the sound. For example, a middle C tone always has the same frequency and wavelength (pitch), but the sound can vary from loud (high amplitude) to soft (low amplitude).

The frequency of electromagnetic radiation is related to its wavelength by

$$\nu\lambda = c$$

where c is the speed of light, 2.998×10^{8} m/s.

Note that because the speed of light, c, in a vacuum is constant, frequency and wavelength are *inversely* related; that is, as one decreases, the other increases. *Short*-wavelength radiation has a *high* frequency.

As Figure 7.1 shows, visible light is only a small portion of the electromagnetic spectrum. Radiation with shorter wavelengths includes ultraviolet radiation (the type

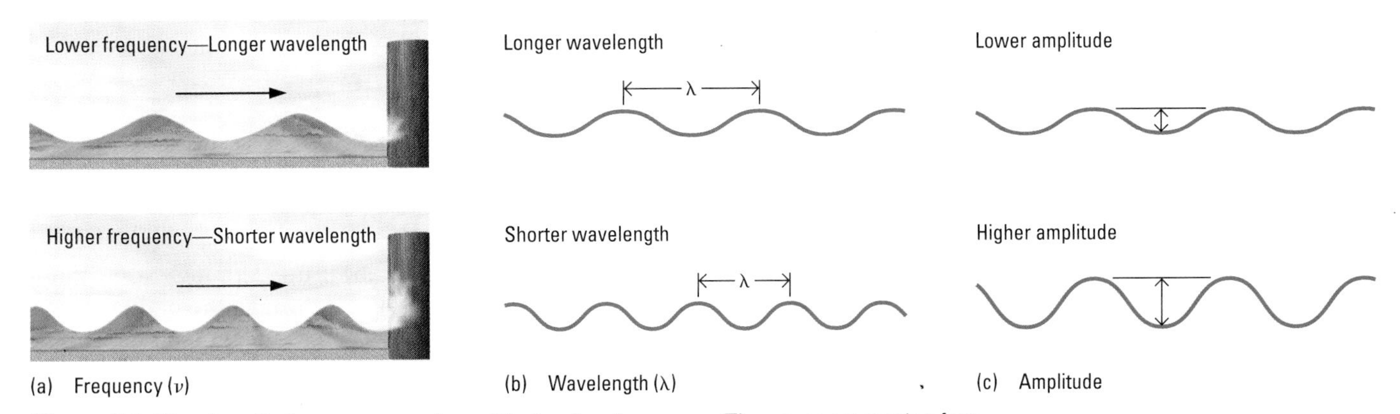

Figure 7.2 Wavelength, frequency, and amplitude of water waves. The waves are moving from left to right toward the post at the same speed. (a) The upper wave has a long wavelength (large λ) and low frequency (the number of times per second the peak of the wave hits the post). The lower wave has a shorter wavelength and a higher frequency (its peaks hit the post more often in a given time). (b) Variation in wavelength. (c) Variation in amplitude.

that leads to sunburn), X-rays, and gamma (γ) rays (emitted in the process of radioactive disintegration of some atoms). Infrared radiation, the type that is sensed as heat from a fire, has wavelengths longer than those of visible light. Longer still is the wavelength of the radiation in a microwave oven or of television and radio transmissions, such as that used for cellular telephones (Table 7.1).

PROBLEM-SOLVING EXAMPLE 7.1 Wavelength and Frequency

Microwaves that heat food in a home microwave oven have a frequency of $2.45 \times 10^9\ s^{-1}$. Calculate the wavelength of this radiation in nanometers, nm.

Answer 1.22×10^8 nm

Strategy and Explanation Rearrange the equation relating frequency, the speed of light, and wavelength to calculate the wavelength of this radiation. Recall that 1 m = 1×10^9 nm.

$$\lambda = \frac{c}{\nu} = \frac{2.998 \times 10^8\ m/s}{2.45 \times 10^9\ s^{-1}} = 0.122\ m$$

Convert meters to nanometers:

$$0.122\ m \times \frac{1 \times 10^9\ nm}{1\ m} = 1.22 \times 10^8\ nm$$

☑ **Reasonable Answer Check** Compare the calculated wavelength (0.122 m) with the wavelength of the microwave region of the electromagnetic spectrum illustrated in Figure 7.1. The calculated wavelength falls within the microwave region and thus the answer is reasonable.

PROBLEM-SOLVING PRACTICE 7.1

A laser developed for use with digital video disc (DVD) players has a wavelength of 405 nm. Calculate the frequency of this radiation in hertz (Hz).

CONCEPTUAL EXERCISE 7.1 Frequency and Wavelength

A fellow chemistry student says that low-frequency radiation is short-wavelength radiation. You disagree. Explain why the other student is wrong.

CONCEPTUAL EXERCISE 7.2 Estimating Wavelengths

The size of a radio antenna is proportional to the wavelength of the radiation. Cellular phones have antennas often less than 0.076 m long, whereas submarine antennas are up to 2000 m long. Which is using higher-frequency radio waves?

7.2 Planck's Quantum Theory

Have you ever looked at the wires in a toaster as they heat up? Although you cannot see the metal atoms in the toaster wires, as electricity flows through them, their atoms gain energy and then emit it as radiation. First the wires emit a slight amount of heat that you can feel (infrared radiation). As the wire gets hotter, it begins to glow, initially emitting red light, and then orange (Figure 7.3a). If a wire gets very hot, it appears almost white. Figure 7.3b shows the spectrum of a typical heated object.

In trying to explain the nature of these emissions from hot objects, late nineteenth-century scientists assumed that vibrating atoms in a hot wire caused the emission of electromagnetic vibrations (light waves). According to the laws of classical physics, the

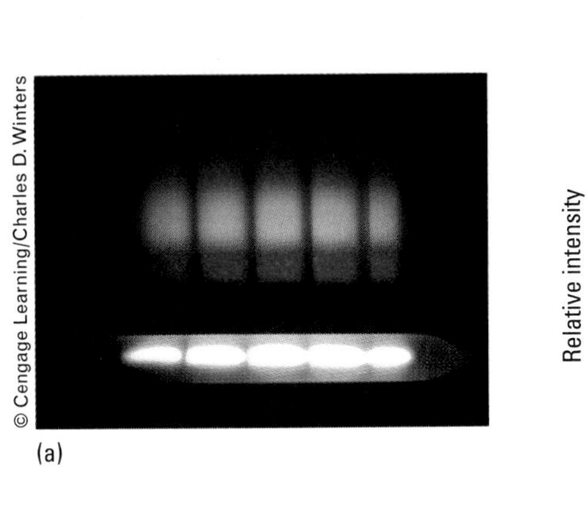

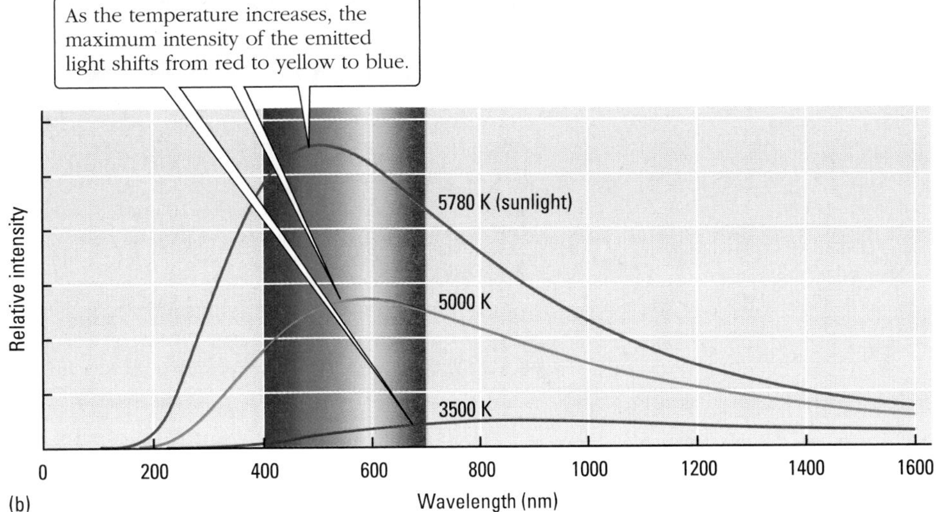

Figure 7.3 Heated objects and temperature. (a) A heated filament. The color of the hot filament and its continuous spectrum change as the temperature is increased. (b) The spectrum of radiation given off by a heated object. At very high temperatures, the heated object becomes "white hot" as all wavelengths of visible light become almost equally intense.

light waves could have any frequency along a continuously varying range. Using such laws, however, the scientists were unable to predict the experimentally observed spectrum.

In 1900, Max Planck (1858–1947) offered an explanation for the spectrum of a heated body. His ideas contained the seeds of a revolution in scientific thought. Planck made what was at that time an incredible assertion, one that departed dramatically from the laws of classical physics. He stated that when an atom in a hot object emits radiation, it does so only in packets having a minimum amount of energy. That is, there must be a small packet of energy such that no smaller quantity can be emitted, just as an atom is the smallest packet of an element. Planck called this packet of energy a **quantum.** He further asserted that the energy of a quantum is proportional to the frequency of the radiation according to the equation

A quantum is the smallest possible unit of a distinct quantity—for example, the smallest possible unit of energy for electromagnetic radiation of a given frequency.

$$E_{\text{quantum}} = h\nu_{\text{radiation}}$$

The proportionality constant h is called **Planck's constant** in his honor; it has a value of 6.626×10^{-34} J·s and relates the frequency of radiation to its energy *per quantum.*

Orange light has a measured frequency of $4.80 \times 10^{14}\ \text{s}^{-1}$. The energy of *one* quantum of orange light is therefore

$$E = h\nu = (6.626 \times 10^{-34}\ \text{J} \cdot \text{s})(4.80 \times 10^{14}\ \text{s}^{-1}) = 3.18 \times 10^{-19}\ \text{J}$$

The theory based on Planck's work is called the **quantum theory.** By using his quantum theory, Planck was able to calculate results that agreed very well with the experimentally measured spectra of heated objects, spectra for which the laws of classical physics had no explanation.

Max Planck won the 1918 Nobel Prize in Physics for his quantum theory.

PROBLEM-SOLVING EXAMPLE 7.2 Calculating Quantum Energies

In the stratosphere, ultraviolet radiation with a frequency of $1.36 \times 10^{15}\ \text{s}^{-1}$ can break C—Cl bonds in chlorofluorocarbons (CFCs), which can lead to stratospheric ozone depletion. Calculate the energy per quantum of this radiation.

Answer 9.01×10^{-19} J

Strategy and Explanation According to Planck's quantum theory, the energy and frequency of radiation are related by $E = h\nu$. Substituting the values for h and ν into the equation, followed by multiplying and canceling the units, gives the correct answer.

$$E = h\nu = (6.626 \times 10^{-34}\ \text{J}\cdot\text{s})(1.36 \times 10^{15}\ \text{s}^{-1}) = 9.01 \times 10^{-19}\ \text{J}$$

PROBLEM-SOLVING PRACTICE 7.2

Which has more energy,

(a) One quantum of microwave radiation or one quantum of ultraviolet radiation?

(b) One quantum of blue light or one quantum of green light?

There is a very important relationship between the energy and wavelength of a quantum of radiation. Since $E = h\nu$ and $\nu = c/\lambda$, then

$$E = \frac{hc}{\lambda}$$

where h is Planck's constant, c is the velocity of light (2.998×10^8 m/s), and λ is the wavelength of the radiation. Note that energy and wavelength are inversely proportional: ***The energy per quantum of radiation increases as the wavelength gets shorter.*** For example, red light and blue light have wavelengths of 656.3 nm and 434.1 nm, respectively. Because of its shorter wavelength, blue light has a higher energy per quantum than red light. We can calculate their different energies by applying the previous equation.

$$E_{\text{red}} = \frac{(6.626 \times 10^{-34}\ \text{J}\cdot\text{s})(2.998 \times 10^8\ \text{m/s})}{(656.3\ \text{nm})(1\ \text{m}/10^9\ \text{nm})} = 3.027 \times 10^{-19}\ \text{J}$$

$$E_{\text{blue}} = \frac{(6.626 \times 10^{-34}\ \text{J}\cdot\text{s})(2.998 \times 10^8\ \text{m/s})}{(434.1\ \text{nm})(1\ \text{m}/10^9\ \text{nm})} = 4.576 \times 10^{-19}\ \text{J}$$

ESTIMATION Turning on the Light Bulb

You turn the switch and a living room lamp comes on. Assume that the lamp emits only yellow light (585 nm), producing 20 J of energy in one second. Approximately how many quanta of yellow light are given off by the lamp in that time?

To find out, we first determine the energy of each 585-nm quantum by using the relationship $E = \frac{hc}{\lambda}$. We approximate by rounding the value of Planck's constant to 7×10^{-34} J·s and that of c, the speed of light, to 3×10^8 m/s. The wavelength of the quanta we round to approximately 600×10^{-9} m; remember we have to convert nanometers to meters: 1 nm = 1×10^{-9} m. Therefore, the energy, E, of each quantum is

$$\frac{(7 \times 10^{-34}\ \text{J}\cdot\text{s})\ (3 \times 10^8\ \text{m/s})}{600 \times 10^{-9}\ \text{m}}$$

which is about 4×10^{-19} J per quantum. Using this value and the total energy generated in one second, we can calculate the number of quanta of yellow light emitted in that time.

$$\text{Number of quanta} = \frac{\text{total energy}}{\text{energy of each quantum}} = \frac{20\ \text{J}}{4 \times 10^{-19}\ \text{J/quantum}} = 5 \times 10^{19}\ \text{quanta}$$

a considerable number, just slightly less than the number of milliliters of water (6×10^{18} mL) in the Atlantic Ocean.

Visit this book's companion website at **www.cengage.com/chemistry/moore** to work an interactive module based on this material.

EXERCISE 7.3 Energy, Wavelength, and Drills

A drilling method uses a microwave drill rather than a mechanical one. The apparatus contains a powerful source of 2.45-GHz microwave radiation. Calculate the wavelength and the energy of the microwaves emitted by such an apparatus.

The Photoelectric Effect

When a theory can accurately predict experimental results, the theory is usually regarded as useful. Planck's quantum theory was not widely accepted at first because of its radical assertion that energy is quantized. The quantum theory of electromagnetic energy was firmly accepted only after Planck's quanta were used by Albert Einstein to explain another phenomenon, the photoelectric effect.

In the early 1900s it was known that certain metals exhibit a **photoelectric effect**: They emit electrons when illuminated by light of certain wavelengths (Figure 7.4a). For each photosensitive metal there is a threshold wavelength below which no photoelectric effect is observed. For example, cesium (Cs) metal emits electrons when illuminated by red light, whereas some metals require yellow light, and others require ultraviolet light to cause emission of electrons. The differences occur because each color of light has a different wavelength and thus a different energy per quantum. Red

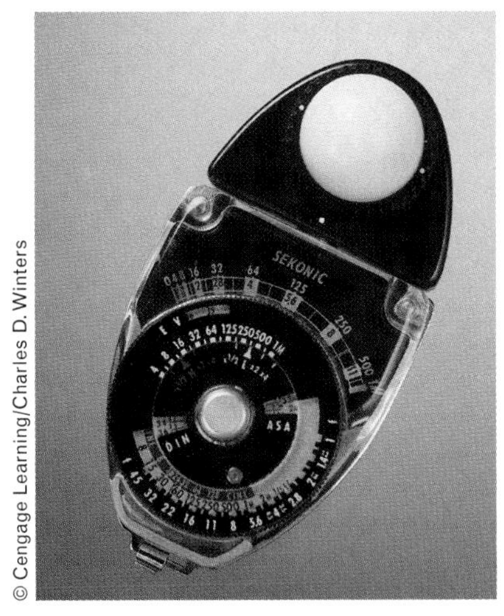

An application of photoelectric effect. The photoelectric effect is used in a light meter's sensors to determine exposure for photographic film.

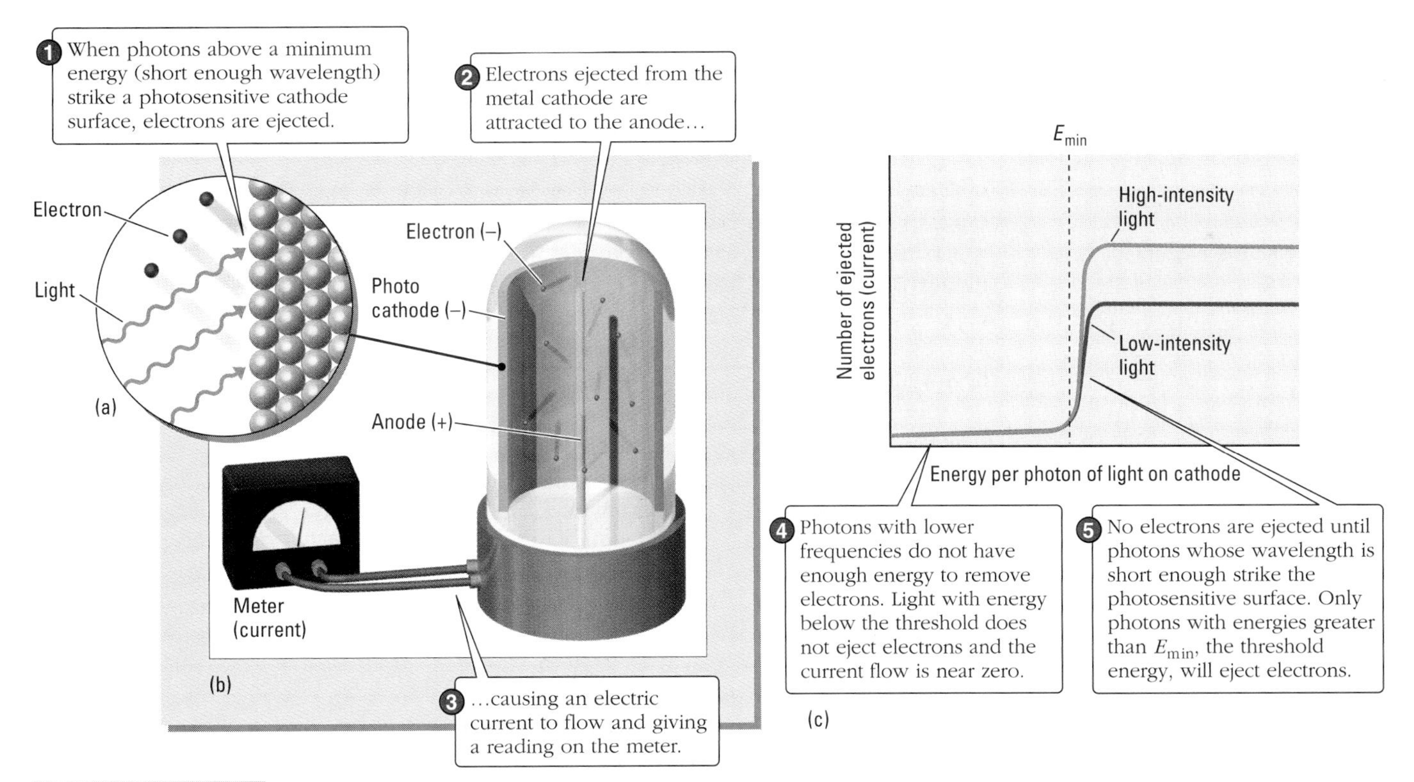

1 When photons above a minimum energy (short enough wavelength) strike a photosensitive cathode surface, electrons are ejected.
2 Electrons ejected from the metal cathode are attracted to the anode…
Electron
Light
(a)
Electron (–)
Photo cathode (–)
Anode (+)
Meter (current)
(b)
3 …causing an electric current to flow and giving a reading on the meter.
E_{min}
High-intensity light
Low-intensity light
Number of ejected electrons (current)
Energy per photon of light on cathode
4 Photons with lower frequencies do not have enough energy to remove electrons. Light with energy below the threshold does not eject electrons and the current flow is near zero.
5 No electrons are ejected until photons whose wavelength is short enough strike the photosensitive surface. Only photons with energies greater than E_{min}, the threshold energy, will eject electrons.
(c)

Active Figure 7.4 **The photoelectric effect.** (a) A photoelectric surface. (b) A photoelectric cell. Only photons with energies greater than the threshold energy will eject electrons from the photosensitive surface. (c) Ejected electrons and energy. Photons with lower energy (left side of Figure 7.4c) do not have enough energy to remove electrons. Note that the *number* of electrons ejected (the current) depends on the light's intensity, not its energy. Visit this book's companion website at **www.cengage.com/chemistry/moore** to test your understanding of the concepts in this figure.

light, for example, has a lower energy (longer wavelength) than yellow light or ultraviolet light. Light with energy that is below the threshold needed to eject electrons from the metal will not cause an electric current to flow, *no matter how bright the light.* Figure 7.4b illustrates how the electrons ejected from the photosensitive cathode are attracted to the anode in the photoelectric cell, causing an electric current to flow, resulting in a display on the meter. Figure 7.4c shows how an electric current suddenly increases when light of sufficient energy per quantum (short enough wavelength) shines on a photosensitive metal.

Albert Einstein explained these experimental observations by assuming that Planck's quanta were *massless* "particles" of light. These became known as **photons** instead of quanta. That is, light could be described as a stream of photons that had particle-like properties as well as wavelike properties. To remove one electron from a photosensitive metal surface requires a certain minimum quantity of energy; we call it E_{min}. Since each photon has an energy given by $E = h\nu$, only photons whose E is greater than E_{min} will have enough energy to knock an electron loose. Thus, no electrons are ejected until photons with high enough frequencies (and short enough wavelengths) strike the photosensitive surface. Photons with lower frequencies (left side of Figure 7.4c) do not have enough energy to remove electrons. This means that if a photosensitive metal requires photons of green light to eject electrons from its surface, then yellow light, red light, or light of any other lower frequency (longer wavelength) will not have sufficient energy to cause the photoelectric effect with that metal. This brilliant deduction about the quantized nature of light and how it relates to light's interaction with matter won Einstein the Nobel Prize for Physics in 1921.

A common misconception is that Einstein won the Nobel Prize for his theory of relativity.

The energies of photons are important for practical reasons. Photons of ultraviolet light can damage skin, while photons of visible light cannot. We use sunblocks containing molecules that selectively absorb ultraviolet photons to protect our skin from solar UV radiation. X-ray photons are even more energetic than ultraviolet photons and can disrupt molecules at the cellular level, causing genetic damage, among other effects. For this reason we try to limit our exposure to X-rays even more than we limit our exposure to ultraviolet light.

Look again at Figure 7.4c. Notice how a higher-intensity light source causes a higher photoelectric current. Higher intensities of ultraviolet light (or more time of exposure) can cause greater damage to the skin than lower intensities (or less exposure time). The same holds true for other high-energy forms of electromagnetic radiation, such as X-rays.

In the quantum theory the intensity of radiation is proportional to the number of photons. The total energy is proportional to the number of photons times the energy per photon.

Developments such as Einstein's explanation of the photoelectric effect led eventually to acceptance of what is referred to as the *dual nature* of light. Depending on the experimental circumstances, visible light and all other forms of electromagnetic radiation appear to have either "wave" or "particle" characteristics. However, both ideas are needed to fully explain light's behavior. Classical wave theory fails to explain the photoelectric effect. But it does explain quite well the *refraction,* or bending, of light by a prism and the diffraction of light by a diffraction grating, a device with a series of parallel, closely spaced small slits. When waves of light pass through such adjacent narrow slits, the waves are scattered so that the emerging light waves spread out, a phenomenon called *diffraction.* The semicircular waves emerging from the narrow slits can either amplify or cancel each other. Such behavior creates a diffraction pattern of dark and bright spots, as seen in Figure 7.5.

Refraction is the bending of light as it crosses the boundary from one medium to another—for example, from air to water. Diffraction is the bending of light around the edges of objects, such as slits in a diffraction grating.

It is important to realize, however, that this "dual nature" description arises because of our attempts to explain observations by using inadequate models. Light is not changing back and forth from being a wave to being a particle, but has a single consistent nature that can be described by modern quantum theory. The dual-nature description arises when we try to explain our observations by using our familiarity with classical models for "wave" or "particle" behavior.

Prior to Einstein's explanation of the photoelectric effect, classical physics considered light as being only wave-like.

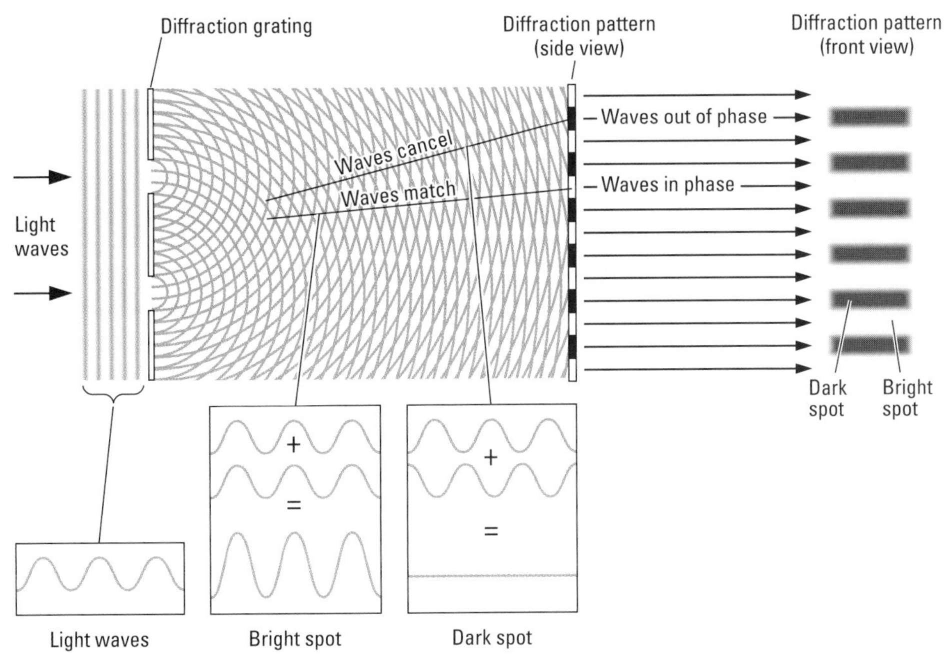

Figure 7.5 Diffraction pattern. Light waves passing through slits of a diffraction grating are scattered. When adjacent waves are in phase, they reinforce each other, producing a bright spot. When the waves are out of phase, they cancel each other and a dark spot occurs.

7.3 The Bohr Model of the Hydrogen Atom

Within just about a decade (1900–1911), three major discoveries were made about the atom and the nature of electromagnetic radiation. The first two, discussed in the previous section, were Max Planck's suggestion that energy was quantized and, five years later, Albert Einstein's application of the quantum idea to explain the photoelectric effect. The third came in 1911 when Ernest Rutherford demonstrated experimentally that atoms contain a dense positive core, the nucleus, surrounded by electrons (⇐ *p. 38*). Niels Bohr linked these three powerful ideas when, in 1913, he used quantum theory to explain the behavior of the electron in a hydrogen atom. In doing so, Bohr developed a mathematical model to explain how excited hydrogen atoms emit or absorb only certain wavelengths of light, a phenomenon that had been unexplained for nearly a century. To understand what Bohr did, we turn first to the visible spectrum.

The various colors of light emitted by neon-type signs are due to emissions from different gases.

The spectrum of white light, such as that from the sun or an incandescent light bulb, consists of a rainbow of colors, as shown in Figure 7.6. This **continuous spectrum** contains light of all wavelengths in the visible region.

If a high voltage is applied to a gaseous element at low pressure, the atoms absorb energy and are excited. The excited atoms then emit, as electromagnetic radiation, the energy that was previously absorbed. A neon advertising sign, in which the neon atoms emit red-orange light, makes commercial use of the light from excited neon atoms. When light from such a source passes through a prism onto a white surface, only a few colored lines are seen in a **line emission spectrum,** characteristic of the element. The line spectra of the visible light emitted by excited atoms of hydrogen, mercury, and neon are shown in Figure 7.7.

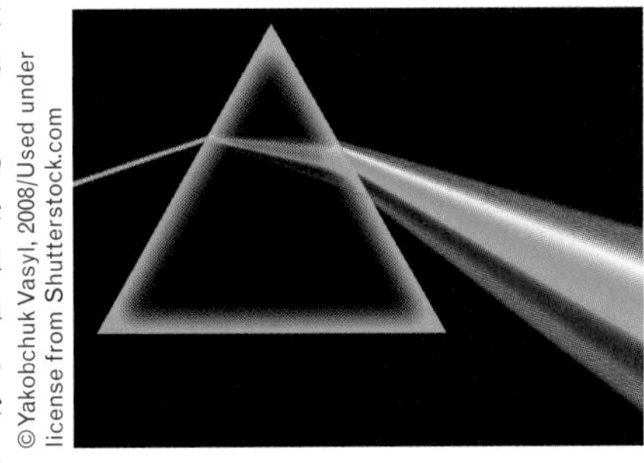

Figure 7.6 A continuous spectrum from white light. When white light is passed through slits to produce a narrow beam and then refracted by a glass prism, the various colors blend smoothly into one another.

Hydrogen has the simplest line emission spectrum. It was first studied in the 1880s. At that time, lines in the visible region of the spectrum and their wavelengths were accurately measured (Figure 7.7). Other series of lines in the ultraviolet region ($\lambda < 400$ nm) and in the infrared region ($\lambda > 700$ nm) were discovered subsequently.

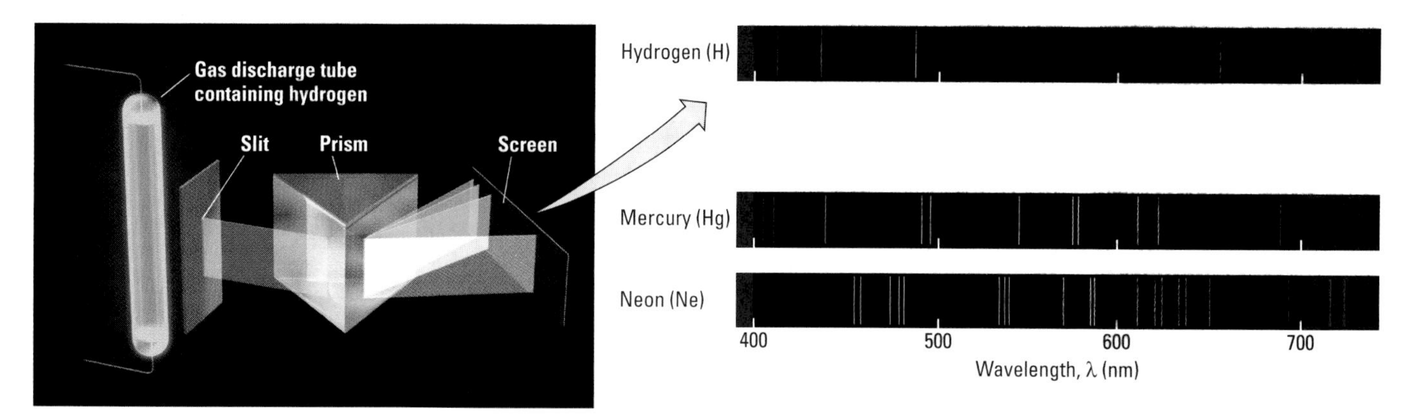

Figure 7.7 Line emission spectra of hydrogen, mercury, and neon. Excited gaseous elements produce characteristic spectra that can be used to identify the elements as well as to determine how much of an element is present in a sample.

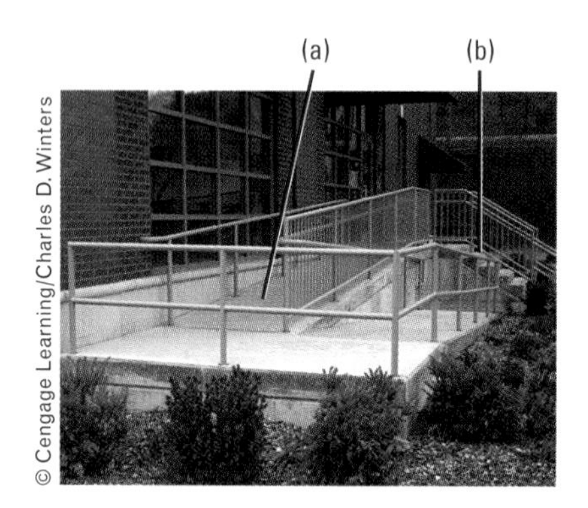

(a) A ramp (not quantized) (b) Stair steps (quantized)

Niels Bohr provided the first explanation of the line emission spectra of atoms by audaciously assuming that the single electron of a hydrogen atom moved in a circular orbit around its nucleus. He then related the energies of the electron to the radius of its orbit, but not by using the classic laws of physics. Such laws allowed the electron to have a wide range of energy. Instead, Bohr introduced quantum theory into his atomic model by invoking Planck's idea that energies are quantized. In the Bohr model, the electron could circle the nucleus in orbits of only certain radii, which correspond to specific energies. Thus, the energy of the electron is "quantized," and the electron is restricted to certain energy levels unless it gains or loses a certain amount of energy. Bohr referred to these energy levels as *orbits* and represented the energy difference between any two adjacent orbits as a specific quantity of energy. Each allowed orbit was assigned an integer, n, known as the **principal quantum number.**

In the Bohr model, the value of n for the possible orbits can be any integer from 1 to infinity (∞). *The energy of the electron and the size of its orbit increase as the value of n increases.* The orbit of lowest energy, $n = 1$, is closest to the nucleus, and the electron of a hydrogen atom is normally in this energy level. Any atom with its electrons arranged to give the lowest total energy is said to be in its **ground state.** Because the positive nucleus attracts the negative electron, an electron must absorb energy to go from a lower energy state (lower n value) to a higher energy state (higher n value), an excited state. Conversely, energy is emitted when the reverse process occurs—an electron going from a higher energy state to a lower energy state (Figure 7.8). Emission lines in the ultraviolet region result from electron transitions from a higher n value down to the $n = 1$ level. Emission lines in the visible region occur due to electron transitions from levels with values of $n > 2$ to the $n = 2$ level. Infrared radiation emission lines result from electron transitions from levels with $n > 3$ or 4 down to the $n = 3$ or $n = 4$ levels. The highest potential energy the electron in a hydrogen atom can have is when sufficient energy has been added to separate the electron from the proton. Bohr designated this situation as zero energy when $n = \infty$ (Figure 7.8). Thus, the potential energy of the electron in $n < \infty$ is negative, as shown in Figure 7.8.

For the hydrogen atom, visible photons are due to n_{high} ($n > 2$) to $n = 2$ transitions.
Ultraviolet photons are due to n_{high} ($n > 1$) to $n = 1$ transitions.
Infrared photons are due to n_{high} ($n > 3$) to $n = 3$ transitions.

Consider this analogy to Bohr's arbitrarily selecting zero as the highest energy for the electron in a hydrogen atom. A book resting on a table is arbitrarily designated as having zero potential energy. As the book falls from the table to a chair, to a stool, and then to the floor, it loses energy so that when it is on the floor, its potential energy is negative with respect to when it was on the table or the chair or the stool. Such is also the case for an electron going from a higher energy level to a lower energy level. The electron must lose energy. Correspondingly, the electron's energy in all its allowed energy states within the atom must be less than zero; that is, the energy must be negative.

When the electron in a hydrogen atom absorbs a quantized amount of energy and moves to an orbit with $n > 1$, the atom is said to be in an **excited state** in which it has

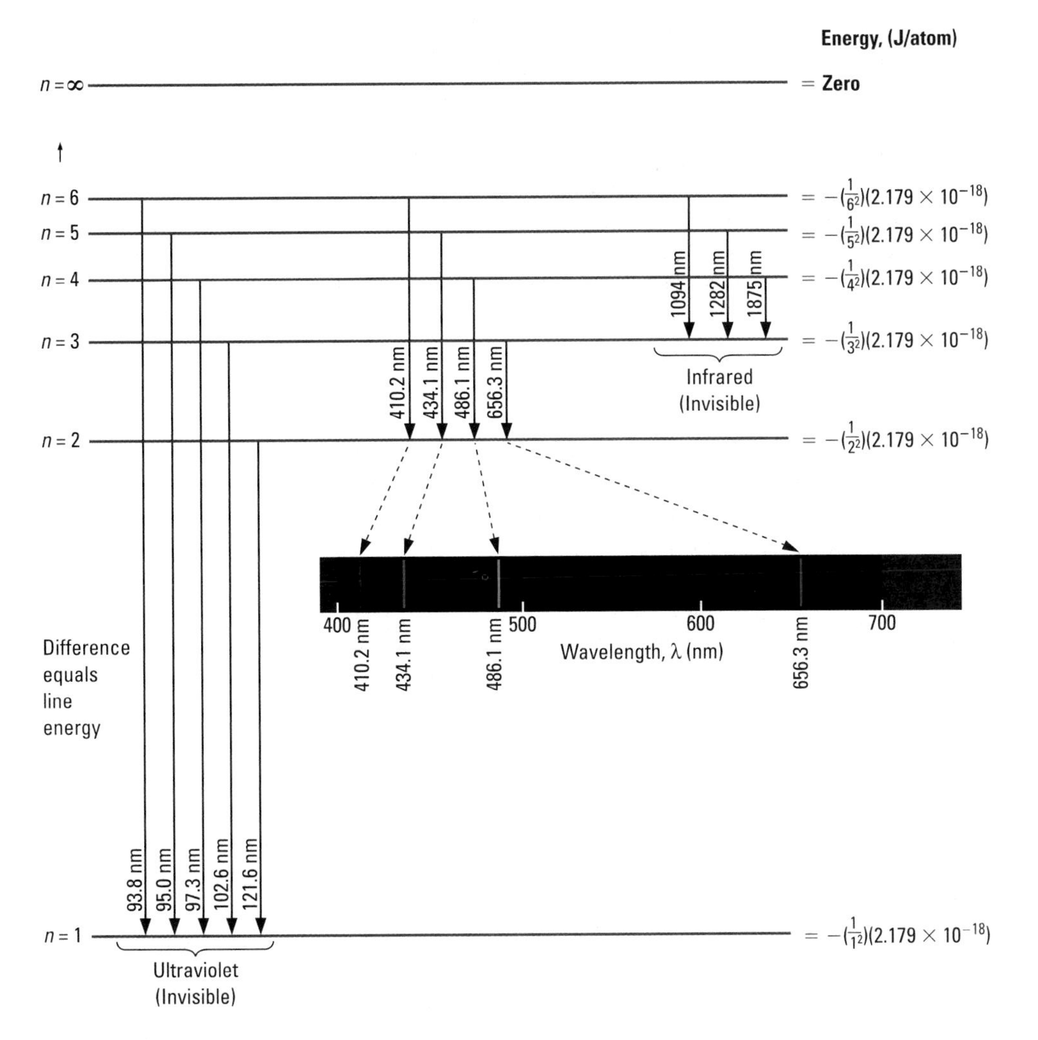

Figure 7.8 Electron transitions in an excited H atom. The lines in the ultraviolet region all result from transitions to the $n = 1$ level. Transitions from levels with values of $n > 2$ to the $n = 2$ level occur in the visible region. Lines in the infrared region result from the transitions from levels with values of $n > 3$ or 4 to the $n = 3$ or 4 levels (only the series for transitions to the $n = 3$ level is shown here).

The Bohr orbit model is analogous to a set of stairs in which the higher stairs are closer together. As you move up and down the stairs, you can stop at any step, *but not between steps* because the steps are quantized; only whole steps are allowed.

more energy than in its ground state. Any excited state of any atom is unstable. When the electron returns to the ground state from an excited state, the electron emits the same quantity of energy it gained when it moved from the ground state to the excited state. The emitted energy is given off as photons of light having an energy of hc/λ. The energy of an emitted photon corresponds to the *difference* between two energy levels of the atom (Figure 7.8). According to Bohr, the light forming the lines in the emission spectrum of hydrogen (Figure 7.7) comes from electrons in hydrogen atoms moving from higher energy orbits to lower energy orbits closer to the nucleus (Figure 7.8).

Electron transitions in the visible region of the spectrum are responsible for the impressive displays of color in fireworks.

PROBLEM-SOLVING EXAMPLE 7.3 Electron Transitions

Four possible electron transitions in a hydrogen atom are given below.

	n_{initial}	n_{final}
(1)	2	5
(2)	5	3
(3)	7	2
(4)	4	6

(a) Which transition(s) represent a loss of energy for the atom?
(b) For which transition does the atom gain the greatest quantity of energy?

(c) Which transition corresponds to emission of the greatest quantity of energy?

Answer

(a) (2) and (3) (b) (1) (c) (3)

Strategy and Explanation First, consider the initial and final energy levels, and then whether the initial level is greater or less than the final one.

(a) Energy is emitted when electrons fall from a higher n level to a lower n level. The $n_7 \rightarrow n_2$ and $n_5 \rightarrow n_3$ transitions release energy.

(b) Energy is gained when electrons go from a lower n level to a higher n level. The $n_2 \rightarrow n_5$ transition requires more energy than the $n_4 \rightarrow n_6$ one.

(c) The $n_7 \rightarrow n_2$ transition is the most energetic (Figure 7.8). It represents the greatest energy difference between energy levels.

PROBLEM-SOLVING PRACTICE 7.3

Identify two electron transitions in a hydrogen atom that would be of greater energy than any of those listed in Problem-Solving Example 7.3.

CONCEPTUAL

EXERCISE 7.4 Many Spectral Lines, but Only One Kind of Atom

The hydrogen atom contains only one electron, but there are several lines in its line emission spectrum (Figure 7.7). How does the Bohr theory explain this?

EXERCISE 7.5 Electron Transitions

In which of these transitions in the hydrogen atom is energy emitted and in which is it absorbed?

(a) $n = 3 \rightarrow n = 2$ (b) $n = 1 \rightarrow n = 4$

(c) $n = 5 \rightarrow n = 3$ (d) $n = 6 \rightarrow n = 2$

Through his calculations, Bohr found that the allowed energies of the electron in a hydrogen atom are restricted by n, the principal quantum number, according to the equation

$$E = -\left(\frac{1}{n^2}\right)2.179 \times 10^{-18}\,\text{J} \qquad (n = 1, 2, 3, \ldots)$$

where the constant, 2.179×10^{-18} J, called the Rydberg constant, can be calculated from theory. The negative sign in the equation arises due to the choice Bohr made to select zero as the energy of a completely separated electron and proton ($n = \infty$). As a consequence of this choice, all energies for the electron in the atom are negative (Figure 7.8).

Division by infinity makes E equal to zero when $n = \infty$.

The energy of a ground-state hydrogen atom ($n = 1$) is

$$E = -\left(\frac{1}{n^2}\right)2.179 \times 10^{-18}\,\text{J} = -\left(\frac{1}{1^2}\right)2.179 \times 10^{-18}\,\text{J} = -2.179 \times 10^{-18}\,\text{J}$$

Think of the atom as the system and everything else as the surroundings (⟵ p. 184).

The difference in energy, ΔE, when the electron moves from its initial state, E_i, to its final state, E_f, can be calculated as $\Delta E = E_f - E_i$. When an electron moves from a higher energy state to a lower energy state, one with a lower n value, energy is emitted and ΔE is negative. When an electron moves from a lower energy state to a higher one, energy must be absorbed (ΔE is positive). These energy differences can be expressed in relation to the change in principal energy levels, where n_i and n_f are the initial and final energy states of the atom, respectively.

$$\Delta E = E_f - E_i = -2.179 \times 10^{-18}\text{ J}\left(\frac{1}{n_f^2} - \frac{1}{n_i^2}\right)$$

If $n_f > n_i$, energy is absorbed; if $n_f < n_i$, energy is emitted.

In the Bohr model, only certain frequencies of light can be absorbed or emitted by an atom. This is because the energy of the photon absorbed or emitted must be the same as the difference in energy, ΔE, of the two energy levels between which the electron moves. If ΔE is positive, then the photon must have transferred its energy to the atom (the photon must have been absorbed). If ΔE is negative, then the photon must have received energy from the atom (the photon must have been emitted). Planck's quantum theory showed that the energy of a photon equals Planck's constant times the frequency,

$$E = h\nu = \frac{hc}{\lambda}$$

This allows us to calculate the frequency and wavelength of the photon absorbed or emitted. Notice that the energy of the photon is always positive, but the sign of the energy change of the atom indicates whether the photon is absorbed (positive change in E_{atom}) or emitted (negative change in E_{atom}).

For example, the frequency of light absorbed in the $n = 2$ to $n = 3$ transition for a hydrogen atom can be calculated by calculating the change in energy of the atom

$$\Delta E = E_f - E_i = -2.179 \times 10^{-18}\text{ J}\left(\frac{1}{n_f^2} - \frac{1}{n_i^2}\right)$$

$$= -2.179 \times 10^{-18}\text{ J}\left(\frac{1}{3^2} - \frac{1}{2^2}\right) = -2.179 \times 10^{-18}\text{ J}\left(\frac{1}{9} - \frac{1}{4}\right)$$

$$= -2.179 \times 10^{-18}\text{ J}(0.1111 - 0.2500) = 3.026 \times 10^{-19}\text{ J}$$

The positive energy change indicates that the atom has absorbed a photon whose energy is 3.026×10^{-19} J. Therefore the frequency can be calculated as

$$\nu = \frac{E_{photon}}{h} = \frac{3.026 \times 10^{-19}\text{ J}}{6.626 \times 10^{-34}\text{ J s}} = 4.567 \times 10^{14}\text{ s}^{-1}$$

The wavelength (λ) of the light absorbed can be obtained from its frequency by using the relationship $\lambda = c/\nu$, where c is the velocity of light (2.998×10^8 m/s). For the $n = 2$ to $n = 3$ transition,

$$\lambda = \frac{2.998 \times 10^8\text{ m/s}}{4.567 \times 10^{14}\text{ s}^{-1}} = 6.565 \times 10^{-7}\text{ m} = 656.5\text{ nm}$$

This is in the red region of visible light (Figure 7.1, *p. 222*).

There is exceptional agreement between the experimentally measured wavelengths and those calculated by the Bohr theory (Table 7.2). Thus, Niels Bohr had tied the unseen (the atom) to the seen (the observable lines of the hydrogen emission spectrum)—a fantastic achievement!

Spectroscopy is the science of measuring spectra. Many kinds of spectroscopy have emerged since the first studies of simple line spectra. Some spectral measurements are done for quantitative analytical purposes; others are done to determine molecular structure.

PROBLEM-SOLVING EXAMPLE 7.4 Electron Transitions

(a) Calculate the frequency and wavelength (nm) corresponding to the $n = 2$ to $n = 5$ transition in a hydrogen atom.

(b) In what region of the electromagnetic spectrum does this transition occur?

AIP Emilio Segre Visual Archives, Physics Today Collection

Niels Bohr 1885–1962

Born in Copenhagen, Denmark, Bohr worked with J. J. Thomson and Ernest Rutherford in England, where he began to develop the ideas that led to the publication of his explanation of atomic spectra. He received the Nobel Prize in Physics in 1922 for this work.

As the director of the Institute of Theoretical Physics in Copenhagen, Bohr was a mentor to many young physicists, seven of whom later received Nobel Prizes for their studies in physics or chemistry, including Werner Heisenberg, Wolfgang Pauli, and Linus Pauling.

Table 7.2 Agreement Between Bohr's Theory and the Lines of the Hydrogen Emission Spectrum*

Changes in Energy Levels	Wavelength Predicted by Bohr's Theory (nm)	Wavelength Determined from Laboratory Measurement (nm)	Spectral Region
$2 \rightarrow 1$	121.6	121.7	Ultraviolet
$3 \rightarrow 2$	656.5	656.3	Visible red
$5 \rightarrow 2$	434.3	434.1	Visible blue
$4 \rightarrow 3$	1876	1876	Infrared

*These lines are typical; other lines could be cited as well, with equally good agreement between theory and experiment. The unit of wavelength is the nanometer (nm), 10^{-9} m.

Answer

(a) $6.906 \times 10^{14}\ s^{-1}$; 434.1 nm (b) Visible region

Strategy and Explanation

(a) We first calculate the change in energy for the $n = 2$ to $n = 5$ transition using the equation given above. Using the calculated ΔE, we can calculate the frequency and convert it into wavelength.

$$\Delta E = E_f - E_i = -2.179 \times 10^{-18}\,J\left(\frac{1}{n_f^2} - \frac{1}{n_i^2}\right)$$

$$= -2.179 \times 10^{-18}\,J\left(\frac{1}{5^2} - \frac{1}{2^2}\right) = -2.179 \times 10^{-18}\,J\left(\frac{1}{25} - \frac{1}{4}\right)$$

$$= -2.179 \times 10^{-18}\,J\,(0.04000 - 0.02500) = 4.576 \times 10^{-19}\,J$$

The frequency can be calculated.

$$\nu = \frac{E_{photon}}{h} = \frac{4.576 \times 10^{-19}\,J}{6.626 \times 10^{-34}\,J\,s} = 6.906 \times 10^{14}\,s^{-1}$$

Convert the frequency to wavelength.

$$\lambda = \frac{c}{\nu} = \frac{2.998 \times 10^{8}\,m/s}{6.906 \times 10^{14}\,s^{-1}} = 4.341 \times 10^{-7}\,m = 434.1\,nm$$

(b) Checking Figure 7.8 shows that electron transitions between any higher n level and the $n = 2$ level emit radiation in the visible region ($\lambda > 400$ nm; $\nu < 7.5 \times 10^{14}\ s^{-1}$).

PROBLEM-SOLVING PRACTICE 7.4

(a) Calculate the frequency and the wavelength of the line for the $n = 6$ to $n = 4$ transition.

(b) Is this wavelength longer or shorter than that of the $n = 7$ to $n = 4$ transition?

CONCEPTUAL **EXERCISE 7.6 Conversions**

Show that the value of the Rydberg constant, 2.179×10^{-18} J, which applies to the emission or absorption of a single photon, is equivalent to 1312 kJ/mol photons.

CHEMISTRY IN THE NEWS

Light Up Your Money

Technological advances in digital printing and graphic arts have led to an increase in counterfeiting of U.S. paper currency, particularly bills worth \$20 or more. The U.S. Treasury Department's Bureau of Engraving and Printing, which produces the banknotes, recently introduced new features to the banknotes to stymie would-be counterfeiters. Among the new features on the authentic bills are inks that shift color when a bill is tilted and the insertion of a narrow plastic security strip that glows when exposed to ultraviolet light. The ultraviolet light is absorbed by molecules in the bills, exciting the molecules to higher energy after which they emit orange (\$10 bills) or green (\$50 bills) light.

AP Photo/Mary Godleski

Source: Rovner, S. L. "Counterfeiting Countermeasures," *Chemical & Engineering News,* June 11, 2007; pp. 30–34.

7.4 Beyond the Bohr Model: The Quantum Mechanical Model of the Atom

The Bohr atomic model was accepted almost immediately. Bohr's success with the hydrogen atom soon led to attempts by him and others to extend the same model to more complex atoms. Before long it became apparent, however, that line spectra for elements other than hydrogen had more lines than could be explained by the simple Bohr model. A very different approach was needed to explain electron behavior in atoms or ions with more than one electron. The new approach again took a radical departure from classical physics.

In 1924, the young physicist Louis de Broglie posed the question: *If light can be viewed in terms of both wave and particle properties, why can't particles of matter, such as electrons, be treated the same way?* And so de Broglie proposed the revolutionary idea that electrons could have wavelike properties. The wavelength (λ) of the electron (or any other particle) would depend on its mass (m) and its velocity (v) according to the relationship

$$\lambda = \frac{h}{mv}$$

where h is Planck's constant. The product of $m \times v$ for any object, including electrons, is the **momentum** of the object. Note that particles of very small mass, such as the electron (mass = 9.11×10^{-28} g), have wavelengths that are measurable. Objects of ordinary size, such as a tennis ball (mass = 56.7 g), have wavelengths too short to be observed.

© Aflo Foto Agency/Photolibrary

A served tennis ball. Even when the ball is served at 125 mi/h, the wavelength of the tennis ball is too short to be observed.

PROBLEM-SOLVING EXAMPLE 7.5 Tennis Balls and Electrons

At Wimbledon, tennis serves routinely reach more than 100 mi/h. Compare the de Broglie wavelength (nm) of an electron moving at a velocity of 5.0×10^6 m/s with that of a tennis ball traveling at 56.0 m/s (125 mi/h). Masses: electron = 9.11×10^{-31} kg; tennis ball = 0.0567 kg.

Answer The wavelength of the electron is *much* longer than that of the tennis ball: electron = 0.15 nm; tennis ball = 2.09×10^{-25} nm.

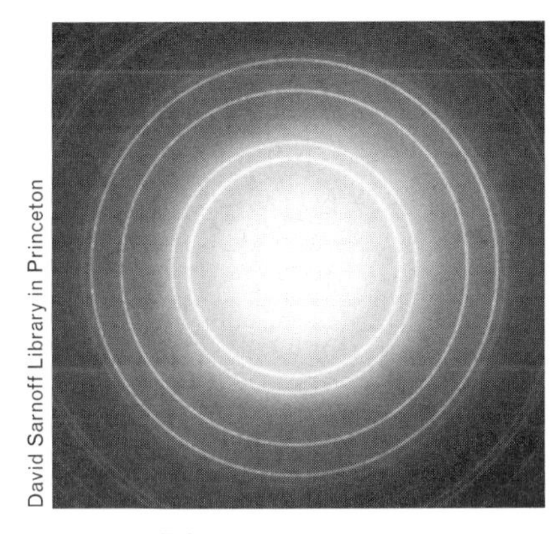

David Sarnoff Library in Princeton

Figure 7.9 Electron diffraction pattern obtained from aluminum foil.

Strategy and Explanation We can substitute the mass and velocity into the de Broglie wave equation to calculate the corresponding wavelength. Planck's constant, h, is

$$6.626 \times 10^{-34}\ \text{J}\cdot\text{s, and } 1\ \text{J} = \frac{1\ \text{kg}\cdot\text{m}^2}{\text{s}^2} \text{ so that } h = 6.626 \times 10^{-34}\ \text{kg}\cdot\text{m}^2\ \text{s}^{-1}.$$

For the electron:

$$\lambda = \frac{6.626 \times 10^{-34}\ \text{kg}\cdot\text{m}^2\ \text{s}^{-1}}{(9.11 \times 10^{-31}\ \text{kg})(5.0 \times 10^6\ \text{m/s})} = 1.5 \times 10^{-10}\ \text{m} \times \frac{1\ \text{nm}}{10^{-9}\ \text{m}} = 0.15\ \text{nm}$$

For the tennis ball:

$$\lambda = \frac{6.626 \times 10^{-34}\ \text{kg}\cdot\text{m}^2\ \text{s}^{-1}}{(0.0567\ \text{kg})(56.0\ \text{m/s})} = 2.09 \times 10^{-34}\ \text{m} \times \frac{1\ \text{nm}}{10^{-9}\ \text{m}} = 2.09 \times 10^{-25}\ \text{nm}$$

The wavelength of the electron is in the X-ray region of the electromagnetic spectrum (Figure 7.1, ⇐ *p. 222*). The wavelength of the tennis ball is far too short to observe.

PROBLEM-SOLVING PRACTICE 7.5

Calculate the de Broglie wavelength of a neutron moving at 10% the velocity of light. The mass of a neutron is 1.67×10^{-24} g.

Many scientists found de Broglie's concept hard to accept, much less believe. But experimental support for de Broglie's hypothesis was soon produced. In 1927, C. Davisson and L. H. Germer, working at the Bell Telephone Laboratories, found that a beam of electrons is diffracted by planes of atoms in a thin sheet of metal foil (Figure 7.9) in the same way that light waves are diffracted by a diffraction grating. Since diffraction is readily explained by the wave properties of light, it followed that electrons also can be described by the equations of waves under some circumstances.

Bryn Lennon/Getty Images, Inc.

Al Bello/Allsport/Getty Images, Inc.

A speeding race car. (a) Photo taken at high shutter speed. (b) Photo taken at low shutter speed.

A few years after de Broglie's hypothesis about the wave nature of the electron, Werner Heisenberg (1901–1976) proposed the **uncertainty principle,** which states that *it is impossible to simultaneously determine the exact position and the exact momentum of an electron.* This limitation is not a problem for a macroscopic object because the energy of photons used to locate such an object does not cause a measurable change in the position or momentum of that object. However, the very act of measurement would affect the position and momentum of the electron because of its very small size and mass. High-energy photons would be required to locate the small electron; when such photons collide with the electron, the momentum of the electron would be changed. If lower-energy photons were used to avoid affecting the momentum, little information would be obtained about the location of the electron. Consider an analogy in photography. If you take a picture at a higher shutter speed setting of a car race, you get a clear picture of the cars but you can't tell how fast they are going or even whether they are moving. With a slow shutter speed, you can tell from the blur of the car images something about the speed and direction, but you have less information about where the car is.

The Heisenberg uncertainty principle illustrated another inadequacy in the Bohr model—its representation of the electron in the hydrogen atom in terms of well-defined orbits about the nucleus. In practical terms, the best we can do is to represent the *probability* of finding an electron of a given energy and momentum within a given space. This probability-based model of the atom is what chemists now use.

In 1926 Erwin Schrödinger (1877–1961) combined de Broglie's hypothesis with classic equations for wave motion. From these and other ideas he derived a new equation called the *wave equation* to describe the behavior of an electron in the hydrogen atom. Solutions to the wave equation are called *wave functions* and are represented by the Greek letter psi, ψ. Wave functions predict the allowed energy states of an electron and the probability of finding that electron in a given region of space. Only certain wave functions are satisfactory for an electron in an atom.

The wave function is a complex mathematical equation that has no direct physical meaning. However, the square of the wave function, ψ^2, can be represented in graphical form as a picture of the three-dimensional region of the atom where an electron with a given energy state is most likely to be found. That is, ψ^2 is related to the probability of finding the electron in a given region of space. This probability, known as the **electron density,** can be visualized as an array of dots, each dot representing the possible location of the electron with a given energy in relation to the nucleus. Figure 7.10a is an electron density distribution, sometimes called an electron cloud picture, for a ground state hydrogen atom. In such a plot, the density of the dots indicates the probability of the electron being there. The cloud of dots is more dense nearer the nucleus than farther from it, indicating that the probability of finding the electron is higher nearer the nucleus than farther from it. The electron-cloud density decreases as the distance from the nucleus increases, illustrating the probability of finding the electron as distance from the nucleus increases.

Electron density is also called probability density.

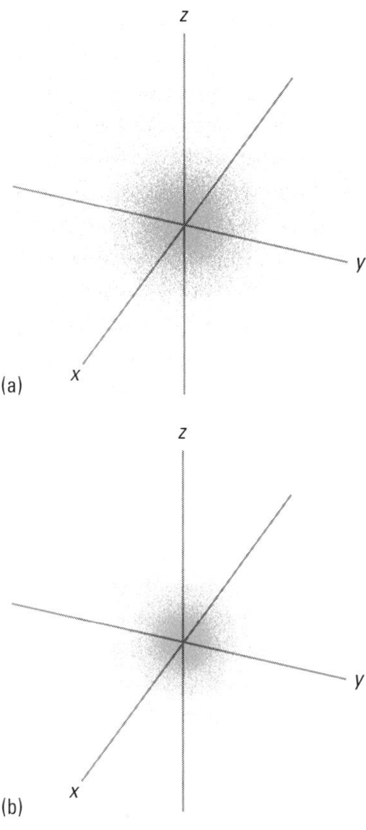

Figure 7.10 (a) Electron density distribution in a ground-state hydrogen atom. (b) A boundary surface.

Solving the Schrödinger wave equation results in a set of wave functions called **orbitals.** Each orbital contains information about the region of space where an electron of a given energy is most likely located because ψ^2 represents electron density. Another way to represent an electron is to draw a surface within which there is a 90% probability that the electron will be found. That is, nine times out of ten an electron will be somewhere inside such a **boundary surface;** there is one chance in ten that the electron will be outside of it (Figure 7.10b). A 100% probability is not chosen because such a surface would have no definite boundary. Consider that a typical dart board has a finite size, and normally, players in a dart game hit the board more than 90% of the time. But if you wanted to be certain that any player, no matter how far away, would be able to hit the board on 100% of his or her throws, then the board would have to be considerably larger. By similar reasoning, a boundary surface in which there would be a 100% probability of finding a given kind of electron would have to be infinitely large.

Note that an *orbital* (quantum mechanical model) is not the same as an *orbit* (Bohr model). In the quantum mechanical model, the principal quantum number, n, is a measure of the most probable distance of the electron from the nucleus, not the radius of a well-defined orbit.

We turn next to describing quantum numbers, which define the orbitals and energy states available to an electron in an atom. Following that, we will relate quantum numbers to the shapes of atomic orbitals.

7.5 Quantum Numbers, Energy Levels, and Atomic Orbitals

In their quantum mechanical calculations, Schrödinger and others found that a set of three integers, called **quantum numbers**—the principal quantum number n and two others, ℓ and m_ℓ—are needed to describe the three-dimensional coordinates of an electron's motion in the orbitals in a hydrogen atom. The need for a fourth quantum number, m_s, was identified in subsequent work by others. Thus, a set of four quantum numbers—n, ℓ, m_ℓ, and m_s—is used to denote the energy and the shape of the electron cloud for each electron.

We can apply quantum numbers to electrons in any atom, not just hydrogen. The quantum numbers, their meanings, and the quantum number notations used for atomic orbitals are given in the following subsections.

First Quantum Number, n: Principal Energy Levels

The principal quantum number, n, is the same n found from spectra (⟸ p. 230).

The first quantum number, n, the principal quantum number, is the most important one in determining the energy of an electron. The quantum number n has only integer values, starting with 1:

$$n = 1, 2, 3, 4, \ldots$$

The value of n corresponds to a **principal energy level** so that an electron is in the first principal energy level when $n = 1$, in the second principal level when $n = 2$, and so on. As n increases, the energy of the electron increases as well, and the electron, on average, is farther away from the nucleus and is less tightly bound to it.

Second Quantum Number, ℓ: Orbital Shapes

Each principal energy level is also known as an electron **shell,** a collection of orbitals with the same principal quantum number value, n. When $n = 1$, there is only one kind of orbital possible. When $n = 2$, two kinds of orbitals are possible; likewise, three kinds are possible when $n = 3$, four when $n = 4$, and so on.

Within each principal energy level (shell) are **subshells.** The $n = 1$ principal energy level has only one subshell; all other energy levels ($n > 1$) have more than one subshell. The subshells are designated by the second quantum number, ℓ (sometimes referred to as the *azimuthal quantum number*). *The shape of the electron cloud corresponding to an orbital is determined by the value of* ℓ.

A subshell is one or more orbitals with the *same n and* ℓ quantum numbers. Thus, the value of ℓ is associated with an n value. The value of ℓ is an integer that ranges from zero to a maximum of $n - 1$:

$$\ell = 0, 1, 2, 3, \ldots, (n - 1)$$

According to this relationship, when $n = 1$, then ℓ must be zero. This indicates that in the first principal energy level ($n = 1$) there is only one subshell, with an ℓ value of 0. The second principal energy level ($n = 2$) has two subshells—one with an ℓ value of 0 and a second with an ℓ value of 1. Continuing in this manner,

$$n = 3; \quad \ell = 0, 1, \text{or } 2 \quad \text{(three subshells)}$$

$$n = 4; \quad \ell = 0, 1, 2 \text{ or } 3 \quad \text{(four subshells)}$$

As n increases, two things happen: (1) orbitals become larger and (2) there is an increase in the number and types of orbitals within a given level. ***The number of orbital types within a principal energy level equals n.*** Thus, in the $n = 3$ level, there are three different types of subshells: $\ell = 0$; $\ell = 1$; and $\ell = 2$. *Atomic orbitals with the same n and* ℓ *values are in the same subshell.*

Rather than using ℓ values, subshells are more commonly designated by letters: *s*, *p*, *d*, or *f*. The first four subshells are known as the *s* subshell, *p* subshell, *d* subshell, and *f* subshell.

The letters *s*, *p*, *d*, and *f* derived historically from spectral lines called *s*harp, *p*rincipal, *d*iffuse, and *f*undamental.

ℓ value	0	1	2	3
Subshell	*s*	*p*	*d*	*f*

A number (the n value) and a letter (*s*, *p*, *d*, or *f*) are used to designate specific subshells.

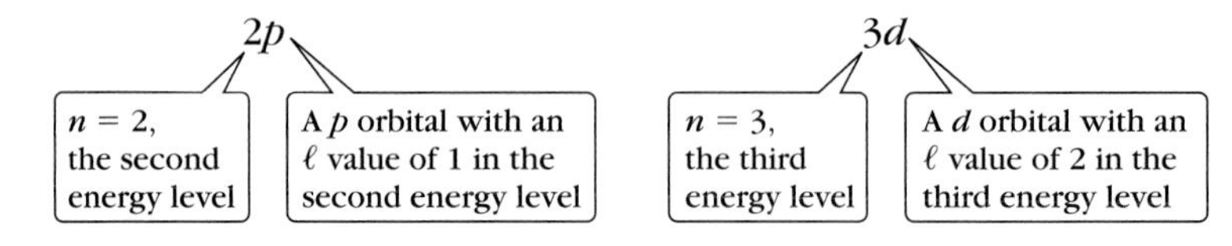

For the first four principal energy levels, these designations are as follows.

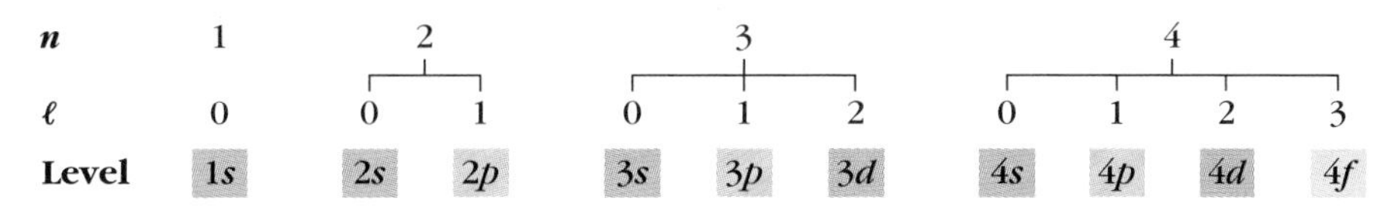

n	1	2		3			4			
ℓ	0	0	1	0	1	2	0	1	2	3
Level	1*s*	2*s*	2*p*	3*s*	3*p*	3*d*	4*s*	4*p*	4*d*	4*f*

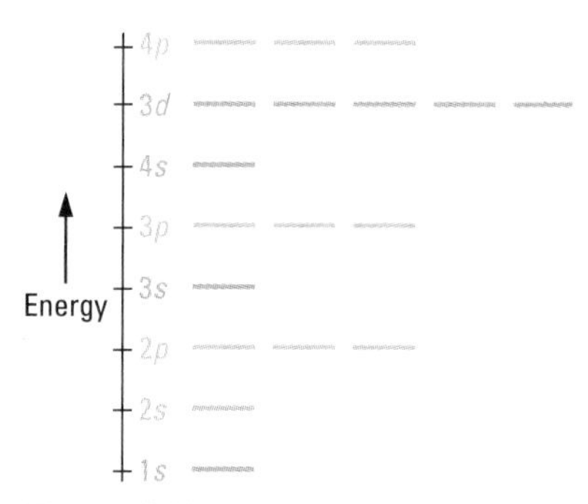

Figure 7.11 The ordering of orbital energy levels in a many-electron atom from the 1*s* level through the 4*p* sublevels. Orbitals in different sublevels of the same level have different energy.

For atoms with two or more electrons, orbitals in different subshells but with the same n value have different energies. The energies of subshells within a given principal energy level always increase in the order $ns < np < nd < nf$. Consequently, a 3*p* subshell has a higher energy than a 3*s* subshell but less energy than a 3*d* subshell, which would have the highest energy in the $n = 3$ level. Figure 7.11 shows the relative energies of orbitals in many-electron atoms through the 4*p* orbitals.

EXERCISE 7.7 Subshell Designations

Give the subshell designation for an electron with these quantum numbers.
(a) $n = 5, \ell = 2$ (b) $n = 4, \ell = 3$ (c) $n = 6, \ell = 1$

EXERCISE 7.8 Subshell Designations

Explain why 3*f* is an incorrect subshell designation. Why is $n = 2, \ell = 2$ incorrect for a subshell?

Third Quantum Number, m_ℓ: Orientation of Atomic Orbitals

The magnetic quantum number, m_ℓ, can have any integer value between ℓ and $-\ell$, including zero. Thus,

$$m_\ell = \ell, (\ell - 1), \ldots, +1, 0, -1, \ldots -(\ell - 1), -\ell$$

For an *s* subshell, which has an ℓ value of zero, m_ℓ has only one value—zero. Therefore, an *s* subshell, regardless of its n value—1*s*, 2*s*, and so on—contains only one orbital, whose m_ℓ value is zero. For a *p* subshell, ℓ equals 1, and so m_ℓ can be +1, 0, or −1. This means that within each *p* subshell there are three *different* types of orbitals: one with $m_\ell = +1$, another with $m_\ell = 0$, and a third with $m_\ell = -1$. In general, for a subshell of quantum number ℓ, there is a total of $2\ell + 1$ orbitals within that subshell. Orbitals within the same *subshell* have essentially the same energy.

Table 7.3 summarizes the relationships among n, ℓ, and m_ℓ. ***The total number of orbitals in a shell equals n^2.*** For example, an $n = 3$ shell has a total of nine orbitals: one 3*s* + three 3*p* + five 3*d*.

In terms of an atomic orbital, the quantum number

- n relates to the orbital's size.
- ℓ relates to the orbital's shape.
- m_ℓ relates to the orbital's orientation.

The m_ℓ value, in conjunction with the ℓ value, is related to the shape and orientation of an orbital in space. The *s* orbitals are spherical; their size enlarges as n increases.

For each n level from $n = 2$ on up, there are three *p* orbitals. The three *p* orbitals are dumbbell-shaped and oriented at right angles to each other, with maximum electron density directed along either the *x*-, *y*-, or *z*-axis. They are sometimes designated as p_x, p_y, and p_z orbitals (Figure 7.14, p. 244.).

Table 7.3 Relationships Among n, ℓ, and m_ℓ for the First Four Principal Energy Levels

n Value	ℓ Value	Subshell Designation	m_ℓ Values	Number of Orbitals in Subshell, $2\ell + 1$	Total Number of Orbitals in Shell, n^2
1	0	1*s*	0	1	1
2	0	2*s*	0	1	
2	1	2*p*	1, 0, −1	3	4
3	0	3*s*	0	1	
3	1	3*p*	1, 0, −1	3	
3	2	3*d*	2, 1, 0, −1, −2	5	9
4	0	4*s*	0	1	
4	1	4*p*	1, 0, −1	3	
4	2	4*d*	2, 1, 0, −1, −2	5	
4	3	4*f*	3, 2, 1, 0, −1, −2, −3	7	16

PROBLEM-SOLVING EXAMPLE 7.6 Quantum Numbers, Subshells, and Atomic Orbitals

Consider the $n = 4$ principal energy level.

(a) Without referring to Table 7.3, predict the number of subshells in this level.

(b) Identify each of the subshells by its number and letter designation (as in 1*s*) and give its ℓ values.

(c) Use the $2\ell + 1$ rule to calculate how many orbitals each subshell has and identify the m_ℓ value for each orbital.

(d) What is the total number of orbitals in the $n = 4$ level?

Answer

(a) Four subshells

(b) 4*s*, 4*p*, 4*d*, and 4*f*; $\ell = 0, 1, 2$, and 3, respectively

(c) One 4*s* orbital, three 4*p* orbitals, five 4*d* orbitals, and seven 4*f* orbitals

(d) 16 orbitals

Strategy and Explanation

(a) There are n subshells in the nth level. Thus, the $n = 4$ level contains four subshells.

(b) The number refers to the principal quantum number, n; the letter is associated with the ℓ quantum number. The four sublevels correspond to the four possible ℓ values:

Sublevels	4*s*	4*p*	4*d*	4*f*
ℓ value	0	1	2	3

(c) There are a total of $2\ell + 1$ orbitals within a sublevel. Only one 4*s* orbital is possible ($\ell = 0$, so m_ℓ must be zero). There are three 4*p* orbitals ($\ell = 1$) with m_ℓ values of 1, 0, or −1. There are five 4*d* orbitals ($\ell = 2$) corresponding to the five allowed values for m_ℓ: 2, 1, 0, −1, and −2. There are seven 4*f* orbitals ($\ell = 3$), each with one of the seven permitted values of m_ℓ: 3, 2, 1, 0, −1, −2, and −3.

(d) The total number of orbitals in a level is n^2. Therefore, the $n = 4$ level has a total of 16 orbitals.

PROBLEM-SOLVING PRACTICE 7.6

(a) Identify the subshell with $n = 6$ and $\ell = 2$.

(b) How many orbitals are in this subshell?

(c) What are the m_ℓ values for these orbitals?

Fourth Quantum Number, m_s: Electron Spin

When spectroscopists more closely studied emission spectra of hydrogen and sodium atoms, they discovered that what were originally thought to be single lines were actually very closely spaced pairs of lines. In 1925 the Dutch physicists George Uhlenbeck and Samuel Goudsmit proposed that the line splitting could be explained by assuming that each electron in an atom can exist in one of two possible spin states. To visualize these states, consider an electron as a charged sphere rotating about an axis through its center (Figure 7.12). Such a spinning charge generates a magnetic field, so that each electron acts like a tiny bar magnet with north and south magnetic poles. Only two directions of spin are possible in relation to the direction of an external magnetic field—clockwise or counterclockwise. Spins in opposite directions produce oppositely directed magnetic fields, which result in two slightly different energies. This slight difference in energy splits the spectral lines into closely spaced pairs.

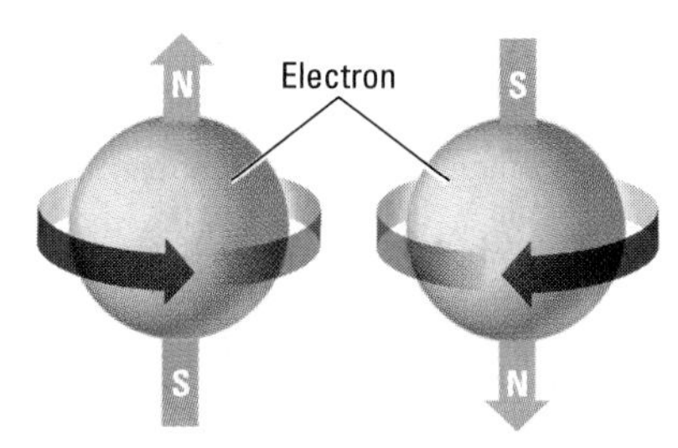

Figure 7.12 Spin directions. The electron can be pictured as though it were a charged sphere spinning about an axis through its center. The electron can have only two directions of spin; any other position is forbidden. Therefore, the spin of the electron is said to be quantized. One direction of spin corresponds with spin quantum number $m_s = +\frac{1}{2}$, the other with $m_s = -\frac{1}{2}$.

Thus, to describe an electron in an atom completely, a fourth quantum number, m_s, called the *spin quantum number,* is needed in conjunction with the other three quantum numbers.

The spin quantum number can have just one of two values: $+\frac{1}{2}$ or $-\frac{1}{2}$. Electrons are said to have *parallel* spins if they have the same m_s quantum number (both $+\frac{1}{2}$ or both $-\frac{1}{2}$). Electrons are said to be *paired* when they are in the same orbital and have opposite spins—one has an m_s of $+\frac{1}{2}$ and the other has a $-\frac{1}{2}$ value.

To make quantum theory consistent with experiment, Wolfgang Pauli stated in 1925 what is now known as the **Pauli exclusion principle:** ***No more than two electrons can occupy the same orbital in an atom, and these electrons must have opposite spins.*** This is equivalent to saying that no two electrons in the same atom can have the same set of four quantum numbers n, ℓ, m_ℓ, and m_s. For example, two electrons can occupy the same $3p$ orbital only if their spins are paired ($+\frac{1}{2}$ and $-\frac{1}{2}$). In such a case, the electrons would have the same first three quantum numbers, for example, 3, 1, +1, but would differ in their m_s values.

EXERCISE 7.9 Quantum Numbers

Give the set of four quantum numbers for each electron in a $3s$ orbital in a sodium atom.

CONCEPTUAL
EXERCISE 7.10 Orbitals and Quantum Numbers

Give sets of four quantum numbers for two electrons that are (a) in the same level and subshell, but in different orbitals, and (b) in the same level, but in different subshells and in different orbitals.

EXERCISE 7.11 Quantum Number Comparisons

Two electrons in the same atom have these sets of quantum numbers: electron$_a$: 3, 1, 0, $+\frac{1}{2}$; electron$_b$: 3, 1, −1, $+\frac{1}{2}$. Show that these two electrons are not in the same orbital. Which subshell are these electrons in?

The restriction that only two electrons can occupy a single orbital has the effect of establishing the maximum number of electrons for each principal energy level, n, as summarized in Table 7.4. Since each orbital can hold a maximum of two paired electrons, the general rule is that ***each principal energy level, n, can accommodate a maximum number of $2n^2$ electrons.*** For example, the $n = 2$ energy level has one s and three p orbitals, each of which can accommodate two paired electrons. Therefore,

The results expressed in this table were predicted by the Schrödinger theory and have been confirmed by experiment.

Table 7.4 Number of Electrons Accommodated in Electron Shells and Subshells

Electron Shell (n)	Number of Subshells Available ($=n$)	Number of Orbitals Available ($=2\ell + 1$)	Number of Electrons Possible in Subshell	Maximum Electrons for nth Shell ($=2n^2$)
1	*s*	1	2	2
2	*s*	1	2	
	p	3	6	8
3	*s*	1	2	
	p	3	6	
	d	5	10	18
4	*s*	1	2	
	p	3	6	
	d	5	10	
	f	7	14	32
5	*s*	1	2	
	p	3	6	
	d	5	10	
	f	7	14	
	*g**	9	18	50
6	*s*	1	2	
	p	3	6	
	d	5	10	
	*f**	7	14	
	*g**	9	18	
	*h**	11	22	72
7	*s*	1	2	

*These orbitals are not used in the ground state of any known element.

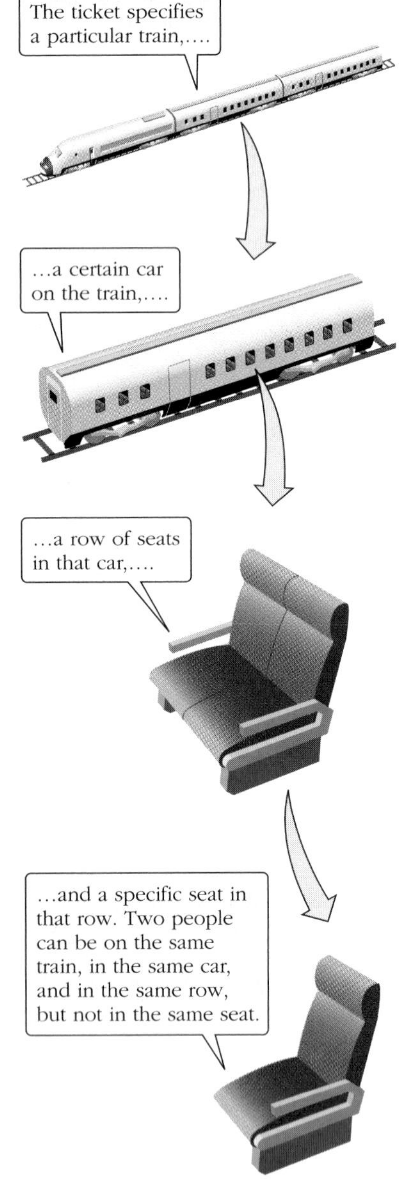
The ticket specifies a particular train,....
...a certain car on the train,....
...a row of seats in that car,....
...and a specific seat in that row. Two people can be on the same train, in the same car, and in the same row, but not in the same seat.

A quantum number analogy. A ticket for a reserved seat on a train is analogous to a set of four quantum numbers, n, ℓ, m_ℓ, and m_s. The ticket specifies a particular train, a certain car on the train, a row of seats in that car, and a specific seat in that row. The row is analogous to an atomic orbital, and the occupants of the row are like two electrons in the same orbital (same train, car, and row), but with opposite spins (different seats).

this level can accommodate a total of eight electrons (one pair in the 2*s* orbital and three pairs, one in each of the three 2*p* orbitals).

In summary,

- ***The principal energy level (shell) has a quantum number n = 1, 2, 3, ...***
- ***Within each principal energy level there are subshells equal in number to n and designated as the s, p, d, and f subshells.***
- ***The number of orbitals in each subshell is $2\ell + 1$: one s orbital ($\ell = 0$), three p orbitals ($\ell = 1$), five d orbitals ($\ell = 2$), and seven f orbitals ($\ell = 3$).***
- ***Within a principal energy level n there are n^2 orbitals.***
- ***Within a principal energy level n there is a maximum of $2n^2$ electrons.***

EXERCISE 7.12 Maximum Number of Electrons

(a) What is the maximum number of electrons in the $n = 3$ level? Identify the orbital of each electron.

(b) What is the maximum number of electrons in the $n = 5$ level? Identify the orbital of each electron. (*Hint:* *g* orbitals follow *f* orbitals.)

CONCEPTUAL
EXERCISE 7.13 *g* Orbitals

For $\ell = 4$, the letter designation is *g*. Using the same reasoning as was developed for *s, p, d,* and *f* orbitals, what should be the *n* value of the first shell that could contain *g* orbitals, and how many *g* orbitals would be in that shell?

7.6 Shapes of Atomic Orbitals

As noted in the previous section, the ℓ and m_ℓ quantum numbers relate to the shapes and spatial orientations of atomic orbitals. We now consider these in more detail.

s Orbitals ($\ell = 0$)

Electron density plots, such as that shown in Figure 7.10, are one way to depict atomic orbitals. In Figure 7.13a, the dot-density diagram of the 1*s* orbital of a hydrogen atom is shown with its electron density decreasing rapidly as the distance from the nucleus increases. The value r_{90} in the figure is the radius of the sphere in which the electron is found 90% of the time. Note from Figure 7.13a that the probability of finding a 1*s* electron is the same *in any direction* at the same distance from the nucleus. Thus, the 1*s* orbital is spherical.

In Figure 7.13a the electron density illustrates that the probability of finding the 1*s* electron in a hydrogen atom varies throughout space. This probability can be represented in another way known as a **radial distribution plot**, as shown in Figure 7.13b. In this case, the *y*-axis represents the probability of finding the electron at a given distance, *r*, from the nucleus (*x*-axis). The probability approaches zero as *r* approaches zero, that is, there is a very small probability of finding the electron at the nucleus. The greatest probability of finding the 1*s* electron in a ground-state hydrogen

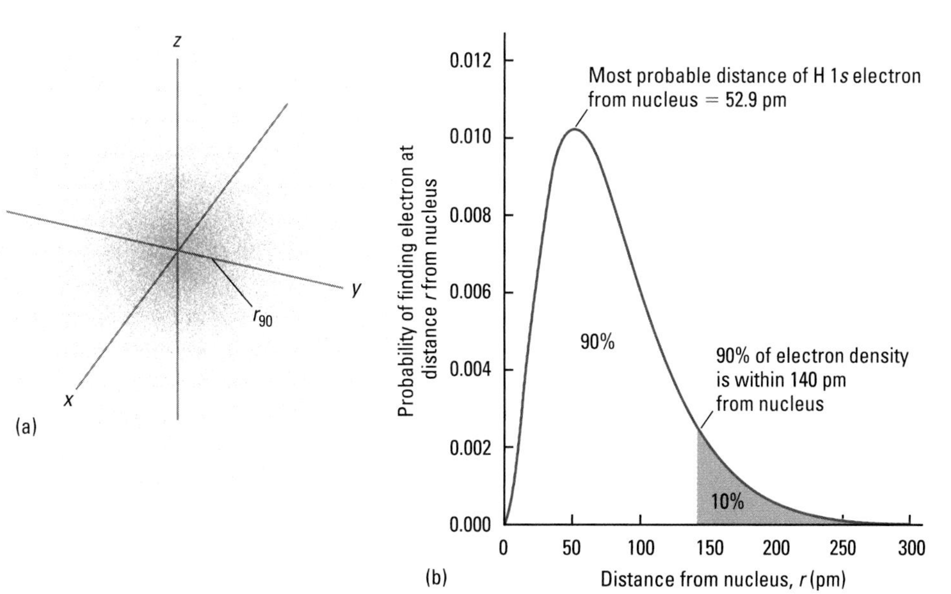

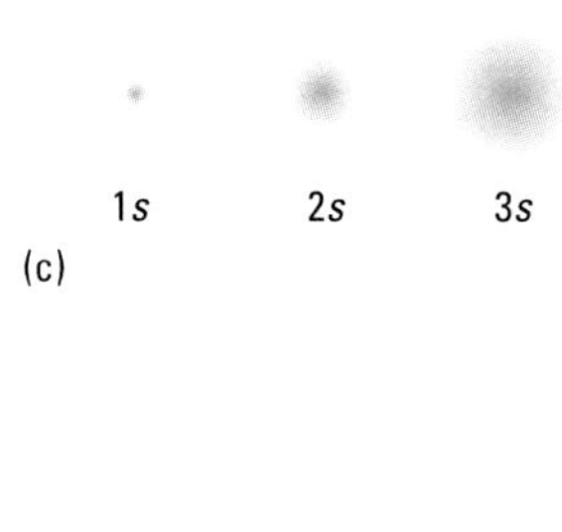

In Figure 7.13b, the *y*-axis probability values equal the square of the wave function, ψ^2, times 4π times the square of the distance *r*: $4\pi r^2\psi^2$.

Figure 7.13 Depicting *s* atomic orbitals. (a) An electron density illustration of a 1*s* orbital. The density of dots represents the probability of finding the electron in any region of the diagram. (b) A radial distribution plot as a function of distance from the nucleus for a 1*s* orbital of hydrogen. (c) The relative sizes of the boundary surfaces of *s* orbitals.

atom is at 0.0529 nm (52.9 pm) from the nucleus. Note from the figure that at larger distances the probability drops very close to zero, but does not quite become zero. This indicates that although it is extremely small, there is a finite probability of the electron being at a considerable distance from the nucleus.

As the n value increases, the shape of s orbitals remains the same; they are all spherical. Their sizes, however, increase as n increases, as shown in Figure 7.13c. For example, the boundary surface of a 3s orbital has a greater volume (and radius) than that of a 2s orbital, which is greater than that of a 1s orbital.

p Orbitals ($\ell = 1$)

Unlike s atomic orbitals, which are all spherical, the p atomic orbitals, those for which $\ell = 1$, are all dumbbell-shaped (Figure 7.14). In a p orbital, there are two lobes with electron density on either side of the nucleus. The p orbitals within a given subshell (same n and ℓ value) differ from each other in their orientation in space; each set of three p orbitals within the same subshell are mutually perpendicular to each other along the x-, y-, and z-axes. These orientations correspond to the three allowed m_ℓ values of +1, 0, and −1.

d Orbitals ($\ell = 2$)

Each energy level with $n \geq 3$ contains a subshell for which $\ell = 2$ consisting of five d atomic orbitals with corresponding m_ℓ values of +2, +1, 0, −1, and −2. The d orbitals consist of two different types of shapes and spatial orientations. Three of them (d_{xz}, d_{yz}, and d_{xy}) each have four lobes that lie in the plane of and between the designated x-, y-, or z-axes; the other pair—the $d_{x^2-y^2}$ and d_{z^2} orbitals—have their principal electron density along the designated axes. The shapes of five d orbitals, along with those of s and p orbitals, are illustrated in Figure 7.15.

We leave the discussion of f orbitals, which are very complex, to subsequent courses.

7.7 Atom Electron Configurations

The complete description of the orbitals occupied by all the electrons in an atom or ion is called its **electron configuration.** The periodic table can serve as a guide to determine the electron configurations of atoms. As you will see, the chemical similarities of elements in the same periodic table group are explained by the similar electron configurations of their atoms.

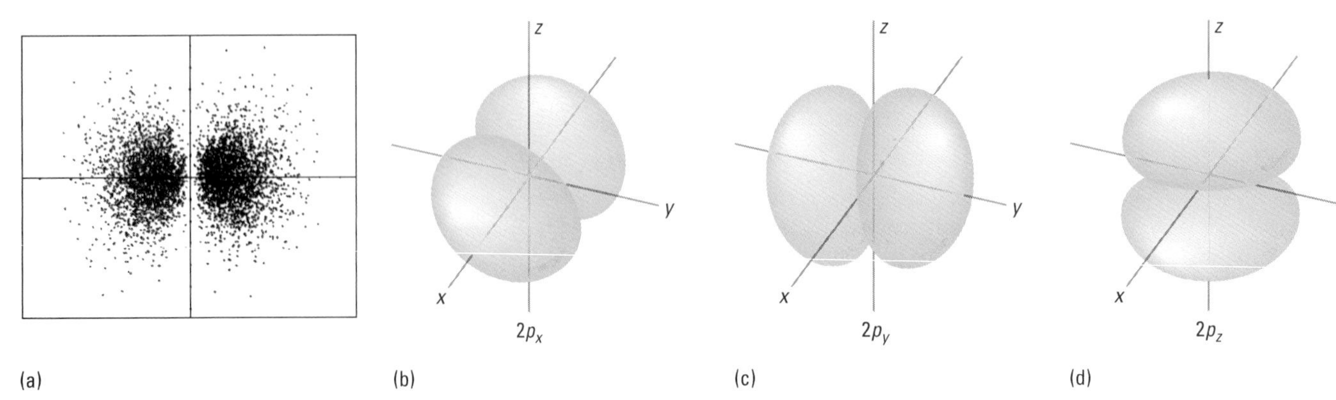

Figure 7.14 Depicting p atomic orbitals. (a) Electron density distribution of a 2p atomic orbital. (b), (c), and (d) Boundary surfaces for the set of three 2p orbitals. (From Zumdahl and Zumdahl, *Chemistry,* 6th ed., Fig. 7.14, p. 310. Copyright © 2002 South-Western, a part of Cengage Learning, Inc. Reproduced by permission. www.cengage.com/permissions.)

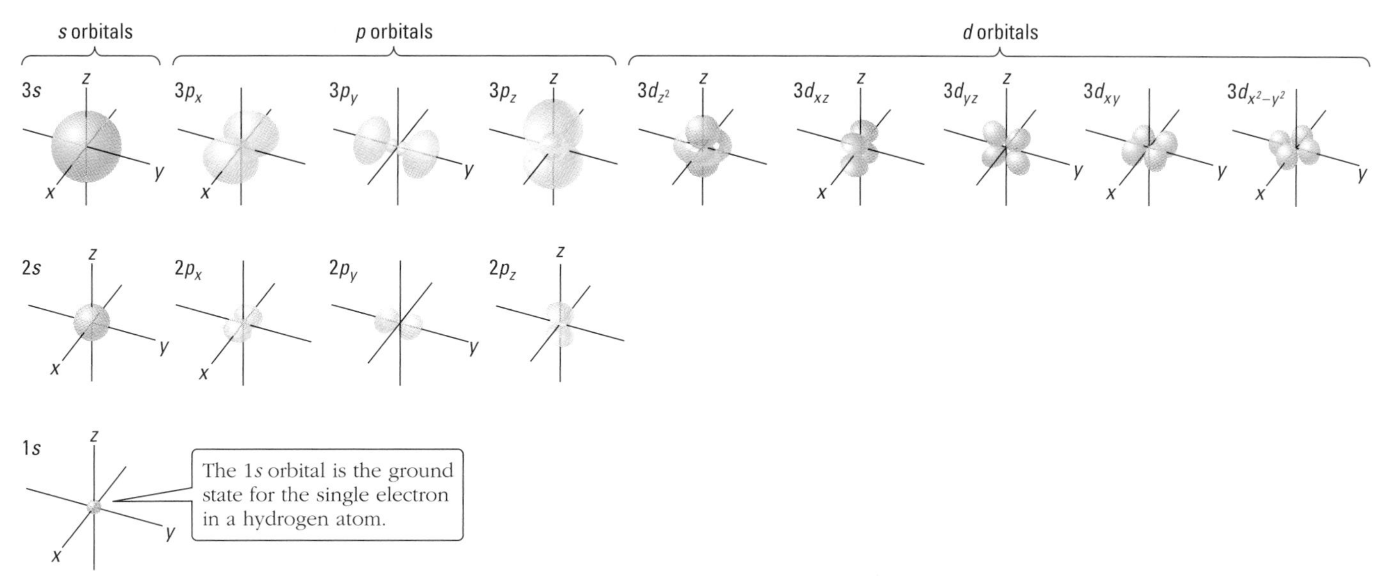

Figure 7.15 Depicting atomic orbitals. Boundary-surface diagrams for electron densities of 1*s*, 2*s*, 2*p*, 3*s*, 3*p*, and 3*d* atomic orbitals.

Electron Configurations of Main Group Elements

The atomic numbers of the elements increase in numerical order throughout the periodic table. As a result, atoms of a particular element each contain one more electron (and proton) than atoms of the preceding element. How do we know which shell and orbital each new electron occupies? An important principle for answering this question is the following: *For an atom in its ground state, electrons are found in the energy shells, subshells, and orbitals that produce the lowest energy for the atom.* In other words, electrons fill orbitals starting with the 1*s* orbital and work upward in the subshell energy order starting with $n = 1$.

To better understand how this filling of orbitals works, consider the experimentally determined electron configurations of the first ten elements, which are written in three different ways in Table 7.5—condensed, expanded, and orbital box diagram. Since electrons assigned to the $n = 1$ shell are closest to the nucleus and therefore lowest in energy, electrons are assigned to it first (H and He). At the left in Table 7.5 (p. 246), the occupied orbitals and the number of electrons in each orbital are represented by this notation:

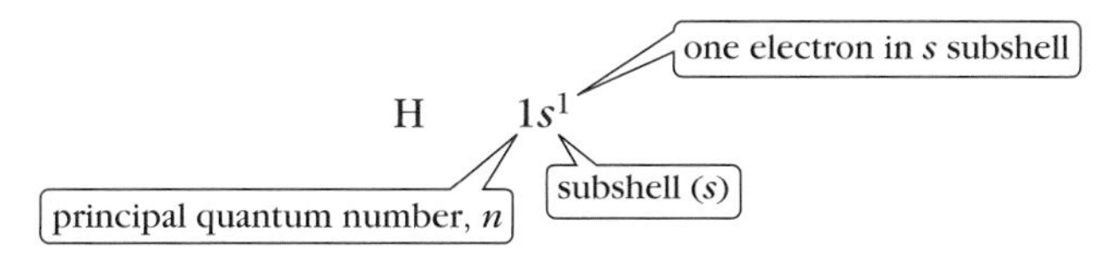

At the right in the table, each occupied orbital is represented by a box in which electrons are shown as arrows: ↑ for a single electron in an orbital and ⇅ for paired electrons in an orbital. The arrow represents spin direction.

In helium the two electrons are paired in the 1*s* orbital so that the lowest energy shell ($n = 1$) is filled. After the $n = 1$ shell is filled, electrons are assigned to the next lowest unoccupied energy level (Table 7.5), the $n = 2$ shell, beginning with lithium. This second shell can hold eight electrons ($2n^2$), and its orbitals are occupied sequentially in the eight elements from lithium to neon. Notice in the periodic table inside the front cover that these are the eight elements of the second period.

As happens in each principal energy level (and each period), the first two electrons fill the *s* orbital. In the second period this occurs in Li ($1s^22s^1$) and Be ($1s^22s^2$),

Table 7.5 Electron Configurations of the First Ten Elements

	Electron Configurations		Orbital Box Diagrams				
	Condensed	*Expanded*	1*s*	2*s*	2*p*		
H	$1s^1$		↑				
He	$1s^2$		↑↓				
Li	$1s^2 2s^1$		↑↓	↑			
Be	$1s^2 2s^2$		↑↓	↑↓			
B	$1s^2 2s^2 2p^1$		↑↓	↑↓	↑		
C	$1s^2 2s^2 2p^2$	$1s^2 2s^2 2p^1 2p^1$	↑↓	↑↓	↑	↑	
N	$1s^2 2s^2 2p^3$	$1s^2 2s^2 2p^1 2p^1 2p^1$	↑↓	↑↓	↑	↑	↑
O	$1s^2 2s^2 2p^4$	$1s^2 2s^2 2p^2 2p^1 2p^1$	↑↓	↑↓	↑↓	↑	↑
F	$1s^2 2s^2 2p^5$	$1s^2 2s^2 2p^2 2p^2 2p^1$	↑↓	↑↓	↑↓	↑↓	↑
Ne	$1s^2 2s^2 2p^6$	$1s^2 2s^2 2p^2 2p^2 2p^2$	↑↓	↑↓	↑↓	↑↓	↑↓

which has the 2*s* orbital filled. The next element is boron, B ($1s^2 2s^2 2p^1$), and the fifth electron goes into a 2*p* orbital. The three 2*p* orbitals are of equal energy, and it does not matter which 2*p* orbital is occupied first. Adding a second *p* electron for the next element, carbon ($1s^2 2s^2 2p^2$), presents a choice. Does the second 2*p* electron in the carbon atom pair with the existing electron in a 2*p* orbital, or does it occupy a 2*p* orbital by itself? It has been shown experimentally that both 2*p* electrons have the same spin. (The electrons are said to have parallel spins.) Hence, they must occupy different 2*p* orbitals; otherwise, they would violate the Pauli exclusion principle. The expanded electron configurations in the middle of Table 7.5 show the locations of the *p* electrons individually in the boron, carbon, and nitrogen atoms' 2*p* orbitals.

Because electrons are negatively charged particles, electron configurations where electrons have parallel spins also minimize electron–electron repulsions, making the total energy of the set of electrons as low as possible. **Hund's rule** summarizes how subshells are filled: The most stable arrangement of electrons in the same subshell has the maximum number of unpaired electrons, all with the same spin. ***Electrons pair only after each orbital in a subshell is occupied by a single electron.*** The general result of Hund's rule is that in *p, d,* or *f* orbitals, each successive electron enters a different orbital of the subshell until the subshell is half-full, after which electrons pair in the orbitals one by one. We can see this in Table 7.5 for the elements that follow boron—carbon, nitrogen, oxygen, fluorine, and neon.

Suppose you need to know the electron configuration of a phosphorus atom. Checking the periodic table (inside the front cover), you find that phosphorus has atomic number 15 and therefore has 15 electrons. The electron configuration can be derived by using the general relationships among periodic table position, shells (n), subshells, and electron configuration as summarized in Figure 7.16. Phosphorus is in the third period ($n = 3$). From Figure 7.16 we see that the $n = 1$ and $n = 2$ shells are filled in the first two periods, giving the first ten electrons in phosphorus the configuration $1s^2 2s^2 2p^6$ (the same as neon). The next two electrons are assigned to the 3*s* orbital, giving $1s^2 2s^2 2p^6 3s^2$ so far. The final three electrons have to be assigned to the 3*p* orbitals, so the electron configuration for phosphorus is

$$\text{P} \quad 1s^2 2s^2 2p^6 3s^2 3p^3$$

The electron configurations of all the elements are given in Appendix D. In that appendix, an abbreviated representation of the configurations called the **noble gas**

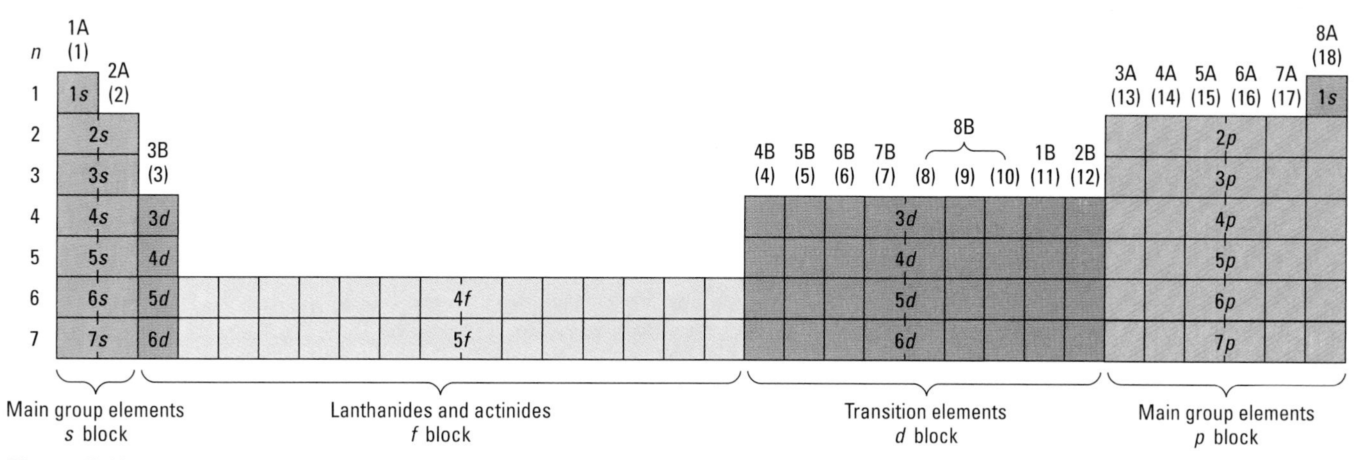

Figure 7.16 Electron configuration and the periodic table. Electron configurations can be generated by assigning electrons to the subshells shown here starting at H and moving through this table in atomic number order until the desired element is reached. The electron configurations in atomic number order are given in Appendix D.

notation is used in which the symbol of the preceding noble gas represents filled subshells. Using the noble gas notation, the electron configuration of phosphorus is [Ne] $3s^23p^3$.

According to Hund's rule, the three electrons in the $3p$ orbitals of the phosphorus atom must be unpaired. To show this, you can write the *expanded* electron configuration

$$\text{P} \quad 1s^22s^22p^22p^22p^23s^23p^13p^13p^1$$

or the orbital box diagram

P	1s	2s	2p	3s	3p
	↑↓	↑↓	↑↓ ↑↓ ↑↓	↑↓	↑ ↑ ↑

Thus, all the electrons are paired except for the three electrons in $3p$ orbitals. These three electrons each occupy different $3p$ orbitals, and they have parallel spins. The partial orbital box diagram of elements in Period 3 is given in Figure 7.17.

Atomic number/ element	Partial orbital box diagram (3s and 3p sublevels only) 3s	3p	Electron configuration	Noble gas notation
$_{11}$Na	↑	[ ][ ][ ]	$[1s^22s^22p^6]\,3s^1$	[Ne] $3s^1$
$_{12}$Mg	↑↓	[ ][ ][ ]	$[1s^22s^22p^6]\,3s^2$	[Ne] $3s^2$
$_{13}$Al	↑↓	[↑][ ][ ]	$[1s^22s^22p^6]\,3s^23p^1$	[Ne] $3s^23p^1$
$_{14}$Si	↑↓	[↑][↑][ ]	$[1s^22s^22p^6]\,3s^23p^2$	[Ne] $3s^23p^2$
$_{15}$P	↑↓	[↑][↑][↑]	$[1s^22s^22p^6]\,3s^23p^3$	[Ne] $3s^23p^3$
$_{16}$S	↑↓	[↑↓][↑][↑]	$[1s^22s^22p^6]\,3s^23p^4$	[Ne] $3s^23p^4$
$_{17}$Cl	↑↓	[↑↓][↑↓][↑]	$[1s^22s^22p^6]\,3s^23p^5$	[Ne] $3s^23p^5$
$_{18}$Ar	↑↓	[↑↓][↑↓][↑↓]	$[1s^22s^22p^6]\,3s^23p^6$	[Ne] $3s^23p^6$

Figure 7.17 Partial orbital diagrams for Period 3 elements.

At this point, you should be able to write electron configurations and orbital box diagrams for main group elements through Ca, atomic number 20, using the periodic table and Figure 7.16 to assist you.

PROBLEM-SOLVING EXAMPLE 7.7 Electron Configurations

Using only Figure 7.16 and the periodic table as a guide, give the complete electron configuration, orbital box diagram, and noble gas notation for vanadium, V.

Answer $1s^22s^22p^63s^23p^63d^34s^2$

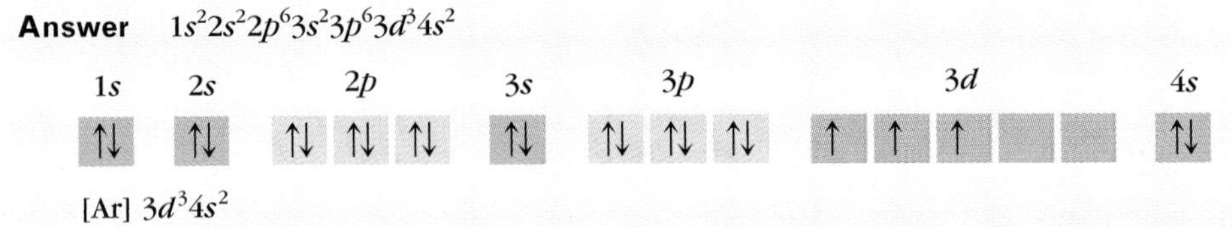

[Ar] $3d^34s^2$

Strategy and Explanation The periodic table shows vanadium in the fourth period with atomic number 23. Therefore, vanadium has 23 electrons. Applying Figure 7.16, the first 18 are represented by $1s^22s^22p^63s^23p^6$, the electron configuration of argon, Ar, the noble gas that precedes vanadium. The last five electrons in vanadium have the configuration $3d^34s^2$. The orbital box diagram shown above follows from the total electron configuration.

PROBLEM-SOLVING PRACTICE 7.7

(a) Write the electron configuration of silicon in the noble gas notation.
(b) Determine how many unpaired electrons a silicon atom has by drawing the orbital box diagram.

EXERCISE 7.14 Highest Energy Electrons in Ground State Atoms

Give the electron configuration of electrons in the highest occupied principal energy level (highest n) in a ground state chlorine atom. Do the same for a selenium atom.

We move now from considering the electron configuration of all electrons in an atom to those electrons of primary importance to the chemical reactivity of an element, its valence electrons, those in the outermost orbitals of the atom.

Valence Electrons

The chemical reactivity of an element is not proportional to the total number of electrons in each of its atoms. If such were the case, chemical reactivity would increase sequentially with increasing atomic number. Instead, chemically similar behavior occurs among elements within a group in the periodic table, and differs among groups. As early as 1902, Gilbert N. Lewis (1875–1946) proposed the idea that electrons in atoms might be arranged in shells, starting close to the nucleus and building outward. Lewis explained the similarity of chemical properties for elements in a given group by assuming that atoms of all elements in that group have the same number of electrons in their outer shell. These electrons are known as **valence electrons** for the main group elements. The electrons in the filled inner shells of these elements are called **core electrons.** For elements in the fourth and higher periods of Groups 3A to 7A, electrons in the filled d subshell are also core electrons. Using this notation, the core electrons of As are represented by [Ar] $3d^{10}$. Examples of core and valence electrons for Na, Si, and As are

Element	Core Electrons	Total Electron Configuration	Valence Electrons	Periodic Group
Na	$1s^2 2s^2 2p^6$, ([Ne])	[Ne] $3s^1$	$3s^1$	1A
Si	$1s^2 2s^2 2p^6$, ([Ne])	[Ne] $3s^2 3p^2$	$3s^2 3p^2$	4A
As	$1s^2 2s^2 2p^6 3s^2 3p^6$, ([Ar])	[Ar] $3d^{10} 4s^2 4p^3$	$4s^2 4p^3$	5A

For transition elements (*d*-block) and *f*-block elements, the *d* and *f* electrons of inner shells may also be valence electrons. (See p. 250.)

While teaching his students about atomic structure, Lewis used the element's symbol to represent the atomic nucleus together with the core electrons. He introduced the practice of representing the valence electrons as dots. The dots are placed around the symbol one at a time until they are used up or until all four sides are occupied; any remaining electron dots are paired with ones already there. The result is a **Lewis dot symbol.** Table 7.6 shows Lewis dot symbols for the atoms of the elements in Periods 2 and 3.

Main-group elements in Groups 1A and 2A are known as ***s*-block elements,** and their valence electrons are *s* electrons (ns^1 for Group 1A, ns^2 for Group 2A). Elements in the main groups at the right in the periodic table, Groups 3A through 8A, are known as ***p*-block elements.** Their valence electrons include the outermost *s* and *p* electrons. Notice in Table 7.6 how the Lewis dot symbols show that *in each A group the number of valence electrons is equal to the group number.*

Main-group elements are those in periodic table groups labeled with A, such as group 1A or 7A (Figure 7.16, p. 247).

PROBLEM-SOLVING EXAMPLE 7.8 Valence Electrons

(a) Using the noble gas notation, write the electron configuration for bromine. Identify its core and valence electrons.
(b) Write the Lewis dot symbol for bromine.

Answer
(a) [Ar] $3d^{10}4s^2 4p^5$ Core electrons are [Ar]$3d^{10}$; valence electrons are $4s^2 4p^5$.
(b) :B̈r̤·

Strategy and Explanation Bromine is element 35, a Group 7A element. Its first 18 electrons are represented by [Ar], as shown in Table 7.6. These, along with the ten 3*d* electrons, make up the core electrons. Because it is in Group 7A, bromine has seven valence electrons, as given by $4s^2 4p^5$ and shown in the Lewis dot symbol.

PROBLEM-SOLVING PRACTICE 7.8

Use the noble gas notation to write electron configurations and Lewis dot symbols for Se and Te. What do these configurations illustrate about elements in the same main group?

Table 7.6 Lewis Dot Symbols for Atoms

	1A ns^1	2A ns^2	3A ns^2np^1	4A ns^2np^2	5A ns^2np^3	6A ns^2np^4	7A ns^2np^5	8A ns^2np^6
Period 2	Li·	·Be·	·Ḃ·	·Ċ̣·	·N̈̇·	:Ö̇·	:F̤̈·	:N̤̈e:
Period 3	Na·	·Mg·	·Ȧl·	·Ṡi̤·	·P̈̇·	:S̈̇·	:C̤̈l·	:Ä̤r:

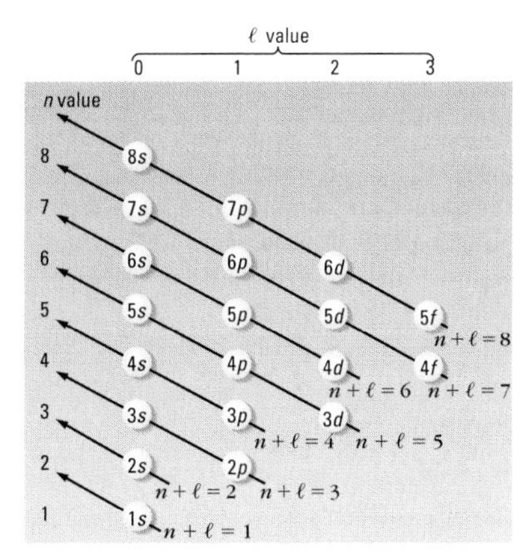

Figure 7.18 Order of subshell filling. As n increases, the energies of electron shells also increase. The energies of subshells within a shell increase with ℓ. Subshells are filled in increasing order of $n + \ell$. When two subshells have the same sum of $n + \ell$, the subshell with the lower n value is filled first.

Electron Configurations and Valence Electrons of Transition Elements

The **transition elements** are those in the B groups in Periods 4 through 7 in the middle of the periodic table. In these metallic elements, a d subshell is being filled as shown in Figure 7.16 (⇐ ***p. 247***). In addition to s and p electrons in the outermost shells of these atoms, electrons in an incompletely filled $(n - 1)d$ subshell are valence electrons.

In each period in which they occur, the transition elements are immediately preceded by two s-block elements. As shown in Figure 7.16, once the $4s$ subshell is filled, the next subshell filled is $3d$ (not $4p$, as you might expect). ***In general, $(n - 1)d$ orbitals are filled after ns orbitals and before filling of np orbitals begins.***

In atoms more complex than hydrogen (multielectron atoms), the subshell energies depend on both n and ℓ, not simply n. Subshells are filled sequentially in order of increasing $n + \ell$. When two subshells have the same $n + \ell$ value, the electrons are first assigned to the subshell with the lower n value (Figure 7.18). For example, the sequence for $3d$, $4s$, and $4p$ subshells for manganese, element 25, is that the $4s$ subshell $(n + \ell = 4 + 0 = 4)$ is filled before the $3d$ subshell $(n + \ell = 3 + 2 = 5)$, which fills before the $4p$ subshell $(n + \ell = 4 + 1 = 5)$. The electron configuration for manganese is [Ar] $3d^5 4s^2$. The $3d$ and $4p$ subshells have the same $n + \ell$ value, but because of its lower n value, the five electrons enter the $3d$ subshell before the $4p$ subshell. Manganese has seven valence electrons.

Occupancy of d orbitals begins with the $n = 3$ level and therefore with the first transition metal, scandium, which has the configuration [Ar] $3d^1 4s^2$. After scandium comes titanium with [Ar] $3d^2 4s^2$ and vanadium with [Ar] $3d^3 4s^2$. We would expect the configuration of the next element, chromium, to be [Ar] $3d^4 4s^2$, but that turns out to be incorrect. Based on spectroscopic and magnetic measurements, the correct configuration is [Ar] $3d^5 4s^1$. This illustrates one of several anomalies that occur in predicting electron configurations for atoms of transition and f subshell-filling elements. When half-filled d subshells are possible, they are sometimes favored (an illustration of Hund's rule). As a result, transition elements that have a half-filled s orbital and half-filled or filled d orbitals have stable electron configurations. Chromium ($s^1 d^5$) and copper ($s^1 d^{10}$) are examples. The electron configuration for copper, the next-to-last element in the first transition series, is [Ar] $3d^{10} 4s^1$ instead of the expected [Ar] $3d^9 4s^2$.

This also occurs with silver and gold, elements that are in the same group as copper.

The number of unpaired electrons for atoms of most transition elements can be predicted according to Hund's rule by placing valence electrons in orbital diagrams. For example, the electron configuration of Co ($Z = 27$) is [Ar] $3d^7 4s^2$, and the number of unpaired electrons is three, corroborated by experimental measurements, and seen from the following orbital box diagram.

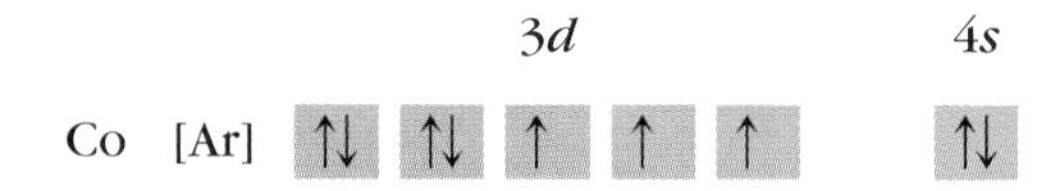

EXERCISE 7.15 Unpaired Electrons

Use orbital box diagrams to determine which chromium ground-state configuration has the greater number of unpaired electrons: [Ar] $3d^4 4s^2$ or [Ar] $3d^5 4s^1$.

Elements with Incompletely Filled *f* Orbitals

In the elements of the sixth and seventh periods, f subshell orbitals exist and can be filled (Figure 7.16). The elements (all metals) for which f subshells are filling are sometimes called the *inner transition* elements or, more usually, *lanthanides* (for lanthanum, the element just before those filling the $4f$ subshell) and *actinides* (for actinium, the element just before those filling the $5f$ subshell). The lanthanides start with lanthanum (La), which has the electron configuration [Xe] $5d^16s^2$. The next element, cerium (Ce), begins a separate row at the bottom of the periodic table, and it is with these elements that *f* orbitals are filled (Appendix D). The electron configuration of Ce is [Xe] $4f^15d^16s^2$. Each of the lanthanide elements, from Ce to Lu, continues to add $4f$ electrons until the seven $4f$ orbitals are filled by 14 electrons in lutetium (Lu, [Xe] $4f^{14}5d^16s^2$). Note that both the $n = 5$ and $n = 6$ levels are partially filled before the $4f$ orbital starts to be occupied.

It is hard to overemphasize how useful the periodic table is as a guide to electron configurations. As another example, using Figure 7.16 as a guide we can find the configuration of Te, element 52. Because tellurium is in Group 6A, by now you should immediately recognize that of its 52 electrons, *six* are *valence electrons* with an outer electron configuration of ns^2np^4. And, because Te is in the fifth period, $n = 5$. Thus, the complete electron configuration is given by starting with the electron configuration of krypton, $[1s^22s^22p^63s^23p^63d^{10}4s^24p^6]$, the noble gas at the end of the fourth period. Then add the filled $4d^{10}$ subshell and the six valence electrons of Te ($5s^25p^4$) to give $1s^22s^22p^63s^23p^63d^{10}4s^24p^64d^{10}5s^25p^4$, or [Kr] $4d^{10}5s^25p^4$. To predict the number of unpaired electrons, look at the outermost subshell's electron configuration, because the inner shells (represented by [Kr] in this case) are completely filled with paired electrons. For Te, [Kr] $4d^{10}5s^25p^25p^15p^1$ indicates two *p* orbitals with unpaired electrons for a total of two unpaired electrons.

7.8 Ion Electron Configurations

In studying ionic compounds *(⇦ p. 71)*, you learned that atoms from Groups 1A through 3A form positive ions (cations) with charges equal to their group numbers—for example, Li^+, Mg^{2+}, and Al^{3+}. Nonmetals in Groups 5A through 7A that form ions do so by adding electrons to form negative ions (anions) with *charges equal to eight minus the A group number.* Examples of such anions are N^{3-}, O^{2-}, and F^-. Here's the explanation. When atoms from *s*- and *p*-block elements form ions, electrons are removed or added so that a noble gas configuration is achieved. Atoms from Groups 1A, 2A, and 3A *lose* 1, 2, or 3 valence electrons to form 1+, 2+, or 3+ ions, respectively; the positive ions have the electron configurations of the *preceding* noble gas. Atoms from Groups 7A, 6A, and some in 5A *gain* 1, 2, or 3 valence electrons to form 1−, 2−, or 3− ions, respectively; the negative ions have the electron configurations of the *next* noble gas. ***Metal atoms lose electrons to form cations with a positive charge equal to the group number; nonmetals gain electrons to form anions with a negative charge equal to the A group number minus eight.*** This relationship holds for Groups 1A to 3A and 5A to 7A. Atoms and ions that have the same electron configuration are said to be **isoelectronic.**

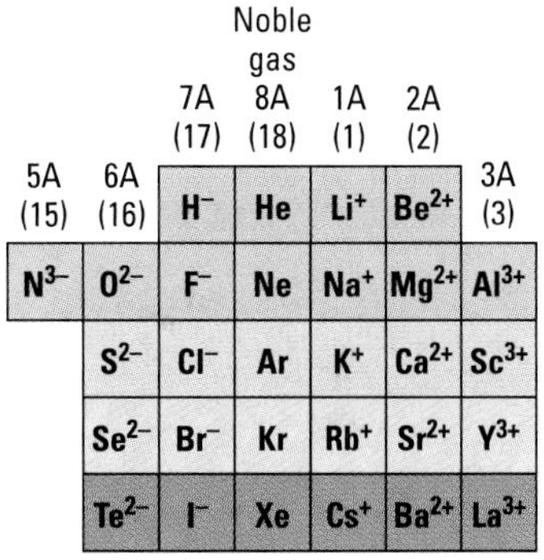

5A (15)	6A (16)	7A (17)	Noble gas 8A (18)	1A (1)	2A (2)	3A (3)
		H^-	He	Li^+	Be^{2+}	
N^{3-}	O^{2-}	F^-	Ne	Na^+	Mg^{2+}	Al^{3+}
	S^{2-}	Cl^-	Ar	K^+	Ca^{2+}	Sc^{3+}
	Se^{2-}	Br^-	Kr	Rb^+	Sr^{2+}	Y^{3+}
	Te^{2-}	I^-	Xe	Cs^+	Ba^{2+}	La^{3+}

Cations, anions, and atoms with ground state noble gas configurations. Atoms and ions shown in the same color are isoelectronic, that is, they have the same electron configuration.

PROBLEM-SOLVING EXAMPLE 7.9 Atoms and Their Ions

Complete this table.

Neutral Atom	Neutral Atom Electron Configuration	Ion	Ion Electron Configuration
Se	________	________	[Kr]
Ba	________	Ba^{2+}	________
Br	________	Br^-	________
________	[Kr] $5s^1$	Rb^+	________
________	[Ne] $3s^23p^3$	________	[Ar]

Answer

Neutral Atom	Neutral Atom Electron Configuration	Ion	Ion Electron Configuration
Se	[Ar] $3d^{10}4s^24p^4$	Se^{2-}	[Kr]
Ba	[Xe] $6s^2$	Ba^{2+}	[Xe]
Br	[Ar] $3d^{10}4s^24p^5$	Br^-	[Kr]
Rb	[Kr] $5s^1$	Rb^+	[Kr]
P	[Ne] $3s^23p^3$	P^{3-}	[Ar]

Strategy and Explanation A neutral Se atom has the electron configuration [Ar] $3d^{10}4s^24p^4$; Se is in Group 6A, so it gains two electrons in the $4p$ sublevel to form Se^{2-} and achieve the noble gas configuration of krypton, Kr (36 electrons). Barium is a Group 2A element and loses the two $6s$ electrons to acquire the electron configuration of xenon (54 electrons), the preceding noble gas. Therefore, Ba is [Xe] $6s^2$ and Ba^{2+} is [Xe]. Bromine is in Group 7A and will gain one electron to form Br^-, which has the electron configuration of krypton, the next noble gas. An electron configuration of [Kr] $5s^1$ indicates 37 electrons in the neutral atom, which is a rubidium atom, Rb. A neutral Rb atom loses the $5s^1$ electron to form an Rb^+ ion, which has a [Kr] configuration. The [Ne] $3s^23p^3$ configuration is for an element with 15 electrons, which is a phosphorus atom, P. By gaining three electrons, a neutral phosphorus atom becomes a P^{3-} ion with the [Ar] configuration.

PROBLEM-SOLVING PRACTICE 7.9

(a) What Period 3 anion with a 3− charge has the [Ar] electron configuration?
(b) What Period 4 cation with a 2+ charge has the electron configuration of argon?

Table 7.7 lists some isoelectronic ions and the noble gas atom that has the same electron configuration. The table emphasizes that ***metal ions are isoelectronic with the preceding noble gas atom, while nonmetal ions have the electron configuration of the next noble gas atom.***

Transition Metal Ions

The closeness in energy of the $4s$ and $3d$ subshells was mentioned earlier in connection with the electron configurations of transition metal atoms. The $5s$ and $4d$ subshells are also close to each other in energy, as are the $6s$, $4f$, and $5d$ subshells and the $7s$, $5f$, and $6d$ subshells. The ns and $(n-1)d$ subshells are so close in energy that once d electrons are added, the $(n-1)d$ subshell becomes slightly lower in energy than the ns subshell. As a result, ***the ns electrons are at higher energy and are always removed before (n−1)d electrons when transition metals form cations***. A simple way to remember this is that when a transition metal atom loses electrons to form an ion, the electrons in the outermost shell are lost first.

Table 7.7 Noble Gas Atoms and Their Isoelectronic Ions

(All $1s^2$)	He, Li^+, Be^{2+}, H^-
(All $1s^22s^22p^6$)	Ne, Na^+, Mg^{2+}, Al^{3+}, F^-, O^{2-}
(All $1s^22s^22p^63s^23p^6$)	Ar, K^+, Ca^{2+}, Cl^-, S^{2-}
(All $1s^22s^22p^63s^23p^63d^{10}4s^24p^6$)	Kr, Rb^+, Sr^{2+}, Br^-, Se^{2-}
(All $1s^22s^22p^63s^23p^63d^{10}4s^24p^64d^{10}5s^25p^6$)	Xe, Cs^+, Ba^{2+}, I^-, Te^{2-}

For example, a nickel atom, Ni: [Ar] $3d^8 4s^2$, will lose its two $4s$ electrons to form a Ni^{2+} ion.

$$Ni \rightarrow Ni^{2+} + 2e^-$$

That two $4s$ electrons are lost, and not two electrons from the $3d$ subshell, is corroborated by the experimental evidence that Ni^{2+} has two unpaired electrons. This is shown by the following box diagrams.

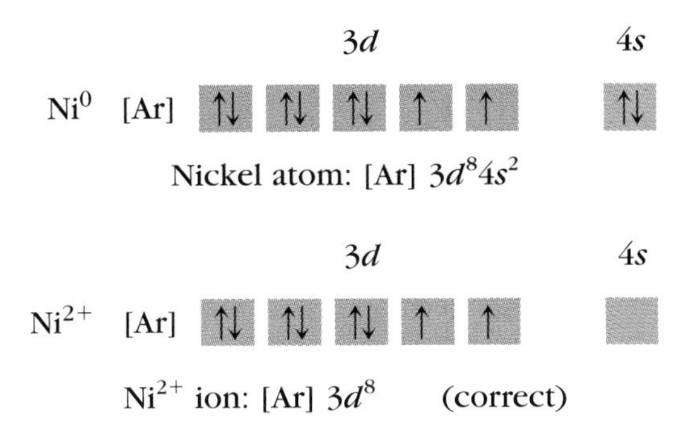

If two electrons had been removed from the $3d$ subshell rather than the $4s$ subshell, a Ni^{2+} ion would have four unpaired electrons, which is inconsistent with experimental evidence.

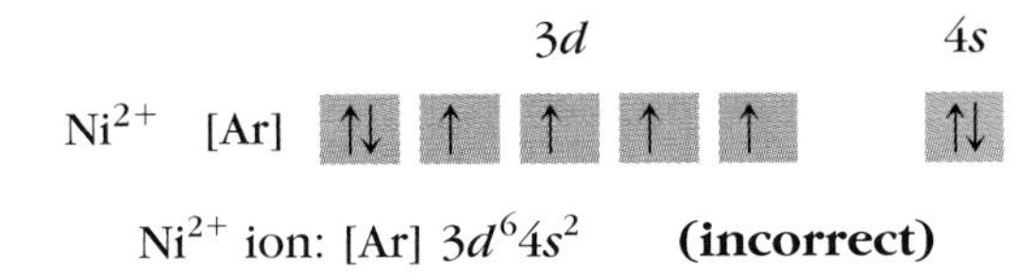

Atoms or ions of inner transition elements (lanthanides and actinides) can have as many as seven unpaired electrons in the f subshell, as occurs in Eu^{2+} and Gd^{3+} ions.

PROBLEM-SOLVING EXAMPLE 7.10 Electron Configurations for Transition Elements and Ions

(a) Write the electron configuration for the Co atom using the noble gas notation. Then draw the orbital box diagram for the electrons beyond the preceding noble gas configuration.
(b) Cobalt commonly exists as 2+ and 3+ ions. How does the orbital box diagram given in part (a) have to be changed to represent the outer electrons of Co^{2+} and Co^{3+}?
(c) How many unpaired electrons do Co, Co^{2+}, and Co^{3+} each have?

Answer

(a)

		3d	4s
[Ar]$3d^7 4s^2$;	[Ar]	↑↓ ↑↓ ↑ ↑ ↑	↑↓

(b) For Co^{2+}, remove the two $4s$ electrons from Co to give

		3d
Co^{2+}	[Ar]	↑↓ ↑↓ ↑ ↑ ↑

For Co^{3+}, remove the two $4s$ electrons and one of the paired $3d$ electrons from Co to give

		3d
Co^{3+}	[Ar]	↑↓ ↑ ↑ ↑ ↑

(c) Co has three unpaired electrons; Co^{2+} has three unpaired electrons; Co^{3+} has four unpaired electrons.

Strategy and Explanation

(a) Co, with an atomic number of 27, is in the fourth period. It has nine more electrons than Ar, with two of the nine in the 4*s* subshell and seven in the 3*d* subshell, so its electron configuration is [Ar] $3d^74s^2$. For the orbital box diagram, all *d* orbitals get one electron before pairing occurs (Hund's rule).

(b) To form Co^{2+}, two electrons are removed from the outermost shell (4*s* subshell). To form Co^{3+} from Co^{2+}, one of the paired electrons is removed from a 3*d* orbital.

(c) Looking at the orbital box diagrams, you should see that the numbers of unpaired electrons are three for Co, three for Co^{2+}, and four for Co^{3+}.

PROBLEM-SOLVING PRACTICE 7.10

Use the electron configuration of a neutral ground state copper atom to explain why copper readily forms the Cu^+ ion.

Paramagnetism: The atoms or ions with magnetic moments are not aligned. If the substance is in a magnetic field, they align with and against the field. Magnetism is weak.

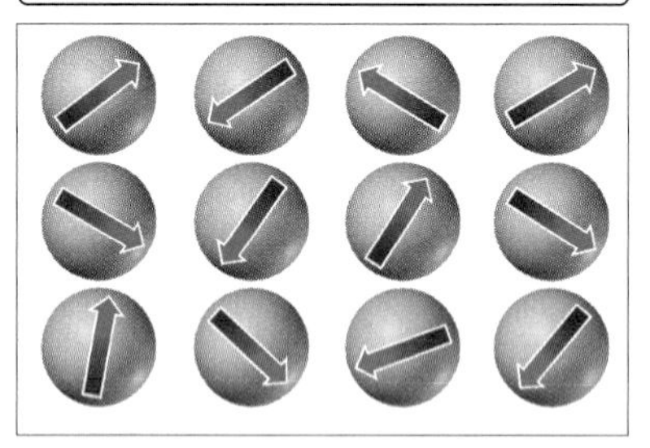

Ferromagnetism: The spins of unpaired electrons in clusters of atoms or ions are aligned in the same direction. In a magnetic field these domains all align and stay aligned when the field is removed.

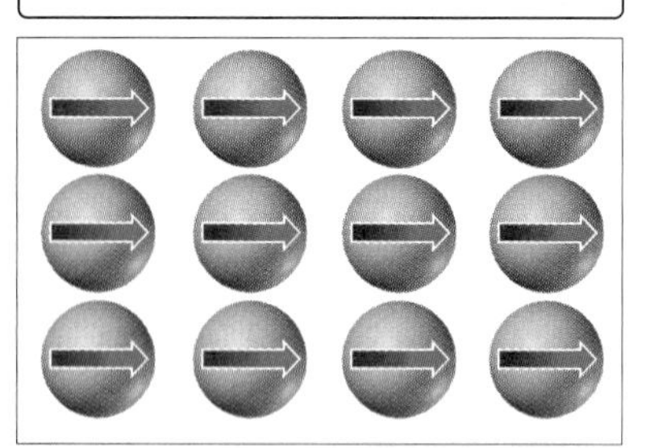

Figure 7.19 Types of magnetic behavior.

Paramagnetism and Unpaired Electrons

The magnetic properties of a spinning electron were described in Section 7.5. In atoms and ions that have filled shells, all the electrons are paired (opposite spins) and their magnetic fields effectively cancel each other. Such substances are called **diamagnetic.** Atoms or ions with unpaired electrons are attracted to a magnetic field; the more unpaired electrons, the greater is the attraction. Such substances are called **paramagnetic.** For example, the greater paramagnetism of chromium with six unpaired electrons rather than four is experimental evidence that Cr has the electron configuration [Ar] $3d^54s^1$, rather than [Ar] $3d^44s^2$.

Ferromagnetic substances are permanent magnets; they retain their magnetism. The magnetic effect in ferromagnetic materials is much larger than that in paramagnetic materials. Ferromagnetism occurs when the spins of unpaired electrons in a cluster of atoms (called a domain) in a solid are aligned in the same direction (Figure 7.19). Only the metals of the iron, cobalt, and nickel subgroups in the periodic table exhibit this property. They are also unique in that, once the domains are aligned in a magnetic field, the metal is permanently magnetized. In such a case, the magnetism can be eliminated only by heating or shaping the metal to rearrange the electron spin domains. Many alloys, such as alnico (an alloy of aluminum, nickel, and cobalt), exhibit greater ferromagnetism than do the pure metals themselves. Some metal oxides, such as CrO_2 and Fe_3O_4, are also ferromagnetic and are used in magnetic recording tape. Computer discs use ferromagnetic materials to store data in a binary code of zeroes and ones. Very small magnetic domains store bits of binary code as a magnetized region that represents a one and an unmagnetized region that represents a zero.

CONCEPTUAL **EXERCISE 7.16 Unpaired Electrons**

The acetylacetonate ion, $acac^-$, which has no unpaired electrons, forms compounds with Fe^{2+} and Fe^{3+} ions. Their formulas are $Fe(acac)_2$ and $Fe(acac)_3$ respectively. Which one will have the greater attraction to a magnetic field? Explain.

7.9 Periodic Trends: Atomic Radii

Using knowledge of electron configurations, we can now answer fundamental questions about why atoms of different elements fit as they do in the periodic table, as well as explain the trends in the properties of the elements in the table.

For atoms that form simple diatomic molecules, such as Cl_2, the **atomic radius** can be defined experimentally by finding the distance between the centers of the two atoms in the molecule. Assuming the atoms to be spherical, one-half of this distance is a good estimate of the atom's radius. In the Cl_2 molecule, the atom-to-atom distance (the distance from the nucleus of one atom to the nucleus of the other) is 200 pm. Dividing by 2 shows that the Cl radius is 100 pm. Similarly, the C—C distance in diamond is 154 pm, so the radius of the carbon atom is 77 pm. To test these estimates, we can add them together to estimate the distance between Cl and C in CCl_4. The estimated distance of 177 pm (100 pm + 77 pm) is in good agreement with the experimentally measured C—Cl distance of 176 pm.

This approach can be extended to other atomic radii. The radii of O, C, and S atoms can be estimated by measuring the O—H, C—Cl, and H—S distances in H_2O, CCl_4, and H_2S, and then subtracting the H and Cl radii found from H_2 and Cl_2. By this and other techniques, a reasonable set of atomic radii for main group elements has been assembled (Figure 7.20).

200 pm

Cl Cl

100 pm

Atomic radius of chlorine. The atomic radius is taken to be one half of the inter-nuclear distance in the Cl_2 molecule.

Atomic Radii of the Main Group Elements

For the main group elements, atomic radii increase going down a group in the periodic table and decrease going across a period (Figure 7.20). These trends reflect three important effects:

- From the top to the bottom of a group in the periodic table, the atomic radii increase because electrons occupy orbitals that are successively larger as the value of n, the principal quantum number, increases.

The main group elements are those in the A groups in the periodic table (Figure 7.16, p. 247).

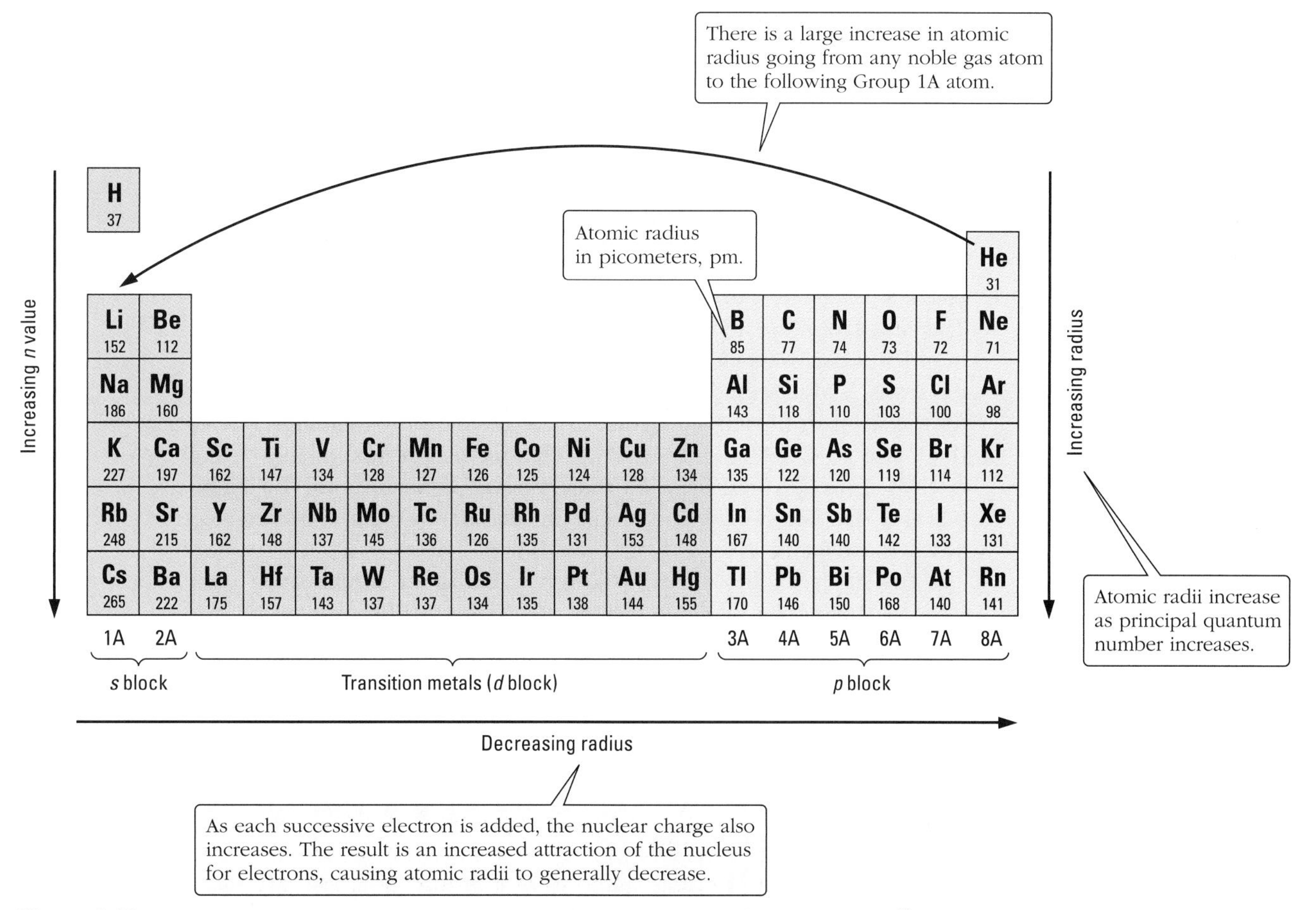

Figure 7.20 Atomic radii of *s*-, *p*-, and *d*-block elements (in picometers, pm; 1 pm = 10^{-12} m). Note the slight change in atomic radii for the transition elements (*d* block) across a period.

- The atomic radii decrease from left to right across a period. The n value of the outermost orbitals stays the same, so we might expect the radii of the occupied orbitals to remain approximately constant. However, in crossing a period, as each successive electron is added, the nuclear charge also increases by the addition of one proton. The result is an increased attraction between the nucleus and electrons that is somewhat stronger than the increasing repulsion between electrons, causing atomic radii to decrease (Figure 7.20).
- There is a large increase in atomic radius going from any noble gas atom to the following Group 1A atom, where the outermost electron is assigned to the next higher energy level (the next larger shell, higher n value). For example, compare the atomic radii of Ne (71 pm) and Na (186 pm) or Ar (98 pm) and K (227 pm).

PROBLEM-SOLVING EXAMPLE 7.11 Atomic Radii and Periodic Trends

Using only a periodic table, list these atoms in order of decreasing size: Br, Cl, Ge, K, S.

Answer K > Ge > Br > S > Cl

Strategy and Explanation Based on the periodic trends in atomic size, we expect the radius of Cl to be smaller than that of Br, which is below it in Group 7A; S in Period 3 should be smaller than Br, which is in Period 4. Because size decreases across a period, we expect Cl to be smaller than S. Of the Period 4 atoms, Ge is earlier in the period than Br and therefore larger than it, but smaller than K, which starts the period. From these trends we can summarize the relative sizes of the radii, from largest to smallest, as K > Ge > Br > S > Cl.

PROBLEM-SOLVING PRACTICE 7.11

Using just a periodic table, arrange these atoms in order of increasing atomic radius: B, Mg, K, Na.

Atomic Radii of Transition Metals

The periodic trend in the atomic radii of main group and transition metal atoms is illustrated in Figure 7.20. The sizes of transition metal atoms change very little across a period, especially beginning at Group 5B (V, Nb, or Ta), because the sizes are determined by the radius of an ns orbital ($n = 4$, 5, or 6) occupied by at least one electron. The variation in the number of electrons occurs instead in the $(n - 1)d$ orbitals. As the number of electrons in these $(n = 1)d$ orbitals increases, they increasingly repel the ns electrons, which would cause the atomic radius to increase. This partly compensates for the stronger nucleus-electron attraction due to the increased nuclear charge across the periods. Consequently, the ns electrons experience only a slightly increasing nuclear attraction, and the radii remain nearly constant until the slight rise at the copper and zinc groups due to the continually increasing electron-to-electron repulsions as the d subshell is filled.

The similar radii of the transition metals and the similar sizes of their ions have an important effect on their chemistry—they tend to be more alike in their properties than other elements in their respective groups. The nearly identical radii of the fifth- and sixth-period transition elements lead to very similar properties and great difficulty in separating the elements from one another. The metals Ru, Os, Rh, Ir, Pd, and Pt, called the "platinum group metals," occur together in nature. Their radii and chemistry are so similar that their minerals are similar and are found in the same geologic zones.

Effective Nuclear Charge

We have seen that the atomic radii of main group elements decrease across a period, as the number of protons and electrons increases with each element. This change in radii can be explained by **effective nuclear charge,** the nuclear positive charge experienced by outer-shell electrons in a many-electron atom. The effective nuclear charge felt by outer-shell electrons is less than the actual nuclear positive charge. Outer electrons are shielded from the full nuclear positive charge by intervening inner electrons, those closer to the nucleus (lower n values). These electrons repel outer electrons and this repulsion partially offsets the attraction of the nucleus for outer-shell electrons. Therefore, $Z^* = Z - \sigma$, where Z^* is the effective nuclear charge, Z is the number of protons in the nucleus (the actual nuclear charge), and σ is the amount of screening, the amount by which the actual nuclear charge is reduced. This decrease is called the **screening effect.** The effective nuclear charge increases across a period and outer electrons are pulled closer to the nucleus causing the decrease of atomic radii across a period.

7.10 Periodic Trends: Ionic Radii

As shown in Figure 7.21, the periodic trends in the **ionic radii** parallel the trends in atomic radii within the same periodic group: ***positive ions of elements in the same group increase in size down the group***—for example, Li^+ to Rb^+ in Group 1A. ***This also occurs with negative ions in the same group*** (see F^- to I^-, Group 7A). But take time to compare the two types of entries in Figure 7.21. When an electron is removed from an atom to form a cation, the size shrinks considerably; ***the radius of a cation is always smaller than that of the atom from which it is derived.*** The radius of Li is 152 pm, whereas that for Li^+ is only 90 pm. This is understandable, because when an electron is removed from a lithium atom, the nuclear charge remains the same (3+), but there are fewer electrons repelling each other. Consequently, the positive nucleus can attract the two remaining electrons more strongly, causing the electrons to contract toward the nucleus. The decrease in ion size is especially great when the electron removed comes from a higher energy level than the new outer electron. This is the case for Li, for which the "old" outer electron was from a 2s orbital and the "new" outer electron in Li^+ is in a 1s orbital.

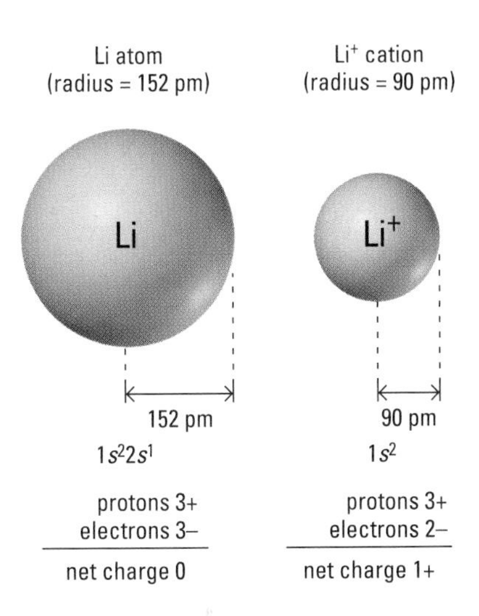

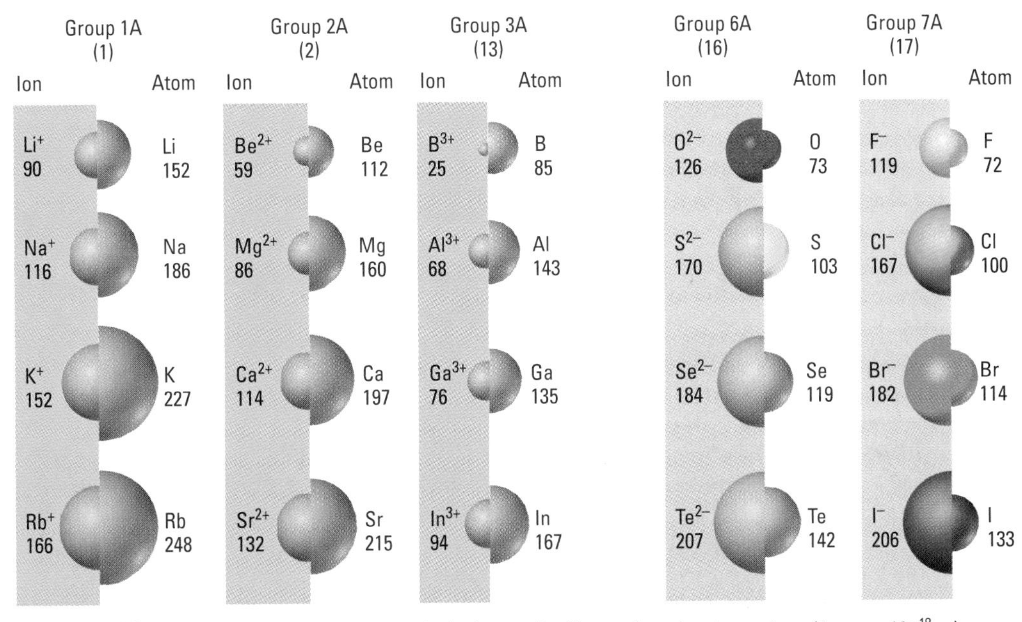

Figure 7.21 Sizes of ions and their neutral atoms. Radii are given in picometers (1 pm = 10^{-12} m).

The shrinkage is also great when two or more electrons are removed—for example, Mg^{2+} and Al^{3+}.

Mg atom (radius = 160 pm) $1s^22s^22p^63s^2$	Mg^{2+} cation (radius = 86 pm) $1s^22s^22p^6$
Al atom (radius 143 pm) $1s^22s^22p^63s^23p^1$	Al^{3+} cation (radius = 68 pm) $1s^22s^22p^6$

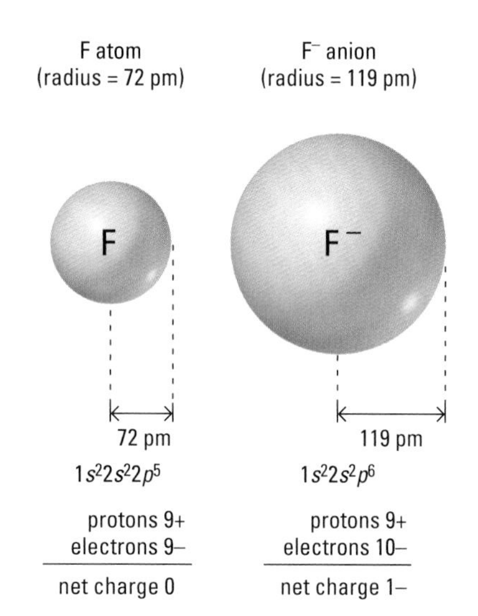

From Figure 7.21 you can see that ***anions are always larger than the atoms from which they are derived.*** Here the argument is the opposite of that used to explain the radii of cations. For anions, the nuclear charge is unchanged, but the added electron(s) introduce new repulsions and the electron clouds swell. The F atom has nine protons and nine electrons. When it forms the F^- anion, the nuclear charge is still 9+ but there are now ten electrons in the anion. The F^- ion (119 pm) is much larger than the F atom (72 pm) because of increased electron-to-electron repulsions.

The oxide ion, O^{2-}, and fluoride ion, F^-, are *isoelectronic;* they both have the same electron configuration (the neon configuration). However, the oxide ion is larger than the fluoride ion because the O^{2-} ion has only eight protons available to attract ten electrons, whereas F^- has more protons (nine) to attract the same number of electrons. The effect of nuclear charge is evident when the sizes of isoelectronic ions across the periodic table are compared. Consider O^{2-}, F^-, Na^+, and Mg^{2+}.

Isoelectronic Ions	O^{2-}	F^-	Na^+	Mg^{2+}
Ionic radius (pm)	126	119	116	86
Number of protons	8	9	11	12
Number of electrons	10	10	10	10

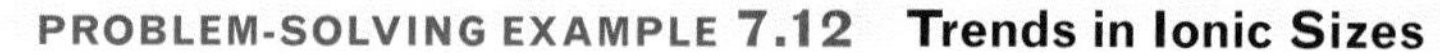

Each ion contains ten electrons. However, the O^{2-} ion has only eight protons in its nucleus to attract these electrons, while F^- has nine, Na^+ has eleven, and Mg^{2+} has twelve. Thus, there is an increasing number of protons to attract the same number of electrons (10). As the electron-to-proton ratio decreases in an isoelectronic series of ions, such as those given, the overall electron cloud is attracted more tightly to the nucleus and the ion size shrinks, as seen for O^{2-} to Mg^{2+}.

PROBLEM-SOLVING EXAMPLE 7.12 Trends in Ionic Sizes

For each of these pairs, choose the smaller atom or ion:

(a) A Cu atom or a Cu^{2+} ion
(b) A Se atom or a Se^{2-} ion
(c) A Cu^+ ion or a Cu^{2+} ion

Answer

(a) Cu^{2+} ion (b) Se atom (c) Cu^{2+} ion

Strategy and Explanation Consider the differences in the sizes of atoms and their corresponding ions based on electron loss or gain.

(a) A cation is smaller than its parent atom; therefore, Cu^{2+} is smaller than a Cu atom.
(b) Anions are larger than their parent atoms, so a Se^{2-} ion is larger than a selenium atom.
(c) Both copper ions contain 26 protons, but Cu^{2+} has one fewer electron than does Cu^+. Correspondingly, the radius of Cu^{2+} is smaller than that of Cu^+.

PROBLEM-SOLVING PRACTICE 7.12

Which of these isoelectronic ions, Ba^{2+}, Cs^+, or La^{3+}, will be (a) the largest? (b) the smallest?

7.11 Periodic Trends: Ionization Energies

An element's chemical reactivity is determined, in part, by how easily valence electrons are removed from its atoms, a process that requires energy. The **ionization energy** of an atom is the energy needed to remove an electron from that atom in the gas phase. For a gaseous sodium atom, the ionization process is

$$\text{Na(g)} \longrightarrow \text{Na}^+\text{(g)} + \text{e}^- \qquad \Delta E = \text{ionization energy (IE)}$$

Because energy is always required to remove an electron, the process is endothermic, and the sign of the ionization energy is always positive. The more difficult an electron is to remove, the greater its ionization energy. ***For s- and p-block elements, the first ionization energy (the energy needed to remove one electron from the neutral atom) generally decreases down a group and increases across a period*** (Figure 7.22). The decrease down a group reflects the increasing radii of the atoms—it is easier to remove a valence electron from a larger atom because the force of attraction between the electron and the nucleus is smaller due to the greater separation. Thus, for example, the ionization energy of Rb (403 kJ/mol) is less than that of K (419 kJ/mol) or Na (496 kJ/mol). Ionization energies increase across a period for the same reason that the radii decrease. Increasing nuclear charge attracts electrons in the same shell more tightly. However, as shown in Figure 7.22, the trend across a given period is not always smooth, as in the second, third, fourth, and fifth periods. The single np electron of the Group 3A element is more easily removed than one of the two ns electrons in the preceding Group 2A element. Another deviation occurs with Group 6A elements (ns^2np^4), which have smaller ionization energies than the Group 5A elements that precede them in the second, third, and fourth periods. Beginning in Group 6A, two electrons are assigned to the same p orbital. Thus, greater electron repulsion is experienced by the fourth p electron, making it easier to remove.

Ionization energies increase much more gradually across the period for the transition and inner transition elements than for the main group elements (Figure 7.22). Just as the atomic radii of transition elements change very little across a period, the ionization energy for the removal of an ns electron also shows small changes, until the d orbitals are filled (Zn, Cd, Hg).

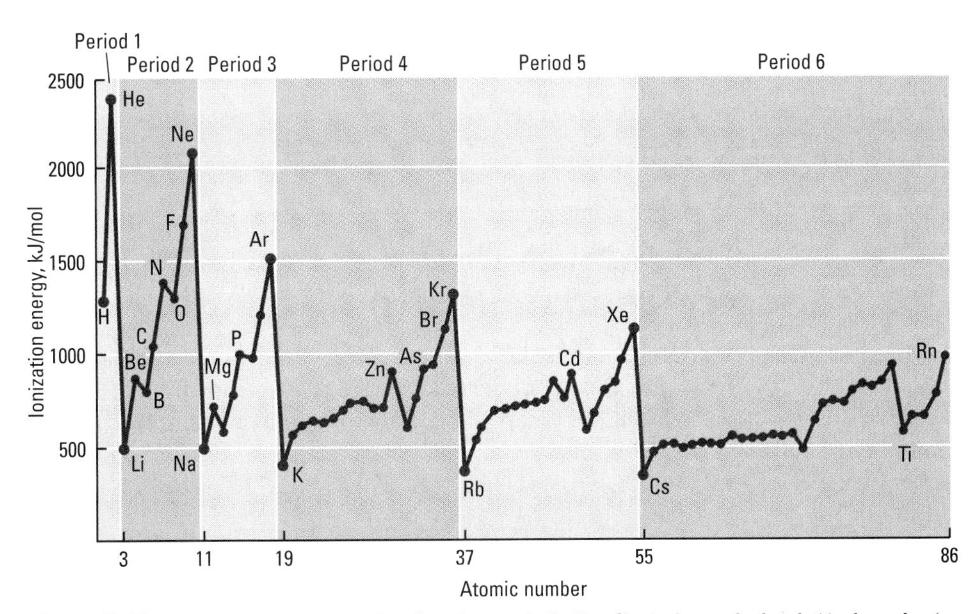

Figure 7.22 First ionization energies for elements in the first six periods plotted against atomic number. *Note* that in each period the alkali metal has the lowest, and the noble gas the highest ionization energy.

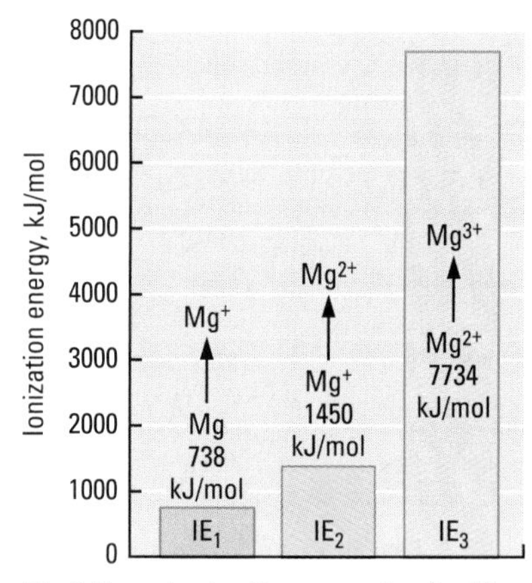

First three ionization energies for Mg.

Every atom except hydrogen has a series of ionization energies, since more than one electron can be removed. For example, the first three ionization energies of Mg in the gaseous state are

$$Mg(g) \longrightarrow Mg^{+}(g) + e^{-} \qquad IE_1 = 738 \text{ kJ/mol}$$
$$1s^22s^22p^63s^2 \qquad 1s^22s^22p^63s^1$$

$$Mg^{+}(g) \longrightarrow Mg^{2+}(g) + e^{-} \qquad IE_2 = 1450 \text{ kJ/mol}$$
$$1s^22s^22p^63s^1 \qquad 1s^22s^22p^6$$

$$Mg^{2+}(g) \longrightarrow Mg^{3+}(g) + e^{-} \qquad IE_3 = 7734 \text{ kJ/mol}$$
$$1s^22s^22p^6 \qquad 1s^22s^22p^5$$

Notice that removing each subsequent electron requires more energy, and the jump from the second (IE_2) to the third (IE_3) ionization energy of Mg is particularly great. The first electron removed from a magnesium atom comes from the 3*s* orbital. The second ionization energy corresponds to removing a 3*s* electron from a Mg^+ ion. As expected, the second ionization energy is higher than the first ionization energy because the electron is being removed from a positive ion, which strongly attracts the electron. The third ionization energy corresponds to removing a 2*p* electron, a core electron, from a filled *p* subshell in Mg^{2+}. The great difference between the second and third ionization energies for Mg is excellent experimental evidence for the existence of electron shells in atoms. Removal of the first core electron requires much more energy than removal of a valence electron. A table of ionization energies of the elements is given in Appendix E.

PROBLEM-SOLVING EXAMPLE 7.13 Ionization Energies

Using only a periodic table, arrange these atoms in order of increasing first ionization energy: Al, Ar, Cl, Na, K, Si.

Answer K < Na < Al < Si < Cl < Ar

Strategy and Explanation Since ionization energy increases across a period, we expect the ionization energies for Ar, Al, Cl, Na, and Si, which are all in Period 3, to be in the order Na < Al < Si < Cl < Ar. Because K is below Na in Group 1A, the first ionization energy of K is less than that of Na. Therefore, the final order is K < Na < Al < Si < Cl < Ar.

PROBLEM-SOLVING PRACTICE 7.13

Use only a periodic table to arrange these elements in decreasing order of their first ionization energy: Na, F, N, P.

7.12 Periodic Trends: Electron Affinities

The **electron affinity** of an element is the energy change when an electron is added to a gaseous atom to form a 1− ion. As the term implies, the electron affinity (EA) is a measure of the attraction an atom has for an additional electron. For example, the electron affinity of fluorine is −328 kJ/mol.

Two sign conventions are used for electron affinity—the one used here and an alternative one. The latter uses the opposite sign—for example, +328 kJ/mol for the electron affinity of fluorine.

$$F(g) + e^{-} \longrightarrow F^{-}(g) \qquad \Delta E = EA = -328 \text{ kJ/mol}$$
$$[He]\ 2s^22p^5 \qquad [He]\ 2s^22p^6$$

This large negative value indicates that fluorine atoms readily accept an electron. As expected, fluorine and the rest of the halogens have large negative electron affinities because, by acquiring an electron, the halogen atoms achieve a stable noble-gas configuration of valence electrons, ns^2np^6 (Table 7.8). Notice from Table 7.8 that electron

Table 7.8 Electron Affinities (kJ/mol)

1A (1)	2A (2)	3A (13)	4A (14)	5A (15)	6A (16)	7A (17)	8A (18)
H −73							He >0
Li −60	Be >0	B −27	C −122	N >0	O −141	F −328	Ne >0
Na −53	Mg >0	Al −43	Si −134	P −72	S −200	Cl −349	Ar >0
K −48	Ca −2	Ga −30	Ge −119	As −78	Se −195	Br −325	Kr >0
Rb −47	Sr −5	In −30	Sn −107	Sb −103	Te −190	I −295	Xe >0

affinities generally become more negative across a period toward the halogen, which is only one electron away from a noble-gas configuration.

Some elements have a positive electron affinity, meaning that the negative ion is less stable than the neutral atom. Such electron affinity values are difficult to obtain experimentally, and the electron affinity is simply indicated as > 0, as in the case of neon.

$$\underset{[\text{He}]\,2s^22p^6}{\text{Ne}} + e^- \longrightarrow \underset{[\text{He}]\,2s^22p^63s^1}{\text{Ne}^-} \qquad \text{EA} > 0 \text{ kJ/mol}$$

The Ne^- ion would revert back to the neutral neon atom and an electron. From their electron configurations we can understand why the Ne^- ion would be less stable than the neutral neon atom; the anion would exceed the stable noble-gas configuration of valence electrons of Ne atoms.

© Cengage Learning/Charles D. Winters

Figure 7.23 The reaction of sodium with chlorine. The reaction liberates heat and light and forms Na^+ and Cl^- ions.

7.13 Energy Considerations in Ionic Compound Formation

Energy is released when an alkali metal and a nonmetal react to form a salt. For example, the vigorous reaction of sodium metal and chlorine gas forms solid sodium chloride, NaCl, and also transfers energy to the surroundings (Figure 7.23).

$$\text{Na(s)} + \tfrac{1}{2}\,\text{Cl}_2\text{(g)} \longrightarrow \text{NaCl(s)} \qquad \Delta H_f^\circ = -410.9 \text{ kJ}$$

As seen in Figure 7.23, energy transfers occur during the formation of an ionic compound such as NaCl.

Let's consider the energy changes in the formation of solid potassium fluoride, formed by the reaction of potassium, a very reactive alkali metal, and fluorine, a highly reactive halogen. Potassium fluoride contains K^+ and F^- ions grouped in a crystalline lattice of alternating cations and anions (Figure 7.24). Potassium atoms have low ionization energy and readily lose their $4s^1$ outer electrons; fluorine atoms have a high electron affinity and readily accept the electrons into $2p$ orbitals. Thus, electron transfer occurs from potassium atoms to fluorine atoms during KF formation.

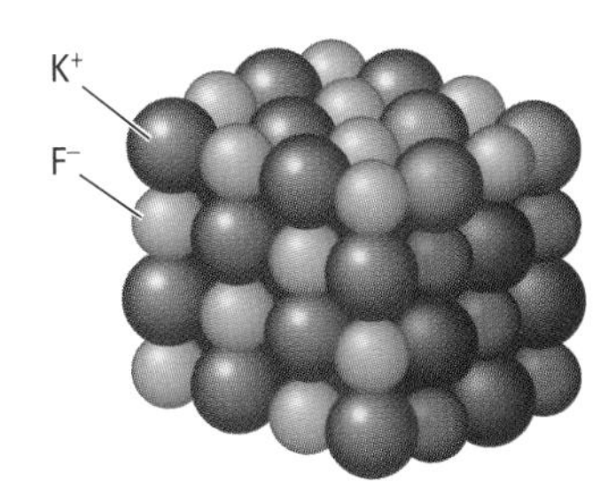

Figure 7.24 A potassium fluoride crystal illustrating the arrangement of K^+ and F^- ions in a lattice structure.

$$\text{K}\ 1s^22s^22p^63s^23p^64s^1 \longrightarrow \text{K}^+\ 1s^22s^22p^63s^23p^6 \qquad \text{Ionization energy} = 419 \text{ kJ}$$

$$\text{F}\ 1s^22s^22p^5 + e^- \longrightarrow \text{F}^-\ 1s^22s^22p^6 \qquad \text{Electron affinity} = -328 \text{ kJ}$$

The transfer of electrons converts potassium and fluorine atoms into K^+ and F^- ions, each of which has an octet of outer electrons, a very stable noble-gas configuration.

Electron configurations: K: [Ar] $4s^1$; F: [He] $2s^2 2p^5$ K^+ [Ar]; F^- [Ne]

The standard molar enthalpy of formation, ΔH_f°, of KF is −567.27 kJ/mol.

$$K(s) + \frac{1}{2} F_2(g) \longrightarrow KF(s) \qquad \Delta H_f^\circ = -567.27 \text{ kJ}$$

This cycle is named for Max Born and Fritz Haber, two Nobel-Prize winning German scientists.

We can evaluate the energy required to form the ions and the solid compound by applying Hess's Law (⇐ ***p. 206***) to a series of five hypothetical steps, known as a **Born-Haber cycle,** as outlined below and illustrated in Figure 7.25. When combined with the thermochemical expression for enthalpy of formation, such a hypothetical multistep sequence allows the calculation of the **lattice energy** of the crystal, the enthalpy of change when 1 mol an ionic solid is formed from its separated gaseous ions.

Step 1: *Convert solid potassium to gaseous potassium atoms.* This is the enthalpy of sublimation, ΔH_1, which is +89 kJ for 1 mol potassium.

$$K(s) \longrightarrow K(g) \qquad \Delta H_1 = +89 \text{ kJ/mol}$$

Step 2: *Dissociate F_2 molecules into F atoms.* One mol of KF contains 1 mol fluorine, so $\frac{1}{2}$ mol of F_2 is required. Breaking the bonds in diatomic fluorine molecules requires energy. This energy, ΔH_2, is +79 kJ, one-half of the bond energy (BE) of F_2, $\frac{1}{2}$(158 kJ/mol).

$$\frac{1}{2} F_2(g) \longrightarrow F(g)$$

$$\Delta H_2 = \frac{1}{2} \text{BE of } F_2 = \frac{1}{2} \text{mol}(+158 \text{ kJ/mol}) = +79 \text{ kJ}$$

Step 3: *Convert gaseous potassium atoms to gaseous potassium ions.* The energy, ΔH_3, required to remove the $4s^1$ electron is the first ionization energy, IE_1, of potassium, +419 kJ, which can be obtained from a table of ionization energies.

$$K(g) \longrightarrow K^+(g) + e^- \qquad \Delta H_3 = IE_1 = +419 \text{ kJ/mol}$$

Step 4: *Form F^- ions from F atoms.* The addition of a sixth electron to the $2p$ subshell completes the octet of valence electrons for F^-. For the formation of 1 mol F^- ions, this energy, ΔH_4, is the electron affinity (EA) of F, −328 kJ, obtained from Table 7.8.

$$F(g) + e^- \longrightarrow F^-(g) \qquad \Delta H_4 = EA = -328 \text{ kJ}$$

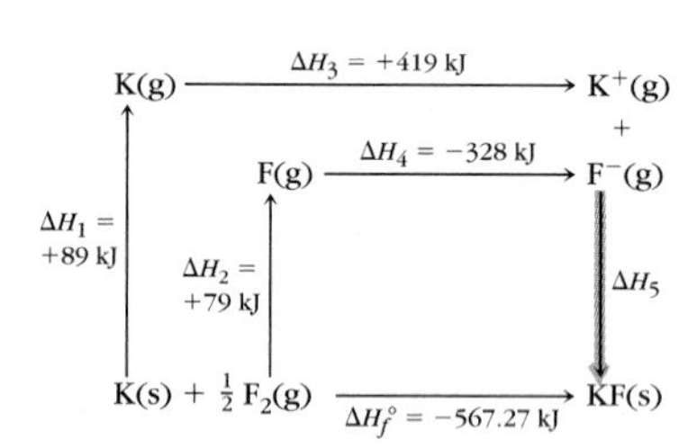

Figure 7.25 Born-Haber cycle for KF.

Step 5: *Consolidate gaseous K^+ and F^- ions into a crystal.* The cations and anions assemble into a crystal, releasing a significant quantity of energy due to the attraction between oppositely charged ions. This energy, ΔH_5, is the lattice energy, the enthalpy of change when 1 mol of an ionic solid is formed from its separated gaseous ions.

$$K^+(g) + F^-(g) \longrightarrow KF(s) \qquad \Delta H_5 = \text{lattice energy of KF}$$

Therefore, the sum of the enthalpy changes in the five steps of the cycle represents the enthalpy change for the overall reaction of solid potassium metal with gaseous fluorine to produce solid potassium fluoride (Figure 7.25). According to Hess's Law, the sum of these enthalpy changes must equal the standard molar enthalpy of formation (ΔH_f°) of KF, which is experimentally measured as −567.27 kJ.

$$\Delta H_f^\circ = -567.72 \text{ kJ} = \Delta H_1 + \Delta H_2 + \Delta H_3 + \Delta H_4 + \Delta H_5$$

Although data for Steps 1–4 can be determined experimentally, lattice energies cannot be measured directly. Instead, we can apply Hess's Law in a Born-Haber cycle to calculate the lattice energy for KF, ΔH_5, using data from Steps 1–4 and the enthalpy of formation, −567.27 kJ.

$$-567.27 \text{ kJ} = 89 \text{ kJ} + 79 \text{ kJ} + 419 \text{ kJ} + (-328 \text{ kJ}) + \Delta H_5$$

$$\Delta H_5 = \text{Lattice energy of KF} = (-567.27 \text{ kJ}) + (-89 \text{ kJ}) + (-79 \text{ kJ}) + (-419 \text{ kJ}) + (328 \text{ kJ}) = -826 \text{ kJ}$$

Notice that the formation of the gaseous cations and anions from the respective elements (Steps 1–4) is not energetically favorable (Figure 7.25). The net enthalpy change for these four steps in the cycle is positive, +259 kJ. The solid compound forms because of the very large lattice energy released during the assembling of the gaseous cations and anions into the solid. It is the lattice energy, not the formation of K^+ and F^- ions, that drives the formation of solid KF (Figure 7.25).

The strength of the ionic bonds in a crystal—its lattice energy—depends on the strength of the electrostatic interactions among the crystal's cations and anions. In a crystal lattice there are attractive forces between oppositely charged ions and repulsive forces between ions of like charge. All these forces obey Coulomb's Law (***p. 74***). Both attractive and repulsive forces are affected in the same proportion by two factors:

- *The sizes of the charges of the cations and anions.* As the charges get larger (from 1+ and 1−, to 2+ and 2−, etc.), both attractive and repulsive forces among the ions increase.
- *The distances among the ions.* The forces increase as the distances decrease. The shortest distance is $r_+ + r_-$, where r_+ is the cation radius and r_- is the anion radius. The distances among ions of like charge increase or decrease in proportion to this shortest distance. The smaller the ionic radii, the shorter the distances among the ions.

Ions are arranged in a crystal so that the attractive forces predominate over the repulsive forces. This makes the energy lower when the ions are in a crystal lattice than when they are completely separated from one another. Because the energy of the crystal lattice is lower than for the separated ions (that is, a mole of KF(s) is lower in energy that a mole of separated K^+ and F^- ions), the lattice energy is negative. The stronger the net ionic attractions are, the more negative the lattice energy is, and the more stable the crystal lattice is. Thus, crystal lattices are more stable for small ions of large charge.

Compounds with high lattice energies (large negative lattice energies) have high melting points. The entries in Table 7.9 exemplify this trend. Consider NaCl and BaO. In both compounds the sum of the ionic radii of the cation and anion is about the same; it is 283 pm for NaCl and 275 pm for BaO. However, the product of the charges differs by a factor of four [(1+) × (1−) for NaCl versus [(+2) × (−2)] for BaO. This charge factor of four is reflected in a lattice energy (−3054 kJ/mol) and a melting point (2193 K) for BaO that are much larger than the lattice energy (−786 kJ/mol) and the melting point of NaCl (1073 K) of NaCl. The effect of cation size can be seen by comparing MgO and BaO in Table 7.9. Because Mg^{2+} (r_+ = 86 pm) is much smaller than Ba^{2+}, the 2+ charge on

Table 7.9 Effect of Ion Size and Charge on Lattice Energy and Melting Point

Compound	Charges of Ions	$r_+ + r_-$	Lattice Energy (kJ/mol)	Melting Point (K)
NaCl	1+, 1−	116 pm + 167 pm = 283 pm	−786	1073
BaO	2+, 2−	149 pm + 126 pm = 275 pm	−3054	2193
MgO	2+, 2−	86 pm + 126 pm = 212 pm	−3791	3073

magnesium ion is more concentrated (higher charge density) than the 2+ charge on barium ion. Consequently, the net attractive force between Mg^{2+} and O^{2-} is greater than the net attractive force between Ba^{2+} and O^{2-}, and the lattice energy of MgO is more negative. This results in a higher melting point for MgO (3073 K) than for BaO (2193 K).

IN CLOSING

Having studied this chapter, you should be able to . . .

- Use the relationships among frequency, wavelength, and the speed of light for electromagnetic radiation (Section 7.1). ■ End-of-chapter Question assignable in OWL: 7
- Explain the relationship between Planck's quantum theory and the photoelectric effect (Section 7.2). ■ Question 16
- Use the Bohr model of the atom to interpret line emission spectra and the energy absorbed or emitted when electrons in atoms change energy levels (Section 7.3). ■ Question 21
- Calculate the frequency, energy, or wavelength of an electron transition in a hydrogen atom and determine in what region of the electromagnetic spectrum the emission would occur (Section 7.3).
- Explain the use of the quantum mechanical model of the atom to represent the energy and probable location of electrons (Section 7.4). ■ Question 34
- Apply quantum numbers (Section 7.5). ■ Question 30
- Understand the spin properties of electrons and how they affect electron configurations and the magnetic properties of atoms (Section 7.5). ■ Question 55
- Describe and explain the relationships among shells, subshells, and orbitals (Section 7.5). ■ Question 32
- Describe how the shapes of *s*, *p*, and *d* orbitals and their sets of the first three quantum numbers differ (Section 7.6).
- Use the periodic table to write the electron configurations of atoms and ions of main group and transition elements (Sections 7.7, 7.8). ■ Questions 39, 40, 43
- Explain variations in valence electrons, electron configurations, ion formation, and paramagnetism of transition metals (Section 7.8). ■ Question 45
- Describe trends in atomic radii, based on electron configurations (Section 7.9). ■ Question 58
- Describe the role of effective nuclear charge in the trend of atomic radii across a period (Section 7.9).
- Describe trends in ionic radii and explain why ions differ in size from their atoms (Section 7.10). ■ Question 60
- Use electron configurations to explain trends in the ionization energies of the elements (Section 7.11). ■ Questions 64
- Describe electron affinity (Section 7.12).
- Describe the relationship between ion formation and the formation of ionic compounds (Section 7.13).
- Discuss the energy changes that occur during the formation of an ionic compound from its elements (Section 7.13).
- Use a Born-Haber cycle to determine the lattice energy of an ionic compound (Section 7.13). ■ Question 67

Selected end-of-chapter Questions may be assigned in OWL.

For quick review, download Go Chemistry mini-lecture flashcard modules (or purchase them at **www.ichapters.com**).

Prepare for an exam with a **Summary Problem** that brings together concepts and problem-solving skills from throughout the chapter. Go to **www.cengage.com/chemistry/moore** to download this chapter's Summary Problem from the Student Companion Site.

KEY TERMS

atomic radius *(Section 7.9)*
Born-Haber cycle *(7.13)*
boundary surface *(7.4)*
continuous spectrum *(7.3)*
core electrons *(7.7)*
diamagnetic *(7.8)*
effective nuclear charge *(7.9)*
electromagnetic radiation *(7.1)*
electron affinity *(7.12)*
electron configuration *(7.7)*
electron density *(7.4)*
excited state *(7.3)*
ferromagnetic *(7.8)*
frequency *(7.1)*
ground state *(7.3)*
Hund's rule *(7.7)*
ionic radii *(7.10)*
ionization energy *(7.11)*
isoelectronic *(7.8)*
lattice energy *(7.13)*
Lewis dot symbol *(7.7)*
line emission spectrum *(7.3)*
momentum *(7.4)*
noble gas notation *(7.7)*
orbital *(7.4)*
paramagnetic *(7.8)*
Pauli exclusion principle *(7.5)*
***p*-block elements** *(7.7)*
photoelectric effect *(7.2)*
photons *(7.2)*
Planck's constant *(7.2)*
principal energy level *(7.5)*
principal quantum number *(7.3)*
quantum *(7.2)*
quantum number *(7.5)*
quantum theory *(7.2)*
radial distribution plot *(7.6)*
***s*-block elements** *(7.7)*
screening effect *(7.9)*
shell *(7.5)*
spectrum *(7.1)*
subshell *(7.5)*
transition elements *(7.7)*
uncertainty principle *(7.4)*
valence electrons *(7.7)*
wavelength *(7.1)*

QUESTIONS FOR REVIEW AND THOUGHT

■ denotes questions assignable in OWL.

Blue-numbered questions have short answers at the back of this book and fully worked solutions in the *Student Solutions Manual.*

Review Questions

1. Light is given off by a sodium- or mercury-containing streetlight when the atoms are excited in some way. The light you see arises for which of these reasons?
   (a) Electrons moving from a given quantum level to one of higher n.
   (b) Electrons being removed from the atom, thereby creating a metal cation.
   (c) Electrons moving from a given quantum level to one of lower n.
   (d) Electrons whizzing about the nucleus in an absolute frenzy.
2. Tell what happens to atomic size and ionization energy across a period and down a group.
3. Why is the radius of Na^+ much smaller than the radius of Na? Why is the radius of Cl^- much larger than the radius of Cl?
4. Write electron configurations to show the first two ionization steps for potassium. Explain why the second ionization energy is much larger than the first.

Topical Questions

Electromagnetic Radiation

5. The regions of the electromagnetic spectrum are shown in Figure 7.1. Answer these questions on the basis of this figure.
   (a) Which type of radiation involves less energy, radio waves or infrared radiation?
   (b) Which radiation has the higher frequency, radio waves or microwaves?
6. The colors of the visible spectrum and the wavelengths corresponding to the colors are given in Figure 7.1.
   (a) Which colors of light involve photons with less energy than yellow light?
   (b) Which color of visible light has photons of greater energy, green or violet?
   (c) Which color of light has the greater frequency, blue or green?
7. ■ Assume that a microwave oven operates at a frequency of $1.00 \times 10^{11}\ s^{-1}$.
   (a) What is the wavelength of this radiation in meters?
   (b) What is the energy in joules per photon?
   (c) What is the energy per mole of photons?
8. IBM scientists have developed a prototype computer chip that operates at 350 GHz (1 GHz = 10^9 Hz).
   (a) What is its wavelength in meters? In nanometers (nm)?
   (b) Calculate its energy in joules.

9. If green light has a wavelength of 495 nm, what is its frequency?
10. Which has more energy,
    (a) One photon of infrared radiation or one photon of microwave radiation?
    (b) One photon of yellow light or one photon of orange light?
11. Stratospheric ozone absorbs damaging UV-C radiation from the sun, preventing the radiation from reaching the earth's surface. Calculate the frequency and energy of UV-C radiation that has a wavelength of 270 nm.
12. Calculate the energy of one photon of X-radiation having a wavelength of 2.36 nm, and compare it with the energy of one photon of orange light (3.18×10^{-19} J).
13. Green light of wavelength 516 nm is absorbed by an atomic gas. Calculate the energy difference between the two quantum states involved with this absorption.

Photoelectric Effect

14. Describe the role Einstein's explanation of the photoelectric effect played in the development of the quantum theory.
15. Light of very long wavelength strikes a photosensitive metallic surface and no electrons are ejected. Explain why increasing the intensity of this light on the metal still will not cause the photoelectric effect.
16. ■ To eject electrons from the surface of potassium metal requires a minimum energy of 3.69×10^{-19} J. When 600.-nm photons shine on a potassium surface, will they cause the photoelectric effect?
17. A bright red light strikes a photosensitive surface and no electrons are ejected, even though dim blue light ejects electrons from the surface. Explain.

Atomic Spectra and the Bohr Atom

18. Which transition involves the emission of less energy in the H atom, an electron moving from $n = 4$ to $n = 3$ or an electron moving from $n = 3$ to $n = 1$? (See Figure 7.8.)
19. For which of these transitions in a hydrogen atom is energy absorbed? Emitted?
    (a) $n = 1$ to $n = 3$ (b) $n = 5$ to $n = 1$
    (c) $n = 2$ to $n = 4$ (d) $n = 5$ to $n = 4$
20. For the transitions in Question 19:
    (a) Which ones involve the ground state?
    (b) Which one involves the greatest energy change?
    (c) Which one absorbs the most energy?
21. ■ If energy is absorbed by a hydrogen atom in its ground state, the atom is excited to a higher energy state. For example, the excitation of an electron from the energy level with $n = 1$ to a level with $n = 4$ requires radiation with a wavelength of 97.3 nm. Which of these transitions would require radiation of a wavelength longer than this? (See Figure 7.8.)
    (a) $n = 2$ to $n = 4$ (b) $n = 1$ to $n = 3$
    (c) $n = 1$ to $n = 5$ (d) $n = 3$ to $n = 5$
22. (a) Calculate the wavelength, in nanometers, of the emission line that results from the $n = 2$ to $n = 1$ transition in hydrogen.
    (b) The emitted radiation is in what region of the electromagnetic spectrum?
23. The Brackett series of emissions has $n_f = 4$.
    (a) Calculate the wavelength, in nanometers, of the $n = 7$ to $n = 4$ electron transition.
    (b) The emitted radiation is in what region of the electromagnetic spectrum?
24. Calculate the energy and the wavelength of the electron transition from $n = 1$ to $n = 4$ in the hydrogen atom.
25. Calculate the energy and wavelength for the electron transition from $n = 2$ to $n = 5$ in the hydrogen atom.

de Broglie Wavelength

26. Calculate the de Broglie wavelength of a 4400-lb sport-utility vehicle moving at 75 miles per hour.
27. Calculate the de Broglie wavelength of an electron moving at 5% the speed of light.

Quantum Numbers

28. Assign a set of four quantum numbers for
    (a) *Each* electron in a nitrogen atom.
    (b) The valence electron in a sodium atom.
    (c) A 3*d* electron in a nickel atom.
29. One electron has the set of quantum numbers $n = 3$, $\ell = 1$, $m_\ell = -1$, and $m_s = +\frac{1}{2}$; another electron has the set $n = 3$, $\ell = 1$, $m_\ell = 1$, and $m_s = +\frac{1}{2}$.
    (a) Could the electrons be in the same atom? Explain.
    (b) Could they be in the same orbital? Explain.
30. ■ Some of these sets of quantum numbers (n, ℓ, m_ℓ, m_s) could not occur. Explain why.
    (a) 2, 1, 2, $+\frac{1}{2}$ (b) 3, 2, 0, $-\frac{1}{2}$
    (c) 1, 0, 0, 1 (d) 3, 3, 2, $-\frac{1}{2}$
    (e) 2, 0, 0, $+\frac{1}{2}$
31. Give the n, ℓ, and m_ℓ values for
    (a) Each orbital in the 6*f* sublevel.
    (b) Each orbital in the $n = 5$ level.

Quantum Mechanics

32. ■ How many subshells are there in the electron shell with the principal quantum number $n = 4$?
33. How many subshells are there in the electron shell with the principal quantum number $n = 5$?
34. ■ Bohr pictured the electrons of the atom as being located in definite orbits about the nucleus, just as planets orbit the sun. Criticize this model in view of the quantum mechanical model.
35. How did the Heisenberg uncertainty principle illustrate the fundamental flaw in Bohr's model of the atom?
36. Which type of orbitals are found in the $n = 3$ shell? How many orbitals altogether are found in this shell?

Electron Configurations

37. ■ Write electron configurations for Mg and Sb atoms.
38. Write electron configurations for these atoms.
    (a) Strontium (Sr), named for a town in Scotland.
    (b) Tin (Sn), a metal used in the ancient world. Alloys of tin (solder, bronze, and pewter) are important.

39. ■ Germanium had not been discovered when Mendeleev formulated his ideas of chemical periodicity. He predicted its existence, however, and germanium was found in 1886 by Winkler. Write the electron configuration of germanium.
40. ■ (a) Which ions in this list are likely to be formed: K^{2+}, Cs^{+}, Al^{4+}, F^{2-}, Se^{2-}?
    (b) Which, if any, of these ions have a noble gas configuration?
41. These ground state orbital diagrams are incorrect. Explain why they are incorrect and how they should be corrected.

    (a) Fe [Ar] 4s: [↑] 3d: [↑↓] [↑↓] [↑↓] [↑] [↑] [↑]

    (b) P [Ne] 2s: [↑↓] 2p: [↑↓] [↑] [ ]

    (c) Sn^{2+} [Kr] 4d: [↑↓] [↑↓] [↑↓] [↑↓] [↑] [↑] 5s: [↑↓] 5p: [↑] [↑] [ ]

42. Write the orbital diagrams for
    (a) A nitrogen atom and a nitride, N^{3-}, ion.
    (b) The 3*p* electrons of a sulfur atom and a sulfide, S^{2-}, ion.
43. ■ How many elements are there in the fourth period of the periodic table? Explain why it is not possible for there to be another element in this period.
44. When transition metals form ions, electrons are lost first from which type of orbital? Why?
45. ■ Give the electron configurations of Mn, Mn^{2+}, and Mn^{3+}. Use orbital box diagrams to determine the number of unpaired electrons for each species.
46. Write the electron configurations of chromium: Cr, Cr^{2+}, and Cr^{3+}. Use orbital box diagrams to determine the number of unpaired electrons for each species.
47. Write electron configurations for these elements.
    (a) Zirconium (Zr). This metal is exceptionally resistant to corrosion and so has important industrial applications. Moon rocks show a surprisingly high zirconium content compared with rocks on earth.
    (b) Rhodium (Rh), used in jewelry and in industrial catalysts.
48. The lanthanides, or rare earths, are only "medium rare." All can be purchased for a reasonable price. Give electron configurations for atoms of these lanthanides.
    (a) Europium (Eu), the most expensive of the rare earth elements; 1 g can be purchased for about $90.
    (b) Ytterbium (Yb). Less expensive than Eu, Yb costs only about $12 per gram. It was named for the village of Ytterby in Sweden, where a mineral source of the element was found.

Valence Electrons

49. Locate these elements in the periodic table, and draw a Lewis dot symbol that represents the number of valence electrons for an atom of each element.
    (a) F (b) In
    (c) Te (d) Cs
50. ■ Locate these elements in the periodic table, and draw a Lewis dot symbol that represents the number of valence electrons for an atom of each element.
    (a) Sr (b) Br
    (c) Ga (d) Sb
51. Give the electron configurations of these ions, and indicate which ones are isoelectronic.
    (a) Na^{+} (b) Al^{3+}
    (c) Cl^{-}
52. ■ Give the electron configurations of these ions, and indicate which ones are isoelectronic.
    (a) Ca^{2+} (b) K^{+}
    (c) O^{2-}
53. What is the electron configuration for
    (a) A bromine atom? (b) A bromide ion?
54. (a) What is the electron configuration for an atom of tin?
    (b) What are the electron configurations for Sn^{2+} and Sn^{4+} ions?

Paramagnetism and Unpaired Electrons

55. ■ (a) In the first transition series (in row four of the periodic table), which elements would you predict to be diamagnetic?
    (b) Which element in this series has the greatest number of unpaired electrons?
56. How do the spins of unpaired electrons from paramagnetic and ferromagnetic materials differ in their behavior in a magnetic field?

Periodic Trends

57. Arrange these elements in order of increasing size: Al, B, C, K, Na. (Try doing it without looking at Figure 7.20 and then check yourself by looking up the necessary atomic radii.)
58. ■ Arrange these elements in order of increasing size: Ca, Rb, P, Ge, Sr. (Try doing it without looking at Figure 7.20 and then check yourself by looking up the necessary atomic radii.)
59. Select the atom or ion in each pair that has the larger radius.
    (a) Cl or Cl^{-} (b) Ca or Ca^{2+}
    (c) Al or N (d) Cl^{-} or K^{+}
    (e) In or Sn
60. ■ Select the atom or ion in each pair that has the smaller radius.
    (a) Cs or Rb (b) O^{2-} or O
    (c) Br or As (d) Ba or Ba^{2+}
    (e) Cl^{-} or Ca^{2+}
61. Write electron configurations to show the first two ionization steps for sodium. Explain why the second ionization energy is much larger than the first.
62. Which of these groups of elements is arranged correctly in order of increasing ionization energy?
    (a) C, Si, Li, Ne (b) Ne, Si, C, Li
    (c) Li, Si, C, Ne (d) Ne, C, Si, Li
63. Rank these ionization energies (IE) in order from the smallest value to the largest value. Briefly explain your answer.
    (a) First IE of Be (b) First IE of Li
    (c) Second IE of Be (d) Second IE of Na
    (e) First IE of K
64. ■ Predict which of these elements would have the greatest difference between the first and second ionization energies: Si, Na, P, Mg. Briefly explain your answer.

65. Explain why nitrogen has a higher first ionization energy than does carbon, the preceding element in the periodic table.
66. The first electron affinity of oxygen is negative, the second is positive. Explain why this change in sign occurs.

Born-Haber Cycle

67. ■ Determine the lattice energy for LiCl(s) given these data: enthalpy of sublimation of Li, 161 kJ; IE_1 for Li, 520 kJ; BE of Cl_2(g), 242 kJ; electron affinity of Cl, −349 kJ; enthalpy of formation of LiCl(s), −408.7 kJ.
68. The lattice energy of KCl(s) is −719 kJ/mol. Use data from the text to calculate the enthalpy of formation of KCl.

General Questions

69. How many *p* orbital electron pairs are there in an atom of selenium (Se) in its ground state?
70. Give the symbol of all the ground state elements that have
    (a) No *p* electrons.
    (b) From two to four *d* electrons.
    (c) From two to four *s* electrons.
71. Give the symbol of the ground state element that
    (a) Is in Group 8A but has no *p* electrons.
    (b) Has a single electron in the 3*d* subshell.
    (c) Forms a 1+ ion with a $1s^22s^22p^6$ electron configuration.
72. Answer these questions about the elements X and Z, which have the electron configurations shown.

$$X = [Kr]\ 4d^{10}5s^1 \qquad Z = [Ar]\ 3d^{10}4s^24p^4$$

    (a) Is element X a metal or a nonmetal?
    (b) Which element has the larger atomic radius?
    (c) Which element would have the greater first ionization energy?
73. (a) Place these in order of increasing radius: Ne, O^{2-}, N^{3-}, F^-.
    (b) Place these in order of increasing first ionization energy: Cs, Sr, Ba.
74. Name the element corresponding to each of these characteristics.
    (a) The element whose atoms have the electron configuration $1s^22s^22p^63s^23p^4$
    (b) The element in the alkaline earth group that has the largest atomic radius
    (c) The element in Group 5A whose atoms have the largest first ionization energy
    (d) The element whose 2^+ ion has the configuration [Kr] $4d^6$
    (e) The element whose neutral atoms have the electron configuration [Ar] $3d^{10}4s^1$
75. The ionization energies for the removal of the first electron from atoms of Si, P, S, and Cl are listed in the following table. Briefly rationalize this trend.

Element	First Ionization Energy (kJ/mol)
Si	780
P	1060
S	1005
Cl	1255

76. Which of these ions are unlikely, and why: Cs^+, In^{4+}, V^{6+}, Te^{2-}, Sn^{5+}, I^-?
77. Rank these in order of increasing first ionization energy: Zn, Ca, Ca^{2+}, Cl^-. Briefly explain your answer.
78. Criticize these statements.
    (a) The energy of a photon is inversely related to its frequency.
    (b) The energy of the hydrogen electron is inversely proportional to its principal quantum number *n*.
    (c) Electrons start to enter the fourth energy level as soon as the third level is full.
    (d) Light emitted by an $n = 4$ to $n = 2$ transition will have a longer frequency than that from an $n = 5$ to $n = 2$ transition.
79. A general chemistry student tells a chemistry classmate that when an electron goes from a 2*d* orbital to a 1*s* orbital, it emits more energy than that for a 2*p* to 1*s* transition. The other student is skeptical and says that such an energy change is not possible and explains why. What explanation was given?
80. Which of these types of radiation—infrared, visible, or ultraviolet—is required to ionize a hydrogen atom? Explain.

Applying Concepts

81. What compound will most likely form between chlorine and element X, if element X has the electronic configuration $1s^22s^22p^63s^1$?
82. Write the formula for the compound that most likely forms between potassium and element Z, if element Z has the electronic configuration $1s^22s^22p^63s^23p^4$.
83. ■ Which of these electron configurations are for atoms in the ground state? In excited states? Which are impossible?
    (a) $1s^22s^1$ (b) $1s^22s^22p^3$
    (c) [Ne] $3s^23p^34s^1$ (d) [Ne] $3s^23p^64s^33d^2$
    (e) [Ne] $3s^23p^64f^4$ (f) $1s^22s^22p^43s^2$
84. Which of these electron configurations are for atoms in the ground state? In excited states? Which are impossible?
    (a) $1s^22s^2$ (b) $1s^22s^23s^1$
    (c) [Ne] $3s^23p^84s^1$ (d) [He] $2s^22p^62d^2$
    (e) [Ar] $4s^23d^3$ (f) [Ne] $3s^22p^54s^1$
85. These questions refer to the graph below.

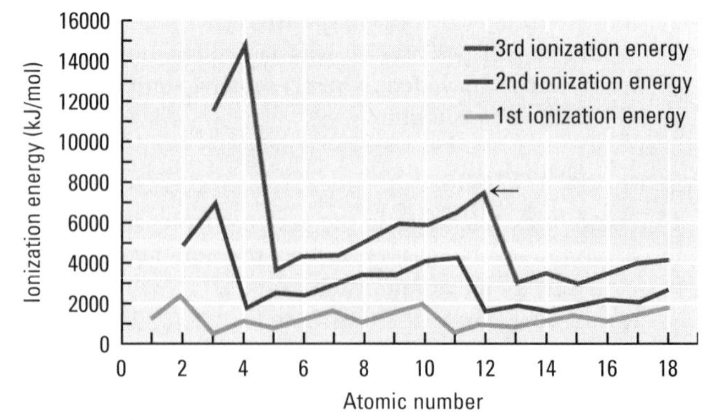

    (a) Based on the graphic data, ionization energies ________ (decrease, increase) left to right and ________ (decrease, increase) top to bottom on the periodic table.
    (b) Which element has the largest first ionization energy?

(c) A plot of the fourth ionization energy versus atomic number for elements 1 through 18 would have peaks at which atomic numbers?
(d) Why is there no third ionization energy for helium?
(e) What is the reason for the large second ionization energy for lithium?
(f) Find the arrow pointing to the third ionization energy curve. Write the equation for the process corresponding to this data point.

86. Use Coulomb's law to predict which substance in each of these pairs has the larger lattice energy.
(a) CaO or KI (b) CaF_2 or BaF_2
(c) KCl or LiBr

More Challenging Questions

87. An element has an ionization energy of 1.66×10^{-18} J. The three longest wavelengths in its absorption spectrum are 253.7 nm, 185.0 nm, and 158.5 nm.
(a) Construct an energy-level diagram similar to Figure 7.8 for this element.
(b) Indicate all possible emission lines. Start from the highest energy level on the diagram.

88. The energy of a photon needed to cause ejection of an electron from a photoemissive metal is expressed as the sum of the binding energy of the photon plus the kinetic energy of the emitted electron. When photons of 4.00×10^{-7} m light strike a calcium metal surface, electrons are ejected with a kinetic energy of 6.3×10^{-20} J.
(a) Calculate the binding energy of the calcium electrons.
(b) The minimum frequency at which the photoelectric effect occurs for a metal is that for which the emitted electron has no kinetic energy. Calculate the minimum frequency required for the photoelectric effect to occur with calcium metal.

89. Calculate the kinetic energy of an electron that is emitted from a strontium metal surface irradiated with photons of 4.20×10^{-7} m light. The binding energy of strontium is 4.39×10^{-19} J. See Question 88 for a statement about binding energy.

90. Ionization energy is the minimum energy required to remove an electron from a ground state atom. According to Niels Bohr, the total energy of an electron in a stable orbit of quantum number n is equal to

$$E_n = -\frac{Z^2}{n^2}(2.18 \times 10^{-18}\ \text{J})$$

where Z is the atomic number. Calculate the ionization energy for the electron in a ground state hydrogen atom.

91. Oxygen atoms are smaller than nitrogen atoms, yet oxygen has a lower first ionization energy than nitrogen. Explain.

92. Beryllium atoms are larger than boron atoms, yet boron has a lower first ionization energy than beryllium. Explain.

93. Suppose two electrons in the same system each have $n = 3$, $\ell = 0$.
(a) How many different electron arrangements would be possible if the Pauli exclusion principle did not apply in this case?
(b) How many would apply if it is operative?

94. Although yet unnamed, element 112 is known and its existence verified.
(a) The chemical properties of the element can be expected to be those of what kind of element—main group, transition metal, lanthanide, or actinide?
(b) Using the noble gas notation, write the electron configuration for a ground state atom of element 112 to corroborate your answer to part (a).

95. You are given the atomic radii of 110 pm, 118 pm, 120 pm, 122 pm, and 135 pm, but do not know to which element (As, Ga, Ge, P, Si) these values correspond. Which must be the value for Ge?

Conceptual Challenge Problems

CP7.A (Section 7.2) Planck stated in 1900 that the energy of a single photon of electromagnetic radiation was directly proportional to the frequency of the radiation ($E = h\nu$). The constant h is known as Planck's constant and has a value of 6.626×10^{-34} J · s. Soon after Planck's statement, Einstein proposed his famous equation ($E = mc^2$), which states that the total energy in any system is equal to its mass times the speed of light squared.

According to the de Broglie relation, what is the apparent mass of a photon emitted by an electron undergoing a change from the second to the first energy level in a hydrogen atom? How does the photon mass compare with the mass of the electron (9.109×10^{-31} kg)?

CP7.B (Section 7.7) When D. I. Mendeleev proposed a periodic law around 1870, he asserted that the properties of the elements are a periodic function of their atomic weights. Later, after H. G. J. Moseley measured the charge on the nuclei of atoms, the periodic law could be revised to state that the properties of the elements are a periodic function of their atomic numbers. What would be another way to define the periodic function that relates the properties of the elements?

CP7.C (Sections 7.3, 7.11) Figure 7.8 shows a diagram of the energy states that an electron can occupy in a hydrogen atom. Use this diagram to show that the first ionization energy for hydrogen, 1312 kJ/mol, is correct.

8

Covalent Bonding

© David DeLossey/Photodisc Green/Getty Images

Human body cells are made up of many thousands of compounds, most of which consist of atoms covalently bonded by sharing electron pairs. The chapter explains aspects of these kinds of bonds—their lengths, strengths, and polarities—and describes some molecular compounds and their covalent bonding.

Atoms of elements are rarely found uncombined in nature. Only the noble gases consist of individual atoms. Most nonmetallic elements consist of molecules, and in a solid metallic element each atom is closely surrounded by eight or twelve neighbors. What makes atoms stick to one another? Interactions among valence electrons of bonded atoms form the glue, but how? In ionic compounds, the transfer of one or more valence electrons from a metal atom to the valence shell of a nonmetal atom produces ions whose opposite charges hold the ions in a crystal lattice. Ionic compounds conduct an electric current when melted due to the mobile cations and anions present. But many other compounds, molecular compounds, do not conduct electricity when in the liquid state and correspondingly do not consist of ions. Examples are carbon monoxide, CO; water, H_2O; methane, CH_4; and the millions of organic compounds.

In these molecular compounds atoms are held together by bonds consisting of one or more pairs of electrons *shared* between the bonded atoms. The attraction of positively charged nuclei for electrons between them pulls the nuclei together. This simple idea can account for the bonding in nearly all molecular compounds, allowing us to correlate their structures with their physical and chemical properties. This chapter describes the bonding found in molecules ranging from simple diatomic gases to hydrocarbons and more complex molecules.

8.1 Covalent Bonding

Previously we have seen that cations and anions form substances such as NaCl (Na^+ and Cl^-) and CaO (Ca^{2+} and O^{2-}) with characteristic properties ***(⇐ p. 71).*** For example, ionic compounds are usually solids at room temperature and when molten they conduct an electric current. Many everyday substances, however, are composed of molecules, not ions. Examples are CO_2, H_2O, CH_4, and $C_6H_{12}O_6$ (glucose). Such molecular compounds have properties very different from those of ionic compounds. The melting points of molecular compounds are lower, and when melted, molecular compounds do not conduct an electric current. In a molecular compound, the atoms that form each molecule are held together in a specific pattern by chemical bonds.

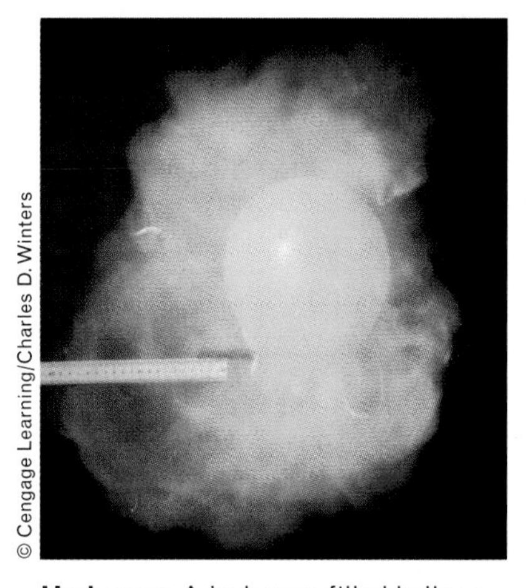

Hydrogen. A hydrogen-filled balloon burning in air forms water.

An explanation of bonding in the simplest molecule H_2 was proposed in 1916 by G. N. Lewis, who suggested that valence electrons rearrange to give noble gas electron configurations when atoms join together chemically. Lewis regarded the lack of reactivity of the noble gases as due to their stable electron configuration, a filled outermost shell. Lewis proposed that an attractive force called a **covalent bond** results when one or more pairs of electrons are shared between the bonded atoms. ***Covalent bonds, which are shared electron pairs, connect the atoms in molecular (covalent) compounds.*** For example, the two H atoms in H_2 and the H and Cl atoms in HCl are held together by covalent bonds.

But why does sharing electrons provide an attractive force between bonded atoms? Consider the formation of the simplest stable molecule, H_2 (Figure 8.1). If the two hydrogen atoms are widely separated, there is little if any interaction between them. When the two atoms get close enough, however, their 1*s* electron clouds overlap. This overlap allows the electron from each atom to be attracted by the other atom's nucleus, causing a net attraction between the two atoms. Experimental data and calculations indicate that an H_2 molecule has its lowest potential energy and is therefore most stable when the nuclei are 74 pm apart (Figure 8.1). At that distance—*the bond length*—the attractive and repulsive electrostatic forces are balanced. If the nuclei get closer than 74 pm, repulsion of each nucleus by the other begins to take over. When the H nuclei are 74 pm apart, it takes 436 kJ of energy to break the hydrogen-to-hydrogen covalent bonds when a mole of gaseous H_2 molecules is converted into isolated H atoms. This energy is the *bond energy* of H_2 ***(⇐ p. 202).***

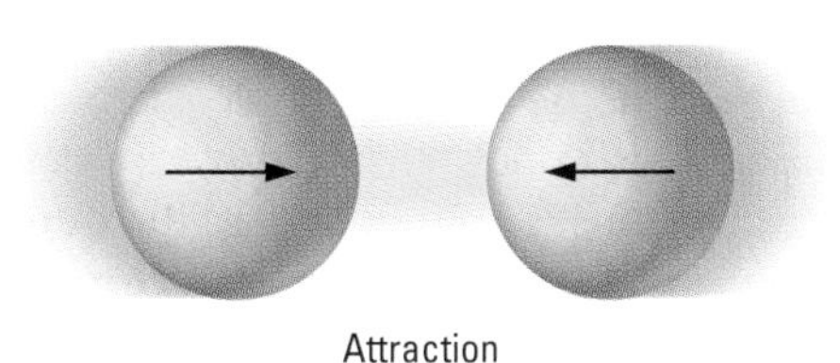

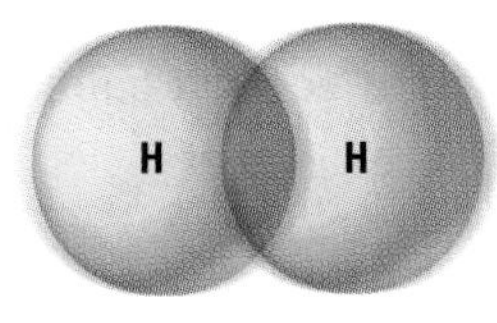

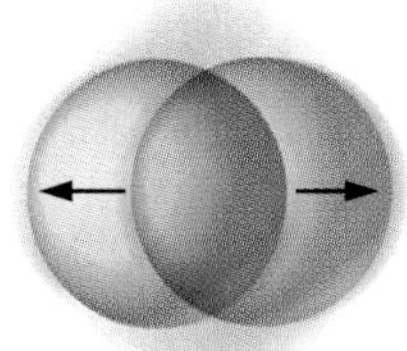

Lowering potential energy is favorable to bond formation.

Potential energy (kJ/mol)
600
0
−436
−600
Higher energy than separated atoms
Lower energy than separated atoms
1 Hydrogen atoms are too close together; nuclei and electrons repel.
2 Atoms are at right distance for stable bond.
3 Atoms are so far apart they do not interact; zero energy.
74 pm
0 74 100 200 300 400 500
Distance between nuclei (pm)

Figure 8.1 H—H bond formation from isolated H atoms. Energy is at a minimum at an internuclear distance of 74 pm, where there is a balance between electrostatic attractions and repulsions.

8.2 Single Covalent Bonds and Lewis Structures

Oesper Collection in the History of Chemistry, University of Cincinnati

Gilbert Newton Lewis
1875–1946

In 1916, G. N. Lewis introduced the theory of the shared electron pair chemical bond in a paper published in the *Journal of the American Chemical Society.* His theory revolutionized chemistry, and it is in honor of this contribution that we refer to "electron dot" structures as Lewis structures. Of particular interest in this text is the extension of his theory of bonding to a generalized theory of acids and bases (Section 15.9).

The shared electron pairs of covalent bonds occupy the same shell as the valence electrons of each atom. For main group elements, the electrons commonly contribute to a noble gas configuration on each atom. Lewis further proposed that by counting the valence electrons of an atom, it would be possible to predict how many bonds that atom can form. ***The number of covalent bonds an atom can form is determined by the number of electrons that the atom must share to achieve a noble gas configuration.***

A **single covalent bond** is formed when two atoms share one pair of electrons. The simplest examples are the bonds in diatomic molecules such as H_2, F_2, and Cl_2, which, like other simple molecules, can be represented by Lewis structures. A **Lewis structure** for a molecule shows all valence electrons as dots or lines that represent covalent bonds.

Lewis structures are drawn by starting with the Lewis dot symbols for the atoms ***(p. 249, Table 7.6)*** and arranging the valence electrons until each atom in the molecule has a noble gas configuration. For example, the Lewis structure for H_2 shows two bonding electrons shared between two hydrogen atoms.

H:H or H—H

The shared electron pair of a *single* covalent bond is often represented by a line instead of a pair of dots. Note that *each* hydrogen atom shares the pair of electrons, thereby achieving the same two-electron configuration as helium, the simplest noble gas.

Atoms with more than two valence electrons achieve a noble gas structure by sharing enough electrons to attain an octet of valence electrons. This is known as the **octet rule:** ***To form bonds, main group elements gain, lose, or share electrons to achieve a stable electron configuration characterized by eight valence elec-***

trons. To obtain the Lewis structure for F_2, for example, we start with the Lewis dot symbol for a fluorine atom. Fluorine, in Group 7A, has seven valence electrons, so there are seven dots, one less than an octet. If each F atom in F_2 shares one valence electron with the other F atom to form a shared electron pair, a single covalent bond forms, and each fluorine atom achieves an octet. ***Shared electrons are counted with each of the atoms in the bond.***

Achieving a noble gas configuration is also referred to as "obeying the octet rule" because all noble gases except helium have eight valence electrons.

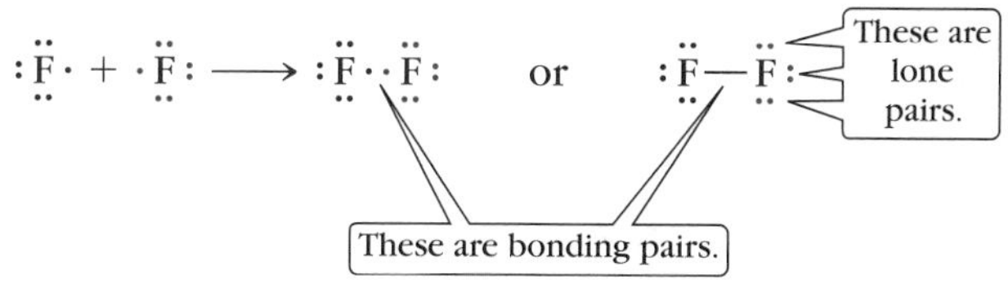

:F̤̈· + ·F̤̈: ⟶ :F̤̈··F̤̈: or :F̤̈—F̤̈:

These are lone pairs.

These are bonding pairs.

A Lewis structure such as that for F_2 shows valence electrons in a molecule as **bonding electrons** (shared electron pairs) and **lone pair electrons** (unshared pairs). In a Lewis structure the atomic symbols, such as F, represent the nucleus and core (nonvalence) electrons of each atom in the molecule. The pair of electrons shared between the two fluorine atoms is a *bonding electron pair.* The other three pairs of electrons on each fluorine atom are *lone pair electrons.* In writing Lewis structures, the bonding pairs of electrons are usually indicated by lines connecting the atoms they hold together; lone pairs are usually represented by pairs of dots.

The term "lone pairs" will be used in this text to refer to unshared electron pairs.

Magnetic measurements (Section 7.7) support the concept that each electron in a pair of electrons (bonding or lone pair) has its spin opposite that of the other electron.

What about Lewis structures for molecules such as H_2O or NH_3? Oxygen (Group 6A) has six valence electrons and must share two electrons to satisfy the octet rule. This can be accomplished by forming covalent bonds with two hydrogen atoms.

H· + H· + ·Ö· forms H:Ö:H or H—Ö—H

Nitrogen (Group 5A) in NH_3 must share three electrons to achieve a noble gas configuration, which can be done by forming covalent bonds with three hydrogen atoms.

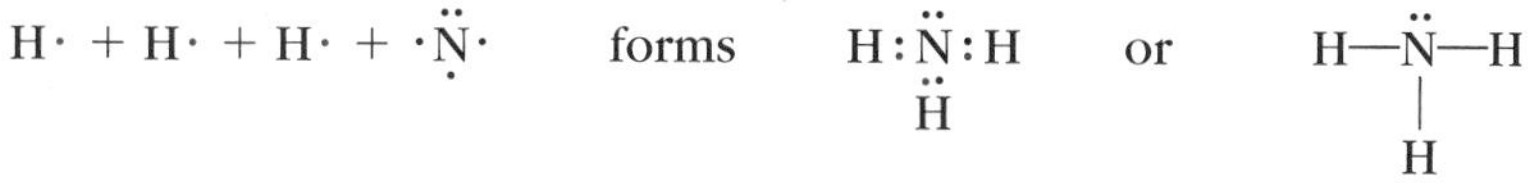

H· + H· + H· + ·N̈· forms H:N̈:H (with :H below) or H—N̈—H (with H bonded below N)

From the Lewis structures of F_2, H_2O, and NH_3, we can make an important generalization: ***The number of electrons that an atom of a main group element must share to achieve an octet equals eight minus its A group number.*** Carbon, for example, which is in Group 4A, needs to share four electrons to reach an octet.

Main group elements are those in the groups labeled "A" in the periodic table inside the front cover of this book.

Group Number	Number of Valence Electrons	Number of Electrons Shared to Complete an Octet (8 − A group number)	Example	
4A	4	4	C in CH_4	H—C—H with H above and below C
5A	5	3	N in NF_3	:F̤̈—N̈—F̤̈: with :F̤̈: below N
6A	6	2	O in H_2O	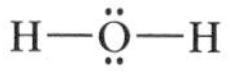 H—Ö—H
7A	7	1	F in HF	H—F̤̈:

Many essential biochemical molecules contain carbon, hydrogen, oxygen, and nitrogen atoms. The structures of these molecules are dictated by the number of bonds C, O, and N need to complete their octets, as pointed out in the table. For example, glycerol, $C_3H_8O_3$,

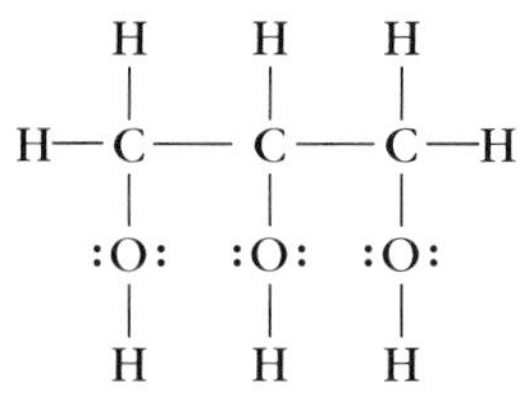

glycerol

is vital to the formation and metabolism of fats. In glycerol, each carbon atom has four shared pairs to complete an octet; each oxygen atom completes an octet by sharing two pairs and having two lone pairs; and hydrogen is satisfied by sharing two electrons. The same type of electron sharing occurs in deoxyribose, $C_5H_{10}O_4$, an essential component of DNA (deoxyribonucleic acid). Four of the carbon atoms are joined in a ring with an oxygen atom.

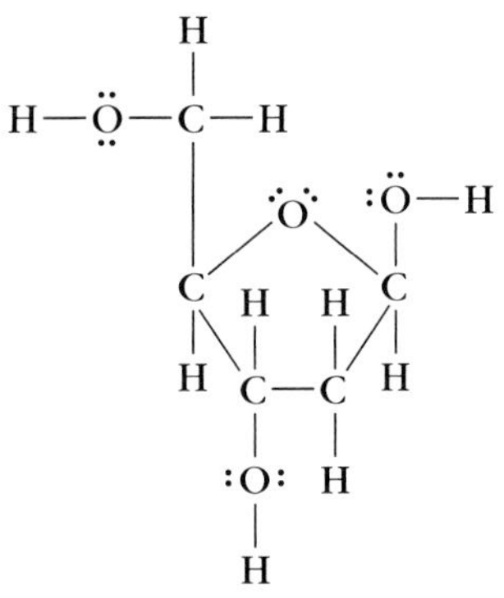

deoxyribose

Guidelines for Writing Lewis Structures

Guidelines have been developed for writing Lewis structures correctly, and we will illustrate using these guidelines to write the Lewis structure for PCl_3.

1. ***Count the total number of valence electrons in the molecule or ion.*** Use the A group number in the periodic table as a guide to indicate the number of valence electrons in each atom. The total of the A group numbers equals the total number of valence electrons of the atoms in a neutral molecule. For a negative ion, add electrons equal to the ion's charge. For a positive ion, subtract the number of electrons equal to the charge. For example, add one electron for the negative charge of OH^- (total of eight valence electrons); subtract one electron for the positive charge of NH_4^+ (total of eight valence electrons: $5 + 4(1) - 1$).

   Because PCl_3 is a neutral molecule, its number of valence electrons is five for P (it is in Group 5A) and seven for *each* Cl (chlorine is in Group 7A): Total number of valence electrons $= 5 + (3 \times 7) = 26$.

2. ***Use atomic symbols to draw a skeleton structure by joining the atoms with shared pairs of electrons (a single line).*** A skeleton structure indicates the attachment of terminal atoms to a central atom. The central atom is usually the one written first in the molecular formula and is the one that can form the most bonds, such as Si in $SiCl_4$ and P in PO_4^{3-}. Hydrogen, oxygen, and the halogens are often terminal atoms. In PCl_3, the central atom is phosphorus, so we draw a skeleton structure with P as the central atom and the three terminal chlorine atoms arranged around it. The three bonding pairs account for six of the total of 26 valence electrons.

Although H is given first in the formulas of H_2O and H_2O_2, for example, it is not the central atom. H is never the central atom in a molecule or ion because hydrogen atoms form only one covalent bond.

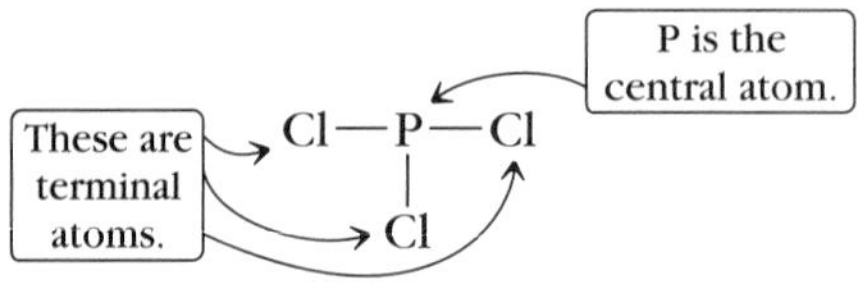

These are terminal atoms.

P is the central atom.

3. ***Place lone pairs of electrons around each atom (except H) to satisfy the octet rule, starting with the terminal atoms.*** Using lone pairs in this way on the Cl atoms accounts for 18 of the remaining 20 valence electrons, leaving two electrons for a lone pair on the P atom. When you check for octets, remember that shared electrons are counted as "belonging" to *each* of the atoms bonded by the

shared pair. Thus, each P—Cl bond has two shared electrons that count for phosphorus and also count for chlorine.

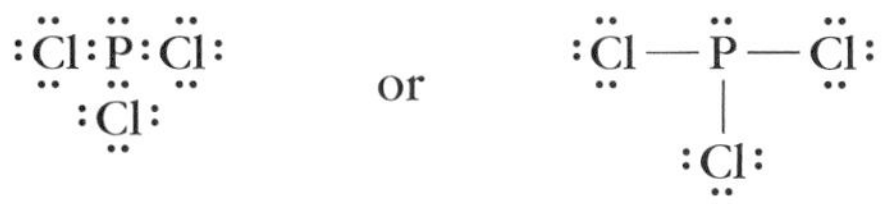

```
 ..  ..  ..              ..    ..    ..
:Cl:P:Cl:        or     :Cl — P — Cl:
 ..  ..  ..              ..    |     ..
    :Cl:                      :Cl:
     ..                        ..
```

Counting dots and lines in the Lewis structure above indicates 26 electrons, accounting for all valence electrons. This is the correct Lewis structure for PCl_3.

The next two steps apply when Steps 1 through 3 result in a structure that does not use all the valence electrons or fails to give an octet of electrons to each atom that should have an octet.

4. ***Place any leftover electrons on the central atom, even if it will give the central atom more than an octet.*** If the central atom is from the third or a higher period, it can accommodate more than an octet of electrons (Section 8.9). Sulfur tetrafluoride, SF_4, is such a molecule. It has a total of 34 valence electrons—six for S (Group 6A) plus seven for each F (Group 7A). In the Lewis structure, each F atom has an octet of electrons giving a total of 32. The remaining two valence electrons are on the central S atom, which, as a third-period element, can accommodate more than an octet of electrons.

```
       ..
      :F:
       |      ..
       :S — F:
  ..  /  |    ..
 :F /    |
  ..    :F:
         ..
```

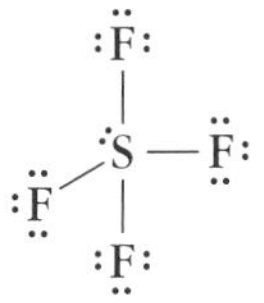

5. ***If the number of electrons around the central atom is less than eight, change single bonds to the central atom to multiple bonds.*** Some atoms can share more than one pair of electrons, resulting in a double covalent bond (two shared pairs) or a triple covalent bond (three shared pairs), known as *multiple bonds* (Section 8.3). Where multiple bonds are needed to complete an octet, use one or more lone pairs of electrons from the terminal atoms to form *double* (two shared pairs) or *triple* (three shared pairs) covalent bonds until the central atom and all terminal atoms have octets. This guideline does not apply to PCl_3, but will be illustrated in Section 8.3.

Double bond: two shared pairs of electrons, as in C=C.

Triple bond: three shared electron pairs, as in C≡N.

We apply these guidelines to write the Lewis structure of phosphate ion, PO_4^{3-}, which has a total of 32 valence electrons (five from P, six from each O, and three for the 3− charge). Phosphorus appears first in the formula and is the central atom. Oxygens are the terminal atoms, giving a skeleton structure of

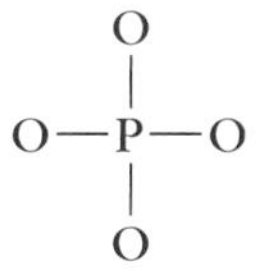

```
       O
       |
   O — P — O
       |
       O
```

The four single P—O bonds account for eight of the valence electrons and provide phosphorus with an octet of electrons. The remaining 24 valence electrons are distributed as three lone pairs around each oxygen to complete their octets.

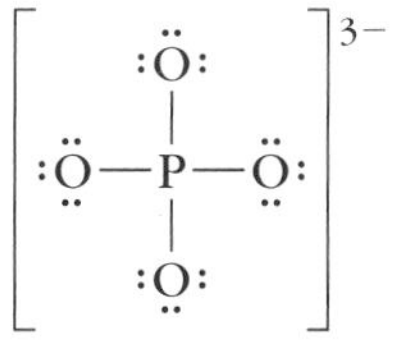

```
 [        ..         ] 3−
 [       :O:         ]
 [  ..    |     ..   ]
 [ :O  —  P  —  O:   ]
 [  ..    |     ..   ]
 [       :O:         ]
 [        ..         ]
```

This is the correct Lewis structure for phosphate ion.

PROBLEM-SOLVING EXAMPLE 8.1 Lewis Structures

Write Lewis structures for these molecules or ions.

(a) HOCl (b) ClO_3^-

(c) SF_6 (d) ClF_4^+

Answer

(a) H—Ö—C̈l: (b) [:Ö—C̈l—Ö: with :Ö: bonded above Cl]$^-$

(c) SF_6 with six :F̤: atoms bonded to S (d) [:F̤—C̈l—F̤: with :F̤: above and below Cl]$^+$

Strategy and Explanation Count up the total valence electrons and distribute them in such a way that each atom (except hydrogen) has an octet of electrons.

(a) There are 14 valence electrons, one from hydrogen (Group 1A), six from oxygen (Group 6A), and seven from chlorine (Group 7A). The central atom is O and the skeleton structure is H—O—Cl. This arrangement uses four of the 14 valence electrons. The remaining ten valence electrons are used by placing three lone pairs on the chlorine and two lone pairs on the oxygen, thereby satisfying the octet rule for oxygen and chlorine; hydrogen needs just two electrons.

(b) There are 26 valence electrons—seven from chlorine and six from each oxygen, plus one for the 1− charge of the ion. Chlorine is the central atom with the three oxygen atoms each single bonded to it. The bonding uses six of the 26 valence electrons, leaving 20 for lone pairs. Placing three lone pairs on each oxygen and a lone pair on chlorine satisfies the octet rule for each atom and accounts for the remaining valence electrons.

(c) Sulfur forms more bonds than fluorine does, so sulfur is the central atom. There are 48 valence electrons: six from sulfur (Group 6A) and seven from each fluorine (Group 7A) for a total of 42 from fluorine. Because sulfur is a Period 3 nonmetal, it can expand its octet to accommodate six fluorine atoms around it, an exception to the octet rule. The 48 valence electrons are distributed as six bonding pairs in S—F bonds and the remaining 36 valence electrons as three lone pairs around each of the six fluorine atoms.

(d) Chlorine, a Period 3 element, can accommodate more than an octet of electrons, appears first in the formula, and is the central atom. There are 34 valence electrons: seven from chlorine, seven for each fluorine, and one subtracted for the 1+ charge. The 34 valence electrons are distributed as four pairs in the four Cl—F bonds and the remaining 26 valence electrons as three lone pairs on each fluorine and one lone pair on the chlorine, making chlorine in this case an exception to the octet rule.

PROBLEM-SOLVING PRACTICE 8.1

Write the Lewis structures for (a) NF_3, (b) N_2H_4, and (c) ClO_4^-.

H—Ö—H

Although Lewis structures are useful for predicting the number of covalent bonds an atom will form, they do not give an accurate representation of where electrons are located in a molecule. Bonding electrons do not stay in fixed positions between nuclei, as Lewis's dots might imply. Instead, quantum mechanics tells us that there is a high probability of finding the bonding electrons between the nuclei. Also, Lewis structures do not convey the shapes of molecules. The angle between the two O—H bonds in a water molecule is not 180°, as the Lewis structure in the margin seems to imply. However, Lewis structures can be used to predict geometries by a method based on the repulsions between valence shell electron pairs (Section 9.1).

Carbon is unique among elements because its atoms form strong bonds with one another as well as with atoms of hydrogen, oxygen, nitrogen, sulfur, and the halogens. The strength of the carbon-carbon bond permits long chains of carbon atoms to form:

```
   |   |   |   |   |   |   |   |   |   |   |   |
 —C—C—C—C—C—C—C—C—C—C—C—C—
   |   |   |   |   |   |   |   |   |   |   |   |
```

Because each carbon atom can form four covalent bonds, such chains contain sites to which other atoms (including more carbon atoms) can bond, leading to the enormous variety of carbon compounds.

Hydrocarbons, compounds of only carbon and hydrogen atoms, have carbon's four valence electrons shared with hydrogen atoms or other carbon atoms. In methane, CH_4, the simplest hydrocarbon, each of carbon's four valence electrons is shared with one electron from each of four hydrogen atoms to form four single covalent C—H bonds. Alkanes, hydrocarbons that contain only C—H and C—C single covalent bonds, are often referred to as **saturated hydrocarbons** because each carbon atom is bonded to a maximum number of hydrogen atoms. The carbon atoms in alkanes with four or more carbon atoms per molecule can be arranged in either a straight chain or a branched chain.

See Table 3.4 for a list of selected alkanes.

```
                                         H
                                         |
    H   H   H   H                   H  H—C—H  H
    |   |   |   |                   |    |    |
 H—C—C—C—C—H                    H—C——C——C—H
    |   |   |   |                   |    |    |
    H   H   H   H                   H    H    H
```

butane
(straight chain)

2-methylpropane
(branched chain)

Bonding in methane can be represented by a Lewis structure or by an *electron density model.* In an electron density model, a ball-and-stick model is surrounded by a space-filling model that represents the distribution of electron density on the surface of the molecule. ***Red indicates regions of higher electron density, and blue indicates regions of lower electron density.***

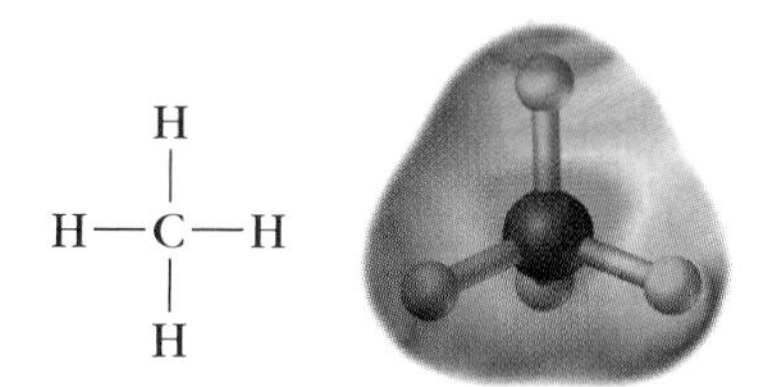

An electron density model of methane.

In addition to straight-chain and branched-chain alkanes, there are *cycloalkanes,* saturated hydrocarbon compounds consisting of carbon atoms joined in rings of —CH_2— units. The cycloalkanes shown below are drawn using a common convention in which only the single bonds between carbon atoms are shown. For example, each line in the drawing of cyclohexane represents a C—C bond. It is assumed that you understand that at the intersection of two lines there is a carbon atom and that each carbon atom forms four bonds. Therefore, neither hydrogen atoms nor bonds to hydrogen atoms are shown. You have to fill in hydrogen atoms and C—H bonds for yourself. The example of cyclopropane, in which a drawing with all bonds and atoms showing is next to a drawing in which only C—C bonds are shown, indicates how this works. The simplest cycloalkane is cyclopropane; other common cycloalkanes include cyclobutane, cyclopentane, and cyclohexane.

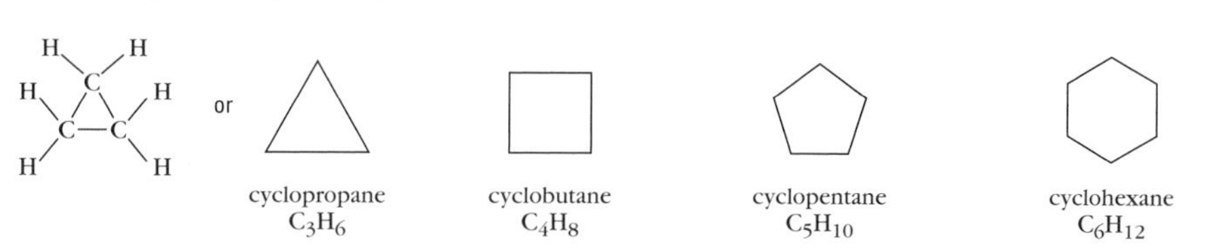

cyclopropane C_3H_6 | cyclobutane C_4H_8 | cyclopentane C_5H_{10} | cyclohexane C_6H_{12}

EXERCISE 8.1 Cyclic Hydrocarbons

Write the Lewis structure and the polygon that represents a molecule of cyclooctane. Write the molecular formula for this compound.

8.3 Multiple Covalent Bonds

A nonmetal atom with fewer than seven valence electrons can form covalent bonds in more than one way. We have seen that the atom can share a single electron with another atom, which can also contribute a single electron, forming a shared electron pair—a *single* covalent bond. But the atom can also share two or three pairs of electrons with another atom, in which case there will be two or three bonds, respectively, between the two atoms. When *two* shared pairs of electrons join the same pair of atoms, the bond is called a **double bond.** When *three* shared pairs are involved, the bond is called a **triple bond.** Double and triple bonds are referred to as **multiple covalent bonds.** Nitrous acid, HNO_2, contains an N=O double bond, and hydrogen cyanide, HCN, a C≡N triple bond.

H—Ö—N̈=Ö: H—C≡N:

nitrous acid hydrogen cyanide

In molecules where there are not enough electrons to complete all octets using just single bonds, one or more lone pairs of electrons from the terminal atoms can be shared with the central atom to form double or triple bonds, so that all atoms have octets of electrons ***(Guideline 5,*** ***p. 275)***. Let's apply this guideline to the Lewis structure for formaldehyde, H_2CO. There is a total of 12 valence electrons (Guideline 1): two from two H atoms (Group 1A), four from the C atom (Group 4A), and six from the O atom (Group 6A). To complete noble gas configurations, H should form one bond, C four bonds, and O two bonds. Because C forms the most bonds, it is the central atom, and we can write this skeleton structure (Guideline 2).

You might have written the skeleton structure O—C—H—H, but remember that H forms only one bond. Another possible skeleton is H—O—C—H, but with this skeleton it is impossible to achieve an octet around carbon without having more than two bonds to oxygen.

H—C—H
|
O

Putting bonding pairs and lone pairs in the skeleton structure according to Guideline 3 yields a structure using all 12 valence electrons in which oxygen has an octet, but carbon does not.

H—C—H
|
:Ö:

We use one of the lone pairs on oxygen as a shared pair with carbon to change the C—O single bond to a C=O double bond (Guideline 5).

H—C—H | :Ö: to form H—C—H |: :O: which is written H—C—H ‖ :O:

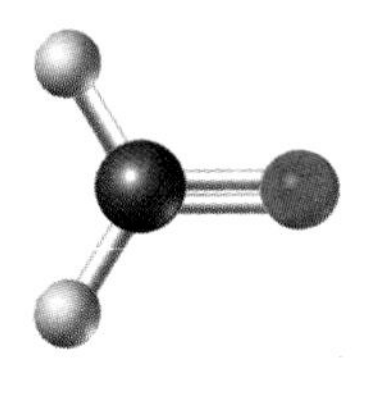

Formaldehyde

This gives carbon and oxygen a share in an octet of electrons, and each hydrogen has a share of two electrons, accounting for all 12 valence electrons and verifying that this is the correct Lewis structure for formaldehyde.

A functional group ***(******p. 67)*** is a distinctive group of atoms in an organic molecule that imparts characteristic chemical properties to the molecule. The C=O combination, called the *carbonyl group,* is one such functional group that is very important in organic and biochemical molecules. The carbonyl-containing —CHO functional group, $-\overset{\overset{\displaystyle O}{\|}}{C}-H$, that is present in formaldehyde is known as the *aldehyde functional group*.

Formaldehyde is the simplest compound with an aldehyde functional group.

As another example of multiple bonds, let's write the Lewis structure for molecular nitrogen, N_2. There is a total of ten valence electrons (five from each N). If two nonbonding pairs of electrons (one pair from each N) become bonding pairs to give a triple bond, the octet rule is satisfied. This is the correct Lewis structure of N_2.

$$\ddot{\underset{..}{N}}-\ddot{\underset{..}{N}} \quad \text{to form} \quad :N\equiv N:$$

CONCEPTUAL
EXERCISE 8.2 Lewis Structures

Why is $:\ddot{\underset{..}{N}}-\ddot{\underset{..}{N}}:$ an incorrect Lewis structure for N_2?

A molecule can have more than one multiple bond, as in carbon dioxide, where carbon is the central atom. There is a total of 16 valence electrons in CO_2, and the skeleton structure uses four of them (two shared pairs):

$$O-C-O$$

Adding lone pairs to give each O an octet of electrons uses up the remaining 12 electrons, but leaves C needing four more valence electrons to complete an octet.

$$:\ddot{\underset{..}{O}}-C-\ddot{\underset{..}{O}}:$$

With no more valence electrons available, the only way that carbon can have four more valence electrons is to use one lone pair of electrons on each oxygen to form a covalent bond to carbon. In this way the 16 valence electrons are accounted for, and each atom has an electron octet.

$$:\ddot{\underset{..}{O}}-C-\ddot{\underset{..}{O}}: \quad \text{forms} \quad :\underset{..}{O}=C=\underset{..}{O}:$$

PROBLEM-SOLVING EXAMPLE 8.2 Lewis Structures

Write Lewis structures for (a) carbon monoxide, CO, an air pollutant; (b) nitrosyl chloride, ClNO, an unstable solid; (c) N_3^-, a polyatomic ion; and (d) HCN, a poison.

Answer

(a) $:C\equiv O:$

(b) $:\ddot{\underset{..}{Cl}}-\ddot{N}=\ddot{O}:$

(c) $[\ddot{\underset{..}{N}}=N=\ddot{\underset{..}{N}}]^-$

(d) $H-C\equiv N:$

Strategy and Explanation Follow the steps given previously to write proper Lewis structures.

(a) The molecule CO contains ten valence electrons (four from C and six from O). We start with a single C—O bond as a skeleton structure.

$$C-O$$

Putting lone pairs around the carbon and oxygen atoms satsfies the octet rule for one of the atoms, but not both. Therefore, lone pairs must become bonding pairs to make up this deficiency and achieve an appropriate Lewis structure.

$$:C-\ddot{\underset{..}{O}}: \quad \text{to form} \quad :C\equiv O:$$

(b) In ClNO, nitrogen is the central atom (it can form more bonds than can oxygen or chlorine) and there are 18 valence electrons (five from N, six from O, and seven from Cl). The skeleton structure is

Cl—N—O

Placing lone pairs on the terminal atoms and a lone pair on the N atom uses the remaining valence electrons, but leaves the N atom without an octet.

:C̤̈l—N̈—Ö:

Converting a lone pair on the oxygen to a bonding pair results in the correct Lewis structure, with each atom having an octet of electrons.

:C̤̈l—N̈=Ö:

(c) There are 16 valence electrons in N_3^- (15 from three N atoms and one for the ion's 1− charge). The skeleton structure with lone pairs on all nitrogen atoms uses all 16 valence electrons.

[N̤̈=N=N̤̈]⁻

is a plausible Lewis structure.

(d) Hydrogen cyanide has a total of ten valence electrons (four from C, five from N, plus one from H.)

H—C≡N:

PROBLEM-SOLVING PRACTICE 8.2

Write Lewis structures for these: (a) nitrosyl ion, NO^+; (b) carbonyl chloride, $COCl_2$.

CONCEPTUAL **EXERCISE** **8.3 Lewis Structures**

Which of these are appropriate Lewis structures and which are not? Explain what is wrong with the incorrect ones.

(a) S bonded to :Ö: (above) and to two :Ö: atoms (b) :F̈=N—C̤̈l: with :C̤̈l: below N (c) H—C=C with H above and below each C (d) :Ö=C—C̤̈l:

H H
| |
C=C
| |
H H
ethylene

H H H
| | |
H—C—C=C
| |
H H
propylene

H H
| |
C=C
| |
H H
unsaturated

H H
| |
H—C—C—H
| |
H H
saturated

8.4 Multiple Covalent Bonds in Hydrocarbons

Carbon atoms are connected by double bonds in some compounds and triple bonds in others as well as by single bonds. **Alkenes** *are hydrocarbons that have one or more carbon-carbon double bonds,* C=C. The general formula for alkenes with one double bond is C_nH_{2n}, where $n = 2, 3, 4$, and so on. The first two members of the alkene series are ethene, $CH_2{=}CH_2$, and propene, $CH_3CH{=}CH_2$, commonly called ethylene and propylene, particularly when referring to the polymers polyethylene and polypropylene.

Alkenes are said to be **unsaturated hydrocarbons.** The carbon atoms connected by double bonds are the *unsaturated sites*—they contain fewer hydrogen atoms than the corresponding saturated alkanes (ethylene, $CH_2{=}CH_2$; ethane, $CH_3{-}CH_3$).

Alkenes are named by using the name of the corresponding alkane (⇐ ***p. 67***) to indicate the number of carbons and the suffix *-ene* to indicate one or more double bonds. The first member, ethylene (ethene), is the most important raw material in the organic chemical industry, where it is used in making polyethylene, antifreeze (ethylene glycol), ethanol, and other chemicals.

CONCEPTUAL EXERCISE **8.4 Alkenes**

(a) Write the molecular formula and structural formula of an alkene with five carbon atoms and one C=C double bond.
(b) How many different alkenes have five carbon atoms and one C=C double bond?

An oxyacetylene torch cutting steel. An oxyacetylene torch cuts through the steel door of a Titan II missile silo that is being retired as part of a disarmament treaty. A mixture of acetylene and oxygen burns with a flame hot enough (3000 °C) to cut steel.

Hydrocarbons with one or more triple bonds, —C≡C—, per molecule are **alkynes.** The general formula for alkynes with one triple bond is C_nH_{2n-2}, where $n = 2, 3, 4$, and so on. The simplest one is ethyne, commonly called acetylene, C_2H_2.

H—C≡C—H

acetylene

Double Bonds and Isomerism

The C=C double bond creates an important difference between alkanes and alkenes—the degree of flexibility of the carbon-carbon bonds in the molecules. The C—C single bonds in alkanes allow the carbon atoms to rotate freely along the C—C bond axis (Figure 8.2). But in alkenes, the C=C double bond prevents such free rotation. This limitation is responsible for the *cis-trans* isomerism of alkenes.

If free rotation could occur around a carbon-carbon double bond, these two molecules would be the same. Therefore, there would only be a single compound.

Two or more compounds with the same molecular formula but different arrangements of atoms are known as isomers (⇐ p. 69). *Cis-trans* isomerism occurs when molecules differ in their arrangement of atoms on either side of a C=C double bond because there is no rotation around the C=C double bond. When two atoms or groups of atoms are attached on the *same* side of the C=C bond, the groups or atoms are said to be *cis* to each other and the compound is the ***cis* isomer;** when two atoms or groups of atoms are on opposite sides, they are *trans* to each other and the compound is the ***trans* isomer.** An example is the *cis* and *trans* isomers of ClHC=CHCl, 1,2-dichloroethene. Because of their different geometries, the two isomers have different physical properties, including melting point, boiling point, and density.

Cis-trans isomerism is also called geometric isomerism.

The 1 and 2 indicate that the two chlorine atoms are attached to the first and second carbon atoms, respectively.

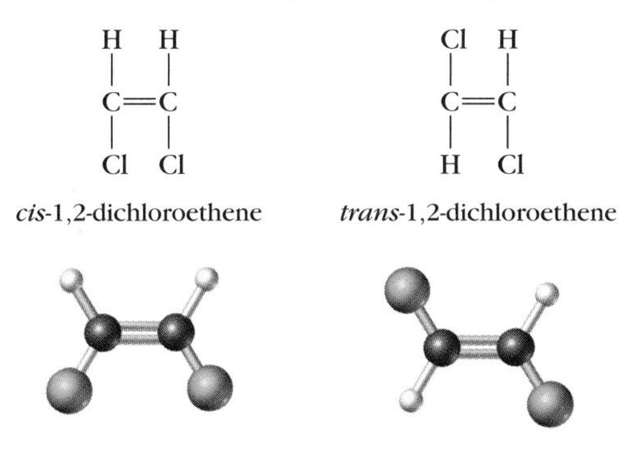
H H
C=C
Cl Cl
cis-1,2-dichloroethene
Cl H
C=C
H Cl
trans-1,2-dichloroethene

The *cis* isomer has two chlorine atoms on the *same* side of the double bond; the *trans* isomer has two chlorine atoms on *opposite* sides of the double bond.

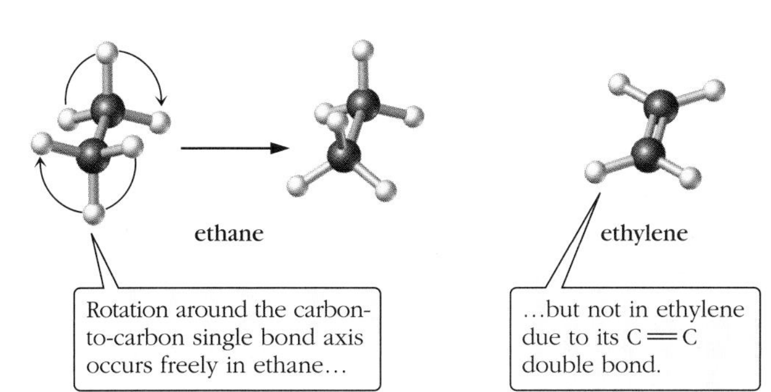
ethane
ethylene
Rotation around the carbon-to-carbon single bond axis occurs freely in ethane...
...but not in ethylene due to its C=C double bond.

Figure 8.2 Nonrotation around C=C. At room temperature, rotation around the carbon-to-carbon axis occurs freely in ethane, but not in ethylene because of its double bond.

Physical Property	*cis*-1,2-dichloroethene	*trans*-1,2-dichloroethene
Melting point	−80.0 °C	−49.8 °C
Boiling point (at 1 atm)	60.1 °C	48.7 °C
Density (at 20 °C)	1.284 g/mL	1.265 g/mL

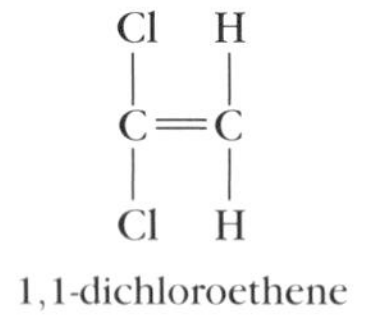

1,1-dichloroethene

Cis-trans **isomerism** in alkenes is possible ***only when each of the carbon atoms connected by the double bond has two different groups attached.*** (For the sake of simplicity, the word "groups" refers to both atoms and groups of atoms.) For example, two chlorine atoms can also bond to the *first* carbon to give 1,1-dichloroethene, which does not have *cis* and *trans* isomers because each carbon atom is attached to two identical atoms (one carbon to two chlorines, the other carbon to two hydrogens).

CONCEPTUAL
EXERCISE 8.5 *Cis-Trans* Isomerism

Maleic acid and fumaric acid are very important molecules that undergo different reactions in metabolism because they are *cis-trans* isomers.

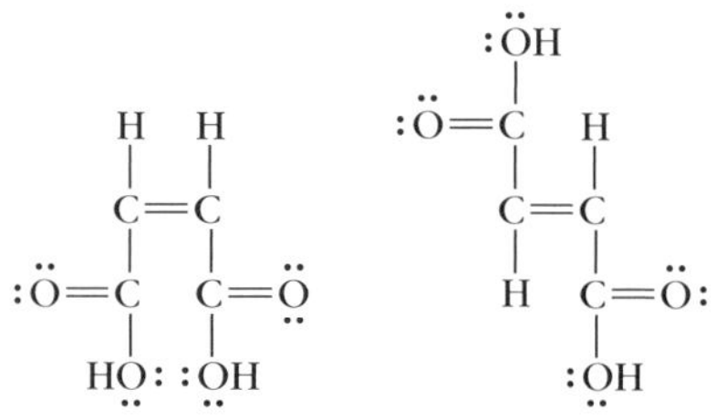

maleic acid

fumaric acid

Identify the *cis* isomer and the *trans* isomer.

When there are four or more carbon atoms in an alkene, the possibility exists for *cis* and *trans* isomers even when only carbon and hydrogen atoms are present. For example, 2-butene has both *cis* and *trans* isomers. (The 2 indicates that the double bond is at the second carbon atom, with the straight carbon chain beginning with carbon 1.) Note the differences in the melting and boiling points of the two compounds, which are due to their structural differences.

Physical Property	***cis*-2-butene**	***trans*-2-butene**
Melting point	−138.9 °C	−105.5 °C
Boiling point (at 1 atm)	3.7 °C	0.88 °C

PROBLEM-SOLVING EXAMPLE 8.3 *Cis* and *Trans* Isomers

Which of these molecules can have *cis* and *trans* isomers? For those that do, write the structural formulas for the two isomers and label them *cis* and *trans*.

(a) $(CH_3)_2C{=}CCl_2$
(b) $CH_3ClC{=}CClCH_3$
(c) $CH_3BrC{=}CClCH_3$
(d) $(CH_3)_2C{=}CBrCl$

Answer (b) and (c)

(b) $H_3C(Cl)C{=}C(CH_3)(Cl)$ *cis*; $(Cl)(H_3C)C{=}C(CH_3)(Cl)$ *trans*

(c) $H_3C(Br)C{=}C(CH_3)(Cl)$ *cis*; $(Br)(H_3C)C{=}C(CH_3)(Cl)$ *trans*

Strategy and Explanation Recall that molecules having two identical groups on a C=C carbon cannot be *cis* and *trans* isomers. Because both of the groups on each carbon are the same, (a) cannot have *cis* and *trans* isomers. In (b), the two —CH_3 groups and the two Cl atoms can both be on the same side of the C=C bond (the *cis* form) or on opposite sides (the *trans* form). The same holds true for the —CH_3 groups in (c). There are two CH_3 groups on the same carbon in (d), so *cis* and *trans* isomers are not possible.

PROBLEM-SOLVING PRACTICE 8.3

Which of these molecules can have *cis* and *trans* isomers? For those that do, write the structural formulas for the two isomers and label them *cis* and *trans*.

(a) $H_3C{-}C(CH_3){=}C(H){-}CH_3$ (b) $H_2C{=}C(H){-}C(H)(H){-}CH_3$ (c) $Br{-}C(H)(H){-}C(Cl){=}C(H){-}CH_3$

8.5 Bond Properties: Bond Length and Bond Energy

Bond Length

The most important factor determining **bond length,** the distance between nuclei of two bonded atoms, is the sizes of the atoms themselves (***p. 254***). The bond length can be considered as the sum of the atomic radii of the two bonded atoms. Bond lengths are given in Table 8.1. As expected, the bond length is greater for larger atoms. Figure 8.3 illustrates the change in atomic size across Periods 2 and 3, and down Groups 4A–6A. Thus, single bonds with carbon increase in length along the series:

C—N < C—C < C—P

Increase in bond length ⟶

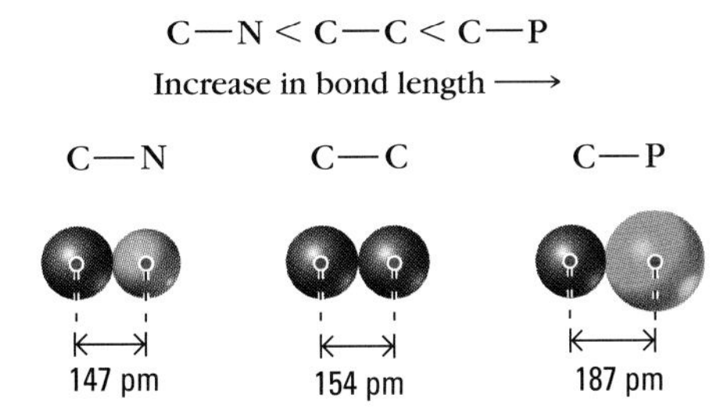

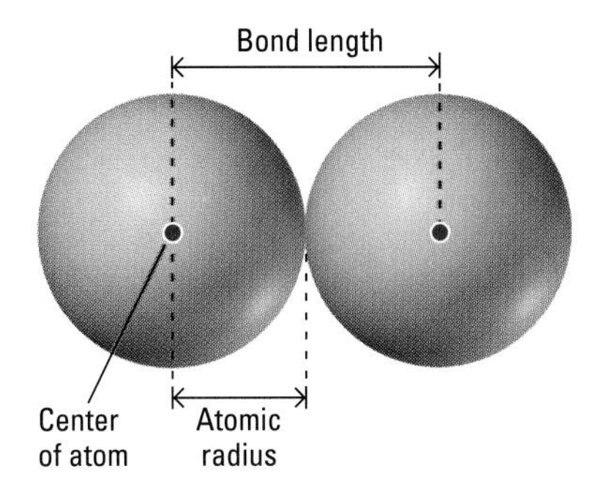

In the case of multiple bonds, a C=O bond is shorter than a C=S bond because S is a larger atom than O. Likewise, a C≡N bond is shorter than a C≡C bond because N is a smaller atom than C (Figure 8.3). Each of these trends can be predicted from the relative sizes shown in Figure 8.3 and is confirmed by the average bond lengths given in Table 8.1.

The effect of bond type is evident when bonds between the same two atoms are compared. For example, structural data show that the bonds become shorter in the series C—O > C=O > C≡O. As the electron density between the atoms increases, the bond lengths decrease because the atoms are pulled together more strongly.

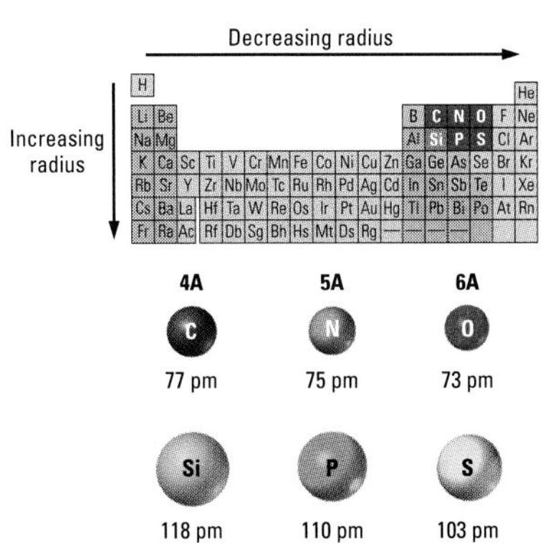

Figure 8.3 Relative atom sizes for second- and third-period elements in Groups 4A, 5A, and 6A.

Table 8.1 Some Average Single and Multiple Bond Lengths (in picometers, pm)*

Single Bonds

	I	Br	Cl	S	P	Si	F	O	N	C	H
H	161	142	127	132	138	145	92	94	98	110	74
C	210	191	176	181	187	194	141	143	147	154	
N	203	184	169	174	180	187	134	136	140		
O	199	180	165	170	176	183	130	132			
F	197	178	163	168	174	181	128				
Si	250	231	216	221	227	234					
P	243	224	209	214	220						
S	237	218	203	208							
Cl	232	213	200								
Br	247	228									
I	266										

Multiple Bonds

Bond	Length	Bond	Length
N=N	120	C=C	134
N≡N	110	C≡C	121
C=N	127	C=O	122
C≡N	115	C≡O	113
O=O (in O_2)	112	N≡O	108
N=O	115		

*1 pm = 10^{-12} m.

Bond lengths are given in picometers (pm) in Table 8.1, but many scientists use nanometers (1 nm = 10^3 pm) or the older unit of Ångströms (Å). 1 Å equals 100 pm. A C—C single bond is 0.154 nm, 1.54 Å, or 154 pm in length.

Bond	Bond length (pm)
C—O	143 pm
C=O	122 pm
C≡O	113 pm

The bond lengths in Table 8.1 are average values, because variations in neighboring parts of a molecule can affect the length of a particular bond. For example, the C—H bond has a length of 105.9 pm in acetylene, HC≡CH, but a length of 109.3 pm in methane, CH_4. Although there can be a variation of as much as 10% from the average values listed in Table 8.1, the average bond lengths are useful for estimating bond lengths and building models of molecules.

PROBLEM-SOLVING EXAMPLE 8.4 Bond Lengths

In each pair of bonds, predict which will be shorter.
(a) P—O or S—O
(b) C≡C or C=C
(c) C=O or C=N

Answer The shorter bonds will be (a) S—O, (b) C≡C, (c) C=O

Strategy and Explanation Apply the periodic trends in atomic radii to determine the relative sizes of the atoms in the question and consequently their relative bond distances.
(a) S—O is shorter than P—O because a P atom is larger than an S atom.
(b) C≡C is shorter than C=C because the more electrons that are shared by atoms, the more closely the atoms are pulled together.
(c) C=O is shorter than C=N because an O atom is smaller than an N atom.

PROBLEM-SOLVING PRACTICE 8.4

Explain the increasing order of bond lengths in these pairs of bonds.
(a) C—S is shorter than C—Si.
(b) C—Cl is shorter than C—Br.
(c) N≡O is shorter than N=O.

Bond Enthalpies

In any chemical reaction, bonds are broken and new bonds are formed. The energy required to break bonds (an *endothermic* reaction) and the energy released when bonds are formed (an *exothermic* reaction) contribute to the enthalpy change for the overall reaction *(⟵ p. 202)*. **Bond enthalpy (bond energy)** *is the enthalpy change that occurs when the bond between two bonded atoms in the gas phase is broken and the atoms are separated completely at constant pressure.*

The two terms "bond enthalpy" and "bond energy" are used interchangeably, although they are not quite equal.

You have seen that as the number of bonding electrons between a pair of atoms increases (single to double to triple bonds), the bond length decreases. It is therefore reasonable to expect that multiple bonds are stronger than single bonds. ***As the electron density between two atoms increases, the bond gets shorter and stronger.*** For example, the bond energy of C=O in CO_2 is 803 kJ/mol and that of C$\equiv$O is 1073 kJ/mol. In fact, the C$\equiv$O triple bond in carbon monoxide is the strongest known covalent bond. Data on the strengths of bonds between atoms in gas phase molecules are summarized in Table 8.2.

The quantity of energy released when 1 mol of a particular bond is made equals that needed when 1 mol of that bond is broken. For example, the H—Cl bond energy is 431 kJ/mol, indicating that 431 kJ must be supplied to break 1 mol H—Cl bonds ($\Delta H° = +431$ kJ). Conversely, when 1 mol H—Cl bonds is formed, 431 kJ is released ($\Delta H° = -431$ kJ).

The data in Table 8.2 can help us understand why an element such as nitrogen, which forms many compounds with oxygen, is unreactive enough to remain in the earth's atmosphere as N_2 molecules even though there is plenty of O_2 to react with. The two N atoms are connected by a very strong bond (946 kJ/mol). Reactions in which N_2 combines with other elements are less likely to occur, because they require breaking this very strong N$\equiv$N bond. This allows us to inhale and exhale N_2 without its undergoing any chemical change. If this were not the case and N_2 reacted readily at body temperature (37 °C) to form oxides and other compounds, there would be severe consequences for us. In fact, life on earth as we know it would not be possible.

We can use data from Table 8.2 to estimate $\Delta H°$ for reactions such as the reaction of hydrogen gas with chlorine gas. Breaking the covalent bond of H_2 requires an input to the system of 436 kJ/mol; breaking the covalent bond in Cl_2 requires 242 kJ/mol.

The bond enthalpies in Table 8.2 are for gas phase reactions. If liquids or solids are involved, there are additional energy transfers for the phase changes needed to convert the liquids or solids to the gas phase. We shall restrict our use of bond enthalpies to gas phase reactions for that reason.

Table 8.2 Average Bond Enthalpies (in kJ/mol)*

Single Bonds

	I	Br	Cl	S	P	Si	F	O	N	C	H
H	299	366	431	347	322	323	566	467	391	416	436
C	213	285	327	272	264	301	486	336	285	356	
N	—	—	193	—	~200	335	272	201	160		
O	201	—	205	—	~340	368	190	146			
F	—	—	255	326	490	582	158				
Si	234	310	391	226	—	226					
P	184	264	319	—	209						
S	—	213	255	226							
Cl	209	217	242								
Br	180	193									
I	151										

Multiple Bonds

N=N	418	C=C	598
N$\equiv$N	946	C$\equiv$C	813
C=N	616	C=O (as in CO_2, O=C=O)	803
C$\equiv$N	866	C=O (as in H_2C=O)	695
O=O (in O_2)	498	C$\equiv$O	1073

*Data from Cotton, F. A., Wilkinson, G., and Gaus, P. L. *Basic Inorganic Chemistry,* 3rd ed. New York: Wiley, 1995, p. 12.

Bond enthalpy is represented by the letter *D* because it refers to *d*issociation of a bond.

In both cases, the sign of the enthalpy change is positive. Forming a covalent bond in HCl transfers 431 kJ/mol out of the system. Since 2 mol HCl bonds are formed, there will be 2 mol × 431 kJ/mol = 862 kJ transferred *out*, which makes the sign of this enthalpy change negative (−862 kJ). If we represent bond enthalpy by the letter *D*, with a subscript to show which bond it refers to, the net transfer of energy is

$$\begin{aligned}\Delta H° &= \{[(1 \text{ mol H—H}) \times D_{\text{H—H}}] + [(1 \text{ mol Cl—Cl}) \times D_{\text{Cl—Cl}}]\} \\ &\quad - [(2 \text{ mol H—Cl}) \times D_{\text{H—Cl}}] \\ &= [(1 \text{ mol H—H})(436 \text{ kJ/mol})] + [(1 \text{ mol Cl—Cl})(242 \text{ kJ/mol})] \\ &\quad - (2 \text{ mol H—Cl})(431 \text{ kJ/mol}) \\ &= -184 \text{ kJ}\end{aligned}$$

This differs only very slightly from the experimentally determined value of −184.614 kJ.

As illustrated in Figure 8.4, we can think of the process in the calculation in terms of breaking all the bonds in each reactant molecule and then forming all the bonds in the product molecules. Each bond enthalpy was multiplied by the number of moles of bonds that were broken or formed. For bonds in reactant molecules, we added the bond enthalpies because breaking bonds is endothermic. For products, we subtracted the bond enthalpies because bond formation is an exothermic process. This is summarized in Equation 8.1.

The Σ (Greek capital letter *sigma*) represents summation. We add the bond enthalpies for all bonds broken, and we subtract the bond enthalpies for all bonds formed. This equation and the values in Table 8.2 allow us to estimate enthalpy changes for a wide variety of gas phase reactions.

$$\begin{aligned}\Delta H° &= \sum[(\text{moles of bonds}) \times D(\text{bonds broken})] \\ &\quad - \sum[(\text{moles of bonds}) \times D(\text{bonds formed})]\end{aligned} \qquad [8.1]$$

There are several important points about the bond enthalpies in Table 8.2:

- The enthalpies listed are often average bond enthalpies and may vary depending on the molecular structure. For example, the enthalpy of a C—H bond is given as

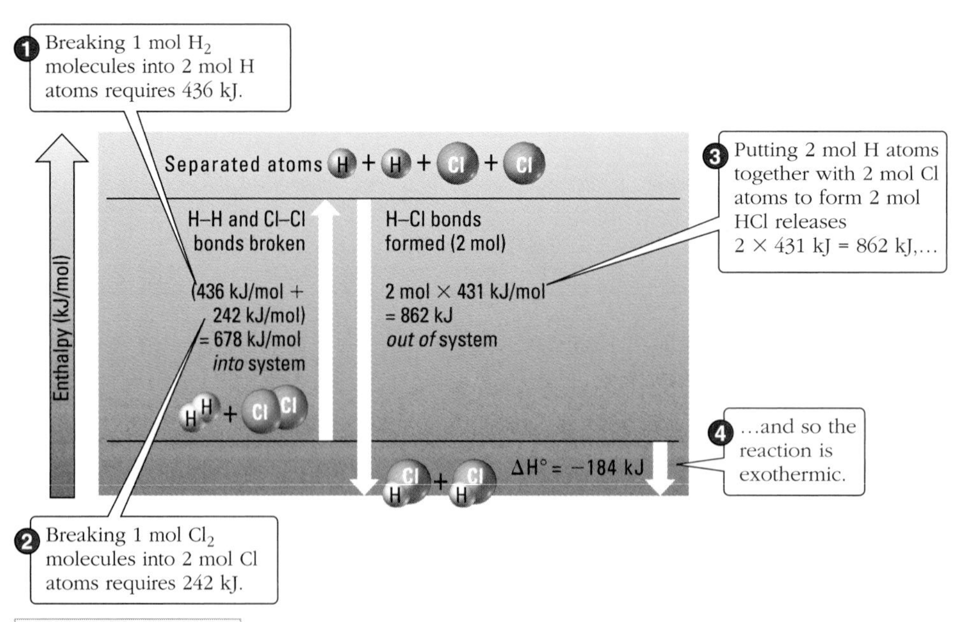

Active Figure 8.4 **Stepwise energy changes in the reaction of hydrogen with chlorine.** The enthalpy change for the reaction is −184 kJ; $\Delta H°$ = 436 kJ + 242 kJ − 862 kJ = −184 kJ. Visit this book's companion website at **www.cengage.com/chemistry/moore** to test your understanding of the concepts in this figure.

413 kJ/mol, but C—H bond strengths are affected by other nearby atoms and bonds in the same molecule. Depending on the structure of the molecule, the energy required to break a mole of C—H bonds may vary by 30 to 40 kJ/mol, so the values in Table 8.2 can be used only to estimate an enthalpy change, not to calculate it exactly.

- The enthalpies in Table 8.2 are for breaking bonds in molecules in the gaseous state. If a reactant or a product is in the liquid or solid state, the energy required to convert it to or from the gas phase will also contribute to the enthalpy change of a reaction and must be accounted for.
- Multiple bonds, shown as double and triple lines between atoms, are listed at the bottom of Table 8.2. In some cases, different enthalpies are given for multiple bonds in specific molecules such as $O_2(g)$ or $CO_2(g)$.

PROBLEM-SOLVING EXAMPLE 8.5 Estimating $\Delta H°$ from Bond Enthalpies

The conversion of diazomethane to ethene and nitrogen is given by this equation:

$$2\ H_2C{=}N{=}\ddot{N} \longrightarrow H_2C{=}CH_2 + 2\ :N{\equiv}N:$$

diazomethane ethene

Using the bond enthalpies in Table 8.2, calculate the $\Delta H°$ for this reaction.

Answer $\Delta H° = -422$ kJ

Strategy and Explanation Begin by using the given Lewis structures, which show that there are two C—H bonds, one C=N bond, and one N=N bond in each diazomethane molecule. In each ethene molecule there are four C—H bonds and one C=C bond; each nitrogen molecule contains one N≡N bond. Bond enthalpies for other than single bonds, such as C=N and C=C, are listed at the bottom of Table 8.2. We can use these data and the balanced chemical equation in the change-in-enthalpy Equation 8.1 for bond breaking and bond making.

$$\Delta H° = \Sigma[(\text{moles of bonds}) \times D(\text{bonds broken})] - \Sigma[(\text{moles of bonds}) \times D(\text{bonds formed})]$$

$$= \{[(4 \text{ mol C—H}) \times D_{C—H}] + [(2 \text{ mol C=N}) \times D_{C=N}] + [(2 \text{ mol N=N}) \times D_{N=N}]\}$$
$$-\{[(4 \text{ mol C—H}) \times D_{C—H}] + [(1 \text{ mol C=C}) \times D_{C=C}] + [(2 \text{ mol N}\equiv\text{N}) \times D_{N\equiv N}]\}$$

$$= \{[(4 \text{ mol C—H}) \times (416 \text{ kJ/mol})] + [(2 \text{ mol C=N}) \times (616 \text{ kJ/mol})]$$
$$+ [(2 \text{ mol N=N}) \times (418 \text{ kJ/mol})]\} - \{[(4 \text{ mol C—H}) \times (416 \text{ kJ/mol})$$
$$+ [(1 \text{ mol C=C}) \times (598 \text{ kJ/mol})]$$
$$+ [(2 \text{ mol N}\equiv\text{N}) \times (946 \text{ kJ/mol})]\}$$
$$= (3732 \text{ kJ} - 4154 \text{ kJ}) = -422 \text{ kJ}$$

☑ **Reasonable Answer Check** Note that the energy for the bond-breaking steps in the reactants (3732 kJ) is less than the energy given off during bond formation in the products (4154 kJ). Therefore, the reaction is exothermic as indicated by the negative $\Delta H°$ value. This means that the bonds in the products are stronger than those of the reactants. Thus, the potential energy of the products is lower than that of the reactants and so $\Delta H°$ will be negative.

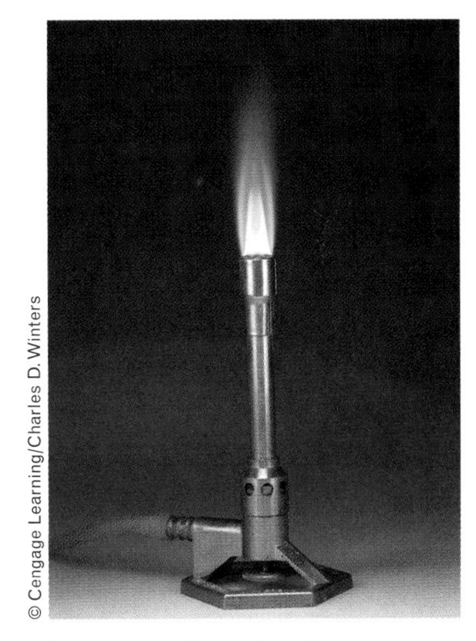

Burning methane in a Bunsen burner. Combustion of methane is highly exothermic, transferring 802 kJ of energy to the surroundings for every mole of CH_4 that burns.

PROBLEM-SOLVING PRACTICE 8.5

Use Equation 8.1 and values from Table 8.2 to estimate the enthalpy change when methane, CH_4, and oxygen combine according to the equation:

$$CH_4(g) + 2\ O_2(g) \longrightarrow CO_2(g) + 2\ H_2O(g)$$

The measured value for the enthalpy of combustion of methane is −802 kJ; the average bond enthalpies used in Problem-Solving Practice 8.5 give a good estimate.

CONCEPTUAL
EXERCISE 8.6 Bond Length

Arrange C=N, C≡N, and C—N in order of decreasing bond length. Is the order for decreasing bond energy the same or the reverse order? Explain.

Oesper Collection in the History of Chemistry, University of Cincinnati

Linus Pauling 1901–1994

The son of a druggist, Linus Pauling completed his Ph.D. in chemistry and then traveled to Europe, where he worked briefly with Erwin Schrödinger and Niels Bohr (see Chapter 7). For his bonding theories and his work with structural aspects of proteins, Pauling was awarded the Nobel Prize in Chemistry in 1954. Shortly after World War II, Pauling and his wife began a crusade to limit nuclear weapons, a crusade culminating in the limited test ban treaty of 1963. For this effort, Pauling was awarded the 1963 Nobel Peace Prize, the first time any person received two unshared Nobel Prizes.

Linus Pauling died in August 1994, leaving a remarkable legacy of breakthrough scientific research and a social consciousness to be remembered in guarding against the possible misapplications of technology.

8.6 Bond Properties: Bond Polarity and Electronegativity

In a molecule such as H_2 or F_2, where both atoms are the same, there is *equal* sharing of the bonding electron pair and the bond is a **nonpolar covalent bond.** When two different atoms are bonded, however, the sharing of the bonding electrons is usually *unequal* and results in a displacement of the bonding electrons toward one of the atoms. If the displacement is complete, electron transfer occurs and the bond is ionic. If the displacement is less than complete, the bonding electrons are shared *unequally,* and the bond is a **polar covalent bond** (Figure 8.5). As you will see in Chapter 9, properties of molecules are dramatically affected by bond polarity.

Linus Pauling, in 1932, first proposed the concept of electronegativity based on an analysis of bond energies. **Electronegativity** represents the ability of an atom in a covalent bond to attract *shared* electrons to itself.

Pauling's electronegativity values (Figure 8.6) are relative numbers with an arbitrary value of 4.0 assigned to fluorine, the most electronegative element. The nonmetal with the next highest electronegativity is oxygen, with a value of 3.5, followed by chlorine and nitrogen, each with the same value of 3.0. Elements with electronegativities of 2.5 or more are all nonmetals in the upper-right corner of the periodic table. By contrast, elements with electronegativities of 1.3 or less are all metals on the left side of the periodic table. These elements are often referred to as the most electropositive elements; *they are the metals that invariably form ionic compounds.* Between these two extremes are most of the remaining metals (largely transition metals) with electronegativities between 1.4 and 1.9, the metalloids with electronegativities between 1.8 and 2.1, and some nonmetals (P, Se) with electronegativities between 2.1 and 2.4.

Electronegativities show a periodic trend (Figure 8.6): Electronegativity *increases* across a period and *decreases* down a group. ***In general, electronegativity increases diagonally upward and to the right in the periodic table.*** Because metals typically lose electrons, they are the least electronegative elements. Nonmetals, which have a tendency to gain electrons, are the most electronegative.

Electronegativity values are approximate and are primarily used to predict the polarity of covalent bonds. Bond polarity is indicated by writing $\delta+$ by the *less* electronegative atom and $\delta-$ by the *more* electronegative atom, where δ stands for partial charge. δ is a fraction—between zero and one. For example, the polar H—Cl bond in hydrogen chloride can be represented as shown on the next page.

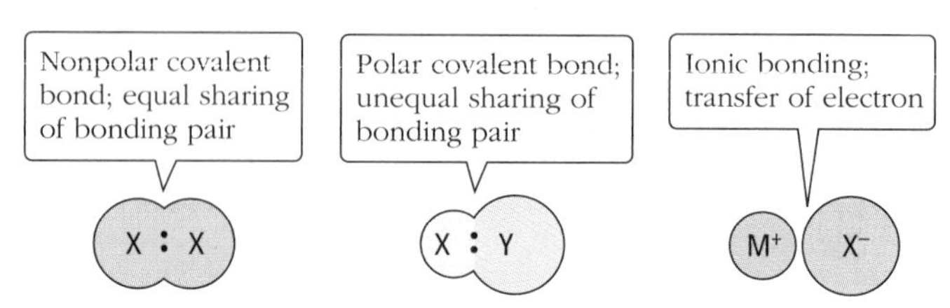

Figure 8.5 **Nonpolar covalent, polar covalent, and ionic bonding.**

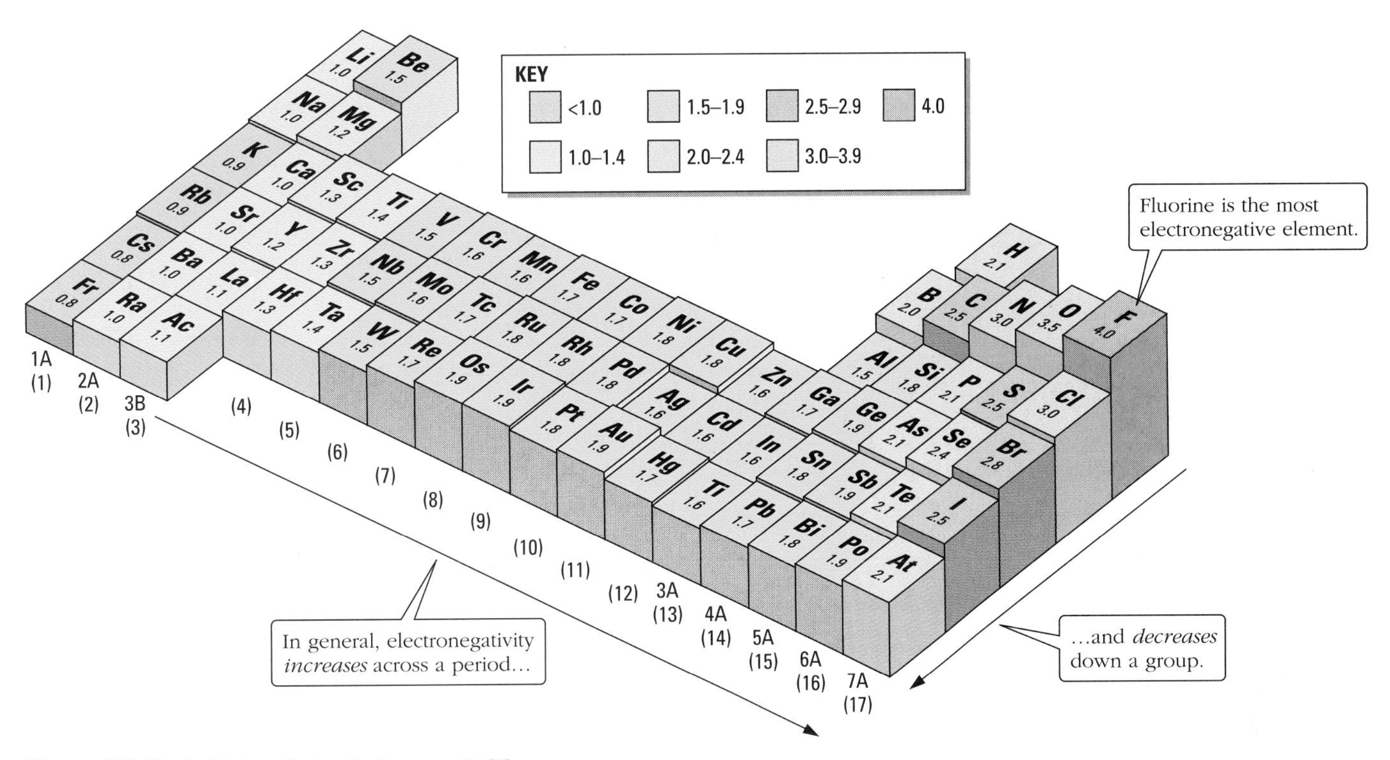

Figure 8.6 Periodic trends in electronegativities.

Except for bonds between identical atoms, all bonds are polar to some extent, and the difference in electronegativity values is a qualitative measure of the degree of polarity. ***The greater the difference in electronegativity between two atoms, the more polar will be the bond between them.*** The change from nonpolar covalent bonds to slightly polar covalent bonds to very polar covalent bonds to ionic bonds can be regarded as a continuum (Figure 8.7). Examples are F_2 (nonpolar), HBr (slightly polar), HF (very polar), and Na^+F^- (ionic).

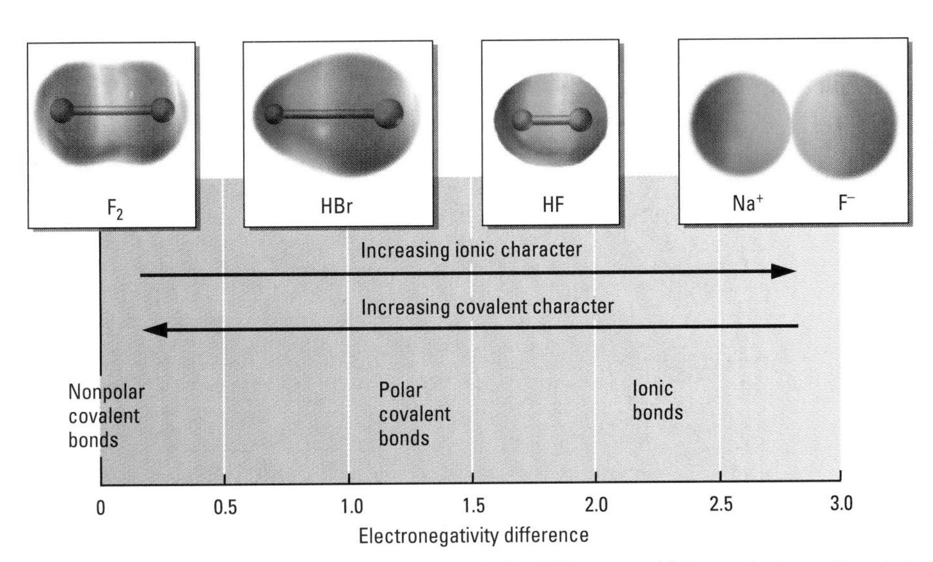

Figure 8.7 Bond character and electronegativity differences. This graph shows the relationship between electronegativity difference and the ionic character of a bond.

$\delta+ \quad \delta-$
H—Cl

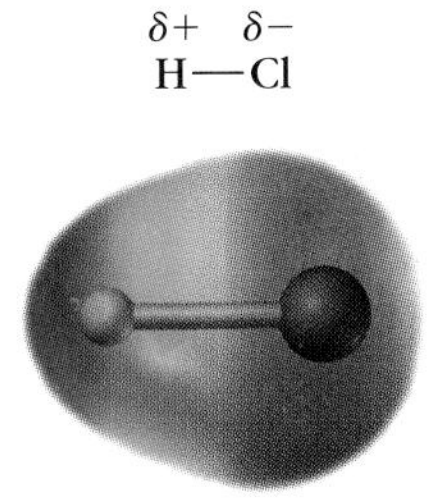

Polar H—Cl bond in hydrogen chloride. Red indicates regions of higher electron density; blue represents regions of lower electron density.

PROBLEM-SOLVING EXAMPLE 8.6 Bond Polarity

For each of these bond pairs, indicate the partial positive and negative atoms and tell which is the more polar bond.

(a) Cl—F and Br—F (b) N—Cl and P—Cl

Answer

(a) $\overset{\delta+}{\text{Br}}$—$\overset{\delta-}{\text{F}}$ (more polar); $\overset{\delta-}{\text{F}}$—$\overset{\delta+}{\text{Cl}}$ (b) $\overset{\delta+}{\text{P}}$—$\overset{\delta-}{\text{Cl}}$ (more polar);

Strategy and Explanation Compare the electronegativities of the two bonded atoms by using Figure 8.6. The atom with the lesser electronegativity will have a partial positive charge and the one with the greater electronegativity will bear a partial negative charge.

(a) The Br—F bond is more polar because the difference in electronegativity is greater between Br and F (1.2) than between Cl and F (1.0). The F atom is the partial negative end in BrF; because F is more electronegative than Cl, the F atom is the partial negative end in ClF.

(b) P—Cl is more polar than N—Cl because P is less electronegative than N, so the electronegativity difference is greater between P and Cl (0.9) than between N and Cl (zero).

PROBLEM-SOLVING PRACTICE 8.6

For each of these pairs of bonds, decide which is the more polar. For each polar bond, indicate the partial positive and partial negative atoms.

(a) B—C and B—Cl (b) N—H and O—H

CONCEPTUAL
EXERCISE 8.7 Bond Types

Use Figures 8.6 and 8.7 to explain why:

(a) NaCl is considered to be an ionic compound rather than a polar covalent compound.

(b) BrF is considered to be a polar covalent compound rather than an ionic compound.

At atmospheric pressure, $AlCl_3$ sublimes at about 180 °C. At roughly this temperature the lattice of Al^{3+} and Cl^- ions changes to Al_2Cl_6 molecules.

In moving across the third period from sodium to sulfur, changes in electronegativity difference cause a significant shift in the properties of compounds of these elements with chlorine. The third period begins with the ionic compounds NaCl and $MgCl_2$; both are crystalline solids at room temperature. The electronegativity differences between the metal and chlorine are 2.0 and 1.8, respectively. Aluminum chloride is less ionic; with an aluminum-chlorine electronegativity difference of 1.5, its bonding is sometimes ionic and sometimes polar covalent. As electronegativity increases across the period, the remaining Period 3 chlorides—$SiCl_4$, PCl_3, and SCl_2—are molecular compounds, with decreasing electronegativity differences between the nonmetal and chlorine from Si to S. This results in a decrease in bond polarity from Si—Cl to P—Cl to S—Cl bonds, culminating in no electronegativity difference in Cl—Cl bonds.

Electronegativities: Si = 1.8; P = 2.1; S = 2.5; Cl = 3.0.

Compound	$SiCl_4$	PCl_3	SCl_2	Cl_2
Bond	Si—Cl	P—Cl	S—Cl	Cl—Cl
Electronegativity difference	1.2	0.9	0.5	0.0

8.7 Formal Charge

Lewis structures depict how valence electrons are distributed in a molecule or ion. For some molecules or ions, more than one Lewis structure can be written, each of which obeys the octet rule. Which of the structures is more correct? How do you decide?

Using formal charge is one way to do so. **Formal charge *is the charge a bonded atom would have if its bonding electrons were shared equally.*** In calculating formal charges, the following assignments are made:

- ***All of the lone pair electrons are assigned to the atom on which they are found.***
- ***Half of the bonding electrons are assigned to each atom in the bond.***
- ***The sum of the formal charges must equal the actual charge: zero for molecules and the ionic charge for an ion.***

Thus, in assigning formal charge to an atom in a Lewis structure,

Formal charge = [*number valence electrons in an atom*] − [(*number lone pair electrons*) + ($\frac{1}{2}$ *number bonding electrons*)]

Applying these rules, we can, for example, calculate the formal charges of the atoms in a cyanate ion, $[:\text{N}{\equiv}\text{C}{-}\ddot{\underset{..}{\text{O}}}:]^-$

	N	C	O
Valence electrons	5	4	6
Lone pair electrons	2	0	6
$\frac{1}{2}$ shared electrons	3	4	1
Formal charge	0	0	−1

Note that the sum of the formal charges equals −1, the charge on the ion. The −1 formal charge is on oxygen, the most electronegative atom in the ion.

It is important to recognize that formal charges do *not* indicate actual charges on atoms. Formal charge is a useful way to determine the most likely structure from among several Lewis structures. In evaluating possible structures with different formal charge distributions, the following principles apply:

- ***Smaller formal charges are more favorable than larger ones.***
- ***Negative formal charges should reside on the more electronegative atoms.***
- ***Like charges should not be on adjacent atoms.***

PROBLEM-SOLVING EXAMPLE 8.7 Formal Charges

(a) Two possible Lewis structures for N_2O are $:\ddot{\text{O}}{=}\text{N}{=}\ddot{\text{N}}:$ and $:\ddot{\underset{..}{\text{O}}}{-}\text{N}{\equiv}\text{N}:$. Determine the formal charges on each atom and the preferred structure.

(b) Determine the formal charges on each atom of these two Lewis structures of dichlorine monoxide, Cl_2O. Which structure is preferred?

Structure 1 $:\ddot{\underset{..}{\text{O}}}{-}\ddot{\underset{..}{\text{Cl}}}{-}\ddot{\underset{..}{\text{Cl}}}:$ Structure 2 $:\ddot{\underset{..}{\text{Cl}}}{-}\ddot{\underset{..}{\text{O}}}{-}\ddot{\underset{..}{\text{Cl}}}:$

Answer

(a) $\overset{0}{:\ddot{\text{O}}}{=}\overset{+1}{\text{N}}{=}\overset{-1}{\ddot{\text{N}}:}$ $\overset{-1}{:\ddot{\underset{..}{\text{O}}}}{-}\overset{+1}{\text{N}}{\equiv}\overset{0}{\text{N}:}$ (preferred)

(b) $\overset{-1}{:\ddot{\underset{..}{\text{O}}}}{-}\overset{+1}{\ddot{\underset{..}{\text{Cl}}}}{-}\overset{0}{\ddot{\underset{..}{\text{Cl}}}:}$ $\overset{0}{:\ddot{\underset{..}{\text{Cl}}}}{-}\overset{0}{\ddot{\underset{..}{\text{O}}}}{-}\overset{0}{\ddot{\underset{..}{\text{Cl}}}:}$ (preferred)

Strategy and Explanation Count the valence electrons, account for the number of lone pair electrons, and assign half the bonding electrons to each atom in the bond. The formal charges should add up to zero for N_2O and in Cl_2O because they are both neutral molecules.

(a) In the first N_2O structure, $:\ddot{O}{=}N{=}\ddot{N}:$, the formal charges for the atoms in order in the structure are 0, +1, and −1. The sum of the formal charges is 0 + (+1) + (−1) = 0. This is as it should be for a neutral molecule. Applying the rules for assigning electrons to the second structure, $:\ddot{\underset{..}{O}}{-}N{\equiv}N:$, the formal charges are (in order) −1, +1, and 0. The sum of the formal charges in this structure is also zero.

	Structure 1			Structure 2		
	O	**N**	**N**	**O**	**N**	**N**
Valence electrons	6	5	5	6	5	5
Lone pair electrons	4	0	4	6	0	2
$\frac{1}{2}$ shared electrons	2	4	2	1	4	3
Formal charge	0	+1	−1	−1	+1	0

Formal charges are low in both structures, but the second structure is preferred because it has the negative charge on the more electronegative atom (O) rather than on the less electronegative N, as in the first structure.

(b)

	Structure 1			Structure 2		
	O	**Cl**	**Cl**	**Cl**	**O**	**Cl**
Valence electrons	6	7	7	7	6	7
Lone pair electrons	6	4	6	6	4	6
$\frac{1}{2}$ shared electrons	1	2	1	1	2	1
Formal charge	−1	+1	0	0	0	0

In both cases, the total formal charge is zero, which it should be for a neutral molecule. Structure 2 is preferred because it has all zero formal charges.

PROBLEM-SOLVING PRACTICE 8.7

A third Lewis structure can be written for N_2O, which also obeys the octet rule. Write this other Lewis structure and determine the formal charges on its atoms.

CONCEPTUAL
EXERCISE 8.8 Formal Charge

Determine the formal charge of each atom in hydrazine, H_2NNH_2.

8.8 Lewis Structures and Resonance

Ozone, O_3, is an unstable, pale blue, diamagnetic gas with a pungent odor. Depending on its location, ozone can be either beneficial or harmful. A low-concentration, but very important, layer of ozone in the stratosphere protects the earth and its inhabitants from intense ultraviolet solar radiation, but ozone pollution in the lower atmosphere causes respiratory problems.

As you have seen, the number of bonding electron pairs between two atoms is important in determining bond length and strength. The experimentally measured lengths for the two oxygen-oxygen bonds in ozone are the same, 127.8 pm, implying that both bonds contain the same number of bond pairs. However, using the guidelines for writing Lewis structures, you might come to a different conclusion. Two possible Lewis structures are

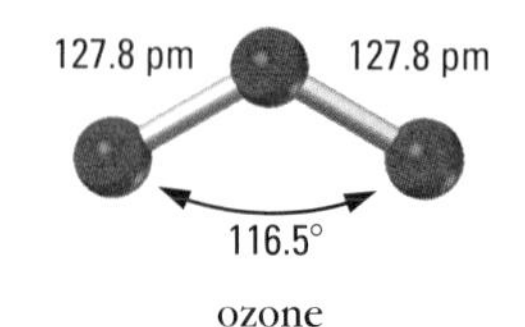

ozone

$$:\ddot{O}{=}\ddot{O}{-}\ddot{\underset{..}{O}}: \quad \text{and} \quad :\ddot{\underset{..}{O}}{-}\ddot{O}{=}\ddot{O}:$$

Each structure shows a double bond on one side of the central O atom and a single bond on the other side. If either one were the actual structure of O_3, then one bond (O═O) should be shorter than the other (O—O), but this is not the case. That the oxygen-to-oxygen bonds in ozone are neither double bonds nor single bonds is supported by the fact that the experimentally determined 127.8 pm bond length is longer than that for O═O (112 pm), but shorter than that for O—O (132 pm). Therefore, no single Lewis structure can be written that is consistent with all of the experimental data. When this situation arises, the concept of *resonance* is invoked to reconcile the experimental observation with two or more Lewis structures for the same molecule. Each of the Lewis structures, called **resonance structures,** is thought of as contributing to the true structure that cannot be written. The actual structure of O_3 is neither of the Lewis structures above, but a composite called a **resonance hybrid.** It is conventional to connect the resonance structures with a double-headed arrow, ↔, to emphasize that the actual bonding is a composite of these structures.

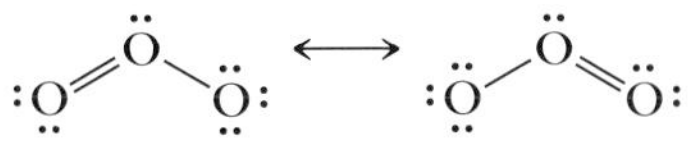

:O═Ö—Ö: ⟷ :Ö—Ö═O:

Resonance structures of ozone

The resonance concept is useful whenever there is a choice about which of two or three atoms contribute lone pairs to achieve an octet of electrons about a central atom by multiple bond formation.

When applying the concept of resonance, keep several important things in mind:

- Lewis structures contributing to the resonance hybrid structure differ only in the assignment of electron pair positions; the positions of the atoms don't change.
- Contributing Lewis structures differ in the number of bond pairs between one or more specific pairs of atoms.
- The resonance hybrid structure represents a single composite structure, not different structures that are continually changing back and forth.

"Resonance" truly is an unfortunate term, because it implies that the molecule somehow "resonates," moving in some way to form different kinds of molecules, which is not true. There is only one kind of ozone molecule.

To further illustrate the use of resonance, consider what happens in writing the Lewis structure of the carbonate ion, CO_3^{2-}, which has 24 valence electrons (four from C, 18 from three O atoms, and two for the 2− charge). Writing the skeleton structure and putting in lone pairs so that each O has an octet uses all 24 electrons but leaves carbon without an octet:

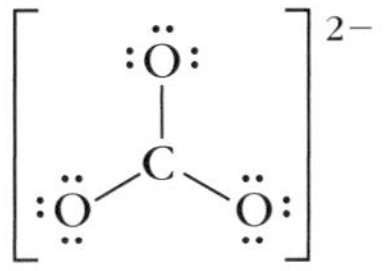

$$\left[\ :\ddot{O}: \text{—C(—}:\ddot{O}:\text{)(—}:\ddot{O}:\text{)}\ \right]^{2-}$$

To give carbon an octet requires changing a single bond to a double bond, which can be done in three equivalent ways:

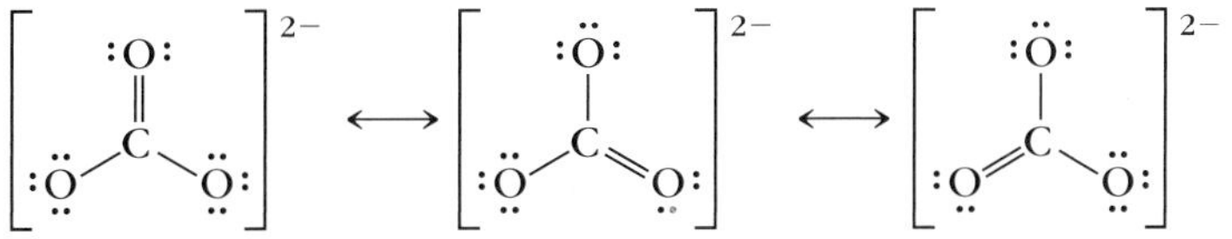

$$[\ :O:═C(\ddot{O})(\ddot{O})\ ]^{2-} \longleftrightarrow [\ \ddot{O}—C(\ddot{O})═O\ ]^{2-} \longleftrightarrow [\ \ddot{O}—C(═O)(\ddot{O})\ ]^{2-}$$

Writing the Lewis structures of oxygen-containing anions often requires using resonance structures.

The three resonance structures contribute to the resonance hybrid. As with ozone, there are no single and double bonds in the carbonate ion. All three carbon-oxygen bond distances are 129 pm, intermediate between the C—O single bond (143 pm) and the C═O double bond (122 pm) distances.

CONCEPTUAL EXERCISE **8.9 Resonance Structures**

The correct Lewis structure for cyanate ion is $\left[:N{\equiv}C{-}\ddot{\underset{\cdot\cdot}{O}}:\right]^-$.

Why is $\left[:\ddot{\underset{\cdot\cdot}{N}}{-}O{\equiv}C:\right]^-$ not a resonance structure for cyanate ion?

To nineteenth-century chemists, the molecular formula of benzene, C_6H_6, implied that benzene was an unsaturated compound because it lacked the ratio of carbon to hydrogen found in saturated noncyclic hydrocarbons, C_nH_{2n+2}. A six-membered ring structure for benzene with alternating C—C and C=C bonds uses all the valence electrons and gives each carbon atom an octet of valence electrons. In 1892 Friedrich A. Kekulé proposed that the ring structure of benzene could be represented by two structures, now called resonance structures, indicated by the double-headed arrow.

Bonding in benzene can be described better using the molecular orbital theory (see Section 8.10).

Neither of these alternating single- and double-bond resonance structures, however, accurately represents benzene. Experimental structural data for benzene indicate a planar, symmetric molecule in which all six carbon-carbon bonds are equivalent. Each carbon-carbon bond is 139 pm long, intermediate between the length of a C—C single bond (154 pm) and a C=C double bond (134 pm). Benzene is a resonance hybrid of these resonance structures—it is a molecule in which the six electrons of the suggested three double bonds are actually delocalized uniformly around the ring, with the six **delocalized electrons** shared equally by all six carbon atoms. When hydrogen and carbon atoms are not shown, the benzene ring is sometimes written as a hexagon with a circle in the middle. The circle represents the six delocalized electrons spread evenly over all of the carbon atoms. Each corner in the hexagon represents one carbon atom and one hydrogen atom, and each line represents a single C—C bond.

Sometimes a dotted circle rather than a solid circle is used to represent the delocalized electrons.

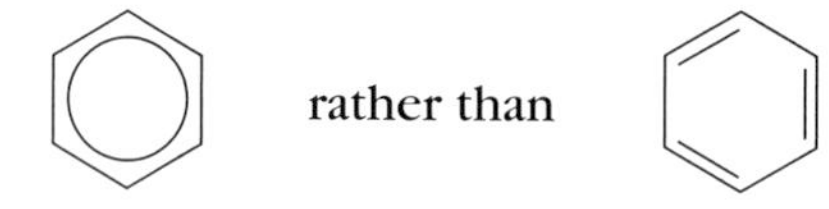

8.9 Exceptions to the Octet Rule

Many molecules and polyatomic ions have structures that are not consistent with the octet rule (***⇐ p. 272***). Consideration of the electron configurations of their central atoms demonstrates why three kinds of exceptions occur: (1) molecules or ions with central atoms having fewer than eight electrons; (2) molecules or ions with an odd number of valence electrons; and (3) molecules or ions with central atoms having more than an octet of electrons.

Fewer Than Eight Valence Electrons

Boron trifluoride, BF_3, is a molecule with less than an octet of valence electrons around the central boron atom. Boron, a Group 3A element, has only three valence electrons; each fluorine contributes seven, for a total of 24 valence electrons. Although the Lewis structure has an octet around each fluorine atom, there are only six electrons around the B atom, an exception to the octet rule.

$$:\ddot{F}: \; | \; B-\ddot{F}: \; | \; :\ddot{F}:$$

A Lewis structure such as

$$\ddot{F}: \; \| \; B-\ddot{F}: \; | \; :\ddot{F}:$$

has a formal charge of +1 on the top F atom and −1 on the B, making it a less likely structure than the one shown in the adjacent text.

Because it lacks an octet around boron, BF_3 is very reactive—for example, it readily combines with NH_3 to form a compound with the formula BF_3NH_3. The bonding between BF_3 and NH_3 can be explained by using the lone pair of electrons on N to form a covalent bond with B in BF_3. In this case, the nitrogen lone pair provides *both* of the shared electrons, resulting in an octet of electrons for both B and N.

$$H_3N: + BF_3 \longrightarrow H_3N-BF_3$$

The type of bond in which both electrons are provided by the same atom is known as a coordinate covalent bond. It is discussed further in Section 15.9.

Odd Number of Valence Electrons

All the molecules we have discussed up to this point have contained only *pairs* of valence electrons. However, a few stable molecules have an odd number of valence electrons. For example, NO has 11 valence electrons, and NO_2 has 17 valence electrons. The most plausible Lewis structures for these molecules are

$$:\dot{N}=\ddot{O} \qquad :\ddot{O}-\ddot{N}=\ddot{O}:$$

NO_2 is a free radical; N_2O_4 is not because it has no unpaired valence electrons.

Atoms and molecules that have an unpaired electron are known as **free radicals.** How do unpaired electrons affect reactivity? Simple free radicals such as atoms of H· and Cl· are very reactive and readily combine with other atoms to give molecules such as H_2, Cl_2, and HCl. Therefore, we would expect free radical molecules to be more reactive than molecules that have all paired electrons, and they are. A free radical either combines with another free radical to form a more stable molecule in which the electrons are paired, or it reacts with other molecules to produce new free radicals. These kinds of reactions are central to the formation of air pollutants. For example, when gaseous NO and NO_2 are released in vehicle exhaust, the colorless NO reacts with O_2 in the air to form brown NO_2. The NO_2 decomposes in the presence of sunlight to give NO and O, both of which are free radicals.

$$:\ddot{O}-\dot{N}=\ddot{O}: \xrightarrow{\text{sunlight}} :\dot{N}=\ddot{O} + \cdot\ddot{O}\cdot$$

The free O atom reacts with O_2 in the air to give ozone, O_3, an air pollutant that affects the respiratory system. Free radicals also have a tendency to combine with themselves to form dimers, substances made from two smaller units. For example, when NO_2 gas is cooled it dimerizes to N_2O_4. As expected, NO and NO_2 are paramagnetic (⇐ *p. 254*) because of their odd numbers of electrons.

Nitrogen dioxide formation. As it emerges from water, colorless nitrogen monoxide (also called nitric oxide) reacts with oxygen to form reddish-brown NO_2.

Third-period elements can also satisfy the octet rule by using just 3*s* and 3*p* orbitals, such as in PF_3.

See Suidan, L., *et al. Journal of Chemical Education* Vol. 72, 1995; p. 583 for a discussion of bonding in exceptions to the octet rule.

More Than Eight Valence Electrons

Exceptions to the octet rule are most common among molecules or ions with an "expanded octet"—that is, more than eight electrons in the valence shell around a central atom. For example, sulfur and phosphorus commonly form stable molecules and ions in which S or P is surrounded by more than an octet of valence electrons. Expanded octets are not found with elements in the first or second period. The octet rule is reliable for predicting stable molecules that contain only H, C, N, O, or F.

Compounds with expanded octets around the central atom occur mainly with central atoms of elements in the third period and beyond. Such central atoms are larger than those from Period 2 and thus, can accommodate more other atoms around them. For example, the Period 3 elements P and S form the known compounds PF_5 and SF_4, but their Period 2 analogs, NF_5 and OF_4, do not exist. Due to its larger size, phosphorus as a central atom can violate the octet rule to form PF_5 and PF_6^-.

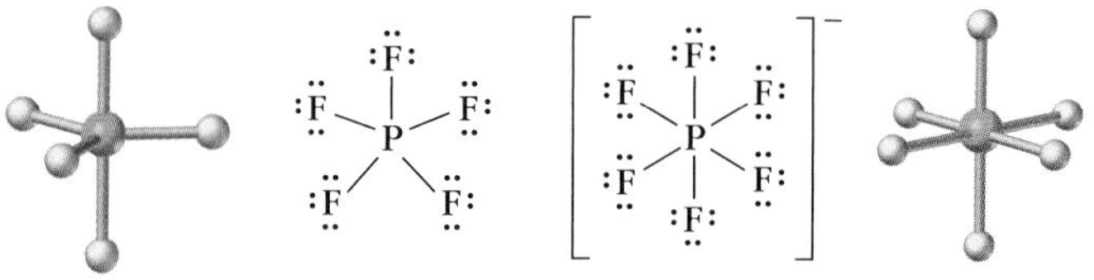

phosphorus pentafluoride

phosphorus hexafluoride ion

Table 8.3 illustrates molecules and ions of central atoms beyond the second period that violate the octet rule.

PROBLEM-SOLVING EXAMPLE 8.8 Exceptions to the Octet Rule

Write the Lewis structure for (a) tellurium tetrabromide, $TeBr_4$; (b) triiodide ion, I_3^-; and (c) boric acid, $B(OH)_3$.

Table 8.3 Lewis Structures for Some Ions and Molecules with More Than Eight Electrons Around the Central Atom*

	Group 4A	Group 5A	Group 6A	Group 7A	Group 8A
Central atoms with five valence pairs		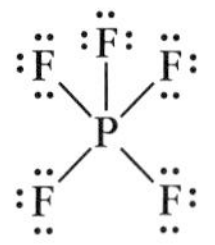	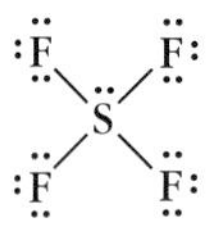	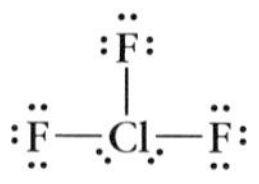	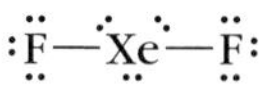
	—	PF_5	SF_4	ClF_3	XeF_2
Bonding pairs	—	5	4	3	2
Lone pairs	—	0	1	2	3
Central atoms with six valence pairs	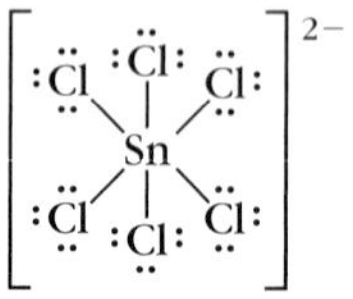	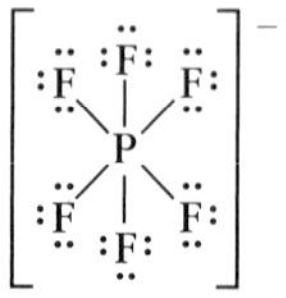		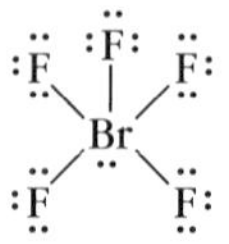	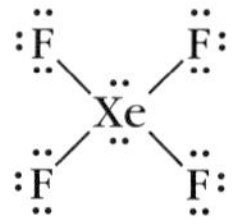
	$SnCl_6^{2-}$	PF_6^-	SF_6	BrF_5	XeF_4
Bonding pairs	6	6	6	5	4
Lone pairs	0	0	0	1	2

*In each case, the numbers of bond pairs and lone pairs about the central atom are given.

Answer

(a)
```
        ..
       :Br:
  ..    ..|     ..
 :Br — Te — Br:
  ..      |     ..
       :Br:
        ..
```

(b) $[:\ddot{I}—\ddot{I}—\ddot{I}:]^-$

(c)
```
H—Ö—B—Ö—H
       |
      :O:
       |
       H
```

Strategy and Explanation

(a) $TeBr_4$ has 34 valence electrons, eight of which are distributed among four Te—Br bonds. Of the remaining 26 lone pair electrons, 24 complete octets for the Br atoms. The other two electrons form a lone pair on Te, which has a total of ten electrons (five pairs: four shared, one unshared) around it, acceptable for a Period 5 element.

(b) There is a total of 22 valence electrons: seven from each iodine atom and one for the 1− charge on the ion. Forming two I—I single bonds with the central I atom and then distributing six of the remaining nine electron pairs as lone pairs on terminal I atoms to satisfy the octet rule uses a total of eight electron pairs. The remaining three electron pairs are placed on the central iodine, which can accommodate more than eight electrons because it is from the fifth period.

(c) $B(OH)_3$ uses six of its 24 valence electrons to form three B—O bonds and an additional six electrons for three O—H bonds. The remaining 12 electrons complete two lone pairs on each oxygen atom, giving each oxygen atom an octet. This leaves boron with only three electron pairs, an exception to the octet rule.

PROBLEM-SOLVING PRACTICE 8.8

Write the Lewis structure for each of these molecules or ions. Indicate which central atoms break the octet rule, and why.

(a) BeF_2 (b) ClO_2 (c) PCl_5
(d) BH_2^+ (e) IF_7

8.10 Molecular Orbital Theory

Earlier in this chapter, we used electron-pair bonds to explain bonding in molecules, accounting qualitatively for the stability of the covalent bond in terms of the overlap of atomic orbitals. Where more than one Lewis structure could be drawn for the same arrangement of atomic nuclei, as in O_3, the concept of resonance (**p. 292**) was used to account for observed properties of molecules.

A major weakness of the Lewis diagram theory presented earlier in this chapter is that the theory does not always correctly predict the magnetic properties of substances. An important example is O_2, which is paramagnetic (Figure 8.8). This means that O_2 molecules must have unpaired electrons. Diatomic oxygen has an even number of valence electrons (12), and the octet rule predicts that all of these electrons should be paired. According to the Lewis diagram theory, O_2 should be diamagnetic, but experimental measurements show that it is not.

Figure 8.8 Paramagnetism of liquid oxygen. Liquid oxygen, the liquid suspended between the poles of the magnet, can be suspended between the poles of a magnet because O_2 is paramagnetic. Paramagnetic substances are attracted into a magnetic field.

Molecular Orbitals

This discrepancy between experiment and theory for O_2 (and other molecules) can be resolved by using an alternative model of covalent bonding, the **molecular orbital** (MO) approach. *Molecular orbital theory* treats bonding in terms of orbitals that can extend over an entire molecule. Thus, the molecular orbitals are not confined to two atoms at a time. The MO approach involves three basic operations.

Step 1: Combine the valence atomic orbitals of all atoms in the molecule to give a new set of molecular orbitals (MOs) that are characteristic of the molecule as a whole. *The number of MOs formed is equal to the number of atomic* *orbitals combined.* For example, when two H atoms combine to form H_2, two *s* orbitals, one from each atom, yield two molecular orbitals. For O_2 there are one *s* orbital and three *p* orbitals in the valence shell of each of the two oxygen atoms. This gives eight atomic orbitals, which combine to give eight MOs.

Step 2: Arrange the MOs in order of increasing energy, as will be illustrated later in this section for second-period elements (Figure 8.11 on p. 300). The relative energies of MOs are usually deduced from experiments involving spectral and magnetic properties.

Step 3: Distribute the *valence electrons* of the molecule among the available MOs, filling the lowest-energy MO first and continuing to build up the electron configuration of the molecule in the same way that electron configurations of atoms are built up **(⇐ *p. 245*)**. As electrons are added,

(a) *Each MO can hold a maximum of two electrons.* In a filled molecular orbital the two electrons have opposed spins, in accordance with the Pauli exclusion principle **(⇐ *p. 241*)**.

(b) *Electrons go into the lowest-energy MO available.* A higher-energy molecular orbital starts to fill only when each molecular orbital below it has its quota of two electrons.

(c) *Hund's rule is obeyed.* When two molecular orbitals of equal energy are available to two electrons, one electron goes into each, giving two half-filled molecular orbitals.

Molecular Orbitals for Diatomic Molecules

To illustrate molecular orbital theory, we apply it to the diatomic molecules of the elements in the first two periods of the periodic table.

Hydrogen and Helium (Combining Two 1*s* Orbitals) When two hydrogen atoms (or two helium atoms) come close together, two 1*s* orbitals overlap and combine to give two MOs. One MO has an energy lower than that of the atomic orbitals from which it was formed; the other MO is of higher energy (Figure 8.9). A molecule that has two electrons in the lower-energy MO is lower in energy (more stable) than the isolated atoms. That lowering of energy corresponds to the bond energy. For that reason the lower-energy MO in Figure 8.9a is called a **bonding molecular orbital.** If electrons are placed in the higher-energy MO, the molecule's energy is higher than the energy of the isolated atoms. This unstable situation is the opposite of bonding; the higher-energy MO is called an **antibonding molecular orbital.**

The electron density in these MOs is shown in Figure 8.9b. The bonding MO has higher electron density between the nuclei than the sum of the individual atomic

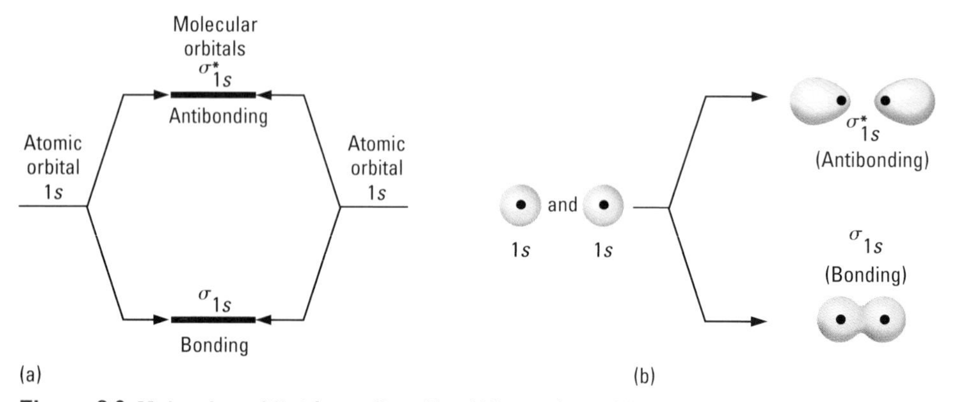

Figure 8.9 Molecular orbital formation. Two MOs are formed by combining two 1*s* atomic orbitals. (a) Energy level diagram. (b) Orbital overlap diagram. (Black dots in (b) represent atomic nuclei.)

orbitals. This attracts the nuclei together and forms the bond. In the antibonding orbital, the electron density between the nuclei is smaller than it would have been in the atoms alone. The positively charged nuclei have less electron "glue" to hold them together, they repel each other, and the molecule flies apart—the opposite of a bond.

When two atomic orbitals of equal energy combine, the resulting MOs lie above and below the energy of the atomic orbitals.

The electron density in both MOs is symmetrical around a line connecting the two nuclei, which means that both are sigma orbitals. In MO notation, the 1*s* bonding orbital is designated as σ_{1s}. The antibonding orbital is given the symbol σ^*_{1s}. An asterisk designates an antibonding molecular orbital.

In the H_2 molecule, there are two 1*s* electrons. They fill the σ_{1s} orbital, giving a single bond. If an He_2 molecule could form, there would be four electrons—two from each atom. These would fill the bonding and antibonding orbitals. One bond and one antibond give a *bond order* (number of bonds) of zero in He_2. The bond order is calculated as

$$\text{Bond order} = \text{number of bonds} = \frac{n_B - n_A}{2}$$

where n_B is the number of electrons in bonding molecular orbitals, n_A is the number of electrons in antibonding molecular orbitals, and dividing by 2 accounts for the fact that two electrons are needed for a bond. In H_2, $n_B = 2$ and $n_A = 0$, so we have one bond. For He_2, $n_B = n_A = 2$, so the number of bonds is zero. The MO theory predicts that the He_2 molecule should not exist, and it does not.

Second-Period Elements (Combining 2*s* and 2*p* Orbitals) Three of the elements in the second period form familiar diatomic molecules: N_2, O_2, and F_2. Less common, but also known, are Li_2, B_2, and C_2, which have been observed as gases. The molecules Be_2 and Ne_2 are either highly unstable or nonexistent. To see what MO theory predicts about the stability of diatomic molecules from the second period, consider the valence atomic orbitals, 2*s* and 2*p*.

Combining two 2*s* atomic orbitals, one from each atom, gives two MOs. These are very similar to the ones shown in Figure 8.9. They are designated as σ_{2s} (sigma, bonding, 2*s*) and σ^*_{2s} (sigma, antibonding, 2*s*).

In an isolated atom, there are three 2*p* orbitals, oriented at right angles to each other. We call these atomic orbitals p_x, p_y, and p_z (Figure 8.10a). Assume that two

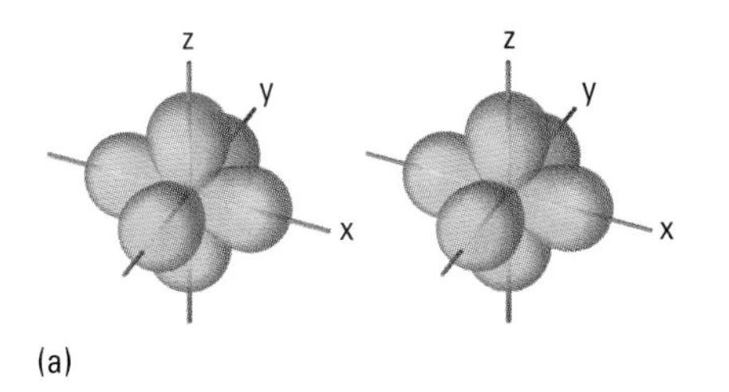

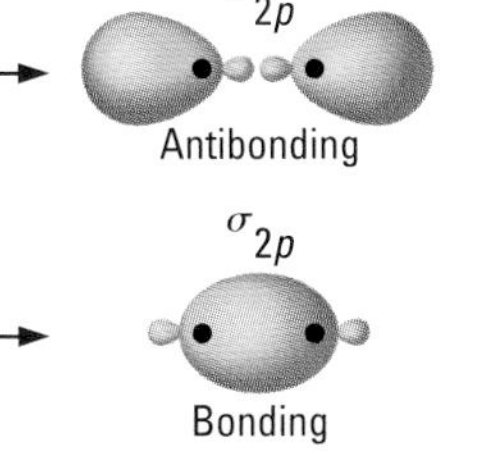

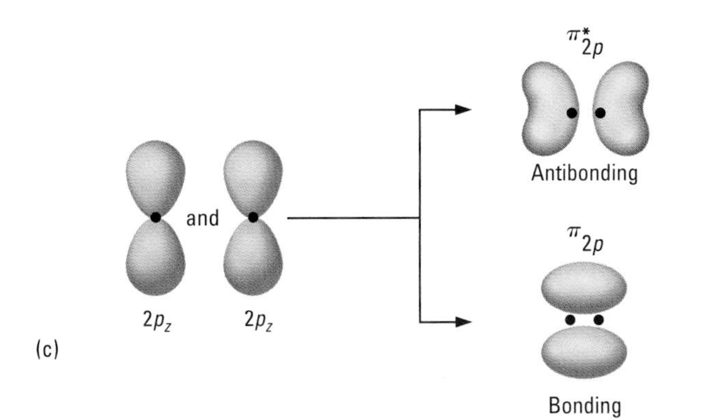

Figure 8.10 Forming molecular orbitals from *p* orbitals. When *p* orbitals (a) from two different atoms overlap, there are two quite different possibilities. If they overlap head to head (b), two σ MOs are produced. If, on the other hand, they overlap side to side (c), two π MOs result. In this example, both p_z and p_y orbitals can overlap to form π MOs. Only p_z overlap is shown. (Black dots represent atomic nuclei.)

If you look down the internuclear axis, a π molecular orbital looks like a *p* atomic orbital.

atoms approach along the *x*-axis. The two p_x atomic orbitals overlap head to head to form two orbitals that are symmetric around the line connecting the two nuclei (Figure 8.10b). That is, they form a sigma bonding orbital, σ_{2p}, and a sigma antibonding orbital, σ^*_{2p}.

The situation is quite different when the p_z orbitals overlap. Because they are oriented parallel to one another, they overlap side to side (Figure 8.10c). The two MOs formed in this case are pi orbitals; one is a bonding MO, π_{2p}, and the other is an antibonding MO, π^*_{2p}. Similarly, the p_y orbitals of the two atoms interact to form another pair of pi MOs, π_{2p} and π^*_{2p}, which are not shown in Figure 8.10.

The relative energies of the MOs available for occupancy by the valence electrons of diatomic molecules formed from second-period atoms are shown in Figure 8.11. To obtain the MO structure of the diatomic molecules of the elements in the second period, we fill the available MOs in order of increasing energy. The results are shown in Table 8.4. Note that MO theory correctly predicts the number of unpaired electrons in each molecule and the bonding in the molecule.

There is also a general correlation between the predicted bond order, $(n_B - n_A)/2$, and the experimental bond energy. The bond order of two for C_2 and O_2 implies a double bond, which is expected to be stronger than the single bonds in Li_2, B_2, and F_2 (bond order = 1). The bond order of three for N_2 (triple bond) implies a still

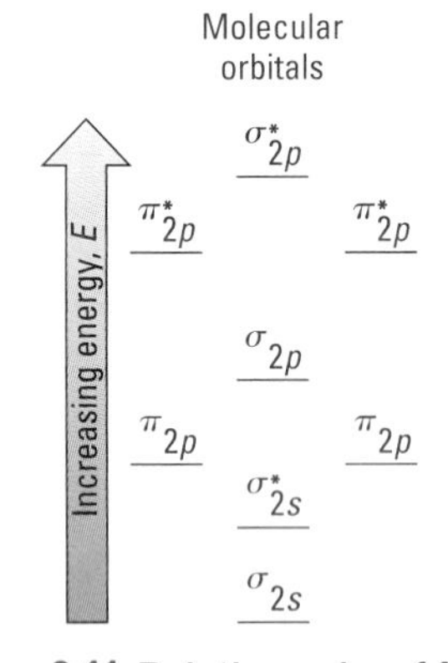

Figure 8.11 Relative order of filling molecular orbitals. The energies of MOs formed by combining 2*s* and 2*p* atomic orbitals increase from bottom to top of the diagram. This order of energies applies to Li_2 through N_2, but for O_2 and F_2 the energy of the σ_{2p} orbital drops below the energy of the two π_{2p} orbitals. This does not affect the filling order, because in O_2 and F_2 the σ_{2p} and π_{2p} orbitals are filled.

Table 8.4 Predicted and Observed Properties of Diatomic Molecules of Second-Period Elements

Occupancy of Orbitals

	σ_{2s}	σ^*_{2p}	π_{2p}	π_{2p}	σ_{2p}	π^*_{2p}	π^*_{2p}	σ^*_{2p}
Li_2	(↑↓)	( )	( )	( )	( )	( )	( )	( )
Be_2	(↑↓)	(↑↓)	( )	( )	( )	( )	( )	( )
B_2	(↑↓)	(↑↓)	(↑)	(↑)	( )	( )	( )	( )
C_2	(↑↓)	(↑↓)	(↑↓)	(↑↓)	( )	( )	( )	( )
N_2	(↑↓)	(↑↓)	(↑↓)	(↑↓)	(↑↓)	( )	( )	( )
O_2	(↑↓)	(↑↓)	(↑↓)	(↑↓)	(↑↓)	(↑)	(↑)	( )
F_2	(↑↓)	(↑↓)	(↑↓)	(↑↓)	(↑↓)	(↑↓)	(↑↓)	( )
Ne_2	(↑↓)	(↑↓)	(↑↓)	(↑↓)	(↑↓)	(↑↓)	(↑↓)	(↑↓)

	Predicted Properties		**Observed Properties**	
	Number of Unpaired e⁻	**Bond Order**	**Number of Unpaired e⁻**	**Bond Energy (kJ/mol)**
Li_2	0	1	0	105
Be_2	0	0	0	Unstable
B_2	2	1	2	289
C_2	0	2	0	598
N_2	0	3	0	946
O_2	2	2	2	498
F_2	0	1	0	158
Ne_2	0	0	0	Nonexistent

stronger bond. A major triumph of MO theory is its ability to explain the properties of O_2. The bond order of two and the presence of two unpaired electrons in the two π^*_{2p} MOs explain how the molecule can have a double bond and at the same time be paramagnetic.

PROBLEM-SOLVING EXAMPLE 8.9 Electron Structure of Peroxide Ion

Using MO theory, predict the bond order and number of unpaired electrons in the peroxide ion, O_2^{2-}.

Answer Bond order = 1; there are no unpaired electrons.

Strategy and Explanation First, find the number of valence electrons. Then, construct an orbital diagram, filling the available MOs (Figure 8.11) in order of increasing energy.

Recall that oxygen is in Group 6A of the periodic table, so an oxygen atom has six valence electrons. The peroxide ion has two oxygen atoms and two extra electrons (2− charge). The total number of valence electrons is 2(6) + 2 = 14. The molecular orbital diagram is

	σ_{2s}	σ^*_{2s}	π_{2p}	π_{2p}	σ_{2p}	π^*_{2p}	π^*_{2p}	σ^*_{2p}
O_2^{2-}	(↑↓)	(↑↓)	(↑↓)	(↑↓)	(↑↓)	(↑↓)	(↑↓)	( )

The bond order is (8 − 6)/2 = 1. There are no unpaired electrons. These conclusions are in agreement with the Lewis structure theory for the peroxide ion:

$$\left[:\ddot{\underset{..}{O}}-\ddot{\underset{..}{O}}:\right]^{2-}$$

PROBLEM-SOLVING PRACTICE 8.9

Use MO theory to predict the bond order and the number of unpaired electrons in the superoxide ion, O_2^-.

Molecular orbital theory can also be applied to predict the properties of molecules containing different kinds of atoms, such as NO. There are 11 valence electrons in this molecule (five from nitrogen and six from oxygen). These 11 valence electrons are placed into molecular orbitals of increasing energy in the molecular orbital diagram:

	σ_{2s}	σ^*_{2s}	π_{2p}	π_{2p}	σ_{2p}	π^*_{2p}	π^*_{2p}	σ^*_{2p}
NO	(↑↓)	(↑↓)	(↑↓)	(↑↓)	(↑↓)	(↑)	( )	( )

There are eight electrons in bonding orbitals (two in σ_{2s}, four in π_{2p}, two in σ_{2p}) and three electrons in antibonding orbitals (two in σ^*_{2s}, one in π^*_{2p}). Therefore there is one unpaired electron (in the π^*_{2p} molecular orbital). The bond order is

$$\text{Number of bonds} = \frac{8-3}{2} = \frac{5}{2} = 2.5$$

Polyatomic Molecules and Ions; Delocalized π Electrons

The bonding in molecules containing more than two atoms can also be described in terms of MOs. We will not attempt to do this; the energy level structure is considerably more complex than that for diatomic molecules. However, one point is worth mentioning: In polyatomic species, *the MOs can be spread over the entire molecule* rather than being localized between two atoms.

The nitrate ion is known from experiment to be triangular planar, with three N—O bonds whose lengths are equal. However, a single Lewis structure for nitrate ion does not predict three equal-length bonds.

Lewis structure theory invokes three contributing resonance structures to explain the fact that the three N—O bonds are identical. In each contributing structure there is a double bond from N to a different O atom. MO theory, on the other hand, considers that the skeleton of the nitrate ion involves three sigma bonds, but the fourth electron pair (the second pair in the double bond) is in a pi orbital that is delocalized; it is shared by all of the atoms in the ion. That is, this pair of electrons is shared equally among the three O atoms. According to MO theory, a similar interpretation applies with all of the resonance hybrids described in this chapter.

IN CLOSING

Having studied this chapter, you should be able to . . .

- Recognize the different types of covalent bonding (Sections 8.1–8.3, 8.8).
- Use Lewis structures to represent covalent bonds in molecules and polyatomic ions (Sections 8.1–8.4, 8.7, 8.9). ■ End-of-chapter Questions assignable in OWL: 13, 17
- Describe multiple bonds in alkenes and alkynes (Section 8.4). ■ Question 24
- Recognize molecules that can have *cis-trans* isomerism (Section 8.4). ■ Question 26
- Predict bond lengths from periodic trends in atomic radii (Section 8.5). ■ Questions 30, 32, 56
- Relate bond energy to bond length (Section 8.5). ■ Question 56
- Use bond enthalpies to calculate the enthalpy of a reaction (Section 8.5). ■ Question 36
- Predict bond polarity from electronegativity trends (Section 8.6). ■ Questions 39, 54
- Use formal charges to compare Lewis structures (8.7). ■ Question 41
- Use resonance structures to model multiple bonding in molecules and polyatomic ions (Section 8.8). ■ Question 45
- Recognize three types of exceptions to the octet rule (Section 8.9). ■ Question 50
- Use molecular orbital theory to explain bonding in diatomic molecules and ions (Section 8.10).

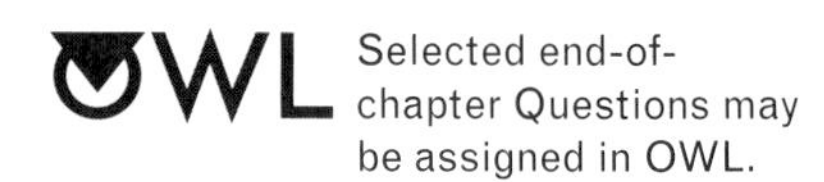

OWL Selected end-of-chapter Questions may be assigned in OWL.

For quick review, download Go Chemistry mini-lecture flashcard modules (or purchase them at **www.ichapters.com**).

Prepare for an exam with a **Summary Problem** that brings together concepts and problem-solving skills from throughout the chapter. Go to **www.cengage.com/chemistry/moore** to download this chapter's Summary Problem from the Student Companion Site.

KEY TERMS

alkenes *(Section 8.4)*

alkynes *(8.4)*

antibonding molecular orbital *(8.10)*

bond enthalpy (bond energy) *(8.5)*

bond length *(8.5)*

bonding electrons *(8.2)*

bonding molecular orbital *(8.10)*

***cis* isomer** *(8.4)*

***cis-trans* isomerism** *(8.4)*

covalent bond *(8.1)*

delocalized electrons *(8.8)*

double bond *(8.3)*

electronegativity *(8.6)*

formal charge *(8.7)*

free radicals *(8.9)*

Lewis structure *(8.2)*

lone pair electrons *(8.2)*

molecular orbital *(8.10)*

multiple covalent bonds *(8.3)*
nonpolar covalent bond *(8.6)*
octet rule *(8.2)*
polar covalent bond *(8.6)*
resonance hybrid *(8.8)*
resonance structures *(8.8)*
saturated hydrocarbons *(8.2)*
single covalent bond *(8.2)*
***trans* isomer** *(8.4)*
triple bond *(8.3)*
unsaturated hydrocarbons *(8.4)*

QUESTIONS FOR REVIEW AND THOUGHT

■ denotes questions assignable in OWL.

Blue-numbered questions have short answers at the back of this book and fully worked solutions in the *Student Solutions Manual.*

Review Questions

1. Explain the difference between an ionic bond and a covalent bond.
2. What kind of bonding (ionic or covalent) would you predict for the products resulting from the following combinations of elements?
   (a) Na + I_2 (b) C + S
   (c) Mg + Br_2 (d) P_4 + Cl_2
3. What characteristics must atoms A and X have if they are able to form a covalent bond A—X with each other? A polar covalent bond with each other?
4. Indicate the difference among alkanes, alkenes, and alkynes by giving the structural formula of a compound in each class that contains three carbon atoms.
5. While sulfur forms the compounds SF_4 and SF_6, no equivalent compounds of oxygen, OF_4 and OF_6, are known. Explain.
6. Which of these molecules have an odd number of valence electrons: NO_2, SCl_2, NH_3, NO_3?
7. Write resonance structures for NO_2^-. Predict a value for the N—O bond length based on bond lengths given in Table 8.1, and explain your answer.
8. Consider these structures for the formate ion, HCO_2^-. Designate which two are resonance structures and which is equivalent to one of the resonance structures.

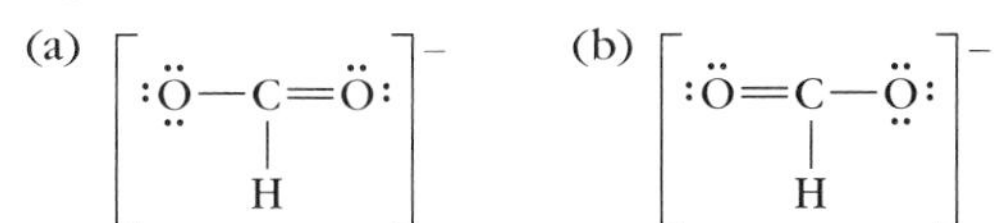
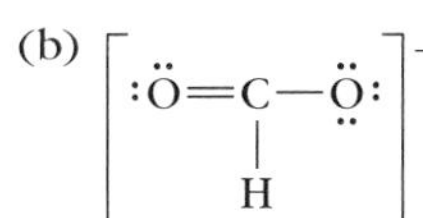

(c) $[\,$:Ö—C=O: (C bonded to H)$\,]^-$

9. Consider a series of molecules in which the C atom is bonded to atoms of second-period elements: C—O, C—F, C—N, C—C, and C—B. Place these bonds in order of increasing bond length.
10. What are the trends in bond length and bond energy for a series of related bonds—for instance, single, double, and triple carbon-to-oxygen bonds?
11. Why is *cis-trans* isomerism not possible for alkynes?

Topical Questions

Lewis Structures

12. Write the Lewis structures of (a) dichlorine monoxide, Cl_2O; (b) hydrogen peroxide, H_2O_2; (c) borohydride ion, BH_4^-; (d) phosphonium ion, PH_4^+; and (e) PCl_5.
13. ■ Write Lewis structures for these molecules or ions.
   (a) ClF (b) H_2Se
   (c) BF_4^- (d) PO_4^{3-}
14. Write Lewis structures for these molecules.
   (a) $CHClF_2$, one of several chlorofluorocarbons that has been used in refrigeration
   (b) Methyl alcohol, CH_3OH
   (c) Methyl amine, CH_3NH_2
15. Write Lewis structures for these molecules or ions.
   (a) CH_3Cl (b) SiO_4^{4-}
   (c) ClF_4^+ (d) C_2H_6
16. Write Lewis structures for these molecules.
   (a) Formic acid, HCOOH, in which atomic arrangement is

   O (bonded to C)
   H—C—O—H

   (b) Acetonitrile, CH_3CN
   (c) Vinyl chloride, CH_2CHCl, the molecule from which PVC plastics are made
17. ■ Write Lewis structures for these molecules.
   (a) Tetrafluoroethylene, C_2F_4, the molecule from which Teflon is made
   (b) Acrylonitrile, CH_2CHCN, the molecule from which Orlon is made
18. Write Lewis structures for:
   (a) N_2^+ (b) XeF_7^-
   (c) tetracyanoethane, C_6N_4
19. Write Lewis structures for:
   (a) Disulfur dichloride, S_2Cl_2
   (b) Nitrous acid, HNO_2, which has no oxygen-to-oxygen bonds
   (c) ClF_6^+
20. Which of these are correct Lewis structures and which are incorrect? Explain what is wrong with the incorrect ones.
   (a) :N=N̈:
   N_2
   (b) 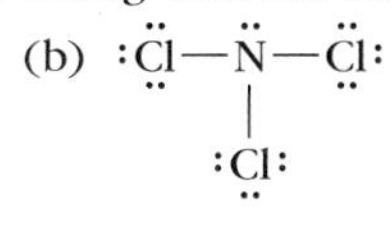
   NCl_3

(c) 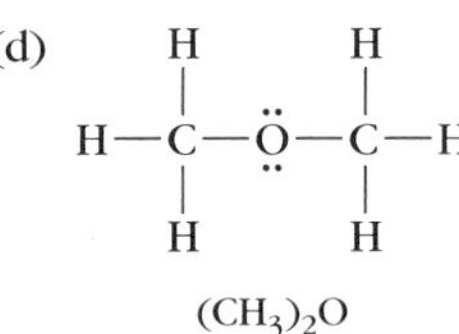

ClO_3^-

(d) H–C(H)(H)–Ö–C(H)(H)–H

$(CH_3)_2O$

(e) $[H–N(H)(H)–H]^+$

21. Which of these are correct Lewis structures and which are incorrect? Explain what is wrong with the incorrect ones.

(a) F:Ö:F
OF_2

(b) :O≡O:
O_2

(c) H:C̈:H:C̤̈l:
CH_3Cl

(d) :Ö: ‖ :C̈l–C–C̈l:
CCl_2O

(e) $[$:Ö–N=Ö:$]^-$
NO_2^-

Bonding in Hydrocarbons

22. Write the structural formulas for all the branched-chain compounds with the molecular formula C_6H_{14}.
23. Write structural formulas for two straight-chain alkenes with the formula C_5H_{10}. Are these the only two structures that meet these specifications?
24. ■ From their molecular formulas, classify each of these straight-chain hydrocarbons as an alkane, an alkene, or an alkyne.
   (a) C_5H_8 (b) $C_{24}H_{50}$ (c) C_7H_{14}
25. From their molecular formulas, classify each of these straight-chain hydrocarbons as an alkyne.
   (a) $C_{21}H_{44}$ (b) C_4H_6 (c) C_8H_{16}
26. ■ In each case, tell whether *cis* and *trans* isomers exist. If they do, write structural formulas for the two isomers and label each *cis* or *trans.*
   (a) Br_2CH_2
   (b) $CH_3CH_2CH{=}CHCH_2CH_3$
   (c) $CH_3CH{=}CHCH_3$
   (d) $CH_2{=}CHCH_2CH_3$
27. ■ Which of these molecules can have *cis* and *trans* isomers? For those that do, write the structural formulas of the two isomers and label each *cis* or *trans.* For those that cannot have these isomers, explain why.
   (a) $CH_3CH_2BrC{=}CBrCH_3$
   (b) $(CH_3)_2C{=}C(CH_3)_2$
   (c) $CH_3CH_2IC{=}CICH_2CH_3$
   (d) $CH_3ClC{=}CHCH_3$
   (e) $(CH_3)_2C{=}CHCH_3$
28. Oxalic acid has this structural formula.

:O: :O:
‖ ‖
HÖ–C–C–ÖH

Is *cis-trans* isomerism possible for oxalic acid? Explain your answer.

29. 2-methyl propene has this Lewis structure.

$CH_3–C(CH_3){=}CH_2$ (CH₃ above C; two H on the right C)

Is *cis-trans* isomerism possible for this molecule? Explain your answer.

Bond Properties

30. ■ For each pair of bonds, predict which will be the shorter.
   (a) B—Cl or Ga—Cl
   (b) C—O or Sn—O
   (c) P—S or P—O
   (d) The C=C or the C=O bond in acrolein
   $H_2C{=}CH{-}C(H){=}O$
31. For each pair of bonds, predict which will be the shorter.
   (a) Si—N or P—O
   (b) Si—O or C—O
   (c) C—F or C—Br
   (d) The C=C or the C≡N bond in acrylonitrile, $H_2C{=}CH{-}C{\equiv}N$
32. ■ Consider the carbon-oxygen bonds in formaldehyde, H_2CO, and in carbon monoxide, CO. In which molecule is the carbon-oxygen bond shorter?
33. Which bond will require more energy to break, the carbon-oxygen bond in formaldehyde, H_2CO, or the CO bond in carbon monoxide, CO?
34. ■ Compare the carbon-oxygen bond lengths in the formate ion, HCO_2^-, and in the carbonate ion, CO_3^{2-}. In which ion is the bond longer? Explain briefly.
35. Compare the nitrogen-oxygen bond lengths in NO_2^+ and in NO_3^-. In which ion are the bonds longer? Explain briefly.

Bond Energies and Enthalpy Changes

36. ■ Estimate $\Delta H°$ for forming 2 mol ammonia from molecular nitrogen and molecular hydrogen. Is this reaction exothermic or endothermic? (N_2 has a triple bond.)
37. Estimate $\Delta H°$ for the conversion of 1 mol carbon monoxide to carbon dioxide by combination with molecular oxygen. Is this reaction exothermic or endothermic? (CO has a triple bond.)

Electronegativity and Bond Polarity

38. For each pair of bonds, indicate the more polar bond and use $\delta+$ or $\delta-$ to show the partial charge on each atom.
   (a) C—O and C—N
   (b) B—O and P—S
   (c) P—H and P—N
   (d) B—H and B—I
39. ■ Given the bonds C—N, C—H, C—Br, and S—O,
   (a) Which atom in each is the more electronegative?
   (b) Which of these bonds is the most polar?

40. For each pair of bonds, identify the more polar one and use δ+ or δ− to indicate the partial charge on each atom.
(a) B—Cl and B—O (b) O—F and O—Se
(c) S—Cl and B—F (d) N—H and N—F

Formal Charge

41. ■ Write correct Lewis structures and assign a formal charge to each atom.
(a) SO_3
(b) C_2N_2 (atoms bonded in the order NCCN)
(c) NO_2^-
42. Write correct Lewis structures and assign a formal charge to each atom.
(a) OCS
(b) HNC (atoms bonded in that order)
(c) CH_3^-
43. Write correct Lewis structures and assign a formal charge to each atom.
(a) CH_3CHO (b) N_3^-
(c) CH_3CN
44. Write correct Lewis structures and assign a formal charge to each atom.
(a) KrF_4 (b) ClO_3^-
(c) SO_2Cl_2

Resonance

45. ■ These have two or more resonance structures. Write all the resonance structures for each.
(a) Nitric acid

H—O—N(=O)—O

(b) Nitrate ion, NO_3^-
46. These have two or more resonance structures. Write all the resonance structures for each molecule or ion.
(a) SO_3 (b) SCN^-
47. Several Lewis structures can be written for perbromate ion, BrO_4^-, the central Br with all single Br—O bonds, or with one, two, or three Br=O double bonds. Draw the Lewis structures of these possible resonance forms, and use formal charges to predict the most plausible one.
48. Use formal charges to predict which of the resonance forms is most plausible for
(a) SO_3 and (b) HNO_3 (See Question 45.)
49. Use formal charges to predict the most plausible Lewis structure for thiosulfate ion, $S_2O_3^{2-}$.

Exceptions to the Octet Rule

50. ■ Write the Lewis structure for each of these molecules or ions.
(a) BrF_5 (b) IF_5 (c) IBr_2^-
51. Write the Lewis structure for each of these molecules or ions.
(a) BrF_3 (b) I_3^- (c) XeF_4

Molecular Orbital Theory

52. Use MO theory to predict the number of electrons in each of the 2*s* and 2*p* molecular orbitals, the number of bonds, and the number of unpaired electrons in:
(a) CO (b) F_2^- (c) NO^-
53. Use MO theory to predict the number of bonds and the number of unpaired electrons in each of these ions:
(a) B_2^+ (b) Li_2^+ (c) O_2^+

General Questions

54. ■ Using just a periodic table (not a table of electronegativities), decide which of these is likely to be the most polar bond. Explain your answer.
(a) C—F (b) S—F
(c) Si—F (d) O—F
55. Is it a good generalization that elements that are close together in the periodic table form covalent bonds, whereas elements that are far apart form ionic bonds? Why or why not?
56. ■ The molecule pictured below is acrylonitrile, the building block of the synthetic fiber Orlon.

H—C(H)=C(H)—C≡N:

(a) Which is the shorter carbon-carbon bond?
(b) Which is the stronger carbon-carbon bond?
(c) Which is the most polar bond and what is the partial negative end of the bond?
57. In nitryl chloride, NO_2Cl, there is no oxygen-oxygen bond. Write a Lewis structure for the molecule. Write any resonance structures for this molecule.
58. Arrange these bonds in order of increasing length (shortest first). List all the factors responsible for each placement: O—H, O—O, Cl—O, O=O, O=C.
59. List the bonds in Question 58 in order of increasing bond *strength*.
60. Judging from the number of carbon and hydrogen atoms in their formulas, which of these formulas represent alkanes? Which do not fall into this category? (It may help to write structural formulas.)
(a) C_8H_{10} (b) $C_{10}H_8$ (c) C_6H_{12}
(d) C_6H_{14} (e) C_8H_{18} (f) C_6H_{10}

Applying Concepts

61. A student drew this incorrect Lewis structure for ClO_3^-. What errors were made when determining the number of valence electrons?

[:Ö—Cl(—:Ö:)—Ö:]⁻

62. This Lewis structure for SF_5^+ is drawn incorrectly. What error was made when determining the number of valence electrons?

$$\left[\begin{array}{c} :\ddot{F}: \\ | \\ :\ddot{F}-\ddot{S}(-\ddot{F}:)(-\ddot{F}:) \\ | \\ :\ddot{F}: \end{array}\right]^+$$

63. Why is this not an example of resonance structures?

$$[:\ddot{N}=S=\ddot{C}:]^- \longleftrightarrow [:\ddot{N}=C-\ddot{\underset{..}{S}}:]^-$$

64. How many bonds would you expect the elements in Groups 3A through 7A to form if they obeyed the octet rule?
65. Elemental phosphorus has the formula P_4. Propose a Lewis structure for this molecule. [*Hints:* (1) Each phosphorus atom is bonded to three other phosphorus atoms. (2) Visualize the structure three-dimensionally, not flat on a page.]
66. ■ The elements As, Br, Cl, S, and Se have electronegativity values of 2.1, 2.4, 2.5, and 3.0, but not in that order. Using the periodic trend for electronegativity, assign the values to the elements. Which assignments are you certain about? Which are you not?
67. A substance is analyzed and found to contain 85.7% carbon and 14.3% hydrogen by weight. A gaseous sample of the substance is found to have a density of 1.87 g/L, and 1 mol of it occupies a volume of 22.4 L. What are two possible Lewis structures for molecules of the compound? (*Hint:* First determine the empirical formula and molar mass of the substance.)
68. When we estimate $\Delta H°$ from bond enthalpies we assume that all bonds of the same type (single, double, triple) between the same two atoms have the same energy, regardless of the molecule in which they occur. The purpose of this problem is to show you that this is only an approximation. You will need these standard enthalpies of formation:

C(g)	$\Delta H° = 716.7$ kJ/mol
CH(g)	$\Delta H° = 596.3$ kJ/mol
CH_2(g)	$\Delta H° = 392.5$ kJ/mol
CH_3(g)	$\Delta H° = 146.0$ kJ/mol
H(g)	$\Delta H° = 218.0$ kJ/mol

(a) What is the average C—H bond energy in methane, CH_4?
(b) Using bond enthalpies, estimate $\Delta H°$ for the reaction

$$CH_4(g) \longrightarrow C(g) + 2\,H_2(g)$$

(c) By heating CH_4 in a flame it is possible to produce the reactive gaseous species CH_3, CH_2, CH, and even carbon atoms, C. Experiments give these values of $\Delta H°$ for the reactions shown:

$$CH_3(g) \longrightarrow C(g) + H_2(g) + H(g) \qquad \Delta H° = 788.7 \text{ kJ}$$
$$CH_2(g) \longrightarrow C(g) + H_2(g) \qquad \Delta H° = 324.2 \text{ kJ}$$
$$CH(g) \longrightarrow C(g) + H(g) \qquad \Delta H° = 338.3 \text{ kJ}$$

For each of these reactions, draw a diagram similar to Figure 8.4. Then calculate the average C—H bond energy in CH_3, CH_2, and CH. Comment on any trends you see.

69. Methylcyanoacrylate is the active ingredient in "super glues." Its Lewis structure is

$$:N\equiv C-C(=CH_2)-C(=\ddot{\underset{..}{O}})-\ddot{\underset{..}{O}}-CH_3$$

In this molecule, which is the:
(a) Weakest carbon-containing bond?
(b) Strongest carbon-containing bond?
(c) Most polar bond?

More Challenging Questions

70. Write the Lewis structure for:
(a) tetraboron tetrachloride, B_4Cl_4
(b) hyponitrous acid, $H_2N_2O_2$
71. Nitrosyl azide, N_4O, is a pale yellow solid first synthesized in 1993. Write the Lewis structure for nitrosyl azide.
72. Write the Lewis structures for:
(a) $(Cl_2PN)_3$ (b) $(Cl_2PN)_4$
73. Write the Lewis structures for:
(a) dithiocyanogen, $(NCS)_2$ (b) tetrahydrofuran, C_4H_8O
74. In sulfuric acid production, sulfur trioxide is combined with sulfuric acid to produce oleum, which contains $H_2S_2O_7$. Write the Lewis structure for this compound.
75. Tetrasulfur tetranitride, S_4N_4, can be converted to disulfur dinitride, S_2N_2. Write the Lewis structures for these compounds.
76. Experimental evidence indicates the existence of HC_3N molecules in interstellar clouds. Write a plausible Lewis structure for this molecule.
77. Molecules of SiC_3 have been discovered in interstellar clouds. Write a plausible Lewis structure for this molecule.
78. Phosphorus and sulfur form a series of compounds, one of which is tetraphosphorus trisulfide. Write the Lewis structure for this compound.
79. Suppose in building up molecular orbitals, the π_{2p} were placed above the σ_{2p}. Prepare a diagram similar to Figure 8.11 based on these changes. For which species in Table 8.4 would this change in relative energies of the MOs affect the prediction of number of bonds and number of unpaired electrons?

Conceptual Challenge Problems

CP8.A (Section 8.2) What deficiency is acknowledged by chemists when they write the formula for silicon dioxide as $(SiO_2)_n$ but write CO_2 for carbon dioxide?

CP8.B (Section 8.6) Without referring to the periodic table, write an argument to predict how the electronegativities of elements change based on the composition of their atoms.

CP8.C (Section 8.6) How would you rebut the statement, "There are no ionic bonds, only polar covalent bonds"?

9

Molecular Structures

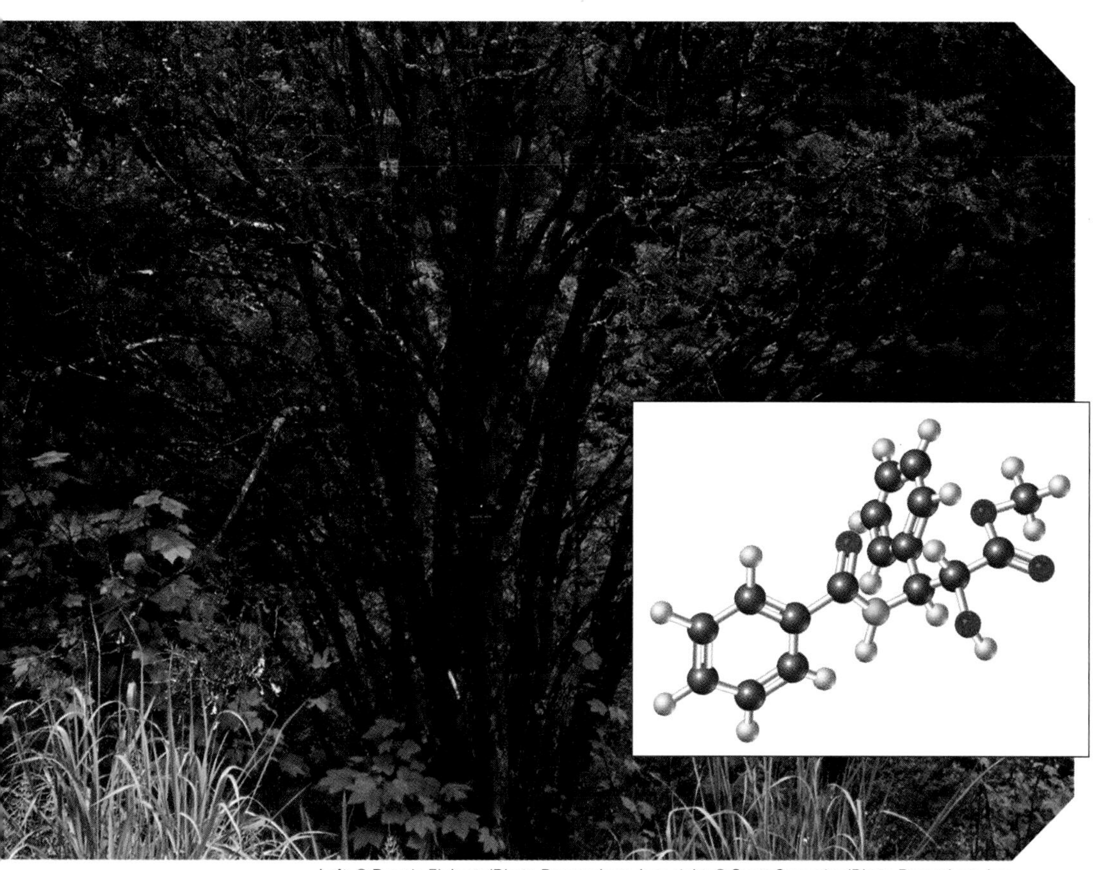

The bark of the Pacific yew tree (above) contains paclitaxel, a highly effective anti-cancer drug. The ball-and-stick model illustrates the shape of part of the paclitaxel molecule. By a detailed study of its molecular shape, researchers were able to determine the mechanism for the drug's anti-cancer activity. This chapter explains aspects of the shapes of molecules using the valence-shell electron-pair repulsion (VSEPR) model.

Although the composition, empirical formula, molecular formula, and Lewis structure of a substance provide important information, they are not sufficient to predict or explain the properties of most molecular compounds. Also important are the arrangement of the atoms and how they occupy three-dimensional space—the shape of a molecule. Because of differences in the sequence in which their atoms are joined, molecules can have the same numbers of the same kinds of atoms, yet have different properties. For example, ethanol (in alcoholic beverages) and dimethyl ether (a refrigerant) have the same molecular formula, C_2H_6O. But the C, H,

Dimethyl ether is a gas at room temperature (approximately 21 °C; ethanol is a liquid at that temperature.

	Ethanol	Dimethyl ether
Melting Pts. °C	−114.1	−141.5
Boiling Pts. °C	78.3	−24.8

Francis H. C. Crick *(right)* and James Watson. Working in the Cavendish Laboratory at Cambridge, England, Watson and Crick built a scale model of the double-helical structure of DNA, based on X-ray data. Knowing distances and angles between atoms, they compared the task to working on a three-dimensional jigsaw puzzle. Watson, Crick, and Maurice Wilkins received the 1962 Nobel Prize for their work relating to the structure of DNA.

Ball-and-stick model kits are available in many campus bookstores. The models are easy to assemble and will help you to visualize the molecular geometries described in this chapter. Ball-and-stick models are relatively inexpensive compared with space-filling models.

and O atoms are arranged so differently in molecules of these two compounds that their melting points differ by 27 °C and their boiling points by 103 °C.

```
     H   H              H       H
     |   |              |  ..   |
 H—C—C—Ö—H          H—C—O—C—H
     |   |  ..          |  ..   |
     H   H              H       H
     ethanol          dimethyl ether
```

The ideas about molecular shape presented in this chapter are crucial to understanding the relationships between the structure and properties of molecules, properties related to the behavior of molecules in living organisms, the design of molecules that are effective drugs, and many other aspects of modern chemistry.

9.1 Using Molecular Models

Molecules are three-dimensional aggregates of atoms, much too small to examine visually directly. Our ability to understand three-dimensional molecular structures is helped by the use of molecular models. Probably the best example of the impact a model can have on the advancement of science is the double helix model of DNA used by James Watson and Francis Crick, which revolutionized the understanding of human heredity and genetic disease.

Watson and Crick's DNA model was a physical model assembled atom by atom, like those for water shown below on the left. The ball-and-stick model uses balls to represent atoms and short rods of wood or plastic to represent bonds. For example, the ball-and-stick model for water has a red ball representing oxygen, with holes at the correct angles connected by sticks to two white balls representing hydrogen atoms. In the space-filling physical model, the atomic models are scaled according to the experimental values for atom sizes, and the links between parts of the model are not visible when the model is assembled. This gives a more accurate representation of the actual space occupied by atoms.

This book uses many computer-generated pictures of molecular models, like those for water shown on the next page. The computer software that generates these pictures relies on the most accurate experimentally derived data on atomic radii, bond lengths, and bond angles.

To convey a three-dimensional perspective for a molecule drawn on a flat surface, such as the page of a book, we can also make a perspective drawing that uses solid wedges (◀) to represent bonds extending in front of the page, and dashed lines (----) or dashed wedges (⋯) to represent bonds behind the page. Bonds that lie in the plane of the page are indicated by a line (—), as illustrated in the following perspective drawing for methane, a tetrahedral molecule:

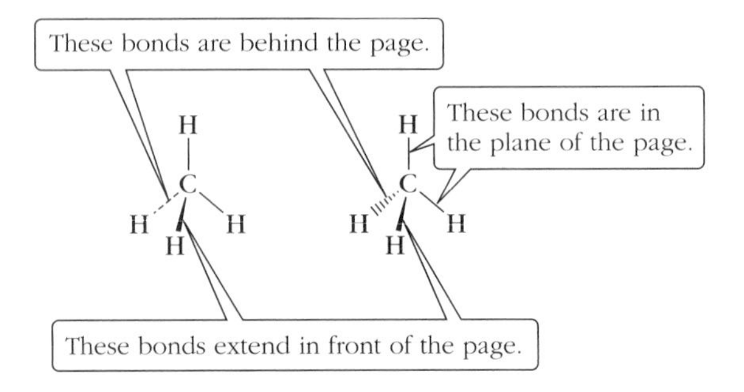
These bonds are behind the page.
These bonds are in the plane of the page.
These bonds extend in front of the page.
H
C
H
H
H

Computers can draw and also rotate molecules so that they can be viewed from any angle, as illustrated in Figure 9.1 for ethane.

Advances in computer graphics have made it possible to draw scientifically accurate pictures of extremely complex molecules, as well as to study interactions between

Photos: © Cengage Learning/Charles D. Winters

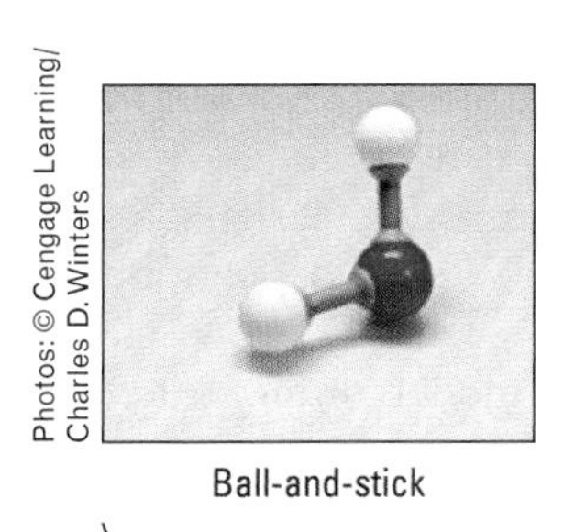
Ball-and-stick

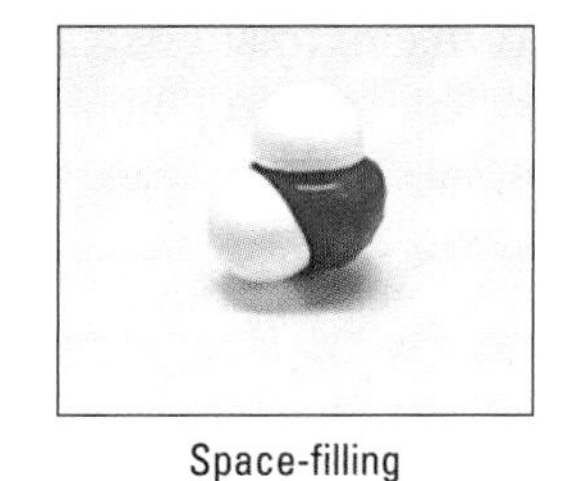
Space-filling

Physical molecular models

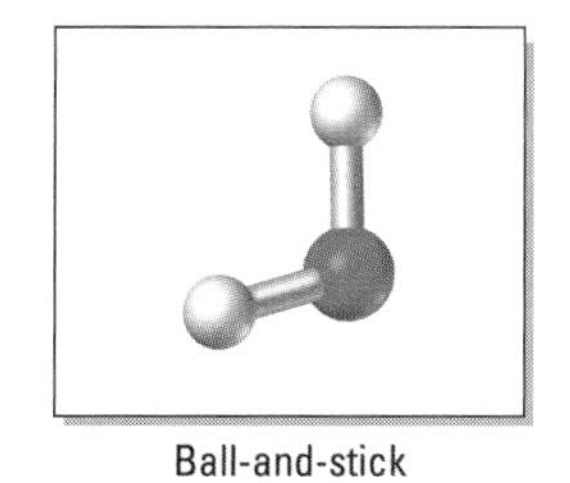
Ball-and-stick

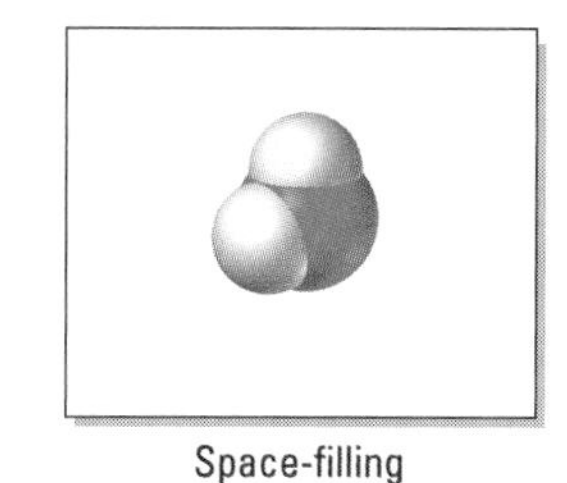
Space-filling

Computer-generated molecular models

molecules. Figure 9.2 is a computer-generated model of hemoglobin, the remarkable and essential protein in the blood that takes up and releases oxygen. This vital protein is made up of two folded sets of pairs of long chains of atoms, indicated by the red ribbons. In a hemoglobin molecule, oxygen binds to Fe^{2+} ions (yellow) at four sites shown in green, red, and blue in Figure 9.2. Molecular graphics programs can rotate such computer-generated models to provide different views of the molecule.

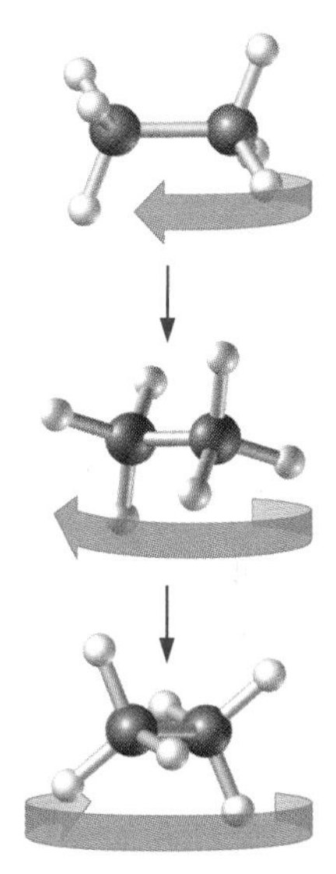
Figure 9.1 Rotation around an axis perpendicular to the C—C bond axis in an ethane molecule, C_2H_6.

9.2 Predicting Molecular Shapes: VSEPR

Because molecular shape is so important to the reactivity and properties of molecules, it is essential to have a simple, reliable method for predicting the shapes of molecules and polyatomic ions. The **valence-shell electron-pair repulsion (VSEPR)** model is such a method. The VSEPR model is based on the idea that repulsions occur among regions of electron density, such as the repulsions among pairs of bonding and lone pair electrons. Such repulsions control the angles between bonds from a central atom to other atoms surrounding it.

How do repulsions among electron pairs result in different shapes? Imagine that the volume of a balloon represents the repulsive force of an electron pair that prevents other electron pairs from occupying the same space. When two or more balloons are tied together at a central point, the cluster of balloons assumes the shapes shown in Figure 9.3. The central point represents the nucleus and the core electrons of a central atom. The arrangements of the balloons are those that minimize interactions among them, and among the electron pairs that the balloons represent.

To minimize their repulsions, electron pairs (regions of electron density) are oriented as far apart as possible. Because electrons are constrained by electrostatic attraction to be near the nucleus, the shapes resulting from minimizing the electron-pair repulsions are those predicted by VSEPR and illustrated in the balloon analogy (Figure 9.3.)

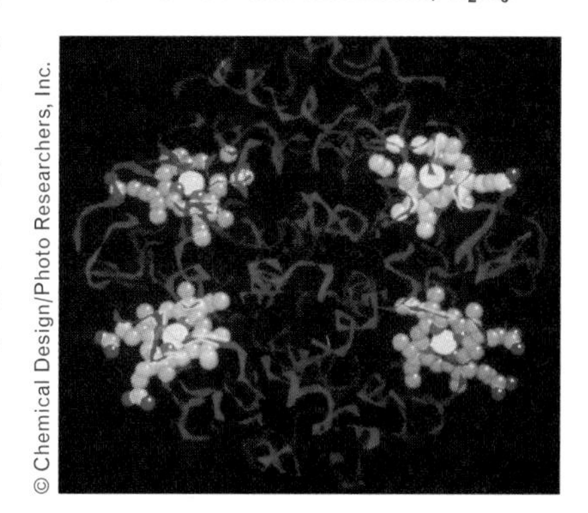
© Chemical Design/Photo Researchers, Inc.

Figure 9.2 Hemoglobin, a protein essential to human life. An Fe^{2+} ion shown in yellow is at the center of each green heme cluster.

Central Atoms with Only Bonding Pairs

The simplest application of VSEPR is to molecules and polyatomic ions having a central atom that has only shared electron pairs to bond it to other atoms (terminal atoms) by only single bonds. In such cases, the central atom has no lone pairs. We can sum-

Photos: © Cengage Learning/Charles D. Winters

Linear

Triangular planar

Tetrahedral

Triangular bipyramidal

Octahedral

Figure 9.3 Balloon models of the geometries predicted by the VSEPR model.

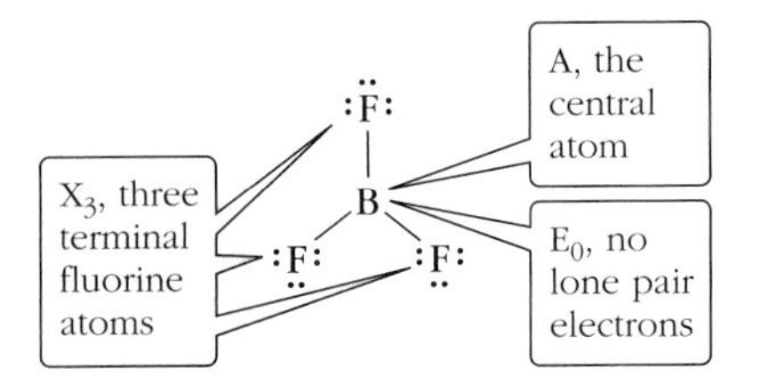

Using the AX_3E_0 designation for boron trifluoride.

The term "molecular shape" is sometimes used rather than "molecular geometry."

The predicted bond angles given in the examples in Figure 9.4 are in agreement with experimental values obtained from structural studies.

methane (CH_4)

Central atoms from Period 2 are too small to accommodate more than four atoms bonded to them.

marize this using an AXE-based shorthand notation: A (the central atom) is bonded to X_n (the n terminal atoms); E_m represents the m lone pairs on the central atom. When there are no lone pairs, the symbolism is E_0. For example, an AX_3E_0 designation denotes a central atom (A) with zero lone pairs (E_0) and whose shared pairs are bonded to three terminal atoms (X_3), such as in BF_3.

Figure 9.4 illustrates the geometries predicted by the VSEPR model for molecules of types AX_2E_0 to AX_6E_0. (Table 9.1 also summarizes this information.)

Notice in Figure 9.4 that there are two categories of geometries—electron-pair geometry and molecular geometry. In a molecule or polyatomic ion, the **electron-pair geometry** is determined by the number and arrangement of the valence electron pairs around the central atom. By contrast, the **molecular geometry** is the arrangement of the atoms (nuclei and core electrons) in space. *For molecules whose central atoms have no lone pairs, the electron-pair geometry and the molecular geometry are the same.* When lone pairs are present on the central atom, the electron-pair and molecular geometries are not the same, as will be discussed shortly. **Bond angles,** as shown in Figure 9.4, are the angles between the bonds of two atoms that are bonded to the same third atom. In a methane molecule, CH_4, for example, all the H—C—H bond angles are 109.5°.

In molecules with *linear* geometry for two bonding pairs and the *triangular planar* geometry for three bonding pairs the central atom has less than an octet of electrons around it (⇐ *p. 294*). The central atom in a *tetrahedral* molecule obeys the octet rule with four bonding pairs (Figure 9.4). The central atoms in *triangular bipyramidal* and *octahedral* molecules do not obey the octet rule because they have five and six bonding pairs, respectively. Hence, triangular bipyramidal and octahedral geometries would be expected only when the central atom is an element in Period 3 or higher (⇐ *p. 294*). The geometries illustrated in Figure 9.4 are by far the most common in molecules and polyatomic ions, and you should be thoroughly familiar with them.

Suppose you want to use VSEPR concepts to predict the shape of SiCl4. First, draw the Lewis structure, with Si as the central atom. From the Lewis structure, you can see that there are four bonding pairs from the terminal Cl atoms to the central Si atom forming four single bonds to Si and no lone pairs on Si. Therefore, SiCl4 is an AX_4E_0 type of molecule. A is Si, X is Cl, and X_4 indicates four Cl atoms bonded to A; E_0 denotes no lone pairs on A (Si), the central atom and you would predict a tetrahedral electron-pair geometry and a tetrahedral molecular geometry for the molecule. (Figure 9.4). This is in agreement with structural studies of SiCl4, which indicate a tetrahedral molecule with all Cl—Si—Cl bond angles 109.5°.

Molecular model	180°	120°	109.5°	90°, 120°	90°
Type	AX_2E_0	AX_3E_0	AX_4E_0	AX_5E_0	AX_6E_0
Electron-pair geometry	Linear	Triangular planar	Tetrahedral	Triangular bipyramidal	Octahedral
Molecular geometry	Linear	Triangular planar	Tetrahedral	Triangular bipyramidal	Octahedral
Example molecule	BeF_2	BF_3	CH_4	PCl_5	SF_6

Figure 9.4 Geometries predicted by the VSEPR model for molecules of types AX_2E_0 through AX_6E_2 that contain only single covalent bonds and no lone pairs on the central atom, A.

Table 9.1 Examples of Electron-Pair Geometries and Molecular Geometries Predicted by the VSEPR Model

Type (X = atoms bonded to central atom A; E = lone pairs on central atom)	Number of X Atoms on Central Atom	Number of Lone Pairs on Central Atom	Electron-Pair Geometry	Molecular Geometry	Example
AX_2E_0	Two	None	Linear	Linear	CO_2, $BeCl_2$
AX_2E_1	Two	One	Triangular planar	Angular (bent)	$SnCl_2$
AX_2E_2	Two	Two	Tetrahedral	Angular (bent)	H_2O, OCl_2
AX_2E_3	Two	Three	Triangular bipyramidal	Linear	XeF_2
AX_3E_0	Three	None	Triangular planar	Triangular planar	BCl_3, CO_3^{2-}
AX_3E_1	Three	One	Tetrahedral	Triangular pyramidal	NCl_3
AX_3E_2	Three	Two	Triangular bipyramidal	T-shaped	ClF_3
AX_4E_0	Four	None	Tetrahedral	Tetrahedral	CH_4, $SiCl_4$
AX_4E_1	Four	One	Triangular bipyramidal	Seesaw	SF_4
AX_4E_2	Four	Two	Octahedral	Square planar	XeF_4
AX_5E_0	Five	None	Triangular bipyramidal	Triangular bipyramidal	PF_5
AX_5E_1	Five	One	Octahedral	Square pyramidal	BrF_5
AX_6E_0	Six	None	Octahedral	Octahedral	SF_6

PROBLEM-SOLVING EXAMPLE 9.1 Molecular Geometry

Use the VSEPR model to predict the electron-pair geometry, molecular geometry, and bond angles of (a) BF_3 and (b) CBr_4.

Answer

(a) Triangular planar electron-pair and molecular geometries with 120° F—B—F bond angles

(b) Tetrahedral electron-pair and molecular geometries with 109.5° Br—C—Br bond angles

Strategy and Explanation The first step in using VSEPR is to write the correct Lewis structure and classify the molecule according to the number of bonding electron pairs and lone pairs on the central atom.

(a) The Lewis structure of BF_3 is

:F̤̈:
| 120°
B
:F̤̈ F̤̈:

BF_3 is an AX_3E_0-type molecule (three bonding pairs, no lone pairs around the central B atom; see Figure 9.4). The molecule has three B—F single bonds around the central boron atom, arranged at 120° angles from each other. Therefore, BF_3 has a triangular planar electron-pair geometry and molecular geometry with all the atoms lying in the same plane.

(b) Being an AX_4E_0-type molecule, CBr_4 will be a tetrahedral molecule with 109.5° Br—C—Br bond angles. The electron pair and molecular geometries are tetrahedral.

109.5°

:Br: — C — :Br: , :Br: , :Br:

PROBLEM-SOLVING PRACTICE 9.1

Identify the electron-pair geometry, the molecular geomety, and the bond angles for BeF_2.

Multiple Bonds and Molecular Geometry

Although double bonds and triple bonds are shorter and stronger than single bonds, this has only a minor effect on predictions of molecular shape. Why? Electron pairs involved in a multiple bond are all shared between the same two nuclei and therefore occupy the same region of space. Because they must remain in that region, two electron pairs in a double bond or three electron pairs in a triple bond are like a single, slightly fatter balloon, rather than two or three balloons. Hence, ***for the purpose of determining molecular geometry, the electron pairs in a multiple bond constitute a single region of electron density as in a single bond.*** For example, compare BeF_2 (Figure 9.4) with CO_2. BeF_2 is a linear molecule with the two Be—F single bonds 180° apart. In CO_2, each C=O double bond is a single region of electron density, just as each Be—F single bond is, so the structure of CO_2 is also linear.

:Ö=C=Ö:

Carbon dioxide is a linear molecule.

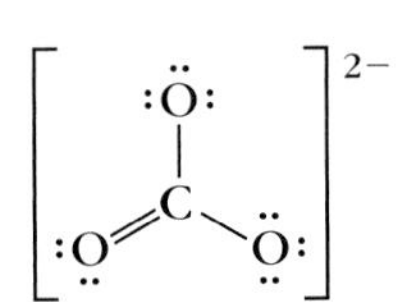

This structural formula is one of three resonance forms for carbonate ion (⇐ p. 293).

When resonance structures are possible, the geometry can be predicted from the resonance hybrid structure. For example, the geometry of the CO_3^{2-} ion is predicted to be triangular planar because the carbon atom has three sets of equivalent bonds and no lone pairs. (Each of the three bonds is intermediate between a single bond and a double bond.) It can be predicted from the Lewis structure.

CONCEPTUAL EXERCISE 9.1 Geometries

Based on the discussion so far, identify a characteristic that is common to all situations where electron-pair geometry and molecular geometry are the same for a molecule or a polyatomic ion.

Central Atoms with Bonding Pairs and Lone Pairs

How does the presence of lone pairs on the central atom affect the geometry of a molecule or polyatomic ion? The easiest way to visualize this situation is to return to the balloon model and notice that the electron pairs on the central atom do not all have to be bonding pairs. We can predict the electron-pair geometry, molecular geometry, and bond angles by applying the VSEPR model to the total number of valence electron pairs—that is, bonding pairs as well as lone pairs around the central atom. We will use NH_3, ammonia, to illustrate guidelines for doing so.

1. ***Draw the Lewis structure.*** The correct Lewis structure for NH_3 is

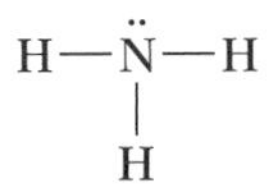

H—N̈—H
|
H

2. ***Determine the number of bonds and the number of lone pairs around each atom.*** In the case of NH_3, there are three bonding pairs and one lone pair; thus, NH_3 is an AX_3E_1-type molecule (Table 9.1). If there were multiple bonds, any electron pairs in a multiple bond would contribute to the molecular geometry as those in a single bond. The central N atom in NH_3 has no multiple bonds, only three H—N single covalent bonds.

In ammonia, A is N, X is H, and X_3 indicates three H atoms bonded to A, the central atom; E_1 denotes one lone pair on A.

3. ***Choose the appropriate electron-pair geometry around each central atom, and then choose the molecular shape that matches the total number of bonds and lone pairs.*** The *electron-pair geometry* around a central atom includes the spatial positions of all bonds and lone pairs. Because there are three bonds and one lone pair for a total of four pairs of electrons, we predict that the electron-pair geometry of NH_3 is tetrahedral. To represent this geometry, draw a tetrahedron with N as the central atom and the three bonding pairs represented by lines, wedges, or dashed lines, since they are single covalent bonds. The lone pair is drawn as a balloon shape to indicate its spatial position in the tetrahedron.

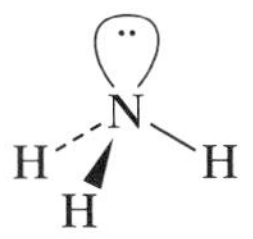

   The positions of the lone pairs are *not* specified when describing the molecular geometry of molecules. The *molecular geometry* of ammonia is described as a triangular pyramid because the three hydrogen atoms form a triangular base with the nitrogen atom at the apex of the pyramid. (This can be seen by covering up the lone pair of electrons and looking at the molecular geometry—the location of the three H atoms and the N atoms.)

4. ***Predict the bond angles, remembering that lone pairs occupy more space around the central atom than do bonding pairs.*** Bonding pairs are concentrated in the bonding region between two atoms by the strong attractive forces of two positive nuclei and are, therefore, relatively compact, or "skinny." For a lone pair, there is only one nucleus attracting the electron pair. As a result, lone pairs are less compact. Using the balloon analogy, a lone pair is like a fatter balloon that takes up more room and squeezes the thinner balloons closer together. The relative strengths of electron-pair repulsions are

   lone pair–lone pair > lone pair–bonding pair > bonding pair–bonding pair

   This listing predicts that lone pairs will force bonding pairs closer together and decrease the angles between the bonding pairs. Recognizing this, we can predict that *bond angles adjacent to lone pairs will be smaller than those predicted for perfect geometric shapes.* The determination of electron-pair geometry and molecular geometry for NH_3 is summarized by this sequence:

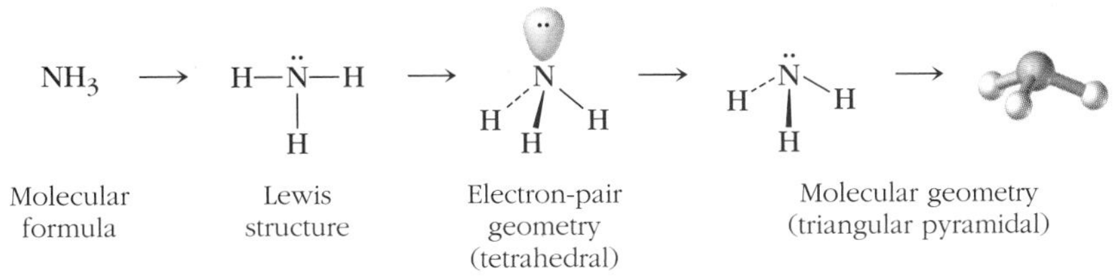

NH_3 → H—N̈—H → (tetrahedral) → (triangular pyramidal) →

Molecular formula | Lewis structure | Electron-pair geometry (tetrahedral) | Molecular geometry (triangular pyramidal)

The success of the VSEPR model in predicting molecular shapes indicates that it is appropriate to account for the effects of lone pairs in this way.

Because the electron-pair geometry is tetrahedral, we would expect the H—N—H bond angles to be 109.5°. However, the experimentally determined bond angles in NH_3 are 107.5°. This is attributed to the bulkier lone pair forcing the bonding pairs closer together, thereby reducing the bond angle from 109.5° to 107.5°. Table 9.1 **(⇐ p. 311)** summarizes the electron-pair geometry and molecular geometry for cases of zero, one, two, or three lone pairs on the central atom.

The effect that increasing numbers of lone pairs on the central atom have on bond angles is shown below for CH_4, NH_3, and H_2O. Note from the series of molecules that as the number of lone pairs on the central atom increases, the bond angle decreases. The three molecules all have tetrahedral electron-pair geometry because each has four electron pairs around the central atom. Each molecule, however, has slightly different bond angles due to the lone pair effect. There are bonding pair–bonding pair repulsions in all three molecules, and also bonding pair–lone pair repulsions in NH_3, and additionally lone pair–lone pair repulsions in water. This results in tetrahedral molecular geometry for CH_4 (109.5°), triangular pyramidal molecular geometry for NH_3 (107.5°), and angular (bent) molecular geometry for H_2O (104.5°).

Sometimes the term "bent" is used rather than "angular" to describe the molecular geometry for water.

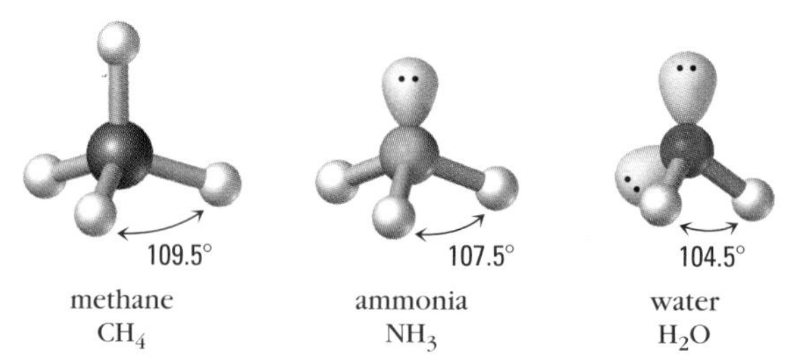

Methane, which has a tetrahedral shape, is the smallest member of the large family of saturated hydrocarbons called alkanes (**p. 272**). It is important to recognize that every carbon atom in an alkane has a tetrahedral environment. For example, notice below that the carbon atoms in propane and in the much longer carbon chain of hexadecane do not lie in a straight line because of the tetrahedral geometry about each carbon atom. The tetrahedron is arguably the most important shape in chemistry because of its predominance in the chemistry of carbon and the carbon-based chains that form the backbone of many biomolecules.

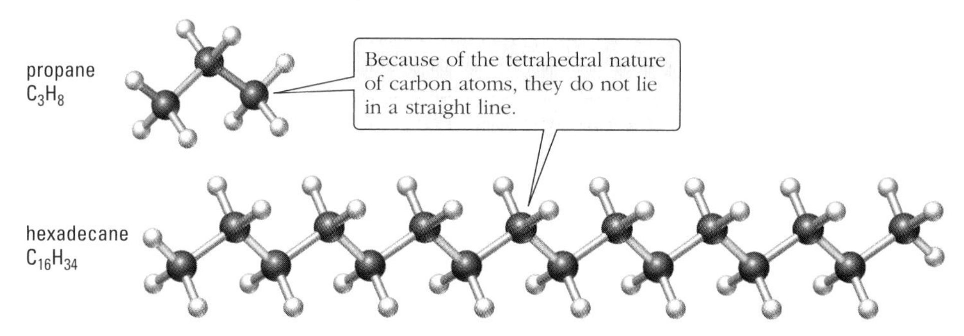

Figure 9.5 gives additional examples of electron-pair and molecular geometries for molecules or ions with three and four electron pairs around the central atom. To check

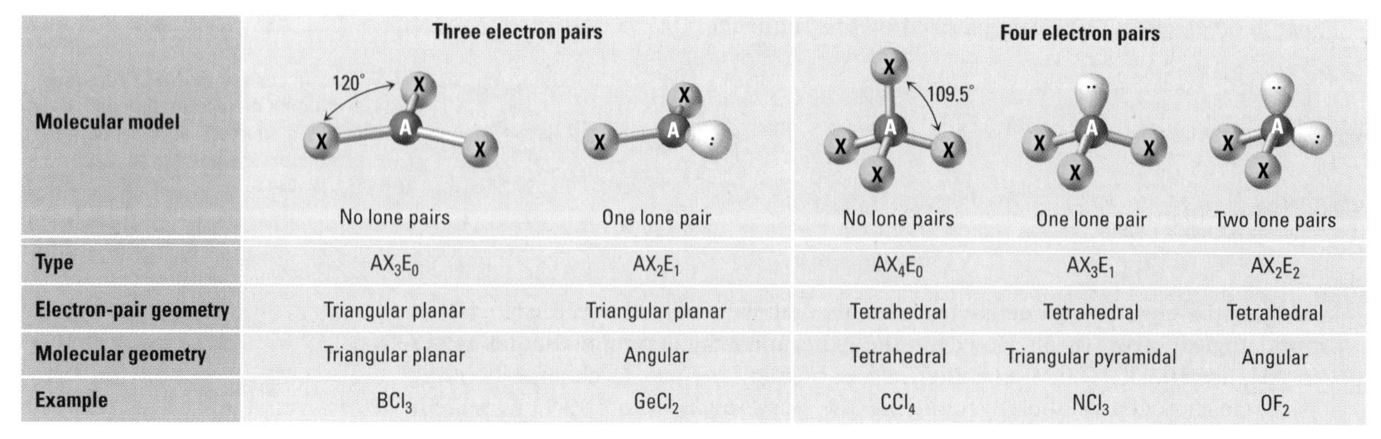

	Three electron pairs		Four electron pairs		
Molecular model	No lone pairs	One lone pair	No lone pairs	One lone pair	Two lone pairs
Type	AX_3E_0	AX_2E_1	AX_4E_0	AX_3E_1	AX_2E_2
Electron-pair geometry	Triangular planar	Triangular planar	Tetrahedral	Tetrahedral	Tetrahedral
Molecular geometry	Triangular planar	Angular	Tetrahedral	Triangular pyramidal	Angular
Example	BCl_3	$GeCl_2$	CCl_4	NCl_3	OF_2

Figure 9.5 Three and four electron pairs around a central atom. Examples are shown of electron-pair geometries and molecular shapes for molecules and polyatomic ions with three and four electron pairs around the central atom.

your understanding of the VSEPR model, try to explain the electron-pair geometry, the molecular geometry, and bond angles given in Figure 9.5.

PROBLEM-SOLVING EXAMPLE 9.2 Molecular Structure

Use the VSEPR model to predict the electron-pair geometry and the molecular geometry of (a) ClF_2^+, (b) $SnCl_3^-$, (c) SeO_3, and (d) CS_2.

Answer

Electron-Pair Geometry	Molecular Geometry
(a) Tetrahedral	Angular
(b) Tetrahedral	Triangular pyramidal
(c) Triangular planar	Triangular planar
(d) Linear	Linear

Strategy and Explanation Writing the correct Lewis structures identifies the central atom in each ion or molecule. The Lewis structures are

(a) $[:\ddot{F}-\ddot{Cl}-\ddot{F}:]^+$ (b) $[:\ddot{Cl}-\ddot{Sn}(-\ddot{Cl}:)-\ddot{Cl}:]^-$ (c) $O=Se(-O)-O$ (d) $:\ddot{S}=C=\ddot{S}:$

(a) The Lewis structure of ClF_2^+ reveals that the central Cl atom has four electron pairs: two bonding pairs to terminal fluorine atoms and two lone pairs. Consequently, the ion is an AX_2E_2 type and has a tetrahedral electron-pair geometry and an angular (bent) molecular geometry. The two lone pairs push the Cl—F bonds closer than the ideal 109.5° angle.
(b) The central tin atom is surrounded by three bonding pairs and one lone pair (AX_3E_1 type). These four electron pairs give a tetrahedral electron-pair geometry. The molecular geometry is triangular pyramidal. The lone pair pushes the Sn—Cl bonds closer together than the purely 109.5° tetrahedral angle expected for four bonding pair–bonding pair repulsions.
(c) There are three resonance structures, which give three equivalant regions of electron density around the Se atom (AX_3E_0 type). Because there are no lone pairs around Se, the electron-pair and molecular geometries are the same, triangular planar.
(d) For determining molecular structure, each C=S double bond is treated as one region of electron density. There are no lone pairs on the central C atom (AX_2E_0 type). Therefore, the electron-pair geometry and molecular geometry are both linear.

PROBLEM-SOLVING PRACTICE 9.2

Use Lewis structures and the VSEPR model to determine electron-pair and molecular geometries for (a) BrO_3^-, (b) SeF_2, and (c) NO_2^-.

Expanded Octets: Central Atoms with Five or Six Electron Pairs

Analysis of electron-pair and molecular geometries becomes more complicated if a central atom has five or six electron pairs, some of which are lone pairs. Let's look first at the entries in Figure 9.6 for the case of five electron pairs. The three angles in the triangular plane are all 120°. The angles between any of the pairs in this plane and an upper or lower pair are only 90°. Thus, the *triangular bipyramidal* structure has two sets of positions that are not equivalent. Because the positions in the triangular plane lie in the equator of an imaginary sphere around the central atom, they are called **equatorial positions.** The north and south poles are called **axial positions.** Each

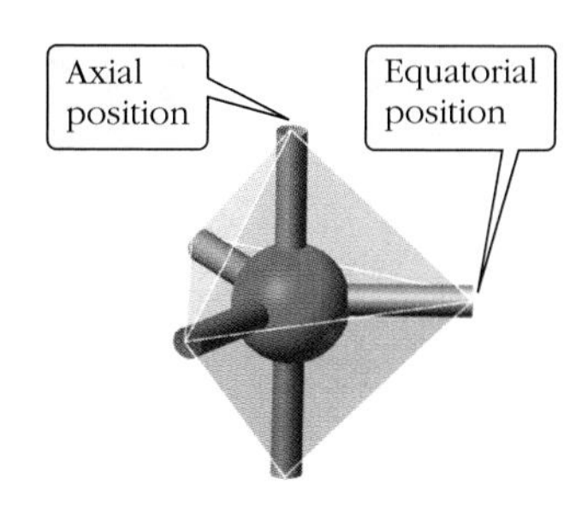

Triangular bipyramidal structure

equatorial position is next to only two electron pairs at 90° angles, the axial ones, while an axial position is next to three electron pairs, the equatorial ones. This means that *any lone pairs, because they are more extensive than bonding pairs, will occupy equatorial positions rather than axial positions.* For example, consider the ClF_3 molecule, which has three bonding pairs and two lone pairs, as shown in Figure 9.6. The two lone pairs in ClF_3 are equatorial; two bonding pairs are axial, with a third occupying an equatorial position, so the molecular geometry is T-shaped (Figure 9.6). (Given our viewpoint of axial positions lying on a vertical line, the T of the molecule is actually on its side ⊢.)

The electron-pair geometry for six electron pairs is an *octahedron* with each angle 90°. Unlike the triangular bipyramid, the octahedron has no distinct axial and equatorial positions; all six positions around the central atom are equivalent. As shown in Figure 9.6, six atoms bonded to the central atom in an octahedron can occupy three axes at right angles to one another (the *x*-, *y*-, and *z*-axes). All six atoms are *equidistant* from the central atom. Thus, if a molecule with octahedral electron-pair geometry has one lone pair on the central atom, it makes no difference which apex the lone pair occupies. An example is BrF_5, whose molecular geometry is *square pyramidal* (Figure 9.6). There is a *square plane* containing the central atom and four of the atoms bonded to it, with the other bonded atom directly above the central atom and equidistant from the other four. If the central atom has two lone pairs, as in ICl_4^-, each of the lone pairs needs as much room as possible. This is best achieved by placing the lone pairs at a 180° angle to each other above and below the square plane that contains the I atom and the four Cl atoms. Thus, the molecular geometry of ICl_4^- is *square planar* (Figure 9.6).

In Figure 9.6, the lone pair in BrF_5 could have been located in any one of the six positions around Br. We arbitrarily chose the "down" position.

An example of the power of the VSEPR model is its use to correctly predict the shape of XeF_4. At one time the noble gases were not expected to form compounds because their atoms have a stable octet of valence electrons ***(⇐ p. 272)***. The synthesis of XeF_4 was a surprise to chemists, but you can use the VSEPR model to predict the correct geometry. The molecule has 36 valence electrons (eight from Xe and seven from each of four F atoms). There are eight electrons in four bonding pairs around Xe,

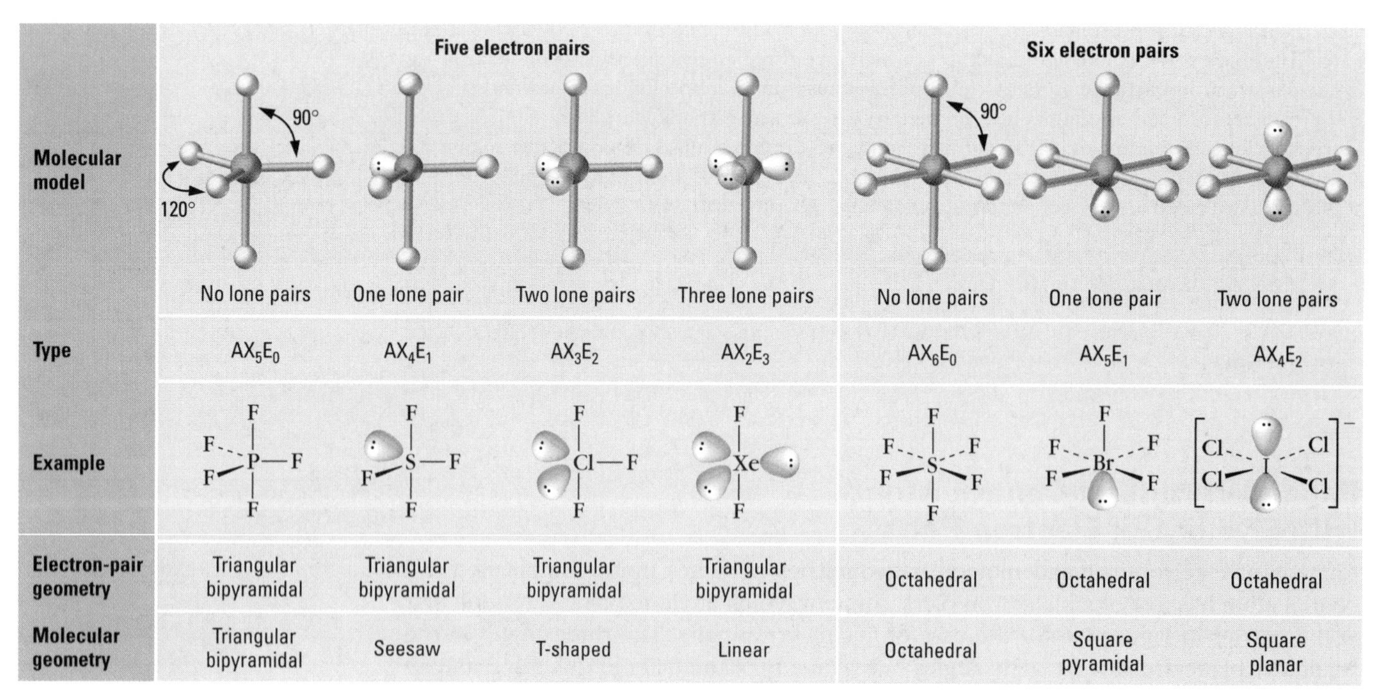

Five electron pairs
Six electron pairs
Molecular model
90°
120°
90°
No lone pairs
One lone pair
Two lone pairs
Three lone pairs
No lone pairs
One lone pair
Two lone pairs
Type
AX5E0
AX4E1
AX3E2
AX2E3
AX6E0
AX5E1
AX4E2
Example
Electron-pair geometry
Triangular bipyramidal
Triangular bipyramidal
Triangular bipyramidal
Triangular bipyramidal
Octahedral
Octahedral
Octahedral
Molecular geometry
Triangular bipyramidal
Seesaw
T-shaped
Linear
Octahedral
Square pyramidal
Square planar

Figure 9.6 Five and six electron pairs around a central atom. Molecules and polyatomic ions with five and six electron pairs around the central atom can have these electron-pair geometries and molecular shapes.

and a total of 24 electrons in the lone pairs on the four F atoms. That leaves four electrons in two lone pairs on the Xe atom.

Because Xe is in Period 5, it can accommodate more than an octet of electrons (⇐ *p. 296*). The total of six electron pairs on Xe leads to a prediction of an octahedral *electron-pair geometry* (AX_4E_2 type). Where do you put the lone pairs? As explained earlier for the ICl_4^- ion, the lone pairs are placed at opposite corners of the octahedron to minimize repulsion by keeping them as far apart as possible. The result is a square planar *molecular geometry* for the XeF_4 molecule (cover the lone pairs in the drawing for ICl_4^- in Figure 9.6 to see this). This shape agrees with experimental structural results. A number of other xenon compounds have been prepared, and the VSEPR model has been useful in predicting their geometries as well.

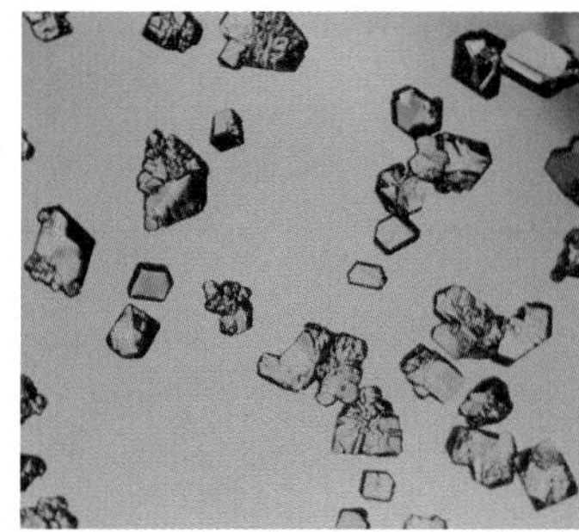

Xenon tetrafluoride crystals.

CONCEPTUAL
EXERCISE 9.2 Two Dissimilar Shapes

Triangular bipyramidal and square pyramidal molecular geometries each have a central atom with five terminal atoms around it, but the molecular shapes are different. Explain how and why these two shapes differ.

Table 9.1 (⇐ *p. 311*) gives additional examples of molecules and ions whose shapes can be predicted by using the VSEPR model. The model also can be used to predict the geometry around atoms in molecules with more than one central atom. Consider, for example, lactic acid, a compound important in carbohydrate metabolism. The molecular structure of lactic acid is

```
    H   H  :O:
    |   |   ||   ..
H — C — C — C — O — H
    |   |        ..
    H  :O:
        |
        H
```

Around C in the —CH_3 group are four bonding pairs and no lone pairs, so its molecular geometry is tetrahedral (109.5° bond angles). The middle carbon also has four bonding pairs associated with it and so also has tetrahedral geometry (109.5° bond angles). The remaining carbon atom has four bonding pairs—two in single bonds and two in a C═O double bond—giving, for molecular geometry purposes, three bonding regions around it and a triangular planar geometry (120° bond angles). The single-bonded oxygens, in each case, have two bonding pairs and two lone pairs, creating an angular molecular geometry around each single-bonded oxygen atom (with a bond angle of less than 109.5°). The combined effect of these geometries is shown in the molecular models below.

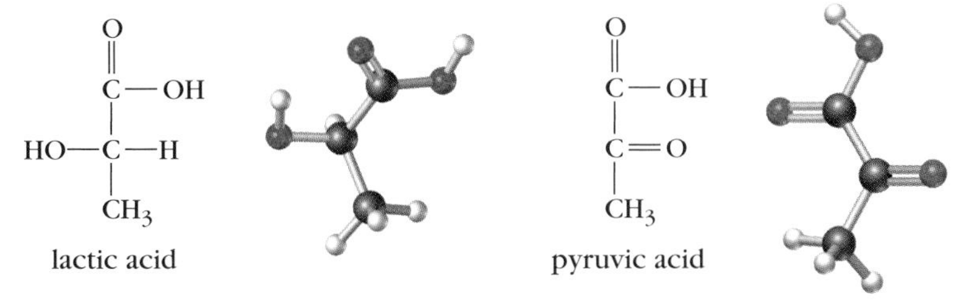

Lactic acid, a waste product of glucose metabolism, builds up rapidly in muscles during strenuous short-term exercise such as sprinting (100 to 800 m) and swimming (100 or 200 m). After the exercise stops, the lactic acid is converted to pyruvic acid, which is removed from the muscles.

PROBLEM-SOLVING EXAMPLE 9.3 VSEPR and Molecular Shape

What are the electron-pair and the molecular geometries of (a) SeF_6, (b) IF_3, (c) TeF_4, and (d) XeF_5^+?

Answer

(a) Octahedral electron-pair and molecular geometries
(b) Triangular bipyramidal electron-pair geometry and T-shaped molecular geometry
(c) Triangular bipyramidal electron-pair geometry and seesaw-shaped molecular geometry
(d) Octahedral electron-pair geometry and square pyramidal molecular geometry

Strategy and Explanation First, draw the Lewis structures, then use them as a guide to determine the electron-pair and molecular geometries. The Lewis structures are

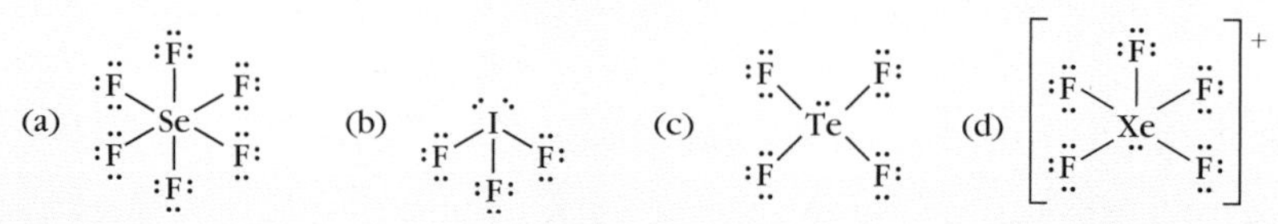

(a) The 48 valence electrons are used for six single bonds from the central Se to the terminal fluorine atoms and three lone pairs around each fluorine. Because SeF_6 is an AX_6E_0-type molecule, there are six bonding pairs around Se and no lone pairs. The electron-pair geometry and the molecular geometry are octahedral.

(b) IF_3 is an AX_3E_2-type molecule. There are five electron pairs around I, giving a triangular bipyramidal electron-pair geometry. As noted in Figure 9.6, the two lone pairs will be in equatorial positions. One bonding pair is equatorial, while the remaining two bonding pairs are axial. The molecule is T-shaped.

(c) In TeF_4, an AX_4E_1-type molecule, the central Te atom is surrounded by five electron pairs, giving it a triangular bipyramidal electron-pair geometry. The lone pair is in an equatorial position, creating a seesaw-shaped molecule (Figure 9.6).

(d) In XeF_5^+, an AX_5E_1-type polyatomic ion, the central Xe atom has a lone pair of electrons plus five bonding pairs to terminal fluorine atoms. The ion has octahedral electron-pair geometry and a square pyramidal molecular geometry (Figure 9.6).

PROBLEM-SOLVING PRACTICE 9.3

What is the electron-pair geometry and the molecular geometry of (a) ClF_2^- and (b) XeO_3?

EXERCISE 9.3 Electron Pairs and Molecular Shapes

Classify each of these species according to its AX_nE_m type:
(a) XeF_2 (b) NI_3 (c) I_3^-

9.3 Orbitals Consistent with Molecular Shapes: Hybridization

Although Lewis structures are helpful in assigning molecular geometries, they indicate nothing about the orbitals occupied by the bonding and lone pair electrons. A theoretical model of covalent bonding, referred to as the **valence bond model,** does so by describing a covalent bond as the result of an overlap of orbitals on each of two bonded atoms. For H_2, for example, the bond is a shared electron pair located in the overlapping *s* atomic orbitals.

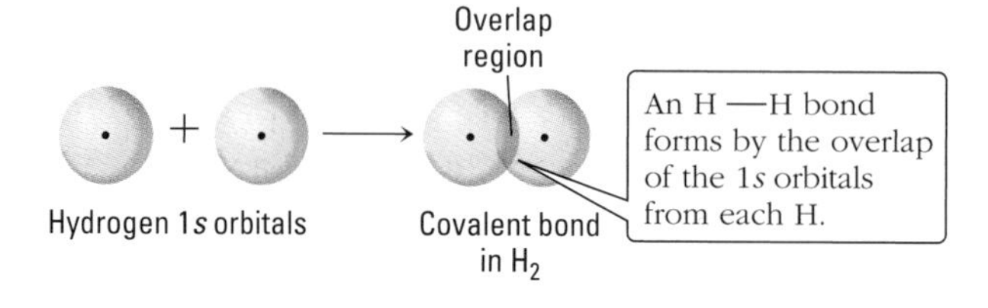
Overlap region
Hydrogen 1s orbitals
Covalent bond in H2
An H—H bond forms by the overlap of the 1s orbitals from each H.

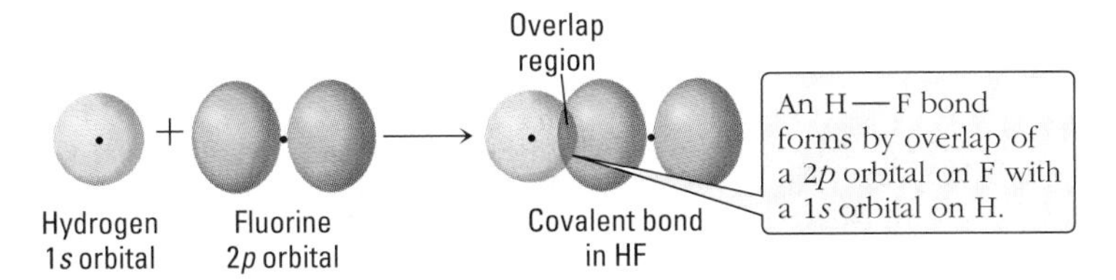

In hydrogen fluoride, HF, the overlap occurs between a $2p$ orbital with a single electron on fluorine and the single electron in the $1s$ orbital of a hydrogen atom.

The simple valence bond model of overlapping s and p orbitals, however, must be modified to account for bonding in molecules with central atoms, such as Be, B, and C. Consider the electron configurations for Be, B, and C atoms:

			1s	2s		2p	
Be	$1s^2 2s^2$		↑↓	↑↓			
B	$1s^2 2s^2 2p^1$		↑↓	↑↓	↑		
C	$1s^2 2s^2 2p^2$	$1s^2 2s^2 2p_x^1 2p_y^1$	↑↓	↑↓	↑	↑	

The simple valence bond model would predict that Be, with an outer s^2 configuration like He, should form no covalent bonds; B, with one unpaired electron ($2p^1$) should form only one bond; and C ($2p^2$) should form only two bonds with its two unpaired $2p$ electrons. But BeF_2, BF_3, and CF_4 all exist, as well as other Be, B, and C compounds in which these atoms have two, three, and four bonds, respectively.

To account for molecules like CH_4, in which the actual molecular geometry is incompatible with simple overlap of s and p orbitals, valence bond theory is modified to include a new kind of atomic orbital. Atomic orbitals of the proper energy and orientation in the same atom are **hybridized,** meaning that the atomic orbitals are combined to form **hybrid orbitals.** The resulting hybrid orbitals all have the same shape. Consequently, they are better able to overlap with bonding orbitals on other atoms. Hybrid atomic orbitals result in more bonds, stronger bonds, or both between atoms than do the unhybridized atomic orbitals from which they are formed. We turn now to a discussion of several types of hybrid orbitals. ***The total number of hybrid orbitals formed is always equal to the number of atomic orbitals that are hybridized.***

Hybridization is described by a combination of mathematical functions that represent *s*, *p*, or *d* orbitals to give new functions that predict shapes of the hybrid orbitals.

Another theory, the *molecular orbital theory,* is also used to describe chemical bonding, as well as the magnetic properties of molecules. Molecular orbital theory was discussed in Section 8.10.

sp Hybrid Orbitals

The simplest hybrid orbitals are ***sp* hybrid orbitals** formed by the combination of one s orbital and one p orbital; these *two* atomic orbitals combine to form *two sp* hybrid orbitals (Figure 9.7).

$$\text{One } s \text{ atomic orbital} + \text{one } p \text{ atomic orbital} \rightarrow \text{two } sp \text{ hybrid orbitals}$$

In BeF_2, for example, a $2s$ and a $2p$ orbital on Be are hybridized to form two sp hybrid orbitals that are 180° from each other. That is, the hybrid orbitals are oriented exactly the same way we predicted earlier for two regions of electron density.

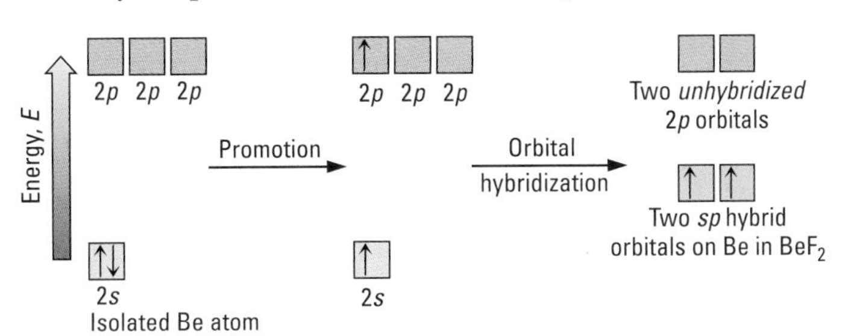

Note that one of the $2s$ electrons is promoted to a $2p$ orbital. The $2s$ and that $2p$ orbital are then hybridized to form *two sp* hybrid orbitals.

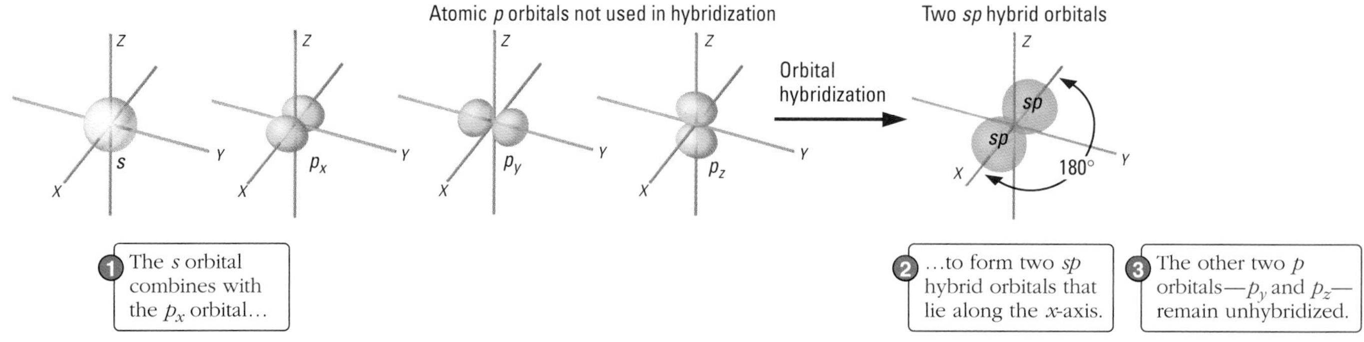

Active Figure 9.7 Formation of two *sp* hybrid orbitals. An *s* orbital combines with a *p* orbital in the same atom, say, p_x, to form two *sp* hybrid orbitals that lie along the *x*-axis. The angle between the two *sp* orbitals is 180°. The other two *p* orbitals remain unhybridized. Visit this book's companion website at **www.cengage.com/chemistry/moore** to test your understanding of the concepts in this figure.

Each *sp* hybrid orbital has one electron that is shared with an electron from a fluorine atom to form two equivalent Be—F bonds. The remaining two 2*p* orbitals in Be are *unhybridized* and are 90° to each other and to the two *sp* hybrid orbitals.

sp^2 Hybrid Orbitals

In BF_3, three atomic orbitals on the central B atom—a 2*s* and two 2*p* orbitals—are hybridized to form *three* **sp^2 hybrid orbitals** (Figure 9.8). The superscript sum (1 + 2 = 3) indicates the number of orbitals that have hybridized—three in this case: one *s* and two *p* orbitals.

$$\text{One } s \text{ atomic orbital} + \text{two } p \text{ atomic orbitals} \rightarrow \text{three } sp^2 \text{ hybrid orbitals}$$

The three sp^2 hybridized orbitals on boron are 120° apart in a plane, each with an electron shared with a fluorine atom electron to form three equivalent B—F bonds. One of the boron 2*p* orbitals remains unhybridized.

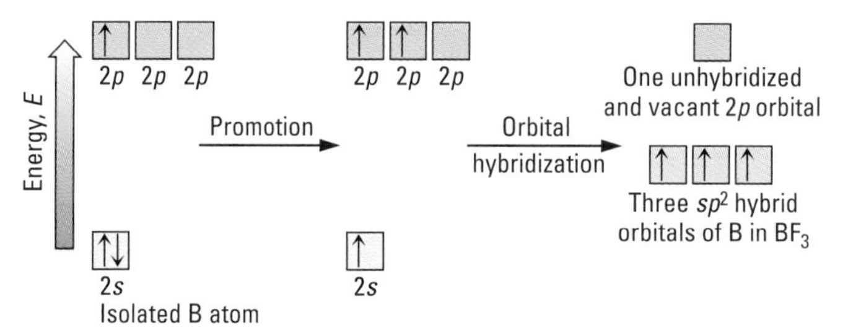

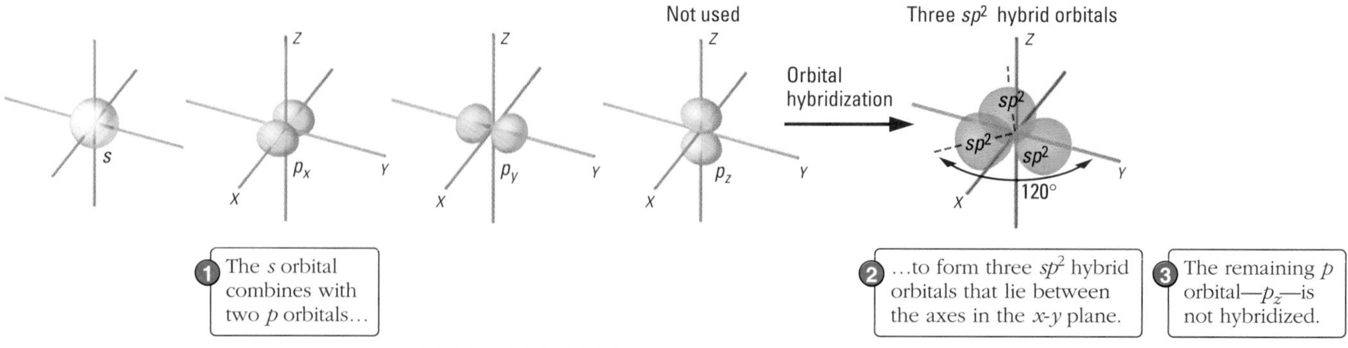

Active Figure 9.8 Formation of three sp^2 hybrid orbitals. An *s* orbital combines with two *p* orbitals in the same atom, say, p_x and p_y, to form three sp^2 hybrid orbitals that lie in the *x*-*y* plane. The angle between the three sp^2 orbitals is 120°. The other *p* orbital is unhybridized. Visit this book's companion website at **www.cengage.com/chemistry/moore** to test your understanding of the concepts in this figure.

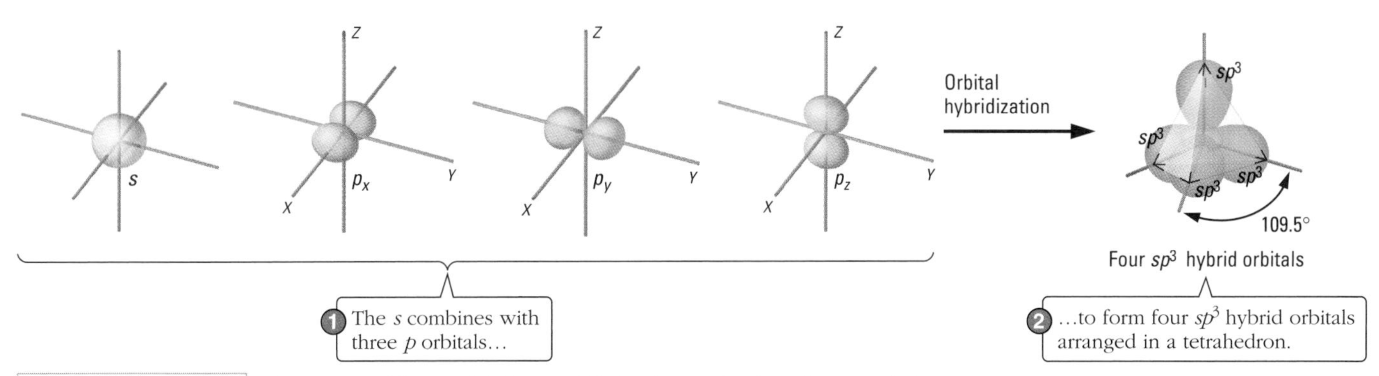

Active Figure 9.9 **Formation of four sp^3 hybrid orbitals.** An *s* orbital combines with three *p* orbitals to form four hybrid sp^3 orbitals that are directed to the corners of a tetrahedron. The angle between the four sp^3 orbitals is 109.5°. The bonds of carbon atoms with four single bonds are described as formed by sp^3 hybrid orbitals. Visit this book's companion website at **www.cengage.com/chemistry/moore** to test your understanding of the concepts in this figure.

sp^3 Hybrid Orbitals

If an *s* and the three *p* orbitals on a central atom are hybridized, *four* hybrid orbitals, called **sp^3 hybrid orbitals,** are formed (Figure 9.9).

One *s* atomic orbital + three *p* atomic orbitals $\longrightarrow$ four sp^3 hybrid orbitals

The four sp^3 hybrid orbitals are equivalent and directed to the corners of a tetrahedron.

In forming sp^3 hybrid orbitals, all of the 2*p* orbitals are hybridized, leaving no unhybridized 2*p* orbitals.

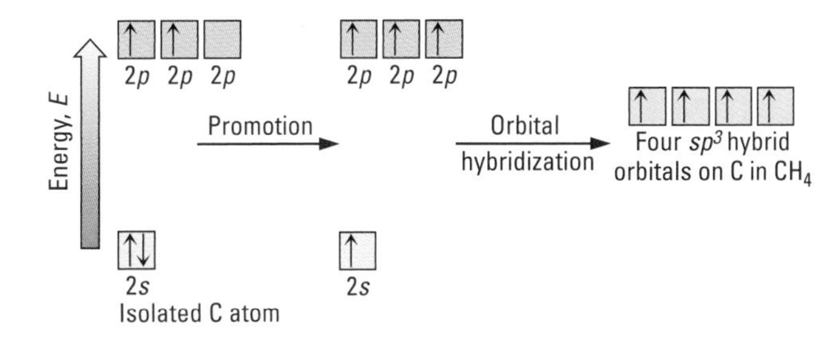

Geometry	Hybrid Orbitals
Linear	Two *sp*
Triangular planar	Three sp^2
Tetrahedral	Four sp^3

The sp^3 hybridization is consistent with the fact that carbon forms four tetrahedral bonds in CF_4 and in all other single-bonded carbon compounds. Overlap of each half-filled sp^3 hybrid orbital on carbon with half-filled orbitals from four fluorine atoms forms four equivalent C—F bonds. The central atoms of Periods 2 and 3 elements that obey the octet rule commonly have sp^3 hybridization.

Because the bond angles in NH_3 and H_2O are close to those in CF_4 and CH_4, this suggests that in general, lone pairs as well as bonding electron pairs can occupy hybrid orbitals. For example, according to the valence bond model, the four electron pairs surrounding the nitrogen atom in ammonia and the oxygen atom in water occupy sp^3 hybrid orbitals like those in CF_4 and CH_4. In NH_3, three shared pairs on nitrogen occupy three of these orbitals, with a lone pair filling the fourth. In H_2O, two of the sp^3 hybrid orbitals on oxygen contain bonding pairs, and two contain lone pairs (Figure 9.10).

It is important for you to recognize that because hybrid orbitals are oriented as far apart as possible to minimize repulsions, they are consistent with the molecular geometries predicted using VSEPR theory. Table 9.2 summarizes information about hybrid orbitals formed from *s* and *p* atomic orbitals.

Bonds in which there is head-to-head orbital overlap so that the electron density of the bond lies along the bonding axis are called **sigma bonds (σ bonds).** The single bonds in diatomic molecules such as H_2 and Cl_2 and also those in the molecules

Only the five valence electrons of N are shown.

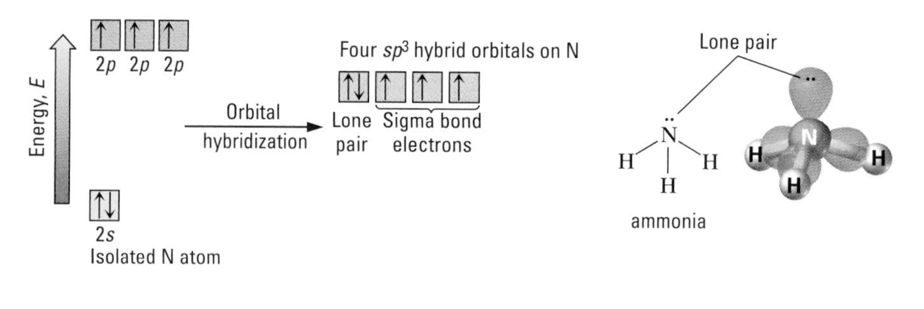

Only the six valence electrons of O are shown.

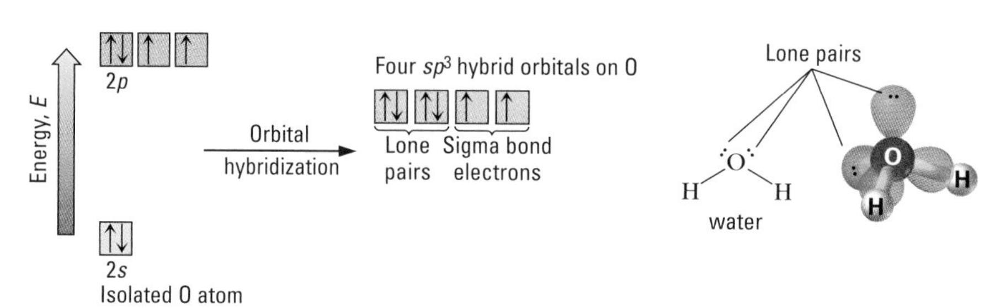

Figure 9.10 Hybridization of nitrogen in ammonia and oxygen in water. Nitrogen and oxygen are both sp^3 hybridized in these ammonia and water molecules. The lone pair electrons are in sp^3 hybrid orbitals.

Table 9.2 Hybrid Orbitals and Their Geometries

	Linear	Trigonal Planar	Tetrahedral
Atomic orbitals mixed	One *s* and one *p*	One *s* and two *p*	One *s* and three *p*
Hybrid orbitals formed	Two *sp*	Three sp^2	Four sp^3
Unhybridized orbitals remaining	Two *p*	One *p*	None

illustrated in Figure 9.10 and BeF_2, BF_3, and CF_4 are all sigma bonds. The N—H and O—H bonds in ammonia and water, respectively, are also sigma bonds.

PROBLEM-SOLVING EXAMPLE 9.4 Hybridization of Atomic Orbitals

Describe the hybridization around the central atom and the bonding in

(a) NCl_3 (b) BCl_3

(c) CH_3OH (hybridization of C and O) (d) $AlCl_4^-$

Answer

(a) sp^3 with three N—Cl sigma bonds (b) sp^2 with three B—Cl sigma bonds (c) sp^3 on C as well as O with three C—H sigma bonds, one C—O sigma bond, and one O—H sigma bond (d) sp^3 with four sigma bonds between Al and Cl atoms

Strategy and Explanation Begin by writing the correct Lewis structure of each compound or ion. Each Lewis structure provides a way to determine the number of bonding pairs and lone pairs around the central atom, a number that indicates how many hybrid orbitals are required.

(a) (b) (c) (d)

(a) There are four electron pairs around nitrogen in NCl_3, a compound analogous to NH_3. Three of the pairs are bonding pairs and the other is a lone pair. Each electron pair is in an sp^3 hybrid orbital. The three N—Cl bonds are sigma bonds formed by the head-to-head overlap of three half-filled sp^3 hybrid orbitals on nitrogen with those of chlorine. The lone pair on nitrogen is in a filled sp^3 hybrid orbital.
(b) BCl_3 is analogous to BF_3 and, therefore, boron has sp^2 orbital hybridization with boron forming three sigma bonds to chlorine.
(c) Carbon forms four sigma bonds in CH_3OH (three to hydrogen atoms and one to oxygen) and so has four bonding pairs, all in sp^3 hybrid orbitals. Oxygen has two sigma bonds formed by the head-to-head overlap of two half-filled sp^3 hybrid orbitals on oxygen with one on carbon and one on hydrogen. The two lone pairs on oxygen are in filled sp^3 hybrid orbitals.
(d) The four sigma bonds in $AlCl_4^-$ are due to four bonding pairs being in sp^3 hybrid orbitals on aluminum that overlap head-to-head with the chlorine atoms.

PROBLEM-SOLVING PRACTICE 9.4

Describe the hybridization around the central atom and the bonding in

(a) $BeCl_2$
(b) NH_4^+
(c) CH_3OCH_3 (hybridization of C and O)
(d) BF_4^-

Hybrid orbitals can also be formed using *d* atomic orbitals in combination with *s* and *p* atomic orbitals. The resulting two additional types of hybrid orbitals are called sp^3d and sp^3d^2 hybrid orbitals. These types of hybrid orbitals will not be discussed in this book.

9.4 Hybridization in Molecules with Multiple Bonds

When using hybrid orbitals to account for molecular shapes, it must be recognized that in a multiple bond, the shared electron pairs do *not* all occupy hybrid orbitals. On a central atom with a multiple bond, the hybrid orbitals contain

- The electron pairs in single bonds to central atoms
- All lone pairs
- Only *one* of the shared electron pairs in a double or triple bond

The electrons in *unhybridized* atomic orbitals are used to form the second bond of a double bond and the second and third bonds in a triple bond. Such bonds, called **pi bonds (π bonds),** are formed by the *sideways* (edgewise) overlap of *parallel* atomic *p* orbitals, as shown in Figure 9.11. In contrast to sigma bonds, in which there is overlap along the bond axis, pi bonds result when *parallel p* orbitals (such as a p_y on one bonding atom and a p_y on the other bonding atom) overlap above and below the bond axis.

It is also possible for *d* orbitals to overlap above and below a bond axis, forming a pi bond.

The bonding in formaldehyde, H_2CO, exemplifies sigma and pi bonding. As shown by the figure in the margin, formaldehyde has two C—H single bonds and one C=O double bond. The triangular planar electron-pair geometry suggests sp^2 hybridization of the carbon atom to supply three sp^2 orbitals for three sigma bonds. As seen in Figure 9.12,

formaldehyde

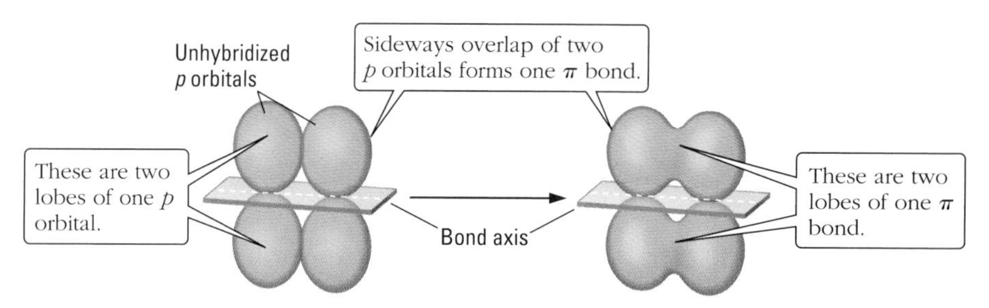

Figure 9.11 Pi bonding. Pi bonding occurs by sideways overlap of two adjacent *p* orbitals to make one pi (π) bond.

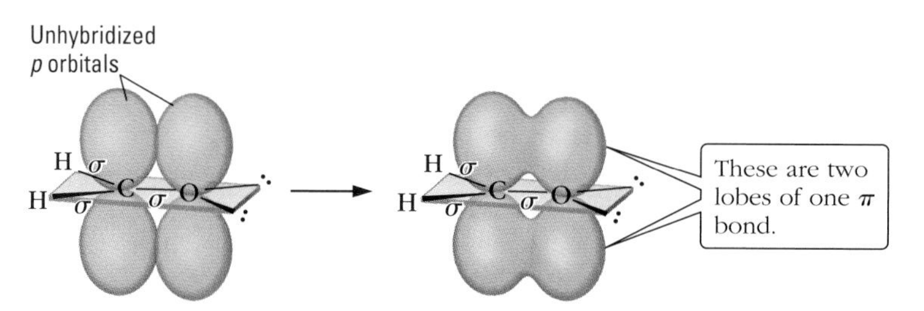

Figure 9.12 Sigma and pi bonding in formaldehyde. In formaldehyde, the C and O atoms are sp^2 hybridized. The C—O sigma bond forms from the end-to-end overlap of two sp^2 hybrid orbitals. The sp^2 hybridization leaves a half-filled unhybridized *p* orbital on each C and O atom. These orbitals overlap edgewise to form the carbon-oxygen pi bond.

two of these sp^2 hybrid orbitals on carbon form two sigma bonds with half-filled 1*s* H orbitals; the third sp^2 hybrid orbital on carbon forms a sigma bond with a half-filled oxygen orbital. The *unhybridized p* orbital on carbon overlaps sideways with an unhybridized *p* orbital on oxygen to form a pi bond, thus completing a double bond between carbon and oxygen.

Bond energies are given in Table 8.2 (← p. 285).

Because pi bonds have less orbital overlap than sigma bonds, pi bonds are generally weaker than sigma bonds. Thus, a C═C double bond is stronger (and shorter) than a C—C single bond, but not twice as strong; correspondingly, a C≡C triple bond, although stronger (and shorter) than a C═C double bond, is not three times stronger than a C—C single bond.

Consider what would happen if one end of a pi bond were to rotate around the sigma bond axis relative to the other end. Because the *p* orbitals need to be parallel to overlap, the pi bond would be broken and the molecule containing the pi bond would be less stable by an amount of energy equal to the strength of the pi bond. The resulting instability means that significant energy is required to rotate around a pi bond. This is the reason that we stated earlier that at room temperature there is no rotation around a double bond. A consequence of this barrier to rotation is *cis-trans* isomerism.

PROBLEM-SOLVING EXAMPLE 9.5 Hybrid Orbitals; Sigma and Pi Bonding

Use hybridized orbitals and sigma and pi bonding to describe bonding in

(a) ethane, C_2H_6 (b) ethylene, C_2H_4 (c) acetylene, C_2H_2

Answer

(a) On each carbon, sp^3 hybridized orbitals form sigma bonds to three H atoms and the other C atom.

(b) On each carbon, sp^2 hybridized orbitals form sigma bonds to two H atoms and to the other C atom; the unhybridized 2*p* orbitals form one pi bond between C atoms.

(c) The *sp* hybridized orbitals on each carbon atom form sigma bonds to one H atom and to the other C; the two unhybridized 2*p* orbitals on each carbon atom form two pi bonds between C atoms.

Strategy and Explanation The Lewis structures can be used to derive the number of sigma and pi bonds, which indicate where and what type of hybridization must occur. The Lewis structures for the three compounds are

```
   H  H          H         H
   |  |           \       /
H—C—C—H           C=C          H—C≡C—H
   |  |           /       \
   H  H          H         H

  ethane         ethylene        acetylene
```

(a) Consistent with the tetrahedral bond angles for single-bonded carbon atoms, each carbon atom in ethane is sp^3 hybridized and all bonds are sigma bonds.
(b) The C=C double bond in ethylene, like the C=O double bond in formaldehyde, requires sp^2 hybridization of carbon. Two of the three sp^2 hybrid orbitals on each carbon form sigma bonds with hydrogens; the third sp^2 hybrid orbital overlaps head-to-head with an sp^2 hybrid orbital on the other carbon, creating a C—C sigma bond. The double bond between the carbon atoms is completed by the sideways overlap of parallel unhybridized *p* orbitals from each carbon to form a pi bond. Thus, the C=C double bond (like all double bonds) consists of a sigma bond and a pi bond.
(c) In acetylene, *sp* hybridization of each carbon atom is indicated by the linear molecular geometry. One of the two *sp* hybrid orbitals on each carbon forms a sigma bond between C and H; the other joins the carbon atoms by a sigma bond. The two unhybridized *p* orbitals on each carbon overlap sideways to form two pi bonds, completing the triple bond. Therefore, the triple bond consists of one sigma and two pi bonds in which the pi bonds are at right angles (90°) to each other (Figure 9.13).

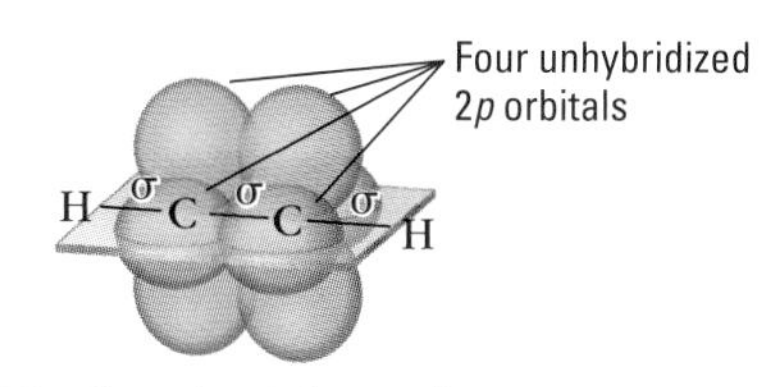

(a) Sigma bonds in acetylene

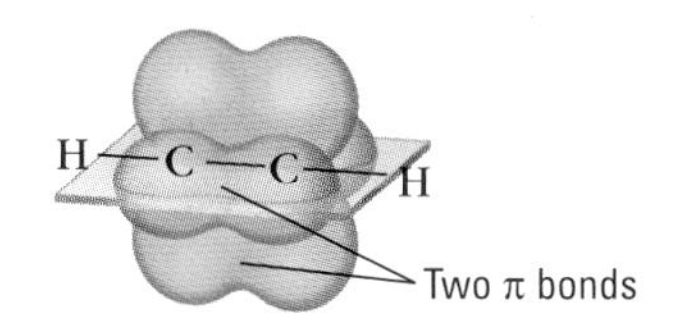

(b) Pi bonds in acetylene

Figure 9.13 Sigma and pi bonding in acetylene. (a) The carbon atoms in acetylene are *sp* hybridized. (b) Thus, each carbon atom contains two unhybridized *p* orbitals. These overlap sideways to form two pi bonds. There is also a carbon-carbon sigma bond formed by the head-to-head overlap of two *sp* hybrid orbitals.

PROBLEM-SOLVING PRACTICE 9.5

Using hybridization and sigma and pi bonding, explain the bonding in (a) HCN and (b) H_2CNH.

CONCEPTUAL
EXERCISE 9.4 Pi Bonding

Explain why pi bonding is not possible for an sp^3 hybridized carbon atom.

9.5 Molecular Polarity

A molecule that has polar bonds (p. 288) may or may not be polar. If the polar bonds are oriented so that their polarities cancel each other, such as in CO_2, a **nonpolar molecule** results. If electron density is concentrated at one end of the molecule, as in HCl, the result is a **polar molecule.** A polar molecule is a *permanent* dipole with a partial negative charge (δ−) where the electron density is concentrated, and a partial positive (δ+) charge at the opposite end, as illustrated in Figure 9.14.

Chlorine gas.

Before examining the factors that determine whether a molecule is polar, let's look at the experimental measurement of the polarity of molecules. Polar molecules experience a force in an electric field that tends to align them with the field (Figure 9.15). In an electric field created by a pair of oppositely charged plates, the partial positive end of each polar molecule is attracted toward the negative plate, and the partial negative end is attracted toward the positive plate. The extent to which the molecules line up with the field depends on their **dipole moment** (μ), which is defined as the product of the magnitude of the partial charges (δ−) and (δ+) times the distance of separation between them. The derived unit of the dipole moment is the coulomb-meter (C·m); a more convenient derived unit is the debye (D), defined as $1\ D = 3.34 \times 10^{-30}$ C · m. Some typical experimental values are listed in Table 9.3. Nonpolar molecules have a zero dipole moment (μ = 0); ***the dipole moments for polar molecules are always greater than zero and increase with greater molecular polarity.***

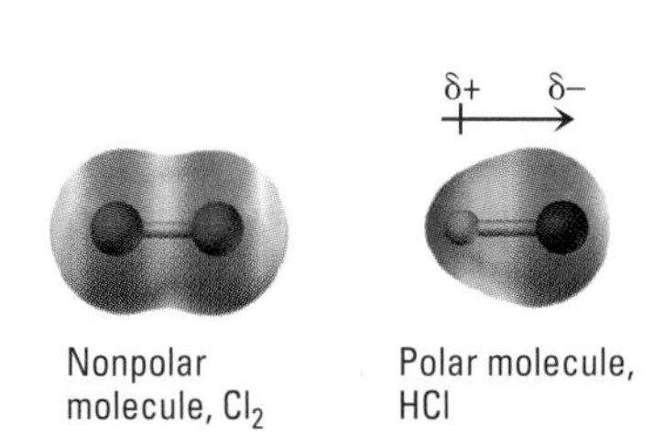

Figure 9.14 Nonpolar (Cl_2) and polar (HCl) molecules. The red color indicates higher electron density; the blue color indicates lower electron density.

Peter Debye 1884–1966

Peter Debye was one of the leading figures in physical chemistry during the early- to mid-twentieth century. His many contributions included the first major X-ray diffraction studies with randomly oriented microcrystals, the seminal theoretical work in conjunction with Hückel on interactions of ionic solutes in solution, and the defining research on molecular polarities. After receiving the 1936 Nobel Prize in Chemistry, Debye left Germany in 1940 due to increasing Nazi influence, accepting an invitation to join the chemistry faculty at Cornell University where he remained until the end of his career.

Table 9.3 Dipole Moments of Selected Molecules

Molecule	Dipole Moment, μ (D)
H_2	0
HF	1.78
HCl	1.07
HBr	0.79
HI	0.38
ClF	0.88
BrF	1.29
BrCl	0.52
H_2O	1.85
H_2S	0.95
CO_2	0
NH_3	1.47
NF_3	0.23
NCl_3	0.39
CH_4	0
CH_3Cl	1.92
CH_2Cl_2	1.60
$CHCl_3$	1.04
CCl_4	0

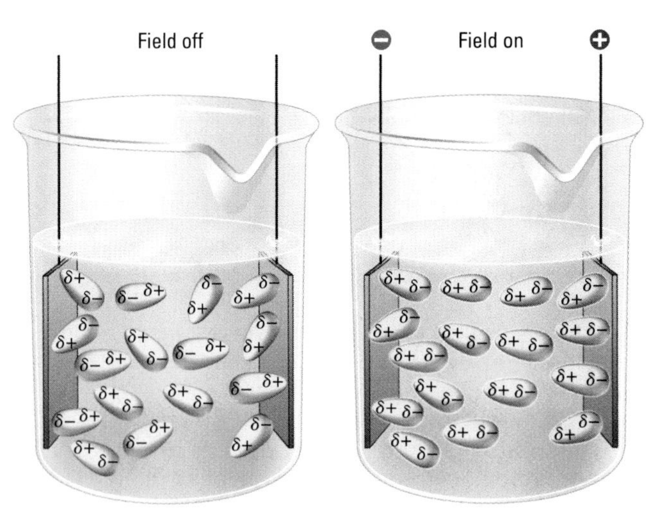

Figure 9.15 Polar molecules in an electric field. Polar molecules experience a force in an electric field that tends to align them so that oppositely charged ends of adjacent molecules are closer to each other.

To predict whether a molecule is polar, we need to consider two factors: (1) whether the molecule has polar bonds, and (2) how those bonds are positioned relative to one another, that is, the molecular geometry. *The net dipole moment of a molecule is the sum of its bond dipoles.* A diatomic molecule is, of course, linear. If its atoms differ in electronegativity, then the bond and the molecule will be polar, with the partial negative charge at the more electronegative atom, as in HCl (Figure 9.14). In polyatomic molecules, the individual bonds can be polar, but because of the molecular shape, there may be no net dipole ($\mu = 0$); if so, the *molecule* is nonpolar. In other nondiatomic molecules, the bond polarity and molecular shape combine to give a net dipole ($\mu > 0$) and a polar molecule. We can correlate the types of molecular geometry with dipole moment by applying a general rule to a molecule of the type AB_n where A is the central atom, B represents a terminal atom or group of atoms, and n is the number of terminal atoms or groups. Such a molecule will *not* be polar if it meets *both* of the following conditions:

- All the terminal atoms or groups are the same, *and*
- All the terminal atoms or groups are symmetrically arranged around the central atom. A symmetric arrangement corresponds with the molecular geometries given in Figure 9.4 ***(p. 310)*** or the AX_5E_0 and AX_6E_0 geometries given in Figure 9.6 ***(p. 316).***

These conditions mean that molecules with the molecular geometries given in Figures 9.4 and 9.6 will never be polar if all their terminal atoms or groups are the same. The dipoles of the individual bonds will cancel due to the symmetry of the molecular structure.

Consider, for example, carbon dioxide, CO_2, a linear triatomic molecule. Each C=O bond is polar because O is more electronegative than C, so O is the partial negative end of the bond dipole. The dipole moment contribution from each bond (the bond dipole) is represented by the symbol ↦, in which the plus sign indicates the partial positive charge and the arrow points to the partial negative end of the bond. We can use the arrows to help estimate whether a molecule is polar. In CO_2, the O atoms are at the same distance from the C atom, both have the same $\delta-$ charge, and they are symmetrically arranged on opposite sides of C. Therefore, their bond dipoles cancel each other, resulting in a molecule with a zero molecular dipole moment. Even though each C=O bond is polar, CO_2 is a nonpolar molecule due to its linear shape.

The situation is different for water, a triatomic, but angular, molecule. Here, both O—H bonds are polar, with the H atoms having the same δ+ charge (Figure 9.16). Note, however, that the two bond dipoles are not symmetrically arranged; they do not point directly toward or away from each other, as in CO_2, but augment each other to give a molecular dipole with a dipole moment of 1.85 D (Table 9.3). Thus, water is a polar molecule.

Using a microwave oven to make popcorn or to heat dinner is an everyday application of the fact that water is a polar molecule. Most foods have a high water content. When water molecules in foods absorb microwave radiation, they rotate, turning so that the dipoles align with the crests and troughs of the oscillating microwave radiation. The radiation generator (called a magnetron) in the oven creates microwave radiation that oscillates at 2.45 GHz (2.45 gigahertz, $2.45 \times 10^9\ s^{-1}$), very nearly the optimum frequency to rotate water molecules, warming them. So, the leftover pizza warms up in a hurry for a late-night snack.

Now consider the differing dipole moments of CF_4 ($\mu = 0$ D) and CF_3H ($\mu = 1.60$ D). Both molecules have the same geometry, with their atoms tetrahedrally arranged around a central carbon atom (Figure 9.16). The terminal F atoms are all the same in CF_4 and thus have the same partial charges. However, the terminal atoms in CF_3H are not the same; F is more electronegative than C, but H is less electronegative than C, so that the bond dipoles reinforce each other in CF_3H. Consequently, CF_4 is a nonpolar molecule and CF_3H is polar.

Nonpolar molecules have a dipole moment of zero.

O=C=O no net dipole
δ− 2δ+ δ−

Combining bond dipoles to obtain a molecular dipole involves addition of vectors (the bond dipoles) to obtain a resultant vector (the molecular dipole).

PROBLEM-SOLVING EXAMPLE 9.6 Molecular Polarity

Are boron trifluoride, BF_3, and dichloromethane, CH_2Cl_2, polar or nonpolar? If polar, indicate the direction of the net dipole.

Answer BF_3 is nonpolar; CH_2Cl_2 is polar, with the partial negative end at the Cl atoms.

Strategy and Explanation To decide whether a molecule is polar requires knowing the geometry of the molecule, the polarity of its bonds, and whether the bonds, if polar, are arranged symmetrically or unsymmetrically. As predicted for a molecule with three electron pairs around the central atom, BF_3 is triangular planar.

F, B, F, F
Bond dipoles cancel; no net dipole

Because F is more electronegative than B, the B—F bonds are polar, with F being the partial negative end. The molecule is nonpolar, though, because the three terminal F atoms are identical, they have the same partial charge, and they are arranged symmetrically around the central B atom. The bond dipole of each B—F is canceled by the bond dipoles of the other two B—F bonds.

H, O, H — Net dipole — H_2O is polar

F, C, F, F, F — No net dipole — CF_4 is nonpolar

H, C, F, F, F — Net dipole — CHF_3 is polar

Figure 9.16 Water ($\mu = 1.85$) and CHF_3 ($\mu = 1.65$) are polar molecules; CF_4 ($\mu = 0$) is a nonpolar molecule.

From the Lewis structure for CH_2Cl_2 in the margin, you might be tempted to say that CH_2Cl_2 is nonpolar because the hydrogens appear across from each other, as do the chlorines. In fact, CH_2Cl_2 is a tetrahedral molecule in which neither hydrogen nor chlorine is directly across from each other.

Net dipole

Because Cl is more electronegative than C, while H is less electronegative, negative charge is drawn away from H atoms and toward Cl atoms. As a result, CH_2Cl_2 has a net dipole (μ = 1.60 D), with the partial negative end at the Cl atoms and the partial positive end between the H atoms.

PROBLEM-SOLVING PRACTICE 9.6

For each of these molecules, decide whether the molecule is polar and, if so, which region is partially positive and which is partially negative.
(a) $BFCl_2$ (b) NH_2Cl (c) SCl_2

EXERCISE 9.5 Dipole Moments

Explain the differences in the dipole moments of
(a) HI (0.38 D) and HBr (0.79 D)
(b) CH_3Cl (1.92 D), CH_3Br (1.81 D), and CH_3I (1.62 D)

9.6 Noncovalent Interactions and Forces Between Molecules

Molecules are attracted to one another. If there were no attractions, then there would be no liquid or solid compounds, only gaseous ones. The attraction between molecules is always the result of an attraction between opposite charges, regardless of whether the charges are permanent or temporary. Polar molecules attract each other due to their permanent dipoles. Even nonpolar molecules, those without a permanent dipole, have some attraction for one another, as you will see in this section.

Atoms within the same molecule are held together by covalent chemical bonds with strengths ranging from 150 to 1000 kJ/mol. Covalent bonds are strong forces. For example, it takes 1656 kJ to break 4 mol C—H covalent bonds and separate the one C atom and four H atoms in all the molecules in 1 mol methane molecules:

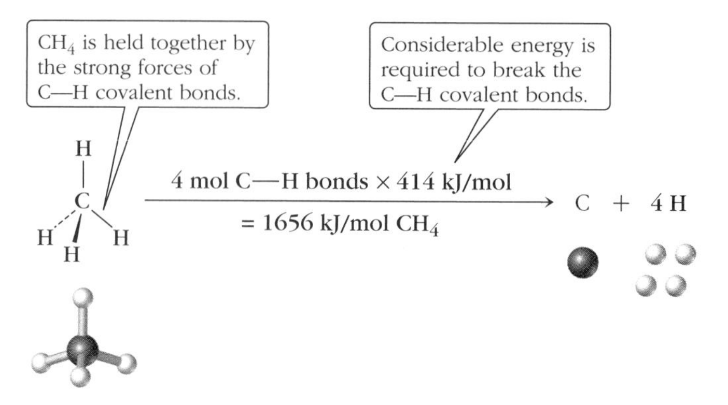
CH4 is held together by the strong forces of C—H covalent bonds.
Considerable energy is required to break the C—H covalent bonds.
4 mol C—H bonds × 414 kJ/mol
= 1656 kJ/mol CH4
C + 4 H

Weaker forces attract one molecule to another molecule. In contrast to the 1656 kJ/mol it takes to atomize methane, only 8.9 kJ is required to separate 1 mol of methane molecules that are close together in liquid methane away from each other to evaporate liquid methane to a gas.

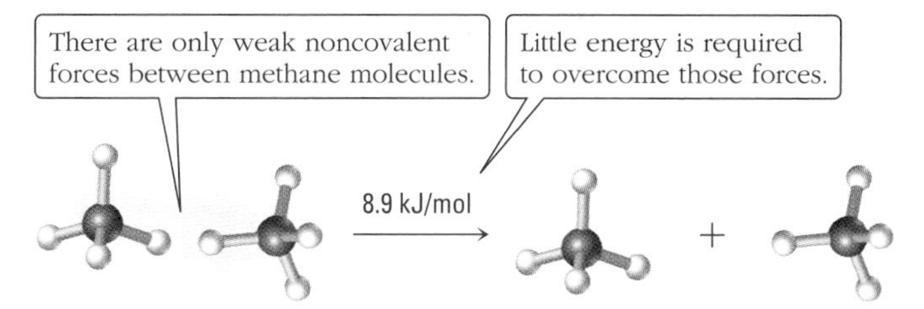
There are only weak noncovalent forces between methane molecules.
Little energy is required to overcome those forces.
8.9 kJ/mol
+

We use the term **noncovalent interactions** to refer to all forces of attraction other than covalent, ionic, or metallic bonding (*metallic bonding* is discussed in Section 11.8). When noncovalent interactions act between one molecule and another, they are referred to as **intermolecular forces.** Because intermolecular forces do not result from the sharing of electron pairs between atoms, they are weaker than covalent bonds. The strengths of noncovalent interactions between molecules (*inter*molecular forces) account for melting points, boiling points, and other properties of molecular substances. Noncovalent interactions between different parts of the same large molecule (*intra*molecular forces) maintain biologically important molecules in the exact shapes required to carry out their functions. Chymotrypsin, a large enzyme, is folded into its biochemically active shape with the assistance of noncovalent interactions between strategically placed atoms. Figure 9.17 shows a ribbon model of chymotrypsin. Red and green ribbons show portions where the protein chain is folded into helical or sheet structures, respectively; the narrow purple ribbons represent simpler, unfolded protein regions. Noncovalent interactions (intermolecular forces) help maintain the shapes of thousands of different kinds of protein molecules in our bodies.

*Intra*molecular means *within* a single molecule; *inter*molecular means *between* two or more separate molecules.

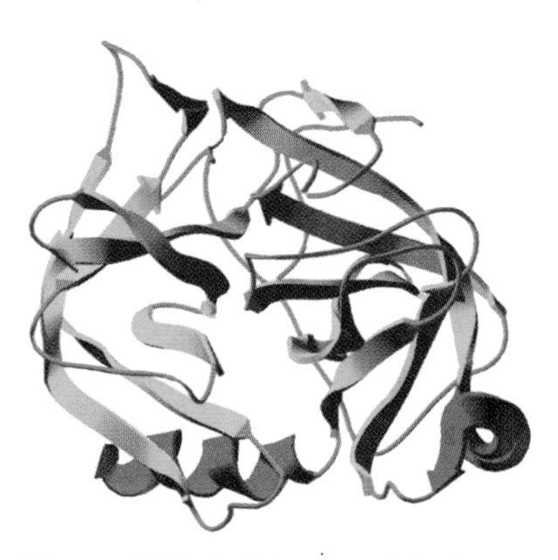

Figure 9.17 A ribbon model of a chymotrypsin molecule. Noncovalent interactions help fold chymotrypsin into its biochemically active shape.

The next few sections explore three types of noncovalent interactions: London forces, dipole-dipole attractions, and hydrogen bonding.

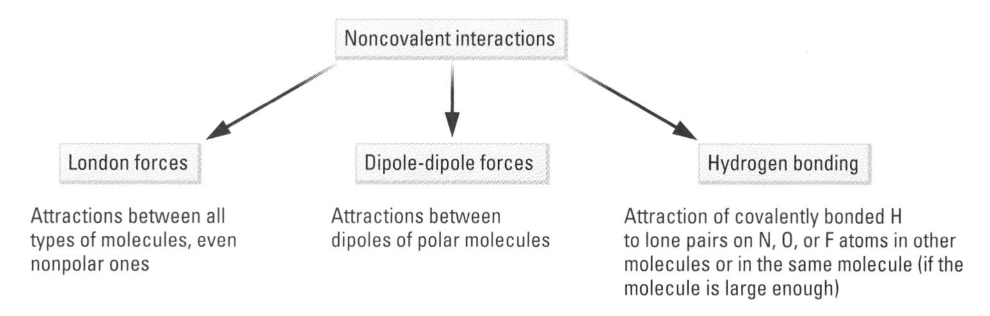
Noncovalent interactions
London forces
Dipole-dipole forces
Hydrogen bonding
Attractions between all types of molecules, even nonpolar ones
Attractions between dipoles of polar molecules
Attraction of covalently bonded H to lone pairs on N, O, or F atoms in other molecules or in the same molecule (if the molecule is large enough)

London forces are named to recognize the work of Fritz London, who extensively studied the origins and nature of such forces. London forces are also called *dispersion forces.*

London Forces

London forces, also known as *dispersion forces,* occur in *all* molecular substances. They result from the attraction between the positive and negative ends of *induced* (nonpermanent) dipoles in adjacent molecules. An **induced dipole** is caused in one molecule when the electrons of a neighboring molecule are momentarily unequally distributed, resulting in a *temporary* dipole in each molecule that attracts the molecules together. Figure 9.18 illustrates how one H_2 molecule with a momentary unevenness in its electron distribution can induce a dipole in a neighboring H_2 molecule. This kind of shift in electron distribution in a molecule is known as **polarization.** Due to polarization, *all* molecules have fleeting, nonpermanent dipoles, even noble gas atoms, molecules of diatomic gases such as oxygen, nitrogen, and chlorine (all of which must be nonpolar), and nonpolar hydrocarbon molecules such as CH_4 and C_2H_6. Such London forces are the *only* noncovalent interactions among nonpolar molecules.

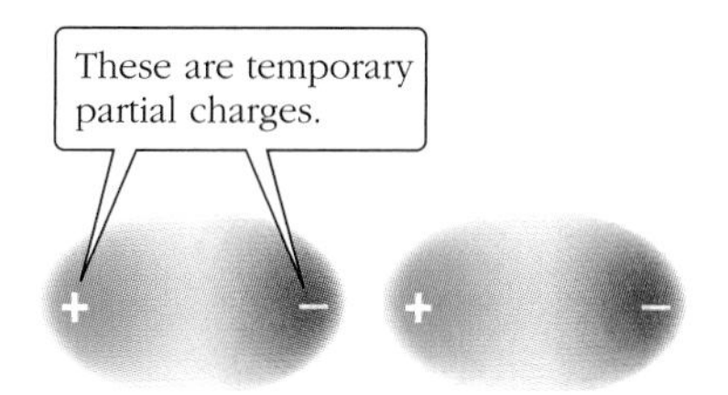
These are temporary partial charges.
+
−
+
−

Figure 9.18 Origin of London forces. Such attractive forces originate when electrons are momentarily distributed unevenly in the molecule, such as the one on the right. The molecule has a temporary positive charge that is close to the molecule on the left. The positive charge attracts electrons in the left-hand molecule, temporarily creating an induced dipole (an unbalanced electron distribution) in that molecule.

Lava lamps take advantage of differences in the intermolecular forces within the red liquid and within the yellow liquid in the lamp and those intermolecular forces between them.

Table 9.4 Effect of Number of Electrons on Boiling Points of Nonpolar Molecular Substances

Noble Gases			Halogens			Hydrocarbons		
	No. e's	bp (°C)		No. e's	bp (°C)		No. e's	bp (°C)
He	2	−269	F_2	18	−188	CH_4	10	−161
Ne	10	−246	Cl_2	34	−34	C_2H_6	18	−88
Ar	18	−186	Br_2	70	59	C_3H_8	26	−42
Kr	36	−152	I_2	106	184	C_4H_{10}*	34	0

*Butane.

London forces range in energy from approximately 0.05 to 40 kJ/mol. Their strength depends on how readily electrons in a molecule can be polarized, which depends on the number of electrons in a molecule and how tightly they are held by nuclear attraction. In general, electrons are more easily polarized when the molecule contains more electrons and the electrons are less tightly attracted to the nucleus; *London forces increase with increased number of electrons in a molecule.* Thus, large molecules with many electrons, such as Br_2 and I_2, are relatively polarizable. In contrast, smaller molecules (F_2, N_2, O_2) are less polarizable because they have fewer electrons.

When we look at the boiling points of several groups of nonpolar molecules (Table 9.4), the effect of the total number of electrons becomes readily apparent: *Boiling points increase as the total number of electrons increases.* (This effect also correlates with molar mass—the heavier an atom or molecule, the more electrons it has.) For a liquid to boil, its molecules must have enough energy to overcome their noncovalent intermolecular attractive forces. Thus, *the boiling point of a liquid depends on the nature and strength of intermolecular forces.* If more energy is required to overcome the intermolecular attractions between molecules of liquid A than the intermolecular attractions between molecules of liquid B, then the boiling point of A will be higher than that of B. Such is the case for Cl_2 (bp = −34 °C) compared with F_2 (bp = −188 °C) due to weaker London forces between diatomic fluorine molecules than between diatomic chlorine molecules. Conversely, weaker intermolecular attractions result in lower boiling points. For example, the boiling point of Br_2 (59 °C) is higher than that of Cl_2 (−34 °C), indicating stronger London forces among Br_2 molecules than among Cl_2 molecules.

Interestingly, molecular shape can also play a role in London forces. Two of the isomers of pentane—straight-chain pentane and 2,2-dimethylpropane (both with the molecular formula C_5H_{12})—differ in boiling point by 27 °C. The linear shape of the pentane molecule allows close contact with adjacent molecules over its entire length, resulting in stronger London forces, while the more compact 2,2-dimethylpropane molecule does not allow as much close contact.

Structure and boiling point. The boiling points of pentane and 2,2-dimethylpropane differ because of differences in their molecular structures even though the total number of electrons in each molecule is the same.

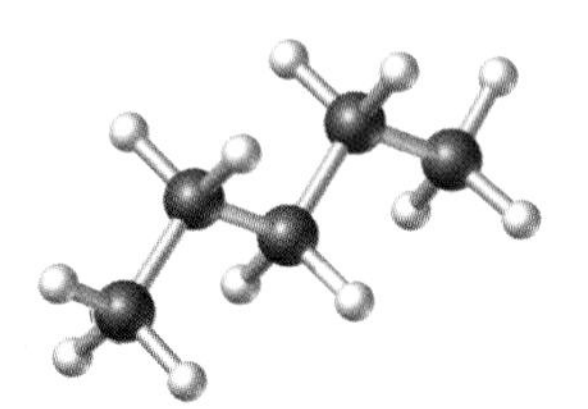
pentane, bp = 36.0 °C

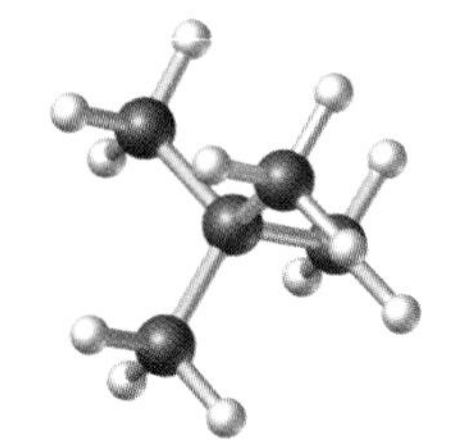
2,2-dimethylpropane, bp = 9.5 °C

Dipole-Dipole Attractions

Polar molecules have *permanent* dipoles that create a **dipole-dipole attraction,** a noncovalent interaction between two *polar* molecules or two polar groups in the same large molecule. Molecules that are dipoles attract each other when the partial positive region of one is close to the partial negative region of another (Figure 9.19).

The boiling points of several nonpolar and polar substances with comparable numbers of electrons, and therefore comparable London forces, are given in Table 9.5. In general, the more polar its molecules, the higher the boiling point of a substance, provided the London forces are similar (molecules with similar numbers of electrons). The lower boiling points of nonpolar substances compared with those of polar substances in Table 9.5 reflect this relationship. Dipole-dipole forces range from 5 to 25 kJ/mol, but London forces (0.05 to 40 kJ/mol) can be stronger. For example, the greater London forces in HI cause it to have a higher boiling point (−36 °C) than HCl (−85 °C), even though HCl is more polar. When London forces are similar, however, a more polar substance will have stronger intermolecular attractions than a less polar one. An example of this is the difference in boiling points of polar ICl with dipole-dipole attractions (97 °C) and nonpolar Br_2 with only London forces (59 °C), even though the substances have approximately the same London forces due to their similar number of electrons (Table 9.5).

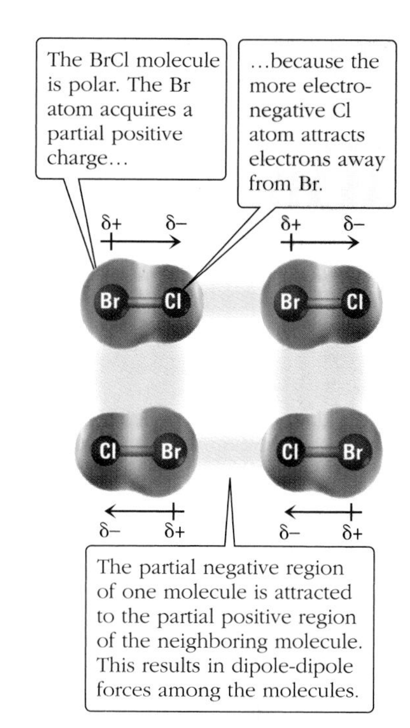

The BrCl molecule is polar. The Br atom acquires a partial positive charge…
…because the more electro-negative Cl atom attracts electrons away from Br.
δ+ δ−
Br Cl
Cl Br
δ− δ+
The partial negative region of one molecule is attracted to the partial positive region of the neighboring molecule. This results in dipole-dipole forces among the molecules.

Figure 9.19 Dipole-dipole attractions between BrCl molecules.

Table 9.5 Numbers of Electrons and Boiling Points of Nonpolar and Polar Substances

Nonpolar Molecules			Polar Molecules		
	No. e's	bp (°C)		No. e's	bp (°C)
SiH_4	18	−112	PH_3	18	−88
GeH_4	36	−90	AsH_3	36	−62
Br_2	70	59	ICl	70	97

CONCEPTUAL EXERCISE 9.6 Dipole-Dipole Forces

Draw a sketch, like that in Figure 9.19, of four CO molecules to indicate dipole-dipole forces among the CO molecules.

PROBLEM-SOLVING EXAMPLE 9.7 Molecular Forces

Which forces must be overcome to

(a) Evaporate gasoline, a mixture of hydrocarbons?
(b) Convert solid carbon dioxide (Dry Ice) into a vapor?
(c) Decompose ammonium nitrate, NH_4NO_3, into N_2 and H_2O?
(d) Boil ClF?
(e) Convert P_4 to P_2?

Answer

(a) and (b) London forces
(c) Covalent bonds between N and H atoms, and O and N atoms; ionic attractions between NH_4^+ and NO_3^- ions.
(d) London forces and dipole-dipole forces
(e) Covalent bonds between phosphorus atoms

Strategy and Explanation Changing the physical state of a molecular substance requires breaking noncovalent intermolecular attractions between molecules. If the molecules are polar, dipole-dipole forces are involved depending on the molecular structure. London forces are present to some extent between all molecules, but are the only

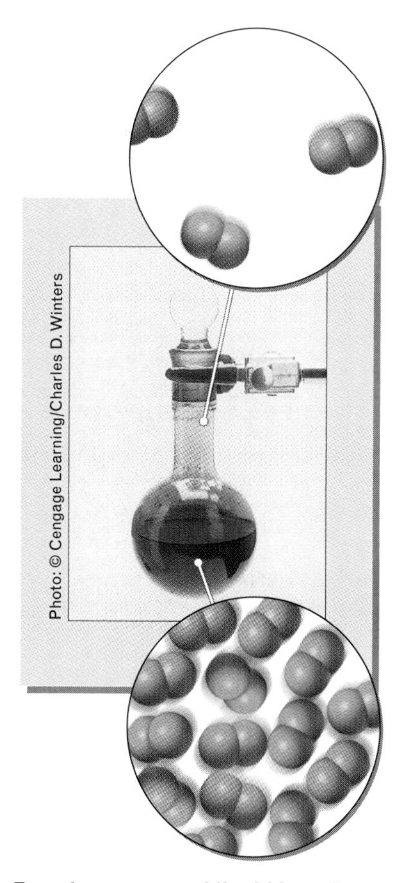

Bromine vapor and liquid bromine.

intermolecular forces between nonpolar molecules. London forces can be significant between large molecules. The decomposition of a molecular substance is drastically different, requiring breaking covalent bonds *within* the molecules.

(a) Gasoline is a mixture of hydrocarbons, which are nonpolar molecules. To evaporate, the hydrocarbon molecules must overcome London forces among them in the liquid to escape from each other and enter the gas phase.

(b) Solid carbon dioxide converts to a vapor when the thermal energy becomes great enough to overcome some of the noncovalent forces of attraction (London forces) among the nonpolar CO_2 molecules. Noncovalent forces still exist in the gaseous CO_2.

(c) Decomposing ammonium nitrate into N_2 and H_2O requires breaking the N—H and N—O covalent bonds in the NH_4^+ and NO_3^- polyatomic ions and rearranging the atoms to form nitrogen and water. These bond-breaking changes involve much greater energy (the bond enthalpy) than that needed to melt the compound, which involves overcoming ionic interactions.

(d) Dipole-dipole forces, like those between BrCl molecules in Figure 9.19, attract the polar ClF molecules to each other in the liquid. London forces are also present. These collective forces must be overcome so that ClF molecules can get away from their neighbors at the surface of the liquid and enter the gaseous state.

(e) Covalent bonds must be broken between phosphorus atoms in P_4.

PROBLEM-SOLVING PRACTICE 9.7

Explain the principal type of forces that must be overcome for

(a) Kr to melt.

(b) Propane to release C and H_2.

Hydrogen Bonds

A hydrogen bond is an especially significant type of noncovalent interaction due to a special kind of dipole-dipole force. A **hydrogen bond** is the attraction between a partially positive hydrogen atom in a molecule and a lone electron pair on a small, very electronegative atom (usually F, O, or N) in another molecule or in a different part of the same molecule. Electron density within molecules shifts toward F, O, or N because of their high electronegativity, giving these atoms *partial negative charges.* As a result, a hydrogen atom bonded to the nitrogen, oxygen, or fluorine atom acquires a *partial positive charge.* In a hydrogen bond, a partially positive hydrogen atom bonded covalently to one of the electronegative atoms (X) is attracted electrostatically to the negative charge of a lone pair on the other atom (Z). Hydrogen bonds are typically shown as dotted lines (· · ·) between the atoms involved:

This represents a hydrogen bond.

X—H · · · :Z—

X = N, O, F Z = N, O, F

Hydrogen bonds can form between two separate molecules or between different parts of the same molecule. These are known as *inter*molecular and *intra*molecular hydrogen bonds, respectively.

The hydrogen bond forms a "bridge" between a hydrogen atom and a highly electronegative atom, which may be in a different molecule or in the same molecule, if the molecule is large enough. This type of bridge from hydrogen to nitrogen and hydrogen to oxygen plays an essential role in determining the folding (three-dimensional structure) of large protein molecules.

This represents a hydrogen bond.

>N̈—H · · · :Ö=C< or —Ö—H · · · :N<

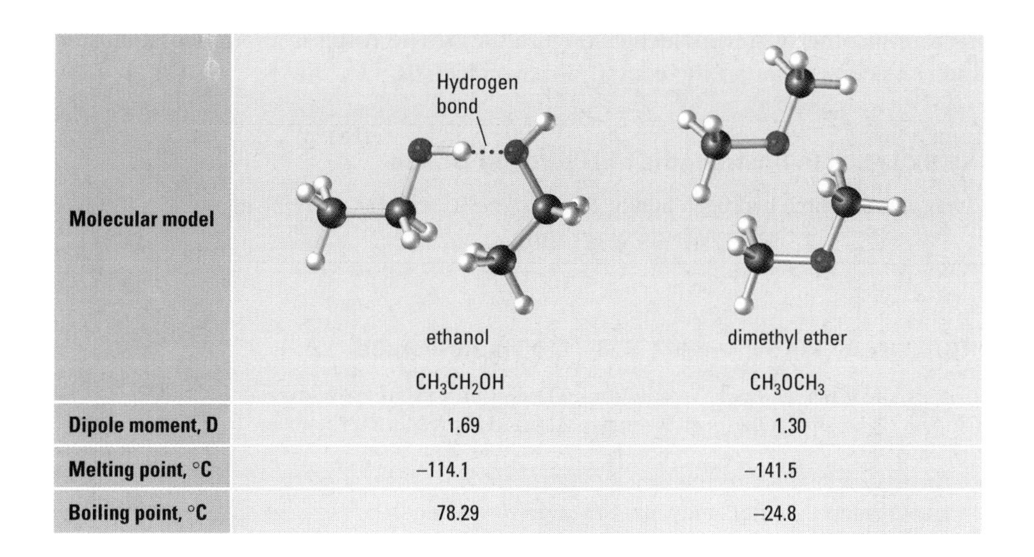

Molecular model	ethanol CH_3CH_2OH	dimethyl ether CH_3OCH_3
Dipole moment, D	1.69	1.30
Melting point, °C	−114.1	−141.5
Boiling point, °C	78.29	−24.8

Noncovalent interactions in ethanol and dimethyl ether. The molecules have the same number of electrons, so London forces are roughly the same. An ethanol molecule has an —OH group, which means that there are both dipole forces and hydrogen bond forces attracting ethanol molecules to each other. A dimethyl ether molecule is polar, but there is no hydrogen bonding, so the noncovalent intermolecular noncovalent forces are weaker than in ethanol.

The greater the electronegativity of the atom connected to H, the greater the partial positive charge on H and hence the stronger the hydrogen bond.

The H atom is very small and its partial positive charge is concentrated in a very small volume, so it can come very close to the lone pair to form an especially strong dipole-dipole force through hydrogen bonding. Hydrogen-bond strengths range from 10 to 40 kJ/mol, less than those of covalent bonds. However, a great many hydrogen bonds often occur in a sample of matter, and the overall effect can be very dramatic. An example of this effect can be seen in the melting and boiling points of ethanol. This chapter began by noting the very different melting and boiling points of ethanol and dimethyl ether, both of which have the same molecular formula, C_2H_6O, and thus the same number of electrons and roughly the same London forces. Both molecules are polar (dipole moment > 0), so there are dipole-dipole forces in each case. The differences in melting and boiling points arise because of hydrogen bonding in ethanol. The O—H bonds in ethanol make intermolecular hydrogen bonding possible, while this is not possible in dimethyl ether because it has no O—H bonds.

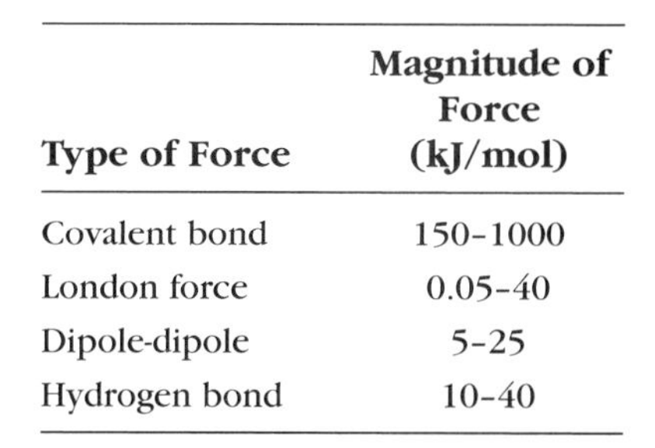

Type of Force	Magnitude of Force (kJ/mol)
Covalent bond	150–1000
London force	0.05–40
Dipole-dipole	5–25
Hydrogen bond	10–40

The hydrogen halides also illustrate the significant effects of hydrogen bonding (Figure 9.20). The boiling point of hydrogen fluoride, HF, the lightest hydrogen halide, is much higher than expected; this is attributed to hydrogen bonding, which does not occur significantly in the other hydrogen halides.

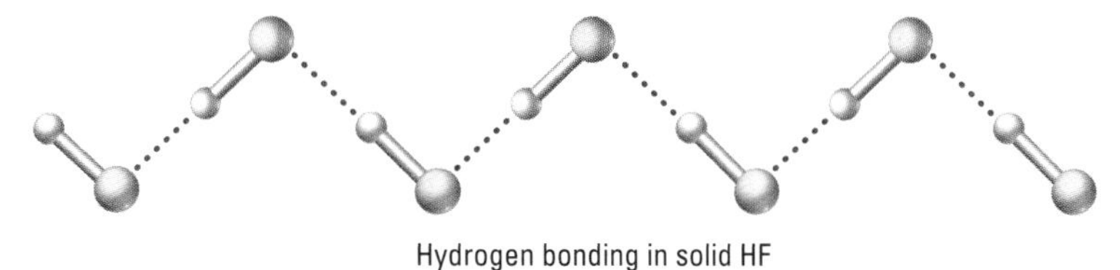

Hydrogen bonding in solid HF

Hydrogen bonding is especially strong among water molecules and is responsible for many of the unique properties of water (Section 11.4). Hydrogen compounds of oxygen's neighbors and family members in the periodic table are gases at room temperature: CH_4, NH_3, H_2S, H_2Se, H_2Te, PH_3, HF, and HCl. But H_2O is a liquid at room temperature, indicating a strong degree of intermolecular attraction. Figure 9.20 shows that the boiling point of H_2O is about 200 °C higher than would be predicted if hydrogen bonding were not present.

In liquid and solid water, where the molecules are close enough to interact, the hydrogen atom on one water molecule is attracted to the lone pair of electrons on the oxygen atom of an adjacent water molecule. Because each hydrogen atom can form a hydrogen bond to an oxygen atom in another water molecule and because each oxygen atom has two lone pairs, every water molecule can participate in four hydrogen

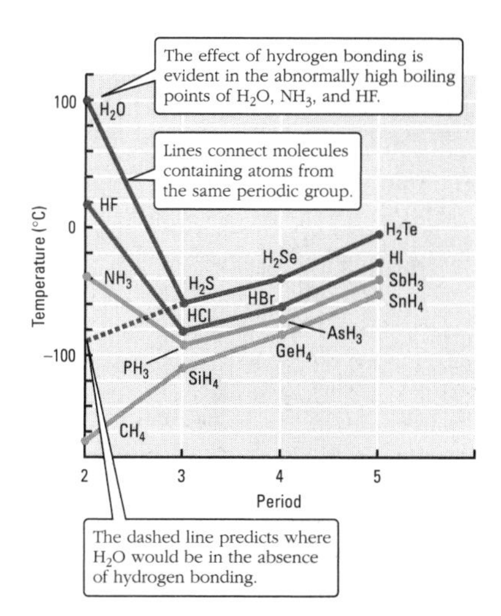

Figure 9.20 Boiling points of some simple hydrogen-containing binary compounds.

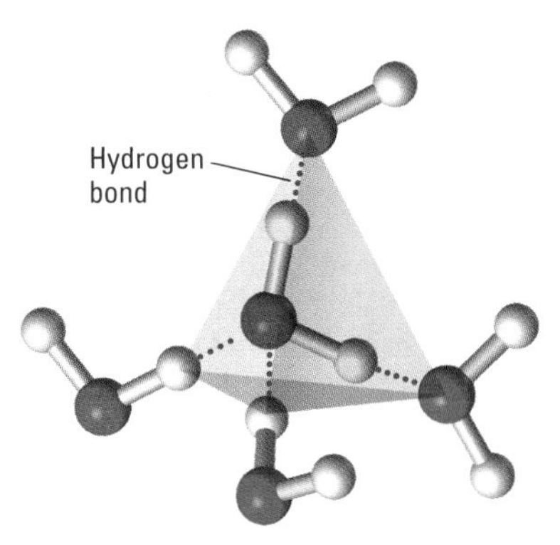

Figure 9.21 Hydrogen bonding between one water molecule and its neighbors. Each water molecule can participate in four hydrogen bonds—one for each hydrogen atom and two through the two lone pairs of oxygen. Since each hydrogen bond is shared between two water molecules, there are two hydrogen bonds per molecule ($\frac{1}{2} \cdot 4$).

bonds to four other water molecules (Figure 9.21). The result is a tetrahedral cluster of water molecules around the central water molecule.

CONCEPTUAL EXERCISE 9.7 Strengths of Hydrogen Bonds

Which of these three hydrogen bonds, F—H · · · F—H, O—H · · · N—C, and N—H · · · O=C, is the strongest? Explain why.

PROBLEM-SOLVING EXAMPLE 9.8 Molecular Forces

What are the types of forces, in addition to London forces, that are overcome in these changes? Using structural formulas, make a sketch representing the major type of force in each case.

(a) The evaporation of liquid ethanol, CH_3CH_2OH

(b) The decomposition of ammonia, NH_3, into N_2 and H_2

(c) The sublimation of iodine, I_2, that is, its conversion from a solid directly into a vapor

(d) The boiling of liquid NH_3

Answer

(a) Hydrogen bonding and dipole forces:

```
   H  H                H  H
   |  |     ..      .. |  |
H—C—C—O—H· · · · · ·:O—C—C—H
   |  |  ..            |  |  |
   H  H  :             H  H  H
         :       H  H
         :       |  |
         H—O—C—C—H
            ..   |  |
                 H  H
```

(b) Covalent bonds between N—H:

```
      ..
      N
2  H / | \ H  ⟶  :N≡N: + 3 H—H
       H
```

(c) London forces between I_2 molecules:

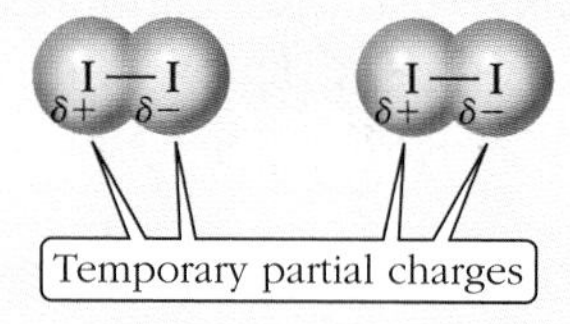

(d) Hydrogen bonding and dipole forces:

```
          ..
          N
       H / | \ H
       :   H
       :
       ..
       N
    H / | \ H···:N        H
        H     H / | \ H···:N<
                    H         H
                           |
                           H
```

Strategy and Explanation In each case identify the force to be overcome. Where the change involves an intermolecular force, consider the particular kind of force.

(a) Ethanol molecules have a hydrogen atom covalently bonded to a highly electronegative oxygen atom with a lone pair of electrons. Therefore, ethanol molecules can hydrogen bond to each other. The bent O—H group is polar also.

(b) This is not a case of overcoming an intermolecular force, but rather an *intra*molecular one, covalent bonds. The decomposition of ammonia involves breaking the N—H covalent bonds in the NH_3 molecule.

Ice crystals.

Sublimation of iodine.

(c) Iodine is composed of nonpolar I_2 molecules, which are held to each other in the solid by London forces. Therefore, these forces must be overcome for the iodine molecules in the solid to escape from one another and become gaseous.

(d) Ammonia molecules are held together in the liquid by hydrogen bonds, which must be broken for the molecules at the surface of the liquid to escape into the vapor. (Figure 9.20 shows the much higher boiling point of NH_3 compared to PH_3.) Because NH_3 is polar, dipole-dipole forces must also be overcome.

PROBLEM-SOLVING PRACTICE 9.8

Decide what types of intermolecular forces are involved in the attraction between molecules of (a) N_2 and N_2, (b) CO_2 and H_2O, and (c) CH_3OH and NH_3.

IN CLOSING

Having studied this chapter, you should be able to . . .

- Recognize the various ways that the shapes of molecules are represented by models and on the printed page (Section 9.1).
- Predict shapes of molecules and polyatomic ions by using the VSEPR model (Section 9.2). ■ End-of-chapter Questions assignable in OWL: 12, 14, 16, 18, 58
- Determine the orbital hybridization of a central atom and the associated molecular geometry (Section 9.3). ■ Questions 24, 30
- Describe in terms of sigma and pi bonding the orbital hybridization of a central atom and the associated molecular geometry for molecules containing multiple bonds (Section 9.4). ■ Question 32
- Describe covalent bonding between two atoms in terms of sigma or pi bonds or both (Section 9.3). ■ Question 34
- Use molecular structure and electronegativities to predict the polarities of molecules (Section 9.5). ■ Questions 36, 38, 42
- Describe the different types of noncovalent interactions and use them to explain melting points and boiling points (Section 9.6). ■ Questions 48, 51

Selected end-of-chapter Questions may be assigned in OWL.

For quick review, download Go Chemistry mini-lecture flashcard modules (or purchase them at **www.ichapters.com**).

Prepare for an exam with a **Summary Problem** that brings together concepts and problem-solving skills from throughout the chapter. Go to **www.cengage.com/chemistry/moore** to download this chapter's Summary Problem from the Student Companion Site.

KEY TERMS

axial positions *(Section 9.2)*

bond angle *(9.2)*

dipole-dipole attraction *(9.6)*

dipole moment *(9.5)*

electron-pair geometry *(9.2)*

equatorial positions *(9.2)*

hybrid orbital *(9.3)*

hybridized *(9.3)*

hydrogen bond *(9.6)*

induced dipole *(9.6)*

intermolecular forces *(9.6)*

London forces *(9.6)*

molecular geometry *(9.2)*

noncovalent interactions *(9.6)*

nonpolar molecule *(9.5)*

pi bond, π bond *(9.4)*

polar molecule *(9.5)*

polarization *(9.6)*

sigma bond, σ bond *(9.3)*

***sp* hybrid orbital** *(9.3)*

sp^2 hybrid orbital *(9.3)*

sp^3 hybrid orbital *(9.3)*

valence bond model *(9.3)*

valence-shell electron-pair repulsion (VSEPR) model *(9.2)*

QUESTIONS FOR REVIEW AND THOUGHT

■ denotes questions assignable in OWL.

Blue-numbered questions have short answers at the back of this book and fully worked solutions in the *Student Solutions Manual.*

Review Questions

1. What is the VSEPR model? What is the physical basis of the model?
2. What is the difference between the electron-pair geometry and the molecular geometry of a molecule? Use the water molecule as an example in your discussion.
3. Designate the electron-pair geometry for each case from two to six electron pairs around a central atom.
4. What are the molecular geometries for each of these?
   (a) H—Ä: (b) H—Ä—H
   (c) 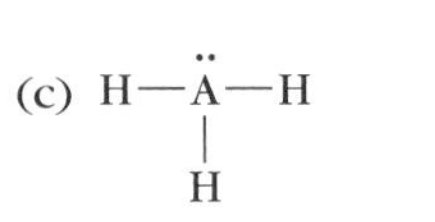 (d) 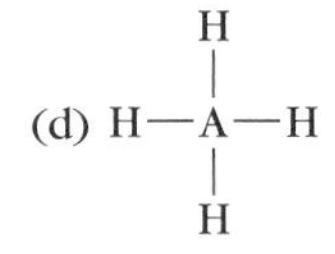
   Give the H—A—H bond angle for each of the last three.
5. If you have three electron pairs around a central atom, how can you have a triangular planar molecule? An angular molecule? What bond angles are predicted in each case?
6. Use VSEPR to explain why ethylene is a planar molecule.
7. How can a molecule with polar bonds be nonpolar? Give an example.

Topical Questions

Molecular Shape

8. All of these molecules have central atoms with only bonding pairs of electrons. After drawing the Lewis structure, identify the molecular shape of each molecule.
   (a) BeH_2 (b) CH_2Cl_2
   (c) BH_3 (d) $SeCl_6$
   (e) PF_5
9. Draw the Lewis structure for each of these molecules or ions. Describe the electron-pair geometry and the molecular geometry.
   (a) NH_2Cl (b) OCl_2
   (c) SCN^- (d) HOF
10. Use the VSEPR model to predict the electron-pair geometry and the molecular geometry of
   (a) PH_4^+ (b) OCl_2
   (c) SO_3 (d) H_2CO
11. In each of these molecules or ions, two oxygen atoms are attached to a central atom. Draw the Lewis structure for each one, and then describe the electron-pair geometry and the molecular geometry. Comment on similarities and differences in the series.
   (a) CO_2 (b) NO_2^-
   (c) SO_2 (d) O_3
   (e) ClO_2^-
12. ■ In each of these ions, three oxygen atoms are attached to a central atom. Draw the Lewis structure for each one, and then describe the electron-pair geometry and the molecular geometry. Comment on similarities and differences in the series.
   (a) BO_3^{3-} (b) CO_3^{2-}
   (c) SO_3^{2-} (d) ClO_3^-
13. These are examples of molecules and ions that do not obey the octet rule. After drawing the Lewis structure, describe the electron-pair geometry and the molecular geometry for each.
   (a) ClF_2^- (b) ClF_3
   (c) ClF_4^- (d) ClF_5
14. ■ These are examples of molecules and ions that do not obey the octet rule. After drawing the Lewis structure, describe the electron-pair geometry and the molecular geometry for each.
   (a) SiF_6^{2-} (b) SF_4
   (c) PF_5 (d) XeF_4
15. Iodine forms three compounds with chlorine: ICl, ICl_3, and ICl_5. Draw the Lewis structures and determine the molecular shapes of these three molecules.
16. ■ Give the approximate values for the indicated bond angles. The figures given are not the actual molecular shapes, but are used only to show the bonds being considered.
   (a) O—S—O angle in SO_2 (b) F—B—F angle in BF_3
   (c) 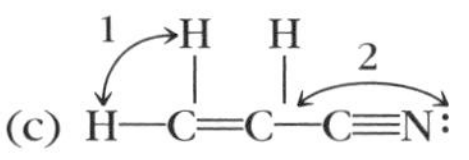 (d) 
17. Give approximate values for the indicated bond angles. The figures given are not the actual molecular shapes, but are used only to show the bonds being considered.
   (a) Cl—S—Cl angle in SCl_2 (b) N—N—O angle in N_2O
   (c) 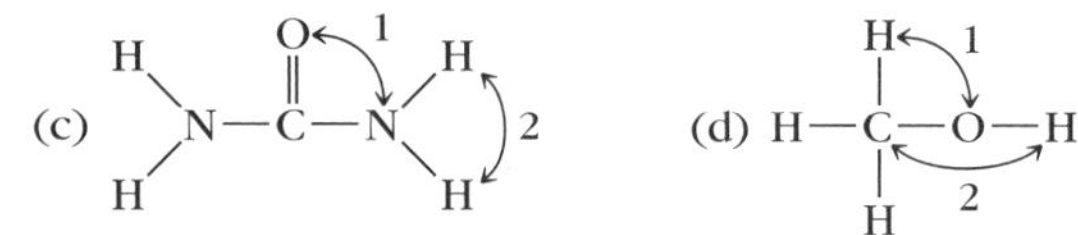 (d) 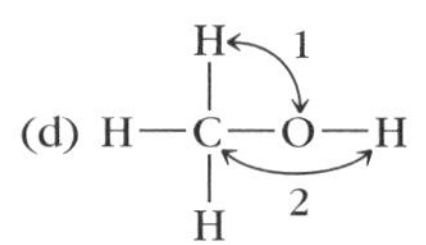
18. ■ Give approximate values for the indicated bond angles.
   (a) F—Se—F angles in SeF_4
   (b) O—S—F angles in SOF_4 (The O atom is in an equatorial position.)
   (c) F—Br—F angles in BrF_5
19. Give approximate values for the indicated bond angles.
   (a) F—S—F angles in SF_6
   (b) F—Xe—F angle in XeF_2
   (c) F—Cl—F angle in ClF_2^-
20. Which would have the greater O—N—O bond angle, NO_2 or NO_2^+? Explain your answer.
21. Compare the F—Cl—F angles in ClF_2^+ and ClF_2^-. From Lewis structures, determine the approximate bond angle in each ion. Explain which ion has the greater angle and why.

Hybridization

22. What designation is used for the hybrid orbitals formed by these combinations of atomic orbitals?
    (a) One *s* and three *p* (b) One *s* and two *p*
23. How many hybrid orbitals are formed in each case in Question 22?
24. ■ Describe the geometry and hybridization of carbon in chloroform, $CHCl_3$.
25. Describe the geometry and hybridization for each inner atom in ethylene glycol, $HOCH_2CH_2OH$, the main component in antifreeze.
26. The hybridization of the two carbon atoms differs in an acetic acid, CH_3COOH, molecule.
    (a) Designate the correct hybridization for each carbon atom in this molecule.
    (b) What is the approximate bond angle around each carbon?
27. The hybridization of the two nitrogen atoms differs in NH_4NO_3.
    (a) Designate the correct hybridization for each nitrogen atom.
    (b) What is the approximate bond angle around each nitrogen?
28. What are the hybridization and approximate bond angles for the N, C, and O atoms in alanine, an amino acid, whose Lewis structure is

```
       H   H  :O:
       |   |   ||   ..
   H—C—C—C—O—H
       |   |        ..
       H   N
          / ..\
        H      H
```

29. Identify the type of hybridization, geometry, and approximate bond angle for each carbon atom in an alkane.
30. ■ (a) Identify the type of hybridization and approximate bond angle for each carbon atom in CH_3CH_2CCH.
    (b) Which is the shortest carbon-to-carbon bond length in this molecule?
    (c) Which is the strongest carbon-to-carbon bond in this molecule?
31. Write the Lewis structure and designate which are sigma and pi bonds in each of these molecules.
    (a) HCN (b) N_2H_2
    (c) HN_3
32. ■ Write the Lewis structure and designate which are sigma and pi bonds in each of these molecules.
    (a) OCS (b) NH_2OH
    (c) CH_2CHCHO (d) $CH_3CH(OH)COOH$
33. Methylcyanoacrylate is the active ingredient in "super glues." Its Lewis structure is

```
  ..
  N≡C        H
      \     /
       C=C
  ..  /     \
 :O=C    H    H
      \   |
       ..O—C—H
       ..  |
           H
```

    (a) How many sigma bonds are in the molecule?
    (b) How many pi bonds are in the molecule?
    (c) What is the hybridization of the carbon atom bonded to nitrogen?
    (d) What is the hybridization of the carbon atom bonded to oxygen?
    (e) What is the hybridization of the double-bonded oxygen?
34. ■ Acrylonitrile is polymerized to manufacture carpets and wigs. Its Lewis structure is

```
               H
               |
 :N≡C—C=C
          |    |
          H    H
```

    (a) Locate the sigma bonds in the molecule. How many are there?
    (b) Locate the pi bonds in the molecule. How many are there?
    (c) What is the hybridization of the carbon atom that is bonded to nitrogen?
    (d) What is the hybridization of the nitrogen atom?
    (e) What is the hybridization of the hydrogen-bearing carbon atoms?

Molecular Polarity

35. Consider these molecules: CH_4, NCl_3, BF_3, CS_2.
    (a) In which compound are the bonds most polar?
    (b) Which compounds in the list are not polar?
36. ■ Consider these molecules: H_2O, NH_3, CO_2, ClF, CCl_4.
    (a) In which compound are the bonds most polar?
    (b) Which compounds in the list are not polar?
    (c) Which atom in ClF has a partial negative charge?
37. Which of these molecules is (are) not polar? Which molecule has the most polar bonds?
    (a) CO (b) PCl_3
    (c) BCl_3 (d) GeH_4
    (e) CF_4
38. ■ Which of these molecules is (are) polar? For each polar molecule, what is the direction of polarity; that is, which is the partial negative end and which is the partial positive end of the molecule?
    (a) CO_2 (b) HBF_2
    (c) CH_3Cl (d) SO_3
39. Which of these molecules have a dipole moment? For each of these polar molecules, indicate the direction of the dipole in the molecule.
    (a) XeF_2 (b) H_2S
    (c) CH_2Cl_2 (d) HCN
40. Explain the differences in the dipole moments of
    (a) BrF (1.29 D) and BrCl (0.52 D)
    (b) H_2O (1.86 D) and H_2S (0.95 D)
41. Which of these molecules has a dipole moment? For each of the polar molecules, indicate the direction of the dipole in the molecule.
    (a) Nitrosyl fluoride, FNO
    (b) Disulfur difluoride, S_2F_2
42. ■ Which of these molecules is polar? For each of the polar molecules, indicate the direction of the dipole in the molecule.
    (a) Hydroxylamine, NH_2OH
    (b) Sulfur dichloride, SCl_2, an unstable, red liquid

Noncovalent Interactions

43. Explain in terms of noncovalent interactions why water and ethanol are miscible, but water and cyclohexane are not.
44. Construct a table covering all the types of noncovalent interactions and comment about the distance dependence for each. (In general, the weaker the force, the closer together the molecules must be to feel the attractive force of nearby molecules.) You should also include an example of a substance that exhibits each type of noncovalent interaction in the table.
45. Explain the trends seen in the diagram below for the boiling points of some main group hydrogen compounds.
    (a) Group IV: CH_4, SiH_4, GeH_4, SnH_4
    (b) Group V: NH_3, PH_3, AsH_3, SbH_3
    (c) Group VI: H_2O, H_2S, H_2Se, H_2Te
    (d) Group VII: HF, HCl, HBr, HI

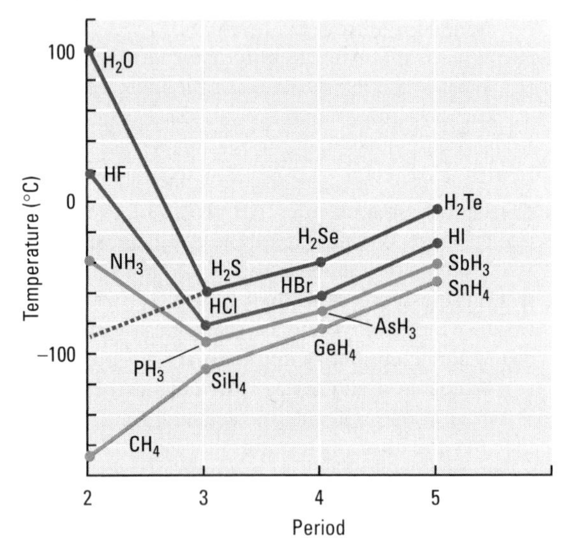

46. Explain why water "beads up" on a freshly waxed car, but not on a dirty, unwaxed car.
47. Explain why water will not remove tar from your shoe, but kerosene will.
48. ■ Which of these will form hydrogen bonds?
    (a) CH_2Br_2 (b) $CH_3OCH_2CH_3$
    (c) H_2NCH_2COOH (d) H_2SO_3
    (e) CH_3CH_2OH
49. The structural formula for vitamin C is

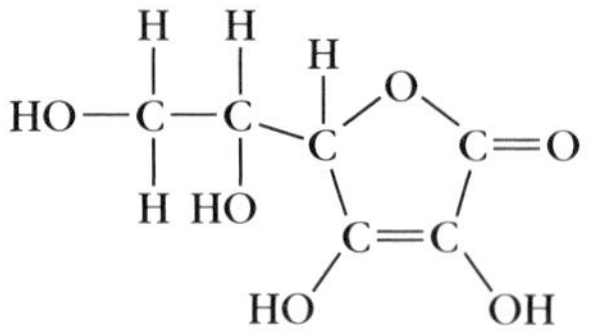

    Give a molecular-level explanation why vitamin C is a water-soluble rather than a fat-soluble vitamin.
50. Which of these would you expect to be most soluble in cyclohexane, C_6H_{12}? The least soluble? Explain your reasoning.
    (a) NaCl (b) CH_3CH_2OH
    (c) C_3H_8
51. ■ What are the types of forces that must be overcome in these changes?
    (a) The sublimation of solid $C_{10}H_8$
    (b) The melting of propane, C_3H_8
    (c) The decomposition of water into H_2 and O_2
    (d) The evaporation of liquid PCl_3
52. The structural formula of alpha-tocopherol, a form of vitamin E, is

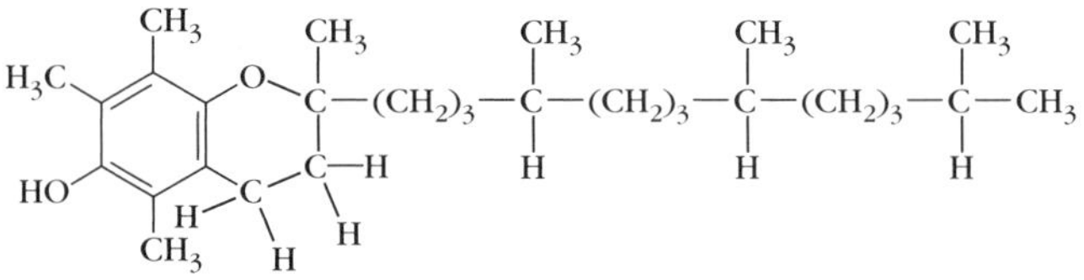

    Give a molecular-level explanation why alpha-tocopherol dissolves in fat, but not in water.

General Questions

53. The formula for nitryl chloride is NO_2Cl. Draw the Lewis structure for the molecule, including all resonance structures. Describe the electron-pair and molecular geometries, and give values for all bond angles.
54. ■ Methylcyanoacrylate is the active ingredient in "super glues." Its Lewis structure is

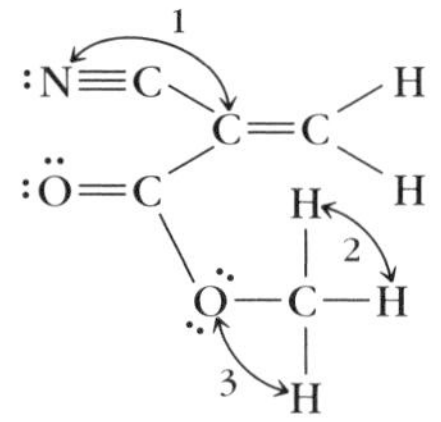

    (a) Give values for the three bond angles indicated.
    (b) Indicate the most polar bond in the molecule.
    (c) Circle the shortest carbon-oxygen bond.
55. Acrylonitrile is polymerized to manufacture carpets and wigs. Its Lewis structure is

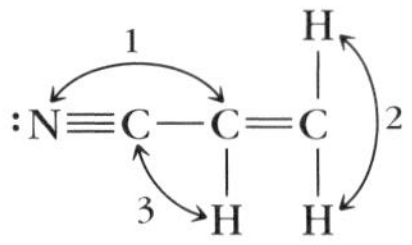

    (a) Give approximate values for the indicated bond angles.
    (b) Which are the most polar bonds in the molecule?
56. In addition to CO and CO_2, there are other carbon oxides. One is tricarbon dioxide, C_3O_2, also called carbon suboxide, a foul-smelling gas.
    (a) Write the Lewis structure of this compound.
    (b) What is the value of the C-to-C-to-O bond angle in carbon suboxide?
    (c) What is the value of the C-to-C-to-C bond angle in tricarbon dioxide?
57. In addition to CO, CO_2, and C_3O_2, there is another molecular oxide of carbon, pentacarbon dioxide, C_5O_2, a yellow solid.
    (a) What is the value of the C-to-C-to-O bond angle in pentacarbon dioxide?
    (b) What is the value of the C-to-C-to-C bond angle in this compound?

58. ■ Each of these molecules has fluorine atoms attached to an atom from Group 1A or 3A to 6A. Draw the Lewis structure for each one and then describe the electron-pair geometry and the molecular geometry. Comment on similarities and differences in the series.
   (a) BF_3 (b) CF_4
   (c) NF_3 (d) OF_2
   (e) HF
59. Sketch the geometry of a carbon-containing molecule or ion in which the angle between two atoms bonded to carbon is
   (a) Exactly 109.5°
   (b) Slightly different from 109.5°
   (c) Exactly 120°
   (d) Exactly 180°
60. How can a diatomic molecule be nonpolar? Polar?
61. How can a molecule have polar bonds yet have a dipole moment of zero?

Applying Concepts

62. Complete this table.

Molecule or Ion	Electron-Pair Geometry	Molecular Geometry
ICl_2^+		
I_3^-		
ICl_3		
ICl_4^-		
IO_4^-		
IF_4^+		
IF_5		
IF_6^+		

63. Complete this table.

Molecule or Ion	Electron-Pair Geometry	Molecular Geometry
SO_2		
SCl_2		
SO_3		
SO_3^{2-}		
SF_4		
SO_4^{2-}		
SF_5^+		
SF_6		

64. What are the types of forces, in addition to London forces, that are overcome in these changes? Using structural formulas, make a sketch representing the major type of force in each case.
   (a) The evaporation of liquid methanol, CH_3OH
   (b) The decomposition of hydrogen peroxide, H_2O_2, into water and oxygen
   (c) The melting of urea, H_2NCONH_2
   (d) The boiling of liquid HCl
65. Name a Group 1A to 8A element that could be the central atom (X) in these compounds.
   (a) XH_3 with one lone pair of electrons
   (b) XCl_3
   (c) XF_5
   (d) XCl_3 with two lone pairs of electrons
66. Name a Group 1A to 8A element that could be the central atom (X) in these compounds.
   (a) XCl_2
   (b) XH_2 with two lone pairs of electrons
   (c) XF_4 with one lone pair of electrons
   (d) XF_4
67. What is the maximum number of water molecules that could hydrogen-bond to an acetic acid molecule? Draw in the water molecules and use dotted lines to show the hydrogen bonds.

```
     H  :O:
     |   ‖   ..
  H—C—C—O—H
     |       ..
     H
```

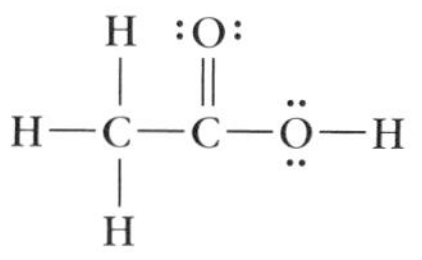

68. What is the maximum number of water molecules that could hydrogen-bond to an ethylamine molecule? Draw in the water molecules and use dotted lines to show the hydrogen bonds.

```
     H   H       H
     |   |  ..  /
  H—C—C—N
     |   |      \
     H   H       H
```

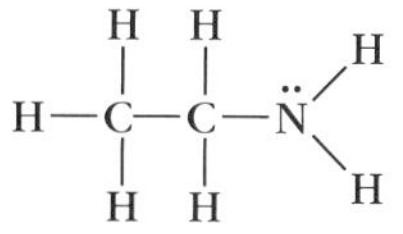

69. These are responses students wrote when asked to give an example of hydrogen bonding. Which are correct?
   (a) H—H···H—H
   (b) 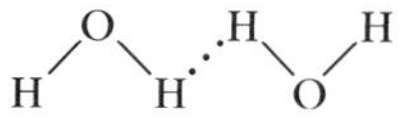 H—O—H···O(H)—H (water dimer)
   (c) H—F
   (d) 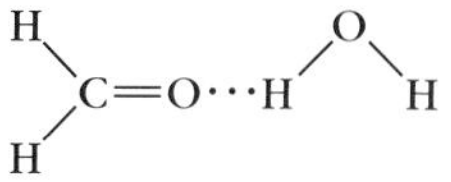 H_2C=O···H—O—H
70. Which of these are examples of hydrogen bonding?
   (a) 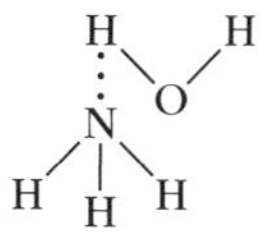 H_3N···H—O—H
   (b) H—H
   (c) 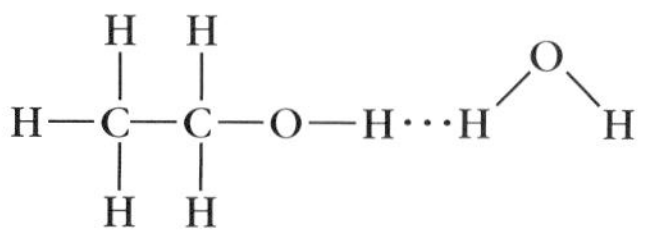 H—C(H)(H)—C(H)(H)—O—H···H—O—H
   (d) 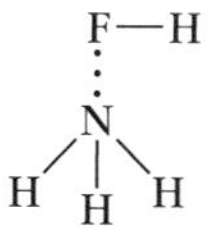 F—H···NH_3

More Challenging Questions

71. There are three isomers with the formula $C_6H_6O_2$. Each isomer contains a benzene ring to which two —OH groups are attached.
    (a) Write the Lewis structures for the three isomers.
    (b) Taking their molecular structure and the likelihood of hydrogen bonding into account, list them in order of increasing melting point.
72. Halomethane, which had been used as an anesthetic, has the molecular formula $CHBrClCF_3$.
    (a) Write the Lewis structure for halomethane.
    (b) Is halomethane a polar molecule? Explain your answer.
    (c) Does hydrogen bonding occur in halomethane? Explain.
73. Ketene, C_2H_2O, is a reactant for synthesizing cellulose acetate, which is used to make films, fibers, and fashionable clothing.
    (a) Write the Lewis structure of ketene. Ketene does not contain an —OH bond.
    (b) Identify the geometry around each carbon atom and all the bond angles in the molecule.
    (c) Identify the hybridization of each carbon and oxygen atom.
    (d) Is the molecule polar or nonpolar? Use appropriate data to support your answer.
74. Gamma hydroxybutyric acid, GHB, infamous as a "date rape" drug, is used illicitly because of its effects on the nervous system. The condensed molecular formula for GHB is $HO(CH_2)_3COOH$.
    (a) Write the Lewis structure for GHB.
    (b) Identify the hybridization of the carbon atom in the CH_2 groups and of the terminal carbon.
    (c) Is hydrogen bonding possible in GHB? If so, write Lewis structures to illustrate the hydrogen bonding.
    (d) Which carbon atoms are involved in sigma bonds? In pi bonds?
    (e) Which oxygen atom is involved in sigma bonds? In pi bonds?
75. There are two compounds with the molecular formula HN_3. One is called hydrogen azide; the other is cyclotriazene.
    (a) Write the Lewis structure for each compound.
    (b) Designate the hybridization of each nitrogen in hydrogen azide.
    (c) What is the hybridization of each nitrogen in cyclotriazene?
    (d) How many sigma bonds are in hydrogen azide? In cyclotriazene?
    (e) How many pi bonds are in hydrogen azide? In cyclotriazene?
    (f) Give approximate values for the N-to-N-to-N bond angles in each molecule.
76. Nitrosyl azide, a yellow solid first synthesized in 1993, has the molecular formula N_4O.
    (a) Write its Lewis structure.
    (b) What is the hybridization on the terminal nitrogen?
    (c) What is the hybridization on the "central" nitrogen?
    (d) Which is the shortest nitrogen-nitrogen bond?
    (e) Give the approximate bond angle between the three nitrogens, beginning with the nitrogen that is bonded to oxygen.
    (f) Give the approximate bond angle between the last three nitrogens, those not involved in bonding to oxygen.
    (g) How many sigma bonds are there? How many pi bonds?
77. ■ Piperine, the active ingredient in black pepper, has this Lewis structure. 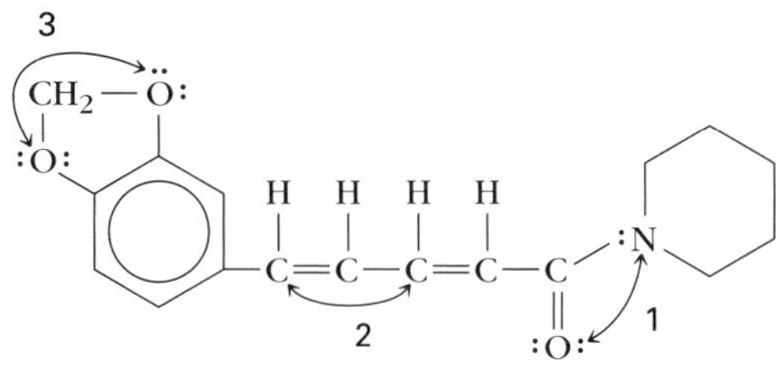
    (a) Give the values for the indicated bond angles.
    (b) What is the hybridization of the nitrogen?
    (c) What is the hybridization of the oxygens?

Conceptual Challenge Problems

CP9.A (Section 9.2) What advantages does the VSEPR model of chemical bonding have compared with the Lewis dot formulas predicted by the octet rule?

CP9.B (Section 9.2) The VSEPR model does not differentiate between single bonds and double bonds for predicting molecular shapes. What experimental evidence supports this?

CP9.C (Section 9.3) What evidence could you present to show that two carbon atoms joined by a single sigma bond are able to rotate about an axis that coincides with the bond, but two carbon atoms bonded by a double bond cannot rotate about an axis along the double bond?

10

Gases and the Atmosphere

© Breitling

The Breitling Orbiter 3 became the first balloon to fly around the world nonstop in March 1999. The 20-day trip covered 28,431 miles. During its flight, the balloon attained an altitude of 36,000 feet and was pushed by the jet stream. The balloon combined an outer hot-air chamber with an inner helium-filled chamber. As the temperature and pressure of the air and the helium changed, the volumes occupied by the gases changed, and the balloon rose and fell in elevation in response to these changes. The black object hanging beneath the gondola is a set of solar panels that supplied the necessary power for telecommunication and navigation instruments during the flight. The Breitling Orbiter 3 gondola in this picture is now on display at the Smithsonian National Air and Space Museum in Washington, DC.

Early chemists studied gases and their chemistry extensively. They carried out reactions that generated gases, bubbled the gases through water into glass containers, and transferred the gases to animal bladders for storage. They mixed gases to see whether the gases would react and, if so, how much of each gas would be consumed or formed. They discovered that gases have many properties in common, much more so than liquids or solids. All gases are transparent, although some are colored. (F_2 is light yellow, Cl_2 is greenish yellow, Br_2 and NO_2 are reddish brown, and I_2 vapor is violet.) All gases are also mobile, that is, they expand to fill all space available, and they mix together physically in any proportions. The volume of a gas sample can be altered by changing its temperature, its pressure, or both. Thus, gas densities (mass/volume) are quite variable, depending on the conditions. At first this might seem to make the study of gases complicated, but it really does not. The way that gas volume depends on temperature and pressure is essentially the same for all gases, so it is easy to formulate an expression that accounts for the dependence.

The earth is surrounded by a layer of gases we call the atmosphere. A vast number of chemical reactions take place in this gas mixture, many of them driven by energy from solar photons. Although many of these reactions are beneficial to the earth's inhabitants, some produce undesirable products. This chapter contains fundamental facts and concepts about gases. The relationships among pressure, temperature, and amount of a gas and its volume are expressed in the ideal gas law. Other quantitative treatments of gases in chemical reactions are introduced, and adjustments to the ideal gas law to account for real gas behaviors are described. The chapter concludes with a discussion of carbon dioxide in the atmosphere and its relationship to the greenhouse effect and global warming.

Compared to the size of the earth, the atmosphere is about the thickness of an apple skin compared to the size of the apple.

NASA

Planet earth.

10.1 The Atmosphere

Earth is enveloped by a few vertical miles of atoms and molecules that compose the atmosphere, the gaseous medium in which we exist, a perfect place to begin the study of gases. Everything on the surface of the earth experiences the atmosphere's pressure. The atmosphere's total mass is approximately 5.3×10^{15} metric tons, a huge figure, but still only about one millionth of the earth's total mass.

A metric ton is 1000 kg.

The 5.3×10^{15} metric tons of the atmosphere has about the same mass as a cube of water 100 km on a side.

The two major chemicals in our atmosphere are nitrogen, a rather unreactive gas, and oxygen, a highly reactive one. In dry air at sea level, nitrogen is the most abundant atmospheric gas, followed by oxygen, and then 13 other gases, each at less than 1% by volume (Table 10.1). For every 100 volume units of air, 21 units are oxygen. When it is pure, oxygen supports very rapid combustion (Figure 10.1). When it is diluted with nitrogen at moderate temperatures such as in the atmosphere, however, oxygen's oxidizing capability is tamed somewhat.

© Cengage Learning/Charles D. Winters

Figure 10.1 Combustion in pure oxygen. When a glowing wood splint is inserted into the mouth of a vessel containing pure oxygen, the high oxygen concentration causes the splint to burn much faster.

Table 10.1 The Composition of Dry Air at Sea Level*

Gas	Percentage by Volume	Gas	Percentage by Volume
Nitrogen	78.084	Krypton	0.0001
Oxygen	20.948	Carbon monoxide†	0.00001
Argon	0.934	Xenon	0.000008
Carbon dioxide†	0.033	Ozone‡	0.000002
Neon	0.00182	Ammonia	0.000001
Hydrogen	0.0010	Nitrogen dioxide‡	0.0000001
Helium	0.00052	Sulfur dioxide‡	0.00000002
Methane†	0.0002		

*The data in this table are for dry air because the percentage of water in air varies. Therefore, water has been omitted even though it is an important component of the atmosphere.
†The greenhouse gases carbon dioxide and methane are discussed in Section 10.9 as they relate to fuels and the burning of fuels for energy production.
‡Trace gases of environmental importance.

Due to the force of gravity, molecules making up the atmosphere are most concentrated near the earth's surface. In addition to the percentages by volume used in Table 10.1, *parts per million (ppm)* and *parts per billion (ppb)* by volume are used to describe the concentrations of components of the atmosphere. Since volume is proportional to the number of molecules, ppm and ppb units also give a ratio of molecules of one kind to another kind. For example, "10 ppm SO_2" means that for every 1 million molecules in air, 10 of them are SO_2 molecules. This may not sound like much until you consider that in just 1 cm^3 of air there are about 2.7×10^{13} million molecules. If this 1 cm^3 of air contains 10 ppm SO_2, then it contains 2.7×10^8 million SO_2 molecules.

To convert percent to ppm, multiply by 10,000. Divide by 10,000 to convert ppm to percent.

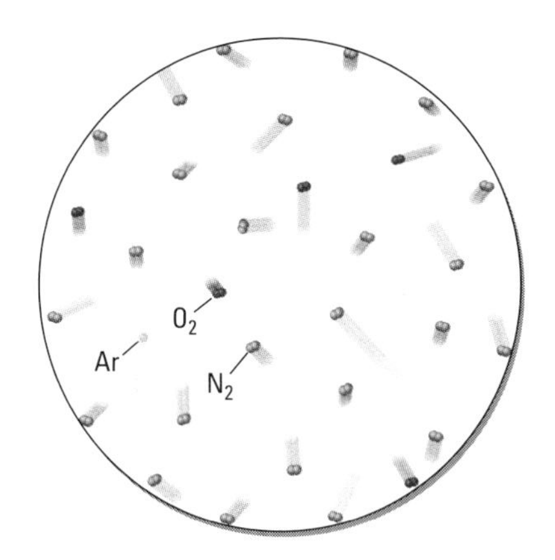

Nanoscale view of air. This instantaneous view of a sample of air at the nanoscale shows nitrogen, oxygen, and argon particles in rapid, random motion.

EXERCISE 10.1 Calculating the Mass of a Gas

Calculate the mass of the SO_2 molecules found in 1 cm^3 of air that contains 10 ppm SO_2.

The Troposphere and the Stratosphere

The earth's atmosphere can be roughly divided into layers. From the surface to an altitude of about 10 km, in the region named the **troposphere,** the temperature of the atmosphere decreases with increasing altitude. In this region the most violent mixing of air and the biggest variations in moisture content and temperature occur. Winds, clouds, storms, and precipitation are the result, the phenomena we know as weather. The troposphere is where we live. A commercial jet airplane flying at an altitude of about 10 km (approximately 33,000 ft) is still in the troposphere, although near its upper limits. The composition of the troposphere is roughly that of dry air near sea level (see Table 10.1), but the concentration of water vapor varies considerably, with an average of about 10 ppm.

Just above the troposphere, from about 12 to 50 km above the earth's surface, is the **stratosphere.** The pressures in the stratosphere are extremely low, and little mixing occurs between the stratosphere and the troposphere. The lower limit of the stratosphere varies from night to day over the globe, and in the polar regions it may be as low as 8 to 9 km above the earth's surface. About 75% of the mass of the atmosphere is in the troposphere, and 99.9% of the atmosphere's mass is below 50 km, in the troposphere and stratosphere.

The troposphere was named by the British meteorologist Sir Napier Shaw from the Greek word *tropos,* meaning "turning."

The stratosphere was named by the French meteorologist Leon Phillipe Treisserenc de Bort, who believed this region consisted of orderly layers with no turbulence or mixing. *Stratum* is a Latin word meaning "layer."

ESTIMATION Thickness of Earth's Atmosphere

In the photograph of earth on page 342, the earth's diameter is about 5 cm. About 99.9% of the atmosphere is contained within 50 km of the earth's surface. To draw a circle around the photo properly relating the diameter of the earth with the thickness of the atmosphere, how thick should the line be drawn?

The diameter of earth is about 7900 miles, which is 12,700 km. Set up an equation that equates the ratio of the diameter of the photo to the diameter of earth with the ratio of the sought line width (x) to the thickness of the atmosphere.

$$\frac{5 \text{ cm}}{12{,}700 \text{ km}} = \frac{x}{50 \text{ km}}$$

$$x = 0.020 \text{ cm} = 0.2 \text{ mm}$$

If the circle representing the earth is 5 cm in diameter, then the line width that properly shows the relative thickness of the atmosphere would be only 0.2 mm, less than the width of the period at the end of this sentence. The proportions are about the same as the skin on an apple in relation to the apple. The atmosphere of earth is really quite thin.

Visit this book's companion website at **www.cengage.com/chemistry/moore** to work an interactive module based on this material.

10.2 Gas Pressure

Many molecular compounds are gases under ordinary conditions of temperature and pressure. The air we breathe consists mostly of N_2 and O_2 molecules. Other elements that exist as gases at normal temperature and pressure include H_2, Cl_2, F_2, and the noble gases (Group 8A). Other common gases include acetylene, C_2H_2; ammonia, NH_3; carbon dioxide, CO_2; methane, CH_4; and sulfur dioxide, SO_2. At sufficiently high temperatures, many liquid or solid compounds become gases.

One of the most important properties of any gas is its pressure. The firmness of a balloon filled with air indicates that the gas inside exerts pressure caused by gas molecules striking the balloon's inner surface with each collision exerting a force on the surface. The force per unit area is the **pressure** of the gas. A gas exerts pressure on every surface it contacts, no matter what the direction of contact.

$$\text{Pressure} = \frac{\text{force}}{\text{area}}$$

A force can accelerate an object, and the force equals the mass of the object times its acceleration.

$$\text{Force} = \text{mass} \times \text{acceleration}$$

The SI units for mass and acceleration are kilograms (kg) and meters per second per second (m/s^2), respectively, so force has the units kg m/s^2. A force of 1 kg m/s^2 is defined as a **newton (N)** in the International System of Units (SI). A pressure of one newton per square meter (N/m^2) is defined as a **pascal (Pa).** Table 10.2 provides units of pressure and conversion factors among units.

1 kPa = 10^3 Pa

The earth's atmosphere exerts pressure on everything with which it comes into contact. Atmospheric pressure can be measured with a **barometer,** which can be made by filling a tube closed at one end with a liquid and then inverting the tube in a dish containing the same liquid. Figure 10.2 shows a mercury barometer—a glass tube filled with mercury, inverted, and placed in a container of mercury. At sea level the height of the mercury column is 760 mm above the surface of the mercury in the dish. The pressure at the bottom of a column of mercury 760 mm tall is balanced by the pressure at the bottom of the column of air surrounding the dish—a column that extends to the top of the atmosphere. Pressure measured with a mercury barometer is usually reported in **millimeters of mercury (mm Hg),** a unit that is also called the **torr** after Evangelista Torricelli, who invented the mercury barometer in 1643. The **standard atmosphere (atm)** is defined as

$$1 \text{ standard atmosphere} = 1 \text{ atm} = 760 \text{ mm Hg (exactly)} = 101.325 \text{ kPa}$$

The pressure of the atmosphere at sea level is about 101,300 Pa (101.3 kPa). A related unit, the **bar,** equal to 100,000 Pa, is sometimes used for atmospheric pressure. The standard thermodynamic properties (← ***p. 210***) are given for a gas pressure of 1 bar.

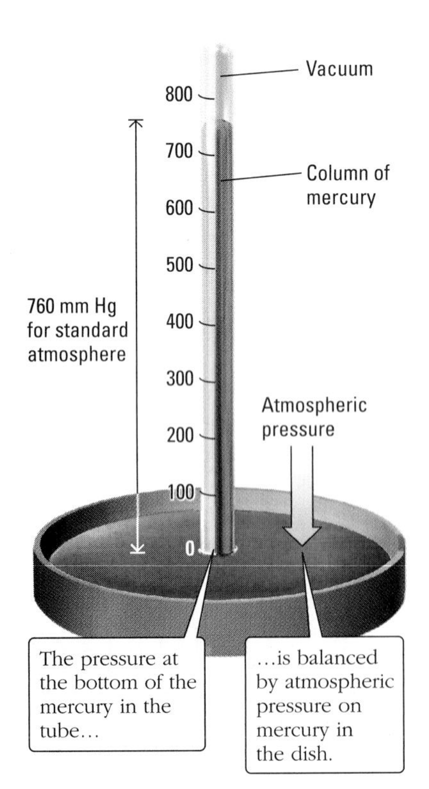

Figure 10.2 A Torricellian barometer.

Table 10.2 Pressure Units

SI Unit: Pascal (Pa)
1 Pa = 1 kg m^{-1} s^{-2} = 1 N/m^2; 1 kPa = 10^3 Pa
Other Common Units
1 bar = 10^5 Pa = 100 kPa; 1 millibar (mbar) = 10^2 Pa
1 atm = 1.01325×10^5 Pa = 101.325 kPa
1 atm = 760 torr = 760 mm Hg (this conversion is exact)*
1 atm = 14.7 lb/inch^2 (psi) = 1.01325 bar

*Exact conversion factors do not limit the number of significant figures in calculations.

PROBLEM-SOLVING EXAMPLE 10.1 Converting Pressure Units

Convert a pressure reading of 0.988 atm to (a) mm Hg, (b) kPa, (c) bar, (d) torr, and (e) psi.

Answer (a) 751. mm Hg (b) 100. kPa (c) 1.00 bar (d) 751. torr (e) 14.5 psi

Strategy and Explanation We use the unit equivalencies shown in Table 10.2 as conversion factors.

(a) mm Hg: $0.988 \text{ atm} \times \dfrac{760. \text{ mm Hg}}{1 \text{ atm}} = 751. \text{ mm Hg}$

(b) kPa: $0.988 \text{ atm} \times \dfrac{101.31 \text{ kPa}}{1 \text{ atm}} = 100. \text{ kPa}$

(c) bar: $0.988 \text{ atm} \times \dfrac{1.013 \text{ bar}}{1 \text{ atm}} = 1.00 \text{ bar}$

(d) torr: $0.988 \text{ atm} \times \dfrac{760. \text{ torr}}{1 \text{ atm}} = 751. \text{ torr}$

(e) psi: $0.988 \text{ atm} \times \dfrac{14.7 \text{ psi}}{1 \text{ atm}} = 14.5 \text{ psi}$

PROBLEM-SOLVING PRACTICE 10.1

A TV weather person says the barometric pressure is "29.5 inches of mercury." What is this pressure in (a) atm, (b) mm Hg, (c) bar, and (d) kPa?

Any liquid can be used in a barometer, but the height of the column depends on the density of the liquid. The atmosphere can support a column of mercury that is 760 mm high, or a column of water almost 34 ft high since water (1.0 g/cm^3) is much less dense than mercury (13.55 g/cm^3). A water barometer would be between 33 and 34 ft high, far too tall to be practical.

10.3 Kinetic-Molecular Theory

All gases have a number of properties in common:

- ***Gases can be compressed.*** We often pump compressed air into an automobile or bicycle tire. Compressed air occupies less volume than the noncompressed air (Figure 10.3).

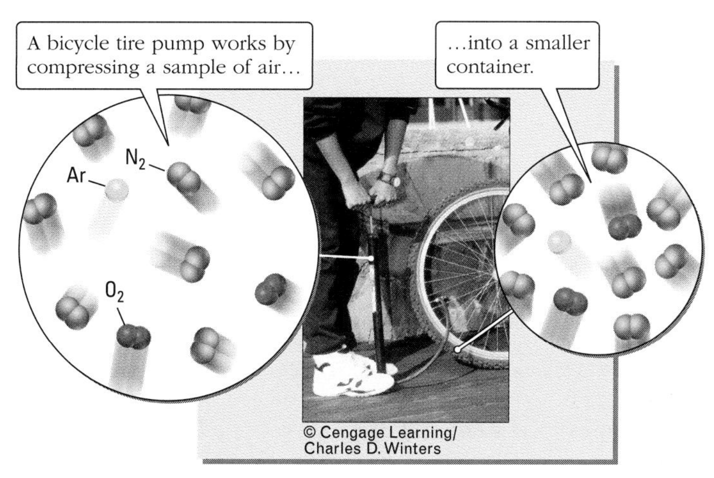

Figure 10.3 Compression of gases.

- ***Gases exert pressure on whatever surrounds them.*** A gas sample inside a balloon or a closed storage cylinder exerts pressure on its surroundings. If the container cannot sustain the pressure, some of the gas sample will escape, perhaps by bursting the container.
- ***Gases expand into whatever volume is available.*** The gaseous contents of a container will escape when it is opened.
- ***Gases mix completely with one another.*** Once gases are mixed, they do not separate spontaneously. The gases in the atmosphere are an example.
- ***Gases are described in terms of their temperature and pressure, the volume occupied, and the amount (numbers of molecules or moles) of gas present.*** For example, a hot gas in a fixed volume exerts a greater pressure than does the same sample of gas when it is cold.

In the kinetic-molecular theory the word "molecule" is taken to include atoms of the monatomic noble gases He, Ne, Ar, Kr, and Xe.

To explain why gases behave as they do, we first look at the nanoscale behavior of gas molecules. The fact that all gases behave in very similar ways can be interpreted by means of the kinetic-molecular theory, a theory that applies to the properties of liquids and solids as well as gases. According to the kinetic-molecular theory, a gas consists of tiny molecules in constant, rapid, random motion. The pressure a gas exerts on the walls of its container results from the continual bombardment of the walls by rapidly moving gas molecules (Figure 10.4).

Four fundamental concepts form the foundation of the kinetic-molecular theory, and a fifth is closely associated with it. Each is consistent with the results of experimental studies of gases.

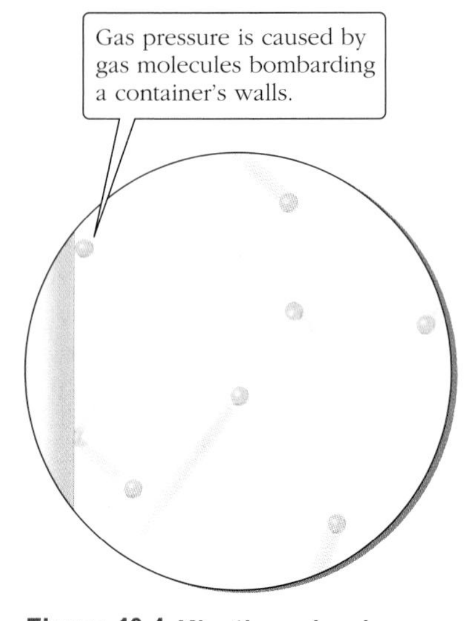

Gas pressure is caused by gas molecules bombarding a container's walls.

Figure 10.4 Kinetic-molecular theory and pressure.

1. ***A gas is composed of molecules whose size is much smaller than the distances between them.*** This concept accounts for the ease with which gases can be compressed and for the fact that gases at ordinary temperature and pressure mix completely with each other. These facts imply that there must be much unoccupied space in gases that provides substantial room for additional molecules in a sample of gas.
2. ***Gas molecules move randomly at various speeds and in every possible direction.*** This concept is consistent with the fact that gases quickly and completely fill any container in which they are placed.
3. ***Except when gas molecules collide, forces of attraction and repulsion between them are negligible.*** This concept is consistent with the fact that all gases behave in the same way, regardless of the types of noncovalent interactions among their molecules.
4. ***When collisions between molecules occur, the collisions are elastic.*** In an elastic collision, the speeds of colliding molecules may change, but the total kinetic energy of two colliding molecules is the same after a collision as before the collision. This concept is consistent with the fact that a gas sample at constant temperature never "runs down," with all molecules falling to the bottom of the container.
5. ***The average kinetic energy of gas molecules is proportional to the absolute temperature.*** Though not part of the kinetic-molecular theory, this useful concept is consistent with the fact that gas molecules escape through a tiny hole faster as the temperature increases and with the fact that rates of chemical reactions are faster at higher temperatures.

Like any moving object, a gas molecule has kinetic energy. An object's kinetic energy, E_k, depends on its mass, m, and its speed or velocity, v, according to the equation ***(⇐ p. 178)***

$$E_k = \frac{1}{2}(\text{mass})(\text{speed})^2 = \frac{1}{2}mv^2$$

All the molecules in a gas are moving, but they do not all move at the same speed, so they do not all have the same kinetic energy. At any given time a few molecules are moving very quickly, a large number are moving at close to the average speed, and a

CHEMISTRY IN THE NEWS

Nitrogen in Tires

Usually, we fill our automobile tires with compressed air because it is cheap. But race cars, jet aircraft, and some heavy trucks use nitrogen. For the ordinary motorist, many tire shops now offer to fill car's tires with nitrogen for about $5.00 each. Is it worth it? What advantages might nitrogen have over air?

Many bogus claims are being made about such fillings. For example, some claim that nitrogen will leak out of your tires more slowly than air because the molecules are bigger, which is false.

Some good reasons for using nitrogen also exist. For example, nitrogen is relatively inert and will not support combustion in case of a fire around the tires. Pure nitrogen contains no oxygen, whereas air does, so corrosion of rims and the insides of tires is said to be decreased. The generation units used to provide the nitrogen produce very dry nitrogen, whereas the air going through a typical gas station compressor contains substantial amounts of moisture, which could aid in corrosion.

From 2006 to 2007, *Consumer Reports* performed a yearlong study of nitrogen in automobile tires, and they concluded that there was little benefit to consumers. Overall, it seems that most professionals recommend using compressed air to fill automobile tires except in specialized applications.

Landing gear on a Boeing 747. Jet aircraft use nitrogen in the tires.

Source: Chemical & Engineering News, October 25, 2004; p. 88; **http://blogs.consumerreports.org/cars/2007/10/tires-nitrogen-.html;** T. Moran, Deflated Hope for Nitrogen, *New York Times,* March 3, 2008.

few may be in the process of colliding with a surface, in which case their speed is momentarily zero. The speed of any individual molecule changes as it collides with and exchanges energy with other molecules.

The relative number of molecules of a gas that have a given speed can be measured experimentally. Figure 10.5 is a graph of the number of molecules plotted versus their speed. The higher a point on the curve, the greater the number of molecules moving at that speed. Notice in the plot that some molecules are moving quickly (have high kinetic energy) and some are moving slowly (have low kinetic energy). The maximum in the distribution curve is the most probable speed. For oxygen gas at 25 °C, for example, the maximum in the curve occurs at a speed of about 400 m/s (1000 mph), and most of the molecules' speeds are in the range from 200 m/s to 700 m/s. Notice that the curves are not symmetric. A consequence of this asymmetry is that the average

Graphs of molecular speeds (or energies) versus numbers of molecules are called Boltzmann distribution curves. They are named after Ludwig Boltzmann (1844–1906), an Austrian physicist who helped develop the kinetic-molecular theory of gases.

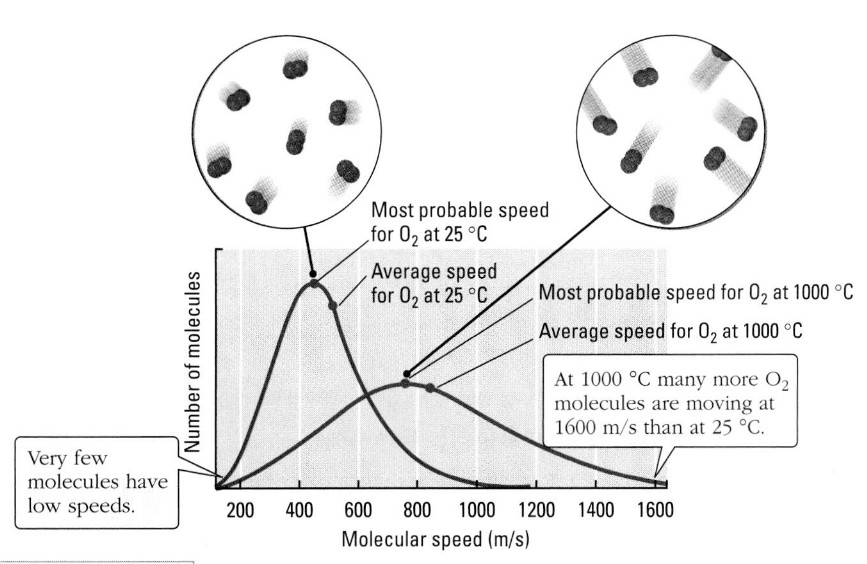

Active Figure 10.5 **Distribution of molecular speeds.** A plot of the relative number of oxygen molecules with a given speed versus that speed. Visit this book's companion website at **www.cengage.com/chemistry/moore** to test your understanding of the concepts in this figure.

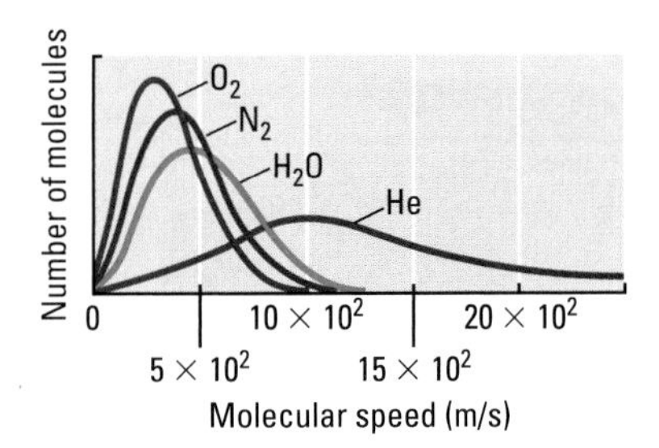

Figure 10.6 The effect of molar mass on the distribution of molecular speeds at a given temperature. On average, heavier molecules move slower than lighter ones.

speed (shown in Figure 10.5 for O_2) is a little faster than the most probable speed. Molecules with speeds higher than the most probable speed outnumber those with speeds lower than the most probable speed, moving the average speed toward a somewhat higher value.

Also notice in Figure 10.5 that as temperature increases, the most probable speed increases, and the number of molecules moving very quickly increases. The areas under the two curves representing the two oxygen samples at different temperatures are the same because the total number of oxygen molecules is the same in both samples.

Since $E_k = \frac{1}{2}mv^2$ and the average kinetic energy of the molecules of any gas is the same at a given temperature, the larger m is, the smaller the average v must be. That is, *heavier molecules have slower average speed than lighter ones*. Figure 10.6 illustrates this relationship. The peaks in the curves for the heavier molecules, O_2 and N_2, occur at a much lower speed than that for the lightest molecule, He. You can also see from the graph that average speeds range from a few hundred to a few thousand meters per second.

PROBLEM-SOLVING EXAMPLE 10.2 Kinetic-Molecular Theory

Consider a sample of oxygen gas, $O_2(g)$, at 25 °C in a rigid container. Use the kinetic-molecular theory to answer these questions:

(a) How would the pressure of the gas change if the temperature were raised to 40 °C?
(b) If some of the O_2 molecules were removed from the sample (at constant T), how would the pressure change?

Answer (a) The pressure would increase. (b) The pressure would decrease.

Strategy and Explanation We use the fundamental concepts of kinetic-molecular theory to answer each question.

(a) Raising the temperature of the O_2 molecules causes them to move at higher average speed and higher average kinetic energy. At the higher temperature the molecules would hit the container walls harder and more often, which would yield a higher pressure.
(b) Molecules in the new sample of O_2, one with fewer gas molecules than originally present but at the same temperature, would still have the same average speed and average energy, so they would hit the walls with the same impact after the removal of part of the sample. However, the gas molecules would hit the walls less frequently, so the pressure would be lower after the removal of some of the gas.

PROBLEM-SOLVING PRACTICE 10.2

(a) How does kinetic-molecular theory explain the change in pressure when a sample of gas has its volume decreased while the temperature remains constant?
(b) How does kinetic-molecular theory explain the change in pressure when gas molecules are added to a sample of gas in a fixed-volume container at constant temperature?

EXERCISE 10.2 Molecular Kinetic Energies

Arrange these gaseous substances in order of increasing average kinetic energy of their molecules at 25 °C: Cl_2, H_2, NH_3, SF_6.

CONCEPTUAL EXERCISE 10.3 Molecular Kinetic Energies

Using Figure 10.6 as a source of information, first draw a plot of the number of molecules versus molecular speed for a sample of helium at 25 °C. Now assume that an equal number of molecules of argon, also at 25 °C, are added to the helium. What would the distribution curve for the mixture of gases look like (one distribution curve for all molecules in the mixture)?

CONCEPTUAL EXERCISE **10.4 Molecular Motion**

Suppose you have two helium-filled balloons of about equal size, put one of them in the freezer compartment of your refrigerator, and leave the other one out in your room. After a few hours you take the balloon from the freezer and compare it to the one left out in the room. Based on the kinetic-molecular theory, what differences would you expect to see (a) immediately after taking the balloon from the freezer and (b) after the cold balloon warms to room temperature?

The dependence of molecular speeds on molecular mass can be used to explain some interesting phenomena. One is **effusion,** the escape of gas molecules from a container through a tiny hole into a vacuum. The rate of effusion of a gas is inversely proportional to the square root of its molar mass: rate $\propto 1/\sqrt{M}$. Therefore, He will effuse much faster than N_2, and a balloon filled with He will deflate faster than one filled with air. A related phenomenon is **diffusion,** the spread of gas molecules of one type through those of another type—for example, a pleasant odor spreading throughout a room. Diffusion is more complicated than effusion because the spreading gas molecules are colliding with the second type of gas molecules. Nevertheless, diffusion follows the same relationship between rate and molar mass. Diffusion was employed in the Manhattan project during World War II to purify uranium for the atomic bomb because the gaseous UF_6 molecules diffuse at slightly different rates due to the difference in masses of the two uranium isotopes, ^{235}U and ^{238}U.

The symbol $\propto$ means "proportional to."

Photos: © Cengage Learning/Charles D. Winters

He (red balloon), with its lower molar mass, effuses through tiny holes in a balloon surface faster than N_2 (blue balloon), with its higher molar mass. The lower frame shows the two balloons when first filled, and the upper frame shows the two balloons after several hours have passed.

10.4 The Behavior of Ideal Gases

The kinetic-molecular theory explains gas behavior on the nanoscale. On the macroscale, gases have been studied for hundreds of years, and the properties that all gases display have been summarized into *gas laws* that are named for their discoverers. Using the variables pressure, volume, temperature, and amount (number of moles), we can write equations that explain how gases behave. A gas that behaves exactly as described by these equations is called an **ideal gas.** At room temperature and atmospheric pressure, most gases behave nearly ideally. However, at pressures much higher than 1 atm or at temperatures just above the boiling point, gases deviate substantially from ideal behavior (Section 10.8). Each of the gas laws can be explained by the kinetic-molecular theory.

The Pressure-Volume Relationship: Boyle's Law

Boyle's law *states that the volume (V) of an ideal gas varies inversely with the applied pressure (P) when temperature (T) and amount (n, moles) are constant.*

$$V \propto \frac{1}{P} \qquad V = \text{constant} \times \frac{1}{P} \qquad PV = \text{constant} \qquad (\text{unchanging } T \text{ and } n)$$

The value of the constant depends on the temperature and the amount of the gas. The inverse relationship between V and P is shown graphically in Figure 10.7a. Plotting V versus $1/P$ yields a linear relationship, as shown in Figure 10.7b. For a gas sample under two sets of pressure and volume conditions (with unchanging T and n), Boyle's law can be written as $P_1V_1 = P_2V_2$.

In terms of the kinetic-molecular theory, a decrease in volume of a gas increases its pressure because there is less room for the gas molecules to move around before they collide with the walls of the container. Thus, there are more frequent collisions with the walls. These collisions produce pressure on the container walls, and more collisions mean a higher pressure. When you pump up a tire with a bicycle pump, the

Boyle's law explains how expanding and contracting your chest cavity (changing its volume) leads to pressure changes, which in turn lead to inhalation and exhalation of air.

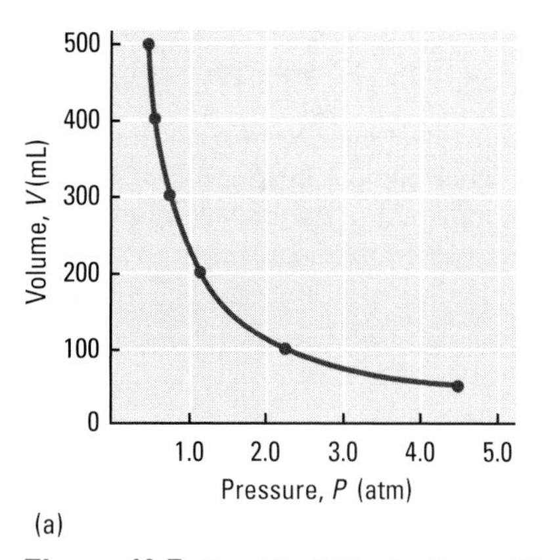

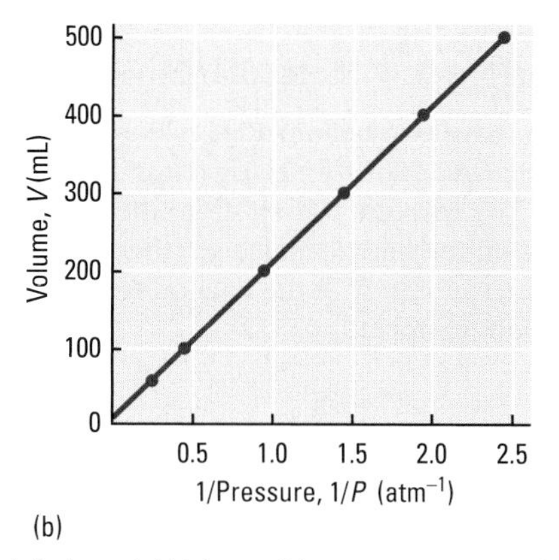

Figure 10.7 Graphical illustration of Boyle's law. (a) Volume (V) versus pressure (P). This curve shows the inverse proportionality between volume and pressure. As pressure increases, volume decreases. (b) V versus $1/P$. A linear plot results.

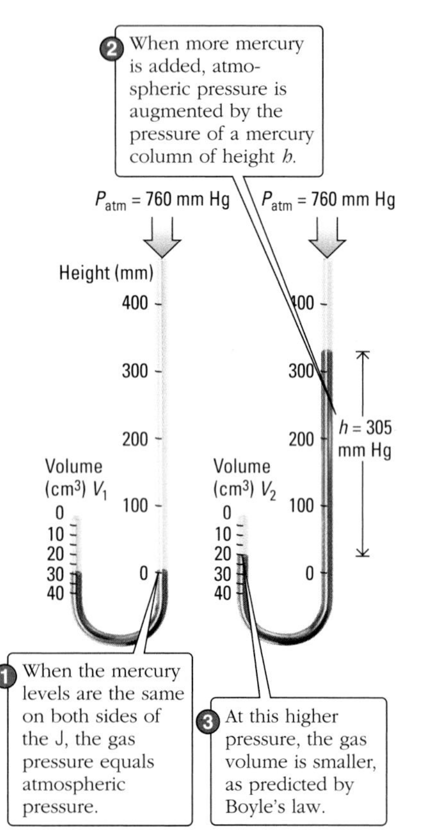

2 When more mercury is added, atmospheric pressure is augmented by the pressure of a mercury column of height h.
P_atm = 760 mm Hg
P_atm = 760 mm Hg
Height (mm)
400
300
200
100
0
Volume (cm³) V₁
Volume (cm³) V₂
h = 305 mm Hg
1 When the mercury levels are the same on both sides of the J, the gas pressure equals atmospheric pressure.
3 At this higher pressure, the gas volume is smaller, as predicted by Boyle's law.

Figure 10.8 Boyle's law. Boyle's experiment showing the compressibility of gases.

gas in the pump is squeezed into a smaller volume by application of pressure. This property is called *compressibility*. In contrast to gases, liquids and solids are only slightly compressible.

Robert Boyle studied the compressibility of gases in 1661 by pouring mercury into a J-shaped tube containing a sample of trapped gas. Each time he added more mercury, at constant temperature, the volume of the trapped gas decreased (Figure 10.8). The mercury additions increased the pressure on the gas and changed the gas volume in a predictable fashion.

CONCEPTUAL EXERCISE 10.5 Visualizing Boyle's Law

Many cars have gas-filled shock absorbers to give the car and its occupants a smooth ride. If a four-passenger car is loaded with four NFL linemen, describe the gas inside the shock absorbers compared with that when the car has no passengers.

The Temperature-Volume Relationship: Charles's Law

***Charles's law** states that the volume (V) of an ideal gas varies directly with absolute temperature (T) when pressure (P) and amount (n, moles) are constant.*

$$V \propto T \qquad V = \text{constant} \times T \qquad \frac{V}{T} = \text{constant} \qquad (\text{unchanging } P \text{ and } n)$$

The value of the constant depends on pressure and the amount of the gas. If the volume, V_1, and temperature, T_1, of a sample of gas are known, then the volume, V_2, at some other temperature, T_2, at the same pressure is given by

$$\frac{V_1}{T_1} = \frac{V_2}{T_2} \qquad (P \text{ and } n \text{ constant})$$

In terms of the kinetic-molecular theory, higher temperature means faster molecular motion and a higher average kinetic energy. The more rapidly moving molecules therefore strike the walls of a container more often, and each collision exerts greater force. For the pressure to remain constant, the volume of the container must expand.

When using the gas law relationships, temperature must be expressed in terms of the **absolute temperature scale.** The zero on this scale is the lowest possible temperature. The unit of the absolute temperature scale is the kelvin, symbol K (with no degree sign). The kelvin is the SI unit of temperature. The relationship between the absolute scale and the Celsius scale is shown in Figure 10.9. The kelvin is the same size as a degree Celsius. Thus, when ΔT is calculated by subtracting one temperature from another, the result is the same on both scales, even though the numbers involved are different. The lowest possible temperature, 0 K, known as **absolute zero,** is -273.15 °C. Temperature on the Celsius scale T(°C) and on the absolute scale T(K) are related as follows:

The absolute temperature scale is also called the **Kelvin temperature scale.** The kelvin is the SI unit of temperature.

$$T(\text{K}) = T(°\text{C}) + 273.15$$

Thus, 25.00 °C (a typical room temperature) is the same as 298.15 K.

In 1787, Jacques Charles discovered that the volume of a fixed quantity of a gas at constant pressure increases with increasing temperature. Figure 10.10 shows how the volume of 1.0 mol H_2 and the volume of 0.55 mol O_2 change with the temperature (pressure remains constant at 1.0 atm). When the plots of volume versus temperature for different gases are extended toward lower temperatures, they all reach zero volume at the same temperature, -273.15 °C, which is 0 K.

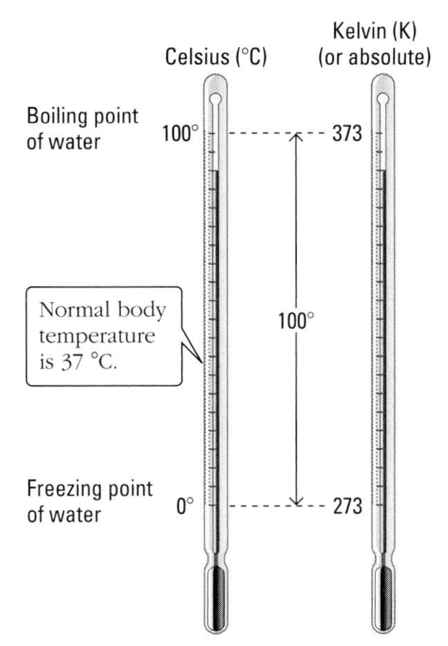

Figure 10.9 Temperature scales. Zero on the absolute scale is the lowest possible temperature (0 K or -273.15 °C).

Suppose you want to calculate the new volume when 450.0 mL of a gas is cooled from 60.0 °C to 20.0 °C, at constant pressure. First, convert the temperatures to kelvins by adding 273.15 to the Celsius values: 60.0 °C becomes 333.2 K and 20.0 °C becomes 293.2 K. Using Charles's law for the two sets of conditions at constant pressure gives

$$V_2 = \frac{V_1 T_2}{T_1} = \frac{450.0 \text{ mL} \times 293.2 \text{ K}}{333.2 \text{ K}} = 396.0 \text{ mL}$$

CONCEPTUAL
EXERCISE 10.6 Visualizing Charles's Law

Consider a collection of gas molecules at temperature T_1. Now increase the temperature to T_2. Use the ideas of the kinetic-molecular theory and explain why the volume would have to be larger if the pressure remains constant. What would have to be done to maintain a constant volume if the temperature increased?

The Amount-Volume Relationship: Avogadro's Law

***Avogadro's law** states that the volume (V) of an ideal gas varies directly with amount (n) when temperature (T) and pressure (P) are constant.*

$$V \propto n \qquad V = \text{constant} \times n \qquad \frac{V}{n} = \text{constant} \qquad (\text{unchanging } T \text{ and } P)$$

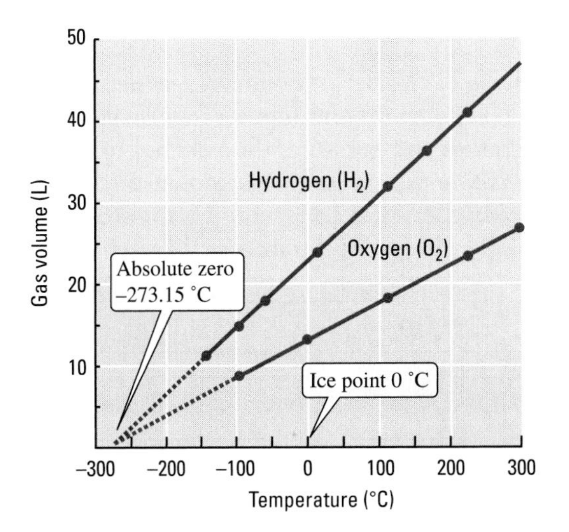

Figure 10.10 Charles's law. The volumes of two different samples of gases decrease with decreasing temperature (at constant pressure and constant molar amount). These graphs (as would those of all gases) intersect the temperature axis at about -273 °C.

The volume of any gas would appear to be zero at -273.15 °C. However, at sufficiently high pressure, all gases liquefy before reaching this temperature.

Jacques Alexandre Cesar Charles 1746–1823

The French scientist Jacques Charles was most famous in his lifetime for his experiments in ballooning. The first such flights were made by the Montgolfier brothers in June 1783 using a balloon made of linen and paper and filled with hot air. In August 1783, Charles filled a silk balloon with hydrogen. Inflating the bag to its final diameter took several days and required nearly 500 lb acid and 1000 lb iron to generate the required volume of hydrogen gas. A huge crowd watched the ascent on August 27, 1783. The balloon stayed aloft for almost 45 min and traveled about 15 miles. When it landed in a village, the people there were so terrified that they tore the balloon to shreds.

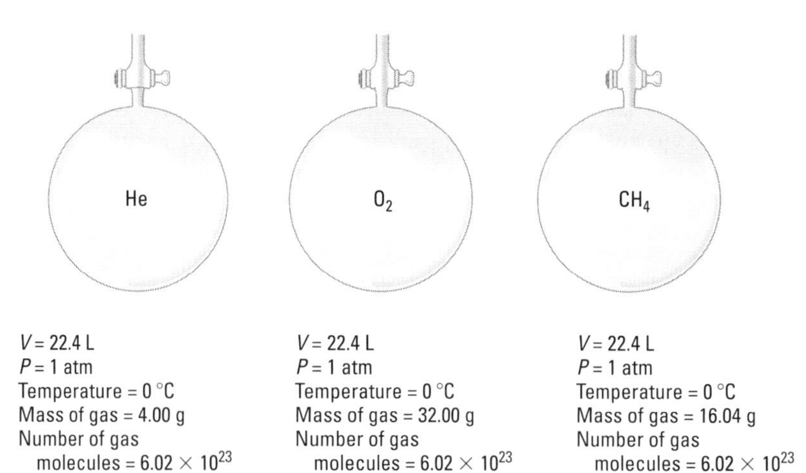

Figure 10.11 Comparison of volumes and amounts of gases illustrating Avogadro's hypothesis. Each gas has the same volume, temperature, and pressure, so each container holds the same number of molecules. Because the molar masses of the three gases are different, the masses of gas in each container are different.

The value of the constant depends on the temperature and the pressure. Avogadro's law means, for example, that at constant temperature and pressure, if the number of moles of gas doubles, the volume doubles. It also means that at the same temperature and pressure, the volumes of two different amounts of gases are related as follows:

$$\frac{V_1}{n_1} = \frac{V_2}{n_2} \qquad (T \text{ and } P \text{ constant})$$

In terms of the kinetic-molecular theory, increasing the number of gas molecules at a constant temperature means that the added molecules have the same average molecular kinetic energy as the molecules to which they were added. As a result, the number of collisions with the container walls increases in proportion to the number of molecules. This would increase the pressure if the volume were held constant. To maintain constant pressure, the volume must increase.

Another way to state the relationship between the amount of gas and its volume is *Avogadro's hypothesis: Equal volumes of gases contain equal numbers of molecules at the same temperature and pressure.* Experiments show that at 0 °C and 1 atm pressure, 22.4 L of any gas will contain 1 mol of the gas (6.02×10^{23} gas molecules). This is illustrated in Figure 10.11.

The Law of Combining Volumes

In 1809 the French scientist Joseph Gay-Lussac (1778–1850) conducted experiments in which he measured the volumes of gases reacting with one another to form gaseous products. He found that, at constant temperature and pressure, the volumes of reacting gases were always in the *ratios of small whole numbers.* This is known as the **law of combining volumes.** For example, the reaction of 2 L H_2 with 1 L O_2 produces 2 L water vapor. Similarly, the reaction of 4 L H_2 with 2 L O_2 produces 4 L water vapor.

	$2\ H_2(g)$	+	$O_2(g)$	→	$2\ H_2O(g)$
Experiment 1	2 L		1 L		2 L
Experiment 2	4 L		2 L		4 L

In 1811 Amadeo Avogadro suggested that Gay-Lussac's observations actually showed that equal volumes of all gases under the same temperature and pressure conditions con-

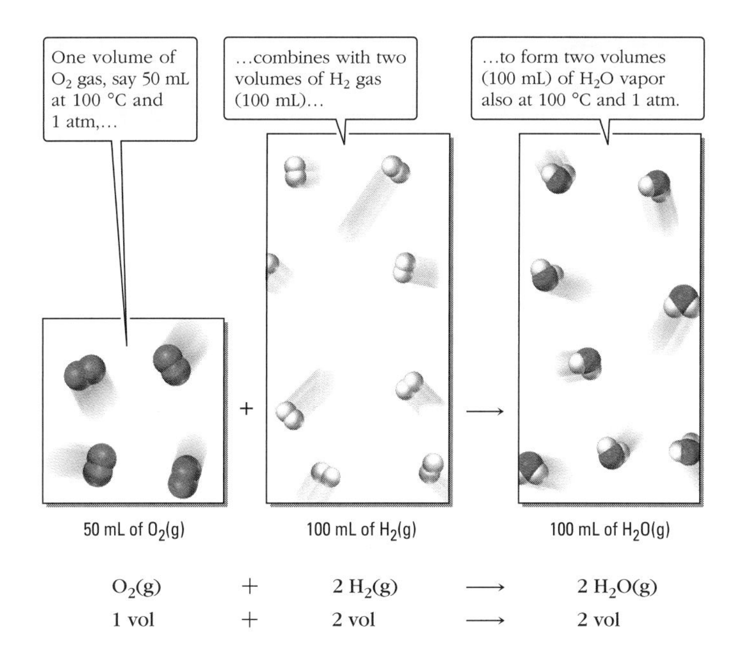
One volume of O_2 gas, say 50 mL at 100 °C and 1 atm,...
...combines with two volumes of H_2 gas (100 mL)...
...to form two volumes (100 mL) of H_2O vapor also at 100 °C and 1 atm.
50 mL of $O_2(g)$
100 mL of $H_2(g)$
100 mL of $H_2O(g)$
$O_2(g)$ + $2\ H_2(g)$ ⟶ $2\ H_2O(g)$
1 vol + 2 vol ⟶ 2 vol

Figure 10.12 Law of combining volumes. When gases at the same temperature and pressure combine with one another, their volumes are in the ratio of small whole numbers.

tain the same number of molecules. Viewed on a molecular scale, if a tiny volume contained only one molecule of a gas, then twice that volume would contain two molecules, and so on (Figure 10.12).

John Dalton (who devised the atomic theory) strongly opposed Avogadro's ideas and never accepted them. It took about 50 years—long after Avogadro and Dalton had died—for Avogadro's explanation of Gay-Lussac's experiments to be generally accepted.

PROBLEM-SOLVING EXAMPLE 10.3 Using Avogadro's Law and the Law of Combining Volumes

Nitrogen and oxygen gases react to form ammonia gas in the reaction

$$N_2(g) + 3\ H_2(g) \longrightarrow 2\ NH_3(g)$$

If 500. mL N_2 at 1 atm and 25 °C were available for reaction, what volume of H_2, at the same temperature and pressure, would be required in the reaction?

Answer 1.50 L H_2

Strategy and Explanation We want to find the volume of H_2, and the information given is the volume of the N_2 at the same temperature and pressure. Therefore, we can solve the problem using the law of combining volumes and the reaction coefficients of 3 for H_2 and 1 for N_2 as the volume ratio.

$$\text{Volume of } H_2 = 500.\text{ mL } N_2 \times \frac{3\text{ mL } H_2}{1\text{ mL } N_2} = 1.50\text{ L } H_2$$

PROBLEM-SOLVING PRACTICE 10.3

Nitrogen monoxide, NO, combines with oxygen to form nitrogen dioxide.

$$2\ NO(g) + O_2(g) \longrightarrow 2\ NO_2(g)$$

If 1.0 L oxygen gas at 30.25 °C and 0.975 atm is used and there is an excess of NO, what volume of NO gas at the same temperature and pressure will be converted to NO_2?

EXERCISE 10.7 Filling Balloons

One hundred balloons of equal volume are filled with a total of 26.8 g helium gas at 23 °C and 748 mm Hg. The total volume of these balloons is 168 L. Next, you are given 150 more balloons of the same size and 41.8 g He gas. The temperature and pressure remain the same. Determine by calculation whether you will be able to fill all the balloons with the He you have available.

The Ideal Gas Law

The three gas laws just discussed focus on the effects of changes in P, T, or n on gas volume:

- ***Boyle's law and pressure*** $(V \propto 1/P);\ V_1P_1 = V_2P_2$
- ***Charles's law and temperature*** $(V \propto T);\ \frac{V_1}{T_1} = \frac{V_2}{T_2}$
- ***Avogadro's law and amount (mol)*** $(V \propto n);\ \frac{V_1}{n_1} = \frac{V_2}{n_2}$

These three gas laws can be combined to give the ideal gas law, which summarizes the relationships among volume, temperature, pressure, and amount for any gas.

$$V \propto \frac{nT}{P} \quad \text{or} \quad PV \propto nT$$

To make this proportionality into an equation, a proportionality constant, R, named the **ideal gas constant,** is used. The equation becomes

$$V = R\frac{nT}{P}$$

and, on rearranging, gives the equation called the **ideal gas law.**

$$PV = nRT$$

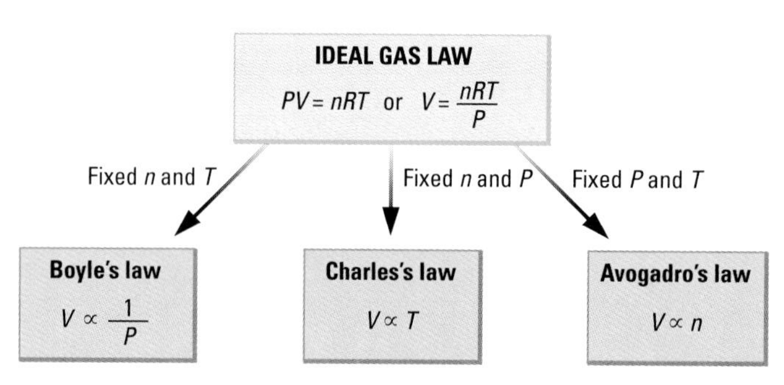

The ideal gas law correctly predicts the amount, pressure, volume, and temperature for samples of most gases at pressures of a few atmospheres or less and at temperatures well above their boiling points. The constant R can be calculated from the experimental fact that at 0 °C and 1 atm the volume of 1 mol gas is 22.414 L. This temperature and pressure are called **standard temperature and pressure (STP),** and the volume is called the **standard molar volume.** Solving the ideal gas law for R, and substituting, gives

$$R = \frac{PV}{nT} = \frac{(22.414\ \text{L})(1\ \text{atm})}{(1\ \text{mol})(273.15\ \text{K})} = 0.082057\ \text{L atm mol}^{-1}\ \text{K}^{-1}$$

which is usually rounded to 0.0821 L atm mol^{-1} K^{-1}. The ideal gas constant has different numerical values in different units, as shown in Table 10.3.

The ideal gas law can be used to calculate P, V, n, or T whenever three of the four variables are known, provided that the conditions of temperature and pressure are not extreme.

Table 10.3 Values of R, in Different Units

$$R = 0.08206\ \frac{\text{L atm}}{\text{mol K}}$$

$$R = 62.36\ \frac{\text{mm Hg L}}{\text{mol K}}$$

$$R = 8.314\ \frac{\text{kPa dm}^3}{\text{mol K}}$$

$$R = 8.314\ \frac{\text{J}}{\text{mol K}}$$

PROBLEM-SOLVING EXAMPLE 10.4 Using the Ideal Gas Law

Calculate the volume that 0.40 g methane, CH_4, occupies at 25 °C and 1.0 atm.

Answer 0.61 L or 6.1×10^2 mL CH_4

Strategy and Explanation The given information is the mass of methane and its temperature (T) and pressure (P). We can convert the mass of methane to the number of moles (n) and then use the ideal gas equation to find the volume (V). When we use $PV = nRT$, we must have all the variables in the same units as the gas constant, $R = 0.0821$ L atm mol^{-1} K^{-1}. That is, temperature must be in kelvins, pressure in atmospheres, volume in liters, and the amount of gas in moles.

We begin by converting the mass of methane to moles.

$$(0.40 \text{ g } CH_4) \times \frac{1 \text{ mol } CH_4}{16.04 \text{ g } CH_4} = 0.0249 \text{ mol } CH_4$$

Next, we convert the temperature to kelvins.

$$(25 + 273.15) \text{ K} = 298 \text{ K}$$

Now we can solve for V in the ideal gas equation. The result will be in liters.

$$V = \frac{nRT}{P} = \frac{(0.0249 \text{ mol})(0.0821 \text{ L atm mol}^{-1} \text{ K}^{-1})(298 \text{ K})}{1.0 \text{ atm}}$$
$$= 0.61 \text{ L or } 6.1 \times 10^2 \text{ mL } CH_4$$

Notice how the units cancel. Also, notice that the answer is given with two significant digits because the mass of methane and the pressure had only two significant digits.

☑ **Reasonable Answer Check** The temperature and pressure are close to STP. The mass of methane is about 2.5% of a mole, so the volume should also be about 2.5% of the volume of a mole of gas (22.4 L × 0.025 = 0.56 L). This approximate result is close to our more exact calculation.

PROBLEM-SOLVING PRACTICE 10.4

What volume will 2.64 mol N_2 occupy at 0.640 atm and 31 °C?

For many calculations involving a gas sample under two sets of conditions, it is convenient to use the ideal gas law in the following manner. For two sets of conditions (n_1, P_1, V_1, and T_1; n_2, P_2, V_2, and T_2), the ideal gas law can be written as

$$R = \frac{P_1V_1}{n_1T_1} \quad \text{and} \quad R = \frac{P_2V_2}{n_2T_2}$$

Since in both sets of conditions the quotient is equal to R, we can set the two quotients equal to each other.

$$\frac{P_1V_1}{n_1T_1} = \frac{P_2V_2}{n_2T_2}$$

When the amount of gas, n, is constant, so that $n_1 = n_2$, this equation simplifies to what is known as the **combined gas law:**

$$\frac{P_1V_1}{T_1} = \frac{P_2V_2}{T_2}$$

PROBLEM-SOLVING EXAMPLE 10.5 Pressure and Volume

Consider a gas sample with a volume of 1.00 L at a pressure of 0.500 atm. What volume will this gas sample occupy if the pressure is increased to 2.00 atm? Assume that the temperature remains constant and the gas sample obeys the ideal gas law.

Answer 0.250 L or 250. mL

Strategy and Explanation The temperature (T) and the amount of gas (n) are constant. The given conditions are $P_1 = 0.500$ atm, $P_2 = 2.00$ atm, and $V_1 = 1.00$ L. We must find V_2, which can be calculated by canceling T in the combined gas law and rearranging to solve for V_2.

$$\frac{P_1V_1}{\cancel{T_1}} = \frac{P_2V_2}{\cancel{T_2}}$$

$$V_2 = \frac{P_1V_1}{P_2} = \frac{(0.500 \text{ atm})(1.00 \text{ L})}{2.00 \text{ atm}} = 0.250 \text{ L} = 250. \text{ mL}$$

☑ **Reasonable Answer Check** The inverse relationship between P and V tells us that V_2 will be smaller than V_1 because P_2 is larger than P_1. The answer is reasonable.

PROBLEM-SOLVING PRACTICE 10.5

At a pressure of exactly 1 atm and some temperature, a gas sample occupies 400. mL. What will be the volume of the gas at the same temperature if the pressure is decreased to 0.750 atm?

PROBLEM-SOLVING EXAMPLE 10.6 The Combined Gas Law

Helium-filled balloons are used to carry scientific instruments high into the atmosphere. Suppose that such a balloon is launched on a summer day when the temperature at ground level is 22.5 °C and the barometer reading is 754 mm Hg. If the balloon's volume is 1.00×10^6 L at launch, what will its volume be at a height of 37 km, where the pressure is 76.0 mm Hg and the temperature is 240. K? (Assume that no He escapes.)

Answer 8.05×10^6 L

Strategy and Explanation Assume that no gas escapes from the balloon. Then only T, P, and V change. Using subscript 1 to indicate initial conditions (at launch) and subscript 2 to indicate final conditions (high in the atmosphere) gives

Initial: $P_1 = 754$ mm Hg $\quad T_1 = (22.5 + 273.15)$ K $= 295.6$ K $\quad V_1 = 1.00 \times 10^6$ L
Final: $P_2 = 76.0$ mm Hg $\quad T_2 = 240.$ K

Solving the combined gas law for V_2 gives

$$V_2 = \frac{P_1V_1T_2}{P_2T_1} = \frac{(754 \text{ mm Hg})(1.00 \times 10^6 \text{ L})(240. \text{ K})}{(76.0 \text{ mm Hg})(295.6 \text{ K})} = 8.05 \times 10^6 \text{ L}$$

Thus, the final volume is about eight times larger. The volume has increased because the pressure has dropped. For this reason, weather balloons are never fully inflated at launch. A great deal of room has to be left so that the helium can expand at high altitudes.

☑ **Reasonable Answer Check** The pressure is dropping by a factor of about 10, and T is changing only 20%, so the volume should increase by a factor of about 20% less than 10, that is, 8, and it does.

PROBLEM-SOLVING PRACTICE 10.6

A small sample of a gas is prepared in the laboratory and found to occupy 21 mL at a pressure of 710. mm Hg and a temperature of 22.3 °C. The next morning the temperature has changed to 26.5 °C, and the pressure is found to be 740. mm Hg. No gas has escaped from the container.

(a) What volume does the sample of gas now occupy?
(b) Assume that the pressure does not change. What volume would the gas occupy at the new temperature?

It is important to remember that Boyle's, Charles's, and Avogadro's laws do not depend on the identity of the gas being studied. These laws reflect properties of all gases and therefore describe the behavior of any gaseous substance, regardless of its identity.

In summary, for problems involving gases, you have your choice of two useful equations:

- ***When three of the variables P, V, n, and T are given, and the value of the fourth variable is needed, use the ideal gas law.***

$$PV = nRT$$

- ***When one set of conditions is given for a single gas sample and one of the variables under a new set of conditions is needed, then use the combined gas law and cancel any of the three variables that do not change.***

$$\frac{P_1V_1}{T_1} = \frac{P_2V_2}{T_2} \qquad (n \text{ constant})$$

CONCEPTUAL EXERCISE 10.8 Predicting Gas Behavior

Name three ways the volume occupied by a sample of gas can be decreased.

10.5 Quantities of Gases in Chemical Reactions

The law of combining volumes and the ideal gas law make it possible to use volumes as well as masses or molar amounts in calculations based on reaction stoichiometry (***← p. 107***). Consider this balanced equation for the combination reaction of solid carbon and gaseous oxygen to produce gaseous carbon dioxide.

$$C(s) + O_2(g) \longrightarrow CO_2(g)$$

Let's look at some of the questions that might be asked about this reaction, which involves one gaseous reactant and one gaseous product. As with any problem to be solved, you have to recognize what information is known and what is needed. Then you can decide which relationship provides the connection between the two.

1. *How many liters of CO_2 are produced from 0.5 L O_2 and excess carbon?* The known and needed information here are both volumes of gases, and the coefficients of the balanced chemical equation represent volumes of gases, so the law of combining volumes leads to the answer.
   Answer: 0.5 L CO_2, because the reaction shows 1 mol CO_2 formed for every mole of O_2 that reacts—a 1:1 ratio.

$$0.5 \text{ L } O_2 \times \frac{1 \text{ vol } CO_2}{1 \text{ vol } O_2} = 0.5 \text{ L } CO_2$$

2. *How many moles of O_2 are required to completely react with 4.00 g C?* Here the known information is the mass of the solid reactant and the needed information is the moles of gaseous reactant. The answer is provided from the balanced reaction, without using any of the gas laws.
   Answer: 0.333 mol O_2, because 4.00 g C is 0.333 mol C, and 1 mol O_2 is required for every 1 mol C.

$$4.00 \text{ g C} \times \frac{1 \text{ mol C}}{12.011 \text{ g C}} \times \frac{1 \text{ mol } O_2}{1 \text{ mol C}} = 0.333 \text{ mol } O_2$$

3. *What volume (in liters) of O_2 at STP is required to react with 12.011 g C?* Because the needed information is the volume of a gaseous reactant, the law of combining volumes can lead to the answer, but first the number of moles of carbon must be found.

Answer: 22.4 L O_2, because 12.011 g C is 1 mol C, which requires 1 mol O_2, and the standard molar volume of any gas at STP is 22.4 L.

$$12.011 \text{ g C} \times \frac{1 \text{ mol C}}{12.011 \text{ g C}} \times \frac{1 \text{ mol } O_2}{1 \text{ mol C}} \times \frac{22.4 \text{ L } O_2}{1 \text{ mol } O_2} = 22.4 \text{ L } O_2$$

4. *What volume (in liters) of O_2 at 747 mm Hg and 21 °C is required to react with 12.011 g C?* Realizing that 12.011 g C is 1 mol C and that 1 mol O_2 will be required, you then use the ideal gas law to calculate the volume occupied by 1 mol O_2 at the temperature and pressure given.
Answer: The units must be consistent with those of R. Convert the temperature to kelvins to get 294 K, and convert the given pressure to atmospheres to get 0.983 atm. Finally, use the ideal gas law to calculate the volume of oxygen required.

$$V_{O_2} = \frac{nRT}{P} = \frac{(1 \text{ mol})(0.0821 \text{ L atm mol}^{-1} \text{ K}^{-1})(294 \text{ K})}{0.983 \text{ atm}} = 24.5 \text{ L}$$

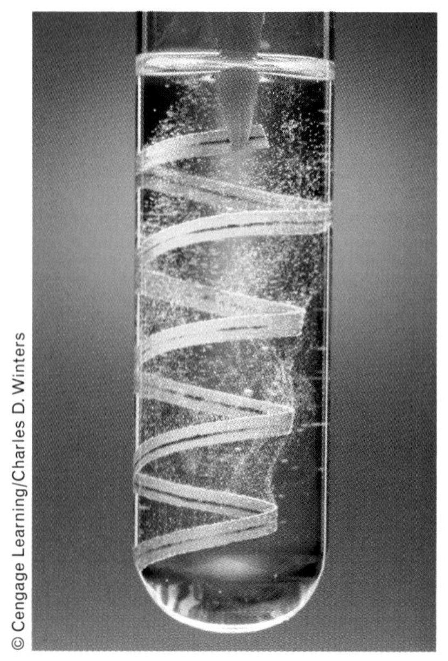

Magnesium reacting with hydrochloric acid.

© Cengage Learning/Charles D. Winters

PROBLEM-SOLVING EXAMPLE 10.7 Gases and Stoichiometry

If 13.8 g Mg is reacted with excess HCl, how many liters of H_2 gas are produced at 28.0 °C and 675 torr?

$$\text{Mg(s)} + 2 \text{ HCl(aq)} \longrightarrow \text{MgCl}_2\text{(aq)} + \text{H}_2\text{(g)}$$

Answer 15.8 L H_2

Strategy and Explanation We start to solve this problem by finding the number of moles of H_2 generated, using the coefficients from the balanced equation and the molar mass of magnesium.

$$13.8 \text{ g Mg} \times \frac{1 \text{ mol Mg}}{24.305 \text{ g Mg}} \times \frac{1 \text{ mol } H_2}{1 \text{ mol Mg}} = 0.568 \text{ mol } H_2$$

Once we know the number of moles of H_2 produced, we can find the volume of H_2 from the ideal gas law, $PV = nRT$. We solve the ideal gas law for V and substitute the values for P, T, and n. When we substitute into the rearranged ideal gas equation, the units of V, P, and T must be compatible with the units of R. This means that we must express the pressure in atmospheres and the temperature in kelvins (28.0 + 273.15 = 301.2 K).

$$P = (675 \text{ mm Hg}) \times \frac{1 \text{ atm}}{760 \text{ mm Hg}} = 0.888 \text{ atm}$$

$$V = \frac{nRT}{P} = \frac{(0.568 \text{ mol } H_2)(0.0821 \text{ L atm mol}^{-1} \text{ K}^{-1})(301.2 \text{ K})}{0.888 \text{ atm}} = 15.8 \text{ L } H_2$$

☑ **Reasonable Answer Check** We start with 13.8/24.3 or about 0.6 mol Mg, so about the same amount of H_2 should be produced. Six-tenths of 22.4 L (the volume of one mole of gas at STP) is about 13.4 L, which is comparable to our more exact answer.

PROBLEM-SOLVING PRACTICE 10.7

Ammonium nitrate, NH_4NO_3, is an explosive that undergoes this decomposition.

$$2 \text{ NH}_4\text{NO}_3\text{(s)} \longrightarrow 4 \text{ H}_2\text{O(g)} + \text{O}_2\text{(g)} + 2 \text{ N}_2\text{(g)}$$

If 10.0 g NH_4NO_3 explodes, how many liters of gas are generated at 25 °C and 1 atm?

Air bags deploying on crash test dummies. Air bags have saved numerous lives in automobile collisions.

©TRL Ltd./Photo Researchers, Inc.

PROBLEM-SOLVING EXAMPLE 10.8 Stoichiometry with Gases—Air Bags

Automobile air bags are filled with N_2 from the decomposition of sodium azide.

$$2 \text{ NaN}_3\text{(s)} \longrightarrow 2 \text{ Na(s)} + 3 \text{ N}_2\text{(g)}$$

How many grams of sodium azide are needed to generate 60.0 L N_2 at 1 atm and 26.0 °C?

Answer 106 g NaN_3

Strategy and Explanation First, we need to calculate the number of moles of N_2 in 60.0 L.

$$n = \frac{PV}{RT} = \frac{(1 \text{ atm})(60.0 \text{ L})}{(0.0821 \text{ L atm mol}^{-1} \text{ K}^{-1})(299.15 \text{ K})} = 2.44 \text{ mol } N_2$$

Now we use the balanced equation to calculate the moles of sodium azide.

$$2.44 \text{ mol } N_2 \times \frac{2 \text{ mol NaN}_3}{3 \text{ mol } N_2} = 1.63 \text{ mol NaN}_3$$

We were asked for the mass of sodium azide, so we convert moles to grams.

$$1.63 \text{ mol NaN}_3 \times \frac{65.01 \text{ g NaN}_3}{1 \text{ mol NaN}_3} = 106 \text{ g NaN}_3$$

☑ **Reasonable Answer Check** Dividing the 106 g NaN_3 by its molar mass of 65 g/mol gives an answer of about 1.6 mol NaN_3. We get $\frac{3}{2}$ as many moles of N_2, or about 2.4 mol N_2, which is about 54 L N_2. This is close to the required volume of N_2.

Actual air bags use additional reactants because the reaction described would produce elemental sodium, which is much too hazardous to remain in a consumer product. However, the actual reactions have nearly the same stoichiometric ratio of NaN_3 to N_2 as the simpler reaction used in this example.

PROBLEM-SOLVING PRACTICE 10.8

Lithium hydroxide is used in spacecraft to absorb the CO_2 exhaled by astronauts.

$$2 \text{ LiOH(s)} + CO_2\text{(g)} \longrightarrow Li_2CO_3\text{(s)} + H_2O(\ell)$$

What volume of CO_2 at 22 °C and 1 atm is absorbed per gram of LiOH?

10.6 Gas Density and Molar Masses

The ideal gas law can be used to relate gas density to molar mass. The definition of density is mass per unit volume (← ***p. 6***). The densities of gases are extremely variable because the volume of a gas sample (but not its mass) varies with temperature and pressure. However, once T and P are specified, the density of a gas can be calculated from the ideal gas law. Additionally, because equal volumes of gas at the same T and P contain equal numbers of molecules, the densities of different gases are directly proportional to their molar masses. As a result, experimental gas densities can be used to determine molar masses.

To derive the relationship between gas density and molar mass, start with the ideal gas law, $PV = nRT$. For any substance, the number of moles (n) equals its mass (m) divided by its molar mass (M), so we will substitute m/M for n in the ideal gas equation:

$$PV = \frac{m}{M}RT$$

Density (d) is mass divided by volume (m/V), and we can rearrange the equation so that m/V is on one side:

$$d = \frac{m}{V} = \frac{PM}{RT}$$

Thus, the density of a gas is directly proportional to its molar mass, M.

Consider the densities of the three pure gases He, O_2, and SF_6. If we take 1 mol of each of these gases at 25 °C and 0.750 atm, we can see from the table below that density (at the same conditions) increases with molar mass.

Gas	Molar Mass (g/mol)	Density (25 °C)
He	4.003	0.123 g/L
O_2	31.999	0.981 g/L
SF_6	146.06	4.48 g/L

Let's use the density equation to compute the approximate density of air at STP. The density of nitrogen at STP is

$$d = \frac{PM}{RT} = \frac{(1\text{ atm})(28.01\text{ g/mol})}{(0.0821\text{ L atm mol}^{-1}\text{ K}^{-1})(273.15\text{ K})} = 1.249\text{ g/L}$$

The density of oxygen at STP is

$$d = \frac{PM}{RT} = \frac{(1\text{ atm})(31.999\text{ g/mol})}{(0.0821\text{ L atm mol}^{-1}\text{ K}^{-1})(273.15\text{ K})} = 1.427\text{ g/L}$$

The density of air is estimated as the weighted average of the densities of nitrogen and oxygen

$$\begin{aligned} d &= (0.80)(\text{density of N}_2) + (0.20)(\text{density of O}_2) \\ &= (0.80)(1.249\text{ g/L}) + (0.20)(1.427\text{ g/L}) = 1.28\text{ g/L} \end{aligned}$$

Air has a density of approximately 1.28 g/L at STP.

EXERCISE 10.9 Calculating Gas Densities

Calculate the densities of Cl_2 and of SO_2 at 25 °C and 0.750 atm. Then calculate the density of Cl_2 at 35 °C and 0.750 atm and the density of SO_2 at 25 °C and 2.60 atm.

CONCEPTUAL EXERCISE 10.10 Comparing Densities

Express the gas density of He (0.123 g/L) in grams per milliliter (g/mL) and compare that value with the density of metallic lithium (0.53 g/mL). The mass of a Li atom (7 amu) is less than twice that of a He atom (4 amu). What do these densities tell you about how closely Li atoms are packed compared with He atoms when each element is in its standard state? To which of the concepts of the kinetic-molecular theory does this comparison apply?

NASA

A weather balloon filled with helium. As it ascends into the atmosphere, does the volume increase or decrease?

PROBLEM-SOLVING EXAMPLE 10.9 Using the Ideal Gas Law to Calculate Molar Mass

Suppose you have a 1.02-g sample of a gas containing C, H, and F atoms, and you suspect that it is $C_2H_2F_4$. The gas sample has a pressure of 750. mm Hg in a 250.-mL vessel at 25.0 °C. From this information, calculate the molar mass of the compound to determine whether your molecular formula is correct.

Answer 101 g/mol. The molecular formula is correct.

Strategy and Explanation Begin by organizing the data and changing units where necessary.

$$V = 0.250\text{ L} \quad P = 750.\text{ mm Hg} \times \frac{1\text{ atm}}{760\text{ mm Hg}} = 0.987\text{ atm} \quad T = 298.1\text{ K}$$

Use the ideal gas law equation to find n.

$$n = \frac{PV}{RT} = \frac{(0.987\text{ atm})(0.250\text{ L})}{(0.0821\text{ L atm mol}^{-1}\text{K}^{-1})(298.1\text{ K})} = 0.0101\text{ mol gas}$$

The mass of the sample is 1.02 g, so the molar mass according to this gas law calculation is

$$\frac{1.02\text{ g}}{0.0101\text{ mol}} = 101.\text{ g/mol}$$

Now sum the atomic weights of the molecular formula, $C_2H_2F_4$. This gives a molar mass of 102.0 g/mol, which is in close agreement with that determined using the ideal gas law.

☑ **Reasonable Answer Check** 250. mL of gas is about 0.250 L/22.4 L = 0.01 mol gas. If 1.0 g of gas is 0.01 mol, then the molar mass is about 1.0 g/0.01 mol = 100 g/mol, which is close to the more accurate answer.

PROBLEM-SOLVING PRACTICE 10.9

A flask contains 1.00 L of a pure gas at 0.850 atm and 20. °C. The mass of the gas is 1.13 g. What is the molar mass of the gas? What is its identity?

10.7 Gas Mixtures and Partial Pressures

Our atmosphere is a mixture of nitrogen, oxygen, argon, carbon dioxide, water vapor, and small amounts of several other gases (⇐ *p. 342,* Table 10.1). What we call atmospheric pressure is the sum of the pressures exerted by all these individual gases. The same is true of every gas mixture. Consider the mixture of nitrogen and oxygen illustrated in Figure 10.13. The pressure exerted by the mixture is equal to the sum of the pressures that the nitrogen alone and the oxygen alone would exert in the same volume at the same temperature and pressure. The pressure of one gas in a mixture of gases is called the **partial pressure** of that gas.

John Dalton was the first to observe that ***the total pressure exerted by a mixture of gases is the sum of the partial pressures of the individual gases in the mixture.*** This statement, known as **Dalton's law of partial pressures,** is a consequence of the fact that gas molecules behave independently of one another. As a demonstration of Dalton's law, consider the three most abundant components of our atmosphere, for which the total number of moles is

$$n_{total} = n_{N_2} + n_{O_2} + n_{Ar}$$

If we replace n in the ideal gas law with n_{total}, the summation of the individual numbers of moles of gases, the equation becomes

$$P_{total}V = n_{total}RT$$

$$P_{total} = \frac{n_{total}RT}{V} = \frac{(n_{N_2} + n_{O_2} + n_{Ar})RT}{V}$$

Expanding the right side of this equation and rearranging gives

$$P_{total} = \frac{n_{N_2}RT}{V} + \frac{n_{O_2}RT}{V} + \frac{n_{Ar}RT}{V} = P_{N_2} + P_{O_2} + P_{Ar}$$

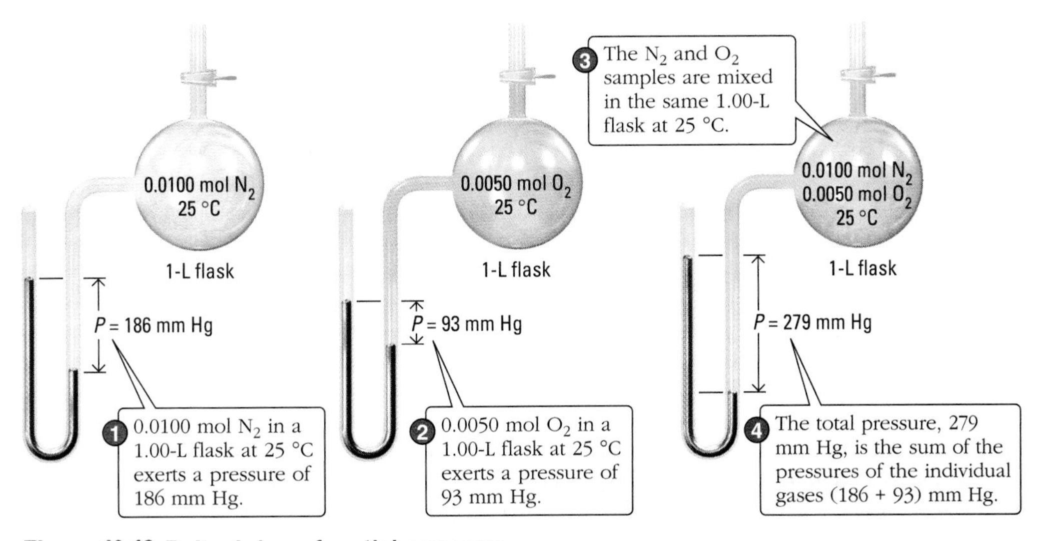

Figure 10.13 Dalton's law of partial pressures.

The quantities P_{N_2}, P_{O_2}, and P_{Ar} are the partial pressures of the three major components of the atmosphere. Dalton's law means that the pressure exerted by the atmosphere is the sum of the pressures due to nitrogen, oxygen, argon, and the other much less abundant components.

We can write a ratio of the partial pressure of one of the components, A, of a gas mixture over the total pressure,

$$\frac{P_A}{P_{total}} = \frac{n_A(RT/V)}{n_{total}(RT/V)}$$

On canceling terms on the right-hand side of this equation, we get

$$\frac{P_A}{P_{total}} = \frac{n_A}{n_{total}}$$

The ratio of the pressures is the same as the ratio of the number of moles of gas A to the total number of moles. This ratio (n_A/n_{total}) is called the **mole fraction** of A and is given the symbol X_A. Hence, rearranging the equation gives

$$P_A = X_A P_{total}$$

In the gas mixture in Figure 10.13, the mole fractions of nitrogen and oxygen are

$$X_{N_2} = \frac{0.010 \text{ mol } N_2}{0.010 \text{ mol } N_2 + 0.0050 \text{ mol } O_2} = 0.67$$

$$X_{O_2} = \frac{0.0050 \text{ mol } O_2}{0.010 \text{ mol } N_2 + 0.0050 \text{ mol } O_2} = 0.33$$

Because these two gases are the only components of the mixture, the sum of the two mole fractions must equal 1:

$$X_{N_2} + X_{O_2} = 0.67 + 0.33 = 1.00$$

An interesting application of partial pressures is the composition of the breathing atmosphere in deep-sea-diving vessels. If normal air at 1 atm pressure, with an oxygen mole fraction of 0.21, is compressed to 2 atm, the partial pressure of oxygen becomes about 0.42 atm. Such high oxygen partial pressures are toxic, so a diluting gas must be added to lower the oxygen partial pressure to near-normal values. Nitrogen gas might seem the logical choice because it is the diluting gas in the atmosphere. The problem is that nitrogen is fairly soluble in the blood and at high concentrations causes *nitrogen narcosis,* a condition similar to alcohol intoxication.

Helium is much less soluble in the blood and is therefore a good substitute for nitrogen in a deep-sea-diving atmosphere. However, using helium leads to interesting side effects. Because He atoms, on average, move faster than the heavier nitrogen molecules at the same temperature (Figure 10.6), they strike a diver's skin more often than would the nitrogen molecules and are therefore more efficient at transferring away energy. This effect causes divers to complain of feeling chilled while breathing a helium/oxygen mixture.

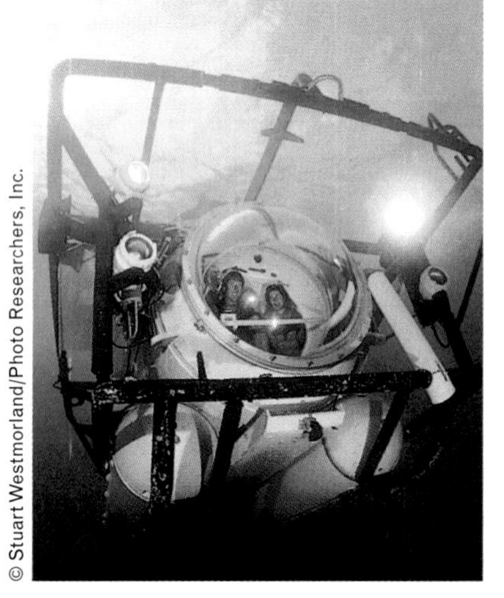

An undersea explorer. Some deep-sea–diving vessels like this one use an atmosphere of oxygen and helium for their occupants.

PROBLEM-SOLVING EXAMPLE 10.10 Calculating Partial Pressures

Halothane, $F_3C—CHBrCl$, is a commonly used surgical anesthetic delivered by inhalation. What is the partial pressure of each gas if 15.0 g halothane gas is mixed with 22.6 g oxygen gas and the total pressure is 862 mm Hg?

Answer $P_{halothane} = 83.8$ mm Hg $\quad P_{O_2} = 778$ mm Hg

Strategy and Explanation To find the partial pressures, we need the mole fraction of each gas. We first calculate the number of moles of each gas, and then we calculate the mole fractions.

$$15.0\text{ g} \times \left(\frac{1\text{ mol halothane}}{197.4\text{ g}}\right) = 0.07600\text{ mol halothane}$$

$$22.6\text{ g} \times \left(\frac{1\text{ mol O}_2}{32.00\text{ g}}\right) = 0.7062\text{ mol O}_2$$

$$X_{\text{halothane}} = \frac{0.07600\text{ mol halothane}}{0.7822\text{ total moles}} = 0.0972$$

Because the sum of the two mole fractions must equal 1.000, the mole fraction of O_2 is 0.903.

$$X_{\text{halothane}} + X_{O_2} = 1.000 = 0.0972 + X_{O_2}$$

$$X_{O_2} = 1.000 - 0.0972 = 0.903$$

Finally, we calculate the partial pressure of each gas.

$$P_{\text{halothane}} = 0.0972 \times P_{\text{total}} = 0.0972 \times (862\text{ mm Hg}) = 83.8\text{ mm Hg}$$

$$P_{O_2} = 0.903 \times P_{\text{total}} = 0.903 \times (862\text{ mm Hg}) = 778\text{ mm Hg}$$

☑ **Reasonable Answer Check** We have a 10:1 mole ratio of O_2 to halothane, so the ratio of partial pressures should be roughly 10:1, and it is.

PROBLEM-SOLVING PRACTICE 10.10

A mixture of 7.0 g N_2 and 6.0 g H_2 is confined in a 5.0-L reaction vessel at 500. °C. Assume that no reaction occurs and calculate the total pressure. Then calculate the mole fraction and partial pressure of each gas.

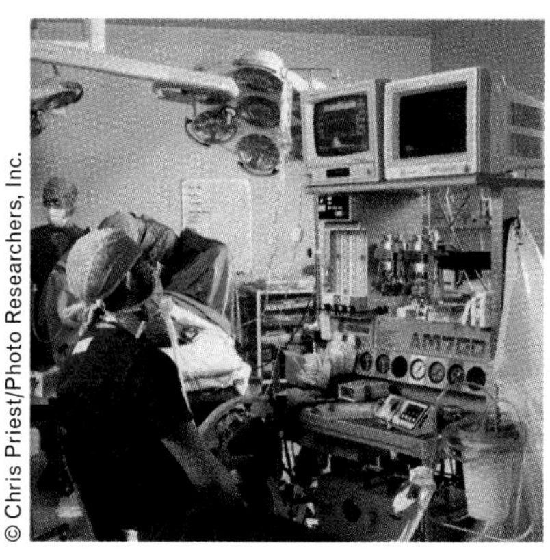

A gas-mixing manifold for anesthesia. An anesthesiologist uses such equipment to prepare a gas mixture to keep a patient unconscious during an operation. By proper mixing, the anesthetic gas can be added slowly to the breathing mixture. Near the end of the operation, the anesthetic gas can be replaced by air of normal composition or by pure oxygen.

CONCEPTUAL
EXERCISE 10.11 Pondering Partial Pressures

What happens to the partial pressure of each gas in a mixture when the volume is decreased by (a) lowering the temperature or (b) increasing the total pressure?

EXERCISE 10.12 Partial Pressures

A 355-mL flask contains 0.146 g neon gas, Ne, and an unknown amount of argon gas, Ar, at 35 °C and a total pressure of 626 mm Hg. How many grams of Ar are in the flask?

Collecting Gases over Water

While studying chemical reactions that produce a gas as a product, it is often necessary to determine the number of moles of the product gas. One convenient way to do this, for gases that are insoluble in water, involves collecting the gas over water (Figure 10.14). The gas bubbles through the water and is collected by displacing the water in an inverted vessel that was initially filled with water. The levels of the water inside and outside the collection vessel are made equal at the end of the experiment. This ensures that the total pressure inside the vessel is equal to the barometric pressure in the laboratory. The volume of gas collected is then determined. Because some of the water evaporates, forming water vapor, the total pressure of the mixture of gases equals the partial pressure of the gas being collected plus the partial pressure of the water vapor. The partial pressure of water in the gaseous mixture is the vapor pressure of water, and it depends on the temperature of the liquid water (Table 10.4).

To calculate the amount of gaseous product, we first subtract the vapor pressure of the water from the total gas pressure (the barometric pressure), which yields the partial pressure of the gaseous product. Substituting this partial pressure and the

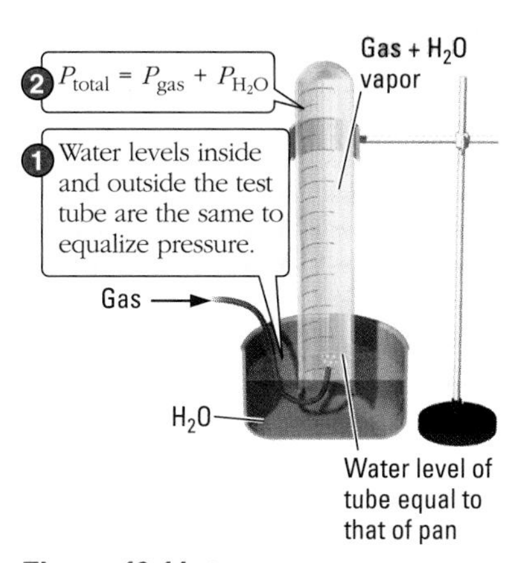

Gas + H_2O vapor
2 $P_{total} = P_{gas} + P_{H_2O}$
1 Water levels inside and outside the test tube are the same to equalize pressure.
Gas
H_2O
Water level of tube equal to that of pan

Figure 10.14 Collecting a gas over water.

Table 10.4 Vapor Pressure of Water at Different Temperatures

T (°C)	P_{water} (mm Hg)
0	4.6
10	9.2
20	17.5
21	18.6
22	19.8
23	21.1
24	22.4
25	23.8
30	31.8
35	42.2
40	55.3
50	92.5
60	149.4
70	233.7
80	355.1
90	525.8
100	760.0

known volume and temperature into the ideal gas law allows the calculation of the number of moles of the gaseous product.

PROBLEM-SOLVING EXAMPLE 10.11 Collecting a Gas over Water

Before electric lamps were available, the reaction between calcium carbide, CaC_2, and water was used to produce acetylene, C_2H_2, to generate a bright flame in miners' lamps.

$$CaC_2(s) + 2\,H_2O(\ell) \longrightarrow C_2H_2(g) + Ca(OH)_2(aq)$$

At a barometric pressure of 745 mm Hg and a temperature of 24.0 °C, 625 mL of gas was collected over water. How many milligrams of acetylene were produced? The vapor pressure of water at 24.0 °C is 22.4 mm Hg.

Answer 635 mg acetylene

Strategy and Explanation We will need to calculate the number of moles and then milligrams of acetylene, so we must first find the partial pressure of the acetylene. The total pressure is 745 mm Hg, and the vapor pressure of water at 24.0 °C is 22.4 mm Hg, so the partial pressure of acetylene is

$$P_{acetylene} = P_{total} - P_{water} = 745 \text{ mm Hg} - 22.4 \text{ mm Hg} = 723 \text{ mm Hg}$$

Now we calculate the molar amount and mass of acetylene

$$n = \frac{PV}{RT} = \frac{(723/760 \text{ atm})(0.625 \text{ L})}{(0.0821 \text{ L atm mol}^{-1}\text{ K}^{-1})(273.15 + 24.0) \text{ K}}$$

$$= 0.0244 \text{ mol acetylene}$$

$$0.0244 \text{ mol} \times 26.04 \text{ g/mol} = 0.635 \text{ g} = 635 \text{ mg acetylene}$$

☑ **Reasonable Answer Check** The quantities entered into the ideal gas equation can be approximated as $n = \frac{(1)(0.6)}{(0.082)(300)} = 0.024$ mol, which is close to our more exact answer.

PROBLEM-SOLVING PRACTICE 10.11

Zinc metal reacts with HCl to produce hydrogen gas, H_2.

$$2\,HCl(aq) + Zn(s) \longrightarrow ZnCl_2(aq) + H_2(g)$$

The H_2 can be collected over water. If you collected 260. mL H_2 at 23 °C, and the total pressure was 740. mm Hg, how many milligrams of H_2 were collected?

H—C≡C—H

acetylene

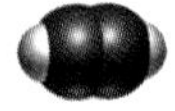

10.8 The Behavior of Real Gases

The ideal gas law provides accurate predictions for the pressure, volume, temperature, and amount of a gas well above its boiling point. At STP (0 °C and 1 atm), most gases deviate only slightly from ideal behavior. At much higher pressures or much lower temperatures, however, the ideal gas law does not work nearly as well. We can illustrate this departure from ideality by plotting PV/nRT for a gas as a function of pressure (Figure 10.15). For one mole of a gas that follows ideal gas behavior, the ratio PV/nRT must equal 1. But Figure 10.15 shows that for real gases, the ratio PV/nRT deviates from 1, first dipping to values lower than 1 and then rising to values higher than 1 as pressure increases. Thus, the measured volume of the real gas is smaller than expected at medium pressures, and it is higher than expected at high pressures.

To see what causes these deviations, we must revisit two of the fundamental concepts of kinetic-molecular theory (p. 345). The theory assumes that the molecules of a gas occupy no volume themselves and that gas molecules do not attract one another.

At STP, the volume occupied by a single molecule is very small relative to its share of the total gas volume. Recall that there are 6.02×10^{23} molecules in a mole and that

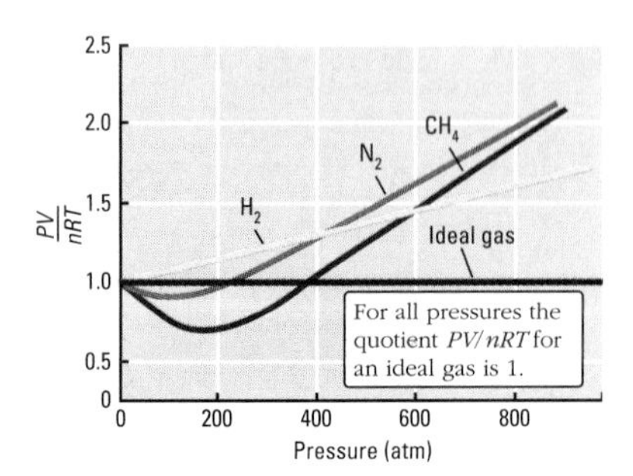

Figure 10.15 The nonideal behavior of real gases compared with that of an ideal gas.

1 mol of a gas occupies about 22.4 L (22.4×10^{-3} m^3) at STP (⇐ p. 352). The volume, V, that each molecule has to move around in is given by

$$V = \frac{22.4 \times 10^{-3}\,\text{m}^3}{6.02 \times 10^{23}\,\text{molecules}} = 3.72 \times 10^{-26}\,\text{m}^3/\text{molecule}$$

The volume of a sphere is given by $\frac{4}{3}\pi r^3$. Solving for r gives

$$r = \sqrt[3]{\frac{3V}{4\pi}}$$

$$= \sqrt[3]{\frac{3(3.72 \times 10^{-26}\,\text{m}^3)}{4(3.14)}}$$

$$r = 2.07 \times 10^{-9}\,\text{m} = 2070\,\text{pm}$$

If this volume is assumed to be a sphere, then the radius, r, of the sphere is about 2000 pm. The radius of the smallest gas molecule, the helium atom, is 31 pm (⇐ p. 255), so a helium atom has a substantial volume of space to move around. Now suppose the pressure is increased significantly, to 1000 atm. The volume available to each molecule is now a sphere only about 200 pm in radius, which means the helium atom has a much smaller volume of space to move around. The volume occupied by the gas molecules themselves relative to the volume of the sphere is no longer negligible. This violates the first concept of the kinetic-molecular theory. The kinetic-molecular theory and the ideal gas law deal with the volume available for the molecules to move around in, not the volume of the molecules themselves. However, the measured volume of the gas must include both. Therefore, at very high pressures, the measured volume will be larger than predicted by the ideal gas law, and the value of PV/nRT will be greater than 1.

At medium pressures, the product PV for a real gas is smaller than the predicted value of nRT for an ideal gas (Figure 10.15). For a given volume of gas, the pressure exerted against the container walls is smaller than expected. Consider a gas molecule that is about to hit the wall of the container, as shown in Figure 10.16. The kinetic-molecular theory assumes that the other molecules exert no forces on the molecule, but in fact such forces do exist (⇐ p. 329). Their influence increases as higher pressures push gas molecules closer to each other. This means that when a molecule is about to hit the wall of the container, the other molecules tend to pull it *away* from the wall, which causes the molecule to hit the wall with less impact. The collision is softer than if there were no attraction among the molecules. Since all collisions with the walls are softer, the internal pressure is less than that predicted by the ideal gas law, and PV/nRT is less than 1. As the external pressure increases, the gas volume decreases, the molecules are squeezed closer together, and the attraction among the molecules grows stronger, which increases this deviation from ideal behavior. Eventually, the pressure gets so high that the loss of empty space becomes the dominating factor, and PV/nRT becomes larger than 1 (Figure 10.15).

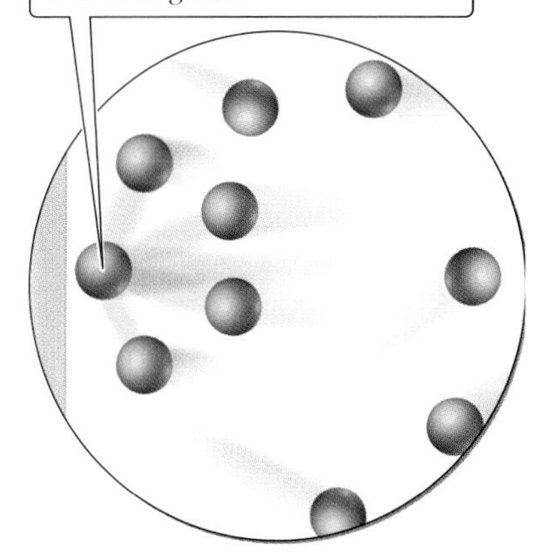

A gas molecule strikes the walls of a container with less force due to the attractive forces between it and its neighbors.

Figure 10.16 Nonideal gas behavior.

The result of these considerations is that the measured pressure for real gases is less than the ideal pressure due to intermolecular interactions, and the measured volume is larger than the ideal volume due to the volume of the molecules themselves. Thus, the ideal pressure is *larger* than the measured pressure, and the ideal volume is *smaller* than the measured volume.

The **van der Waals equation** predicts the behavior of n moles of a real gas quantitatively by taking these two effects into account. It assumes that $P_{\text{ideal}} V_{\text{ideal}} = nRT$, but expresses the left-hand side in terms of *measured* P and V.

$$\left(P_{\text{measured}} + \frac{n^2 a}{V^2_{\text{measured}}}\right)(V_{\text{measured}} - nb) = nRT$$

Correction for molecular attraction (adjusts measured P up)

Correction for volume of molecules (adjusts measured V down)

The measured volume is decreased by the factor nb, which accounts for the volume occupied by n moles of the gas molecules themselves. The van der Waals constant b has units of L/mol, and it is larger for larger molecules. The measured pressure is increased by the factor n^2a/V^2, which accounts for the attractive forces between molecules (Figure 10.16). The van der Waals constant a has units of L^2 atm/mol^2. The a pressure correction term has this form because the attractive forces being accounted

Table 10.5 Van der Waals Constants for Some Common Gases

Gas	$a\left(\frac{\text{L}^2\,\text{atm}}{\text{mol}^2}\right)$	$b\left(\frac{\text{L}}{\text{mol}}\right)$
He	0.034	0.0237
Ne	0.211	0.0171
Ar	1.35	0.0322
H_2	0.244	0.0266
N_2	1.39	0.0391
O_2	1.36	0.0318
Cl_2	6.49	0.0562
CO_2	3.59	0.0427
CH_4	2.25	0.0428
NH_3	4.17	0.0371
H_2O	5.46	0.0305

for depend on interactions between *pairs* of molecules; the a term is therefore proportional to the square of the number of molecules per unit volume, $(n/V)^2$. The value of a shows how strongly the gas molecules attract one another; the stronger the intermolecular interactions, the larger the value of a. The constants a and b are different for each gas (Table 10.5) and must be determined experimentally. In general, the values of a and b increase as the gas molecule's size increases and as the molecule's complexity increases.

Note that for an ideal gas, where the kinetic-molecular theory of gases holds, the attractive forces between molecules are negligible, so $a \approx 0$, and $P_{measured} = P_{ideal}$. The volume of the molecules themselves is negligible compared with the container volume, so $b \approx 0$ and $V_{measured} = V_{ideal}$. Therefore, for ordinary conditions of P and T, the van der Waals equation simplifies to the ideal gas law.

10.9 Atmospheric Carbon Dioxide, the Greenhouse Effect, and Global Warming

The greenhouse effect derives its name by analogy with a botanical greenhouse. However, the warming in a botanical greenhouse is much more dependent on the glass reducing convection than on blocking infrared radiation from leaving through the glass.

An important function of our atmosphere is to moderate the earth's surface temperature. Most of the energy that heats the earth comes from the sun as electromagnetic radiation. Some of the electromagnetic radiation from the sun is reflected by the atmosphere back into space, and some is absorbed by the atmosphere. The remainder reaches the earth, warming its surface and oceans. The warmed surfaces then reradiate this energy into the troposphere as infrared radiation (**p. 222**). Carbon dioxide, water vapor, methane, and ozone all absorb radiation in various portions of the infrared region. By absorbing this reradiated energy they warm the atmosphere, creating what is called the **greenhouse effect** (Figure 10.17). Thus, all four are "greenhouse gases." Such gases constitute an absorbing blanket that reduces the quantity of energy radiated back into space. Thanks to the greenhouse effect, the earth's average temperature is a com-

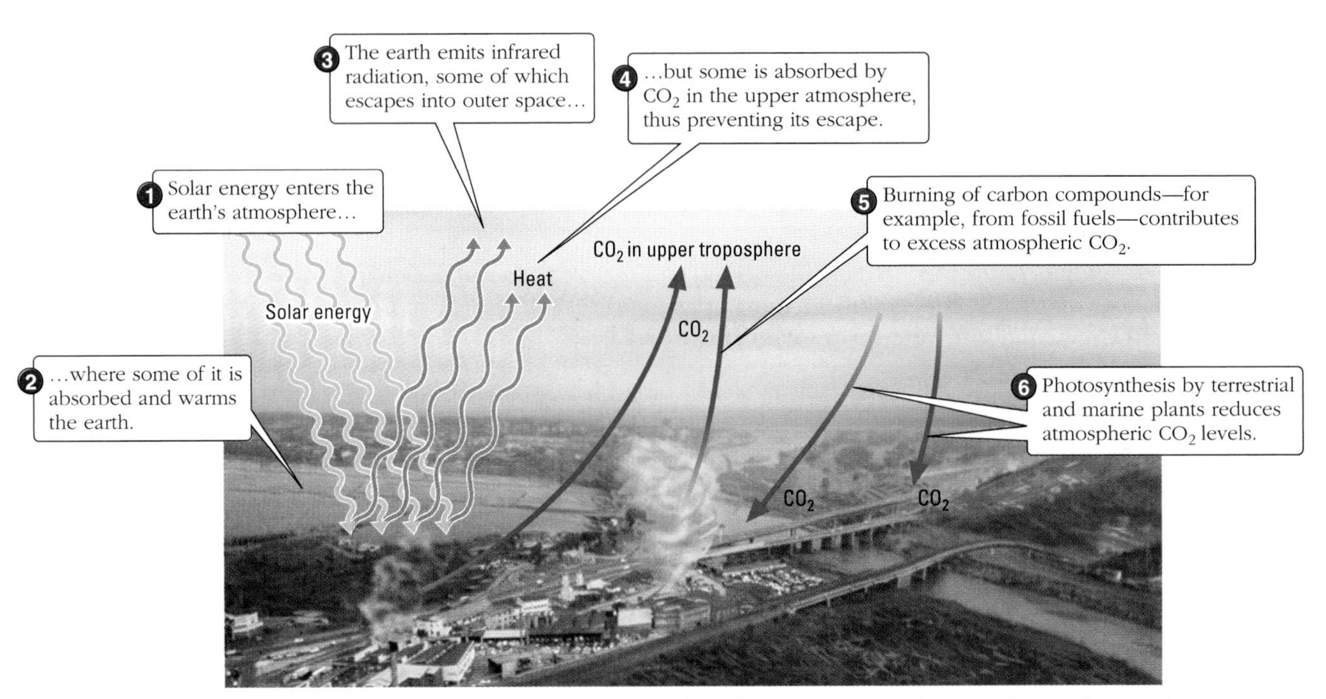

Figure 10.17 The greenhouse effect. Greenhouse gases form an effective barrier that prevents some heat from escaping the earth's surface. Without the greenhouse effect, earth's average temperature would be much lower.

fortable 15 °C (59 °F). By comparison, the moon, with no moderating atmosphere, has a surface mean temperature that fluctuates between approximately 107 °C at daytime and −153 °C at night.

Earth has such a vast reservoir of water in the oceans that human activity has a negligible influence on the concentration of water vapor in the atmosphere. In addition, methane is produced by natural processes in such large quantities that human contributions are negligible. Ozone is present in such small concentrations that its contribution to the greenhouse effect is small. So among the four greenhouse gases, most attention is focused on CO_2.

Carbon dioxide is a greenhouse gas because it absorbs infrared radiation, causing C=O bonds in the molecule to stretch and bend. Stretching and compressing the C=O bonds requires more energy (shorter wavelength) than does bending them. Thus, C=O stretching and compressing occurs when infrared radiation with a wavelength of 4.257 μm is absorbed, whereas the C=O bending vibrations occur at lower energy (longer wavelength) when the molecule absorbs 15.000-μm infrared radiation.

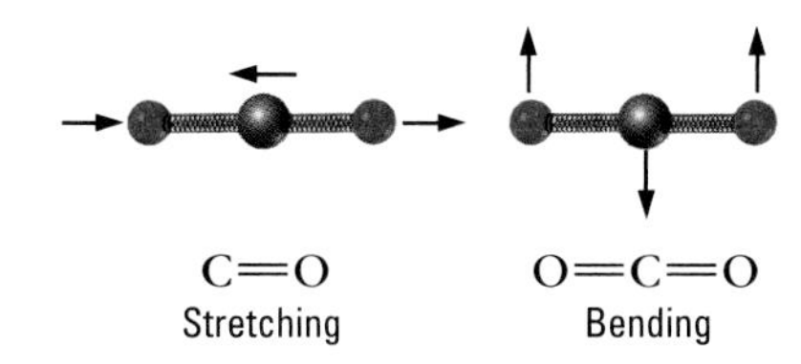

$1\ \mu m = 1 \times 10^{-6}\ m = 1 \times 10^{3}\ nm$

Recall from Section 7.1 that energy and wavelength are inversely related.

Each year combustion of fossil fuels (coal, petroleum, natural gas) worldwide puts billions of metric tons of carbon into the atmosphere as CO_2. About 45% is removed from the atmosphere by natural processes—some by plants during photosynthesis and the rest by dissolving in rainwater and the oceans to form hydrogen carbonates and carbonates.

$$CO_2(g) + 2\ H_2O(\ell) \longrightarrow H_3O^+(aq) + HCO_3^-(aq)$$

$$HCO_3^-(aq) + H_2O(\ell) \longrightarrow H_3O^+(aq) + CO_3^{2-}(aq)$$

The other 55% of the carbon dioxide from fossil fuel combustion remains in the atmosphere, increasing the global CO_2 concentration.

Worldwide, nearly one third of all atmospheric CO_2 is released as a by-product from fossil fuel–burning electric power plants.

CONCEPTUAL
EXERCISE 10.13 Sources of CO_2

List as many natural sources of CO_2 as you can. List as many sources of CO_2 from human activities as you can.

Without human influences, the flow of carbon dioxide among the air, plants, animals, and the oceans would be roughly balanced. In 1750, during the preindustrial era, CO_2 concentration in the atmosphere was 277 parts per million (ppm). During the next 130 years, as the Industrial Revolution progressed, the concentration increased to 291 ppm, a 5% increase. Since 1900, however, the rate of increase of CO_2 has reflected the rapid increase in the use of fossil fuel combustion for industrial and domestic purposes, especially for motor vehicle transportation. Between 1958 and 2006, the atmospheric concentration of CO_2 increased from 315 to 379 ppm, an increase of 20% (Figure 10.18). Population pressure is also contributing heavily to increased CO_2 concentrations. In the Amazon region of Brazil, for example, forests are being cut and burned to create cropland. This activity places a double burden on the natural CO_2 cycle, since there are fewer trees to use the CO_2 in photosynthesis and, at the same time, CO_2 is added to the atmosphere during burning.

Parts per million (ppm) is a convenient way to express low concentrations. One ppm means one part of something in one million parts. A CO_2 concentration of 360 ppm means that for every million molecules of air, 360 are CO_2 molecules.

Expectations are that the atmospheric CO_2 concentration will continue to increase at the current rate of about 1.5 ppm per year, due principally to industrialization, agricultural production, and an expanding global population. As a result, according to sophisticated computer models of climate change, the CO_2 concentration is projected to reach 550 ppm, double its preindustrial value, between 2030 and 2050. The increase may perhaps not be so great if fuel costs increase substantially as fossil fuel supplies get tighter.

Clearing a Brazilian rain forest.

The Intergovernmental Panel on Climate Change (IPCC) is a prestigious multinational group of climate scientists organized by the United Nations. In the *IPCC Fourth*

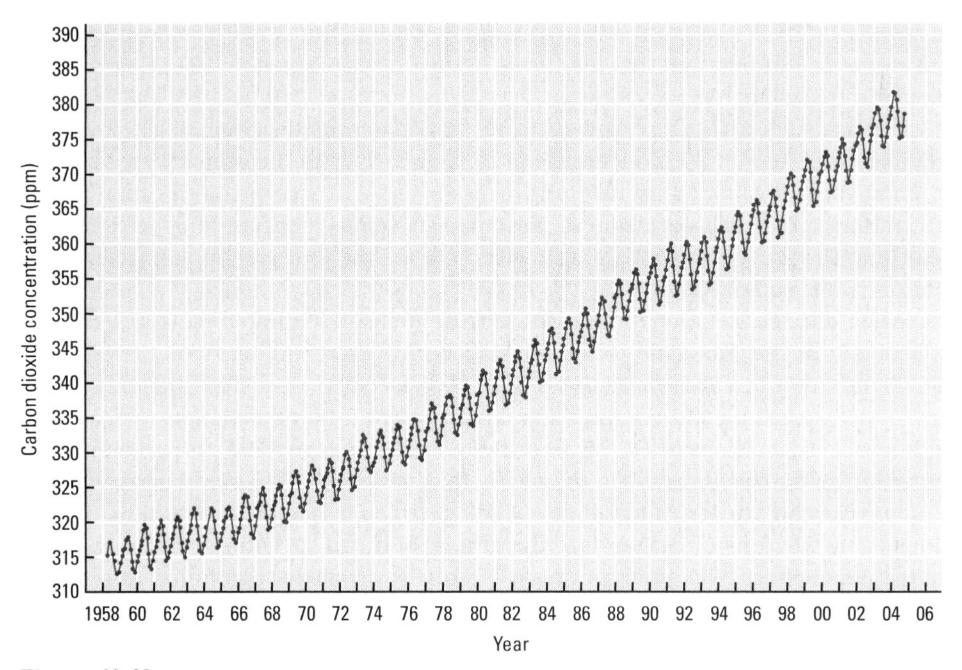

Figure 10.18 Atmospheric carbon dioxide concentration measured at Mauna Loa, Hawaii. These data have been carefully collected by a standardized methodology for 47 years, and they are the longest continuous record of atmospheric CO_2 concentrations available. The Mauna Loa record shows a 20% increase in CO_2 concentration, from 316 ppm to 379 ppm, over the period 1959 to 2006. *Source:* **http://cdiac.esd.ornl.gov/trends/co2/sio-mlo.htm.**

Assessment Report of 2007, the IPCC noted that, "Atmospheric concentrations of CO_2 (379 ppm) . . . in 2006 exceed by far the natural range over the last 650,000 years." The IPCC noted earlier that nearly 75% of anthropogenic CO_2 emissions in the past 20 years are due to the burning of fossil fuels.

To see how easily everyday activities affect the quantity of CO_2 being released into the atmosphere, consider a round-trip flight from New York to Los Angeles. Each passenger pays for about 200 gal jet fuel, which weighs 1400 lb. When burned, each pound of jet fuel produces about 3.14 lb carbon dioxide. So 4400 lb, or 2 metric tons, of carbon dioxide are produced per passenger during that trip.

CONCEPTUAL
EXERCISE 10.14 Annual CO_2 Changes

During a year, atmospheric CO_2 concentration fluctuates, building up to a high value, then dropping to a low value before building up again (Figure 10.18). Explain what causes this fluctuation and when during the year the high and the low occur in the Northern Hemisphere.

Global warming is an average increase in the temperature of the atmosphere. It is attributed to an increased greenhouse effect due to increasing concentrations of CO_2 and other greenhouse gases. Figure 10.19 shows global surface temperature change over the past 160 years, during which the average global surface temperature rose 0.6 to 0.8 °C. There is a parallel relationship between changes in global surface temperature and atmospheric CO_2 concentration: *As atmospheric CO_2 levels have increased, average global temperatures have increased.*

Many computer models predict that increasing atmospheric CO_2 to 600 ppm will increase average global temperature, and most climate experts who do global warming research agree. What is uncertain is the extent of the temperature increase, with

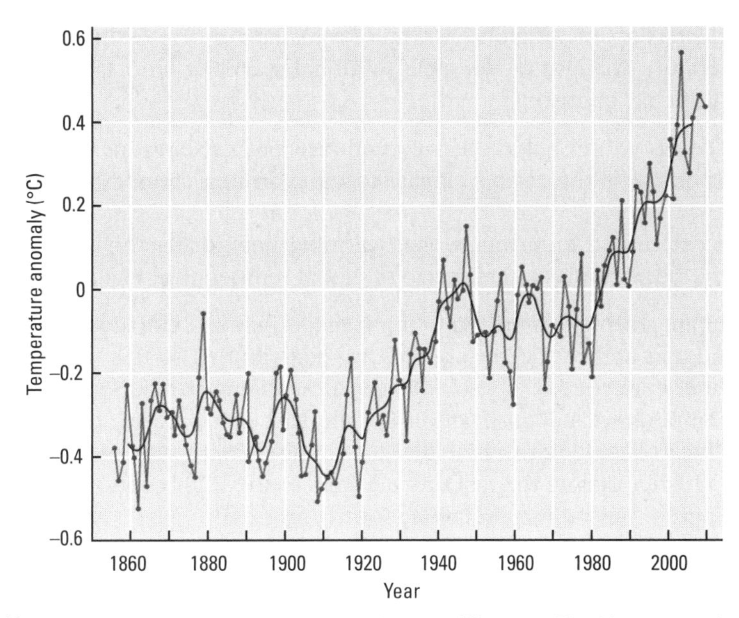

Figure 10.19 Global temperature change over the past 160 years. The blue points show the annual average temperatures, and the red line shows the five-year average temperatures. Temperature anomaly refers to the annual temperature minus the average temperature during the period 1961–1990. The global average temperature has risen approximately 0.6 to 0.8 °C over the past 160 years. *Source:* **http://en.wikipedia.org/wiki/Global_warming.**

estimates varying from 1.5 to 4.5 °C (2.7 to 8.1 °F). The IPCC estimates a 1.0 to 3.5 °C (2 to 6 °F) increase by the year 2100. Warming by as little as 1.5 °C would produce the warmest climate seen on earth in the past 6000 years; an increase of 4.5 °C would produce world temperatures higher than any since the Mesozoic era, the time of the dinosaurs.

A 1.5 °C rise in global temperature is a significant change, requiring a very large input of energy.

The 2007 Nobel Peace Prize was awarded to former Vice President Al Gore and the IPCC "for their efforts to build up and disseminate greater knowledge about man-made climate change."

Many scientists are concerned that rising temperatures are causing ice caps in Greenland and Antarctica to melt more rapidly, raising the sea level. In conjunction with this, atmospheric currents will change and produce significant changes in weather and agricultural productivity. Since the 1940s, average summertime temperatures in Antarctica have increased 2.5 °C to just above 0 °C.

Computer models used to predict future global temperature changes have become more sophisticated and accurate over the past decade. When factors in addition to greenhouse gases, such as the presence of aerosols (tiny, suspended particles) and changes in the sun's irradiance are taken into account, the computer models predict temperature changes close to what has been observed since 1860. Such work adds credibility to the predictions of temperature increases of 1.5 to 4.5 °C due to global warming.

Global warming has become an acknowledged worldwide problem. Concerns about global warming prompted delegates from 159 countries to convene in 1997 in Kyoto, where the Kyoto Protocols were negotiated. Goals were set to reduce greenhouse gas emissions below 1990 levels. Enough countries ratified these protocols to make them official in February 2005. The only industrialized country that did not ratify them was the United States. The Kyoto protocols are due to expire in 2012. In December 2007 delegates from around the world met in Bali, Indonesia, to negotiate a successor to the Kyoto Protocols.

The most obvious measure to reduce global warming is to control anthropogenic CO_2 emissions throughout the world. Given our dependence on fossil fuels, however, getting and keeping these emissions under control will be very difficult. One direct way to decrease the quantity of atmospheric CO_2 is to plant more trees and to replace those in deforested areas.

The projected extent of global warming remains a thorny issue. The uncertainty in the temperature changes attributable to global warming is related to three fundamental, but difficult, questions:

1. To what levels will CO_2 and other greenhouse gases rise during the next decades?
2. How responsive is the earth's climate to the warming created by the greenhouse effect?
3. To what extent is the current global warming being caused by human activities, or could it be part of a natural cycle of global temperature changes?

Most climate scientists worldwide think that enhanced global warming is already occurring and that anthropogenic sources are contributing to this warming. In 2006, an expert panel convened by the U.S. National Academy of Sciences, the Committee of Surface Temperature Reconstruction for the Last 2,000 Years, produced a report that stated, "It can be said with a high level of confidence that global mean surface temperature was higher during the last few decades of the 20th century than during any comparable period during the preceding four centuries."

IN CLOSING

Having studied this chapter, you should be able to . . .

- Describe the components of the atmosphere (Section 10.1). ■ End-of-chapter Question assignable in OWL: 8
- Describe the properties of gases (Section 10.2). ■ Questions 11, 66
- State the fundamental concepts of the kinetic-molecular theory and use them to explain gas behavior (Section 10.3). ■ Question 17
- Solve problems using the appropriate gas laws (Section 10.4). ■ Questions 22, 25, 29, 31
- Calculate the quantities of gaseous reactants and products involved in chemical reactions (Section 10.5). ■ Questions 38, 40, 42
- Apply the ideal gas law to finding gas densities and molar masses (Section 10.6). ■ Questions 44, 46
- Perform calculations using partial pressures of gases in mixtures (Section 10.7). ■ Questions 49, 51
- Describe the differences between real and ideal gases (Section 10.8). ■ Question 58
- Relate atmospheric CO_2 concentration to the greenhouse effect and to enhanced global warming (Section 10.9). ■ Question 60

Selected end-of-chapter Questions may be assigned in OWL.

For quick review, download Go Chemistry mini-lecture flashcard modules (or purchase them at **www.ichapters.com**).

Prepare for an exam with a **Summary Problem** that brings together concepts and problem-solving skills from throughout the chapter. Go to **www.cengage.com/chemistry/moore** to download this chapter's Summary Problem from the Student Companion Site.

KEY TERMS

absolute temperature scale *(Section 10.4)*

absolute zero *(10.4)*

Avogadro's law *(10.4)*

bar *(10.2)*

barometer *(10.2)*

Boyle's law *(10.4)*

Charles's law *(10.4)*

combined gas law *(10.4)*

Dalton's law of partial pressures *(10.7)*

diffusion *(10.3)*

effusion *(10.3)*

ideal gas *(10.4)*

ideal gas constant *(10.4)*

ideal gas law *(10.4)*

global warming *(10.9)*

greenhouse effect *(10.9)*

Kelvin temperature scale *(10.4)*

law of combining volumes *(10.4)*

millimeters of mercury (mm Hg) *(10.2)*

mole fraction *(10.7)*

newton (N) *(10.2)*

partial pressure *(10.7)*

pascal (Pa) *(10.2)*

pressure *(10.2)*

standard atmosphere (atm) *(10.2)*

standard molar volume *(10.4)*

standard temperature and pressure (STP) *(10.4)*

stratosphere *(10.1)*

torr *(10.2)*

troposphere *(10.1)*

van der Waals equation *(10.8)*

QUESTIONS FOR REVIEW AND THOUGHT

■ denotes questions assignable in OWL.

Blue-numbered questions have short answers at the back of this book and fully worked solutions in the *Student Solutions Manual.*

Review Questions

1. What are the conditions represented by STP?
2. What is the volume occupied by 1 mol of an ideal gas at STP?
3. State Avogadro's law. Explain why two volumes of hydrogen react with one volume of oxygen to form two volumes of steam.
4. State Dalton's law of partial pressures. If the air we breathe is 78% N_2 and 22% O_2 on a mole basis, what is the mole fraction of O_2? What is the partial pressure of O_2 if the total pressure is 720 mm Hg?
5. Explain Boyle's law on the basis of the kinetic-molecular theory.
6. Explain why gases at low temperature and high pressure do not obey the ideal gas equation as well as gases at high temperature and low pressure.
7. Gaseous water and carbon dioxide each absorb infrared radiation. Does either of them absorb ultraviolet radiation? Explain your answer.

Topical Questions

The Atmosphere

8. ■ Convert all of the "percentage by volume" figures in Table 10.1 into (a) parts per million and (b) parts per billion. Which atmospheric gases are present at concentrations of less than 1 ppb? Between 1 ppb and 1 ppm? Greater than 1 ppm?
9. The mass of the earth's atmosphere is 5.3×10^{15} metric tons. The atmospheric abundance of helium is 0.7 ppm when expressed as a fraction by *weight* instead of by *volume,* as in Table 10.1. How many metric tons of helium are there in the atmosphere? How many moles of helium is this?
10. ■ Sulfur is about 2.5% of the mass of coal, and when coal is burned the sulfur is all converted to SO_2. In one year, 3.1×10^9 metric tons of coal was burned worldwide. How many tons of SO_2 were added to the atmosphere? How many tons of SO_2 are currently in the atmosphere? (*Note:* The weight fraction of SO_2 in air is 0.4 ppb.)

Properties of Gases

11. ■ Gas pressures can be expressed in units of mm Hg, atm, torr, and kPa. Convert these pressure values.
    (a) 720. mm Hg to atm (b) 1.25 atm to mm Hg
    (c) 542. mm Hg to torr (d) 740. mm Hg to kPa
    (e) 700. kPa to atm
12. Convert these pressure values.
    (a) 120. mm Hg to atm (b) 2.00 atm to mm Hg
    (c) 100. kPa to mm Hg (d) 200. kPa to atm
    (e) 36.0 kPa to atm (f) 600. kPa to mm Hg
13. Mercury has a density of 13.96 g/cm^3. A barometer is constructed using an oil with a density of 0.75 g/cm^3. If the atmospheric pressure is 1.0 atm, what will be the height in meters of the oil column in the barometer?
14. A vacuum pump is connected to the top of an upright tube whose lower end is immersed in a pool of mercury. How high will the mercury rise in the tube when the pump is turned on?

Kinetic-Molecular Theory

15. List the five basic concepts of the kinetic-molecular theory. Which assumption is incorrect at very high pressures? Which one is incorrect at low temperatures? Which assumption is probably most nearly correct?
16. You are given two flasks of equal volume. Flask A contains H_2 at 0 °C and 1 atm pressure. Flask B contains CO_2 gas at 0 °C and 2 atm pressure. Compare these two samples with respect to each of these properties.
    (a) Average kinetic energy per molecule
    (b) Average molecular velocity
    (c) Number of molecules
17. ■ Place these gases in order of increasing average molecular speed at 25 °C: Kr, CH_4, N_2, CH_2Cl_2.
18. Arrange these four gases in order of increasing average molecular speed at 25 °C: Cl_2, F_2, N_2, O_2.
19. If equal amounts of the four inert gases Ar, Ne, Kr, and Xe are released at the same time at the end of a long tube, which gas will reach the end of the tube first?

Gas Behavior and the Ideal Gas Law

20. How many moles of CO are present in 1.0 L air at STP that contains 950 ppm CO?
21. A sample of gaseous CO exerts a pressure of 45.6 mm Hg in a 56.0-L flask at 22 °C. If the gas is released into a 2.70×10^4-L room, what is the partial pressure of the CO in the room at 22 °C?

22. ■ A sample of a gas has a pressure of 100. mm Hg in a 125-mL flask. If this gas sample is transferred to another flask with a volume of 200. mL, what will be the new pressure? Assume that the temperature remains constant.
23. A sample of gas has a pressure of 62 mm Hg in a 100-mL flask. This sample of gas is transferred to another flask, where its pressure is 29 mm Hg. What is the volume of the new flask? (The temperature does not change.)
24. A sample of gas at 30. °C has a pressure of 2.0 atm in a 1.0-L container. What pressure will it exert in a 4.0-L container? The temperature does not change.
25. ■ Suppose you have a sample of CO_2 in a gas-tight syringe with a movable piston. The gas volume is 25.0 mL at a room temperature of 20. °C. What is the final volume of the gas if you hold the syringe in your hand to raise the gas temperature to 37. °C?
26. A balloon is inflated with helium to a volume of 4.5 L at 23 °C. If you take the balloon outside on a cold day (−10. °C), what will be the new volume of the balloon?
27. A sample of gas has a volume of 2.50 L at a pressure of 670. mm Hg and a temperature of 80. °C. If the pressure remains constant but the temperature is decreased, the gas occupies 1.25 L. What is this new temperature, in degrees Celsius?
28. A sample of 9.0 L CO_2 at 20 °C and 1 atm pressure is cooled so that it occupies a volume of 8.0 L at some new temperature. The pressure remains constant. What is the new temperature, in kelvins?
29. ■ A sample of gas occupies 754 mL at 22 °C and a pressure of 165 mm Hg. What is its volume if the temperature is raised to 42 °C and the pressure is raised to 265 mm Hg? (The number of moles does not change.)
30. A balloon is filled with helium to a volume of 1.05×10^3 L on the ground, where the pressure is 745 mm Hg and the temperature is 20. °C. When the balloon ascends to a height of 2 miles, where the pressure is only 600. mm Hg and the temperature is −33 °C, what is the volume of the helium in the balloon?
31. ■ What is the pressure exerted by 1.55 g Xe gas at 20. °C in a 560-mL flask?
32. A 1.00-g sample of water is allowed to vaporize completely inside a 10.0-L container. What is the pressure of the water vapor at a temperature of 150. °C?
33. Which of these gas samples contains the largest number of molecules and which contains the smallest?
    (a) 1.0 L H_2 at STP
    (b) 1.0 L N_2 at STP
    (c) 1.0 L H_2 at 27 °C and 760. mm Hg
    (d) 1.0 L CO_2 at 0 °C and 800. mm Hg

Quantities of Gases in Chemical Reactions

34. When a commercial drain cleaner containing sodium hydroxide and small pieces of aluminum is poured into a clogged drain, this reaction occurs:

$$2\,Al(s) + 2\,NaOH(aq) + 6\,H_2O(\ell) \longrightarrow 2\,NaAl(OH)_4(aq) + 2\,H_2(g)$$

If 6.5 g Al and excess NaOH are reacted, what volume of H_2 gas measured at 742 mm Hg and 22.0 °C will be produced?

35. If 2.7 g Al metal is reacted with excess HCl, how many liters of H_2 gas are produced at 25 °C and 1.00 atm pressure?
36. The yeast in rising bread dough converts sugar (sucrose, $C_{12}H_{22}O_{11}$) into carbon dioxide. A popular recipe for two loaves of French bread requires 1 package of yeast and $\frac{1}{4}$ teaspoon (about 2.4 g) of sugar. What volume of CO_2 at STP is produced by the complete conversion of this quantity of sucrose into CO_2 by the yeast? Compare this volume with the typical volume of two loaves of bread.
37. Water can be made by combining gaseous O_2 and H_2. If you begin with 1.5 L $H_2(g)$ at 360. mm Hg and 23 °C, what volume in liters of $O_2(g)$ will you need for complete reaction if the O_2 gas is also measured at 360. mm Hg and 23 °C?
38. ■ Gaseous silane, SiH_4, ignites spontaneously in air according to the equation

$$SiH_4(g) + 2\,O_2(g) \longrightarrow SiO_2(s) + 2\,H_2O(g)$$

If 5.2 L SiH_4 is treated with O_2, what volume in liters of O_2 is required for complete reaction? What volume of H_2O vapor is produced? Assume all gases are measured at the same temperature and pressure.

39. Hydrogen can be made in the "water gas reaction."

$$C(s) + H_2O(g) \longrightarrow H_2(g) + CO(g)$$

If you begin with 250. L gaseous water at 120. °C and 2.00 atm and excess C(s), what mass in grams of H_2 can be prepared?

40. ■ If boron hydride, B_4H_{10}, is treated with pure oxygen, it burns to give B_2O_3 and H_2O.

$$2\,B_4H_{10}(s) + 11\,O_2(g) \longrightarrow 4\,B_2O_3(s) + 10\,H_2O(g)$$

If a 0.050-g sample of the boron hydride burns completely in O_2, what will be the pressure of the gaseous water in a 4.25-L flask at 30. °C?

41. Metal carbonates decompose to the metal oxide and CO_2 on heating according to this general equation.

$$M_x(CO_3)_y(s) \longrightarrow M_xO_y(s) + y\,CO_2(g)$$

You heat 0.158 g of a white, solid carbonate of a Group 2A metal and find that the evolved CO_2 has a pressure of 69.8 mm Hg in a 285-mL flask at 25 °C. What is the molar mass of the metal carbonate?

42. Nickel carbonyl, $Ni(CO)_4$, can be made by the room-temperature reaction of finely divided nickel metal with gaseous CO. This is the basis for purifying nickel on an industrial scale. If you have CO in a 1.50-L flask at a pressure of 418 mm Hg at 25.0 °C, what is the maximum mass in grams of $Ni(CO)_4$ that can be made?
43. Zinc reacts with aqueous sulfuric acid to form hydrogen gas.

$$Zn(s) + H_2SO_4(aq) \longrightarrow ZnSO_4(aq) + H_2(g)$$

In an experiment, 201 mL wet H_2 is collected over water at 27 °C and a barometric pressure of 733 torr. The vapor pressure of water at 27 °C is 26.74 torr. How many grams of H_2 are produced?

Gas Density and Molar Mass

44. ■ What is the molar mass of a gas that has a density of 5.75 g/L at STP?
45. To find the volume of a flask, it is first evacuated so it contains no gas. Next, 4.4 g CO_2 is introduced into the

flask. On warming to 27 °C, the gas exerts a pressure of 730. mm Hg. What is the volume of the flask in milliliters?

46. ■ What mass of helium in grams is required to fill a 5.0-L balloon to a pressure of 1.1 atm at 25 °C?
47. Consider two 5.0-L containers, each filled with gas at 25 °C. One container is filled with helium and the other with N_2. The density of gas in the two containers is the same. What is the relationship between the pressures in the two containers?
48. Forty miles above the earth's surface the temperature is −23 °C, and the pressure is only 0.20 mm Hg. What is the density of air (molar mass = 29.0 g/mol) at this altitude?

Partial Pressures of Gases

49. ■ What is the total pressure exerted by a mixture of 1.50 g H_2 and 5.00 g N_2 in a 5.00-L vessel at 25 °C?
50. At 298 K, a 750-mL vessel contains equimolar amounts of O_2, H_2, and He at a total pressure of 3.85 atm. What is the partial pressure of the H_2 gas?
51. ■ A sample of the atmosphere at a total pressure of 740. mm Hg is analyzed to give these partial pressures: $P(N_2)$ = 575 mm Hg; $P(Ar)$ = 6.9 mm Hg; $P(CO_2)$ = 0.2 mm Hg; $P(H_2O)$ = 4.0 mm Hg. No other gases except O_2 have appreciable partial pressures.
    (a) What is the partial pressure of O_2?
    (b) What is the mole fraction of each gas?
    (c) What is the composition of each component of this sample in percentage by volume? Compare your results with those of Table 10.1.
52. Gaseous CO exerts a pressure of 45.6 mm Hg in a 56.0-L tank at 22.0 °C. If this gas is released into a room with a volume of 2.70×10^4 L, what is the partial pressure of CO (in mm Hg) in the room at 22 °C?
53. Three flasks are connected as shown in the figure. The starting conditions are shown.
    (a) What is the final pressure inside the system when all the stopcocks are open?
    (b) What is the partial pressure of each of the three gases? Assume that the connecting tube has negligible volume. Assume no change in *T*.

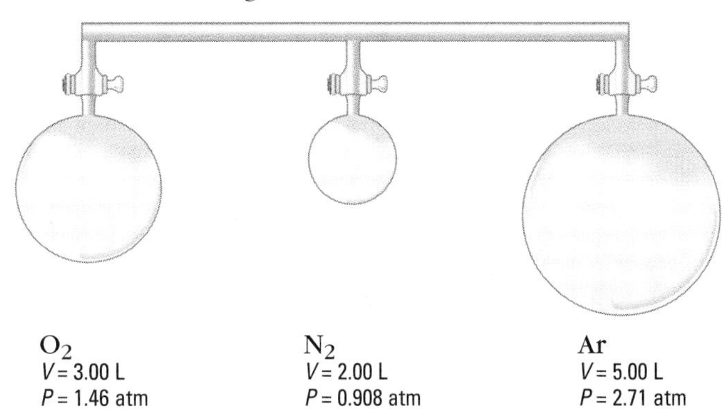

54. Acetylene can be made by reacting calcium carbide with water.

$$CaC_2(s) + 2\,H_2O(\ell) \longrightarrow C_2H_2(g) + Ca(OH)_2(aq)$$

Assume that you place 2.65 g CaC_2 in excess water and collect the acetylene over water. The volume of the acetylene and water vapor is 795 mL at 25.0 °C and a barometric pressure of 735.2 mm Hg. Calculate the percent yield of acetylene. The vapor pressure of water at 25 °C is 23.8 mm Hg.

55. Potassium chlorate, $KClO_3$, can be decomposed by heating.

$$2\,KClO_3(s) \longrightarrow 2\,KCl(s) + 3\,O_2(g)$$

If 465 mL gas was collected over water at a total pressure of 750. mm Hg and a temperature of 23 °C, how many grams of O_2 were collected?

The Behavior of Real Gases

56. From the density of liquid water and its molar mass, calculate the volume that 1 mol liquid water occupies. If water were an ideal gas at STP, what volume would a mole of water vapor occupy? Can we achieve the STP conditions for water vapor? Why or why not?
57. At high temperatures and low pressures, gases behave ideally, but as the pressure is increased the product *PV* becomes greater than the product *nRT*. Give a molecular-level explanation of this fact.
58. ■ At low temperatures and very low pressures, gases behave ideally, but as the pressure is increased the product *PV* becomes less than the product *nRT*. Give a molecular-level explanation of this fact.
59. The densities of liquid noble gases and their normal boiling points are given in the following table.

Gas	Normal Boiling Point (K)	Liquid Density (g/cm³)
He	4.2	0.125
Ne	27.1	1.20
Ar	87.3	1.40
Kr	120.	2.42
Xe	165	2.95

Calculate the volume occupied by 1 mol of each of these liquids. Comment on any trend that you see. What is the volume occupied by exactly 1 mol of each of these substances as an ideal gas at STP? On the basis of these calculations, which gas would you expect to show the largest deviations from ideality at room temperature?

Greenhouse Gases and Global Warning

60. ■ What is the difference between the greenhouse effect and global warming? How are they related?
61. What are the four greenhouse gases, and why are they called that?
62. Carbon dioxide is known to be a major contributor to the greenhouse effect. List some of its sources in our atmosphere and some of the processes that remove it. Currently, which predominates—the production of CO_2 or its removal?
63. Name a favorable effect of the global increase of CO_2 in the atmosphere.

General Questions

64. HCl can be made by the direct reaction of H_2 and Cl_2 in the presence of light. Assume that 3.0 g H_2 and 140. g Cl_2 are mixed in a 10-L flask at 28 °C.
Before the reaction:
(a) What are the partial pressures of the two reactants?
(b) What is the total pressure due to the gases in the flask?
After the reaction:
(c) What is the total pressure in the flask?
(d) What reactant remains in the flask? How many moles of it remain?
(e) What are the partial pressures of the gases in the flask?
(f) What will be the pressure inside the flask if the temperature is increased to 40. °C?

65. Worldwide, about 100. million metric tons of H_2S are produced annually from sources that include the oceans, bogs, swamps, and tidal flats. One of the major sources of SO_2 in the atmosphere is the oxidation of H_2S, produced by the decay of organic matter. The reaction in which H_2S molecules are oxidized to SO_2 involves O_3. Write an equation showing that one molecule of each reactant combines to form two product molecules, one of them being SO_2. Then calculate the annual production in tons of H_2SO_4, assuming all of this SO_2 is converted to sulfuric acid.

Applying Concepts

66. Which graph below would best represent the distribution of molecular speeds for the gases acetylene, C_2H_2, and N_2? Both gases are in the same flask with a total pressure of 750. mm Hg. The partial pressure of N_2 is 500. mm Hg.

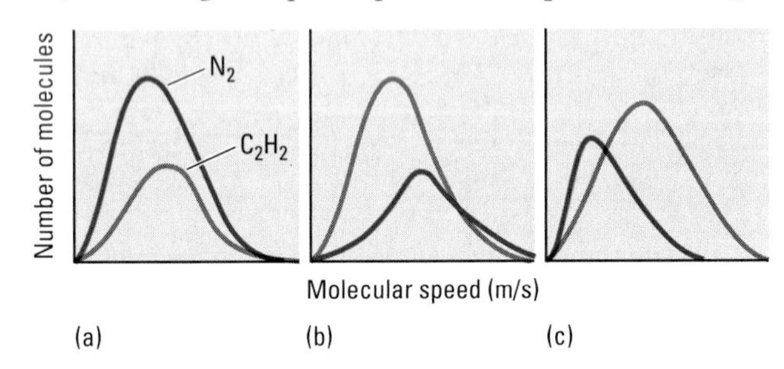

67. Draw a graph representing the distribution of molecular speeds for the gases ethane, C_2H_6, and F_2 when both are in the same flask with a total pressure of 720. mm Hg and a partial pressure of 540. mm Hg for F_2.

68. This drawing represents a gas collected in a syringe (the needle end was sealed after collecting) at room temperature and pressure. Redraw the syringe and gas to show what it would look like under the following conditions. Assume that the plunger can move freely but no gas can escape.
(a) The temperature of the gas is decreased by one half.
(b) The pressure of the gas is decreased to one half of its initial value.
(c) The temperature of the gas is tripled and the pressure is doubled.

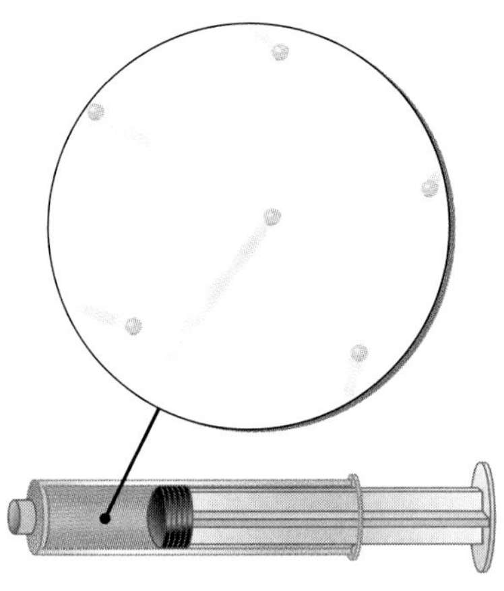

69. A gas phase reaction takes place in a syringe at a constant temperature and pressure. If the initial volume is 40. cm^3 and the final volume is 60. cm^3, which of these general reactions took place? Explain your reasoning.
(a) $A(g) + B(g) \rightarrow AB(g)$
(b) $2\ A(g) + B(g) \rightarrow A_2B(g)$
(c) $2\ AB_2(g) \rightarrow A_2(g) + 2\ B_2(g)$
(d) $2\ AB(g) \rightarrow A_2(g) + B_2(g)$
(e) $2\ A_2(g) + 4\ B(g) \rightarrow 4\ AB(g)$

70. The gas molecules in the box below undergo a reaction at constant temperature and pressure.

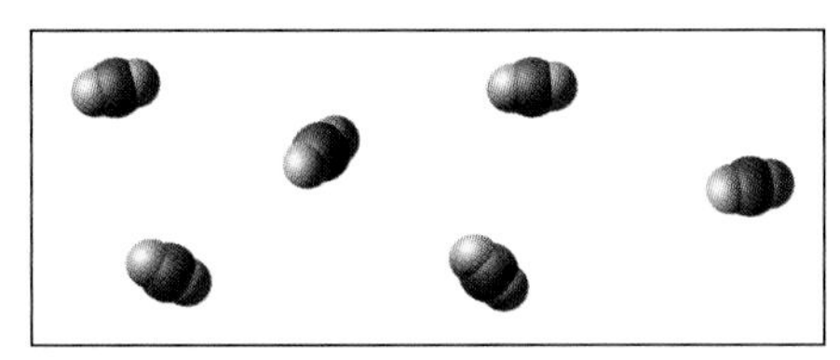

If the initial volume is 1.8 L and the final volume is 0.9 L, which of the boxes below could be the products of the reaction? Explain your reasoning.

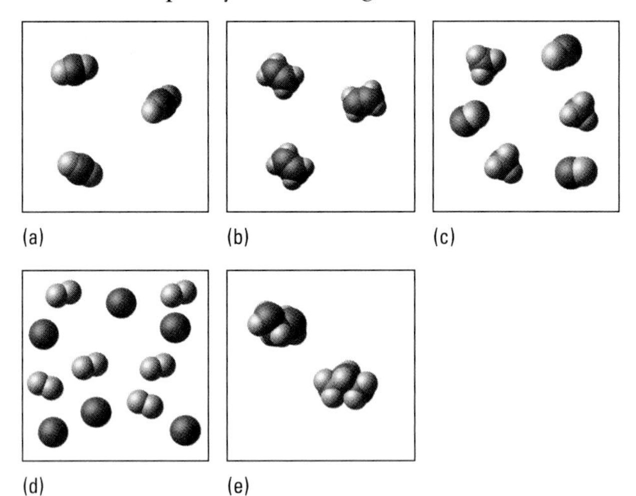

71. A substance is analyzed and found to contain 85.7% carbon and 14.3% hydrogen by mass. A gaseous sample of the substance is found to have a density of 1.87 g/L at STP.
(a) What is the molar mass of the compound?
(b) What are the empirical and molecular formulas of the compound?

(c) What are two possible Lewis structures for molecules of the compound?

72. A compound consists of 37.5% C, 3.15% H, and 59.3% F by mass. When 0.298 g of the compound is heated to 50. °C in an evacuated 125-mL flask, the pressure is observed to be 750. mm Hg. The compound has three isomers.
(a) What is the molar mass of the compound?
(b) What are the empirical and molecular formulas of the compound?
(c) Draw the Lewis structure for each isomer of the compound.

More Challenging Questions

73. What is the total pressure in an 8.7-L vessel that contains 3.44 g carbon dioxide, 1.88 g methane, and 4.28 g Ar, all at 37 °C?

74. An ideal gas was contained in a glass vessel of unknown volume with a pressure of 0.960 atm. Some of the gas was withdrawn from the vessel and used to fill a 25.0-mL glass bulb to a pressure of 1.00 atm. The pressure of the gas remaining in the vessel of unknown volume was 0.882 atm. All the measurements were done at the same temperature. What is the volume of the vessel?

75. You are holding two balloons, an orange balloon and a blue balloon, both at the same temperature and pressure. The orange balloon is filled with neon gas and the blue balloon is filled with argon gas. The orange balloon has twice the volume of the blue balloon. What is the mass ratio of Ne to Ar in the two balloons?

76. A sealed bulb filled with a gas is immersed in an ice bath, and its pressure is measured as 444 mm Hg. Then the same bulb is placed in a heated oven, and the pressure is measured as 648 mm Hg. What is the temperature of the oven in °C?

77. Nitroglycerine, the explosive ingredient in dynamite, decomposes violently when shocked.

$$4\ C_3H_5(NO_3)_3(\ell) \longrightarrow 12\ CO_2(g) + 10\ H_2O(\ell) + 6\ N_2(g) + O_2(g)$$

(a) How many total moles of gas are produced when 1.00 kg nitroglycerine explodes?
(b) What volume would the gases occupy at 1.0 atm pressure and 25 °C?
(c) At $P = 1.00$ atm, what would be the partial pressure of each product gas?

78. A container of gas has a pressure of 550. torr. A chemical change then occurs that consumes half of the molecules present at the start and produces two new molecules for each three consumed. What is the new pressure in the container? T and V are unchanged.

79. Imagine a tiny gas bubble with a volume of exactly one cubic millimeter being formed at the bottom of a lake at a pressure of 4.4 atm. When the bubble reaches the surface of the lake, where the pressure is 740. torr, what will its volume be? Assume that the temperature is constant.

80. Propane, $CH_3CH_2CH_3$, is a gaseous fuel that can be liquified when put under modest pressure.
(a) How many moles of propane can be stored in a container with a volume of 150. L at a pressure of 3.00 atm and a temperature of 25 °C?
(b) How many moles of liquid propane can be stored in a vessel of the same volume? (Liquid propane's density is approximately 0.60 g/mL.)

81. The effects of intermolecular interactions on gas properties depend on T and P. Are these effects more or less significant when these changes occur? Why?
(a) A container of gas is compressed to a smaller volume at constant temperature.
(b) A container of gas has more gas added into the same volume at constant temperature.
(c) The gas in a container of variable volume is heated at constant pressure.

Conceptual Challenge Problems

CP10.A (Section 10.1) How would you quickly estimate whether the mass of the earth's atmosphere is 5.3×10^{15} metric tons?

CP10.B (Section 10.4) Under what conditions would you expect to observe that the pressure of a confined gas at constant temperature and volume is *not* constant?

CP10.C (Section 10.4) Suppose that the gas constant, R, were defined as 1.000 L atm mol^{-1} deg^{-1} where the "deg" referred to a newly defined Basic temperature scale. What would be the melting and boiling temperatures of water in degrees Basic (°B)?

11

Liquids, Solids, and Materials

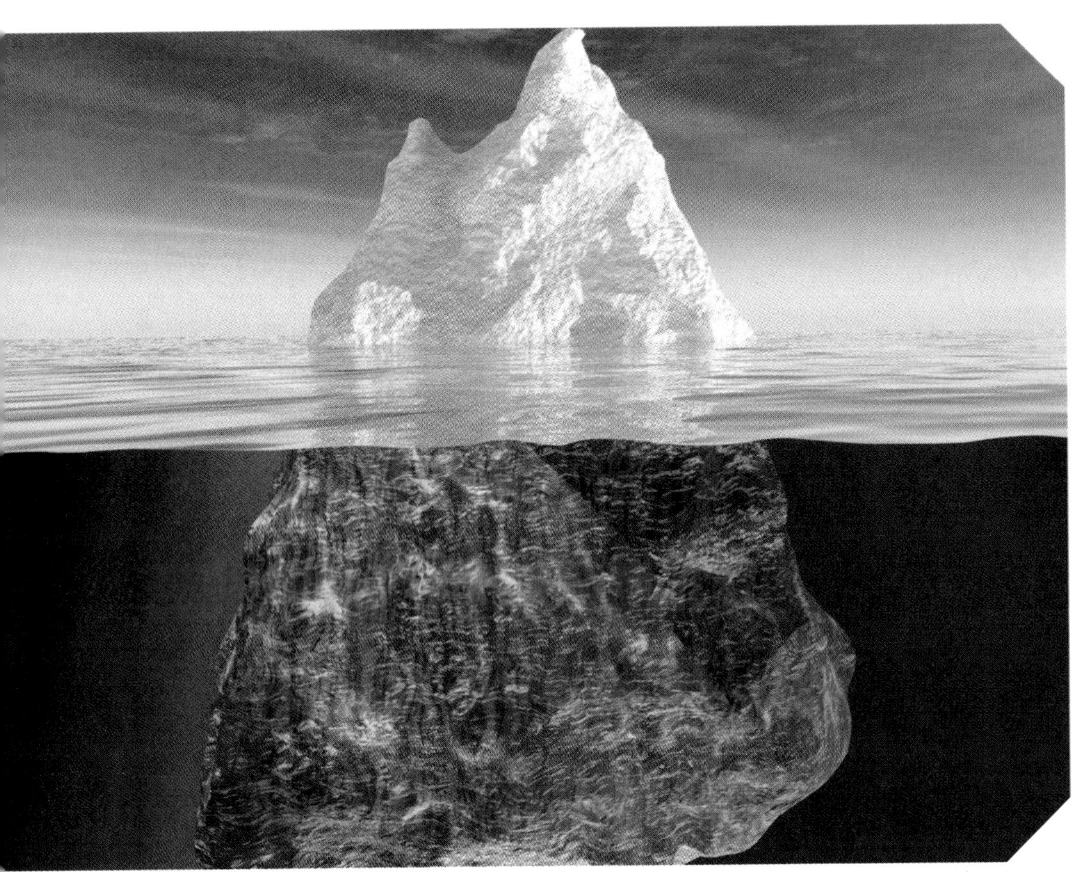

An iceberg floats with about 10% of its mass above the surface of the surrounding water. The hydrogen bonding of water molecules in ice makes the solid less dense than liquid water, which is unusual. This property of water, however, is crucial for life on earth.

The properties of liquids and solids, like those of gases, can be understood on the molecular, nanoscale level. In common liquids and solids—water, salt crystals, metals—the molecules, ions, or atoms are close enough to have strong interactions with each other. The strength of these interactions accounts for the properties of liquids and solids. For example, the strong repulsions when the molecules, ions, or atoms in a liquid or solid are forced very close together account for the fact that solids and liquids can be compressed very little—in contrast to gases. Liquids are like gases in that their molecules, ions (in the case of molten ionic compounds), or atoms can move freely. Liquids are therefore able to flow and fill a container to a certain level. Solids, on the other hand, are composed of molecules, ions, or atoms in relatively fixed positions, which explains the fact that solids have definite shapes. When the molecules, ions, or atoms making up a solid have enough energy to overcome the attractive forces holding them in place, the solid melts or sublimes. Some solids decompose when heated.

Sublimation is the direct conversion of a solid to its vapor without becoming a liquid.

In molecular liquids and solids, molecules having sufficient energy can leave the surface of the liquid or solid and enter the gaseous state. In the reverse process, as energy is removed, gases condense into liquids, and liquids solidify. All of these properties of solids and liquids have great practical importance.

© Cengage Learning/Charles D. Winters

The hydraulic lifts are raised and lowered by changes in the pressure on the hydraulic fluid in the cylinders. The lack of compressibility of the fluid makes this application possible.

11.1 The Liquid State

At low enough temperatures, gases condense to liquids. Condensation occurs when most molecules no longer have enough kinetic energy to overcome their intermolecular attractions *(⇐ p. 328)*. Most liquids, such as water, alcohol, and gasoline, are substances whose condensation temperatures are above room temperature. Some liquids, such as molten salts and liquid polymers, have no corresponding vapor state so condensation of their vapors has no meaning.

In a liquid, the molecules are much closer together than are those in a gas. Nevertheless, the molecules remain mobile enough that the liquid flows. At the nanoscale, liquids have a regular structure only in very small regions, and most of the molecules continue to move about randomly. Because confined liquids are difficult to compress and their molecules move in all directions, they can transmit applied pressure equally in all directions. This property has a practical application in the hydraulic fluids that operate mechanical devices such as the hydraulic lifts that raise and lower autos in repair shops.

The resistance of a liquid to flow is called **viscosity.** Water flows smoothly and quickly (low viscosity), while motor oil and honey flow more slowly (higher viscosity; Figure 11.1). The viscosity of a liquid is related to its intermolecular forces, which determine how easily the molecules can move past each other, and also the degree to which the molecules can become entangled. Viscosity decreases as temperature increases because the molecules have more kinetic energy at higher temperature and the attractive intermolecular forces are more easily overcome. Examples of common materials whose viscosity decreases as their temperature rises include motor oil, cooking oil, and honey.

Unlike gases, liquids have *surface properties.* Molecules beneath the surface of the liquid are completely surrounded by other liquid molecules and experience intermolecular attractions in all directions. By contrast, molecules at the liquid surface have intermolecular interactions only with molecules below or beside them, but not above them (Figure 11.2). This unevenness of attractive forces at the liquid surface causes the

© Cengage Learning/Charles D. Winters

Figure 11.1 Viscosity. Honey is viscous, so it builds up rather than spreading out, as less viscous water would.

Photo: © Cengage Learning/Charles D. Winters

Fewer forces act on surface molecules.

More forces act on molecules completely surrounded by other molecules.

Figure 11.2 Surface tension. Surface tension, the energy required to increase the surface area of a liquid, arises from the difference between the forces acting on a molecule within the liquid and those acting on a molecule at the surface of the liquid.

The chemicals known as surfactants (soaps and detergents are examples) can dissolve in water and dramatically lower its surface tension. When this happens, water becomes "wetter" and does a better job of cleaning. You can see the effect of a surfactant by comparing how water alone beads on a surface such as the waxed hood of a car and how a soap solution fails to bead on the same surface.

Surface tension of water supports a water bug.

Table 11.1 Surface Tensions of Some Liquids

Substance	Formula	Surface Tension (J/m^2 at 20 °C)
Octane	C_8H_{18}	2.16×10^{-2}
Ethanol	CH_3CH_2OH	2.23×10^{-2}
Chloroform	$CHCl_3$	2.68×10^{-2}
Benzene	C_6H_6	2.85×10^{-2}
Water	H_2O	7.29×10^{-2}
Mercury	Hg	46×10^{-2}

surface to contract. The energy required to expand a liquid surface is called its **surface tension,** and it is higher for liquids that have stronger intermolecular attractions. For example, water's surface tension is high compared with those of most other liquids (Table 11.1) because extensive hydrogen bonding causes the water molecules to attract each other strongly (*p. 332*). The surface tension of water, 7.29×10^{-2} J/m^2, is the quantity of energy required to increase the surface area of water by 1 m^2 at 20 °C. The very high surface tension of mercury (six times that of water) is due to the much stronger metallic bonding (Section 11.8) that holds Hg atoms together in the liquid.

The teardrop shape of raindrops you see in illustrations of rain bears little resemblance to actual raindrops, which are spherical. Air pressure begins to distort the sphere into an egg shape for 2- to 3-mm-radius raindrops. Larger raindrops assume other shapes, but they never look like teardrops.

Even though their densities are greater than that of water, water bugs can walk on the surface of water and small metal objects can float. Surface tension prevents the objects from breaking through the surface and sinking. Surface tension also accounts for the nearly spherical shape of rain droplets, as well as the rounded shape of water droplets that bead up on waxy surfaces such as leaves. A sphere has less surface area per unit volume than any other shape, so a spherical rain droplet has fewer surface H_2O molecules, minimizing the surface energy.

CONCEPTUAL **EXERCISE** **11.1 Explaining Differences in Surface Tension**

At 20 °C, chloroform, $CHCl_3$, has a surface tension of 2.68×10^{-2} J/m^2, while bromoform, $CHBr_3$, has a higher surface tension of 4.11×10^{-2} J/m^2. Explain this observation in terms of intermolecular forces.

CONCEPTUAL **EXERCISE** **11.2 Predicting Surface Tension**

Predict which substance listed in Table 11.1 has a surface tension most similar to that of

(a) Glycerol, $C_3H_8O_3$, which has three —OH groups in each molecule

(b) Decane, $C_{10}H_{22}$

Explain your answer in each case.

When a glass tube with a small diameter is put into water, the water rises in the tube due to **capillary action.** The glass walls of the tube are largely silicon dioxide, SiO_2, so the water molecules form hydrogen bonds to the oxygen atoms of the glass. This attractive force is stronger than the attractive forces among water molecules (also hydrogen bonding), so water creeps up the wall of the tube. Simultaneously, the surface tension of the water tries to keep the water's surface area small. The combination of the forces raises the water level in the tube. The liquid surface—concave upward for water—is called a **meniscus** (Figure 11.3a). The water rises in the tube until the force of the water-to-wall hydrogen bonds is balanced by the pull of gravity. Capillary action

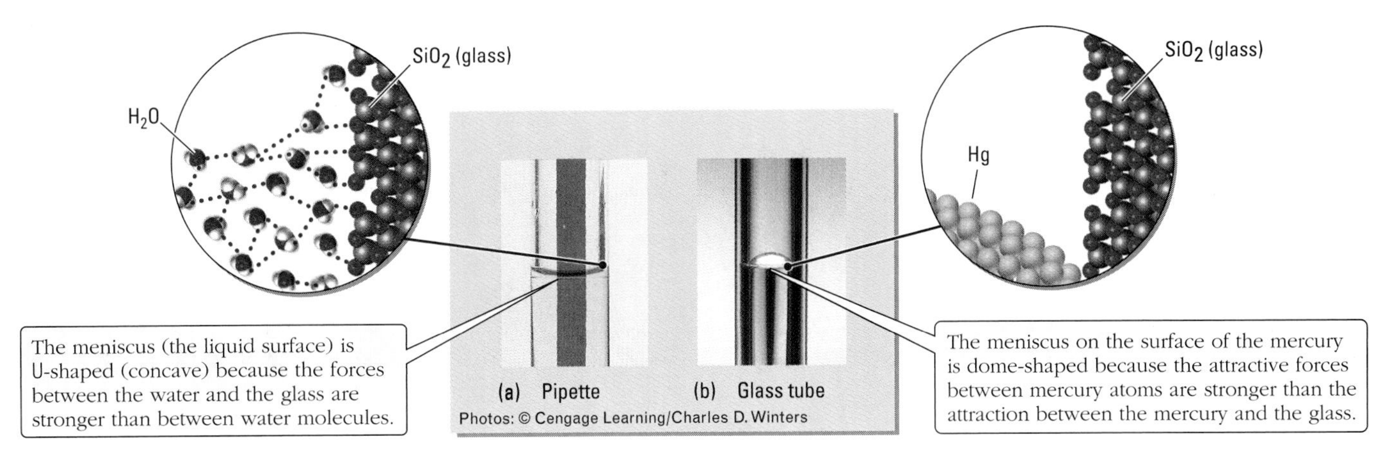

SiO_2 (glass)
H_2O
SiO_2 (glass)
Hg
The meniscus (the liquid surface) is U-shaped (concave) because the forces between the water and the glass are stronger than between water molecules.
(a) Pipette
(b) Glass tube
Photos: © Cengage Learning/Charles D. Winters
The meniscus on the surface of the mercury is dome-shaped because the attractive forces between mercury atoms are stronger than the attraction between the mercury and the glass.

Figure 11.3 Noncovalent forces at surfaces. (a) The meniscus of water in a pipette is concave upward. (b) The meniscus of mercury in a glass tube is dome-shaped (convex upward).

is crucial to plant life because it helps water, with its dissolved nutrients, to move upward through plant tissues against the force of gravity.

Mercury behaves oppositely from water. Mercury in a small-diameter tube has a dome-shaped (convex upward) meniscus because the attractive forces between mercury atoms are stronger than the attraction between the mercury and the glass wall (Figure 11.3b).

11.2 Vapor Pressure

The tendency of a liquid to vaporize, its **volatility,** increases as temperature increases. Everyday experiences such as heating water or soup in a pan on a stove, or evaporation of rain from hot pavement, demonstrate this phenomenon. Conversely, the volatility of a liquid is lower at lower temperatures.

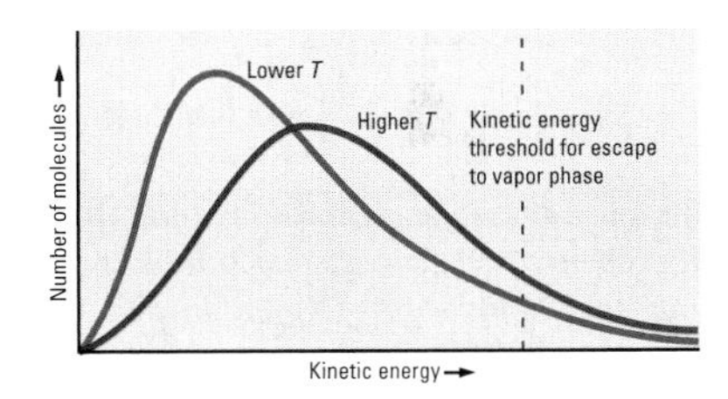

Figure 11.4 Kinetic energy and evaporation. At any given temperature, a fraction of the molecules at the surface of a liquid have sufficient energy to escape to the vapor phase. When the temperature of the liquid is raised, a larger fraction of the molecules have energies greater than the kinetic energy threshold, so the vapor pressure increases.

The change of volatility with temperature can be explained by considering what is happening at the nanoscale. The molecules in a liquid have varying kinetic energies and speeds (Figure 11.4). At any time, some fraction of the molecules have sufficient energy to escape from the liquid into the gas phase. When the temperature of the liquid is raised, a larger number of molecules exceeds the energy threshold, and evaporation proceeds more rapidly. Thus, hot water evaporates more quickly than cold water.

A liquid in an open container will eventually evaporate completely because air currents and diffusion take away most of the gas phase molecules before they can reenter the liquid phase. In a closed container, however, no molecules can escape. If a liquid is injected into an evacuated, closed container, the rate of vaporization (number of molecules entering the gas phase per unit time) will at first far exceed the rate of condensation. Over time, the pressure of the gas above the liquid will increase as the number of gas-phase molecules increases. Eventually the system will attain a state of *dynamic equilibrium,* in which molecules are entering and leaving the liquid state at equal rates. At this point the pressure of the gas will no longer increase; this pressure is known as the **equilibrium vapor pressure** (or just the **vapor pressure**) of the liquid.

An evacuated container is specified so that only the pure liquid is present.

As shown in Figure 11.5 for three liquids, the vapor pressure of a liquid increases with increasing temperature. The differences in vapor pressures among the three liquids at the same temperature are due to differences in the strengths of the intermolecular interactions that bind their molecules together in the liquid phase. Consider the vapor pressures of ethyl alcohol and water at 60 °C (see Figure 11.5). At that temperature, ethyl alcohol has a higher vapor pressure, which means that it has weaker

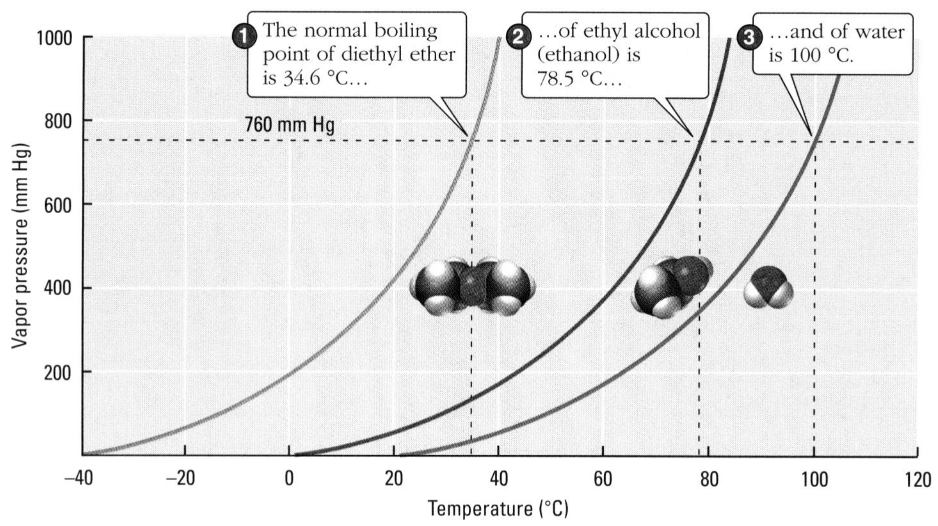

Figure 11.5 Vapor pressure curves for diethyl ether, $C_2H_5OC_2H_5$; ethyl alcohol, C_2H_5OH; and water. Each curve represents the conditions of T and P where the two phases (pure liquid and its vapor) are in equilibrium. Each compound is a liquid in the temperature and pressure region to the left of its curve, and each compound is a vapor for all temperatures and pressures to the right of its curve.

intermolecular interactions than those in water. Water's stronger intermolecular interactions (due to more hydrogen bonds) require a higher temperature to disrupt them and volatilize liquid water *(⇐ p. 332)*.

To shorten cooking times, one can use a pressure cooker, a sealed pot (with a relief valve for safety) that allows water vapor to build up pressures slightly greater than the external atmospheric pressure. At the higher pressure, the boiling point of water is higher, and foods cook faster.

If a liquid is placed in an open container and heated, a temperature eventually is reached at which the vapor pressure of the liquid is equal to the atmospheric pressure. Below this temperature, only molecules at the surface of the liquid can go into the gas phase. But when the vapor pressure equals the atmospheric pressure, the liquid begins vaporizing throughout. Bubbles of vapor form and immediately rise to the surface due to their lower density. The liquid is said to be **boiling** (Figure 11.6). The temperature at which the equilibrium vapor pressure equals the atmospheric pressure is the

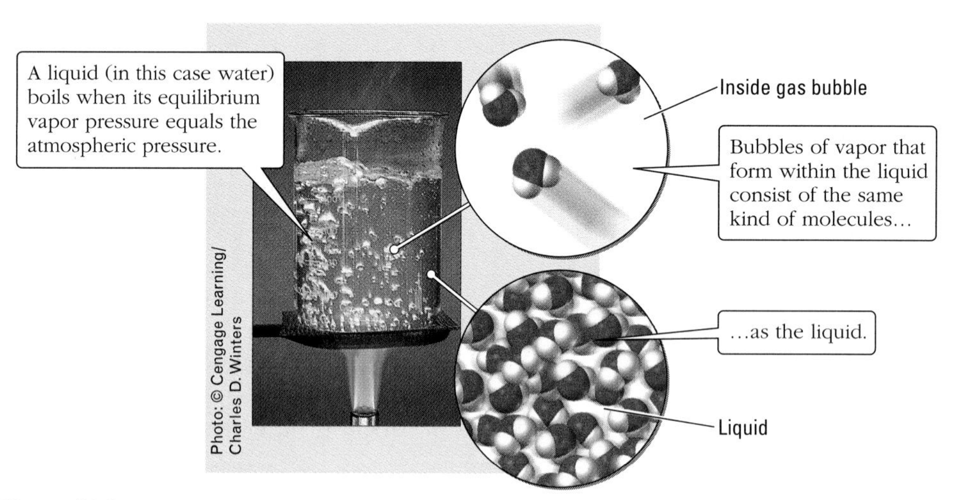

Figure 11.6 Boiling liquid. From Figure 11.5, we see that the lower the atmospheric pressure, the lower the vapor pressure at which boiling can occur. It takes longer to hard-boil an egg high in the mountains, where the atmospheric pressure is lower, than it does at sea level, because water at a higher elevation boils at a lower temperature. In Salt Lake City, Utah (elevation 4390 ft), where the average barometric pressure is about 650 mm Hg, water boils at about 95 °C.

boiling point of the liquid. When the atmospheric pressure is 1 atm (760 mm Hg), the temperature is designated the **normal boiling point.** The temperatures where the vapor pressure curves cross the 760 mm Hg (1 atm) line are the normal boiling points of the three liquids (Figure 11.5).

CONCEPTUAL
EXERCISE 11.3 Estimating Boiling Points

Use Figure 11.5 to estimate the boiling points of these liquids: (a) ethyl alcohol at 400 mm Hg; (b) diethyl ether at 200 mm Hg; (c) water at 400 mm Hg.

CONCEPTUAL
EXERCISE 11.4 Explaining Bubbles

One of your classmates believes that the bubbles in a boiling liquid are air bubbles. Explain to him what is wrong with that idea and what the bubbles actually are. Suggest an experiment to show that the bubbles are not air.

Clausius-Clapeyron Equation

As seen in Figure 11.5, plotting the vapor pressure of a substance versus its temperature results in a curved line. The relationship between vapor pressure and temperature is given by the **Clausius-Clapeyron equation:**

$$\ln P = \frac{-\Delta H_{vap}}{RT} + C$$

which relates the natural logarithm of the vapor pressure P of the liquid to the absolute temperature T, the universal gas constant R (8.314 J mol^{-1} K^{-1}), the molar enthalpy of vaporization ΔH_{vap} (J/mol), and a constant C that is characteristic of the liquid. The form of the Clausius-Clapeyron equation shows that a graph of ln P versus $1/T$ is a straight line with its slope equal to $-\Delta H_{vap}/R$ and a y-axis intercept of C. That is, ln P equals C in the limit when T is very large and $1/T$ tends toward zero. Figure 11.7 shows a plot of ln P versus $1/T$ for ethanol. A comparison of the plots for ethanol in Figures 11.5 and 11.7 illustrates the two different ways of expressing the relationship between T and P.

Note that R is in units of J mol^{-1} K^{-1} here.

The Clausius-Clapeyron equation can be recast for two sets of pressure and temperature of a liquid to give

$$\ln\left(\frac{P_2}{P_1}\right) = \frac{-\Delta H_{vap}}{R}\left[\frac{1}{T_2} - \frac{1}{T_1}\right]$$

which can be used to calculate ΔH_{vap} from the two sets of data: T_1, T_2 and P_1, P_2. If ΔH_{vap} is known, then this form of the equation can be used to find the vapor pressure of a liquid at a new temperature given its vapor pressure at some other temperature.

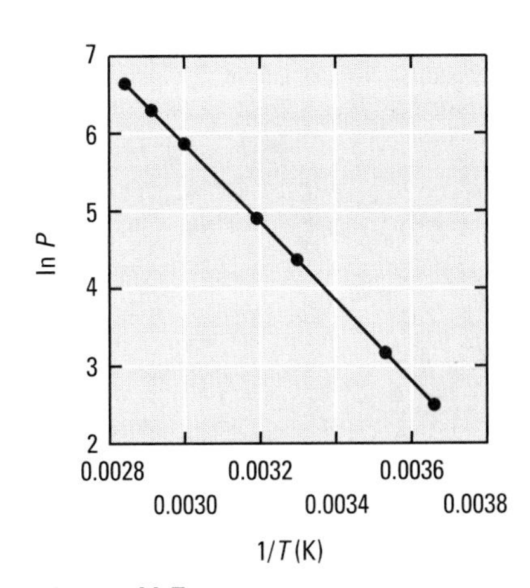

Figure 11.7 Clausius-Clapeyron equation plot for ethanol. Plotting ln P versus $1/T$ yields a straight line whose slope is directly proportional to ΔH_{vap}.

PROBLEM-SOLVING EXAMPLE 11.1 Clausius-Clapeyron Equation

Ethanol has a vapor pressure of 44.22 mm Hg at 20.0 °C and 350.8 mm Hg at 60.0 °C. What is ΔH_{vap} for ethanol?

Answer 42.04 kJ/mol

Strategy and Explanation Use the Clausius-Clapeyron equation in the form that contains the four known quantities: T_1, T_2, P_1, and P_2. The temperatures must be converted to kelvins. Substitute the given values into the equation and solve for ΔH_{vap}.

$$\ln\left(\frac{350.8 \text{ mm Hg}}{44.22 \text{ mm Hg}}\right) = \frac{-\Delta H_{vap}}{8.314 \text{ J mol}^{-1} \text{ K}^{-1}}\left[\frac{1}{333.15 \text{ K}} - \frac{1}{293.15 \text{ K}}\right]$$

$$\ln(7.933) = \frac{-\Delta H_{vap}}{8.314 \text{ J mol}^{-1} \text{ K}^{-1}}(-4.096 \times 10^{-4} \text{ K}^{-1})$$

$$\Delta H_{vap} = \frac{\ln(7.933) \times 8.314 \text{ J mol}^{-1} \text{ K}^{-1}}{4.096 \times 10^{-4} \text{ K}^{-1}} = 42037 \text{ J/mol} = 42.04 \text{ kJ/mol}$$

PROBLEM-SOLVING PRACTICE 11.1

At 30.0 °C the vapor pressure of carbon tetrachloride is 143.0 mm Hg. What is its vapor pressure at 60.0 °C? The ΔH_{vap} of carbon tetrachloride is 32.1 kJ/mol.

Vapor Pressure of Water and Relative Humidity

Water has an appreciable vapor pressure in the temperature range 10 °C to 25 °C, normal outdoor temperatures. Therefore, water vapor is present in the atmosphere at all times. The equilibrium vapor pressure depends on the temperature as shown in Table 10.4 *(⇐ p. 363)*. The vapor pressure of water in the atmosphere is expressed as the **relative humidity,** which is the ratio of the actual partial pressure of the water vapor in the atmosphere, P_{H_2O}, to the equilibrium partial pressure of water vapor at the relevant temperature, $P^{\circ}_{H_2O}$. The factor of 100 changes a fraction into a percentage.

$$\text{Relative humidity} = \frac{P_{H_2O}}{P^{\circ}_{H_2O}} \times 100\%$$

As an example, consider a typical warm day with an outdoor temperature of 25 °C. The equilibrium water vapor pressure at this temperature is 23.8 mm Hg (see Table 10.4). If the actual partial pressure of water vapor were 17.8 mm Hg, then the relative humidity would be

$$\text{Relative humidity} = \frac{17.8 \text{ mm Hg}}{23.8 \text{ mm Hg}} \times 100\% = 75\%$$

If the temperature were lowered (which would lower the equilibrium vapor pressure of water) and the partial pressure of water vapor remained constant, the relative humidity would rise. Eventually, it would reach 100%. The temperature at which the actual partial pressure of water vapor equals the equilibrium vapor pressure is the **dew point.** At this temperature, water vapor will condense as water droplets to form fog or dew. When the dew point rises above about 20 °C we consider it to be somewhat muggy; when it rises above about 25 °C we consider the day to be extremely uncomfortable. In desert conditions the relative humidity can fall to less than 10%; this, too, can be uncomfortable for some people.

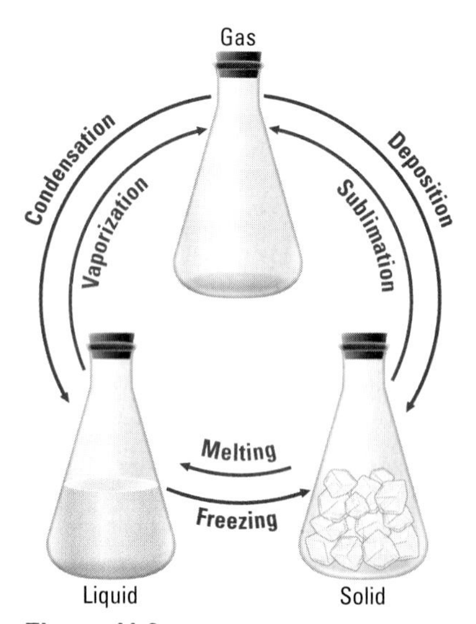

Figure 11.8 The three states of matter and their six interconversions. The red arrows indicate endothermic conversions—processes that require an input of energy. The blue arrows indicate exothermic conversions—processes that transfer energy to the surroundings.

11.3 Phase Changes: Solids, Liquids, and Gases

A substance in one state of matter—solid, liquid, gas—can change to another. Water evaporates from the street after a rain shower. Ice cubes melt in your cold drink, and water vapor in the air condenses to water droplets on the outside wall of the glass. Each of the three states of matter can be converted to the other two states through six important processes (Figure 11.8).

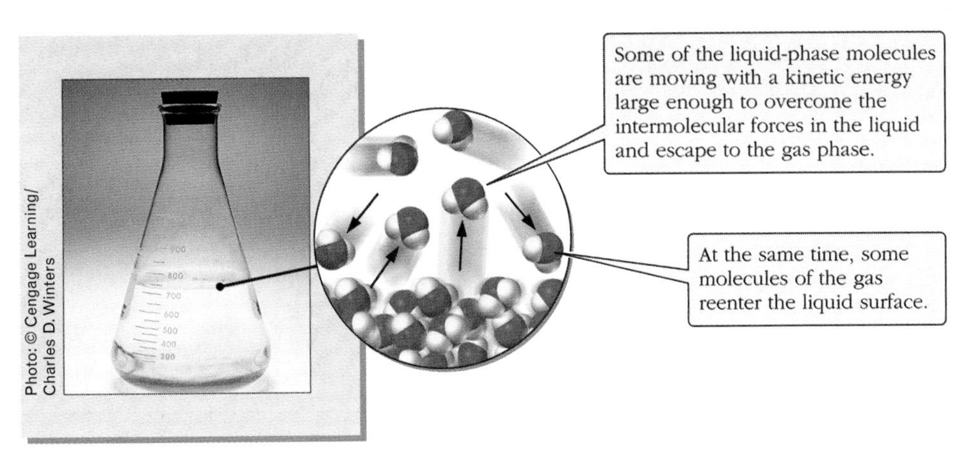

Some of the liquid-phase molecules are moving with a kinetic energy large enough to overcome the intermolecular forces in the liquid and escape to the gas phase.
At the same time, some molecules of the gas reenter the liquid surface.
Photo: © Cengage Learning/ Charles D. Winters

Figure 11.9 Molecules in the liquid and gas phases. All the molecules in both phases are moving, although the distances traveled in the liquid before collision with another molecule are much smaller.

Vaporization and Condensation

Like molecules in a gas, molecules in a liquid are in constant motion and have a range of kinetic energies, as shown in Figure 11.4. A fraction of the molecules in the liquid have enough kinetic energy to overcome the intermolecular attractive forces among the liquid molecules *(⇐ p. 379)*. If such a molecule is at the surface of the liquid and moving in the right direction, it will leave the liquid phase and enter the gas phase (Figure 11.9). This process is called **vaporization** or **evaporation.**

As the high-energy molecules leave the liquid, they take some energy with them. Therefore, vaporization is *endothermic;* the enthalpy change is called the enthalpy of vaporization *(⇐ p. 194)*. Vaporization increases with temperature because an increasingly larger fraction of the molecules have enough energy to vaporize.

Enthalpy of vaporization is also referred to as heat of vaporization.

CONCEPTUAL
EXERCISE 11.5 Evaporative Cooling

In some countries where electric refrigeration is not readily available, drinking water is chilled by placing it in porous clay water pots. Water slowly passes through the clay, and when it reaches the outer surface, it evaporates. Explain how this process cools the water inside.

When a molecule in the gas phase hits the liquid's surface, it can transfer some of its energy to the other liquid particles and remain in the liquid, a process called **condensation.** The overall effect of molecules reentering the liquid phase is a release of thermal energy, making condensation an exothermic process. The heat transferred out of the system upon condensation is equal to the heat transferred into the system upon vaporization.

$$\text{Liquid} \underset{\text{heat of condensation}}{\overset{\text{heat of vaporization}}{\rightleftharpoons}} \text{Gas} \qquad \Delta H_{\text{vaporization}} = -\Delta H_{\text{condensation}}$$

The thermal energy absorbed during evaporation is sometimes called the latent heat of vaporization. It is somewhat dependent on the temperature: at 100 °C, ΔH°_{vap} (H_2O) = 40.7 kJ/mol, but at 25 °C the value is 44.0 kJ/mol.

For example, the quantity of enthalpy required to completely vaporize 1 mol liquid water once it has reached the boiling point of 100 °C at 1 bar (the *molar enthalpy of vaporization*) and the quantity of enthalpy released when 1 mol water vapor at 100 °C condenses to liquid water at 100 °C (the *molar enthalpy of condensation*) have the same values but opposite signs

$$H_2O(\ell) \rightarrow H_2O(g) \qquad \Delta H^\circ = \Delta H_{vap} = +40.7 \text{ kJ/mol}$$

$$H_2O(g) \rightarrow H_2O(\ell) \qquad \Delta H^\circ = \Delta H_{cond} = -40.7 \text{ kJ/mol}$$

Table 11.2 Molar Enthalpies of Vaporization and Boiling Points for Some Common Substances

Substance	Number of Electrons	ΔH°_{vap} (kJ/mol)*	Boiling Point (°C)†
Noble gases			
He	2	0.08	−269.0
Ne	10	1.8	−246.0
Ar	18	6.5	−185.9
Xe	54	12.6	−107.1
Nonpolar molecules			
H_2	2	0.90	−252.8
O_2	16	6.8	−183.0
F_2	18	6.54	−188.1
Cl_2	34	20.39	−34.6
Br_2	70	29.54	59.6
CH_4 (methane)	10	8.9	−161.5
$CH_3—CH_3$ (ethane)	18	15.7	−88.6
$CH_3—CH_2—CH_3$ (propane)	26	19.0	−42.1
$CH_3—CH_2—CH_2—CH_3$ (butane)	34	24.3	−0.5
Polar molecules			
HF	10	25.2	19.7
HCl	18	17.5	−84.8
HBr	36	19.3	−66.5
HI	54	21.2	−35.1
NH_3	10	25.1	−33.4
H_2O	10	40.7	100.0
SO_2	32	26.8	−10.0

*ΔH°_{vap} is given at the normal boiling point of the liquid.
†Boiling point is the temperature at which vapor pressure = 760 mm Hg.

Table 11.2 illustrates the influence of noncovalent forces on enthalpies of vaporization and boiling points. In the series of nonpolar molecules and noble gases, the increasing noncovalent forces (*p. 328*) with increasing molecular size and numbers of electrons are shown by the increasing ΔH°_{vap} and boiling point values. Comparison of HF and H_2O with CH_4, which have the same number of electrons, shows the effect of molecular polarity as well as hydrogen bonding.

PROBLEM-SOLVING EXAMPLE 11.2 Enthalpy of Vaporization

To vaporize 50.0 g carbon tetrachloride, CCl_4, requires 9.69 kJ at its normal boiling point of 76.7 °C and 1 bar pressure. What is the molar enthalpy of vaporization of CCl_4 under these conditions?

Answer 29.8 kJ/mol

Strategy and Explanation The molar mass of CCl_4 is 153.8 g/mol, so

$$50.0 \text{ g CCl}_4 \times \frac{1 \text{ mol CCl}_4}{153.8 \text{ g CCl}_4} = 0.325 \text{ mol CCl}_4$$

The enthalpy change per mole of CCl_4 is given by

$$\frac{9.69 \text{ kJ}}{0.325 \text{ mol}} = 29.8 \text{ kJ/mol}$$

☑ **Reasonable Answer Check** We have approximately one third mol CCl_4, so the enthalpy required to vaporize this amount of CCl_4 should be approximately one third of the molar enthalpy of vaporization, and it is.

PROBLEM-SOLVING PRACTICE 11.2

Using data from Table 11.2, calculate the thermal energy transfer required to vaporize 0.500 mol Br_2 at its normal boiling point.

PROBLEM-SOLVING EXAMPLE 11.3 Enthalpy of Vaporization

You put 1.00 L water (about 4 cups) in a pan at 100 °C, and the water evaporates. How much thermal energy must have been transferred (at 1 bar) to the water for all of it to vaporize? (The density of liquid water at 100 °C is 0.958 g/mL.)

Answer 2.16×10^3 kJ

Strategy and Explanation You need three pieces of information to solve this problem:

(a) ΔH°_{vap} for water, which is 40.7 kJ/mol at 100 °C (Table 11.2).

(b) The density of water, which is needed because ΔH°_{vap} has units of kJ/mol, so you must first find the mass of water and then the molar amount.

(c) The molar mass of water, 18.02 g/mol, which is needed to convert grams to moles.

From the density of water, a volume of 1.00 L (or 1.00×10^3 cm^3) is equivalent to 958. g. Therefore,

$$\text{Heat required for vaporization} = 958. \text{ g } H_2O \times \frac{1 \text{ mol } H_2O}{18.02 \text{ g } H_2O} \times \frac{40.7 \text{ kJ}}{1 \text{ mol}} = 2160 \text{ kJ}$$

This enthalpy change of 2160 kJ is equivalent to about one third to one half of the caloric energy in the daily food intake of an average person in the United States.

PROBLEM-SOLVING PRACTICE 11.3

A rainstorm deposits 2.5×10^{10} kg of rain. Using the heat of vaporization of water at 25 °C of 44.0 kJ/mol, calculate the quantity of thermal energy, in joules, transferred when this much rain forms. Is this process exothermic or endothermic?

CONCEPTUAL **EXERCISE 11.6 Understanding Boiling Points**

(a) Chlorine and bromine are both diatomic. Explain the difference in their boiling points.

(b) Methane and ammonia have the same number of electrons. Explain the difference in their boiling points.

Melting and Freezing

When a solid is heated, its temperature increases until the solid begins to melt. Unless the solid decomposes first, a temperature is reached at which the kinetic energies of the molecules or ions are sufficiently high that the intermolecular interactions in the solid are no longer strong enough to keep the particles in their fixed positions. The solid's structure collapses, and the solid melts (Figure 11.10). This temperature is the *melting point* of the solid. Melting requires transfer of energy, the *enthalpy of fusion,* from the surroundings into the system, so it is always endothermic. The *molar enthalpy of fusion* (*p. 192*) is the quantity of enthalpy required to melt 1 mol of

Some solids do not have measurable melting points, and some liquids do not have measurable boiling points, because increasing temperature causes them to decompose before they melt or boil. Making peanut brittle provides an example: Melted sucrose chars before it boils, but it produces great-tasting candy when done correctly.

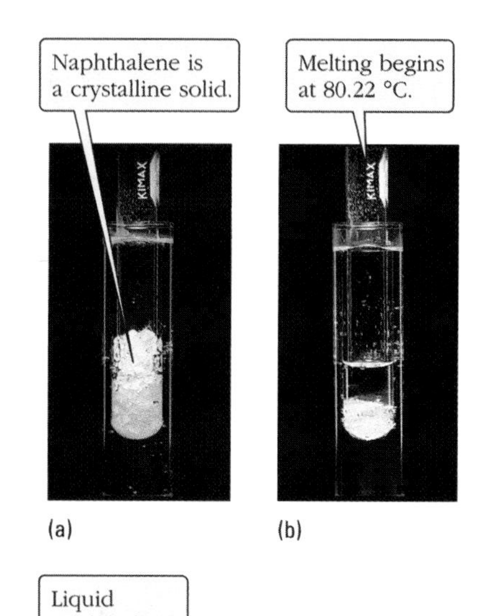
Naphthalene is a crystalline solid.
Melting begins at 80.22 °C.
KIMAX
(a)
(b)

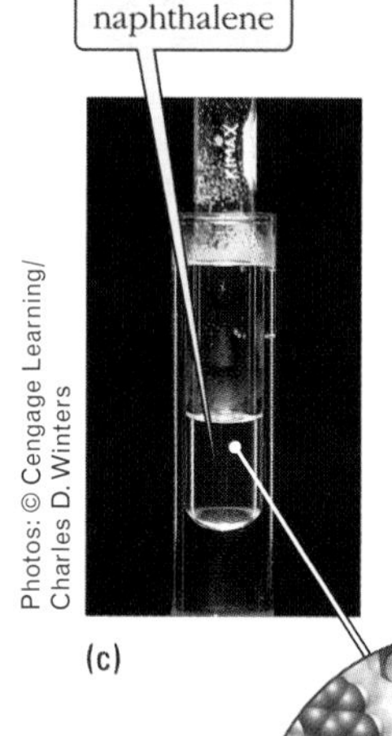
Liquid naphthalene
(c)

Figure 11.10 The melting of naphthalene, $C_{10}H_8$, at 80.22 °C.

Table 11.3 Melting Points and Enthalpies of Fusion of Some Solids

Solid	Melting Point (°C)	Enthalpy of Fusion (kJ/mol)	Type of Intermolecular Forces
Molecular solids: Nonpolar molecules			
O_2	−248	0.445	These molecules have only London forces (that increase with the number of electrons).
F_2	−220	1.020	
Cl_2	−103	6.406	
Br_2	−7.2	10.794	
Molecular solids: Polar molecules			
HCl	−114	1.990	All of these molecules have London forces enhanced significantly by dipole-dipole forces. H_2O also has significant hydrogen bonding.
HBr	−87	2.406	
HI	−51	2.870	
H_2O	0	6.020	
H_2S	−86	2.395	
Ionic solids			
NaCl	800	30.21	All ionic solids have strong attractions between oppositely charged ions.
NaBr	747	25.69	
NaI	662	21.95	

a pure solid. Solids with high enthalpies of fusion usually melt at high temperatures, and solids with low enthalpies of fusion usually melt at low temperatures. The reverse of melting—called *solidification, freezing,* or **crystallization**—is always an exothermic process. The *molar enthalpy of crystallization* has the same magnitude as the molar enthalpy of fusion, but the opposite sign.

$$\text{Solid} \underset{\text{enthalpy of crystallization}}{\overset{\text{enthalpy of fusion}}{\rightleftharpoons}} \text{Liquid} \qquad \Delta H_{\text{fusion}} = -\Delta H_{\text{crystallization}}$$

Table 11.3 lists melting points and enthalpies of fusion for examples of three classes of compounds: (a) nonpolar molecular solids, (b) polar molecular solids, some capable of hydrogen bonding, and (c) ionic solids. Solids composed of low-molecular-weight nonpolar molecules have the lowest melting temperatures, because their intermolecular attractions are weakest. These molecules are held together by London forces only (⇐ *p. 329*), and they form solids with melting points so low that we seldom encounter them in the solid state at normal temperatures. Melting points and enthalpies of fusion of nonpolar molecular solids increase with increasing number of electrons (which corresponds to increasing molar mass) as the London forces become stronger. The ionic compounds in Table 11.3 have the highest melting points and enthalpies of fusion because of the very strong ionic bonding that holds the oppositely charged ions together in the solid. The polar molecular solids have intermediate melting points.

PROBLEM-SOLVING EXAMPLE 11.4 Enthalpy of Fusion

The molar enthalpy of fusion of NaCl is 30.21 kJ/mol at its melting point. How much thermal energy will be absorbed when 10.00 g NaCl melts?

Answer 5.169 kJ

Strategy and Explanation The molar mass of NaCl is 58.443 g/mol, so

$$10.00 \text{ g NaCl} \times \frac{1 \text{ mol NaCl}}{58.443 \text{ g NaCl}} = 0.1711 \text{ mol NaCl}$$

Now we can use the molar enthalpy of fusion to calculate the thermal energy needed to melt this sample of NaCl.

$$0.1711 \text{ mol NaCl} \times \frac{30.21 \text{ kJ}}{\text{mol NaCl}} = 5.169 \text{ kJ}$$

☑ **Reasonable Answer Check** We have approximately one sixth mol NaCl, so the energy required to melt this amount of NaCl should be approximately one sixth of the molar enthalpy of fusion, and it is.

PROBLEM-SOLVING PRACTICE 11.4

Calculate the thermal energy transfer required to melt 0.500 mol NaI at its normal melting point.

EXERCISE 11.7 Heat Liberated upon Crystallization

Which would liberate more thermal energy, the crystallization of 2 mol liquid bromine or the crystallization of 1 mol liquid water?

Sublimation and Deposition

Atoms or molecules can escape directly from the solid to the gas phase, a process known as **sublimation.** The enthalpy change is the *enthalpy of sublimation.* The reverse process, in which a gas is converted directly to a solid, is called **deposition.** The enthalpy change for this exothermic process (the enthalpy of deposition) has the same magnitude as the enthalpy of sublimation, but the opposite sign.

$$\text{Solid} \underset{\text{enthalpy of deposition}}{\overset{\text{enthalpy of sublimation}}{\rightleftharpoons}} \text{Gas} \qquad \Delta H_{\text{sublimation}} = -\Delta H_{\text{deposition}}$$

Figure 11.11 Dry Ice. In this photo the cold vapors of CO_2 cause moisture, which is seen as wispy white clouds, to condense. Being more dense than air at room temperature, the CO_2 vapors glide slowly downward toward the table top.

One common substance that sublimes at normal atmospheric pressure is solid carbon dioxide (Dry Ice), which has a vapor pressure of 1 atm at −78 °C (Figure 11.11). Having such a high vapor pressure below its melting point causes solid carbon dioxide to sublime rather than melt. Because of the high vapor pressure of the solid, liquid carbon dioxide can exist only at pressures much higher than 1 atm.

Have you noticed that snow outdoors and ice cubes in a frost-free refrigerator slowly disappear even if the temperature never gets above freezing? The enthalpy of sublimation of ice is 51 kJ/mol, and its vapor pressure at 0 °C is 4.60 mm Hg. Therefore, ice sublimes readily in dry air when the partial pressure of water vapor is below 4.60 mm Hg (Figure 11.12). Given enough air passing over it, a sample of ice will sublime completely, leaving no trace behind. In a frost-free refrigerator, a current of dry air periodically blows across any ice formed in the freezer compartment, taking away water vapor (and hence the ice) without warming the freezer enough to thaw the food.

"Freezer burn" refers to loss of water by the process of sublimation from food stored in a freezer.

In the reverse of sublimation, atoms or molecules in the gas phase can deposit (solidify directly) on the surface of a solid. Deposition is used to form thin coatings of metal atoms on surfaces. Audio CD discs and CD-ROM discs, for example, have shiny metallic surfaces of deposited aluminum or gold atoms. To make such discs, a metal filament is heated in a vacuum to a temperature at which metal atoms begin to sublime rapidly off the surface. The plastic compact disc is cooler than the filament, so the metal atoms in the gas phase quickly deposit on the cool surface. The purpose of the metal coating is to provide a reflective surface for the laser beam that reads the pits and lands (unpitted areas) containing the digital audio or data information.

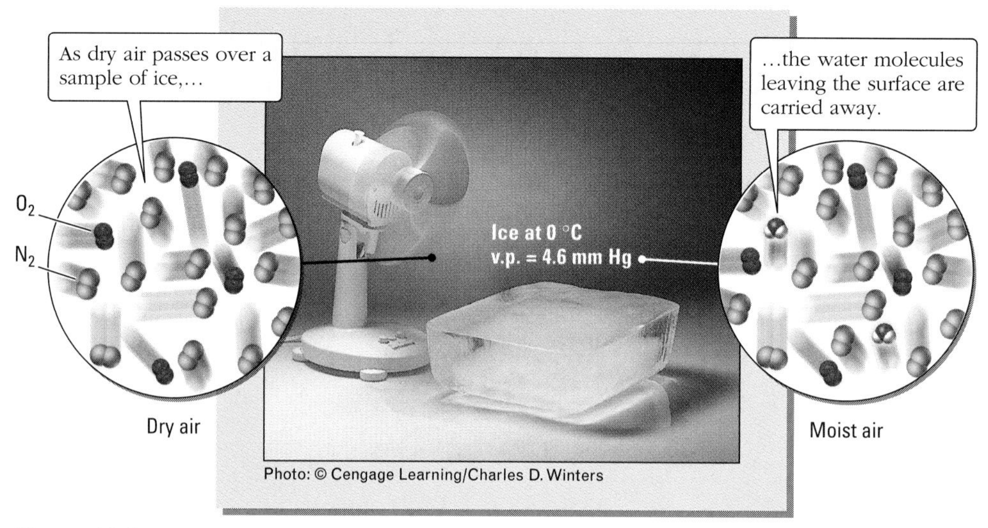

Figure 11.12 **Ice subliming.**

CONCEPTUAL **EXERCISE** **11.8 Frost-Free Refrigeration**

Sometimes, because of high humidity, a frost-free refrigerator doesn't work as efficiently as it should. Explain why.

CONCEPTUAL **EXERCISE** **11.9 Purification by Sublimation**

Sublimation is an excellent means of purification for compounds that will readily sublime. Explain how purification by sublimation works at the nanoscale.

Heating Curve

Heating a solid or liquid increases its temperature as long as its phase does not change and it does not decompose. The size of this temperature increase is governed by the quantity of energy added, the mass of the substance, and its specific heat capacity (***p. 186***). The temperature rises until a phase-transition temperature is reached—for example, 100 °C for water. At this point, the temperature no longer rises. Instead, the energy goes into changing the phase of the substance. Once all of the substance has changed phase, the addition of still more energy causes the temperature to rise. A plot of the temperature of a substance versus the energy added is called a **heating curve** (Figure 11.13).

For example, consider how much energy is required to heat 100. g water from −20 °C to 120 °C, as illustrated in Figure 11.13. The constants used to construct the heating curve are as follows: specific heat capacity of liquid water = 4.184 J g^{-1} °C^{-1}; specific heat capacity of solid water = 2.06 J g^{-1} °C^{-1}; specific heat capacity of water vapor = 1.84 J g^{-1} °C^{-1}; ΔH°_{fusion} = 6.020 kJ/mol; $\Delta H^\circ_{vaporization}$ = 40.7 kJ/mol.

The heating curve has five distinct parts. Three parts relate to heating the water in its three states (red lines). Two parts relate to phase changes (horizontal blue lines).

1. *Heat the ice from −20 °C to 0 °C (A → B).*

$$H_2O(s) \text{ at } -20\ °C \rightarrow H_2O(s) \text{ at } 0\ °C$$

$$\Delta H^\circ = (100.\ g)(2.06\ J\ g^{-1}\ °C^{-1})(20\ °C) = 4120\ J = 4.12\ kJ$$

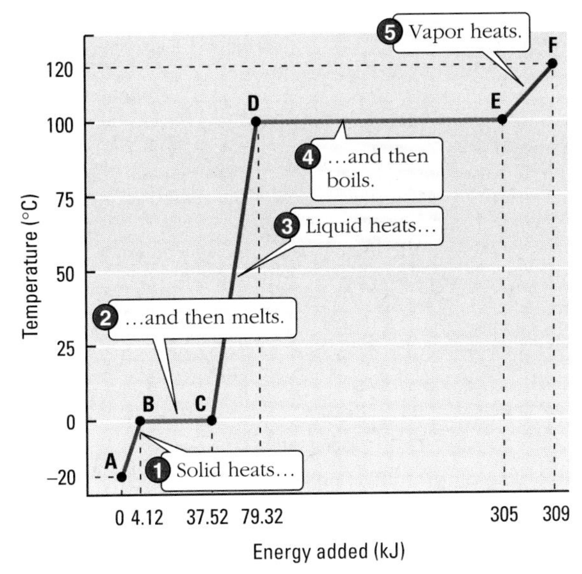

Figure 11.13 Heating curve for 100. g water. (Horizontal axis not to scale.)

Notice that there is no temperature change during a phase change.

2. *Melt the ice at 0 °C (B → C).*

$$H_2O(s) \text{ at } 0\ °C \rightarrow H_2O(\ell) \text{ at } 0\ °C$$

$$\Delta H° = 100.\ g \times \frac{1\ mol}{18.015\ g} \times \frac{6.020\ kJ}{1\ mol} = 33.4\ kJ$$

3. *Heat the water from 0 °C to 100 °C (C → D).*

$$H_2O(\ell) \text{ at } 0\ °C \rightarrow H_2O(\ell) \text{ at } 100\ °C$$

$$\Delta H° = (100.\ g)(4.184\ J\ g^{-1}\ °C^{-1})(100\ °C) = 41800\ J = 41.8\ kJ$$

4. *Boil the water at 100 °C (D → E).*

$$H_2O(\ell) \text{ at } 100\ °C \rightarrow H_2O(g) \text{ at } 100\ °C$$

$$\Delta H° = 100.\ g \times \frac{1\ mol}{18.015\ g} \times \frac{40.7\ kJ}{1\ mol} = 226.\ kJ$$

5. *Heat the water vapor from 100 °C to 120 °C (E → F).*

$$H_2O(g) \text{ at } 100\ °C \rightarrow H_2O(g) \text{ at } 120\ °C$$

$$\Delta H° = (100.\ g)(1.84\ J\ g^{-1}\ °C^{-1})(20\ °C) = 3680\ J = 3.68\ kJ$$

The total energy required to complete the transformation is 309. kJ, the sum of the five steps. Notice that the largest portion of the energy, 226. kJ (73%), goes into vaporizing water at 100 °C to steam at 100 °C.

Phase Diagrams

All three phases (states of matter) and the six interconversions among them shown in Figure 11.8 *(p. 382)* can be represented in a **phase diagram** (Figure 11.14). The pressure-temperature values at which each phase exists are also shown on the phase diagram. Each of the three phases—solid, liquid, gas—is represented. Each solid line in the diagram represents the conditions of temperature and pressure at which equilibrium exists between the two phases on either side of the line. The point at which all three phases are in equilibrium is the **triple point.** Every pure substance that exists in all three phases has a characteristic phase diagram.

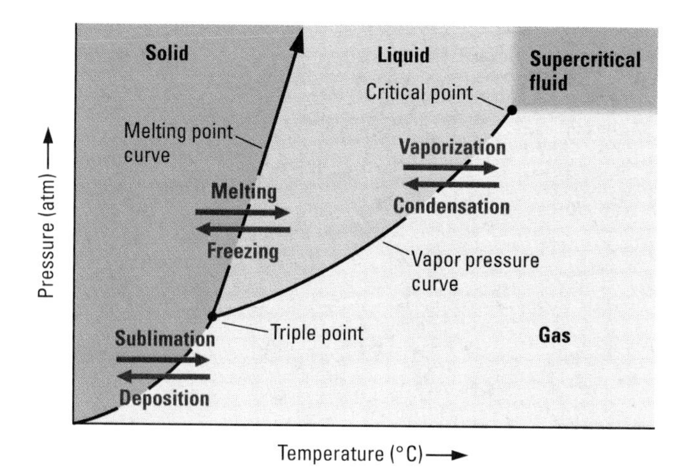

Figure 11.14 Generic phase diagram. The phase diagram shows the pressure-temperature regions in which the substance is solid, liquid, or gas (vapor). It also shows the melting point curve, the vapor pressure curve, and the six interconversions among the three phases.

The phase diagram for water is shown in Figure 11.15. The three shaded regions represent the three phases. The temperature scale has been exaggerated to better illustrate some of the features of water's phase diagram. The temperatures and pressures along the line AD represent conditions at which liquid water and gaseous water are in equilibrium. This line, the *vapor pressure curve,* is the same one shown in Figure 11.5 (⇐ *p. 380*). The line AC, the *melting point curve,* represents the solid/liquid (ice/water) equilibrium, and the line BA represents the solid/gaseous (ice/water vapor) equilibrium. The triple point for water occurs at $P = 4.58$ mm Hg and $T = 0.01$ °C.

You are most likely familiar with the common situation in which ice and liquid water are in equilibrium because of keeping food cool in an ice chest. If you fill an ice chest with ice, some of the ice melts to produce liquid. (Thermal energy is slowly transferred into the ice chest, no matter how well it is insulated.) The ice-liquid water mixture remains at approximately 0 °C until all of the ice has melted. This equilibrium between solid water and liquid water is represented in the phase diagram (Figure 11.15) by point C, at which $P = 760$ mm Hg and $T = 0$ °C.

On a very cold day, ice can have a temperature well below 0 °C. Look along the temperature axis of the phase diagram for water and notice that below 1 atm and

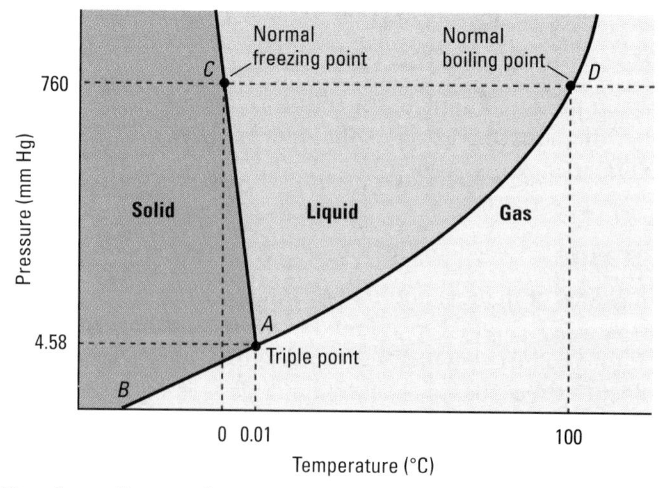

Figure 11.15 The phase diagram for water. The temperature and pressure scales are nonlinear for emphasis. Line AD is the vapor pressure curve, and line AC is the melting point curve. (Axes not to scale.)

0 °C the only equilibrium possible is between ice and water vapor. As a result, if the partial pressure of water in the air is low enough, ice will sublime on a cold day. Sublimation, which is endothermic, takes place more readily when solar radiation provides energy to warm the ice. The sublimation of ice allows snow or ice to gradually disappear even though the temperature does not climb above freezing.

The phase diagram for water is unusual in that the melting point curve *AC* slopes in the opposite direction from that seen for almost every other substance. The right-to-left, or negative, slope of the solid/liquid water equilibrium line is a consequence of the lower density of ice compared to that of liquid water (which is discussed in the next section). Thus, when ice and water are in equilibrium, one way to melt the ice is to apply greater pressure. This is evident from Figure 11.15. If you start at the normal freezing point (0 °C, 1 atm) and increase the pressure, you will move into the area of the diagram that corresponds to liquid water.

Ice skating was long thought to provide a practical example of this property of water. One's body weight applied to a very small area of a skate blade was thought to cause sufficient pressure to form liquid water to make the ice slippery. However, more recent investigations show that the water molecules on the surface of ice are actually in a mobile state, resembling their mobility in liquid water, due to "surface melting." Surface melting is an intrinsic property of the ice surface, even at tens of degrees Celsius below the usual melting point. It arises due to the fact that the surface water molecules have interactions with fewer neighbors than do those molecules deeper in the solid. The ice lattice becomes less and less ordered closer to the surface because the surface molecules have the fewest hydrogen bonds holding them in place. That is, these molecules vibrate around their lattice positions more than the molecules deeper in the solid. At temperatures well below the melting point, the surface molecules begin to take on the characteristics of a liquid. Ice becomes much more slippery when a thin coating of liquid water is present. This thin coat of liquid water acts as a lubricant to help the skates move easily over the ice (Figure 11.16). Although increasing the pressure on ice does cause it to melt, the effect is too small to contribute significantly in the context of ice skating.

The phase diagram for CO_2 (Figure 11.17) is similar to that of water, except for one very important difference—the solid/liquid equilibrium line has a positive slope, as it does for most substances. Solid CO_2 is more dense than liquid CO_2. Notice also that the values of the pressure and temperature axes in the phase diagram for CO_2 are much different from those in the diagram for water. For pressures below about 5 atm, the only equilibrium that can exist is between solid and gaseous CO_2. This means solid

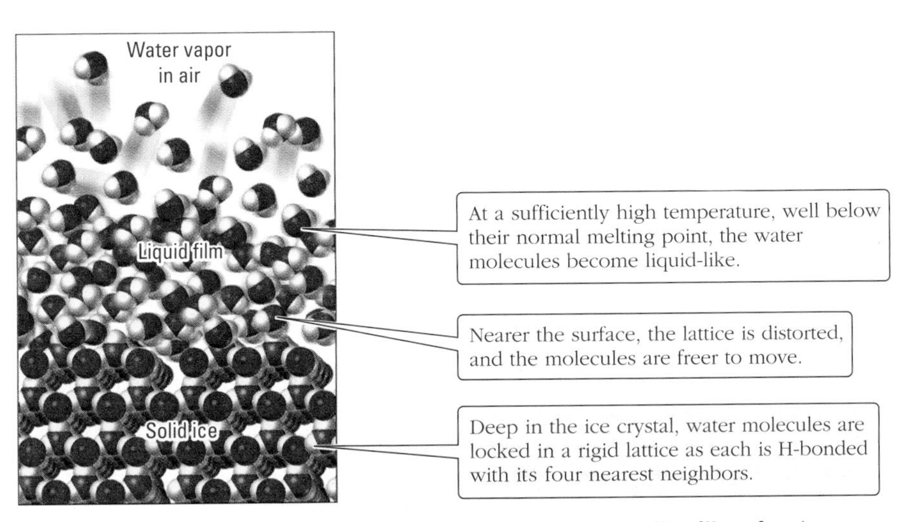

Figure 11.16 **The surface of ice is slick due to a liquid-like film of water.**

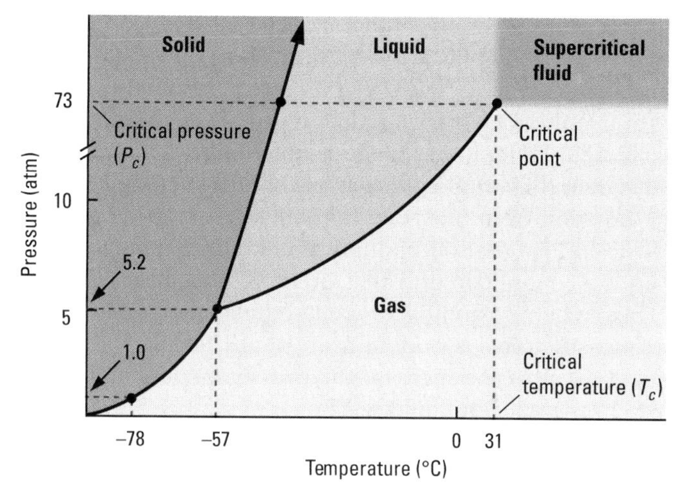

Figure 11.17 The phase diagram for carbon dioxide. (Axes are not to scale.)

CO_2 sublimes when heated. Liquid CO_2 can be produced only at pressures above 5 atm. The temperature range for liquid CO_2 is narrower than that for water. Tanker trucks marked "liquid carbonic" carry liquid CO_2, a convenient source of CO_2 gas for making carbonated beverages.

PROBLEM-SOLVING EXAMPLE 11.5 Phase Diagrams

Use the phase diagram for CO_2 (Figure 11.17) to answer these questions.

(a) What is the temperature and pressure at the triple point?

(b) Starting at the triple point, what phase exists when the pressure is held constant and the temperature is increased to 65 °C?

(c) Starting at $T = -70$ °C and $P = 4$ atm, what phase change occurs when the pressure is held constant and the temperature is increased to -30 °C?

(d) Starting at $T = -30$ °C and $P = 5$ atm, what phase change occurs when the temperature is held constant and the pressure is increased to 10 atm?

Answer (a) $T = -57$ °C, $P = 5.2$ atm (b) Gas (c) Solid to gas (sublimation) (d) Gas to liquid (condensation)

Strategy and Explanation In each case, consult the location on the phase diagram that corresponds to the given temperature and pressure conditions.

(a) The phase diagram shows that the triple point occurs at a temperature of -57 °C and a pressure of 5.2 atm.

(b) Increasing the temperature and holding the pressure constant means moving somewhat to the right from the triple point. This is the gas region of the phase diagram.

(c) At the starting temperature and pressure, the CO_2 is a solid. Increasing the temperature while holding the pressure constant causes the solid to sublime and change directly to the gas phase.

(d) At the starting temperature and pressure, the CO_2 is a gas. Increasing the pressure while holding the temperature constant causes the gas to condense to the liquid phase.

PROBLEM-SOLVING PRACTICE 11.5

On the phase diagram for CO_2 (Figure 11.17), the point that corresponds to a temperature of -13 °C and a pressure of 7.5 atm is on the vapor pressure curve where liquid phase and gas phase are in equilibrium. What phase of CO_2 exists when the pressure remains the same and the temperature is increased by several degrees?

EXERCISE 11.10 Using a Phase Diagram

Consider the phase diagram for CO_2 (Figure 11.17). A curve connects the triple point and the critical point; what is the name of this curve? What phase transition occurs when you traverse this curve at its midpoint going upward, that is, increasing the pressure at constant temperature? Downward? What phase transition occurs when you traverse this curve at its midpoint going from left to right, that is, increasing the temperature at constant pressure? Right to left?

Critical Temperature and Pressure

On the phase diagram for CO_2, notice the upper point where the liquid/gas equilibrium line terminates (Figure 11.17). For CO_2 the conditions at this critical point are 73 atm and 31 °C. For water the termination of the liquid/gas curve occurs at a pressure of 217.7 atm and a temperature of 374.0 °C (Figure 11.15). These conditions at the critical point are called the **critical pressure** (P_c) and **critical temperature** (T_c). At any temperature above T_c the molecules of a substance have sufficient kinetic energy to overcome any attractive forces, and no amount of pressure can cause the substance to act like a liquid again. Above T_c and P_c the substance becomes a **supercritical fluid,** which has a density characteristic of a liquid but the flow properties of a gas, thereby enabling it to diffuse through many substances easily.

The term "fluid" describes a substance that will flow. It is used for both liquids and gases.

Supercritical CO_2 is present in a fire extinguisher under certain conditions. As the phase diagram shows, at any temperature below 31 °C, the pressurized CO_2 (usually at about 73 atm) is a liquid that can slosh around in the container. On a hot day, however, the CO_2 becomes a supercritical fluid and cannot be heard sloshing. In fact, under these conditions the only way to know whether CO_2 is in the extinguisher—without discharging it—is to weigh it and compare its mass with the mass of the empty container, which is usually indicated on a tag attached to the fire extinguisher.

Supercritical fluids are excellent solvents. Because supercritical CO_2 is nonpolar and is therefore a good solvent for nonpolar substances, it is used for dry cleaning. Supercritical CO_2 is also used to extract caffeine from coffee beans and for extracting toxic components from hazardous industrial wastes.

Coffee beans.

CONCEPTUAL
EXERCISE 11.11 The Behavior of CO_2

If liquid CO_2 is slowly released to the atmosphere from a cylinder, what state will the CO_2 be in? If the liquid is suddenly released, as in the discharge of a CO_2 fire extinguisher, why is solid CO_2 seen? Can you explain this phenomenon on the basis of the phase diagram alone, or do you need to consider other factors?

11.4 Water: An Important Liquid with Unusual Properties

Earth is sometimes called the blue planet because the large quantities of water on its surface make it look blue from outer space. Three quarters of the globe is covered by oceans, and vast ice sheets cover the poles. Large quantities of water are present in soils and rocks on the surface. Water is essential to almost every form of life, has played a key role in human history, and is a significant factor in weather and climate. Water's importance is linked to its unique properties (Table 11.4). Ice floats on water; in contrast, when most substances freeze, the solid sinks in its liquid. More thermal energy must be transferred to melt ice, heat water, and vaporize water than to accomplish

Table 11.4 Unusual Properties of Water

Property	Comparison with Other Substances	Importance in Physical and Biological Environment
Specific heat capacity ($4.18 \text{ J g}^{-1} {}^\circ\text{C}^{-1}$)	Highest of all liquids and solids except NH_3	Moderates temperature in the environment and in organisms; climate affected by movement of water (e.g., Gulf Stream)
Heat of fusion (333 J/g)	Highest of all molecular solids except NH_3	Freezing water releases large quantity of thermal energy; used to save crops from freezing by spraying them with liquid water
Heat of vaporization (2250 J/g)	Highest of all molecular substances	Condensation of water vapor in clouds releases large quantities of thermal energy, fueling storms
Surface tension ($7.3 \times 10^{-2} \text{ J/m}^2$)	Highest of all molecular liquids	Contributes to capillary action in plants; causes formation of spherical droplets; supports insects on water surfaces
Thermal conductivity ($0.6 \text{ J s}^{-1} \text{ m}^{-1} {}^\circ\text{C}^{-1}$)	Highest of all molecular liquids	Provides for transfer of thermal energy within organisms; rapidly cools organisms immersed in cold water, causing hypothermia

these changes for the same mass of almost any other substance. Water has the largest thermal conductivity and the highest surface tension of any molecular substance in the liquid state. In other words, this most common of substances in our daily lives has properties that are highly unusual. These properties are crucial for the welfare of our planet and our species.

"If there is magic on this planet, it is contained in water. . . . Its substance reaches everywhere."
—Loren Eiseley

Most of water's unusual properties can be attributed to its molecules' unique capacity for hydrogen bonding. As seen from Figure 9.21 (⇐ *p. 334*), one water molecule can participate in four hydrogen bonds to other water molecules. When liquid water freezes, a three-dimensional network of water molecules forms to accommodate the maximum hydrogen bonding. In the crystal lattice of ice, the oxygen atoms lie at the corners of puckered, six-sided rings. Considerable open space is left within the rings, forming empty channels that run through the entire crystal lattice, lowering its density.

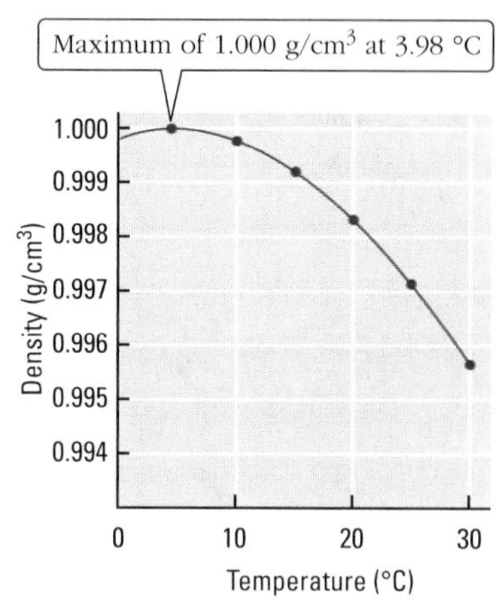

Figure 11.18 The density of liquid water between 0 °C and 30 °C. The density is largest at 1.000 g/cm^3 at 3.98 °C. The density of ice at 0 °C is 0.917 g/cm^3.

When ice melts to form liquid water, approximately 15% of the hydrogen bonds are broken and the rigid ice lattice collapses. This makes the density of liquid water greater than that of ice at the melting point. The density of ice at 0 °C is 0.917 g/cm^3, and that of liquid water at 0 °C is 0.998 g/cm^3. The density difference is not large, but it is enough so that ice floats on the surface of the liquid. This explains why about 90% of an iceberg is submerged and about 10% is above water.

As the liquid water is warmed further, more hydrogen bonds break, and more empty space is filled by water molecules. The density continues to increase until a temperature of 3.98 °C and a density of 1.000 g/cm^3 are reached. As the temperature rises beyond 3.98 °C, increased molecular motion causes the molecules to push each other aside more vigorously, the empty space between molecules increases, and the density decreases by about 0.0001 g/cm^3 for every 1 °C temperature rise above 3.98 °C (Figure 11.18).

Because of this unusual variation of density with temperature, when water in a lake is cooled to 3.98 °C, the higher density causes the cold water to sink to the bottom. Water cooled below 3.98 °C is less dense and stays on the surface, where it can be cooled even further. Consequently, the water at the bottom of the lake remains at

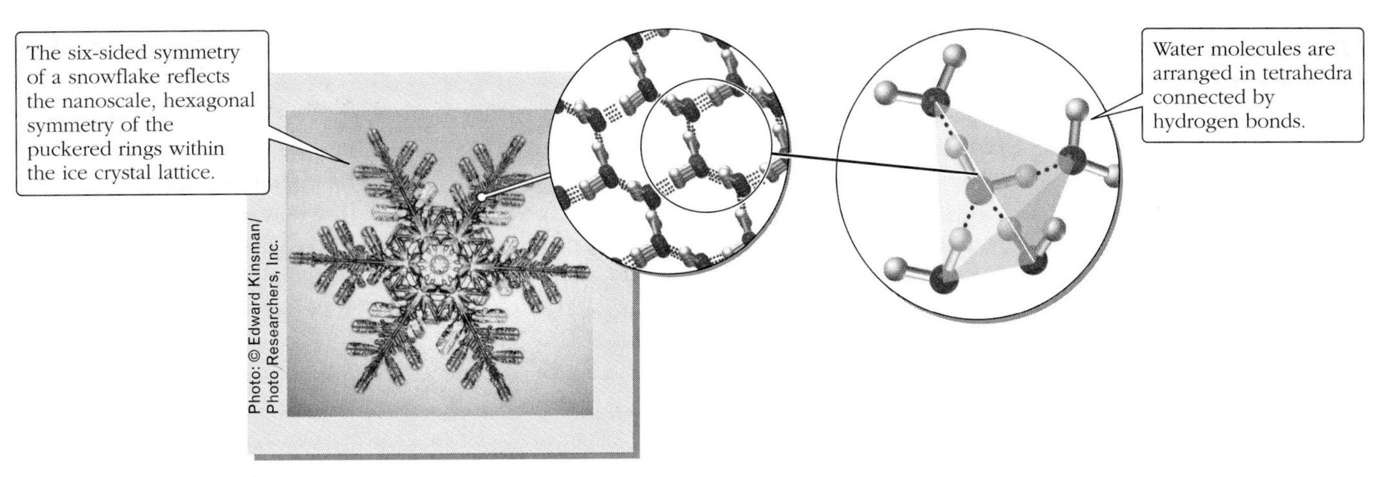
The six-sided symmetry of a snowflake reflects the nanoscale, hexagonal symmetry of the puckered rings within the ice crystal lattice.
Water molecules are arranged in tetrahedra connected by hydrogen bonds.
Photo: © Edward Kinsman/ Photo Researchers, Inc.

Open-cage structure of ice.

3.98 °C, while that on the surface freezes. Ice on the surface insulates the remaining liquid water from the cold air, and, unless the lake is quite shallow, not all of the water freezes. This allows fish and other organisms to survive without being frozen in the winter. When water at 3.98 °C sinks to the bottom of a lake in the fall, it carries with it dissolved oxygen. Nutrients from the bottom are brought to the surface by the water it displaces. This is called "turnover" of the lake. The same thing happens in the spring, when the ice melts and water on the surface warms to 3.98 °C. Spring and fall turnovers are essential to maintain nutrient and oxygen levels required by fish and other lake-dwelling organisms.

When water is heated, the increased molecular motion breaks additional hydrogen bonds. The strength of these intermolecular forces requires that considerable energy be transferred to raise the temperature of 1 g of water by 1 °C. That is, water's specific heat capacity is quite large. Many hydrogen bonds are broken when liquid water vaporizes, because the molecules are completely separated. This gives rise to water's very large enthalpy of vaporization, which is a factor in humans' ability to regulate their body temperature by evaporation of sweat. As we have already described, hydrogen bonds are broken when ice melts, and this requires a large enthalpy of fusion. Its larger-than-normal enthalpy changes upon vaporization and freezing, together with its large specific heat capacity, allow water to moderate climate and influence weather by a much larger factor than other liquids could. In the vicinity of a large body of water, summer temperatures do not get as high and winter temperatures do not get as low (at least until the water freezes over) as they do far away from water ***(⇐ p. 187)***. Seattle is farther north than Minneapolis, for example, but Seattle has a much more moderate climate because it borders the Pacific Ocean and Puget Sound.

11.5 Types of Solids

The relationship of nanoscale structure to macroscale properties is a central theme of this text. Nowhere is the influence of the nanoscale arrangement of atoms, molecules, or ions on properties more evident than in the study of solids. Chemists, in collaboration with physicists, engineers, and other scientists, explore such relationships as they work to create new and useful materials. The nature of solid substances is determined by the type of forces holding them together. *Ionic solids,* such as common table salt, NaCl, are held together by electrostatic interactions between cations and anions. *Molecular solids,* such as water in the form of ice, are held together by intermolecular interactions such as London forces, dipole-dipole forces, and hydrogen bonding. *Network solids,* such as diamond, are bonded into infinite molecules by covalent bonds between

Table 11.5 Structures and Properties of Various Types of Solid Substances

Type	Examples	Structural Units
Ionic (*p. 71*)	NaCl, K_2SO_4, $CaCl_2$, $(NH_4)_3PO_4$	Positive and negative ions (some polyatomic); no discrete molecules
Metallic (Section 11.8)	Iron, silver, copper, other metals and alloys	Metal atoms (positive metal ions surrounded by an electron sea)
Molecular (*p. 63*)	H_2, O_2, I_2, H_2O, CO_2, CH_4, CH_3OH, CH_3COOH	Molecules with covalent bonds
Network (Section 11.7)	Graphite, diamond, quartz, feldspars, mica	Atoms held in an infinite one-, two-, or three-dimensional network
Amorphous (glassy)	Glass, polyethylene, nylon	Covalently bonded networks of atoms or collections of large molecules with short-range order only

atoms. Table 11.5 summarizes the characteristics and physical properties of the major types of solid substances. By classifying a substance as one of these types of solid, you will be able to form a reasonably good idea of what general physical properties to expect, even for a substance that you have never encountered before.

A solid is rigid, having its own shape rather than assuming the shape of its container as a liquid does. Solids have varying degrees of hardness that depend on the kinds of atoms in the solid and the types of forces that hold the atoms, molecules, or ions of the solid together. For example, talc (soapstone, Figure 11.19a), which is used as a lubricant and in talcum powder, is one of the softest solids known. At the atomic level, talc consists of layered sheets containing silicate and magnesium ions. Attractive forces between these sheets are very weak, so one sheet of talc can slide along another and be removed easily from the rest. In contrast, diamond (Figure 11.19b) is one of the hardest solids known. In diamond, each carbon atom is covalently bonded to four neighbors in a very symmetrical tetrahedral arrangement. Each of those neighbor atoms is in turn strongly bonded to four other carbon atoms, and so on throughout the solid (a network solid, Section 11.7). Because of the number and strength of the bonds holding each carbon atom to its neighbors, diamond is so hard that it can scratch or cut almost any other solid. For this reason diamonds are used in cutting tools and abrasives, which are more important commercially than diamonds used as gemstones.

(a)

(b)

Figure 11.19 (a) Talc. Although they do not appear to be soft, these talc crystals are so soft that they can be crushed between one's fingers. **(b) Diamond.** One of the hardest substances known.

Although all solids consist of atoms, molecules, or ions in relatively immobile positions, some solids exhibit a greater degree of regularity than others. In **crystalline solids,** the long-range ordered arrangement of the individual particles is reflected in the planar faces and sharp angles of the crystals. Salt crystals, minerals, gemstones, and ice are examples of crystalline solids. **Amorphous solids** are somewhat like liquids in that they exhibit very little long-range order, yet they are hard and have definite shapes. Ordinary glass is an amorphous solid, as are organic polymers such as polyethylene and polystyrene. The remainder of this chapter is devoted to explaining the nanoscale structures that give rise to the properties of solids summarized in Table 11.5.

PROBLEM-SOLVING EXAMPLE 11.6 Types of Solids

What types of solids are these substances?

(a) Sucrose, $C_{12}H_{22}O_{11}$ (table sugar), has a melting point of about 185 °C. It has poor electrical conductance both as a solid and as a liquid.

(b) Solid Na_2SO_4 has a melting point of 884 °C and has low electrical conductivity that increases dramatically when the solid melts.

Forces Holding Units Together	Typical Properties
Ionic bonding; attractions among charges on positive and negative ions	Hard; brittle; high melting point; poor electrical conductor as solid, good as molten liquid; often water-soluble
Metallic bonding; electrostatic attraction among metal ions and electrons	Malleable; ductile; good electrical conductor in solid and liquid; good heat conductor; wide range of hardness and melting point
London forces, dipole-dipole forces, hydrogen bonds	Low to moderate melting point and boiling point; soft; poor electrical conductor in solid and liquid
Covalent bonds (directional electron-pair bonds)	Wide range of hardness and melting point (three-dimensional bonding > two-dimensional bonding > one-dimensional bonding); poor electrical conductor with some exceptions
Covalent bonds (directional electron-pair bonds)	Noncrystalline; wide temperature range for melting; poor electrical conductor, with some exceptions

Answer (a) Molecular solid (b) Ionic solid

Strategy and Explanation

(a) Sucrose molecules contain covalently bonded atoms. Sucrose's properties correspond to those of a molecular solid since it has poor electrical conductance both as a solid and as a liquid.

(b) Sodium sulfate is composed of Na^+ and SO_4^{2-} ions. The given properties are consistent with those of ionic solids. Ionic solids have high melting points and are poor conductors as solids but good conductors as liquids.

PROBLEM-SOLVING PRACTICE 11.6

What types of solids are these substances?

(a) The hydrocarbon decane, $C_{10}H_{22}$, has a melting point of -31 °C and is a poor electrical conductor.

(b) Solid $MgCl_2$ has a melting point of 780 °C and conducts electricity only when melted.

CONCEPTUAL EXERCISE 11.12 Lead into Gold?

Imagine a sample of gold and a sample of lead, each with a smoothly polished surface. The surfaces of these two samples are placed in contact with one another and held in place, under pressure, for about one year. After that time the two surfaces are analyzed. The gold surface is tested for the presence of lead, and the lead surface is tested for the presence of gold. Predict what the outcome of these two tests will be and explain what has happened, if anything.

11.6 Crystalline Solids

The beautiful regularity of ice crystals, crystalline salts, and gemstones suggests that they must have some *internal* regularity. Toward the end of the eighteenth century, scientists found that shapes of crystals can be used to identify minerals. The angles at which crystal faces meet are characteristic of a crystal's composition. The shape of each crystalline solid reflects the shape of its **crystal lattice**—the orderly, repeating arrangement of atoms, molecules, or ions that shows the position of each individual particle. In such a lattice each atom, molecule, or ion is surrounded by neighbors in exactly the same arrangement. Each crystal is built up from a three-dimensional repetition of the same pattern, which gives the crystal long-range order throughout.

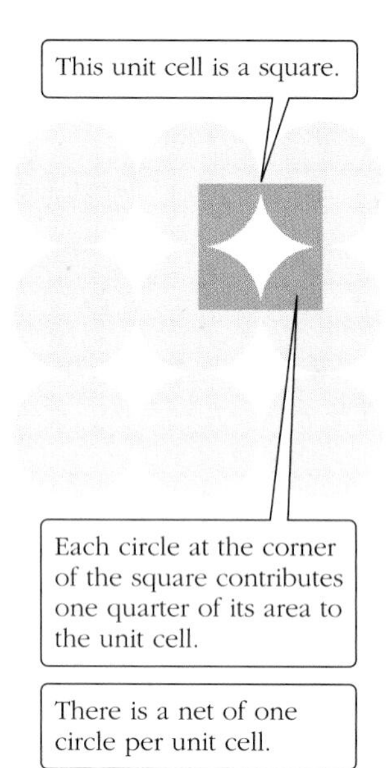
This unit cell is a square.
Each circle at the corner of the square contributes one quarter of its area to the unit cell.
There is a net of one circle per unit cell.

Figure 11.20 Unit cell for a two-dimensional solid made from flat, circular objects, such as coins.

Unit Cells

A convenient way to describe and classify the repeating pattern of atoms, molecules, or ions in a crystal is to define a small segment of a crystal lattice as a **unit cell**—a small part of the lattice that, when repeated along the directions defined by its edges, reproduces the entire crystal structure. To help understand the idea of the unit cell, look at the simple two-dimensional array of circles shown in Figure 11.20. The same size circle is repeated over and over, but a circle is not a proper unit cell because it gives no indication of its relationship to all the other circles. A better choice is to recognize that the centers of four adjacent circles lie at the corners of a square and to draw four lines connecting those centers. A square unit cell results. Figure 11.20 shows this simple unit cell superimposed on the circles. As you look at the unit cell drawn in darker blue, notice that each of four circles contributes one quarter of itself to the unit cell, so a net of one circle is located within the unit cell. When this unit cell is repeated by moving the square parallel to its edges (that is, when unit cells are placed next to and above and below the first one), the two-dimensional lattice results. Notice that the corners of a unit cell are equivalent to each other and that collectively they define the crystal lattice.

The three-dimensional unit cells from which all known crystal lattices can be constructed fall into only seven categories. These seven types of unit cells have edges of different relative lengths that meet at different angles. We will limit our discussion here to one of these types—cubic unit cells composed of atoms or monatomic ions. Such unit cells are quite common in nature and are simpler and easier to visualize than the other types of unit cells. The principles illustrated, however, apply to all unit cells and all crystal structures, including those composed of polyatomic ions and large molecules.

Cubic unit cells have edges of equal length that meet at 90° angles. There are three types: *primitive (simple) cubic (psc), body-centered cubic (bcc),* and *face-centered cubic (fcc)* (Figure 11.21). Many metals and ionic compounds crystallize in cubic unit cells.

psc: 1 atom/unit cell
bcc: 2 atoms/unit cell
fcc: 4 atoms/unit cell

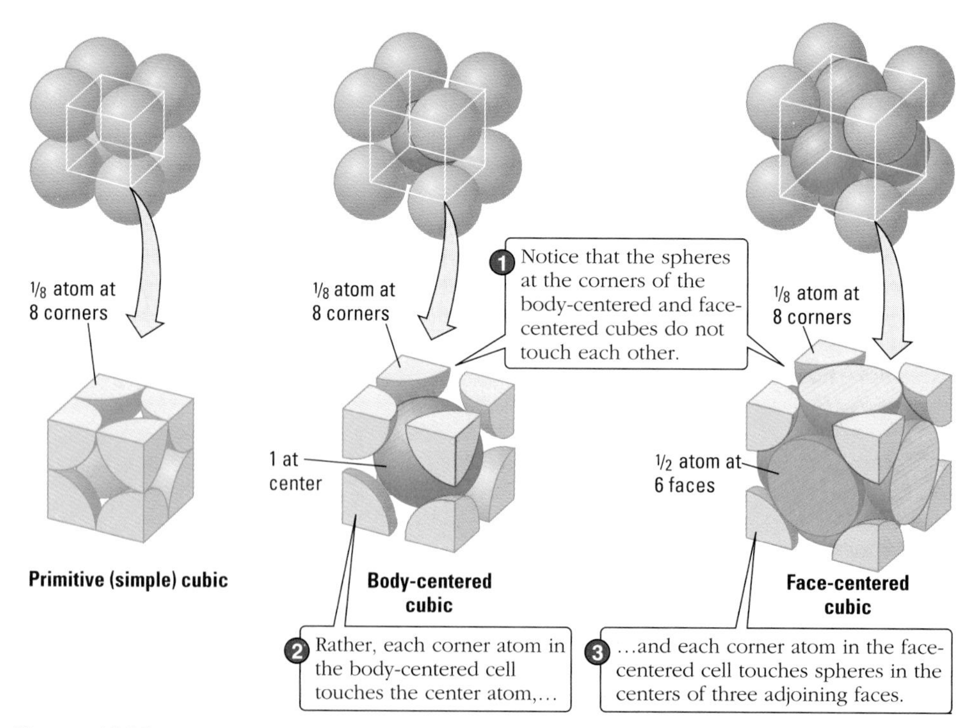
1/8 atom at 8 corners
1/8 atom at 8 corners
1 Notice that the spheres at the corners of the body-centered and face-centered cubes do not touch each other.
1/8 atom at 8 corners
1 at center
1/2 atom at 6 faces
Primitive (simple) cubic
Body-centered cubic
Face-centered cubic
2 Rather, each corner atom in the body-centered cell touches the center atom,...
3 ...and each corner atom in the face-centered cell touches spheres in the centers of three adjoining faces.

Figure 11.21 The three different types of cubic unit cells. The top row shows the lattice points of the three cells superimposed on space-filling spheres centered on the lattice points. The bottom row shows the three cubic unit cells with only the parts of each atom that are within the unit cell.

In metals, all three types of cubic unit cells have identical atoms centered at each corner of the cube. When the cubes pack into three-dimensional space, an atom at a corner is shared among eight cubes (Figure 11.21); thus only one eighth of each corner atom is actually within the unit cell. Since a cube has eight corners and since one eighth of the atom at each corner belongs to the unit cell, the net result is $8 \times \frac{1}{8} = 1$ *atom per primitive (simple) cubic unit cell.*

In the bcc unit cell an additional atom is at the center of the cube and lies entirely within the unit cell. This atom, combined with the net of one atom from the corners, gives a total of *two atoms per body-centered cubic unit cell.*

In the face-centered cubic unit cell six atoms lie in the centers of the faces of the cube. One half of each of these atoms belongs to the unit cell (Figure 11.21). In this case there is a net result of $6 \times \frac{1}{2} = 3$ atoms per unit cell, plus the net of one atom contributed by the corners, for a total of *four atoms per face-centered cubic unit cell.* The number of atoms per unit cell helps to determine the density of a solid.

Examples of metals with cubic unit cells:

Primitive (simple) cubic: Po

Body-centered cubic: alkali metals, V, Cr, Mo, W, Fe

Face-centered cubic: Ni, Cu, Ag, Au, Al, Pb

EXERCISE 11.13 Counting Atoms in Unit Cells of Metals

Crystalline polonium has a primitive (simple) cubic unit cell, lithium has a body-centered cubic unit cell, and calcium has a face-centered cubic unit cell. How many Po atoms belong to one unit cell? How many Li atoms belong to one unit cell? How many Ca atoms belong to one unit cell? Draw each unit cell. Indicate on your drawing what fraction of each atom lies within the unit cell.

Closest Packing of Spheres

In crystalline solids, the atoms, molecules, or ions are usually arranged as closely as possible so that their interactions are maximized, which results in a stable structure. This arrangement is most easily illustrated for metals, in which the individual particles are identical atoms that can be represented as spheres. The arrangement known as **closest packing** can be built up as a series of layers of atoms. In each layer of equal-sized spheres, as shown in Figure 11.22, part 1, each sphere is surrounded by six neighbors. A second layer can be put on top of the first layer so the spheres nestle into the depressions of the first layer (Figure 11.22, part 2). Then a third layer can be added. There are two ways to add the third layer, each of which yields a different structure, as shown in Figure 11.22, parts 3 and 4. If the third layer is directly above the first layer, an *ababab* arrangement results, called a **hexagonal close-packed** structure. In this *ababab* structure, the centers of the spheres of the third layer are directly above the centers of the spheres of the first layer. If the third layer is *not* directly above the first layer, an *abcabc* arrangement results, giving a **cubic close-packed** structure. In the *abcabc* structure, the centers of the spheres of the third layer are directly above the holes between the spheres of the first layer, and the centers of the spheres of the fourth layer are directly above the centers of the spheres of the first layer. The unit cell for a cubic close-packed structure is face-centered cubic (fcc).

In *ababab* packing, the third layer repeats the first layer; in *abcabc* packing, the fourth layer repeats the first layer.

In each of these two close-packed structures, each sphere has 12 nearest-neighbor spheres that are touching it. In each of these arrangements, 74% of the total volume of the structure is occupied by spheres and 26% is empty. Compare these percentages with those for the body-centered cubic structure, which is 68% occupied, and for the simple cubic structure, which is only 52% occupied.

Recently, a rigorous mathematical proof was developed that showed the impossibility of packing spheres to get a packing fraction greater than 74%.

The stacked fruits in markets are placed in closest packed arrangements for efficiency.

Unit Cells and Density

The information given so far about unit cells allows us to check whether a proposed unit cell is reasonable. Because a unit cell can be replicated to give the entire crystal lattice, the unit cell should have the same density as the crystal. As an example, we will consider platinum, which has a face-centered cubic unit cell.

Each layer is formed by packing the spheres as closely as possible—each sphere touches six others.

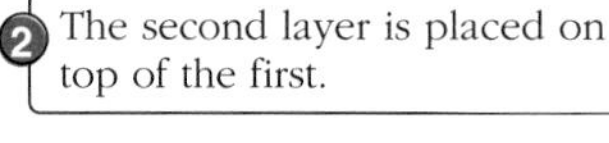

The second layer is placed on top of the first.

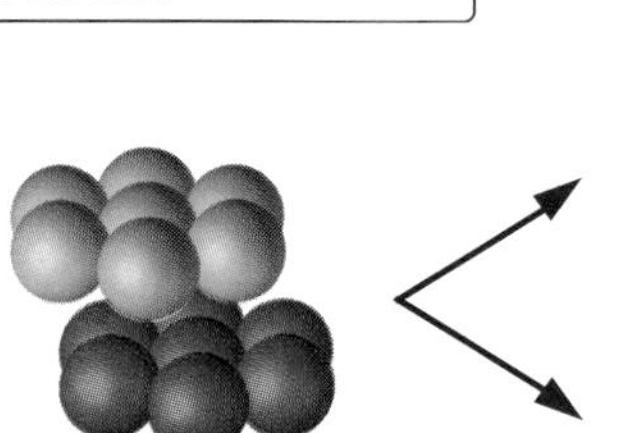

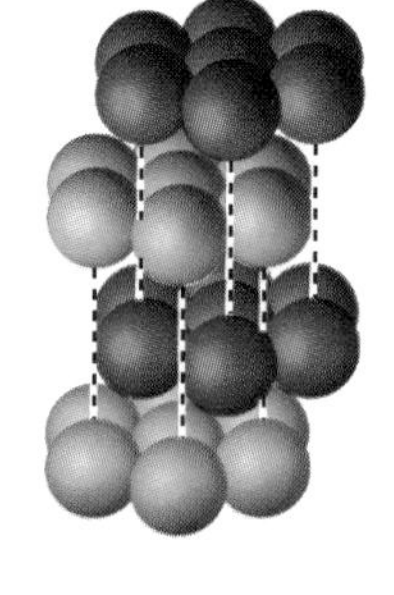

The third-layer atoms are placed directly above the atoms in the first layer, forming an *ababab* arrangement.

The third-layer atoms are placed directly above the holes in the first layer, forming an *abcabc* arrangement.

Figure 11.22 Closest packing of spheres. Hexagonal close-packed structure (part 3) and cubic close-packed structure (part 4).

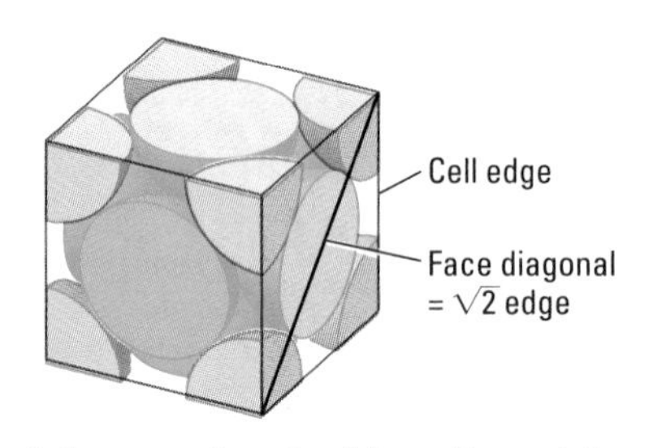

A face-centered cubic unit crystal.

PROBLEM-SOLVING EXAMPLE 11.7 Unit Cell Dimension, Type, and Density

Platinum crystals are fcc, with a density of 21.45 g/cm^3. Platinum has a molar mass of 195.08 g/mol, and platinum atoms have a 139-pm radius. What is the length of a unit cell edge? Is this consistent with Pt being fcc?

Answer 392 pm. Yes, it is consistent with fcc.

Strategy and Explanation Starting with the radius of a Pt atom, we can calculate the diagonal length of the unit cell, the length of an edge of the unit cell, and the unit cell's volume. From the mass of Pt atoms per unit cell and the unit cell's volume we can calculate the density and compare this to the known density.

The mass of one Pt atom is

$$\frac{195.08 \text{ g Pt}}{1 \text{ mol Pt}} \times \frac{1 \text{ mol Pt}}{6.022 \times 10^{23} \text{ atoms Pt}} = 3.239 \times 10^{-22} \text{ g}$$

In an fcc unit cell there are four atoms per unit cell, so the mass of Pt atoms per unit cell is

$$4 \times 3.239 \times 10^{-22} \text{ g} = 1.296 \times 10^{-21} \text{ g Pt per unit cell}$$

Now we calculate the volume of the unit cell. To start we calculate the face diagonal length (see figure in margin).

$$\text{Face diagonal length} = 4 \times \text{Pt atom radius} = 4 \times 139 \text{ pm} = 556 \text{ pm}$$

For an fcc unit cell, the relationship between the edge length of the unit cell and its face diagonal length is (see figure in margin)

$$(\text{Diagonal length})^2 = \text{edge}^2 + \text{edge}^2 = 2(\text{edge})^2$$

so

$$\text{Edge} = \frac{\text{diagonal length}}{\sqrt{2}} = \frac{556 \text{ pm}}{1.414} = 393 \text{ pm}$$

The volume of the unit cell is

$$V = \ell^3 = (393 \text{ pm})^3 = 6.07 \times 10^{-23} \text{ cm}^3$$
$$(1 \text{ pm} = 10^{-12} \text{ m} = 10^{-10} \text{ cm})$$

The density is the mass of Pt atoms per unit cell divided by the unit cell's volume.

$$d = \frac{m}{V} = \frac{1.296 \times 10^{-21} \text{ g Pt}}{6.07 \times 10^{-23} \text{ cm}^3} = 21.4 \text{ g/cm}^3$$

This density value compares well with the given value, and the calculation shows that the data are consistent with an fcc unit cell for the Pt crystal.

☑ **Reasonable Answer Check** The density, atomic weight, unit cell edge length and diagonal length, and Pt atom radius are all in agreement.

PROBLEM-SOLVING PRACTICE 11.7

Gold crystals have a bcc structure. The radius of a gold atom is 144 pm, and the atomic weight of gold is 196.97 amu. Calculate the density of gold.

For each of the three cubic unit cells, there is a close relationship between the radius of the atom forming the unit cell and the unit cell size. For the primitive (simple) cubic unit cell, radii of two atoms form the edge of the unit cell. For the face-centered cubic unit cell, one complete atom diameter and radii of two other atoms form the face diagonal of the unit cell. For the body-centered cubic unit cell, one complete atom diameter and radii of two other atoms form the body diagonal of the unit cell. To summarize:

Primitive (simple) cubic: $2 \times$ (atomic radius) = edge
Face-centered cubic: $4 \times$ (atomic radius) = edge $\times \sqrt{2}$
Body-centered cubic: $4 \times$ (atomic radius) = edge $\times \sqrt{3}$

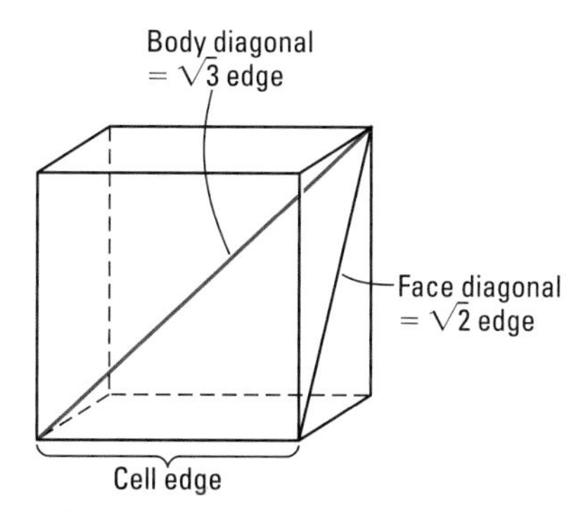

Dimensions of a cubic unit cell.

Ionic Crystal Structures

The crystal structures of many ionic compounds can be described as primitive (simple) cubic or face-centered cubic lattices of spherical negative ions, with smaller positive ions occupying spaces (called holes) among the larger negative ions. The number and locations of the occupied holes are the keys to understanding the relation between the lattice structure and the formula of an ionic compound. The simplest example is an ionic compound in which the hole in a primitive (simple) cubic unit cell (Figure 11.21) is occupied. The ionic compound cesium chloride, CsCl, has such a structure. In it, each cube of Cl^- ions has a Cs^+ ion at its center (Figure 11.23). The spaces occupied by the Cs^+ ions are called cubic holes, and each Cs^+ has eight nearest-neighbor Cl^- ions.

Remember that negative ions are usually larger than positive ions. Therefore, building an ionic crystal is a lot like placing marbles (smaller positive ions) in the spaces among ping-pong balls (larger negative ions).

It is not appropriate to describe the CsCl structure as bcc. In a bcc lattice, the species at the center of the unit cell must be the same as the species at the corner of the cell. In the CsCl unit cell described here, there are chloride ions at the corners and a cesium ion at the center. Another, equally valid, unit cell for CsCl has cesium ions at the corners and a chloride ion at the center.

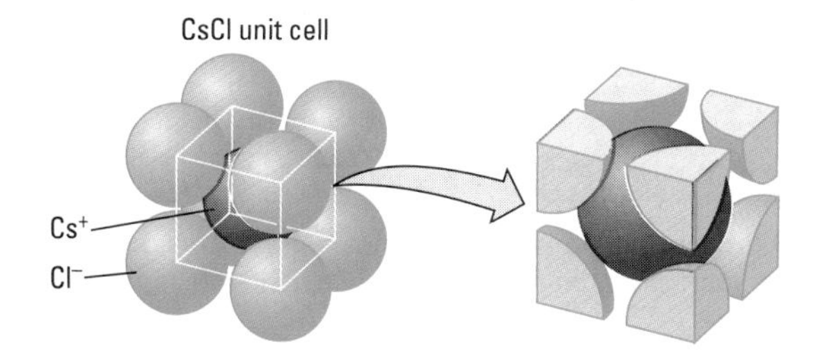

Figure 11.23 Unit cell of the cesium chloride (CsCl) crystal lattice.

Since the edge of a cube in a cubic lattice is surrounded by four cubes, one fourth of a spherical ion at the midpoint of an edge is within any one of the cubes.

The structure of sodium chloride, NaCl, is a very common ionic crystal lattice (Figures 11.24 and 11.25). If you look carefully at Figure 11.24, it is possible to determine the number of Na^+ and Cl^- ions in the NaCl unit cell. There is one eighth of a Cl^- ion at each corner of the unit cell and one half of a Cl^- ion in the middle of each face. The total number of Cl^- ions within the unit cell is

$\frac{1}{8}$ Cl^- per corner × 8 corners = 1 Cl^-

$\frac{1}{2}$ Cl^- per face × 6 faces = 3 Cl^-

Total of 4 Cl^- in a unit cell

There is one fourth of a Na^+ at the midpoint of each edge and a whole Na^+ in the center of the unit cell. For Na^+ ions, the total is

$\frac{1}{4}$ Na^+ per edge × 12 edges = 3 Na^+

1 Na^+ per center × 1 center = 1 Na^+

Total of 4 Na^+ in a unit cell

Thus, the unit cell contains four Na^+ and four Cl^- ions. This result agrees with the formula of NaCl for sodium chloride.

As shown in Figure 11.25, the NaCl crystal lattice consists of an fcc lattice of the larger Cl^- ions, in which Na^+ ions occupy so-called *octahedral* holes—octahedral because each Na^+ ion is surrounded by six Cl^- ions at the corners of an octahedron. Likewise, each Cl^- ion is surrounded by six Na^+ ions. Figure 11.25 also shows a space-filling model of the NaCl lattice, in which each ion is drawn to scale based on its ionic radius.

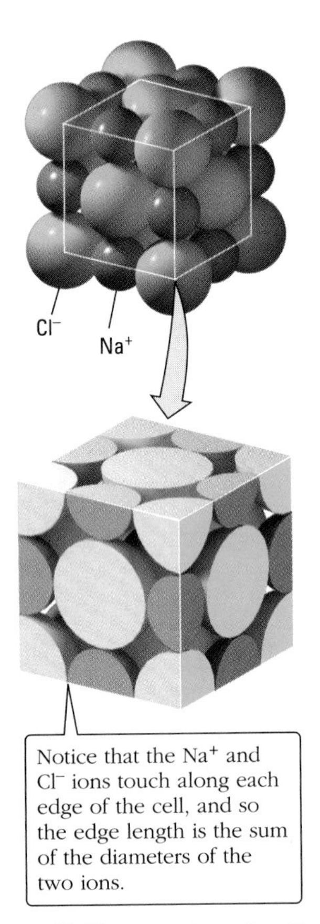

Cl⁻
Na⁺
Notice that the Na⁺ and Cl⁻ ions touch along each edge of the cell, and so the edge length is the sum of the diameters of the two ions.

Figure 11.24 An NaCl unit cell.

CONCEPTUAL **EXERCISE 11.14 Formulas and Unit Cells**

Cesium chloride has a cubic unit cell, as seen in Figure 11.23. Show that the formula for the salt must be CsCl.

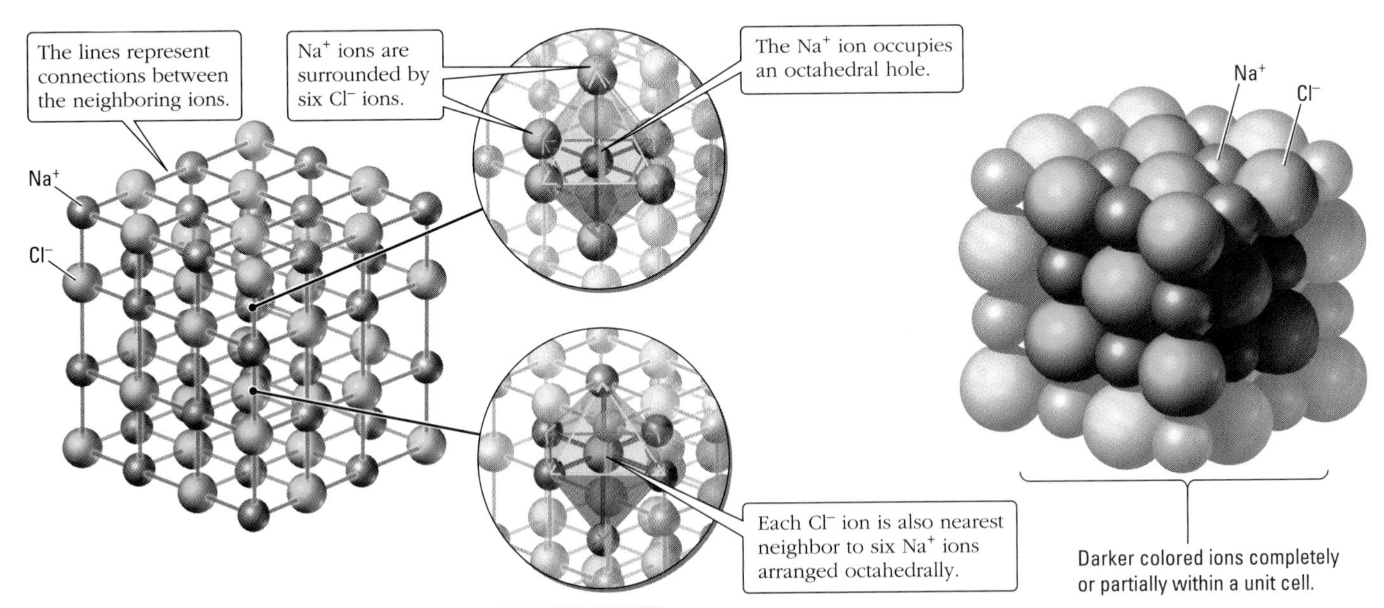

Darker colored ions completely or partially within a unit cell.

Active Figure 11.25 The NaCl crystal lattice. Visit this book's companion website at **www.cengage.com/chemistry/moore** to test your understanding of the concepts in this figure.

PROBLEM-SOLVING EXAMPLE 11.8 Calculating the Volume and Density of a Unit Cell

The unit cell of NaCl is shown in Figure 11.24. The ionic radii for Na^+ and Cl^- are 116 pm and 167 pm, respectively. Calculate the density of NaCl.

Answer 2.14 g/cm^3

Strategy and Explanation To calculate the density, we need to determine the mass and the volume of the unit cell. The mass can be found from the number of formula units in a unit cell and the molar mass of NaCl. The volume can be found from the length of an edge of the cubic unit cell. As seen in Figure 11.24, the Na^+ and Cl^- ions touch along the edge of the unit cell. Thus, the edge length is equal to two Cl^- radii plus two Na^+ radii.

$$\text{Edge} = 167 \text{ pm} + (2 \times 116 \text{ pm}) + 167 \text{ pm} = 566 \text{ pm}$$

The volume of the cubic unit cell is the cube of the edge length.

$$\text{Volume of unit cell} = (\text{edge})^3 = (566 \text{ pm})^3 = 1.81 \times 10^8 \text{ pm}^3$$

Converting this to cm^3 gives

$$1.81 \times 10^8 \text{ pm}^3 \times \left(\frac{10^{-10} \text{ cm}}{\text{pm}}\right)^3 = 1.81 \times 10^{-22} \text{ cm}^3$$

Next we can calculate the mass of a unit cell and divide it by the volume to get the density. With four NaCl formula units per unit cell,

$$4 \text{ NaCl formula units} \times \frac{58.44 \text{ g}}{\text{mol NaCl}} \times \frac{1 \text{ mol}}{6.022 \times 10^{23} \text{ formula units}} = 3.88 \times 10^{-22} \text{ g}$$

This means the density of a NaCl unit cell is

$$\frac{3.88 \times 10^{-22} \text{ g}}{1.81 \times 10^{-22} \text{ cm}^3} = 2.14 \text{ g/cm}^3$$

☑ **Reasonable Answer Check** The experimental density is 2.164 g/cm^3, which is in reasonable agreement with the calculated value. Remember, however, that all experiments have uncertainties associated with them. The density of NaCl calculated from unit cell dimensions could easily have given a value closer to the experimental density if the tabulated radii for the Na^+ and Cl^- ions were slightly smaller. There is, of course, a slight uncertainty in the published radii of all ions.

The general relationships among the unit cell type, ion or atom size, and density for most metallic or ionic solids are as follows:

Mass of 1 formula unit (e.g., NaCl) × number of formula units per unit cell (4) = mass of unit cell

Mass of unit cell ÷ unit cell volume (= edge^3 for NaCl) = density

PROBLEM-SOLVING PRACTICE 11.8

KCl has the same crystal structure as NaCl. Calculate the volume of the unit cell for KCl, given that the ionic radii are K^+ = 152 pm and Cl^- = 167 pm. Compute the density of KCl. Which has the larger unit cell, NaCl or KCl?

11.7 Network Solids

A number of solids are composed of nonmetal atoms connected by a network of covalent bonds. Such **network solids** really consist of huge molecules in which all the atoms are connected to all the others via such a network. Separate small molecules do not exist in a network solid.

The most important network solids are the *silicates.* Many bonding patterns exist among the silicates, but extended arrays of covalently bonded silicon and oxygen atoms are common, such as quartz, SiO_2.

Graphite, Diamond, and Fullerenes

Graphite, diamond, and fullerenes are allotropes of carbon (⇐ ***p. 23***). Graphite's name comes from the Greek *graphein,* meaning "to write," because one of its earliest uses was for writing on parchment. Artists today still draw with charcoal, an impure

The distance between graphite planes is more than twice the distance between the nearest carbon atoms within a plane.

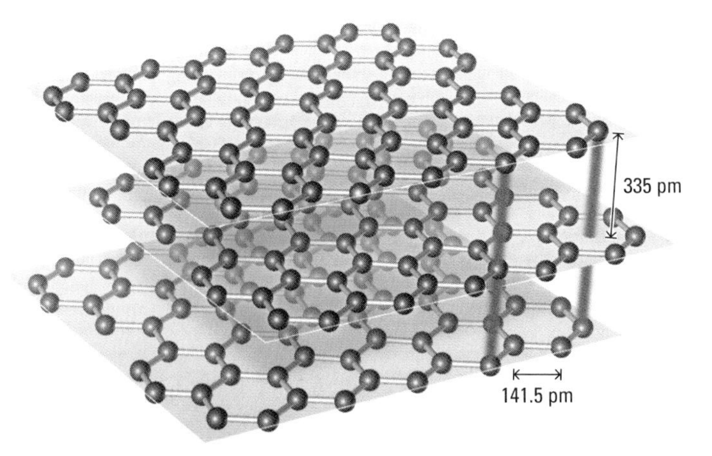

Figure 11.26 The structure of graphite. Three of the many layers of six-membered carbon rings are shown. These layers can slide past one another relatively easily, making graphite a good lubricant. In addition, some of the carbon valence electrons in the layers are delocalized, allowing graphite to be a good conductor of electricity parallel to the layers.

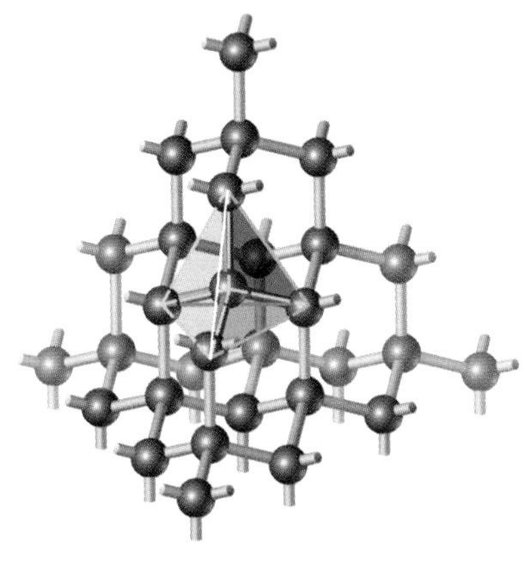

Figure 11.27 The structure of diamond. Each carbon atom is covalently bonded to four other carbon atoms in an extended tetrahedral arrangement.

Synthetic diamonds were first produced in the 1950s.

The C_{60} fullerene molecule, known as a buckyball.

form of graphite, and we write with pencil leads that contain graphite. Graphite is an example of a *planar network solid* (Figure 11.26). Each carbon atom is covalently bonded to three other carbon atoms. The planes consist of six-membered rings of carbon atoms. Each hexagon shares all six of its sides with other hexagons around it, forming a *two-dimensional network* resembling chicken wire. Some of the bonding electrons are able to move freely around this network, so graphite readily conducts electricity. Within a plane, there are strong covalent bonds between carbon atoms, but attractions between the planes are caused by London forces and hence are weaker. Because of this, the planes can easily slip across one another, which makes graphite an excellent solid lubricant for uses such as in locks, where greases and oils are undesirable.

The structure of diamond is also built of six-membered carbon rings, with each carbon atom bonded to four others by single covalent bonds. This structure forms a *three-dimensional network* (Figure 11.27). Because of the tetrahedral arrangement of bonds around each carbon atom, the six-membered rings in the diamond structure are puckered. As a result, diamond (3.51 g/cm^3) is much denser than graphite (2.22 g/cm^3). Also, because its valence electrons are localized in covalent bonds between carbon atoms, diamond does not conduct electricity. Diamond is one of the hardest materials and also one of the best conductors of heat known. It is also transparent to visible, infrared, and ultraviolet radiation.

Buckyballs, the prototype of the third allotrope of carbon, were discovered in the 1980s. Since then, a variety of carbon structures have been made, all classified as **fullerenes.** The prototype structure, C_{60}, is a sphere composed of pentagons and hexagons of carbon atoms similar to a soccer ball. Larger cage-like structures have also been synthesized. In fullerenes, each carbon atom has three carbon atom neighbors. Each carbon atom has sp^2 hybridization and has one unhybridized p orbital. Electrons can be delocalized over these p orbitals, just as they can in graphite. Unlike graphite, each carbon atom and its three bonds are not quite planar, resulting in the curved fullerene structures. Another fullerene geometry is elongated tubes of carbon atoms, called **nanotubes** because of their small size. Nanotubes have been formed with a single wall of carbon atoms similar to graphite rolled into a tube, resembling rolled-up chicken wire. The nanotube is capped at each end with a truncated buckyball (Figure 11.28). Nanoscale materials such as nanotubes have physical and chemical properties different from bulk materials. Scientists are now exploring the possible useful properties of nanotubes and other similar materials.

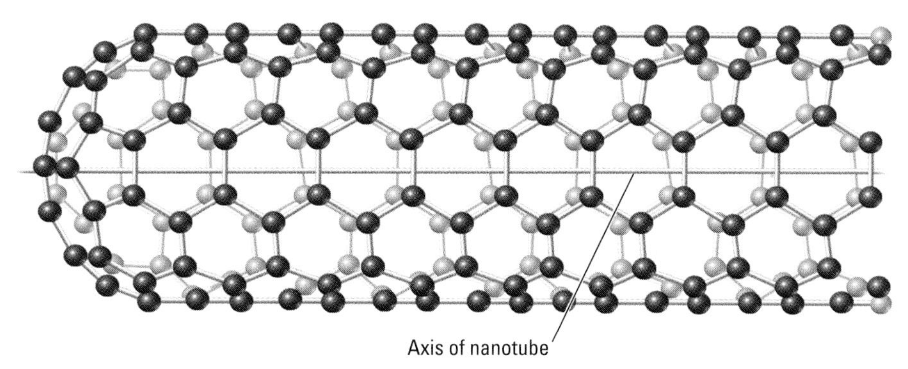

Figure 11.28 A single-walled carbon nanotube composed of hexagons of carbon atoms.

11.8 Metals, Semiconductors, and Insulators

All metals are solids at room temperature except mercury (m.p. −38.8 °C), and they exhibit common properties that we call metallic.

- **High electrical conductivity** Metal wires are used to carry electricity from power plants to homes and offices because electrons in metals are highly mobile.
- **High thermal conductivity** We learn early in life not to touch any part of a hot metal pot because it will transfer heat rapidly and painfully.
- **Ductility and malleability** Most metals are easily drawn into wire (ductility) or hammered into thin sheets (malleability); some metals (gold, for example) are more easily formed into shapes than others.
- **Luster** Polished metal surfaces reflect light. Most metals have a silvery-white color because they reflect all wavelengths equally well.
- **Insolubility in water and other common solvents** No metal dissolves in water, but a few (mainly from Groups 1A and 2A) react with water to form hydrogen gas and solutions of metal hydroxides.

As you will see, these properties are explained by the kinds of bonding that hold atoms together in metals.

The enthalpies of fusion and melting points of metals vary greatly. Low melting points correlate with low enthalpies of fusion, which implies weaker attractive forces holding the metal atoms together. Mercury, which is a liquid at room temperature, has an enthalpy of fusion of only 2.3 kJ/mol. The alkali metals and gallium also have very low enthalpies of fusion and notably low melting points (Table 11.6). Compare these values with those in Table 11.3 for nonmetals *(**p. 384)**.*

Figure 11.29 shows the relative enthalpies of fusion of the metals related to their positions in the periodic table. The transition metals, especially those in the third transition series, have very high melting points and extraordinarily high enthalpies of fusion. Tungsten (W) has the highest melting point (3410 °C) of all the metals, and among all the elements it is second only to carbon as graphite, which has a melting point of 3550 °C. Pure tungsten is used in incandescent light bulbs as the filament, the wire that glows white-hot. No other material has been found to be better since the invention of commercially successful, long-lived light bulbs in 1908 by Thomas Edison and his co-workers.

Table 11.6 Enthalpies of Fusion and Melting Points of Some Metals

Metal	ΔH°_{fusion} (kJ/mol)	Melting Point (°C)
Hg	2.3	−38.8
Ga	7.5	29.78*
Na	2.59	97.9
Li	3.0	180.5
Al	10.7	660.4
U	12.6	1132.1
Fe	13.8	1535.0
Ti	20.9	1660.1
Cr	16.9	1857.0
W	35.2	3410.1

*This means that gallium metal will melt in the palm of your hand from the warmth of your body (37 °C). It happens that gallium is a liquid over the largest range of temperature of any metal. Its boiling point is approximately 2250 °C.

PROBLEM-SOLVING EXAMPLE 11.9 Calculating Enthalpies of Fusion

Use data from Table 11.6 to calculate the thermal energy transfer required to melt 10.0 g chromium at its melting point.

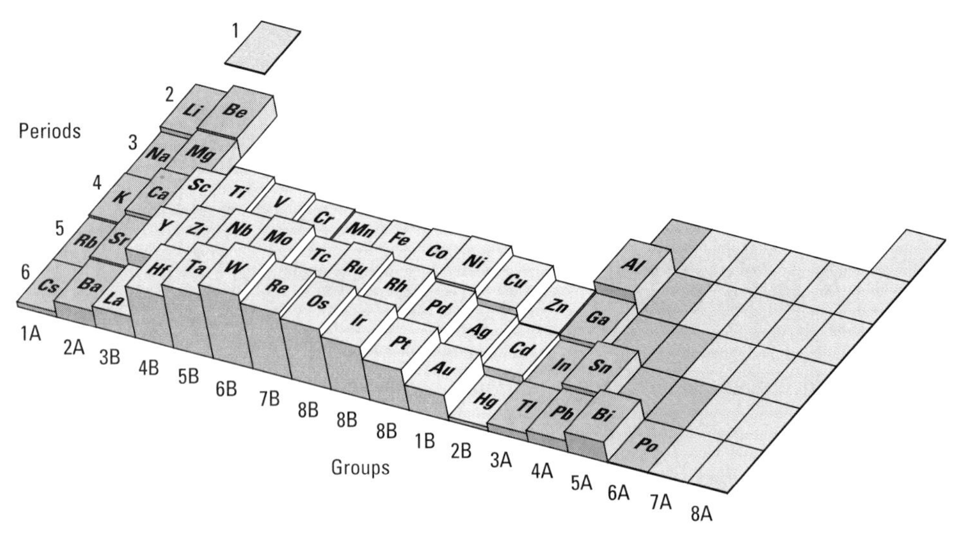

Figure 11.29 Relative enthalpies of fusion for the metals in the periodic table. See Table 11.6 for some numerical values of enthalpies of fusion.

Answer 3.25 kJ

Strategy and Explanation First, we determine how many moles of chromium are present. Then we calculate the energy required from the ΔH°_{fusion} value in Table 11.6.

$$10.0 \text{ g Cr} \times \frac{1 \text{ mol Cr}}{52.00 \text{ g Cr}} = 0.192 \text{ mol Cr}$$

$$0.192 \text{ mol Cr} \times \frac{16.9 \text{ kJ}}{1 \text{ mol Cr}} = 3.25 \text{ kJ}$$

☑ **Reasonable Answer Check** We have about two tenths of a mole of chromium, so the energy required should be about two tenths of the molar enthalpy of fusion, about 17 kJ/mol, or about 3.4 kJ, which compares well with our more accurate answer.

PROBLEM-SOLVING PRACTICE 11.9

Use data from Table 11.6 to calculate the heat transfer required to melt 1.45 g aluminum.

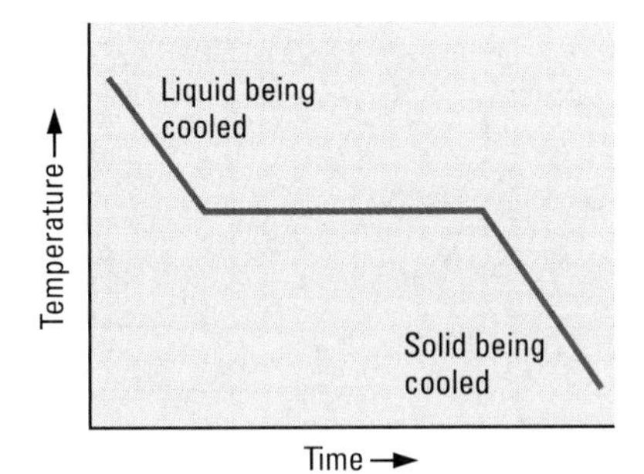

CONCEPTUAL EXERCISE 11.15 Cooling a Liquid Metal Until It Solidifies

When a liquid metal is cooled at a constant rate to the temperature at which it solidifies, and the solid is then cooled to an even lower temperature, the "cooling curve"—a plot of temperature against time—looks like the chart to the left. Account for the shape of this curve. Would all substances exhibit similar curves?

CONCEPTUAL EXERCISE 11.16 Heats of Fusion and Electronic Configuration

Look in Appendix D and compare the electron configurations shown there with the heats of fusion for the metals shown in Table 11.6. Is there any correlation between these configurations and this property? Does strength of attraction among metal atoms correlate with number of valence electrons? Explain.

Electrons in Metals, Semiconductors, and Insulators

Metals behave as though metal cations exist in a "sea" of mobile electrons—the valence electrons of all the metal atoms. **Metallic bonding** is the nondirectional attraction between positive metal ions and the surrounding sea of negative charge (valence electrons). Each metal ion has a large number of near neighbors. The valence electrons are spread throughout the metal's crystal lattice, holding the positive metal ions together. When an electric field is applied to a metal, these valence electrons move toward the positive end of the field, and the metal conducts electricity.

The mobile valence electrons provide a uniform charge distribution in a metal lattice, so the positions of the positive ions can be changed without destroying the attractions among positive ions and electrons. Thus, most metals can be bent and drawn into wire. Conversely, when we try to deform an ionic solid, which consists of a lattice of positive and negative ions, the crystal usually shatters because the balance of positive ions surrounded by negative ions, and vice versa, is disrupted.

To visualize how bonding electrons behave in a metal, first consider the arrangement of electrons in an individual atom far enough away from any neighbor so that no bonding occurs. In such an atom the electrons occupy orbitals that have definite energy levels. In a large number of separated, identical atoms, all of the energy levels are identical. If the atoms are brought closer together, however, they begin to influence one another. The identical energy levels shift up or down and become bands of energy levels characteristic of the large collections of metal atoms (Figure 11.30). An **energy band** is a large group of orbitals whose energies are closely spaced and whose average energy is the same as the energy of the corresponding orbital in an individual atom. In some cases, energy bands for different types of electrons (*s*, *p*, *d*, and so on) overlap; in other cases there is a gap between different energy bands.

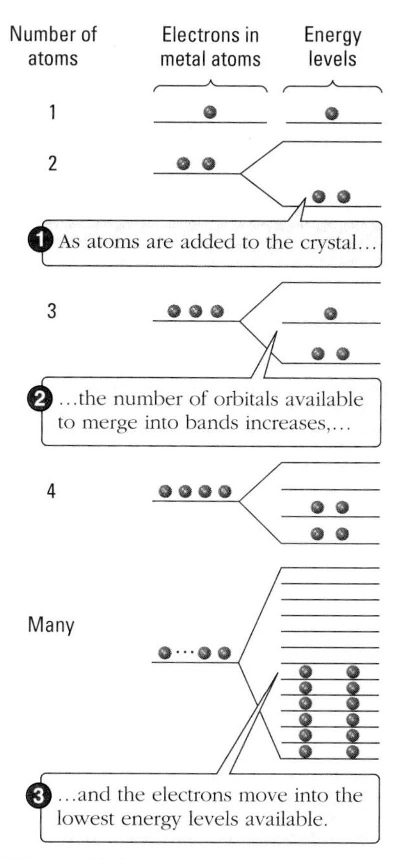

Figure 11.30 Formation of bands of electron orbitals in a metal crystal.

Within each band, electrons fill the lowest energy orbitals much as electrons fill orbitals in atoms or molecules. The number of electrons in a given energy band depends on the number of metal atoms in the crystal. In considering conductivity and other metallic properties, it is usually necessary to consider only valence electrons, as other electrons all occupy completely filled bands in which two electrons occupy every orbital. In these bands, no electron can move from one orbital to another, because there is no empty spot for it.

A band containing the valence electrons is called the **valence band**. If the valence band is partially filled, it requires little added energy to excite a valence electron to a slightly higher energy orbital. Such a small increment of energy can be provided by applying an electric field, for example. The presence of low-energy, empty orbitals into which electrons can move allows the electrons to be mobile and to conduct an electric current (Figure 11.31).

Another band containing higher energy orbitals (the **conduction band**) exists at an average energy above the valence band. In a metal, the valence band and the conduction band overlap, so electrons can move from the valence band to the conduction band freely; this ability explains metals' electrical conductivity. Such a metal with overlapping valence and conduction bands is a **conductor.**

When the conduction band is close in energy to the valence band, the electrons can absorb a wide range of wavelengths in the visible region of the spectrum. As the excited electrons fall back to their lower energy states, they emit their extra energy as visible light, producing the luster characteristic of metals.

The energy band theory also explains why some solids are **insulators,** which do not conduct electricity. In an insulator the valence band and conduction band do not overlap. Rather, there is a large band gap between them (Figure 11.31). Very few electrons have enough energy to move across the large gap from a filled lower energy band to an empty higher energy band, so no current flows through an insulator when an external electric field is applied.

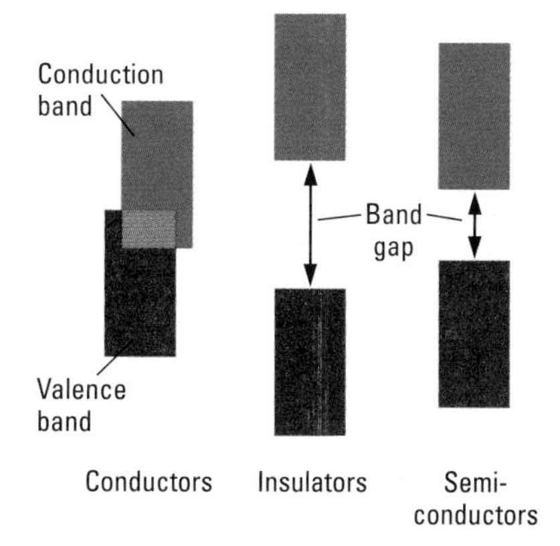

Figure 11.31 Differences in the energy bands of available orbitals in conductors, insulators, and semiconductors.

Ceramic insulators are used on high-voltage electrical transmission lines.

In a **semiconductor** a very *narrow* energy gap separates the valence band and the conduction band (Figure 11.31). At quite low temperatures, electrons remain in the filled lower energy valence band, and semiconductors are not good conductors. At higher temperatures, or when an electric field is applied, some electrons have enough energy to jump across the band gap into the conduction band. This allows an electric current to flow. This property of semiconductors—to switch from insulator to conductor with the application of an external electric field—is the basis for the operation of transistors, the cornerstone of modern electronics.

IN CLOSING

Having studied this chapter, you should be able to . . .

- Explain the properties of surface tension, capillary action, vapor pressure, and boiling point, and describe how these properties are influenced by intermolecular forces (Sections 11.1, 11.2). ■ End-of-chapter Questions assignable in OWL: 19, 22, 37, 77
- Calculate the energy transfers associated with vaporization and fusion (Section 11.3). ■ Questions 15, 25, 27, 64
- Describe the phase changes that occur among solids, liquids, and gases (Section 11.3). ■ Questions 31, 33, 35
- Use phase diagrams to predict what happens when temperature and pressure are changed for a sample of matter (Section 11.3). ■ Questions 38, 40, 70
- Understand critical temperature and critical pressure (Section 11.3).
- Describe and explain the unusual properties of water (Section 11.4).
- Differentiate among the major types of solids (Section 11.5). ■ Questions 44, 45
- Do calculations based on knowledge of simple unit cells and the dimensions of atoms and ions that occupy positions in those unit cells (Section 11.6). ■ Question 49
- Explain the bonding in network solids and how it results in their properties (Section 11.7). ■ Question 56
- Explain metallic bonding and how it results in the properties of metals and semiconductors (Section 11.8). ■ Question 59

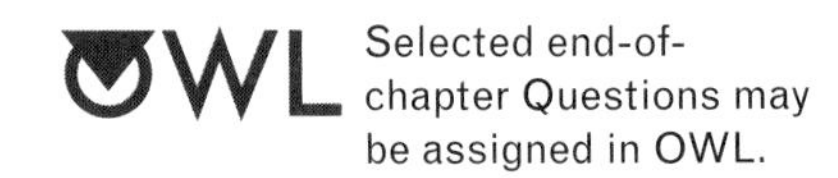

Selected end-of-chapter Questions may be assigned in OWL.

For quick review, download Go Chemistry mini-lecture flashcard modules (or purchase them at **www.ichapters.com**).

Prepare for an exam with a **Summary Problem** that brings together concepts and problem-solving skills from throughout the chapter. Go to **www.cengage.com/chemistry/moore** to download this chapter's Summary Problem from the Student Companion Site.

KEY TERMS

amorphous solids *(Section 11.5)*

boiling *(11.2)*

boiling point *(11.2)*

capillary action *(11.1)*

Clausius-Clapeyron equation *(11.2)*

closest packing *(11.6)*

condensation *(11.3)*

conduction band *(11.8)*

conductor *(11.8)*

critical pressure (P_c) *(11.3)*

critical temperature (T_c) *(11.3)*

crystal lattice *(11.6)*

crystalline solids *(11.5)*

crystallization *(11.3)*

cubic close packing *(11.6)*

cubic unit cell *(11.6)*

deposition *(11.3)*

dew point *(11.2)*

energy band *(11.8)*

equilibrium vapor pressure *(11.2)*

evaporation *(11.3)*

fullerenes *(11.7)*

heating curve *(11.3)*

hexagonal close packing *(11.6)*

insulator *(11.8)*

meniscus *(11.1)*

metallic bonding *(11.8)*

nanotube *(11.7)*
network solids *(11.7)*
normal boiling point *(11.2)*
phase diagram *(11.3)*
relative humidity *(11.2)*
semiconductor *(11.8)*
sublimation *(11.3)*
supercritical fluid *(11.3)*
surface tension *(11.1)*
triple point *(11.3)*
unit cell *(11.6)*
valence band *(11.8)*
vapor pressure *(11.2)*
vaporization *(11.3)*
viscosity *(11.1)*
volatility *(11.2)*

QUESTIONS FOR REVIEW AND THOUGHT

■ denotes questions assignable in OWL.

Blue-numbered questions have short answers at the back of this book and fully worked solutions in the *Student Solutions Manual.*

Review Questions

1. List the concepts of the kinetic-molecular theory that apply to liquids.
2. What causes surface tension in liquids? Name a substance that has a very high surface tension. What kinds of intermolecular forces account for the high value?
3. Define boiling point and normal boiling point.
4. What is the heat of crystallization of a substance, and how is it related to the substance's heat of fusion?
5. What is sublimation?
6. Which of these processes are endothermic?
   (a) Condensation (b) Melting (c) Evaporation
   (d) Sublimation (e) Deposition (f) Freezing
7. What is the unit cell of a crystal?
8. Assuming the same substance could form crystals with its atoms or ions in either simple cubic packing or hexagonal closest packing, which form would have the higher density? Explain.

Topical Questions

The Liquid State

9. Explain on the molecular scale the processes of condensation and vaporization.
10. How would you convert a sample of liquid to vapor without changing the temperature?
11. What is the heat of vaporization of a liquid? How is it related to the heat of condensation of that liquid? Using the idea of intermolecular attractions, explain why the process of vaporization is endothermic.
12. The substances Ne, HF, H_2O, NH_3, and CH_4 all have the same number of electrons. In a thought experiment, you can make HF from Ne by removing a single proton a short distance from the nucleus and having the electrons follow the new arrangement of nuclei so as to make a new chemical bond. You can do the same for each of the other compounds. (Of course, none of these thought experiments can actually be done because of the enormous energies required to remove protons from nuclei.) For all of these substances, make a plot of (a) the boiling point in kelvins versus the number of hydrogen atoms and (b) the molar heat of vaporization versus the number of hydrogen atoms. Explain any trend that you see in terms of intermolecular forces.
13. How much heat energy transfer is required to vaporize 1.0 metric ton of ammonia? (1 metric ton = 10^3 kg.) The ΔH_{vap} for ammonia is 25.1 kJ/mol.
14. The chlorofluorocarbon CCl_3F has an enthalpy of vaporization of 24.8 kJ/mol. To vaporize 1.00 kg of the compound, how much heat energy transfer is required?
15. ■ The molar enthalpy of vaporization of methanol is 38.0 kJ/mol at 25 °C. How much heat energy transfer is required to convert 250. mL of the alcohol from liquid to vapor? The density of CH_3OH is 0.787 g/mL at 25 °C.
16. Some camping stoves contain liquid butane, C_4H_{10}. They work only when the outside temperature is warm enough to allow the butane to have a reasonable vapor pressure (so they are not very good for camping in temperatures below about 0 °C). Assume the enthalpy of vaporization of butane is 24.3 kJ/mol. If the camp stove fuel tank contains 190. g liquid C_4H_{10}, how much heat energy transfer is required to vaporize all of the butane?
17. Mercury is a highly toxic metal. Although it is a liquid at room temperature, it has a high vapor pressure and a low enthalpy of vaporization (294 J/g). What quantity of heat energy transfer is required to vaporize 0.500 mL of mercury at 357 °C, its normal boiling point? The density of Hg(ℓ) is 13.6 g/mL. Compare this heat energy transfer with the amount needed to vaporize 0.500 mL water. See Table 11.2 for the molar enthalpy of vaporization of H_2O.
18. Rationalize the observation that 1-propanol, $CH_3CH_2CH_2OH$, has a boiling point of 97.2 °C, whereas a compound with the same empirical formula, ethyl methyl ether, $CH_3CH_2OCH_3$, boils at 7.4 °C.
19. ■ Briefly explain the variations in the boiling points in the table following. In your discussion be sure to mention the types of intermolecular forces involved.

Compound	Boiling Point (°C)
NH_3	−33.4
PH_3	−87.5
AsH_3	−62.4
SbH_3	−18.4

Vapor Pressure

20. Give a molecular-level explanation of why the vapor pressure of a liquid increases with temperature.
21. Methanol, CH_3OH, has a normal boiling point of 64.7 °C and a vapor pressure of 100 mm Hg at 21.2 °C. Formaldehyde, $H_2C{=}O$, has a normal boiling point of −19.5 °C and a vapor pressure of 100 mm Hg at −57.3 °C. Explain why these two compounds have different boiling points and require different temperatures to achieve the same vapor pressure.
22. ■ The vapor pressure curves for four substances are shown in the plot. Which one of these four substances will have the greatest intermolecular attractive forces at 25 °C? Explain your answer.

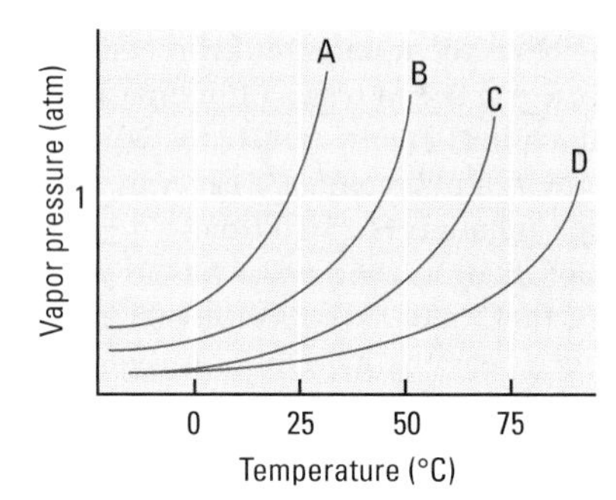

23. The highest mountain in the western hemisphere is Mt. Aconcagua, in the central Andes of Argentina (22,834 ft). If atmospheric pressure decreases at a rate of 3.5 millibar every 100 ft, estimate the atmospheric pressure at the top of Mt. Aconcagua, and then estimate from Figure 11.5 the temperature at which water would boil at the top of the mountain.
24. A liquid has a ΔH_{vap} of 38.7 kJ/mol and a boiling point of 110 °C at 1 atm pressure. What is the vapor pressure of the liquid at 97 °C?
25. ■ A liquid has a ΔH_{vap} of 44.0 kJ/mol and a vapor pressure of 370 mm Hg at 90 °C. What is the vapor pressure of the liquid at 130 °C?
26. The vapor pressure of ethanol, C_2H_5OH, at 50.0 °C is 233 mm Hg, and its normal boiling point at 1 atm is 78.3 °C. What is the ΔH_{vap} of ethanol?
27. ■ What would the ΔH_{vap} be for a substance whose vapor pressure doubled when its temperature was raised from 70.0 °C to 80.0 °C?

Phase Changes: Solids, Liquids, and Gases

28. What does a high melting point and a high enthalpy of fusion tell you about a solid (its bonding or type)?
29. Which would you expect to have the higher enthalpy of fusion, N_2 or I_2? Explain your choice.
30. The enthalpy of fusion for H_2O is about 2.5 times larger than the enthalpy of fusion for H_2S. What does this say about the relative strengths of the forces between the molecules in these two solids? Explain.
31. ■ What is the total quantity of heat energy transfer required to change 0.50 mol ice at −5 °C to 0.50 mol steam at 100 °C?
32. How much thermal energy is needed to melt a 36.00-g ice cube that is initially at −10 °C and bring it to room temperature (20 °C)? The solid ice and liquid water have heat capacities of 2.06 J g^{-1} °C^{-1} and 4.184 J g^{-1} °C^{-1}, respectively. The enthalpy of fusion for solid ice is 6.02 kJ/mol and the enthalpy of vaporization of liquid water is 40.7 kJ/mol.
33. ■ The chlorofluorocarbon CCl_2F_2 was once used as a refrigerant. What mass of this substance must evaporate to freeze 2.0 mol water initially at 20 °C? The enthalpy of vaporization for CCl_2F_2 is 289 J/g. The enthalpy of fusion for solid ice is 6.02 kJ/mol and specific heat capacity for liquid water is 4.184 J g^{-1} °C^{-1}.
34. The ions of NaF and MgO all have the same number of electrons, and the internuclear distances are about the same (235 pm and 212 pm). Why, then, are the melting points of NaF and MgO so different (992 °C and 2642 °C, respectively)?
35. ■ For the pair of compounds LiF and CsI, tell which compound is expected to have the higher melting point, and briefly explain why.
36. Which of these substances has the highest melting point? The lowest melting point? Explain your choice briefly.
    (a) LiBr (b) CaO
    (c) CO (d) CH_3OH
37. ■ Which of these substances has the highest melting point? The lowest melting point? Explain your choice briefly.
    (a) SiC (b) I_2
    (c) Rb (d) $CH_3CH_2CH_2CH_3$
38. ■ In this phase diagram, make these identifications:

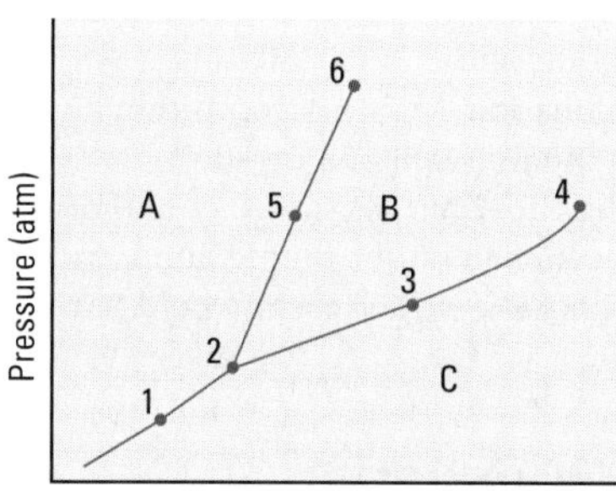

    (a) What phase is present in region A? Region B? Region C?
    (b) What phases are in equilibrium at point 1? Point 2? Point 3? Point 5?
39. From memory, sketch the phase diagram of water. Label all the regions as to the physical state of water. Draw either horizontal (constant pressure) or vertical (constant temperature) paths (i.e., lines with arrows indicating a direction) for these changes of state:
    (a) Sublimation (b) Condensation to a liquid
    (c) Melting (d) Vaporization
    (e) Crystallization
40. ■ Consult the phase diagram of CO_2 in Figure 11.17. What phase or phases are present under these conditions:
    (a) T = −70 °C and P = 1.0 atm
    (b) T = −40 °C and P = 15.5 atm
    (c) T = −80 °C and P = 4.7 atm

Types of Solids

41. Classify each of these solids as ionic, metallic, molecular, network, or amorphous.
 (a) KF (b) I_2
 (c) SiO_2 (d) BN
42. Classify each of these solids as ionic, metallic, molecular, network, or amorphous.
 (a) Tetraphosphorus decaoxide
 (b) Brass
 (c) Graphite
 (d) Ammonium phosphate
43. On the basis of the description given, classify each of these solids as molecular, metallic, ionic, network, or amorphous, and explain your reasoning.
 (a) A brittle, yellow solid that melts at 113 °C; neither its solid nor its liquid conducts electricity
 (b) A soft, silvery solid that melts at 40 °C; both its solid and its liquid conduct electricity
 (c) A hard, colorless, crystalline solid that melts at 1713 °C; neither its solid nor its liquid conducts electricity
 (d) A soft, slippery solid that melts at 63 °C; neither its solid nor its liquid conducts electricity
44. ■ On the basis of the description given, classify each of these solids as molecular, metallic, ionic, network, or amorphous, and explain your reasoning.
 (a) A soft, slippery solid that has no definite melting point but decomposes at temperatures above 250 °C; the solid does not conduct electricity
 (b) Violet crystals that melt at 114 °C and whose vapor irritates the nose; neither the solid nor the liquid conducts electricity
 (c) Hard, colorless crystals that melt at 2800 °C; the liquid conducts electricity, but the solid does not
 (d) A hard solid that melts at 3410 °C; both the solid and the liquid conduct electricity
45. ■ What type of solid exhibits each of these sets of properties?
 (a) Melts below 100 °C and is insoluble in water
 (b) Conducts electricity only when melted
 (c) Insoluble in water and conducts electricity
 (d) Noncrystalline and melts over a wide temperature range

Crystalline Solids

46. Each diagram below represents an array of like atoms that would extend indefinitely in two dimensions. Draw a two-dimensional unit cell for each array. How many atoms are in each unit cell?

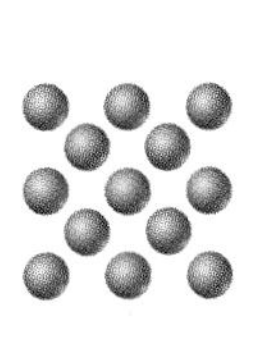
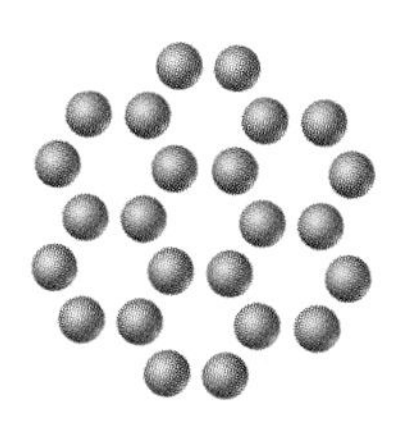
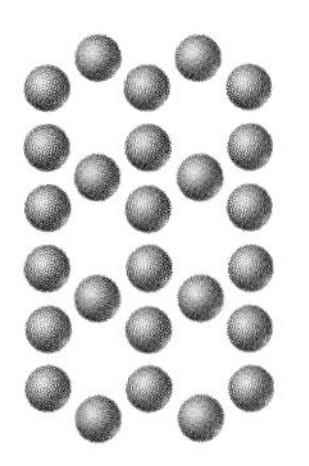

47. Name and draw the three cubic unit cells. Describe their similarities and differences.
48. Explain how the volume of a primitive (simple) cubic unit cell is related to the radius of the atoms in the cell.
49. ■ Solid xenon forms crystals with a face-centered unit cell that has an edge of 620 pm. Calculate the atomic radius of xenon.
50. Gold (atomic radius = 144 pm) crystallizes in an fcc unit cell. What is the length of a side of the cell?
51. Using the NaCl structure shown in Figure 11.25, how many unit cells share each of the Na^+ ions in the front face of the unit cell? How many unit cells share each of the Cl^- ions in this face?
52. The ionic radii of Cs^+ and Cl^- are 181 and 167 pm, respectively. What is the length of the body diagonal in the CsCl unit cell? What is the length of the side of this unit cell? (See Figure 11.23.)
53. A primitive (simple) cubic unit cell is formed so that the spherical atoms or ions just touch one another along the edge. Prove mathematically that the percentage of empty space within the unit cell is 47.6%. (The volume of a sphere is $\frac{4}{3}\pi r^3$, where r is the radius of the sphere.)
54. Metallic lithium has a body-centered cubic structure, and its unit cell is 351 pm along an edge. Lithium iodide has the same crystal lattice structure as sodium chloride. The cubic unit cell is 600 pm along an edge.
 (a) Assume that the metal atoms in lithium touch along the body diagonal of the cubic unit cell, and estimate the radius of a lithium atom.
 (b) Assume that in lithium iodide the I^- ions touch along the face diagonal of the cubic unit cell and that the Li^+ and I^- ions touch along the edge of the cube; calculate the radius of an I^- ion and of an Li^+ ion.
 (c) Compare your results in parts (a) and (b) for the radius of a lithium atom and a lithium ion. Are your results reasonable? If not, how could you account for the unexpected result? Could any of the assumptions that were made be in error?

Network Solids

55. Explain why diamond is denser than graphite.
56. ■ Determine, by looking up data in a reference such as the *Handbook of Chemistry and Physics,* whether the examples of network solids given in the text are soluble in water or other common solvents. Explain your answer in terms of the chemical bonding in network solids.
57. Explain why diamond is an electrical insulator and graphite is an electrical conductor.

Metals, Semiconductors, and Insulators

58. What is the principal difference between the orbitals that electrons occupy in individual, isolated atoms and the orbitals they occupy in solids?
59. ■ In terms of band theory, what is the difference between a conductor and an insulator? Between a conductor and a semiconductor?
60. Name three properties of metals, and explain them by using a theory of metallic bonding.
61. Which substance has the greatest electrical conductivity? The smallest electrical conductivity? Explain your choice briefly.
 (a) Si (b) Ge
 (c) Ag (d) P_4

62. Which substance has the greatest electrical conductivity? The smallest electrical conductivity? Explain your choices briefly.
(a) RbCl(ℓ) (b) NaBr(s)
(c) Rb (d) Diamond

General Questions

63. The chlorofluorocarbon CCl_2F_2 was once used in air conditioners as the heat transfer fluid. Its normal boiling point is −30 °C, and its enthalpy of vaporization is 165 J g^{-1}. The gas and the liquid have specific heat capacities of 0.61 J g^{-1} $°C^{-1}$ and 0.97 J g^{-1} $°C^{-1}$, respectively. How much heat is evolved when 10.0 g CCl_2F_2 is cooled from 40 °C to −40 °C?

64. ■ Liquid ammonia, $NH_3(\ell)$, was used as a refrigerant fluid before the discovery of the chlorofluorocarbons and is still widely used today. Its normal boiling point is −33.4 °C, and its enthalpy of vaporization is 23.5 kJ/mol. The gas and liquid have specific heat capacities of 2.2 J g^{-1} K^{-1} and 4.7 J g^{-1} K^{-1}, respectively. How much heat transfer is required to 10.0 kg liquid ammonia to raise its temperature from −50.0 °C to −33.4 °C, and then to 0.0 °C?

65. Potassium chloride and rubidium chloride both have the sodium chloride structure. Experiments indicate that their cubic unit cell dimensions are 629 pm and 658 pm, respectively.
(i) One mol KCl and 1 mol RbCl are ground together in a mortar and pestle to a very fine powder, and two cubic unit cells are measured—one with an edge length of 629 pm and one with an edge length of 658 pm. Call this Sample 1.
(ii) One mol KCl and 1 mol RbCl are heated until the entire mixture is molten and then cooled to room temperature. A single cubic unit cell is observed with an edge length of roughly 640 pm. Call this Sample 2.
(a) Suppose that Samples 1 and 2 were analyzed for their chloride content. What fraction of each sample is chloride? Could the samples be distinguished by means of chemical analysis?
(b) Interpret the two results (i and ii) in terms of the structures of the crystal lattices of Samples 1 and 2.
(c) What chemical formula should you write for Sample 1? For Sample 2?
(d) Suppose that you dissolved 1.00 g Sample 1 in 100 mL water in a beaker and did the same with 1.00 g Sample 2. Which sample would conduct electricity better, or would both be the same? What ions would be present in each solution at what concentrations?

66. Sulfur dioxide, SO_2, is found in polluted air.
(a) What type of forces are responsible for binding SO_2 molecules to one another in the solid or liquid phase?
(b) Using the information below, place the compounds listed in order of increasing intermolecular attractions.

Compound	Normal Boiling Point (°C)
SO_2	−10
NH_3	−33.4
CH_4	−161.5
H_2O	100

Applying Concepts

67. Refer to Figure 11.5 when answering these questions.
(a) What is the equilibrium vapor pressure for ethyl alcohol at room temperature?
(b) At what temperature does diethyl ether have an equilibrium vapor pressure of 400 mm Hg?
(c) If a pot of water were boiling at a temperature of 95 °C, what would be the atmospheric pressure?
(d) At 200 mm Hg and 60 °C, which of the three substances are gases?
(e) If you put an equal number of drops of each substance on your hand, which would immediately evaporate, and which would remain as a liquid?
(f) Which of the three substances has the greatest intermolecular attractions?

68. The normal boiling point of SO_2 is 263.1 K and that of NH_3 is 239.7 K. At −40 °C, would you predict that ammonia has a vapor pressure greater than, less than, or equal to that of sulfur dioxide? Explain.

69. Butane, C_4H_{10}, is a gas at room temperature; however, if you look closely at a butane lighter you see it contains liquid butane. How is this possible?

70. ■ Examine the nanoscale diagrams and the phase diagram below. Match each particulate diagram (1 through 7) to its corresponding point (A through H) on the phase diagram.

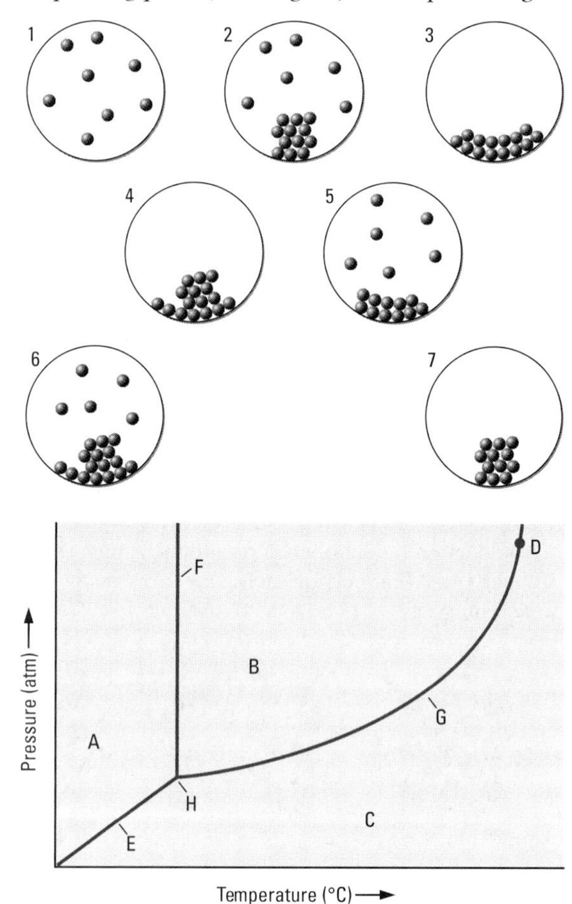

71. Consider the phase diagram below. Draw corresponding heating curves for T_1 to T_2 at pressures P_1 and P_2. Label each phase and phase change on your heating curves.

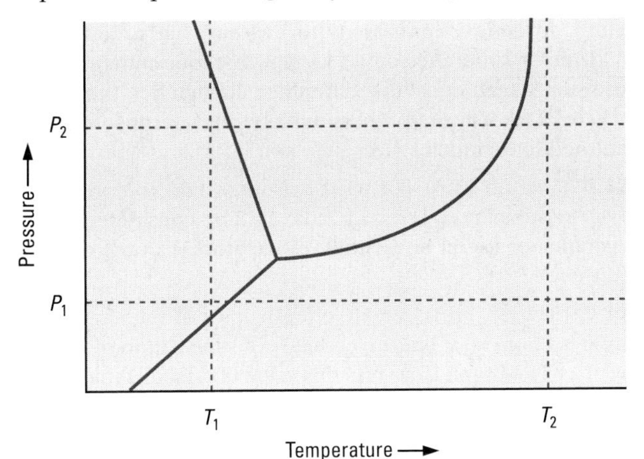

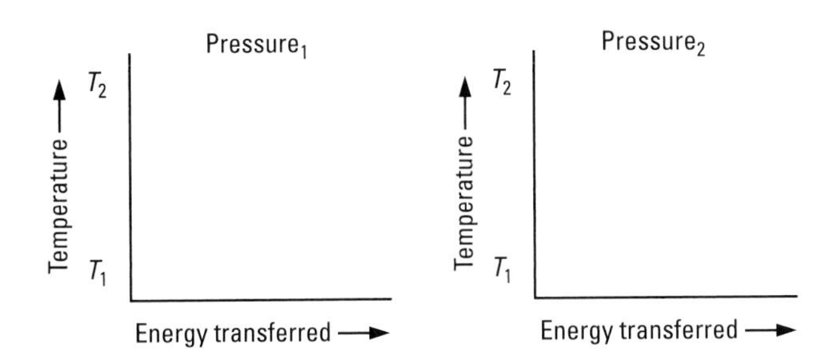

More Challenging Questions

72. If you get boiling water at 100 °C on your skin, it burns. If you get 100 °C steam on your skin, it burns much more severely. Explain why this is so.

73. If water at room temperature is placed in a flask that is connected to a vacuum pump and the vacuum pump then lowers the pressure in the flask, we observe that the volume of the water has decreased and the remaining water has turned into ice. Explain what has happened.

74. We hear reports from weather forecasters of "relative humidity," the ratio of the partial pressure of water in the air to the equilibrium vapor pressure of water at the same temperature. (The vapor pressure of water at 32.2 °C is 36 mm Hg.)
   (a) On a sticky, humid day, the relative humidity may reach 90% with a temperature of 90. °F (32.2 °C). What is the partial pressure of water under these conditions? How many moles per liter are present in the air? How many water molecules per cm^3 are present in such air?
   (b) On a day in a desert, the relative humidity may be 5.% with the same temperature of 90. °F. How many water molecules per cm^3 are present in such air?

75. (a) In the diagram below, for a substance going from point F (initial state) to point G (final state), what changes in phase occur?
   (b) What does point A represent?
   (c) What does the curve from point A to point B represent?

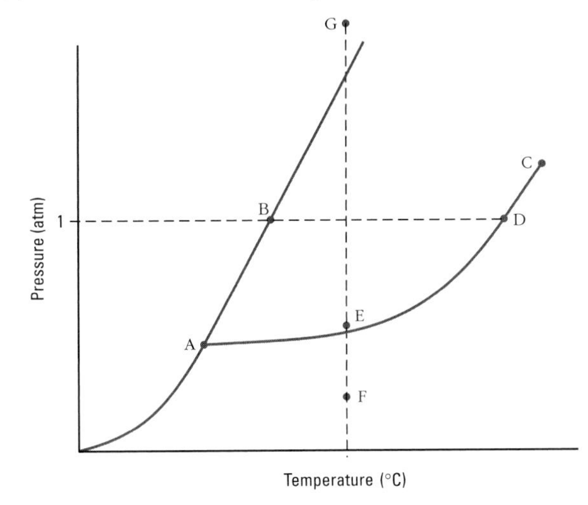

76. Suppose that liquid A has stronger intermolecular forces than liquid B at room temperature.
   (a) Which substance will have the greater surface tension?
   (b) Which substance will have the greater vapor pressure?
   (c) Which substance will have the greater viscosity?

77. ■ Use the vapor pressure curves shown in the figure below for methyl ethyl ether, $CH_3OCH_2CH_3$; carbon disulfide, CS_2; and benzene, C_6H_6, to answer these questions.
   (a) What is the vapor pressure of methyl ethyl ether at 0 °C?
   (b) Which of these three liquids has the strongest intermolecular attractions?
   (c) At what temperature does benzene have a vapor pressure of 600 mm Hg?
   (d) What are the normal boiling points of these three liquids?

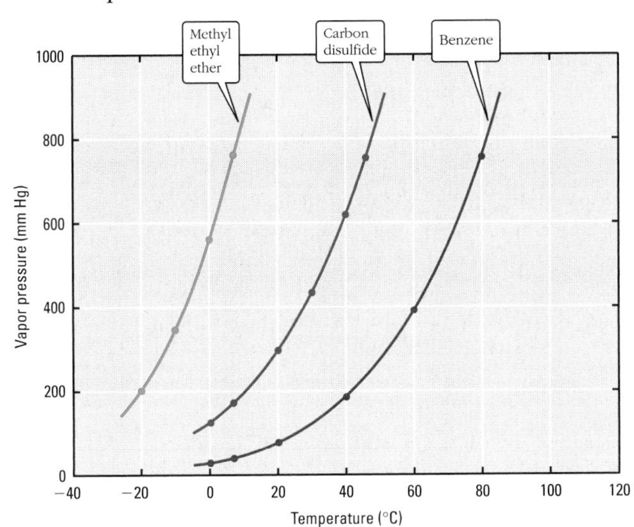

78. Will a closed container of water at 70 °C or an open container of water at the same temperature cool faster? Explain why.
79. Calculate the boiling point of water at 24 mm Hg. (The ΔH_{vap} of water is 40.7 kJ/mol.)
80. Solid lithium has a body-centered cubic unit cell with the length of the edge of 351 pm at 20 °C. What is the density of lithium at this temperature?
81. Tungsten has a body-centered cubic unit cell and an atomic radius of 139 pm. What is the density of solid tungsten?
82. Copper is an important metal in the U.S. economy. Most of it is mined in the form of the mineral chalcopyrite, $CuFeS_2$.
    (a) To obtain one metric ton (1000. kilograms) of copper metal, how many metric tons of chalcopyrite would you have to mine?
    (b) If the sulfur in chalcopyrite is converted to SO_2, how many metric tons of the gas would you get from one metric ton of chalcopyrite?
    (c) Copper crystallizes as a face-centered cubic lattice. Knowing that the density of copper is 8.95 g/cm^3, calculate the radius of the copper atom.

Conceptual Challenge Problems

CP11.A (Section 11.2) In Section 11.2 you read that the enthalpy of vaporization of water "is somewhat dependent on the temperature." At 100 °C this value is 40.7 kJ/mol, but at 25 °C it is 44.0 kJ/mol, a difference of 3.3 kJ/mol. List four enthalpy changes whose sum would equal this difference. Remember, the sum of the changes for a cyclic process must be zero because the system is returned to its initial state.

CP11.B (Section 11.3) For what reasons would you propose that two of the substances listed in Table 11.2 be considered better refrigerants for use in household refrigerators than the others listed there?

CP11.C (Section 11.3) A table of enthalpies of sublimation is not given in Section 11.3, but the enthalpy of sublimation of ice at 0 °C is given as 51 kJ/mol. How was this value obtained? Tables 11.2 and 11.3 list the enthalpies of vaporization and fusion, respectively, for several substances. Determine from data in these tables the ΔH_{sub} for ice. Using the same method, estimate the enthalpies of sublimation of HBr and HI at their melting points.

12

Chemical Kinetics: Rates of Reactions

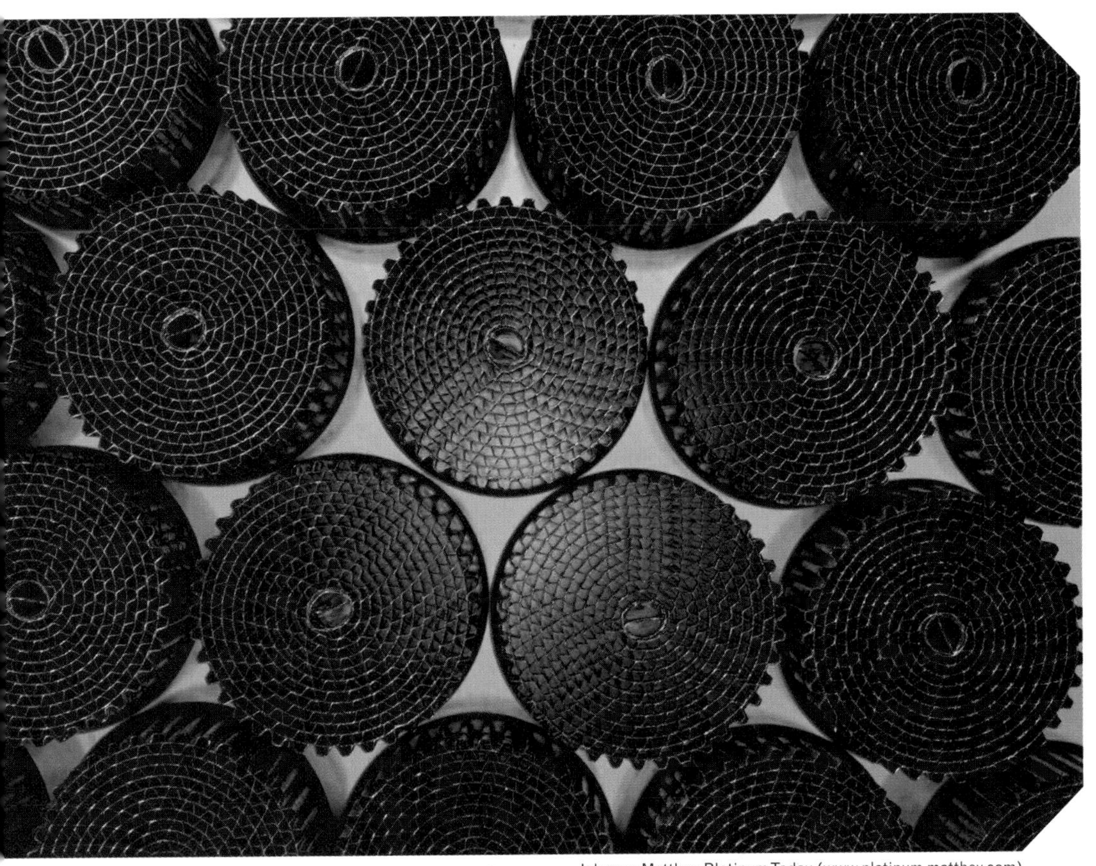

Johnson Matthey Platinum Today (www.platinum.matthey.com)

Catalysis is an exciting and profitable field of chemistry that affects our lives every day. In the chemical industry, catalysts speed up reactions that produce new substances such as ammonia fertilizer. In the pharmaceutical industry, catalysts enable reactions that produce medicines. Catalytic converters, such as the ones for motorcycles shown here, remove noxious gases from automobile exhaust, thereby reducing air pollution. In our bodies, enzyme catalysts turn reactions on and off by changing their rates, thereby maintaining proper concentrations of many substances and transmitting nerve impulses.

Turn on the valve of a Bunsen burner in your laboratory, bring up a lighted match, and a rapid combustion reaction begins with a whoosh:

$$CH_4(g) + 2\ O_2(g) \longrightarrow CO_2(g) + 2\ H_2O(g) \qquad \Delta H^\circ = -802.34\ \text{kJ}$$

What would happen if you didn't put a lighted match in the methane-air stream? Nothing obvious. At room temperature the reaction of methane with oxygen is so slow that the two potential reactants can be mixed in a closed flask and stored unreacted for centuries. These facts about combustion of methane lie within the realm of **chemical kinetics**—*the study of the speeds of reactions and the nanoscale pathways or rearrangements by which atoms and molecules are transformed from reactants to products.*

Chemical kinetics is extremely important, because knowing about kinetics enables us to control many kinds of reactions in addition to combustion. In pharmaceutical chemistry an important problem is devising drugs that remain in their active form long enough to get to the site in the body where they are intended to act. Consequently, it is important to know whether a drug will react with other substances in the body and how long it will take to do so. In environmental chemistry, there was more than a decade of controversy over whether stratospheric ozone is being depleted by chlorofluorocarbons. Much of this hinged on verifying the sequence and rates of reactions by which stratospheric ozone is produced and consumed. Their careful studies of such reactions led to a Nobel Prize in Chemistry for Sherwood Rowland, Mario Molina, and Paul Crutzen.

This chapter focuses on the factors that affect the speeds of reactions, the nanoscale basis for understanding those factors, and their importance in modern society, from industrial plants to cars to the cells of our bodies.

12.1 Reaction Rate

For a chemical reaction to occur, reactant molecules must come together so that their atoms can be exchanged or rearranged. Atoms and molecules are more mobile in the gas phase or in solution than in the solid phase, so reactions are often carried out in a mixture of gases or among solutes in a solution. For a **homogeneous reaction,** one in which reactants and products are all in the same phase (gas or solution, for example), four factors affect the speed of a reaction:

- The *properties* of reactants and products—in particular, their molecular structure and bonding
- The *concentrations* of the reactants and sometimes the products
- The *temperature* at which the reaction occurs
- The *presence* of a catalyst and, if one is present, its concentration

A catalyst (described in Section 12.8) speeds up a reaction but undergoes no *net* chemical change itself. The catalytic converter in an automobile speeds up reactions that remove pollutants from the exhaust gases.

Many important reactions, including the ones in catalytic converters that remove air pollutants from automobile exhaust, are **heterogeneous reactions.** They take place at a surface—at an interface between two different phases (solid and gas, for example). The speed of a heterogeneous reaction depends on the four factors listed above as well as on the area and nature of the surface at which the reaction occurs. For example, very finely divided metal powder can burn very rapidly, whereas a pile of powder with much less surface exposed to oxygen in the air is difficult to ignite (Figure 12.1). The much more rapid reaction when greater surface is exposed has been responsible for explosions in grain elevators and sugar mills where finely divided, combustible solids in the air are exposed to a spark or flame.

Recall that the Greek letter Δ (delta) means that a change in some quantity has been measured (**← p. 184**). As usual, Δ means to subtract the initial value of the quantity from the final value.

Change in concentration is used (rather than change in total amount of reactant) because using change in concentration makes the rate independent of the volume of the reaction mixture.

The speed of any process is expressed as its **rate,** which is *the change in some measurable quantity per unit of time.* A car's rate of travel, for example, is found by measuring the change in its position, Δx, during a given time interval, Δt. Suppose you are driving on an interstate highway. If you pass mile marker 43 at 2:00 PM and mile marker 173 at 4:00 PM, $\Delta x = (173 - 43)$ mi $= 130$ mi and $\Delta t = 2.00$ h. You are traveling at an average rate of $\Delta x/\Delta t = 65$ mi/h. For a chemical process, the **reaction rate** is defined as *the change in concentration of a reactant or product per unit time.* (Time can be measured in seconds, hours, days, or whatever unit is most convenient for the speed of the reaction.)

(a)

(b)

Figure 12.1 Combustion of iron powder. (a) The very finely divided metal powder burns slowly in a pile where only a small surface area is exposed to air. (b) Spraying the same powder through the air greatly increases the exposed surface. When the powder enters a flame, combustion is rapid—even explosive.

As an example of measurements made in chemical kinetics, consider Figure 12.2, which shows the violet-colored dye crystal violet reacting with aqueous sodium hydroxide to form a colorless product. The dye's violet color disappears over time, and the intensity of color can be used to determine the concentration of the dye. Crystal violet consists of polyatomic positive ions that we abbreviate as Cv^+ and chloride ions, Cl^-. Its formula is therefore represented by CvCl. The Cv^+ ions cause the beautiful violet color. In aqueous solution they combine with hydroxide ions, OH^-, to form a colorless, uncharged product abbreviated as CvOH. The reaction can be represented by the net ionic equation

$$Cv^+(aq) + OH^-(aq) \longrightarrow CvOH(aq) \qquad [12.1]$$

Concentrations in moles per liter are represented by square brackets surrounding the formula of the substance. For example, the concentration of crystal violet cation is represented as $[Cv^+]$.

The rate at which this reaction occurs can be calculated by dividing the change in concentration of crystal violet cation, $\Delta[Cv^+]$, by the elapsed time, Δt. For example, if the concentration of crystal violet is measured at some time t_1 to give $[Cv^+]_1$, and

Figure 12.2 Disappearance of a dye. Violet-colored crystal violet dye in aqueous solution reacts with aqueous sodium hydroxide, which converts the dye into a colorless product. The intensity of the solution's color decreases and eventually the color disappears. The rate of the reaction can be determined by repeated, simultaneous measurements of the intensity of color and the time. From the intensity of color the concentration of dye can be calculated, so concentration can be determined as a function of time.

Table 12.1 Concentration-Time Data for Reaction of Crystal Violet with 0.10 M NaOH(aq) at 23 °C

Time, t (s)	Concentration of Crystal Violet Cation, $[Cv^+]$ (mol/L)	Average Rate (mol L^{-1} s^{-1})
0.0	5.000×10^{-5}	
		13.2×10^{-7}
10.0	3.680×10^{-5}	
		9.70×10^{-7}
20.0	2.710×10^{-5}	
		7.20×10^{-7}
30.0	1.990×10^{-5}	
		5.30×10^{-7}
40.0	1.460×10^{-5}	
		3.82×10^{-7}
50.0	1.078×10^{-5}	
		2.85×10^{-7}
60.0	0.793×10^{-5}	
		1.82×10^{-7}
80.0	0.429×10^{-5}	
		0.985×10^{-7}
100.0	0.232×10^{-5}	

the measurement is repeated at a subsequent time t_2 to give $[Cv^+]_2$, then the rate of reaction is

$$\text{Rate of change of concentration of } Cv^+ = \frac{\text{change in concentration of } Cv^+}{\text{elapsed time}}$$

$$= \frac{\Delta [Cv^+]}{\Delta t} = \frac{[Cv^+]_2 - [Cv^+]_1}{t_2 - t_1}$$

The experimentally measured concentration of crystal violet as a function of time is shown in Table 12.1. Because the concentration of crystal violet decreases as time increases, $[Cv^+]_2$ is smaller than $[Cv^+]_1$. Therefore $\Delta[Cv^+]/\Delta t$ is negative. By convention, reaction rate is defined as a positive quantity. Therefore, for the crystal violet reaction the rate is defined as

The negative sign converts a negative $\Delta[Cv^+]/\Delta t$ value to a positive one.

$$\text{Reaction rate} = -\frac{\Delta [Cv^+]}{\Delta t}$$

Table 12.1 also shows calculated values of reaction rates for each time interval. Because calculating the rate involves dividing a concentration difference by a time difference, the units of reaction rate are units of concentration divided by units of time, in this case mol/L divided by s, mol L^{-1} s^{-1}.

PROBLEM-SOLVING EXAMPLE 12.1 Calculating Average Rates

Using the data in the first two columns of Table 12.1, calculate $\Delta[Cv^+]$, Δt, and the average rate of reaction for each time interval given. Use the numbers given in the third column to check your results. (A good way to do so is to use a computer spreadsheet program.)

Answer See the third column in Table 12.1.

Strategy and Explanation The time interval from 80.0 s to 100.0 s provides an example of the calculation.

$$\Delta [Cv^+] = 0.232 \times 10^{-5} \text{ mol/L} - 0.429 \times 10^{-5} \text{ mol/L} = -0.197 \times 10^{-5} \text{ mol/L}$$

$$\Delta t = 100.0 \text{ s} - 80.0 \text{ s} = 20.0 \text{ s}$$

$$\text{Rate} = -\frac{\Delta [Cv^+]}{\Delta t} = -\frac{(-0.197 \times 10^{-5} \text{ mol/L})}{20.0 \text{ s}} = 0.985 \times 10^{-7} \text{ mol L}^{-1} \text{ s}^{-1}$$

Do the other calculations in a similar way.

☑ **Reasonable Answer Check** The rates of reaction should be positive numbers and should have units of concentration divided by time (such as mol L^{-1} s^{-1}). Both of these conditions are met.

PROBLEM-SOLVING PRACTICE 12.1

For the reaction of crystal violet with NaOH(aq), the measured rate of reaction is 1.27×10^{-6} mol L^{-1} s^{-1} when the concentration of crystal violet cation is 4.13×10^{-5} mol/L.

(a) Estimate how long it will take for the concentration of crystal violet to drop from 4.30×10^{-5} mol/L to 3.96×10^{-5} mol/L.

(b) Could you use the same method to make an accurate estimate of how long it would take for the concentration of crystal violet to drop from 4.30×10^{-5} mol/L to 0.43×10^{-5} mol/L? Explain why or why not.

EXERCISE 12.1 Rates of Reaction

(a) From data in Table 12.1, calculate the rate of reaction for each time interval: (i) from 40.0 s to 60.0 s; (ii) from 20.0 s to 80.0 s; (iii) from 0.0 to 100.0 s.

(b) Use all of the data in the first two columns of Table 12.1 to draw a graph with time on the horizontal (x) axis and concentration on the vertical (y) axis. Draw a smooth curve through the data. On the graph, draw lines that correspond to $\Delta[Cv^+]/\Delta t$ for each interval.

(c) Why is the rate not the same for each time interval in part (a), even though the average time for each interval is 50.0 s? (That is, for interval i, the average time is (40.0 s + 60.0 s)/2 = 50.0 s.) Write an explanation of the reason for a friend who is taking this course, and ask your friend to evaluate what you have written.

Reaction Rates and Stoichiometry

From the stoichiometry of the crystal violet reaction (Reaction 12.1), it is clear that for every mole of crystal violet that reacts, a mole of CvOH product is formed, because the coefficients of Cv^+(aq), OH^-(aq), and CvOH(aq) are the same. Thus the rate of appearance of CvOH(aq) equals the rate of disappearance of Cv^+(aq). In many reactions, however, the coefficients are not all the same. For example, in the reaction

$$2\,NO_2(g) \longrightarrow 2\,NO(g) + O_2(g) \qquad [12.2]$$

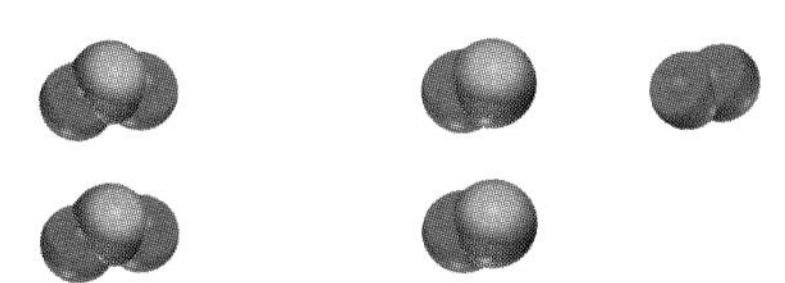

for every mole of O_2 formed, two moles of NO_2 react. Thus the rate of disappearance of NO_2 is twice as great as the rate of appearance of O_2. We would like to define the rate of reaction in a way that does not depend on which substance's concentration change is measured. If we multiply $-\dfrac{\Delta[NO_2]}{\Delta t}$ by $\frac{1}{2}$ in the definition of the rate, then the rate is the same whether expressed in terms of $[O_2]$ or $[NO_2]$. Notice that $\frac{1}{2}$ is the reciprocal of the stoichiometric coefficient of NO_2 in Equation 12.2. For the general reaction equation

$$a\,A + b\,B \longrightarrow c\,C + d\,D$$

where A, B, C, and D represent formulas of substances and a, b, c, and d are coefficients, the reaction rate can be defined uniformly in terms of each substance as

$$\text{Rate} = -\frac{1}{a}\frac{\Delta[A]}{\Delta t} = -\frac{1}{b}\frac{\Delta[B]}{\Delta t} = \frac{1}{c}\frac{\Delta[C]}{\Delta t} = \frac{1}{d}\frac{\Delta[D]}{\Delta t} \qquad [12.3]$$

Because reaction rates are related to Δconcentration/Δt as shown in Equation 12.3, it is important to know the exact chemical equation for which a rate is reported. If the coefficients in the equation are changed, for example, by doubling all of them, then the definition of reaction rate also changes.

That is, *the rate of change in concentration of any of the reactants or products is multiplied by the reciprocal of the stoichiometric coefficient to find the rate of reaction*. Because the concentrations of reactants decrease with time, their rates of change are given negative signs.

PROBLEM-SOLVING EXAMPLE 12.2 Rates and Stoichiometry

Equation 12.2 shows the decomposition of $NO_2(g)$ to form NO(g) and $O_2(g)$.

(a) Define the rate of reaction in terms of the rate of change in concentration of each reactant and product.

(b) If the rate of appearance of $O_2(g)$ is 0.023 mol L^{-1} s^{-1}, what is the rate of disappearance of $NO_2(g)$?

Answer

(a) $\text{Rate} = \frac{1}{2}\frac{\Delta[\text{NO}]}{\Delta t} = \frac{\Delta[\text{O}_2]}{\Delta t} = -\frac{1}{2}\frac{\Delta[\text{NO}_2]}{\Delta t}$

(b) 0.046 mol L^{-1} s^{-1}

Strategy and Explanation

(a) Define the rate of reaction as change in concentration divided by time elapsed (Δt). Multiply the rate for each substance by the reciprocal of the stoichiometric coefficient, and place a negative sign in front of the rate for each reactant.

(b) From the coefficients, 2 mol NO forms for every 1 mol O_2. Therefore, the rate of formation of NO is twice the rate of formation of O_2, which equals the rate of reaction. Algebraically,

$$\frac{1}{2}\frac{\Delta[\text{NO}]}{\Delta t} = \frac{1}{1}\frac{\Delta[\text{O}_2]}{\Delta t}$$

and

$$\frac{\Delta[\text{NO}]}{\Delta t} = 2 \times \frac{\Delta[\text{O}_2]}{\Delta t} = 2 \times 0.023 \text{ mol L}^{-1}\text{ s}^{-1} = 0.046 \text{ mol L}^{-1}\text{ s}^{-1}$$

PROBLEM-SOLVING PRACTICE 12.2

For the reaction

$$4\text{ NO}_2(g) + \text{O}_2(g) \longrightarrow 2\text{ N}_2\text{O}_5(g)$$

(a) Express the rate of formation of N_2O_5 in terms of the rate of disappearance of O_2.

(b) If the rate of disappearance of O_2 is 0.0037 mol L^{-1} s^{-1}, what is the rate of disappearance of NO_2?

Average Rate and Instantaneous Rate

A reaction rate calculated from a change in concentration divided by a change in time is called the **average reaction rate** over the time interval from which it was calculated. For example, the average reaction rate at 23 °C for the crystal violet reaction over the interval from 0.0 to 10.0 s is 13.2×10^{-7} mol L^{-1} s^{-1}. The data in Table 12.1 indicate that as the concentration of Cv^+ decreases, the average rate also decreases. Most reactions are like this: The average rate becomes smaller as the concentration of one or more reactants decreases. Your results in Exercise 12.1 should have been different for each range of time over which you calculated. Because the average reaction rate changes over time, the rate you calculate depends on when, and for what range of time, you calculate. If you want to know the rate that corresponds to a particular concentration of Cv^+ (and therefore to a particular time after the reaction began), the average rate is not appropriate, because it depends on the size of the time interval.

The **instantaneous reaction rate** is *the rate at a particular time after a reaction has begun*. To obtain it, the rate must be calculated over a very small interval

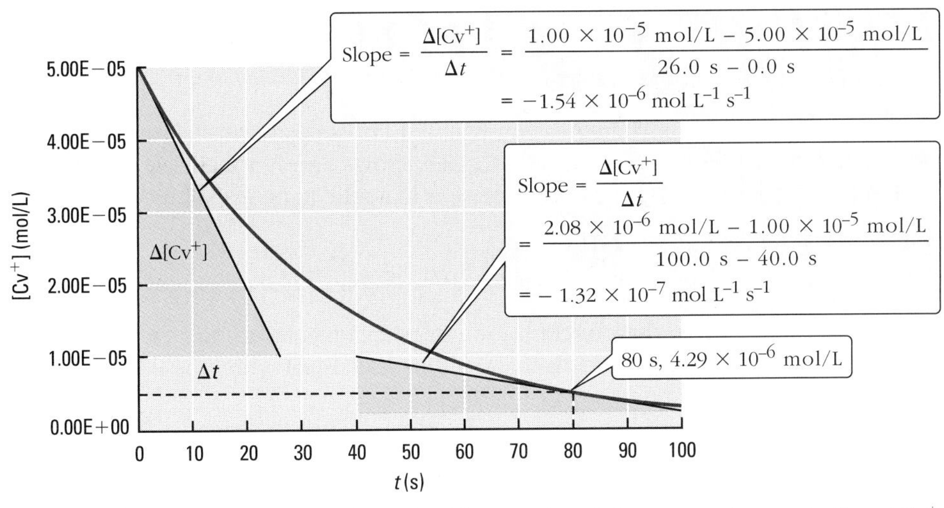

Figure 12.3 Instantaneous reaction rates. The experimentally measured concentration of Cv^+ is plotted as a function of time during the reaction in which Cv^+ reacts with OH^- in aqueous solution. The slopes at time 0 s and at 80 s are indicated on the graph. From these slopes the instantaneous rates 1.54×10^{-6} mol L^{-1} s^{-1} and 1.32×10^{-7} mol L^{-1} s^{-1} can be obtained.

around the time or concentration for which the rate is desired. For example, to calculate the rate at which Cv^+ is disappearing when its concentration is 4.29×10^{-6} mol/L, you would need to calculate $\Delta[Cv^+]/\Delta t$ at exactly 80 s from the start of the reaction. A good way to do so is shown in Figure 12.3. ***The instantaneous rate is the slope of a line tangent to the concentration-time curve at the point corresponding to the specified concentration and time.*** For a particular concentration of the same reactant at the same temperature and the same concentrations of other species, the instantaneous rate has a specific value. As you saw in Exercise 12.1c, the value of the average rate depends on the size of the Δt used to calculate the average rate.

If you are familiar with calculus, then you may recognize that in the limit of very small time intervals, $\Delta[A]/\Delta t$, where A represents a substance, becomes the same as the derivative of concentration with respect to time. That is,

$$\lim_{\Delta t \to 0} \frac{\Delta[A]}{\Delta t} = \frac{d[A]}{dt}$$

This also means that the rate of reaction at any time can be found from the *slope* (at that time) of the tangent to a curve of concentration versus time, such as the curve in Figure 12.3. Appendix A.8 discusses how to determine the slope and intercept of a graph. The slope of a tangent to the graph at a given point is referred to as the derivative of the graph at that point and can be obtained using scientific graphing programs.

CONCEPTUAL
EXERCISE 12.2 Instantaneous Rates

Instantaneous rates for the reaction of hydroxide ion with Cv^+ can be determined from the slope of the curve in Figure 12.3 at various concentrations. They are

(1) At 4.0×10^{-5} mol/L, rate = 12.3×10^{-7} mol L^{-1} s^{-1}
(2) At 3.0×10^{-5} mol/L, rate = 9.25×10^{-7} mol L^{-1} s^{-1}
(3) At 2.0×10^{-5} mol/L, rate = 6.16×10^{-7} mol L^{-1} s^{-1}
(4) At 1.5×10^{-5} mol/L, rate = 4.60×10^{-7} mol L^{-1} s^{-1}
(5) At 1.0×10^{-5} mol/L, rate = 3.09×10^{-7} mol L^{-1} s^{-1}

(a) What is the relationship between the rates in (1) and (3)? Between (2) and (4)? Between (3) and (5)?
(b) What is the relationship between the concentrations in each of these cases?
(c) Is the rate of the reaction proportional to the concentration of Cv^+?

CONCEPTUAL
EXERCISE 12.3 Graphing Concentrations versus Time

Consider the decomposition of $N_2O_5(g)$,

$$2\,N_2O_5(g) \longrightarrow 4\,NO_2(g) + O_2(g)$$

Assume that the initial concentration of $N_2O_5(g)$ is 0.0200 mol/L and that none of the products are present. Make a graph that shows concentrations of $N_2O_5(g)$, $NO_2(g)$, and $O_2(g)$ as a function of time, all on the same set of axes and roughly to scale.

(a)

(b)

Figure 12.4 Reaction of aqueous potassium permanganate with aqueous hydrogen peroxide. The rate of the reaction of potassium permanganate with hydrogen peroxide depends on the permanganate concentration. With dilute $KMnO_4$ the reaction is slow (a); it is more rapid in more concentrated $KMnO_4$ (b). (In both cases the temperature is the same, so the difference in rate must be due to concentration of permanganate.)

12.2 Effect of Concentration on Reaction Rate

The rates of most reactions change when reactant concentrations change, just as we found for the crystal violet reaction. Figure 12.4 shows another example. The oxidation of hydrogen peroxide by permanganate ion in acidic aqueous solution

$$2\,MnO_4^-(aq) + 5\,H_2O_2(aq) + 6\,H_3O^+(aq) \longrightarrow 2\,Mn^{2+}(aq) + 5\,O_2(g) + 14\,H_2O(\ell)$$

is visibly more rapid when the concentration of permanganate is higher. One goal of chemical kinetics is to find out whether a reaction speeds up when the concentration of a reactant is increased and, if so, by how much.

The Rate Law

How the concentration of a reactant affects the rate can be determined by performing a series of experiments in which the concentration of that reactant is varied systematically (and temperature is held constant). Alternatively, a single experiment can be done in which concentration is determined continuously as a function of time. The latter approach gave the data for crystal violet shown in Table 12.1 and Figure 12.3, which you analyzed in Problem-Solving Example 12.1 and Exercise 12.2. You should have discovered that if the concentration of crystal violet cation is halved, the reaction rate is also halved. If the concentration of crystal violet cation is doubled, then the reaction rate is doubled. This leads to the expression

$$\text{Rate} \propto [\text{Cv}^+]$$

It says that the rate is directly proportional to (symbol $\propto$) the concentration of one of the reactants, crystal violet cation.

This proportionality can be changed to a mathematical equation by including a proportionality constant, k. *A mathematical equation that summarizes the relationship between reactant concentration and reaction rate* is called a **rate law** (or *rate equation*). For the crystal violet reaction the rate law is

$$\text{Rate} = k \times [\text{Cv}^+]$$

The proportionality constant, k, is called the **rate constant.** The rate constant is independent of concentration, but it has different values at different temperatures, usually becoming larger as temperature increases. The rate constant applies only to the specific reaction being studied and it applies at a specific temperature. Thus the chemical equation and the temperature for the reaction should be given along with the rate constant. In this case we write

$$\text{Cv}^+(aq) + \text{OH}^-(aq) \longrightarrow \text{CvOH}(aq) \qquad k = 3.07 \times 10^{-2}\ \text{s}^{-1}\ (\text{at } 25\ °\text{C})$$

Determining Rate Laws from Initial Rates

An advantage of measuring initial rates is that the concentrations of products are low early in the process. As a reaction proceeds, more and more products are formed. In some cases products can alter the rate; comparing initial rates with rates when products are present can reveal such a complication.

The relation between rate and concentration (the rate law) must be determined experimentally. One way to do so was illustrated in Exercise 12.3, but it is difficult to determine rates from tangents to a curve such as that in Figure 12.3. Another way is to measure initial rates. The **initial rate** of a reaction is *the instantaneous rate determined at the very beginning of the reaction*. A good approximation to the initial rate is to calculate $-\Delta[\text{reactant}]/\Delta t$ after no more than 2% of the limiting reactant has been consumed.

Many reactions can be started by mixing two different solutions or two different gas samples. Usually the concentrations of the reactants are known before they are mixed, so the initial rate corresponds to a known set of reactant concentrations. Several experiments can then be done in which initial concentrations are varied, and the change in the reaction rate can be correlated with changes in the concentration of each reactant. As an example, consider the reaction of a base with methyl acetate, CH_3COOCH_3, which produces acetate ion and methanol.

$$\underset{\text{methyl acetate}}{CH_3\overset{\overset{O}{\|}}{C}—O—CH_3} + \underset{\text{hydroxide ion}}{OH^-} \longrightarrow \underset{\text{acetate ion}}{CH_3\overset{\overset{O}{\|}}{C}—O^-} + \underset{\text{methanol}}{CH_3OH} \qquad [12.4]$$

To control for the effect of temperature on rate, several experiments were done at the same temperature:

Experiment	Initial Concentration (mol/L) $[CH_3COOCH_3]$	Initial Concentration (mol/L) $[OH^-]$	Initial Rate (mol L^{-1} s^{-1})
1	0.040	0.040	0.00022
	no change	× 2	× 2
2	0.040	0.080	0.00045
	× 2	no change	× 2
3	0.080	0.080	0.00090

Notice that in Experiments 1 and 2 the initial concentration of methyl acetate is the same. In Experiments 2 and 3 the initial concentration of hydroxide is the same.

To determine the rate law, compare two experiments in which only a single initial concentration changed. In Experiments 1 and 2, the $[CH_3COOCH_3]$ remained constant and the $[OH^-]$ doubled. The rate also doubled, which means that the rate is *directly proportional* to the $[OH^-]$. In Experiments 2 and 3, the $[OH^-]$ remained the same, the $[CH_3COOCH_3]$ doubled, and the rate doubled, indicating that the rate is also proportional to the $[CH_3COOCH_3]$. Therefore, the experimental data show that the rate is proportional to the *product* of the two concentrations, and the rate law is

$$\text{Rate} = k\,[CH_3COOCH_3]\,[OH^-]$$

This equation also tells us that doubling both initial concentrations at the same time would cause the rate to go up by a factor of 4, which it does from Experiment 1 to Experiment 3.

Another way to approach this problem involves proportions. As before, choose two experiments in which one concentration did not change. Then calculate the ratio of the other concentrations and the ratio of rates. For the methyl acetate reaction, using Experiments 1 and 2 where the $[CH_3COOCH_3]$ was constant,

$$\frac{[OH^-]_2}{[OH^-]_1} = \frac{0.080\text{ M}}{0.040\text{ M}} = 2.0 \quad \text{and} \quad \frac{\text{rate}_2}{\text{rate}_1} = \frac{0.00045\text{ mol L}^{-1}\text{ s}^{-1}}{0.00022\text{ mol L}^{-1}\text{ s}^{-1}} = 2.0$$

it is clear that both the concentrations and the rates change in the same proportion. This same method could be applied to analyze results of Experiments 2 and 3, where the initial $[OH^-]$ was constant and the initial $[CH_3COOCH_3]$ changed.

Once the rate law is known, a value for k, the rate constant, can be found by substituting rate and initial concentration data for any one experiment into the rate law.

For example, a value of k for the methyl acetate-hydroxide ion reaction could be obtained from data for the first experiment,

$$\underbrace{0.00022 \text{ mol L}^{-1} \text{ s}^{-1}}_{\text{rate}} = k\underbrace{(0.040 \text{ mol/L})}_{[CH_3COOCH_3]}\underbrace{(0.040 \text{ mol/L})}_{[OH^-]}$$

$$k = \frac{0.00022 \text{ mol L}^{-1} \text{ s}^{-1}}{(0.040 \text{ mol/L})(0.040 \text{ mol/L})} = 0.14 \text{ L mol}^{-1} \text{ s}^{-1}$$

A more precise value for k can be obtained by using all available experimental data—that is, by calculating a k for each experiment and then averaging the k values to obtain an overall result.

EXERCISE 12.4 Rates and Concentrations

Use the rate law for the reaction of methyl acetate with OH^- to predict the effect on the rate of reaction if the concentration of methyl acetate is doubled and the concentration of hydroxide ions is halved.

For the reaction as written in Problem-Solving Example 12.3,

$$\text{Rate} = -\frac{1}{2}\frac{\Delta[NO]}{\Delta t} = -\frac{\Delta[Cl_2]}{\Delta t} = +\frac{1}{2}\frac{\Delta[NOCl]}{\Delta t}$$

PROBLEM-SOLVING EXAMPLE 12.3 Rate Law from Initial Rates

Initial rates $\left(-\frac{\Delta[Cl_2]}{\Delta t}\right)$ for the reaction of nitrogen monoxide and chlorine

$$2\ NO(g) + Cl_2(g) \longrightarrow 2\ NOCl(g)$$

were measured at 27 °C starting with various concentrations of NO and Cl_2. These data were collected.

Experiment	Initial Concentrations (mol/L) [NO]	[Cl_2]	Initial Rate (mol L^{-1} s^{-1})
1	0.020	0.010	8.27×10^{-5}
2	0.020	0.020	1.65×10^{-4}
3	0.020	0.040	3.31×10^{-4}
4	0.040	0.020	6.60×10^{-4}
5	0.010	0.020	4.10×10^{-5}

(a) What is the rate law?
(b) What is the value of the rate constant k?

Answer

(a) Rate = $k[Cl_2][NO]^2$
(b) $k = 2.1 \times 10^1 \text{ L}^2 \text{ mol}^{-2} \text{ s}^{-1}$

Strategy and Explanation (a) Analyze data from experiments in which one concentration remains the same. In Experiments 1, 2, and 3, the concentration of NO is constant, while the Cl_2 concentration increases from 0.010 to 0.020 to 0.040 mol/L. Each time [Cl_2] is doubled, the initial rate also doubles. For example, when [Cl_2] is doubled from 0.020 to 0.040 mol/L in Experiments 2 and 3, the initial rate doubles from 1.65×10^{-4} to 3.31×10^{-4} mol L^{-1} s^{-1}. Therefore the initial rate is directly proportional to [Cl_2].

In Experiments 2, 4, and 5, [Cl_2] is constant, while [NO] varies. In Experiments 2 and 4, [NO] is doubled, but the initial rate increases by a factor of 4, or 2^2.

$$\frac{\text{Experiment 4 rate}}{\text{Experiment 2 rate}} = \frac{6.60 \times 10^{-4} \text{ mol L}^{-1} \text{ s}^{-1}}{1.65 \times 10^{-4} \text{ mol L}^{-1} \text{ s}^{-1}} = \frac{4}{1} = \frac{2^2}{1}$$

This same result is found in Experiments 2 and 5. Thus the initial rate is proportional to the *square* of [NO]. Therefore, the rate law is

$$\text{Rate} = k[Cl_2][NO]^2$$

When comparing one experiment with another, as is done in the Strategy and Explanation of Problem-Solving Example 12.3, it is usually convenient to put the larger rate in the numerator. Otherwise fractions are obtained instead of whole numbers.

(b) Once the rate law is known, the rate constant k can be calculated. For Experiment 1, for example,

$$8.27 \times 10^{-5} \text{ mol L}^{-1} \text{ s}^{-1} = k(0.010 \text{ mol/L})(0.020 \text{ mol/L})^2$$

$$k = \frac{8.27 \times 10^{-5} \text{ mol L}^{-1} \text{ s}^{-1}}{(0.010 \text{ mol/L})(0.020 \text{ mol/L})^2} = 2.1 \times 10^1 \text{ mol}^{-2} \text{ L}^2 \text{ s}^{-1}$$

For Experiments 2, 3, 4, and 5, the rate constants are 2.1×10^1, 2.1×10^1, 2.1×10^1, and 2.0×10^1 L^2 mol^{-2} s^{-1}, respectively. The average of these values is the rate constant from this series of experiments, 2.1×10^1 L^2 mol^{-2} s^{-1}. It can be used to calculate the rate for any set of NO and Cl_2 concentrations at 27 °C.

For this reaction the rate is proportional to one concentration and to the square of another. The rate constant equals the rate (units of mol L^{-1} s^{-1}) divided by three concentration terms multiplied together (units of mol^3 L^{-3}). Thus, the rate constant has units of

$$\frac{\text{mol L}^{-1}\text{ s}^{-1}}{\text{mol}^3 \text{ L}^{-3}} = \text{L}^2 \text{mol}^{-2} \text{s}^{-1}.$$

☑ **Reasonable Answer Check** The five calculated k values are nearly equal. If the rate law were incorrect, or if an error were made in one or more calculations, some k values would be quite different from the others.

PROBLEM-SOLVING PRACTICE 12.3

At 23 °C, these data were collected for the crystal violet reaction

$$Cv^+(aq) + OH^-(aq) \longrightarrow CvOH(aq)$$

Experiment	Initial Concentrations (mol/L)		Initial Rate
	$[Cv^+]$	$[OH^-]$	(mol L^{-1} s^{-1})
1	4.3×10^{-5}	0.10	1.3×10^{-6}
2	2.2×10^{-5}	0.10	6.7×10^{-7}
3	1.1×10^{-5}	0.10	3.3×10^{-7}

(a) Is it possible to determine the complete rate law from the data given? Why or why not?

(b) Assume that the rate does not depend on the concentration of hydroxide ion. What is the rate law?

(c) Calculate the rate constant, again assuming that the rate does not depend on the concentration of hydroxide ion.

(d) Calculate the initial rate of reaction when the concentration of crystal violet cation is 0.00045 M and the concentration of hydroxide ion is 0.10 M. Report your results in mol L^{-1} s^{-1}.

(e) Calculate the rate when the concentration of Cv^+ is half the initial value of 0.00045 M.

EXERCISE 12.5 Determining the Rate Law Using Logarithms

For the reaction in Problem-Solving Example 12.3, assume that the rate law is of the form Rate $= k[\text{NO}]^x[\text{Cl}_2]^y$. Show mathematically that by taking logarithms of both sides of the rate law and comparing Experiments 2 and 4, where the concentration of chlorine is the same, x, is given by

$$x = \log\left(\frac{\text{Rate}_4}{\text{Rate}_2}\right) \Big/ \log\left(\frac{[\text{NO}]_4}{[\text{NO}]_2}\right)$$

12.3 Rate Law and Order of Reaction

For many (but not all) homogeneous reactions, the rate law has the general form

$$\text{Rate} = k[\text{A}]^m[\text{B}]^n \ldots$$

where concentrations of substances, [A], [B], . . . are raised to powers, m, n, ... The substances A, B, . . . might be reactants, products, or catalysts. The exponents m, n, . . .

are usually positive whole numbers but might be negative numbers or fractions. These exponents define the **order of the reaction** with respect to each reactant. If n is 1, for example, the reaction is first-order with respect to B; if m is 2, then the reaction is second-order with respect to A. The sum of m and n (plus the exponents on any other concentration terms in the rate equation) gives the **overall reaction order.** (The reaction in Problem-Solving Example 12.3 is first-order in Cl_2, second-order in NO, and third-order overall.) A very important point to remember is that ***the rate law and reaction orders must be determined experimentally; they cannot be predicted from stoichiometric coefficients in the balanced overall chemical equation.***

None of the reactions in Problem-Solving Example 12.4 has a rate law that can be derived correctly from the stoichiometric equation. For example, H_2 has a coefficient of 2 in Reaction (a), but the rate law involves $[H_2]$ to the first power.

PROBLEM-SOLVING EXAMPLE 12.4 Reaction Order and Rate Law

For each reaction and experimentally determined rate law listed below, determine the order with respect to each reactant and the overall order.

(a) $2\ NO(g) + 2\ H_2(g) \longrightarrow N_2(g) + 2\ H_2O(g)$ Rate $= k[NO]^2[H_2]$

(b) $14\ H_3O^+(aq) + 2\ HCrO_4^-(aq) + 6\ I^-(aq) \longrightarrow 2\ Cr^{3+}(aq) + 3\ I_2(aq) + 22\ H_2O(\ell)$
Rate $= k[HCrO_4^-][I^-]^2[H_3O^+]^2$

(c) *cis*-2-butene(g) $\longrightarrow$ *trans*-2-butene(g) (catalytic concentration of I_2 present)
Rate $= k$[*cis*-2-butene]$[I_2]^{1/2}$

Answer

(a) First-order in H_2, second-order in NO, third-order overall
(b) First-order in $HCrO_4^-$, second-order in I^-, second-order in H_3O^+, fifth-order overall
(c) First-order in *cis*-2-butene, 0.5-order in I_2, 1.5-order overall

Strategy and Explanation Use the exponents in the rate law—not the stoichiometric coefficients—to determine the order.

(a) The rate law has a single term that is raised to the first power and another term that is squared, so the reaction is first-order in H_2, second-order in NO, and third-order overall.
(b) The rate law contains three terms. Since the $HCrO_4^-$ term is raised to the first power, the reaction is first-order in $HCrO_4^-$. The other two terms are squared, so the reaction is second-order in I^- and second-order in H_3O^+. The exponents sum to five, so the reaction is fifth-order overall.
(c) In this case the rate of reaction depends on the concentration of the reactant and also on the square root ($\frac{1}{2}$ power) of the concentration of a catalyst, I_2. The reaction is therefore first-order in *cis*-2-butene, 0.5-order in I_2, and 1.5-order overall.

PROBLEM-SOLVING PRACTICE 12.4

In Problem-Solving Example 12.3 the rate law for reaction of NO with Cl_2 was found to be

$$2\ NO(g) + Cl_2(g) \longrightarrow 2\ NOCl(g) \qquad \text{Rate} = k[NO]^2[Cl_2]$$

(a) What is the order of the reaction with respect to NO? With respect to Cl_2?
(b) Suppose that you triple the concentration of NO and simultaneously decrease the concentration of Cl_2 by a factor of 8. Will the reaction be faster or slower under the new conditions? How much faster or slower? (Assume that the temperature is the same in both sets of conditions.)

We have already seen that the rate law allows us to calculate the rate of reaction from the concentration of the reactants (and perhaps other substances). The integrated rate law allows us to calculate the concentration of a reactant (or perhaps another substance) as a function of time.

The Integrated Rate Law

Another approach to experimental determination of the rate law and rate constant for a reaction uses calculus to derive what is called the integrated rate law. As an example of the integrated rate law method, suppose that we have a hypothetical reaction in which a single substance A reacts to form products.

$$A \longrightarrow \text{products}$$

First-Order Reaction If the rate law is first-order, then

$$\text{Rate} = -\frac{\Delta[\text{A}]}{\Delta t} = k[\text{A}]$$

This expression can be transformed, using calculus, to the integrated first-order rate law,

$$\ln[\text{A}]_t = -kt + \ln[\text{A}]_0 \qquad [12.5]$$

where $[\text{A}]_t$ represents the concentration of A at time t, $[\text{A}]_0$ represents the initial concentration of A (when $t = 0$), and ln represents the natural logarithm function. (Logarithms are discussed in Appendix A.6.)

Equation 12.5 has the same form as the general equation for a straight line, $y = mx + b$, in which m is the slope and b is the y-intercept.

y-axis variable | slope | x-axis variable | y-intercept

$$\ln[\text{A}]_t = -kt + \ln[\text{A}]_0$$

$$y = mx + b \qquad [12.5]$$

If the reaction is actually first-order, then a graph of ln[A] on the vertical (y) axis versus t on the horizontal (x) axis should be a straight line. A linear graph, such as the one in Figure 12.5a, is evidence that the reaction is first-order.

You do not have to know calculus to use the results that constitute the integrated rate law method. If you do know calculus, however, you will be able to derive the results for yourself.

Using calculus,

$$-\frac{d[\text{A}]}{dt} = k[\text{A}] \quad \text{and} \quad \frac{d[\text{A}]}{[\text{A}]} = -k\,dt$$

$$\int_{[\text{A}]_0}^{[\text{A}]_t} \frac{d[\text{A}]}{[\text{A}]} = -k\int_0^t dt$$

$$\ln[\text{A}]_t - \ln[\text{A}]_0 = -k(t_t - t_0)$$

If the reaction starts at time t_0, then $t_t - t_0 = t$, the elapsed time, and $\ln[\text{A}]_t - \ln[\text{A}]_0 = -kt$ or $\ln[\text{A}]_t = -kt + \ln[\text{A}]_0$.

Second-Order Reaction For the same reaction

$$\text{A} \longrightarrow \text{products}$$

suppose that the rate depends on the square of the concentration of the reactant; that is, suppose the rate law is second-order.

$$\text{Rate} = k[\text{A}]^2$$

The integrated rate law derived using calculus is

$$\frac{1}{[\text{A}]_t} = kt + \frac{1}{[\text{A}]_0}$$

This equation is also of the form $y = mx + b$. If a reaction is second-order, a graph of $1/[\text{A}]_t$ versus t will be linear with slope $= k$ and y-intercept $= 1/[\text{A}]_0$. Such a straight-line graph is evidence that a reaction is second-order. An example graph is shown in Figure 12.5b.

Using calculus,

$$-\frac{d[\text{A}]}{dt} = k[\text{A}]^2 \quad \text{and} \quad \frac{d[\text{A}]}{[\text{A}]^2} = -k\,dt$$

$$\int_{[\text{A}]_0}^{[\text{A}]_t} \frac{d[\text{A}]}{[\text{A}]} = -k\int_0^t dt$$

$$-\frac{1}{[\text{A}]_t} - \left(-\frac{1}{[\text{A}]_0}\right) = -k(t_t - t_0)$$

If the reaction starts at time t_0, then $t_t - t_0 = t$, the elapsed time, and

$$\frac{1}{[\text{A}]_t} = kt + \frac{1}{[\text{A}]_0}$$

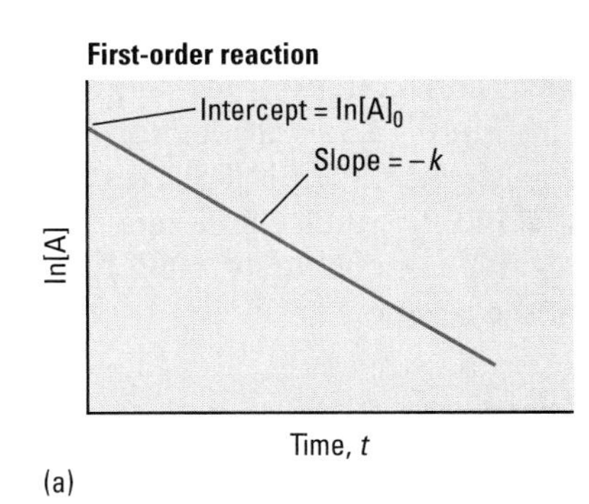

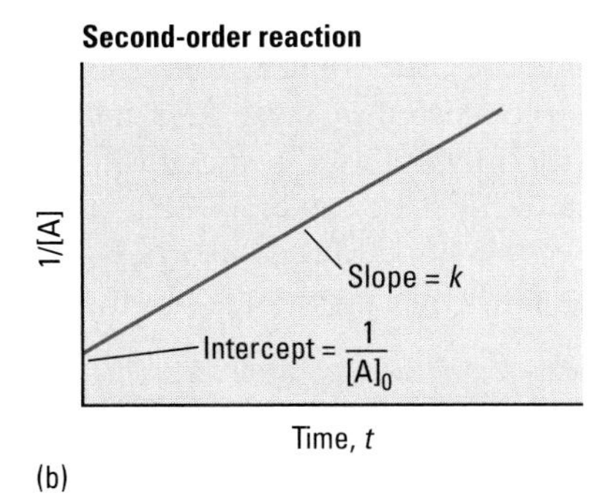

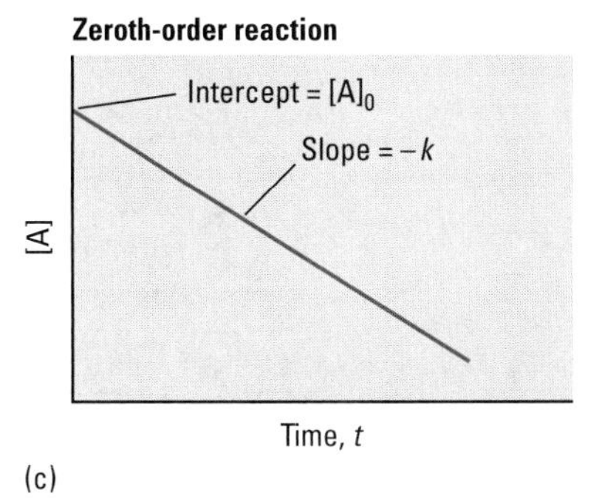

Figure 12.5 First-order, second-order, and zeroth-order plots. (a) If a reaction is first-order in reactant A, plotting $\ln[\text{A}]_t$ versus t gives a straight line. (b) If a reaction is second-order in reactant A, plotting $1/[\text{A}]_t$ versus t gives a straight line. (c) If a reaction is zeroth-order in reactant A, plotting $[\text{A}]_t$ versus t gives a straight line.

Table 12.2 Integrated Rate Laws

Order	Rate Equals	Integrated Rate Law*	Straight-line Plot	Slope of Plot	Units of k
0	$k[A]^0 = k$	$[A]_t = -kt + [A]_0$	$[A]_t$ vs t	$-k$	conc. time^{-1}
1	$k[A]$	$\ln[A]_t = -kt + \ln[A]_0$	$\ln[A]_t$ vs t	$-k$	time^{-1}
2	$k[A]^2$	$\frac{1}{[A]_t} = kt + \frac{1}{[A]_0}$	$\frac{1}{[A]_t}$ vs t	k	conc.$^{-1}$ time^{-1}

*In the table, $[A]_0$ indicates the initial concentration of substance A, that is, the concentration of A at $t = 0$, the time when the reaction was started.

Zeroth-Order Reaction There are a few reactions for which the rate does not depend on the concentration of a reactant at all. These are called zeroth-order reactions, because the rate law can be written as a rate constant k times a concentration to the zeroth power. (Recall that anything raised to the zeroth power equals 1.)

$$\text{Rate} = -\frac{\Delta[A]}{\Delta t} = k[A]^0 = k$$

This rate law says that for a zeroth-order reaction the rate is always the same no matter what the concentration of reactant. The rate is equal to the rate constant k. For a zeroth-order reaction you can derive the integrated form without calculus. Simply use algebra to rearrange the rate law just given.

$$\Delta[A] = -k\Delta t$$
$$[A]_t - [A]_0 = -k(t_t - t_0) = -kt$$
$$[A]_t = -kt + [A]_0$$

Again, the equation is of the form $y = mx + b$. If a reaction is zeroth-order, graphing A_t versus t gives a straight line with slope $= -k$ and y-intercept $= [A]_0$, as shown in Figure 12.5c.

To summarize these three situations, a rate law that involves powers of the reactant concentration can be written as

$$\text{Rate} = -\frac{\Delta[A]}{\Delta t} = k[A]^m$$

where m is the order of the reaction as defined earlier. The integrated rate law depends on the value of m, and the results for m equal to 0, 1, and 2 are given in Table 12.2.

To determine the order of reaction, then, we collect concentration-time data and make the three plots listed in Table 12.2 and shown in Figure 12.5. Only one (or perhaps none) of the plots will be a straight line. If one is straight, then it indicates the order, and the rate constant can be calculated from its slope. The units of the rate constant are those of the slope of the line; they are given in the last column of Table 12.2. The units of the rate constant depend on the order of the reaction.

Cyclopentene has the structure

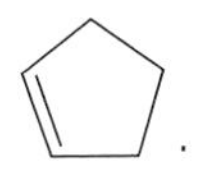

PROBLEM-SOLVING EXAMPLE 12.5 Reaction Order and Rate Constant from Integrated Rate Law

These data were obtained for decomposition of cyclopentene at 825 K.

$$C_5H_8(g) \longrightarrow C_5H_6(g) + H_2(g)$$

Time (s)	$[C_5H_8]$ (mol/L)	Time (s)	$[C_5H_8]$ (mol/L)
0	0.0200	300	0.0084
20	0.0189	400	0.0063
50	0.0173	500	0.0047
100	0.0149	700	0.0027
200	0.0112	1000	0.0011

Obtain the order of the reaction and the rate constant.

Answer The reaction is first-order in C_5H_8; $k = 2.88 \times 10^{-3}\ s^{-1}$.

Strategy and Explanation Use the integrated rate law method, making graphs to test for zeroth-, first-, and second-order. Such plots are shown in Figure 12.6. The zeroth-order and second-order plots are curved, while the first-order plot is a straight line. Thus the reaction must be first-order. From the first-order plot, calculate the slope using the points marked on the graph as open circles.

$$\text{Slope} = \frac{\{-6.56 - (-4.15)\}}{\{910 - 74\}\ s} = -2.88 \times 10^{-3}\ s^{-1}$$

The slope of $-2.88 \times 10^{-3}\ s^{-1}$ is the negative of the rate constant (see Table 12.2), which means that $k = 2.88 \times 10^{-3}\ s^{-1}$.

☑ **Reasonable Answer Check** The units are s^{-1}, which corresponds to the reciprocal time units indicated in Table 12.2 for a first-order rate constant.

It is important to use two points on the straight line through the experimental data (open circles in Figure 12.6b), not two of the data points themselves, when you calculate the slope. Making a graph is similar to averaging, because the straight line and its slope are based on all ten points in the data table, not just two of them.

PROBLEM-SOLVING PRACTICE 12.5

Use the concentration-time data in Table 12.1 for the reaction of Cv^+ with OH^- (⇐ *p. 418*) to deduce the order of the reaction with respect to Cv^+. Determine the rate constant. (Assume that $[OH^-]$ is constant during the reaction.)

Calculating Concentration or Time from Rate Law

Once the rate law has been determined experimentally, it provides a way to calculate the concentration of a reactant or product at any time after the reaction has begun. All that is needed is the integrated rate law (from Table 12.2), the value of the rate constant, and the initial concentration of reactant or product. These are related by the equations given in Table 12.2.

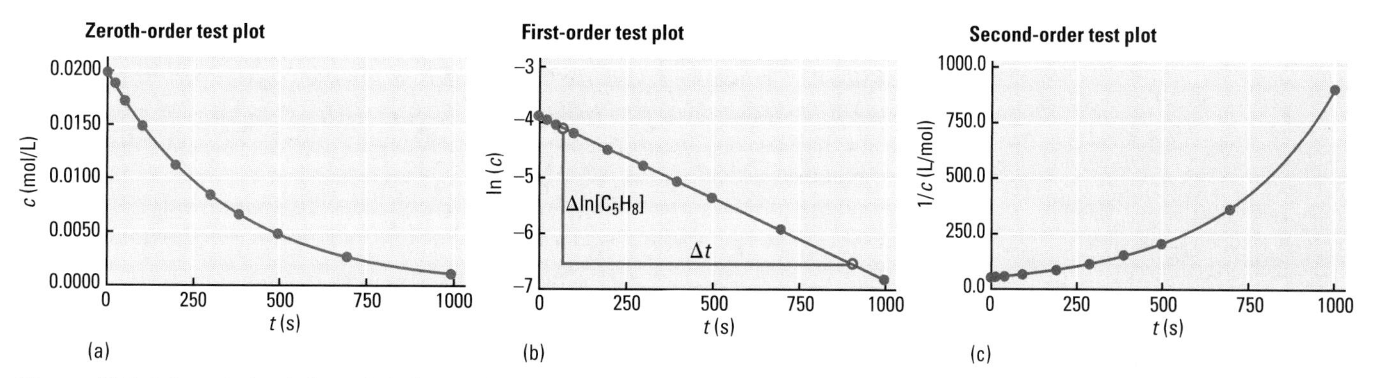

Figure 12.6 Integrated rate law plots for cyclopentene decomposition reaction. (a) Zeroth-order, (b) first-order, and (c) second-order plots for the decomposition reaction of cyclopentene in aqueous solution at 825 K.

PROBLEM-SOLVING EXAMPLE 12.6 Calculating Concentrations

The first-order rate constant is 1.87×10^{-3} min^{-1} at 37 °C (body temperature) for reaction of cisplatin, a cancer chemotherapy agent, with water. The reaction is

$$\text{cisplatin} + H_2O \longrightarrow \text{cisplatinOH}_2^+ + Cl^-$$

Suppose that the concentration of cisplatin in the bloodstream of a cancer patient is 4.73×10^{-4} mol/L. What will the concentration be exactly 24 hours later?

Answer 3.20×10^{-5} mol/L

Strategy and Explanation Assume that the rate of reaction in the blood is the same as in water. Let [cisplatin] represent the concentration of cisplatin at any time and $[\text{cisplatin}]_0$ represent the initial concentration. The reaction is first-order, so from Table 12.2,

$$\ln[\text{cisplatin}] = -kt + \ln[\text{cisplatin}]_0$$

Mathematical operations involving logarithms and exponentials (antilogarithms) are discussed in Appendix A.6.

Because the solution to this example involves a ratio of concentrations in which the units divide out, the same approach can be taken in problems that involve the number of moles, the mass, or the number of atoms or molecules at two different times.

Rearrange the equation algebraically to

$$\ln[\text{cisplatin}] - \ln[\text{cisplatin}]_0 = -kt$$

$$\ln\left\{\frac{[\text{cisplatin}]}{[\text{cisplatin}]_0}\right\} = -kt$$

$$\frac{[\text{cisplatin}]}{[\text{cisplatin}]_0} = \text{anti}\ln(-kt) = e^{-kt}$$

$$[\text{cisplatin}] = [\text{cisplatin}]_0\, e^{-kt}$$

$$[\text{cisplatin}] = (4.73 \times 10^{-4}\ \text{mol/L})\, e^{-(1.87\times 10^{-3}\ \text{min}^{-1}\times 24\ \text{h})}$$

$$[\text{cisplatin}] = (4.73 \times 10^{-4}\ \text{mol/L})\, e^{-(1.87\times 10^{-3}\ \text{min}^{-1}\times 24\ \text{h}\times 60\ \text{min h}^{-1})}$$

$$[\text{cisplatin}] = (4.73 \times 10^{-4}\ \text{mol/L})(6.77 \times 10^{-2})$$

$$= 3.20 \times 10^{-5}\ \text{mol/L}$$

☑ **Reasonable Answer Check** If the drug is to be effective, it almost certainly needs to be in the body for some time—several minutes to hours or more. The concentration has dropped to a little less than 10% of its initial value in 24 hours, which is reasonable.

PROBLEM-SOLVING PRACTICE 12.6

The first-order rate constant for decomposition of an insecticide in the environment is 3.43×10^{-2} d^{-1}. How long does it take for the concentration of insecticide to drop to $\frac{1}{10}$ of its initial value?

Half-Life

The **half-life** of a reaction, $t_{1/2}$, is *the time required for the concentration of a reactant A to fall to one half of its initial value.* That is, $[A]_{t_{1/2}} = \frac{1}{2}[A]_0$. For a first-order reaction the half-life has the same value, no matter what the initial concentration is. For other reaction orders this is not true.

Remember that Equation 12.5 was derived for a reaction of the form A ⟶ products. Had the equation been 2 A ⟶ products, that is, had the coefficient of A been 2 in the chemical equation, the result would be slightly different.

The half-life is related to the first-order rate constant. To see how, use algebra to rearrange Equation 12.5 (which was associated with a reaction A ⟶ products):

$$\ln[A]_t = -kt + \ln[A]_0 \qquad [12.5]$$

$$\ln[A]_t - \ln[A]_0 = -kt$$

$$\ln[A]_{t_{1/2}} - \ln[A]_0 = -kt_{1/2} \qquad [12.6]$$

Because $[A]_{t_{1/2}} = \frac{1}{2}[A]_0$, Equation 12.5 can be rewritten as

$$\ln([A]_0/2) - \ln[A]_0 = -kt_{1/2}$$

and so

$$-kt_{1/2} = \ln[A]_0 - \ln(2) - \ln[A]_0 = -\ln 2$$

$$t_{1/2} = \frac{-\ln 2}{-k} = \frac{0.693}{k} \qquad [12.7]$$

This means that measuring the half-life of a first-order reaction determines the rate constant, and vice versa. Radioactive decay (Section 19.4) is a first-order process, and half-life is typically used to report the rate of decay of radioactive nuclei.

PROBLEM-SOLVING EXAMPLE 12.7 Half-Life and Rate Constant

In Problem-Solving Example 12.5, the rate constant for decomposition of cyclopentene at 825 K was found to be $2.88 \times 10^{-3}\ s^{-1}$. The reaction gave a linear first-order plot. What is the half-life in seconds for this reaction?

Answer 241 s

Strategy and Explanation The reaction is first-order and is of the form A $\longrightarrow$ products, so Equation 12.7 can be used.

$$t_{1/2} = \frac{0.693}{k} = \frac{0.693}{2.88 \times 10^{-3}\ s^{-1}} = 2.41 \times 10^2\ s = 241\ s$$

☑ **Reasonable Answer Check** The rate constant is about $3 \times 10^{-3}\ s^{-1}$. If the concentration of cyclohexene were 1.0 mol L^{-1}, then the reaction rate would be $0.003\ s^{-1} \times 1.0$ mol L^{-1} = 0.003 mol $L^{-1}\ s^{-1}$. This means that 0.003 mol L^{-1} would react every second. For the concentration to drop to half of 1.0 mol L^{-1}, the change in concentration would be 0.500 mol L^{-1}. Therefore, it would take at least $\frac{0.500\ \text{mol L}^{-1}}{0.003\ \text{mol L}^{-1}\ \text{s}^{-1}} = 167$ s for the concentration to drop to half its initial value, and the half-life should be at least 167 s. (Because the rate decreases as concentration decreases, it will take longer than 167 s.) A half-life of 241 s is reasonable.

PROBLEM-SOLVING PRACTICE 12.7

From Figure 12.3 determine the time required for the concentration of Cv^+ to fall to one half the initial value. Verify that the same period is required for the concentration to fall from one half to one fourth the initial value. From this half-life, calculate the rate constant.

ESTIMATION Pesticide Decay

There are usually several different ways that a pesticide can decompose in an ecosystem. For this reason, it is difficult to define an accurate rate of decomposition and even more difficult to define a rate law. Often it is assumed that decomposition is first-order, and an approximate half-life is reported.

Organochlorine pesticides such as DDT, lindane, and dieldrin may have half-lives as long as 10 years in the environment. The maximum contaminant level (MCL) for lindane is 0.2 ppb (parts per billion). Suppose that an ecosystem has been contaminated with lindane at a concentration of 200 ppb. How long would you have to wait before it would be safe to enter the ecosystem without protection from the pesticide?

The level of contamination is 200 ppb/0.2 ppb = 1000 times the MCL. Presumably it would be safe to wait until the level had dropped to 1/1000 of its initial value. You can estimate how long this would be by using powers of 2, because the number of half-lives required is n, where $(1/2)^n = 1/1000$. That is, as soon as 2^n exceeds 1000, you have waited long enough. Computer scientists, who deal with binary arithmetic, can easily tell you that $2^{10} = 1024$, so $n = 10$. You can verify this using your calculator's y^x-key, or you could simply raise 2 to a power until a value greater than 1000 was calculated. If you did not have a calculator, you could multiply $2 \times 2 \times 2 \ldots$ in your head until you had enough factors to multiply out to a number bigger than 1000. Ten half-lives means 10×10 years, or 100 years, so after 100 years the ecosystem would be free of significant lindane contamination.

Visit this book's companion website at **www.cengage.com/chemistry/moore** to work an interactive module based on this material.

12.4 A Nanoscale View: Elementary Reactions

Macroscale experimental observations reveal that reactant concentrations, temperature, and catalysts can affect reaction rates. But how can we interpret such observations in terms of nanoscale models? We will use the *kinetic-molecular theory of matter,* which was first introduced in Section 1.8 ***(⟵ p. 17)*** and developed further in Section 10.3 ***(⟵ p. 345),*** together with the ideas about molecular structure developed in Chapters 8 and 9. These concepts provide a good basis for understanding how atoms and molecules move and chemical bonds are made or broken during the very short time it takes for reactant molecules to be converted into product molecules.

According to kinetic-molecular theory, molecules are in constant motion. In a gas or liquid they bump into one another; in a solid they vibrate about specific locations. Molecules also rotate, flex, or vibrate around or along the bonds that hold the atoms together. These motions produce the transformations of molecules that occur during chemical reactions. It turns out that ***there are only two important types of molecular transformations:*** unimolecular and bimolecular. In a **unimolecular reaction** *the structure of a single particle (molecule or ion) rearranges to produce a different particle or particles.* A unimolecular reaction might involve breaking a bond and forming two new molecules, or it might involve rearrangement of one isomeric structure into another. In a **bimolecular reaction** *two particles (atoms, molecules, or ions) collide and rearrange into products.* In a bimolecular reaction new bonds may be formed between the reactant particles, and existing bonds may be broken. Sometimes the two particles combine to form a new, larger one. Sometimes two or more new molecules are formed from the original two.

All chemical reactions can be understood in terms of simple reactions such as those just described. Very complicated reactions can be built up from combinations of unimolecular and bimolecular reactions, just as complicated compounds can be built from chemical elements. For example, hundreds of such reactions are needed to understand how smog is produced in a city such as Los Angeles. Like the chemical elements, the simplest nanoscale reactions are building blocks, so they are referred to as **elementary reactions.** ***The equation for an elementary reaction shows exactly which molecules, atoms, or ions take part in the elementary reaction.*** The next two sections describe the two important types of elementary reactions in more detail.

Remember the color codes for atoms:

carbon,	black	nitrogen,	blue
hydrogen,	white	chlorine,	green
oxygen,	red	iodine,	violet

EXERCISE 12.6 Unimolecular and Bimolecular Reactions

For each of the nanoscale molecular diagrams below, write a balanced equation using chemical formulas. Which of the reactions are unimolecular reactions? Which are bimolecular?

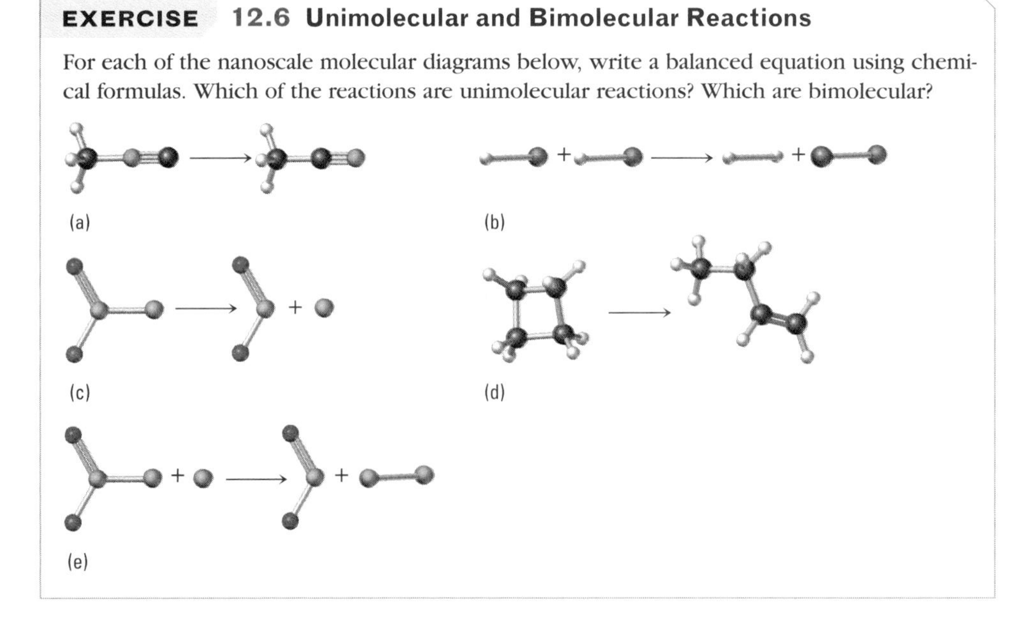

Unimolecular Reactions

An example of a unimolecular reaction is the conversion of *cis*-2-butene to *trans*-2-butene.

$$\underset{cis\text{-2-butene}}{H_3C(H)C{=}C(H)CH_3}\,(g) \longrightarrow \underset{trans\text{-2-butene}}{H_3C(H)C{=}C(H)CH_3}\,(g)$$

Cis-2-butene and *trans*-2-butene are *cis-trans* stereoisomers (**p. 281**). The difference between the two molecules is the orientation of the methyl groups, which are on the same side of the double bond in the *cis* structure and on opposite sides in the *trans* structure. If we could grab one end of the molecule and twist it 180° around the axis of the double bond, we would get the other molecule. Thus, it is a reasonable hypothesis that the molecular pathway by which *cis*-2-butene changes to *trans*-2-butene involves twisting the molecule around the double bond. The angle of twist around the double-bond axis measures the progress of the reaction on the nanoscale. The greater the angle, the less the molecule is like *cis*-2-butene and the more it is like *trans*-2-butene, until an angle of 180° is reached and it has become *trans*-2-butene.

Such a twist requires that the reactant molecule have sufficient energy. Chemical bonds are like springs. They can be stretched, twisted, and bent, but these changes raise the potential energy. Consequently, some kinetic energy must be converted to potential energy when one end of the *cis*-2-butene molecule twists relative to the other, just as it would if a spring were twisted. At room temperature most of the molecules do not have enough energy to twist far enough to change *cis*-2-butene into *trans*-2-butene. Therefore, *cis*-2-butene can be kept in a sealed flask at room temperature for a long time without any appreciable quantity of *trans*-2-butene being formed. However, as the temperature is raised, more and more molecules have sufficient energy to react, and the reaction gets faster and faster.

Figure 12.7 shows a plot of potential energy versus the angle of twist in *cis*- and *trans*-2-butene. The potential energy is 435×10^{-21} J higher when one end of a *cis*-2-butene molecule is twisted by 90° from the initial flat molecule. This is similar to the increased potential energy that an object such as a car has at the top of a hill compared with its energy at the bottom. Just as a car cannot reach the top of a hill unless it has enough energy, a molecule cannot reach the top of the "hill" for a reaction unless it has enough energy. Notice that the top of the hill can be approached from either side, and from the top a twisted molecule can go downhill energetically to either the *cis* or the *trans* form. *The structure at the top of an energy diagram like this one* is called the **transition state** or **activated complex.** In this case it is a molecule that has been twisted so that the methyl groups are at a 90° angle.

Since molecules are very small, the energy required to twist one *cis*-2-butene molecule is very small. However, if we wanted to twist a mole of molecules all at once, it would take a lot of energy. The energy required to reach the top of the "hill" is often reported per mole of molecules—that is, as $(435 \times 10^{-21}$ J/molecule$) \times (1$ kJ/10^3J$) \times (6.022 \times 10^{23}$ molecules/mol$) =$ 262 kJ/mol.

CONCEPTUAL
EXERCISE 12.7 Transition State

Methyl isonitrile reacts to form acetonitrile in a single-step elementary reaction.

$$CH_3NC(g) \longrightarrow CH_3CN(g)$$

During the reaction the nitrogen atom and one of the carbon atoms exchange places, but the rest of the molecule is unchanged. Suggest a structure for the transition state for this reaction. Draw this structure as a Lewis structure and as a ball-and-stick molecular model.

Figure 12.7 is similar to Figure 6.15 (p. 201), which showed the energy change when bonds were broken and formed as H_2 and Cl_2 changed into HCl.

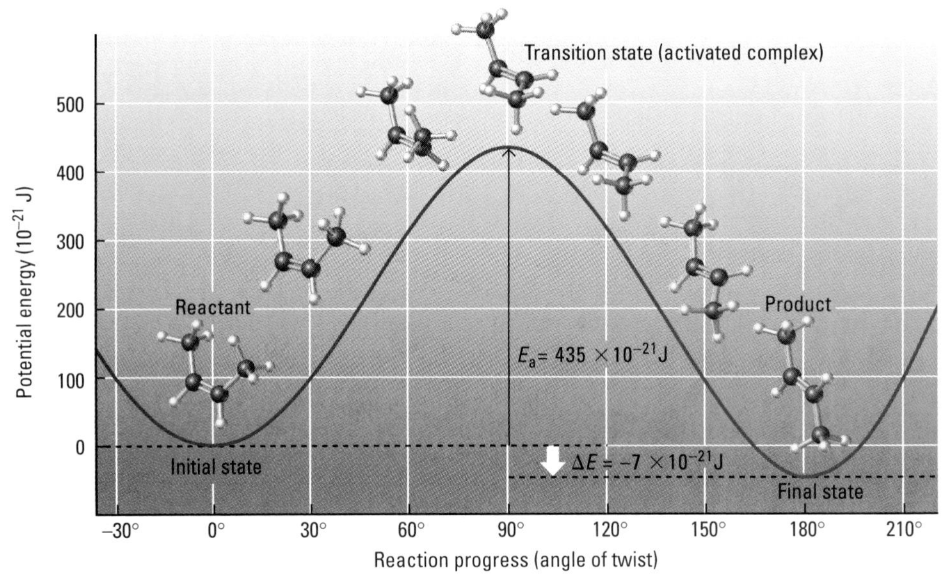

Active Figure 12.7 Energy diagram for the conversion of *cis*-2-butene to *trans*-2-butene. Twisting one end of the molecule with respect to the other requires an increase in potential energy because the double bond between the two central C atoms is like a spring and resists twisting. Twisting in either direction requires the same increase in potential energy, so the energy for +30° equals that for −30°. When the angle between the ends of the molecule is 90°, the potential energy has risen by 435×10^{-21} J. A molecule of *cis*-2-butene must have at least this quantity of energy before it can twist past 90° to 180°, which converts it to *trans*-2-butene. The 90° twisted structure at the top of the diagram is called the *transition state* or *activated complex*. The progress of the reaction (the change in the structure of this single molecule) is measured by the angle of twist. Visit this book's companion website at **www.cengage.com/chemistry/moore** to test your understanding of the concepts in this figure.

The generalization that higher activation energy results in slower reaction applies best if the reactions are similar. For example, it applies to a group of reactions that all involve twisting around a double bond. It also applies to reactions that involve collisions of one molecule with each of a group of similar molecules. It would be less applicable if we were comparing one reaction that involved collision of two molecules with another reaction that involved twisting around a bond.

Almost every chemical reaction has an energy barrier that must be surmounted as reactant molecules change into product molecules. The heights of such barriers vary greatly—from almost zero to hundreds of kilojoules per mole. *At a given temperature, the higher the energy barrier, the slower the reaction.* The minimum energy required to surmount the barrier is called the **activation energy, E_a**, for the reaction. For the *cis*-2-butene ⟶ *trans*-2-butene reaction the activation energy is 435×10^{-21} J/molecule, or 262 kJ/mol (see Figure 12.7).

Another interesting relationship shown in Figure 12.7 connects kinetics and thermodynamics. The energy of the product, one molecule of *trans*-2-butene, is 7×10^{-21} J *lower* than that of the reactant, one molecule of *cis*-2-butene. This means that the *cis* ⟶ *trans* reaction is *exothermic* by 7×10^{-21} J/molecule, which translates to 4 kJ/mol. Also, *cis*-2-butene is higher in energy by 7×10^{-21} J/molecule, so the reverse reaction requires that 4 kJ/mol be absorbed from the surroundings; it is *endothermic*. The height of the energy hill that must be climbed when the reverse reaction occurs is $(435 + 7) \times 10^{-21}$ J/molecule or 442×10^{-21} J/molecule (266 kJ/mol). Thus, the activation energy for the forward reaction is 4 kJ/mol less than that for the reverse reaction. For almost all reactions the activation energy for a forward reaction will differ from the activation energy of the reverse reaction, and the difference is $\Delta E°$ for the reaction.

The actual relation is $\Delta E° = E_a$(forward) − E_a(reverse). Since $\Delta E°$ differs from $\Delta H°$ only when there is a change in volume of the reaction system (under constant pressure), the difference in activation energies is often equated with the enthalpy change (p. 192).

Bimolecular Reactions

An example of a bimolecular process is the reaction of iodide ion, I^-, with methyl bromide, CH_3Br, in aqueous solution.

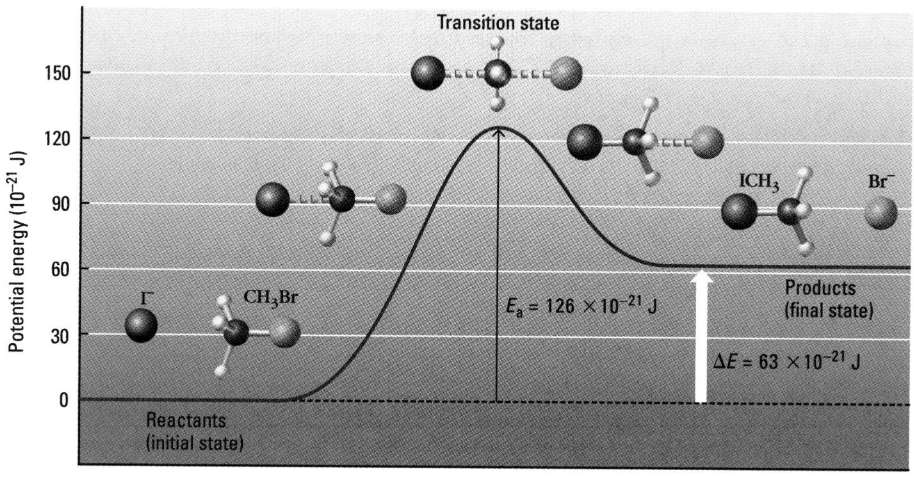

Figure 12.8 Energy diagram for iodide–methyl bromide reaction. During collision of an iodide ion with a methyl bromide molecule, a new iodine-carbon bond forms at the same time that the carbon-bromine bond is breaking. Forming the new bond lowers the potential energy, which otherwise would be raised a lot by breaking the carbon-bromine bond. This results in a lower activation energy and a faster reaction than would otherwise occur. In this case reaction progress is measured in terms of stretching of the carbon-bromine bond and formation of the iodine-carbon bond.

$$I^-(aq) + CH_3Br(aq) \longrightarrow ICH_3(aq) + Br^-(aq)$$

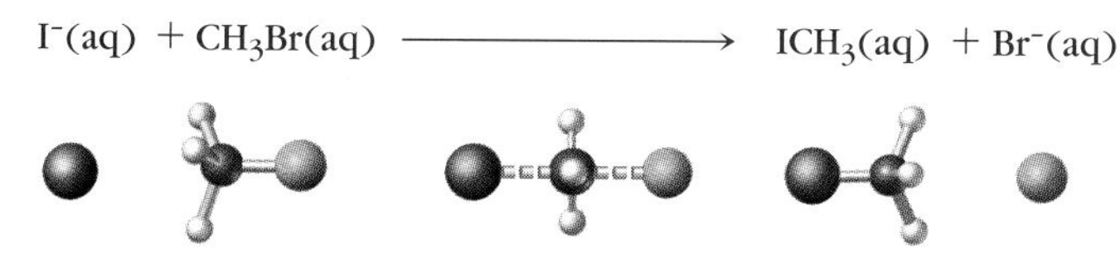

Here the equation for the elementary reaction shows that an iodide ion must collide with a methyl bromide molecule for the reaction to occur. The carbon-bromine bond does not break until after the iodine-carbon bond has begun to form. This makes sense, because just breaking a carbon-bromine bond would require a large increase in potential energy. Partially forming a carbon-iodine bond while the other bond is breaking lowers the potential energy. This helps keep the activation energy hill low. Figure 12.8 shows the energy-versus-reaction progress diagram for this reaction.

The methyl bromide molecule has a tetrahedral shape that is distorted because the Br atom is much larger than the H atoms. Numerous experiments suggest that the reaction occurs most rapidly in solution when the I^- ion approaches the methyl bromide from the side of the tetrahedron opposite the bromine atom. That is, approach to only one of the four sides of CH_3Br can be effective, which limits reaction to only one fourth of all the collisions at most. This factor of one fourth is called a **steric factor** because it depends on the three-dimensional shapes of the reacting molecules. For molecules much more complicated than methyl iodide, such geometry restraints mean that only a very small fraction of the total collisions can lead to reaction. No wonder some chemical reactions are slow.

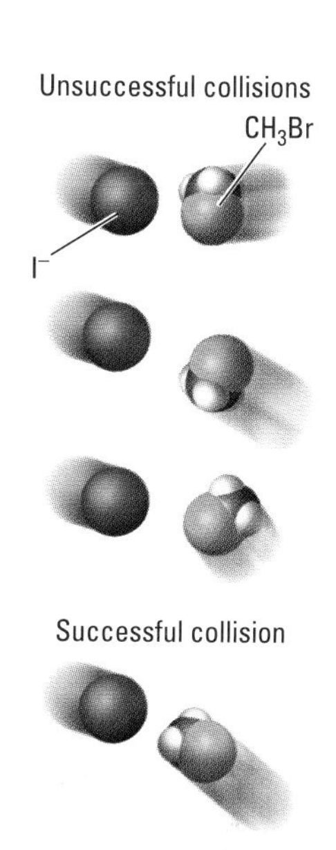

Unsuccessful collisions. In the first three collisions shown, the iodide ion does not approach the methyl bromide molecule from the side opposite the bromine atom. None of these collisions is as likely to result in a reaction as is the collision shown at the bottom.

The word *steric* comes from the same root as the prefix *stereo-*, which means "three-dimensional."

PROBLEM-SOLVING EXAMPLE 12.8 Reaction Energy Diagrams

A reaction by which ozone is destroyed in the stratosphere is

$$O_3(g) + O(g) \longrightarrow 2\ O_2(g)$$

(O represents atomic oxygen, which is formed in the stratosphere when photons of ultraviolet light from the sun split oxygen molecules in two.) The activation energy for ozone destruction is 19 kJ/mol of O_3 consumed. Use standard enthalpies of formation from

Remember that because there are equal numbers of moles of gas phase reactants and products, there is no volume change at constant temperature, and $\Delta E° = \Delta H°$ for this reaction (⟵ p. 192).

Appendix J to calculate the enthalpy change for this reaction. Then construct an energy diagram for the reaction. Draw vertical arrows to indicate the sizes of $\Delta H°$, E_a(forward), and E_a(reverse) for the reaction.

Answer $\Delta H° = \Delta E° = -392$ kJ/mol of O_3 consumed, E_a(forward) = 19 kJ/mol O_3 consumed, E_a(reverse) = 411 kJ/mol O_3 formed; the energy diagram is shown below.

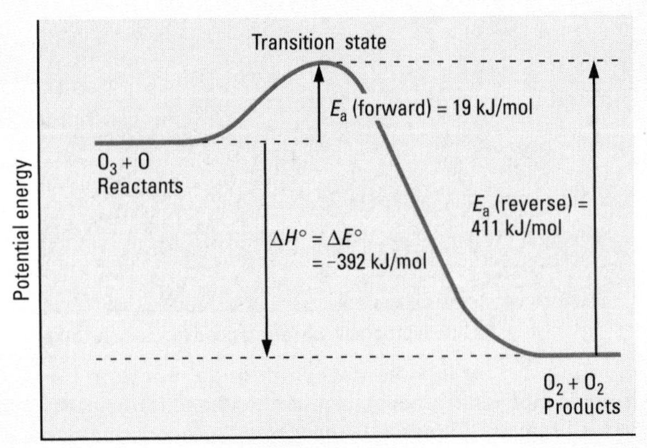

Strategy and Explanation Standard enthalpies of formation are 249.2 kJ/mol for ozone, 0 kJ/mol for diatomic oxygen, and 142.7 kJ/mol for atomic oxygen. Using these enthalpy values, we get $\Delta H° = 0 - (249.2 + 142.7)$ kJ/mol = −391.9 kJ/mol of O_3 consumed. The negative sign of $\Delta H°$ indicates that the reaction is exothermic, so the products must be lower in energy than the reactants by 391.9 kJ/mol. Since E_a(forward) = 19 kJ/mol, the transition state must be this much higher in energy than the reactants. Thus, the first two arrows on the left side of the diagram can be drawn. Then the third arrow (from products to the transition state) can be drawn. It indicates that

$$E_a(\text{reverse}) = -\Delta H° + E_a(\text{forward}) = (391.9 + 19)\ \text{kJ/mol} = 411\ \text{kJ/mol}$$

PROBLEM-SOLVING PRACTICE 12.8

For the reaction

$$Cl_2(g) + 2\,NO(g) \longrightarrow 2\,NOCl(g)$$

the activation energy is 18.9 kJ/mol. For the reaction

$$2\,NOCl(g) \longrightarrow 2\,NO(g) + Cl_2(g)$$

the activation energy is 98.1 kJ/mol. Draw a diagram similar to Figure 12.7 for the reaction of NO with Cl_2 to form NOCl. Is this reaction exothermic or endothermic? Explain.

Ahmed H. Zewail 1946–

For his studies of the transition states of chemical reactions using femtosecond spectroscopy, Ahmed H. Zewail of the California Institute of Technology received the 1999 Nobel Prize in Chemistry. Zewail, who holds joint Egyptian and U.S. citizenship, pioneered the use of extremely short laser pulses—on the order of femtoseconds (10^{-15} s)—to study chemical kinetics. His technique has been called the world's fastest camera, and his research has enhanced understanding of many reactions, among them those involving rhodopsin and vision.

CONCEPTUAL EXERCISE 12.8 Successful and Unsuccessful Collisions

The reaction

$$2\,NOCl(g) \longrightarrow 2\,NO(g) + Cl_2(g)$$

occurs in a single bimolecular step. Draw at least four possible ways that two NOCl molecules could collide, and rank them in order of greatest likelihood that a collision will be successful in producing products.

12.5 Temperature and Reaction Rate: The Arrhenius Equation

The most common way to speed up a reaction is to increase the temperature. A mixture of methane and air can be ignited by a lighted match, which raises the temperature of the mixture of reactants. This increases the reaction rate, the thermal energy

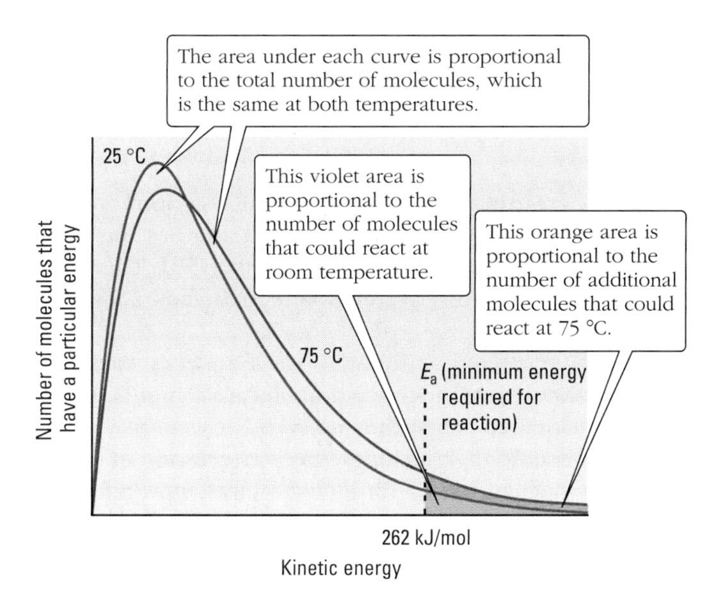

Figure 12.9 Energy distribution curves. The vertical axis gives the number of molecules that have the energy shown on the horizontal axis. Assume that a molecule reacts if it has more energy than the activation energy. The number of reactive molecules is given by the area under each curve to the right of the activation energy (262 kJ/mol). The two curves show that at 75 °C many more molecules have energies of 262 kJ/mol or higher than at 25 °C. (This graph is not to scale. Actually, 262 kJ/mol would be much farther to the right on the graph.)

evolved maintains the high temperature, and the reaction continues at a rapid rate. Reactions that speed up when the temperature is raised must slow down when the temperature is lowered. Foods are stored in refrigerators or freezers because the reactions in cells of microorganisms that produce spoilage occur more slowly at the lower temperature.

As a rough rule of thumb, the reaction rate increases by a factor of 2 to 4 for each 10-K rise in temperature.

Reaction rates increase with temperature because at a higher temperature a greater fraction of reactant molecules has enough energy to surmount the activation energy barrier. Consider again the conversion of *cis*- to *trans*-2-butene (Figure 12.7, *p. 434*). You learned in Section 10.3 (*p. 345*) that gas phase molecules are constantly in motion and have a wide distribution of speeds and energies. At room temperature relatively few *cis*-2-butene molecules have sufficient energy to surmount the energy barrier. However, as the temperature goes up, the number of molecules that have enough energy goes up rapidly, so the reaction rate increases rapidly.

The number of *cis*-2-butene molecules that have a given energy is shown by the curves in Figure 12.9. One curve is for 25 °C; the other is for 75 °C. The higher a point is on either curve, the greater is the number of molecules that have the energy corresponding to that point. The areas under the two curves are the same and represent the same total number of molecules. With a 50 °C rise in temperature, the number of molecules whose energy exceeds the activation energy is much higher, and so is the reaction rate.

A reaction is faster at a higher temperature because its rate constant is larger. That is, ***a rate constant is constant only for a given reaction at a given temperature.*** For example, for the reaction of iodide ion with methyl bromide, the data shown in Table 12.3 are found for the rate constant at different temperatures.

Table 12.3 Temperature Dependence of Rate Constant for Iodide Plus Methyl Bromide Reaction

T(K)	k (L mol^{-1} s^{-1})	T(K)	k (L mol^{-1} s^{-1})	T(K)	k (L mol^{-1} s^{-1})
273	4.18×10^{-5}	310	2.31×10^{-3}	350	6.80×10^{-2}
280	9.68×10^{-5}	320	5.82×10^{-3}	360	1.41×10^{-1}
290	2.00×10^{-4}	330	1.39×10^{-2}	370	2.81×10^{-1}
300	8.60×10^{-4}	340	3.14×10^{-2}		

Exponential curves also represent the growth of populations (such as human population) over time and are important in many other scientific fields. Logarithms and exponentials are discussed in Appendix A.6.

R is the constant found in the ideal gas law, $PV = nRT$ (p. 354). Here it is expressed in units of J mol^{-1} K^{-1} instead of L atm mol^{-1} K^{-1}, which is the reason that the numerical value is not 0.0821 L atm mol^{-1} K^{-1}.

When the data from Table 12.3 are graphed (Figure 12.10), it is obvious that the rate constant increases very rapidly as temperature increases. A graph that is shaped like this is called an exponential curve. It can be represented by the equation

$$k = Ae^{-E_a/RT} \qquad \text{[Arrhenius equation]}$$

where A is the **frequency factor,** e is the base of the natural logarithm system (2.718 . . .), E_a is the activation energy, R is the gas law constant and has the value 8.314 J mol^{-1} K^{-1}, and T is the absolute (Kelvin) temperature (p. 350). This equation is called the **Arrhenius equation** after its discoverer, Svante Arrhenius, a Swedish chemist.

The Arrhenius equation can be interpreted as follows. The frequency factor, A, depends on how often molecules collide when all concentrations are 1 mol/L and on whether the molecules are properly oriented when they collide. For example, in the case of the iodide-methyl bromide reaction, A includes the steric factor of $\frac{1}{4}$ that resulted because only one of the four sides of the CH_3Br molecule was appropriate for iodide to approach. The rest of the equation, $e^{-E_a/RT}$, gives the fraction of all the reactant molecules that have sufficient energy to surmount the activation energy barrier.

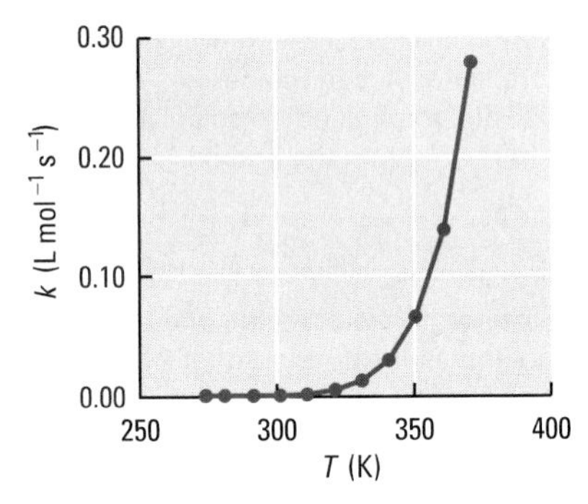

Figure 12.10 Effect of temperature on rate constant. The rate constant for the reaction of iodide ion with methyl bromide in aqueous solution is plotted as a function of temperature. The rate constant increases very rapidly with temperature, and the shape of the curve is characteristic of exponential increase.

Determining Activation Energy

The activation energy and frequency factor can be obtained from experimental measurements of rate constants as a function of temperature (such as those in Table 12.3). When a large number of experimental data pairs are given, a graph is usually a good way of obtaining information from the data. This is easier to do if the graph is linear. The activation energy equation can be modified by taking natural logarithms of both sides so that its graph is linear.

$$k = Ae^{-E_a/RT}$$

$$\ln(k) = \ln(A) + \ln(e^{-E_a/RT}) = \ln(A) + (-E_a/RT)$$

Rearranging this equation gives the equation of a straight line.

$$\ln(k) = -\left(\frac{E_a}{R} \times \frac{1}{T}\right) + \ln(A)$$

$$y = (m \times x) + b$$

That is, if we graph ln(k) on the vertical (y) axis and $1/T$ on the horizontal (x) axis, the result should be a straight line whose slope is $-E_a/R$ and whose y-intercept is ln(A). For the data in Table 12.3, such a graph is shown in Figure 12.11. It is linear, its slope is -9.29×10^3 K, and the y-intercept is 23.85. Since the slope $= -E_a/R$, the activation energy can be calculated as

$$E_a = -(\text{slope}) \times R = -(-9.29 \times 10^3 \text{ K})\left(\frac{8.314 \text{ J}}{\text{K mol}}\right)\left(\frac{1 \text{ kJ}}{1000 \text{ J}}\right) = 77.2 \text{ kJ/mol}$$

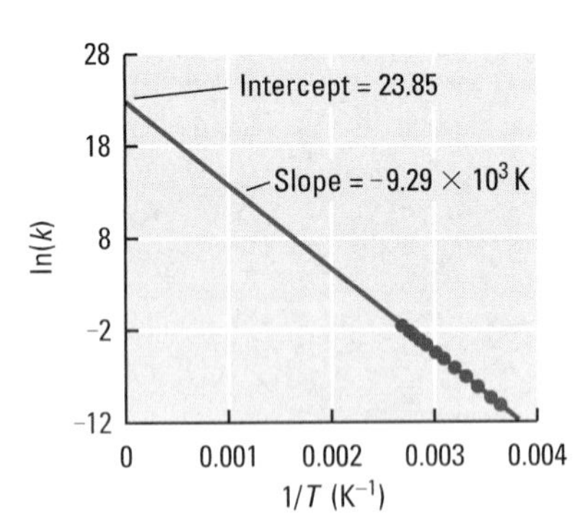

Figure 12.11 Determining activation energy graphically. A graph of ln(k) versus $1/T$ gives a straight line for the iodide–methyl bromide reaction. The activation energy can be obtained from the slope of the line, and the frequency factor from its y-intercept. Because the line through the experimental data has to be extrapolated a long way to reach the y-intercept, determining A accurately requires measuring k over a wide range of temperatures.

The vertical axis plots ln(k), and k has units of L mol^{-1} s^{-1}. At the y-intercept, ln(A) equals ln(k), and therefore A must have the same units as k. Since ln(A) = 23.85, the frequency factor, A, is

$$e^{23.85} \text{ L mol}^{-1} \text{ s}^{-1} = 2.28 \times 10^{10} \text{ L mol}^{-1} \text{ s}^{-1}$$

The Arrhenius equation can be used to calculate the rate constant at any temperature. For example, the rate constant for the reaction of iodide ion with methyl bromide can be calculated at 50. °C by substituting the temperature (in kelvins), the frequency factor, the activation energy, and the constant R into the equation.

$$k = Ae^{-E_a/RT} = (2.28 \times 10^{10} \text{ L mol}^{-1} \text{ s}^{-1})e^{(-77{,}200 \text{ J/mol})/(8.314 \text{ J K}^{-1} \text{ mol}^{-1})(273.15+50.) \text{ K}}$$
$$= (2.28 \times 10^{10} \text{ L mol}^{-1} \text{ s}^{-1})e^{-28.7} = 7.56 \times 10^{-3} \text{ L mol}^{-1} \text{ s}^{-1}$$

As a means of calculating rate constants, the Arrhenius equation works best within the range of temperatures over which the activation energy and frequency factor were determined. (For the reaction of iodide with methyl bromide, that range was 273 to 370 K.)

PROBLEM-SOLVING EXAMPLE 12.9 Temperature Dependence of Rate Constant

The experimental rate constant for the reaction of iodide ion with methyl bromide is 7.70×10^{-3} L mol^{-1} s^{-1} at 50 °C and 4.25×10^{-5} L mol^{-1} s^{-1} at 0 °C. Calculate the frequency factor and activation energy.

Answer $A = 1.66 \times 10^{10}$ L mol^{-1} s^{-1}; $E_a = 7.63 \times 10^4$ J mol^{-1}

Strategy and Explanation This problem could be solved by making a graph of ln(k) versus $1/T$, but with only two data pairs, a graph is not the best way. Instead, write two equations, one for each data pair.

$$k_1 = Ae^{-E_a/RT_1}$$

$$k_2 = Ae^{-E_a/RT_2}$$

Now divide the first equation by the second to eliminate A.

$$\frac{k_1}{k_2} = \frac{Ae^{-E_a/RT_1}}{Ae^{-E_a/RT_2}} = e^{-E_a/RT_1} \times e^{+E_a/RT_2}$$

$$\frac{k_1}{k_2} = e^{\frac{E_a}{R}\left(\frac{1}{T_2} - \frac{1}{T_1}\right)}$$

Next take the natural logarithm of both sides.

$$\ln\left(\frac{k_1}{k_2}\right) = \frac{E_a}{R}\left(\frac{1}{T_2} - \frac{1}{T_1}\right)$$

This equation can be solved for E_a.

$$E_a = \frac{R \ln\left(\frac{k_1}{k_2}\right)}{\frac{1}{T_2} - \frac{1}{T_1}} = \frac{(8.314 \text{ J mol}^{-1} \text{ K}^{-1})\ln\left(\frac{7.70 \times 10^{-3}}{4.25 \times 10^{-5}}\right)}{\frac{1}{273.15 \text{ K}} - \frac{1}{323.15 \text{ K}}}$$

$$= \frac{43.23 \text{ J mol}^{-1} \text{ K}^{-1}}{5.66 \times 10^{-4} \text{ K}^{-1}} = 7.63 \times 10^4 \text{ J mol}^{-1}$$

Finally, using the calculated value of E_a, solve one of the rate constant expressions for A.

$$k_1 = Ae^{-E_a/RT_1}$$

$$A = \frac{k_1}{e^{-E_a/RT_1}} = k_1 e^{E_a/RT_1}$$

$$= (7.70 \times 10^{-3} \text{ L mol}^{-1} \text{ s}^{-1})e^{\left(\frac{7.63 \times 10^4 \text{ J mol}^{-1}}{(8.314 \text{ J mol}^{-1} \text{ K}^{-1})(323.15 \text{ K})}\right)}$$

$$= (7.70 \times 10^{-3} \text{ L mol}^{-1} \text{ s}^{-1})(2.156 \times 10^{12}) = 1.66 \times 10^{10} \text{ L mol}^{-1} \text{ s}^{-1}$$

PROBLEM-SOLVING PRACTICE 12.9

Calculate the rate constant for the reaction of iodide ion with methyl bromide at a temperature of 75 °C.

The Arrhenius equation enables us to calculate the rate constant as a function of temperature, and the rate law shows how the rate depends on concentration. To obtain a single equation that summarizes the effects of both temperature and concentration,

Recall that the collision frequency contribution to the frequency factor is for 1 M concentrations. This equation summarizes the effects of both temperature and concentration on rate of a reaction. The temperature effect depends primarily on the large increase in the number of sufficiently energetic collisions as the temperature increases, which shows up as larger values of k at higher temperatures. The effect of concentration is clearly indicated by the concentration terms in the rate law. If the rate law is known for a reaction, and if both the A and E_a values are known, then the rate can be calculated over a wide range of conditions.

substitute k from the Arrhenius equation into the rate law for the iodide–methyl bromide reaction, giving

$$\text{Rate} = k \times [\text{I}^-] \times [\text{CH}_3\text{Br}]$$
$$\text{Rate} = A \times e^{-E_a/RT} \times [\text{I}^-] \times [\text{CH}_3\text{Br}] \qquad [12.8]$$

Collision frequency × steric factor | Fraction of sufficiently energetic molecules | Concentrations of colliding molecules

EXERCISE 12.9 Activation Energy and Experimental Data

The frequency factor A is 6.31×10^8 L mol^{-1} s^{-1} and the activation energy is 10. kJ/mol for the gas phase reaction

$$\text{NO(g)} + \text{O}_3\text{(g)} \longrightarrow \text{NO}_2\text{(g)} + \text{O}_2\text{(g)}$$

which is important in the chemistry of stratospheric ozone depletion.

(a) Calculate the rate constant for this reaction at 370. K.

(b) Assuming that this is an elementary reaction, calculate the rate of the reaction at 370. K if [NO] = 0.0010 M and $[\text{O}_3]$ = 0.00050 M.

12.6 Rate Laws for Elementary Reactions

An elementary reaction is a one-step process whose equation describes which nanoscale particles break apart, rearrange their positions, or collide to make a reaction occur. Therefore it is possible to figure out what the rate law and reaction order are for an elementary reaction, without doing an experiment. By contrast, when an equation represents a reaction that we do not understand at the nanoscale, rate laws and reaction orders must be determined experimentally ***(⇐ p. 422)***. The macroscale rate law can then be used to help develop a hypothesis about how a particular reaction takes place at the nanoscale.

Rate Law for a Unimolecular Reaction

In Section 12.4 we used the isomerization of *cis*-2-butene as an example of a reaction in which a single reactant molecule was converted to a product molecule or molecules—a unimolecular reaction ***(⇐ p. 433)***.

$$cis\text{-2-butene} \longrightarrow trans\text{-2-butene}$$

Suppose that the fraction of molecules that have enough energy to react is 0.1%, or 0.001. If there are 10,000 molecules in a given volume, then 0.001 × 10,000 gives only 10 that have enough energy to react. If there are twice as many molecules in the same volume—that is, 20,000 molecules—then 0.001 × 20,000 gives 20 with enough energy to react, and the number reacting per unit volume (the rate) is twice as great.

Suppose a flask contains 0.0050 mol/L of *cis*-2-butene vapor at room temperature. The molecules have a wide range of energies, but only a few of them have enough energy at this temperature to get over the activation energy barrier. Thus, during a given period only a few molecules twist sufficiently to become *trans*-2-butene. Now suppose that we double the concentration of *cis*-2-butene in the flask to 0.0100 mol/L, while keeping the temperature the same. The fraction of molecules with enough energy to cross over the barrier remains the same. However, as there are now twice as many molecules, twice as many must be crossing the barrier in any given time. Therefore, the rate of the *cis* ⟶ *trans* reaction is twice as great. That is, the reaction rate is proportional to the concentration of *cis*-2-butene, and the rate law must be

$$\text{Rate} = k[cis\text{-2-butene}]$$

In the general case of any unimolecular elementary reaction,

$$\text{A} \longrightarrow \text{products} \qquad \text{the rate law is} \qquad \text{Rate} = k[\text{A}]$$

For any unimolecular reaction the nanoscale mechanism predicts that a first-order rate law will be observed in a macroscale laboratory experiment.

Rate Law for a Bimolecular Reaction

A good example of a reaction in which two molecules collide [a bimolecular reaction (p. 434)] is the gas phase reaction of nitrogen monoxide and ozone that was mentioned in Exercise 12.9.

$$NO(g) + O_3(g) \longrightarrow NO_2(g) + O_2(g)$$

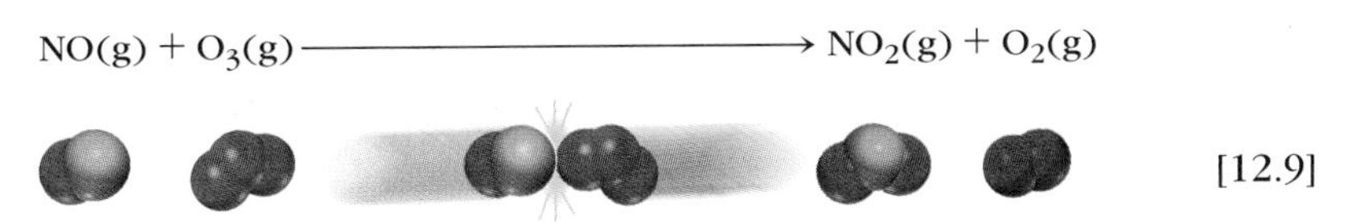

[12.9]

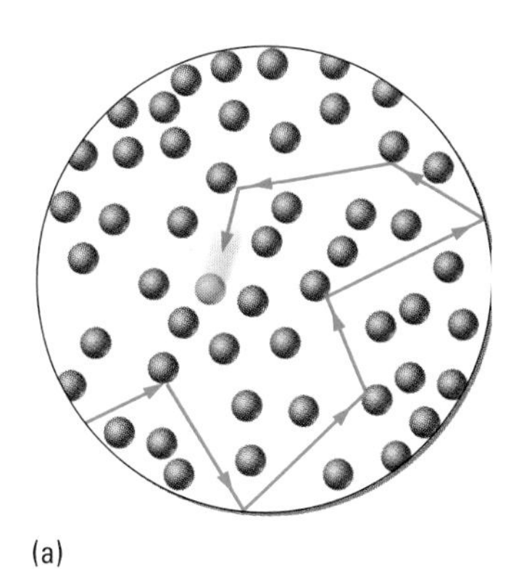

(a)

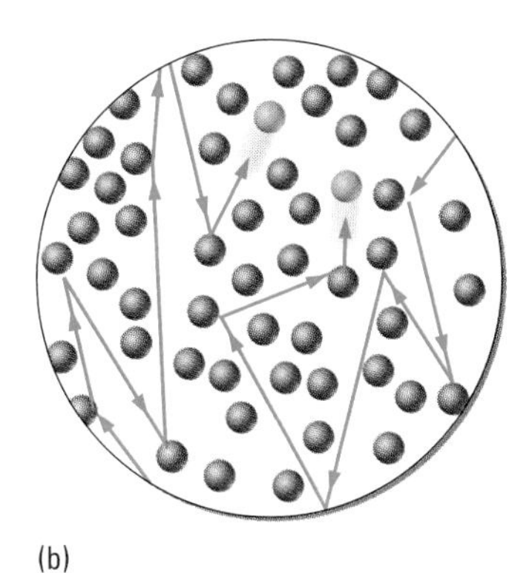

(b)

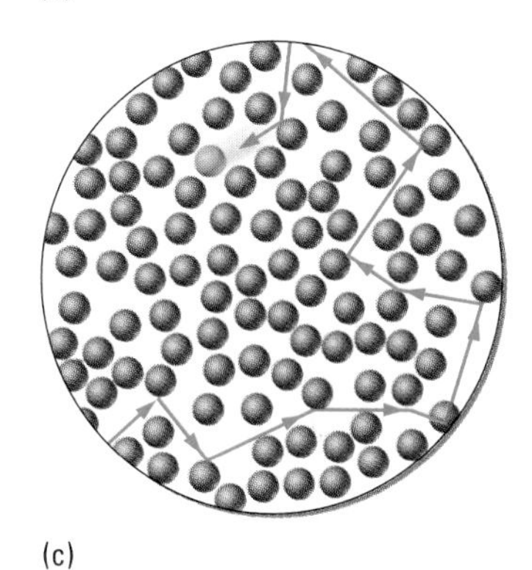

(c)

Figure 12.12 Effect of concentration on frequency of bimolecular collisions. (a) A single green molecule moves among 50 purple molecules and collides with five of them per second. (b) Two green molecules now move among 50 purple molecules, and there are 10 green-purple collisions per second. (c) If the number of purple molecules is doubled to 100, the frequency of green-purple collisions is also doubled, to 10 per second. The number of collisions is proportional to *both* the concentration of green molecules *and* the concentration of purple molecules. (In a real sample both the green and the purple molecules would be moving, but in this diagram the motion of the purple molecules is not shown.)

Here the equation shows that the elementary reaction involves the collision of one NO molecule and one O_3 molecule. Since the molecules must collide to exchange atoms, the rate depends on the number of collisions per unit time.

Figure 12.12a represents one NO molecule (the green ball) and many O_3 molecules (the purple balls) in a tiny region within a flask where Reaction 12.9 is taking place. In a given time, the NO molecule collides with five O_3 molecules. If the concentration of NO molecules is doubled to two NO molecules in the same portion of the flask (Figure 12.12b), *each* NO molecule collides with five different O_3 molecules. Doubling the concentration of NO has doubled the number of collisions. This also doubles the rate, because the rate is proportional to the number of collisions (Equation 12.8, p. 440). The number of collisions also doubles when the O_3 concentration is doubled (Figure 12.12c). Thus, the rate law for this reaction must be

$$\text{Rate} = k[NO][O_3]$$

This description of the NO + O_3 reaction applies in general to bimolecular elementary reactions, even if the two molecules that must collide are of the same kind. That is, for the elementary reaction

$$A + B \longrightarrow \text{products} \qquad \text{the rate law is} \qquad \text{Rate} = k[A][B]$$

and for the elementary reaction

$$A + A \longrightarrow \text{products} \qquad \text{the rate law is} \qquad \text{Rate} = k[A]^2$$

For the NO + O_3 reaction the experimentally determined rate law is the same as the one we just derived by assuming the reaction occurs in one step. The equation for the reaction is

$$NO(g) + O_3(g) \longrightarrow NO_2(g) + O_2(g)$$

and the experimental rate law is

$$\text{Rate} = k[NO][O_3]$$

This experimental observation suggests, but does not prove, that the reaction does take place in a single step. Other experimental evidence also suggests that this reaction is bimolecular.

In contrast, for the decomposition of hydrogen peroxide the equation is

$$2\,H_2O_2(aq) \longrightarrow 2\,H_2O(\ell) + O_2(g)$$

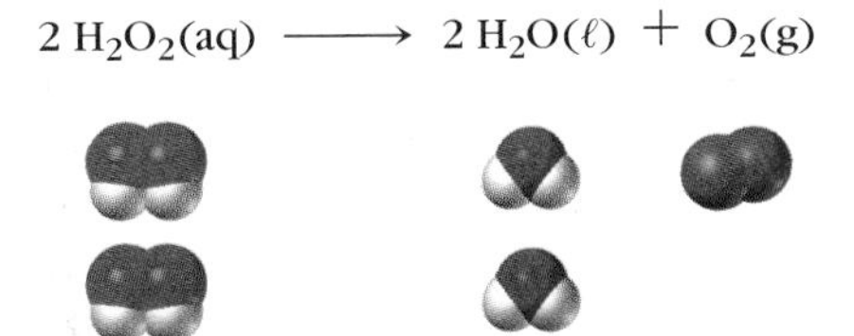

and the experimental rate law is Rate = $k[H_2O_2]$. This rate law proves that this reaction *cannot* occur in a single step that involves collision of two H_2O_2 molecules. A single-step bimolecular reaction would have a second-order rate law, but the observed rate law is first-order. This means that more than a single elementary step is needed when hydrogen peroxide decomposes.

CONCEPTUAL
EXERCISE 12.10 Rate Law and Elementary Reactions

For the reaction

$$NO_2(g) + CO(g) \longrightarrow NO(g) + CO_2(g)$$

the experimentally determined rate law is

$$\text{Rate} = k[NO_2]^2$$

Does this reaction occur in a single step? Explain why or why not.

12.7 Reaction Mechanisms

Most chemical reactions do not take place in a single step. Instead, they involve a sequence of unimolecular or bimolecular elementary reactions. For each elementary reaction in the sequence we can write an equation. A set of such equations is called a **reaction mechanism** (or just a *mechanism*). For example, iodide ion can be oxidized by hydrogen peroxide in acidic solution to form iodine and water according to this overall equation:

$$2\,I^-(aq) + H_2O_2(aq) + 2\,H_3O^+(aq) \longrightarrow I_2(aq) + 4\,H_2O(\ell) \qquad [12.10]$$

When the acid concentration is between 10^{-3} M and 10^{-5} M, experiments show that the rate law is

$$\text{Rate} = k[I^-][H_2O_2]$$

The reaction is first-order in the concentrations of I^- and H_2O_2, and second-order overall.

Looking at Equation 12.10 for the oxidation of iodide ion by hydrogen peroxide, you might think that two iodide ions, one hydrogen peroxide molecule, and two hydronium ions would all have to come together at once. However, the rate law corresponds with a bimolecular collision of I^- and H_2O_2. It is highly unlikely that at the same time five ions or molecules would all be at the same place, be properly oriented, and have enough energy to react. Instead, chemists who have studied this reaction propose that initially one H_2O_2 (HOOH) molecule and one I^- ion come together:

Step 1: $HOOH + I^- \xrightarrow{\text{slow}} OH^- + HOI$

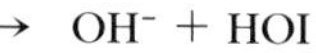

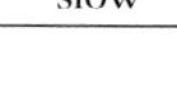

This first step forms hypoiodous acid, HOI, and hydroxide ion, both known substances. The HOI then reacts with another I^- to form the product I_2:

Step 2: $HOI + I^- \xrightarrow{\text{fast}} OH^- + I_2$

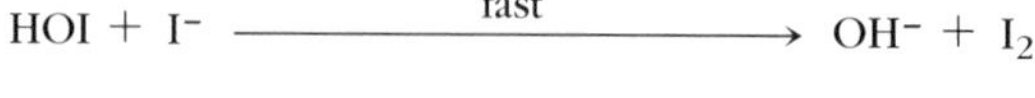

In each of Steps 1 and 2, a hydroxide ion was produced. Since the solution is acidic, these OH^- ions react immediately with H_3O^+ ions to form water.

Step 3: $2\,OH^- + 2\,H_3O^+ \xrightarrow{\text{fast}} 4\,H_2O$

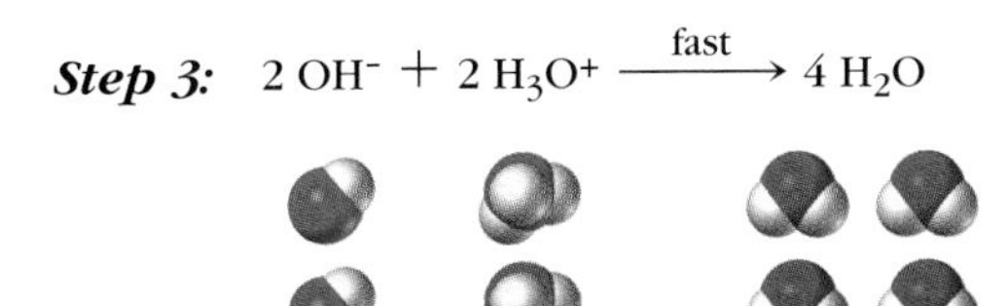

Each of the three steps in this mechanism is an elementary reaction. Each has its own activation energy, E_a, and its own rate constant, k. When the three steps are summed (by putting all the reactants on the left, putting all the products on the right, and eliminating formulas that appear as both reactants and products), the overall stoichiometric equation (Equation 12.10) is obtained. ***Any valid mechanism must consist of a series of unimolecular or bimolecular elementary reaction steps that sum to the overall reaction.***

Step 1: $HOOH + I^- \xrightarrow{\text{slow}} \cancel{HOI} + \cancel{OH^-}$

Step 2: $\cancel{HOI} + I^- \xrightarrow{\text{fast}} I_2 + \cancel{OH^-}$

Step 3: $\cancel{2\,OH^-} + 2\,H_3O^+ \xrightarrow{\text{fast}} 4\,H_2O$

Overall: $2\,I^- + HOOH + 2\,H_3O^+ \longrightarrow I_2 + 4\,H_2O$ [12.10]

Step 1 of the mechanism is slow, while Steps 2 and 3 are fast. Step 1 is called the **rate-limiting step;** because it is the slowest in the sequence, it limits the rate at which I_2 and H_2O can be produced. Steps 2 and 3 are rapid and therefore not rate-limiting. As soon as some HOI and OH^- are produced by Step 1, they are transformed into I_2 and H_2O by Steps 2 and 3. ***The rate of the overall reaction is limited by, and equal to, the rate of the slowest step in the mechanism.***

Step 1 is a bimolecular elementary reaction. Therefore its rate must be first-order in HOOH and first-order in I^-. The mechanism predicts that the rate law should be

$$\text{Reaction rate} = k[\text{HOOH}][\text{I}^-]$$

which agrees with the experimentally observed rate law. ***A valid mechanism must correctly predict the experimentally observed rate law.***

An analogy to the rate-limiting or rate-determining step is that no matter how quickly you shop in the supermarket, it seems that the time it takes to get out of the store depends on the rate at which you move through the checkout line.

EXERCISE 12.11 Rate Law for an Elementary Reaction

What is the rate law for Step 2 of the mechanism for the reaction of hydrogen peroxide and iodide ion?

The species HOI and OH^- are produced in Step 1 and used up in Step 2 or 3. In a mechanism, atoms, molecules, or ions that are produced in one step and consumed in a later step (or later steps) are called **reaction intermediates** (or just **intermediates**). Very small concentrations of HOI and OH^- are produced while the reaction is going on. Once the HOOH, the I^-, or both are used up, the intermediates HOI and OH^- are consumed by Steps 2 and 3 and disappear. HOI and OH^- are crucial to the reaction mechanism, but neither of them appears in the overall stoichiometric equation. If an experimenter is proficient enough to demonstrate that a particular intermediate was present, this provides additional evidence that a mechanism involving that intermediate is the correct one.

If significant concentrations of an intermediate build up while a reaction is occurring, then reactants may disappear faster than products are formed (because buildup of the intermediate stores some of the used reactant before it is converted to final product). In such a case, the definition of reaction rate given in Equation 12.3 will not be correct. One way to detect formation of an intermediate is to notice that products are not formed as fast as reactants are used up.

In summary, a valid reaction mechanism should

- Consist of a series of unimolecular or bimolecular elementary reactions;
- Consist of reaction steps that sum to the overall reaction equation; and
- Correctly predict the experimentally observed rate law.

Kinetics and Mechanism

Studying the kinetics of a chemical reaction involves collecting data on the concentrations of reactants as a function of time. From such data the rate law for the reaction and a rate constant can usually be obtained. The reaction can also be studied at several different temperatures to determine its activation energy. This allows us to predict how fast the macroscale reaction will be under a variety of experimental conditions, but it does not provide definitive information about the nanoscale mechanism by which the reaction takes place. A reaction mechanism is an educated guess—a hypothesis—about the way the reaction occurs. If the mechanism predicts correctly the overall stoichiometry of the reaction and the experimentally determined rate law, then it is a reasonable hypothesis. However, it is impossible to prove for certain that a mechanism is correct. Sometimes several mechanisms can agree with the same set of experiments. This is what makes kinetic studies one of the most interesting and rewarding areas of chemistry, but it also can provoke disputes among scientists who favor different possible mechanisms for the same reaction.

PROBLEM-SOLVING EXAMPLE 12.10 Rate Law and Reaction Mechanism

The gas phase reaction between nitrogen dioxide and fluorine,

$$2\ NO_2(g) + F_2(g) \longrightarrow 2\ NO_2F(g)$$

is found experimentally to obey the rate law

$$\text{Rate} = k[NO_2][F_2]$$

Decide which of these mechanisms is compatible with the rate law.

(a) $NO_2 + NO_2 \longrightarrow N_2O_4$ slow
$N_2O_4 + F_2 \longrightarrow 2\ NO_2F$ fast

(b) $NO_2 + F_2 \longrightarrow NO_2F_2$ slow
$NO_2F_2 + F_2 \longrightarrow NO_2F_4$ fast

(c) $NO_2 + F_2 \longrightarrow NO_2F + F$ slow
$NO_2 + F \longrightarrow NO_2F$ fast

(d) $2\ NO_2 + F_2 \longrightarrow 2\ NO_2F$

Answer Mechanism (c) is compatible with the rate law and stoichiometry.

Strategy and Explanation Examine each mechanism to see whether it (1) consists only of unimolecular and bimolecular steps, (2) agrees with the overall stoichiometry, and (3) predicts the experimental rate law. Eliminate those that do not. The remaining mechanism(s) may be correct.

Mechanism (a) has a slow, bimolecular first step. This implies a rate law, Rate = $k[NO_2]^2$, which does not match the observed rate law and eliminates it from consideration.

Mechanism (b) does not sum to the overall stoichiometry, and so it can be eliminated.

Mechanism (d) involves simultaneous collision of three molecules: two NO_2 and one F_2. It can be eliminated because it does not consist of unimolecular or bimolecular steps. It also predicts a third-order rate law, which differs from the experimental rate law.

Mechanism (c) satisfies all three criteria. Atomic fluorine, F, is an intermediate.

PROBLEM-SOLVING PRACTICE 12.10

The Raschig reaction produces the industrially important reducing agent hydrazine, N_2H_4, from ammonia, NH_3, and hypochlorite ion, OCl^-, in basic aqueous solution. A proposed mechanism is

Step 1: $NH_3(aq) + OCl^-(aq) \xrightarrow{\text{slow}} NH_2Cl(aq) + OH^-(aq)$

Step 2: $NH_2Cl(aq) + NH_3(aq) \xrightarrow{\text{fast}} N_2H_5^+(aq) + Cl^-(aq)$

Step 3: $N_2H_5^+(aq) + OH^-(aq) \xrightarrow{\text{fast}} N_2H_4(aq) + H_2O(\ell)$

(a) What is the overall stoichiometric equation?
(b) Which step is rate-limiting?
(c) What reaction intermediates are involved?
(d) What rate law is predicted by this mechanism?

12.8 Catalysts and Reaction Rate

Raising the temperature increases a reaction rate because it increases the fraction of molecules that are energetic enough to surmount the activation energy barrier. Increasing reactant concentrations can also increase the rate because it increases the number of molecules per unit volume. A third way to increase reaction rates is to add a catalyst ***(⟸ p. 416)***.

For example, an aqueous solution of hydrogen peroxide can decompose to water and oxygen.

$$2\,H_2O_2(aq) \longrightarrow O_2(g) + 2\,H_2O(\ell)$$

At room temperature the rate of the decomposition reaction is exceedingly slow. If the hydrogen peroxide solution is stored in a cool, dark place in a clean plastic container, it is stable for months. However, in the presence of a manganese salt, an iodide-containing salt, or a biological catalyst *(an enzyme;* see p. 448), the decomposition reaction can occur quite rapidly (Figure 12.13a).

Ammonium nitrate is used as fertilizer and is stable at room temperature. At higher temperatures and in the presence of chloride ion as a catalyst, however, ammonium nitrate can explode with tremendous force (Figure 12.13b). Approximately 600 people were killed in Texas City, Texas, in 1947 when workers tried to use salt water (which contains ~0.5 M Cl^-) to extinguish a fire in the hold of the ship *Grandcamp* and the ammonium nitrate cargo exploded (Figure 12.13c).

How does a catalyst or an enzyme help a reaction to go faster? It does so by participating in the reaction mechanism. That is, *the mechanism for a catalyzed reaction is different from the mechanism of the same reaction without the catalyst.* The rate-limiting step in the catalyzed mechanism has a lower activation energy and therefore is faster than the slow step for the uncatalyzed reaction. To see how this works, let us again consider conversion of *cis-* to *trans-*2-butene in the gas phase.

```
H3C      CH3            H       CH3
   \    /                \     /
    C=C        ---->      C=C          Rate = k [cis-2-butene]
   /    \                /     \
  H      H            H3C       H
 cis-2-butene          trans-2-butene
```

If a trace of gaseous molecular iodine, I_2, is added to a sample of *cis*-2-butene, the iodine accelerates the change to *trans*-2-butene. The iodine is neither consumed nor produced in the overall reaction, so it does not appear in the overall balanced equation. However, because the reaction rate depends on the concentration of I_2, there is a term involving concentration of I_2 in the rate law for the catalyzed reaction.

$$\text{Rate} = k[cis\text{-2-butene}][I_2]^{1/2}$$

The exponent of $\frac{1}{2}$ for the concentration of I_2 in the rate law indicates the square root of the concentration. A square root dependence usually means that only half a molecule—in this case a single iodine atom—is involved in the mechanism.

(a)

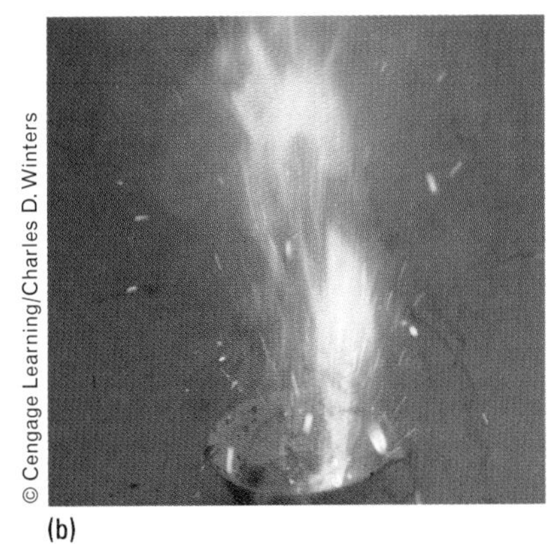

(b)

(c)

Figure 12.13 Catalysis in action. (a) A 30% aqueous solution of H_2O_2 is dropped onto a piece of liver. The liquid foams as H_2O_2 rapidly decomposes to O_2 and H_2O. Liver contains an enzyme that catalyzes the decomposition of H_2O_2. (b) Laboratory-scale explosion of a sample of ammonium nitrate, NH_4NO_3, catalyzed by chloride ion. (c) Scene following explosion of the ship *Grandcamp* in Texas City, Texas, April 16, 1947.

The rate of the conversion of *cis*- to *trans*-2-butene changes because the presence of I_2 somehow changes the reaction mechanism. The best hypothesis is that iodine molecules first dissociate to form iodine atoms.

***Step 1:* I_2 dissociation**

$$\tfrac{1}{2}[\, I_2(g) \longrightarrow 2\, I(g)\,]$$

(This equation is multiplied by $\frac{1}{2}$ because only one of the two I atoms from the I_2 molecule is needed in subsequent steps of the mechanism.) An iodine atom then attaches to the *cis*-2-butene molecule, breaking half of the double bond between the two central carbon atoms and allowing the ends of the molecule to twist freely relative to each other.

***Step 2:* Attachment of I atom to *cis*-2-butene**

cis-2-butene

***Step 3:* Rotation around the C—C bond**

***Step 4:* Loss of an I atom and reformation of the carbon-carbon double bond**

trans-2-butene

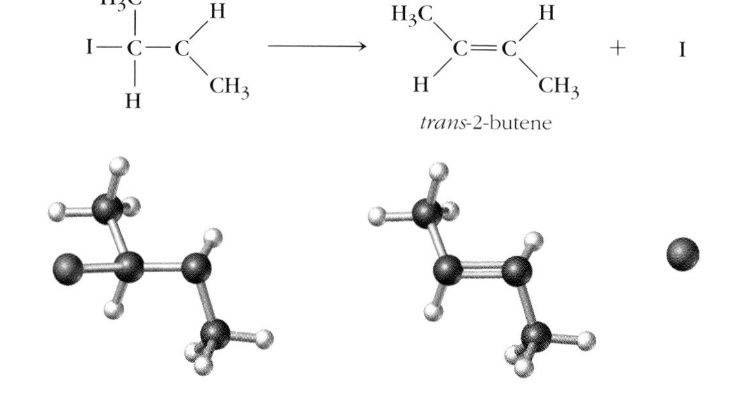

After the new double bond forms to give *trans*-2-butene and the iodine atom falls away, two iodine atoms come together to regenerate molecular iodine.

Step 5: **I_2 regeneration**

$$\frac{1}{2}[\,2\,I(g) \longrightarrow I_2(g)\,]$$

There are five important points concerning this mechanism.

- The I_2 dissociates to atoms and then reforms. To an "outside" observer the concentration of I_2 is unchanged; I_2 is not involved in the balanced stoichiometric equation even though it has appeared in the mechanism. *This is generally true of catalysts.*
- Figure 12.14 shows that the activation energy barrier is significantly lower for the catalyzed reaction (because the mechanism is different). Consequently the reaction rate is much faster. Dropping the activation energy from 262 kJ/mol for the uncatalyzed reaction to 115 kJ/mol for the catalyzed process makes the catalyzed reaction 10^{15} times faster at a temperature of 500. K.
- The catalyzed mechanism has five reaction steps, and its energy-versus-reaction progress diagram (Figure 12.14) has five energy barriers (five humps appear in the curve).
- The catalyst I_2 and the reactant *cis*-2-butene are both in the gas phase during the reaction. When a catalyst is present in the same phase as the reacting substance or substances, it is called a **homogeneous catalyst.**
- Although the mechanism is different, the initial and final energies for the catalyzed reaction are the same as for the uncatalyzed reaction. This means that ΔE and ΔH are the same for the catalyzed as for the uncatalyzed reaction.

Because $k = Ae^{-E_a/RT}$,

$$\frac{k_2}{k_1} = \frac{Ae^{-E_{a_2}/RT}}{Ae^{-E_{a_1}/RT}} = e^{-(E_{a_2} - E_{a_1})/RT}$$

$$= e^{\left(\frac{(262{,}000 - 115{,}000)\ \text{J mol}^{-1}}{8.314\ \text{J K}^{-1}\ \text{mol}^{-1} \times 500.\ \text{K}}\right)}$$

$$= e^{3.536 \times 10^{1}} = 2.28 \times 10^{15}$$

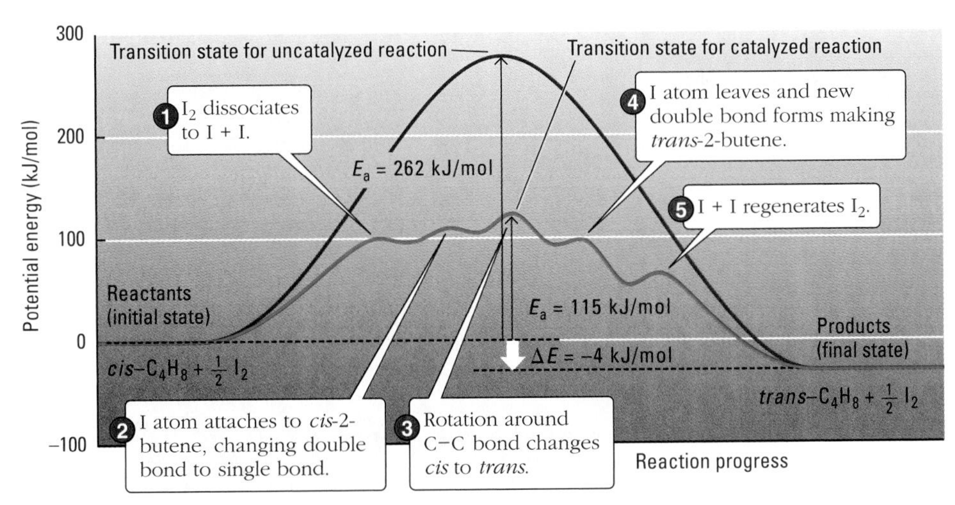

Figure 12.14 Energy diagrams for catalyzed and uncatalyzed reactions. A catalyst accelerates a reaction by altering the mechanism so that the activation energy is reduced. With a smaller barrier to overcome, more reactant molecules have enough energy to cross the barrier, and reaction occurs more readily. (The steps involved are described in the text.) Notice that the shape of the barrier has changed because the mechanism has changed. This changes the activation energy, but not ΔE for the reaction.

EXERCISE 12.12 Catalysis

The oxidation of thallium(I) ion by cerium(IV) ion in aqueous solution has the equation

$$2\ Ce^{4+}(aq) + Tl^{+}(aq) \longrightarrow 2\ Ce^{3+}(aq) + Tl^{3+}(aq)$$

The accepted mechanism for this reaction is

Step 1: $Ce^{4+}(aq) + Mn^{2+}(aq) \longrightarrow Ce^{3+}(aq) + Mn^{3+}(aq)$

Step 2: $Ce^{4+}(aq) + Mn^{3+}(aq) \longrightarrow Ce^{3+}(aq) + Mn^{4+}(aq)$

Step 3: $Mn^{4+}(aq) + Tl^{+}(aq) \longrightarrow Mn^{2+}(aq) + Tl^{3+}(aq)$

(a) Verify that this mechanism predicts the overall reaction.
(b) Identify all intermediates in this mechanism.
(c) Identify the catalyst in this mechanism.
(d) Suppose that the first step in this mechanism is rate-limiting. What would the rate law be?

12.9 Enzymes: Biological Catalysts

Your body is a chemical factory of cells that can manufacture a broad range of compounds that are needed so that you can move, breathe, digest food, see, hear, smell, and even think. But did you ever consider how the reactions that make those compounds are controlled? And how they can all occur reasonably quickly at the relatively low body temperature of 37 °C? Oxidation of glucose powers all the systems of your body, but you would not want it to take place at the temperature it does when cellulose (which consists of many glucose molecules linked in a long, molecular chain) in wood burns in a fireplace. The chemical reactions of your body are catalyzed by enzymes. An **enzyme** is *a highly efficient catalyst for one or more chemical reactions in a living system.* The presence or absence of appropriate enzymes turns these reactions on or off by speeding them up or slowing them down. This allows your body to maintain nearly constant temperature and nearly constant concentrations of a variety of molecules and ions, an absolute necessity if you are to continue functioning.

Enzymes are usually proteins, large molecules made up of many individual units called amino acids. The amino acids are linked into long molecular chains that can fold back upon themselves in the same way as a metal chain that has been piled on the floor becomes tangled. Most enzymes are globular proteins, in which one or more long chains of amino acids fold into a nearly spherical shape. The shape of a globular protein is determined by noncovalent interactions among the amino acid components (hydrogen bonds, attractions of opposite ionic charges, dipole-dipole and ion-dipole forces), a few weak covalent bonds, and the fact that nonpolar (hydrophobic) amino acid side groups congregate in the middle of the molecule, avoiding the surrounding aqueous solution.

The weak covalent bonds in some globular proteins are disulfide bonds. They occur between sulfur atoms in side chains of the amino acid cysteine. A cysteine side chain at one point in the protein can become bonded to a cysteine side chain much farther along the protein backbone.

Enzymes are among the most effective catalysts known. They can increase reaction rates by factors of 10^9 to 10^{19}. For example, essentially every collision of the enzyme carbonic anhydrase with a carbonic acid molecule results in decomposition, and the enzyme can decompose about 36 million H_2CO_3 molecules every minute.

$$H_2CO_3(aq) \xrightarrow{\text{carbonic anhydrase}} CO_2(g) + H_2O(\ell)$$

Most enzymes are highly specific catalysts. Some act on only one or two of the hundreds of different substances found in living cells. For example, carbonic anhydrase catalyzes only the decomposition of carbonic acid. Other enzymes can speed up several reactions, but usually these reactions are all of the same type.

Enzyme Activity and Specificity

A **substrate** is a molecule whose reaction is catalyzed by an enzyme. In some cases there may be more than one substrate, as when an enzyme catalyzes transfer of a group of atoms from one molecule to another. Enzyme catalysis is extremely effective and specific because the structure of the enzyme is finely tuned to minimize the activation energy barrier. Usually one part of the enzyme molecule, called the **active site,** interacts with the substrate via the same kinds of noncovalent attractions that hold the enzyme in its globular structure. The nanoscale structure of an enzyme's active site is specifically suited to attract and bind a substrate molecule and to help the substrate react.

When a substrate binds to an enzyme, both molecular structures can change. Each structure adjusts to fit closely with the other, and the structures become complementary. The change in shape of either the enzyme, the substrate, or both molecules when they bind is called **induced fit.** Enzymes catalyze reactions of only certain molecules because the structures of most molecules are not close enough to the structure of the active site for an induced fit to occur. The induced fit of a substrate to an enzyme also can lower the activation energy for a reaction. For example, it may distort the substrate and stretch a bond that will be broken in the desired reaction. A schematic example of how this can work is shown in Figure 12.15.

To summarize, enzymes are extremely effective as biological catalysts for several reasons:

- Enzymes bring substrates into close proximity and hold them there while a reaction occurs.
- Enzymes hold substrates in the shape that is most effective for reaction.
- Enzymes can act as acids and bases during reaction, donating or accepting hydrogen ions from the substrate quickly and easily.
- The potential energy of a bond distorted by the induced fit of the substrate to the enzyme is already partway up the activation energy hill that must be surmounted for reaction to occur.
- Enzymes sometimes contain metal ions that are needed to help catalyze oxidation-reduction reactions.

Enzyme Kinetics

An enzyme changes the mechanism of a reaction, as does any catalyst. The first step in the mechanism for any enzyme-catalyzed reaction is binding of the substrate and the enzyme, which is referred to as formation of an **enzyme-substrate complex.** Representing enzyme by E, substrate by S, and products by P, we can write a single-step uncatalyzed mechanism and a two-step enzyme-catalyzed mechanism as follows.

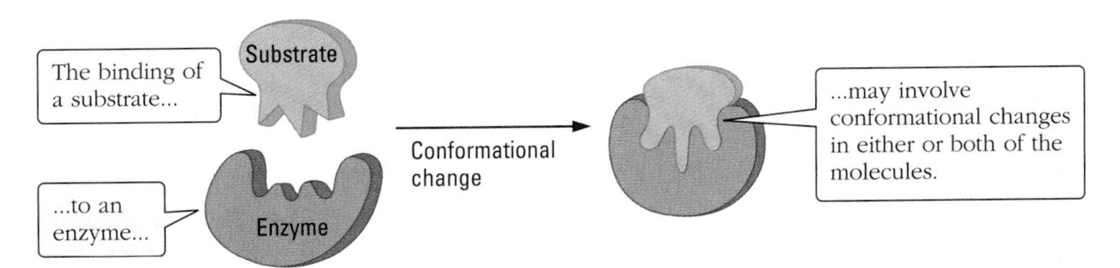

Figure 12.15 Induced fit of substrate to enzyme. Binding of a substrate to an enzyme may involve changing the shape of either or both molecules, thereby inducing them to fit together. In some cases a substrate molecule may be stretched or strained, helping bonds to break and reaction to occur.

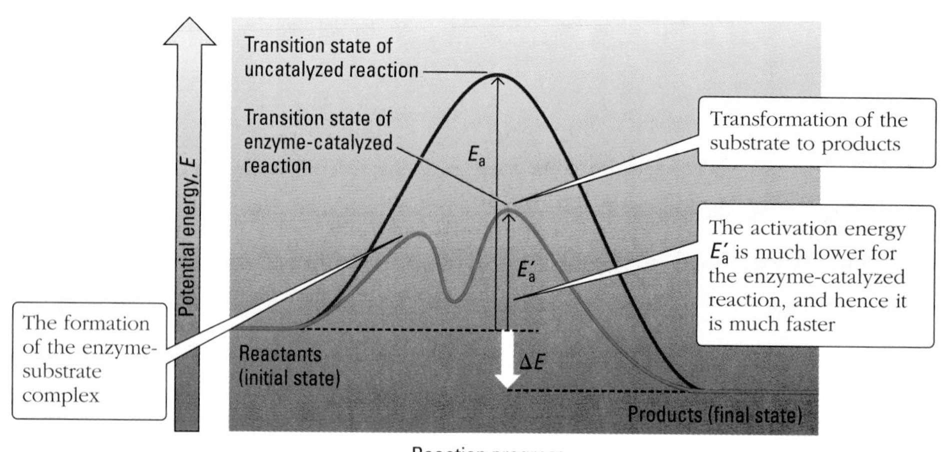

Figure 12.16 Energy diagram for enzyme-catalyzed reaction. The red curve is the energy profile for a typical reaction in a living system with no enzyme present. The green energy profile is drawn to the same scale for the same reaction with enzyme catalysis.

Uncatalyzed mechanism: $S \longrightarrow P$

Enzyme-catalyzed mechanism:

Step 1 (fast): $S + E \rightleftharpoons ES$ (formation of enzyme-substrate complex)

Step 2 (slow): $ES \longrightarrow P + E$ (formation of products and regeneration of enzyme)

That the enzyme is a catalyst is evident from the fact that it is a reactant in the first step and is regenerated in the second. This mechanism applies to nearly all enzyme-catalyzed reactions. Because the second step is slow, the enzyme-substrate complex can often separate and reform S + E before it reacts to form products. This possibility is indicated by the double arrow in the first step.

Because of the noncovalent interactions between enzyme and substrate, the activation energy is significantly lower for the enzyme-catalyzed reaction than it would be for the uncatalyzed process. This situation is shown in Figure 12.16. Even at temperatures only a little above room temperature, significant numbers of molecules have enough energy to surmount this lower barrier. Thus, enzyme-catalyzed reactions can occur reasonably quickly at body temperature.

12.10 Catalysis in Industry

An expert in the field of industrial chemistry has said that every year more than one trillion dollars' worth of goods is manufactured with the aid of manmade catalysts. Without them, fertilizers, pharmaceuticals, fuels, synthetic fibers, solvents, and detergents would be in short supply. Indeed, 90% of all manufactured items use catalysts at some stage of production. The major areas of catalyst use are in petroleum refining, industrial production of chemicals, and environmental controls. In this section we provide a few examples of the many important industrial reactions that depend on catalysis.

Many industrial reactions use **heterogeneous catalysts.** Such catalysts are present in a different phase from that of the reactants being catalyzed. Usually the catalyst is a solid and the reactants are in the gaseous or liquid phase. Heterogeneous catalysts are used in industry because they are more easily separated from the products and leftover reactants than are homogeneous catalysts. Catalysts for chemical processing are generally metal-based and often contain precious metals such as platinum and palladium. In the United States more than $600 million worth of such catalysts are used annually by the chemical processing industry, almost half of them in the preparation of polymers.

Manufacture of Acetic Acid

The importance of acetic acid, CH_3COOH, in the organic chemicals industry is comparable to that of sulfuric acid in the inorganic chemicals industry; annual production of acetic acid in the United States exceeds 5 billion pounds. Acetic acid is used widely in industry to make plastics and synthetic fibers, as a fungicide, and as the starting material for preparing many dietary supplements. One way of synthesizing the acid is an excellent example of homogeneous catalysis: Rhodium(III) iodide is used to speed up the combination of carbon monoxide and methyl alcohol, both inexpensive chemicals, to form acetic acid.

$$\underset{\text{methyl alcohol}}{CH_3OH} + \underset{\text{carbon monoxide}}{CO} \xrightarrow{\text{RhI}_3 \text{ catalyst}} \underset{\text{acetic acid}}{CH_3C(=O){-}OH}$$

The role of the rhodium(III) iodide catalyst in this reaction is to bring the reactants together and allow them to rearrange to form the products. Carbon monoxide and the methyl group from the alcohol become attached to the rhodium atom, which helps transfer the methyl group to the CO. After this rearrangement, the intermediate reacts with solvent water to form acetic acid and the catalyst is regenerated.

Controlling Automobile Emissions

A major use of catalysts is in *emissions control* for both automobiles and power plants. This market uses very large quantities of platinum group metals: platinum, palladium, rhodium, and iridium. In 2006, more than 130,00 kg platinum and more than 128,000 kg palladium were sold worldwide for automotive uses. Demand was also high for rhodium for this same purpose (about 26,000 kg). All three metals are also used in chemical processing as catalysts, and the petroleum industry uses platinum and rhodium to catalyze refining processes.

The purpose of the catalysts in the exhaust system of an automobile is to ensure that the combustion of carbon monoxide and hydrocarbons is complete (Figure 12.17).

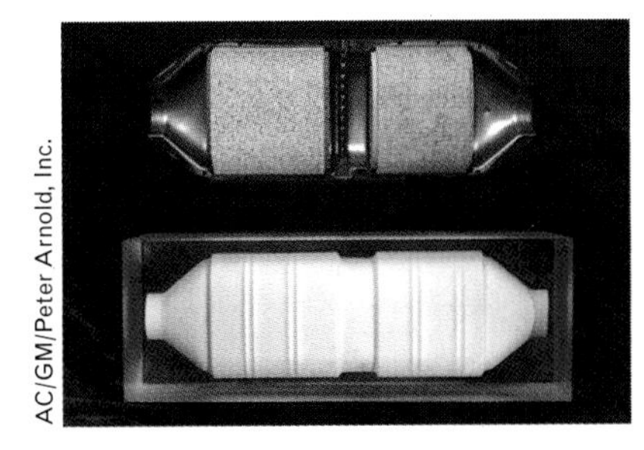

AC/GM/Peter Arnold, Inc.

Figure 12.17 Automobile catalytic converter. Catalytic converters are standard equipment on the exhaust systems of all new automobiles. This one contains two catalysts: One converts nitrogen monoxide to nitrogen and the other converts carbon monoxide and hydrocarbons to carbon dioxide and water.

$$2\,CO(g) + O_2(g) \xrightarrow{\text{Pt-NiO catalyst}} 2\,CO_2(g)$$

$$2\,C_8H_{18}(g) + 25\,O_2(g) \xrightarrow{\text{Pt-NiO catalyst}} 16\,CO_2(g) + 18\,H_2O(g)$$

2,2,4-trimethylpentane, a component of gasoline

and to convert nitrogen oxides to molecules that are less harmful to the environment. At the high temperature of combustion, some N_2 from air reacts with O_2 to give NO, a serious air pollutant. Thermodynamic data show that nitrogen monoxide is unstable and should revert to N_2 and O_2. But remember that thermodynamics says nothing about rate. Unfortunately, the rate of reversion of NO to N_2 and O_2 is slow. Fortunately, catalysts have been developed that greatly speed this reaction.

$$2\,NO(g) \xrightarrow{\text{catalyst}} N_2(g) + O_2(g)$$

The role of the heterogeneous catalyst in the preceding reactions is probably to weaken the bonds of the reactants and to assist in product formation. For example, Figure 12.18 shows how NO molecules can dissociate into N and O atoms on the surface of a platinum metal catalyst.

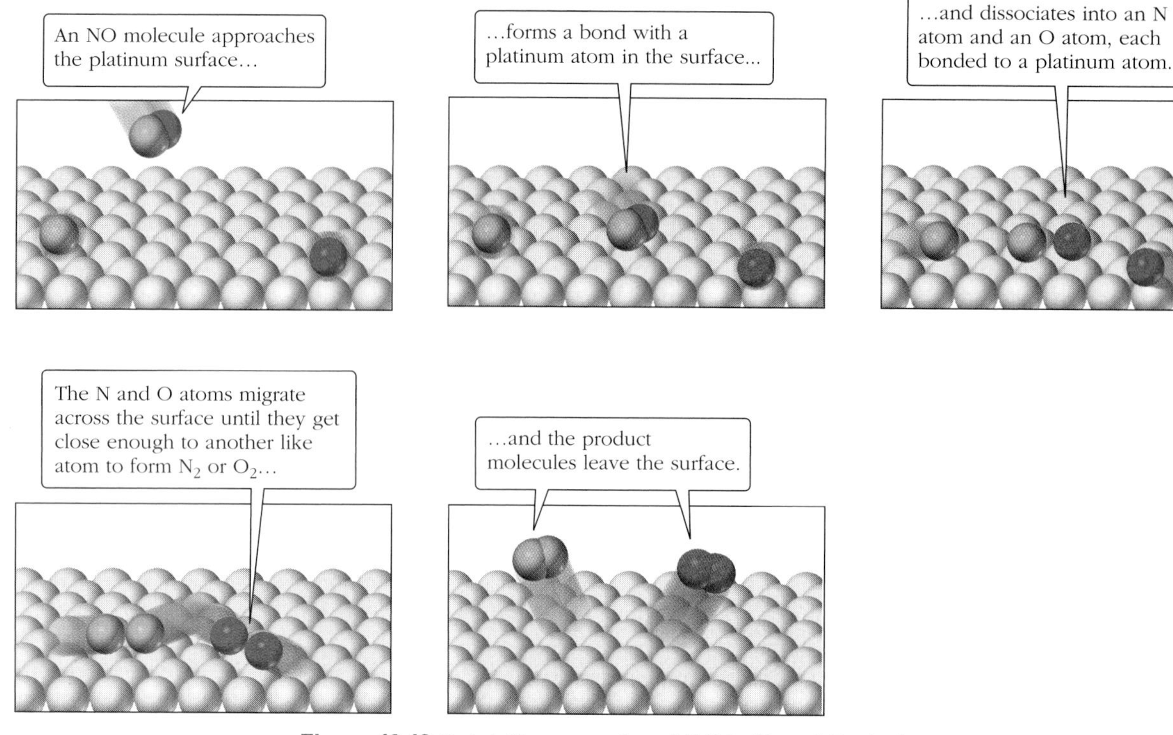

Figure 12.18 Catalytic conversion of NO to N_2 and O_2. A platinum surface can speed conversion of NO to N_2 and O_2 by helping to dissociate NO into N atoms and O atoms, which then travel across the surface and combine to form N_2 and O_2 molecules. The entire process of interaction with the surface and dissociation takes about 1.7×10^{-12} s.

CONCEPTUAL
EXERCISE 12.13 Catalysis

Which of these statements is (are) true? If any are false, change the wording to make them true.

(a) The concentration of a homogeneous catalyst may appear in the rate law.
(b) A catalyst is always consumed in the overall reaction.
(c) A catalyst must always be in the same phase as the reactants.

IN CLOSING

Having studied this chapter, you should be able to . . .

- Define reaction rate and calculate average rates (Section 12.1).
- Describe the effect that reactant concentrations have on reaction rate, and determine rate laws and rate constants from initial rates (Section 12.2). ■ End-of-chapter Questions assignable in OWL: 15, 20
- Determine reaction orders from a rate law, and use the integrated rate law method to obtain orders and rate constants (Section 12.3). ■ Question 21
- Calculate concentration at a given time, time to reach a certain concentration, and half-life for a first-order reaction (Section 12.3). ■ Questions 25, 27
- Define and give examples of unimolecular and bimolecular elementary reactions (Section 12.4). ■ Question 29

Selected end-of-chapter Questions may be assigned in OWL.

For quick review, download Go Chemistry mini-lecture flashcard modules (or purchase them at **www.ichapters.com**).

- Show by using an energy profile what happens as two reactant molecules interact to form product molecules (Section 12.4). ■ Question 43
- Define activation energy and frequency factor, and use them to calculate rate constants and rates under different conditions of temperature and concentration (Section 12.5). ■ Question 40
- Derive rate laws for unimolecular and bimolecular elementary reactions (Section 12.6). ■ Question 47
- Define reaction mechanism and identify rate-limiting steps and intermediates (Section 12.7). ■ Questions 49, 51
- Given several reaction mechanisms, decide which is (are) in agreement with experimentally determined stoichiometry and rate law (Section 12.7). ■ Question 53
- Explain how a catalyst can speed up a reaction, and draw energy profiles for catalyzed and uncatalyzed reaction mechanisms (Section 12.8). ■ Question 56
- Define the terms *enzyme, substrate,* and *active site,* and identify similarities and differences between enzyme-catalyzed reactions and uncatalyzed reactions (Section 12.9). ■ Question 59
- Describe two important industrial processes that depend on catalysts (Section 12.10).

Prepare for an exam with a **Summary Problem** that brings together concepts and problem-solving skills from throughout the chapter. Go to **www.cengage.com/chemistry/moore** to download this chapter's Summary Problem from the Student Companion Site.

KEY TERMS

activated complex *(Section 12.4)*
activation energy (E_a) *(12.4)*
active site *(12.9)*
Arrhenius equation *(12.5)*
average reaction rate *(12.1)*
bimolecular reaction *(12.4)*
chemical kinetics *(Introduction)*
elementary reaction *(12.4)*
enzyme *(12.9)*
enzyme-substrate complex *(12.9)*
frequency factor *(12.5)*
half-life *(12.3)*
heterogeneous catalyst *(12.10)*
heterogeneous reaction *(12.1)*
homogeneous catalyst *(12.8)*
homogeneous reaction *(12.1)*
induced fit *(12.9)*
initial rate *(12.2)*
instantaneous reaction rate *(12.1)*
intermediate *(12.7)*
order of reaction *(12.3)*
overall reaction order *(12.3)*
rate *(12.1)*
rate constant *(12.2)*
rate law *(12.2)*
rate-limiting step *(12.7)*
reaction intermediate *(12.7)*
reaction mechanism *(12.7)*
reaction rate *(12.1)*
steric factor *(12.4)*
substrate *(12.9)*
transition state *(12.4)*
unimolecular reaction *(12.4)*

QUESTIONS FOR REVIEW AND THOUGHT

■ denotes questions assignable in OWL.

Blue-numbered questions have short answers at the back of this book and fully worked solutions in the *Student Solutions Manual.*

Review Questions

1. Which of these is appropriate for determining the rate law for a chemical reaction?
   (a) Theoretical calculations based on balanced equations
   (b) Measuring the rate of the reaction as a function of the concentrations of the reacting species
   (c) Measuring the rate of the reaction as a function of temperature
2. Name at least three factors that affect the rate of a chemical reaction.
3. Define the terms "enzyme," "substrate," and "active site," and give an example of each kind of molecule.
4. Explain the difference between a homogeneous and a heterogeneous catalyst. Give an example of each.
5. Define the terms "unimolecular elementary reaction" and "bimolecular elementary reaction," and give an example of each.
6. Define the terms "activation energy" and "frequency factor." Write an equation that relates activation energy and frequency factor to reaction rate.

Topical Questions

Reaction Rate

7. Consider the dissolving of sugar as a simple process in which kinetics is important. Suppose that you dissolve an equal mass of each kind of sugar listed. Which dissolves the fastest? Which dissolves the slowest? Explain why in terms of rates of heterogeneous reactions. (If you are not sure which is fastest or slowest, try them all.)
   (a) Rock candy sugar (large sugar crystals)
   (b) Sugar cubes
   (c) Granular sugar
   (d) Powdered sugar
8. A cube of aluminum 1.0 cm on each edge is placed into 9 M NaOH(aq), and the rate at which H_2 gas is given off is measured.
   (a) By what factor will this reaction rate change if the aluminum cube is cut exactly in half and the two halves are placed in the solution? Assume that the reaction rate is proportional to the surface area, and that all of the surface of the aluminum is in contact with the NaOH(aq).
   (b) If you had to speed up this reaction as much as you could without raising the temperature, what would you do to the aluminum?
9. Using data given in the table for the reaction

$$N_2O_5 \longrightarrow 2\ NO_2 + \tfrac{1}{2}\ O_2$$

calculate the average rate of reaction during each of these intervals:
(a) 0.00 to 0.50 h (b) 0.50 to 1.0 h
(c) 1.0 to 2.0 h (d) 2.0 to 3.0 h
(e) 3.0 to 4.0 h (f) 4.0 to 5.0 h

Time (h)	$[N_2O_5]$ (mol/L)	Time (h)	$[N_2O_5]$ (mol/L)
0.00	0.849	3.00	0.352
0.50	0.733	4.00	0.262
1.00	0.633	5.00	0.196
2.00	0.472		

10. Using data from Question 9, calculate the average rate over the interval 0 to 5.0 h. Compare your result with the average rates over the intervals 1.0 to 4.0 h and 2.0 to 3.0 h, all of which have the same midpoint (2.5 h from the start).
11. Using all your calculated rates from Question 9,
   (a) Show that the reaction obeys the rate law

$$\text{Rate} = -\frac{\Delta[N_2O_2]}{\Delta t} = k[N_2O_5]$$

   (b) Evaluate the rate constant k as an average of the values obtained for the six intervals.
12. Using the rate law and the rate constant you calculated in Question 11, calculate the reaction rate exactly 2.5 h from the start. Do your results from this and Question 10 agree with the statement in the text that the smaller the time interval, the more accurate the average rate?
13. For the reaction

$$2\ NO_2(g) \longrightarrow 2\ NO(g) + O_2(g)$$

make qualitatively correct plots of the concentrations of $NO_2(g)$, NO(g), and $O_2(g)$ versus time. Draw all three graphs on the same axes, assume that you start with $NO_2(g)$ at a concentration of 1.0 mol/L, and assume that the reaction is first-order. Explain how you would determine, from these plots,
   (a) The initial rate of the reaction
   (b) The final rate (i.e., the rate as time approaches infinity)
14. For the reaction

$$O_3(g) + O(g) \longrightarrow 2\ O_2(g)$$

make qualitatively correct plots of the concentrations of $O_3(g)$, O(g), and $O_2(g)$ versus time. Draw all three graphs on the same axes, assume that you start with $O_3(g)$ and O(g), each at a concentration of 1.0 μmol/L, and assume that the reaction is second-order. Explain how you would determine, from these plots,
   (a) The initial rate of the reaction
   (b) The final rate (i.e., the rate as time approaches infinity)

Effect of Concentration on Reaction Rates

15. ■ The reaction of CO(g) + $NO_2(g)$ is second-order in NO_2 and zeroth-order in CO at temperatures less than 500 K.
   (a) Write the rate law for the reaction.
   (b) How will the reaction rate change if the NO_2 concentration is halved?
   (c) How will the reaction rate change if the concentration of CO is doubled?
16. Nitrosyl bromide, NOBr, is formed from NO and Br_2.

$$2\ NO(g) + Br_2(g) \longrightarrow 2\ NOBr(g)$$

Experiment shows that the reaction is first-order in Br_2 and second-order in NO.
   (a) Write the rate law for the reaction.
   (b) If the concentration of Br_2 is tripled, how will the reaction rate change?
   (c) What happens to the reaction rate when the concentration of NO is doubled?
17. For the reaction of $Pt(NH_3)_2Cl_2$ with water,

$$Pt(NH_3)_2Cl_2 + H_2O \longrightarrow Pt(NH_3)_2(H_2O)Cl^+ + Cl^-$$

the rate law is rate = $k[Pt(NH_3)_2Cl_2]$ with $k = 0.090\ h^{-1}$.
   (a) Calculate the initial rate of reaction when the concentration of $Pt(NH_3)_2Cl_2$ is
   (i) 0.010 M (ii) 0.020 M (iii) 0.040 M
   (b) How does the rate of disappearance of $Pt(NH_3)_2Cl_2$ change with its initial concentration?
   (c) How is this related to the rate law?
   (d) How does the initial concentration of $Pt(NH_3)_2Cl_2$ affect the rate of appearance of Cl^- in the solution?
18. Methyl acetate, CH_3COOCH_3, reacts with base to break one of the C—O bonds.

$$CH_3\overset{\overset{\displaystyle O}{\|}}{C}—O—CH_3(aq) + OH^-(aq) \longrightarrow CH_3\overset{\overset{\displaystyle O}{\|}}{C}—O^-(aq) + HO—CH_3(aq)$$

The rate law is rate = $k[CH_3COOCH_3][OH^-]$ where k = 0.14 L mol^{-1} s^{-1} at 25 °C.

(a) What is the initial rate at which the methyl acetate disappears when both reactants, CH_3COOCH_3 and OH^-, have a concentration of 0.025 M?

(b) How rapidly (i.e., at what rate) does the methyl alcohol, CH_3OH, initially appear in the solution?

19. Measurements of the initial rate of reaction between two compounds, triphenylmethyl hexachloroantimonate, **I**, and bis-(9-ethyl-3-carbazolyl)methane, **II**, in 1,2-dichloroethane at 40 °C yielded these data:

Initial Concentration × 10^5 (mol/L)		Initial Rate × 10^9 (mol L^{-1} s^{-1})
[I]	[II]	
1.65	10.6	1.50
14.9	10.6	17.7
14.9	7.10	11.2
14.9	3.52	6.30
14.9	1.76	3.10
4.97	10.6	4.52
2.48	10.6	2.70

(a) Derive the rate law for this reaction.

(b) Calculate the rate constant k and express it in appropriate units.

20. ■ For the reaction

$$2\ NO(g) + 2\ H_2(g) \longrightarrow N_2(g) + 2\ H_2O(g)$$

these data were obtained at 1100 K:

[NO] (mol/L)	$[H_2]$ (mol/L)	Initial Rate (mol L^{-1} s^{-1})
5.00×10^{-3}	2.50×10^{-3}	3.0×10^{-3}
15.0×10^{-3}	2.50×10^{-3}	9.0×10^{-3}
15.0×10^{-3}	10.0×10^{-3}	3.6×10^{-2}

(a) What is the order with respect to NO? With respect to H_2?

(b) What is the overall order?

(c) Write the rate law.

(d) Calculate the rate constant.

(e) Calculate the initial rate of this reaction at 1100 K when $[NO] = [H_2] = 8.0 \times 10^{-3}$ mol L^{-1}.

Rate Law and Order of Reaction

21. ■ For each of these rate laws, state the reaction order with respect to the hypothetical substances A and B, and give the overall order.

(a) Rate = $k[A][B]^3$ (b) Rate = $k[A][B]$
(c) Rate = $k[A]$ (d) Rate = $k[A]^3[B]$

22. The reaction

$$2\ NO(g) + 2\ H_2(g) \longrightarrow N_2(g) + 2\ H_2O(g)$$

is found to be first-order in $H_2(g)$. Which rate equation cannot be correct?

(a) Rate = $k[NO]^2[H_2]$ (b) Rate = $k[H_2]$
(c) Rate = $k[NO]^2[H_2]^2$

23. The bromination of acetone is catalyzed by acid.

$$CH_3COCH_3(aq) + Br_2(aq) + H_2O(\ell) \xrightarrow{\text{acid catalyst}} CH_3COCH_2Br(aq) + H_3O^+(aq) + Br^-(aq)$$

The rate of disappearance of bromine was measured for several different initial concentrations of acetone, bromine, and hydronium ion.

Initial Concentration (mol/L)			Initial Rate of Change of $[Br_2]$ (mol L^{-1} s^{-1})
$[CH_3COCH_3]$	$[Br_2]$	$[H_3O^+]$	
0.30	0.05	0.05	5.7×10^{-5}
0.30	0.10	0.05	5.7×10^{-5}
0.30	0.05	0.10	12.0×10^{-5}
0.40	0.05	0.20	31.0×10^{-5}
0.40	0.05	0.05	7.6×10^{-5}

(a) Deduce the rate law for the reaction and give the order with respect to each reactant.

(b) What is the numerical value of k, the rate constant?

(c) If $[H_3O^+]$ is maintained at 0.050 M, whereas both $[CH_3COCH_3]$ and $[Br_2]$ are 0.10 M, what is the rate of the reaction?

24. One of the major eye irritants in smog is formaldehyde, CH_2O, formed by reaction of ozone with ethylene.

$$C_2H_4(g) + O_3(g) \longrightarrow 2\ CH_2O(g) + \tfrac{1}{2}\ O_2(g)$$

These data were collected:

Initial Concentration (mol/L)		Initial Rate of Formation of CH_2O (mol L^{-1} s^{-1})
$[O_3]$	$[C_2H_4]$	
0.50×10^{-7}	1.0×10^{-8}	1.0×10^{-12}
1.5×10^{-7}	1.0×10^{-8}	3.0×10^{-12}
1.0×10^{-7}	2.0×10^{-8}	4.0×10^{-12}

(a) Determine the rate law for the reaction using the data in the table.

(b) What is the reaction order with respect to O_3? What is the order with respect to C_2H_4?

(c) Calculate the rate constant, k.

(d) What is the rate of reaction when $[C_2H_4]$ and $[O_3]$ are both 2.0×10^{-7} M?

25. ■ If the initial concentration of the reactant in a first-order reaction A ⟶ products is 0.64 mol/L and the half-life is 30. s,

(a) Calculate the concentration of the reactant 60 s after initiation of the reaction.

(b) How long would it take for the concentration of the reactant to drop to one-eighth its initial value?

(c) How long would it take for the concentration of the reactant to drop to 0.040 mol L^{-1}?

26. If the initial concentration of the reactant in a first-order reaction A $\longrightarrow$ products is 0.50 mol/L and the half-life is 400 s,
(a) Calculate the concentration of the reactant 1600 s after initiation of the reaction.
(b) How long would it take for the concentration of the reactant to drop to one-sixteenth its initial value?
(c) How long would it take for the concentration of the reactant to drop to 0.062 mol L^{-1}?

27. ■ The compound SO_2Cl_2 decomposes in a first-order reaction

$$SO_2Cl_2(g) \longrightarrow SO_2(g) + Cl_2(g)$$

that has a half-life of 1.47×10^4 s at 600 K. If you begin with 1.6×10^{-3} mol of pure SO_2Cl_2 in a 2.0-L flask, at what time will the amount of SO_2Cl_2 be 1.2×10^{-4} mol?

28. The hypothetical compound A decomposes in a first-order reaction that has a half-life of 2.3×10^2 s at 450. °C. If the initial concentration of A is 4.32×10^{-2} M, how long will it take for the concentration of A to drop to 3.75×10^{-3} M?

A Nanoscale View: Elementary Reactions

29. ■ Which of these reactions are unimolecular and elementary, which are bimolecular and elementary, and which are not elementary?
(a) $CH_4(g) + 2\ O_2(g) \longrightarrow CO_2(g) + 2\ H_2O(g)$
(b) $O_3(g) + O(g) \longrightarrow 2\ O_2(g)$
(c) $Mg(s) + 2\ H_2O(\ell) \longrightarrow H_2(g) + Mg(OH)_2(s)$
(d) $O_3(g) \longrightarrow O_2(g) + O(g)$

30. Which of these reactions are unimolecular and elementary, which are bimolecular and elementary, and which are not elementary?
(a) $HCl(g) + H_2O(g) \longrightarrow H_3O^+(g) + Cl^-(g)$
(b) $I^-(g) + CH_3Cl(g) \longrightarrow ICH_3(g) + Cl^-(g)$
(c) $C_2H_6(g) \longrightarrow C_2H_4(g) + H_2(g)$
(d) $N_2(g) + 3\ H_2(g) \longrightarrow 2\ NH_3(g)$
(e) $O_2(g) + O(g) \longrightarrow O_3(g)$

31. Assume that each gas phase reaction occurs via a single bimolecular step. For which reaction would you expect the steric factor to be more important? Why?

$$Cl + O_3 \longrightarrow ClO + O_2 \quad \text{or} \quad NO + O_3 \longrightarrow NO_2 + O_2$$

32. Assume that each gas phase reaction occurs via a single bimolecular step. For which reaction would you expect the steric factor to be more important? Why?

$$H_2C{=}CH_2 + H_2 \longrightarrow H_3C{-}CH_3 \quad \text{or}$$
$$(CH_3)_2C{=}CH_2 + HBr \longrightarrow (CH_3)_2CBr{-}CH_3$$

Temperature and Reaction Rate

33. From Problem-Solving Example 12.8 (p. 435), where the energy profile of the ozone plus atomic oxygen reaction was derived, obtain the activation energy. Then determine the ratio of the reaction rate for this reaction at 50. °C to the reaction rate at room temperature (25 °C). Assume that the initial concentrations are the same at both temperatures.

34. A chemical reaction has an activation energy of 30. kJ/mol. If you had to slow down this reaction a thousandfold by cooling it from room temperature (25 °C), what would the temperature be?

35. For the reaction of iodine atoms with hydrogen molecules in the gas phase, these rate constants were obtained experimentally.

$$2\ I(g) + H_2(g) \longrightarrow 2\ HI(g)$$

T (K)	10^{-5} k ($L^2\ mol^{-2}\ s^{-1}$)
417.9	1.12
480.7	2.60
520.1	3.96
633.2	9.38
666.8	11.50
710.3	16.10
737.9	18.54

(a) Calculate the activation energy and frequency factor for this reaction.
(b) Estimate the rate constant of the reaction at 400.0 K.

36. Make an Arrhenius plot and calculate the activation energy for the gas phase reaction

$$2\ NOCl(g) \longrightarrow 2\ NO(g) + Cl_2(g)$$

T (K)	Rate Constant (L $mol^{-1}\ s^{-1}$)
400.	6.95×10^{-4}
450.	1.98×10^{-2}
500.	2.92×10^{-1}
550.	2.60
600.	16.3

37. For the gas phase reaction

$$CH_3CH_2I(g) \longrightarrow CH_2CH_2(g) + HI(g)$$

the activation energy E_a is 221 kJ/mol and the frequency factor A is $1.2 \times 10^{14}\ s^{-1}$. If the concentration of CH_3CH_2I is 0.012 mol/L, what is the rate of reaction at
(a) 400. °C? (b) 800. °C?

38. For the gas phase reaction

$$cis\text{-CHClCHCl}(g) \longrightarrow trans\text{-CHClCHCl}(g)$$

the activation energy E_a is 234 kJ/mol and the frequency factor A is $6.3 \times 10^{12}\ s^{-1}$. If the concentration of *cis*-CHClCHCl is 0.0043 mol/L, what is the rate of reaction at
(a) 400. °C? (b) 800. °C?

39. For the reaction

$$N_2O_5(g) \longrightarrow 2\ NO_2(g) + \tfrac{1}{2}\ O_2(g)$$

the rate constant k at 25 °C is $3.46 \times 10^{-5}\ s^{-1}$ and at 55 °C it is $1.5 \times 10^{-3}\ s^{-1}$. Calculate the activation energy, E_a.

40. ■ For a hypothetical reaction the rate constant is $4.63 \times 10^{-3}\ s^{-1}$ at 25 °C and $2.37 \times 10^{-2}\ s^{-1}$ at 43 °C. Calculate the activation energy.

Rate Laws for Elementary Reactions

41. For the hypothetical reaction A + B $\longrightarrow$ C + D, the activation energy is 32 kJ/mol. For the reverse reaction (C + D $\longrightarrow$ A + B), the activation energy is 58 kJ/mol. Is the reaction A + B $\longrightarrow$ C + D exothermic or endothermic?

42. Use the diagram to answer these questions.

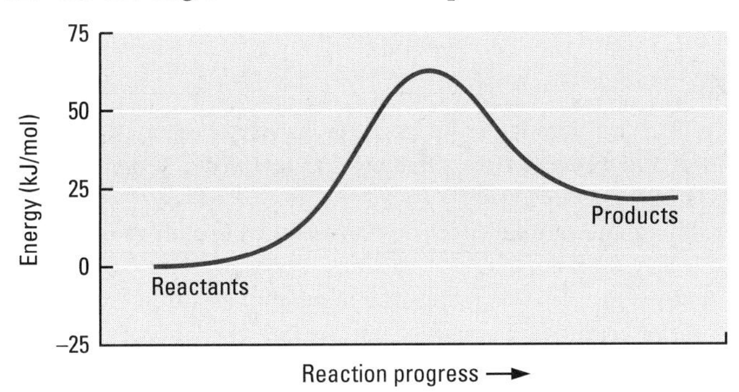

(a) Is the reaction exothermic or endothermic?
(b) What is the approximate value of ΔE for the forward reaction?
(c) What is the activation energy in each direction?
(d) A catalyst is found that lowers the activation energy of the reaction by about 10 kJ/mol. How will this catalyst affect the rate of the reverse reaction?

43. ■ Draw an energy versus reaction progress diagram (similar to the one in Question 42) for each of the reactions whose activation energy and enthalpy change are given below.
(a) $\Delta H° = -145\ kJ\ mol^{-1}$; $E_a = 75\ kJ\ mol^{-1}$
(b) $\Delta H° = -70\ kJ\ mol^{-1}$; $E_a = 65\ kJ\ mol^{-1}$
(c) $\Delta H° = 70\ kJ\ mol^{-1}$; $E_a = 85\ kJ\ mol^{-1}$

44. Draw an energy versus reaction progress diagram (similar to the one in Question 42) for each of the reactions whose activation energy and enthalpy change are given below.
(a) $\Delta H° = 105\ kJ\ mol^{-1}$; $E_a = 175\ kJ\ mol^{-1}$
(b) $\Delta H° = -43\ kJ\ mol^{-1}$; $E_a = 95\ kJ\ mol^{-1}$
(c) $\Delta H° = 15\ kJ\ mol^{-1}$; $E_a = 55\ kJ\ mol^{-1}$

45. Which of the reactions in Question 43 would be expected to (a) Occur fastest? (b) Occur slowest? (Assume equal temperatures, equal concentrations, equal frequency factors, and the same rate law for all reactions.)

46. Which of the reactions in Question 44 would be expected to (a) Occur fastest? (b) Occur slowest? (Assume equal temperatures, equal concentrations, equal frequency factors, and the same rate law for all reactions.)

47. ■ Assuming that each reaction is elementary, predict the rate law.
(a) $NO(g) + NO_3(g) \longrightarrow 2\ NO_2(g)$
(b) $O(g) + O_3(g) \longrightarrow 2\ O_2(g)$
(c) $(CH_3)_3CBr(aq) \longrightarrow (CH_3)_3C^+(aq) + Br^-(aq)$
(d) $2\ HI(g) \longrightarrow H_2(g) + I_2(g)$

48. Assuming that each reaction is elementary, predict the rate law.
(a) $Cl(g) + ICl(g) \longrightarrow I(g) + Cl_2(g)$
(b) $Cl(g) + H_2(g) \longrightarrow HCl(g) + H(g)$
(c) $2\ NO_2(g) \longrightarrow N_2O_4(g)$
(d) Cyclopropane(g) $\longrightarrow$ propene(g)

Reaction Mechanisms

49. ■ Experiments show that the reaction of nitrogen dioxide with fluorine

Overall reaction: $2\ NO_2(g) + F_2(g) \longrightarrow 2\ FNO_2(g)$

has the rate law

$$\text{Initial reaction rate} = k[NO_2][F_2]$$

and the reaction is thought to occur in two steps.

Step 1: $NO_2(g) + F_2(g) \longrightarrow FNO_2(g) + F(g)$

Step 2: $NO_2(g) + F(g) \longrightarrow FNO_2(g)$

(a) Show that the sum of this sequence of reactions gives the balanced equation for the overall reaction.
(b) Which step is rate-determining?

50. Nitrogen monoxide is reduced by hydrogen to give water and nitrogen

$$2\ H_2(g) + 2\ NO(g) \longrightarrow N_2(g) + 2\ H_2O(g)$$

and one possible mechanism for this reaction is a sequence of three elementary steps.

Step 1 (fast): $2\ NO(g) \rightleftharpoons N_2O_2(g)$

Step 2 (slow): $N_2O_2(g) + H_2(g) \longrightarrow N_2O_2(g) + H_2O(g)$

Step 3 (fast): $N_2O(g) + H_2(g) \longrightarrow N_2(g) + H_2O(g)$

Show that the sum of these steps gives the net reaction.

51. ■ For the reaction mechanism

$$H_3C-C(=O)-O-CH_3 + H_3O^+ \rightleftharpoons H_3C-C(=\overset{+}{O}H)-O-CH_3 + H_2O \qquad \text{fast}$$

$$H_3C-C(=\overset{+}{O}H)-O-CH_3 + H_2O \rightleftharpoons H_3C-C(OH)(\overset{+}{O}H_2)-O-CH_3 \qquad \text{slow}$$

$$H_3C-C(OH)(\overset{+}{O}H_2)-O-CH_3 \rightleftharpoons H_3C-C(OH)(OH)-\overset{+}{O}(H)-CH_3 \qquad \text{fast}$$

$$H_3C-C(OH)(OH)-\overset{+}{O}(H)-CH_3 + H_2O \rightleftharpoons H_3C-C(=O)(OH) + O(H)-CH_3 + H_3O^+ \qquad \text{fast}$$

(a) Write the chemical equation for the overall reaction.
(b) Is there a catalyst involved in this reaction? If so, what is it?
(c) Identify all intermediates in the reaction.

52. For the reaction mechanism

$$A + B \rightleftharpoons C \qquad \text{fast}$$

$$C + A \longrightarrow 2\ D \qquad \text{slow}$$

(a) Write the chemical equation for the overall reaction.
(b) Is there a catalyst involved in this reaction? If so, what is it?
(c) Identify all intermediates in the reaction.

53. ■ For the reaction

$$CH_3-C(CH_3)_2-Br + OH^- \longrightarrow CH_3-C(CH_3)_2-OH + Br^-$$

the rate law is

$$\text{Rate} = k[(CH_3)_3CBr]$$

Identify each mechanism that is compatible with the rate law.

(a) $(CH_3)_3CBr \longrightarrow (CH_3)_3C^+ + Br^-$ slow
$(CH_3)_3C^+ + OH^- \longrightarrow (CH_3)_3COH$ fast
(b) $(CH_3)_3CBr + OH^- \longrightarrow (CH_3)_3COH + Br^-$

Catalysts and Reaction Rate

54. Which of these statements is (are) true?
 (a) The concentration of a homogeneous catalyst may appear in the rate law.
 (b) A catalyst is always consumed in the reaction.
 (c) A catalyst must always be in the same phase as the reactants.
 (d) A catalyst can change the course of a reaction and allow different products to be produced.
55. Hydrogenation reactions—processes in which H_2 is added to a molecule—are usually catalyzed. An excellent catalyst is a very finely divided metal suspended in the reaction solvent. Tell why finely divided rhodium, for example, is a much more efficient catalyst than a small block of the metal that has the same mass.
56. ■ Which of these reactions appear to involve a catalyst? In those cases where a catalyst is present, tell whether it is homogeneous or heterogeneous.
 (a) $CH_3CO_2CH_3(aq) + H_2O(\ell) \longrightarrow CH_3COOH(aq) + CH_3OH(aq)$
 Rate $= k[CH_3CO_2CH_3][H_3O^+]$
 (b) $H_2(g) + I_2(g) \longrightarrow 2\ HI(g)$
 Rate $= k[H_2][I_2]$
 (c) $2\ H_2(g) + O_2(g) \longrightarrow 2\ H_2O(g)$
 Rate $= k[H_2][O_2]$ (area of Pt surface)
 (d) $H_2(g) + CO(g) \longrightarrow H_2CO(g)$
 Rate $= k[H_2]^{1/2}[CO]$
57. In acid solution, methyl formate forms methyl alcohol and formic acid.

$$\underset{\text{methyl formate}}{HCO_2CH_3(aq)} + H_2O(\ell) \longrightarrow \underset{\text{formic acid}}{HCOOH(aq)} + \underset{\text{methyl alcohol}}{CH_3OH(aq)}$$

 The rate law is as follows: Rate $= k[HCO_2CH_3][H_3O^+]$. Why does H_3O^+ appear in the rate law but not in the overall equation for the reaction?

Enzymes: Biological Catalysts

58. Write a one- or two-sentence definition in your own words for each term:

active site	induced fit
enzyme	protein
enzyme-substrate complex	substrate
globular protein	

59. ■ When enzymes are present at very low concentration, their effect on reaction rate can be described by first-order kinetics. By what factor does the rate of an enzyme-catalyzed reaction change when the enzyme concentration is changed from 1.5×10^{-7} to 4.5×10^{-6} M?
60. When substrates are present at relatively high concentration and are catalyzed by enzymes, the effect on reaction rate of changing substrate concentration can be described by zeroth-order kinetics. By what factor does the rate of an enzyme-catalyzed reaction change when the substrate concentration is changed from 1.5×10^{-3} to 4.5×10^{-2} M?

Catalysis in Industry

61. Why are homogeneous catalysts harder to separate from products and leftover reactants than are heterogeneous reactants?
62. In an automobile catalytic converter the catalysis is accomplished on a surface consisting of platinum and other precious metals. The metals are deposited as a thin layer on a honeycomb-like ceramic support (see the photo).

Johnson Matthey Platinum Today (www.platinum.matthey.com)

 (a) Why is the ceramic support arranged in the honeycomb geometry?
 (b) Why are the metals deposited on the ceramic surface instead of being used as strips or rods?

General Questions

63. Draw a reaction energy diagram for an exothermic process. Mark the positions of reactants, products, and activated complex. Indicate the activation energies of the forward and reverse processes and explain how ΔE for the reaction can be calculated from the diagram.
64. Draw a reaction energy diagram for an endothermic process. Mark the positions of reactants, products, and activated complex. Indicate the activation energies of the forward and reverse processes and explain how ΔE for the reaction can be calculated from the diagram.
65. Indicate whether each of these statements is true or false. Change the wording of each false statement to make it true.
 (a) It is possible to change the rate constant for a reaction by changing the temperature.
 (b) The reaction rate remains constant as a first-order reaction proceeds at a constant temperature.
 (c) The rate constant for a reaction is independent of reactant concentrations.
 (d) As a second-order reaction proceeds at a constant temperature, the rate constant changes.
66. Nitrogen monoxide can be reduced with hydrogen.

$$2\ H_2(g) + 2\ NO(g) \longrightarrow 2\ H_2O(g) + N_2(g)$$

 Experiment shows that when the concentration of H_2 is halved, the reaction rate is halved. Furthermore, raising the concentration of NO by a factor of 3 raises the rate by a factor of 9. Write the rate equation for this reaction.
67. For the reaction of NO and O_2 at 660 K,

$$2\ NO(g) + O_2(g) \longrightarrow 2\ NO_2(g)$$

Concentration (mol/L)		Rate of Disappearance of NO (mol L^{-1} s^{-1})
[NO]	[O_2]	
0.010	0.010	2.5×10^{-5}
0.020	0.010	1.0×10^{-4}
0.010	0.020	5.0×10^{-5}

(a) Determine the order of the reaction for each reactant.
(b) Write the rate equation for the reaction.
(c) Calculate the rate constant.
(d) Calculate the rate when [NO] = 0.025 mol/L and [O_2] = 0.050 mol/L.
(e) If O_2 disappears at a rate of 1.0×10^{-4} mol L^{-1} s^{-1}, what is the rate at which NO disappears? What is the rate at which NO_2 is forming?

68. Nitryl fluoride is an explosive compound that can be made by oxidizing nitrogen dioxide with fluorine:

$$2\,NO_2(g) + F_2(g) \longrightarrow 2\,NO_2F(g)$$

Several kinetics experiments, all done at the same temperature and involving formation of nitryl fluoride are summarized in this table:

Experiment	Initial Concentration (mol/L)			Initial Rate
	[NO_2]	[F_2]	[NO_2F]	(mol L^{-1} s^{-1})
1	0.0010	0.0050	0.0020	2.0×10^{-4}
2	0.0020	0.0050	0.0020	4.0×10^{-4}
3	0.0020	0.0020	0.0020	1.6×10^{-4}
4	0.0020	0.0020	0.0010	1.6×10^{-4}

(a) Write the rate law for the reaction.
(b) What is the order of the reaction with respect to each reactant and each product?
(c) Calculate the rate constant k and express it in appropriate units.

Applying Concepts

69. The graph shows the change in concentration as a function of time for the reaction

$$2\,H_2O_2(g) \longrightarrow 2\,H_2O(g) + O_2(g)$$

What do each of the curves A, B, and C represent?

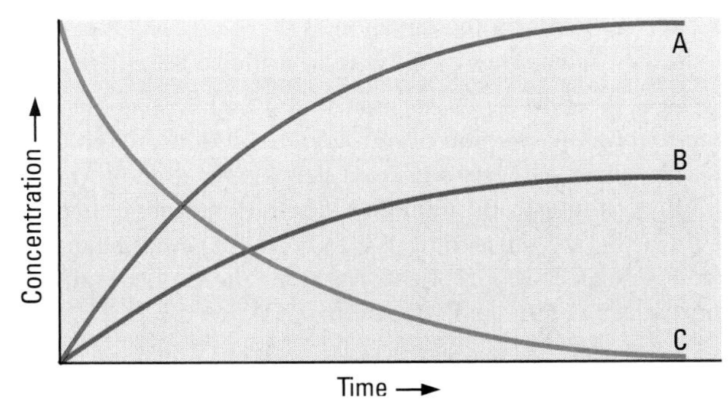

70. Draw a graph similar to the one in Question 69 for the reaction

$$2\,N_2O_5(g) \longrightarrow 4\,NO_2(g) + O_2(g)$$

71. The picture below is a "snapshot" of the reactants at time = 0 for the reaction

$$H_2(g) + I_2(g) \longrightarrow 2\,HI(g)$$

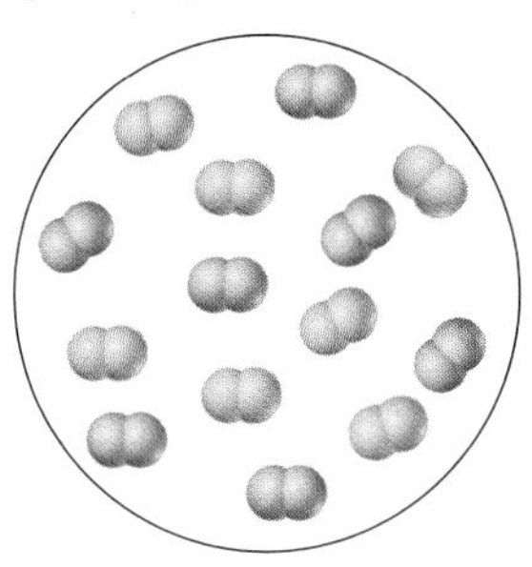

Suppose the reaction is carried out at two different temperatures and that another snapshot is taken after a constant time has elapsed.

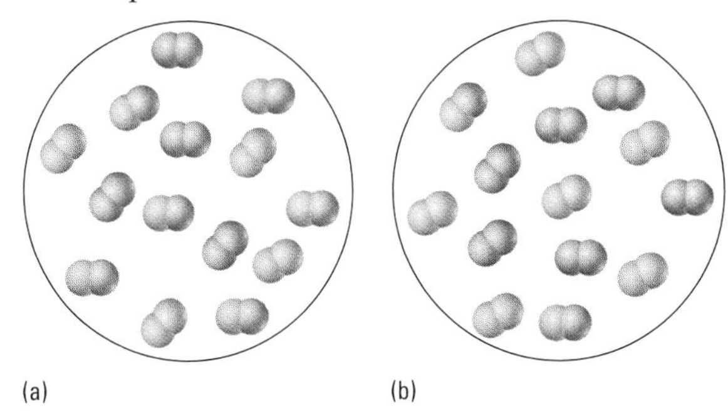

(a) (b)

Which of these two snapshots corresponds to the lower temperature reaction condition?

72. Consider Question 71 again, only this time a catalyst is used instead of a lower temperature. Which of the two snapshots corresponds to the presence of a catalyst?

73. Initial rates for the reaction A + B + C $\longrightarrow$ D + E were measured with various concentrations of A, B, and C as represented in the pictures below. Based on these data, what is the rate law?

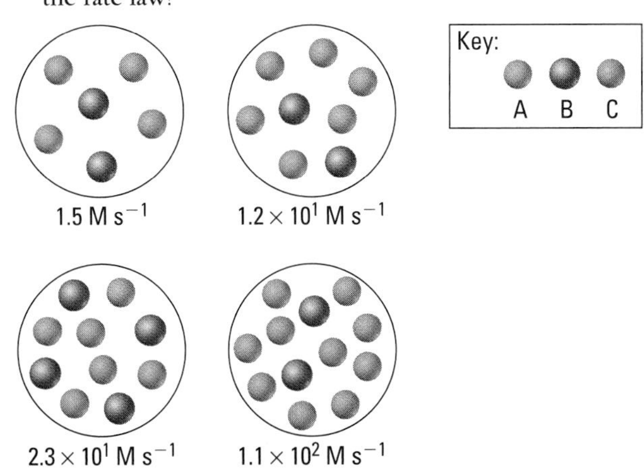

74. The rate of decay of a radioactive solid is independent of the temperature of that solid—at least for temperatures easily obtained in the laboratory. What does this observation imply about the activation energy for this process?

75. Platinum metal is used as a catalyst in the decomposition of NO(g) into N_2(g) and O_2(g). A graph of the rate of the reaction as a function of NO concentration is shown below. Explain why the rate stops increasing and levels out.

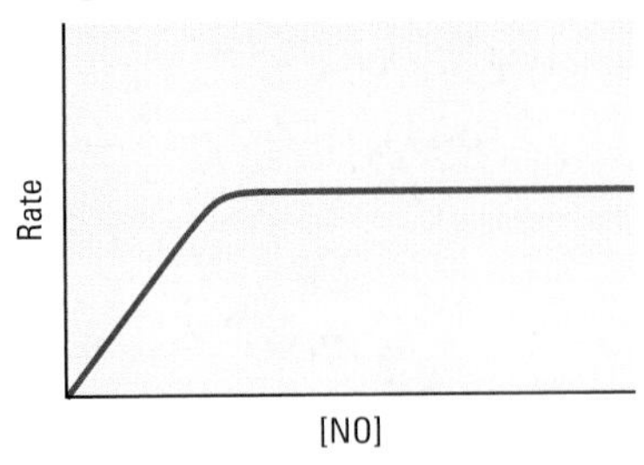

More Challenging Questions

76. Measurements of the initial rate of hydrolysis of benzenesulfonyl chloride in aqueous solution at 15 °C in the presence of fluoride ion yielded the results in the table for a fixed concentration of benzenesulfonyl chloride of 2×10^{-4} M. The reaction rate is known to be proportional to the concentration of benzenesulfonyl chloride.

$[F^-] \times 10^2$ (mol/L)	Initial Rate $\times 10^7$ (mol L^{-1} s^{-1})
0	2.4
0.5	5.4
1.0	7.9
2.0	13.9
3.0	20.2
4.0	25.2
5.0	32.0

Note that some reaction must be occurring in the absence of any fluoride ion, because at zero concentration of fluoride the rate is not zero. This residual rate should be subtracted from each observed rate to give the rate of the reaction being studied.
(a) Derive the complete rate law for the reaction.
(b) Calculate the rate constant k and express it in appropriate units.

77. The rate constant for decomposition of azomethane at 425 °C is 0.68 s^{-1}.

$$CH_3N{=}NCH_3(g) \longrightarrow N_2(g) + C_2H_6(g)$$

(a) Based on the units of the rate constant, is the reaction zeroth-, first-, or second-order?
(b) If 2.0 g azomethane is placed in a 2.0-L flask and heated to 425 °C, what mass of azomethane remains after 5.0 s?
(c) How long does it take for the mass of azomethane to drop to 0.24 g?
(d) What mass of nitrogen would be found in the flask after 0.5 s of reaction?

78. Cyclopropane isomerizes to propene when heated. Rate constants for the reaction

$$\text{cyclopropane} \longrightarrow \text{propene}$$

are 1.10×10^{-4} s^{-1} at 470. °C and 1.02×10^{-3} s^{-1} at 510. °C.
(a) What is the activation energy, E_a, for this reaction?
(b) At 500. °C and [cyclopropane] = 0.10 M, how long does it take for the concentration to drop to 0.023 M?

79. When heated, cyclobutane, C_4H_8, decomposes to ethylene, C_2H_4.

$$C_4H_8(g) \longrightarrow 2\ C_2H_4(g)$$

The activation energy, E_a, for this reaction is 262 kJ/mol.
(a) If the rate constant $k = 0.032\ s^{-1}$ at 800. K, what is the value of k at 900. K?
(b) At 850. K and an initial concentration of cyclobutane of 0.0427 M, what is the concentration of cyclobutane after 2 h?

80. For each reaction listed with its rate law, propose a reasonable mechanism.
(a) $CH_3CO_2CH_3(aq) + H_2O(\ell) \longrightarrow CH_3COOH(aq) + CH_3OH(aq)$
Rate = $k[CH_3CO_2CH_3][H_3O^+]$
(b) $H_2(g) + I_2(g) \longrightarrow 2\ HI(g)$
Rate = $k[H_2][I_2]$
(c) $2\ H_2(g) + O_2(g) \longrightarrow 2\ H_2O(g)$
Rate = $k[H_2][I_2]$ (area of Pt surface)

81. When a molecule of *cis*-2-butene has rotated 90° around the axis through the two central carbon atoms, the π bond between these atoms is essentially broken. Use bond energies to estimate the activation energy for rotation around the π bond in ethylene, C_2H_4. Compare your result with the activation energy value in Figure 12.7.

82. Consider the reaction mechanism for iodine-catalyzed isomerization of *cis*-2-butene presented in Section 12.8. Use bond enthalpies (Table 8.2) to estimate the energy change for each step in the mechanism. (In Step 3 assume that rotation around the single bond requires about 5 kJ/mol.) Do your estimates agree with the energy values for the intermediates in Figure 12.14?

Conceptual Challenge Problems

CP12.A (Section 12.5) A rule of thumb is that for a typical reaction, if concentrations are unchanged, a 10-K rise in temperature increases the reaction rate by two to four times. Use an average increase of three times to answer the questions below.
(a) What is the approximate activation energy of a "typical" chemical reaction at 298 K?
(b) If a catalyst increases a chemical reaction's rate by providing a mechanism that has a lower activation energy, then what change do you expect a 10-K increase in temperature to make in the rate of a reaction whose uncatalyzed activation energy of 75 kJ/mol has been lowered to one-half this value (at 298 K) by addition of a catalyst?

CP12.B (Section 12.7) A sentence in an introductory chemistry textbook reads, "Dioxygen reacts with itself to form trioxygen, ozone, according to the equation, $3\ O_2 \longrightarrow 2\ O_3$." As a student of chemistry, what would you write to criticize this sentence?

CP12.C (Section 12.7) A classmate consults you about a problem concerning the reaction of nitrogen monoxide and dioxygen in the gas phase. She has been told that the reaction is second-order in nitrogen monoxide and first-order in dioxygen; hence, the rate law may be written as rate = $k[NO]^2[O_2]$. She has been asked to propose a mechanism for this reaction. She proposes that the mechanism is this single equation:

$$NO + NO + O_2 \longrightarrow NO_2 + NO_2$$

She asks your opinion about whether this is correct. What should you tell her to explain why the answer is correct or incorrect?

13

Chemical Equilibrium

© Ken Lucas/Visuals Unlimited

This pinecone fish, *Monocentris gloriamaris,* has a bioluminescent area on its lower jaw. The bioluminescent area glows orange in the daytime and blue-green at night. It apparently helps to attract zooplankton upon which the pinecone fish feeds. In this daytime photograph, the orange area is clearly visible below and between the eye and the mouth. The bioluminescent area consists of a colony of rod-shaped bacteria, *Vibrio fischeri,* that are common in marine environments around the world. Colonies of *Vibrio fischeri* become luminescent only when they reach a minimum population and the bacteria in the colony sense that a sufficiently large population has been reached. This biological sensing depends on a chemical equilibrium involving acylhomoserine lactone, a substance that signals to one bacterium that others are nearby (see p. 496).

Chemical reactions that involve only pure solids or pure liquids are simpler than those that occur in the gas phase or in a solution. Either no reaction occurs between the solid and liquid reactants, or a reaction occurs in which at least one reactant is completely converted into products. [If one or more of the reactants are present in excess, those reactants will be left over, but the limiting reactant

(⇐ *p. 115*) will be completely reacted away.] This happens because the concentration of a pure solid or a pure liquid does not change during a reaction, provided the temperature remains constant. If the initial concentrations of reactants are large enough to cause the reaction to occur, those same concentrations will be present throughout the reaction, and the reaction will not stop until the limiting reactant is used up.

When a reaction occurs in the gas phase or in a solution, concentrations of reactants decrease as the reaction takes place (⇐ *p. 420*). Eventually the concentrations decrease to the point at which conversion of reactants to products is no longer favored. Then the concentrations of reactants and products stop changing, but none of the concentrations has become zero. At least a tiny bit (and often a lot) of each reactant and each product is present in the reaction mixture. Because the concentrations have stopped changing, it is often relatively easy to measure them, thus providing quantitative information about *how much* product can be obtained from the reaction. It is also possible to predict how changes in temperature, pressure, and concentrations will affect the quantity of product produced. This kind of information, combined with what you learned in Chapter 12 about factors that affect the rates of chemical reactions, enables us to predict which reactions will be useful for manufacturing a broad range of substances that enhance our quality of life.

As an example of the importance of such information, consider ammonia. The United States uses approximately 36 billion pounds of liquefied ammonia per year, mostly as fertilizer to provide nitrogen needed to support growth of a broad range of crops. Therefore, ammonia is a very important factor in providing people with food. Ammonia is synthesized directly from nitrogen and hydrogen by the *Haber-Bosch process* (Section 13.8).

$$N_2(g) + 3\,H_2(g) \rightleftharpoons 2\,NH_3(g) \qquad \Delta H = -92.2\text{ kJ}$$

The chemists and chemical engineers who operate ammonia manufacturing plants do their best to obtain the maximum quantity of ammonia with the minimum input of reactants and the minimum consumption of energy resources. The German chemist Fritz Haber won the 1918 Nobel Prize for research that showed how to determine the best conditions for carrying out this reaction. The German engineer Carl Bosch received the Nobel Prize in 1931 (together with Friedrich Bergius) for his pioneering chemical engineering work that enabled large-scale ammonia synthesis to be successful.

In this chapter you will learn the same principles that Haber used. With them you will be able to make both qualitative and quantitative predictions about how much product will be formed under a given set of reaction conditions.

13.1 Characteristics of Chemical Equilibrium

When the concentrations of reactants stop decreasing and the concentrations of products stop increasing, we say that a chemical reaction has reached equilibrium. In a **chemical equilibrium,** *there are finite concentrations of reactants and products, and these concentrations remain constant.* An equilibrium reaction always results in smaller amounts of products than the theoretical yield predicts (⇐ *p. 121*), and sometimes hardly any products are produced. The concentrations of reactants and products at equilibrium provide a quantitative way of determining how successful a reaction has been. *When products predominate over reactants*, the reaction is **product-favored.** *When the equilibrium mixture consists mostly of reactants* with very little product, the reaction is **reactant-favored.**

EXERCISE 13.1 Concentrations of Pure Solids and Liquids

The introduction to this chapter states that at a given temperature the concentration of a pure solid or liquid does not depend on the quantity of substance present. Verify this assertion by calculating the concentration (in mol/L) of these solids and liquids at 20 °C. Obtain densities from Table 1.1 *(⇐ p. 8)*.

(a) Aluminum (b) Benzene (c) Water (d) Gold

Equilibrium Is Dynamic

When equilibrium is reached and concentrations of reactants and products remain constant, it appears that a chemical reaction has stopped. This is true only of the net, macroscopic reaction, however. On the nanoscale both forward and reverse reactions continue, but *the rate of the forward reaction exactly equals the rate of the reverse reaction*. To emphasize that *chemical equilibrium involves a balance between opposite reactions*, it is often referred to as a **dynamic equilibrium,** and an equilibrium reaction is usually written with a double arrow ($\rightleftharpoons$) between reactants and products.

A good example of chemical equilibrium is provided by weak acids *(⇐ p. 143)*, which ionize only partially in water.

$$\underset{\text{acetic acid}}{CH_3COOH(aq)} + \underset{\text{water}}{H_2O(\ell)} \rightleftharpoons \underset{\text{acetate ion}}{CH_3COO^-(aq)} + \underset{\text{hydronium ion}}{H_3O^+(aq)}$$

After equilibrium has been reached at room temperature, more than 90% of the acetic acid remains in molecular form (CH_3COOH) and the equilibrium concentrations of acetate ions and hydronium ions are each less than one tenth the concentration of acetic acid molecules. Nevertheless, both the forward and reverse reactions continue. Evidence to support this idea can be obtained by adding a tiny quantity of sodium acetate in which radioactive carbon-14 has been substituted into the CH_3COO^- ion to give $^{14}CH_3COO^-$. Almost immediately the radioactivity can be found in acetic acid molecules as well. This would not happen if the reaction had come to a halt, but it does happen because the reverse reaction,

$$^{14}CH_3COO^-(aq) + H_3O^+(aq) \longrightarrow {}^{14}CH_3COOH(aq) + H_2O(\ell)$$

is still taking place. To a macroscopic observer nothing seems to be happening because the reverse reaction and the forward reaction are occurring at equal rates and there is no *net* change in the concentrations of reactants or products.

The *equi* in the word *equilibrium* means "equal." It refers to equal rates of forward and reverse reactions, not to equal quantities or concentrations of the substances involved. The *librium* part of the word comes from *libra*, meaning "balance." Chemical equilibrium is an equal balance between two reaction rates.

A set of double arrows, $\rightleftharpoons$, in an equation indicates a dynamic equilibrium in which forward and reverse reactions are occurring at equal rates; it also indicates that the reaction should be thought of in terms of the concepts of chemical equilibrium.

Equilibrium Is Independent of Direction of Approach

Another important characteristic of chemical equilibrium is that, *for a specific reaction at a specific temperature, the equilibrium state will be the same, no matter what the direction of approach to equilibrium*. As an example, consider again the synthesis of ammonia from N_2 and H_2.

$$N_2(g) + 3\,H_2(g) \rightleftharpoons 2\,NH_3(g)$$

Suppose that you introduce 1.0 mol $N_2(g)$ and 3.0 mol $H_2(g)$ into an empty (evacuated), closed 1.00-L container at 472 °C. Some (but not all) of the N_2 reacts with the H_2 to form NH_3. After equilibrium is established, you would find that the concentration of H_2 has fallen from its initial value of 3.0 mol/L to an equilibrium value of 0.89 mol/L. You would also find equilibrium concentrations of 0.30 mol/L for N_2 and 1.4 mol/L for NH_3.

Now consider a second experiment at the same temperature in which you introduce 2.0 mol $NH_3(g)$ into an empty 1.00-L container at 472 °C. The 2.0 mol NH_3

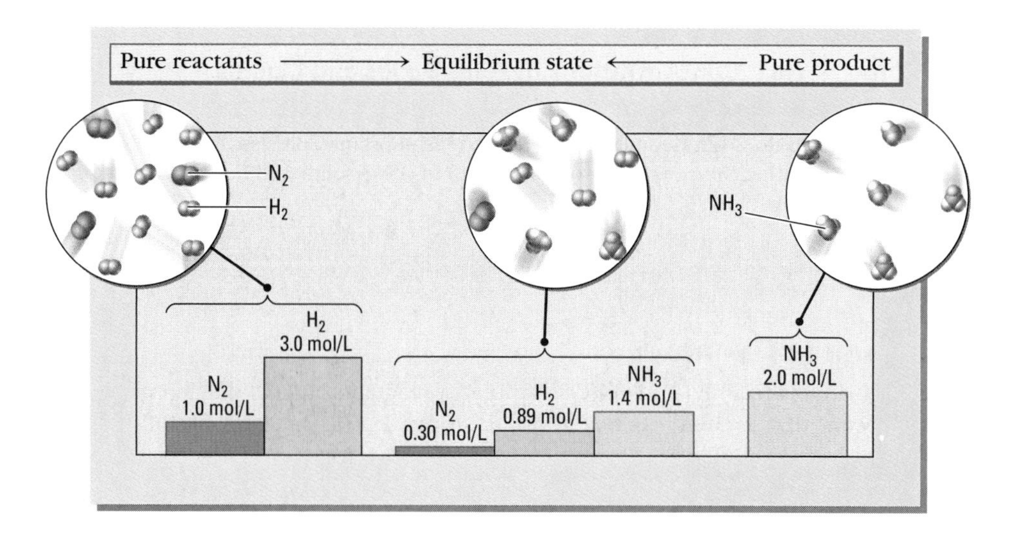

Figure 13.1 Reactants, products, and equilibrium. In the ammonia synthesis reaction, as in any equilibrium, it is possible to start with reactants or products and achieve the same equilibrium state. Here 1.0 mol N_2 reacts with 3.0 mol H_2 in a 1.0-L container to give the same equilibrium concentrations as when 2.0 mol NH_3 is introduced into the same container at the same temperature (472 °C).

consists of 2.0 mol N atoms and 6.0 mol H atoms—the same number of N atoms and H atoms contained in the 1.0 mol N_2 and 3.0 mol H_2 used in the first experiment. Because there is only NH_3 present initially, the reverse reaction occurs, producing some N_2 and H_2. Measuring the concentrations at equilibrium reveals that the concentration of NH_3 dropped from the initial 2.0 mol/L to 1.4 mol/L, and the concentrations of N_2 and H_2 built up to 0.30 mol/L and 0.89 mol/L. These equilibrium concentrations are the same as those achieved in the first experiment. Thus, *whether you start with reactants or products, the same equilibrium state is achieved*—as long as the number of atoms of each type, the volume of the container, and the temperature remain the same. This is shown schematically in Figure 13.1.

Catalysts Do Not Affect Equilibrium Concentrations

Another important characteristic of chemical equilibrium is that *if a catalyst is present, the same equilibrium state will be achieved, but more quickly*. A catalyst speeds up the forward reaction, but it also speeds up the reverse reaction. The overall effect is to produce exactly the same concentrations at equilibrium, whether or not a catalyst is in the reaction mixture. A catalyst can be used to speed up production of products in an industrial process, but it will not result in greater equilibrium concentrations of products, nor will it reduce the concentration of product present when the system reaches equilibrium.

CONCEPTUAL
EXERCISE 13.2 Recognizing an Equilibrium State

A mixture of hydrogen gas and oxygen gas is maintained at 25 °C for 1 year. On the first day of each month the mixture is sampled and the concentrations of hydrogen and oxygen measured. In every case they are found to be 0.50 mol/L H_2 and 0.50 mol/L O_2; that is, the concentrations do not change over a long period. Is this mixture at equilibrium? If you think not, how could you do an experiment to prove it?

13.2 The Equilibrium Constant

Consider again the isomerization reaction of *cis*-2-butene to *trans*-2-butene, whose rate we discussed previously (*p. 433*).

cis-2-butene (H_3C–CH=CH–CH_3) $\rightleftharpoons$ *trans*-2-butene (H_3C–CH=CH–CH_3)

At 500 K the reaction reaches an equilibrium in which the concentration of *trans*-2-butene is 1.65 times the concentration of *cis*-2-butene. This is a single-step, elementary process (⇐ *p. 433*). Both the forward and reverse reactions in this system involve only a single molecule and, therefore, are unimolecular and first-order (⇐ *p. 440*). Thus, the rate equations for forward and reverse reactions can be derived from the reaction equations.

$$\text{Rate}_{\text{forward}} = k_{\text{forward}}[\textit{cis}\text{-2-butene}] \qquad \text{Rate}_{\text{reverse}} = k_{\text{reverse}}[\textit{trans}\text{-2-butene}]$$

Suppose that we start with 0.100 mol *cis*-2-butene in a 5-L closed flask at 500 K. The *cis*-2-butene begins to react at a rate given by the forward rate equation. Initially no *trans*-2-butene is present, so the initial rate of the reverse reaction is zero. As the forward reaction proceeds, *cis*-2-butene is converted to *trans*-2-butene. The concentration of *cis*-2-butene decreases, so the forward rate decreases. As soon as some *trans*-2-butene has formed, the reverse reaction begins. As the concentration of *trans*-2-butene builds up, the reverse rate gets faster. The forward rate slows down (and the reverse rate speeds up) until the two rates are equal. At this time equilibrium has been achieved, and in the macroscopic system no further change in concentrations will be observed (Figure 13.2).

On the nanoscale, when equilibrium has been achieved, both reactions are still occurring, but the forward and reverse rates are equal. Therefore we can equate the two rates to give

$$\text{Rate}_{\text{forward}} = \text{Rate}_{\text{reverse}}$$

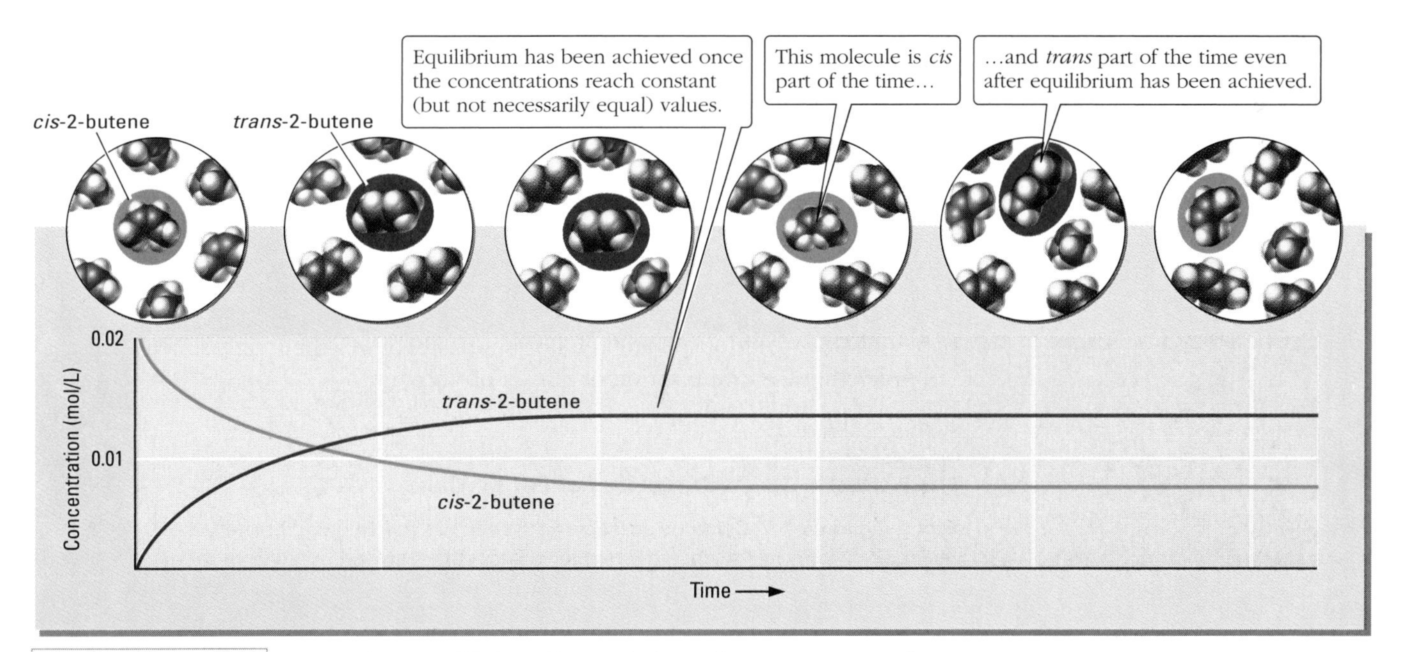

Active Figure 13.2 **Approach to equilibrium.** The graph shows the concentrations of *cis*-2-butene and *trans*-2-butene as the *cis* compound reacts to form the *trans* compound at 500 K. The nanoscale diagrams above the graph show "snapshots" of the composition of a tiny portion of the reaction mixture. The same molecule has been circled in each diagram. Visit this book's companion website at **www.cengage.com/chemistry/moore** to test your understanding of the concepts in this figure.

It is important to distinguish equilibrium concentrations, which do not change over time, from the changing concentrations of reactants and products before equilibrium is achieved. Later in this chapter we will use the notation "conc. X" to indicate the concentration of X in a system that is not at equilibrium. This distinction is important because only equilibrium concentrations can be calculated from equilibrium constants.

In Chapter 12 we noted that the rate law for an overall chemical reaction cannot be derived from the balanced equation, but must be determined experimentally by kinetic studies. This is not true of the equilibrium constant expression. When a reaction takes place by a mechanism that consists of a sequence of steps, the equilibrium constant expression can be obtained by multiplying together the rate constants for the forward reactions in all steps and then dividing by the rate constants for the reverse reactions in all steps. This process gives the same equilibrium constant expression that can be obtained from the coefficients of the balanced overall equilibrium equation.

It turns out that when concentrations are large enough, the value of the equilibrium "constant" expressed in terms of concentrations does not remain constant, even at the same temperature. This happens because of noncovalent interactions among molecules, and especially ions, that cause them to behave differently as their concentrations become larger. To deal with this behavior, true equilibrium constants must be expressed in terms of corrected concentrations that are called *activities*.

and, by substituting from the two previous rate equations,

$$k_{\text{forward}}[\textit{cis}\text{-2-butene}] = k_{\text{reverse}}[\textit{trans}\text{-2-butene}]$$

In this and subsequent equations in this chapter, square brackets designate *equilibrium* concentrations. The equation can be rearranged so that both rate constants are on one side and both equilibrium concentrations are on the other side.

$$\frac{k_{\text{forward}}}{k_{\text{reverse}}} = \frac{[\textit{trans}\text{-2-butene}]}{[\textit{cis}\text{-2-butene}]}$$

This expression shows that the ratio of equilibrium concentrations is equal to a ratio of rate constants. Because a ratio of two constants is also a constant, the ratio of equilibrium concentrations must also be constant. We call this ratio K_c, where the capital letter K is used to distinguish it from the rate constants, k_{forward} and k_{reverse}, and the subscript c indicates that it is a ratio of equilibrium *concentrations*. At 500 K the experimental value of K_c is 1.65 for the *cis* $\rightleftharpoons$ *trans* butene reaction.

$$K_c = \frac{k_{\text{forward}}}{k_{\text{reverse}}} = \frac{[\textit{trans}\text{-2-butene}]}{[\textit{cis}\text{-2-butene}]} = 1.65 \qquad \text{(at 500 K)}$$

Because the values of the rate constants vary with temperature *(⇐ p. 438)*, the value of K_c also varies with temperature. For the butene isomerization reaction, K_c is 1.47 at 600 K and 1.36 at 700 K.

*An **equilibrium constant** is a quotient of equilibrium concentrations of product and reactant substances that has a constant value for a given reaction at a given temperature.* It is given the symbol K_c. On the next page we show how to derive a mathematical expression for the equilibrium constant directly from the chemical equation for any equilibrium process. The mathematical expression is called an **equilibrium constant expression.**

Equilibrium constants can be used to answer three important questions about a reaction:

- When equilibrium has been achieved, do products predominate over reactants?
- Given initial concentrations of reactants and products, in which direction will the reaction go to achieve equilibrium?
- What concentrations of reactants and products are present at equilibrium?

If a reaction moves quickly to equilibrium, you can use equilibrium constants to determine the composition soon after reactants are mixed. Equilibrium constants are less valuable for slow reactions. Until equilibrium has been reached, only kinetics is capable of predicting the composition of a reaction mixture.

CONCEPTUAL **EXERCISE** **13.3 Properties of Equilibrium**

After a mixture of *cis*-2-butene and *trans*-2-butene has reached equilibrium at 600 K, where $K_c = 1.47$, half of the *cis*-2-butene is suddenly removed. Answer these questions:

(a) Is the new mixture at equilibrium? Explain why or why not.
(b) In the new mixture, which rate is faster, *cis* → *trans* or *trans* → *cis*? Or are both rates the same?
(c) In an equilibrium mixture, which concentration is larger, *cis*-2-butene or *trans*-2-butene?
(d) If the concentration of *cis*-2-butene at equilibrium is 0.10 mol/L, what will be the concentration of *trans*-2-butene?

Writing Equilibrium Constant Expressions

The equilibrium constant expression for any reaction has concentrations of products in the numerator and concentrations of reactants in the denominator. Each concentration is raised to the power of its stoichiometric coefficient in the balanced equation. *The only concentrations that appear in an equilibrium constant expression are those of gases and of solutes in dilute solutions*, because these are the only concentrations that can change as a reaction occurs. Concentrations of pure solids, pure liquids, and solvents in dilute solutions ***do not*** appear in equilibrium constant expressions.

To illustrate how this works, consider the general equilibrium reaction

$$a\,\mathrm{A} + b\,\mathrm{B} \rightleftharpoons c\,\mathrm{C} + d\,\mathrm{D} \qquad [13.1]$$

By convention we write the equilibrium constant expression for this reaction as

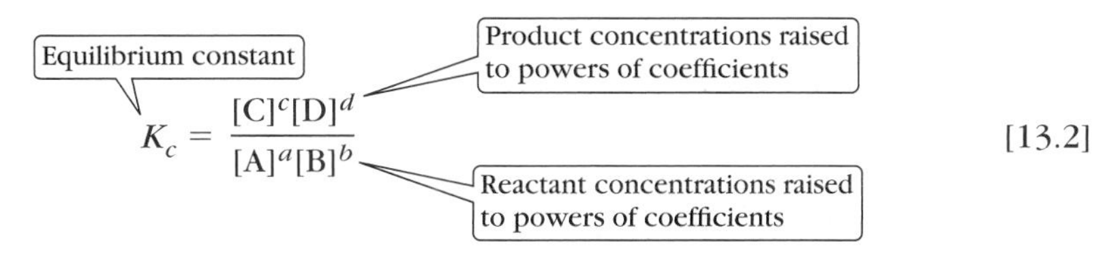

$$K_c = \frac{[\mathrm{C}]^c[\mathrm{D}]^d}{[\mathrm{A}]^a[\mathrm{B}]^b} \qquad [13.2]$$

Let's apply these ideas to the combination of nitrogen and oxygen gases to form nitrogen monoxide.

$$N_2(g) + O_2(g) \rightleftharpoons 2\,NO(g)$$

The reaction of nitrogen with oxygen occurs in automobile engines and other high-temperature combustion processes where air is present.

Because all substances in the reaction are gaseous, all concentrations appear in the equilibrium constant expression. The concentration of the product NO(g) is in the numerator and is squared because of the coefficient 2 in the balanced chemical equation. The concentrations of the reactants $N_2(g)$ and $O_2(g)$ are in the denominator. Each is raised to the first power because each has a coefficient of 1.

$$\text{Equilibrium constant} = K_c = \frac{[NO]^2}{[N_2][O_2]}$$

Equilibria Involving Pure Liquids and Solids

As another example, consider the combustion of solid yellow sulfur, which consists of S_8 molecules. The combustion reaction produces sulfur dioxide gas.

$$\tfrac{1}{8}\,S_8(s) + O_2(g) \rightleftharpoons SO_2(g)$$

This reaction occurs whenever a material that contains sulfur burns in air, and it is responsible for a good deal of sulfur dioxide air pollution. It is also the first reaction in a sequence by which sulfur is converted to sulfuric acid, the number-one industrial chemical in the world.

Because sulfur is a solid, the number of molecules per unit volume is fixed by the density of sulfur at any given temperature (see Exercise 13.1, *p. 463*). Therefore, the sulfur concentration is not changed either by reaction or by addition or removal of solid sulfur. It is an experimental fact that, as long as some solid sulfur is present, the equilibrium concentrations of O_2 and SO_2 are not affected by changes in the amount of sulfur. Therefore, the equilibrium constant expression for this reaction is properly written as

$$K_c = \frac{[SO_2(g)]}{[O_2(g)]} \qquad \text{At 25 °C, } K_c = 4.2 \times 10^{52}$$

It is very important to remember that concentrations of pure solids, pure liquids, and solvents for dilute solutions do not appear in the equilibrium constant expression. Any substance designated as (s) or (ℓ) in the equilibrium equation does not appear in the equilibrium constant expression.

Equilibria in Dilute Solutions

Consider an aqueous solution of the weak base ammonia, which contains a small concentration of ammonium ions and hydroxide ions because ammonia reacts with water.

$$NH_3(aq) + H_2O(\ell) \rightleftharpoons NH_4^+(aq) + OH^-(aq)$$

If the concentration of ammonia molecules (and consequently of ammonium ions and hydroxide ions) is small, the number of water molecules per unit volume remains essentially the same as in pure water. Because the molar concentration of water is effectively constant for reactions involving dilute solutions, the concentration of water should not be included in the equilibrium constant expression. Thus, for this reaction we write

$$K_c = \frac{[NH_4^+][OH^-]}{[NH_3]}$$

and the concentration of water is not included in the denominator. At 25 °C, $K_c = 1.8 \times 10^{-5}$ for reaction of ammonia with water.

Notice that no units were included in the equilibrium constant for the ammonia ionization reaction. There are two concentrations in the numerator and only one in the denominator, which ought to give units of mol/L. However, if we always express concentrations in mol/L, the units of the equilibrium constant can be figured out from the equilibrium constant expression. Therefore it is customary to omit the units from the equilibrium constant value, even if the equilibrium constant should have units. We follow that custom in this book.

PROBLEM-SOLVING EXAMPLE 13.1 Writing Equilibrium Constant Expressions

Write an equilibrium constant expression for each chemical equation.

(a) $4\,NO_2(g) + O_2(g) + 2\,H_2O(g) \rightleftharpoons 4\,HNO_3(g)$
(b) $BaSO_4(s) \rightleftharpoons Ba^{2+}(aq) + SO_4^{2-}(aq)$
(c) $6\,NH_3(aq) + Ni^{2+}(aq) \rightleftharpoons Ni(NH_3)_6^{2+}(aq)$
(d) $2\,BaO_2(s) \rightleftharpoons 2\,BaO(s) + O_2(g)$
(e) $NH_3(g) \rightleftharpoons NH_3(\ell)$

Answer

(a) $K_c = \dfrac{[HNO_3]^4}{[NO_2]^4[O_2][H_2O]^2}$

(b) $K_c = [Ba^{2+}][SO_4^{2-}]$

(c) $K_c = \dfrac{[Ni(NH_3)_6^{2+}]}{[NH_3]^6[Ni^{2+}]}$

(d) $K_c = [O_2]$

(e) $K_c = \dfrac{1}{[NH_3]}$

Strategy and Explanation Concentrations of products go in the numerator of the fraction, and concentrations of reactants go in the denominator. Each concentration is raised to the power of the stoichiometric coefficient of the species. In part (b), one species, $BaSO_4(s)$, is a pure solid and does not appear in the expression. In part (d), two species, $BaO_2(s)$ and $BaO(s)$, are solids and do not appear. In part (e), one species, $NH_3(\ell)$, is a pure liquid and does not appear.

PROBLEM-SOLVING PRACTICE 13.1

Write an equilibrium constant expression for each equation.

(a) $CaCO_3(s) \rightleftharpoons CaO(s) + CO_2(g)$
(b) $HCl(g) + LiH(s) \rightleftharpoons H_2(g) + LiCl(s)$
(c) $CH_4(g) + H_2O(g) \rightleftharpoons CO(g) + 3\,H_2(g)$
(d) $CN^-(aq) + H_2O(\ell) \rightleftharpoons HCN(aq) + OH^-(aq)$

Equilibrium Constant Expressions for Related Reactions

Consider the equilibrium involving nitrogen, hydrogen, and ammonia.

$$N_2(g) + 3\,H_2(g) \rightleftharpoons 2\,NH_3(g) \qquad K_{c_1} = 3.5 \times 10^8 \quad \text{(at 25 °C)}$$

We could also write the equation so that 1 mol NH_3 is produced.

$$\tfrac{1}{2}\,N_2(g) + \tfrac{3}{2}\,H_2(g) \rightleftharpoons NH_3(g) \qquad K_{c_2} = ?$$

Coefficients half as big

Is the value of the equilibrium constant, K_{c_2}, for the second equation the same as the value of the equilibrium constant, K_{c_1}, for the first equation? To see the relation between K_{c_1} and K_{c_2}, write the equilibrium constant expression for each balanced equation.

Concentrations raised to powers half as big

$$K_{c_1} = \frac{[NH_3]^2}{[N_2][H_2]^3} = 3.5 \times 10^8 \quad \text{and} \quad K_{c_2} = \frac{[NH_3]}{[N_2]^{1/2}[H_2]^{3/2}} = ?$$

This makes it clear that K_{c_1} is the square of K_{c_2}; that is, $K_{c_1} = (K_{c_2})^2$. Therefore, the answer to our question is

$$K_{c_2} = (K_{c_1})^{1/2} = (3.5 \times 10^8)^{1/2} = 1.9 \times 10^4$$

Whenever the stoichiometric coefficients of a balanced equation are multiplied by some factor, the equilibrium constant for the new equation (K_{c_2} in this case) is the old equilibrium constant (K_{c_1}) raised to the power of the multiplication factor.

What is the value of K_{c_3}, the equilibrium constant for the decomposition of ammonia to the elements, which is the reverse of the first equation?

$$2\,NH_3(g) \rightleftharpoons N_2(g) + 3\,H_2(g) \qquad K_{c_3} = ? = \frac{[N_2][H_2]^3}{[NH_3]^2}$$

Concentration of NH_3 is in denominator

K_{c_3} is the reciprocal of K_{c_1}; that is, $K_{c_3} = 1/K_{c_1} = 1/(3.5 \times 10^8) = 2.9 \times 10^{-9}$. ***The equilibrium constant for a reaction and that for its reverse are the reciprocals of one another.*** If a reaction has a very large equilibrium constant, the reverse reaction will have a very small one. That is, if a reaction is strongly product-favored, then its reverse is strongly reactant-favored. In the case of the production of ammonia from its elements at room temperature, the forward reaction has a large equilibrium constant (3.5×10^8). As expected, the reverse reaction, decomposition of ammonia to its elements, has a small equilibrium constant (2.9×10^{-9}).

EXERCISE 13.4 Manipulating Equilibrium Constants

The balanced equation for conversion of oxygen to ozone has a very small value of K_c.

$$3\,O_2(g) \rightleftharpoons 2\,O_3(g) \qquad K_c = 6.25 \times 10^{-58}$$

(a) What is the value of K_c if the equation is written as

$$\tfrac{3}{2}\,O_2(g) \rightleftharpoons O_3(g)$$

(b) What is the value of K_c for the conversion of ozone to oxygen?

$$2\,O_3(g) \rightleftharpoons 3\,O_2(g)$$

Equilibrium Constant for a Reaction That Combines Two or More Other Reactions

If two chemical equations can be combined to give a third, the equilibrium constant for the combined reaction can be obtained from the equilibrium constants for the two original reactions. For example, air pollution is produced when nitrogen monoxide forms from nitrogen and oxygen and then combines with additional oxygen to form nitrogen dioxide.

(1) $N_2(g) + O_2(g) \rightleftharpoons 2\ NO(g)$ $$K_{c_1} = \frac{[NO]^2}{[N_2][O_2]}$$

(2) $2\ NO(g) + O_2(g) \rightleftharpoons 2\ NO_2(g)$ $$K_{c_2} = \frac{[NO_2]^2}{[NO]^2[O_2]}$$

The sum of these two equations is

Sum of Equations 1 and 2

(3) $N_2(g) + 2\ O_2(g) \rightleftharpoons 2\ NO_2(g)$

Product of equilibrium constants K_{c_1} and K_{c_2}

$$K_{c_3} = \frac{[NO_2]^2}{[N_2][O_2]^2} = \frac{[NO]^2}{[N_2][O_2]} \times \frac{[NO_2]^2}{[NO]^2[O_2]} = K_{c_1} \times K_{c_2}$$

Because K_{c_1} and K_{c_2} were known experimentally and could be used to calculate K_{c_3}, there is no need to measure K_{c_3} experimentally.

That is, ***if two chemical equations can be summed to give a third, the equilibrium constant for the overall equation equals the product of the two equilibrium constants for the equations that were summed.*** This is a powerful tool for obtaining equilibrium constants without having to measure them experimentally for each individual reaction.

PROBLEM-SOLVING EXAMPLE 13.2 Manipulating Equilibrium Constants

Given these equilibrium reactions and constants,

(1) $\frac{1}{4}\ S_8(s) + 2\ O_2(g) \rightleftharpoons 2\ SO_2(g)$ $K_{c_1} = 1.86 \times 10^{105}$

(2) $\frac{1}{8}\ S_8(s) + \frac{3}{2}\ O_2(g) \rightleftharpoons SO_3(g)$ $K_{c_2} = 1.77 \times 10^{53}$

calculate the equilibrium constant for this reaction, which is important in the formation of acid rain air pollution.

$2\ SO_2(g) + O_2(g) \rightleftharpoons 2\ SO_3(g)$ $$K_{c_3} = \frac{[SO_3]^2}{[SO_2]^2[O_2]}$$

Answer 16.8

Strategy and Explanation Compare the given equations with the target equation to see how each given equation needs to be manipulated. The target equation has SO_2 on the left, so Equation 1 needs to be reversed and we need to take the reciprocal of the equilibrium constant. The target equation has SO_3 on the right. Equation 2 need not be reversed, but each coefficient must be multiplied by 2, which means squaring K_{c_2}. Once this has been done, the two new equations sum to the target equation, and we can multiply their equilibrium constants.

(1)′ $2\ SO_2(g) \rightleftharpoons \frac{1}{4}\ S_8(s) + 2\ O_2(g)$ $$K'_{c_1} = \frac{1}{1.86 \times 10^{105}} = 5.38 \times 10^{-106}$$

(2)′ $\frac{1}{4}\ S_8(s) + 3\ O_2(g) \rightleftharpoons 2\ SO_3(g)$ $K'_{c_2} = (1.77 \times 10^{53})^2 = 3.13 \times 10^{106}$

$2\ SO_2(g) + O_2(g) \rightleftharpoons 2\ SO_3(g)$ $K_{c_3} = K'_{c_1} \times K'_{c_2} = 16.8$

☑ **Reasonable Answer Check** For Equation 1 the equilibrium constant was quite large, so for the reverse reaction it should be quite small, which it is. Also check the order of magnitude of the answer by checking the powers of 10. In the square the power of 10 should be doubled, which it is. And the sum of the powers of 10 for the two equilibrium constants that were multiplied should be close to the power of 10 in the answer, which it is.

PROBLEM-SOLVING PRACTICE 13.2

When carbon dioxide dissolves in water it reacts to produce carbonic acid, $H_2CO_3(aq)$, which can ionize in two steps.

$$H_2CO_3(aq) + H_2O(aq) \rightleftharpoons HCO_3^-(aq) + H_3O^+(aq) \qquad K_{c_1} = 4.2 \times 10^{-7}$$

$$HCO_3^-(aq) + H_2O(aq) \rightleftharpoons CO_3^{2-}(aq) + H_3O^+(aq) \qquad K_{c_2} = 4.8 \times 10^{-11}$$

Calculate the equilibrium constant for the reaction

$$H_2CO_3(aq) + 2\,H_2O(aq) \rightleftharpoons CO_3^{2-}(aq) + 2\,H_3O^+(aq)$$

Equilibrium Constants in Terms of Pressure

In a constant-volume system, when the concentration of a gas changes, the partial pressure (p. 361) of the gas also changes. This follows from the ideal gas equation

$$PV = nRT$$

Solving for the partial pressure P_A of a gaseous substance, A,

[A] = n_A/V, the number of moles of A per unit volume

$$P_A = \frac{n_A}{V}RT = [A]RT \qquad [13.3]$$

Equation 13.3 allows us to express the equilibrium constant for the general reaction in Equation 13.1 (p. 467) in a form similar to Equation 13.2, but in terms of partial pressures as

$$K_P = \frac{P_C^c \times P_D^d}{P_A^a \times P_B^b} \qquad [13.4]$$

Product pressures raised to powers of coefficients

Reactant pressures raised to powers of coefficients

The subscript on K_P indicates that the equilibrium constant has been expressed in terms of partial pressures. For some gas phase equilibria $K_c = K_P$; for many others it does not. Therefore it is useful to be able to relate one type of equilibrium constant to the other. This can be done by combining Equations 13.2, 13.3, and 13.4 to give

Because K_c is related to K_P for the same gas phase reaction, either can be used to calculate the composition of an equilibrium mixture. Most examples in this chapter involve K_c, but the same rules apply to solving problems with K_P.

$\Delta n = c + d - a - b$ is the number of moles of gaseous products minus the number of moles of gaseous reactants

$$K_P = \frac{P_C^c \times P_D^d}{P_A^a \times P_B^b} = \frac{\{[C]RT\}^c\{[D]RT\}^d}{\{[A]RT\}^a\{[B]RT\}^b} = \frac{[C]^c[D]^d}{[A]^a[B]^b}(RT)^{c+d-a-b} = K_c(RT)^{\Delta n} \qquad [13.3]$$

As an example of this relation, consider the equilibrium

$$2\,NOCl(g) \rightleftharpoons 2\,NO(g) + Cl_2(g) \qquad K_c = 4.02 \times 10^{-2}\ \text{mol/L at 298 K}$$

For this reaction $\Delta n = 2 + 1 - 2 = 3 - 2 = 1$, because there are three moles of gas phase product molecules and only two moles of gas phase reactants. Therefore, for this reaction

$$K_P = K_c \times (RT)^1$$
$$= (4.02 \times 10^{-2}\ \text{mol/L}) \times (0.0821\ \text{L atm K}^{-1}\ \text{mol}^{-1}) \times (298\ \text{K}) = 0.984\ \text{atm}$$

EXERCISE 13.5 Relating K_c and K_P

For each of these reactions, calculate K_P from K_c.

(a) $N_2(g) + 3\,H_2(g) \rightleftharpoons 2\,NH_3(g)$ $\quad K_c = 3.5 \times 10^8$ at 25 °C

(b) $2\,H_2(g) + O_2(g) \rightleftharpoons 2\,H_2O(g)$ $\quad K_c = 3.2 \times 10^{81}$ at 25 °C

(c) $N_2(g) + O_2(g) \rightleftharpoons 2\,NO(g)$ $\quad K_c = 1.7 \times 10^{-3}$ at 2300 K

(d) $2\,NO_2(g) \rightleftharpoons N_2O_4(g)$ $\quad K_c = 1.7 \times 10^2$ at 25 °C

13.3 Determining Equilibrium Constants

To determine the value of an equilibrium constant it is necessary to know all of the equilibrium concentrations that appear in the equilibrium constant expression. This is most commonly done by allowing a system to reach equilibrium and then measuring the **equilibrium concentration** of one or more of the reactants or products. Algebra and stoichiometry are then used to obtain K_c.

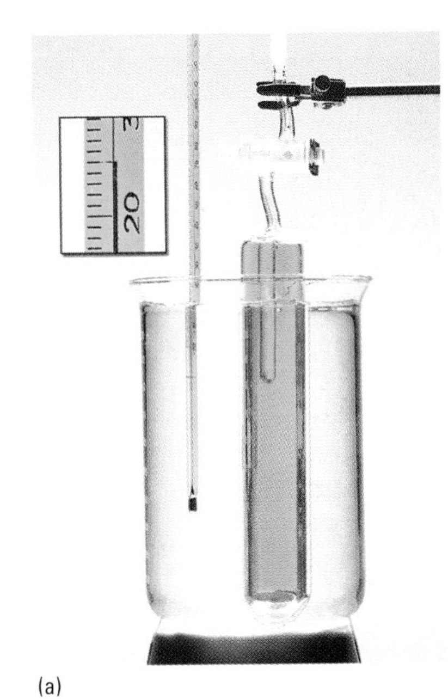

(a)

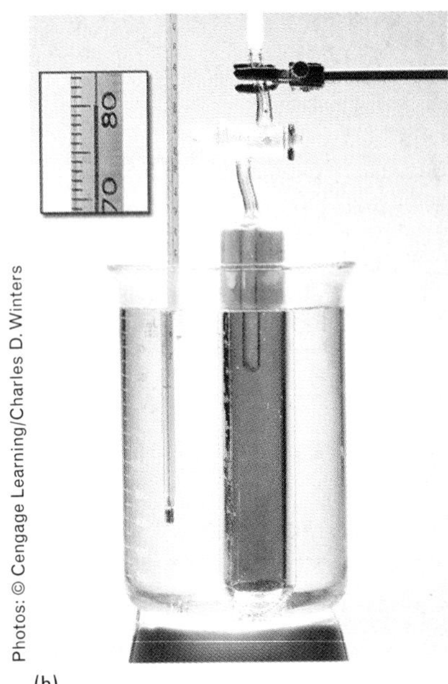

(b)

Figure 13.3 Formation of NO_2 from N_2O_4. Sealed glass tubes containing an equilibrium mixture of dinitrogen tetraoxide (colorless) and nitrogen dioxide (red-brown) are shown immersed in water (a) at room temperature and (b) at 80 °C. Notice the much darker color at the higher temperature, indicating that some dinitrogen tetraoxide has reacted to form nitrogen dioxide. The intensity of color can be used to measure the concentration of nitrogen dioxide.

Reaction Tables, Stoichiometry, and Equilibrium Concentrations

A systematic approach to calculations involving equilibrium constants involves making a table that shows initial conditions, changes that take place when a reaction occurs, and final (equilibrium) conditions. As an example, consider the colorless gas dinitrogen tetraoxide, $N_2O_4(g)$. When heated it dissociates to form red-brown $NO_2(g)$ according to the equation

$$N_2O_4(g) \rightleftharpoons 2\ NO_2(g)$$

Suppose that 2.00 mol $N_2O_4(g)$ is placed into an empty 5.00-L flask and heated to 407 K. Almost immediately a dark red-brown color appears, indicating that much of the colorless gas has been transformed into NO_2 (Figure 13.3). By measuring the intensity of color, it can be determined that the concentration of NO_2 at equilibrium is 0.525 mol/L. To use this information to calculate the equilibrium constant, follow these steps:

1. ***Write the balanced equation for the equilibrium reaction. From it derive the equilibrium constant expression.*** The balanced equation and equilibrium constant expression are

$$N_2O_4(g) \rightleftharpoons 2\ NO_2(g) \qquad K_c = \frac{[NO_2]^2}{[N_2O_4]}$$

2. ***Set up a table containing initial concentration, change in concentration, and equilibrium concentration for each substance included in the equilibrium constant expression. Enter all known information into this reaction table.*** In this case the number of moles of $N_2O_4(g)$ and the volume of the flask were given, so we first calculate the initial concentration of $N_2O_4(g)$ as (conc. N_2O_4) $= \frac{2.00\text{ mol}}{5.00\text{ L}} = 0.400$ mol/L. Because the flask initially contained no NO_2, the initial concentration of NO_2 is zero. After the reaction took place and equilibrium was reached, the equilibrium concentration of NO_2 was measured as 0.525 mol/L. The reaction table looks like this:

	$N_2O_4(g)$	$\rightleftharpoons$	$2\ NO_2(g)$
Initial concentration (mol/L)	0.400		0
Change as reaction occurs (mol/L)	______		______
Equilibrium concentration (mol/L)	______		0.525

3. ***Use x to represent the change in concentration of one substance. Use the stoichiometric coefficients in the balanced equilibrium equation to calculate the other changes in terms of x.*** When the reaction proceeds from left to right, the concentrations of reactants decrease. Therefore the change in concentration of a reactant is negative. The concentrations of products increase, so the change in concentration of a product is positive. Usually it is best to begin with the reactant or product column that contains the most information. In this case that is the NO_2 column, where both initial and equilibrium concentrations are known. Therefore, we let x represent the unknown change in concentration of NO_2.

	$N_2O_4(g)$	$\rightleftharpoons$	$2\ NO_2(g)$
Initial concentration (mol/L)	0.400		0
Change as reaction occurs (mol/L)	______		x
Equilibrium concentration (mol/L)	______		0.525

A gas or a solution is colored if it absorbs visible light. The greater the concentration of the colored substance is, the more light is absorbed. An instrument known as a spectrophotometer can measure absorbance of light, thereby determining the concentration.

Next, we use the mole ratio from the balanced equation to find the change in concentration of N_2O_4 in terms of x.

$$\Delta(\text{conc. } N_2O_4) = \frac{x \text{ mol } NO_2 \text{ formed}}{L} \times \frac{1 \text{ mol } N_2O_4 \text{ reacted}}{2 \text{ mol } NO_2 \text{ formed}}$$
$$= \tfrac{1}{2}x \text{ mol } N_2O_4 \text{ reacted/L}$$

For every 1 mol N_2O_4 that decomposes, 2 mol NO_2 forms. Therefore there is a $\frac{1}{2}$:1 mol ratio of N_2O_4:NO_2.

The sign of the change in concentration of N_2O_4 is *negative*, because the concentration of N_2O_4 *decreases*. The table becomes

	$N_2O_4(g)$	$\rightleftharpoons$	$2\ NO_2(g)$
Initial concentration (mol/L)	0.400		0
Change as reaction occurs (mol/L)	$-\frac{1}{2}x$		x
Equilibrium concentration (mol/L)	______		0.525

4. ***From initial concentrations and the changes in concentrations, calculate the equilibrium concentrations in terms of* x *and enter them in the table.*** The concentration of N_2O_4 at equilibrium, $[N_2O_4]$, is the sum of the initial 0.40 mol/L of N_2O_4 and the change due to reaction $-\frac{1}{2}x$ mol/L; that is, $[N_2O_4] = (0.40 - \frac{1}{2}x)$ mol/L. Similarly, the equilibrium concentration of NO_2 (which is already known to be 0.525 mol/L) is $0 + x$, and the table becomes

	$N_2O_4(g)$	$\rightleftharpoons$	$2\ NO_2(g)$
*I*nitial concentration (mol/L)	0.400		0
*C*hange as reaction occurs (mol/L)	$-\frac{1}{2}x$		x
*E*quilibrium concentration (mol/L)	$0.400 - \frac{1}{2}x$		$0.525 = 0 + x$

Tables like this one are often called ICE tables from the initial letters of the labels on the rows: *I*nitial, *C*hange, and *E*quilibrium.

5. ***Use the simplest possible equation to solve for* x*. Then use* x *to calculate the unknown you were asked to find.*** (Usually the unknown is K_c or a concentration.) In this case the simplest equation to solve for x is the last entry in the table, $0.525 = 0 + x$, and it is easy to see that $x = 0.525$. Calculate $[N_2O_4] = (0.400 - \frac{1}{2}x)$ mol/L $= (0.400 - \frac{1}{2} \times 0.525)$ mol/L $= 0.138$ mol/L. The problem stated that $[NO_2] = 0.525$ mol/L, so K_c is given by

$$K_c = \frac{[NO_2]^2}{[N_2O_4]} = \frac{(0.525 \text{ mol/L})^2}{(0.138 \text{ mol/L})} = 2.00 \qquad \text{(at 407 K)}$$

6. ***Check your answer to make certain it is reasonable.*** In this case the equilibrium concentration of product is larger than the concentration of reactant. Since the products are in the numerator of the equilibrium constant expression, we expect a value greater than 1, which is what we calculated.

PROBLEM-SOLVING EXAMPLE 13.3 Determining an Equilibrium Constant Value

Consider the gas phase reaction

$$H_2(g) + I_2(g) \rightleftharpoons 2\ HI(g)$$

Note that products form at the expense of reactants.

Suppose that a flask containing H_2 and I_2 has been heated to 425 °C and the initial concentrations of H_2 and I_2 were each 0.0175 mol/L. With time, the concentrations of H_2 and I_2 decline and the concentration of HI increases. At equilibrium [HI] = 0.0276 mol/L. Use this experimental information to calculate the equilibrium constant.

Answer $K_c = 56$

Strategy and Explanation Use the information given and follow the six steps. In the table below, each entry has been color-coded to match the numbered steps.

1. ***Write the balanced equation and equilibrium constant expression.***

$$H_2(g) + I_2(g) \rightleftharpoons 2\ HI(g) \qquad K_c = \frac{[HI]^2}{[H_2][I_2]}$$

2. ***Construct a reaction (ICE) table (see below) and enter known information.***
3. ***Represent changes in concentration in terms of* x.**

The best choice is to enter x in the third column, because both initial and equilibrium concentrations of HI are known, and this provides a simple way to calculate x. Next, derive the rest of the concentration changes in terms of x. If the concentration of HI increases by a given quantity, the mole ratios say that the concentrations of H_2 and I_2 must decrease only half as much:

$$\frac{x\ \text{mol HI produced}}{\text{L}} \times \frac{1\ \text{mol } H_2\ \text{consumed}}{2\ \text{mol HI produced}} = \frac{1}{2}x\ \text{mol/L } H_2\ \text{consumed}$$

Because the coefficients of H_2 and I_2 are equal, each of their concentrations decreases by $\frac{1}{2}x$ mol/L. The entries in the table for change in concentration of H_2 and I_2 are negative, because their concentrations decrease.

4. ***Calculate equilibrium concentrations and enter them in the table.***

Color-coded entries in the table in this example correspond with color-coded steps in the explanation section above.

	$H_2(g)$	+	$I_2(g)$	$\rightleftharpoons$	$2\ HI(g)$
Initial concentration (mol/L)	0.0175		0.0175		0
Change as reaction occurs (mol/L)	$-\frac{1}{2}x$		$-\frac{1}{2}x$		x
Equilibrium concentration (mol/L)	$0.0175 - \frac{1}{2}x$		$0.0175 - \frac{1}{2}x$		$0.0276 = 0 + x$

5. ***Solve the simplest equation for* x.** The last row and column in the table contains $0.0276 = 0 + x$, which gives $x = 0.0276$. Substitute this value into the other two equations in the last row of the table to get the equilibrium concentrations, and substitute them into the equilibrium constant expression:

$$K_c = \frac{[HI]^2}{[H_2][I_2]} = \frac{(0.0276)^2}{(0.0175 - [\frac{1}{2} \times 0.0276])(0.0175 - [\frac{1}{2} \times 0.0276])}$$

$$= \frac{(0.0276)^2}{(0.0037)(0.0037)} = 56 \qquad \text{(at 424 °C)}$$

☑ **Reasonable Answer Check** The equilibrium constant is larger than 1, so there should be more products than reactants when equilibrium is reached. The equilibrium concentration of the product (0.0276 mol/L) is larger than those of the reactants (0.0037 mol/L each).

PROBLEM-SOLVING PRACTICE 13.3

Saying that 2.96% of the acetic acid molecules have ionized means that at equilibrium the concentration of acetate ions is 2.96/100 = 0.0296 times the initial concentration of acetic acid molecules.

Measuring the conductivity of an aqueous solution in which 0.0200 mol CH_3COOH has been dissolved in 1.00 L of solution shows that 2.96% of the acetic acid molecules have ionized to CH_3COO^- ions and H_3O^+ ions. Calculate the equilibrium constant for ionization of acetic acid and compare your result with the value given in Table 13.1.

Experimentally determined equilibrium constants for a few reactions are given in Table 13.1. These reactions occur to widely differing extents, as shown by the wide range of K_c values.

Table 13.1 Selected Equilibrium Constants at 25 °C

Reaction	K_c	K_P
Gas phase reactions		
$\frac{1}{8}S_8(s) + O_2(g) \rightleftharpoons SO_2(g)$	4.2×10^{52}	4.2×10^{52}
$2\ H_2(g) + O_2(g) \rightleftharpoons 2\ H_2O(g)$	3.2×10^{81}	1.3×10^{80}
$N_2(g) + 3\ H_2(g) \rightleftharpoons 2\ NH_3(g)$	3.5×10^{8}	5.8×10^{5}
$N_2(g) + O_2(g) \rightleftharpoons 2\ NO(g)$	4.5×10^{-31}	4.5×10^{-31}
	1.7×10^{-3} (at 2300 K)	
$H_2(g) + I_2(g) \rightleftharpoons 2\ HI(g)$	2.5×10^{1}	2.5×10^{1}
$2\ NO_2(g) \rightleftharpoons N_2O_4(g)$	1.7×10^{2}	7.0
$CH_4(g) + H_2O(g) \rightleftharpoons CO(g) + 3\ H_2(g)$	2.0×10^{-28}	1.25×10^{-25}
cis-2-butene(g) $\rightleftharpoons$ *trans*-2-butene(g)	3.2	3.2
Weak acids and bases		
Formic acid		
$HCOOH(aq) + H_2O(\ell) \rightleftharpoons H_3O^+(aq) + HCOO^-(aq)$	1.8×10^{-4}	—
Acetic acid		
$CH_3COOH(aq) + H_2O(\ell) \rightleftharpoons H_3O^+(aq) + CH_3COO^-(aq)$	1.8×10^{-5}	—
Carbonic acid		
$H_2CO_3(aq) + H_2O(\ell) \rightleftharpoons H_3O^+(aq) + HCO_3^-(aq)$	4.2×10^{-7}	—
Ammonia (weak base)		
$NH_3(aq) + H_2O(\ell) \rightleftharpoons NH_4^+(aq) + OH^-(aq)$	1.8×10^{-5}	—
Very slightly soluble solids		
$CaCO_3(s) \rightleftharpoons Ca^{2+}(aq) + CO_3^{2-}(aq)$	3.8×10^{-9}	—
$AgCl(s) \rightleftharpoons Ag^+(aq) + Cl^-(aq)$	1.8×10^{-10}	—
$AgI(s) \rightleftharpoons Ag^+(aq) + I^-(aq)$	1.5×10^{-16}	—

13.4 The Meaning of the Equilibrium Constant

The numerical size of the equilibrium constant tells how far a reaction has proceeded by the time equilibrium has been achieved. In addition, it can be used to calculate how much product will be present at equilibrium. There are three important cases to consider.

Case 1. $K_c \gg 1$: *Reaction is strongly product-favored; equilibrium concentrations of products are much greater than equilibrium concentrations of reactants.*

The symbol $\gg$ means "much greater than."

A large value of K_c *means that reactants have been converted almost entirely to products when equilibrium has been achieved.* That is, the products are strongly favored over the reactants. An example is the reaction of NO(g) with $O_3(g)$, which is one way that ozone is destroyed in the stratosphere.

$$NO(g) + O_3(g) \rightleftharpoons NO_2(g) + O_2(g) \qquad K_c = \frac{[NO_2][O_2]}{[NO][O_3]} = 6 \times 10^{34} \qquad \text{(at 25 °C)}$$

The very large value of K_c tells us that if 1 mol each of NO and O_3 are mixed in a flask at 25 °C and allowed to come to equilibrium, $[NO_2][O_2] \gg [NO][O_3]$. Virtually none of the reactants will remain, and essentially only NO_2 and O_2 will be found in the flask. For practical purposes, this reaction goes to completion, and it would not be necessary to use the equilibrium constant to calculate the quantities of products that would be

The reaction goes to essentially completion when equilibrium has been reached. This could take a long time if the reaction is slow.

obtained. The simpler methods developed in Chapter 4 (⇐ *p. 110*) would work just fine in this case.

The symbol << means "much less than."

Case 2. K_c << 1: *Reaction is strongly reactant-favored; equilibrium concentrations of reactants are greater than equilibrium concentrations of products.*

Conversely, *an extremely small* K_c *means that when equilibrium has been achieved, very little of the reactants have been transformed into products*. The reactants are favored over the products at equilibrium. An example is

$$3\ O_2(g) \rightleftharpoons 2\ O_3(g) \qquad K_c = \frac{[O_3]^2}{[O_2]^3} = 6.25 \times 10^{-58} \quad (\text{at } 25\ ^\circ\text{C})$$

This means that $[O_3]^2 << [O_2]^3$ and if O_2 is placed in a flask at 25 °C, *very* little O_3 will be found when equilibrium is achieved. The concentration of O_2 would remain essentially unchanged. In the terminology of Chapter 5, we would write "N.R." and say that no reaction occurs.

The symbol ≅ means "approximately equal to."

Case 3. K_c ≅ 1: *Equilibrium mixture contains significant concentrations of reactants and products; calculations are needed to determine equilibrium concentrations.*

If K_c *is neither extremely large nor extremely small, the equilibrium constant must be used to calculate how far a reaction proceeds toward products*. In contrast with the reactions in Case 1 and Case 2, dissociation of dinitrogen tetraoxide has neither a very large nor a very small equilibrium constant. At 391 K the value is 1.00, which means that significant concentrations of both N_2O_4 and NO_2 are present at equilibrium.

Recall that the equation for dissociation of dinitrogen tetraoxide is

$$N_2O_4(g) \rightleftharpoons 2\ NO_2(g)$$

$$K_c = 1.00 = \frac{[NO_2]^2}{[N_2O_4]}, \text{ so } [NO_2]^2 = [N_2O_4] \quad (\text{at } 391\text{ K}) \qquad [13.6]$$

What range of equilibrium constants represents this middle ground depends on how small a concentration is significant and on the form of the equilibrium constant. If the concentrations of N_2O_4 and NO_2 at equilibrium are both 1.0 mol/L, then the ratio of $[NO_2]^2/[N_2O_4]$ does equal the K_c value of 1.00 at 391 K. But what if the concentrations were much smaller? Would they still be equal? You can verify this by using Equation 13.6. If the numeric value of the equilibrium concentration of NO_2 is 0.01, then the concentration of N_2O_4 must be $(0.01)^2$, which equals 0.0001. Thus, even though $K_c = 1.00$, the concentration of one substance can be much larger than the concentration of the other. This happens because there is a squared term in the numerator of the equilibrium constant expression and a term to a different power (the first power) in the denominator. Whenever the total of the exponents in the numerator differs from the total in the denominator, it becomes very difficult to say whether the concentrations of the products will exceed those of the reactants without doing a calculation.

By contrast, if the total of the exponents is the same, then if $K_c > 1$, products predominate over reactants, and if $K_c < 1$, reactants predominate over products. Examples in which this is true include

$$cis\text{-2-butene}(g) \rightleftharpoons trans\text{-2-butene}(g) \qquad K_c = \frac{[trans]}{[cis]} = 3.2 \quad (\text{at } 25\ ^\circ\text{C})$$

and

$$2\ HI(g) \rightleftharpoons H_2(g) + I_2(g) \qquad K_c = \frac{[H_2][I_2]}{[HI]^2} = 0.040 \quad (\text{at } 25\ ^\circ\text{C})$$

Figure 13.4 diagrams reactant and product concentrations as a function of K_c for the isomerization of *cis*-2-butene in the gas phase.

You might wonder whether reactant-favored systems in which small quantities of products form are important. Many are. Examples include the acids and bases listed in Table 13.1. For acetic acid, the acidic ingredient in vinegar, the reaction is

$$CH_3COOH(aq) + H_2O(\ell) \rightleftharpoons H_3O^+(aq) + CH_3COO^-(aq)$$

$$K_c = \frac{[H_3O^+][CH_3COO^-]}{[CH_3COOH]} = 1.8 \times 10^{-5} \quad \text{(at 25 °C)}$$

The value of K_c for acetic acid is small, and at equilibrium the concentrations of the products (acetate ions and hydronium ions) are small relative to the concentration of the reactant (acetic acid molecules). This confirms that acetic acid is a weak acid. Nevertheless, vinegar tastes sour because a small percentage of the acetic acid molecules react with water to produce $H_3O^+(aq)$.

If the form of the equilibrium constant is the same for two or more different reactions, then the degree to which each of those reactions is product-favored can be compared quantitatively in a very straightforward way. For example, the equilibrium reactions are similar and the equilibrium constant expressions all have the same form for formic acid, acetic acid, and carbonic acid:

$$K_c = \frac{[H_3O^+][\text{anion}^-]}{[\text{acid}]}$$

where anion$^-$ is $HCOO^-$, CH_3COO^-, or HCO_3^-, and acid is HCOOH, CH_3COOH, or H_2CO_3. From data in Table 13.1, we can say that formic acid is stronger (has a larger K_c value) than acetic acid and carbonic acid is the weakest of the three acids.

If a reaction has a large tendency to occur in one direction, then the reverse reaction has little tendency to occur. This means that the equilibrium constant for the reverse of a strongly product-favored reaction will be extremely small. Table 13.1 shows that combustion of hydrogen to form water vapor has an enormous equilibrium constant (3.2×10^{81}). This reaction is strongly product-favored. We say that it goes to completion.

The reverse reaction, decomposition of water to its elements,

$$2\,H_2O(g) \rightleftharpoons 2\,H_2(g) + O_2(g) \qquad K_c = \frac{[H_2]^2[O_2]}{[H_2O]^2} = 3.1 \times 10^{-82} \quad \text{(at 25 °C)}$$

is strongly reactant-favored, as indicated by the *very* small value of K_c.

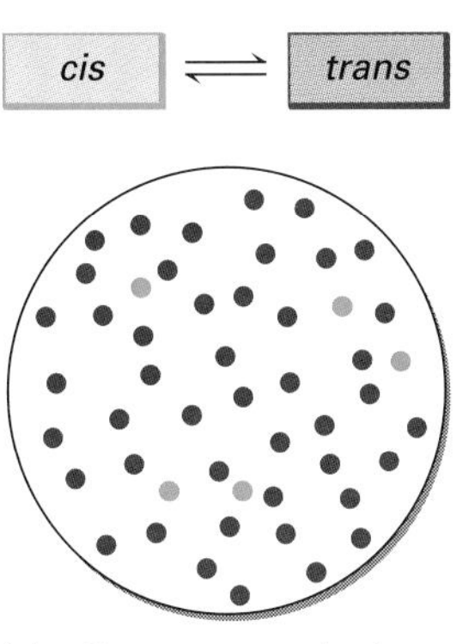

(a) $K_c = 9$; *trans* predominates

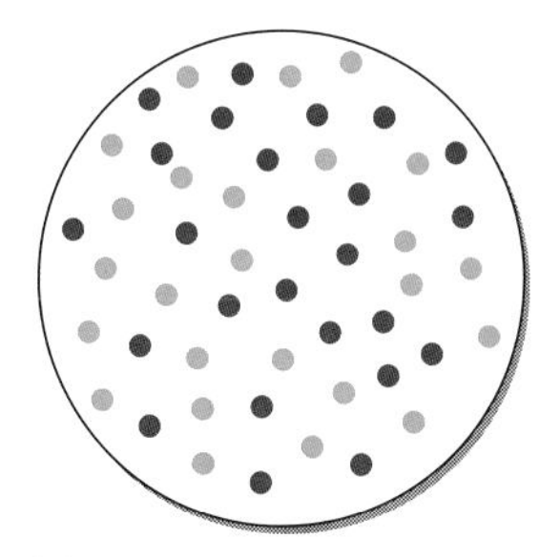

(b) $K_c = 1$

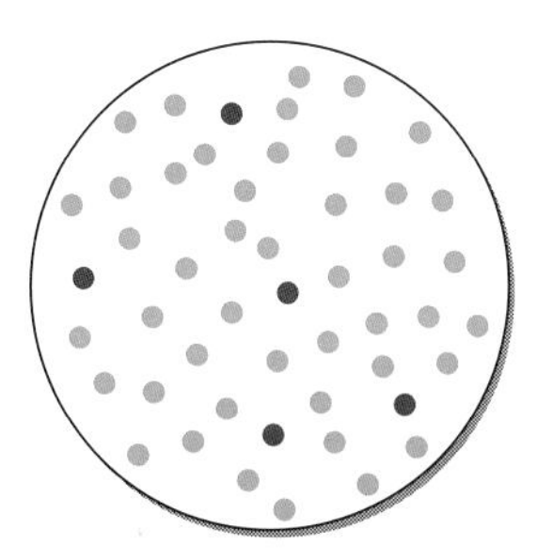

(c) $K_c = 1/9$; *cis* predominates

Figure 13.4 Equilibrium constants and concentrations of reactants and products. In the system *cis*-2-butene(g) $\rightleftharpoons$ *trans*-2-butene(g), the equilibrium constant decreases as temperature increases. (a) At a low temperature, when $K_c = 9$, the ratio [*trans*]/[*cis*] is 9/1 or 45/5. (b) At 1000 K, when $K_c = 1$, the ratio is 1/1 or 25/25. (c) At a much higher temperature, when $K_c = 0.111 = 1/9$, the ratio is 1/9 or 5/45.

PROBLEM-SOLVING EXAMPLE 13.4 Using Equilibrium Constants

Use equilibrium constants (Table 13.1) to predict which of the reactions below will be product-favored at 25 °C. Place all of the reactions in order from most reactant-favored to least reactant-favored.

(a) $NH_3(aq) + H_2O(\ell) \rightleftharpoons NH_4^+(aq) + OH^-(aq)$
(b) $HCOOH(aq) + H_2O(\ell) \rightleftharpoons H_3O^+ + HCOO^-(aq)$
(c) $N_2O_4(g) \rightleftharpoons 2\,NO_2(g)$

Answer All reactions are reactant-favored. The order from most reactant-favored to least reactant-favored is (a), (b), (c).

Strategy and Explanation Check whether the reactions all have equilibrium constant expressions of the same form. If they do, then the smaller the equilibrium constant is, the less product-favored (more reactant-favored) the reaction is. The equilibrium constant expressions are all of the form

$$K_c = \frac{[\text{product 1}][\text{product 2}]}{[\text{reactant}]}$$

because $H_2O(\ell)$ does not appear in the expressions for (a) and (b). The equilibrium constants for reactions (a) and (b) are 1.8×10^{-5} and 1.8×10^{-4}, respectively. The equilibrium constant for reaction (c) is not given in Table 13.1, but K_c for the reverse reaction is given as 1.7×10^2. Because the reaction is reversed it is necessary to take the reciprocal (p. 469), which gives an equilibrium constant for reaction (c) of 5.8×10^{-3}. Therefore the most reactant-favored reaction (smallest K_c) is (a), the next smallest K_c is for reaction (b), and the largest K_c is for reaction (c), which is the least reactant-favored.

PROBLEM-SOLVING PRACTICE 13.4

Suppose that solid AgCl and AgI are placed in 1.0 L water in separate beakers.

(a) In which beaker would the silver ion concentration, $[Ag^+]$, be larger?

(b) Does the volume of water in which each compound dissolves affect the equilibrium concentration?

CONCEPTUAL EXERCISE 13.6 Manipulating Equilibrium Constants

The equilibrium constant is 1.8×10^{-5} for reaction of ammonia with water.

$$NH_3(aq) + H_2O(\ell) \rightleftharpoons NH_4^+(aq) + OH^-(aq)$$

(a) Is the equilibrium constant large or small for the reverse reaction, the reaction of ammonium ion with hydroxide ion to give ammonia and water?

(b) What is the value of K_c for the reaction of ammonium ions with hydroxide ions?

(c) What does the value of this equilibrium constant tell you about the extent to which a reaction can occur between ammonium ions and hydroxide ions?

(d) Predict what would happen if you added a 1.0 M solution of ammonium chloride to a 1.0 M solution of sodium hydroxide. What observations might allow you to test your prediction in the laboratory?

13.5 Using Equilibrium Constants

Because equilibrium constants have numeric values, they can be used to predict quantitatively in which direction a reaction will proceed and how far it will go.

Predicting the Direction of a Reaction

Recall that the metric prefix "m" means $\frac{1}{1000} = 10^{-3}$. Therefore, 1 mmol = 1×10^{-3} mol.

Suppose that you have a mixture of 50. mmol $NO_2(g)$ and 100. mmol $N_2O_4(g)$ at 25 °C in a container with a volume of 10. L. Is the system at equilibrium? If not, in which direction will it react to achieve equilibrium? A useful way to approach such questions is to use the **reaction quotient**, Q, which *has the same mathematical form as the equilibrium constant expression but is a ratio of actual concentrations in the mixture,* instead of equilibrium concentrations. For the reaction

$$2\,NO_2(g) \rightleftharpoons N_2O_4(g) \qquad K_c = \frac{[N_2O_4]}{[NO_2]^2} = 1.7 \times 10^2 \quad (\text{at } 25\ °C)$$

$$Q = \frac{(\text{conc. } N_2O_4)}{(\text{conc. } NO_2)^2} = \frac{(100. \times 10^{-3}\ \text{mol}/10.\ \text{L})}{(50. \times 10^{-3}\ \text{mol}/10.\ \text{L})^2} = \frac{1.0 \times 10^{-2}}{(5.0 \times 10^{-3})^2} = 4.0 \times 10^2$$

In the expression for Q, we have used (conc. N_2O_4) to represent the *actual* concentration of N_2O_4 at a given time. We use $[N_2O_4]$ to represent the equilibrium concentration of N_2O_4. When the reaction is at equilibrium (conc. N_2O_4) = $[N_2O_4]$.

- *If Q is equal to* K_c, *then the reaction is at equilibrium.* The concentrations will not change.
- *If Q is less than* K_c, *then the concentrations of products are not as large as they would be at equilibrium.* The reaction will proceed from left to right to increase the product concentrations until they reach their equilibrium values.

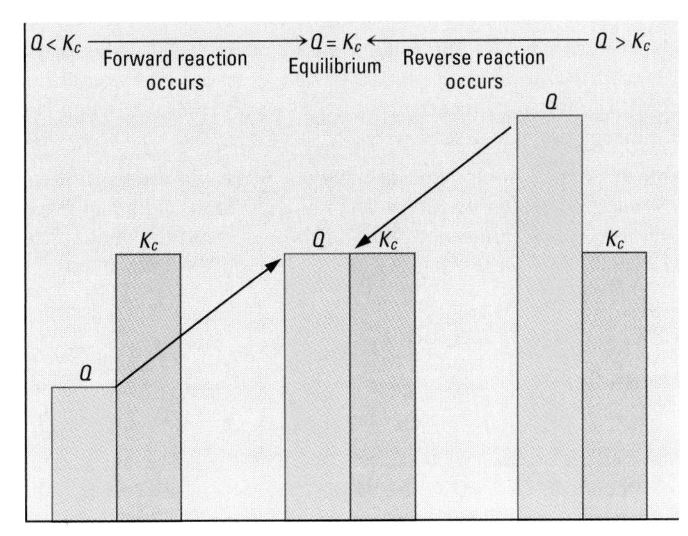

Figure 13.5 Predicting direction of a reaction. The relative sizes of the reaction quotient, Q, and the equilibrium constant, K_c, determine in which direction a mixture of substances will react to achieve equilibrium. The rule is

$Q < K_c$, reaction goes in forward direction (→)
$Q = K_c$, reaction at equilibrium
$Q > K_c$, reaction goes in reverse direction (←)

The reaction will proceed until $Q = K_c$.

- *If Q is greater than* K_c, *then the product concentrations are higher than they would be at equilibrium.* The reaction will proceed from right to left, increasing reactant concentrations, until equilibrium is achieved.

These relationships are shown schematically in Figure 13.5. In the case of the NO_2/N_2O_4 mixture described above, Q is greater than K_c, so, to establish equilibrium, some N_2O_4 will react to form NO_2. As the reverse reaction takes place, Q becomes smaller and eventually becomes equal to K_c.

PROBLEM-SOLVING EXAMPLE 13.5 Predicting Direction of Reaction

Consider this equilibrium, which is used industrially to generate hydrogen gas.

$$CH_4(g) + H_2O(g) \rightleftharpoons CO(g) + 3\,H_2(g) \qquad K_c = 0.94 \text{ at } 25\ ^\circ C$$

If 1.0 mol CH_4, 1.0 mol H_2O, 2.0 mol H_2, and 0.50 mol CO are mixed in a 10.0-L container at 25 °C, will the concentration of H_2O be greater or less than 0.10 mol/L when equilibrium is reached?

Answer Less than 0.10 mol/L

Strategy and Explanation Calculate the initial concentration of each gas and thus evaluate Q. Then compare Q with K_c.

$$(\text{conc. } CH_4) = \frac{1.0 \text{ mol}}{10.0 \text{ L}} = 0.10 \text{ mol/L} \qquad (\text{conc. } H_2) = \frac{2.0 \text{ mol}}{10.0 \text{ L}} = 0.20 \text{ mol/L}$$

$$(\text{conc. } H_2O) = \frac{1.0 \text{ mol}}{10.0 \text{ L}} = 0.10 \text{ mol/L} \qquad (\text{conc. CO}) = \frac{0.50 \text{ mol}}{10.0 \text{ L}} = 0.050 \text{ mol/L}$$

Because the units are defined by the equilibrium constant expression, leave them out.

$$Q = \frac{(\text{conc. CO})(\text{conc. } H_2)^3}{(\text{conc. } CH_4)(\text{conc. } H_2O)} = \frac{(0.050)(0.20)^3}{(0.10)(0.10)} = 0.040$$

which is much smaller than 0.94, the value of K_c. Because $Q < K_c$, the forward reaction—reaction of CH_4 with H_2O to form CO and H_2—will occur until the equilibrium concentrations are reached. The initial concentration of H_2O was 0.10 mol/L; when H_2O reacts with CH_4, the H_2O concentration decreases.

☑ **Reasonable Answer Check** Because all concentrations are fractions, multiplying concentrations will give smaller fractions. The concentrations are all approximately 10^{-1}, but the numerator has concentration to the fourth power and the denominator to the second power overall. This means that Q should be much less than 1, and it is.

PROBLEM-SOLVING PRACTICE 13.5

For the equilibrium

$$2\,SO_2(g) + O_2(g) \rightleftharpoons 2\,SO_3(g) \qquad K_c = 245 \quad \text{(at 1000 K)}$$

the equilibrium concentrations are $[SO_2] = 0.102$, $[O_2] = 0.0132$, and $[SO_3] = 0.184$. Suppose that the concentration of SO_2 is suddenly doubled. Calculate Q and use it to show that the forward reaction would take place to reach a new equilibrium.

CONCEPTUAL EXERCISE 13.7 Reaction Quotient and Pressure Equilibrium Constant

Is it possible to apply the idea of the reaction quotient to gas phase reactions where the equilibrium constant is given in terms of pressure (as in Equation 13.4, *p. 471*)? Define Q for such a reaction and give an appropriate set of rules by which you can predict in which direction a gas phase reaction will go to achieve equilibrium.

Calculating Equilibrium Concentrations

Equilibrium constants from Table 13.1 (or one of the appendixes, or a reference compilation) can be used to calculate how much product is formed and how much of the reactants remain once a system has reached equilibrium. To verify our earlier statement that if an equilibrium constant is very large, essentially all of the reactants are converted to products (***p. 475***), consider the reaction

$$\tfrac{1}{8}\,S_8(s) + O_2(g) \rightleftharpoons SO_2(g) \qquad K_c = 4.2 \times 10^{52} \quad \text{(at 25 °C)}$$

Suppose we place 4.0 mol O_2 and a large excess of sulfur in an empty 1.00-L flask and allow the system to reach equilibrium. We can calculate the quantity of O_2 left and the quantity of SO_2 formed at equilibrium by summarizing information in a table. (Because S_8 is a solid and does not appear in the equilibrium constant expression, we do not need any entries under S_8 in the table.)

	$\frac{1}{8}S_8(s)$	+	$O_2(g)$	$\rightleftharpoons$	$SO_2(g)$
Initial concentration (mol/L)			4.0		0.0
Change as reaction occurs (mol/L)			$-x$		$+x$
Equilibrium concentration (mol/L)			$4.0 - x$		$0.0 + x$

We know the concentrations of reactant and product before the reaction, but we do not know how many moles per liter of O_2 are consumed during the reaction, and so we designate this as x mol/L. (There is a minus sign in the table because O_2 is consumed.) Since the mole ratio is (1 mol SO_2)/(1 mol O_2), we know that x mol/L SO_2 is formed when x mol/L O_2 is consumed. To calculate the concentration of O_2 we take what was present initially (4.0 mol/L) minus what was consumed in the reaction

(x mol/L). The equilibrium concentration of SO_2 must be the initial concentration (0 mol/L) plus what was formed by the reaction (x mol/L). Putting these values into the equilibrium constant expression, we have

$$K_c = \frac{[SO_2]}{[O_2]} = \frac{x}{4.0 - x} = 4.2 \times 10^{52}$$

Solving algebraically for x (and following the usual rules for significant figures), we find

$$x = (4.2 \times 10^{52})(4.0 - x)$$
$$x = (16.8 \times 10^{52}) - (4.2 \times 10^{52})x$$
$$x + (4.2 \times 10^{52})x = 16.8 \times 10^{52}$$

Notice that $x + (4.2 \times 10^{52})x = (1 + 4.2 \times 10^{52})x$, which to a very good approximation is equal to $(4.2 \times 10^{52})x$. Thus,

$$x = \frac{16.8 \times 10^{52}}{4.2 \times 10^{52}} = 4.0$$

Because 4.2×10^{52} is so much larger than 1, adding 1 to it makes no appreciable change in the very large number.

The equilibrium concentration of SO_2 is $x = 4.0$ mol/L and that of O_2 is $(4.0 - x)$ mol/L, or 0 mol/L. That is, within the precision of our calculation, all the O_2 has been converted to SO_2. As stated earlier, a very large K_c value (4.2×10^{52} in this case) implies that essentially all of the reactants have been converted to products. The reaction is strongly product-favored and goes to completion, so the calculation could have been done using the methods in Section 4.4 ***(⟵ p. 110)***.

The fact that samples of sulfur at 25 °C can be exposed to oxygen in the air for long periods without being converted to sulfur dioxide shows the importance of chemical kinetics. This reaction is very slow at room temperature, so only a faint odor of sulfur dioxide is noticeable in the vicinity of solid sulfur.

PROBLEM-SOLVING EXAMPLE 13.6 Calculating Equilibrium Concentrations

When colorless hydrogen iodide gas is heated to 745 K, a beautiful purple color appears. This shows that some iodine gas has been formed, which shows that some HI has been decomposed to its elements.

$$2\,HI(g) \rightleftharpoons H_2(g) + I_2(g) \qquad K_c = 0.0200 \quad \text{(at 745 K)}$$

Suppose that a mixture of 1.00 mol HI(g) and 1.00 mol $H_2(g)$ is sealed into a 10.0-L flask and heated to 745 K. What will be the concentrations of all three substances when equilibrium has been achieved?

Answer [HI] = 0.096 M, $[I_2]$ = 0.0018 M, $[H_2]$ = 0.102 M

Strategy and Explanation Follow the usual steps ***(⟵ p. 472)***. The balanced equation was given, and the equilibrium constant expression is given on the next page. Because amounts are given instead of concentrations, divide each number of moles by the volume of the flask to get (conc. HI) = (conc. H_2) = 1.00 mol/10.0 L = 0.100 mol/L. Because no I_2 is present at the beginning, $Q = 0$. This is much less than K_c, even though K_c is small. Therefore the reaction must go from left to right, so it makes sense to let x mol/L be the concentration of I_2 when equilibrium is reached. Since the coefficients of H_2 and I_2 are the same, if x mol/L of I_2 is produced, then x mol/L of H_2 must also be produced. Therefore the change in concentration of both H_2 and I_2 is $+x$ mol/L. Because the coefficient of HI is twice the coefficient of I_2, twice as many moles of HI must react as moles of I_2 formed; the change in concentration of HI is $-2x$ mol/L. The reaction table is then

Iodine vapor has a beautiful purple color.

	2 HI(g)	⇌	H_2(g)	+	I_2(g)
Initial concentration (mol/L)	0.100		0.100		0.000
Change as reaction occurs (mol/L)	$-2x$		$+x$		$+x$
Equilibrium concentration (mol/L)	$(0.100 - 2x)$		$(0.100 + x)$		x

In solving this problem we might have chosen $-x$ to be the change in concentration of HI, in which case the changes in concentrations of H_2 and I_2 would each have been $+\frac{1}{2}x$.

In some cases an equation for K_c like this one will have an expression involving x that is a perfect square. In such cases it is safe to take the square root of both sides of the equation and then solve for x.

Now write the equilibrium constant expression in terms of the equilibrium concentrations calculated in the third row of the table.

$$K_c = \frac{[H_2][I_2]}{[HI]^2} = \frac{(0.100 + x)x}{(0.100 - 2x)^2} = 0.0200$$

To solve the equation directly, multiply out the numerator and denominator to obtain

$$\frac{0.100x + x^2}{0.0100 - 0.400x + 4x^2} = 0.0200$$

Multiply both sides by the denominator and then multiply out the terms. This gives

$$0.100x + x^2 = 0.0200 \times (0.0100 - 0.400x + 4x^2)$$
$$0.100x + x^2 = 0.000200 - 0.00800x + 0.0800x^2$$

Collecting terms in x^2 and x, we have

$$0.9200x^2 + 0.10800x - 0.000200 = 0$$

This is a quadratic equation of the form $ax^2 + bx + c = 0$, where $a = 0.9200$, $b = 0.10800$, and $c = -0.000200$. The equation can be solved using the quadratic formula (Appendix A.7).

$$x = \frac{-b \pm \sqrt{b^2 - 4ac}}{2a} = \frac{-0.10800 \pm \sqrt{0.10800^2 - 4 \times 0.9200 \times (-0.000200)}}{2 \times 0.9200}$$

$$x = \frac{-0.10800 + 0.11135}{1.840} = 1.83 \times 10^{-3}$$

The other root, $x = -0.119$, can be eliminated because it would result in a negative concentration of I_2 at equilibrium, which is clearly impossible. From the equilibrium concentration row of the table,

$$[HI] = (0.100 - 2x)\text{mol/L} = 0.100 - (2 \times 1.83 \times 10^{-3}) = 0.0963 \text{ mol/L}$$
$$[H_2] = (0.100 + x)\text{mol/L} = 0.100 + (1.83 \times 10^{-3}) = 0.102 \text{ mol/L}$$
$$[I_2] = x \text{ mol/L} = 0.00183 \text{ mol/L}$$

☑ **Reasonable Answer Check** Solving this type of problem involves a lot of algebra, and it would be easy to make a mistake. It is very important to check the result by substituting the equilibrium concentrations into the equilibrium constant expression and verifying that the correct value of K_c (0.0200) results.

$$K_c = \frac{[H_2][I_2]}{[HI]^2} = \frac{(0.102)(0.00183)}{(0.0963)^2} = 0.0201$$

which is acceptable agreement because it differs from K_c by 1 in the last significant figure.

PROBLEM-SOLVING PRACTICE 13.6

Obtain the equilibrium constant for dissociation of dinitrogen tetraoxide to form nitrogen dioxide from Table 13.1. If 1.00 mol N_2O_4 and 0.500 mol NO_2 are initially placed in a container with a volume of 4.00 L, calculate the concentrations of $N_2O_4(g)$ and $NO_2(g)$ present when equilibrium is achieved at 25 °C.

13.6 Shifting a Chemical Equilibrium: Le Chatelier's Principle

Le Chatelier is pronounced "luh SHOT lee ay."

Suppose you are an environmental engineer, biologist, or geologist, and you have just measured the concentration of hydronium ion, H_3O^+, in a lake. You know that the H_3O^+ ions are involved in many different equilibrium reactions in the lake. How can you predict the influence of changing conditions? For example, what happens if there is a large increase in acid rainfall that has a hydronium ion concentration different from that of the lake? Or what happens if lime (calcium oxide), a strong base, is added to the lake? These questions and many others like them can be answered qualitatively by applying a useful guideline known as **Le Chatelier's principle:** ***If a system is at equilibrium and the conditions are changed so that it is no longer at equilibrium, the system will react to reach a new equilibrium in a way that partially counteracts the change.*** To adjust to a change, a system reacts in either the forward direction (producing more products) or the reverse direction (producing more reactants) until a new equilibrium state is achieved. Le Chatelier's principle applies to changes in the concentrations of reactants or products that appear in the equilibrium constant expression, the pressure or volume of a gas phase equilibrium, and the temperature. Changing conditions, thereby changing the equilibrium concentrations of reactants and products, is called **shifting an equilibrium.** If the reaction occurs in the *forward direction,* we say that the equilibrium reaction has *shifted to the right*. If the system reacts in the *reverse direction,* the reaction has *shifted to the left*.

Henri Le Chatelier (1850–1936) was a French chemist who, as a result of his studies of the chemistry of cement, developed his ideas about how altering conditions affects an equilibrium system.

Changing Concentrations of Reactants or Product

If the concentration of a reactant or a product that appears in the equilibrium constant expression is changed, a system can no longer be at equilibrium because Q must have a different value from K_c. The equilibrium will shift to use up a substance that was added, or to replenish a substance that was removed. The shift will occur in whichever direction causes Q to approach the value of K_c.

- If the concentration of a reactant is increased, the system will react in the forward direction.
- If the concentration of a reactant is decreased, the system will react in the reverse direction.
- If the concentration of a product is increased, the system will react in the reverse direction.
- If the concentration of a product is decreased, the system will react in the forward direction.

To see why this happens, consider the simple equilibrium discussed in Section 13.2.

$$cis\text{-2-butene(g)} \rightleftharpoons trans\text{-2-butene(g)} \qquad K_c = \frac{[trans]}{[cis]} = 1.5 \quad \text{(at 600 K)}$$

Suppose that 2 mmol *cis*-2-butene is placed into a 1.0-L container at 1000 K. The forward reaction will be faster than the reverse reaction until the concentration of *trans*-2-butene builds up to 1.5 times the concentration of *cis*-2-butene. Then equilibrium is achieved. Now suppose that half of the *cis*-2-butene is instantaneously removed from the container (see Exercise 13.3, ⇐ ***p. 466***). Because the concentration of *cis*-2-butene suddenly drops to half its former value, the forward reaction rate will also drop to half its former value. The reverse rate will not be affected, because the concentration of *trans*-2-butene has not changed. This means that the reverse reaction is twice as fast as the forward reaction, and *cis*-2-butene molecules are being formed twice as fast as they are reacting away. Therefore the concentration of *cis*-2-butene will increase and the concentration of *trans*-2-butene will decrease until the forward and reverse rates are again equal. The graph in Figure 13.6 illustrates this situation.

As long as the temperature remains the same, the value of the equilibrium constant also remains the same. Adding or removing a reactant or a product does not change the equilibrium constant value. If the concentration of the substance added or removed appears in the equilibrium constant expression, however, the value of Q changes. Because Q no longer equals K_c, the system is no longer at equilibrium and must react to achieve a new equilibrium.

The effect of changing concentration has many important consequences. For example, when the concentration of a substance needed by your body falls slightly, several enzyme-catalyzed chemical equilibria shift so as to increase the concentration of the essential substance. In industrial processes, reaction products are often continuously removed. This shifts one or more equilibria to produce more products and thereby maximize the yield of the reaction.

In nature, slight changes in conditions are responsible for effects such as the formation of limestone stalactites and stalagmites in caves and the crust of limestone that slowly develops in a tea kettle if you boil hard water in it. Both of these examples

Hard water contains dissolved CO_2 and metal ions such as Ca^{2+} and Mg^{2+}.

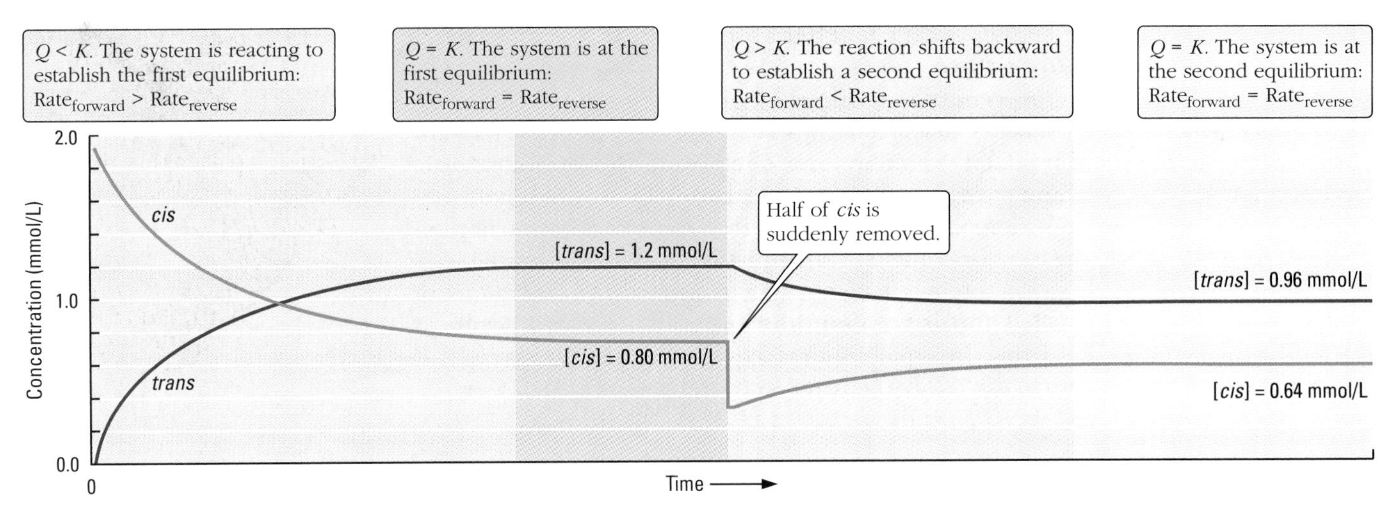

Figure 13.6 Approach to new equilibrium after a change in conditions.

involve calcium carbonate. Limestone, a form of calcium carbonate, $CaCO_3$, is present in underground deposits, a leftover of the ancient oceans from which it precipitated long ago. Limestone dissolves when it reacts with an aqueous solution of CO_2.

$$CaCO_3(s) + CO_2(aq) + H_2O(\ell) \rightleftharpoons Ca^{2+}(aq) + 2\,HCO_3^-(aq) \qquad [13.7]$$

If groundwater that is saturated with CO_2 encounters a bed of limestone below the surface of the earth, the forward reaction can occur until equilibrium is reached, and the water subsequently contains significant concentrations of aqueous Ca^{2+} and HCO_3^- ions in addition to dissolved CO_2. These ions, often accompanied by $Mg^{2+}(aq)$, constitute hard water.

As with all equilibria, a reverse reaction is occurring in addition to the forward reaction. This reverse reaction can be demonstrated by mixing aqueous solutions of $CaCl_2$ and $NaHCO_3$ (salts containing the Ca^{2+} and HCO_3^- ions) in an open beaker (Figure 13.7). You will eventually see bubbles of CO_2 gas and a precipitate of solid $CaCO_3$. Because the beaker is open to the air, any gaseous CO_2 that escapes the solution is

© Cengage Learning/Charles D. Winters
1 These two salt solutions provide Ca^{2+} and HCO_3^- ions, which are products in the net ionic equation for the reaction of calcium carbonate with carbon dioxide.
2 Bringing the reaction products together by mixing the two solutions shifts the equilibrium toward the reactants.
3 The CO_2 produced by the reverse reaction bubbles out of the solution into the air, thereby reducing a reactant concentration in solution and shifting the equilibrium to the left.
4 Therefore the reverse reaction continues until almost all of the Ca^{2+} and HCO_3^- ions have reacted to form CO_2, $CaCO_3$, and water.

Figure 13.7 Reaction of $CaCl_2(aq)$ with $NaHCO_3(aq)$.

swept away. Its removal reduces the concentration of $CO_2(aq)$ and causes the equilibrium to shift to the left to produce more CO_2. Eventually all of the dissolved Ca^{2+} and HCO_3^- ions are converted to gaseous CO_2, solid $CaCO_3$, and water.

Suppose that water containing dissolved CO_2, Ca^{2+}, and HCO_3^- contacts the air in a cave (or hard water contacts air in your tea kettle). Carbon dioxide bubbles out of the solution, the concentration of CO_2 decreases on the reactant side of Equation 13.8, and the equilibrium shifts toward the reactants. The reverse reaction forms $CO_2(aq)$, compensating partially for the reduced concentration of $CO_2(aq)$. Some of the calcium ions and hydrogen carbonate ions combine, and some $CaCO_3(s)$ precipitates (Figure 13.8) as a beautiful formation in the cave (or as scale in your kettle).

Figure 13.8 Stalactites in a limestone cave. Stalactites hang from the ceilings of caves. Stalagmites grow from the floors of caves up toward the stalactites. Both consist of limestone, $CaCO_3$. The process that produces these lovely formations is an excellent example of chemical equilibrium.

CONCEPTUAL **EXERCISE** **13.8 Effect of Adding a Substance**

Solid phosphorus pentachloride decomposes when heated to form gaseous phosphorus trichloride and gaseous chlorine. Write the equation for the equilibrium that is set up when solid phosphorus pentachloride is introduced into a container, the container is evacuated and sealed, and the solid is heated. Once the system has reached equilibrium at a given temperature, what will be the effect on the equilibrium of each of these changes?

(a) Adding chlorine to the container
(b) Adding phosphorus trichloride to the container
(c) Adding a small quantity of phosphorus pentachloride to the container

Changing Volume or Pressure in Gaseous Equilibria

One way to change the pressure of a gaseous equilibrium mixture is to keep the volume constant and to add or remove one or more of the substances whose concentrations appear in the equilibrium constant expression. The effect of adding or removing a substance has just been discussed. We consider here other ways of changing pressure or volume.

The pressures of all substances in a gaseous equilibrium can be changed by changing the volume of the container. Consider the effect of tripling the pressure on the equilibrium

$$N_2O_4(g) \rightleftharpoons 2\,NO_2(g) \qquad K_c = \frac{[NO_2]^2}{[N_2O_4]}$$

by reducing the volume of the container to one third of its original value (at constant temperature). This situation is shown in Figure 13.9 (p. 486). Decreasing the volume increases the pressures of N_2O_4 and NO_2 to three times their equilibrium values. Decreasing the volume also increases the concentrations of N_2O_4 and NO_2 to three times their equilibrium values. Because $[NO_2]$ is squared in the equilibrium constant expression but $[N_2O_4]$ is not, tripling both concentrations increases the numerator of Q by $3^2 = 9$ but increases the denominator by only 3.

$$Q = \frac{(\text{conc. } NO_2)^2}{(\text{conc. } N_2O_4)} = \frac{(3 \times [NO_2])^2}{(3 \times [N_2O_4])} = \frac{9}{3} \times \frac{[NO_2]^2}{[N_2O_4]} = 3 \times K_c$$

Because Q is larger than K_c under the new conditions, the reaction should produce more reactant; that is, the equilibrium should shift to the left.

The same prediction is made using Le Chatelier's principle: The reaction should shift to partially compensate for the increase in pressure. That means decreasing the pressure, which can happen if the total number of gas phase molecules decreases. In the case of the N_2O_4/NO_2 equilibrium, the reverse reaction should occur, because one N_2O_4 molecule is produced for every two NO_2 molecules that react. A shift to the left reduces the number of gas phase molecules and hence the pressure.

Remember that the pressure of an ideal gas is proportional to the number of moles of gas, and therefore to the number of molecules of gas (⟵ ***p. 354***).

CHEMISTRY IN THE NEWS

Bacteria Communicate Chemically

Stories about fish fingers that glowed in the dark after being stored too long in a dorm-room fridge sound like urban legends, but they may well be true. Certain bacteria, such as *Vibrio fischeri*, are bioluminescent when a colony gets big enough, and these bacteria are present in many fish. As a fish decomposes, the bacteria grow rapidly and may produce a green glow.

Rapid growth of many forms of bacteria involves communication among individual bacterial cells. Only after a colony reaches a certain size and the bacteria in the colony realize it has reached that size is there really rapid growth. In this group state, pathogenic bacteria initiate the majority of infections in plants and animals. How do bacteria sense when they have reached an adequate group size? And could bacterial infections be treated by preventing bacteria from realizing they actually are in a group? Answers to these questions could be very beneficial to humankind, and scientists are working to find them.

Gram-negative bacteria communicate by secreting so-called signal molecules that can diffuse out of one bacterial cell and into another. In the case of *V. fischeri*, the signal molecule is an acyl-homoserine lactone (AHL) and it is produced by a protein designated LuxI. The AHL molecules diffuse out of the cell in which they were produced and can enter other *V. fischeri* cells, if any are nearby. When the signal molecule enters a cell, it can bind with another protein, LuxR, in an equilibrium process:

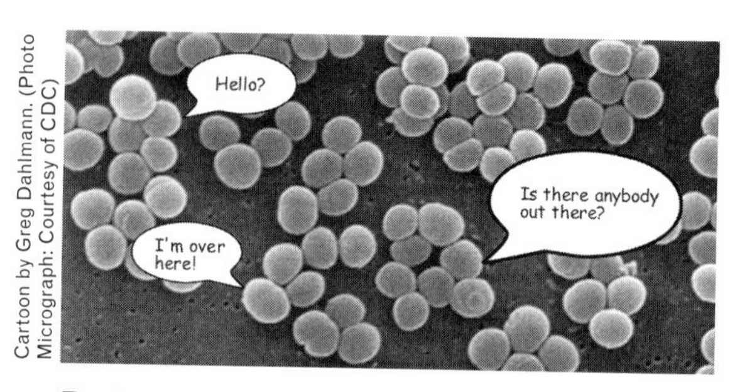

Bacterial communication.

$$\text{LuxR} + \text{AHL} \rightleftharpoons \text{LuxR—AHL}$$

If only a few *V. fischeri* cells are close together, most of the AHL diffuses away, and its concentration remains small. But once a bacterial colony reaches a certain size, most of the cells are surrounded by other cells, all of which are secreting AHL. The increased concentration of AHL shifts the equilibrium to the right. This forms sizable concentrations of the LuxR-AHL combination, which binds to DNA and switches on genes that encode the enzyme luciferase. In the presence of O_2, luciferase produces luminescence like that of a firefly. The LuxR-AHL combination also switches on genes that code for formation of LuxI, which produces more AHL. Therefore, AHL is said to *autocatalyze* its own synthesis, and this explains why the colony of bacteria suddenly grows very rapidly (and produces the green glow) after an initial period of slow growth.

This process is referred to as *quorum sensing,* because it requires a significant number of bacteria (a quorum) to initiate the autocatalysis of formation of AHL. Many bacteria display quorum-sensing behavior, and often their rapid growth has much worse consequences than a green-glowing fish stick. Quorum-sensing behavior is thought to account for more than 50% of all crop diseases and as much as 80% of all human bacterial infections. Therefore scientists are

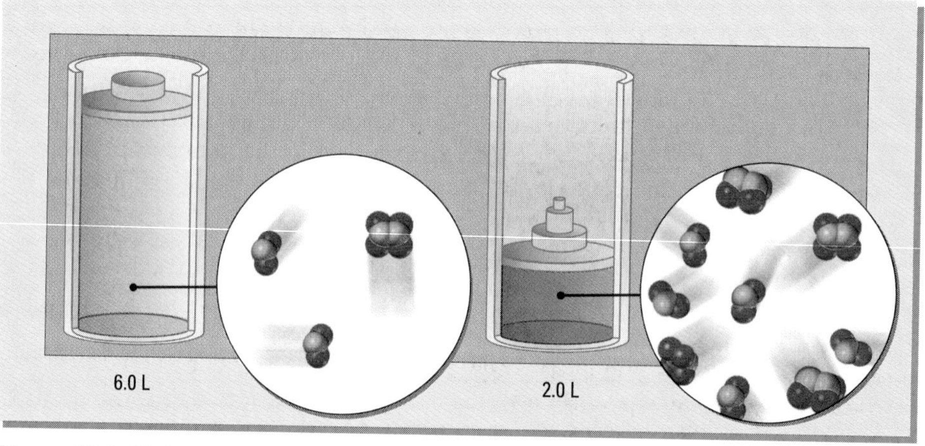

Figure 13.9 Shifting an equilibrium by changing pressure and volume. If the volume of an equilibrium mixture of NO_2 and N_2O_4 is decreased from 6.0 to 2.0 L, the equilibrium shifts toward the smaller number of molecules in the gas phase (the N_2O_4 side) to partially compensate for the increased pressure. When the new equilibrium is achieved, the concentrations of both NO_2 and N_2O_4 have increased, but the NO_2 concentration is less than three times as great, while the N_2O_4 concentration is more than three times as great. The total pressure goes from 1.00 to 2.62 atm (rather than to 3.00 atm, which was the new pressure before the equilibrium shifted).

Photo by Jason Varney Photography, Courtesy of Helen Blackwell

Helen Blackwell.

looking for a way to prevent quorum-sensing equilibria from shifting right and forming the enzyme–signal molecule combination. Substances that could do this are called quorum-sensing antagonists. They have great potential to alleviate problems of bacterial resistance to antibiotics and could bring great benefit to humankind.

One way to find quorum-sensing antagonists is being explored by the research group of Professor Helen Blackwell of the University of Wisconsin-Madison. Blackwell and her co-workers make many different compounds whose molecular structures are closely related and test those compounds for their ability to inhibit quorum-sensing behavior in bacteria. To help make the compounds Blackwell uses a laboratory version of a microwave oven such as those found in kitchens to heat food. Microwave irradiation heats reactant chemicals quickly and results in greater yields of higher-purity products. In a single day Blackwell's research group can prepare as many as 100–500 related compounds. These can subsequently be tested for activity, and the rate at which new biologically active compounds can be discovered is greatly accelerated. Recently Blackwell and her students have identified quorum-sensing antagonists that are among the most potent discovered to date.

Knowledge of quorum sensing also allows chemists to enhance the growth of bacteria, and in some cases that may be a good thing. For example, changing the concentrations of autocatalyst substances might be a way of controlling the differentiation of stem cells into different type of tissues. Or, a system can be imagined in which bacterial populations could be maintained at a desired level. A quorum-sensing signal molecule could turn on a suicide gene if the colony got too large, but when the population decreased the signal concentration would become smaller, thus turning off the gene and not completely wiping out the colony. Exploring the reasons for that green glow in the refrigerator might have incredibly beneficial consequences.

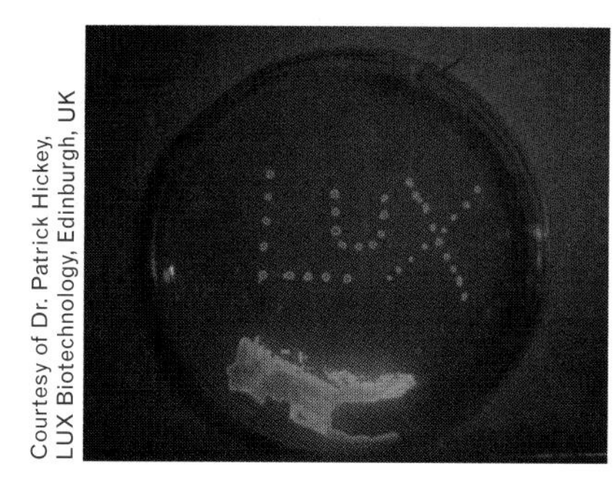

Courtesy of Dr. Patrick Hickey, LUX Biotechnology, Edinburgh, UK

Glowing *Vibrio fischeri.*

Sources: Based on information from *New York Times*, January 17, 2006; *Wisconsin State Journal*, November 4, 2007, *The Health Show*, April 21, 2005, **http://www.healthshow.org/archive/week_2005_04_17.shtml;** Blackwell laboratory, **http://www.chem.wisc.edu/blackwell/index.htm;** and *New York Times*, February 27, 2001.

However, consider the situation with respect to another equilibrium we have already mentioned in Problem-Solving Example 13.3 (**p. 473**). At 425 °C all substances are in the gas phase.

$$2\,HI(g) \rightleftharpoons H_2(g) + I_2(g)$$

Suppose that the pressure of this system were tripled by reducing its volume to one third of the original volume. What would happen to the equilibrium? In this case, all the concentrations triple, but because equal numbers of moles of gaseous substances appear on both sides of the equation, Q still has the same numeric value as K_c. That is, the system is still at equilibrium, and no shift occurs. Thus, *changing pressure by changing the volume shifts an equilibrium only if the sum of the coefficients for gas phase reactants is different from the sum of the coefficients for gas phase products.*

Finally, consider what happens if the pressure of the N_2O_4/NO_2 equilibrium system is increased by adding an inert gas such as nitrogen while retaining exactly the same volume. The total pressure of the system would increase, but since neither the amounts of N_2O_4 and NO_2 nor the volume would change, the concentrations of N_2O_4 and NO_2 would remain the same. Q still equals K_c, the system is still at equilibrium, and no shift occurs (or needs to). Thus, changing the pressure of an equilibrium system must change the concentration of at least one substance in the equilibrium constant expression if a shift in the equilibrium is to occur.

CONCEPTUAL EXERCISE **13.9 Changing Volume Does Not Always Shift an Equilibrium**

In Problem-Solving Example 13.6 you found that for the reaction

$$2\ HI(g) \rightleftharpoons H_2(g) + I_2(g) \qquad K_c = 0.0200 \quad (\text{at } 745\ K)$$

the equilibrium concentrations were $[HI] = 0.0963$ mol/L, $[H_2] = 0.102$ mol/L, and $[I_2] = 0.00183$ mol/L. Use algebra to show that if each of these concentrations is tripled by reducing the volume of this equilibrium system to one third of its initial value, the system is still at equilibrium, and therefore the pressure change causes no shift in the equilibrium.

EXERCISE **13.10 Effect of Changing Volume**

Verify the statement in the text that, for the N_2O_4/NO_2 equilibrium system, decreasing the volume to one third of its original value increases the equilibrium concentration of N_2O_4 by more than a factor of 3 while it increases the equilibrium concentration of NO_2 by less than a factor of 3. Start with the equilibrium conditions you calculated in Problem-Solving Practice 13.6. Then decrease the volume of the system from 4.00 L to 1.33 L, and calculate the new concentrations of N_2O_4 and NO_2 assuming the shift in the equilibrium has not yet taken place. Set up the usual table of initial concentrations, change, and equilibrium concentrations, and calculate the concentrations at the new equilibrium.

Changing Temperature

When temperature changes, the values of most equilibrium constants also change, and systems will react to achieve new equilibria consistent with new values of K_c. You can make a qualitative prediction about the effect of temperature on an equilibrium if you know whether the reaction is exothermic or endothermic. As an example, consider the endothermic, gas phase reaction of N_2 with O_2 to give nitrogen monoxide, NO.

$$N_2(g) + O_2(g) \rightleftharpoons 2\ NO(g) \qquad \Delta H^\circ = 180.5\ \text{kJ}$$

$$K_c = 4.5 \times 10^{-31} \quad \text{at } 298\ K$$

$$K_c = \frac{[NO]^2}{[N_2][O_2]} \qquad K_c = 6.7 \times 10^{-10} \quad \text{at } 900\ K$$

$$K_c = 1.7 \times 10^{-3} \quad \text{at } 2300\ K$$

In this case the equilibrium constant increases very significantly as the temperature increases. At 298 K the equilibrium constant is so small that essentially no reaction occurs. Suppose that a room-temperature equilibrium mixture were suddenly heated to 2300 K. What would happen? The equilibrium should shift to partially compensate for the temperature increase. This happens if the reaction shifts in the endothermic direction, since that change would involve a transfer of energy into the reaction system, cooling the surroundings. For the $N_2 + O_2$ reaction the forward process is endothermic ($\Delta H^\circ > 0$), so at the higher temperature N_2 and O_2 should react to produce more NO. At the new equilibrium, the concentration of NO should be higher and the concentrations of N_2 and O_2 should be lower. Since this makes the numerator in the K_c expression bigger and the denominator smaller, K_c should be larger at the higher temperature. This outcome corresponds with the experimental result.

The effect on K_c of raising the temperature of the $N_2 + O_2$ reaction leads us to a general conclusion: ***For an endothermic reaction, an increase in temperature always means an increase in K_c; the reaction will become more product-favored at higher temperatures.*** When equilibrium is achieved at the higher

ESTIMATION Generating Gaseous Fuel

The reaction of coke (mainly carbon) with steam is called the water-gas reaction. It was used for many years to generate gaseous fuel from coal. The thermochemical expression is

$$C(s) + H_2O(g) \rightleftharpoons CO(g) + H_2(g) \qquad \Delta H° = 131.293 \text{ kJ}$$

The equilibrium constant K_P has the value 9.5×10^{-17} at 298 K, 1.9×10^{-7} at 500 K, 2.60×10^{-2} at 800 K, 18.8 at 1200 K, and 500. at 1600 K. Suppose that you have an equilibrium mixture in which the partial pressure of steam is 1.00 atm. Estimate the temperature at which the partial pressures of CO and H_2 would also equal 1 atm. (The reaction would need to be carried out at temperatures roughly this high to produce appreciable quantities of products.)

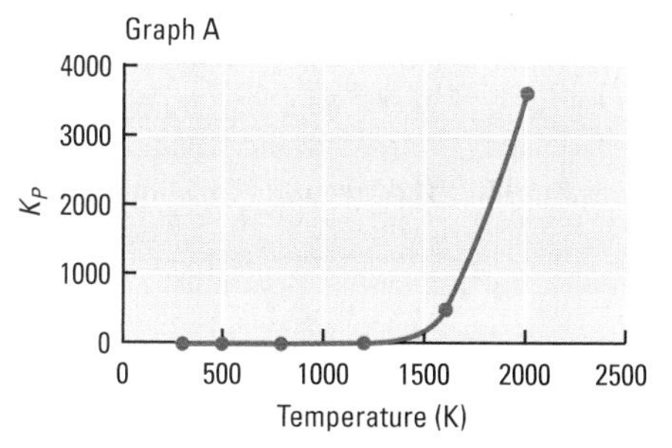

A good way to make this estimation is to use a spreadsheet program to plot the data and see which temperature corresponds to $K_P = 1$. This has been done in Graph A. It is clear that the range of values for the equilibrium constant is so wide that an accurate temperature cannot be read corresponding to $K_P = 1$. An estimate would be easier to make if the graph were a straight line. Because the values of K_P rise very rapidly, it is possible that taking the logarithm of each K_P value would generate a linear graph. This has been done in Graph B, where the natural logarithm of the equilibrium constant, $\ln(K_P)$, has been plotted on the vertical axis. Since $\ln(K_P) = 0$ when $K_P = 1$, we are looking for the temperature at which the graph crosses zero on the vertical axis. Although the graph is not linear, the appropriate temperature can be read from the graph and is slightly less than 1000 K.

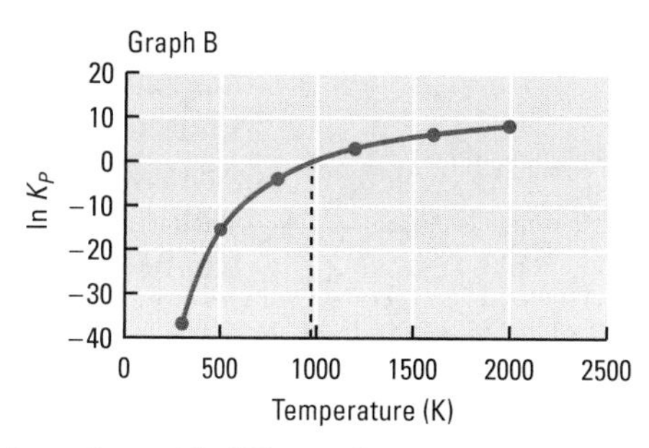

Experimenting with different functions of K_P on the vertical axis and different functions of T on the horizontal axis reveals that a linear graph is obtained when $\ln(K_P)$ is plotted against $1/T$, as in Graph C. (Note that the spreadsheet graph displays too many significant figures.) From this graph the equation of the straight line is found to be

$$\ln(K_P) = (-1.58 \times 10^4 \text{ K})\left(\frac{1}{T}\right) + 16.1$$

Substituting $K_P = 1$, that is, $\ln(K_P) = 0$, we find that

$$T = \frac{1.58 \times 10^4 \text{ K}}{16.1} = 980 \text{ K}$$

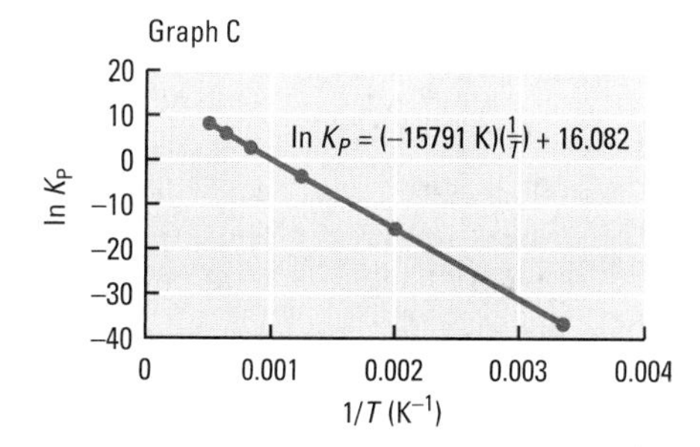

Visit this book's companion website at **www.cengage.com/chemistry/moore** to work an interactive module based on this material.

temperature, the concentration of products is greater and that of the reactants is smaller. Likewise, and as illustrated by Problem-Solving Example 13.7, the opposite is true for an exothermic reaction: ***For an exothermic reaction, an increase in temperature always means a decrease in K_c; the reaction will become less product-favored at higher temperatures.***

The effect of temperature on the reaction of N_2 with O_2 has important consequences. This reaction produces NO in earth's atmosphere when lightning suddenly raises the temperature of the air along its path. Because the reverse reaction is slow at room temperature, and because after the lightning bolt is over, the air rapidly cools to normal temperatures, much of the NO that is produced does not react to re-form N_2 and O_2 as it would at equilibrium. This situation provides one natural mechanism by which nitrogen in the air can be converted into a form that can be used by plants. (Converting nitrogen into a useful form is called nitrogen fixation.) Humans have tried to use the same kind of process to produce NO and from it HNO_3 for use in manufac-

Another consequence of the shift toward products in the $N_2 + O_2$ reaction at high temperatures is that automobile engines emit small concentrations of NO. The NO is rapidly oxidized to brown NO_2 in the air above cities, and the NO_2 in turn produces many further reactions that create air pollution problems. One of the functions of catalytic converters in automobiles is to speed up reduction of these nitrogen oxides back to elemental nitrogen.

turing fertilizer. About a century ago scientists and many in the general public were worried that the earth's farmland could not grow enough food to support a growing population. Consequently, strenuous efforts were made to adjust the conditions of the $N_2 + O_2$ reaction so that significant yields of NO could be obtained. At the end of the nineteenth century, a chemical plant at Niagara Falls, New York (where there was plentiful electric power), operated an electric arc process for fixing nitrogen for several years. This electric arc plant was important because it was the first attempt to deal with the limitations on plant growth caused by lack of sufficient nitrogen in soils that had been heavily farmed.

PROBLEM-SOLVING EXAMPLE 13.7 Le Chatelier's Principle

Consider an equilibrium mixture of nitrogen, hydrogen, and ammonia in which the reaction is

$$N_2(g) + 3\,H_2(g) \rightleftharpoons 2\,NH_3(g) \qquad \Delta H^\circ = -92.2 \text{ kJ} \quad \text{at } 25\ ^\circ\text{C}$$

For each of the changes listed below, tell whether the value of K_c increases or decreases, and tell whether more NH_3 or less NH_3 is present at the new equilibrium established after the change.

(a) More H_2 is added (at a constant temperature of 25 °C and constant volume).
(b) The temperature is increased.
(c) The volume of the container is doubled (at constant temperature).

Answer

(a) K_c stays the same; more NH_3 is present.
(b) K_c decreases; less NH_3 is present.
(c) K_c stays the same; less NH_3 is present.

Strategy and Explanation

(a) Since the temperature does not change, the value of K_c does not change. Adding a reactant to the equilibrium mixture will shift the equilibrium toward the product, producing more NH_3. This can be seen in another way by considering the reaction quotient.

$$Q = \frac{(\text{conc. } NH_3)^2}{(\text{conc. } N_2)(\text{conc. } H_2)^3}$$

When more H_2 is added, the denominator becomes larger. This makes Q smaller than K_c, which predicts that the reaction should produce more product; that is, some of the added H_2 reacts with N_2 to make more NH_3. Notice also that the concentration of N_2 decreases, because some reacts with the H_2, but the concentration of H_2 in the new equilibrium is still greater than it was before.

(b) The reaction is exothermic. Increasing the temperature shifts the equilibrium in the endothermic direction—that is, to the left (toward the reactants). This shift leads to a decrease in the NH_3 concentration, an increase in the concentrations of H_2 and N_2, and a decrease in the value of K_c.

(c) Since the temperature is constant, the value of K_c must be constant. Doubling the volume should cause the reaction to shift toward a greater number of moles of gaseous substance—that is, toward the left. Doubling the volume would normally halve the pressure, but the shift of the equilibrium partially compensates for this effect, and the final equilibrium will be at a pressure somewhat more than half the pressure of the initial equilibrium.

PROBLEM-SOLVING PRACTICE 13.7

Consider the equilibrium between N_2O_4 and NO_2 in a closed system.

$$N_2O_4(g) \rightleftharpoons 2\,NO_2(g)$$

Draw Lewis structures for the molecules involved in this equilibrium. Based on the bonding in the molecules, predict whether the reaction is exothermic or endothermic; hence, predict whether the concentration of N_2O_4 is larger in an equilibrium system at 25 °C or at 80 °C. Verify your prediction by looking at Figure 13.3.

CONCEPTUAL
EXERCISE 13.11 Summarizing Le Chatelier's Principle

Construct a table to summarize your understanding of Le Chatelier's principle. Consider these changes in conditions:

(a) Addition of a reactant
(b) Removal of a reactant
(c) Addition of a product
(d) Removal of a product
(e) Increasing pressure by decreasing volume
(f) Decreasing pressure by increasing volume
(g) Increasing temperature
(h) Decreasing temperature

For each of these changes in conditions, indicate (1) how the reaction system changes to achieve a new equilibrium, (2) in which direction the equilibrium reaction shifts, and (3) whether the value of K_c changes and, if so, in which direction. For some of these changes there are qualifications. For example, increasing pressure by decreasing volume does not always shift an equilibrium. List as many of these qualifications as you can.

13.7 Equilibrium at the Nanoscale

In Section 13.2 (**p. 464**), we used the isomerization of *cis*-2-butene to show that both forward and reverse reactions occur simultaneously in an equilibrium system.

$$\textit{cis}\text{-2-butene(g)} \rightleftharpoons \textit{trans}\text{-2-butene(g)} \qquad K_c = \frac{[\textit{trans}]}{[\textit{cis}]} = 2.0 \qquad \text{(at 415 K)}$$

Because the equilibrium constant is 2.0, there are twice as many *trans* molecules as *cis* molecules in an equilibrium mixture at 415 K. In other words, two thirds of the molecules have the *trans* structure and one third have the *cis* structure. Because the molecules are continually reacting in both the forward and the reverse directions, another way to think about this situation is that each molecule is *trans* two thirds of the time and *cis* one third of the time.

Based on its molecular structure, you might think that a 2-butene molecule ought to be just as likely to be *cis* as *trans*, so why is the *trans* isomer favored at 415 K? In Figure 12.7 (**p. 434**) we noted that one molecule of the *trans* isomer is 7×10^{-21} J lower in energy than one molecule of the *cis* isomer, and this difference is all important. For a mole of molecules, the difference in energy is $(-7 \times 10^{-21}\ \text{J})(6.022 \times 10^{23}\ \text{mol}^{-1}) = -4 \times 10^3\ \text{J/mol} = -4\ \text{kJ/mol}$, giving the thermochemical expression

$$\textit{cis}\text{-2-butene(g)} \rightleftharpoons \textit{trans}\text{-2-butene(g)} \qquad \Delta H^\circ = -4\ \text{kJ}$$

Consider the rate constants for the forward and reverse reactions in the isomerization of 2-butene. Based on Figure 12.7, E_a(forward) = 262 kJ/mol and E_a(reverse) = 266 kJ/mol. This means that the rate constant for the reverse reaction will be smaller than the rate constant for the forward reaction. At equilibrium the forward and reverse rates are equal, so

Larger k value and smaller concentration

Smaller k value and larger concentration

$$k_{\text{forward}} \times [\textit{cis}] = k_{\text{reverse}} \times [\textit{trans}]$$

If k_{reverse} is smaller than k_{forward}, then the concentration of *trans* must be larger than the concentration of *cis*, or the rates would not be equal. Because *cis*-2-butene is 4 kJ/mol higher in energy than *trans*-2-butene, it occurs only half as often as *trans*-2-butene at 415 K. We can generalize that ***in an equilibrium system, molecules that are higher in energy occur less often.*** We shall refer to this tendency as the energy effect on the position of equilibrium.

A second energy factor also affects how large an equilibrium constant is. It depends on how spread out in space the total energy of products is compared to the total energy of reactants and can be illustrated by the dissociation of dinitrogen tetraoxide.

$$N_2O_4(g) \rightleftharpoons 2\ NO_2(g) \qquad \Delta H^\circ = 57.2\ kJ$$

In a constant-pressure system (such as a reaction at atmospheric pressure), the two moles of product molecules occupy twice the volume of the one mole of reactant molecules. This means that the energy of the NO_2 molecules is spread over twice the volume that would have been occupied by the N_2O_4 molecules. A thermodynamic quantity called **entropy** provides a quantitative measure of how much energy is spread out when something happens. Entropy will be defined more completely in Section 17.3. For now it suffices to say that ***if there are more product molecules than reactant molecules, entropy favors the products in an equilibrium system.*** Although it depends on spreading out of energy, we shall refer to this tendency as the entropy effect on the position of equilibrium.

Despite the fact that the entropy factor favors NO_2 in the dissociation of N_2O_4, when data from Table 13.1 are used to calculate K_c at 25 °C (298 K) the result is $1/(1.7 \times 10^2) = 5.9 \times 10^{-3}$. This is because the dissociation of N_2O_4 is endothermic ($\Delta H^\circ = 57.2$ kJ). The enthalpy of 2 mol NO_2 is 57.2 kJ higher than the enthalpy of 1 mol N_2O_4. From our first generalization that molecules with more energy occur less often, we expect that 2 NO_2 is less likely than N_2O_4 because of the energy effect. Therefore the equilibrium constant should be smaller than the entropy argument predicts. *Both the energy effect and the entropy effect must be taken into account to predict an equilibrium constant value.*

$N_2O_4(g) \rightleftharpoons 2\ NO_2(g)$

T (K)	K_c
298	5.9×10^{-3}
350	1.3×10^{-1}
400	1.5×10^{0}
500	4.6×10^{1}
600	4.6×10^{2}

These ideas can also help us to understand the temperature dependence of the equilibrium constant. K_c values for the dissociation of N_2O_4 are given in the table in the margin. As the temperature rises, K_c becomes much larger. At high temperatures the molecules have lots of energy, and when the volume doubles, much more energy is dispersed. The energy difference between reactants and products (ΔH) is smaller relative to the average energy per molecule. ***The higher the temperature is, the less important the energy effect (ΔH) becomes, and the more the entropy effect (dispersal of energy) determines the position of equilibrium.***

We have shown that if the number of gas phase molecules increases when a reaction occurs, then the products will be favored by entropy. This is not the only way for the products of a reaction to have greater entropy than the reactants, but it is one of the most important. In Chapter 17 we will show how entropy can be measured and tabulated. That discussion will enable us to use enthalpy changes and entropy changes for a reaction to calculate its equilibrium constant at a variety of temperatures.

PROBLEM-SOLVING EXAMPLE 13.8 Energy and Entropy Effects on Equilibria

For the equilibrium

$$CH_4(g) + H_2O(g) \rightleftharpoons CO(g) + 3\ H_2(g)$$

(a) Estimate whether the entropy increases, decreases, or remains the same when products form.
(b) Does the entropy effect favor reactants or products?
(c) Use data from Appendix J to calculate ΔH°.
(d) Does the energy effect favor reactants or products?
(e) Is the reaction likely to be product-favored at high temperatures? Why or why not?

Answer

(a) Entropy increases
(b) Entropy favors products
(c) $\Delta H^\circ = 206.10$ kJ
(d) Reactants

(e) Yes. Entropy favors products and the energy effect is less important at high temperatures.

Strategy and Explanation Consider how many gas phase reactants and products there are and apply the ideas about entropy in the previous paragraphs.

(a) Because there are 4 mol gas phase products and only 2 mol gas phase reactants, the products will have higher entropy.

(b) The products have higher entropy, and the products are favored.

(c) $\Delta H^\circ = \Sigma\{(\text{moles product})\Delta H_f^\circ(\text{product})\} - \Sigma\{(\text{moles reactant})\Delta H_f^\circ(\text{reactant})\}$

$= \{(1\text{ mol})(-110.525\text{ kJ/mol})\}$

$- \{(1\text{ mol})(-74.81\text{ kJ/mol}) + (1\text{ mol})(-241.818\text{ kJ/mol})\}$

$= -110.525\text{ kJ} - (-316.628\text{ kJ})$

$= 206.10\text{ kJ}$

(d) Because the reactants are lower in energy, they are favored by the energy effect.

(e) The reaction is product-favored at high temperatures because the energy effect favoring reactants becomes less important, and there is a relatively large entropy effect (four product molecules for every two reactant molecules in the gas phase).

☑ **Reasonable Answer Check** In part (c) it is reasonable that the reaction is endothermic, because six bonds are broken and only four are formed when reactant molecules are changed into product molecules.

PROBLEM-SOLVING PRACTICE 13.8

For the reaction

$$2\ SOCl_2(g) + O_2(g) \rightleftharpoons 2\ SO_2(g) + 2\ Cl_2(g)$$

(a) Does the entropy effect favor products?

(b) Does the energy effect favor products?

(c) Will there be a greater concentration of $SO_2(g)$ at high temperature or at low temperature? Explain.

13.8 Controlling Chemical Reactions: The Haber-Bosch Process

The principles that allow us to control a reaction are based on our understanding of both equilibrium systems and the rates of chemical reactions. Some generalizations about equilibrium systems are

- ***A product-favored reaction has an equilibrium constant larger than 1.***
- ***If a reaction is exothermic, this energy factor favors the products.***
- ***If there is an increase in entropy when a reaction occurs, this entropy factor favors the products.***
- ***Product-favored reactions at low temperatures are usually exothermic.***
- ***Product-favored reactions at high temperatures are usually ones in which the entropy increases (energy becomes more dispersed).***

Using these general rules about equilibria, we can often predict whether a reaction is capable of yielding products. But it is also important that those products be produced rapidly. Recall these useful generalizations about reaction rates from Chapter 12.

- ***Reactions in the gas phase or in solution, where molecules of one reactant are completely mixed with molecules of another, occur more rapidly than do reactions between pure liquids or solids that do not dissolve in one another (⇐ p. 416).***
- ***Reactions occur more rapidly at high temperatures than at low temperatures (⇐ p. 436).***

For gases, higher concentration corresponds to higher partial pressure.

Because finely divided solids react more rapidly, coal is ground to a powder before it is burned to generate electricity.

The current interest in biotechnology is largely driven by the fact that naturally occurring enzymes are among the most effective catalysts known *(⇐ p. 448)*.

It is estimated that 40% to 60% of the nitrogen in the average human body has come from ammonia produced by the Haber-Bosch process and that increased agricultural productivity resulting from this process supports about 40% of the world's population.

- ***Reactions are faster when the reactant concentrations are high than when they are low (⇐ p. 422).***
- ***Reactions between a solid and a gas, or between a solid and something dissolved in solution, are usually much faster when the solid particles are as small as possible (⇐ p. 416).***
- ***Reactions are faster in the presence of a catalyst. Often the right catalyst makes the difference between success and failure in industrial chemistry (⇐ p. 445).***

One of the best examples of the application of the principles of chemical reactivity is the chemical reaction used for the synthesis of ammonia from its elements. Even though the earth is bathed in an atmosphere that is about 80% N_2 gas, nitrogen cannot be used by most plants until it has been fixed—that is, converted into biologically useful forms. Although nitrogen fixation is done naturally by organisms such as cyanobacteria and some field crops such as alfalfa and soybeans, most plants cannot fix N_2. They must instead obtain nitrogen from cyanobacteria, some other organism, or fertilizer. Proper fertilization is especially important for recently developed varieties of wheat, corn, and rice that have resulted in much improved food production.

Direct combination of nitrogen and oxygen was used at the beginning of the twentieth century to provide fertilizer, but this process was not very efficient ***(⇐ p. 490)***. A much better way of manufacturing ammonia was devised by Fritz Haber and Carl Bosch, who chose the *direct synthesis of ammonia from its elements* as the basis for an industrial process.

$$N_2(g) + 3\,H_2(g) \rightleftharpoons 2\,NH_3(g)$$

Oesper Collection in the History of Chemistry, University of Cincinnati

Fritz Haber 1868–1934

The industrial chemical process by which ammonia is manufactured was developed by Fritz Haber, a chemist, and Carl Bosch, an engineer. Haber's studies in the early 1900s revealed that direct ammonia synthesis should be possible. In 1914 the engineering problems were solved by Bosch. Haber's contract with the manufacturer of ammonia called for him to receive 1 pfennig (one hundredth of a German mark—similar to a penny) per kilogram of ammonia, and he soon became not only famous but also rich. In 1918 he was awarded a Nobel Prize for the ammonia synthesis, but the choice was criticized because of his role in developing the use of poison gases for Germany during World War I.

CONCEPTUAL **EXERCISE** **13.12 Ammonia Synthesis**

For the ammonia synthesis reaction, predict

(a) Whether the reaction is exothermic or endothermic.
(b) Whether the reaction product is favored by entropy.
(c) Whether the reaction produces more products at low or high temperatures.
(d) What would happen if you tried to increase the rate of the reaction by increasing the temperature.

At first glance this reaction might seem to be a poor choice. Hydrogen is available naturally only in combined form—for example, in water or hydrocarbons—meaning that hydrogen must be extracted from these compounds at considerable expense in energy resources and money. As you discovered in Problem-Solving Practice 13.8, the ammonia synthesis reaction becomes less product-favored at higher temperatures. But higher temperatures are needed for ammonia to be produced fast enough for the process to be efficient and economical. Nonetheless, the **Haber-Bosch process** (shown schematically in Figure 13.10) has been so well developed that ammonia is inexpensive (less than $600 per ton). For this reason it is widely used as a fertilizer and is often among the "top five" chemicals produced in the United States. Annual U.S. production of NH_3 by the Haber-Bosch process is approximately 18 billion pounds and about 12 billion pounds are imported each year.

Both the thermodynamics and the kinetics of the direct synthesis of ammonia have been carefully studied and fine-tuned by industry so that the maximum yield of product is obtained in a reasonable time and at a reasonable cost of both money and energy resources.

- The reaction is exothermic, and there is a decrease in entropy when it takes place. Therefore this reaction is predicted to be product-favored at low temperatures, but reactant-favored at high temperatures. (You should have made this prediction in

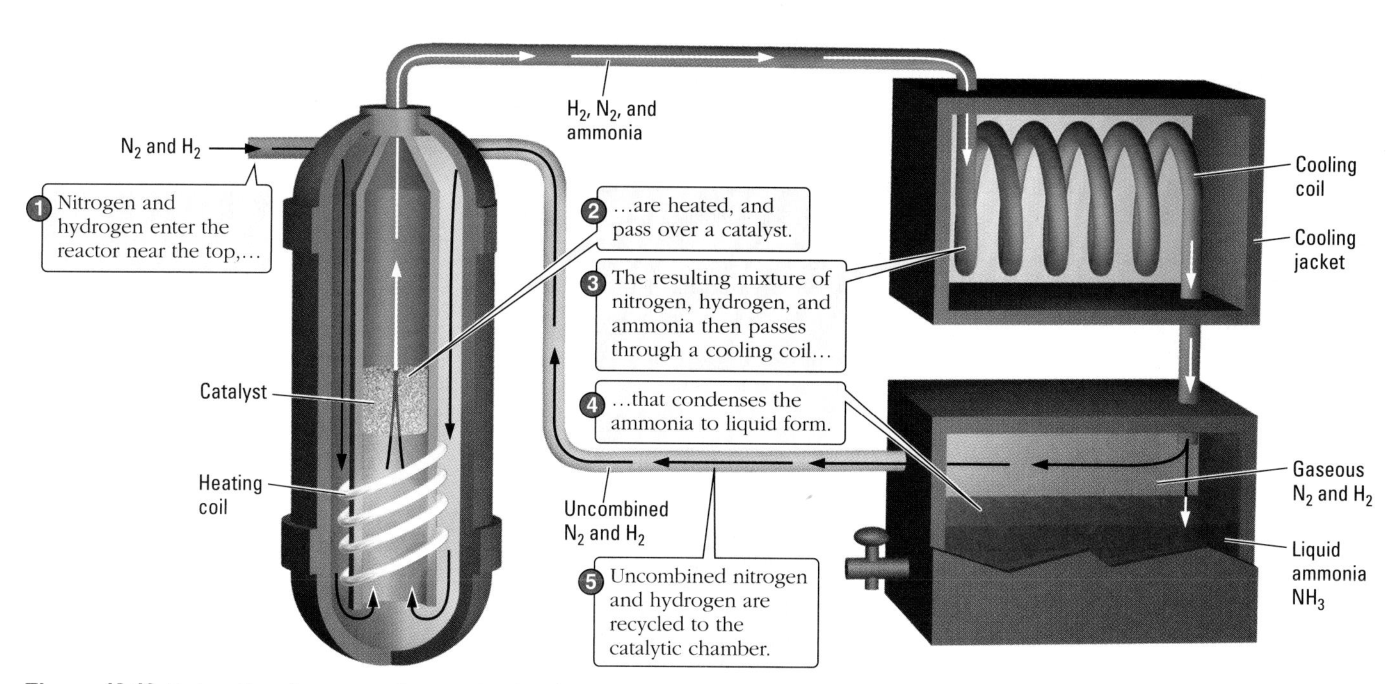

Figure 13.10 Haber-Bosch process for synthesis of ammonia (schematic).

Problem-Solving Practice 13.8, and it is in accord with Le Chatelier's principle for an exothermic process.)

- To increase the equilibrium concentration of NH_3, the reaction is carried out at high pressure (200 atm). This does not change the value of K_c, or K_P, but an increase in pressure can be compensated for by converting N_2 and H_2 to NH_3; 2 mol NH_3(g) exerts less pressure than a total of 4 mol gaseous reactants [N_2(g) + 3 H_2(g)] in the same-sized container.
- Ammonia is continually liquefied and removed from the reaction vessel, which reduces the concentration of the product of the reaction and shifts the equilibrium toward the right.
- The reaction is quite slow at room temperature, so the temperature must be raised to increase the rate. Although the rate increases with increasing temperature, the equilibrium constant declines. Thus, the faster the reaction, the smaller the yield.
- The temperature cannot be raised too much in an attempt to increase the rate, but a rate increase can be achieved with a catalyst. An effective catalyst for the Haber-Bosch process is Fe_3O_4 mixed with KOH, SiO_2, and Al_2O_3. Since the catalyst is not effective below about 400 °C, the optimum temperature, considering all the factors controlling the reaction, is about 450 °C.

Making predictions about chemical reactivity is part of the challenge, the adventure, and the art of chemistry. Many chemists enjoy the challenge of making useful new materials, which usually means choosing to make them by reactions that we believe will be product-favored and reasonably rapid. Such predictions are based on the ideas outlined in this chapter and Chapter 12.

IN CLOSING

Having studied this chapter, you should be able to . . .

- Recognize a system at equilibrium and describe the properties of equilibrium systems (Section 13.1).
- Describe the dynamic nature of equilibrium and the changes in concentrations of reactants and products that occur as a system approaches equilibrium (Sections 13.1, 13.2). ■ End-of-chapter Question assignable in OWL: 7
- Write equilibrium constant expressions, given balanced chemical equations (Section 13.2). ■ Question 11
- Obtain equilibrium constant expressions for related reactions from the expression for one or more known reactions (Section 13.2). ■ Question 13
- Calculate K_P from K_c or K_c from K_P for the same equilibrium (Section 13.2).
- Calculate a value of K_c for an equilibrium system, given information about initial concentrations and equilibrium concentrations (Section 13.3). ■ Question 21
- Make qualitative predictions about the extent of reaction based on equilibrium constant values—that is, predict whether a reaction is product-favored or reactant-favored based on the size of the equilibrium constant (Section 13.4). ■ Question 23
- Calculate concentrations of reactants and products in an equilibrium system if K_c and initial concentrations are known (Section 13.5). ■ Questions 27, 33, 67
- Use the reaction quotient Q to predict in which direction a reaction will go to reach equilibrium (Section 13.5). ■ Questions 31, 32
- Use Le Chatelier's principle to show how changes in concentrations, pressure or volume, and temperature shift chemical equilibria (Section 13.6). ■ Question 39
- Use the change in enthalpy and the change in entropy qualitatively to predict whether products are favored over reactants (Section 13.7). ■ Questions 45, 55
- List the factors affecting chemical reactivity, and apply them to predicting optimal conditions for producing products (Section 13.8).

Selected end-of-chapter Questions may be assigned in OWL.

For quick review, download Go Chemistry mini-lecture flashcard modules (or purchase them at **www.ichapters.com**).

Prepare for an exam with a **Summary Problem** that brings together concepts and problem-solving skills from throughout the chapter. Go to **www.cengage.com/chemistry/moore** to download this chapter's Summary Problem from the Student Companion Site.

KEY TERMS

chemical equilibrium *(Section 13.1)*

dynamic equilibrium *(13.1)*

entropy *(13.7)*

equilibrium concentration *(13.3)*

equilibrium constant, *K* *(13.2)*

equilibrium constant expression *(13.2)*

Haber-Bosch process *(13.8)*

Le Chatelier's principle *(13.6)*

product-favored *(13.1)*

reactant-favored *(13.1)*

reaction quotient, *Q* *(13.5)*

shifting an equilibrium *(13.6)*

QUESTIONS FOR REVIEW AND THOUGHT

■ denotes questions assignable in OWL.

Blue-numbered questions have short answers at the back of this book and fully worked solutions in the *Student Solutions Manual*.

Review Questions

1. Define the terms *chemical equilibrium* and *dynamic equilibrium.*
2. If an equilibrium is product-favored, is its equilibrium constant large or small with respect to 1? Explain.
3. List three characteristics that you would need to verify in order to determine that a chemical system is at equilibrium.
4. The decomposition of ammonium dichromate,

$$(NH_4)_2Cr_2O_7(s),$$

yields nitrogen gas, water vapor, and solid chromium(III) oxide. The reaction is exothermic. In a closed container this process reaches an equilibrium state. Write a balanced equation for the equilibrium reaction. How is the equilibrium affected if

(a) More ammonium dichromate is added to the equilibrium system?
(b) More water vapor is added?
(c) More chromium(III) oxide is added?

© Cengage Learning/Charles D. Winters

Decomposition of $(NH_4)_2Cr_2O_7$.

5. For the equilibrium reaction in Question 4, write the expression for the equilibrium constant.
   (a) How would this equilibrium constant change if the total pressure on the system were doubled?
   (b) How would the equilibrium constant change if the temperature were increased?
6. Indicate whether each statement below is true or false. If a statement is false, rewrite it to produce a closely related statement that is true.
   (a) For a given reaction, the magnitude of the equilibrium constant is independent of temperature.
   (b) If there is an increase in entropy and a decrease in enthalpy when reactants in their standard states are converted to products in their standard states, the equilibrium constant for the reaction will be negative.
   (c) The equilibrium constant for the reverse of a reaction is the reciprocal of the equilibrium constant for the reaction itself.
   (d) For the reaction

$$H_2O_2(\ell) \rightleftharpoons H_2O(\ell) + \tfrac{1}{2}O_2(g)$$

   the equilibrium constant is one-half the magnitude of the equilibrium constant for the reaction

$$2\,H_2O_2(\ell) \rightleftharpoons 2\,H_2O(\ell) + O_2(g)$$

Topical Questions

Characteristics of Chemical Equilibrium

7. ■ Think of an experiment you could do to demonstrate that the equilibrium

$$2\,NO_2(g) \rightleftharpoons N_2O_4(g)$$

   is a dynamic process in which the forward and reverse reactions continue to occur after equilibrium has been achieved. Describe how such an experiment might be carried out.
8. Discuss this statement: "No true chemical equilibrium can exist unless reactant molecules are constantly changing into product molecules, and vice versa."

The Equilibrium Constant

9. Consider the gas phase reaction of $N_2 + O_2$ to give 2 NO and the reverse reaction of 2 NO to give $N_2 + O_2$, discussed in Section 13.2. An equilibrium mixture of NO, N_2, and O_2 at 5000. K that contains equal concentrations of N_2 and O_2 has a concentration of NO about half as great. Make qualitatively correct plots of the concentrations of reactants and products versus time for these two processes, showing the initial state and the final dynamic equilibrium state. Assume a temperature of 5000. K. Don't do any calculations—just sketch how you think the plots will look.
10. After 0.1 mol of pure *cis*-2-butene is allowed to come to equilibrium with *trans*-2-butene in a closed, 5.0-L flask at 25 °C, another 0.1 mol *cis*-2-butene is suddenly added to the flask.
    (a) Is the new mixture at equilibrium? Explain why or why not.
    (b) In the new mixture, immediately after addition of the *cis*-2-butene, which rate is faster: *cis* → *trans* or *trans* → *cis*? Or are both rates the same?
    (c) After the second 0.1 mol *cis*-2-butene has been added and the system is at equilibrium, if the concentration of *trans*-2-butene is 0.01 mol/L, what is the concentration of *cis*-2-butene?
11. ■ Write the equilibrium constant expression for each reaction.
    (a) $2\,H_2O_2(g) \rightleftharpoons 2\,H_2O(g) + O_2(g)$
    (b) $PCl_3(g) + Cl_2(g) \rightleftharpoons PCl_5(g)$
    (c) $SiO_2(s) + 3\,C(s) \rightleftharpoons SiC(s) + 2\,CO(g)$
    (d) $H_2(g) + \frac{1}{8}S_8(s) \rightleftharpoons H_2S(g)$
12. Write the equilibrium constant expression for each reaction.
    (a) $3\,O_2 \rightleftharpoons 2\,O_3(g)$
    (b) $SiH_4(g) + 2\,O_2(g) \rightleftharpoons SiO_2(s) + 2\,H_2O(g)$
    (c) $MgO(s) + SO_2(g) + \frac{1}{2}O_2(g) \rightleftharpoons MgSO_4(s)$
    (d) $2\,PbS(s) + 3\,O_2(g) \rightleftharpoons 2\,PbO(s) + 2\,SO_2(g)$
13. ■ Consider these two equilibria involving $SO_2(g)$ and their corresponding equilibrium constants.

$$SO_2(g) + \tfrac{1}{2}O_2(g) \rightleftharpoons SO_3(g) \qquad K_{c_1}$$
$$2\,SO_3(g) \rightleftharpoons 2\,SO_2(g) + O_2(g) \qquad K_{c_2}$$

    Which of these expressions correctly relates K_{c_1} to K_{c_2}?
    (a) $K_{c_2} = K_{c_1}^2$ (b) $K_{c_2}^2 = K_{c_1}$
    (c) $K_{c_2} = 1/K_{c_1}$ (d) $K_{c_2} = K_{c_1}$
    (e) $K_{c_2} = 1/K_{c_1}^2$
14. The reaction of hydrazine, N_2H_4, with chlorine trifluoride, ClF_3, has been used in experimental rocket motors.

$$N_2H_4(g) + \tfrac{4}{3}ClF_3(g) \rightleftharpoons 4\,HF(g) + N_2(g) + \tfrac{2}{3}Cl_2(g)$$

    How is the equilibrium constant, K_P, for this reaction related to K_P' for the reaction written this way?

$$3\,N_2H_4(g) + 4\,ClF_3(g) \rightleftharpoons 12\,HF(g) + 3\,N_2(g) + 2\,Cl_2(g)$$

    (a) $K_P = K_P'$ (b) $K_P = 1/K_P'$
    (c) $K_P^3 = K_P'$ (d) $K_P = (K_P')^3$
    (e) $3K_P = K_P'$

■ in OWL Blue-numbered questions answered at the back of this book

15. For each reaction in Question 11, write the equilibrium constant expression for K_P.
16. For each reaction in Question 12, write the equilibrium constant expression for K_P.

Determining Equilibrium Constants

17. Isomer A is in equilibrium with isomer B, as in the reaction

$$A(g) \rightleftharpoons B(g)$$

Three experiments are done, each at the same temperature, and equilibrium concentrations are measured. For each experiment, calculate the equilibrium constant, K_c.
(a) [A] = 0.74 mol/L, [B] = 0.74 mol/L
(b) [A] = 2.0 mol/L, [B] = 2.0 mol/L
(c) [A] = 0.01 mol/L, [B] = 0.01 mol/L

18. This reaction was examined at 250 °C.

$$PCl_5(g) \rightleftharpoons PCl_3(g) + Cl_2(g)$$

At equilibrium, $[PCl_5] = 4.2 \times 10^{-5}$ M, $[PCl_3] = 1.3 \times 10^{-2}$ M, and $[Cl_2] = 3.9 \times 10^{-3}$ M. Calculate the equilibrium constant K_c for the reaction.

19. At high temperature, hydrogen and carbon dioxide react to give water and carbon monoxide.

$$H_2(g) + CO_2(g) \rightleftharpoons H_2O(g) + CO(g)$$

Laboratory measurements at 986 °C show that there is 0.11 mol each of CO and water vapor and 0.087 mol each of H_2 and CO_2 at equilibrium in a 1.0-L container. Calculate the equilibrium constant K_P for the reaction at 986 °C.

20. Carbon dioxide reacts with carbon to give carbon monoxide according to the equation

$$C(s) + CO_2(g) \rightleftharpoons 2\ CO(g)$$

At 700. °C, a 2.0-L flask is found to contain at equilibrium 0.10 mol CO, 0.20 mol CO_2, and 0.40 mol C. Calculate the equilibrium constant K_P for this reaction at the specified temperature.

21. ■ An equilibrium mixture contains 3.00 mol CO, 2.00 mol Cl_2, and 9.00 mol $COCl_2$ in a 50.-L reaction flask at 800. K. Calculate the value of the equilibrium constant K_c for the reaction

$$CO(g) + Cl_2(g) \rightleftharpoons COCl_2(g)$$

at this temperature.

22. At 667 K, HI is found to be 11.4% dissociated into its elements.

$$2\ HI(g) \rightleftharpoons H_2(g) + I_2(g)$$

If 1.00 mol HI is placed in a 1.00-L container and heated to 667 K, calculate (a) the equilibrium concentration of all three substances and (b) the value of K_c for this equilibrium at this temperature.

The Meaning of the Equilibrium Constant

23. ■ Using the data of Table 13.1, predict which of these reactions will be product-favored at 25 °C. Then place all the reactions in order from most reactant-favored to most product-favored.
(a) $2\ NH_3(g) \rightleftharpoons N_2(g) + 3\ H_2(g)$
(b) $NH_4^+(aq) + OH^-(aq) \rightleftharpoons NH_3(aq) + H_2O(\ell)$
(c) $2\ NO(g) \rightleftharpoons N_2(g) + O_2(g)$

24. Using the data of Table 13.1, predict which of these reactions will be product-favored at 25 °C. Then place all the reactions in order from most reactant-favored to most product-favored.
(a) $2\ NO_2(g) \rightleftharpoons N_2O_4(g)$
(b) $H_2CO_3(aq) + H_2O(\ell) \rightleftharpoons HCO_3^-(aq) + H_3O^+(aq)$
(c) $AgI(s) \rightleftharpoons Ag^+(aq) + I^-(aq)$

Using Equilibrium Constants

25. The hydrocarbon C_4H_{10} can exist in two forms: butane and 2-methylpropane. The value of K_c for conversion of butane to 2-methylpropane is 2.5 at 25 °C.

$$CH_3-CH_2-CH_2-CH_3 \rightleftharpoons H-\underset{CH_3}{\overset{CH_3}{\underset{|}{\overset{|}{C}}}}-CH_3$$

butane 2-methylpropane

(a) Suppose that the initial concentrations of both butane and 2-methylpropane are 0.100 mol/L. Make up a table of initial concentrations, change in concentrations, and equilibrium concentrations for this reaction.
(b) Write the equilibrium constant expression in terms of x, the change in the concentration of butane, and then solve for x.
(c) If you place 0.017 mol butane in a 0.50-L flask at 25 °C, what will be the equilibrium concentrations of the two isomers?

26. A mixture of butane and 2-methylpropane at 25 °C has [butane] = 0.025 mol/L and [2-methylpropane] = 0.035 mol/L. Is this mixture at equilibrium? If the *equilibrium* concentration of butane is 0.025 mol/L, what must [2-methylpropane] be at equilibrium? (See the reaction and K_c value in Question 25.)

27. ■ Hydrogen gas and iodine gas react via the equation

$$H_2(g) + I_2(g) \rightleftharpoons 2\ HI(g) \qquad K_c = 76 \quad (\text{at } 600.\ K)$$

If 0.050 mol HI is placed in an empty 1.0-L flask at 600. K, what are the equilibrium concentrations of HI, I_2, and H_2?

28. Consider the equilibrium

$$N_2(g) + O_2(g) \rightleftharpoons 2\ NO(g)$$

At 2300 K the equilibrium constant $K_c = 1.7 \times 10^{-3}$. If 0.15 mol NO(g) is placed into a 10.0-L flask and heated to 2300 K, what are the equilibrium concentrations of all three substances?

29. The equilibrium constant K_c for the reaction

$$H_2(g) + I_2(g) \rightleftharpoons 2\ HI(g)$$

has the value 50.0 at 745 K.
(a) When 1.00 mol I_2 and 3.00 mol H_2 are allowed to come to equilibrium at 745 K in a flask of volume 10.00 L, what amount (in moles) of HI will be produced?
(b) What amount of HI is produced in a 5.00-L flask?
(c) What total amount of HI is present at equilibrium if an additional 3.00 mol H_2 is added to the 10.00-L flask?

30. The equilibrium constant K_c for the *cis-trans* isomerization of gaseous 2-butene has the value 1.50 at 580. K.

$$\underset{cis}{H(H_3C)C{=}C(H)CH_3} \rightleftharpoons \underset{trans}{H(H_3C)C{=}C(CH_3)H}$$

(a) Is the reaction product-favored at 580. K?
(b) Calculate the amount (in moles) of *trans* isomer produced when 1 mol *cis*-2-butene is heated to 580. K in the presence of a catalyst in a flask of volume 1.00 L and reaches equilibrium.
(c) What would the answer be if the flask had a volume of 10.0 L?

31. ■ Consider the equilibrium at 25 °C

$$2\ SO_3(g) \rightleftharpoons 2\ SO_2(g) + O_2(g) \quad K_c = 3.58 \times 10^{-3}$$

Suppose that 0.15 mol SO_3(g), 0.015 mol SO_2(g), and 0.0075 mol O_2(g) are placed into a 10.0-L flask at 25 °C.
(a) Is the system at equilibrium?
(b) If the system is not at equilibrium, in which direction must the reaction proceed to reach equilibrium? Explain your answer.

32. Consider the equilibrium at 25 °C

$$CH_4(g) + H_2O(g) \rightleftharpoons CO(g) + 3\ H_2(g) \quad K_c = 9.4 \times 10^{-1}$$

Suppose that 0.25 mol CH_4(g), 0.15 mol H_2O(g), 0.25 mol CO(g), and 0.15 mol H_2(g) are placed into a 10.0-L flask at 25 °C.
(a) Is the system at equilibrium?
(b) If the system is not at equilibrium, in which direction must the reaction proceed to reach equilibrium? Explain your answer.

33. ■ Consider the equilibrium

$$N_2(g) + O_2(g) \rightleftharpoons 2\ NO(g)$$

At 2300 K the equilibrium constant $K_c = 1.7 \times 10^{-3}$. Suppose that 0.015 mol NO(g), 0.25 mol N_2(g), and 0.25 mol O_2(g) are placed into a 10.0-L flask and heated to 2300 K.
(a) Is the system at equilibrium?
(b) If not, in which direction must the reaction proceed to reach equilibrium?
(c) Calculate the equilibrium concentrations of all three substances.

34. Consider the equilibrium

$$H_2(g) + I_2(g) \rightleftharpoons 2\ HI(g)$$

At 745 K the equilibrium constant $K_c = 50.0$. Suppose that 0.75 mol HI(g), 0.025 mol H_2(g), and 0.025 mol I_2(g) are placed into a 20.0-L flask and heated to 745 K.
(a) Is the system at equilibrium?
(b) If not, in which direction must the reaction proceed to reach equilibrium?
(c) Calculate the equilibrium concentrations of all three substances.

Shifting a Chemical Equilibrium: Le Chatelier's Principle

35. Consider this equilibrium, established in a 2.0-L flask at 25 °C:

$$N_2O_4(g) \rightleftharpoons 2\ NO_2(g) \qquad \Delta H° = +57.2\ kJ$$

What will happen to the concentration of N_2O_4 if the temperature is increased? Explain your choice.
(a) It will increase. (b) It will decrease.
(c) It will not change.
(d) It is not possible to tell from the information provided.

36. Solid barium sulfate is in equilibrium with barium ions and sulfate ions in solution.

$$BaSO_4(s) \rightleftharpoons Ba^{2+}(aq) + SO_4^{2-}(aq)$$

What will happen to the barium ion concentration if more solid $BaSO_4$ is added to the flask? Explain your choice.
(a) It will increase. (b) It will decrease.
(c) It will not change.
(d) It is not possible to tell from the information provided.

37. Hydrogen, bromine, and HBr in the gas phase are in equilibrium in a container of fixed volume.

$$H_2(g) + Br_2(g) \rightleftharpoons 2\ HBr(g) \qquad \Delta H° = -103.7\ kJ$$

How will each of these changes affect the indicated quantities? Write "increase," "decrease," or "no change."

Change	$[Br_2]$	[HBr]	K_c	K_P
Some H_2 is added to the container.	____	____	____	____
The temperature of the gases in the container is increased.	____	____	____	____
The pressure of HBr is increased.	____	____	____	____

38. Nitrogen, oxygen, and nitrogen monoxide are in equilibrium in a container of fixed volume.

$$N_2(g) + O_2(g) \rightleftharpoons 2\ NO(g) \qquad \Delta H° = 180.5\ kJ$$

How will each of these changes affect the indicated quantities? Write "increase," "decrease," or "no change."

Change	$[N_2]$	[NO]	K_c	K_P
Some NO is added to the container.	____	____	____	____
The temperature of the gases in the container is decreased.	____	____	____	____
The pressure of N_2 is decreased.	____	____	____	____

39. ■ The equilibrium constant K_c for this reaction is 0.16 at 25 °C, and the standard enthalpy change is 16.1 kJ.

$$2\ NOBr(g) \rightleftharpoons 2\ NO(g) + Br_2(\ell)$$

Predict the effect of each of these changes on the position of the equilibrium; that is, state which way the equilibrium

will shift (left, right, or no change) when each of the following changes is made.
(a) Adding more Br_2 (b) Removing some NOBr
(c) Lowering the temperature

40. The formation of hydrogen sulfide from the elements is exothermic.

$$H_2(g) + \frac{1}{8} S_8(s) \rightleftharpoons H_2S(g) \qquad \Delta H° = -20.6 \text{ kJ}$$

Predict the effect of each of these changes on the position of the equilibrium; that is, state which way the equilibrium will shift (left, right, or no change) when each of the following changes is made.
(a) Adding more sulfur (b) Adding more H_2
(c) Raising the temperature

41. Consider the equilibrium

$$PbCl_2(s) \rightleftharpoons Pb^{2+}(aq) + 2\ Cl^-(aq)$$

(a) Will the equilibrium concentration of aqueous lead(II) ion increase, decrease, or remain the same if some solid NaCl is added to the flask?
(b) Make a graph like the one in Figure 13.6 to illustrate what happens to each of the concentrations after the NaCl is added.

42. Consider the transformation of butane into 2-methylpropane (see Question 25). The system is originally at equilibrium at 25 °C in a 1.0-L flask with [butane] = 0.010 M and [2-methylpropane] = 0.025 M. Suppose that 0.0050 mol of 2-methylpropane is suddenly added to the flask, and the system shifts to a new equilibrium.
(a) What is the new equilibrium concentration of each gas?
(b) Make a graph like the one in Figure 13.6 to show how the concentrations of the isomers change when the 2-methylpropane is added.

Equilibrium at the Nanoscale

43. For each of these reactions at 25 °C, indicate whether the entropy effect, the energy effect, both, or neither favors the reaction.
(a) $N_2(g) + 3\ F_2(g) \rightleftharpoons 2\ NF_3(g) \qquad \Delta H° = -249$ kJ
(b) $N_2F_4(g) \rightleftharpoons 2\ NF_2(g) \qquad \Delta H° = 93.3$ kJ
(c) $N_2(g) + 3\ Cl_2(g) \rightleftharpoons 2\ NCl_3(g) \qquad \Delta H° = 460$ kJ

44. For each of these chemical reactions, predict whether the equilibrium constant at 25 °C is greater than 1 or less than 1, or state that insufficient information is available. Also indicate whether each reaction is product-favored or reactant-favored.
(a) $2\ NaCl(s) \rightleftharpoons 2\ Na(s) + Cl_2(g) \qquad \Delta H° = -823$ kJ
(b) $2\ CO(g) + O_2(g) \rightleftharpoons 2\ CO_2(g) \qquad \Delta H° = -566$ kJ
(c) $3\ CO_2(g) + 4\ H_2O(g) \rightleftharpoons C_3H_8(g) + 5\ O_2(g)$
$\Delta H° = 2045$ kJ

Controlling Chemical Reactions: The Haber-Bosch Process

45. ■ Although ammonia is made in enormous quantities by the Haber-Bosch process, sulfuric acid is made in even greater quantities by the *contact process*. A simplified version of this process can be represented by these three reactions.

$$S(s) + O_2(g) \rightleftharpoons SO_2(g)$$
$$2\ SO_2(g) + O_2(g) \rightleftharpoons 2\ SO_3(g)$$
$$SO_3(g) + H_2O(\ell) \rightleftharpoons H_2SO_4(\ell)$$

(a) Use data from Appendix J to calculate $\Delta H°$ for each reaction.
(b) Which reactions are exothermic? Which are endothermic?
(c) In which of the reactions does entropy increase? In which does it decrease? In which does it stay about the same?
(d) For which reaction(s) do low temperatures favor formation of products?

46. Lime, CaO(s), can be produced by heating limestone, $CaCO_3(s)$, to cause a decomposition reaction.
(a) Write a balanced equation for the reaction.
(b) Predict the sign of the enthalpy change for the reaction.
(c) From the data in Appendix J, calculate $\Delta H°$ for this reaction at 25 °C to verify or contradict your prediction in part (b).
(d) Predict the entropy effect for this reaction.
(e) Is the reaction favored by entropy, energy, both, or neither?
(f) Explain in terms of Le Chatelier's principle why limestone must be heated to make lime.

General Questions

47. Write equilibrium constant expressions, in terms of reactant and product concentrations, for each of these reactions.

$2\ O_3(g) \rightleftharpoons 3\ O_2(g) \qquad K_c = 7 \times 10^{56}$
$2\ NO_2(g) \rightleftharpoons N_2O_4(g) \qquad K_c = 1.7 \times 10^2$
$HCOO^-(aq) + H^+(aq) \rightleftharpoons HCOOH(aq) \qquad K_c = 5.6 \times 10^3$
$Ag^+(aq) + I^-(aq) \rightleftharpoons AgI(s) \qquad K_c = 6.7 \times 10^{15}$

Assume that all gases and solutes have initial concentrations of 1.0 mol/L. Then let the *first* reactant in each reaction change its concentration by $-x$.
(a) Using the reaction table (ICE table) approach, write equilibrium constant expressions in terms of the unknown variable x for each reaction.
(b) Which of these expressions yield quadratic equations?
(c) How would you go about solving the others for x?

48. Many common nonmetallic elements exist as diatomic molecules at room temperature. When these elements are heated to 1500. K, the molecules break apart into atoms. A general equation for this type of reaction is

$$E_2(g) \rightleftharpoons 2\ E(g)$$

where E stands for each element. Equilibrium constants for dissociation of these molecules at 1500. K are

Species	K_c	Species	K_c
Br_2	8.9×10^{-2}	H_2	3.1×10^{-10}
Cl_2	3.4×10^{-3}	N_2	1×10^{-27}
F_2	7.4	O_2	1.6×10^{-11}

(a) If 1.00 mol of each diatomic molecule is placed in a separate 1.0-L container and heated to 1500. K, what is the equilibrium concentration of the atomic form of each element at 1500. K?

(b) From these results, predict which of the diatomic elements has the lowest bond dissociation energy, and compare your results with thermochemical calculations and with Lewis structures.

49. A small sample of *cis*-dichloroethene in which one carbon atom is the radioactive isotope ^{14}C is added to an equilibrium mixture of the *cis* and *trans* isomers at a certain temperature. Eventually, 40% of the radioactive molecules are found to be in the *trans* configuration at any given time.
(a) What is the value of K_c for the *cis* $\rightleftharpoons$ *trans* equilibrium?
(b) What would have happened if a small sample of radioactive *trans* isomer had been added instead of the *cis* isomer?

50. In a 0.0020 M solution of acetic acid, any acetate species spends 9% of its time as an acetate ion, CH_3COO^-, and the remaining 91% of its time as acetic acid, CH_3COOH. What is the value of K_c for the reaction?

$$CH_3COOH + H_2O \rightleftharpoons CH_3COO^- + H_3O^+$$

51. The following amounts of HI, H_2, and I_2 are introduced into a 10.00-L reaction flask and heated to 745 K.

	n_{HI} (mol)	n_{H_2} (mol)	n_{I_2} (mol)
Case a	1.0	0.10	0.10
Case b	10.	1.0	1.0
Case c	10.	10.	1.0
Case d	5.62	0.381	1.75

The equilibrium constant for the reaction

$$2\,HI(g) \rightleftharpoons H_2(g) + I_2(g)$$

has the value 0.0200 at 745 K. In which cases will the concentration of HI increase as equilibrium is attained, and in which cases will the concentration of HI decrease?

52. The following amounts of CO(g), H_2O(g), CO_2(g), and H_2(g) are introduced into a 10.00-L reaction flask and heated to 745 K.

	n_{CO} (mol)	n_{H_2O} (mol)	n_{CO_2} (mol)	n_{H_2O} (mol)
Case a	1.0	0.10	0.10	0.10
Case b	10.	1.0	1.0	1.0
Case c	10.	10.	1.0	1.0
Case d	5.62	0.381	1.75	1.75

The equilibrium constant for the reaction

$$CO(g) + H_2O(g) \rightleftharpoons CO_2(g) + H_2(g)$$

has the value $K_c = 4.00$ at 500 K. For which cases will the concentration of CO increase as equilibrium is attained, and in which cases will the concentration of CO decrease?

Applying Concepts

53. Suppose that you have heated a mixture of *cis*- and *trans*-2-pentene to 600. K, and after 1 h you find that the composition is 40% *cis*. After 4 h the composition is found to be 42% *cis*, and after 8 h it is 42% *cis*. Next, you heat the mixture to 800. K and find that the composition changes to 45% *cis*. When the mixture is cooled to 600. K and allowed to stand for 8 h, the composition is found to be 42% *cis*. Is this system at equilibrium at 600. K? Or would more experiments be needed before you could conclude that it was at equilibrium? If so, what experiments would you do?

54. For the reaction

$$cis\text{-2-butene} \rightleftharpoons trans\text{-2-butene}$$

K_c is 1.65 at 500. K, 1.47 at 600. K, and 1.36 at 700. K. Predict whether the conversion from the *cis* to the *trans* isomer of 2-butene is exothermic or endothermic.

Use this table to answer Questions 55 and 56.

Equilibrium Constants K_c for Some *Cis-Trans* Interconversions

$$(H)(R)C{=}C(H)(R)\ [cis] \rightleftharpoons (H)(R)C{=}C(R)(H)\ [trans]$$

Temperature (*K*)	R is F	R is Cl	R is CH_3
500.	0.420	0.608	1.65
600.	0.491	0.678	1.47
700.	0.549	0.732	1.36

55. ■ Based on the data in the table and these reaction equations,
(i) *cis*-(R is F) $\rightleftharpoons$ *trans*-(R is F)
(ii) *trans*-(R is Cl) $\rightleftharpoons$ *cis*-(R is Cl)
(iii) *cis*-(R is CH_3) $\rightleftharpoons$ *trans*-(R is CH_3)
(a) Which reaction is most exothermic?
(b) Which reaction is most endothermic?

56. Based on the data in the table and these reaction equations,
(i) *cis*-(R is F) $\rightleftharpoons$ *trans*-(R is F)
(ii) *trans*-(R is Cl) $\rightleftharpoons$ *cis*-(R is Cl)
(iii) *cis*-(R is CH_3) $\rightleftharpoons$ *trans*-(R is CH_3)
(a) Which reaction is most product-favored at 500. K?
(b) Which reaction is most reactant-favored at 700. K?

57. Figure 13.3 shows the equilibrium mixture of N_2O_4 and NO_2 at two different temperatures. Imagine that you can shrink yourself down to the size of the molecules in the two glass tubes and observe their behavior for a short period of time. Write a brief description of what you observe in each of the flasks.

58. Imagine yourself to be the size of ions and molecules inside a beaker containing this equilibrium mixture with a K_c greater than 1.

$$\underset{\text{pink}}{Co(H_2O)_6^{2+}(aq)} + 4\,Cl^-(aq) \rightleftharpoons \underset{\text{blue}}{CoCl_4^{2-}(aq)} + 6\,H_2O(\ell)$$

Write a brief description of what you observe around you before and after additional water is added to the mixture.

59. Which of the diagrams represent equilibrium mixtures for the reaction

$$A_2(g) + B_2(g) \rightleftharpoons 2\ AB(g)$$

at temperatures where $10^2 > K_c > 0.1$?

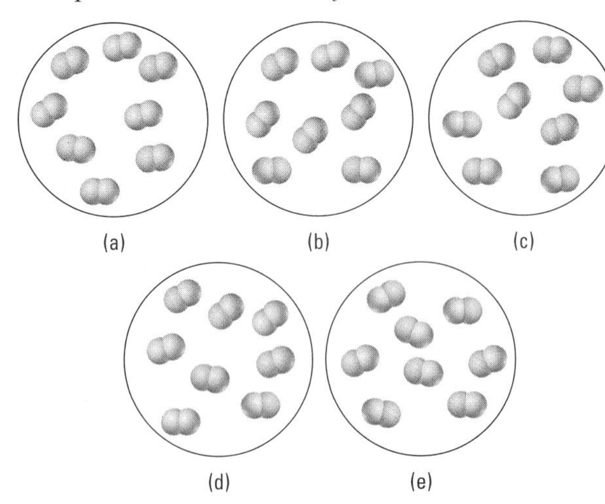

60. Which diagram in Question 59 best represents an equilibrium mixture with an equilibrium constant of
(a) 0.44? (b) 4.0? (c) 36?

61. Draw a nanoscale (particulate) level diagram for an equilibrium mixture of

$$CO(g) + H_2O(g) \rightleftharpoons CO_2(g) + H_2(g) \qquad K_c = 4.00$$

62. The diagram below represents an equilibrium mixture for the reaction

$$N_2(g) + O_2(g) \rightleftharpoons 2\ NO(g)$$

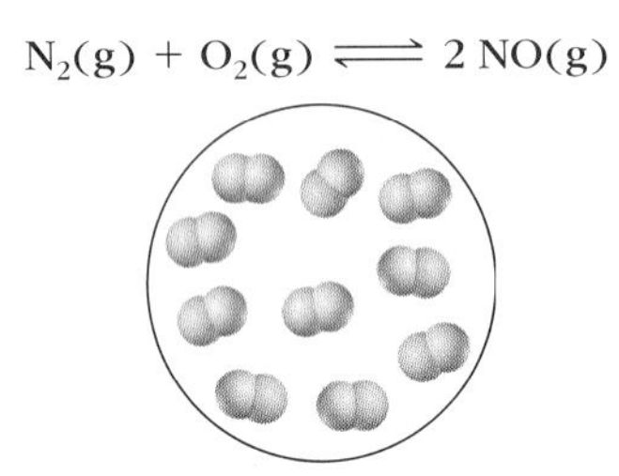

What is the equilibrium constant?

63. A sample of solid benzoic acid, a carboxylic acid, is in equilibrium with an aqueous solution of benzoic acid. A tiny quantity of D_2O, water containing the isotope ^2H, deuterium, is added to the solution. The solution is allowed to stand at constant temperature for several hours, after which some of the solid benzoic acid is removed and analyzed. The benzoic acid is found to contain a tiny quantity of deuterium, D, and the formula of the deuterium-containing molecules is C_6H_5COOD. Explain how this can happen.

64. In a second experiment with benzoic acid (see Question 63), a tiny quantity of water that contains the isotope ^{18}O is added to a saturated solution of benzoic acid in water. When some of the solid benzoic acid is analyzed, no ^{18}O is found in the benzoic acid. Compare this situation with the experiment involving deuterium, and explain how the results of the two experiments can differ as they do.

More Challenging Questions

65. A sample of nitrosyl bromide is heated to 100. °C in a 10.00-L container to decompose it partially according to the equation

$$2\ NOBr(g) \rightleftharpoons 2\ NO(g) + Br_2(g)$$

The container is found to contain 6.44 g NOBr, 3.15 g NO, and 8.38 g Br_2 at equilibrium.
(a) Find the value of K_c at 100. °C.
(b) Find the total pressure exerted by the mixture of gases.
(c) Calculate K_P for this reaction at 100. °C.

66. Exactly 5.0 mol ammonia was placed in a 2.0-L flask that was then heated to 473 K. When equilibrium was established, 0.2 mol nitrogen had been formed according to the decomposition reaction

$$2\ NH_3(g) \rightleftharpoons N_2(g) + 3\ H_2(g)$$

(a) Calculate the value of the equilibrium constant K_c for this reaction at 473 K.
(b) Calculate the total pressure exerted by the mixture of gases inside the 2.0-L flask at this temperature.
(c) Calculate K_P for this reaction at 473 K.

67. ■ The equilibrium constant K_c has a value of 3.30 at 760. K for the decomposition of phosphorus pentachloride,

$$PCl_5(g) \rightleftharpoons PCl_3(g) + Cl_2(g)$$

(a) Calculate the equilibrium concentrations of all three species arising from the decomposition of 0.75 mol PCl_5 in a 5.00-L vessel.
(b) Calculate the equilibrium concentrations of all three species resulting from an initial mixture of 0.75 mol PCl_5 and 0.75 mol PCl_3 in a 5.00-L vessel.

68. A 1.00-mol sample of CO_2 is heated to 1000. K with excess solid graphite in a container of volume 40.0 L. At 1000. K, K_c is 2.11×10^{-2} for the reaction

$$C(graphite) + CO_2(g) \rightleftharpoons 2\ CO(g)$$

(a) What is the composition of the equilibrium mixture at 1000. K?
(b) If the volume of the flask is changed and a new equilibrium established in which the amount of CO_2 is equal to the amount of CO, what is the new volume of the flask? (Assume the temperature remains 1000. K.)

69. Predict whether the equilibria listed below will be shifted to the left or the right when the following changes occur: (i) the temperature is increased; (ii) the pressure is decreased; (iii) more of the substance indicated by a colored box is added.
(a) $C(s) + H_2O(g) \rightleftharpoons CO(g) + H_2(g)$ $\Delta H°$ (298 K) = +131.3 kJ mol^{-1}
(b) $3\ Fe(s) + 4\ H_2O(g) \rightleftharpoons Fe_3O_4(s) + 4\ H_2(g)$ $\Delta H°$ (298 K) = −149.9 kJ mol^{-1}
(c) $C(s) + CO_2(g) \rightleftharpoons 2\ CO(g)$ $\Delta H°$ (298 K) = +172.5 kJ mol^{-1}
(d) $N_2O_4(g) \rightleftharpoons 2\ NO_2(g)$ $\Delta H°$ (298 K) = +54.8 kJ mol^{-1}

70. Predict whether the equilibria listed below will be shifted to the left or the right when the following changes are made: (i) the temperature is increased; (ii) the pressure is

decreased; (iii) more of the substance indicated by a colored box is added.

(a) $N_2(g) + O_2(g) \rightleftharpoons 2\ NO(g)$
$\Delta H°$ (298 K) = +180.0 kJ mol^{-1}

(b) $CH_4(g) + 2\ O_2(g) \rightleftharpoons CO_2(g) + 2\ H_2O(g)$
$\Delta H°$ (298 K) = −802.3 kJ mol^{-1}

(c) $CaCO_3(s) \rightleftharpoons CaO(s) + CO_2(g)$
$\Delta H°$ (298 K) = +177.9 kJ mol^{-1}

71. When a mixture of hydrogen and bromine is maintained at normal atmospheric pressure and heated above 200. °C, the hydrogen and bromine react to form hydrogen bromide and a gas-phase equilibrium is established.
(a) Write a balanced chemical equation for the equilibrium reaction.
(b) Use bond enthalpies from Table 8.2 to estimate the enthalpy change for the reaction.
(c) Based on your answers to parts (a) and (b), which is more important in determining the position of this equilibrium, the entropy effect or the energy effect?
(d) In which direction will the equilibrium shift as the temperature increases above 200. °C? Explain.
(e) Suppose that the pressure were increased to triple its initial value. In which direction would the equilibrium shift?
(f) Why is the equilibrium not established at room temperature?

72. A sample of pure SO_3 weighing 0.8312 g was placed in a 1.00-L flask and heated to 1100. K to decompose it partially.

$$2\ SO_3(g) \rightleftharpoons 2\ SO_2(g) + O_2(g)$$

If a total pressure of 1.295 atm was developed, find the value of K_c for this reaction at this temperature.

Conceptual Challenge Problems

Conceptual Challenge Problems CP13.A, CP13.B, and CP13.C are related to the information in this paragraph. Aqueous iron(III) ions, $Fe^{3+}(aq)$, are nearly colorless. If their concentration is 0.001 M or lower, a person cannot detect their color. Thiocyanate ions, $SCN^-(aq)$, are colorless also, but monothiocyanatoiron(III) ions, $Fe(SCN)^{2+}(aq)$, can be detected at very low concentrations because of their color. These ions are light amber in very dilute solutions, but as their concentration increases, the color intensifies and appears blood-red in more concentrated solutions. Suppose you prepared a stock solution by mixing equal volumes of 1.0×10^{-3} M solutions of both iron(III) nitrate and potassium thiocyanate solutions. The equilibrium reaction is

$$\underset{\text{colorless}}{Fe^{3+}(aq)} + \underset{\text{colorlesss}}{SCN^-(aq)} \rightleftharpoons \underset{\text{amber}}{Fe(SCN)^{2+}(aq)}$$

CP13.A (Section 13.1) Describe how you would use 5-mL samples of the stock solution and additional solutions of 0.010 M $Fe^{3+}(aq)$ and 0.010 M $SCN^-(aq)$ to show experimentally that the reaction between $Fe^{3+}(aq)$ and $SCN^-(aq)$ does not go to completion but instead reaches an equilibrium state in which appreciable quantities of reactants and product are present. (Refer to the first paragraph for further information.)

CP13.B (Section 13.1) Suppose that you added 1 drop of 0.010 M $Fe^{3+}(aq)$ to a 5-mL sample of the stock solution, followed by 10 drops of 0.010 M $SCN^-(aq)$. You treated a second 5-mL sample of the stock solution by first adding 10 drops of 0.010 M $SCN^-(aq)$, followed by 1 drop of $Fe^{3+}(aq)$. How would the color intensity of these two solutions compare after the same quantities of the same solutions were added in reverse order? (Refer to the first paragraph for further information.)

CP13.C (Section 13.6) Predict what will happen if you put a 5-mL sample of the stock solution (described in the first paragraph) in a hot water bath. Predict what will happen if it is placed in an ice bath.

CP13.D (Section 13.4) Consider the equilibrium reaction between dioxygen and trioxygen (ozone). What is the minimum volume of air (21% dioxygen by volume) at 1.00 atm and 25 °C that you would predict to have at least one molecule of trioxygen, if the only source of trioxygen were its formation from dioxygen and if the atmospheric system were at equilibrium?

$$3\ O_2(g) \rightleftharpoons 2\ O_3(g) \qquad K_c = 6.3 \times 10^{-58} \quad \text{(at 25 °C)}$$

(The volume of 1 mol air at 1 atm and 25 °C is 24.45 L.)

14

The Chemistry of Solutes and Solutions

© Karl Weatherly/Photodisc Green/Getty

Oceans are vast aqueous solutions containing Na^+, Mg^{2+}, and Cl^- ions plus other solutes that give ocean water its characteristic salinity, vapor pressure, and freezing point. This chapter considers the macroscale to nanoscale links that govern how solutes dissolve in solvents, how to express solute concentration in solution, and how such concentration affects the freezing point and boiling point of solutions.

Every day we all encounter many solutions, such as soft drinks, juices, coffee, and gasoline. A *solution* is a homogeneous mixture of two or more substances **(← p. 12)**. The component present in greatest amount is usually called the *solvent;* the other components are *solutes* **(← p. 161)**. In sweetened iced tea, for example, water is the solvent; sugar and soluble extracts of tea are the solutes.

Although solids dissolved in liquids or homogeneous mixtures of liquids are the most common types of solutions, other kinds are possible as well, encompassing the three physical states of matter (Table 14.1). Chemistry often focuses on liquid solutions, and in particular on those in which water is the solvent (aqueous solutions).

Much of this chapter is devoted to aqueous solutions because water is the most important solvent on our planet.

In this chapter we will explore in some detail the macroscale to nanoscale connections between solutes and solvents to answer questions such as:

- Why does a particular solvent readily dissolve one kind of solute, but not another? Water, for example, dissolves NaCl, but does not dissolve gasoline.
- In what ways can the concentration of a dissolved solute be expressed?
- Is thermal energy released or absorbed when a solute dissolves?
- How do factors such as temperature or pressure changes affect the solubility of a solute in a given solvent? Why, for example, does a cold, carbonated beverage become "flat" when it is opened and warmed to room temperature?

To answer these questions, we will apply to the interactions among solute and solvent molecules what you learned from Chapter 9 about the noncovalent forces that act between molecules—London forces, dipole-dipole forces, and hydrogen bonding (⇐ *p. 332*). We will also utilize the thermodynamic and equilibrium principles presented in Chapters 6 and 13. These principles are important to understand the effect that solutes have on the vapor pressures, melting points, and boiling points of solvents.

Table 14.1 Types of Solutions

Type of Solution	Example
Gas in gas	Air—a mixture principally of N_2 and O_2 but containing other gases as well
Gas in liquid	Carbonated beverages (CO_2 in water)
Gas in solid	Hydrogen in palladium metal
Liquid in liquid	Motor oil—a mixture of liquid hydrocarbons; ethanol in water
Solid in liquid	The oceans (dissolved Na^+, Cl^-, and other ions)
Solid in solid	Bronze (copper and tin); pewter (tin, antimony, and lead)

14.1 Solubility and Intermolecular Forces

In every solution, the interplay between solute and solvent particles determines whether a solute will dissolve in a particular solvent and how much is dissolved.

Solute-Solvent Interactions

An old adage says that "oil and water don't mix." Chemists use a similar saying about solubility: "Like dissolves like," where "like" refers to solutes and solvents whose molecules attract each other by similar types of noncovalent intermolecular forces. Such forces exist between solute particles, as well as between solvent particles. Dissolving a solute in a solvent is favored when the solute-solvent intermolecular forces are stronger than the solute-solute or the solvent-solvent intermolecular forces.

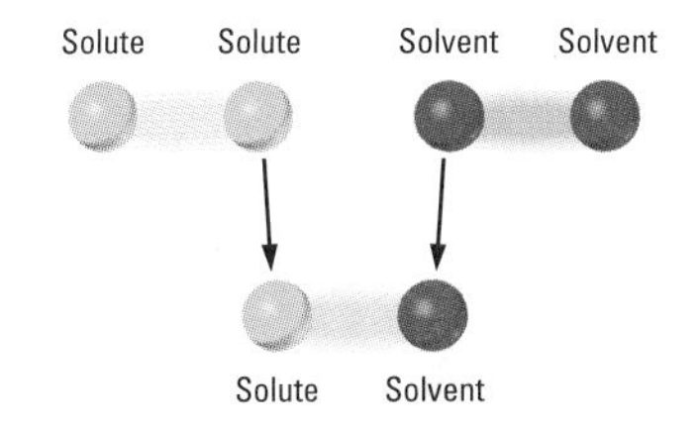

Consider, for example, dissolving the hydrocarbon octane, C_8H_{18}, in carbon tetrachloride, CCl_4 (Figure 14.1a). Both are nonpolar liquids that dissolve in each other because of the similar London forces between molecules in each compound. Liquids such as these that dissolve in each other in any proportion are said to be **miscible.** In contrast, gasoline, a mixture of nonpolar hydrocarbons, is not miscible with water, a polar substance (Figure 14.1b). The nonpolar hydrocarbons cannot hydrogen-bond to water molecules, but rather stay attracted to each other through London forces (solute-solute attractions); the water molecules remain hydrogen-bonded to each other (solvent-solvent attractions). Liquids such as gasoline and water, with noncovalent attractions so different between their molecules that they do not dissolve in each other, are described as **immiscible.**

Alaskan oil spill. This oil spill in Prince William Sound, Alaska, is a large-scale example of the nonsolubility of crude oil, a nonpolar material, in water, a polar substance. The oil floats on top of the water, but does not dissolve in it.

The differing solubilities of various alcohols in water further illustrate the "like dissolves like" principle and the role of noncovalent intermolecular forces. Simple, low-molar-mass alcohols dissolve in water due to hydrogen bonding between water molecules and the —OH groups of the alcohol molecules. The hydrogen bonding forces in the solution are the same forces as those in pure water or pure ethanol. The hydrogen bonding between ethanol and water is illustrated in Figure 14.2c.

Alcohol molecules contain a polar portion, the —OH group, and a nonpolar portion, the hydrocarbon part. The polar region is *hydrophilic* ("water loving"); any polar part of a molecule will be hydrophilic because of its attraction to polar water molecules. The nonpolar hydrocarbon region is *hydrophobic* ("water hating").

There is a dye in the gasoline to make it a different color.

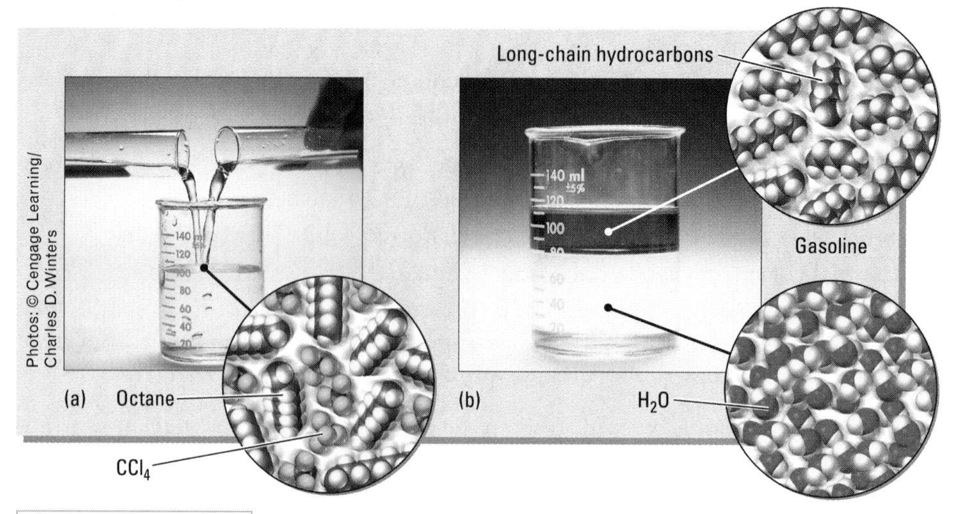

Active Figure 14.1 **(a) Miscible and (b) immiscible liquids.** When carbon tetrachloride and octane, both colorless, clear liquids, are mixed (a), each dissolves completely in the other, and there is no sign of an interface or boundary between them. (b) When gasoline and water are put together, the mixture remains as two distinct layers. Visit this book's companion website at **www.cengage.com/chemistry/moore** to test your understanding of the concepts in this figure.

In a low-molar-mass alcohol such as methanol or ethanol (Figure 14.2), hydrogen bonding of the —OH group with water molecules is stronger than the London forces between the nonpolar hydrocarbon-like parts of the alcohol molecules. As the hydrophobic hydrocarbon chain of the alcohol increases in length, the molecular structure of the alcohol becomes less and less like that of water and more and more like that of a hydrocarbon. The London forces between the nonpolar hydrocarbon portion of the alcohol molecules become stronger, so a point is reached at which the

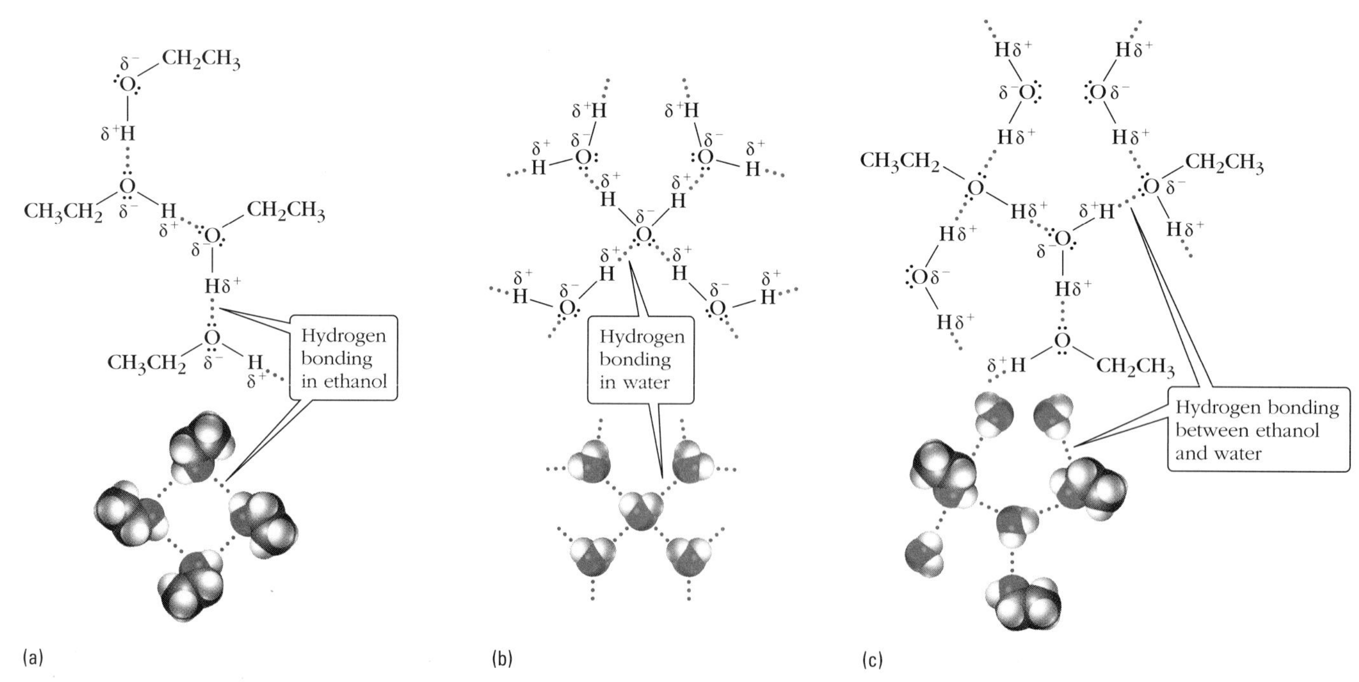

Figure 14.2 **Hydrogen bonding.** Hydrogen bonding (a) among ethanol molecules (solute-solute attraction); (b) among water molecules (solvent-solvent attraction); and (c) among ethanol and water molecules (solute-solvent attraction).

Table 14.2 Solubilities of Some Alcohols in Water

Name	Formula	Solubility in Water (g/100 g H_2O at 20 °C)	Name	Formula	Solubility in Water (g/100 g H_2O at 20 °C)
Methanol	CH_3OH	Miscible	1-Butanol	$CH_3(CH_2)_3OH$	7.9
Ethanol	CH_3CH_2OH	Miscible	1-Pentanol	$CH_3(CH_2)_4OH$	2.7
1-Propanol	$CH_3(CH_2)_2OH$	Miscible	1-Hexanol	$CH_3(CH_2)_5OH$	0.6

London forces between the hydrocarbon portion of alcohol molecules (solute-solute attraction) become sufficiently large that the water solubility of the alcohol becomes small, such as that of 1-hexanol compared with ethanol (Table 14.2). The hydrocarbon portion becomes a relatively larger portion of the molecule from methanol to 1-hexanol compared with the —OH group of the alcohol. Thus, methanol and ethanol, with just one and two carbon atoms, respectively, are infinitely soluble in water, whereas alcohols with more than six carbon atoms per molecule are virtually insoluble in water.

We can summarize the principle of "like dissolves like" as follows:

- ***Substances with similar noncovalent forces are likely to be soluble in each other.***
- ***Solutes do not readily dissolve in solvents whose noncovalent forces are quite different from their own.***
- ***Stronger solute-solvent attractions favor solubility. Stronger solute-solute or solvent-solvent attractions reduce solubility.***

CONCEPTUAL
EXERCISE 14.1 Predicting Water Solubility

How could the data in Table 14.2 be used to predict the solubility in water of 1-octanol, $CH_3(CH_2)_7OH$, or 1-decanol, $CH_3(CH_2)_9OH$?

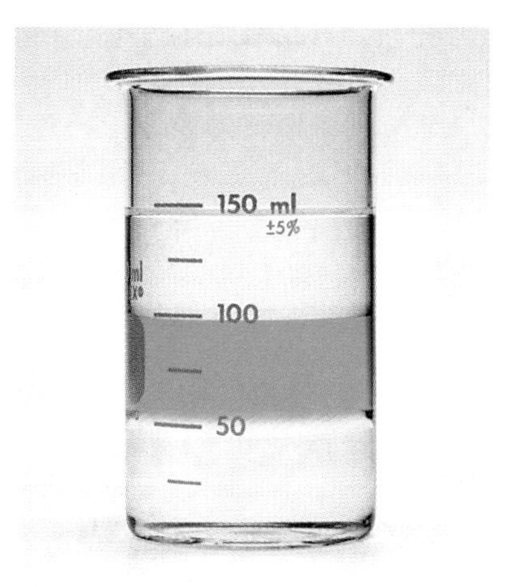

(a)

PROBLEM-SOLVING EXAMPLE 14.1 Predicting Solubilities

A beaker initially contains three layers (Figure 14.3): In part (a) the top layer is colorless heptane, C_7H_{16} (density = 0.684 g/mL); the middle layer is a green solution of $NiCl_2$ in water (density = 1.10 g/mL); the bottom layer is colorless carbon tetrachloride, CCl_4 (density = 1.59 g/mL). After the liquids are mixed, two layers remain—a lower, colorless layer and an upper green-colored layer. Using the principle of "like dissolves like," explain what resulted when the layers were mixed.

Answer The nonpolar liquids carbon tetrachloride and heptane dissolved in each other; neither the nickel chloride nor the water dissolved in the nonpolar liquids.

Strategy and Explanation To apply the "like dissolves like" principle requires identifying the polarity and intermolecular forces of the substances in the beaker. Hexane is a hydrocarbon and thus nonpolar with London forces between hexane molecules. Likewise, carbon tetrachloride is a nonpolar molecule (no *net* dipole even though each C—Cl bond is polar) due to the symmetry of its tetrahedral shape (⇐ *p. 310*). Nickel chloride is an ionic solid that dissolves in water due to the attraction of polar water molecules for the Ni^{2+} cations and Cl^- anions in the compound. Thus, there is little attraction for the polar aqueous solution to mix with either of the nonpolar liquids above and below the aqueous layer. Once the mixture is stirred to mix the components, the nonpolar liquids dissolve in each other to form a bottom layer denser than the upper aqueous layer containing dissolved $NiCl_2$.

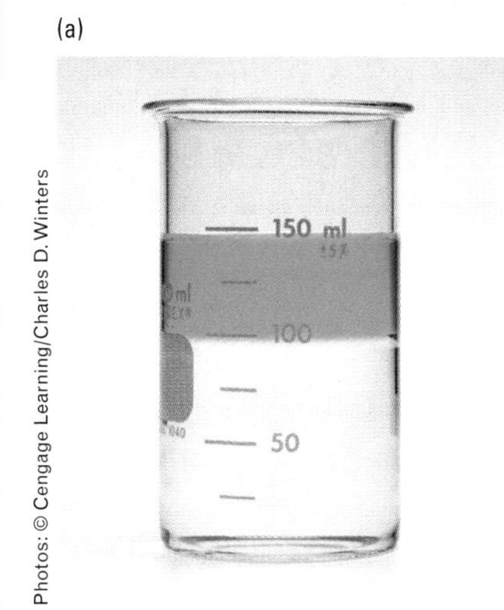

(b)

Figure 14.3 Miscibility of liquids. (a) This mixture was prepared by carefully layering three liquids of differing densities: a bottom layer of colorless carbon tetrachloride; a middle layer of green aqueous nickel chloride; and an upper layer of colorless heptane. (b) The mixture after the components have been mixed.

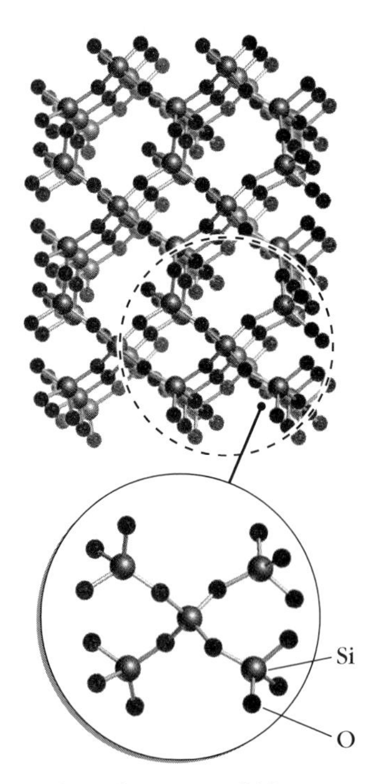

A portion of a quartz, SiO_2, structure. The structure is based on SiO_4 tetrahedra linked through shared oxygen atoms.

PROBLEM-SOLVING PRACTICE 14.1

Using the principle of "like dissolves like," predict whether
(a) Ethylene glycol, $HOCH_2CH_2OH$, dissolves in gasoline
(b) Molecular iodine dissolves in carbon tetrachloride
(c) Motor oil dissolves in carbon tetrachloride
Explain your predictions.

Solids such as quartz, SiO_2, that are held together by an extensive network of covalent bonds are generally insoluble in polar or nonpolar solvents. In quartz, the silicon and oxygen atoms form SiO_4 tetrahedra linked through covalent bonds to shared oxygen atoms. These strong covalent bonds are not broken by weaker attractions to solvent molecules. Thus, quartz (and sand derived from it) is insoluble in water or any other solvent at room temperature. Sandy beaches do not dissolve in ocean water.

The solubility of a substance in water or in the triglycerides—nonpolar fats or oily substances in our bodies—plays an important role in our body chemistry. Vitamins, for example, are either water-soluble or fat-soluble. Vitamins A, D, E, and K are known as fat-soluble vitamins because they dissolve in triglycerides. All the other vitamins are water-soluble. The major significance of this difference is the danger of overdosing on fat-soluble vitamins because they are stored in fatty tissues and may accumulate to harmful levels. By contrast, overdosing on water-soluble vitamins is difficult and uncommon because these vitamins are not stored in the body and any excess of them is excreted in urine.

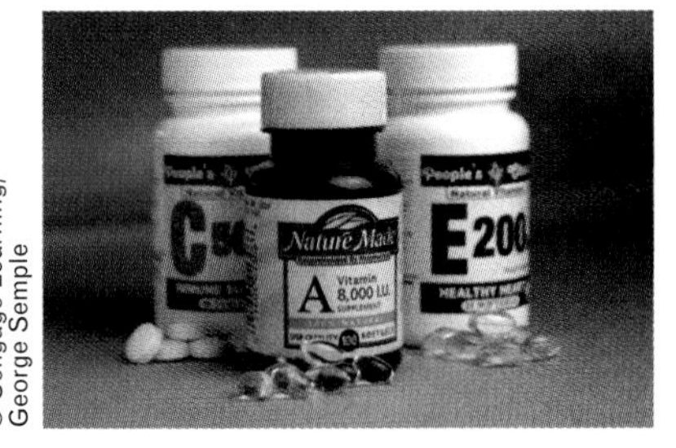

Vitamins are fat-soluble (A, D, E, K) or water-soluble (the rest).

PROBLEM-SOLVING EXAMPLE 14.2 Solubility and Noncovalent Forces

Use the structural formulas of niacin (nicotinic acid) and vitamin A to determine which vitamin is more soluble in water and which is more soluble in fat.

niacin (nicotinic acid)

vitamin A

Answer Niacin is water-soluble; vitamin A is fat-soluble.

Strategy and Explanation Use the structural formulas to determine whether any water-soluble or fat-soluble portions exist in the molecules.

Niacin is water-soluble because its —COOH group and nitrogen atom can hydrogen-bond with water molecules.

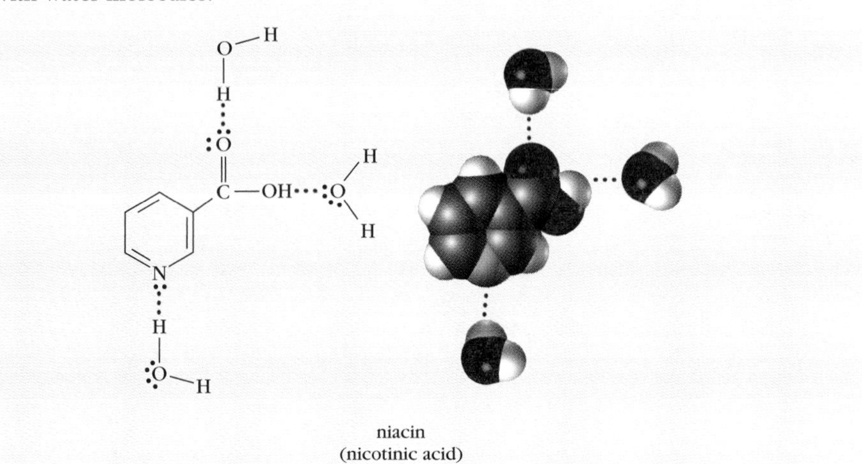

niacin (nicotinic acid)

Niacin dissolves because the solute-solvent interaction between niacin and water is stronger than hydrogen bonding between water molecules or the dipole-dipole attractions between niacin molecules.

Vitamin A has an —OH group that can hydrogen bond with water. However, as is the case with other long-chain alcohols, the hydrocarbon portion of the molecule is large enough to make the molecule hydrophobic. In this case, London forces between the hydrocarbon portions are more important than the hydrogen bonding, and vitamin A is insoluble in water. On the other hand, the extended hydrocarbon portion of the molecule is soluble in the long-chain hydrocarbon portion of fats.

PROBLEM-SOLVING PRACTICE 14.2

Explain why vitamin C, which has the structure shown, is water-soluble.

14.2 Enthalpy, Entropy, and Dissolving Solutes

A solution forms when atoms, molecules, or ions of one kind mix with atoms, molecules, or ions of a different kind. For the solution to form, intermolecular forces attracting solvent molecules to each other, and such forces attracting solute molecules or ions to each other must be overcome (*steps a* and *b* in Figure 14.4). In *step (a)* the enthalpy of the collection of solvent molecules increases due to their separation, resulting in an endothermic process with a positive ΔH ($\Delta H > 0$). Likewise, this must also occur in *step (b)* for the separation of solute ions or molecules. When the solute and solvent particles mix (*step c*), they attract each other and there is a decrease in enthalpy; ΔH is negative ($\Delta H < 0$) and energy is released in this exothermic step. If the enthalpy of the final solution is lower than that of the initial solute and solvent, as shown in Figure 14.4 (*left*), a net release of energy to the surroundings occurs and the solution-making process is *exothermic*. When the enthalpy of the final solution is greater than the enthalpies of the initial solute and solvent, the process is *endothermic* and the surroundings transfer energy to the system (Figure

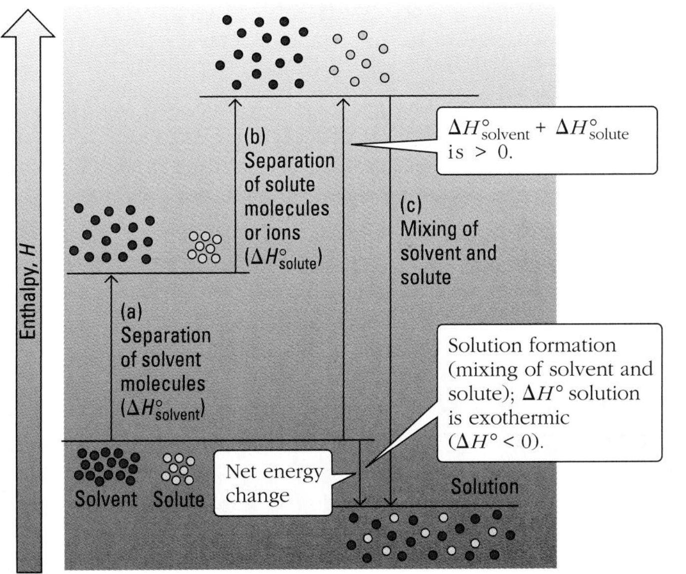

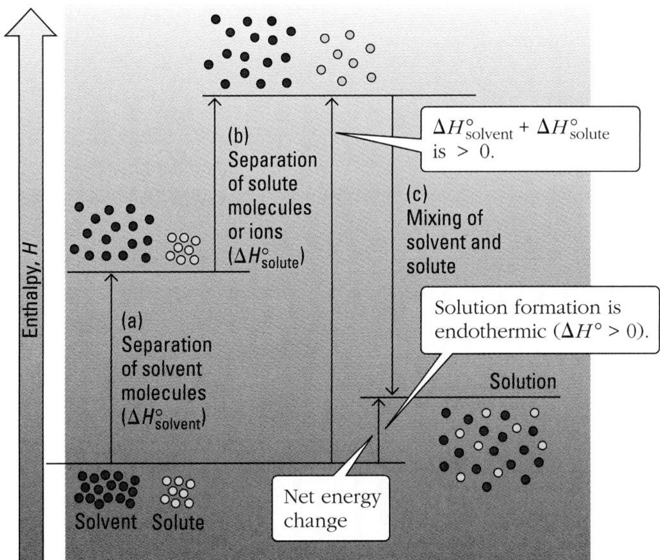

Figure 14.4 The solution-making process.

14.4, *right*). The net energy change, either exothermic or endothermic, is called the **enthalpy of solution, ΔH_{soln}**.

The enthalpy of solution is also known as the heat of solution.

$$\text{Enthalpy of solution} = \Delta H_{soln} = \Delta H_{step\ a} + \Delta H_{step\ b} + \Delta H_{step\ c}$$

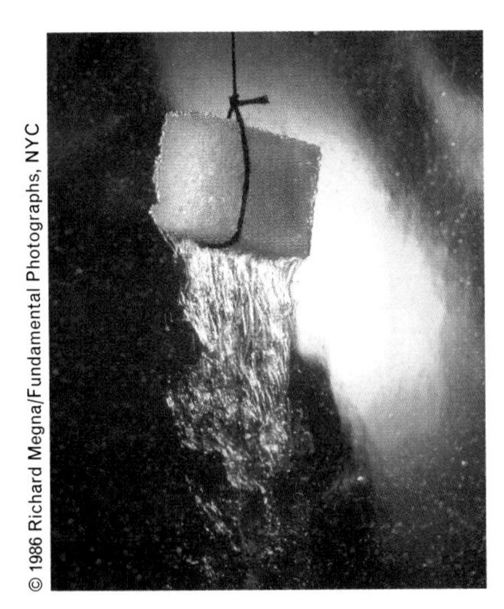

A solute (sugar) dissolving in a solvent (water). As the solute dissolves, concentration gradients are observed as wavy lines.

After the solute and solvent mix, the solvent-solute forces (indicated by the magnitude of $\Delta H_{step\ c}$) may not be strong enough to overcome the solute-solute and solvent-solvent attractions ($\Delta H_{step\ a} + \Delta H_{step\ b}$). Under these conditions, dissolving is not favored by the enthalpy effect (⇐ **p. 492**).

When solutes and solvents mix to form solutions, solute and solvent molecules usually spread out over a larger volume. This spreads their energy over a larger volume and results in an increase in entropy (⇐ **p. 492**). As described in Section 13.7, *entropy* (symbolized by S) is a measure of the energy dispersal of a system; an increase in entropy ($\Delta S > 0$) indicates a process where energy is spread out more, and dispersal of energy corresponds with dispersal of matter. *Processes in which the entropy of the system increases tend to be product-favored.* Therefore, in many cases when a solute dissolves, a large entropy increase occurs as solvent and solute molecules mix. When the enthalpy of solution is rather small, entropy is the most important factor behind solution formation, such as when octane and carbon tetrachloride are mixed. In other cases, even though the enthalpy of solution is positive and significantly large ($\Delta H_{soln} > 0$), the entropy increase is large enough to cause the solution to form. This is true for some ionic solutes, such as ammonium nitrate, NH_4NO_3 (Section 14.3). Because molecules in gases are always much more dispersed than liquids, there is a significant decrease in entropy ($\Delta S < 0$) as gas solute molecules are brought closer together between solvent molecules.

14.3 Solubility and Equilibrium

Some solutes dissolve to a much greater extent than others in the same mass of the same solvent. The **solubility** of a solute is the maximum quantity of solute that dissolves in a given quantity of solvent at a particular temperature (Figure 14.5). A solution can be described as saturated, unsaturated, or supersaturated depending on the quantity of solute that is dissolved in it. A **saturated solution,** as its name implies, is one whose solute concentration equals its solubility (all points *along* the curve in Figure 14.5). When a saturated solution forms, there is a *dynamic equilibrium* between undissolved and dissolved solute.

Remember that the concentration of a pure substance (solute or solvent) does not appear in the K_c expression.

$$\text{Solute} + \text{solvent} \rightleftharpoons \text{solute (in solution)} \qquad K_c = [\text{solute (in solution)}]$$

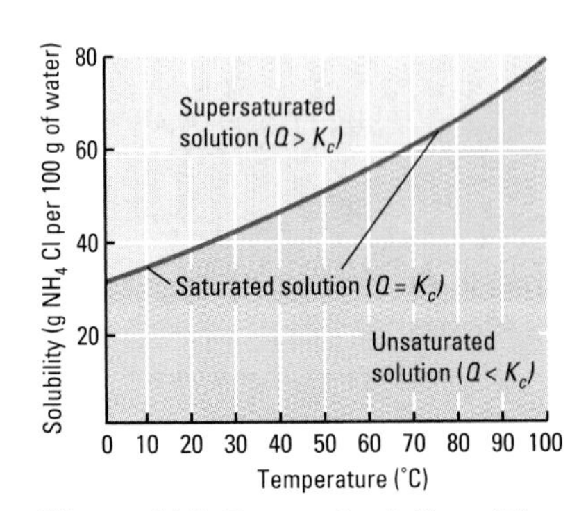

Figure 14.5 Types of solutions. The solubility of solid ammonium chloride and three types of ammonium chloride solutions—saturated (the curve), unsaturated (below the curve), and supersaturated (above the curve).

Some solute molecules or ions are mixing with solvent molecules and going into solution, while others are separating from solvent molecules and entering the pure solute phase. Both processes are going on simultaneously at identical rates. When a solid dissolves in a liquid, for example, the quantity of solid is observed to decrease until the solution becomes saturated. A saturated solution is in equilibrium with its solute and solvent, and in this case $Q = K_c$ (⇐ **p. 478**). Once equilibrium is reached the quantity of solid solute (and the quantity of dissolved solute) stops changing.

An **unsaturated solution** is one in which the solute concentration is less than its solubility; that is, an unsaturated solution can accommodate additional solute at a given temperature (region *under* the curve in Figure 14.5). Thus, for an unsaturated solution, $Q < K_c$, and more solute can dissolve. If solute continues to be added to an unsaturated solution at a given temperature, a point is reached where the solution becomes saturated, at which point $Q = K_c$.

For some solutes, there is a third case. It is possible to prepare solutions that contain *more* than the equilibrium concentration of solute at a given temperature; such a solution is **supersaturated** (the region *above* the curve in Figure 14.5). For example, a supersaturated solution of ammonium chloride can be made by first making a saturated solution at 90 °C and stirring for some time. This is done by adding 72 g of solid NH_4Cl to 100 g water, heating the solution to 90 °C. At that point the solution is saturated. To make the supersaturated solution, the temperature of the saturated solution is lowered very slowly to 25 °C. If this is done carefully, none of the NH_4Cl will crystallize out of solution, and the resulting solution will be supersaturated, holding more dissolved NH_4Cl than a saturated solution at that temperature; 72 g of the solute is dissolved at 25 °C, whereas the solution should hold a maximum of 40 g at that temperature (Figure 14.5). In equilibrium terms, $Q > K_c$, and excess solute should crystallize out of solution, but it is very slow to do so. Some supersaturated solutions, such as honey, can remain so for days or months. To form a crystal, several ions or molecules must move into an arrangement much like the appropriate crystal lattice positions, and it can take a long time for such an alignment to occur by chance. However, precipitation of a solid from a supersaturated solution occurs rapidly if a tiny crystal of the solute is added to the solution (Figure 14.6). The lattice of the added crystal provides a template onto which more ions or molecules can be added.

Unsaturated solution: $Q < K_c$; saturated solution: $Q = K_c$; supersaturated solution: $Q > K_c$.

Fudge is made using a supersaturated sugar solution. In smooth fudge, the sugar remains uncrystallized; poor-quality fudge has a gritty texture because it contains crystallized excess sugar.

Sometimes other actions, such as stirring a supersaturated solution or scratching the inner walls of its container, will cause solute to precipitate rapidly.

EXERCISE 14.2 Crystallizing Out of Solution

Using Figure 14.5, how many grams of excess NH_4Cl would crystallize out of the supersaturated NH_4Cl solution described previously at 25 °C?

EXERCISE 14.3 Solubility

Refer to Figure 14.5 to determine whether each of these NH_4Cl solutions in 100 g H_2O is unsaturated, saturated, or supersaturated:

(a) 30 g NH_4Cl at 70 °C (b) 60 g NH_4Cl at 60 °C
(c) 50 g NH_4Cl at 50 °C

Figure 14.6 Crystallization from a supersaturated solution. Sodium acetate, $NaCH_3COO$, easily forms supersaturated solutions in water. (a) The solution looks ordinary, but it is supersaturated, holding more dissolved sodium acetate than a saturated solution at that temperature. (b) After a tiny seed crystal of sodium acetate is added to the supersaturated solution, some of the excess dissolved sodium acetate immediately begins to crystallize. Very soon, numerous solid sodium acetate crystals can be seen.

Dissolving Ionic Solids in Liquids

Sodium chloride is an ionic compound whose crystal lattice consists of Na^+ and Cl^- ions in a cubic array *(⇐ p. 402)*. Strong electrostatic attractions between oppositely charged ions hold the ions tightly in the lattice. The enthalpy change when 1 mol of Na^+ and Cl^- ions is completely separated from a crystal lattice is referred to as the lattice energy of the ionic compound *(⇐ p. 262)*. The large lattice energy of NaCl (788 kJ/mol) accounts for sodium chloride's high melting point (800 °C). It is possible for solvent molecules to attract Na^+ ions away from Cl^- ions in a crystal, but they have to be the right kind of solvent molecules. Trying to dissolve NaCl (or any other ionic compound) with carbon tetrachloride or hexane (both nonpolar solvents) is a futile exercise because nonpolar molecules have very little attraction for ions. On the other hand, water dissolves NaCl.

Water is a good solvent for an ionic compound because water molecules are small and highly polar *(⇐ p. 327)*. As shown in Figure 14.7, the partially negative oxygen atoms of water molecules are attracted to positive ions and help pull them away from the crystal lattice, while the partially positive hydrogen atoms of other water molecules are attracted to the negative ions in the lattice and help pull them away from the lattice. This process, in which water molecules surround positive and negative ions, is called **hydration** (Figure 14.8). Energy known as the *enthalpy of hydration* is released when these new attractions form between ions and water molecules as they mix and get close enough to one another. Energy is always required to separate the ions to overcome their attraction, and energy is always released when ions become hydrated because noncovalent interactions are being formed. Whether dissolving a particular ionic compound is exothermic (ΔH_{soln} is negative) or endothermic (ΔH_{soln} is positive) depends on the relative sizes of the lattice energy *(⇐ p. 262)* of the ionic

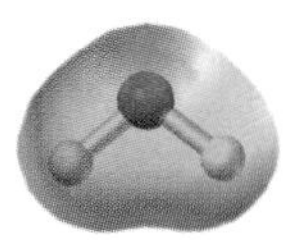

Electron density of a water molecule. The red area has high electron density and partial negative charge; the blue area has low electron density and partial positive charge.

Enthalpies of Hydration of Selected Ions (kJ/mol)

Cations	
H^+	−1130
Li^+	−558
Na^+	−444
Mg^{2+}	−2003
Ca^{2+}	−1557
Al^{3+}	−2537
Anions	
F^-	−483
Cl^-	−340
Br^-	−309

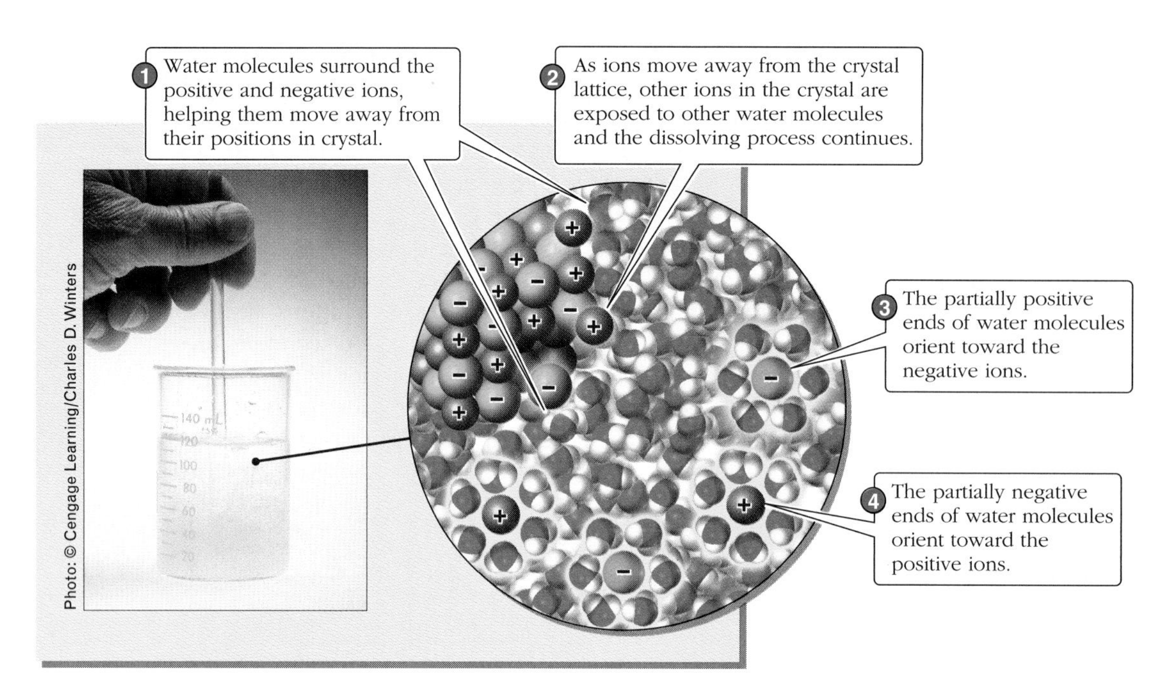

Figure 14.7 Water dissolving an ionic solid. Water molecules surround the positive (gray) and negative (green) ions, helping them move away from their positions in the crystal.

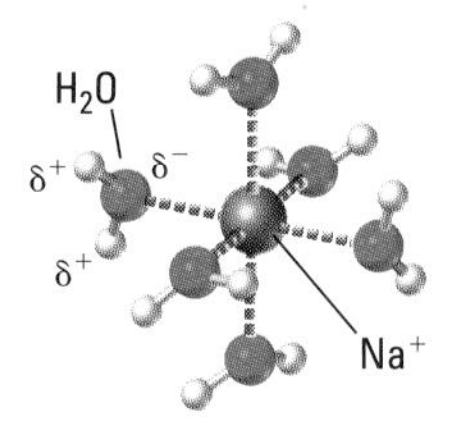

Figure 14.8 Hydration of a sodium ion. The arrangement of water molecules around this Na^+ ion is highly ordered.

compound and the hydration enthalpies of its positive and negative ions. The relationship between the enthalpy of solution, the lattice energy of the ionic compound, and the enthalpies of hydration of the ions is

$$\Delta H_{soln} = -\text{lattice energy} + \Delta H_{hydration}\,(\text{cations}) + \Delta H_{hydration}\,(\text{anions})$$

Figure 14.9 shows how lattice energy and the enthalpies of hydration combine to give the enthalpy of solution.

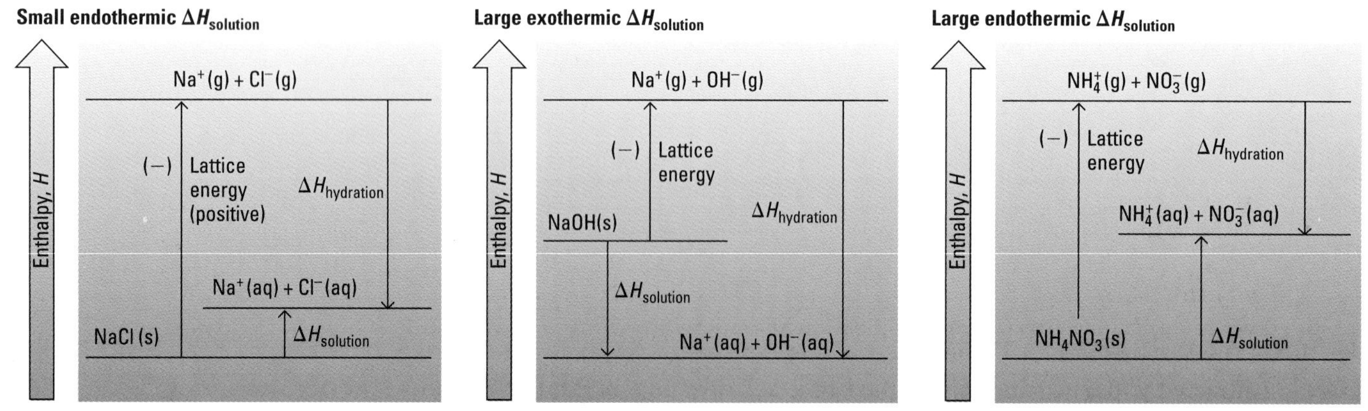

Figure 14.9 Enthalpies of solution of three different ionic compounds dissolving in equal volumes of water. (*Left*) For NaCl, the lattice energy that must be overcome is larger than the energy released on hydration of the ions. This causes the ΔH_{soln} of NaCl to be positive (endothermic). (*Center*) The lattice energy for NaOH is much smaller than the energy released when the ions become hydrated. Consequently, the ΔH_{soln} of NaOH is negative (highly exothermic). (*Right*) For NH_4NO_3, the enthalpy of hydration of the ions is much smaller than the lattice energy, so the resulting ΔH_{soln} has a large positive value (highly endothermic). NH_4NO_3 can therefore be dissolved in water to make a cold pack cold.

Practical applications of endothermic and exothermic dissolution include cold packs containing NH_4NO_3 used to treat athletic injuries and hot packs containing $CaCl_2$ used to warm foods.

The solubility rules for ionic compounds in water (⇐ p. 137) remind us that not all ionic compounds are highly water-soluble in spite of the strong attractions between water molecules and ions. For some ionic compounds, the lattice energy is so large that water molecules cannot effectively pull ions away from the lattice. As a result, such compounds have large positive enthalpies of solution and usually are only slightly soluble.

Energy is always released when particles that attract each other get closer together, so the lattice energy is always negative. Energy is always required to separate particles that attract each other.

Although generally useful, Le Chatelier's principle does not always correctly predict how the solubility of ionic solutes changes with temperature.

Entropy and the Dissolving of Ionic Compounds in Water

The spreading of energy introduced when a crystal lattice breaks down, and the spreading of energy introduced by the mixing of ions with solvent molecules, both favor the dissolving process ($\Delta S > 0$). This entropy increase is counteracted by the ordering of solvent molecules around the ions ($\Delta S < 0$) (Figure 14.8). For 1+ and 1− charged ions, the overall entropy change is positive, and dissolving is favored. For some salts that contain 2+ or 3+ ions, the charges on the ions are so large and the ions so small that water molecules are aligned in a highly organized manner around the ions. When a large number of water molecules are locked into place by this strong hydration, the entropy of solution may be negative, which does not favor solubility. Calcium oxide (only 0.131 g CaO dissolves in 100 mL of water at 10 °C) and aluminum oxide Al_2O_3 (insoluble) exemplify this effect.

Commercial cold and hot packs. (*Top*) When the desired cooling is needed, ammonium nitrate in the inner container of this cold pack is brought into contact with water in the outer container by breaking the seal of the inner container. (*Bottom*) This hot pack releases heat when the inner pouch containing either $CaCl_2$ or $MgSO_4$ is punctured and the compound dissolves in the water in the outer container.

14.4 Temperature and Solubility

Solubility of Gases

To understand how temperature affects gas solubility, we can apply Le Chatelier's principle (⇐ p. 482) to the equilibrium between a pure solute gas, the solvent, and a saturated solution of the gas.

$$\text{Gas} + \text{solvent} \rightleftharpoons \text{saturated solution} \qquad \text{Usually, } \Delta H_{soln} < 0 \text{ (exothermic)}$$

When a gas dissolves to form a saturated liquid solution, the process is almost always exothermic. Gas molecules that were relatively far apart are brought much closer in the solution to other molecules that attract them, lowering the potential energy of the system and releasing some energy to the surroundings.

If the temperature of a solution of a gas in a liquid increases, the equilibrium shifts in the direction that partially counteracts the temperature rise. That is, the equilibrium shifts to the left in the preceding equation. Thus, a dissolved gas becomes less soluble with increasing temperature. Conversely, cooling a solution of a gas that is at equilibrium with undissolved gas will cause the equilibrium to shift to the right in the direction that liberates heat, so more gas dissolves. This is illustrated in Figure 14.10 with data for the solubility of oxygen in water.

Cooler water in contact with the atmosphere contains more dissolved oxygen at equilibrium than water at a higher temperature. For this reason fish seek out cooler (usually deeper) waters in the summer. Fish have an easier time obtaining oxygen when its concentration in the water is higher. The decrease in gas solubility as temperature increases makes *thermal pollution* a problem for aquatic life in rivers and streams. Natural heating of water by sunlight and by warmer air can usually be accommodated. But excess heat from extended periods of very hot weather or from sources such as industrial facilities and electrical power plants can warm the water sufficiently that the concentration of dissolved oxygen is reduced to the point where some species of fish die.

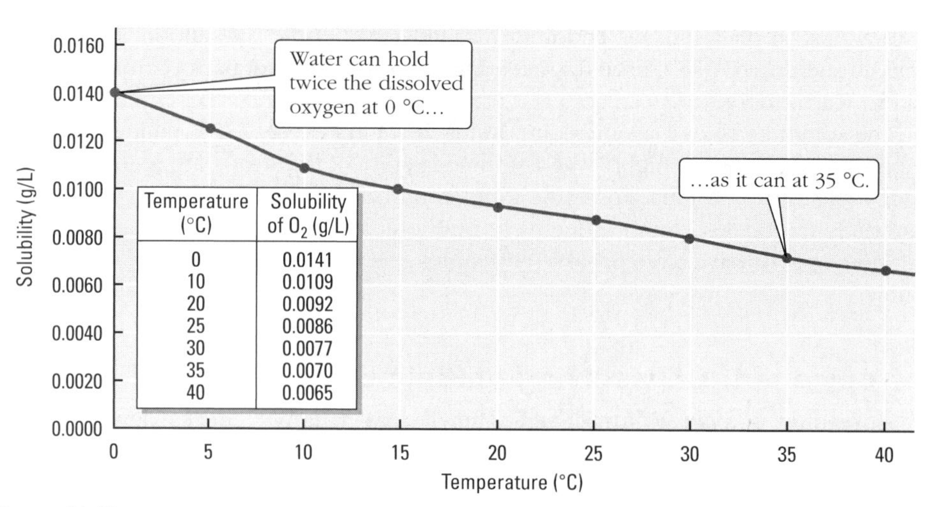

Temperature (°C)	Solubility of O_2 (g/L)
0	0.0141
10	0.0109
20	0.0092
25	0.0086
30	0.0077
35	0.0070
40	0.0065

Figure 14.10 Solubility of oxygen in water at various temperatures. The solubility of oxygen, like that of other gases, decreases with increasing temperature.

A fish kill caused by a lack of dissolved oxygen.

CONCEPTUAL
EXERCISE 14.4 Carbonated Beverages

Explain on a molecular basis why a carbonated beverage goes "flat" once it is opened and warms to room temperature.

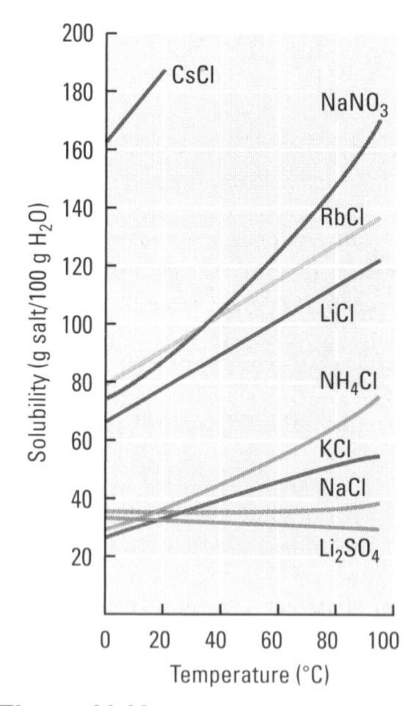

Figure 14.11 Solubility of ionic compounds and temperature. The solubility of most ionic compounds in water depends on the temperature.

Solubility of Solids

Common experience tells us that more sugar (sucrose) can dissolve in hot coffee or hot tea than in cooler coffee or tea. This is an example of the fact that the aqueous solubility of most solid solutes, including ionic compounds, increases with increasing temperature (Figure 14.11). Although predictions of solubility based on enthalpies of solution usually work, there are notable exceptions, such as Li_2SO_4.

CONCEPTUAL
EXERCISE 14.5 Temperature and Solubility

If a substance has a positive enthalpy of solution, which would likely cause more of it to dissolve, hot solvent or cold solvent? Explain.

14.5 Pressure and Dissolving Gases in Liquids: Henry's Law

Pressure does not measurably affect the solubilities of solids or liquids in liquid solvents, but *the solubility of any gas in a liquid increases as the partial pressure of the gas increases.* A dynamic equilibrium is established when a gas is in contact with a liquid—the rate at which gas molecules enter the liquid phase equals the rate at which gas molecules escape from the liquid. If the pressure is increased, gas molecules strike the surface of the liquid more often, increasing the rate of dissolution of the gas. A new equilibrium is established when the rate of escape increases to match the rate of dissolution. The rate of escape is first-order *(⇐ p. 426)* in concentration of solute, so a higher rate of gas escape requires a higher concentration of solute gas molecules—that is, a higher gas solubility.

The partial pressure of a gas in a mixture of gases is the pressure that a pure sample of the gas would exert if it occupied the same volume as the mixture. Partial pressure is proportional to the mole fraction of the gas *(⇐ p. 362).*

The relationship between gas pressure and solubility is known as **Henry's law:**

$$S_g = k_H P_g$$

where S_g is the solubility of the gas in the liquid, and P_g is the pressure of the gas above the solution (or the partial pressure of the gas if the solution is in contact with a mixture of gases). The value of k_H, known as the Henry's law constant, depends on the identities of both the solute and the solvent and on the temperature (Table 14.3). It has units of moles per liter per millimeter of mercury (mm Hg). The behavior of a carbonated drink when the cap is opened is an everyday illustration of the solubility of gases in liquids under pressure. The drink fizzes when opened because the partial pressure of CO_2 over the solution drops, the solubility of the gas decreases, and dissolved gas escapes from the solution.

Table 14.3 Henry's Law Constants, 25 °C; Water Is the Solvent

Gas	k_H mol L^{-1} mm Hg^{-1}
N_2	8.42×10^{-7}
O_2	1.66×10^{-6}
CO_2	4.4×10^{-5}
He	4.6×10^{-7}
Air	8.6×10^{-4}

PROBLEM-SOLVING EXAMPLE 14.3 Using Henry's Law

The "bends" is a condition that can occur in scuba divers due to the solubility of N_2 in the blood. The Henry's law constant for N_2 in water at 25 °C is 8.42×10^{-7} mol L^{-1} mm Hg^{-1}. Assume that the solubility of N_2 in blood is approximately that of N_2 in water. The partial pressure of N_2 in the atmosphere is 0.78 atm. Calculate
(a) the mass of dissolved N_2 in 1 L blood at 25 °C.
(b) the volume (mL) of N_2 dissolved in 1 L blood at 25 °C.

Answer (a) 1.40×10^{-2} g N_2 (b) 16 mL of N_2

Strategy and Explanation The solubility of N_2 in blood (g/L) can be calculated, but first the solubility must be calculated in terms of molarity (mol/L) (***p. 161***) using Henry's law, and then the result is converted to grams by using the molar mass. The volume (mL) of N_2 in 1 L blood can then be calculated using the ideal gas law.

(a) We assume that the dissolved N_2 is in equilibrium with nitrogen in the air. Converting atm to mm Hg we have

$$0.78 \text{ atm} \times \frac{760 \text{ mm Hg}}{1 \text{ atm}} = 593 \text{ mm Hg}$$

Use Henry's law to calculate the solubility of N_2, $S_{N_2} = k_H P_{N_2}$, in mol/L:

$$S_{N_2} = k_H P_{N_2} = (8.42 \times 10^{-7} \text{ mol L}^{-1} \text{ mm Hg}^{-1})(593 \text{ mm Hg}) = 4.99 \times 10^{-4} \text{ mol/L}$$

$$\text{mass of } N_2 = (4.99 \times 10^{-4} \text{ mol/L})\left(\frac{28.02 \text{ g}}{1 \text{ mol}}\right) = 1.40 \times 10^{-2} \text{ g}_{N_2}/\text{L}$$

(b) Apply the ideal gas law to calculate the volume (mL) of the 4.99×10^{-4} mol N_2 dissolved in 1 L blood. Remember that the temperature must be in kelvins.

$$V = \frac{nRT}{P} = \frac{(4.99 \times 10^{-4} \text{ mol})(0.0821 \text{ L atm mol}^{-1} \text{ K}^{-1})(298 \text{ K})}{0.78 \text{ atm}}$$

$$= 0.0157 \text{ L} = 16 \text{ mL}$$

This is a significant volume of nitrogen gas dissolved in one liter of blood, which can cause serious pain when the gaseous nitrogen is released from the blood when the pressure drops as the scuba diver rises to the surface of the water.

PROBLEM-SOLVING PRACTICE 14.3

The Henry's law constant for oxygen in water at 25 °C is 1.66×10^{-6} mol L^{-1} mm Hg^{-1}. Suppose that a trout stream at 25 °C is in equilibrium with air at normal atmospheric pressure. What is the concentration of O_2 in this stream? Express the result in milligrams per liter (mg/L). The mole fraction of oxygen in air is 0.21.

Henry's law. The greater the partial pressure of CO_2 over the soft drink in the bottle, the greater the concentration of dissolved CO_2. When the bottle is opened, the partial pressure of CO_2 drops and CO_2 bubbles out of the solution.

14.6 Solution Concentration: Keeping Track of Units

The terms *unsaturated, saturated*, and *supersaturated* describe a solution with respect to the quantity of solute in a given quantity of solvent at a certain temperature. Sometimes the terms *concentrated* or *dilute* are also used to describe a solution. Although useful, these are all qualitative descriptors of solution concentration, and it is often important to specify the concentration of a solute in a solution more precisely. Several concentration units are used to do so, including mass fraction, weight percent, molarity, and molality. Often it is useful to be able to quantitatively describe quantities of solute across a wide range of concentrations, from rather large to very small. For example, a variety of units are used to express the concentrations of dilute solutions containing unwanted—even potentially harmful—ionic solutes, such as those containing lead, mercury, selenium, and nitrate, and organic compounds in drinking water. These units are also useful clinically in discussing solute concentrations in blood or urine.

Mass Fraction and Weight Percent

The **mass fraction** of a solute is the fraction of the total mass of the solution that a solute contributes—that is, the mass of a single solute divided by the total mass of all the solutes and the solvent. Mass fraction is commonly expressed as a percentage and called **weight percent,** which is the mass fraction multiplied by 100%. This is the same as the mass of solute in exactly 100 g of solution. For example, the mass fraction of sucrose in a solution consisting of 25.0 g sucrose, 10.0 g fructose, and 300. g water is

The symbol % means "per hundred," that is, divided by 100. Thus, 100% is $\frac{100}{100} = 1$.

$$\text{Mass fraction of sucrose} = \frac{\text{mass of sucrose}}{\text{mass of sucrose} + \text{mass of fructose} + \text{mass of water}}$$

$$= \frac{25.0\text{ g}}{25.0\text{ g} + 10.0\text{ g} + 300.\text{ g}} = 0.0746$$

The weight percent of sucrose in the solution is (0.0746) × 100% = 7.46%.

Weight percent can be thought of as parts per hundred.

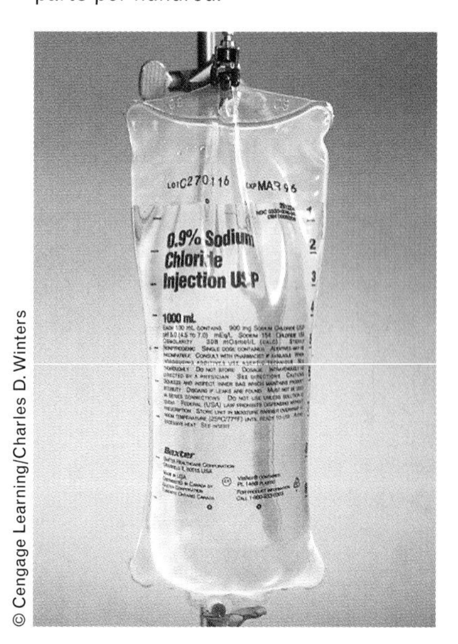

Sterile saline solution. Saline solutions like this one are routinely given to patients who have lost body fluids.

PROBLEM-SOLVING EXAMPLE 14.4 Mass Fraction and Weight Percent

Sterile saline solutions containing NaCl in water are often used in medicine. What is the weight percent of NaCl in a solution made by dissolving 4.6 g NaCl in 500. g pure water?

Answer 0.91%

Strategy and Explanation We apply the definitions of mass fraction and weight percent to the solute and solution stated in the problem.

$$\text{Mass fraction of NaCl} = \frac{4.6\text{ g NaCl}}{4.6\text{ g NaCl} + 500.\text{ g H}_2\text{O}} = 0.0091$$

$$\text{Weight percent of NaCl} = \text{mass fraction of NaCl} \times 100\%$$

$$= 0.0091 \times 100\% = 0.91\%$$

☑ **Reasonable Answer Check** The mass fraction is about 5 g NaCl in 500 g solution, or about 0.01, which is close to the more accurate result.

PROBLEM-SOLVING PRACTICE 14.4

What is the weight percent of glucose in a solution containing 21.5 g glucose, $C_6H_{12}O_6$, in 750. g pure water?

EXERCISE 14.6 Mass Fraction and Weight Percent

Ringer's solution is used in physiology experiments. One liter of the solution contains 6.5 g NaCl, 0.20 g $NaHCO_3$, 0.10 g $CaCl_2$, and 0.10 g KCl. Assume that one liter of water has been used to prepare the solution and that the density of water is 1.00 g/mL.

(a) Calculate the mass fraction and weight percent of $NaHCO_3$ in the solution.
(b) Which solute has the lowest weight percent in the solution?

The density of liquid water is essentially 1 g/mL, 10^3 g/L, or 1 kg/L.

Parts per Million, Billion, and Trillion

Solutes in very dilute solutions have very low mass fractions. Consequently, the mass fraction in such solutions is often expressed in **parts per million** (abbreviated ppm). One part per million is equivalent to one gram of solute per one million grams of solution, or proportionally, one milligram of solute per one thousand grams of solution (1 mg/kg). Thus, a commercial bottled water with a calcium ion concentration of 66 ppm contains 66 mg Ca^{2+} per 1000 g water, essentially one liter. For even smaller mass fractions, **parts per billion** (1 ppb = one microgram, μg, of solute per thousand grams of solution) and **parts per trillion** (1 ppt = one nanogram, ng, of solute in one thousand grams of solution) are often used. As the names imply, a mass fraction converts to parts per billion by multiplying by 10^9 ppb and to parts per trillion by multiplying by 10^{12} ppt.

1 ppm is equivalent to one penny in $10,000; 1 ppb is one penny in $10,000,000.

1 μg (microgram) = 10^{-6} g; 1 ng (nanogram) = 10^{-9} g.

PROBLEM-SOLVING EXAMPLE 14.5 ppm, ppb, and Mass Fraction

Most of the wells in the country of Bangladesh are contaminated with arsenic, a toxic material in sufficient dose, with levels well above the World Health Organization's (WHO) guideline maximum value of 0.010 mg As/L. The WHO estimates that 28–35 million Bangladesh citizens have been exposed to high arsenic levels by drinking water from wells in which the arsenic concentration in the water is at least 0.050 mg/L.

(a) Calculate the mass of arsenic per liter of solution in the 0.050 mg/L solution. Assume that the solution is almost entirely water and the density of the solution is 1.0 g/mL.
(b) Calculate the concentration of arsenic in ppb.

Answer (a) 5.0×10^{-5} g arsenic/L solution (b) 50. ppb

Strategy and Explanation

(a) Assuming that the solution is almost entirely water, 1 L of the solution has a mass of 1.0×10^3 g. The mass of arsenic in this 1-L water sample is calculated from its mass fraction, expressed as the ratio of grams of arsenic per 10^6 g of solution.

$$\left(\frac{0.050 \text{ g arsenic}}{1 \times 10^6 \text{ g solution}}\right)\left(\frac{1.0 \times 10^3 \text{ g solution}}{\text{L solution}}\right) = 5.0 \times 10^{-5} \text{ g arsenic/L solution}$$

(b) This answer can be converted to ppb by first converting it to micrograms of arsenic per liter.

$$\left(\frac{5.0 \times 10^{-5} \text{ g arsenic}}{1 \text{ L}}\right)\left(\frac{1 \ \mu\text{g}}{1 \times 10^{-6} \text{ g}}\right) = 50. \ \mu\text{g arsenic/L}$$

Thus, a mass fraction of 5.0×10^{-2} g arsenic per 10^6 g of solution (0.050 ppm) corresponds to a concentration of 50. μg arsenic/L, which is the equivalent of 50. ppb.

To express the mass fraction as parts per billion, multiply it by 10^9 ppb. This makes a very small number bigger and easier to handle.

☑ **Reasonable Answer Check** Used correctly, the conversion factors change 0.050 mg/L directly to 5.0×10^{-5} g arsenic/L and 50 ppb arsenic.

PROBLEM-SOLVING PRACTICE 14.5

Drinking water often contains small concentrations of selenium (Se). If a sample of water contains 30 ppb Se, how many micrograms of Se are present in 100. mL of this water?

At the height of the Roman Empire, worldwide lead production was about 80,000 tons per year. Today it is about 3 million tons annually. Lead was first used for water pipes in ancient Rome. The Latin name for lead, *plumbum*, gave us the word "plumber."

CONCEPTUAL
EXERCISE 14.7 Lead in Drinking Water

One drinking-water sample has a lead concentration of 20 ppb; another has a lead concentration of 0.003 ppm.

(a) Which sample has the higher lead concentration?

(b) The current EPA acceptable limit for lead in drinking water is 0.015 mg/L. Compare each of the water sample's lead concentration with the acceptable limit.

EXERCISE 14.8 Striking It Rich in the Oceans?

The concentration of gold in seawater is about 1×10^{-3} ppm. The earth's oceans contain 3.5×10^{20} gal of seawater. Approximately how many pounds of gold are in the oceans? 1 gal = 3.785 L; 1 lb = 454 g.

Molarity

As defined in Section 5.6 (*p. 161*), the *molarity* of a solution is

$$\text{Molarity} = \frac{\text{number of moles of solute}}{\text{number of liters of solution}}$$

Multiplying the volume of a solution (L) by its molarity yields the number of moles of solute in that volume of solution. For example, the number of moles of KNO_3 in 250. mL of 0.0200 M KNO_3 is

$$0.250\text{ L} \times \left(\frac{0.0200\text{ mol KNO}_3}{1\text{ L}}\right) = 5.00 \times 10^{-3}\text{ mol KNO}_3$$

from which the number of grams can be determined using the molar mass of potassium nitrate.

$$5.00 \times 10^{-3}\text{ mol KNO}_3 \times \frac{101.1\text{ g KNO}_3}{1\text{ mol KNO}_3} = 0.506\text{ g KNO}_3$$

Thus, to make 250. mL of 0.0200 M KNO_3 solution, you would weigh 0.506 g KNO_3, put it into a 250-mL volumetric flask, and add to it sufficient water to bring the volume of the solution to 250. mL (*p. 162*).

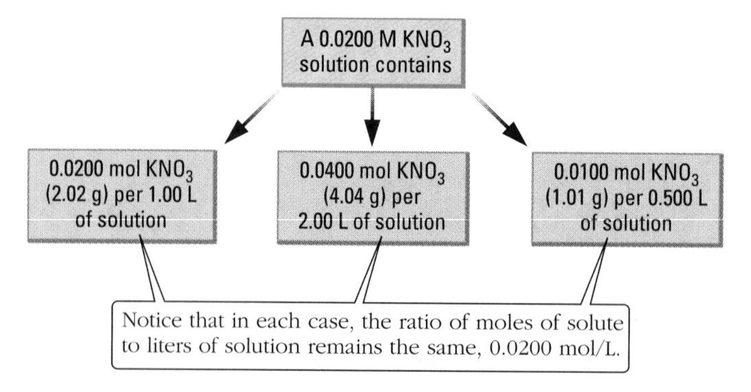
A 0.0200 M KNO3 solution contains
0.0200 mol KNO3 (2.02 g) per 1.00 L of solution
0.0400 mol KNO3 (4.04 g) per 2.00 L of solution
0.0100 mol KNO3 (1.01 g) per 0.500 L of solution
Notice that in each case, the ratio of moles of solute to liters of solution remains the same, 0.0200 mol/L.

PROBLEM-SOLVING EXAMPLE 14.6 Molarity

(a) Calculate the mass in grams of $NiCl_2$ needed to prepare 500. mL of 0.125 M $NiCl_2$.

(b) How many milliliters of this solution are required to prepare 250. mL of 0.0300 M $NiCl_2$?

Answer (a) 8.10 g $NiCl_2$ (b) 60.0 mL

Strategy and Explanation To solve this problem we apply the concept of molarity in part (a) and use the dilution relationship (⟵ *p. 163*) in part (b).

(a) To find the number of grams of $NiCl_2$, the number of moles of $NiCl_2$ must first be calculated. This can be done by multiplying the volume (L) times the molarity.

$$0.500 \text{ L} \times \frac{0.125 \text{ mol NiCl}_2}{1 \text{ L}} = 0.0625 \text{ mol NiCl}_2$$

The number of grams of $NiCl_2$ can be determined from the number of moles by using the molar mass of $NiCl_2$, 129.6 g/mol.

$$0.0625 \text{ mol NiCl}_2 \times \frac{129.6 \text{ g NiCl}_2}{1 \text{ mol NiCl}_2} = 8.10 \text{ g NiCl}_2$$

(b) A more concentrated solution is being diluted to a less concentrated one. The number of moles of $NiCl_2$ remain the same; only the volume of solution changes. To calculate the volume of the more concentrated (undiluted) solution required we can use the relation

$$\text{Molarity}_{\text{conc.}} \times V_{\text{conc.}} = \text{Molarity}_{\text{dil}} \times V_{\text{dil}}$$

in which $\text{Molarity}_{\text{conc.}}$ and $V_{\text{conc.}}$ represent the molarity and volume of the initial (undiluted) solution and $\text{Molarity}_{\text{dil}}$ and V_{dil} are the molarity and volume of the final (diluted) solution (⟵ *p. 163*). In this case,

$$\text{Molarity}_{\text{conc.}} = 0.125 \text{ M} \qquad V_{\text{conc.}} = \text{volume to be determined}$$

$$\text{Molarity}_{\text{dil}} = 0.0300 \text{ M} \qquad V_{\text{dil}} = 250.\text{ mL} = 0.250 \text{ L}$$

Solving for $V_{\text{conc.}}$

$$V_{\text{conc.}} = \frac{\text{Molarity}_{\text{dil}} \times V_{\text{dil}}}{\text{Molarity}_{\text{conc.}}} = \frac{(0.0300 \text{ M})(0.250 \text{ L})}{0.125 \text{ M}} = 0.0600 \text{ L} = 60.0 \text{ mL}$$

☑ **Reasonable Answer Check** The molar mass of $NiCl_2$ is approximately 130 g/mol, so 1 L of a 0.125 M $NiCl_2$ solution would contain about 0.125 mol or 16 g $NiCl_2$. One half of a liter (500 mL) of that solution would contain one half as much $NiCl_2$, or about 8 g, which is close to the calculated value of 8.10 g. The diluted solution is 0.0300 M, which is only about one fourth the undiluted concentration. Therefore, in the dilution the volume must change by a factor of about 4, from 60.0 mL to 250. mL.

PROBLEM-SOLVING PRACTICE 14.6

(a) Calculate the mass in grams of NaBr needed to prepare 250. mL of 0.0750 M NaBr.
(b) How many mL of this solution are required to prepare 500. mL of 0.00150 M NaBr?

PROBLEM-SOLVING EXAMPLE 14.7 Molarity and ppm

Seawater contains 19,000 ppm Cl^- making chloride the most abundant anion in the oceans. Assume that the density of seawater is 1.03 g/mL. Calculate
(a) the mass in grams of chloride ion per liter of seawater.
(b) the molarity of chloride in seawater.

Answer (a) 20. g Cl^-/L seawater (b) 0.56 mol Cl^-/L seawater

Strategy and Explanation

(a) Calculate the mass of Cl^- per liter using the definition of ppm and the density of seawater.

$$\frac{(1.9 \times 10^4 \text{ g Cl}^-)}{(1 \times 10^6 \text{ g seawater})} \times \frac{(1.03 \times 10^3 \text{ g seawater})}{(1 \text{ L seawater})} = 20.\text{ g Cl}^-/\text{L seawater}$$

(b) Calculate the molarity of chloride ion by converting grams to moles.

$$\frac{(20.\text{ g Cl}^-)}{(1 \text{ L seawater})} \times \frac{(1 \text{ mol Cl}^-)}{(35.5 \text{ g Cl}^-)} = 0.56 \text{ mol Cl}^-/\text{L seawater}$$

☑ **Reasonable Answer Check** There is about 0.02 g Cl^- per gram of seawater. A liter of seawater is about 1000 g seawater, which contains about 20 g Cl^-; this is close to the calculated answer.

PROBLEM-SOLVING PRACTICE 14.7

The concentration of magnesium ion in seawater is 0.0556 M making magnesium the second most abundant cation in oceans. Express the Mg^{2+} concentration as (a) mass fraction and (b) ppm. Assume that the density of seawater is 1.03 g/mL.

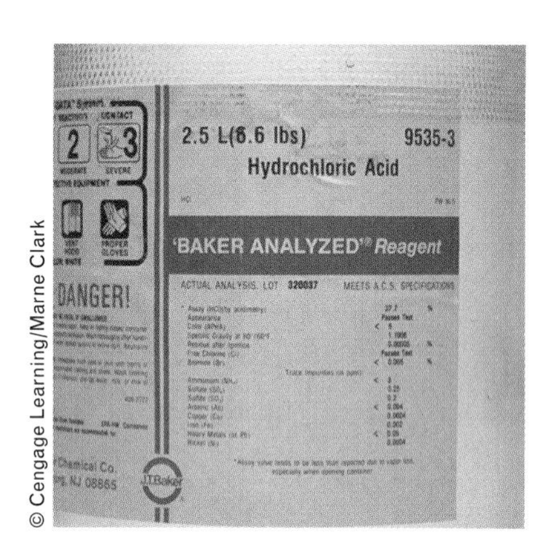

A bottle of commercial concentrated hydrochloric acid.

PROBLEM-SOLVING EXAMPLE 14.8 Weight Percent and Molarity

Hydrochloric acid is sold as a concentrated aqueous solution of HCl with a density of 1.18 g/mL. The concentrated acid contains 38% HCl by mass. Calculate the molarity of hydrochloric acid in this solution.

Answer 12.4 M

Strategy and Explanation To calculate the molarity, the number of moles of HCl and the volume of the solution in liters must be determined. The 38.0% hydrochloric acid solution means that 100. g solution contains 38.0 g HCl. Therefore,

$$\text{Amount (moles) of HCl} = 38.0 \text{ g HCl} \times \frac{1 \text{ mol HCl}}{36.5 \text{ g HCl}} = 1.04 \text{ mol HCl}$$

The molarity of the solution can be calculated using the number of moles of HCl and the density of the solution to determine its volume.

$$\text{Molarity} = \frac{1.04 \text{ mol}}{100. \text{ g solution}} \times \frac{1.18 \text{ g solution}}{1 \text{ mL solution}} \times \frac{1000 \text{ mL solution}}{1 \text{ L solution}}$$

$$= 12.4 \frac{\text{mol}}{\text{L}} = 12.4 \text{ M}$$

☑ **Reasonable Answer Check** The molar mass of HCl, 36.5 g/mol, is close to the number of grams of HCl in 100 g solution. The density of the solution is about 1, so there is about 1 mol HCl in 100 mL of this solution. Consequently, there are approximately 10 mol HCl in 1 L solution—approximately a 10 M HCl solution. The calculated answer is a bit higher because we rounded the actual density (1.18 g/mL) to 1.

PROBLEM-SOLVING PRACTICE 14.8

The density of a commercial 30.0% hydrogen peroxide, H_2O_2, solution is 1.11 g/mL at 25 °C. Calculate the molarity of hydrogen peroxide in this solution.

Molarity and molality. The photo shows a 0.10 molal solution (0.10 mol/kg) of potassium chromate (*flask at right*) and a 0.10 molar solution (0.10 mol/L) of potassium chromate (*flask at left*). Each solution contains 0.10 mol (19.4 g) of yellow K_2CrO_4, shown in the dish at the front. The 0.10 molar (0.10 mol/L) solution on the left was made by placing the solid in the flask and adding enough water to make 1.0 L solution. The 0.10 molal (0.10 mol/kg) solution on the right was made by placing the solid in the flask and adding 1000 g (1 kg) water. Adding 1 kg water produces a solution that has a volume greater than 1 L.

Molality

Solute concentration can also be expressed in terms of the moles of solute in relation to the mass of the solvent. **Molality** (abbreviated m), is defined as the amount (moles) of solute per *kilogram* of *solvent* (not solution).

$$\text{Molality of solute A} = m_A = \frac{\text{moles of solute A}}{\text{kilograms of solvent}}$$

For example, we can calculate the molality of a solution prepared by dissolving 0.413 g methanol, CH_3OH, in 1.50×10^3 g water by determining the number of moles of methanol solute and the number of kilograms of solvent, water.

$$0.413 \text{ g methanol} \times \frac{1 \text{ mol methanol}}{32.042 \text{ g methanol}} = 0.0129 \text{ mol methanol}$$

The mass of water is 1.50×10^3 g, which is 1.50 kg. Substituting these values into the molality equation gives

$$\text{Molality} = m = \frac{0.0129 \text{ mol methanol}}{1.50 \text{ kg water}} = 0.00860 \text{ mol/kg}$$

Molality will be used in Section 14.7 when we deal with the effect of solute concentration on decreasing the freezing point and raising the boiling point of a solution.

PROBLEM-SOLVING EXAMPLE 14.9 Molarity and Molality

Automobile lead storage batteries contain an aqueous solution of sulfuric acid. The solution has a density of 1.230 g/mL at 25 °C and contains 368. g H_2SO_4 per liter.

Calculate (a) the molarity and (b) the molality of sulfuric acid in the solution.

Answer (a) Molarity = 3.75 mol H_2SO_4/L (b) Molality = 4.35 mol H_2SO_4/kg solvent

Strategy and Explanation

(a) Determine the number of moles of solute in a liter of solution and calculate the molarity.

$$368.\text{ g } H_2SO_4 \times \frac{1 \text{ mol } H_2SO_4}{98.1 \text{ g } H_2SO_4} = 3.75 \text{ mol } H_2SO_4$$

This is the amount of solute in 1 L solution; the molarity is 3.75 mol/L, 3.75 M.

(b) To calculate the molality we first need to determine the mass (kg) of solvent (water) and then divide that into the moles of solute. The density of the solution is 1.230 g/mL, which is equivalent to 1.230×10^3 g solution/L; 1 L solution contains 1.230×10^3 g of the water and sulfuric acid mixture, 368. g of which is sulfuric acid. Therefore, the mass of water can be determined.

$$1230 \text{ g solution} - 368 \text{ g } H_2SO_4 = 862 \text{ g water} = 0.862 \text{ kg water}$$

We can now calculate the molality of sulfuric acid in the solution.

$$\text{Molality of } H_2SO_4 = \frac{\text{moles of } H_2SO_4}{\text{kg of solvent}} = \frac{3.75 \text{ mol } H_2SO_4}{0.862 \text{ kg solvent}}$$
$$= 4.35 \text{ mol } H_2SO_4\text{/kg solvent}$$

☑ **Reasonable Answer Check** The molar mass of sulfuric acid is about 100 g/mol and so 368 g of the acid would be about 3.7 mol. There is about 4 mol of the acid in about a kilogram of water, so the molality is about 4, close to the more accurate answer of 4.35 mol/kg. Because the density of the solution is not 1.00 g/ml, the mass of solvent and the volume of the solution are not equal, so the molarity and the molality differ.

PROBLEM-SOLVING PRACTICE 14.9

Calculate the molarity and the molality of NaCl in a 20% aqueous NaCl solution whose density is 1.148 g/mL at 25 °C.

Molality and molarity are *not* the same, although the difference becomes negligibly small for dilute solutions, those less than 0.01 mol/L.

Table 14.4 Comparison of Concentration Expressions

Concentration Expression	Units of Concentration
Mass fraction	None
Weight percent	Percent
Parts per million, parts per billion, parts per trillion	ppm, ppb, ppt
Molarity	$\frac{\text{Moles solute}}{\text{liter solution}}$
Molality	$\frac{\text{Moles solute}}{\text{kg solvent}}$

Table 14.4 summarizes the types of concentration units we have used so far, comparing their units.

CONCEPTUAL EXERCISE 14.9 Molality and Molarity

(a) What information is required to calculate the molality of a solution?

(b) What information is needed to calculate the molarity of a solution if the solution's composition is given in weight percent?

14.7 Vapor Pressures, Boiling Points, Freezing Points, and Osmotic Pressures of Solutions

Up to this point, we have discussed solutions in terms of the nature of the solute and the nature of the solvent. Some properties of solutions do not depend on the nature of the solute or solvent, but rather depend only on the *number* of dissolved solute particles—ions or molecules—per unit volume.

In liquid solutions, solute molecules or ions disrupt solvent-solvent noncovalent attractions, causing changes in solvent properties that depend on these attractions. For example, when solute is added to a solvent, the freezing point of the resulting solution is lower and its boiling point is higher than that of the pure solvent. How much the properties of the solution differ from those of the pure solvent depends only on the concentration of the solute particles. **Colligative properties** of solutions are those that *depend only on the concentration of solute particles* (ions or molecules) in the solution, regardless of what kinds of particles are present. We will consider four colligative properties: vapor pressure lowering, boiling point elevation, freezing point depression, and osmotic pressure. These are all quite common and important in the world around us.

Vapor Pressure Lowering

In a closed container a dynamic equilibrium exists between a pure liquid and its vapor—the rate at which molecules escape the liquid phase equals the rate at which vapor phase molecules return to the liquid. This equilibrium gives rise to a vapor pressure that depends on the temperature (⇐ *p. 379*). But the vapor pressure of a pure liquid differs from that of the liquid when a solute has been dissolved in it. Compare a small portion of the liquid/vapor boundary for pure water with that for seawater (mainly an aqueous sodium chloride solution), as shown at the molecular scale in Figure 14.12. For an aqueous solution such as seawater, in which sodium ions and chlo-

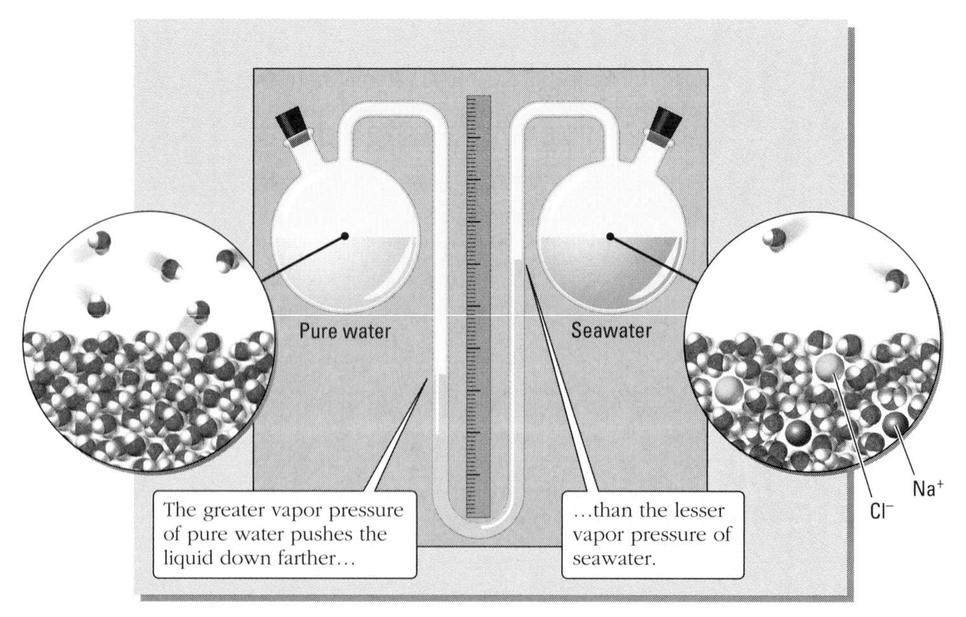
Pure water
Seawater
The greater vapor pressure of pure water pushes the liquid down farther...
...than the lesser vapor pressure of seawater.
Na^+
Cl^-

Figure 14.12 The vapor pressure of pure water and that of seawater. Seawater is an aqueous solution of NaCl and many other salts. The vapor pressure over an aqueous solution is not as great as that over pure water at the same temperature.

ride ions (and many other kinds of ions and molecules) are present, the vapor pressure of water is lower than for a sample of pure water.

The vapor pressure of any pure solvent will be lowered by the addition of a nonvolatile solute to the solvent. How much a dissolved solute lowers the vapor pressure of a solvent depends on the solute concentration and is expressed by **Raoult's law:**

Raoult's law works best with dilute solutions. Deviations from Raoult's law occur when solute-solvent intermolecular forces are either much weaker or much stronger than the solvent-solvent and solute-solute intermolecular forces.

$$P_1 = X_1 P_1^0$$

where P_1 is the vapor pressure of the solvent over the *solution,* P_1^0 is the vapor pressure of the *pure* solvent at the same temperature, and X_1 is the mole fraction of *solvent* in the solution. For example, suppose you want to calculate the vapor pressure over a sucrose solution at 25 °C in which the mole fraction of water is 0.986. The vapor pressure of pure water at 25 °C is 23.76 mm Hg. From these data, the vapor pressure of water over the solution can be calculated using Raoult's law.

$$P_{water} = (X_{water})(P^0_{water}) = (0.986)(23.76 \text{ mm Hg}) = 23.42 \text{ mm Hg}$$

The vapor pressure has been lowered by (23.76 − 23.42) mm Hg = 0.34 mm Hg. Therefore, the vapor pressure of water over this solution is only 98.6% that of pure water at 25 °C.

Raoult's law can also be applied to solutions in which the solvent and the solute are both volatile so that an appreciable amount of each can be in the vapor above the solution. We will not consider such cases.

PROBLEM-SOLVING EXAMPLE 14.10 Raoult's Law

Ethylene glycol, $HOCH_2CH_2OH$, is used as an antifreeze. What is the vapor pressure of water above a solution of 100.0 mL ethylene glycol and 100.0 mL water at 90 °C? Densities: ethylene glycol, 1.15 g/mL; water, 1.00 g/mL. The vapor pressure of pure water at 90 °C is 525.8 mm Hg.

ethylene glycol

$HOCH_2CH_2OH$

Answer 394 mm Hg

Strategy and Explanation To determine the vapor pressure of water over the solution using Raoult's law, we must first calculate the mole fraction of water in the solution. For the mole fraction, the milliliters of ethylene glycol and water must be converted to grams and then to moles.

$$100.0 \text{ mL eth. gly.} \times \frac{1.15 \text{ g eth. gly.}}{1.00 \text{ mL eth. gly.}} \times \frac{1 \text{ mol eth. gly.}}{62.0 \text{ g eth. gly.}} = 1.86 \text{ mol eth. gly.}$$

$$100.0 \text{ mL water} \times \frac{1.00 \text{ g water}}{1.00 \text{ mL water}} \times \frac{1 \text{ mol water}}{18.0 \text{ g water}} = 5.56 \text{ mol water}$$

$$X_{water} = \frac{5.56}{5.56 + 1.86} = \frac{5.56}{7.42} = 0.749$$

Mole fraction of A, X_A,

$$= \frac{\text{moles of A}}{\text{total number of moles}}$$

$$= \frac{\text{moles A}}{\text{moles A} + \text{moles B} + \cdots}$$

Applying Raoult's law:

$$P_{water} = (X_{water})(P^0_{water}) = (0.749)(525.8 \text{ mm Hg}) = 394 \text{ mm Hg}$$

The dissolved ethylene glycol lowers the vapor pressure of water to 394 mm Hg, a 25% decrease.

☑ **Reasonable Answer Check** Because there are nearly three times the number of moles of water as there are moles of ethylene glycol, the mole fraction of water should be greater than the mole fraction of ethylene glycol, and it is (mole fraction of ethylene glycol = 1 − 0.749 = 0.251). Therefore, the vapor pressure of water over the solution should be about 75% that of pure water (394/525.8 × 100%), which it is. The answer is reasonable.

PROBLEM-SOLVING PRACTICE 14.10

The vapor pressure of an aqueous solution of urea, CH_4N_2O, is 291.2 mm Hg. The vapor pressure of water at that temperature is 355.1 mm Hg. Calculate the mole fraction of each component.

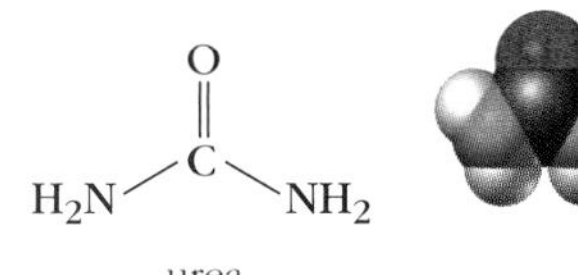

urea

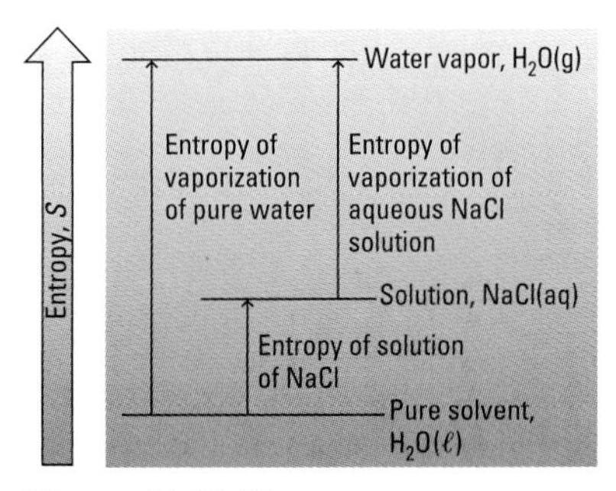

Figure 14.13 Vapor pressure lowering and entropies of solution and vaporization. The entropy of vaporization of pure water is greater than the entropy of solution.

EXERCISE 14.10 Vapor Pressure of a Mixture

Calculate the vapor pressure of water over a solution containing 50.0 g sucrose, $C_{12}H_{22}O_{11}$, and 100.0 g water at 45 °C. The vapor pressure of pure water at this temperature is 71.88 mm Hg.

Entropy plays a role in vapor pressure lowering. Consider the entropy change for the vaporization of pure water with that for the vaporization of a corresponding quantity of water from a sodium chloride solution (Figure 14.13). Very few sodium or chloride ions escape from the solution and the vapor in equilibrium with the salt water consists almost entirely of water molecules. Thus, the entropy of a given amount of the vapor is approximately the same in both cases (pure water and salt water). The entropy within the salt solution, however, is greater than that in pure water.

Now consider what happens when water vaporizes from pure water and from the salt solution. As with any change from a liquid to a gas, there is a significant increase in entropy of the water vaporizing from either source (pure water or salt solution) (⇐ *p. 523*). But the entropy of vaporization is not the same in each case. As illustrated in Figure 14.13, the entropy of vaporization is larger for the water vapor from pure water than from the salt solution because the entropy of pure water was already smaller than that of the salt solution. The main point is that a bigger entropy increase corresponds to a more product-favored process. The result is that the vaporization from pure water creates a higher pressure of water vapor (that is, more water molecules per unit volume) in equilibrium with pure water than does the vaporization of water from a salt solution.

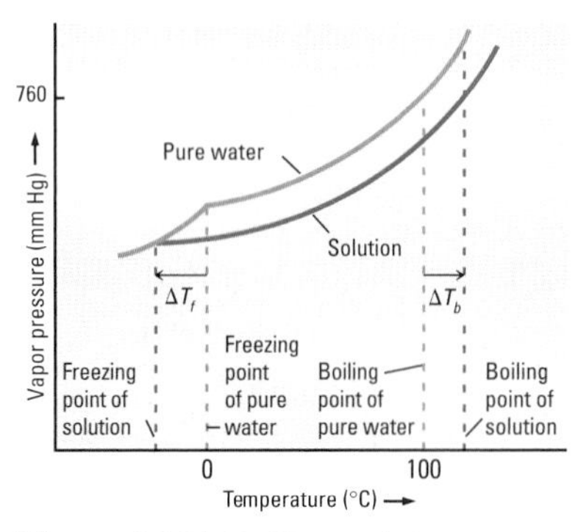

Figure 14.14 Boiling point elevation (ΔT_b) and freezing point lowering (ΔT_f) for aqueous solutions. Addition of solute to a pure solvent raises its boiling point and lowers its freezing point.

The units of the molal boiling point elevation constant are abbreviated °C kg mol^{-1}.

Boiling Point Elevation

As a result of vapor pressure lowering, the vapor pressure of an aqueous solution of a nonvolatile solute at 100 °C is less than 760 mm Hg (1 atm). For the solution to boil, it must be heated *above* 100 °C. The **boiling point elevation,** ΔT_b, is the difference between the normal boiling point of water and the higher boiling point of an aqueous solution of a nonvolatile nonelectrolyte solute (Figure 14.14).

The increase in boiling point is proportional to the concentration of the solute, expressed as molality, and can be calculated from this relationship.

$$\Delta T_b = T_b\,(\text{solution}) - T_b\,(\text{solvent}) = K_b m_{\text{solute}}$$

where ΔT_b is the boiling point of solution minus the boiling point of pure solvent. The value of K_b, the *molal boiling point elevation constant* of the *solvent,* depends only on the solvent. For example, K_b for water is 0.52 °C kg mol^{-1}; that of benzene is 2.53 °C kg mol^{-1}.

Molality, rather than molarity, is used in boiling point elevation determinations because the molality of a solution does not change with temperature changes, but the molarity of a solution does. Molality is based on the masses of solute *and* solvent, which are unaffected by temperature changes. The volume of a solution, which is used in molarity, expands or contracts when the solution is heated or cooled.

EXERCISE 14.11 Calculating the Boiling Point of a Solution

The boiling point elevation constant for benzene is 2.53 °C kg mol^{-1}. The boiling point of pure benzene is 80.10 °C. If a solute's concentration in benzene is 0.10 mol/kg, what will be the boiling point of the solution?

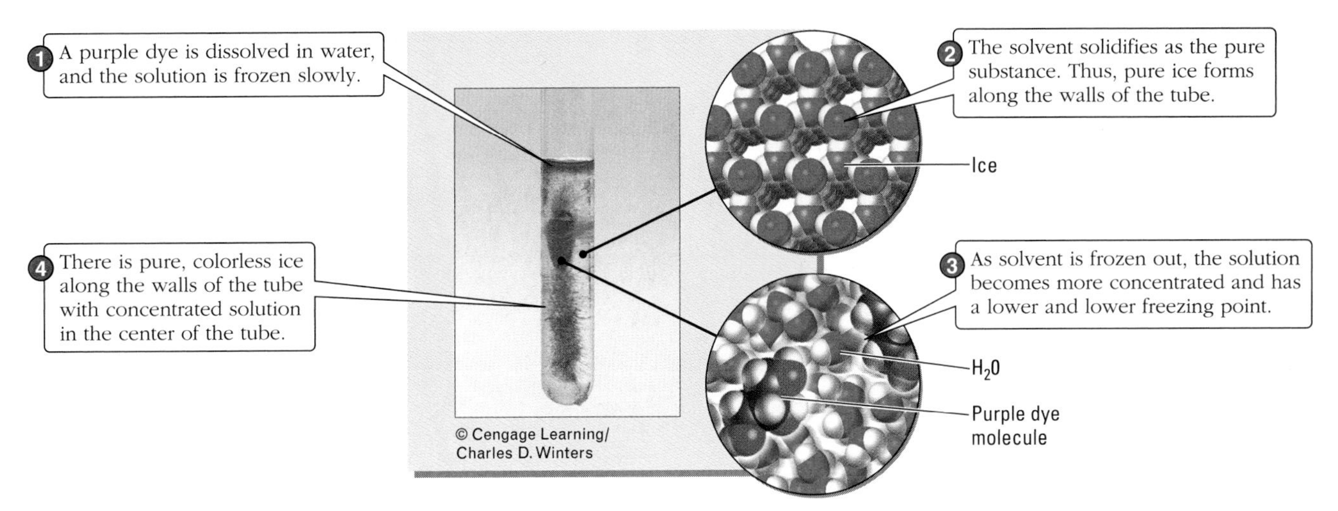

Figure 14.15 Solvent freezing.

Freezing Point Lowering

A pure liquid begins to freeze when the temperature is lowered to the substance's freezing point and the first few molecules cluster together into a crystal lattice forming a tiny quantity of solid. As long as both solid and liquid phases are present and the temperature is at the freezing point, there is a dynamic equilibrium in which the rate of crystallization equals the rate of melting. When a *solution* freezes, a few molecules of solvent cluster together to form pure solid *solvent* (Figure 14.15), and a dynamic equilibrium is set up between solution and solid solvent.

The molecules or ions in the liquid in contact with the frozen solvent in a freezing solution are not all solvent molecules. This causes a slower rate at which particles move from solution to solid than the rate in the pure liquid solvent. To achieve dynamic equilibrium, a correspondingly slower rate of escape of molecules from the solid crystal lattice must occur. According to the kinetic-molecular theory, this slower rate occurs at a lower temperature, so the freezing point of the solution is lower than that of the pure liquid solvent (Figure 14.14).

The **freezing point lowering,** ΔT_f, is proportional to the concentration of the solute (molality) in the same way as the boiling point elevation.

$$\Delta T_f = K_f m_{\text{solute}}$$

As with K_b, the *molal freezing point constant* of the solvent, K_f, depends only on the solvent and not the type of solute. For water, the freezing point constant is 1.86 °C kg mol^{-1}; by comparison, that of benzene is 5.10 °C kg mol^{-1} and that of cyclohexane is 20.2 °C kg mol^{-1}.

Using ethylene glycol, $HOCH_2CH_2OH$, a relatively nonvolatile alcohol, in automobile cooling systems is a practical application of boiling point elevation and freezing point lowering. Ethylene glycol raises the boiling temperature of the coolant mixture of ethylene glycol and water to a level that prevents engine overheating in hot weather. Ethylene glycol also lowers the freezing point of the coolant, thereby keeping the solution from freezing in the winter.

During winter in cold climates, road salt, principally sodium chloride, is applied liberally to roads to lower the freezing point of ice and snow, making it easier to clear them from the roads.

Another practical application of freezing point lowering is adding salt, NaCl, to ice when making homemade ice cream. Lowering the freezing temperature of the ice-salt water mixture freezes the ice cream more quickly.

PROBLEM-SOLVING EXAMPLE 14.11 Boiling Point Elevation and Freezing Point Lowering

Calculate the boiling and freezing points of an aqueous solution containing 39.5 g ethylene glycol, $HOCH_2CH_2OH$, dissolved in 750. mL water. Assume the density of water to be 1.00 g/mL; its K_b = 0.52 °C kg mol^{-1} and its K_f = 1.86 °C kg mol^{-1}.

Answer Boiling point = 100.44 °C; freezing point = −1.58 °C

Strategy and Explanation To use the equations for freezing point and boiling point changes, the molality of the solution must be determined. The molar mass of ethylene glycol is 62.07 g/mol, and the number of moles of ethylene glycol is

$$39.5\text{ g} \times \frac{1\text{ mol}}{62.07\text{ g}} = 0.636\text{ mol}$$

The mass of solvent is

$$750.\text{ mL} \times \frac{1.00\text{ g}}{\text{mL}} \times \frac{1\text{ kg}}{10^3\text{ g}} = 0.750\text{ kg}$$

The molality of the solution is

$$\frac{0.636\text{ mol}}{0.750\text{ kg}} = 0.848\text{ mol/kg}$$

The boiling point elevation is

$$\Delta T_b = (0.52\text{ °C kg mol}^{-1})(0.848\text{ mol/kg}) = 0.44\text{ °C}$$

Therefore, the solution boils at 100.44 °C.

The freezing point lowering is

$$\Delta T_f = (1.86\text{ °C kg mol}^{-1})(0.848\text{ mol/kg}) = 1.58\text{ °C}$$

The freezing point is lowered by 1.58 °C and so the solution freezes at −1.58 °C.

☑ **Reasonable Answer Check** There is a little more than half a mole of ethylene glycol in three-fourths kilogram of solvent, so the molality should be less than 1 mol/kg, which it is. Because the concentration is less than 1 mol/kg, the boiling point should be raised less than 0.52 °C and the freezing point should be lowered less than 1.86 °C, which they are. The answers are reasonable.

PROBLEM-SOLVING PRACTICE 14.11

A water tank contains 6.50 kg water. Will the addition of 1.20 kg ethylene glycol be sufficient to prevent the solution from freezing if the temperature drops to −25. °C?

EXERCISE 14.12 Protection Against Freezing

Suppose that you are closing a cabin in the north woods for the winter and you do not want the water in the toilet tank to freeze. You know that the temperature might get as low as −30. °C, and you want to protect about 4.0 L water in the toilet tank from freezing. What volume of ethylene glycol (density = 1.113 g/mL; molar mass = 62.1 g/mol) should you add to the 4.0 L water?

Ethylene glycol is toxic and should not be allowed to get into drinking water supplies.

PROBLEM-SOLVING EXAMPLE 14.12 Molar Mass from Freezing Point Lowering

A researcher synthesizes a new compound and uses freezing point depression measurements to determine the molar mass of the compound. The researcher dissolves 1.50 g compound in 75.0 g cyclohexane, which has a freezing point of 6.50 °C and a freezing point depression constant of 20.2 °C kg mol^{-1}. The freezing point of the solution is measured as 2.70 °C. Use these data to calculate the molar mass of the new compound.

Answer 106 g/mol

Strategy and Explanation To calculate the molar mass, we need to calculate the change in freezing point and the molality of the solution. The change in the freezing point, ΔT_f, is 6.50 °C − 2.70 °C = 3.80 °C. Rearranging the freezing point lowering equation, we can solve for the molality of the solution.

$$\text{Molality} = \frac{\Delta T_f}{K_f} = \frac{3.80\ °\text{C}}{20.2\ °\text{C kg mol}^{-1}} = 0.188 \text{ mol/kg}$$

From the molality we can calculate the molar mass of the new compound from the relation

$$\text{Molality, } m = \frac{\text{moles solute}}{\text{kg solvent}} = \frac{\text{grams solute/molar mass}}{\text{kg solvent}}$$

This equation can be solved for the molar mass, g/mol.

$$\text{Molar mass} = \frac{\text{grams solute}}{(\text{molality})(\text{kg solvent})} = \frac{1.50 \text{ g}}{(0.188)(0.0750) \text{ mol}} = 106 \text{ g/mol}$$

☑ **Reasonable Answer Check** About 1.5 g compound in 0.0750 kg solvent, equivalent to 20 g compound per kilogram of solvent, forms a 0.188 mol/kg solution. Thus, a solution containing 1 mol compound per kilogram would contain 20 g/kg solvent × $\frac{1 \text{ kg solvent}}{0.188 \text{ mol}}$, which equals 106 g/mol.

PROBLEM-SOLVING PRACTICE 14.12

A student determines the freezing point to be 5.15 °C for a solution made from 0.180 g of a nonelectrolyte in 50.0 g benzene. Calculate the molar mass of the solute. The freezing point constant of benzene is 5.10 °C kg mol^{-1}; its freezing point is 5.50 °C.

Colligative Properties of Electrolytes

Experimentally, the vapor pressures of 1 M aqueous solutions of sucrose, NaCl, and $CaCl_2$ are all less than that of water at the same temperature. This is to be expected because solutes lower the vapor pressure of the pure solvent. However,

$$\text{vp pure water} > \text{vp 1 M sucrose} > \text{vp 1 M NaCl} > \text{vp 1 M CaCl}_2$$

Because colligative properties of dilute solutions are proportional to the concentration of solute *particles,* this vapor pressure order is not surprising. Because of their dissociation, electrolytes such as NaCl and $CaCl_2$ contribute more particles per mole than do nonelectrolytes such as sucrose or ethanol. Whereas 1 mol sucrose contributes 1 mol of particles (sucrose molecules) to solution, 1 mol NaCl contributes 2 mol of particles (1 mol Na^+ and 1 mol Cl^-), and 1 mol $CaCl_2$ produces 3 mol of particles (1 mol Ca^{2+} and 2 mol Cl^-). Therefore, electrolytes have a greater effect on vapor pressure and boiling point than nonelectrolytes do.

For solutions of electrolytes, the boiling point elevation and freezing point lowering equations can be written as

$$\Delta T_b = K_b\, m_{\text{solute}}\, i_{\text{solute}} \qquad \Delta T_f = K_f\, m_{\text{solute}}\, i_{\text{solute}}$$

The i_{solute} factor gives the *number of particles per formula unit of solute.* It is called the van't Hoff factor, named after Jacobus Henricus van't Hoff (1852–1911). The value of i is 1 for nonelectrolytes because these molecular solutes, such as ethanol, sucrose, benzene, and carbon tetrachloride, do *not* dissociate in solution. For soluble ionic solutes (strong electrolytes), i equals the number of ions per formula unit of the ionic compound. In extremely dilute solutions $i_{\text{solute}} = 2$ for NaCl and $i_{\text{solute}} = 3$ for calcium chloride. The actual i_{solute} value must be determined experimentally. The theoretical i value assumes that the ions act independently in solution, which is achieved only in extremely dilute solutions where the interaction between cations and anions is minimal. In more concentrated solutions, cations and anions interact and $i_{\text{expt}} < i_{\text{theor}}$. For example, in aqueous $MgSO_4$ solutions, $i_{\text{theor}} = 2$, whereas in 0.50 M $MgSO_4$, $i_{\text{expt}} = 1.07$; in 0.005 M $MgSO_4$, $i_{\text{expt}} = 1.72$.

Edgar Fahs Smith Memorial Collection/University of Pennsylvania

Jacobus Henricus van't Hoff 1852–1911

Van't Hoff was one of the founders of physical chemistry. While still a graduate student, he proposed an explanation of optical isomerism based on the tetrahedral nature of the carbon atom. Physical chemistry is the branch of chemistry that applies the laws of physics to the understanding of chemical phenomena. Van't Hoff conducted seminal experimental studies in chemical kinetics, chemical equilibrium, osmotic pressure, and chemical affinity. Van't Hoff received the first Nobel Prize in Chemistry (1901) for his fundamental discoveries in physical chemistry, including his work on the colligative properties of solutions.

16908

Salt helps to lower the melting point of snow and ice, facilitating their removal from roads.

A buildup of dissolved NaCl or $CaCl_2$ along roads is environmentally hazardous because the excessive Cl^- concentration is harmful to roadside plants.

Another practical application of freezing point lowering can be seen in areas where winter weather produces lots of frozen precipitation. To remove snow and particularly ice, roads and walkways are often salted. Although sodium chloride is usually used, calcium chloride is particularly good for this purpose because it has three ions per formula unit and dissolves exothermically. Not only is the freezing point of water lowered, but the enthalpy of solution helps melt the ice.

CONCEPTUAL
EXERCISE 14.13 Freezing Point Lowering

The freezing point of a 2.0 mol/kg $CaCl_2$ solution is measured as −4.78 °C. Calculate the *i* factor and use it to approximate the degree of dissociation of $CaCl_2$ in this solution.

Osmotic Pressure of Solutions

A *membrane* is a thin layer of material that allows molecules or ions to pass through it. A **semipermeable membrane** allows only certain kinds of molecules or ions to pass through while excluding others (Figure 14.16). Examples of semipermeable membranes are animal bladders, cell membranes in plants and animals, and cellophane, a polymer derived from cellulose. When two solutions containing the same solvent are separated by a membrane permeable only to solvent molecules, osmosis will occur. **Osmosis** is *the movement of a solvent through a semipermeable membrane from a region of lower solute concentration (higher solvent concentration) to a region of higher solute concentration (lower solvent concentration).* The **osmotic pressure** of a solution is *the pressure that must be applied to the solution to stop osmosis from a sample of pure solvent.*

Consider the osmosis example shown in Figure 14.17. A 5% aqueous sugar solution is placed in a bag attached to a glass tube. The bag is made of a semipermeable membrane that allows water but not sugar molecules to pass through it. When the bag is submerged in pure water, water flows into the bag by osmosis and raises the liquid level in the tube. When the bag is first submerged, more collisions of solvent mole-

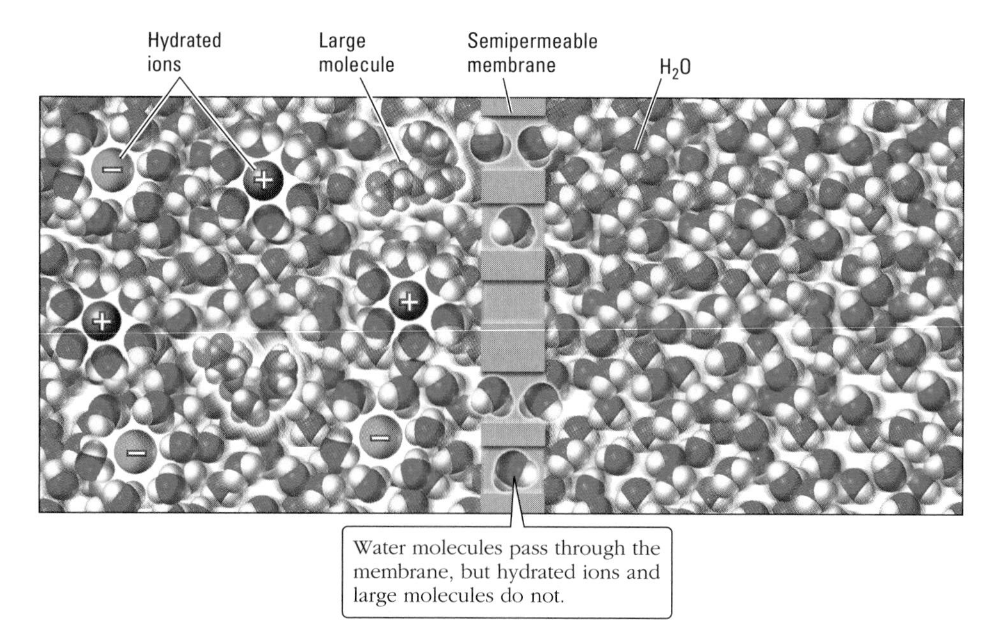
Hydrated ions
Large molecule
Semipermeable membrane
H_2O
Water molecules pass through the membrane, but hydrated ions and large molecules do not.

Figure 14.16 Osmotic flow of a solvent through a semipermeable membrane to a solution. The semipermeable membrane is shown acting as a size-selective sieve. Many membranes operate in different ways, but the ultimate effect is the same.

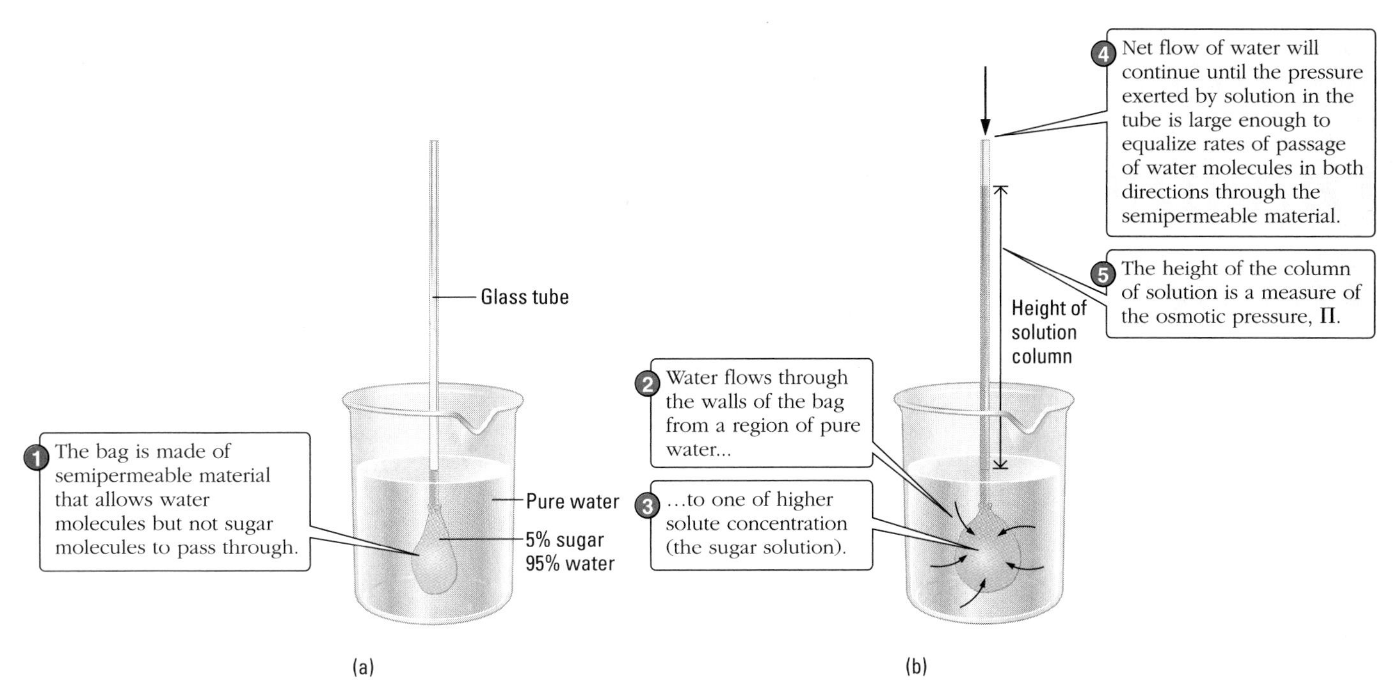

Figure 14.17 Demonstration of osmotic pressure.

cules per unit area of the membrane occur on the pure solvent side than there are on the solution side (where there are fewer solvent molecules per unit volume). Hence, water moves through the membrane from the beaker, where water is in greater concentration, into the solution in the bag, where the water concentration is lower. As this continues, the number of water molecules in the solution increases, the number of collisions of water molecules within the solution increases, and water rises in the tube as pressure builds up in the bag. A dynamic equilibrium is achieved when the pressure in the bag equals the osmotic pressure, at which point the rate of passing water molecules is the same in both directions. The height of the water column then remains unchanged, and it is a measure of the osmotic pressure.

Osmotic pressure—like vapor pressure lowering, boiling point elevation, and freezing point lowering—results from the unequal rates at which solvent molecules pass through an interface or boundary. In the case of evaporation and boiling, it is the solution/vapor interface; for freezing, it is the solution/solid interface. The semipermeable membrane is the interface for osmosis.

All colligative properties can be understood in terms of differences in entropy between a pure solvent and a solution. This is perhaps most easily seen in the case of osmosis. When solvent and solute molecules mix, entropy usually increases. If pure solvent is added to a solution, a higher entropy state will be achieved as solvent and solute molecules diffuse among one another to form a more dilute solution. Unless strong noncovalent intermolecular forces exist between the solute and solvent, there will be a negligible enthalpy change, so the increase in entropy makes mixing of solvent and solution a product-favored process. A semipermeable membrane prevents solute molecules from passing into pure solvent, so the only way mixing can occur (and entropy can increase) is for solvent to flow into the solution, and it does.

The more concentrated the solution, the more product-favored the mixing is and the greater is the pressure required to prevent it. Osmotic pressure (Π) is proportional to the molarity of the solution, c,

$$\Pi = cRTi$$

where R is the gas constant, T is the absolute temperature (in kelvins), and i is the number of particles per formula unit of solute.

The osmotic pressure equation is similar to the ideal gas law equation, $PV = nRT$, which can be rearranged to $P = (n/V)RT = cRT$, where n/V is the molar concentration of the gas.

Even though the solution concentration is small, osmotic pressure can be quite large. For example, the osmotic pressure of a 0.020 M solution of a nonelectrolyte solute ($i = 1$) at 25 °C is

$$\Pi = cRTi = \left(\frac{0.020 \text{ mol}}{\text{L}}\right)(0.0821 \text{ L atm mol}^{-1} \text{ K}^{-1})(298 \text{ K})(1) = 0.49 \text{ atm}$$

This pressure would support a water column more than 15 ft high. One way to determine osmotic pressure is to measure the height of a column of solution in a tube, as shown in Figure 14.17. Heights of a few centimeters can be measured accurately, so quite small concentrations can be determined by osmotic pressure experiments. If the mass of solute dissolved in a measured volume of solution is known, it is possible to calculate the molar mass of the solute by using the definition of molar concentration, $c = n/V$ = amount (mol)/volume (L). Osmotic pressure is especially useful in studying large molecules whose molar mass is difficult to determine by other means.

Freezing point lowering and boiling point elevation measurements can also be used to find the molar mass in the same manner as shown in Problem-Solving Example 14.13 for osmotic pressure measurements.

PROBLEM-SOLVING EXAMPLE 14.13 Molar Mass from Osmotic Pressure

A solution containing 2.50 g of a nonelectrolyte polymer, a long-chain molecule, dissolved in 150. mL solution has an osmotic pressure of 1.25×10^{-2} atm at 25 °C. Calculate the molar mass of the polymer.

Answer 3.26×10^{4} g/mol

Strategy and Explanation We can calculate the molarity of the solution using the osmotic pressure equation. Because the polymer is a nonelectrolyte, $i = 1$.

$$c = \frac{\Pi}{RTi} = \frac{1.25 \times 10^{-2} \text{ atm}}{(0.0821 \text{ L atm mol}^{-1} \text{ K}^{-1})(298 \text{ K})(1)} = 5.11 \times 10^{-4} \text{ mol/L}$$

The volume of solution is 0.150 L, so the number of moles of polymer is

$$\text{Amount polymer} = (0.150 \text{ L})(5.11 \times 10^{-4} \text{ mol/L}) = 7.67 \times 10^{-5} \text{ mol}$$

This is the number of moles in 2.50 g of the polymer; therefore the average molar mass of the polymer is

$$\text{Average molar mass of polymer} = \frac{2.50 \text{ g polymer}}{7.67 \times 10^{-5} \text{ mol polymer}} = 3.26 \times 10^{4} \text{ g/mol}$$

☑ **Reasonable Answer Check** Because the molarity is low, the molar mass of the polymer must be relatively large to create an osmotic pressure of 1.25×10^{-2} atm.

PROBLEM-SOLVING PRACTICE 14.13

The osmotic pressure of a solution of 5.0 g of horse hemoglobin (a protein) in 1.0 L water is 1.8×10^{-3} atm at 25 °C. What is the molar mass of the hemoglobin?

Blood and other fluids inside living cells contain many different solutes, and the osmotic pressures of these solutions play an important role in the distribution and balance of solutes within the body. Dehydrated patients are often given water and nutrients intravenously. However, pure water cannot simply be dripped into a patient's veins. The water would flow into the red blood cells by osmosis, causing them to burst (Figure 14.18c). A solution that causes this condition is called a **hypotonic** solution. To prevent cells from bursting, an **isotonic** (or iso-osmotic) intravenous solution must be used. Such a solution has the same total concentration of solutes and therefore the same osmotic pressure as the patient's blood (Figure 14.18a). A solution of 0.9% sodium chloride is isotonic with fluids inside cells in the body.

If an intravenous solution more concentrated than the solution inside a red blood cell were added to blood, the cell would lose water to the solution and shrivel up. A solution that causes this condition is a **hypertonic** solution (Figure 14.18b). Cell-

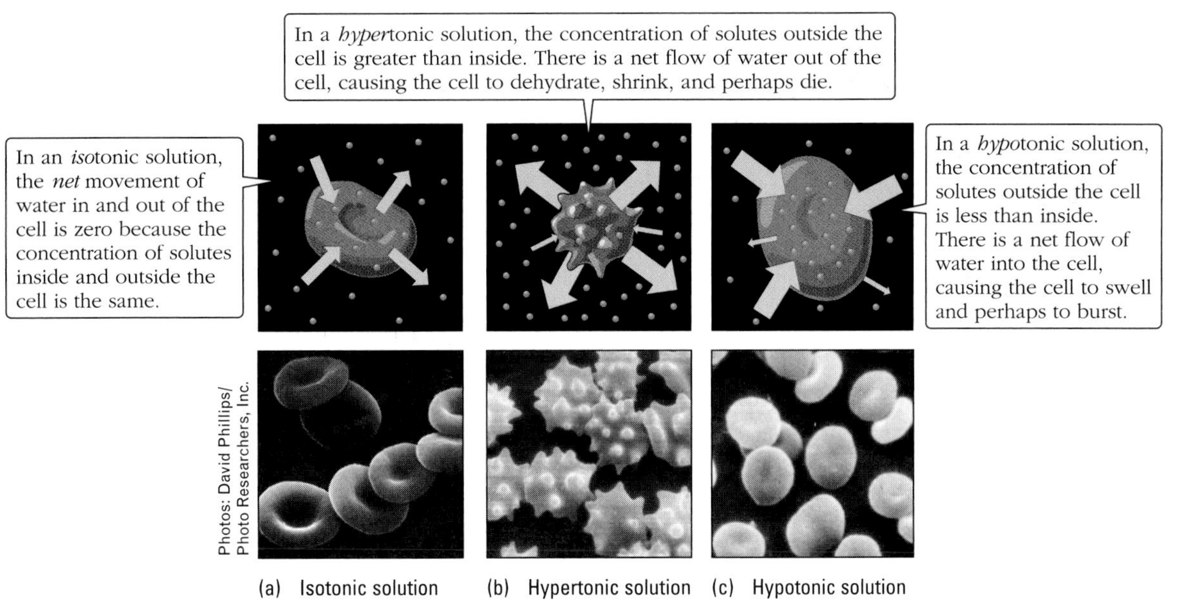

In a *hyper*tonic solution, the concentration of solutes outside the cell is greater than inside. There is a net flow of water out of the cell, causing the cell to dehydrate, shrink, and perhaps die.
In an *iso*tonic solution, the *net* movement of water in and out of the cell is zero because the concentration of solutes inside and outside the cell is the same.
In a *hypo*tonic solution, the concentration of solutes outside the cell is less than inside. There is a net flow of water into the cell, causing the cell to swell and perhaps to burst.
Photos: David Phillips/ Photo Researchers, Inc.
(a) Isotonic solution
(b) Hypertonic solution
(c) Hypotonic solution

Figure 14.18 Osmosis and the living cell.

shriveling by osmosis happens when vegetables or meats are cured in *brine,* a concentrated solution of NaCl. If you put a fresh cucumber into brine, water will flow out of its cells and into the brine, leaving behind a shriveled vegetable. With proper spices added to the brine, a cucumber will become a tasty pickle.

Courtesy of IDE Technologies

This reverse osmosis plant in Israel can annually convert 100 million cubic meters of salt water to fresh water.

Reverse Osmosis

Reverse osmosis occurs when pressure greater than the osmotic pressure is applied and solvent is made to flow through a semipermeable membrane from a concentrated solution to a dilute solution. In effect, the semipermeable membrane serves as a filter with very tiny pores through which only the solvent can pass. Reverse osmosis can be used to remove small molecules or ions to obtain highly purified water. Seawater contains a high concentration of dissolved salts; its osmotic pressure is 24.8 atm. If a pressure greater than 24.8 atm is applied to a chamber containing seawater, water molecules can be forced to flow from seawater through a semipermeable membrane to a region containing purer water (Figure 14.19). Pressures up to 100 atm are used to

Small reverse osmosis units are used to make the ultrapure water for some "spotless" car washes.

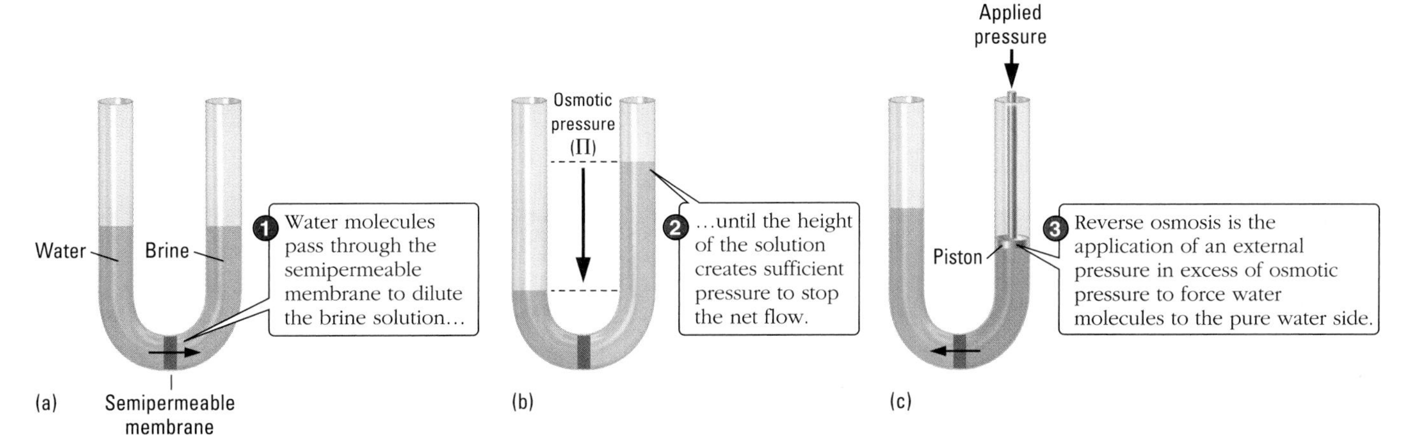

2 ...until the height of the solution creates sufficient pressure to stop the net flow.

Figure 14.19 Normal and reverse osmosis. Normal osmosis is represented in (a) and (b). Reverse osmosis is represented in (c).

Brackish water is a mixture of salt water and fresh water.

Table 14.5 Ions Present in Seawater at 100 ppm or More

Ion	Mass Fraction g/kg	ppm
Cl^-	19.35	19,350
Na^+	10.76	10,760
SO_4^{2-}	2.710	2710
Mg^{2+}	1.290	1290
Ca^{2+}	0.410	410
K^+	0.400	400
HCO_3^-, CO_3^{2-}	0.106	106
Total	35.026	35,026

provide reasonable rates of seawater purification. Seawater, which contains upward of 35,000 ppm of dissolved salts (Table 14.5), can be purified by reverse osmosis to between 400 and 500 ppm of solutes, which is well within the World Health Organization's limits for drinking water. Large reverse osmosis plants in places like the Persian Gulf countries and Florida can purify more than 100 million gallons of water per day. Nearly 50% of Saudi Arabia's fresh water is provided by the world's largest desalination plant at Jubai. Reverse osmosis purifies nearly 15 million gallons of brackish underground water daily for the city of Cape Coral, Florida. That facility is one of more than 100 in the state of Florida.

IN CLOSING

OWL Selected end-of-chapter Questions may be assigned in OWL.

For quick review, download Go Chemistry mini-lecture flashcard modules (or purchase them at **www.ichapters.com**).

Prepare for an exam with a **Summary Problem** that brings together concepts and problem-solving skills from throughout the chapter. Go to **www.cengage.com/chemistry/moore** to download this chapter's Summary Problem from the Student Companion Site.

Having studied this chapter, you should be able to . . .

- Describe how liquids, solids, and gases dissolve in a solvent (Section 14.1). ■ End-of-chapter Question assignable in OWL: 1
- Predict solubility based on properties of solute and solvent (Section 14.1). ■ Questions 18, 21
- Interpret the dissolving of solutes in terms of enthalpy and entropy changes (Section 14.2). ■ Question 20
- Differentiate among unsaturated, saturated, and supersaturated solutions (Section 14.2). ■ Question 65
- Describe how ionic compounds dissolve in water (Section 14.3).
- Predict how temperature affects the solubility of ionic compounds (Section 14.4).
- Predict the effects of temperature (Section 14.4) and pressure on the solubility of gases in liquids (Section 14.5). ■ Question 7
- Describe the compositions of solutions in terms of weight percent, mass fraction, parts per million, parts per billion, parts per trillion, molarity, and molality (Section 14.6). ■ Questions 33, 37
- Interpret vapor pressure lowering in terms of Raoult's law (Section 14.7). ■ Question 46
- Use molality to calculate the colligative properties: freezing point lowering and boiling point elevation (Section 14.7). ■ Questions 50, 53, 73
- Differentiate the colligative properties of nonelectrolytes and electrolytes (Section 14.7). ■ Question 42
- Explain the phenomena of osmosis and reverse osmosis, and calculate osmotic pressure (Section 14.7). ■ Question 55

KEY TERMS

boiling point elevation *(Section 14.7)*
colligative properties *(14.7)*
enthalpy of solution *(14.2)*
freezing point lowering *(14.7)*
Henry's law *(14.5)*
hydration *(14.3)*
hypertonic *(14.7)*
hypotonic *(14.7)*
immiscible *(14.1)*
isotonic *(14.7)*
mass fraction *(14.6)*
miscible *(14.1)*
molality *(14.6)*
osmosis *(14.7)*
osmotic pressure *(14.7)*
parts per billion *(14.6)*
parts per million *(14.6)*
parts per trillion *(14.6)*
Raoult's law *(14.7)*
reverse osmosis *(14.7)*
saturated solution *(14.3)*
semipermeable membrane *(14.7)*
solubility *(14.3)*
supersaturated solution *(14.3)*
unsaturated solution *(14.3)*
weight percent *(14.6)*

QUESTIONS FOR REVIEW AND THOUGHT

■ denotes questions assignable in OWL.

Blue-numbered questions have short answers at the back of this book and fully worked solutions in the *Student Solutions Manual.*

Review Questions

1. ■ Which of these general types of substances would you expect to dissolve readily in water?
   (a) Alcohols (b) Hydrocarbons
   (c) Metals (d) Nonpolar molecules
   (e) Polar molecules (f) Salts
2. Explain on a molecular basis why the components of blended motor oils remain dissolved and do not separate.
3. Explain why gasoline and motor oil are miscible, as in the fuel mixtures used in two-cycle lawn mower engines.
4. Describe the differences among solutions that are unsaturated, saturated, and supersaturated in terms of amount of solute.
5. Describe the differences among unsaturated, saturated, and supersaturated solutions in terms of Q and K_c.
6. In general, how does the water solubility of most ionic compounds change as the temperature is increased?
7. ■ How does the solubility of gases in liquids change with increased temperature? Explain why.
8. Which is the highest solute concentration: 50 ppm, 500 ppb, or 0.05% by weight?
9. Estimate your concentration on campus in parts per million and parts per thousand.
10. Define molality. How does it differ from molarity?
11. Explain the difference between the mass fraction and the mole fraction of solute in a solution.
12. Explain why the vapor pressure of a solvent is lowered by the presence of a nonvolatile solute.
13. Why is a higher temperature required for boiling a solution containing a nonvolatile solute than for boiling the pure solvent?
14. Which would have the lowest freezing point?
    (a) A 1.0 mol/kg NaCl solution in water
    (b) A 1.0 mol/kg $CaCl_2$ solution in water
    (c) A 1.0 mol/kg methanol solution in water
    Explain your choice.
15. Explain the difference between (a) a hypotonic and an isotonic solution; (b) an isotonic and a hypertonic solution.
16. Explain how reverse osmosis works.

Topical Questions

How Substances Dissolve

17. Explain why some liquids are miscible in each other while other liquids are immiscible. Using only three liquids, give an example of a miscible pair and an immiscible pair.
18. ■ Why would the same solid readily dissolve in one liquid and be almost insoluble in another liquid? Give an example of such behavior.
19. ■ A saturated solution of NH_4Cl was prepared by adding solid NH_4Cl to water until no more solid NH_4Cl would dissolve. The resulting mixture felt very cold and had a layer of undissolved NH_4Cl on the bottom. When the mixture reached room temperature, no solid NH_4Cl was present. Explain what happened. Was the solution still saturated?
20. ■ The lattice energy of $CaCl_2$ is −2258 kJ/mol, and its enthalpy of hydration is +2175 kJ/mol. Is the process of dissolving $CaCl_2$ in water endothermic or exothermic?
21. ■ Simple acids such as formic acid, HCOOH, and acetic acid, CH_3COOH, are very soluble in water; however, fatty acids such as stearic acid, $CH_3(CH_2)_{16}COOH$, and palmitic acid, $CH_3(CH_2)_{14}COOH$, are water-insoluble. Based on what you know about the solubility of alcohols, explain the solubility of these organic acids.
22. If a solution of a certain salt in water is saturated at some temperature and a few crystals of the salt are added to the

solution, what do you expect will happen? What happens if the same quantity of the same salt crystals is added to an unsaturated solution of the salt? What would you expect to happen if the temperature of this second salt solution is slowly lowered?

23. Describe what happens when an ionic solid dissolves in water. Sketch an illustration that includes at least three positive ions, three negative ions, and a dozen or so water molecules in the vicinity of the ions.

24. The partial pressure of O_2 in your lungs varies from 25 mm Hg to 40 mm Hg. What concentration of O_2 (in grams per liter) can dissolve in water at 37 °C when the O_2 partial pressure is 40. mm Hg? The Henry's law constant for O_2 at 37 °C is 1.5×10^{-6} mol L^{-1} mm Hg^{-1}.

25. The Henry's law constant for nitrogen in blood serum is approximately 8×10^{-7} mol L^{-1} mm Hg^{-1}. What is the N_2 concentration in a diver's blood at a depth where the total pressure is 2.5 atm? The air the diver is breathing is 78% N_2 by volume.

Concentration Units

26. What mass (in grams) of sucrose is in 1.0 kg of a 0.25% sucrose solution?

27. How many grams of ethanol are in 750. mL of a 12% ethanol solution? (Assume its density is the same as that of water.)

28. A sample of lead-based paint is found to contain 60.5 ppm lead. The density of the paint is 8.0 lb/gal. What mass of lead (in grams) would be present in 50. gal of this paint?

29. A paint contains 200. ppm lead. Approximately what mass of lead (in grams) will be in 1.0 cm^2 of this paint (density = 8.0 lb/gal) when 1 gal is uniformly applied to 500. ft^2 of a wall?

30. Calculate the mass in grams of solute needed to prepare each of these solutions.
    (a) 250. mL of 0.50 M NaCl
    (b) 0.50 L of 0.15 M sucrose, $C_{12}H_{22}O_{11}$
    (c) 200. mL of 0.20 M $NaHCO_3$

31. Calculate the mass in grams of solute required to prepare each of these solutions.
    (a) 750. mL of 4.00 M NH_4Cl
    (b) 1.50 L of 0.750 M KCl
    (c) 150. mL of 0.350 M Na_2SO_4

32. Calculate the molarity of the solute in a solution containing
    (a) 14.2 g KCl in 250. mL solution
    (b) 5.08 g K_2CrO_4 in 150. mL solution
    (c) 0.799 g $KMnO_4$ in 400. mL solution
    (d) 15.0 g $C_6H_{12}O_6$ in 500. mL solution

33. ■ Calculate the molarity of the solute in a solution containing
    (a) 6.18 g $MgNH_4PO_4$ in 250. mL solution
    (b) 16.8 g $NaCH_3COO$ in 300. mL solution
    (c) 2.50 g CaC_2O_4 in 750. mL solution
    (d) 2.20 g $(NH_4)_2SO_4$ in 400. mL solution

34. Concentrated nitric acid is a 70.0% solution of nitric acid, HNO_3, in water. The density of the solution is 1.41 g/mL at 25 °C. What is the molarity of nitric acid in this solution?

35. Concentrated sulfuric acid has a density of 1.84 g/cm^3 and is 18 M. What is the weight percent of H_2SO_4 in the solution?

36. Consider a 13.0% solution of sulfuric acid, H_2SO_4, whose density is 1.090 g/mL.
    (a) Calculate the molarity of this solution.
    (b) To what volume should 100. mL of this solution be diluted to prepare a 1.10 M solution?

37. ■ You want to prepare a 1.0 mol/kg solution of ethylene glycol, $C_2H_4(OH)_2$, in water. What mass of ethylene glycol do you need to mix with 950. g water?

38. A 23.2% by weight aqueous solution of sucrose has a density of 1.127 g/mL. Calculate the molarity of sucrose in this solution.

39. Calculate the number of grams of KI required to prepare 100. mL of 0.0200 M KI. How many milliliters of this solution are required to produce 250. mL of 0.00100 M KI?

40. Calculate the molality of a solution made by dissolving 6.58 g NaCl in 250. mL water.

Colligative Properties

41. Place these aqueous solutions in order of increasing boiling point.
    (a) 0.10 mol KCl/kg
    (b) 0.10 mol glucose/kg
    (c) 0.080 mol $MgCl_2$/kg

42. ■ List these aqueous solutions in order of decreasing freezing point.
    (a) 0.10 mol methanol/kg
    (b) 0.10 mol KCl/kg
    (c) 0.080 mol $BaCl_2$/kg
    (d) 0.040 mol Na_2SO_4/kg
    (Assume that all of the salts dissociate completely into their ions in solution.)

43. Calculate the boiling point at 760 mm Hg and the freezing point of these solutions.
    (a) 20.0 g citric acid, $C_6H_8O_7$, in 100.0 g water
    (b) 3.00 g CH_3I in 20.0 g benzene (K_b benzene = 2.53 °C kg mol^{-1}; K_f benzene = 5.10 °C kg mol^{-1})
    (The freezing point of benzene is 5.50 °C; its normal boiling point is 80.10 °C.)

44. Calculate the freezing and boiling points (at 760 mm Hg) of a solution of 4.00 g urea, $CO(NH_2)_2$, dissolved in 75.0 g water.

45. Calculate the mass in grams of urea that must be added to 150. g water to give a solution whose vapor pressure is 2.5 mm Hg less than that of pure water at 40 °C (vp H_2O at 40 °C = 55.34 mm Hg).

46. ■ At 60 °C the vapor pressure of pure water is 149.44 mm Hg and that above an aqueous sucrose, $C_{12}H_{22}O_{12}$, solution is 119.55 mm Hg. Calculate the mole fraction of water and the mass in grams of sucrose in the solution if the mass of water is 150. g.

47. At 760 mm Hg, a solution of 5.52 g glycerol in 40.0 g water has a boiling point of 100.777 °C. Calculate the molar mass of glycerol.

48. The boiling point of benzene is increased by 0.65 °C when 5.0 g of an unknown organic compound (a nonelectrolyte) is dissolved in 100. g benzene. Calculate the approximate molar mass of the organic compound. (K_b benzene = 2.53 °C kg mol^{-1}.)

49. The freezing point of *p*-dichlorobenzene is 53.1 °C, and its K_f is 7.10 °C kg mol^{-1}. A solution of 1.52 g of the drug sulfanilamide in 10.0 g *p*-dichlorobenzene freezes at 46.7 °C. What is the molar mass of sulfanilamide?

50. ■ You add 0.255 g of an orange crystalline compound with an empirical formula of $C_{10}H_8Fe$ to 11.12 g benzene. The boiling point of the solution is 80.26 °C. The normal boiling point of benzene is 80.10 °C and its K_b = 2.53 °C kg mol^{-1}. What are the molar mass and molecular formula of the compound?
51. If you use only water and pure ethylene glycol, $HOCH_2CH_2OH$, in your car's cooling system, what mass (in grams) of the glycol must you add to each quart of water to give freezing protection down to −31.0 °C?
52. A 1.00 mol/kg aqueous sulfuric acid solution, H_2SO_4, freezes at −4.04 °C. Calculate i, the van't Hoff factor, for sulfuric acid in this solution.
53. ■ Some ethylene glycol, $HOCH_2CH_2OH$, was added to your car's cooling system along with 5.0 kg water.
    (a) If the freezing point of the solution is −15.0 °C, what mass (in grams) of the glycol must have been added?
    (b) What is the boiling point of the coolant mixture?
54. The molar mass of a polymer was determined by measuring the osmotic pressure, 7.6 mm Hg, of a solution containing 5.0 g of the polymer dissolved in 1.0 L benzene. What is the molar mass of the polymer? Assume a temperature of 298.15 K.
55. ■ The osmotic pressure at 25 °C is 1.79 atm for a solution prepared by dissolving 2.50 g sucrose, empirical formula $C_{12}H_{22}O_{11}$, in enough water to give a solution volume of 100 mL. Use the osmotic pressure equation to show that the empirical formula for sucrose is the same as its molecular formula.

General Questions

56. A chemistry classmate tells you that a supersaturated solution is also saturated. Is the student correct? What would you tell the student about her/his statement?
57. In *The Rime of the Ancient Mariner* the poet Samuel Taylor Coleridge wrote, ". . . Water, water, everywhere/And all the boards did shrink. . . ." Explain this effect in terms of osmosis.
58. A 10.0 M aqueous solution of NaOH has a density of 1.33 g/cm^3 at 20 °C. Calculate the weight percent of NaOH in the solution.
59. Concentrated aqueous ammonia is 14.8 M and has a density of 0.90 g/cm^3. Calculate the weight percent of NH_3 in the solution.
60. (a) What is the molality of a solution made by dissolving 115.0 g ethylene glycol, $HOCH_2CH_2OH$, in 500. mL water? The density of water at this temperature is 0.978 g/mL.
    (b) What is the molarity of the solution?
61. The solubility of NaCl in water at 100 °C is 39.1 g/100. g of water. Calculate the boiling point of a saturated solution of NaCl.
62. Aluminum chloride reacts with phosphoric acid to give aluminum phosphate, $AlPO_4$, which is used industrially as the basis of adhesives, binders, and cements.
    (a) Write a balanced equation for the reaction of aluminum chloride and phosphoric acid.
    (b) If you begin with 152. g aluminum chloride and 3.00 L of 0.750 M phosphoric acid, what mass of $AlPO_4$ can be isolated?
    (c) Use the solubility table (Table 5.1) to predict the solubility of $AlPO_4$.

Applying Concepts

63. Using these symbols,

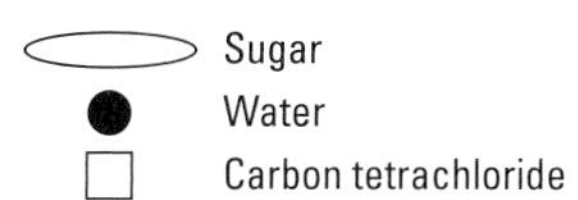

draw nanoscale diagrams for the contents of a beaker containing
    (a) Water and sugar
    (b) Carbon tetrachloride and sugar
64. Using these symbols,

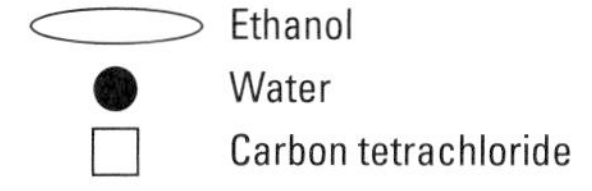

draw nanoscale diagrams for the contents of a beaker containing
    (a) Water and ethanol
    (b) Water and carbon tetrachloride
65. ■ Refer to Figure 14.11 to determine whether these situations would result in an unsaturated, saturated, or supersaturated solution.
    (a) 40. g NH_4Cl is added to 100. g H_2O at 80 °C.
    (b) 100. g LiCl is dissolved in 100. g H_2O at 30 °C.
    (c) 120. g $NaNO_3$ is added to 100. g H_2O at 40 °C.
    (d) 50. g Li_2SO_4 is dissolved in 200. g H_2O at 50 °C.
66. Refer to Figure 14.11 to determine whether these situations would result in an unsaturated, saturated, or supersaturated solution.
    (a) 120. g RbCl is added to 100. g H_2O at 50 °C.
    (b) 30. g KCl is dissolved in 100. g H_2O at 70 °C.
    (c) 20. g NaCl is dissolved in 50. g H_2O at 60 °C.
    (d) 150. g CsCl is added to 100. g H_2O at 10 °C.
67. Complete this table.

Compound	Mass of Compound	Mass of Water	Mass Fraction of Solute	Weight Percent of Solute	ppm of Solute
Lye	______	125 g	0.375	______	______
Glycerol	33 g	200. g	______	______	______
Acetylene	0.0015 g	______	______	0.0009%	______

68. Complete this table.

Compound	Mass of Compound	Mass of Water	Mass Fraction	Weight Percent	ppm
Table salt	52 g	175 g	______	______	______
Glucose	15 g	______	______	______	7×10^4
Methane	______	100. g	______	0.0025%	______

69. In your own words, explain why
    (a) Seawater has a lower freezing point than fresh water.
    (b) Salt is added to the ice in an ice cream maker to freeze the ice cream faster.
70. Criticize these statements.
    (a) A saturated solution is always a concentrated one.
    (b) A 0.10 mol/kg sucrose solution and a 0.10 mol KCl/kg solution have the same osmotic pressure.

More Challenging Questions

71. In chemical research, newly synthesized compounds are often sent to commercial laboratories for analysis that determines the weight percent of C and H by burning the compound and collecting the evolved CO_2 and H_2O. The molar mass is determined by measuring the osmotic pressure of a solution of the compound. Calculate the empirical and molecular formulas of a compound, C_xH_yCr, given this information:
    (a) The compound contains 73.94% C and 8.27% H; the remainder is chromium.
    (b) At 25 °C, the osmotic pressure of 5.00 mg of the unknown dissolved in 100. mL of chloroform solution is 3.17 mm Hg.
72. An osmotic pressure of 5.15 atm is developed by a solution containing 4.80 g dioxane (a nonelectrolyte) dissolved in 250. mL water at 15.0 °C. The empirical formula of dioxane is C_2H_4O. Use the osmotic pressure data to show that the empirical formula and the molecular formula of dioxane are not the same.
73. ■ The "proof" of an alcohol-containing beverage is twice the volume percentage of ethanol, C_2H_5OH (density = 0.789 g/mL), in the beverage. A bottle of 100-proof vodka is left outside where the temperature is −15 °C.
    (a) Will the vodka freeze?
    (b) What is the boiling point of the vodka?
74. A martini is a 5-oz (142-g) cocktail containing 30% by mass of alcohol. When the martini is consumed, about 15% of it passes directly into the bloodstream (blood volume = 7.0 L in an adult). Consider an adult who drinks two martinis with lunch. Estimate the blood alcohol concentration in this person after the two martinis have been consumed. An adult with a blood alcohol concentration of 3.0×10^{-3} g/mL or more is considered intoxicated. Is the person intoxicated?
75. Osmosis is responsible for sap rising in trees. Calculate the approximate height to which it could rise if the sap were 0.13 M in sugar and the dissolved solids in the water outside the tree were at a concentration of 0.020 M. *Hint:* A column of liquid exerts a pressure directly proportional to its density.
76. The osmotic pressure of human blood at 37 °C is 7.7 atm. Calculate what the molarity of a glucose solution should be if it is to be safely administered intravenously.
77. A 0.109 mol/kg aqueous formic acid solution, HCOOH, freezes at −0.210 °C. Calculate the percent dissociation of formic acid.
78. A 0.63% by weight aqueous tin(II) fluoride, SnF_2, solution is used as an oral rinse in dentistry to decrease tooth decay.
    (a) Calculate the SnF_2 concentration in ppm and ppb.
    (b) What is the molarity of SnF_2 in the solution?
    (c) Large quantities of tin are produced commercially by the reduction of cassiterite, SnO_2, the principal tin ore, with carbon.

$$SnO_2(s) + 2\,C(s) \longrightarrow Sn(s) + 2\,CO(g)$$

    Once purified, the tin is reacted with hydrogen fluoride vapor to produce SnF_2. Consider that one metric ton of cassiterite ore is reduced with sufficient carbon to tin metal in 80% yield. The tin is purified and reacted with sufficient hydrogen fluoride to produce SnF_2 in 94% yield. Calculate how many 250.-mL bottles of 0.63% SnF_2 solution could be prepared from the one metric ton of cassiterite by these steps.

Conceptual Challenge Problems

CP14.A (Section 14.6) Concentrations expressed in units of parts per million and parts per billion often have no meaning for people until they relate these small and large numbers to their own experiences.
(a) What time in seconds is 1 ppm of a year?
(b) What time in seconds is 1 ppb of a 70-year lifetime?

CP14.B (Section 14.4) Bodies of water with an abundance of nutrients that support a blooming growth of plants are said to be eutrophoric. In general, fish do not thrive for long in eutrophoric waters because little oxygen is available for them. Suppose someone asked you why this was true, given the fact that growing plants produce oxygen as a product of photosynthesis. How would you respond to this person's inquiry?

CP14.C (Section 14.7) Suppose that you want to produce the lowest temperature possible by using ice, sodium chloride, and water to chill homemade ice cream made in a 1.5-L metal cylinder surrounded by a coolant held in a wooden bucket. You have all the ice, salt, and water you want. How would you plan to do this?

15

Acids and Bases

© Cengage Learning/Charles D. Winters

Tomato juice has a pH of 4.0 as indicated by the pH meter. Measurement of the acidity or alkalinity of substances, expressed as pH, is important in commerce, medicine, and studies of the environment. Acids, bases, and the measurement and uses of pH are discussed in this chapter.

It is difficult to overstate the importance of acids and bases. Aqueous solutions, which abound in our environment and in all living organisms, are almost always acidic or basic to some degree. Photosynthesis and respiration, the two most important biological processes on earth, depend on acid-base reactions. Carbon dioxide, CO_2, is the most important acid-producing compound in nature. Rainwater is generally slightly acidic because of dissolved CO_2, and acid rain results from further acidification of rainwater by acids formed by the reaction of gaseous pollutants SO_2 and NO_2 with water vapor. The oceans are slightly basic, as are many ground and surface waters. Natural waters can also be acidic; the more acidic the water, the more easily metals such as lead can be dissolved from water pipes or soldered joints.

Because of their importance in much of chemistry and biochemistry, the properties of acids and bases have been studied extensively. In 1677 Antoine Lavoisier proposed that all acids contain oxygen, making them acidic; he even derived the name *oxygen* from Greek words meaning "acid former." But in 1808 it was discovered that the gaseous compound HCl, which dissolves in water to give hydrochloric acid, contains only hydrogen and chlorine. It later became clear that hydrogen, not oxygen, is common to all acids in aqueous solution. It was also shown that aqueous solutions of both acids and bases conduct an electrical current because acids and bases are electrolytes—they release ions into solution ***(⟵ p. 143).*** In 1887 the Swedish chemist Svante Arrhenius proposed that acids *ionize in aqueous solution to produce hydrogen ions (protons) and anions;* bases *ionize to produce hydroxide ions and cations.* We now rely on a more general acid-base concept: ***hydronium ions, H_3O^+, are responsible for the properties of acidic aqueous solutions, and hydroxide ions, OH^-, are responsible for the properties of basic aqueous solutions.***

Because of their small size and high charge density, hydrogen ions (protons) are always associated with water molecules in aqueous solution and are usually represented as H_3O^+, the hydronium ion.

$$H^+ + H_2O \longrightarrow H_3O^+$$

15.1 The Brønsted-Lowry Concept of Acids and Bases

A major problem with the Arrhenius acid-base concept is that certain substances, such as ammonia, NH_3, produce basic solutions and react with acids, yet contain no hydroxide ions. In 1923 J. N. Brønsted in Denmark and T. M. Lowry in England independently proposed a new way of defining acids and bases in aqueous solutions:

- **Brønsted-Lowry acids** are hydrogen ion donors.
- **Brønsted-Lowry bases** are hydrogen ion acceptors.

According to the Brønsted-Lowry concept, an acid can donate a hydrogen ion, H^+, to another substance, while a base can accept an H^+ ion from another substance. In **acid-base reactions,** acids *donate* H^+ ions, and bases *accept* them. To accept an H^+ and serve as a Brønsted-Lowry base, a molecule or ion must have an *unshared pair of electrons.* For example, in aqueous solutions, H^+ ions from acids such as nitric acid, HNO_3, react with water molecules, which accept protons to form hydronium ions, H_3O^+.

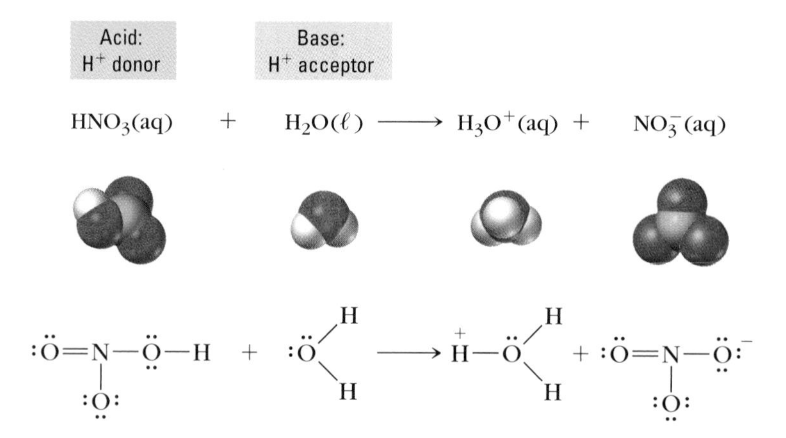

In reacting with an acid, water acts as a Brønsted-Lowry base by using an unshared electron pair to accept the H^+. Because nitric acid is a strong acid, it is 100% ionized in aqueous solution, and the equation above is strongly product-favored.

An ionic substance that dissolves in water is a strong electrolyte ***(⟵ p. 82).*** A strong acid ionizes completely and is a strong electrolyte.

In contrast, a weak acid does *not* ionize *completely* and therefore is a weak electrolyte; at equilibrium appreciable quantities of un-ionized acid molecules are pres-

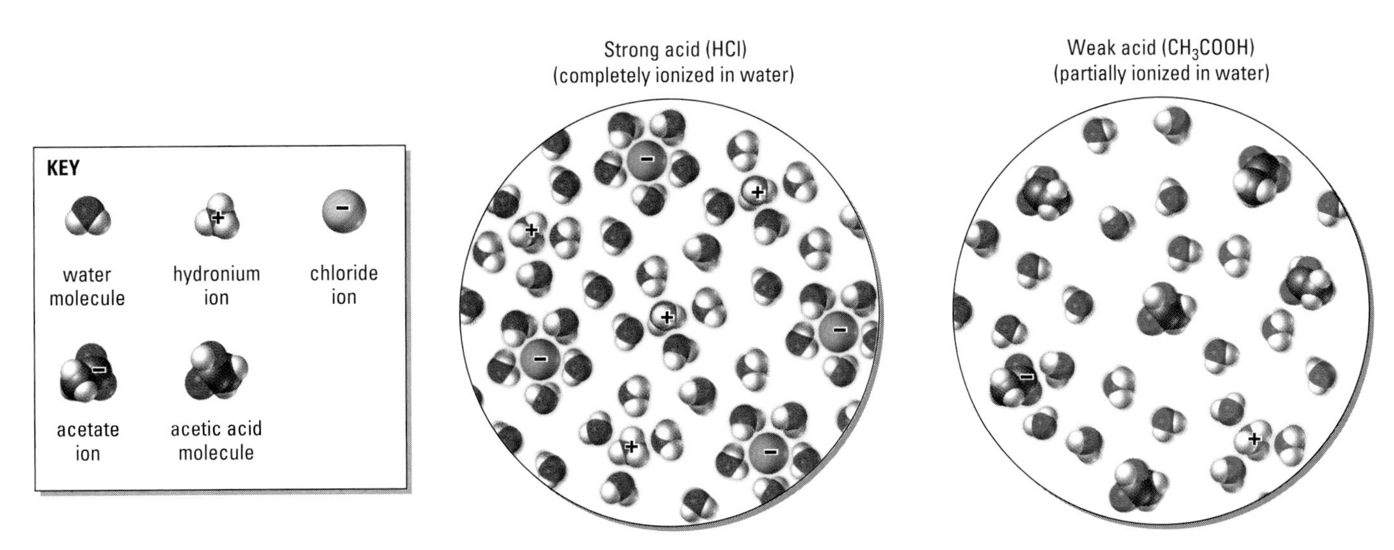

Ionization of acids in water. A strong acid such as hydrochloric acid, HCl, is completely ionized in water; a weak acid such as acetic acid, CH_3COOH, is only partially ionized in water.

ent. For example, hydrofluoric acid, HF, is a weak acid, and its ionization in water is written as

$$\underset{\text{hydrofluoric acid, a weak acid}}{HF(aq)} + H_2O(\ell) \rightleftharpoons H_3O^+(aq) + F^-(aq)$$

The double arrow indicates an equilibrium between the reactants and the products (⇐ *p. 462*). Because ionization of HF is much less than 100%, this means that at equilibrium most HF molecules are intact, un-ionized; there are relatively few hydronium and fluoride ions in solution. Thus, the ionization of *weak* acids is a reactant-favored process.

Ammonia, NH_3, is a base because it accepts an H^+ from water (an acid) to form an ammonium ion, NH_4^+. Water, having donated an H^+, is converted into a hydroxide ion, OH^-.

Ammonia, NH_3, is a compound consisting of electrically neutral molecules; the ammonium ion, NH_4^+, is a polyatomic ion.

Base: H^+ acceptor		Acid: H^+ donor			
$NH_3(g)$	+	$H_2O(\ell)$	$\rightleftharpoons$	$NH_4^+(aq)$	+ $OH^-(aq)$

$$H-\ddot{N}(H)_2 + H-\ddot{\underset{..}{O}}-H \rightleftharpoons [NH_4]^+ + :\ddot{\underset{..}{O}}-H^-$$

Ammonia establishes an equilibrium with water, ammonium ions, and hydroxide ions and is therefore a weak base (⇐ *p. 144*). At equilibrium the solution contains far more ammonia molecules than ammonium and hydroxide ions. Therefore, ammonium hydroxide is not an appropriate name for an aqueous solution of ammonia.

CONCEPTUAL
EXERCISE 15.1 Brønsted-Lowry Acids and Bases

Identify each molecule or ion as a Brønsted-Lowry acid or base.

(a) HBr (b) Br^- (c) HNO_2 (d) CH_3NH_2

CONCEPTUAL
EXERCISE 15.2 Using Le Chatelier's Principle

Use Le Chatelier's principle (⟸ *p. 482*) to explain why a larger percentage of NH_3 will react with water in a very dilute solution than in a less dilute solution.

Water's Role as Acid or Base

In aqueous solution, all Brønsted-Lowry acids and bases react with water molecules. As we have seen, a water molecule *accepts* an H^+ *from* an acid such as nitric acid, while a water molecule *donates* an H^+ *to* a base such as an ammonia molecule. According to the Brønsted-Lowry definitions, water serves as a base (an H^+ acceptor) when an acid is present and as an acid (an H^+ donor) when a base is present. Therefore, water displays both acid and base properties—*it can donate or accept H^+ ions,* depending on the circumstances. The general reactions of water with acids (HA) and molecular bases (B) are

Water acting as a base

$$\underset{\text{acid}}{HA} + \underset{\text{base}}{H_2O} \rightleftharpoons H_3O^+ + A^-$$

Water acting as an acid

If the base is an anion, B^-, it accepts H^+ from water to form BH.

$$\underset{\text{base}}{B} + \underset{\text{acid}}{H_2O} \rightleftharpoons BH^+ + OH^-$$

A substance, like water, that can donate or accept H^+ is said to be **amphiprotic.**

EXERCISE 15.3 Acids and Bases

Complete these equations. (*Hint:* CH_3NH_2 and $(CH_3)_2NH$ are amines, compounds related to ammonia, so they are bases.)

(a) $HCN + H_2O \longrightarrow$ (b) $HBr + H_2O \longrightarrow$

(c) $CH_3NH_2 + H_2O \longrightarrow$ (d) $(CH_3)_2NH + H_2O \longrightarrow$

Only the hydrogen in the —COOH group is acidic; the other hydrogen atoms in acetic acid are not.

```
     H  :O:
     |   ||
  H—C—C—Ö—H
     |       ↑
     H
```

nonacidic hydrogens — acidic hydrogen

Conjugate Acid-Base Pairs

Whenever an acid donates H^+ to a base, a new acid and a new base are formed. We illustrate this using the reaction between acetic acid, CH_3COOH, and water (Active Figure p. 541). Acetic acid is an H^+ ion donor (an acid), and water is an H^+ ion acceptor (a base). The products of the reaction are a new acid, H_3O^+, and a new base, CH_3COO^-. In the reverse reaction, H_3O^+ acts as an acid (H^+ donor), and acetate ion as a base (H^+ acceptor). The structures of CH_3COOH and CH_3COO^- differ from one another by only a single H^+, just as the structures of H_2O and H_3O^+ do.

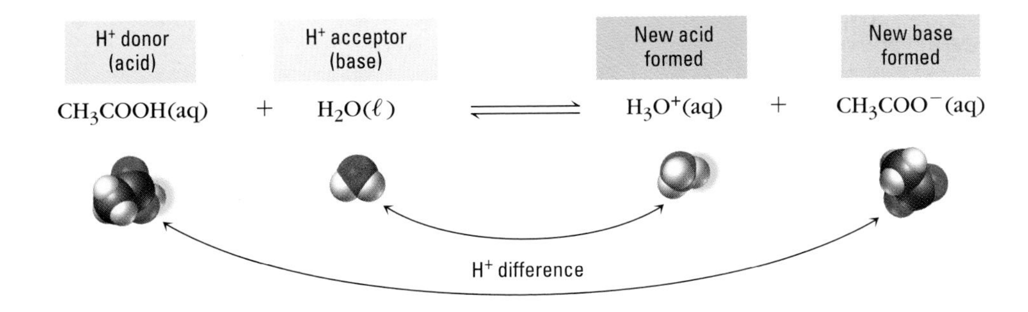

Active Figure **Conjugate acid-base pairs.** Transfer of an H^+ converts a conjugate acid into its conjugate base. Visit this book's companion website at **www.cengage.com/chemistry/moore** to test your understanding of the concepts in this figure.

In the illustration, acetic acid and acetate ion are a conjugate acid-base pair.

A pair of molecules or ions *related to each other by the loss or gain of a single H^+* is called a **conjugate acid-base pair.** Every Brønsted-Lowry acid has its conjugate base, and every Brønsted-Lowry base has its conjugate acid.

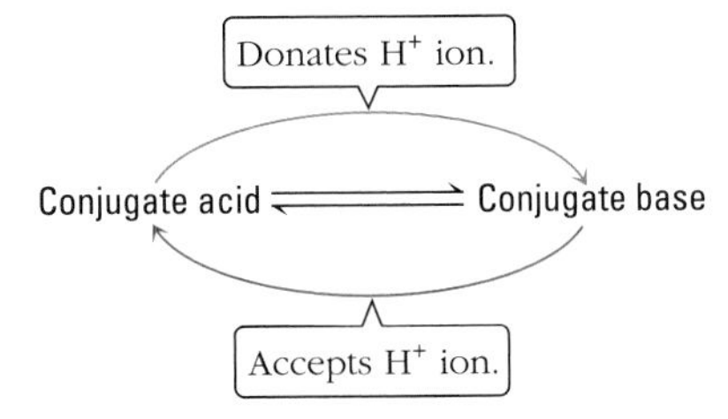

Removing an H^+ ion from the acid forms the conjugate base making the charge of the remaining portion of the acid one unit more negative. For example, hydrofluoric acid, HF, has the F^- (fluoride) ion as its conjugate base; HF and F^- are a conjugate acid-base *pair*.

ACID
Donates $H^+ \longrightarrow$

$$HF(aq) + H_2O(\ell) \rightleftharpoons H_3O^+(aq) + F^-(aq)$$

ACID base acid BASE

BASE
$\longleftarrow$ Accepts H^+

In the forward reaction, HF is a Brønsted-Lowry acid. It donates an H^+ to water, which, by accepting the H^+, acts as a Brønsted-Lowry base. In the reverse reaction, fluoride ion is the Brønsted-Lowry base, accepting an H^+ from H_3O^+, the acid. As noted in the equation, there are two conjugate acid-base pairs: (1) HF and F^- and (2) H_2O and H_3O^+. *One member of a conjugate acid-base pair is always a reactant and the other is always a product; they are never both products or both reactants.*

The conjugate acid formula can be derived from the formula of its conjugate base by adding an H^+ ion to the conjugate base and making the charge of the resulting conjugate acid one unit more positive than the conjugate base. Therefore, the conjugate acid of Cl^- ion is HCl; the conjugate acid of NH_3 is NH_4^+.

BASE
Accepts $H^+ \longrightarrow$

$$NH_3(aq) + H_2O(\ell) \rightleftharpoons NH_4^+(aq) + OH^-(aq)$$

BASE acid ACID base

ACID
$\longleftarrow$ Donates H^+

PROBLEM-SOLVING EXAMPLE 15.1 Conjugate Acid-Base Pairs

Complete this table by identifying the correct conjugate acid or conjugate base.

Acid	Its Conjugate Base	Base	Its Conjugate Acid
HCOOH	________	CN^-	________
H_2S	________	________	HSO_4^-
PH_4^+	________	________	H_2SO_3
________	ClO^-	S^{2-}	________

Answer

Acid	Its Conjugate Base	Base	Its Conjugate Acid
HCOOH	$HCOO^-$	CN^-	HCN
H_2S	HS^-	SO_4^{2-}	HSO_4^-
PH_4^+	PH_3	HSO_3^-	H_2SO_3
HClO	ClO^-	S^{2-}	HS^-

Strategy and Explanation In each of these cases, apply the relationship

$$\text{Conjugate acid} \underset{\text{BASE} \longleftarrow \text{Accepts H}^+}{\overset{\text{ACID Donates H}^+ \longrightarrow}{\rightleftharpoons}} \text{Conjugate base}$$

The conjugate acid can be identified by adding H^+ to the conjugate base; the conjugate base forms by loss of H^+ from the conjugate acid. For example, because CN^- has no H^+ to donate, it must be a base—the conjugate base of HCN, its conjugate acid. Likewise, HSO_3^- is the conjugate base of its conjugate acid, H_2SO_3. The other conjugate acid-base pairs can be worked out similarly.

PROBLEM-SOLVING PRACTICE 15.1

Complete the table.

Acid	Its Conjugate Base	Base	Its Conjugate Acid
$H_2PO_4^-$	________	________	HPO_4^{2-}
________	H^-	NH_2^-	________
HSO_3^-	________	ClO_4^-	________
HF	________	________	HBr

EXERCISE 15.4 HSO_4^- as a Base

(a) Write the equation for HSO_4^- ion acting as a base in water.
(b) Identify the conjugate acid-base pairs.
(c) Is HSO_4^- amphiprotic? Explain your answer.

Relative Strengths of Acids and Bases

Strong acids are better H^+ ion donors than weak acids. Correspondingly, strong bases are better H^+ ion acceptors than weak bases. Thus, *stronger acids have weaker con-*

jugate bases and weaker acids have stronger conjugate bases. For example, compare HCl, a strong acid, with HF, a weak acid.

$$HCl(aq) + H_2O(\ell) \longrightarrow H_3O^+(aq) + Cl^-(aq)$$

$$HF(aq) + H_2O(\ell) \rightleftharpoons H_3O^+(aq) + F^-(aq)$$

In a dilute solution (<1.0 M) the ionization of HCl is virtually 100%; essentially all of the HCl molecules react with water to form H_3O^+ and Cl^-. The Cl^- ion exhibits virtually no tendency to accept H^+ from H_3O^+. On the other hand, the reverse of the ionization of HF is significant; F^- ion readily accepts H^+ from H_3O^+, and hydrofluoric acid is a weak acid that is mainly un-ionized.

By measuring the extent to which various acids donate H^+ ions to water, chemists have developed an extensive tabulation of the relative strengths of acids and their conjugate bases. An abbreviated table is given in Figure 15.1, where the strongest acids are at the top left and the weakest bases at the top right. The weakest acids are at the bottom left, with the strongest bases at the bottom right. From Figure 15.1 we can draw two important generalizations regarding conjugate acid-base pairs:

- ***As acid strength decreases, conjugate base strength increases; the weaker the acid, the stronger its conjugate base.***
- ***As base strength decreases, conjugate acid strength increases; the weaker the base, the stronger its conjugate acid.***

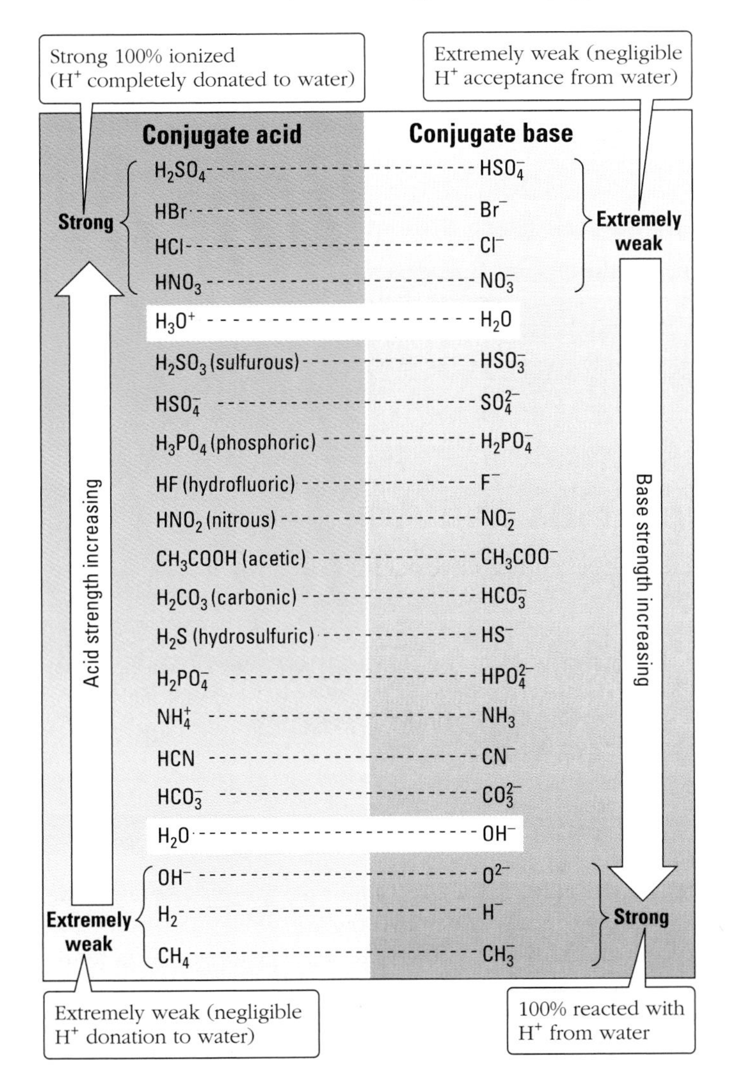

Figure 15.1 Relative strengths of conjugate acids and bases in water.

Knowing the relative acid and base strengths of the reactants, we can predict the direction of an acid-base reaction. ***The stronger acid and the stronger base will always react to form a weaker conjugate base and a weaker conjugate acid.*** Strong Brønsted-Lowry bases such as hydride ion, H^-, sulfide ion, S^{2-}, oxide ion, O^{2-}, amide ion, NH_2^-, and hydroxide ion, OH^-, readily accept H^+ ions, while weaker Brønsted-Lowry bases do so less readily. For example, the reaction of calcium hydride, CaH_2, with water is highly exothermic because of the extremely strong basic properties of the hydride ion, H^-, which avidly accepts H^+ from water to produce H_2, the extremely weak conjugate acid of hydride ion. In this reaction, hydride ion is a stronger base than OH^-, and water is a stronger acid than H_2, so the forward reaction is favored.

The strongly basic properties of hydride ion. The reaction of calcium hydride with water is highly exothermic due to the strongly basic hydride ion.

$$H^-(aq) + H_2O(\ell) \longrightarrow H_2(g) + OH^-(aq)$$

A great many weak bases are anions, such as CN^- (cyanide), F^- (fluoride), and CH_3COO^- (acetate).

We can apply information from Figure 15.1 to consider whether the forward or the reverse reaction is favored in an equilibrium. For example, consider the equilibrium

$$HSO_4^-(aq) + CO_3^{2-}(aq) \rightleftharpoons HCO_3^-(aq) + SO_4^{2-}(aq)$$

From Figure 15.1 we note that HSO_4^- is a stronger acid than HCO_3^-, and CO_3^{2-} is a stronger base than SO_4^{2-}. Since acid-base reactions favor going from the stronger to the weaker member of each conjugate acid-base pair, the forward reaction is favored and H^+ will be transferred from HSO_4^- to CO_3^{2-} to form bicarbonate, HCO_3^-, and sulfate, SO_4^{2-}, ions.

CONCEPTUAL
EXERCISE 15.5 Conjugate Acid-Base Strength

Use Figure 15.1 to predict whether the forward or reverse reaction is favored for the equilibrium

$$CH_3COOH(aq) + SO_4^{2-}(aq) \rightleftharpoons HSO_4^-(aq) + CH_3COO^-(aq)$$

15.2 Carboxylic Acids and Amines

Many weak acids such as acetic, lactic, and pyruvic acids are organic acids.

$$CH_3{-}\overset{\overset{\displaystyle O}{\|}}{C}{-}OH \qquad CH_3\overset{\overset{\displaystyle OH}{|}}{\underset{\underset{\displaystyle H}{|}}{C}}{-}\overset{\overset{\displaystyle O}{\|}}{C}{-}OH \qquad CH_3\overset{\overset{\displaystyle O}{\|}}{C}{-}\overset{\overset{\displaystyle O}{\|}}{C}{-}OH$$

acetic acid — lactic acid — pyruvic acid

Carboxylic acid group

These acids all contain the carboxylic acid, —COOH, functional group. Although carboxylic acid molecules usually contain many other hydrogen atoms, only the hydrogen atom bound to the oxygen atom of the carboxylic acid group is sufficiently positive to be donated as an H^+ ion in aqueous solution. The two highly electronegative oxygen atoms of the carboxylic acid group pull electron density away from the hydrogen atom. As a result, the —O—H bond of the acid is even more polar, and its hydrogen atom more acidic. The C—H bonds in organic acids are relatively nonpolar

Note that although the carboxylic acid group is abbreviated —COOH, the oxygen atoms are not bonded to each other, but to carbon.

and strong, and these hydrogen atoms are *not* acidic, as can be seen with butanoic acid.

$$H-\underset{H}{\overset{H}{C}}-\underset{H}{\overset{H}{C}}-\underset{H}{\overset{H}{C}}-\overset{O}{\overset{\|}{C}}-O-H$$

nonacidic hydrogens   acidic hydrogen

butanoic acid

Anions formed by loss of an H^+ from a —COOH group, such as acetate ion, CH_3COO^-, from acetic acid, CH_3COOH, are stabilized by resonance *(p. 292)*.

$CH_3-C(=O)-OH$ — electron-attracting oxygen atom; acidic hydrogen

acetic acid

$CH_3-C(=O)-O^- \longleftrightarrow CH_3-C(-O^-)=O$

acetate ion

Amines, such as methylamine, CH_3NH_2, are compounds that, like ammonia, have a nitrogen atom with three of its valence electrons in covalent bonds and an unshared electron pair on the nitrogen atom. The lone pair of electrons can accept an H^+, and so, like ammonia, amines such as methylamine react as weak bases with water, accepting an H^+ from water.

$$CH_3-\ddot{N}H_2(aq) + H_2O(\ell) \rightleftharpoons CH_3-NH_3^+(aq) + OH^-(aq)$$

$CH_3-\ddot{N}H_2$

methylamine

CONCEPTUAL
EXERCISE 15.6 Piperidine, an Analog of Ammonia

Write the equation for the reaction of piperidine, a component of pepper, with (a) water and with (b) hydrochloric acid.

piperidine

15.3 The Autoionization of Water

Even highly purified water conducts a very small electrical current, which indicates that pure water contains a very small concentration of ions. These ions are formed when water molecules react to produce hydronium ions and hydroxide ions in a process called **autoionization.**

$$H_2O(\ell) + H_2O(\ell) \rightleftharpoons H_3O^+(aq) + OH^-(aq)$$

BASE acid ACID base

In this reaction, one water molecule serves as an H^+ acceptor (base) while the other is an H^+ donor (acid). The equilibrium between the water molecules and the hydronium and hydroxide ions is very reactant-favored. Therefore, the concentrations of these ions in pure water are *very* low. Nevertheless, autoionization of water is very important in understanding how acids and bases function in aqueous solutions. As in the case of any equilibrium reaction, an equilibrium constant expression can be written for the autoionization of water.

$$2\,H_2O(\ell) \rightleftharpoons H_3O^+(aq) + OH^-(aq) \qquad K_w = [H_3O^+][OH^-]$$

Like all equilibrium constant expressions, the one for K_w includes concentrations of solutes but not the concentration of the solvent, which in this case is water.

Table 15.1 Temperature Dependence of K_w for Water

T (°C)	K_w
10	0.29×10^{-14}
15	0.45×10^{-14}
20	0.68×10^{-14}
25	1.01×10^{-14}
30	1.47×10^{-14}
50	5.48×10^{-14}

This equilibrium constant K_w is known as the **ionization constant for water.** From electrical conductivity measurements of pure water, we know that $[H_3O^+] = [OH^-] = 1.0 \times 10^{-7}$ M at 25 °C. Hence

$$K_w = [H_3O^+][OH^-] = (1.0 \times 10^{-7})(1.0 \times 10^{-7}) = 1.0 \times 10^{-14} \quad \text{(at 25 °C)}$$

The equation $K_w = [H_3O^+][OH^-]$ ***applies to pure water and all aqueous solutions.*** Like other equilibrium constants, the value of K_w is temperature-dependent (Table 15.1).

According to the K_w expression, the product of the hydronium ion concentration times the hydroxide ion concentration will always remain the same at a given temperature. If the hydronium ion concentration increases (because an acid was added to the water, for example), then the hydroxide ion concentration must decrease, and vice versa. The equation also tells us that if we know the concentration of one, we can calculate that of the other.

In aqueous solutions, the H_3O^+ and OH^- concentrations are inversely related; as one increases, the other must decrease. Their product always equals 1.0×10^{-14} at 25 °C.

The relative concentrations of H_3O^+ and OH^- also indicate the acidic, neutral, or basic nature of the aqueous solution. For any aqueous solution there are three possibilities.

Neutral solution:	$[H_3O^+] = [OH^-]$	both equal to 1.0×10^{-7} M at 25 °C
Acidic solution:	$[H_3O^+] > 1.0 \times 10^{-7}$ M	$[OH^-] < 1.0 \times 10^{-7}$ M
Basic solution:	$[H_3O^+] < 1.0 \times 10^{-7}$ M	$[OH^-] > 1.0 \times 10^{-7}$ M

When a solution has equal concentrations of $[H_3O^+]$ and $[OH^-]$, it is said to be **neutral.** If either an acid or a base is added to a neutral solution, the autoionization equilibrium between H_3O^+ and OH^- will be disturbed. Recall that according to Le Chatelier's principle (⟽ ***p. 482***), an equilibrium shifts in such a way as to offset the effect of any disturbance. When an acid is added, the concentration of H_3O^+ ions increases. To oppose this increase, some added H_3O^+ ions react with OH^- ions in water to form H_2O, thereby reducing the $[OH^-]$. When equilibrium is re-established, $[H_3O^+] > [OH^-]$ and the solution is **acidic;** however, the mathematical product $[H_3O^+][OH^-]$ is still equal to 1.0×10^{-14} at 25 °C. Similarly, if a base is added to water, some of the added OH^- ions react with H_3O^+ ions in water to form H_2O, thereby decreasing the $[H_3O^+]$. When equilibrium is re-established, $[H_3O^+] < [OH^-]$ and the solution is **basic;** $[H_3O^+][OH^-]$ still equals 1.0×10^{-14}.

The term *alkaline* is also used to describe basic solutions.

PROBLEM-SOLVING EXAMPLE 15.2 Hydronium and Hydroxide Concentrations

Calculate:

(a) The hydroxide ion concentration at 25 °C in 0.10 M HCl, a strong acid.

(b) The hydronium ion concentration at 25 °C in 0.010 M KOH, a strong base.

Based on your calculations, explain why the 0.10 M HCl solution is acidic and the 0.010 M KOH solution is basic.

Answer (a) $[OH^-] = 1.0 \times 10^{-13}$ M (b) $[H_3O^+] = 1.0 \times 10^{-12}$ M

The 0.10 M HCl solution is acidic because its hydronium concentration is greater than its hydroxide concentration: $[H_3O^+] > [OH^-]$. The 0.010 M KOH solution is basic because $[OH^-] > [H_3O^+]$.

Strategy and Explanation In each case, the relationship $[H_3O^+][OH^-] = 1.0 \times 10^{-14}$ applies.

(a) Being a strong acid, hydrochloric acid is 100% ionized and so the $[H_3O^+]$ is 1.0×10^{-1} M. The $[OH^-]$ can readily be calculated.

$$[H_3O^+][OH^-] = 1.0 \times 10^{-14} = (1.0 \times 10^{-1}\ \text{M})\ [OH^-]$$

$$[OH^-] = \frac{1.0 \times 10^{-14}}{1.0 \times 10^{-1}} = 1.0 \times 10^{-13}\ \text{M}$$

(b) The strong base KOH is completely ionized and the $[OH^-]$ is 1.0×10^{-2} M. The hydronium concentration can be calculated directly:

$$[H_3O]^+ = \frac{1.0 \times 10^{-14}}{[OH^-]} = \frac{1.0 \times 10^{-14}}{1.0 \times 10^{-2}} = 1.0 \times 10^{-12}\ M$$

PROBLEM-SOLVING PRACTICE 15.2

Which is more acidic, a solution whose H_3O^+ concentration is 5.0×10^{-4} M or one that has an OH^- concentration of 3.0×10^{-8} M?

PROBLEM-SOLVING EXAMPLE 15.3 $[H_3O^+]$ and $[OH^-]$ Concentrations

Calculate the hydroxide ion concentration at 25 °C in 6.0 M HNO_3 and the hydronium ion concentration in 6.0 M NaOH.

Answer $[OH^-]$ of 6.0 M HNO_3 = 1.7×10^{-15} M; $[H_3O^+]$ of 6.0 M NaOH = 1.7×10^{-15} M

Strategy and Explanation Assume that nitric acid and sodium hydroxide, both strong electrolytes, are 100% ionized (Figure 15.1). Therefore, a 6.0 M nitric acid solution has $[H_3O^+] = 6.0$ M and its $[OH^-]$ can be calculated.

$$[H_3O^+][OH^-] = (6.0)[OH^-] = 1.0 \times 10^{-14}$$

$$[OH^-] = \frac{1.0 \times 10^{-14}}{6.0} = 1.7 \times 10^{-15}\ M$$

Because NaOH is a strong base, 6.0 M sodium hydroxide has $[OH^-] = 6.0$ M.

$$[H_3O^+][OH^-] = [H_3O^+](6.0) = 1.0 \times 10^{-14}$$

$$[H_3O^+] = \frac{1.0 \times 10^{-14}}{6.0} = 1.7 \times 10^{-15}\ M$$

☑ **Reasonable Answer Check** Note that at the high hydronium ion concentration of 6.0 M nitric acid, the hydroxide ion concentration is very, very low, which is to be expected of a highly acidic solution. In contrast, the hydronium ion concentration is exceedingly low in 6.0 M NaOH, a highly basic solution.

PROBLEM-SOLVING PRACTICE 15.3

Which is more basic, a solution whose H_3O^+ concentration is 2.0×10^{-5} M or one that has an OH^- concentration of 5.0×10^{-9} M?

15.4 The pH Scale

The $[H_3O^+]$ and $[OH^-]$ in an aqueous solution vary widely depending on the acid or base present and its concentration. In general, the $[H_3O^+]$ or $[OH^-]$ in aqueous solutions can range from about 10 mol/L down to about 10^{-15} mol/L.

Because these concentrations can be so small, they have very large negative exponents. It is more convenient to express these concentrations in terms of logarithms. We define the **pH** of a solution as ***the negative of the base 10 logarithm (log) of the hydronium ion concentration (mol/L).***

$$\mathbf{pH = -\log[H_3O^+]}$$

The *negative* logarithm of the small concentration values is used since it gives a positive pH value. Thus, the pH of pure water at 25 °C is given by

$$pH = -\log[1.0 \times 10^{-7}] = -[\log(1.0) + \log(10^{-7})]$$
$$= -[0 + (-7.00)] = 7.00$$

The definition $pH = -\log[H_3O^+]$ is accurate only for small concentrations of hydronium ions. A more accurate definition is $pH = -\log a_{H_3O^+}$, where $a_{H_3O^+}$ represents the *activity* of hydronium ions. Activity represents an effective concentration that has been corrected for noncovalent interactions among ions and molecules in solutions. For examples where the definition involving concentration fails, see McCarty, C. G.; Vitz, E. *Journal of Chemical Education*, Vol. 83, 2006; pp. 752–757. A complete definition of pH is quite complicated (see http://www.iupac.org/goldbook/P04524.pdf and Galster, H. *pH Measurement: Fundamentals, Methods, Applications, Instrumentation.* New York: VCH, 1991).

The p in pH is derived from French "puissance" meaning "power." Thus, pH is the "power of hydrogen."

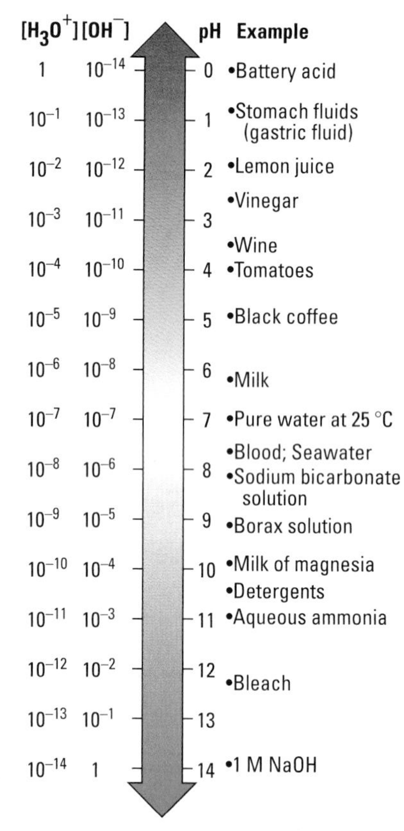

Figure 15.2 The pH of aqueous solutions. The relationship of pH to the concentrations of H_3O^+ and OH^- (in moles/liter at 25 °C) is shown. The pH values of some common substances are also included in the diagram.

In terms of pH, for aqueous solutions at 25 °C we can write

Neutral solution	pH = 7.00
Acidic solution	pH < 7.00
Basic (alkaline) solution	pH > 7.00

Figure 15.2 shows the pH values along with the corresponding H_3O^+ and OH^- concentrations of some common solutions. Notice that, for example, $-\log(1 \times 10^{-x}) = x$ as seen in these examples from Figure 15.2.

$$\text{Lemon juice:}\quad [H_3O^+] = 1 \times 10^{-2}\ M;\quad pH = -\log(1 \times 10^{-2}) = 2$$

$$\text{Black coffee:}\quad [H_3O^+] = 1 \times 10^{-5}\ M;\quad pH = -\log(1 \times 10^{-5}) = 5$$

Keep in mind that a change of *one* pH unit represents a *ten-fold* change in H_3O^+ concentration, a change of two pH units represents a 100-fold change, and so on. Thus, according to Figure 15.2, the $[H_3O^+]$ in lemon juice (pH = 2) is 1000 times *greater* than that in black coffee (pH = 5).

For solutions in which $[H_3O^+]$ or $[OH^-]$ has a value other than an exact power of 10 ($1, 1 \times 10^{-1}, 1 \times 10^{-2}, \ldots$) a calculator is convenient for finding the pH. For example, the pH is 2.35 for a solution that contains 0.0045 mol of the strong acid HNO_3 per liter.

$$pH = -\log(4.5 \times 10^{-3}) = 2.35$$

See Appendix A.6 for more about using logarithms.

See Appendix A.3 for a discussion of significant figures.

The digits to the left of the decimal point in a pH represent a power of 10. Only the digits to the right of the decimal point are significant. In Problem-Solving Example 15.4, where $pH = -\log(2.95 \times 10^{-2}) = -\log(2.95) + (-\log 10^{-2}) = -0.470 + 2.000 = 1.530$, there are three significant figures in 1.530, the result, because there are three significant figures in 2.95.

Notice that, as in the case of equilibrium constants, the concentration units of mol/L are ignored when the logarithm is taken. It is not possible to take the logarithm of a unit.

PROBLEM-SOLVING EXAMPLE 15.4 Calculating pH from $[H_3O^+]$

Calculate the pH of an aqueous HNO_3 solution that has a volume of 250. mL and contains 0.4649 g HNO_3.

Answer 1.530 (*Note:* This pH has been calculated to three significant figures, those to the right of the decimal point. In actual measurements, pH values are seldom measurable to this degree of accuracy.)

Strategy and Explanation Nitric acid is a strong acid, so every mole of HNO_3 that dissolves produces a mole of H_3O^+ and a mole of NO_3^-. First, determine the number of moles of HNO_3 and then calculate the H_3O^+ concentration.

$$0.465\text{ g }HNO_3 \times \frac{1\text{ mol }HNO_3}{63.012\text{ g }HNO_3} = 0.007380\text{ mol }HNO_3$$

$$[H_3O^+] = \frac{0.007380\text{ mol }HNO_3}{0.250\text{ L}} = 0.0295\text{ M}$$

Then, express this concentration as pH.

$$pH = -\log(2.95 \times 10^{-2}) = 1.530$$

☑ **Reasonable Answer Check** If the $[H_3O^+]$ were 0.10 M, the pH would be 1.00; the pH would be 2.00 for an H_3O^+ concentration of 0.010 M. Therefore, a solution with an H_3O^+ concentration of 0.0295 M, which is between these two values, should have a pH between 1.00 and 2.00, which it does.

PROBLEM-SOLVING PRACTICE 15.4

Calculate the pH of a 0.040 M NaOH solution.

The calculation done in Problem-Solving Example 15.4 can be reversed; the hydronium ion concentration of a solution can be calculated from its pH value as shown in Problem-Solving Example 15.5.

PROBLEM-SOLVING EXAMPLE 15.5 Calculating $[H_3O^+]$ from pH

A hospital patient's blood sample has a pH of 7.40. K_w at body temperature (37 °C) = 2.4×10^{-14}.
(a) Calculate the sample's H_3O^+ concentration. (b) What is its OH^- concentration?
(c) Is the sample acidic, neutral, or basic?

Answer (a) 4.0×10^{-8} M (b) 6.0×10^{-7} M (c) Basic

Strategy and Explanation Use the pH value to find the hydronium ion concentration. From that concentration, calculate the OH^- concentration from the $[H_3O^+][OH^-] = 1.0 \times 10^{-14}$ relationship.

(a) Substituting into the definition of pH,

$$-\log[H_3O^+] = 7.40, \quad \text{so} \quad \log[H_3O^+] = -7.40$$

By the rules of logarithms, $10^{\log(x)} = x$, so we can write $10^{\log[H_3O^+]} = 10^{-pH} = [H_3O^+]$. Finding $[H_3O^+]$ therefore requires finding the antilogarithm of -7.40 (Appendix A.6).

$$[H_3O^+] = 10^{-7.40} = 4.0 \times 10^{-8}\ M$$

(b) Rearrange the $[H_3O^+][OH^-] = K_w = 2.4 \times 10^{-14}$ equation to solve for $[OH^-]$.

$$[OH^-] = \frac{2.4 \times 10^{-14}}{[H_3O^+]} = \frac{2.4 \times 10^{-14}}{4.0 \times 10^{-8}} = 6.0 \times 10^{-7}\ M$$

(c) Because the pH is greater than 7.0, the sample is basic.

☑ **Reasonable Answer Check** Because the pH of 7.40 is slightly above 7.00, the hydroxide ion concentration should be a bit higher than 1.6×10^{-7} M, that of a neutral solution, which it is. Therefore, the sample is slightly basic.

PROBLEM-SOLVING PRACTICE 15.5

In a hospital laboratory the pH of a bile sample is measured as 7.90.
(a) What is the H_3O^+ concentration? (b) Is the sample acidic or basic?

Arnold Beckman 1900–2004

When he invented the first electronic pH meter in 1934, Arnold Beckman revolutionized pH measurement. Beckman, a professor at the California Institute of Technology at the time, developed the instrument in response to a request from the California Fruit Growers' Association for a quicker, more accurate way to measure the acidity of lemon juice. He went on to found the highly successful Beckman Instrument Company, a firm that invented the first widely used infrared and ultraviolet spectrophotometers and other laboratory instruments. The Arnold and Mabel Beckman Foundation has donated millions of dollars to advance chemical research and education nationwide.

CONCEPTUAL EXERCISE 15.7 pH of Solutions of Different Acids

Would the pH of a 0.1 M solution of the strong acid HNO_3 be the same as the pH of a 0.1 M solution of the strong acid HCl? Explain.

The OH^- concentration can also be expressed in exponential terms as pOH.

$$pOH = -\log[OH^-]$$

The $[OH^-]$ of pure water at 25 °C is 1.0×10^{-7} M, and therefore its pOH is

$$pOH = -\log(1 \times 10^{-7}) = -(-7.00) = 7.00$$

Because the values of $[H_3O^+]$ and $[OH^-]$ are related by the K_w expression, for all aqueous solutions at 25 °C, we can write

$$K_w = [H_3O^+][OH^-] = 1.0 \times 10^{-14}$$

This equation can be rewritten by taking $-\log$ of each side

$$-\log K_w = -\log[H_3O^+] + (-\log[OH^-]) = -\log(1.0 \times 10^{-14})$$

or

$$pK_w = pH + pOH = 14.00$$

$-\log K_w = pK_w$;
$-\log[H_3O^+] = pH$;
$-\log[OH^-] = pOH$

The relation between pH and pOH can be used to find one value when the other is known. A 0.0010 M solution of the strong base NaOH, for example, has an OH^- concentration of 0.0010 M and a pOH given by

$$pOH = -\log(1.0 \times 10^{-3}) = 3.00$$

Knowing the pH, then the pOH is just 14.00 − pH; knowing the pOH, the pH is 14.00 − pOH.

and therefore

$$\text{pH} = 14.00 - \text{pOH} = 14.00 - 3.00 = 11.00$$

EXERCISE 15.8 pOH and pH

Which solution is more basic, one that has a pH of 5.5 or one with a pOH of 8.5? What is the H_3O^+ concentration in each solution?

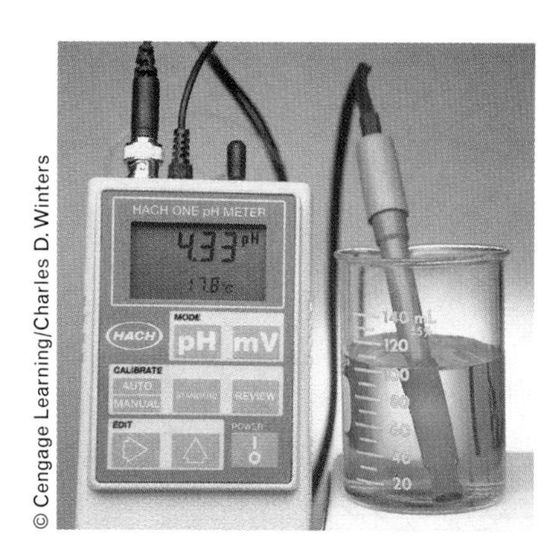

Figure 15.3 A pH meter. A pH meter can quickly and accurately determine the pH of a sample.

Measuring pH

The pH of a solution is readily measured using a pH meter (Figure 15.3). This device consists of a pair of electrodes (often in one probe) that detect the H_3O^+ concentration of the test solution, convert it into an electrical signal, and display it directly as the pH value. The meter is initially calibrated using standard solutions of known pH. The pH of body fluids, soil, environmental and industrial samples, and other substances can be measured easily and accurately with a pH meter.

Acid-base indicators are a much older and less precise (but more convenient) method to determine the pH of a sample. Such indicators are substances that change color within a narrow pH range, generally 1 to 2 pH units, by the loss or gain of an H^+ ion that changes the indicator's molecular structure so that it absorbs light in different regions of the visible spectrum. The indicator is one color at a lower pH (its "acid" form), and a different color at a higher pH (its "base" form). Consider the indicator bromthymol blue (Figure 15.4). At or below pH 6 it is yellow (its acid form); at pH 8 and above it is blue (its base form). Between pH 6 and 7, the indicator changes from pure yellow to a yellow-green color. At pH 7, it is a mixture of 50% yellow and 50% blue, so it appears green. As the pH changes from 7 to 8, the color becomes pure blue. Thus, the pH of a sample that turns bromthymol blue to green has a pH of about 7. If the indicator color is blue, the pH of the sample is at least 8, and it could be much higher.

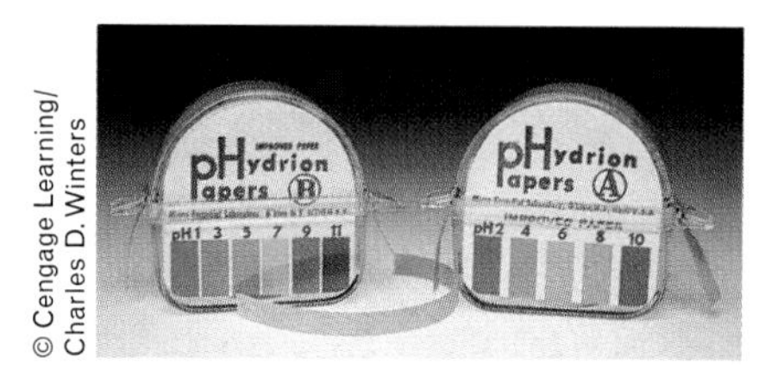

Indicator paper strips. Strips of paper impregnated with indicator are used to find an approximate pH.

Strips of paper impregnated with acid-base indicators are also used to test the pH of many substances. The color of the paper after it has been dampened by the solution to be tested is compared with a set of colors at known pHs.

ESTIMATION Using an Antacid

Estimate how many Rolaids antacid tablets it would take to neutralize the acidity in one glass (250 mL) of a regular cola drink. Assume the pH of the cola is 3.0. One Rolaids tablet contains 334 mg $NaAl(OH)_2CO_3$.

With a pH of 3.0, the cola has 1×10^{-3} mol acid per liter of cola, so 0.250 L cola has one fourth that much acid, or about 3×10^{-4} mol acid. To neutralize this amount of acid requires 3×10^{-4} mol base (1 mol base for every 1 mol acid). There are two bases in Rolaids—hydroxide ions and carbonate ions. Each mole of hydroxide neutralizes 1 mol acid, and each mole of carbonate neutralizes 2 mol acid.

$$H^+(aq) + OH^-(aq) \longrightarrow H_2O(\ell)$$

$$2\,H^+(aq) + CO_3^{2-}(aq) \longrightarrow H_2O(\ell) + CO_2(g)$$

Because each mole of $NaAl(OH)_2CO_3$ contains 2 mol OH^- ions and 1 mol CO_3^{2-} ions, 1 mol $NaAl(OH)_2CO_3$ neutralizes 4 mol acid. The molar mass of $NaAl(OH)_2CO_3$ is 144 g/mol, so one Rolaids tablet contains about 0.002 mol of the antacid.

$$\frac{0.334 \text{ g antacid}}{1 \text{ antacid tablet}} \times \frac{1 \text{ mol antacid}}{144 \text{ g antacid}} = 0.00232 \text{ mol antacid/tablet}$$

This tablet can neutralize four times that many moles of acid, or about 0.01 mol acid. To neutralize the 3×10^{-4} mol acid in the cola requires about 0.03 tablet.

$$3 \times 10^{-4} \text{ mol antacid} \times \frac{1 \text{ tablet}}{0.01 \text{ mol acid}} \approx 0.03 \text{ tablet}$$

It would take only a small portion of a tablet to do the job.

Visit this book's companion website at **www.cengage.com/chemistry/moore** to work an interactive module based on this material.

15.5 Ionization Constants of Acids and Bases

You learned earlier that the greater the value of the equilibrium constant for a reaction, the more product-favored that reaction is. In an acid-base reaction, the stronger the reactant acid and the reactant base, the more product-favored the reaction (p. 461). Consequently, the magnitude of equilibrium constants can give us an idea about the relative strengths of weak acids and bases. For example, *the larger the equilibrium constant for an acid's ionization, the stronger the acid.*

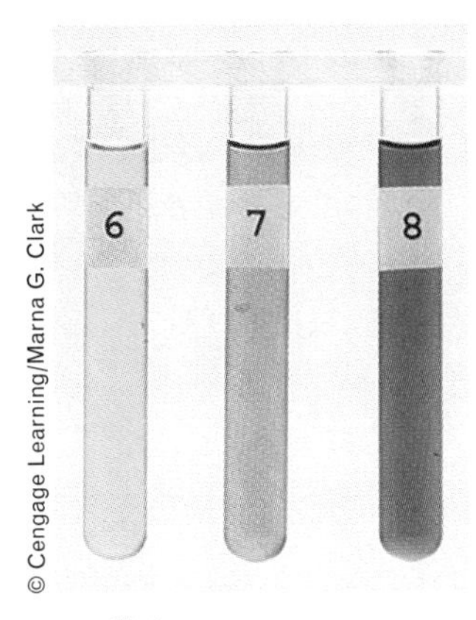

Figure 15.4 Bromthymol blue indicator. At or below a pH of 6 the indicator is yellow. At pH 7 it is pale green, and at pH 8 and above, the color is blue.

Acid Ionization Constants

An ionization equation for the transfer to water of H^+ from any acid represented by the general formula HA is

$$\underset{\text{conjugate acid}}{HA(aq)} + H_2O(\ell) \rightleftharpoons H_3O^+(aq) + \underset{\text{conjugate base}}{A^-(aq)}$$

The corresponding **acid ionization constant expression** is

$$K_a = \frac{[H_3O^+][\text{conjugate base}]}{[\text{un-ionized conjugate acid}]} = \frac{[H_3O^+][A^-]}{[HA]}$$

In the acid ionization constant expression, the *equilibrium* concentrations of conjugate base and hydronium ion appear in the numerator; the *equilibrium* concentration of *un-ionized* conjugate acid appears in the denominator. As with other equilibrium constant expressions, pure solids and liquids, such as water, are not included.

The equilibrium constant K_a is called the **acid ionization constant.** As more acid ionizes, the [HA] denominator term in the acid ionization constant expression gets smaller as the numerator terms increase. Consequently, the ratio gets larger. For strong acids such as hydrochloric acid, the equilibrium is so product-favored that the acid ionization constant value is much larger than 1. In contrast with strong acids, weak acids such as acetic acid ionize to a much smaller extent, establishing equilibria in which significant concentrations of un-ionized weak acid molecules are still present in the solution. All weak acids have K_a values less than 1 because the ionization of a weak acid is reactant-favored; weak acids are weak electrolytes.

Acid ionization constants are also called acid dissociation constants.

The larger its acid ionization constant, the stronger the acid.

The common strong acids are hydrochloric HCl, nitric HNO_3, sulfuric H_2SO_4, perchloric $HClO_4$, hydrobromic HBr, and hydroiodic HI.

PROBLEM-SOLVING EXAMPLE 15.6 Acid Ionization Constant Expressions

Write the ionization equation and the ionization constant expression for these weak acids.

(a) HF (b) HBrO (c) $H_2PO_4^-$

Answer

(a) $HF(aq) + H_2O(\ell) \rightleftharpoons H_3O^+(aq) + F^-(aq)$ $\quad K_a = \frac{[H_3O^+][F^-]}{[HF]}$

(b) $HBrO(aq) + H_2O(\ell) \rightleftharpoons H_3O^+(aq) + BrO^-(aq)$ $\quad K_a = \frac{[H_3O^+][BrO^-]}{[HBrO]}$

(c) $H_2PO_4^-(aq) + H_2O(\ell) \rightleftharpoons H_3O^+(aq) + HPO_4^{2-}(aq)$ $\quad K_a = \frac{[H_3O^+][HPO_4^{2-}]}{[H_2PO_4^-]}$

Strategy and Explanation In each case, the ionization equation represents the transfer of an H^+ ion from an acid to water, creating a hydronium ion and the conjugate base of the acid. In the K_a expression, the product concentrations at equilibrium are divided by the reactant concentrations at equilibrium; $[H_2O]$ is not included. For example, in the case of HF:

$$HF(aq) + H_2O(\ell) \rightleftharpoons H_3O^+(aq) + F^-(aq) \qquad K_a = \frac{[H_3O^+][F^-]}{[HF]}$$

Remember that the solvent (water, in this case) does not appear in the equilibrium constant expression because it is a liquid and a pure substance.

PROBLEM-SOLVING PRACTICE 15.6

Write the ionization equation and ionization constant expression for each of these weak acids:

(a) Hydrazoic acid, HN_3 (b) Formic acid, HCOOH

(c) Chlorous acid, $HClO_2$

That weak acids are only slightly ionized can be shown by measuring the pH of their aqueous solutions. The pH of a 0.10 M acetic acid solution is 2.88, which corresponds to an H_3O^+ concentration of only 1.3×10^{-3} M. Compare this value with the 0.10 M concentration of H_3O^+ ions in a 0.10 M solution of HCl, a strong acid that has pH = 1.00. In a 0.10 M acetic acid solution, only 1.3% of the initial concentration of acetic acid is ionized:

Stronger acid than CH_3COOH — Stronger base than H_2O

$$CH_3COOH(aq) + H_2O(\ell) \rightleftharpoons H_3O^+(aq) + CH_3COO^-(aq)$$

$$\% \text{ ionization} = \frac{[H_3O^+] \text{ at equilibrium}}{\text{initial un-ionized acid conc.}} \times 100\% = \frac{1.3 \times 10^{-3}}{1.0 \times 10^{-1}} \times 100\% = 1.3\%$$

Therefore, almost 99% of the acetic acid remains in the un-ionized molecular form, CH_3COOH. This is why weak acids (and bases) are weak electrolytes.

In an acetic acid solution, or an aqueous solution of any weak acid, two different bases compete for H^+ ions that can be donated from two different acids. In the equation above, the two bases are water and acetate ion; the two acids are acetic acid and hydronium ion. Since acetic acid is a weak acid, its K_a is much less than 1 and the equilibrium favors the reactants. The acetate ion must be a stronger H^+ acceptor than the water molecule. Another way of looking at the same reaction is that the hydronium ion must be a stronger H^+ donor than the acetic acid molecule. Both of these statements are true. Recall from Section 15.1 that acid-base reactions favor going from the stronger to the weaker member of each conjugate acid-base pair. Thus, the acetic acid equilibrium is reactant-favored; a significant concentration of un-ionized acetic acid molecules is present in solution at equilibrium.

Base Ionization Constants

A general equation analogous to that for the donation of H^+ to water by acids can be written for the *acceptance* of an H^+ *from* water by a molecular base, B, to form its conjugate acid, BH^+.

$$\underset{\text{conjugate base}}{B(aq)} + H_2O(\ell) \rightleftharpoons \underset{\text{conjugate acid}}{BH^+(aq)} + OH^-(aq)$$

If the base B were NH_3, then BH^+ would be NH_4^+.

The corresponding equilibrium constant expression is

$$K_b = \frac{[\text{conjugate acid}][OH^-]}{[\text{conjugate base}]} = \frac{[BH^+][OH^-]}{[B]}$$

The equilibrium constant K_b is called the **base ionization constant,** a term that can be misleading. Notice from the chemical equation above that the base does not ionize. Rather, K_b and its equilibrium constant expression refer to *the reaction in which a base forms its conjugate acid by removing an H^+ ion from water.*

When the base is an anion, A^- (such as the anion of a weak acid), the general equation is

$$\underset{\text{conjugate base}}{A^-(aq)} + H_2O(\ell) \rightleftharpoons \underset{\text{conjugate acid}}{HA(aq)} + OH^-(aq)$$

If the base A^- were CH_3COO^-, then HA would be CH_3COOH. The corresponding **base ionization constant expression** is

$$K_b = \frac{[\text{conjugate acid}][OH^-]}{[\text{conjugate base}]} = \frac{[HA][OH^-]}{[A^-]}$$

The *magnitude of the K_b value indicates the extent to which the base removes H^+ ions from water to produce OH^- ions.* The larger the base ionization constant, K_b, the stronger the base, the more product-favored the H^+ transfer reaction from water, and the greater the OH^- concentration produced. For a strong base, the base ionization constant is greater than 1. For a weak base, the ionization constant is less than 1, sometimes considerably less than 1, because at equilibrium there is a significant concentration of unreacted weak conjugate base and a much smaller concentration of its conjugate acid and OH^- ions in solution.

PROBLEM-SOLVING EXAMPLE 15.7 Base Ionization

For each of these weak bases, write the equation for the reaction of the base with water and the companion K_b expression.

(a) C_5H_5N (b) NH_2OH (c) F^-

Answer

(a) $C_5H_5N(aq) + H_2O(\ell) \rightleftharpoons C_5H_5NH^+(aq) + OH^-(aq)$ $\quad K_b = \frac{[C_5H_5NH^+][OH^-]}{[C_5H_5N]}$

(b) $NH_2OH(aq) + H_2O(\ell) \rightleftharpoons NH_3OH^+(aq) + OH^-(aq)$ $\quad K_b = \frac{[NH_3OH^+][OH^-]}{[NH_2OH]}$

(c) $F^-(aq) + H_2O(\ell) \rightleftharpoons HF(aq) + OH^-(aq)$ $\quad K_b = \frac{[HF][OH^-]}{[F^-]}$

Strategy and Explanation The general reaction is the same for each of the bases: The base removes an H^+ from water to form the corresponding conjugate acid. In the first two parts, the base is a neutral molecule to which an H^+ is added, forming a positively charged conjugate acid. In part (c), the H^+ adds to a negatively charged ion, F^-, resulting in a conjugate acid, HF, with no net charge.

PROBLEM-SOLVING PRACTICE 15.7

Write the ionization equation and the K_b expression for these weak bases.

(a) CH_3NH_2 (b) Phosphine, PH_3 (c) NO_2^-

Values of Acid and Base Ionization Constants

Table 15.2 summarizes the ionization constants for a number of acids and their conjugate bases. The ionization constants for strong acids (those above H_3O^+ in Table 15.2) and strong bases (those below OH^- in Table 15.2) are too large to be measured easily. Fortunately, because their ionization reactions are virtually complete, these K_a and K_b values are hardly ever needed. For weak acids, K_a values show relative strengths quantitatively; for weak bases, K_b values do the same.

Table 15.2 Ionization Constants for Some Acids and Their Conjugate Bases at 25 °C

Acid Name	Acid	$K_a = \frac{[H_3O^+][\text{conj base}]}{[\text{conj acid}]}$	Base Name	Base	$K_b = \frac{[\text{conj acid}][OH^-]}{[\text{conj base}]}$
Perchloric acid	$HClO_4$	Large	Perchlorate ion	ClO_4^-	Very small
Sulfuric acid	H_2SO_4	Large	Hydrogen sulfate ion	HSO_4^-	Very small
Hydrochloric acid	HCl	Large	Chloride ion	Cl^-	Very small
Nitric acid	HNO_3	≈ 20	Nitrate ion	NO_3^-	$\approx 5 \times 10^{-16}$
Hydronium ion	H_3O^+	1.0	Water	H_2O	1.0×10^{-14}
Sulfurous acid	H_2SO_3	1.2×10^{-2}	Hydrogen sulfite ion	HSO_3^-	8.3×10^{-13}
Hydrogen sulfate ion	HSO_4^-	1.2×10^{-2}	Sulfate ion	SO_4^{2-}	8.3×10^{-13}
Phosphoric acid	H_3PO_4	7.5×10^{-3}	Dihydrogen phosphate ion	$H_2PO_4^-$	1.3×10^{-12}
Hydrofluoric acid	HF	7.2×10^{-4}	Fluoride ion	F^-	1.4×10^{-11}
Nitrous acid	HNO_2	4.5×10^{-4}	Nitrite ion	NO_2^-	2.2×10^{-11}
Formic acid	$HCOOH$	1.8×10^{-4}	Formate ion	$HCOO^-$	5.6×10^{-11}
Benzoic acid	C_6H_5COOH	6.3×10^{-5}	Benzoate ion	$C_6H_5COO^-$	1.6×10^{-10}
Acetic acid	CH_3COOH	1.8×10^{-5}	Acetate ion	CH_3COO^-	5.6×10^{-10}
Propanoic acid	CH_3CH_2COOH	1.4×10^{-5}	Propanoate ion	$CH_3CH_2COO^-$	7.1×10^{-10}
Carbonic acid	H_2CO_3	4.2×10^{-7}	Hydrogen carbonate ion	HCO_3^-	2.4×10^{-8}
Hydrogen sulfide	H_2S	1×10^{-7}	Hydrogen sulfide ion	HS^-	1×10^{-7}
Dihydrogen phosphate ion	$H_2PO_4^-$	6.2×10^{-8}	Hydrogen phosphate ion	HPO_4^{2-}	1.6×10^{-7}
Hydrogen sulfite ion	HSO_3^-	6.2×10^{-8}	Sulfite ion	SO_3^{2-}	1.6×10^{-7}
Hypochlorous acid	$HClO$	3.5×10^{-8}	Hypochlorite ion	ClO^-	2.9×10^{-7}
Boric acid	$B(OH)_3(H_2O)$	7.3×10^{-10}	Tetrahydroxoborate ion	$B(OH)_4^-$	1.4×10^{-5}
Ammonium ion	NH_4^+	5.6×10^{-10}	Ammonia	NH_3	1.8×10^{-5}
Hydrocyanic acid	HCN	4.0×10^{-10}	Cyanide ion	CN^-	2.5×10^{-5}
Hydrogen carbonate ion	HCO_3^-	4.8×10^{-11}	Carbonate ion	CO_3^{2-}	2.1×10^{-4}
Hydrogen phosphate	HPO_4^{2-}	3.6×10^{-13}	Phosphate ion	PO_4^{3-}	2.8×10^{-2}
Water	H_2O	1.0×10^{-14}	Hydroxide ion	OH^-	1.0
Hydrogen sulfide ion	HS^-	1×10^{-19}	Sulfide ion	S^{2-}	1×10^{5}
Ethanol	C_2H_5OH	Very small	Ethoxide ion	$C_2H_5O^-$	Large
Ammonia	NH_3	Very small	Amide ion	NH_2^-	Large
Hydrogen	H_2	Very small	Hydride ion	H^-	Large
Methane	CH_4	Very small	Methide ion	CH_3^-	Large

Increasing Acid Strength (upward) | Increasing Base Strength (downward)

The smaller the K_a value, the weaker the acid; the smaller the K_b value, the weaker the base.

Consider acetic acid and boric acid. Boric acid is below acetic acid in Table 15.2, so boric acid must be a weaker acid than acetic acid; the K_a values tell us how much weaker. The K_a for boric acid is 7.3×10^{-10}; that for acetic acid is 1.8×10^{-5}, which shows that boric acid is somewhat more than 10^4 times weaker than acetic acid. In fact, boric acid is such a weak acid that a dilute solution of it can be used safely as an eyewash. Don't try that with acetic acid!

CONCEPTUAL **EXERCISE** **15.9 Acid Strengths**

The K_a of lactic acid is 1.5×10^{-4}; that of pyruvic acid is 3.2×10^{-3}.

(a) Which of these acids is the stronger acid?
(b) Which acid's ionization reaction is more reactant-favored?

K_a Values for Polyprotic Acids

So far we have concentrated on **monoprotic acids** such as hydrogen fluoride, HF, hydrogen chloride, HCl, and nitric acid, HNO_3—acids that can donate a single H^+ per molecule.

Some acids, called **polyprotic acids,** can donate more than one H^+ per molecule. These include sulfuric acid, H_2SO_4, carbonic acid, H_2CO_3, and phosphoric acid, H_3PO_4. Oxalic acid, $H_2C_2O_4$ or HOOC—COOH, and other organic acids with two or more carboxylic acid (—COOH) groups are also polyprotic acids (Table 15.3).

sulfuric acid

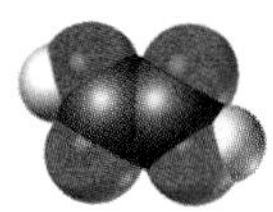
oxalic acid

phosphoric acid

In aqueous solution, a polyprotic acid donates its H^+ ions to water molecules in a stepwise manner. In the first step for sulfuric acid, hydrogen sulfate ion, HSO_4^-, is formed. Sulfuric acid is a strong acid, so this first ionization is essentially complete.

$$\underset{\text{ACID}}{H_2SO_4(aq)} + \underset{\text{base}}{H_2O(\ell)} \longrightarrow \underset{\text{acid}}{H_3O^+(aq)} + \underset{\text{BASE}}{HSO_4^-(aq)}$$

Hydrogen sulfate ion is the conjugate base of sulfuric acid.

In the next ionization step, hydrogen sulfate ion acting as an acid donates an H^+ ion to another water molecule. In this case, hydrogen sulfate ion is a weak acid ($K_a < 1$) and, as with other weak acids, an equilibrium is established. Sulfate ion, SO_4^{2-}, is the conjugate base of HSO_4^-, its conjugate acid.

$$\underset{\text{ACID}}{HSO_4^-(aq)} + \underset{\text{base}}{H_2O(\ell)} \rightleftharpoons \underset{\text{acid}}{H_3O^+(aq)} + \underset{\text{BASE}}{SO_4^{2-}(aq)}$$

Many chemical reactions occur in steps that can be represented by individual equations. Sometimes only the overall equation is written.

CONCEPTUAL
EXERCISE 15.10 Explaining Acid Strengths

Look at the charge on the hydrogen sulfate ion. What does the charge have to do with the fact that this ion is a weaker acid than H_2SO_4?

The successive K_a values for the ionization of a polyprotic acid decrease by a factor of 10^4 to 10^5, indicating that each ionization step occurs to a lesser extent than the one before it. The $H_2PO_4^-$ ion ($K_a = 6.2 \times 10^{-8}$) is a much weaker acid than phosphoric acid ($K_a = 7.5 \times 10^{-3}$), and the HPO_4^{2-} ion ($K_a = 3.6 \times 10^{-13}$) is an even weaker acid than $H_2PO_4^-$. The K_a values indicate that it is more difficult to remove H^+ from a negatively charged $H_2PO_4^-$ ion than from a neutral H_3PO_4 molecule and even more difficult to remove H^+ from a doubly negative HPO_4^{2-} ion.

EXERCISE 15.11 Polyprotic Acids

Write equations for the stepwise ionization in aqueous solution of (a) oxalic acid and (b) citric acid. (Formulas for these acids are given in Table 15.3.)

Table 15.3 Polyprotic Acids

Acid Form	Conjugate Base Form
H_2S (hydrosulfuric acid)	HS^- (hydrogen sulfide or bisulfide ion)
H_3PO_4 (phosphoric acid)	$H_2PO_4^-$ (dihydrogen phosphate ion)
$H_2PO_4^-$ (dihydrogen phosphate ion)	HPO_4^{2-} (monohydrogen phosphate ion)
H_2CO_3 (carbonic acid)	HCO_3^- (hydrogen carbonate or bicarbonate ion)
$H_2C_2O_4$ (oxalic acid)	$HC_2O_4^-$ (hydrogen oxalate ion)
$C_3H_5(COOH)_3$ (citric acid)	$C_3H_5(COOH)_2COO^-$ (monocitrate ion)

There are polyprotic bases, such as CO_3^{2-}, that can accept more than one H^+ per molecule of base. We will not discuss polyprotic bases here.

15.6 Molecular Structure and Acid Strength

If all acids donate H^+ ions, why are some acids strong while others are weak? Why do K_a values cover such a broad range? To answer these questions we turn to the relationship between an acid's strength and its molecular structure. In doing so, we will consider a wide range of acids, from simple binary ones like HBr to structurally more complex ones containing oxygen, carbon, and other elements.

Factors Affecting Acid Strength

All acids have their acidic hydrogen bonded to some other atom, call it A, which can be bonded to other atoms as well. The H—A bond must be broken for the acid to transfer its hydrogen as an H^+ to water, and that will occur only if the H—A bond is sufficiently polar.

$$\overset{\delta^+ \quad \delta^-}{\overset{\longleftrightarrow}{\text{H—A}}}$$

An H—A bond with very little polarity, such as the H—C bond in methane, CH_4, makes the hydrogens nonacidic, and methane is not a significant H^+ donor to water.

The simplest case of an acid is a *binary acid,* such as HBr, one that contains only hydrogen and one other element. In this case, A is bromine.

Blue represents high partial positive charge; red indicates high partial negative charge.

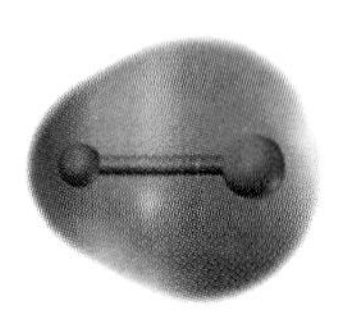

HBr

$$\overset{\delta^+ \quad \delta^-}{\overset{\longleftrightarrow}{\text{H—Br}}}$$

The H—Br bond is polar and HBr is a strong acid ($K_a \approx 10^8$). For acids with A atoms from the same group in the periodic table, for example HF, HCl, HBr, and HI, the H—A bond energies determine the relative acid strengths for binary acids. *As H—A bond energies decrease down a group, the H—A bond weakens, and binary acid strengths increase.*

Strengths of Oxoacids

Acids in which the acidic hydrogen is bonded directly to oxygen in an H—O— bond are called **oxoacids.** Three of the strong acids—nitric, HO—NO_2, perchloric, HO—ClO_3, and sulfuric, $(HO)_2$—SO_2—are oxoacids. Like other oxoacids, they have at least one hydrogen atom bonded to an oxygen and have the general formula

$$\text{H—O—Z}\begin{smallmatrix}/\\—\\\backslash\end{smallmatrix}$$

$$\text{H—}\ddot{\underset{\cdot\cdot}{\text{O}}}\text{—N}\begin{smallmatrix}\diagup \ddot{\text{O}}: \\ \diagdown\!\!\diagdown \ddot{\text{O}}:\end{smallmatrix}$$

nitric acid; Z is nitrogen.

The nature of Z and other atoms that may be attached to it are important in determining the strength of the H—O bond and thus the strength of an oxoacid. In general, *acid strength decreases with the decreasing electronegativity of Z.* This is reflected in the differences among the K_a values of HOCl, HOBr, and HOI, as the electronegativity of the halogen decreases from chlorine (3.0) to bromine (2.8) to iodine (2.5).

Acid:	HOCl	HOBr	HOI
K_a:	3.5×10^{-8}	2.5×10^{-9}	2.3×10^{-11}

The number of oxygen atoms attached to Z also significantly affects the strength of the H—O bond and oxoacid strength: ***The acid strength increases as the number of oxygen atoms attached to Z increases.*** The terminal oxygen atoms (those not in an H—O bond) are sufficiently electronegative, along with Z, to withdraw electron density from the H—O bond. This weakens that bond, promoting the transfer of an H^+ ion to water. The more terminal oxygen atoms present, the greater the electron density shift and the greater the acid strength. A particularly striking example of this trend is seen with the oxoacids of chlorine from the weakest, hypochlorous acid, HOCl, to the strongest, perchloric acid, $HOClO_3$.

HOCl	HOClO	$HOClO_2$	$HOClO_3$
hypochlorous acid	chlorous acid	chloric acid	perchloric acid
K_a: 3.5×10^{-8}	1.1×10^{-2}	$\approx 10^3$	$\approx 10^8$

To be a strong acid, an inorganic oxoacid must have at least two more oxygen atoms than acidic hydrogen atoms in the molecule. Thus, sulfuric acid is a strong acid. In contrast, the inorganic oxoacid phosphoric acid, H_3PO_4, has only four oxygen atoms for three hydrogen atoms and is a weak acid.

15.7 Problem Solving Using K_a and K_b

Calculations with K_a or K_b follow the same patterns as those of other equilibrium calculations illustrated earlier (**p. 472**). Similar important relationships apply to these calculations.

- *Starting with only reactants, equilibrium can be achieved only if some amount of the reactants is converted to products; that is,* ***products are formed at the expense of reactants.***
- *The chemical equilibrium equation for the ionization of the acid or the reaction of the base with water is the basis for the acid ionization or base ionization constant expression, respectively.*
- *The concentrations in the acid ionization or base ionization expression, expressed as molarity (mol/L), are those* ***at equilibrium.***
- *The magnitude of the K_a or K_b value indicates how far the forward reaction occurs at equilibrium (K_a: H^+ donation by an acid* ***to*** *water; K_b: H^+ gain by a base* ***from*** *water).*

There are several experimental methods for determining acid or base ionization constants. The simplest is based on measuring the pH of an acid solution of known concentration. If both the acid concentration and the pH are known, the K_a for the acid can be calculated, as illustrated in Problem-Solving Example 15.8.

PROBLEM-SOLVING EXAMPLE 15.8 K_a from pH

What is the K_a of butanoic acid, $CH_3CH_2CH_2COOH$, a weak organic acid, if a 0.025 M butanoic acid solution has a pH of 3.21 at 25 °C?

Answer 1.6×10^{-5}

Strategy and Explanation We can determine the hydronium concentration from the pH. The hydronium concentration equals the butanoate ion, $CH_3CH_2CH_2COO^-$, concentration because the acid ionizes according to the balanced equation

$$CH_3CH_2CH_2COOH(aq) + H_2O(\ell) \rightleftharpoons H_3O^+(aq) + CH_3CH_2CH_2COO^-(aq)$$

Using the definition of pH, the equilibrium concentration of H_3O^+ is calculated to be 0.00062 M; $[H_3O^+] = 10^{-pH} = 10^{-3.21} = 0.00062$ M.

Butanoic acid dissociates to give equal concentrations of hydronium ions and butanoate ions, in this case 0.00062 M. At equilibrium the concentration of un-ionized butanoic acid equals the original concentration minus the amount that dissociated. The equilibrium concentrations of the species are represented by using a reaction (ICE) table.

	$CH_3CH_2CH_2COOH$	H_3O^+	$CH_3CH_2COO^-$
Initial concentration (mol/L)	0.025	1.0×10^{-7} (from water)*	0
Change as reaction occurs (mol/L)	−0.00062	+0.00062	+0.00062
Equilibrium concentration (mol/L)	0.025 − 0.00062	0.00062	0.00062

*This concentration can be ignored because it is so small.

From the measured pH we have calculated H_3O^+ to be 6.2×10^{-4} mol/L, which is also the $CH_3CH_2CH_2COO^-$ concentration at equilibrium because the ions are formed in equal amounts as butanoic acid ionizes. The concentration of the un-ionized acid at equilibrium is 0.025 − 0.00062 = 0.0244 M. Using these values, we can now calculate K_a for butanoic acid.

$$K_a = \frac{[H_3O^+][CH_3CH_2CH_2COO^-]}{[CH_3CH_2CH_2COOH]} = \frac{[0.00062][0.00062]}{0.0244} = 1.6 \times 10^{-5}$$

☑ **Reasonable Answer Check** This K_a is small, indicating that butanoic acid is a weak acid, as reflected by the fact that a 0.025 M butanoic acid solution has an $[H_3O^+]$ of just 0.00062 M. Thus, the answer makes sense; butanoic acid is only slightly ionized. Butanoic acid is similar to acetic acid in strength, as expected from its similar structure, and a 0.025 M solution has a pH identical to that of 0.025 M acetic acid (pH = 3.21).

PROBLEM-SOLVING PRACTICE 15.8

Lactic acid is a monoprotic acid that occurs naturally in sour milk and also forms by metabolism in the human body. A 0.10 M aqueous solution of lactic acid, $CH_3CH(OH)COOH$, has a pH of 2.43. What is the value of K_a for lactic acid? Is lactic acid stronger or weaker than propanoic acid?

Acid-base ionization constants such as those in Table 15.2 can be used to calculate the pH of a solution of a weak acid or a weak base from its concentration.

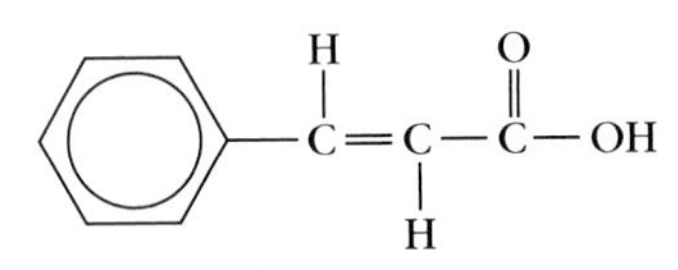

trans-cinnamic acid

PROBLEM-SOLVING EXAMPLE 15.9 pH from K_a

(a) Calculate the pH of a 0.020 M solution of *trans*-cinnamic acid, whose $K_a = 3.6 \times 10^{-5}$ at 25 °C. For simplicity, we will symbolize *trans*-cinnamic acid as HtCA.

(b) What percent of the acid has ionized in this solution?

Answer (a) pH = 3.07 (b) 4.3% ionized

Strategy and Explanation (a) We first must relate the information to the ionization of HtCA to release hydronium ions and *trans*-cinnamate ions, (tCA^-), into solution. Start by writing the equilibrium equation and equilibrium constant expression for *trans*-cinnamic acid.

$$HtCA(aq) + H_2O(\ell) \rightleftharpoons H_3O^+(aq) + tCA^-(aq) \qquad K_a = \frac{[H_3O^+][tCA^-]}{[HtCA]}$$

Next, define equilibrium concentrations and organize the known information in the usual ICE table. In this case, let x equal the H_3O^+ concentration at equilibrium. At equilibrium, the tCA^- ion concentration is also equal to x because the reaction produces H_3O^+ and tCA^- in equal amounts.

	HtCA	H_3O^+	tCA^-
*I*nitial concentration (mol/L)	0.020	1.0×10^{-7} (from water)*	0
*C*hange as reaction occurs (mol/L)	$-x$	$+x$	$+x$
*E*quilibrium concentration (mol/L)	$0.020 - x$	x	x

*This concentration can be ignored because it is so small.

The equilibrium constant expression can be rewritten using the values from the table where all equilibrium concentrations are defined in terms of the single unknown, x.

$$K_a = 3.6 \times 10^{-5} = \frac{[H_3O^+][tCA^-]}{[HtCA]} = \frac{[x][x]}{0.020 - x}$$

Because K_a is very small, the reaction is reactant-favored and therefore, not very much product will form. Consequently, at equilibrium the concentrations of H_3O^+ and tCA^- will be very small. Therefore, x must be quite small compared with 0.020. When x is subtracted from 0.020, the result will still be almost exactly 0.020, and so we can approximate $0.020 - x$ as 0.020 to get

$$\frac{x^2}{0.020} = 3.6 \times 10^{-5}$$

Solving for x gives

$$x = \sqrt{(3.6 \times 10^{-5})(0.020)} = \sqrt{7.2 \times 10^{-7}} = 8.5 \times 10^{-4} = [H_3O^+]$$

$$pH = -\log[H_3O^+] = -\log(8.5 \times 10^{-4}) = 3.07$$

The solution is acidic.

(b) The ionization of the acid is the major source of H_3O^+ ions (the concentration from water is insignificant). Therefore, the percent ionization is calculated by comparing the H_3O^+ concentration at equilibrium with the initial concentration of the un-ionized acid.

$$\% \text{ ionization} = \frac{[H_3O^+]}{(HtCA)_{initial}} \times 100\% = \frac{8.5 \times 10^{-4}}{0.020} \times 100\% = 4.3\%$$

☑ **Reasonable Answer Check** Its K_a value of 3.6×10^{-5} indicates that *trans*-cinnamic acid is a weak acid, similar to acetic acid ($K_a = 1.8 \times 10^{-5}$) in strength and should be only slightly ionized. That 0.020 M *trans*-cinnamic acid is only 4.3% ionized and has a $[H_3O^+]$ of 8.5×10^{-4} M and a pH of 3.07 is reasonable. To check if the approximation was valid, substitute the equilibrium values into the equilibrium constant expression. The calculated result should equal the K_a.

PROBLEM-SOLVING PRACTICE 15.9

(a) What is the pH of a 0.015 M solution of hydrazoic acid, HN_3 ($K_a = 1.9 \times 10^{-5}$), at 25 °C?
(b) What percent of the acid has ionized in this solution?

H—N̈=N=N̈:

hydrazoic acid

An analogous calculation can be done to find the pH of a solution of a weak base, such as piperidine, a component of pepper.

PROBLEM-SOLVING EXAMPLE 15.10 pH of a Weak Base from K_b

Piperidine, $C_5H_{11}N$, a nitrogen-containing base analogous to ammonia, has a $K_b = 1.3 \times 10^{-3}$. Calculate the OH^- concentration and the pH of a 0.025 M solution of piperidine.

piperidine, $C_5H_{11}N$

Piperidinium ion, $C_5H_{11}NH^+$, is analogous to ammonium ion, NH_4^+.

Answer $[OH^-] = 5.1 \times 10^{-3}$ M; pH = 11.71

Strategy and Explanation We use the balanced equilibrium equation to provide the K_b expression. Using the value of K_b we can calculate $[OH^-]$ and then pOH from the OH^- concentration. Knowing the pOH, the pH can then be derived from the relation pH + pOH = 14.

Piperidine reacts with water to transfer H^+ ions from water to form hydroxide ions and piperidinium ions, $C_5H_{11}NH^+$, according to the equation

$$C_5H_{11}N(aq) + H_2O(\ell) \rightleftharpoons C_5H_{11}NH^+(aq) + OH^-(aq)$$

$$K_b = \frac{[C_5H_{11}NH^+][OH^-]}{[C_5H_{11}N]} = 1.3 \times 10^{-3}$$

A table can be set up like the one in Problem-Solving Example 15.9, letting x be the concentration of OH^- and of piperidinium ions at equilibrium because the forward reaction produces them in equal amounts. The equilibrium concentration of *unreacted* piperidine will be its initial concentration, 0.025 mol/L, minus x, the amount per liter that has reacted with water.

	$C_5H_{11}N$	$C_5H_{11}NH^+$	OH^-
*I*nitial concentration (mol/L)	0.025	0	1.0×10^{-7}*
*C*hange as reaction occurs (mol/L)	$-x$	$+x$	$+x$
*E*quilibrium concentration (mol/L)	$(0.025 - x)$	x	x

*The low concentration can be ignored, as it was in the K_a calculations.

Substitution into the base ionization constant expression gives

$$K_b = \frac{[C_5H_{11}NH^+][OH^-]}{[C_5H_{11}N]} = \frac{x^2}{0.025 - x} = 1.3 \times 10^{-3}$$

When $\frac{x}{\text{initial conc.}} \times 100\% > 5\%$, the x term cannot be dropped from the denominator term.

Generally in K_a and K_b calculations, if $\frac{x}{\text{initial concentration}} \times 100\% > 5\%$, the x term cannot be dropped from the denominator in the equilibrium constant expression and the quadratic equation is used. As seen below, in this case x is not negligible compared with 0.025 because piperidine reacts with water sufficiently and so x, the OH^- concentration, must be found by using the quadratic formula.

Multiplying out the terms gives Equation A.

$$x^2 = (0.025 - x)(1.3 \times 10^{-3}) = 3.3 \times 10^{-5} - (1.3 \times 10^{-3}x) \quad \text{(A)}$$

Rearranging Equation A into the quadratic form $ax^2 + bx + c = 0$ then gives Equation B.

Appendix A.6 reviews the use of the quadratic equation.

$$x^2 + (1.3 \times 10^{-3}x) - (3.3 \times 10^{-5}) = 0 \quad \text{(B)}$$

Solving for x using the quadratic formula gives

$$x = \frac{-(1.3 \times 10^{-3}) \pm \sqrt{(1.3 \times 10^{-3})^2 - (4 \times 1)(-3.3 \times 10^{-5})}}{2(1)}$$

$$= \frac{-(1.3 \times 10^{-3}) \pm \sqrt{1.34 \times 10^{-4}}}{2}$$

$$= \frac{-(1.3 \times 10^{-3}) \pm (1.16 \times 10^{-2})}{2} = \frac{1.03 \times 10^{-2}}{2} = 5.1 \times 10^{-3}$$

Therefore x, the OH^- concentration, equals 5.1×10^{-3}. (The negative root in the solution of the quadratic equation is disregarded because concentration cannot be negative; you can't have less than 0 mol/L of a substance.)

Note that $\frac{5.1 \times 10^{-3}}{0.025} \times 100\% = 20\%$, which is far greater than 5%, so the quadratic equation was necessary in this case; the approximation of

$$\frac{x^2}{0.025 - x} \approx \frac{x^2}{0.025}$$

would not have given the correct answer.

The pOH can be calculated from the OH^- concentration.

$$\text{pOH} = -\log(5.1 \times 10^{-3}) = 2.29$$

$$\text{pH} = 14.00 - \text{pOH} = 14.00 - 2.29 = 11.71$$

Therefore, piperidine reacts sufficiently with water to generate a fairly basic solution.

☑ **Reasonable Answer Check** Both the initial concentration and the K_b of piperidine are small. Consequently, the pH should be less than that of a 0.025 M solution of a strong base like NaOH, which would be 12.40.

$$[OH^-] = 0.025 \text{ M; pOH} = -\log(0.025) = 1.60\text{; pH} = 12.40$$

A pH of 11.71 for 0.025 M piperidine is reasonable.

PROBLEM-SOLVING PRACTICE 15.10

Calculate the OH^- concentration and the pH of a 0.015 M solution of cyclohexylamine, $C_6H_{11}NH_2$. $K_b = 4.6 \times 10^{-4}$.

cyclohexylamine

Relationship between K_a and K_b Values

The right-hand side of Table 15.2 (⇐ *p. 554*) gives K_b values for the conjugate base of each acid. Try an experiment with these data: Multiply a few of the K_a values by K_b values for their conjugate bases. What do you find? Within a very small error you ought to find that $K_a \times K_b = 1.0 \times 10^{-14}$. This value is the same as K_w, the autoionization constant for water. To see why, multiply the equilibrium constant expressions for K_a and K_b.

$$K_a \times K_b = \left(\frac{[H_3O^+][A^-]}{[HA]}\right)\left(\frac{[HA][OH^-]}{[A^-]}\right)$$

Canceling like terms in the numerator and denominator of this expression gives

$$K_a \times K_b = \left(\frac{[H_3O^+]\cancel{[A^-]}}{\cancel{[HA]}}\right)\left(\frac{\cancel{[HA]}[OH^-]}{\cancel{[A^-]}}\right) = [H_3O^+][OH^-] = K_w$$

This relation shows that if you know K_a for an acid, you can find K_b for its conjugate base by using K_w. Furthermore, the larger the K_a, the smaller the K_b, and vice versa (because when multiplied they always have to give the same product, K_w). For example, K_a for HCN is 4.0×10^{-10}. The value of K_b for the conjugate base, CN^-, is

$$K_b\,(\text{for } CN^-) = \frac{K_w}{K_a\,(\text{for HCN})} = \frac{1.0 \times 10^{-14}}{4.0 \times 10^{-10}} = 2.5 \times 10^{-5}$$

HCN has a relatively small K_a and lies fairly far down in Table 15.2, which means it is a very weak acid. However, CN^- is a fairly strong weak base; its K_b of 2.5×10^{-5} is nearly the same as the K_b for ammonia (1.8×10^{-5}), making CN^- a slightly stronger base than ammonia. In general, if $K_a > K_b$, the acid is stronger than its conjugate base. Alternatively, if $K_b > K_a$, the conjugate base is stronger than its conjugate acid. For example, hypochlorite ion, OCl^- ($K_b = 2.9 \times 10^{-7}$), is a stronger base than hypochlorous acid, HOCl, is an acid ($K_a = 3.5 \times 10^{-8}$).

EXERCISE 15.12 K_b from K_a

Phenol, or carbolic acid, C_6H_5OH, is a weak acid, $K_a = 1.3 \times 10^{-10}$. Calculate K_b for the phenolate ion, $C_6H_5O^-$. Which base in Table 15.2 is closest in strength to the phenolate ion? How did you make your choice?

15.8 Acid-Base Reactions of Salts

An exchange reaction between an acid and a base produces a salt plus water ***(← p. 145)***. The salt's positive ion comes from the base and its negative ion comes from the acid. In the case of a metal hydroxide as a base, the salt-forming general reaction is

$$\underset{\text{acid}}{HX(aq)} + \underset{\text{base}}{MOH(aq)} \longrightarrow \underset{\text{salt}}{MX(aq)} + HOH(\ell)$$

Now that you know more about the Brønsted-Lowry acid-base concept and the strengths of acids and bases, it is useful to consider acid-base reactions and salt formation in more detail.

Salts of Strong Bases and Strong Acids

Strong acids react with strong bases to form *neutral* salts. Consider the reaction of the strong acid HCl with the strong base NaOH to form the salt NaCl. If the amounts of HCl and NaOH are in the correct stoichiometric ratio (1 mol HCl per 1 mol NaOH), this reaction occurs with the complete neutralization of the acidic properties of HCl and the basic properties of NaOH. The reaction can be described first by an overall equation, then by a complete ionic equation, and finally by a net ionic equation ***(← p. 140)***. Each of these equations contains useful information.

Overall equation	$HCl(aq) + NaOH(aq)$	$\longrightarrow NaCl(aq) + H_2O(\ell)$
Complete ionic equation	$H_3O^+(aq) + Cl^-(aq) + Na^+(aq) + OH^-(aq)$	$\longrightarrow Na^+(aq) + Cl^-(aq) + 2\,H_2O(\ell)$
Net ionic equation	$H_3O^+(aq) + OH^-(aq)$	$\longrightarrow H_2O(\ell) + H_2O(\ell)$
	ACID base	BASE acid

The overall equation shows the substances that were dissolved or that could be recovered at the end of the reaction. The complete ionic equation indicates all of the ions that are present before and after reaction. The net ionic equation emphasizes that a Brønsted-Lowry acid (H_3O^+) is reacting with a Brønsted-Lowry base (OH^-); the spectator ions, Na^+ and Cl^-, are omitted. This reaction goes to completion because H_3O^+ is a strong acid, OH^- is a strong base, and water is a very weak acid and a very weak base.

The resulting solution contains only sodium ions and chloride ions, with some more water molecules than before. Its properties are the same as if it had been prepared by simply dissolving some NaCl(s) in water. It has a neutral pH because it contains no significant concentrations of acids or bases. The Cl^- ion is the conjugate base of a strong acid and hence is such a weak base that it does not react with water. The Na^+ ion also does not react as either an acid or a base with water. Examples of some other salts of this type are given in Table 15.4. These salts all form neutral solutions.

Table 15.4 Some Salts Formed by Neutralization of Strong Acids with Strong Bases

	Base		
Acid	**NaOH**	**KOH**	**$Ba(OH)_2$**
	Salts	**Salts**	**Salts**
HCl	NaCl	KCl	$BaCl_2$
HNO_3	$NaNO_3$	KNO_3	$Ba(NO_3)_2$
H_2SO_4	Na_2SO_4	K_2SO_4	$BaSO_4$
$HClO_4$	$NaClO_4$	$KClO_4$	$Ba(ClO_4)_2$

Salts of Strong Bases and Weak Acids

Strong bases react with weak acids to form *basic* salts. Suppose, for example, that 0.010 mol NaOH is added to 0.010 mol of the weak acid acetic acid in 1 L of solution. The three equations are

$$CH_3COOH(aq) + NaOH(aq) \longrightarrow NaCH_3COO(aq) + H_2O(\ell) \quad \text{Overall equation}$$

$$CH_3COOH(aq) + Na^+(aq) + OH^-(aq) \longrightarrow Na^+(aq) + CH_3COO^-(aq) + H_2O(\ell) \quad \text{Complete ionic equation}$$

$$\underset{\text{weak acid}}{CH_3COOH(aq)} + \underset{\text{strong base}}{OH^-(aq)} \longrightarrow \underset{\text{base}}{CH_3COO^-(aq)} + \underset{\text{acid}}{H_2O(\ell)} \quad \text{Net ionic equation}$$

In this case, acetate ion, a weaker base than OH^-, has been formed by the reaction. Therefore, the solution is slightly basic (pH > 7), even though exactly the stoichiometric amount of acetic acid was added to the sodium hydroxide. The reaction that makes the solution basic is the reaction of water with acetate ion, a weak Brønsted-Lowry base.

$$CH_3COO^-(aq) + H_2O(\ell) \rightleftharpoons CH_3COOH(aq) + OH^-(aq)$$

This is a **hydrolysis** reaction, one in which a water molecule is split—in this case, into an H^+ ion and an OH^- ion. An H^+ ion is donated to the acetate ion to form acetic acid. The extent of hydrolysis is determined by the value of K_b for acetate ion.

The term "hydrolysis" is derived from *hydro*, meaning "water," and *lysis*, meaning "to break apart."

All of the weak bases in Table 15.2, except for the very weak bases (NO_3^-, Cl^-, HSO_4^-, and ClO_4^-) above water in the next to last column, undergo hydrolysis reactions in aqueous solution. The larger the K_b value of a base, the more basic the solutions it produces. The pH of a solution of a salt of a strong base and a weak acid can be calculated from K_b, as shown in Problem-Solving Example 15.11.

As K_b of the base increases, the pH of the solution increases.

PROBLEM-SOLVING EXAMPLE 15.11 pH of a Salt Solution

Sodium benzoate, $NaC_7H_5O_2$, is used as a preservative in foods. Calculate the pH of a 0.025 M solution of $NaC_7H_5O_2$ ($K_b = 1.6 \times 10^{-10}$).

Answer 8.30

Strategy and Explanation To calculate the pH we must first calculate the hydroxide ion concentration in the solution. Sodium benzoate is a basic salt that could be synthesized from a strong base, NaOH, and a weak acid, benzoic acid, $HC_7H_5O_2$. The Na^+ ion does not react with water, but benzoate ion, $C_7H_5O_2^-$, the conjugate base of a weak acid, $HC_7H_5O_2$, reacts with water to produce a basic solution.

$$C_7H_5O_2^-(aq) + H_2O(\ell) \rightleftharpoons HC_7H_5O_2(aq) + OH^-(aq)$$

$$K_b = 1.6 \times 10^{-10} = \frac{[\text{conjugate acid}][OH^-]}{[\text{conjugate base}]} = \frac{[HC_7H_5O_2][OH^-]}{[C_7H_5O_2^-]}$$

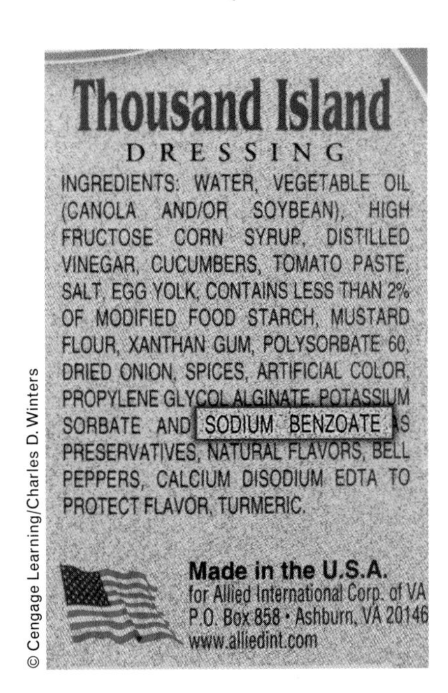

The concentrations of benzoate ion, benzoic acid, and hydroxide ion initially and at equilibrium are summarized in the following table. We let x be equal to the equilibrium concentration of OH^- as well as that of $C_7H_5O_2^-$, because they are formed in equal amounts.

	$C_7H_5O_2^-$	$HC_7H_5O_2$	OH^-
Initial concentration (mol/L)	0.025	0	1.0×10^{-7} (from water)
Change as reaction occurs (mol/L)	$-x$	$+x$	$+x$
Equilibrium concentration (mol/L)	$0.025 - x$	x	x

Benzoate ion has a very small K_b (1.6×10^{-10}) and thus is a very weak base, so it is safe to assume that x will be negligibly small compared with 0.025, and $0.025 - x \approx 0.025$.

$$K_b = 1.6 \times 10^{-10} \approx \frac{x^2}{0.025}$$

Solving for x gives $x = 2.0 \times 10^{-6}$, which is equal to the hydroxide and benzoate ion concentrations. Because $0.025 - 2.0 \times 10^{-6} = 0.025$ (using the significant figures rules), our assumption that x is negligible compared with 0.025 is justified. Therefore, at equilibrium

$$OH^- = [HC_7H_5O_2] = 2.0 \times 10^{-6} \text{ mol/L}; [C_7H_5O_2^-] = 0.025 \text{ mol/L}$$

Finally, the pH of the solution can be calculated.

$$K_w = [H_3O^+][OH^-] = [H_3O^+](2.0 \times 10^{-6}) = 1.0 \times 10^{-14}$$

$$[H_3O^+] = \frac{1.0 \times 10^{-14}}{2.0 \times 10^{-6}} = 5.0 \times 10^{-9}$$

$$\text{pH} = -\log(5.0 \times 10^{-9}) = 8.30$$

As expected, the solution is basic.

☑ **Reasonable Answer Check** The reaction of benzoate ion with water produces hydroxide ions in addition to those from the dissociation of water. The excess hydroxide ions cause the solution to become basic, as indicated by the pH greater than 7. This is expected because the salt is formed from a strong base and a weak acid.

PROBLEM-SOLVING PRACTICE 15.11

Sodium carbonate is an environmentally benign paint stripper. It is water-soluble, and carbonate ion is a strong enough base to loosen paint so it can be scraped off. What is the pH of a 1.0 M solution of Na_2CO_3?

CONCEPTUAL
EXERCISE 15.13 pH of Soap Solutions

Ordinary soaps are often sodium salts of fatty acids, which are weak organic acids. Would you expect the pH of a soap solution to be greater than or less than 7? Explain your answer.

Salts of Weak Bases and Strong Acids

When a weak base reacts with a strong acid, the resulting salt solution is *acidic*. The conjugate acid of the weak base determines the pH of the solution. For example, suppose equal volumes of 0.10 M NH_3 and 0.10 M HCl are mixed. The reaction, shown in overall, complete ionic, and net ionic equations, is

$$NH_3(aq) + HCl(aq) \longrightarrow NH_4Cl(aq)$$

$$NH_3(aq) + H_3O^+(aq) + Cl^-(aq) \longrightarrow NH_4^+(aq) + Cl^-(aq) + H_2O(\ell)$$

$$\underset{\text{weak base}}{NH_3(aq)} + \underset{\text{strong acid}}{H_3O^+(aq)} \longrightarrow \underset{\text{acid}}{NH_4^+} + \underset{\text{base}}{H_2O(\ell)}$$

As soon as it is formed, the weak acid NH_4^+ reacts with water and establishes an equilibrium. The resulting solution is slightly acidic, because the reaction produces hydronium ions.

$$NH_4^+(aq) + H_2O(\ell) \rightleftharpoons NH_3(aq) + H_3O^+(aq)$$

Strong base + strong acid yields a neutral salt (solution pH = 7.0)

Strong base + weak acid yields a basic salt (solution pH > 7.0)

Strong acid + weak base yields an acidic salt (solution pH < 7.0)

PROBLEM-SOLVING EXAMPLE 15.12 pH of Another Salt Solution

Ammonium nitrate, NH_4NO_3, is a salt used in fertilizers and in making matches. The salt is made by the reaction of ammonia, NH_3, and nitric acid, HNO_3. Calculate the pH of a 0.15 M solution of ammonium nitrate; K_a (NH_4^+) = 5.6×10^{-10}.

Answer 5.04

Strategy and Explanation You should first recognize that ammonium nitrate is the salt of a weak base and a strong acid so the solution will be acidic, with a pH less than 7.0. Because ammonium ion is a weak acid, ammonium ions react with water to form ammonia and hydronium ions, making the solution acidic.

$$NH_4^+(aq) + H_2O(\ell) \rightleftharpoons NH_3(aq) + H_3O^+(aq)$$

The following table summarizes the concentrations of ammonium ion, ammonia, and hydronium ions initially and at equilibrium. We let x be equal to the equilibrium concentration of H_3O^+ as well as that of NH_3 because they are formed in equal amounts.

	NH_4^+	NH_3	H_3O^+
*I*nitial concentration (mol/L)	0.15	0	1.0×10^{-7} (from water)
*C*hange as reaction occurs (mol/L)	$-x$	$+x$	$+x$
*E*quilibrium concentration (mol/L)	$0.15 - x$	x	x

As in prior problems, we substitute the equilibrium values into the K_a expression.

$$K_a = \frac{[NH_3][H_3O^+]}{[NH_4^+]} = \frac{(x)(x)}{(0.15 - x)} = 5.6 \times 10^{-10}$$

Because ammonium ion is such a weak acid, as indicated by its very low K_a, we can simplify the equation

$$K_a = \frac{(x)(x)}{(0.15)} \approx 5.6 \times 10^{-10}$$

and solve for x, the H_3O^+ concentration.

$$x^2 = (0.15)(5.6 \times 10^{-10}); x = 9.2 \times 10^{-6} = [H_3O^+]$$

The pH can be calculated from the hydronium ion concentration.

$$pH = -\log[H_3O^+] = -\log(9.2 \times 10^{-6}) = 5.04$$

☑ **Reasonable Answer Check** As predicted, the pH of the solution is less than 7.0, which it should be because of the release of hydronium ions by the forward reaction.

PROBLEM-SOLVING PRACTICE 15.12

Calculate the pH of a 0.10 M solution of ammonium chloride, NH_4Cl.

Salts of Weak Bases and Weak Acids

What is the pH of a solution of a salt such as NH_4F containing an acidic cation and a basic anion? The salt could be formed by the reaction of a weak acid and a weak base. There are two possible reactions that can determine the pH of the NH_4F solution: formation of H_3O^+ by H^+ transfer from the cation; and formation of OH^- by hydrolysis of the anion.

$$NH_4^+(aq) + H_2O(\ell) \rightleftharpoons H_3O^+(aq) + NH_3(aq) \qquad K_a(NH_4^+) = 5.6 \times 10^{-10}$$

$$F^-(aq) + H_2O(\ell) \rightleftharpoons HF(aq) + OH^-(aq) \qquad K_b(F^-) = 1.4 \times 10^{-11}$$

Since $K_a(NH_4^+) > K_b(F^-)$, the reaction of ammonium ions with water to produce hydronium ions is the more favorable reaction causing the resulting solution to be slightly acidic.

In general, the K_a of the weak acid and the K_b of the weak base need to be considered to determine whether the aqueous solution of a salt of a weak acid and weak base will be acidic or basic.

CONCEPTUAL **EXERCISE 15.13 Hydrolysis of a Salt of a Weak Acid and a Weak Base**

Name a salt of a weak acid and a weak base where $K_a = K_b$. What should the pH of a solution of this salt be?

Table 15.5 summarizes the acid-base behavior of many different ions in aqueous solution.

These generalizations can be made about acid-base neutralization reactions in aqueous solution and the pH of the resulting salt solutions.

- Solution of strong acid + solution of strong base $\longrightarrow$ salt solution with pH = 7 (neutral)

Table 15.5 Acid-Base Properties of Typical Ions in Aqueous Solution

	Neutral		Basic			Acidic
Anions	Cl^-	NO_3^-	CH_3COO^-	CN^-	SO_4^{2-}	HSO_4^-
	Br^-	ClO_4^-	$HCOO^-$	PO_4^{3-}	HPO_4^{2-}	$H_2PO_4^-$
	I^-		CO_3^{2-}	HCO_3^-	SO_3^{2-}	HSO_3^-
			S^{2-}	HS^-	ClO^-	
			F^-	NO_2^-		
Cations	Li^+	Mg^{2+}		*None*		Al^{3+}
	Na^+	Ca^{2+}				NH_4^+
	K^+	Ba^{2+}				

- Solution of strong acid + solution of weak base ⟶ salt solution with pH < 7 (acidic)
- Solution of weak acid + solution of strong base ⟶ salt solution with pH > 7 (basic)
- Solution of weak acid + solution of weak base ⟶ salt solution with pH determined by relative strengths of conjugate base and conjugate acid formed

15.9 Lewis Acids and Bases

In 1923 when Brønsted and Lowry independently proposed their acid-base concept, Gilbert N. Lewis also was developing a new concept of acids and bases. By the early 1930s Lewis had proposed definitions of acids and bases that are more general than those of Brønsted and Lowry because they are based on sharing of electron pairs rather than on H^+ ion transfers. A **Lewis acid** is *a substance that can accept a pair of electrons to form a new bond*, and a **Lewis base** is *a substance that can donate a pair of electrons to form a new bond*. Those definitions mean that in the Lewis sense, an acid-base reaction occurs when a molecule (or ion) that has a lone pair of electrons that can be donated (a Lewis base) reacts with a molecule (or ion) that can accept an electron pair (a Lewis acid). In general, Lewis acids are cations or neutral molecules with an available, empty orbital; Lewis bases are anions or neutral molecules with a lone pair of electrons. When both electrons in an electron-pair bond were originally associated with one of the bonded atoms (the Lewis base), the bond is called a coordinate covalent bond.

A + :B ⟶ A—B

Lewis Acid (electron pair acceptor)
Lewis Base (electron pair donor)
Coordinate covalent bond

A simple example of a Lewis acid-base reaction is the formation of a hydronium ion from H^+ and water. The H^+ ion has no electrons, while the water molecule has two lone pairs of electrons on the oxygen atom. One of the lone pairs can be shared between H^+ and oxygen, thus forming an O—H bond.

$$H^+ + H_2O \longrightarrow H_3O^+$$

hydronium ion

A Brønsted-Lowry base (H^+ ion acceptor) must also be a Lewis base by donating an electron pair to bond with the H^+.

Positive Metal Ions as Lewis Acids

All metal cations are potential Lewis acids. Not only do the positively charged metal cations attract electrons but all such cations also have at least one empty orbital. This empty orbital can accommodate an electron pair donated by a base, thereby forming a **coordinate covalent bond.** Consequently, metal ions readily form coordination complexes and also are hydrated in aqueous solution. When a metal ion becomes hydrated, one of the lone pairs on the oxygen atom in each of several water molecules forms a coordinate covalent bond to the metal ion; the metal ion acts as a Lewis acid, and water acts as a Lewis base. The combination of a metal ion and a Lewis base forms a **complex ion.** For example, $[Ag(NH_3)_2]^+$, is a complex ion in which a silver ion (Lewis acid) is bonded to two ammonia molecules (Lewis base) through donation of an electron pair from each ammonia.

$$Ag^+(aq) + 2 :NH_3(aq) \longrightarrow [H_3N:Ag:NH_3]^+(aq)$$

Lone pairs of electrons on water molecules form coordinate covalent bonds with the metal ion.

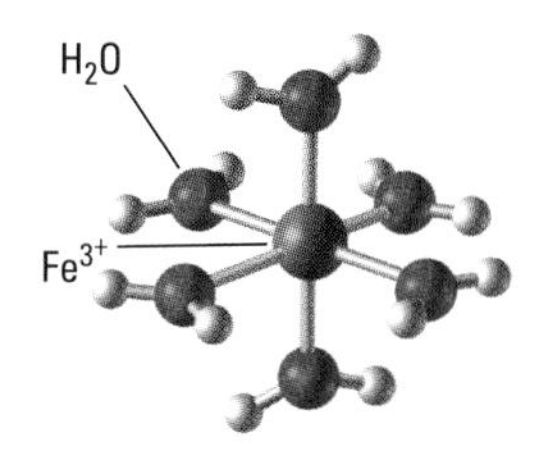

Water molecules bonded to an Fe^{3+} ion in $[Fe(H_2O)_6]^{3+}$. Square brackets in the formula indicate that ions or molecules within the brackets are bonded to the metal ion.

Indeed, this complex is so stable that the very water-insoluble compound AgCl can be dissolved in aqueous ammonia.

$$AgCl(s) + 2\,{:}NH_3(aq) \longrightarrow [H_3N{:}Ag{:}NH_3]^+(aq) + Cl^-(aq)$$

The hydroxide ion (OH^-) is an excellent Lewis base and so it binds readily to metal cations to give metal hydroxides. An important feature of the chemistry of many metal hydroxides is that they are **amphoteric,** meaning that they can react as both a base and an acid (Table 15.6). The amphoteric aluminum hydroxide, for example, behaves as a Lewis acid when it dissolves in a basic solution by forming a complex ion containing one additional OH^- ion, a Lewis base.

$$Al(OH)_3(s) + OH^-(aq) \rightleftharpoons [Al(OH)_4]^-(aq)$$

This reaction is shown in Figure 15.5 (b). The same compound behaves as a Brønsted-Lowry base when it reacts with a Brønsted-Lowry acid, as seen in Figure 15.5 (c).

$$Al(OH)_3(s) + 3\,H_3O^+(aq) \rightleftharpoons Al^{3+}(aq) + 6\,H_2O(\ell)$$

Neutral Molecules as Lewis Acids

Lewis's ideas about acids and bases account nicely for the fact that oxides of nonmetals behave as acids. Two important examples are carbon dioxide and sulfur dioxide, whose Lewis structures are

$$:\ddot{O}{=}C{=}\ddot{O}: \qquad :\ddot{O}{=}\ddot{S}{-}\ddot{\underset{..}{O}}: \;\longleftrightarrow\; :\ddot{\underset{..}{O}}{-}\ddot{S}{=}\ddot{O}:$$

carbon dioxide — sulfur dioxide

Table 15.6 Some Common Amphoteric Metal Hydroxides

Hydroxide	Reaction as a Brønsted-Lowry Base	Reaction as a Lewis Acid
$Al(OH)_3$	$Al(OH)_3(s) + 3\,H_3O^+(aq) \longrightarrow Al^{3+}(aq) + 6\,H_2O(\ell)$	$Al(OH)_3(s) + OH^-(aq) \longrightarrow [Al(OH)_4]^-(aq)$
$Zn(OH)_2$	$Zn(OH)_2(s) + 2\,H_3O^+(aq) \longrightarrow Zn^{2+}(aq) + 4\,H_2O(\ell)$	$Zn(OH)_2(s) + 2\,OH^-(aq) \longrightarrow [Zn(OH)_4]^{2-}(aq)$
$Sn(OH)_4$	$Sn(OH)_4(s) + 4\,H_3O^+(aq) \longrightarrow Sn^{4+}(aq) + 8\,H_2O(\ell)$	$Sn(OH)_4(s) + 2\,OH^-(aq) \longrightarrow [Sn(OH)_6]^{2-}(aq)$
$Cr(OH)_3$	$Cr(OH)_3(s) + 3\,H_3O^+(aq) \longrightarrow Cr^{3+}(aq) + 6\,H_2O(\ell)$	$Cr(OH)_3(s) + OH^-(aq) \longrightarrow [Cr(OH)_4]^-(aq)$

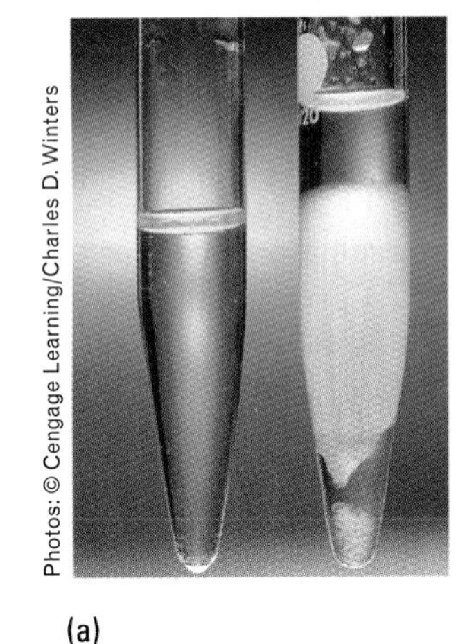

(a)

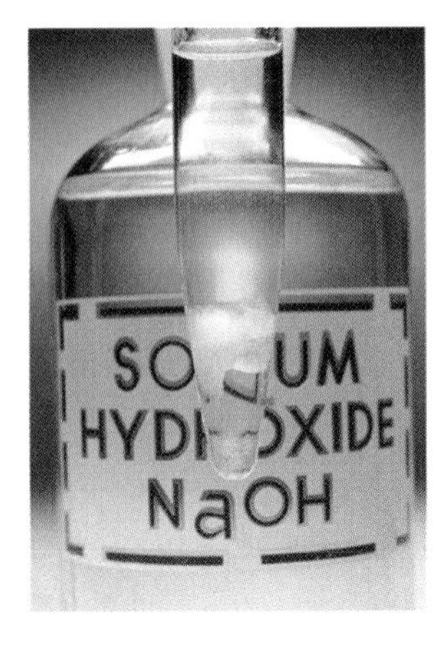

(b)

(c)

Photos: © Cengage Learning/Charles D. Winters

Figure 15.5 The amphoteric nature of $Al(OH)_3$. (a) Adding aqueous ammonia to a solution of Al^{3+} (left test tube) causes formation of a precipitate of $Al(OH)_3$ (right test tube). (b) Adding a strong base, NaOH, to the $Al(OH)_3$ dissolves the precipitate. Here the aluminum hydroxide acts as a Lewis acid toward the Lewis base OH^- and forms a soluble salt of the complex ion $[Al(OH)_4]^-$. (c) If we begin again with freshly precipitated $Al(OH)_3$, it dissolves as strong acid, HCl, is added. In this case $Al(OH)_3$ acts as a Brønsted-Lowry base and forms a soluble aluminum salt and water.

In each case, there is a double bond; an "extra" pair of electrons is being shared between an oxygen atom and the central atom. Because oxygen is highly electronegative, electrons in these bonds are attracted away from the less electronegative central atom, which becomes slightly positively charged. This makes the central atom a likely site to attract a pair of electrons. A Lewis base such as OH^- can bond to the carbon atom in CO_2 to give bicarbonate ion, HCO_3^-. This bonding displaces one double-bond pair of electrons back onto an oxygen atom.

$$:\ddot{O}{=}C{=}\ddot{O}: + :\ddot{O}{-}H^- \longrightarrow :\ddot{O}{=}C(\ddot{O}{-}H)(\ddot{O}:^-)$$

bicarbonate ion

Carbon dioxide from the air can react to form sodium carbonate around the mouth of a bottle of sodium hydroxide. Sulfur dioxide can react similarly with hydroxide ion to form HSO_3^- ion.

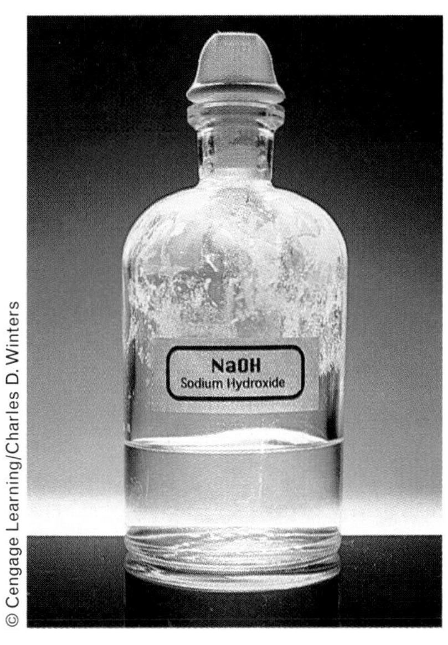

Carbon dioxide in air reacts with spilled base such as NaOH to form Na_2CO_3. If the mouth of a glass-stoppered bottle such as the one shown here is not routinely cleaned, the sodium carbonate formed can virtually cement the top of the bottle to the neck, making it difficult to open the bottle.

EXERCISE 15.14 Lewis Acids and Bases

Predict whether each of these is a Lewis acid or a Lewis base. (*Hint:* Drawing a Lewis structure for a molecule or ion is often helpful in making such a prediction.)

(a) PH_3 (b) BCl_3 (c) H_2S (d) NO_2 (e) Ni^{2+} (f) CO

IN CLOSING

Having studied this chapter, you should be able to . . .

- Describe water's role in aqueous acid-base chemistry (Section 15.1). ■ End-of-chapter Questions assignable in OWL: 8, 12
- Identify the conjugate base of an acid, the conjugate acid of a base, and the relationship between conjugate acid and base strengths (Section 15.1). ■ Question 20
- Recognize how amines act as bases and how carboxylic acids ionize in aqueous solution (Section 15.2). ■ Question 12
- Use the autoionization of water and show how this equilibrium takes place in aqueous solutions of acids and bases (Section 15.3).
- Classify an aqueous solution as acidic, neutral, or basic based on its concentration of H_3O^+ or OH^- and its pH or pOH (Section 15.4). ■ Question 30
- Calculate pH (or pOH) given $[H_3O^+]$, or $[OH^-]$ given pH (or pOH) (Section 15.5). ■ Questions 24, 26
- Estimate acid and base strengths from K_a and K_b values (Section 15.5). ■ Question 34
- Write the ionization steps of polyprotic acids (Section 15.5). ■ Question 22
- Describe the relationships between acid strength and molecular structure (Section 15.6). ■ Question 73
- Calculate pH from K_a or K_b values and solution concentration (Section 15.7). ■ Question 40
- Describe the hydrolysis of salts in aqueous solution (Section 15.8). ■ Question 50
- Recognize Lewis acids and bases and describe how they react (Section 15.9). ■ Questions 52, 54

Selected end-of-chapter Questions may be assigned in OWL.

For quick review, download Go Chemistry mini-lecture flashcard modules (or purchase them at **www.ichapters.com**).

Prepare for an exam with a **Summary Problem** that brings together concepts and problem-solving skills from throughout the chapter. Go to **www.cengage.com/chemistry/moore** to download this chapter's Summary Problem from the Student Companion Site.

KEY TERMS

acid-base reaction *(Section 15.1)*

acid ionization constant, K_a *(15.5)*

acid ionization constant expression *(15.5)*

acidic solution *(15.3)*

amines *(15.2)*

amphiprotic *(15.1)*

amphoteric *(15.9)*

autoionization *(15.3)*

base ionization constant, K_b *(15.5)*

base ionization constant expression *(15.5)*

basic solution *(15.3)*

Brønsted-Lowry acid *(15.1)*

Brønsted-Lowry base *(15.1)*

complex ion *(15.9)*

conjugate acid-base pair *(15.1)*

coordinate covalent bond *(15.9)*

hydrolysis *(15.8)*

ionization constant for water, K_w *(15.3)*

Lewis acid *(15.9)*

Lewis base *(15.9)*

monoprotic acids *(15.5)*

neutral solution *(15.3)*

oxoacids *(15.6)*

pH *(15.4)*

polyprotic acids *(15.5)*

QUESTIONS FOR REVIEW AND THOUGHT

■ denotes questions assignable in OWL.

Blue-numbered questions have short answers at the back of this book and fully worked solutions in the *Student Solutions Manual*.

Review Questions

1. Explain in your own words what 100% ionization means.
2. Write the chemical equation for the autoionization of water. Write the equilibrium constant expression for this reaction. What is the value of the equilibrium constant at 25 °C? What is this constant called?
3. When OH^- is the base in a conjugate acid-base pair, the acid is ________; when OH^- is the acid, the base is ________.
4. Dissolving ammonium bromide in water gives an acidic solution. Write a balanced equation showing how that can occur.
5. Solution A has a pH of 8 and solution B a pH of 10. Which has the greater hydronium ion concentration? How many times greater is its concentration?
6. Contrast the main ideas of the Brønsted-Lowry and Lewis acid-base concepts. Name and write the formula for a substance that behaves as a Lewis acid but not as a Brønsted-Lowry acid.

Topical Questions

The Brønsted-Lowry Concept of Acids and Bases

7. Write an equation to describe the proton transfer that occurs when each of these acids is added to water.
   (a) HBr (b) CF_3COOH
   (c) HSO_4^- (d) HNO_2
8. ■ Write an equation to describe the proton transfer that occurs when each of these acids is added to water.
   (a) HCO_3^- (b) HCl
   (c) CH_3COOH (d) HCN
9. Write an equation to describe the proton transfer that occurs when each of these acids is added to water.
   (a) HClO (b) CH_3CH_2COOH
   (c) $HSeO_3^-$ (d) HO_2^-
10. Write an equation to describe the proton transfer that occurs when each of these acids is added to water.
    (a) HIO (b) $CH_3(CH_2)_4COOH$
    (c) HOOCCOOH (d) $CH_3NH_3^+$
11. Write an equation to describe the proton transfer that occurs when each of these bases is added to water.
    (a) H^- (b) HCO_3^- (c) NO_2^-
12. ■ Write an equation to describe the proton transfer that occurs when each of these bases is added to water.
    (a) HSO_4^- (b) CH_3NH_2
    (c) I^- (d) $H_2PO_4^-$
13. Write an equation to describe the proton transfer that occurs when each of these bases is added to water.
    (a) PO_4^{3-} (b) SO_3^{2-} (c) HPO_4^{2-}
14. Write an equation to describe the proton transfer that occurs when each of these bases is added to water.
    (a) AsO_4^{3-} (b) S^{2-} (c) N_3^-
15. Write the formula and name for the conjugate partner for each acid or base.
    (a) CN^- (b) SO_4^{2-}
    (c) HS^- (d) S^{2-}
    (e) HSO_3^- (f) HCOOH (formic acid)
16. Write the formula and name for the conjugate partner for each acid or base.
    (a) HI (b) NO_3^-
    (c) CO_3^{2-} (d) H_2CO_3
    (e) HSO_4^- (f) SO_3^{2-}
17. Which are conjugate acid-base pairs?
    (a) H_2O and H_3O^+ (b) H_3O^+ and OH^-
    (c) NH_2^- and NH_4^+ (d) NH_3 and NH_4^+
    (e) O^{2-} and H_2O
18. Which are conjugate acid-base pairs?
    (a) NH_2^- and NH_4^+ (b) NH_3 and NH_2^-
    (c) H_3O^+ and H_2O (d) OH^- and O^{2-}
    (e) H_3O^+ and OH^-

19. Identify the acid and the base that are reactants in each equation; identify the conjugate base and conjugate acid on the product side of each equation.
    (a) $HI(aq) + H_2O(\ell) \longrightarrow H_3O^+(aq) + I^-(aq)$
    (b) $OH^-(aq) + NH_4^+(aq) \longrightarrow H_2O(\ell) + NH_3(aq)$
    (c) $NH_3(aq) + H_2CO_3(aq) \longrightarrow NH_4^+(aq) + HCO_3^-(aq)$
20. ■ Identify the acid and the base that are reactants in each equation; identify the conjugate base and conjugate acid on the product side of the equation.
    (a) $HS^-(aq) + H_2O(\ell) \longrightarrow H_2S(aq) + OH^-(aq)$
    (b) $S^{2-}(aq) + NH_4^+(aq) \longrightarrow NH_3(g) + HS^-(aq)$
    (c) $HCO_3^-(aq) + HSO_4^-(aq) \longrightarrow H_2CO_3(aq) + SO_4^{2-}(aq)$
    (d) $NH_3(aq) + NH_2^-(aq) \longrightarrow NH_2^-(aq) + NH_3(aq)$
21. Write stepwise equations for protonation or deprotonation of each of these polyprotic acids and bases in water.
    (a) H_2SO_3 (b) S^{2-}
    (c) $NH_3CH_2COOH^+$ (glycinium ion, a diprotic acid)
22. ■ Write stepwise equations for protonation or deprotonation of each of these polyprotic acids and bases in water.
    (a) CO_3^{2-} (b) H_3AsO_4
    (c) $NH_2CH_2COO^-$ (glycinate ion, a diprotic base)

pH Calculations

23. The pH of a popular soft drink is 3.30. What is its hydronium ion concentration? Is the drink acidic or basic?
24. ■ Milk of magnesia, $Mg(OH)_2$, has a pH of 10.5. What is the hydronium ion concentration of the solution? Is this solution acidic or basic?
25. What is the pH of a 0.0013 M solution of HNO_3? What is the pOH of this solution?
26. ■ What is the pH of a solution that is 0.025 M in NaOH? What is the pOH of this solution?
27. The pH of a $Ba(OH)_2$ solution is 10.66 at 25 °C. What is the hydroxide ion concentration of this solution? If the solution volume is 250. mL, how many grams of $Ba(OH)_2$ must have been used to make this solution?
28. A 1000.-mL solution of hydrochloric acid has a pH of 1.3. How many grams of HCl are dissolved in the solution?
29. Make these interconversions. In each case tell whether the solution is acidic or basic.

	pH	$[H_3O^+]$ (M)	$[OH^-]$ (M)
(a)	1.00	______	______
(b)	10.5	______	______
(c)	______	1.8×10^{-4}	______
(d)	______	______	2.3×10^{-5}

30. ■ Make these interconversions. In each case tell whether the solution is acidic or basic.

	pH	$[H_3O^+]$ (M)	$[OH^-]$ (M)
(a)	______	6.1×10^{-7}	______
(b)	______	______	2.2×10^{-9}
(c)	4.67	______	______
(d)	______	2.5×10^{-2}	______
(e)	9.12	______	______

Acid-Base Strengths

31. Write ionization equations and ionization constant expressions for these acids and bases.
    (a) CH_3COOH (b) HCN
    (c) SO_3^{2-} (d) PO_4^{3-}
    (e) NH_4^+ (f) H_2SO_4
32. Write ionization equations and ionization constant expressions for these acids and bases.
    (a) F^- (b) NH_3
    (c) H_2CO_3 (d) H_3PO_4
    (e) CH_3COO^- (f) S^{2-}
33. Which solution will be more acidic?
    (a) 0.10 M H_2CO_3 or 0.10 M NH_4Cl
    (b) 0.10 M HF or 0.10 M $KHSO_4$
    (c) 0.1 M $NaHCO_3$ or 0.1 M Na_2HPO_4
    (d) 0.1 M H_2S or 0.1 M HCN
34. ■ Which solution will be more basic?
    (a) 0.10 M NH_3 or 0.10 M NaF
    (b) 0.10 M K_2S or 0.10 M K_3PO_4
    (c) 0.10 M $NaNO_3$ or 0.10 M $NaCH_3COO$
    (d) 0.10 M NH_3 or 0.10 M KCN

Using K_a and K_b

35. A 0.015 M solution of cyanic acid has a pH of 2.67. What is the ionization constant, K_a, of the acid?
36. What is the K_a of butyric acid if a 0.025 M butyric acid solution has a pH of 3.21?
37. What are the equilibrium concentrations of H_3O^+, acetate ion, and acetic acid in a 0.20 M aqueous solution of acetic acid, CH_3COOH?
38. The ionization constant of a very weak acid, HA, is 4.0×10^{-9}. Calculate the equilibrium concentrations of H_3O^+, A^-, and HA in a 0.040 M solution of the acid.
39. (a) What is the pH of a 0.050 M solution of benzoic acid, C_6H_5COOH ($K_a = 6.3 \times 10^{-5}$ at 25 °C)?
    (b) What percent of the acid has ionized in this solution?
40. ■ The pH of a 0.10 M solution of propanoic acid, CH_3CH_2COOH, a weak organic acid, is measured at equilibrium and found to be 2.93 at 25 °C. Calculate the K_a of propanoic acid.
41. The weak base methylamine, CH_3NH_2, has $K_b = 5.0 \times 10^{-4}$. It reacts with water according to the equation

$$CH_3NH_2(aq) + H_2O(\ell) \rightleftharpoons CH_3NH_3^+(aq) + OH^-(aq)$$

    What is the pH of a 0.23 M methylamine solution?
42. Calculate the pH of a 0.12 M aqueous solution of the base aniline, $C_6H_5NH_2$; $K_b = 4.2 \times 10^{-10}$.
43. Amantadine, $C_{10}H_{15}NH_2$, is used in the treatment of Parkinson's disease. For amantadine: $K_a = 7.9 \times 10^{-11}$.
    (a) Write the chemical equation for the reaction of amantadine with water.
    (b) Calculate the pH of a 0.0010 M aqueous solution of amantadine at 25 °C.
44. Lactic acid, $C_3H_6O_3$, occurs in sour milk as a result of the metabolism of certain bacteria. What is the pH of a solution of 56. mg lactic acid in 250. mL water? K_a for lactic acid is 1.4×10^{-4}.

Acid-Base Reactions

45. Complete each of these reactions by filling in the blanks. Predict whether each reaction is product-favored or reactant-favored, and explain your reasoning.
    (a) ______(aq) + Br^-(aq) $\rightleftharpoons$ NH_3(aq) + HBr(aq)
    (b) CH_3COOH(aq) + CN^-(aq) $\rightleftharpoons$ ______(aq) + HCN(aq)
    (c) ______(aq) + $H_2O(\ell)$ $\rightleftharpoons$ NH_3(aq) + OH^-(aq)
46. ■ Complete each of these reactions by filling in the blanks. Predict whether each reaction is product-favored or reactant-favored, and explain your reasoning.
    (a) ______(aq) + HSO_4^-(aq) $\rightleftharpoons$ HCN(aq) + SO_4^{2-}(aq)
    (b) H_2S(aq) + $H_2O(\ell)$ $\rightleftharpoons$ H_3O^+(aq) + ______(aq)
    (c) H^-(aq) + $H_2O(\ell)$ $\rightleftharpoons$ OH^-(aq) + ______(g)
47. For each salt, predict whether an aqueous solution will have a pH less than, equal to, or greater than 7. Explain your answer.
    (a) $NaHSO_4$ (b) NH_4Br (c) $KClO_4$
48. For each salt, predict whether an aqueous solution will have a pH less than, equal to, or greater than 7. Explain your answer.
    (a) $AlCl_3$ (b) Na_2S (c) $NaNO_3$
49. For each salt, predict whether an aqueous solution will have a pH less than, equal to, or greater than 7. Explain your answer.
    (a) NaH_2PO_4 (b) NH_4NO_3 (c) $SrCl_2$
50. ■ For each salt, predict whether an aqueous solution will have a pH less than, equal to, or greater than 7. Explain your answer.
    (a) Na_2HPO_4 (b) $(NH_4)_2S$ (c) KCH_3COO

Lewis Acids and Bases

51. Which of these is a Lewis acid? A Lewis base?
    (a) NH_3 (b) $BeCl_2$ (c) BCl_3
52. ■ Which of these is a Lewis acid? A Lewis base?
    (a) O^{2-} (b) CO_2 (c) H^-
53. Identify the Lewis acid and the Lewis base in each reaction.
    (a) $H_2O(\ell)$ + SO_2(aq) $\longrightarrow$ H_2SO_3(aq)
    (b) H_3BO_3(aq) + OH^-(aq) $\longrightarrow$ $B(OH)_4^-$(aq)
    (c) Cu^{2+}(aq) + 4 NH_3(aq) $\longrightarrow$ $[Cu(NH_3)_4]^{2+}$(aq)
    (d) 2 Cl^-(aq) + $SnCl_2$(aq) $\longrightarrow$ $SnCl_4^{2-}$(aq)
54. ■ Identify the Lewis acid and the Lewis base in each reaction.
    (a) I_2(s) + I^-(aq) $\longrightarrow$ I_3^-(aq)
    (b) SO_2(g) + BF_3(g) $\longrightarrow$ O_2SBF_3(s)
    (c) Au^+(aq) + 2 CN^-(aq) $\longrightarrow$ $[Au(CN)_2]^-$(aq)
    (d) CO_2(g) + $H_2O(\ell)$ $\longrightarrow$ H_2CO_3(aq)

General Questions

55. Classify each of these as a strong acid, weak acid, strong base, weak base, amphiprotic substance, or neither acid nor base.
    (a) HCl (b) NH_4^+ (c) H_2O
    (d) CH_3COO^- (e) CH_4 (f) CO_3^{2-}
56. Classify each of these as a strong acid, weak acid, strong base, weak base, amphiprotic substance, or neither acid nor base.
    (a) CH_3COOH (b) Na_2O (c) H_2SO_4
    (d) NH_3 (e) $Ba(OH)_2$ (f) $H_2PO_4^-$
57. Several acids and their respective equilibrium constants are:

$$HF(aq) + H_2O(\ell) \rightleftharpoons H_3O^+(aq) + F^-(aq) \quad K_a = 7.2 \times 10^{-4}$$

$$HS^-(aq) + H_2O(\ell) \rightleftharpoons H_3O^+(aq) + S^{2-}(aq) \quad K_a = 8 \times 10^{-18}$$

$$CH_3COOH(aq) + H_2O(\ell) \rightleftharpoons H_3O^+(aq) + CH_3COO^-(aq) \quad K_a = 1.8 \times 10^{-5}$$

    (a) Which is the strongest acid? Which is the weakest acid?
    (b) Which acid has the weakest conjugate base?
    (c) Which acid has the strongest conjugate base?
58. State whether equal molar amounts of these would have a pH equal to 7, less than 7, or greater than 7.
    (a) A weak base and a strong acid react.
    (b) A strong base and a strong acid react.
    (c) A strong base and a weak acid react.
59. Sulfurous acid, H_2SO_3, is a weak diprotic acid (K_{a_1} = 1.2 × 10^{-2}, K_{a_2} = 6.2 × 10^{-8}). What is the pH of a 0.45 M solution of H_2SO_3? (Assume that only the first ionization is important in determining pH.)
60. Ascorbic acid (vitamin C, $C_6H_8O_6$) is a diprotic acid (K_{a_1} = 7.9 × 10^{-5}, K_{a_2} = 1.6 × 10^{-12}). What is the pH of a solution that contains 5.0 mg of the acid per mL of water? (Assume that only the first ionization is important in determining pH.)
61. (a) Solid ammonium bromide, NH_4Br, is added to water. Will the resulting solution be acidic, basic, or neutral? Use a chemical equation to explain your answer.
    (b) An aqueous ammonium bromide solution has a pH of 5.50. Calculate the ammonium bromide concentration of this solution.
62. Sodium hypochlorite, NaOCl, is used as a source of chlorine in some laundry bleaches, swimming pool disinfectants, and water treatment plants. Calculate the pH of a 0.010 M solution of NaOCl (K_b = 2.9 × 10^{-7}).

Applying Concepts

63. When a 0.1 M aqueous ammonia solution is tested with a conductivity apparatus (← p. 83), the bulb glows dimly. When a 0.1 M hydrochloric acid solution is tested, the bulb glows brightly. As water is added to each of the solutions, would you expect the bulb to glow brighter, stop glowing, or stay the same? Explain your reasoning.
64. The diagrams below are nanoscale representations of different acids.

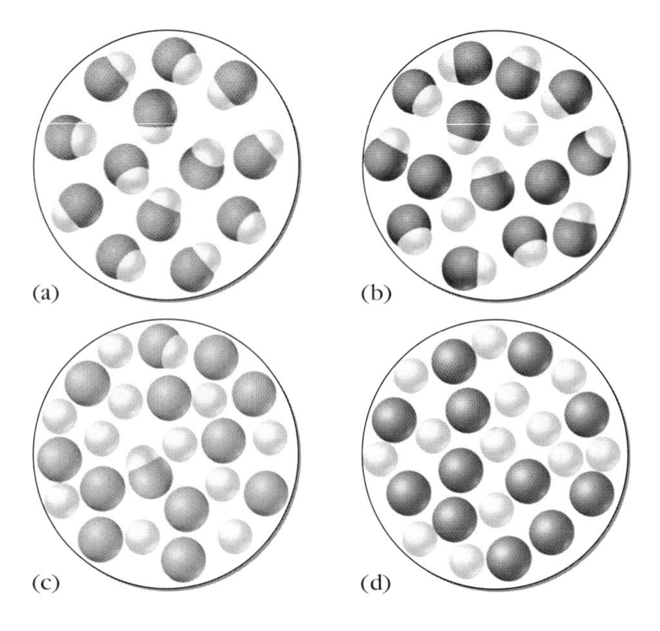

(a) Which diagram best represents hydrochloric acid? (The yellow spheres are H^+ ions and the other spheres are Cl^- ions.)

(b) Which diagram best represents acetic acid? (The yellow spheres are H^+ ions and the other spheres are CH_3COO^- ions.)

65. When asked to identify the conjugate acid-base pairs in the reaction

$$HCO_3^-(aq) + HSO_4^-(aq) \rightleftharpoons H_2CO_3(aq) + SO_4^{2-}(aq)$$

a student incorrectly wrote: "HCO_3^- is a base and HSO_4^- is its conjugate acid. H_2CO_3 is an acid and SO_4^- is its conjugate base." Write a brief explanation to the student telling why the answer is incorrect.

66. Explain how the Arrhenius acid-base theory and the Brønsted-Lowry theory of acids and bases are explained by the Lewis acid-base theory.

67. Calculate the K_b for the conjugate base of gallic acid, found in tea; $K_a = 3.9 \times 10^{-5}$. Identify a base from Table 15.2 with a K_b close to that of the conjugate base of gallic acid.

68. Trichloroacetic acid, CCl_3COOH, has an acid dissociation constant of 3.0×10^{-1} at 25 °C.
(a) Calculate the pH of a 0.0100 M trichloroacetic acid solution.
(b) How many times greater is the H_3O^+ concentration in this solution than in a 0.0100 M acetic acid?

69. Explain why a 0.1 M NH_4NO_3 aqueous solution has a pH of 5.4.

More Challenging Questions

70. A person claimed that his stomach ruptured when he took a teaspoonful of baking soda in a glass of water to relieve heartburn after a full meal ($\frac{1}{2}$ tsp = 2.5 g $NaHCO_3$). Assume that the pH of stomach acid is 1 and that the stomach has a volume of 1 L when expanded fully. Body temperature is 37 °C. What volume of carbon dioxide gas was generated by the reaction of baking soda with stomach acids? Might his stomach have ruptured from this volume of CO_2?

71. A chilled carbonated beverage is opened and warmed to room temperature. Will the pH change and, if so, how will it change?

72. For an experiment a student needs a pH = 9.0 solution. He plans to make the solution by diluting 6.0 M HCl until the H_3O^+ concentration equals 1.0×10^{-9} M.
(a) Will this plan work? Explain your answer.
(b) Will diluting 1.00 M NaOH work? Explain your answer.

73. ■ Explain why $BrNH_2$ is a weaker base than ammonia, NH_3. Which will have the smaller K_b value? Will $ClNH_2$ be a stronger or weaker base than $BrNH_2$? Explain your answer.

74. ■ Aniline has a $pK_a = 4.63$; 3-bromoaniline has a $pK_a = 3.58$; $pK_a = -\log(K_a)$. Which is the stronger base? Explain your answer.

75. It is determined that 0.1 M solutions of the sodium salts NaM, NaQ, and NaZ have pH values of 7.0, 8.0, and 9.0, respectively. Arrange the acids HM, HQ, and HZ in order of decreasing strength. Where possible, find the K_a values of these acids.

76. Use the data in Appendix J to calculate the enthalpy change for the reaction of:
(a) 100. mL of 0.100 M NaOH with 100. mL of 1.00 M HCl.
(b) 25.0 mL of 1.00 M H_2SO_4 with 75.0 mL of 1.0 M NaOH.

77. Hydrogen peroxide, HOOH, is a powerful oxidizing agent that, in diluted form, is used as an antiseptic. For HOOH: $K_a = 2.5 \times 10^{-12}$; $K_b = 4.0 \times 10^{-3}$.
(a) Is hydrogen peroxide a strong or weak acid?
(b) Is OOH^- a strong or weak base?
(c) What is the relationship between HOOH and OOH^-?
(d) Calculate the pH of 0.100 M aqueous hydrogen peroxide.

78. *para*-Bromoaniline is used in dye production. Two different pK_b values have been reported for this compound: 10.25 and 9.98; $pK_b = -\log(K_b)$. Using these values, calculate how different the pH values are for a 0.010 M solution of *para*-bromoaniline.

Br — NH_2

79. At 25 °C, a 0.1% aqueous solution of adipic acid, $C_5H_9O_2COOH$, has a pH of 3.2. A saturated solution of the acid, which contains 1.44 g acid per 100. mL of solution, has a pH = 2.7. Calculate the percent dissociation of adipic acid in each solution.

80. Niacin, a B vitamin, can act as an acid and as a base.
(a) Write chemical equations showing the action of niacin as an acid in water. Symbolize niacin as HNc.
(b) Write chemical equations showing the action of niacin as a base in water.
(c) A 0.020 M aqueous solution of niacin has a pH = 3.26. Calculate the K_a of niacin.
(d) Calculate the K_b of niacin.
(e) Write the chemical equation that accompanies the K_b expression for niacin.

81. At normal body temperature, 37 °C, $K_w = 2.5 \times 10^{-14}$.
(a) Calculate the pH of a neutral solution at this temperature.
(b) The pH of blood at this temperature ranges from 7.35 to 7.45. Assuming a pH of 7.40, calculate the H_3O^+ concentration of the blood at this temperature.
(c) By how much does the H_3O^+ concentration differ at this temperature than at 25 °C?

Conceptual Challenge Problems

CP15.A (Section 15.4) Is it possible for an aqueous solution to have a pH of 0 or even less than 0? Explain your answer mathematically as well as practically based on what you know about acid solubilities.

CP15.B (Section 15.4) What is the pH of water at 200 °C? Liquid water this hot would have to be under a pressure greater than 1.0 atm and might be found in a pressurized water reactor located in a nuclear power plant.

CP15.C (Section 15.5) Develop a set of rules by which you could predict the pH for solutions of strong or weak acids and strong or weak bases without using a calculator. Your predictions need to be accurate to ±1 pH units. Assume that you know the concentration of the acid or base and that for the weak acids and bases you can look up the pK_a ($-\log K_a$) or K_a values. What rules would work to predict pH?

16

Additional Aqueous Equilibria

George D. Lepp/Corbis

Precipitation reactions occur not only in the laboratory, but also on a massive scale in nature, as seen by the deposition of the mineral travertine (calcium carbonate) at Mammoth Hot Springs in Yellowstone National Park (USA). Calcium ions and hydrogen carbonate ions dissolved in the hot spring water react to form calcium carbonate (travertine), which precipitates from solution. In this chapter we will consider factors affecting the solubility of ionic compounds, as well as quantitative aspects of ionic compound solubilities, and acid-base reactions such as those in buffers and in acid-base titrations.

In environments as different as the interior of red blood cells, coral reefs, in ocean waters, and clouds high above the earth, important interactions occur among solutes in aqueous solution. This chapter extends the discussion of aqueous solutions begun in Chapters 14 and 15 (conjugate acid-base behavior, acid-base neutralization, the link between solubility and precipitation) to the quantitative aspects of dealing with (1) buffers, which are combinations of a weak acid and its conjugate base or a weak base and its conjugate acid; (2) acid-base titrations, which are neutralization reactions of an acid with a base; and (3) equilibria associated with solutions of slightly soluble salts.

16.1 Buffer Solutions

Adding a small amount of acid or base to pure water radically changes the pH. Consider what happens if 0.010 mol HCl is added to 1 L pure water. The pH changes from 7.00 to 2.00 because $[H_3O^+]$ changes from 1.0×10^{-7} M to 1.0×10^{-2} M. This pH change represents a *100,000-fold increase* in H_3O^+ concentration. Similarly, if 0.010 mol NaOH is added to 1 L pure water, the pH goes from 7.00 to 12.00, a *100,000-fold decrease* in $[H_3O^+]$ and a *100,000-fold increase* in $[OH^-]$. Most aquatic organisms could not survive such dramatic pH changes; the organisms can survive only within a narrow pH range. For example, if acid rain lowers the pH of a lake or stream sufficiently, fish such as trout may die.

Unlike pure water and aqueous solutions of NaOH or HCl, there are aqueous solutions that maintain a relatively constant pH when limited amounts of base or acid are added to them. Such solutions contain a **buffer**—*a chemical system that resists change in pH*—and are referred to as **buffer solutions.** For example, a solution that contains 0.50 mol acetic acid, CH_3COOH, and 0.50 mol sodium acetate, $NaCH_3COO$, in 1.0 L of solution is a buffer with a pH of 4.74. When 0.010 mol of a strong acid is added to 1.0 L of this buffer solution, the pH changes from 4.74 to 4.72, only 0.02 pH unit; adding 0.010 mol of strong base to this buffer changes the pH from 4.74 to 4.76 (Figure 16.1). These are only slight pH changes, clearly much less than those that occur when the same number of moles of HCl or NaOH are added to pure water, as described above. How do the sodium acetate-acetic acid buffer and other buffers offset such additions of acid or base without significant change in pH?

Buffer Action

To maintain a relatively constant pH, a buffer must contain *a weak acid that can react with added base,* and the buffer also must contain *a weak base that can react with added acid.* In addition, it is necessary that the acid and base components of a buffer

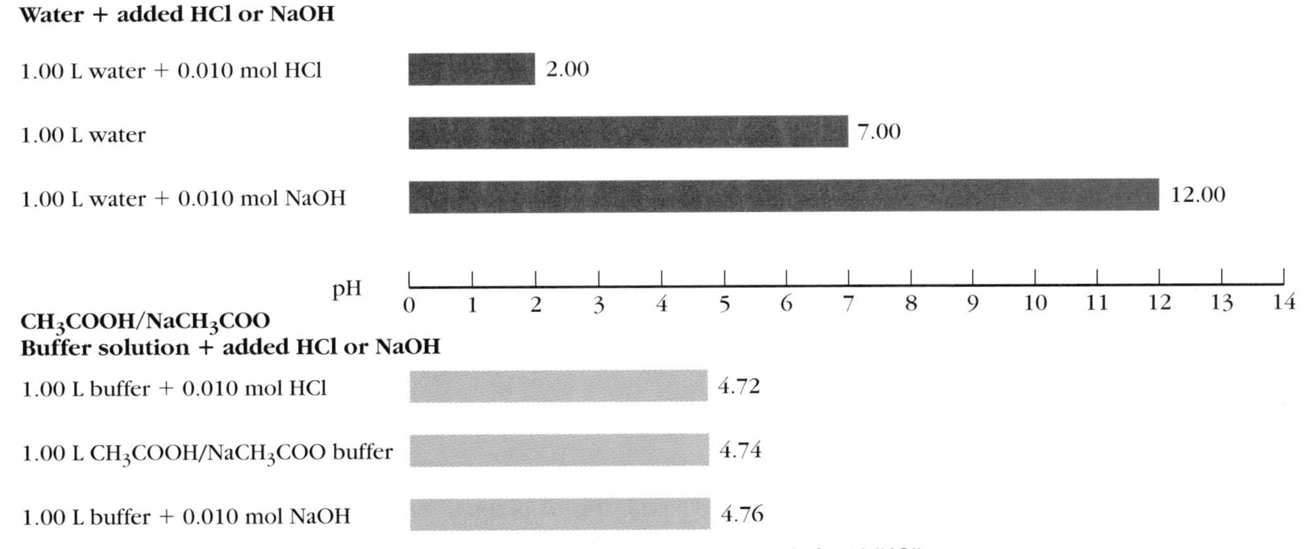

Figure 16.1 pH changes. The pH of water changes dramatically when 0.010 mol of acid (HCl) or base (NaOH) is added to it. By contrast, the pH of a sodium acetate–acetic acid buffer containing 0.50 mol acetic acid and 0.50 mol sodium acetate remains relatively constant when 0.010 mol HCl or NaOH is added to it. Only a slight change in pH occurs.

solution not react with each other. A conjugate acid-base pair, such as acetic acid and acetate ion (from sodium acetate), satisfies this requirement. In a conjugate pair, if the acid and base react with each other they just produce conjugate base and conjugate acid—no observable change occurs. For example, acetic acid reacts with acetate ion to form acetate ion and acetic acid.

$$\underset{\text{CONJ ACID}}{CH_3COOH(aq)} + \underset{\text{conj base}}{CH_3COO^-(aq)} \rightleftharpoons \underset{\text{CONJ BASE}}{CH_3COO^-(aq)} + \underset{\text{conj acid}}{CH_3COOH(aq)}$$

A buffer usually consists of approximately equal quantities of a weak acid and its conjugate base, or a weak base and its conjugate acid.

The blood of mammals is an aqueous solution that maintains a constant pH. To see how a buffer works, consider human blood, whose normal pH is 7.40 ± 0.05. If the pH decreases below 7.35, a condition known as *acidosis* occurs; increasing the pH above 7.45 causes *alkalosis.* Both conditions can be life-threatening. Acidosis, for example, causes a decrease in oxygen transport by hemoglobin and depresses the central nervous system, leading in extreme cases to coma and death by creating weak and irregular cardiac contractions—symptoms of heart failure. To prevent such problems, your body must keep the pH of your blood nearly constant.

The term "acidosis" is used in this context even though the pH is not less than 7.0.

Carbon dioxide provides the most important blood buffer (but not the only one). In solution, CO_2 reacts with water to form H_2CO_3, which ionizes to produce H_3O^+ and HCO_3^- ions. The equilibria are

$$CO_2(aq) + H_2O(\ell) \rightleftharpoons H_2CO_3(aq)$$

$$H_2CO_3(aq) + H_2O(\ell) \rightleftharpoons H_3O^+(aq) + HCO_3^-(aq)$$

Since H_2CO_3 is a weak acid and HCO_3^- is its conjugate weak base, they constitute a buffer. The normal concentrations of H_2CO_3 and HCO_3^- in blood are 0.0025 M and 0.025 M, respectively—a 1:10 ratio. As long as the ratio of H_2CO_3 to HCO_3^- concentrations remains about 1 to 10, the pH of the blood remains near 7.4. (We will calculate this pH in Problem-Solving Example 16.1.)

If a strong base such as NaOH is added to this buffer, carbonic acid in the buffer will react with the added OH^-.

$$H_2CO_3(aq) + OH^-(aq) \longrightarrow HCO_3^-(aq) + H_2O(\ell)$$

$$K = \frac{1}{K_b(HCO_3^-)} = \frac{1}{2.4 \times 10^{-8}} = 4.2 \times 10^7$$

Here the equilibrium constant K equals $1/K_b$ of hydrogen carbonate ion, because the reaction is the reverse of the reaction of hydrogen carbonate ion with H_2O. Since OH^- is the strongest base that can exist in water solution, the reaction of hydroxide ions with carbonic acid is essentially complete, as indicated by the very large K value of 4.2×10^7.

If a strong acid such as HCl is added to this buffer, the HCO_3^- ion—the conjugate base in the buffer—will react with the hydronium ions from HCl.

$$HCO_3^-(aq) + H_3O^+(aq) \longrightarrow H_2CO_3(aq) + H_2O(\ell)$$

$$K = \frac{1}{K_a(H_2CO_3)} = \frac{1}{4.2 \times 10^{-7}} = 2.4 \times 10^6$$

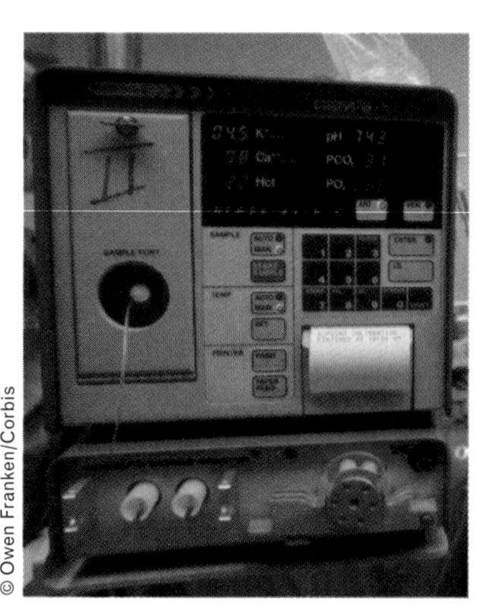

A blood gas analyzer. This instrument measures CO_2 level in blood, blood pH, and oxygen level. These values are related and must be within a narrow range for good health.

In this case, the equilibrium constant K equals $1/K_a$ of carbonic acid, because the reaction is the reverse of the ionization of carbonic acid. Since the H_3O^+ ion is such a strong acid, the reaction between HCO_3^- and H_3O^+ is essentially complete and K is large, 2.4×10^6.

In many buffers, the ratio of conjugate acid:conjugate base concentrations is about 1:1. In blood, however, the $[HCO_3^-]/[H_2CO_3]$ ratio is about 10 to 1, with good reason. There are more acidic byproducts of metabolism in the blood that must be neutralized than there are basic ones.

CONCEPTUAL
EXERCISE 16.1 Possible Buffers?

Could a solution of equimolar amounts of HCl and NaCl be a buffer? What about a solution of equimolar amounts of KOH and KCl? Explain each of your answers.

The pH of Buffer Solutions

The pH of a buffer solution can be calculated in two ways: (1) by using $[H_3O^+]$ in the K_a expression if the K_a and the concentrations of the conjugate acid and the conjugate base are known (Problem-Solving Example 16.1), and (2) by using the Henderson-Hasselbalch equation (Problem-Solving Example 16.2).

PROBLEM-SOLVING EXAMPLE 16.1 The pH of a Buffer from K_a

To mimic a blood buffer, a scientist prepared 1.000 L buffer containing 0.0025 mol carbonic acid and 0.025 mol hydrogen carbonate ion. Calculate the pH of the buffer. The K_a of carbonic acid is 4.2×10^{-7}.

Answer 7.38

Strategy and Explanation Carbonic acid (conjugate acid) and hydrogen carbonate ion (conjugate base) are a conjugate acid–conjugate base pair. To calculate the pH, we use the ionization constant equation for carbonic acid and its K_a value to find $[H_3O^+]$ and then pH.

$$H_2CO_3(aq) + H_2O(\ell) \rightleftharpoons H_3O^+(aq) + HCO_3^-(aq)$$

$$K_a = 4.2 \times 10^{-7} = \frac{[H_3O^+][\text{conj base}]}{[\text{conj acid}]} = \frac{[H_3O^+][HCO_3^-]}{[H_2CO_3]}$$

We can set up an ICE table such as we have done for other equilibrium calculations.

	H_2CO_3	+	H_2O	$\rightleftharpoons$	H_3O^+	+	HCO_3^-
*I*nitial concentration (mol/L)	0.0025				≈ 0		0.025
*C*hange as reaction occurs (mol/L)	$-x$				$+x$		$+x$
*E*quilibrium concentration (mol/L)	$0.0025 - x$				$+x$		$0.025 + x$

Given the small value of K_a, we will assume that $x \ll 0.0025$ and thus, $0.0025 - x \approx 0.0025$ and $0.025 + x \approx 0.025$. These values can be substituted into the K_a expression to calculate $[H_3O^+]$ and then pH.

$$K_a = 4.2 \times 10^{-7} = \frac{x(0.025 + x)}{(0.0025 - x)} \approx \frac{x(0.025)}{(0.0025)}$$

$$x = [H_3O^+] = \frac{(4.2 \times 10^{-7})(0.0025)}{0.025} = 4.2 \times 10^{-8}$$

$$pH = -\log(4.2 \times 10^{-8}) = 7.38$$

Note that $x = H_3O^+ = 4.2 \times 10^{-8}$ M, which is $\ll 0.0025$. Therefore, the assumption that $x \ll 0.0025$ was valid.

☑ **Reasonable Answer Check** From the K_a expression, we see that when $[HCO_3^-] = [H_2CO_3]$, the $[H_3O^+] = K_a$. But, in this example, the concentration of HCO_3^-, the conjugate base, is ten times that of H_2CO_3, the conjugate acid. Therefore, the $[H_3O^+]$ should be ten times less than the K_a, which it is.

PROBLEM-SOLVING PRACTICE 16.1

Calculate the pH of blood containing 0.0020 M carbonic acid and 0.025 M hydrogen carbonate ion.

The *Henderson-Hasselbalch equation* can be used conveniently to calculate the pH of a buffer containing known concentrations of conjugate base and conjugate acid. The equation can also be applied to determine the ratio of conjugate base to conjugate acid concentrations needed to achieve a buffer of a given pH. This equation is derived by writing the acid ionization constant expression for a weak acid, HA, and solving for $[H_3O^+]$; A^- is the conjugate base of HA.

$$HA(aq) + H_2O(\ell) \rightleftharpoons H_3O^+(aq) + A^-(aq)$$

$$K_a = \frac{[H_3O^+][\text{conj base}]}{[\text{conj acid}]} = \frac{[H_3O^+][A^-]}{[HA]}$$

$$[H_3O^+] = \frac{K_a[\text{conj acid}]}{[\text{conj base}]} = \frac{K_a[HA]}{[A^-]}$$

The next steps convert $[H_3O^+]$ to pH. Taking the base 10 logarithm of each side of this equation gives

$$\log[H_3O^+] = \log K_a + \log\frac{[\text{conj acid}]}{[\text{conj base}]} = \log K_a + \log\frac{[HA]}{[A^-]}$$

Multiplying both sides of the equation by -1 and using the relation $-\log(x) = \log(1/x)$, we get

$$-\log[H_3O^+] = -\log K_a + \log\frac{[\text{conj base}]}{[\text{conj acid}]} = -\log K_a + \log\frac{[A^-]}{[HA]}$$

Because $-\log(x) = \log\left(\frac{1}{x}\right)$, $-\log\frac{[\text{conj acid}]}{[\text{conj base}]} = \log\frac{[\text{conj base}]}{[\text{conj acid}]}$ and $-\log\frac{[HA]}{[A^-]} = \log\frac{[A^-]}{[HA]}$.

Using the definition of pH and defining $-\log K_a$ as pK_a (analogous to the definition of pH), the equation becomes the Henderson-Hasselbalch equation.

$$pH = pK_a + \log\frac{[\text{conj base}]}{[\text{conj acid}]} = pK_a + \log\frac{[A^-]}{[HA]}$$

Henderson-Hasselbalch equation

Using the Henderson-Hasselbalch equation to find the blood pH calculated in Problem-Solving Example 16.1 gives the same results.

$$pH = pK_a + \log\frac{[\text{conj base}]}{[\text{conj acid}]} = -\log K_a + \log\frac{[HCO_3^-]}{[H_2CO_3]}$$

$$= -\log(4.2 \times 10^{-7}) + \log\left(\frac{0.0250}{0.00250}\right)$$

$$= 6.38 + \log(10) = 6.38 + 1.00 = 7.38$$

Note that since the concentration of conjugate base is ten times that of the conjugate acid, the pH should be greater than the pK_a by 1 pH unit, and it is.

It is important to note that the Henderson-Hasselbalch equation has two important limits for when it can be applied to buffer solutions:

1. The value of the $\frac{[\text{conj base}]}{[\text{conj acid}]}$ ratio must be between 0.1 and 10.
2. The [conj base] and [conj acid] must *each* exceed the K_a by a factor of 100 or more.

EXERCISE 16.2 Blood pH

Calculate the pH of blood containing 0.0025 M HPO_4^{2-}(aq) and 0.0015 M $H_2PO_4^-$(aq). K_a of $H_2PO_4^- = 6.2 \times 10^{-8}$. (Assume that this is the only blood buffer.)

PROBLEM-SOLVING EXAMPLE 16.2 A Buffer Solution and Its pH

Calculate the pH of a buffer containing 0.050 mol/L pyruvic acid, $CH_3COCOOH$, and 0.060 mol/L sodium pyruvate, $Na^+CH_3COCOO^-$. K_a of pyruvic acid $= 3.2 \times 10^{-3}$.

Answer 2.57

Strategy and Explanation We will solve for the pH in two ways. First, we will use the K_a expression to find the H_3O^+ concentration from which the pH can be calculated. Second, we will use the Henderson-Hasselbalch equation to determine pH.

Using the K_a expression, we first write the chemical equation for the equilibrium of pyruvic acid with pyruvate ions, and its corresponding K_a expression.

$$CH_3COCOOH(aq) + H_2O(\ell) \rightleftharpoons H_3O^+(aq) + CH_3COCOO^-(aq)$$

$$K_a = 3.2 \times 10^{-3} = \frac{[H_3O^+][CH_3COCOO^-]}{[CH_3COCOOH]} = \frac{[H_3O^+](0.060)}{(0.050)}$$

Rearranging the equation to calculate $[H_3O^+]$ and then the pH gives

$$[H_3O^+] = \frac{3.2 \times 10^{-3}(0.050)}{0.060} = 2.67 \times 10^{-3}$$

$$pH = -\log(2.67 \times 10^{-3}) = 2.57$$

Applying the Henderson-Hasselbalch equation:

$$pH = pK_a + \log\frac{[\text{conj base}]}{[\text{conj acid}]} = -\log(3.2 \times 10^{-3}) + \log\frac{[CH_3COCOO^-]}{[CH_3COCOOH]}$$

$$pH = 2.49 + \log\left(\frac{0.060}{0.050}\right) = 2.49 + \log(1.2) = 2.49 + 0.079 = 2.57$$

☑ **Reasonable Answer Check** The concentration of conjugate base is a bit greater than the concentration of conjugate acid. Therefore, the pH should be slightly higher than the pK_a, which it is. The answer is reasonable.

If the conjugate base and conjugate acid concentrations were equal, the concentration ratio would be one and its log would be zero. In such a case, $pH = pK_a$, which is not the case here.

PROBLEM-SOLVING PRACTICE 16.2

Calculate the ratio of $[HPO_4^{2-}]$ to $[H_2PO_4^-]$ in blood at a normal pH of 7.40. Assume that this is the only buffer system present.

From the Henderson-Hasselbalch equation note that when the concentrations of conjugate base and conjugate acid are equal,

$$\frac{[\text{conj base}]}{[\text{conj acid}]} = \frac{[A^-]}{[HA]} = 1 \qquad \log\frac{[\text{conj base}]}{[\text{conj acid}]} = \log\frac{[A^-]}{[HA]} = \log(1) = 0$$

and so $pH = pK_a$. Thus, ***a buffer's pH equals the pK_a of its weak acid when the concentrations of the acid and its conjugate base in the buffer are equal.***

A buffer for maintaining a desired pH can therefore be chosen easily by examining pK_a values, which are often tabulated along with K_a values. *Choose a conjugate acid-base pair whose conjugate acid has a pK_a near the desired pH.* Table 16.1 lists pK_a values of several common acids that could be used with their conjugate bases to prepare buffers over the pH range from 4 to 10.

To have comparable quantities of both acid and conjugate base in a buffer solution, the ratio of conjugate base to conjugate acid cannot get much smaller than 1:10 or much bigger than 10:1. ***The pH range of a buffer is limited to about one pH unit above***

Notice that as K_a decreases, the pK_a increases; therefore, the weaker the acid, the larger its pK_a.

Table 16.1 Buffer Systems That Are Useful at Various pH Values*

Desired pH	Weak Conjugate Acid	Weak Conjugate Base	K_a of Weak Conjugate Acid	pK_a
4	Lactic acid, $CH_3CHOHCOOH$	Lactate ion, $CH_3CHOHCOO^-$	1.4×10^{-4}	3.85
5	Acetic acid, CH_3COOH	Acetate ion, CH_3COO^-	1.8×10^{-5}	4.74
6	Carbonic acid, H_2CO_3	Hydrogen carbonate ion, HCO_3^-	4.2×10^{-7}	6.38
7	Dihydrogen phosphate ion, $H_2PO_4^-$	Monohydrogen phosphate ion, HPO_4^{2-}	6.2×10^{-8}	7.21
8	Hypochlorous acid, HClO	Hypochlorite ion, ClO^-	3.5×10^{-8}	7.46
9	Ammonium ion, NH_4^+	Ammonia, NH_3	5.6×10^{-10}	9.25
10	Hydrogen carbonate ion, HCO_3^-	Carbonate ion, CO_3^{2-}	4.8×10^{-11}	10.32

*Adapted from Masterton, W. L., and Hurley, C. N. *Chemistry—Principles and Reactions,* 4th ed. Philadelphia: Harcourt College Publishers, 2001; p. 416.

or below the p*K*a of the conjugate acid. In the carbonic acid/hydrogen carbonate case, for example, that would be a pH from 5.38 to 7.38 because $pK_a = -\log K_a = -\log 4.2 \times 10^{-7} = 6.38$. Other conjugate acid-base pairs, such as those in Table 16.1, can be used to prepare buffers with much different pH ranges, as determined by the pK_a value of the acid in the buffer (Table 16.1).

PROBLEM-SOLVING EXAMPLE 16.3 Selecting an Acid-Base Pair for a Buffer Solution of Known pH

You are doing an experiment that requires a buffer solution with a pH of 4.00. Available to you are solutions needed to make buffers with these acid-base pairs: CH_3COOH/CH_3COO^-; H_2CO_3/HCO_3^-; and $CH_3CHOHCOOH/CH_3CHOHCOO^-$. Which acid-base pair should you use to make a pH 4.00 buffer solution, and what molar ratio of the compounds should you use? Use Table 16.1 for K_a and pK_a values.

Answer The $CH_3CHOHCOOH/CH_3CHOHCOO^-$ conjugate acid-base pair with a molar ratio of 0.10 mol/L lactic acid to 0.14 mol/L lactate would work.

Strategy and Explanation Because you want to have a buffer with a pH of 4.00, you need an acid-base pair whose conjugate acid has a pK_a near the desired pH of 4.00. You can use Table 16.1 to evaluate and select the conjugate acid-base pair whose pK_a is closest to pH = 4.00. Once that acid-base pair has been selected, you can use the Henderson-Hasselbalch equation to calculate the necessary conjugate acid and conjugate base concentrations to give a buffer with pH 4.00.

The acid with the pK_a (3.85) closest to the target pH is lactic acid, $CH_3CHOHCOOH$; the pK_a values of the other acids are not close enough to 4.00.

Substituting this pK_a and the lactic acid–lactate conjugate acid-base pair concentration terms into the Henderson-Hasselbalch equation gives

$$pH = 4.00 = 3.85 + \log \frac{[CH_3CHOHCOO^-]}{[CH_3CHOHCOOH]}$$

$$\log \frac{[CH_3CHOHCOO^-]}{[CH_3CHOHCOOH]} = 4.00 - 3.85 = 0.15; \quad \frac{[CH_3CHOHCOO^-]}{[CH_3CHOHCOOH]} = 10^{0.15} = 1.41$$

This means that the required concentration of lactate ion, $CH_3CHOHCOO^-$, will be roughly 1.41 times that of lactic acid, $CH_3CHOHCOOH$. The buffer could be made using lactic acid and a soluble lactate salt, such as sodium lactate, $Na^+CH_3CHOHCOO^-$. If the concentration of lactic acid is 0.10 mol/L, the required concentration of sodium lactate is 1.41(0.10 mol/L) = 0.14 mol/L.

☑ **Reasonable Answer Check** The target pH is 4.00. We can see from the Henderson-Hasselbalch equation that with equal concentrations of conjugate acid and base, the pH would equal the pK_a, 3.85. Therefore, to reach the target pH of 4.00, the concentration of lactate ion, the conjugate base, must be greater than that of lactic acid, the conjugate acid. Adding more conjugate base than conjugate acid raises the pH of the buffer pair from 3.85 to 4.00.

PROBLEM-SOLVING PRACTICE 16.3

Use the data in Table 16.1 to select a conjugate acid-base pair you could use to make buffer solutions having each of these hydrogen ion concentrations.

(a) 3.2×10^{-4} M (b) 5.0×10^{-5} M
(c) 7.0×10^{-8} M (d) 6.0×10^{-11} M

EXERCISE 16.3 Making a Buffer Solution

Use data from Table 16.1 to calculate the molar ratio of sodium acetate and acetic acid needed to make a buffer of pH 4.68.

EXERCISE 16.4 Buffers and pH

Use data from Table 16.1 to calculate the pH of these buffers.

(a) H_2CO_3 (0.10 M)/HCO_3^- (0.25 M) (b) $H_2PO_4^-$ (0.10 M)/HPO_4^{2-} (0.25 M)

Addition of Acid or Base to a Buffer

When acid (H_3O^+) is added to a buffer, the acid reacts with the conjugate base of the buffer to form its conjugate acid:

Conjugate base in buffer + H_3O^+ added $\longrightarrow$
conjugate acid of the buffer base + water

When base (OH^-) is added to a buffer, the base reacts with the conjugate acid of the buffer, which is converted into its conjugate base:

Conjugate acid in buffer + OH^- added $\longrightarrow$
conjugate base of the buffer acid + water

Figure 16.2 summarizes these relationships. For a buffer made from acetic acid and sodium acetate, the changes are

$$\underset{\text{Conjugate base of the buffer}}{CH_3COO^-(aq)} + \underset{\text{added}}{H_3O^+(aq)} \longrightarrow \underset{\text{Conjugate acid of the buffer}}{CH_3COOH(aq)} + H_2O(\ell)$$

$$\underset{\text{Conjugate acid of the buffer}}{CH_3COOH(aq)} + \underset{\text{added}}{OH^-(aq)} \longrightarrow \underset{\text{Conjugate base of the buffer}}{CH_3COO^-(aq)} + H_2O(\ell)$$

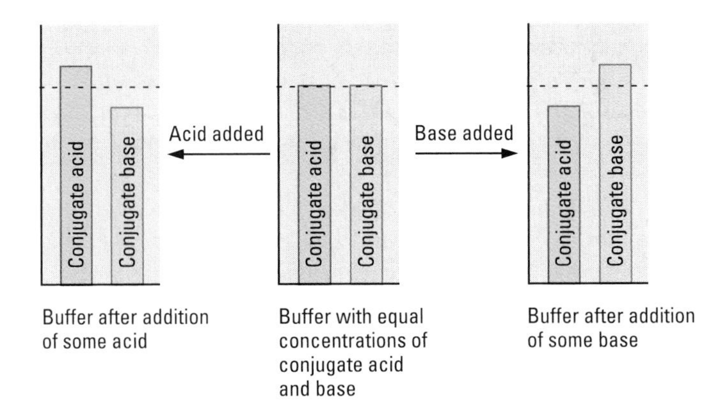

Figure 16.2 The effects of adding acid or base to a buffer. When H_3O^+ or OH^- ions are added to a buffer, the amounts of conjugate acid and conjugate base in the buffer change.

The presence of a conjugate acid together with its conjugate base can form a buffer, but not in every case; it depends on the amounts of conjugate acid and conjugate base. Consider these three examples:

(a) The combination of 0.010 mol $NaHCO_3$ with 0.010 mol Na_2CO_3 in 1.00 L solution. In this case, the conjugate acid bicarbonate ion, HCO_3^-, and the conjugate base carbonate ion, CO_3^{2-}, are in equal concentrations, so the solution is a buffer.

(b) The combination of 0.045 mol ammonia, NH_3, with 0.025 mol HCl in 1.00 L solution.

$$NH_3(aq) + HCl(q) \longrightarrow NH_4Cl(aq)$$

In this case, there are 0.045 mol of the weak base ammonia, NH_3; 0.025 mol of it reacts with 0.025 mol HCl to form 0.025 mol NH_4Cl, leaving 0.020 mol NH_3 of the original 0.045 mol unreacted. The reaction produces 0.025 mol NH_4Cl, which provides 0.025 mol ammonium ion, NH_4^+, the conjugate acid. Thus, the solution contains 0.020 mol conjugate base (NH_3) and 0.025 mol conjugate acid (NH_4^+); the solution is a buffer.

(c) The combination of 0.010 mol acetic acid, CH_3COOH, the conjugate acid, with 0.010 mol NaOH in 1.00 L solution.

$$CH_3COOH(aq) + NaOH(aq) \longrightarrow NaCH_3COO(aq) + HOH(\ell)$$

Here, 0.010 mol acetic acid reacts with 0.010 mol NaOH to form 0.010 mol sodium acetate, $NaCH_3COO$. Essentially all of the acetic acid has reacted, leaving no conjugate acid for a conjugate acid/base pair with acetate ion, CH_3COO^-. Therefore, the solution is not a buffer.

In Case (c) a small amount of acetic acid would be formed by the hydrolysis of sodium acetate, but the amount is not sufficient to act significantly as part of a buffer pair with acetate ion.

The pH Change on Addition of an Acid or a Base to a Buffer

As has been discussed, the pH of a buffer changes when acid or base is added to it due to shifts in the concentration of conjugate acid and conjugate base in the buffer. The extent of the pH change depends on the amount of acid or base added and on the amounts of conjugate acid or conjugate base remaining in the buffer to offset additional amounts of acid or base to be added. As shown in Problem-Solving Example 16.4, the change in pH of a buffer when acid or base is added to it can be calculated in two ways: (1) by using the K_a expression in conjunction with the K_a value, or (2) by using the Henderson-Hasselbalch equation.

PROBLEM-SOLVING EXAMPLE 16.4 pH Changes in a Buffer

A buffer is prepared by adding 0.15 mol lactic acid, $CH_3CHOHCOOH$, and 0.20 mol sodium lactate, $Na^+CH_3CHOHCOO^-$, to sufficient water to make 1.00 L of buffer solution. The K_a of lactic acid is 1.4×10^{-4}.

(a) Calculate the pH of the buffer.
(b) Calculate the pH of the buffer after 0.050 mol HCl has been added (neglect volume changes).
(c) Calculate the pH of the buffer after 0.10 mol NaOH has been added to the original buffer (neglect volume changes).

Answer (a) 3.97 (b) 3.73 (c) 4.63

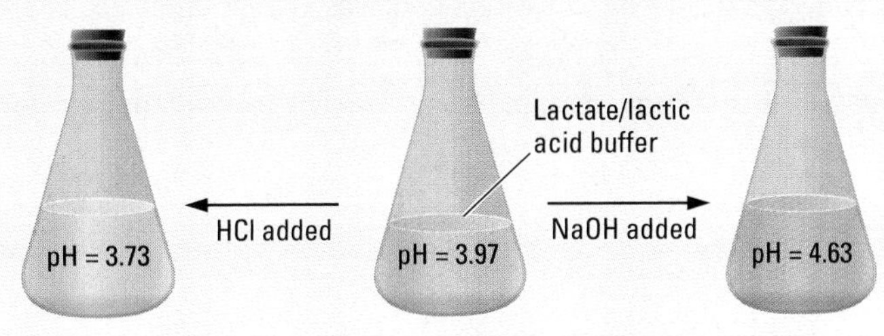

Comparison of pH of lactate/lactic acid buffer after addition of HCl or NaOH.

Strategy and Explanation These calculations can be done using the K_a expression to calculate hydronium ion concentration, from which the pH can be obtained. The Henderson-Hasselbalch equation could also be used.

(a) The chemical equation for the equilibrium and the equilibrium constant expression are

$$CH_3CHOHCOOH(aq) + H_2O(\ell) \rightleftharpoons H_3O^+(aq) + CH_3CHOHCOO^-(aq)$$

$$K_a = 1.4 \times 10^{-4} = \frac{[H_3O^+][CH_3CHOHCOO^-]}{[CH_3CHOHCOOH]}$$

Solving for $[H_3O^+]$ gives the value of 1.1×10^{-4} M and a corresponding pH of 3.97.

$$[H_3O^+] = 1.4 \times 10^{-4}\frac{[CH_3CHOHCOOH]}{[CH_3CHOHCOO^-]} = 1.4 \times 10^{-4}\frac{(0.15\text{ M})}{(0.20\text{ M})} = 1.1 \times 10^{-4}\text{ M}$$

$$pH = -\log(1.1 \times 10^{-4}) = 3.97$$

Using the Henderson-Hasselbalch equation to calculate the pH begins with finding the pK_a,

$$pK_a = -\log K_a = -\log(1.4 \times 10^{-4}) = 3.85$$

and then calculating the pH,

$$pH = 3.85 + \log\frac{(0.20\text{ mol/L lactate})}{(0.15\text{ mol/L lactic acid})} = 3.85 + \log(1.33) = 3.85 + 0.12 = 3.97$$

(b) The 0.050 mol HCl added reacts with 0.050 mol lactate ion to form 0.050 mol lactic acid and water, which adds to the 0.15 mol lactic acid originally in the buffer.

$$\underset{\text{0.050 mol from buffer}}{CH_3CHOHCOO^-(aq)} + \underset{\text{0.050 mol added}}{H_3O^+(aq)} \longrightarrow \underset{\text{0.050 mol formed in buffer}}{CH_3CHOHCOOH(aq)} + H_2O(\ell)$$

This changes the original lactate/lactic acid ratio, so the pH changes because of the added HCl:

	Moles Lactic Acid	Moles Lactate
Before reaction	0.15	0.20
After reaction	0.15 + 0.050	0.20 − 0.050

$$1.4. \times 10^{-4} = \frac{[H_3O^+][CH_3CHOHCOO^-]}{[CH_3CHOHCOOH]}$$

$$= [H_3O^+]\frac{(0.20 - 0.050\text{ mol/L})}{(0.15 + 0.050\text{ mol/L})}$$

$$= [H_3O^+](0.75\text{ M})$$

$$\begin{array}{c} \text{H} \quad \text{OH} \quad \text{O} \\ \text{H—C—C—C—OH} \\ \text{H} \quad \text{H} \end{array}$$

lactic acid

In this case, there is 1.00 L of buffer solution so that the number of moles is the same as the number of moles per liter, the molarity.

$$[H_3O^+] = \frac{1.4 \times 10^{-4}}{0.75} = 1.9 \times 10^{-4}$$

$$pH = -\log(1.9 \times 10^{-4}) = 3.73$$

The pH drops from 3.97 to 3.73. Note that 0.15 mol lactate remains to react with additional acid that might be added.

The Henderson-Hasselbalch equation could also be used to calculate the pH directly.

$$pH = 3.85 + \log \frac{[\text{lactate}]}{[\text{lactic acid}]} = 3.85 + \log \frac{(0.20 - 0.050 \text{ mol/L})}{(0.15 + 0.050 \text{ mol/L})}$$

$$= 3.85 + \log\left(\frac{0.15}{0.20}\right) = 3.85 + \log(0.75) = 3.85 + (-0.12) = 3.73$$

(c) To offset the addition of 0.10 mol NaOH requires 0.10 mol lactic acid.

$$\underset{\substack{0.10 \text{ mol} \\ \text{from buffer}}}{CH_3CHOHCOOH(aq)} + \underset{\substack{0.10 \text{ mol} \\ \text{added}}}{OH^-(aq)} \longrightarrow \underset{\substack{0.10 \text{ mol} \\ \text{formed in buffer}}}{CH_3CHOHCOO^-(aq)} + H_2O(\ell)$$

	Moles Lactic Acid	**Moles Lactate**
Before reaction	0.15	0.20
After reaction	0.15 − 0.10 = 0.05	0.20 + 0.10 = 0.30

Addition of NaOH changes the initial lactate/lactic acid ratio from $\frac{0.20}{0.15} = 1.33$ to $\frac{0.30}{0.05} = 6.0$ and a pH change occurs. We can calculate the pH by using the K_a expression to find the hydronium ion concentration, taking into account the changes in lactate and lactic acid concentrations due to the addition of 0.10 mol base.

$$K_a = 1.4 \times 10^{-4} = \frac{[H_3O^+][CH_3CHOHCOO^-]}{[CH_3CHOHCOOH]}$$

$$= \frac{[H_3O^+](0.20 + 0.10 \text{ mol/L})}{(0.15 - 0.10 \text{ mol/L})}$$

$$= [H_3O^+](6.0)$$

$$[H_3O^+] = \frac{1.4 \times 10^{-4}}{6.0} = 2.33 \times 10^{-5}; \; pH = -\log(2.33 \times 10^{-5}) = 4.63$$

The pH rises from the original value of 3.97 to 4.63. This change can also be determined using the Henderson-Hasselbalch equation.

$$pH = 3.85 + \log \frac{[\text{lactate}]}{[\text{lactic acid}]} = 3.85 + \log \frac{(0.20 + 0.10 \text{ mol/L})}{(0.15 - 0.10 \text{ mol/L})}$$

$$= 3.85 + \log\left(\frac{0.30}{0.05}\right) = 3.85 + \log(6) = 3.85 + 0.78 = 4.63$$

☑ **Reasonable Answer Check** The pH of the initial buffer (3.97) is reasonable because the concentration of conjugate base is slightly greater than that of the conjugate acid; therefore, the ratio is greater than 1 and the log of the ratio is positive, so the pH of the buffer should be slightly greater than the pK_a of the acid (3.85). When acid is added to the buffer, we expect that if the pH changes, it should decrease, and it does—from 3.97 to 3.73. The pH change that occurs when NaOH is added is greater than that for the addition of HCl because more moles of base than acid were added. But in either case, the buffer worked because the changes were less than 1 pH unit; the answers are reasonable.

PROBLEM-SOLVING PRACTICE 16.4

For the initial lactate-lactic acid buffer given above, calculate the pH when these samples are added to it; assume there is no volume change.
(a) 0.075 mol HCl (b) 0.025 mol NaOH

CONCEPTUAL **EXERCISE 16.5 Blood Buffer Reaction**

If an abnormally high CO_2 concentration is present in blood, which phosphorus-containing ion—$H_2PO_4^-$ or HPO_4^{2-}—can counteract the presence of excess CO_2? Explain your answer.

Buffer Capacity

The *amounts* of conjugate acid and conjugate base in the buffer solution determine the **buffer capacity**—the quantity of acid or base added to the buffer that the buffer can accommodate without undergoing significant pH change (more than 1 pH unit). When nearly all of the conjugate acid in a buffer has reacted with added base, adding a little more base can increase the pH significantly, because there is almost no conjugate acid left in the buffer to neutralize the added base. Similarly, if enough acid is added to a buffer to react with all of the buffer's conjugate base and excess acid remains, the pH will decrease significantly. In either case, the buffer capacity has been exceeded. For example, 1 L of a buffer solution that is 0.25 M in CH_3COOH and 0.25 M in CH_3COO^- contains 0.25 mol CH_3COOH and 0.25 mol CH_3COO^-. This buffer can accommodate the addition of up to 0.25 mol H_3O^+ or OH^-, at which point it has used up its buffer capacity. Thus, the initial buffer cannot accommodate the addition of 0.30 mol of strong acid or 0.30 mol of strong base without undergoing a major change in pH. Such additions would use up all of the buffer's conjugate base or all of its conjugate acid, respectively, and exceed the buffer's capacity. The pH would drop or rise accordingly.

When the ratio of conj base:conj acid changes to 1:10, the pH *decreases* by 1 unit. When the ratio changes to 10:1, the pH *increases* by 1 unit.

PROBLEM-SOLVING EXAMPLE 16.5 Buffer Capacity

A buffer is prepared using 0.25 mol $H_2PO_4^-$ and 0.15 mol HPO_4^{2-} in 500. mL of solution.
(a) Will the buffer capacity be exceeded if 6.2 g KOH are added to it? What will be the pH of the new solution?
(b) Will the buffer capacity be exceeded if 23.0 mL of 6.0 M HCl are added to the original buffer? What is the resulting pH?

Answer (a) No, the buffer capacity will not be exceeded; pH = 7.48. (b) Yes, the buffer capacity is exceeded; pH = 5.61.

Strategy and Explanation The initial pH of the buffer can be calculated using the Henderson-Hasselbalch equation. From Table 16.1, the pK_a of $H_2PO_4^-$ is 7.21.

$$pH = 7.21 + \log\frac{[HPO_4^{2-}]}{[H_2PO_4^-]} = 7.21 + \log\frac{(0.15/0.500)}{(0.25/0.500)}$$

$$= 7.21 + \log(0.60) = 7.21 - 0.22 = 6.99$$

(a) The 6.2 g KOH is 0.11 mol KOH, which contributes 0.11 mol OH^- to be neutralized by the buffer.

$$6.2 \text{ g KOH}\left(\frac{1 \text{ mol KOH}}{56.1 \text{ g KOH}}\right) = 0.11 \text{ mol KOH}$$

The 0.11 mol OH^- added to the buffer is neutralized by reacting with 0.11 mol $H_2PO_4^-$, the conjugate acid of the buffer, to form 0.11 mol HPO_4^{2-}.

$$\underset{\substack{0.11\text{ mol}\\\text{from buffer}}}{H_2PO_4^-(aq)} + \underset{\substack{0.11\text{ mol}\\\text{from KOH}}}{OH^-} \longrightarrow \underset{\substack{0.11\text{ mol}\\\text{formed}}}{HPO_4^{2-}(aq)} + H_2O(\ell)$$

The reaction changes the amounts of HPO_4^{2-} and $H_2PO_4^-$ remaining in the buffer:

	Before Reaction	After Reaction
Moles HPO_4^{2-}	0.15	$0.15 + 0.11 = 0.26$
Moles $H_2PO_4^-$	0.25	$0.25 - 0.11 = 0.14$

Because there is still some $H_2PO_4^-$ remaining (0.14 mol), the buffer's capacity was not exceeded by adding 6.2 g KOH. The pH after the KOH addition is

$$pH = 7.21 + \log\frac{[HPO_4^{2-}]}{[H_2PO_4^-]} = 7.21 + \log\frac{(0.26/0.500)}{(0.14/0.500)}$$
$$= 7.21 + \log(1.9) = 7.21 + 0.27 = 7.48$$

(b) Adding 0.0230 L of 6.0 M HCl provides $(0.0230\text{ L})\dfrac{6.0\text{ mol }H_3O^+}{\text{L}} = 1.4 \times 10^{-1}$ mol

H_3O^+, which reacts with 1.4×10^{-1} mol HPO_4^{2-} to form 1.4×10^{-1} mol $H_2PO_4^-$.

$$\underset{\substack{0.14\text{ mol}\\\text{from buffer}}}{HPO_4^{2-}(aq)} + \underset{\substack{0.14\text{ mol}\\\text{from HCl}}}{H_3O^+(aq)} \longrightarrow \underset{\substack{0.14\text{ mol}\\\text{formed}}}{H_2PO_4^-(aq)} + H_2O(\ell)$$

	Before Reaction	After Reaction
Moles HPO_4^{2-}	0.15	$0.15 - 0.14 = 0.01$
Moles $H_2PO_4^-$	0.25	$0.25 + 0.14 = 0.39$

Thus, the composition of the buffer changes. The number of moles of HPO_4^{2-} *decreases* to 0.01 mol; the moles of $H_2PO_4^-$ *increase* from 0.25 to 0.39. The volume of the resulting solution is 0.523 L due to the addition of 0.0230 L of 6.0 M HCl. The pH after addition of HCl is

$$pH = 7.21 + \log\frac{[HPO_4^{2-}]}{[H_2PO_4^-]} = 7.21 + \log\frac{(0.01/0.523)}{(0.39/0.523)}$$
$$= 7.21 + \log(0.026) = 7.21 - 1.6 = 5.6$$

The pH changed by 1.3 units (from 6.99 to 5.6) showing that the buffer capacity was exceeded by the addition of the HCl. The final conjugate base:conjugate acid ratio was 0.026, smaller than 1:10.

☑ **Reasonable Answer Check**

(a) Addition of base to a buffer should increase the pH to some extent depending on the amount of base added. The amount of KOH (0.11 mol) added was less than the amount of conjugate acid (0.25 mol) in the buffer. The conjugate base:conjugate acid ratio changed from 0.60 to 1.9; this change did not exceed ten and therefore, the buffer's capacity was not exceeded. The pH change should be less than 1 pH unit, which it is.

(b) The amount of acid added (0.14 mol) lowered the amount of conjugate base remaining in the buffer to 0.01 mol, while the amount of conjugate acid increased to 0.39 mol, thereby making the conjugate acid:conjugate base ratio greater than ten. Consequently, the pH rose by more than one pH unit and the buffer capacity was exceeded.

PROBLEM-SOLVING PRACTICE 16.5

Calculate the minimum mass (grams) of KOH that would have to be added to the initial buffer in Problem-Solving Example 16.5a to exceed its buffer capacity.

16.2 Acid-Base Titrations

In Section 5.8 (⇐ ***p. 168***) acid-base titrations were described as a method by which the concentration of an acid or a base could be determined. The apparatus usually used for this type of titration is shown in Figure 16.3. An acid-base titration is carried out by slowly adding a measurable amount of an aqueous solution of a base or acid whose concentration is known to a known volume of an aqueous acid or base whose concentration is to be determined. For example, a standardized base could be added from a buret to a known volume of acid whose concentration is to be determined, as in Figure 16.3. A *standard solution* (⇐ ***p. 168***), one whose concentration is known accurately, is added from the buret, a device that allows the required volume of a solution to be measured accurately. The solution in the buret is known as the **titrant.** The *equivalence point* (⇐ ***p. 168***) is reached when the stoichiometric amount of titrant has been added, the amount that exactly neutralizes the acid or base being titrated.

Detection of the Equivalence Point

A method is needed in a titration to detect the equivalence point. This can be done by using a pH meter, which electronically monitors the pH of the solution as the titration proceeds. Alternatively, the color change of an acid-base indicator can be used to detect when sufficient titrant has been added (Figure 16.4). The **end point** of a titration occurs when the indicator changes color. The goal is to use an indicator that gives an end point very close to the equivalence point.

Acid-base indicators, described in Section 15.4 (⇐ ***p. 550***), are typically weak organic acids (HIn) that differ in color from their conjugate bases (In^-).

$$\underset{\text{Color 1}}{HIn(aq)} + H_2O(\ell) \rightleftharpoons H_3O^+(aq) + \underset{\text{Color 2}}{In^-(aq)}$$

(a) (b) (c)

Figure 16.3 An acid-base titration setup for titrating an acid sample with NaOH as the titrant. (a) A buret calibrated in 0.1-mL divisions contains an NaOH solution of known concentration. (b) An acid-base indicator is added to the acid solution before the titration begins. The NaOH solution is added slowly from the buret into the acid solution to be titrated. (c) The indicator changes color when the end point is reached.

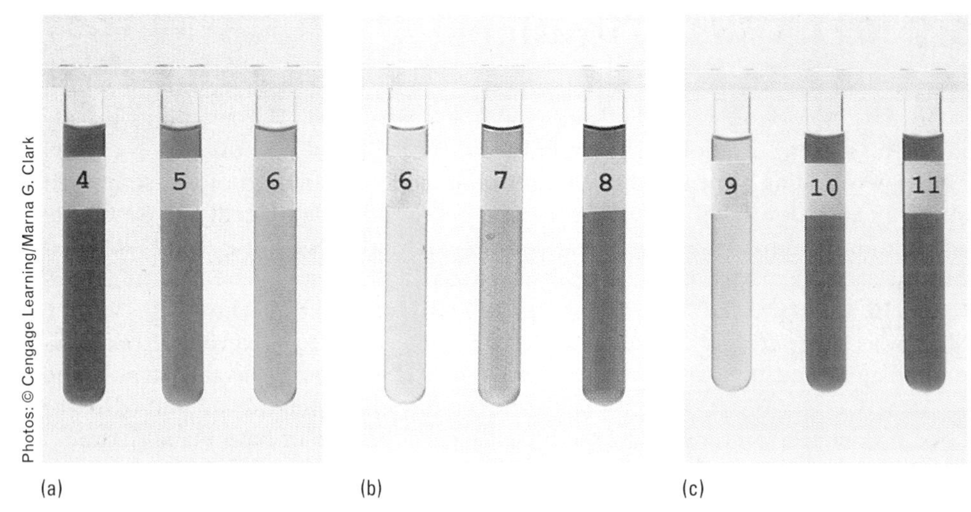

Figure 16.4 Acid-base indicators. Acid-base indicators are compounds that change color in a particular pH range. (a) Methyl red is red at pH 4 or lower, orange at pH 5, and yellow at pH 6.3 and higher. (b) Bromthymol blue changes from yellow to blue as the pH changes from 6 to 8. (c) Phenolphthalein is colorless below a pH of 8.3 and changes to red between 8.3 and 11.

Removal of an H^+ from the indicator molecule changes the structure of the indicator molecule so that it absorbs light in a different region of the visible spectrum, thus changing the color of the indicator.

The color observed for the indicator during an acid-base titration depends on the $\frac{[HIn]}{[In^-]}$ ratio, for which three cases apply:

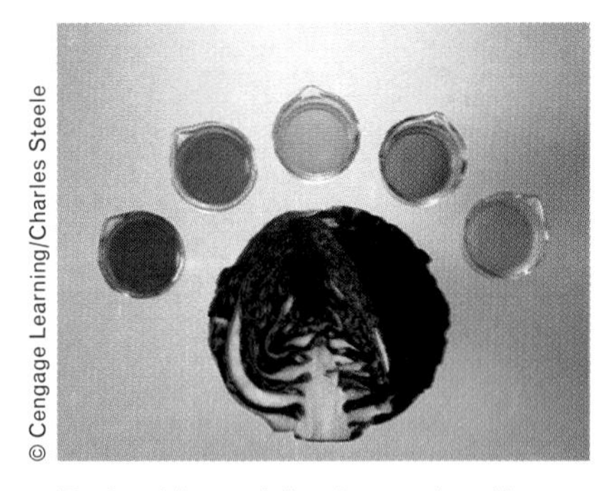

Red cabbage juice is a naturally occurring acid-base indicator. From left to right are solutions of pH 1, 4, 7, 10, and 13.

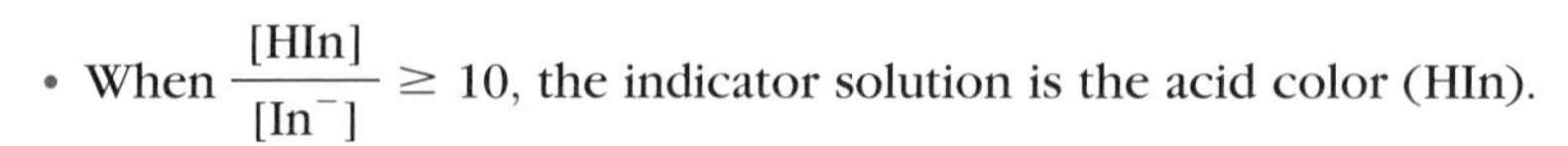

- When $\frac{[HIn]}{[In^-]} \geq 10$, the indicator solution is the acid color (HIn).
- When $\frac{[HIn]}{[In^-]} \leq 0.1$, the indicator solution is the conjugate base color (In^-).
- When $\frac{[HIn]}{[In^-]} \approx 1$, the indicator solution color is intermediate between the acid and the conjugate base colors.

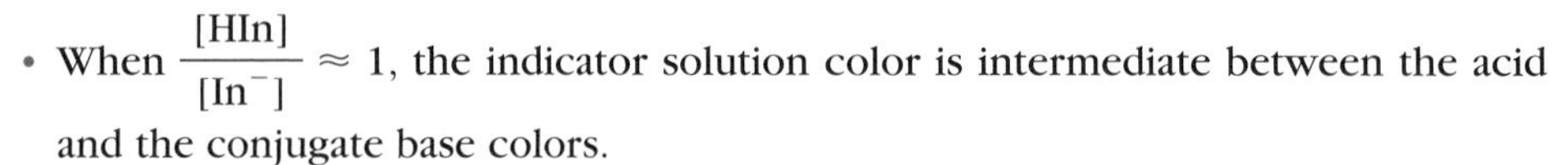

Bromthymol blue, for example, changes from yellow to blue as the pH changes from 6 to 8 (Figure 16.4b).

We will consider in some detail three types of acid-base titrations: (1) a strong acid, HCl, titrated by a strong base, NaOH; (2) a weak acid, CH_3COOH, titrated by a strong base, NaOH; and (3) a weak base, NH_3, titrated by a strong acid, HCl. For each titration we will examine its **titration curve,** a graph of pH as a function of the volume of titrant added. In each case, we will be interested in the pH particularly at four stages of the titration:

- Prior to the addition of titrant
- After addition of titrant, but prior to the equivalence point
- At the equivalence point
- After the equivalence point

Titration of a Strong Acid with a Strong Base

The titration curve for the titration of 50.0 mL of 0.100 M HCl with 0.100 M NaOH, the titrant, is given in Figure 16.5. The titration of a strong acid with a strong base produces a neutral salt with a pH = 7.0 at the equivalence point.

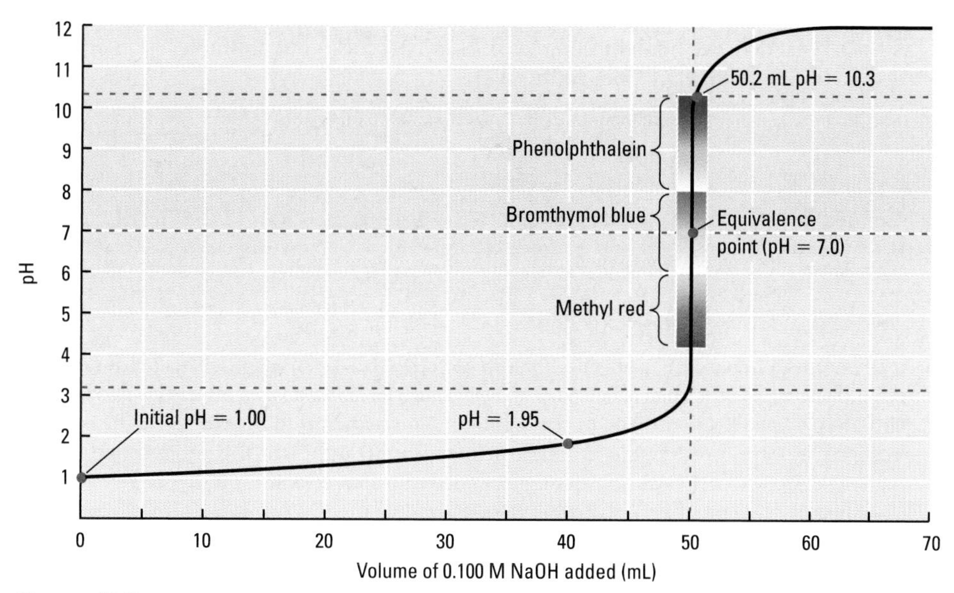

Figure 16.5 Curve for titration of 0.100 M HCl with 0.100 M NaOH. This strong acid reacts with this strong base to form a solution with a pH of 7.0 at the equivalence point.

In general, prior to the equivalence point in a strong acid-strong base titration, the $[H_3O^+]$ can be calculated for any volume of base added by the relation

$$[H_3O^+] = \frac{\text{original moles acid} - \text{total moles base added}}{\text{volume acid (L)} + \text{volume base added (L)}}$$

This equation assumes that the added volumes equal the exact volume of the mixed solutions.

Problem-Solving Example 16.6 illustrates calculations for the four points marked on the curve.

PROBLEM-SOLVING EXAMPLE 16.6 Titration of HCl with NaOH

A 0.100 M NaOH solution is used to titrate 50.0 mL of 0.100 M HCl. Calculate the pH of the solution at these four points:

(a) Before any titrant is added
(b) After 40.0 mL of titrant has been added
(c) After 50.0 mL of NaOH has been added. What indicator—methyl red, bromthymol blue, or phenolphthalein—can be used to detect the equivalence point? See Figure 16.4.
(d) After 50.2 mL of NaOH has been added

Answer

(a) 1.00 (b) 1.954
(c) 7.00; methyl red, bromthymol blue, or phenolphthalein
(d) 10.3

Strategy and Explanation Initially only hydrochloric acid is present, so the pH is dictated by its hydronium ion concentration (part a). As the titration proceeds, the H_3O^+ concentration decreases as NaOH is added. In a strong acid-strong base titration, the $[H_3O^+]$ can be calculated prior to the equivalence point for any volume of base added by the equation given previously. We will apply that equation in part (b).

(a) Because HCl is a strong acid, the initial H_3O^+ concentration is 0.100 M and the pH is $-\log(0.100) = 1.00$. (We will round the pH values to two significant figures. See Appendix A.6 for treatment of significant figures when using logarithms.)

(b) The initial 50.0-mL solution of acid contains

$$(0.0500 \text{ L})(0.100 \text{ mol/L}) = 5.00 \times 10^{-3} \text{ mol } H_3O^+$$

As NaOH is added, the number of moles of H_3O^+ decreases due to the reaction of added OH^- ions with H_3O^+ ions in the acid.

$$\underset{\text{from HCl}}{H_3O^+(aq)} + \underset{\text{from NaOH}}{OH^-(aq)} \longrightarrow H_2O(\ell)$$

After 40.0 mL of 0.100 M NaOH is added to the original 50.0 mL of 0.100 M HCl, the $[H_3O^+]$ is

$$[H_3O^+] = \frac{(5.00 \times 10^{-3}) - (4.00 \times 10^{-3})}{0.0500 \text{ L} + 0.0400 \text{ L}} = 1.11 \times 10^{-2} \text{ M}; \qquad pH = 1.954$$

(c) At the equivalence point, 50.0 mL of 0.100 M NaOH has been added. This amounts to $(0.0500 \text{ L})(0.100 \text{ mol OH}^-/\text{L}) = 5.00 \times 10^{-3}$ mol OH^-, which exactly neutralizes the 5.00×10^{-3} mol H_3O^+ initially in the solution. No residual acid or excess NaOH is present. The NaCl produced is a neutral salt, so the pH is 7.00 at the equivalence point. Because the pH rises so rapidly near the equivalence point, methyl red, bromthymol blue, or phenolphthalein can be used as the indicator in a strong acid–strong base titration (Figure 16.5).

(d) Adding 50.2 mL of 0.100 M NaOH to the solution puts 5.02×10^{-3} mol OH^- into the solution: $(0.0502 \text{ L})(0.100 \text{ mol OH}^-/\text{L}) = 5.02 \times 10^{-3}$ mol OH^-. As seen in part (c), 5.00×10^{-3} mol OH^- neutralized all of the HCl in the initial sample. Therefore, the additional 0.02×10^{-3} mol OH^-, now in 100.2 mL of solution, is not neutralized. The pH of the solution is

$$[OH^-] = \frac{0.02 \times 10^{-3} \text{ mol OH}^-}{0.1002 \text{ L}} = 2.0 \times 10^{-4} \text{ M}$$

$$pOH = -\log(2.0 \times 10^{-4}) = 3.70; \qquad pH = 14.00 - pOH = 14.00 - 3.70 = 10.3$$

Notice that the addition of just 0.2 mL of excess NaOH to the now *unbuffered* solution (part(c)) dramatically raises the pH, as seen from the titration curve (Figure 16.5).

A volume of 0.2 mL is approximately 4 drops.

PROBLEM-SOLVING PRACTICE 16.6

For the HCl-NaOH titration described above, calculate the pH when these volumes of NaOH have been added:

(a) 10.0 mL (b) 25.00 mL (c) 45.0 mL (d) 50.5 mL

CONCEPTUAL
EXERCISE 16.6 Titration Curve

Draw the titration curve for the titration of 50.0 mL of 0.100 M NaOH using 0.100 M HCl as the titrant.

Titration of a Weak Acid with a Strong Base

As noted in Section 15.8 (p. 562), the reaction of a weak acid, such as acetic acid, with a strong base, like NaOH, produces a salt—sodium acetate in this case—that has a basic anion. Like other basic anions, the acetate ions react with water to produce hydroxide ions due to the hydrolysis reaction:

$$CH_3COO^-(aq) + H_2O(\ell) \rightleftharpoons CH_3COOH(aq) + OH^-(aq)$$

As a result, when a weak acid is titrated with a strong base, the pH of the solution at the equivalence point will be greater than 7 due to hydrolysis of the basic anion formed by the titration reaction. The titration curve in Figure 16.6 for the titration of 50.0 mL of 0.100 M acetic acid with 0.100 M NaOH represents this type of titration, and Problem-Solving Example 16.7 illustrates the calculations associated with the titration curve.

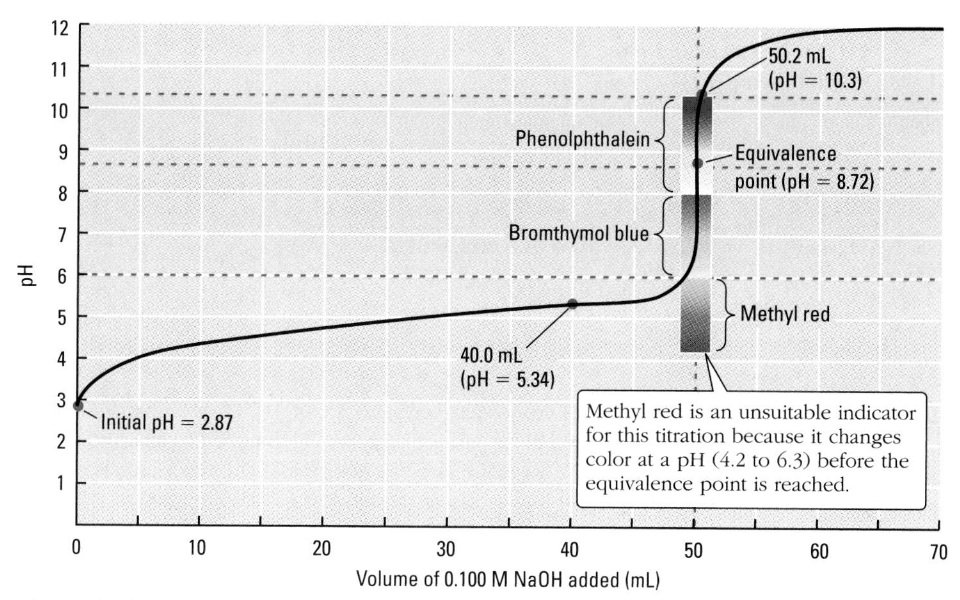

Figure 16.6 **Curve for titration of 50.0 mL of 0.100 M acetic acid with 0.100 M NaOH.**

Notice in Figure 16.6 that the initial pH of 0.100 M acetic acid (2.87) is higher than that of the 0.100 M HCl (1.00) in Figure 16.5. This is to be expected because acetic acid is a much weaker acid than HCl. Acetic acid is only slightly ionized (K_a of acetic acid = 1.8×10^{-5}), and the pH of 0.100 M acetic acid is 2.78, larger than $-\log(10^{-1}) = 1$ (*p. 550*). Also notice from Figure 16.6 that the rapidly rising portion of the titration curve near the equivalence point is shorter than it is for the NaOH-HCl titration (Figure 16.5). The equivalence point in Figure 16.6 occurs at pH = 8.72, making methyl red an unsuitable indicator because its color changes (pH 4 to 6.3) before the equivalence point is reached. Bromthymol blue or phenolphthalein can be used.

PROBLEM-SOLVING EXAMPLE 16.7 Titration of CH_3COOH with NaOH

A 0.100 M NaOH solution is used to titrate 50.0 mL of 0.100 M acetic acid. Calculate the pH of the solution at these three points:
(a) After 40.0 mL of titrant has been added
(b) After 50.0 mL of NaOH has been added
(c) After 50.2 mL of NaOH has been added

Answer (a) 5.34 (b) 8.72 (c) 10.30

Strategy and Explanation

(a) The acetic acid sample contains (0.0500 L)(0.100 mol acetic acid/L) = 5.00×10^{-3} mol acetic acid. Adding 40.0 mL of 0.100 M NaOH puts 4.00×10^{-3} mol OH^- ions into the solution, which neutralizes 4.00×10^{-3} mol acetic acid and forms 4.00×10^{-3} mol acetate ions; 1.00×10^{-3} mol acetic acid remains un-neutralized.

$$CH_3COOH(aq) + OH^-(aq) \longrightarrow CH_3COO^-(aq) + H_2O(\ell)$$

5.00×10^{-3} mol in acid soln	4.00×10^{-3} mol added	4.00×10^{-3} mol formed

The total volume of the solution is now 90.0 mL and the concentrations are

$$\text{Acetic acid} = \frac{1.00 \times 10^{-3}\text{ mol}}{0.0900\text{ L}} = 0.0111\text{ M}$$

$$\text{Acetate ion} = \frac{4.00 \times 10^{-3}\text{ mol}}{0.0900\text{ L}} = 0.0444\text{ M}$$

The pH can also be calculated using the Henderson-Hasselbalch equation.

The pH can be calculated using the K_a value and expression for acetic acid to solve for $[H_3O^+]$ and then pH.

$$K_a = \frac{[H_3O^+][CH_3COO^-]}{[CH_3COOH]} = 1.8 \times 10^{-5}$$

$$= \frac{[H_3O^+][0.00400 \text{ mol}/0.0900 \text{ L}]}{[0.00100 \text{ mol}/0.0900 \text{ L}]}$$

$$= [H_3O^+](4.00)$$

$$[H_3O^+] = \frac{1.8 \times 10^{-5}}{4.00} = 4.5 \times 10^{-6}; \qquad \text{pH} = -\log(4.5 \times 10^{-6}) = 5.34$$

(b) At this point, 5.00×10^{-3} mol OH^- has been added to 5.0×10^{-3} mol acetic acid initially present, so the stoichiometric amount of base has been added to exactly neutralize the acid in the sample. This is the equivalence point. The reaction has produced 5.00×10^{-3} mol acetate ion, whose concentration is 0.0500 M.

$$\frac{5.00 \times 10^{-3} \text{ mol acetate}}{0.100 \text{ L solution}} = 0.0500 \text{ M}$$

The pH at the equivalence point is governed by the hydrolysis of acetate ion:

$$CH_3COO^-(aq) + H_2O(\ell) \rightleftharpoons CH_3COOH(aq) + OH^-(aq)$$

We can use the K_b expression to calculate $[OH^-]$ and from it the pH. The K_b for acetate ion can be calculated from K_w and the K_a for acetic acid.

$$K_b = \frac{K_w}{K_a} = \frac{1.0 \times 10^{-14}}{1.8 \times 10^{-5}} = 5.6 \times 10^{-10}$$

Substituting into the K_b expression we let $x = [OH^-] = [CH_3COOH]$.

$$K_b = 5.6 \times 10^{-10} = \frac{[CH_3COOH][OH^-]}{[CH_3COO^-]} = \frac{x^2}{0.0500 - x}$$

Because K_b is small, we can approximate $0.0500 - x$ to be 0.0500. Solving for x,

$$K_b = 5.6 \times 10^{-10} = \frac{x^2}{0.0500}; \qquad x = 5.3 \times 10^{-6} = [OH^-]$$

which converts to a pOH of 5.28 and a pH of 8.72. The pH at this equivalence point is in marked contrast to the equivalence point pH of 7.00 for the NaOH-HCl titration, where neutral NaCl was the titration product.

(c) The pH beyond the equivalence point is controlled by the OH^- concentration from excess NaOH, which is greater than the OH^- contributed by the hydrolysis of acetate ion. Therefore, the calculation for the pH beyond the equivalence point is like that for the NaOH-HCl titration with excess NaOH.

$$\text{Final } OH^- \text{ concentration} = \frac{0.02 \times 10^{-3} \text{ mol of } OH^-}{0.1002 \text{ L}} = 2.00 \times 10^{-4} \text{ M}$$

$$\text{pOH} = -\log(2.00 \times 10^{-4}) = 3.70; \qquad \text{pH} = 14.00 - \text{pOH} = 14.00 - 3.70 = 10.30$$

PROBLEM-SOLVING PRACTICE 16.7

Calculate the pH when these volumes of 0.100 M NaOH have been added when titrating 50.0 mL of 0.100 M acetic acid:

(a) 10.0 mL (b) 25.00 mL (c) 45.0 mL (d) 51.0 mL

As seen from Figures 16.5 and 16.6 and their associated Problem-Solving Examples, there are differences in the titration curves for a strong base with a strong acid or a weak acid of equal concentration. Comparing Figures 16.5 and 16.6 we see in particular that

- Before the titration, the initial pH of the solution is higher for the weak acid.
- Very near the equivalence point, the length of the rapid rise of the curve is shorter for the weak acid.
- The pH at the equivalence point is higher for the weak acid titration.

A shorter rise in pH means that greater care is needed to select an appropriate indicator.

Polyprotic acids—those with more than one ionizable hydrogen *(p. 555)*—react stepwise when titrated with bases, one step for each ionizable hydrogen. If the K_a values of the ionizable forms of the acid are sufficiently different, the titration curve has an equivalence point for each ionizable hydrogen removed from the acid by the titration.

EXERCISE 16.7 Titration of Acetic Acid with NaOH

Use the K_a expression and value for acetic acid to calculate the pH after 30.0 mL of 0.100 M NaOH has been added to 50.0 mL of 0.100 M acetic acid.

CONCEPTUAL EXERCISE 16.8 Shape of the Titration Curve

Explain why the NaOH-acetic acid titration curve in Figure 16.6 has a relatively flat region between ~10.0 and ~40.0 mL of NaOH added.

Titration of a Weak Base with a Strong Acid

The titration of the weak base NH_3 with the strong acid HCl has the titration curve shown in Figure 16.7. The reaction produces ammonium chloride, NH_4Cl. Notice that because NH_3 is a weak base, the starting pH is greater than 7.0, but less than it would be for 0.100 M NaOH (13.00). Also notice that the pH at the equivalence point, 5.28, is less than 7.0 because ammonium chloride is an acidic salt.

The pH of 0.100 M NaOH = $-\log\left(\frac{10^{-14}}{10^{-1}}\right) = 13$

$$NH_4^+(aq) + H_2O(\ell) \rightleftharpoons NH_3(aq) + H_3O^+(aq)$$

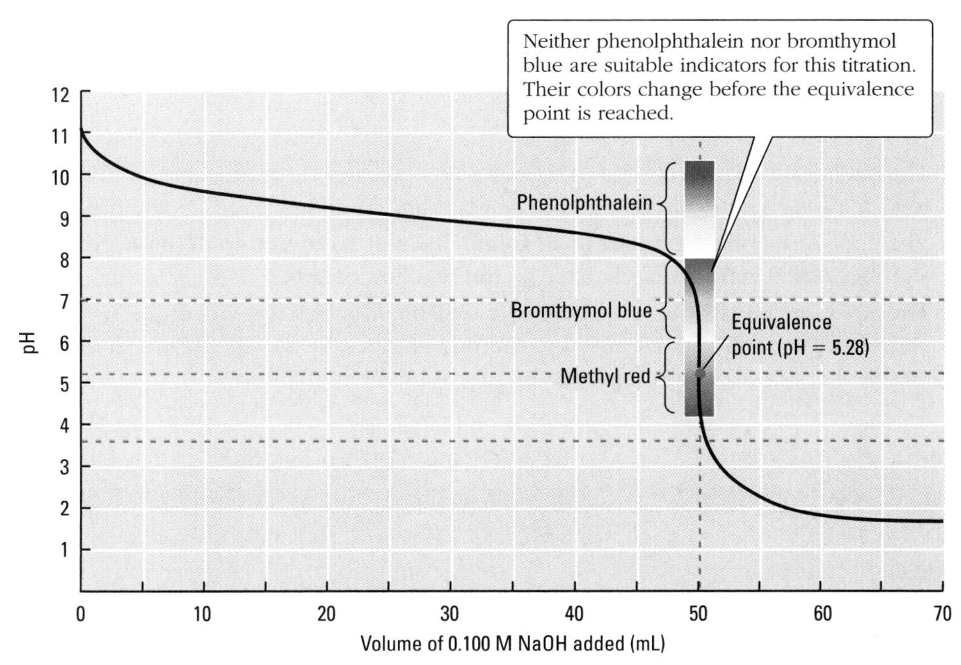

Figure 16.7 The titration of a weak base, NH_3, by a strong acid, HCl. This titration curve is for the titration of 50.0 mL of 0.100 M NH_3 with 0.100 M HCl. It is essentially the inverse of the curve for the titration of a weak acid by a strong base, Figure 16.6.

Beyond the equivalence point the pH continues to drop as excess acid is added. Although methyl red is a suitable indicator for this titration, phenolphthalein or bromthymol blue are not suitable because their color changes occur well before the equivalence point.

16.3 Solubility Equilibria and the Solubility Product Constant, K_{sp}

Many ionic compounds, such as NH_4Cl and $NaNO_3$, are very soluble in water. Many other ionic compounds, however, are only modestly or slightly water-soluble; they produce *saturated* solutions of 0.001 M or less, far less in some cases. Consider the case of a saturated aqueous solution of silver bromide, AgBr, a light-sensitive ionic compound used in photographic film. When sufficient AgBr is added to water, some of it dissolves to form a saturated solution, and some undissolved AgBr is present. The dissolved AgBr forms aqueous Ag^+ and Br^- ions in solution that are in equilibrium with the undissolved solid AgBr.

$$AgBr(s) \rightleftharpoons Ag^+(aq) + Br^-(aq)$$

The solubility product constant is commonly just called the solubility product.

This balanced chemical equation represents a solubility equilibrium. As with other equilibria *(⇐ p. 465)*, we can derive an equilibrium constant from the chemical equation. In this case, the equilibrium constant is called the **solubility product constant, K_{sp}.** The magnitude of K_{sp} indicates the extent to which the solid solute dissolves to give ions in solution (Table 16.2). To evaluate the equilibrium constant, we first must write a **solubility product constant expression.** For the chemical equation given above, the solubility product constant expression is

$$K_{sp} = [Ag^+][Br^-]$$

In general, the balanced chemical equation for dissolving a slightly soluble salt with the general formula A_xB_y is

$$A_xB_y(s) \rightleftharpoons x\,A^{n+}(aq) + y\,B^{m-}(aq)$$

This results in the general K_{sp} expression

$$K_{sp} = [A^{n+}]^x[B^{m-}]^y$$

A solubility product expression has the same general form that other equilibrium constant expressions have, except there is no denominator in the K_{sp} expression, because the reactant is always a pure solid.

Notice that

- The chemical equation related to the solubility product constant expression is written for a solid solute compound as a reactant and its aqueous ions as the products.
- The concentration of the pure solid solute reactant is omitted from the K_{sp} expression because it remains unchanged as the reaction occurs.
- The K_{sp} value equals the product of the *equilibrium* molar concentrations of the cation and the anion, each raised to the power given by the coefficient in the balanced chemical equation representing the solubility equilibrium.

PROBLEM-SOLVING EXAMPLE 16.8 Writing K_{sp} Expressions

Write the K_{sp} expression for each of these slightly soluble salts:

(a) $Fe(OH)_2$ (b) MgC_2O_4 (c) Ag_3PO_4

Answer

(a) $K_{sp} = [Fe^{2+}][OH^-]^2$ (b) $K_{sp} = [Mg^{2+}][C_2O_4^{2-}]$ (c) $K_{sp} = [Ag^+]^3[PO_4^{3-}]$

Strategy and Explanation To write the correct K_{sp} expression we first must write the balanced equation for the dissociation of the solute, and then substitute the products in it appropriately into the K_{sp} expression.

(a) The equilibrium reaction for the solubility of $Fe(OH)_2$ in water is

$$Fe(OH)_2(s) \rightleftharpoons Fe^{2+}(aq) + 2\,OH^-(aq)$$

In this example, the cation:anion relationship is 2:1; for every Fe^{2+} ion there are two OH^- ions produced, so the hydroxide ion concentration is raised to the power of 2 in the K_{sp} expression: $K_{sp} = [Fe^{2+}][OH^-]^2$.

(b) The equilibrium reaction for the solubility of MgC_2O_4 in water is

$$MgC_2O_4(s) \rightleftharpoons Mg^{2+}(aq) + C_2O_4^{2-}(aq)$$

Since the equilibrium chemical equation shows the cations and anions in a one-to-one relationship, the equilibrium constant expression is $K_{sp} = [Mg^{2+}][C_2O_4^{2-}]$.

(c) The equilibrium reaction for the solubility of Ag_3PO_4 in water is

$$Ag_3PO_4(s) \rightleftharpoons 3\,Ag^+(aq) + PO_4^{3-}(aq)$$

Since three Ag^+ ions are produced for every PO_4^{3-} ion, the K_{sp} expression is written $K_{sp} = [Ag^+]^3[PO_4^{3-}]$.

PROBLEM-SOLVING PRACTICE 16.8

Write the K_{sp} expression for each of these slightly soluble salts:

(a) CuBr (b) HgI_2 (c) $SrSO_4$

Solubility and K_{sp}

The solubility of a sparingly soluble solute and its solubility product constant, K_{sp}, are not the same thing, but they are related. The solubility is the amount of solute per unit volume of solution (mol/L) that dissolves to form a *saturated* solution. On the other hand, the solubility product constant is the equilibrium constant for the chemical equilibrium that exists between a solid ionic solute and its ions in a saturated solution. If the equilibrium concentrations of the ions are known, they can be used to calculate the K_{sp} value for the solute. For example, in a saturated AgCl solution at 10 °C, the molar concentrations of Ag^+ and Cl^- each are experimentally determined to be 6.3×10^{-6} M. This means that K_{sp} at 10 °C is

$$K_{sp} = [Ag^+][Cl^-] = [6.3 \times 10^{-6}][6.3 \times 10^{-6}] = 4.0 \times 10^{-11}$$

The K_{sp} values for selected ionic compounds at 25 °C are listed in Table 16.2. A more extensive listing is in Appendix H. In general, solutes with very low solubility have very small K_{sp} values.

Table 16.2 K_{sp} Values for Some Slightly Soluble Salts

Compound	K_{sp} at 25 °C
AgBr	3.3×10^{-13}
AuBr	5.0×10^{-17}
$AuBr_3$	4.0×10^{-36}
CuBr	5.3×10^{-9}
Hg_2Br_2*	1.3×10^{-22}
$PbBr_2$	6.3×10^{-6}
AgCl	1.8×10^{-10}
AuCl	2.0×10^{-13}
$AuCl_3$	3.2×10^{-25}
CuCl	1.9×10^{-7}
Hg_2Cl_2*	1.1×10^{-18}
$PbCl_2$	1.7×10^{-5}
AgI	1.5×10^{-13}
AuI	1.6×10^{-13}
AuI_3	1.0×10^{-46}
CuI	5.1×10^{-12}
Hg_2I_2*	4.5×10^{-29}
HgI_2	4.0×10^{-29}
PbI_2	8.7×10^{-9}
Ag_2SO_4	1.7×10^{-5}
$BaSO_4$	1.1×10^{-10}
$PbSO_4$	1.8×10^{-8}
Hg_2SO_4*	6.8×10^{-7}
$SrSO_4$	2.8×10^{-7}

*These compounds contain the diatomic ion Hg_2^{2+}.

PROBLEM-SOLVING EXAMPLE 16.9 Solubility and K_{sp}

The K_{sp} of $BaSO_4$ is 1.1×10^{-10} at 25 °C. Calculate the solubility of $BaSO_4$, expressing the result in molarity, moles per liter.

Answer 1.0×10^{-5} M

Strategy and Explanation The solubility product constant expression for barium sulfate is derived from the chemical equation

$$BaSO_4(s) \rightleftharpoons Ba^{2+}(aq) + SO_4^{2-}(aq) \qquad K_{sp} = [Ba^{2+}][SO_4^{2-}] = 1.1 \times 10^{-10}$$

The ionization of solid $BaSO_4$ forms Ba^{2+} ions and SO_4^{2-} ions in equal amounts. Therefore, if we let S equal the solubility of $BaSO_4$, then at equilibrium the concentration of Ba^{2+} and SO_4^{2-} ions will each be S.

$$1.1 \times 10^{-10} = (S)(S) = S^2$$

$$S = \sqrt{1.1 \times 10^{-10}} = 1.0 \times 10^{-5}\ \text{M}$$

Consequently, the aqueous solubility of $BaSO_4$ at 25 °C is 1.0×10^{-5} M.

PROBLEM-SOLVING PRACTICE 16.9

The K_{sp} of AgBr at 100 °C is 5×10^{-10}. Calculate the solubility of AgBr at that temperature in moles per liter.

A note of caution is in order here. It might seem perfectly straightforward to calculate the solubility of an ionic compound from its K_{sp}, the calculation just completed in Problem-Solving Example 16.9, or to do the reverse, that is, calculate the K_{sp} from the solubility. Doing so, however, will often lead to incorrect answers. This approach is too simplified and overlooks several complicating factors. One is that there are ionic solids, such as $PbCl_2$, that dissociate stepwise, so that $PbCl^+$ ions as well as Pb^{2+} and Cl^- are present in a $PbCl_2$ solution. Also, ion pairs such as $PbCl^+Cl^-$ can exist, reducing the concentrations of unassociated Pb^{2+} and Cl^-. In addition, the solubilities of some solutes, such as metal hydroxides, depend on the acidity or alkalinity of the solution. Also, solutes containing anions such as CO_3^{2-} and PO_4^{3-} that react with water are more soluble than predicted by their K_{sp} values. Solubilities calculated from K_{sp} values, and K_{sp} values calculated from solubilities, best agree with the experimentally measured solubilities of compounds with 1+ and 1− charged ions and ions that do not react with water.

PROBLEM-SOLVING EXAMPLE 16.10 Solubility and K_{sp}

A saturated aqueous solution of lead(II) iodate, $Pb(IO_3)_2$, contains 3.1×10^{-5} M Pb^{2+} at 25 °C. Assuming that the ions do not react with water, calculate the K_{sp} of lead(II) iodate at that temperature.

Answer $K_{sp} = 1.2 \times 10^{-13}$

Strategy and Explanation We must write the chemical equation for the dissociation of the solute, derive the K_{sp} expression correctly from it, and substitute concentrations of Pb^{2+} and IO_3^- ions into the expression to obtain K_{sp}. In the saturated solution, the chemical equilibrium is

$$Pb(IO_3)_2(s) \rightleftharpoons Pb^{2+}(aq) + 2\,IO_3^-(aq)$$

and the solubility product expression is $K_{sp} = [Pb^{2+}][IO_3^-]^2$.

We next determine the *molar* concentrations of Pb^{2+} and IO_3^-. From the balanced chemical equation we see that when dissociation of the solid lead(II) iodate occurs, one mole of lead ions is produced for two moles of iodate ions. Since the Pb^{2+} concentration is given as 3.1×10^{-5} M, the iodate ion concentration is 6.2×10^{-5} M, twice that of lead ion. The K_{sp} value can be calculated directly from these equilibrium concentrations and the solubility product equilibrium constant expression.

You can review molarity calculations in Section 5.7.

$$K_{sp} = [Pb^{2+}][IO_3^-]^2 = (3.1 \times 10^{-5})(6.2 \times 10^{-5})^2 = 1.2 \times 10^{-13}$$

☑ **Reasonable Answer Check** Given the fact that the iodate ion concentration is twice that of the lead ion concentration, substituting these values correctly into the proper K_{sp} expression gives the calculated answer, which is reasonable.

PROBLEM-SOLVING PRACTICE 16.10

A saturated solution of silver oxalate, $Ag_2C_2O_4$, contains 6.9×10^{-5} M $C_2O_4^{2-}$ at 25 °C. Calculate the K_{sp} of silver oxalate at that temperature, assuming that the ions do not react with water.

CONCEPTUAL
EXERCISE 16.9 Solubility and Le Chatelier's Principle

At 25 °C, 0.014 g calcium carbonate dissolves in 100 mL water. Two equilibria are present in this solution.

(a) $CaCO_3(s) \rightleftharpoons Ca^{2+}(aq) + CO_3^{2-}(aq)$

(b) $CO_3^{2-}(aq) + H_2O(\ell) \rightleftharpoons HCO_3^-(aq) + OH^-(aq)$

Suppose reaction (b) occurs to an appreciable extent. Use Le Chatelier's principle (⇐ *p. 482*) to predict how the extent of reaction (b) will affect the solubility of $CaCO_3$.

16.4 Factors Affecting Solubility

The aqueous solubility of ionic compounds is affected by a number of factors, some of which have already been mentioned—temperature (⇐ *p. 513*), the formation of ion pairs, and competing equilibria. In this section we will consider four additional factors affecting the aqueous solubility of ionic compounds:

- The effect of acids and pH
- The presence of common ions
- The formation of complex ions
- Amphoterism

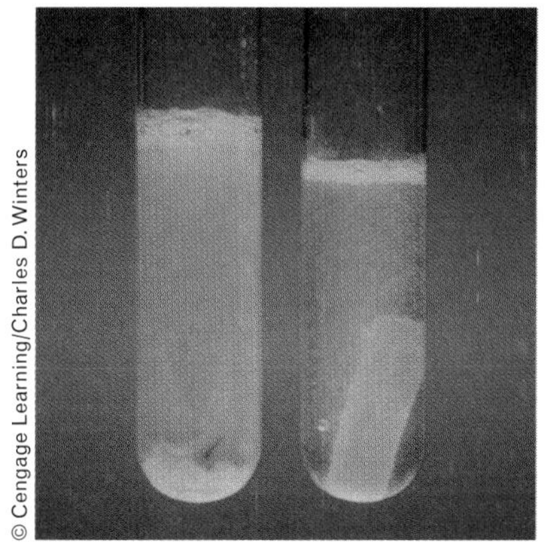
© Cengage Learning/Charles D. Winters

Reaction of calcium carbonate with acid. A piece of chalk (calcium carbonate) is dissolved by reacting it with hydrochloric acid. Bubbles of CO_2 gas can be seen being formed by the reaction.

pH and Dissolving Slightly Soluble Salts Using Acids

As noted earlier in the discussion of solubility rules, many salts are only slightly soluble in water (⇐ *p. 137*). An acid can dissolve an insoluble salt containing a moderately basic ion. As an example, consider calcium carbonate, $CaCO_3$, which is found in minerals such as limestone and marble. $CaCO_3$ is not very soluble in pure water.

$$\text{(a) } CaCO_3(s) \rightleftharpoons Ca^{2+}(aq) + CO_3^{2-}(aq) \qquad K_{sp} = 8.7 \times 10^{-9}$$

Since the solubility of calcium carbonate is so low, the equilibrium concentrations of Ca^{2+} and CO_3^{2-} must also be small. However, if acid is added to the solution, calcium carbonate will dissolve and CO_2 will be released from the solution. Adding acid adds hydronium ions, which react with carbonate and hydrogen carbonate ions.

$$\text{(b) } CO_3^{2-}(aq) + H_3O^+(aq) \rightleftharpoons HCO_3^-(aq) + H_2O(\ell)$$

$$\text{(c) } HCO_3^-(aq) + H_3O^+(aq) \rightleftharpoons H_2CO_3(aq) + H_2O(\ell)$$

$$\text{(d) } H_2CO_3(aq) \rightleftharpoons CO_2(g) + H_2O(\ell) \qquad K \approx 10^5$$

Reaction (b) is the reaction of a fairly strong weak base, CO_3^{2-}, with a strong acid, so nearly all of the carbonate is converted to hydrogen carbonate ion. Reaction (c) produces a product, carbonic acid, that is unstable and breaks down to CO_2 gas and water. Reaction (d) is a very product-favored reaction, as indicated by the large K value, $\approx 10^5$.

Reactions (a) through (d) are linked through carbonate, hydrogen carbonate, and carbonic acid. As CO_2 gas escapes from the solution, the H_2CO_3 concentration decreases. This shifts Reaction (c) to the right, which decreases the concentration of HCO_3^-. This in turn shifts Reaction (b) to the right, decreasing the concentration of CO_3^{2-} to an even lower value. To oppose this decrease in carbonate ion concentration, Reaction (a) shifts to the right, and more $CaCO_3(s)$ dissolves. Because the acidity of the solution determines the positions of equilibria (b) and (c), small changes in pH can cause limestone and marble to dissolve or precipitate (⇐ *p. 484*). Acid rain can dissolve a marble statue; it can also dissolve underground limestone deposits, creat-

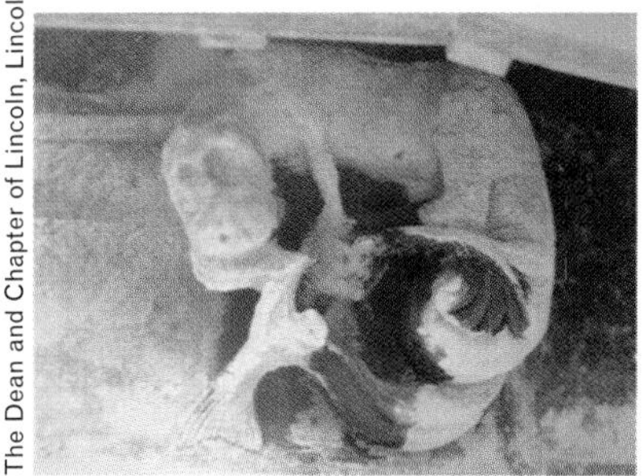
The Dean and Chapter of Lincoln, Lincoln Cathedral

Damage caused by acid rain. The erosion of limestone figures by acid rain over the years is readily apparent in these photos of limestone statuary at Lincoln Cathedral (UK). The upper photo was taken in 1910; the lower one in 1984.

ing massive cave formations. Impressive stalactite and stalagmite formations in caves result from such changes. In addition, limestone has precipitated as layers of sedimentary rock on the ocean floor where a slight increase in the pH of seawater has occurred.

In general, ***insoluble salts containing anions that are Brønsted-Lowry bases dissolve in acidic solutions (those of low pH).*** This rule covers carbonates, sulfides (which produce $H_2S(g)$), phosphates, and other anions listed as bases in Table 15.5 ***(⇐ p. 566)***. The principal exceptions to this rule are a few sulfides, such as HgS, CuS, and CdS, that have extremely low solubilities and therefore do not dissolve even when the pH is extremely low.

In contrast, an insoluble salt such as AgCl, which contains the conjugate base of a strong acid, is not soluble in strongly acidic solution because Cl^- is a very weak base and so does not react with H_3O^+.

Solubility and the Common Ion Effect

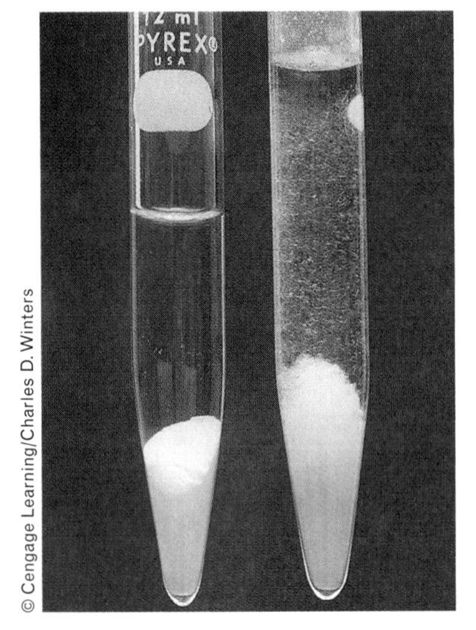

The common ion effect. The tube on the left contains a saturated solution of silver acetate, $AgCH_3COO$. When 1 M $AgNO_3$ is added to the tube, the equilibrium

$$AgCH_3COO(s) \rightleftharpoons Ag^+(aq) + CH_3COO^-(aq)$$

shifts to the left due to the addition of Ag^+ ions, as evidenced by the tube on the right, where more solid silver acetate has formed.

It is often desirable to remove a particular ion from solution by forming a precipitate of one of its insoluble compounds. For example, barium ions readily absorb X-rays and so are quite effective in making the intestinal tract visible when X-ray photographs are taken. But barium ions are poisonous and must not be allowed to dissolve in body fluids. Barium sulfate, an insoluble compound, can be ingested and used as an X-ray absorber, but both physician and patient want to be certain that no harmful amounts of barium ions will be in solution.

The solubility of $BaSO_4$ in water at 25 °C is 1.0×10^{-5} mol/L, which means that the concentration of Ba^{2+} ions is 1.0×10^{-5} M.

$$BaSO_4(s) \rightleftharpoons Ba^{2+}(aq) + SO_4^{2-}(aq)$$

The concentration of aqueous Ba^{2+} ions can be reduced by adding a soluble sulfate salt, such as Na_2SO_4. The solubility of $BaSO_4$ decreases because of the increased concentration of SO_4^{2-} ions due to the sulfate ions from Na_2SO_4. The sulfate ion is called a "common ion" because it is common to both substances dissolved in the solution—barium sulfate and sodium sulfate. The common ion shifts the preceding equilibrium to the left by what is called the **common ion effect.** *The presence of a second solute that provides a common ion lowers the solubility of an ionic compound.*

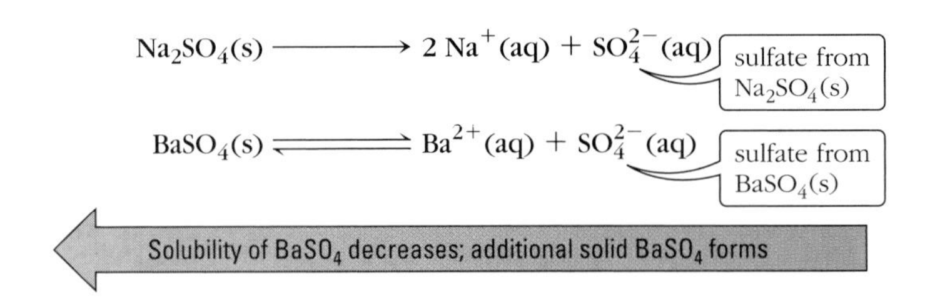

The common ion effect can be interpreted by using Le Chatelier's principle ***(⇐ p. 482)***. Adding sodium sulfate to the barium sulfate solution causes a stress on the equilibrium (additional sulfate ions), which shifts to offset the added sulfate ions. To use up some of them, the equilibrium shifts to the left, removing an equal concentration of Ba^{2+} ions from solution as well, to form solid $BaSO_4$. The outcome is that less $BaSO_4$ is dissolved, lowering its solubility in the presence of sulfate, the common ion.

CONCEPTUAL
EXERCISE 16.10 Common Ion Effect

Consider 0.0010 M solutions of these sparingly soluble solutes in equilibrum with their ions.

$$BaSO_4(s) \rightleftharpoons Ba^{2+}(aq) + SO_4^{2-}(aq)$$
$$AgI(s) \rightleftharpoons Ag^{+}(aq) + I^{-}(aq)$$
$$PbI_2(s) \rightleftharpoons Pb^{2+}(aq) + 2\,I^{-}(aq)$$

(a) Using Le Chatelier's principle, explain what would likely happen if a saturated solution of sodium iodide were added to the last two solutions.
(b) What would likely occur if a saturated solution of potassium sulfate were added to the $BaSO_4$ solution?

PROBLEM-SOLVING EXAMPLE 16.11 pH and Common Ion Effect

Manganese(II) hydroxide, $Mn(OH)_2$, is sparingly soluble in water: K_{sp} of $Mn(OH)_2$ = 4.6×10^{-14} at 25 °C. Calculate the solubility of manganese(II) hydroxide at that temperature in (a) pure water and (b) at a pH of 11.00.

Answer (a) 2.3×10^{-5} M (b) 4.6×10^{-8} M

Strategy and Explanation In part (a) we apply solubility equilibrium concepts to the equilibrium between a solute and its ions in solution in the case where no common ion is present. In part (b) where the initial solution has a pH of 11.00, there will be a significant hydroxide ion concentration before any solid $Mn(OH)_2$ has been added to the solution. After the addition, there are two possible sources of hydroxide ion, the common ion: (1) the initial solution and (2) that resulting from the dissolution of solid manganese(II) hydroxide.

(a) The chemical equilibrium is

$$Mn(OH)_2(s) \rightleftharpoons Mn^{2+}(aq) + 2\,OH^{-}(aq)$$

and the equilibrium constant expression is $K_{sp} = [Mn^{2+}][OH^{-}]^2 = 4.6 \times 10^{-14}$. Let S equal the solubility of $Mn(OH)_2$. From the solubility equation we see that the concentration of Mn^{2+} is S, that is, S moles per liter of Mn^{2+} for each mole per liter of $Mn(OH)_2$ that dissolves. Hydroxide ion concentration is $2S$ (Mn^{2+} and OH^{-} are in a 1:2 mole ratio). We can summarize the equilibrium concentrations in a table.

	$Mn(OH)_2(s) \rightleftharpoons$	$Mn^{2+}(aq)$ +	$2\,OH^{-}(aq)$
*I*nitial concentration (mol/L)		0	0*
*C*hange as reaction occurs (mol/L)		$+S$	$+2S$
*E*quilibrium concentration (mol/L)		$+S$	$+2S$

*We assume that the contribution of OH^{-} from the ionization of water is negligible.

We substitute into the K_{sp} expression

$$K_{sp} = [Mn^{2+}][OH^{-}]^2 = (S)(2S)^2 = 4.6 \times 10^{-14}$$

and solve for S, the solubility of $Mn(OH)_2$.

$$(S)(2S)^2 = 4S^3 = 4.6 \times 10^{-14}$$

$$S = [Mn^{2+}] = \sqrt[3]{4.6 \times 10^{-14}/4} = 2.3 \times 10^{-5}\ M$$

Therefore, the solubility of $Mn(OH)_2$ equals 2.3×10^{-5} M, the Mn^{2+} concentration; the OH^{-} concentration is $2S$, which is 4.6×10^{-5} M.

(b) Before any $Mn(OH)_2$ is added to the solution, the pH is 11.00. Thus, initially the $[H_3O^+] = 1.0 \times 10^{-11}$ M and the $[OH^-] = \dfrac{1.0 \times 10^{-14}}{1.0 \times 10^{-11}} = 1.0 \times 10^{-3}$ M. When $Mn(OH)_2(s)$ is added, let x equal the concentration of $Mn(OH)_2$ that dissolves, thus forming x mol/L of Mn^{2+} and $2x$ mol/L of OH^-. In the solution there are two possible sources of hydroxide ions: (1) the initial solution in which the hydroxide concentration is 1.0×10^{-3} M and (2) the hydroxide ions arising from the dissociation of dissolved $Mn(OH)_2$. Correspondingly, the OH^- concentration at equilibrium will be $1.0 \times 10^{-3} + 2x$.

	$Mn(OH)_2(s) \rightleftharpoons$	$Mn^{2+}(aq)$	+	$2\ OH^-(aq)$
*I*nitial concentration (mol/L)		0		1.0×10^{-3}
*C*hange as addition of $Mn(OH)_2$ (mol/L)		$+x$		$+2x$
*E*quilibrium concentration (mol/L)		$+x$		$1.0 \times 10^{-3} + 2x$

To calculate the solubility of $Mn(OH)_2$ we substitute these into the K_{sp} expression.

$$K_{sp} = [Mn^{2+}][OH^-]^2 = (x)(1.0 \times 10^{-3} + 2x)^2 = 4.6 \times 10^{-14}$$

The solubility of $Mn(OH)_2$ in pure water is very slight, as calculated in part (a). In addition, the dissolution will be suppressed by the presence of hydroxide ions (the common ion) in the solution at pH 11.00, so we will assume that $2x \ll 1.0 \times 10^{-3}$, which simplifies the equation to

$$(x)(1.0 \times 10^{-3})^2 \approx 4.6 \times 10^{-14}$$

$$x \approx \frac{4.6 \times 10^{-14}}{1.0 \times 10^{-6}} = 4.6 \times 10^{-8}\ \text{M}$$

Thus, the presence of hydroxide as the common ion decreased the solubility of $Mn(OH)_2$ from that in pure water, 2.3×10^{-5} M, to 4.6×10^{-8} M in a starting solution of pH 11.00.

☑ **Reasonable Answer Check** In part (a) the solubility is described by the expression $(S)(2S)^2 = 4S^3 = 4.6 \times 10^{-14}$. The answer calculated using this expression is reasonable considering how very small the K_{sp} is. We can check the approximation we made in part (b) by substituting the approximate value of x into the actual expression $K_{sp} = (x)(1.0 \times 10^{-3} + 2x)^2$. Doing so we find that the product $(x)(1.0 \times 10^{-3} + 2x)^2$ equals the given K_{sp} value, so the approximation is legitimate.

PROBLEM-SOLVING PRACTICE 16.11

Calculate the solubility of $PbCl_2$ in (a) pure water and (b) 0.20 M NaCl. K_{sp} of $PbCl_2 = 1.7 \times 10^{-5}$.

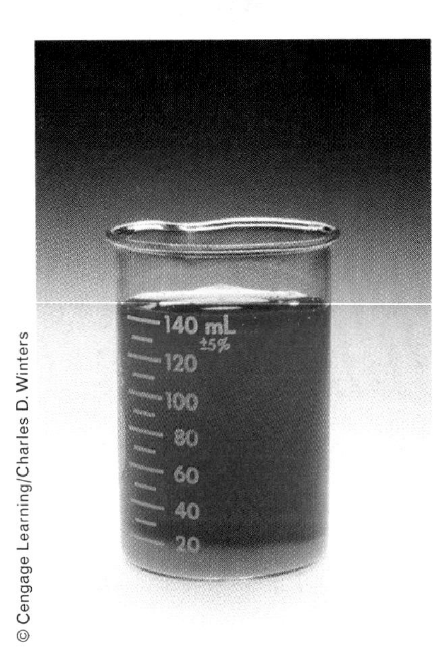

The addition of aqueous ammonia to aqueous Cu^{2+} ions forms the intense deep-blue/purple $[Cu(NH_3)_4]^{2+}$ complex ion.

Complex Ion Formation

As pointed out in Section 15.9 *(p. 567)*, all metal cations are potential Lewis acids because they can accept an electron pair donated by a Lewis base to form a complex ion. The reaction of Cu^{2+} ions with NH_3 is typical.

$$\underset{\text{Lewis acid}}{Cu^{2+}(aq)} + \underset{\text{Lewis base}}{4\ NH_3(aq)} \rightleftharpoons \underset{\text{complex ion}}{[Cu(NH_3)_4]^{2+}(aq)}$$

Many metal salts that are insoluble in water are brought into solution by complex ion formation with Lewis bases such as $S_2O_3^{2-}$, NH_3, OH^-, and CN^-. For example, the

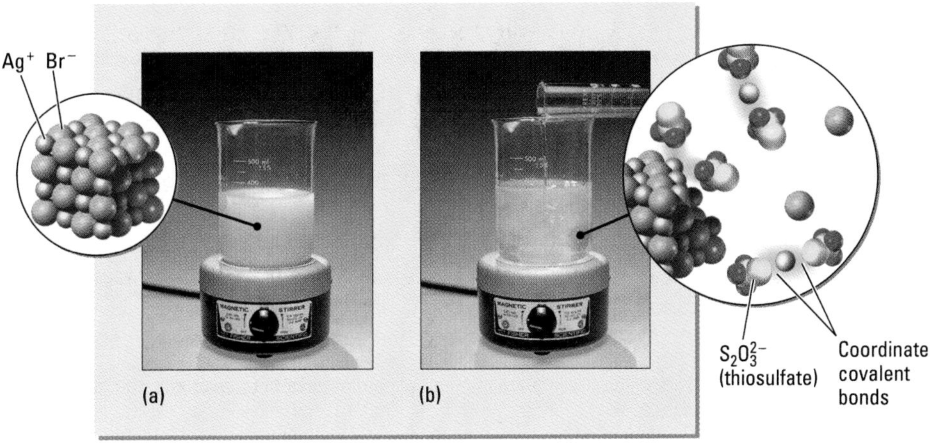

Photos: © Cengage Learning/Charles D. Winters

Active Figure 16.8 **Sodium thiosulfate dissolves silver bromide.** (a) Silver bromide (white solid) is insoluble in water. (b) When aqueous sodium thiosulfate is added, the AgBr dissolves. Water molecules have been omitted from the illustration for simplicity's sake. Visit this book's companion website at **www.cengage.com/chemistry/moore** to test your understanding of the concepts in this figure.

solubility of AgBr is very low in water, 1.35×10^{-4} g/L, equivalent to 7.19×10^{-7} M, but AgBr dissolves readily in a sodium thiosulfate ($Na_2S_2O_3$) solution due to the formation of the $[Ag(S_2O_3)_2]^{3-}$ complex ion (Figure 16.8).

$$AgBr(s) + 2\ S_2O_3^{2-}(aq) \rightleftharpoons [Ag(S_2O_3)_2]^{3-}(aq) + Br^-(aq)$$

The dissolving of AgBr in this way can be considered as the sum of two reactions—the solubility equilibrium of aqueous AgBr and the formation of the complex ion.

$$AgBr(s) \rightleftharpoons Ag^+(aq) + Br^-(aq)$$

$$Ag^+(aq) + 2\ S_2O_3^{2-}(aq) \rightleftharpoons [Ag(S_2O_3)_2]^{3-}(aq)$$

$$\textit{Net reaction}: AgBr(s) + 2\ S_2O_3^{2-}(aq) \longrightarrow [Ag(S_2O_3)_2]^{3-}(aq) + Br^-(aq)$$

This reaction is commercially important for removing unreacted AgBr from photographic film, fixing the image.

The extent to which complex ion formation occurs can be evaluated from the magnitude of the equilibrium constant for the formation of the complex ion, K_f, called the **formation constant.** For example, the formation constant for $[Ag(S_2O_3)_2]^{3-}$ is 2.0×10^{13}.

$$Ag^+(aq) + 2\ S_2O_3^{2-}(aq) \rightleftharpoons [Ag(S_2O_3)_2]^{3-}(aq)$$

$$K_f = \frac{[[Ag(S_2O_3)_2]^{3-}]}{[Ag^+][S_2O_3^{2-}]^2} = 2.0 \times 10^{13}$$

Formation constants for some metal complex ions are given in Table 16.3.

Table 16.3 Formation Constants for Some Complex Ions in Aqueous Solution

Formation Equilibrium	K_f
$Ag^+ + 2\ CN^- \rightleftharpoons [Ag(CN)_2]^-$	5.6×10^{18}
$Ag^+ + 2\ S_2O_3^{2-} \rightleftharpoons [Ag(S_2O_3)_2]^{3-}$	2.0×10^{13}
$Ag^+ + 2\ NH_3 \rightleftharpoons [Ag(NH_3)_2]^+$	1.6×10^{7}
$Al^{3+} + 4\ OH^- \rightleftharpoons [Al(OH)_4]^-$	7.7×10^{33}
$Au^+ + 2\ CN^- \rightleftharpoons [Au(CN)_2]^-$	2.0×10^{38}
$Cd^{2+} + 4\ CN^- \rightleftharpoons [Cd(CN)_4]^{2-}$	1.3×10^{17}
$Cd^{2+} + 4\ NH_3 \rightleftharpoons [Cd(NH_3)_4]^{2+}$	1.0×10^{7}
$Co^{2+} + 6\ NH_3 \rightleftharpoons [Co(NH_3)_6]^{2+}$	7.7×10^{4}
$Cu^{2+} + 4\ NH_3 \rightleftharpoons [Cu(NH_3)_4]^{2+}$	6.8×10^{12}
$Fe^{2+} + 6\ CN^- \rightleftharpoons [Fe(CN)_6]^{4-}$	7.7×10^{36}
$Hg^{2+} + 4\ Cl^- \rightleftharpoons [HgCl_4]^{2-}$	1.2×10^{15}
$Ni^{2+} + 6\ NH_3 \rightleftharpoons [Ni(NH_3)_6]^{2+}$	5.6×10^{8}
$Zn^{2+} + 4\ OH^- \rightleftharpoons [Zn(OH)_4]^{2-}$	2.9×10^{15}
$Zn^{2+} + 4\ NH_3 \rightleftharpoons [Zn(NH_3)_4]^{2+}$	2.9×10^{9}

PROBLEM-SOLVING EXAMPLE 16.12 Solubility and Complex Ion Formation

The K_{sp} of AgCl is 1.8×10^{-10}. The K_f of $[Ag(CN)_2]^-$ is 5.6×10^{18}. Use these data to show that AgCl will dissolve in aqueous NaCN.

Answer The net equilibrium constant is 1.0×10^9, which indicates that dissolving AgCl by complex ion formation is highly favored and AgCl will dissolve in NaCN.

Strategy and Explanation To answer this question requires using the solubility product constant for AgCl and the formation constant for $[Ag(CN)_2]^-$ to determine the overall equilibrium constant for the reaction of CN^- with AgCl. The magnitude of an equilibrium constant indicates whether a reaction is product-favored; a $K \gg 1$ indicates a very product-favored reaction (⇐ *p. 475*).

The net reaction for dissolving AgCl by $[Ag(CN)_2]^-$ complex ion formation is the sum of the K_{sp} and K_f equations.

$$AgCl(s) \rightleftharpoons Ag^+(aq) + Cl^-(aq) \qquad K_{sp} = 1.8 \times 10^{-10}$$

$$Ag^+(aq) + 2\,CN^-(aq) \rightleftharpoons [Ag(CN)_2]^-(aq) \qquad K_f = 5.6 \times 10^{18}$$

Net reaction: $AgCl(s) + 2\,CN^-(aq) \rightleftharpoons [Ag(CN)_2]^-(aq) + Cl^-(aq)$

Therefore, the equilibrium constant for the net reaction is the product of K_{sp} and K_f : $K_{net} = K_{sp} \times K_f = (1.8 \times 10^{-10})(5.6 \times 10^{18}) = 1.0 \times 10^9$. Because K_{net} is much greater than 1, the net reaction is product-favored, and AgCl is much more soluble in a NaCN solution than it is in water.

PROBLEM-SOLVING PRACTICE 16.12

The K_{sp} of AgCl is 1.8×10^{-10}. The K_f of $[Ag(S_2O_3)_2]^{3-}$ is 2.0×10^{13}. Use these data to show that dissolving AgCl by $[Ag(S_2O_3)_2]^{3-}$ complex ion formation is a product-favored process.

Amphoterism

The majority of metal hydroxides are insoluble in water, but many dissolve in highly acidic or basic solutions. This is because these hydroxides are *amphoteric;* that is, they can react with both H_3O^+ ions and OH^- ions (⇐ *p. 568*). Aluminum hydroxide, $Al(OH)_3$, is an example of an amphoteric hydroxide (Figure 16.9). When it reacts with acid, aluminum hydroxide dissolves by acting as a base, donating OH^- ions to react with hydronium ions from the acid to form water.

$$Al(OH)_3(s) + 3\,H_3O^+(aq) \longrightarrow Al^{3+}(aq) + 6\,H_2O(\ell)$$

In highly basic solutions, $Al(OH)_3$ is dissolved through complex ion formation. In this case, Al^{3+} ions act as a Lewis acid by accepting an electron pair from OH^- ions, a Lewis base, to form $[Al(OH)_4]^-$.

$$Al(OH)_3(s) + OH^-(aq) \longrightarrow [Al(OH)_4]^-(aq)$$

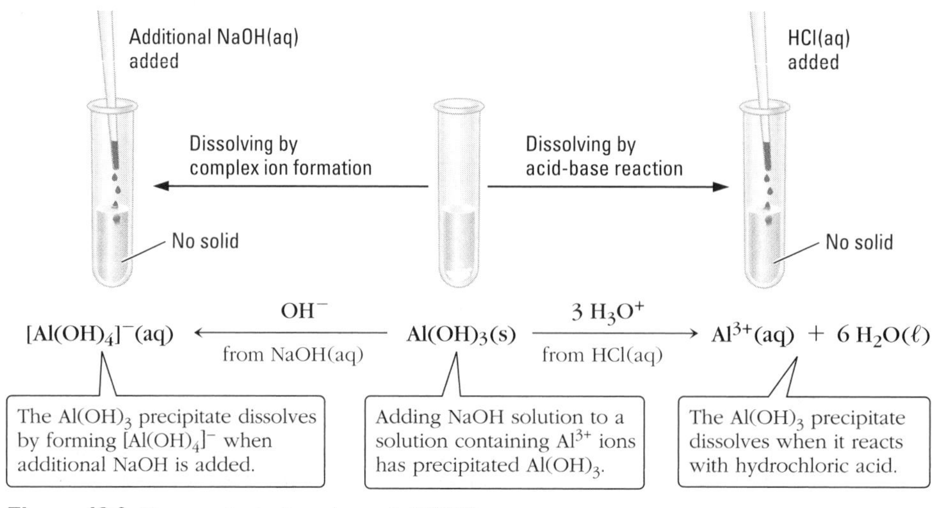

Figure 16.9 The amphoteric nature of $Al(OH)_3$.

16.5 Precipitation: Will It Occur?

Earlier, when writing net ionic equations, we used the solubility rules to predict whether a precipitate will form when ions in two solutions are mixed. Those rules apply to situations where the ions involved are at concentrations of 0.1 M or greater. If the ion concentrations are considerably less than 0.1 M, precipitation may or may not occur. The result depends on the concentrations of the ions in the resulting solution and the K_{sp} value for any precipitate that might form.

For example, AgBr might precipitate when a water-soluble silver salt, such as $AgNO_3$, is added to an aqueous solution of a bromide salt, such as KBr. The net ionic equation for the reaction is

$$Ag^+(aq) + Br^-(aq) \rightleftharpoons AgBr(s)$$

This is the reverse of the equation for K_{sp} of AgBr:

$$AgBr(s) \rightleftharpoons Ag^+(aq) + Br^-(aq)$$

To determine whether a precipitate will form, we compare the magnitude of the **ion product, *Q*,** with that of the solubility product constant, K_{sp}. The Q expression has the same form as that for K_{sp}. For Q, however, the *original* concentrations are used, not those at equilibrium as in K_{sp}. For AgBr the two expressions are

$$Q = (\text{conc. } Ag^+)(\text{conc. } Br^-) \qquad K_{sp} = [Ag^+][Br^-]$$

When the value of Q is compared with that of the K_{sp}, three cases are possible (Figure 16.10).

1. ***$Q < K_{sp}$* *The solution is unsaturated and no precipitate forms.*** In this case, the solution contains ions at a concentration lower than required for equilibrium with the solid. An equilibrium is not established between a solid solute and its ions because no solid solute is present; more solute can be added to the solution before precipitation occurs. If solid were present, more solid would dissolve.
2. ***$Q > K_{sp}$* *The solution contains a higher concentration of ions than it can hold at equilibrium; that is, the solution is supersaturated.*** To reach equilibrium, a precipitate forms, decreasing the concentration of ions until the ion product equals the K_{sp}.
3. ***$Q = K_{sp}$* *The solution is saturated with ions and is at equilibrium and at the point of precipitation.***

Precipitation of lead iodide. Adding a drop of aqueous potassium iodide solution to an aqueous lead(II) nitrate solution precipitates yellow lead(II) iodide. Potassium nitrate, a soluble salt, remains dissolved in the solution.

Q is the reaction quotient introduced in Chapter 13 (p. 478).

The bracket notation, [ ], represents molarity at *equilibrium*.

Consider the case of two solutions, each made by combining $Pb(NO_3)_2$ and Na_2SO_4 solutions. In one case (solution 1) when the solutions are mixed the initial concentrations of Pb^{2+} and SO_4^{2-} are each 1.0×10^{-4} M. In the other case (solution 2) these concentrations are each 2.0×10^{-4}, twice that of the first solution. In each of the two solutions, the products are $NaNO_3$ and $PbSO_4$. The solubility rules indicate that $NaNO_3$ is soluble and remains in solution as Na^+ and NO_3^- ions, whereas $PbSO_4$ is insoluble. Will a precipitate of $PbSO_4$ form in either or both solutions? We can determine this by using Q and K_{sp} for $PbSO_4$; K_{sp} of $PbSO_4 = 1.8 \times 10^{-8}$.

$$\begin{aligned} Q \text{ of solution 1} &= (\text{conc. } Pb^{2+})(\text{conc. } SO_4^{2-}) \\ &= (1.0 \times 10^{-4}\text{ M})(1.0 \times 10^{-4}\text{ M}) = 1.0 \times 10^{-8} < K_{sp} = 1.8 \times 10^{-8} \end{aligned}$$

$$\begin{aligned} Q \text{ of solution 2} &= (\text{conc. } Pb^{2+})(\text{conc. } SO_4^{2-}) \\ &= (2.0 \times 10^{-4}\text{ M})(2.0 \times 10^{-4}\text{ M}) = 4.0 \times 10^{-8} > K_{sp} = 1.8 \times 10^{-8} \end{aligned}$$

Since Q of solution 1 is less than the K_{sp}, no precipitate will form. In contrast, Q of solution 2 exceeds the K_{sp}, and precipitation will occur.

If their solubilities are sufficiently different, ionic compounds can be precipitated selectively from solution. The more soluble compound remains in solution as the less soluble one starts to precipitate. For example, silver chloride, AgCl, and silver

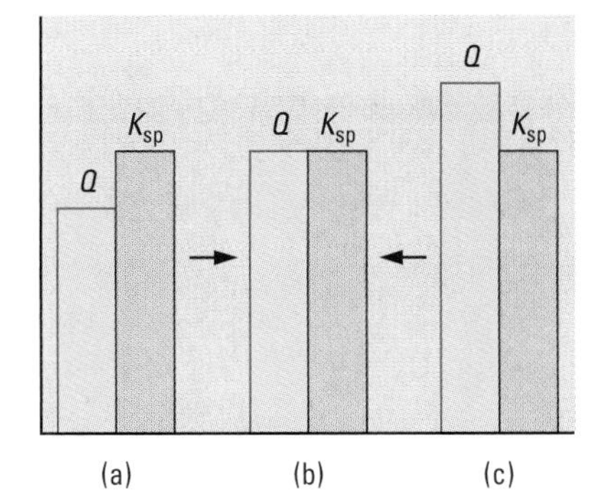

Figure 16.10 Predicting precipitation. (a) When $Q < K_{sp}$, the solution is unsaturated, and no precipitation occurs. (b) $Q = K_{sp}$: the solution is saturated and just at the point of precipitation. (c) $Q > K_{sp}$: the solution is supersaturated, and precipitation occurs until $Q = K_{sp}$.

chromate, Ag_2CrO_4, are each only slightly soluble in water. The solubilities of these two solutes differ enough, however, that when they are both in the same solution, one can be precipitated from solution, leaving the other behind in solution.

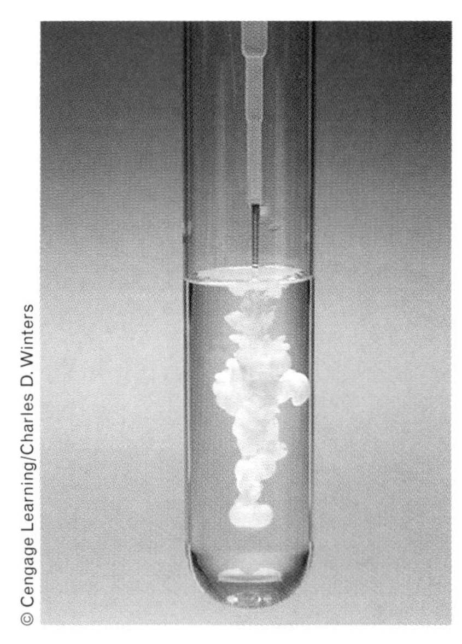

Precipitation of lead sulfate. The addition of sufficiently high concentrations of Pb^{2+} and SO_4^{2-} ions causes $Q > K_{sp}$ and lead sulfate ($PbSO_4$) precipitates.

PROBLEM-SOLVING EXAMPLE 16.13 Selective Precipitation

Consider a solution containing 0.020 M Cl^- and 0.010 M CrO_4^{2-} ions to which Ag^+ ions are added slowly. Which precipitate forms first—AgCl or Ag_2CrO_4? K_{sp} AgCl = 1.8×10^{-10}; K_{sp} Ag_2CrO_4 = 9×10^{-12}.

Answer AgCl precipitates first.

Strategy and Explanation To answer this question we first find the minimum Ag^+ concentration required to precipitate each compound, which is the molar concentration product of its ions that just barely exceeds the K_{sp}. To precipitate AgCl,

$$K_{sp} \text{ of AgCl} = [Ag^+][Cl^-] = 1.8 \times 10^{-10}$$

$$[Ag^+] = \frac{1.8 \times 10^{-10}}{[Cl^-]} = \frac{1.8 \times 10^{-10}}{2.0 \times 10^{-2}} = 9.0 \times 10^{-9}\text{ M}$$

An Ag^+ concentration of slightly greater than 9.0×10^{-9} M will precipitate some AgCl from the solution. To precipitate Ag_2CrO_4,

$$K_{sp} \text{ of } Ag_2CrO_4 = [Ag^+]^2[CrO_4^-] = 9 \times 10^{-12}$$

$$[Ag^+]^2 = \frac{9 \times 10^{-12}}{[CrO_4^{2-}]} = \frac{9 \times 10^{-12}}{1.0 \times 10^{-2}} = 9 \times 10^{-10}\text{ M}; \qquad [Ag^+] = 3 \times 10^{-5}\text{ M}$$

Silver chromate will precipitate when the Ag^+ concentration slightly exceeds 3×10^{-5} M. Because a *much* smaller concentration of Ag^+, 9.0×10^{-9} M, is required to precipitate AgCl, it will precipitate before Ag_2CrO_4.

When the Ag^+ concentration exceeds 3×10^{-5} M, Q = (conc. Ag^+) × (conc. CrO_4^{2-}) > K_{sp} Ag_2CrO_4, and Ag_2CrO_4 will precipitate.

☑ **Reasonable Answer Check** The Ag^+ concentration required to precipitate Ag_2CrO_4 is approximately 10,000 times greater than that for AgCl (3×10^{-5} M versus 9×10^{-9} M). Therefore, the answer is reasonable; AgCl will precipitate first. In fact, the difference is so great that essentially all of the AgCl will precipitate before Ag_2CrO_4 precipitation begins.

PROBLEM-SOLVING PRACTICE 16.13

Hydrochloric acid is slowly added to a solution that is 0.10 M in Pb^{2+} and 0.01 M in Ag^+. Which precipitate is formed first, AgCl or $PbCl_2$?

Kidney stones. These kidney stones were surgically removed from a patient.

Kidney Stones—Common Ion Effect and Le Chatelier's Principle

Certain combinations of ions passing through the kidneys can precipitate as kidney stones, which can become large enough to be extremely painful or even life-threatening. Kidney stones usually consist of insoluble calcium and magnesium compounds, or a mixture of them. For example, the equilibrium for calcium oxalate is $CaC_2O_4(s) \rightleftharpoons Ca^{2+}(aq) + C_2O_4^{2-}(aq)$. High intake of foods rich in calcium or oxalate ions can cause a rise in the urinary concentration of either ion (or both) sufficient to make $Q > K_{sp}$ so that the equilibrium shifts to the left, precipitating calcium oxalate as a kidney stone. Thus, foods rich in Ca^{2+} such as dairy products, or those high in $C_2O_4^{2-}$, such as chocolate or spinach, can trigger kidney stone formation through the common ion effect.

Leaving chocolate out of a diet seems far more punishment than forgoing spinach.

IN CLOSING

Having studied this chapter, you should be able to . . .

- Explain how buffers maintain pH, how to calculate their pH, how they are prepared, and the importance of buffer capacity (Section 16.1). ■ End-of-chapter Questions assignable in OWL: 14, 23
- Use the Henderson-Hasselbalch equation or the K_a expression to calculate the pH of a buffer and the pH change after acid or base has been added to the buffer (Section 16.1). ■ Questions 21, 25
- Interpret acid-base titration curves and calculate the pH of the solution at various stages of the titration (Section 16.2). ■ Questions 28, 30, 72
- Relate a K_{sp} expression to its chemical equation (Section 16.3). ■ Question 41
- Use the solubility of a slightly soluble solute to calculate its solubility product (Section 16.3).
- Describe the factors affecting the aqueous solubility of ionic compounds (Section 16.4). ■ Questions 51, 53
- Apply Le Chatelier's principle to the common ion effect (Section 16.4).
- Use the solubility product to calculate the solubility of a sparingly soluble solute in pure water and in the presence of a common ion (Section 16.4). ■ Question 46
- Describe the effect of complex ion formation on the solubility of a sparingly soluble ionic compound (Section 16.4).
- Relate Q, the ion product, to K_{sp} to determine whether precipitation will occur (Section 16.5).
- Predict which of two sparingly soluble ionic solutes will precipitate first (Section 16.5). ■ Question 83

Selected end-of-chapter Questions may be assigned in OWL.

For quick review, download Go Chemistry mini-lecture flashcard modules (or purchase them at **www.ichapters.com**).

Prepare for an exam with a **Summary Problem** that brings together concepts and problem-solving skills from throughout the chapter. Go to **www.cengage.com/chemistry/moore** to download this chapter's Summary Problem from the Student Companion Site.

KEY TERMS

buffer *(Section 16.1)*

buffer capacity *(16.1)*

buffer solution *(16.1)*

common ion effect *(16.4)*

end point *(16.2)*

formation constant, K_f *(16.4)*

Henderson-Hasselbalch equation *(16.1)*

ion product, Q *(16.5)*

solubility product constant, K_{sp} *(16.3)*

solubility product constant expression *(16.3)*

titrant *(16.2)*

titration curve *(16.2)*

QUESTIONS FOR REVIEW AND THOUGHT

■ denotes questions assignable in OWL.

Blue-numbered questions have short answers at the back of this book and fully worked solutions in the *Student Solutions Manual.*

Review Questions

1. What is meant by the term "buffer capacity"?
2. Which would form a buffer?
   (a) HCl and CH_3COOH (b) NaH_2PO_4 and Na_2HPO_4
   (c) H_2CO_3 and $NaHCO_3$
3. Which would form a buffer?
   (a) NaOH and NaCl (b) NaOH and NH_3
   (c) Na_3PO_4 and Na_2HPO_4
4. Briefly describe how a buffer solution can control the pH of a solution when strong acid is added and when strong base is added. Use NH_3/NH_4Cl as an example of a buffer and HCl and NaOH as the strong acid and strong base.
5. What is the difference between the end point and the equivalence point in an acid-base titration?
6. What are the characteristics of a good acid-base indicator?

7. A strong acid is titrated with a strong base, such as KOH. Describe the changes in the composition of the solution as the titration proceeds: prior to the equivalence point, at the equivalence point, and beyond the equivalence point.
8. Repeat the description for Question 8, but use a weak acid rather than a strong one.
9. Use Le Chatelier's principle to explain why $PbCl_2$ is less soluble in 0.010 M $Pb(NO_3)_2$ than in pure water.
10. Describe what a complex ion is and give an example.
11. What is amphoterism?
12. Describe how the solubility of a sparingly soluble metal hydroxide can be changed.

Topical Questions

Buffer Solutions

13. Many natural processes can be studied in the laboratory but only in an environment of controlled pH. Which of these combinations would be the best choice to buffer the pH at approximately 7?
    (a) H_3PO_4/NaH_2PO_4
    (b) NaH_2PO_4/Na_2HPO_4
    (c) Na_2HPO_4/Na_3PO_4
14. ■ Which of these combinations would be the best to buffer the pH at approximately 9?
    (a) $CH_3COOH/NaCH_3COO$
    (b) $HCl/NaCl$
    (c) NH_3/NH_4Cl
15. Without doing calculations, determine the pH of a buffer made from equimolar amounts of these acid-base pairs.
    (a) Nitrous acid and sodium nitrite
    (b) Ammonia and ammonium chloride
    (c) Formic acid and potassium formate
16. Without doing calculations, determine the pH of a buffer made from equimolar amounts of these acid-base pairs.
    (a) Phosphoric acid and sodium dihydrogen phosphate
    (b) Sodium monohydrogen phosphate and sodium dihydrogen phosphate
    (c) Sodium phosphate and sodium monohydrogen phosphate
17. Select from Table 16.1 a conjugate acid-base pair that would be suitable for preparing a buffer solution whose concentration of hydronium ions is
    (a) 4.5×10^{-3} M (b) 5.2×10^{-8} M
    (c) 8.3×10^{-6} M (d) 9.7×10^{-11} M
    Explain your choices.
18. Select from Table 16.1 a conjugate acid-base pair that would be suitable for preparing a buffer solution with pH equal to
    (a) 3.45 (b) 5.48
    (c) 8.32 (d) 10.15
    Explain your choices.
19. A buffer solution can be made from benzoic acid, C_6H_5COOH, and sodium benzoate, NaC_6H_5COO. How many grams of the acid would you have to mix with 14.4 g of the sodium salt to have a liter of a solution with a pH of 3.88?
20. You dissolve 0.425 g NaOH in 2.00 L of a solution that originally had $[H_2PO_4^-] = [HPO_4^{2-}] = 0.132$ M. Calculate the resulting pH.
21. ■ A buffer solution is prepared by adding 0.125 mol ammonium chloride to 500. mL of 0.500 M aqueous ammonia. What is the pH of the buffer? If 0.0100 mol HCl gas is bubbled into 500. mL of the buffer, what is the new pH of the solution?
22. If added to 1 L of 0.20 M acetic acid, CH_3COOH, which of these would form a buffer?
    (a) 0.10 mol $NaCH_3COO$ (b) 0.10 mol NaOH
    (c) 0.10 mol HCl (d) 0.30 mol NaOH
    Explain your answers.
23. ■ If added to 1 L of 0.20 M NaOH, which of these would form a buffer?
    (a) 0.10 mol acetic acid (b) 0.30 mol acetic acid
    (c) 0.20 mol HCl (d) 0.10 mol $NaCH_3COO$
    Explain your answers.
24. Calculate the pH change when 10.0 mL of 0.10 M NaOH is added to 90.0 mL pure water, and compare the pH change with that when the same amount of NaOH solution is added to 90.0 mL of a buffer consisting of 1.0 M NH_3 and 1.0 M NH_4Cl. Assume that the volumes are additive. K_b of NH_3 = 1.8×10^{-5}.
25. ■ Calculate the pH change when 1.0 mL of 1.0 M NaOH is added to 0.100 L of a solution of
    (a) 0.10 M acetic acid and 0.10 M sodium acetate.
    (b) 0.010 M acetic acid and 0.010 M sodium acetate.
    (c) 0.0010 M acetic acid and 0.0010 M sodium acetate.
26. Calculate the pH change when 1.0 mL of 1.0 M HCl is added to 0.100 L of a solution of
    (a) 0.10 M acetic acid and 0.10 M sodium acetate.
    (b) 0.010 M acetic acid and 0.010 M sodium acetate.
    (c) 0.0010 M acetic acid and 0.0010 M sodium acetate.
27. A buffer consists of 0.20 M propanoic acid (K_a = 1.4×10^{-5}) and 0.30 M sodium propanoate.
    (a) Calculate the pH of this buffer.
    (b) Calculate the pH after the addition of 1.0 mL of 0.10 M HCl to 0.010 L of the buffer.
    (c) Calculate the pH after the addition of 3.0 mL of 1.0 M HCl to 0.010 L of the buffer.

Titrations and Titration Curves

28. ■ The titration curves for two acids with the same base are

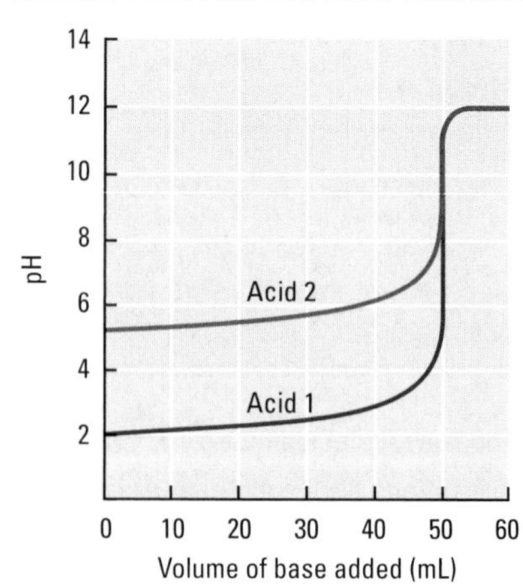

    (a) Which is the curve for the weak acid? Explain your choice.
    (b) Give the approximate pH at the equivalence point for the titration of each acid.
    (c) Explain why the pH at the equivalence point differs for each acid.

(d) Explain why the starting pH values of the two acids differ.
(e) Which indicator could be used for the titration of Acid 1 and for the titration of Acid 2? Explain your choices.

29. Sketch the titration curve for the titration of 20.0 mL of a 0.100 M solution of a strong acid by a 0.100 M weak base; that is, the base is the titrant. In particular, note the pH of the solution:
(a) Prior to the titration
(b) When half the required volume of titrant has been added
(c) At the equivalence point
(d) 10. mL beyond the equivalence point.

30. ■ Consider all acid-base indicators discussed in this chapter. Which of these indicators would be suitable for the titration of each of these?
(a) NaOH with $HClO_4$ (b) Acetic acid with KOH
(c) NH_3 solution with HBr (d) KOH with HNO_3
Explain your choices.

31. Which of the acid-base indicators discussed in this chapter would be suitable for the titration of
(a) HNO_3 with KOH (b) KOH with acetic acid
(c) HCl with NH_3 (d) KOH with HNO_2
Explain your answers.

32. ■ It required 22.6 mL of 0.0140 M $Ba(OH)_2$ solution to titrate a 25.0-mL sample of HCl to the equivalence point. Calculate the molarity of the HCl solution.

33. It took 12.4 mL of 0.205 M H_2SO_4 solution to titrate 20.0 mL of a sodium hydroxide solution to the equivalence point. Calculate the molarity of the original NaOH solution.

34. Vitamin C is a monoprotic acid. To analyze a vitamin C capsule weighing 0.505 g by titration took 24.4 mL of 0.110 M NaOH. Calculate the percentage of vitamin C, $C_6H_8O_6$, in the capsule. Assume that vitamin C is the only substance in the capsule that reacts with the titrant.

35. An acid-base titration was used to find the percentage of $NaHCO_3$ in 0.310 g of a powdered commercial product used to relieve upset stomachs. The titration required 14.3 mL of 0.101 M HCl to titrate the powder to the equivalence point. Assume that the $NaHCO_3$ in the powder is the only substance that reacted with the titrant. Calculate the percentage of $NaHCO_3$ in the powder.

36. What volume of 0.150 M HCl is required to titrate to the equivalence point each of these samples?
(a) 25.0 mL of 0.175 M KOH
(b) 15.0 mL of 6.00 M NH_3
(c) 15.0 mL of propylamine, $CH_3CH_2CH_2NH_2$, which has a density of 0.712 g/mL
(d) 40.0 mL of 0.0050 M $Ba(OH)_2$

37. What volume of 0.225 M NaOH is required to titrate to the equivalence point each of these samples?
(a) 20.0 mL of 0.315 M HBr
(b) 30.0 mL of 0.250 M $HClO_4$
(c) 6.00 g of concentrated acetic acid, CH_3COOH, which is 99.7% pure

38. The titration of 50.00 mL of 0.150 NaOH with 0.150 M HCl is carried out in a chemistry laboratory. Calculate the pH of the solution after these volumes of the titrant have been added:
(a) 0.00 mL (b) 25.00 mL (c) 49.9 mL
(d) 50.00 mL (e) 50.1 mL (f) 75.00 mL
Use the results of your calculations to plot a titration curve for this titration. On the curve indicate the position of the equivalence point.

39. The titration of 50.00 mL of 0.150 HCl with 0.150 M NaOH is carried out in a chemistry laboratory. Calculate the pH of the solution after these volumes of the titrant have been added:
(a) 0.00 mL (b) 25.00 mL (c) 49.9 mL
(d) 50.00 mL (e) 50.1 mL (f) 75.00 mL
Use the results of your calculations to plot a titration curve for this titration. On the curve indicate the position of the equivalence point.

Solubility Product

40. Write a balanced chemical equation for the equilibrium occurring when each of these solutes is added to water, then write the K_{sp} expression for each solute.
(a) Lead(II) carbonate (b) Nickel(II) hydroxide
(c) Strontium phosphate (d) Mercury(I) sulfate

41. ■ Write a balanced chemical equation for the equilibrium occurring when each of these solutes is added to water, then write the K_{sp} expression.
(a) Iron(II) carbonate (b) Silver sulfate
(c) Calcium phosphate (d) Mn(II) hydroxide

42. The solubility of silver chromate, Ag_2CrO_4, in water is 2.7×10^{-3} g/100. mL. Calculate the K_{sp} of silver chromate. Assume that there are no other reactions but the K_{sp} reaction.

43. Calculate the K_{sp} of HgI_2 given that its solubility in water is 2.2×10^{-10} M. Assume that there are no other reactions but the K_{sp} reaction.

44. The solubility of $PbCl_2$ in water is 1.62×10^{-2} M. Calculate the K_{sp} of $PbCl_2$. Assume that there are no other reactions but the K_{sp} reaction.

45. In a saturated CaF_2 solution at 25°C, the calcium concentration is analyzed to be 9.1 mg/L. Use this value to calculate the K_{sp} of CaF_2 assuming that the solute dissociates completely into Ca^{2+} and F^- ions and that neither ion reacts with water.

Common Ion Effect

46. ■ The K_{sp} of $ZnCO_3$ is 1.5×10^{-11}. Calculate the solubility of $ZnCO_3$ in
(a) Water (b) 0.050 M $Zn(NO_3)_2$
(c) 0.050 M K_2CO_3

47. Calculate the solubility (mol/L) of $SrSO_4$ ($K_{sp} = 3.1 \times 10^{-7}$) in 0.010 M Na_2SO_4.

48. Iron(II) hydroxide, $Fe(OH)_2$, has a solubility in water of 6.0×10^{-1} mg/L at a given temperature.
(a) Calculate the K_{sp} of iron(II) hydroxide.
(b) Calculate the hydroxide concentration needed to precipitate Fe^{2+} ions such that no more than 1.0 μg Fe^{2+} per liter remains in the solution.

49. The solubility of $Mg(OH)_2$ in water is approximately 9 mg/L at a given temperature.
(a) Calculate the K_{sp} of magnesium hydroxide.
(b) Calculate the hydroxide concentration needed to precipitate Mg^{2+} ions such that no more than 5.0 μg Mg^{2+} per liter remains in the solution.

50. What is the Cl^- concentration (in mol/L) in a solution that is 0.05 M in $AgNO_3$ and contains some undissolved AgCl?

Factors Affecting the Solubility of Sparingly Soluble Solutes

51. ■ What is the maximum concentration of Zn^{2+} in a solution of pH 10.00?
52. Determine the maximum concentration of Mn^{2+} in two solutions with pH as follows:
    (a) 7.81 (b) 11.15
53. ■ Hydrochloric acid is added to dissolve 5.00 g $Mg(OH)_2$ in a liter of water. To what value must the pH be adjusted to do so?
54. When a few drops of 1×10^{-5} M $AgNO_3$ are added to 0.01 M NaCl, a white precipitate immediately forms. When a few drops of 1×10^{-5} M $AgNO_3$ are added to 5 M NaCl, no precipitate forms. Explain these observations.

Complex Ion Formation

55. For these complex ions, write the chemical equation for the formation of the complex ion, and write its formation constant expression.
    (a) $[Ag(CN)_2]^-$ (b) $[Cd(NH_3)_4]^{2+}$
56. For these complex ions, write the chemical equation for the formation of the complex ion and write its formation constant expression.
    (a) $[Co(Cl)_6]^{3-}$ (b) $[Zn(OH)_4]^{2-}$
57. ■ Calculate how many moles of $Na_2S_2O_3$ must be added to dissolve 0.020 mol AgBr in 1.0 L water.
58. Gaseous ammonia is added to a 0.063 M solution of $AgNO_3$ until the aqueous ammonia concentration rises to 0.18 M. Calculate the concentrations of $[Ag(NH_3)_2]^+$ and Ag^+ in the solution.
59. Write chemical equations to illustrate the amphoteric behavior of
    (a) $Zn(OH)_2$ (b) $Sb(OH)_3$
60. Write chemical equations to illustrate the amphoteric behavior of
    (a) $Cr(OH)_3$ (b) $Sn(OH)_2$

General Questions

61. A buffer solution was prepared by adding 4.95 g sodium acetate, $NaCH_3COO$, to 250. mL of 0.150 M acetic acid, CH_3COOH.
    (a) What ions and molecules are present in the solution? List them in order of decreasing concentration.
    (b) What is the resulting pH of the buffer?
    (c) What is the pH of 100. mL of the buffer solution if you add 80. mg NaOH? (Assume negligible change in volume.)
    (d) Write a net ionic equation for the reaction that occurs to change the pH.
62. Calculate the relative concentrations of *o*-ethylbenzoic acid ($pK_a = 3.79$) and potassium *o*-ethylbenzoate that are needed to prepare a pH = 4.0 buffer.
63. Calculate the relative concentrations of aniline ($pK_b = 9.42$) and anilinium chloride that are required to prepare a buffer of pH 5.00.
64. Which of these buffers has the greater resistance to change in pH?
    (a) Conjugate acid concentration = 0.100 M = conjugate base concentration
    (b) Conjugate acid concentration = 0.300 M = conjugate base concentration
    Explain your answer.
65. (a) Calculate the pH of a 0.050 M solution of HF.
    (b) What is the pH of the solution if you add 1.58 g NaF to 250. mL of the 0.050 M solution?
66. When 40.00 mL of a weak monoprotic acid solution is titrated with 0.100 M NaOH, the equivalence point is reached when 35.00 mL base have been added. After 20.00 mL NaOH solution has been added, the titration mixture has a pH of 5.75. Calculate the ionization constant of the acid.
67. Each of the solutions given in the table has equal volume and the same concentration, 0.1 M.

Acid	pH	Acid	pH
HCl	1.1	Acetic	2.9
Formic	2.3	HCN	5.1

Which solution requires the greatest volume of 0.1 M NaOH to titrate to the equivalence point? Explain your answer.

68. At 20.°C, 2.03 g $CaSO_4$ dissolve per liter of water. From these data calculate the K_{sp} of calcium sulfate at 20.°C.

Applying Concepts

69. The average normal concentration of Ca^{2+} in urine is 5.33 g/L.
    (a) What concentration of oxalate is needed to precipitate calcium oxalate to initiate formation of a kidney stone? K_{sp} of calcium oxalate = 2.3×10^{-9}.
    (b) What minimum phosphate concentration would it take to precipitate a calcium phosphate kidney stone? K_{sp} of calcium phosphate = 1×10^{-25}.
70. Explain why even though an aqueous acetic acid solution contains acetic acid and acetate ions, it cannot be a buffer.
71. Vinegar must contain at least 4% acetic acid (0.67 M). A 5.00-mL sample of commercial vinegar required 33.5 mL of 0.100 M NaOH to reach the equivalence point. Does the vinegar meet the legal limit of 4% acetic acid?
72. ■ An unknown acid is titrated with base, and the pH is 3.64 at the point where exactly half of the acid in the original sample has been neutralized. What is the value of the ionization constant of the acid?
73. When asked to prepare a carbonate buffer with a pH = 10, a lab technician wrote this equation to determine the ratio of weak acid to conjugate base needed:

$$10 = 10.32 + \log \frac{[HCO_3^-]}{[H_2CO_3]}$$

What is wrong with this setup? If the technician prepared a solution containing equimolar concentrations of HCO_3^- and CO_3^{2-}, what would be the pH of the resulting buffer?

74. When you hold your breath, carbon dioxide gas is trapped in your body. Does this increase or decrease your blood pH? Does it lead to acidosis or alkalosis? Explain your answers.

75. Apatite, $Ca_5(PO_4)_3OH$, is the mineral in teeth.

$$Ca_5(PO_4)_3OH(s) \rightleftharpoons 5\ Ca^{2+}(aq) + 3\ PO_4^{3-}(aq) + OH^-(aq)$$

(a) On a chemical basis explain why drinking milk strengthens young children's teeth.
(b) Sour milk contains lactic acid. Not removing sour milk from the teeth of young children can lead to tooth decay. Use chemical principles to explain why.

More Challenging Questions

76. You are given four different aqueous solutions and told that they each contain NaOH, Na_2CO_3, $NaHCO_3$, or a mixture of these solutes. You do some experiments and gather these data about the samples.

Sample A: Phenolphthalein is colorless in the solution.
Sample B: The sample was titrated with HCl until the pink color of phenolphthalein disappeared, then methyl orange was added to the solution. The solution became a pink color. Methyl orange changes color from pH 3.01 (red) to pH 4.4 (orange).
Sample C: Equal volumes of the sample were titrated with standardized acid. Using phenolphthalein as an indicator required 15.26 mL of standardized acid to change the phenolphthalein color; it required 17.90 mL for a color change using methyl orange as the indicator.
Sample D: Two equal volumes of the sample were titrated with standardized HCl. Using phenolphthalein as the indicator, it took 15.00 mL of acid to reach the equivalence point; using methyl orange as the indicator required 30.00 mL HCl to achieve neutralization.

Identify the solute in each of the solutions.

77. An aqueous solution contains these ions:

Ion	Concentration (M)
Br^-	0.0010
Cl^-	0.0010
CrO_4^{2-}	0.0010

A solution of silver nitrate is added dropwise to the aqueous solution. Which silver salt will precipitate first? Last?

78. You have 1.00 L of 0.10 M formic acid, HCOOH, whose $K_a = 1.8 \times 10^{-4}$. You want to bubble into the formic acid solution sufficient HCl gas to decrease the pH of the formic acid solution by 1.0 pH unit. Calculate the volume of HCl (liters) that must be used at STP to bring about the desired change in pH. Assume no volume change has occurred in the solution due to the addition of HCl gas.

79. One liter (1.0 L) of distilled water has an initial pH of 5.6 because it is in equilibrium with carbon dioxide in the atmosphere. One drop of concentrated hydrochloric acid, 12 M HCl, is added to the distilled water. Calculate the pH of the resulting solution. 20 drops = 1.0 mL.

80. A 0.0010 M aqueous solution of anisic acid, $C_8H_8O_3$, is prepared and 100 mL of it is left in an uncovered beaker. After some time, 50. mL water has evaporated from the solution in the beaker (assume no solute evaporated). The K_a of anisic acid is 3.38×10^{-5}.
(a) Calculate the pH of the initial solution and the pH after evaporation has occurred.
(b) Calculate the degree of dissociation of anisic acid in each of the solutions.

81. You want to prepare a pH 4.50 buffer using sodium acetate and glacial acetic acid. You have on hand 300 mL of 0.100 M sodium acetate. How many grams of glacial acetic acid (d = 1.05 g/mL) should you use to prepare the buffer?

82. The Lewis structures and the titration curves for the amino acids glutamic acid and lysine are given below.

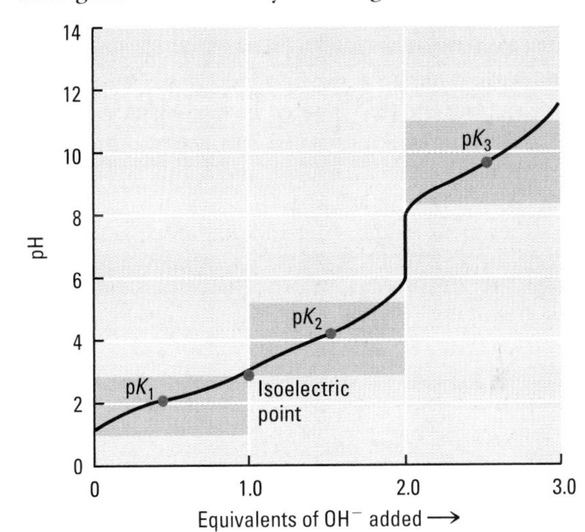

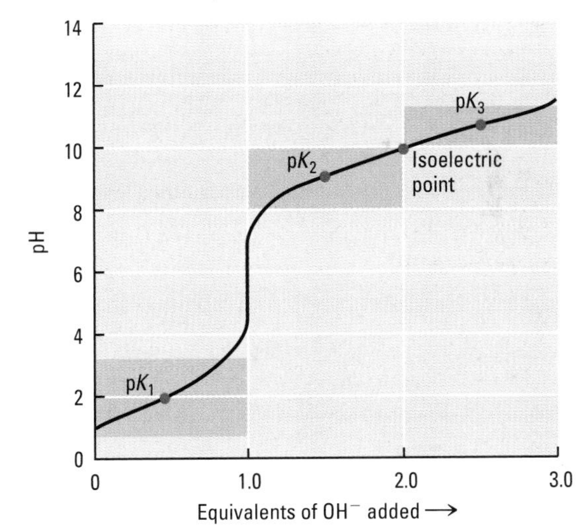

$$H_2N{-}CH_2CH_2CH_2CH_2{-}\underset{\large NH_2}{\overset{\large H}{C}}{-}\overset{\large O}{\overset{\|}{C}}{-}OH$$

lysine

$$HO{-}\overset{\large O}{\overset{\|}{C}}{-}CH_2CH_2{-}\underset{\large NH_2}{\overset{\large H}{C}}{-}\overset{\large O}{\overset{\|}{C}}{-}OH$$

glutamic acid

(a) Using the Lewis structures given, write the structural formulas for the forms of each amino acid corresponding to points on the titration curve when 0, 1.0, 2.0, and 3.0 equivalents of OH^- have been added. An equivalent in this case is the amount of base added that neutralizes an amount of H_3O^+ ions equivalent to the amount of amino acid present.

(b) The isoelectric point is the pH at which an amino acid has no net charge. Write the structural formulas of glutamic acid and lysine at their isoelectric points.

(c) What relationship exists between the conjugate acid and conjugate base forms of these amino acids at each of the pK points noted on the titration curve?

83. ■ A 1.00-L solution contains 0.010 M F^- and 0.010 M SO_4^{2-}. Solid barium nitrate is slowly added to the solution.
(a) Calculate the $[Ba^{2+}]$ when $BaSO_4$ begins to precipitate.
(b) Calculate the $[Ba^{2+}]$ when BaF_2 starts to precipitate.
Assume no volume change occurs. K_{sp} values: $BaSO_4$ = 1.1×10^{-11}; $BaF_2 = 1.7 \times 10^{-6}$.

84. An experiment requires the addition of 0.075 mol gaseous NH_3 to 1.0 L of 0.025 M $Mg(NO_3)_2$. Ammonium chloride, NH_4Cl, is added prior to the addition of the NH_3 to prevent precipitation of $Mg(OH)_2$. Calculate the minimum number of grams of ammonium chloride that must be added. K_{sp} of $Mg(OH)_2 = 2.1 \times 10^{-5}$.

Conceptual Challenge Problems

CP16.A (Section 16.2) Suppose you were asked on a laboratory test to outline a procedure to prepare a buffered solution of pH 8.0 using hydrocyanic acid, HCN. You realize that a pH of 8.0 is basic, and you find that the K_a of hydrocyanic acid is 4.0×10^{-10}. What is your response?

CP16.B (Sections 16.3 and 16.4) Barium sulfate is swallowed to enhance X-ray studies of the gastrointestinal tract.

(a) Calculate the solubility of Ba^{2+} in mol/L in pure water. K_{sp} of $BaSO_4 = 1.1 \times 10^{-10}$.

(b) In the stomach, the HCl concentration is 0.10 M. Calculate the solubility of Ba^{2+} in mol/L in this solution. K_a of $HSO_4^- =$ 1.2×10^{-2}.

(c) The two calculations in parts (a) and (b) were done using data at 25 °C. Repeat part (b) at 37 °C, body temperature. At that temperature, K_{sp} of $BaSO_4 = 1.5 \times 10^{-10}$; K_a of $HSO_4^- =$ 7.1×10^{-3}.

Thermodynamics: Directionality of Chemical Reactions

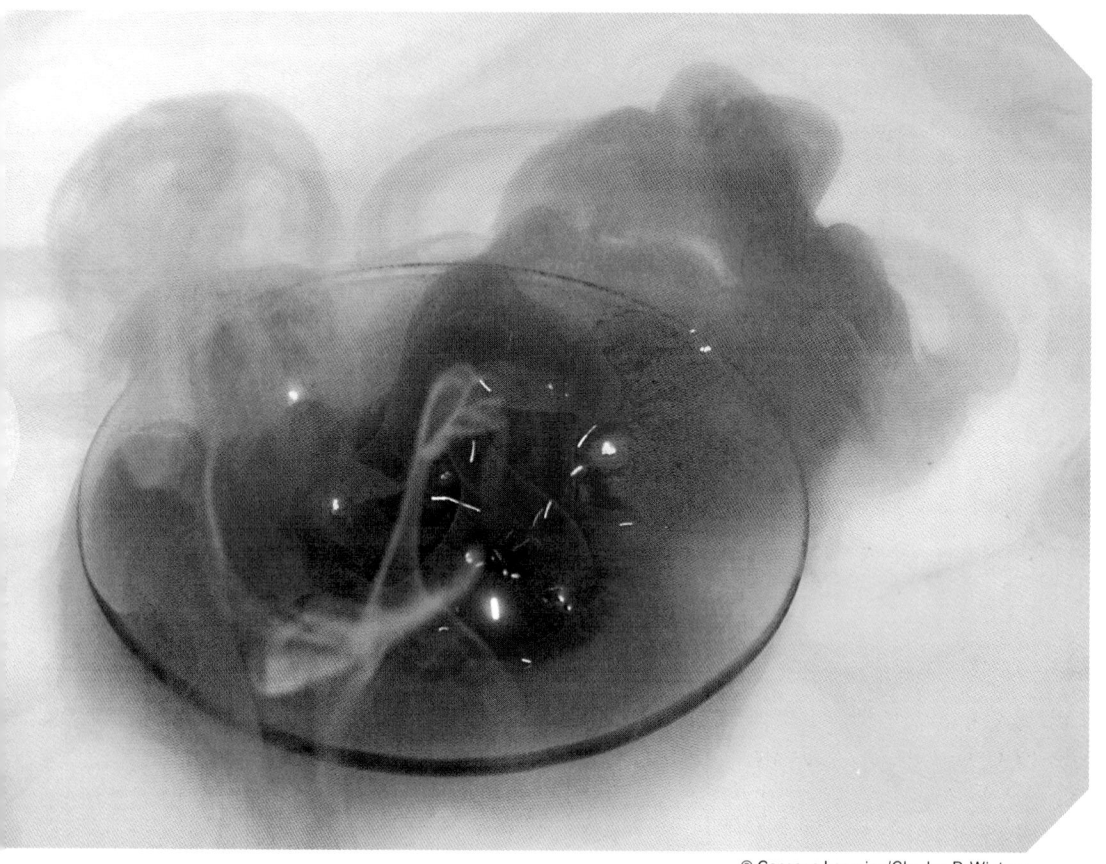

© Cengage Learning/Charles D. Winters

When liquid bromine is poured onto small pieces of aluminum, nothing appears to happen at first. Eventually the mixture becomes hot and a violent exothermic reaction occurs. Sparks fly, and you see flames. The reaction begins slowly as soon as the bromine contacts the aluminum, and its rate increases as the temperature rises. The reaction continues until either the bromine or the aluminum has been completely consumed. Chemical thermodynamics provides methods for predicting whether a reaction will be product-favored, like this one.

Many chemical reactions behave as the reaction of bromine and aluminum does (see photos in margin on the next page). They begin when the reactants come into contact and continue until at least one reactant, (the limiting reactant, ⟵ ***p. 115***), is completely converted to products. Other reactions, such as the rusting of iron at room temperature, happen much more slowly, but reactants are still converted completely to products. After many years and enough flaking of hydrated iron(III) oxide (rust) from its surface, a piece of iron exposed to air will rust away. Still

Because air can slowly oxidize hydrocarbons to alkenes, which can polymerize and form insoluble gums, gasoline is usually drained from lawn mowers or cars that are going to be stored for many months, or else an antioxidant (such as N-nitrosodiethylamine) is added.

other reactions are even slower at room temperature. Gasoline reacts so slowly with air at room temperature that it can be stored safely for long periods, although it may go bad after a very long time. However, if its temperature is raised by a spark or flame, gasoline vapor burns rapidly and is essentially completely converted to CO_2 and H_2O.

By contrast with the reaction of bromine and aluminum, many chemical reactions do not occur by themselves. For example, table salt, NaCl, does not of its own accord decompose into sodium and chlorine. Neither does water change into hydrogen and oxygen all by itself. These reactions take place only if another process occurs simultaneously and transfers energy to them. (A significant portion of world energy resources is used to cause desirable reactions to occur—reactions that transform inexpensive, readily available substances into new substances with more useful properties, such as plastics and medicines.) It is important to differentiate between a reaction that is so slow that it *appears* not to occur, such as air oxidation of gasoline, and one that *cannot* take place of its own accord, such as decomposition of sodium chloride. The principles of chemical kinetics ***(⇐ Chapter 12, p. 415)*** can be applied to find ways to speed up a slow reaction, but they are of no use in dealing with one that cannot occur by itself.

In Chapter 6 ***(⇐ p. 197)*** you learned that thermal energy is transferred when most reactions occur. You also learned how to predict whether a reaction is exothermic or endothermic and how to calculate what quantity of energy transfer takes place as a reaction occurs. In this chapter you will learn how thermodynamics helps us to predict what will happen when potential reactants are mixed. Will most or all of the reactants be converted to products, as in the case of bromine and aluminum? Will some be converted? Or virtually none?

Photos: © Cengage Learning/Charles D. Winters

The product-favored reaction of bromine with aluminum.

17.1 Reactant-Favored and Product-Favored Processes

In Chapter 13 ***(⇐ p. 462)*** we introduced the idea that a chemical process can be described as reactant-favored or product-favored. When products predominate over reactants, we designate the reaction as a *product-favored process.* Examples include the reaction of bromine with aluminum, rusting of iron, and combustion of gasoline. If a process is product-favored, most or all of the reactants will eventually be converted to products without continuous outside intervention, although "eventually" may mean a very, very long time.

Other reactions have virtually no tendency to occur by themselves. Examples include the reactions for which we wrote "N.R." for "no reaction" in Chapter 5 ***(⇐ p. 135)***. For example, nitrogen and oxygen have coexisted in the earth's atmosphere for at least a billion years without significant concentrations of nitrogen oxides such as N_2O, NO, or NO_2 building up. Similarly, deposits of salt, NaCl(s), have existed on earth for millions of years without forming the elements Na(s) and $Cl_2(g)$. If, when equilibrium has been reached, reactants predominate over products, we categorize a chemical reaction as a *reactant-favored process.*

The term "product-favored" designates reactions that many scientists refer to as "spontaneous"; many people use the two terms interchangeably. We prefer "product-favored" because some reactions do begin spontaneously, but produce only tiny quantities of products when equilibrium is reached. "Product-favored" describes clearly a situation in which products predominate over reactants. Also, the nonscientific usage of "spontaneous" implies a rapid change; if the rate of a product-favored reaction is very slow, the reaction does not appear spontaneous at all.

A reactant-favored process is always *exactly the opposite* of a product-favored process. For example, the equation

$$2\,Na(s) + Cl_2(g) \longrightarrow 2\,NaCl(s)$$

describes a product-favored reaction, because sodium metal and chlorine gas react readily to produce salt. However, if we had written the same equation in the reverse direction

$$2\,NaCl(s) \longrightarrow 2\,Na(s) + Cl_2(g)$$

the system would be designated as reactant-favored. This equation represents decomposition of sodium chloride to form sodium and chlorine, a reaction that does not

occur of its own accord. The designations "product-favored" and "reactant-favored" indicate the direction in which a chemical reaction will take place—either forward or backward based on a given equation.

Unless there is some continuous outside intervention, a reactant-favored process does not produce large quantities of products. What do we mean by continuous outside intervention? Usually it is some flow of energy. For example, if enough energy is provided to a sample of air to keep it at a very high temperature, small but significant quantities of NO can be formed from the N_2 and O_2. Such high temperatures are found in lightning bolts and in combustion chambers of electric power generating plants and automobile engines. A power plant or a large number of automobiles can produce enough NO and other nitrogen oxides to cause significant air pollution problems. Salt can be decomposed to its elements by continuously heating it to keep it molten and passing a direct electric current through it to separate the ions, carry out oxidation and reduction, and form the elements.

Although earth's atmosphere is 78% N_2 and 21% O_2, the concentration of N_2O, the most abundant oxide of nitrogen in the atmosphere, is more than two million times smaller than the concentration of N_2.

The air in the immediate vicinity of a lightning bolt can be heated enough to cause a small fraction of the nitrogen and oxygen to combine to form NO, but this reaction takes place only while the lightning is present. A similar reaction can occur in the engine of an automobile, but again only a small fraction of the air is converted to nitrogen oxides and only while the temperature remains high.

$$2\ NaCl(\ell) \xrightarrow{\text{electricity}} 2\ Na(\ell) + Cl_2(g)$$

In each case, a reactant-favored process can be forced to produce products if sufficient electrical energy is continuously supplied. This is in contrast to the situation for a product-favored process such as combustion of gasoline, which requires only a brief spark to initiate the reaction. Once started, gasoline combustion continues of its own accord without an additional supply of energy from outside.

EXERCISE 17.1 Reactant-Favored and Product-Favored Processes

Write a chemical equation for each process described below, and classify each as reactant-favored or product-favored.

(a) A puddle of water evaporates on a summer day.
(b) Silicon dioxide (sand) decomposes to the elements silicon and oxygen.
(c) Paper, which is mainly cellulose $(C_6H_{10}O_5)_n$, burns at a temperature of 451 °F.
(d) A pinch of sugar dissolves in water at room temperature.

17.2 Chemical Reactions and Dispersal of Energy

The fundamental rule that governs whether a process is product-favored is that ***energy will spread out (disperse) unless it is hindered from doing so.*** A simple example is the one-way transfer of energy from a hotter sample to a colder sample **(⟵ p. 183)**. As a hot frying pan on a stove cools to room temperature, thermal energy that was concentrated in the pan spreads out over the atoms, molecules, and ions in the stove, the pan, and the surrounding air. When thermal equilibrium is reached, the room has become slightly warmer—energy has been dispersed and the process is product-favored.

Most exothermic reactions are product-favored at room temperature for a similar reason. When an exothermic reaction takes place, energy is transferred from the system to the surroundings. (See, for example, the reaction of bromine with aluminum in the photo at the beginning of this chapter.) Chemical potential energy that has been stored in bonds between relatively few atoms, ions, and molecules (the reactants) spreads over many more atoms, ions, and molecules as the surroundings (as well as the products) are heated. Therefore it is usually true that after an exothermic reaction, energy is more dispersed than it was before.

A chemical reaction system is usually defined as the collection of atoms that make up the reactants. These same atoms also make up the products, but there they are bonded in a different way. Everything else is designated as the surroundings ***(⟵ p. 184).***

Probability and Dispersal of Energy

Dispersal of energy occurs because the probability is much higher that energy will be spread over many particles than that it will be concentrated in a few. To better understand energy dispersal and probability, consider the hypothetical case of a very small sample of matter consisting of two atoms, A and B. Suppose that this sample contains two units of energy, each designated by *. The energy can be distributed over the two atoms in three ways: Atom A could have both units of energy, atom A and atom B could each have one unit, or atom B could have both units. Designate these three situations as

$$A^{**} \quad A^{*}B^{*} \quad B^{**}$$

Now suppose that atoms A and B come into contact with two other atoms, C and D, that have no energy. There are ten possibilities for distributing the two units of energy over four atoms.

$$A^{**} \quad A^{*}B^{*} \quad A^{*}C^{*} \quad A^{*}D^{*} \quad B^{**} \quad B^{*}C^{*} \quad B^{*}D^{*} \quad C^{**} \quad C^{*}D^{*} \quad D^{**}$$

The low probability that a lot of energy will be associated with only a few particles makes a substance with a lot of chemical potential energy valuable. Humans call substances such as coal, oil, and natural gas "energy resources" and sometimes fight wars over them because of their concentrated energy.

Only three of these cases (A^{**}, $A^{*}B^{*}$, and B^{**}) have all the energy in atoms A and B, which was the initial situation. When all four atoms are in contact, there are seven chances out of ten that some energy will have transferred from A and B to C and D. Thus, there is a probability of $7/10 = 0.70$ that the energy will become spread out over more than just the two atoms A and B.

CONCEPTUAL
EXERCISE 17.2 Probability of Energy Dispersal

Suppose that you have three units of energy to distribute over two atoms, A and B. Designate each possible arrangement. Now suppose that atoms A and B come into contact with three more atoms, C, D, and E. From the possible arrangements of energy over the five atoms, calculate the probability that all the energy will remain confined to atoms A and B.

For the same amount of substance or number of particles, higher thermal energy corresponds to higher temperature (*p. 184*). Therefore, a substance at a higher temperature has greater energy per particle on average. Dispersal of energy over a larger number of particles corresponds to transfer of energy from a substance at a higher temperature to another substance at a lower temperature.

The probability that energy will become dispersed becomes overwhelming when large numbers of atoms or molecules are involved. For example, suppose that atoms A and B had been brought into contact with a mole of other atoms. There would still be only three arrangements in which all the energy was associated with atoms A and B, but there would be many, many more arrangements (more than 10^{47}) in which all the energy was associated with other atoms. In such a case it is essentially certain that energy will be transferred. ***If energy can be dispersed over a very much larger number of particles, it will be.***

Dispersal of Energy Accompanies Dispersal of Matter

Energy becomes more dispersed when a system consisting of atoms or molecules expands to occupy a larger volume. This kind of energy dispersal is associated with a characteristic property of gases: A gas expands until it fills a container. Recall that molecules in the gas phase are essentially independent of one another and that only weak forces attract the molecules together (*p. 346*). Suppose a sample of bromine gas at room temperature is confined within one flask that is connected through a tube with a barrier to a second flask of equal size from which all gas molecules have been removed (Figure 17.1). What happens if the barrier is removed? The confined bromine expands to fill the vacuum in the second flask.

The conclusions drawn about dispersal of bromine molecules on this page apply in general to particles of a gas, whether they are atoms or molecules.

That dispersal of atoms and molecules also involves dispersal of energy seems obvious, because energy accompanies the particles as they disperse. However, dispersal of energy refers to spreading of energy over a greater number of different *energy levels (quantum levels)* (*p. 229*) of the atoms and molecules. When the volume of a gas-phase system increases, the energy levels associated with the motion of each atom or molecule get closer together; that is, there are more levels within the same range of energies. Therefore, at a given temperature, more different energy levels are accessible in the larger volume than in the smaller volume.

When matter spreads out, the number of ways of arranging the energy associated with that matter increases. The energy is more dispersed, and therefore the spreading out of matter is product-favored. This basic idea applies to mixing of different gases and dissolving of one liquid in another, as well as to expansion of a gas. We have already mentioned it in connection with dissolving of solids ***(p. 514)***, and it is also useful in understanding colligative properties ***(p. 522)***.

Expansion of a gas is product-favored, so the opposite process—compression of a gas—should be reactant-favored. If we wanted to reverse the expansion of a gas by concentrating all the particles into a smaller volume, a continuous outside influence such as a pump would be required—the pump could do work on the gas to force it into a less probable arrangement in which energy was less dispersed. The work done by the pump would be stored in the gas and could later be used for some other purpose.

To summarize, any physical or chemical process in which energy is concentrated in a few energy levels in the initial state and energy is dispersed over many energy levels in the final state is product-favored. Two important situations where this is true are

- an exothermic reaction, which disperses potential energy of chemical bonds to thermal energy of a much larger number of atoms or molecules, and
- a process where matter spreads out, which disperses energy as well as matter.

If both of these situations apply to a reaction, then it will definitely be product-favored, because the final distribution of energy will be more probable. On the other hand, a process that spreads out neither energy nor matter will be reactant-favored—the initial substances will remain no matter how long we wait. If one of these situations applies but not the other, then quantitative information is needed to decide which effect is greater. The remainder of this chapter develops that quantitative information.

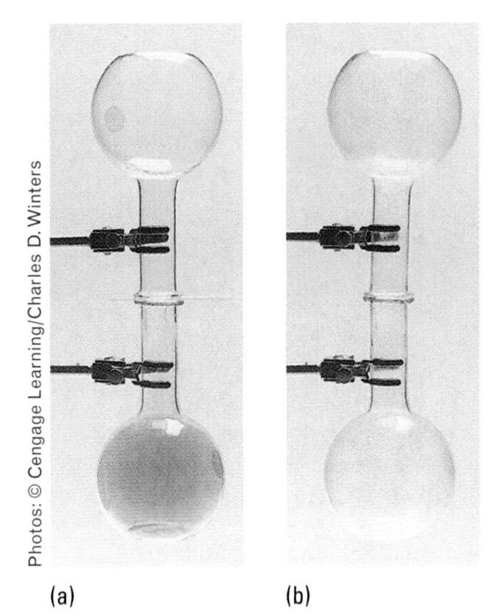

(a) (b)

Figure 17.1 Expansion of a gas. A gas will expand to fill any container. (a) Bromine vapor is confined in the lower flask. There is a vacuum in the upper flask. (b) When the barrier between the flasks is removed, the bromine molecules rapidly rush into flask B, and eventually bromine is evenly distributed throughout both flasks.

Atoms or molecules of a material that is a solid at room temperature (like the glass flask) do not disperse, because there are strong attractive forces between them. Their tendency to disperse becomes more obvious if the temperature is raised so that they can vaporize.

17.3 Measuring Dispersal of Energy: Entropy

The nanoscale dispersal of energy in a sample of matter is measured by a thermodynamic quantity called *entropy*, symbolized by *S* ***(p. 492)***. Entropy changes can be measured with a calorimeter ***(p. 203)***, the same instrument used to measure the enthalpy change when a reaction occurs. For a process that takes place at constant temperature and pressure, the entropy change can be calculated by dividing the thermal energy transferred, q_{rev}, by the absolute temperature, T,

$$\Delta S = S_{final} - S_{initial} = \frac{q_{rev}}{T} \qquad [17.1]$$

The symbol Δ was defined in Section 6.2 ***(p. 184)***. The quantity of energy transferred by heating, q, was defined in Section 6.3 ***(p. 186)***.

The subscript "rev" has been added to q to indicate that the equation applies only to processes that can be reversed by a very small change in conditions. An example of such a process is melting of ice at 0 °C and normal atmospheric pressure. If the temperature is just a tiny bit below 0 °C, so that energy is transferred from the water to its surroundings, the water will freeze. If the temperature is a tiny bit above 0 °C, the ice will melt. *Any process for which a very small change in conditions can reverse its direction* is called a **reversible process.**

The absolute temperature scale is also called the *Kelvin temperature scale or the thermodynamic temperature scale.* It was defined in Section 10.4 ***(p. 350)***. The Kelvin scale should be used in all thermodynamic calculations involving temperature.

Another case where Equation 17.1 can be used is when energy is transferred to or from a thermal reservoir at constant temperature and pressure. A thermal reservoir is any large sample of matter, such as everything in your dorm room or the surroundings of a chemical reaction. For example, when a cup of hot coffee cools to room temper-

Oesper Collection in the History of Chemistry, University of Cincinnati

Ludwig Boltzmann 1844–1906

An Austrian physicist, Ludwig Boltzmann gave us the useful interpretation of entropy as probability. Engraved on his tombstone in Vienna is the equation

$$S = k \log W$$

It relates entropy, S, and thermodynamic probability, W—the number of different nanoscale arrangements of energy that correspond to a given macroscale system. The proportionality constant, k, is called Boltzmann's constant, and log stands for the natural logarithm (in modern symbolism, ln).

Entropy changes are usually reported in units of joules per kelvin (J/K), whereas enthalpy changes are usually given in kilojoules (kJ). This means that you need to be careful about the units to avoid being wrong by a factor of 1000.

ature, the quantity of energy transferred is small enough that the temperature of the thermal reservoir (your room) hardly changes. This small quantity of energy could be transferred into or out of your room reversibly. Therefore, to calculate the entropy change of your room you could measure room temperature (in kelvins), calculate how much energy transferred to the room as the coffee cooled, and divide energy transfer by temperature. You could not do the same calculation to determine the entropy change of the coffee, because its temperature changed during the process.

PROBLEM-SOLVING EXAMPLE 17.1 Calculating Entropy Change

The melting point of glacial acetic acid (pure acetic acid) is 16.6 °C at 1 bar. Calculate $\Delta S°$ for the process

$$CH_3COOH(s) \longrightarrow CH_3COOH(\ell)$$

given that the molar enthalpy of fusion of acetic acid is 11.53 kJ/mol. Express your answer in units of joules per kelvin.

Answer 39.79 J/K

Strategy and Explanation Melting acetic acid at its melting point of 16.6 °C is reversible, so use the equation $\Delta S = q_{rev}/T$. The pressure is 1 bar, the standard-state pressure (⇐ p. 208), so $\Delta S° = \Delta S$. Remember that for a constant-pressure process, thermal energy transfer, q_{rev}, is the same as the enthalpy change, $\Delta H°$ (⇐ p. 191). Therefore, the entropy change can be calculated from the enthalpy of fusion and the temperature.

$$\Delta S° \text{ (melting acetic acid)} = \frac{q_{rev}}{T} = \frac{\Delta H° \text{ (melting acetic acid)}}{T} = \frac{\Delta H_{fusion}}{T}$$

$$= \frac{11.53 \text{ kJ}}{(16.6 + 273.15) \text{ K}} = 3.979 \times 10^{-2} \text{ kJ/K} = 39.79 \text{ J/K}$$

☑ **Reasonable Answer Check** The result is positive, meaning that entropy increased when solid acetic acid was converted to liquid. Since larger entropy corresponds to greater spreading out of energy, and since molecules move greater distances and in more different ways in a liquid than in a solid (⇐ p. 17), this is reasonable.

PROBLEM-SOLVING PRACTICE 17.1

A chemical reaction transfers 30.8 kJ to a thermal reservoir that has a temperature of 45.3 °C before and after the energy transfer. Calculate the entropy change for the thermal reservoir.

CONCEPTUAL
EXERCISE 17.3 The Importance of Absolute Temperature

Consider what would happen if the Celsius temperature scale were used when calculating entropy change by means of $\Delta S = q_{rev}/T$. Suppose, for example, that energy were transferred reversibly to $H_2O(s)$ at a temperature 10° below its melting point and we wanted to calculate the entropy change. Would the value calculated from the Celsius temperature agree with the fact that transfer of thermal energy to a sample always increases its entropy?

Absolute Entropy Values

In Chapter 6 we mentioned that there is no way to measure the total energy content of a sample of matter. Therefore, to summarize a large number of calorimetric measurements of enthalpy changes, we tabulated standard molar enthalpies of formation in Table 6.2 (⇐ p. 210). The standard enthalpy of formation is the difference between the enthalpy of a substance in its standard state and the enthalpies of the elements that make up that substance, all in their standard states. The elements have enthalpy values, but we do not know what they are. Therefore, the standard enthalpies of formation of elements are by definition set to zero.

For entropy the situation is simpler, because it is possible to define conditions for which it is logical to assume the entropy of a substance has its lowest possible value—namely, zero. Measuring ΔS for a change from those zero-entropy conditions tells us the absolute entropy value for the substance under new conditions. This follows from the definition $\Delta S = S_{final} - S_{initial}$, because if $S_{initial} = 0$, then $\Delta S = S_{final}$, the absolute entropy of the substance. Because decreasing temperature corresponds to decreasing molecular motion, the minimum possible temperature can reasonably be expected to correspond to minimum motion and thus minimum dispersal of energy. Thus it is logical to assume that a perfect crystal of a substance at 0 K has an entropy value of zero. [In a perfect crystal every nanoscale particle is in exactly the right position in the crystal lattice (*p. 397*), and there are no empty spaces or discontinuities.] To calculate the entropy change, start as close as possible to absolute zero, successively introduce small quantities of energy, and calculate ΔS from the equation $\Delta S = q_{rev}/T$ for each increment of energy at each temperature. Then sum these entropy changes to give the total (or absolute) entropy of a substance at any desired temperature.

The results of such measurements for several substances at 298.15 K are given in Table 17.1. These are standard molar entropy values, and so they apply to 1 mol of each substance at the standard pressure of 1 bar and the specified temperature of 25 °C. The units are joules per kelvin per mole (J K^{-1} mol^{-1}). Because there is a real zero on the entropy scale, the values in Table 17.1 are not measured relative to elements in their most stable form under standard-state conditions. Therefore, absolute entropies can be determined for elements as well as compounds. ***The standard molar entropy of a substance at temperature* T *is a sum of the quantities of energy that must be dispersed in that substance at successive temperatures up to* T*, that is, it is* ΔS *from 0 K to* T.**

Though it is impossible to cool anything all the way to absolute zero, it is possible to get very close. Temperatures of a few nanokelvins can be achieved in a Bose-Einstein condensate—the coldest thing known to science. (See **http://www.nobel.se/physics/laureates/2001/index.html.**)

The idea that a perfect crystal of any substance at 0 K has minimum entropy is called the **third law of thermodynamics.** Even if absolute zero cannot be achieved, there are ways of estimating how much energy has been dispersed in a substance near 0 K. Thus, accurate entropy values can be obtained for many substances.

The process of introducing small quantities of energy, calculating an entropy increase, and then summing these small entropy increases is actually done by measuring the heat capacity of a substance as a function of temperature and then using integral calculus to calculate the integral of the function q_{rev}/T between the limits of 0 K and the desired temperature.

Qualitative Guidelines for Entropy

Some useful guidelines can be drawn from the data given in Table 17.1.

- ***Entropies of gases are usually much larger than those of liquids, which in turn are usually larger than those of solids.*** In a solid the particles can vibrate only around their lattice positions. When a solid melts, its particles can move around more freely, and molar entropy increases. When a liquid vaporizes,

For a more detailed look at estimating entropy changes, see Craig, N. C. *Journal of Chemical Education*, Vol. 80, 2003; pp. 1432–1436.

Table 17.1 Some Standard Molar Entropy Values at 298.15 K*

Compound or Element	Entropy, $S°$ (J K^{-1} mol^{-1})	Compound or Element	Entropy, $S°$ (J K^{-1} mol^{-1})	Compound or Element	Entropy, $S°$ (J K^{-1} mol^{-1})
C(graphite)	5.740	$Br_2(\ell)$	152.231	Ca(s)	41.42
C(g)	158.096	$I_2(s)$	116.135	NaF(s)	51.5
$CH_4(g)$	186.264	Ar(g)	154.7	MgO(s)	26.94
$CH_3CH_3(g)$	229.60	$H_2(g)$	130.684	NaCl(s)	72.13
$CH_3CH_2CH_3(g)$	269.9	$N_2(g)$	191.61	KOH(s)	78.9
$CH_3OH(\ell)$	126.8	$O_2(g)$	205.138	$MgCO_3(s)$	65.7
CO(g)	197.674	$NH_3(g)$	192.45	$NH_4NO_3(s)$	151.08
$CO_2(g)$	213.74	HCl(g)	186.908	NaCl(aq)	115.5
$F_2(g)$	202.78	$H_2O(g)$	188.825	$NH_4NO_3(aq)$	259.8
$Cl_2(g)$	223.066	$H_2O(\ell)$	69.91	KOH(aq)	91.6

*Data from Wagman, D. D., Evans, W. H., Parker, V. B., Schumm, R. H., Halow, I., Bailey, S. M., Churney, K. L., and Nuttall, R. "The NBS Tables of Chemical Thermodynamic Properties." *Journal of Physical and Chemical Reference Data,* Vol. 11, Suppl. 2, 1982.

Particles in a gas have no fixed positions. They move in different directions and at different speeds.

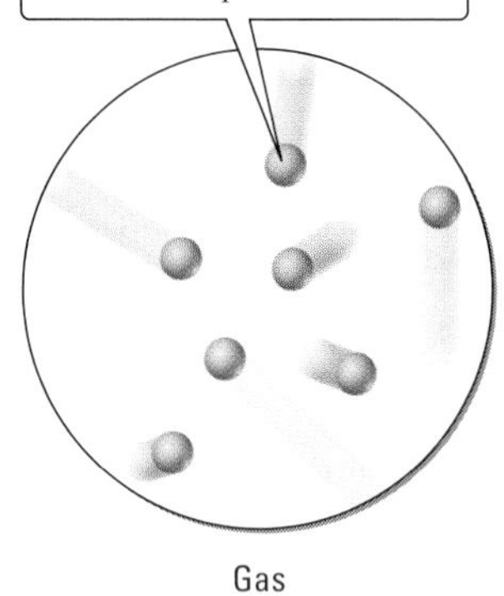

Gas

Particles in a liquid are closer together than in a gas. They have no fixed positions and move in different directions and at different speeds.

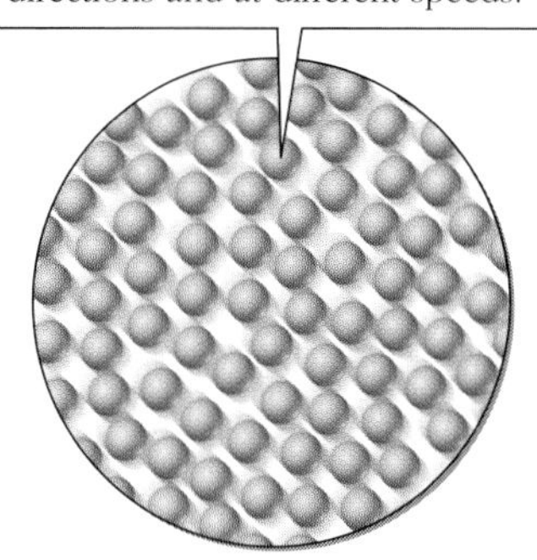

Liquid

Particles in a solid are constrained to vibrate about specific positions.

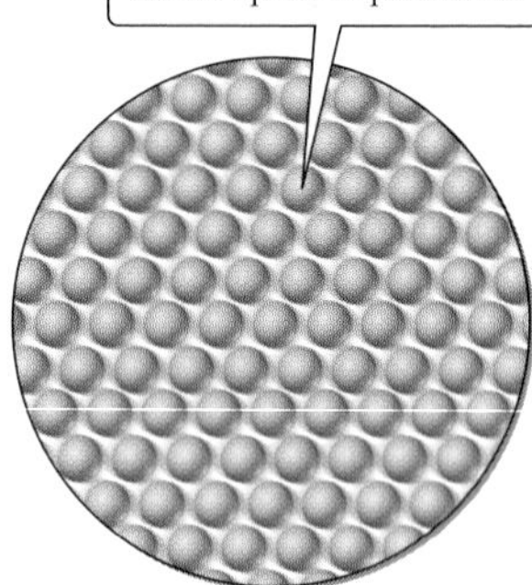

Solid

Figure 17.2 Entropies of solid, liquid, and gas phases. The entropy of each phase is related to the freedom of motion of the particles. (The figure is not drawn to scale.)

the position restrictions due to forces between the particles nearly disappear, and another large entropy increase occurs (Figure 17.2). For example, the entropies (in $J\ K^{-1}\ mol^{-1}$) of the halogens $I_2(s)$, $Br_2(\ell)$, and $Cl_2(g)$ are 116.1, 152.2, and 223.1, respectively. Similarly, the entropies of C(s, graphite) and C(g) are 5.7 and 158.1.

- ***Entropies of more complex molecules are larger than those of simpler molecules,*** especially in a series of closely related compounds. In a more complicated molecule there are more ways for the atoms to move about in three-dimensional space, and hence there is greater entropy. For example, the entropies (in $J\ K^{-1}\ mol^{-1}$) of the gases methane, CH_4; ethane, CH_3CH_3; and propane, $CH_3CH_2CH_3$, are 186.26, 229.6, and 269.9, respectively. For atoms or molecules of similar molar mass, compare the gases Ar, CO_2, and $CH_3CH_2CH_3$, which have entropies of 154.7, 213.74, and 269.9, respectively (Figure 17.3).
- ***Entropies of ionic solids that have similar formulas are larger when the attractions among the ions are weaker.*** The weaker such forces are, the easier it is for ions to vibrate about their lattice positions and the greater the entropy is. The entropy of NaF(s) is $51.5\ J\ K^{-1}\ mol^{-1}$, and that of MgO(s) is $26.94\ J\ K^{-1}\ mol^{-1}$; Na^+ and F^-, with unit positive and negative charges, attract each other less than Mg^{2+} and O^{2-}, each of which has two units of charge ***(⇐ p. 75);*** therefore NaF(s) has higher entropy. NaF(s) and NaCl(s) have entropies of 51.5 and $72.13\ J\ K^{-1}\ mol^{-1}$. Chloride ions, Cl^-, are larger than fluoride ions, F^-, and attractions are smaller when the ions are farther apart.
- ***Entropy usually increases when a pure liquid or solid dissolves in a solvent.*** Energy usually becomes more dispersed when different kinds of molecules mix together and occupy a larger volume (Figure 17.4). An example is $NH_4NO_3(s)$ and $NH_4NO_3(aq)$ with standard molar entropies of $151.08\ J\ K^{-1}\ mol^{-1}$ and $259.8\ J\ K^{-1}\ mol^{-1}$, respectively. Some ionic compounds dissolving in water are exceptions to this generalization because the ions are strongly hydrated.
- ***Entropy decreases when a gas dissolves in a liquid.*** Although gas molecules are dispersed among solvent molecules in solution, the very large entropy of the gas phase is lost when the widely separated gas particles become crowded together with solvent particles in the liquid solution (Figure 17.5).

Name	Propane	Carbon dioxide	Argon
Molecular model			
Entropy ($J\ K^{-1}\ mol^{-1}$)	269.9	213.74	154.7
Molar mass (g/mol)	44.1	44.0	39.9

Figure 17.3 Entropy and molecular structure. Three groups of particles of similar molar mass are shown. In propane, $CH_3CH_2CH_3$, many different conformations (different ways to arrange the atoms relative to one another within a molecule) and many different bond stretching and bending vibrations are possible ***(⇐ p. 317).*** In CO_2 there is a single conformation and many fewer vibrations. The individual Ar atoms can move about, but there are no conformations or vibrations possible.

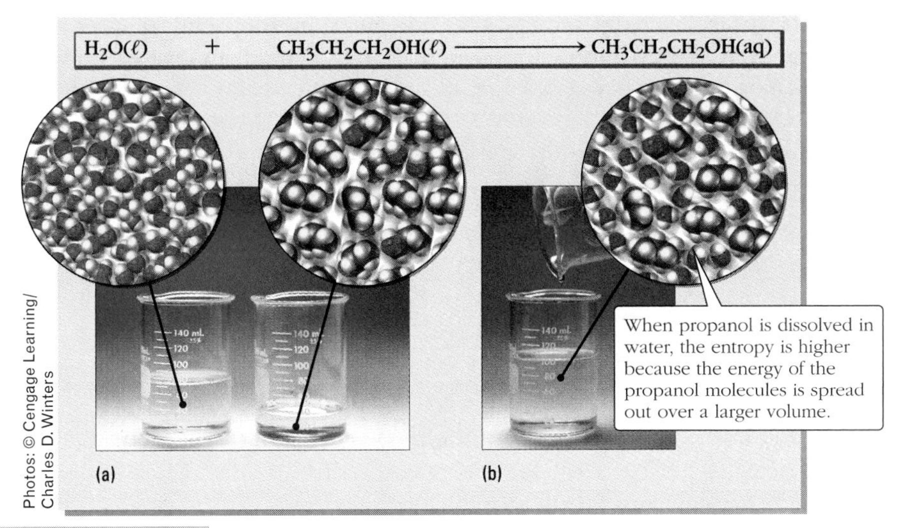

$H_2O(\ell)$ + $CH_3CH_2CH_2OH(\ell)$ ⟶ $CH_3CH_2CH_2OH(aq)$
When propanol is dissolved in water, the entropy is higher because the energy of the propanol molecules is spread out over a larger volume.
(a)
(b)
Photos: © Cengage Learning/Charles D. Winters

Active Figure 17.4 **Entropy and dissolving.** There is usually an increase in entropy when a solid or liquid dissolves in a liquid solvent, because, as solute particles become dispersed among solvent particles, the energy of motion of the particles is spread out over a larger volume. Visit this book's companion website at **www.cengage.com/chemistry/moore** to test your understanding of the concepts in this figure.

Figure 17.5 **Entropy of solution of a gas.** The very large entropy of the gas exceeds that of the solution. Even though particles are dispersed among each other in the liquid solution, the gas particles are much more widely spread out and have much higher entropy.

PROBLEM-SOLVING EXAMPLE 17.2 Relative Entropy Values

For each pair of substances below, predict which has greater entropy and give a reason for your choice. (Assume 1-mol samples at 25 °C and 1 bar.)

(a) $H_2C{=}CH_2(g)$ or $CH_3CH_2CH_2CH_3(g)$ (b) $CO_2(aq)$ or $CO_2(g)$
(c) LiF(s) or RbBr(s) (d) $N_2(\ell)$ or $N_2(s)$

Answer

(a) $CH_3CH_2CH_2CH_3(g)$ (b) $CO_2(g)$
(c) RbBr(s) (d) $N_2(\ell)$

Strategy and Explanation Use the rules given on p. 617 and 618.

(a) Larger, more complex molecules have greater entropy than similar smaller ones, so the entropy of $CH_3CH_2CH_2CH_3(g)$ is greater.
(b) The molecules of a gas are free to move, rotate, and vibrate, so when a gas dissolves in a liquid, the entropy decreases; therefore $CO_2(g)$ has greater entropy.
(c) These are ionic solids, both with 1+ and 1− ions. The attractive forces are greater the closer the ions are to each other, and the ions are smaller in LiF ***(see table of ionic radii, ⟵ p. 258),*** so RbBr has greater entropy.
(d) Entropy increases from solid to liquid to gas for the same substance, so $N_2(\ell)$ has greater entropy.

PROBLEM-SOLVING PRACTICE 17.2

In each case, predict which of the two substances has greater entropy, assuming 1-mol samples at 25 °C and 1 bar. Check your prediction by looking up each substance's absolute entropy in Table 17.1.

(a) C(g) or C(s, graphite) (b) Ca(s) or Ar(g) (c) KOH(s) or KOH(aq)

When ionic solids that consist of very small ions (such as Li^+) or of ions that carry two or more units of charge (such as Mg^{2+} or Al^{3+}) dissolve in water, there is often a decrease in entropy. This happens because very small ions and highly charged ions strongly attract water dipoles. The water molecules surrounding such ions are held in a relatively rigid structure and are no longer as free to move and rotate as they were in pure water. Although the ions become dispersed among the water molecules, the reduced dispersion of energy among the water molecules results in an overall decrease in entropy.

Predicting Entropy Changes

The general guidelines about entropy can be used to predict whether an increase or decrease in entropy will occur when reactants are converted to products. For both of the processes

$$H_2O(s) \longrightarrow H_2O(\ell) \quad \text{and} \quad H_2O(\ell) \longrightarrow H_2O(g)$$

an entropy increase is expected. Water molecules in the solid phase are more restricted and their energy is more localized than in the liquid, and water molecules in the liquid cannot move, rotate, or vibrate as freely as they can in the gas. This is confirmed by the data in Table 17.1, where $S°$ ($H_2O(g)$) > $S°$ ($H_2O(\ell)$).

Predicting entropy changes for chemical processes is usually easier than predicting enthalpy changes. For gas phase reactions, the enthalpy-change guideline is that having more bonds or stronger bonds (or both) in the products gives a negative $\Delta H°$ (⟵ ***p. 201***); however, a table of bond enthalpies is usually needed to tell which bonds are stronger.

For the decomposition of iron(III) oxide to its elements,

$$2\,Fe_2O_3(s) \longrightarrow 4\,Fe(s) + 3\,O_2(g)$$

an increase in entropy is also predicted, because 3 mol gaseous oxygen is present in the products and the reactant is a solid. This is confirmed by the experimental $\Delta S°$, 551.7 J/K. Because gases have much higher entropy than solids or liquids, gaseous substances are most important in determining entropy changes.

An example where a decrease in entropy can be predicted is

$$2\,CO(g) + O_2(g) \longrightarrow 2\,CO_2(g)$$

Here there is 3 mol gaseous substance (2 mol CO and 1 mol O_2) at the beginning but only 2 mol gaseous substance at the end of the reaction. Two moles of gas almost always contains less entropy than three moles of gas, so $\Delta S°$ is negative (experimentally, $\Delta S° = -173.0$ J/K). Another example in which entropy decreases is the process

$$Ag^+(aq) + Cl^-(aq) \longrightarrow AgCl(s)$$

Predicting an entropy decrease when ions precipitate from aqueous solution does not always work, especially when the ions have more than single positive or negative charges, such as Mg^{2+} or SO_4^{2-}. The higher the charge on an ion the more tightly water-molecule dipoles are attracted to it. When water molecules are tightly attracted around an ion their motion is restricted and entropy decreases.

Here the reactant ions are free to move about among water molecules in aqueous solution, but those same ions are held in a crystal lattice in the solid, a situation with greater constraint and thus less spreading out of energy.

CONCEPTUAL
EXERCISE 17.4 Predicting Entropy Changes

For each process, tell whether entropy increases or decreases, and explain how you arrived at your prediction.

(a) $2\,CO_2(g) \longrightarrow 2\,CO(g) + O_2(g)$ (b) $NaCl(s) \longrightarrow NaCl(aq)$

(c) $MgCO_3(s) \xrightarrow{\text{heat}} MgO(s) + CO_2(g)$

17.4 Calculating Entropy Changes

The standard molar entropy values given in Table 17.1 can be used to calculate entropy changes for physical and chemical processes. Assume that each reactant and each product is at the standard pressure of 1 bar and at the temperature given (298.15 K). The number of moles of each substance is specified by its stoichiometric coefficient in the equation for the process. Multiply the entropy of each product substance by the number of moles of that product and add the entropies of all products. Calculate the total entropy of the reactants in the same way and subtract it from the total entropy of the products. This is summarized in the equation

$$\Delta S° = \Sigma\{(\text{moles product}) \times S°(\text{product})\} - \Sigma\{(\text{moles reactant}) \times S°(\text{reactant})\}$$

Notice that the equation for calculating $\Delta S°$ has the same form as that for calculating $\Delta H°$ for a reaction (⟵ ***p. 211***).

Note that this calculation gives the entropy change for the chemical reaction *system* only. It tells whether the energy of the atoms that make up the system is more dispersed or less dispersed after the reaction. It does not account for any entropy change in the surroundings.

PROBLEM-SOLVING EXAMPLE 17.3 Calculating an Entropy Change from Tabulated Values

The reaction

$$CO(g) + 2\,H_2(g) \longrightarrow CH_3OH(\ell)$$

is being evaluated as a possible way to manufacture liquid methanol, $CH_3OH(\ell)$, for use in motor fuel. Calculate $\Delta S°$ for the reaction.

Answer $\Delta S° = -332.3 \text{ J/K}$

Strategy and Explanation Use information in Table 17.1, subtracting the entropies of the reactants from the entropy of the product. Because all substances, including elements, have nonzero absolute entropy values, elements as well as compounds must be included.

$$\Delta S° = \Sigma\{(\text{moles of product}) \times S°(\text{product})\} - \Sigma\{(\text{moles of reactant}) \times S°(\text{reactant})\}$$
$$= (1 \text{ mol}) \times S°[CH_3OH(\ell)] - \{(1 \text{ mol}) \times S°[CO(g)] + (2 \text{ mol}) \times S°[H_2(g)]\}$$
$$= (1 \text{ mol}) \times (126.8 \text{ J K}^{-1} \text{ mol}^{-1})$$
$$- \{(1 \text{ mol}) \times (197.7 \text{ J K}^{-1} \text{ mol}^{-1}) + (2 \text{ mol}) \times (130.7 \text{ J K}^{-1} \text{ mol}^{-1})\}$$
$$= -332.3 \text{ J/K}$$

☑ **Reasonable Answer Check** There is a decrease in entropy, which is reasonable because 3 mol gaseous reactants is converted to 1 mol liquid phase product. The product molecule is more complicated, but the fact that it is a liquid makes its entropy much smaller than that of 1 mol gaseous carbon monoxide and 2 mol gaseous hydrogen.

PROBLEM-SOLVING PRACTICE 17.3

Use absolute entropies from Table 17.1 to calculate the entropy change for each of these processes, thereby verifying the predictions made in Conceptual Exercise 17.4.

(a) $2\ CO_2(g) \longrightarrow 2\ CO(g) + O_2(g)$ (b) $NaCl(s) \longrightarrow NaCl(aq)$

(c) $MgCO_3(s) \xrightarrow{\text{heat}} MgO(s) + CO_2(g)$

17.5 Entropy and the Second Law of Thermodynamics

A great deal of experience with many chemical reactions and other processes in which energy is transformed and transferred is consistent with the conclusion that *whenever a product-favored chemical or physical process occurs, energy becomes more dispersed or disordered.* This is summarized in the **second law of thermodynamics,** which states that the ***total entropy of the universe (a system plus its surroundings) is continually increasing.*** Evaluating whether the total entropy increases during a proposed chemical reaction allows us to predict whether reactants will form appreciable quantities of products.

Predicting whether a reaction is product-favored can be done in three steps:

1. Calculate how much entropy changes as a result of transfer of energy between system and surroundings ($\Delta S_{surroundings}$).
2. Calculate how much entropy changes as a result of dispersal of energy within the system (ΔS_{system}).
3. Add these two results to get $\Delta S_{universe} = \Delta S_{system} + \Delta S_{surroundings}$.

Steps 1 and 2 can be carried out by assuming that reactants under standard conditions of 1 bar and a specified temperature are converted to products under the same standard conditions.

Let us apply these steps to the reaction

$$CO(g) + 2\ H_2(g) \longrightarrow CH_3OH(\ell)$$

for which we calculated the entropy change in Problem-Solving Example 17.3. If the reaction is product-favored, it would be a good way to produce methanol for use as automotive fuel. The reactants can be obtained from plentiful resources: coal and water. We base our prediction upon having as reactants 1 mol CO(g) and 2 mol $H_2(g)$ and as product 1 mol liquid methanol, with all substances at 1 bar and 298.15 K (25 °C). Then data from Table 6.2 (*p. 210*) and Table 17.1 (or Appendix J) apply.

If the entropy of the universe is predicted to be higher after the product has been produced, then the reaction is product-favored under these conditions and might be useful. If not, perhaps some other conditions could be used, or perhaps we should consider some other reaction altogether.

We will consider the effect of changing temperature later in this chapter. The effects of changing other conditions can often be predicted qualitatively by using Le Chatelier's principle (⇐ p. 482).

Step 1: ***Calculate the reaction's dispersal or concentration of energy to or from the surroundings by calculating $\Delta H°$ and assuming that this quantity of thermal energy is transferred reversibly.*** The entropy change for the surroundings can be calculated as

$$\Delta S°_{\text{surroundings}} = \frac{q_{\text{rev}}}{T} = \frac{\Delta H_{\text{surroundings}}}{T} = \frac{-\Delta H°_{\text{system}}}{T} = \frac{-\Delta H°}{T}$$

The minus sign in this equation comes from the fact that the direction of energy transfer for the surroundings is opposite from the direction of energy transfer for the system. For an exothermic reaction (negative $\Delta H°$) there will be an increase in entropy of the surroundings, a fact that we have already mentioned. For the proposed methanol-producing reaction, $\Delta H°$ (calculated from data in Appendix J) is -128.1 kJ, so the entropy change is

$$\Delta S°_{\text{surroundings}} = \frac{-(-128.1 \text{ kJ})}{298 \text{ K}} \times \frac{1000 \text{ J}}{\text{kJ}} = 430.\text{ J/K}$$

Step 2: ***Calculate the entropy change for dispersal of energy within the system*** (that is, for the atoms involved in the methanol-producing reaction). This entropy change can be evaluated from the absolute entropies of the products and reactants as described in the previous section and has already been calculated in Problem-Solving Example 17.3 to be

$$\Delta S°_{\text{system}} = \Delta S° = -332.3 \text{ J/K}$$

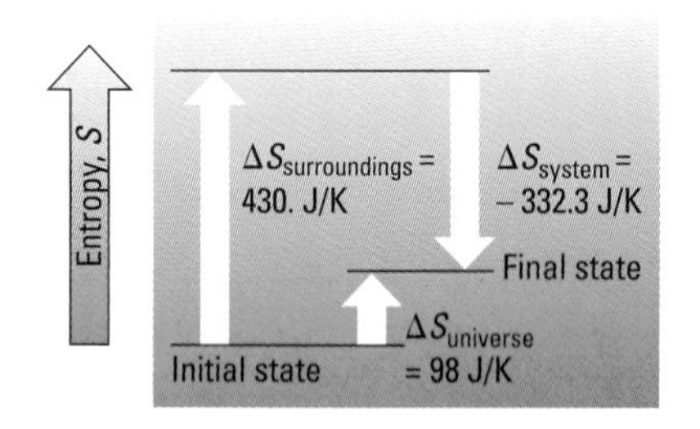

Entropy diagram for combination of carbon monoxide and hydrogen to form methanol.

Step 3: ***Calculate the total entropy change for the system and the surroundings, $\Delta S°_{\text{universe}}$.*** Because the universe includes both system and surroundings, $\Delta S°_{\text{universe}}$ is the sum of the entropy change for the system and the entropy change for the surroundings. (We assume that nothing else but our reaction happens, so there are no other entropy changes.) This total entropy change is

$$\Delta S°_{\text{universe}} = \Delta S°_{\text{surroundings}} + \Delta S°_{\text{system}} = \frac{-\Delta H°_{\text{system}}}{T} + \Delta S°_{\text{system}} \qquad [17.2]$$

$$= (430. - 332.3) \text{ J/K} = 98 \text{ J/K}$$

Combination of carbon monoxide and hydrogen to form methanol increases the entropy of the universe. Because the process is product-favored, it could be useful for manufacturing methanol.

CONCEPTUAL
EXERCISE 17.5 Effect of Temperature on Entropy Change

The reaction of carbon monoxide with hydrogen to form methanol is quite slow at room temperature. As a general rule, reactions go faster at higher temperatures. Suppose that you tried to speed up this reaction by increasing the temperature.

(a) Assuming that $\Delta H°$ does not change very much as the temperature changes, what effect would increasing the temperature have on $\Delta S°_{\text{surroundings}}$?

(b) Assuming that $\Delta S°$ for a reaction system does not change much as the temperature changes, what effect would increasing the temperature have on $\Delta S°_{\text{universe}}$?

PROBLEM-SOLVING EXAMPLE 17.4 Calculating $\Delta S^\circ_{universe}$

When gasoline burns, one reaction is combustion of octane in air.

$$2\ C_8H_{18}(g) + 25\ O_2(g) \longrightarrow 16\ CO_2(g) + 18\ H_2O(\ell)$$

Calculate $\Delta S^\circ_{universe}$ for this reaction, thereby confirming that the reaction is product-favored. (Assume that reactants and products are at 298.15 K and 1 bar; use data from Appendix J.)

Answer 35,591 J/K

Strategy and Explanation A balanced chemical equation for the reaction is given. Calculate ΔH° and ΔS° by subtracting the sum of reactant values from the sum of product values.

$$\begin{aligned}\Delta H^\circ &= \Sigma\{(\text{moles of product}) \times \Delta H_f^\circ\ (\text{product})\} \\ &\quad - \Sigma\{(\text{moles of reactant}) \times \Delta H_f^\circ\ (\text{reactant})\} \\ &= \{16 \times (-393.509\ \text{kJ}) + 18 \times (-285.830\ \text{kJ})\} - \{2 \times (-208.447\ \text{kJ}) + 25 \times 0\ \text{kJ}\} \\ &= -11{,}024.2\ \text{kJ}\end{aligned}$$

$$\begin{aligned}\Delta S^\circ &= \Sigma\{(\text{moles of product}) \times S^\circ\ (\text{product})\} - \Sigma\{(\text{moles of reactant}) \times S^\circ\ (\text{reactant})\} \\ &= \{16 \times (213.74\ \text{J/K}) + 18 \times (69.91\ \text{J/K}\} - \{2 \times (466.835\ \text{J/K}) \\ &\quad + 25 \times (205.138\ \text{J/K})\} \\ &= -1383.9\ \text{J/K}\end{aligned}$$

Now use Equation 17.2 to calculate $\Delta S^\circ_{universe}$.

$$\begin{aligned}\Delta S^\circ_{universe} &= \Delta S_{surroundings} + \Delta S_{system} = \frac{-\Delta H^\circ}{T} + \Delta S^\circ \\ &= \frac{-(-11{,}024.2\ \text{kJ})}{298.15\ \text{K}} + (-1383.9\ \text{J/K}) \\ &= 36.975\ \text{kJ/K} - 1383.9\ \text{J/K} \\ &= 36{,}975\ \text{J/K} - 1383.9\ \text{J/K} = 35{,}591\ \text{J/K}\end{aligned}$$

Because $\Delta S^\circ_{universe}$ is positive, the reaction is product-favored.

Notice that for $-\Delta H^\circ/T$ the units were kilojoules per kelvin (kJ/K), while for ΔS° the units were joules per kelvin (J/K). Therefore it was necessary to convert kilojoules to joules by multiplying the first term by 1000 J/kJ.

☑ **Reasonable Answer Check** A positive result is reasonable, because you know that gasoline burns in air, which means that the reaction is product-favored. Even though the entropy change for the system is unfavorable, the reaction is highly product-favored because it is strongly exothermic.

PROBLEM-SOLVING PRACTICE 17.4

Use data from Appendix J to determine whether the synthesis of ammonia from nitrogen and hydrogen is product-favored at 298.15 K and 1 bar.

CONCEPTUAL EXERCISE 17.6 Variation of ΔH° and ΔS° with Temperature

Suppose that the combustion of gaseous octane is carried out at 150 °C.

(a) How would this affect the chemical equation for the reaction?

(b) What effect would the change in the chemical equation have on ΔH° and ΔS° for the reaction?

(c) When is it definitely *not* safe to assume that ΔH° and ΔS° for a reaction will be almost the same over a broad range of temperatures?

Predictions of the sort we have just made by calculating $\Delta S^\circ_{universe}$ can also be made qualitatively, without calculating, if we know whether a reaction is exothermic and if

Table 17.2 Predicting Whether a Reaction Is Product-Favored (at constant T and P)

Sign of ΔH_{system}	Sign of ΔS_{system}	Product-Favored?
Negative (exothermic)	Positive	Yes
Negative (exothermic)	Negative	Yes at low T; no at high T
Positive (endothermic)	Positive	No at low T; yes at high T
Positive (endothermic)	Negative	No

A product-favored reaction. Carbonates react rapidly with acid to produce carbon dioxide and water.

we can predict whether energy is dispersed within the system when the reaction takes place. ***A reaction is certain to be product-favored if it is exothermic and the entropy of the product atoms, molecules, and ions is greater than the entropy of the reactants ($\Delta S_{system} > 0$). Also, a reaction is certainly not product-favored if it is endothermic and there is a decrease in entropy for the system.*** There are two other possible cases, as indicated in Table 17.2, but they are more difficult to predict without quantitative information.

As examples, consider the reactions of carbonates with acids (***p. 150***). These reactions are product-favored because they are exothermic and produce gases. Reaction of limestone with hydrochloric acid is typical.

$$CaCO_3(s) + 2\,HCl(aq) \longrightarrow CaCl_2(aq) + H_2O(\ell) + CO_2(g) \qquad \text{(exothermic)}$$

Similarly, combustion reactions of hydrocarbons such as butane, $CH_3CH_2CH_2CH_3$, are product-favored because they are exothermic and produce a larger number of gas phase product molecules than there were gas-phase reactant molecules.

$$2\,CH_3CH_2CH_2CH_3(g) + 13\,O_2(g) \longrightarrow 8\,CO_2(g) + 10\,H_2O(g) \qquad \text{(exothermic)}$$

But what about a reaction such as the production of ethylene, $CH_2{=}CH_2$, from ethane, CH_3CH_3? Although entropy is predicted to increase (one gas phase molecule forms two), the reaction is very endothermic. (That the reaction is endothermic might be predicted on the basis of a decrease from seven bonds in the reactant molecule to six bonds—five single, one double—in the products.)

$$CH_3CH_3(g) \longrightarrow H_2(g) + CH_2{=}CH_2(g) \qquad \Delta H° = +137\text{ kJ};\ \Delta S° = +121\text{ J/K}$$

At 25 °C:

$$\Delta S°_{surroundings} = \frac{-137\text{ kJ}}{298\text{ K}} = -460\text{ J/K}$$

$$\Delta S°_{system} = 121\text{ J/K}$$

$$\Delta S°_{universe} = (-460 + 121)\text{ J/K} = -339\text{ J/K}$$

At 1000 °C:

$$\Delta S°_{surroundings} = \frac{-137\text{ kJ}}{(1000 + 273)\text{ K}} = -108\text{ J/K}$$

$$\Delta S°_{system} = 121\text{ J/K}$$

$$\Delta S°_{universe} = 13\text{ J/K}$$

Enthalpy change predicts that this process is reactant-favored, while entropy change predicts the opposite. Which is more important? It depends on the temperature.

Calculating $\Delta S°_{surroundings}$ ($= -\Delta H°/T$) requires dividing the enthalpy change by the temperature. Because $\Delta H°$ stays pretty much the same at different temperatures, the higher the temperature, the smaller the absolute value of $\Delta S°_{surroundings}$ ($= -\Delta H°/T$). At room temperature $\Delta S°_{surroundings}$ is usually bigger in absolute value than $\Delta S°_{system}$, so exothermic reactions are expected to be product-favored and endothermic reactions (like this one) are expected to be reactant-favored. The ethylene-producing reaction is reactant-favored at 25 °C, because $\Delta S°_{universe} = -339$ J/K. To make a successful industrial process, chemical engineers have designed plants that carry out this reaction at about 1000 °C. At this higher temperature, $\Delta S°_{surroundings}$ is smaller in magnitude than $\Delta S°_{system}$. Thus $\Delta S°_{universe} = 13$ J/K, and products are predicted to predominate over reactants.

EXERCISE 17.7 Predicting the Direction of a Reaction

Using data from Appendix J, complete the table and then classify each reaction into one of the four types in Table 17.2. Predict whether each reaction is product-favored or reactant-favored at room temperature.

Reaction	$\Delta H°$, 298 K (kJ)	$\Delta S°$, 298 K (J/K)
(a) $C_2H_4(g) + 3\ O_2(g) \longrightarrow 2\ H_2O(\ell) + 2\ CO_2(g)$	______	______
(b) $2\ Fe_2O_3(s) + 3\ C(graphite) \longrightarrow 4\ Fe(s) + 3\ CO_2(g)$	______	______
(c) $C(graphite) + O_2(g) \longrightarrow CO_2(g)$	______	______
(d) $2\ Ag(s) + 3\ N_2(g) \longrightarrow 2\ AgN_3(s)$	______	______

In Chapter 13 (p. 492) we stated that if a reaction is exothermic or involves an increase in entropy of the system, this favors the products. The entropy effect becomes more important the higher the temperature is.

EXERCISE 17.8 Product- or Reactant-Favored?

(a) Is the combination reaction of hydrogen gas and chlorine gas to give hydrogen chloride gas (at 1 bar) predicted to be product-favored or reactant-favored at 298 K?
(b) What is the value for $\Delta S°_{universe}$ for the reaction in part (a)?

Josiah Willard Gibbs
1838–1903

Gibbs was the son of a Yale professor of sacred literature, and like many of his forebears he studied at Yale, where he received his Ph.D. in 1863. His interest in both mathematics and engineering were reflected in the title of his thesis, "On the Form of the Teeth of Wheels in Spur Gearing." He spent three years in Europe, studying in Paris, Berlin, and Heidelberg, before becoming the first professor of mathematical physics at Yale. His work was not well known until long after he published it, because his published articles were compactly written and very abstract. They were also profound and influential, providing the basis for chemical thermodynamics. (See **http://www.aip.org/history/gap/Gibbs/Gibbs.html** for a more detailed biography.)

17.6 Gibbs Free Energy

Calculations of the sort done in the previous section would be simpler if we did not have to separately evaluate the entropy change of the surroundings from T and a table of ΔH_f° values and the entropy change of the system from a table of $S°$ values. To simplify such calculations, a new thermodynamic function was defined by J. Willard Gibbs (1838–1903). It is now called **Gibbs free energy** and it is given the symbol G. Gibbs defined his free energy so that $\Delta G_{system} = -T\Delta S_{universe}$. Because of the minus sign, if the entropy of the universe increases, the Gibbs free energy of the system must decrease. That is, ***a decrease in Gibbs free energy of a system is characteristic of a process that is product-favored at constant temperature and pressure.***

In the previous section we showed that the total entropy change accompanying a chemical reaction carried out at constant temperature and pressure is

$$\Delta S_{universe} = \Delta S_{surroundings} + \Delta S_{system} = \frac{-\Delta H_{system}}{T} + \Delta S_{system}$$

Combining this algebraically with Gibbs's definition of free energy, we have

$$\Delta G_{system} = -T\Delta S_{universe} = -T\left[\frac{-\Delta H_{system}}{T} + \Delta S_{system}\right] = \Delta H_{system} - T\Delta S_{system}$$

or, under standard-state conditions.

$$\Delta G° = \Delta H° - T\Delta S° \qquad [17.3]$$

This equation summarizes the ideas about chemical equilibrium that were developed in Chapter 13 (p. 493). A *negative* value of $\Delta G°$ indicates that a reaction is *product-favored*, and the equation says that two conditions will make $\Delta G°$ more negative: (1) if the reaction is exothermic, $\Delta H°$ will be negative, thereby favoring the products; and (2) if the products have greater entropy than the reactants, then $\Delta S°$ will

For $\Delta H°$, $\Delta S°$, and $\Delta G°$ you can assume that the values apply to the system—that is, $\Delta G° = \Delta G°_{system}$—unless a subscript is attached to indicate that a value is for the surroundings.

be positive and the $-T\Delta S°$ term will be negative, which favors the products. Because $\Delta S°$ is multiplied by T, the entropy of the system is more important at higher temperatures.

CONCEPTUAL **EXERCISE** 17.9 **Predicting Whether a Process Is Product-Favored**

Make a table similar to Table 17.2, but add a new column for the sign of $\Delta G°$. Based on the value of $\Delta G°$, predict whether the reaction is product-favored. If there is insufficient information, indicate whether the products would be favored more at high temperatures or at low temperatures. Check your results against Table 17.2.

It is important to realize that $\Delta G°$ varies significantly as the temperature changes (because of the $-T\Delta S°$ term). Therefore, values of $\Delta G°$ calculated from Equation 17.4 apply only to the temperature specified in the table of $\Delta G_f°$ values. Appendix J specifies a temperature of 25 °C.

The Gibbs free energy change provides a way of predicting whether a reaction will be product-favored that depends only on the system—that is, the chemical substances undergoing reaction. Therefore, we can tabulate values of the standard Gibbs free energy of formation, $\Delta G_f°$, for a variety of substances, and from them calculate

$$\Delta G° = \Sigma\{(\text{moles of product}) \times \Delta G_f°(\text{product})\} - \Sigma\{(\text{moles of reactant}) \times \Delta G_f°(\text{reactant})\} \quad [17.4]$$

for a great many reactions. The calculation is similar to using $\Delta H_f°$ values from Table 6.2 or Appendix J to calculate $\Delta H°$ for a reaction (← ***p. 210***). As was the case for $\Delta H_f°$ values, there are no $\Delta G_f°$ values for elements in their standard states, because forming an element from itself constitutes no change at all. Appendix J contains a table that includes $\Delta G_f°$ values for many compounds.

PROBLEM-SOLVING EXAMPLE 17.5 **Using Standard Gibbs Free Energies of Formation**

Calculate the standard Gibbs free energy change for the combustion of octane using values of $\Delta G_f°$ from Appendix J. Assume that the initial and final states have the same temperature and pressure so that Equation 17.4 applies.

Answer $-10{,}611$ kJ

Strategy and Explanation Write a balanced equation for the combustion reaction and look up $\Delta G_f°$ values in Appendix J.

$$2\,C_8H_{18}(g) + 25\,O_2(g) \longrightarrow 18\,H_2O(\ell) + 16\,CO_2(g)$$

$\Delta G_f°$(kJ/mol): 16.72 0 −237.1 −394.4

(Notice that elements in their standard states have $\Delta G_f° = 0$, just as they have $\Delta H_f° = 0$.)

Now calculate

$$\begin{aligned}\Delta G° &= \Sigma\{(\text{moles of product}) \times \Delta G_f°\ (\text{product})\}\\ &\quad - \Sigma\{(\text{moles of reactant}) \times \Delta G_f°\ (\text{reactant})\}\\ &= \{(18\text{ mol} \times \Delta G_f°\ [H_2O(g)]\} + \{16\text{ mol} \times \Delta G_f°[CO_2(g)]\}\\ &\quad - \{2\text{ mol} \times \Delta G_f°[C_8H_{18}(g)]\} - \{25\text{ mol} \times \Delta G_f°[O_2(g)]\}\\ &= \{18\text{ mol} \times (-237.1\text{ kJ/mol})\} + \{16\text{ mol} \times (-394.4\text{ kJ/mol})\}\\ &\quad - \{2\text{ mol} \times (16.72\text{ kJ/mol})\} - \{25\text{ mol} \times (0\text{ kJ/mol})\}\\ &= -10{,}611\text{ kJ}\end{aligned}$$

The Gibbs free energy change, $\Delta G°$, is a large negative number, indicating that the reaction is product-favored under standard-state conditions.

☑ **Reasonable Answer Check** A large, negative Gibbs free energy change is reasonable for a combustion reaction, which, once initiated, occurs of its own accord.

PROBLEM-SOLVING PRACTICE 17.5

In the text we concluded that the reaction to produce methanol from CO and H_2 is product-favored.

$$CO(g) + 2\,H_2(g) \longrightarrow CH_3OH(\ell)$$

(a) Verify this result by calculating $\Delta G°$ from $\Delta H°$ and $\Delta S°$ for the system. Use values of $\Delta H_f°$ and $S°$ from Appendix J.
(b) Compare your result in part (a) with the calculated value of $\Delta G°$ obtained from $\Delta G_f°$ values from Appendix J.
(c) Is the sign of $\Delta G°$ positive or negative? Is the reaction product-favored? At all temperatures?

The Effect of Temperature on Reaction Direction

Many reactions are product-favored at some temperatures and reactant-favored at other temperatures. Thus, it might be possible to make such a reaction produce products by increasing or decreasing the temperature. There is a simple, approximate way to estimate the temperature at which a reactant-favored process becomes product-favored.

The effect of temperature on reaction direction can be seen in Equation 17.2 (p. 622). Because $\Delta S_{surroundings} = -\Delta H_{system}/T$ and ΔH is nearly constant over a broad range of temperatures, the larger T is the smaller (and therefore less important) the entropy change of the surroundings becomes relative to the entropy change of the system. At high temperatures ΔS_{system} governs the directionality of the reaction.

In Exercises 17.5 and 17.6 (p. 622 and p. 623) we developed the idea that $\Delta H°$ and $\Delta S°$ have nearly constant values over a broad range of temperatures, provided that each of the substances involved in a chemical reaction remains in the same state of matter (solid, liquid, or gas). Because $\Delta H°$ and $\Delta S°$ are nearly constant, the T on the right-hand side of the equation $\Delta G° = \Delta H° - T\Delta S°$ implies that $\Delta G°$ must vary with temperature. It also implies that if we know $\Delta H°$ and $\Delta S°$ at one temperature, we can estimate $\Delta G°$ over a range of temperatures. For example, suppose we want to know whether the reaction

$$2\,HgO(s) \longrightarrow 2\,Hg(\ell) + O_2(g)$$

will produce products at a temperature of 350. °C and a pressure of 1 bar. To find out, calculate $\Delta H°$ and $\Delta S°$ at 298 K and then estimate $\Delta G°$, assuming that $\Delta H°$ and $\Delta S°$ have the same values at 350. °C (623 K) that they do at 298 K. The boiling point of mercury is 356 °C, and mercury(II) oxide does not melt until well above 500. °C, so the substances are all in the same states at 350. °C that they were at 25 °C.

$$\begin{aligned}
\Delta S°(298.15\text{ K}) &= \{2\text{ mol} \times S°[Hg(\ell)] + (1\text{ mol}) \times S°[O_2(g)]\} \\
&\quad - \{2\text{ mol} \times S°[HgO(s)]\} \\
&= 2 \times 76.02\text{ J/K} + 205.138\text{ J/K} - 2 \times 70.29\text{ J/K} = 216.60\text{ J/K} \\
\Delta H°(298.15\text{ K}) &= \{2\text{ mol} \times \Delta H_f°[Hg(\ell)] + 1\text{ mol} \times \Delta H_f°[O_2(g)]\} \\
&\quad - \{2\text{ mol} \times \Delta H_f°[HgO(s)]\} \\
&= 0\text{ kJ} + 0\text{ kJ} - 2 \times (-90.83)\text{ kJ} = 181.66\text{ kJ} \\
\Delta G°(623\text{ K}) &= \Delta H°(298.15\text{ K}) - T \times \Delta S°(298.15\text{ K}) \\
&= 181.66\text{ kJ} - 623\text{ K} \times 216.60\text{ J/K} \\
&= 181.66\text{ kJ} - 134{,}942\text{ J} = 181.66\text{ kJ} - 134.9\text{ kJ} = 47\text{ kJ}
\end{aligned}$$

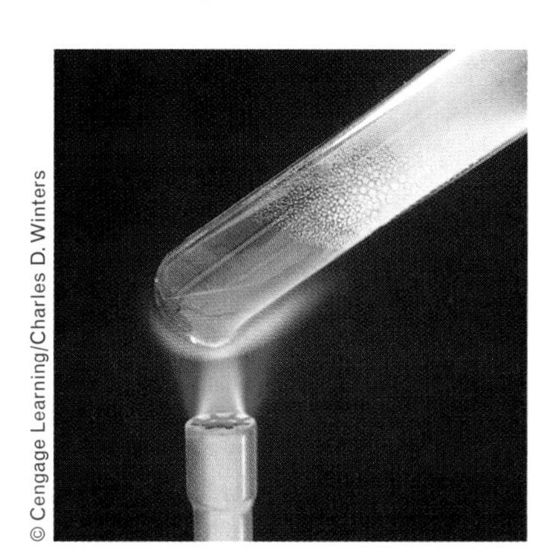

© Cengage Learning/Charles D. Winters

Figure 17.6 Decomposition of HgO(s). When heated, red mercury(II) oxide decomposes to liquid mercury metal and oxygen gas. Shiny droplets of mercury can be seen where they have candensed in the cooler part of the test tube.

Note that 134,942 J has been converted to 134.9 kJ in the last step.

Because $\Delta G°$ is positive, the reaction is not product-favored at 623 K. But at 623 K $\Delta G°$ has a smaller positive value than at 298 K, where it has the value $\Delta G° = -(2\text{ mol}) \times \Delta G_f°(HgO[s]) = +117.1$ kJ. That is, the reaction is more product-favored at 623 K than it is at room temperature. Because the calculated values of $\Delta H°$ and $\Delta S°$ are both positive, the reaction is expected to be product-favored at high temperatures but not at low temperatures. Heating to an even higher temperature than 623 K does decompose mercury(II) oxide (see Figure 17.6).

Joseph Priestley's discovery of oxygen in 1774 involved heating mercury(II) oxide to decompose it.

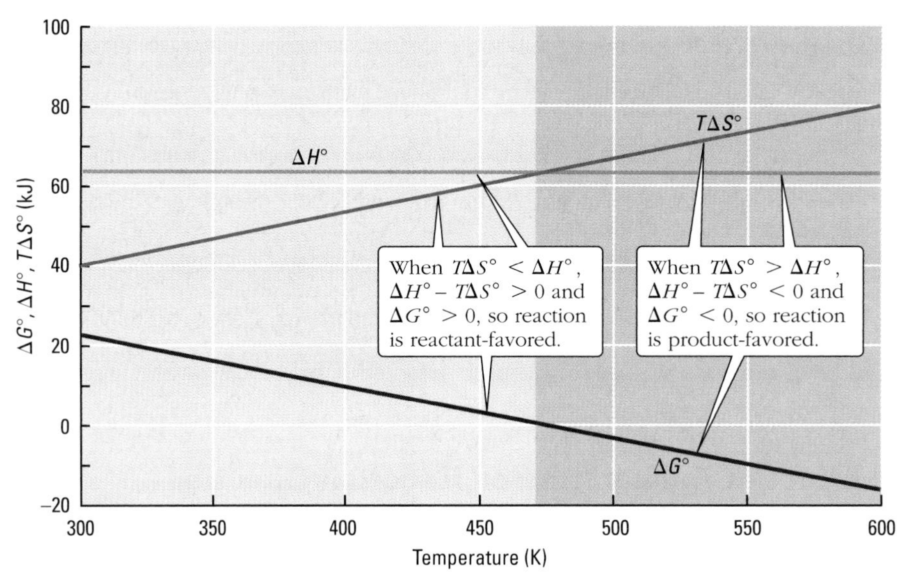

Figure 17.7 Effect of temperature on reaction spontaneity. The two terms that contribute to the Gibbs free energy change, $\Delta H°$ and $T\Delta S°$, are plotted as a function of temperature for the reaction of silver ore, $Ag_2O(s)$, to form silver metal and oxygen gas.

CONCEPTUAL
EXERCISE 17.10 High-Temperature Decomposition

Suppose that a sample of HgO(s) is heated above the boiling point of mercury (356 °C).

(a) Could you use the same method to estimate the Gibbs free energy change that was used in the preceding paragraph? Why or why not? (*Hint:* Write the equation for the process that would occur if the temperature were 400 °C.)

(b) At 400 °C, would you expect the reaction to be more or less product-favored than it was at 350 °C? Give two reasons for your choice.

For reactions such as decomposition of mercury(II) oxide that are reactant-favored at low temperatures and product-favored at high temperatures, it is possible to calculate the minimum temperature to which the system must be heated to make it product-favored. Below that temperature $\Delta G°$ is positive, and above that temperature $\Delta G°$ is negative. Therefore, $\Delta G°$ must equal zero at the desired temperature. Because $\Delta G° = \Delta H° - T\Delta S°$ (Equation 17.3), we can set $\Delta H° - T\Delta S° = 0$. Solving for T gives

$$T(\text{at which } \Delta G° \text{changes sign}) = \frac{\Delta H°}{\Delta S°} \quad [17.5]$$

As an example, the contributions of $\Delta H°$ and $-T\Delta S°$ to $\Delta G°$ for the decomposition of silver(I) oxide are shown graphically in Figure 17.7. Equation 17.5 also applies to reactions that are product-favored at low temperatures and reactant-favored at high temperatures. For such reactions Equation 17.5 gives the temperature *below* which the reaction is product-favored. Heating the system above this temperature will probably result in insufficient quantities of products being produced.

PROBLEM-SOLVING EXAMPLE 17.6 Effect of Temperature on Gibbs Free Energy Change

Calculate the temperature to which silver ore, $Ag_2O(s)$, must be heated to decompose the ore to oxygen gas and solid silver metal.

Answer 468 K (195 °C)

Strategy and Explanation Begin by writing the equation for the reaction, and use values from Appendix J to calculate $\Delta H°$ and $\Delta S°$ at 25 °C. Then use Equation 17.5 to calculate the desired temperature.

$$2\,Ag_2O(s) \longrightarrow 4\,Ag(s) + O_2(g)$$

$$\Delta H° = (4\text{ mol}) \times (0\text{ kJ/mol}) + (1\text{ mol}) \times (0\text{ kJ/mol}) - (2\text{ mol}) \times (-31.05\text{ kJ/mol}) = 62.10\text{ kJ}$$

$$\Delta S° = (4\text{ mol}) \times (42.55\text{ J K}^{-1}\text{ mol}^{-1}) + (1\text{ mol}) \times (205.138\text{ J K}^{-1}\text{ mol}^{-1}) - (2\text{ mol}) \times (121.3\text{ J K}^{-1}\text{ mol}^{-1}) = 132.7\text{ J/K}$$

Because both $\Delta H°$ and $\Delta S°$ are positive, the reaction will become more product-favored as temperature increases.

$$T = \frac{62.1\text{ kJ}}{132.7\text{ J/K}} \times \frac{1000\text{ J}}{1\text{ kJ}} = \frac{62.1\text{ kJ}}{0.1327\text{ kJ/K}} = 467.9\text{ K} = 468\text{ K}$$

Note that $\Delta H°$ has units of kJ, whereas $\Delta S°$ has units of J/K; this necessitates a unit conversion so that the energy units can cancel.

☑ **Reasonable Answer Check** The reaction produces a gas and a solid from a solid, so the entropy change should be positive. Silver is one of the few metals that can be found as the element in nature, so its oxide probably does not require an extremely high temperature to decompose; 468 K (195 °C) is a reasonable answer. Check that each reactant and each product is in the same physical state at 195 °C as it was at room temperature. The *CRC Handbook* gives the melting point of silver as 962 °C and reports that silver(I) oxide decomposes before it melts, so the approximation of nearly constant $\Delta S°$ and $\Delta H°$ is valid.

PROBLEM-SOLVING PRACTICE 17.6

For the reaction

$$2\,CO(g) + O_2(g) \longrightarrow 2\,CO_2(g)$$

(a) Predict the temperature at which the reaction changes from being product-favored to being reactant-favored.

(b) If you wanted this reaction to produce CO_2(g), what temperature conditions would you choose?

17.7 Gibbs Free Energy Changes and Equilibrium Constants

In Chapter 13 you learned that if a chemical reaction has an equilibrium constant greater than 1, then the reaction is product-favored. If the equilibrium constant is less than 1, then the reaction is reactant-favored. In the preceding section you learned that if the Gibbs free energy change for a process is negative, the process is product-favored; if the Gibbs free energy change is positive, the process is reactant-favored. Because both the Gibbs free energy and the equilibrium constant tell us whether a reaction is product-favored, it is reasonable to conclude that there these two seemingly different quantities are related. That conclusion is correct. The relationship is

$$\Delta G° = -RT \ln K° \qquad [17.6]$$

where $\Delta G°$ is the standard Gibbs free energy change for the reaction, R is the ideal gas constant, T is the absolute temperature, and $K°$ is the **standard equilibrium constant.** The expression for $K°$ is similar to the equilibrium constant as defined in Chapter 13 (← ***p. 466***) except that each concentration is divided by the standard-state concentration of 1 mol/L and each pressure is divided by the standard-state pressure of 1 bar. That is, for the reaction

$$a\,A + b\,B \rightleftharpoons c\,C + d\,D$$

Relation between $\Delta G°$ and $K°$ at 25 °C

$\Delta G°$ (kJ/mol)	$K°$
200	9×10^{-36}
100	3×10^{-18}
10	2×10^{-2}
1	7×10^{-1}
0	1
−1	1.5
−10	6×10^{1}
−100	3×10^{17}
−200	1×10^{35}

$K°$	$\Delta G°$ ($-RT \ln K°$)	Product-Favored?
<1	Positive	No
>1	Negative	Yes
=1	0	Neither

For formic acid in water, the equilibrium constant is K_a, a ratio of concentrations (p. 550).

the standard equilibrium constant in terms of pressures is

$$K° = \frac{\left(\frac{P_C}{P°}\right)^c \left(\frac{P_D}{P°}\right)^d}{\left(\frac{P_A}{P°}\right)^a \left(\frac{P_B}{P°}\right)^b}$$

where $P°$ is the standard-state pressure of 1 bar. Even if the equilibrium constant has units (of concentration or pressure), $K°$ is unitless. If the reaction occurs in solution, $K°$ has the same form (and the same value) as the concentration equilibrium constant. For gases, $K°$ relates pressures, not concentrations, and has the same value as K_P.

Regardless of the choice of standard state, Equation 17.6 indicates that the Gibbs free energy change for a reaction is the negative of a constant times the temperature times the natural logarithm of the equilibrium constant. If $K°$ is larger than 1, then $\ln K°$ is positive and $\Delta G°$ will be negative because of the minus sign. Both of these conditions, a negative $\Delta G°$ and $K° > 1$, indicate that the reaction is *product-favored* under standard-state conditions. Conversely, if $K° < 1$, then $\ln K°$ is negative and $\Delta G°$ must be positive, indicating a *reactant-favored* system.

$\Delta G°$ is the difference in Gibbs free energy between products in their standard states and reactants in their standard states. For example, for ionization of formic acid in water,

$$\underset{1\ \text{mol/L}}{HCOOH(aq)} \rightleftharpoons \underset{1\ \text{mol/L}}{HCOO^-(aq)} + \underset{1\ \text{mol/L}}{H^+(aq)} \qquad \Delta G° = 21.3\ \text{kJ}$$

the Gibbs free energy change of 21.3 kJ is for converting 1 mol HCOOH(aq) at a concentration of 1 mol/L into 1 mol $HCOO^-$(aq) and 1 mol H^+(aq), each at a concentration of 1 mol/L. Since $\Delta G°$ is positive, we predict that the process will be reactant-favored. This agrees with the fact that formic acid is a weak acid, is only slightly ionized, and therefore has an equilibrium constant much smaller than 1.

PROBLEM-SOLVING EXAMPLE 17.7 Gibbs Free Energy and Equilibrium Constant

In the preceding paragraph you learned that $\Delta G° = 21.3$ kJ at 25 °C for the reaction

$$HCOOH(aq) \rightleftharpoons HCOO^-(aq) + H^+(aq)$$

Use this information to calculate the equilibrium constant, K_a, for ionization of formic acid in aqueous solution at 25 °C.

Answer $K_a = K_c = K° = 1.8 \times 10^{-4}$

Strategy and Explanation The relation between Gibbs free energy and $K°$ was given in Equation 17.6 as

$$\Delta G° = -RT \ln K°$$

To obtain $K°$ from $\Delta G°$, we first divide both sides of the equation by $-RT$:

$$-\Delta G°/RT = \ln K°$$

Next, we make use of the properties of logarithms (which are discussed in Appendix A.6). Since ln represents a logarithm to the base e, we can remove the logarithm function by using each side of the equation as an exponent of e.

$$e^{-\Delta G°/RT} = e^{\ln K°} = K°$$

Now we can substitute the known values into the equation.

$$K° = e^{-\Delta G°/RT} = e^{-(21.3\ \text{kJ/mol})(1000\ \text{J/kJ})/(8.314\ \text{J K}^{-1}\ \text{mol}^{-1})(298\ \text{K})} = 1.8 \times 10^{-4}$$

Thus, the positive value of $\Delta G°$ results in a value of $K°$ less than 1 and, indeed, indicates a reactant-favored system. Because this reaction occurs in aqueous solution, the standard states of reactants and products involve concentrations. Therefore $K° = K_c = K_a$.

Make certain to check units in calculations like this one. Standard Gibbs free energy changes involve kilojoules, and the gas constant R involves joules, so a unit conversion is needed.

☑ **Reasonable Answer Check** Formic acid is a weak acid and is not expected to have a very large equilibrium constant. The value calculated appears to be reasonable.

PROBLEM-SOLVING PRACTICE 17.7

For each of the following reactions, evaluate $K°$ at 298 K from the standard free energy change. If necessary, obtain data from Appendix J to calculate $\Delta G°$. Check your results against the K_c and K_P values in Table 13.1 (p. 475). For which of these reactions is $K_c = K°$?

(a) $CaCO_3(s) \rightleftharpoons Ca^{2+}(aq) + CO_3^{2-}(aq)$

(b) $H_2CO_3(aq) \rightleftharpoons HCO_3^-(aq) + H^+(aq)$

(c) $2\ NO_2(g) \rightleftharpoons N_2O_4(g)$

Remember also that $\Delta G°$ can be calculated from the equation

$$\Delta G° = \Delta H° - T\Delta S°$$

If we know or can estimate changes in enthalpy and entropy for a reaction, then we can calculate or estimate the standard Gibbs free energy change and hence the equilibrium constant. And because $\Delta H°$ and $\Delta S°$ have nearly constant values over a wide range of temperatures, we can estimate equilibrium constants at a variety of temperatures, not just at 25 °C. Problem-Solving Example 17.8 shows how to do this.

Assuming constant values of $\Delta H°$ and $\Delta S°$ over a broad range of temperatures does not work well for reactions involving ions in aqueous solution, such as (a) and (b) in Problem-Solving Practice 17.7, because the extent of hydration of the ions varies with temperature.

PROBLEM-SOLVING EXAMPLE 17.8 Estimating $K°$ at Different Temperatures

Use data from Appendix J to obtain values of $\Delta H°$ and $\Delta S°$ for the reaction

$$N_2(g) + O_2(g) \rightleftharpoons 2\ NO(g)$$

From these data estimate the value of $\Delta G°$ and hence the value of $K°$ at (a) 298 K, (b) 1000. K, and (c) 2300. K.

This reaction is of great importance because it can take place to a significant extent in high-temperature combustion processes. If the temperature is high enough, nitrogen and oxygen form the air pollutant NO.

Answer

(a) $\Delta G° = 173.1$ kJ; $K° = 4.5 \times 10^{-31}$ (b) $\Delta G° = 155.7$ kJ; $K° = 7.3 \times 10^{-9}$

(c) $\Delta G° = 123.5$ kJ; $K° = 1.57 \times 10^{-3}$

Strategy and Explanation At each temperature, use the Gibbs equation to calculate $\Delta G° = \Delta H° - T\Delta S°$. Then calculate $K°$ as was done in Problem-Solving Example 17.7. Part (c) is done below to illustrate the calculations:

$$\Delta G° = \Delta H° - T\Delta S° = 180{,}500\ \text{J} - (2300.\ \text{K})(24.772\ \text{J/K})$$

$$= 1.235 \times 10^5\ \text{J} = 123.5\ \text{kJ}$$

$$K° = e^{-\Delta G°/RT} = e^{-(1.235\times10^5\ \text{J})/(8.314\ \text{J/K})(2300.\ \text{K})} = 1.57 \times 10^{-3}$$

☑ **Reasonable Answer Check** The reaction is endothermic, so Le Chatelier's principle predicts that the equilibrium will shift toward products as the temperature increases. This is reflected by the increasing values of $K°$ as the temperature rises.

PROBLEM-SOLVING PRACTICE 17.8

For the ammonia synthesis reaction,

$$N_2(g) + 3\ H_2(g) \rightleftharpoons 2\ NH_3(g)$$

estimate the equilibrium constant at (a) 298. K, (b) 450. K, and (c) 800. K.

17.8 Gibbs Free Energy and Maximum Work

An important interpretation of the Gibbs free energy is that **ΔG *represents the maximum work that can be done by a product-favored system on its surroundings under conditions of constant temperature and pressure.* ΔG *also represents***

The work represented by ΔG does not include work done by expansion of a system, pushing back the atmosphere.

the minimum work that must be done to cause a reactant-favored process to occur. Consider the product-favored reaction of hydrogen with oxygen to form liquid water under standard conditions.

$$2\ H_2(g) + O_2(g) \rightleftharpoons 2\ H_2O(\ell) \qquad \Delta G° = -474.258\ kJ$$

This thermochemical expression tells us that for every 2 mol $H_2O(\ell)$ produced, as much as 474.258 kJ of work could be done. The negative sign of $\Delta G°$ tells us that the work is done on the surroundings. (Because the system has less Gibbs free energy after the reaction than before it, the surroundings will have more energy.) Even if the reactants and the products are not at standard pressure or concentration, ΔG still equals $-w_{max}$, the maximum work the system can do on its surroundings.

$$\Delta G = -w_{max}\ \text{(work done on the surroundings)} \qquad [17.7]$$

It is important to remember that w_{max} is the maximum work the system can do on the *surroundings*. Therefore the sign of w_{max} is opposite to that of ΔG.

Because transformations of energy from one form to another are not 100% efficient, we seldom observe anything close to the maximum quantity of work given by the value of $\Delta G°$.

Now consider the decomposition of water to form hydrogen and oxygen, which is the reverse of the previous reaction.

$$2\ H_2O(\ell) \rightleftharpoons 2\ H_2(g) + O_2(g) \qquad \Delta G° = 474.258\ kJ$$

The positive value of $\Delta G°$ indicates that this process is reactant-favored. Because the Gibbs free energy of the products is greater than the Gibbs free energy of the reactant, at least 474.258 kJ must be supplied for every 2 mol $H_2O(\ell)$ that decomposes. This 474.258 kJ is the minimum work that must be done to change 2 mol liquid water into hydrogen gas and oxygen gas. One way to supply this work is to use a direct electric current to carry out electrolysis of the water. In general, a continuous supply of energy is required for a reactant-favored process, such as decomposition of liquid water, to continue.

PROBLEM-SOLVING EXAMPLE 17.9 Gibbs Free Energy Change and Maximum Work

Use data from Appendix J to predict whether each reaction is product-favored or reactant-favored at 25 °C and 1 bar. For each product-favored reaction, calculate the maximum work the reaction could do. For each reactant-favored process, calculate the minimum work needed to cause it to occur.

(a) $2\ Al_2O_3(s) \longrightarrow 4\ Al(s) + 3\ O_2(g)$ (b) $Cl_2(g) + Mg(s) \longrightarrow MgCl_2(s)$

Answer

(a) Reactant-favored; at least 3164.6 kJ must be supplied

(b) Product-favored; can do up to 591.79 kJ of useful work

Strategy and Explanation Use data from Appendix J to calculate $\Delta G°$ for each reaction. If $\Delta G°$ is negative, the process is product-favored, and the value of $\Delta G°$ gives the maximum work that can be done. If $\Delta G°$ is positive, the process is reactant-favored, and the value tells the minimum work that has to be done to force the reaction to occur.

(a) $\Delta G° = 0 + 0 - 2(-1582.3)\ kJ = 3164.6\ kJ$; at least 3164.6 kJ is required.

(b) $\Delta G° = -591.79\ kJ - 0 - 0 = -591.79\ kJ$; up to 591.79 kJ useful work can be done.

☑ **Reasonable Answer Check** Reaction (a) is decomposition of an oxide to a metal and oxygen. Because metals are good reducing agents and oxygen is a strong oxidizing agent, the reverse of this reaction is likely to be product-favored, which would make Reaction (a) reactant-favored. This result agrees with the calculation. Reaction (b) is combination of an alkaline earth element with a halogen, which should form a stable ionic compound. Therefore Reaction (b) should be product-favored, which agrees with the calculation. In both cases the value of $\Delta G°$ is large, which also would be expected based on the arguments just given.

PROBLEM-SOLVING PRACTICE 17.9

Predict whether each reaction is reactant-favored or product-favored at 298 K and 1 bar, and calculate the minimum work that would have to be done to force it to occur, or the maximum work that could be done by the reaction.

(a) $2\ CO_2(g) \longrightarrow 2\ CO(g) + O_2(g)$ (b) $4\ Fe(s) + 3\ O_2(g) \longrightarrow 2\ Fe_2O_3(s)$

Coupling Reactant-Favored Processes with Product-Favored Processes

A dead car battery will not charge itself. The process that takes place when a battery is charged is reactant-favored. But a battery can be charged if it is connected to a charger that is, in turn, powered by electricity generated in a power plant that burns coal. Coal, which is mainly carbon, burns in air according to the equation

$$C(s) + O_2(g) \longrightarrow CO_2(g) \qquad \Delta G^\circ = -394.4 \text{ kJ}$$

If enough coal is burned, the large negative Gibbs free energy change for its combustion more than offsets the positive Gibbs free energy change of the battery-charging process. An overall decrease in Gibbs free energy occurs, even though the battery-charging part we are interested in has an increase. Once a battery has been charged, the charging reaction's reverse (which is product-favored) can supply electricity to start a car's engine or play its radio. Some of the Gibbs free energy lost when the coal was burned has been stored in the car's battery for use later.

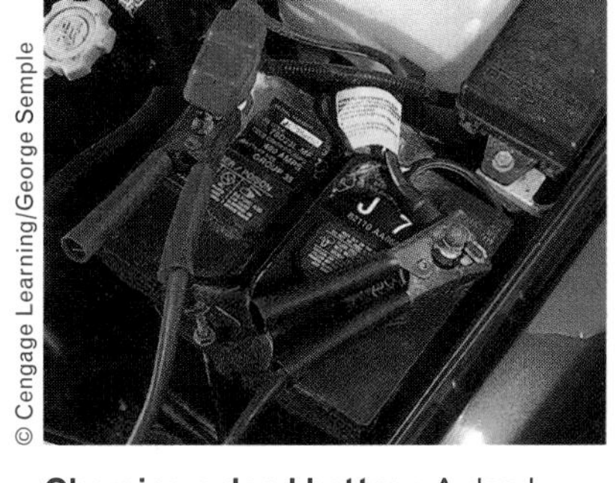

© Cengage Learning/George Semple

Charging a dead battery. A dead battery can be charged by using electricity from a power plant to cause a reactant-favored process to occur. After the battery has been charged, the reverse of that reactant-favored process (a product-favored process) can generate electricity to start the car.

Charging a battery is an example of coupling a product-favored reaction with a reactant-favored process to cause the latter to take place. Both processes occur at the same time and in a way that allows the Gibbs free energy released by the product-favored reaction to be used by the reactant-favored reaction. Other examples include obtaining aluminum or iron from their ores; synthesizing large, complicated molecules from simple reactants to make medicines, plastics, and other useful materials; and keeping a house cool when the outside temperature is above 100 °F. All involve decreasing entropy in the region of our interest, but all can be made to occur provided that there is a larger increase in entropy at a power plant or somewhere else.

The Gibbs free energy change indicates a chemical reaction's capacity to drive a different reactant-favored system to produce products. The word "free" in the name indicates not "zero cost," but rather "available." ***Gibbs free energy is available to do useful tasks that would not happen on their own.*** Another way of saying this is that Gibbs free energy is a measure of the *quality* of the energy contained in a chemical reaction system. If it contains a lot of Gibbs free energy, a chemical system can do a lot of useful work for us; the energy is of high quality—potentially useful to humankind. When the system's reactants are transformed into products, that available free energy can do useful work, but only if the reaction is coupled to some other, reactant-favored process we want to carry out. If systems are not coupled, then the free energy released by a reaction will be wasted as thermal energy. All living things depend on coupling of chemical reactions to provide for their energy needs.

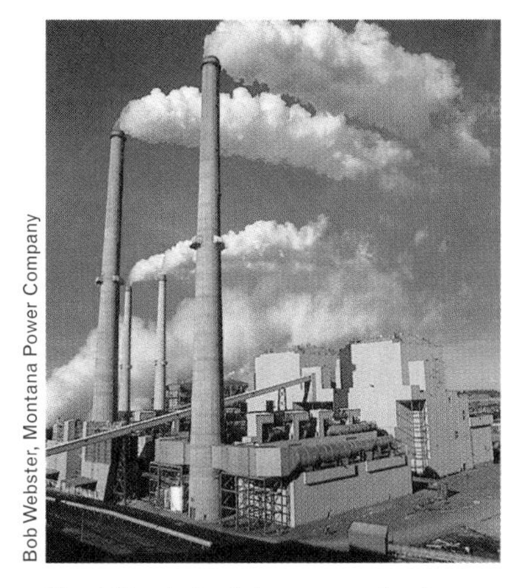

Bob Webster, Montana Power Company

Coal-fired electric power plant.

CONCEPTUAL
EXERCISE 17.11 Coupling Reactions

One way to produce iron metal is to reduce iron(III) oxide with aluminum. This is called the thermite reaction. You can think of the reaction as occurring in two steps. The first is the loss of oxygen from iron(III) oxide,

(i) $Fe_2O_3(s) \longrightarrow 2\,Fe(s) + \frac{3}{2}\,O_2(g)$

The second is the combination of aluminum with the oxygen,

(ii) $2\,Al(s) + \frac{3}{2}\,O_2(g) \longrightarrow Al_2O_3(s)$

(a) Calculate the enthalpy, entropy, and Gibbs free energy changes for each step. Decide whether each step is product- or reactant-favored. Comment on the signs of ΔH°, ΔS°, and ΔG° for each step.

(b) What is the overall net reaction that occurs when aluminum is combined with iron(III) oxide? What are the enthalpy, entropy, and Gibbs free energy changes for the overall reaction? Is it product- or reactant-favored? Comment on the signs of ΔH°, ΔS°, and ΔG° for the overall reaction.

(c) Discuss briefly how coupling Reaction (i) with Reaction (ii) affects our ability to obtain iron metal from iron(III) oxide by reacting it with aluminum.

(d) Suggest a reaction other than oxidation of aluminum that might be used to reduce iron(III) oxide to iron. Test your selection by calculating the Gibbs free energy change for the coupled system.

© Cengage Learning/Charles Steele

Thermite reaction. Exercise 17.11 deals with the reaction of aluminum with iron(III) oxide. The reaction is very product-favored and releases a large quantity of Gibbs free energy.

Diamond: a material resource. Energy resources are like other natural resources in that they contain high-quality, concentrated energy. An analogy is a material resource such as a diamond, which is pure carbon with the atoms bonded so that each is surrounded tetrahedrally by four others. A diamond is valuable because it consists of a single crystal with atoms arranged in a specific way. If you ground a diamond into dust and spread the carbon it was made of over the area of a city block, the carbon would be nearly worthless, because it would require tremendous expense to collect the carbon and convert it back to diamond. Similarly, an energy resource is valuable not for the energy it contains, but because that energy is concentrated and available to do useful work.

17.9 Conservation of Gibbs Free Energy

When a ton of coal is burned its energy has not been used up. The law of conservation of energy (⇐ *p. 180*) summarizes many experiments whose results verify that energy cannot be destroyed. When coal is burned in a power plant, its chemical energy is changed to an equal quantity of energy in other forms. These are electrical energy, which can be very useful, and thermal energy in the gases going up the smokestack and in the immediate surroundings of the plant, which is much less useful. However, an *energy resource* has been used up: the coal's ability to store energy and release it to do work. When coal burns in air, some of the Gibbs free energy that was in the coal and the oxygen that combined with it has been used up. This fact is indicated by the negative value of $\Delta G°$ for the combustion of coal. The same is true of any other product-favored reaction.

What we commonly refer to as **energy conservation** *is actually conservation of useful energy: Gibbs free energy.* Energy conservation does not mean conserving energy—nature takes care of conserving energy automatically. But nature does not automatically conserve Gibbs free energy. Substances with high Gibbs free energies are energy resources, and it is their *useful* energy that we must take pains to conserve. Once a product-favored reaction with a negative $\Delta G°$ has taken place, it cannot be reversed, thereby restoring the Gibbs free energy of its reactants, without coupling the reverse reaction with some other product-favored reaction. That is, once we have used an energy resource, it cannot be restored, except by using some other energy resource. Analysis of chemical systems in terms of Gibbs free energy can lead to important insights into how energy resources can be conserved effectively.

By comparing Gibbs free energy changes calculated using the equations in this chapter with the actual loss of Gibbs free energy in industrial processes, environmentalists and industrialists can suggest ways to minimize loss of Gibbs free energy. For example, there is a very large quantity of Gibbs free energy stored in aluminum metal and oxygen gas compared with aluminum ore, Al_2O_3. This can be seen from the thermochemical expression

$$2\,Al_2O_3(s) \longrightarrow 4\,Al(s) + 3\,O_2(g) \qquad \Delta G° = 3164.6\text{ kJ}$$

ESTIMATION Gibbs Free Energy and Automobile Travel

Given that $\Delta G° = -5295.74$ kJ per mole of octane burned, estimate the quantity of Gibbs free energy consumed when a typical car makes a 1000-mile round trip on interstate highways.

Assume that the typical car averages 20 miles per gallon and that combustion of gasoline can be approximated by combustion of octane. Because the trip is a round trip, the car ends up exactly where it started out, which means that it has done no useful thermodynamic work. Therefore all of the Gibbs free energy released by combustion of the fuel is lost. The combustion reaction is

$$C_8H_{18}(\ell) + \tfrac{25}{2}\,O_2(g) \longrightarrow 8\,CO_2(g) + 9\,H_2O(\ell) \qquad \Delta G° = -5295.74\text{ kJ}$$

Fuel economy of 20 miles per gallon means that five gallons of fuel will be used in 100 miles and 50 gallons in 1000 miles. One gallon is four quarts and a quart is about a liter, so the volume of octane is about 4 × 50 = 200 L. The density of gasoline must be less than that of water, because gasoline floats on water. Assume that it is about 80% as big. Then the density is 0.8 g/mL or 800 g/L. The 200 L of fuel weighs about 200 × 800 = 160,000 g.

The molar mass of octane, C_8H_{18}, is about 8 × 12 + 18 = 114 g/mol. To make the arithmetic easier, round this value to 100 g/mol. Then 160,000 g of octane corresponds to 1600 mol of octane. The Gibbs free energy released by combustion is about 5000 kJ for every mole of octane burned. Since 1600 × 5000 = 8,000,000 kJ, about 8 million kJ of useful energy is consumed for every 1000 miles a car is driven. Most of us drive ten times that far every year, and there are a lot of cars in the United States, so the energy resources consumed by automobile travel are huge.

Visit this book's companion website at **www.cengage.com/chemistry/moore** to work an interactive module based on this material.

which shows that the Gibbs free energy of 4 mol Al(s) and 3 mol O_2(g) is 3164.6 kJ higher than the Gibbs free energy of 2 mol Al_2O_3(s). If 4 mol Al(s) is oxidized to aluminum oxide, 3164.6 kJ of Gibbs free energy is lost—energy that was expended to manufacture the aluminum is wasted if the aluminum is oxidized. It is not surprising, then, that major programs for recycling aluminum operate throughout the United States. A similar statement can be made about almost every metal: Once reduced from their ores, metals are storehouses of Gibbs free energy that should be maintained in their reduced forms to avoid repeating the expenditure of Gibbs free energy needed to separate them from chemical combination with oxygen.

There is a lot of energy in the molecules of a bathtub full of water at room temperature, but that energy cannot easily be used to generate electricity. The energy is not useful in the sense that energy stored in a fuel and oxygen is useful.

It is important to recognize that completely eliminating consumption of Gibbs free energy is impossible. Whenever anything happens, whether a chemical reaction or a physical process, the final state must have less Gibbs free energy than was available initially. This is the same as saying that the entropy of the universe must have increased during the change. This statement is true of any system in which the initial substances are changed into something new—any product-favored system. Thus, losses of Gibbs free energy are inevitable. The aim of energy conservation is to minimize—not eliminate—them. This can be done by maximizing the efficiency of coupling reactions that release Gibbs free energy to processes we want to cause to occur. The ideas of thermodynamics help us figure out how to accomplish that goal and are the most powerful tool we have for conserving energy while maintaining a high standard of living.

17.10 Thermodynamic and Kinetic Stability

Chemists often say that substances are "stable," but what exactly does that statement mean? Usually it means that the substance in question does not decompose or react with other substances that normally come in contact with it. Most chemists, for example, would say that the aluminum can that holds the soda you drink is stable. It will be around for quite a long time. The fact that aluminum cans do not decompose rapidly is one of the reasons you are encouraged to recycle them instead of throwing them away. Some aluminum cans have emerged almost unchanged from landfills after 40 or 50 years.

Strictly speaking, there are two kinds of stability. We discussed one of them earlier in this chapter. A substance is *thermodynamically stable* if it does not undergo product-favored reactions. Such reactions disperse energy and increase the entropy of their surroundings. Although we just said it was stable, the aluminum in a soda can is actually *thermodynamically unstable* compared with its oxide, because its reaction with oxygen in air has a negative Gibbs free energy change.

$$4\,Al(s) + 3\,O_2(g) \longrightarrow 2\,Al_2O_3(s) \qquad \Delta G^\circ = -3164.6\text{ kJ}$$

The aluminum exhibits a different kind of stability—it is ***kinetically stable.*** Although it has the potential to undergo a product-favored oxidation reaction, this reaction proceeds so slowly that the can remains essentially unchanged for a long time. This happens because a thin coating of aluminum oxide forms on the surface of the aluminum and prevents oxygen from reaching the rest of the aluminum atoms below the can's surface. If we grind the aluminum into a fine powder and throw it into a flame, the powder will burn and the evolved heat will lead to an entropy increase in the little piece of the universe around the burning metal. An aluminum can is stable (does not oxidize away) because the oxidation is slow (kinetics), not because formation of the oxide would not occur of its own accord (thermodynamics).

Another substance that is *thermodynamically unstable* but *kinetically stable* is diamond. If you look up the data in Appendix J, you will find that the conversion of diamond to graphite has a negative Gibbs free energy change. But diamonds don't change into graphite. Engagement rings contain diamonds precisely because the diamond (like the love it represents) is expected to last for a long time. It does so because there is a very high activation energy barrier ***(p. 435)*** for the change from the diamond structure to the graphite structure. When a chemist says something is stable, it usually means that it is kinetically stable—only an activation energy barrier prevents it from reacting fast enough for us to see a change.

PROBLEM-SOLVING EXAMPLE 17.10 Thermodynamic and Kinetic Stability

Whenever air is heated to a very high temperature, the reaction between nitrogen and oxygen to form nitrogen monoxide occurs. It is an important source of nitrogen-containing air pollutants that can be formed in the cylinders of an automobile engine.

(a) Write a balanced equation with minimum whole-number coefficients for the equilibrium reaction of N_2 with O_2 to form NO.

(b) Is this reaction product-favored at room temperature? That is, is NO thermodynamically stable compared to N_2 and O_2?

(c) Estimate the temperature at which the standard equilibrium constant for this reaction equals 1.

(d) If NO is formed at high temperature in an automobile engine, why does it not all change back to N_2 and O_2 when the mixture of gases enters the exhaust system and its temperature falls?

(e) How might the concentration of NO in automobile exhaust be reduced?

Answer

(a) $N_2(g) + O_2(g) \rightleftharpoons 2\ NO(g)$ (b) No (c) 7301 K

(d) The reverse reaction is too slow. (e) Use a suitable catalyst.

Strategy and Explanation

(a) See answer.

(b) Calculate $\Delta G°$ at 25 °C using data from Appendix J.
$\Delta G° = 2\{\Delta G_f°[NO(g)]\} = 2(86.55\ kJ) = 173.10\ kJ$. Since $\Delta G° > 0$, the reaction is not product-favored.

(c) If $K° = 1$, then $\Delta G° = -RT \ln K° = -RT \ln(1) = 0$.
Because $\Delta G° = \Delta H° - T\Delta S° = 0$, $\Delta H° = T\Delta S°$ and $T = \Delta H°/\Delta S°$.
Using data from Appendix J gives $\Delta H° = 2\{\Delta H_f°\ [NO(g)]\} = 2(90.25\ kJ) = 180.50\ kJ$, and

$$\begin{aligned}\Delta S° &= 2\{S°[NO(g)]\} - \{S°[N_2(g)] + S°[O_2(g)]\}\\ &= 2(210.76\ J/K) - 191.66\ J/K - 205.138\ J/K = 24.722\ J/K\end{aligned}$$

Therefore $T = 180.50\ kJ/24.722\ J/K = 180{,}500\ J/24.722\ J/K = 7301\ K$.

(d) When the mixture of gases, which contains some NO as well as N_2 and O_2, leaves the cylinder of the automobile engine and enters the exhaust system, it cools very rapidly to a temperature below 500 K. The reverse reaction should occur, according to thermodynamics, but it does not. The activation energy for the decomposition of NO is quite high, because NO contains a double bond, and it is very difficult to separate the two atoms (which must be done to form N_2 and O_2). Therefore, the reaction rate is greatly affected by temperature. At low temperatures the reverse reaction is very slow, so significant concentrations of NO exist in automobile exhaust, even though N_2 and O_2 are thermodynamically stable compared to NO.

(e) With a suitable catalyst, decomposition of NO to its elements can take place at appreciable rates even at relatively low temperatures. Catalytic converters are installed in the exhaust systems of cars partly to reduce the concentration of NO. Because N_2 and O_2 are more stable thermodynamically, it is reasonable to use a catalyst to speed up their formation.

☑ **Reasonable Answer Check** It is reasonable that $\Delta G°$ for the reaction of N_2 with O_2 is positive, because N_2 and O_2 are the principal components of the atmosphere where their partial pressures are close to standard pressure, and they do not react with each other. It is reasonable that $\Delta S°$ for the reaction is small and positive. The total number of gas phase molecules does not change, but the product molecules have two different atoms, and both reactant molecules have two atoms that are the same, making the product molecules slightly more probable. It is reasonable that the reaction is endothermic, because the reactant molecules have a triple bond and a double bond, and the product molecules have two double bonds. The bonds broken are therefore expected to be stronger than the bonds formed.

PROBLEM-SOLVING PRACTICE 17.10

All of these substances are stable with respect to decomposition to their elements at 25 °C. Which are kinetically stable and which are thermodynamically stable?

(a) $MgO(s)$ (b) $N_2H_4(\ell)$ (c) $C_2H_6(g)$ (d) $N_2O(g)$

Finally, think about whether you yourself are stable (thermodynamically or kinetically). From a thermodynamic standpoint, most of the substances you are made of are unstable with respect to oxidation to carbon dioxide, water, and other substances. That is, based on Gibbs free energy changes, most of the substances that you are made of should undergo product-favored reactions that would completely destroy them. Your proteins, fats, carbohydrates, and even DNA should spontaneously change into much smaller, simpler molecules. Fortunately for you, the reactions by which this change would happen are very slow at room temperature and body temperature. Only when enzymes catalyze those reactions do they occur with reasonable speed. It is the combination of thermodynamic instability and kinetic stability that allows those enzymes to control the reactions in your body or in any living organism. Were it not for the kinetic stability of a wide variety of substances, everything would be quickly converted to a small number of very thermodynamically stable substances. Life and the environment as we know them would then be impossible.

The roles of thermodynamics and kinetics in determining chemical reactivity can be summarized by saying that ***thermodynamics tells whether a reaction can produce predominantly products under standard conditions and, if it does, how much useful work can be accomplished by coupling the reaction to another process.*** If a reaction involves dilution of substances in the gas phase or in solution, ***thermodynamics tells the value of the standard equilibrium constant and allows quantitative prediction of how much product is formed. Chemical kinetics tells how fast a given reaction goes and indicates how we can control the rate of reaction.*** Together, thermodynamics and kinetics provide the intellectual foundation on which modern chemical industries are based and the principles upon which fundamental understanding of physiology and medicine depends.

IN CLOSING

Having studied this chapter, you should be able to . . .

- Understand and be able to use the terms "product-favored" and "reactant-favored" (Section 17.1).
- Explain why there is a higher probability that energy will be dispersed than that it will be concentrated in a small number of nanoscale particles (Section 17.2). ■ End-of-chapter Question assignable in OWL: 14
- Calculate the entropy change for a process occurring at constant temperature (Section 17.3). ■ Question 24
- Use qualitative rules to predict the sign of the entropy change for a process (Section 17.3). ■ Question 18
- Calculate the entropy change for a chemical reaction, given a table of standard molar entropy values for elements and compounds (Section 17.4).
- Use entropy and enthalpy changes to predict whether a reaction is product-favored (Section 17.5). ■ Question 30

Selected end-of-chapter Questions may be assigned in OWL.

For quick review, download Go Chemistry mini-lecture flashcard modules (or purchase them at **www.ichapters.com**).

Prepare for an exam with a **Summary Problem** that brings together concepts and problem-solving skills from throughout the chapter. Go to **www.cengage.com/chemistry/moore** to download this chapter's Summary Problem from the Student Companion Site.

- Describe the connection between enthalpy and entropy changes for a reaction and the Gibbs free energy change; use this relation to estimate quantitatively how temperature affects whether a reaction is product-favored (Section 17.6). ■ Questions 32, 38, 93, 99
- Calculate the Gibbs free energy change for a reaction from values given in a table of standard molar Gibbs free energies of formation (Section 17.6). ■ Questions 40, 46, 48, 54
- Relate the Gibbs free energy change and the standard equilibrium constant for the same reaction and be able to calculate one from the other (Section 17.7). ■ Questions 56, 62, 78
- Describe how a reactant-favored system can be coupled to a product-favored system so that a desired reaction can be carried out (Section 17.8). ■ Question 66
- Explain the relationship between Gibbs free energy and energy conservation (Sections 17.8, 17.9).
- Distinguish between thermodynamic stability and kinetic stability and describe the effect of each on whether a reaction is useful in producing products (Section 17.10). ■ Question 74

KEY TERMS

energy conservation *(Section 17.9)*

Gibbs free energy *(17.6)*

reversible process *(17.3)*

second law of thermodynamics *(17.5)*

standard equilibrium constant *(17.7)*

third law of thermodynamics *(17.3)*

QUESTIONS FOR REVIEW AND THOUGHT

■ denotes questions assignable in OWL.

Blue-numbered questions have short answers at the back of this book and fully worked solutions in the *Student Solutions Manual.*

Review Questions

1. Define the terms "product-favored system" and "reactant-favored system." Give one example of each.
2. What are the two ways that a final chemical state of a system can be more probable than its initial state?
3. Define the term "entropy," and give an example of a sample of matter that has zero entropy. What are the units of entropy? How do they differ from the units of enthalpy?
4. State five useful qualitative rules for predicting entropy changes when chemical or physical changes occur.
5. State the second law of thermodynamics.
6. In terms of values of $\Delta H°$ and $\Delta S°$, under what conditions can you be sure that a reaction is product-favored? When can you be sure that it is not product-favored?
7. Define the Gibbs free energy change of a chemical reaction in terms of its enthalpy and entropy changes. Why is the Gibbs free energy change especially useful in predicting whether a reaction is product-favored?
8. Why are materials whose reactions release large quantities of Gibbs free energy useful to society? Give two examples of such materials.
9. Describe two ways to cause reactant-favored reactions to form products.

Topical Questions

Reactant-Favored and Product-Favored Processes

10. For each process, write a chemical equation and classify the process as reactant-favored or product-favored.
    (a) Water decomposes to its elements, hydrogen and oxygen.
    (b) Gasoline spilled on the ground evaporates (use octane, C_8H_{18}, to represent gasoline).
    (c) Sugar dissolves in water at room temperature.
11. For each process, write a chemical equation and classify the process as reactant-favored or product-favored.
    (a) Carbon dioxide gas decomposes to its elements, carbon and oxygen.
    (b) The steel (mostly iron) body of an automobile rusts.
    (c) Gasoline reacts with oxygen to form carbon dioxide and water (use octane, C_8H_{18}, to represent gasoline).

Chemical Reactions and Dispersal of Energy

12. Suppose you flip a coin.
    (a) What is the probability that the coin will come up heads?
    (b) What is the probability that it will come up tails?
    (c) If you flip the coin 100 times, what is the most likely number of heads and tails you will see?
13. Suppose you make a tetrahedron and put numbers 1, 2, 3, and 4 on each of the four sides. You toss the tetrahedron in the air and observe it after it comes to rest.
    (a) What is the probability that the tetrahedron will come to rest with the numbers 2, 3, and 4 visible?
    (b) What is the probability that the tetrahedron will come to rest with the numbers 1, 2, and 3 visible?
    (c) If you toss the tetrahedron 100 times, what is the most likely number of times you will see a 1 after it comes to rest?
14. ■ Consider two equal-sized flasks connected as in shown in the figure.

    (a) Suppose you put one molecule inside. What is the probability that the molecule will be in flask A? What is the probability that it will be in flask B?
    (b) If you put 100 molecules into the two-flask system, what is the most likely arrangement of molecules? Which arrangement has the highest entropy?
15. Suppose you have four identical molecules labeled 1, 2, 3, and 4. Draw 16 simple two-flask diagrams as in the figure for Question 14, and draw all possible arrangements of the four molecules in the two flasks. How many of these arrangements have two molecules in each flask? How many have no molecules in one flask? From these results, what is the most probable arrangement of molecules? Which arrangement has the highest entropy?

Measuring Dispersal of Energy: Entropy

16. For each process, tell whether the entropy change of the system is positive or negative.
    (a) Water vapor (the system) deposits as ice crystals on a cold windowpane.
    (b) A can of carbonated beverage loses its fizz. (Consider the beverage but not the can as the system. What happens to the entropy of the dissolved gas?)
    (c) A glassblower heats glass (the system) to its softening temperature.
17. For each process, tell whether the entropy change of the system is positive or negative.
    (a) Water boils.
    (b) A teaspoon of sugar dissolves in a cup of coffee. (The system consists of both sugar and coffee.)
    (c) Calcium carbonate precipitates out of water in a cave to form stalactites and stalagmites. (Consider only the calcium carbonate to be the system.)
18. ■ For each pair of items, tell which has the higher entropy, and explain why.
    (a) Item 1, a sample of solid CO_2 at −78 °C, or item 2, CO_2 vapor at 0 °C
    (b) Item 1, solid sugar, or item 2, the same sugar dissolved in a cup of tea
    (c) Item 1, a 100-mL sample of pure water and a 100-mL sample of pure alcohol, or item 2, the same samples of water and alcohol after they had been poured together and stirred
19. For each pair of items, tell which has the higher entropy, and explain why.
    (a) Item 1, a sample of pure silicon (to be used in a computer chip), or item 2, a piece of silicon having the same mass but containing a trace of some other element, such as B or P
    (b) Item 1, an ice cube at 0 °C, or item 2, the same mass of liquid water at 0 °C
    (c) Item 1, a sample of pure I_2 solid at room temperature, or item 2, the same mass of iodine vapor at room temperature
20. ■ Comparing the formulas or states for each pair of substances, tell which you would expect to have the higher entropy per mole at the same temperature, and explain why.
    (a) NaCl(s) or CaO(s)
    (b) Cl_2(g) or P_4(g)
    (c) NH_4NO_3(s) or NH_4NO_3(aq)
21. Comparing the formulas or states for each pair of substances, tell which you would expect to have the higher entropy per mole at the same temperature, and explain why.
    (a) CH_3NH_2(g) or $(CH_3)_2NH$(g)
    (b) Au(s) or Hg(ℓ)
    (c) Kr(g) or C_6H_{14}(g)
22. Without consulting a table of standard molar entropies, predict whether ΔS°_{system} will be positive or negative for each of these reactions.
    (a) $2\ CO(g) + O_2(g) \longrightarrow 2\ CO_2(g)$
    (b) $2\ H_2(g) + O_2(g) \longrightarrow 2\ H_2O(\ell)$
    (c) $2\ O_3(g) \longrightarrow 3\ O_2(g)$
23. Without consulting a table of standard molar entropies, predict whether ΔS°_{system} will be positive or negative for each of these reactions.
    (a) $2\ NH_3(g) \longrightarrow N_2(g) + 3\ H_2(g)$
    (b) $2\ Na(s) + Cl_2(g) \longrightarrow 2\ NaCl(s)$
    (c) $H_2(g) + I_2(s) \longrightarrow 2\ HI(g)$

Calculating Entropy Changes

24. ■ Calculate the entropy change, ΔS°, for the vaporization of ethanol, C_2H_5OH, at the boiling point of 78.3 °C. The heat of vaporization of the alcohol is 39.3 kJ/mol.
    $$C_2H_5OH(\ell) \longrightarrow C_2H_5OH(g) \qquad \Delta S^\circ = ?$$
25. Diethyl ether, $(C_2H_5)_2O$, was once used as an anesthetic. What is the entropy change, ΔS°, for the vaporization of ether if its heat of vaporization is 26.0 kJ/mol at the boiling point of 35.0 °C?
26. The standard molar entropy of methanol vapor, CH_3OH(g), is 239.8 J K^{-1} mol^{-1}.
    (a) Calculate the entropy change for the vaporization of 1 mol methanol (use data from Table 17.1 or Appendix J).

(b) Calculate the enthalpy of vaporization of methanol, assuming that $\Delta S°$ doesn't depend on temperature and taking the boiling point of methanol to be 64.6 °C.

27. The standard molar entropy of iodine vapor, $I_2(g)$, is 260.7 J K^{-1} mol^{-1} and the standard molar enthalpy of formation is 62.4 kJ/mol.
(a) Calculate the entropy change for vaporization of 1 mol of solid iodine (use data from Table 17.1 or Appendix J).
(b) Calculate the enthalpy change for sublimation of iodine.
(c) Assuming that $\Delta S°$ does not change with temperature, estimate the temperature at which iodine would sublime (change directly from solid to gas).

28. Check your predictions in Question 22 by calculating the entropy change for each reaction. Standard molar entropies not in Table 17.1 can be found in Appendix J.

29. Check your predictions in Question 23 by calculating the entropy change for each reaction. Standard molar entropies not in Table 17.1 can be found in Appendix J.

Entropy and the Second Law of Thermodynamics

30. ■ Calculate $\Delta S°_{system}$ at 25 °C for the reaction

$$C_2H_4(g) + H_2O(g) \longrightarrow C_2H_5OH(\ell)$$

Can you tell from the result of this calculation whether this reaction is product-favored? If you cannot tell, what additional information do you need? Obtain that information and determine whether the reaction is product-favored.

31. Calculate $\Delta S°_{system}$ at 25 °C for the reaction

$$C_6H_6(\ell) + 4\ H_2(g) \longrightarrow C_6H_{14}(\ell)$$

Can you tell from the result of this calculation whether this reaction is product-favored? If you cannot tell, what additional information do you need? Obtain that information and determine whether the reaction is product-favored.

32. Is this reaction predicted to favor the products at low temperatures, at high temperatures, or both? Explain your answer briefly.

$$Mg(s) + \tfrac{1}{2}\ O_2(g) \longrightarrow MgO(s) \qquad \Delta H° = -601.70\ kJ$$

33. Is this reaction predicted to favor the products at low temperatures, at high temperatures, or both? Explain your answer briefly.

$$MgCO_3(s) \longrightarrow MgO(s) + CO_2(g) \qquad \Delta H° = 116.48\ kJ$$

34. Sodium reacts violently with water according to the equation

$$Na(s) + H_2O(\ell) \longrightarrow NaOH(aq) + \tfrac{1}{2}\ H_2(g)$$

(a) Predict the signs of $\Delta H°$ and $\Delta S°$ for the reaction.
(b) Verify your predictions with calculations.

35. Once ignited, magnesium reacts vigorously with oxygen in air according to the equation

$$2\ Mg(s) + O_2(g) \longrightarrow 2\ MgO(s)$$

(a) Predict the signs of $\Delta H°$ and $\Delta S°$ for the reaction.
(b) Verify your predictions with calculations.

36. Hydrogen burns in air with considerable heat transfer to the surroundings. Consider the decomposition of water to gaseous hydrogen and oxygen. Without doing any calculations, and basing your prediction on the enthalpy change and the entropy change, is this reaction product-favored at 25 °C? Explain your answer briefly.

37. Hydrogen gas combines with chlorine gas in an exothermic reaction to form HCl(g). Consider the decomposition of gaseous hydrogen chloride to hydrogen and chlorine. Without doing any calculations, and basing your prediction on the enthalpy change and the entropy change, is this reaction product-favored at 25 °C? Explain your answer briefly.

38. ■ For each reaction, calculate $\Delta H°$ and $\Delta S°$ and predict whether the reaction is always product-favored, product-favored only at low temperatures, product-favored only at high temperatures, or never product-favored.
(a) $Fe_2O_3(s) + 2\ Al(s) \longrightarrow 2\ Fe(s) + Al_2O_3(s)$
(b) $N_2(g) + 2\ O_2(g) \longrightarrow 2\ NO_2(g)$

39. For each reaction, calculate $\Delta H°$ and $\Delta S°$ and predict whether the reaction is always product-favored, product-favored only at low temperatures, product-favored only at high temperatures, or never product-favored.
(a) $C_6H_{12}O_6(s) + 6\ O_2(g) \longrightarrow 6\ CO_2(g) + 6\ H_2O(\ell)$
(b) $MgO(s) + C(graphite) \longrightarrow Mg(s) + CO(g)$

Gibbs Free Energy

40. Determine whether the combustion of ethane, C_2H_6, is product-favored at 25 °C.

$$C_2H_6(g) + \tfrac{7}{2}\ O_2(g) \longrightarrow 2\ CO_2(g) + 3\ H_2O(\ell)$$

(a) Calculate $\Delta S_{universe}$. Required values of $\Delta H_f°$ and S° are in Appendix J.
(b) Verify your result by calculating the value of $\Delta G°$ for the reaction.
(c) Do your calculated answers in parts (a) and (b) agree with your preconceived idea of this reaction?

41. The reaction of magnesium with water can be used as a means for heating food.

$$Mg(s) + 2\ H_2O(\ell) \longrightarrow Mg(OH)_2(s) + H_2(g)$$

Determine whether this reaction is product-favored at 25 °C.
(a) Calculate $\Delta S_{universe}$. See Appendix J for the needed data.
(b) Verify your result by calculating $\Delta G°$ for the reaction.

42. Add a column for the sign of the Gibbs free energy to Table 17.2 (**p. 624**). For the first and last lines in the table, tell whether ΔG is positive or negative.

43. Based on your table from Question 42, when ΔH_{system} and ΔS_{system} are both negative, is ΔG is positive or negative or does the sign depend on temperature? If the sign depends on temperature, does the reaction become product-favored at high or low temperatures?

44. Use a mathematical equation to show how the statement leads to the conclusion cited: If a reaction is exothermic (negative ΔH) and if the entropy of the system increases (positive ΔS), then ΔG must be negative, and the reaction will be product-favored.

45. Use a mathematical equation to show how the statement leads to the conclusion cited: If ΔH and ΔS have the same sign, then the magnitude of T determines whether ΔG will be negative and whether the reaction will be product-favored.

46. ■ Predict whether the reaction below is product-favored or reactant-favored by calculating $\Delta G°$ from the entropy and enthalpy changes for the reaction at 25 °C.

$$H_2(g) + CO_2(g) \longrightarrow H_2O(g) + CO(g)$$
$$\Delta H° = 41.17\ kJ \qquad \Delta S° = 42.08\ J/K$$

47. Predict whether this reaction is product-favored at 25 °C by calculating the change in standard Gibbs free energy from the entropy and enthalpy changes.

$$H_2(g) + I_2(g) \rightleftharpoons 2\ HI(g)$$
$$\Delta H° = 52.96\ kJ \qquad \Delta S° = 166.4\ J/K$$

48. ■ Use data from Appendix J to calculate $\Delta G°$ for each reaction at 25 °C. Which are product-favored?
(a) $C_2H_2(g) + H_2(g) \longrightarrow C_2H_4(g)$
(b) $2\ SO_3(g) \longrightarrow 2\ SO_2(g) + O_2(g)$
(c) $4\ NH_3(g) + 5\ O_2(g) \longrightarrow 4\ NO(g) + 6\ H_2O(g)$

49. Evaluate $\Delta H°$ for each reaction in Question 48. Use your results to calculate standard molar entropies at 25.00 °C for
(a) $C_2H_2(g)$ (b) $SO_3(g)$
(c) $NO(g)$

50. ■ If a system falls within the second or third category in Table 17.2 (p. 624), then there must be a temperature at which it shifts from being reactant-favored to being product-favored. For each reaction, obtain data from Appendix J and calculate what that temperature is.
(a) $CO(g) + 2\ H_2(g) \rightleftharpoons CH_3OH(\ell)$
(b) $2\ Fe_2O_3(s) + 3\ C(graphite) \rightleftharpoons 4\ Fe(s) + 3\ CO_2(g)$

51. If a system falls within the second or third category in Table 17.2 (p. 624) then there must be a temperature at which it shifts from being reactant-favored to being product-favored. For each reaction, obtain data from Appendix J and calculate what that temperature is.
(a) $2\ H_2O(g) \rightleftharpoons 2\ H_2(g) + O_2(g)$
(b) $N_2(g) + 3\ H_2(g) \rightleftharpoons 2\ NH_3(g)$

52. Many metal carbonates can be decomposed to the metal oxide and carbon dioxide by heating.

$$CaCO_3(s) \longrightarrow CaO(s) + CO_2(g)$$

(a) What are the enthalpy, entropy, and Gibbs free energy changes for this reaction at 25.00 °C?
(b) Is it product-favored or reactant-favored?
(c) Based on the signs of $\Delta H°$ and $\Delta S°$, predict whether the reaction is product-favored at all temperatures.
(d) Predict the lowest temperature at which appreciable quantities of products can be obtained.

53. Some metal oxides, such as lead(II) oxide, can be decomposed to the metal and oxygen simply by heating.

$$PbO(s) \longrightarrow Pb(s) + \tfrac{1}{2}O_2(g)$$

(a) Is the decomposition of lead(II) oxide product-favored at 25 °C? Explain.
(b) If not, can it become so if the temperature is raised?
(c) As the temperature increases, at what temperature does the reaction first become product-favored?

54. ■ Use the thermochemical expression

$$CaC_2(s) + 2\ H_2O(\ell) \longrightarrow C_2H_2(g) + Ca(OH)_2(aq)$$
$$\Delta G° = -119.282\ kJ$$

and data from Appendix J to calculate $\Delta G_f°$ for $Ca(OH)_2(aq)$ at 25 °C.

55. Use the thermochemical expression

$$PCl_3(g) + Cl_2(g) \longrightarrow PCl_5(g) \qquad \Delta G° = -37.2\ kJ$$

and data from Appendix J to calculate $\Delta G_f°$ for $PCl_5(g)$.

Gibbs Free Energy Changes and Equilibrium Constants

56. ■ Use data from Appendix J to obtain the equilibrium constant K_P for each reaction at 298.15 K.
(a) $2\ HCl(g) \rightleftharpoons H_2(g) + Cl_2(g)$
(b) $N_2(g) + O_2(g) \rightleftharpoons 2\ NO(g)$

57. Use data from Appendix J to obtain the equilibrium constant K_P for each of these reactions at 298 K.
(a) $CH_4(g) + 2\ O_2(g) \rightleftharpoons CO_2(g) + 2\ H_2O(g)$
(b) $2\ NO_2(g) \rightleftharpoons N_2O_4(g)$

58. Ethylene reacts with hydrogen to produce ethane.

$$H_2C{=}CH_2(g) + H_2(g) \rightleftharpoons H_3C{-}CH_3(g)$$

(a) Using the data in Appendix J, calculate $\Delta G°$ for the reaction at 25 °C. Is the reaction predicted to be product-favored under standard conditions?
(b) Calculate K_P from $\Delta G°$. Comment on the connection between the sign of $\Delta G°$ and the magnitude of K_P.

59. Use the data in Appendix J to calculate $\Delta G°$ and K_P at 25 °C for the reaction

$$2\ HBr(g) + Cl_2(g) \rightleftharpoons 2\ HCl(g) + Br_2(\ell)$$

Comment on the connection between the sign of $\Delta G°$ and the magnitude of K_P.

60. For each reaction, estimate $K°$ at the temperature indicated.
(a) $2\ H_2(g) + O_2(g) \rightleftharpoons 2\ H_2O(g)$ at 800. K
(b) $2\ SO_2(g) + O_2(g) \rightleftharpoons 2\ SO_3(g)$ at 500. K
(c) $2\ HF(g) \rightleftharpoons H_2(g) + F_2(g)$ at 2000. K

61. For each reaction, estimate $K°$ at the temperature indicated.
(a) $H_2(g) + I_2(g) \rightleftharpoons 2\ HI(g)$ at 500. K
(b) $N_2(g) + 3\ H_2(g) \rightleftharpoons 2\ NH_3(g)$ at 400. K
(c) $CO(g) + 3\ H_2(g) \rightleftharpoons CH_4(g) + H_2O(g)$ at 800. K

62. ■ For each reaction, an equilibrium constant at 298 K is given. Calculate $\Delta G°$ for each reaction.
(a) $Br_2(\ell) + H_2(g) \rightleftharpoons 2\ HBr(g)$ $\quad K_P = 4.4 \times 10^{18}$
(b) $H_2O(\ell) \rightleftharpoons H_2O(g)$ $\quad K_P = 3.17 \times 10^{-2}$
(c) $N_2(g) + 3\ H_2(g) \rightleftharpoons 2\ NH_3(g)$ $\quad K_c = 3.5 \times 10^8$

63. For each reaction, an equilibrium constant at 298 K is given. Calculate $\Delta G°$ for each reaction.
(a) $\tfrac{1}{8}S_8(s) + O_2(g) \rightleftharpoons SO_2(g)$ $\quad K_P = 4.2 \times 10^{52}$
(b) $2\ H_2(g) + O_2(g) \rightleftharpoons 2\ H_2O(g)$ $\quad K_c = 3.3 \times 10^{81}$
(c) $CH_4(g) + H_2O(g) \rightleftharpoons CO(g) + 3\ H_2(g)$ $\quad K_c = 9.4 \times 10^{-1}$

Gibbs Free Energy and Maximum Work

64. ■ Which of these reactions are capable of being harnessed to do useful work at 298 K and 1 bar? Which require that work be done to make them occur?
(a) $2\ C_6H_6(\ell) + 15\ O_2(g) \longrightarrow 12\ CO_2(g) + 6\ H_2O(g)$
(b) $2\ NF_3(g) \longrightarrow N_2(g) + 3\ F_2(g)$
(c) $TiO_2(s) \longrightarrow Ti(s) + O_2(g)$

65. Which of these reactions are capable of being harnessed to do useful work at 298 K and 1 bar? Which require that work be done to make them occur?
(a) $Al_2O_3(s) \longrightarrow 2\ Al(s) + \tfrac{3}{2}O_2(g)$
(b) $2\ CO(g) + O_2(g) \longrightarrow 2\ CO_2(g)$
(c) $C_2H_6(g) \longrightarrow C_2H_4(g) + H_2(g)$

66. ■ For each of the reactions in Question 64 that requires work to be done, calculate the minimum mass of graphite that would have to be oxidized to $CO_2(g)$ to provide the necessary work.
67. For each of the reactions in Question 65 that requires work to be done, calculate the minimum mass of hydrogen gas that would have to be burned to form water vapor to provide the necessary work.
68. Titanium is obtained from its ore, $TiO_2(s)$, by heating the ore in the presence of chlorine gas and coke (carbon) to produce gaseous titanium(IV) chloride and carbon monoxide.
    (a) Write a balanced equation for this process.
    (b) Calculate $\Delta H°$, $\Delta S°$, and $\Delta G°$ for the reaction.
    (c) Is this reaction product-favored or reactant-favored at 25 °C?
    (d) Does the reaction become more product-favored or more reactant-favored as the temperature increases?
69. To obtain a metal from its ore, the decomposition of the metal oxide to form the metal and oxygen is often coupled with oxidation of coke (carbon) to carbon monoxide. For each metal oxide listed, write a balanced equation for the decomposition of the oxide and for the overall reaction when the decomposition is coupled to oxidation of coke to carbon monoxide. Calculate the overall value of $\Delta G°$ for each coupled reaction at 25 °C. Which of the metals could be obtained from these ores at 25 °C by this method?
    (a) $CuO(s)$ (b) $Ag_2O(s)$
    (c) $HgO(s)$ (d) $MgO(s)$
    (e) $PbO(s)$
70. From which of the metal oxides in Question 69 could the metal be obtained by coupling reduction of the oxide with oxidation of coke to carbon monoxide at 800 °C?
71. From which of the metal oxides in Question 69 could the metal be obtained by coupling reduction of the oxide with oxidation of coke to carbon monoxide at 1500 °C?

Conservation of Gibbs Free Energy

72. What are the resources human society uses to supply Gibbs free energy?
73. For one day, keep a log of all the activities you undertake that consume Gibbs free energy. Distinguish between Gibbs free energy provided by nutrient metabolism and that provided by other energy resources.

Thermodynamic and Kinetic Stability

74. ■ Billions of pounds of acetic acid are made each year, much of it by the reaction of methanol with carbon monoxide.

$$CH_3OH(\ell) + CO(g) \longrightarrow CH_3COOH(\ell)$$

    (a) By calculating the standard Gibbs free energy change, $\Delta G°$, for this reaction, show that it is product-favored.
    (b) Determine the standard Gibbs free energy change, $\Delta G°$, for the reaction of acetic acid with oxygen to form gaseous carbon dioxide and liquid water.
    (c) Based on this result, is acetic acid thermodynamically stable compared with $CO_2(g)$ and $H_2O(\ell)$?
    (d) Is acetic acid kinetically stable compared with $CO_2(g)$ and $H_2O(\ell)$?
75. Actually, the carbon in $CO_2(g)$ is thermodynamically unstable with respect to the carbon in calcium carbonate (limestone). Verify this by determining the standard Gibbs free energy change for the reaction of lime, $CaO(s)$, with $CO_2(g)$ to make $CaCO_3(s)$.

General Questions

Reaction	Chemical Equation	K_c	$\Delta H°$ (kJ)
1	$CH_3OH(g) + H_2(g) \rightleftharpoons CH_4(g) + H_2O(g)$	3.6×10^{20}	−115.4
2	$Mg(OH)_2(s) \rightleftharpoons MgO(s) + H_2O(g)$	1.24×10^{-5}	81.1
3	$2\ CH_4(g) \rightleftharpoons C_2H_6(g) + H_2(g)$	9.5×10^{-13}	64.9
4	$2\ H_2(g) + CO(g) \rightleftharpoons CH_3OH(g)$	3.76	−90.7
5	$H_2(g) + Br_2(g) \rightleftharpoons 2\ HBr(g)$	1.9×10^{24}	−103.7

76. The table above provides data at 25 °C for five reactions. For which (if any) of the reactions 1 through 5 is
    (a) K_P greater than K_c?
    (b) the reaction product-favored?
    (c) there only a single concentration in the K_c expression?
    (d) there an increase in the concentrations of products when the temperature increases?
    (e) there a change in the sign of $\Delta G°$ if water is liquid instead of gas?
77. The table above provides data at 25 °C for five reactions. For which (if any) of the reactions 1 through 5 is
    (a) K_P less than K_c?
    (b) there a decrease in the concentrations of products when the pressure increases?
    (c) the value of $\Delta S°$ positive?
    (d) the sign of $\Delta G°$ dependent on temperature?
78. ■ Consider the gas phase decomposition of sulfur trioxide to sulfur dioxide and oxygen.
    (a) Calculate $\Delta G°$ for the reaction at 25 °C.
    (b) Is the reaction product-favored under standard conditions at 25 °C?
    (c) If the reaction is not product-favored at 25 °C, is there a temperature at which it will become so?
    (d) Estimate K_P for the reaction at 1500. °C.
    (e) Estimate K_c for the reaction at 1500. °C.
79. The Haber process for the synthesis of ammonia involves the reaction

$$N_2(g) + 3\ H_2(g) \rightleftharpoons 2\ NH_3(g)$$

Using data from Appendix J, estimate the amount (in moles) of $NH_3(g)$ that would be produced from 1 mol $N_2(g)$ and 3 mol $H_2(g)$ once equilibrium is reached at 450 °C and a total pressure of 1000. atm.
80. Mercury is a poison, and its vapor is readily absorbed through the lungs. Therefore it is important that the partial pressure of mercury be kept as low as possible in any area where people could be exposed to it (such as a dentist's office). The relevant equilibrium reaction is

$$Hg(\ell) \rightleftharpoons Hg(g)$$

For Hg(g), $\Delta H_f^\circ = 61.4$ kJ/mol, $S^\circ = 175.0 \text{ J K}^{-1} \text{ mol}^{-1}$, and $\Delta G_f^\circ = 31.8$ kJ/mol. Use data from Appendix J and these values to evaluate the vapor pressure of mercury at different temperatures. (Remember that concentrations of pure liquids and solids do not appear in the equilibrium constant expression, and for gases K° involves pressures in bars.)

(a) Calculate ΔG° for vaporization of mercury at 25 °C.
(b) Write the equilibrium constant expression for vaporization of mercury.
(c) Calculate K° for this reaction at 25 °C.
(d) What is the vapor pressure of mercury at 25 °C?
(e) Estimate the temperature at which the vapor pressure of mercury reaches 10 mm Hg.

Applying Concepts

81. A friend says that the boiling point of water is twice that of cyclopentane, which boils at 50 °C. Write a brief statement about the validity of this observation.
82. Using the second law of thermodynamics, explain why it is very difficult to unscramble an egg. Who was Humpty-Dumpty? Why did his moment of glory illustrate the second law of thermodynamics?
83. Appendix J lists standard molar entropies S°, not standard entropies of formation ΔS_f°. Why is this possible for entropy but not for internal energy, enthalpy, or Gibbs free energy?
84. When calculating ΔS° from S° values, it is necessary to look up all substances, including elements in their standard state, such as $O_2(g)$, $H_2(g)$, and $N_2(g)$. When calculating ΔH° from ΔH_f° values, however, elements in their standard state can be ignored. Why is the situation different for S° values?
85. Explain why the entropy of the system increases when solid NaCl dissolves in water.
86. Explain how the entropy of the universe increases when an aluminum metal can is made from aluminum ore. The first step is to extract the ore, which is primarily a form of Al_2O_3, from the ground. After it is purified by freeing it from oxides of silicon and iron, aluminum oxide is changed to the metal by an input of electrical energy.

$$2\,Al_2O_3(s) \xrightarrow{\text{electrical energy}} 4\,Al(s) + 3\,O_2(g)$$

87. Suppose that at a certain temperature T a chemical reaction is found to have a standard equilibrium constant K° of 1.0. Indicate whether each statement is true or false and explain why.
(a) The enthalpy change for the reaction, ΔH°, is zero.
(b) The entropy change for the reaction, ΔS°, is zero.
(c) The Gibbs free energy change for the reaction, ΔG°, is zero.
(d) ΔH° and ΔS° have the same sign.
(e) $\Delta H^\circ/T = \Delta S^\circ$ at the temperature T.
88. How can kinetically stable substances exist at all, if they are not thermodynamically stable?
89. Criticize this statement: Provided it occurs at an appreciable rate, any chemical reaction for which $\Delta G < 0$ will proceed until all reactants have been converted to products.
90. Reword the statement in Question 89 so that it is always true.

More Challenging Questions

91. Calculate the entropy change for formation of exactly 1 mol of each of these gaseous hydrocarbons under standard conditions from carbon (graphite) and hydrogen. What trend do you see in these values? Does ΔS° increase or decrease on adding H atoms?
(a) acetylene, $C_2H_2(g)$ (b) ethylene, $C_2H_4(g)$
(c) ethane, $C_2H_6(g)$
92. Calcium hydroxide, $Ca(OH)_2(s)$, can be dehydrated to form lime, CaO, by heating. Without doing any calculations, and basing your prediction on the enthalpy change and the entropy change, is this reaction product-favored at 25 °C? Explain your answer briefly.
93. ■ Octane is the product of adding hydrogen to 1-octene.

$$\underset{\text{1-octene}}{C_8H_{16}(g)} + H_2(g) \longrightarrow \underset{\text{octane}}{C_8H_{18}(g)}$$

The enthalpies of formation are

$$\Delta H_f^\circ[C_8H_{16}(g)] = -82.93 \text{ kJ/mol}$$
$$\Delta H_f^\circ[C_8H_{18}(g)] = -208.45 \text{ kJ/mol}$$

Predict whether this reaction is product-favored or reactant-favored at 25 °C and explain your reasoning.
94. This is a group project: Estimate or look up, to the nearest order of magnitude,
(a) the number of kilograms of CH_3OH made each year
(b) the number of kilograms of CO in the entire atmosphere
(c) the number of kilograms of CH_3COOH made each year
(d) the number of kilograms of H_2O on earth
(e) the number of kilograms of CO_2 in the atmosphere
What do these facts tell you about the difference between kinetic stability and thermodynamic stability?
95. Nitric oxide and chlorine combine at 25 °C to produce nitrosyl chloride, NOCl.

$$2\,NO(g) + Cl_2(g) \longrightarrow 2\,NOCl(g)$$

(a) Calculate the equilibrium constant K_P for the reaction.
(b) Is the reaction product-favored or reactant-favored?
(c) Calculate the equilibrium constant K_c for the reaction.
96. Hydrogen for use in the Haber-Bosch process for ammonia synthesis is generated from natural gas by the reaction

$$CH_4(g) + H_2O(g) \rightleftharpoons CO(g) + 3\,H_2(g)$$

(a) Calculate ΔG° for this reaction at 25 °C.
(b) Calculate K_P for the reaction at 25 °C.
(c) Is the reaction product-favored under standard conditions? If not, at what temperature will it become so?
(d) Estimate K_c for the reaction at 1000. K.
97. It would be very useful if we could use the inexpensive carbon in coal to make more complex organic molecules such as gaseous or liquid fuels. The formation of methane from coal and water is reactant-favored and thus cannot occur unless there is some energy transfer from outside. This problem examines the feasibility of other reactions using coal and water.
(a) Write three balanced equations for the reactions of coal (carbon) and steam to make ethane gas, $C_2H_6(g)$, propane gas, $C_3H_8(g)$, and liquid methanol, $CH_3OH(\ell)$, with carbon dioxide as a by-product.

(b) Using the data in Appendix J, calculate $\Delta H°$, $\Delta S°$, and $\Delta G°$ for each reaction, and then comment on whether any of them would be a feasible way to make the stated products.

98. You are exploring the marketing possibilities of a scheme by which every family in the United States produces enough water for its own needs by the combustion of hydrogen and oxygen. Would the release of Gibbs free energy from the combination of hydrogen and oxygen be sufficient to supply the family's energy needs? Do not try to collect the actual data you would use, but define the problem well enough so that someone else could collect the necessary data and do the calculations that would be needed.

99. ■ Without consulting tables of $\Delta H_f°$, $S°$, or $\Delta G_f°$ values, predict which of these reactions is
(i) always product-favored.
(ii) product-favored at low temperatures, but not product-favored at high temperatures.
(iii) not product-favored at low temperatures, but product-favored at high temperatures.
(iv) never product-favored.

(a) $2\ NO_2(g) \longrightarrow N_2O_4(g)$
(b) $C_5H_{12}(g) + 8\ O_2(g) \longrightarrow 5\ CO_2(g) + 6\ H_2O(g)$
(c) $P_4(g) + 10\ F_2(g) \longrightarrow 4\ PF_5(g)$
[*Hint:* Use the qualitative rules regarding bond enthalpies in Section 6.7 (⇐ ***p. 202***) to predict the sign of $\Delta H°$.]

Conceptual Challenge Problems

CP17.A (Section 17.2) Suppose that you are invited to play a game as either the "player" or the "house." A pair of dice is used to determine the winner. Each die is a cube having a different number, one through six, showing on each face. The player rolls two dice and sums the numbers showing on the top side of each die to determine the number rolled. Obviously, the number rolled has a minimum value of 2 (both dice showing a 1) and a maximum of 12 (both dice showing a 6). The player begins the game with his or her initial roll of the dice. If the player rolls a 7 or an 11, he or she wins on the first roll and the house loses. If the player does not roll a 7 or an 11 on the initial roll, then whatever number was rolled is called the point, and the player must roll again. For the player to win, he or she must roll the point again before either a 7 or an 11 is rolled. Should the player roll a 7 or an 11 before rolling the point a second time, the house wins. Which would you choose to be, player or house? Explain clearly in terms of the probabilities of rolling the dice why you chose the role you did.

CP17.B (Section 17.3) When thermal energy is transferred to a substance at its standard melting point or boiling point, the substance melts or vaporizes, but its temperature does not change while it is doing so. It is clear then that temperature cannot be a measure of "how much energy is in a sample of matter" or the "intensity of energy in a sample of matter." In "Qualitative Guidelines for Entropy" (⇐ ***p. 617***) we noted that atoms and molecules are not stationary, but rather are in constant motion. When heated, their motion increases. If this is true, what can you infer that temperature measures about a sample of matter?

CP17.C (Section 17.9) Suppose that you are a member of an environmental group and have been assigned to evaluate various ways of delivering milk to consumers with respect to Gibbs free energy conservation. Think of all the ways that milk could be delivered, the kinds of containers that could be used, and the ways they could be transported. Consider whether the containers could be reused (refilled) or recycled. Define the problem in terms of the kinds of information you would need to collect, how you would analyze the information, and which criteria you would use to decide which systems are more efficient in use of Gibbs free energy. Do not try to collect the actual data you would use, but define the problem well enough so that someone could collect the necessary data based on your statement of the problem.

CP17.D (Section 17.10) Consider planet earth as a thermodynamic system. Is earth thermodynamically or kinetically stable? Discuss your choice, providing as many arguments as you can to support it.

18

Electrochemistry and Its Applications

© Enigma/Alamy

Personal electronic devices, such as this iPhone, depend on advanced types of batteries to supply their electrical power needs (lithium-ion batteries power iPhones). Dependable, long-lasting, light weight batteries are important for many of the modern conveniences we use every day such as cellular telephones, calculators, flashlights, computers, CD and MP3 players, cordless tools, and hybrid cars. In this chapter we discuss electrochemical reactions and their extraordinary range of applications.

Many of our modern devices are powered by batteries—portable storage devices for electrochemical energy that is produced by product-favored redox reactions. Making these reactions do useful work is the goal of significant chemical research. What chemistry goes on inside a battery? How does it produce electricity? Is this chemistry similar to or different from the other kinds of reactions you have studied?

Oxidation-reduction (redox) reactions, such as those in batteries, are an important class of chemical reactions *(⟸ p. 151)*. Redox reactions involve the transfer of electrons from one atom, molecule, or ion to another. **Electrochemistry** is the study of the relationship between electron flow and redox reactions, that is, the relationship between electricity and chemical changes. Applications of electrochemistry are numerous and important. In electrochemical cells (commonly called batteries), electrons from a product-favored redox reaction are released and transferred as an electrical current through an external circuit. The current depends on how much of the reactants are converted to products, and how fast they are converted. We rely on batteries to power many useful devices, including MP3 players, cellular telephones, calculators, flashlights, laptop computers, heart pacemakers, and cordless tools.

The voltage produced by an electrochemical reaction depends on the oxidizing agents and reducing agents used as reactants. A knowledge of the strengths of oxidizing and reducing agents is used in the design of better batteries. Product-favored electrochemical reactions, however, are not always beneficial. Corrosion of iron, for example, is a product-favored redox reaction. Damage to materials as a result of corrosion is very costly, so preventing corrosion is an important goal. By contrast, electroplating and electrolysis are applications of reactant-favored redox reactions. In an electrolysis cell, an external energy source creates an electrical current that forces a reactant-favored process to occur. Electrolysis is important in the production of many products, such as aluminum metal and the chlorine used to disinfect water supplies.

18.1 Redox Reactions

Redox reactions form a large class of chemical reactions in which the reactants can be atoms, ions, or molecules. How do you know when a reaction involves oxidation-reduction?

- ***By identifying the presence of strong oxidizing or reducing agents as reactants (Table 5.3;*** ***⇐ p. 154).***
- ***By recognizing a change in oxidation number (******⇐ p. 155******).*** This means you have to determine the oxidation number of each element as it appears in a reactant or a product.
- ***By recognizing the presence of an uncombined element as a reactant or product.*** Producing a free element or incorporating one into a compound almost always results in a change in oxidation number.

An uncombined element is always assigned an oxidation number of 0.

You may want to review the definitions of oxidation and reduction in Section 5.3 and the rules for assigning oxidation numbers in Section 5.4.

To briefly review the definitions of oxidation and reduction, consider the displacement reaction between magnesium, a relatively reactive metal, and hydrochloric acid. The oxidation numbers of the species are shown above their symbols.

H^+ is reduced

$$\overset{0}{Mg(s)} + 2\,\overset{+1\,-1}{HCl(aq)} \longrightarrow \overset{+2\;-1}{MgCl_2(aq)} + \overset{0}{H_2(g)}$$

Mg is oxidized

The presence of the uncombined elements Mg and H_2 indicates a redox reaction, as do the changes in oxidation number. Mg(s) is oxidized, indicated by an *increase* in its oxidation number (from 0 to +2). Hydrogen ions from HCl are reduced to H_2, as shown by a *decrease* in oxidation number (from +1 to 0). Magnesium metal is the reducing agent because it causes the hydrogen ions in HCl to be reduced (gain electrons). In the process of giving electrons to HCl, metallic Mg is oxidized to Mg^{2+} ions. Hydrochloric acid is the oxidizing agent because it causes magnesium metal to be oxidized (lose electrons). Note that in this and all redox reactions, ***the oxidizing agent is reduced, and the reducing agent is oxidized.*** Oxidation and reduction *always* occur together, with one reactant acting as the oxidizing agent and another acting as the reducing agent. Oxidizing and reducing agents are *always* reactants, never products.

Oxidation: loss of electron(s) and increase in oxidation number

Reduction: gain of electron(s) and decrease in oxidation number

Redox reactions such as the one between Mg and HCl involve complete loss or gain of electrons by the reacting species—the type of reaction that is utilized in electrochemistry. A flow of electrons through the reaction system to make a complete electrical circuit is necessary for the reaction to proceed.

PROBLEM-SOLVING EXAMPLE 18.1 Identifying Oxidizing and Reducing Agents in Redox Reactions

In the thermite reaction, iron(III) oxide and aluminum metal react to give iron metal and aluminum oxide:

$$Fe_2O_3(s) + 2\ Al(s) \longrightarrow 2\ Fe(s) + Al_2O_3(s)$$

Is this a redox reaction? Identify the oxidation numbers of all the atoms that change oxidation number. What gets reduced? What gets oxidized? What is the oxidizing agent? What is the reducing agent?

Answer Yes, this is a redox reaction. Oxidation number changes: Fe, +3 to 0; Al, 0 to +3. The iron in the iron(III) oxide is reduced to metallic iron. The aluminum metal is oxidized to Al^{3+} ions. The oxidizing agent is the Fe^{3+} ions in iron(III) oxide and the reducing agent is the aluminum metal.

Strategy and Explanation First, determine the oxidation number of each element on the reactant side of the equation. Oxygen in compounds is normally −2 (← *p. 154*). Since the sum of the oxidation numbers of all the atoms in a formula must equal the charge on the formula, each iron in Fe_2O_3 is +3 and each oxygen is −2. The oxidation number for metallic Al is 0, as it is for all uncombined elements.

On the product side, the Al in Al_2O_3 has an oxidation number of +3. Iron is now in its uncombined form, so its oxidation number is now 0.

Thus, the elements that change oxidation number are

$$\overset{+3}{Fe}_2O_3(s) + 2\ \overset{0}{Al}(s) \longrightarrow 2\ \overset{0}{Fe}(s) + \overset{+3}{Al}_2O_3(s)$$

The oxidation state of the iron atoms decreased, while the oxidation state of the aluminum atoms increased. Thus, each iron in Fe_2O_3 has been reduced (+3 to 0) and aluminum has been oxidized (0 to +3). Consequently, Fe^{3+} ions in Fe_2O_3 are the oxidizing agent and Al metal is the reducing agent. Oxygen is neither reduced nor oxidized in this reaction; its oxidation number remains −2.

The thermite reaction in progress.

PROBLEM-SOLVING PRACTICE 18.1

Give the oxidation number for each atom and identify the oxidizing and reducing agents in these balanced chemical equations.

(a) $2\ Fe(s) + 3\ Cl_2(g) \longrightarrow 2\ FeCl_3(s)$

(b) $2\ H_2(g) + O_2(g) \longrightarrow 2\ H_2O(\ell)$

(c) $Cu(s) + 2\ NO_3^-(aq) + 4\ H_3O^+(aq) \longrightarrow Cu^{2+}(aq) + 2\ NO_2(g) + 6\ H_2O(\ell)$

(d) $C(s) + O_2(g) \longrightarrow CO_2(g)$

(e) $6\ Fe^{2+}(aq) + Cr_2O_7^{2-}(aq) + 14\ H_3O^+(aq) \longrightarrow 6\ Fe^{3+}(aq) + 2\ Cr^{3+}(aq) + 21\ H_2O(\ell)$

18.2 Using Half-Reactions to Understand Redox Reactions

Consider the redox reaction between zinc metal and copper(II) ions shown in Figure 18.1. The net ionic equation is

$$Zn(s) + Cu^{2+}(aq) \longrightarrow Zn^{2+}(aq) + Cu(s)$$

Zinc metal is oxidized to Zn^{2+} ions, and Cu^{2+} ions are reduced to copper metal.

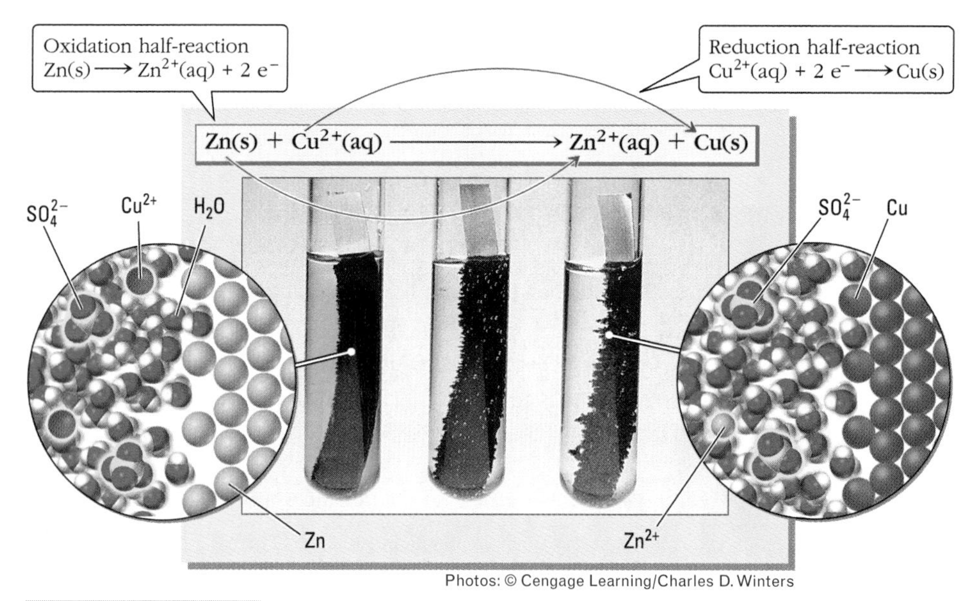

Photos: © Cengage Learning/Charles D. Winters

Active Figure 18.1 **An oxidation-reduction reaction.** A strip of zinc is placed in a solution of copper(II) sulfate (*left*). The zinc reacts with the copper(II) ions to produce copper metal (the brown-colored deposit on the zinc strip) and zinc ions in solution.

$$Zn(s) + Cu^{2+}(aq) \longrightarrow Zn^{2+}(aq) + Cu(s)$$

As copper metal accumulates on the zinc strip, the blue color due to the aqueous copper ions gradually fades (*middle* and *right*) as Cu^{2+} ions are reduced to metallic copper. The zinc ions in aqueous solution are colorless. Visit this book's companion website at **www.cengage.com/chemistry/moore** to test your understanding of the concepts in this figure.

To see more clearly how electrons are transferred, this overall reaction can be thought of as the result of two simultaneous **half-reactions:** one half-reaction for the oxidation of Zn and one half-reaction for the reduction of Cu^{2+} ions. The oxidation half-reaction

$$Zn(s) \longrightarrow Zn^{2+}(aq) + 2\,e^-$$

Note how the sum of the charges on the left side of the reaction equals the sum of the charges on the right side, even for half-reactions.

shows that each zinc atom loses two electrons when it is oxidized to a Zn^{2+} ion. These two electrons are accepted by a Cu^{2+} ion in the reduction half-reaction,

$$Cu^{2+}(aq) + 2\,e^- \longrightarrow Cu(s)$$

As Cu^{2+} ions are converted to Cu(s) in this half-reaction, the blue color of the solution becomes less intense and metallic copper forms on the zinc surface.

The net reaction is the sum of the oxidation and reduction half-reactions.

$$\begin{array}{rll} Zn(s) & \longrightarrow Zn^{2+}(aq) + 2\,e^- & \text{(oxidation half-reaction)} \\ Cu^{2+}(aq) + 2\,e^- & \longrightarrow Cu(s) & \text{(reduction half-reaction)} \\ \hline Zn(s) + Cu^{2+}(aq) & \longrightarrow Zn^{2+}(aq) + Cu(s) & \text{(net reaction)} \end{array}$$

Notice that no electrons appear in the equation for the net reaction because the number of electrons produced by the oxidation half-reaction must equal the number of electrons gained by the reduction half-reaction. This must always be true in a net reaction. Otherwise, electrons would be created or destroyed, violating the laws of conservation of mass and conservation of electrical charge.

Consider another example, as shown in Figure 18.2. A piece of metallic copper screen is immersed in a solution of silver nitrate. As the reaction proceeds, the colorless solution gradually turns blue, and fine, silvery, hair-like crystals form on the copper screen. Knowing that Cu^{2+} ions in aqueous solution appear blue, we can conclude that the copper metal is being oxidized to Cu^{2+}. Reduction must also be taking place, so it is reasonable to conclude that the hair-like crystals of silver result from the reduction of Ag^+ ions to metallic silver. The two half-reactions are

$$Cu(s) \longrightarrow Cu^{2+}(aq) + 2\,e^- \qquad \text{(oxidation half-reaction)}$$
$$Ag^+(aq) + e^- \longrightarrow Ag(s) \qquad \text{(reduction half-reaction)}$$

In this case, two electrons are produced in the oxidation half-reaction, but only one is needed for the reduction half-reaction. *One* atom of copper provides enough electrons (two) to reduce *two* Ag^+ ions, so the reduction half-reaction must occur twice every time the oxidation half-reaction occurs once. To indicate this relationship, we multiply the reduction half-reaction by 2.

$$2\,Ag^+(aq) + 2\,e^- \longrightarrow 2\,Ag(s) \qquad \text{(reduction half-reaction)} \times 2$$

Adding this reduction half-reaction to the oxidation half-reaction gives the net equation

$$\begin{array}{rcl} 2\,Ag^+(aq) + 2\,e^- & \longrightarrow & 2\,Ag(s) \\ Cu(s) & \longrightarrow & Cu^{2+}(aq) + 2\,e^- \\ \hline Cu(s) + 2\,Ag^+(aq) & \longrightarrow & Cu^{2+}(aq) + 2\,Ag(s) \end{array}$$

The method shown here is a general one. A net equation can always be generated by writing oxidation and reduction half-reactions, using coefficients to adjust the half-reaction equations so that the number of electrons released by the oxidation equals the number gained by the reduction, and then adding the two half-reactions to give the equation for the net reaction.

Figure 18.2 Copper metal screen in a solution of $AgNO_3$. The blue color intensifies as more copper is oxidized to aqueous Cu^{2+} ion. Ag^+ ions are reduced to silver metal.

PROBLEM-SOLVING EXAMPLE 18.2 Determining Half-Reactions from Net Redox Reactions

Aluminum will undergo a redox reaction with an acid such as HCl to produce $Al^{3+}(aq)$ and $H_2(g)$.

(unbalanced equation) $Al(s) + H^+(aq) \longrightarrow Al^{3+}(aq) + H_2(g)$

Write the oxidation half-reaction and the reduction half-reaction equations, and combine them to give the balanced equation for the net reaction.

Answer
Oxidation half-reaction: $Al(s) \longrightarrow Al^{3+}(aq) + 3\,e^-$
Reduction half-reaction: $2\,H^+(aq) + 2\,e^- \longrightarrow H_2(g)$
Net reaction: $2\,Al(s) + 6\,H^+(aq) \longrightarrow 2\,Al^{3+}(aq) + 3\,H_2(g)$

Strategy and Explanation To identify the half-reactions, we must first identify the species whose oxidation number increases (it is oxidized) and the species whose oxidation number decreases (it is reduced). Aluminum's oxidation number increases from 0 to +3, so it is oxidized.

$$Al(s) \longrightarrow Al^{3+}(aq) + 3\,e^-$$

This half-reaction must have three electrons on the right side to balance the 3+ charge of the aluminum ion and yield equal charges on the right side and the left side of the half-reaction.

Hydrogen's oxidation number decreases from +1 to 0 so it is reduced.

$$2\,H^+(aq) + 2\,e^- \longrightarrow H_2(g)$$

The half-reaction must have two electrons on the left side to balance the charge of the two hydrogen ions.

Notice that these two half-reactions contain different numbers of electrons. The two half-reactions are multiplied by 2 and 3, respectively, so that six e^- appear in each half-reaction.

$$\begin{array}{lll} 2\,[Al(s) \longrightarrow Al^{3+}(aq) + 3\,e^-] & \text{gives} & 2\,Al(s) \longrightarrow 2\,Al^{3+}(aq) + 6\,e^- \\ 3\,[2\,H^+(aq) + 2\,e^- \longrightarrow H_2(g)] & \text{gives} & 6\,H^+(aq) + 6\,e^- \longrightarrow 3\,H_2(g) \\ \hline \text{Net reaction:} & & 2\,Al(s) + 6\,H^+(aq) \longrightarrow 2\,Al^{3+}(aq) + 3\,H_2(g) \end{array}$$

The net reaction is the sum of these two half-reactions multiplied by the proper whole numbers to make the number of electrons produced by the oxidation half-reaction equal to the number of electrons gained in the reduction half-reaction.

PROBLEM-SOLVING PRACTICE 18.2

Write oxidation and reduction half-reactions for these net redox equations. Show that their sum is the net reaction.

(a) $Cd(s) + Cu^{2+}(aq) \longrightarrow Cu(s) + Cd^{2+}(aq)$

(b) $Zn(s) + 2\ H^{+}(aq) \longrightarrow Zn^{2+}(aq) + H_2(g)$

(c) $2\ Al(s) + 3\ Zn^{2+}(aq) \longrightarrow 2\ Al^{3+}(aq) + 3\ Zn(s)$

Balancing Redox Equations Using Half-Reactions

All of the equations in Problem-Solving Practice 18.2 are balanced. While these particular redox equations could be easily balanced by inspection, this is not always the case. Equations for redox reactions often involve water, hydronium ions, or hydroxide ions as reactants or products. It is difficult to tell by observing the unbalanced equation how many H_2O, H_3O^+, and OH^- are involved, or whether they will be reactants, or products, or even present at all. Fortunately, there are systematic ways to figure this out.

Oxalic acid, HOOC—COOH, is the simplest organic acid containing two carboxylic acid groups.

Balancing Redox Equations in Acidic Solutions Consider the reaction of oxalic acid with permanganate ion in an acidic solution. The products are manganese(II) ions and carbon dioxide, so the *unbalanced* equation is

$$\text{(unbalanced equation)} \quad \underset{\text{oxalic acid}}{H_2C_2O_4(aq)} + \underset{\text{permanganate ion}}{MnO_4^-(aq)} \longrightarrow Mn^{2+}(aq) + CO_2(g)$$

If you try to balance this equation by trial and error, you will almost certainly have a hard time balancing hydrogen and oxygen. You have probably already noticed that no hydrogen-containing species appears on the product side of the unbalanced equation. Because the reaction takes place in an aqueous acidic solution, water and hydronium ions are involved. Generating the balanced equation for a reaction like this one is best done by following a systematic approach, a series of steps. In each step you must use what you know about the oxidation and reduction half-reactions, as well as conservation of matter and conservation of electrical charge. Problem-Solving Example 18.3 illustrates the steps that produce a balanced equation for a redox reaction occurring in acidic solution.

PROBLEM-SOLVING EXAMPLE 18.3 Balancing Redox Equations for Reactions in Acidic Solutions

Balance the previous equation for the oxidation of oxalic acid in an acidic permanganate solution.

Answer $5\ H_2C_2O_4(aq) + 6\ H_3O^+(aq) + 2\ MnO_4^-(aq) \longrightarrow 10\ CO_2(g) + 2\ Mn^{2+}(aq) + 14\ H_2O(\ell)$

Strategy and Explanation Follow a systematic approach listed in this series of steps to balance the equation for this reaction (and all similar redox reactions).

Step 1: *Recognize whether the reaction is an oxidation-reduction process. If it is, then determine what is reduced and what is oxidized.* This is a redox reaction because the oxidation number of Mn changes from +7 in MnO_4^- to +2 in Mn^{2+}, so the Mn in MnO_4^- is reduced. The oxidation number of each C changes from +3 in $H_2C_2O_4$ to +4 in CO_2, so the C in $H_2C_2O_4$ is oxidized. The oxidation numbers of H (+1) and O (−2) are unchanged.

Step 2: *Break the overall unbalanced equation into half-reactions.*

$$H_2C_2O_4(aq) \longrightarrow CO_2(g) \qquad \text{(oxidation half-reaction)}$$

$$MnO_4^-(aq) \longrightarrow Mn^{2+}(aq) \qquad \text{(reduction half-reaction)}$$

Step 3: *Balance the atoms in each half-reaction.* First balance all atoms except for O and H, then balance O by adding H_2O and balance H by adding H^+. (Hydroxide ion, OH^-, cannot be used here because the reaction occurs in an acidic solution and the OH^- concentration is very low.)

In acidic solution, balance O by adding H_2O, and balance H by adding H^+.

Oxalic acid half-reaction: First, balance the carbon atoms in the half-reaction.

$$H_2C_2O_4(aq) \longrightarrow 2\,CO_2(g)$$

This step balances the O atoms as well (no H_2O needed here), so only H atoms remain to be balanced. Because the product side is deficient by two H, we put $2\,H^+$ there.

$$H_2C_2O_4(aq) \longrightarrow 2\,CO_2(g) + 2\,H^+(aq) \qquad \text{(oxalic acid half-reaction)}$$

Strictly speaking, we ought to use H_3O^+ instead of H^+, but this would result in adding water molecules to each side of the equation, which is rather cumbersome. It is simpler to add H^+ now and add the water molecules at the end.

Permanganate half-reaction: The Mn atoms are already balanced, but the oxygen atoms are not balanced until H_2O is added. Adding $4\,H_2O$ on the product side takes care of the needed oxygen atoms.

$$MnO_4^-(aq) \longrightarrow Mn^{2+}(aq) + 4\,H_2O(\ell)$$

Now there are 8 H atoms on the right and none on the left. To balance hydrogen atoms, $8\,H^+$ are placed on the left side of the half-reaction.

$$8\,H^+(aq) + MnO_4^-(aq) \longrightarrow Mn^{2+}(aq) + 4\,H_2O(\ell) \qquad \text{(permanganate half-reaction)}$$

Step 4: *Balance the half-reactions for charge using electrons (e^-).* The oxalic acid half-reaction has a net charge of 0 on the left side and 2+ on the right. The reactants have lost two electrons. To show this fact, $2\,e^-$ must appear on the right side.

$$H_2C_2O_4(aq) \longrightarrow 2\,CO_2(g) + 2\,H^+(aq) + 2\,e^-$$

This confirms that $H_2C_2O_4$ is the reducing agent (it loses electrons and is oxidized). The loss of two electrons is also in keeping with the increase in the oxidation number of each of two C atoms by 1, from +3 to +4. The $2\,e^-$ also balance the charge on the product side of the equation.

The MnO_4^- half-reaction has a charge of 7+ on the left and 2+ on the right. Therefore, to achieve a net 2+ charge on each side, $5\,e^-$ must appear on the left. The gain of electrons shows that MnO_4^- is the oxidizing agent; it is reduced.

$$5\,e^- + 8\,H^+(aq) + MnO_4^-(aq) \longrightarrow Mn^{2+}(aq) + 4\,H_2O(\ell)$$

Step 5: *Multiply the half-reactions by appropriate factors so that the oxidation half-reaction produces as many electrons as the reduction half-reaction accepts.* In this case, one half-reaction involves two electrons, and the other half-reaction involves five electrons. It takes ten electrons to balance each half-reaction. The oxalic acid half-reaction must be multiplied by 5, and the permanganate half-reaction by 2.

$$5\,[H_2C_2O_4(aq) \longrightarrow 2\,CO_2(g) + 2\,H^+(aq) + 2\,e^-]$$

$$2\,[5\,e^- + 8\,H^+(aq) + MnO_4^-(aq) \longrightarrow Mn^{2+}(aq) + 4\,H_2O(\ell)]$$

Step 6: *Add the half-reactions to give the net reaction and cancel equal amounts of reactants and products that appear on both sides of the arrow.*

$$5\,H_2C_2O_4(aq) \longrightarrow 10\,CO_2(g) + 10\,H^+(aq) + 10\,e^-$$

$$10\,e^- + 16\,H^+(aq) + 2\,MnO_4^-(aq) \longrightarrow 2\,Mn^{2+}(aq) + 8\,H_2O(\ell)$$

$$5\,H_2C_2O_4(aq) + 16\,H^+(aq) + 2\,MnO_4^-(aq) \longrightarrow 10\,CO_2(g) + 10\,H^+(aq) + 2\,Mn^{2+}(aq) + 8\,H_2O(\ell)$$

Since $16\,H^+$ appear on the left and $10\,H^+$ appear on the right, $10\,H^+$ are canceled, leaving $6\,H^+$ on the left.

$$5\ H_2C_2O_4(aq) + 6\ H^+(aq) + 2\ MnO_4^-(aq) \longrightarrow 10\ CO_2(g) + 2\ Mn^{2+}(aq) + 8\ H_2O(\ell)$$

Step 7: *Check the balanced net equation to make sure both atoms and charge are balanced.*

Atom balance: Each side of the equation has 2 Mn, 28 O, 10 C, and 16 H atoms.

Charge balance: Each side has a net charge of 4+.
On the left side, $(6 \times 1+) + (2 \times 1-) = 4+$.

On the right side, $2(2+) = 4+$.

Step 8: *Add enough water molecules to both sides of the equation to convert all H^+ to H_3O^+.* In this case, six water molecules are needed ($6\ H_2O + 6\ H^+ \rightarrow 6\ H_3O^+$). Six water molecules are added to each side of the equation, which increases the total to 14 on the product side.

$$5\ H_2C_2O_4(aq) + 6\ H_3O^+(aq) + 2\ MnO_4^-(aq) \longrightarrow 10\ CO_2(g) + 2\ Mn^{2+}(aq) + 14\ H_2O(\ell)$$

Step 9: *Check the final results to make sure both atoms and charges are balanced.* The net equation is balanced. The net charges on each side of the reaction are the same, and the numbers of atoms of each kind on each side of the reaction are equal.

PROBLEM-SOLVING PRACTICE 18.3

Balance this equation for the reaction of Zn with $Cr_2O_7^{2-}$ in acidic aqueous solution.

$$Zn(s) + Cr_2O_7^{2-}(aq) \longrightarrow Cr^{3+}(aq) + Zn^{2+}(aq)$$

CONCEPTUAL
EXERCISE 18.1 Electrons Lost Equal Electrons Gained

Why must the number of electrons lost always equal the number gained in a redox reaction?

Balancing Redox Equations in Basic Solutions For redox reactions that occur in basic solutions, the final electrochemical reaction must be completed with water and OH^- ions rather than water and H_3O^+ ions that we used for acidic solutions. During the balancing process, the half-reactions can be balanced as if they occurred in acidic solution. Then, at the end of the series of steps, the H^+ ions can be neutralized by adding an equal number of OH^- ions to both sides of the electrochemical equation and canceling, if necessary, water molecules that appear as both reactants and products. Problem-Solving Example 18.4 illustrates how to balance a redox reaction in basic solution.

PROBLEM-SOLVING EXAMPLE 18.4 Balancing Redox Equations for Reactions in Basic Solutions

In a nickel-cadmium (nicad) battery, cadmium metal forms $Cd(OH)_2$ and Ni_2O_3 forms $Ni(OH)_2$ in an alkaline solution. Write the balanced net equation for this reaction.

Answer $Cd(s) + Ni_2O_3(s) + 3\ H_2O(\ell) \longrightarrow Cd(OH)_2(s) + 2\ Ni(OH)_2(s)$

Strategy and Explanation Use the systematic approach given below.

Step 1: *Recognize whether the reaction is an oxidation-reduction process. Then determine what is reduced and what is oxidized.* This is a redox reaction because the oxidation number of Cd changes from 0 in Cd metal to +2 in $Cd(OH)_2$, so Cd metal is oxidized. The oxidation number of each Ni changes from +3 in Ni_2O_3 to +2 in $Ni(OH)_2$, so the Ni is reduced.

Step 2: *Break the overall unbalanced equation into half-reactions.*

$$Cd(s) \longrightarrow Cd(OH)_2(s) \qquad \text{(oxidation half-reaction)}$$
$$Ni_2O_3(s) \longrightarrow Ni(OH)_2(s) \qquad \text{(reduction half-reaction)}$$

Step 3: *Balance the atoms in each half-reaction.* First, balance all atoms except the O and H atoms; do them last. Balance each half-reaction as if it were in an *acidic* solution to start. Balance O by adding H_2O and balance H by adding H^+. We will revert to using OH^- ions characteristic of basic solutions later in the process.

In the Cd half-reaction, the Cd atoms are balanced. Adding two water molecules on the left balances the two O atoms on the right, but this leaves four H atoms on the left and only two on the right. Adding two H^+ ions on the right balances H atoms in this half-reaction.

$$2\,H_2O(\ell) + Cd(s) \longrightarrow Cd(OH)_2(s) + 2\,H^+(aq)$$

For the Ni_2O_3 half-reaction, a coefficient of 2 is needed for $Ni(OH)_2$ because there are two Ni atoms on the left. To balance the four O atoms and four H atoms now on the right with the three O atoms on the left requires one water molecule and two H^+ ions on the left.

$$2\,H^+(aq) + H_2O(\ell) + Ni_2O_3(s) \longrightarrow 2\,Ni(OH)_2(s)$$

Step 4: *Balance the half-reactions for charge using electrons.* The Cd half-reaction produces 2 e^- as a product.

$$2\,H_2O(\ell) + Cd(s) \longrightarrow Cd(OH)_2(s) + 2\,H^+(aq) + 2\,e^- \qquad \text{(balanced)}$$

The Ni_2O_3 half-reaction requires 2 e^- as a reactant.

$$2\,H^+(aq) + H_2O(\ell) + Ni_2O_3(s) + 2\,e^- \longrightarrow 2\,Ni(OH)_2(s) \qquad \text{(balanced)}$$

Step 5: *Multiply the half-reactions by appropriate factors so that the reducing agent produces as many electrons as the oxidizing agent accepts.* The Cd half-reaction produces two electrons, and the Ni_2O_3 half-reaction accepts two, so the electrons are balanced.

Step 6: *Because H^+ does not exist at any appreciable concentration in a basic solution, remove H^+ by adding an appropriate amount of OH^- to both sides of the equation. H^+ and OH^- react to form H_2O.* In the Cd half-reaction, add two OH^- ions to each side to get

$$2\,OH^-(aq) + 2\,H_2O(\ell) + Cd(s) \longrightarrow Cd(OH)_2(s) + 2\,H_2O(\ell) + 2\,e^-$$

On the product side, two OH^- ions plus two H^+ ions form two H_2O molecules. For the Ni_2O_3 half-reaction, add two OH^- ions to each side to get

$$3\,H_2O(\ell) + Ni_2O_3(s) + 2\,e^- \longrightarrow 2\,Ni(OH)_2(s) + 2\,OH^-(aq)$$

Step 7: *Add the half-reactions to give the net reaction, and cancel reactants and products that appear on both sides of the reaction arrow.*

$$\cancel{2\,OH^-(aq)} + \cancel{2\,H_2O(\ell)} + Cd(s) \longrightarrow Cd(OH)_2(s) + \cancel{2\,H_2O(\ell)} + 2\,e^-$$
$$3\,H_2O(\ell) + Ni_2O_3(s) + 2\,e^- \longrightarrow 2\,Ni(OH)_2(s) + \cancel{2\,OH^-(aq)}$$
$$\overline{Cd(s) + Ni_2O_3(s) + 3\,H_2O(\ell) \longrightarrow Cd(OH)_2(s) + 2\,Ni(OH)_2(s)}$$

Step 8: *Check the final results to make sure both atoms and charge are balanced.* The net equation is balanced. In the final equation, there are no net charges on either side of the reaction arrow, and the numbers of atoms of each kind on each side of the reaction arrow are equal.

PROBLEM-SOLVING PRACTICE 18.4

In a basic solution, aluminum metal forms $Al(OH)_4^-$ ion as the metal reduces NO_3^- ion to NH_3. Write the balanced net equation for this reaction using the steps outlined in Problem-Solving Example 18.4.

18.3 Electrochemical Cells

Strictly speaking, many devices we call batteries consist of several voltaic cells connected together, but the term "battery" has taken on the same meaning as "voltaic cell."

Voltaic cells are sometimes referred to as *galvanic cells* in recognition of the work of Luigi Galvani, who discovered that a frog's leg placed in salt water would twitch when it was touched simultaneously by two dissimilar metals.

To identify the anode and cathode, remember that **O**xidation takes place at the **A**node (both words begin with vowels), and **R**eduction takes place at the **C**athode (both words begin with consonants).

On a flashlight battery, the anode is marked "−" because an oxidation reaction produces electrons that make the anode negative. Conversely, the cathode is marked "+" because a reduction reaction consumes electrons, leaving the metal electrode positive.

In a redox reaction, electrons are *transferred* from one kind of atom, molecule, or ion to another. It is easy to see by the color changes in the two redox reactions shown in Figures 18.1 and 18.2 that these two reactions favor the formation of products—as soon as the reactants are mixed, changes take place. All product-favored reactions release Gibbs free energy (***p. 625***), energy that can do useful work. An electrochemical cell is a way to capture that useful work as electrical work.

In an **electrochemical cell,** an oxidizing agent and a reducing agent pair are arranged in such a way that they can react only if electrons flow through an outside conductor. Such electrochemical cells are also known as **voltaic cells** or **batteries.** Figure 18.3 diagrams how a voltaic cell can be made from the Zn/Cu^{2+} reaction shown previously in Figure 18.1. The two half-reactions occur in separate beakers, each of which is called a **half-cell.** When Zn atoms are oxidized, the electrons that are given up by zinc metal pass through the wire and a light bulb to the copper metal. There the electrons are available to reduce Cu^{2+} ions from the solution. The metallic zinc and copper strips are called electrodes. An **electrode** conducts electrical current (electrons) into or out of something, in this case, a solution. An electrode is most often a metal plate or wire, but it can also be a piece of graphite or another conductor. The electrode where oxidation occurs is the **anode;** the electrode where reduction takes place is the **cathode.**

The voltaic cell is named after the Italian scientist Alessandro Volta, who, in about 1800, constructed the first electrochemical cell, a stack of alternating zinc and silver disks separated by pieces of paper soaked in salt water (an electrolyte). Later, Volta

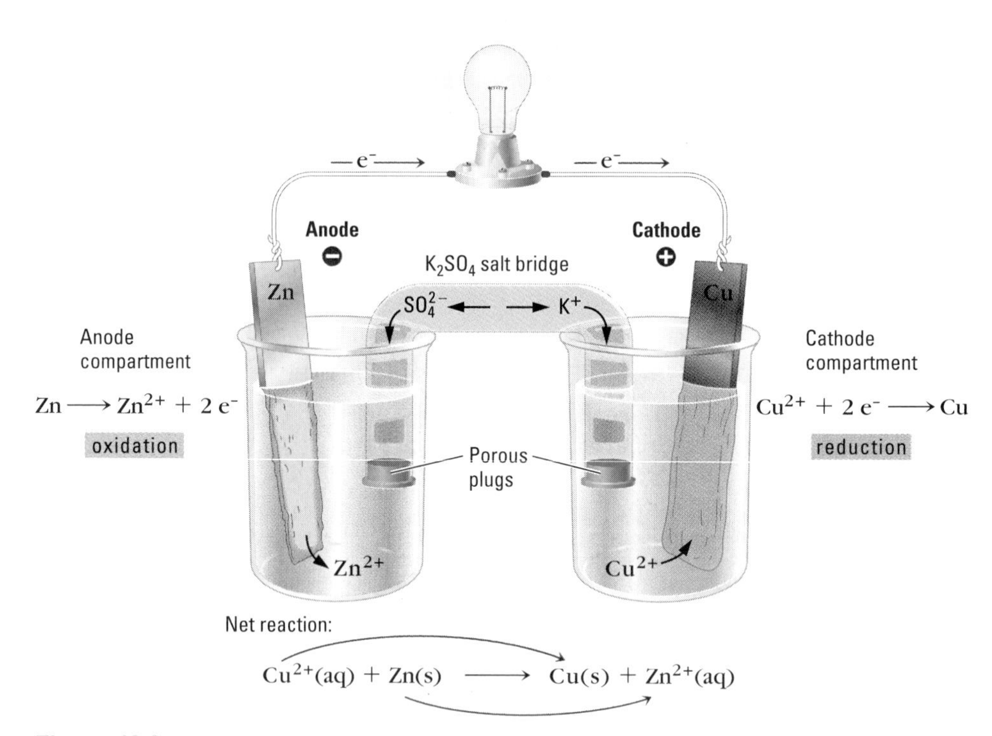

Figure 18.3 A simple electrochemical cell. The cell consists of a zinc electrode in a solution containing Zn^{2+} ions (*left*), a copper electrode in a solution containing Cu^{2+} ions (*right*), and a salt bridge that allows ions to flow into and out of the two solutions. When the two metal electrodes are connected by a conducting circuit, electrons flow through the wires and light bulb from the zinc electrode, where zinc is oxidized, to the copper electrode, where copper ions from the solution are reduced.

showed that any two different metals and an electrolyte could be used to make a battery. Figure 18.4 shows a cell constructed by sticking a strip of zinc metal and a strip of copper metal into a grapefruit. In this case, the acidic solution of the grapefruit juice is the electrolyte.

In the electrochemical cell diagrammed in Figure 18.3, electrons are released at the anode by the oxidation half-reaction

$$Zn(s) \longrightarrow Zn^{2+}(aq) + 2\,e^- \qquad \text{(anode reaction)}$$

They then flow from the anode through the filament in the bulb, causing it to glow, and eventually travel to the cathode, where they react with copper(II) ions in the reduction half-reaction

$$Cu^{2+}(aq) + 2\,e^- \longrightarrow Cu(s) \qquad \text{(cathode reaction)}$$

© Cengage Learning/Charles D. Winters

Figure 18.4 A grapefruit battery. A voltaic cell can be made by inserting zinc and copper electrodes into a grapefruit. A potential of 0.95 V is obtained. (The water and citric acid of the fruit allow for ion conduction between electrodes.) This cell is more complicated than the one in Figure 18.3. To learn how the grapefruit battery works, see Goodisman, J. *Journal of Chemical Education,* Vol. 78, 2001; p. 516.

If nothing else but electron flow took place, the concentration of Zn^{2+} ions in the anode compartment would increase as zinc metal is oxidized, building up positive charge in the solution. The concentration of Cu^{2+} ions in the cathode compartment would decrease as Cu^{2+} ions are reduced to copper metal. This makes that solution less positive due to the decrease of positive charge as Cu^{2+} ions are reduced. Due to charge imbalance, the flow of electrons in the wires would very quickly stop. For the cell to work, there must be a way for the positive charge buildup in the anode compartment to be balanced by addition of negative ions or removal of positive ions, and vice versa for the cathode compartment.

The charge buildup can be avoided by using a salt bridge to connect the two compartments. A **salt bridge** is a solution of a salt (K_2SO_4 in Figure 18.3) arranged so that the bulk of that solution cannot flow into the cell solutions, but the ions (K^+ and SO_4^{2-}) can pass freely. As electrons flow through the wire from the zinc electrode to the copper electrode, negative ions (SO_4^{2-}) move from the salt bridge into the anode compartment solution and positive ions (K^+) move in the opposite direction into the cathode compartment solution. In general, anions from the salt bridge flow into the anode cell, and cations from the salt bridge flow into the cathode cell. This flow of ions completes the electrical circuit, allowing current to flow. If the salt bridge is removed from this battery, the flow of electrons will stop.

All voltaic cells and batteries operate in a similar fashion and have these characteristics:

- The oxidation-reduction reaction that occurs must favor the formation of products.
- There must be an external circuit through which electrons flow.
- There must be a salt bridge, porous barrier, or some other means of allowing ions in the salt bridge to flow into the electrode compartments to offset charge buildup.

In commercial batteries, the salt bridge is often a porous membrane.

The components of an electrochemical cell are summarized in Figure 18.5.

PROBLEM-SOLVING EXAMPLE 18.5 Electrochemical Cells

A simple voltaic cell is assembled with Fe(s) and $Fe(NO_3)_2(aq)$ in one compartment and Cu(s) and $Cu(NO_3)_2(aq)$ in the other compartment. An external wire connects the two electrodes, and a salt bridge containing $NaNO_3$ connects the two solutions. The net reaction is

$$Fe(s) + Cu^{2+}(aq) \longrightarrow Cu(s) + Fe^{2+}(aq)$$

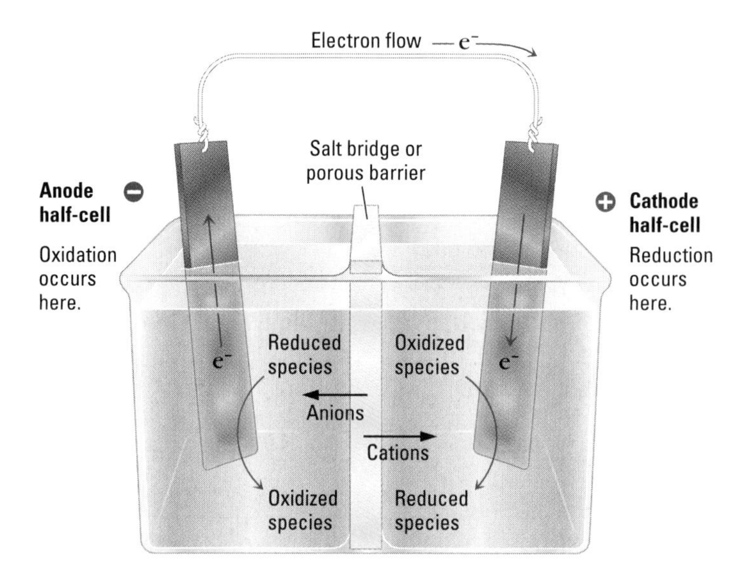

Figure 18.5 Summary of the terminology used in voltaic cells. Oxidation occurs at the anode, and reduction occurs at the cathode. Electrons move from the negative electrode (anode) to the positive electrode (cathode) through the external wire. The electrical circuit is completed in the solution by the movement of ions—anions move from the salt bridge compartment to the anode compartment, and cations move from the salt bridge compartment to the cathode compartment. The half-cells can be separated by either a salt bridge or a porous barrier.

This voltaic cell is shown without an electrical device in the external part of the circuit for simplicity.

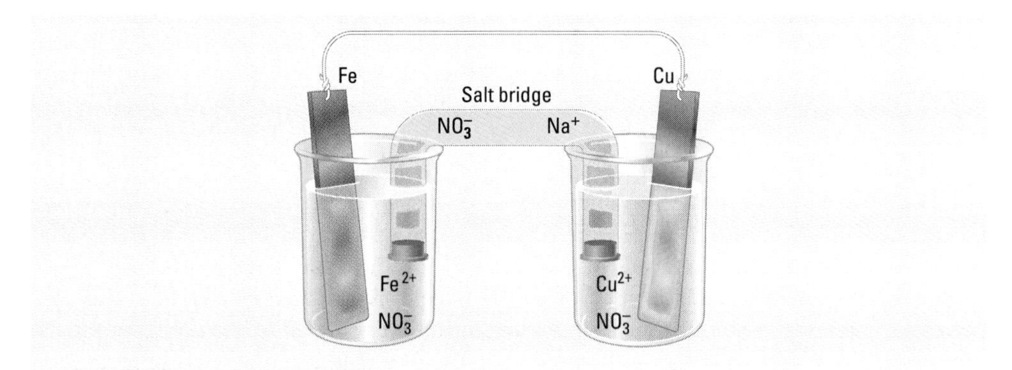

What is the reaction at the anode? What is the reaction at the cathode? What is the direction of electron flow in the external wire? What is the direction of ion flow in the salt bridge? Draw a cell diagram, indicating the anode, the cathode, and the directions of electron flow and ion flow.

Answer

Anode reaction: $Fe(s) \longrightarrow Fe^{2+}(aq) + 2\ e^-$

Cathode reaction: $Cu^{2+}(aq) + 2\ e^- \longrightarrow Cu(s)$

The electrons flow through the wire from the anode to the cathode. Nitrate ions move from the salt bridge into the anode compartment. Sodium ions move from the salt bridge into the cathode compartment. The completed cell diagram is:

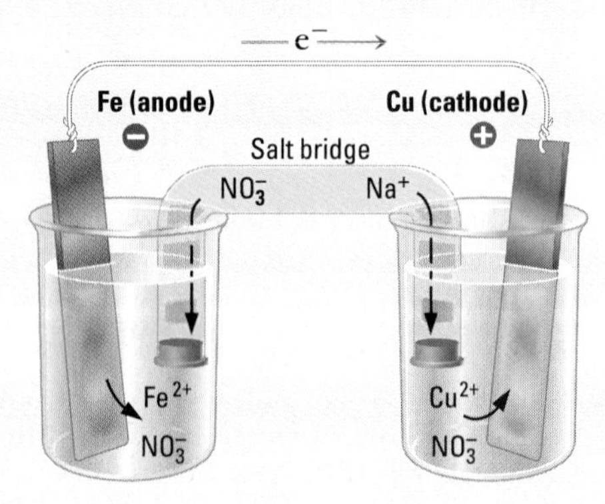

Strategy and Explanation The net reaction shows that iron is being oxidized from $Fe(s)$ to $Fe^{2+}(aq)$ and that $Cu^{2+}(aq)$ is being reduced to $Cu(s)$. We need to decide at which electrodes these reactions occur. Since Fe metal is being oxidized (increase in oxidation number), the electrode in the $Fe(s)/Fe^{2+}(aq)$ compartment is the anode. The electrode in the $Cu(s)/Cu^{2+}(aq)$ compartment must be the cathode because the Cu^{2+} ions are being reduced to Cu metal (decrease in oxidation number).

The half-reactions are

$$Fe(s) \longrightarrow Fe^{2+}(aq) + 2\,e^- \qquad \text{(oxidation—anode)}$$
$$Cu^{2+}(aq) + 2\,e^- \longrightarrow Cu(s) \qquad \text{(reduction—cathode)}$$

Electrons flow *from* their source (the oxidation of the Fe at the iron electrode which is the anode) through the wire *to* the electrode where they reduce Cu^{2+} ions to copper metal (the Cu electrode which is the cathode). Because positive Fe^{2+} ions are being produced in the anode compartment, negative NO_3^- ions in the salt bridge move into the anode compartment from the salt bridge to balance the overall charge. Because Cu^{2+} ions are being removed from the cathode compartment, Na^+ ions move into the cathode compartment from the salt bridge to replace the charge of the Cu^{2+} ions reduced to Cu metal. The external part of the electrical circuit is shown in the margin.

Anode $\xrightarrow{e^-\text{ flow}}$ Cathode
⊖ ⊕
oxidation (loss of e^- s) $Fe \rightarrow Fe^{2+}$
reduction (gain of e^- s) $Cu^{2+} \rightarrow Cu$

PROBLEM-SOLVING PRACTICE 18.5

A voltaic cell is assembled to use this net reaction.

$$Ni(s) + 2\,Ag^+(aq) \longrightarrow Ni^{2+}(aq) + 2\,Ag(s)$$

(a) Write half-reactions for this cell, and indicate which is the oxidation reaction and which is the reduction reaction.
(b) Name the electrodes at which these reactions take place.
(c) What is the direction of flow of electrons in an external wire connected between the electrodes?
(d) If a salt bridge connecting the two electrode compartments contains KNO_3, what is the direction of flow of the K^+ ions and the NO_3^- ions?

A shorthand notation has been developed to represent an electrochemical cell. For the cell shown in Figure 18.3 with the redox reaction

$$Zn(s) + Cu^{2+}(aq) \longrightarrow Zn^{2+}(aq) + Cu(s)$$

the representation is

anode electrode / salt bridge / cathode electrode

$$Zn(s)\,|\,Zn^{2+}(aq)\,||\,Cu^{2+}(aq)\,|\,Cu(s)$$

anode (Zn(s) | Zn²⁺(aq)) — cathode (Cu²⁺(aq) | Cu(s))

The anode half-cell is represented on the left, and the cathode half-cell is represented on the right. The electrodes are written on the extreme left (anode, Zn) and extreme right (cathode, Cu) of the notation. The single vertical lines denote boundaries between phases, and the double vertical lines denote the salt bridge and the separation between half-cells. Within each half-cell the reactants are written first, followed by the products. The electron flow is from left to right (anode to cathode).

CONCEPTUAL EXERCISE 18.2 Battery Design

Devise an internal on-off switch for a battery that would not be a part of the flow of electrons.

18.4 Electrochemical Cells and Voltage

Because electrons flow from the anode to the cathode of an electrochemical cell, they can be thought of as being "driven" or "pushed" by an **electromotive force (emf).** The emf is produced by the difference in electrical potential energy between the two electrodes. Just as water flows downhill in response to a difference in gravitational potential energy, so an electron moves from an electrode of higher electrical potential energy to an electrode of lower potential energy. The moving water can do work, as can moving electrons; for example, they could run a motor.

This is similar to comparing the quantity of work a few drops of water can do when falling 100 m with that possible when a few tons of water fall the same distance.

The quantity of electrical work done is proportional to the number of electrons that go from higher to lower potential energy as well as to the size of the potential energy difference.

$$\text{Electrical work} = \text{charge} \times \text{potential energy difference}$$

or

$$\text{Electrical work} = \text{number of electrons} \times \text{potential energy difference}$$

$$\frac{1.6022 \times 10^{-19}\ \text{C}}{1\ e^-} \times 6.24 \times 10^{18}\ e^- = 1\ \text{C}$$

Electrical charge is measured in coulombs. The charge on a single electron is very small (1.6022×10^{-19} C), so it takes 6.24×10^{18} electrons to produce 1 coulomb of charge. A **coulomb, C,** is the quantity of charge that passes a fixed point in an electrical circuit when a current of 1 ampere flows for 1 second. The **ampere, A,** is the unit of electrical current.

$$1\ \text{coulomb} = 1\ \text{ampere} \times 1\ \text{second}$$

When a single electron moves through a potential of 1 V, the work done is one electron-volt, abbreviated eV.

Electrical potential energy difference is measured in volts. The **volt, V,** is defined such that one joule of work is performed when one coulomb of charge moves through a potential difference of one volt:

$$1\ \text{volt} = \frac{1\ \text{joule}}{1\ \text{coulomb}} \quad \text{or} \quad 1\ \text{joule} = 1\ \text{volt} \times 1\ \text{coulomb}$$

The electromotive force of an electrochemical cell, commonly called its **cell voltage,** shows how much work a cell can produce for each coulomb of charge that the chemical reaction produces.

© Cengage Learning/Charles D. Winters

Figure 18.6 Dry cell batteries. The larger batteries are capable of more work since they contain more oxidizing and reducing agents.

The voltage of an electrochemical cell depends on the temperature and substances that make up the cell, including their pressures if they are gases or their concentrations if they are solutes in solution. The quantity of charge (coulombs) depends on how much of each substance reacts. Look at the 1.5-V batteries shown in Figure 18.6. They have the same voltage because they have electrodes with the same potential difference between them. Yet a larger battery is capable of far more work than a smaller one, because it contains larger quantities of oxidizing and reducing agents. In this section and the next, we consider how cell voltage depends on the materials from which a cell is made. In Section 18.5, we will return to the question of how much electrical work a cell can do.

CONCEPTUAL **EXERCISE** **18.3 Electrical Charges**

Which has the larger charge, a coulomb of charge or Avogadro's number of electrons?

Cell Voltage

It is important that the voltmeter have very high electrical resistance so that very little chemical reaction occurs.

A cell's voltage is readily measured by inserting a voltmeter into the circuit. Because the voltage varies with concentrations, **standard conditions** are defined for voltage measurements. These are the same as those used for $\Delta H°$ (⇐ ***p. 195***)***:*** All reactants and products must be present as pure solids, pure liquids, gases at 1 bar pressure, or solutes at 1 M concentration. Voltages measured under these conditions are **standard voltages,** symbolized by $E°$**.** Unless specified otherwise, all values of $E°$ are for 25 °C (298 K). By definition, *cell voltages for product-favored electrochemical reactions are positive.* For example, the standard cell voltage for the product-favored Zn/Cu^{2+} cell discussed earlier is + 1.10 V at 25 °C.

Since every redox reaction can be thought of as the sum of two half-reactions, it is convenient to assign a voltage to every possible half-reaction. Then the cell voltage for any net reaction can be obtained by using the standard voltages of the half-reactions that occur at the cathode and the anode.

$$\boxed{E°_{cell} = E°_{cathode} - E°_{anode}}$$

The two quantities on the right side of the equation, $E°_{cathode}$ and $E°_{anode}$, are the values for the half-reactions *written as reduction half-reactions.* ***If*** $E°_{cell}$ ***is positive, the reaction is product-favored. If*** $E°_{cell}$ ***is negative, the reaction is reactant-favored.*** However, because only *differences* in potential energy can be measured, it is not possible to measure the voltage for a single half-reaction. Instead, one half-reaction is chosen as the standard, and then all others are compared to it. The half-reaction chosen as the standard is the one that occurs at the **standard hydrogen electrode,** in which hydrogen gas at a pressure of 1 bar is bubbled over a platinum electrode immersed in 1 M aqueous acid at 25 °C (Figure 18.7).

$$2\,H_3O^+(aq, 1\,M) + 2\,e^- \longrightarrow H_2(g, 1\,bar) + 2\,H_2O(\ell)$$

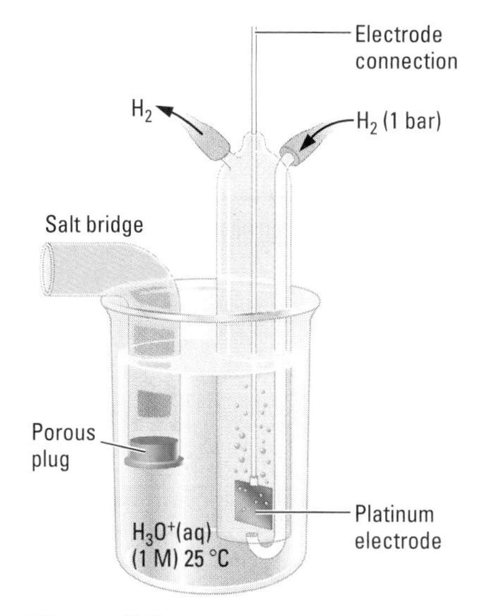

Figure 18.7 The standard hydrogen electrode. Hydrogen gas at 1 bar pressure bubbles over an inert platinum electrode that is immersed in a solution containing exactly 1 M H_3O^+ ions at 25 °C. The potential for this electrode is defined as exactly 0 V.

A voltage of exactly 0 V is *assigned* to this half-cell. In a cell that combines another half-reaction with the standard hydrogen electrode, the overall cell voltage is determined by the difference between the potentials of the two electrodes. Because the potential of the hydrogen electrode is assigned a value of 0, the overall cell voltage equals the voltage of the other electrode.

When the standard hydrogen electrode is paired with a half-cell that contains a better reducing agent than H_2, $H_3O^+(aq)$ is reduced to H_2.

H_3O^+ reduced: $2\,H_3O^+(aq) + 2\,e^- \longrightarrow H_2(g, 1\,bar) + 2\,H_2O(\ell)$ $\qquad E° = 0.0\text{ V}$

When the standard hydrogen electrode is paired with a half-cell that contains a better oxidizing agent than H_3O^+, then H_2 is oxidized to H_3O^+.

H_2 oxidized: $H_2(g, 1\,bar) + 2\,H_2O(\ell) \longrightarrow 2\,H_3O^+(aq, 1\,M) + 2\,e^-$ $\qquad E° = 0.0\text{ V}$

The reaction that occurs at the standard hydrogen electrode can occur in either direction. In either case, the standard hydrogen electrode, by definition, has a potential of 0 V.

Figure 18.8 diagrams a cell in which one compartment contains a standard hydrogen electrode and the other contains a zinc metal electrode immersed in a 1 M solution of Zn^{2+}. The voltmeter connected between the two electrodes measures the difference in electrical potential energy. For this cell the voltage is +0.76 V. After the cell generates current for a time, the zinc electrode decreases in mass as zinc metal is oxidized to $Zn^{2+}(aq)$. Therefore, the Zn electrode must be the anode; that is, oxidation takes place at this electrode. The hydrogen electrode must be the cathode, where reduction is taking place. The cell reaction is the difference of the half-cell reactions.

$$\begin{array}{rll} Zn(s) & \longrightarrow Zn^{2+}(aq, 1\,M) + 2\,e^- & E°_{anode} = ? \\ 2\,H_3O^+(aq, 1\,M) + 2\,e^- & \longrightarrow H_2(g, 1\,bar) + 2\,H_2O(\ell) & E°_{cathode} = 0\text{ V} \\ \hline Zn(s) + 2\,H_3O^+(aq, 1\,M) & \longrightarrow Zn^{2+}(aq, 1\,M) + H_2(g, 1\,bar) + 2\,H_2O(\ell) & \\ & & E°_{cell} = +0.76\text{ V (cell reaction)} \end{array}$$

The voltmeter tells us that the potential difference between the two electrodes is +0.76 V. Using the equation

$$E°_{cell} = E°_{cathode} - E°_{anode}$$

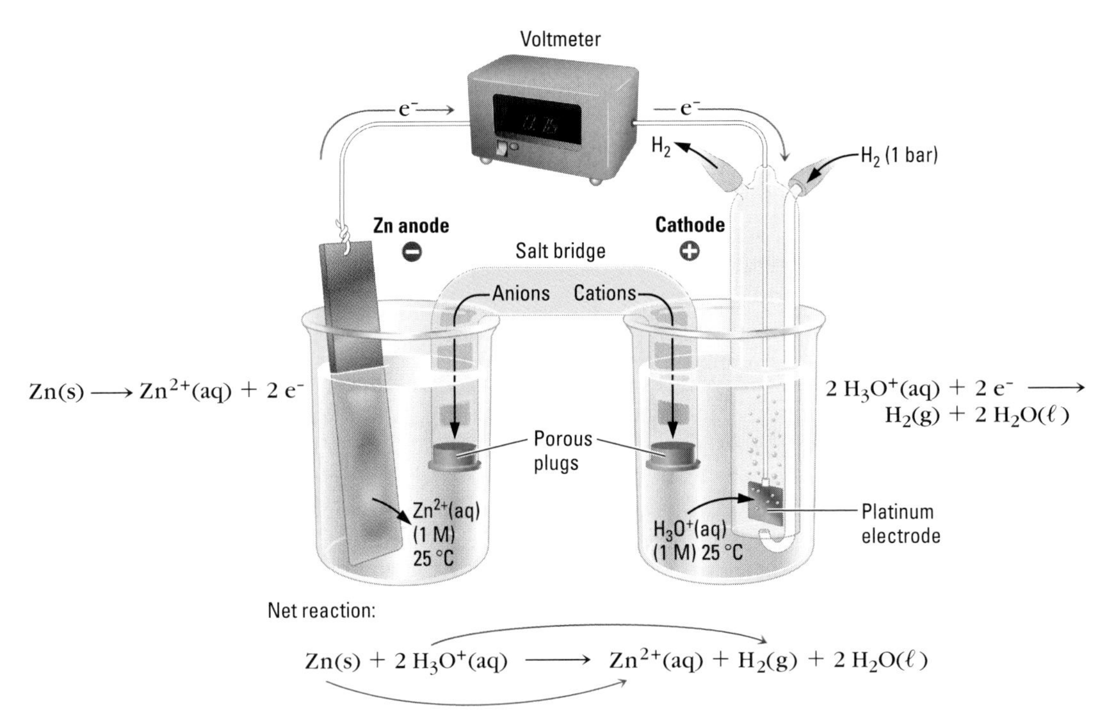

Figure 18.8 An electrochemical cell using a Zn^{2+}/Zn(s) half-cell and a standard hydrogen electrode. In this cell, the zinc electrode is the anode and the standard hydrogen electrode is the cathode. The cell voltage is +0.76 V. Zinc is the reducing agent and is oxidized to Zn^{2+}; H_3O^+ is the oxidizing agent and is reduced to H_2. In the standard hydrogen electrode, reaction occurs only where the three phases—gas, solution, and solid electrode—are in contact. The platinum electrode does not undergo any chemical change, and in the cell pictured here the cathodic half-cell reaction is $2\ H_3O^+(aq) + 2\ e^- \rightarrow H_2(g) + 2\ H_2O(\ell)$. (When the standard hydrogen electrode is the anode, the half-cell reaction is $H_2(g) + 2\ H_2O(\ell) \rightarrow 2\ H_3O^+(aq) + 2\ e^-$.)

we have

$$0.76\ V = 0\ V - E^\circ_{anode}$$
$$E^\circ_{anode} = E^\circ_{Zn(s)/Zn^{2+}(aq,\ 1\ M)} = -0.76\ V$$

Thus, by using the standard hydrogen electrode, it is possible to assign a standard reduction potential E° value of -0.76 V to the $Zn(s)/Zn^{2+}$(aq, 1 M) electrode.

CONCEPTUAL **EXERCISE** **18.4 What Is Going On Inside the Electrochemical Cell?**

Devise an experiment that would show that Zn is being oxidized in the electrochemical cell shown in Figure 18.8.

The convention of assigning voltages to half-reactions is similar to the convention of tabulating standard enthalpies of formation; in both cases a relatively small table of data can provide information about a large number of different reactions.

The half-cell potentials of many different half-reactions can be measured by comparing them with the standard hydrogen electrode. For example, in a cell consisting of the Cu^{2+}/Cu half-cell connected to a standard hydrogen electrode, the mass of the copper electrode increases and the voltmeter reads +0.34 V (Figure 18.9). This means that the reactions are

$$H_2(g,\ 1\ bar) + 2\ H_2O(\ell) \longrightarrow 2\ H_3O^+(aq,\ 1\ M) + 2\ e^- \qquad E^\circ_{anode} = 0\ V$$
$$Cu^{2+}(aq,\ 1\ M) + 2\ e^- \longrightarrow Cu(s) \qquad E^\circ_{cathode} = ?$$

$$H_2(g,\ 1\ bar) + Cu^{2+}(aq,\ 1\ M) + 2\ H_2O(\ell) \longrightarrow 2\ H_3O^+(aq,\ 1\ M) + Cu(s) \qquad E^\circ_{cell} = +0.34\ V$$

$$E^\circ_{cell} = E^\circ_{cathode} - E^\circ_{anode} = E^\circ_{Cu(s)/Cu^{2+}(aq,\ 1\ M)} - 0\ V = +0.34\ V$$

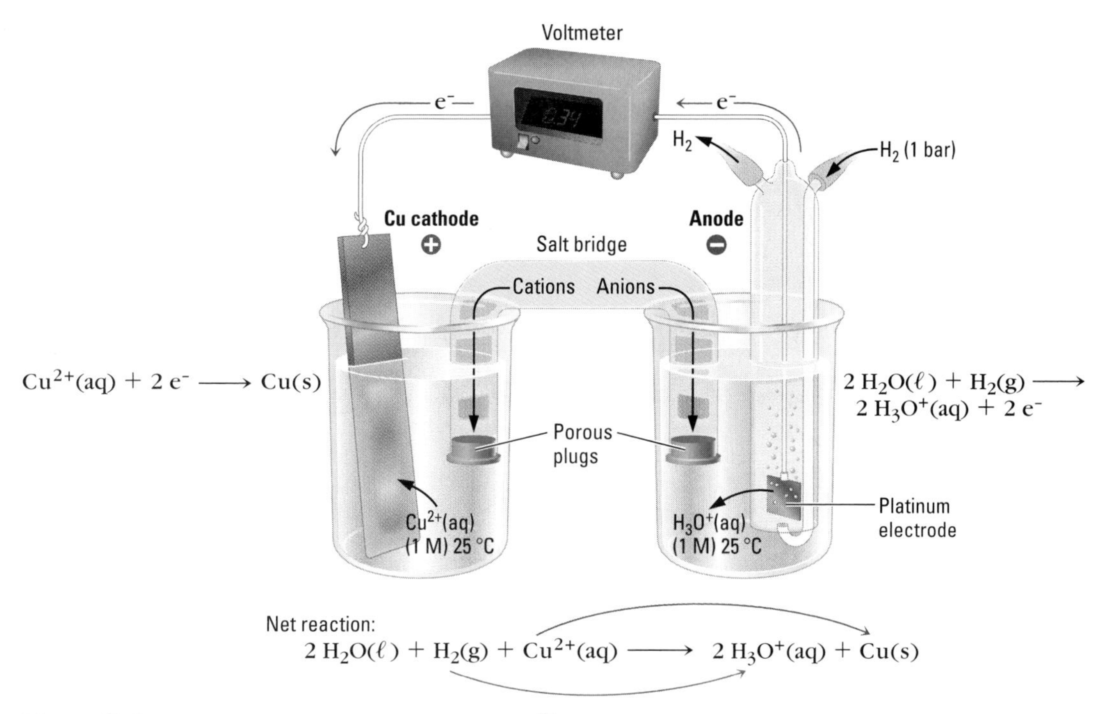

Figure 18.9 An electrochemical cell using the Cu^{2+}/Cu half-cell and the standard hydrogen electrode. A voltage of +0.34 V is produced. In this cell, Cu^{2+} ions are reduced to form Cu metal, and H_2 is oxidized at the standard hydrogen electrode. The reaction at the standard hydrogen electrode is the opposite of that shown in Figure 18.8.

The half-cell potential for the Cu^{2+} (aq, 1 M) + 2 e⁻ ⟶ Cu(s) reduction half-reaction must be +0.34 V. Note that in this cell the standard hydrogen electrode is the anode.

We can now return to the first electrochemical cell we looked at, in which Zn reduces Cu^{2+} ions to Cu. Using the potentials for the half-reactions, we can write

$$\begin{array}{rll} Zn(s) \longrightarrow Zn^{2+}(aq, 1\ M) + 2\ e^- & & E^\circ_{anode} = -0.76\ V \\ Cu^{2+}(aq, 1\ M) + 2\ e^- \longrightarrow Cu(s) & & E^\circ_{cathode} = +0.34\ V \\ \hline Zn(s) + Cu^{2+}(aq, 1\ M) \longrightarrow Zn^{2+}(aq, 1\ M) + Cu(s) & & E^\circ_{cell} = ? \end{array}$$

When we combine the *standard reduction potentials* of the two half-reactions, we have the measured potential for the cell reaction.

$$E^\circ_{cell} = E^\circ_{cathode} - E^\circ_{anode} = (+0.34\ V) - (-0.76\ V) = +1.10\ V$$

The experimentally measured potential for this cell is 1.10 V, confirming that half-cell potentials measured with the standard hydrogen electrode can be subtracted to obtain overall cell potentials.

PROBLEM-SOLVING EXAMPLE 18.6 Determining a Half-Cell Potential

The voltaic cell shown in the drawing below generates a potential of $E^\circ = 0.36$ V under standard conditions at 25 °C. The net cell reaction is

$$Zn(s) + Cd^{2+}(aq, 1\ M) \longrightarrow Zn^{2+}(aq, 1\ M) + Cd(s)$$

The standard half-cell potential for Zn(s)/Zn^{2+}(aq, 1 M) is −0.76 V.

(a) Determine which electrode is the anode and which is the cathode.

(b) Show the direction of electron flow through the circuit outside the cell, and complete the cell diagram.

For simplicity, this voltaic cell is shown without an electrical device in the external circuit. To generate its standard voltage, a high resistance would have to be in the circuit.

(c) Calculate the standard potential for the half-cell $Cd^{2+}(aq) + 2\ e^- \longrightarrow Cd(s)$.

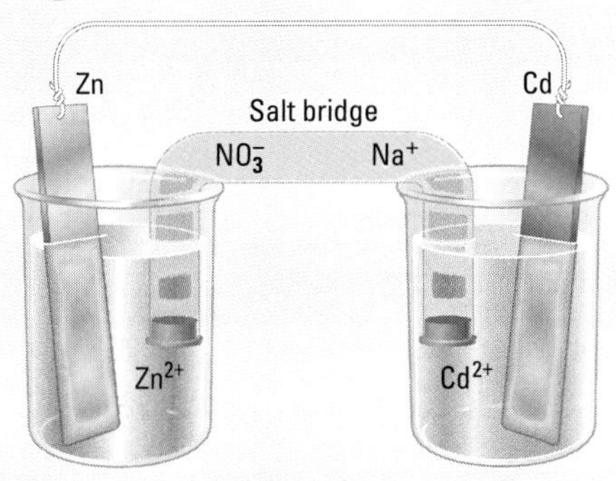

Answer

(a) Zinc is the anode, and cadmium is the cathode.
(b) The completed cell diagram is shown below.
(c) The standard cell potential is −0.40 V.

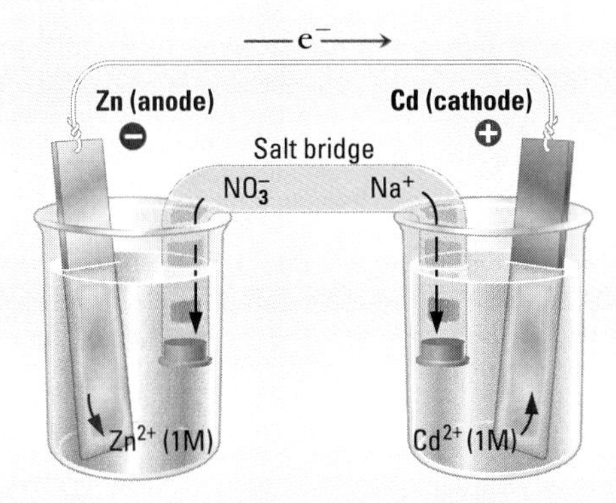

Strategy and Explanation The electrode where oxidation occurs is the anode. Because Zn(s) is oxidized to $Zn^{2+}(aq)$, the Zn electrode is the anode. Cadmium(II) ions are reduced at the Cd electrode, so it is the cathode.

The net cell potential and the potential for the $Zn(s)/Zn^{2+}(aq, 1\ M)$ half-cell are known, so the value of $E°$ for $Cd^{2+}(aq, 1\ M) + 2\ e^- \longrightarrow Cd(s)$ can be calculated.

$$Zn(s) \longrightarrow Zn^{2+}(aq, 1\ M) + 2\ e^- \qquad E°_{anode} = -0.76\ V\ (anode)$$

$$Cd^{2+}(aq, 1\ M) + 2\ e^- \longrightarrow Cd(s) \qquad E°_{cathode} = ?\ V\ (cathode)$$

$$Zn(s) + Cd^{2+}(aq, 1\ M) \longrightarrow Zn^{2+}(aq, 1\ M) + Cd(s) \qquad E°_{cell} = +0.36\ V$$

Using $E°_{cell} = E°_{cathode} - E°_{anode}$, we can solve for $E°_{cathode}$.

$$E°_{cathode} = E°_{cell} + E°_{anode} = 0.36\ V + (-0.76\ V) = -0.40\ V$$

At 25 °C, the value of $E°$ for the $Cd^{2+}(aq) + 2\ e^- \longrightarrow Cd(s)$ half-reaction is −0.40 V.

PROBLEM-SOLVING PRACTICE 18.6

Given that the reaction of aqueous copper(II) ions with iron metal has $E°_{cell} = +0.78$ V, what is the value of $E°$ for the half-cell $Fe(s) \rightarrow Fe^{2+}(aq) + 2\ e^-$?

$$Fe(s) + Cu^{2+}(aq, 1\ M) \longrightarrow Fe^{2+}(aq, 1\ M) + Cu(s) \qquad E°_{cell} = +0.78\ V$$

18.5 Using Standard Reduction Potentials

The results of a great many measurements of cell potentials such as the ones just described are summarized as **standard reduction potentials** in Table 18.1. A much longer and more complete list of standard reduction potentials appears in Appendix I. The values reported in the tables are called standard reduction potentials because they

Table 18.1 Standard Reduction Potentials in Aqueous Solution at 25 °C*

Reduction Half-Reaction		$E°$ (V)
$F_2(g) + 2\,e^-$	$\longrightarrow 2\,F^-(aq)$	+2.87
$H_2O_2(aq) + 2\,H_3O^+(aq) + 2\,e^-$	$\longrightarrow 4\,H_2O(\ell)$	+1.77
$PbO_2(s) + SO_4^{2-}(aq) + 4\,H_3O^+(aq) + 2\,e^-$	$\longrightarrow PbSO_4(s) + 6\,H_2O(\ell)$	+1.685
$MnO_4^-(aq) + 8\,H_3O^+(aq) + 5\,e^-$	$\longrightarrow Mn^{2+}(aq) + 12\,H_2O(\ell)$	+1.51
$Au^{3+}(aq) + 3\,e^-$	$\longrightarrow Au(s)$	+1.50
$Cl_2(g) + 2\,e^-$	$\longrightarrow 2\,Cl^-(aq)$	+1.358
$Cr_2O_7^{2-}(aq) + 14\,H_3O^+(aq) + 6\,e^-$	$\longrightarrow 2\,Cr^{3+}(aq) + 21\,H_2O(\ell)$	+1.33
$O_2(g) + 4\,H_3O^+(aq) + 4\,e^-$	$\longrightarrow 6\,H_2O(\ell)$	+1.229
$Br_2(\ell) + 2\,e^-$	$\longrightarrow 2\,Br^-(aq)$	+1.066
$NO_3^-(aq) + 4\,H_3O^+(aq) + 3\,e^-$	$\longrightarrow NO(g) + 6\,H_2O(\ell)$	+0.96
$OCl^-(aq) + H_2O(\ell) + 2\,e^-$	$\longrightarrow Cl^-(aq) + 2\,OH^-(aq)$	+0.89
$Hg^{2+}(aq) + 2\,e^-$	$\longrightarrow Hg(\ell)$	+0.855
$Ag^+(aq) + e^-$	$\longrightarrow Ag(s)$	+0.7994
$Hg_2^{2+}(aq) + 2\,e^-$	$\longrightarrow 2\,Hg(\ell)$	+0.789
$Fe^{3+}(aq) + e^-$	$\longrightarrow Fe^{2+}(aq)$	+0.771
$I_2(s) + 2\,e^-$	$\longrightarrow 2\,I^-(aq)$	+0.535
$O_2(g) + 2\,H_2O(\ell) + 4\,e^-$	$\longrightarrow 4\,OH^-(aq)$	+0.403
$Cu^{2+}(aq) + 2\,e^-$	$\longrightarrow Cu(s)$	+0.337
$Sn^{4+}(aq) + 2\,e^-$	$\longrightarrow Sn^{2+}(aq)$	+0.15
$2\,H_3O^+(aq) + 2\,e^-$	$\longrightarrow H_2(g) + 2\,H_2O(\ell)$	0.00
$Sn^{2+}(aq) + 2\,e^-$	$\longrightarrow Sn(s)$	−0.14
$Ni^{2+}(aq) + 2\,e^-$	$\longrightarrow Ni(s)$	−0.25
$PbSO_4(s) + 2\,e^-$	$\longrightarrow Pb(s) + SO_4^{2-}(aq)$	−0.356
$Cd^{2+}(aq) + 2\,e^-$	$\longrightarrow Cd(s)$	−0.403
$Fe^{2+}(aq) + 2\,e^-$	$\longrightarrow Fe(s)$	−0.44
$Zn^{2+}(aq) + 2\,e^-$	$\longrightarrow Zn(s)$	−0.763
$2\,H_2O(\ell) + 2\,e^-$	$\longrightarrow H_2(g) + 2\,OH^-(aq)$	−0.8277
$Al^{3+}(aq) + 3\,e^-$	$\longrightarrow Al(s)$	−1.66
$Mg^{2+}(aq) + 2\,e^-$	$\longrightarrow Mg(s)$	−2.37
$Na^+(aq) + e^-$	$\longrightarrow Na(s)$	−2.714
$K^+(aq) + e^-$	$\longrightarrow K(s)$	−2.925
$Li^+(aq) + e^-$	$\longrightarrow Li(s)$	−3.045

*In volts (V) versus the standard hydrogen electrode.

If a half-reaction is written as an oxidation, the voltage of the oxidation half-reaction is the same in magnitude but reversed in sign compared to the standard reduction potential.

Standard reduction potential:

$$Cu^{2+}(aq) + 2\,e^- \longrightarrow Cu(s) \quad E° = +0.337\,V$$

Reverse reaction:

$$Cu(s) \longrightarrow Cu^{2+}(aq) + 2\,e^- \quad E° = -0.337\,V$$

are the potentials, reported as voltages, that are measured for a cell in which a half-reaction *occurs as a reduction* when paired with the standard hydrogen electrode. Here are some important points to notice about Table 18.1:

1. ***Each half-reaction is written as a reduction.*** Thus, the species on the left-hand side of each half-reaction is in a higher oxidation state, and the species on the right-hand side is in a lower oxidation state.
2. ***Each half-reaction can occur in either direction.*** A given substance can react at the anode or the cathode, depending on the conditions. For example, we have already seen cases in which H_2 is oxidized to H_3O^+ and others in which H_3O^+ is reduced to H_2 by different reactants.
3. ***The more positive the value of the standard reduction potential, E°, the more easily the substance on the left-hand side of a half-reaction can be***

F_2 is always reduced and Li(s) is always oxidized.

reduced. When a substance is easy to reduce, it is a strong oxidizing agent. (Recall that an oxidizing agent is reduced when it oxidizes something else.) Thus, F_2(g) is the best oxidizing agent in the table, and Li^+ is the poorest oxidizing agent in the table. Other strong oxidizing agents are at the top left of the table:

$$H_2O_2(aq),\ PbO_2(s),\ Au^{3+}(aq),\ Cl_2(g),\ O_2(g)$$

4. ***The less positive the value of the standard reduction potential, E°, the less likely the reaction will occur as a reduction, and the more likely an oxidation (the reverse reaction) will occur.*** The farther down we go in the table, the better the reducing (electron donating) ability of the atom, ion, or molecule on the right. Thus, Li(s) is the strongest reducing agent in the table, and F^- is the weakest reducing agent in the table. Other strong reducing agents are alkali and alkaline earth metals and hydrogen at the lower right of the table.
5. ***Under standard conditions, any species on the left of a half-reaction will oxidize any species that is below it on the right side of the table.*** For example, we can apply this rule to predict that Fe^{3+}(aq) will oxidize Al(s), $Br_2(\ell)$ will oxidize Mg(s), and Na^+(aq) will oxidize Li(s). The net reaction is found by adding the half-reactions, and the cell voltage can be calculated from the $E^\circ_{cell} = E^\circ_{cathode} - E^\circ_{anode}$ equation, as illustrated below.

$$Br_2(\ell) + 2\,e^- \longrightarrow 2\,Br^-(aq) \qquad E^\circ_{cathode} = +1.07\text{ V}$$
$$Mg(s) \longrightarrow Mg^{2+}(aq) + 2\,e^- \qquad E^\circ_{anode} = -2.37\text{ V}$$
$$Br_2(\ell) + Mg(s) \longrightarrow Mg^{2+}(aq) + 2\,Br^-(aq) \qquad E^\circ_{cell} = +3.45\text{ V}$$
$$E^\circ_{cell} = E^\circ_{cathode} - E^\circ_{anode} = 1.07\text{ V} - (-2.37\text{ V}) = +3.44\text{ V}$$

A positive cell potential denotes a product-favored reaction.
6. ***Electrode potentials depend on the nature and concentration of reactants and products, but not on the quantity of each that reacts.*** Changing the stoichiometric coefficients for a half-reaction does *not* change the value of E°. For example, the reduction of Fe^{3+} has an E° of +0.771 V whether the reaction is written as

$$Fe^{3+}(aq, 1\text{ M}) + e^- \longrightarrow Fe^{2+}(aq, 1\text{ M}) \qquad E^\circ = +0.771\text{ V}$$

or

$$2\,Fe^{3+}(aq, 1\text{ M}) + 2\,e^- \longrightarrow 2\,Fe^{2+}(aq, 1\text{ M}) \qquad E^\circ = +0.771\text{ V}$$

In this respect, E° values differ from ΔH° and ΔG° values, which do depend on the coefficients in thermochemical equations.

This fact about half-cell potentials seems unusual at first, but consider that a half-cell voltage is energy per unit charge (1 volt = 1 joule/1 coulomb). When a half-reaction is multiplied by some number, both the energy and the charge are multiplied by that number. Thus the ratio of the energy to the charge (voltage) does not change.

Using the preceding guidelines and the table of standard reduction potentials, we will make some *predictions* about whether reactions will occur and then check our results by calculating E°_{cell}.

PROBLEM-SOLVING EXAMPLE 18.7 Predicting Redox Reactions

(a) Will zinc metal react with a 1 M Ag^+(aq) solution? If so, what is E° for the reaction?
(b) Will a 1 M Fe^{2+}(aq) solution react with metallic tin? If so, what is E° for the reaction?

Answer

(a) Yes. $E^\circ_{cell} = +1.56$ V
(b) No. E°_{cell} is negative.

Strategy and Explanation We will answer the questions by referring to Table 18.1 and comparing the positions of the reactants there.

(a) $Ag^+(aq)$ is above metallic zinc in Table 18.1, so it is a better oxidizing agent, and we predict that it can oxidize zinc, causing metallic zinc atoms to form $Zn^{2+}(aq)$ ions. To be certain, we combine the half-cell reactions to give the net equation. We subtract the half-cell potentials, which yields a positive $E°_{cell}$, so this reaction is product-favored, as we predicted from Table 18.1.

$$\begin{array}{lll} Zn(s) \longrightarrow Zn^{2+}(aq, 1\ M) + 2\ e^- & & E°_{anode} = -0.763\ V \\ 2\ [Ag^+(aq, 1\ M) + e^- \longrightarrow Ag(s)] & & E°_{cathode} = +0.80\ V \\ \hline Zn(s) + 2\ Ag^+(aq, 1\ M) \longrightarrow Zn^{2+}(aq, 1\ M) + 2\ Ag(s) & & E°_{cell} = ? \end{array}$$

$$E°_{cell} = E°_{cathode} - E°_{anode} = +0.80\ V - (-0.763\ V) = +1.56\ V$$

Note that the half-reaction voltages are not multiplied by the balancing coefficients.

The positive value for $E°_{cell}$ shows that this is a product-favored reaction.

(b) We evaluate the reaction between $Fe^{2+}(aq)$ and metallic Sn the same way. $Fe^{2+}(aq)$ is on the left in Table 18.1 below Sn(s), which is on the right. Therefore, $Fe^{2+}(aq)$ is not a strong enough oxidizing agent to oxidize Sn(s), and we predict that this reaction will not occur. The combined half-reactions are

$$\begin{array}{lll} Sn(s) \longrightarrow Sn^{2+}(aq) + 2e^- & & E°_{anode} = -0.14\ V \\ Fe^{2+}(aq) + 2\ e^- \longrightarrow Fe(s) & & E°_{cathode} = -0.44\ V \\ \hline Sn(s) + Fe^{2+}(aq) \longrightarrow Sn^{2+}(aq) + Fe(s) & & E°_{cell} = ? \end{array}$$

$$E°_{cell} = E°_{cathode} - E°_{anode} = -0.44\ V - (-0.14\ V) = -0.30\ V$$

The negative value for $E°_{cell}$ shows that this process is reactant-favored, and it will not form appreciable quantities of products under standard conditions. In fact, iron metal will reduce Sn^{2+} ($E°_{cell} = +0.30$ V), the reverse of the net reaction shown above.

PROBLEM-SOLVING PRACTICE 18.7

Look at Table 18.1 and determine which two half-reactions would produce the largest value of $E°_{cell}$. Write the two half-reactions and the overall cell reaction, and give the $E°$ for the reaction.

CONCEPTUAL EXERCISE 18.5 Using $E°$ Values

Transporting chemicals is of great practical and economic importance. Suppose that you have a large volume of 1 M mercury(II) chloride solution, $HgCl_2$, that needs to be transported. A driver brings a tanker truck made of aluminum to the loading dock. Will it be okay to load the truck with your solution? Explain your answer fully.

CONCEPTUAL EXERCISE 18.6 Predicting Redox Reactions Using $E°$ Values

Consider these reduction half-reactions:

Half-Reaction	$E°$ (V)
$Cl_2(g) + 2\ e^- \longrightarrow 2\ Cl^-(aq)$	+1.36
$I_2(s) + 2\ e^- \longrightarrow 2\ I^-(aq)$	+0.535
$Pb^{2+}(aq) + 2\ e^- \longrightarrow Pb(s)$	−0.126
$V^{2+}(aq) + 2\ e^- \longrightarrow V(s)$	−1.18

(a) Which is the weakest oxidizing agent?
(b) Which is the strongest oxidizing agent?
(c) Which is the strongest reducing agent?
(d) Which is the weakest reducing agent?
(e) Will Pb(s) reduce $V^{2+}(aq)$ to V(s)?
(f) Will $I_2(s)$ oxidize $Cl^-(aq)$ to $Cl_2(g)$?
(g) Name the molecules or ions in the above reactions that can be reduced by Pb(s).

18.6 $E°$ and Gibbs Free Energy

The sign of $E°_{cell}$ indicates whether a redox reaction is product-favored (positive $E°$) or reactant-favored (negative $E°$). Earlier you learned another way to decide whether a reaction is product-favored: The change in standard Gibbs free energy, $\Delta G°$, must be negative (← *p. 625*). Since both $E°_{cell}$ and $\Delta G°$ tell something about whether a reaction will occur, it should be no surprise that the two quantities are related.

The "free" in Gibbs free energy indicates that it is energy available to do work. The energy available for electrical work from an electrochemical cell can be calculated by multiplying the quantity of electrical charge transferred times the cell voltage, $E°$. The quantity of charge is given by the number of moles of electrons transferred in the overall reaction, n, multiplied by the number of coulombs per mole of electrons.

$$\text{Quantity of charge} = \text{moles of electrons} \times \text{coulombs per mole of electrons}$$

The charge on 1 mol of electrons can be calculated from the charge on one electron and Avogadro's number.

$$\text{Charge on 1 mol of e}^- = \left(\frac{1.60218 \times 10^{-19}\text{ C}}{\text{e}^-}\right)\left(\frac{6.02214 \times 10^{23}\text{ e}^-}{1\text{ mol e}^-}\right)$$
$$= 9.6485 \times 10^4\text{ C/mol e}^-$$

Michael Faraday 1791–1867

Science Photo Library/Photo Researchers, Inc.

As an apprentice to a London bookbinder, Michael Faraday became fascinated by science when he was a boy. At age 22 he was appointed as a laboratory assistant at the Royal Institution and became its director within 12 years. A skilled experimenter in chemistry and physics, he made many important discoveries, the most significant of which was electromagnetic induction, the basis of modern electromagnetic technology. Faraday built the first electric motor, generator, and transformer. A popular speaker and educator, he also performed chemical and electrochemical experiments, and he first synthesized benzene.

The quantity 9.6485×10^4 C/mol of electrons (commonly rounded to 96,500 C/mol of electrons) is known as the **Faraday constant, *F*,** in honor of Michael Faraday, who first explored the quantitative aspects of electrochemistry.

The electrical work that can be done by a cell is equal to the Faraday constant (F) multiplied by the number of moles of electrons transferred (n) and by the cell voltage ($E°_{cell}$).

$$\text{Electrical work} = nFE°_{cell}$$

Unlike the cell voltage, the electrical work a cell can do *does* depend on the quantity of reactants in the cell reaction. More reactants mean more moles of electrons transferred and hence more work. Equating the electrical work of a cell at standard conditions with $\Delta G°$, we get

$$\Delta G° = -nFE°_{cell}$$

The negative sign on the right side of the equation accounts for the fact that $\Delta G°$ *is always negative for a product-favored process, but* $E°_{cell}$ *is always positive for a product-favored process.* Thus, these values must have opposite signs.

Using this equation we can calculate $\Delta G°$ for the Cu^{2+}/Zn cell. This value represents the maximum work that the cell can do. The reaction is

$$Cu^{2+}(aq) + Zn(s) \longrightarrow Cu(s) + Zn^{2+}(aq) \qquad E°_{cell} = +1.10\text{ V}$$

so 2 mol electrons are transferred per mole of copper ions reduced. The Gibbs free energy change when this quantity of reactants is converted is

$$\Delta G° = -\left(\frac{2\text{ mol e}^-}{\text{transferred}}\right)\left(\frac{9.65 \times 10^4\text{ C}}{\text{mol e}^-}\right)\left(\frac{1\text{ J}}{1\text{ V} \times 1\text{ C}}\right)\left(\frac{1\text{ kJ}}{10^3\text{ J}}\right)(+1.10\text{ V})$$
$$= -212\text{ kJ}$$

The positive $E°_{cell}$ and the negative Gibbs free energy change values indicate a very product-favored reaction.

PROBLEM-SOLVING EXAMPLE 18.8 Determining $E°_{cell}$ and $\Delta G°$

Consider the redox reaction

$$Zn^{2+}(aq) + H_2(g) + 2\,H_2O(\ell) \longrightarrow Zn(s) + 2\,H_3O^+(aq)$$

Use the standard reduction potentials in Table 18.1 to calculate $E°_{cell}$ and $\Delta G°$ and to determine whether the reaction as written favors product formation.

Answer $E°_{cell} = -0.763$ V; $\Delta G° = 147$ kJ. The reaction as written is not product-favored.

Strategy and Explanation The first step is to write the half-reactions for the oxidation and reduction that occur and to use their standard reduction potentials from Table 18.1.

Reduction: $Zn^{2+}(aq) + 2\,e^- \longrightarrow Zn(s)$ $\quad E°_{cathode} = -0.763$ V

Oxidation: $H_2(g) + 2\,H_2O(\ell) \longrightarrow 2\,H_3O^+(aq) + 2\,e^-$ $\quad E°_{anode} = 0$ V

We obtain the $E°_{cell}$ for the reaction from the standard reduction potentials for the two half-reactions.

$$E°_{cell} = E°_{cathode} - E°_{anode} = -0.763\text{ V} - 0\text{ V} = -0.763\text{ V}$$

Since $E°_{cell}$ is negative, the reaction is not product-favored in the direction written. Zn^{2+} will not oxidize H_2O. The reverse reaction is product-favored; that is, zinc metal reacts with acid.

From the calculated $E°_{cell}$ we can calculate $\Delta G°$.

$$\Delta G° = -nFE°_{cell}$$

$$= -(2\text{ mol e}^-) \times \left(\frac{9.65 \times 10^4\text{ C}}{1\text{ mol e}^-}\right)\left(\frac{1\text{ J}}{1\text{ V} \times 1\text{ C}}\right)(-0.763\text{ V})$$

$$= 1.47 \times 10^5\text{ J} = 147\text{ kJ}$$

The positive value for $\Delta G°$ also shows that the reaction as written is not product-favored.

☑ **Reasonable Answer Check** The given oxidation is the half-reaction at the standard hydrogen electrode with a potential of zero, so the overall reaction is governed by the Zn reduction. The negative cell potential and the positive $\Delta G°$ are consistent with the reaction being reactant-favored as written.

PROBLEM-SOLVING PRACTICE 18.8

Using standard reduction potentials, determine whether this reaction is product-favored as written.

$$Hg^{2+}(aq) + 2\,I^-(aq) \longrightarrow Hg(\ell) + I_2(s)$$

$\Delta G°$, $E°_{cell}$, and $K°$

We have seen that the standard Gibbs free energy change is directly proportional to the $E°_{cell}$ for an electrochemical cell at standard conditions

$$\Delta G° = -nFE°_{cell}$$

Recall from Chapter 17 (p. 630) that the standard Gibbs free energy change is directly proportional to the logarithm of the equilibrium constant of a reaction.

$$\Delta G° = -RT \ln K°$$

Putting these two equations together yields

$$-nFE°_{cell} = -RT \ln K°$$

which, when solved for $E°_{cell}$, yields

$$E°_{cell} = \frac{RT}{nF} \ln K°$$

Thus by measuring $E°_{cell}$, the values of $\Delta G°$ and $K°$ can be calculated. The relationships linking $\Delta G°$, $E°_{cell}$, and $K°$ are summarized in Figure 18.10.

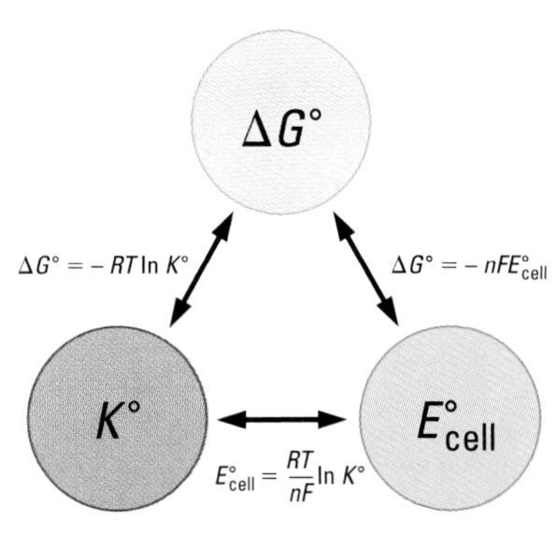

Figure 18.10 The relationships linking $\Delta G°$, $E°_{cell}$, and $K°$. Given any one of the values, the other two can be calculated.

The E°_{cell} expression can be simplified by substituting numerical values for R (8.314 J mol^{-1} K^{-1}) and F (96,485 J V^{-1} mol^{-1}) and assuming standard-state temperature (298 K). For n electrons transferred we get

$$E^\circ_{cell} = \frac{RT}{nF}\ln K^\circ = \frac{(8.314\ \text{J mol}^{-1}\ \text{K}^{-1})(298\ \text{K})}{n\ (96{,}485\ \text{J V}^{-1}\text{mol}^{-1})} = \frac{0.0257\ \text{V}}{n}\ln K^\circ$$

$2.303 \log(x) = \ln(x)$

Changing from natural logarithms to base-10 logarithms (multiplying by 2.303) yields

$$E^\circ_{cell} = \frac{0.0592\ \text{V}}{n}\log K^\circ \qquad \text{and} \qquad \log K^\circ = \frac{nE^\circ_{cell}}{0.0592\ \text{V}} \qquad \text{(at 298 K)}$$

These equations hold for standard states of all reactants and products.

PROBLEM-SOLVING EXAMPLE 18.9 Equilibrium Constant for a Redox Reaction

Calculate the equilibrium constant K_c for the reaction

$$Fe(s) + Cd^{2+}(aq) \rightleftharpoons Fe^{2+}(aq) + Cd(s)$$

using the standard reduction potentials listed in Table 18.1.

Answer $K_c = K^\circ = 22$

Strategy and Explanation We first need to calculate E°_{cell}. To do so we separate the reaction into its two half-reactions.

$$Fe(s) \longrightarrow Fe^{2+}(aq) + 2\,e^- \qquad E^\circ_{anode} = -0.44\ \text{V}$$

$$Cd^{2+}(aq) + 2\,e^- \longrightarrow Cd(s) \qquad E^\circ_{cathode} = -0.40\ \text{V}$$

$$Fe(s) + Cd^{2+}(aq) \longrightarrow Fe^{2+}(aq) + Cd(s) \qquad E^\circ_{cell} = +0.04\ \text{V}$$

Two moles of electrons are transferred.

$$\log K^\circ = \frac{nE^\circ_{cell}}{0.0592\ \text{V}} = \frac{(2)(0.04\ \text{V})}{0.0592\ \text{V}} = 1.35 \qquad \text{and} \qquad K = 10^{1.35} = 22$$

The K° value is larger than 1, which shows that the reaction is product-favored as written. Since the reaction occurs in aqueous solution, $K_c = K^\circ = 22$.

☑ **Reasonable Answer Check** The value for E°_{cell} is positive, which indicates a product-favored reaction, as does $K^\circ > 1$.

PROBLEM-SOLVING PRACTICE 18.9

Using the standard reduction potentials listed in Table 18.1, calculate the equilibrium constant for the reaction

$$I_2(s) + Sn^{2+}(aq) \longrightarrow 2\,I^-(aq) + Sn^{4+}(aq)$$

18.7 Effect of Concentration on Cell Potential

The electrochemical cells discussed previously, all voltaic cells, are based on product-favored chemical reactions. As the reactions proceed in such an electrochemical cell, reactants are converted to products, so the concentrations of the species change continuously. As the reactant concentrations decrease, the voltage produced by the cell drops. The voltage finally reaches zero when the reactants and products are at equilibrium.

E°_{cell} is the voltage at standard-state conditions; E_{cell} is the voltage at nonstandard conditions.

We can relate the voltage of a voltaic cell to the concentration of the reactants and products of its chemical reaction. The relation is

$$E_{cell} = E^\circ_{cell} - \frac{RT}{nF}\ln Q$$

where Q is the reaction quotient. Q has the same form as the equilibrium constant, but it refers to a reaction mixture at a given instant in time, that is, a reaction mixture not necessarily in equilibrium. This equation is called the **Nernst Equation.** We can change from natural logarithms to base-10 logarithms (replace ln Q by 2.303 log Q) to get

In the Nernst equation, n is the number of moles of electrons transferred in the *balanced equation* of the process.

$$E_{cell} = E^\circ_{cell} - \frac{2.303\,RT}{nF}\log Q$$

We can simplify this expression further by substituting numerical values for R (8.314 J mol^{-1} K^{-1}) and F (96,485 J V^{-1} mol^{-1}) and by assuming 25 °C (298 K) to get

$$E_{cell} = E^\circ_{cell} - \frac{(2.303)(8.314\text{ J mol}^{-1}\text{ K}^{-1})(298\text{ K})}{n(96{,}485\text{ J V}^{-1}\text{ mol}^{-1})}\log Q$$

$$E_{cell} = E^\circ_{cell} - \frac{0.0592\text{ V}}{n}\log Q \qquad (T = 298\text{ K})$$

If all the concentrations in Q are equal to 1 (which is the standard state), then $Q = 1$ and $\log(1) = 0$, so the Nernst equation reduces to $E_{cell} = E^\circ_{cell}$.

The Nernst equation can be used to calculate the voltage produced by an electrochemical cell under nonstandard conditions. It can also be used to calculate the concentration of a reactant or product in an electrochemical reaction from the measured value of the voltage produced.

PROBLEM-SOLVING EXAMPLE 18.10 Using the Nernst Equation

Consider this electrochemical reaction:

$$Zn(s) + Ni^{2+}(aq) \longrightarrow Zn^{2+}(aq) + Ni(s)$$

The standard cell potential $E^\circ_{cell} = 0.51$ V. Find the voltage if the Ni^{2+} concentration is 5.0 M and the Zn^{2+} concentration is 0.050 M.

Answer 0.57 V

Strategy and Explanation We use the Nernst equation to solve the problem. Two moles of electrons are transferred from 1 mol Zn to 1 mol Ni^{2+}, giving $n = 2$. At 298 K,

$$E^\circ_{cell} = 0.51\text{ V} - \frac{0.0592\text{ V}}{2}\log\frac{(\text{conc. }Zn^{2+})}{(\text{conc. }Ni^{2+})}$$

$$E^\circ_{cell} = 0.51\text{ V} - \frac{0.0592\text{ V}}{2}\log\left(\frac{0.050}{5.0}\right) = 0.51\text{ V} - \frac{0.0592\text{ V}}{2}\log(10^{-2})$$

$$= 0.51\text{ V} - \frac{0.0592\text{ V}}{2}(-2.00) = 0.57\text{ V}$$

☑ **Reasonable Answer Check** The reactant concentration of Ni^{2+} is 5.0 M, larger than the standard-state value of 1.0 M, and the product concentration of Zn^{2+} is 0.050 M, smaller than the standard-state value. Each of these departures from standard-state conditions tends to make the voltage under these conditions slightly larger than the standard cell potential ($E^\circ_{cell} = 0.51$ V), and it is.

PROBLEM-SOLVING PRACTICE 18.10

What would the voltage in the chemical system above become if (conc. Zn^{2+}) = 3.0 M and (conc. Ni^{2+}) = 0.010 M?

Used "dead" batteries.

18.8 Common Batteries

Voltaic cells include the convenient, portable sources of energy that we call *batteries.* Some batteries, such as the common flashlight battery, consist of a single cell while others, such as automobile batteries, contain multiple cells. Batteries can be classified

Wilson Greatbatch 1919–

In the early 1960s Wilson Greatbatch had an idea of how a battery might be used to help an ailing heart keep pumping. His story, told in his own words, is fascinating:

I quit all my jobs, and with two thousand dollars I went out in the barn in the back of my house and built 50 pacemakers in two years. I started making the rounds of all the doctors in Buffalo who were working in this field, and I got consistently negative results. The answer I got was, well, these people all die in a year, you can't do much for them. . . . When I first approached Dr. Shardack with the idea of the pacemaker, he alone thought that it really had a future. He said, "You know—if you can do that—you can save a thousand lives a year."

After the first ten years, we were still only getting one or two years out of pacemakers . . . and the failure mechanism was always the battery. The human body is a very hostile environment. . . . You're trying to run things in a warm salt water environment. . . . So we started looking around for new power sources. And we finally wound up with this lithium battery. It really revolutionized the pacemaker business. The doctors have told me that the introduction of the lithium battery was more significant than the invention of the pacemaker in the first place.

Source: *The World of Chemistry* video, Program 15, The Annenberg/CPB Collection.

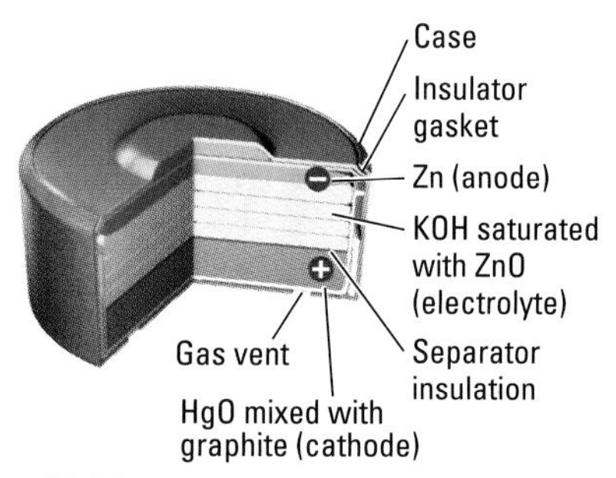

Figure 18.11 Mercury battery. The reducing agent is zinc and the oxidizing agent is mercury(II) oxide.

as primary or secondary depending on whether the reactions at the anode and cathode can be easily reversed. In a **primary battery** the electrochemical reactions cannot easily be reversed, so when the reactants are used up the battery is "dead" and must be discarded. In contrast, a **secondary battery** (sometimes called a storage battery or a rechargeable battery) uses an electrochemical reaction that can be reversed, so this type of battery can be recharged.

Primary Batteries

An important primary battery is the *alkaline battery,* which produces 1.54 V. The anode reaction is the oxidation of zinc under alkaline (pH >7) conditions.

$$Zn(s) + 2\,OH^-(aq) \longrightarrow ZnO(aq) + H_2O(\ell) + 2\,e^- \qquad \text{(anode, oxidation)}$$

The electrons that pass through the external circuit are consumed by reduction of manganese(IV) oxide at the cathode.

$$MnO_2(s) + H_2O(\ell) + e^- \longrightarrow MnO(OH)(s) + OH^-(aq) \qquad \text{(cathode, reduction)}$$

In the *mercury battery* (Figure 18.11), the oxidation of zinc is again the anode reaction. The cathode reaction is the reduction of mercury(II) oxide.

$$HgO(s) + H_2O(\ell) + 2\,e^- \longrightarrow Hg(\ell) + 2\,OH^-(aq)$$

The voltage of this battery is about 1.35 V. Mercury batteries are used in calculators, watches, hearing aids, cameras, and other devices in which small size is an advantage. However, mercury and its compounds are poisonous, so proper disposal of mercury batteries is necessary.

Mercury batteries are hermetically sealed to prevent leakage of mercury and should never be heated. Heating increases the pressure of mercury vapor within the battery, ultimately causing the battery to explode.

Secondary Batteries

Secondary batteries are rechargeable because, as they discharge, the oxidation products remain at the anode and the reduction products remain at the cathode. As a result, if the direction of electron flow is reversed, the anode and cathode reactions are reversed and the reactants are regenerated. Under favorable conditions, secondary batteries may be discharged and recharged hundreds or even thousands of times. Examples of secondary batteries include automobile batteries, nicad (nickel-cadium) batteries, and lithium-ion batteries.

Lead-Acid Storage Batteries The familiar automobile battery, the lead-acid storage battery, is a secondary battery consisting of six cells, each containing porous metallic lead electrodes and lead(IV) oxide electrodes immersed in aqueous sulfuric acid (Figure 18.12). When this battery produces an electric current, metallic lead is oxidized to lead(II) sulfate at the anode, and lead(IV) oxide is reduced to lead(II) sulfate at the cathode.

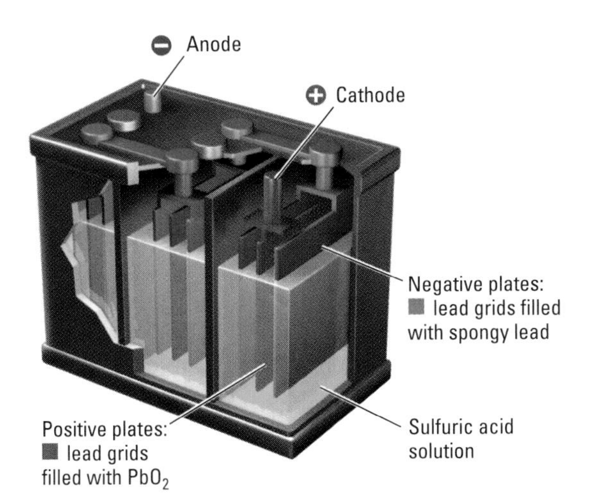

Figure 18.12 Lead-acid storage battery. The anodes are lead grids filled with spongy lead. The cathodes are lead grids filled with lead(IV) oxide, PbO_2. Each cell produces a potential of about 2 V. Six cells connected in series produce the desired overall battery voltage.

$$Pb(s) + HSO_4^-(aq) + H_2O(\ell) \longrightarrow PbSO_4(s) + H_3O^+(aq) + 2\,e^- \qquad E^\circ_{anode} = -0.356\ V$$

$$PbO_2(s) + 3\,H_3O^+(aq) + HSO_4^-(aq) + 2\,e^- \longrightarrow PbSO_4(s) + 5\,H_2O(\ell) \qquad E^\circ_{cathode} = 1.685\ V$$

$$Pb(s) + PbO_2(s) + 2\,H_3O^+(aq) + 2\,HSO_4^-(aq) \longrightarrow 2\,PbSO_4(s) + 4\,H_2O(\ell) \qquad E^\circ_{cell} = +2.041\ V$$

The voltage from the six cells connected in series in a typical automobile battery gives a total of 12 V.

To understand why the lead storage battery is rechargeable, consider that the lead sulfate formed at both electrodes is an insoluble compound that mostly *stays on the electrode surface.* As a result, it remains available for the reverse reaction. To recharge a secondary battery, a source of direct electrical current is supplied so that electrons are forced to flow in the direction opposite from when the battery was discharging. This causes the overall battery reaction to be reversed and regenerates the reactants that originally produced the battery's voltage and current. For the lead-acid storage battery, the overall redox reaction is

$$Pb(s) + PbO_2(s) + 2\,HSO_4^-(aq) + 2\,H_3O^+(aq) \rightleftharpoons 2\,PbSO_4(s) + 4\,H_2O(\ell)$$

Discharging battery produces electricity (forward); Charging battery requires electricity from an external source (reverse)

The lead-acid battery was first described to the French Academy of Sciences in 1860 by Gaston Planté.

Normal charging of an automobile lead-acid storage battery occurs during driving. In addition to reversing the overall battery reaction, charging reduces hydronium ion at the cathode and oxidizes water at the anode.

Reduction of hydronium ion: $4\,H_3O^+(aq) + 4\,e^- \longrightarrow 2\,H_2(g) + 4\,H_2O(\ell)$

Oxidation of water: $6\,H_2O(\ell) \longrightarrow O_2(g) + 4\,H_3O^+(aq) + 4\,e^-$

These reactions produce a hydrogen-oxygen mixture inside the battery, which, if accidentally ignited, can explode. Therefore, no sparks or open flames should be brought near a lead-acid storage battery, even the sealed kind.

When a car with a dead battery is jump-started, the last jumper cable connection should be to the car's frame—well away from the battery—to avoid igniting any H_2 in the battery with a spark.

As the battery operates, sulfuric acid is consumed in both the anode and the cathode reactions, thereby decreasing the concentration of the sulfuric acid electrolyte.

Before the introduction of modern sealed automotive batteries, the density of this battery acid was routinely measured to indicate the state of charge of the battery. The density of the battery acid decreases as the battery discharges. Consequently, the lower the density, the lower the charge of the battery. In modern sealed batteries, it is difficult to gain access to the acid to measure its density.

The number of electrons that a battery can move from the anode to the cathode is proportional to the amount of reactants involved.

The lead-acid storage battery is relatively inexpensive, reliable, and simple, and it has an adequate lifetime. High weight is its major fault. A typical automobile battery contains about 15 to 20 kg lead, which is required to provide the large number of electrons needed to start an automobile engine, especially on a cold morning. Another problem with lead batteries is that lead mining and manufacturing and disposal of the used batteries can contaminate air and groundwater. Auto batteries should be recycled by companies equipped with the proper safeguards.

© Cengage Learning/Charles D. Winters

Figure 18.13 Nickel-cadmium (nicad) batteries in a battery charger.

Nickel-Cadmium and Nickel-Metal Hydride Batteries Nickel-cadmium (nicad) secondary batteries are lightweight, can be quite small, and produce a constant voltage until completely discharged, which makes them useful in cordless appliances, video camcorders, portable radios, and other applications (Figure 18.13).

Nicad batteries can be recharged because the reaction products are insoluble hydroxides that remain at the electrode surfaces. The anode reaction during the discharge cycle is the oxidation of cadmium, and the cathode reaction is the reduction of nickel oxyhydroxide, NiO(OH).

$$\mathrm{Cd(s) + 2\,OH^-(aq) \longrightarrow Cd(OH)_2(s) + 2\,e^-} \qquad E^\circ_{\text{anode}} = -0.809\ \mathrm{V}$$

$$\mathrm{2[NiO(OH)(s) + H_2O(\ell) + e^- \longrightarrow Ni(OH)_2(s) + OH^-(aq)]} \qquad E^\circ_{\text{cathode}} = 0.490\ \mathrm{V}$$

$$\mathrm{Cd(s) + 2\,NiO(OH)(s) + 2\,H_2O(\ell) \longrightarrow Cd(OH)_2(s) + 2\,Ni(OH)_2(s)} \qquad E^\circ_{\text{cell}} = 1.299\ \mathrm{V}$$

Like mercury batteries, nicad batteries should be disposed of properly because of the toxicity of cadmium and its compounds.

Another nickel battery that uses the same cathode reaction but a different anode reaction is the *nickel-metal hydride* (NiMH) *battery*, which eliminates the use of cadmium. These batteries are used in portable power tools, cordless shavers, and photoflash units. Here the anode is a metal (M) alloy, often nickel or a rare earth, in a basic electrolyte (KOH). The anode reaction oxidizes hydrogen absorbed in the metal alloy, and water is produced.

$$\mathrm{MH(s) + OH^-(aq) \longrightarrow M(s) + H_2O(\ell) + e^-} \qquad \text{(anode reaction)}$$

The overall reaction is

$$\mathrm{MH(s) + NiO(OH)(s) \longrightarrow M(s) + Ni(OH)_2(s)} \qquad E_{\text{cell}} = 1.4\ \mathrm{V}$$

EXERCISE 18.7 Recharging a Nicad Battery

Write the electrode reactions that take place when a nicad battery is recharged; identify the anode and cathode reactions.

© Cengage Learning/Charles D. Winters

Figure 18.14 Lithium-ion battery. It finds many uses in which a high energy density and low mass are desired.

Lithium-Ion Batteries Lithium-ion batteries (Figure 18.14) benefit from the low density and high reducing strength of lithium metal (Table 18.1). The anode in such a battery is made of lithium metal that has been mixed with a conducting carbon polymer. The polymer has tiny spaces in its structure that can hold the lithium atoms as well as the lithium ions formed by the oxidation reaction.

$$\mathrm{Li(s)\ (in\ polymer) \longrightarrow Li^+(in\ polymer) + e^-} \qquad \text{(anode reaction)}$$

The cathode also contains lithium ions, but in the lattice of a metal oxide such as CoO_2. This oxide lattice, like the carbon-polymer electrode, has holes in it that accommodate Li^+ ions. The reduction reaction is

$$Li^+(\text{in } CoO_2) + e^- + CoO_2 \longrightarrow LiCoO_2 \qquad \text{(cathode reaction)}$$

The overall reaction in the lithium-ion battery is therefore

$$Li(s) + CoO_2(s) \longrightarrow LiCoO_2(s) \qquad E_{cell} = 3.4\ V$$

Lithium-ion batteries have a large voltage (3.4 V per cell) and very high energy output for their mass. They can be recharged many hundreds of times. Because of these desirable characteristics, lithium-ion batteries are used in cellular telephones, laptop computers, and digital cameras.

CONCEPTUAL **EXERCISE** **18.8 Emergency Batteries**

You are stranded on an island and need to communicate your location to receive help. You have a battery-powered radio transmitter, but the lead batteries are discharged. There is a swimming pool nearby and you find a tank of chlorine gas and some plastic tubing that can withstand being oxidized by chlorine. Devise a battery that might be used to power the radio using these items.

CHEMISTRY IN THE NEWS

Hybrid Cars

A number of hybrid cars are now for sale in the United States and Japan, and more are to be offered soon. Hybrid cars have two propulsion systems: an electric motor and a gasoline engine. The energy to power such a car comes from gasoline. The electricity comes ultimately from its gasoline engine, which charges the battery that is used to run the electric motor. However, the gasoline engine in a hybrid car is smaller than that in a normal car, and the gasoline engine switches off when the hybrid car is stopped or cruising at low speed. Therefore, fuel efficiency is high and hybrid cars get up to 60 mpg in city driving, twice as much as the gasoline mileage of non-hybrid, conventional cars.

Overall, the hybrid car is much more energy efficient than a conventional car. When the Toyota Prius starts up, the electric motor is used, but when the car is accelerating, and the demand for power is high, both the electric motor and the gasoline engine are used. At speeds of less than about 20 mph, the electric motor alone provides the propulsion, so hybrid cars get their best gasoline mileage in city traffic. The Prius cruises using both propulsion systems, although some of the energy from the engine is used to charge the batteries using the motor-generator. When going downhill, the Prius turns off the gasoline engine. Furthermore, when the brakes are applied, the motor-generator converts some of the kinetic energy of the car into electricity, charging the batteries, and saving energy wasted in a conventional car.

The batteries that power the motor are nickel–metal hydride (NiMH) batteries that are charged by the gasoline engine during normal driving or as the car goes downhill, so the car never needs to be plugged in to be recharged. Eventually, the batteries need to be replaced.

Because a hybrid car does not use the gasoline engine all of the time, it produces much less exhaust, both polluting gases and carbon dioxide, than a conventional car. For example, the Toyota Prius has such low tailpipe emissions that it qualifies for the California Air Resources Board's stringent Super-Ultra-Low-Emission Vehicle class. Due to their excellent gas mileage, demand for hybrid cars is projected to rise as emission standards grow ever stricter and gasoline prices rise.

Toyota Prius, a hybrid car.

Sources: Reisch, M. "Thirst for Power." *Chemical & Engineering News,* July 9, 2007; Maynard, M. "Getting to Green." *New York Times,* October 24, 2007.

18.9 Fuel Cells

A **fuel cell** is an electrochemical cell that converts the chemical energy of fuels directly into electricity. It functions somewhat like a battery, but in contrast to a battery, its reactants are continually supplied from an external reservoir. The best-known fuel cell is the alkaline fuel cell, which is used in the Space Shuttle. Alkaline fuel cells have been used since the 1960s by NASA to power electrical systems on spacecraft, but they are very expensive and unlikely to be commercialized. They require very pure hydrogen and oxygen to operate and are susceptible to contamination. Therefore, alternative fuel cells have been developed.

Proton-Exchange Membrane Fuel Cell

The hydrogen-oxygen fuel cell produces electrical power from the oxidation of hydrogen in an electrochemical cell. The two half-reactions are

$$H_2 \longrightarrow 2\,H^+ + 2\,e^- \qquad \text{(anode, oxidation)}$$

$$\tfrac{1}{2}O_2 + 2\,H^+ + 2\,e^- \longrightarrow H_2O \qquad \text{(cathode, reduction)}$$

Hydrogen gas is pumped to the anode, and oxygen gas or air is pumped to the cathode (Figure 18.15). The graphite electrodes are surrounded by catalysts containing platinum. The two electrodes are separated by a semipermeable membrane—a proton-exchange membrane (PEM), which is a thin plastic sheet—that allows the passage of H^+ ions but not electrons. Instead, the electrons must pass through an external circuit, where they can be used to perform work. The electrochemical reactions are aided by platinum catalysts on both sides of the PEM membrane that are in contact with the anode and the cathode. The H^+ ions produced at the anode pass through the PEM to the cathode. At the cathode, hydroxide ions produced by the reduction of O_2 react with the protons to produce water. The overall reaction of the fuel cell

$$H_2(g) + \tfrac{1}{2}O_2(g) \longrightarrow H_2O(\ell)$$

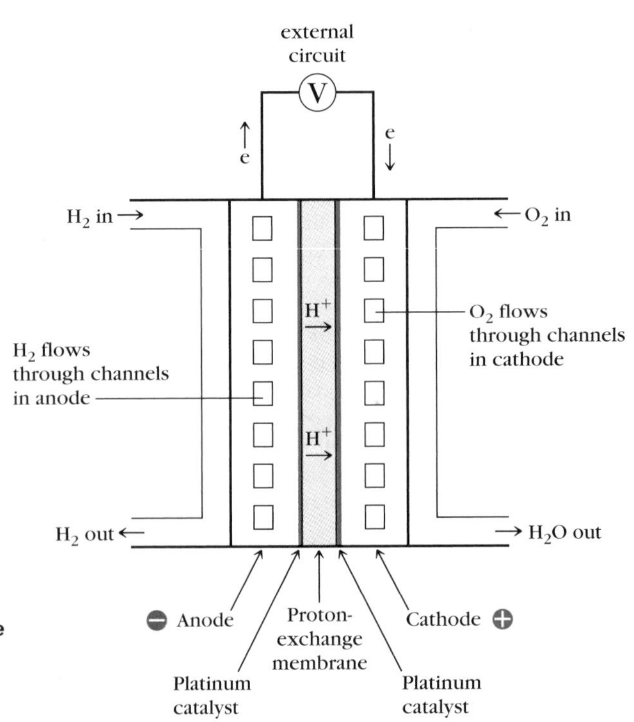

Figure 18.15 A proton-exchange membrane H_2/O_2 fuel cell. H_2 is oxidized in the anode chamber. O_2 is reduced in the cathode chamber.

produces approximately 0.7 V. To get a higher voltage, fuel cells are stacked and connected in series.

The pure hydrogen needed by the PEM fuel cell poses a problem for its more general use. Pure hydrogen is flammable and is difficult to store and distribute. Other hydrogen sources are more attractive. A device called a *reformer* can turn a hydrocarbon (such as natural gas) or alcohol (such as methanol) fuel into hydrogen, which can then be fed to the fuel cell. However, a fuel cell with a reformer has lower overall efficiency and produces products in addition to water.

Hydrogen does not occur in nature as H_2. Therefore, H_2 fuel has to be manufactured. Currently, most H_2 is produced as a by-product of petroleum refining or by treating methane with steam (⇐ ***p. 492***).

18.10 Electrolysis—Causing Reactant-Favored Redox Reactions to Occur

Reactant-favored redox systems (for which $\Delta G°$ is positive) can be made to produce products if electrons are forced into the electrochemical system from an external source of electrical current, such as a battery. This process, called **electrolysis,** provides a way to carry out reactant-favored electrochemical reactions that will not take place by themselves. Electrolytic processes are important in our economy being used in the production and purification of many metals, including copper and aluminum and in electroplating processes that produce a thin coating of metal on many different kinds of items.

Lysis means "splitting," so electrolysis means "splitting with electricity." Electrolysis reactions are chemical reactions caused by the flow of a direct current.

Like voltaic cells, electrolysis cells contain electrodes in contact with a conducting medium and an external circuit. As in a voltaic cell, the electrode where reduction takes place in an electrolysis cell is called the cathode, and the electrode where oxidation takes place is called the anode. The electrodes in electrolysis cells are often inert, and their function is to furnish a path for electrons to enter and leave the cell. In contrast to voltaic cells, however, the external circuit connected to an electrolysis cell must contain a direct current *source* of electrons. A battery can serve as a source of electrical current when an electrolysis is carried out on a small scale. The battery forces electrons at a high enough voltage into one of the electrodes (which becomes negative) and removes electrons from the other electrode (which becomes positive). There is often no need for a physical separation of the two electrode reactions, so there is usually no salt bridge. The conducting medium in contact with the electrodes is often the same for both electrodes, and it can be a molten salt or an aqueous solution.

An example of electrolysis is the decomposition of molten sodium chloride. In this process, a pair of electrodes dips into pure sodium chloride that has been heated above its melting temperature (Figure 18.16). In the molten liquid, Na^+ and Cl^- ions are free to move. The Na^+ ions are attracted to the negative electrode, and the Cl^- ions are attracted to the positive electrode. Reduction of Na^+ ions to Na atoms occurs at the cathode (negative electrode). Oxidation of Cl^- ions to chlorine molecules occurs at the anode (positive electrode).

$$2\,Na^+(\text{in melt}) + 2\,e^- \longrightarrow 2\,Na(\ell) \qquad \text{(cathode, reduction)}$$
$$2\,Cl^-(\text{in melt}) \longrightarrow Cl_2(g) + 2\,e^- \qquad \text{(anode, oxidation)}$$
$$2\,Na^+(\text{in melt}) + 2\,Cl^-(\text{in melt}) \longrightarrow 2\,Na(\ell) + Cl_2(g) \qquad \text{(net cell reaction)}$$

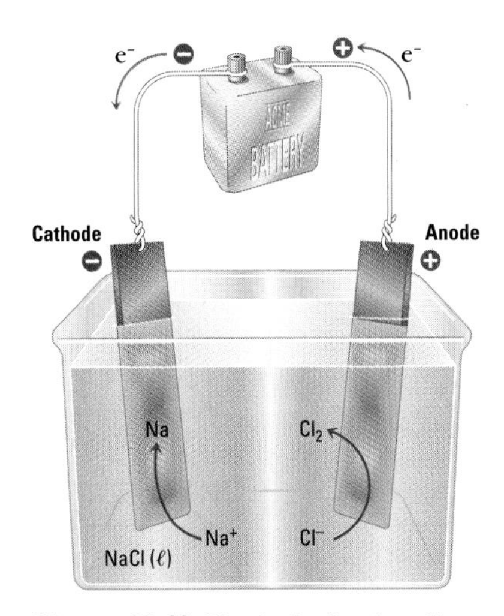

Figure 18.16 Electrolysis of molten sodium chloride.

The electrolysis of molten salts is an energy-intensive process because energy is needed to melt the salt as well as to cause the anode and cathode reactions to take place.

What happens if we pass a direct current through an *aqueous solution* of a salt, such as potassium iodide, KI, rather than through the molten salt? To predict the outcome of the electrolysis we must first decide what in the solution can be oxidized and reduced. For KI(aq), the solution contains K^+ ions, I^- ions, and H_2O molecules. Potassium is already in its highest common oxidation state in K^+, so it cannot be oxidized.

However, both the I^- ion and the H_2O could be oxidized. The possible anode half-reaction oxidations and their standard reduction potentials are

$$2\,I^-(aq) \longrightarrow I_2(s) + 2\,e^- \qquad E^\circ_{anode} = 0.535\text{ V}$$

$$6\,H_2O(\ell) \longrightarrow O_2(g) + 4\,H_3O^+(aq) + 4\,e^- \qquad E^\circ_{anode} = 1.229\text{ V}$$

*Whenever two or more electrochemical reactions are possible at the same electrode, you can use Table 18.1 (**← p. 663**) to decide which reaction is more likely to occur under standard-state conditions.* Considering the two possible anode reactions, item 4 in the discussion of Table 18.1 indicates that the less positive the standard reduction potential, the more likely a half-reaction is to occur as an oxidation. That is, the farther toward the bottom of Table 18.1 a half-reaction is, the more likely it is to occur as an oxidation. Oxidation of I^- to I_2 has a less positive standard reduction potential ($E^\circ = +0.535$ V) than does H_2O ($E^\circ = +1.229$ V), so oxidation of iodide ion to iodine is the more likely anode reaction.

Since I^- is the lowest common oxidation state of iodine, there are only two species that can be reduced at the cathode: K^+ ions and water molecules. The possible cathode half-reaction reductions and their standard reduction potentials are

$$K^+(aq) + e^- \longrightarrow K(s) \qquad E^\circ_{cathode} = -2.925\text{ V}$$

$$2\,H_2O(\ell) + 2\,e^- \longrightarrow H_2(g) + 2\,OH^-(aq) \qquad E^\circ_{cathode} = -0.8277\text{ V}$$

Point 3 in the discussion of Table 18.1 states that the more positive the standard reduction potential, the more easily a substance on the left-hand side of a half-reaction can be reduced. Since E° for H_2O is more positive (less negative) than E° for K^+, H_2O is more likely to be reduced. Therefore, in aqueous KI solution, the overall reaction and cell potential are

$$E^\circ_{cell} = E^\circ_{cathode} - E^\circ_{anode} = -0.8277\text{ V} - (0.535\text{ V}) = -1.363\text{ V}$$

The negative value of E°_{cell} indicates that the reaction is not product-favored and that an external energy source is needed for the reaction to occur.

$$2\,I^-(aq) + 2\,H_2O(\ell) \longrightarrow I_2(s) + H_2(g) + 2\,OH^-(aq) \qquad E^\circ_{cell} = -1.363\text{ V}$$

An experiment in which a direct electrical current is passed through aqueous KI (Figure 18.17) shows that this prediction is correct. At the anode (on the right in Figure 18.17a), the I^- ions are oxidized to I_2, which produces a yellow-brown color in the solution. At the cathode, water is reduced to gaseous hydrogen and aqueous hydroxide ions. The formation of excess OH^- ions is shown by the pink color of the phenolphthalein indicator that has been added to the solution.

When electrolysis is carried out by passing a direct electrical current through an aqueous solution, the electrode reactions most likely to take place are those that require the least voltage, that is, the half-reactions that combine to give the least negative overall cell voltage. This means that in aqueous solution the following conditions apply.

1. ***A metal ion or other species can be reduced if it has a reduction potential more positive than −0.8 V, the potential for reduction of water.*** If a species has a reduction potential more negative than −0.8 V, then water will preferentially be reduced to $H_2(g)$ and OH^- ions. Table 18.1 shows that many metal ions are in this category. These include Na^+, K^+, Mg^{2+}, and Al^{3+}. Consequently, producing these metals from their ions requires electrolysis of a molten salt with no water present.
2. ***A species can be oxidized in aqueous solution if it has a reduction potential less positive than 1.2 V, the potential for reduction of $O_2(g)$ to water.*** If the reduction potential is less positive than for reduction of water, then oxidation of the species on the right-hand side of a half-reaction is more likely than oxidation of water. Most of the half-equations in Table 18.1 are in this category. If a species has a reduction potential more positive than 1.2 V (that is, if its half-

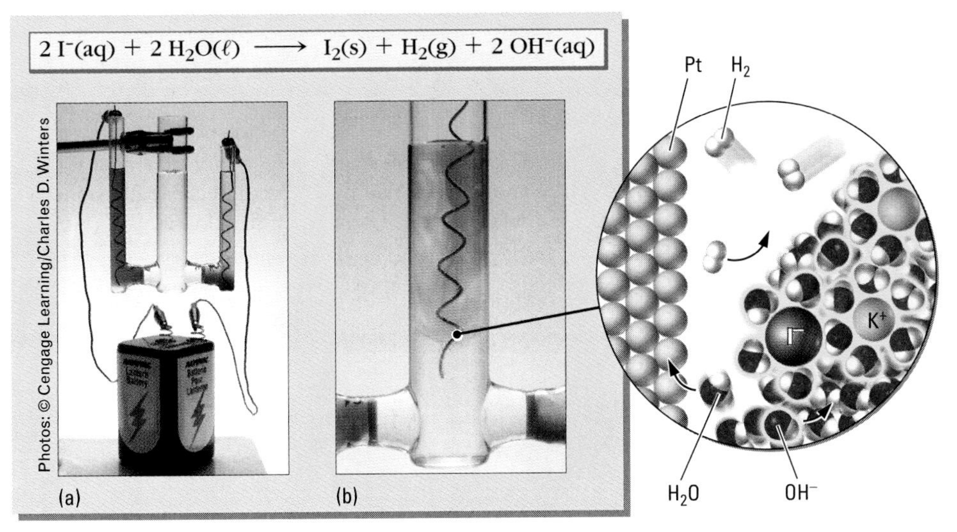

Figure 18.17 The electrolysis of aqueous potassium iodide. (a) Aqueous KI is found in all three compartments of the cell, and both electrodes are platinum. At the positive electrode, or anode (*right*), the I^- ion is oxidized to iodine, which gives the solution a yellow-brown color.

$$2\,I^-(aq) \longrightarrow I_2(aq) + 2\,e^-$$

At the negative electrode, or cathode (*left*), water is reduced, and the presence of OH^- ion is indicated by the pink color of the acid-base indicator, phenolphthalein.

$$2\,H_2O(\ell) + 2\,e^- \longrightarrow H_2(g) + 2\,OH^-(aq)$$

(b) A close-up of the cathode of a different cell running the same reaction. Bubbles of H_2 and evidence of OH^- generation at the electrode are readily apparent.

reaction is above the water-oxygen half-reaction in Table 18.1), water will be oxidized preferentially. For example, $F^-(aq)$ cannot be oxidized electrolytically to $F_2(g)$ because water will be oxidized to $O_2(g)$ instead.

An additional important factor is that concentrations and pressure are usually not the standard values.

The voltage that must be applied to an electrolysis cell is always somewhat greater than the voltage calculated from standard reduction potentials. An *overvoltage* is required, which is an additional voltage needed to overcome limitations in the electron transfer rate at the interface between electrode and solution. Redox reactions that involve the formation of O_2 or H_2 are especially prone to having large overvoltages. Since overvoltages cannot be predicted accurately, the only way to determine with certainty which half-reaction will occur in an electrolysis cell when two possible reactions have similar standard reduction potentials is to perform the experiment.

PROBLEM-SOLVING EXAMPLE 18.11 Electrolysis of Aqueous NaOH

Predict the results of passing a direct electrical current through an aqueous solution of NaOH. Calculate the cell potential.

Answer The net cell reaction is $2\,H_2O(\ell) \rightarrow 2\,H_2(g) + O_2(g)$. Hydrogen is produced at the cathode and oxygen is produced at the anode. The cell potential is -1.23 V.

Strategy and Explanation First, list all the species in the solution: Na^+, OH^-, and H_2O. Next, use Table 18.1 to decide which species can be oxidized and which can be reduced, and note the standard reduction potential of each possible half-reaction.

Reductions:

$$Na^+(aq) + e^- \longrightarrow Na(s) \qquad E^\circ_{cathode} = -2.71\text{ V}$$

$$2\,H_2O(\ell) + 2\,e^- \longrightarrow H_2(g) + 2\,OH^-(aq) \qquad E^\circ_{cathode} = -0.83\text{ V}$$

Oxidations:

$$4\,OH^-(aq) \longrightarrow O_2(g) + 2\,H_2O(\ell) + 4\,e^- \qquad E^\circ_{anode} = +0.40\text{ V}$$

$$6\,H_2O(\ell) \longrightarrow O_2(g) + 4\,H_3O^+(aq) + 4\,e^- \qquad E^\circ_{anode} = +1.229\text{ V}$$

Water will be reduced to H_2 at the cathode because the standard reduction potential for this half-reaction is more positive. At the anode, OH^- will be oxidized because the standard reduction potential is smaller than that for water. The net cell reaction is

$$2\,H_2O(\ell) \longrightarrow 2\,H_2(g) + O_2(g)$$

and the cell potential under standard conditions is

$$E^\circ_{cell} = E^\circ_{cathode} - E^\circ_{anode} = (-0.83\text{ V}) - (+0.40\text{ V}) = -1.23\text{ V}$$

PROBLEM-SOLVING PRACTICE 18.11

Predict the results of passing a direct electrical current through (a) molten NaBr, (b) aqueous NaBr, and (c) aqueous $SnCl_2$.

Electrolysis of Brine: The Chlor-Alkali Process

Chlorine is produced industrially by the electrolysis of aqueous sodium chloride in the **chlor-alkali process;** the alkali produced by this process is sodium hydroxide. More than 12 million tons of chlorine are produced annually in the United States, along with 8.8 million tons of sodium hydroxide. These large amounts testify to the usefulness of these two products. The oxidizing and bleaching ability of chlorine is utilized in many industrial and everyday applications, and this element is a raw material in the manufacture of chlorine-containing chemicals. Sodium hydroxide is the base of choice in many industrial chemistry applications because it is inexpensive, and so it is used widely to produce soaps, detergents, and other compounds.

The chlor-alkali process electrolyzes brine (saturated aqueous NaCl), as illustrated in Figure 18.18. Chloride ions are oxidized at the anode, and water is reduced at the cathode. The anode and cathode compartments are separated by a special polymeric membrane that allows only cations to pass through it. The brine solution is added to the anode compartment, and sodium ions pass through the membrane into the cathode compartment. The half-reactions are

The reduction potential of Na^+ is more negative than that of water, so water, not sodium ions, is reduced.

$$2\,Cl^-(aq) \longrightarrow Cl_2(g) + 2\,e^- \qquad \text{(anode, oxidation)}$$

$$2\,H_2O(\ell) + 2\,e^- \longrightarrow 2\,OH^-(aq) + H_2(g) \qquad \text{(cathode, reduction)}$$

$$2\,Cl^-(aq) + 2\,H_2O(\ell) \longrightarrow Cl_2(g) + 2\,OH^-(aq) + H_2(g) \qquad \text{(net cell reaction)}$$

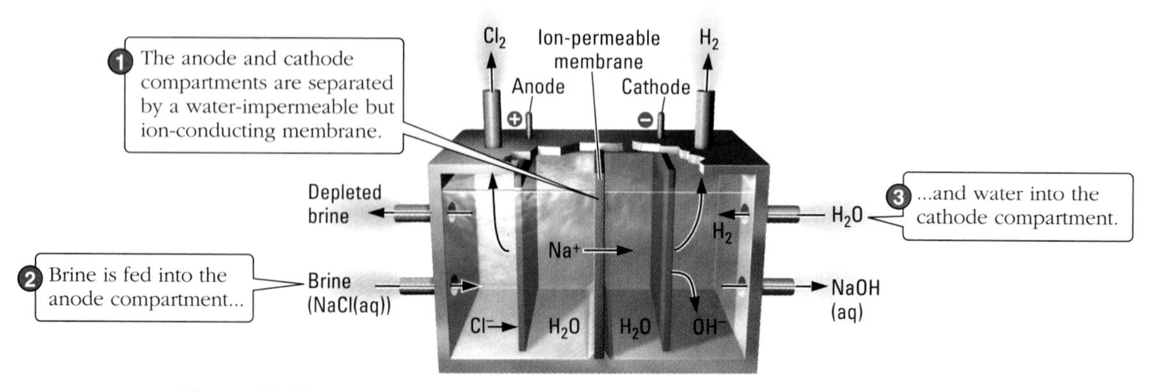

Figure 18.18 A membrane cell used in the chlor-alkali process.

The anode is specially treated titanium, and the cathode is stainless steel or nickel. The membrane is not permeable to water and acts as a salt bridge. Thus, as chloride ions are oxidized in the anode compartment, sodium ions must migrate from there to the cathode compartment to maintain charge balance. The resulting NaOH solution in the cathode compartment is 21% to 30% NaOH by weight.

CONCEPTUAL
EXERCISE 18.9 Making F_2 Electrolytically

In 1886 Moissan was the first to prepare F_2 by the electrolysis of F^- ions. He electrolyzed KF dissolved in pure HF. No water was present, so only F^- ions were available at the anode. What was produced at the cathode? Write the half-equations for the oxidation and the reduction reactions, and then write the net cell reaction.

18.11 Counting Electrons

When a direct electric current is passed through an aqueous solution of the soluble salt $AgNO_3$, metallic silver is produced at the cathode. One mole of electrons is required for every mole of Ag^+ reduced.

$$Ag^+(aq) + e^- \longrightarrow Ag(s)$$

If a copper(II) salt in aqueous solution were reduced, 2 mol electrons would be required to produce 1 mol metallic copper from 1 mol copper(II) ions.

$$Cu^{2+}(aq) + 2\,e^- \longrightarrow Cu(s)$$

Large electric currents, like those needed to run a hair dryer or refrigerator, are measured in amperes (C/s). Smaller currents, in the milliampere (mA) range, are more commonly used in laboratory electrolysis experiments. 1 mA = 10^{-3} A.

Each of these balanced half-reactions is like any other balanced chemical equation. That is, each illustrates the fact that both matter and charge are conserved in chemical reactions. Thus, if you could measure the number of moles of electrons flowing through an electrolysis cell, you would know the number of moles of silver or copper produced. Conversely, if you knew the amount of silver or copper produced, you could calculate the number of moles of electrons that had passed through the circuit.

The number of moles of electrons transferred during a redox reaction is usually determined experimentally by measuring the current flowing in the external electrical circuit during a given time. The product of the current (measured in amperes, A) and the time interval (in seconds, s) equals the electric charge (in coulombs, C) that flowed through the circuit.

$$\text{Charge} = \text{current} \times \text{time}$$

$$1 \text{ coulomb} = 1 \text{ ampere} \times 1 \text{ second}$$

The Faraday constant (96,500 C/mol of electrons; *p. 666*) can then be used to find the number of moles of electrons from a known number of coulombs of charge. This information is of practical significance in chemical analysis and synthesis.

Figure 18.19 shows the relationship between quantity of charge used and the quantities of substances that are oxidized or reduced during electrolysis.

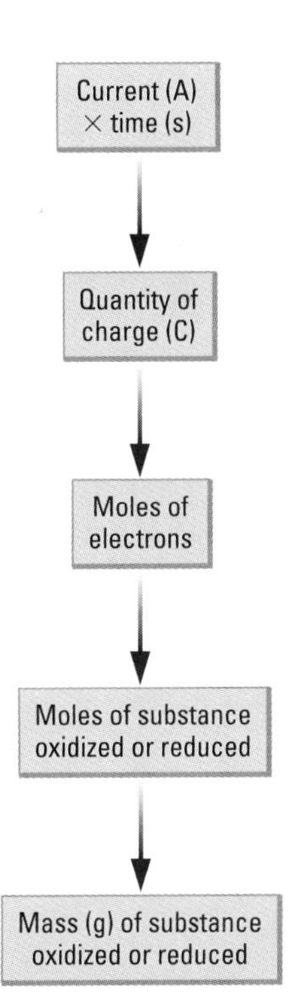

Figure 18.19 Calculation steps for electrolysis. These steps relate the quantity of electrical charge used in electrolysis to the amounts of substances oxidized or reduced.

PROBLEM-SOLVING EXAMPLE 18.12 Using the Faraday Constant

How many grams of copper will be deposited at the cathode of an electrolysis cell if an electric current of 15 mA is applied for 1.0 h through an aqueous solution containing excess Cu^{2+} ions?

Answer 0.018 g Cu

Chris Sharp/Photo Researchers, Inc.

A copper electrorefining facility showing the refined copper plated on cathodes.

Strategy and Explanation We use the strategy presented in Figure 18.19. First, we write and balance the relevant half-reaction that occurs at the cathode.

$$Cu^{2+}(aq) + 2\,e^- \rightarrow Cu(s)$$

Then, we calculate the quantity of charge transferred.

$$\text{Charge} = 15 \times 10^{-3}\,\text{A} \times 3600\,\text{s} = 15 \times 10^{-3}\,\text{C/s} \times 3600.\,\text{s} = 54\,\text{C}$$

Finally, we determine the mass of copper deposited.

$$(54\,\text{C})\left(\frac{1\text{ mol e}^-}{9.65 \times 10^4\text{ C}}\right)\left(\frac{1\text{ mol Cu}}{2\text{ mol e}^-}\right)\left(\frac{63.5\text{ g Cu}}{1\text{ mol Cu}}\right) = 0.018\text{ g Cu}$$

PROBLEM-SOLVING PRACTICE 18.12

In the commercial production of sodium metal by electrolysis, the cell operates at 7.0 V and a current of 25×10^3 A. What mass of metallic sodium can be produced in 1.0 h?

CONCEPTUAL
EXERCISE 18.10 How Many Faradays?

Which would require more Faradays of electricity?

(a) Making 1 mol Al from Al^{3+}
(b) Making 2 mol Na from Na^+
(c) Making 2 mol Cu from Cu^{2+}

EXERCISE 18.11 NaOH Production

A chlor-alkali membrane cell operates at 2.0×10^4 A for 100. hours. How many tons of NaOH are produced in this time?

© Cengage Learning/Charles D. Winters

Electrolysis of water. A very dilute solution of sulfuric acid is electrolyzed to produce H_2 at the cathode (*left*) and O_2 at the anode (*right*).

Electrolytic Production of Hydrogen

Hydrogen can be produced by the electrolysis of water to which a drop or two of sulfuric acid has been added to make the solution conductive. The overall electrochemical reaction is

$$2\,H_2O(\ell) \longrightarrow 2\,H_2(g) + O_2(g) \qquad E^\circ_{cell} = -1.24\text{ V}$$

Oxygen is produced at the anode and hydrogen at the cathode. The minimum voltage required for this reaction is 1.24 V (J/C), but in practice overvoltage requires a higher voltage of about 2 V.

Let's consider how much electrical energy would be required to produce 1.00 kg of gaseous H_2 (about 11,200 L at STP) and at what cost. First we calculate the required charge in coulombs by using the Faraday constant; then we use the fact that 1 joule = 1 volt × 1 coulomb.

The reduction half-reaction shows that 2 mol electrons produces 1 mol (2.02 g) $H_2(g)$.

$$2\,H_3O^+(aq) + 2\,e^- \longrightarrow H_2(g) + 2\,H_2O(\ell)$$

The amount (number of moles) of electrons required to produce 1.00 kg H_2 is calculated as follows:

$$1.00\text{ kg H}_2 \times \left(\frac{1 \times 10^3\text{ g}}{\text{kg}}\right)\left(\frac{1\text{ mol H}_2}{2.016\text{ g H}_2}\right)\left(\frac{2\text{ mol e}^-}{1\text{ mol H}_2}\right) = 9.92 \times 10^2\text{ mol e}^-$$

Now we can calculate the charge using the Faraday constant.

$$(9.92 \times 10^2\text{ mol e}^-) \times \left(\frac{9.65 \times 10^4\text{ C}}{1\text{ mol e}^-}\right) = 9.57 \times 10^7\text{ C}$$

The energy (in joules) can be calculated from the charge and the cell voltage.

$$\text{Energy} = \text{charge} \times \text{voltage} = (9.57 \times 10^7\ \text{C})(1.24\ \text{J/C}) = 1.19 \times 10^8\ \text{J}$$

We convert joules to kilowatt-hours (kWh), which is the unit we see when we pay the electric bill.

The kilowatt-hour (kWh) is a unit of energy: 1 kWh = 3.60×10^6 J.

$$1.19 \times 10^8\ \text{J} \times \frac{1\ \text{kWh}}{3.60 \times 10^6\ \text{J}} = 33.1\ \text{kWh}$$

At a rate of 10 cents per kilowatt-hour, the production of 1.00 kg hydrogen would cost $3.31.

Hydrogen holds great promise as a fuel in our economy because it is a gas and can be transported through pipelines, it burns without producing pollutants, and it could be used in fuel cells to generate electricity on demand. Both water and sulfuric acid are in plentiful supply. The major problem with producing hydrogen in quantities large enough to meet the nation's energy demands is finding a cheap enough source of electricity. Another challenging problem is the development of economical, reliable, and safe methods for hydrogen storage.

EXERCISE 18.12 Calculations Based on Electrolysis

In the production of aluminum metal, Al^{3+} is reduced to Al metal using currents of about 50,000 A and a low voltage of about 4.0 V. How much energy (in kilowatt-hours) is required to produce 2000. metric tons of aluminum metal?

CONCEPTUAL
EXERCISE 18.13 How Many Joules?

Think of a battery you just purchased at the store as an energy source that can deliver some number of joules. Name the two pieces of information you need to calculate the number of joules this battery can deliver. Which one is obviously available as you read the label on the battery? Devise a means of determining the other information needed.

Electroplating

If a metal or other electrical conductor serves as the cathode in an electrolysis cell, the metal can be plated with another metal to decorate it or protect it against corrosion. To plate an object with copper, for example, we have only to make the object's surface conducting and use the object as the cathode in an electrolysis cell containing a solution of a soluble copper salt as a source of Cu^{2+} ions. The object will become coated with metallic copper, and the coating will thicken as the electrolysis continues and electrons reduce more Cu^{2+} ions to Cu atoms. If the plated object is a metal, it will conduct electricity by itself. If the object is a nonmetal, its surface can be lightly dusted with graphite powder to make it conducting.

Don Smetzer

Oscar is gold-plated. The Oscar award and most gold jewelry are made by plating a thin coating of gold onto a base metal.

Precious metals such as gold are often plated onto cheaper metals such as copper to make jewelry. If the current and duration of the plating reaction are known, it is possible to calculate the mass of gold that will be reduced onto the cathode surface. For example, suppose the object to be plated is immersed in an aqueous solution of $AuCl_3$ and is made a cathode by connecting it to the negative pole of a battery. The circuit is completed by immersing an inert electrode connected to the positive battery pole in the solution, and gold is reduced at the cathode for 60. min at a current of 0.25 A. The reduction half-reaction is

$$Au^{3+}(aq) + 3\ e^- \longrightarrow Au(s)$$

ESTIMATION The Cost of Aluminum in a Beverage Can

How much does it cost to generate the mass of aluminum (14 g) in one beverage can? The aluminum in these cans is produced by reducing Al^{3+} to Al(s). The reaction is run commercially at 50,000 A and 4 V (4 J/C), 1 kWh of electricity costs about 10 cents, and 1 kWh = 3.60×10^6 J.

The charge needed to generate 14 g Al is

$$(14\text{ g Al})\left(\frac{1\text{ mol Al}}{26.98\text{ g Al}}\right)\left(\frac{3\text{ mol e}^-}{1\text{ mol Al}}\right)\left(\frac{96{,}500\text{ C}}{1\text{ mol e}^-}\right) = 1.5 \times 10^5\text{ C}$$

The quantity of energy used is

$$(1.5 \times 10^5\text{ C})(4\text{ J/C})\left(\frac{1\text{ kWh}}{3.60 \times 10^6\text{ J}}\right) = 0.17\text{ kWh}$$

The cost of 0.17 kWh at 10 cents per kWh is 1.7 cents. Thus, the mass of Al in one beverage can could be generated by electrolysis for less than 2 cents.

It is estimated that recycling of aluminum for cans requires overall less than 1% of the cost of mining and processing aluminum ore into aluminum metal, an energy-intensive and costly process.

Visit this book's companion website at **www.cengage.com/chemistry/moore** to work an interactive module based on this material.

The mass of gold that is reduced is calculated by

$$(0.25\text{ C/s})(60.\text{ min})\left(\frac{60\text{ s}}{1\text{ min}}\right)\left(\frac{1\text{ mol e}^-}{9.65 \times 10^4\text{ C}}\right)\left(\frac{1\text{ mol Au}}{3\text{ mol e}^-}\right)\left(\frac{197.\text{ g Au}}{1\text{ mol Au}}\right) = 0.61\text{ g Au}$$

Assuming that gold is selling for \$900 per ounce, that's about \$19.35 worth of gold.

EXERCISE 18.14 Electroplating Silver

Calculate the mass of silver that could be plated from solution with a current of 0.50 A for 20. min. The cathode reaction is $Ag^+(aq) + e^- \rightarrow Ag(s)$.

© George B. Diebold/Corbis

Figure 18.20 Rusting. The formation of rust destroys the structural integrity of objects made of iron and steel. Given time, this chain will completely rust away.

18.12 Corrosion—Product-Favored Redox Reactions

Corrosion is the oxidation of a metal that is exposed to the environment. Visible corrosion on the steel supports of a bridge, for example, indicates possible structural failure. Corrosion reactions are invariably product-favored, which means that $E°$ for the reaction is positive and $\Delta G°$ is negative. Corrosion of iron, for example, takes place quite readily and is difficult to prevent. It produces the red-brown substance we call rust, which is hydrated iron(III) oxide ($Fe_2O_3 \cdot x\,H_2O$, where x varies from 2 to 4). The rust that forms when iron corrodes does not adhere to the surface of the metal, so it can easily flake off and expose fresh metal surface to corrosion (Figure 18.20). The corrosion of aluminum, a metal that is even more reactive than iron, is also very product-favored. The aluminum oxide that forms as a result of corrosion adheres tightly as a thin coating on the surface of the metal, creating a protective coating that prevents further corrosion.

For corrosion of a metal (M) to occur, the metal must have an anodic area where the oxidation can occur. The general reaction is

Anode reaction: $$M(s) \longrightarrow M^{n+} + n\,e^-$$

There must also be a cathodic area where electrons are consumed. Frequently, the cathode reactions are reductions of oxygen or water.

Cathode reactions:

$$O_2(g) + 2\ H_2O(\ell) + 4\ e^- \longrightarrow 4\ OH^-(aq)$$
$$2\ H_2O(\ell) + 2\ e^- \longrightarrow 2\ OH^-(aq) + H_2(g)$$

Anodic areas may occur at cracks in the oxide coating that protects the surfaces of many metals or around impurities. Cathodic areas may occur at the metal oxide coating, at less reactive metallic impurity sites, or around other metal compounds trapped at the surface, such as sulfides or carbides.

The other requirements for corrosion are an electrical connection between the anode and the cathode and an electrolyte in contact with both anode and cathode. Both requirements are easily fulfilled—the metal itself is the conductor, and ions dissolved in moisture from the environment provide the electrolyte.

Corrosion is so commonplace that about 25% of the annual steel production in the United States is used to replace material lost to corrosion.

In the corrosion of iron, the anodic reaction is the oxidation of metallic iron (Figure 18.21). If both water and O_2 gas are present, the cathode reaction is the reduction of oxygen, giving the net reaction

$$2\ [Fe(s) \longrightarrow Fe^{2+}(aq) + 2\ e^-] \quad \text{(anode reaction)}$$
$$O_2(g) + 2\ H_2O(\ell) + 4\ e^- \longrightarrow 4\ OH^-(aq) \quad \text{(cathode reaction)}$$
$$2\ Fe(s) + O_2(g) + 2\ H_2O(\ell) \longrightarrow \underset{\text{iron(II) hydroxide}}{2\ Fe(OH)_2(s)}$$

In the presence of an ample supply of oxygen and water, as in open air or flowing water, the iron(II) hydroxide is oxidized to the hydrated red-brown iron(III) oxide (Figure 18.20). This hydrated iron oxide is the familiar rust you see on iron and steel objects and the substance that colors the water red in some mountain streams and home water pipes. Rust is easily removed from the metal surface by mechanical shaking, rubbing, or even the action of rain or freeze-thaw cycles, thus exposing more iron at the surface and allowing the objects to eventually deteriorate completely.

Other substances in air and water can hasten corrosion. Metal salts, such as the chlorides of sodium and calcium from sea air or from salt spread on roadways in the winter, function as salt bridges between anodic and cathodic regions, thus speeding up corrosion reactions.

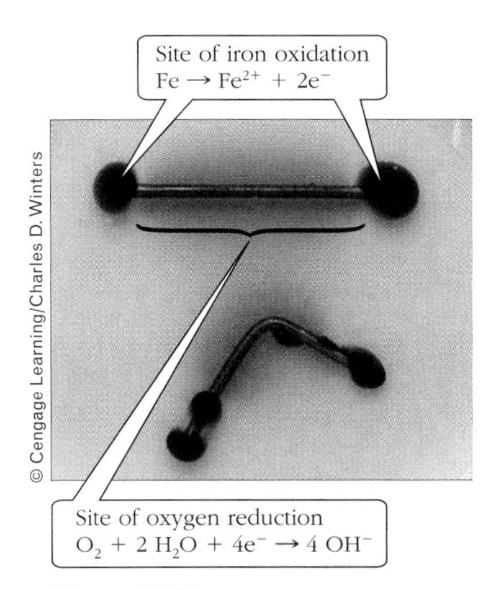

Figure 18.21 Corroding iron nails. Two nails were placed in an agar gel, which also contained the indicator phenolphthalein and $[Fe(CN)_6]^{3-}$. The nails began to corrode and produced Fe^{2+} ions at the tip and where the nail is bent. (These points of stress corrode more quickly.) These points are the anode, as indicated by the formation of the blue-colored compound called Prussian blue, $Fe_3[Fe(CN)_6]_2$. The remainder of the nail is the cathode, since oxygen is reduced in water to give OH^-. The presence of OH^- ions causes the phenolphthalein to turn pink.

CONCEPTUAL
EXERCISE 18.15 Do All Metals Corrode?

Do all metals corrode as readily as iron and aluminum? Name three metals that you would expect to corrode about as readily as iron and aluminum, and name three metals that do not corrode readily. Name a use for each of the three noncorroding metals. Explain why metals fall into these two groups.

Corrosion Protection

How can metal corrosion be prevented? The general approaches are to (1) inhibit the anodic process, (2) inhibit the cathodic process, or (3) do both. The most common method is **anodic inhibition,** which directly limits or prevents the oxidation half-reaction by painting the metal surface, coating it with grease or oil, or allowing a thin film of metal oxide to form. More recently developed methods of anodic protection are illustrated by the following reaction, which occurs when the surface is treated with a solution of sodium chromate.

$$2\ Fe(s) + 2\ Na_2CrO_4(aq) + 2\ H_2O(\ell) \longrightarrow Fe_2O_3(s) + Cr_2O_3(s) + 4\ NaOH(aq)$$

The surface iron is oxidized by the chromate salt to give iron(III) and chromium(III) oxides. These form a coating that is impervious to O_2 and water, and further atmospheric oxidation is inhibited.

Cathodic protection is accomplished by forcing the metal to become the cathode instead of the anode. Usually, this goal is achieved by attaching another, more readily oxidized metal to the metal being protected. The best example involves galvanized iron, iron that has been coated with a thin film of zinc (Figure 18.22). The standard reduction potential for zinc is considerably more negative than that for iron (Zn is lower in Table 18.1 than Fe), so zinc is more easily oxidized. Therefore, the zinc metal film is oxidized before any of the iron and the zinc coating forms a *sacrificial anode.* In addition, when the zinc is corroded, $Zn(OH)_2$ forms an insoluble film on the surface (K_{sp} of $Zn(OH)_2 = 4.5 \times 10^{-17}$) that further slows corrosion.

Galvanized objects. A thin coating of zinc helps prevent the oxidation of iron.

CONCEPTUAL EXERCISE 18.16 Corrosion Rates

Rank these environments in terms of their relative rates of corrosion of iron. Place the fastest first. Explain your answers.

(a) Moist clay
(b) Sand by the seashore
(c) The surface of the moon
(d) Desert sand in Arizona

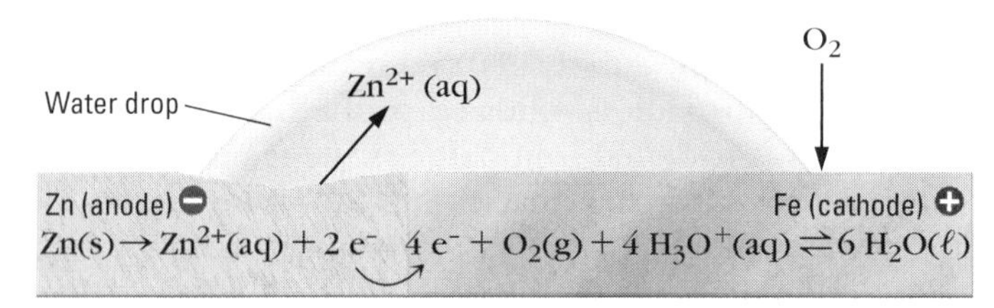

Figure 18.22 Cathodic protection of an iron-containing object. The iron is coated with a film of zinc, a metal more easily oxidized than iron. The zinc acts as the anode and forces iron to become the cathode, thereby preventing the corrosion of the iron.

IN CLOSING

Having studied this chapter, you should be able to . . .

- Identify the oxidizing and reducing agents in a redox reaction (Section 18.1).
  ■ End-of-chapter Question assignable in OWL: 6
- Write equations for oxidation and reduction half-reactions, and use them to balance the net equation (Section 18.2). ■ Questions 8, 10
- Identify and describe the functions of the parts of an electrochemical cell; describe the direction of electron flow outside the cell and the direction of the ion flow inside the cell (Section 18.3). ■ Question 16
- Describe how standard reduction potentials are defined and use them to predict whether a reaction will be product-favored as written (Sections 18.4, 18.5).
  ■ Questions 21, 25, 27
- Calculate $\Delta G°$ from the value of $E°$ for a redox reaction (Section 18.6).
  ■ Questions 29, 33, 35
- Explain how product-favored electrochemical reactions can be used to do useful work, and list the requirements for using such reactions in rechargeable batteries (Section 18.6).

Selected end-of-chapter Questions may be assigned in OWL.

For quick review, download Go Chemistry mini-lecture flashcard modules (or purchase them at **www.ichapters.com**).

Prepare for an exam with a **Summary Problem** that brings together concepts and problem-solving skills from throughout the chapter. Go to **www.cengage.com/chemistry/moore** to download this chapter's Summary Problem from the Student Companion Site.

- Explain how the Nernst equation relates concentrations of redox reactants to E_{cell} (Section 18.7). ■ Question 37
- Use the Nernst equation to calculate the potentials of cells that are not at standard conditions (Section 18.7). ■ Question 40
- Describe the chemistry of the mercury battery and the lead-acid storage battery (Section 18.8).
- Describe how a fuel cell works, and indicate how it differs from a battery (Section 18.9). ■ Question 46
- Use standard reduction potentials to predict the products of electrolysis of an aqueous salt solution (Section 18.10). ■ Question 51
- Calculate the quantity of product formed at an electrode during an electrolysis reaction, given the current passing through the cell and the time during which the current flows (Section 18.11). ■ Questions 52, 60
- Explain how electroplating works (Section 18.11).
- Describe what corrosion is and how it can be prevented by cathodic protection (Section 18.12).

KEY TERMS

ampere, A *(Section 18.4)*
anode *(18.3)*
anodic inhibition *(18.12)*
battery *(18.3)*
cathode *(18.3)*
cathodic protection *(18.12)*
cell voltage *(18.4)*
chlor-alkali process *(18.10)*
corrosion *(18.12)*
coulomb, C *(18.4)*
electrochemical cell *(18.3)*
electrochemistry *(Introduction)*
electrode *(18.3)*
electrolysis *(18.10)*
electromotive force (emf) *(18.4)*
Faraday constant, *F* *(18.6)*
fuel cell *(18.9)*
half-cell *(18.3)*
half-reaction *(18.2)*
Nernst equation *(18.7)*
primary battery *(18.8)*
salt bridge *(18.3)*
secondary battery *(18.8)*
standard conditions *(18.4)*
standard hydrogen electrode *(18.4)*
standard reduction potentials *(18.5)*
standard voltages, $E°$ *(18.4)*
volt, V *(18.4)*
voltaic cell *(18.3)*

QUESTIONS FOR REVIEW AND THOUGHT

■ denotes questions assignable in OWL.

Blue-numbered questions have short answers at the back of this book and fully worked solutions in the *Student Solutions Manual.*

Review Questions

1. Describe the principal parts of an electrochemical cell by drawing a hypothetical cell, indicating the cathode, the anode, the direction of electron flow outside the cell, and the direction of ion flow within the cell.
2. Explain how product-favored electrochemical reactions can be used to do useful work.
3. Explain how reactant-favored electrochemical reactions can be induced to make products.
4. Explain how electroplating works.
5. Tell whether each of these statements is true or false. If false, rewrite it to make it a correct statement.
   (a) Oxidation always occurs at the anode of an electrochemical cell.
   (b) The anode of a battery is the site of reduction and is negative.
   (c) Standard conditions for electrochemical cells are a concentration of 1.0 M for dissolved species and a pressure of 1 bar for gases.
   (d) The potential of a cell does not change with temperature.
   (e) All product-favored oxidation-reduction reactions have a standard cell voltage $E°_{cell}$, with a negative sign.

Topical Questions

Redox Reactions

6. ■ In each of these reactions assign oxidation numbers to all species, and tell which substance is oxidized and which is reduced. Tell which is the oxidizing agent and which is the reducing agent.
   (a) $2\ Al(s) + 3\ Cl_2(g) \longrightarrow 2\ AlCl_3(s)$
   (b) $8\ H_3O^+(aq) + MnO_4^-(aq) + 5\ Fe^{2+}(aq) \longrightarrow 5\ Fe^{3+}(aq) + Mn^{2+}(aq) + 12\ H_2O(\ell)$
   (c) $FeS(s) + 3\ NO_3^-(aq) + 4\ H_3O^+(aq) \longrightarrow 3\ NO(g) + SO_4^{2-}(aq) + Fe^{3+}(aq) + 6\ H_2O(\ell)$
7. In each of these reactions assign oxidation numbers to all species, and tell which substance is oxidized and which is reduced. Tell which is the oxidizing agent and which is the reducing agent.
   (a) $Fe(s) + Br_2(\ell) \longrightarrow FeBr_2(s)$
   (b) $8\ HI(aq) + H_2SO_4(aq) \longrightarrow H_2S(aq) + 4\ I_2(s) + 4\ H_2O(\ell)$
   (c) $H_2O_2(aq) + 2\ Fe^{2+}(aq) + 2\ H_3O^+(aq) \longrightarrow 2\ Fe^{3+}(aq) + 4\ H_2O(\ell)$

Using Half-Reactions to Understand Redox Reactions

8. ■ Write half-reactions for these changes:
   (a) Oxidation of zinc to Zn^{2+} ions
   (b) Reduction of H_3O^+ ions to hydrogen gas
   (c) Reduction of Sn^{4+} ions to Sn^{2+} ions
   (d) Reduction of chlorine to Cl^- ions
   (e) Oxidation of sulfur dioxide to sulfate ions in acidic solution
9. Write half-reactions for these changes:
   (a) Reduction of MnO_4^- ion to Mn^{2+} ion in acid solution
   (b) Reduction of $Cr_2O_7^{2-}$ ion to Cr^{3+} ion in acid solution
   (c) Oxidation of hydrogen gas to H_3O^+ ions
   (d) Reduction of hydrogen peroxide to water in acidic solution
   (e) Oxidation of nitric oxide to nitrogen monoxide in acidic solution
10. ■ Balance this redox reaction in a basic solution:
    $$Zn(s) + NO_3^-(aq) \longrightarrow Zn(OH)_4^{2-}(aq) + NH_3(aq)$$
11. Balance this redox reaction in a basic solution:
    $$NO_2^-(aq) + Al(s) \longrightarrow NH_3(aq) + Al(OH)_4^-(aq)$$
12. Balance these redox reactions, and identify the oxidizing agent and the reducing agent.
    (a) $CO(g) + O_3(g) \longrightarrow CO_2(g)$
    (b) $H_2(g) + Cl_2(g) \longrightarrow HCl(g)$
    (c) $H_2O_2(aq) + Ti^{2+}(aq) \longrightarrow H_2O(\ell) + Ti^{4+}(aq)$ in acidic solution
    (d) $Cl^-(aq) + MnO_4^-(aq) \longrightarrow Cl_2(g) + MnO_2(s)$ in acidic solution
    (e) $FeS_2(s) + O_2(g) \longrightarrow Fe_2O_3(s) + SO_2(g)$
    (f) $O_3(g) + NO(g) \longrightarrow O_2(g) + NO_2(g)$
    (g) $Zn(Hg)$ (amalgam) $+ HgO(s) \longrightarrow ZnO(s) + Hg(\ell)$ in basic solution (This is the reaction in a mercury battery.)
13. Balance these redox reactions, and identify the oxidizing agent and the reducing agent.
    (a) $FeO(s) + O_3(g) \longrightarrow Fe_2O_3(s)$
    (b) $P_4(s) + Br_2(\ell) \longrightarrow PBr_5(\ell)$
    (c) $H_2O_2(aq) + Co^{2+}(aq) \longrightarrow H_2O(\ell) + Co^{3+}(aq)$ in acidic solution
    (d) $Cl^-(aq) + Cr_2O_7^{2-}(aq) \longrightarrow Cl_2(g) + Cr^{3+}(aq)$ in acidic solution
    (e) $CuFeS_2(s) + O_2(g) \longrightarrow Cu_2S(s) + FeO(s) + SO_2(g)$
    (f) $H_2CO(g) + O_2(g) \longrightarrow CO_2(g) + H_2O(\ell)$
    (g) $C_3H_8(g) + O_2(g) \longrightarrow CO_2(g) + H_2O(\ell)$ in acidic solution (This is the reaction in a propane fuel cell.)

Electrochemical Cells

14. For the redox reaction $Cu^{2+}(aq) + Zn(s) \longrightarrow Cu(s) + Zn^{2+}(aq)$, why can't you generate electric current by placing a piece of copper metal and a piece of zinc metal in a solution containing $CuCl_2(aq)$ and $ZnCl_2(aq)$?
15. Explain the function of a salt bridge in an electrochemical cell.
16. ■ A voltaic cell is assembled with Pb(s) and $Pb(NO_3)_2(aq)$ in one compartment and Zn(s) and $ZnCl_2(aq)$ in the other. An external wire connects the two electrodes, and a salt bridge containing KNO_3 connects the two solutions.
    (a) In the product-favored reaction, zinc metal is oxidized to Zn^{2+}. Write a balanced net ionic equation for this reaction.
    (b) Which half-reaction occurs at each electrode? Which is the anode and which is the cathode?
    (c) Draw a diagram of the cell, indicating the direction of electron flow outside the cell and of ion flow within the cell.
17. A voltaic cell is assembled with Sn(s) and $Sn(NO_3)_2(aq)$ in one compartment and Ag(s) and $AgNO_3(aq)$ in the other. An external wire connects the two electrodes, and a salt bridge containing KNO_3 connects the two solutions.
    (a) In the product-favored reaction, Ag^+ is reduced to silver metal. Write a balanced net ionic equation for this reaction.
    (b) Which half-reaction occurs at each electrode? Which is the anode and which is the cathode?
    (c) Draw a diagram of the cell, indicating the direction of electron flow outside the cell and of ion flow within the cell.

Electrochemical Cells and Voltage

18. ■ You light a 25-W light bulb with the current from a 12-V lead-acid storage battery. After 1.0 h of operation, how much energy has the light bulb utilized? How many coulombs have passed through the bulb? Assume 100% efficiency. (A watt is the transfer of 1 J of energy in 1 s.)
19. Copper can reduce silver ion to metallic silver, a reaction that could, in principle, be used in a battery.
    $$Cu(s) + 2\ Ag^+(aq) \longrightarrow Cu^{2+}(aq) + 2\ Ag(s)$$
    (a) Write equations for the half-reactions involved.
    (b) Which half-reaction is an oxidation and which is a reduction? Which half-reaction occurs in the anode compartment and which takes place in the cathode compartment?

20. Chlorine gas can oxidize zinc metal in a reaction that has been suggested as the basis of a battery. Write the half-reactions involved. Label which is the oxidation half-reaction and which is the reduction half-reaction.

Using Standard Reduction Potentials

21. ■ What is the strongest oxidizing agent in Table 18.1? What is the strongest reducing agent? What is the weakest oxidizing agent? What is the weakest reducing agent?
22. Using the reduction potentials in Table 18.1 or Appendix I, place these elements in order of increasing ability to function as reducing agents:
(a) Cl_2 (b) Fe
(c) Ag (d) Na
(e) H_2
23. Using the reduction potentials in Table 18.1, place these elements in order of increasing ability to function as oxidizing agents:
(a) O_2 (b) H_2O_2
(c) $PbSO_4$ (d) H_2O
24. One of the most energetic redox reactions is that between F_2 gas and lithium metal.
(a) Write the half-reactions involved. Label which is the oxidation half-reaction and which is the reduction half-reaction.
(b) According to data from Table 18.1, what is E°_{cell} for this reaction?
25. ■ Calculate the value of E°_{cell} for each of these reactions. Decide whether each is product-favored.
(a) $I_2(s) + Mg(s) \longrightarrow Mg^{2+}(aq) + 2\,I^-(aq)$
(b) $Ag(s) + Fe^{3+}(aq) \longrightarrow Ag^+(aq) + Fe^{2+}(aq)$
(c) $Sn^{2+}(aq) + 2\,Ag^+(aq) \longrightarrow Sn^{4+}(aq) + 2\,Ag(s)$
(d) $2\,Zn(s) + O_2(g) + 2\,H_2O(\ell) \longrightarrow 2\,Zn^{2+}(aq) + 4\,OH^-(aq)$
26. Consider these half-reactions:

Half-Reaction	E° (V)
$Au^{3+}(aq) + 3\,e^- \longrightarrow Au(s)$	1.50
$Pt^{2+}(aq) + 2\,e^- \longrightarrow Pt(s)$	1.2
$Co^{2+}(aq) + 2\,e^- \longrightarrow Co(s)$	−0.28
$Mn^{2+}(aq) + 2\,e^- \longrightarrow Mn(s)$	−1.18

(a) Which is the weakest oxidizing agent?
(b) Which is the strongest oxidizing agent?
(c) Which is the strongest reducing agent?
(d) Which is the weakest reducing agent?
(e) Will Co(s) reduce $Pt^{2+}(aq)$ to Pt(s)?
(f) Will Pt(s) reduce $Co^{2+}(aq)$ to Co(s)?
(g) Which ions can be reduced by Co(s)?
27. ■ Consider these half-reactions:

Half-Reaction	E° (V)
$Ce^{4+}(aq) + e^- \longrightarrow Ce^{3+}(aq)$	1.61
$Ag^+(aq) + e^- \longrightarrow Ag(s)$	0.80
$Hg_2^{2+}(aq) + 2\,e^- \longrightarrow 2\,Hg(\ell)$	0.79
$Sn^{2+}(aq) + 2\,e^- \longrightarrow Sn(s)$	−0.14
$Ni^{2+}(aq) + 2\,e^- \longrightarrow Ni(s)$	−0.25
$Al^{3+}(aq) + 3\,e^- \longrightarrow Al(s)$	−1.66

(a) Which is the weakest oxidizing agent?
(b) Which is the strongest oxidizing agent?
(c) Which is the strongest reducing agent?
(d) Which is the weakest reducing agent?
(e) Will Sn(s) reduce $Ag^+(aq)$ to Ag(s)?
(f) Will $Hg(\ell)$ reduce $Sn^{2+}(aq)$ to Sn(s)?
(g) Name the ions that can be reduced by Sn(s).
(h) Which metals can be oxidized by $Ag^+(aq)$?

E° and Gibbs Free Energy

28. For each of the reactions in Question 25, compute the Gibbs free energy change, ΔG°.
29. ■ Hydrazine, N_2H_4, can be used as the reducing agent in a fuel cell.
$$N_2H_4(aq) + O_2(aq) \longrightarrow H_2(g) + 2\,H_2O(\ell)$$
(a) If ΔG° for the reaction is −598 kJ, calculate the value of E° expected for the reaction.
(b) Suppose the equation is written with all coefficients doubled. Determine ΔG° and E° for this new reaction.
30. The standard cell potential for the oxidation of Mg by Br_2 is 3.45 V.
$$Br_2(\ell) + Mg(s) \longrightarrow Mg^{2+}(aq) + 2\,Br^-(aq)$$
(a) Calculate ΔG° for this reaction.
(b) Suppose the equation is written with all coefficients doubled. Determine ΔG° and E° for this new reaction.
31. The standard cell potential, E°, for the reaction of Zn(s) and $Cl_2(g)$ is 2.12 V. Write the chemical equation for the reaction of 1 mol zinc. What is the standard Gibbs free energy change, ΔG°, for this reaction?
32. What is the equilibrium constant K_c and ΔG° for the reaction between Cd(s) and $Cu^{2+}(aq)$?
33. ■ What is the equilibrium constant K_c and ΔG° for the reaction between $I_2(s)$ and $Br^-(aq)$?
34. Consider a voltaic cell with the following reaction. As the cell reaction proceeds, what happens to the values of E_{cell}, ΔG, and K_c? Explain your answers.
$$Cu^{2+}(aq, 1\ M) + Zn(s) \longrightarrow Cu(s) + Zn^{2+}(aq, 1\ M) \qquad E^\circ_{cell} = 1.10\ V$$
35. ■ Estimate the equilibrium constant K_c for this reaction.
$$Ni(s) + Co^{2+}(aq) \rightleftharpoons Ni^{2+}(aq) + Co(s) \qquad E^\circ_{cell} = +0.046\ V$$

Effect of Concentration on Cell Potential

36. Consider the voltaic cell
$$Zn(s) + Cd^{2+}(aq) \longrightarrow Zn^{2+}(aq) + Cd(s)$$
operating at 298 K.
(a) What is the E°_{cell} for this cell?
(b) If $E_{cell} = 0.390$ and (conc. Cd^{2+}) = 2.00 M, what is (conc. Zn^{2+})?
(c) If (conc. Cd^{2+}) = 0.068 M and (conc. Zn^{2+}) = 1.00 M, what is E_{cell}?
37. ■ Consider the voltaic cell
$$2\,Ag^+(aq) + Cd(s) \longrightarrow 2\,Ag(s) + Cd^{2+}(aq)$$
operating at 298 K.
(a) What is the E°_{cell} for this cell?
(b) If (conc. Cd^{2+}) = 2.0 M and (conc. Ag^+) = 0.25 M, what is E_{cell}?

(c) If E_{cell} = 1.25 V and (conc. Cd^{2+}) = 0.100 M, what is (conc. Ag^+)?

38. Consider the reaction

$$H_2(g) + Sn^{4+}(aq) \longrightarrow 2\ H^+(aq) + Sn^{2+}(aq)$$

operating at 298 K.

(a) What is the $E°_{cell}$ for this cell?

(b) What is the E_{cell} for P_{H_2} = 1.0 bar, (conc. Sn^{2+}) = 6.0 × 10^{-4} M, (conc. Sn^{4+}) = 5.0 × 10^{-4} M, and pH = 3.60?

39. What is the potential of an electrode made from zinc metal immersed in a solution where (conc. Zn^{2+}) = 0.010 M?

40. ■ For a voltaic cell with the reaction

$$Pb(s) + Sn^{2+}(aq) \longrightarrow Pb^{2+}(aq) + Sn(s)$$

at what ratio of concentrations of lead and tin ions will E_{cell} = 0?

Common Batteries

41. What are the advantages and disadvantages of lead-acid storage batteries?

42. ■ Nicad batteries are rechargeable and are commonly used in cordless appliances. Although such batteries actually function under basic conditions, imagine an electrochemical cell using this setup.

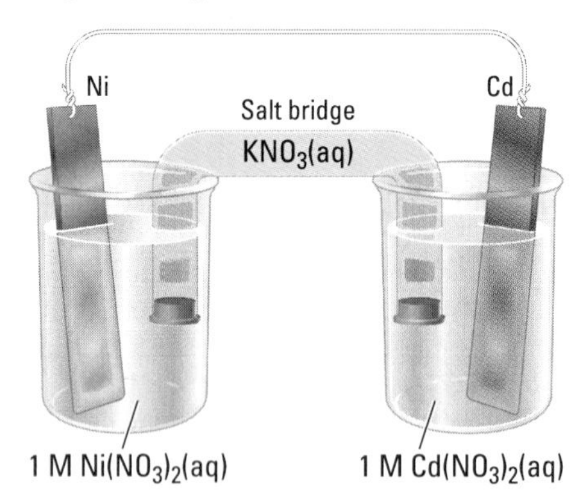

(a) Write a balanced net ionic equation depicting the reaction occurring in the cell.

(b) What is oxidized? What is reduced? What is the reducing agent and what is the oxidizing agent?

(c) Which is the anode and which is the cathode?

(d) What is $E°$ for the cell?

(e) What is the direction of electron flow in the external wire?

(f) If the salt bridge contains KNO_3, toward which compartment will the NO_3^- ions migrate?

43. Consider the nicad cell in Question 42.

(a) If the concentration of Cd^{2+} is reduced to 0.010 M, and (conc. Ni^{2+}) = 1.0 M, will the cell emf be smaller or larger than when the concentration of Cd^{2+}(aq) was 1.0 M? Explain your answer in terms of Le Chatelier's principle.

(b) Begin with 1.0 L of each of the solutions, both initially 1.0 M in dissolved species. Each electrode weighs 50.0 g at the start. If 0.050 A is drawn from the battery, how long can it last?

Fuel Cells

44. How does a fuel cell differ from a battery?

45. Describe the principal parts of an H_2/O_2 fuel cell. What is the reaction at the cathode? At the anode? What is the product of the fuel cell reaction?

46. ■ Hydrazine, N_2H_4, has been proposed as the fuel in a fuel cell in which oxygen is the oxidizing agent. The reactions are

$$N_2H_4(aq) + 4\ OH^-(aq) \longrightarrow N_2(g) + 4\ H_2O(\ell) + 4\ e^-$$
$$O_2(g) + 2\ H_2O(\ell) + 4\ e^- \longrightarrow 4\ OH^-(aq)$$

(a) Which reaction occurs at the anode and which at the cathode?

(b) What is the net cell reaction?

(c) If the cell is to produce 0.50 A of current for 50.0 h, what mass in grams of hydrazine must be present?

(d) What mass in grams of O_2 must be available to react with the mass of N_2H_4 determined in part (c)?

Electrolysis—Reactant-Favored Redox Reactions

47. Consider the electrolysis of water in the presence of very dilute H_2SO_4. What species is produced at the anode? At the cathode? What are the relative amounts of the species produced at the two electrodes?

48. From Table 18.1 write down all of the aqueous metal ions that can be reduced by electrolysis to the corresponding metal.

49. From Table 18.1 write down all of the aqueous species that can be oxidized by electrolysis, and determine the products.

50. What are the products of the electrolysis of a 1 M aqueous solution of NaBr? What species are present in the solution? What is formed at the cathode? What is formed at the anode?

51. ■ For each of these solutions, tell what reactions take place at the anode and at the cathode during electrolysis.

(a) $NiBr_2$(aq) (b) NaI(aq)
(c) $CdCl_2$(aq) (d) CuI_2(aq)
(e) MgF_2(aq) (f) HNO_3(aq)

Counting Electrons

52. ■ A current of 0.015 A is passed through a solution of $AgNO_3$ for 155 min. What mass of silver is deposited at the cathode?

53. A current of 2.50 A is passed through a solution of $Cu(NO_3)_2$ for 2.00 h. What mass of copper is deposited at the cathode?

54. A current of 0.0125 A is passed through a solution of $CuCl_2$ for 2.00 h. What mass of copper is deposited at the cathode and what volume of Cl_2 gas (in mL at STP) is produced at the anode?

55. The vanadium(II) ion can be produced by electrolysis of a vanadium(III) salt in solution. How long must you carry out an electrolysis if you wish to convert completely 0.125 L of 0.0150 M V^{3+}(aq) to V^{2+}(aq) using a current of 0.268 A?

56. The reactions occurring in a lead-acid storage battery are given in Section 18.8. A typical battery might be rated at 50. ampere-hours (A-h). This means that it has the capacity to deliver 50. A for 1.0 h or 1.0 A for 50. h. If it does deliver 1.0 A for 50. h, what mass of lead would be consumed?

57. An effective battery can be built using the reaction between Al metal and O_2 from the air. If the Al anode of this battery consists of a 3-oz piece of aluminum (84 g), for how many hours can the battery produce 1.0 A of electricity?

58. Assume that the anode reaction for the lithium battery is

$$Li(s) \longrightarrow Li^+(aq) + e^-$$

and the anode reaction for the lead-acid storage battery is

$$Pb(s) + HSO_4^-(aq) + H_2O(\ell) \longrightarrow PbSO_4(s) + 2\ e^- + H_3O^+(aq)$$

Compare the masses of metals consumed when each of these batteries supplies a current of 1.0 A for 10. min.

59. A hydrogen-oxygen fuel cell operates on the simple reaction

$$2\ H_2(g) + O_2(g) \longrightarrow 2\ H_2O(\ell)$$

If the cell is designed to produce 1.5 A of current, how long can it operate if there is an excess of oxygen and only sufficient hydrogen to fill a 1.0-L tank at 200. bar pressure at 25 °C?

60. ■ How long would it take to electroplate a metal surface with 0.500 g nickel metal from a solution of Ni^{2+} with a current of 4.00 A?

61. How much current is required to electroplate a metal surface with 0.400 g chromium metal from a solution of Cr^{3+} in 1.00 h?

Corrosion—Product-Favored Redox Reactions

62. Explain how rust is formed from iron materials by corrosion.

63. Why does iron corrode faster in salt water than in fresh water?

64. What common metal does not corrode readily under normal conditions?

65. Why does coating a steel object with chromium stop corrosion of the iron?

66. Explain how galvanizing iron stops corrosion of the underlying iron.

General Questions

67. A 12-V automobile battery consists of six cells of the type described in Section 18.8. The cells are connected in series so that the same current flows through all of them. Calculate the theoretical minimum electrical potential difference needed to recharge an automobile battery. (Assume standard-state concentrations.) How does this compare with the maximum voltage that could be delivered by the battery? Assuming that the lead plates in an automobile battery each weigh 2.50 kg and that sufficient PbO_2 is available, what is the maximum possible work that could be obtained from the battery?

68. Three electrolytic cells are connected in series, so that the same current flows through all of them for 20. min. In cell A, 0.0234 g Ag plates from a solution of $AgNO_3(aq)$; cell B contains $Cu(NO_3)_2(aq)$; cell C contains $Al(NO_3)_3(aq)$. What mass of Cu will plate in cell B? What mass of Al will plate in cell C?

69. Fluorinated organic compounds are important commercially; they are used as herbicides, flame retardants, and fire-extinguishing agents, among other things. A reaction such as

$$CH_3SO_2F + 3\ HF \longrightarrow CF_3SO_2F + 3\ H_2$$

is actually carried out electrochemically in liquid HF as the solvent.

(a) Draw the structural formula for CH_3SO_2F. (S is the "central" atom with the O atoms, F atom, and CH_3 group bonded to it.) What is the geometry around the S atom? What are the O—S—O and O—S—F bond angles?

(b) If you electrolyze 150. g CH_3SO_2F, how many grams of HF are required and how many grams of each product can be isolated?

(c) Is H_2 produced at the anode or the cathode of the electrolysis cell?

(d) A typical electrolysis cell operates at 8.0 V and 250 A. How many kilowatt-hours of energy does one such cell consume in 24 h?

Applying Concepts

70. Four metals A, B, C, and D exhibit these properties:

(a) Only A and C react with 1.0 M HCl to give H_2 gas.

(b) When C is added to solutions of ions of the other metals, metallic A, B, and D are formed.

(c) Metal D reduces B^{n+} ions to give metallic B and D^{n+} ions.

On the basis of this information, arrange the four metals in order of increasing ability to act as reducing agents.

71. The table below lists the cell potentials for the ten possible electrochemical cells assembled from the elements A, B, C, D, and E and their respective ions in solutions. Using the data in the table, establish a standard reduction potential table similar to Table 18.1. Assign a reduction potential of 0.00 V to the element that falls in the middle of the series.

	A(s) in A^{n+}(aq)	**B(s) in B^{n+}(aq)**
E(s) in E^{n+}(aq)	+0.21 V	+0.68 V
D(s) in D^{n+}(aq)	+0.35 V	+1.24 V
C(s) in C^{n+}(aq)	+0.58 V	+0.31 V
B(s) in B^{n+}(aq)	+0.89 V	—

	C(s) in C^{n+}(aq)	**D(s) in D^{n+}(aq)**
E(s) in E^{n+}(aq)	+0.37 V	+0.56 V
D(s) in D^{n+}(aq)	+0.93 V	—
C(s) in C^{n+}(aq)	—	—
B(s) in B^{n+}(aq)	—	—

72. When this electrochemical cell runs for several hours, the green solution gets lighter and the yellow solution gets darker.

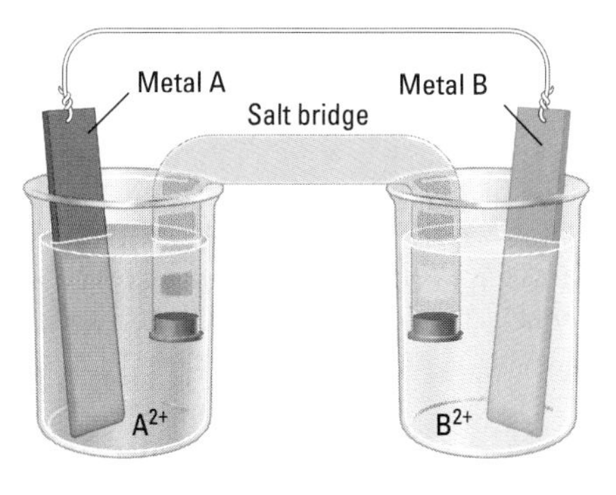

(a) What is oxidized, and what is reduced?
(b) What is the oxidizing agent, and what is the reducing agent?
(c) What is the anode, and what is the cathode?
(d) Write equations for the half-reactions.
(e) Which metal gains mass?
(f) What is the direction of the electron transfer through the external wire?
(g) If the salt bridge contains $KNO_3(aq)$, into which solution will the K^+ ions migrate?

73. An electrolytic cell is set up with Cd(s) in $Cd(NO_3)_2(aq)$ and Zn(s) in $Zn(NO_3)_2(aq)$. Initially both electrodes weigh 5.00 g. After running the cell for several hours the electrode in the left compartment weighs 4.75 g.
(a) Which electrode is in the left compartment?
(b) Does the mass of the electrode in the right compartment increase, decrease, or stay the same? If the mass changes, what is the new mass?
(c) Does the mass of the solution in the right compartment increase, decrease, or stay the same?
(d) Does the volume of the electrode in the right compartment increase, decrease, or stay the same? If the volume changes, what is the new volume? (The density of Cd is 8.65 g/cm^3.)

74. When H_2O_2 is mixed with Fe^{2+}, which redox reaction will occur—the oxidation of Fe^{2+} to Fe^{3+} or the reduction of Fe^{2+} to Fe? What are the $E°_{cell}$ values for the electrochemical cells corresponding to the two reactions?

75. The permanganate ion MnO_4^- can be reduced to the manganese(II) ion Mn^{2+} in aqueous acidic solution, and the reduction potential for this half-cell reaction is 1.52 V. If this half-cell is combined with a Zn^{2+}/Zn half-cell to form a galvanic cell at standard conditions,
(a) Write the chemical equation for the half-reaction occurring at the anode.
(b) Write the chemical equation for the half-reaction occurring at the cathode.
(c) Write the overall balanced equation for the reaction.
(d) Calculate the cell voltage.

More Challenging Questions

76. Fluorine, F_2, is made by the electrolysis of anhydrous HF.

$$2\ HF(\ell) \longrightarrow H_2(g) + F_2(g)$$

Typical electrolysis cells operate at 4000 to 6000 A and 8 to 12 V. A large-scale plant can produce about 9.0 metric tons of F_2 gas per day.
(a) What mass in grams of HF is consumed?
(b) Using the conversion factor of 3.60×10^6 J/kWh, how much energy in kilowatt-hours is consumed by a cell operating at 6.0×10^3 A at 12 V for 24 h?

77. What reaction would take place if a 1.0 M solution of $Cr_2O_7^{2-}$ was added to a 1.0 M solution of HBr?

78. If Cl_2 and Br_2 are added to an aqueous solution that contains Cl^- and Br^-, what product-favored reaction will occur?

79. This reaction occurs in a cell with $H_2(g)$ pressure of 1.0 atm and (conc. Cl^-) = 1.0 M at 25 °C; the measured E_{cell} = 0.34 V. What is the pH of the solution?

$$2\ H_2O(\ell) + 2\ H_2(g) + 2\ AgCl(s) \longrightarrow 2\ H_3O^+(aq) + 2\ Cl^-(aq) + 2\ Ag(s)$$

80. An electric current of 2.00 A was passed through a platinum salt solution for 3.00 hours, and 10.9 g of metallic platinum was formed at the cathode. What is the charge on the platinum ions in the solution?

81. E_{cell} = 0.010 V for a galvanic cell with this reaction at 25 °C.

$$Sn(s) + Pb^{2+}(aq) \longrightarrow Sn^{2+}(aq) + Pb(s)$$

(a) What is the equilibrium constant K_c for the reaction?
(b) If a solution with (conc. Pb^{2+}) = 1.1 M had excess tin metal added to it, what would the equilibrium concentration of Sn^{2+} and Pb^{2+} be?

82. You wish to electroplate a copper surface having an area of 1200 mm^2 with a 1.0-μm-thick coating of silver from a solution of $Ag(CN)_2^-$ ions. If you use a current of 150.0 mA, what electrolysis time should you use? The density of metallic silver is 10.5 g/cm^3.

83. A student wanted to measure the copper(II) concentration in an aqueous solution. For the cathode half-cell she used a silver electrode with a 1.00 M solution of $AgNO_3$. For the anode half-cell she used a copper electrode dipped into the aqueous sample. If the cell gave E_{cell} = 0.62 V at 25 °C, what was the copper(II) concentration of the solution?

84. In a mercury battery, the anode reaction is

$$Zn(s) + 2\ OH^-(aq) \longrightarrow ZnO(aq) + H_2O(\ell) + 2\ e^-$$

and the cathode reaction is

$$HgO(s) + H_2O(\ell) + 2\ e^- \longrightarrow Hg(\ell) + 2\ OH^-(aq)$$

The cell potential is 1.35 V. How many hours can such a battery provide power at a rate of 4.0×10^{-4} watt (1 watt = 1 J s^{-1}) if 1.25 g HgO is available?

Conceptual Challenge Problems

CP18.A (Section 18.6) Most automobiles run on internal combustion engines, in which the energy used to run the vehicle is obtained from the combustion of gasoline. The main component of gasoline is octane, C_8H_{18}. An automobile manufacturer has recently announced a chemical method for generating hydrogen gas from gasoline and proposes to develop a car in which an H_2/O_2 fuel cell powers an electric propulsion motor, thus eliminating the internal combustion engine with its problems (for example, the generation of unwanted by-products that pollute the air). The hydrogen for the fuel cell would be directly generated from gasoline on board the vehicle. There are two steps in this hydrogen generation process:
(i) Partial oxidation of octane by oxygen to carbon monoxide and hydrogen
(ii) Combination of carbon monoxide with additional gaseous water to form carbon dioxide and more hydrogen (the water-gas shift reaction)

(a) Write the chemical equation for the complete combustion of 1 mol octane.
(b) Write balanced chemical equations for the two-step hydrogen generation process. How many moles of H_2 are produced per mole of octane? (Remember that water is a reactant in the two-step process.)
(c) By combining these equations, show that the net *overall* reaction is the same as in the combustion of octane.
(d) Assuming that the entire Gibbs free energy change of the H_2/O_2 fuel cell reaction is available for use by the electric

propulsion motor, calculate the energy produced by a fuel cell when it consumes all of the hydrogen produced from 1 mol of octane. Compare this energy with the Gibbs free energy change for the combustion of 1 mol of octane. (*Note:* The Gibbs free energy of formation, ΔG_f°, for $C_8H_{18}(\ell)$ is 6.14 kJ/mol.)

CP18.B (Section 18.4) People obtain energy by oxidizing food. Glucose is a typical foodstuff. This carbohydrate is oxidized to water and carbon dioxide.

$$C_6H_{12}O_6(aq) + 6\ O_2(g) \longrightarrow 6\ CO_2(g) + 6\ H_2O(\ell)$$

The heat of combustion of glucose is 2.80×10^3 kJ/mol, which means that as glucose is oxidized, its electrons lose 2.80×10^3 kJ/mol as they give up potential energy in a complicated series of chemical steps.

(a) Assume that a person requires 2400 food Calories per day and that this energy is obtained from the oxidation of glucose. How much O_2 must a person breathe each day to react with this much glucose?

(b) Each mole of O_2 requires 4 mol electrons, regardless of whether the O atoms become part of CO_2 or H_2O. What would be the average electric current (C/s) in a human body using the above amount of energy described in part (a) per day?

(c) Use the answer from part (b) and calculate the electrical potential this current flows through in a day to produce the 2400 food Calories. (1 Calorie = 4.18 kJ)

CP18.C (Section 18.10) A piece of chromium metal is attached to a battery and dipped into 50 mL of 0.3 M KOH solution in a 250-mL beaker. A stainless steel electrode is connected to the other electrode of the battery and immersed in the same solution. A steady current of 0.50 A is maintained for exactly 2 hours. Several samples of a gas formed at the stainless steel electrode during the electrolysis are captured, and all are found to ignite in air. After the electrolysis, the chromium electrode is weighed and found to have decreased in weight by 0.321 g. The mass of the stainless steel electrode does not change.

After electrolysis, the KOH solution is neutralized with nitric acid to a pH of slightly less than 7, then is heated and reacted with 0.151 M lead(II) nitrate solution. As the lead(II) nitrate solution is added, a yellow precipitate quickly forms from the hot solution. The formation of precipitate stops after 40.4 mL of the lead(II) nitrate solution has been added. The yellow solid is then filtered, dried, and weighed. Its mass is 1.97 g.

(a) How much electrical charge passes through the cell?

(b) How many moles of Cr react?

(c) What is the oxidation state of the Cr after reacting?

(d) Assuming that the yellow compound that precipitates from the solution during the titration contains both Pb and Cr, what do you conclude to be the ratio of the numbers of atoms of Pb and Cr?

(e) If the yellow compound contains an element other than Pb and Cr, what is it and how much is in the compound? What is the formula for the yellow compound?

19

Nuclear Chemistry

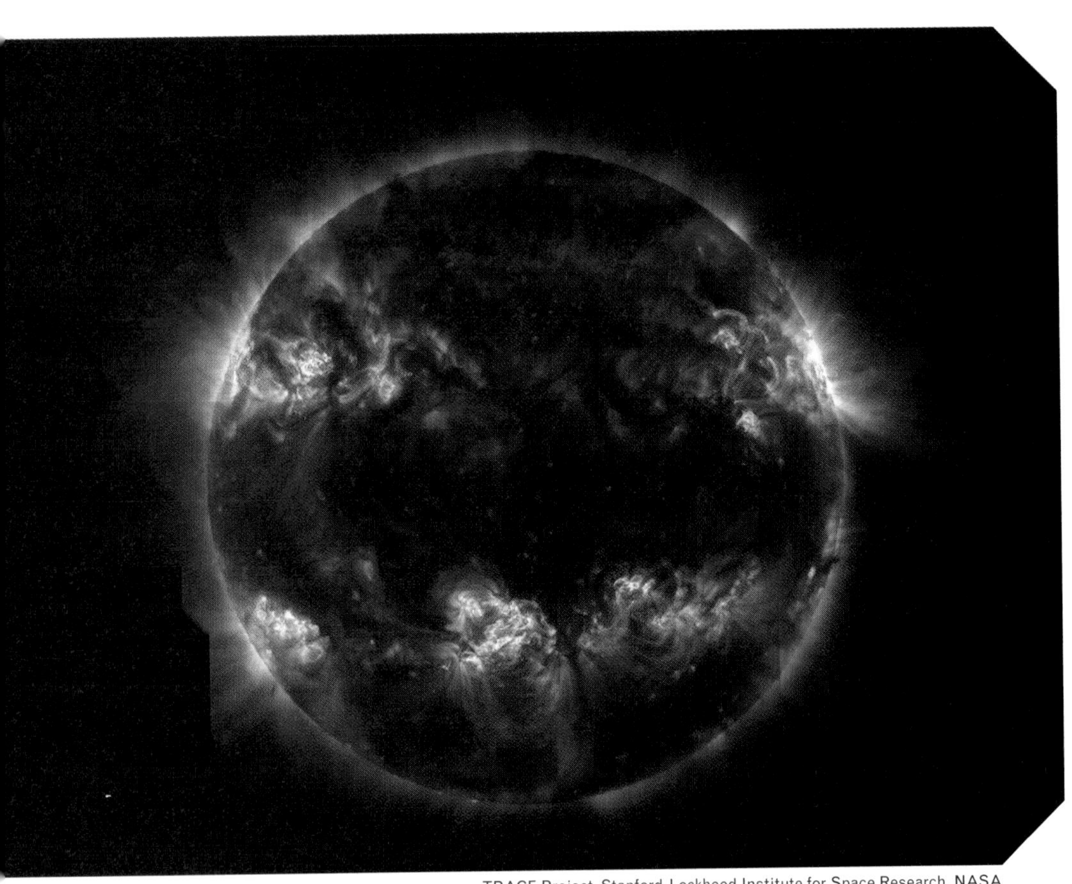

TRACE Project, Stanford-Lockheed Institute for Space Research, NASA

Nuclear fusion reactions are the source of energy in the sun, whose surface is shown here. Such reactions occurring in the sun are the ultimate source of almost all of the energy available on earth. During fusion, light nuclei such as $^{1}_{1}H$ combine to form heavier nuclei such as $^{4}_{2}He$, and tremendous energy is released. Temperatures of 10^6 to 10^7 K are required for fusion to occur.

Nuclear chemistry, a subject that bridges chemistry and physics, has a significant impact on our society. No matter what your reason for taking a college course in chemistry—to prepare for a career in one of the sciences or simply to gain knowledge as a concerned citizen—you should know about nuclear chemistry. Radioactive isotopes are widely used in medicine. Your room may be protected by a smoke detector that uses a radioactive element as part of its sensor, and research in all fields of science employs radioactive elements and their compounds. More than 30 nations use commercial nuclear reactors to produce electricity. This chapter considers several aspects of nuclear chemistry: changes in atomic nuclei and their effects, fission and fusion of nuclei and the energy that can be derived from such changes, and units used to measure radioactivity.

19.1 The Nature of Radioactivity

In 1896 French physicist Antoine Henri Becquerel made an unexpected observation that led to the discovery of radioactivity. In a dark drawer, he stored a photographic plate wrapped in black paper along with a uranium salt. When the image of the uranium salt appeared on the plate, he realized that radiation from the uranium salt had darkened the plate even though it had not been exposed to sunlight. He later found that uranium metal produced the same effect. Failing to find the reason for this effect, he gave the project to his graduate student Marie Curie, who, along with her husband Pierre, a physicist, studied the phenomenon intensively and termed it *radioactivity*.

One of Marie Curie's first findings was to confirm Becquerel's observation that uranium metal itself was radioactive and that the degree to which a uranium-containing sample was radioactive depended on the percentage of uranium present. When she tested pitchblende, a common ore containing uranium and other metals (such as lead, bismuth, and copper), she was surprised to find that it was even more radioactive than pure uranium. Only one explanation was possible: pitchblende contained an element (or elements) more radioactive than uranium. Eventually, the Curies discovered an element they named *polonium* after Marie's homeland of Poland. They also discovered radium, another highly radioactive element.

For more on experiments done by Becquerel and the Curies, see Walton, H. F. *Journal of Chemical Education*, Vol. 69, 1992; p. 10.

At about the same time in England, Sir J. J. Thomson and his graduate student Ernest Rutherford were studying the radiation from uranium and thorium. Rutherford found that "There are present at least two distinct types of radiation—one that is readily absorbed, which will be termed for convenience alpha (α) radiation, and the other of a more penetrative character, which will be termed beta (β) radiation." **Alpha radiation,** he discovered, was composed of particles that, when passed through an electric field, were attracted to the negative side of the field ***(⇐ p. 36).*** Indeed, his later studies showed these **alpha (α) particles** to be helium nuclei, ${}^4_2He^{2+}$, which were ejected at high speeds from a radioactive element (Table 19.1). Alpha particles have limited penetrating power and can be absorbed by skin, clothing, or several sheets of paper.

Encouraged by the Curies, Becquerel returned to the study of radiation. He found that the radiation from uranium was affected by magnetic fields and consisted of two kinds of particles, which we now know to be alpha and beta particles.

In the same experiment with electric fields, Rutherford found that **beta radiation** must be composed of negatively charged particles, since the beam of beta radiation was attracted to the electrically positive plate of an electric field. Later work by Becquerel showed that these particles have an electric charge and mass equal to those of an electron. Thus, **beta (β) particles** are electrons ejected at high speeds from radioactive nuclei. They are more penetrating than alpha particles (Table 19.1), and a $\frac{1}{8}$-inch-thick piece of aluminum will to stop them. Beta particles can penetrate 1 to 2 cm into living bone or tissue.

A third type of radiation was later discovered by P. Villard, a Frenchman, who named it **gamma (γ) radiation**, using the third letter in the Greek alphabet in keeping with

Table 19.1 Characteristics of α, β, and γ Emissions

Name	Symbol	Charge	Mass (g/particle)	Penetrating Power*
Alpha	${}^4_2He^{2+}$, ${}^4_2\alpha$, ${}^4_2He$	2+	6.65×10^{-24}	0.03 mm
Beta	${}^{0}_{-1}e$, ${}^{0}_{-1}\beta$	1−	9.11×10^{-28}	2 mm
Gamma	${}^0_0\gamma$, γ	0	0	10 cm

*Distance at which half the radiation has been stopped by water.

Rutherford's scheme. Unlike alpha and beta particles, which are particulate in nature, gamma rays are a form of electromagnetic radiation and are not affected by an electric field. Gamma radiation is the most penetrating, and it can pass completely through the human body. Thick layers of lead or concrete are required to stop a beam of gamma rays.

19.2 Nuclear Reactions

Equations for Nuclear Reactions

Ernest Rutherford found that radium not only emits alpha particles but also produces the radioactive gas radon in the process. Such observations led Rutherford and Frederick Soddy, in 1902, to propose the revolutionary theory that *radioactivity is the result of a natural change of a radioactive isotope of one element into an isotope of a different element.* In such changes, called **nuclear reactions** or *transmutations,* an unstable nucleus (the *parent nucleus*) spontaneously emits radiation and is converted (decays) into a more stable nucleus of a different element (the *daughter product*). Thus, a nuclear reaction results in a change in atomic number and, in some cases, a change in mass number as well. For example, the reaction of radium studied by Rutherford can be written as

When a radioactive atom decays, the emission of a charged particle leaves behind a charged atom. Thus, when radium-226 decays, it gives a helium-4 cation ($^{4}_{2}He^{2+}$) and a radon-222 anion ($^{222}_{86}Rn^{2-}$). By convention, the ion charges are not shown in balanced nuclear equations.

$$^{226}_{88}Ra \longrightarrow ^{4}_{2}He + ^{222}_{86}Rn$$

In this representation, the subscripts are the atomic numbers (number of protons) and the superscripts are the mass numbers (sum of protons plus neutrons).

In a chemical change, the atoms in molecules and ions are rearranged, but atoms are neither created nor destroyed; the number of atoms remains the same. Similarly, in nuclear reactions the total number of nuclear particles, or **nucleons** (protons and neutrons), remains the same. The essence of nuclear reactions, however, is that one nucleon can change into a different nucleon along with the release of energy. A proton can change to a neutron or a neutron can change to a proton, but the total number of nucleons remains the same. Therefore, ***the sum of the mass numbers of reacting nuclei must equal the sum of the mass numbers of the nuclei produced.*** Furthermore, because electrical charge cannot be created or destroyed, ***the sum of the atomic numbers of the products must equal the sum of the atomic numbers of the reactants.*** These principles can be verified for the nuclear equations that follow.

Recall that the atomic number is the number of protons in an atom's nucleus (that is, the total positive charge on the nucleus), and the mass number is the sum of protons and neutrons in a nucleus.

Alpha and Beta Particle Emission

One way a radioactive isotope can decay is to eject an alpha particle from the nucleus. This is illustrated by the radium-226 reaction above and by the conversion of uranium-234 to thorium-230 by alpha emission.

	$^{234}_{92}U$	$\longrightarrow$	$^{4}_{2}He$	+	$^{230}_{90}Th$
	uranium-234 (parent nucleus)		alpha particle		thorium-230 (daughter product)
Mass number:	234	$\longrightarrow$	4	+	230
Atomic number:	92	$\longrightarrow$	2	+	90

In alpha emission, *the atomic numbers of the parent nucleus and the daughter nucleus differ by two units and the mass numbers differ by four units for each alpha particle emitted.*

Emission of a beta particle is another way for a radioactive isotope to decay. For example, loss of a beta particle by uranium-239 (parent nucleus) to form neptunium-239 (daughter product) is represented by

Note that in beta decay the mass number is unchanged; the atomic number increases by one unit.

	$^{239}_{92}U$	$\longrightarrow$	$^{0}_{-1}e$	+	$^{239}_{93}Np$
	uranium-239		beta particle		neptunium-239
Mass number:	239	$\longrightarrow$	0	+	239
Atomic number:	92	$\longrightarrow$	−1	+	93

How does a nucleus, composed only of protons and neutrons, increase its number of protons by ejecting an electron during beta emission? It is generally accepted that a series of reactions is involved, but the net process is

$$\underset{\text{neutron}}{{}^{1}_{0}\text{n}} \longrightarrow \underset{\text{electron}}{{}^{0}_{-1}\text{e}} + \underset{\text{proton}}{{}^{1}_{1}\text{p}}$$

where we use the symbol p for a proton and n for a neutron. In this process, a neutron is converted to a proton and a beta particle is released. Therefore, *the ejection of a beta particle always means that a different element is formed because a neutron has been converted into a proton. The new element (daughter product) has an atomic number one unit greater (one additional proton) than that of the decaying (parent) nucleus.* The mass number does not change, however, because no proton or neutron has been emitted.

If a neutron changes to a proton, conservation of charge requires that a negative particle (a beta particle) be created.

In many cases, the emission of an alpha or beta particle results in the formation of a product nucleus that is also unstable and therefore radioactive. The new radioactive product may undergo a number of successive transformations until a stable, nonradioactive nucleus is finally produced. Such a series of reactions is called a **radioactive series.** One such series begins with uranium-238 and ends with lead-206, as illustrated in Figure 19.1. The first step in the series is

A nucleus formed as a result of alpha or beta emission is often in an excited state and therefore emits a gamma ray.

$${}^{238}_{92}\text{U} \longrightarrow {}^{4}_{2}\text{He} + {}^{234}_{90}\text{Th}$$

The final step, the conversion of polonium-210 to lead-206, is

$${}^{210}_{84}\text{Po} \longrightarrow {}^{4}_{2}\text{He} + {}^{206}_{82}\text{Pb}$$

PROBLEM-SOLVING EXAMPLE 19.1 Radioactive Series

An intermediate species in the uranium-238 decay series shown in Figure 19.1 is polonium-218. It emits an alpha particle, followed by emission of a beta particle, followed by the emission of a beta particle. Write the nuclear equations for these three reactions.

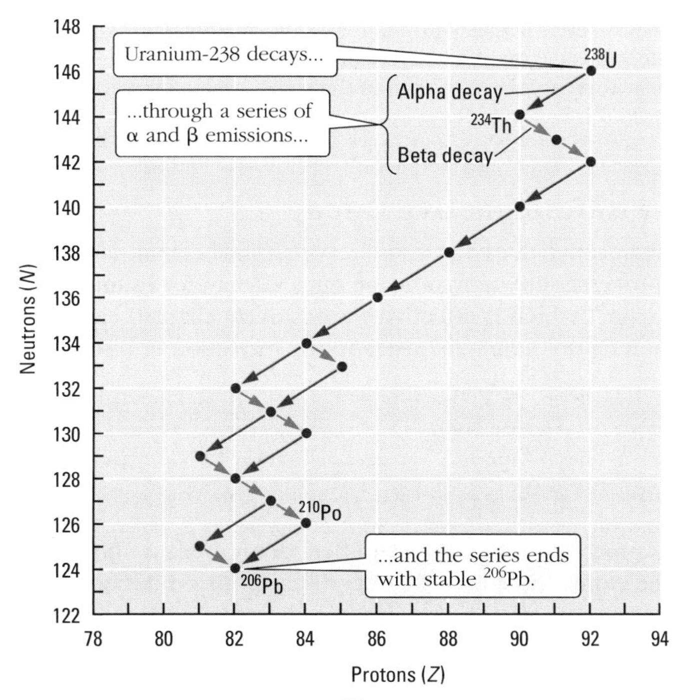

Figure 19.1 The ^{238}U decay series.

Answer $^{218}_{84}\text{Po} \longrightarrow {}^{4}_{2}\text{He} + {}^{214}_{82}\text{Pb}$

$^{214}_{82}\text{Pb} \longrightarrow {}^{0}_{-1}\text{e} + {}^{214}_{83}\text{Bi}$

$^{214}_{83}\text{Bi} \longrightarrow {}^{0}_{-1}\text{e} + {}^{214}_{84}\text{Po}$

Strategy and Explanation The starting point of these linked reactions is polonium-218. When it emits an alpha particle, the atomic number decreases by two and the mass number decreases by four to produce lead-214.

$$^{218}_{84}\text{Po} \longrightarrow {}^{4}_{2}\text{He} + {}^{214}_{82}\text{Pb}$$

polonium-218 lead-214

When lead-214 emits a beta particle, the atomic number increases by one and the mass number remains constant to produce bismuth-214.

$$^{214}_{82}\text{Pb} \longrightarrow {}^{0}_{-1}\text{e} + {}^{214}_{83}\text{Bi}$$

lead-214 bismuth-214

Bismuth-214 emits a beta particle, so the atomic number increases by one and the mass number remains constant to produce polonium-214.

$$^{214}_{83}\text{Bi} \longrightarrow {}^{0}_{-1}\text{e} + {}^{214}_{84}\text{Po}$$

bismuth-214 polonium-214

☑ **Reasonable Answer Check** An alpha emission decreases the atomic number by two and decreases the mass number by four. Each beta emission increases the atomic number by one and leaves the mass number unchanged. Therefore, one alpha and two beta emissions would leave the atomic number unchanged and decrease the mass number by four, which is what our systematic analysis found.

PROBLEM-SOLVING PRACTICE 19.1

(a) Write an equation showing the emission of an alpha particle by an isotope of neptunium, $^{237}_{93}\text{Np}$, to produce an isotope of protactinium.
(b) Write an equation showing the emission of a beta particle by sulfur-35, $^{35}_{16}\text{S}$, to produce an isotope of chlorine.

EXERCISE 19.1 Radioactive Decay Series

The actinium decay series begins with uranium-235, $^{235}_{92}\text{U}$, and ends with lead-207, $^{207}_{82}\text{Pb}$. The first five steps involve the successive emission of α, β, α, α, and β particles. Identify the radioactive isotope produced in each of the steps, beginning with uranium-235.

Other Types of Radioactive Decay

In addition to radioactive decay by emission of alpha, beta, or gamma radiation, other nuclear decay processes are known. Some nuclei decay, for example, by emission of a **positron,** $^{0}_{+1}\text{e}$ or β^{+}, which is effectively a positively charged electron. For example, positron emission by polonium-207 leads to the formation of bismuth-207.

The positron was discovered by Carl Anderson in 1932. It is sometimes called an "antielectron," one of a group of particles that have become known as "antimatter." Contact between an electron and a positron leads to mutual annihilation of both particles with production of two high-energy photons (gamma rays). This process is the basis of positron emission tomography (PET) scanning to detect tumors.

	$^{207}_{84}\text{Po}$	$\longrightarrow$	$^{0}_{+1}\text{e}$	+	$^{207}_{83}\text{Bi}$
	polonium-207		positron		bismuth-207
Mass number:	207	$\longrightarrow$	0	+	207
Atomic number:	84	$\longrightarrow$	+1	+	83

Notice that this process is the opposite of beta decay, because positron decay leads to a *decrease* in the atomic number. Like beta decay, positron decay does not change the mass number because no proton or neutron is ejected.

In electron capture, an electron and a proton combine to form a neutron. Because no proton or neutron is emitted, the mass number of the nucleus is unchanged.

Another nuclear process is **electron capture,** in which the atomic number is reduced by one but the mass number remains unchanged. In this process an inner-shell electron (for example, a 1*s* electron) is captured by the nucleus.

	$^{7}_{4}Be$ beryllium-7	+	$^{0}_{-1}e$ electron	⟶	$^{7}_{3}Li$ lithium-7
Mass number:	7	+	0	⟶	7
Atomic number:	4	+	−1	⟶	3

In the old nomenclature of atomic physics, the innermost shell (n = 1 principal quantum number) was called the K-shell, so the electron capture mechanism is sometimes called *K-capture*.

In summary, radioactive nuclei can decay in four ways, as summarized in the figure at right.

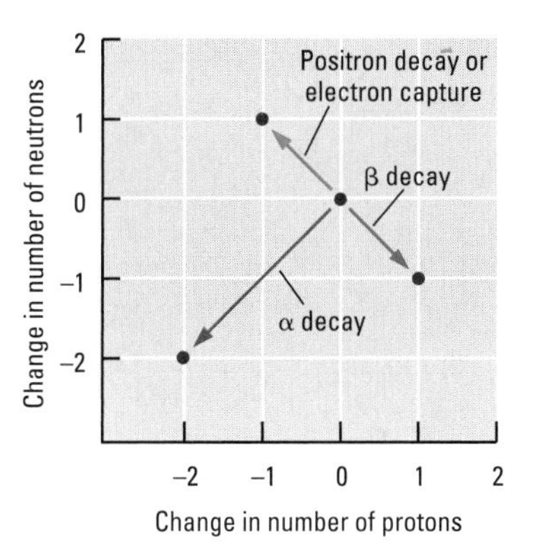

Effects of four radioactive decay processes. The chart shows the changes in the number of protons and neutrons during alpha decay, beta decay, positron emission, and electron capture.

PROBLEM-SOLVING EXAMPLE 19.2 Nuclear Equations

Complete these nuclear equations by filling in the missing symbol, mass number, and atomic number of the product species.

(a) $^{18}_{9}F \longrightarrow ^{18}_{8}O +$ ____   (b) $^{26}_{13}Al + ^{0}_{-1}e \longrightarrow$ ____

(c) $^{208}_{79}Au \longrightarrow ^{208}_{80}Hg +$ ____   (d) $^{218}_{84}Po \longrightarrow ^{4}_{2}He +$ ____

Answer

(a) $^{0}_{+1}e$ (b) $^{26}_{12}Mg$ (c) $^{0}_{-1}e$ (d) $^{214}_{82}Pb$

Strategy and Explanation In each case we deduce the missing species by comparing the atomic numbers and mass numbers before and after the reaction.

(a) The missing particle has a mass number of zero and a charge of +1, so it must be a positron, $^{0}_{+1}e$. When the positron is included in the equation, the atomic mass is 18 on each side, and the atomic numbers sum to 9 on each side.

(b) The missing nucleus must have a mass number of 26 + 0 = 26 and an atomic number of 13 − 1 = 12, so it is $^{26}_{12}Mg$.

(c) The missing particle has a mass number of zero and a charge of −1, so it must be a beta particle, $^{0}_{-1}e$.

(d) The missing nucleus has a mass number of 218 − 4 = 214 and an atomic number of 84 − 2 = 82, so it is $^{214}_{82}Pb$.

PROBLEM-SOLVING PRACTICE 19.2

Complete these nuclear equations by filling in the missing symbol, mass number, and atomic number of the product species.

(a) $^{11}_{6}C \longrightarrow ^{11}_{5}B + ?$ (b) $^{35}_{16}S \longrightarrow ^{35}_{17}Cl + ?$

(c) $^{30}_{15}P \longrightarrow ^{0}_{+1}e + ?$ (d) $^{22}_{11}Na \longrightarrow ^{0}_{-1}e + ?$

EXERCISE 19.2 Nuclear Reactions

Aluminum-26 can undergo either positron emission or electron capture. Write the balanced nuclear equation for each case.

19.3 Stability of Atomic Nuclei

The naturally occurring isotopes of elements from hydrogen to bismuth are shown in Figure 19.2, where the radioactive isotopes are represented by orange circles and the stable (nonradioactive) isotopes are represented by purple and green circles. It is surprising that so few stable isotopes exist. Why are there not hundreds more? To investigate this question, we will systematically examine the elements, starting with hydrogen.

In its simplest and most abundant form, hydrogen has only one nuclear particle, a single proton. In addition, the element has two other well-known isotopes: nonradioactive deuterium, with one proton and one neutron, $^{2}_{1}H$ = D, and radioactive tritium, with one proton and two neutrons, $^{3}_{1}H$ = T. Helium, the next element, has two protons

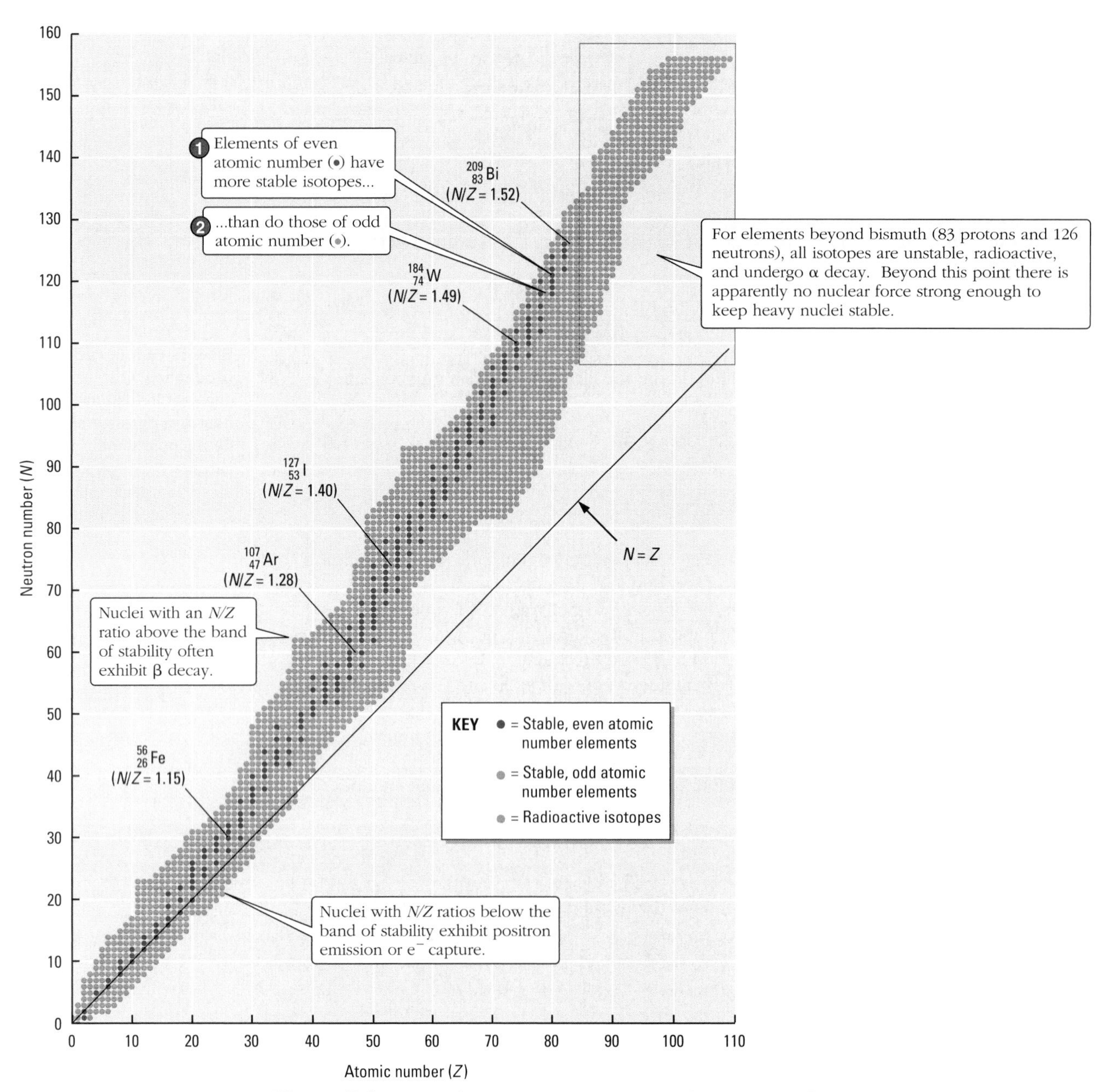

Figure 19.2 A plot of neutrons versus protons for the nuclei from hydrogen (Z = 1) through bismuth (Z = 83). A narrow band of stability is apparent. The N/Z values for some example stable nuclei are shown.

and two neutrons in its most stable isotope. At the end of the actinide series is element 103, lawrencium, one isotope of which has 154 neutrons and a mass number of 257. From hydrogen to lawrencium, except for $^{1}_{1}H$ and $^{3}_{2}He$, *the mass numbers of stable isotopes are always at least twice as large as the atomic number.* In other words, except for $^{1}_{1}H$ and $^{3}_{2}He$, every isotope of every element has a nucleus containing *at least* one neutron for every proton. Apparently the tremendous *repulsive* forces between the positively charged protons in the nucleus are moderated by the presence of neutrons, which have no electrical charge. Figure 19.2 illustrates a number of principles:

1. For light elements up to Ca (Z = 20), the stable isotopes usually have equal numbers of protons and neutrons, or perhaps one more neutron than protons. Examples include $^{7}_{3}Li$, $^{12}_{6}C$, $^{16}_{8}O$, and $^{32}_{16}S$.

2. Beyond calcium the neutron/proton ratio becomes increasingly greater than 1. The band of stable isotopes deviates more and more from the line $N = Z$ (number of neutrons = number of protons). It is evident that more neutrons are needed for nuclear stability in the heavier elements. For example, whereas one stable isotope of Fe has 26 protons and 30 neutrons ($N/Z = 1.15$), one stable isotope of platinum has 78 protons and 117 neutrons ($N/Z = 1.50$).
3. For elements beyond bismuth-209 (83 protons and 126 neutrons), all nuclei are unstable and radioactive. Furthermore, the rate of disintegration becomes greater the heavier the nucleus. For example, half of a sample of $^{238}_{92}U$ disintegrates in 4.5 billion years, whereas half of a sample of $^{256}_{103}Lr$ decays in only 28 seconds.
4. A careful analysis of Figure 19.2 reveals additional interesting features. First, elements with an even atomic number have a greater number of stable isotopes than do those with an odd atomic number. Second, stable isotopes usually have an even number of neutrons. For elements with an odd atomic number, the most stable isotope has an even number of neutrons. In fact, of the nearly 300 stable isotopes represented in Figure 19.2, roughly 200 have an even number of neutrons *and* an even number of protons. Only about 120 have an odd number of either protons or neutrons. Only four isotopes ($^{2}_{1}H$, $^{6}_{3}Li$, $^{10}_{5}B$, and $^{14}_{7}N$) have odd numbers of *both* protons and neutrons.

The Band of Stability and Type of Radioactive Decay

The narrow band of stable isotopes in Figure 19.2 (the purple and green circles) is sometimes called the *peninsula of stability* in a "sea of instability." Any nucleus not on this peninsula (the orange circles) will decay in such a way that the nucleus can come ashore on the peninsula. The chart can help us predict what type of decay will be observed.

The nuclei of all elements beyond Bi ($Z = 83$) are unstable—that is, radioactive—and most decay by *alpha particle emission.* For example, americium, the radioactive element used in smoke alarms, decays in this manner.

$$^{243}_{95}Am \longrightarrow {}^{4}_{2}He + {}^{239}_{93}Np$$

Beta emission occurs in isotopes that have too many neutrons to be stable—that is, isotopes above the peninsula of stability in Figure 19.2. When beta decay converts a neutron to a proton and an electron (beta particle), which is then ejected, the atomic number increases by one, and the mass number remains constant.

$$^{60}_{27}Co \longrightarrow {}^{0}_{-1}e + {}^{60}_{28}Ni$$

Conversely, lighter nuclei—below the peninsula of stability—that have too few neutrons attain stability by *positron emission* or by *electron capture,* because these processes convert a proton to a neutron in one step.

$$^{13}_{7}N \longrightarrow {}^{0}_{+1}e + {}^{13}_{6}C$$

$$^{41}_{20}Ca + {}^{0}_{-1}e \longrightarrow {}^{41}_{19}K$$

Decay by these two routes is observed for elements with atomic numbers ranging from 4 to greater than 100; as Z increases, electron capture becomes more likely than positron emission.

PROBLEM-SOLVING EXAMPLE 19.3 Nuclear Stability

For each of these unstable isotopes, write a nuclear equation for its probable mode of decay.

(a) Silicon-32, $^{32}_{14}Si$ (b) Titanium-43, $^{43}_{22}Ti$

(c) Plutonium-239, $^{239}_{94}Pu$ (d) Manganese-56, $^{56}_{25}Mn$

Answer

(a) $^{32}_{14}\text{Si} \longrightarrow {}^{0}_{-1}\text{e} + {}^{32}_{15}\text{P}$ (b) $^{43}_{22}\text{Ti} \longrightarrow {}^{0}_{+1}\text{e} + {}^{43}_{21}\text{Sc}$ or $^{43}_{22}\text{Ti} + {}^{0}_{-1}\text{e} \longrightarrow {}^{43}_{21}\text{Sc}$

(c) $^{239}_{94}\text{Pu} \longrightarrow {}^{4}_{2}\text{He} + {}^{235}_{92}\text{U}$ (d) $^{56}_{25}\text{Mn} \longrightarrow {}^{0}_{-1}\text{e} + {}^{56}_{26}\text{Fe}$

Strategy and Explanation Note the ratio of protons to neutrons. If there are excess neutrons, beta emission is probable. If there are excess protons, either electron capture or positron emission is probable. If the atomic number is greater than 83, then alpha emission is probable.

(a) Silicon-32 has excess neutrons, so beta decay is expected.
(b) Titanium-43 has excess protons, so either positron emission or electron capture is probable.
(c) Plutonium-239 has an atomic number greater than 83, so alpha decay is probable.
(d) Manganese-56 has excess neutrons, so beta decay is expected.

PROBLEM-SOLVING PRACTICE 19.3

For each of these unstable isotopes, write a nuclear equation for its probable mode of decay.

(a) $^{42}_{19}\text{K}$ (b) $^{234}_{92}\text{U}$ (c) $^{20}_{9}\text{F}$

Binding Energy

As demonstrated by Ernest Rutherford's alpha particle scattering experiment (*p. 38*), the nucleus of the atom is extremely small. Yet the nucleus can contain up to 83 protons before becoming unstable, suggesting that there must be a very strong short-range binding force that can overcome the electrostatic repulsive force of a number of protons packed into such a tiny volume. A measure of the force holding the nucleus together is the nuclear **binding energy.** This energy (E_b) is defined as the *negative* of the energy change (ΔE) that would occur if a nucleus were formed directly from its component protons and neutrons. For example, if a mole of protons and a mole of neutrons directly formed a mole of deuterium nuclei, the energy change would be more than 200 million kJ, the equivalent of exploding 73 tons of TNT.

The nuclear binding energy is similar to the bond energy for a chemical bond (*p. 285*) in that the binding energy is the energy that must be supplied to separate all of the particles (protons and neutrons) that make up the atomic nucleus and the bond energy is the energy that must be supplied to separate one mole of two bonded atoms. In both cases the energy change is positive, because work must be done to separate the particles.

$$^{1}_{1}\text{H} + {}^{1}_{0}\text{n} \longrightarrow {}^{2}_{1}\text{H} \qquad \Delta E = -2.15 \times 10^{8}\ \text{kJ}$$

$$\text{Binding energy} = E_b = -\Delta E = 2.15 \times 10^{8}\ \text{kJ}$$

This nuclear synthesis reaction is highly exothermic (so E_b is very positive), an indication of the strong attractive forces holding the nucleus together. The deuterium nucleus is more stable than an isolated proton and an isolated neutron.

The enormous energy released during the formation of atomic nuclei occurs because the mass of a nucleus is always slightly less than the sum of the masses of its constituent protons and neutrons.

$$\underset{1.007825\ \text{g/mol}}{^{1}_{1}\text{H}} + \underset{1.008665\ \text{g/mol}}{^{1}_{0}\text{n}} \longrightarrow \underset{2.01410\ \text{g/mol}}{^{2}_{1}\text{H}}$$

Change in mass $= \Delta m =$ mass of product $-$ sum of masses of reactants

$$\Delta m = 2.01410\ \text{g/mol} - 1.008665\ \text{g/mol} - 1.007825\ \text{g/mol}$$
$$\Delta m = 2.01410\ \text{g/mol} - 2.016490\ \text{g/mol}$$
$$\Delta m = -0.00239\ \text{g/mol} = -2.39 \times 10^{-6}\ \text{kg/mol}$$

The mass difference, Δm, is released as energy, which we describe as the binding energy.

The relationship between mass and energy is contained in Albert Einstein's 1905 theory of special relativity, which holds that mass and energy are simply different manifestations of the same quantity. Einstein stated that the energy of a body is equivalent to its mass times the square of the speed of light, $E = mc^2$. To calculate the energy change in a process in which the mass has changed, the equation becomes $\Delta E = (\Delta m)c^2$. For the formation of 1 mol deuterium nuclei from 1 mol protons and 1 mol neutrons, we have

1 J = 1 kg m^2 s^{-2}

$$\Delta E = (-2.39 \times 10^{-6}\,\text{kg})(3.00 \times 10^{8}\,\text{m/s})^2$$
$$= -2.15 \times 10^{11}\,\text{J} = -2.15 \times 10^{8}\,\text{kJ}$$

This is the value of ΔE given at the beginning of this section for the change in energy when a mole of protons and a mole of neutrons form a mole of deuterium nuclei.

Consider another example, the formation of a helium-4 nucleus from two protons and two neutrons.

$$2\,{}^{1}_{1}\text{H} + 2\,{}^{1}_{0}\text{n} \longrightarrow {}^{4}_{2}\text{He} \qquad E_b = +2.73 \times 10^{9}\,\text{kJ/mol}\ {}^{4}_{2}\text{He nuclei}$$

This binding energy, E_b, is very large, even larger than that for deuterium. To compare nuclear stabilities more directly, nuclear scientists generally calculate the **binding energy per nucleon.** Each ${}^{4}_{2}$He nucleus contains four nucleons—two protons and two neutrons. Therefore, 1 mol ${}^{4}_{2}$He atoms contains 4 mol nucleons.

$$E_b \text{ per mol nucleons} = \frac{2.73 \times 10^{9}\,\text{kJ}}{\text{mol}\ {}^{4}_{2}\text{He nuclei}} \times \frac{1\,\text{mol}\ {}^{4}_{2}\text{He nuclei}}{4\,\text{mol nucleons}}$$
$$= 6.83 \times 10^{8}\,\text{kJ/mol nucleons}$$

The greater the binding energy per nucleon, the greater the stability of the nucleus. The binding energies per nucleon are known for a great number of nuclei and are plotted as a function of mass number in Figure 19.3. It is very interesting and important that the point of maximum stability occurs in the vicinity of iron-56, ${}^{56}_{26}$Fe. This means that *all nuclei are thermodynamically unstable with respect to iron-56.* That is, very heavy nuclei can split, or *fission,* to form smaller, more stable nuclei with atomic numbers nearer to that of iron, while simultaneously releasing enormous quantities of energy (Section 19.5). In contrast, two very light nuclei can come together and undergo *nuclear fusion* exothermically to form heavier nuclei (Section 19.6). Because of its high nuclear stability, *iron is the most abundant of the heavier elements in the universe.*

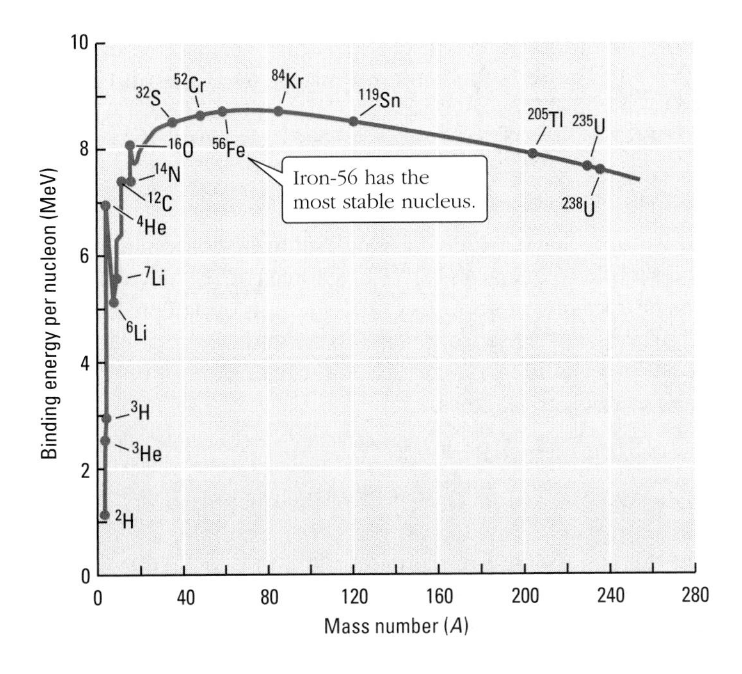

Figure 19.3 Binding energy per nucleon. The values plotted were derived by calculating the binding energy per nucleon in million electron volts (MeV) for the most abundant isotope of each element from hydrogen to uranium (1 MeV = 1.602×10^{-13} J). The nuclei at the top of the curve are most stable.

EXERCISE 19.3 Binding Energy

Calculate the binding energy, in kJ/mol, for the formation of lithium-6.

$$3\,{}^{1}_{1}\text{H} + 3\,{}^{1}_{0}\text{n} \longrightarrow {}^{6}_{3}\text{Li}$$

The necessary masses are ${}^{1}_{1}\text{H} = 1.00783$ g/mol, ${}^{1}_{0}\text{n} = 1.00867$ g/mol, and ${}^{6}_{3}\text{Li} = 6.015125$ g/mol. Is the binding energy greater than or less than that for helium-4? Compare the binding energy per nucleon of lithium-6 and helium-4. Which nucleus is more stable?

CONCEPTUAL
EXERCISE 19.4 Binding Energy

By interpreting the shape of the curve in Figure 19.3, determine which is more exothermic per gram—fission or fusion. Explain your answer.

19.4 Rates of Disintegration Reactions

Cobalt-60 is radioactive and is used as a source of β particles and γ rays to treat malignancies in the human body. One-half of a sample of cobalt-60 will change via beta decay into nickel-60 in a little more than five years (Table 19.2). On the other hand, copper-64, which is used in the form of copper acetate to detect brain tumors, decays much more rapidly; half of the radioactive copper decays in slightly less than 13 hours. These two radioactive isotopes differ greatly in their rates of decay.

Table 19.2 Half-Lives of Some Common Radioactive Isotopes

Decay Process	Half-Life
${}^{238}_{92}\text{U} \longrightarrow {}^{234}_{90}\text{Th} + {}^{4}_{2}\text{He}$	4.15×10^{9} yr
${}^{3}_{1}\text{H} \longrightarrow {}^{3}_{2}\text{He} + {}^{0}_{-1}\text{e}$	12.3 yr
${}^{14}_{6}\text{C} \longrightarrow {}^{14}_{7}\text{N} + {}^{0}_{-1}\text{e}$	5730 yr
${}^{131}_{53}\text{I} \longrightarrow {}^{131}_{54}\text{Xe} + {}^{0}_{-1}\text{e}$	8.04 d
${}^{123}_{53}\text{I} + {}^{0}_{-1}\text{e} \longrightarrow {}^{123}_{52}\text{Te}$	13.2 h
${}^{57}_{24}\text{Cr} \longrightarrow {}^{57}_{25}\text{Mn} + {}^{0}_{-1}\text{e}$	21 s
${}^{28}_{15}\text{P} \longrightarrow {}^{28}_{14}\text{Si} + {}^{0}_{+1}\text{e}$	0.270 s
${}^{90}_{38}\text{Sr} \longrightarrow {}^{90}_{39}\text{Y} + {}^{0}_{-1}\text{e}$	28.8 yr
${}^{60}_{27}\text{Co} \longrightarrow {}^{60}_{28}\text{Ni} + {}^{0}_{-1}\text{e}$	5.26 yr

Half-Life

The relative instability of a radioactive isotope is expressed as its **half-life,** the time required for one half of a given quantity of the isotope to undergo radioactive decay. In terms of reaction kinetics (***p. 427***), radioactive decay is a first-order reaction. Therefore, the rate of decay is given by the first-order rate law equation

$$\ln[\text{A}]_t = -kt + \ln[\text{A}]_0$$

where $[\text{A}]_0$ is the initial concentration of isotope A, $[\text{A}]_t$ is the concentration of A after time t has passed, and k is the first-order rate constant. Because radioactive decay is first-order, the half-life ($t_{1/2}$) of an isotope is the same no matter what the initial concentration. It is given by

$$t_{1/2} = \frac{\ln 2}{k} = \frac{0.693}{k}$$

The relationship $t_{1/2} = \frac{0.693}{k}$ was introduced in the context of first-order kinetics of reactions (***p. 431***).

As illustrated by Table 19.2, isotopes have widely varying half-lives. Some take years, even millennia, for half of the sample to decay (${}^{238}\text{U}$, ${}^{14}\text{C}$), whereas others decay to half the original number of atoms in fractions of seconds (${}^{28}\text{P}$). The unit of half-life is whatever time unit is most appropriate—anything from years to seconds.

As an example of the concept of half-life, consider the decay of plutonium-239, an alpha-emitting isotope formed in nuclear reactors.

$${}^{239}_{94}\text{Pu} \longrightarrow {}^{4}_{2}\text{He} + {}^{235}_{92}\text{U}$$

The half-life of plutonium-239 is 24,400 years. Thus, half of the quantity of ${}^{239}_{94}\text{Pu}$ present at any given time will disintegrate every 24,400 years. For example, if we begin with 1.00 g ${}^{239}_{94}\text{Pu}$, 0.500 g of the isotope will remain after 24,400 years. After 48,800 years (two half-lives), only half of the 0.500 g, or 0.250 g, will remain. After 73,200 years (three half-lives), only half of the 0.250 g will still be present, or 0.125 g. The amounts of $^{239}_{94}Pu$ present at various times are illustrated in Figure 19.4. All radioactive isotopes follow this type of decay curve.

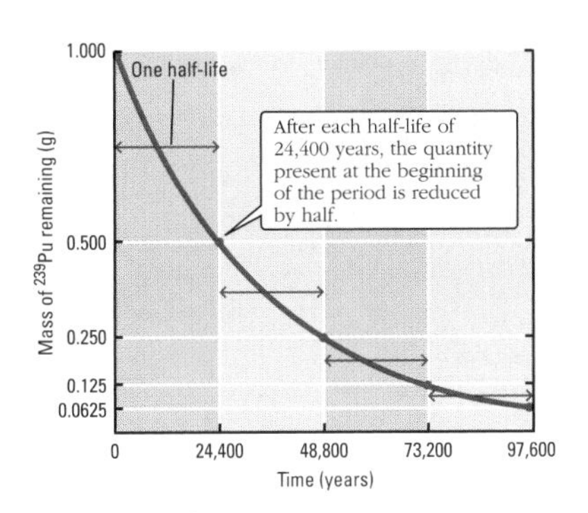

Figure 19.4 The decay of 1.00 g plutonium-239.

PROBLEM-SOLVING EXAMPLE 19.4 Half-Life

Iodine-131, used to treat hyperthyroidism, has a half-life of 8.04 days.

$$^{131}_{53}I \longrightarrow {}^{131}_{54}Xe + {}^{0}_{-1}e \qquad t_{1/2} = 8.04 \text{ days}$$

If you have a sample containing 10.0 μg of iodine-131, what mass of the isotope will remain after 32.2 days?

Answer 0.0625 μg

Strategy and Explanation First, we find the number of half-lives in the given 32.2-day time period. Since the half-life is 8.04 days, the number of half-lives is

$$32.2 \text{ days} \times \frac{1 \text{ half-life}}{8.04 \text{ days}} = 4.00 \text{ half-lives}$$

This means that the initial quantity of 10.0 μg is reduced by half four times.

$$10.0\ \mu g \times 1/2 \times 1/2 \times 1/2 \times 1/2 = 10.0\ \mu g \times 1/16 = 0.0625\ \mu g$$

After 32.2 days, only one sixteenth of the original ^{131}I remains.

☑ **Reasonable Answer Check** After the passage of four half-lives, the remaining ^{131}I should be a small fraction of the starting amount, and it is.

PROBLEM-SOLVING PRACTICE 19.4

Strontium-90 is a radioisotope ($t_{1/2}$ = 29 years) produced in atomic bomb explosions. Its long half-life and tendency to concentrate in bone marrow by replacing calcium make it particularly dangerous to people and animals.

(a) The isotope decays with loss of a β particle. Write a balanced equation showing the other product of decay.

(b) A sample of the isotope emits 2000 β particles per minute. How many half-lives and how many years are necessary to reduce the emission to 125 β particles per minute?

EXERCISE 19.5 Half-Lives

The radioactivity of formerly highly radioactive isotopes is essentially negligible after ten half-lives. What percentage of the original radioisotope remains after this amount of time (ten half-lives)?

Rate of Radioactive Decay

To determine the half-life of a radioactive element, its *rate of decay,* that is, the number of atoms that disintegrate in a given time—per second, per hour, per year—must be measured.

Radioactive decay is a first-order process ***(⟵ p. 427)***, with a rate that is directly proportional to the number of radioactive atoms present (N). This proportionality is expressed as a rate law (Equation 19.1) in which A is the **activity** of the sample—the number of disintegrations observed per unit time—and k is the first-order rate constant or *decay constant* characteristic of that radioisotope.

$$A = kN \qquad [19.1]$$

Suppose the activity of a sample is measured at some time t_0 and then measured again after a few minutes, hours, or days. If the initial activity is A_0 at t_0, then a second measurement at a later time t will detect a smaller activity A. Using Equation 19.1, the ratio of the activity A at some time t to the activity at the beginning of the experiment (A_0)

must be equal to the ratio of the number of radioactive atoms N that are present at time t to the number present at the beginning of the experiment (N_0).

$$\frac{A}{A_0} = \frac{kN}{kN_0} \quad \text{or} \quad \frac{A}{A_0} = \frac{N}{N_0}$$

Thus, either A/A_0 or N/N_0 expresses the fraction of radioactive atoms remaining in a sample after some time has elapsed.

The activity of a sample can be measured with a device such as a Geiger counter (Figure 19.5). It detects radioactive emissions as they ionize a gas to form free electrons and cations that can be attracted to a pair of electrodes. In the Geiger counter, a metal tube is filled with low-pressure argon gas. The inside of the tube acts as the cathode. A thin wire running through the center of the tube acts as the anode. When radioactive emissions enter the tube through the thin window at the end, they collide with argon atoms; these collisions produce free electrons and argon cations. As the free electrons accelerate toward the anode, they collide with other argon atoms to generate more free electrons. The free electrons all go to the anode, and they constitute a pulse of current. This current pulse is counted, and the rate of pulses per unit time is the output of the Geiger counter.

The *curie* was named for Pierre Curie by his wife Marie; the *becquerel* honors Henri Becquerel.

The unit for curie and becquerel is s^{-1} because each is a number (of disintegrations) per second.

The **curie, Ci,** is commonly used as a unit of activity. One curie represents a decay rate of 3.7×10^{10} disintegrations per second (s^{-1}), which is the decay rate of 1 g radium. One millicurie (mCi) $= 10^{-3}$ Ci $= 3.7 \times 10^{7}\ s^{-1}$. Another unit of radioactivity is the **becquerel, Bq;** 1 becquerel is equal to one nuclear disintegration per second (1 Bq $= 1\ s^{-1}$).

The change in activity of a radioactive sample over a period of time, or the fraction of radioactive atoms still present in a sample after some time has elapsed, can be calculated using the integrated rate equation for a first-order reaction

$$\ln A = -kt + \ln A_0$$

which can be rearranged to

Equation 19.2 can be derived from Equation 19.1 using calculus.

$$\ln \frac{A}{A_0} = -kt \qquad [19.2]$$

where A/A_0 is the ratio of activities at time t. Equation 19.2 can also be stated in terms of the fraction of radioactive atoms present in the sample after some time, t, has passed.

$$\ln \frac{N}{N_0} = -kt \qquad [19.3]$$

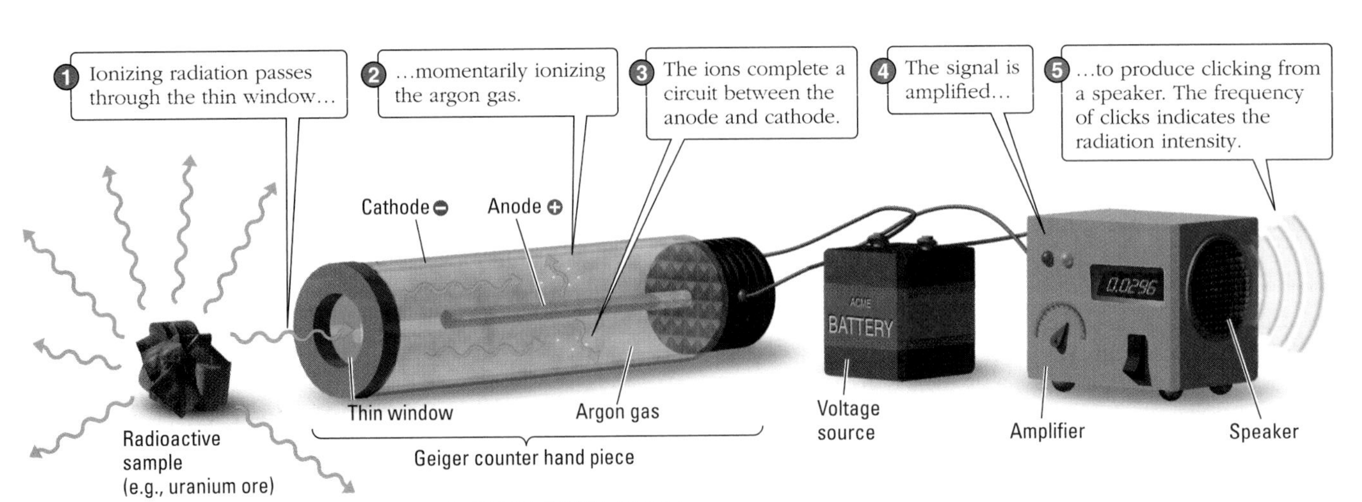
1 Ionizing radiation passes through the thin window...
2 ...momentarily ionizing the argon gas.
3 The ions complete a circuit between the anode and cathode.
4 The signal is amplified...
5 ...to produce clicking from a speaker. The frequency of clicks indicates the radiation intensity.
Cathode
Anode
BATTERY
0.0296
Radioactive sample (e.g., uranium ore)
Thin window
Argon gas
Geiger counter hand piece
Voltage source
Amplifier
Speaker

Active Figure 19.5 **A Geiger counter.** Visit this book's companion website at **www.cengage.com/chemistry/moore** to test your understanding of the concepts in this figure.

In words, Equation 19.3 says

$$\text{Natural logarithm}\left(\frac{\text{number of radioactive atoms at time } t}{\text{number of radioactive atoms at start of experiment}}\right)$$
$$= \text{natural logarithm(fraction of radioactive atoms remaining at time } t)$$
$$= -(\text{decay constant})(\text{time})$$

Notice the negative sign in Equation 19.3. The ratio N/N_0 is less than 1 because N is always less than N_0. This means that the logarithm of N/N_0 is negative, and the other side of the equation has a compensating negative sign because k and t are always positive.

As radioactive atoms decay, N becomes a smaller and smaller fraction of N_0.

As we have seen, the half-life of an isotope is inversely proportional to the first-order rate constant k:

$$t_{1/2} = \frac{0.693}{k}$$

Thus, the half-life can be found by calculating k from Equation 19.3 using N and N_0 from laboratory measurements over the time period t, as illustrated in Problem-Solving Example 19.5.

PROBLEM-SOLVING EXAMPLE 19.5 Half-Life

A sample of ^{24}Na initially undergoes 3.50×10^4 disintegrations per second (s^{-1}). After 24.0 h, its disintegration rate has fallen to $1.16 \times 10^4\ s^{-1}$. What is the half-life of ^{24}Na?

Answer 15.0 h

Strategy and Explanation We use Equation 19.2 relating activity (disintegration rate) at time zero and time t with the decay constant k. The experiment provided us with A, A_0, and the time.

$$\ln\left(\frac{1.16 \times 10^4\ s^{-1}}{3.50 \times 10^4\ s^{-1}}\right) = \ln(0.331) = -k(24.0\ h)$$

$$k = -\frac{\ln(0.331)}{24\ h} = -\left(\frac{-1.104}{24.0\ h}\right) = 0.0460\ h^{-1}$$

From k we can determine $t_{1/2}$.

$$t_{1/2} = \frac{0.693}{k} = \frac{0.693}{0.0460\ h^{-1}} = 15.0\ h$$

☑ **Reasonable Answer Check** The activity (disintegration rate) fell to between one half and one quarter of its initial value in 24.0 h, so the half-life must be less than 24.0 h, and this agrees with our more accurate calculation.

PROBLEM-SOLVING PRACTICE 19.5

The decay of iridium-192, a radioisotope used in cancer radiation therapy, has a rate constant of $9.3 \times 10^{-3}\ d^{-1}$.

(a) What is the half-life of ^{192}Ir?

(b) What fraction of an ^{192}Ir sample remains after 100 days?

Carbon-14 Dating

In 1946 Willard Libby developed a technique for measuring the age of archaeological objects using radioactive carbon-14. Carbon is an important building block of all living systems, and all organisms contain the three isotopes of carbon: ^{12}C, ^{13}C, and ^{14}C. The first two isotopes are stable (nonradioactive) and have been present for billions of years. Carbon-14, however, is radioactive and decays to nitrogen-14 by beta emission.

Willard Libby won the 1960 Nobel Prize in Chemistry for his discovery of radiocarbon dating.

$$^{14}_{6}C \longrightarrow {}^{0}_{-1}e + {}^{14}_{7}N$$

© Bettmann/Corbis

Willard Libby and his apparatus for carbon-14 dating.

The half-life of ^{14}C is known by experiment to be 5.73×10^3 years. The number of carbon-14 atoms (N) in a carbon-containing sample can be measured from the activity of the sample (A). If the number of carbon-14 atoms originally in the sample (N_0) can be determined, or if the initial activity (A_0) can be determined, the age of the sample can be found from Equation 19.2 or 19.3.

This method of age determination clearly depends on knowing how much ^{14}C was originally in the sample. The answer to this question comes from work by physicist Serge Korff, who discovered in 1929 that ^{14}C is continually generated in the upper atmosphere. High-energy cosmic rays collide with gas molecules in the upper atmosphere and cause them to eject neutrons. These free neutrons collide with nitrogen atoms to produce carbon-14.

$$^{14}_{7}N + ^{1}_{0}n \longrightarrow ^{14}_{6}C + ^{1}_{1}H$$

Throughout the *entire* atmosphere, only about 7.5 kg ^{14}C is produced per year. However, this relatively small quantity of radioactive carbon is incorporated into CO_2 and other carbon compounds and then distributed worldwide as part of the carbon cycle. The continual formation of ^{14}C, transfer of the isotope within the oceans, atmosphere, and biosphere, and decay of living matter keep the supply of ^{14}C constant.

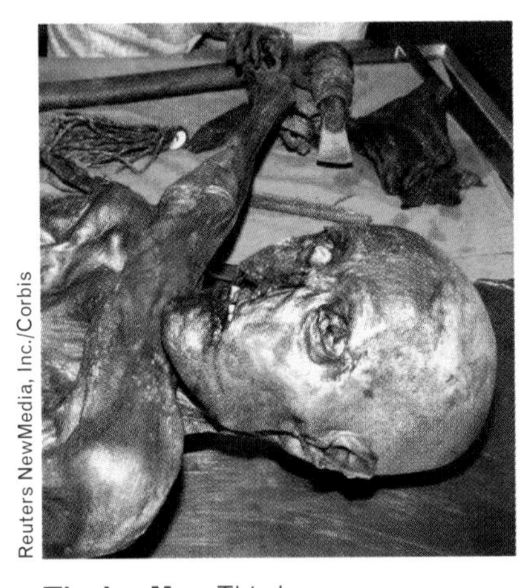

Reuters NewMedia, Inc./Corbis

The Ice Man. This human mummy was found in 1991 in glacial ice high in the Alps. Carbon-14 dating determined that he lived about 5300 years ago. The mummy is exhibited at the South Tyrol Archaeological Museum in Bolzano, Italy.

Plants absorb carbon dioxide from the atmosphere and convert it into food via photosynthesis. In this way, the ^{14}C becomes incorporated into living tissue, where radioactive ^{14}C atoms and nonradioactive ^{12}C atoms in CO_2 chemically react in the same way. The beta decay activity of carbon-14 in *living* plants and in the air is constant at 15.3 disintegrations per minute per gram ($min^{-1}\ g^{-1}$) of carbon. When a plant dies, however, carbon-14 disintegration continues *without the ^{14}C being replaced.* Consequently, the ^{14}C activity of the dead plant material decreases with the passage of time. The smaller the activity of carbon-14 in the plant, the longer the period of time between the death of the plant and the present. Assuming that ^{14}C activity in living plants was about the same hundreds or thousands of years ago as it is now, measurement of the ^{14}C beta activity of an artifact can be used to date an article containing carbon. The slight fluctuations of the ^{14}C activity in living plants for the past several thousand years have been measured by studying growth rings of long-lived trees, and the carbon-14 dates of objects can be corrected accordingly.

The time scale accessible to carbon-14 dating is determined by the half-life of ^{14}C. Therefore, this method for dating objects can be extended back approximately 50,000 years. This span of time is almost nine half-lives, during which the number of disintegrations per minute per gram of carbon would fall by a factor of about $(\frac{1}{2})^9 = 1.95 \times 10^{-3}$, from about 15.3 $min^{-1}\ g^{-1}$ to about 0.030 $min^{-1}\ g^{-1}$, which is a disintegration rate so low that it is difficult to measure accurately.

PROBLEM-SOLVING EXAMPLE 19.6 Carbon-14 Dating

Charcoal fragments found in a prehistoric cave in Lascaux, France, had a measured disintegration rate of 2.40 $min^{-1}\ g^{-1}$ carbon. Calculate the approximate age of the charcoal.

Answer 15,300 years old

Strategy and Explanation We will use Equation 19.2 to solve the problem

$$\ln\left(\frac{A}{A_0}\right) = -kt$$

where A is proportional to the known activity of the charcoal (2.40 $min^{-1}\ g^{-1}$) and A_0 is proportional to the activity of the carbon-14 in living material (15.3 $min^{-1}\ g^{-1}$). We first need to calculate k, the rate constant, using the half-life of carbon-14, 5.73×10^3 yr.

$$k = \frac{0.693}{t_{1/2}} = \frac{0.693}{5.73 \times 10^3\ \text{yr}} = 1.21 \times 10^{-4}\ \text{yr}^{-1}$$

© images-of-france/Alamy

Prehistoric cave paintings from Lascaux, France.

Now we are ready to calculate the time, t.

$$\ln\left(\frac{2.40\ \text{min}^{-1}\ \text{g}^{-1}}{15.3\ \text{min}^{-1}\ \text{g}^{-1}}\right) = -kt$$

$$\ln(0.1569) = -(1.21 \times 10^{-4}\ \text{yr}^{-1})t$$

$$t = \frac{1.852}{1.21 \times 10^{-4}\ \text{yr}^{-1}} = 1.53 \times 10^{4}\ \text{yr}$$

Thus, the charcoal is approximately 15,300 years old.

☑ **Reasonable Answer Check** The disintegration rate has fallen a factor of six from the rate for living material, so more than two but less than three half-lives have elapsed. This agrees with our calculated result.

PROBLEM-SOLVING PRACTICE 19.6

Tritium, ^{3}H ($t_{1/2}$ = 12.3 yr), is produced in the atmosphere and incorporated in living plants in much the same way as ^{14}C. Estimate the age of a sealed sample of Scotch whiskey that has a tritium content 0.60 times that of the water in the area where the whiskey was produced.

EXERCISE 19.6 Radiochemical Dating

The radioactive decay of uranium-238 to lead-206 provides a method of radiochemically dating ancient rocks by using the ratio of ^{206}Pb atoms to ^{238}U atoms in a sample. Using this method, a moon rock was found to have a $^{206}Pb/^{238}U$ ratio of 100/109, that is, 100 ^{206}Pb atoms for every 109 ^{238}U atoms. No other lead isotopes were present in the rock, indicating that all of the ^{206}Pb was produced by ^{238}U decay. Estimate the age of the moon rock. The half-life of ^{238}U is 4.51×10^{9} years.

CONCEPTUAL EXERCISE 19.7 Radiochemical Dating

Ethanol, C_2H_5OH, is produced by the fermentation of grains or by the reaction of water with ethylene, which is made from petroleum. The alcohol content of wines can be increased fraudulently beyond the usual 12% available from fermentation by adding ethanol produced from ethylene. How can carbon dating techniques be used to differentiate the ethanol sources in these wines?

19.5 Nuclear Fission

In 1938 the nuclear chemists Otto Hahn and Fritz Strassman were confounded when they isolated barium from a sample of uranium that had been bombarded with neutrons. Further work by Lise Meitner, Otto Frisch, Niels Bohr, and Leo Szilard confirmed that the bombarded uranium-235 nucleus had formed barium by the capture of a neutron and had undergone **nuclear fission,** a nuclear reaction in which the bombarded nucleus splits into two lighter nuclei (Figure 19.6).

$$^{235}_{92}U + ^{1}_{0}n \longrightarrow ^{236}_{92}U \longrightarrow ^{141}_{56}Ba + ^{92}_{36}Kr + 3\ ^{1}_{0}n \qquad \Delta E = -2 \times 10^{10}\ \text{kJ}$$

This nuclear equation shows that bombardment with a single neutron produces three neutrons among the products. The fact that the fission reaction produces more neutrons than are required to begin the process is important. Each of the product neutrons is capable of inducing another fission reaction, so three neutrons would induce

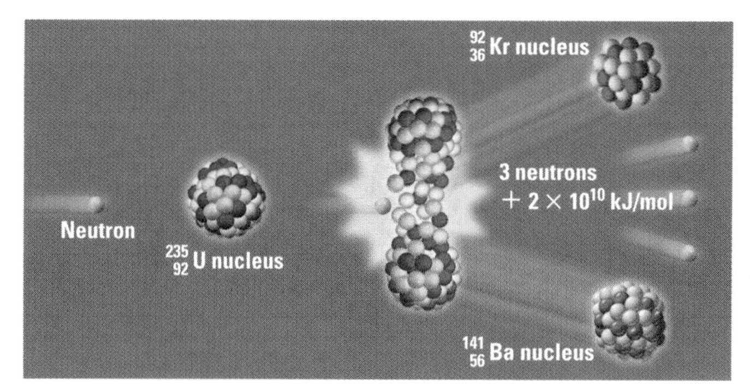

Figure 19.6 The fission of a $^{235}_{92}U$ nucleus from its bombardment with a neutron.

three fissions, which would release nine neutrons to induce nine more fissions, from which 27 neutrons are obtained, and so on. Since the neutron-induced fission of uranium-235 is extremely rapid, this sequence of reactions can lead to an explosive chain reaction, as illustrated in Figure 19.7.

Nuclear fission of uranium-235 produces a variety of products. Thirty-four elements have been detected among the fission products, including those shown in Figure 19.7.

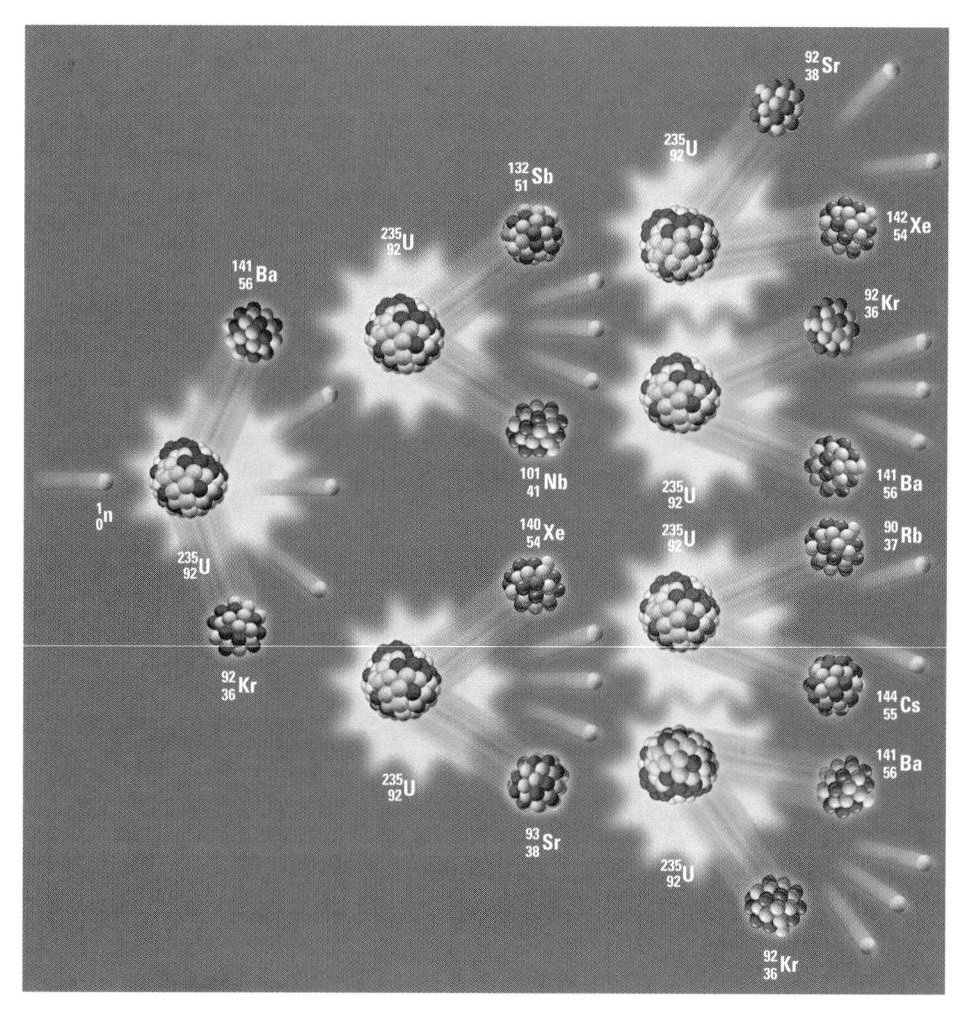

Figure 19.7 A self-propagating nuclear chain reaction initiated by capture of a neutron. The nuclear fission of uranium-235 produces a variety of products. Thirty-four elements have been detected among the fission products, including those shown here. Each fission event produces two lighter nuclei plus two or three neutrons.

If the quantity of uranium-235 is small, then most of the neutrons escape without hitting a nucleus and so few neutrons are captured by ^{235}U nuclei that the chain reaction cannot be sustained. In an atomic bomb, two small masses of uranium-235, neither capable of sustaining a chain reaction, are brought together rapidly to form one mass capable of supporting a chain reaction. An atomic explosion results. The minimum mass of fissionable material required for a self-sustaining chain reaction is termed the **critical mass.**

Nuclear Reactors

Nuclear fission reactions are the energy source in nuclear power plants. The nuclear reactor fuel is fissionable material, for example, uranium-235. The rate of fission in a **nuclear reactor** is controlled by limiting the number and energy of neutrons available so that energy derived from fission can be used safely as a heat source for a nuclear power plant (Figure 19.8). In a nuclear reactor, the rate of fission is controlled by inserting cadmium rods or other neutron absorbers into the reactor. The rods absorb the neutrons that would otherwise propagate fission reactions. The rate of the fission reaction can be increased or decreased by withdrawing or inserting the control rods, respectively. The materials that control the numbers of neutrons (by absorbing them) or control their energy (by absorbing some of their energy) are known as *moderators.*

The uranium fuel in the reactor core is uranium dioxide (UO_2) pellets, each about the size of the eraser on a pencil. The pellets are placed end-to-end in metal alloy tubes, which are then grouped into stainless steel–clad bundles. As pointed out earlier, once a fission reaction is started, it can be sustained as a chain reaction. However, a source of neutrons is needed to initiate the chain reaction. One means of generating these initial neutrons is a nuclear reaction source, such as beryllium-9, and a heavy, alpha-emitting element, such as plutonium or americium. The heavy element emits alpha particles; when they strike a beryllium-9 nucleus, the two nuclei combine to form a carbon-12 nucleus and a neutron is emitted.

$$^{238}_{94}\text{Pu} \longrightarrow {}^{4}_{2}\text{He} + {}^{234}_{92}\text{U}$$

$$^{4}_{2}\text{He} + {}^{9}_{4}\text{Be} \longrightarrow {}^{12}_{6}\text{C} + {}^{1}_{0}\text{n}$$

These neutrons then initiate the nuclear fission of uranium-235 in the reactor core.

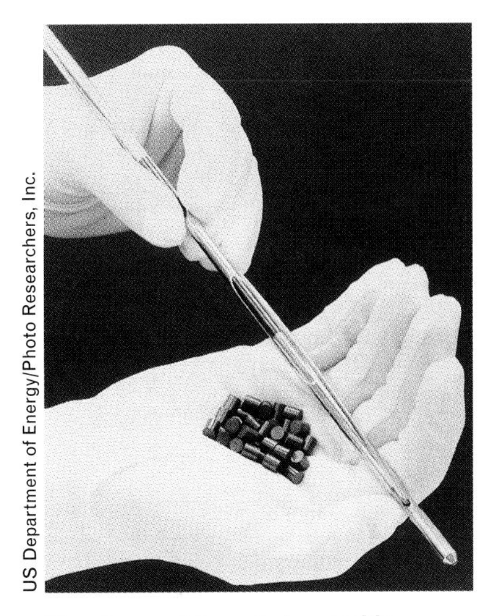
US Department of Energy/Photo Researchers, Inc.

Uranium oxide pellets used in nuclear fuel rods.

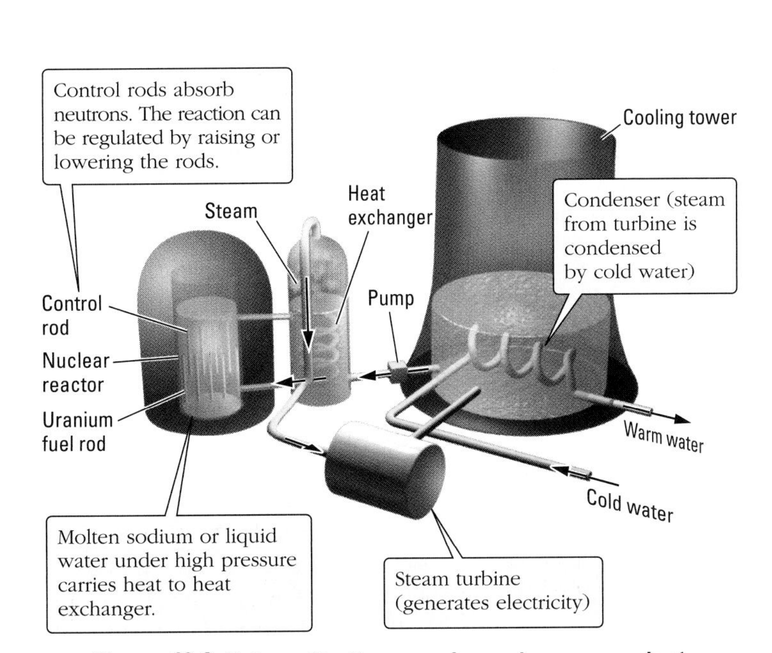

Figure 19.8 Schematic diagram of a nuclear power plant.

Conventional (non-nuclear) power plants burn fossil fuel to generate the heat to produce steam to drive the turbine.

Cooling towers are also used by fossil fuel–burning power plants.

Uranium-238 can fission, but only when bombarded by fast neutrons, unlike those in nuclear reactors. Thus, we consider uranium-238 to be nonfissionable in the context of nuclear reactors.

Tom Brakefield/Getty Images

A nuclear power plant with two prominent cooling towers and two nuclear containment buildings (the domed buildings).

Nuclear fuel rods that have become depleted of U-235 are known as *spent* fuel.

A sample is considered to be of weapons-grade quality only if its ^{235}U content is greater than 90%. Even in reactors using weapons-grade quality, the ^{235}U is still too dispersed to produce uncontrolled fission.

The tremendous heat generated by the fission reaction heats the primary coolant, a substance with a very high heat capacity, usually water. The primary coolant is at a pressure of more than 150 atm, so it does not boil, even though the temperature is higher than its normal boiling point (**p. 381**). The hot primary coolant is pumped in a closed loop from the reaction vessel to a heat exchanger and back again to the reaction vessel. Heat transfer to water that generates steam in the heat exchanger lowers the temperature of the primary coolant, which returns to the reactor to be heated again. This closed loop links the nuclear reactor and the rest of the power plant.

Water in the heat exchanger is sometimes referred to as the secondary coolant. As the water in the heat exchanger is vaporized to steam, the steam strikes the large turbine blades, causing the turbine to spin. The turbine shaft is connected to a large metal rod in the generator, which is surrounded by a magnetic field. The rapid spinning of the turbine shaft in a magnetic field produces electricity.

After striking the turbine blades, the steam must be condensed so that the heating/cooling cycle can be repeated to create additional electricity. To do so, cooling water is pumped from a neighboring river or lake to the secondary coolant loop. Enormous amounts of outside cooling water are needed to condense the vast quantity of steam produced by such power plants. For example, the nuclear power reactor at the Entergy Arkansas Unit 1 uses 750,000 gal of cooling water per minute. Having picked up heat from the secondary coolant, the cooling water must then be cooled itself before being returned to its source. Such cooling is done in many nuclear power plants by passing the water through large concave evaporative cooling towers, which are often mistaken for the nuclear reactors themselves.

Not all nuclei can be made to fission on colliding with a neutron, but ^{235}U and ^{239}Pu are both fissionable isotopes. Natural uranium contains an average of only 0.72% of the fissionable ^{235}U isotope. More than 99% of the natural element is nonfissionable uranium-238. Since the percentage of natural ^{235}U is too small to sustain a chain reaction, uranium for nuclear power fuel must be enriched to about 3% uranium-235. To accomplish this, some of the ^{238}U isotope in a sample is effectively discarded, thereby raising the concentration of ^{235}U. If sufficient fissionable uranium-235 is present (a critical mass), it can capture enough neutrons to sustain the fission chain reaction. Approximately one third of the fuel rods in a nuclear reactor are replaced annually because fission by-products absorb neutrons, reducing the efficiency of the fission reactions.

Because the mass of uranium-235 in the fuel rods of a nuclear power plant is lower than the critical mass needed for an atomic bomb, the reactor core *cannot* undergo an uncontrolled chain reaction to convert the reactor into an atomic bomb.

Nuclear fission fuels have extremely large energy density. For example, fission of 1.0 kg (2.2 lb) uranium-235 releases 9.0×10^{13} J, the equivalent of exploding 33,000 tons (33 kilotons) TNT. Each UO_2 fuel pellet used in a nuclear reactor has the energy equivalent to burning 136 gallons oil, 2.5 tons wood, or 1 ton coal.

Courtesy of Department of Energy

These stainless steel canisters (2 feet in diameter and 10 feet tall) hold high-level radioactive waste that has been vitrified into a glassy solid.

EXERCISE 19.8 Energy of Nuclear Fission

Burning 1.0 kg high-grade coal produces 2.8×10^4 kJ, whereas fission of 1.0 mol uranium-235 generates 2.1×10^{10} kJ. How many metric tons of coal (1 metric ton = 10^3 kg) are needed to produce the same quantity of energy as that released by the fission of 1.0 kg uranium-235? (Assume that the processes have equal efficiency.)

There is, of course, substantial controversy surrounding the use of nuclear power plants, and not just in the United States. Their proponents argue that the health of our economy and our standard of living depend on inexpensive, reliable, and safe sources of energy. Just within the past few years the demand for electric power has at times exceeded the supply in the United States, and many believe nuclear power plants should be built to meet that demand. Nuclear power plants can be the source of "clean" energy, in that they do not pollute the atmosphere with ash, smoke, or oxides of sulfur, nitrogen, or carbon as coal-fired plants do. In addition, nuclear plants help to ensure that our supplies of fossil fuels will not be depleted as quickly in the near future, and they reduce our dependency on buying such fuels from other countries. Currently, 104 nuclear plants operate in the United States. Approximately 440 nuclear power plants worldwide in 31 countries produce about 17% of the world's electricity. Currently, 32 nuclear power plants are under construction worldwide, none in the United States. The nuclear plants in the United States supply about 19% of our nation's electric energy (Figure 19.9).

No new nuclear power plants have been built in the United States since 1979—the year of the accident at the Three Mile Island nuclear power plant near Harrisburg, Pennsylvania. One problem associated with nuclear power plants is the highly radioactive fission products in the spent fuel. In the United States tens of thousands of tons of spent fuel waste are being stored, and the amount is growing steadily. Although some of the products are put to various uses, many are unsuitable as a fuel or for other purposes. Because these products are often highly radioactive and some have long half-lives (plutonium-239, $t_{1/2} = 24{,}400$ yr), proper disposal of this high-level nuclear waste poses an enormous problem. One approach is that high-level radioactive wastes can be encased in a glassy material having a volume of about 2 m^3 per reactor per year. In 1996 the Department of Energy Savannah River site near Augusta, Georgia, began encapsulating radioactive waste in glass, a process called vitrification, in which a mixture of glass particles and radioactive waste is heated to 1200 °C. The molten mixture is poured into stainless steel canisters, cooled, and stored. Eventually, such high-level nuclear wastes may be stored underground in geological formations, such as salt deposits, that are known to be stable for hundreds of millions of years. A site at Yucca Mountain, Nevada, is the designated national repository for high-level nuclear waste, such as that from spent nuclear reactor cores. No such nuclear waste has yet been stored at that site.

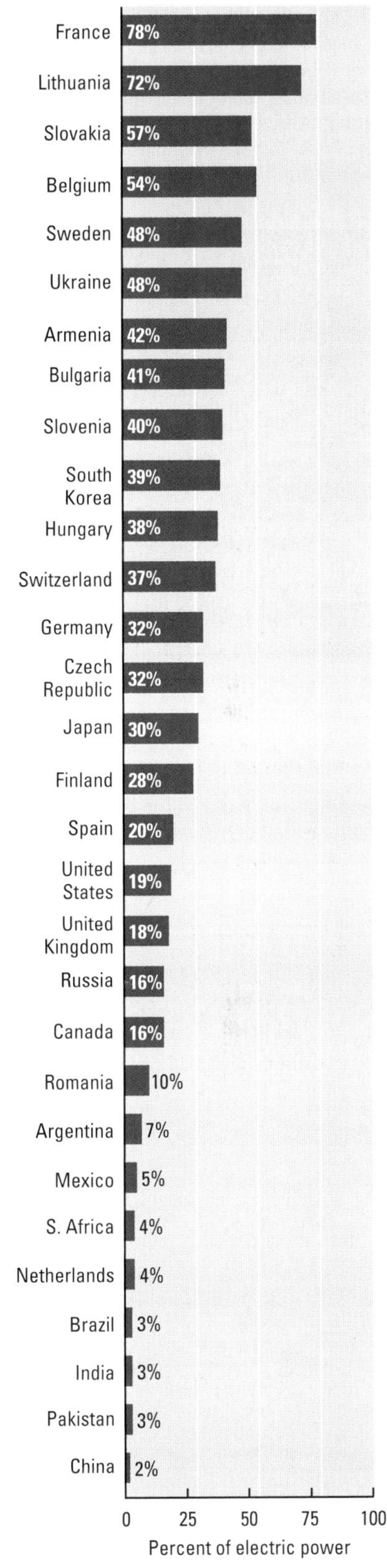

Figure 19.9 The approximate fraction of electricity generated by nuclear power in various countries. About 19% of the electricity in the United States is produced by nuclear power.

EXERCISE 19.9 Radioactive Decay of Fission Products

Unlike the 1979 incident at Three Mile Island, the accident at the Chernobyl nuclear plant in the former Soviet Union in 1986 released significant quantities of radioisotopes into the atmosphere. One of those radioisotopes was strontium-90 ($t_{1/2} = 29.1$ yr). What fraction of strontium-90 released at that time remains?

EXERCISE 19.10 Nuclear Waste

Cesium-137 ($t_{1/2} = 30.2$ yr) is produced by ^{235}U fission. If ^{137}Cs is part of nuclear waste stored deep underground, how long will it take for the initial ^{137}Cs activity when it was first buried to drop (a) by 60%? and (b) by 90%?

19.6 Nuclear Fusion

Tremendous amounts of energy are generated when two light nuclei combine to form a heavier nucleus in a reaction called **nuclear fusion.** One of the best examples is the fusion of hydrogen nuclei (protons) to produce a helium nucleus.

$$4\,^1_1H \longrightarrow\, ^4_2He + 2\,^{\ 0}_{+1}e \qquad \Delta E = -2.5 \times 10^9 \text{ kJ}$$

The helium nucleus produced by this reaction is more stable (has higher binding energy per nucleon) than the reactant hydrogen nuclei, as shown in Figure 19.3

The activation energy (⇐ p. 434) for nuclear fusion is very large because there is very strong repulsion between two positive nuclei when they are brought very close together—Coulomb's law. Thus, very high temperatures are needed for fusion to occur.

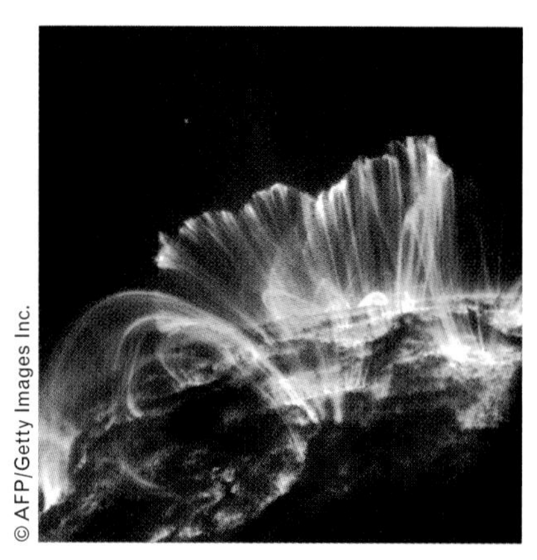

© AFP/Getty Images Inc.

Nuclear fusion reactions power the sun, whose surface is shown here.

Containment is one of the biggest problems in developing controlled nuclear fusion.

(⇐ p. 701). This fusion reaction is the main source of the energy from our sun and other stars, and it is the beginning of the synthesis of the elements in the universe. Temperatures of 10^6 to 10^7 K, found in the core and radioactive zone of the sun, are required to bring the positively charged ${}^1_1H$ nuclei together with enough kinetic energy to overcome nuclear repulsions and react.

Deuterium can also be fused to give helium-3,

$${}^2_1H + {}^2_1H \longrightarrow {}^3_2He + {}^1_0n \qquad \Delta E = -3.2 \times 10^8 \text{ kJ}$$

which can undergo further fusion with a proton to give helium-4.

$${}^1_1H + {}^3_2He \longrightarrow {}^4_2He + {}^{0}_{+1}e \qquad \Delta E = -1.9 \times 10^9 \text{ kJ}$$

Each of these reactions releases an enormous quantity of energy, so it has been the dream of many nuclear physicists to harness them to provide energy for the people of the world.

Development of nuclear fusion as a commercial energy source is appealing because hydrogen isotopes are available (from water), and fusion products are usually nonradioactive or have short half-lives, which eliminates the problems associated with the disposal of high-level radioactive fission reactor products. However, controlling a nuclear fusion reaction for peaceful commercial uses has proven to be extraordinarily difficult. Three critical requirements must be met to achieve controlled fusion. First, the temperature must be high enough for fusion to occur sufficiently rapidly. The fusion of deuterium and tritium, for example, requires a temperature of at least 100 million K. Second, the plasma must be confined long enough to generate a net output of energy. Third, the energy must be recovered in some usable form.

Magnetic "bottles" (enclosures in space bounded by magnetic fields) have confined the plasma so that controlled fusion has been achieved. But the energy generated by the fusion has been less than that required to produce the magnetic bottle and control the fusion reaction. Using more energy to produce less energy is not a commercially appealing investment. Thus, commercial fusion reactors are not likely in the near future without a dramatic breakthrough in fusion technology.

EXERCISE 19.11 Nuclear Fusion

Complete the equations for these nuclear fusion reactions.

(a) ${}^7_3Li + {}^1_1H \longrightarrow {}^1_0n + \_\_\_\_$

(b) ${}^2_1H + \_\_\_\_ \longrightarrow {}^4_2He + {}^1_1H$

Hydrogen bombs are based on fusion. At the very high temperatures that allow fusion reactions to occur rapidly, atoms do not exist as such. Instead, there is a **plasma,** which consists of unbound nuclei and electrons. To achieve the high temperatures required for the fusion reaction of the hydrogen bomb, a fission bomb (atomic bomb) is first set off. In one type of hydrogen bomb, lithium-6 deuteride (LiD, a solid salt) is placed around a ${}^{235}U$ or ${}^{239}Pu$ fission bomb, and the fission reaction is set off to initiate the process. Lithium-6 nuclei absorb neutrons produced by the fission and split into tritium and helium.

$${}^1_0n + {}^6_3Li \longrightarrow {}^3_1H + {}^4_2He$$

The temperature reached by the fission of uranium or plutonium is high enough to bring about the fusion of tritium and deuterium, accompanied by the release of 1.7×10^9 kJ per mole of ${}^3H$. A 20-megaton hydrogen bomb usually contains about 300 lb LiD, as well as a considerable mass of plutonium and uranium.

19.7 Nuclear Radiation: Effects and Units

The use of nuclear energy and radiation is a double-edged sword that carries both risks and benefits. It can be used to harm (nuclear armaments) or to cure (radioisotopes in medicine).

Alpha, beta, and gamma radiation disrupt normal cell processes in living organisms by interacting with key biomolecules, breaking their covalent bonds, and producing energetic free radicals and ions that can lead to further disruptive reactions. The potential for serious radiation damage to humans is well known. The biological effects of the atomic bombs exploded at Hiroshima and Nagasaki, Japan, at the close of World War II in 1945, have been well documented. However, controlled exposure to nuclear radiation can be beneficial in destroying malignant tissue, as in radiation therapy for treating some cancers.

All technologies carry risks as well as benefits. In the 1800s, railroads were new and the poet William Wordsworth wrote of their risks and benefits in terms of "Weighing the mischief with the promised gain. . . ."

Radiation Units

The SI unit of radioactivity is the becquerel; 1 Bq = 1 s^{-1}. Another common unit of radioactivity is the curie; 1 Ci = 3.70×10^{10} s^{-1}. However, to measure the effects of radiation on tissue, units are needed for radiation dose that take into account the energy absorbed by tissue when radiation passes into it.

To quantify radiation and its effects, particularly on humans, several units have been developed. The SI unit of absorbed radiation dose is the **gray, Gy,** which is equal to the absorption of 1 J per kilogram of material; 1 Gy = 1 J/kg. The **rad** (*r*adiation *a*bsorbed *d*ose) also measures the quantity of radiation energy *absorbed;* 1 rad represents a dose of 1.00×10^{-2} J absorbed per kilogram of material. Thus, 1 Gy = 100 rad. Another unit is the **roentgen, R,** which corresponds to the deposition of 93.3×10^{-7} J per gram of tissue.

The roentgen is named to honor Wilhelm Röntgen, the German physicist who discovered X-rays.

The biological effects of radiation per rad or gray differ with the type of radiation, which can be quantified more generally using the **rem** (*r*oentgen *e*quivalent in *m*an).

$$\text{Effective dose in rems} = \text{quality factor} \times \text{dose in rads}$$

The quality factor depends on the type of radiation and other factors. It is arbitrarily set as 1 for beta and gamma radiation. It is between 10 and 20 for alpha particles, depending on total dose, dose rate, and type of tissue. Since one rem is a large quantity of radiation, the millirem (mrem) is commonly used (1 mrem = 10^{-3} rem). The SI unit of effective dose is the **sievert, Sv,** which is defined similarly to the rem, except that the absorbed dose is in grays, not rads. Consequently, 1 Sv = 100 rem.

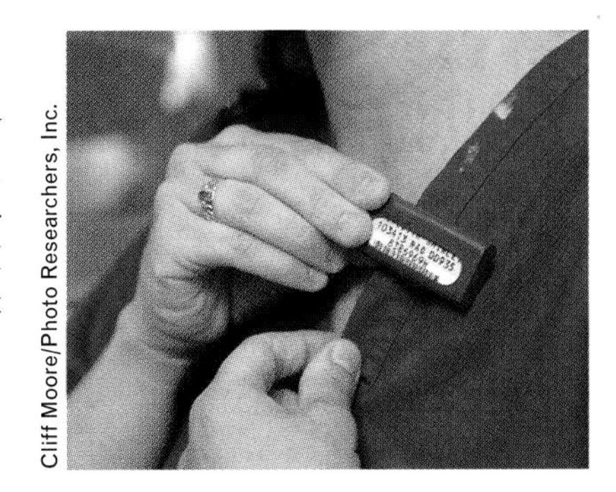

Cliff Moore/Photo Researchers, Inc.

Individuals who work where there is potential danger from exposure to excessive nuclear radiation wear film badges to monitor their radiation dose.

Background Radiation

Humans are constantly exposed to natural and artificial **background radiation,** estimated to be collectively about 360 mrem per year (Figure 19.10), well below 500 mrem, the federal government's background radiation standard for the general public. Note in Figure 19.10 that *most* background radiation, about 300 mrem per year (82%), comes from *natural* background radiation sources: cosmic radiation and radioactive elements and minerals found naturally in the earth, air, and materials around and within us. The remaining 18% comes from artificial sources.

Cosmic radiation, emitted by the sun and other stars, continually bombards the earth and accounts for about 8% of natural background radiation. The remainder comes from radioactive isotopes such as ^{40}K. Potassium is present to the extent of about 0.3 g per kilogram of soil and is essential to all living organisms. We all acquire some radioactive potassium from the foods we eat. For example, a hamburger contains 960 mg of ^{40}K, giving off 29 disintegrations per second (s^{-1}); a hot dog contains 200 mg of ^{40}K and gives off 6 s^{-1}; a serving of french fries has 650 mg ^{40}K and gives

There are no observable physiologic effects from a single dose of radiation less than 25 rem (25×10^3 mrem).

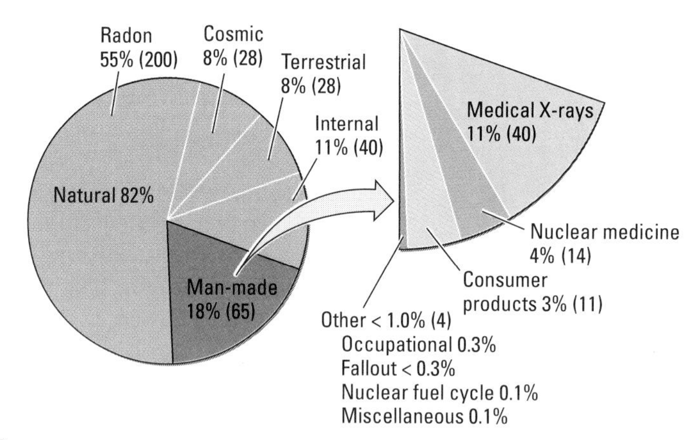

Figure 19.10 Sources of average background radiation exposure in the United States. The sources are expressed as percentages of the total, as well as in millirems per year, the values in parentheses. As seen from the figure, background radiation from natural sources far exceeds that from artificial sources.

off 20 s^{-1}. Other radioactive elements found in some abundance on the earth include thorium-232 and uranium-238. Approximately 8% of the natural background radiation arises from Th-232 and U-238 in rocks and soil. Thorium, for example, is found to the extent of 12 g per 1000 kg soil. Most natural background radiation comes from radon, a by-product of radium decay, as discussed in the next subsection.

On average, roughly 15% of our annual exposure comes from medical procedures such as diagnostic X-rays and the use of radioactive compounds to trace the body's functions. Consumer products account for 3% of our total annual exposure. Contrary to popular belief, less than 1% comes from sources such as the radioactive fission products from testing nuclear explosives in the atmosphere, nuclear power plants and their wastes, nuclear weapons manufacture, and nuclear fuel processing.

Burning fossil fuels (coal and oil) releases into the atmosphere considerable quantities of naturally occurring radioactive isotopes originally in the fossil fuel. Thus, fossil fuel plants add significantly to the background radiation. Far more thorium and uranium are released annually into the atmosphere from fossil fuel–burning plants than from nuclear power plants. The emission from nuclear power plants is essentially zero.

A recently published report by the National Research Council concludes that there is no safe level of radiation for humans, although the risks of low-dose radiation are small. The researchers studied doses of radiation of 0.1 Sv (10,000 mrem) or less, which is much greater than the 360 mrem per year natural background level discussed above. Rules for nuclear workers and others who are systematically exposed to elevated radiation levels may be affected by this type of research finding.

Radon

As shown in Figure 19.10, radon accounts for 55% of natural background radiation. Radon is a chemically inert gas in the same periodic table group as helium, neon, argon, and krypton. Radon-222 is produced in the decay series of uranium-238. Other isotopes of radon are products of other decay series. Although chemically inert, radon is problematic because it is a radioactive gas that can be inhaled.

An online radiation calculator is available at **http://www.epa.gov/radiation/understand/calculate.html.**

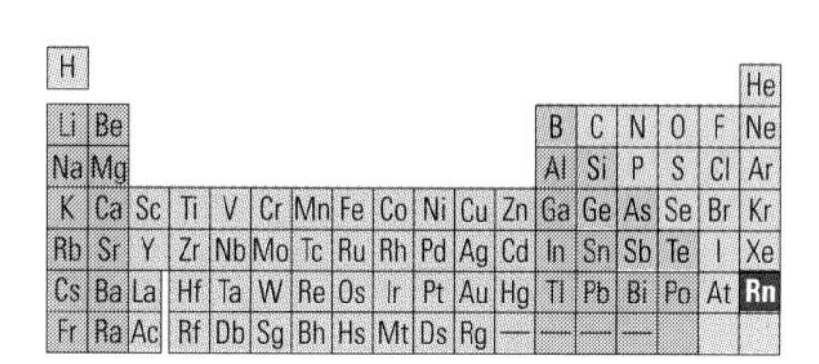

It should be kept in mind that radon occurs naturally in our environment. Because it comes from natural uranium deposits, the quantity of radon depends on the nature of the rocks and soil in a given locality. Furthermore, since the gas is chemically inert, it is not trapped by chemical processes in the soil or water. Thus, it can seep up from the ground and into underground mines or into homes through pores in concrete block walls, cracks in basement floors or walls, and around pipes. Radon-222 decays to give polonium-218, a radioactive, heavy metal element that is not a gas and is not chemically inert.

If radon is inhaled, some of the inhaled radon will decay to polonium in the lungs, and ^{218}Po will be generated there.

$$^{222}_{86}Rn \longrightarrow {}^{4}_{2}He + {}^{218}_{84}Po \qquad t_{1/2} = 3.82 \text{ days}$$

$$^{218}_{84}Po \longrightarrow {}^{4}_{2}He + {}^{214}_{82}Pb \qquad t_{1/2} = 3.10 \text{ minutes}$$

ESTIMATION Counting Millirems: Your Radiation Exposure

In 1987 the Committee on Biological Effects of Ionizing Radiation of the National Academy of Sciences issued a report that contained a survey for an individual to evaluate his or her exposure to ionizing radiation. The table below is adapted from this report and updated. By adding up your exposure, you can compare your annual dose to the U.S. annual average of 360 mrem.

Source: Based on the BEIR Report III. National Academy of Sciences, Committee on Biological Effects of Ionizing Radiation. *The Effects on Populations of Exposure to Low Levels of Ionizing Radiation.* Washington, DC: National Academy of Sciences, 1987.

Common Sources of Radiation		Your Annual Dose (mrem)
Where you live	**Location:** Cosmic radiation at sea level	27
	For your elevation (in feet), add this number of mrem	
	Elevation 1000: 2; 2000: 5; 3000: 9; 4000: 15; 5000: 21; 6000: 29; 7000: 40; 8000: 53; 9000: 70	
	Ground: U.S. average	26
	Radon: U.S. average	200
	House construction: For stone, concrete, or masonry building, add 7; for wood, add 30	
What you eat, drink, and breathe	**Radioisotopes** in the body from	
	Food, air, water: U.S. average	40
	Weapons test fallout	4
How you live	**X-ray and radiopharmaceutical diagnosis**	
	Number of chest X-rays × 10	
	Number of lower gastrointestinal tract X-rays × 500	
	Number of radiopharmaceutical examinations × 300	
	(Average dose to total U.S. population = 53 mrem)	
	Jet plane travel: For each 2500 miles add 1 mrem	
	TV viewing: Number of hours per day × 0.15	
How close you live to a nuclear plant	**At site boundary:** Average number of hours per day × 0.2	
	One mile away: Average number of hours per day × 0.02	
	Five miles away: Average number of hours per day × 0.002	
	Over five miles away: None	
	Note: Maximum allowable dose determined by "as low as reasonably achievable" (ALARA) criteria established by the U.S. Nuclear Regulatory Commission. Experience shows that your actual dose is substantially less than these limits.	
	Your total annual dose in mrem:	

Elevation table:

Elevation	*mrem*	*Elevation*	*mrem*	*Elevation*	*mrem*
1000	2	4000	15	7000	40
2000	5	5000	21	8000	53
3000	9	6000	29	9000	70

Compare your annual dose with the U.S. annual average of 360 mrem.

Visit this book's companion website at **www.cengage.com/chemistry/moore** to work an interactive module based on this material.

Polonium-218 can lodge in lung tissues, where it undergoes alpha decay to give lead-214, itself a radioactive isotope. The range of an alpha particle is quite small, perhaps 0.7 mm (about the thickness of a sheet of paper). However, this is approximately the thickness of the epithelial cells of the lungs, so the radiation can damage these tissues and induce lung cancer.

Most homes in the United States are believed to have some level of radon gas. There is currently a great deal of controversy over the level of radon in a house that is considered safe. Estimates indicate that only about 6% of U.S. homes have radon levels above 4 picocuries per liter (pCi/L) of air, the action level standard set by the U.S. Environmental Protection Agency. Some believe that 1.5 pCi/L is more likely the average level and that only about 2% of the homes will contain more than 8 pCi/L. To test for the presence of radon, you can purchase testing kits of various kinds. If your home shows higher levels of radon gas than 4 pCi/L, you should probably have it tested further. Perhaps corrective actions are needed, such as sealing cracks around the foundation and in the basement, but keep in mind the relative risks involved. A 1.5 pCi/L level of radon leads to a lung cancer risk about the same as the risk of your dying in an accident in your home.

1 picocurie, pCi = 10^{-12} Ci.

A commercially available kit to test for radon gas in a residence.

EXERCISE 19.12 Radon Levels

Calculate how long it will take for the activity of a radon-222 sample ($t_{1/2}$ = 3.82 days) initially at 8 pCi to drop

(a) To 4 pCi, the EPA action level.

(b) To 1.5 pCi, approximately the U.S. average.

IN CLOSING

Having studied this chapter, you should be able to . . .

- Characterize the three major types of radiation observed in radioactive decay: alpha (α), beta (β), and gamma (γ) (Section 19.1). ■ End-of-chapter Question assignable in OWL: 11
- Write a balanced equation for a nuclear reaction or transmutation (Section 19.2). ■ Questions 13, 17
- Decide whether a particular radioactive isotope will decay by α, β, or positron emission or by electron capture (Sections 19.2, 19.3). ■ Question 20
- Calculate the binding energy for a particular isotope and understand what this energy means in terms of nuclear stability (Section 19.3). ■ Question 22
- Use Equation 19.3, $\ln\frac{N}{N_0} = -kt$, which relates (through the decay constant k) the time period over which a sample is observed (t) to the number of radioactive atoms present at the beginning (N_0) and end (N) of the time period (Section 19.4). ■ Questions 27, 31
- Calculate the half-life of a radioactive isotope ($t_{1/2}$) from the activity of a sample at two times, or use the half-life to find the time required for an isotope to decay to a particular activity (Section 19.4). ■ Questions 35, 67
- Describe nuclear chain reactions, nuclear fission, and nuclear fusion (Sections 19.5, 19.6).
- Describe the basic functioning of a nuclear power reactor (Section 19.5).
- Describe some sources of background radiation and the units used to measure radiation (Section 19.7).

Selected end-of-chapter Questions may be assigned in OWL.

For quick review, download Go Chemistry mini-lecture flashcard modules (or purchase them at **www.ichapters.com**).

Prepare for an exam with a **Summary Problem** that brings together concepts and problem-solving skills from throughout the chapter. Go to **www.cengage.com/chemistry/moore** to download this chapter's Summary Problem from the Student Companion Site.

KEY TERMS

activity *(Section 19.4)*
alpha (α) particles *(19.1)*
alpha radiation *(19.1)*
background radiation *(19.7)*
becquerel, Bq *(19.4)*
beta (β) particles *(19.1)*
beta radiation *(19.1)*
binding energy *(19.3)*
binding energy per nucleon *(19.3)*
critical mass *(19.5)*
curie, Ci *(19.4)*
electron capture *(19.2)*
gamma (γ) radiation *(19.1)*
gray, Gy *(19.7)*
half-life *(19.4)*
nuclear fission *(19.5)*
nuclear fusion *(19.6)*
nuclear reactions *(19.2)*
nuclear reactor *(19.5)*
nucleons *(19.2)*
plasma *(19.6)*
positron *(19.2)*
rad *(19.7)*
radioactive series *(19.2)*
rem *(19.7)*
roentgen, R *(19.7)*
sievert, Sv *(19.7)*

QUESTIONS FOR REVIEW AND THOUGHT

■ denotes questions assignable in OWL.

Blue-numbered questions have short answers at the back of this book and fully worked solutions in the *Student Solutions Manual.*

Review Questions

1. Complete the tables.

	Symbol	Mass	Charge
α particle	______	______	______
β particle	______	______	______
γ radiation	______	______	______

	Ionizing Power	Penetrating Power
α particle	______	______
β particle	______	______
γ radiation	______	______

2. Compare nuclear and chemical reactions in terms of changes in reactants, type of products formed, and conservation of matter and energy.
3. What is meant by the "band of stability"?
4. What is the binding energy of a nucleus?
5. If the mass number of an isotope is much greater than twice the atomic number, what type of radioactive decay might you expect?
6. If the number of neutrons in an isotope is much less than the number of protons, what type of radioactive decay might you expect?
7. Define critical mass and chain reaction.
8. What is the difference between nuclear fission and nuclear fusion? Illustrate your answer with an example of each.
9. Use the World Wide Web to locate the nuclear reactor power plant nearest to your college residence. Do you consider it to pose a threat to your health and safety? If so, why? If not, why not?
10. Name at least two uses of radioactive isotopes (outside of their use in power reactors and weapons).

Topical Questions

Nuclear Reactions

11. ■ By what processes do these transformations occur?
    (a) Thorium-230 to radium-226
    (b) Cesium-137 to barium-137
    (c) Potassium-38 to argon-38
    (d) Zirconium-97 to niobium-97
12. By what processes do these transformations occur?
    (a) Uranium-238 to thorium-234
    (b) Iodine-131 to xenon-131
    (c) Nitrogen-13 to carbon-13
    (d) Bismuth-214 to polonium-214
13. ■ Fill in the mass number, atomic number, and symbol for the missing particle in each nuclear equation.
    (a) $^{242}_{94}Pu \longrightarrow {}^{4}_{2}He + \_\_\_\_$
    (b) $\_\_\_\_ \longrightarrow {}^{32}_{16}S + {}^{0}_{-1}e$
    (c) $^{252}_{98}Cf + \_\_\_\_ \longrightarrow 3\,{}^{1}_{0}n + {}^{259}_{103}Lr$
    (d) $^{55}_{26}Fe + \_\_\_\_ \longrightarrow {}^{55}_{25}Mn$
    (e) $^{15}_{8}O \longrightarrow \_\_\_\_ + {}^{0}_{+1}e$
14. Fill in the mass number, atomic number, and symbol for the missing particle in each nuclear equation.
    (a) $\_\_\_\_ \longrightarrow {}^{22}_{10}Ne + {}^{0}_{+1}e$
    (b) $^{122}_{53}I \longrightarrow {}^{122}_{54}Xe + \_\_\_\_$
    (c) $^{210}_{84}Po \longrightarrow \_\_\_\_ + {}^{4}_{2}He$
    (d) $^{195}_{79}Au + \_\_\_\_ \longrightarrow {}^{195}_{78}Pt$
    (e) $^{241}_{94}Pu + {}^{16}_{8}O \longrightarrow 5\,{}^{1}_{0}n + \_\_\_\_$

15. Write a balanced nuclear equation for each word statement.
    (a) Magnesium-28 undergoes β emission.
    (b) When uranium-238 is bombarded with carbon-12, four neutrons are emitted and a new element forms.
    (c) Hydrogen-2 and helium-3 react to form helium-4 and another particle.
    (d) Argon-38 forms by positron emission.
    (e) Platinum-175 forms osmium-171 by spontaneous radioactive decay.
16. Write a balanced nuclear equation for each word statement.
    (a) Einsteinium-253 combines with an alpha particle to form a neutron and a new element.
    (b) Nitrogen-13 undergoes positron emission.
    (c) Iridium-178 captures an electron to form a stable nucleus.
    (d) A proton and boron-11 fuse, forming three identical particles.
    (e) Nobelium-252 and six neutrons form when carbon-12 collides with a transuranium isotope.
17. ■ One radioactive series that begins with uranium-235 and ends with lead-207 undergoes this sequence of emission reactions: α, β, α, β, α, α, α, α, β, β, α. Identify the radioisotope produced in each of the *first five steps.*
18. One radioactive series that begins with uranium-235 and ends with lead-207 undergoes this sequence of emission reactions: α, β, α, β, α, α, α, α, β, β, α. Identify the radioisotope produced in each of the *last six steps.*
19. Radon-222 is unstable, and its presence in homes may constitute a health hazard. It decays by this sequence of emissions: α, α, β, β, α, β, β, α. Write out the sequence of nuclear reactions leading to the final product nucleus, which is stable.

Nuclear Stability

20. ■ Write a nuclear equation for the type of decay each of these unstable isotopes is most likely to undergo.
    (a) Neon-19 (b) Thorium-230
    (c) Bromine-82 (d) Polonium-212
21. Write a nuclear equation for the type of decay each of these unstable isotopes is most likely to undergo.
    (a) Silver-114 (b) Sodium-21
    (c) Radium-226 (d) Iron-59
22. ■ Boron has two stable isotopes, ^{10}B (abundance = 19.78%) and ^{11}B (abundance = 80.22%). Calculate the binding energies per nucleon of these two nuclei and compare their stabilities.

$$5\,^{1}_{1}H + 5\,^{1}_{0}n \longrightarrow {}^{10}_{5}B$$
$$5\,^{1}_{1}H + 6\,^{1}_{0}n \longrightarrow {}^{11}_{5}B$$

The required masses (in g/mol) are $^{1}_{1}H$ = 1.00783; $^{1}_{0}n$ = 1.00867; $^{10}_{5}B$ = 10.01294; and $^{11}_{5}B$ = 11.00931.
23. Calculate the binding energy in kJ per mole of P for the formation of $^{30}_{15}P$ and $^{31}_{15}P$.

$$15\,^{1}_{1}H + 15\,^{1}_{0}n \longrightarrow {}^{30}_{15}P$$
$$15\,^{1}_{1}H + 16\,^{1}_{0}n \longrightarrow {}^{31}_{15}P$$

Which is the more stable isotope? The required masses (in g/mol) are $^{1}_{1}H$ = 1.00783; $^{1}_{0}n$ = 1.00867; $^{30}_{15}P$ = 29.97832; and $^{31}_{15}P$ = 30.97376.
24. The most abundant isotope of uranium is U-238, which has an isotopic mass of 238.0508 g/mol. What is its nuclear binding energy in kJ/mol and binding energy per nucleon?
25. What is the nuclear binding energy in kJ/mol and binding energy per nucleon of chlorine-35, which has an isotopic mass of 34.9689 g/mol?
26. What is the nuclear binding energy and binding energy per nucleon in kJ/mol of iodine-127, which has an isotopic mass of 126.9004 g/mol?

Rates of Disintegration Reactions

27. ■ Sodium-24 is a diagnostic radioisotope used to measure blood circulation time. How much of a 20-mg sample remains after 1 day and 6 hours if sodium-24 has $t_{1/2}$ = 15 hours?
28. Iron-59 in the form of iron(II) citrate is used in iron metabolism studies. Its half-life is 45.6 days. If you start with 0.56 mg iron-59, how much would remain after 1 year?
29. Iodine-131 is used in the form of sodium iodide to treat cancer of the thyroid.
    (a) The isotope decays by ejecting a β particle. Write a balanced equation to show this process.
    (b) The isotope has a half-life of 8.05 days. If you begin with 25.0 mg of radioactive $Na^{131}I$, what mass remains after 32.2 days?
30. Phosphorus-32 is used in the form of $Na_2H^{32}PO_4$ in the treatment of chronic myeloid leukemia, among other things.
    (a) The isotope decays by emitting a β particle. Write a balanced equation to show this process.
    (b) The half-life of ^{32}P is 14.3 days. If you begin with 9.6 mg of radioactive $Na_2H^{32}PO_4$, what mass remains after 28.6 days?
31. ■ What is the half-life of a radioisotope if it decays to 12.5% of its radioactivity in 12 years?
32. After 2 hours, tantalum-172 has $\frac{1}{16}$ of its initial radioactivity. How long is its half-life?
33. Radioisotopes of iodine are widely used in medicine. For example, iodine-131 ($t_{1/2}$ = 8.05 days) is used to treat thyroid cancer. If you ingest a sample of $Na^{131}I$, how much time is required for the isotope to decrease to 5.0% of its original activity?
34. The noble gas radon has been the focus of much attention because it may be found in homes. Radon-222 emits α particles and has a half-life of 3.82 days.
    (a) Write a balanced equation to show this process.
    (b) How long does it take for a sample of radon to decrease to 10.0% of its original activity?
35. ■ A sample of wood from a Thracian chariot found in an excavation in Bulgaria has a ^{14}C activity of 11.2 disintegrations per minute per gram. Estimate the age of the chariot and the year it was made. ($t_{1/2}$ for ^{14}C is 5.73×10^3 years, and the activity of ^{14}C in living material is 15.3 disintegrations per minute per gram.)
36. A piece of charred bone found in the ruins of a Native American village has a $^{14}C/^{12}C$ ratio of 0.72 times that found in living organisms. Calculate the age of the bone fragment. (See Question 35 for required data on carbon-14.)
37. How long will it take for a sample of plutonium-239 with a half-life of 2.4×10^4 years to decay to 1% of its original activity?

38. A 1.00-g sample of wood from an archaeological site gave 4100 disintegrations of ^{14}C in a 10-hour measurement. In the same time, a 1.00-g modern sample gave 9200 disintegrations. What is the age of the wood?

Nuclear Fission and Fusion

39. Name the fundamental parts of a nuclear fission reactor and describe their functions.
40. Explain why it is easier for a nucleus to capture a neutron than to force a nucleus to capture a proton.
41. ■ What is the missing product in each of these fission equations?
   (a) $^{235}_{92}U + ^{1}_{0}n \longrightarrow \_\_\_\_ + ^{93}_{38}Sr + 3\,^{1}_{0}n$
   (b) $^{235}_{92}U + ^{1}_{0}n \longrightarrow \_\_\_\_ + ^{132}_{51}Sb + 3\,^{1}_{0}n$
   (c) $^{235}_{92}U + ^{1}_{0}n \longrightarrow \_\_\_\_ + ^{141}_{56}Ba + 3\,^{1}_{0}n$
42. Explain why no commercial fusion reactors are in operation today.
43. The average energy output of a barrel of oil is 5.9×10^6 kJ/barrel. Fission of 1 mol ^{235}U releases 2.1×10^{10} kJ of energy. Calculate the number of barrels of oil needed to produce the same energy as 1.0 lb ^{235}U.
44. A concern in the nuclear power industry is that, if nuclear power becomes more widely used, there may be serious shortages in worldwide supplies of fissionable uranium. One solution is to build breeder reactors that manufacture more fuel than they consume. One such cycle works as follows:
   (i) A ^{238}U nucleus collides with a neutron to produce ^{239}U.
   (ii) ^{239}U decays by β emission ($t_{1/2}$ = 24 minutes) to give an isotope of neptunium.
   (iii) The neptunium isotope decays by β emission to give a plutonium isotope.
   (iv) The plutonium isotope is fissionable. On its collision with a neutron, fission occurs and gives energy, at least two neutrons, and other nuclei as products.

   Write an equation for each of these steps, and explain how this process can be used to breed more fuel than the reactor originally contained and still produce energy.

Effects of Nuclear Radiation

45. Two common units of radiation used in newspaper and news magazine articles are the rad and rem. What does each measure? Which would you use in an article describing the damage an atomic bomb would inflict on a human population? What relationship does the gray have with these units?
46. Which electrical power plant—fossil fuel or nuclear—exposes a community to more nuclear radiation? Explain why.
47. Explain how our own bodies are sources of nuclear radiation.
48. What is the source of radiation exposure during jet plane travel?

General Questions

49. Complete these nuclear equations.
   (a) $^{214}Bi \longrightarrow \_\_\_\_ + ^{214}Po$
   (b) $4\,^{1}_{1}H \longrightarrow \_\_\_\_$ + 2 positrons
   (c) ^{249}Es + neutron $\longrightarrow$ 2 neutrons + ____ + ^{161}Gd
   (d) $^{220}Rn \longrightarrow \_\_\_\_$ + alpha particle
   (e) ^{68}Ge + electron $\longrightarrow \_\_\_\_$
50. Complete these nuclear equations.
   (a) ____ + neutron $\longrightarrow$ 2 neutrons + $^{137}Tc + ^{97}Zr$
   (b) $^{45}Ti \longrightarrow \_\_\_\_$ + positron
   (c) ____ $\longrightarrow$ beta particle + ^{59}Co
   (d) ^{24}Mg + neutron $\longrightarrow$ ____ + proton
   (e) ^{131}Cs + ____ $\longrightarrow ^{131}Xe$
51. Radioactive nitrogen-13 has a half-life of 10 minutes. After an hour, how much of this isotope remains in a sample that originally contained 96 mg?
52. The half-life of molybdenum-99 is 67.0 hours. How much of a 1.000-mg sample of ^{99}Mo is left after 335 hours? How many half-lives did it undergo?
53. The oldest known fossil cells were found in South Africa. The fossil has been dated by the reaction

   $$^{87}Rb \longrightarrow ^{87}Sr + ^{0}_{-1}e \qquad t_{1/2} = 4.9 \times 10^{10} \text{ years}$$

   If the ratio of the present quantity of ^{87}Rb to the original quantity is 0.951, calculate the age of the fossil cells.
54. Cobalt-60 is a therapeutic radioisotope used in treating certain cancers. If a sample of cobalt-60 initially disintegrates at a rate of $4.3 \times 10^6\ s^{-1}$ and after 21.2 years the rate has dropped to $2.6 \times 10^5\ s^{-1}$, what is the half-life of cobalt-60?
55. On December 2, 1942, the first man-made self-sustaining nuclear fission chain reactor was operated by Enrico Fermi and others under the University of Chicago stadium. In June 1972, natural fission reactors, which operated billions of years ago, were discovered in Oklo, Gabon. At present, natural uranium contains 0.72% ^{235}U. How many years ago did natural uranium contain 3.0% ^{235}U, sufficient to sustain a natural reactor? ($t_{1/2}$ for $^{235}U = 7.04 \times 10^8$ years.)

Applying Concepts

56. During the Three Mile Island incident, people in central Pennsylvania were concerned that strontium-90 (a beta emitter) released from the reactor could become a health threat (it did not). Where would this isotope collect in the body? If so, what types of problems could it cause?
57. ■ Classify the isotopes ^{17}Ne, ^{20}Ne, and ^{23}Ne as stable or unstable. What type of decay would you expect the unstable isotope(s) to have?
58. The following demonstration was carried out to illustrate the concept of a nuclear chain reaction. Explain the connections between the demonstration and the reaction.

   Eighty mousetraps are arranged side by side in eight rows of ten traps each. Each trap is set with two rubber stoppers for bait. A small plastic mouse is tossed into the middle of the traps, setting off one trap, which in turn sets off two traps and so on until all the traps are sprung.
59. Most students have no trouble understanding that 1.5 g of a 24-g sample of a radioisotope would remain after 8 h if it had $t_{1/2}$ = 2 h. What they don't always understand is where the other 22.5 g went. How would you explain this disappearance to another student?

60. Nuclear chemistry is a topic that raises many debatable issues. Briefly discuss your views regarding:
    (a) Twice a year the general public is allowed to visit the Trinity Site in Alamogordo, New Mexico, where the first atomic bomb was tested. If you had the opportunity to do so, would you visit the site? Explain your answer.
    (b) Now that the Cold War has ended, should the United States continue to stockpile nuclear weapons? Explain your answer.

More Challenging Questions

61. All radioactive decays are first-order. Why is this so?
62. An average 70.0-kg adult contains about 170 g potassium. Potassium-40, with a relative abundance of 0.0118%, undergoes beta decay with a half-life of 1.28×10^9 years. What is the total activity (dps) due to beta decay of ^{40}K for this average person?
63. If the earth receives 7.0×10^{14} kJ/s of energy from the sun, what mass of solar material is lost per hour to supply this amount of energy?
64. A sample of the alpha emitter ^{222}Ra had an initial activity A_0 of 7.00×10^4 Bq. After 10.0 days its activity A had fallen to 1.15×10^4 Bq. Calculate the decay constant and half-life of radon-222.
65. When a bottle of wine was analyzed for its tritium (^3H) content, it was found to contain 1.45% of the tritium originally present when the wine was produced. How old is the bottle of wine? ($t_{1/2}$ of ^3H = 12.3 years.)
66. A chemist is setting up an experiment using ^{47}Ca, which has a half-life of 4.5 days. He needs 10.0 μg of the calcium. How many μg of $^{47}CaCO_3$ must he order if the delivery time is 50 hours?
67. ■ To determine the age of the charcoal found at an archaeological site, this sequence of experiments was done: The charcoal sample was burned in oxygen and the CO_2 obtained was captured by bubbling it through lime water, $Ca(OH)_2$, to form a precipitate of $CaCO_3$. This precipitate was filtered, dried, and weighed. A sample of 1.14 g $CaCO_3$ produced 2.17×10^{-2} Bq from carbon-14. Modern carbon produces 15.3 disintegrations min^{-1} g^{-1} of carbon. What is the age of the charcoal? The half-life of carbon-14 is 5730 years.

Conceptual Challenge Problems

CP19.A (Section 19.4) The half-life for the alpha decay of uranium-238 to thorium-234 is 4.5×10^9 years, which happens to be the estimated age of the earth.
(a) How many atoms were decaying per second in a 1.0-g sample of uranium-238 that existed 1.0×10^6 years ago?
(b) How would you find the number of atoms now decaying per second in this sample?

CP19.B (Section 19.4) If the earth is 4.5×10^9 years old and the amount of radioactivity in a sample becomes smaller with time, how is it possible for there to be any radioactive elements left on earth that have half-lives less than a few million years?

CP19.C (Section 19.4) Using experiments based on a sample of living wood, a nuclear chemist estimates that the uncertainty of her measurements of the carbon-14 radioactivity in the sample is 1.0%. The half-life of carbon-14 is 5730 years.
(a) How long must a sample of wood be separated from a living tree before the chemist's radioactivity measurements on the sample provide evidence for the time when it died?
(b) Suppose that the chemist's uncertainty in the radioactivity of carbon-14 continues to be 1.0% of the radioactivity of living wood. How long must a sample of wood be dead before the chemist's measurements support the claim that the time since the wood was separated from the tree is not changing?

CP19.D (Section 19.7) You have read that alpha radiation is the least penetrating type of radiation, followed by beta radiation. Gamma radiation penetrates matter well, and thick samples of matter are required to contain gamma radiation. Knowing these facts, what can you correctly deduce about the harmful effects of these three types of radiation on living tissue?

CP-19.E (Section 19.7) Death will likely occur within weeks for a 150-lb person who receives 500,000 mrem of radiation over a short time, an exposure that is 1000 times the federal government's standard for 1 year (500 mrem/yr). A student realizes that 500,000 mrem is 500 rem, and that 500 rem has the effect of depositing 317 J of energy on the body of the 150-lb person. The student is puzzled. How can the deposition of only 317 J of energy from nuclear radiation—much less energy than that deposited by cooling a cup of coffee 1 °C within a person's body—have such a disastrous effect on the person?

APPENDIX

A

Problem Solving and Mathematical Operations

In this book we have provided many illustrations of problem solving and many problems for practice. Some are numerical problems that must be solved by mathematical calculations. Others are conceptual problems that must be solved by applying an understanding of the principles of chemistry. Often, it is necessary to use chemical concepts to relate what we know about matter at the nanoscale to the properties of matter at the macroscale. The problems throughout this book are representative of the kinds of problems that chemists and other scientists must regularly solve to pursue their goals, although our problems are often not as difficult as those encountered in the real world.

Problem solving is not a simple skill that can be mastered with a few hours of study or practice. Because there are many different kinds of problems and many different kinds of people who are problem solvers, no hard and fast rules are available that are guaranteed to lead you to solutions. The general guidelines presented in this appendix are, however, helpful in getting you started on any kind of problem and in checking whether your answers are correct. The problem-solving skills you develop in a chemistry course can later be applied to difficult and important problems that may arise in your profession, your personal life, or the society in which you live.

In getting a clear picture of a problem and asking appropriate questions regarding the problem, you need to keep in mind all the principles of chemistry and other subjects that you think may apply. In many real-life problems, not enough information is available for you to arrive at an unambiguous solution; in such cases, try to look up or estimate what is needed and then forge ahead, noting assumptions you have made. Often the hardest part is deciding which principle or idea is most likely to help solve the problem and what information is needed. To some degree this can be a matter of luck or chance. Nevertheless, in the words of Louis Pasteur, "In the field of observation chance only favors those minds which have been prepared." The more practice you have had, and the more principles and facts you can keep in mind, the more likely you are to be able to solve the problems that you face.

A.1 General Problem-Solving Strategies

"The mere formulation of a problem is often far more essential than its solution. To raise new questions, new possibilities, to regard old problems from a new angle, requires creative imagination and marks real advances in science."
—Albert Einstein

1. **Define the problem.** Carefully review the information contained in the problem. What is the problem asking you to find? What key principles are involved? What known information is necessary for solving the problem and what information is there just to place the question in context? Organize the information to see what is necessary and to see the relationships among the known data. Try writing the information in an organized way. If the information is numerical, be sure to include proper units. Can you picture the situation under consideration? Try sketching it and including any relevant dimensions in the sketch.
2. **Develop a plan.** Have you solved a problem of this type before? If you recognize the new problem as similar to ones you know how to solve, you can use the same method that worked before. Try reasoning backward from the units of what is being sought. What data are needed to find an answer in those units?

Can the problem be broken down into smaller pieces, each of which can be solved separately to produce information that can be assembled to solve the entire problem? When a problem can be divided into simpler problems, it often helps to write down a plan that lists the simpler problems and the order in which those problems must be put together to arrive at an overall solution. Many major problems in chemical research have to be solved in this way. In problems in this book we have mostly provided the needed numerical data, but in the laboratory, the first aspect of solving a problem is often devising experiments to gather the data or searching databases to find needed information.

If you are still unsure about what to do, do something anyway. It may not be the right thing to do, but as you work on it, the way to solve the problem may become apparent, or you may see what is wrong with your initial approach, thereby making clearer what a good plan would be.

3. **Execute the plan.** Carefully write down each step of a mathematical problem, being sure to keep track of the units. Do the units cancel to give you the answer in the desired units? Don't skip steps. Don't do any except the simplest steps in your head. Once you've written down the steps for a mathematical problem, check what you've written—is it all correct? Students often say they got a problem wrong because they "made a stupid mistake." Teachers—and textbook authors—make mistakes, too. These errors usually arise because they don't take the time to write down the steps of a problem clearly and correctly. In solving a mathematical problem, remember to apply the principles of dimensional analysis and significant figures. Dimensional analysis is introduced in Sections 1.4 ***(⇐ p. 5)*** and 2.3 ***(⇐ p. 40);*** it is reviewed below. Section 2.4 ***(⇐ p. 44)*** and Appendix A.3 ***(p. A.5)*** introduce significant figures.

4. ✓ **Check the answer to see whether it is reasonable.** As a final check of your solution to any problem, ask yourself whether the answer is reasonable: Are the units of a numerical answer correct? Is a numerical answer of about the right size? Don't just copy a result from your calculator without thinking about whether it makes sense.

Suppose you have been asked to convert 100. yards to a distance in meters. Using dimensional analysis and some well-known factors for converting from the English system to the metric system, you could write

$$100.\text{ yd} \times \frac{3\text{ ft}}{1\text{ yd}} \times \frac{12\text{ in.}}{1\text{ ft}} \times \frac{2.54\text{ cm}}{1\text{ in.}} \times \frac{1\text{ m}}{100\text{ cm}} = 91.4\text{ m}$$

To check that a distance of 91.4 m is about right, recall that a yard is a little shorter than a meter. Therefore 100 yd should be a little less than 100 m. If you had mistakenly divided instead of multiplied by 3 ft/yd in the first step, your final answer would have been a little more than 10 m. This is equivalent to only about 30 ft, and you probably know a 100-yd football field is longer than that.

A.2 Numbers, Units, and Quantities

Many scientific problems require you to use mathematics to calculate a result or draw a conclusion. Therefore, knowledge of mathematics and its application to problem solving is important. However, one aspect of scientific calculations is often absent from pure mathematical work: Science deals with *measurements* in which an unknown quantity is compared with a standard or unit of measure. For example, using a balance to determine the mass of an object involves comparing the object's mass with standard masses, usually in multiples or fractions of one gram; the result is reported as some number of grams, say 4.357 g. *Both the number and the unit are important.* If the result had been 123.5 g, this would clearly be different, but a result of 4.357 oz (ounces) would also be different, because the unit "ounce" is different from the unit "gram." A *result that describes the quantitative measurement of a property,* such as 4.357 g, is called a *quantity* (or physical quantity), and it consists of a number and a unit. Chemical problem solving requires calculating with quantities. Notice that whether a quantity is large or small depends on the units as well as the number; the two quantities 123.5 g and 4.357 oz, for example, represent the *same* mass.

A quantity is always treated as though the number and the units are multiplied together; that is, 4.357 g can be handled mathematically as $4.357 \times \text{g}$. Using this simple rule, you will see that calculations involving quantities follow the normal rules of algebra and arithmetic: $5\text{ g} + 7\text{ g} = 5 \times \text{g} + 7 \times \text{g} = (5 + 7) \times \text{g} = 12\text{ g}$; or $6\text{ g} \div 2\text{ g} = (6\text{ g})/(2\text{ g}) = 3$. (Notice that in the second calculation the unit g appears in both the numerator and the denominator and cancels out, leaving a pure number, 3.) Treating units as algebraic entities has the advantage that *if a calculation is set up correctly, the units will cancel or multiply together so that the final result has appropriate units.* For example, if you measured the size of a sheet of paper and found it to be 8.5 in. by 11 in., the area A of the sheet could be calculated as area = length × width = 11 in. × 8.5 in. = 94 in.^{2}, or 94 square inches. If a calculation is set up incorrectly, the units of the result will be inappropriate. Using units to check whether a calculation has been properly set up is called ***dimensional analysis*** **(⇐ p. 8)**.

This idea of using algebra on units as well as numbers is useful in all kinds of situations. For example, suppose you are having a party for some friends who like pizza. A large pizza consists of 12 slices and costs \$10.75. You expect to need 36 slices of pizza and want to know how much you will have to spend. A strategy for solving the problem is first to figure out how many pizzas you need and then to figure out the cost in dollars. This solution could be diagrammed as

$$\text{Slices} \xrightarrow[\text{step 1}]{\text{slices per pizza}} \text{pizzas} \xrightarrow[\text{step 2}]{\text{dollars per pizza}} \text{dollars}$$

Step 1: Find the number of pizzas required by dividing the number of slices per pizza into the number of slices, thus converting "units" of slices to "units" of pizzas:

$$\text{Number of pizzas} = 36 \text{ slices} \left(\frac{1 \text{ pizza}}{12 \text{ slices}}\right) = 3 \text{ pizzas}$$

If you had multiplied the number of slices times the number of slices per pizza, the result would have been labeled pizzas × slices^2, which does not make sense. In other words, the labels indicate whether multiplication or division is appropriate.

Strictly speaking, slices and pizzas are not units in the same sense that a gram is a unit. Nevertheless, labeling things this way will often help you keep in mind what a number refers to—pizzas, slices, or dollars in this case.

Step 2: Find the total cost by multiplying the cost per pizza by the number of pizzas needed, thus converting "units" of pizzas to "units" of dollars:

$$\text{Total price} = 3 \text{ pizzas} \left(\frac{\$10.75}{1 \text{ pizza}}\right) = \$32.25$$

Notice that in each step you have multiplied by a factor that allowed the initial units to cancel algebraically, giving the answer in the desired units. A factor such as (1 pizza/12 slices) or (\$10.75/pizza) is referred to as a ***proportionality factor*** **(⇐ p. 8)**. This name indicates that it comes from a proportion. For instance, in the pizza problem you could set up the proportion

$$\frac{x \text{ pizzas}}{36 \text{ slices}} = \frac{1 \text{ pizza}}{12 \text{ slices}} \quad \text{or} \quad x \text{ pizzas} = 36 \text{ slices}\left(\frac{1 \text{ pizza}}{12 \text{ slices}}\right) = 3 \text{ pizzas}$$

A proportionality factor such as (1 pizza/12 slices) is also called a ***conversion factor,*** which indicates that it converts one kind of unit or label to another; in this case the label "slices" is converted to the label "pizzas."

Many everyday scientific problems involve proportionality. For example, the bigger the volume of a solid or liquid substance, the bigger its mass. When the volume is zero, the mass is also zero. These facts indicate that mass, m, is directly proportional to volume, V, or, symbolically,

$$m \propto V$$

where the symbol $\propto$ means "is proportional to." Whenever a proportion is expressed this way, it can also be expressed as an equality by using a proportionality constant—for example,

$$m = d \times V$$

In this case the proportionality constant, d, is called the density of the substance. This equation embodies the definition of density as mass per unit volume, since it can be rearranged algebraically to

$$d = \frac{m}{V}$$

As with any algebraic equation involving three variables, it is possible to calculate any one of the three quantities m, V, or d, provided the other two are known. If density is wanted, simply use the definition of mass per unit volume; if mass or volume is to be calculated, the density can be used as a proportionality factor.

Suppose that you are going to buy a ton of gravel and want to know how big a bin you will need to store it. You know the mass of gravel and want to find the volume of the bin; this implies that density will be useful. If the gravel is primarily limestone, you can assume that its density is about the same as for limestone and look it up. Limestone has the chemical formula $CaCO_3$, and its density is 2.7 kg/L. However, these mass units are different from the units for mass of gravel—namely, tons. Therefore you need to recall or look up the mass of 1 ton (exactly 2000 pounds [lb]) and the fact that there are 2.20 lb per kilogram. This provides enough information to calculate the volume needed. Here is a "roadmap" plan for the calculation:

$$\text{Mass of gravel in tons} \xrightarrow[\text{step1}]{\text{change units}} \text{mass of gravel in kilograms} \xrightarrow[\text{step 2}]{\text{density}} \text{volume of bin}$$

Step 1: Figure out how many kilograms of gravel are in a ton.

$$m_{\text{gravel}} = 1 \text{ ton} = 2000 \text{ lb} = 2000 \text{ lb}\left(\frac{1 \text{ kg}}{2.20 \text{ lb}}\right) = 909 \text{ kg}$$

The fact that there are 2.20 pounds per kilogram implies two proportionality factors: (2.20 lb/1 kg) and (1 kg/2.20 lb). The latter was used because it results in appropriate cancellation of units.

Step 2: Use the density to calculate the volume of 909 kg of gravel.

$$V_{\text{gravel}} = \frac{m_{\text{gravel}}}{d_{\text{gravel}}} = \frac{909 \text{ kg}}{2.7 \text{ kg/L}} = 909 \text{ kg}\left(\frac{1 \text{ L}}{2.7 \text{ kg}}\right) = 340 \text{ L}$$

In this step we used the definition of density, solved algebraically for volume, substituted the two known quantities into the equation, and calculated the result. However, it is quicker simply to remember that mass and volume are related by a proportionality factor called density and to use the units of the quantities to decide whether to multiply or divide by that factor. In this case we divided mass by density because the units kilograms canceled, leaving a result in liters, which is a unit of volume.

Also, it is quicker and more accurate to solve a problem like this one by using a single setup. Then all the calculations can be done at once, and no intermediate results need to be written down. The "roadmap" plan given above can serve as a guide to the single-setup solution, which looks like this:

$$V_{\text{gravel}} = 1 \text{ ton}\left(\frac{2000 \text{ lb}}{1 \text{ ton}}\right)\left(\frac{1 \text{ kg}}{2.20 \text{ lb}}\right)\left(\frac{1 \text{ L}}{2.7 \text{ kg}}\right) = 340 \text{ L}$$

To calculate the result, then, you would enter 2000 on your calculator, divide by 2.20, and divide by 2.7. Such a setup makes it easy to see what to multiply and divide by, and the calculation goes more quickly when it can be entered into a calculator all at once.

The liter is not the most convenient volume unit for this problem, however, because it does not relate well to what we want to find out—how big a bin to make. A liter is about the same volume as a quart, but whether you are familiar with liters, quarts, or both, 300 of them is not easy to visualize. Let's convert liters to something we can understand better. A liter is a volume equal to a cube one tenth of a meter (1 dm) on a side; that is, $1\text{L} = 1 \text{ dm}^3$. Consequently,

$$340 \text{ L} = 340 \text{ L}\left(\frac{1 \text{ dm}^3}{1 \text{ L}}\right)\left(\frac{1 \text{ m}}{10 \text{ dm}}\right)^3 = 340 \text{ dm}^3\left(\frac{1 \text{ m}^3}{1000 \text{ dm}^3}\right) = 0.34 \text{ m}^3$$

Thus, the bin would need to have a volume of about one third of a cubic meter; that is, it could be a meter wide, a meter long, and about a third of a meter high and it would hold the ton of gravel.

One more thing should be noted about this example. We don't need to know the volume of the bin very precisely, because being off a bit will make very little difference; it might mean getting a little too much wood to build the bin, or not making the bin quite big enough and having a little gravel spill out, but this isn't a big deal. In other cases, such as calculating the quantity of fuel needed to get a space shuttle into orbit, being off by a few percent could be a life-or-death matter. Because it is important to know how precise data are and to be able to evaluate how important precision is, scientific results usually indicate their precision. The simplest way to do so is by means of significant figures.

A.3 Precision, Accuracy, and Significant Figures

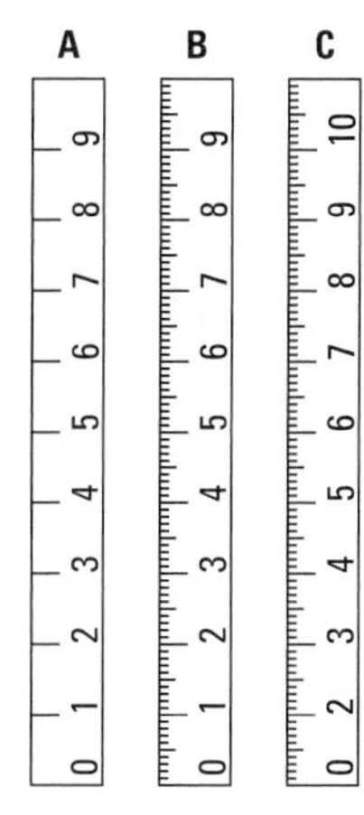

Three rulers.

The **precision** of a measurement indicates how well several determinations of the same quantity agree. Some devices can make more precise measurements than others. Consider rulers A and B shown in the margin. Ruler A is marked every centimeter and ruler B is marked every millimeter. Clearly you could measure at least to the nearest millimeter with ruler B, and probably you could estimate to the nearest 0.2 mm or so. However, using ruler A you could probably estimate only to the nearest millimeter or so. We say that ruler B allows more precise measurement than ruler A. If several different people used ruler B to measure the length of an iron bar, their measurements would probably agree more closely than if they used ruler A.

Precision is also illustrated by the results of throwing darts at a bull's-eye (Figure A.1). In part (a), the darts are scattered all over the board; the dart thrower was apparently not very skillful (or threw the darts from a long distance away from the board), and the precision of their placement on the board is low. This is analogous to the results that would be obtained by several people using ruler A. In part (b), the darts are all clustered together, indicating much better reproducibility on the part of the thrower—that is, greater precision. This is analogous to measurements that might be made using ruler B.

Notice also that in Figure A.1b every dart has come very close to the bull's-eye; we describe this result by saying that the thrower has been quite **accurate**—the average of all throws is very close to the accepted position, the bull's-eye. Figure A.1c illustrates that it is possible to be precise without being accurate—the dart thrower has consistently missed the bull's-eye, but all darts are clustered very precisely around the wrong point on the board. This third case is like an experiment with some flaw (either in its design or in a measuring device) that causes all results to differ from the correct value by the same quantity. An example is measuring length with ruler C, on which the scale is incorrectly labeled. The precision (reproducibility) of measurements with ruler C might be quite good, but the accuracy would be poor because all items longer than 1 cm would be off by a centimeter.

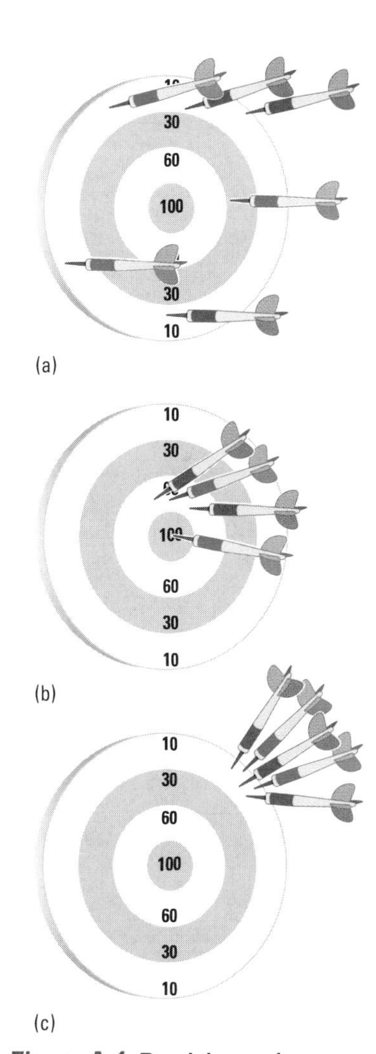

Figure A.1 Precision and accuracy. (a) Poor precision. (b) Good precision and good accuracy. (c) Good precision and poor accuracy.

In the laboratory we attempt to set up experiments so that the greatest possible accuracy can be obtained. As a further check on accuracy, results may be compared among different laboratories so that any flaw in experimental design or measurement can be detected. For each individual experiment, several measurements are usually made and their precision is determined. In most cases, better precision is taken as an indication of better experimental work, and it is necessary to know the precision to compare results among different experimenters. If two different experimenters both had results like those in Figure A.1a, for example, their average values could differ quite a lot before they would say that their results did not agree within experimental error.

In most experiments several different kinds of measurements must be made, and some can be done more precisely than others. It is common sense that *a calculated result can be no more precise than the least precise piece of information that went into the calculation.* This is where the rules for significant figures come in. In the last example in the preceding section, the quantity of gravel was described as "a ton." Usually gravel is measured by weighing an empty truck, putting some gravel in the truck, weighing the truck again, and subtracting the weight of the truck from the weight of the truck plus gravel. The quantity of gravel is not adjusted if there is a bit too much or a bit too little, because that would be a lot of trouble. You might end up with as much as 2200 pounds or as little as 1800 pounds, even though you asked for a ton. In terms of significant figures this would be expressed as 2.0×10^3 lb.

The quantity 2.0×10^3 lb is said to have two significant figures; it designates a quantity in which the 2 is taken to be exactly right but the 0 is not known precisely. (In this case, the number could be as large as 2.2 or as small as 1.8, so the 0 obviously is not exactly right.) In general, in a number that represents a scientific measurement, the last digit on the right is taken to be inexact, but all digits farther to the left are assumed to be exact. When you do calculations using such numbers, you must follow some simple rules so that your results will reflect the precision of all the measurements that go into the calculations. Here are the rules:

Rule 1: To determine the number of significant figures in a measurement, read the number from left to right and count all digits, starting with the first digit that is *not* zero.

For a number written in scientific notation, all digits are significant.

The number π is now known to 1,011,196,691 digits. It is doubtful that you will need that much precision in this course—or ever.

Example	Number of Significant Figures
1.23 g	3
0.00123 g	3; the zeros to the left of the 1 simply locate the decimal point. The number of significant figures is more obvious if you write numbers in scientific notation; thus, $0.00123 = 1.23 \times 10^{-3}$.
2.0 g and 0.020 g	2; both have two significant digits. When a number is greater than 1, *all zeros to the right of the decimal point are significant.* For a number less than 1, only zeros to the right of the first significant digit are significant.
100 g	1; in numbers that do not contain a decimal point, "trailing" zeros may or may not be significant. To eliminate possible confusion, the practice followed in this book is to include a decimal point if the zeros are significant. Thus, 100. has three significant digits, while 100 has only one. Alternatively, we write in scientific notation 1.00×10^2 (three significant digits) or 1×10^2 (one significant digit). For a number written in scientific notation, all digits preceding the 10^x term are significant.
100 cm/m	Infinite number of significant figures, because this is a defined quantity. There are *exactly* 100 centimeters in one meter.
$\pi = 3.1415926\ldots$	The value of π is known to a greater number of significant figures than any data you will ever use in a calculation.

Rule 2: When adding or subtracting, the number of decimal places in the answer should be equal to the number of decimal places in the number with the *fewest* places. Suppose you add three numbers:

0.12	2 significant figures	2 decimal places
1.6	2 significant figures	1 decimal place
+ 10.976	5 significant figures	3 decimal places
12.696		

This sum should be reported as 12.7, a number with one decimal place, because 1.6 has only one decimal place.

Rule 3: In multiplication or division, the number of significant figures in the answer should be the same as that in the quantity with the fewest significant figures.

$$\frac{0.1208}{0.0236} = 0.512 \text{ or, in scientific notation, } 5.12 \times 10^{-1}$$

Since 0.0236 has only three significant figures, while 0.01208 has four, the answer is limited to three significant figures.

Rule 4: When a number is rounded (the number of digits is reduced), the last digit retained is increased by 1 only if the following digit is 5 or greater. If the following digit is exactly 5 (or 5 followed by zeros), then increase the last retained digit by 1 if it is *odd* or leave the last digit unchanged if it is *even.* Thus, both 18.35 and 18.45 are rounded to 18.4.

Full Number	Number Rounded to Three Significant Figures
12.696	12.7
16.249	16.2
18.350	18.4
18.351	18.4

One last word regarding significant figures and calculations. In working problems on a pocket calculator, you should do the calculation using all the digits allowed by the calculator and round only at the end of the problem. Rounding in the middle can introduce small errors (called *rounding errors* or *round-off errors*). If your answers do not quite agree with those in the Appendices of this book, rounding errors may be the source of the disagreement.

Now let us consider a problem that is of practical importance and that makes use of all the rules. Suppose you discover that young children are eating chips of paint that flake off a wall in an old house. The paint contains 200. ppm lead (200. milligrams of Pb per kilogram of paint). Suppose that a child eats five such chips. How much lead has the child gotten from the paint?

The ppm unit stands for "parts per million." If a substance is present with a concentration of 1 ppm, there is 1 gram of the substance in 1 million grams of the sample.

As stated, this problem does not include enough information for a solution to be obtained; however, some reasonable assumptions can be made, and they can lead to experiments that could be used to obtain the necessary information. The statement does not say how big the paint chips are. Let's assume that they are 1.0 cm by 1.0 cm so that the area is 1.0 cm^2. Then eating five chips means eating 5.0 cm^2 of paint. (This assumption could be improved by measuring similar chips from the same place.) Since the concentration of lead is reported in units of mass of lead per mass of paint, we need to know the mass of 5.0 cm^2 of paint. This could be determined by measuring the areas of several paint chips and determining the mass of each. Suppose that the results of such measurements were those given in the table.

Mass of Chip (mg)	Area of Chip (cm^2)	Mass per Unit Area (mg/cm^2)
29.6	2.34	12.65
21.9	1.73	12.66
23.6	1.86	12.69

$$\text{Average mass per unit area} = \frac{(12.65 + 12.66 + 12.69)\ \text{mg/cm}^2}{3}$$

$$= 12.67\ \text{mg/cm}^2 = 12.7\ \text{mg/cm}^2$$

The average has been rounded to three significant figures because each experimentally measured mass and area has three significant figures. (Notice that more than three significant figures were kept in the intermediate calculations so as not to lose precision.) Now we can use this information to calculate how much lead the child has consumed.

$$m_{\text{paint}} = 5.0\ \text{cm}^2\ \text{paint}\left(\frac{12.7\ \text{mg paint}}{1\ \text{cm}^2\ \text{paint}}\right)\left(\frac{1\ \text{g}}{1000\ \text{mg}}\right)\left(\frac{1\ \text{kg}}{1000\ \text{g}}\right)$$

$$= 6.35 \times 10^{-5}\ \text{kg paint}$$

$$m_{\text{Pb}} = 6.35 \times 10^{-5}\ \text{kg paint}\left(\frac{200.\ \text{mg Pb}}{1\ \text{kg paint}}\right) = 1.27 \times 10^{-2}\ \text{mg Pb}$$

$$= 1.3 \times 10^{-2}\ \text{mg Pb} = 0.013\ \text{mg Pb}$$

The final result was rounded to two significant figures because there were only two significant figures in the initial 5.0 cm^2 area of the paint chip. This is quite adequate precision, however, for you to determine whether this quantity of lead is likely to harm the child.

The methods of problem solving presented here have been developed over time and represent a good way of keeping track of the precision of results, the units in which those results were obtained, and the correctness of calculations. These methods are not the only way that such goals can be achieved, but they do work well. We recommend that you include units in all calculations and check that they cancel appropriately. It is also important not to overstate the precision of results by keeping too many significant figures. By solving many problems, you should be able to develop your problem-solving skills so that they become second nature and you can use them without thinking about the mechanics. You can then devote all your thought to the logic of a problem solution.

A.4 Electronic Calculators

The directions for calculator use in this section are given for calculators using "algebraic" logic. Such calculators are the most common type used by students in introductory courses. For calculators using RPN logic (such as those made by Hewlett-Packard), the procedure will differ slightly.

The advent of inexpensive electronic calculators has made calculations in introductory chemistry much more straightforward. You are well advised to purchase a calculator that has the capability of performing calculations in scientific notation, has both base 10 and natural logarithms, and is capable of raising any number to any power and of finding any root of any number. In the discussion below, we will point out in general how these functions of your calculator can be used. You should practice using your calculator to carry out arithmetic operations and make certain that you are able to use all of its functions correctly.

Although electronic calculators have greatly simplified calculations, they have also forced us to focus again on significant figures. A calculator easily handles eight or more significant figures, but real laboratory data are rarely known to this precision. Therefore, if you have not already done so, review Appendix A.3 on significant figures, precision, and rounding numbers.

The mathematical skills required to read and study this textbook successfully involve algebra, some geometry, scientific notation, logarithms, and solving quadratic equations. The next three sections review the last three of these topics.

A.5 Exponential or Scientific Notation

In exponential or scientific notation, a number is expressed as a product of two numbers: $N \times 10^n$. The first number, N, is called the digit term and is a number between 1 and 10. The second number, 10^n, the exponential term, is some integer power of 10. For example, 1234 would be written in scientific notation as 1.234×10^3 or 1.234 multiplied by 10 three times.

$$1234 = 1.234 \times 10^1 \times 10^1 \times 10^1 = 1.234 \times 10^3$$

Conversely, a number less than 1, such as 0.01234, would be written as 1.234×10^{-2}. This notation tells us that 1.234 should be divided twice by 10 to obtain 0.01234.

$$0.01234 = \frac{1.234}{10^1 \times 10^1} = 1.234 \times 10^{-1} \times 10^{-1} = 1.234 \times 10^{-2}$$

Some other examples of scientific notation follow:

$10,000 = 1 \times 10^4$ | $12,345 = 1.2345 \times 10^4$
$1000 = 1 \times 10^3$ | $1234 = 1.234 \times 10^3$
$100 = 1 \times 10^2$ | $123 = 1.23 \times 10^2$
$10 = 1 \times 10^1$ | $12 = 1.2 \times 10^1$
$1 = 1 \times 10^0$ | (any number to the zeroth power = 1)
$1/10 = 1 \times 10^{-1}$ | $0.12 = 1.2 \times 10^{-1}$
$1/100 = 1 \times 10^{-2}$ | $0.012 = 1.2 \times 10^{-2}$
$1/1000 = 1 \times 10^{-3}$ | $0.0012 = 1.2 \times 10^{-3}$
$1/10,000 = 1 \times 10^{-4}$ | $0.00012 = 1.2 \times 10^{-4}$

When converting a number to scientific notation, notice that the exponent n is positive if the number is greater than 1 and negative if the number is less than 1. The value of n is the number of places by which the decimal was shifted to obtain the number in scientific notation.

$$1\,2\,3\,4\,5. = 1.2345 \times 10^4$$

Decimal shifted 4 places to the left. Therefore,* n *is positive and equal to 4.

$$0.0\,0\,1\,2 = 1.2 \times 10^{-3}$$

Decimal shifted 3 places to the right. Therefore,* n *is negative and equal to 3.

If you wish to convert a number in scientific notation to the usual form, the procedure above is simply reversed.

$$6\,.\,2\,7\,3 \times 10^2 = 627.3$$

Decimal point shifted 2 places to the right, because* n *is positive and equal to 2.

$$0\,0\,6.273 \times 10^{-3} = 0.006273$$

Decimal point shifted 3 places to the left, because* n *is negative and equal to 3.

To enter a number in scientific notation into a calculator, first enter the number itself. Then press the EE (Enter Exponent) key followed by *n*, the power of 10. For example, to enter 6.022×10^{23}, you press these keys in succession:

6 . 0 2 2 EE 2 3 .

(Do not enter the number using the multiplication key, the number 10, and the EE key. This will result in a number that is 10 times bigger than you want. For example, pressing these keys

6 . 0 2 2 × 1 0 EE 2 3

enters the number $6.022 \times 10 \times 10^{23} = 6.022 \times 10^{24}$, because EE 2 3 means 10^{23}.)

There are two final points concerning scientific notation. First, if you are used to working on a computer you may be in the habit of writing a number such as 1.23×10^3 as 1.23E3, or 6.45×10^{-5} as 6.45E-5. Second, some electronic calculators allow you to convert numbers readily to scientific notation. If you have such a calculator, you can change a number shown in the usual form to scientific notation by pressing an appropriate key or keys.

Usually you will handle numbers in scientific notation with a calculator. In case you need to work without a calculator, however, the next few sections describe pencil-and-paper calculations as well as calculator methods.

1. Adding and Subtracting

When adding or subtracting numbers in scientific notation without using a calculator, first convert the numbers to the same powers of 10. Then add or subtract the digit terms.

$$(1.234 \times 10^{-3}) + (5.623 \times 10^{-2}) = (0.1234 \times 10^{-2}) + (5.623 \times 10^{-2})$$
$$= 5.746 \times 10^{-2}$$

$$(6.52 \times 10^{2}) - (1.56 \times 10^{3}) = (6.52 \times 10^{2}) - (15.6 \times 10^{2})$$
$$= -9.1 \times 10^{2}$$

In this calculation, the result has only two significant figures, although each of the original numbers had three. Subtracting two numbers that are nearly the same can reduce the number of significant figures appreciably.

2. Multiplying

The digit terms are multiplied in the usual manner, and the exponents are added algebraically. The result is expressed with a digit term with only one nonzero digit to the left of the decimal.

$$(1.23 \times 10^{3})(7.60 \times 10^{2}) = (1.23 \times 7.60)(10^{3} \times 10^{2}) = (1.23)(7.60) \times 10^{3+2}$$
$$= 9.35 \times 10^{5}$$

$$(6.02 \times 10^{23})(2.32 \times 10^{-2}) = (6.02)(2.32) \times 10^{23-2}$$
$$= 13.966 \times 10^{21} = 1.3966 \times 10^{22}$$
$$= 1.40 \times 10^{22} \quad \text{(rounded to 3 significant figures)}$$

3. Dividing

The digit terms are divided in the usual manner, and the exponents are subtracted algebraically. The quotient is written with one nonzero digit to the left of the decimal in the digit term.

$$\frac{7.60 \times 10^{3}}{1.23 \times 10^{2}} = \frac{7.60}{1.23} \times 10^{3-2} = 6.18 \times 10^{1}$$

$$\frac{6.02 \times 10^{23}}{9.10 \times 10^{-2}} = \frac{6.02}{9.10} \times 10^{(23)-(-2)} = 0.662 \times 10^{25} = 6.62 \times 10^{24}$$

4. Raising Numbers in Scientific Notation to Powers

When raising a number in scientific notation to a power, treat the digit term in the usual manner. The exponent is then multiplied by the number indicating the power.

$$(1.25 \times 10^{3})^2 = (1.25)^2 \times (10^{3})^2 = (1.25)^2 \times 10^{3\times 2} = 1.5625 \times 10^{6} = 1.56 \times 10^{6}$$

$$(5.6 \times 10^{-10})^3 = (5.6)^3 \times 10^{(-10)\times 3} = 175.6 \times 10^{-30} = 1.8 \times 10^{-28}$$

Electronic calculators usually have two methods of raising a number to a power. To square a number, enter the number and then press the "x^2" key. To raise a number to any power, use the "y^x" key. For example, to raise 1.42×10^2 to the 4th power, that is, to find $(1.42 \times 10^2)^4$,

(a) Enter 1.42×10^2.
(b) Press "y^x."
(c) Enter 4 (this should appear on the display).
(d) Press "=" and $4.0658689... \times 10^8$ will appear on the display. (The number of digits depends on the calculator.)

As a final step, express the number in the correct number of significant figures (4.07×10^8 in this case).

5. Taking Roots of Numbers in Scientific Notation

Unless you use an electronic calculator, the number must first be put into a form in which the exponential is exactly divisible by the root. The root of the digit term is found in the usual way, and the exponent is divided by the desired root.

$$\sqrt{3.6 \times 10^7} = \sqrt{36 \times 10^6} = \sqrt{36} \times \sqrt{10^6} = 6.0 \times 10^3$$

$$\sqrt[3]{2.1 \times 10^{-7}} = \sqrt[3]{210 \times 10^{-9}} = \sqrt[3]{210} \times \sqrt[3]{10^{-9}} = 5.9 \times 10^{-3}$$

To take a square root on an electronic calculator, enter the number and then press the "$\sqrt{x}$" key. To find a higher root of a number, such as the fourth root of 5.6×10^{-10},

On some calculators, Steps (a) and (c) may be interchanged.

(a) Enter the number, 5.6×10^{-10} in this case.
(b) Press the "$\sqrt[x]{y}$" key. (On most calculators, the sequence you actually use is to press "2ndF" and then "$\sqrt[x]{y}$." Alternatively, you may have to press "INV" and then "y^x".)
(c) Enter the desired root, 4 in this case.
(d) Press "=". The answer here is 4.8646×10^{-3} or 4.9×10^{-3}.

A general procedure for finding any root is to use the "y^x" key. For a square root, x is 0.5 (or $\frac{1}{2}$), whereas it is 0.33 (or $\frac{1}{3}$) for a cube root, 0.25 (or $\frac{1}{4}$) for a fourth root, and so on.

A.6 Logarithms

There are two types of logarithms used in this text: common logarithms (abbreviated log), whose base is 10, and natural logarithms (abbreviated ln), whose base is e (=2.7182818284).

Logarithms to the base 10 are needed when dealing with pH.

$$\log x = n \qquad \text{where } x = 10^n$$

$$\ln x = m \qquad \text{where } x = e^m$$

Most equations in chemistry and physics were developed in natural or base e logarithms, and this practice is followed in this text. The relation between log and ln is

$$\ln x = 2.303 \log x$$

Aside from the different bases of the two logarithms, they are used in the same manner. What follows is largely a description of the use of common logarithms.

A common logarithm is the power to which you must raise 10 to obtain the number. For example, the log of 100 is 2, since you must raise 10 to the power 2 to obtain 100. Other examples are

$$\log 1000 = \log (10^3) = 3$$

$$\log 10 = \log (10^1) = 1$$

$$\log 1 = \log (10^0) = 0$$

$$\log 1/10 = \log (10^{-1}) = -1$$

$$\log 1/10{,}000 = \log (10^{-4}) = -4$$

To obtain the common logarithm of a number other than a simple power of 10, use an electronic calculator. For example,

$$\log 2.10 = 0.3222, \text{ which means that } 10^{0.3222} = 2.10$$

$$\log 5.16 = 0.7126, \text{ which means that } 10^{0.7126} = 5.16$$

$$\log 3.125 = 0.49485, \text{ which means that } 10^{0.49485} = 3.125$$

To check this result on your calculator, enter the number and then press the "log" key.

To obtain the natural logarithm of the numbers above, use a calculator having this function. Enter each number and press "ln".

$$\ln 2.10 = 0.7419, \text{ which means that } e^{0.7419} = 2.10$$

$$\ln 5.16 = 1.6409, \text{ which means that } e^{1.6409} = 5.16$$

To find the common logarithm of a number greater than 10 or less than 1 with a log table, first express the number in scientific notation. Then find the log of each part of the number and add the logs. For example,

Nomenclature of Logarithms: The number to the left of the decimal in a logarithm is called the *characteristic,* and the number to the right of the decimal is called the *mantissa.*

$$\log 241 = \log (2.41 \times 10^2) = \log 2.4 + \log 10^2$$
$$= 0.382 + 2 = 2.382$$

$$\log 0.00573 = \log (5.73 \times 10^{-3}) = \log 5.73 + \log 10^{-3}$$
$$= 0.758 + (-3) = -2.242$$

Significant Figures and Logarithms

The mantissa (digits to the right of the decimal point in the logarithm) should have as many significant figures as the number whose log was found. (So that you could more clearly see the result obtained with a calculator or a table, this rule was not strictly followed until the last two examples.)

Obtaining Antilogarithms

If you are given the logarithm of a number and need to find the number from it, you need to obtain the "antilogarithm" or "antilog" of the number. There are two common procedures used by electronic calculators to do this:

Procedure A	Procedure B
(a) Enter the log or ln (a number).	(a) Enter the log or ln (a number).
(b) Press 2ndF.	(b) Press INV.
(c) Press 10^x or e^x.	(c) Press log or ln x.

Test one or the other of these procedures with the following examples.

EXAMPLE 1 Find the number whose log is 5.234.

Recall that $\log x = n$, where $x = 10^n$. In this case $n = 5.234$. Enter that number in your calculator and find the value of 10^n, the antilog. In this case,

$$10^{5.234} = 10^{0.234} \times 10^5 = 1.71 \times 10^5$$

Notice that the characteristic (5) sets the decimal point; it is the power of 10 in the exponential form. The mantissa (0.234) gives the value of the number x. Thus, if you use a log table to find x, you need only look up 0.234 in the table and see that it corresponds to 1.71.

EXAMPLE 2 Find the number whose log is −3.456.

$$10^{-3.456} = 10^{0.544} \times 10^{-4} = 3.50 \times 10^{-4}$$

Notice here that -3.456 must be expressed as the sum of -4 and $+0.544$.

Mathematical Operations Using Logarithms

Because logarithms are exponents, operations involving them follow the same rules as the use of exponents. Thus, multiplying two numbers can be done by adding logarithms.

$$\log xy = \log x + \log y$$

For example, we multiply 563 by 125 by adding their logarithms and finding the antilogarithm of the result.

$$\log 563 = 2.751$$
$$\log 125 = 2.097$$
$$\log (563 \times 125) = 2.751 + 2.097 = 4.848$$
$$563 \times 125 = 10^{4.848} = 10^4 \times 10^{0.848} = 7.05 \times 10^4$$

One number (x) can be divided by another (y) by subtraction of their logarithms.

$$\log \frac{x}{y} = \log x - \log y$$

For example, to divide 125 by 742,

$$\log 125 = 2.097$$
$$\log 742 = 2.870$$
$$\log (125/742) = 2.097 - 2.870 = -0.773$$
$$125/742 = 10^{-0.773} = 10^{0.227} \times 10^{-1} = 1.69 \times 10^{-1}$$

Similarly, powers and roots of numbers can be found using logarithms.

$$\log x^y = y(\log x)$$
$$\log \sqrt[y]{x} = \log x^{1/y} = \frac{1}{y} \log x$$

As an example, find the fourth power of 5.23. First find the log of 5.23 and then multiply it by 4. The result, 2.874, is the log of the answer. Therefore, find the antilog of 2.874.

$$(5.23)^4 = ?$$
$$\log (5.23)^4 = 4 \log 5.23 = 4(0.719) = 2.874$$
$$(5.23)^4 = 10^{2.874} = 748$$

As another example, find the fifth root of 1.89×10^{-9}.

$$\sqrt[5]{1.89 \times 10^{-9}} = (1.89 \times 10^{9})^{1/5} = ?$$
$$\log (1.89 \times 10^{-9})^{1/5} = \frac{1}{5} \log (1.89 \times 10^{-9}) = \frac{1}{5}(-8.724) = -1.745$$

The answer is the antilog of -1.745.

$$(1.89 \times 10^{-9})^{1/5} = 10^{-1.745} = 1.80 \times 10^{-2}$$

A.7 Quadratic Equations

Algebraic equations of the form $ax^2 + bx + c = 0$ are called **quadratic equations.** The coefficients a, b, and c may be either positive or negative. The two roots of the equation may be found using the *quadratic formula.*

$$x = \frac{-b \pm \sqrt{b^2 - 4ac}}{2a}$$

As an example, solve the equation $5x^2 - 3x - 2 = 0$. Here $a = 5$, $b = -3$, and $c = -2$. Therefore,

$$x = \frac{3 \pm \sqrt{(-3^2) - 4(5)(-2)}}{2(5)}$$

$$= \frac{3 \pm \sqrt{9 - (-40)}}{10} = \frac{3 \pm \sqrt{49}}{10} = \frac{3 \pm 7}{10}$$

$$x = 1 \textit{ and } -0.4$$

How do you know which of the two roots is the correct answer? You have to decide in each case which root has physical significance. However, it is usually true in this course that negative values are not significant.

Many calculators have a built-in quadratic-equation solver. If your calculator offers this capability, it will be convenient to use it, and you should study the manual until you know how to use your calculator to obtain the roots of a quadratic equation.

When you have solved a quadratic expression, you should always check your values by substituting them into the original equation. In the example above, we find that $5(1)^2 - 3(1) - 2 = 0$ and that $5(-0.4)^2 - 3(-0.4) - 2 = 0$.

You will encounter quadratic equations in the chapters on chemical equilibria, particularly in Chapters 13, 15, and 16. Here you may be faced with solving an equation such as

$$1.8 \times 10^{-4} = \frac{x^2}{0.0010 - x}$$

This equation can certainly be solved by using the quadratic formula or your calculator (to give $x = 3.4 \times 10^{-4}$). However, you may find the method of successive approximations to be especially convenient. Here you begin by making a reasonable approximation of x. This approximate value is substituted into the original equation, and this expression is solved to give what is hoped to be a more correct value of x. This process is repeated until the answer converges on a particular value of x—that is, until the value of x derived from two successive approximations is the same.

Step 1: Assume that x is so small that $(0.0010 - x) \approx 0.0010$. This means that

$$x^2 = 1.8 \times 10^{-4}(0.0010)$$

$$x = 4.2 \times 10^{-4} \text{ (to 2 significant figures)}$$

Step 2: Substitute the value of x from Step 1 into the denominator (but not the numerator) of the original equation and again solve for x.

$$x^2 = (1.8 \times 10^{-4})(0.0010 - 0.00042)$$

$$x = 3.2 \times 10^{-4}$$

Step 3: Repeat Step 2 using the value of x found in that step.

$$x = \sqrt{1.8 \times 10^{-4}(0.0010 - 0.00032)} = 3.5 \times 10^{-4}$$

Step 4: Continue by repeating the calculation, using the value of x found in the previous step.

Step 5. $x = \sqrt{1.8 \times 10^{-4}(0.0010 - 0.00034)} = 3.4 \times 10^{-4}$

Here we find that iterations after the fourth step give the same value for x, indicating that we have arrived at a valid answer (and the same one obtained from the quadratic formula).

Some final thoughts on using the method of successive approximations: First, in some cases this method does not work. Successive steps may give answers that are random or that diverge from the correct value. For quadratic equations of the form $K = x^2/(C - x)$, the method of approximations will work only as long as $K < 4C$ (assuming one begins with $x = 0$ as the first guess; that is, $K \approx x^2/C$). This will always be true for weak acids and bases.

Second, values of K in the equation $K = x^2/(C - x)$ are usually known only to two significant figures. Therefore, we are justified in carrying out successive steps until two answers are the same to two significant figures.

Finally, if your calculator does not automatically obtain roots of quadratic equations, we highly recommend this method. If your calculator has a memory function, successive approximations can be carried out easily and very rapidly. Even without a memory function, the method of successive approximations is much quicker than solving a quadratic equation by hand.

A.8 Graphing

Mass (g)	Volume (mL)
2.03	0.76
5.27	1.95
9.57	3.54
11.46	4.25
14.96	5.55
18.02	6.68
21.83	8.10
25.17	9.32
30.08	11.14
32.84	12.17
36.27	13.43
36.77	13.62
39.36	14.58

When analyzing experimental data, chemists and other scientists often graph the data to see whether the data agree with a mathematical equation. If the equation does fit the data (often indicated by a linear graph), then the graph can be used to obtain numerical values (parameters) that can be used in the equation to predict information not specifically included in the data set. For example, suppose that you measured the masses and the volumes of several samples of aluminum. Your data set might look like the table in the margin. When these data are plotted with volume along the horizontal (x) axis and mass on the vertical (y) axis, a graph like this one can be obtained.

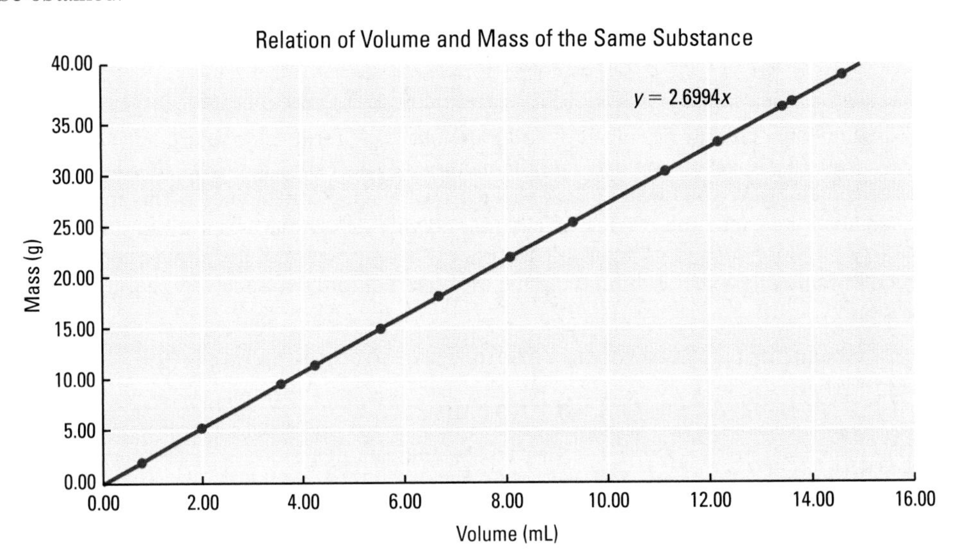

Notice the features of this graph. It has a title that describes its contents, it has a label for each axis that specifies the quantity plotted and the units used, it has equally spaced grid lines in the vertical and horizontal directions, and the numbers associated with those grid lines have the same difference between successive grid lines. (The differences are 2.00 mL on the horizontal axis and 5.00 g on the vertical axis.) The experimental points are indicated by circular markers, and a straight line has been drawn through the points. Because the line passes through all of the points and through the origin (point 0, 0), the data can be represented by the equation $y = 2.70x$, where 2.70 is the slope of the line. (On the graph, the slope is indicated by 2.6994; in the equation, it has been rounded to 2.70 because some of the data have only three significant figures.)

The slope of a graph such as this one can be obtained by choosing two points on the straight line (not two data points). The points should be far apart to obtain the most precise result. The slope is then given by the change in the y-axis variable divided by the change in the x-axis variable from the first point to the second. An example is shown on the graph below. From the calculation of the slope, it is more obvious that the slope has only three significant figures. For this graph, the slope represents mass divided by volume—that is, density. A good way to measure density is to measure the mass and volume of several samples of the same substance and then plot the data. The resulting graph should be linear and should pass very nearly through the origin. The density can be obtained from its slope. When a graph passes through the origin, we say that the y-axis variable is directly proportional to the x-axis variable.

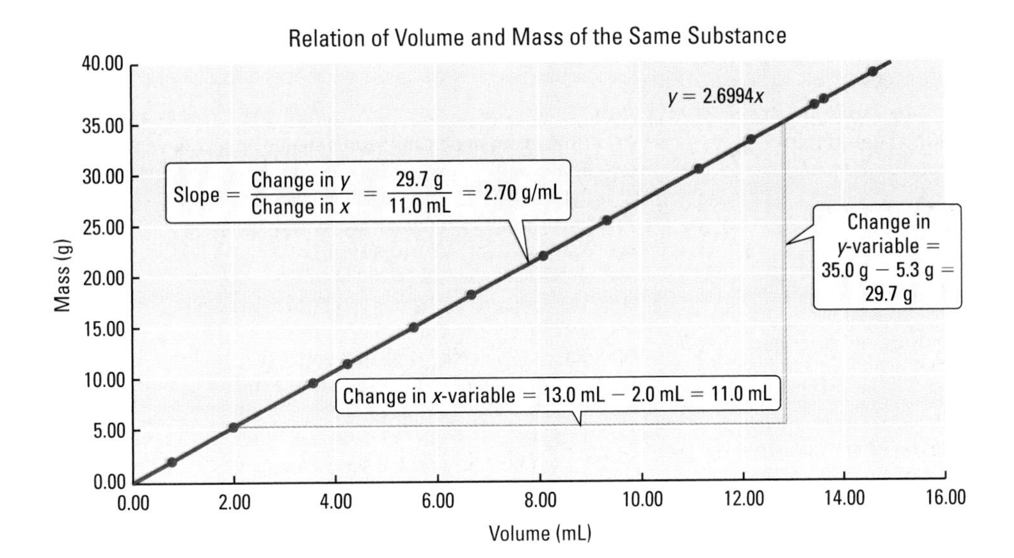

When a graph does not pass through the origin, there is an intercept as well as a slope. The intercept is the value of the *y*-axis variable when the *x*-axis variable is zero. A good example is the relation between Celsius and Fahrenheit temperatures. Suppose you have measured a series of temperatures using both a Celsius thermometer and a Fahrenheit thermometer. Your data set might look something like the table in the margin.

When graphed, these data look like this.

Temperature (°C)	Temperature (°F)
−5.23	22.586
13.54	56.372
32.96	91.328
48.34	119.012
63.59	146.462
74.89	166.802
88.02	190.436
105.34	221.612

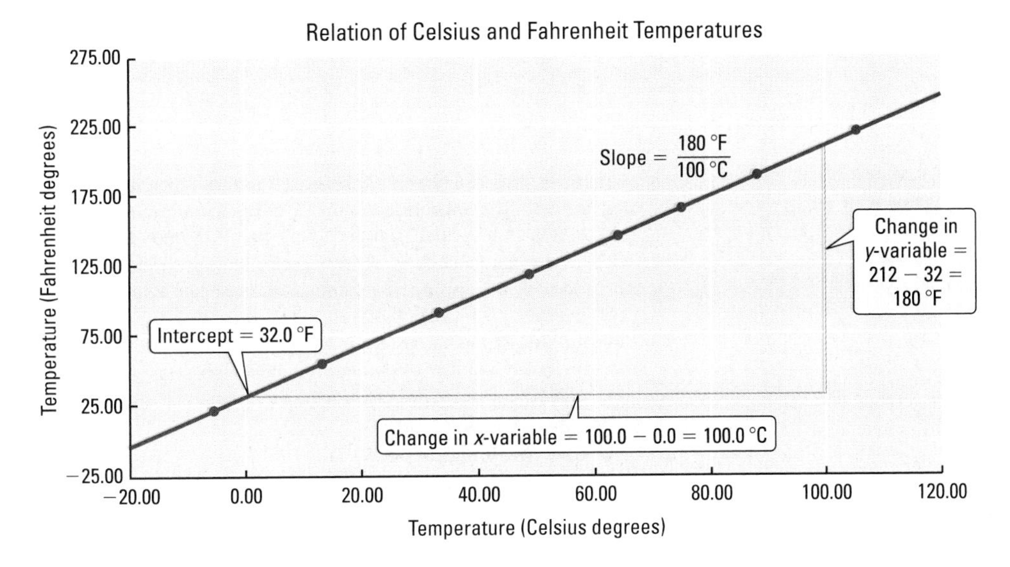

In this case, although the data are on a straight line, that line does not pass through the origin. Instead, it intersects the vertical line corresponding to $x = 0$ (0 °C) at a value of 32 °F. This makes sense, because the normal freezing point of water is 0 °C, which is the same temperature as 32 °F. This value, 32 °F, is the intercept.

If we determine the slope by starting at the intercept value of 32 °F and going to 212° F, the normal boiling point of water, the change in temperature on the Fahrenheit scale is 180 °F. The corresponding change on the Celsius scale is 100 °C, from which we can determine that the slope is 180 °F/100 °C. Thus the equation relating Celsius and Fahrenheit temperatures is

$$y \quad = \quad m \quad\quad x \quad + \quad b$$

$$\text{temperature } ^\circ\text{F} = \frac{180\ ^\circ\text{F}}{100\ ^\circ\text{C}} \times \text{temperature } ^\circ\text{C} + 32\ ^\circ\text{F}$$

QUESTIONS FOR APPENDIX A

Blue-numbered questions have short answers at the back of this book and fully worked solutions in the *Student Solutions Manual.*

General Problem-Solving Strategies

1. List four steps that can be used for guidance in solving problems. Choose a problem that you are interested in solving, and apply the four steps to that problem.
2. You are asked to study a lake in which fish are dying and determine the cause of their deaths. Suggest three things you might do to define this problem.
3. You have calculated the area in square yards of a carpet whose dimensions were originally given to you as 96 in by 72 in. Suggest at least one way to check that the results of your calculation are reasonable. (12 in = 1 ft; 3 ft = 1 yd)

Numbers, Units, and Quantities

4. When a calculation is done and the units for the answer do not make sense, what can you conclude about the solution to the problem? What would you do if you were faced with a situation like this?
5. The term "quantity" (or "physical quantity") has a specific meaning in science. What is that meaning, and why is it important?
6. To measure the length of a pencil, you use a tape measure calibrated in inches with marks every sixteenth of an inch. Which of these results would be a suitable record of your observation? Why would the other results be unsuitable? (1/16 in = 0.0625 in)
   (a) 8.38 ft (b) 8.38 m (c) 8.38 in
   (d) 8.38 (e) 0.698 ft
7. To measure the inseam length of a pair of slacks, you use a tape measure calibrated in centimeters with marks every tenth of a centimeter. Which of these results would be a suitable record of your observation? Why would the other results be unsuitable?
   (a) 75.0 cm (b) 75.0 m (c) 75.0
   (d) 75.0 in (e) 750 mm
8. What is wrong with each calculation? How would you carry out each calculation correctly? What is the correct result for each calculation?
   (a) $4.32 \text{ g} + 5.63 \text{ g} = 9.95 \text{ g}^2$
   (b) $5.23 \text{ g} \times \dfrac{4.87 \text{ g}}{1.00 \text{ mL}} = 25.5 \text{ g}$
   (c) $3.57 \text{ cm}^3 \times \left(\dfrac{1 \text{ m}}{100 \text{ cm}}\right) = 3.57 \times 10^{-2} \text{ m}^3$
9. What is wrong with each calculation? How would you carry out each calculation correctly? What is the correct result for each calculation?
   (a) $7.86 \text{ g} - 5.63 \text{ g} = 2.23 \text{ g}^2$
   (b) $7.37 \text{ mL} \times \dfrac{1.00 \text{ mL}}{2.23 \text{ g}} = 3.30 \text{ mL}$
   (c) $9.26 \text{ m}^3 \times \left(\dfrac{100 \text{ cm}}{1 \text{ m}}\right) = 9.26 \times 10^2 \text{ m}^3$

Precision, Accuracy, and Significant Figures

10. These measurements were reported for the length of an eight-foot pole: 95.31 in; 96.44 in; 96.02 in; 95.78 in; 95.94 in (1 ft = 12 in).
    (a) Based on these measurements, what would you report as the length of the pole?
    (b) How many significant figures should appear in your result?
    (c) Assuming that the pole is exactly eight feet long, is the result accurate?
11. These measurements were reported for the mass of a sample in the laboratory: 32.54 g; 32.67 g; 31.98 g; 31.76 g; 32.05 g.
    (a) Based on these measurements, what would you report as the mass of the sample?
    (b) How many significant figures should appear in your result?
    (c) The sample is exactly balanced by a weight known to have a mass of 35.0 g. Is the result accurate?
12. How many significant figures are in each quantity?
    (a) 3.274 g (b) 0.0034 L
    (c) 43,000 m (d) 6200. ft
13. How many significant figures are in each quantity?
    (a) 0.2730 g (b) 8.3 g/mL
    (c) 300 m (d) 2030.0 dm^3
14. Round each number to four significant figures.
    (a) 43.3250 (b) 43.3165 (c) 43.3237
    (d) 43.32499 (e) 43.3150 (f) 43.32501
15. Round each number to three significant figures.
    (a) 88.3520 (b) 88.365 (c) 88.45
    (d) 88.5500 (e) 88.2490 (f) 88.4501
16. Evaluate each expression and report the result to the appropriate number of significant figures.
    (a) $\dfrac{4.47}{0.3260}$ (b) $\dfrac{4.03 + 3.325}{29.75}$
    (c) $\dfrac{8.234}{5.673 - 4.987}$
17. Evaluate each expression and report the result to the appropriate number of significant figures.
    (a) $\dfrac{4.47}{0.3260}$ (b) $\dfrac{4.03 + 3.325}{29.75}$
    (c) $\dfrac{8.234}{5.673 - 4.987}$

Exponential or Scientific Notation

18. Without using a calculator, express each number in scientific notation and with the appropriate number of significant figures.
    (a) 76,003 (b) 0.00037 (c) 34,000
19. Without using a calculator, express each number in scientific notation and with the appropriate number of significant figures.
    (a) 49,002 (b) 0.0234 (c) 23,400
20. Evaluate each expression using your calculator and report the result to the appropriate number of significant figures with appropriate units.
    (a) $\dfrac{0.7346}{304.2}$
    (b) $\dfrac{(3.45 \times 10^{-3})(1.83 \times 10^{12})}{23.4}$
    (c) $3.240 - 4.33 \times 10^{-3}$
    (d) $(4.87 \text{ cm})^3$
21. Evaluate each expression using your calculator and report the result to the appropriate number of significant figures with appropriate units.
    (a) $\dfrac{893.0}{0.2032}$
    (b) $\dfrac{(5.4 \times 10^3)(8.36 \times 10^{-12})}{5.317 \times 10^{-3}}$
    (c) $3.240 \times 10^5 - 8.33 \times 10^3$
    (d) $(4.87 \text{ cm} + 7.33 \times 10^{-1} \text{ cm})^3$

Logarithms

22. Use your calculator to find the logarithm of each number and report the logarithm to the appropriate number of significant figures.
    (a) $\log(0.7327)$ (b) $\ln(34.5)$
    (c) $\log(6.022 \times 10^{23})$ (d) $\ln(6.022 \times 10^{23})$
    (e) $\log\left(\dfrac{8.34 \times 10^{-5}}{2.38 \times 10^3}\right)$

23. Use your calculator to find the logarithm of each number and report the logarithm to the appropriate number of significant figures.
    (a) log(54.3) (b) ln(0.0345)
    (c) $\log(4.344 \times 10^{-3})$ (d) $\ln(8.64 \times 10^{4})$
    (e) $\ln\left(\dfrac{4.33 \times 10^{24}}{8.32 \times 10^{-2}}\right)$
24. Use your calculator to evaluate each expression and report the result to the appropriate number of significant figures.
    (a) antilog(0.7327) (b) antiln(34.5)
    (c) $10^{2.043}$ (d) $e^{3.20 \times 10^{-4}}$
    (e) exp(4.333/3.275)
25. Use your calculator to evaluate each expression and report the result to the appropriate number of significant figures.
    (a) antilog(87.2) (b) antiln(0.0034)
    (c) $e^{2.043}$ (d) $10^{(3.20 \times 10^{-4})}$
    (e) $\exp(4.3 \times 10^{3}/8.314)$

Quadratic Equations

26. Find the roots of each quadratic equation.
    (a) $3.27x^2 + 4.32x - 2.83 = 0$
    (b) $x^2 + 4.32 = 4.57x$
27. Find the roots of each quadratic equation.
    (a) $8.33x^2 - 2.32x - 7.53 = 0$
    (b) $4.3x^2 - 8.37 = -2.22x$

Graphing

28. Graph these data involving mass and volume, label the graph appropriately, and determine whether m is directly proportional to V.

V (mL)	m (g)
0.347	0.756
1.210	2.638
2.443	5.326
7.234	15.76
11.43	24.90

29. Graph these data for heat transfers during a reaction, label the graph appropriately, and determine whether the heat evolved is directly proportional to the amount of reactant consumed.

Amount of Reactant (mol)	Heat Evolved (J)
94.2	43.2
70.7	32.5
65.7	30.1
34.2	15.7
54.3	24.9

APPENDIX B

Units, Equivalences, and Conversion Factors

B.1 Units of the International System (SI)

The metric system was begun by the French National Assembly in 1790 and has undergone many modifications since its inception. The International System of Units, or *Système International* (SI), which represents an extension of the metric system, was adopted by the 11th General Conference on Weights and Measures in 1960. It is constructed from seven base units, each of which represents a particular physical quantity (Table B.1). More information about the SI is available at **http://physics.nist.gov/cuu/Units/index.html**.

The first five units listed in Table B.1 are particularly useful in chemistry. They are defined as follows:

1. The *meter* is the length of the path traveled by light in a vacuum during a time interval of 1/299,792,458 of a second.
2. The *kilogram* represents the mass of a platinum-iridium block kept at the International Bureau of Weights and Measures in Sevres, France.
3. The *second* is the duration of 9,192,631,770 periods of a certain line in the microwave spectrum of cesium-133.
4. The *kelvin* is 1/273.16 of the temperature interval between absolute zero and the triple point of water (the temperature at which liquid water, ice, and water vapor coexist).
5. The *mole* is the amount of substance that contains as many elementary entities (atoms, molecules, ions, or other particles) as there are atoms in exactly 0.012 kg of carbon-12 (12 g of ^{12}C atoms).

Decimal fractions and multiples of metric and SI units are designated by using the **prefixes** listed in Table B.2. The prefix *kilo-*, for example, means that a unit is multiplied by 10^3.

$$1 \text{ kilogram} = 1 \times 10^3 \text{ grams} = 1000 \text{ grams}$$

The prefix *centi-* means that the unit is multiplied by the factor 10^{-2}.

$$1 \text{ centigram} = 1 \times 10^{-2} \text{ gram} = 0.01 \text{ gram}$$

Table B.1 SI Fundamental Units

Physical Quantity	Name of Unit	Symbol
Length	Meter	m
Mass	Kilogram	kg
Time	Second	s
Temperature	Kelvin	K
Amount of substance	Mole	mol
Electric current	Ampere	A
Luminous intensity	Candela	cd

The prefixes are added to give units of a magnitude appropriate to what is being measured. The distance from New York to London (5.6×10^3 km = 5600 km) is much easier to

Table B.2 Prefixes for Metric and SI Units*

Factor	Prefix	Symbol	Factor	Prefix	Symbol
10^{18}	exa-	E	10^{-1}	*deci-*	d
10^{15}	peta-	P	10^{-2}	*centi-*	c
10^{12}	tera-	T	10^{-3}	*milli-*	m
10^{9}	giga-	G	10^{-6}	*micro-*	μ
10^{6}	mega-	M	10^{-9}	*nano-*	n
10^{3}	*kilo-*	k	10^{-12}	*pico-*	p
10^{2}	hecto-	h	10^{-15}	femto-	f
10^{1}	deka-	da	10^{-18}	atto-	a

*The prefixes most commonly used in chemistry are shown in italics.

comprehend measured in kilometers than in meters (5.6×10^6 m = 5,600,000 m). Following Table B.2 is a list of units for measuring very small and very large distances.

attometer (am)	0.000000000000000001 meter
femtometer (fm)	0.000000000000001 meter
picometer (pm)	0.000000000001 meter
nanometer (nm)	0.000000001 meter
micrometer (μm)	0.000001 meter
millimeter (mm)	0.001 meter
centimeter (cm)	0.01 meter
decimeter (dm)	0.1 meter
meter (m)	1 meter
dekameter (dam)	10 meters
hectometer (hm)	100 meters
kilometer (km)	1000 meters
megameter (Mm)	1,000,000 meters
gigameter (Gm)	1,000,000,000 meters
terameter (Tm)	1,000,000,000,000 meters
petameter (Pm)	1,000,000,000,000,000 meters
exameter (Em)	1,000,000,000,000,000,000 meters

In the International System of Units, all physical quantities are represented by appropriate combinations of the base units listed in Table B.1. The result is a derived unit for each kind of measured quantity. The most common derived units are listed in Table B.3. It is easy to see that the

Table B.3 Derived SI Units

Physical Quantity	Name of Unit	Symbol	Definition	Expressed in Fundamental Units
Area	Square meter	m^2	—	
Volume	Cubic meter	m^3	—	
Density	Kilogram per cubic meter	kg/m	—	
Force	Newton	N	$\frac{(\text{kilogram})(\text{meter})}{(\text{second})^2}$	kg m/s^2
Pressure	Pascal	Pa	$\frac{(\text{newton})}{(\text{meter})^2}$	$N/m^2 = kg\ m^{-1}\ s^{-2}$
Energy	Joule	J	$\frac{(\text{kilogram})(\text{meter})^2}{(\text{second})^2}$	$kg\ m^2\ s^{-2}$
Electric charge	Coulomb	C	(ampere)(second)	A s
Electric potential difference	Volt	V	$\frac{(\text{joule})}{(\text{ampere})(\text{second})}$	$J\ A^{-1}\ s^{-1} = kg\ m^2\ s^{-3}\ A^{-1}$

derived unit for area is length × length = meter × meter = square meter, m^2, or that the derived unit for volume is length × length × length = meter × meter × meter = cubic meter, m^3. More complex derived units are arrived at by a similar kind of combination of units. Units such as the joule, which measures energy, have been given simple names that represent the combination of fundamental units by which they are defined.

B.2 Conversion of Units for Physical Quantities

The result of a measurement is a physical quantity, which consists of a number and a unit. Algebraically, a physical quantity can be treated as if the number is multiplied by the unit. To convert a physical quantity from one unit of measure to another requires a conversion factor (proportionality factor) based on equivalences between units of measure such as those given in Table B.4. (See Appendix A.2 for more about physical quantities and proportionality factors.) Each equivalence provides two conversion factors that are the reciprocals of each other. For example, the equivalence between a quart and a liter, 1 quart = 0.9463 liter, gives

$$\frac{1\ \text{quart}}{0.9463\ \text{liter}}$$ There is 1 quart per 0.9463 liter.

$$\frac{0.9463\ \text{liter}}{1\ \text{quart}}$$ There is 0.9463 liter per 1 quart.

The method of canceling units described in Appendix A.2 provides the basis for choosing which conversion factor is needed: It is always the one that allows the unit being converted to be canceled and leaves the new unit uncanceled.

To convert 2 quarts to liters:

$$2\ \text{quarts} \times \frac{0.9463\ \text{liter}}{1\ \text{quart}} = 1.893\ \text{liters}$$

To convert 2 liters to quarts:

$$2\ \text{liters} \times \frac{1\ \text{quart}}{0.9463\ \text{liter}} = 2.113\ \text{quarts}$$

Because of the definitions of Celsius degrees and Fahrenheit degrees, conversions between these temperature scales are a bit more complicated. Both units are based on the properties of water. The Celsius unit is defined by assigning 0 °C as the freezing point of pure water and 100 °C as its boiling point, when the pressure is exactly 1 atm. The size of the Fahrenheit degree is equally arbitrary. Fahrenheit defined 0 °F as the freezing point of a solution in which he had dissolved the maximum quantity of ammonium chloride (because this was the lowest temperature he could reproduce reliably), and he intended 100 °F to be the normal human body temperature (but this value turned out to be 98.6 °F). Today, the reference points are set at exactly 32 °F and 212 °F (the freezing and boiling points of pure water, at 1 atm). The number of units between these two Fahrenheit temperatures is 180 °F. Thus, the Celsius degree is almost twice as large as the Fahrenheit degree; it takes only 5 Celsius degrees to cover the same temperature range as 9 Fahrenheit degrees.

To be entirely correct, we must specify that pure water boils at 100 °C and freezes at 0 °C only when the pressure of the surrounding atmosphere is 1 atm.

$$\frac{100\ °\text{C}}{180\ °\text{F}} = \frac{5\ °\text{C}}{9\ °\text{F}}$$

This relationship is the basis for converting a temperature on one scale to a temperature on the other. If t_C is the numerical value of the temperature in °C and t_F is the numerical value of the temperature in °F, then

$$t_C = \left(\tfrac{5}{9}\right)(t_F - 32)$$
$$t_F = \left(\tfrac{9}{5}\right)t_C + 32$$

Table B.4 Common Units of Measure

Mass and Weight

1 pound = 453.59 grams = 0.45359 kilogram
1 kilogram = 1000 grams = 2.205 pounds
1 gram = 10 decigrams = 100 centigrams = 1000 milligrams
1 gram = 6.022×10^{23} atomic mass units
1 atomic mass unit = 1.6605×10^{-24} grams
1 short ton = 2000 pounds = 907.2 kilograms
1 long ton = 2240 pounds
1 metric tonne = 1000 kilograms = 2205 pounds

Length

1 inch = 2.54 centimeters (exactly)
1 mile = 5280 feet = 1.609 kilometers
1 yard = 36 inches = 0.9144 meter
1 meter = 100 centimeters = 39.37 inches = 3.281 feet = 1.094 yards
1 kilometer = 1000 meters = 1094 yards = 0.6215 miles
1 Angstrom = 1.0×10^{-8} centimeters = 0.10 nanometers = 100 picometers
= 1.0×10^{-10} meters = 3.937×10^{-9} inches

Volume

1 quart = 0.9463 liters
1 liter = 1.0567 quarts
1 liter = 1 cubic decimeter = 10^3 cubic centimeters = 10^{-3} cubic meters
1 milliliter = 1 cubic centimeters = 0.001 liters = 1.056×10^{-3} quarts
1 cubic foot = 28.316 liters = 29.924 quarts = 7.481 gallons

Force and Pressure

1 atmosphere = 760.0 millimeters of mercury = 1.01325×10^5 pascals
= 14.70 pounds per square inch
1 bar = 10^5 pascals = 0.98692 atmospheres
1 torr = 1 millimeter of mercury
1 pascal = $1 \text{ kg m}^{-1} \text{ s}^{-2} = 1 \text{ N/m}^2$

Energy

1 joule = 1×10^7 ergs
1 thermochemical calorie = 4.184 joules = 4.184×10^7 ergs
= 4.129×10^{-2} liter-atmospheres
= 2.612×10^{19} electron volts
1 erg = 1×10^{-7} joules = 2.3901×10^{-8} calories
1 electron volt = 1.6022×10^{-19} joules = 1.6022×10^{-12} ergs = 96.85 kJ/mol*
1 liter-atmosphere = 24.217 calories = 101.32 joules = 1.0132×10^9 ergs
1 British thermal unit = 1055.06 joules = 1.05506×10^{10} ergs = 252.2 calories

Temperature

0 K = −273.15 °C
If T_K is the numerical value of the temperature in kelvins, t_C is the numerical value of the temperature in °C, and t_F is the numerical value of the temperature in °F, then

$$T_K = t_C + 273.15$$
$$t_C = \left(\tfrac{5}{9}\right)(t_F - 32)$$
$$t_F = \left(\tfrac{9}{5}\right)t_C + 32$$

*The other units in this line are per particle and must be multiplied by 6.022×10^{23} to be strictly comparable.

For example, to show that your normal body temperature of 98.6 °F corresponds to 37.0 °C, use the first equation.

$$t_C = \left(\tfrac{5}{9}\right)(t_F - 32) = \left(\tfrac{5}{9}\right)(98.6 - 32) = \left(\tfrac{9}{5}\right)(66.6) = 37.0$$

Thus, body temperature in °C = 37.0 °C.

Laboratory work is almost always done using Celsius units, and we rarely need to make conversions to and from Fahrenheit degrees. It is best to try to calibrate your senses to Celsius units; to help you do so, it is useful to know that water freezes at 0 °C, a comfortable room temperature is about 22 °C, your body temperature is 37 °C, and the hottest water you could leave your hand in for some time is about 60 °C.

QUESTIONS FOR APPENDIX B

Blue-numbered questions have short answers at the back of this book and fully worked solutions in the *Student Solutions Manual.*

Units of the International System

1. Which SI unit accompanied by which prefix would be most convenient for describing each quantity?
   (a) Mass of this book
   (b) Volume of a glass of water
   (c) Thickness of this page
2. Which SI unit accompanied by which prefix would be most convenient for describing each quantity?
   (a) Distance from New York to San Francisco
   (b) Mass of a glass of water
   (c) Area of this page
3. Explain the difference between an SI base (fundamental) unit and a derived SI unit.
4. What is the official SI definition of the mole? Describe in your own words each part of the definition and explain why each part of the definition is important.

Conversion of Units for Physical Quantities

5. Express each quantity in SI base (fundamental) units. Use exponential (scientific) notation whenever it is needed.
   (a) 475 pm (b) 56 Gg (c) 4.28 μA
6. Express each quantity in SI base (fundamental) units. Use exponential (scientific) notation whenever it is needed.
   (a) 32.5 ng (b) 56 Mm (c) 439 pm
7. Express each quantity in SI base (fundamental) units. Use exponential (scientific) notation whenever it is needed.
   (a) 8.7 nm^2 (b) 27.3 aJ (c) 27.3 μN
8. Express each quantity in SI base (fundamental) units. Use exponential (scientific) notation whenever it is needed.
   (a) 56.3 cm^3 (b) 5.62 MJ (c) 33.4 kV
9. Express each quantity in the units indicated. Use scientific notation.
   (a) 1.00 kg in pound
   (b) 2.45 ton in kilograms
   (c) 1 L in cubic inches (in^3)
   (d) 1 atm in pascals and in bars
10. Express each quantity in the units indicated. Use scientific notation.
    (a) 24.3 amu in grams
    (b) 87.3 mL in cubic feet (ft^3)
    (c) 24.7 dg in ounces
    (d) 1.02 bar in millimeters of mercury (mm Hg) and in torrs
11. Express each temperature in Fahrenheit degrees.
    (a) 37 °C (b) −23.6 °C (c) −40.0 °C
12. Express each temperature in Celsius degrees.
    (a) 180. °F (b) −40.0 °F (c) 28.3 °F

Physical Constants* and Sources of Data

Quantity	Symbol	Traditional Units	SI Units
Acceleration of gravity	g_n	$980.665\ cm/s^2$	$9.806\ 65\ m/s^2$
Atomic mass unit ($\frac{1}{12}$ the mass of ^{12}C atom)	amu or u	$1.660\ 54 \times 10^{-24}$ g	$1.660\ 54 \times 10^{-27}$ kg
Avogadro constant	N_A, L	$6.022\ 141\ 79\ (30) \times 10^{23}$ particles/mol	$6.022\ 142 \times 10^{23}$ particles/mol
Bohr radius	a_o	0.529 177 2108 Å	$5.291\ 772\ 108 \times 10^{-11}$ m
Boltzmann constant	k	$1.380\ 650\ 5 \times 10^{-16}$ erg/K	$1.380\ 650\ 5 \times 10^{-23}$ J/K
Charge-to-mass ratio of electron	e/m	$-1.758\ 820\ 12 \times 10^{8}$ C/g	$-1.758\ 820\ 12 \times 10^{11}$ C/kg
Elementary charge (electron or proton charge)	e	$1.602\ 176\ 487\ (40) \times 10^{-19}$ C	$1.602\ 176\ 53 \times 10^{-19}$ C
Electron rest mass	m_e	$9.109\ 382\ 15\ (15) \times 10^{-28}$ g	$9.109\ 3826 \times 10^{-31}$ kg
Faraday constant	F	96 485.3383 C/mol e^-	96 485.3383 C/mol e^-
		23.06 kcal V^{-1} mol^{-1}	96 485 J V^{-1} mol^{-1}
Gas constant	R	0.082 057 L atm mol^{-1} K^{-1}	8.314 472 dm^3 Pa mol^{-1} K^{-1}
		1.987 cal mol^{-1} K^{-1}	8.314 472 J mol^{-1} K^{-1}
Molar volume (STP)	V_m	22.414 L/mol	$22.414 \times 10^{-3}\ m^3/mol$
			22.414 dm^3/mol
Neutron rest mass	m_n	$1.674\ 927\ 2811\ (84) \times 10^{-24}$ g	$1.674\ 927\ 28 \times 10^{-27}$ kg
		1.008 664 amu	
Planck's constant	h	$6.626\ 0693 \times 10^{-27}$ erg s	$6.626\ 0693 \times 10^{-34}$ J s
Proton rest mass	m_p	$1.672\ 621\ 637\ (83) \times 10^{-24}$ g	$1.672\ 621\ 71 \times 10^{-27}$ kg
		1.007 276 amu	
Rydberg constant	R_∞	$3.289\ 841\ 960 \times 10^{15}\ s^{-1}$	$1.097\ 373\ 156\ 8525 \times 10^{7}\ m^{-1}$
		$2.179\ 871\ 90 \times 10^{-11}$ erg	$2.179\ 871\ 90 \times 10^{-18}$ J
Velocity of light (in a vacuum)	c, c_0	$2.997\ 924\ 58 \times 10^{10}$ cm/s	$2.997\ 924\ 58 \times 10^{8}$ m/s
		186 282 mile/s	

*Data from the National Institute for Standards and Technology reference on constants, units, and uncertainty (in parentheses), **http://physics.nist.gov/cuu/Constants/Index.html.**

Online Sources

- Finding Chemical Data in Web and Library Sources. University of Adelaide Library
  http://www.library.adelaide.edu.au/guide/sci/Chemistry/propindex.html
- SIRCh: Selected Internet Resources for Chemistry
  http://www.indiana.edu/~cheminfo/cis_ca.html
  SIRCh: Physical Property Information
  http://www.indiana.edu/~cheminfo/ca_ppi.html
- Thermodex. University of Texas at Austin.
  http://thermodex.lib.utexas.edu/
- How Many? A Dictionary of Units of Measurement.
  http://www.unc.edu/~rowlett/units/index.html
- NIST Chemistry Web Book
  http://webbook.nist.gov/chemistry

Print Sources

- Lide, David R., ed. *CRC Handbook of Chemistry and Physics,* 88th edition, Boca Raton, FL: CRC Press, 2007–2008.
- Budavari, Susan; O'Neil, Maryadele J.; Smith, Ann; Heckelman, Patricia E., eds. *The Merck Index: An Encyclopedia of Chemicals, Drugs, and Biologicals,* 14th edition, Rahway, NJ: Merck & Co., 2006.
- Speight, James, ed. *Lange's Handbook of Chemistry,* 16th edition, New York: McGraw-Hill, 2004.
- Perry, Robert H., Green, Don W., eds. *Perry's Chemical Engineer's Handbook,* 8th edition, New York: McGraw-Hill, 2007.
- Zwillinger, Daniel, ed. *CRC Standard Mathematical Tables and Formulae,* 31st edition, Boca Raton, FL: CRC Press, 2003.
- Lewis, Richard J., Sr. *Hawley's Condensed Chemical Dictionary* (with CD-ROM), 15th edition, New York: Wiley, 2007.

APPENDIX D

Ground-State Electron Configurations of Atoms

Z	Element	Configuration	Z	Element	Configuration	Z	Element	Configuration
1	H	$1s^1$	40	Zr	[Kr] $4d^2 5s^2$	78	Pt	[Xe] $4f^{14} 5d^9 6s^1$
2	He	$1s^2$	41	Nb	[Kr] $4d^4 5s^1$	79	Au	[Xe] $4f^{14} 5d^{10} 6s^1$
3	Li	[He] $2s^1$	42	Mo	[Kr] $4d^5 5s^1$	80	Hg	[Xe] $4f^{14} 5d^{10} 6s^2$
4	Be	[He] $2s^2$	43	Tc	[Kr] $4d^5 5s^2$	81	Tl	[Xe] $4f^{14} 5d^{10} 6s^2 6p^1$
5	B	[He] $2s^2 2p^1$	44	Ru	[Kr] $4d^7 5s^1$	82	Pb	[Xe] $4f^{14} 5d^{10} 6s^2 6p^2$
6	C	[He] $2s^2 2p^2$	45	Rh	[Kr] $4d^8 5s^1$	83	Bi	[Xe] $4f^{14} 5d^{10} 6s^2 6p^3$
7	N	[He] $2s^2 2p^3$	46	Pd	[Kr] $4d^{10}$	84	Po	[Xe] $4f^{14} 5d^{10} 6s^2 6p^4$
8	O	[He] $2s^2 2p^4$	47	Ag	[Kr] $4d^{10} 5s^1$	85	At	[Xe] $4f^{14} 5d^{10} 6s^2 6p^5$
9	F	[He] $2s^2 2p^5$	48	Cd	[Kr] $4d^{10} 5s^2$	86	Rn	[Xe] $4f^{14} 5d^{10} 6s^2 6p^6$
10	Ne	[He] $2s^2 2p^6$	49	In	[Kr] $4d^{10} 5s^2 5p^1$	87	Fr	[Rn] $7s^1$
11	Na	[Ne] $3s^1$	50	Sn	[Kr] $4d^{10} 5s^2 5p^2$	88	Ra	[Rn] $7s^2$
12	Mg	[Ne] $3s^2$	51	Sb	[Kr] $4d^{10} 5s^2 5p^3$	89	Ac	[Rn] $6d^1 7s^2$
13	Al	[Ne] $3s^2 3p^1$	52	Te	[Kr] $4d^{10} 5s^2 5p^4$	90	Th	[Rn] $6d^2 7s^2$
14	Si	[Ne] $3s^2 3p^2$	53	I	[Kr] $4d^{10} 5s^2 5p^5$	91	Pa	[Rn] $5f^2 6d^1 7s^2$
15	P	[Ne] $3s^2 3p^3$	54	Xe	[Kr] $4d^{10} 5s^2 5p^6$	92	U	[Rn] $5f^3 6d^1 7s^2$
16	S	[Ne] $3s^2 3p^4$	55	Cs	[Xe] $6s^1$	93	Np	[Rn] $5f^4 6d^1 7s^2$
17	Cl	[Ne] $3s^2 3p^5$	56	Ba	[Xe] $6s^2$	94	Pu	[Rn] $5f^6 7s^2$
18	Ar	[Ne] $3s^2 3p^6$	57	La	[Xe] $5d^1 6s^2$	95	Am	[Rn] $5f^7 7s^2$
19	K	[Ar] $4s^1$	58	Ce	[Xe] $4f^1 5d^1 6s^2$	96	Cm	[Rn] $5f^7 6d^1 7s^2$
20	Ca	[Ar] $4s^2$	59	Pr	[Xe] $4f^3 6s^2$	97	Bk	[Rn] $5f^9 7s^2$
21	Sc	[Ar] $3d^1 4s^2$	60	Nd	[Xe] $4f^4 6s^2$	98	Cf	[Rn] $5f^{10} 7s^2$
22	Ti	[Ar] $3d^2 4s^2$	61	Pm	[Xe] $4f^5 6s^2$	99	Es	[Rn] $5f^{11} 7s^2$
23	V	[Ar] $3d^3 4s^2$	62	Sm	[Xe] $4f^6 6s^2$	100	Fm	[Rn] $5f^{12} 7s^2$
24	Cr	[Ar] $3d^5 4s^1$	63	Eu	[Xe] $4f^7 6s^2$	101	Md	[Rn] $5f^{13} 7s^2$
25	Mn	[Ar] $3d^5 4s^2$	64	Gd	[Xe] $4f^7 5d^1 6s^2$	102	No	[Rn] $5f^{14} 7s^2$
26	Fe	[Ar] $3d^6 4s^2$	65	Tb	[Xe] $4f^9 6s^2$	103	Lr	[Rn] $5f^{14} 6d^1 7s^2$
27	Co	[Ar] $3d^7 4s^2$	66	Dy	[Xe] $4f^{10} 6s^2$	104	Rf	[Rn] $5f^{14} 6d^2 7s^2$
28	Ni	[Ar] $3d^8 4s^2$	67	Ho	[Xe] $4f^{11} 6s^2$	105	Db	[Rn] $5f^{14} 6d^3 7s^2$
29	Cu	[Ar] $3d^{10} 4s^1$	68	Er	[Xe] $4f^{12} 6s^2$	106	Sg	[Rn] $5f^{14} 6d^4 7s^2$
30	Zn	[Ar] $3d^{10} 4s^2$	69	Tm	[Xe] $4f^{13} 6s^2$	107	Bh	[Rn] $5f^{14} 6d^5 7s^2$
31	Ga	[Ar] $3d^{10} 4s^2 4p^1$	70	Yb	[Xe] $4f^{14} 6s^2$	108	Hs	[Rn] $5f^{14} 6d^6 7s^2$
32	Ge	[Ar] $3d^{10} 4s^2 4p^2$	71	Lu	[Xe] $4f^{14} 5d^1 6s^2$	109	Mt	[Rn] $5f^{14} 6d^7 7s^2$
33	As	[Ar] $3d^{10} 4s^2 4p^3$	72	Hf	[Xe] $4f^{14} 5d^2 6s^2$	110	Ds	[Rn] $5f^{14} 6d^8 7s^2$
34	Se	[Ar] $3d^{10} 4s^2 4p^4$	73	Ta	[Xe] $4f^{14} 5d^3 6s^2$	111	Rg	[Rn] $5f^{14} 6d^9 7s^2$
35	Br	[Ar] $3d^{10} 4s^2 4p^5$	74	W	[Xe] $4f^{14} 5d^4 6s^2$	112	—	[Rn] $5f^{14} 6d^{10} 7s^2$
36	Kr	[Ar] $3d^{10} 4s^2 4p^6$	75	Re	[Xe] $4f^{14} 5d^5 6s^2$	113	—	[Rn] $5f^{14} 6d^{10} 7s^2 7p^1$
37	Rb	[Kr] $5s^1$	76	Os	[Xe] $4f^{14} 5d^6 6s^2$	114	—	[Rn] $5f^{14} 6d^{10} 7s^2 7p^2$
38	Sr	[Kr] $5s^2$	77	Ir	[Xe] $4f^{14} 5d^7 6s^2$	115	—	[Rn] $5f^{14} 6d^{10} 7s^2 7p^3$
39	Y	[Kr] $4d^1 5s^2$						

Naming Hydrocarbons

The systematic nomenclature for organic compounds was proposed by the International Union of Pure and Applied Chemistry (IUPAC). The IUPAC set of rules provides different names for the more than 10 million known organic compounds, and allows names to be assigned to new compounds as they are synthesized. Many organic compounds also have *common* names. Usually the common name came first and is widely known. Many consumer products are labeled with the common name, and when only a few isomers are possible, the common name adequately identifies the product for the consumer. However, as illustrated in Section 3.4 (⇐ ***p. 69***), a system of common names quickly fails when several structural isomers are possible.

Hydrocarbons

The name of each member of the hydrocarbon classes has two parts. The first part, called the prefix (*meth-*, *eth-*, *prop-*, *but-*, and so on), reflects the number of carbon atoms. When more than four carbons are present, the Greek or Latin number prefixes are used: *pent-*, *hex-*, *hept-*, *oct-*, *non-*, and *dec-*. The second part of the name, called the suffix, tells the class of hydrocarbon. Alkanes have carbon-carbon single bonds, alkenes have carbon-carbon double bonds, and alkynes have carbon-carbon triple bonds.

Unbranched Alkanes and Alkyl Groups

The names of the first 20 unbranched (straight-chain) alkanes are given in Table E.1.

Alkyl groups are named by dropping *-ane* from the parent alkane and adding *-yl* (see Table 3.5 for examples).

Branched-Chain Alkanes

The rules for naming branched-chain alkanes are as follows:

1. *Find the longest continuous chain of carbon atoms; it determines the parent name for the compound.* For example, the following compound has two methyl groups attached to a *heptane* parent; the longest continuous chain contains seven carbon atoms.

$$CH_3CH_2CH_2\underset{\displaystyle CH_3}{\underset{|}{C}}HCH_2\underset{\displaystyle CH_3}{\underset{|}{C}}HCH_3$$

Table E.1 Names of Unbranched Alkanes

CH_4	Methane	$C_{11}H_{24}$	Undecane
C_2H_6	Ethane	$C_{12}H_{26}$	Dodecane
C_3H_8	Propane	$C_{13}H_{28}$	Tridecane
C_4H_{10}	Butane	$C_{14}H_{30}$	Tetradecane
C_5H_{12}	Pentane	$C_{15}H_{32}$	Pentadecane
C_6H_{14}	Hexane	$C_{16}H_{34}$	Hexadecane
C_7H_{16}	Heptane	$C_{17}H_{36}$	Heptadecane
C_8H_{18}	Octane	$C_{18}H_{38}$	Octadecane
C_9H_{20}	Nonane	$C_{19}H_{40}$	Nonadecane
$C_{10}H_{22}$	Decane	$C_{20}H_{42}$	Eicosane

The longest continuous chain may not be obvious from the way the formula is written, especially for the straight-line format that is commonly used. For example, the longest continuous chain of carbon atoms in the following chain is *eight*, not *four* or *six*.

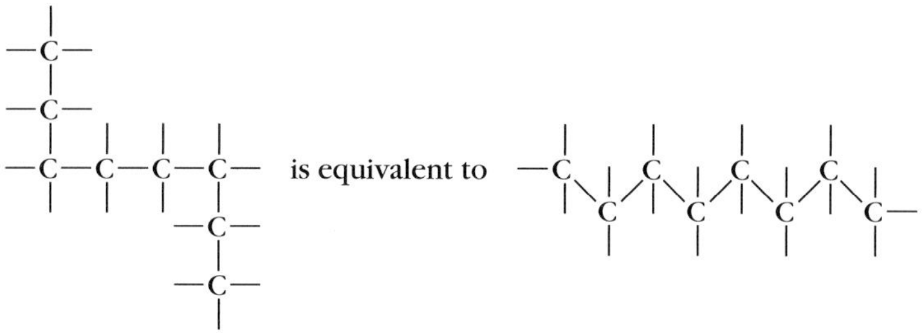

is equivalent to

2. *Number the longest chain beginning with the end of the chain nearest the branching. Use these numbers to designate the location of the attached group. When two or more groups are attached to the parent, give each group a number corresponding to its location on the parent chain.* For example, the name of

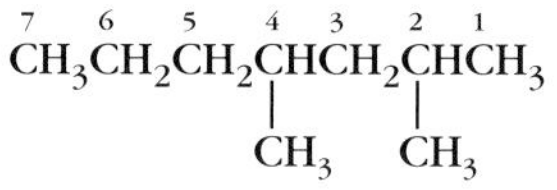

$$\overset{7}{C}H_3\overset{6}{C}H_2\overset{5}{C}H_2\overset{4}{C}H(CH_3)\overset{3}{C}H_2\overset{2}{C}H(CH_3)\overset{1}{C}H_3$$

is 2,4-dimethylheptane. The name of the compound below is 3-methylheptane, not 5-methylheptane or 2-ethylhexane.

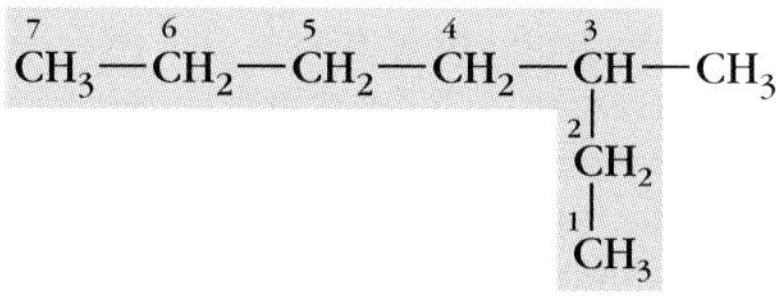

$$\overset{7}{C}H_3-\overset{6}{C}H_2-\overset{5}{C}H_2-\overset{4}{C}H_2-\overset{3}{C}H(-CH_3)-\overset{2}{C}H_2-\overset{1}{C}H_3$$

3. *When two or more substituents are identical, indicate this by the use of the prefixes* di-, tri-, tetra, *and so on. Positional numbers of the substituents should have the smallest possible sum.*

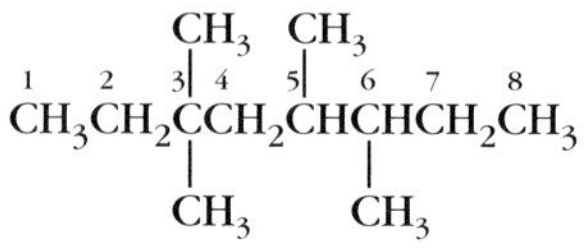

$$\overset{1}{C}H_3\overset{2}{C}H_2\overset{3}{C}(CH_3)_2\overset{4}{C}H_2\overset{5}{C}H(CH_3)\overset{6}{C}H(CH_3)\overset{7}{C}H_2\overset{8}{C}H_3$$

The correct name of this compound is 3,3,5,6-tetramethyloctane.

4. *If there are two or more different groups, the groups are listed alphabetically.*

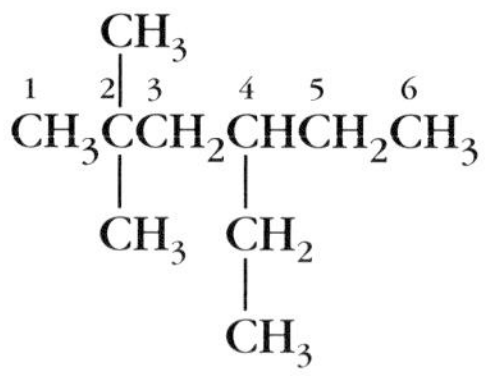

$$\overset{1}{C}H_3\overset{2}{C}(CH_3)_2\overset{3}{C}H_2\overset{4}{C}H(CH_2CH_3)\overset{5}{C}H_2\overset{6}{C}H_3$$

The correct name of this compound is 4-ethyl-2,2-dimethylhexane. Note that the prefix *di-* is ignored in determining alphabetical order.

Alkenes

Alkenes are named by using the prefix to indicate the number of carbon atoms and the suffix *-ene* to indicate one or more double bonds. The systematic names for the first two members of the alkene series are *ethene* and *propene*.

$$CH_2{=}CH_2 \qquad CH_3CH{=}CH_2$$

When groups, such as methyl or ethyl, are attached to carbon atoms in an alkene, the longest hydrocarbon chain is numbered from the end that will give the double bond the lowest number, and then numbers are assigned to the attached groups. For example, the name of

$$\begin{array}{l} \quad\quad\; CH_3 \\ \quad\;\; ^{5}\quad ^{4}| \quad ^{3} \qquad ^{2} \quad ^{1} \\ CH_3CHCH{=}CHCH_3 \end{array}$$

is 4-methyl-2-pentene. See Section 8.4 for a discussion of *cis-trans* isomers of alkenes.

Alkynes

The naming of alkynes is similar to that of alkenes, with the lowest number possible being used to locate the triple bond. For example, the name of

$$\begin{array}{l} \qquad\qquad\;\; CH_3 \\ ^{1} \quad ^{2} \quad ^{3\,4}| \quad ^{5} \\ CH_3C{\equiv}CCHCH_3 \end{array}$$

is 4-methyl-2-pentyne.

APPENDIX F

Ionization Constants for Weak Acids at 25 °C

Acid	Formula and Ionization Equation	K_a
Acetic	$CH_3COOH + H_2O \rightleftharpoons H_3O^+ + CH_3COO^-$	1.8×10^{-5}
Arsenic	$H_3AsO_4 + H_2O \rightleftharpoons H_3O^+ + H_2AsO_4^-$	$K_1 = 2.5 \times 10^{-4}$
	$H_2AsO_4^- + H_2O \rightleftharpoons H_3O^+ + HAsO_4^{2-}$	$K_2 = 5.6 \times 10^{-8}$
	$HAsO_4^{2-} + H_2O \rightleftharpoons H_3O^+ + AsO_4^{3-}$	$K_3 = 3.0 \times 10^{-13}$
Arsenous	$H_3AsO_3 + H_2O \rightleftharpoons H_3O^+ + H_2AsO_3^-$	$K_1 = 6.0 \times 10^{-10}$
	$H_2AsO_3^- + H_2O \rightleftharpoons H_3O^+ + HAsO_3^{2-}$	$K_2 = 3.0 \times 10^{-14}$
Benzoic	$C_6H_5COOH + H_2O \rightleftharpoons H_3O^+ + C_6H_5COO^-$	6.3×10^{-5}
Boric	$B(OH)_3(H_2O) + H_2O \rightleftharpoons H_3O^+ + B(OH)_4^-$	7.3×10^{-10}
Carbonic	$H_2CO_3 + H_2O \rightleftharpoons H_3O^+ + HCO_3^-$	$K_1 = 4.2 \times 10^{-7}$
	$HCO_3^- + H_2O \rightleftharpoons H_3O^+ + CO_3^{2-}$	$K_2 = 4.8 \times 10^{-11}$
Citric	$H_3C_6H_5O_7 + H_2O \rightleftharpoons H_3O^+ + H_2C_6H_5O_7^-$	$K_1 = 7.4 \times 10^{-3}$
	$H_2C_6H_5O_7^- + H_2O \rightleftharpoons H_3O^+ + HC_6H_5O_7^{2-}$	$K_2 = 1.7 \times 10^{-5}$
	$HC_6H_5O_7^{2-} + H_2O \rightleftharpoons H_3O^+ + C_6H_5O_7^{3-}$	$K_3 = 4.0 \times 10^{-7}$
Cyanic	$HOCN + H_2O \rightleftharpoons H_3O^+ + OCN^-$	3.5×10^{-4}
Formic	$HCOOH + H_2O \rightleftharpoons H_3O^+ + HCOO^-$	1.8×10^{-4}
Hydrazoic	$HN_3 + H_2O \rightleftharpoons H_3O^+ + N_3^-$	1.9×10^{-5}
Hydrocyanic	$HCN + H_2O \rightleftharpoons H_3O^+ + CN^-$	4.0×10^{-10}
Hydrofluoric	$HF + H_2O \rightleftharpoons H_3O^+ + F^-$	7.2×10^{-4}
Hydrogen peroxide	$H_2O_2 + H_2O \rightleftharpoons H_3O^+ + HO_2^-$	2.4×10^{-12}
Hydrosulfuric	$H_2S + H_2O \rightleftharpoons H_3O^+ + HS^-$	$K_1 = 1 \times 10^{-7}$
	$HS^- + H_2O \rightleftharpoons H_3O^+ + S^{2-}$	$K_2 = 1 \times 10^{-19}$
Hypobromous	$HOBr + H_2O \rightleftharpoons H_3O^+ + OBr^-$	2.5×10^{-9}
Hypochlorous	$HOCl + H_2O \rightleftharpoons H_3O^+ + OCl^-$	3.5×10^{-8}
Nitrous	$HNO_2 + H_2O \rightleftharpoons H_3O^+ + NO_2^- H_3O^+$	4.5×10^{-4}
Oxalic	$H_2C_2O_4 + H_2O \rightleftharpoons H_3O^+ + HC_2O_4^-$	$K_1 = 5.9 \times 10^{-2}$
	$HC_2O_4^- + H_2O \rightleftharpoons H_3O^+ + C_2O_4^{2-}$	$K_2 = 6.4 \times 10^{-5}$
Phenol	$HC_6H_5O + H_2O \rightleftharpoons H_3O^+ + C_6H_5O^-$	1.3×10^{-10}
Phosphoric	$H_3PO_4 + H_2O \rightleftharpoons H_3O^+ + H_2PO_4^-$	$K_1 = 7.5 \times 10^{-3}$
	$H_2PO_4^- + H_2O \rightleftharpoons H_3O^+ + HPO_4^{2-}$	$K_2 = 6.2 \times 10^{-8}$
	$HPO_4^{2-} + H_2O \rightleftharpoons H_3O^+ + PO_4^{3-}$	$K_3 = 3.6 \times 10^{-13}$
Phosphorous	$H_3PO_3 + H_2O \rightleftharpoons H_3O^+ + H_2PO_3^-$	$K_1 = 1.6 \times 10^{-2}$
	$H_2PO_3^- + H_2O \rightleftharpoons H_3O^+ + HPO_3^{2-}$	$K_2 = 7.0 \times 10^{-7}$
Selenic	$H_2SeO_4 + H_2O \rightleftharpoons H_3O^+ + HSeO_4^-$	K_1 = very large
	$HSeO_4^- + H_2O \rightleftharpoons H_3O^+ + SeO_4^{2-}$	$K_2 = 1.2 \times 10^{-2}$
Selenous	$H_2SeO_3 + H_2O \rightleftharpoons H_3O^+ + HSeO_3^-$	$K_1 = 2.7 \times 10^{-3}$
	$HSeO_3^- + H_2O \rightleftharpoons H_3O^+ + SeO_3^{2-}$	$K_2 = 2.5 \times 10^{-7}$
Sulfuric	$H_2SO_4 + H_2O \rightleftharpoons H_3O^+ + HSO_4^-$	K_1 = very large
	$HSO_4^- + H_2O \rightleftharpoons H_3O^+ + SO_4^{2-}$	$K_2 = 1.2 \times 10^{-2}$
Sulfurous	$H_2SO_3 + H_2O \rightleftharpoons H_3O^+ + HSO_3^-$	$K_1 = 1.7 \times 10^{-2}$
	$HSO_3^- + H_2O \rightleftharpoons H_3O^+ + SO_3^{2-}$	$K_2 = 6.4 \times 10^{-8}$
Tellurous	$H_2TeO_3 + H_2O \rightleftharpoons H_3O^+ + HTeO_3^-$	$K_1 = 2 \times 10^{-3}$
	$HTeO_3^- + H_2O \rightleftharpoons H_3O^+ + TeO_3^{2-}$	$K_2 = 1 \times 10^{-8}$

APPENDIX G

Ionization Constants for Weak Bases at 25 °C

Base	Formula and Ionization Equation	K_b
Ammonia	$NH_3 + H_2O \rightleftharpoons NH_4^+ + OH^-$	1.8×10^{-5}
Aniline	$C_6H_5NH_2 + H_2O \rightleftharpoons C_6H_5NH_3^+ + OH^-$	4.2×10^{-10}
Dimethylamine	$(CH_3)_2NH + H_2O \rightleftharpoons (CH_3)_2NH_2^+ + OH^-$	7.4×10^{-4}
Ethylenediamine	$(CH_2)_2(NH_2)_2 + H_2O \rightleftharpoons (CH_2)_2(NH_2)_2H^+ + OH^-$	$K_1 = 8.5 \times 10^{-5}$
	$(CH_2)_2(NH_2)_2H^+ + H_2O \rightleftharpoons (CH_2)_2(NH_2)_2H_2^{2+} + OH^-$	$K_2 = 2.7 \times 10^{-8}$
Hydrazine	$N_2H_4 + H_2O \rightleftharpoons N_2H_5^+ + OH^-$	$K_1 = 8.5 \times 10^{-7}$
	$N_2H_5^+ + H_2O \rightleftharpoons N_2H_6^{2+} + OH^-$	$K_2 = 8.9 \times 10^{-16}$
Hydroxylamine	$NH_2OH + H_2O \rightleftharpoons NH_3OH^+ + OH^-$	6.6×10^{-9}
Methylamine	$CH_3NH_2 + H_2O \rightleftharpoons CH_3NH_3^+ + OH^-$	5.0×10^{-4}
Pyridine	$C_5H_5N + H_2O \rightleftharpoons C_5H_5NH^+ + OH^-$	1.5×10^{-9}
Trimethylamine	$(CH_3)_3N + H_2O \rightleftharpoons (CH_3)_3NH^+ + OH^-$	7.4×10^{-5}

APPENDIX H

Solubility Product Constants for Some Inorganic Compounds at 25 °C*

Substance	K_{sp}
Aluminum Compounds	
$AlAsO_4$	1.6×10^{-16}
$Al(OH)_3$	1.9×10^{-33}
$AlPO_4$	1.3×10^{-20}
Barium Compounds	
$Ba_3(AsO_4)_2$	1.1×10^{-13}
$BaCO_3$	8.1×10^{-9}
$BaC_2O_4 \cdot 2\,H_2O$†	1.1×10^{-7}
$BaCrO_4$	2.0×10^{-10}
BaF_2	1.7×10^{-6}
$Ba(OH)_2 \cdot 8\,H_2O$†	5.0×10^{-3}
$Ba_3(PO_4)_2$	1.3×10^{-29}
$BaSeO_4$	2.8×10^{-11}
$BaSO_3$	8.0×10^{-7}
$BaSO_4$	1.1×10^{-10}
Bismuth Compounds	
$BiOCl$	7.0×10^{-9}
$BiO(OH)$	1.0×10^{-12}
$Bi(OH)_3$	3.2×10^{-40}
BiI_3	8.1×10^{-19}
$BiPO_4$	1.3×10^{-23}
Cadmium Compounds	
$Cd_3(AsO_4)_2$	2.2×10^{-32}
$CdCO_3$	2.5×10^{-14}
$Cd(CN)_2$	1.0×10^{-8}
$Cd_2[Fe(CN)_6]$	3.2×10^{-17}
$Cd(OH)_2$	1.2×10^{-14}
Calcium Compounds	
$Ca_3(AsO_4)_2$	6.8×10^{-19}
$CaCO_3$	3.8×10^{-9}
$CaCrO_4$	7.1×10^{-4}
$CaC_2O_4 \cdot H_2O$†	2.3×10^{-9}
CaF_2	3.9×10^{-11}
$Ca(OH)_2$	7.9×10^{-6}
$CaHPO_4$	2.7×10^{-7}
$Ca(H_2PO_4)_2$	1.0×10^{-3}
$Ca_3(PO_4)_2$	1.0×10^{-25}

Substance	K_{sp}
$CaSO_3 \cdot 2\,H_2O$†	1.3×10^{-8}
$CaSO_4 \cdot 2\,H_2O$†	2.4×10^{-5}
Chromium Compounds	
$CrAsO_4$	7.8×10^{-21}
$Cr(OH)_3$	6.7×10^{-31}
$CrPO_4$	2.4×10^{-23}
Cobalt Compounds	
$Co_3(AsO_4)_2$	7.6×10^{-29}
$CoCO_3$	8.0×10^{-13}
$Co(OH)_2$	2.5×10^{-16}
$Co(OH)_3$	4.0×10^{-45}
Copper Compounds	
$CuBr$	5.3×10^{-9}
$CuCl$	1.9×10^{-7}
$CuCN$	3.2×10^{-20}
$Cu_2O(Cu^+ + OH^-)$‡	1.0×10^{-14}
CuI	5.1×10^{-12}
$CuSCN$	1.6×10^{-11}
$Cu_3(AsO_4)_2$	7.6×10^{-36}
$CuCO_3$	2.5×10^{-10}
$Cu_2[Fe(CN)_6]$	1.3×10^{-16}
$Cu(OH)_2$	1.6×10^{-19}
Gold Compounds	
$AuBr$	5.0×10^{-17}
$AuCl$	2.0×10^{-13}
AuI	1.6×10^{-23}
$AuBr_3$	4.0×10^{-36}
$AuCl_3$	3.2×10^{-25}
$Au(OH)_3$	1×10^{-53}
AuI_3	1.0×10^{-46}
Iron Compounds	
$FeCO_3$	3.5×10^{-11}
$Fe(OH)_2$	7.9×10^{-15}
FeS	4.9×10^{-18}
$Fe_4[Fe(CN)_6]_3$	3.0×10^{-41}
$Fe(OH)_3$	6.3×10^{-38}

Substance	K_{sp}
Lead Compounds	
$Pb_3(AsO_4)_2$	4.1×10^{-36}
$PbBr_2$	6.3×10^{-6}
$PbCO_3$	1.5×10^{-13}
$PbCl_2$	1.7×10^{-5}
$PbCrO_4$	1.8×10^{-14}
PbF_2	3.7×10^{-8}
$Pb(OH)_2$	2.8×10^{-16}
PbI_2	8.7×10^{-9}
$Pb_3(PO_4)_2$	3.0×10^{-44}
$PbSeO_4$	1.5×10^{-7}
$PbSO_4$	1.8×10^{-8}
Magnesium Compounds	
$Mg_3(AsO_4)_2$	2.1×10^{-20}
$MgCO_3 \cdot 3\,H_2O$†	4.0×10^{-5}
MgC_2O_4	8.6×10^{-5}
MgF_2	6.4×10^{-9}
$MgNH_4PO_4$	2.5×10^{-12}
Manganese Compounds	
$Mn_3(AsO_4)_2$	1.9×10^{-11}
$MnCO_3$	1.8×10^{-11}
$Mn(OH)_2$	4.6×10^{-14}
$Mn(OH)_3$	$\sim 1 \times 10^{-36}$
Mercury Compounds	
Hg_2Br_2	1.3×10^{-22}
Hg_2CO_3	8.9×10^{-17}
Hg_2Cl_2	1.1×10^{-18}
Hg_2CrO_4	5.0×10^{-9}
Hg_2I_2	4.5×10^{-29}
$Hg_2O \cdot H_2O\ (Hg_2^{2+} + 2\,OH^-)$†‡	1.6×10^{-23}
Hg_2SO_4	6.8×10^{-7}
$Hg(CN)_2$	3.0×10^{-23}
$Hg(OH)_2$	2.5×10^{-26}
HgI_2	4.0×10^{-29}
Nickel Compounds	
$Ni_3(AsO_4)_2$	1.9×10^{-26}
$NiCO_3$	6.6×10^{-9}
$Ni(CN)_2$	3.0×10^{-23}
$Ni(OH)_2$	2.8×10^{-16}
Silver Compounds	
Ag_3AsO_4	1.1×10^{-20}
$AgBr$	3.3×10^{-13}
Ag_2CO_3	8.1×10^{-12}
$AgCl$	1.8×10^{-10}
Ag_2CrO_4	9.0×10^{-12}
$AgCN$	1.2×10^{-16}
$Ag_4[Fe(CN)_6]$	1.6×10^{-41}
$Ag_2O\ (Ag^+ + OH^-)$‡	2.0×10^{-8}
AgI	1.5×10^{-16}
Ag_3PO_4	8.9×10^{-17}
Ag_2SO_3	1.5×10^{-14}
Ag_2SO_4	1.7×10^{-5}
$AgSCN$	1.0×10^{-12}
Strontium Compounds	
$Sr_3(AsO_4)_2$	1.3×10^{-18}
$SrCO_3$	9.4×10^{-10}
$SrC_2O_4 \cdot 2\,H_2O$†	5.6×10^{-8}
$SrCrO_4$	3.6×10^{-5}
$Sr(OH)_2 \cdot 8\,H_2O$†	3.2×10^{-4}
$Sr_3(PO_4)_2$	1.0×10^{-31}
$SrSO_3$	4.0×10^{-8}
$SrSO_4$	2.8×10^{-7}
Tin Compounds	
$Sn(OH)_2$	2.0×10^{-26}
SnI_2	1.0×10^{-4}
$Sn(OH)_4$	1×10^{-57}
Zinc Compounds	
$Zn_3(AsO_4)_2$	1.1×10^{-27}
$ZnCO_3$	1.5×10^{-11}
$Zn(CN)_2$	8.0×10^{-12}
$Zn_3[Fe(CN)_6]$	4.1×10^{-16}
$Zn(OH)_2$	4.5×10^{-17}
$Zn_3(PO_4)_2$	9.1×10^{-33}

*No metallic sulfides are listed in this table because sulfide ion is such a strong base that the usual solubility product equilibrium equation does not apply. See Myers, R. J. *Journal of Chemical Education,* Vol. 63, 1986; pp. 687–690.

†Since $[H_2O]$ does not appear in equilibrium constants for equilibria in aqueous solution in general, it does *not* appear in the K_{sp} expressions for hydrated solids.

‡Very small amounts of these oxides dissolve in water to give the ions indicated in parentheses. Solid hydroxides of these metal ions are unstable and decompose to oxides as rapidly as they are formed.

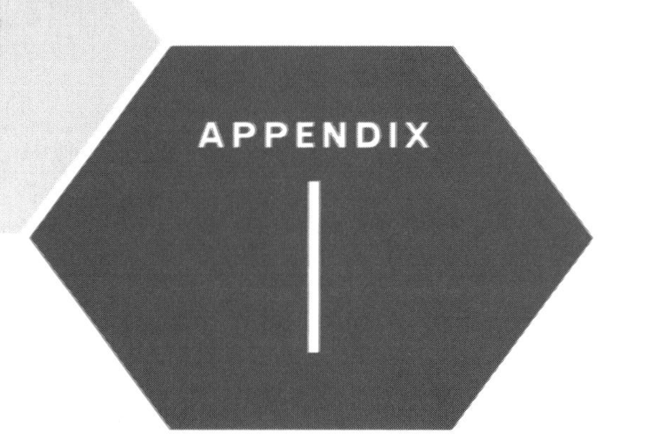

Standard Reduction Potentials in Aqueous Solution at 25 °C

Acidic Solution	Standard Reduction Potential, $E°$ (volts)
$F_2(g) + 2e^- \longrightarrow 2\,F^-(aq)$	2.87
$Co^{3+}(aq) + e^- \longrightarrow Co^{2+}(aq)$	1.82
$Pb^{4+}(aq) + 2\,e^- \longrightarrow Pb^{2+}(aq)$	1.8
$H_2O_2(aq) + 2\,H_3O^+(aq) + 2\,e^- \longrightarrow 4\,H_2O(\ell)$	1.77
$NiO_2(s) + 4\,H_3O^+(aq) + 2\,e^- \longrightarrow Ni^{2+}(aq) + 6\,H_2O(\ell)$	1.7
$PbO_2(s) + SO_4^{2-}(aq) + 4\,H_3O^+(aq) + 2\,e^- \longrightarrow PbSO_4(s) + 6\,H_2O(\ell)$	1.685
$Au^+(aq) + e^- \longrightarrow Au(s)$	1.68
$2\,HClO(aq) + 2\,H_3O^+(aq) + 2\,e^- \longrightarrow Cl_2(g) + 4\,H_2O(\ell)$	1.63
$Ce^{4+}(aq) + e^- \longrightarrow Ce^{3+}(aq)$	1.61
$NaBiO_3(s) + 6\,H_3O^+(aq) + 2\,e^- \longrightarrow Bi^{3+}(aq) + Na^+(aq) + 9\,H_2O(\ell)$	~ 1.6
$MnO_4^-(aq) + 8\,H_3O^+(aq) + 5\,e^- \longrightarrow Mn^{2+}(aq) + 12\,H_2O(\ell)$	1.51
$Au^{3+}(aq) + 3\,e^- \longrightarrow Au(s)$	1.50
$2\,ClO_3^-(aq) + 12\,H_3O^+(aq) + 10\,e^- \longrightarrow Cl_2(g) + 18\,H_2O(\ell)$	1.47
$BrO_3^-(aq) + 6\,H_3O^+(aq) + 6\,e^- \longrightarrow Br^-(aq) + 9\,H_2O(\ell)$	1.44
$Cl_2(g) + 2\,e^- \longrightarrow 2\,Cl^-(aq)$	1.358
$Cr_2O_7^{2-}(aq) + 14\,H_3O^+(aq) + 6\,e^- \longrightarrow 2\,Cr^{3+}(aq) + 21\,H_2O(\ell)$	1.33
$N_2H_5^+(aq) + 3\,H_3O^+(aq) + 2\,e^- \longrightarrow 2\,NH_4^+(aq) + 3\,H_2O(\ell)$	1.24
$MnO_2(s) + 4\,H_3O^+(aq) + 2\,e^- \longrightarrow Mn^{2+}(aq) + 6\,H_2O(\ell)$	1.23
$O_2(g) + 4\,H_3O^+(aq) + 4\,e^- \longrightarrow 6\,H_2O(\ell)$	1.229
$Pt^{2+}(aq) + 2\,e^- \longrightarrow Pt(s)$	1.2
$IO_3^-(aq) + 6\,H_3O^+(aq) + 5\,e^- \longrightarrow \frac{1}{2}\,I_2(aq) + 9\,H_2O(\ell)$	1.195
$ClO_4^-(aq) + 2\,H_3O^+(aq) + 2\,e^- \longrightarrow ClO_3^-(aq) + 3\,H_2O(\ell)$	1.19
$Br_2(\ell) + 2\,e^- \longrightarrow 2\,Br^-(aq)$	1.066
$AuCl_4^-(aq) + 3\,e^- \longrightarrow Au(s) + 4\,Cl^-(aq)$	1.00
$Pd^{2+}(aq) + 2\,e^- \longrightarrow Pd(s)$	0.987
$NO_3^-(aq) + 4\,H_3O^+(aq) + 3\,e^- \longrightarrow NO(g) + 6\,H_2O(\ell)$	0.96
$NO_3^-(aq) + 3\,H_3O^+(aq) + 2\,e^- \longrightarrow HNO_2(aq) + 4\,H_2O(\ell)$	0.94
$2\,Hg^{2+}(aq) + 2\,e^- \longrightarrow Hg_2^{2+}(aq)$	0.920
$Hg^{2+}(aq) + 2\,e^- \longrightarrow Hg(\ell)$	0.855
$Ag^+(aq) + e^- \longrightarrow Ag(s)$	0.7994
$Hg_2^{2+}(aq) + 2\,e^- \longrightarrow 2\,Hg(\ell)$	0.789
$Fe^{3+}(aq) + e^- \longrightarrow Fe^{2+}(aq)$	0.771
$SbCl_6^-(aq) + 2\,e^- \longrightarrow SbCl_4^-(aq) + 2\,Cl^-(aq)$	0.75
$[PtCl_4]^{2-}(aq) + 2\,e^- \longrightarrow Pt(s) + 4\,Cl^-(aq)$	0.73
$O_2(g) + 2\,H_3O^+(aq) + 2\,e^- \longrightarrow H_2O_2(aq) + 2\,H_2O(\ell)$	0.682
$[PtCl_6]^{2-}(aq) + 2\,e^- \longrightarrow [PtCl_4]^{2-}(aq) + 2\,Cl^-(aq)$	0.68
$H_3AsO_4(aq) + 2\,H_3O^+(aq) + 2\,e^- \longrightarrow H_3AsO_3(aq) + 3\,H_2O(\ell)$	0.58
$I_2(s) + 2\,e^- \longrightarrow 2\,I^-(aq)$	0.535

Acidic Solution	**Standard Reduction Potential, $E°$ (volts)**
$TeO_2(s) + 4\,H_3O^+(aq) + 4\,e^- \longrightarrow Te(s) + 6\,H_2O(\ell)$	0.529
$Cu^+(aq) + e^- \longrightarrow Cu(s)$	0.521
$[RhCl_6]^{3-}(aq) + 3\,e^- \longrightarrow Rh(s) + 6\,Cl^-(aq)$	0.44
$Cu^{2+}(aq) + 2\,e^- \longrightarrow Cu(s)$	0.337
$Hg_2Cl_2(s) + 2\,e^- \longrightarrow 2\,Hg(\ell) + 2\,Cl^-(aq)$	0.27
$AgCl(s) + e^- \longrightarrow Ag(s) + Cl^-(aq)$	0.222
$SO_4^{2-}(aq) + 4\,H_3O^+(aq) + 2\,e^- \longrightarrow SO_2(g) + 6\,H_2O(\ell)$	0.20
$SO_4^{2-}(aq) + 4\,H_3O^+(aq) + 2\,e^- \longrightarrow H_2SO_3(aq) + 5\,H_2O(\ell)$	0.17
$Cu^{2+}(aq) + e^- \longrightarrow Cu^+(aq)$	0.153
$Sn^{4+}(aq) + 2\,e^- \longrightarrow Sn^{2+}(aq)$	0.15
$S(s) + 2\,H_3O^+(aq) + 2\,e^- \longrightarrow H_2S(aq) + 2\,H_2O(\ell)$	0.14
$AgBr(s) + e^- \longrightarrow Ag(s) + Br^-(aq)$	0.0713
$2\,H_3O^+(aq) + 2\,e^- \longrightarrow H_2(g) + 2\,H_2O(\ell)$ (reference electrode)	0.0000
$N_2O(g) + 6\,H_3O^+(aq) + 4\,e^- \longrightarrow 2\,NH_3OH^+(aq) + 5\,H_2O(\ell)$	−0.05
$Pb^{2+}(aq) + 2\,e^- \longrightarrow Pb(s)$	−0.126
$Sn^{2+}(aq) + 2\,e^- \longrightarrow Sn(s)$	−0.14
$AgI(s) + e^- \longrightarrow Ag(s) + I^-(aq)$	−0.15
$[SnF_6]^{2-}(aq) + 4\,e^- \longrightarrow Sn(s) + 6\,F^-(aq)$	−0.25
$Ni^{2+}(aq) + 2\,e^- \longrightarrow Ni(s)$	−0.25
$Co^{2+}(aq) + 2\,e^- \longrightarrow Co(s)$	−0.28
$Tl^+(aq) + e^- \longrightarrow Tl(s)$	−0.34
$PbSO_4(s) + 2\,e^- \longrightarrow Pb(s) + SO_4^{2-}(aq)$	−0.356
$Se(s) + 2\,H_3O^+(aq) + 2\,e^- \longrightarrow H_2Se(aq) + 2\,H_2O(\ell)$	−0.40
$Cd^{2+}(aq) + 2\,e^- \longrightarrow Cd(s)$	−0.403
$Cr^{3+}(aq) + e^- \longrightarrow Cr^{2+}(aq)$	−0.41
$Fe^{2+}(aq) + 2\,e^- \longrightarrow Fe(s)$	−0.44
$2\,CO_2(g) + 2\,H_3O^+(aq) + 2\,e^- \longrightarrow (COOH)_2(aq) + 2\,H_2O(\ell)$	−0.49
$Ga^{3+}(aq) + 3\,e^- \longrightarrow Ga(s)$	−0.53
$HgS(s) + 2\,H_3O^+(aq) + 2\,e^- \longrightarrow Hg(\ell) + H_2S(g) + 2\,H_2O(\ell)$	−0.72
$Cr^{3+}(aq) + 3\,e^- \longrightarrow Cr(s)$	−0.74
$Zn^{2+}(aq) + 2\,e^- \longrightarrow Zn(s)$	−0.763
$Cr^{2+}(aq) + 2\,e^- \longrightarrow Cr(s)$	−0.91
$Mn^{2+}(aq) + 2\,e^- \longrightarrow Mn(s)$	−1.18
$V^{2+}(aq) + 2\,e^- \longrightarrow V(s)$	−1.18
$Zr^{4+}(aq) + 4\,e^- \longrightarrow Zr(s)$	−1.53
$Al^{3+}(aq) + 3\,e^- \longrightarrow Al(s)$	−1.66
$H_2(g) + 2\,e^- \longrightarrow 2\,H^-(aq)$	−2.25
$Mg^{2+}(aq) + 2\,e^- \longrightarrow Mg(s)$	−2.37
$Na^+(aq) + e^- \longrightarrow Na(s)$	−2.714
$Ca^{2+}(aq) + 2\,e^- \longrightarrow Ca(s)$	−2.87
$Sr^{2+}(aq) + 2\,e^- \longrightarrow Sr(s)$	−2.89
$Ba^{2+}(aq) + 2\,e^- \longrightarrow Ba(s)$	−2.90
$Rb^+(aq) + e^- \longrightarrow Rb(s)$	−2.925
$K^+(aq) + e^- \longrightarrow K(s)$	−2.925
$Li^+(aq) + e^- \longrightarrow Li(s)$	−3.045

Basic Solution	**Standard Reduction Potential, $E°$ (volts)**
$ClO^-(aq) + H_2O(\ell) + 2e^- \longrightarrow Cl^-(aq) + 2OH^-(aq)$	0.89
$OOH^-(aq) + H_2O(\ell) + 2e^- \longrightarrow 3OH^-(aq)$	0.88
$2NH_2OH(aq) + 2e^- \longrightarrow N_2H_4(aq) + 2OH^-(aq)$	0.74
$ClO_3^-(aq) + 3H_2O(\ell) + 6e^- \longrightarrow Cl^-(aq) + 6OH^-(aq)$	0.62
$MnO_4^-(aq) + 2H_2O(\ell) + 3e^- \longrightarrow MnO_2(s) + 4OH^-(aq)$	0.588
$MnO_4^-(aq) + e^- \longrightarrow MnO_4^{2-}(aq)$	0.564
$NiO_2(s) + 2H_2O(\ell) + 2e^- \longrightarrow Ni(OH)_2(s) + 2OH^-(aq)$	0.49
$Ag_2CrO_4(s) + 2e^- \longrightarrow 2Ag(s) + CrO_4^{2-}(aq)$	0.446
$O_2(g) + 2H_2O(\ell) + 4e^- \longrightarrow 4OH^-(aq)$	0.40
$ClO_4^-(aq) + H_2O(\ell) + 2e^- \longrightarrow ClO_3^-(aq) + 2OH^-(aq)$	0.36
$Ag_2O(s) + H_2O(\ell) + 2e^- \longrightarrow 2Ag(s) + 2OH^-(aq)$	0.34
$2NO_2^-(aq) + 3H_2O(\ell) + 4e^- \longrightarrow N_2O(g) + 6OH^-(aq)$	0.15
$N_2H_4(aq) + 2H_2O(\ell) + 2e^- \longrightarrow 2NH_3(aq) + 2OH^-(aq)$	0.10
$[Co(NH_3)_6]^{3+}(aq) + e^- \longrightarrow [Co(NH_3)_6]^{2+}(aq)$	0.10
$HgO(s) + H_2O(\ell) + 2e^- \longrightarrow Hg(\ell) + 2OH^-(aq)$	0.0984
$O_2(g) + H_2O(\ell) + 2e^- \longrightarrow OOH^-(aq) + OH^-(aq)$	0.076
$NO_3^-(aq) + H_2O(\ell) + 2e^- \longrightarrow NO_2^-(aq) + 2OH^-(aq)$	0.01
$MnO_2(s) + 2H_2O(\ell) + 2e^- \longrightarrow Mn(OH)_2(s) + 2OH^-(aq)$	−0.05
$CrO_4^{2-}(aq) + 4H_2O(\ell) + 3e^- \longrightarrow Cr(OH)_3(s) + 5OH^-(aq)$	−0.12
$Cu(OH)_2(s) + 2e^- \longrightarrow Cu(s) + 2OH^-(aq)$	−0.36
$Fe(OH)_3(s) + e^- \longrightarrow Fe(OH)_2(s) + OH^-(aq)$	−0.56
$2H_2O(\ell) + 2e^- \longrightarrow H_2(g) + 2OH^-(aq)$	−0.8277
$2NO_3^-(aq) + 2H_2O(\ell) + 2e^- \longrightarrow N_2O_4(g) + 4OH^-(aq)$	−0.85
$Fe(OH)_2(s) + 2e^- \longrightarrow Fe(s) + 2OH^-(aq)$	−0.877
$SO_4^{2-}(aq) + H_2O(\ell) + 2e^- \longrightarrow SO_3^{2-}(aq) + 2OH^-(aq)$	−0.93
$N_2(g) + 4H_2O(\ell) + 4e^- \longrightarrow N_2H_4(aq) + 4OH^-(aq)$	−1.15
$[Zn(OH)_4]^{2-}(aq) + 2e^- \longrightarrow Zn(s) + 4OH^-(aq)$	−1.22
$Zn(OH)_2(s) + 2e^- \longrightarrow Zn(s) + 2OH^-(aq)$	−1.245
$[Zn(CN)_4]^{2-}(aq) + 2e^- \longrightarrow Zn(s) + 4CN^-(aq)$	−1.26
$Cr(OH)_3(s) + 3e^- \longrightarrow Cr(s) + 3OH^-(aq)$	−1.30
$SiO_3^{2-}(aq) + 3H_2O(\ell) + 4e^- \longrightarrow Si(s) + 6OH^-(aq)$	−1.70

APPENDIX J

Selected Thermodynamic Values*

Species	ΔH_f° (298.15 K) (kJ/mol)	S° (298.15 K) (J K^{-1} mol^{-1})	ΔG_f° (298.15 K) (kJ/mol)
Aluminum			
Al(s)	0	28.275	0
Al^{3+}(aq)	−531	−321.7	−485
$AlCl_3$(s)	−704.2	110.67	−628.8
Al_2O_3(s, corundum)	−1675.7	50.92	−1582.3
Argon			
Ar(g)	0	154.843	0
Ar(aq)	−12.1	59.4	16.4
Barium			
$BaCl_2$(s)	−858.6	123.68	−810.4
BaO(s)	−553.5	70.42	−525.1
$BaSO_4$(s)	−1473.2	132.2	−1362.2
$BaCO_3$(s)	−1216.3	112.1	85.35
Beryllium			
Be(s)	0	9.5	0
$Be(OH)_2$(s)	−902.5	51.9	−815
Bromine			
Br(g)	111.884	175.022	82.396
Br_2(ℓ)	0	152.231	0
Br_2(g)	30.907	245.463	3.110
Br_2(aq)	−2.59	130.5	3.93
Br^-(aq)	−121.55	82.4	−103.96
BrCl(g)	14.64	240.10	−0.98
BrF_3(g)	−255.6	292.53	−229.43
HBr(g)	−36.40	198.695	−53.45
Calcium			
Ca(s)	0	41.42	0
Ca(g)	178.2	158.884	144.3
Ca^{2+}(g)	1925.9	—	—
Ca^{2+}(aq)	−542.83	−53.1	−553.58
CaC_2(s)	−59.8	69.96	−64.9
$CaCO_3$(s, calcite)	−1206.92	92.9	−1128.79
$CaCl_2$(s)	−795.8	104.6	−748.1
CaF_2(s)	−1219.6	68.87	−1167.3
CaH_2(s)	−186.2	42	−147.2
CaO(s)	−635.09	39.75	−604.03
CaS(s)	−482.4	56.5	−477.4

*Taken from Wagman, D. D., Evans, W. H., Parker, V. B., Schumm, R. H., Halow, I., Bailey, S. M., Churney, K. L., and Nuttall, R. The NBS Tables of Chemical Thermodynamic Properties. *Journal of Physical and Chemical Reference Data,* Vol. 11, Suppl. 2, 1982.

Species	ΔH_f° (298.15 K) (kJ/mol)	S° (298.15 K) (J K^{-1} mol^{-1})	ΔG_f° (298.15 K) (kJ/mol)
$Ca(OH)_2$(s)	−986.09	83.39	−898.49
$Ca(OH)_2$(aq)	−1002.82	−74.5	−868.07
$CaSO_4$(s)	−1434.11	106.7	−1321.79
Carbon			
C(s, graphite)	0	5.74	0
C(s, diamond)	1.895	2.377	2.9
C(g)	716.682	158.096	671.257
CCl_4(ℓ)	−135.44	216.4	−65.21
CCl_4(g)	−102.9	309.85	−60.59
$CHCl_3$(ℓ)	−134.47	201.7	−73.66
$CHCl_3$(g)	−103.14	295.71	−70.34
CH_4 (g, methane)	−74.81	186.264	−50.72
C_2H_2 (g, ethyne)	226.73	200.94	209.2
C_2H_4 (g, ethene)	52.26	219.56	68.15
C_2H_6 (g, ethane)	−84.68	229.6	−32.82
C_3H_8 (g, propane)	−103.8	269.9	−23.49
C_4H_{10} (g, butane)	−126.148	310.227	−16.985
C_6H_6(ℓ, benzene)	49.03	172.8	124.5
C_6H_{14}(ℓ, hexane)	−198.782	296.018	−4.035
C_8H_{18} (g, octane)	−208.447	466.835	16.718
C_8H_{18}(ℓ, octane)	−249.952	361.205	6.707
CH_3OH(ℓ, methanol)	−238.66	126.8	−166.27
CH_3OH(g, methanol)	−200.66	239.81	−161.96
CH_3OH(aq, methanol)	−245.931	133.1	−175.31
C_2H_5OH(ℓ, ethanol)	−277.69	160.7	−174.78
C_2H_5OH(g, ethanol)	−235.1	282.7	−168.49
C_2H_5OH(aq, ethanol)	−288.3	148.5	−181.64
$C_6H_{12}O_6$ (s, glucose)	−1274.4	235.9	−917.2
CH_3COO^-(aq)	−486.01	86.6	−369.31
CH_3COOH(aq)	−485.76	178.7	−396.46
CH_3COOH(ℓ)	−484.5	159.8	−389.9
CO(g)	−110.525	197.674	−137.168
CO_2(g)	−393.509	213.74	−394.359
H_2CO_3(aq)	−699.65	187.4	−623.08
HCO_3^-(aq)	−691.99	91.2	−586.77
CO_3^{2-}(aq)	−677.14	−56.9	−527.81
$HCOO^-$(aq)	−425.55	92.0	−351.0
HCOOH(aq)	−425.43	163	−372.3
HCOOH(ℓ)	−424.72	128.95	−361.35
CS_2(g)	117.36	237.84	67.12
CS_2(ℓ)	89.70	151.34	65.27
$COCl_2$(g)	−218.8	283.53	−204.6
Cesium			
Cs(s)	0	85.23	0
Cs^+(g)	457.964	—	—
CsCl(s)	−443.04	101.17	−414.53
Chlorine			
Cl(g)	121.679	165.198	105.68
Cl^-(g)	−233.13	—	—
Cl^-(aq)	−167.159	56.5	−131.228

Species	ΔH_f° (298.15 K) (kJ/mol)	S° (298.15 K) (J K^{-1} mol^{-1})	ΔG_f° (298.15 K) (kJ/mol)
$Cl_2(g)$	0	223.066	0
$Cl_2(aq)$	−23.4	121	6.94
$HCl(g)$	−92.307	186.908	−95.299
$HCl(aq)$	−167.159	56.5	−131.228
$ClO_2(g)$	102.5	256.84	120.5
$Cl_2O(g)$	80.3	266.21	97.9
$ClO^-(aq)$	−107.1	42.0	−36.8
$HClO(aq)$	−120.9	142.	−79.9
$ClF_3(g)$	−163.2	281.61	−123.0
Chromium			
$Cr(s)$	0	23.77	0
$Cr_2O_3(s)$	−1139.7	81.2	−1058.1
$CrCl_3(s)$	−556.5	123	−486.1
Copper			
$Cu(s)$	0	33.15	0
$CuO(s)$	−157.3	42.63	−129.7
$CuCl_2(s)$	−220.1	108.07	−175.7
$CuSO_4(s)$	−771.36	109.	−661.8
Fluorine			
$F_2(g)$	0	202.78	0
$F(g)$	78.99	158.754	61.91
$F^-(g)$	−255.39	—	—
$F^-(aq)$	−332.63	−13.8	−278.79
$HF(g)$	−271.1	173.779	−273.2
HF(aq, un-ionized)	−320.08	88.7	−296.82
HF(aq, ionized)	−332.63	−13.8	−278.79
Hydrogen †			
$H_2(g)$	0	130.684	0
$H_2(aq)$	−4.2	57.7	17.6
$HD(g)$	0.318	143.801	−1.464
$D_2(g)$	0	144.960	0
$H(g)$	217.965	114.713	203.247
$H^+(g)$	1536.202	—	—
$H^+(aq)$	0	0	0
$OH^-(aq)$	−229.994	−10.75	−157.244
$H_2O(\ell)$	−285.83	69.91	−237.129
$H_2O(g)$	−241.818	188.825	−228.572
$H_2O_2(\ell)$	−187.78	109.6	−120.35
$H_2O_2(aq)$	−191.17	143.9	−134.03
$HO_2^-(aq)$	−160.33	23.8	−67.3
$HDO(\ell)$	−289.888	79.29	−241.857
$D_2O(\ell)$	−294.600	75.94	−243.439
Iodine			
$I_2(s)$	0	116.135	0
$I_2(g)$	62.438	260.69	19.327
$I_2(aq)$	22.6	137.2	16.40
$I(g)$	106.838	180.791	70.25
$I^-(g)$	−197	—	—

Species	ΔH_f° (298.15 K) (kJ/mol)	S° (298.15 K) (J K^{-1} mol^{-1})	ΔG_f° (298.15 K) (kJ/mol)
I^-(aq)	−55.19	111.3	−51.57
I_3^-(aq)	−51.5	239.3	−51.4
HI(g)	26.48	206.594	1.70
HI(aq, ionized)	−55.19	111.3	−51.57
IF(g)	−95.65	236.17	−118.51
ICl(g)	17.78	247.551	−5.46
ICl_3(s)	−89.5	167.4	−22.29
ICl(ℓ)	−23.89	135.1	−13.58
IBr(g)	40.84	258.773	3.69
Iron			
Fe(s)	0	27.78	0
FeO(s, wustite)	−266.27	57.9	−245.12
Fe_2O_3(s, hematite)	−824.2	87.4	−742.2
Fe_3O_4(s, magnetite)	−1118.4	146.4	−1015.4
$FeCl_2$(s)	−341.79	117.95	−302.3
$FeCl_3$(s)	−399.49	142.3	−344
FeS_2 (s, pyrite)	−178.2	52.93	−166.9
$Fe(CO)_5$(ℓ)	−774	338.1	−705.3
Lead			
Pb(s)	0	64.81	0
$PbCl_2$(s)	−359.41	136	−314.1
PbO(s, yellow)	−217.32	68.7	−187.89
PbS(s)	−100.4	91.2	−98.7
Lithium			
Li(s)	0	29.12	0
Li^+(g)	685.783	—	—
LiOH(s)	−484.93	42.8	−438.95
LiOH(aq)	−508.48	2.8	−450.58
LiCl(s)	−408.701	59.33	−384.37
Magnesium			
Mg(s)	0	32.68	0
Mg^{2+}(aq)	−466.85	−138.1	−454.8
$MgCl_2$(g)	−400.4	—	—
$MgCl_2$(s)	−641.32	89.62	−591.79
$MgCl_2$(aq)	−801.15	−25.1	−717.1
MgO(s)	−601.70	26.94	−569.43
$Mg(OH)_2$(s)	−924.54	63.18	−833.51
MgS(s)	−346	50.33	−341.8
$MgSO_4$(s)	−1284.9	91.6	−1170.6
$MgCO_3$(s)	−1095.8	65.7	−1012.1
Mercury			
Hg(ℓ)	0	76.02	0
$HgCl_2$(s)	−224.3	146	−178.6
HgO(s, red)	−90.83	70.29	−58.539
HgS(s, red)	−58.2	82.4	−50.6

Species	ΔH_f° (298.15 K) (kJ/mol)	S° (298.15 K) (J K^{-1} mol^{-1})	ΔG_f° (298.15 K) (kJ/mol)
Nickel			
Ni(s)	0	29.87	0
NiO(s)	−239.7	37.99	−211.7
$NiCl_2$(s)	−305.332	97.65	−259.032
Nitrogen			
N_2(g)	0	191.61	0
N_2(aq)	−10.8	—	—
N(g)	472.704	153.298	455.563
NH_3(g)	−46.11	192.45	−16.45
NH_3(aq)	−80.29	111.3	−26.50
NH_4^+(aq)	−132.51	113.4	−79.31
N_2H_4(ℓ)	50.63	121.21	149.34
NH_4Cl(s)	−314.43	94.6	−202.87
NH_4Cl(aq)	−299.66	169.9	−210.52
NH_4NO_3(s)	−365.56	151.08	−183.87
NH_4NO_3(aq)	−339.87	259.8	−190.56
NO(g)	90.25	210.761	86.55
NO_2(g)	33.18	240.06	51.31
N_2O(g)	82.05	219.85	104.20
N_2O_4(g)	9.16	304.29	97.89
N_2O_4(ℓ)	−19.50	209.2	97.54
NOCl(g)	51.71	261.69	66.08
HNO_3(ℓ)	−174.10	155.60	−80.71
HNO_3(g)	−135.06	266.38	−74.72
HNO_3(aq)	−207.36	146.4	−111.25
NO_3^-(aq)	−205.0	146.4	−108.74
NF_3(g)	−124.7	260.73	−83.2
Oxygen †			
O_2(g)	0	205.138	0
O_2(aq)	−11.7	110.9	16.4
O(g)	249.170	161.055	231.731
O_3(g)	142.7	238.93	163.2
OH^-(aq)	−229.994	−10.75	−157.244
Phosphorus			
P_4(s, white)	0	164.36	0
P_4(s, red)	−70.4	91.2	−48.4
P(g)	314.64	163.193	278.25
PH_3(g)	5.4	310.23	13.4
PCl_3(g)	−287	311.78	−267.8
PCl_3(ℓ)	−319.7	217.1	−272.3
PCl_5(s)	−443.5	—	—
P_4O_{10}(s)	−2984	228.86	−2697.7
H_3PO_4(s)	−1279	110.5	−1119.1
Potassium			
K(s)	0	64.18	0
KF(s)	−567.27	66.57	−537.75

Species	ΔH_f° (298.15 K) (kJ/mol)	S° (298.15 K) (J K^{-1} mol^{-1})	ΔG_f° (298.15 K) (kJ/mol)
KCl(s)	−436.747	82.59	−409.14
KCl(aq)	−419.53	159.0	−414.49
KBr(s)	−393.798	95.90	−380.66
KI(s)	−327.900	106.32	−324.892
$KClO_3$(s)	−397.73	143.1	−296.25
KOH(s)	−424.764	78.9	−379.08
KOH(aq)	−482.37	91.6	−440.5
Silicon			
Si(s)	0	18.83	0
$SiBr_4$(ℓ)	−457.3	277.8	−443.8
SiC(s)	−65.3	16.61	−62.8
$SiCl_4$(g)	−657.01	330.73	−616.98
SiH_4(g)	34.3	204.62	56.9
SiF_4(g)	−1614.94	282.49	−1572.65
SiO_2(s, quartz)	−910.94	41.84	−856.64
Silver			
Ag(s)	0	42.55	0
Ag^+(aq)	105.579	72.68	77.107
Ag_2O(s)	−31.05	121.3	−11.2
AgCl(s)	−127.068	96.2	−109.789
AgI(s)	−61.84	115.5	−66.19
AgN_3(s)	620.60	99.22	591.0
$AgNO_3$(s)	−124.39	140.92	−33.41
$AgNO_3$(aq)	−101.8	219.2	−34.16
Sodium			
Na(s)	0	51.21	0
Na(g)	107.32	153.712	76.761
Na^+(g)	609.358	—	—
Na^+(aq)	−240.12	59.0	−261.905
NaF(s)	−573.647	51.46	−543.494
NaF(aq)	−572.75	45.2	−540.68
NaCl(s)	−411.153	72.13	−384.138
NaCl(g)	−176.65	229.81	−196.66
NaCl(aq)	−407.27	115.5	−393.133
NaBr(s)	−361.062	86.82	−348.983
NaBr(aq)	−361.665	141.4	−365.849
NaI(s)	−287.78	98.53	−286.06
NaI(aq)	−295.31	170.3	−313.47
NaOH(s)	−425.609	64.455	−379.484
NaOH(aq)	−470.114	48.1	−419.15
$NaClO_3$(s)	−365.774	123.4	−262.259
$NaHCO_3$(s)	−950.81	101.7	−851.0
Na_2CO_3(s)	−1130.68	134.98	−1044.44
Na_2SO_4(s)	−1387.08	149.58	−1270.16

Species	ΔH_f° (298.15 K) (kJ/mol)	S° (298.15 K) (J K^{-1} mol^{-1})	ΔG_f° (298.15 K) (kJ/mol)
Sulfur			
S(s, monoclinic)	0.33	—	—
S(s, rhombic)	0	31.80	0
S(g)	278.805	167.821	238.250
S^{2-}(aq)	33.1	−14.6	85.8
S_2Cl_2(g)	−18.4	331.5	−31.8
SF_6(g)	−1209.	291.82	−1105.3
SF_4(g)	−774.9	292.03	−731.3
H_2S(g)	−20.63	205.79	−33.56
H_2S(aq)	−39.7	121	−27.83
HS^-(aq)	−17.6	62.8	12.08
SO_2(g)	−296.830	248.22	−300.194
SO_3(g)	−395.72	256.76	−371.06
$SOCl_2$(g)	−212.5	309.77	−198.3
SO_4^{2-}(aq)	−909.27	20.1	−744.53
H_2SO_4(ℓ)	−813.989	156.904	−690.003
H_2SO_4(aq)	−909.27	20.1	−744.53
HSO_4^-(aq)	−887.34	131.8	−755.91
Tin			
Sn(s, white)	0	51.55	0
Sn(s, gray)	−2.09	44.14	0.13
$SnCl_2$(s)	−325.1	—	—
$SnCl_4$(ℓ)	−511.3	258.6	−440.1
$SnCl_4$(g)	−471.5	365.8	−432.2
SnO_2(s)	−580.7	52.3	−519.6
Titanium			
Ti(s)	0	30.63	0
$TiCl_4$(ℓ)	−804.2	252.34	−737.2
$TiCl_4$(g)	−763.2	354.9	−726.7
TiO_2(s)	−939.7	49.92	−884.5
Uranium			
U(s)	0	50.21	0
UO_2(s)	−1084.9	77.03	−1031.7
UO_3(s)	−1223.8	96.11	−1145.9
UF_4(s)	−1914.2	151.67	−1823.3
UF_6(g)	−2147.4	377.9	−2063.7
UF_6(s)	−2197.0	227.6	−2068.5
Zinc			
Zn(s)	0	41.63	0
$ZnCl_2$(s)	−415.05	111.46	−369.398
ZnO(s)	−348.28	43.64	−318.3
ZnS(s, sphalerite)	−205.98	57.7	−201.29

†Many hydrogen-containing and oxygen-containing compounds are listed only under other elements; for example, HNO_3 appears under nitrogen.

Answers to Problem-Solving Practice Problems

Chapter 1

1.1 (1) *Define the problem:* You are asked to find the volume of the sample, and you know the mass.

(2) *Develop a plan:* Density relates mass and volume and is the appropriate conversion factor, so look up the density in a table. Volume is proportional to mass, so the mass has to be either multiplied by the density or multiplied by the reciprocal of the density. Use the units to decide which.

(3) *Execute the plan:* According to Table 1.1, the density of benzene is 0.880 g/mL. Setting up the calculation so that the unit (grams) cancels gives

$$4.33\ \text{g} \times \frac{1\ \text{mL}}{0.880\ \text{g}} = 4.92\ \text{mL}$$

Notice that the result is expressed to three significant figures, because both the mass and the density had three significant figures.

(4) *Check your answer:* Because the density is a little less than 1.00 g/mL, the volume in milliliters should be a little larger than the mass in grams. The calculated answer, 4.92 mL, is a little larger than the mass, 4.33 g.

1.2 Substance A must be a mixture since some of it dissolves and some, substance B, does not.

Substance C is the soluble portion of substance A. Since all of substance C dissolves in water there is no way to determine how many components it has. Additionally, it is not possible to determine whether the one or more components themselves are elements or compounds. Therefore it is not possible to say whether C is an element, a compound, or a mixture.

The only thing we know about substance B is that it is insoluble in water. We do not know whether it is one insoluble substance, or more than one insoluble substance. Additionally, we do not know whether the substance or substances of B are elements or compounds. Therefore it is not possible to say whether B is an element, a compound, or a mixture.

1.3 Oxygen is O_2; ozone is O_3. Oxygen is a colorless, odorless gas; ozone is a pale blue gas with a pungent odor.

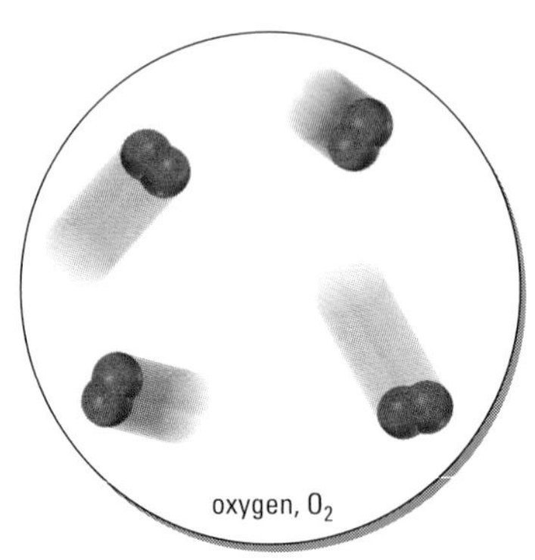

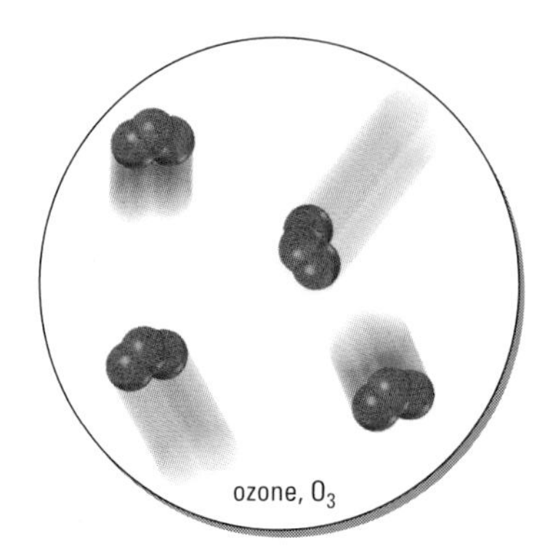

Chapter 2

2.1 (a) 10 gal = 40 qt. There are 1.0567 quarts per liter, so

$$40\ \text{qt} \times \frac{1\ \text{L}}{1.0567\ \text{qt}} = 37.9\ \text{L or } 38\ \text{L}$$

(b) $100\ \text{yds} \times \frac{3\ \text{ft}}{\text{yd}} \times \frac{12\ \text{in}}{\text{ft}} \times \frac{2.54\ \text{cm}}{\text{in}} \times \frac{1\ \text{m}}{100\ \text{cm}} = 91.5\ \text{m}$

2.2 (a) 1 lb = 453.59 g, so 5 lb = 2268. g

(b) $3\ \text{pt} \times \frac{1\ \text{qt}}{2\ \text{pt}} \times \frac{1\ \text{L}}{1.057\ \text{qt}} \times \frac{1000\ \text{ml}}{1\ \text{L}} = 1420\ \text{ml}$

(c) $\frac{1.420\ \text{L}}{5\ \text{L}} \times 100\% = 28\%$

2.3 Work with the numerator first: 165 mg is 0.165 g. Work with the denominator next: 1 dL = 0.1 L. Therefore, the concentration is

$$\frac{0.165\ \text{g}}{0.1\ \text{L}} = 1.65\ \text{g/L}.$$

2.4 (a) 0.00602 g — 3 sf
(b) 22.871 mg — 5 sf
(c) 344. °C — 3 sf
(d) 100.0 mL — 4 sf
(e) 0.00042 m — 2 sf
(f) 0.002001 L — 4 sf

2.5 (a) $244.2 + 0.1732 = 244.4$
(b) $6.19 \times 5.2222 = 32.3$
(c) $\frac{7.2234 - 11.3851}{4.22} = -0.986$

2.6 (a) A phosphorus atom ($Z = 15$) with 16 neutrons has $A = 31$.
(b) A neon-22 atom has $A = 22$ and $Z = 10$, so the number of electrons must be 10 and the number of neutrons must be $A - Z = 22 - 10 = 12$ neutrons.
(c) The periodic table shows us that the element with 82 protons is lead. The atomic weight of this isotope of lead is 82 + 125 = 207, so the correct symbol is $^{207}_{82}Pb$.

2.7 The magnesium isotope with 12 neutrons has 12 protons, so $Z = 12$ and the notation is $^{24}_{12}Mg$; the isotope with 13 neutrons has $Z = 12$ and $^{25}_{12}Mg$; and the isotope with 14 neutrons has $Z = 12$ and $^{26}_{12}Mg$.

2.8 75 g wire × (fraction Ni) = g Ni, so 75 g Ni × 0.80 = 60 g Ni. For Cr we have 75 g Cr × 0.20 = 15 g Cr. Or, we could have solved for the mass of Cr from 75 g wire − 60 g Ni = 15 g Cr.

2.9 (a) 1.00 mg Mo = 1.00×10^{-3} g Mo

$$1.00 \times 10^{-3}\ \text{g Mo} \times \frac{1\ \text{mol Mo}}{95.94\ \text{g Mo}} = 1.04 \times 10^{-5}\ \text{mol Mo}$$

(b) $5.00 \times 10^{-3}\ \text{mol Au} \times \frac{196.97\ \text{g Au}}{1\ \text{mol Au}} = 0.985\ \text{g Au}$

Chapter 3

3.1 (a) $C_{10}H_{11}O_{13}N_5P_3$ (b) $C_{18}H_{27}O_3N$ (c) $C_2H_2O_4$

3.2 (a) Sulfur dioxide
(b) Boron trifluoride
(c) Carbon tetrachloride

3.3 Yes, a similar trend would be expected. The absolute values of the boiling points of the chlorine-containing compounds would be different from their alkane parents, but the differences between successive chlorine-containing compounds should follow a similar trend as for the alkanes themselves.

3.4 (a) A Ca^{4+} charge is unlikely because calcium is in Group 2A, the elements of which lose two electrons to form 2+ ions.

(b) Cr^{2+} is possible because chromium is a transition metal ion that forms 2+ and 3+ ions.

(c) Strontium is a Group 2A metal and forms 2+ ions; thus, a Sr^- ion is highly unlikely.

3.5 (a) CH_4 is formed from two nonmetals and is molecular.
(b) $CaBr_2$ is formed from a metal and a nonmetal, so it is ionic.
(c) $MgCl_2$ is formed from a metal and a nonmetal, so it is ionic.
(d) PCl_3 is formed from two nonmetals and is molecular.
(e) KCl is formed from a metal and a nonmetal and is ionic.

3.6 (a) $In_2(SO_3)_3$ contains two In^{3+} and three SO_3^{2-} ions. There are 14 atoms in this formula unit.
(b) $(NH_4)_3PO_4$ contains three ammonium ions, NH_4^+, and one phosphate ion, PO_4^{3-}, collectively containing 20 atoms.

3.7 (a) One Mg^{2+} ion and two Br^- ions
(b) Two Li^+ ions and one CO_3^{2-} ion
(c) One NH_4^+ ion and one Cl^- ion
(d) Two Fe^{3+} ions and three SO_4^{2-} ions
(e) CuCl and $CuCl_2$

3.8 (a) KNO_2 is potassium nitrite.
(b) $NaHSO_3$ is sodium hydrogen sulfite or sodium bisulfite.
(c) $Mn(OH)_2$ is manganese(II) hydroxide.
(d) $Mn_2(SO_4)_3$ is manganese(III) sulfate.
(e) Ba_3N_2 is barium nitride.
(f) LiH is lithium hydride.

3.9 (a) KH_2PO_4 (b) CuOH (c) NaClO
(d) NH_4ClO_4 (e) $CrCl_3$ (f) $FeSO_3$

3.10 (a) The molar mass of $K_2Cr_2O_7$ is 294.2 g/mol.

$$\frac{12.5\text{ g}}{294.2\text{ g/mol}} = 4.25 \times 10^{-2}\text{ mol}$$

(b) The molar mass of $KMnO_4$ is 158.0 g/mol.

$$\frac{12.5\text{ g}}{158.0\text{ g/mol}} = 7.91 \times 10^{-2}\text{ mol}$$

(c) The molar mass of $(NH_4)_2CO_3$ is 96.1 g/mol.

$$\frac{12.5\text{ g}}{96.1\text{ g/mol}} = 1.30 \times 10^{-1}\text{ mol}$$

3.11 (a) The molar mass of sucrose, $C_{12}H_{22}O_{11}$, is 342.3 g/mol.

$$5.0 \times 10^{-3}\text{ mol sucrose} \times \frac{342.3\text{ g sucrose}}{1\text{ mol sucrose}} = 1.7\text{ g sucrose}$$

(b) $$3.0 \times 10^{-6}\text{ mol ACTH} \times \frac{4600\text{ g ACTH}}{1\text{ mol ACTH}} = 1.4 \times 10^{-2}\text{ g ACTH}$$

$$1.4 \times 10^{-2}\text{ g} \times \frac{10^3\text{ mg}}{1\text{ g}} = 14.\text{ mg ACTH}$$

3.12 (a) The molar mass of cholesterol is 386.7 g/mol.

$$\frac{10.0\text{ g}}{386.7\text{ g/mol}} = 2.59 \times 10^{-2}\text{ mol}$$

The molar mass of $Mn_2(SO_4)_3$ is 398.1 g/mol.

$$\frac{10.0\text{ g}}{398.1\text{ g/mol}} = 2.51 \times 10^{-2}\text{ mol}$$

(b) The molar mass of K_2HPO_4 is 174.2 g/mol.

$$0.25\text{ mol} \times \frac{174.2\text{ g}}{1\text{ mol}} = 44.\text{ g}$$

The molar mass of caffeine is 194.2 g/mol.

$$0.25\text{ mol} \times \frac{194.2\text{ g}}{1\text{ mol}} = 49.\text{ g}$$

3.13 The mass of Si in 1 mol SiO_2 is 28.0855 g. The mass of O in 1 mol SiO_2 is 31.9988 g.

$$\%\text{ Si in SiO}_2 = \frac{28.0855\text{ g}}{60.08\text{ g}} \times 100\% = 46.74\%\text{ Si}$$

$$\%\text{ O in SiO}_2 = \frac{31.9988\text{ g}}{60.08\text{ g}} \times 100\% = 53.26\%\text{ O}$$

3.14 The molar mass of hydrated nickel chloride is

$$58.69\text{ g/mol} + 2(35.45\text{ g/mol}) + 12(1.008\text{ g/mol}) + 6(16.00\text{ g/mol}) = 237.69\text{ g/mol.}$$

The percentages by weight for each element are found from the ratios of the mass of each element in 1 mole of hydrated nickel chloride to the molar mass of hydrated nickel chloride:

$$\frac{58.69\text{ g Ni}}{237.69\text{ g hydrated nickel chloride}} \times 100\% = 24.69\%\text{ Ni}$$

$$\frac{70.90\text{ g Cl}}{237.69\text{ g hydrated nickel chloride}} \times 100\% = 29.83\%\text{ Cl}$$

$$\frac{12.096\text{ g H}}{237.69\text{ g hydrated nickel chloride}} \times 100\% = 5.089\%\text{ H}$$

$$\frac{96.00\text{ g O}}{237.69\text{ g hydrated nickel chloride}} \times 100\% = 40.39\%\text{ O}$$

3.15 A 100-g sample of the phosphorus oxide contains 43.64 g P and 56.36 g O.

$$43.64\text{ g P} \times \frac{1\text{ mol P}}{30.9738\text{ g P}} = 1.409\text{ mol P}$$

$$56.36\text{ g O} \times \frac{1\text{ mol O}}{15.9994\text{ g O}} = 3.523\text{ mol O}$$

The mole ratio is

$$\frac{3.523\text{ mol O}}{1.409\text{ mol P}} = \frac{2.500\text{ mol O}}{1.000\text{ mol P}}$$

There are 2.500 oxygen atoms for every phosphorus atom. Thus, the empirical formula is P_2O_5. The molar mass corresponding to this empirical formula is

$$\left(2\text{ mol P} \times \frac{30.9738\text{ g P}}{1\text{ mol P}}\right) + \left(5\text{ mol O} \times \frac{15.994\text{ g O}}{1\text{ mol O}}\right) = 141.9\text{ g/mol}$$

The known molar mass is 283.89 g/mol. The molar mass is twice as large as the empirical formula mass, so the molecular formula of the oxide is P_4O_{10}.

3.16 Find the number of moles of each element in 100.0 g of vitamin C.

$$40.9\text{ g C} \times \frac{1\text{ mol C}}{12.011\text{ g C}} = 3.405\text{ mol C}$$

$$4.58\text{ g H} \times \frac{1\text{ mol H}}{1.0079\text{ g H}} = 4.544\text{ mol H}$$

$$54.5\text{ g O} \times \frac{1\text{ mol O}}{15.9994\text{ g O}} = 3.406\text{ mol O}$$

Find the mole ratios.

$$\frac{4.544\text{ mol H}}{3.406\text{ mol O}} = \frac{1.334\text{ mol H}}{1.000\text{ mol O}}$$

The same ratio holds for H to C. Using whole numbers, we have $C_3H_4O_3$ for the empirical formula. The empirical formula weight is

(3)(12.011) + (4)(1.0079) + (3)(15.9994) = 88.06 g. The molar mass, however, is 176.13 g/mol, so the molecular formula must be twice the empirical formula: $C_6H_8O_6$.

Chapter 4

4.1 (a) N_2; combination (b) O_2; combination
(c) N_2; decomposition

4.2 (a) Decomposition reaction:

$$2\ Al(OH)_3(s) \longrightarrow Al_2O_3(s) + 3\ H_2O(g)$$

(b) Combination reaction:

$$Na_2O(s) + H_2O(\ell) \longrightarrow 2\ NaOH(aq)$$

(c) Combination reaction: $S_8(s) + 24\ F_2(g) \rightarrow 8\ SF_6(g)$
(d) Exchange reaction:

$$3\ NaOH(aq) + H_3PO_4(aq) \longrightarrow Na_3PO_4(aq) + 3\ H_2O(\ell)$$

(e) Displacement:

$$3\ C(s) + Fe_2O_3(s) \longrightarrow 3\ CO(g) + 2\ Fe(\ell)$$

4.3 (a) $2\ Cr(s) + 3\ Cl_2 \rightarrow 2\ CrCl_3(s)$
(b) $As_2O_3(s) + 3\ H_2(g) \rightarrow 2\ As(s) + 3\ H_2O(\ell)$

4.4 (a) $C_2H_5OH + 3\ O_2 \rightarrow 2\ CO_2 + 3\ H_2O$
(b) $C_2H_5OH + 2\ O_2 \rightarrow 2\ CO + 3\ H_2O$

4.5 0.433 mol hematite needs 0.433 × 3 = 1.30 mol CO. Molar mass of CO is 28.01, so 1.30 mol × 28.01 g/mol = 36.4 g CO.

4.6 (a) $$0.300\text{ mol cassiterite} \times \frac{1\text{ mol Sn}}{1\text{ mol cassiterite}} \times \frac{118.7\text{ g Sn}}{1\text{ mol Sn}} = 35.6\text{ g Sn}$$

(b) $$35.6\text{ g Sn} \times \frac{1\text{ mol Sn}}{118.7\text{ g Sn}} \times \frac{2\text{ mol C}}{1\text{ mol Sn}} \times \frac{12.01\text{ g C}}{1\text{ mol C}} = 7.20\text{ g C}$$

4.7 (a) $$57.\text{ g C} \times \frac{1\text{ mol C}}{12.01\text{ g C}} \times \frac{1\text{ mol }O_2}{2\text{ mol C}} \times \frac{32.0\text{ g }O_2}{1\text{ mol }O_2} = 76.\text{ g }O_2$$

(b) $$57.\text{ g C} \times \frac{1\text{ mol C}}{12.01\text{ g C}} \times \frac{2\text{ mol CO}}{2\text{ mol C}} \times \frac{28.0\text{ g CO}}{1\text{ mol CO}} = 1.3 \times 10^2\text{ g CO}$$

4.8 $$6.46\text{ g }MgCl_2 \times \frac{1\text{ mol }MgCl_2}{95.2104\text{ g }MgCl_2} = 6.78 \times 10^{-2}\text{ mol }MgCl_2$$

The same number of moles of Mg as $MgCl_2$ are involved, so

$$6.78 \times 10^{-2}\text{ mol Mg} \times \frac{24.3050\text{ g Mg}}{1\text{ mol Mg}} = 1.65\text{ g Mg}$$

$$\frac{1.65\text{ g Mg}}{1.72\text{-g sample}} \times 100\% = 95.9\%\text{ Mg in sample}$$

4.9 $$0.75\text{ mol }CO_2 \times \frac{1\text{ mol }(NH_2)_2CO}{1\text{ mol }CO_2} = 0.75\text{ mol }(NH_2)_2CO$$

4.10 (a) $CS_2(\ell) + 3\ O_2(g) \rightarrow CO_2(g) + 2\ SO_2(g)$
(b) Determine the quantity of CO_2 produced by each reactant; the limiting reactant produces the lesser quantity.

$$3.5\text{ g }CS_2 \times \frac{1\text{ mol }CS_2}{76.0\text{ g }CS_2} \times \frac{1\text{ mol }CO_2}{1\text{ mol }CS_2} \times \frac{44.01\text{ g }CO_2}{1\text{ mol }CO_2} = 2.0\text{ g }CO_2$$

$$17.5\text{ g }O_2 \times \frac{1\text{ mol }O_2}{31.998\text{ g }O_2} \times \frac{1\text{ mol }CO_2}{3\text{ mol }O_2} \times \frac{44.01\text{ g }CO_2}{1\text{ mol }CO_2} = 8.02\text{ g }CO_2$$

Therefore, CS_2 is the limiting reagent.

(c) The yield of SO_2 must be calculated using the limiting reagent, CS_2.

$$3.5\text{ g }CS_2 \times \frac{1\text{ mol }CS_2}{76.0\text{ g }CS_2} \times \frac{2\text{ mol }SO_2}{1\text{ mol }CS_2} \times \frac{64.1\text{ g }SO_2}{1\text{ mol }SO_2} = 5.9\text{ g }SO_2$$

4.11 Find the number of moles of each reactant.

$$100.\text{ g }SiCl_4 \times \frac{1\text{ mol }SiCl_4}{169.90\text{ g }SiCl_4} = 0.589\text{ mol }SiCl_4$$

$$100.\text{ g Mg} \times \frac{1\text{ mol Mg}}{24.3050\text{ g Mg}} = 4.11\text{ mol Mg}$$

Find the mass of Si produced, based on the mass available of each reactant.

$$0.589\text{ mol }SiCl_4 \times \frac{1\text{ mol Si}}{1\text{ mol }SiCl_4} \times \frac{28.0855\text{ g Si}}{1\text{ mol Si}} = 16.5\text{ g Si}$$

$$4.11\text{ mol Mg} \times \frac{1\text{ mol Si}}{2\text{ mol Mg}} \times \frac{28.0855\text{ g Si}}{1\text{ mol Si}} = 57.7\text{ g Si}$$

Thus, $SiCl_4$ is the limiting reactant, and the mass of Si produced is 16.5 g.

4.12 To make 1.0 kg CH_3OH with 85% yield will require using enough reactant to produce 1000/0.85, or 1.2×10^3 g CH_3OH.

$$1.2 \times 10^3\text{ g }CH_3OH \times \frac{1\text{ mol}}{32.042\text{ g}} = 37.5\text{ mol }CH_3OH$$

$$37.5\text{ mol }CH_3OH \times \frac{2\text{ mol }H_2}{1\text{ mol }CH_3OH} = 75.0\text{ mol }H_2$$

$$75.0\text{ mol }H_2 \times \frac{2.0158\text{ g }H_2}{1\text{ mol }H_2} = 1.5 \times 10^2\text{ g }H_2$$

4.13 Calculate the mass of Cu_2S you should have produced and compare it with the amount actually produced.

$$2.50\text{ g Cu} \times \frac{1\text{ mol Cu}}{63.546\text{ g Cu}} = 3.93 \times 10^{-2}\text{ mol Cu}$$

$$3.93 \times 10^{-2}\text{ mol Cu} \times \frac{8\text{ mol }Cu_2S}{16\text{ mol Cu}} = 1.97 \times 10^{-2}\text{ mol }Cu_2S$$

$$1.97 \times 10^{-2}\text{ mol }Cu_2S \times \frac{159.16\text{ g }Cu_2S}{1\text{ mol }Cu_2S} = 3.14\text{ g }Cu_2S$$

$$\frac{2.53\text{ g}}{3.14\text{ g}} \times 100\% = 80.6\%\text{ yield was obtained}$$

Your synthesis met the standard.

4.14 (a) $$491\text{ mg }CO_2 \times \frac{1\text{ g }CO_2}{10^3\text{ mg }CO_2} \times \frac{1\text{ mol }CO_2}{44.01\text{ g }CO_2} \times \frac{1\text{ mol C}}{1\text{ mol }CO_2} = 1.116 \times 10^{-2}\text{ mol C}$$

$$1.116 \times 10^{-2}\text{ mol C} \times \frac{12.01\text{ g C}}{1\text{ mol C}} = 0.1340\text{ g C} = 134.0\text{ mg C}$$

$$100\text{ mg }H_2O \times \frac{1\text{ g }H_2O}{10^3\text{ mg }H_2O} \times \frac{1\text{ mol }H_2O}{18.02\text{ g }H_2O} \times \frac{2\text{ mol H}}{1\text{ mol }H_2O} = 1.110 \times 10^{-2}\text{ mol H}$$

$$1.110 \times 10^{-2}\text{ mol H} \times \frac{1.008\text{ g H}}{1\text{ mol H}} = 1.119 \times 10^{-2}\text{ g H} = 11.2\text{ mg H}$$

The mass of oxygen in the compound

= total mass − (mass C + mass H)

= 175 mg − (134.0 mg C + 11.2 mg H) = 29.8 mg O

The moles of oxygen are

$$29.8 \text{ mg O} \times \frac{1 \text{ g O}}{10^3 \text{ mg O}} \times \frac{1 \text{ mol O}}{16.00 \text{ g O}} = 1.862 \times 10^{-3} \text{ mol O}$$

The empirical formula can be derived from the mole ratios of the elements.

$$\frac{1.116 \times 10^{-2} \text{ mol C}}{1.862 \times 10^{-3} \text{ mol O}} = 5.993 \text{ mol C/mol O}$$

$$\frac{1.110 \times 10^{-2} \text{ mol H}}{1.862 \times 10^{-3} \text{ mol O}} = 5.961 \text{ mol H/mol O}$$

$$\frac{1.862 \times 10^{-3} \text{ mol O}}{1.862 \times 10^{-3} \text{ mol O}} = 1.000 \text{ mol O}$$

The empirical formula of phenol is C_6H_6O.

(b) The molar mass is needed to determine the molecular formula.

4.15 $$0.569 \text{ g Sn} \times \frac{1 \text{ mol Sn}}{118.7 \text{ g Sn}} = 4.794 \times 10^{-3} \text{ mol Sn}$$

$$2.434 \text{ g } I_2 \times \frac{1 \text{ mol } I_2}{253.81 \text{ g } I_2} \times \frac{2 \text{ mol I}}{1 \text{ mol } I_2} = 1.918 \times 10^{-2} \text{ mol I}$$

$$\frac{1.918 \times 10^{-2} \text{ mol I}}{4.794 \times 10^{-3} \text{ mol Sn}} = \frac{4.001 \text{ mol I}}{1.000 \text{ mol Sn}}$$

Therefore, the empirical formula is SnI_4.

Chapter 5

5.1 (a) NaF is soluble.
(b) $Ca(CH_3COO)_2$ is soluble.
(c) $SrCl_2$ is soluble.
(d) MgO is not soluble.
(e) $PbCl_2$ is not soluble.
(f) HgS is not soluble.

5.2 (a) This exchange reaction forms insoluble nickel hydroxide and aqueous sodium chloride.

$$NiCl_2(aq) + 2\ NaOH(aq) \longrightarrow Ni(OH)_2(s) + 2\ NaCl(aq)$$

(b) This is an exchange reaction that forms aqueous potassium bromide and a precipitate of calcium carbonate.

$$K_2CO_3(aq) + CaBr_2(aq) \longrightarrow CaCO_3(s) + 2\ KBr(aq)$$

5.3 (a) $BaCl_2(aq) + Na_2SO_4(aq) \rightarrow BaSO_4(s) + 2\ NaCl(aq)$

$$Ba^{2+}(aq) + SO_4^{2-}(aq) \longrightarrow BaSO_4(s)$$

(b) $(NH_4)_2S(aq) + FeCl_2(aq) \rightarrow FeS(s) + 2\ NH_4Cl(aq)$

$$Fe^{2+}(aq) + S^{2-}(aq) \rightarrow FeS(s)$$

5.4 Any of the strong acids in Table 5.2 would also be strong electrolytes. Any of the weak acids or bases in Table 5.2 would be weak electrolytes. Any organic compound that yields no ions on dissolution would be a nonelectrolyte.

5.5 $H_3PO_4(aq) + 3\ NaOH(aq) \rightarrow Na_3PO_4(aq) + 3\ H_2O(\ell)$

5.6 (a) Sulfuric acid and magnesium hydroxide
(b) Carbonic acid and strontium hydroxide

5.7

$$2\ HCN(aq) + Ca(OH)_2(aq) \longrightarrow Ca(CN)_2(aq) + 2\ H_2O(\ell)$$

$$2\ HCN(aq) + Ca^{2+}(aq) + 2\ OH^-(aq) \longrightarrow Ca^{2+}(aq) + 2\ CN^-(aq) + 2\ H_2O(\ell)$$

$$HCN(aq) + OH^-(aq) \longrightarrow CN^-(aq) + H_2O(\ell)$$

5.8 The oxidation numbers of Fe and Sb are 0 (Rule 1). The oxidation numbers in Sb_2S_3 are +3 for Sb^{3+} and −2 for S^{2-} (Rules 2 and 4). The oxidation numbers in FeS are +2 for Fe^{2+} and −2 for S^{2-} (Rules 2 and 4).

5.9 In the reaction $PbO(s) + CO(g) \rightarrow Pb(s) + CO_2(g)$, Pb^{2+} is reduced to Pb; Pb^{2+} is the oxidizing agent. C^{+2} is oxidized to C^{+4}; C^{2+} is the reducing agent.

5.10 Reactions (a) and (b) will occur. Aluminum is above copper and chromium in Table 5.5; therefore, aluminum will be oxidized and acts as the reducing agent in reactions (a) and (b). In reaction (a), Cu^{2+} is reduced, and Cu^{2+} is the oxidizing agent. Cr^{3+} is the oxidizing agent in reaction (b) and is reduced to Cr metal. Reactions (c) and (d) do not occur because Pt cannot reduce H^+, and Au cannot reduce Ag^+.

5.11 $$36.0 \text{ g } Na_2SO_4 \times \frac{1 \text{ mol } Na_2SO_4}{142.0 \text{ g } Na_2SO_4} = 0.254 \text{ mol } Na_2SO_4$$

$$\text{Molarity} = \frac{0.254 \text{ mol}}{0.750 \text{ L}} = 0.339 \text{ molar}$$

5.12 $$V(\text{conc}) = \frac{0.150 \text{ molar} \times 0.050 \text{ L}}{0.500 \text{ molar}} = 0.015 \text{ L} = 15 \text{ mL}$$

5.13 (a) 1.00 L of 0.125 M Na_2CO_3 contains 0.125 mol Na_2CO_3.

$$0.125 \text{ mol} \times \frac{105.99 \text{ g}}{1 \text{ mol}} = 13.2 \text{ g } Na_2CO_3$$

Prepare the solution by adding 13.2 g Na_2CO_3 to a volumetric flask, dissolving it and mixing thoroughly, and adding sufficient water until the solution volume is 1.0 L.

(b) Use water to dilute a specific volume of the 0.125 M solution to 100 mL.

$$V(\text{conc}) = \frac{0.0500 \text{ M} \times 0.100 \text{ L}}{0.125 \text{ M}} = 0.0400 \text{ L} = 40.0 \text{ mL of } 0.125 \text{ M solution}$$

Therefore, put 40.0 mL of the more concentrated solution into a container and add water until the solution volume equals 100 mL.

(c) 500 mL of 0.215 M $KMnO_4$ contains 1.70 g $KMnO_4$.

$$0.500 \text{ L} \times \frac{0.0215 \text{ mol } KMnO_4}{1 \text{ L}} = 0.01075 \text{ mol } KMnO_4$$

$$0.01075 \text{ mol } KMnO_4 \times \frac{158.0 \text{ g } KMnO_4}{1 \text{ mol } KMnO_4} = 1.70 \text{ g } KMnO_4$$

Put 1.70 g $KMnO_4$ into a container and add water until the solution volume is 500 mL.

(d) Dilute the more concentrated solution by adding sufficient water to 52.3 mL of 0.0215 M $KMnO_4$ until the solution volume is 250 mL.

$$V(\text{conc}) = \frac{0.00450 \text{ M} \times 0.250 \text{ L}}{0.0215 \text{ M}} = 0.0523 \text{ L} = 52.3 \text{ mL}$$

5.14 $$1.2 \times 10^{10} \text{ kg NaOH} \times \frac{1 \text{ mol NaOH}}{0.040 \text{ kg NaOH}} \times \frac{2 \text{ mol NaCl}}{2 \text{ mol NaOH}} \times \frac{58.5 \text{ g NaCl}}{1 \text{ mol NaCl}} \times \frac{1 \text{ L brine}}{360 \text{ g NaCl}} = 4.9 \times 10^{10} \text{ L}$$

5.15 $H_2SO_4(aq) + 2\ NaOH(aq) \rightarrow Na_2SO_4(aq) + 2\ H_2O(\ell)$

$$\text{Moles NaOH} = (0.0413 \text{ L})(0.100 \text{ M}) = 0.00413 \text{ mol NaOH}$$

$$\text{Moles } H_2SO_4 = \tfrac{1}{2} \times 0.00413 = 0.002065 \text{ mol } H_2SO_4$$

$$\text{Molarity} = \frac{0.002065 \text{ mol}}{0.0200 \text{ L}} = 0.103 \text{ M } H_2SO_4$$

Chapter 6

6.1 (a) $$160 \text{ Cal} \times \frac{1000 \text{ cal}}{\text{Cal}} \times \frac{4.184 \text{ J}}{\text{cal}} = 6.69 \times 10^5 \text{ J}$$

(b) 75 W = 75 J/s;
$75 \text{ J/s} \times 3.0 \text{ h} \times 60 \text{ min/h} \times 60 \text{ s/min} = 8.1 \times 10^5 \text{ J}$

(c) $16 \text{ kJ} \times \frac{1 \text{ kcal}}{4.184 \text{ kJ}} = 3.8 \text{ kcal}$

6.2 $\Delta E = -2400 \text{ J} = q + w = -1.89 \text{ kJ} + w$
$w = -2400 \text{ J} + 1.89 \text{ kJ} = -2.4 \text{ kJ} + 1.89 \text{ kJ} = -0.5 \text{ kJ}$

6.3 $q = c \times m \times \Delta T = c \times m \times (T_{\text{final}} - T_{\text{initial}})$

$$T_{\text{final}} = T_{\text{initial}} + \frac{q}{c \times m} = 5\ °\text{C} + \frac{24{,}100 \text{ J}}{(0.902 \text{ J g}^{-1}\ °\text{C}^{-1})(250.\text{ g})}$$
$$= 5\ °\text{C} + 106.8\ °\text{C} = 112\ °\text{C}$$

6.4 $q_{\text{water}} = -q_{\text{iron}}$
$(4.184 \text{ J/g °C})(1000.\text{ g})(32.8\ °\text{C} - 20.0\ °\text{C})$
$= -(0.451 \text{ J/g °C})(400.\text{ g})(32.8\ °\text{C} - T_i)$
$T_i = (297 + 32.8)\ °\text{C} = 330.\ °\text{C}$

6.5 $1.00 \text{ g K(s)} \times \frac{1 \text{ mL}}{0.86 \text{ g}} = 1.16 \text{ mL};$

$1.00 \text{ g K}(\ell) \times \frac{1 \text{ mL}}{0.82 \text{ g}} = 1.22 \text{ mL}$

The change in volume is $(1.22 - 1.16) \text{ mL} = 0.06 \text{ mL}$.

$$w = 0.06 \text{ mL} \times \frac{0.10 \text{ J}}{1 \text{ mL}} = 6 \times 10^{-3} \text{ J}$$
$$\Delta H = \frac{14.6 \text{ cal}}{1 \text{ g}} \times 1.00 \text{ g} \times \frac{4.184 \text{ J}}{1 \text{ cal}} = 61.1 \text{ J}$$
$$\Delta E = \Delta H + w = 61.1 \text{ J} + (6 \times 10^{-3} \text{ J}) = 61.1 \text{ J}$$

6.6 (a) $10.0 \text{ g I}_2 \times \frac{1 \text{ mol I}_2}{253.8 \text{ g I}_2} \times \frac{62.4 \text{ kJ}}{1 \text{ mol I}_2} = 2.46 \text{ kJ}$

(b) $3.42 \text{ g I}_2 \times \frac{1 \text{ mol I}_2}{253.8 \text{ g I}_2} \times \frac{-62.4 \text{ kJ}}{1 \text{ mol I}_2} = -0.841 \text{ kJ} = -841 \text{ J}$

(c) This process is the reverse of the one in part (a), so $\Delta H°$ is negative. Thus the process is exothermic. The quantity of energy transferred is 841 J.

6.7 The equation as written involves CO_2 as a reactant, but the question asks for CO_2 as a product. Therefore we will have to change the sign of $\Delta H°$. In addition, since the question asks about production of 4 mol CO_2, we must multiply $\Delta H°$ for the endothermic decomposition by 4. So for the production of 4 mol CO_2,

$$\Delta H° = (4) \times (662.8 \text{ kJ}) = 2651 \text{ kJ}$$

6.8 According to the thermochemical expression, the reaction is endothermic, so 285.8 kJ of energy is transferred into the system per mole of $H_2O(\ell)$ decomposed. Thus,

$$12.6 \text{ g H}_2\text{O} \times \frac{1 \text{ mol H}_2\text{O}}{18.02 \text{ g H}_2\text{O}} \times \frac{285.8 \text{ kJ}}{1 \text{ mol H}_2\text{O}} = 200.\text{ kJ}$$

6.9 $\Delta T = (25.43 - 20.64)\ °\text{C} = 4.79\ °\text{C}$

$$\Delta E_{\text{calorimeter}} = \frac{877 \text{ J}}{°\text{C}} \times 4.79\ °\text{C} = 4.200 \times 10^3 \text{ J} = 4.200 \text{ kJ}$$
$$\Delta E_{\text{water}} = 832 \text{ g} \times \frac{4.184 \text{ J}}{\text{g °C}} \times 4.79\ °\text{C} = 16.67 \text{ kJ}$$
$$\Delta E_{\text{reaction}} = -(q_{\text{calorimeter}} + q_{\text{water}}) = -(4.200 + 16.67) \text{ kJ} = -20.87 \text{ kJ}$$
$$20.87 \text{ kJ} \times \frac{1 \text{ kcal}}{4.184 \text{ kJ}} \times \frac{1 \text{ Cal}}{1 \text{ kcal}} = 4.99 \text{ Cal}$$

Since metabolizing the Fritos chip corresponds to oxidizing it, the result of 4.99 Cal verifies the statement that one chip provides 5 Cal.

6.10 The total volume of the initial solutions is 200. mL, which corresponds to 200. g of solution. The quantities of reactants are 0.10 mol $H^+(aq)$ and 0.050 mol $OH^-(aq)$, so 0.050 mol H_2O is formed.

$$0.050 \text{ mol H}_2\text{O} \times \frac{-58.6 \text{ kJ}}{1 \text{ mol H}_2\text{O}} = -2.9 \text{ kJ}$$

Since $\Delta H°$ is negative, energy is transferred to the water, and its temperature will rise.

$$\Delta T = \frac{q}{c \times m} = \frac{2.93 \times 10^3 \text{ J}}{(4.184 \text{ J g}^{-1}\ °\text{C}^{-1})(200.\text{ g})} = 3.5\ °\text{C}$$

The final temperature will be $(20.4 + 3.5)\ °\text{C} = 23.9\ °\text{C}$.

6.11 The balanced chemical equation is

$$2\ Fe_3O_4(s) \rightarrow 6\ FeO(s) + O_2(g)$$

The equations given in the problem can be arranged so that when added they give this balanced equation:

$2 \times [Fe_3O_4(s) \rightarrow 3\ Fe(s) + 2\ O_2(g)]$ $\quad \Delta H° = 2(1118.4 \text{ kJ})$

$6 \times [Fe(s) + \frac{1}{2} O_2(g) \rightarrow FeO(s)]$ $\quad \Delta H° = 6(-272.0 \text{ kJ})$

So

$$\Delta H° = 2236.8 \text{ kJ} - 1632.0 \text{ kJ} = 604.8 \text{ kJ}$$

6.12 (a) $\frac{1}{2} N_2(g) + \frac{3}{2} H_2(g) \rightarrow NH_3(g)$ $\quad \Delta H_f° = -46.11 \text{ kJ/mol}$

(b) $C(\text{graphite}) + \frac{1}{2} O_2(g) \rightarrow CO(g)$ $\quad \Delta H_f° = -110.525 \text{ kJ/mol}$

6.13 For the reaction given,

$$\Delta H° = \{6 \text{ mol } CO_2(g)\} \times \Delta H_f°\{CO_2(g)\} + \{5 \text{ mol } H_2O(g)\} \times \Delta H_f°\{H_2O(g)\} - (2 \text{ mol } \{C_3H_5(NO_3)_3(\ell)\} \times \Delta H_f°\{C_3H_5(NO_3)_3(\ell)\}$$
$$= \{6(-393.509) + 5(-241.818) - 2(-364)\} \text{ kJ}$$
$$= -2.84 \times 10^3 \text{ kJ}$$

For 10.0 g nitroglycerin (nitro),

$$q = 10.0 \text{ g} \times \frac{1 \text{ mol nitro}}{227.09 \text{ g}} \times \frac{-2.84 \times 10^3 \text{ kJ}}{2 \text{ mol nitro}} = -62.5 \text{ kJ}$$

(The 2 mol nitro in the last factor comes from the coefficient of 2 associated with nitroglycerin in the chemical equation.)

6.14 $SO_2(g) + \frac{1}{2} O_2(g) \rightarrow SO_3(g)$ $\quad \Delta H° = ?$

$$\Delta H° = \Delta H_f°\{SO_3(g)\} - [\Delta H_f°\{SO_2(g)\} + \tfrac{1}{2}\Delta H_f°\{O_2(g)\}]$$
$$= -395.72 \text{ kJ} - (-296.830 + 0) \text{ kJ} = -98.89 \text{ kJ}$$

Chapter 7

7.1 $\nu = \frac{c}{\lambda} = \frac{2.998 \times 10^8 \text{ m/s}}{4.05 \times 10^{-7} \text{ m}} = 7.40 \times 10^{14} \text{ s}^{-1}$ or 7.40×10^{14} Hz

7.2 (a) One photon of ultraviolet radiation has more energy because ν is larger in the UV spectral region than in the microwave region.
(b) One photon of blue light has more energy because the blue portion of the visible spectrum has a higher frequency than the green portion of the visible spectrum.

7.3 Any from $n_{\text{hi}} > 8$ to $n_{\text{lo}} = 2$

7.4 (a) $\Delta E = E_f - E_i = -2.179 \times 10^{-18} \text{ J}\left(\frac{1}{n_f^2} - \frac{1}{n_i^2}\right)$

$$= -2.179 \times 10^{-18} \text{ J}\left(\frac{1}{4^2} - \frac{1}{6^2}\right)$$
$$= -2.179 \times 10^{-18} \text{ J}\left(\frac{1}{16} - \frac{1}{36}\right)$$
$$= -2.179 \times 10^{-18} \text{ J}(0.6250 - 0.02778) = -7.565 \times 10^{-20} \text{ J}$$

The negative value of the energy change indicates that energy is emitted.

$$\nu = \frac{E_{photon}}{h} = \frac{7.565 \times 10^{-19}\ \text{J}}{6.626 \times 10^{-34}\ \text{J s}} = 1.141 \times 10^{14}\ \text{s}^{-1}$$

$$\lambda = \frac{c}{\nu} = \frac{2.998 \times 10^{8}\ \text{m/s}}{1.141 \times 10^{14}\ \text{s}^{-1}} = 2.628 \times 10^{-6}\ \text{m} = 2.628 \times 10^{3}\ \text{nm}$$

(b) Longer than the $n = 7$ to $n = 4$ transition.

7.5 $$\lambda = \frac{h}{mv} = \frac{6.626 \times 10^{-34}\ \text{J}\cdot\text{s}}{1.67 \times 10^{-27}\ \text{kg} \times 2.998 \times 10^{7}\ \text{m/s}}$$
$$= 1.32 \times 10^{-14}\ \text{m}$$

7.6 (a) 6d (b) 5 (c) 2, 1, 0, −1, −2

7.7 (a) [Ne] $3s^2 3p^2$ (b) [Ne] 3s [↑↓] 3p [↑] [↑] [ ]

7.8 The electron configurations for :Se: and :Te: are [Ar] $4s^2 3d^{10} 4p^4$ and [Kr] $5s^2 4d^{10} 5p^4$, respectively. Elements in the same main group have similar electron configurations.

7.9 (a) P^{3-} (b) Ca^{2+}

7.10 The ground state Cu atom has a configuration [Ar] $4s^1 3d^{10}$. When it loses one electron, it becomes the Cu^+ ion with configuration [Ar] $3d^{10}$. There is an added stability for the completely filled set of 3*d* orbitals.

7.11 B < Mg < Na < K

7.12 (a) Cs^+ (b) La^{3+}

7.13 F > N > P > Na

Chapter 8

8.1 (a) 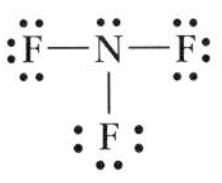 (b) 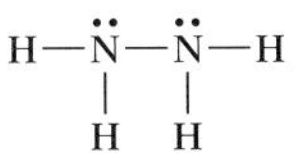

(c) 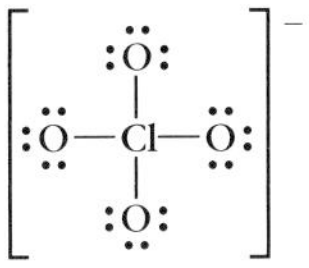

8.2 (a) [:N≡O:]$^+$ (b) 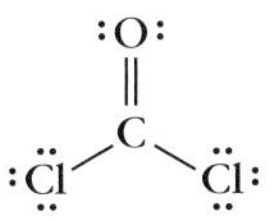

8.3 Only (c) can have geometric isomers. Molecules in (a) and (b) each have the same two groups on one of the double-bonded carbons.

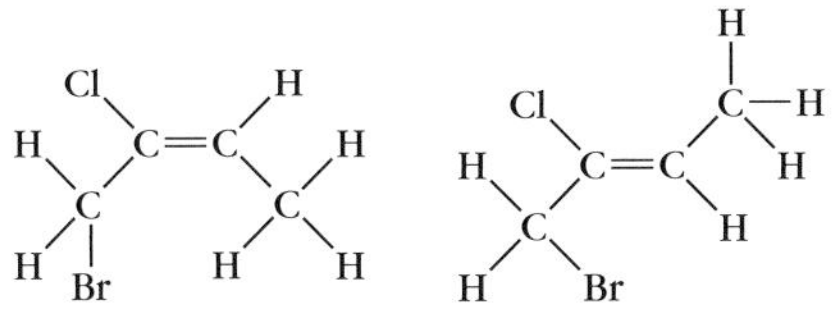

cis *trans*

8.4 (a) Si is a larger atom than S. (b) Br is a larger atom than Cl.
(c) The greater electron density in the triple bond brings the N≡O atoms closer together than the smaller electron density in the N═O double bond does.

8.5
$$\Delta H = [(4\ \text{mol C—H})(416\ \text{kJ/mol}) + (2\ \text{mol O═O})(498\ \text{kJ/mol})]$$
$$- [(4\ \text{mol O—H})(467\ \text{kJ/mol}) + (2\ \text{mol C═O})(803\ \text{kJ/mol})]$$
$$= (2660\ \text{kJ}) - (3474\ \text{kJ}) = -814\ \text{kJ}$$

8.6 (a) B—Cl is more polar; $\overset{\delta+}{\text{B}}$—$\overset{\delta-}{\text{Cl}}$; O—H is more polar; $\overset{\delta-}{\text{O}}$—$\overset{\delta+}{\text{H}}$.

8.7 The other Lewis structure is :O≡N—N̈:

	O	N	N
Valence electrons	6	5	5
Lone pair electrons	2	0	6
$\frac{1}{2}$ shared electrons	3	4	1
Formal charge	+1	+1	−2

8.8 (a) :F̤—Be—F̤: (b) :Ö—C̈l—Ö:

(c) 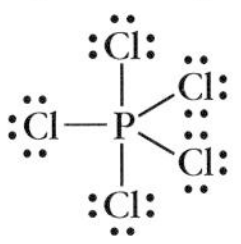 (d) [H—B—H]$^+$

(e) 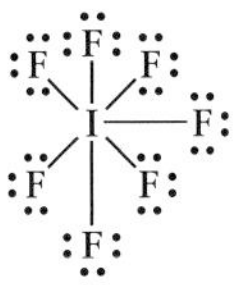

BeF_2—not an octet around the central Be atom; ClO_2—an odd number of electrons around Cl; PCl_5—more than four electron pairs around the central phosphorus atom; BH_2^+—only two electron pairs around the central B atom; IF_7—iodine has seven shared electron pairs.

8.9 Bond order = (8 − 5)/2 = 1.5. There is one unpaired electron.

Chapter 9

9.1 There are two Be—F bonds in BeF_2 and no lone pairs on beryllium. Therefore, the electron-pair and the molecular geometry are the same, linear, with 180° Be—F bond angles.

9.2

Central Atom (underlined)	Bond Pairs	Lone Pairs	Electron-Pair Geometry	Molecular Shape
$\underline{Br}O_3^-$	3	1	Tetrahedral	Triangular pyramid
$\underline{Se}F_2$	2	2	Tetrahedral	Angular
$\underline{N}O_2^-$	3	1	Triangular planar	Angular

9.3 (a) ClF_2^-: triangular bipyramidal electron-pair geometry and linear molecular geometry
(b) XeO_3: tetrahedral electron-pair geometry and triangular pyramidal molecular geometry

9.4 (a) $BeCl_2$: *sp* hybridization (two bonding electron pairs around the central Be atom), linear geometry

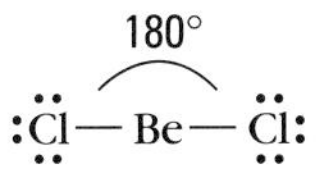

(b) The central N atom has four bonding pairs and no lone pairs in sp^3 hybridized orbitals on N giving tetrahedral electron-pair and molecular geometries.

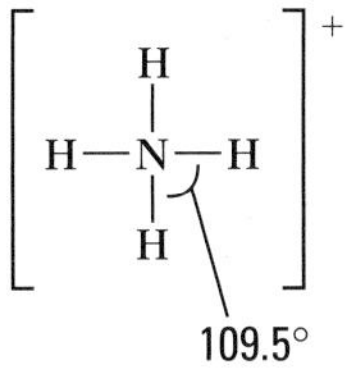

(c) Each carbon is sp^3 hybridized with no lone pairs, so the electron-pair and molecular geometries are both tetrahedral. The sp^3 hybridized oxygen atom has two bonding pairs, each in single bonds to a carbon atom plus two lone pairs giving it a tetrahedral electron-pair geometry and an angular molecular geometry.

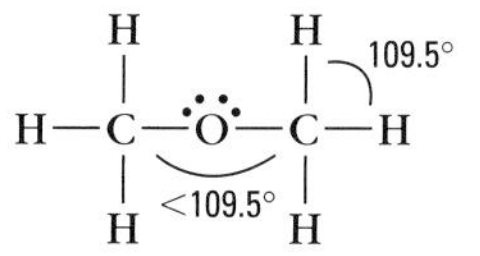

```
    H       H   109.5°
    |       |
H—C—Ö—C—H
    |  <109.5°  |
    H       H
```

(d) The central boron atom has four single bonds, one to each fluorine atom, and no lone pairs resulting in tetrahedral electron-pair and molecular geometries. The boron is sp^3 hybridized to accommodate the four bonding pairs.

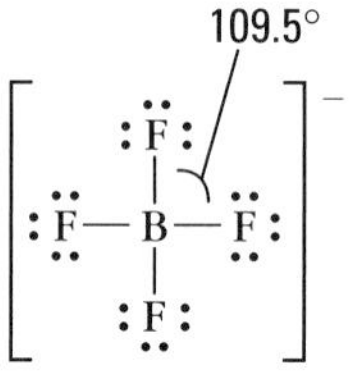

```
        109.5°
  [     :F:      ]⁻
  [      |       ]
  [ :F—B—F:      ]
  [      |       ]
  [     :F:      ]
```

9.5 (a) In HCN, the *sp* hybridized carbon atom is sigma bonded to H and to N, as well as having two pi bonds to N. The sigma and two pi bonds form the C≡N triple bond. The nitrogen is *sp* hybridized with a sigma and two pi bonds to carbon; a lone pair is in the nonbonding *sp* hybrid orbital on N.

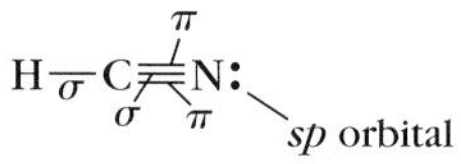

```
       π
H—C≡N:
 σ  σ π   sp orbital
```

(b) The double-bonded carbon and nitrogen are both sp^2 hybridized. The sp^2 hybrid orbitals on C form sigma bonds to H and to N; the unhybridized *p* orbital on C forms a pi bond with the unhybridized *p* orbital on N. The sp^2 hybrid orbitals on N form sigma bonds to carbon and to H; the N lone pair is in the nonbonding sp^2 hybrid orbital.

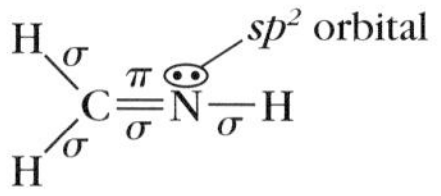

```
H  σ         sp² orbital
  \    π
   C=N—H
  /  σ   σ
H  σ
```

9.6 (a) $BFCl_2$ is a triangular planar molecule with polar B—F and B—Cl bonds. The molecule is polar because the B—F bond is more polar than the B—Cl bonds, resulting in a net dipole.

(b) NH_2Cl is a triangular pyramidal molecule with polar N—H bonds. (N—Cl is a nonpolar bond; N and Cl have the same electronegativity.) It is a polar molecule because the N—H dipoles do not cancel and produce a net dipole.

(c) SCl_2 is an angular polar molecule. The polar S—Cl bond dipoles do not cancel each other because they are not symmetrically arranged due to the two lone pairs on S.

9.7 (a) London forces between Kr atoms must be overcome for krypton to melt.

(b) The C—H covalent bonds in propane must be broken to form C and H atoms; the H atoms covalently bond to form H_2.

9.8 (a) London forces occur between N_2 molecules.

(b) CO_2 is nonpolar, and London forces occur between it and polar water molecules.

(c) London forces occur between the two molecules, but the principal intermolecular forces are the hydrogen bonds between the H on NH_3 with the lone pairs on the OH oxygen, and the hydrogen bonds between the H on the oxygen in CH_3OH and lone pair on nitrogen in NH_3.

Chapter 10

10.1 (a) Pressure in atm =

$$29.5 \text{ in Hg} \times \frac{1 \text{ atm}}{76.0 \text{ cm Hg}} \times \frac{2.54 \text{ cm}}{1 \text{ in}} = 0.986 \text{ atm}$$

(b) Pressure in mm Hg $= 29.5 \text{ in Hg} \times \frac{25.4 \text{ mm}}{1 \text{ in}} = 749 \text{ mm Hg}$

(c) Pressure in bar =

$$29.5 \text{ in Hg} \times \frac{1.013 \text{ bar}}{760 \text{ mm Hg}} \times \frac{25.4 \text{ mm}}{1 \text{ in}} = 0.999 \text{ bar}$$

(d) Pressure in kPa =

$$29.5 \text{ in Hg} \times \frac{101.3 \text{ kPa}}{760 \text{ mm Hg}} \times \frac{25.4 \text{ mm}}{1 \text{ in}} = 99.9 \text{ kPa}$$

10.2 (a) The temperature remains constant, so the average energy of the gas molecules remains constant. If the volume is decreased, then the gas molecules must hit the walls more frequently, and the pressure is increased.

(b) The temperature remains constant, so the average energy of the gas molecules remains constant. The addition of more molecules within a fixed volume must mean that the molecules hit walls more frequently, so the pressure is increased.

10.3 Volume of NO gas $= 1.0 \text{ L } O_2 \times \frac{2 \text{ L NO}}{1 \text{ L } O_2} = 2.0 \text{ L NO}$

10.4 $V = \frac{nRT}{P}$

$$= \frac{(2.64 \text{ mol})(0.0821 \text{ L atm mol}^{-1} \text{ K}^{-1})(304 \text{ K})}{0.640 \text{ atm}} = 103 \text{ L}$$

10.5 $V_2 = \frac{P_1V_1}{P_2} = \frac{(1.00 \text{ atm})(400. \text{ mL})}{0.750 \text{ atm}} = 533 \text{ mL}$

10.6 (a) $V_2 = \frac{P_1V_1T_2}{P_2T_1}$

$$= \frac{(710 \text{ mm Hg})(21. \text{ mL})(299.6 \text{ K})}{(740 \text{ mm Hg})(295.4 \text{ K})} = 20. \text{ mL}$$

(b) $V_2 = \frac{(21. \text{ mL})(299.6 \text{ K})}{(295.4 \text{ K})} = 21. \text{ mL}$

10.7 $\frac{10.0 \text{ g } NH_4NO_3}{80.043 \text{ g/mol}} = 0.1249 \text{ mol } NH_4NO_3$

$$0.1249 \text{ mol } NH_4NO_3 \times \frac{7 \text{ mol product gases}}{2 \text{ mol } NH_4NO_3} = 0.437 \text{ mol produced}$$

$$V = \frac{(0.437 \text{ mol})(0.0821 \text{ L atm mol}^{-1} \text{ K}^{-1})(298 \text{ K})}{1 \text{ atm}} = 10.7 \text{ L}$$

10.8 $\frac{1.0 \text{ g LiOH}}{23.94 \text{ g/mol}} = 0.0418 \text{ mol LiOH}$

$$0.0418 \text{ mol LiOH} \times \frac{1 \text{ mol } CO_2}{2 \text{ mol LiOH}} = 0.0209 \text{ mol } CO_2$$

$$V = \frac{(0.0209 \text{ mol})(0.0821 \text{ L atm mol}^{-1} \text{ K}^{-1})(295 \text{ K})}{1 \text{ atm}} = 0.51 \text{ L } CO_2$$

10.9 Amount of gas, $n = \frac{PV}{RT}$

$$= \frac{(0.850 \text{ atm})(1.00 \text{ L})}{(0.0821 \text{ L atm mol}^{-1} \text{ K}^{-1})(293 \text{ K})} = 0.0353 \text{ mol}$$

Molar mass $= \frac{1.13 \text{ g}}{0.0353 \text{ mol}} = 32.0 \text{ g/mol}$

The gas is probably oxygen.

10.10 Amount of $N_2 = 7.0 \text{ g } N_2 \times \frac{1 \text{ mol } N_2}{28.10 \text{ g } N_2} = 0.25 \text{ mol } N_2$

Amount of $H_2 = 6.0 \text{ g } H_2 \times \frac{1 \text{ mol } H_2}{2.02 \text{ g } H_2} = 3.0 \text{ mol } H_2$

Total number of moles = 3.0 + 0.25 = 3.25 mol

$$X_{N_2} = \frac{0.25 \text{ mol}}{3.25 \text{ mol}} = 0.077 \qquad X_{H_2} = \frac{3.0 \text{ mol}}{3.25 \text{ mol}} = 0.92$$

$$P_{N_2} = \frac{(0.25 \text{ mol})(0.0821 \text{ L atm mol}^{-1} \text{ K}^{-1})(773 \text{ K})}{5.0 \text{ L}} = 3.2 \text{ atm}$$

$$P_{H_2} = \frac{(3.0 \text{ mol})(0.0821 \text{ L atm mol}^{-1} \text{ K}^{-1})(773 \text{ K})}{5.0 \text{ L}} = 38 \text{ atm}$$

10.11 $P_{HCl} = P_{total} - P_{water} = 740 \text{ mm Hg} - 21 \text{ mm Hg} = 719 \text{ mm Hg}$

$$n = \frac{PV}{RT} = \frac{(719/760 \text{ atm})(0.260 \text{ L})}{(0.0821 \text{ L atm mol}^{-1} \text{ K}^{-1})(296 \text{ K})}$$

$$= 0.0101 \text{ mol } H_2$$

$$0.0101 \text{ mol} \times 2.0158 \text{ g/mol} = 0.0204 \text{ g} = 20.4 \text{ mg } H_2$$

Chapter 11

11.1 $\ln\left(\frac{P_2}{143.0 \text{ torr}}\right) = \frac{3.21 \times 10^4 \text{ J/mol}}{8.31 \text{ J mol}^{-1} \text{ K}^{-1}}\left[\frac{1}{303.15 \text{ K}} - \frac{1}{333.15 \text{ K}}\right]$

$$\ln\left(\frac{P_2}{143.0 \text{ torr}}\right) = 1.1473$$

$$\ln P_2 = -1.1473 - \ln(143.0)$$

$$P_2 = 450. \text{ torr}$$

11.2 From Table 11.2, $\Delta H_{vap}(Br_2)$ is 29.54 kJ/mol at its normal boiling point.
So 29.54 kJ/mol × 0.500 mol = 14.77 kJ.

11.3 Heat $= 2.5 \times 10^{10} \text{ kg } H_2O \times \frac{10^3 \text{ g}}{1 \text{ kg}} \times \frac{1 \text{ mol } H_2O}{18.02 \text{ g } H_2O}$

$$\times \frac{-44.0 \text{ kJ}}{\text{mol}} = -6.1 \times 10^{13} \text{ kJ}$$

This process is exothermic as water vapor condenses, forming rain.

11.4 (21.95)/2 = 10.98 kJ

11.5 The gas phase.

11.6 (a) Solid decane is a molecular solid.
(b) Solid $MgCl_2$ is composed of Mg^{2+} and Cl^- ions and is an ionic solid.

11.7 There are two atoms per bcc unit cell. The diagonal of the bcc unit cell is four times the radius of the atoms in the unit cell, so, solving for the edge,

$$\text{Edge} = \frac{4 \times 144 \text{ pm}}{\sqrt{3}} = 333 \text{ pm}$$

$$\text{Density} = \frac{\text{mass}}{\text{volume}}$$

$$= \frac{(2 \text{ Au atoms})(196.97 \text{ g Au}/6.022 \times 10^{23} \text{ Au atoms})}{[(333 \text{ pm})(1 \text{ m}/10^{12} \text{ pm})(10^2 \text{ cm/m})]^3}$$

$$= 17.7 \text{ g/cm}^3$$

11.8 The edge of the KCl unit cell would be 2 × 152 pm + 2 × 167 pm = 638 pm.
The unit cell of KCl is larger than that of NaCl.

Volume of the unit cell = $(638 \text{ pm})^3$

$$= 2.60 \times 10^{18} \text{ pm}^3 \times \left(\frac{10^{-10} \text{ cm}}{\text{pm}}\right)^3$$

$$= 2.60 \times 10^{-22} \text{ cm}^3$$

$$D = \frac{m}{V}$$

$$= \frac{(4 \text{ formula units KCl}) \times 74.55 \text{ g KCl}/6.022 \times 10^{23} \text{ formula units}}{2.60 \times 10^{-22} \text{ cm}^3}$$

$$D = 1.91 \text{ g/cm}^3$$

11.9 Energy transfer required

$$= 1.45 \text{ g Al} \times \frac{1 \text{ mol Al}}{26.98 \text{ g Al}} \times \frac{10.7 \text{ kJ}}{\text{mol}} = 0.575 \text{ kJ}$$

Chapter 12

12.1 (a) Rate $= \frac{-\Delta[Cv^+]}{\Delta t} = 1.27 \times 10^{-6} \text{ mol L}^{-1} \text{ s}^{-1}$

$$\Delta t = \frac{-\Delta[Cv^+]}{1.27 \times 10^{-6} \text{ mol L}^{-1} \text{ s}^{-1}}$$

$$= \frac{(4.30 \times 10^{-5} - 3.96 \times 10^{-5}) \text{ mol/L}}{1.27 \times 10^{-6} \text{ mol L}^{-1} \text{ s}^{-1}}$$

$$= 2.7 \text{ s}$$

(b) No. The rate of reaction depends on the concentration of Cv^+ and, therefore, becomes slower as the reaction progresses. Therefore, the method used in part (a) works only over a small range of concentrations.

12.2 (a) The balanced chemical equation shows that for every mole of O_2 consumed two moles of N_2O_5 are produced. Therefore, the rate of formation of N_2O_5 is twice the rate of disappearance of O_2.
(b) Four moles of NO_2 are consumed for every mole of O_2 consumed. Therefore, if O_2 is consumed at the rate of 0.0037 mol L^{-1} s^{-1} the rate of disappearance of NO_2 is four times this rate.

$$4 \times (0.0037 \text{ mol L}^{-1} \text{ s}^{-1}) = 0.015 \text{ mol L}^{-1} \text{ s}^{-1}$$

12.3 (a) The effect of $[OH^-]$ on the rate of reaction cannot be determined, because the $[OH^-]$ is the same in all three experiments.

(b) Rate $= k[Cv^+]$

(c) $k_1 = \frac{\text{rate}}{[Cv^+]} = \frac{1.3 \times 10^{-6} \text{ mol L}^{-1} \text{ s}^{-1}}{4.3 \times 10^{-5} \text{ mol/L}}$

$$= 3.0 \times 10^{-2} \text{ s}^{-1}$$

$$k_2 = 3.0 \times 10^{-2} \text{ s}^{-1}$$

$$k_3 = 3.0 \times 10^{-2} \text{ s}^{-1}$$

$$k = \frac{k_1 + k_2 + k_3}{3} = 3.0 \times 10^{-2} \text{ s}^{-1}$$

(d) Rate $= k[Cv^+]$

$$= (3.0 \times 10^{-2} \text{ s}^{-1})(0.00045 \text{ mol/L})$$

$$= 1.4 \times 10^{-5} \text{ mol L}^{-1} \text{ s}^{-1}$$

(e) Rate $= (3.0 \times 10^{-2} \text{ s}^{-1})(0.5 \times 0.00045 \text{ mol/L})$

$$= 6.8 \times 10^{-6} \text{ mol L}^{-1} \text{ s}^{-1}$$

12.4 (a) The order of the reaction with respect to each chemical is the exponent associated with the concentration of that chemical. So the reaction is second-order with respect to NO and it is first-order with respect to Cl_2.
(b) Tripling the concentration of NO will make the reaction go $3^2 = 9$ times faster but decreasing the concentration of Cl_2 by a factor of 8 will make the reaction go at 1/8 the initial rate. If these two changes are made simultaneously the relevant factor will be 9/8 = 1.13, so the reaction will occur 13% faster than it did under the initial conditions.

12.5 Make three plots of the data.

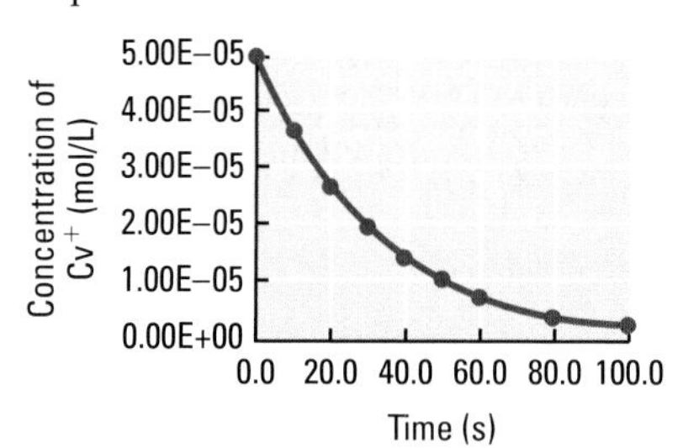

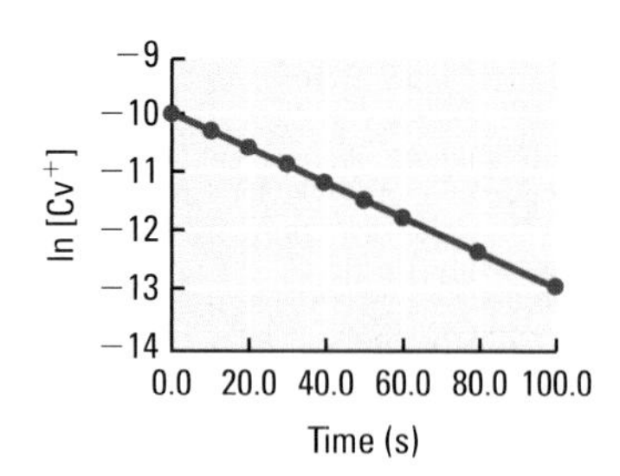

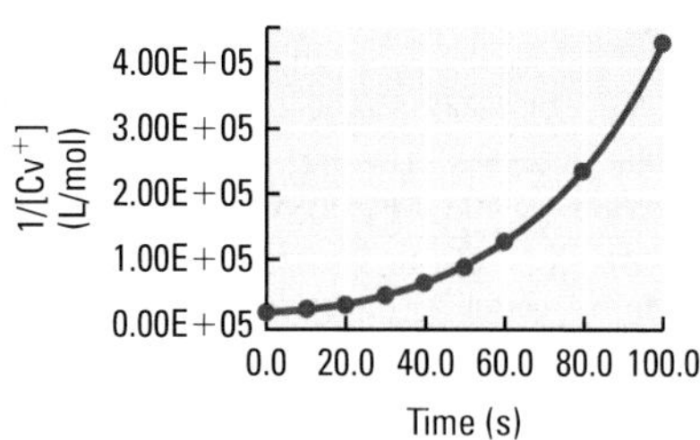

The first-order plot is a straight line and the other plots are curved, so the reaction is first-order. The slope of the first-order plot is $-0.0307\ s^{-1}$, so $k = -\text{slope} = 0.031\ s^{-1}$.

12.6 Use the integrated first-order rate law from Table 12.2.

$$\ln[A]_t = -kt + \ln[A]_0$$

$$\ln\frac{[A]_t}{[A]_0} = -kt$$

$$t = -\frac{1}{k}\ln\frac{[A]_t}{[A]_0} = -\left(\frac{1}{3.43 \times 10^{-2}\ d^{-1}}\right)\ln\left(\frac{0.1}{1.0}\right)$$

$$= -(29.15\ d)(-2.303) = 67.1\ d$$

12.7 In Figure 12.3 the $[Cv^+]$ falls from 5.00×10^{-5} M to 2.5×10^{-5} M in 23 s. The $[Cv^+]$ falls from 2.5×10^{-5} M to 1.25×10^{-5} M between 23 s and 46 s. The two times are equal, so $t_{1/2} = 23$ s.

$$k = \frac{0.693}{t_{1/2}} = \frac{0.693}{23\ s} = 3.0 \times 10^{-2}\ s^{-1}$$

12.8

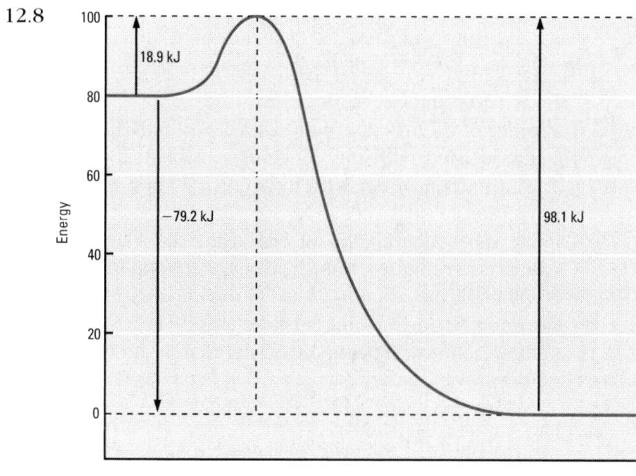

$Cl_2(g) + 2\ NO(g) \rightarrow 2\ NOCl(g)$

Since the energy of the products is less than that of reactants, the reaction is exothermic.

12.9 Obtain the value $E_a = 76.3$ kJ/mol from the discussion and analysis of the data in Figure 12.10 and the value $k = 4.18 \times 10^{-5}\ L\ mol^{-1}\ s^{-1}$ at 273 K from Table 12.3.

$$\frac{k_1}{k_2} = e^{\left[\frac{E_a}{R}\left(\frac{1}{T_2} - \frac{1}{T_1}\right)\right]} = e^{\left[\frac{76{,}300\ J/mol}{8.314\ J\ K^{-1}\ mol^{-1}}\left(\frac{1}{348\ K} - \frac{1}{273\ K}\right)\right]}$$

$$= e^{-7.245} = 7.138 \times 10^{-4}$$

$$k_2 = \frac{k_1}{7.138 \times 10^{-4}} = \frac{4.18 \times 10^{-5}\ L\ mol^{-1}\ s^{-1}}{7.138 \times 10^{-4}}$$

$$= 5.86 \times 10^{-2}\ L\ mol^{-1}\ s^{-1}$$

12.10 (a) $2\ NH_3(aq) + OCl^-(aq) \rightarrow N_2H_4(aq) + Cl^-(aq) + H_2O(\ell)$
(b) Step 1
(c) NH_2Cl, OH^-, $N_2H_5^+$
(d) Rate = rate of step 1 = $k\ [NH_3][OCl^-]$

Chapter 13

13.1 (a) $K_c = [CO_2(g)]$ (b) $K_c = \frac{[H_2]}{[HCl]}$

(c) $K_c = \frac{[CO][H_2]^3}{[CH_4][H_2O]}$ (d) $K_c = \frac{[HCN][OH^-]}{[CN^-]}$

13.2 $K_c = K_{c_1} \times K_{c_2} = (4.2 \times 10^{-7})(4.8 \times 10^{-11})$
$= 2.0 \times 10^{-17}$

13.3 $[CH_3COO^-] = \frac{2.96}{100} \times 0.0200\ mol/L = 5.92 \times 10^{-4}$ M

$[H_3O^+] = 5.92 \times 10^{-4}$ M

$$[CH_3\ COOH] = \frac{100 - 2.96}{100} \times 0.0200\ mol/L$$
$$= 1.94 \times 10^{-2}\ M$$

$$K_c = \frac{[H_3O^+][CH_3COO^-]}{[CH_3COOH]} = \frac{(5.92 \times 10^{-4})(5.92 \times 10^{-4})}{1.94 \times 10^{-2}}$$
$$= 1.81 \times 10^{-5}$$

The result agrees with the value in Table 13.1.

13.4 (a) $K_c(AgCl) = 1.8 \times 10^{-10}$; $K_c(AgI) = 1.5 \times 10^{-16}$. Because $K_c(AgI) < K_c(AgCl)$, the concentration of silver ions is larger in the beaker of AgCl.
(b) Unless all of the solid AgCl or AgI dissolves (which would mean that there was no equilibrium reaction), the concentrations at equilibrium are independent of the volume.

13.5 $$Q = \frac{(\text{conc. } SO_3)^2}{(\text{conc. } SO_2)^2(\text{conc. } O_2)}$$
$$= \frac{(0.184)^2}{(0.102 \times 2)^2(0.0132)} = 61.6$$

Since $Q < K_c$, the forward reaction should occur.

13.6 The reaction is the reverse of the one in Table 13.1, so

$$K_c = \frac{1}{1.7 \times 10^2} = 5.9 \times 10^{-3}$$

$$Q = \frac{(\text{conc. } NO_2)^2}{(\text{conc. } N_2O_4)} = \frac{\left(\frac{0.500\ mol}{4.00\ L}\right)^2}{\left(\frac{1.00\ mol}{4.00\ L}\right)} = 6.25 \times 10^{-2}$$

Because $Q > K_c$, the reaction should go in the reverse direction. Therefore, let x be the change in concentration of N_2O_4, giving the ICE table:

	N_2O_4	$\rightleftharpoons$	$2\ NO_2$
Initial concentration (mol/L)	$\frac{1.00}{4.00} = 0.250$		$\frac{0.500}{4.00} = 0.125$
Change as reaction occurs (mol/L)	x		$-2x$
Equilibrium concentration (mol/L)	$0.250 + x$		$0.125 - 2x$

$$K_c = 5.9 \times 10^{-3} = \frac{(0.125 - 2x)^2}{0.250 + x}$$

$$= \frac{(1.56 \times 10^{-2}) - 0.500x + 4x^2}{0.250 - x}$$

$$(1.48 \times 10^{-3}) + (5.9 \times 10^{-3})x = (1.56 \times 10^{-2}) - 0.500x + 4x^2$$

$$4x^2 - 0.5059x + (1.412 \times 10^{-2}) = 0$$

$$x = \frac{-(-0.5059) \pm \sqrt{(-0.5059)^2 - 4 \times 4 \times 1.412 \times 10^{-2}}}{2 \times 4}$$

$$= \frac{0.5059 \pm \sqrt{3.001 \times 10^{-2}}}{8}$$

$$= \frac{0.5059 \pm 0.1732}{8}$$

$$x = 8.49 \times 10^{-2} \text{ or } x = 4.16 \times 10^{-2}$$

If $x = 8.49 \times 10^{-2}$, then $[N_2O_4] = 0.250 + x = 0.335$ and $[NO_2] = 0.125 - 2x = 0.125 - (2 \times 7.86 \times 10^{-2}) = -0.0448$. A negative concentration is impossible, so x must be 4.16×10^{-2}. Then

$$[N_2O_4] = 0.250 + 0.0416 = 0.292$$

$$[NO_2] = 0.125 - (2 \times 0.0416) = 0.0418$$

As predicted by Q, the reverse reaction has occurred, and the concentration of N_2O_4 has increased.

13.7 $O_2N{-}NO_2$ and $\cdot NO_2 + \cdot NO_2$ (structures: O=N(–O)–N(=O)–O and O=N•(–O) + •N(=O)–O)

Because a bond is broken and because bond breaking is always endothermic (**p. 202**), the reaction must be endothermic. Increasing temperature shifts the equilibrium in the endothermic direction. Figure 13.3 shows that at a higher temperature there is a greater concentration of brown NO_2.

13.8 (a) There are more moles of gas phase reactants than products, so entropy favors the reactants.
(b) Data from Appendix J show that $\Delta H° = -168.66$ kJ. The reaction is exothermic, so the energy effect favors the products.
(c) As T increases the reaction shifts in the endothermic direction, which is toward reactants. The entropy effect also becomes more important at high T, and it favors reactants. There is a greater concentration of SO_2 at low temperature.

Chapter 14

14.1 (a) Ethylene glycol molecules are polar and attracted to each other by dipole-dipole attractions and hydrogen bonding. They will not dissolve in gasoline, a nonpolar substance.
(b) Molecular iodine and carbon tetrachloride are nonpolar; therefore iodine should dissolve readily in carbon tetrachloride.
(c) Motor oil contains a mixture of nonpolar hydrocarbons that will dissolve in carbon tetrachloride, a nonpolar solvent.

14.2 The —OH groups attached to the ring and to the side chain of vitamin C hydrogen bond to water molecules. The oxygen atoms in the ring also form hydrogen bonds to water.

14.3 Use Henry's law, the Henry's law constant, and the fact that air has a 0.21 mole fraction of oxygen.

$$\text{Pressure of } O_2 = (1.0 \text{ atm})\left(\frac{760 \text{ mm Hg}}{1 \text{ atm}}\right)(0.21) = 160. \text{ mm Hg}$$

$$S_g = k_H P_g = \left(1.66 \times 10^{-6} \frac{\text{mol/L}}{\text{mm Hg}}\right)(160. \text{ mm Hg}) = 2.66 \times 10^{-4} \text{ mol/L}$$

$$(2.66 \times 10^{-4} \text{ mol/L})\left(\frac{32.00 \text{ g } O_2}{1 \text{ mol } O_2}\right) = 0.0085 \text{ g/L or } 8.5 \text{ mg/L}$$

14.4 Total mass is $750 + 21.5 = 771.5$ g.

$$\text{Weight percent glucose} = \frac{21.5 \text{ g}}{771.5 \text{ g}} \times 100\% = 2.79\%$$

14.5
$$\left(\frac{30 \text{ g Se}}{10^9 \text{ g } H_2O}\right)\left(\frac{1 \text{ g } H_2O}{1 \text{ mL } H_2O}\right)\left(\frac{10^6\ \mu\text{g Se}}{1 \text{ g Se}}\right) = 3.0 \times 10^{-2}\ \mu\text{g Se/mL } H_2O$$

Se in 100 mL of water

$$= \left(\frac{3.0 \times 10^{-2} \mu\text{g Se}}{1 \text{ mL } H_2O}\right)(100 \text{ mL } H_2O) = 3.0\ \mu\text{g Se}$$

14.6 (a) $$(0.250 \text{ L})\left(\frac{0.0750 \text{ mol NaBr}}{1 \text{ L}}\right)\left(\frac{102.9 \text{ g NaBr}}{1 \text{ mol NaBr}}\right) = 1.93 \text{ g NaBr}$$

(b) $$V_c = \frac{M_d \times V_d}{M_c} = \frac{(0.00150 \text{ M})(0.500 \text{ L})}{0.0750 \text{ M}} = 0.0100 \text{ L} = 10.0 \text{ mL}$$

14.7 (a) $$\left[\left(\frac{0.0556 \text{ mol } Mg^{2+}}{1 \text{ L}}\right)\left(\frac{24.3 \text{ g } Mg^{2+}}{1 \text{ mol } Mg^{2+}}\right)\right] \Big/ 1.03 \times 10^{-3} \text{ g solution} = 0.00131 \frac{\text{g } Mg^{2+}}{\text{g solution}}$$

(b) 0.00131×10^6 ppm $= 1310$ ppm

14.8 $$\text{Molarity } H_2O_2 = \frac{\text{moles } H_2O_2}{\text{L solution}}$$

$$= \frac{30.0 \text{ g } H_2O_2}{100. \text{ g solution}} \times \frac{1.11 \text{ g solution}}{1 \text{ mL solution}} \times \frac{10^3 \text{ mL solution}}{1 \text{ L solution}} \times \frac{1 \text{ mol } H_2O_2}{34.0 \text{ g } H_2O_2}$$

$$= \frac{3.33 \times 10^4 \text{ mol } H_2O_2}{3.40 \times 10^3 \text{ L}} = 9.79 \text{ M}$$

14.9 $$\text{Molarity, M} = \left(\frac{20 \text{ g NaCl}}{100 \text{ g solution}}\right)\left(\frac{1.148 \text{ g solution}}{1 \text{ mL solution}}\right) \times \left(\frac{1 \text{ mol NaCl}}{58.5 \text{ g NaCl}}\right)\left(\frac{10^3 \text{ mL}}{1 \text{ L}}\right) = 3.9 \text{ mol NaCl/L}$$

$$\text{Molality, } m = \left(\frac{20 \text{ g NaCl}}{0.0080 \text{ kg } H_2O}\right)\left(\frac{1 \text{ mol NaCl}}{58.5 \text{ g NaCl}}\right) = 4.3 \text{ mol NaCl/kg } H_2O$$

14.10 $$P_{water} = (X_{water})(P°_{water})$$

$$291.2 \text{ mm Hg} = (X_{water})(355.1 \text{ mm Hg})$$

$$X_{water} = \frac{291.2 \text{ mm Hg}}{355.1 \text{ mm Hg}} = 0.8201$$

$$X_{urea} = 1.000 - X_{water} = 1.000 - 0.8201 = 0.1799$$

14.11 $$\text{Molality of solution} = \left(\frac{1.20 \text{ kg ethylene glycol}}{6.50 \text{ kg } H_2O}\right)\left(\frac{1000 \text{ g}}{\text{kg}}\right)\left(\frac{1 \text{ mol ethylene glycol}}{62.068 \text{ g ethylene glycol}}\right) = 2.97 \text{ mol/kg}$$

Next, calculate the freezing point depression of a 2.97 mol/kg solution.

$$\Delta T_f = (1.86\ °\text{C kg mol}^{-1})(2.97 \text{ mol/kg}) = 5.52\ °\text{C}$$

This solution will freeze at -5.52 °C, so this quantity of ethylene glycol will not protect the 6.5 kg water in the tank if the temperature drops to -25 °C.

14.12 F.P. benzene = 5.50 °C; $\Delta T_f = 5.50\ °\text{C} - 5.15\ °\text{C} = 0.35\ °\text{C}$.

$$\text{Molality} = \frac{\Delta T_f}{K_f} = \frac{0.35\ °\text{C}}{5.10\ °\text{C kg mol}^{-1}} = 0.0686 \text{ mol/kg}$$

$$\frac{0.0686 \text{ mol solute}}{1 \text{ kg benzene}} \times 0.0500 \text{ kg benzene} = 0.00343 \text{ mol solute}$$

$$\frac{0.180 \text{ g solute}}{0.00343 \text{ mol solute}} = 52.5 \text{ g/mol}$$

14.13 Hemoglobin is a molecular substance so the factor i is 1.

$$c = \frac{\Pi}{RTi} = \frac{1.8 \times 10^{-3}\ \text{atm}}{(0.0821\ \text{L atm mol}^{-1}\ \text{K}^{-1})(298\ \text{K})(1)}$$
$$= 7.36 \times 10^{-5}\ \text{mol/L}$$
$$\text{MM} = \frac{5.0\ \text{g}}{7.36 \times 10^{-5}\ \text{mol}} = 6.8 \times 10^{4}\ \text{g/mol}$$

Chapter 15

15.1

Acid	Its Conjugate Base	Base	Its Conjugate Acid
$H_2PO_4^-$	HPO_4^{2-}	PO_4^{3-}	HPO_4^{2-}
H_2	H^-	NH_2^-	NH_3
HSO_3^-	SO_3^{2-}	ClO_4^-	$HClO_4$
HF	F^-	Br^-	HBr

15.2 $[OH^-] = 3.0 \times 10^{-8}$ M;

$$[H_3O^+] = \frac{1.0 \times 10^{-14}}{3.0 \times 10^{-8}} = 3.3 \times 10^{-7}\ \text{M}$$

Therefore, the solution whose $[H_3O^+]$ is 5.0×10^{-4} M is more acidic.

15.3 The 2.0×10^{-5} M H^+ solution is more acidic.

15.4 In a 0.040 M solution of NaOH, the $[OH^-]$ is 0.040 M because the NaOH is 100% dissociated; pH = 12.60.

15.5 (a) $[H_3O^+] = 10^{-7.90} = 1.3 \times 10^{-8}$ M

(b) A pH of 7.90 is basic.

15.6 (a) $HN_3(aq) \rightleftharpoons H^+(aq) + N_3^-(aq)$ $\quad K_a = \dfrac{[H^+][N_3^-]}{[HN_3]}$

(b) $HCOOH(aq) \rightleftharpoons H^+(aq) + HCOO^-(aq)$

$$K_a = \frac{[H^+][HCOO^-]}{[HCOOH]}$$

(c) $HClO_2(aq) \rightleftharpoons H^+(aq) + ClO_2^-(aq)$

$$K_a = \frac{[H^+][ClO_2^-]}{[HClO_2]}$$

15.7 (a) $CH_3NH_2(aq) + H_2O(\ell) \rightleftharpoons CH_3NH_3^+(aq) + OH^-(aq)$

$$K_b = \frac{[CH_3NH_3^+][OH^-]}{[CH_3NH_2]}$$

(b) $PH_3(aq) + H_2O(\ell) \rightleftharpoons PH_4^+(aq) + OH^-(aq)$

$$K_b = \frac{[PH_4^+][OH^-]}{[PH_3]}$$

(c) $NO_2^-(aq) + H_2O(\ell) \rightleftharpoons HNO_2(aq) + OH^-(aq)$

$$K_b = \frac{[HNO_2][OH^-]}{[NO_2^-]}$$

15.8 Setting up a small table for lactic acid, HLa:

	HLa + H_2O $\rightleftharpoons$	H_3O^+ +	La^-
Initial concentration (mol/L)	0.10	10^{-7}	0
Concentration change due to reaction (mol/L)	$-x$	$+x$	$+x$
Equilibrium concentration (mol/L)	$0.10 - x$	x	x

But $x = [H_3O^+] = 10^{-\text{pH}} = 10^{-2.43} = 3.7 \times 10^{-3}$. Substituting in the K_a expression,

$$K_a = \frac{[H_3O^+][La^-]}{[HLa]} = \frac{(3.7 \times 10^{-3})^2}{0.10 - (3.7 \times 10^{-3})}$$
$$= \frac{1.4 \times 10^{-5}}{0.1} = 1.4 \times 10^{-4}$$

Lactic acid is a stronger acid than propionic acid, with a K_a of 1.4×10^{-5}.

15.9 (a) Using the same methods as shown in the example,

$$\frac{x^2}{0.015} = 1.9 \times 10^{-5}$$

Solving for x, which is $[H_3O^+]$, we get

$$x = \sqrt{(1.9 \times 10^{-5})(0.015)} = 5.3 \times 10^{-4} = [H_3O^+].$$

So the pH of this solution is $-\log(5.3 \times 10^{-4}) = 3.28$.

(b) $$\%\ \text{ionization} = \frac{[H_3O^+]}{[HN_3]_{\text{initial}}} \times 100\% = \frac{5.4 \times 10^{-4}}{0.015} \times 100\% = 3.6\%$$

15.10 In such cases use the K_b expression and value to calculate $[OH^-]$ and then pOH from $[OH^-]$. Calculate pH from 14 − pOH.

$$C_6H_{11}NH_2(aq) + H_2O(\ell) \rightleftharpoons C_6H_{11}NH_3^+(aq) + OH^-(aq)$$
$$K_b = \frac{[C_6H_{11}NH_3^+][OH^-]}{[C_6H_{11}NH_2]} = 4.6 \times 10^{-4}$$
$$K_b = \frac{[C_6H_{11}NH_3^+][OH^-]}{[C_6H_{11}NH_2]} = \frac{x^2}{0.015 - x} = 4.6 \times 10^{-4}$$
$$x^2 = (0.015 - x)(4.6 \times 10^{-4}) = (6.9 \times 10^{-6}) - (4.6 \times 10^{-4}x)$$

Solve the quadratic equation for x.

$$x = \frac{4.8 \times 10^{-3}}{2} = 2.4 \times 10^{-3} = [OH^-]$$
$$\text{pOH} = -\log(2.4 \times 10^{-3}) = 2.62; \quad \text{pH} = 14.00 - 2.62 = 11.38$$

15.11 Using the same methods as those used in the example, letting $x = [OH^-]$ and $[HCO_3^-]$, and using the value of 2.1×10^{-4} for K_b for CO_3^{2-}, we get

$$\frac{x^2}{1.0} = 2.1 \times 10^{-4} \qquad x = \sqrt{2.1 \times 10^{-4}} = 1.45 \times 10^{-2}$$

pOH = 1.84 and pH = 12.16

15.12 NH_4Cl dissolves by dissociating into NH_4^+ and Cl^- ions. The ammonium ions react with water to produce an acidic solution.

$$NH_4^-(aq) - H_2O(\ell) \rightleftharpoons H_3O^-(aq) + NH_3(aq)$$

$$\text{The } K_a \text{ of } NH_4^+ = \frac{1.0 \times 10^{-14}}{1.8 \times 10^{-5}} = \frac{K_w}{K_b};$$
$$K_a = 5.6 \times 10^{-10} = \frac{[H_3O^+][NH_3]}{[NH_4^+]}$$

	NH_4^+ + H_2O $\rightleftharpoons$	H_3O^+ +	NH_3
Initial	0.10	0	0
Change	$-x$	$+x$	$+x$
Equilibrium	$0.10 - x$	x	x

$$5.6 \times 10^{-10} = \frac{(x)(x)}{(0.10 - x)}$$

Assume $0.10 - x \approx 0.10$ because K_a is so small. Thus, $(5.6 \times 10^{-10})(0.10) = x^2$; $x = [H_3O^+] = 7.48 \times 10^{-6}$ M.

$\text{pH} = -\log[H_3O^+] = -\log(7.48 \times 10^{-6}) = 5.16$

$$NH_4Cl(s) \rightleftharpoons NH_4^+(aq) + Cl^-(aq)$$

Chapter 16

16.1 $$K_a = \frac{[H^+][HCO_3^-]}{[H_2CO_3]} = \frac{H^+(0.025)}{(0.0020)} = [H^+] \times 12.5$$
$$= 4.2 \times 10^{-7}$$
$$[H^+] = \frac{4.2 \times 10^{-7}}{12.5} = 3.4 \times 10^{-8}$$
$$\text{pH} = -\log(3.4 \times 10^{-8}) = 7.47$$

16.2 $7.40 = 7.21 + \log(\text{ratio}) = 7.21 + \log\dfrac{[HPO_4^{2-}]}{[H_2PO_4^-]}$

$$\log\frac{[HPO_4^{2-}]}{[H_2PO_4^-]} = 7.40 - 7.21 = 0.19$$

$$\frac{[HPO_4^{2-}]}{[H_2PO_4^-]} = 10^{0.19} = 1.5$$

Therefore, $[HPO_4^{2-}] = 1.5 \times [H_2PO_4^-]$.

16.3 (a) Lactic acid-lactate (b) Acetic acid-acetate
(c) Hypochlorous acid-hypochlorite or $H_2PO_4^-$-HPO_4^{2-}
(d) CO_3^{2-}-HCO_3^-

16.4 (a) 0.075 mol HCl converts 0.075 mol of lactate to lactic acid (0.075 mol).

$$pH = 3.85 + \log\frac{(0.20 - 0.075)}{(0.15 + 0.075)} = 3.85 + \log\frac{(0.125)}{(0.225)}$$

$$= 3.85 + \log(0.556) = 3.85 + (-0.25) = 3.60$$

(b) 0.025 mol NaOH converts 0.025 mol of lactic acid to 0.025 mol of lactate.

$$pH = 3.85 + \log\frac{(0.20 + 0.025)}{(0.15 - 0.025)} = 3.85 + \log\frac{(0.225)}{(0.125)}$$

$$= 3.85 + \log(1.8) = 3.85 + (0.26) = 4.11$$

16.5 The buffer capacity will be exceeded when just over 0.25 mol KOH is added, which will have reacted with the 0.25 mol $H_2PO_4^-$.

$0.25\text{ mol OH}^- = 0.25\text{ mol KOH} \times \dfrac{56\text{ g KOH}}{1\text{ mol KOH}} = 14\text{ g}$. Thus, slightly more than 14 g KOH will exceed the buffer capacity.

16.6 (a) $[H_3O^+] = \dfrac{(5.00 \times 10^{-3}) - (1.00 \times 10^{-3})}{0.0500 + 0.0100}$

$$= \frac{4.00 \times 10^{-3}}{0.0600} = 6.67 \times 10^{-2}\text{ M}$$

$$pH = 1.176 = 1.18$$

(b) $[H_3O^+] = \dfrac{(5.00 \times 10^{-3}) - 0.00250}{0.0500 + 0.0250}$

$$= \frac{2.50 \times 10^{-3}}{0.0750} = 3.33 \times 10^{-2}\text{ M}$$

$$pH = -\log(3.33 \times 10^{-2}) = 1.48$$

(c) $[H_3O^+] = \dfrac{(5.00 \times 10^{-3}) - 0.00450}{0.0500 + 0.0450}$

$$= \frac{5.00 \times 10^{-4}}{0.0950} = 5.26 \times 10^{-3}$$

$$pH = -\log(5.26 \times 10^{-3}) = 2.28$$

(d) $[OH^-] = \dfrac{0.05 \times 10^{-3}\text{ mol}}{0.0500\text{ L} + 0.0505\text{ L}}$

$$= 5.0 \times 10^{-4}\text{ mol/L}$$

$$pOH = -\log(5.0 \times 10^{-4}) = 3.30$$

$$pH = 14.00 - 3.30 = 10.70$$

16.7 (a) Adding 10.0 mL of 0.100 M NaOH is adding (0.100 mol/L) (0.0100 L) = 0.00100 mol OH^-, which neutralizes 0.00100 mol acetic acid, converting it to 0.00100 mol acetate ion.

$$pH = pH + \log\frac{[\text{acetate}]}{[\text{acetic acid}]}$$

$$= 4.74 + \log\frac{(0.00100/0.0600)}{(0.00400/0.0600)}$$

$$= 4.74 + \log(0.25) = 4.74 + (-0.602) = 4.14$$

(b) $pH = 4.74 + \log\dfrac{(0.00250/0.0750)}{(0.00250/0.0750)}$

$$= 4.74 + \log(1) = 4.74$$

(c) $pH = 4.74 + \log\dfrac{(0.00450/0.0950)}{(0.00050/0.0950)}$

$$= 4.74 + \log(9) = 4.74 + 0.95 = 5.70$$

(d) $[OH^-] = \dfrac{0.10 \times 10^{-3}\text{ mol}}{0.0500\text{ L} + 0.0510\text{ L}}$

$$= 9.9 \times 10^{-4}\text{ mol/L}$$

$$pOH = -\log(9.9 \times 10^{-4}) = 3.00$$

$$pH = 14.00 - 3.00 = 11.00$$

16.8 (a) $K_{sp} = [Cu^+][Br^-]$ (b) $K_{sp} = [Hg^{2+}][I^-]^2$
(c) $K_{sp} = [Sr^{2+}][SO_4^{2-}]$

16.9 $AgBr(s) \rightleftharpoons Ag^+(aq) + Br^-(aq)$
$K_{sp} = [Ag^+][Br^-] = 5 \times 10^{-10} = x^2$; x = solubility
$x = \sqrt{5 \times 10^{-10}} \cong 2 \times 10^{-5}$

16.10 $Ag_2C_2O_4(s) \rightleftharpoons 2\,Ag^+(aq) + C_2O_4^{2-}(aq)$
$K_{sp} = [Ag^+]^2[C_2O_4^{2-}]$
$K_{sp} = [Ag^+]^2[C_2O_4^{2-}] = (1.4 \times 10^{-4})^2(6.9 \times 10^{-5})$
$= 1.4 \times 10^{-12}$

16.11 (a) $PbCl_2(s) \rightleftharpoons Pb^{2+}(aq) + 2\,Cl^-(aq)$
$K_{sp} = [Pb^{2+}][Cl^-]^2 = 1.7 \times 10^{-5}$. Let S equal the solubility of lead chloride, which equals $[Pb^{2+}]$

$$K_{sp} = 1.7 \times 10^{-5} = (S)(2S)^2 = 4S^3;\ S = \sqrt[3]{\frac{1.7 \times 10^{-5}}{4}}$$

$$= (4.25 \times 10^{-6})^{1/3} = 1.6 \times 10^{-2}\text{ mol/L}$$

(b) Let solubility of $PbCl_2 = [Pb^{2+}] = S$; $[Cl^-] = 0.20$ M

$[Pb^{2+}] = S = \dfrac{1.7 \times 10^{-5}}{(0.20)^2} = 4.3 \times 10^{-4}$ mol/L This is less than that in pure water due to the common ion effect of the presence of chloride ion.

16.12 $AgCl(s) \rightleftharpoons Ag^+(aq) + Cl^-(aq)$ $K_{sp} = 1.8 \times 10^{-10}$
$Ag^+(aq) + 2\,S_2O_3^{2-}(aq) \rightleftharpoons [Ag(S_2O_3)_2]^{3-}(aq)$ $K_f = 2.0 \times 10^{13}$

Net reaction:
$AgCl(s) + 2\,S_2O_3^{2-}(aq) \rightleftharpoons [Ag(S_2O_3)_2]^{3-}(aq) + Cl^-(aq)$

Therefore, the equilibrium constant for the net reaction is the product of $K_{sp} \times K_f$: $K_{net} = K_{sp} \times K_f = (1.8 \times 10^{-10})(2.0 \times 10^{13}) = 3.6 \times 10^3$. Because K_{net} is much greater than 1, the net reaction is product-favored, and AgCl is much more soluble in a $Na_2S_2O_3$ solution than it is in water.

16.13 AgCl will precipitate first. $[Cl^-]$ needed to precipitate AgCl:

$$[Cl^-] = \frac{1.8 \times 10^{-10}}{1.0 \times 10^{-2}} = 1.8 \times 10^{-8}\text{ M}$$

$[Cl^-]$ needed to precipitate $PbCl_2$:

$$[Cl^-] = \sqrt{\frac{1.7 \times 10^{-5}}{1.0 \times 10^{-1}}} = 1.3 \times 10^{-2}\text{ M}$$

Chapter 17

17.1 $\Delta S = q_{rev}/T = (30.8 \times 10^3\text{ J})/(273.15 + 45.3)\text{ K}$

$$= (30.8 \times 10^3\text{ J})/(318.45\text{ K})$$

$$\Delta S = 96.7\text{ J/K}$$

17.2 (a) C(g) has higher $S°$, 158.096 J K^{-1} mol^{-1}, versus 5.740 J K^{-1} mol^{-1} for C (graphite).
(b) Ar(g) has higher $S°$, 154.7 J K^{-1} mol^{-1}, versus 41.42 J K^{-1} mol^{-1} for Ca(s).

(c) KOH(aq) has higher $S°$, 91.6 J K^{-1} mol^{-1}, versus 78.9 J K^{-1} mol^{-1} for KOH(s).

17.3 (a) $\Delta S° = 2 \text{ mol CO(g)} \times S°(\text{CO(g)}) + 1 \text{ mol } O_2(g) \times S°(O_2(g)) - 2 \text{ mol } CO_2(g) \times S°(CO_2(g))$
$= \{2 \times (197.674) + (205.138) - 2 \times (213.74)\}$ J/K
$= 173.01$ J/K

(b) $\Delta S° = 1 \text{ mol NaCl(aq)} \times S°(\text{NaCl(aq)}) - 1 \text{ mol NaCl(s)} \times S°(\text{NaCl(s)})$
$= (115.5 - 72.13)$ J/K $= 43.4$ J/K

(c) $\Delta S° = 1 \text{ mol MgO(s)} \times S°(\text{MgO(s)}) + 1 \text{ mol } CO_2(g) \times S°(CO_2(g)) - 1 \text{ mol } MgCO_3(s) \times S°(MgCO_3(s))$
$= (26.94 + 213.74 - 65.7)$ J/K
$= 175.0$ J/K

17.4 $N_2(g) + 3\,H_2(g) \rightarrow 2\,NH_3(g)$

$$\Delta H° = 2 \text{ mol } NH_3 \times \Delta H_f°(NH_3(g)) = 2(-46.11) \text{ kJ} = -92.22 \text{ kJ}$$

$$\Delta S° = 2 \text{ mol } NH_3 \times S°(NH_3(g)) - 1 \text{ mol } N_2(g) \times S°(N_2(g)) - 3 \text{ mol } H_2 \times S°(H_2(g))$$
$$= 2(192.45) \text{ J/K} - (191.61) \text{ J/K} - 3(130.684) \text{ J/K}$$
$$= -198.76 \text{ J/K}$$

$$\Delta S°_{universe} = \frac{-\Delta H°}{T} + \Delta S° = \frac{92.2 \text{ kJ}}{298.15 \text{ K}} + (-198.76 \text{ J/K})$$
$$= \frac{92{,}200 \text{ J}}{298.15 \text{ K}} - 198.76 \text{ J/K} = 110.5 \text{ J/K}$$

The process is product-favored.

17.5 (a) $\Delta H° = \{(-238.66) - (-110.525)\}$ kJ $= -128.14$ kJ
$\Delta S° = \{(126.8) - 197.674 - 2 \times 130.684\}$ J/K
$= -332.2$ J/K
$\Delta G° = \Delta H° = T\Delta S°$
$= -128.14 \times 10^3$ J $- 298.15$ K $\times (-332.2$ J/K)
$= -29.09 \times 10^3$ J $= -29.09$ kJ

(b) $\Delta G° = [-166.27 - (-137.168)]$ kJ $= -29.10$ kJ. The two results agree.

(c) $\Delta G°$ is negative. The reaction is product-favored at 298.15 K. Because $\Delta S°$ is negative, at very high temperatures the reaction will become reactant-favored.

17.6 (a) $T = \Delta H°/\Delta S° = (-565{,}968 \text{ J})/(-173.01 \text{ J/K}) = 3271$ K

(b) The reaction is exothermic and therefore is product-favored at temperatures lower than 3271 K.

17.7 (a) $\Delta G° = \{-553.04 - 527.81 - (-1128.79)\}$ kJ
$= 47.94$ kJ
$K° = e^{-\Delta G°/RT} = e^{-(47.94 \text{ kJ/mol})/(8.314 \text{ J K}^{-1} \text{ mol}^{-1})(298 \text{ K})}$
$= e^{-(47{,}940 \text{ J/mol})/(8.314 \text{ J K}^{-1} \text{ mol}^{-1})(298 \text{ K})}$
$= e^{-19.35} = 3.9 \times 10^{-9}$ (close to K_c)

(b) $K° = e^{-14.68} = 4.2 \times 10^{-7}$ (agrees with K_c)

(c) $K° = e^{-(-1.909)} = 6.75$ (agrees with K_P)

For reactions (a) and (b), $K_c = K°$.

17.8 (a) At 298 K,
$\Delta G° = 2 \times (-16.45)$ kJ $= -32.9$ kJ
$K° = e^{-(-32{,}900 \text{ J})/(8.314 \text{ J K}^{-1} \text{ mol}^{-1})(298 \text{ K})} = e^{13.28} = 5.8 \times 10^5$

(b) At 450. K,
$\Delta G° = \Delta H° - T\Delta S° = -92.22$ kJ $- (450.)(-0.19876)$ kJ
$= -2.78$ kJ
$K° = e^{-(-2780 \text{ J})/(8.314 \text{ J K}^{-1})(450. \text{ K})} = 2.10$

(c) At 800. K,
$\Delta G° = -92.22$ kJ $- (800.)(-0.19876)$ kJ $= 66.79$ kJ
$K° = e^{-(66{,}790 \text{ J})/(8.314 \text{ J K}^{-1} \text{ mol}^{-1})(800. \text{ K})} = 4.3 \times 10^{-5}$

17.9 (a) $\Delta G° = 2(-137.168)$ kJ $- 2(-394.359)$ kJ $= 514.382$ kJ. The reaction is reactant-favored, and at least 514.382 kJ of work must be done to make it occur.

(b) $\Delta G° = 2(-742.2)$ kJ $= -1484.4$ kJ. The reaction is product-favored and could do up to 1484.4 kJ of useful work.

17.10 (a) $\Delta G_f°(\text{MgO(s)}) = -569.43$ kJ, so formation of MgO(s) is product-favored and MgO(s) is thermodynamically stable.

(b) $\Delta G_f°(N_2H_4(\ell)) = 149.34$ kJ; kinetically stable.

(c) $\Delta G_f°(C_2H_6(g)) = -32.82$ kJ; thermodynamically stable.

(d) $\Delta G_f°(N_2O(g)) = 104.20$ kJ; kinetically stable.

Chapter 18

18.1 Reducing agents are indicated by "red" and oxidizing agents are indicated by "ox." Oxidation numbers are shown above the symbols for the elements.

(a) $2\,\underset{red}{\overset{0}{Fe}}(s) + 3\,\underset{ox}{\overset{0}{Cl_2}}(g) \rightarrow 2\,\overset{+3}{Fe}\overset{-1}{Cl_3}(s)$

(b) $2\underset{red}{\overset{0}{H_2}}(g) + \underset{ox}{\overset{0}{O_2}}(g) \rightarrow 2\,\overset{+1}{H_2}\overset{-2}{O}(\ell)$

(c) $\underset{red}{\overset{0}{Cu}}(S) + 2\,\underset{ox}{\overset{+5}{N}\overset{-2}{O_3^-}}(aq) + 4\,\overset{+1}{H_3}\overset{-2}{O^+}(aq) \rightarrow \overset{+2}{Cu^{2+}}(aq) + 2\,\overset{+4}{N}\overset{-2}{O_2}(g) + 6\,\overset{+1}{H_2}\overset{-2}{O}(\ell)$

(d) $\underset{red}{\overset{0}{C}}(s) + \underset{ox}{\overset{0}{O_2}}(g) \rightarrow \overset{+4}{C}\overset{-2}{O_2}(g)$

(e) $6\,\underset{red}{\overset{+2}{Fe^{2+}}}(aq) + \underset{ox}{\overset{+6}{Cr_2}\overset{-2}{O_7^{2-}}}(aq) + 14\,\overset{+1}{H_3}\overset{-2}{O^+}(aq) \rightarrow 6\,\overset{+3}{Fe^{3+}}(aq) + 2\,\overset{+3}{Cr^{3+}}(aq) + 21\,\overset{+1}{H_2}\overset{-2}{O}(\ell)$

18.2 (a)
Ox: $Cd(s) \rightarrow Cd^{2+}(aq) + 2\,e^-$
Red: $Cu^{2+}(aq) + 2\,e^- \rightarrow Cu(s)$
Net: $Cd(s) + Cu^{2+}(aq) \rightarrow Cd^{2+}(aq) + Cu(s)$

(b)
Ox: $Zn(s) \rightarrow Zn^{2+}(aq) + 2\,e^-$
Red: $2\,H_3O^+(aq) + 2\,e^- \rightarrow H_2(g) + 2\,H_2O(\ell)$
Net: $Zn(s) + 2\,H_3O^+(aq) \rightarrow Zn^{2+}(aq) + H_2(g) + 2\,H_2O(\ell)$

(c)
Ox: $2\,Al(s) \rightarrow 2\,Al^{3+}(aq) + 6\,e^-$
Red: $3\,Zn^{2+}(aq) + 6\,e^- \rightarrow 3\,Zn(s)$
Net: $2\,Al(s) + 3\,Zn^{2+}(aq) \rightarrow 2\,Al^{3+}(aq) + 3\,Zn(s)$

18.3 ***Step 1.*** This is an oxidation-reduction reaction. It is obvious that Zn is oxidized by its change in oxidation state.

Step 2. The half-reactions are

$Zn(s) \longrightarrow Zn^{2+}(aq)$ (This is the oxidation reaction.)

$Cr_2O_7^{2-}(aq) \longrightarrow 2\,Cr^{3+}(aq)$ (This is the reduction reaction.)

Step 3. Balance the atoms in the half-reactions. The atoms are balanced in the Zn half-reaction. We need to add water and H^+ in the $Cr_2O_7^{2-}$ half-reaction. Fourteen H^+ ions are required on the right to combine with the seven O atoms.

$$Cr_2O_7^{2-}(aq) + 14\,H^+(aq) \longrightarrow 2\,Cr^{3+}(aq) + 7\,H_2O(\ell)$$

Step 4. Balance the half-reactions for charge. Write the Zn half-reaction as

$$Zn(s) \longrightarrow Zn^{2+}(aq) + 2\,e^-$$

and write the $Cr_2O_7^{2-}$ half-reaction as

$$Cr_2O_7^{2-}(aq) + 14\,H^+(aq) + 6\,e^- \longrightarrow 2\,Cr^{3+}(aq) + 7\,H_2O(\ell)$$

Step 5. Multiply the half-reactions by factors to make the number of electrons gained equal to the number lost.

$$3\,[Zn(s) \longrightarrow Zn^{2+}(aq) + 2\,e^-]$$
$$1\,[Cr_2O_7^{2-}(aq) + 14\,H^+(aq) + 6\,e^- \longrightarrow 2\,Cr^{3+}(aq) + 7\,H_2O(\ell)]$$

Step 6. Add the two half-reactions, canceling the electrons.

$$3\ Zn(s) \longrightarrow 3\ Zn^{2+}(aq) + 6\ e^-$$

$$Cr_2O_7^{2-}(aq) + 14\ H^+(aq) + 6\ e^- \longrightarrow 2\ Cr^{3+}(aq) + 7\ H_2O(\ell)$$

$$Cr_2O_7^{2-}(aq) + 3\ Zn(s) + 14\ H^+(aq) \longrightarrow 2\ Cr^{3+}(aq) + 3\ Zn^{2+}(aq) + 7\ H_2O(\ell)$$

Step 7. Everything checks.
Step 8. Water was added in Step 3. The balanced equation is

$$Cr_2O_7^{2-}(aq) + 3\ Zn(s) + 14\ H_3O^+(aq) \longrightarrow 2\ Cr^{3+}(aq) + 3\ Zn^{2+}(aq) + 21\ H_2O(\ell)$$

18.4 ***Step 1.*** This is an oxidation-reduction reaction. The wording of the question says Al reduces NO_3^- ion. Al is oxidized.
Step 2. The half-reactions are:

$Al(s) \longrightarrow Al(OH)_4^-(aq)$ (This is the oxidation reaction.)

$NO_3^-(aq) \longrightarrow NH_3(aq)$ (This is the reduction reaction.)

Step 3. Balance the atoms in the half-reactions. For the Al half-reaction, add four H^+ ions on the right and four water molecules on the left.

$$Al(s) + 4\ H_2O(\ell) \longrightarrow Al(OH)_4^- + 4\ H^+(aq)$$

For the NO_3^- half-reaction,

$$NO_3^-(aq) + 9\ H^+(aq) \longrightarrow NH_3(aq) + 3\ H_2O(\ell)$$

Step 4. Balance the half-reactions for charge. Put 3 e^- on the right in the Al half-reaction

$$Al(s) + 4\ H_2O(\ell) \longrightarrow Al(OH)_4^- + 4\ H^+(aq) + 3\ e^-$$

and put 8 e^- on the left side of the NO_3^- half-reaction.

$$NO_3^-(aq) + 9\ H^+(aq) + 8\ e^- \longrightarrow NH_3(aq) + 3\ H_2O(\ell)$$

Step 5. Multiply the half-reactions by factors to make the electrons gained equal to those lost.

$$8[Al(s) + 4\ H_2O(\ell) \longrightarrow Al(OH)_4^- + 4\ H^+(aq) + 3\ e^-]$$

$$3[NO_3^-(aq) + 9\ H^+(aq) + 8\ e^- \longrightarrow NH_3(aq) + 3\ H_2O(\ell)]$$

Step 6. Remove $H^+(aq)$ ions by adding an appropriate amount of OH^-. For the Al half-reaction, add 32 OH^- ions to get

$$8\ Al(s) + 32\ OH^-(aq) + 32\ H_2O(\ell) \longrightarrow 8\ Al(OH)_4^- + 32\ H_2O + 24\ e^-$$

For the NO_3^- half-reactions, add 27 OH^- ions to get

$$3\ NO_3^-(aq) + 27\ H_2O + 24\ e^- \longrightarrow 3\ NH_3(aq) + 9\ H_2O(\ell) + 27\ OH^-(aq)$$

Step 7. Add both half-reactions and cancel the electrons.

$$8\ Al(s) + 32\ OH^-(aq) + 32\ H_2O(\ell) \longrightarrow 8\ Al(OH)_4^- + 32\ H_2O + 24\ e^-$$

$$3\ NO_3^-(aq) + 27\ H_2O + 24\ e^- \longrightarrow 3\ NH_3(aq) + 9\ H_2O(\ell) + 27\ OH^-(aq)$$

$$3\ NO_3^-(aq) + 8\ Al(s) + 59\ H_2O(\ell) + 32\ OH^-(aq) \longrightarrow 8\ Al(OH)_4^-(aq) + 3\ NH_3(aq) + 27\ OH^-(aq) + 41\ H_2O(\ell)$$

Step 8. Make a final check. Since there are OH^- ions and water molecules on both sides of the equation, cancel them out. This gives the final balanced equation.

$$3\ NO_3^-(aq) + 8\ Al(s) + 18\ H_2O(\ell) + 5\ OH^-(aq) \longrightarrow 8\ Al(OH)_4^-(aq) + 3\ NH_3(aq)$$

(This is a fairly complicated equation to balance. If you balanced this one with a minimum of effort, your understanding of balancing redox equations is rather good. If you had to struggle with one or more of the steps, go back and repeat them.)

18.5 (a) $Ni(s) \rightarrow Ni^{2+}(aq) + 2\ e^-$ (This is the oxidation half-reaction.)

$2\ Ag^+(aq) + 2\ e^- \rightarrow 2\ Ag(s)$ (This is the reduction half-reaction.)

(b) The oxidation of Ni takes place at the anode and the reduction of Ag^+ takes place at the cathode.
(c) Electrons would flow through an external circuit from the anode (where Ni is oxidized) to the cathode (where Ag^+ ions are reduced).
(d) Nitrate ions would flow through the salt bridge to the anode compartment. Potassium ions would flow into the cathode compartment.

18.6 Oxidation half-reaction: $Fe(s) \rightarrow Fe^{2+}(aq, 1\ M) + 2\ e^-$ (anode)

Reduction half-reaction: $Cu^{2+}(aq, 1\ M) + 2\ e^- \rightarrow Cu(s)$ (cathode)

$$E^\circ_{cell} = +0.78\ V = E^\circ_{cathode} - E^\circ_{anode}$$

Since $E^\circ_{cathode} = +0.34$ V, E°_{anode} must be -0.44 V.

18.7 $F_2(g) + 2\ e^- \longrightarrow 2\ F^-(aq)$ $\quad E^\circ_{cathode} = +2.87$ V

$2\ Li(s) \longrightarrow 2\ Li^+(aq) + 2\ e^-$ $\quad E^\circ_{anode} = -3.045$ V

$2\ Li(s) + F_2(g) \longrightarrow 2\ Li^+(aq) + 2\ F^-(aq)$ $\quad E^\circ_{cell} = +5.91$ V

$$E^\circ_{cell} = E^\circ_{cathode} - E^\circ_{anode} = +2.87 - (-3.045) = +5.91\ V$$

18.8 The two half-reactions are

$Hg^{2+}(aq) + 2\ e^- \longrightarrow 2\ Hg(\ell)$ $\quad E^\circ_{cathode} = +0.855$ V

$2\ I^-(aq) \longrightarrow I_2(s) + 2\ e^-$ $\quad E^\circ_{anode} = +0.535$ V

$$E^\circ_{cell} = E^\circ_{cathode} - E^\circ_{anode} = +0.855 - 0.535 = +0.320\ V$$

The reaction is product-favored as written.

18.9 We first need to calculate E°_{cell}, and to do this we break the reaction into two half-reactions.

Ox: $Sn^{2+}(aq) \longrightarrow Sn^{4+}(aq) + 2\ e^-$ $\quad E^\circ_{anode} = +0.15$ V

Red: $I_2(s) + 2\ e^- \longrightarrow 2\ I^-(aq)$ $\quad E^\circ_{cathode} = +0.535$ V

$I_2(s) + Sn^{2+}(aq) \longrightarrow 2\ I^-(aq) + Sn^{4+}(aq)$ $\quad E^\circ_{cell} = +0.385$ V

$$E^\circ_{cell} = E^\circ_{cathode} - E^\circ_{anode} = +0.535 - 0.15 = +0.385\ V$$

E°_{cell} is 0.385 V, and 2 mol of electrons are transferred.

$$\log K = \frac{nE^\circ\ V}{0.0592\ V} = \frac{2 \times 0.385\ V}{0.0592\ V}$$

$$= 13.00 \text{ and } K = 1 \times 10^{13}$$

The large value of K indicates that the reaction is strongly product-favored as written.

18.10 $$E_{cell} = 0.51\ V - \left(\frac{0.0592\ V}{2} \times \log\frac{3}{0.010}\right)$$

$$= 0.51\ V - (0.0296\ V \times \log(300))$$
$$= 0.51\ V - 0.073\ V = +0.44\ V$$

18.11 (a) The net cell reaction would be

$$2\ Na^+ + 2\ Br^- \longrightarrow 2\ Na + Br_2$$

Sodium ions would be reduced at the cathode and bromide ions would be oxidized at the anode.
(b) H_2 would be produced at the cathode for the same reasons given in Problem-Solving Example 18.8. That reaction is

$$2\ H_2O(\ell) + 2\ e^- \longrightarrow H_2(g) + 2\ OH^-(aq)$$

At the anode, two reactions are possible: the oxidation of water and the oxidation of Br^- ions.

$6\ H_2O(\ell) \longrightarrow O_2(g) + 4\ H_3O^+(aq) + 4\ e^-$ $\quad E^\circ = 1.229$ V

$2\ Br^-(aq) \longrightarrow Br_2(\ell) + 2\ e^-$ $\quad E^\circ = 1.08$ V

Bromide ions will be oxidized to Br_2 because that potential is smaller. The net cell reaction is

$$2\,H_2O(\ell) + 2\,Br^-(aq) \longrightarrow Br_2(\ell) + H_2(g) + 2\,OH^-(aq)$$

(c) Sn metal will be formed at the cathode because its reduction potential (−0.14 V) is less negative than the potential for the reduction of water. O_2 will form at the anode because the $E°$ value for the oxidation of water is smaller than the $E°$ value for the oxidation of Cl^-. The net cell reaction is

$$2\,Sn^{2+}(aq) + 6\,H_2O(\ell) \longrightarrow 2\,Sn(s) + O_2(g) + 4\,H_3O^+(aq)$$

18.12 First, calculate the quantity of charge:

$$\text{Charge} = (25 \times 10^3\ \text{A})(1\ \text{h})\left(\frac{60\ \text{s}}{1\ \text{min}}\right)\left(\frac{60\ \text{min}}{1\ \text{h}}\right) = 9.0 \times 10^7\ \text{A} \cdot \text{s} = 9.0 \times 10^7\ \text{C}$$

Then calculate the mass of Na:

$$\text{Mass of Na} = (9.0 \times 10^7\ \text{C}) \times \left(\frac{1\ \text{mol e}^-}{96{,}500\ \text{C}}\right)\left(\frac{1\ \text{mol Na}}{1\ \text{mol e}^-}\right)\left(\frac{22.99\ \text{g Na}}{1\ \text{mol Na}}\right) = 2.1 \times 10^4\ \text{g Na}$$

Chapter 19

19.1 (a) $^{237}_{93}Np \rightarrow\ ^{4}_{2}He +\ ^{233}_{91}Pa$ (b) $^{35}_{16}S \rightarrow\ ^{0}_{-1}e +\ ^{35}_{17}Cl$

19.2 (a) $^{11}_{6}C \rightarrow\ ^{11}_{5}B +\ ^{0}_{1}e$ (b) $^{35}_{16}S \rightarrow\ ^{35}_{17}Cl +\ ^{0}_{-1}e$
(c) $^{30}_{15}P \rightarrow\ ^{0}_{+1}e +\ ^{30}_{14}Si$ (d) $^{22}_{11}Na \rightarrow\ ^{0}_{-1}e +\ ^{22}_{12}Mg$

19.3 (a) $^{42}_{19}K \rightarrow\ ^{0}_{-1}e +\ ^{42}_{20}Ca$ (b) $^{234}_{92}U \rightarrow\ ^{4}_{2}He +\ ^{230}_{90}Th$
(c) $^{20}_{9}F \rightarrow\ ^{0}_{-1}e +\ ^{20}_{10}Ne$

19.4 (a) $^{90}_{38}Sr \rightarrow\ ^{0}_{-1}e +\ ^{90}_{39}Y$
(b) It takes 4 half-lives (4 × 29 y = 116 y) for the activity to decrease to 125 beta particles emitted per minute:

Number of Half-lives	Change of Activity	Total Elapsed Time (y)
1	2000 to 1000	29
2	1000 to 500	58
3	500 to 250	87
4	250 to 125	116

19.5 (a) $t_{1/2} = \dfrac{0.693}{9.3 \times 10^{-3}\ \text{d}^{-1}} = 75\ \text{d}$

(b) ln(fraction remaining) $= -k \times t = -(9.3 \times 10^{-3}\ \text{d}^{-1}) \times (100\ \text{d}) = -0.930$
Fraction of iridium-192 remaining $= e^{-0.930} = 0.39$. Therefore, 39% of the original iridium-192 remains.

19.6 $\ln(0.60) = -0.510 = -k \times t$

$$k = \frac{0.693}{t_{1/2}} = \frac{0.693}{12.3\ \text{y}} = 0.0563\ \text{y}^{-1}$$

$$t = \frac{-0.510}{-0.0563\ \text{y}^{-1}} = 9.1\ \text{y}$$

Answers to Exercises

Chapter 1

1.1 (a) These temperatures can be compared to the boiling point of water, 212 °F or 100 °C. So 110 °C is a higher temperature than 180 °F.
(b) These temperatures can be compared to normal body temperature, 98.6 °F or 37.0 °C. So 36 °C is a lower temperature than 100 °F.
(c) This temperature can be compared to normal body temperature, 37.0 °C. Since body temperature is above the melting point, gallium held in one's hand will melt.

1.2 Reference to the figure on page 7 indicates that kerosene is the top layer, vegetable oil is the middle layer, and water is the bottom layer.
(a) Since the least dense liquid will be the top layer and the densest liquid will be the bottom layer, the densities increase in the order kerosene, vegetable oil, water.
(b) If vegetable oil is added to the tube, the top and bottom layers will remain the same, but the middle layer will become larger.
(c) If kerosene is now added to the tube the top layer will grow, but the middle and bottom layers will remain the same. The order of levels will *not change*. Density does not depend on the quantity of material present. So no matter how much of each liquid is present, the densities increase in the order kerosene, vegetable oil, water.

1.3 (a) Properties: blue (qualitative), melts at 99 °C (quantitative)
Change: melting
(b) Properties: white, cubic (both qualitative)
Change: none
(c) Properties: mass of 0.123 g, melts at 327 °C (both quantitative)
Change: melting
(d) Properties: colorless, vaporizes easily (both qualitative), boils at 78 °C, density of 0.789 g/mL (both quantitative)
Changes: vaporizing, boiling

1.4 Physical change: boiling water
Chemical changes: combustion of propane, cooking the egg

1.5 (a) Homogeneous mixture (solution)
(b) Heterogeneous mixture (contains carbon dioxide gas bubbles in a solution of sugar and other substances in water)
(c) Heterogeneous mixture of dirt and oil
(d) Element; diamond is pure carbon.
(e) Modern quarters (since 1965) are composed of a pure copper core (that can be seen when they are viewed side-on) and an outer layer of 75% Cu, 25% Ni alloy, so they are heterogeneous matter. Pre-1965 quarters are fairly pure silver.
(f) Compound; contains carbon, hydrogen, and oxygen

1.6 (a) Energy from the sun warms the ice and the water molecules vibrate more; eventually they break away from their fixed positions in the solid and liquid water forms. As the temperature of the liquid increases, some of the molecules have enough energy to become widely separated from the other molecules, forming water vapor (gas).
(b) Some of the water molecules in the clothes have enough speed and energy to escape from the liquid state and become water vapor; these molecules are carried away from the clothes by breezes or air currents. Eventually nearly every water molecule in the clothes vaporizes, and the clothes become dry.
(c) Water molecules from the air come into contact with the cold glass, and their speeds are decreased, allowing them to become liquid. As more and more molecules enter the liquid state, droplets form on the glass.
(d) Some water molecules escape from the liquid state, forming water vapor. As more and more molecules escape, the ratio of sugar molecules to water molecules becomes larger and larger, and eventually some sugar molecules start to stick together. As more and more sugar molecules stick to each other, a visible crystal forms. Eventually all of the water molecules escape, leaving sugar crystals behind.

1.7 (a) Tellurium, Te, earth (Latin *tellus* means earth); uranium, U, for Uranus; neptunium, Np, for Neptune; and plutonium, Pu, for Pluto. (Mercury, like the planet Mercury, is named for a Roman god.)
(b) Californium, Cf
(c) Curium, Cm, for Marie Curie; and meitnerium, Mt, for Lise Meitner
(d) Scandium, Sc, for Scandinavia; gallium, Ga, for France (Latin *Gallia* means France); germanium, Ge, for Germany; ruthenium, Ru, for Russia; europium, Eu, for Europe; polonium, Po, for Poland; francium, Fr, for France; americium, Am, for America; californium, Cf, for California
(e) H, He, C, N, O, F, Ne, P, S, Cl, Ar, Se, Br, Kr, I, Xe, At, Rn

1.8 (a) Elements that consist of diatomic molecules are H, N, O, F, Cl, Br, and I; At is radioactive and there is probably less than 50 mg of naturally occurring At on earth, but it does form diatomic molecules; H, N, O, plus group 7A.
(b) Metalloids are B, Si, Ge, As, Sb, and Te; along a zig-zag line from B to Te.

1.9 Tin and lead are two different elements; allotropes are two different forms of the same element, so tin and lead are not allotropes.

Chapter 2

2.1 The movement of the comb though your hair removes some electrons, leaving slight charges on your hair and the comb. The charges must sum to zero; therefore, one must be slightly positive and one must be slightly negative, so they attract each other.

2.2 (a) A nucleus is about one ten-thousandth as large as an atom, so $100 \text{ m} \times (1 \times 10^{-4}) = 1 \times 10^{-2} \text{ m} = 1 \text{ cm}$. (b) Many everyday objects are about 1 cm in size—for example, a grape.

2.3 The statement is wrong because two atoms that are isotopes always have the same number of protons. It is the number of neutrons that varies from one isotope of an element to another.

2.4 Atomic weight of lithium
$= (0.07500)(6.015121 \text{ amu}) + (0.9250)(7.016003 \text{ amu})$
$= 0.451134 \text{ amu} + 6.489802 \text{ amu}$
$= 6.940936 \text{ amu, or } 6.941 \text{ amu}$

2.5 Because the most abundant isotope is magnesium-24 (78.70%), the atomic weight of magnesium is closer to 24 than to 25 or 26, the mass numbers of the other magnesium isotopes, which make up approximately 21% of the remaining mass. The simple arithmetic average is $(24 + 25 + 26)/3 = 25$, which is larger than the atomic weight. In the arithmetic average, the relative abundance of each magnesium isotope is 33%, far less than the actual percent abundance of magnesium-24, and much more than the natural percent abundances of magnesium-25 and magnesium-26.

2.6 There is no reasonable pair of values of the mass numbers for Ga that would have an average value of 69.72.

2.7 Start by calculating the number of moles in 10.00 g of each element.

$$10.00 \text{ g Li} \times \frac{1 \text{ mol Li}}{6.941 \text{ g Li}} = 1.441 \text{ mol Li}$$

$$10.00 \text{ g Ir} \times \frac{1 \text{ mol Ir}}{192.22 \text{ g Ir}} = 0.05202 \text{ mol Ir}$$

Multiply the number of moles of each element by Avogadro's number.

$$1.441 \text{ mol Li} \times 6.022 \times 10^{23} \text{ atoms/mol} = 8.678 \times 10^{23} \text{ atoms Li}$$

$$0.05202 \text{ mol Ir} \times 6.022 \times 10^{23} \text{ atoms/mol} = 3.133 \times 10^{22} \text{ atoms Ir}$$

Find the difference.
$(8.678 \times 10^{23}) - (0.3133 \times 10^{23})$
$= 8.365 \times 10^{23}$ more atoms of Li than Ir

Chapter 3

3.1 Propylene glycol structural formula:

```
    H  OH OH
    |  |  |
 H—C—C—C—H
    |  |  |
    H  H  H
```

Condensed formula:

```
   OHOH
   |  |
CH3CHCH2
```

Molecular formula: $C_3H_8O_2$

3.2 (a) CS_2 (b) PCl_3 (c) SBr_2 (d) SeO_2 (e) OF_2 (f) XeO_3

3.3 (a) $C_{16}H_{34}$ and $C_{28}H_{58}$ (b) $C_{14}H_{30}$, 14 carbon atoms and 30 hydrogen atoms

3.4 The structural and condensed formulas for three constitutional isomers of five-carbon alkanes (pentanes) are

```
    H  H  H  H  H
    |  |  |  |  |
 H—C—C—C—C—C—H      CH3CH2CH2CH2CH3
    |  |  |  |  |
    H  H  H  H  H

    H  H  H  H
    |  |  |  |
 H—C—C—C—C—H         CH3CH2CHCH3
    |  |  |  |                 |
    H  H  |  H                CH3
       H—C—H
          |
          H

          H
          |
       H—C—H
    H     |     H             CH3
    |     |     |              |
 H—C————C————C—H        CH3—C—CH3
    |     |     |              |
    H     |     H             CH3
       H—C—H
          |
          H
```

3.5 The compound is a solid at room temperature and is soluble in water, so it is likely to be an ionic compound.

3.6 (a) 174.18 g/mol (b) 386.66 g/mol (c) 398.07 g/mol (d) 194.19 g/mol

3.7 The statement is true. Because both compounds have the same formula, they have the same molar mass. Thus, 100 g of each compound contains the same number of moles.

3.8 Epsom salt is $MgSO_4 \cdot 7H_2O$, which has a molar mass of 246 g/mol.

$$20 \text{ g} \times \frac{1 \text{ mol}}{246 \text{ g}} = 8.1 \times 10^{-2} \text{ mol Epsom salt}$$

3.9 (a) SF_6 molar mass is 146.06 g/mol; 1.000 mol SF_6 contains 32.07 g S and 18.9984 × 6 = 113.99 g F. The mass percents are

$$\frac{32.07 \text{ g S}}{146.06 \text{ g SF}_6} \times 100\% = 21.96\% \text{ S}$$

$$100.00\% - 21.96\% = 78.04\% \text{ F}$$

(b) $C_{12}H_{22}O_{11}$ has a molar mass of 342.3 g/mol; 1.000 mol $C_{12}H_{22}O_{11}$ contains

$$12.011 \times 12 = 144.13 \text{ g C}$$

$$1.0079 \times 22 = 22.174 \text{ g H}$$

$$15.9994 \times 11 = 175.99 \text{ g O}$$

The mass percents of the three elements are

$$\frac{144.13 \text{ g C}}{342.3 \text{ g}} \times 100\% = 42.12\% \text{ C}$$

$$\frac{22.174 \text{ g H}}{342.3 \text{ g}} \times 100\% = 6.478\% \text{ H}$$

$$\frac{175.99 \text{ g O}}{342.3 \text{ g}} \times 100\% = 51.41\% \text{ O}$$

(c) $Al_2(SO_4)_3$ molar mass is 342.15 g/mol; 1.000 mol $Al_2(SO_4)_3$ contains

$$26.9815 \text{ g/mol} \times 2 \text{ mol} = 53.96 \text{ g Al}$$

$$32.066 \text{ g/mol} \times 3 \text{ mol} = 96.20 \text{ g S}$$

$$15.9994 \text{ g/mol} \times 12 \text{ mol} = 192.0 \text{ g O}$$

The mass percents of the three elements are

$$\frac{53.96 \text{ g Al}}{342.15 \text{ g}} \times 100\% = 15.77\% \text{ Al}$$

$$\frac{96.20 \text{ g S}}{342.15 \text{ g}} \times 100\% = 28.12\% \text{ S}$$

$$\frac{192.0 \text{ g O}}{342.15 \text{ g}} \times 100\% = 56.12\% \text{ O}$$

(d) $U(OTeF_5)_6$ molar mass is 1669.6 g/mol; 1.000 mol $U(OTeF_5)_6$ contains 238.0289 g of U and

$$15.9994 \text{ g/mol} \times 6 \text{ mol} = 96.00 \text{ g O}$$

$$127.60 \text{ g/mol} \times 6 \text{ mol} = 765.6 \text{ g Te}$$

$$18.9984 \text{ g/mol} \times 30 \text{ mol} = 570.0 \text{ g F}$$

The mass percents of the four elements are

$$\frac{238.0289 \text{ g U}}{1669.6 \text{ g}} \times 100\% = 14.256\% \text{ U}$$

$$\frac{96.00 \text{ g O}}{1669.6 \text{ g}} \times 100\% = 5.750\% \text{ O}$$

$$\frac{765.6 \text{ g Te}}{1669.6 \text{ g}} \times 100\% = 45.86\% \text{ Te}$$

$$\frac{570.0 \text{ g F}}{1669.6 \text{ g}} \times 100\% = 34.14\% \text{ F}$$

3.10 (a) Carbon, nitrogen, oxygen, phosphorus, hydrogen, selenium, sulfur, Cl, Br, I (b) Calcium and magnesium (c) Chloride, bromide, and iodide (d) Iron, copper, zinc, vanadium (also chromium, manganese, cobalt, nickel, molybdenum, and cadmium)

Chapter 4

4.1 One mol of methane reacts with 2 mol oxygen to produce 1 mol carbon dioxide and 2 mol water.

4.2 (a) The total mass of reactants {4 Fe(s) + 3 O_2(g)} must equal the total mass of products {2 Fe_2O_3(s)}, which is 2.50 g.
(b) The stoichiometric coefficients are 4, 3, and 2.
(c)
$$1.000 \times 10^4 \text{ O atoms} \times \frac{1 \text{ O}_2 \text{ molecule}}{2 \text{ O atoms}} \times \frac{4 \text{ Fe atoms}}{3 \text{ O}_2 \text{ molecules}} = 6.667 \times 10^3 \text{ Fe atoms}$$

4.3 (a) Not balanced; the number of oxygen atoms do not match.
(b) Not balanced; the number of bromine atoms do not match.
(c) Not balanced; the number of sulfur atoms do not match.

4.4 (a) To predict the product of a combination reaction between two elements, we need to know the ion that will be formed by each element when combined. (b) For calcium, Ca^{2+} ions are formed, and for fluorine, F^- ions are formed. (c) The product is CaF_2.

4.5 (a) Magnesium chloride, $MgCl_2$
(b) Magnesium oxide, MgO, and carbon dioxide, CO_2

4.6 $\frac{2 \text{ mol Al}}{3 \text{ mol Br}_2}, \frac{2 \text{ mol Al}}{1 \text{ mol Al}_2\text{Br}_6}, \frac{3 \text{ mol Br}_2}{1 \text{ mol Al}_2\text{Br}_6}$, and their reciprocals

4.7 $0.300 \text{ mol CH}_4 \times \frac{2 \text{ mol H}_2\text{O}}{1 \text{ mol CH}_4} = 0.600 \text{ mol H}_2\text{O}$

$0.600 \text{ mol H}_2\text{O} \times \frac{18.02 \text{ g H}_2\text{O}}{1 \text{ mol H}_2\text{O}} = 10.8 \text{ g H}_2\text{O}$

4.8 (a) $300. \text{ g urea} \times \frac{1 \text{ mol urea}}{60.06 \text{ g urea}} \times \frac{2 \text{ mol NH}_3}{1 \text{ mol urea}} \times \frac{17.03 \text{ g NH}_3}{1 \text{ mol NH}_3} = 170. \text{ g NH}_3$

$100. \text{ g H}_2\text{O} \times \frac{1 \text{ mol H}_2\text{O}}{18.02 \text{ g H}_2\text{O}} \times \frac{2 \text{ mol NH}_3}{1 \text{ mol H}_2\text{O}} \times \frac{17.03 \text{ g NH}_3}{1 \text{ mol NH}_3} = 189. \text{ g HN}_3$

Therefore, urea is the limiting reactant.

(b) 176. g NH_3

$300. \text{ g urea} \times \frac{1 \text{ mol urea}}{60.06 \text{ g urea}} \times \frac{1 \text{ mol H}_2\text{O}}{1 \text{ mol urea}} \times \frac{44.01 \text{ g CO}_2}{1 \text{ mol CO}_2} = 220. \text{ g CO}_2$

(c) $300. \text{ g urea} \times \frac{1 \text{ mol urea}}{60.06 \text{ g urea}} \times \frac{1 \text{ mol H}_2\text{O}}{1 \text{ mol urea}} \times \frac{18.02 \text{ g H}_2\text{O}}{1 \text{ mol H}_2\text{O}} = 90.0 \text{ g H}_2\text{O}$

100. g − 90.0 g = 10. g H_2O remains

4.9 (1) Impure reactants; (2) Inaccurate weighing of reactants and products

4.10 Assuming that the nicotine is pure, weigh a sample of nicotine and burn the sample. Separately collect and weigh the carbon dioxide and water generated, and calculate the moles and grams of carbon and hydrogen collected. By mass difference, determine the mass of nitrogen in the original sample, then calculate the moles of nitrogen. Calculate the mole ratios of carbon, hydrogen, and nitrogen in nicotine to determine its empirical formula.

Chapter 5

5.1 It is possible for an exchange reaction to form two different precipitates—for example, the reaction between barium hydroxide and iron(II) sulfate:

$$Ba(OH)_2(aq) + FeSO_4(aq) \rightarrow BaSO_4(s) + Fe(OH)_2(s)$$

5.2 $H_3PO_4(aq) \rightleftharpoons H_2PO_4^-(aq) + H^+(aq)$
$H_2PO_4^-(aq) \rightleftharpoons HPO_4^{2-}(aq) + H^+(aq)$
$HPO_4^{2-}(aq) \rightleftharpoons PO_4^{3-}(aq) + H^+(aq)$

5.3 (a) Hydrogen ions and perchlorate ions:

$$HClO_4(aq) \rightarrow H^+(aq) + ClO_4^-(aq)$$

(b) $Ca(OH)_2(aq) \rightarrow Ca^{2+}(aq) + 2\,OH^-(aq)$

5.4 (a) $H^+(aq) + Cl^-(aq) + K^+(aq) + OH^-(aq) \rightarrow H_2O(\ell) + K^+(aq) + Cl^-(aq)$
$H^+(aq) + OH^-(aq) \rightarrow H_2O(\ell)$
(b) $2\,H^+(aq) + SO_4^{2-}(aq) + Ba^{2+}(aq) + 2\,OH^-(aq) \rightarrow 2\,H_2O(\ell) + BaSO_4(s)$
$H^+(aq) + OH^-(aq) \rightarrow H_2O(\ell)$
$Ba^{2+}(aq) + SO_4^{2-}(aq) \rightarrow BaSO_4(s)$
(c) $2\,CH_3\,COOH(aq) + Ca^{2+}(aq) + 2\,OH^-(aq) \rightarrow Ca^{2+}(aq) + 2\,CH_3COO^-(aq) + 2\,H_2O(\ell)$
$CH_3\,COOH(aq) + OH^-(aq) \rightarrow H_2O(\ell) + CH_3COO^-(aq)$

5.5 $Al(OH)_3(s) + 3\,H^+(aq) + 3\,Cl^-(aq) \rightarrow 3\,H_2O(\ell) + Al^{3+}(aq) + 3\,Cl^-(aq)$
$H^+(aq) + OH^-(aq) \rightarrow H_2O(\ell)$

5.6 (a) The products are aqueous sodium sulfate, water, and carbon dioxide gas.

$Na_2CO_3(aq) + H_2SO_4(aq) \rightarrow Na_2SO_4(aq) + H_2O(\ell) + CO_2(g)$
$2\,H^+(aq) + CO_3^{2-}(aq) \rightarrow H_2O(\ell) + CO_2(g)$

(b) The products are aqueous iron(II) chloride and hydrogen sulfide gas.

$FeS(s) + 2\,HCl(aq) \rightarrow FeCl_2(aq) + H_2S(g)$
$2\,H^+(aq) + S^{2-}(aq) \rightarrow H_2S(g)$

(c) The products are aqueous potassium chloride, water, and sulfur dioxide gas.
$K_2SO_3(aq) + 2\,HCl(aq) \rightarrow 2\,KCl(aq) + H_2O(\ell) + SO_2(g)$
$2\,H^+(aq) + SO_3^{2-}(aq) \rightarrow H_2O(\ell) + SO_2(g)$

5.7 (a) Gas-forming reaction; the products are aqueous nickel sulfate, water, and carbon dioxide gas.
$NiCO_3(s) + H_2SO_4(aq) \rightarrow NiSO_4(aq) + H_2O(\ell) + CO_2(g)$
$NiCO_3(s) + 2\,H^+(aq) \rightarrow Ni^{2+}(aq) + H_2O(\ell) + CO_2(g)$

(b) Acid-base reaction; nitric acid reacts with strontium hydroxide, a base, to produce water and strontium nitrate, a salt.
$2\,HNO_3(aq) + Sr(OH)_2(s) \rightarrow Sr(NO_3)_2(aq) + 2\,H_2O(\ell)$
$Sr(OH)_2(s) + 2\,H^+(aq) \rightarrow Sr^{2+}(aq) + 2\,H_2O(\ell)$

(c) Precipitation reaction; aqueous sodium chloride and insoluble barium oxalate are produced.
$BaCl_2(aq) + Na_2C_2O_4(aq) \rightarrow BaC_2O_4(s) + 2\,NaCl(aq)$
$Ba^{2+}(aq) + C_2O_4^{2-}(aq) \rightarrow BaC_2O_4(s)$

(d) Precipitation and gas-forming reaction; lead sulfate precipitates and carbon dioxide gas is released.
$PbCO_3(aq) + H_2SO_4(aq) \rightarrow PbSO_4(s) + H_2O(\ell) + CO_2(g)$
$Pb^{2+}(aq) + SO_4^{2-}(aq) \rightarrow PbSO_4(s)$
$2\,H^+(aq) + CO_3^{2-}(aq) \rightarrow H_2O(\ell) + CO_2(g)$

5.8 In the reaction $2\,Ca(s) + O_2(g) \rightarrow 2\,CaO(s)$, Ca loses electrons, is oxidized, and is the reducing agent; O gains electrons, is reduced, and is the oxidizing agent.

5.9 $Cl_2(g) + Ca(s) \rightarrow CaCl_2(s)$. $Cl_2(g)$ is the oxidizing agent.

5.10 (a) This is not a redox reaction. Nitric acid is a strong oxidizing agent, but here it serves as an acid.
(b) In this redox reaction, chromium metal (Cr) is oxidized (loses electrons) to form Cr^{3+} ions in Cr_2O_3; oxygen (O_2) is reduced (gains electrons) to form oxide ions, O^{2-}. Oxygen is the oxidizing agent, and chromium is the reducing agent.
(c) This is an acid-base reaction, but not a redox reaction; there are no strong oxidizing or reducing agents present.
(d) Copper is oxidized and chlorine is reduced in this redox reaction, in which copper is the reducing agent and chlorine is the oxidizing agent. The equations are $Cu \rightarrow Cu^{2+} + 2e^-$ and $Cl_2 + 2e^- \rightarrow 2\,Cl^-$.

5.11 (a) Carbon in oxalate ion, $C_2O_4^{2-}$ (oxidation state = +3), is oxidized to oxidation state +4 in CO_2.
(b) Carbon is reduced from +4 in CCl_2F_2 to 0 in C(s).

5.12 (a) $CH_3CH_2OH(\ell) + 3\ O_2(g) \rightarrow 3\ H_2O(\ell) + 2\ CO_2(g)$; redox
(b) $2\ Fe(s) + 6\ HNO_3(aq) \rightarrow 2\ Fe(NO_3)_3(aq) + 3\ H_2(g)$; redox
(c) $AgNO_3(aq) + KBr(aq) \rightarrow AgBr(s) + KNO_3(aq)$; not redox

5.13 Molar mass of cholesterol = 386.7 g/mol

$$240\text{ mg} \times \frac{1\text{ g}}{10^3\text{ mg}} \times \frac{1\text{ mol}}{386.7\text{ g}} = 6.21 \times 10^{-4}\text{ mol cholesterol}$$

$$\frac{6.21 \times 10^{-4}\text{ mol}}{0.100\text{ L}} = 6.2 \times 10^{-3}\text{ M}$$

5.14 (a) $$6.37\text{ g Al(NO}_3)_3 \times \frac{1\text{ mol Al(NO}_3)_3}{213.0\text{ g Al(NO}_3)_3} = 0.0299\text{ mol Al(NO}_3)_3;$$

$$\frac{0.0299\text{ mol}}{0.250\text{ L}} = 0.120\text{ M Al(NO}_3)_3$$

(b) Molarity: Al^{3+} = 0.120; NO_3^- = 3(0.120) = 0.360

5.15 If the description of solution preparation is always worded in terms of adding enough solvent to make a specific volume of solution, then any possible expansion or contraction has no effect on the molarity of the solution. The denominator of the definition of molarity is liters of *solution.*

5.16 The moles of HCl in the concentrated solution are given by (6.0 mol/L)(0.100 L) = 0.60 mol HCl. The moles of HCl in the dilute solution are given by (1.20 mol/L)(0.500 L) = 0.600 mol HCl.

5.17 The molarity could be increased by evaporating some of the solvent.

5.18 $$0.0193\text{ L} \times \frac{0.200\text{ mol AgNO}_3}{1\text{ L}} \times \frac{1\text{ mol Ag}^+}{1\text{ mol AgNO}_3} \times \frac{1\text{ mol Cl}^-}{1\text{ mol Ag}^+} \times \frac{1\text{ mol NaCl}}{1\text{ mol Cl}^-} = 3.86 \times 10^{-3}\text{ mol NaCl}$$

$$\frac{3.86 \times 10^{-3}\text{ mol NaCl}}{0.0250\text{ L}} = 0.154\text{ M NaCl}$$

Chapter 6

6.1 You transfer some mechanical energy to the ball to accelerate it upward. The ball's potential energy increases the higher it gets, but its kinetic energy decreases by an equal quantity, and eventually it stops rising and begins to fall. As it falls, some of the ball's potential energy changes to kinetic energy, and the ball goes faster and faster until it hits the floor. When the ball hits the floor, some of its kinetic energy is transferred to the atoms, molecules, or ions that make up the floor, causing them to move faster. Eventually all of the ball's kinetic energy is transferred, and the ball stops moving. The nanoscale particles in the floor (and some in the air that the ball fell through) are moving faster on average, and the temperature of the floor (and the air) is slightly higher. The energy has spread out over a much larger number of particles.

6.2

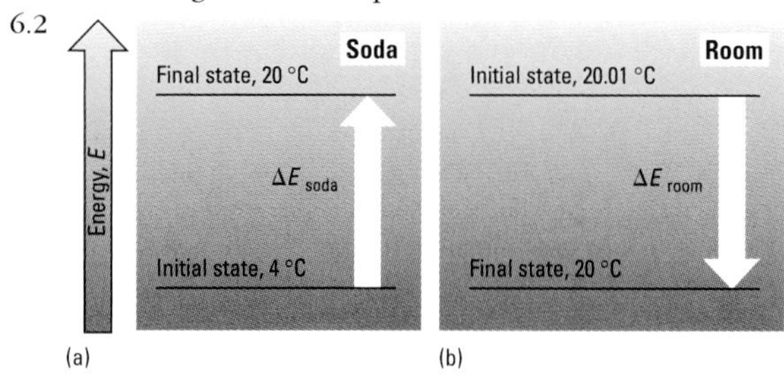

6.3 (a) Heat transfer = $(25.0\text{ °C} - 1.0\text{ °C})\left(\frac{1.5\text{ kJ}}{1.0\text{ °C}}\right) = 36\text{ kJ}$

(b) The system is the can and the liquid it contains.

(c) The surroundings includes the air and other materials in contact with the can, or close to the can.

(d) ΔE is negative because the system transferred energy to the surroundings as it cooled; $\Delta E = -36$ kJ.

(e)

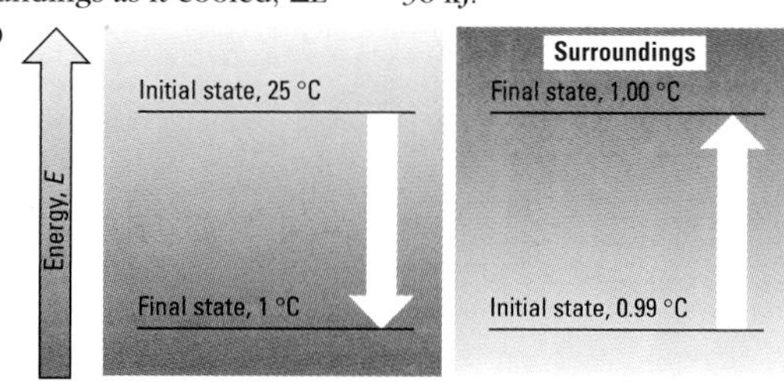

The surroundings are warmed very slightly, say from 0.99 °C to 1.00 °C, by the heat transfer from the can of soda.

6.4 The same quantity of energy is transferred out of each beaker and the mass of each sample is the same. Therefore the sample with the smaller specific heat capacity will cool more. Look up the specific heat capacities in Table 6.1. Because glass has a larger specific heat capacity than carbon, the carbon will cool more and therefore will have the lower temperature.

6.5 The calculation for Al is given as an example.

$$\text{Molar heat capacity} = \frac{0.902\text{ J}}{\text{g °C}} \times \frac{26.98\text{ g}}{1\text{ mol}} = 24.3\text{ J mol}^{-1}\text{ °C}^{-1}$$

Metal	Molar Heat Capacity (J mol^{-1} °C^{-1})	Metal	Molar Heat Capacity (J mol^{-1} °C^{-1})
Al	24.3	Cu	24.5
Fe	25.2	Au	25.2

The molar heat capacities of most metals are close to 25 J mol^{-1} °C^{-1}. This rule does not work for ethanol or other compounds listed in Table 6.2.

6.6 (a) Since the heat of vaporization is almost seven times larger than the heat of fusion, the temperature stays constant at 100 °C almost seven times longer than it stays constant at 0 °C. It stays constant at 0 °C for slightly less time than it takes to heat the water from 0 °C to 100 °C (see graph). Because the heating is at a constant rate, time is proportional to quantity of energy transferred.

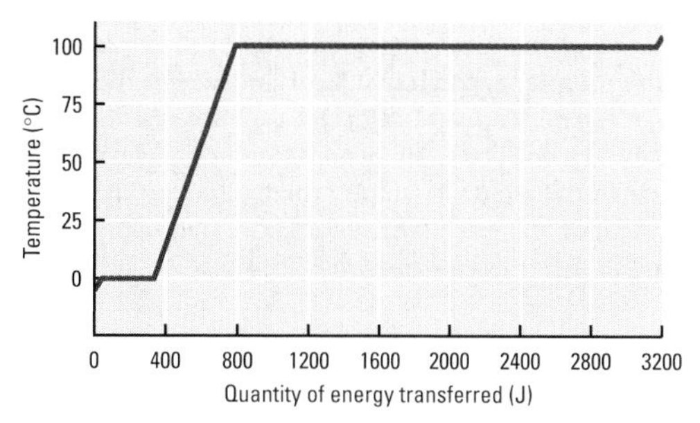

(b) The mass of water is half as great as in part (a), so each process takes half as long. A graph to the same scale as in part (a) begins at 105 °C and reaches −5 °C with half the quantity of energy transferred.

6.7 Heat of fusion: 237 g × 333 J/g = 78.9 kJ
Heating liquid: 237 g × 4.184 J g^{-1} °C^{-1} × 100.0 °C = 99.2 kJ
Heat of vaporization: 237 g × 2260 J/g = 536 kJ
Total heating = (78.9 + 99.2 + 536) kJ = 714 kJ

6.8 The direction of energy transfer is indicated by the sign of the enthalpy change. Transfer to the system corresponds to a positive enthalpy change.

6.9 Because of heats of fusion and heats of vaporization, the enthalpy change is different when a reactant or product is in a different state.

6.10 When 1.0 mol of H_2 reacts (Equation 6.3), $\Delta H = -241.8$ kJ. When half that much H_2 reacts, ΔH is half as great; that is, 0.50 × (−241.8) = −120.9 kJ.

6.11 $\dfrac{-92.22\text{ kJ}}{1\text{ mol }N_2(g)}$ $\dfrac{-92.22\text{ kJ}}{3\text{ mol }H_2(g)}$ $\dfrac{-92.22\text{ kJ}}{2\text{ mol }NH_3(g)}$

$\dfrac{1\text{ mol }N_2(g)}{-92.22\text{ kJ}}$ $\dfrac{3\text{ mol }H_2(g)}{-92.22\text{ kJ}}$ $\dfrac{2\text{ mol }NH_3(g)}{-92.22\text{ kJ}}$

6.12 The reaction used must be exothermic. Because it can be started by opening the package, it probably involves oxygen from the air, and the sealed package prevents the reaction from occurring before it is needed. Many metals can be oxidized easily and exothermically. The reaction of iron with oxygen is a good candidate.

6.13 Yes, it would violate the first law of thermodynamics. According to the supposition, we could create energy by starting with 2 mol HCl, breaking all the molecules apart, recombining the atoms to form 1 mol H_2 and 1 mol Cl_2, and then reacting the H_2 and Cl_2 to give 2 HCl.

$$2\,HCl \longrightarrow \text{atoms} \longrightarrow H_2 + Cl_2 \qquad \Delta H° = +185\text{ kJ}$$
$$H_2 + Cl_2 \longrightarrow 2\,HCl \qquad \Delta H° = -190\text{ kJ}$$

The net effect of these two processes is that there is still 2 mol HCl, but 5 kJ of energy has been created. This is impossible according to the first law of thermodynamics.

6.14 (a) In the reaction $2\,HF \rightarrow H_2 + F_2$ there are two bonds in the two reactant molecules and two bonds in the two product molecules. Since the reaction is endothermic, the bonds in the reactant molecules must be stronger than in the products.
(b) For the reaction $2\,H_2O \rightarrow 2\,H_2 + O_2$, there are four bonds in the two reactant molecules but only three bonds in the three product molecules. The reaction is endothermic because more bonds are broken than are formed.

6.15 $C_6H_{12}O_6(s) + 6\,O_2(g) \rightarrow 6\,CO_2(g) + 6\,H_2O(\ell)$
Because the volume of any ideal gas is proportional to the amount (moles) of gas, and because there are 6 mol of gaseous reactant and 6 mol of gaseous product, there will be very little change in volume. Almost no work will be done, and $\Delta H \cong \Delta E$.

6.16 In the Problem-Solving Example 6.10 the reaction produced 0.25 mol NaCl and the heat transfer associated with the reaction caused 500 mL of solution to warm by 7 °C.
(a) Here the reaction produces 0.20 mol NaCl by neutralizing 0.20 mol NaOH and the heat transfer warms 400. mL of water. So there is less heat transfer but it will be heating a smaller volume. The quantity of reaction is $\frac{0.20\text{ mol}}{0.25\text{ mol}} = 0.80$ as much, so the heat transfer is 0.80 as much. The quantity of water to be heated is $\frac{400.\text{mL}}{500.\text{mL}} = 0.800$ as much. Therefore the combined effect on temperature is the same as in Problem-Solving Example 6.10, and the temperature change associated with this process will also be 7 °C.
(b) Here the limiting reactant is 0.10 mol NaOH. (Only half of the H_2SO_4 is used up.) The heat transfer from the reaction warms 200. mL of water. So there is less heat transfer but it will heat a smaller volume. The quantity of reaction is $\frac{0.10\text{ mol}}{0.25\text{ mol}} = 0.40$ as much, so the heat transfer is 0.40 as much. The quantity of water is $\frac{200.}{400.} = 0.40$ as much. Therefore once again the combined effect on temperature is the same as in Problem-Solving Example 6.10, and the temperature change associated with this process will also be 7 °C.

6.17 $N_2(g) \rightarrow N_2(g)$
(a) The product is the same as the reactant, so there is no change—nothing happens.
(b) Since product and reactant are the same, $\Delta H = 0$.

Chapter 7

7.1 Wavelength and frequency are inversely related. Therefore, low-frequency radiation has long-wavelength radiation.

7.2 Cellular phones use higher frequency radio waves.

7.3 $E = h\nu;\quad \lambda = \dfrac{hc}{E}$

$$E = h\nu = (6.626 \times 10^{-34}\text{ J}\cdot\text{s})(2.45 \times 10^{9}\text{ s}^{-1})$$
$$= 1.62 \times 10^{-24}\text{ J}$$
$$\lambda = \frac{(6.626 \times 10^{-34}\text{ J}\cdot\text{s})(2.998 \times 10^{8}\text{ m/s})}{1.62 \times 10^{-24}\text{ J}} = 1.23 \times 10^{-1}\text{ m}$$

7.4 In a sample of excited hydrogen gas there are many atoms, and each can exist in one of the excited states possible for hydrogen. The observed spectral lines are a result of all the possible transitions of all these hydrogen atoms.

7.5 (a) Emitted (b) Absorbed
(c) Emitted (d) Emitted

7.6 $(2.179 \times 10^{-18}\text{ J/photon}) \times \left(\dfrac{1\text{ kJ}}{10^3\text{ J}}\right)\left(\dfrac{6.022 \times 10^{23}\text{ photons}}{1\text{ mol}}\right)$
$= 1312\text{ kJ/mol photons}$

7.7 (a) $5d$ (b) $4f$ (c) $6p$

7.8 The $n = 3$ level can have only three types of sublevels—s, p, and d. The $n = 2$ level can have only s and p sublevels, not d sublevels ($l = 2$).

7.9 3, 0, 0, $+\frac{1}{2}$; 3, 0, 0 $-\frac{1}{2}$

7.10 (a) For example 3, 1, +1, $+\frac{1}{2}$ and 3, 1, 0, $+\frac{1}{2}$
(b) 3, 1, 1, $+\frac{1}{2}$ and 3, 0, 0, $-\frac{1}{2}$

7.11 Electron a is in the $3p_y$ orbital. Electron b is in the $3p_z$ orbital.

7.12 (a) The maximum number of electrons in the $n = 3$ level is 18 (2 electrons per orbital). The orbitals would be designated $3s$, $3p_x$, $3p_y$, $3p_z$, $3d_{z^2}$, $3d_{xy}$, $3d_{yz}$, $3d_{xz}$, and $3d_{x^2-y^2}$.
(b) The maximum number of electrons in the $n = 5$ level is 50. The orbitals would be designated $5s$, $5p_x$, $5p_y$, $5p_z$, $5d_{yz}$, $5d_{xz}$, $5d_{z^2}$, $5d_{xy}$, $5d_{x^2-y^2}$, and the seven $5f$ orbitals and the nine $5g$, which are not designated by name in the text.

7.13 The first shell that could contain g orbitals would be the $n = 5$ shell. There would be nine g orbitals.

7.14 For the chlorine atom, $n = 3$, and there are seven electrons in this highest energy level. The configuration is

3s: [↑↓] 3p: [↑↓][↑↓][↑].

For the selenium atom, the highest energy level is $n = 4$, and there are six electrons in the $n = 4$ level. The configuration is

4s: [↑↓] 4p: [↑↓][↑][↑].

7.15 The $[Ar]3d^44s^2$ configuration for chromium has four unpaired electrons, and the $[Ar]3d^54s^1$ configuration has six unpaired electrons.

7.16 The $Fe(acac)_2$ contains an Fe^{2+} ion with a $3d$ electron configuration of [↑↓][↑][↑][↑][↑]. This configuration has four unpaired electrons. The compound $Fe(acac)_3$ contains an Fe^{3+} ion, with a $3d$ electron configuration of [↑][↑][↑][↑][↑]. This configuration has five unpaired electrons. The $Fe(acac)_3$, with more unpaired electrons per molecule, would be attracted more strongly into a magnetic field.

Chapter 8

8.1 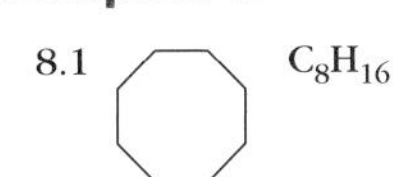 C_8H_{16}

8.2 N_2 has only 10 valence electrons. The Lewis structure shown has 14 valence electrons.

8.3 None of the structures are correct. (a) is incorrect because sulfur does not have an octet of electrons (it has only six); (b) is incorrect because, although it shows the correct number of valence electrons (26), there is a double bond between F and N rather than a single bond with a lone pair on N; (c) is incorrect because the left carbon has five bonds; (d) is incorrect because COCl should have 17 valence electrons, not 18 as shown.

8.4 (a) C_5H_{10} (b) Two

8.5

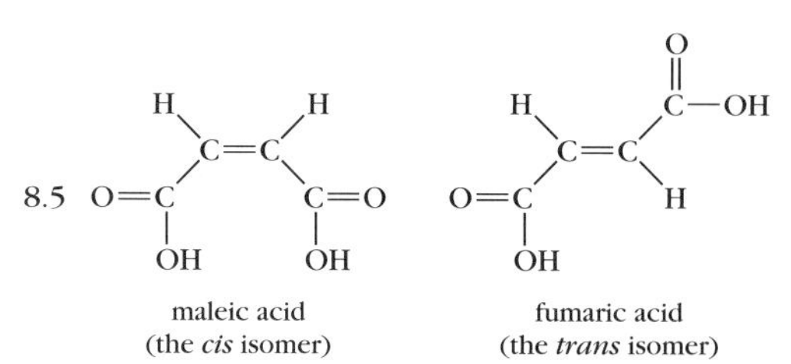

8.6 C—N > C=N > C≡N. The order of decreasing bond energy is the reverse order: C≡N > C=N > C—N. See Tables 8.1 and 8.2.

8.7 (a) The electronegativity difference between sodium and chlorine is 2.0, sufficient to cause electron transfer from sodium to chlorine to form Na^+ and Cl^- ions. Molten NaCl conducts an electric current, indicating the presence of ions.
(b) There is an electronegativity difference of 1.2 in BrF, which is sufficient to form a polar covalent bond, but not great enough to cause electron transfer leading to ion formation.

8.8 The Lewis structure of hydrazine is

H—N̈—N̈—H
   | |
   H   H

	H	H	N	N	H	H
Valence electrons	1	1	5	5	1	1
Lone pair electrons	0	0	2	2	0	0
$\frac{1}{2}$ shared electrons	1	1	3	3	1	1
Formal charge	0	0	0	0	0	0

8.9 Atoms cannot be rearranged to derive a resonance structure. There is no N-to-O bond in cyanate ion; therefore, such an arrangement cannot be a resonance structure of cyanate ion.

Chapter 9

9.1 When the central atom has no lone pairs

9.2 The triangular bipyramidal shape has three of the five pairs situated in equatorial positions 120° apart and the remaining two pairs in axial positions. The square pyramidal shape has four of the atoms bonded to the central atom in a square plane, with the other bonded atom directly above the central atom and equidistant from the other four.

9.3 (a) AX_2E_3 (b) AX_3E_1 (c) AX_2E_3

9.4 Pi bonding is not possible for a carbon atom with sp^3 hybridization because it has no unhybridized $2p$ orbitals. All of its $2p$ orbitals have been hybridized.

9.5 (a) Bromine is more electronegative than iodine, and the H—Br bond is more polar than the H—I bond.
(b) Chlorine is more electronegative than the other two halogens; therefore, the C—Cl bond is more polar than the C—Br and C—I bonds.

9.6
$\delta^+\ \delta^-$ $\delta^+\ \delta^-$
:C≡O:- - - -:C≡O: - - - - dipole-dipole forces
$\delta^-\ \delta^+$ $\delta^-\ \delta^+$
:O≡C:- - - -:O≡C:

9.7 The F—H · · · F—H hydrogen bond is the strongest because the electronegativity difference between H and F produces a more polar F—H bond than does the lesser electronegativity difference between O and H or N and H in the O—H or N—H bonds.

Chapter 10

10.1 $(2.7 \times 10^{14}\text{ molecules}) \times \left(\dfrac{64.06\text{ g }SO_2}{6.02 \times 10^{23}\text{ molecules}}\right) = 2.9 \times 10^{-8}\text{ g }SO_2$

10.2 All have the same kinetic energy at the same temperature.

10.3 For a sample of helium, the plot would look like the curve marked He in Figure 10.6. When an equal number of argon molecules, which are heavier, are added to the helium, the distribution of molecular speeds would look like the sum of the curves marked He and O_2 in Figure 10.7, except that the curve for Ar would have its peak a little to the left of the O_2 curve.

10.4 (a) The balloon placed in the freezer will be smaller than the one kept at room temperature because its sample of helium is colder.
(b) Upon warming, the helium balloon that had been in the freezer will be either the same size as the balloon kept at room temperature or perhaps slightly larger because there is a greater chance that He atoms leaked out of the room temperature balloon during the time the other balloon was kept in the freezer. This would be caused by the faster-moving He atoms in the room temperature balloon having more chances to escape from tiny openings in the balloon's walls.

10.5 The gas in the shock absorbers will be more highly compressed. The gas molecules will be closer together. The gas molecules will collide with the walls of the shock absorber more often, and the pressure exerted will be larger.

10.6 Increasing the temperature of a gas causes the gas molecules to move faster, on average. This means that each collision with the container walls involves greater force, because on average, a molecule is moving faster and hits the wall harder. If the container remained the same (constant volume), there would also be more collisions with the container wall because faster-moving molecules would hit the walls more often. Increasing the volume of the container, on the other hand, requires that the faster-moving molecules must travel a greater distance before they strike the container walls. Increasing the volume enough would just balance the greater numbers of harder collisions caused by increased temperature. To maintain a constant volume requires that the pressure increases to match the greater pressure due to more and harder collisions of gas molecules with the walls.

10.7 Using the ratio of 100 balloons/26.8 g He, calculate the number of balloons 41.8 g He can fill.

$$\text{Balloons} = (41.8\text{ g He})\left(\frac{100\text{ balloons}}{26.8\text{ g He}}\right) = 155\text{ balloons}$$

This much He will fill more balloons than needed.

10.8 1. Increase the pressure.
2. Decrease the temperature.
3. Remove some of the gas by reaction to form a nongaseous product.

10.9 Density of Cl_2 at 25 °C and 0.750 atm $= \dfrac{PM}{RT}$

$$= \frac{(0.750\text{ atm})(70.906\text{ g/mol})}{(0.0821\text{ L atm mol}^{-1}\text{ K}^{-1})(298\text{ K})} = 2.17\text{ g/L}$$

Density of SO_2 at 25 °C and 0.750 atm $= \dfrac{PM}{RT}$

$$= \frac{(0.750\text{ atm})(64.06\text{ g/mol})}{(0.0821\text{ L atm mol}^{-1}\text{ K}^{-1})(298\text{ K})} = 1.96\text{ g/L}$$

$$\text{Density of } Cl_2 \text{ at } 35\ °C \text{ and } 0.750 \text{ atm} = \frac{PM}{RT}$$

$$= \frac{(0.750 \text{ atm})(70.906 \text{ g/mol})}{(0.0821 \text{ L atm mol}^{-1} \text{ K}^{-1})(308 \text{ K})} = 2.10 \text{ g/L}$$

$$\text{Density of } SO_2 \text{ at } 25\ °C \text{ and } 2.60 \text{ atm} = \frac{PM}{RT}$$

$$= \frac{(2.60 \text{ atm})(64.06 \text{ g/mol})}{(0.0821 \text{ L atm mol}^{-1} \text{ K}^{-1})(298 \text{ K})} = 6.81 \text{ g/L}$$

10.10 Density of He = 1.23×10^{-4} g/mL
Density of Li = 0.53 g/mL
Since the density of He is so much less than that of Li, the atoms in a sample of He must be much farther apart than the atoms in a sample of Li. This idea is in keeping with the general principle of the kinetic-molecular theory that the particles making up a gas are far from one another.

10.11 (a) If lowering the temperature causes the volume to decrease, by $PV = nRT$, the pressure can be assumed to be constant. The value of n is unchanged. Since both P and n remain unchanged, the partial pressures of the gases in the mixture remain unchanged.
(b) When the total pressure of a gas mixture increases, the partial pressure of each gas in the mixture increases because the partial pressure of each gas in the mixture is the product of the mole fraction for that gas and the total pressure.

10.12 We can calculate the total number of moles of gas in the flask from the given information.

$$n = \frac{PV}{RT} = \frac{\left(\frac{626}{760}\right) \text{atm } (0.355 \text{ L})}{(0.0821 \text{ L atm mol}^{-1} \text{ K}^{-1})(308 \text{ K})} = 0.01156 \text{ mol gas}$$

The number of moles of Ne is

$$0.146 \text{ g Ne} \times \frac{1 \text{ mol Ne}}{20.18 \text{ g/mol}} = 0.007235 \text{ mol Ne}$$

We find the number of moles of Ar by subtraction.

$$0.01156 \text{ mol gas} - 0.007235 \text{ mol Ne} = 0.004325 \text{ mol Ar}$$

$$0.004325 \text{ mol Ar} \times 39.948 \text{ g/mol} = 0.173 \text{ g Ar}$$

10.13 Natural sources: animal respiration, forest fires, decay of cellulose materials, partial digestion of carbohydrates, volcanoes. Human sources: burning fossil fuels, burning agricultural wastes and refined cellulose products such as paper, decay of carbon compounds in landfills.

10.14 The fluctuations occur because of the seasons. Photosynthesis, which uses CO_2, is greatest during the spring and summer, accounting for lower CO_2 levels.

Chapter 11

11.1 The London forces are greater between bromoform molecules than between chloroform molecules because the bromoform molecules have more electrons. This stronger intermolecular attraction causes the $CHBr_3$ molecules to exhibit a greater surface tension. (The dipole in each molecule contributes less than the London forces to the intermolecular attractions.)

11.2 (a) Water and glycerol would have similar surface tensions because of extensive hydrogen bonding. (b) Octane and decane would have similar surface tensions because both are alkane hydrocarbons.

11.3 (a) 62 °C (b) 0 °C (c) 80 °C

11.4 Bubbles form within a boiling liquid when the vapor pressure of the liquid equals the pressure of the surroundings of the liquid sample. The bubbles are actually filled with vapor of the boiling liquid. One way to prove this would be to trap some of these bubbles and allow them to condense. They would condense to form the liquid that had boiled.

11.5 The evaporating water carries with it thermal energy from the water inside the pot. In addition, a large quantity of thermal energy is required to cause the water to evaporate. Much of this thermal energy comes from the water inside the pot.

11.6 (a) Bromine molecules have more electrons than chlorine molecules. Therefore, bromine molecules are held together by stronger intermolecular attractions.
(b) Ammonia molecules are attracted to one another by hydrogen bonds. This causes ammonia to have a higher boiling point than that of methane, which has no hydrogen bonding.

11.7 Two moles of liquid bromine crystallizing liberates 21.59 kJ of energy. One mole of liquid water crystallizing liberates 6.02 kJ of energy.

11.8 High humidity conditions make the evaporation of water or the sublimation of ice less favorable. Under these conditions, the sublimation of ice required to make the frost-free refrigerator work is less favorable, so the defrost cycle is less effective.

11.9 The impurity molecules are less likely to be converted from the solid phase to the vapor phase. This causes them to be left behind as the molecules that sublime go into the gas phase and then condense at some other place. The molecules that condense are almost all of the same kind, so the sublimed sample is much purer than the original.

11.10 The curve is the vapor pressure curve. Upward: condensation. Downward: vaporization. Left to right: vaporization. Right to left: condensation.

11.11 (a) If liquid CO_2 is slowly released from a cylinder of CO_2, gaseous CO_2 is formed. The temperature remains constant (at room temperature) because there is time for energy to be transferred from the surroundings to separate the CO_2 molecules from their intermolecular attractions. This can be seen from the phase diagram as the phase changes from liquid to vapor as the pressure decreases.
(b) If the pressure is suddenly released, the attractive forces between a large number of CO_2 molecules must be overcome, which requires energy. This energy comes from the surroundings as well as from the CO_2 molecules themselves, causing the temperature of both the surroundings and the CO_2 molecules to decrease. On the phase diagram for CO_2, a decrease in both temperature and pressure moves into a region where only solid CO_2 exists.

11.12 It is predicted that a small concentration of gold will be found in the lead and that a small concentration of lead will be found in the gold. This will occur because of the movement of the metal atoms with time, as predicted by the kinetic molecular theory.

11.13 One Po atom belongs to its unit cell. Two Li atoms belong to its unit cell. Four Ca atoms belong to its unit cell.

Simple cubic

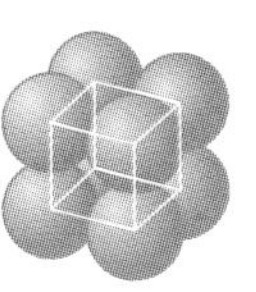

= 1 atom
Each of the 8 atoms contributes 1/8 to unit cell

Body-centered cubic

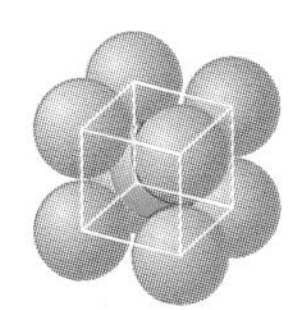

= 2 atoms
8/8 from corner atoms
+ 1 atom at center

Face-centered cubic

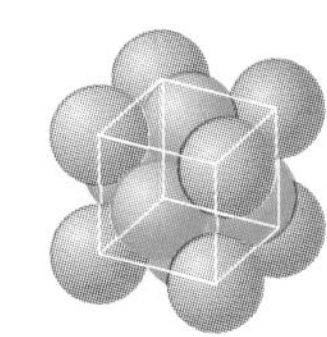

= 4 atoms
8/8 from corner atoms
+ 6/2 from each atom on the 6 faces contributing 1/2 atom

11.14 Each Cs^+ ion at the center of the cube has eight Cl^- ions as its neighbors. One eighth of each Cl^- ion belongs to that Cs^+ ion. So the formula for this salt must be a 1 : 1 ratio of Cs^+ ions to Cl^- ions, or CsCl.

11.15 Cooling a liquid above its freezing point causes the temperature to decrease. When the liquid begins to solidify, energy is released as atoms, molecules, or ions move closer together to form in the solid crystal lattice. This causes the temperature to remain constant until all the molecules in the liquid have positioned themselves in the lattice. Further cooling then causes the temperature to decrease. The shape of this curve is common to all substances that can exist as liquids.

11.16 Increasing strength of metallic bonding is related to increasing numbers of valence electrons. In the transition metals, the presence of d-orbital electrons causes stronger metallic bonding. Beyond a half-filled set of d-orbitals, however, extra electrons have the effect of decreasing the strength of metallic bonding.

Chapter 12

12.1 (a)

i. $$\text{Rate} = -\frac{\Delta[\text{Cv}^+]}{\Delta t} = -\frac{(0.793 \times 10^{-5} - 1.46 \times 10^{-5})\ \text{mol/L}}{(60.0 - 40.0)\ \text{s}} = \frac{6.67 \times 10^{-6}\ \text{mol/L}}{20.0\ \text{s}} = 3.3 \times 10^{-7}\ \text{mol L}^{-1}\ \text{s}^{-1}$$

ii. $$\text{Rate} = -\frac{(0.429 \times 10^{-5} - 2.71 \times 10^{-5})\ \text{mol/L}}{(80.0 - 20.0)\ \text{s}} = 3.8 \times 10^{-7}\ \text{mol L}^{-1}\ \text{s}^{-1}$$

iii. $$\text{Rate} = -\frac{(0.232 \times 10^{-5} - 5.00 \times 10^{-5})\ \text{mol/L}}{(100.0 - 0.0)\ \text{s}} = 4.8 \times 10^{-7}\ \text{mol L}^{-1}\ \text{s}^{-1}$$

(b)

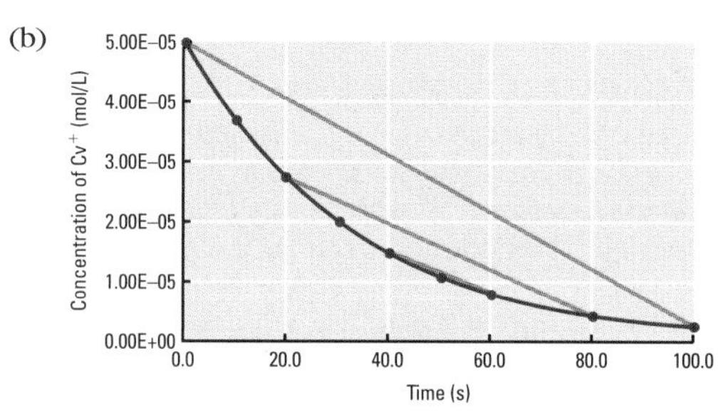

(c) The rate is faster when the concentration of Cv^+ is larger. As the reaction takes place, the rate gets slower. There is a much larger change in concentration from 0 s to 50 s than from 50 s to 100 s. Therefore, the $\Delta\ [\text{Cv}^+]$ is more than three times bigger for the time range from 20 s to 80 s than it is for the time range from 40 s to 60 s, even though Δt is exactly three times larger.

12.2 (a) Rate (1) is twice rate (3); rate (2) is twice rate (4); rate (3) is twice rate (5).
(b) In each case, the $[\text{Cv}^+]$ is twice as great when the rate is twice as great.
(c) Yes, the rate doubles when $[\text{Cv}^+]$ doubles.

12.3

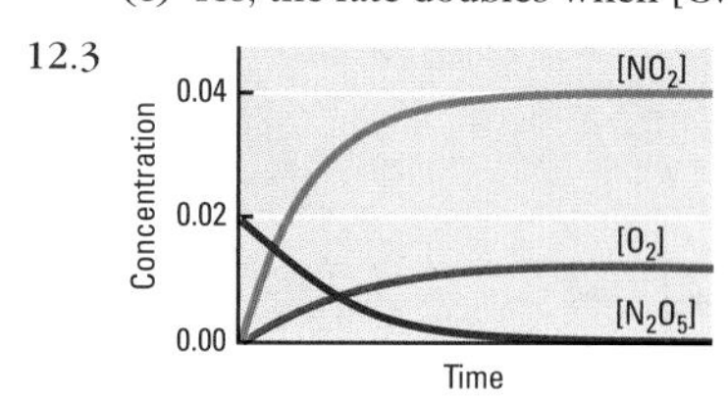

12.4 $\text{Rate}_1 = k[\text{CH}_3\text{COOCH}_3][\text{OH}^-]$
$\text{Rate}_2 = k(2[\text{CH}_3\text{COOCH}_3])(\tfrac{1}{2}[\text{OH}^-]) = k[\text{CH}_3\text{COOCH}_3][\text{OH}^-]$
The rate is unchanged.

12.5

$$\text{Rate} = k[\text{NO}]^x[\text{Cl}_2]^y$$

Taking the log of both sides and applying to experiments 4 and 2 gives

$$\log(\text{Rate}_4) = \log(k) + x \log[\text{NO}]_4 + y \log[\text{Cl}_2]_4$$
$$\log(\text{Rate}_2) = \log(k) + x \log[\text{NO}]_2 + y \log[\text{Cl}_2]_2$$

Recognizing that $[\text{Cl}_2]_4 = [\text{Cl}_2]_2$ and then subtracting the second equation from the first gives

$$\log(\text{Rate}_4) - \log(\text{Rate}_2) = x \log[\text{NO}]_4 - x \log[\text{NO}]_2$$

Using the properties of logarithms gives

$$\log\{\text{Rate}_4/\text{Rate}_2\} = x\{\log[\text{NO}]_4/[\text{NO}]_2\}$$

So

$$x = \log\{\text{Rate}_4/\text{Rate}_2\}/\{\log[\text{NO}]_4/[\text{NO}]_2\} \quad \text{QED}$$

Now we can insert the actual numbers from experiments 4 and 2:

$$x = [\log(6.60 \times 10^{-4}/1.65 \times 10^{-4})]/\log(0.04/0.02)$$
$$x = \log 4/\log 2 = 0.602/0.301 = 2$$

12.6 (a) $\text{CH}_3\text{NC} \rightarrow \text{CH}_3\text{CN}$ unimolecular
(b) $2\ \text{HI} \rightarrow \text{H}_2 + \text{I}_2$ bimolecular
(c) $\text{NO}_2\text{Cl} \rightarrow \text{NO}_2 + \text{Cl}$ unimolecular
(d) $\text{C}_4\text{H}_8 \rightarrow \text{C}_4\text{H}_8$ unimolecular
(e) $\text{NO}_2\text{Cl} + \text{Cl} \rightarrow \text{NO}_2 + \text{Cl}_2$ bimolecular

12.7 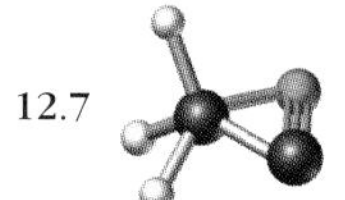 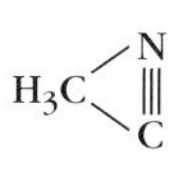

The Lewis structure has five pairs of electrons around one C atom, which would not be stable. However, this is a transition state, which is, by definition, unstable.

12.8 (1) :Ö=N̈ ... N̈=Ö: with :C̈l::C̈l: bridging

(2) :Ö=N̈–:C̈l: :C̈l:–N̈=Ö:

(3) :C̈l–N̈=Ö::Ö=N̈–C̈l:

(4) :C̈l–N̈=Ö: :Ö=N̈–C̈l:

(1) and (2) are much more likely to result in a reaction than are (3) and (4).

12.9 (a) $k = Ae^{-E_a/RT}$

$$= (6.31 \times 10^8\ \text{L mol}^{-1}\ \text{s}^{-1})\, e^{\frac{-10{,}000\ \text{J/mol}}{(8.314\ \text{J K}^{-1}\ \text{mol}^{-1})(370\ \text{K})}}$$

$$k = 2.4 \times 10^7\ \text{L mol}^{-1}\ \text{s}^{-1}$$

(b) $\text{Rate} = k[\text{NO}][\text{O}_3]$

$$= (2.4 \times 10^7\ \text{L mol}^{-1}\ \text{s}^{-1})(1.0 \times 10^{-3}\ \text{mol/L}) \times (5.0 \times 10^{-4}\ \text{mol/L})$$

$$= 12\ \text{mol L}^{-1}\ \text{s}^{-1}$$

12.10 The reaction does not occur in a single step. If it did, the rate law would be $\text{Rate} = k[\text{NO}_2][\text{CO}]$.

12.11 $\text{Rate} = k[\text{HOI}][\text{I}^-]$

12.12 (a)
$\text{Ce}^{4+} + \text{Mn}^{2+} \rightarrow \text{Ce}^{3+} + \text{Mn}^{3+}$
$\text{Ce}^{4+} + \text{Mn}^{3+} \rightarrow \text{Ce}^{3+} + \text{Mn}^{4+}$
$\text{Mn}^{4+} + \text{Tl}^{+} \rightarrow \text{Mn}^{2+} + \text{Tl}^{3+}$

$2\ \text{Ce}^{4+} + \text{Tl}^{+} \rightarrow 2\ \text{Ce}^{3+} + \text{Tl}^{3+}$

(b) Intermediates are Mn^{3+} and Mn^{4+}.
(c) The catalyst is Mn^{2+}.
(d) $\text{Rate} = k[\text{Ce}^{4+}][\text{Mn}^{2+}]$

12.13 (a) The concentration of a homogeneous catalyst *must* appear in the rate law.
(b) A catalyst does not appear in the equation for an overall reaction.
(c) A *homogeneous* catalyst must always be in the same phase as the reactants.

Chapter 13

13.1 (a) $(\text{conc. Al}) = \frac{2.70\text{ g}}{\text{mL}} \times \frac{1000\text{ mL}}{1\text{ L}} \times \frac{1\text{ mol}}{26.98\text{ g}}$
$= 100.\text{ mol/L}$

(b) $(\text{conc. benzene}) = \frac{0.880\text{ g}}{\text{mL}} \times \frac{1000\text{ mL}}{1\text{ L}} \times \frac{1\text{ mol}}{78.11\text{ g}}$
$= 11.3\text{ mol/L}$

(c) $(\text{conc. water}) = \frac{0.998\text{ g}}{\text{mL}} \times \frac{1000\text{ mL}}{1\text{ L}} \times \frac{1\text{ mol}}{18.02\text{ g}}$
$= 55.4\text{ mol/L}$

(d) $(\text{conc. Au}) = \frac{19.32\text{ g}}{\text{mL}} \times \frac{1000\text{ mL}}{1\text{ L}} \times \frac{1\text{ mol}}{196.97\text{ g}}$
$= 98.1\text{ mol/L}$

13.2 The mixture is not at equilibrium, but the reaction is so slow that there is no change in concentrations. You could show that the system was not at equilibrium by providing a catalyst or by raising the temperature to speed up the reaction.

13.3 (a) The new mixture is not at equilibrium because the quotient (conc. *trans*)/(conc. *cis*) no longer equals the equilibrium constant. Because (conc. *cis*) was halved, the quotient is twice K_c.
(b) The rate *trans* → *cis* remains the same as before, because (conc. *trans*) did not change. The rate *cis* → *trans* is only half as great, because (conc. *cis*) is half as great as at equilibrium.
(c) At 600 K, K_c is 1.47. Thus, [*trans*] = 1.47[*cis*].
(d) 0.15 mol/L

13.4 (a) If the coefficients of an equation are halved, the numerical value for the new equilibrium constant is the square root of the previous equilibrium constant. So the new equilibrium constant is $(6.25 \times 10^{-58})^{1/2} = 2.50 \times 10^{-29}$.
(b) If a chemical equation is reversed, the value for the new equilibrium constant is the reciprocal of the previous equilibrium constant. So the new equilibrium constant is $1/(6.25 \times 10^{-58}) = 1.60 \times 10^{57}$.

13.5 (a) $K_P = K_c \times (RT)^{\Delta n} = (3.5 \times 10^8\text{ L}^2\text{ mol}^{-2}) \times \{(0.082057\text{ L atm K}^{-1}\text{ mol}^{-1})(298)\}^{-2}$
$= 5.8 \times 10^5\text{ atm}^{-2}$
(b) $K_P = (3.2 \times 10^{81}\text{ L mol}^{-1}) \times \{(0.082057\text{ L atm mol}^{-1}\text{ K}^{-1})(298\text{ K})\}^{-1}$
$= 1.3 \times 10^{80}\text{ atm}^{-1}$
(c) $K_P = K_c = 1.7 \times 10^{-3}$
(d) $K_P = (1.7 \times 10^2\text{ L mol}^{-1}) \times \{(0.082057\text{ L atm mol}^{-1}\text{ K}^{-1})(298\text{ K})\}^{-1}$
$= 6.9\text{ atm}^{-1}$

13.6 (a) Because K_c for the forward reaction is small, K_c' for the reverse reaction is large.
(b) $K_c' = \frac{1}{K_c} = \frac{1}{1.8 \times 10^{-5}} = 5.6 \times 10^4$
(c) Ammonium ions and hydroxide ions should react, using up nearly all of whichever is the limiting reactant.
(d) $NH_4^+(aq) + OH^-(aq) \rightleftharpoons NH_3(aq) + H_2O(\ell)$
$NH_3(aq) \rightleftharpoons NH_3(g)$
You might detect the odor of $NH_3(g)$ above the solution. A piece of moist red litmus paper above the solution would turn blue.

13.7 Q should have the same mathematical form as K_P, so for the general Equation 13.1

$$Q_P = \frac{P_C^c \times P_D^d}{P_A^a \times P_B^b}$$

The rules for Q_P and K_P are analogous to those for Q and K_c:
If $Q_P > K_P$, then the reverse reaction occurs.
If $Q_P = K_P$, then the system is at equilibrium.
If $Q_P < K_P$, the forward reaction occurs.

13.8 $PCl_5(s) \rightleftharpoons PCl_3(g) + Cl_2(g)$
(a) Adding Cl_2 shifts the equilibrium to the left.
(b) Adding PCl_3 to the container shifts the equilibrium to the left.
(c) Because $PCl_5(s)$ does not appear in the K_c expression, adding some will not affect the equilibrium.

13.9 $Q = \frac{(\text{conc. H}_2)(\text{conc. I}_2)}{(\text{conc. HI})^2} = \frac{(0.102 \times 3)(0.0018 \times 3)}{(0.096 \times 3)^2}$
$= \frac{(0.102)(0.0018)(9)}{(0.096)^2(9)} = 0.020 = K_c$
Since $Q = K_c$, the system is at equilibrium under the new conditions. No shift is needed and none occurs.

13.10 From Problem-Solving Practice 13.6, $[NO_2] = 0.0418$ mol/L and $[N_2O_4] = 0.292$ mol/L.
Decreasing the volume from 4.00 to 1.33 L increases the concentrations as

$$(\text{conc. NO}_2) = 0.0418 \times \frac{4.00}{1.33} = 0.1257\text{ mol/L}$$

$$(\text{conc. N}_2\text{O}_4) = 0.292 \times \frac{4.00}{1.33} = 0.8782\text{ mol/L}$$

$$Q = \frac{(\text{conc. NO}_2)^2}{(\text{conc. N}_2\text{O}_4)} = \frac{(0.1257)^2}{0.8782} = 1.80 \times 10^{-2}$$

which is greater than K_c. Therefore, the equilibrium should shift to the left. Let x be the change in concentration of NO_2, (which should be negative).

	N_2O_4	$\rightleftharpoons$	$2\ NO_2$
Initial concentration (mol/L)	0.8782		0.1257
Change as reaction occurs (mol/L)	$-\frac{1}{2}x$		x
New equilibrium concentration (mol/L)	$0.8782 - \frac{1}{2}x$		$0.1257 + x$

$$K_c = 5.9 \times 10^{-3} = \frac{(0.1257 + x)^2}{0.8782 - 0.500x}$$

$$= \frac{(1.580 \times 10^{-2}) + (2.514 \times 10^{-1})x + x^2}{0.8782 - 0.500x}$$

$$5.18 \times 10^{-3} - (2.95 \times 10^{-3})x = (1.580 \times 10^{-2}) + (2.514 \times 10^{-1})x + x^2$$

$$x^2 + 0.2544x + (1.062 \times 10^{-2}) = 0$$

$$x = \frac{-0.2544 \pm \sqrt{6.472 \times 10^{-2} - (4 \times 1 \times 1.062 \times 10^{-2})}}{2}$$

$$x = \frac{-0.2544 \pm 0.1491}{2}$$

$x = -0.0526$ or $x = -0.2018$

The second root is mathematically reasonable, but results in a negative value for the $[NO_2]$. The new equilibrium concentrations are

$$[NO_2] = 0.1257 - 0.0526 = 0.0730\text{ mol/L}$$

$$[N_2O_4] = 0.8782 - \tfrac{1}{2}(-0.0526) = 0.9045\text{ mol/L}$$

Compared to the initial equilibrium, the concentrations have changed by

$$NO_2: \frac{0.0730}{0.0418} = 1.75 \qquad N_2O_4: \frac{0.945}{0.292} = 3.10$$

The concentration of N_2O_4 did increase by more than a factor of 3. The concentration of NO_2 increased by 1.75, which is less than a factor of 3 increase.

13.11

	How Reaction System Changes	Equilibrium Shifts	Change in K_c?
(a) Add reactant	Some reactants consumed	To right	No
(b) Remove reactant	More reactants formed	To left	No
(c) Add product	More reactants formed	To left	No
(d) Remove product	More products formed	To right	No
(e) Increase P by decreasing V	Total pressure decreases	Toward fewer gas molecules	No
(f) Decrease P by increasing V	Total pressure increases	Toward more gas molecules	No
(g) Increase T	Heat transfer into system	In endothermic direction	Yes
(h) Decrease T	Heat transfer out of system	In exothermic direction	Yes

If a substance is added or removed, the equilibrium is affected only if the substance's concentration appears in the equilibrium constant expression, or if its addition or removal changes concentrations that appear in the equilibrium constant expression.
Changing pressure by changing volume affects an equilibrium only for gas phase reactions in which there is a difference in the number of moles of gaseous reactants and products.

13.12 (a) The reaction is exothermic. $\Delta H° = -46.11$ kJ.
(b) The reaction is not favored by entropy.
(c) The reaction produces more products at low temperatures.
(d) If you increase T the reaction will go faster, but a smaller amount of products will be produced.

Chapter 14

14.1 The data in Table 14.2 indicate that the solubility of alcohols decreases as the hydrocarbon chain lengthens. Thus, 1-octanol is less soluble in water than 1-heptanol and 1-decanol should be even less soluble than 1-octanol.

14.2 34 g NH_4Cl would crystallize from solution at 20 °C.

14.3 (a) Unsaturated (b) Supersaturated (c) Saturated

14.4 The solubility of CO_2 decreases with increasing temperature, and the beverage loses its carbonation, causing it to go "flat."

14.5 Hot solvent would cause more of the solute to dissolve because Le Chatelier's principle states that at higher temperature an equilibrium will shift in the endothermic direction.

14.6 (a) Mass fraction of $NaHCO_3$

$$= \frac{0.20}{1000 + 6.5 + 0.20 + 0.1 + 0.10} = 2.0 \times 10^{-4}$$

$$\text{Wt. fraction} = 2.0 \times 10^{-4} \times 100\% = 2.0 \times 10^{-2}$$

(b) KCl and $CaCl_2$ each have the lowest mass fraction, 0.015.

14.7 (a) The 20-ppb sample has the higher lead concentration. (The other sample is 3 ppb lead.)
(b) 0.015 mg/L is equivalent to 0.015 ppm, which is 15 ppb. The 20-ppb sample exceeds the EPA limit; the 3-ppb sample does not.

14.8 $$3.5 \times 10^{20}\text{ gal} \times \frac{3.785\text{ L}}{\text{gal}} \times \frac{1 \times 10^{-3}\text{ mg Au}}{1\text{ L}} = 1 \times 10^{18}\text{ mg Au} = 1 \times 10^{15}\text{ g Au}$$

$$1 \times 10^{15}\text{ g Au} \times \frac{1\text{ lb Au}}{454\text{ g Au}} = 2 \times 10^{12}\text{ lb Au}$$

14.9 (a) Moles of solute and kilograms of solvent
(b) Molar mass of the solute and the density of the solution

14.10 $$50.0\text{ g sucrose} \times \left(\frac{1\text{ mol sucrose}}{342\text{ g sucrose}}\right) = 0.146\text{ mol sucrose}$$

$$100.0\text{ g water} \times \left(\frac{1\text{ mol water}}{18.0\text{ g water}}\right) = 5.56\text{ mol water}$$

$$X_{H_2O} = \frac{5.56}{5.56 + 0.146} = \frac{5.56}{5.706} = 0.974$$

Applying Raoult's law:

$$P_{water} = (X_{water})(P^0_{water}) = (0.974)(71.88\text{ mm Hg}) = 70.0\text{ mm Hg}$$

14.11 $\Delta T_b = (2.53\text{ °C kg mol}^{-1})(0.10\text{ mol/kg}) = 0.25$ °C. The boiling point of the solution is 80.10 °C + 0.25 °C = 80.35 °C.

14.12 First, calculate the required molality of the solution that would have a freezing point of −30 °C.

$$\Delta T_f = -30\text{ °C} = (-1.86\text{ °C}\cdot\text{kg mol}^{-1}) \times m$$

$$m = \frac{-30\text{ °C}}{-1.86\text{ °C kg mol}^{-1}} = 16.1\text{ mol/kg}$$

To protect 4.0 kg of water from this freezing temperature, you would need 4.0 kg × 16.1 mol/kg of ethylene glycol, or 65 mol.

$$65\text{ mol}\left(\frac{62.1\text{ g}}{\text{mol}}\right)\left(\frac{1\text{ mL}}{1.113\text{ g}}\right)\left(\frac{1\text{ L}}{1000\text{ mL}}\right) = 3.6\text{ L}$$

14.13 $\Delta T_f = K_f \times m \times i$;

$$4.78\text{ °C} = (1.86\text{ °C kg mol}^{-1})(2.0\text{ mol/kg})(i)$$

$$i = \frac{4.78\text{ °C}}{(2.0\text{ mol/kg})(1.86\text{ °C kg mol}^{-1})} = 1.28$$

Degree of dissociation: If completely dissociated, 1 mol $CaCl_2$ should yield 3 mol ions

$$CaCl_2 \longrightarrow Ca^{2+} + 2\,Cl^-$$

and i should be 3. The degree of dissociation in this solution is $\frac{1.28}{3} = 0.427 \approx 43\%$.

Chapter 15

15.1 (a) Acid (b) Base
(c) Acid (d) Base

15.2 More water molecules are available per NH_3 molecule in a very dilute solution of NH_3.

15.3 (a) $H_3O^+(aq) + CN^-(aq)$
(b) $H_3O^+(aq) + Br^-(aq)$
(c) $CH_3NH_3^+(aq) + OH^-(aq)$
(d) $(CH_3)_2NH_2^+(aq) + OH^-(aq)$

15.4 (a) $HSO_4^-(aq) + H_2O(\ell) \rightarrow H_2SO_4(aq) + OH^-(aq)$
(b) H_2SO_4 is the conjugate acid of HSO_4^-; water is the conjugate acid of OH^-.
(c) Yes; it can be an H^+ donor or acceptor.

15.5 The reverse reaction is favored because HSO_4^- is a stronger acid than CH_3COOH.

15.6 (a) [structure of piperidine (aq)] $+ H_2O \rightleftharpoons$ [structure of piperidinium ion]$^+$(aq) $+ OH^-(aq)$

(b) H (aq) + $HCl(aq) \longrightarrow$ [ ]$^{+}$ (aq) + $Cl^{-}(aq)$

15.7 The pH values of 0.1 M solutions of these two strong acids would be essentially the same since they both are 100% ionized, resulting in $[H_3O^+]$ values that are the same.

15.8 Because pH + pOH = 14.0, both solutions have a pOH of 8.5. The $[H_3O^+] = 10^{-pH} = 10^{-5.5} = 3.16 \times 10^{-6}$ M.

15.9 Pyruvic acid is the stronger acid, as indicated by its larger K_a value. Lactic acid's ionization reaction is more reactant-favored (less acid ionizes).

15.10 Being negatively charged, the HSO_4^- ion has a lesser tendency to lose a positively charged proton because of the electrostatic attractions of opposite charges.

15.11 (a) Step 1: $HOOC{-}COOH(aq) + H_2O(\ell) \rightleftharpoons H_3O^+(aq) + HOOC{-}COO^-(aq)$

Step 2: $HOOC{-}COO^-(aq) + H_2O(\ell) \rightleftharpoons H_3O^+(aq) + {}^-OOC{-}COO^-(aq)$

(b) Step 1: $C_3H_5(COOH)_3(aq) + H_2O \rightleftharpoons H_3O^+(aq) + C_3H_5(COOH)_2COO^-(aq)$

Step 2: $C_3H_5(COOH)_2COO^-(aq) + H_2O \rightleftharpoons H_3O^+(aq) + C_3H_5(COOH)(COO)_2^{2-}(aq)$

Step 3: $C_3H_5(COOH)(COO)_2^{2-}(aq) + H_2O \rightleftharpoons H_3O^+(aq) + C_3H_5(COO)_3^{3-}(aq)$

15.12 $K_b = \dfrac{1.0 \times 10^{-14}}{K_a} = \dfrac{1.0 \times 10^{-14}}{1.3 \times 10^{-10}} = 7.7 \times 10^{-5}$; carbonate ion; by comparing K_b values.

15.13 The pH of soaps is >7 due to the reaction with water of the conjugate base in the soap to form a basic solution.

15.14 Ammonium acetate. The pH of a solution of this salt will be 7.

15.15 (a) Lewis base (b) Lewis acid
(c) Lewis acid and base (d) Lewis acid
(e) Lewis acid (f) Lewis base

Chapter 16

16.1 HCl and NaCl: no; has no significant H^+ acceptor (Cl^- is a very poor base).
KOH and KCl: no; has no H^+ donor.

16.2 $\text{pH} = 7.21 + \log\dfrac{(0.0025)}{(0.00015)} = 7.21 + \log(1.67)$
$= 7.21 + 0.22 = 7.43$

16.3 $\text{pH} = pK_a + \log\dfrac{[\text{acetate}]}{[\text{acetic acid}]}$

$4.68 = 4.74 + \log\dfrac{[\text{acetate}]}{[\text{acetic acid}]}$

$\log\dfrac{[\text{acetate}]}{[\text{acetic acid}]} = 4.68 - 4.74 = -0.06;$

$\dfrac{[\text{acetate}]}{[\text{acetic acid}]} = 10^{-0.06} = 0.86$

Therefore, [acetate] = 0.86 × [acetic acid].

16.4 (a) $\text{pH} = 6.38 + \log\dfrac{[0.25]}{[0.10]} = 6.38 + 0.398 = 6.78$

(b) $\text{pH} = 7.21 + \log\dfrac{[HPO_4^{2-}]}{[H_2PO_4^-]}$

$= 7.21 + \log\dfrac{(0.25)}{(0.10)} = 7.21 + 0.398 = 7.61$

16.5 Since CO_2 reacts to form an acid, H_2CO_3, the phosphate ion that is the stronger base, HPO_4^{2-}, will be used to counteract its presence.

16.6

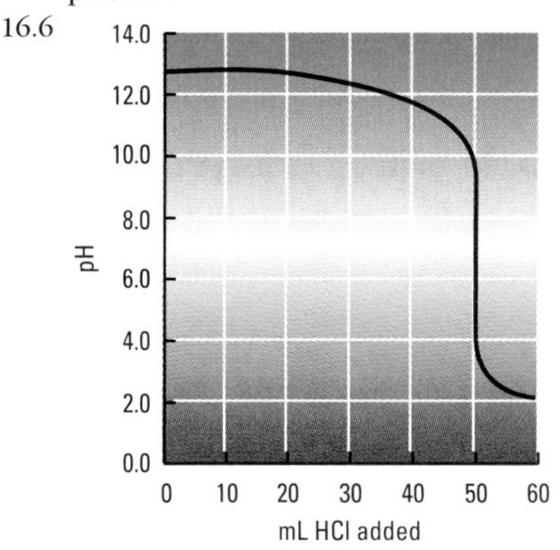

16.7 The addition of 30.0 mL of 0.100 M NaOH neutralizes 30.0 mL of 0.100 M acetic acid, forming 0.0030 mol of acetate ions, which is in 80.0 mL of solution. There is (0.0200 L)(0.100 M) = 0.00200 mol of acetic acid that is unreacted.

$$K_a = 1.8 \times 10^{-5} = \frac{[H^+][C_2H_3O_2^-]}{[HC_2H_3O_2]}$$

$$= \frac{[H^+] \times \left(\dfrac{0.00300 \text{ mol}}{0.0800 \text{ L}}\right)}{\left(\dfrac{0.00200 \text{ mol}}{0.0800 \text{ L}}\right)}$$

$$1.8 \times 10^{-5} = \frac{[H^+] \times (0.0375)}{(0.025)} = [H^+] \times 1.5$$

$$[H^+] = \frac{1.8 \times 10^{-5}}{1.5} = 1.2 \times 10^{-5}$$

$$\text{pH} = -\log(1.2 \times 10^{-5}) = 4.92$$

16.8 As NaOH is added, it reacts with acetic acid to form sodium acetate. After 20.0 mL NaOH has been added, just less than half of the acetic acid has been converted to sodium acetate; when 30.0 mL NaOH has been added, just over half of the acetic acid has been neutralized. Thus, after 20.0 mL and 30.0 mL base have been added, the solution contains approximately equal amounts of acetic acid and acetate ion, its conjugate base, which acts as a buffer.

16.9 Because Reaction (b) occurs to an appreciable extent, CO_3^{2-} is used as it forms by Reaction (a), causing additional $CaCO_3(s)$ to dissolve.

16.10 (a) The excess iodide would create a stress on the equilibrium and shift it to the left; some AgI and some PbI_2 would precipitate from solution.
(b) The added SO_4^{2-} would cause the precipitation of $BaSO_4$.

Chapter 17

17.1 (a) $H_2O(\ell) \rightarrow H_2O(g)$ Product-favored
(b) $SiO_2(s) \rightarrow Si(s) + O_2(g)$ Reactant-favored
(c) $(C_6H_{10}O_5)_n(s) + 6n\ O_2(g) \rightarrow 6n\ CO_2(g) + 5n\ H_2O(g)$ Product-favored
(d) $C_{12}H_{22}O_{11}(s) \rightarrow C_{12}H_{22}O_{11}(aq)$ Product-favored

17.2 A*** A**B* A*B** B***
If C, D, and E are added, there are many more arrangements in addition to these:

A*B*C*	A*B*D*	A*B*E*	A*C*D*	A*C*E*
A*D*E*	B*C*D*	B*C*E*	B*D*E*	C*D*E*
A**C*	A**D*	A**E*	B**C*	B**D*
B**E*	C**A*	C**B*	C**D*	C**E*
D**A*	D**B*	D**C*	D**E*	E**A*
E**B*	E**C*	E**D*	C***	D***
E***				

There are 35 possible arrangements, but only 4 of them have the energy confined to atoms A and B. The probability that all energy remains with A and B is thus 4/35 = 0.114, or a little more than 11%.

17.3 Using Celsius temperature and $\Delta S = q_{rev}/T$, if the temperature were -10 °C, the value of ΔS would be negative, in disagreement with the fact that transfer of energy to a sample should increase molecular motion and, hence, entropy.

17.4 (a) The reactant is a gas. The products are also gases, but the number of molecules has increased, so entropy is greater for products. (Entropy increases.)
(b) The reactant is a solid. The product is a solution. Mixing sodium and chloride ions among water molecules results in greater entropy for the product. (Entropy increases.)
(c) The reactant is a solid. The products are a solid and a gas. The much larger entropy of the gas results in greater entropy for the products. (Entropy increases.)

17.5 (a) Because $\Delta S_{surroundings} = -\Delta H/T$ at a given temperature, the larger the value of T, the smaller the value of $\Delta S_{surroundings}$.
(b) If ΔS_{system} does not change much with temperature, then $S_{universe}$ must also get smaller. In this case, because ΔS_{system} is negative, $\Delta S_{universe}$ would become negative at a high enough temperature.

17.6 (a) The reaction would have gaseous water as a product.
(b) Both $\Delta H°$ and $\Delta S°$ would change. $\Delta H° = -10{,}232.197$ kJ and $\Delta S° = 756.57$ J/K.
(c) If any of the reactants or products change to a different phase (s, ℓ, or g) over the range of temperature, $\Delta H°$ and $\Delta S°$ will change significantly at the temperature of the phase transition.

17.7

Reaction	$\Delta H°$, 298 K (kJ)	$\Delta S°$, 298 K (J/K)
(a)	−1410.94	−267.67
(b)	467.87	560.32
(c)	−393.509	2.862
(d)	1241.2	−461.50

Reaction (a) is product-favored at low T (room temperature) and reactant-favored at high T.
Reaction (b) is reactant-favored at low T, product-favored at high T.
Reaction (c) is product-favored at all values of T.
Reaction (d) is reactant-favored at all values of T.

17.8 (a)
$$\Delta S°_{system} = 2\text{ mol HCl(g)} \times S°(\text{HCl[g]}) - 1\text{ mol H}_2\text{[g]} \times S°(\text{H}_2\text{[g]}) - 1\text{ mol Cl}_2\text{[g]} \times S°(\text{Cl}_2\text{[g]})$$
$$= (2 \times 186.908 - 130.684 - 223.066)\text{ J/K}$$
$$= 20.055\text{ J/K}$$
$$\Delta S°_{surroundings} = -\Delta H°/T = -\Delta H°/T$$
$$= -[2\text{ mol HCl(g)} \times (-92.307\text{ kJ/mol})]/298.15\text{ K}$$
$$= 619.20\text{ J/K}$$
(b) $\Delta S°_{universe} = (619.20 + 20.066)\text{ J/K} = 639.27\text{ J/K}$

(Product-favored)

17.9

Sign of $\Delta H°$	Sign of $\Delta S°$	Sign of $\Delta G°$	Product-favored?
Negative (exothermic)	Positive	Negative	Yes
Negative (exothermic)	Negative	Depends on T	Yes at low T; no at high T
Positive (endothermic)	Positive	Depends on T	No at low T; yes at high T
Positive (endothermic)	Negative	Positive	No

17.10 (a) At 400 °C the equation is
$$2\text{ HgO(s)} \longrightarrow 2\text{ Hg(g)} + \text{O}_2\text{(g)}$$
Because Hg(g) is a product, instead of Hg(ℓ), both $\Delta H°$ and $\Delta S°$ will have significantly different values above 356 °C from their values below 356 °C. Therefore, the method of estimating $\Delta G°$ would not work above 356 °C.
(b) At 400 °C the entropy change should be more positive, which would make the reaction more product-favored. Because $\Delta H°$ and $\Delta S°$ are both positive, the reaction is product-favored at high temperatures.

17.11 (a)
$$\Delta S°(\text{i}) = \{2 \times (27.78) + \tfrac{3}{2} \times (205.138) - (87.40)\}\text{ J/K} = 275.86\text{ kJ}$$
$$\Delta H°(\text{i}) = -\Delta H_f°(\text{Fe}_2\text{O}_3\text{[s]}) = 824.2\text{ kJ}$$
$$\Delta G°(\text{i}) = -\Delta G_f°(\text{Fe}_2\text{O}_3\text{[s]}) = 742.2\text{ kJ}$$
$$\Delta S°(\text{ii}) = \{50.92 - 2 \times (28.3) - \tfrac{3}{2} \times (205.138)\}\text{ J/K} = -313.4\text{ kJ}$$
$$\Delta H°(\text{ii}) = \Delta H_f°(\text{Al}_2\text{O}_3\text{[s]}) = -1675.7\text{ kJ}$$
$$\Delta G°(\text{ii}) = \Delta G_f°(\text{Al}_2\text{O}_3\text{[s]}) = -1582.3\text{ kJ}$$
Step (i) is reactant-favored. Step (ii) is product-favored.
(b) Net reaction
$Fe_2O_3(s) + 2\,Al(s) \rightarrow 2\,Fe(s) + Al_2O_3(s)$
$\Delta S° = 275.86\text{ J/K} + (-313.4\text{ J/K}) = -37.5\text{ J/K}$
$\Delta H° = 824.2\text{ kJ} + (-1675.7\text{ kJ}) = -851.5\text{ kJ}$
$\Delta G° = 742.2\text{ kJ} + (-1582.3\text{ kJ}) = -840.1\text{ kJ}$
The net reaction has negative $\Delta G°$ and is therefore product-favored. For the *net* reaction, $\Delta S°$, $\Delta H°$, and $\Delta G°$ are all negative.
(c) If the two reactions are coupled, it is possible to obtain iron from iron(III) oxide even though that reaction is not product-favored by itself. The large negative $\Delta G°$ for formation of $Al_2O_3(s)$ makes the overall $\Delta G°$ negative for the coupled reactions.
(d) $Mg(s) + \tfrac{1}{2}O_2(g) \rightarrow MgO(s)$
$$\Delta G° = \Delta G_f°(\text{MgO[s]}) = -569.43\text{ kJ}$$
Coupling the reactions, we have
$$\text{Fe}_2\text{O}_3\text{(s)} \longrightarrow 2\text{ Fe(s)} + \tfrac{3}{2}\text{O}_2\text{(g)} \qquad \Delta G_1° = 742.2\text{ kJ}$$
$$3 \times (\text{Mg[s]} + \tfrac{1}{2}\text{O}_2\text{[g]} \longrightarrow \text{MgO[s]}) \qquad \Delta G_2° = 3(-569.43)\text{ kJ} = -1708.29\text{ kJ}$$
$$\text{Fe}_2\text{O}_3\text{(s)} + 3\text{ Mg(s)} \longrightarrow 2\text{ Fe(s)} + 3\text{ MgO(s)} \qquad \Delta G_3° = -966.1\text{ kJ}$$

Chapter 18

18.1 This is an application of the law of conservation of matter. If the number of electrons gained were different from the number of electrons lost, some electrons must have been created or destroyed.

18.2 Removal of the salt bridge would effectively switch off the flow of electricity from the battery.

18.3 Avogadro's number of electrons is 96,500 coulombs of charge, so it is 96,500 times as large as one coulomb of charge.

18.4 The zinc anode could be weighed before the battery was put into use. After a period of time, the zinc anode could be dried and reweighed. A loss in weight would be interpreted as being caused by the loss of Zn atoms from the surface through oxidation.

18.5 No, because Hg^{2+} ions can oxidize Al metal to Al^{3+} ions. The net cell reaction is
$$2\text{ Al(s)} + 3\text{ Hg}^{2+}\text{(aq)} \longrightarrow 2\text{ Al}^{3+}\text{(aq)} + 3\text{ Hg}(\ell)$$
$$E_{cell} = +2.51\text{ V}$$

18.6 For this table,
(a) V^{2+} ion is the weakest oxidizing agent.
(b) Cl_2 is the strongest oxidizing agent.
(c) V is the strongest reducing agent.
(d) Cl^- is the weakest reducing agent.
(e) No, E_{cell} for that reaction would be <0.
(f) No, E_{cell} for that reaction would be <0.
(g) Pb can reduce I_2 and Cl_2.

18.7 During charging, the reactions at each electrode are reversed. At the electrode that is normally the anode, the charging reaction is

$$Cd(OH)_2(s) + 2\,e^- \longrightarrow Cd(s) + 2\,OH^-(aq)$$

This is reduction, so this electrode is now a cathode.
At the electrode that is normally the cathode, the charging reaction is

$$Ni(OH)_2 + OH^-(aq) \longrightarrow NiO(OH)(s) + H_2O(\ell) + e^-$$

This is oxidation, so this electrode is now an anode.

18.8 Remove the lead cathodes and as much sulfuric acid as you can from the discharged battery. Find some steel and construct a battery with Cl_2 gas flowing across a piece of steel. The two half-reactions would be

$$Cl_2(g) + 2\,e^- \longrightarrow 2\,Cl^-(aq) \qquad +1.36\text{ V}$$

$$Pb(s) + SO_4^{2-}(aq) \longrightarrow PbSO_4(s) + 2\,e^- \qquad +0.356\text{ V}$$

$$E_{cell} = 1.36 + 0.356 = 1.71\text{ V}$$

18.9 Potassium metal was produced at the cathode.
Oxidation reaction: $2\,F^-\text{ (molten)} \rightarrow F_2(g) + 2\,e^-$
Reduction reaction: $2\,(K^+[\text{molten}] + e^- \rightarrow K[\ell])$
Net cell reaction:
$2\,K^+\text{ (molten)} + 2\,F^-\text{ (molten)} \rightarrow 2\,K(\ell) + F_2(g)$

18.10 Reaction (c) making 2 mol of Cu from Cu^{2+} would require 4 Faradays of electricity. Two F are required for part (b), and 3 F are required for part (a).

18.11 $C = 2.00 \times 10^4\text{ A} \times 100.\text{ h} \times \dfrac{3600\text{ s}}{\text{h}} = 7.20 \times 10^9\text{ C}$

$$7.20 \times 10^9\text{ C} \times \frac{1\text{ mol e}^-}{9.65 \times 10^4\text{ C}} \times \frac{1\text{ mol NaOH}}{1\text{ mol e}^-} \times \frac{40.00\text{ g NaOH}}{1\text{ mol NaOH}} \times \frac{1\text{ lb}}{454\text{ g}} \times \frac{1\text{ ton}}{2000\text{ lb}} = 3.29\text{ tons NaOH}$$

18.12 First, calculate how many coulombs of electricity are required to make this much aluminum.

$$(2000.\text{ ton Al})\left(\frac{2000\text{ lb Al}}{1\text{ ton Al}}\right)\left(\frac{454.6\text{ g Al}}{1\text{ lb Al}}\right)\left(\frac{1\text{ mol Al}}{26.982\text{ g Al}}\right) \times \left(\frac{3\text{ mol e}^-}{1\text{ mol Al}}\right)\left(\frac{96{,}500\text{ C}}{1\text{ mol e}^-}\right) = 1.950 \times 10^{13}\text{ C}$$

Next, using the product of charge and voltage, calculate how many joules are required; then convert to kilowatt-hours.

$$\text{Energy} = (1.950 \times 10^{13}\text{ C})(4.0\text{ V})\left(\frac{1\text{ J}}{1\text{ C} - \times 1\text{ V}}\right) \times \left(\frac{1\text{ kWh}}{3.60 \times 10^6\text{ J}}\right) = 2.2 \times 10^7\text{ kWh}$$

18.13 To calculate how much energy is stored in a battery, you need the voltage and the number of coulombs of charge the battery can provide. The voltage is generally given on the battery label. To determine the number of coulombs available, you would have to disassemble the battery and determine the masses of the chemicals at the cathode and anode.

18.14 $$(0.50\text{ A})(20.\text{ min})\left(\frac{60\text{ s}}{1\text{ min}}\right)\left(\frac{1\text{ C}}{1\text{ A s}}\right)\left(\frac{1\text{ mol e}^-}{96{,}500\text{ C}}\right) \times \left(\frac{1\text{ mol Ag}}{1\text{ mol e}^-}\right)\left(\frac{107.9\text{ g Ag}}{1\text{ mol Ag}}\right) = 0.67\text{ g Ag}$$

18.15 No, not all metals corrode as easily. Three metals that would corrode about as readily as Fe and Al are Zn, Mg, and Cd. Three metals that do not corrode as readily as Fe and Al are Cu, Ag, and Au. These three metals are used in making coins and jewelry. Metals fall into these two broad groups because of their relative ease of oxidation compared with the oxidation of H_2. In Table 18.1, you can see this breakdown easily.

18.16 (b) > (a) > (d) > (c)
Sand by the seashore, (b), would contain both moisture and salts, which would aid corrosion. Moist clay, (a), would contain water but less dissolved salts. If an iron object were embedded within the clay, its impervious nature might prevent oxygen from getting to the iron, which would also lower the rate of corrosion. Desert sand in Arizona, (d), would be quite dry, and this low-moisture environment would not lead to a rapid rate of corrosion. On the moon, (c), there would be a lack of moisture and oxygen. This would lead to a very low rate of corrosion.

Chapter 19

19.1 $^{235}_{92}U \rightarrow {}^4_2He + {}^{231}_{90}Th$
$^{231}_{90}Th \rightarrow {}^{0}_{-1}e + {}^{231}_{91}Pa$
$^{231}_{91}Pa \rightarrow {}^4_2He + {}^{227}_{89}Ac$
$^{227}_{89}Ac \rightarrow {}^4_2He + {}^{223}_{87}Fr$
$^{223}_{87}Fr \rightarrow {}^{0}_{-1}e + {}^{223}_{88}Ra$

19.2 $^{26}_{13}Al \rightarrow {}^{0}_{+1}e + {}^{26}_{12}Mg$
$^{26}_{13}Al + {}^{0}_{-1}e \rightarrow {}^{26}_{12}Mg$

19.3 Mass difference $= \Delta m = -0.03438$ g/mol
$\Delta E = (-3.438 \times 10^{-5}\text{ kg/mol})(2.998 \times 10^8\text{ m/s})^2$
$= -3.090 \times 10^{12}\text{ J/mol} = -3.090 \times 10^8\text{ kJ/mol}$
E_b per nucleon $= 5.150 \times 10^8$ kJ/nucleon
E_b for ^6Li is smaller than E_b for ^4He; therefore, helium-4 is more stable than lithium-6.

19.4 From the graph it can be seen that the binding energy per nucleon increases more sharply for the fusion of lighter elements than it does for heavy elements undergoing fission. Therefore, fusion is more exothermic per gram than fission.

19.5 $(\frac{1}{2})^{10} = 9.8 \times 10^{-4}$; this is equivalent to 0.098% of the radioisotope remaining.

19.6 All the lead came from the decay of ^{238}U; therefore, at the time the rock was dated, $N = 100$ and $N_0 = 109$. The decay constant, k, can be determined:

$$k = \frac{0.693}{4.51 \times 10^9\text{ y}} = 1.54 \times 10^{-10}\text{ y}^{-1}$$

The age of the rock (t) can be calculated using Equation 19.3:

$$\ln\frac{109}{209} = -(1.54 \times 10^{-10}\text{ y}^{-1}) \times t$$

$$t = 4.22 \times 10^9\text{y}$$

19.7 Ethylene is derived from petroleum, which was formed millennia ago. The half-life of ^{14}C is 5730 y, and thus much of ethylene's ^{14}C would have decayed and would be much less than that of the ^{14}C alcohol produced by fermentation.

19.8 Burning a metric ton of coal produces 2.8×10^7 kJ of energy.

$$\left(\frac{2.8 \times 10^4\text{ kJ}}{1.0\text{ kg}}\right)\left(\frac{10^3\text{ kg}}{\text{metric ton}}\right) = 2.8 \times 10^7\text{ kJ of energy}$$

The fission of 1.0 kg of ^{235}U produces

$$\frac{2.1 \times 10^{10}\text{ kJ}}{0.235\text{ kg }^{235}U} = 8.93 \times 10^{10}\text{ kJ}$$

It would require burning 3.2×10^3 metric tons of coal to equal the amount of energy from 1.0 kg of ^{235}U:

$$8.93 \times 10^{10}\text{ kJ from }^{235}U \times \frac{1\text{ metric ton coal}}{2.9 \times 10^7\text{ kJ}} = 3.2 \times 10^3\text{metric tons}$$

19.9 $k = \dfrac{0.693}{29.1 \text{ y}} = 2.38 \times 10^{-2} \text{ y}^{-1}$

$\ln(\text{fraction}) = -(2.38 \times 10^{-2} \text{ y}^{-1})$ (22 y, as of 2008) $= -0.5236$
fraction $= e^{-0.5236} = 0.592 = 59.2\%$

19.10 $k = \dfrac{0.693}{30.2 \text{ y}} = 2.29 \times 10^{-2} \text{ y}^{-1}$

(a) 60% drop in activity; 40% activity remaining
$\ln(0.40) = -0.916 = -(2.29 \times 10^{-2} \text{ y}^{-1}) \times t$

$$t = \frac{-0.916}{-2.29 \times 10^{-2} \text{ y}^{-1}} = 40 \text{ y}$$

(b) 90% drop in activity, 10% remains
$\ln(0.10) = -2.30 = -(2.29 \times 10^{-2} \text{ y}^{-1}) \times t$

$$t = \frac{-2.30}{-2.29 \times 10^{-2} \text{ y}^{-1}} = 100 \text{ y}$$

19.11 (a) ${}^{7}_{3}\text{Li} + {}^{1}_{1}\text{H} \rightarrow {}^{1}_{0}\text{n} + {}^{7}_{4}\text{Be}$
(b) ${}^{2}_{1}\text{H} + {}^{3}_{2}\text{He} \rightarrow {}^{4}_{2}\text{He} + {}^{1}_{1}\text{H}$

19.12 $k = \dfrac{0.693}{3.82 \text{ d}} = 0.181 \text{ d}^{-1}$

(a) The drop from 8 to 4 pCi represents one half-life, 3.82 days.
(b) $\ln\left(\dfrac{1.5}{8}\right) = -1.67 = -(0.181 \text{ d}^{-1}) \times t$

$$t = \frac{-1.67}{-0.181 \text{ d}^{-1}} = 9.25 \text{ d}$$

Answers to Selected Questions for Review and Thought

Chapter 1

8. (a) Quantitative (b) Qualitative (c) Qualitative
(d) Quantitative and qualitative (e) Qualitative
10. Sulfur is a pale yellow, powdery solid. Bromine is a dark, red-brown liquid and a red-brown gas that fills the upper part of the flask. Both the melting point and the boiling point of sulfur must be above room temperature. The boiling point, but not the melting point, of bromine must be above room temperature. Both substances are colored. Most of their other properties appear to be different.
12. The liquid will boil because your body temperature of 37 °C is above the boiling point of 20 °C.
14. Copper
16. Aluminum
18. (a) Physical (b) Chemical
(c) Chemical (d) Physical
20. (a) Chemical (b) Chemical (c) Physical
22. Heterogeneous; use a magnet.
24. (a) Homogeneous (b) Heterogeneous
(c) Heterogeneous (d) Heterogeneous
26. (a) A compound that decomposed
(b) A compound that decomposed
28. (a) Heterogeneous mixture (b) Pure compound
(c) Heterogeneous mixture (d) Homogeneous mixture
30. (a) No (b) Maybe
32. The macroscopic world; a parallelepiped shape; the atom crystal arrangement is a parallelepiped shape.
34. Carbon dioxide molecules are crowded in the unopened can. When the can is opened, the molecules quickly escape through the hole.
36. When sucrose is heated, the motion of the atoms increases. Only when that motion is extreme enough, will the bonds in the sucrose break, allowing for the formation of new bonds to produce the "caramelization" products.
38. (a) 3.275×10^4 m (b) 3.42×10^4 nm (c) 1.21×10^{-3} μm
40. Because atoms in the starting materials must all be accounted for in the substances produced, and because the mass of each atom does not change, there would be no change in the mass.
42. (Remember, you are instructed to use your own words to answer this question.) Consider two compounds that both contain the same two elements. In each compound, the proportion of these two elements is a whole-number integer ratio. Because they are different compounds, these ratios must be different. If you pick a sample of each of these compounds such that both samples contain the same number of atoms of the first element, and then you count the number of atoms of the second type, you will find that a small integer relationship exists between the number of atoms of the second type in the first compound and the number of atoms of the second type in the second compound.
44. Many responses are equally valid here. Common examples given here: (a) iron, Fe; gold, Au (b) carbon, C; hydrogen, H
(c) boron, B; silicon, Si (d) nitrogen, N_2; oxygen, O_2.
46.

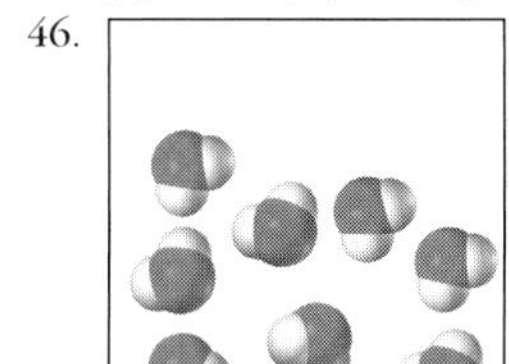

(a) H_2O

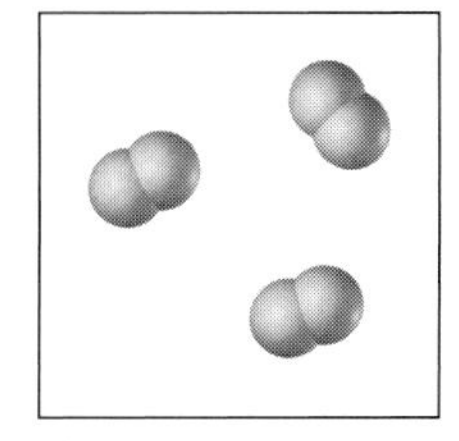

(b) N_2

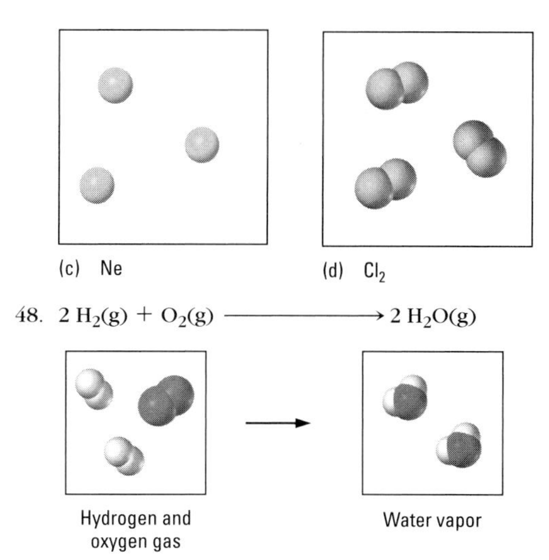

50. (a) Mass is quantitative and related to a physical property. Colors are qualitative and related to physical properties. Reaction is qualitative and related to a chemical property.
(b) Mass is quantitative and related to a physical property. The fact that a chemical reaction occurs between substances is qualitative information and related to a chemical property.
52. In solid calcium, smaller radius atoms are more closely packed, making a smaller volume. In solid potassium, larger radius atoms are less closely packed, making a larger volume.
54. (a) Bromobenzene (b) Gold (c) Lead
56. (a) 2.7×10^2 mL ice
(b) Bulging, cracking, deformed, or broken
58. (a) Water layer on top of bromobenzene layer.
(b) If it is poured slowly and carefully, ethanol will float on top of the water and slowly dissolve in the water. Both ethanol and water will float on the bromobenzene.
(c) Stirring will speed up dissolving of ethanol and water in each other. After stirring only two layers will remain.
60. Drawing (b)
62. 6.02×10^{-29} m^3
64. (a) Gray and blue (b) Lavender (c) Orange
66. It is difficult to prove that something cannot be broken down.
68. (a) Nickel, lead, and magnesium (b) Titanium
70. Obtain four or more lemons. Keep one unaltered to be the "control" case. Perform designated tasks to others, including applying both tasks to the same lemon; then juice all of them, recording results, such as juice volume and ease of task. Repeat with more lemons to achieve better reliability. Hypothesis: Disrupting the "juice sacks" inside the pulp helps to release the juice more easily.

Chapter 2

5. 40,000 cm
8. 1.97×10^3 cm^3, 1.97 L
9. 1550 in^2
11. 2.8×10^9 m^3
13. (a) 4 (b) 3 (c) 4
(d) 4 (e) 3
15. (a) 4.33×10^{-4} (b) 4.47×10^1
(c) 2.25×10^1 (d) 8.84×10^{-3}
17. (a) 1.9 g/mL (b) 218.4 cm^3
(c) 0.0217 (d) 5.21×10^{-5}
19. 80.1% silver, 19.9% copper
21. 245 g sulfuric acid
25. 27 protons, 27 electrons, and 33 neutrons
27. 78.92 amu/atom

29. (a) 20 e^-, 20 p^+, 20 n^0 (b) 50 e^-, 50 p^+, 69 n^0
(c) 94 e^-, 94 p^+, 150 n^0

31.

Z	A	Number of Neutrons	Element
35	81	46	Br
46	108	62	Pd
77	192	115	Ir
63	151	88	Eu

33. $^{18}_{9}X$, $^{20}_{9}X$, and $^{15}_{9}X$

35. (0.07500 × 6.015121 amu/atom ^6Li) + (0.9250 × 7.016003 amu/atom ^7Li) = 6.941 amu/atom

37. 60.12% ^{69}Ga, 39.87% ^{71}Ga

39. ^7Li

41. (a) 27 g B (b) 0.48 g O_2
(c) 6.98×10^{-2} g Fe (d) 2.61×10^3 g H

43. (a) 1.9998 mol Cu (b) 0.499 mol Ca
(c) 0.6208 mol Al (d) 3.1×10^{-4} mol K
(e) 2.1×10^{-5} mol Am

45. 4.131×10^{23} Cr atoms

47. 1.055×10^{-22} g Cu

51. Five; nonmetal: carbon (C), metalloids: silicon (Si) and germanium (Ge), and metals: tin (Sn) and lead (Pb).

53. (a) I (b) In (c) Ir (d) Fe

55. (a) Mg (b) Na (c) C (d) S
(e) I (f) Mg (g) Kr (h) O
(i) Ge [Other answers are possible for (a), (b), and (i).]

58. (a) 0.197 nm (b) 197 pm

59. (a) 0.178 nm^3 (b) 1.78×10^{-22} cm^3

61. 89 tons/yr

62. ^{39}K

64. 0.038 mol

65. $5,070

66. 3.4 mol, 2.0×10^{24} atoms

67. (a) Not possible (b) Possible (c) Not possible
(d) Not possible (e) Possible (f) Not possible

69. (a) Same (b) Second (c) Same
(d) Same (e) Same (f) Second
(g) Same (h) First (i) Second
(j) First

71. (a) 270 mL (b) No

75. (a) Se (b) ^{39}K (c) ^{79}Br (d) ^{20}Ne

Chapter 3

5. (a) BrF_3 (b) XeF_2 (c) P_2F_4 (d) $C_{15}H_{32}$ (e) N_2H_4

7. (a) C_6H_6 (b) $C_6H_8O_6$

9. (a) 1 calcium atom, 2 carbon atoms, and 4 oxygen atoms
(b) 8 carbon atoms and 8 hydrogen atoms
(c) 2 nitrogen atoms, 8 hydrogen atoms, 1 sulfur atom, and 4 oxygen atoms
(d) 1 platinum atom, 2 nitrogen atoms, 6 hydrogen atoms, and 2 chlorine atoms
(e) 4 potassium atoms, 1 iron atom, 6 carbon atoms, and 6 nitrogen atoms

11. (a) Same number of atoms of each kind
(b) Different bonding arrangements

13. (a) +2 (b) +2 (c) +2 or +3 (d) +3

15. (c) and (d) are correct formulas. (a) $AlCl_3$ (b) NaF

17. (a) 1 Pb^{2+} and 2 NO_3^- (b) 1 Ni^{2+} and 1 CO_3^{2-}
(c) 3 NH_4^+ and 1 PO_4^{3-} (d) 2 K^+ and 1 SO_4^{2-}

19. (a) $Ni(NO_3)_2$ (b) $NaHCO_3$ (c) LiClO
(d) $Mg(ClO_3)_2$ (e) $CaSO_3$

21. (b), (c), and (e) are ionic.

23. (a) $(NH_4)_2CO_3$ (b) CaI_2 (c) $CuBr_2$ (d) $AlPO_4$

25. (a) Potassium sulfide (b) Nickel(II) sulfate (c) Ammonium phosphate (d) Aluminum hydroxide (e) Cobalt(III) sulfate

27. (a) K^+ and OH^- (b) K^+ and SO_4^{2-}
(c) Na^+ and NO_3^- (d) NH_4^+ and Cl^-

29. (a) and (d)

31.

	CH_3OH	Carbon
No. of moles	One	One
No. of molecules or atoms	6.022×10^{23} molecules	6.022×10^{23} atoms
Molar mass	32.0417 g/mol	12.0107 g/mol

	Hydrogen	Oxygen
No. of moles	Four	One
No. of molecules or atoms	2.409×10^{24} atoms	6.022×10^{23} atoms
Molar mass	4.0316 g/mol	15.9994 g/mol

33. (a) 159.688 g/mol (b) 67.806 g/mol (c) 44.0128 g/mol
(d) 197.905 g/mol (e) 176.1238 g/mol

35. (a) 0.0312 mol (b) 0.0101 mol (c) 0.0125 mol
(d) 0.00406 mol (e) 0.00599 mol

37. (a) 151.1622 g/mol (b) 0.0352 mol (c) 25.1 g

39. 1.2×10^{24} molecules

40. (a) 0.250 mol CF_3CH_2F
(b) 6.02×10^{23} F atoms

41. (a) 239.3 g/mol PbS, 86.60% Pb, 13.40% S
(b) 30.0688 g/mol C_2H_6, 79.8881% C, 20.1119% H
(c) 60.0518 g/mol CH_3CO_2H, 40.0011% C, 6.7135% H, 53.2854% O
(d) 80.0432 g/mol NH_4NO_3, 34.9979% C, 5.0368% H, 59.9654% O

43. 245.745 g/mol, 25.858% Cu, 22.7992% N, 5.74197% H, 13.048% S, 32.5528% O

45. One

49. $C_4H_4O_4$

50. $C_2H_3OF_3$

52. KNO_3

55. $C_5H_{14}N_2$

56. $x = 7$

61. (a) (i) Chlorine tribromide (ii) Nitrogen trichloride
(iii) Calcium sulfate (iv) Heptane (v) Xenon tetrafluoride
(vi) Oxygen difluoride (vii) Sodium iodide (viii) Aluminum sulfide
(ix) Phosphorus pentachloride (x) Potassium phosphate
(b) (iii), (vii), (viii), and (x)

63. (a) CO_2F_2 (b) CO_2F_2

68.

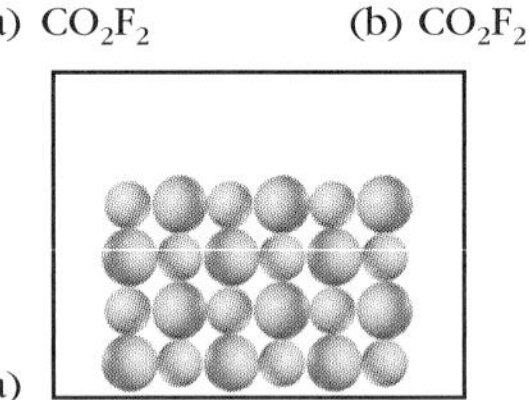

(a)

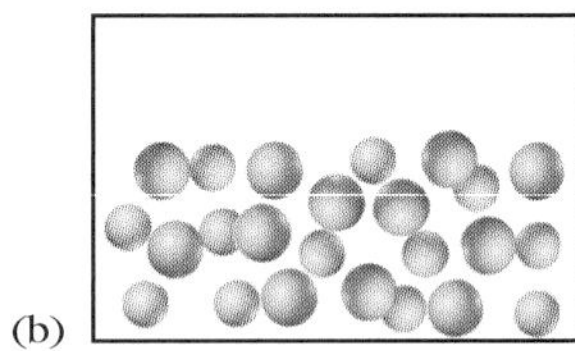

(b)

70. (a) Three (b) Three pairs are identical:

$CH_3—CH_2—CH_2—OH$ and $HO—CH_2—CH_2$ (with CH_3 attached to the last CH_2)

$CH_3—CH—CH_3$ (with OH on CH) and $HO—CH—CH_3$ (with CH_3 on CH)

$CH_3—O—CH_2—CH_3$ and $CH_3—CH_2—O—CH_3$

71. Tl_2CO_3, Tl_2SO_4

73. (a) Perbromate, bromate, bromite, hypobromite
(b) Selenate, selenite

76. (a) 0.0130 mol Ni (b) NiF_2
(c) Nickel(II) fluoride

77. CO_2

79. 5.0 lb N, 4.4 lb P, 4.2 lb K

Chapter 4

6.

	NH_3	O_2
No. molecules	4	5
No. atoms	16	10
No. moles of molecules	4	5
Mass	68.1216 g	159.9940 g
Total mass of reactants	228.1156 g	

	NO	H_2O
No. molecules	4	6
No. atoms	8	18
No. moles of molecules	4	6
Mass	120.0244 g	108.0912 g
Total mass of products	228.1156 g	

8. Equation (b)
11. (a) Combination (b) Decomposition (c) Exchange (d) Displacement
13. (a) Decomposition (b) Displacement (c) Combination (d) Exchange
15. (a) $2\ Mg(s) + O_2(g) \rightarrow 2\ MgO(s)$, magnesium oxide
(b) $2\ Ca(s) + O_2(g) \rightarrow 2\ CaO(s)$, calcium oxide
(c) $4\ In(s) + 3\ O_2(g) \rightarrow 2\ In_2O_3(s)$, indium oxide
17. (a) $UO_2(s) + 4\ HF(\ell) \rightarrow UF_4(s) + 2\ H_2O(\ell)$
(b) $B_2O_3(s) + 6\ HF(\ell) \rightarrow 2\ BF_3(s) + 3\ H_2O(\ell)$
(c) $BF_3(g) + 3\ H_2O(\ell) \rightarrow 3\ HF(\ell) + H_3BO_3(s)$
19. (a) $H_2NCl(aq) + 2\ NH_3(g) \rightarrow NH_4Cl(aq) + N_2H_4(aq)$
(b) $(CH_3)_2N_2H_2(\ell) + 2\ N_2O_4(g) \rightarrow 3\ N_2(g) + 4\ H_2O(g) + 2\ CO_2(g)$
(c) $CaC_2(s) + 2\ H_2O(\ell) \rightarrow Ca(OH)_2(s) + C_2H_2(g)$
21. (a) $C_6H_{12}O_6 + 6\ O_2 \rightarrow 6\ CO_2 + 6\ H_2O$
(b) $C_5H_{12} + 8\ O_2 \rightarrow 5\ CO_2 + 6\ H_2O$
(c) $2\ C_7H_{14}O_2 + 19\ O_2 \rightarrow 14\ CO_2 + 14\ H_2O$
(d) $C_2H_4O_2 + 2\ O_2 \rightarrow 2\ CO_2 + 2\ H_2O$
23. 50.0 mol HCl
25. 1.1 mol O_2, 35 g O_2, 1.0×10^2 g NO_2
27. (a) 12.7 g Cl_2 (b) 0.179 mol $FeCl_2$, 22.7 g $FeCl_2$ expected
29.

$(NH_4)_2PtCl_6$	Pt	HCl
12.35 g	5.428 g	5.410 g
0.02782 mol	0.02782 mol	0.1484 mol

31. 2.0 mol, 36.0304 g
32. 0.699 g Ga and 0.751 g As
34. (a) 699 g (b) 526 g
35. $BaCl_2$, 1.12081 g $BaSO_4$
37. (a) Cl_2 is limiting. (b) 5.08 g Al_2Cl_6
(c) 1.67 g Al unreacted
38. (a) CO (b) 1.3 g H_2 (c) 85.2 g
39. 0 mol CaO, 0.19 mol NH_4Cl, 2.00 mol H_2O, 4.00 mol NH_3, 2.00 mol $CaCl_2$
42. 56.0%
44. 8.8%
48. SO_3
50. CH
52. $C_3H_6O_2$
54. 21.6 g N_2
56. Element (b)
57. 12.5 g $Pt(NH_3)_2Cl_2$
58. SiH_4
62. Equation (b)
63. When the metal mass is less than 1.2 g, the metal is the limiting reactant. When the metal mass is greater than 1.2 g, the bromine is the limiting reactant.
65. $H_2(g) + 3\ Fe_2O_3(s) \rightarrow H_2O(\ell) + 2\ Fe_3O_4(s)$
67. 86.3 g
70. 44.9 amu
72. 0 g $AgNO_3$, 9.82 g Na_2CO_3, 6.79 g Ag_2CO_3, 4.19 g $NaNO_3$
74. 99.7% CH_3OH, 0.3% C_2H_5OH
75. (a) $C_9H_{11}NO_4$ (b) $C_9H_{11}NO_4$

Chapter 5

7. (a) soluble, Fe^{2+} and ClO_4^- (b) soluble, Na^+ and SO_4^{2-} (c) soluble, K^+ and Br^- (d) soluble, Na^+ and CO_3^{2-}
9. (a) soluble, K^+ and HPO_4^{2-} (b) soluble, Na^+ and ClO^- (c) soluble, Mg^{2+} and Cl^- (d) soluble, Ca^{2+} and OH^- (e) soluble, Al^{3+} and Br^-
11. (a) $MnCl_2(aq) + Na_2S(aq) \rightarrow MnS(s) + 2\ NaCl(aq)$
(b) NP
(c) NP
(d) $Hg(NO_3)_2(aq) + Na_2S(aq) \rightarrow HgS(s) + 2\ NaNO_3(aq)$
(e) $Pb(NO_3)_2(aq) + 2\ HCl(aq) \rightarrow PbCl_2(s) + 2\ HNO_3(aq)$
(f) $BaCl_2(aq) + H_2SO_4(aq) \rightarrow BaSO_4(s) + 2\ HCl(aq)$
13. (a) CuS insoluble;
$Cu^{2+} + H_2S(aq) \rightarrow CuS(s) + 2\ H^+$;
spectator ion is Cl^-.
(b) $CaCO_3$ insoluble;
$Ca^{2+} + CO_3^{2-} \rightarrow CaCO_3(s)$; spectator ions are K^+ and Cl^-.
(c) AgI insoluble;
$Ag^+ + I^- \rightarrow AgI(s)$; spectator ions are Na^+ and NO_3^-.
15. (a) $Zn(s) + 2\ HCl(aq) \rightarrow H_2(g) + ZnCl_2(aq)$
$Zn(s) + 2\ H^+(aq) + 2\ Cl^-(aq) \rightarrow H_2(g) + Zn^{2+}(aq) + 2\ Cl^-(aq)$
$Zn(s) + 2\ H^+(aq) \rightarrow H_2(g) + Zn^{2+}(aq)$
(b) $Mg(OH)_2(s) + 2\ HCl(aq) \rightarrow MgCl_2(aq) + 2\ H_2O(\ell)$
$Mg(OH)_2(s) + 2\ H^+(aq) + 2\ Cl^-(aq) \rightarrow Mg^{2+}(aq) + 2\ Cl^-(aq) + 2\ H_2O(\ell)$
$Mg(OH)_2(s) + 2\ H^+(aq) \rightarrow Mg^{2+}(aq) + 2\ H_2O(\ell)$
(c) $2\ HNO_3(aq) + CaCO_3(s) \rightarrow Ca(NO_3)_2(aq) + H_2O(\ell) + CO_2(g)$
$2\ H^+(aq) + 2\ NO_3^-(aq) + CaCO_3(s) \rightarrow Ca^{2+}(aq) + 2\ NO_3^-(aq) + H_2O(\ell) + CO_2(g)$
$2\ H^+(aq) + CaCO_3(s) \rightarrow Ca^{2+}(aq) + H_2O(\ell) + CO_2(g)$
(d) $4\ HCl(aq) + MnO_2(s) \rightarrow MnCl_2(aq) + Cl_2(g) + 2\ H_2O(\ell)$
$4\ H^+(aq) + 4\ Cl^-(aq) + MnO_2(s) \rightarrow Mn^{2+}(aq) + 2\ Cl^-(aq) + Cl_2(g) + 2\ H_2O(\ell)$
$4\ H^+(aq) + 2\ Cl^-(aq) + MnO_2(s) \rightarrow Mn^{2+}(aq) + Cl_2(g) + 2\ H_2O(\ell)$
17. (a) $Ca(OH)_2(s) + 2\ HNO_3(aq) \rightarrow Ca(NO_3)_2(aq) + 2\ H_2O(\ell)$
$Ca(OH)_2(s) + 2\ H^+(aq) + 2\ NO_3^-(aq) \rightarrow Ca^{2+}(aq) + 2\ NO_3^-(aq) + 2\ H_2O(\ell)$
$Ca(OH)_2(s) + 2\ H^+(aq) \rightarrow Ca^{2+}(aq) + 2\ H_2O(\ell)$
(b) $BaCl_2(aq) + Na_2CO_3(aq) \rightarrow BaCO_3(s) + 2\ NaCl(aq)$
$Ba^{2+}(aq) + 2\ Cl^-(aq) + 2\ Na^+(aq) + CO_3^{2-}(aq) \rightarrow BaCO_3(s) + 2\ Na^+(aq) + 2\ Cl^-(aq)$
$Ba^{2+}(aq) + CO_3^{2-}(aq) \rightarrow BaCO_3(s)$
(c) $2\ Na_3PO_4(aq) + 3\ Ni(NO_3)_2(aq) \rightarrow Ni_3(PO_4)_2(s) + 6\ NaNO_3(aq)$
$6\ Na^+(aq) + 2\ PO_4^{3-}(aq) + 3\ Ni^{2+}(aq) + 6\ NO_3^-(aq) \rightarrow Ni_3(PO_4)_2(s) + 6\ Na^+(aq) + 6\ NO_3^-(aq)$
$2\ PO_4^{3-}(aq) + 3\ Ni^{2+}(aq) \rightarrow Ni_3(PO_4)_2(s)$
19. $CdCl_2(aq) + 2\ NaOH(aq) \longrightarrow Cd(OH)_2(s) + 2\ NaCl(aq)$
$Cd^{2+}(aq) + 2\ Cl^-(aq) + 2\ Na^+(aq) + 2\ OH^-(aq) \longrightarrow Cd(OH)_2(s) + 2\ Na^+(aq) + 2\ Cl^-(aq)$
$Cd^{2+}(aq) + 2\ OH^-(aq) \longrightarrow Cd(OH)_2(s)$
21. (a) Base, strong, K^+ and OH^-
(b) Base, strong, Mg^{2+} and OH^-
(c) Acid, weak, small amounts of H^+ and ClO^-
(d) Acid, strong, H^+ and Br^-
(e) Base, strong, Li^+ and OH^-
(f) Acid, weak, small amounts of H^+, HSO_3^-, and SO_3^{2-}

23. (a) Acid: HNO_2, base: NaOH
complete ionic form:
$HNO_2(aq) + Na^+ + OH^- \longrightarrow H_2O(\ell) + Na^+ + NO_2^-$
net ionic form:
$HNO_2(aq) + OH^- \longrightarrow H_2O(\ell) + NO_2^-$
(b) Acid: H_2SO_4 base: $Ca(OH)_2$
complete ionic and net ionic forms:
$H^+ + HSO_4^- + Ca(OH)_2(s) \longrightarrow 2\,H_2O(\ell) + CaSO_4(s)$
(c) Acid: HI, base: NaOH
complete ionic form:
$H^+ + I^- + Na^+ + OH^- \longrightarrow H_2O(\ell) + Na^+ + I^-$
net ionic form: $H^+ + OH^- \longrightarrow H_2O(\ell)$
(d) Acid: H_3PO_4, base: $Mg(OH)_2$
complete ionic and net ionic forms:
$2\,H_3PO_4(aq) + 3\,Mg(OH)_2(s) \longrightarrow 6\,H_2O(\ell) + Mg_3(PO_4)_2(s)$
25. (a) Precipitation reaction; products are NaCl and MnS;
$MnCl_2(aq) + Na_2S(aq) \rightarrow 2\,NaCl(aq) + MnS(s)$ (b) Precipitation reaction; products are NaCl and $ZnCO_3$;
$Na_2CO_3(aq) + ZnCl_2(aq) \rightarrow 2\,NaCl(aq) + ZnCO_3(s)$
(c) Gas-forming reaction; products are $KClO_4$, H_2O, and CO_2;
$K_2CO_3(aq) + 2\,HClO_4(aq) \rightarrow 2\,KClO_4(aq) + H_2O(\ell) + CO_2(g)$
27. (a) Ox. # O = −2, Ox. # S = +6
(b) Ox. # O = −2, Ox. # H = +1, Ox. # N = +5
(c) Ox. # K = +1, Ox. # O = −2, Ox. # Mn = +7
(d) Ox. # O = −2, Ox. # H = +1
(e) Ox. # Li = +1, Ox. # O = −2, Ox. # H = +1
(f) Ox. # Cl = −1, Ox. # H = +1, Ox. # C = 0
29. (a) S: +6, O: −2 (b) N: +5, O: −2
(c) Mn: +7, O: −2 (d) Cr: +3, O: −2, H: +1
(e) P: +5, O: −2, H: +1 (f) S: +2, O: −2
32. Only reaction (b) is an oxidation-reduction reaction; oxidation numbers change. Reaction (a) is precipitation and reaction; (c) is acid-base neutralization.
34. Substances (b), (c), and (d)
36. (a) $CO_2(g)$ or $CO(g)$ (b) $PCl_3(g)$ or $PCl_5(g)$
(c) $TiCl_2(s)$ or $TiCl_4(s)$ (d) $Mg_3N_2(s)$
(e) $Fe_2O_3(s)$ (f) $NO_2(g)$
38. This is just an example:
$Fe(s) + 2\,HCl(aq) \longrightarrow FeCl_2(aq) + H_2(g)$
(a) Fe is oxidized. (b) HCl is reduced.
(c) The reducing agent is Fe(s).
(d) The oxidizing agent is HCl(aq).
(Other answers are possible.)
40. (a) NR (b) NR (c) NR
(d) $Au^{3+}(aq) + 3\,Ag(s) \longrightarrow 3\,Ag^+(aq) + Au(s)$
42. (a) 0.254 M Na_2CO_3 (b) 0.508 M Na^+, 0.254 M CO_3^{2-}
44. 0.494 g $KMnO_4$
46. 5.08×10^3 mL
48. 0.0150 M $CuSO_4$
50. 0.205 g Na_2CO_3
52. 121 mL HNO_3 solution
54. 0.18 g AgCl, NaCl, 0.0080 M NaCl
56. 1.192 M HCl
59. (a) $(NH_4)_2S(aq) + Hg(NO_3)_2(aq) \rightarrow HgS(s) + 2\,NH_4NO_3(aq)$
(b) Reactants: ammonium sulfide, mercury(II) nitrate; products: mercury(II) sulfide, ammonium nitrate
(c) $S^{2-}(aq) + Hg^{2+}(aq) \longrightarrow HgS(s)$
(d) Precipitation reaction
60. (a) Combination reaction; product: $H_2SO_4(aq)$
(b) Combination reaction; product: $SrH_2(s)$
(c) Displacement reaction; products: $MgSO_4(aq)$ and $H_2(g)$
(d) Exchange (precipitation) reaction; products: $Ag_3PO_4(s)$ and $NaNO_3(aq)$
(e) Decomposition and gas-forming reaction; products: CaO(s), $H_2O(\ell)$, and $CO_2(g)$
(f) Oxidation-reduction reaction; products: $Fe^{2+}(aq)$ and $Sn^{4+}(aq)$
61. (a) $NH_3(aq)$, H_2O (b) $CH_3CO_2H(aq)$, H_2O
(c) Na^+, OH^-, H_2O (d) H^+, Br^-, H_2O
63. (c) is a redox reaction. The oxidizing agent is Ti. The reducing agent is Mg.
64. Case 1 (a) Before: clear colorless solution; after: solid at the bottom of beaker with colorless solution above it
(b)

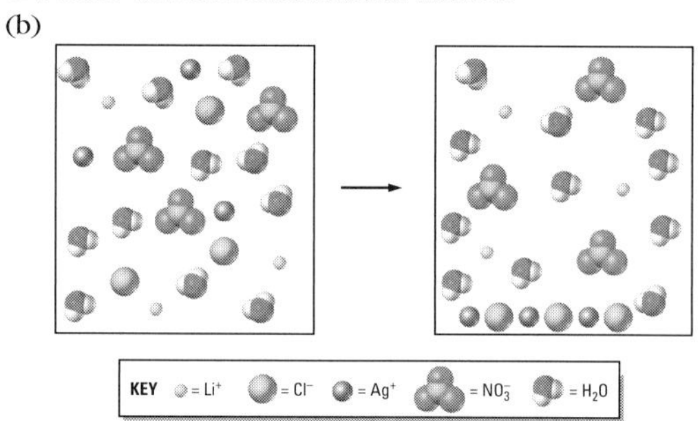

(c) $Li^+ + Cl^- + Ag^+ + NO_3^- \rightarrow Li^+ + AgCl(s) + NO_3^-$
Case 2 (a) Before and after: clear colorless solutions
(b)

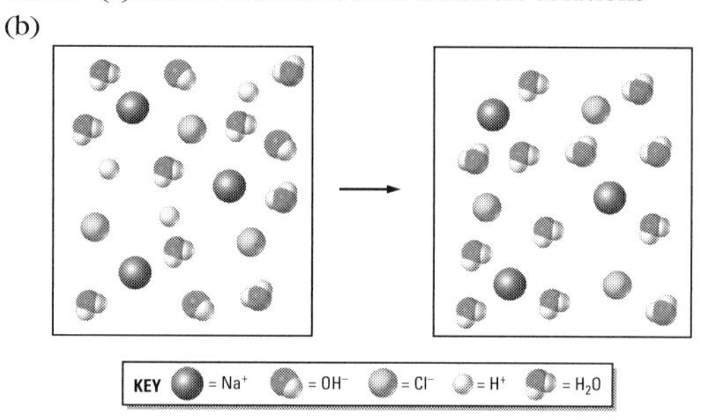

(c) $Na^+ + OH^- + H^+ + Cl^- \rightarrow Na^+ + H_2O(\ell) + Cl^-$
66. (a) Combine $Ba(OH)_2(aq)$ and $H_2SO_4(aq)$
(b) Combine $Na_2SO_4(aq)$ and $Ba(NO_3)_2(aq)$
(c) Combine $BaCO_3(s)$ and $H_2SO_4(aq)$
68. (d)
70. (a) and (d) are correct.
72. (a) Groups C and D: $Ag^+ + Cl^- \rightarrow AgCl(s)$, Groups A and B: $Ag^+ + Br^- \rightarrow AgBr(s)$
(b) Different silver halide produced in Group C and D than in Groups A and B
(c) Bromide is heavier than chloride.
73. 104 g/mol
76. 2.26 g, 1.45 g
77. 0.0154 M $CaSO_4$; 0.341 g $CaSO_4$ undissolved
78. 184 mL
80. 6.28% impurity

Chapter 6

7. (a) 399 Cal (b) 5.0×10^6 J/day
9. 3×10^8 J
11. (a) The chemical potential energy of the atoms in the match, fuse, and fuel are converted to thermal energy (due to the combustion reaction), potential energy (as the rocket's altitude increases), and light energy (colorful sparkles). (b) The chemical potential energy of the atoms in the fuel is converted to thermal energy (due to the combustion reaction), some of which is converted to kinetic energy (for the movement of the vehicle).
13. (a) The system: NH_4Cl; the surroundings: anything not NH_4Cl, including the water. (b) To study the release of energy during the phase change of this chemical, we must isolate it and see how it

interacts with the surroundings. (c) The system's interaction with the surroundings causes heat to be transferred into the system and from the surroundings. There is no material transfer in this process, but there is a change in the specific interaction between the water and system. (d) Endothermic

15. $\Delta E = +32$ J
17. Process (a) requires more energy than (b).
19. It takes less time to raise the Cu sample to body temperature.
21. More energy (1.48×10^6 J) is absorbed by the water than by the ethylene glycol (9.56×10^5 J).
23. Gold
25. Positive; Negative
27. 4.13×10^5 J
29. 5.00×10^5 J
31.

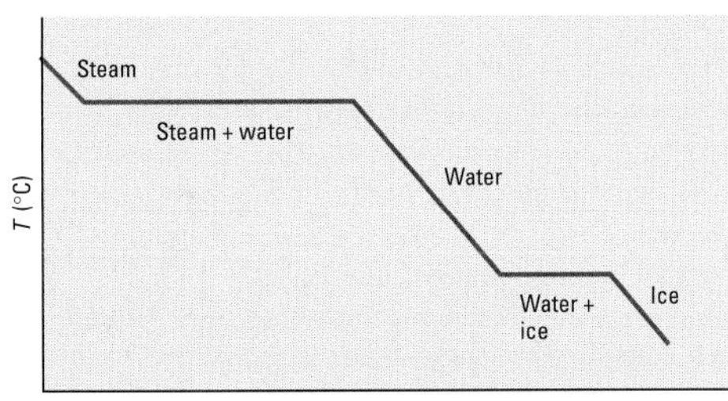

33. Endothermic
35. (a) −210 kJ (b) −33 kJ
37. (a) $\frac{1}{2}$ $C_8H_{18}(\ell) + \frac{25}{4} O_2(g) \rightarrow 4\ CO_2(g) + \frac{9}{2} H_2O(\ell)$ $\Delta H° = -2748.0$ kJ
    (b) $100\ C_8H_{18}(\ell) + 1250\ O_2(g) \rightarrow 800\ CO_2(g) + 900\ H_2O(\ell)$ $\Delta H° = -5.4960 \times 10^5$ kJ
    (c) $C_8H_{18}(\ell) + \frac{25}{2} O_2(g) \rightarrow 8\ CO_2(g) + 9\ H_2O(\ell)$ $\Delta H° = -5496.0$ kJ
39. $\frac{464.8 \text{ kJ}}{1 \text{ mol CaO}}, \frac{464.8 \text{ kJ}}{3 \text{ mol C}}, \frac{464.8 \text{ kJ}}{1 \text{ mol CaC}_2}, \frac{464.8 \text{ kJ}}{1 \text{ mol CO}}, \frac{1 \text{ mol CaO}}{464.8 \text{ kJ}},$
    $\frac{3 \text{ mol C}}{464.8 \text{ kJ}}, \frac{1 \text{ mol CaC}_2}{464.8 \text{ kJ}}, \frac{1 \text{ mol CO}}{464.8 \text{ kJ}}$
41. $\Delta H = -1450$ kJ/mol
43. 6×10^4 kJ released
45. HF
47. For $H_2 + F_2 \longrightarrow 2$ HF reaction: (a) 594 kJ (b) −1132 kJ (c) −538 kJ
    For $H_2 + Cl_2 \longrightarrow 2$ HCl reaction: (a) 678 kJ (b) −862 kJ (c) −184 kJ (d) Reaction to form HF is most exothermic.
49. 18 °C
51. 6.6 kJ
53. $\Delta H_f°(SrCO_3) = -1220.$ kJ/mol
55. (a) $2\ Al(s) + \frac{3}{2} O_2(g) \rightarrow Al_2O_3(s)$ $\Delta H° = -1675.7$ kJ
    (b) $Ti(s) + 2\ Cl_2(g) \rightarrow TiCl_4(\ell)$ $\Delta H° = -804.2$ kJ
    (c) $N_2(g) + 2\ H_2(g) + \frac{3}{2} O_2(g) \rightarrow NH_4NO_3(s)$ $\Delta H° = -365.56$ kJ
57. (a) $\Delta H° = 1372.5$ kJ (b) Endothermic
59. 41.2 kJ evolved
61. Gold reaches 100 °C first.
63. $\Delta H_f°(B_2H_6) = 36$ kJ/mol
65. $\Delta H_f°(C_2H_4Cl_2(\ell)) = -165.2$ kJ/mol
67. (a) 36.03 kJ evolved (b) 1.18×10^4 kJ evolved
69. Step 1: −137.23 kJ, Step 2: 275.341 kJ, Step 3: 103.71 kJ
    $H_2O\ (g) \longrightarrow H_2(g) + \frac{1}{2} O_2(g)$
    $\Delta H° = 241.82$ kJ, endothermic.
71. Substance A
73. Greater; a larger mass of water will contain a larger quantity of thermal energy at a given temperature.
75. The given reaction produces 2 mol SO_3. Formation enthalpy from Table 6.2 is for the production of 1 mol SO_3.
77. $\Delta H_f°(OF_2) = 18$ kJ/mol
79. CH_4, 50.014 kJ/g; C_2H_6, 47.484 kJ/g; C_3H_8, 46.354 kJ/g; C_4H_{10}, 45.7140 kJ/g; $CH_4 > C_2H_6 > C_3H_8 > C_4H_{10}$
81. (a) 26.6 °C (Above $C_6H_8O_6$ masses of 8.81 g, NaOH is the limiting reactant.)
    (b) $C_6H_8O_6$ limits in Experiments 1–3 and NaOH limits in Experiments 4 and 5.
    (c) Ascorbic acid has one hydrogen ion; equal quantities of reactant in Experiment 3 at stoichiometric equivalence point.

Chapter 7

5. (a) Radio waves have less energy than infrared.
   (b) Microwaves are higher frequency than radio waves.
7. (a) 3.00×10^{-3} m (b) 6.63×10^{-23} J/photon
   (c) 39.9 J/mol
9. 6.06×10^{14} Hz
11. 1.1×10^{15} Hz, 7.4×10^{-19} J/photon
12. X-ray (8.42×10^{-17} J/photon) energy is larger than that of orange light (3.18×10^{-19} J/photon).
15. Photons of this light are too low in energy. Increasing the intensity only increases the number of photons, not their individual energy.
16. No
19. (a) Absorbed (b) Emitted (c) Absorbed
    (d) Emitted
21. (a), (b), (d)
25. 4.576×10^{-19} J absorbed, 434.0 nm
27. 0.05 nm
30. (a) Cannot occur, m_ℓ too large (b) Can occur
    (c) Cannot occur, m_s cannot be 1, here.
    (d) Cannot occur, ℓ must be less than n. (e) Can occur
32. Four subshells
34. Electrons do not follow simple paths as planets do.
35. Orbits have predetermined paths—position and momentum are both exactly known at all times. Heisenberg's uncertainty principle says that we cannot know both simultaneously.
36. d, p, and s orbitals; nine orbitals total
39. ${}_{32}$Ge: $1s^22s^22p^63s^23p^63d^{10}4s^24p^2$
41. (a) $4s$ orbital must be full.
    (b) Orbital labels must be 3, not 2; electrons in $3p$ subshell should be in separate orbitals with parallel spin.
    (c) $4d$ orbitals must be completely filled before $5p$ orbitals start filling.
43. 18 elements, all possible orbital electron combinations are already used
45. Mn: $1s^22s^22p^63s^23p^63d^54s^2$; it has 5 unpaired electrons:

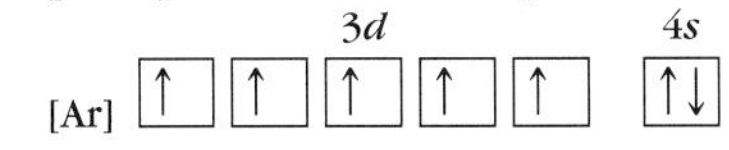

Mn^{2+}: $1s^22s^22p^63s^23p^63d^5$; it has 5 unpaired electrons:

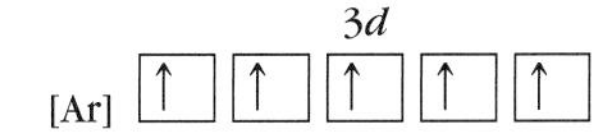

Mn^{3+}: $1s^22s^22p^63s^23p^63d^4$; it has 4 unpaired electrons:

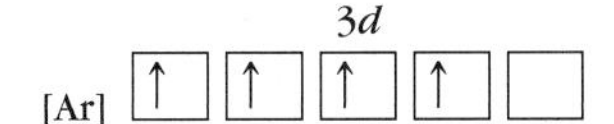

48. (a) ${}_{63}$Eu: [Xe]$4f^76s^2$ (b) ${}_{70}$Yb: [Xe]$4f^{14}6s^2$

50. (a) ·Sr· (b) :B̤r· (c) ·Ġa· (d) ·S̈b·
52. (a) [Ar] (b) [Ar] (c) [Ne]
Ca^{2+} and K^{+} are isoelectronic.
54. $_{50}Sn$: [Kr]$4d^{10}5s^{2}5p^{2}$
$_{50}Sn^{2+}$: [Kr]$4d^{10}5s^{2}$
$_{50}Sn^{4+}$: [Kr]$4d^{10}$
56. In both paramagnetic and ferromagnetic substances, atoms have unpaired spins and thus are attracted to magnets. Ferromagnetic substances retain their aligned spins after an external magnetic field has been removed, so they can function as magnets. Paramagnetic substances lose their aligned spins after a time and, therefore, cannot be used as permanent magnets.
58. P < Ge < Ca < Sr < Rb
60. (a) Rb smaller (b) O smaller (c) Br smaller
(d) Ba^{2+} smaller (e) Ca^{2+} smaller
62. (c)
64. Na; it must have one valence electron.
66. Adding a negative electron to a negatively charged ion requires additional energy to overcome the coulombic charge repulsion.
67. −862 kJ
69. Seven pairs
71. (a) He (b) Sc (c) Na
74. (a) Sulfur (b) Radium (c) Nitrogen
(d) Ruthenium (e) Copper
76. In^{4+}, Fe^{6+}, and Sn^{5+}; very high successive ionization energies
78. (a) Directly related, not inversely related
(b) Inversely proportional to *the square of* the principle quantum number, not inversely proportional to the principle quantum number
(c) Before, not as soon as
(d) Wavelength, not frequency
80. Ultraviolet; 91.18 nm or shorter wavelength is needed to ionize hydrogen.
81. XCl
83. (a) Ground state (b) Could be ground state or excited state
(c) Excited state (d) Impossible
(e) Excited state (f) Excited state
85. (a) Increase, decrease
(b) Helium
(c) 5 and 13
(d) He has only two electrons; there are no electrons left.
(e) The first electron is a valence electron, but the second electron is a core electron.
(f) $Mg^{2+}(g) \rightarrow Mg^{3+}(g) + e^{-}$
88. (a) 4.34×10^{-19} J (b) 6.54×10^{14} Hz
90. 2.18×10^{-18} J
91. To remove an electron from a N atom requires disrupting a half-filled subshell (p^3), which is relatively stable, so the ionization of O requires less energy than the ionization of N.
94. (a) Transition metals (b) [Rn]$7s^{2}5f^{14}6d^{10}$

Chapter 8

13. (a) :C̤l̈—F̤̈: (b) H—S̈e—H
(c) [:F̤̈—B(—F̤̈:)(—F̤̈:)—F̤̈:]⁻ (d) [:Ö—P(—Ö:)(—Ö:)—Ö:]³⁻
15. (a) :C̤l̈—C(H)(H)—H (b) [:Ö—Si(—Ö:)(—Ö:)—Ö:]⁴⁻
(c) [:F̤̈—C̤l(—F̤̈:)(—F̤̈:)—F̤̈:]⁺ (d) H—C(H)(H)—C(H)(H)—H
17. (a) (:F̤̈)₂C=C(F̤̈:)₂ (b) H₂C=C(H)—C≡N:
21. (a) Incorrect; the F atoms are missing electrons.
(b) Incorrect; the structure has 10 electrons, but needs 12 electrons.
(c) Incorrect. The structure has three too many electrons; Carbon has nine electrons. The single electron should be deleted. Oxygen has ten electrons; delete one pair of electrons.
(d) Incorrect; one hydrogen atom has more than two electrons. One hydrogen atom is completely missing.
(e) Incorrect; the structure has 16 electrons, but needs 18 electrons. The N atom doesn't follow the octet rule. It needs another pair of electrons in the form of a lone pair.
22. Four branched-chain compounds

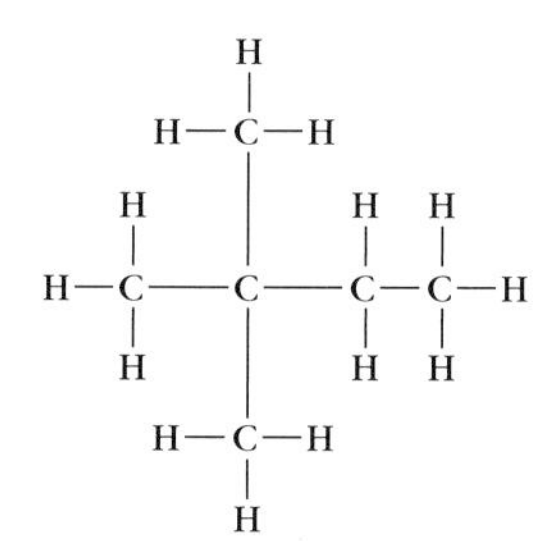

24. (a) Alkyne (b) Alkane (c) Alkene
26. (a) No
(b) Yes
cis: $CH_3CH_2(H)C=C(H)CH_2CH_3$ *trans*: $CH_3CH_2(H)C=C(H)CH_2CH_3$
(c) Yes
cis: $CH_3(H)C=C(H)CH_3$ *trans*: $CH_3(H)C=C(H)CH_3$
(d) No
28. No; free rotation about the single C—C bond prevents it.
30. (a) B—Cl (b) C—O (c) P—O (d) C=O
32. CO
34. CO_3^{2-} has longer C—O bonds.
36. −92 kJ; the reaction is exothermic.
39. (a) N, C, Br, O
(b) S—O is the most polar.

41. (a) 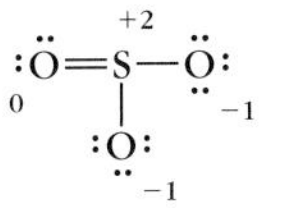 :Ö=S—Ö: with :Ö: below S; formal charges: O 0, S +2, O −1, O −1 (b) 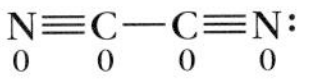 :N≡C—C≡N: (formal charges 0 0 0 0)
(c) [:Ö=N̈—Ö:]⁻ (formal charges 0, 0, −1)

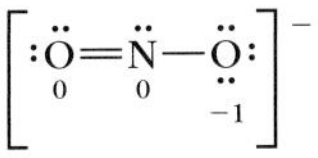

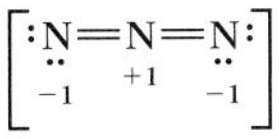

43. (a) 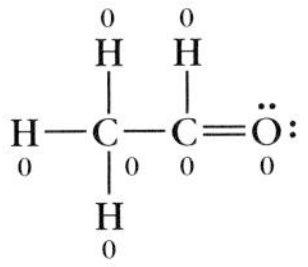 H—C(—H)(—H)—C(—H)=Ö: (formal charges: H 0, H 0, H 0, C 0, C 0, O 0, H 0) (b) [:N̈=N=N̈:]⁻ (formal charges −1, +1, −1)
(c) H—C(—H)(—H)—C≡N: (formal charges: H 0, H 0, H 0, C 0, C 0, N 0)

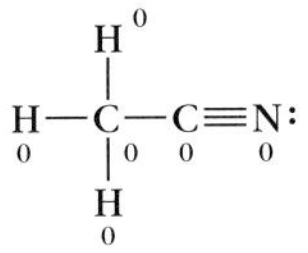

45. (a) H—Ö—N(=Ö:)(—Ö:) ⟷ H—Ö—N(—Ö:)(=Ö:) ⟷ H—O=N(—Ö:)(—Ö:)

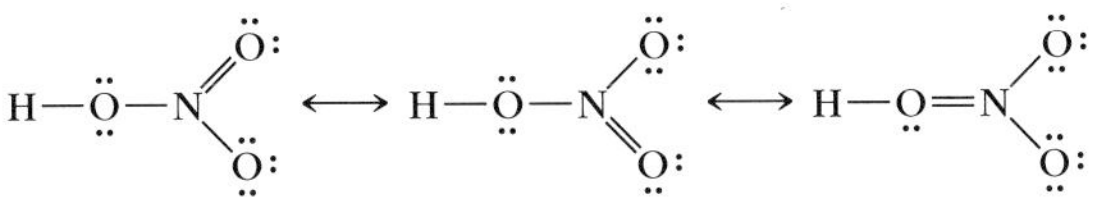

(b) [:Ö=N(—Ö:)(—Ö:)]⁻ ⟷ [:Ö—N(=Ö:)(—Ö:)]⁻ ⟷ [:Ö—N(—Ö:)(=Ö:)]⁻

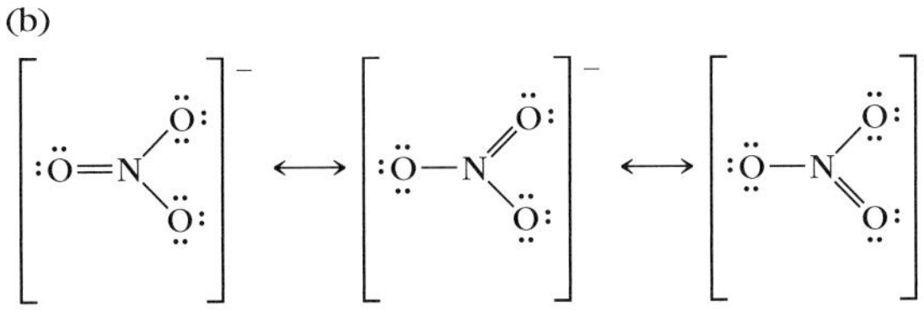

47. 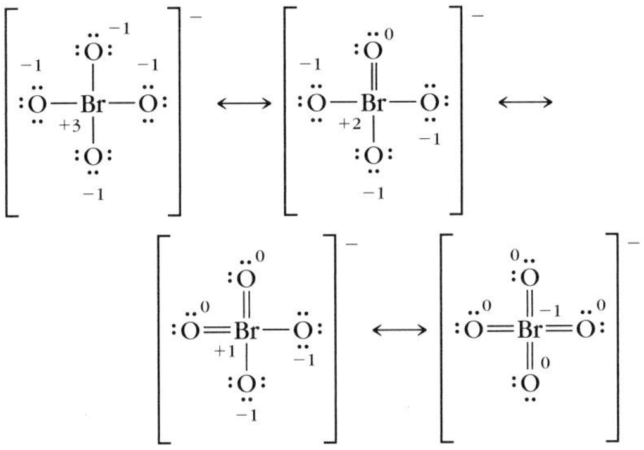 [:Ö—Br(—Ö:)(—Ö:)]⁻ (−1, +3, −1, −1, −1) ⟷ [:Ö—Br(=Ö)(—Ö:)(—Ö:)]⁻ (−1, +2, 0, −1, −1) ⟷
[:Ö=Br(=Ö)(—Ö:)(—Ö:)]⁻ (0, +1, 0, −1, −1) ⟷ [:Ö=Br(=Ö)(=Ö)=Ö:]⁻ (0, −1, 0, 0, 0) Most plausible

50. (a) 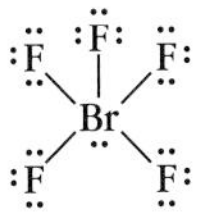 :F̈—Br(—F̈:)(—F̈:)(—F̈:)(—F̈:) (b) 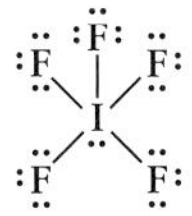 :F̈—I(—F̈:)(—F̈:)(—F̈:)(—F̈:) (c) 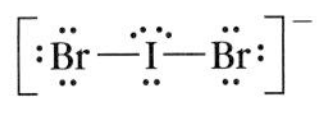 [:Br̈—Ï—Br̈:]⁻

54. Si—F; Si is in the third period and less electronegative than C, S, or O.

55. Yes; "close" elements have similar electronegativities, therefore covalent bonds. "Far apart" elements have different electronegativities, therefore ionic bonds.

56. (a) The C=C is shorter. (b) The C=C is stronger.
(c) $\overset{\delta^+}{C}\equiv\overset{\delta^-}{N}$

59. O—O < Cl—O < O—H < O=O < O=C

60. (d) and (e) are alkanes, (a), (b), (c), and (f) are not.

62. The student forgot to subtract one electron for the positive charge.

63. Atoms are not bonded to the same atoms.

66. Cl: 3.0, S: 2.5, Br: 2.5, Se: 2.4, As: 2.1

69. (a) C—O (b) C≡N (c) C—O

70. (a) 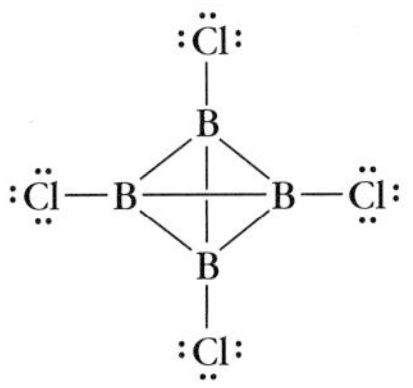 :C̈l: —B; :C̈l—B—B—C̈l: ; B— :C̈l: (four B atoms bonded to each other, each B bonded to one :C̈l:)
(b) H—Ö—N̈=N̈—Ö—H

72. (a) 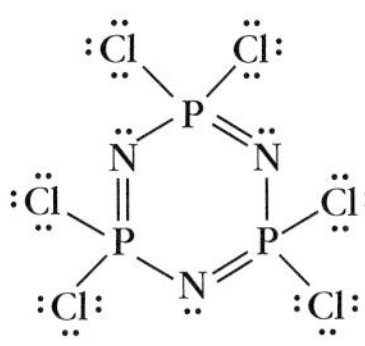 :C̈l, :C̈l: on P; P=N, N—P; :C̈l—P, P—C̈l:; :C̈l:, :C̈l: on P; P—N=P (six-membered P–N ring)
(b) 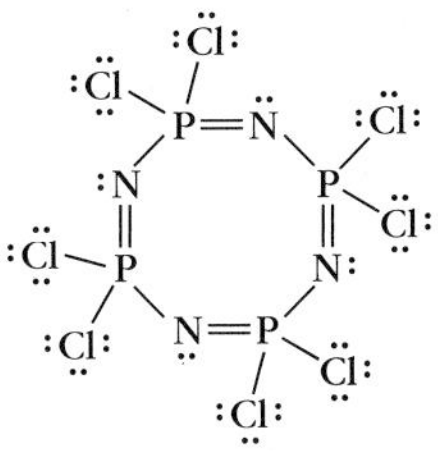 :C̈l:, :C̈l—P=N—P(:C̈l:)(—C̈l:); :N, P=N:; :C̈l—P, N=P; :C̈l:, :C̈l:, :C̈l: (eight-membered P–N ring)

74. 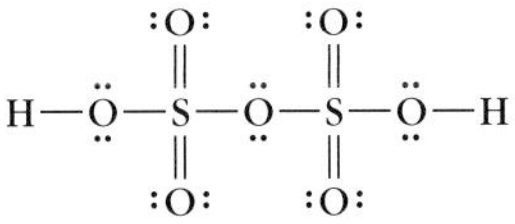 H—Ö—S(=O:)(=O:)—Ö—S(=O:)(=O:)—Ö—H

76. H—C≡C—C≡N:

78. 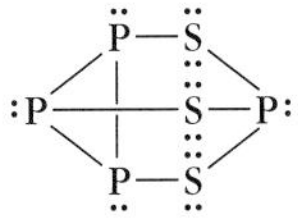 :P—P̈—S̈; :P—S̈—P:; P̈—S̈ (P_4S_3 cage)

Chapter 9

8. (a) H—Be—H Linear (b) 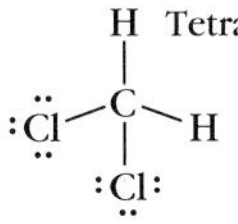 :C̈l—C(—H)(—H)—C̈l: Tetrahedral
(c) 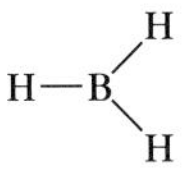 H—B(—H)—H Triangular planar (d) 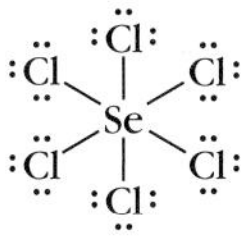 :C̈l:, :C̈l, C̈l:, Se, :C̈l, C̈l:, :C̈l: Octahedral
(e) 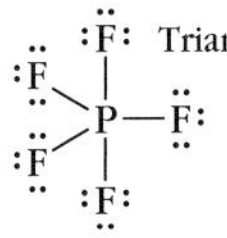 :F̈:, :F̈, P—F̈:, :F̈, :F̈: Triangular bipyramidal

10. (a) Both tetrahedral (b) Tetrahedral, angular (109.5°)
(c) Both triangular planar (d) Both triangular planar

12. (a) Both triangular planar (b) Both triangular planar
(c) Tetrahedral, triangular pyramidal
(d) Tetrahedral, triangular pyramidal
Three atoms bonded to the central atom gives triangular shape; however, structures with 26 valence electrons are triangular pyramidal and structures with 24 valence electrons are triangular planar.

14. (a) Both octahedral
(b) Triangular bipyramidal and seesaw
(c) Both triangular bipyramidal
(d) Octahedral and square planar

16. (a) 120° (b) 120°
(c) H—C—H angle is 120°, C—C—N angle is 180°
(d) H—O—N angle = 109.5°, O—N—O angle = 120°

18. (a) 90°, 120°, and 180° (b) 120° and 90° (c) 90° and 180°

20. The O—N—O angle in NO_2^+ is larger than in NO_2.

22. (a) sp^3 (b) sp^2

24. Tetrahedral, sp^3 hybridized carbon atom

28. The N atom is sp^3 hybridized with 109.5° angles. The first two carbons are sp^3 hybridized with 109.5° angles. The third carbon is sp^2

hybridized with 120° angles. The single-bonded oxygen is sp^3 hybridized with 109.5° angles. The double-bonded O is sp^2 hybridized with 120° angles.

30. (a) The first two carbon atoms are sp^3 hybridized with 109.5° angles. The third and fourth carbon atoms are sp hybridized with 180° angles.
   (b) C≡C (c) C≡C

32. 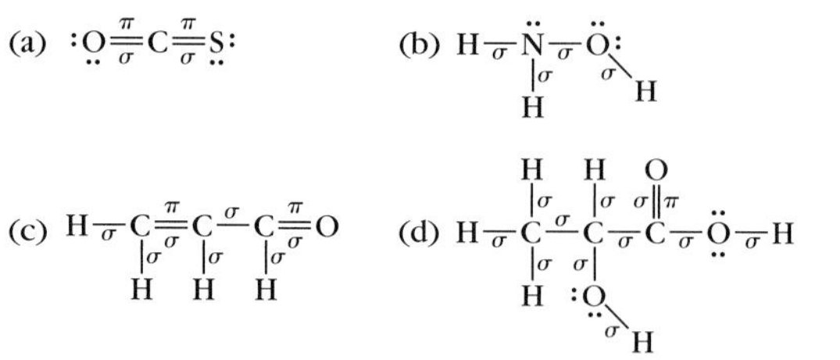 (a) :O=C=S: (σ and π bonds labeled) (b) H–N(–H)–Ö–H (all σ bonds)
   (c) H–C(H)=C(H)–C(H)=O (σ and π bonds labeled) (d) H–C(H)(H)–C(H)(:O–H)–C(=O)–Ö–H (σ and π bonds labeled)

34. (a) Six (b) Three (c) sp (d) sp
   (e) Both sp^2

36. (a) H_2O (b) CO_2 and CCl_4 (c) F

38. Molecules (b) and (c) are polar; HBF_2 has the F atoms on the partial negative end and the H atom on the partial positive end; CH_3Cl has the Cl atom on the partial negative end and the H atoms on the partial positive end.

40. (a) The Br—F bond has a larger electronegativity difference.
   (b) The H—O bond has a larger electronegativity difference.

42. 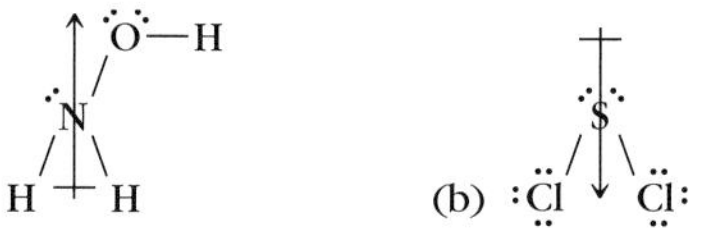 (a) H–N(–H)–Ö–H (with dipole arrow) (b) :Cl–S–Cl: (with dipole arrow)

44.

Interaction	Distance	Example
Ion-ion	Longest range	Na^+ interaction with Cl^-
Ion-dipole	Long range	Na^+ ions in H_2O
Dipole-dipole	Medium range	H_2O interaction with H_2O
Dipole-induced dipole	Short range	H_2O interaction with Br_2
Induced dipole-induced dipole	Shortest range	Br_2 interaction with Br_2

46. Wax molecules interact using London forces. Water molecules interact using hydrogen bonding. The water molecules interact much more strongly with other water molecules; hence, beads form as the water tries to avoid contact with the surface of the wax.

48. (c), (d), and (e)

49. Vitamin C is capable of forming hydrogen bonds with water.

51. (a) London forces (b) London forces
   (c) Intramolecular (covalent) forces
   (d) Dipole-dipole force

54. (a) (1) NCC = angle 180°, (2) HCH angle = 109.5°,
   (3) COC angle = 109.5°
   (b) C=O (c) C=O

56. (a) Ö=C=C=C=Ö (b) 180° (c) 180°

61. If the polarity of the bonds exactly cancel each other, a molecule will be nonpolar.

65. (a) Nitrogen (b) Boron (c) Phosphorus
   (d) Iodine (There are other right answers.)

67. Five

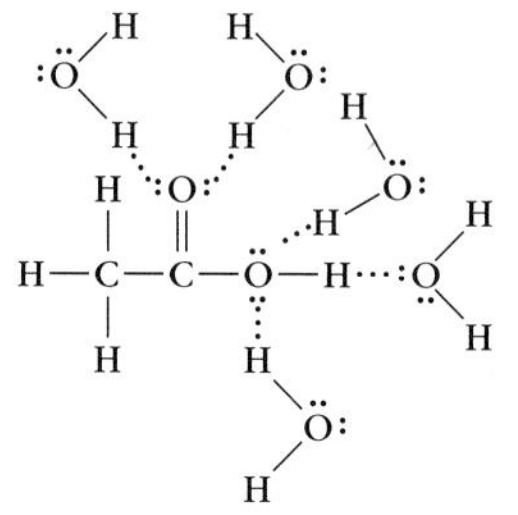

(Structure: H–C(H)(H)–C(=O)–Ö–H hydrogen-bonded to five water molecules)

69. Diagram (d) is correct.

71. (a)

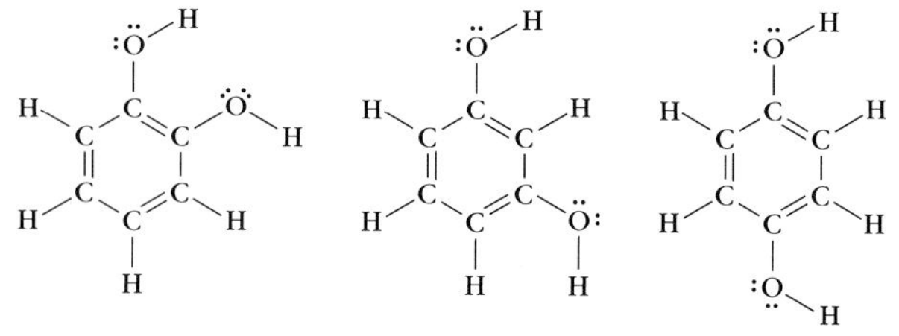

(i) (ii) (iii) (structures of hydroxyphenol isomers)

   (b) (i) < (ii) < (iii)

73. (a) 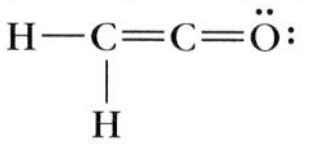 H—C(H)=C=Ö:
   (b) 1st C: triangular planar, 2nd C: linear, H—C—C angle = 120°, H—C—H angle = 120°, C—C—O angle = 180°
   (c) sp^2, sp, sp^2 (d) Polar

75. (a) 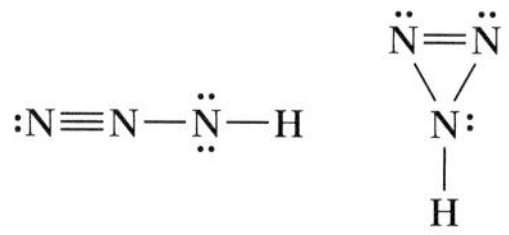 :N≡N—N̈—H and cyclic N=N–N(H) structure
   (b) sp, sp, sp^2 (c) sp^2, sp^2, sp^3
   (d) 3, 4 (e) 2, 1
   (f) molecule 1: 180°; molecule 2: 60°

77. (a) Angle 1: 120°, angle 2: 120°, angle 3: 109.5° (b) sp^3
   (c) For O with two single bonds, sp^3; for O with one double bond, sp^2

Chapter 10

8.

Molecule	ppm	ppb	
N_2	780,840	780,840,000	↑
O_2	209,480	209,480,000	
Ar	9,340	9,340,000	
CO_2	330	330,000	
Ne	18.2	18,200	>1 ppm
H_2	10.	10,000	
He	5.2	5,200	
CH_4	2	2,000	↓
Kr	1	1,000	↑
CO	0.1	100	between
Xe	0.08	80	1 ppm
O_3	0.02	20	and
NH_3	0.01	10	1 ppb
NO_2	0.001	1	↓
SO_2	0.0002	0.2	<1 ppb

10. 1.5×10^8 metric tons SO_2, 2×10^6 metric tons SO_2

11. (a) 0.947 atm (b) 950. mm Hg (c) 542 torr
   (d) 98.7 kPa (e) 6.91 atm

13. 14 m

15. I. A gas is composed of molecules whose size is much smaller than the distances between them.
   II. Gas molecules move randomly at various speeds and in every possible direction.
   III. Except when molecules collide, forces of attraction and repulsion between them are negligible.
   IV. When collisions occur, they are elastic.
   V. The average kinetic energy of gas molecules is proportional to the absolute temperature.

   Assumption I will become false at very high pressures. Assumptions III and IV will become false at very high pressures or very low temperatures. Assumption II is most likely to always be correct.

17. Slowest CH_2Cl_2 < Kr < N_2 < CH_4 fastest

19. Ne will arrive first.

20. 4.2×10^{-5} mol CO

22. 62.5 mm Hg
25. 26.5 mL
27. −96 °C
29. 501 mL
31. 0.507 atm
33. Largest number in (d); smallest number in (c)
34. 6.0 L H_2
36. 1.9 L CO_2; about half the volume is CO_2.
38. 10.4 L O_2, 10.4 L H_2O
40. 21 mm Hg
42. 1.44 g $Ni(CO)_4$
44. 130. g/mol
45. 2.7×10^3 mL
47. P_{He} is 7.000 times greater than P_{N_2}.
48. 3.7×10^{-4} g/L
49. $P_{tot} = 4.51$ atm
51. (a) 154 mm Hg
    (b) $X_{N_2} = 0.777$, $X_{O_2} = 0.208$, $X_{Ar} = 0.0093$, $X_{H_2O} = 0.0053$, $X_{CO_2} = 0.0003$
    (c) 77.7% N_2, 20.8% O_2, 0.93% Ar, 0.03% CO_2, 0.54% H_2O. This sample is wet. Table 1.3 gives percentages for dry air, so the percentages are slightly different due to the water in this sample.
53. (a) $P_{tot} = 1.98$ atm
    (b) $P_{O_2} = 0.438$ atm, $P_{N_2} = 0.182$ atm, $P_{Ar} = 1.36$ atm
56. 18. mL $H_2O(\ell)$, 22.4 L $H_2O(g)$; No, because the vapor pressure of water at 0 °C is 4.6 mm Hg; we cannot achieve 1 atm pressure for this gas at this temperature.
58. Molecular attractions become larger at higher pressures. Molecules hitting the walls hit them with somewhat less force due to the opposing attractions of other molecules.
60. Greenhouse effect = trapping of heat by atmospheric gases. Global warming = increase of the average global temperature. Global warming is caused by an increase in the amount of greenhouse gases in the atmosphere.
62. Examples of CO_2 sources: animal respiration, burning fossil fuels and other plant materials, decomposition of organic matter, etc. Examples of CO_2 removal: photosynthesis in plants, being dissolved in rain water, incorporation into carbonate and bicarbonate compounds in the oceans, etc. Currently, CO_2 production exceeds CO_2 removal.
64. (a) Before: $P_{H_2} = 3.7$ atm; $P_{Cl_2} = 4.9$ atm
    (b) Before: $P_{tot} = 8.6$ atm (c) After: $P_{tot} = 8.6$ atm
    (d) Cl_2; 0.5 moles remain (e) $P_{HCl} = 7.4$ atm; $P_{Cl_2} = 1.2$ atm
    (f) $P = 8.9$ atm
67.

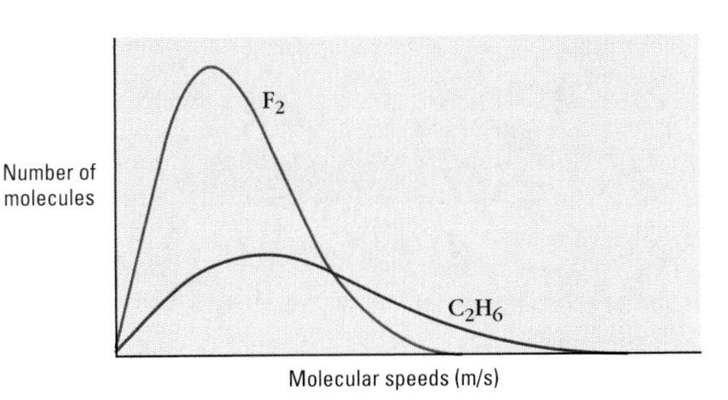

68. For reference, the initial state looks like this:

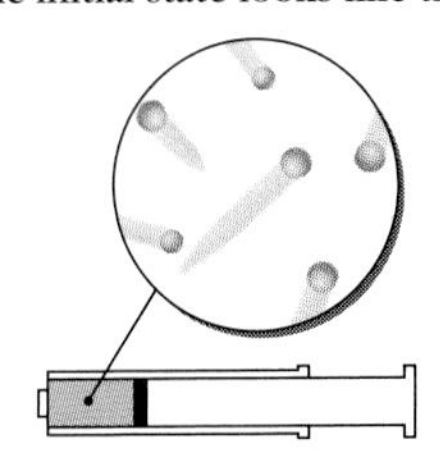

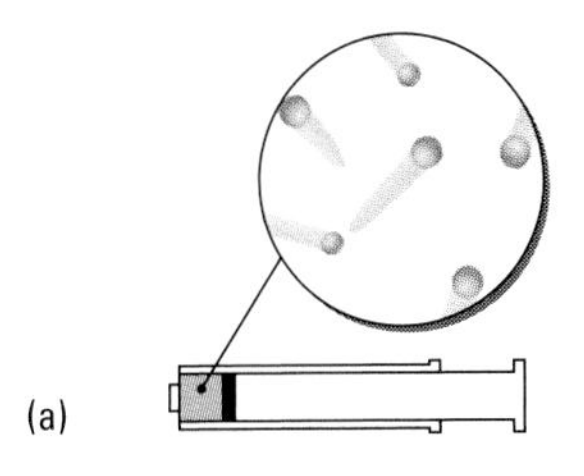

(a)

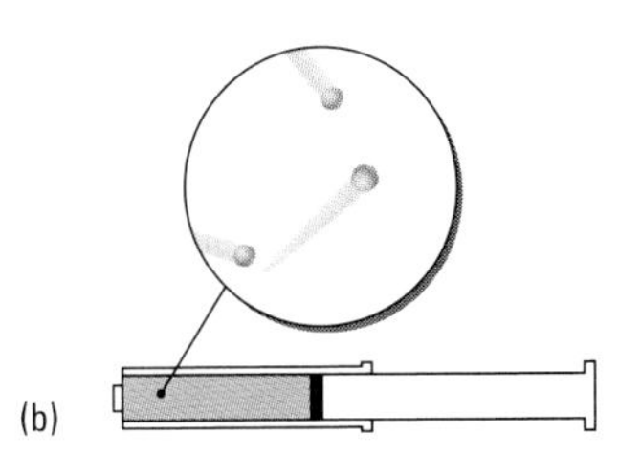

(b)

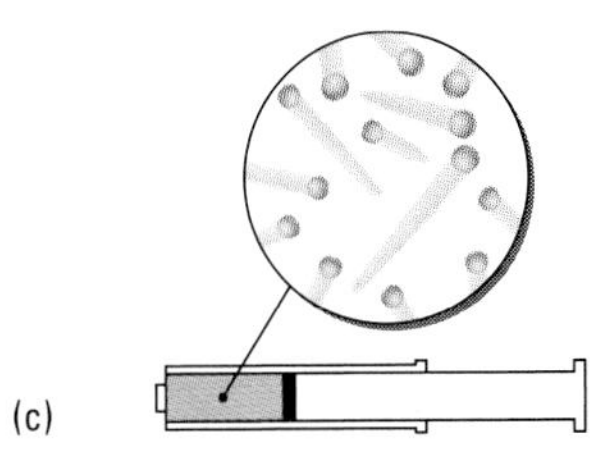

(c)

70. Box (b). The initial-to-final volume ratio is 2:1, so, for every two molecules of gas reactants, there must be one molecule of gas products. 6 reactant molecules must produce 3 product molecules.
72. (a) 64.1 g/mol
    (b) Empirical formula = CHF; molecular formula = $C_2H_2F_2$
    (c)

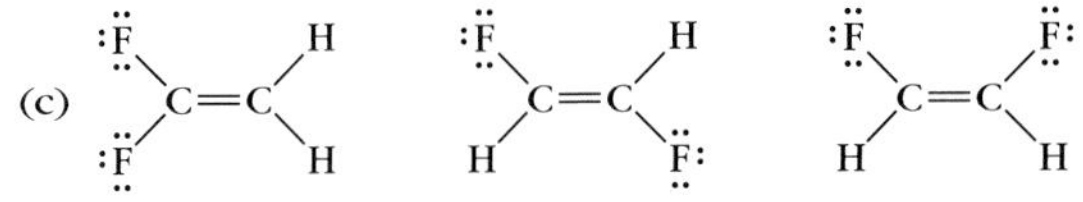

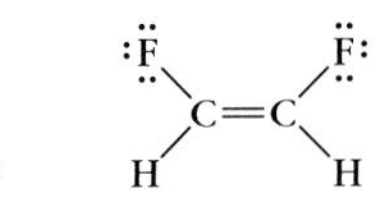

73. $P_{tot} = 0.88$ atm
75. $\frac{m_{Ne}}{m_{Ar}} = 0.2$
77. (a) 29.1 mol CO_2; 14.6 mol N_2; 2.43 mol O_2
    (b) 1.1×10^3 L
    (c) $P_{N_2} = 0.317$ atm; $P_{CO_2} = 0.631$ atm; $P_{O_2} = 0.0527$ atm
78. 458 torr
79. 4.5 mm^3
81. (a) More significant, because of more collisions
    (b) More significant, because of more collisions
    (c) Less significant, because the molecules will move faster

Chapter 11

10. Reduce the pressure
13. 1.5×10^6 kJ
15. 233 kJ

17. 2.00 kJ required for Hg is more than 1.13 kJ required for H_2O sample. Hg has a much greater mass.
19. NH_3 has a relatively large boiling point because the molecules interact using relatively strong hydrogen-bonding intermolecular forces. The increase in the boiling points of the series PH_3, AsH_3, and SbH_3 is related to the increasing London dispersion intermolecular forces experienced, due to the larger central atom in the molecule. (Size: P < As < Sb)
21. Methanol molecules are capable of hydrogen bonding, whereas formaldehyde molecules use dipole-dipole forces to interact. Molecules experiencing stronger intermolecular forces (such as methanol here) will have higher boiling points and lower vapor pressures compared to molecules experiencing weaker intermolecular forces (such as formaldehyde here).
23. 0.21 atm, approximately 57 °C
25. 1600 mm Hg
27. $\Delta H_{vap} = 70.$ kJ/mol
28. The interparticle forces in the solid are very strong.
30. The intermolecular forces between the molecules of H_2O in the solid (hydrogen bonding) are stronger than the intermolecular forces between the molecules of H_2S (dipole-dipole).
31. 27 kJ
33. 51.9 g CCl_2F_2
35. LiF, because the ions are smaller, making the charges more localized and closer together, causing a higher coulombic interaction between the ions.
37. The highest melting point is (a). The extended network of covalent bonds in SiC are the strongest interparticle forces of the listed choices. The lowest melting point is (d). Both I_2 and $CH_3CH_2CH_2CH_3$ interact using the weakest intermolecular forces—London dispersion forces—but large I_2 has many more electrons, so its London forces are relatively stronger.
40. (a) Gas phase
    (b) Liquid phase
    (c) Solid phase
42. (a) Molecular (b) Metallic
    (c) Network (d) Ionic
44. (a) Amorphous; it decomposes before melting and does not conduct electricity. (b) Molecular; low melting point and nonconducting (c) Ionic; high melting point and only liquid (molten ions) conduct electricity. (d) Metallic; both solid and liquid conduct electricity.
45. (a) Molecular (b) Ionic
    (c) Metallic or network (d) Amorphous
47. See Figure 11.21 and its description
49. 220 pm
52. Diagonal is 696 pm, side length is 401 pm.
54. (a) 152 pm
    (b) Radius of I^- = 212 pm and radius of Li^+ = 88.0 pm
    (c) It is reasonable that the atom is larger than the cation. The assumption that anions touch anions seems unreasonable; there would be some repulsion and probably a small gap or distortion.
55. Carbon atoms in diamond are sp^3 hybridized and are tetrahedrally bonded to four other carbon atoms. Carbon atoms in pure graphite are sp^2 hybridized and bonded with a triangle planar shape to other carbon atoms. These bonds are partially double bonded so they are shorter than the single bonds in diamond. However, the planar sheets of sp^2 hybridized carbon atoms are only weakly attracted by intermolecular forces to adjacent layers, so these interplanar distances are much longer than the C—C single bonds. The net result is that graphite is less dense than diamond.
57. Diamond is an electrical insulator because all the electrons are in single bonds that are shared between two specific atoms and cannot move around. However, graphite is a good conductor of electricity because its electrons are delocalized in conjugated double bonds that allow the electrons to move easily through the graphite sheets.
59. In a conductor, the valence band is only partially filled, whereas, in an insulator, the valence band is completely full, the conduction band is empty, and there is a wide energy gap between the two. In a semiconductor, the gap between the valence band and the conduction band is very small so that electrons are easily excited into the conduction band.
61. Substance (c), Ag, has the greatest electrical conductivity because it is a metal. Substance (d), P_4, has the smallest electrical conductivity because it is a nonmetal. (The other two are metalloids.)
64. 780 kJ, 1.6×10^4 kJ
66. (a) Dipole-dipole forces and London forces
    (b) $CH_4 < NH_3 < SO_2 < H_2O$
67. (a) Approx. 80 mm Hg (b) Approx. 18 °C
    (c) Approx. 640 mm Hg (d) Diethyl ether and ethanol
    (e) Diethyl ether evaporates immediately. Ethanol and water remain liquid.
    (f) Water
69. The butane in the lighter is under great enough pressure so that the vapor pressure of butane at room temperature is less than the pressure inside the light. Hence, it exists as a liquid.
70. 1 and C, 2 and E, 3 and B, 4 and F, 5 and G, 6 and H, 7 and A
72. Vapor-phase water condenses on contact with the skin and the condensation is exothermic, which imparts more energy to the skin.
75. (a) Condensation, freezing (b) Triple point
    (c) Melting point curve
77. (a) Approx. 560 mm Hg (b) Benzene
    (c) 73 °C
    (d) Methyl ethyl ether, approx. 7 °C; carbon disulfide, approx. 47 °C; benzene, approx. 81 °C
79. 22.2 °C
80. 0.533 g/cm^3

Chapter 12

7. (d) dissolves fastest; (a) dissolves slowest. Rate of dissolving is larger when the grains of sugar are smaller, because there is more surface contact with the solvent.
9. (a) 0.23 mol/L·h (b) 0.20 mol/L·h (c) 0.161 mol/L·h
   (d) 0.12 mol/L·h (e) 0.090 mol/L·h (f) 0.066 mol/L·h
11. (a) Calculate the average concentration for each time interval and plot average concentration vs. average rate:

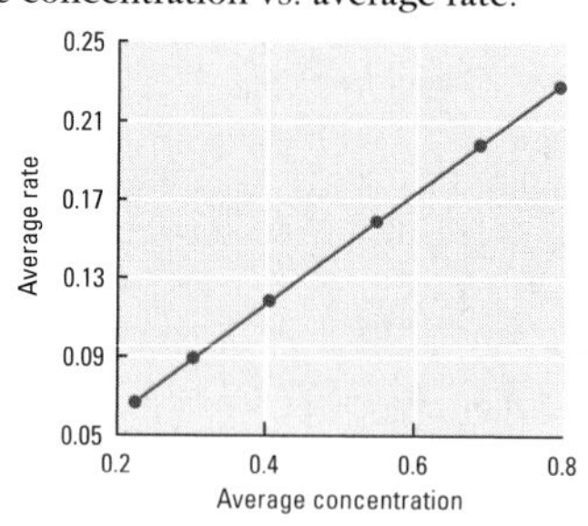

The linear relationship shows: Rate = $k[N_2O_5]$.
(b) $k = 0.29\ h^{-1}$

13.

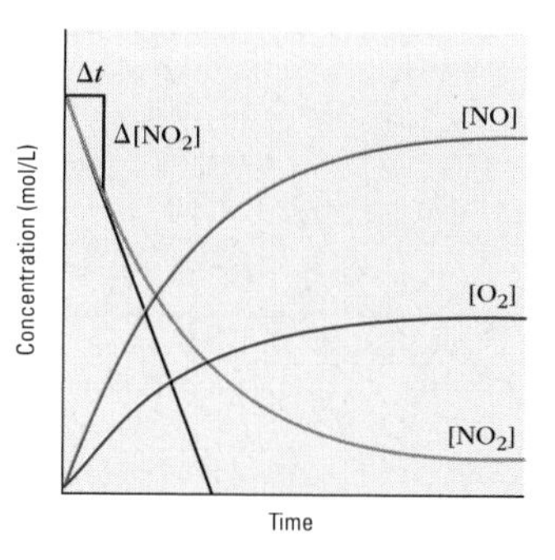

(a) To calculate initial rate, obtain values for Δt and Δ(concentration) near initial time, where the curve can still be approximated by a straight line (the black line on the graph) then divide Δ(concentration) by Δt and change the sign to make it positive, e.g.,

$$\text{Initial rate} = \frac{-\Delta[NO_2]}{\Delta t}$$

(b) The curves on the graph become horizontal after a very long time. The final rate will be zero, when $\Delta[NO_2] = 0$.

15. (a) Rate $= k[NO_2]^2$
(b) The rate will be one fourth as fast.
(c) The rate is unchanged.

17. (a) (i) 9.0×10^{-4} M/h, (ii) 1.8×10^{-3} M/h, (iii) 3.6×10^{-3} M/h
(b) If the initial concentration of $Pt(NH_3)_2Cl_2$ is high, the rate of disappearance of $Pt(NH_3)_2Cl_2$ is high. If the initial concentration of $Pt(NH_3)_2Cl_2$ is low, the rate of disappearance of $Pt(NH_3)_2Cl_2$ is low. The rate of disappearance of $Pt(NH_3)_2Cl_2$ is directly proportional to $[Pt(NH_3)_2Cl_2]$.
(c) The rate law shows direct proportionality between rate and $[Pt(NH_3)_2Cl_2]$.
(d) When the initial $[Pt(NH_3)_2Cl_2]$ is high, the rate of appearance of Cl^- is high. When the initial concentration is low, the rate of appearance of Cl^- is low. The rate of appearance of Cl^- is directly proportional to $[Pt(NH_3)_2Cl_2]$.

19. (a) Rate $= k[I][II]$
(b) $k = 1.04 \frac{L}{mol \cdot s}$

20. (a) The order with respect to NO is one and with respect to H_2 is one.
(b) The reaction is second-order.
(c) Rate $= k[NO][H_2]$
(d) $k = 2.4 \times 10^2$ L mol^{-1} s^{-1}
(e) Initial rate $= 1.5 \times 10^{-2}$ mol L^{-1} s^{-1}

21. (a) First-order in A and third-order in B, fourth-order overall
(b) First-order in A and first-order in B, second-order overall
(c) First-order in A and zero-order in B, first-order overall
(d) Third-order in A and first-order in B, fourth-order overall

23. (a) Rate $= k[CH_3COCH_3][H_3O^+]$; first-order in both H_3O^+ and CH_3COCH_3, and zero-order in Br_2
(b) $4 \times 10^{-3} \frac{L}{mol \cdot s}$ (c) $2 \times 10^{-5} \frac{mol}{L \cdot s}$

25. (a) 0.16 mol/L (b) 90. s (c) 120 s

27. $t = 5.49 \times 10^4$ s

29. (a) Not elementary (b) Bimolecular and elementary
(c) Not elementary (d) Unimolecular and elementary

31. NO + O_3, NO is an asymmetric molecule and Cl is a symmetric atom.

33. $E_a = 19$ kJ/mol, Ratio $= 1.8$

35. (a) $E_a = 22.2$ kJ/mol, $A = 6.66 \times 10^7 \frac{L^2}{mol^2 \cdot s}$
(b) $k = 8.39 \times 10^4 \frac{L^2}{mol^2 \cdot s}$

37. (a) $8 \times 10^{-4} \frac{mol}{L \cdot s}$ (b) $3 \times 10^1 \frac{mol}{L \cdot s}$

39. 3×10^2 kJ/mol

41. Exothermic

43. (a)

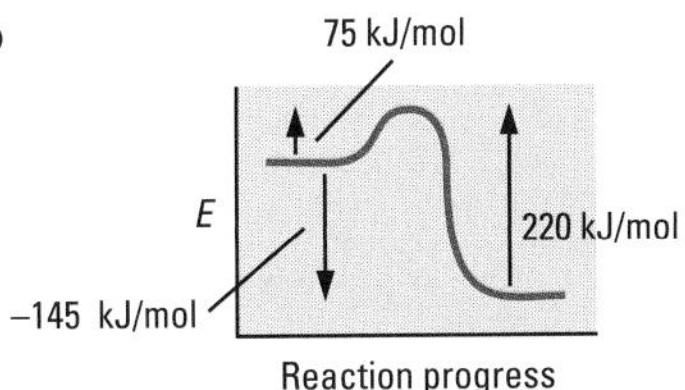

(b)

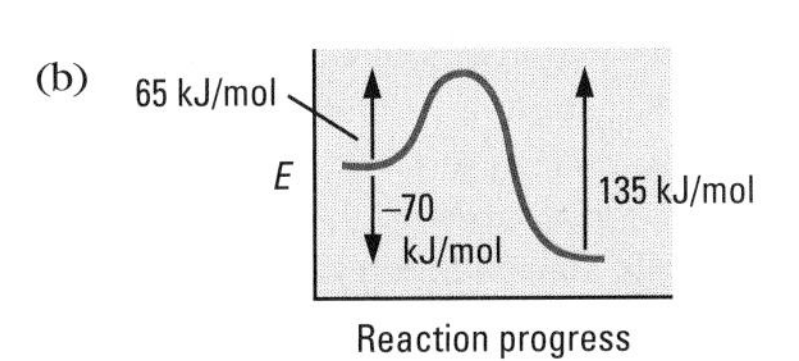

(c)

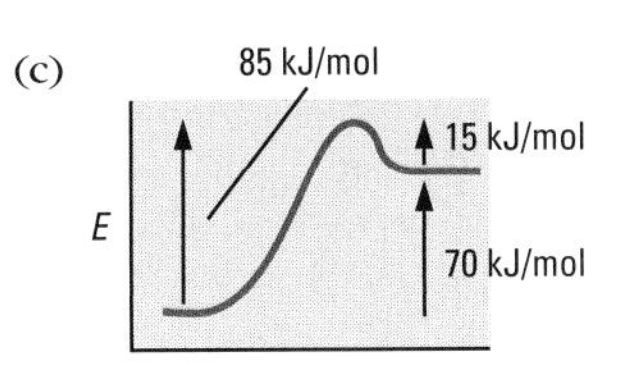

44. (a)

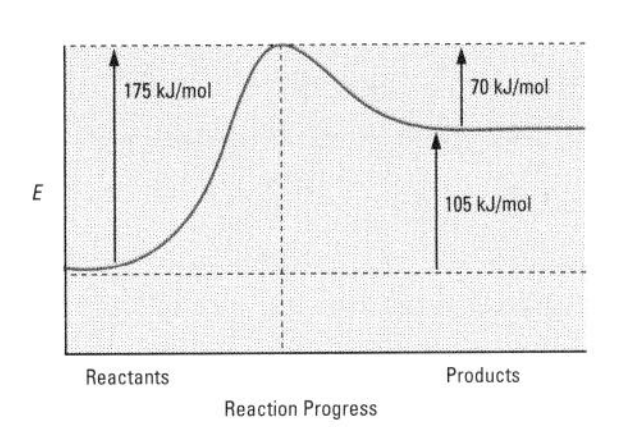

(b)

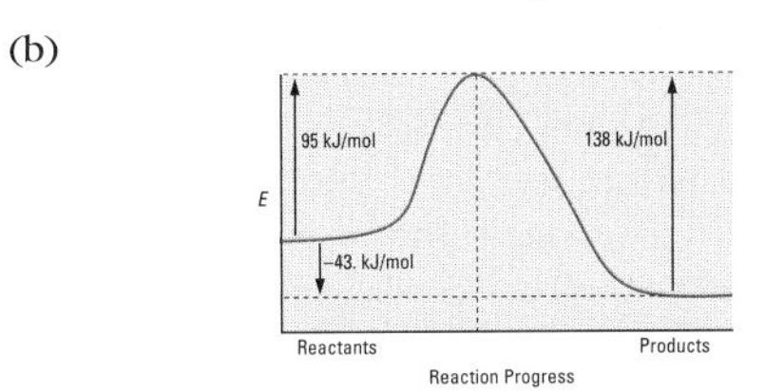

(c)

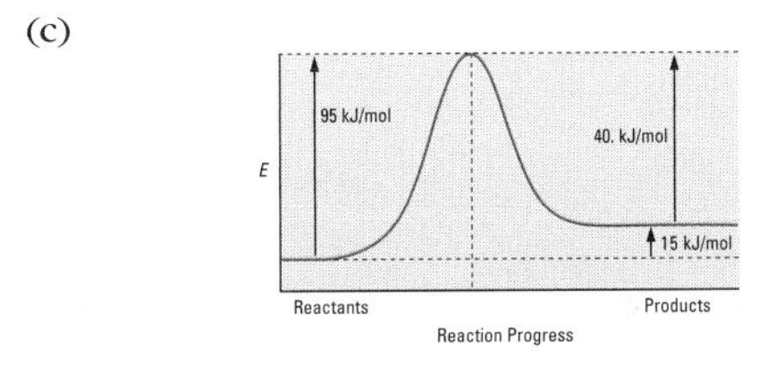

45. (a) Reaction (b) (b) Reaction (c)

47. (a) Rate $= k[NO][NO_3]$ (b) Rate $= k[O][O_3]$
(c) Rate $= k[(CH_3)_3CBr]$ (d) Rate $= k[HI]^2$

49. (a)
$$NO_2 + F_2 \longrightarrow FNO_2 + \cancel{F}$$
$$+\ NO_2 + \cancel{F} \longrightarrow FNO_2$$
$$\overline{2\,NO_2 + F_2 \longrightarrow 2\,FNO_2}$$
(b) The first step is rate-determining.

51. (a) $CH_3COOCH_3 + H_2O \rightleftharpoons CH_3COOH + CH_3OH$
(b) Catalyst: H_3O^+
(c) Intermediates: $H_3C(OH)OCH_3^+$, $H_3C(H_2O)(OH)OCH_3^+$, $H_3C(OH)_2OHCH_3^+$, and H_2O.

53. Only mechanism (a) is compatible with the observed rate law.

54. (a) is true.

56. (a) and (c); (a) homogeneous (c) heterogeneous

58. Enzyme: A protein that catalyzes a biological reaction.
Substrate: The reactant of a reaction catalyzed by an enzyme.
Active site: That part of an enzyme where the substrate binds to the enzyme in preparation for conversion into products.
Proteins: Large polypeptide molecules that serve as structural and functional molecules in living organisms.
Enzyme-substrate complex: A loose affiliation, usually involving intermolecular forces, between the enzyme and the substrate.
Induced fit: When the natural shape of either the enzyme or the substrate changes slightly in the formation of the enzyme-substrate complex.

Globular proteins: Proteins whose structure allows them to fold into tight, compact spherical shapes, in contrast to fibrous proteins which are long and slender.

59. 30. times faster

62. (a) To maximize the surface area and increase contact with the heterogeneous catalyst
(b) Catalysis happens only at the surface of the metal; strips or rods with less surface area would be less efficient and more costly.

63.

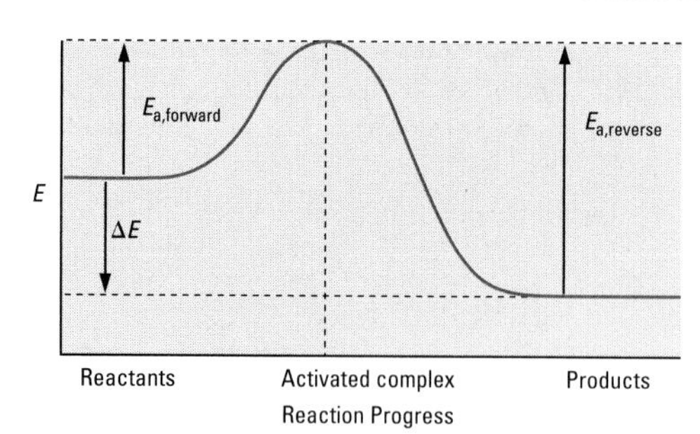

$\Delta E = E_{a,\ forward} - E_{a,\ reverse}$

65. (a) True
(b) False. "The reaction rate decreases as a first-order reaction proceeds at a constant temperature."
(c) True
(d) False. "As a second-order reaction proceeds at a constant temperature, the rate constant does not change."
(Other corrections for (b) and (d) are also possible.)

66. Rate $= k[H_2][NO]^2$

67. (a) NO is second-order, O_2 is first-order.
(b) Rate $= k[NO]^2[O_2]$ (c) $25\ \dfrac{L^2}{mol^2 \cdot s}$
(d) $7.8 \times 10^{-4}\ \dfrac{mol}{L \cdot s}$ (e) $-\dfrac{\Delta[NO]}{\Delta t} = 2.0 \times 10^{-4}\ \dfrac{mol}{L \cdot s}$, $+\dfrac{\Delta[NO_2]}{\Delta t} = 2.0 \times 10^{-4}\ \dfrac{mol}{L \cdot s}$

69. Curve A represents $[H_2O(g)]$ increase with time, Curve B represents $[O_2(g)]$ increase with time, and Curve C represents $[H_2O_2(g)]$ decrease with time.

71. Snapshot (b)

73. Rate $= k[A]^3[B][C]^2$

74. E_a is very, very small—approximately zero.

76. (a) Rate $= 2.4 \times 10^{-7}$ mol L^{-1} s^{-1} $+ k[BSC][F^-]$
(b) 0.3 L mol^{-1} s^{-1}

78. (a) $E_a = 3 \times 10^2$ kJ (b) $t = 2 \times 10^3$ s

80. Note: there are other correct answers to this question; the following are examples:

(a) $CH_3CO_2CH_3 + H_3O^+ \longrightarrow CH_3COHOCH_3 + H_2O$ slow
$CH_3COHOCH_3 + H_2O \longrightarrow CH_3COH(OH_2)OCH_3$ fast
$CH_3COH(OH_2)OCH_3 \longrightarrow CH_3C(OH)_2 + CH_3OH$ fast
$CH_3C(OH)_2 + H_2O \longrightarrow CH_3COOH + H_3O^+$ fast

(b) $H_2 + I_2 \longrightarrow 2\ HI$ slow

(c) $H_2 + Pt(s) \longrightarrow PtH_2$ fast
$PtH_2 + I_2 \longrightarrow PtH_2I_2$ slow
$PtH_2I_2 + O_2 \longrightarrow PtI_2O + H_2O$ fast
$PtI_2O + H_2 \longrightarrow Pt(s) + I_2 + H_2O$ fast

81. Estimated $E_a = 402 \times 10^{-21}$ J/molecule; it is the same to one significant figure as the E_a given in Figure 12.7.

Chapter 13

7. There are many answers to this question; this is an example: Prepare a sample of N_2O_4 in which the N atoms are the heavier isotopes ^{15}N. Introduce the heavy isotope of N_2O_4 into an equilibrium mixture of N_2O_4 and NO_2. Use spectroscopic methods, such as infrared spectroscopy to observe the distribution of the radioisotope among the reactants and products.

9.

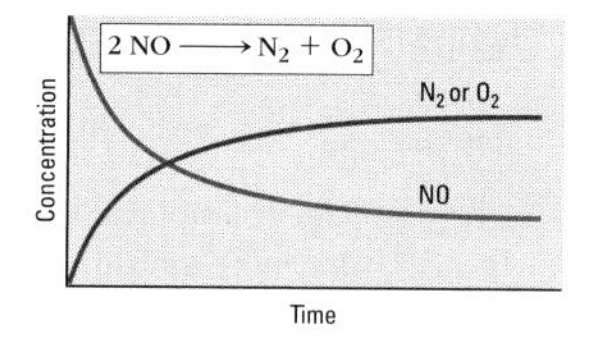

11. (a) $K_c = \dfrac{[H_2O]^2[O_2]}{[H_2O_2]^2}$ (b) $K_c = \dfrac{[PCl_5]}{[PCl_3][Cl_2]}$
(c) $K_c = [CO]^2$ (d) $K_c = \dfrac{[H_2S]}{[H_2]}$

13. Equation (e)

15. (a) $K_P = \dfrac{P^2_{H_2O}P_{O_2}}{P^2_{H_2O_2}}$ (b) $K_P = \dfrac{P_{PCl_5}}{P_{PCl_3}P_{Cl_2}}$
(c) $K_P = P^2_{CO}$ (d) $K_P = \dfrac{P_{H_2S}}{P_{H_2}}$

17. (a) $K_c = 1.0$ (b) $K_c = 1.0$ (c) $K_c = 1$

19. $K_p = 1.6$

21. $K_c = 75$

23. Reactions (b) and (c) are product-favored. Most reactant-favored is (a), then (b), and then (c).

25. (a)

	butane ⇌	2-methylpropane
Conc. initial	0.100 mol/L	0.100 mol/L
Change conc.	$-x$	$+x$
Equilibrium conc.	$0.100 - x$	$0.100 + x$

(b) $K_c = \dfrac{[\text{2-methylpropane}]}{[\text{butane}]} = 2.5 = \dfrac{0.100 + x}{0.100 - x}$ $x = 0.043$
(c) [2-methylpropane] = 0.024 mol/L, [butane] = 0.010 mol/L

27. $[HI] = 4 \times 10^{-2}$ M; $[H_2] = [I_2] = 5 \times 10^{-3}$ M

29. (a) 1.94 mol HI (b) 1.92 mol HI (c) 1.98 mol HI

31. (a) No (b) Proceed toward products

33. (a) No (b) Proceed toward reactants
(c) $[N_2] = [O_2] = 0.002$ M; $[NO] = 0.061$ M

35. Choice (b); heat energy is absorbed, and increasing the temperature drives the reaction toward products.

37.

Change	$[Br_2]$	[HBr]	K_c	K_p
Some H_2 is added to the container.	Decrease	Increase	No change	No change
Temperature of the gases in the container is increased.	Increase	Decrease	Decrease	Decrease
The pressure of HBr is increased.	Increase	Increase	No change	No change

39. (a) No change (b) Left (c) Left

41. (a) Decrease
(b)

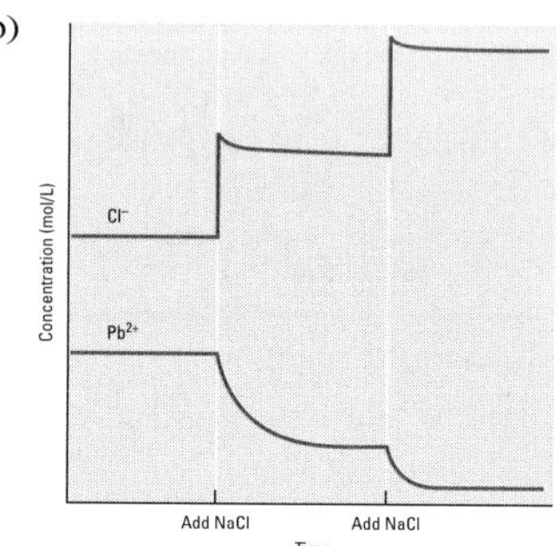

43. (a) Energy effect (b) Entropy effect (c) Neither
45. (a) First reaction: $\Delta H = -296.830$ kJ, second reaction: $\Delta H = -197.78$ kJ, third reaction: $\Delta H = -132.44$ kJ
(b) All three are exothermic, none are endothermic.
(c) None of the reactions have entropy increase, the second and third reactions have entropy decrease, and the first reaction entropy doesn't change.
(d) All three reactions favor products more at low temperatures.
48. (a)

Species	Br_2	Cl_2	F_2	H_2	N_2	O_2
$[E]$ (mol/L)	0.28	0.057	1.44	1.76×10^{-5}	4×10^{-14}	4.0×10^{-6}

(b) F_2. (At this temperature, the lowest bond energy is predicted from the reaction that gives the most products.) Compared to Table 8.2, the product production decreases as the bond energy increases: 158 kJ F_2, 193 kJ Br_2, 242 kJ Cl_2, 436 kJ H_2, 498 kJ O_2, 946 kJ N_2. Lewis structures of F_2, Br_2, Cl_2, H_2 have a single bond and more products are produced than O_2 with double bond and N_2 with triple bonds.
49. (a) $K_c = 0.67$ (b) Same result
51. Cases a and b will have decreased [HI] at equilibrium, and Cases c and d will have increased [HI].
53. It is at equilibrium at 600. K. No more experiments are needed.
55. (a) (iii) (b) (i)
57. In the warmer sample, the molecules would be moving faster and more NO_2 molecules would be seen. In the cooler sample, the molecules would be moving somewhat slower and fewer NO_2 molecules would be seen. In both samples, the molecules are moving very fast. The average speed of gas molecules is commonly hundreds of miles per hour. In both samples, one would see a dynamic equilibrium with some N_2O_4 molecules decomposing and some NO_2 molecules reacting with each other, at equal rates.
59. Diagrams (b), (c), and (d)
61. The following is an example of a diagram. (Other answers are possible.)

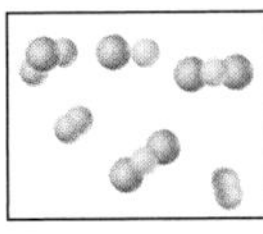

63. Dynamic equilibria with small values of K introduce a small amount of D^+ ions in place of H^+ ions in the place of the acidic hydrogen.
65. (a) $K_c = 0.0168$ (b) 0.661 atm (c) 0.514
67. (a) $[PCl_3] = [Cl_2] = 0.14$ mol/L, $[PCl_5] = 0.005$ mol/L
(b) $[PCl_3] = 0.29$ mol/L, $[Cl_2] = 0.14$ mol/L, $[PCl_5] = 0.01$ mol/L
69. (a) (i) Right (ii) Right (iii) Left
(b) (i) Left (ii) No shift (iii) Right
(c) (i) Right (ii) Right (iii) No shift
(d) (i) Right (ii) Right (iii) Left
71. (a) $H_2(g) + Br_2(g) \rightleftharpoons 2\ HBr\ (g)$ (b) -103 kJ
(c) Energy effect (d) To the left (e) No effect
(f) The reactants must achieve activation energy. At $T < 200°$, reactants do not have enough energy.

Chapter 14

18. If the solid interacts with the solvent using similar (or stronger) intermolecular forces, it will dissolve readily. If the solute interacts with the solvent using different intermolecular forces than those experienced in the solvent, it will be almost insoluble. For example, consider dissolving an ionic solid in water and oil. The interactions between the ions in the solid and water are very strong, since ions would be attracted to the highly polar water molecule; hence, the solid would have a high solubility. However, the ions in the solid interact with each other much more strongly than the London dispersion forces experienced between the non-polar hydrocarbons in the oil; hence, the solid would have a low solubility. (Other examples exist.)
19. The dissolving process was endothermic, so the temperature dropped as more solute was added. The solubility of the solid at the lower temperature is lower, so some of the solid did not dissolve. As the solution warmed up, however, the solubility increased again. What remained of the solid dissolved. The solution was saturated at the lower temperature, but is no longer saturated at the current temperature.
21. When an organic acid has a large (non-polar) piece, it interacts primarily using London dispersion intermolecular forces. Since water interacts via hydrogen-bonding intermolecular forces, it would rather interact with itself than with the acid. Hence, the solubility of the large organic acids drops, and some are completely insoluble.
23. The positive H end of the very polar water molecule interacts with the negative ions. The negative O end of the very polar water molecule interacts with the positive ions.

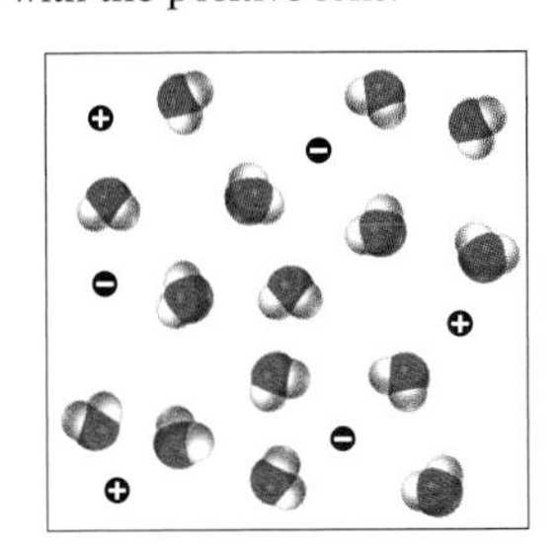

25. 1×10^{-3} M
27. 90. g ethanol
29. 1.6×10^{-6} g Pb
31. (a) 160. g NH_4Cl (b) 83.9 g KCl (c) 7.46 g Na_2SO_4
32. (a) 0.762 M (b) 0.174 M
(c) 0.0126 M (d) 0.167 M
33. (a) 0.180 M (b) 0.683 M
(c) 0.0260 M (d) 0.0416 M
35. 96% H_2SO_4
37. 59 g
42. (a) < (d) < (b) < (c)
44. $T_f = -1.65$ °C, $T_b = 100.46$ °C
46. $X_{H_2O} = 0.79999$, 712 g sucrose
48. 1.9×10^2 g/mol
50. 3.6×10^2 g/mol; $C_{20}H_{16}Fe_2$
53. (a) 2.5 kg (b) 104.2 °C
57. Water in the cells of the wood leaked out, since the osmotic pressure inside the cells was less than that of the seawater in which the wood was sitting.
59. 28% NH_3
63. (a)

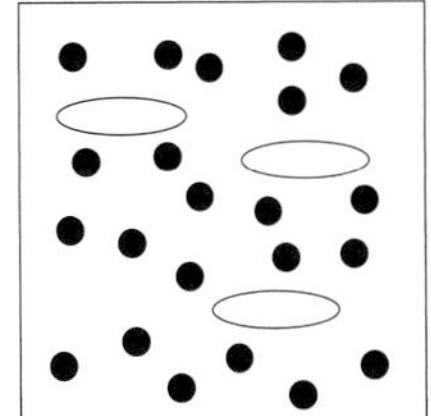

(b)

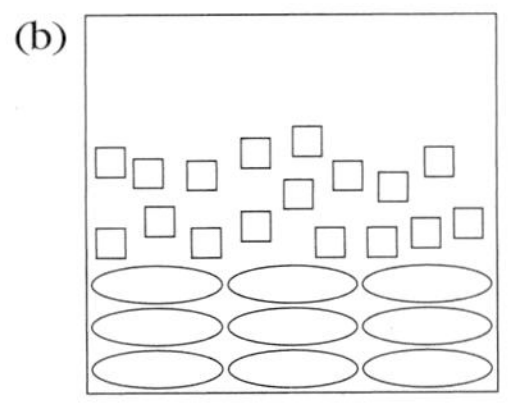

65. (a) Unsaturated (b) Supersaturated
(c) Supersaturated (d) Unsaturated
67.

Compound	Mass of Compound	Mass of Water	Mass Fraction of Solute	Weight Percent of Solute	Conc of Solute
Lye	75.0 g	125 g	0.375	37.5%	3.75×10^5 ppm
Glycerol	33 g	200. g	0.14	14%	1.4×10^5 ppm
Acetylene	0.0015	2×10^2 g	0.000009	0.0009%	9 ppm

69. (a) Seawater contains more dissolved solutes than fresh water. The presence of a solute lowers the freezing point. That means a lower temperature is required to freeze the seawater than to freeze fresh water.
(b) Salt added to a mixture of ice and water will lower the freezing point of the water. If the ice cream is mixed at a lower temperature, its temperature will drop faster; hence, it will freeze faster.
71. The empirical and molecular formulas are both $C_{18}H_{24}Cr$.
73. (a) No (b) 108.9°
75. 28 m
76. 0.30 M
78. (a) 6300 ppm, 6300000 ppb
(b) 0.040 M (c) 4.99×10^5 bottles

Chapter 15

8. (a) $HCO_3^- + H_2O \rightleftharpoons CO_3^{2-} + H_3O^+$
(b) $HCl + H_2O \rightleftharpoons Cl^- + H_3O^+$
(c) $CH_3COOH + H_2O \rightleftharpoons CH_3COO^- + H_3O^+$
(d) $HCN + H_2O \rightleftharpoons CN^- + H_3O^+$
10. (a) $HIO + H_2O \rightleftharpoons IO^- + H_3O^+$
(b) $CH_3(CH_2)_4COOH + H_2O \rightleftharpoons CH_3(CH_2)_4COO^- + H_3O^+$
(c) $HOOCCOOH + H_2O \rightleftharpoons HOOCCOO^- + H_3O^+$
$HOOCCOO^- + H_2O \rightleftharpoons {}^-OOCCOO^- + H_3O^+$
(d) $CH_3NH_3^+ + H_2O \rightleftharpoons CH_3NH_2 + H_3O^+$
12. (a) $HSO_4^- + H_2O \leftrightarrows H_2SO_4 + OH^-$
(b) $CH_3NH_2 + H_2O \rightleftharpoons CH_3NH_3^+ + OH^-$
(c) $I^- + H_2O \leftrightarrows HI + OH^-$
(d) $H_2PO_4^- + H_2O \rightleftharpoons H_3PO_4 + OH^-$
16. (a) I^-, iodide, conjugate base
(b) HNO_3, nitric acid, conjugate acid
(c) HCO_3^-, hydrogen carbonate ion, conjugate acid
(d) HCO_3^-, hydrogen carbonate ion, conjugate base
(e) SO_4^{2-}, sulfate ion, conjugate base and H_2SO_4, sulfuric acid, conjugate acid
(f) HSO_3^-, hydrogen sulfite ion, conjugate acid
18. Pairs (b), (c), and (d)
20. (a) Reactant acid = H_2O, reactant base = HS^-, product conjugate acid = H_2S, product conjugate base = OH^-
(b) Reactant acid = NH_4^+, reactant base = S^{2-}, product conjugate acid = HS^-, product conjugate base = NH_3
(c) Reactant acid = HSO_4^-, reactant base = HCO_3^-, product conjugate acid = H_2CO_3, product conjugate base = SO_4^{2-}
(d) Reactant acid = NH_3, reactant base = NH_2^-, product conjugate base = NH_2^-, product conjugate acid = NH_3
22. (a) $CO_3^{2-} + H_2O \rightleftharpoons HCO_3^- + OH^-$
$HCO_3^- + H_2O \rightleftharpoons H_2CO_3 + OH^-$
(b) $H_3AsO_4 + H_2O \rightleftharpoons H_2AsO_4^- + H_3O^+$
$H_2AsO_4^- + H_2O \rightleftharpoons HAsO_4^{2-} + H_3O^+$
$HAsO_4^{2-} + H_2O \rightleftharpoons AsO_4^{3-} + H_3O^+$
(c) $NH_2CH_2COO^- + H_2O \rightleftharpoons {}^+NH_3CH_2COO^- + OH^-$
${}^+NH_3CH_2COO^- + H_2O \rightleftharpoons {}^+NH_3CH_2COOH + OH^-$
24. 3×10^{-11} M, basic
26. pH = 12.40, pOH = 1.60
28. 5×10^{-2} M, 2 g HCl
30.

pH	$[H_3O^+]$ (M)	$[OH^-]$ (M)	Acidic or Basic
(a) 6.21	6.1×10^{-7}	1.6×10^{-8}	Acidic
(b) 5.34	4.5×10^{-6}	2.2×10^{-9}	Acidic
(c) 4.67	2.1×10^{-5}	4.7×10^{-10}	Acidic
(d) 1.60	2.5×10^{-2}	4.0×10^{-13}	Acidic
(e) 9.12	7.6×10^{-10}	1.3×10^{-5}	Basic

32. (a) $F^-(aq) + H_2O(\ell) \rightleftharpoons HF(aq) + OH^-(aq)$ $K = \frac{[HF][OH^-]}{[F^-]}$
(b) $NH_3(aq) + H_2O(\ell) \rightleftharpoons NH_4^+(aq) + OH^-(aq)$ $K = \frac{[NH_4^+][OH^-]}{[NH_3]}$
(c) $H_2CO_3(aq) + H_2O(\ell) \rightleftharpoons HCO_3^-(aq) + H_3O^+(aq)$ $K = \frac{[HCO_3^-][H_3O^+]}{[H_2CO_3]}$
(d) $H_3PO_4(aq) + H_2O(\ell) \rightleftharpoons H_2PO_4^-(aq) + H_3O^+(aq)$ $K = \frac{[H_2PO_4^-][H_3O^+]}{[H_3PO_4]}$
(e) $CH_3COO^-(aq) + H_2O(\ell) \rightleftharpoons CH_3COOH(aq) + OH^-(aq)$ $K = \frac{[CH_3COOH][OH^-]}{[CH_3COO^-]}$
(f) $S^{2-}(aq) + H_2O(\ell) \rightleftharpoons HS^-(aq) + OH^-(aq)$ $K = \frac{[HS^-][OH^-]}{[S^{2-}]}$
34. (a) NH_3 (b) K_2S (c) $NaCH_3COO$ (d) KCN
36. $K_a = 1.6 \times 10^{-5}$
38. $[H_3O^+] = [A^-] = 1.3 \times 10^{-5}$ M, [HA] = 0.040 M
40. $K_a = 1.4 \times 10^{-5}$
42. 8.85
43. (a) $C_{10}H_{15}NH_2 + H_2O \rightleftharpoons C_{10}H_{15}NH_3^+ + OH^-$
(b) 10.47
44. 3.28
46. (a) CN^-, product-favored (b) HS^-, reactant-favored
(c) $H_2(g)$, reactant-favored
48. (a) pH < 7 (b) pH > 7 (c) pH = 7
50. (a) pH > 7 (b) pH > 7 (c) pH > 7
52. All three are Lewis bases; CO_2 is a Lewis acid.
54. (a) I_2 is a Lewis acid and I^- is a Lewis base.
(b) BF_3 is a Lewis acid and SO_2 is a Lewis base.
(c) Au^+ is a Lewis acid and CN^- is a Lewis base.
(d) CO_2 is a Lewis acid and H_2O is a Lewis base.
56. (a) Weak acid (b) Strong base (c) Strong acid
(d) Weak base (e) Strong base (f) Amphiprotic
58. (a) Less than 7 (b) Equal to 7 (c) Greater than 7
60. 2.85
62. 9.73
65. Conjugates must differ by just one H^+.
66. Arrhenius Theory: Electron pairs on the solvent water molecules (Lewis base) form a bond with the hydrogen ion (Lewis acid) producing aqueous H_3O^+ ions. Electron pairs on the OH^- ions (Lewis base) form a bond with the hydrogen ion (Lewis acid) in the solvent water molecule, producing aqueous OH^- ions.
Brønsted-Lowry Theory: The H^+ ion from the Brønsted-Lowry acid is bonded to a Brønsted-Lowry base using an electron pair on the base. The electron-pair acceptor, the H1 ion, is the Lewis acid and the electron-pair donor is the Lewis base.
68. (a) 2.01 (b) 23 times
70. 0.76 L; Probably not
71. Yes, pH increases
73. Br has a higher electronegativity than H. $ClNH_2$ is weaker, because Cl has a higher electronegativity than Br.
75. Strongest HM > HQ > HZ weakest; $K_{a,HZ} = 1 \times 10^{-5}$, $K_{a,HQ} = 1 \times 10^{-3}$, $K_{a,HM} = 1 \times 10^{-1}$ or larger
77. (a) Weak (b) Weak
(c) Acid/base conjugates (d) 6.26
79. Dilute solution is 9%, saturated solution is 2%.

Chapter 16

14. (c)
16. (a) pH = 2.1 (b) pH = 7.21 (c) pH = 12.46
18. (a) Lactic acid and lactate ion
(b) Acetic acid and acetate ion

(c) HClO and ClO^-
(d) HCO_3^- and CO_3^{2-}
pK_a closest to given pH.
21. pH = 9.55, pH = 9.51
23. Sample (b). Only (b) results in a solution containing a conjugate acid/base pair.
25. (a) ΔpH = 0.1 (b) ΔpH = 3.8 (c) ΔpH = 7.25
27. (a) pH = 5.02 (b) pH = 4.99 (c) pH = 4.06
30. (a) Bromothymol blue
(b) Phenolphthalein
(c) Methyl red
(d) Bromothymol blue. Suitable pH color changes
32. 0.0253 M HCl
34. 93.6%
36. (a) 29.2 mL (b) 600. mL (c) 1.20 L (d) 2.7 mL
39. (a) pH = 0.824 (b) pH = 1.30 (c) pH = 3.8
(d) pH = 7.000 (e) pH = 10.2 (f) 12.48

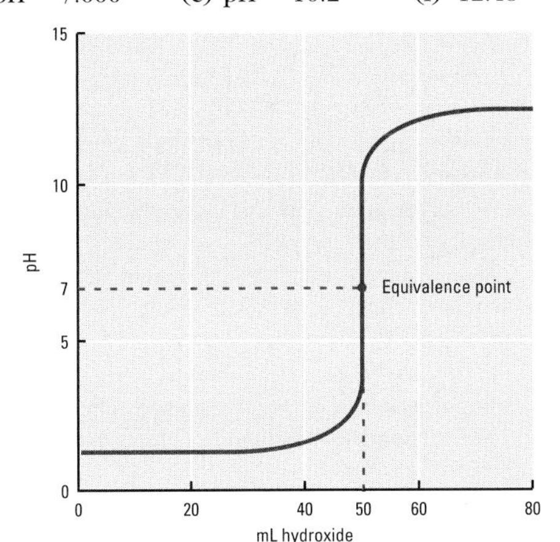

41. (a) $FeCO_3(s) \rightleftharpoons Fe^{2+}(aq) + CO_3^{2-}(aq)$ $K_{sp} = [Fe^{2+}][CO_3^{2-}]$
(b) $Ag_2SO_4(s) \rightleftharpoons 2\,Ag^+(aq) + SO_4^{2-}(aq)$ $K_{sp} = [Ag^+]^2[SO_4^{2-}]$
(c) $Ca_3(PO_4)_2(s) \rightleftharpoons 3\,Ca^{2+}(aq) + 2\,PO_4^{3-}(aq)$
$K_{sp} = [Ca^{2+}]^3[PO_4^{3-}]^2$
(d) $Mn(OH)_2(s) \rightleftharpoons Mn^{2+}(aq) + 2\,OH^-(aq)$
$K_{sp} = [Mn^{2+}][OH^-]^2$
42. $K_{sp} = 2.2 \times 10^{-12}$
44. $K_{sp} = 1.7 \times 10^{-5}$
47. 3.1×10^{-5} mol/L
49. (a) $K_{sp} = 1 \times 10^{-11}$ (b) $[OH^-]$ must be 0.0085 M or higher.
51. 4.5×10^{-9} M or lower
53. pH = 9.0
55. (a) $Ag^+(aq) + 2\,CN^-(aq) \rightleftharpoons [Ag(CN)_2]^-(aq)$

$$K_f = \frac{[[Ag(CN)_2]^-]}{[Ag^+][CN^-]^2}$$

(b) $Cd^{2+}(aq) + 4\,NH_3(aq) \rightleftharpoons [Cd(NH_3)_4]^{2+}(aq)$

$$K_f = \frac{[[Cd(NH_3)_4]^{2+}]}{[Cd^{2+}][NH_3]^4}$$

57. 0.0078 mol or more
59. (a) $Zn(OH)_2(s) + 2\,H_3O^+(aq) \rightarrow Zn^{2+}(aq) + 4\,H_2O(\ell)$
$Zn(OH)_2(s) + 2\,OH^-(aq) \rightarrow [Zn(OH)_4]^{2-}(aq)$
(b) $Sb(OH)_3(s) + 3\,H_3O^+(aq) \rightarrow Sb^{3+}(aq) + 6\,H_2O(\ell)$
$Sb(OH)_3(s) + OH^-(aq) \rightarrow [Sb(OH)_4]^-(aq)$
61. (a) H_2O, CH_3COO^-, Na^+, CH_3COOH, H_3O^+, OH^-
(b) pH = 4.95
(c) pH = 5.05
(d) $CH_3COOH(aq) + H_2O(\ell) \rightleftharpoons CH_3COO^-(aq) + H_3O^+(aq)$
62. Ratio = 1.6
66. $K_a = 3.5 \times 10^{-6}$
70. The tiny amount of base (CH_3COO^-) present is insufficient to prevent the pH from changing dramatically if a strong acid is introduced into the solution.
72. $K_a = 2.3 \times 10^{-4}$
74. Blood pH decreases; acidosis
75. (a) Adding Ca^{2+} drives the reaction more toward reactants, making more apatite.
(b) Acid reacts with OH^-, removing a product and driving the reaction toward the products, causing apatite to decompose.
76. Sample A: $NaHCO_3$; Sample B: NaOH; Sample C: Mixture of NaOH and Na_2CO_3 and/or $NaHCO_3$; Sample D: Na_2CO_3
79. 3.22
81. 3.2 g

Chapter 17

10. (a) $2\,H_2O(\ell) \rightarrow 2\,H_2(g) + O_2(g)$, reactant-favored
(b) $C_8H_{18}(\ell) \rightarrow C_8H_{18}(g)$, product-favored
(c) $C_{12}H_{22}O_{11}(s) \rightarrow C_{12}H_{22}O_{11}(aq)$, product-favored
12. (a) $\frac{1}{2}$ (b) $\frac{1}{2}$ (c) 50 of each
14. (a) Probability of $\frac{1}{2}$ in Flask A; probability of $\frac{1}{2}$ in flask B
(b) 50 in flask A and 50 in flask B
16. (a) Negative (b) Positive (c) Positive
18. (a) Item 2 (b) Item 2 (c) Item 2
20. (a) NaCl(s) (b) $P_4(g)$ (c) $NH_4NO_3(aq)$
22. (a) Negative (b) Negative (c) Positive
24. 112 J K^{-1} mol^{-1}
26. (a) +113.0 J/K (b) +38.17 J/K
28. (a) −173.01 J K^{-1} (b) −326.69 J K^{-1} (c) 137.55 J K^{-1}
30. $\Delta S° = -247.7$ J/K; Cannot tell without $\Delta H°$ also, since that is needed to calculate $\Delta G°$. $\Delta H° = -88.13$ kJ and $\Delta G° = -14.3$ kJ, so it is product-favored at given *T*.
32. Product-favored at low temperatures; the exothermicity is sufficient to favor products if the temperature is low enough to overcome the decrease in entropy.
34. (a) Enthalpy change (ΔH) is negative; entropy change (ΔS) is positive.
(b) $\Delta H° = -184.28$ kJ; $\Delta S° = -7.7$ J K^{-1}; not positive, as predicted. The enthalpy change is negative, as predicted in (a). But the entropy change is negative, not as predicted in (a). The aqueous solute must have sufficient order to compensate for the high disorder of the gas. The value of −7.7 J K^{-1} is very small.
36. Entropy increase is insufficient to drive this highly endothermic reaction to form products without assistance from the surroundings at this temperature; reaction is reactant-favored.
38. (a) $\Delta S° = -37.52$ J/K, $\Delta H° = -851.5$ kJ, product-favored at low temperature
(b) $\Delta S° = -21.77$ J/K, $\Delta H° = 66.36$ kJ, never product-favored
40. (a) $\Delta S_{univ} = 4.92 \times 10^3$ J/K (b) -1.47×10^3 kJ
(c) Yes; ethane is used as a fuel; hence, we might expect that its combustion reaction is product-favored.
42. First line: Sign of $\Delta G°_{system}$ is negative; last line: Sign of $\Delta G°_{system}$ is positive.
44. $\Delta G° = \Delta H° - T\Delta S°$ Here $\Delta H°$ is negative, and $\Delta S°$ is positive, so $\Delta G° = -|\Delta H°| - |T\Delta S°| = -(|\Delta H°| + |T\Delta S°|) < 0$.
46. $\Delta G° = 28.63$ kJ; reactant-favored
48. (a) −141.05 kJ (b) 141.73 kJ (c) −959.43 kJ
Reactions (a) and (c) are product-favored.
50. (a) 385.7 K (b) 835.1 K
52. (a) $\Delta H° = 178.32$ kJ, $\Delta S° = 160.6$ J/K, $\Delta G° = 130.5$ kJ
(b) Reactant-favored
(c) No; it is only product-favored at high temperatures.
(d) 1110. K
54. $\Delta G_f°(Ca(OH)_2) = -867.8$ kJ/mol
56. (a) $K = 4 \times 10^{-34}$ (b) $K = 5 \times 10^{-31}$
58. (a) $\Delta G° = -100.97$ kJ, product-favored
(b) $K = 5 \times 10^{17}$. When $\Delta G°$ is negative, *K* is larger than 1.

62. (a) $\Delta G° = -106$ kJ/mol (b) $\Delta G° = 8.55$ kJ/mol
(c) $\Delta G° = -33.8$ kJ/mol
64. (a) can be harnessed; (b) and (c) require work to be done.
66. Reaction 64(b), 5.068 g graphite oxidized; reaction 64(c), 26.94 g graphite oxidized
68. (a) $2\ C(s) + 2\ Cl_2(g) + TiO_2(s) \rightarrow TiCl_4(g) + 2\ CO(g)$
(b) $\Delta H°$ -44.6 kJ; $\Delta S° = 242.7$ J K^{-1}; $\Delta G° = -116.5$ kJ
(c) Product-favored (d) More reactant-favored
70. CuO, Ag_2O, HgO, and PbO
72. The combustions of coal, petroleum, and natural gas are the most common sources used to supply free energy. We also use solar and nuclear energy as well as the kinetic energy of wind and water. (There may be other answers.)
74. (a) $\Delta G° = -86.5$ kJ, product-favored
(b) $\Delta G° = -873.1$ kJ (c) No (d) Yes
76. (a) Reaction 2 (b) Reactions 1 and 5
(c) Reaction 2 (d) Reactions 2 and 3
(e) None of them
78. (a) $\Delta G° = 141.73$ kJ (b) No (c) Yes
(d) $K_P = 1 \times 10^4$ (e) $K_c = 7 \times 10^1$
80. (a) $\Delta G° = 31.8$ kJ (b) $K_P = P_{Hg(g)}$
(c) $K° = 2.7 \times 10^{-6}$ (d) $P_{Hg(g)} = 2.7 \times 10^{-6}$ atm
(e) $T = 450$ K
82. Scrambled is a very disordered state for an egg. The second law of thermodynamics says that the more disordered state is the more probable state. Putting the delicate tissues and fluids back where they were before the scrambling occurred would take a great deal of energy. Humpty Dumpty is a fictional character who was also an egg. He fell off a wall. A very probable result of that fall is for an egg to become scrambled. The story goes on to tell that all the energy of the king's horses and men was not sufficient to put Humpty together again.
83. Absolute entropies can be determined because the minimum value of $S°$ is zero at $T = 0$ K. It is not possible to define conditions for a specific minimum value for internal energy, enthalpy, or Gibbs free energy of a substance, so relative quantities must be used.
85. NaCl, in an orderly crystal structure, and pure water, with only O—H hydrogen bonding interactions in the liquid state, are far more ordered than the dispersed hydrated sodium and chloride ions interacting with the water molecules.
87. (a) False (b) False (c) True
(d) True (e) True
89. $\Delta G < 0$ means products are favored; however, the equilibrium state will always have some reactants present, too. To get all the reactants to go away requires the removal of the products from the reactants, so the reaction continues forward.
91. (a) 58.78 J/K (b) -53.29 J/K (c) -173.93 J/K
Adding more hydrogen makes the $\Delta S°$ more negative.
93. Product favored; $\Delta H°$ is negative. Assuming the $S°$ of C_8H_{16} and C_8H_{18} are approximately the same value, $\Delta S°$ is negative, but not large enough to give $\Delta G°$ a different sign; hence, $\Delta G°$ is negative, also.
95. (a) $K_P = 1.5 \times 10^7$ (b) Product-favored (c) $K_c = 3.7 \times 10^8$
97. (a) $7\ C(s) + 6\ H_2O(g) \rightarrow 2\ C_2H_6(g) + 3\ CO_2(g)$
$5\ C(s) + 4\ H_2O(g) \rightarrow C_3H_8(g) + 2\ CO_2(g)$
$3\ C(s) + 4\ H_2O(g) \rightarrow 2\ CH_3OH(\ell) + CO_2(g)$
(b) For C_2H_6, $\Delta H° = 101.02$ kJ, $\Delta G° = 122.71$ kJ, $\Delta S° = -72.71$ J/K
For C_3H_8, $\Delta H° = 76.5$ kJ, $\Delta G° = 102.08$ kJ, $\Delta S° = -86.6$ J/K
For CH_3OH, $\Delta H° = 96.44$ kJ, $\Delta G° = 187.39$ kJ, $\Delta S° = -305.2$ J/K
None of these are feasible. $\Delta G°$ is positive. In addition, $\Delta H°$ is positive and $\Delta S°$ is negative, suggesting that there is no temperature at which the products would be favored.
99. (a) (ii) (b) (i) (c) (ii)

Chapter 18

8. (a) $Zn(s) \rightarrow Zn^{2+}(aq) + 2\ e^-$
(b) $2\ H_3O^+(aq) + 2\ e^- \rightarrow 2\ H_2O(\ell) + H_2(g)$
(c) $Sn^{4+}(aq) + 2\ e^- \rightarrow Sn^{2+}(aq)$
(d) $Cl_2(g) + 2\ e^- \rightarrow 2\ Cl^-(aq)$
(e) $6\ H_2O(\ell) + SO_2(g) \rightarrow SO_4^{2-}(aq) + 4\ H_3O^+(aq) + 2\ e^-$
10. $4\ Zn(s) + 7\ OH^-(aq) + NO_3^-(aq) + 6\ H_2O(\ell) \rightarrow 4\ Zn(OH)_4^{2-} + NH_3(aq)$
12. (a) $3\ CO(g) + O_3(g) \rightarrow 3\ CO_2(g)$
O_3 is the oxidizing agent; CO is the reducing agent.
(b) $H_2(g) + Cl_2(g) \rightarrow 2\ HCl(g)$
Cl_2 is the oxidizing agent; H_2 is the reducing agent.
(c) $H_2O_2(aq) + Ti^{2+}(aq) + 2\ H_3O^+(aq) \rightarrow 4\ H_2O(\ell) + Ti^{4+}(aq)$
H_2O_2 is the oxidizing agent; Ti^{2+} is the reducing agent.
(d) $2\ MnO_4^-(aq) + 6\ Cl^-(aq) + 8\ H_3O^+(aq) \rightarrow 2\ MnO_2(s) + 3\ Cl_2(g) + 12\ H_2O(\ell)$
MnO_4^- is the oxidizing agent; Cl^- is the reducing agent.
(e) $4\ FeS_2(s) + 11\ O_2(g) \rightarrow 2\ Fe_2O_3(s) + 8\ SO_2(g)$
O_2 is the oxidizing agent; FeS_2 is the reducing agent.
(f) $O_3(g) + NO(g) \rightarrow O_2(g) + NO_2(g)$
O_3 is the oxidizing agent; NO is the reducing agent.
(g) $Zn(Hg)(amalgam) + HgO(s) \rightarrow ZnO(s) + 2\ Hg(\ell)$
HgO is the oxidizing agent; Zn(Hg) is the reducing agent.
14. The generation of electricity occurs when electrons are transmitted through a wire from the metal to the cation. Here, the transfer of electrons would occur directly from the metal to the cation and the electrons would not flow through any wire.
16. (a) $Zn(s) + Pb^{2+}(aq) \rightarrow Zn^{2+}(aq) + Pb(s)$
(b) Oxidation of zinc occurs at the anode. The reduction of lead ion occurs at the cathode. The anode is metallic zinc. The cathode is metallic lead.
(c)

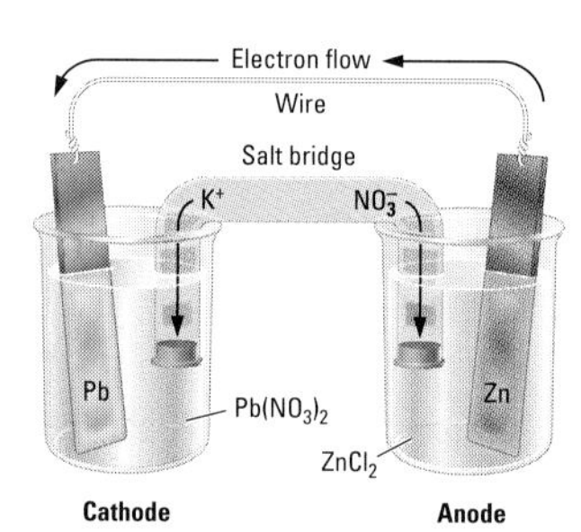

18. 7.5×10^3 C
19. (a) $Cu(s) \rightarrow Cu^{2+}(aq) + 2\ e^-$
$Ag^+(aq) + e^- \rightarrow Ag(s)$
(b) The copper half-reaction is oxidation and it occurs in the anode compartment. The silver half-reaction is reduction and it occurs in the cathode compartment.
21. Li is the strongest reducing agent and Li^+ is the weakest oxidizing agent. F_2 is the strongest oxidizing agent and F^- is the weakest reducing agent.
23. Worst oxidizing agent (d) < (c) < (a) < (b) best oxidizing agent
25. (a) 2.91 V (b) -0.028 V
(c) 0.65 V (d) 1.16 V
Reactions (a), (c), and (d) are product-favored.
27. (a) Al^{3+} (b) Ce^{4+} (c) Al
(d) Ce^{3+} (e) Yes (f) No
(g) Mercury(I) ion, silver ion, and cerium(IV) ion
(h) Hg, Sn, Ni, Al
29. (a) 1.55 V (b) -1196 kJ, 1.55 V
31. -409 kJ
33. 1×10^{-18}, 102 kJ
35. $K° = 4 \times 10^1$

37. (a) 1.20 V (b) 1.16 V (c) 3 M
40. (conc. Pb^{2+})/(conc. Sn^{2+}) = 0.3
42. (a) Ni^{2+} (aq) + Cd(s) $\rightarrow$ Ni(s) + Cd^{2+} (aq)
 (b) Cd is oxidized, Ni^{2+} is reduced, Ni^{2+} is the oxidizing agent, and Cd is the reducing agent.
 (c) Metallic Cd is the anode and metallic Ni is the cathode.
 (d) 0.15 V
 (e) Electrons flow from the Cd electrode to the Ni electrode.
 (f) Toward the anode compartment
44. A fuel cell has a continuous supply of reactants and will be useable for as long as the reactants are supplied. A battery contains all the reactants of the reaction. Once the reactants are gone, the battery is no longer useable.
46. (a) The N_2H_4 half-reaction occurs at the anode and the O_2 half-reaction occurs at the cathode.
 (b) $N_2H_4(g) + O_2(g) \rightarrow N_2(g) + 2\,H_2O(\ell)$
 (c) 7.5 g N_2H_4 (d) 7.5 g O_2
47. $O_2(g)$ produced at anode; $H_2(g)$ produced at cathode; 2 mol H_2 per mol O_2
48. Au^{3+}, Hg^{2+}, Ag^{+}, Hg_2^{2+}, Fe^{3+}, Cu^{2+}, Sn^{4+}, Sn^{2+}, Ni^{2+}, Cd^{2+}, Fe^{2+}, Zn^{2+}
50. H_2, Br_2, and OH^- are produced. After the reaction is complete, the solution contains H_2O, Na^+, OH^-, a small amount of dissolved Br_2 (though it has low solubility in water), and a very small amount of H_3O^+. H_2 is formed at the cathode. Br_2 is formed at the anode.
52. 0.16 g Ag
53. 5.93 g Cu
56. 1.9×10^2 g Pb
58. 0.043 g Li; 0.64 g Pb
60. 6.85 min
63. Ions increase the electrolytic capacity of the solution.
65. Chromium is highly resistant to corrosion and protects iron in steel from oxidizing.
68. 0.00689 g Cu, 0.00195 g Al
70. Worst reducing agent B < D < A < C best reducing agent
72. (a) B is oxidized and A^{2+} is reduced.
 (b) A^{2+} is the oxidizing agent and B is the reducing agent.
 (c) B is the anode and A is the cathode.
 (d) $A^{2+} + 2e^- \rightarrow A$
 $B \rightarrow B^{2+} + 2e^-$
 (e) A gains mass.
 (f) Electrons flow from B to A.
 (g) K^+ ions will migrate toward the A^{2+} solution.
76. (a) 9.5×10^6 g HF (b) 1.7×10^3 kWh
78. $Cl_2(g) + 2\,Br^-(aq) \rightarrow Br_2(\ell) + 2\,Cl^-(aq)$
80. 4+
82. 75 s
83. 5×10^{-6} M Cu^{2+}

Chapter 19

11. (a) α emission (b) β emission (c) Electron capture or positron emission (d) β emission
13. (a) $^{238}_{92}U$ (b) $^{32}_{15}P$ (c) $^{10}_{5}B$ (d) $^{0}_{-1}e$
 (e) $^{15}_{7}N$
15. (a) $^{28}_{12}Mg \rightarrow {}^{28}_{13}Al + {}^{0}_{-1}e$
 (b) $^{238}_{92}U + {}^{12}_{6}C \rightarrow 4\,{}^{1}_{0}n + {}^{246}_{98}Cf$
 (c) $^{2}_{1}H + {}^{3}_{2}He \rightarrow {}^{4}_{2}He + {}^{1}_{1}H$
 (d) $^{38}_{19}K \rightarrow {}^{38}_{18}Ar + {}^{0}_{+1}e$
 (e) $^{175}_{78}Pt \rightarrow {}^{4}_{2}He + {}^{171}_{76}Os$
17. $^{231}_{90}Th$, $^{231}_{91}Pa$, $^{227}_{89}Ac$, $^{227}_{90}Th$, $^{223}_{88}Ra$
19. $^{222}_{86}Rn \rightarrow {}^{4}_{2}He + {}^{218}_{84}Po$
 $^{218}_{84}Po \rightarrow {}^{4}_{2}He + {}^{214}_{82}Pb$
 $^{214}_{82}Pb \rightarrow {}^{0}_{-1}e + {}^{214}_{83}Bi$
 $^{214}_{83}Bi \rightarrow {}^{0}_{-1}e + {}^{214}_{84}Po$
 $^{214}_{84}Po \rightarrow {}^{4}_{2}He + {}^{210}_{82}Pb$
 $^{210}_{82}Pb \rightarrow {}^{0}_{-1}e + {}^{210}_{83}Bi$
 $^{210}_{83}Bi \rightarrow {}^{0}_{-1}e + {}^{210}_{84}Po$
 $^{210}_{84}Po \rightarrow {}^{4}_{2}He + {}^{206}_{82}Pb$
20. (a) $^{19}_{10}Ne \rightarrow {}^{19}_{9}F + {}^{0}_{+1}e$ (b) $^{230}_{90}Th \rightarrow {}^{0}_{-1}e + {}^{230}_{91}Pa$
 (c) $^{82}_{35}Br \rightarrow {}^{0}_{-1}e + {}^{82}_{36}Kr$ (d) $^{212}_{84}Po \rightarrow {}^{4}_{2}He + {}^{208}_{82}Pb$
22. The binding energy per nucleon of ^{10}B is 6.252×10^8 kJ/mol nucleon. The binding energy per nucleon of ^{11}B is 6.688×10^8 kJ/mol nucleon. ^{11}B is more stable than ^{10}B because its binding energy is larger.
24. -1.7394×10^{11} kJ/mol ^{238}U; 7.3086×10^8 kJ/mol nucleon
27. 5 mg
29. (a) $^{131}_{53}I \rightarrow {}^{131}_{54}Xe + {}^{0}_{-1}e$ (b) 1.56 mg
31. 4.0 y
33. 34.8 d
35. 2.58×10^3 y; Approx. 570 B.C.
37. 1.6×10^5 y
39. Cadmium rods (a neutron absorber to control the rate of the fission reaction), uranium rods (a source of fuel, since uranium is a reactant in the nuclear equation), and water (used for cooling by removing excess heat and used in steam/water cycle for the production of turning torque for the generator)
41. (a) $^{140}_{54}Xe$ (b) $^{101}_{41}Nb$ (c) $^{92}_{36}Kr$
43. 6.9×10^3 barrels
45. A rad is a measure of the amount of radiation absorbed. A rem includes a quality factor that better describes the biological impact of a radiation dose. The unit rem would be more appropriate when talking about the effects of an atomic bomb on humans. The unit gray (Gy) is 100 rad.
47. Since most elements have some proportion of unstable isotopes that decay and we are composed of these elements (e.g., ^{14}C), our bodies emit radiation particles.
49. (a) $^{0}_{-1}e$ (b) $^{4}_{2}He$ (c) $^{87}_{35}Br$ (d) $^{216}_{84}Po$ (e) $^{68}_{31}Ga$
51. 2 mg
53. 3.6×10^9 y
57. ^{20}Ne is stable. The ^{17}Ne is likely to decay by positron emission, increasing the ratio of neutrons to protons. The ^{23}Ne is likely to decay by beta emission, decreasing the ratio of neutrons to protons.
59. A nuclear reaction occurred, making products. Therefore, some of the lost mass is found in the decay particles, if the decay is alpha or beta decay, and almost all of the rest is found in the element produced by the reaction.
61. They are first-order because the rate of decay depends only on the quantity of particles decaying.
63. 2.8×10^4 kg
65. 75.1 y
67. 3.92×10^3 y

Appendix A

4. The calculation is incomplete or incorrect. Check for incomplete unit conversions (e.g., cm^3 to L, g to kg, mm Hg to atm, etc.) and check to see that the numerator and denominator of unit factors are placed such that the unwanted units can cancel (e.g., $\frac{g}{mL}$ or $\frac{mL}{g}$).
6. Answer (c) gives the properly reported observation. Answers (a) and (b) do not have the proper units. Answer (d) is incomplete, with no units. Answer (e) shows the conversion of the observed measurement to new units, making (e) the result of a calculation, not the observed measurement.

8. (a) The units should not be squared. To do this correctly, add values with common units (g) and give the answer the same units (g). The result is 9.95 g.
(b) The unit factor is not set up to cancel unwanted units (g); the units reported (g) are not the units resulting from this calculation ($\frac{g^2}{mL}$). To do this correctly, $5.23\text{ g} \times \frac{1.00\text{ mL}}{4.87\text{ g}} = 1.07\text{ mL}$.
(c) The unit factor must be cubed to cancel all unwanted units; the units reported (m^3) are not the units resulting from this calculation ($m \cdot cm^2$). To do this correctly, $3.57\text{ cm}^3 \times (\frac{1\text{ m}}{100\text{ cm}})^3 = 3.57 \times 10^{-6}\text{ m}^3$

10. (a) 95.9 ± 0.59 in
(b) Three, because uncertainty is in the tenths place
(c) The result is accurate (i.e., the true answer, exactly 8 ft or 96 inches, is within the range of the uncertainty, though not very precise.

12. (a) 4 (b) 2 (c) 2 (d) 4
14. (a) 43.32 (b) 43.32 (c) 43.32 (d) 43.32 (e) 43.32 (f) 43.32
16. (a) 13.7 (b) 0.247 (c) 12.0
18. (a) 7.6003×10^4 (b) 3.7×10^{-4} (c) 3.4×10^4
20. (a) 2.415×10^{-3} (b) 2.70×10^8 (c) 3.236 (d) 116 cm^3
22. (a) −0.1351 (b) 3.541 (c) 23.7797 (d) 54.7549 (e) −7.455
24. (a) 5.404 (b) 1×10^{15} (c) 110. (d) 1.000320 (e) 3.75
26. (a) 0.480, −1.80 (b) 1.34, 3.23

28.

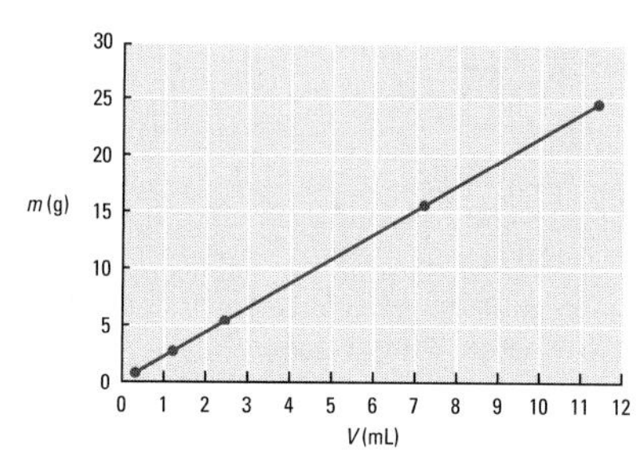

Yes, m is proportional to V.

Appendix B

1. (a) Kilogram, no additional prefix
(b) cubic meter, prefix nano-
(c) meter, prefix milli-
3. SI base (fundamental) units are set by convention; derived units are based on the fundamental units.
5. (a) 4.75×10^{-10} m (b) 5.6×10^7 kg (c) 4.28×10^{-6} A
7. (a) $8.7 \times 10^{-18}\text{ m}^2$ (b) 2.73×10^{-17} J (c) 2.73×10^{-5} N
9. (a) 2.20 lb (b) 2.22×10^3 kg (c) 60 in^3 (d) 1×10^5 Pa and 1 bar
11. (a) 99 °F (b) −10.5 °F (c) −40.0 °F

Glossary

absolute temperature scale (Kelvin temperature scale) A temperature scale on which the zero is the lowest possible temperature and the degree is the same size as the Celsius degree.

absolute zero Lowest possible temperature, 0 K.

acid (Arrhenius) A substance that increases the concentration of hydronium ions, H_3O^+, in aqueous solution. (See also **Brønsted-Lowry acid, Lewis acid.**)

acid ionization constant (K_a) The equilibrium constant for the reaction of a weak acid with water to produce hydronium ions and the conjugate base of the weak acid.

acid ionization constant expression Mathematical expression in which the product of the equilibrium concentrations of hydronium ion and conjugate base is divided by the equilibrium concentration of the un-ionized conjugate acid.

acidic solution An aqueous solution in which the concentration of hydronium ion exceeds the concentration of hydroxide ion.

actinides The elements after actinium in the seventh period; in actinides the 5*f* subshell is being filled.

activated complex A molecular structure corresponding to the maximum of a plot of energy versus reaction progress; also known as the transition state.

activation energy (E_a) The potential energy difference between reactants and activated complex; the minimum energy that reactant molecules must have to be converted to product molecules.

active site The part of an enzyme molecule that binds the substrate to help it to react.

activity (A) A measure of the rate of nuclear decay, given as disintegrations per unit time.

actual yield The quantity of a reaction product obtained experimentally; less than the theoretical yield.

alcohol An organic compound containing a hydroxyl group (—OH) covalently bonded to a saturated carbon atom.

aldehyde An organic compound characterized by a carbonyl group in which the carbon atom is bonded to a hydrogen atom; a molecule containing the —CHO functional group.

alkali metals The Group 1A elements in the periodic table (except hydrogen).

alkaline earth metals The elements in Group 2A of the periodic table.

alkane Any of a class of hydrocarbons characterized by the presence of only single carbon-carbon bonds.

alkene Any of a class of hydrocarbons characterized by the presence of a carbon-carbon double bond.

alkyl group A fragment of an alkane structure that results from the removal of a hydrogen atom from the alkane.

alkyne Any of a class of hydrocarbons characterized by the presence of a carbon-carbon triple bond.

allotropes Different forms of the same element that exist in the same physical state under the same conditions of temperature and pressure.

alpha (α) particles Positively charged (2+) particles ejected at high speeds from certain radioactive nuclei; the nuclei of helium atoms.

alpha radiation Radiation composed of alpha particles (helium nuclei).

amine An organic compound containing an $—NH_2$, —NHR, or $—NR_2$ functional group.

amorphous solid A solid whose constituent nanoscale particles have no long-range repeating structure.

amount A measure of the number of elementary entities (such as atoms, ions, molecules) in a sample of matter compared with the number of elementary entities in exactly 0.012 kg pure ^{12}C. Also called molar amount.

ampere The SI unit of electrical current; involves the flow of one coulomb of charge per second.

amphoteric Refers to a substance that can act as either an acid or a base.

anion An ion with a negative electrical charge.

anode The electrode of an electrochemical cell at which oxidation occurs.

anodic inhibition The prevention of oxidation of an active metal by painting it, coating it with grease or oil, or allowing a thin film of metal oxide to form.

antibonding molecular orbital A higher-energy molecular orbital that, if occupied by electrons, does not result in attraction between the atoms.

aqueous solution A solution in which water is the solvent.

Arrhenius equation Mathematical relation that gives the temperature dependence of the reaction rate constant; $k = Ae^{-E_a/RT}$.

atom The smallest particle of an element that can be involved in chemical combination with another element.

atom economy The fraction of atoms of starting materials incorporated into the desired final product in a chemical reaction.

atomic mass units (amu) The unit of a scale of relative atomic masses of the elements; 1 amu = 1/12 the mass of a six-proton, six-neutron carbon atom.

atomic number The number of protons in the nucleus of an atom of an element.

atomic radius One-half the distance between the nuclei centers of two like atoms in a molecule.

atomic structure The identity and arrangement of subatomic particles in an atom.

atomic weight The average mass of an atom in a representative sample of atoms of an element.

autoionization The equilibrium reaction in which water molecules react with each other to form hydronium ions and hydroxide ions.

average reaction rate A reaction rate calculated from a change in concentration divided by a change in time.

Avogadro's law The volume of a gas, at a given temperature and pressure, is directly proportional to the amount of gas.

Avogadro's number The number of particles in a mole of any substance (6.022×10^{23}).

axial position(s) Positions above and below the equatorial plane in a triangular bipyramidal structure.

background radiation Radiation from natural and synthetic radioactive sources to which all members of a population are exposed.

balanced chemical equation A chemical equation that shows equal numbers of atoms of each kind in the products and the reactants.

bar A pressure unit equal to 100,000 Pa.

barometer A device for measuring atmospheric pressure.

base (Arrhenius) Substance that increases concentration of hydroxide ions, OH^-, in aqueous solution. (See also **Brønsted–Lowry base, Lewis base.**)

base ionization constant (K_b) The equilibrium constant for the reaction of a weak base with water to produce hydroxide ions and the conjugate acid of the weak base.

base ionization constant expression Mathematical expression in which the product of the equilibrium concentrations of hydroxide ion and conjugate acid is divided by the equilibrium concentration of the conjugate base.

basic solution An aqueous solution in which the concentration of hydroxide ion is greater than the concentration of hydronium ion.

battery (voltaic cell) An electrochemical cell (or group of cells) in which a product-favored oxidation-reduction reaction is used to produce an electric current.

becquerel A unit of radioactivity equal to 1 nuclear disintegration per second.

beta particles Electrons ejected from certain radioactive nuclei.

beta radiation Radiation composed of electrons.

bimolecular reaction An elementary reaction in which two particles must collide for products to be formed.

binary molecular compound A molecular compound whose molecules contain atoms of only two elements.

binding energy The energy required to separate all nucleons in an atomic nucleus.

binding energy per nucleon The energy per nucleon required to separate all nucleons in an atomic nucleus.

boiling The process whereby a liquid vaporizes throughout when its vapor pressure equals atmospheric pressure.

boiling point The temperature at which the equilibrium vapor pressure of a liquid equals the external pressure on the liquid.

boiling-point elevation A colligative property; the difference between the normal boiling point of a pure solvent and the higher boiling point of a solution in which a nonvolatile nonelectrolyte solute is dissolved in that solvent.

bond Attractive force between two atoms holding them together, for example, as part of a molecule.

bond angle The angle between the bonds to two atoms that are bonded to the same third atom.

bond enthalpy (bond energy) The change in enthalpy when a mole of chemical bonds of a given type is broken, separating the bonded atoms; the atoms and molecules must be in the gas phase.

bond length The distance between the centers of the nuclei of two bonded atoms.

bonding electrons Electron pairs shared in covalent bonds.

bonding molecular orbital A lower-energy molecular orbital that can be occupied by bonding electrons.

bonding pair A pair of valence electrons that are shared between two atoms.

Born-Haber cycle A stepwise thermochemical cycle in which the constituent elements are converted to ions and combined to form an ionic compound.

boundary surface A surface within which there is a specified probability (often 90%) that an electron will be found.

Boyle's law The volume of a confined ideal gas varies inversely with the applied pressure, at constant temperature and amount of gas.

Brønsted-Lowry acid A hydrogen ion donor.

Brønsted-Lowry acid-base reaction A reaction in which an acid donates a hydrogen ion and a base accepts the hydrogen ion.

Brønsted-Lowry base A hydrogen ion acceptor.

buckyball Buckminsterfullerene; an allotrope of carbon consisting of molecules in which 60 carbon atoms are arranged in a cage-like structure consisting of five-membered rings sharing edges with six-membered rings.

buffer See **buffer solution.**

buffer capacity The quantity of acid or base a buffer can accommodate without a significant pH change (more than one pH unit).

buffer solution A solution that resists changes in pH when limited amounts of acids or bases are added; it contains a weak acid and its conjugate base, or a weak base and its conjugate acid.

calorie (cal) A unit of energy equal to 4.184 J. Approximately 1 cal is required to raise the temperature of 1 g of liquid water by 1 °C.

Calorie (Cal) A unit of energy equal to 4.184 kJ = 1 kcal. (See also **kilocalorie.**)

calorimeter A device for measuring the quantity of thermal energy transferred during a chemical reaction or some other process.

capillary action The process whereby a liquid rises in a small-diameter tube due to noncovalent interactions between the liquid and the tube's material.

carbonyl group An organic functional group consisting of carbon bonded to two other atoms and double bonded to oxygen; $>C{=}O$.

carboxylic acid An organic compound characterized by the presence of the carboxyl group ($-COOH$).

catalyst A substance that increases the rate of a reaction but is not consumed in the overall reaction.

cathode The electrode of an electrochemical cell at which reduction occurs.

cathodic protection A process of protecting a metal from corrosion whereby it is made the cathode by connecting it electrically to a more reactive metal.

cation An ion with a positive electrical charge.

cell voltage The electromotive force of an electrochemical cell; the quantity of work a cell can produce per coulomb of charge that the chemical reaction produces.

Celsius temperature scale A scale defined by the freezing (0 °C) and boiling (100 °C) points of pure water, at 1 atm.

change of state A physical process in which one state of matter is changed into another (such as melting a solid to form a liquid).

Charles's law The volume of an ideal gas at constant pressure and amount of gas varies directly with its absolute temperature.

chemical change (chemical reaction) A process in which substances (reactants) change into other substances (products) by rearrangement, combination, or separation of atoms.

chemical compound (compound) A pure substance (e.g., sucrose or water) that can be decomposed into two or more different pure substances; homogeneous, constant-composition matter that consists of two or more chemically combined elements.

chemical element (element) A substance (e.g., carbon, hydrogen, or oxygen) that cannot be decomposed into two or more new substances by chemical or physical means.

chemical equilibrium A state in which the concentrations of reactants and products remain constant because the rates of forward and reverse reactions are equal.

chemical formula (formula) A notation combining element symbols and numerical subscripts that shows the relative numbers of each kind of atom in a molecule or formula unit of a substance.

chemical kinetics The study of the speeds of chemical reactions and the nanoscale pathways or rearrangements by which atoms, ions, and molecules are converted from reactants to products.

chemical periodicity, law of Law stating that the properties of the elements are periodic functions of atomic number.

chemical property Describes the kinds of chemical reactions that chemical elements or compounds can undergo.

chemical reaction (chemical change) A process in which substances (reactants) change into other substances (products) by rearrangements, combination, or separation of atoms.

chemistry The study of matter and the changes it can undergo.

chlor-alkali process Electrolysis process for producing chlorine and sodium hydroxide from aqueous sodium chloride.

***cis* isomer** The isomer in which two like substituents are on the same side of a carbon-carbon double bond, the same side of a ring of carbon atoms, or the same side of a complex ion.

***cis-trans* isomerism** A form of stereoisomerism in which the isomers have the same molecular formula and the same atom-to-atom bonding sequence, but the atoms differ in the location of pairs of substituents on the same side or on opposite sides of a molecule.

Clausius-Clapeyron equation Equation that gives the relationship between vapor pressure and temperature.

closest packing Arranging atoms so that they are packed into the minimum volume.

coefficients (stoichiometric coefficients) The multiplying numbers assigned to the formulas in a chemical equation in order to balance the equation.

colligative properties Properties of solutions that depend only on the concentration of solute particles in the solution, not on the nature of the solute particles.

combination reaction A reaction in which two reactants combine to give a single product.

combined gas law A form of the ideal gas law that relates the P, V, T of a given amount of gas before and after a change: $P_1V_1/T_1 = P_2V_2/T_2$.

combining volumes, law of At constant temperature and pressure, the volumes of reacting gases are always in the ratios of small whole numbers.

combustion analysis A quantitative method to obtain percent composition data for compounds that can burn in oxygen.

combustion reaction A reaction in which an element or compound burns in air or oxygen.

common ion effect Shift in equilibrium position that results from addition of an ion identical to one in the equilibrium.

complex ion An ion with several molecules or ions connected to a central metal ion by coordinate covalent bonds.

compound See **chemical compound.**
compressibility The property of a gas that allows it to be compacted into a smaller volume by application of pressure.
concentration The relative quantities of solute and solvent in a solution.
condensation Process whereby a molecule in the gas phase enters the liquid phase.
condensed formula A chemical formula of an organic compound indicating how atoms are grouped together in a molecule.
conduction band In a solid, an energy band that contains electrons of higher energy than those in the valence band.
conductor A material that conducts electric current; has an overlapping valence band and conduction band.
conjugate acid-base pair A pair of molecules or ions related to one another by the loss and gain of a single hydrogen ion.
conservation of energy, law of (first law of thermodynamics) Law stating that energy can be neither created nor destroyed—the total energy of the universe is constant.
conservation of mass, law of Law stating that there is no detectable change in mass during an ordinary chemical reaction.
constant composition, law of Law stating that a chemical compound always contains the same elements in the same proportions by mass.
constitutional isomers (structural isomers) Compounds with the same molecular formula that differ in the order in which their atoms are bonded together.
continuous spectrum A spectrum consisting of all possible wavelengths.
conversion factor (proportionality factor) A relationship between two measurement units derived from the proportionality of one quantity to another; e.g., density is the conversion factor between mass and volume.
coordinate covalent bond A chemical bond in which both of the two electrons forming the bond were originally associated with the same one of the two bonded atoms.
core electrons The electrons in the filled inner shells of an atom.
corrosion Oxidation of a metal exposed to the environment.
coulomb The unit of electrical charge equal to the quantity of charge that passes a fixed point in an electrical circuit when a current of one ampere flows for one second.
Coulomb's law Law that represents the force of attraction between two charged particles; $F = k(q_1q_2/d^2)$.
covalent bond Interatomic attraction resulting from the sharing of electrons between two atoms.
critical mass The minimum quantity of fissionable material needed to support a self-sustaining chain reaction.
critical pressure The vapor pressure of a liquid at its critical temperature.
critical temperature The temperature above which there is no distinction between liquid and vapor phases.
crystal lattice The ordered, repeating arrangement of ions, molecules, or atoms in a crystalline solid.
crystalline solids Solids with an ordered arrangement of atoms, molecules, or ions that results in planar faces and sharp angles of the crystals.
crystallization The process in which mobile atoms, molecules, or ions in a liquid or solution convert into a crystalline solid.
cubic close packing The three-dimensional structure that results when atoms or ions are closest packed in the *abcabc* arrangement.
cubic unit cell A unit cell with equal-length edges that meet at 90° angles.
curie (Ci) A unit of radioactivity equal to 3.7×10^{10} disintegrations per second.

Dalton's law of partial pressures The total pressure exerted by a mixture of gases is the sum of the partial pressures of the individual gases in the mixture.
decomposition reaction A reaction in which a compound breaks down chemically to form two or more simpler substances.
delocalized electrons Electrons, such as in benzene, that are spread over several atoms in a molecule or polyatomic ion.
density The ratio of the mass of an object to its volume.
deposition The process of a gas converting directly to a solid.
dew point Temperature at which the actual partial pressure of water vapor equals the equilibrium vapor pressure.
diamagnetic Describes atoms or ions in which all the electrons are paired in filled shells so their magnetic fields effectively cancel each other.
diatomic molecule A molecule that contains two atoms.
dietary minerals Essential elements that are not carbon, hydrogen, oxygen, or nitrogen.
diffusion Spread of gas molecules of one type through those of another type.
dimensional analysis A method of using units in calculations to check for correctness.
dipole moment The product of the magnitude of the partial charges ($\delta+$ and $\delta-$) of a molecule times the distance of separation between the charges.
dipole-dipole attraction The noncovalent force of attraction between any two polar molecules or polar regions in the same large molecule.
displacement reaction A reaction in which one element reacts with a compound to form a new compound and release a different element.
dissociation The separation of cations from anions when an ionic solid dissolves in water.
double bond A bond formed by sharing two pairs of electrons between the same two atoms.
dynamic equilibrium A balance between opposing reactions occurring at equal rates.
effective nuclear charge The nuclear positive charge experienced by outer-shell electrons in a many-electron atom.
effusion Escape of gas molecules from a container through a tiny hole into a vacuum.
electrochemical cell A combination of anode, cathode, and other materials arranged so that a product-favored oxidation-reduction reaction can cause a current to flow or an electric current can cause a reactant-favored redox reaction to occur.
electrochemistry The study of the relationship between electron flow and oxidation-reduction reactions.
electrode A device such as a metal plate or wire that conducts electrons into and out of a system.
electrolysis The use of electrical energy to produce a chemical change.
electrolyte A substance that ionizes or dissociates when dissolved in water to form an electrically conducting solution.
electromagnetic radiation Radiation that consists of oscillating electric and magnetic fields that travel through space at the same rate (the speed of light: 186,000 miles/s or 10^8 m/s in a vacuum).
electromotive force (emf) The difference in electrical potential energy between the two electrodes in an electrochemical cell, measured in volts.
electron A negatively charged subatomic particle that occupies most of the volume of an atom.
electron affinity The energy change when a mole of electrons is added to a mole of atoms in the gas phase.
electron capture A radioactive decay process in which one of an atom's inner-shell electrons is captured by the nucleus, which decreases the atomic number by 1.
electron configuration The complete description of the orbitals occupied by all the electrons in an atom or ion.
electron density The probability of finding an electron within a tiny volume in an atom; determined by the square of the wave function.
electron-pair geometry The geometry around a central atom including the spatial positions of bonding and lone electron pairs.
electronegativity A measure of the ability of an atom in a molecule to attract bonding electrons to itself.
element (chemical element) A substance (e.g., carbon, hydrogen, and oxygen) that cannot be decomposed into two or more new substances by chemical or physical means.
elementary reaction A nanoscale reaction whose equation indicates exactly which atoms, ions, or molecules collide or change as the reaction occurs.
empirical formula A formula showing the simplest possible ratio of atoms of elements in a compound.
end point The point at which the indicator changes color during a titration.

endothermic (process) A process in which thermal energy must be transferred into a thermodynamic system in order to maintain constant temperature.

energy The capacity to do work.

energy band In a solid, a large group of orbitals whose energies are closely spaced; in an atomic solid the average energy of a band equals the energy of the corresponding orbital in an individual atom.

energy conservation The conservation of useful energy, that is, of Gibbs free energy.

enthalpy change (ΔH) The quantity of thermal energy transferred when a process takes place at constant temperature and pressure.

enthalpy of fusion The enthalpy change when a substance melts; the quantity of energy that must be transferred when a substance melts at constant temperature and pressure.

enthalpy of solution The quantity of thermal energy transferred when a solution is formed at constant T and P.

enthalpy of sublimation The enthalpy change when a solid sublimes; the quantity of energy, at constant pressure, that must be transferred to cause a solid to vaporize.

enthalpy of vaporization The enthalpy change when a substance vaporizes: the quantity of energy that must be transferred when a liquid vaporizes at constant temperature and pressure.

entropy A measure of the number of ways energy can be distributed in a system; a measure of the dispersal of energy in a system.

enzyme A highly efficient biochemical catalyst for one or more reactions in a living system.

enzyme-substrate complex The combination formed by the binding of an enzyme with a substrate through noncovalent forces.

equatorial position Position lying on the equator of an imaginary sphere around a triangular bipyramidal molecular or ionic structure.

equilibrium concentration The concentration of a substance (usually expressed as molarity) in a system that has reached the equilibrium state.

equilibrium constant (K) A quotient of equilibrium concentrations of product and reactant substances that has a constant value for a given reaction at a given temperature.

equilibrium constant expression The mathematical expression associated with an equilibrium constant.

equilibrium vapor pressure Pressure of the vapor of a substance in equilibrium with its liquid or solid in a closed container.

equivalence point The point in a titration at which a stoichiometrically equivalent amount of one substance has been added to another substance.

evaporation The process of conversion of a liquid to a gas.

exchange reaction A reaction in which cations and anions that were partners in the reactants are interchanged in the products.

excited state The unstable state of an atom or molecule in which at least one electron does not have its lowest possible energy.

exothermic Refers to a process in which thermal energy must be transferred out of a thermodynamic system in order to maintain constant temperature.

Faraday constant (F) The quantity of electric charge on one mole of electrons, 9.6485×10^4 C/mol.

ferromagnetic A substance that contains clusters of atoms with unpaired electrons whose magnetic spins become aligned, causing permanent magnetism.

first law of thermodynamics (law of conservation of energy) Energy can neither be created nor destroyed—the total energy of the universe is constant.

formal charge The charge a bonded atom would have if its electrons were shared equally.

formation constant (K_f) The equilibrium constant for the formation of a complex ion.

formula (chemical formula) A notation combining element symbols and numerical subscripts that shows the relative numbers of each kind of atom in a molecule or formula unit of a substance.

formula unit The simplest cation-anion grouping represented by the formula of an ionic compound; also the collection of atoms represented by any formula.

formula weight The sum of the atomic weights in amu of all the atoms in a compound's formula.

free radical An atom, ion, or molecule that contains one or more unpaired electrons; usually highly reactive.

freezing-point lowering A colligative property; the difference between the freezing point of a pure solvent and the freezing point of a solution in which a nonvolatile nonelectrolyte solute is dissolved in the solvent.

frequency The number of complete traveling waves passing a point in a given period of time (cycles per second).

frequency factor The factor (A) in the Arrhenius equation that depends on how often molecules collide when all concentrations are 1 mol/L and on whether the molecules are properly oriented to react when they collide.

fuel cell An electrochemical cell that converts the chemical energy of fuels directly into electricity.

fullerenes Allotropic forms of carbon that consist of many five- and six-membered rings of carbon atoms sharing edges.

functional group An atom or group of atoms that imparts characteristic properties and defines a given class of organic compounds (e.g., the —OH group is present in all alcohols).

galvanized Has a thin coating of zinc metal that forms an oxide coating impervious to oxygen, thereby protecting a less active metal, such as iron, from corrosion.

gamma radiation Radiation composed of highly energetic photons.

gas A phase or state of matter in which a substance has no definite shape and has a volume determined by the volume of its container.

Gibbs free energy A thermodynamic function that decreases for any product-favored system. For a process at constant temperature and pressure, $\Delta G = \Delta H - T\Delta S$.

global warming Increase in temperature at earth's surface as a result of the greenhouse effect amplified by increasing concentrations of carbon dioxide and other greenhouse gases.

gram(s) The basic unit of mass in the metric system; equal to 1×10^{-3} kg.

gray The SI unit of absorbed radiation dose equal to the absorption of 1 joule per kilogram of material.

greenhouse effect Atmospheric warming caused when atmospheric carbon dioxide, water vapor, methane, ozone, and other greenhouse gases absorb infrared radiation reradiated from earth.

ground state The state of an atom or molecule in which all of the electrons are in their lowest possible energy levels.

groups The vertical columns of the periodic table of the elements.

Haber-Bosch process The process developed by Fritz Haber and Carl Bosch for the direct synthesis of ammonia from its elements.

half-cell One half of an electrochemical cell in which only the anode or cathode is located.

half-life, $t_{1/2}$ The time required for the concentration of one reactant to reach half its original value; radioactivity—the time required for the activity of a radioactive sample to reach half of its original value.

half-reaction A reaction that represents either an oxidation or a reduction process.

halide ion A monatomic ion (1−) of a halogen.

halogens The elements in Group 7A of the periodic table.

heat (heating) The energy-transfer process between two samples of matter at different temperatures.

heat capacity The quantity of energy that must be transferred to an object to raise its temperature by 1 °C.

heat of See **enthalpy of.**

heating curve A plot of the temperature of a substance versus the quantity of energy transferred to it by heating.

Henderson-Hasselbalch equation The equation describing the relationships among the pH of a buffer solution, the pK_a of the acid, and the concentrations of the acid and its conjugate base.

Henry's law A mathematical expression for the relationship of gas pressure and solubility; $S_g = k_H P_g$.

Hess's law If two or more chemical equations can be combined to give another equation, the enthalpy change for that equation will be the sum of the enthalpy changes for the equations that were combined.

heterogeneous catalyst A catalyst that is in a different phase from that of the reaction mixture.

heterogeneous mixture A mixture in which components remain separate and can be observed as individual substances or phases.

heterogeneous reaction A reaction that takes place at an interface between two phases, solid and gas for example.

hexagonal close packing The three-dimensional structure that results when layers of atoms in a solid are closest packed in the *ababab* arrangement.

homogeneous catalyst A catalyst that is in the same phase as that of the reaction mixture.

homogeneous mixture A mixture of two or more substances in a single phase that is uniform throughout.

homogeneous reaction A reaction in which the reactants and products are all in the same phase.

Hund's rule Electrons pair only after each orbital in a subshell is occupied by a single electron.

hybrid orbitals Orbitals formed by combining atomic orbitals of appropriate energy and orientation.

hybridized Refers to atomic orbitals of proper energy and orientation that have combined to form hybrid orbitals.

hydrate A solid compound that has a stoichiometric amount of water molecules bonded to metal ions or trapped within its crystal lattice.

hydration The binding of one or more water molecules to an ion or molecule within a solution or within a crystal lattice.

hydrocarbon An organic compound composed only of carbon and hydrogen.

hydrogen bond Noncovalent interaction between a hydrogen atom and a very electronegative atom to produce an unusually strong dipole-dipole force.

hydrolysis A reaction in which a bond is broken by reaction with a water molecule and the —H and —OH of the water add to the atoms of the broken bond.

hydronium ion H_3O^+; the simplest proton-water complex; responsible for acidity.

hydroxide ion OH^- ion; bases increase the concentration of hydroxide ions in solution.

hypertonic Refers to a solution having a higher concentration of nanoscale particles and therefore a higher osmotic pressure than another solution.

hypothesis A tentative explanation for an observation and a basis for experimentation.

hypotonic Refers to a solution having a lower solute concentration of nanoscale particles and therefore a lower osmotic pressure than another solution.

ideal gas A gas that behaves exactly as described by the ideal gas law, and by Boyle's, Charles's, and Avogadro's laws.

ideal gas constant The proportionality constant, *R*, in the equation $PV = nRT$; $R = 0.0821$ L atm mol^{-1} K^{-1} = 8.314 J K^{-1} mol^{-1}.

ideal gas law A law that relates pressure, volume, amount (moles), and temperature for an ideal gas; the relationship expressed by the equation $PV = nRT$.

immiscible Describes two liquids that form two separate phases when mixed because each is only slightly soluble in the other.

induced dipole A temporary dipole created by a momentary uneven distribution of electrons in a molecule or atom.

induced fit The change in the shape of an enzyme, its substrate, or both when they bind.

initial rate The instantaneous rate of a reaction determined at the very beginning of the reaction.

inorganic compound A chemical compound that is not an organic compound; usually of mineral or nonbiological origin.

insoluble Describes a solute, almost none of which dissolves in a solvent.

instantaneous reaction rate The rate at a particular time after a reaction has begun.

insulator A material that has a large energy gap between fully occupied and empty energy bands, and does not conduct electricity.

intermolecular forces Noncovalent attractions between separate molecules.

internal energy The sum of the individual energies (kinetic and potential) of all of the nanoscale particles (atoms, molecules, or ions) in a sample of matter.

ion An atom or group of atoms that has lost or gained one or more electrons so that it is no longer electrically neutral.

ion product (*Q*) A value found from an expression with the same mathematical form as the solubility product expression (K_{sp}) but using the actual concentrations rather than equilibrium concentrations of the species involved. (See **reaction quotient.**)

ionic bonding Forces of attraction between cations and anions in an ionic compound.

ionic compound A compound that consists of positive and negative ions (cations and anions).

ionic hydrate Ionic compounds that incorporate water molecules in the ionic crystal lattice.

ionic radius Radius of an anion or cation in an ionic compound.

ionization constant for water (K_w) The equilibrium constant that is the mathematical product of the hydronium ion concentration and the concentration of hydroxide ion in any aqueous solution; $K_w = 1 \times 10^{-14}$ at 25 °C.

ionization energy The energy needed to remove a mole of electrons from a mole of atoms in the gas phase.

isoelectronic Refers to atoms and ions that have identical electron configurations.

isomers Compounds that have the same molecular formula but different arrangements of atoms.

isotonic Refers to a solution having the same concentration of nanoscale particles and therefore the same osmotic pressure as another solution.

isotopes Forms of an element composed of atoms with the same atomic number but different mass numbers owing to a difference in the number of neutrons.

joule (J) A unit of energy equal to 1 kg m^2/s^2. The kinetic energy of a 2-kg object traveling at a speed of 1 m/s.

Kelvin temperature scale (See also **absolute temperature scale.**) A temperature scale on which the zero is the lowest possible temperature and the degree is the same size as a Celsius degree.

ketone An organic compound characterized by the presence of a carbonyl group in which the carbon atom is bonded to two other carbon atoms ($R_2C{=}O$).

kilocalorie (kcal or Cal) (See also **calorie.**) A unit of energy equal to 4.184 kJ. Approximately 1 kcal (1 Cal) is required to raise the temperature of 1 kg of liquid water by 1 °C. The food Calorie.

kinetic energy Energy that an object has because of its motion. Equal to $\frac{1}{2}mv^2$, where *m* is the object's mass and *v* is its velocity.

kinetic-molecular theory The theory that matter consists of nanoscale particles that are in constant, random motion.

lanthanides The elements after lanthanum in the sixth period in which the 4*f* subshell is being filled.

lattice energy Enthalpy of formation of 1 mol of an ionic solid from its separated gaseous ions.

law A statement that summarizes a wide range of experimental results and has not been contradicted by experiments.

law of chemical periodicity Law stating that the properties of the elements are periodic functions of atomic number.

law of combining volumes At constant temperature and pressure, the volumes of reacting gases are always in the ratios of small whole numbers.

law of conservation of energy (first law of thermodynamics) Law stating that energy can be neither created nor destroyed—the total energy of the universe is constant.

law of conservation of mass Law stating that there is no detectable change in mass in an ordinary chemical reaction.

law of constant composition Law stating that a chemical compound always contains the same elements in the same proportions by mass.

law of multiple proportions When two elements A and B can combine in two or more ways, the mass ratio A : B in one compound is a small-whole-number multiple of the mass ratio A : B in the other compound.

Le Chatelier's principle If a system is at equilibrium and the conditions are changed so that it is no longer at equilibrium, the system will react to give a new equilibrium in a way that partially counteracts the change.

Lewis acid A molecule or ion that can accept an electron pair from another atom, molecule, or ion to form a new bond.

Lewis base A molecule or ion that can donate an electron pair to another atom, molecule, or ion to form a new bond.

Lewis dot symbol An atomic symbol with dots representing valence electrons.

Lewis structure Structural formula for a molecule that shows all valence electrons as dots or as lines that represent covalent bonds.

limiting reactant The reactant present in limited supply that controls the amount of product formed in a reaction.

line emission spectrum A spectrum produced by excited atoms and consisting of discrete wavelengths of light.

linear Molecular geometry in which there is a 180° angle between bonded atoms.

liquid A phase of matter in which a substance has no definite shape but a definite volume.

London forces Forces resulting from the attraction between positive and negative regions of momentary (induced) dipoles in neighboring molecules.

lone-pair electrons Paired valence electrons unused in bond formation; also called nonbonding pairs.

major minerals Dietary minerals present in humans in quantities greater than 100 mg per kg of body weight.

macroscale Refers to samples of matter that can be observed by the unaided human senses; samples of matter large enough to be seen, measured, and handled.

main-group elements Elements in the eight A groups to the left and right of the transition elements in the periodic table; the *s*- and *p*-block elements.

mass A measure of an object's resistance to acceleration.

mass fraction The ratio of the mass of one component to the total mass of a sample.

mass number The number of protons plus neutrons in the nucleus of an atom of an element.

mass percent The mass fraction multiplied by 100%.

matter Anything that has mass and occupies space.

melting point The temperature at which the structure of a solid collapses and the solid changes to a liquid.

meniscus A concave or convex surface that forms on a liquid as a result of the balance of noncovalent forces in a narrow container.

metal An element that is malleable, ductile, forms alloys, and conducts an electric current.

metal activity series A ranking of relative reactivity of metals in displacement and other kinds of reactions.

metallic bonding In solid metals, the nondirectional attraction between positive metal ions and the surrounding sea of negatively charged electrons.

metalloid An element that has some typically metallic properties and other properties that are more characteristic of nonmetals.

methyl group A $—CH_3$ group.

metric system A decimalized measurement system.

microscale Refers to samples of matter so small that they have to be viewed with a microscope.

millimeters of mercury (mm Hg) A unit of pressure related to the height of a column of mercury in a mercury barometer (760 mm Hg = 1 atm = 101.3 kPa).

mineral A naturally occurring inorganic compound with a characteristic composition and crystal structure.

miscible Describes two liquids that will dissolve in each other in any proportion.

model A mechanical or mathematical way to make a theory more concrete, such as a molecular model.

molality (m) A concentration term equal to the molar amount of solute per kilogram of solvent.

molar amount See **amount.**

molar enthalpy of fusion The energy transfer required to melt 1 mol of a pure solid.

molar heat capacity The quantity of energy that must be transferred to 1 mol of a substance to increase its temperature by 1 °C.

molar mass The mass in grams of 1 mol of atoms, molecules, or formula units of one kind, numerically equal to the atomic or molecular weight in amu.

molar solubility The solubility of a solute in a solvent, expressed in moles per liter.

molarity Solute concentration expressed as the molar amount of solute per liter of solution.

mole (mol) The amount of substance that contains as many elementary particles as there are atoms in exactly 0.012 kg of carbon-12 isotope.

mole fraction (X) The ratio of number of moles of one component to the total number of moles in a mixture of substances.

mole ratio (stoichiometric factor) A mole-to-mole ratio relating the molar amount of a reactant or product to the molar amount of another reactant or product.

molecular compound A compound composed of atoms of two or more elements chemically combined in molecules.

molecular formula A formula that expresses the number of atoms of each type within one molecule of a substance.

molecular geometry The three-dimensional arrangement of atoms in a molecule.

molecular orbitals Orbitals extending over an entire molecule generated by combining atomic orbitals.

molecular weight The sum of the atomic weights of all the atoms in a substance's formula.

molecule The smallest particle of an element or compound that exists independently, and retains the chemical properties of that element or compound.

momentum The product of the mass (m) times the velocity (v) of an object in motion.

monatomic ion An ion consisting of one atom bearing an electrical charge.

monoprotic acid An acid that can donate a single hydrogen ion per molecule.

multiple covalent bonds Double or triple covalent bonds.

multiple proportions, law of When two elements A and B can combine in two or more ways, the mass ratio A : B in one compound is a small-whole-number multiple of the mass ratio A : B in the other compound.

nanoscale Refers to samples of matter (e.g., atoms and molecules) whose normal dimensions are in the 1–100 nanometer range.

nanotubes Members of the family of fullerenes in which graphite-like layers of carbon atoms form cylindrical shapes.

Nernst equation The equation relating the potential of an electrochemical cell to the concentrations of the chemical species involved in the oxidation-reduction reactions occurring in the cell.

net ionic equation A chemical equation in which only those molecules or ions undergoing chemical changes in the course of the reaction are represented.

network solid A solid consisting of one huge molecule in which all atoms are connected via a network of covalent bonds.

neutral solution A solution containing equal concentrations of H_3O^+ and OH^-; a solution that is neither acidic nor basic.

neutron An electrically neutral subatomic particle found in the nucleus.

newton (N) The SI unit of force; equal to 1 kg times an acceleration of 1 m/s^2; 1 kg m/s^2.

nitrogen fixation The conversion of atmospheric nitrogen (N_2) to nitrogen compounds utilizable by plants or industry.

noble gas notation An abbreviated electron configuration of an element in which filled inner shells are represented by the symbol of the preceding noble gas in brackets. For Al, this would be $[Ne]3s^23p^1$.

noble gases Gaseous elements in Group 8A; the least reactive elements.

noncovalent interactions All forces of attraction other than covalent, ionic, or metallic bonding.

nonelectrolyte A substance that dissolves in water to form a solution that does not conduct electricity.

nonmetal Element that does not have the chemical and physical properties of a metal.

nonpolar covalent bond A bond in which the electron pair is shared equally by the bonded atoms.

nonpolar molecule A molecule that is not polar either because it has no polar bonds or because its polar bonds are oriented symmetrically so that they cancel each other.

normal boiling point The temperature at which the vapor pressure of a liquid equals 1 atm.
nuclear decay Spontaneous emission of radioactivity by an unstable nucleus that is converted into a more stable nucleus.
nuclear fission The highly exothermic process by which very heavy fissionable nuclei split to form lighter nuclei.
nuclear fusion The highly exothermic process by which very light nuclei combine to form heavier nuclei.
nuclear reaction A process in which one or more atomic nuclei change into one or more different nuclei.
nuclear reactor A container in which a controlled nuclear reaction takes place.
nucleon A nuclear particle, either a neutron or a proton.
nucleus (atomic) The tiny central core of an atom; contains protons and neutrons. (There are no neutrons in hydrogen-1.)

octahedral Molecular geometry of six groups around a central atom in which all groups are at angles of 90° to other groups.
octet rule In forming bonds, many main group elements gain, lose, or share electrons to achieve a stable electron configuration characterized by eight valence electrons.
orbital A region of an atom or molecule within which there is a significant probability that an electron will be found.
orbital shape The shape of an electron density distribution determined by an orbital.
order of reaction The reaction rate dependency on the concentration of a reactant or product, expressed as an exponent of a concentration term in the rate equation.
ores Minerals containing a sufficiently high concentration of an element to make its extraction profitable.
organic compound A compound of carbon with hydrogen, possibly also oxygen, nitrogen, sulfur, phosphorus, or other elements.
osmosis The movement of a solvent (water) through a semipermeable membrane from a region of lower solute concentration to a region of higher solute concentration.
osmotic pressure (Π) The pressure that must be applied to a solution to stop osmosis from a sample of pure solvent.
overall reaction order The sum of the exponents for all concentration terms in the rate equation.
oxidation The loss of electrons by an atom, ion, or molecule, leading to an increase in oxidation number.
oxidation number (oxidation state) The hypothetical charge an atom would have if all bonds to that atom were completely ionic.
oxidation-reduction reaction (redox reaction) A reaction involving the transfer of one or more electrons from one species to another so that oxidation numbers change.
oxides Compounds of oxygen combined with another element.
oxidized The result when an atom, molecule, or ion loses one or more electrons.
oxidizing agent The substance that accepts electron(s) and is reduced in an oxidation-reduction reaction.
oxoacids Acids in which the acidic hydrogen is bonded directly to an oxygen atom.
oxoanion A polyatomic anion that contains oxygen.

***p*-block elements** Main-group elements in Groups 3A through 8A whose valence electrons consist of outermost *s* and *p* electrons.
paramagnetic Refers to atoms, molecules, or ions that are attracted to a magnetic field because they have unpaired electrons in incompletely filled electron subshells.
partial pressure The pressure that one gas in a mixture of gases would exert if it occupied the same volume at the same temperature as the mixture.
parts per billion (ppb) One part in one billion (10^9) parts.
parts per million (ppm) One part in one million (10^6) parts.
parts per trillion (ppt) One part in one trillion (10^{12}) parts.
pascal (Pa) The SI unit of pressure; 1 Pa = 1 N/m^2.
Pauli exclusion principle An atomic principle that states that, at most, two electrons can be assigned to the same orbital in the same atom or molecule, and these two electrons must have opposite spins.
percent abundance The percentage of atoms of a particular isotope in a natural sample of a pure element.
percent composition by mass The percentage of the mass of a compound represented by each of its constituent elements.
percent yield The ratio of actual yield to theoretical yield, multiplied by 100%.
periodic table A table of elements arranged in order of increasing atomic number so that those with similar chemical and physical properties fall in the same vertical groups.
periods The horizontal rows of the periodic table of the elements.
pH The negative logarithm of the hydronium ion concentration ($-\log [H_3O^+]$).
phase Any of the three states of matter: gas, liquid, solid. Also, one of two or more solid-state structures of the same substance, such as iron in a body-centered cubic or face-centered cubic structure.
phase change A physical process in which one state or phase of matter is changed into another (such as melting a solid to form a liquid).
phase diagram A diagram showing the relationships among the phases of a substance (solid, liquid, and gas), at different temperatures and pressures.
photoelectric effect The emission of electrons by some metals when illuminated by light of certain wavelengths.
photon A massless particle of light whose energy is given by $h\nu$, where ν is the frequency of the light and h is Planck's constant.
physical changes Changes in the physical properties of a substance, such as the transformation of a solid to a liquid.
physical properties Properties (e.g., melting point or density) that can be observed and measured without changing the composition of a substance.
pi (π) bond A bond formed by the sideways overlap of parallel *p* orbitals.
Planck's constant The proportionality constant, h, that relates energy of a photon to its frequency. The value of h is 6.626×10^{-34} J · s.
plasma A state of matter consisting of unbound nuclei and electrons.
polar covalent bond A covalent bond between atoms with different electronegativities; bonding electrons are shared unequally between the atoms.
polar molecule A molecule that is polar because it has polar bonds arranged so that electron density is concentrated at one end of the molecule.
polarization The induction of a temporary dipole in a molecule or atom by shifting of electron distribution.
polyatomic ion An ion consisting of more than one atom.
polyprotic acids Acids that can donate more than one hydrogen ion per molecule.
positron A nuclear particle having the same mass as an electron, but a positive charge.
potential energy Energy that an object has because of its position.
precipitate An insoluble product of an exchange reaction in aqueous solution.
pressure The force exerted on an object divided by the area over which the force is exerted.
primary battery A voltaic cell (or battery of cells) in which the oxidation and reduction half-reactions cannot easily be reversed to restore the cell to its original state.
principal energy level An energy level containing orbitals with the same quantum number ($n = 1, 2, 3 \ldots$).
principal quantum number An integer assigned to each of the allowed main electron energy levels in an atom.
product A substance formed as a result of a chemical reaction.
product-favored system A system in which, when a reaction appears to be over, products predominate over reactants.
proportionality factor (conversion factor) A relationship between two measurement units derived from the proportionality of one quantity to another; e.g., density is the conversion factor between mass and volume.
proton A positively charged subatomic particle found in the nucleus.

qualitative In observations, nonnumerical experimental information, such as a description of color or texture.

quantitative Numerical information, such as the mass or volume of a substance, expressed in appropriate units.

quantum The smallest possible unit of a distinct quantity; for example, the smallest possible unit of energy for electromagnetic radiation of a given frequency.

quantum theory The theory that energy comes in very small packets (quanta); this is analogous to matter occurring in very small particles—atoms.

rad A unit of radioactivity; a measure of the energy of radiation absorbed by a substance, 1.00×10^{-2} J per kilogram.

radial distribution plot A graph showing the probability of finding an electron as a function of distance from the nucleus of an atom.

radioactive series A series of nuclear reactions in which a radioactive isotope undergoes successive nuclear transformations resulting ultimately in a stable, nonradioactive isotope.

radioactivity The spontaneous emission of energy and/or subatomic particles by unstable atomic nuclei; the energy or particles so emitted.

Raoult's law A mathematical expression for the vapor pressure of the solvent in a solution; $P_1 = X_1 P_1^0$.

rate The change in some measurable quantity per unit time.

rate constant (*k*) A proportionality constant relating reaction rate and concentrations of reactants and other species that affect the rate of a specific reaction.

rate law (rate equation) A mathematical equation that summarizes the relationship between concentrations and reaction rate.

rate-limiting step The slowest step in a reaction mechanism.

reactant A substance that is initially present and undergoes change in a chemical reaction.

reactant-favored system A system in which, when a reaction appears to be over, reactants predominate over products.

reaction intermediate An atom, molecule, or ion produced in one step and used in a later step in a reaction mechanism; does not appear in the equation for the overall reaction.

reaction mechanism A sequence of unimolecular and bimolecular elementary reactions by which an overall reaction may occur.

reaction quotient (*Q*) A value found from an expression with the same mathematical form as the equilibrium constant expression but with the actual concentrations in a mixture not at equilibrium.

reaction rate The change in concentration of a reactant or product per unit time.

redox reaction (oxidation-reduction reaction) A reaction involving the transfer of one or more electrons from one species to another so that oxidation numbers change.

reduced The result when an atom, molecule, or ion gains one or more electrons.

reducing agent The atom, molecule, or ion that donates electron(s) and is oxidized in an oxidation-reduction reaction.

reduction The gain of electrons by an atom, ion, or molecule, leading to a decrease in its oxidation number.

relative humidity In the atmosphere, the ratio of actual partial pressure to equilibrium vapor pressure of water at the prevailing temperature.

rem A unit of radioactivity; 1 rem has the physiological effect of 1 roentgen of radiation.

resonance hybrid The actual structure of a molecule that can be represented by more than one Lewis structure.

resonance structures The possible structures of a molecule for which more than one Lewis structure can be written, differing by the arrangement of electrons but having the same arrangement of atomic nuclei.

reverse osmosis Application of pressure greater than the osmotic pressure to cause solvent to flow through a semipermeable membrane from a concentrated solution to a solution of lower solute concentration.

reversible process A process for which a very small change in conditions will cause a reversal in direction.

roentgen (R) A unit of radioactivity; 1 R corresponds to deposition of 93.3×10^{-7} J per gram of tissue.

***s*-block elements** Main-group elements in Groups 1A and 2A whose valence electrons are *s* electrons.

salt An ionic compound whose cation comes from a base and whose anion comes from an acid.

salt bridge A device for maintaining balance of ion charges in the compartments of an electrochemical cell.

saturated hydrocarbon Hydrocarbon in which carbon atoms are bonded to the maximum number of hydrogen atoms.

saturated solution A solution in which the concentration of solute is the concentration that would be in equilibrium with undissolved solute at a given temperature.

screening effect Reduction of the effective attraction between nucleus and valence electrons as a result of repulsion of the outer valence electrons by electrons in inner shells.

second law of thermodynamics The total entropy of the universe (the system and surroundings) is continually increasing. In any product-favored system, the entropy of the universe is greater after a reaction than it was before.

secondary battery A voltaic cell (or battery of cells) in which the oxidation and reduction half-reactions can be reversed to restore the cell to its original state after discharge.

semiconductor Material with a narrow energy gap between the valence band and the conduction band; a conductor when an electric field or a higher temperature is applied.

semipermeable membrane A thin layer of material through which only certain kinds of molecules can pass.

shell A collection of orbitals with the same value of the principal quantum number, *n*.

shifting an equilibrium Changing the conditions of an equilibrium system so that the system is no longer at equilibrium and there is a net reaction in either the forward or reverse direction until equilibrium is reestablished.

sievert The SI unit of effective dose of absorbed radiation, 1 Sv = 100 rem.

sigma (σ) bond A bond formed by head-to-head orbital overlap along the bond axis.

single covalent bond A bond formed by sharing one pair of electrons between the same two atoms.

solid A state of matter in which a substance has a definite shape and volume.

solubility The maximum amount of solute that will dissolve in a given volume of solvent at a given temperature when pure solute is in equilibrium with the solution.

solubility product constant (K_{sp}) An equilibrium constant that is the product of concentrations of ions in a solution in equilibrium with a solid ionic compound.

solubility product expression Molar concentrations of a cation and anion, each raised to a power equal to its coefficient in the balanced chemical equation for the solubility equilibrium.

solute The material dissolved in a solution.

solution A homogeneous mixture of two or more substances in a single phase.

solvent The medium in which a solute is dissolved to form a solution.

***sp* hybrid orbitals** Orbitals of the same atom formed by the combination of one *s* orbital and one *p* orbital.

sp^2 hybrid orbitals Orbitals of the same atom formed by the combination of one *s* orbital and two *p* orbitals.

sp^3 hybrid orbitals Orbitals of the same atom formed by the combination of one *s* orbital and three *p* orbitals.

specific heat capacity The quantity of energy that must be transferred to 1 g of a substance to increase its temperature by 1 °C.

spectator ion An ion that is present in a solution in which a reaction takes place, but is not involved in the net process.

spectrum A plot of the intensity of light (photons per unit time) as a function of the wavelength or frequency of light.

standard atmosphere (atm) A unit of pressure; 1 atm = 101.325 kPa = 1.01325 bar = 760 mm Hg exactly.

standard-state conditions These are 1 bar pressure for all gases, 1 M concentration for all solutes, at a specified temperature.

standard enthalpy change The enthalpy change when a process occurs with reactants and products all in their standard states.

standard equilibrium constant ($K°$) An equilibrium constant in which each concentration (or pressure) is divided by the standard-state concentration (or pressure); if concentrations are expressed in moles per liter (or pressures in bars) then the concentration (or pressure) equilibrium constant equals the standard equilibrium constant $\Delta G° = -RT \ln K°$.

standard hydrogen electrode The electrode against which standard reduction potentials are measured, consisting of a platinum electrode at which 1 M hydronium ion is reduced to hydrogen gas at 1 bar.

standard molar enthalpy of formation The standard enthalpy change for forming 1 mol of a compound from its elements, with all substances in their standard states.

standard molar volume The volume occupied by exactly 1 mol of an ideal gas at standard temperature (0 °C) and pressure (1 atm), equal to 22.414 L.

standard reduction potential ($E°$) The potential of an electrochemical cell when a given electrode is paired with a standard hydrogen electrode under standard conditions.

standard state The most stable form of a substance in the physical state in which it exists at 1 bar and a specified temperature.

standard solution A solution whose concentration is known accurately.

standard temperature and pressure (STP) A temperature of 0 °C and a pressure of 1 atm.

standard voltages Electrochemical cell voltages measured under standard conditions.

state function A property whose value is invariably the same if a system is in the same state.

steric factor A factor in the expression for rate of reaction that reflects the fact that some three-dimensional orientations of colliding molecules are more likely to result in reaction than others.

stoichiometric coefficients The multiplying numbers assigned to the species in a chemical equation in order to balance the equation.

stoichiometric factor (mole ratio) A factor relating number of moles of a reactant or product to number of moles of another reactant or product.

stoichiometry The study of the quantitative relations between amounts of reactants and products in chemical reactions.

stratosphere The region of the atmosphere approximately 12 to 50 km above sea level.

strong acid An acid that ionizes completely in aqueous solution.

strong base A base that ionizes completely in aqueous solution.

strong electrolyte An electrolyte that consists solely of ions in aqueous solution.

structural formulas Formulas written to show how atoms in a molecule or polyatomic ion are connected to each other.

structural isomers (constitutional isomers) Compounds with the same molecular formula that differ in the order in which their atoms are bonded together.

sublimation Conversion of a solid directly to a gas with no formation of liquid.

subshell A group of atomic orbitals with the same n and ℓ quantum numbers.

substance Matter of a particular kind; each substance, when pure, has a well-defined composition and a set of characteristic properties that differ from the properties of any other substance.

substrate A molecule or molecules whose reaction is catalyzed by an enzyme.

supercritical fluid A substance above its critical temperature and pressure; has density characteristic of a liquid, but flow properties of a gas.

supersaturated solution A solution that temporarily contains more solute per unit volume than a saturated solution at a given temperature.

surface tension The energy required to overcome the attractive forces between molecules at the surface of a liquid.

surroundings Everything that can exchange energy with a thermodynamic system.

system In thermodynamics, that part of the universe that is singled out for observation and analysis. The region of primary concern.

temperature The physical property of matter that determines whether one object can heat another.

tetrahedral Molecular geometry of four atoms or groups of atoms around a central atom with bond angles of 109.5°.

theoretical yield The maximum quantity of product theoretically obtainable from a given quantity of reactant in a chemical reaction.

theory A unifying principle that explains a body of facts and the laws based on them.

thermal equilibrium The condition of equal temperatures achieved between two samples of matter that are in contact.

thermochemical expression A balanced chemical equation, including specification of the states of matter of reactants and products, together with the corresponding value of the enthalpy change.

thermodynamics The science of heat, work, and the transformations of each into the other.

third law of thermodynamics A perfect crystal of any substance at 0 K has the lowest possible entropy.

titrant The solution being added from a buret to another solution during a titration.

titration A procedure whereby a substance in a standard solution reacts with a known stoichiometry with a substance whose concentration is to be determined.

titration curve A plot of the progress of a titration as a function of the volume of titrant added.

torr A unit of pressure equivalent to 1 mm Hg.

trace elements (See also **major minerals.**) The dietary minerals that are present in smaller concentrations than the major minerals, sometimes far smaller concentrations.

***trans* isomer** The isomer in which two like substituents are on opposite sides of a carbon-carbon double bond, a ring of carbon atoms, or a coordination complex.

transition elements Elements that lie in rows 4 through 7 of the periodic table in which d or f subshells are being filled; comprising scandium through zinc, yttrium through cadmium, lanthanum through mercury, and actinium and elements of higher atomic number.

transition state A molecular structure corresponding to the maximum of a plot of energy versus reaction progress; also known as the activated complex.

triangular bipyramidal Molecular geometry of five groups around a central atom in which three groups are in equatorial positions and two are in axial positions.

triangular planar Molecular geometry of three groups at the corners of an equilateral triangle around a central atom at the center of the triangle.

triple bond A bond formed by sharing three pairs of electrons between two atoms.

triple point The point on a temperature/pressure phase diagram of a substance where solid, liquid, and gas phases are all in equilibrium.

troposphere The lowest region of the atmosphere, extending from the Earth's surface to an altitude of about 12 km.

uncertainty principle The statement that it is impossible to determine simultaneously the exact position and the exact momentum of an electron.

unimolecular reaction A reaction in which the rearrangement of the structure of a single molecule produces the product molecule or molecules.

unit cell A small portion of a crystal lattice that can be replicated in each of three directions to generate the entire lattice.

unsaturated hydrocarbon A hydrocarbon containing double or triple carbon-carbon bonds.

unsaturated solution A solution that contains a smaller concentration of solute than the concentration of a saturated solution at a given temperature.

valence band In a solid, an energy band (group of closely spaced orbitals) that contains valence electrons.

valence bond model A theoretical model that describes a covalent bond as resulting from an overlap of one orbital on each of the bonded atoms.

valence electrons Electrons in an atom's highest occupied principal shell and in partially filled subshells of lower principal shells; electrons available to participate in bonding.

valence-shell electron-pair repulsion model (VSEPR) A simple model used to predict the shapes of molecules and polyatomic ions based on repulsions between bonding pairs and lone pairs around a central atom.

van der Waals equation An equation of state for gases that takes into account the volume occupied by molecules and noncovalent attractions between molecules:

$$\left[P + a\left(\frac{n}{V}\right)^2\right][V - bn] = nRT.$$

vaporization The change of a substance from the liquid to the gas phase.

vapor pressure The pressure of the vapor of a substance in equilibrium with its liquid or solid in a sealed container.

viscosity The resistance of a liquid to flow.

volatility The tendency of a liquid to vaporize.

volt (V) Electrical potential energy difference defined so that 1 joule (work) is performed when 1 coulomb (charge) moves through 1 volt (potential difference).

voltaic cell An electrochemical cell in which a product-favored oxidation-reduction reaction is used to produce an electric current.

water of hydration The water molecules trapped within the crystal lattice of an ionic hydrate or coordinated to a metal ion in a crystal lattice or in solution.

wave functions Solutions to the Schrödinger wave equation that describe the behavior of an electron in an atom.

wavelength The distance between adjacent crests (or troughs) in a wave.

weak acid An acid that is only partially ionized in aqueous solution.

weak base A base that is only partially ionized in aqueous solution.

weak electrolyte An electrolyte that is only partially ionized in aqueous solution.

weight percent A mass fraction expressed as a percent by multiplying by 100%; used for elemental composition of a compound, composition of a solute in solution.

work (working) A mechanical process that transfers energy to or from an object.

Index

Page numbers in **bold face** refer to key term definitions
Page numbers in *italics* refer to images or diagrams
Page numbers followed by a "t" indicate a table

Useful Conversion Factors and Relationships

Length

SI unit: meter (m)

1 kilometer = 1000. meters
= 0.62137 mile
1 meter = 100. centimeters
1 centimeter = 10. millimeters
1 nanometer = 1×10^{-9} meter
1 picometer = 1×10^{-12} meter
1 inch = 2.54 centimeter (exactly)
1 Ångström = 1×10^{-10} meter

Mass

SI unit: kilogram (kg)

1 kilogram = 1000. grams
1 gram = 1000. milligrams
1 pound = 453.59237 grams = 16 ounces
1 ton = 2000. pounds

Volume

SI unit: cubic meter (m^3)

1 liter (L) = 1×10^{-3} m^3
= 1 dm^3 = 1000. cm^3
= 1.056710 quarts
1 gallon = 4 quarts

Energy

SI unit: joule (J)

1 joule = 1 kg m^2 s^{-2}
= 0.23901 calorie
= 1 C × 1 V
1 calorie = 4.184 joules

Pressure

SI unit: pascal (Pa)

1 pascal = 1 N/m^2
= 1 kg m^{-1} s^{-1}
1 atmosphere = 101.325 kilopascals, kPa
= 760. mm Hg = 760. torr
= 14.70 lb/in.^{2}
1 bar = 1×10^5 Pa
1 kilopascal = 1×10^3 Pa

Temperature

SI unit: kelvin (K)

If T_K is the numeric value of the temperature in kelvins, t_C the numeric value of the temperature in °C, and t_F the numeric value of the temperature in °F, then

$$T_K = t_C + 273.15$$
$$t_C = (\tfrac{5}{9})(t_F - 32)$$
$$t_F = (\tfrac{9}{5})t_C + 32$$

Location of Useful Tables and Figures

Atomic, Ionic, and Molecular Properties

Atomic radii	Figure 7.20
Bond energies	Table 8.2
Bond lengths	Table 8.1
Electron affinities	Table 7.8
Electron configurations	Table 7.5, Figure 7.16, Appendix D
Electronegativity	Figures 8.6, 8.7
Ionic radii	Figure 7.21
Ionization energies	Figure 7.22
Molecular geometries	Table 9.1
Properties of molecular and ionic compounds	Table 3.9
Properties of solids	Table 11.5

Thermodynamic Properties

Specific heat capacities	Table 6.1
Standard molar enthalpies of formation	Table 6.2, App. J
Standard molar entropies	Table 17.1, App. J
Enthalpy, entropy, Gibbs free energy	Appendix J
Equilibrium constants	Table 13.1 (Also see Appendices F, G, and H)

Acids, Bases, and Salts

Acid-base properties of ions in solution	Table 15.5
Common acids and bases	Table 5.2
Common polyatomic ions	Table 3.7
Solubility guidelines for ionic compounds	Table 5.1
Common monatomic ions	Figure 3.2
Buffer systems	Table 16.1
Acid and base ionization constants	Table 15.2 (See also Appendices F, G)
Solubility product constants	Appendix H

Other Useful Tables and Figures

Common oxidizing and reducing agents	Table 5.3
Electromagnetic spectrum	Figure 7.1
Composition of dry air	Table 10.1
Standard reduction potentials	Table 18.1 (See also Appendix I)